evolve
Anatomy & Physiology Online

GETTING STARTED

If your course is being led by an instructor:

1. System Your instructor will provide information about the system on which your course is being hosted. Evolve® courses can be run on a variety of systems and your instructor will decide which one is right for this course.

2. Username & Password Your instructor will also provide you with the username and password needed to access the system where this course is located.

3. Login Instructions If your instructor's course is being hosted on the Evolve Learning System, please go to *http://evolve.elsevier.com/student* Enter your username and password in the **Login to My Evolve** area and click the **Login** button. You will be taken to your personalized **My Evolve** page where your course will be listed in the **My Course** module.

If your course is on a different system, your instructor will provide information about how to log in.

4. Access Code The first time you access this course, you will need the access code located below, regardless of which system is hosting the course. When you are prompted, enter the code exactly as it appears in this book.

If you plan to take the course on your own:

(**Note:** By taking the course independently, you will not have any instructor to help you with the course. You will have 12 months from the date you are enrolled to complete the course.)

1. System All independent learners are enrolled in a course hosted on the Evolve Learning System.

2. Self-Enrollment

a. Go to: *http://evolve.elsevier.com/Thibodeau*

b. Under the **Online Course** heading, click on the **Self-Study Student? Enroll Here** option. This will launch the enrollment wizard for your course.

c. Complete the enrollment wizard. During this process you will create an Evolve username and password. You will also be asked to provide identifying information about yourself and will need to provide the access code provided below.

d. Once the wizard has been completed you will be able to log in to your Evolve account and begin your Online Course immediately.

TECHNICAL REQUIREMENTS

To use an Evolve Online Course, you will need access to a computer that is connected to the Internet and equipped with web browser software that supports frames. For optimal performance, it is recommended that you have speakers and use a high-speed Internet connection. However, slower dial-up modems (56K minimum) are acceptable.

Screen Settings

For best results, the resolution of your computer monitor should be set at a minimum of 800 x 600. The number of colors displayed should be set to "thousands or higher" (High Color or 16 bit) or "millions of colors" (True Color or 24 bit). To set the resolution:

Windows

1. From the **Start** menu, select **Settings** and **Control Panel**.
2. Double-click on the **Display** icon.
3. Click on the **Settings** tab.
4. In the **Screen area** use the slider bar to select **800 by 600 pixels**.
5. In the **Colors** drop down menu, click on the arrow to show more settings.
6. Click on **High Color (16 bit)** or **True Color (24 bit)**.
7. Click on **Apply**.
8. Click on **OK**.
9. You may be asked to verify the setting changes. Click **Yes**.
10. You may be asked to restart your computer to accept the setting changes.

Click **Yes**.

Macintosh

1. Select the **Monitors** control panel.
2. Select **800 x 600** (or similar) from the **Resolution** area.
3. Select **Thousands** or **Millions** from the **Color Depth** area.

Web Browsers

Supported web browsers include Microsoft Internet Explorer (IE) version 6.0 or higher, Netscape Navigator version 7.1 or higher, and Mozilla 1.4 or higher. If you use America Online (AOL) for Web access, you will need AOL version 4.0 or higher **and** one of the browsers listed above. Earlier versions of AOL and Internet Explorer will not run the course properly and you will have difficulty accessing many features.

For best results with AOL:

- Connect to the Internet using AOL version 4.0 or higher.
- Open a private chat within AOL. (This allows the AOL client to remain open, without asking if you wish to disconnect while minimized.)
- Minimize AOL.
- Launch one of the recommended browsers.

Whichever browser you use, the browser preferences must be set to enable cookies as well as Java/JavaScript, and the cache must be set to reload every time.

Enable Cookies

Browser - Internet Explorer (IE) 6.0 or higher
1. Select **Tools**.
2. Select **Internet Options**.
3. Select **Privacy** tab.
4. Use the slider (slide down) to **Accept All Cookies**.
5. Click **OK**.
-OR-
4. Click the **Advanced** button.
5. Click the check box next to **Override Automatic Cookie Handling**.
6. Click the **Accept** radio buttons under **First-party Cookies** and **Third-party Cookies**.
7. Click **OK**.

Browser - Netscape 7.1 or higher
1. Select **Edit**.
2. Select **Preferences**.
3. Select **Privacy & Security**.
4. Select **Cookies**.
5. Select **Enable All Cookies**.

Browser - Mozilla 1.4 or higher
1. Select **Tools**.
2. Select **Privacy**.
3. Expand the **Cookies** section and check the following box: Allow sites to set cookies.

Enable Java

Browser - Internet Explorer (IE) 6.0 or higher
1. Select **Tools -> Internet Options**.
2. Select **Advanced** tab.
3. Scroll down the list until you see the **Java (Sun)** section and select the box that appears below it.

Browser – Netscape 7.1 or higher
1. Select **Edit -> Preferences**.
2. Select **Advanced**.
3. Select **Scripts & Plugins**.
4. Make sure the **Navigator** box is checked to **Enable JavaScript**.
5. Click **OK**.

Browser - Mozilla 1.4 or higher
1. Select **Tools**.
2. Select **Web Features**.
3. Select the boxes next to **Enable Java** and **Enable Javascript**.

Set Cache to Always Reload a Page

Browser - Internet Explorer (IE) 6.0 or higher
1. Select **Tools -> Internet Options**.
2. Select **General** tab.
3. Go to the **Temporary Internet Files** and click the **Settings** button.
4. Select the radio button for **Every visit to the page** and click **OK** when complete.

Browser - Netscape 7.1 or higher
1. Select **Edit -> Preferences**.
2. Select **Advanced**.
3. Select **Cache**.
4. Select the **Every time I view the page** radio button.
5. Click **OK**.

Browser - Mozilla 1.4 or higher
1. Select **Tools**.
2. Select **Privacy**.
3. Expand the **Cache** section and designate a disk space number if one isn't in place already.

Plug-Ins

Adobe Acrobat Reader—With the free Acrobat Reader software you can view and print Adobe PDF files. Many Evolve products offer documents in this format, including student and instructor manuals, checklists, and more.
Download at: *http://www.adobe.com*

Apple QuickTime—Install this to hear word pronunciations, heart and lung sounds, and many other interesting audio clips within Evolve Online Courses.
Download at: *http://www.apple.com*

Macromedia Flash Player—This player will enhance your viewing of many Evolve web pages as well as educational short-form to long-form animation within the Evolve Learning System.
Download at: *http://www.macromedia.com*

Macromedia Shockwave Player—Shockwave is best for viewing the many interactive learning activities within Evolve Online Courses.
Download at: *http://www.macromedia.com*

Microsoft Word Viewer—With this viewer, Microsoft Word users can share documents with others who don't have Word software. Users without Word can then open and view Word documents. Many Evolve products have test banks, student and instructor manuals, and other documents available for download and viewing on your local computer.
Download at: *http://www.microsoft.com*

Microsoft PowerPoint Viewer—This viewer makes it possible for you to view PowerPoint presentations even if you don't have PowerPoint software. Many Evolve products have slides available for download and viewing on your local computer.
Download at: http://www.microsoft.com

SUPPORT INFORMATION

Technical support is available via calling the USA helpdesk on 001–314–995–3200. Before calling, please have details of your browser and version (i.e. Netscape 7.1, Internet Explorer 6.0); operating system and version (i.e. Windows XP, Macintosh OSX); Course ID; Product Title. Alternatively, customers can e-mail *evolve-support@elsevier.com*

Support Information is also available on the Evolve Portal (*http://evolve.elsevier.com*) including:
- Guided Tours
- Tutorials
- Frequently Asked Questions (FAQ)
- Online Copies of Course User Guides
- And much more!

PASS CODE

MWW9FWH6YS38   MWW9FWH6YH8F

ANATOMY

PHYSIOLOGY

Gar
Chancellor Eme
Univers

Kevi
Professor of Life Sciences
St. Charles Community College
Cottleville, Missouri

Adjunct Assistant Professor of Physiology
St. Louis University Medical School
St. Louis, Missouri

Foreword by

Dr John C Atherton
Senior Lecturer in Medical Education and Physiology,
School of Medicine, University of Keele, UK

Dr Helen L Atherton
Lecturer in Nursing, School of Healthcare, University of Leeds, UK

MOSBY

ELSEVIER

MOSBY
ELSEVIER

11830 Westline Industrial Drive
St. Louis, Missouri 63146

ANATOMY & PHYSIOLOGY

ISBN: 0-7234-3448-4
ISBN-13: 978-0-7234-3448-1

Copyright © 2007, 2003, 1999, 1996, 1993, 1987 by Mosby, an affiliate of Mosby, Inc.

This edition published under licence by Elsevier Ltd for distribution outside the USA and Canada.

ISBN: 0-7234-3448-4
ISBN-13: 978-0-7234-3448-1

Executive Editor: Thomas J. Wilhelm
Managing Editor: Jeff Downing
Associate Developmental Editor: Jennifer Stoces
Editorial Assistant: Carlie Bliss
Publishing Services Manager: Deborah L. Vogel
Senior Project Manager: Steve Ramay
Design Manager: Mark Oberkrom

Printed in China

Last digit is the print number: 9 8 7 6 5 4 3 2 1

Foreword

Attainment and maintenance of good health is essential for both day-to-day functioning of individuals and the achievement of long-term goals and aspirations; it is central to an individual's holistic wellbeing. Good physical health is a state that cannot necessarily be achieved in isolation from others and it would be unusual for us not to seek assistance from one or more healthcare professionals during our lifetime. In doing so we rely on the fact that such professionals will have an accurate and detailed knowledge of what constitutes normal and abnormal functioning of the body, a knowledge that underpins healthcare interventions to effectively meet our needs.

Anatomy & Physiology, an already established text, provides basic anatomical and physiological knowledge for healthcare studies by focusing primarily on developing an understanding of the complexity of how the body works and emphasising the integration of the body systems to sub-serve homeostasis. Written and compiled in a student friendly manner, *Anatomy & Physiology* clearly responds to the need for teaching and learning to be individually driven and enjoyable processes. There is obvious recognition through the structure of this text and accompanying resources that teaching and learning are no longer best served by the traditional didactic teaching methods of higher educational settings, but rather that learning should be a flexible, self-directed and mobile process - essential features in view of the broadening access to healthcare studies for students with either no biological background or no recent experience in biology. There is a clear demand therefore for texts and additional resources of this kind to make learning an exciting, interesting and interactive process.

Anatomy & Physiology achieves this demand with a text that contains good illustrations, uses flow diagrams and summary tables to highlight the integration between the various body systems, and includes transparent coloured overlays to build a picture of the human body and the internal relations of organs and tissues that is almost three dimensional. Another attractive feature of *Anatomy & Physiology* is the successful attempt to apply the content of each chapter to the clinical setting by using both highlighted information and including sections on the mechanism of disease and clinical scenarios to provide the relevance of the basic science. Highlighted information on diagnostic investigations and relevance of information to sports science and medicine add to the reader's interest. Glossaries of terms – the language of science and the language of medicine – facilitate understanding of complex words and, equally important in today's climate of self-directed learning, the pronunciation of these words.

Anatomy & Physiology comes with a package that includes excellent support for many learning strategies: the innovative facility to download the entire book (*Anatomy & Physiology e-book*) to provide an easily portable version which can be annotated and highlighted to create notes and used to rapidly access specified topics; access to a colouring book to support picture-based learning (*Body Spectrum Electronic Coloring Book* on the EVOLVE website); and tutorial-guided instruction to advise and guide the user through the text (*A&P Online*). The opportunity to self-assess and monitor progress is provided in all aspects of the package: *Quick Check Questions* throughout the text; multiple-choice questions relevant to the clinical scenarios; tutor-guided learning and specific chapters; and more advance understanding questions that are chapter specific. The multiple choice questions range from assessment of basic principles to assessment of an advanced understanding and clinical relevance. The highlighting of key issues contained in each chapter provides an important revision tool.

Anatomy & Physiology is an invaluable text and provides a unique learning support network for all students studying in the healthcare professions.

Dr John C Atherton
Senior Lecturer in Medical Education and Physiology,
School of Medicine, University of Keele, UK

Dr Helen L Atherton
Lecturer in Nursing, School of Healthcare,
University of Leeds, UK

About the Authors

Gary Thibodeau has been teaching anatomy and physiology for more than three decades. Since 1975, *Anatomy & Physiology* has been a logical extension of his interest and commitment to education. Gary's teaching style encourages active interaction with students, and he uses a wide variety of teaching methodologies—a style that has been incorporated into every aspect of this edition. He is considered a pioneer in the introduction of collaborative learning strategies to the teaching of anatomy and physiology. Recent conferral of Emeritus status in the University of Wisconsin System has provided him with additional time to interact with students and teachers across the country and around the world. His focus continues to be successful student-centered learning—leveraged by text, Web-based, and ancillary teaching materials. Over the years, his success as a teacher has resulted in numerous awards from both students and professional colleagues. Gary is active in numerous professional organizations including the Human Anatomy and Physiology Society (HAPS), The American Association of Anatomists, and the American Association of Clinical Anatomists. His biography is included in numerous publications, including *Who's Who in America; Who's Who in American Education; Outstanding Educators in America; American Men and Women of Science;* and *Who's Who in Medicine and Healthcare.* While earning master's degrees in both zoology and pharmacology, as well as a PhD in physiology, Gary says that he became "fascinated by the connectedness of the life sciences." That fascination has led to this edition's unifying themes that focus on how each concept fits into the "Big Picture" of the human body.

To my parents M.A. Thibodeau and Florence Thibodeau, who had a deep respect for education at all levels and who truly believed that you never give up being a student.

To my wife Emogene, an ever-generous and uncommonly discerning critic, for her love, support, and encouragement over the years.

To my children Douglas and Beth for making it all worthwhile.

Gary A. Thibodeau

Kevin Patton has taught anatomy and physiology to high school, community college, and university students from various backgrounds for twenty-five years. This experience has helped him produce a text that will be easier to understand for all students. He has earned several citations for teaching anatomy and physiology and in 1993 was awarded the Missouri Governor's Award for Excellence in Teaching. "One thing I've learned," says Kevin, "is that most of us learn scientific concepts more easily when we can *see* what's going on." His talent for using imagery to teach is evident throughout this edition, with its expanded visual resources. Kevin's interest in promoting excellence in teaching anatomy and physiology has led him to take an active role in the Human Anatomy and Physiology Society (HAPS), currently serving as President Emeritus. Like Gary, Kevin found that the work that led him to a PhD in vertebrate anatomy and physiology instilled in him an appreciation for the "Big Picture" of human structure and function.

To my family and friends who never let me forget the joys of discovery, adventure, and good humor.

To the many teachers who taught me more by who they were than by what they said.

To my students who help me keep the joy of learning fresh and exciting.

Kevin T. Patton

Contributors

We gratefully acknowledge the following individuals for their contributions to this text: **ED CALCATERRA, BS, MEd,** Instructor, DeSmet Jesuit High School, Creve Coeur, Missouri; **PAUL KRIEGER, MS,** Professor of Biology, Grand Rapids Community College, Grand Rapids, Michigan; **DANIEL J. MATUSIAK, BS, MA, EdD,** Adjunct Professor, St. Charles Community College; **IZAK PAUL, PhD,** Biological Sciences, Mount Royal College, Calgary, Alberta

Reviewers

Mohammed Abbas, PhD, Wayne County Community College; **Laura Anderson,** Elk County Catholic High School; **Bert Atsma,** Union County College; **John Bagdade,** Northwestern University; **Brenda Blackwelder,** Central Piedmont Community College; **Richard Blonna,** William Paterson College; **Claude Boucheix,** INSERM; **Charles T. Brown,** Barton County Community College; **Laurence Campbell,** Florida Southern College; **Roger Carroll,** University of Tennessee School of Medicine; **Pattie Clark,** Abraham Baldwin College; **Richard Cohen,** Union County College; **Harry W. Colvin, Jr.,** University of California–Davis; **Dorwin Coy,** University of North Florida; **Douglas M. Dearden,** General College of University of Minnesota; **Cheryl Donlon,** Northeast Iowa Community College; **J. Paul Ellis,** St. Louis Community College; **Cammie Emory,** Bossier Parish Community College; **Julie Fiez,** Washington University School of Medicine; **Debbie Gantz,** Mississippi Delta Community College; **Becky Gesler,** Spalding University; **Norman Goldstein,** California State University–Hayward; **John Goudie,** Kalamazoo Area Mathematics & Science Center; **Charles J. Grossman,** Xavier University; **Monica L. Hall-Woods,** St. Charles Community College; **Rebecca Halyard,** Clayton State College; **Ann T. Harmer,** Orange Coast College; **Linden C. Haynes,** Hinds Community College; **Lois Jane Heller,** University of Minnesota School of Medicine; **Lee E. Henderson,** Prairie View A&M University; **Paula Holloway,** Ohio University; **Julie Hotz-Siville,** Mt. San Jacinto College; **Gayle Dranch Insler,** Adelphi University; **Patrick Jackson,** Canadian Memorial Chiropractic College; **Carolyn Jaslow,** Rhodes College; **Gloria El Kammash,** Wake Technical Community College; **Murray Kaplan,** Iowa State University; **Kathy Kath,** Henry Ford Hospital School of Radiologic Technology; **Johanna Krontiris-Litowitz,** Youngstown State University; **William Langley,** Butler County Community College; **Clifton Lewis,** Wayne County Community College; **Jerri Lindsey,** Tarrant County Junior College; **Eddie Lunsford,** Southwestern Community College; **Bruce Luxon,** University of Texas Medical Branch; **Melanie S. MacNeil,** Brock University; **Susan Marshall,** St. Louis University School of Medicine; **Gary Massaglia,** Elk County Christian High School; **Jeff Mellenthin,** The Methodist Debakey Heart Center; **Lanette Meyer,** Regis University/Denver Children's Hospital; **Donald Misumi,** Los Angeles Trade–Technical Center; **Susan Moore,** New Hampshire Community Technical College; **Rose Morgan,** Minot State University; **Jeremiah Morrissey,** Washington University School of Medicine; **Robert Earl Olsen,** Briar Cliff College; **Juanelle Pearson,** Spalding University; **Carolyn Jean Rivard,** Fanshawe College of Applied Arts and Technology; **Mary F. Ruh,** St. Louis University School of Medicine; **Jenny Sarver,** Sarver Chiropractic; **Henry M. Seidel, MD,** The Johns Hopkins University School of Medicine; **Gerry Silverstein,** University of Vermont–Burlington; **Charles Singhas,** East Carolina University; **Marci Slusser,** Reading Area Community College; **Paul Keith Small,** Eureka College; **Aleta Sullivan,** Pearl River Community College; **Kathleen Tatum,** Iowa State University; **Reid Tatum,** St. Martin's Episcopal School; **Kent R. Thomas, Ph.D.,** Wichita State University; **Todd Thuma,** Macon College; **Stuart Tsubota,** St. Louis University; **Judith B. Van Liew,** State University of New York College at Buffalo; **Gordon Wardlaw,** Ohio State University; **Cheryl Wiley,** Andrews University; **Clarence C. Wolfe,** Northern Virginia Community College

ANATOMICAL DIRECTIONS

Directional Terms	Definition	Example of Usage
Left	To the left of the body (not *your* left, the subject's)	The stomach is to the *left* of the liver.
Right	To the right of the body or structure being studied	The *right* kidney is damaged.
Lateral	Toward the side; away from the midsagittal plane	The eyes are *lateral* to the nose.
Medial	Toward the midsagittal plane; away from the side	The eyes are *medial* to the ears.
Anterior	Toward the front of the body	The nose is on the *anterior* of the head.
Posterior	Toward the back (rear)	The heel is *posterior* to the head.
Superior	Toward the top of the body	The shoulders are *superior* to the hips.
Inferior	Toward the bottom of the body	The stomach is *inferior* to the heart.
Dorsal	Along (or toward) the vertebral surface of the body	Her scar is along the *dorsal* surface.
Ventral	Along (toward) the belly surface of the body	The navel is on the *ventral* surface.
Caudad (caudal)	Toward the tail	The neck is *caudad* to the skull.
Cephalad	Toward the head	The neck is *cephalad* to the tail.
Proximal	Toward the trunk (describes relative position in a limb or other appendage)	The joint is *proximal* to the toenail.
Distal	Away from the trunk or point of attachment	The hand is *distal* to the elbow.
Visceral	Toward an internal organ; away from the outer wall (describes positions inside a body cavity)	This organ is covered with the *visceral* layer of the membrane.
Parietal	Toward the wall; away from internal structures	The abdominal cavity is lined with the *parietal* peritoneal membrane.
Deep	Toward the inside of a part; away from the surface	The thigh muscles are *deep* to the skin.
Superficial	Toward the surface of a part; away from the inside	The skin is a *superficial* organ.
Medullary	Refers to an inner region, or *medulla*	The *medullary* portion of the organ contains nerve tissue.
Cortical	Refers to an outer region or *cortex*	The *cortical* area produces hormones.
Ipsilateral	On the same side (of the body) as	The left knee is *ipsilateral* to the left ankle.
Contralateral	On the opposite side of the body	The left knee is *contralateral* to the right knee.

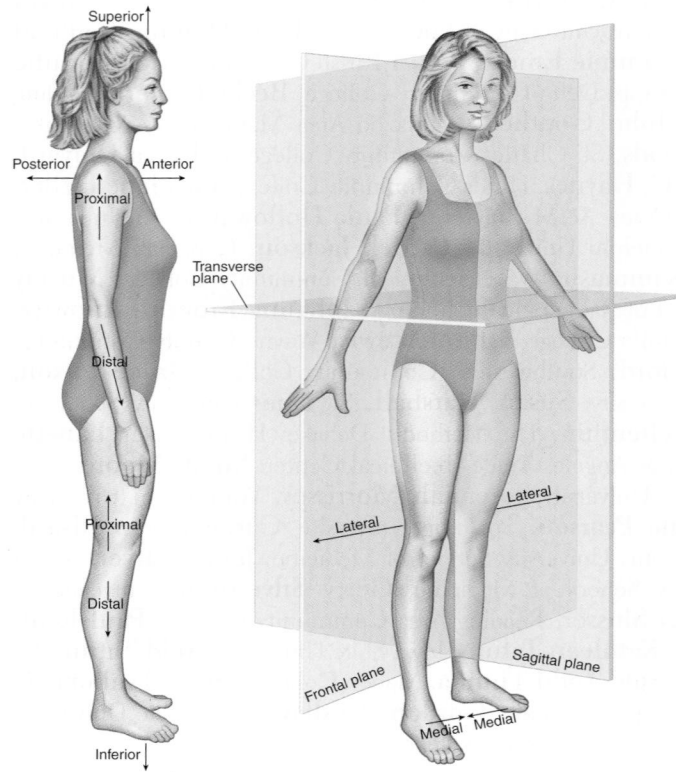

To make the reading of anatomical figures a little easier, an *anatomical compass* is used throughout this book. On many figures, you will notice a small compass rosette similar to those on geographical maps. Rather than being labeled N, S, E, and W, the anatomical rosette is labeled with abbreviated anatomical directions.

A = Anterior
D = Distal
I = Inferior
L (opposite R) = Left
L (opposite M) = Lateral
M = Medial
P (opposite A) = Posterior
P (opposite D) = Proximal
R = Right
S = Superior

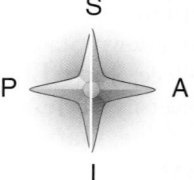

Preface

Success in both teaching and learning is, in many ways, determined by how effective we are in transforming information into knowledge. This is especially true in scientific disciplines, such as anatomy and physiology, where both student and teacher continue to be confronted with an enormous accumulation of factual information. *Anatomy & Physiology* is intended to help transform that information into a manageable knowledge base by effective use of unifying themes and by focusing on the significant and on what is truly relevant in both disciplines.

This textbook is intended for use as both a teaching and a learning tool. It was written to help students unify information, stimulate critical thinking, and hopefully, acquire a taste for knowledge about the wonders of the human body. This is a textbook that will help students avoid becoming lost in a maze of facts in a complex learning environment. It will encourage them to explore, to question, and to look for relationships, not only between related facts in a single discipline, but also between fields of academic inquiry and personal experience. It is our hope that *Anatomy & Physiology* will help both students and teachers transform information into knowledge.

This new edition of the text has been extensively revised. New and improved pedagogical features, a larger and more open page design, many additional full-color illustrations, and the addition of carefully selected new information in both anatomy and physiology provide an accurate and up-to-date presentation. We have retained the basic philosophy of personal and interactive teaching that characterized previous editions. In addition, essential, accurate, and current information continues to be presented in a comfortable writing style. As in previous editions, emphasis is placed on concepts rather than descriptions and the "connectedness" of human structure and function is repeatedly reinforced by unifying themes.

UNIFYING THEMES

Anatomy and physiology encompass a body of knowledge that is large and complex. Students are faced with the need to know and understand a multitude of individual structures and functions that constitute a bewildering array of seemingly disjointed information. Ultimately, the student of anatomy and physiology must be able to "pull together" this information to view the body as a whole—to see the "Big Picture." If a textbook is to be successful as a teaching tool in such a complex learning environment, it must help unify information, stimulate critical thinking, and motivate students to master a new vocabulary.

To accomplish this synthesis of information, unifying themes are required. In addition, a mechanism to position and implement these themes must be an integral part of each chapter. Unit One begins with "Seeing the Big Picture," an overview that encourages students to place individual structures or functions into an integrated and multifunctional framework. Then, throughout the book, the specific information presented is highlighted in a special "Big Picture" section so that it can be viewed as an integral component of a single multifaceted organism.

Anatomy & Physiology is dominated by two major unifying themes: (1) the complementarity of normal structure and function and (2) homeostasis. The student is shown, in every chapter of the book, how organized anatomical structures of a particular size, shape, form, or placement serve unique and specialized functions. The integrating principle of homeostasis is used to show how the "normal" interaction of structure and function is achieved and maintained by dynamic counterbalancing forces within the body. Repeated emphasis of these principles encourages students to integrate otherwise isolated factual information into a cohesive and understandable whole. "The Big Picture" summarizes the larger interaction between structures and functions of the different body systems. As a result, anatomy and physiology emerge as living and dynamic topics of personal interest and importance to students.

AIMS OF THE REVISION

As in past editions, our revision efforts focused on identifying the need for new or revised information and for additional visual presentations that would better serve to clarify important, yet sometimes difficult, content areas.

In this new edition, we have included information on new concepts in cell biology and genetics that have already become important in the practical applications of anatomy and physiology. For example, this edition now features information on the rapidly growing area of RNA interference (RNAi) and its therapeutic applications. We have also included new introductory level information on gene chips, electrocardiography, mucosal immunity, the extracellular matrix (ECM), skin biology, nerve signaling and brain function, hormonal and local regulation, hemodynamics, regulation of digestion, and many other subject areas.

One of the most apparent changes that you will notice in this new edition is the larger width of the pages. This has enabled us to make the page design more open and easier to use. The larger page size also allows us to enlarge some of the figures to make their details easier to see. And, perhaps most importantly, the larger page size has allowed us to add many more illustrations to the book. In nearly every chapter, concepts that in previous editions were discussed but not illustrated are now complemented by images that are carefully designed to make each concept even more clear for the reader—especially the visual learner. A larger array of tables that help students visually organize important concepts complements the expanded art program to provide a multisensory learning tool.

Another of our aims in this edition is to help the students master the language of anatomy and physiology more easily. To this end, we have added a new *Language of Science and Medicine* section to Chapter 1 and have added two new word lists to every chapter of the book. As described later in this preface, both word lists feature a pronunciation guide for each term.

As teachers of anatomy and physiology, we know that to be effective a text must be readable, and it must challenge and excite the student. This text remains one that students will read—one designed to help the teacher teach and the student learn. To accomplish this end, we facilitated the comprehension of difficult material for students with thorough, consistent, and nonintimidating explanations that are free of unnecessary terminology and extraneous information. This easy access to complex ideas remains the single most striking hallmark of our textbook.

Illustrations and Design

A major strength of this text has always been the exceptional quality, accuracy, and beauty of the illustration program. As stated earlier, this edition features a large number of new illustrations and photographs. We worked very closely with professional scientific illustrators to provide attractive and colorful images that clearly and accurately portray the major concepts of anatomy and physiology. Many new photographs and medical images of human body structures have been added to a number of chapters, as well as to the new *Brief Atlas of the Human Body* packaged with this book.

The truest test of any illustration is how effectively it can complement and strengthen the written information in the text and how successfully it can be used by the student as a learning tool. Each illustration is explicitly referred to in the text and is designed to support the text discussion. Careful attention has been paid to placement and sizing of the illustrations to maximize usefulness and clarity. As noted earlier, the larger page size in this new edition permits more and larger illustrations. Each figure and all labels are relevant to and consistent with the text discussion. Each illustration has a boldface title for easy identification. Most illustrations also include a brief explanation that guides the student through the image as a complement to the text narrative.

The artistically drawn full-color artwork is both aesthetically pleasing and functional. Color is used to highlight specific structures in drawings to help organize or highlight complex material in illustrated tables or conceptual flow charts. The text is also filled with dissection photographs, exceptional light micrographs, and scanning (SEM) and transmission (TEM) electron micrographs, some of which are new to this edition. In addition, examples of medical imagery, including CT scans, PET scans, MRIs, and x-ray photographs, are used throughout the text to show structural detail, explain medical procedures, or enhance the understanding of differences that distinguish pathological conditions from normal structure and function. All illustrations used in the text are an integral part of the learning process and should be carefully studied by the student.

LEARNING AIDS

Anatomy & Physiology is a student-oriented text. Written in a readable style, the text is designed with many different pedagogical aids to motivate and maintain interest. The special features and learning aids listed below are intended to facilitate learning and retention of information in the most effective and efficient manner. No textbook can replace the direction and stimulation provided by an enthusiastic teacher to a curious and involved student. However, a full complement of innovative pedagogical aids that are carefully planned and implemented can contribute a great deal to the success of a text as a learning tool. An excellent textbook can and should be enjoyable to read and should be helpful to both student and teacher. We hope you agree that the learning aids in *Anatomy & Physiology* successfully meet the high expectations we have set.

Unit Introductions

Each of the six major units of the text begins with a brief overview statement. The general content of the unit is discussed, and the chapters and their topics are listed. Before beginning the study of material in a new unit, students are encouraged to scan the introduction and each of the chapter outlines in the unit to understand the relationship and "connectedness" of the material to be studied. Each unit has a color-coded tab at the outside edge of every page to help you quickly find the information you need.

Chapter Learning Aids

- **Chapter Outline**—*summarizes the contents of a chapter at a glance.*

An overview outline introduces each chapter and enables the student to preview the content and direction of the chapter at the major concept level before beginning the detailed reading. Page references enable students to quickly locate topics in the chapter.

- **Language of Science**—*introduces you to new scientific terms in the chapter.*

A comprehensive list of new terms is presented at the beginning of the chapter. Each term in the list has an easy-to-use pronunciation guide to help the learner easily "own" the word by being able to say it. Literal translations of the term or its word parts are included to help students learn how to deduce the meaning of new terms themselves. The listed terms are defined in the text body, where they appear in **boldface** type, and may also be found in the glossary at the back of the book. The **boldface** type feature enables students to scan the text for new words before beginning their first detailed reading of the material so they may read without having to disrupt the flow to grapple with new, multisyllabic words or phrases.

The *Language of Science* word list includes terms related to the essential anatomy and physiology presented in the chapter. Another word list near the end of the chapter, a feature described below as the *Language of Medicine,* is an inventory of all the new clinical terms introduced in the chapter.

- **Color-Coded Illustrations**—*help beginning students appreciate the "Big Picture" of human structure and function.*

A special feature of the illustrations in this text is the careful and consistent use of color to identify important structures and substances that recur throughout the book. Consistent use of a color key helps beginning students appreciate the "Big Picture" of human structure and function each time they see a familiar structure in a new illustration. For an explanation of the color scheme, see the color key on page xvi, preceding the Unit One introductory text.

-  **Directional Rosettes**—*help students learn the orientation of anatomical structures.*

Where appropriate, small orientation diagrams and directional rosettes are included as part of an illustration to help students locate a structure with reference to the body as a whole or orient a small structure in a larger view.

-  **Quick Check questions**—*test your knowledge of material just read.*

Short objective-type questions are located immediately following major topic discussions throughout the body of the text. These questions cover important information presented in the preceding section. Students unable to answer the questions should reread that section before proceeding. This feature therefore enhances reading comprehension. New to this edition is a renumbering of the *Quick Check* items by chapter and a numerical listing of their answers on the Evolve site (http://evolve.elsevier.com/thibodeau).

- **Cycle of Life**—*describes major changes that occur over a person's lifetime.*

In many body systems, changes in structure and function are frequently related to a person's age or state of development. In appropriate chapters of the text, these changes are highlighted in this special section.

-  **The Big Picture**—*explains the interactions of the system discussed in a particular chapter with the body as a whole.*

This helps students relate information about body structures or functions that are discussed in the chapter to the body as a whole. *The Big Picture* feature helps you improve critical thinking by focusing on how structures and functions relate to one another on a global basis.

- **Mechanisms of Disease**—*helps you understand the basic principles of human structure and function by showing you what happens when things go wrong.*

Examples of pathology, or disease, are included in many chapters of the book to stimulate student interest and to help students understand that the disease process is a disruption in homeostasis, a breakdown of normal integration of form and function. The intent of the *Mechanisms of Disease* section is to reinforce the normal structures and mechanisms of the body while highlighting the general causes of disorders for a particular body system.

- **Language of Medicine**—*introduces you to new clinical terms in the chapter.*

A brief list of clinical terms is presented near the end of each chapter. As in the *Language of Science* list at the beginning of the chapter, each term has a phonetic pronunciation guide and translations of word parts. The listed terms are defined in the text body, where they appear in **boldface** type.

-  **Case Study**—*challenges you with "real-life" clinical situations so you can creatively apply what you have learned.*

Every chapter has a case study preceding the chapter summary. The case study consists of a description of a clinical situation and a series of multiple-choice questions that require the student to use critical thinking skills to determine the answers.

- **Chapter Summary**—*outlines essential information in a way that helps you organize your study.*

Detailed end-of-chapter summaries provide excellent guides for students as they review the text materials before examinations. Many students also find the summaries to be useful as a chapter preview in conjunction with the chapter outline.

- **Review Questions**—*help you determine whether you have mastered the important concepts of each chapter.*

Review questions at the end of each chapter give students practice in using a narrative format to discuss the concepts presented in the chapter.

- **Critical Thinking Questions**—*actively engage and challenge you to evaluate and synthesize the chapter content.*

Critical thinking questions require students to use their higher-level reasoning skills and demonstrate their understanding of, not just their repetition of, complex concepts.

Answers for all of the case study questions and also the review and critical thinking questions are in the *Instructor's Resource Manual* and *Instructor's Electronic Resource (CD-ROM)* that accompanies *Anatomy & Physiology.* Teachers can then choose to use the questions as homework assignments or include them on tests.

Boxed Information

As always, we made every effort to update factual information and to incorporate the most current anatomy and physiology research findings in this edition. Although there continues to be an incredible explosion of knowledge in the life sciences, not all new information is appropriate for inclusion in a fundamental level textbook. Therefore we were selective in choosing new clinical, pathological, or special-interest material to include in this edition. This text remains focused on normal anatomy and physiology. The addition of new boxed content is intended to stimulate student interest and provide examples that reinforce the immediate personal relevance of anatomy and physiology as important disciplines for study.

Each of the various categories of boxed information has a distinct design and color scheme to make it easily recognizable.

- **Generic Boxes**—*provide an expanded explanation of specific chapter content.*

Many chapters contain boxed essays, occasionally clinical in nature, that expand on or relate to material covered in the text. Examples of subjects include the RNA revolution, inflammation, and the enteric nervous system.

-  **Health Matters**—*present current information on diseases, disorders, treatments, and other health issues related to normal structure and function.*

These boxes contain information related to health issues or clinical applications. In some instances, examples of structural anomalies or pathophysiology are presented. Information of this type is often useful in helping students understand the mechanisms involved in maintaining the "normal" interaction of structure and function.

- **Diagnostic Study**—*keep you abreast of developments in diagnosing diseases and disorders.*

These boxes deal with specific diagnostic tests used in clinical medicine or research. Specific blood chemistry analysis, medical imagery, and ECG interpretation are examples.

-  **FYI**—*give you more in-depth information on interesting topics mentioned in the text.*

Topics of current interest such as new advances in anatomy and physiology research are covered in these "for your information" boxes.

-  **Sports and Fitness**—*highlight sports-related topics.*

Exercise physiology, sports injury, and physical education applications are highlighted in these boxes.

- **Career Choices**—*highlight individuals in health-related careers.*

A *Career Choices* box appears at the end of each unit. These boxes describe some of the diverse opportunities currently available in health-related occupations and also demonstrate the importance of how an understanding of anatomy and physiology will be useful to students in their futures.

Glossary

A comprehensive glossary of terms is located at the end of the text. Accurate, concise definitions and phonetic pronunciation guides are provided. An audio glossary is also available on the Evolve site—**http://evolve.elsevier.com/thibodeau**—with definitions and audio pronunciations for most of the key terms in the text.

LEARNING SUPPLEMENTS FOR STUDENTS
Brief Atlas of Human Anatomy

A new full-color *Brief Atlas of Human Anatomy* containing cadaver dissections, osteology, organ casts, histology specimens, and surface anatomy photographs is packaged with every copy of this edition of *Anatomy & Physiology*. This new work serves as a handy reference for students as they study the human body in class and in the laboratory—and even later on in clinical and career contexts.

Clear View of the Human Body

We are particularly excited to present a full-color, semi-transparent model of the body called the *Clear View of the Human Body.* Found near the end of Chapter 1, this feature permits the virtual dissection of male and female human bodies along several different planes of the body. Developed by Kevin Patton and Paul Krieger, this new type of tool helps learners assimilate their knowledge of the complex structure of the human body. It also provides a unique learning tool that helps students visualize human anatomy in the manner of today's clinical and athletic body imaging technology.

EVOLVE—http://evolve.elsevier.com/thibodeau

This new edition of *Anatomy & Physiology* is supported by an expanded multimedia EVOLVE website, featuring:

- Answers to all of the **Quick Check** questions found in the textbook.

- **Online appendices,** including:
  - Metric Measurements and Their Equivalents
  - Determining the Potential Osmotic Pressure of a Solution
  - Clinical and Laboratory Values
  - Conversion Factors to International System of Units (SI Units)
  - Nobel Prize Winners in Medicine or Physiology

- **Online Tutoring** offers you one-on-one expert assistance from an experienced mentor.

- An interactive **AudioGlossary,** with definitions and pronunciations for over 1000 key terms from the textbook.

- The **Body Spectrum** electronic coloring book, which offers dozens of anatomy illustrations that can be colored online or printed out and colored by hand.

- Over 500 **Student Post-Test** questions that allow you to get instant feedback on what you've learned in each chapter.

- State-of-the-art 3-D animations, which show and describe physiological processes by body system.

- **WebLinks** provide students with access to hundreds of important sites simply by clicking on a subject in the book's table of contents.

- **Expanded index** with over 10,000 citations and page references.

- Access to your **eBook** (see information below and at the back of this textbook).

You can visit the EVOLVE site by pointing your browser to **http://evolve.elsevier.com/thibodeau.**

eBook

By using the pin code found at the back of this book, you can visit the Evolve site and download an electronic version of *Anatomy & Physiology.* Powered by a reader that allows you to search the entire textbook by keyword or topic, and take notes electronically, you now have the tools to create your own custom study guide.

Survival Guide for Anatomy & Physiology

The *Survival Guide for Anatomy & Physiology*, written by Kevin Patton, is an easy-to-read and easy-to-understand brief handbook to help you achieve success in your anatomy and physiology course. Read with greater comprehension using the 10 survival skills, study more effectively, prepare for tests and quizzes, and tap into all of the information resources at your disposal. It also includes a Quick Reference filled with illustrations, tables, and diagrams that convey all of the important facts and concepts students need to know to succeed in an anatomy and physiology course.

Study Guide

The *Study Guide*, written by Linda Swisher, is a valuable student workbook that provides the reinforcement and practice necessary for students to succeed in their study of anatomy and physiology. Important concepts from the text are reinforced through *Concept Reviews*, organized by objectives and referenced to the text. *Clinical Challenges* apply the material to "real-life" situations. Matching, completion, and illustration labeling exercises are provided for every chapter.

TEACHING SUPPLEMENTS FOR INSTRUCTORS

Instructor's Electronic Resource CD-ROM

The *Instructor's Electronic Resource CD-ROM* was written and developed specifically for this new edition of *Anatomy & Physiology*. It includes an **Instructor's Resource Manual** (also available in a print version) for the textbook with critical thinking questions, learning objectives and activities, teaching tips for the text and CD, synopses of difficult concepts, and clinical applications exercises. The new CD-ROM also includes a **Computerized Test Bank** with almost 6000 multi-choice, true/false, short answer, and challenge questions (which you can also import into your **Classroom Performance System** to quickly assess student comprehension and monitor your classroom's response).

The IER also features a separate CD-ROM that contains an **Electronic Image Collection** to accompany *Anatomy & Physiology*, featuring hundreds of full-color illustrations and photographs, with labels that you can turn off and on.

ACKNOWLEDGMENTS

Over the years, many people have contributed to the development and success of *Anatomy & Physiology*. We extend our thanks and deep appreciation to all the students and classroom instructors who have provided us with helpful suggestions. We also thank the many contributors who have, over the last several editions, provided us with extraordinary insights and useful features that we have added to our textbook.

Dan Matusiak and Izak Paul helped us produce our new word lists, a huge task, and we appreciate their help. Paul Krieger helped us design the *Clear View of the Human Body*, for which we are grateful. We thank the late Stanley L. Erlandsen and Jean E. Magney for allowing us to borrow a number of wonderful histology micrographs from their book *Color Atlas of Histology*. Thanks also to Jean Proehl, Monteo Myers, Mark Alderman, Dominic Steward, Craig Huard, and Christina Zaleski for their insights in the Career Choices boxes.

To those at Elsevier who put their best efforts into producing this edition, we are indebted. This new edition, and its expanded library of ancillary resources, would not have been possible without the nearly superhuman efforts of Tom Wilhelm, Executive Editor, and Jeff Downing, Managing Editor. In addition, we are grateful to Sally Schrefer, Executive Vice President, and Robin Carter, Executive Publisher, for their continuing guidance and support. And where the rubber meets the road, we were fortunate to have a wonderful team of professionals working with us to keep it all on track and moving along: Jennifer Stoces, Associate Developmental Editor; Carlie Bliss, Editorial Assistant; Deborah Vogel, Publishing Services Manager; Steve Ramay, Senior Project Manager; Mark Oberkrom, Designer; and Tricia Schroeder, Marketing Manager.

Gary A. Thibodeau
Kevin T. Patton

Contents

COLOR KEY

BIOCHEMISTRY

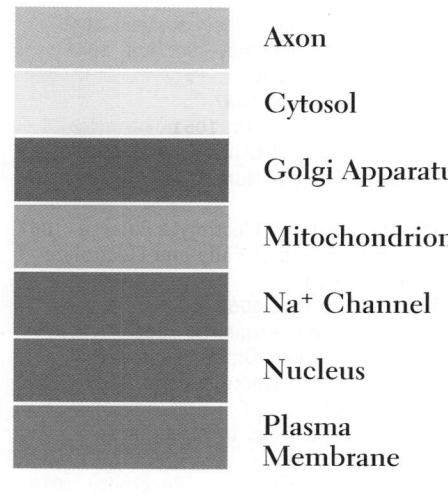

Carbon

Chloride

Energy (ATP)

Hydrogen

Nitrogen

Oxygen

Potassium

Sodium

Sulfur

CELLULAR STRUCTURES

Axon

Cytosol

Golgi Apparatus

Mitochondrion

Na⁺ Channel

Nucleus

Plasma Membrane

OTHER STRUCTURES

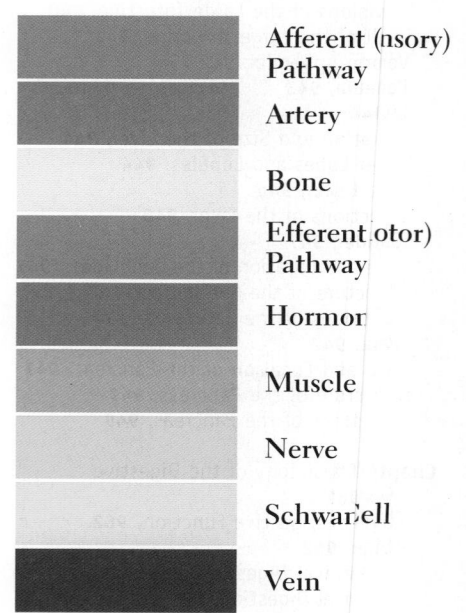

Afferent (nsory) Pathway

Artery

Bone

Efferent otor) Pathway

Hormor

Muscle

Nerve

Schwar ell

Vein

The Body as a Whole

The five chapters in Unit 1 "set the stage" for the study of human anatomy and physiology. They provide the unifying information required to understand the "connectedness" of human structure and function and to understand how organized anatomical structures of a particular size, shape, form, or placement serve unique and specialized functions. Note that the illustration selected to introduce this unit shows the body not as an assemblage of isolated parts but as an integrated whole.

In Chapter 1 the concept of levels of organization in the body is presented, and the unifying theme of homeostasis is introduced to explain how the interaction of structure and function at chemical, organelle, cellular, tissue, organ, and system levels is achieved and maintained by dynamic counterbalancing forces within the body. The material presented in Chapter 2—The Chemical Basis of Life—provides an understanding of the basic chemical interactions that influence the control, integration, and regulation of these counterbalancing forces in every organ system of the body.

Unit 1 concludes with information that builds on the organizational and biochemical information presented in the first two chapters. The structure and function of cells presented in Chapters 3 and 4 explain why physiologists often state that "all body functions are cellular functions." Grouping similar cells into functioning tissues is accomplished in Chapter 5. Subsequent chapters of the text will focus on the remaining organ systems of the body.

SEEING THE BIG PICTURE

SEEING THE BIG PICTURE

Before reading this introduction, you probably spent a few minutes flipping through this book. Naturally, you are curious about your course in human anatomy and physiology, and you wanted to see what lay ahead. It is more than that. You are curious about the human body—about yourself, really. We all have that desire to learn more about how our bodies are put together and how all the parts work. Unlike many other people, though, you now have the opportunity to gain an understanding of the underlying scientific principles of human structure and function.

To truly understand the nature of the human body requires an ability to appreciate "the parts" and "the whole" at the same time. As you flipped through this book for the first time, you probably looked at many different body parts. Some were microscopic—such as muscle cells—and some were very large—such as arms and legs. In looking at these parts, however, you gained very little insight about how they worked together to allow you to sit here, alive and breathing, and read and comprehend these words.

Think about it for a moment. What does it take to be able to read these words and understand them? You might begin by thinking about the eye. How do all of its many intricate parts work together to form an image? The eye is not the only organ you are using right now. What about the bones, joints, and muscles you are using to hold the book, to turn the pages, and to move your eyes as they scan this paragraph? Let's not forget the nervous system. The brain, spinal cord, and nerves are receiving information from the eyes, evaluating it, and using it to coordinate the muscle movements. The squiggles we call letters are being interpreted near the top of the brain to form complex ideas. In short, you are *thinking* about what you are reading.

However, we still have not covered everything. How are you getting the energy to operate your eyes, muscles, brain, and nerves? Ener-getic chemical reactions inside each cell of these organs require oxygen and nutrients captured by the lungs and digestive tract and delivered by the heart and blood vessels. These chemical reactions produce wastes that are handled by the liver, kidneys, and other organs. All these functions must be coordinated, a feat accomplished by regulation of body organs by hormones, nerves, and other mechanisms.

Learning to name the various body parts, to describe their detailed structure, and to explain the mechanisms underlying their functions is an essential step that leads to the goal of understanding the human body. To actually reach that goal, however, you must be able to draw together isolated facts and concepts. In other words, understanding the nature of individual body parts is meaningless if you do not understand how the parts work together in a living, *whole* person.

Many textbooks are written like reference books—encyclopedias, for example. They provide detailed descriptions of the structure and function of individual body parts, often in logical groupings, while rarely stopping to step back and look at the whole person. In this book, however, we have incorporated the "whole body" aspect into our discussion of every major topic. In chapter and unit introductions, in appropriate paragraphs within each section, and in specific sections near the end of each chapter, we have stepped back from the topic at hand and refocused attention to the broader view.

We are confident that our "whole body" approach will help you put each new fact or concept you learn into its proper place within a larger framework of understanding. You may also better appreciate why it is important to learn some detailed facts that may at first seem to have no practical value to you. When you have finished learning the many details covered in this course, however, you will have also gained a more complete understanding of the essential nature of the human body.

SEEING THE BIG PICTURE—cont'd

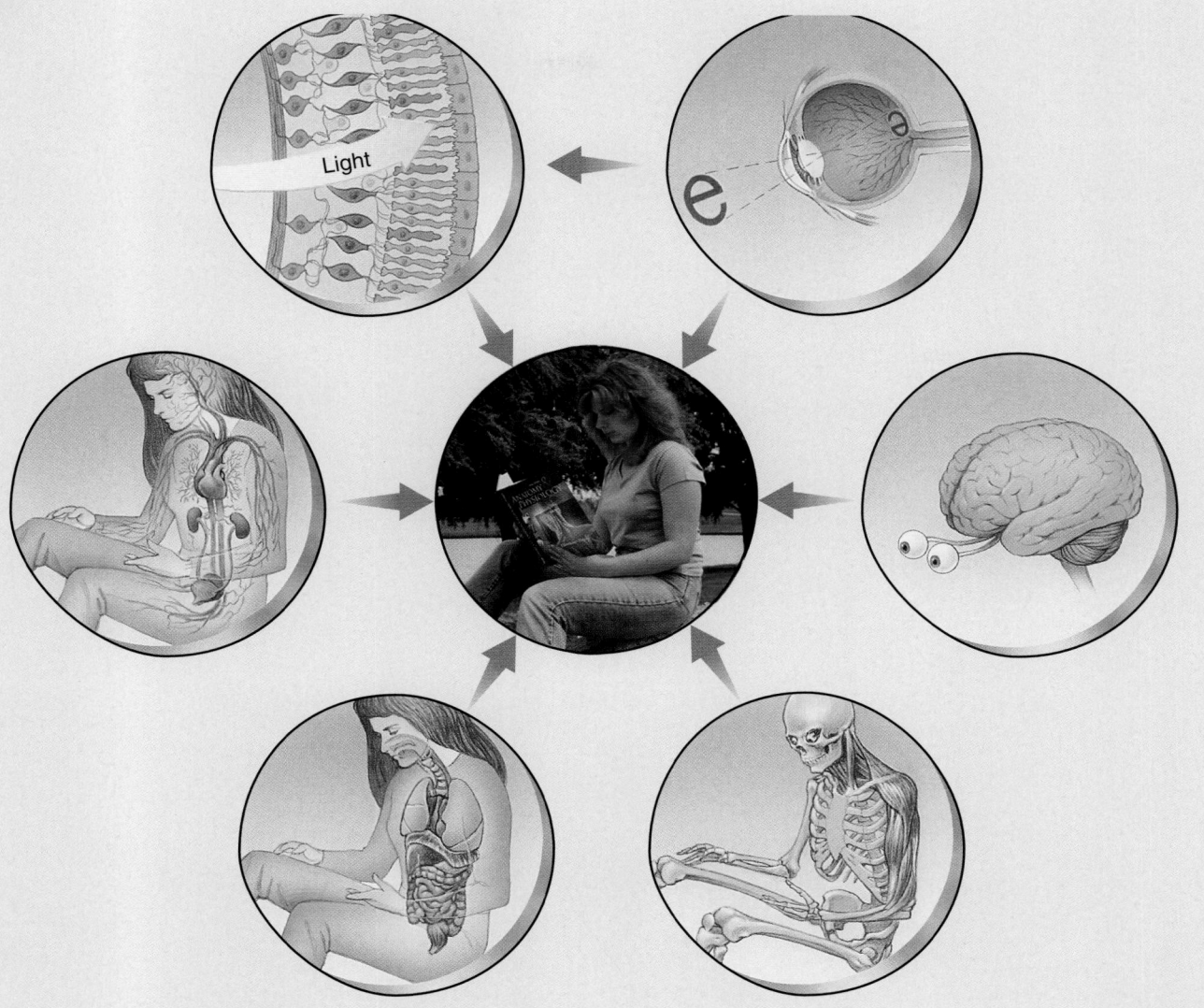

The "big picture." What does it take for your body to sit and read these words? The coordination of your muscles, bones, and ligaments is needed to hold the book in your hands, to sit up without falling, and to hold your head and eyes in position. Your brain and nerves process the image of the words received by the eyes. Your digestive and other organs provide the nutrients—and thus the energy—needed by the muscles, bones, brain, and other organs. The urinary and respiratory systems remove the wastes produced by this conversion of food to energy. You can see that it takes a coordinated effort by nearly every small part of your body to simply read this book! This is the *big picture*—putting all your knowledge of bits and pieces into the larger picture of understanding how the body works as a whole.

CHAPTER 1

Organization of the Body

LANGUAGE OF SCIENCE

abdominopelvic cavity (ab-DOM-i-no-PEL-vik KAV-ih-tee) [abdomin- belly, -pelv- basin, cav cavity]

anatomical position (an-ah-TOM-i-kal po-ZISH-un) [ana- not or without, -tom- cut, posit place]

anatomy (ah-NAT-o-mee) [ana- not or without, -tomy a cut]

anterior (an-TEER-ee-or) [anter front]

apical (AY-pik-al) [apic tip]

autopoiesis (aw-toe-poy-EE-sis) [auto- self, -poiesis making]

basal (BAY-sal) [bas foundation]

bilateral symmetry (bye-LAT-er-al SIM-e-tree) [bi- two, -lateral side, sym- union, -metr- measure]

body planes (BOD-ee playnz)

cadavers (kah-DAV-erz) [cadaver dead body]

cells (sellz) [cell storeroom]

cell theory (sell THEE-o-ree) [cell storeroom, theor look at]

central (SEN-tral) [centr center]

cortical (KOR-tik-al) [cortic bark]

coronal plane (ko-RO-nal plane) [corona crown]

deep

distal (DIS-tal) [dista distant]

dorsal cavity (DOR-sal KAV-ih-tee) [dors back, cav cavity]

eponyms (EP-o-nimz) [epo- above, -nym name]

extrinsic control (eks-TRIN-sik kon-TROL) [extr- outside or beyond, -insic beside]

feedback control loop

homeostasis (ho-me-o-STAY-sis) [homeo- same or equal, -stasis standing still]

hypothesis (hye-POTH-eh-sis) [hypo- under or below, -thesis placing or proposition]

inferior (in-FEER-ee-or) [infer lower]

intracellular control (in-tra-SELL-yoo-lar kon-TROL) [intra- inside or within, -cell storeroom]

intrinsic control (in-TRIN-sik kon-TROL) [intr- inside or within, -insic beside]

lateral (LAT-er-al) [later side]

lumen (LOO-men) [lumen light] pl. lumina

Cont'd on p. 32

You have just begun the study of one of nature's most wondrous structures—the human body. **Anatomy** (ah-NAT-oh-mee) and **physiology** (fiz-ee-OL-oh-jee) are branches of biology that are concerned with the form and functions of the body. Anatomy is the study of body structure, whereas physiology deals with body function—that is, how the body parts work to support life. As you learn about the complex interdependence of structure and function in the human body, you become, in a very real sense, the subject of your own study.

Regardless of your field of study or your future career goals, acquiring and using information about your body structure and functions will enable you to live a more knowledgeable, involved, and healthy life in this science-conscious age. Your study of anatomy and physiology provides a unique and fascinating understanding of self, and this knowledge allows for more active and informed participation in your own personal health care decisions. If you are pursuing a health-related or athletic career, your study of anatomy and physiology takes on added significance. It provides the necessary concepts you will need in your professional courses and clinical experiences.

SCIENCE AND SOCIETY

Before we get to the details, we should emphasize that everything you will read in this book is in the context of a broad field of inquiry called *science*. Science is a style of inquiry that attempts to understand nature in a rational, logical manner. Using detailed observations and vigorous tests, or *experiments*, scientists weed out each idea or **hypothesis** (hye-PAHTH-es-iss) until a reasonable conclusion can be made. Rigorous experiments that eliminate any influences or biases not being directly tested are called *controlled* experiments. If the results of observations and experiments are repeatable, they may verify a hypothesis and eventually lead to enough confidence in the concept to call it a *theory*. Theories in which scientists have an unusually high level of confidence are sometimes called *laws*. Experiments may disprove a hypothesis, a result that often leads to the formation of new hypotheses (hye-PAHTH-es-eez) to be tested.

Figure 1-1 summarizes some of the basic concepts of how new scientific principles are developed. As you can see, science is a dynamic process of getting closer and closer to the truth about nature, including the nature of the human body. Science is definitely not the set of unchanging facts that many people in our culture often assume.

We should also take this opportunity to point out the social and cultural context of the science presented in this book. Scientists drive the process of science, but our culture drives the kinds of questions we ask about nature and how we attempt to answer them. For example, cutting apart human *cadavers* (dead bodies) has not always been an acceptable activity in all cultures. Today we are faced with a debate in our culture about the acceptability of using live animals in scientific experiments. Because our culture does not accept most experiments with living humans, we have until now often done our tests with animals that are similar to humans. In fact, most of the theories presented in this book are based on animal experimentation, but cultural influences are pulling scientists in experimental directions they otherwise may not have taken.

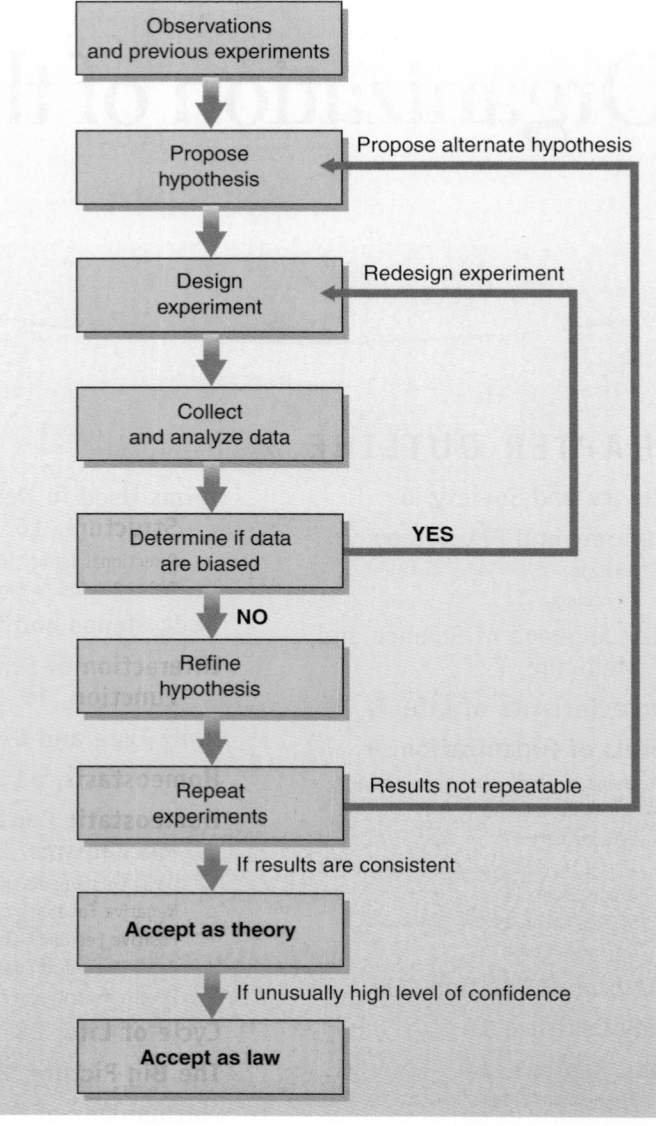

Figure 1-1 *The scientific method.* This flow chart summarizes the classic ideal of how new principles of science are developed. Initial observations or results from other experiments may lead to the formation of a new hypothesis. As more testing is performed to eliminate outside influences or biases and ensure consistent results, scientists begin to have more confidence in the principle and call it a theory or law.

Similarly, science affects culture. Recent advances in understanding human genes and technological advances in our ability to use tissues from human embryos to treat devastating diseases have sparked new debates in how our culture defines what it means to be a human being.

As you study the concepts presented in this book, keep in mind that they are not set in stone. Science is a rapidly changing set of ideas and processes that not only is influenced by our cultural biases but also affects our cultural awareness of who we are.

Before you move on in this chapter, take a moment to go online and visit http://nobelprize.org/medicine. There you will find a table listing most of the Nobel Prizes awarded in the last century in the area the Nobel Committee calls "medicine or physiology." How many women do you see? When they do appear on the list, are they nearer the beginning or the end of the list? Do you think this fact reflects a changing cultural framework for science? Look at the kinds of discoveries for which the prizes were awarded. Do you think these discoveries have affected our culture?

ANATOMY AND PHYSIOLOGY

Anatomy

Anatomy is often defined as study of the structure of an organism and the relationships of its parts. The word anatomy is derived from two Greek words (*ana*, "up," and *temos* or *tomos*, "cutting"). Students of anatomy still learn about the structure of the human body by literally cutting it apart. This process, called *dissection*, remains a principal technique used to isolate and study the structural components or parts of the human body.

Biology is defined as the scientific study of life. Both anatomy and physiology are subdivisions of this very broad area of inquiry. Just as biology can be subdivided into specific areas for study, so can anatomy and physiology. For example, the term *gross anatomy* is used to describe the study of body parts visible to the naked eye. Before discovery of the microscope, anatomists had to study human structure with only the eye during dissection. These early anatomists could make only a *gross*, or whole, examination, as you can see in Figure 1-2. With the use of modern microscopes, many anatomists now specialize in *microscopic anatomy*, including the study of cells, called *cytology* (sye-TOL-oh-jee), and tissues, called *histology* (his-TOL-oh-jee). Other branches of anatomy include the study of human growth and development (*developmental anatomy*) and the study of diseased body structures (*pathological anatomy*). In the chapters that follow you will study the body by systems—a process called *systemic anatomy*. Systems are groups of organs that have a common function, such as the bones in the skeletal system and the muscles in the muscular system.

Physiology

Physiology is the science that deals with the functions of the living organism and its parts. The term is a combination of two Greek words (*physis*, "nature," and *logos*, "words or study"). Simply stated, it is the study of physiology that helps us understand how the body works. Physiologists attempt to discover and understand the intricate control systems that permit the body to operate and survive in an often hostile environment. As a scientific discipline, physiology can be subdivided according to (1) the type of organism involved, such as human physiology or plant physiology; (2) the organizational level studied, such as molecular or cellular physiology; or (3) a specific or *systemic* function being studied, such as neurophysiology, respiratory physiology, or cardiovascular physiology.

In the chapters that follow, both anatomy and physiology are studied by specific organ systems. This unit begins with an overview of the body as a whole. In subsequent chapters the body is dissected and studied, both structurally (anatomy) and functionally (physiology), into "levels of organization" so that its component parts can be more easily understood and then "fit together" into a living and integrated whole. It is knowledge of

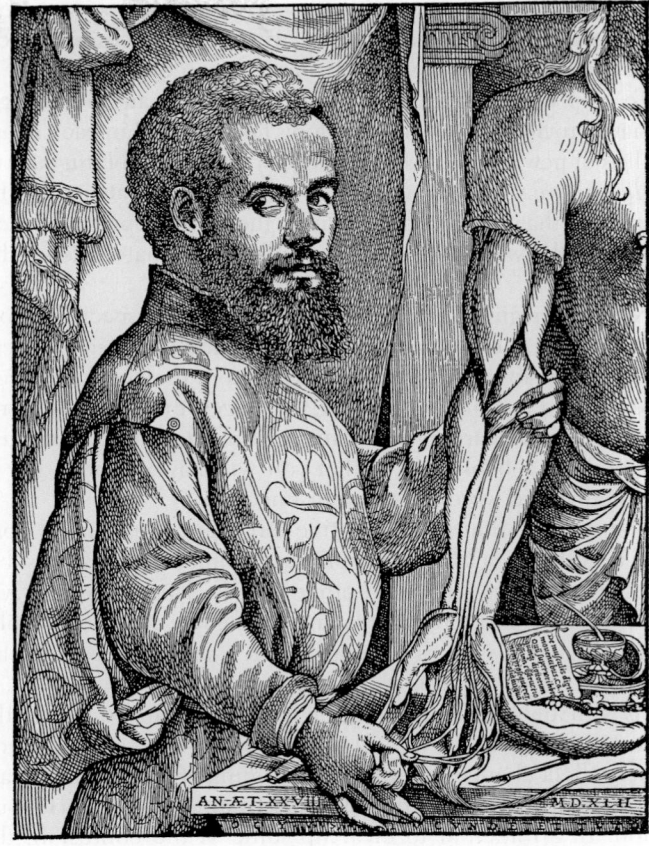

Figure 1-2 *Gross anatomy.* This famous woodcut of a gross dissection appeared in the world's first anatomy textbook, *De Corporis Humani Fabrica (On the Structure of the Human Body)* in 1543. This woodcut features the book's author, Andreas Vesalius, who is considered to be the founder of modern anatomy. The body being dissected is called a *cadaver*.

anatomy and physiology that allows us to understand how nerve impulses travel from one part of the body to another; how muscles contract; how light energy can be transformed into visual images; how we breathe, digest food, reproduce, excrete wastes, and sense changes in our environment; and even how we think and reason.

 QUICK CHECK

1. Describe how science develops new principles.
2. Define anatomy and physiology.
3. List the three ways in which physiology can be subdivided as a scientific discipline.
4. What name is used to describe the study of the body that considers groups of organs that have a common function?

THE LANGUAGE OF SCIENCE AND MEDICINE

You may have noticed by now that many scientific terms, such as anatomy and physiology, are made up of non-English word parts. Many such terms make up the core of the language used to communicate ideas in science and medicine. Learning in science thus begins with learning a new vocabulary, just as it does when you learn a new language as you visit a new region of the world.

To help you learn the vocabulary of anatomy and physiology, we have provided several helpful tools for you. Within each chapter, lists of new terms titled *Language of Science* and *Language of Medicine* give you each new key (boldface) term that you will be learning in that chapter. Each term in the list has a pronunciation guide and the word parts (and their meanings) that make up the term.

Near the beginning of the book, there is also a special removable section called Language of Science and Medicine. Take a moment now to locate it and remove it from the book. After you have finished reading this chapter, we suggest that you quickly review the tips for learning and using scientific language. Then use the insert as a bookmark so that you will have a handy list of commonly used word parts at your fingertips.

You will see that most scientific terms are made up of word parts from Latin or Greek. That is because most scientists first began corresponding with each other in these languages, as it was commonly the first foreign language learned by educated people. Other languages such as German, French, and Japanese are also sources of some scientific word parts.

As with any language, scientific language changes constantly. This is useful because we often need to fine-tune our terminology to reflect changes in our understanding of science and to accommodate new discoveries. But it also sometimes leads to confusion. In an attempt to clear up some of the confusion, the International Federation of Associations of Anatomists (IFAA) formed a worldwide committee to publish a list of "universal" anatomical terminology. The list was published in 1998 as *Terminologia Anatomica*. Although there remain some alternate (and newer) terms used in anatomy, the list is a useful standard reference. *Terminologia Anatomica* lists each term in Latin and English (based on the Latin form), along with a reference number. In our textbook we have used the English terms from this list as our standard reference, but we do occasionally refer to the pure Latin form or an alternate term when appropriate for beginning students.

One of the basic principles of the *Terminologia Anatomica* is the avoidance of **eponyms** (EP-oh-nimz), or terms that are based on a person's name. Instead, a more descriptive Latin-based term is always preferred. Thus the term *eustachian tube* (named after the famed Italian anatomist Eustachius) is now replaced with the more descriptive *auditory tube*. Likewise, the *islets of Langerhans* are now simply *pancreatic islets*.

Although the IFAA plans to eventually produce similar standard lists of terminology related to cells, tissues, and embryological development, there are no standard lists of physiological terms. However, many principles used in anatomical terminology are used in physiology. For example, most terms are in English but are based on Latin or Greek word parts. And, as in anatomy, eponyms are less favored than descriptive terms.

If you focus on learning the new words as you begin each new topic, as though you are in a foreign land and need to pick up a few phrases to get by, you will find your study of anatomy and physiology easy and enjoyable.

CHARACTERISTICS OF LIFE

Anatomy and physiology are important disciplines in biology—the study of life. But what is life? What is the quality that distinguishes a vital and functional being from a dead body? We know that a living organism is endowed with certain characteristics not associated with inorganic matter. However, it is sometimes hard to find a single criterion to define life. One could say that living organisms are self-organizing or self-maintaining and nonliving structures are not. This concept is called **autopoiesis** (aw-toh-poy-EE-sis), which literally means "self making." Another idea, called the **cell theory,** states that any independent structure made up of one or more microscopic units called *cells* is a living organism.

Instead of trying to find a single difference that separates living and nonliving things, scientists sometimes define life by listing what are often called *characteristics of life*. Lists of characteristics of life may differ from one physiologist to the next, depending on the type of organism being studied and the way in which life functions are grouped and defined. Attributes that characterize life in bacteria, plants, or animals may vary.

Characteristics of life that are considered most important in humans are described as follows:

Responsiveness. Responsiveness or irritability is that characteristic of life that permits an organism to sense, monitor, and respond to changes in its external environment. Withdrawing from a painful stimulus, such as a pinprick, is an example of responsiveness.

Conductivity. Conductivity refers to the capacity of living cells and tissues to selectively transmit or propagate a wave of excitation from one point to another within the body. Responsiveness and conductivity are highly developed in nerve and muscle cells in living organisms.

Growth. Growth occurs as a result of a normal increase in the size or number of cells. In most instances, it produces an increase in size of the individual, or a particular organ or part, but little change in the shape of the organism as a whole or the part affected.

Respiration. Respiration involves processes that result in the absorption, transport, utilization, or exchange of respiratory gases (oxygen and carbon dioxide) between an organism and its environment. The exchange of gases may occur between the blood and individual body cells (internal respiration) as the cells use nutrients to produce energy or between the blood and air in the lungs (external respiration).

Digestion. Digestion is the process by which complex food products are broken down into simpler substances that can be absorbed and used by individual body cells.

Absorption. Absorption refers to the movement of digested nutrients through the wall of the digestive tube into the body fluids for transport to cells for use.

Secretion. Secretion is the production and delivery of specialized substances, such as digestive juices and hormones, for diverse body functions.

Excretion. Excretion refers to the removal of waste products produced during many body functions, including the breakdown and use of nutrients in the cell. Carbon dioxide is a gaseous waste that is excreted during respiration.

Circulation. Circulation refers to the movement of body fluids and many other substances, such as nutrients, hormones, and waste products, from one body area to another.

Reproduction. Reproduction involves the formation of a new individual and also the formation of new cells (through cell division) in the body to permit growth, wound repair, and replacement of dead or aging cells on a regular basis.

Each characteristic of life is related to the sum total of all the physical and chemical reactions occurring in the body. The term **metabolism** is used to describe these various processes. They include the steps involved in the breakdown of nutrient materials to produce energy and the transformation of one material into another. For example, if we eat and absorb more sugar than needed for the body's immediate energy requirements, it is converted into an alternate form, such as fat, that can be stored in the body. Metabolic reactions are also required for making complex compounds out of simpler ones, as in tissue growth, wound repair, or manufacture of body secretions.

Each characteristic of life—its functional manifestation in the body, its integration with other body functions and structures, and its mechanism of control—is the subject of study in subsequent chapters of the text.

 QUICK CHECK

5. What is an *eponym*?
6. What single criterion might be used to define life?
7. Define the term *metabolism* as it applies to the characteristics of life.

LEVELS OF ORGANIZATION

Before you begin the study of the structure and function of the human body and its many parts, it is important to think about how the parts are organized and how they might logically fit together and function effectively. The differing levels of organization that influence body structure and function are illustrated in Figure 1-3.

Chemical Level—Basis for Life

Note that organization of the body begins at the chemical level (see Figure 1-3). There are more than 100 different chemical building blocks of nature, called *atoms*—tiny spheres of matter so small they are invisible. Every material thing in our universe, including the human body, is composed of atoms. Combinations of atoms form larger chemical groupings, called *molecules*. Molecules, in turn, often combine with other atoms and molecules to form larger and more complex chemicals, called *macromolecules*.

The unique and complex relationships that exist between atoms, molecules, and macromolecules in living material form a gel-like material made of fluids, particles, and membranes called *cytoplasm* (SYE-toh-plaz-em)—the essential material of human life. Unless proper relationships between chemical elements are maintained, death results. Maintaining the type of chemical or-

ganization in cytoplasm required for life requires the expenditure of energy. In Chapter 2, important information related to the chemistry of life is discussed in more detail.

Organelle Level

Chemical structures may be organized within larger units called *cells* to form various structures called **organelles** (or-gah-NELLZ), the next level of organization (see Figure 1-3). An organelle may be defined as a structure made of molecules organized in such a way that it can perform a specific function. Organelles are the "tiny organs" that allow each cell to live. Organelles cannot survive outside the cell, but without organelles a cell could not survive either.

Dozens of different kinds of organelles have been identified. A few examples include *mitochondria* (my-toe-KON-dree-ah), the "power houses" of cells that provide the energy needed by the cell to carry on day-to-day functioning, growth, and repair; *Golgi* (GOL-jee) *apparatus*, which provides a "packaging" service to the cell by storing material for future internal use or for export from the cell; and *endoplasmic reticulum*, the network of transport channels within the cell that act as "highways" for the movement of chemicals. Chapter 3 contains a complete discussion of organelles and their functions.

Cellular Level

The characteristics of life ultimately result from a hierarchy of structure and function that begins with the organization of atoms, molecules, and macromolecules. Continued organization resulting in organelles is the next step. However, in the view of the anatomist, the most important function of the chemical and organelle levels of organization is to furnish the basic building blocks and specialized structures required for the next higher level of body structure—*the cellular level*. **Cells** are the smallest and most numerous structural units that possess and exhibit the basic characteristics of living matter. How many cells are there in the body? One estimate places the number of cells in a 150-pound adult human body at 100,000,000,000,000. In case you cannot translate this number—1 with 14 zeroes after it—it is 100 trillion! or 100,000 billion! or 100 million million!

Each cell is surrounded by a membrane and is characterized by a single nucleus surrounded by cytoplasm that includes the numerous organelles required for specialized activity. Although all cells have certain features in common, they specialize or *differentiate* to perform unique functions. Fat cells, for example, are structurally modified to permit the storage of fats, whereas other cells, such as cardiac muscle cells, are specialized to contract (see Figure 1-3). Muscle, bone, nerve, and blood cells are other examples of structurally and functionally specialized cells.

Tissue Level

The next higher level of organization beyond the cell is the *tissue level* (see Figure 1-3). Tissues represent another step in the progressive organization of living matter. By definition, a **tissue** is a group of a great many similar cells that all developed together from the same part of the embryo and are all specialized to perform a certain function. Tissue cells are surrounded by varying

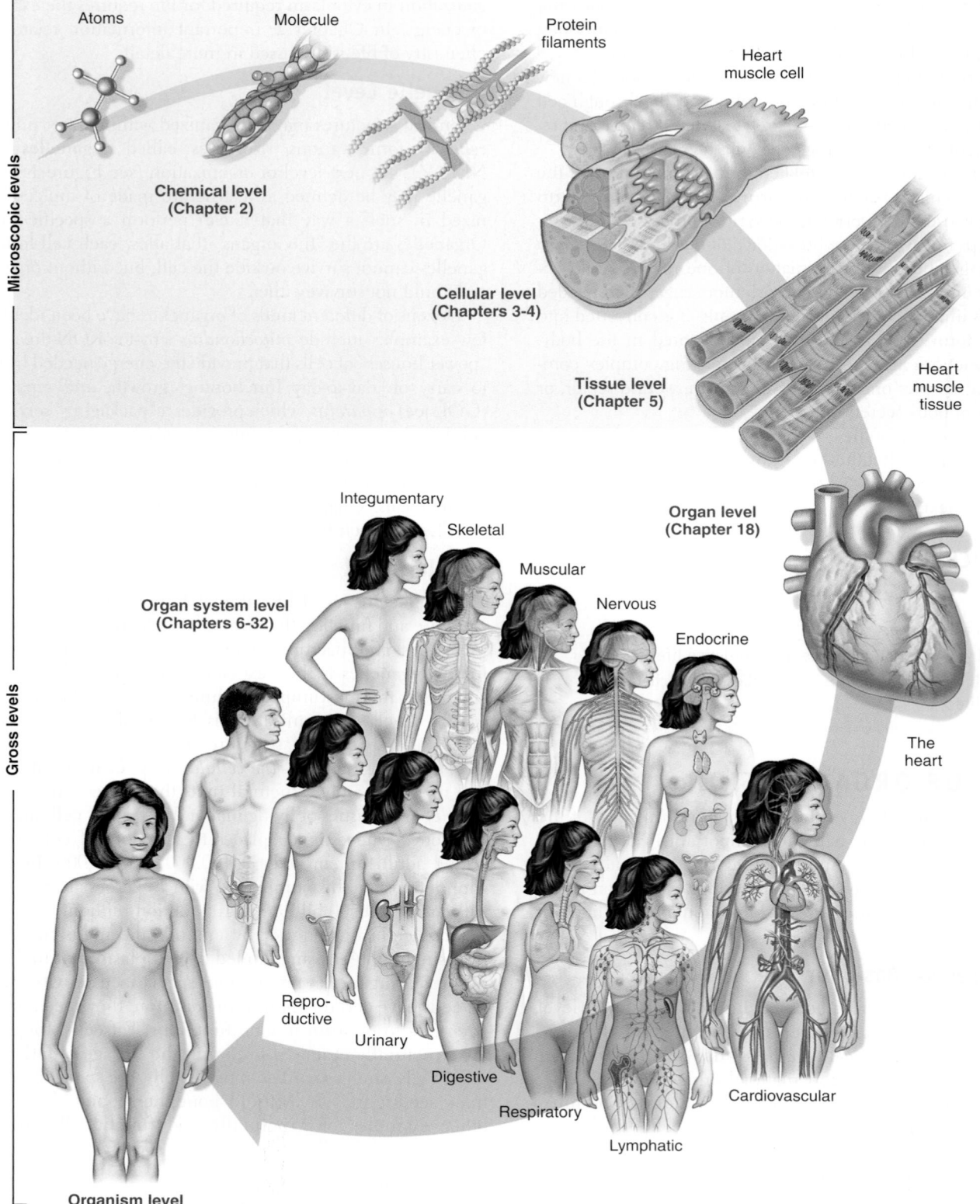

Figure 1-3 *Levels of organization.* The smallest parts of the body are the atoms that make up the chemicals, or molecules, of the body. Molecules, in turn, make up microscopic parts called *organelles* that fit together to form each cell of the body. Groups of similar cells are called *tissues,* which combine with other tissues to form individual organs. Groups of organs that work together are called *systems.* All the systems of the body together make up an individual organism. Knowledge of the different levels of organization will help you understand the basic concepts of human anatomy and physiology.

amounts and kinds of nonliving, intercellular substances, or the matrix. Tissues are the "fabric" of the body.

There are four major or principal tissue types: *epithelial, connective, muscle,* and *nervous*. Considering the complex nature of the human body, this is a surprisingly short list of major tissues. Each of the four major tissues, however, can be subdivided into several specialized subtypes. Together the body tissues are able to meet all the structural and functional needs of the body.

The tissue used as an example in Figure 1-3 is a specialized type of muscle called *cardiac* muscle. Note how the cells are branching and interconnected. The details of tissue structure and function are covered in Chapter 5.

Organ Level

Organs are more complex units than tissues are. An **organ** is defined as a structure made up of several different kinds of tissues so arranged that, together, they can perform a special function. If tissues are the "fabric" of the body, an organ is like an item of clothing with a specific function made up of different fabrics. The heart is an example of the *organ level*: muscle and specialized connective tissues give it shape, specialized epithelial tissues line

the cavities, or chambers, and nervous tissues permit control of muscular contraction.

Tissues seldom exist in isolation. Instead, joined together, they form organs that represent discrete, but functionally complex operational units. Each organ has a unique shape, size, appearance, and placement in the body, and each can be identified by the pattern of tissues that form it. The lungs, heart, brain, kidneys, liver, and spleen are all examples of organs.

System Level

Systems are the most complex of the organizational units of the body. The **system** level of organization involves varying numbers and kinds of organs so arranged that, together, they can perform complex functions for the body. Eleven major systems compose the human body: integumentary, skeletal, muscular, nervous, endocrine, circulatory, lymphatic/immune, respiratory, digestive, urinary, and reproductive. Systems that work together to accomplish the general needs of the body are summarized in Table 1-1.

Take a few minutes to read through Table 1-1. The left column points out that several different systems often work together to accomplish some overall goal. For example, the first three systems listed

Table 1-1	Body Systems (With Unit and Chapter References)		
FUNCTIONAL CATEGORY	**SYSTEM**	**PRINCIPAL ORGANS**	**PRIMARY FUNCTIONS**
Support and movement (Unit 1)	Integumentary (Chapter 6)	Skin	Protection, temperature regulation, sensation
	Skeletal (Chapters 7-9)	Bones, ligaments	Support, protection, movement, mineral and fat storage, blood production
	Muscular (Chapters 10-11)	Skeletal muscles, tendons	Movement, posture, heat production
Communication, control, and integration (Unit 2)	Nervous (Chapters 12-15)	Brain, spinal cord, nerves, sensory organs	Control, regulation, and coordination of other systems, sensation, memory
	Endocrine (Chapter 16)	Pituitary gland, adrenals, pancreas, thyroid, parathyroids, and other glands	Control and regulation of other systems
Transportation and defense (Unit 3)	Cardiovascular (Chapters 17-19)	Heart, arteries, veins, capillaries	Exchange and transport of materials
	Lymphatic (Chapters 20-22)	Lymph nodes, lymphatic vessels, spleen, thymus, tonsils	Immunity, fluid balance
Respiration, nutrition, and excretion (Unit 4)	Respiratory (Chapters 23-24)	Lungs, bronchial tree, trachea, larynx, nasal cavity	Gas exchange, acid-base balance
	Digestive (Chapters 25-27)	Stomach, small and large intestines, esophagus, liver, mouth, pancreas	Breakdown and absorption of nutrients, elimination of waste
	Urinary (Chapters 28-30)	Kidneys, ureters, bladder, urethra	Excretion of waste, fluid and electrolyte balance, acid-base balance
Reproduction and development (Unit 5)	Reproductive (Chapters 31-34)	Male: Testes, vas deferens, prostate, seminal vesicles, penis	Reproduction, continuity of genetic information, nurturing of offspring
		Female: Ovaries, fallopian tubes, uterus, vagina, breasts	

(integumentary, skeletal, muscular) make up the framework of the body and therefore provide support and movement. Notice also that this table corresponds to the organization of this book. Once we get to the system level of organization, we will study each system one by one, chapter by chapter. To help you navigate through the book, we have organized the chapters into units of several systems each—units that group the systems by common or overlapping functions.

You are probably aware that some systems can be grouped together or split apart if you find that more useful to you. For example, because both the skeletal and muscular systems work together to produce athletic movements, an athletic trainer may study them together as the "skeletomuscular system." A physical therapist may also include concepts of nervous control of movement and study the "neuroskeletomuscular system." On the other hand, a neurologist may find it useful to keep in mind a distinction between the sensory nervous system and the motor nervous system. In any case, the idea of levels of organization is universal, and once you know how it works, you can adapt it to suit your own changing needs. The plan of 11 major systems is widely used among biologists, so we will use it as the basis of our study too.

Organism Level

The living human **organism** is certainly more than the sum of its parts. It is a marvelously integrated assemblage of interactive structures that is able to survive and flourish in an often hostile environment. Not only can the human body reproduce itself (and its genetic information) and maintain ongoing repair and replacement of worn or damaged parts, but it can also maintain—in a constant and predictable way—an incredible number of variables required to lead healthy, productive lives.

We are able to maintain a "normal" body temperature and fluid balance under widely varying environmental extremes; we maintain constant blood levels of many important chemicals and nutrients; we experience effective protection against disease, elimination of waste products, and coordinated movement; and we correctly and quickly interpret sound, visual images, and other external stimuli with great regularity. These are a few examples of how the different levels of organization in the human organism permit expression of the characteristics associated with life.

As you study the structure and function of the human body, it is too easy to think of each part or function in isolation from the body as a whole. Always remember that you are ultimately dealing with information related to the entire human organism—not information limited to an understanding of the structure and function of a single organelle, cell, tissue, organ, or organ system. Do not limit your learning to the memorization of facts. Instead, integrate and conceptualize factual information so that your understanding of human structure and function is related not to a part of the body but to the body as a whole.

QUICK CHECK

8. List the seven levels of organization.
9. Identify three organelles.
10. List the four major tissue types.
11. List the 11 major organ systems.

ANATOMICAL POSITION

Discussions about the body, how it moves, its posture, or the relationship of one area to another, assume that the body as a whole is in a specific position called the **anatomical position.** In this reference position the body is in an erect, or standing, posture with the arms at the sides and palms turned forward (Figure 1-4). The head and feet are also pointing forward. The anatomical position is a reference position that gives meaning to the directional terms used to describe the body parts and regions.

Bilateral symmetry is one of the most obvious of the external organizational features in humans. The figure shown in Figure 1-4 is divided by a line into bilaterally symmetrical sides. To say that humans are bilaterally symmetrical simply means that the right and left sides of the body are mirror images of each other and only one plane can divide the body into left and right sides. One of the most important features of bilateral symmetry is balanced proportions. There is a remarkable correspondence in size and shape when comparing similar anatomical parts or external areas on opposite sides of the body.

The terms *ipsilateral* and *contralateral* are often used to identify the placement of one body part with respect to another on the same or opposite side of the body. These terms are used most frequently in describing injury to an extremity. *Ipsilateral* simply means "on the same side," and *contralateral* means "on the op-

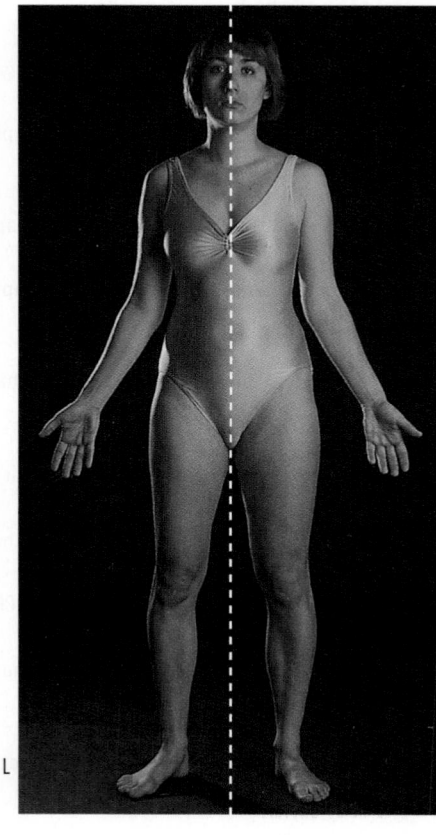

Figure 1-4 *Anatomical position and bilateral symmetry.* In the anatomical position, the body is in an erect, or standing, posture with the arms at the sides and palms forward. The head and feet are also pointing forward. The *dotted line* shows the axis of the body's bilateral symmetry. As a result of this organizational feature, the right and left sides of the body are mirror images of each other.

posite side." Injuries to an arm or leg require careful comparison of the injured with the noninjured side. Minimal swelling or deformity on one side of the body is often apparent only to a trained observer who compares a suspected area of injury with its corresponding part on the opposite side of the body. If the right knee were injured, for example, the left knee would be designated the *contralateral* knee.

BODY CAVITIES

The body, contrary to its external appearance, is not a solid structure. It contains two major cavities that are, in turn, subdivided and contain compact, well-ordered arrangements of internal organs. Although not all anatomists use these terms, it is useful to think of the two major hollow areas as the *ventral* and *dorsal* body cavities. The location and outlines of the body cavities are illustrated in Figure 1-5.

The **ventral cavity** includes the *thoracic* or *chest cavity* and the *abdominopelvic cavity*. The thoracic cavity includes a right and a left *pleural cavity* and a midportion called the *mediastinum* (mee-dee-ass-TI-num). Fibrous tissue forms a wall around the mediastinum that completely separates it from the right pleural cavity,

in which the right lung lies, and from the left pleural cavity, in which the left lung lies. Thus, the only organs in the thoracic cavity that are not located in the mediastinum are the lungs. Organs that are located in the mediastinum are the following: the heart (enclosed in its pericardial cavity), the trachea and right and left bronchi, the esophagus, the thymus, various blood vessels (e.g., thoracic aorta, superior vena cava), the thoracic duct and other lymphatic vessels, various lymph nodes, and nerves (such as the phrenic and vagus nerves). The **abdominopelvic cavity** has an upper portion, the *abdominal cavity*, and a lower portion, the *pelvic cavity*. The abdominal cavity contains the liver, gallbladder, stomach, pancreas, intestines, spleen, kidneys, and ureters. The bladder, certain reproductive organs (uterus, uterine tubes, and ovaries in females; prostate gland, seminal vesicles, and part of the vas deferens in males), and part of the large intestine (namely, the sigmoid colon and rectum) lie in the pelvic cavity (Table 1-2).

The **dorsal cavity** includes the *cranial* and *spinal* cavities. The cranial cavity lies in the skull and houses the brain. The spinal cavity lies in the spinal column and houses the spinal cord (see Figure 1-5).

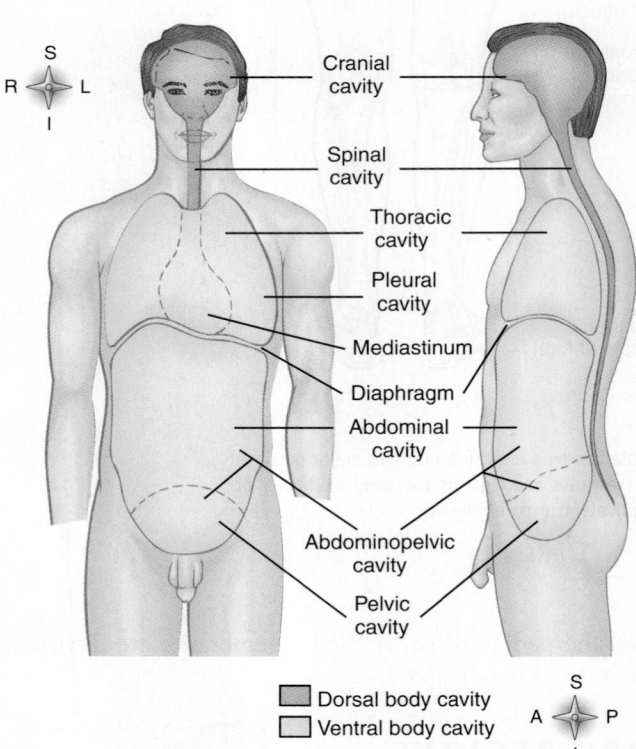

Figure 1-5 *Major body cavities.* The dorsal body cavity is in the dorsal (back) part of the body and is subdivided into a cranial cavity above and a spinal cavity below. The ventral body cavity is on the ventral (front) side of the trunk and is subdivided into the thoracic cavity above the diaphragm and the abdominopelvic cavity below the diaphragm. The thoracic cavity is subdivided into the mediastinum in the center and pleural cavities to the sides. The abdominopelvic cavity is subdivided into the abdominal above the pelvis and the pelvic cavity within the pelvis.

Table 1-2	Organs in Ventral Body Cavities	
AREAS	**ORGANS**	
Thoracic Cavity		
Right pleural cavity Mediastinum	Right lung (in pleural cavity) Heart (in pericardial cavity) Trachea Right and left bronchi Esophagus Thymus gland Aortic arch and thoracic aorta Venae cavae Various lymph nodes and nerves Thoracic duct	
Left pleural cavity	Left lung (in pleural cavity)	
Abdominopelvic Cavity		
Abdominal cavity	Liver Gallbladder Stomach Pancreas Intestines Spleen Kidneys Ureters	
Pelvic cavity	Urinary bladder Female reproductive organs Uterus Uterine tubes Ovaries Male reproductive organs Prostate gland Seminal vesicles Parts of vas deferens Part of large intestine, namely, sigmoid colon and rectum	

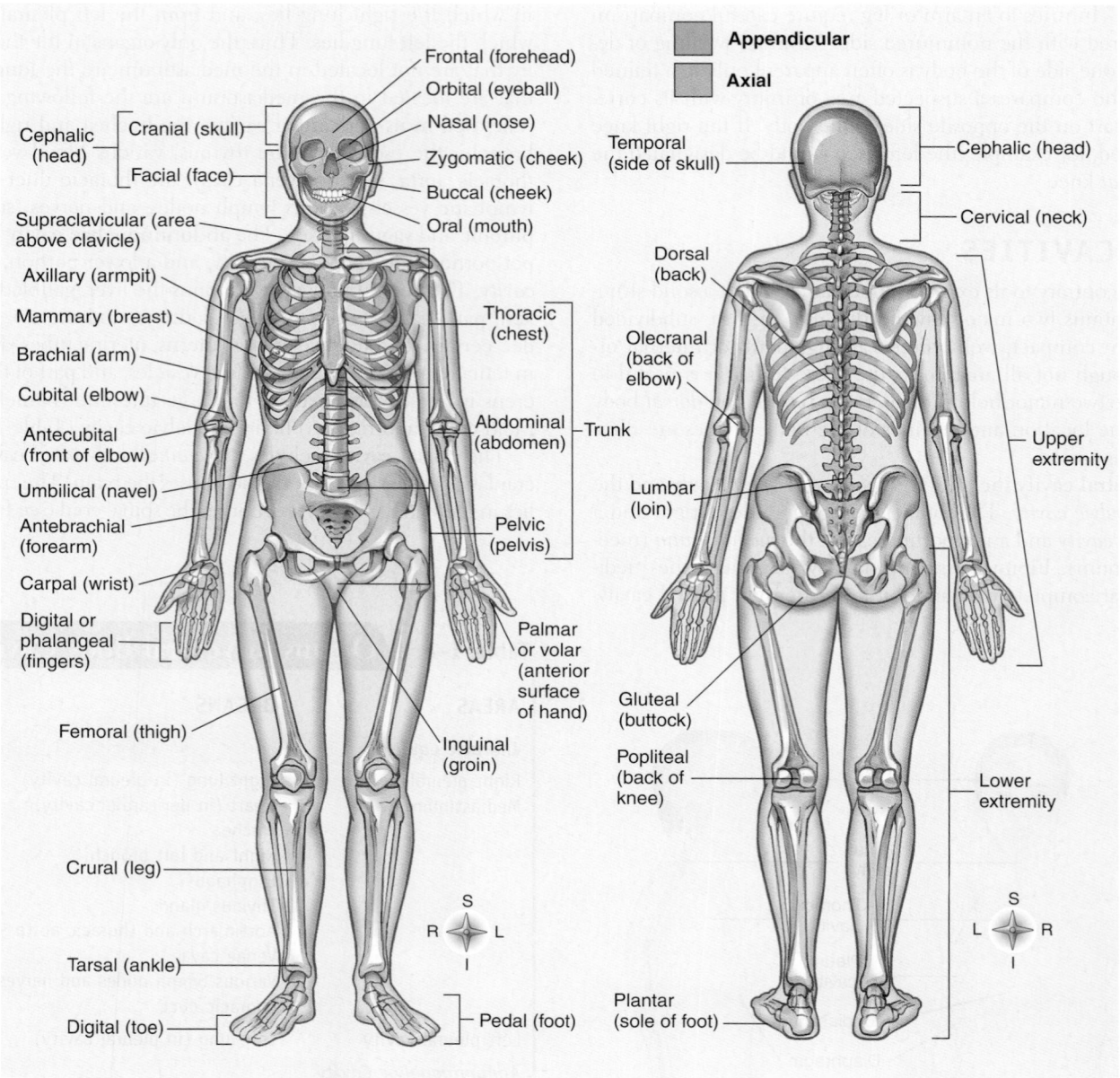

Figure 1-6 *Specific body regions.* Note that the body as a whole can be subdivided into two major portions: axial (along the middle, or *axis,* of the body) and appendicular (the arms and legs, or *appendages*). Names of specific body regions follow the Latin form, with the English equivalent in parentheses.

The thin filmy membranes that line body cavities or cover the surfaces of organs within body cavities also have special names. The term **parietal** refers to the actual wall of a body cavity or the lining membrane that covers its surface. **Visceral** refers not to the wall or lining of a body cavity but to the thin membrane that covers the organs, or **viscera,** within a cavity.

The membrane lining the inside of the abdominal cavity, for example, is called the *parietal peritoneum.* The membrane that covers the organs within the abdominal cavity is called the *visceral peritoneum.* You can see in Figure 1-6 that there is a space or opening between the two membranes in the abdomen. This is

called the *peritoneal cavity.* Body membranes are discussed in greater detail in Chapter 5.

BODY REGIONS

Identification of an object begins with overall recognition of its structure and form. Initially, it is in this way that the human form can be distinguished from other creatures or objects. Recognition occurs as soon as you can identify the overall shape and basic outline. For more specific identification to occur, details of size, shape, and appearance of individual body areas must be described. Individuals differ in over-

Table 1-3	Latin-Based Descriptive Terms for Body Regions*		
BODY REGION	**AREA OR EXAMPLE**	**BODY REGION**	**AREA OR EXAMPLE**
Abdominal (ab-DOM-in-al)	Anterior torso below diaphragm	**Mammary** (MAM-er-ee)	Breast
Acromial (ah-KRO-mee-al)	Shoulder	**Manual** (MAN-yoo-al)	Hand
Antebrachial (an-tee-BRAY-kee-al)	Forearm	**Mental** (MEN-tal)	Chin
		Nasal (NAY-zal	Nose
Antecubital (an-tee-KYOO-bi-tal)	Depressed area just in front of elbow	**Navel** (NAY-val)	Area around navel, or umbilicus
		Occipital (ok-SIP-i-tal)	Back of lower part of skull
Axillary (AK-si-lair-ee)	Armpit (axilla)	**Olecranal** (o-LECK-ra-nal)	Back of elbow
Brachial (BRAY-kee-al)	Upper part of arm	**Oral** (OR-al)	Mouth
Buccal (BUK-al)	Cheek (inside)	**Orbital** or **ophthalmic** (OR-bi-tal or op-THAL-mik)	Eyes
Calcaneal (cal-CANE-ee-al)	Heel of foot		
Carpal (KAR-pal)	Wrist	**Otic** (O-tick)	Ear
Cephalic (se-FAL-ik)	Head	**Palmar** (PAHL-mar)	Palm of hand
Cervical (SER-vi-kal)	Neck	**Patellar** (pa-TELL-er)	Front of knee
Coxal (COX-al)	Hip	**Pedal** (PED-al)	Foot
Cranial (KRAY-nee-al)	Skull	**Pelvic** (PEL-vik)	Lower portion of torso
Crural (KROOR-al)	Leg	**Perineal** (pair-i-NEE-al)	Area (perineum) between anus and genitals
Cubital (KYOO-bi-tal)	Elbow		
Cutaneous (kyoo-TANE-ee-us)	Skin (or body surface)	**Plantar** (PLAN-tar)	Sole of foot
Digital (DIJ-i-tal)	Fingers or toes	**Pollex** (POL-ex)	Thumb
Dorsal (DOR-sal)	Back or top	**Popliteal** (pop-li-TEE-al)	Area behind knee
Facial (FAY-shal)	Face	**Pubic** (PYOO-bik)	Pubis
Femoral (FEM-or-al)	Thigh	**Supraclavicular** (soo-pra-cla-VIK-yoo-lar)	Area above clavicle
Frontal (FRON-tal)	Forehead		
Gluteal (GLOO-tee-al)	Buttock	**Sural** (SUR-al)	Calf
Hallux (HAL-luks)	Great toe	**Tarsal** (TAR-sal)	Ankle
Inguinal (ING-gwi-nal)	Groin	**Temporal** (TEM-por-al)	Side of skull
Lumbar (LUM-bar)	Lower part of back between ribs and pelvis	**Thoracic** (tho-RAS-ik)	Chest
		Zygomatic (zye-go-MAT-ik)	Cheek

*The left column lists English adjectives based on Latin terms that describe the body parts listed in English in the right column.

all appearance because specific body areas, such as the face or torso, have unique identifying characteristics. Detailed descriptions of the human form require that specific regions be identified and appropriate terms be used to describe them (see Figure 1-6 and Table 1-3).

The body as a whole can be subdivided into two major portions or components: axial and appendicular. The axial portion of the body consists of the head, neck, and torso, or trunk; the appendicular portion consists of the upper and lower extremities and their connections to the axial portion. Each major area is subdivided as shown in Figure 1-6. Note, for example, that the torso is composed of the thoracic, abdominal, and pelvic areas and the upper extremity is divided into arm, forearm, wrist, and hand components. Although most terms used to describe gross body regions are familiar, misuse is common. The term *leg* is a good example. To an anatomist, "leg" refers to the area of the lower extremity between the knee and ankle and *not* to the entire lower extremity.

Abdominal Regions

For convenience in locating abdominal organs, anatomists divide the abdomen like a tic-tac-toe grid into nine imaginary regions. The following is a list of the nine regions (Figure 1-7) identified from right to left and from top to bottom:

1. Right hypochondriac region
2. Epigastric region
3. Left hypochondriac region
4. Right lumbar region
5. Umbilical region
6. Left lumbar region
7. Right iliac (inguinal) region
8. Hypogastric region
9. Left iliac (inguinal) region

The most superficial organs located in each of the nine abdominal regions are shown in Figure 1-7. In the right hypochon-

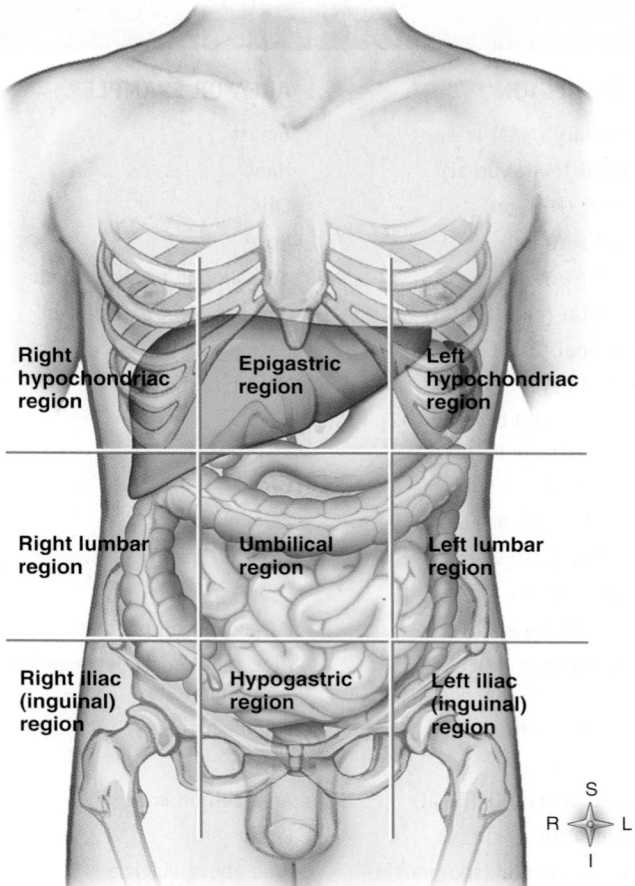

Figure 1-7 *Nine regions of the abdominopelvic cavity.* Only the most superficial structures of the internal organs are shown here.

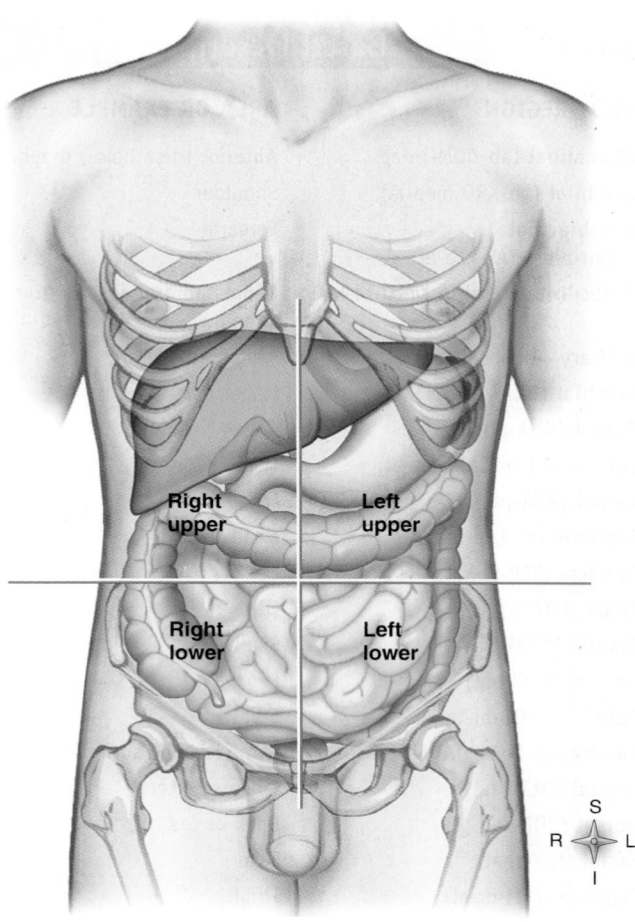

Figure 1-8 *Division of the abdomen into four quadrants.* The diagram shows the relationship of internal organs to the four abdominopelvic quadrants.

driac region the right lobe of the liver and the gallbladder are visible. In the epigastric area, parts of the right and left lobes of the liver and a large portion of the stomach can be seen. Viewed superficially, only a small portion of the stomach and large intestine is visible in the left hypochondriac area. Note that the right lumbar region includes parts of the large and small intestines (see Figure 1-7). The superficial organs seen in the umbilical region include a portion of the transverse colon and loops of the small intestine. Additional loops of the small intestine and a part of the colon can be seen in the left lumbar region. The right iliac region contains the cecum and parts of the small intestine. Only loops of the small intestine, the urinary bladder, and the appendix are seen in the hypogastric region. The left iliac region shows portions of the colon and the small intestine.

Abdominopelvic Quadrants

Physicians and other health professionals frequently divide the abdomen into four quadrants to describe the site of abdominopelvic pain or locate some type of internal pathology such as a tumor or abscess (Figure 1-8). A horizontal and vertical line passing through the umbilicus (navel) divides the abdomen into *right* and *left upper quadrants* and *right* and *left lower quadrants* (see Figure 1-8).

 QUICK CHECK

12. Define the term *anatomical position* and explain its importance.
13. Name the two major subdivisions of the body as a whole.
14. Identify the two major body cavities and the subdivisions of each.
15. List the nine abdominal regions and four abdominopelvic quadrants.

TERMS USED IN DESCRIBING BODY STRUCTURE

Directional Terms

To minimize confusion when discussing the relationship between body areas or the location of a particular anatomical structure, specific terms must be used. When the body is in the anatomical position, the following directional terms can be used to describe the location of one body part with respect to another (Figure 1-9).

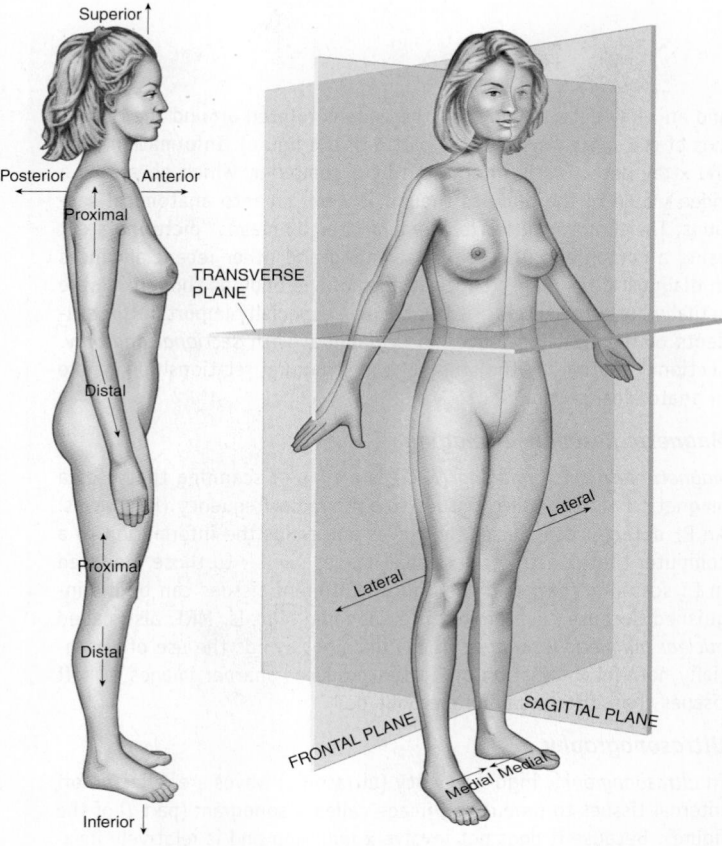

Figure 1-9 Directions and planes of the body.

low it, and the bone of the upper part of the arm is deep to the muscles that surround and cover it. Refer often to the table of anatomical directions on the inside front cover. It is intended to serve as a useful and ready reference for review.

Terms Related to Organs

When discussing anatomical relationships among organs in a system or region, or anatomical relationships within an organ, additional terms are often useful.

Lumen. Many organs of the body are hollow, such as the stomach, small intestine, blood vessels, urinary organs, and so on. The hollow area of any of these organs is called the **lumen.** The term *luminal* means "of or near the lumen."

Central and *peripheral.* **Central** means "near the center"; **peripheral** means "around the boundary." For example, the central nervous system includes the brain and spinal cord, which are near the center of the body. The peripheral nervous system, on the other hand, includes the nerves of the muscles, skin, and other organs that are nearer the periphery, or outer boundaries, of the body.

Medullary and *cortical.* **Medullary** refers to an inner region of an organ; **cortical** refers to an outer region or layer of an organ. For example, the inner region of the kidney is the *medulla,* and any structures there are described as medullary. Similarly, the term cortical describes structures found in the outer layer of kidney tissue (the *cortex* of the kidney).

Basal and *apical.* Some organs, such as the heart and each lung, are somewhat cone-shaped. Thus, we borrow terms that describe the point or *apex* of a cone and the flat part or *base* of a cone. **Basal** refers to the base or widest part of an organ; **apical** refers to the narrow tip of an organ. For example, in the heart the term apical refers to the "point" of the heart that rests on the diaphragm. Basal and apical may also refer to individual cells: the apical surface faces the lumen of a hollow organ, and the basal surface of the cell faces away from the lumen.

Many of the more common directional terms that you will use in this course are listed in a handy table inside the front cover of the book.

BODY PLANES AND SECTIONS

The transparent glasslike plates in Figure 1-9 dividing the body into parts represent cuts, or *sections,* that can be made along a particular axis, or line of orientation, called a *plane.* There are three major **body planes** that lie at right angles to each other. They are called the *sagittal* (SAJ-i-tal), *coronal* (koh-ROH-nal), and *transverse* (or *horizontal*) planes. Literally hundreds of sections can be made in each plane, and each section made is named after the particular plane along which it occurs. For example, the transverse plane in Figure 1-9 is shown dividing the individual into upper and lower parts at about the level of the umbilicus. Many other transverse sections are possible in parallel transverse planes. A transverse section through the knee would amputate the lower extremity at that joint, and a transverse section through the neck would result in decapitation. Box 1-1 shows how these planes are sometimes used in medical imaging.

Superior and *inferior.* **Superior** means "toward the head," and inferior means "toward the feet." Superior also means "upper" or "above," and **inferior** means "lower" or "below." For example, the lungs are located superior to the diaphragm, whereas the stomach is located inferior to it.

Anterior and *posterior.* **Anterior** means "front" or "in front of"; **posterior** means "back" or "in back of." In humans—who walk in an upright position—*ventral* (toward the belly) can be used in place of anterior, and *dorsal* (toward the back) can be used for posterior. For example, the nose is on the anterior surface of the body, and the shoulder blades are on its posterior surface.

Medial and *lateral.* **Medial** means "toward the midline of the body"; **lateral** means "toward the side of the body, or away from its midline." For example, the great toe is at the medial side of the foot, and the little toe is at its lateral side. The heart lies medial to the lungs, and the lungs lie lateral to the heart.

Proximal and *distal.* **Proximal** means "toward or nearest the trunk of the body, or nearest the point of origin of one of its parts"; **distal** means "away from or farthest from the trunk or the point of origin of a body part." For example, the elbow lies at the proximal end of the lower part of the arm, whereas the hand lies at its distal end.

Superficial and *deep.* **Superficial** means "nearer the surface"; **deep** means "farther away from the body surface." For example, the skin of the arm is superficial to the muscles be-

BOX 1-1
Diagnostic Study

Medical Imaging of the Body

Cadavers (preserved human bodies used for scientific study) can be cut into sagittal, frontal, or transverse sections for easy viewing of internal structures, but living bodies, of course, cannot. This fact has been troublesome for medical professionals who must determine whether internal organs are injured or diseased. In some cases the only sure way to detect a *lesion* or variation from normal is extensive *exploratory surgery*. Fortunately, recent advances in medical imaging allow physicians to visualize internal structures of the body without risking the trauma or other complications associated with extensive surgery. Some of the more widely used techniques are briefly described here.

Radiography

Radiography, or *x-ray photography,* is the oldest and still most widely used method of noninvasive imaging of internal body structures. In this method, energy in the x band of the radiation spectrum is beamed through the body to photographic film (part *A* of the figure). The x-ray photograph shows the outlines of bones and other dense structures that partially absorb the x-rays. In *fluoroscopy*, a phosphorescent screen sensitive to x-rays is used instead of photographic film. A visible image is formed on the screen as x-rays passing through the subject cause the screen to glow. Fluoroscopy allows a medical professional to view the internal structures of the subject's body as it moves. One way to make soft, hollow structures such as blood vessels or digestive organs more visible is to use *radiopaque* contrast media. Substances that absorb x-rays, such as barium sulfate, are injected or swallowed to fill the hollow organ of interest. As the screen in part *A* of the figure depicts, the hollow organ shows up as distinctly as a dense bone.

Computed Tomography

A newer variation of traditional x-ray photography is called *computed tomography (CT),* or computed axial tomography (CAT) scanning. In this method, a device with an x-ray source on one side of the body and an x-ray detector on the other side is rotated around the central axis of the subject's body (see part *B* of the figure). Information from the x-ray detectors is interpreted by a computer, which generates a video image of the body as though it were cut into anatomical sections. The term *computed tomography* literally means "picturing a cut using a computer." Because CT scanning and other recent advances in diagnostic imaging produce images of the body as though it were actually cut into sections, it has become especially important for students of the health sciences to be familiar with *sectional anatomy*. Sectional anatomy is the study of the structural relationships visible in anatomical sections.

Magnetic Resonance Imaging

Magnetic resonance imaging (MRI) is a type of scanning that uses a magnetic field to induce tissues to emit radio frequency (RF) waves. An RF detector coil senses the waves and sends the information to a computer that constructs sectional images similar to those produced in CT scanning (part *C* of the figure). Different tissues can be distinguished because each emits different radio signals. MRI, also called *nuclear magnetic resonance* (NMR) imaging, avoids the use of potentially harmful x radiation and often produces sharper images of soft tissues than other imaging methods do.

Ultrasonography

In *ultrasonography*, high-frequency (ultrasonic) waves are reflected off internal tissues to produce an image called a sonogram (part *D* of the figure). Because it does not involve x radiation and is relatively inexpensive and easy to use, ultrasonography has been used extensively—especially in studying maternal or fetal structures in pregnant women. However, the image produced is not nearly as clear or sharp as in MRI, CT scanning, or traditional radiography.

Variations of these and other technological advances that have improved the ability to study the structure and functions of the human body are discussed in later chapters.

Read the following definitions and identify each term in Figure 1-9:

Sagittal plane. A lengthwise plane running from front to back; divides the body or any of its parts into right and left sides. If a sagittal section is made in the exact midline, resulting in equal and symmetrical right and left halves, the plane is called a *midsagittal* plane (see Figure 1-9).

Coronal plane. A lengthwise plane running from side to side; divides the body or any of its parts into anterior and posterior portions; also called a *frontal* plane.

Transverse plane. A crosswise plane; divides the body or any of its parts into upper and lower parts; also called a *horizontal* plane.

Figure 1-10 shows the organs of the abdominal cavity as they would appear in the transverse, or horizontal, plane or "cut" through the abdomen represented in Figure 1-9. In addition to the actual photograph, a simplified line diagram helps in identifying the primary organs. Note that organs near the bottom of the photo or line drawing are in a posterior position. The cut vertebra of the spine, for example, can be identified in its position behind, or posterior, to the stomach. The kidneys are located on either side of the vertebra—they are *lateral* and the vertebra is *medial*. To make the reading of anatomical figures a little easier, an *anatomical rosette* is used throughout this book. On many figures, you will notice a small compass rosette similar to those on geographical maps. Rather than being labeled N, S, E, and W, the anatomical rosette is labeled with abbreviated anatomical directions:

A	Anterior
D	Distal
I	Inferior
L (opposite M)	Lateral
L (opposite R)	Left
M	Medial
P (opposite A)	Posterior
P (opposite D)	Proximal
R	Right
S	Superior

For your convenience, the compass rosette, its possible directions, a helpful diagram of the planes and directions of the body, and a summary table are found on the inside front cover of this book. Refer to it frequently until you are familiar enough with anatomy to do without it.

BOX 1-1
Diagnostic Study—cont'd

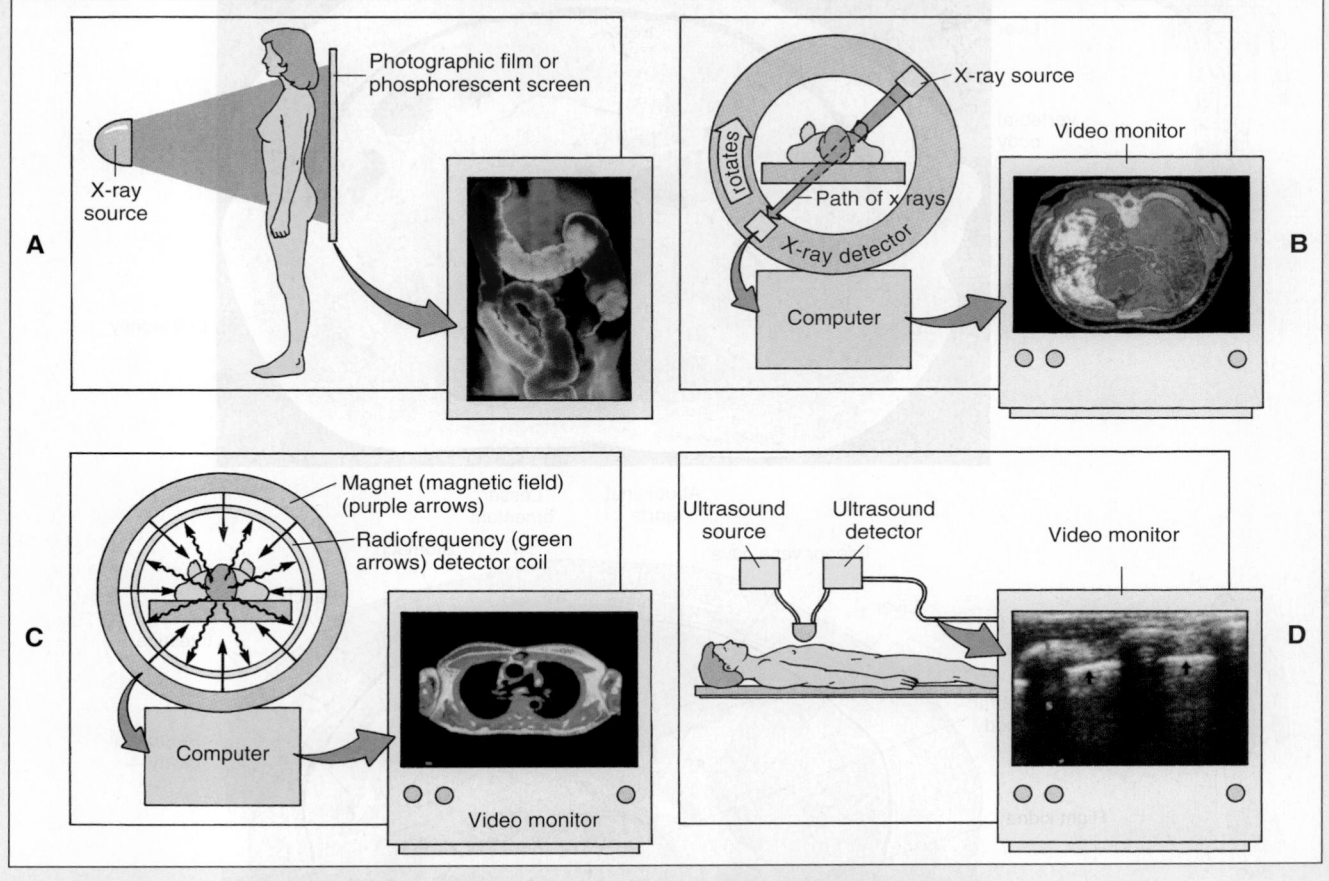

Types of medical imaging. A, Radiography, or x-ray photography. **B,** Computed tomography (CT). **C,** Magnetic resonance imaging (MRI). **D,** Ultrasonography.

INTERACTION OF STRUCTURE AND FUNCTION

One of the most unifying and important concepts of the study of anatomy and physiology is the principle of *complementarity of structure and function.* In the chapters that follow, you will note again and again that anatomical structures seem "designed" to perform specific functions. Each structure has a particular size, shape, form, or placement in the body that makes it especially efficient at performing a unique and specialized activity.

The relationships between the levels of structural organization will take on added meaning as you study the various organ systems in the chapters that follow. For example, as you study the respiratory system in Chapter 24, you will learn about a special chemical substance secreted by cells in the lungs that help to keep tiny air sacs in these organs from collapsing during respiration. Hereditary material called *DNA* (a macromolecule) "directs" the differentiation of specialized cells in the lungs during development so that they can effectively contribute to respiratory function. As a result of DNA activity, special chemicals are produced, cells are modified, and tissues appear that are

uniquely suited to this organ system. The cilia (organelles), which cover the exposed surface of cells that form the tissues lining the respiratory passageways, help trap and eliminate inhaled contaminants such as dust. The structures of the respiratory tubes and lungs assist in efficient and rapid movement of air and also make possible the exchange of critical respiratory gases such as oxygen and carbon dioxide between the air in the lungs and the blood. Working together as the respiratory system, specialized chemicals, organelles, cells, tissues, and organs supply every cell of our body with necessary oxygen and constantly remove carbon dioxide.

Structure determines function, and function influences the actual anatomy of an organism over time. Understanding this fact helps students better understand the mechanisms of disease and the structural abnormalities often associated with pathology. Current research in the study of human biology is now focused in large part on integration, interaction, development, modification, and control of functioning body structures.

By applying the principle of complementarity of structure and function as you study the structural and functional levels of the body's organization in each organ system, you will be able to in-

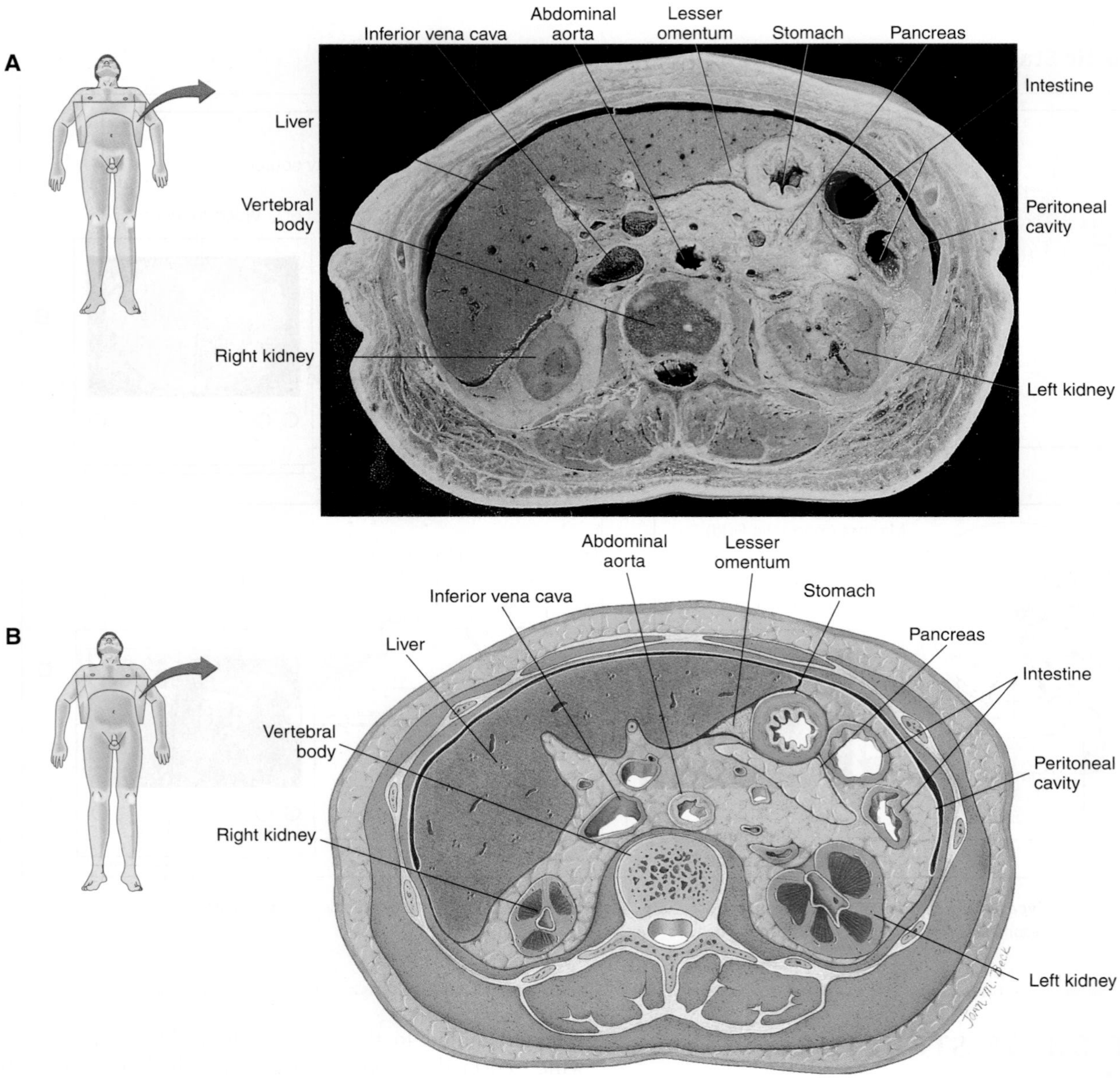

A

Inferior vena cava
Abdominal aorta
Lesser omentum
Stomach
Pancreas
Intestine
Liver
Peritoneal cavity
Vertebral body
Right kidney
Left kidney

B

Abdominal aorta
Lesser omentum
Inferior vena cava
Stomach
Liver
Pancreas
Intestine
Vertebral body
Peritoneal cavity
Right kidney
Left kidney

Figure 1-10 *Transverse section of the abdomen.* **A,** A transverse, or horizontal, plane through the abdomen shows the position of various organs within the cavity. **B,** A drawing of the photograph helps clarify the photo. Compare these views with the medical images shown in Box 1-1, Figures *B* and *C,* on p. 19.

tegrate otherwise isolated factual information into a cohesive and understandable whole. A memorized set of individual and isolated facts is soon forgotten—the parts of an anatomical structure that can be related to its function are not.

QUICK CHECK

16. Define complementarity of structure and function.
17. Give an example of how the chemical macromolecule DNA can have an influence on body structure.
18. Discuss how structure relates to function at the tissue level of organization in the respiratory system.

BODY TYPE AND DISEASE

The concept of body types is a good example of how structure and function are interrelated. The term **somatotype** is used to describe a particular category of body build, or physique. Although the human body comes in many sizes and shapes, every individual can be classified as belonging to one of three basic body types, or somatotypes. The names used to describe these body types are

Endomorph: heavy, rounded physique characterized by large accumulations of fat in the trunk and thighs

Mesomorph: muscular physique

Ectomorph: thin, fragile physique characterized by little body fat accumulation

Figure 1-11 shows extreme examples of the three somatotypes. By carefully studying the body build of numerous individuals, scientists

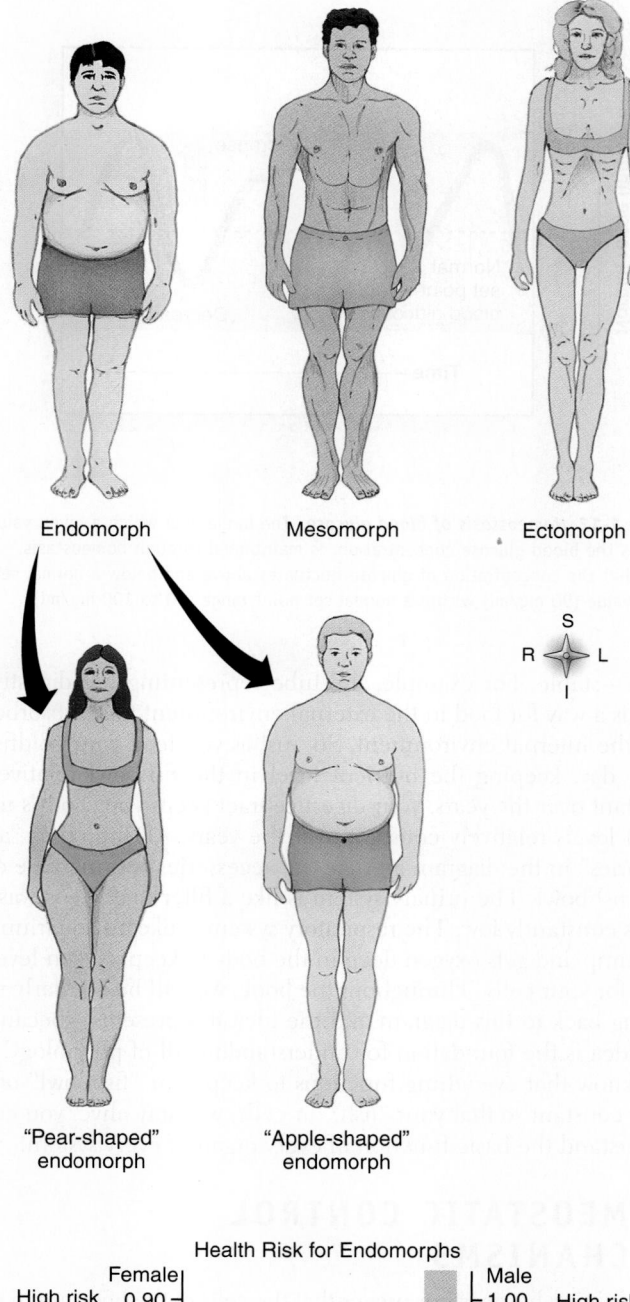

Endomorph   Mesomorph   Ectomorph

"Pear-shaped" endomorph   "Apple-shaped" endomorph

Health Risk for Endomorphs

	Female	Male	
High risk	0.90	1.00	High risk
	0.85	0.95	
	0.80	0.90	
	0.75	0.85	
Low risk	0.70	0.80	Low risk

Pear shape   Apple shape

Figure 1-11 Body types and health risks.

know, for example, that knowledge of physique can provide health care professionals and educators with vital information useful in such areas as disease screening procedures, programs designed to identify individuals who may be at risk for certain diseases, and prediction of performance capability in selected physical education programs.

Researchers have discovered that individuals (especially endomorphs) who have large waistlines and are "apple-shaped," or fattest in the abdomen, have a greater risk for heart disease, stroke, high blood pressure, and diabetes than do individuals with a lower "pear-shaped" distribution pattern of fat in the hips, thighs, and buttocks. Breast cancer in postmenopausal women is also associated with storage of fat in the abdomen and upper body area (apple shape). In both sexes, endomorphic individuals of the same height and weight but with a lower, or pear-shaped, body fat distribution pattern contracted these diseases more frequently than mesomorphs and ectomorphs but less frequently than endomorphs with an apple-shaped, or high body fat distribution pattern.

The *waist-to-hip* ratio is used to assess apple- or pear-shaped body fat distribution patterns and is considered a valuable clinical tool to assist health care professionals in evaluating the risk of individuals developing many diseases. To obtain the ratio, simply divide the waist measurement by the hip size. A ratio greater than 1.0 for men and 0.9 for women places the person at higher risk for diseases associated with a high body, or apple-shaped, fat distribution pattern.

The graph illustrates the increasing health risks for both men and women in whom an apple shape develops as the waist-to-hip ratio increases. This is especially true if the individual is not physically fit and leads a sedentary lifestyle.

QUICK CHECK

19. Contrast each term in these pairs: superior/inferior, anterior/posterior, medial/lateral, dorsal/ventral.
20. How is anatomical left different from your left?
21. List and define the three major planes that are used to divide the body into parts.
22. Explain how an anatomical rosette is used in anatomical illustrations.
23. How might one's body type influence health risks?

HOMEOSTASIS

More than a century ago a great French physiologist, Claude Bernard (1813-1878), made a remarkable observation. He noted that body cells survived in a healthy condition only when the temperature, pressure, and chemical composition of their fluid environment remained relatively constant. He called the environment of cells the *internal environment*, or *milieu intérieur*. Bernard realized that although many elements of the external environment in which we live are in a constant state of change, important elements of the internal environment, such as body temperature, remain remarkably stable. For example, Bernard's neighbor, who travels from his fireplace in Paris to the snowy slopes of the Alps in January, is exposed to dramatic changes in air temperature within a few hours. Fortunately, in a healthy individual, body temperature will remain at or very near normal regard-

have found that the basic components that determine the different categories of physique occur in varying degrees in every person—both men and women. Only in very rare instances does an individual show almost total dominance by a single somatotype component.

Until recently the concept of somatotype was considered largely "historical" and of relatively little practical importance. However, new research findings have rekindled interest in this area. We now less of temperature changes that may occur in the external environment. Just as the external environment surrounding the body as a whole is subject to change, so too is the fluid environment surrounding each body cell. The remarkable fluid that bathes each cell contains literally dozens of different substances. Good health, indeed life itself, depends on the correct and constant amount of each substance in the blood and other body fluids. The precise and constant chemical composition of the internal environment must be maintained within very narrow limits ("normal ranges"), or sickness and death will result.

In 1932 a famous American physiologist, Walter B. Cannon, suggested the name **homeostasis** (hoh-mee-oh-STAY-sis) for the relatively constant states maintained by the body. Homeostasis is a key word in modern physiology. It comes from two Greek words (*homoios*, "the same," and *stasis*, "standing"). "Standing or staying the same," then, is the literal meaning of homeostasis. In his classic publication titled *The Wisdom of the Body*, Cannon advanced one of the most unifying and important themes of physiology. He suggested that every regulatory mechanism of the body existed to maintain homeostasis, or constancy, of the body's internal fluid environment. However, as Cannon emphasized, homeostasis does not mean something set and immobile that stays exactly the same all the time. In his words, homeostasis "means a condition that may vary, but which is relatively constant." It is the maintenance of relatively constant internal conditions despite changes in either the internal or the external environment that characterizes homeostasis. For example, even if external temperatures vary, homeostasis of body temperature means that it remains relatively constant at about 37° C (98.6° F), although it may vary slightly above or below that point and still be "normal." The fasting concentration of blood glucose, an important nutrient, can also vary somewhat and still remain within normal limits (Figure 1-12). This normal reading or range of normal is called the **set point** or **set point range**. A value between 80 and 100 mg of glucose per milliliter of blood, depending on dietary intake and timing of meals, is typical. Although levels of the important gases oxygen and carbon dioxide also vary with the respiratory rate, these substances, like body temperature and blood glucose levels, must be maintained within very narrow limits.

Specific regulatory mechanisms are responsible for adjusting body systems to maintain homeostasis. This ability of the body to "self-regulate," or "return to normal" to maintain homeostasis, is a critically important concept in modern physiology and also serves as a basis for understanding mechanisms of disease. Each cell of the body, each tissue, and each organ system plays an important role in homeostasis. Each of the diverse regulatory systems described in subsequent chapters of the text is explained as a function of homeostasis. You will learn how specific regulatory activities such as temperature control or carbon dioxide elimination are accomplished. In addition, an understanding of the relationship of homeostasis to healthy survival helps explain why such mechanisms are necessary.

Take a moment to study Figure 1-13. This diagram is a classic way of envisioning the idea of the body as "bag of fluid." The fluid inside the bag is our internal environment, and it is this fluid that must be kept at a relatively constant temperature, glucose level, and so on, if the cells that make up the body are to survive. It is like a big, walking fish bowl, and our cells are the fish. All the little tubes and gizmos you see in Figure 1-13 are the systems that keep the "water in the fishbowl"—your internal fluid environ-

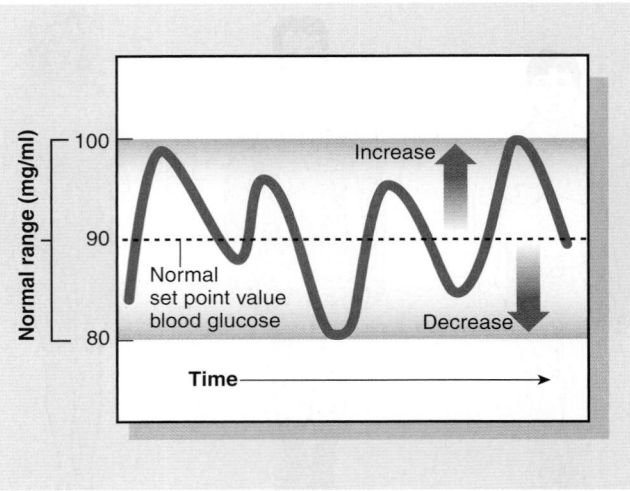

Figure 1-12 *Homeostasis of blood glucose.* The range over which a given value, such as the blood glucose concentration, is maintained through homeostasis. Note that the concentration of glucose fluctuates above and below a normal set point value (90 mg/ml) within a normal set point range (80 to 100 mg/ml).

ment—stable. For example, the tube representing the digestive tract is a way for food in the external environment to be absorbed into the internal environment. So, just as you feed your goldfish every day, keeping the nutrient level in the fishbowl relatively constant over the years, your digestive tract keeps your body's nutrient levels relatively constant over the years. All the other "accessories" in the diagram are like the accessories you may use on your fishbowl. The urinary system is like a filter that keeps waste levels constantly low. The respiratory system is like an aquarium's air pump and gets oxygen deep in the body to keep oxygen levels high for your cells. Throughout the book, we will be regularly referring back to this diagram and the idea it represents—because this idea is the foundation for understanding all of physiology. If you know that everything functions to keep your "fishbowl" of a body constant so that your "fish," or cells, will stay alive, you can understand the basic function of every organ of every system!

HOMEOSTATIC CONTROL MECHANISMS

Maintaining homeostasis means that the cells of the body are in an environment that meets their needs and permits them to function normally under changing external conditions. Processes for maintaining or restoring homeostasis are known as *homeostatic control mechanisms*. They involve virtually all of the body's organs and systems. If circumstances occur that require changes or more active regulation in some aspect of the internal environment, the body must have appropriate control mechanisms available that respond to these changing needs and then restore and maintain a healthy internal environment. For example, exercise increases the need for oxygen and results in accumulation of the waste product carbon dioxide. By increasing our breathing rate above its average of about 17 breaths per minute, we can maintain an adequate blood oxygen level and also increase the elimination of carbon dioxide. When exercise stops, the need for an increased respiratory rate no longer exists and the frequency of breathing returns to normal.

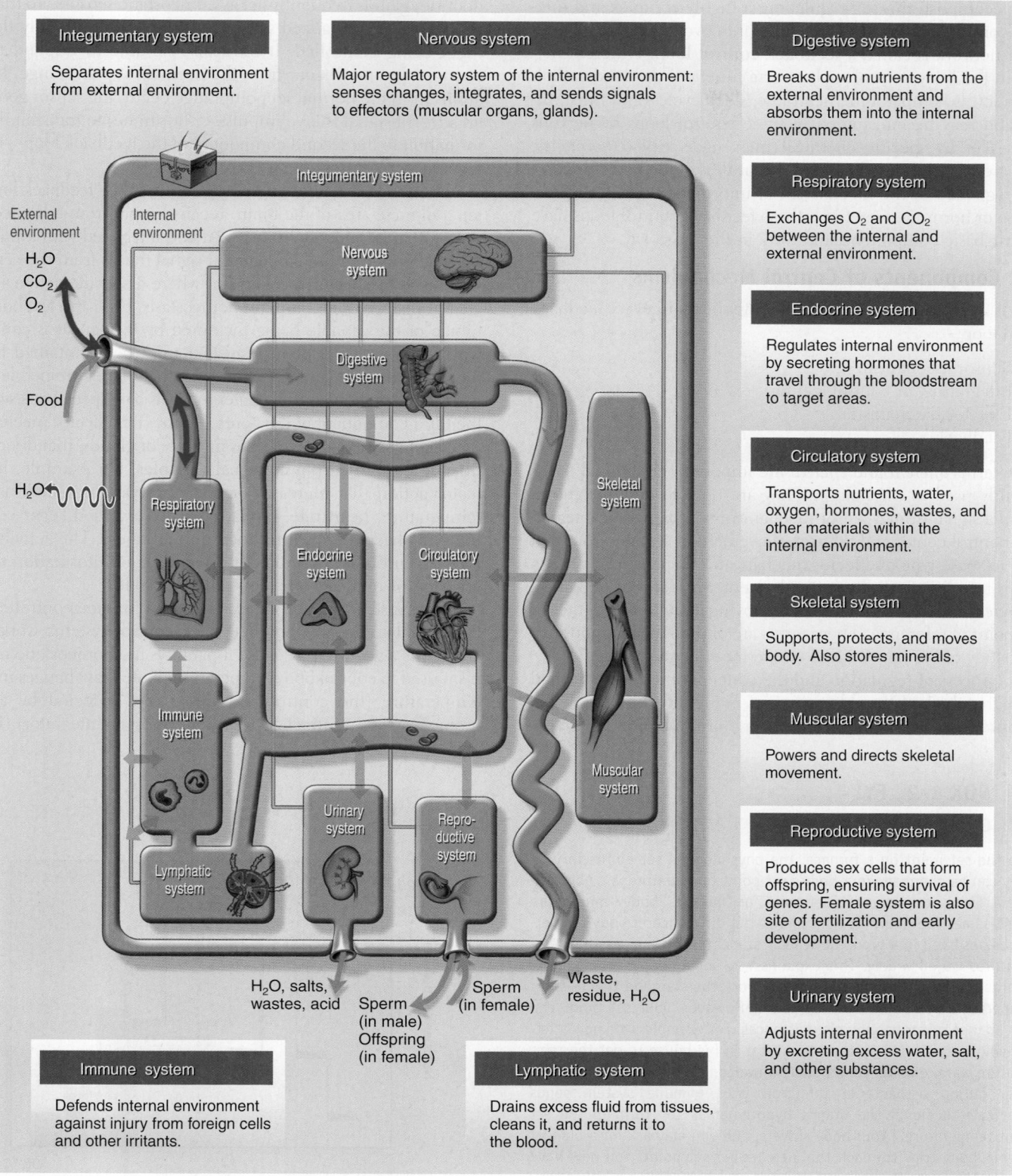

Integumentary system

Separates internal environment from external environment.

Nervous system

Major regulatory system of the internal environment: senses changes, integrates, and sends signals to effectors (muscular organs, glands).

Digestive system

Breaks down nutrients from the external environment and absorbs them into the internal environment.

Respiratory system

Exchanges O_2 and CO_2 between the internal and external environment.

Endocrine system

Regulates internal environment by secreting hormones that travel through the bloodstream to target areas.

Circulatory system

Transports nutrients, water, oxygen, hormones, wastes, and other materials within the internal environment.

Skeletal system

Supports, protects, and moves body. Also stores minerals.

Muscular system

Powers and directs skeletal movement.

Reproductive system

Produces sex cells that form offspring, ensuring survival of genes. Female system is also site of fertilization and early development.

Urinary system

Adjusts internal environment by excreting excess water, salt, and other substances.

Immune system

Defends internal environment against injury from foreign cells and other irritants.

Lymphatic system

Drains excess fluid from tissues, cleans it, and returns it to the blood.

Integumentary system

External environment

H_2O
CO_2
O_2

Internal environment

Food

H_2O

Nervous system

Digestive system

Respiratory system

Endocrine system

Circulatory system

Skeletal system

Muscular system

Immune system

Urinary system

Repro-ductive system

Lymphatic system

H_2O, salts, wastes, acid

Sperm (in male) Offspring (in female)

Sperm (in female)

Waste, residue, H_2O

Figure 1-13 *Diagram of the body's internal environment.* The human body is like a bag of fluid separated from the external environment. Tubes, such as the digestive tract and respiratory tract, bring the external environment to deeper parts of the bag where substances may be absorbed into the internal fluid environment or excreted into the external environment. All the "accessories" somehow help maintain a constant environment inside the bag that allows the cells that live there to survive.

To accomplish this self-regulation, a highly complex and integrated communication control system or network is required. This type of network is called a **feedback control loop.** Different networks in the body control such diverse functions as blood carbon dioxide levels, temperature, heart rate, sleep cycles, and thirst. Information may be transmitted in these control loops by nervous impulses or by specific chemical messengers called *hormones,* which are secreted into the blood. Regardless of the body function being regulated or the mechanism of information transfer (nerve impulse or hormone secretion), these feedback control loops have the same basic components and work in the same way.

Basic Components of Control Mechanisms

There is a minimum of four basic components in every feedback control loop:

1. Sensor mechanism
2. Integrating, or control, center
3. Effector mechanism
4. Feedback

The terms *afferent* and *efferent* are important directional terms frequently used in physiology. They are used to describe movement of a signal from a sensor mechanism to a particular integrating or control center and, in turn, movement of a signal from that center to some type of effector mechanism. *Afferent* means that a signal is traveling toward a particular center or point of reference, and *efferent* means that the signal is moving away from a center or other point of reference. These terms are of particular importance in the study of the nervous and endocrine systems in Unit 3.

The process of regulation and the concept of return to normal require that the body be able to "sense" or identify the variable being controlled. Sensory nerve cells or hormone-producing (endocrine) glands frequently act as homeostatic sensors. To function in this way a specialized sensor must be able to identify the element being controlled. It must also be able to respond to any changes that may occur from the normal set point range. If deviations from the normal set point range occur, the sensor generates an afferent signal (nerve impulse or hormone) to transmit that information to the second component of the feedback loop—the integration or control center (Box 1-2).

When the integration or control center of the feedback loop (often a discrete area of the brain) receives input from a homeostatic sensor, that information is analyzed and integrated with input from other sensors, and then an efferent signal travels from the center to some type of effector mechanism, where a specific action is initiated, if necessary, to maintain homeostasis. First, the level or magnitude of the variable being measured by the sensor is compared with the normal set point level that must be maintained for homeostasis. If significant deviation from that predetermined level exists, the integration/control center sends its own specialized signal to the third component of the control loop—the effector mechanism.

Effectors are organs, such as muscles or glands, that directly influence controlled physiological variables. For example, it is effector action that increases or decreases variables such as body temperature, heart rate, blood pressure, or blood sugar concentration to keep them within their normal range. The activity of effectors is ultimately regulated by feedback of information regarding their own effects on a controlled variable.

Many instructors use the example of a furnace controlled by a thermostat to explain how feedback control systems work. This analogy is a good one because it parallels the homeostatic mechanism used to control body temperature. Note that changes in room temperature (the controlled variable) are detected by a thermometer (sensor) attached to the thermostat (integrator) (Figure

BOX 1-2: FYI
Changing the Set Point

Like the set point on a furnace, the physiological set points in your body can be changed. Your body's set point temperature is a good example. First, not everyone's set point, or "normal," body temperature is the same. The figure at right shows the difference in body temperatures observed in a group of healthy students. You can see that temperatures varied widely. This explains why some people are comfortable at a temperature that is too cold for others around them—their temperature set point must be lower. However, your set point can also change under varying circumstances. For example, we know a fellow who once turned up the thermostat in his house to get the temperature warm enough to get some unwelcome visitors to leave. Likewise, during a bacterial infection your immune system sends chemicals to signal the brain's hypothalamus to "turn up the set point temperature." Your body shivers, and you may ask for a blanket as your body tries to reach this new higher set point. You now have a fever. The bacteria that have invaded your body did so because they liked the temperature of your body. During a fever your body becomes uncomfortably warm for the bacteria, and they slow down their reproductive rate, which slows the infection. At the same time, the warmer temperature helps improve the immune system's function in dealing with the bacteria. After the infection is over, the hypothalamus returns to its usual set point and your fever goes away.

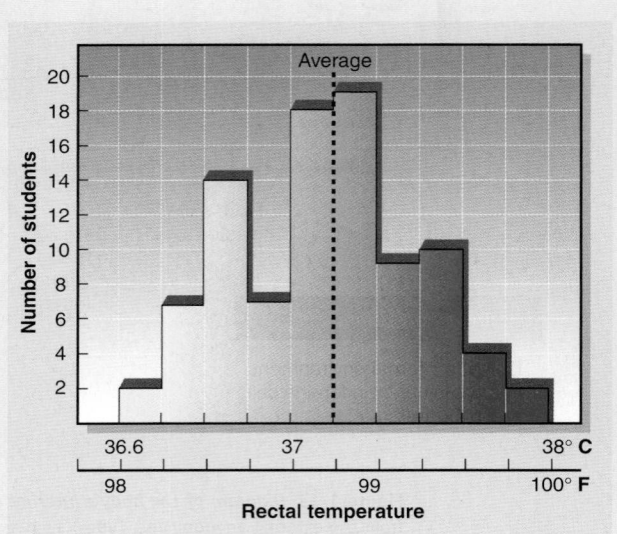

Range of normal body temperatures. In a well-controlled experiment, a group of healthy students show a wide range of normal rectal temperatures. The average (mean) temperature of the group is 37.1° C (98.8° F).

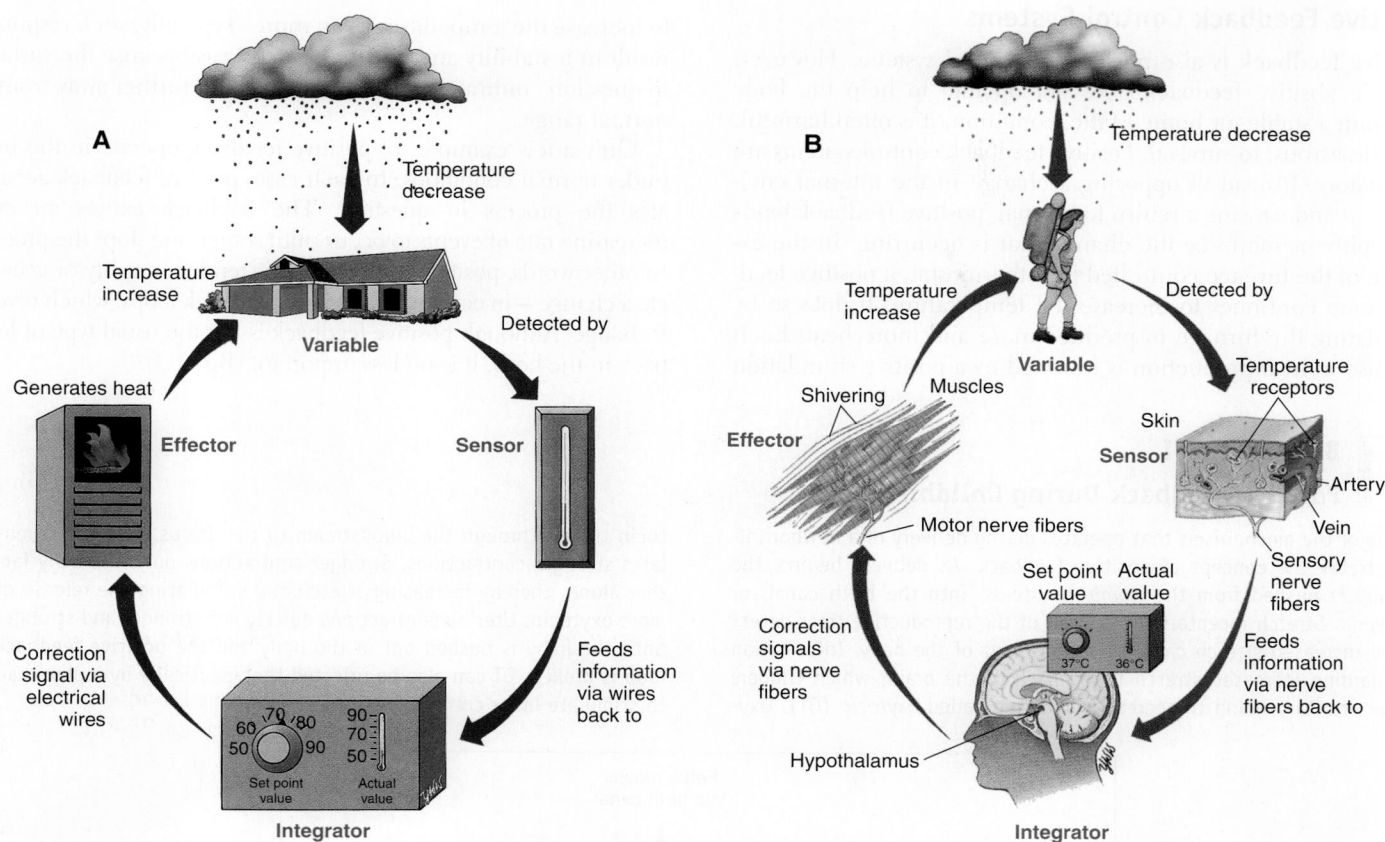

Figure 1-14 *Basic components of homeostatic control mechanisms.* **A,** Heat regulation by a furnace controlled by a thermostat. **B,** Homeostasis of body temperature. Note that in both examples A and B a stimulus (drop in temperature) activates a sensor mechanism (thermostat or body temperature receptor) that sends input to an integrating, or control, center (on-off switch or hypothalamus), which then sends input to an effector mechanism (furnace or contracting muscle). The resulting heat that is produced maintains the temperature in a "normal range." Feedback of effector activity to the sensor mechanism completes the control loop.

1-14, *A*). The thermostat contains a switch that controls the furnace (effector). When cold weather causes a decrease in room temperature, the change is detected by the thermometer and relayed to the thermostat. The thermostat compares the actual room temperature with the set point temperature. After the integrator determines that the actual temperature is too low, it sends a "correction" signal by switching on the furnace. The furnace produces heat and thus increases room temperature back toward normal. As the room temperature increases above normal, feedback information from the thermometer causes the thermostat to switch off the furnace. Thus by intermittently switching the furnace off and on, a relatively constant room temperature can be maintained.

Body temperature can be regulated in much the same way as room temperature is regulated by the furnace system just described (see Figure 1-14, *B*). Here, sensory receptors in the skin and blood vessels act as sensors by monitoring body temperature. When cold weather causes the body temperature to decrease, feedback information is relayed through the nerves to the "thermostat" in a part of the brain called the *hypothalamus* (hye-poh-THAL-ah-muss). The hypothalamic integrator compares the actual body temperature with the "built-in" set point body temperature and subsequently sends a nerve signal to effectors. In this example, the skeletal muscles act as effectors by shivering and thus producing heat. Shivering increases body temperature back

to normal, at which point it stops as a result of feedback information that causes the hypothalamus to shut off its stimulation of the skeletal muscles. More specifics of body temperature control are discussed in Chapter 6.

The impact of effector activity on sensors may be positive or negative. Therefore, homeostatic control mechanisms are categorized as negative or positive feedback systems. By far the most important and numerous of the homeostatic control mechanisms are negative feedback systems.

Negative Feedback Control Systems

The example of temperature regulation by action of a thermostatically regulated furnace is a classic example of **negative feedback.** Negative feedback control systems are inhibitory. They oppose a change (such as a drop in temperature) by creating a response (production of heat) that is opposite in direction to the initial disturbance (fall in temperature below a normal set point). All negative feedback mechanisms in the body respond in this way regardless of the variable being controlled. They produce an action that is opposite to the change that activated the system. It is important to emphasize that negative feedback control systems *stabilize* physiological variables. They keep variables from straying too far outside their normal ranges. Negative feedback systems are responsible for maintaining a constant internal environment.

Positive Feedback Control Systems

Positive feedback is also possible in control systems. However, because positive feedback does not operate to help the body maintain a stable, or homeostatic, condition, it is often harmful, even disastrous, to survival. Positive feedback control systems are stimulatory. Instead of opposing a change in the internal environment and causing a return to normal, positive feedback tends to amplify or reinforce the change that is occurring. In the example of the furnace controlled by a thermostat, a positive feedback loop continues to increase the temperature. It does so by stimulating the furnace to produce more and more heat. Each increase in heat production is followed by a positive stimulation to increase the temperature even more. Typically, such responses result in instability and disrupt homeostasis because the variable in question continues to deviate further and further away from its normal range.

Only a few examples of positive feedback operate in the body under normal conditions. In each case, positive feedback accelerates the process in question. The feedback causes an ever-increasing rate of events to occur until something stops the process. In other words, positive feedback loops tend to amplify or accelerate a change—in contrast to negative feedback loops, which reverse a change. Although positive feedback is not the usual type of feedback in the body, it is no less important (Box 1-3).

BOX 1-3: FYI

Positive Feedback During Childbirth

One of the mechanisms that operates during delivery of a newborn illustrates the concept of positive feedback. As delivery begins, the baby is pushed from the *womb,* or *uterus,* into the birth canal, or *vagina*. Stretch receptors in the wall of the reproductive tract detect the increased stretch caused by movement of the baby. Information regarding increased stretch is fed back to the brain, which triggers the pituitary gland to secrete a hormone called *oxytocin* (OT). Oxytocin travels through the bloodstream to the uterus, where it stimulates stronger contractions. Stronger contractions push the baby farther along, thereby increasing stretch and stimulating the release of more oxytocin. Uterine contractions quickly get stronger and stronger until the baby is pushed out of the body and the positive feedback loop is broken. OT can also be injected therapeutically by a physician to stimulate labor contractions.

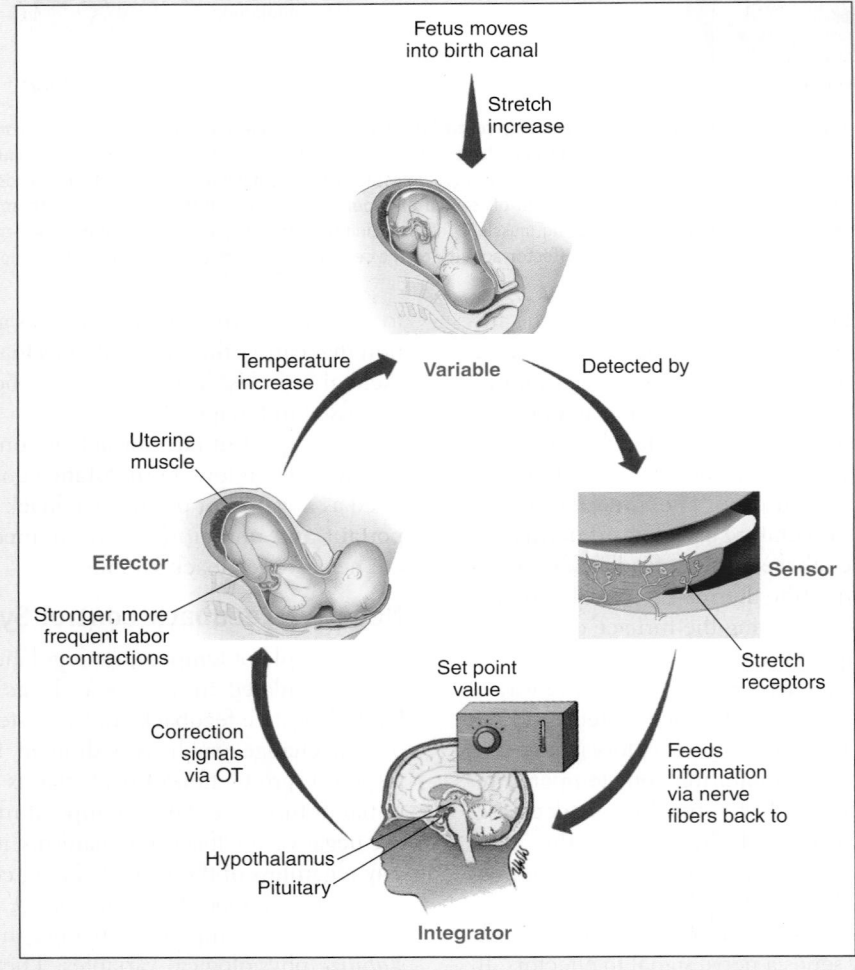

Positive feedback loop.

Events that lead to a simple sneeze, the birth of a baby, an immune response to an infection, or the formation of a blood clot are all examples of positive feedback.

Feed-Forward in Control Systems

As you study the complexity of control systems throughout the body, you will no doubt run into cases of *feed-forward* in control systems. Feed-forward is the concept that information may flow ahead to another process to trigger a change in anticipation of an event that will follow. For example, when you eat a meal, the stomach stretches and this triggers stretch sensors in the stomach wall. As you would expect, the stretch sensors trigger a feedback response that causes the release of digestive juices and contraction of stomach muscles. This is normal negative feedback because secretion and muscle activity eventually get rid of the food and bring the stretch of the stomach back down to normal. It will continue as long as there is food to stretch the stomach. At the same time, the stretch stimulus is triggering the small intestine and related organs to increase secretion there as well—*before* the food has arrived. In other words, information from one feedback loop (in the stomach) has leaped ahead to the next logical feedback loop (in the intestines) to get the second loop ready ahead of time. Feed-forward causes a feedback loop to anticipate a stimulus (in this case, stretch of the intestinal wall caused by food moving down from the stomach) before it actually happens.

In summary, many homeostatic mechanisms operate on the negative feedback principle. They are activated, or turned on, by changes in the environment that surround every body cell. Negative feedback systems are inhibitory. They reverse the change that initially activated the homeostatic mechanism. By reversing the initial change, a homeostatic mechanism tends to maintain or restore internal constancy. Occasionally, a positive (stimulatory) feedback mechanism helps promote survival. Such positive or stimulatory feedback systems may be required to bring specific body functions to swift completion (see Box 1-3). Feed-forward occurs when sensory information "jumps ahead" to a feedback loop to get it started before the stimulus actually changes the controlled physiological variable.

Levels of Control

One of the first principles that will occur to you as you study human physiology is that the functions of cells, tissues, organs, and systems are integrated into a coordinated whole. This is accomplished by many different feedback loops and feed-forward systems operating at many different levels of organization within the body (Figure 1-15).

Intracellular control mechanisms operate at the cell level. These mechanisms regulate functions within the cell, often by means of genes and enzymes. The role of genes and enzymes will be discussed further in Chapter 4.

Intrinsic control mechanisms operate at the tissue and organ levels. Sometimes also called local control or *autoregulation*, intrinsic mechanisms often make use of chemical signals. For example, prostaglandins are molecules sent as signals to other nearby cells. Intrinsic regulation may also be "built in" the tissue or organ. For instance, when cardiac muscle in the heart is stretched, the muscle automatically contracts with more force. Prostaglandins will be discussed further in Chapters 2 and 16. Many other examples of intrinsic control will be discussed throughout the book.

Extrinsic control means "outside" control and operates at the system and organism levels. Extrinsic control usually involves nervous and endocrine (hormonal) regulation. It is called "extrinsic" control because the nerve signals and hormones originate outside the controlled organ. Nervous regulation will be introduced in more detail in Chapters 12 through 15 and endocrine regulation will be introduced in Chapter 16.

QUICK CHECK

24. Define the term *homeostasis*.
25. List the basic components of every feedback control system.
26. Explain the mechanism of action of negative and positive feedback control systems.
27. Contrast *intracellular, intrinsic,* and *extrinsic* levels of control.

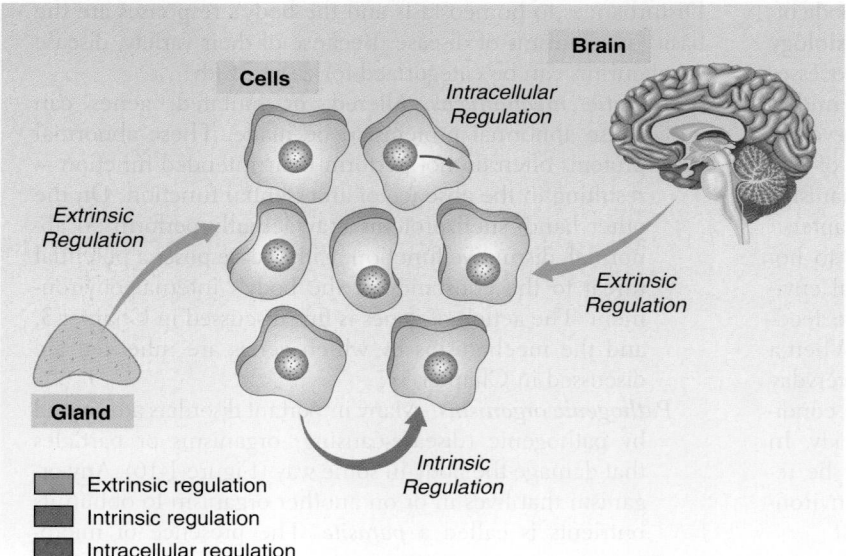

Figure 1-15 *Levels of control.* The many complex processes of the body are coordinated at many levels: intracellular (within cells), intrinsic (within tissues/organs), and extrinsic (organ to organ).

Cycle of Life

Life Span Considerations

An important generalization about body structure is that every organ, regardless of location or function, undergoes change over the years. In general, the body performs its functions least well at both ends of life—in infancy and in old age. Organs develop and grow during the years before maturity, and body functions gradually become more and more efficient and effective. In a healthy young adult all body systems are mature and fully operational. Homeostatic mechanisms tend to function most effectively during this period of life to maintain the constancy of one's internal environment.

After maturity, effective repair and replacement of the body's structural components often decrease. The term **atrophy** is used to describe the wasting effects of advancing age. In addition to structural atrophy, the functioning of many physiological control mechanisms also decreases and becomes less precise with advancing age. The changes in functions that occur during the early years are called *developmental processes*. Those that occur during the late years are called *aging processes*. The study of aging processes and other changes that occur in our lives as we get older is called *gerontology*. Many specific age changes are noted in the chapters that follow.

THE BIG PICTURE

Organization of the Body

Ultimately, your success in the study of anatomy and physiology, your ability to see the "big picture," will require understanding, synthesis, and integration of structural information and functional concepts. After you have completed your study of the individual organ systems of the body presented in the chapters that follow, you must be able to reassemble the parts and view the body in a holistic, integrated way.

The body is truly more than the sum of the parts, and understanding the connectedness of human structure and function is the real challenge—and the greatest reward—in the study of anatomy and physiology. Your ability to integrate otherwise isolated factual information about bones, muscles, nerves, and blood vessels, for example, will allow you to view anatomical components of the body and their specialized functions in a more cohesive and understandable way. This chapter introduces the principle of homeostasis as the glue that integrates and explains how the normal interaction of structure and function is achieved and maintained and how a breakdown of this integration results in disease. Furthermore, it provides the basis for understanding and integrating the body of knowledge, both factual and conceptual, that anatomy and physiology encompass.

Mastery of any academic discipline or achieving success in any health care–related work environment requires the ability to communicate effectively. The ability to understand and appropriately use the vocabulary of anatomy and physiology allows you to accurately describe the body itself, the orientation of the body in its surrounding environment, and the relationships that exist between its component parts in both health and disease. This chapter provides you with information necessary to be successful in seeing the "big picture" as you master the details of each organ system, which although presented separately in subsequent chapters of the text, are in reality part of a marvelously integrated whole.

Mechanisms of Disease

SOME GENERAL CONSIDERATIONS

A clearer understanding of the normal function of the body often comes from our study of disease (Box 1-4). **Pathophysiology** is the organized study of the underlying physiological processes associated with disease. Pathophysiologists attempt to understand the mechanisms of a disease and its course of development, or *pathogenesis*. Near the end of each chapter of this book we briefly describe some important disease mechanisms that illustrate the normal functions described in that chapter.

Many diseases are best understood as disturbances to homeostasis, the relative constancy of the body's internal environment. If homeostasis is disturbed, various negative feedback mechanisms usually return the body to normal. When a disturbance goes beyond the normal fluctuation of everyday life, we can say that a disease condition exists. In acute conditions the body recovers its homeostatic balance quickly. In chronic diseases a normal state of balance may never be restored. If the disturbance keeps the body's internal environment too far from normal for too long, death may result.

Basic Mechanisms of Disease

Disturbances to homeostasis and the body's responses are the basic mechanisms of disease. Because of their variety, disease mechanisms can be categorized for easier study:

Genetic mechanisms. Altered, or mutated, genes can cause abnormal proteins to be made. These abnormal proteins often do not perform their intended function—resulting in the absence of an essential function. On the other hand, such proteins may actually perform an abnormal, disruptive function. Either case poses a potential threat to the constancy of the body's internal environment. The action of genes is first discussed in Chapter 3, and the mechanisms by which genes are inherited are discussed in Chapter 34.

Pathogenic organisms. Many important disorders are caused by pathogenic (disease-causing) organisms or particles that damage the body in some way (Figure 1-16). Any organism that lives in or on another organism to obtain its nutrients is called a *parasite*. The presence of micro-

Mechanisms of Disease—cont.

BOX 1-4: HEALTH MATTERS
Disease Terminology

Everyone is interested in **pathology**—the study of disease. Researchers want to know the scientific basis of abnormal conditions. Health practitioners want to know how to prevent and treat various diseases. Every one of us, when we suffer from the inevitable head cold or something more serious, want to know what is going on and how best to deal with it. Pathology has its own terminology, as in any specialized field. Just as with other scientific terms, most disease-related terms are derived from Latin and Greek word parts. For example, *patho-* comes from the Greek word for disease *(pathos)* and is used in many terms, including "pathology" itself.

Disease conditions are usually *diagnosed* or identified by **signs** and **symptoms.** Signs are objective abnormalities that can be seen or measured by someone other than the patient, whereas symptoms are the subjective abnormalities that are felt only by the patient. Although *sign* and *symptom* are distinct terms, we often use them interchangeably. A **syndrome** is a collection of different signs and symptoms that occur together. When signs and symptoms appear suddenly, persist for a short time, and then disappear, we say that the disease is **acute.** On the other hand, diseases that develop slowly and last for a long time (perhaps for life) are labeled **chronic** diseases. The term *subacute* refers to diseases with characteristics somewhere between acute and chronic.

The study of all the factors involved in causing a disease is its **etiology.** The etiology of a skin infection often involves a cut or abrasion and subsequent invasion and growth of a bacterial colony. Diseases with undetermined causes are said to be **idiopathic. Communicable** diseases are those that can be transmitted from one person to another.

The term *etiology* refers to the theory of a disease's cause, but the actual pattern of a disease's development is called its **pathogenesis.** The common cold, for example, begins with a *latent,* or "hidden," stage during which the cold virus establishes itself in the patient. No signs of the cold are yet evident. In infectious diseases, the latent stage is also called **incubation.** The cold may then manifest itself as a mild nasal drip and trigger a few sneezes. It subsequently progresses to its full fury and continues for a few days. After the cold infection has run its course, a period of *convalescence,* or recovery, occurs. During this stage, body functions return to normal. Some chronic diseases, such as cancer, exhibit a temporary reversal that seems to be a recovery. Such reversal of a chronic disease is called a **remission.** If a remission is permanent, we say that the person is "cured."

Epidemiology is the study of the occurrence, distribution, and transmission of diseases in human populations. A disease that is native to a local region is called an **endemic** disease. If the disease spreads to many individuals at the same time, the situation is called an **epidemic. Pandemics** are epidemics that affect large geographic regions, perhaps spreading worldwide. Because of the speed and availability of modern air travel, pandemics are more common than they once were. Almost every flu season we see a new strain of influenza virus quickly spreading from continent to continent.

Names of specific diseases are often descriptive, such as *rheumatoid arthritis* (meaning "autoimmune inflammation of joints"). Some disease names are eponyms, with a person's name incorporated into the term, as in *Parkinson disease (PD)*. Notice that we follow the American Medical Association (AMA) format for eponyms, which prohibits adding the 's to a person's name. Notice also that some disease names are abbreviated with an acronym such as *DMD* (for *Duchenne Muscular Dystrophy*).

---

scopic or larger parasites may interfere with normal body functions of the *host* and cause disease. Besides parasites, there are organisms that poison or otherwise damage the human body to cause disease. Some of the major pathogenic organisms and particles include the following:

Prions (proteinaceous infectious particles) are proteins that convert normal proteins of the nervous system into abnormal proteins, thereby causing loss of nervous system function. The abnormal form of the protein may also be inherited. Prions are a newly discovered type of pathogen, and not much is known about how the prion works to cause such diseases as bovine spongiform encephalopathy (BSE; "mad cow disease") or variant Creutzfeldt-Jakob disease (vCJD).

Viruses are intracellular parasites that consist of a DNA or RNA core surrounded by a protein coat and, sometimes, a lipoprotein envelope. They are particles that invade human cells and cause them to produce viral components. Sometimes, the term *virion* is used for the complete virus particle as it exists outside the host cell.

Bacteria are tiny, primitive cells that lack nuclei. They cause infection by parasitizing tissues or otherwise disrupting normal function (Box 1-5).

Fungi are simple organisms similar to plants but lack the chlorophyll pigments that allow plants to make their own food. Because they cannot make their own food, fungi must parasitize other tissues, including those of the human body.

Protozoa are protists, one-celled organisms larger than bacteria whose DNA is organized into a nucleus. Many types of protozoa parasitize human tissues.

Pathogenic animals are large, multicellular organisms such as insects and worms. Such animals can parasitize human tissues, bite or sting, or otherwise disrupt normal body function.

Examples of infections or other conditions caused by pathogenic organisms are given in many chapters throughout this book.

Tumors and cancer. Abnormal tissue growths, or neoplasms, can cause various physiological disturbances, as described in Chapter 5.

Mechanisms of Disease — cont.

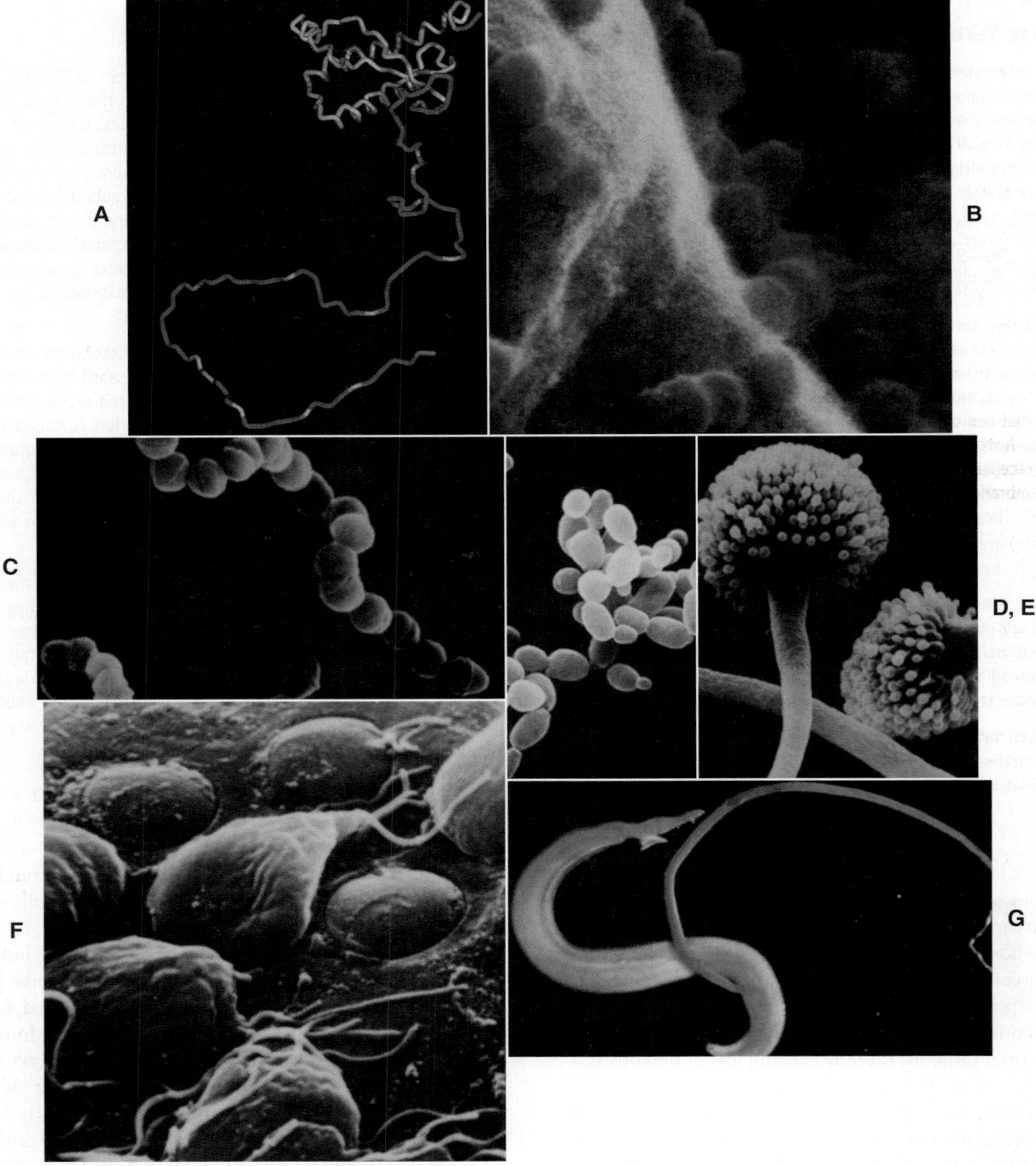

Figure 1-16 *Pathogenic organisms.* **A,** Prion (cause of variant Creutzfeldt-Jakob disease). **B,** Viruses (the human immunodeficiency virus [HIV] that causes AIDS). **C,** Bacteria (*Streptococcus* bacteria that cause strep throat and other infections). **D,** Fungi (yeast cells that commonly infect the urinary and reproductive tracts). **E,** Fungi (the mold that causes aspergillosis). **F,** Protozoans (the flagellated cells that cause traveler's diarrhea). **G,** Pathogenic animals (the parasitic worms that cause snail fever).

Mechanisms of Disease—cont.

Disease as a Weapon

World events have shown us that the intentional transmission of disease can be used as a weapon of terror. Anthrax, a bacterial infection caused by *Bacillus anthracis,* is an example of a pathogen that has been intentionally distributed to otherwise healthy victims in acts of bioterrorism. This particular bacterium ordinarily affects plant-eating animals such as sheep, cattle, and goats, often killing them.

The anthrax bacterium can assume the form of a *spore* that is resistant to heat, drying, and chemicals but later becomes active to cause infection. Rarely, humans inhale some anthrax spores or get the spores in an open cut when handling infected animals or their hides. The inhaled form may be fatal if not treated quickly with antibiotics. The cutaneous (skin) form is less serious and characterized by a reddish brown patch on the skin that ulcerates and forms a dark, nearly black scab (see the figure), followed by muscle pain, internal hemorrhage (bleeding), headache, fever, nausea, and vomiting. Anthrax causes disease by releasing a toxin that latches onto receptors on the cells of the host, punches a hole in the cell's membrane, and inserts a portion of the toxin called "lethal factor" that destroys proteins in the cell and kills it.

If the infection is discovered before the anthrax bacteria have time to make large amounts of toxin, antibiotics such as doxycycline and ciprofloxacin can cure anthrax. Scientists are also working to perfect drugs that imitate the cell's receptors and thus "gum up" the toxin with fake receptors before it can attack cells. Vaccines are available, but they must be given long before possible exposure to the spores.

Anthrax spores have been refined for military purposes, even though this is now outlawed by various treaties, and have been used by terrorists to cause panic among civilian populations. Other bacteria such as *Yersinia pestis* (plague), viruses such as smallpox, and a variety of genetically engineered forms of known pathogens may also be added to the arsenal of the terrorist.

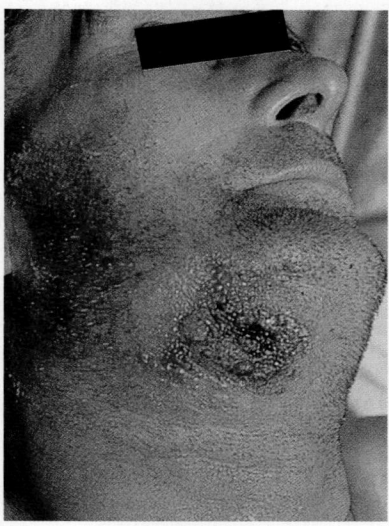

Cutaneous anthrax.

Physical and chemical agents. Agents such as toxic or destructive chemicals, extreme heat or cold, mechanical injury, and radiation can each affect the normal homeostasis of the body. Examples of healing of tissues damaged by physical agents are discussed in Chapters 4, 5, 6, and other chapters.

Malnutrition. Insufficient or imbalanced intake of nutrients causes various diseases; these are outlined in Chapters 26 and 27.

Autoimmunity. Some diseases result from the immune system attacking one's own body *(autoimmunity)* or from other mistakes or overreactions of the immune response. Autoimmunity, literally "self-immunity," is discussed in Chapter 21 with other disturbances of the immune system.

Inflammation. The body often responds to disturbances with an *inflammatory response.* The inflammatory response, which is described in Chapter 5, is a normal mechanism that usually speeds recovery from an infection or injury. However, when the inflammatory response occurs at inappropriate times or is abnormally prolonged or severe, normal tissues may become damaged. Thus some disease symptoms are caused by the inflammatory response.

Degeneration. By means of many still unknown processes, tissues sometimes break apart or *degenerate.* Although a normal consequence of aging, degeneration of one or more tissues resulting from disease can occur at any time. The degeneration of tissues associated with aging is discussed in nearly every chapter of this book.

Risk Factors

Other than direct causes or disease mechanisms, certain predisposing conditions may exist that make a disease more likely to develop. Usually called *risk factors,* they often do not actually cause a disease but just put one "at risk" for it. Risk factors can combine and increase a person's chance for contracting a specific disease even more. Some of the major types of risk factors are as follows:

Genetic factors. There are several types of genetic risk factors. Sometimes an inherited trait puts one at greater than normal risk for development of a specific disease. For example, light-skinned people are more at risk for certain forms of skin cancer than dark-skinned people are. This occurs because light-skinned people have less pigment in their skin to protect them from cancer-causing ultraviolet radiation (see Chapter 6). Membership in a certain ethnic group, or *gene pool,* involves the "risk" of inheriting a disease-causing gene that is common in that gene pool. For example, certain Africans and their descendants are at greater than average risk of inheriting *sickle cell anemia*—a deadly blood disorder.

Age. Biological and behavioral variations during different phases of the human life cycle put us at greater risk for certain diseases at certain times in our life. For example, middle ear infections are more common in infants than in adults because of the difference in ear structure at different ages.

Mechanisms of Disease—cont.

Lifestyle. The way we live and work can put us at risk for some diseases. People whose work or personal activity puts them in direct sunlight for long periods have a greater chance for development of skin cancer because this puts them in more frequent contact with ultraviolet radiation from the sun. Some researchers believe that the high-fat, low-fiber diet common among people in the "developed" nations increases the risk for certain types of cancer.

Stress. Physical, psychological, or emotional stress can put one at risk for problems such as chronic high blood pressure (hypertension), peptic ulcers, and headaches. Conditions caused by psychological factors are sometimes called *psychogenic* (mind-caused) disorders. Chapter 22 discusses the concept of stress and its effect on health.

Environmental factors. Although environmental factors such as climate and pollution can actually cause injury or disease, some environmental situations simply put us at greater risk for getting certain diseases. For example, because some parasites survive only in tropical environments, we are not at risk if we live in a temperate climate.

Microorganisms. Different types of pathological organisms, such as viruses and bacteria, are now suspected of being "infectious cofactors" in the development of certain noninfectious diseases that in the past were not considered to result directly from their presence in the body. For example, we now have very strong evidence to link infections caused by hepatitis B virus with liver cancer and human papillomavirus with cervical cancer. We also know that the bacterium *Helicobacter pylori*, which causes ulcers, is in some way also a factor in the development of certain types of stomach cancer.

Preexisting conditions. A preexisting condition can adversely affect our capacity to defend ourselves against an entirely different condition or disease. Thus the *primary* (preexisting) condition can put a person at risk for development of a *secondary* condition. For example, in individuals with AIDS the primary condition is characterized by a suppressed immune system. As a result, secondary or "opportunistic" infections such as pneumonia often develop.

LANGUAGE OF SCIENCE *(Cont'd from page 5)*

medial (MEE-dee-al) [*media* middle]

medullary (MED-oo-lar-ee) [*medulla* marrow]

metabolism (me-TAB-o-lizm) [*metabol-* change, *-ism* condition]

negative feedback

organ [*organ* instrument]

organelles (or-gah-NELLZ) [*organ-* instrument, *-elle* small]

organism (OR-gah-niz-im) [*organ-* instrument, *-ism* condition]

parietal (pah-RYE-ih-tal) [*parie* wall]

pathology (pah-THOL-oh-jee) [*patho-* disease, *-ology* words (study of)]

peripheral (pe-RIF-er-al) [*peri-* around, *-phera* boundary]

physiology (fiz-ee-OL-o-jee) [*physio-* nature (function), *-ology* words (study of)]

positive feedback

posterior (pos-TEER-ee-or) [*poster-* behind]

proximal (PROK-si-mal) [*proxima-* near]

sagittal plane (SAJ-i-tal plane) [*sagitta* arrow]

set point

somatotype (so-MAT-o-type) [*soma-* body, *-type* mark]

superficial (soo-per-FISH-al) [*super-* over or above, *-fici* face]

superior (soo-PEER-ee-or) [*super-* over or above]

system (SIS-tem) [*system* organized whole]

tissue (TISH-yoo) [*tissu* fabric]

transverse plane (TRANS-vers plane) [*trans-* across or through, *-vers* turn]

ventral cavity (VEN-tral KAV-ih-tee) [*ventr-* belly, *cav-* cavity]

viscera, visceral (VISS-er-ah) (VISS-er-al) [*visc* internal organ] sing. viscus

LANGUAGE OF MEDICINE

acute (ah-KYOOT) [*acut* sharp]

atrophy (AT-ro-fee) [*a-* without, *-troph* nourishment]

bacteria (bak-TEE-ree-ah) [*bacterium* small staff] pl. bacteria

chronic (KRON-ik) [*chron-* time, *-ic* pertaining to]

communicable (kom-MYOO-ni-kah-bil) [*communi* common]

endemic (en-DEM-ik) [*en-* in, *-dem-* people, *-ic* pertaining to]

epidemic (ep-i-DEM-ik) [*epi-* upon, *-dem-* people, *-ic* pertaining to]

epidemiology (EP-i-dee-mee-OL-o-jee) [*epi-* upon, *-dem-* people, *-ology* words (study of)]

etiology (e-tee-OL-o-jee) [*etio-* cause, *-ology* words (study of)]

fungi (FUNG-eye) [*fungus* mushroom] pl. fungi

idiopathic (id-ee-o-PATH-ik) [*idio-* peculiar, *-path-* disease, *-ic* pertaining to]

incubation (in-kyoo-BAY-shun) [*in-* in or on, *-cuba-* to lie, *-tion* condition of]

pandemics (pan-DEM-iks) [*pan-* all, *-dem-* people, *-ic* pertaining to]

pathogenesis (path-o-JEN-e-sis) [*patho-* disease, *-genesis* origin]

pathogenic animals (path-o-JEN-ik AN-im-alz) [*patho-* disease, *-gen-* to produce, *-ic* condition of]

pathophysiology (path-o-fiz-ee-OL-o-jee) [*patho-* disease, *-physio-* nature (function), *-ology* words (study of)]

prions (PREE-ahnz) [condensed from *proteinaceous infectious particle*]

protozoa (pro-toe-ZO-ah) [*proto-* first, *-zoan* animal] pl. protozoa

remission (ree-MISH-un) [*re-* back or again, *-miss-* send, *-sion* condition of]

signs (synz) [*sign* mark]

symptoms (SIMP-tums) [*sym-* together, *-tom* to fall]

syndrome (SIN-drome) [*syn-* together, *-drome* course]

viruses (VYE-rus-ez) [*virus* poison]

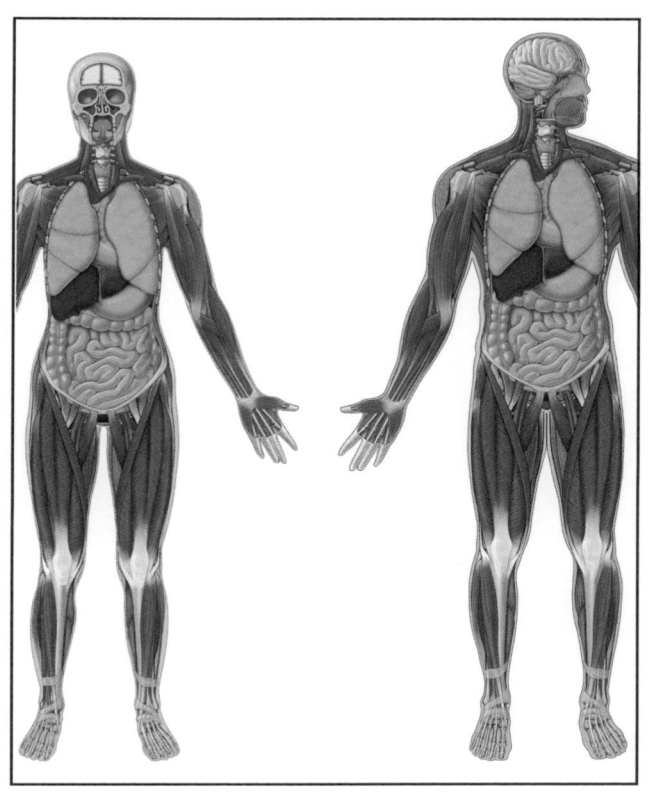

Clear View
of the
Human Body

**Developed by: KEVIN PATTON
and PAUL KRIEGER**

Illustrated by: Dragonfly Media Group

INTRODUCTION

A complete understanding of human anatomy and physiology requires an appreciation for how structures within the body relate to one another. Such appreciation for anatomical structure has become especially important in the 21st century with the explosion in the use of diverse methods of medical imaging that rely on the ability to interpret sectional views of the human body.

The best way to develop your understanding of overall anatomical structure is to carefully dissect a large number of male and female human cadavers—then have those dissected specimens handy while reading and learning about each system of the body. Obviously, such multiple dissections and constant access to specimens are impractical for nearly everyone. However, the experience of a simple dissection can be approximated by layering several partially transparent, two-dimensional anatomical diagrams in a way that allows a student to "virtually" dissect the human body simply by paging through the layers.

This **Clear View of the Human Body** provides a handy tool for dissecting simulated male and female bodies. It also provides views of several different parts of the human body in a variety of cross sections. The many different anterior and posterior views also give you a perspective on body structure that is not available with ordinary anatomical diagrams. This Clear View is an always-available tool to help you learn the three-dimensional structure of the body in a way that allows you to see how they relate to each other in a complete body. It will always be right here in your text-book, so place a bookmark here and refer to the Clear View frequently as you study each of the systems of the human body.

HINTS FOR USING THE CLEAR VIEW OF THE BODY

1. Starting at the first page of the Clear View, slowly lift the page as you look at the anterior view of the male and female bodies. You will see deeper structures appear, as if you had dissected the body. As you lift each successive layer of images, you will be looking at deeper and deeper body structures. A key to the labels is found in the gray sidebar.
2. Starting with the second section of the Clear View, notice that you are looking at the posterior aspect of the male and female body. Lift each layer from the left edge to reveal body structures in successive layers from the back to the front. This very unique view will help you understand structural relationships even better.
3. On each page of the Clear View, look at the horizontal section represented in the sidebar. The section you are looking at on any one page is from the location shown in the larger diagram as a red line. In other words, if you cut the body at the red line and tilted the upper part of the body toward you, you would see what is shown in the section diagram. Notice that each section has its own labeling system that is separate from the labels used in the larger images.

KEY

1. Epicranius m.
2. Temporalis m.
3. Orbicularis oculi m.
4. Masseter m.
5. Orbicularis oris m.
6. Pectoralis major m.
7. Serratus anterior m.
8. Basilic vein
9. Brachial fascia
10. Cephalic vein
11. Rectus sheath
12. Linea alba
13. Rectus abdominis m.
14. Umbilicus
15. Abdominal oblique m., external
16. Abdominal oblique m., internal
17. Transverse abdominis m.
18. Inguinal ring, external
19. Fossa ovalis
20. Fascia of the thigh
21. Great saphenous vein
22. Parietal bone
23. Frontal bone
24. Temporal bone
25. Zygomatic bone
26. Maxilla
27. Mandible
28. Sternocleidomastoid m.
29. Sternohyoid muscle
30. Omohyoid muscle
31. Deltoid m.
32. Pectoralis minor m.
33. Sternum
34. Rib (costal) cartilage
35. Rib
36. Greater omentum
37. Frontal lobe
38. Parietal lobe
39. Temporal lobe
40. Cerebellum
41. Nasal septum
42. Brachiocephalic vein
43. Superior vena cava
44. Thymus gland
45. Right lung
46. Left lung
47. Pericardium
48. Liver
49. Gall bladder
50. Stomach
51. Transverse colon
52. Small intestines
53. Biceps brachii m.
54. Brachioradialis m.
55. Adductor longus m.
56. Sartorius m.
57. Quadriceps femoris m.
58. Patellar ligament
59. Tibialis anterior m.
60. Sup. extensor retinaculum
61. Inf. extensor retinaculum
62. Cerebrum of brain
63. Cerebellum
64. Brain stem
65. Maxillary sinus
66. Nasal cavity
67. Tongue
68. Thyroid gland
69. Heart
70. Hepatic veins
71. Esophagus
72. Spleen
73. Celiac artery
74. Portal vein
75. Duodenum
76. Pancreas
77. Mesenteric artery
78. Ascending colon
79. Transverse colon
80. Descending colon
81. Sigmoid colon
82. Mesentery
83. Appendix
84. Inguinal ligament
85. Pubic symphysis
86. Extensor carpi radialis m.
87. Pronator teres m.
88. Flexor carpi radialis m.
89. Flexor digitorum profundus m.
90. Quadraceps femoris m.
91. Extensor digitorum longus m.
92. Thyroid cartilage
93. Trachea
94. Aortic arch
95. Right lung
96. Left lung
97. Pulmonary artery
98. Right atrium
99. Right ventricle
100. Left atrium
101. Left ventricle
102. Coracobrachialis m.
103. Inferior vena cava
104. Descending aorta
105. Right kidney
106. Left kidney
107. Right ureter
108. Rectum
109. Urinary bladder
110. Prostate gland
111. Iliac artery and vein
112. Uterus
113. Parietal bone
114. Frontal sinus
115. Sphenoidal sinus
116. Occipital bone
117. Palatine process
118. Cervical vertebrae
119. Corpus callosum
120. Thalamus
121. Trapezius m.
122. Acromion process
123. Coracoid process
124. Humerus
125. Subscapularis m.
126. Deltoid m. (cut)
127. Triceps m.
128. Brachialis m.
129. Brachioradialis m.
130. Radius
131. Ulna
132. Diaphragm
133. Thoracic duct
134. Quadratus lumborum m.
135. Psoas m.
136. Lumbar vertebrae
137. Iliacus m.
138. Gluteus medius m.
139. Iliofemoral ligament
140. Sacral nerves
141. Sacrum
142. Coccyx
143. Femur
144. Vastus lateralis m.
145. Femoral artery and vein
146. Adductor magnus m.
147. Patella
148. Fibula
149. Tibia
150. Fibularis longus m.
151. Spinal cord
152. Nerve root
153. Platysma m.
154. Splenius capitis m.
155. Levator scapulae m.
156. Rhomboideus m.
157. Infraspinatis m.
158. Teres major m.
159. Lumbodorsal fascia
160. Erector spinae m.
161. Serratus post. inf. m.
162. Latissimus dorsi m.
163. Gluteus medius m.
164. Gluteus maximus m.
165. Iliotibial tract
166. Flexor carpi ulnaris m.
167. Extensor carpi ulnaris m.
168. Extensor digitorum m.
169. Carpal ligament, dorsal
170. Interosseus m.
171. Gluteus minimus m.
172. Piriformis m.
173. Gemellus sup. m.
174. Obturator internus m.
175. Gemellus inf. m.
176. Quadratus femoris m.
177. Biceps femoris m.
178. Gastrocnemius m.
179. Calcaneal (Achilles) tendon
180. Calcaneus bone
181. Subcutaneous fat
182. Corpus spongiosum
183. Corpora cavernosa
184. Umbilical ligaments
185. Epigastric artery and vein
186. Right testis
187. Transverse thoracic m.
188. Parietal pleura
189. Common bile duct
190. Lesser omentum
191. Flexor digitorum profundus
192. Epiglottis

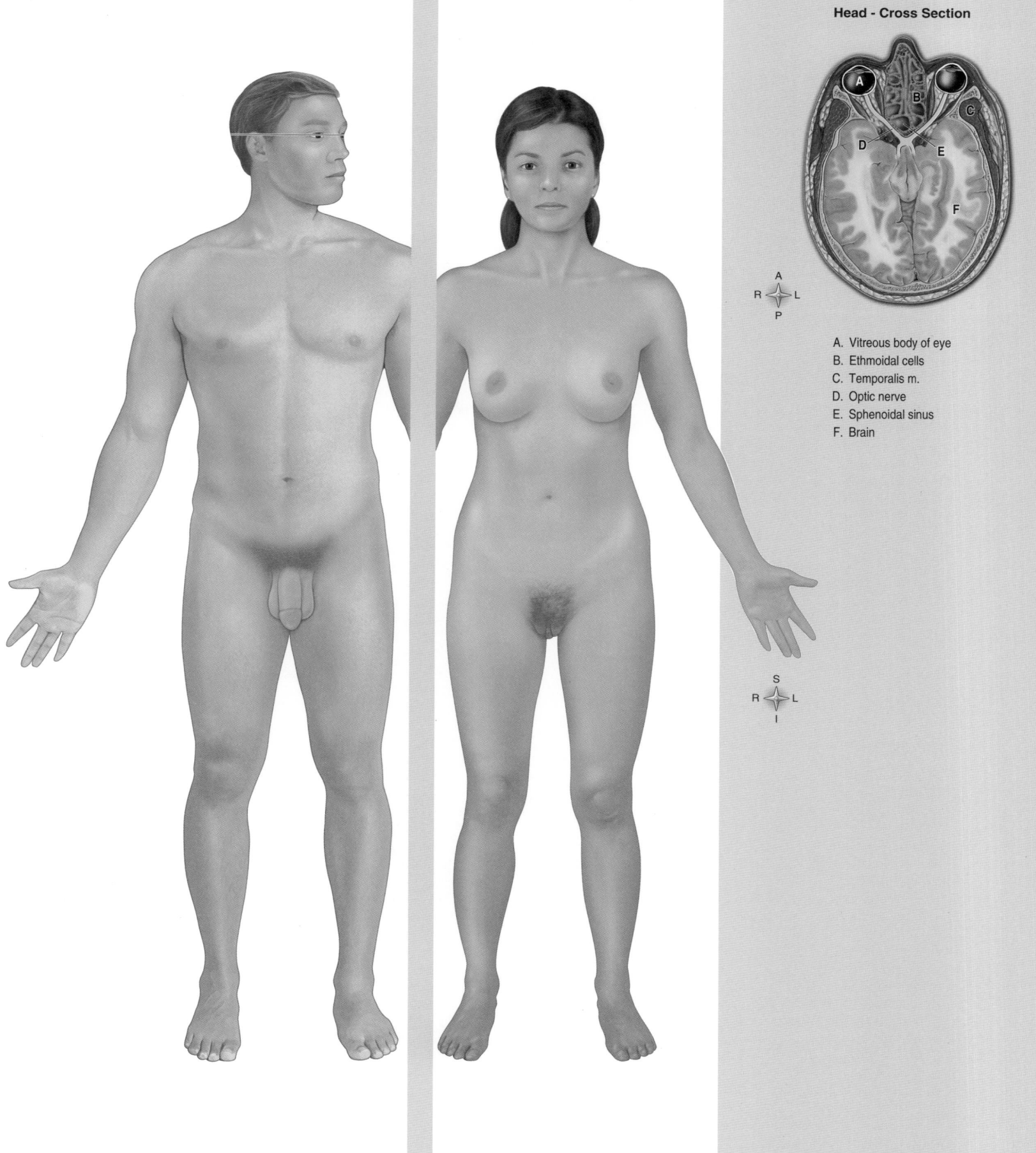

Head - Cross Section

A. Vitreous body of eye
B. Ethmoidal cells
C. Temporalis m.
D. Optic nerve
E. Sphenoidal sinus
F. Brain

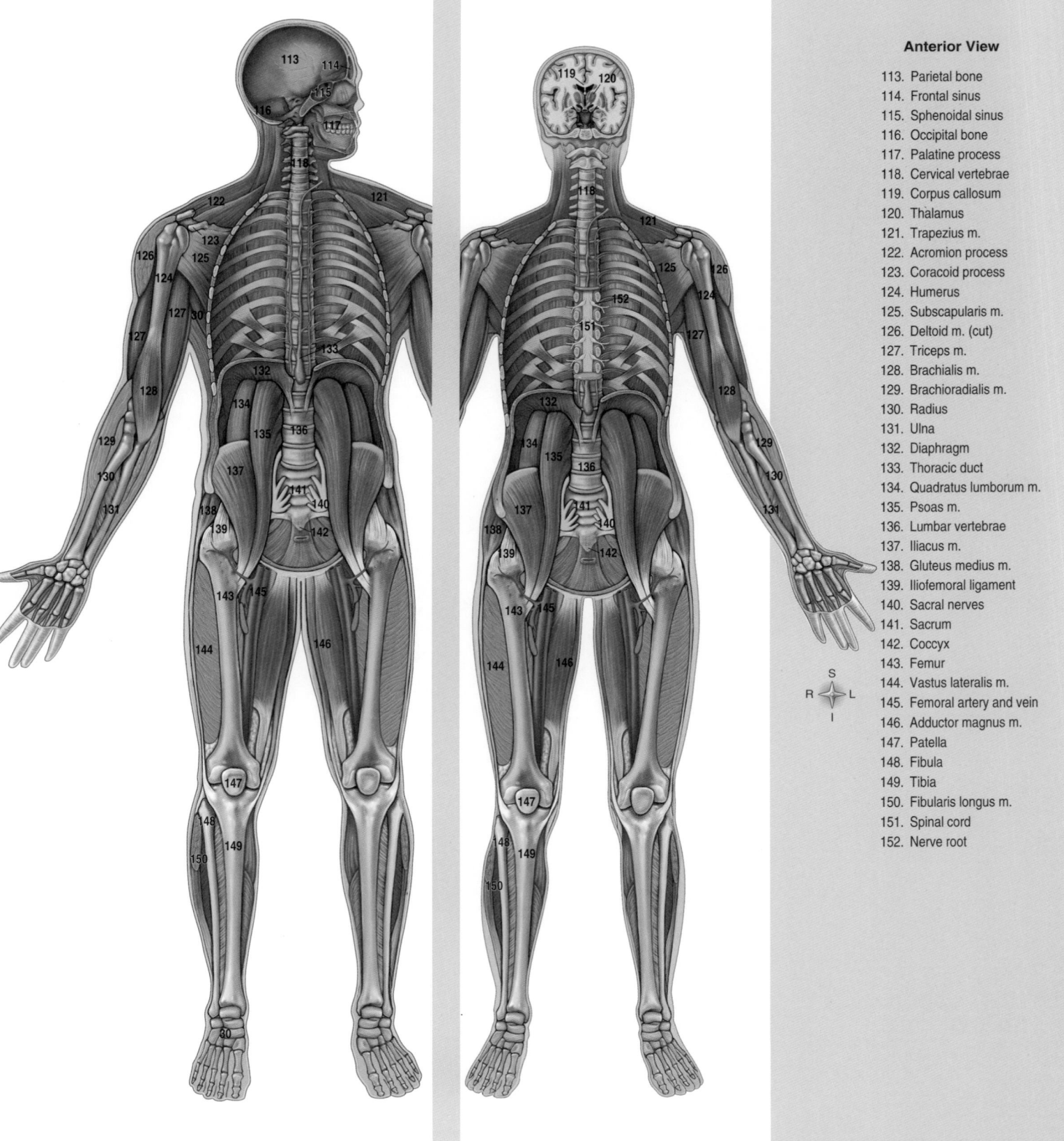

TERMS USED IN DESCRIBING BODY STRUCTURE

A. Directional terms (Figure 1-9)
  1. Superior
  2. Inferior
  3. Anterior (ventral)
  4. Posterior (dorsal)
  5. Medial
  6. Lateral
  7. Proximal
  8. Distal
  9. Superficial
  10. Deep
B. Terms related to organs
  1. Lumen (luminal)
  2. Central
  3. Peripheral
  4. Medullary (medulla)
  5. Cortical (cortex)
  6. Apical (apex)
  7. Basal (base)
C. Many directional terms are listed inside the front cover of the book

BODY PLANES AND SECTIONS (FIGURES 1-9 AND 1-10)

A. Planes are lines of orientation along which cuts or sections can be made to divide the body, or a body part, into smaller pieces
B. There are three major planes, which lie at right angles to each other:
  1. Sagittal plane front to back so that sections through this plane divide the body (or body part) into right and left sides
    a. If section divides the body (or part) into symmetrical right and left halves, the plane is called midsagittal or median sagittal
  2. Frontal (coronal) plane runs lengthwise (side to side) and divides the body (or part) into anterior and posterior portions
  3. Transverse (horizontal) plane is a "crosswise" plane—it divides the body (or part) into upper and lower parts

INTERACTION OF STRUCTURE AND FUNCTION

A. Complementarity of structure and function is an important and unifying concept in the study of anatomy and physiology
B. Anatomical structures often seem "designed" to perform specific functions because of their unique size, shape, form, or body location
C. Understanding the interaction of structure and function assists in the integration of otherwise isolated factual information

BODY TYPE AND DISEASE (FIGURE 1-11)

A. Somatotype—category of body build or physique
B. Endomorph—heavy, rounded physique with accumulation of fat
  1. "Apple-shaped" endomorph has more accumulation of fat in the waist than the hip

    a. Waist-to-hip ratio higher than 0.9 for women and 1.0 for men
    b. Higher risk for health problems than "pear shape"
  2. "Pear-shaped" endomorph has more accumulation of fat in the hips than in the waist
C. Mesomorph—muscular physique
D. Ectomorph—thin, often fragile physique with little fat

HOMEOSTASIS (FIGURE 1-14)

A. Term homeostasis coined by American physiologist Walter B. Cannon
B. *Homeostasis* is the term used to describe the relatively constant states maintained by the body—internal environment around body cells remains constant
C. Body adjusts important variables from a normal "set point" in an acceptable or normal range
D. Examples of homeostasis
  1. Temperature regulation
  2. Regulation of blood carbon dioxide level
  3. Regulation of blood glucose level

HOMEOSTATIC CONTROL MECHANISMS

A. Devices for maintaining or restoring homeostasis by self-regulation through feedback control loops
B. Basic components of control mechanisms
  1. Sensor mechanism—specific sensors detect and react to any changes from normal
  2. Integrating, or control, center—information is analyzed and integrated, and then if needed, a specific action is initiated
  3. Effector mechanism—effectors directly influence controlled physiological variables
  4. Feedback—process of information about a variable constantly flowing back from the sensor to the integrator
C. Negative feedback control systems
  1. Are inhibitory
  2. Stabilize physiological variables
  3. Produce an action that is opposite to the change that activated the system
  4. Are responsible for maintaining homeostasis
  5. Are much more common than positive feedback control systems
D. Positive feedback control systems
  1. Are stimulatory
  2. Amplify or reinforce the change that is occurring
  3. Tend to produce destabilizing effects and disrupt homeostasis
  4. Bring specific body functions to swift completion
E. Feed-forward occurs when information flows ahead to another process or feedback loop to trigger a change in anticipation of an event that will follow
F. Levels of control (Figure 1-15)
  1. Intracellular control
    a. Regulation within cells
    b. Genes or enzymes can regulate cell processes

2. Intrinsic control (autoregulation)
   a. Regulation within tissues or organs
   b. May involve chemical signals
   c. May involve other "built in" mechanisms
3. Extrinsic control
   a. Regulation from organ to organ
   b. May involve nerve signals
   c. May involve endocrine signals (hormones)

CYCLE OF LIFE: LIFE SPAN CONSIDERATIONS

A. Structure and function of body undergo changes over the early years (developmental processes) and late years (aging processes)
B. Infancy and old age are periods when the body functions least well
C. Young adulthood is period of greatest homeostatic efficiency
D. Atrophy—term to describe the wasting effects of advancing age

REVIEW QUESTIONS

1. Define the terms *anatomy* and *physiology*.
2. List and briefly describe the levels of organization that relate the structure of an organism to its function. Give examples characteristic of each level.
3. Give examples of each system level of organization in the body and briefly discuss the function of each.
4. What is meant by the term *anatomical position?* How do the specific anatomical terms of position or direction relate to this body orientation?
5. What is *bilateral symmetry?* What terms are used to identify placement of one body part with respect to another on the same or opposite sides of the body?
6. What does the term *somatotype* mean? Name the three major somatotype categories and briefly describe the general characteristics of each.
7. Define briefly each of the following terms: anterior, distal, sagittal plane, medial, dorsal, coronal plane, organ, parietal peritoneum, superior, tissue.

8. Locate the mediastinum.
9. What does the term *homeostasis* mean? Illustrate some generalizations about body function using homeostatic mechanisms as examples.
10. Define *homeostatic control mechanisms* and feedback control loops.
11. Identify the three basic components of a control loop.
12. Discuss in general terms the principle of *complementarity of structure and function.*
13. Briefly describe medical imaging techniques that allow physicians to examine internal structures in a noninvasive manner.
14. List the major types of risk factors that may increase a person's chance of a specific disease developing.

CRITICAL THINKING QUESTIONS

1. Each characteristic of life is related to body metabolism. Explain how digestion, circulation, and growth are metabolically related.
2. What diseases may result in a patient with an endomorph somatotype and a waist-to-hip ratio of 1:2?
3. An x-ray technician has been asked to take x-rays of the entire large intestine (colon), including the appendix. Which of the nine abdominopelvic regions must be included in the x-ray?
4. Body cavities can be subdivided into smaller and smaller sections. Identify, from largest to smallest, the cavities in which the urinary bladder can be placed.
5. When driving in traffic, it is important to stay in your own lane. If you see that you are drifting out of your lane, your brain tells your arms and hands to move in such a way that you get back in your lane. Identify the three components of a control loop in this example. Explain why this would be a negative feedback loop.

The Chemical Basis of Life

LANGUAGE OF SCIENCE

acid (AS-id) [*acid* sour]

adenosine triphosphate (ATP) (ah-DEN-o-sen
try-FOS-fate) [blend of *adenine* and *ribose, tri-* three,
-phosph- phosphorus, *-ate* oxygen]

amino acids (ah-MEE-no AS-ids) [*amino* NH_2, *acid*
sour]

anabolism (ah-NAB-o-lizm) [*anabol-* to build up, *-ism*
action]

atoms (AT-oms) [*atom* indivisible]

bases (BAYS-es) [*bas* foundation]

buffers (BUFF-ers) [*buffe* to cushion]

carbohydrate (kar-bo-HYE-drate) [*carbo-* carbon,
-hydr- hydrogen, *-ate* oxygen]

catabolism (kah-TAB-o-lizm) [*catabol-* to throw
down, *-ism* action]

compounds (KOM-pownds) [*compon* to assemble]

covalent bond (ko-VAYL-ent bond) [*co-* with, *-valen-*
power, *bond* band]

decomposition reaction (dee-KAHM-poh-sih-shun
ree-AK-shun) [*de-* opposite of, *-compo-* to assemble,
-tion process, *re-* again, *-action* action]

dehydration synthesis (dee-hye-DRAY-shun SIN-
the-sis) [*de-* from, *-hydrat-* water, *-tion* process,
synthesis putting together]

denatures (de-NAYT-shurs) [*de-* remove, *-nature*
nature]

electrolytes (e-LEK-tro-lites) [*electro-* electricity, *-lyt-*
soluble]

elements (EL-em-ents) [*element* first principle]

energy level [*en-* in, *-erg* work]

enzymes (EN-zimes) [*en-* in, *-zyme* ferment]

exchange reactions [*ex-* from, *-change* to change, *re-*
again, *-action* action]

functional groups (FUNK-shun-al groops) [*function-*
to perform, *-al* pertaining to]

functional proteins (FUNK-shun-al PRO-teens)
[*function-* to perform, *-al* pertaining to, *prote-* primary,
-in neutral substance]

high-energy bonds [*en-* in, *-erg* work, *bond* band]

hydrolysis (hye-DROL-i-sis) [*hydro-* water, *-lysis*
loosening]

ions (EYE-ons) [*ion* to go]

Cont'd on p. 71

natomy and physiology are subdivisions of biology—the study of life. To best understand the characteristics of life, what living matter is, how it is organized, and what it can do, we must appreciate and understand certain basic principles of chemistry that apply to the life process.

Life itself depends on proper levels and proportions of chemical substances in the cytoplasm of cells. The various structural levels of organization described in Chapter 1 are ultimately based on the existence and interrelationships of atoms and molecules. Chemistry, like biology, is a very broad scientific discipline. It deals with the structure, arrangement, and composition of substances and the reactions they undergo. Just as biology may be subdivided into many subdisciplines or branches, like anatomy and physiology, chemistry may also be divided into specialized areas. *Biochemistry* is the specialized area of chemistry that deals with living organisms and life processes. It deals directly with the chemical composition of living matter and the processes that underlie such life activities as growth, muscle contraction, and transmission of nervous impulses.

Modern biochemistry is in reality many disciplines. It is closely related to the other life sciences and to modern medicine. Biochemists use many different chemical, physical, biological, nutritional, and immunological techniques to probe life processes at every level of organization. An understanding of homeostatic processes and control mechanisms is in many cases dependent on a knowledge of basic chemistry and on certain facts and concepts in biochemistry.

BASIC CHEMISTRY
Elements and Compounds

Chemists use the term *matter* to describe in a general sense all of the materials or substances around us. Anything that has mass and occupies space is matter.

Substances are either **elements** or **compounds.** An element is said to be "pure." That is, it cannot be broken down or decomposed into two or more different substances. Pure oxygen is a good example of an element. In most living material, elements do not exist alone in their pure state. Instead, two or more elements are joined to form chemical combinations called *compounds.* Compounds can be broken down or decomposed into the elements that are contained within them. Water is a compound (H_2O). It can be broken down into atoms of hydrogen and atoms of oxygen in a 2:1 ratio.

Other examples of elements include phosphorus, copper, and nitrogen. For convenience in writing chemical formulas and in other types of notation, chemists assign a symbol to each element, usually the first letter or two of the English or Latin name of the element: P, phosphorus; Cu, copper (Latin *cuprum*); N, nitrogen (Figure 2-1). Note in Table 2-1 that 26 elements are listed as being present in the human body. Although all are important, 11 are called *major elements.* Four of these major elements—carbon, oxygen, hydrogen, and nitrogen—make up about 96% of the material in the human body. The 15 remaining elements are present in amounts that are less than 0.1% of body weight and are called *trace elements.* It is important to note, however, that the unique "aliveness" of a living organism does not depend on a single ele-

ment or mixture of elements but on the complexity, organization, and interrelationships of all elements required for life.

QUICK CHECK

1. What is biochemistry?
2. What is the difference between an element and a compound?
3. What elements make up 96% of the material in the human body?

Atoms

The most important of all chemical theories was advanced in 1805 by the English chemist John Dalton. He proposed the concept that matter is composed of **atoms** (from the Greek *atomos,* "indivisible"). His idea was revolutionary and yet simple—that all matter, regardless of the form it may assume (liquid, gas, or solid), is composed of units he called *atoms.*

Dalton conceived of atoms as solid, indivisible particles, and for about 100 years this was believed to be true. We now know that atoms are divisible into even smaller or *subatomic* particles, some of which exist in a "cloud" surrounding a dense central core called a *nucleus.* More than 100 million atoms of even very dense and heavy substances, if lined up, would measure barely an inch and would consist mostly of empty space! Our knowledge about the number and nature of subatomic particles and the central nucleus around which they move continues to grow as a result of ongoing research.

Atomic Structure

Atoms contain several different kinds of smaller or subatomic particles that are found in either a central *nucleus* or its surrounding *electron cloud* or field. Figure 2-2, A, shows an atomic model of carbon illustrating the most important types of subatomic particles:
Protons (p^+)
Neutrons (n^0)
Electrons (e^-)
Note that the carbon atom in Figure 2-2 has a central corelike *nucleus.* It is located deep inside the atom and is made up of six positively charged *protons* (+ or *p*) and six uncharged *neutrons (n).* Note also that the nucleus is surrounded by a "cloud" or "field" of six negatively charged electrons (− or *e*). Because protons are positively charged and neutrons are neutral, the nucleus of an atom bears a positive electrical charge equal to the number of protons that are present in it. Electrons move around the atom's nucleus in what can be represented as an electron "cloud" or field (Figure 2-2, B). The number of negatively charged electrons moving around an atom's nucleus equals the number of positively charged protons in the nucleus. The opposite charges therefore cancel or neutralize each other, and atoms are electrically neutral particles.

Atomic Number and Atomic Weight

Elements differ in their chemical and physical properties because of differences in the number of protons in their atomic nuclei. The number of protons in an atom's nucleus, called its *atomic number,* is therefore critically important—it identifies the kind of element it is. Look again at the elements important in living or-

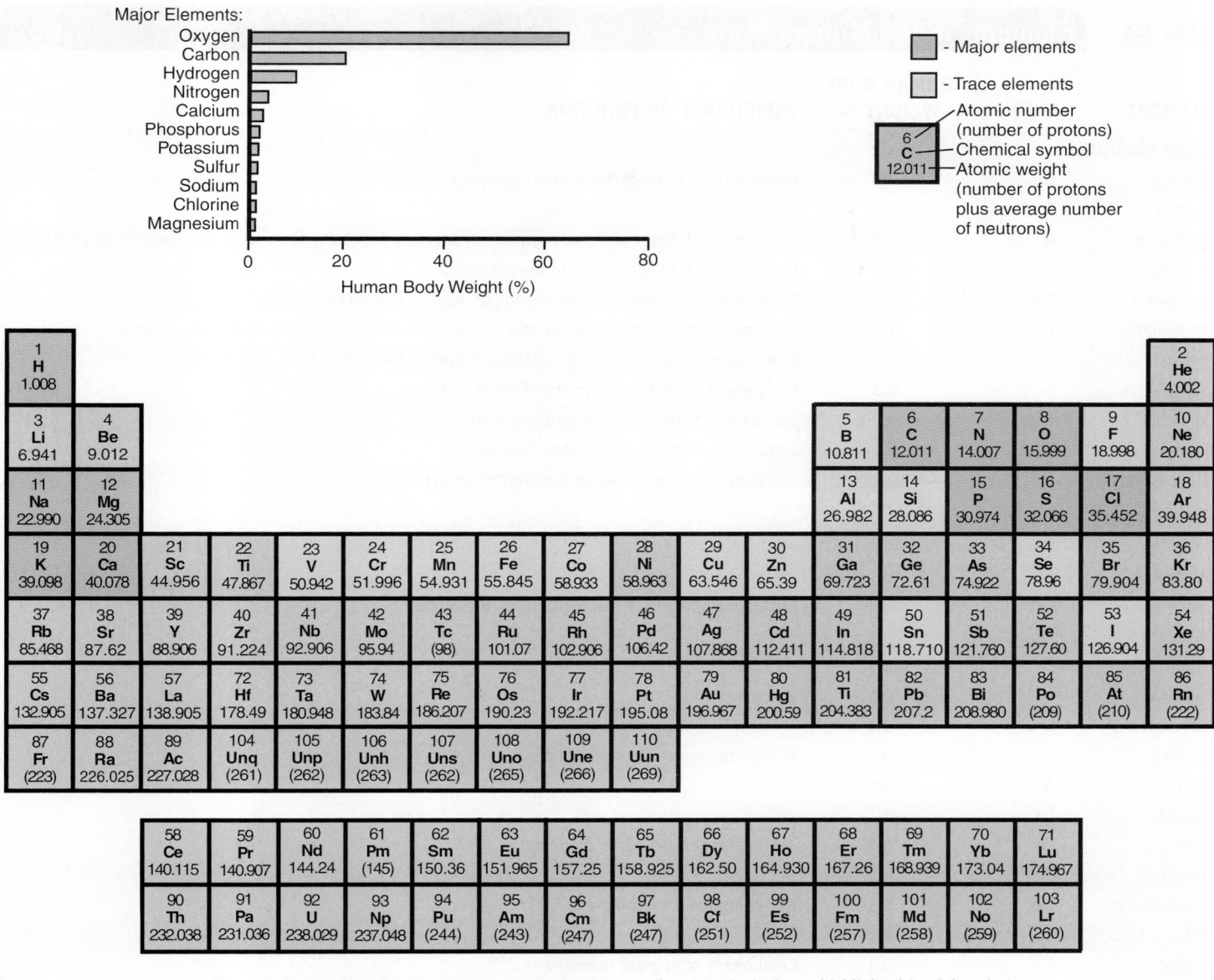

Figure 2-1 *Periodic table of elements.* The major elements found in the body are highlighted in pink and are listed in the small graph that shows their relative abundance in the body. The trace elements, found in very tiny quantities in the body are highlighted in *orange*.

ganisms listed in Table 2-1. Each element is identified by its symbol and atomic number. Hydrogen, for example, has an atomic number of 1; this means that all hydrogen atoms—and only hydrogen atoms—have one proton in their nucleus. All carbon atoms—and only carbon atoms—contain six protons and have an atomic number of 6. All oxygen atoms, and only oxygen atoms, have eight protons and an atomic number of 8. In short, each element is identified by its own unique number of protons, that is, by its own unique atomic number. If two atoms contain a different number of protons, they necessarily have different atomic numbers and are different elements.

There are 92 elements that occur naturally on earth. Because each element is characterized by the number of protons in its atoms (atomic number), there are atoms that contain from 1 to 92

protons. Additional elements have been discovered as a result of sophisticated research in the area of particle physics. At least 110 elements are now known to science.

The term *atomic weight* refers to the mass of a single atom. It equals the number of protons plus the number of neutrons in the atom's nucleus. The weight of electrons is, for practical purposes, negligible. Because protons and neutrons weigh almost exactly the same, the equation for determining atomic weight is as follows:

$$\text{Atomic weight} = (p + n)$$

The largest naturally occurring atom is uranium. It has an atomic weight of 238, with a nucleus containing 92 protons and 146 neutrons. In contrast, hydrogen, which has only one proton and no neutrons in its nucleus, has an atomic weight of 1.

Table 2-1	Elements in the Human Body		
ELEMENT	**SYMBOL**	**HUMAN BODY WEIGHT %**	**IMPORTANCE OR FUNCTION**
Major Elements			
Oxygen	O	65.0	Necessary for cellular respiration; component of water
Carbon	C	18.5	Backbone of organic molecules
Hydrogen	H	9.5	Component of water and most organic molecules; necessary for energy transfer and respiration
Nitrogen	N	3.3	Component of all proteins and nucleic acids
Calcium	Ca	1.5	Component of bones and teeth; triggers muscle contraction
Phosphorus	P	1.0	Principal component in the backbone of nucleic acids; important in energy transfer
Potassium	K	0.4	Principal positive ion within cells; important in nerve function
Sulfur	S	0.3	Component of many energy-transferring enzymes
Sodium	Na	0.2	Important positive ion surrounding cells
Chlorine	Cl	0.2	Important negative ion surrounding cells
Magnesium	Mg	0.1	Component of many energy-transferring enzymes
Trace Elements			
Silicone	Si	<0.1	—
Aluminum	Al	<0.1	—
Iron	Fe	<0.1	Critical component of hemoglobin in the blood
Manganese	Mn	<0.1	—
Fluorine	F	<0.1	—
Vanadium	V	<0.1	—
Chromium	Cr	<0.1	—
Copper	Cu	<0.1	Key component of many enzymes
Boron	B	<0.1	—
Cobalt	Co	<0.1	—
Zinc	Zn	<0.1	Key component of some enzymes
Selenium	Se	<0.1	—
Molybdenum	Mo	<0.1	Key component of some enzymes
Tin	Sn	<0.1	—
Iodine	I	<0.1	Component of thyroid hormone

Energy Levels

The total number of electrons in an atom equals the number of protons in its nucleus (see Figure 2-2). These electrons are known to exist in regions surrounding the atom's nucleus.

No single model of the atom sufficiently explains all we know about atomic structure. However, two simple models of atoms may be useful here to begin our discussion.

The cloud model suggests that any one electron cannot be exactly located at a specific point at any particular time. This concept is called a probability distribution and refers to the probability of finding an electron at any specific location outside the nucleus. Earlier models based on the work of a Danish physicist, Niels Bohr, who won the 1922 Nobel Prize in Physics for his groundbreaking contributions, suggested that electrons moved in regular patterns around the nucleus much like the planets in our solar system move around the sun. A simplified version of the Bohr model of the atom (see Figure 2-2, A) is perhaps most use-

ful in visualizing the structure of atoms as they enter into chemical reactions.

In the Bohr model, the electrons are shown in shells or concentric circles. The different shells show the relative distances of the electrons from the nucleus. The electrons surrounding the atom's nucleus are seen in this model as existing in simple rings or shells. Each ring represents a different **energy level,** and each can hold only a certain maximum number of electrons (Figure 2-3). The number and arrangement of electrons orbiting in an atom's energy levels are important because they determine whether the atom is chemically active.

In chemical reactions between atoms it is the electrons in the outermost energy level that participate in the formation of chemical bonds. In each energy level, electrons tend to group in pairs. As a rule, an atom can be listed as chemically inert and unable to react with another atom if its outermost energy level has four pairs of electrons, or eight. Such an atom is said to have a stable elec-

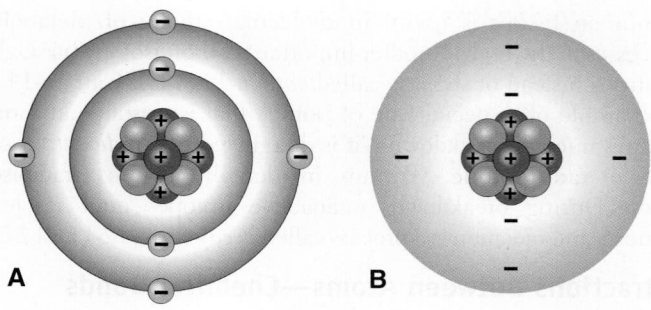

Figure 2-2 *Models of the atom.* The nucleus—protons (+) and neutrons—is at the core. Electrons inhabit outer regions called electron *shells* or energy levels **(A)** or probability distributions called electron *clouds* **(B).** This is a carbon atom, a fact that is determined by the number of its protons. All carbon atoms (and only carbon atoms) have six protons. (Not all of the protons in the nucleus are visible in this illustration.)

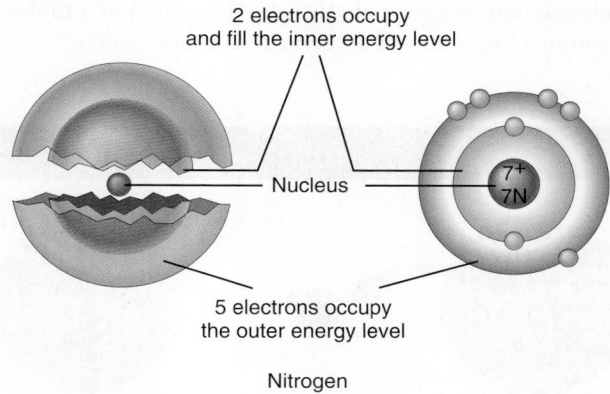

2 electrons occupy
and fill the inner energy level

Nucleus

5 electrons occupy
the outer energy level

Nitrogen

Figure 2-3 *Energy levels (electron shells) surrounding the nucleus of an atom.* Each concentric shell represents a different electron energy level.

tron configuration. The pairing of electrons is important. If the outer energy level contains single, unpaired electrons, the atom will be chemically active. Atoms with fewer or more than eight electrons in the outer energy level will attempt to lose, gain, or share electrons with other atoms to achieve stability. This tendency is called the **octet rule.** This rule holds true except for atoms that are limited to a single energy level that is filled by a maximum of two electrons. For example, hydrogen has but one electron in its single energy level. It therefore has an incomplete energy level with an unpaired electron. The result is a highly reactive tendency of hydrogen to enter into many chemical reactions. Helium, however, has two electrons in its single energy level. Because this is the maximum number for this energy level, no chemical activity is possible, and no naturally occurring compound containing helium exists.

The atoms shown in Figure 2-4 illustrate several of the most important facts related to energy levels. Note that even in the hydrogen atom with its very basic structure, positive and negative charges balance. However, its single energy level contains only one electron because the hydrogen nucleus contains only one proton. As a result of the unpaired electron, hydrogen is chemically active. In contrast, the helium atom has a full outer energy

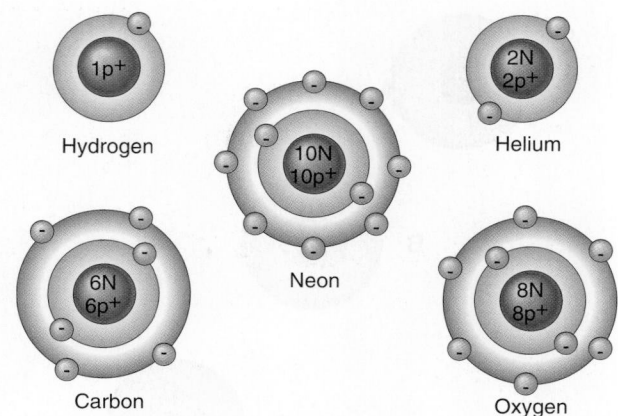

Figure 2-4 *Energy levels (shells) of five common elements.* All atoms are balanced with respect to positive and negative charges. In atoms with a single energy level, two electrons are required for stability. Hydrogen with its single electron is reactive, whereas helium with its full energy level is not. In atoms with more than one energy level, eight electrons in the outermost energy level are required for stability. Neon is stable because its outer energy level has eight electrons. Oxygen and carbon, with six and four electrons, respectively, in their outer energy levels, are chemically active.

level and is therefore inactive, or *inert*, as is neon. With only four electrons and six electrons in the outer energy levels of carbon and oxygen, respectively, these elements will react chemically because they do not satisfy the octet rule.

 QUICK CHECK

4. List and define the three most important types of subatomic particles.
5. How are the atomic number and atomic weight of an atom defined?
6. What is an energy level?
7. Explain what is meant by the "octet rule."

Isotopes

All atoms of the same element contain the same number of protons but do not necessarily contain the same number of neutrons. **Isotopes** of an element contain the same number of protons, but different numbers of neutrons. Isotopes have the same basic chemical properties as any other atom of the same element, and they also have the same atomic number. However, because they have a different number of neutrons, they differ in atomic weight. Usually a hydrogen atom has only one proton and no neutrons (atomic number, 1; atomic weight, 1). Figure 2-5 illustrates this most common type of hydrogen and two of its isotopes. Note that the isotope of hydrogen called *deuterium* (^2H) has one proton and one neutron (atomic weight, 2). *Tritium* (^3H) is the isotope of hydrogen that has one proton and two neutrons (atomic weight, 3).

The atomic nuclei of more than 99% of all carbon atoms in nature have six protons and six neutrons (atomic number, 6; atomic weight, 12). An important isotope of carbon has 7 neutrons instead of six and is called *carbon-13* (^{13}C). Carbon-13 makes up about 1% of the world's carbon atoms. The presence of carbon-13

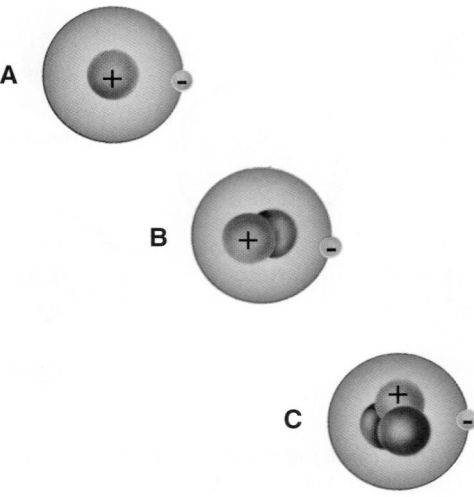

Figure 2-5 *Structure of hydrogen and two of its isotopes.* **A,** The most common form of hydrogen. **B,** An isotope of hydrogen called *deuterium (^2H)*. **C,** The hydrogen isotope *tritium (^3H)*. Note that isotopes of an element differ only in the number of its neutrons.

in human tissues is useful in molecular studies of metabolic processes in the body. Another important carbon isotope has eight neutrons instead of six; it is called *carbon-14* (^{14}C). Carbon-14 is an example of a special type of isotope that is unstable and undergoes nuclear breakdown—it is designated as a *radioactive isotope*, or **radioisotope**. (Tritium, incidentally, is also a radioisotope.) During breakdown, radioactive isotopes emit nuclear particles and radiation—a process called *decay* (Boxes 2-1 and 2-2).

Attractions Between Atoms—Chemical Bonds

Interactions between two or more atoms occur largely as a result of activity between electrons in their outermost energy level. The result, called a *chemical reaction*, most often involves unpaired electrons.

Ultimately, in atoms with fewer or more than eight electrons in the outer energy level, reactions will occur that result in the loss, gain, or sharing of one atom's unpaired electrons with those of another atom to satisfy the octet rule for both atoms. The result of such reactions between atoms is the formation of a **molecule**. For example, two atoms of oxygen can combine with one carbon

---

BOX 2-1 **Radioactivity**

Radioactivity is the emission of radiation from an atom's nucleus. Alpha particles, beta particles, and gamma rays are the three kinds of radiation. *Alpha particles* are relatively heavy particles consisting of two protons plus two neutrons. They shoot out of a radioactive atom's nucleus at a reported speed of 18,000 miles per second. *Beta particles* are electrons formed in a radioactive atom's nucleus by one of its neutrons breaking down into a proton and an electron. The proton remains behind in the nucleus, and the electron is ejected from it as a beta particle. Beta particles, because they are electrons, are much smaller than alpha particles, which consist of two protons and two neutrons. In addition, beta particles travel at a much greater speed than alpha particles do. *Gamma rays* are electromagnetic radiation, a form of light energy.

How does radioactivity change an atom? To find out how the emission of an alpha particle changes the nucleus of a radium atom, examine part *A* in the figure. Note that after the ejection of an alpha particle, the atom's nucleus contains fewer protons and neutrons. Some basic principles about atoms, you will recall, are as follows: All atoms of the same element contain the same number of protons; atoms that contain different numbers of protons are therefore different elements. After an alpha particle is ejected from the nucleus of a radium atom, the atom contains 86 instead of 88 protons and has been changed into an atom of radon. Part *B* of the figure shows how the emission of a beta particle changes an atom's nucleus: iodine loses a neutron and gains a proton to become xenon. Radioactivity usually changes the chemical identity of an atom. It can transform an atom of one element into an atom of a different element by changing the number of protons in the atom's nucleus.

Our bodies are continually exposed to low levels of radiation in the environment. When alpha or beta particles or gamma rays score direct hits on atoms in living cells, they ionize the atoms by knocking electrons out of their outer energy levels. The effect of ionization may injure, kill, or change cells. Knowledge of this fact underlies the use of radiation therapy to kill cancer cells, but radiation can also have an opposite effect. It can lead to *radiation sickness,* a condition that can be mild to severe, or even fatal, depending on the level of

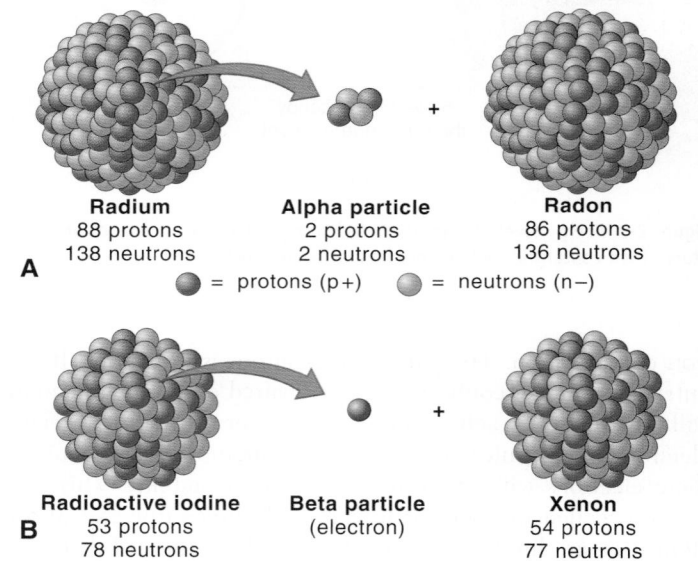

Radioactive changes in the nucleus of an atom. **A,** Emission of an alpha particle. **B,** Emission of a beta particle.

radiation exposure and the length of time exposed. For example, we know that many people who survive atomic bomb explosions or are exposed to high levels of radiation from industrial sources or radiological weapons have leukemia at a much higher rate than expected. However, even those exposed to lower doses of ionizing radiation can have mild symptoms of damage such as headache, nausea and vomiting, appetite loss, and diarrhea. Exposure to lower levels of radiation for a longer period may even lead to effects similar to a single, high-level exposure: sterility, damage to fetal development, cancer (including leukemia), cataracts, hair loss, and skin damage.

BOX 2-2: HEALTH MATTERS
Radon

Radon is a radioactive gas that is invisible and odorless. It is produced as a result of decay of radioactive elements occurring naturally in the soil (see Box 2-1). Radon is a known human carcinogen and is responsible for almost 15,000 lung cancer deaths each year in the United States. Once produced in the soil, the gas can enter and then concentrate in home basements and other building areas with poor ventilation.

Radon gas can be detected by inexpensive and readily available test kits that measure radioactivity in units called *picocuries* (pCi). Levels in excess of 4 pCi per liter (pCi/L) are considered sufficient to warrant intervention. Generally, increasing ventilation in affected areas by installing exhaust fans or sealing basement floors and walls will solve most problems. Exposure to radon gas at elevated levels, especially for prolonged periods, is a serious and potentially lethal health hazard.

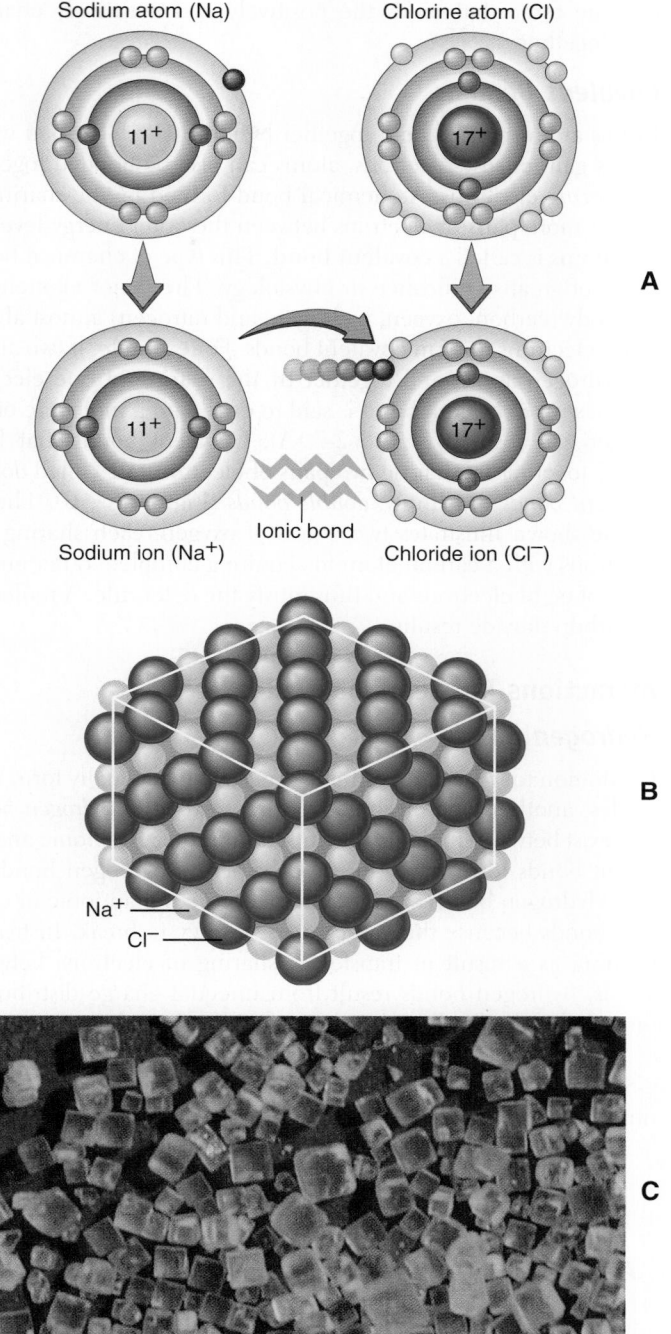

Figure 2-6 *Example of an ionic bond.* **A,** Energy-level models show the steps involved in forming an ionic bond between atoms of sodium and chlorine. Sodium "donates" an electron to chlorine, thereby forming a positive sodium ion and a negative chloride ion. The electrical attraction between the now oppositely charged ions forms an ionic bond. **B,** The space-filling model shows a crystal of sodium chloride (table salt) in typical cube-shaped formation. **C,** Photomicrograph showing cubic crystals of sodium chloride after the removal of water.

atom to form molecular carbon dioxide, or CO_2. If atoms of more than one element combine, the result, as defined earlier, is a compound. In other words, oxygen exists as a molecule (O_2) and is an element. Water exists as a molecule (H_2O) and is a compound. Reactions that hold atoms together do so by the formation of *chemical bonds*. There are two types of chemical bonds that unite atoms into molecules: ionic (or electrovalent) bonds and covalent bonds.

Ionic Bonds

A chemical bond formed by the transfer of electrons from one atom to another is called an *ionic,* or *electrovalent, bond.* Such a bond occurs as a result of the attraction between atoms that have become electrically charged by the loss or gain of electrons. When dissolved in water (Figure 2-6), such atoms are called **ions.** It is important to remember that ions can be positively or negatively charged and that ions with opposite charges are attracted to each other.

Note in Figure 2-6, *A,* that in the outer energy level of the sodium atom there is a single unpaired electron. If this electron were "lost," the outer ring would be stable because it would have a full outer octet (four pairs of electrons). The loss of the electron would result in the formation of a sodium ion (Na^+) with a positive charge. This is because there is now one more proton ($+$) than electron ($-$). The chlorine atom, in contrast, has one unpaired electron plus three paired electrons, or a total of seven electrons, in its outer energy level. By the addition of another electron, chlorine would satisfy the octet rule—its outer energy level would have a full complement of four paired electrons. The addition of another electron would result in the formation of a negatively charged chloride ion (Cl^-). A chemical reaction is set to occur. Sodium transfers or donates its one unpaired electron to chlorine and becomes a positively charged sodium ion (Na^+). Chlorine accepts the electron from sodium and pairs it with its one unpaired electron, thereby filling its outer energy level with the maximum of four electron pairs and becoming a negatively charged chloride ion (Cl^-). The positively charged sodium ion

(Na^+) is attracted to the negatively charged chloride ion (Cl^-), and the formation of NaCl, ordinary table salt, results. This chemical reaction illustrates ionic or electrovalent bonding. The electron transfer changed the two atoms of the elements sodium and chlorine into ions. The **ionic bond** is simply the strong electrostatic force that binds the positively and negatively charged ions together.

Covalent Bonds

Just as atoms can be held together by ionic bonds formed when atoms gain or lose electrons, atoms can also be bonded together by *sharing* electrons. A chemical bond formed by the sharing of one or more pairs of electrons between the outer energy levels of two atoms is called a **covalent bond.** This type of chemical bonding is of great significance in physiology. The major elements of the body (carbon, oxygen, hydrogen, and nitrogen) almost always share electrons to form covalent bonds. For example, if two atoms of hydrogen are bound together by the sharing of one electron pair, a *single* covalent bond is said to exist, and a molecule of hydrogen gas results (Figure 2-7, *A*). Covalent bonds that bind atoms together by sharing *two* pairs of electrons are called *double covalent bonds* or, simply, *double bonds* (Figure 2-7, *B*). The example shown illustrates two atoms of oxygen, each sharing two electrons with a carbon atom to acquire a complete outer energy level of eight electrons and thus satisfy the octet rule. A molecule of carbon dioxide results.

Attractions Between Molecules

Hydrogen Bonds

In addition to ionic and covalent bonds, which actually form molecules, another type of attractive force, called a *hydrogen bond*, can exist between biologically important molecules. Ionic and covalent bonds form new molecules, whereas hydrogen bonds do not. Hydrogen bonds are much weaker forces than ionic or covalent bonds because they require less energy to break. Instead of forming as a result of transfer or sharing of electrons between atoms, hydrogen bonds result from unequal charge distribution on a molecule. Such molecules are said to be **polar.**

Water is a good example of a polar molecule. Note in Figure 2-8 that although an atom of water is electrically neutral (the number of negative charges equals the number of positive charges), it has a partial positive charge (the hydrogen side) and a partial negative charge (the oxygen side); that is, it has a positive pole and a negative pole. The partial charges result from the electrons having a higher probability of being found nearer the highly positive oxygen nucleus than either hydrogen nucleus. Thus water is said to be "polar" because it has regions with different partial charges. Hydrogen bonds serve to weakly attach the partially negative (oxygen) side of one water molecule to the partially positive (hydrogen) side of an adjacent water molecule. Figure 2-9 illustrates hydrogen bonding between water molecules. Depending on how many of these hydrogen bonds are intact at one instant, the water may be either liquid (few bonds) or solid (many bonds). If the water molecules are too far apart to form any hydrogen bonds, then the water is a gas, such as steam.

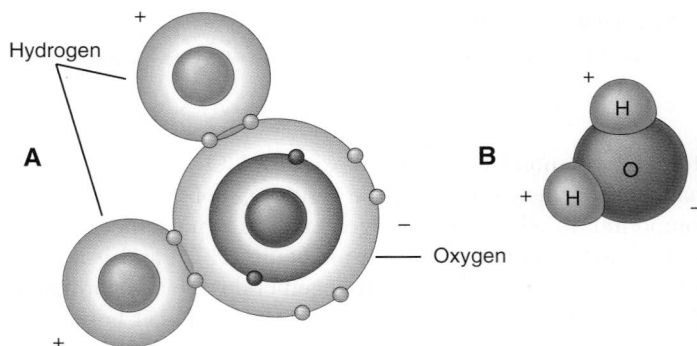

Figure 2-8 *Water is a polar molecule.* The polar nature of water is represented in an energy-level model **(A)** and a space-filling model **(B).** The two hydrogen atoms are nearer one end of the molecule and give that end a partial positive charge. The "oxygen end" of the molecule attracts the electrons more strongly and thus has a partial negative charge.

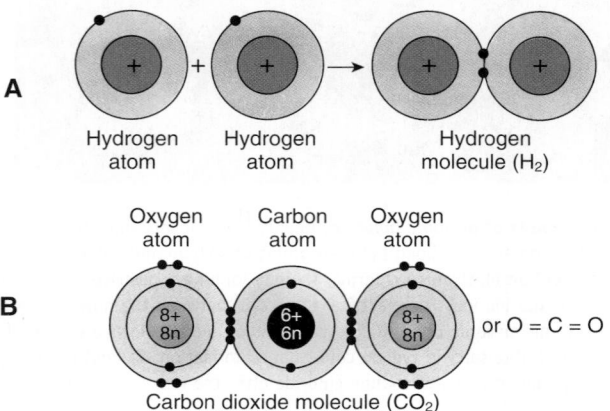

Figure 2-7 *Types of covalent bonds.* **A,** A single covalent bond formed by the sharing of one electron pair between two atoms of hydrogen results in a molecule of hydrogen gas. **B,** A double covalent bond (double bond) forms by the sharing of *two* pairs of electrons between two atoms. In this case, two double bonds form—one between carbon and *each* of the two oxygen atoms.

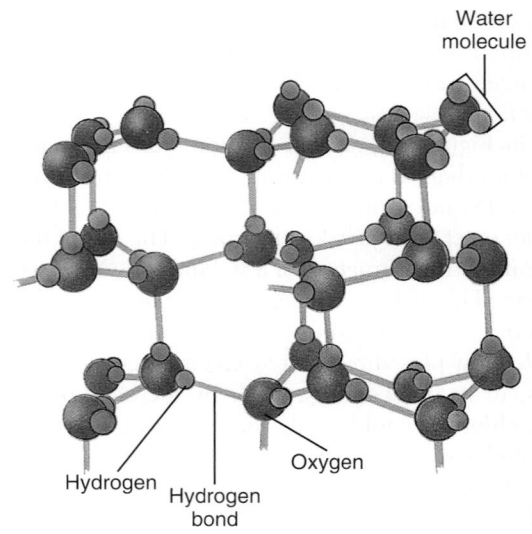

Figure 2-9 *Hydrogen bonds between water molecules.* Hydrogen bonds serve to weakly attach the negative *(oxygen)* side of one water molecule to the positive *(hydrogen)* side of an adjacent water molecule. This diagram depicts many bonded molecules, as one would expect in ice (solid water). In liquid water, there would be fewer intact bonds at any single moment.

The ability of water molecules to form hydrogen bonds with each other accounts for many of the unique properties of water that make it an ideal medium for the chemistry of life. Hydrogen bonds are also important in maintaining the three-dimensional structure of proteins and nucleic acids, also described later in the chapter.

Other Weak Forces

Other weak forces sometimes attract molecules to each other, even if only temporarily. Shifts in the locations of electrons within each molecule result in fleeting changes in the partial electrical charge of some regions of the molecule. This change in electrical charge may result in attraction to oppositely charge regions of another molecule. In the scope of our course, however, we will not concern ourselves with the different varieties of weak forces or how they are produced. It is important to know only that they exist and sometimes play an underlying role in holding the material of the body together.

Chemical Reactions

Chemical reactions involve interactions between atoms and molecules, which in turn involves the formation or breaking of chemical bonds. Three basic types of chemical reactions that you will learn to recognize as you study physiology are the following:

1. Synthesis reactions
2. Decomposition reactions
3. Exchange reactions

To the chemist, reactions can be symbolized by variations on a simple formula. In *synthesis* (from the Greek *syn*, "together," and *thesis*, "putting") *reactions*, two or more substances called *reactants* combine to form a different, more complex substance called a *product*. The process can be summarized by the following formula:

$$A + B \xrightarrow{\text{Energy}} AB$$
$$\text{(Reactants)} \qquad \text{(Product)}$$

Synthesis reactions result in the formation of new bonds, and energy is required for the reaction to occur and the new product to form. Many such reactions occur in the body. Every cell, for example, combines amino acid molecules as reactants to form complex protein compounds as products. The ability of the body to synthesize new tissue in wound repair is a good example of this type of reaction.

Decomposition reactions result in the breakdown of a complex substance into two or more simpler substances. In this type of reaction, chemical bonds are broken and energy is released. Energy can be released in the form of heat, or it can be captured for storage and future use. Decomposition reactions can be summarized by the following formula:

$$AB \longrightarrow A + B + \text{Energy}$$

Decomposition reactions occur when a complex nutrient is broken down in a cell to release energy for other cellular functions. The products of such a reaction are ultimately waste products. Decomposition and synthesis are opposites. Synthesis builds up; decomposition breaks down. Synthesis forms chemical bonds; decomposition breaks chemical bonds. Decomposition and synthesis reactions are often coupled with one another in such a way that the energy released by a decomposition reaction can be used to drive a synthesis reaction.

The nature of **exchange reactions** permits two different reactants to exchange components and, as a result, form two new products. An exchange reaction is often symbolized by the following formula:

$$AB + CD \longrightarrow AD + CB$$

Exchange reactions break down, or decompose, two compounds and, in exchange, synthesize two new compounds. Certain exchange reactions take place in the blood. One example is the reaction between lactic acid and sodium bicarbonate. The decomposition of both substances is exchanged for the synthesis of sodium lactate and carbonic acid. These changes can be seen more easily in the following equation:

$$H \cdot Lactate + NaHCO_3 \longrightarrow Na \cdot Lactate + H \cdot HCO_3$$

The formula "H · Lactate" represents lactic acid; "$NaHCO_3$" is the formula for sodium bicarbonate; "Na · Lactate" represents sodium lactate; and "H · HCO_3" represents carbonic acid.

Reversible reactions, as the name suggests, proceed in both directions. A great many synthesis, decomposition, or exchange reactions are reversible, and a number of them are cited in later chapters of this book. An arrow pointing in both directions is used to denote a reversible reaction:

$$A + B \rightleftharpoons AB$$

 QUICK CHECK

8. List the two types of chemical bonds between atoms and explain how they are formed.
9. What type of bonds attract one molecule to another?
10. Diagram the three basic types of chemical reactions.

METABOLISM

The term **metabolism** is used to describe all the chemical reactions that occur in body cells. The important topics of nutrition and metabolism will be discussed fully in Chapter 27. Nutrition and metabolism are described together because the total of all the chemical reactions or *metabolic activity* occurring in cells is associated with the use the body makes of foods after they have been digested, absorbed, and circulated to cells. The terms *catabolism* and *anabolism* are used to describe the two major types of metabolic activity. **Catabolism** describes chemical reactions—usually hydrolysis reactions—that break down larger food molecules into smaller chemical units and, in so doing, *release energy*. The release of energy is related to the disruption of chemical bonds. This breakdown of bonds in the chemical compounds contained in the foods and beverages that we consume provides the energy to power all of our activities. The released energy is in the form of electrons that are freed as the bonds are broken. (Remember that energy *is*

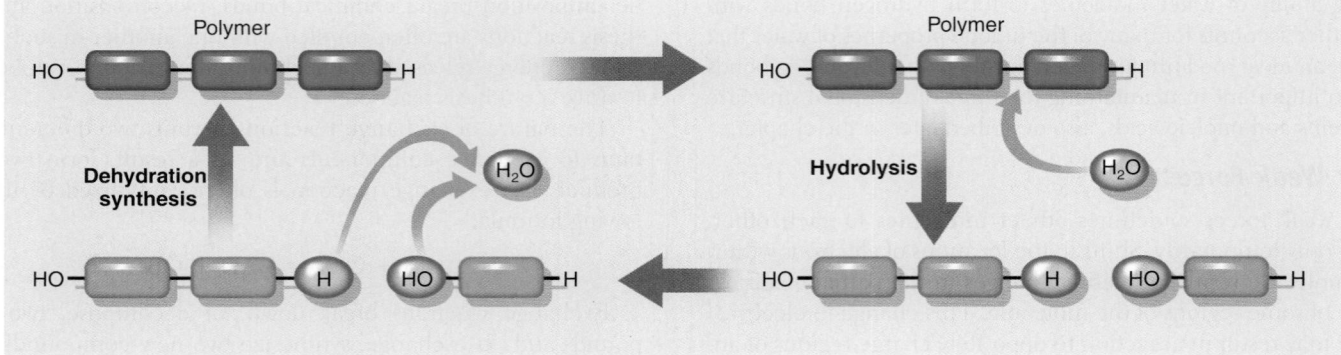

Figure 2-10 *Metabolic reactions.* Hydrolysis is a catabolic reaction that adds water to break down large molecules into smaller molecules, or subunits. Dehydration synthesis is an anabolic reaction that operates in the reverse fashion: small molecules are assembled into large molecules by removing water. Note that specific examples of dehydration synthesis are shown in Figures 2-15 and 2-21.

electrons and that electrons form bonds.) Basic chemical building blocks, such as fatty acids and amino acids, although present in our foods, are rarely available in ready-to-use form. Instead, they must be obtained by catabolic breakdown reactions from more complex nutrients. **Anabolism** involves the many chemical reactions—usually dehydration synthesis reactions—that build larger and more complex chemical molecules from smaller subunits (Figure 2-10). Anabolic chemical reactions *require energy*—energy in the form of **adenosine triphosphate (ATP)**.

Catabolism

Catabolism consists of chemical reactions that not only break down relatively complex compounds into simpler ones but also release energy from them. This breakdown process represents a type of chemical reaction called *hydrolysis* (see Figure 2-10). As a result of hydrolysis occurring during catabolism, a water molecule is added to break a larger compound into smaller subunits. For example, hydrolysis of a fat molecule breaks it down into its subunits—glycerol and fatty acid molecules; a disaccharide such as sucrose breaks down into its monosaccharide subunits—glucose and fructose; and the subunits of protein hydrolysis are amino acids. Ultimately, catabolism reactions will further degrade these building blocks of food compounds—glycerol, fatty acids, monosaccharides, and amino acids—into the end products carbon dioxide, water, and other waste products. During this process, energy is released. Some of the energy released by catabolism is heat energy, the heat that keeps our bodies warm. However, more than half the energy released is immediately recaptured and transferred to a molecule called *ATP*, which transfers the energy to cell components to do work. ATP will be discussed in more detail later in this chapter.

Anabolism

Anabolism is the term used to describe chemical reactions that join simple molecules together to form more complex biomolecules, notably, carbohydrates, lipids, proteins, and nucleic acids.

Literally thousands of anabolic reactions take place continually in the body. The type of chemical reaction responsible for this joining together of smaller units to form larger molecules is called *dehydration synthesis* (see Figure 2-10). It is a key reaction during anabolism. As a result of dehydration synthesis, water is removed as smaller subunits are fused together. The process requires energy, which is transferred from ATP molecules. Anabolism-type reactions join monosaccharide units to form larger carbohydrates, fuse amino acids into peptide chains, and form fat molecules from glycerol and fatty acid subunits.

 QUICK CHECK

11. What does the term *metabolism* mean?
12. What is the difference between *anabolism* and *catabolism*?
13. What is the role of *ATP* in the body?

ORGANIC AND INORGANIC COMPOUNDS

In living organisms, there are two kinds of compounds: *organic* and *inorganic*. Organic compounds are generally defined as compounds composed of molecules that contain carbon–carbon (C–C) covalent bonds or carbon–hydrogen (C–H) covalent bonds—or both kinds of bonds. Few inorganic compounds have carbon atoms in them, and none have C–C or C–H bonds. Organic molecules are generally larger and more complex than inorganic molecules. The term **functional groups** is often used to describe specialized arrangements of atoms attached to the carbon core of many organic molecules. Different functional groups confer unique chemical properties. Examples of important functional groups are illustrated in Figure 2-11. The human body has inorganic and organic compounds because both are equally important to the chemistry of life. We will discuss the chemistry of inorganic compounds first and then move on to some of the more important types of organic compounds.

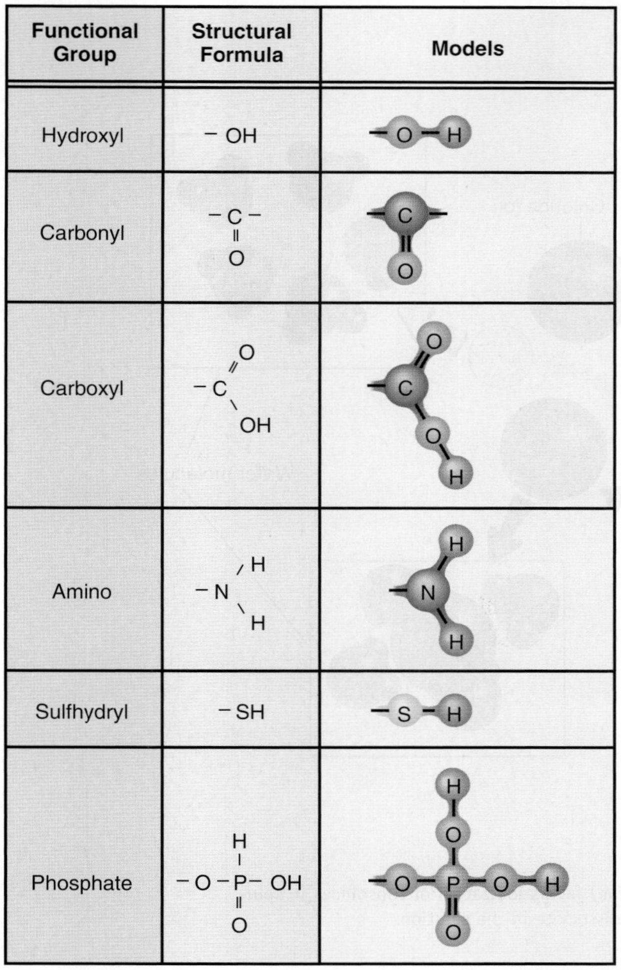

Functional Group	Structural Formula	Models
Hydroxyl	– OH	
Carbonyl	– C – ‖ O	
Carboxyl	– C ⫽O \OH	
Amino	– N ⁄H \H	
Sulfhydryl	– SH	
Phosphate	H | – O – P – OH ‖ O	

Figure 2-11 *The principal functional chemical groups.* Each functional group confers specific chemical properties on the molecules that possess them.

INORGANIC MOLECULES
Water

Water has been called the "cradle of life" because all living organisms require water to survive. Each body cell is bathed in fluid, and it is only in this precisely regulated and homeostatically controlled environment that cells can function. In addition to water surrounding the cell, the basic substance of each cell, *cytoplasm*, is itself largely water. Water is certainly the body's most abundant and important compound. Fifty percent or more of a normal adult's body weight is water, which serves a host of vital functions. Because of water's pervasive importance in all living organisms, an understanding of the basics of water chemistry is important. In a very real sense, "water chemistry" forms the basis for the chemistry of life.

Properties of Water

The chemist views water as a simple and stable compound. It has an atomic structure that results from the combination of two covalent bonds between a single oxygen atom and two hydrogen atoms.

Recall that water molecules are polar molecules and interact with one another because they have a partial positive charge at one end and a partial negative charge at their other end. Take a moment to review Figure 2-8. This simple chemical property, called *polarity*, allows water to act as a very effective *solvent*. Proper functioning of a cell requires the presence of many chemical substances. Many of these compounds are quite large and must be broken into smaller and more reactive particles (ions) for reactions to occur. Because of its polar nature, water tends to ionize substances in solution and surround any molecule that has an electrical charge (Figure 2-12). The fact that so many substances dissolve in water is of utmost importance in the life process.

The critical role that water plays as a solvent permits the *transportation* of many essential materials within the body. By dissolving food substances in blood, for instance, water enables these materials to enter and leave the blood capillaries in the digestive organs and eventually enter cells in every area of the body. In turn, waste products are transported from where they are produced to excretory organs for elimination from the body.

Another important function of water stems from the fact that water both absorbs and gives up heat slowly. These properties of water enable it to maintain a relatively constant temperature. This allows the body, which has a large water content, to resist sudden changes in temperature. Chemists describe this property by saying that water has a high *specific heat*; that is, water can lose and gain large amounts of heat with little change in temperature. As a result, excess body heat produced by the contraction of muscles during exercise, for example, can be transported by blood to the body surface and dissipated into the environment with little actual change in core temperature.

Chemists and biologists recognize water's high *heat of vaporization* as another important physical quality. This characteristic requires the absorption of significant amounts of heat to change water from a liquid to a gas. The energy is required to break the many hydrogen bonds that hold adjacent water molecules together in the liquid state. Thus, the body can dissipate excess heat and maintain a normal temperature by evaporation of water (sweat) from the skin surface whenever excess heat is being produced.

Understanding and appreciating the importance of water in the life process are critical. Water does more than act as a solvent, produce ionization, and facilitate chemical reactions. It has essential chemical roles of its own in addition to the many important physical qualities it brings to body function (Table 2-2). It plays a key role in such processes as cell permeability, active transport of materials, secretion, and membrane potential, to name a few. The requirement of the body to maintain homeostasis is stressed in each chapter of the text. As you progress from system to system in the chapters that follow, the importance of water in almost every regulatory and control mechanism studied will be a constant that should not go unnoticed.

Oxygen and Carbon Dioxide

Oxygen (O_2) and carbon dioxide (CO_2) are important inorganic substances that are closely related to cellular respiration. Molecular oxygen in the body is present as two oxygen atoms joined by a double covalent bond. Oxygen is required to complete the de-

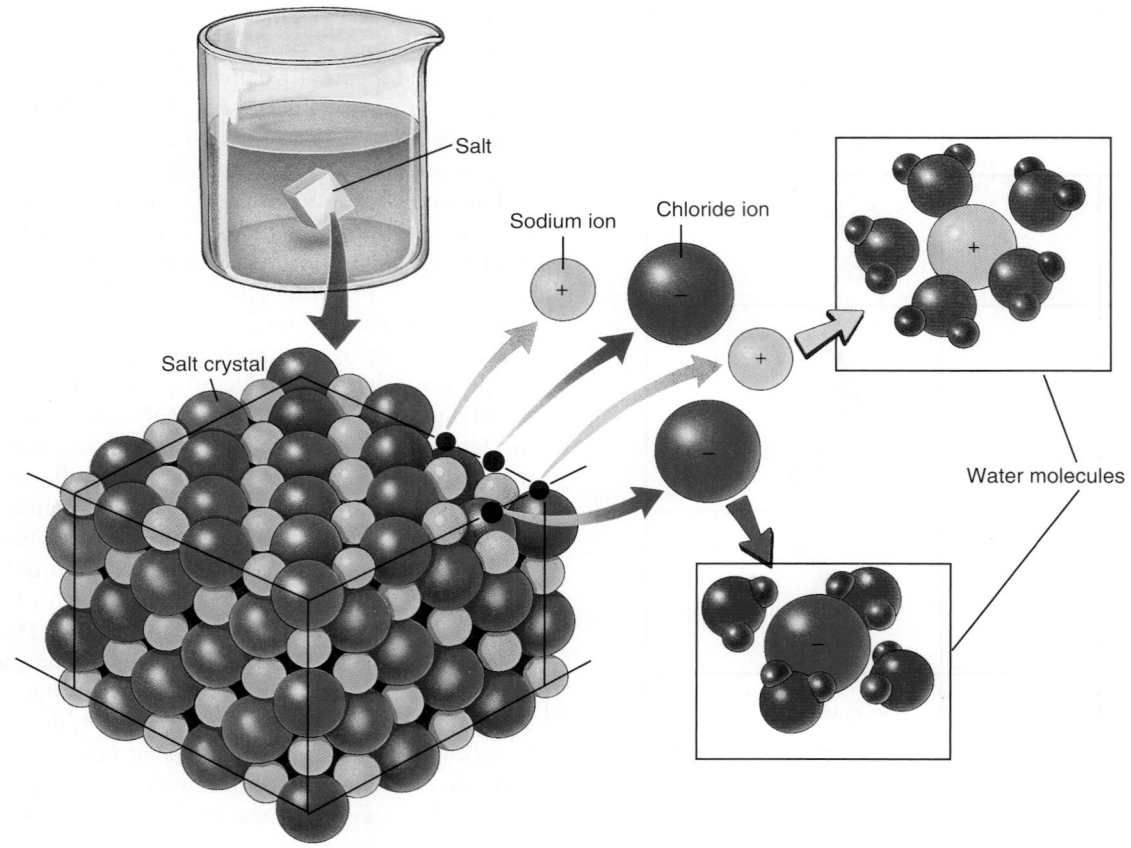

Figure 2-12 *Water as a solvent.* The polar nature of water *(blue)* favors ionization of substances in solution. Sodium (Na^+) ions *(pink)* and chloride (Cl^-) ions *(green)* dissociate in the solution.

Table 2-2	Properties of Water		
PROPERTY	**DESCRIPTION**		**EXAMPLE OF BENEFIT TO THE BODY**
Strong polarity	Polar water molecules attract other polar compounds, which causes them to dissociate		Many kinds of molecules can dissolve in cells, thereby permitting a variety of chemical reactions and allowing many substances to be transported
High specific heat	Hydrogen bonds absorb heat when they break and release heat when they form, thereby minimizing temperature changes		Body temperature stays relatively constant
High heat of vaporization	Many hydrogen bonds must be broken for water to evaporate		Evaporation of water in perspiration cools the body
Cohesion	Hydrogen bonds hold molecules of water together		Water works as lubricant or cushion to protect against damage from friction or trauma

composition reactions required for the release of energy from nutrients burned by the cell. Carbon dioxide is considered one of a group of very simple carbon-containing inorganic compounds. It is an important exception to the "rule of thumb" that inorganic substances do not contain carbon. Like oxygen, carbon dioxide is involved in cellular respiration. It is produced as a waste product during the breakdown of complex nutrients and also serves an important role in maintaining the appropriate acid-base balance in the body.

Electrolytes

Other inorganic substances include acids, bases, and salts. These substances belong to a large group of compounds called **electrolytes** (e-LEK-troh-lites). Electrolytes are substances that break up, or *dissociate*, in solution to form charged particles, or ions. Ions with a positive charge are called *cations*, and those with a negative charge are called *anions*. Figure 2-12 shows the way in which water molecules work to dissociate a typical electrolyte, sodium chloride (NaCl), into Na^+ cations and Cl^- anions.

Acids and Bases

Acids and bases are common and very important chemical substances in the body. Early chemists categorized acids and bases by such characteristics as taste or the ability to change the color of certain dyes. Acids, for example, taste sour and bases taste bitter. The dye *litmus* will turn blue in the presence of a base and red when exposed to an acid. These and other observations illustrate a fundamental point, namely, that acids and bases are chemical opposites. Although acids and bases dissociate in solution, both release different types of ions. The unique chemical properties of acids and bases when they are in solution are perhaps the best way to differentiate them.

Acids

By definition, an **acid** is any substance that will release a hydrogen ion (H^+) when in solution. A hydrogen ion is simply a bare proton—the nucleus of a hydrogen atom. Therefore, acids are frequently called *proton donors*. It is the concentration of hydrogen ions that accounts for the chemical properties of acids. The level of "acidity" of a solution depends on the number of hydrogen ions a particular acid will release.

One point should be understood about water. Water molecules dissociate continually in a reversible reaction to form hydrogen ions (H^+) and hydroxide ions (OH^-):

$$H_2O \rightleftharpoons H^+ + OH^-$$

Recall from our discussion of ionic bonds (pp. 43 to 44) that having a single unpaired electron in the outer energy level makes an atom unstable and that losing that electron results in a more stable structure. This is precisely the reason why dissociation of water occurs. In pure water, the balance between these two ions is equal. However, when an acid such as hydrochloric acid (HCl) dissociates into H^+ and Cl^-, it shifts the H^+/OH^- balance in favor of excess H^+ ions, thus increasing the level of acidity. The more hydrogen ions (H^+) produced, the stronger the acid.

A *strong acid* is an acid that completely, or almost completely, dissociates to form H^+ ions. A *weak acid*, on the other hand, dissociates very little and therefore produces few excess H^+ ions in solution. There are many important acids in the body, and they perform many functions. Hydrochloric acid, for example, is the acid produced in the stomach to aid the digestive process.

Bases

Bases, or *alkaline* compounds, are electrolytes that when dissociated in solution, shift the H^+/OH^- balance in favor of OH^-. This can be accomplished by increasing the number of hydroxide ions (OH^-) in solution or decreasing the number of H^+ ions present. The fact that bases will combine with or accept H^+ ions (protons) is the reason the term *proton acceptor* is used to describe these substances. The dissociation of a common base, sodium hydroxide, yields the cation Na^+ and the OH^- anion.

Like acids, bases are classified as strong or weak, depending on how readily and completely they dissociate into ions. Important bases in the body, such as the bicarbonate ion (HCO_3^-), play critical roles in the transportation of respiratory gases and in the elimination of waste products from the body.

The pH Scale

The term **pH** is literally an abbreviation for a phrase meaning "the power of hydrogen" and is used to mean the H^+ ion concentration of a solution. As you can see at the left side of Figure 2-13, the pH value is the negative of the base-10 logarithm of the H^+ ion concentration. pH indicates the degree of *acidity* or *alkalinity* of a solution. As the concentration of H^+ ions increases, the pH goes down and the solution becomes more acidic; a decrease in H^+ ion concentration makes the solution more alkaline and the pH goes up. A pH of 7 indicates neutrality (equal amounts of H^+ and OH^-), a pH of less than 7 indicates acidity (more H^+ than OH^-), and a pH greater than 7 indicates alkalinity (more OH^- than H^+). The overall pH range is often expressed numerically on a logarithmic scale of 1 to 14. Keep in mind that a change of 1 pH unit on this type of scale represents a 10-fold difference in actual concentration of H^+ ions (Box 2-3).

Buffers

The normal pH range of blood and other body fluids is extremely narrow. For example, venous blood (pH 7.36) is only slightly more acidic than arterial blood (pH 7.41). The difference results primarily from carbon dioxide entering venous blood as a waste product of cellular metabolism. Carbon dioxide is carried as carbonic acid (H_2CO_3) and therefore lowers the pH of venous blood. More than 30 *liters* of carbonic acid is transported in venous blood each day and eliminated as carbon dioxide by the lungs, and yet 1 liter of venous blood contains only about 1/100,000,000 g more H^+ ions than 1 liter of arterial blood does! The incredible constancy of the pH homeostatic mechanism is due to the presence of substances, called **buffers**, that minimize changes in the concentrations of H^+ and OH^- ions in our body fluids. Buffers are said to act as a "reservoir" for H^+ ions. They donate, or remove, H^+ ions to a solution if that becomes necessary to maintain a constant pH. Examples of important buffer systems and specifics of buffer action will be discussed in Chapter 30.

Salts

A salt is any compound that results from the chemical interaction of an acid and a base. Salts, like acids and bases, are electrolyte compounds and dissociate in solution to form positively and negatively charged ions. Ions exist in solution. If the water is removed, the ions will crystallize and form salt. When mixed and allowed to react, the positive ion (cation) of a base and the negative ion (anion) of an acid will join to form a salt and additional water in the manner of a typical exchange reaction. The reaction between an acid and base to form a salt and water is called a *neutralization reaction*:

(A B	+	C D	$\longrightarrow$	C B	+	A D)
HCl	+	NaOH	$\longrightarrow$	NaCl	+	H_2O
Acid		Base		Salt		Water
(hydrochloric		(sodium		(sodium		
acid)		hydroxide)		chloride)		

Note that the sodium and the chloride join to form the salt whereas the hydroxide ion "accepts" or combines with a hydrogen ion to form water.

The sources of many of the major and trace mineral elements listed in Table 2-1 are inorganic salts, which are common in many body fluids and specialized tissues such as bone. These elements

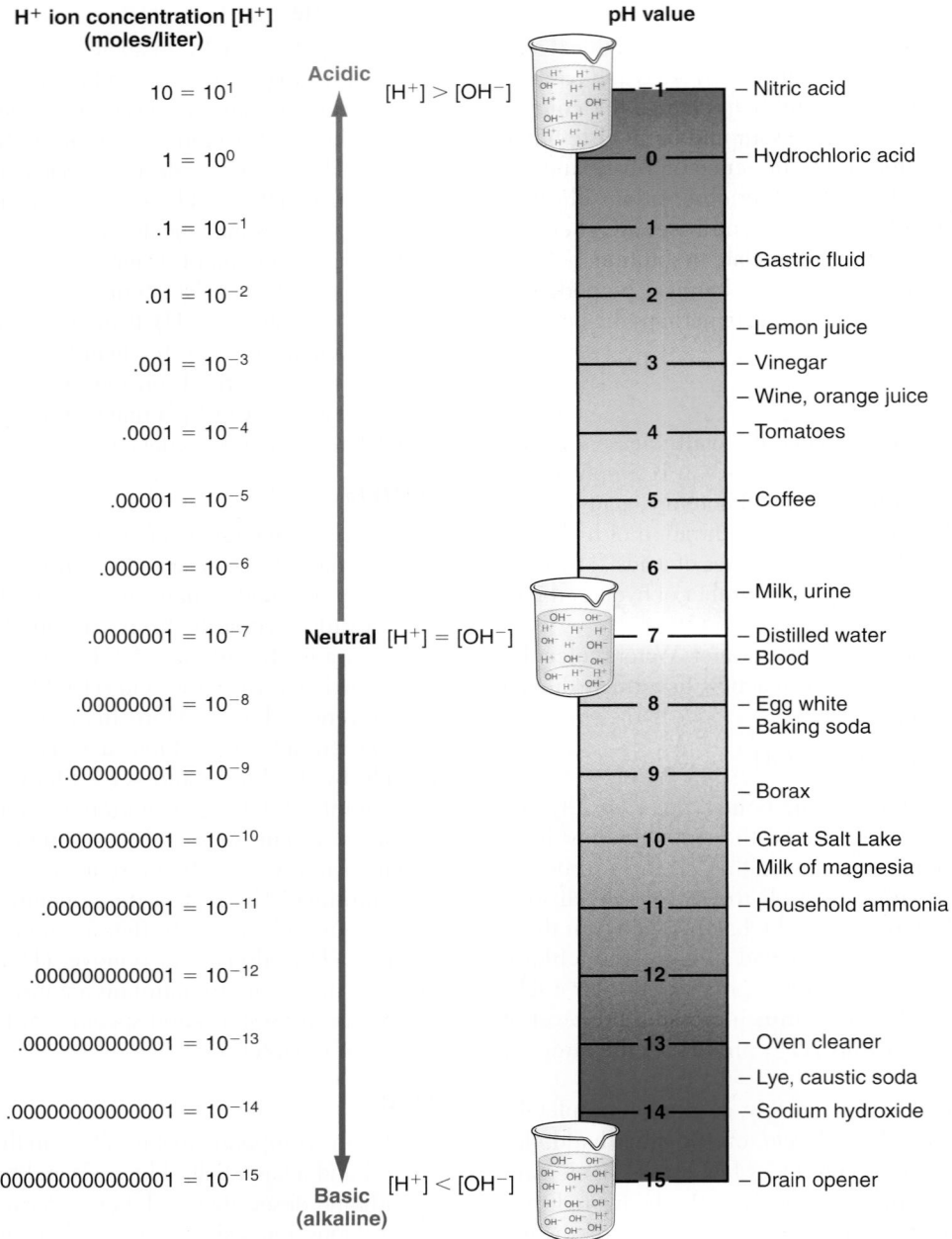

H⁺ ion concentration [H⁺]
(moles/liter)

pH value

$10 = 10^1$

$1 = 10^0$

$.1 = 10^{-1}$

$.01 = 10^{-2}$

$.001 = 10^{-3}$

$.0001 = 10^{-4}$

$.00001 = 10^{-5}$

$.000001 = 10^{-6}$

$.0000001 = 10^{-7}$

$.00000001 = 10^{-8}$

$.000000001 = 10^{-9}$

$.0000000001 = 10^{-10}$

$.00000000001 = 10^{-11}$

$.000000000001 = 10^{-12}$

$.0000000000001 = 10^{-13}$

$.00000000000001 = 10^{-14}$

$.000000000000001 = 10^{-15}$

Acidic $[H^+] > [OH^-]$

Neutral $[H^+] = [OH^-]$

Basic $[H^+] < [OH^-]$
(alkaline)

-1 — Nitric acid

0 — Hydrochloric acid

1

— Gastric fluid

2

— Lemon juice

3 — Vinegar
— Wine, orange juice

4 — Tomatoes

5 — Coffee

6

— Milk, urine

7 — Distilled water
— Blood

8 — Egg white
— Baking soda

9
— Borax

10 — Great Salt Lake
— Milk of magnesia

11 — Household ammonia

12

13 — Oven cleaner
— Lye, caustic soda

14 — Sodium hydroxide

15 — Drain opener

Figure 2-13 *The pH scale.* Note that as the concentration of H⁺ increases, the solution becomes increasingly acidic and the pH value decreases. As the H⁺ concentration decreases, the pH value increases, and the solution becomes more and more basic, or alkaline. (The scale on the left side of the diagram shows the actual concentrations of H⁺ in moles per liter, or molar concentration, as an ordinary number and expressed as an exponent [logarithm] of 10. You can see that the pH scale is simply the negative of the exponent of 10.)

often exert their full physiological effects only when present as charged atoms or ions in solution.

The proper amount and concentration of such mineral salt electrolytes as potassium (K^+), calcium (Ca^{++}), and sodium (Na^+) are required for proper functioning of nerves and for contraction of muscle tissue. See Chapter 29 for specific homeostatic control mechanisms that regulate electrolyte balance in blood and other body fluids. Table 2-3 lists several inorganic salts that on dissociation in body fluids, contribute important electrolytes required for numerous body functions.

QUICK CHECK

14. Discuss the properties of water that make it so important in living organisms.

15. What is an electrolyte?

16. How do acids and bases react with each other when in solution?

17. What is pH?

BOX 2-3: FYI
The pH Unit

A pH of 7, for example, means that a solution contains 10^{-7} g of hydrogen ions per liter. Translating this logarithm into a number, a pH of 7 means that a solution contains 0.0000001 (that is, 1/10,000,000) g of hydrogen ions per liter. A solution of pH 6 contains 0.000001 (1/1,000,000) g of hydrogen ions per liter, and one of pH 8 contains 0.00000001 (1/100,000,000) g of hydrogen ions per liter. Note that a solution with pH 7 contains 10 times as many hydrogen ions as a solution with pH 8 and that pH decreases as the hydrogen ion concentration increases.

Table 2-3	Inorganic Salts Important in Body Functions	
INORGANIC SALT	**FORMULA**	**ELECTROLYTES**
Sodium chloride	NaCl	$Na^+ + Cl^-$
Calcium chloride	$CaCl_2$	$Ca^{++} + 2Cl^-$
Magnesium chloride	$MgCl_2$	$Mg^{++} + 2Cl^-$
Sodium bicarbonate	$NaHCO_3$	$Na^+ + HCO_3$
Potassium chloride	KCl	$K^+ + Cl^-$
Sodium sulfate	Na_2SO_4	$2Na^+ + SO_4^=$
Calcium carbonate	$CaCO_3$	$Ca^{++} + CO_3^=$
Calcium phosphate	$Ca_3(PO_4)_2$	$3Ca^{++} + 2PO_4$

ORGANIC MOLECULES

The term *organic* is used to describe the enormous number of compounds that contain carbon—specifically C–C or C–H bonds. Recall that carbon atoms have only four electrons in their outer energy level (see Figure 2-2, A); it requires four electrons to satisfy the octet rule. As a result, each carbon atom can join with four other atoms to form literally thousands of molecules of varying size and shape. Although some organic molecules are small and have only one or two functional groups, the large *macromolecules* often have many functional groups attached to one another or to other chemical compounds (Figure 2-14). In the human body, the following four major groups of organic substances are the most important:

1. Carbohydrates
2. Proteins
3. Lipids
4. Nucleic acids, nucleotides, and related molecules

Figure 2-14 shows examples of the four major organic substances represented by three-dimensional models. Many macromolecules are composed of basic building blocks, such as glucose or amino acids, that are joined in chains of varying length by covalent bonds. Table 2-4 identifies important biological molecules, including macromolecules and combined forms, shows the type of subunit present, gives a typical function, and lists one or more examples of each. Refer to this table as you read about the large biological molecules in the paragraphs that follow.

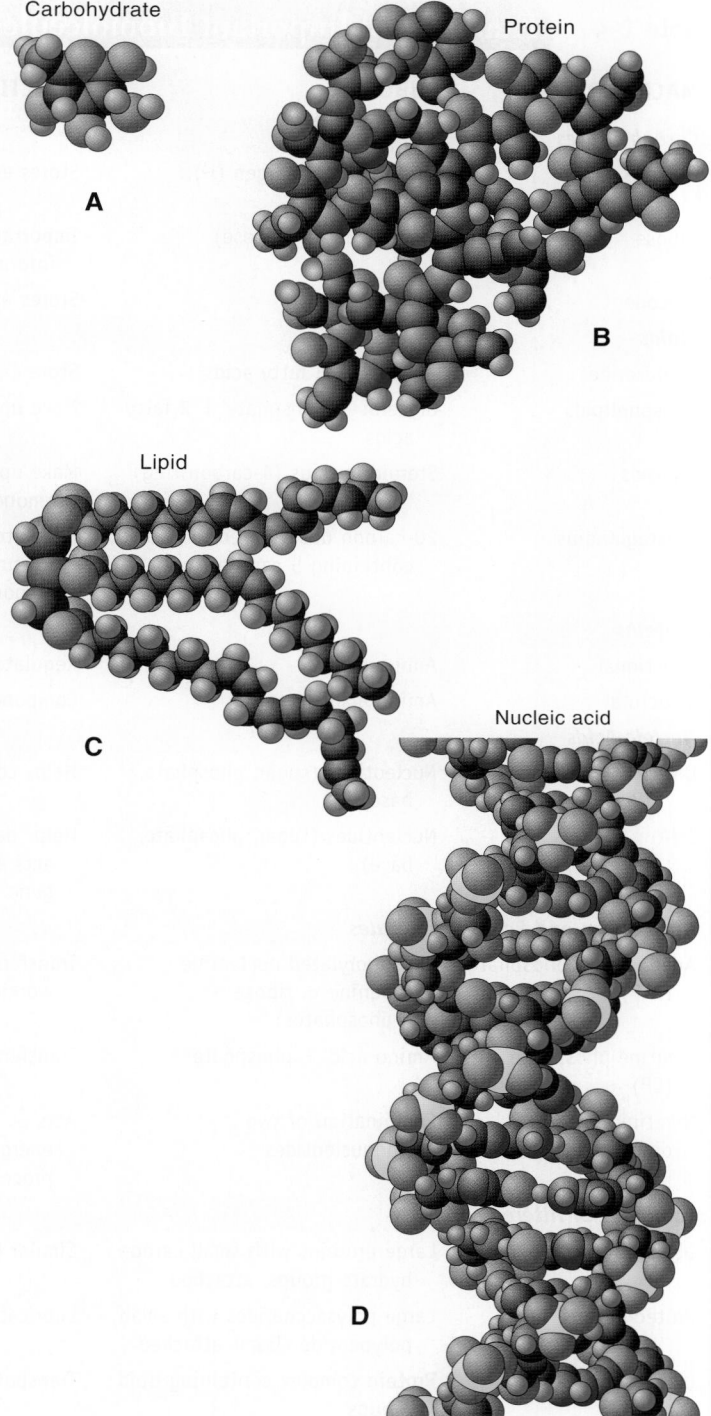

Figure 2-14 *Important organic molecules.* Molecular models showing examples of the four major groups of organic substances: **A,** carbohydrate; **B,** protein; **C,** lipid; **D,** nucleic acid.

Carbohydrates

All **carbohydrate** compounds contain the elements carbon, hydrogen, and oxygen, with the carbon atoms linked to one another to form chains of varying length. Carbohydrates include the substances commonly called *sugars* and *starches* and represent the primary source of chemical energy needed by every body cell. In

Table 2-4   Examples of Important Biomolecules

MACROMOLECULE	SUBUNIT	FUNCTION	EXAMPLE
Carbohydrates			
Glucose	Carbon (C), hydrogen (H), oxygen (O)	Stores energy	Blood glucose
Ribose	Simple sugar (pentose)	Important in expression of hereditary information	Component of RNA
Glycogen	Glucose	Stores energy	Liver glycogen
Lipids			
Triglycerides	Glycerol + 3 fatty acids	Store energy	Body fat
Phospholipids	Glycerol + phosphate + 2 fatty acids	Make up cell membranes	Plasma membrane of cell
Steroids	Steroid nucleus (4-carbon ring)	Make up cell membranes Hormone synthesis	Cholesterol, various steroid hormones Estrogen
Prostaglandins	20-carbon unsaturated fatty acid containing 5-carbon ring	Regulate hormone action; enhance immune system; affect inflammatory response	Prostaglandin E, prostaglandin A
Proteins			
Functional	Amino acids	Regulate chemical reactions	Hemoglobin, antibodies, enzymes
Structural	Amino acids	Component of body support tissues	Muscle filaments, tendons, ligaments
Nucleic Acids			
DNA	Nucleotides (sugar, phosphate, base)	Helps code hereditary information	Chromatin, chromosomes
RNA	Nucleotides (sugar, phosphate, base)	Helps decode hereditary information; acts as "RNA enzyme"; silencing of gene expression	Transfer RNA (tRNA), messenger RNA (mRNA), double-strand RNA (dsRNA)
Nucleotides and Related Molecules			
Adenosine triphosphate (ATP)	Phosphorylated nucleotide (adenine + ribose + 3 phosphates)	Transfers energy from fuel molecules to working molecules	ATP present in every cell of the body
Creatine phosphate (CP)	Amino acid + phosphate	Transfers energy from fuel to ATP	CP present in muscle fiber as "backup" to ATP
Nicotinic adenine dinucleotide (NAD)	Combination of two ribonucleotides	Acts as coenzyme to transfer high-energy particles from one chemical process to another	NAD present in every cell of the body
Combined or Altered Forms			
Glycoproteins	Large proteins with small carbohydrate groups, attached	Similar to functional proteins	Some hormones, antibodies, enzymes, cell membrane components
Proteoglycans	Large polysaccharides with small polypeptide chains attached	Lubrication; increases thickness of fluid	Component of mucous fluid and many tissue fluids in the body
Lipoproteins	Protein complex containing lipid groups	Transport lipids in the blood	LDLs (low-density lipoproteins); HDLs (high-density lipoproteins)
Glycolipids	Lipid molecule with attached carbohydrate group	Component of cell membranes	Component of membranes of nerve cells
Ribonucleoprotein	Combination of RNA nucleotide and protein	Enzyme-like actions such as splicing mRNA	Small nuclear ribonucleoproteins (snRNPs or "snurps") that make up the spliceosome structure in a cell

addition, carbohydrates serve a structural role as components of such critically important molecules as RNA and DNA, which are involved in cell reproduction and protein synthesis.

As a group, carbohydrates are divided into three types or classes that are characterized by the length of their carbon chains. The three types are called

1. Monosaccharides (simple sugars)
2. Disaccharides (double sugars)
3. Polysaccharides (complex sugars)

Monosaccharides

Monosaccharides, or simple sugars, have short carbon chains. The most important simple sugar is *glucose*. It is a six-carbon sugar with the formula $C_6H_{12}O_6$. The chemical formula indicates that each molecule of glucose contains 6 atoms of carbon, 12 atoms of hydrogen, and 6 atoms of oxygen. Because it has six carbon atoms it is called a *hexose* (*hexa*, "six"). Glucose is present in the dry state as a "straight chain" but forms a "cyclic" compound when dissolved in water. In Figure 2-15 the straight chain and cyclic arrangements are shown with a three-dimensional model of the molecule. However, it is important to remember that all forms of glucose, represented in models or illustrations, are the same molecule.

In addition to glucose, other important hexoses, or six-carbon simple sugars, include fructose and galactose. Not all monosaccharides, however, are hexoses. Some are *pentoses* (*penta*, "five"), so named because they contain five carbon atoms. *Ribose* and *deoxyribose* are pentose monosaccharides of great importance in the body—they will be covered further when we discuss nucleic acids later in this chapter. Like all monosaccharides, ribose and deoxyribose are sugars—but strange sugars in that they are not sweet.

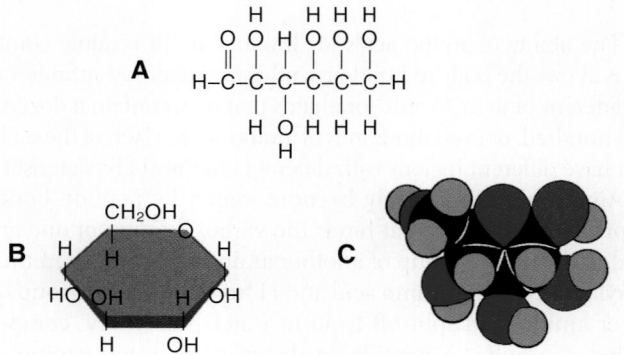

Figure 2-15 *Structure of glucose.* **A,** Straight chain, or linear model, of glucose. **B,** Ring model representing glucose in solution. **C,** Three-dimensional, or space-filling, model of glucose.

Disaccharides and Polysaccharides

Substances classified as *disaccharides* (double sugars) or *polysaccharides* (complex sugars) are carbohydrates composed of two or more simple sugars that are bonded together through a synthesis reaction that involves the removal of water. Sucrose (table sugar), maltose, and lactose are all disaccharides. Each consists of two monosaccharides linked together. Figure 2-16 shows the formation of sucrose from glucose and fructose. Note that a hydrogen atom from the glucose molecule combines with a hydroxyl group (OH) from the fructose molecule to form water, with an oxygen atom left to bind the two subunits together.

Polysaccharides consist of many monosaccharides chemically joined to form straight or branched chains. Once again, water is removed as the many monosaccharide subunits are joined. Any large molecule made up of many identical small molecules is called a *polymer*. Polysaccharides are polymers of monosaccharides. *Glycogen* is sometimes referred to as *animal starch*. It is the main polysaccharide in the body and has an estimated molecular weight of several million—truly a macromolecule.

 QUICK CHECK

18. List the four major groups of organic substances.
19. Identify the most important monosaccharide, or simple sugar.
20. Identify a carbohydrate polymer and explain how it is formed.

Proteins

All **proteins** are characterized by the presence of four elements: carbon, oxygen, hydrogen, and nitrogen. A number of more specialized proteins also contain sulfur, iron, and phosphorus. Proteins (from the Greek *proteios*, "of the first rank") are the most abundant of the carbon-containing, or organic, compounds in the body, and as their name implies, their functions are of first-rank importance (Table 2-5). Protein molecules are giant sized; that is, they are macromolecules.

The many roles played by proteins in the body can be divided into two broad categories: *structural* and *functional*. Structural proteins form the structure of the cells, tissues, and organs of the body. Various unique shapes and compositions such as flexible strands, elastic strands, and waterproof layers allow structural proteins to form the many different building blocks of the body. Functional proteins are chemists. The unique shape of each functional protein allows it to fit with certain other chemicals and cause some change in the molecules. For example, **enzymes** are functional

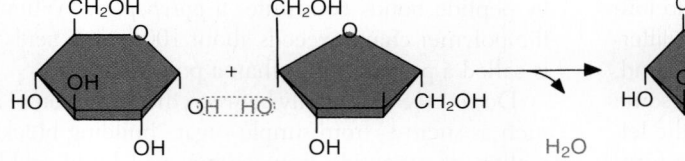
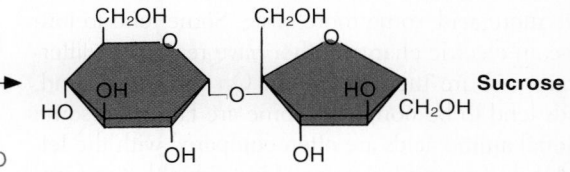

Figure 2-16 *Formation of sucrose.* Glucose and fructose are joined in a synthesis reaction that involves the removal of water.

Table 2-5	Major Functions of Human Protein Compounds
FUNCTION	**EXAMPLE**
Provide structure	Structural proteins include keratin of skin, hair, and nails; parts of cell membranes; tendons
Catalyze chemical reactions	Lactase (enzyme in intestinal digestive juice) catalyzes chemical reaction that changes lactose to glucose and galactose
Transport substances in blood	Proteins classified as albumins combine with fatty acids to transport them in form of lipoproteins
Communicate information to cells	Insulin, a protein hormone, serves as chemical message from islet cells of the pancreas to cells all over the body
Act as receptors	Binding sites of certain proteins on surfaces of cell membranes serve as receptors for insulin and various other hormones
Defend body against many harmful agents	Proteins called *antibodies* or *immunoglobulins* combine with various harmful agents to render agents them harmless
Provide energy	Proteins can be metabolized for energy

proteins that bring molecules together or split them apart in chemical reactions. Protein hormones such as insulin trigger chemical changes in cells to produce the hormone's effects.

You now realize, of course, that it is the *shape* of a protein that determines how it performs. The main principle in understanding how proteins work is that form and function go hand in hand—the right shape for the right job.

Compared with water with a molecular weight of 18, giant protein molecules may weigh in at several million! However, all protein molecules, regardless of size, have a similar basic structure. They are chainlike polymers composed of multiple subunits, or building blocks, linked end to end. The building blocks of all proteins are called *amino acids.*

Amino Acids

The elements that make up a protein molecule are bonded together to form chemical units called **amino acids.** Proteins are composed of 20 commonly occurring amino acids, and nearly all of the 20 amino acids are usually present in every protein. Of these 20, 8 are known as *essential amino acids.* They cannot be produced by the body and must be included in the diet. The 12 remaining *nonessential amino acids* can be produced from other amino acids or from simple organic molecules readily available to the body cells. The basic structural formula for an amino acid is shown in Figure 2-17. As you can see, it consists of a *carbon atom* (called the *alpha* carbon) to which are bonded an *amino group* (NH$_2$), a *carboxyl group* (COOH), a *hydrogen atom,* and a *side chain,* or group of elements designated by the letter *R.* It is this side chain that constitutes the unique, identifying part of an amino acid. The 20 amino acids that make up human proteins are shown in Figure 2-18. You can easily see that each individual amino acid has its own chemical nature. Some are more acid, some more basic. Some tend to ionize and thus have an electric charge. Other have regions of different partial charges and are therefore polar. On the other hand, some amino acids tend to be nonpolar. Some are large and some are small. Individual amino acids are often compared with the letters of the alphabet. Just as combinations of individual letters form word combinations, different amino acids form protein chains. Think of amino acids as the alphabet of proteins.

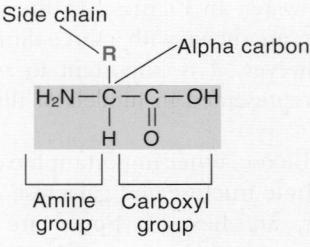

Figure 2-17 *Basic structural formula for an amino acid.* Note relationship of the side chain *(R),* amine group, and carboxyl group to the alpha carbon. The amine group (NH$_2$) is depicted in the figures as H$_2$N to show that the *nitrogen* atom of the group bonds to the alpha carbon.

The ability of amino acids to "link up" in all possible combinations allows the body to build or synthesize an almost infinite variety of different protein "words" or chains that may contain a dozen, several hundred, or even thousands of amino acids. Each of these chains can have different regions with different chemical characteristics.

Amino acids frequently become joined by peptide bonds. A **peptide bond** is one that binds the carboxyl group of one amino acid to the amino group of another amino acid. OH from the carboxyl group of one amino acid and H from the amino group of another amino acid split off to form water plus a new compound called a *peptide.* A peptide made up of only two amino acids linked by a peptide bond is a *dipeptide.* A *tripeptide* consists of three amino acids linked by two bonds. The linkage of four amino acids by these peptide bonds is shown in Figure 2-19, A. A long sequence or chain of amino acids—usually 100 or more—linked by peptide bonds constitutes a *polypeptide.* When the length of the polymer chain exceeds about 100 amino acids, the molecule is called a protein rather than a polypeptide.

Do you see a similarity between the formation of a disaccharide, such as sucrose, from simple sugar "building blocks" and the formation of a peptide from amino acid building blocks? In both processes, subunits are joined together, resulting in the loss of a water molecule. Such processes are called **dehydration synthesis** *re-*

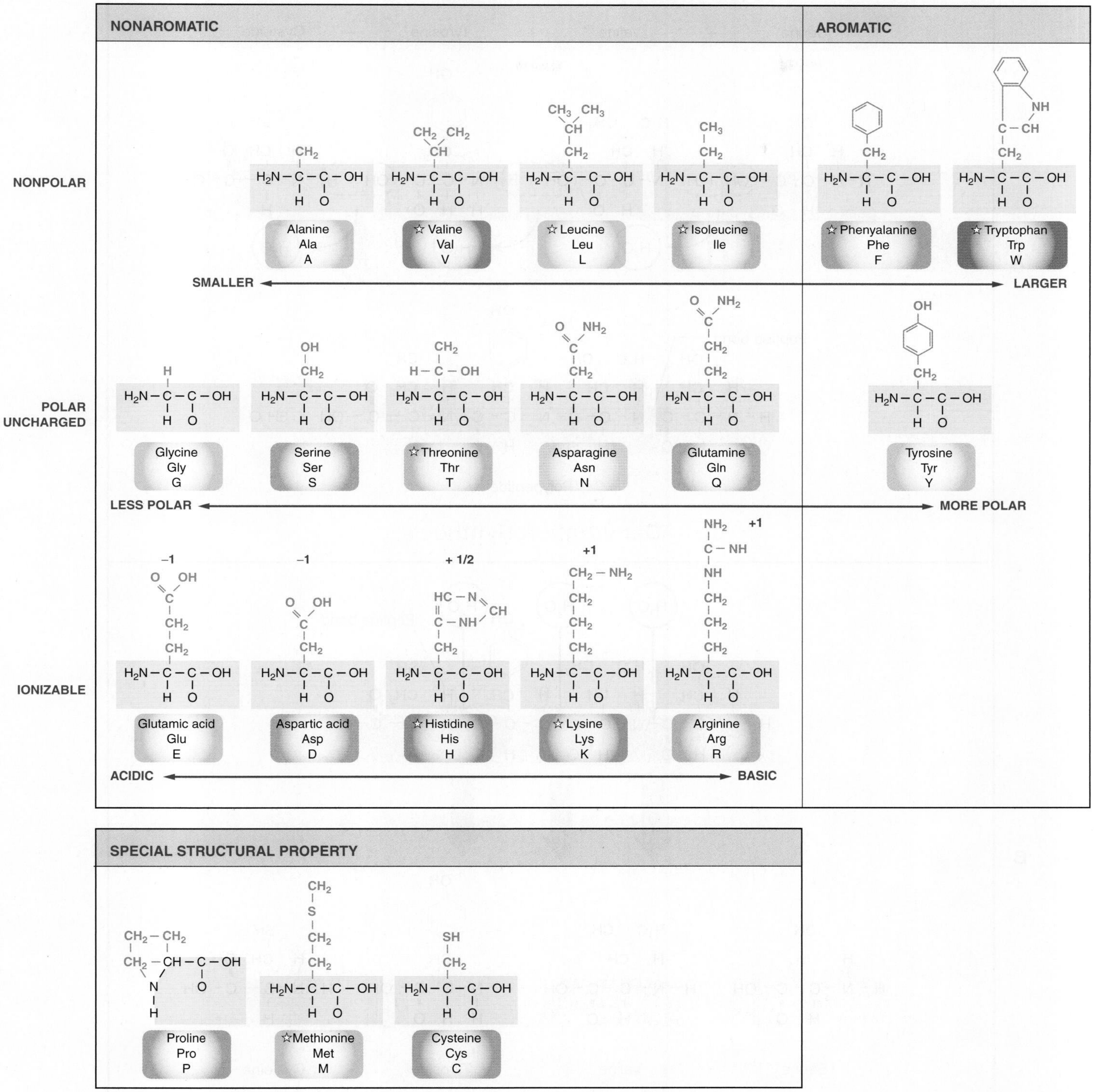

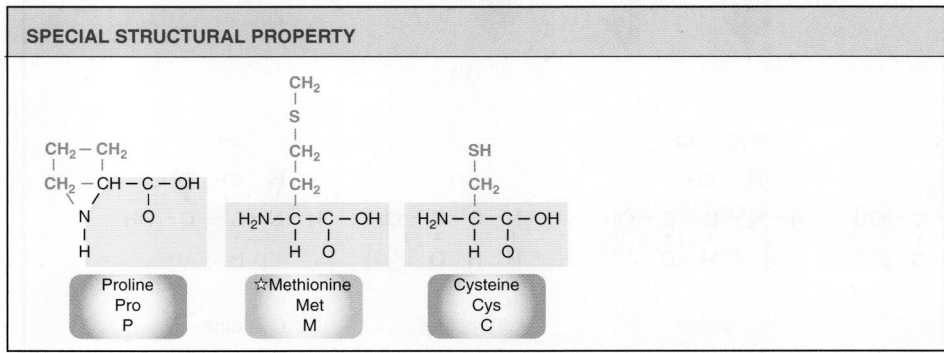

Figure 2-18 *The 20 amino acids of the body.* The full name for each is given, followed by the three-letter abbreviation and the one-letter symbol. The structural formulas show that each amino acid has the same chemical backbone *(yellow highlight)* but differs from the others in the side, or R, group that it possesses (red). Essential amino acids are indicated by a star. Notice the great variety of sizes among the amino acids, the different polarities or charges, the different level of acidity, and the other differing chemical characteristics. Like different kinds of blocks in a set of toy blocks, this makes it possible to build different proteins with a wide variety of chemical characteristics and functions.

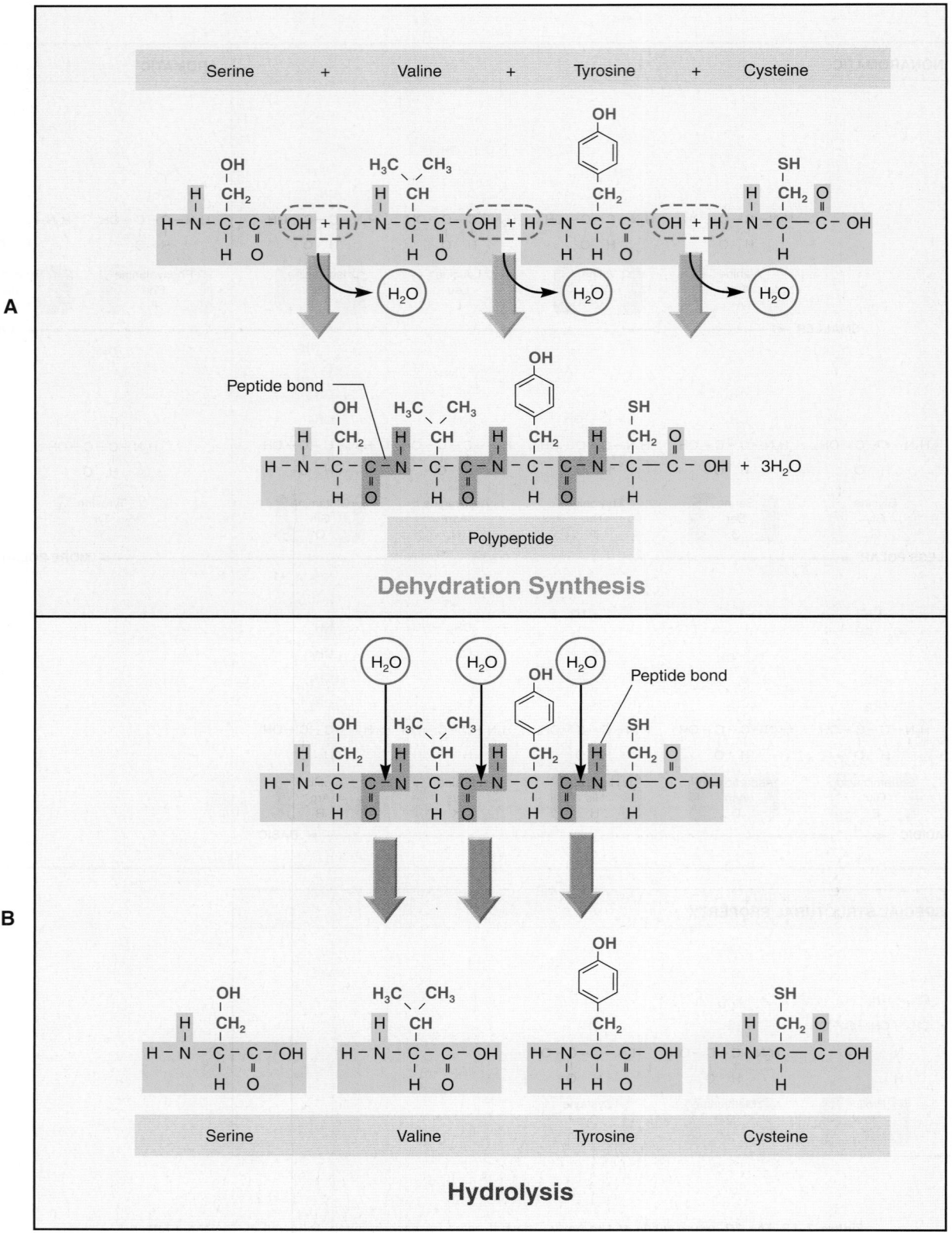

Figure 2-19 *Formation (dehydration synthesis) and decomposition (hydrolysis) of a polypeptide.* **A,** Linkage of four amino acids by three peptide bonds resulting in the dehydration synthesis of a polypeptide chain and three molecules of water. **B,** Decomposition (hydrolysis) reaction resulting from the addition of three molecules of water. Peptide bonds are broken and individual amino acids are released.

actions and are very common in living organisms. Repetitive dehydration synthesis reactions involving the addition of simple sugars will result in the formation of a polysaccharide polymer. If amino acids are linked together in this fashion by formation of peptide bonds, the resulting polymer is a polypeptide (Figure 2-19, *A*).

A decomposition reaction called **hydrolysis** requires the addition of a water molecule to break a bond. During hydrolysis of a peptide chain, the peptide linkages between adjacent amino acids in the sequence are broken by the addition of water, and individual amino acids are released (Figure 2-19, *B*).

Levels of Protein Structure

Biochemists often describe four levels of increasing complexity in protein organization:

1. Primary (first level)
2. Secondary (second level)
3. Tertiary (third level)
4. Quaternary (fourth level)

The four levels of protein structure are illustrated in Figure 2-20.

The *primary structure* of a protein refers simply to the number, kind, and sequence of amino acids that make up the polypeptide chain. The hormone of the human parathyroid gland, parathyroid hormone (PTH), is a primary structure protein—it consists of only one polypeptide chain of 84 amino acids.

Most polypeptides do not exist as a straight chain. Instead, they show a *secondary structure* in which the chains are coiled or bent into pleated sheets. The most common type of coil takes a clockwise direction and is called an *alpha helix*. In this type of secondary structure, the coils of the protein chain resemble a spiral staircase, with the coils stabilized by hydrogen bonds between successive turns of the spiral. This stabilizing function of hydrogen bonding in protein structure is critical.

Just as a primary structure polypeptide chain can pleat or bend into a helical secondary structure, so too can a secondary structure protein chain undergo other contortions and be further twisted so that a globular-shaped *tertiary structure* protein is formed. In this structure, the polypeptide chain is so twisted that its coils touch one another in many places, and "spot welds," or interlocking connections, occur. These linkages are strong covalent bonds between amino acid units that exist in the same chain (Box 2-4). In addition, hydrogen bonds and other weak forces also help stabilize the twisted and convoluted loops of the structure. The red muscle protein myoglobin, which is discussed in Chapter 11 is an example of a tertiary structure protein.

A *quaternary structure* protein is one that contains clusters of more than one polypeptide chain, all linked together into one giant molecule. Antibody molecules that protect us from disease (see Chapter 21) and hemoglobin molecules in red blood cells (see Chapter 17) are examples.

A group of proteins called *chaperones*, which are present in every body cell, acts to direct the steps required for proteins to fold into the twisted and convoluted shape required for them to function properly (Box 2-5). Some of these chaperone proteins are called *chaperonins*. Inappropriate folding of some proteins is known to be associated with certain diseases. The critically important chemical reactions that permit chaperonins to organize proteins into the different organizational levels required for a particular function can occur only within a very narrow pH range. Maintaining acid-base balance and normal pH in body cells and fluids will be discussed in depth in Chapter 30.

Protein molecules are highly organized and show a very definite relationship between their structural appearance and their function. For example, the strong **structural proteins** found in tendons and ligaments are linear, or threadlike, insoluble, and very stable molecules (Figure 2-21). In contrast, **functional pro-**

BOX 2-4: FYI
Disulfide Linkages

Hair contains a fibrous threadlike protein called *keratin* that is rich in the sulfur-containing amino acid cysteine. The protein chains in keratin are linked in numerous places by sulfur-to-sulfur bonds that form between sulfur-containing amino acids, such as cysteine, in adjacent hair fibrils in each hair shaft. The number and placement of these bonds, called *disulfide linkages,* are determined by heredity and partly explain the "natural" straightness or curl of hair seen in every individual.

The object of a "permanent wave" (at least to a chemist!) is to change the arrangement of these bonds by breaking the naturally occurring disulfide linkages and then causing them to re-form in another pattern. In a permanent wave, strong chemicals are applied to the hair that break the existing or "natural" disulfide linkages. The hair is then curled on some type of roller or straightened, and another chemical is applied that causes the disulfide bridges to become reestablished in the new or reoriented configuration. Such treatment must be repeated as new hair growth occurs.

Disulfide linkage. The yellow highlighted area shows a disulfide linkage in a folded protein. Other forces maintaining the folded protein structure are also shown.

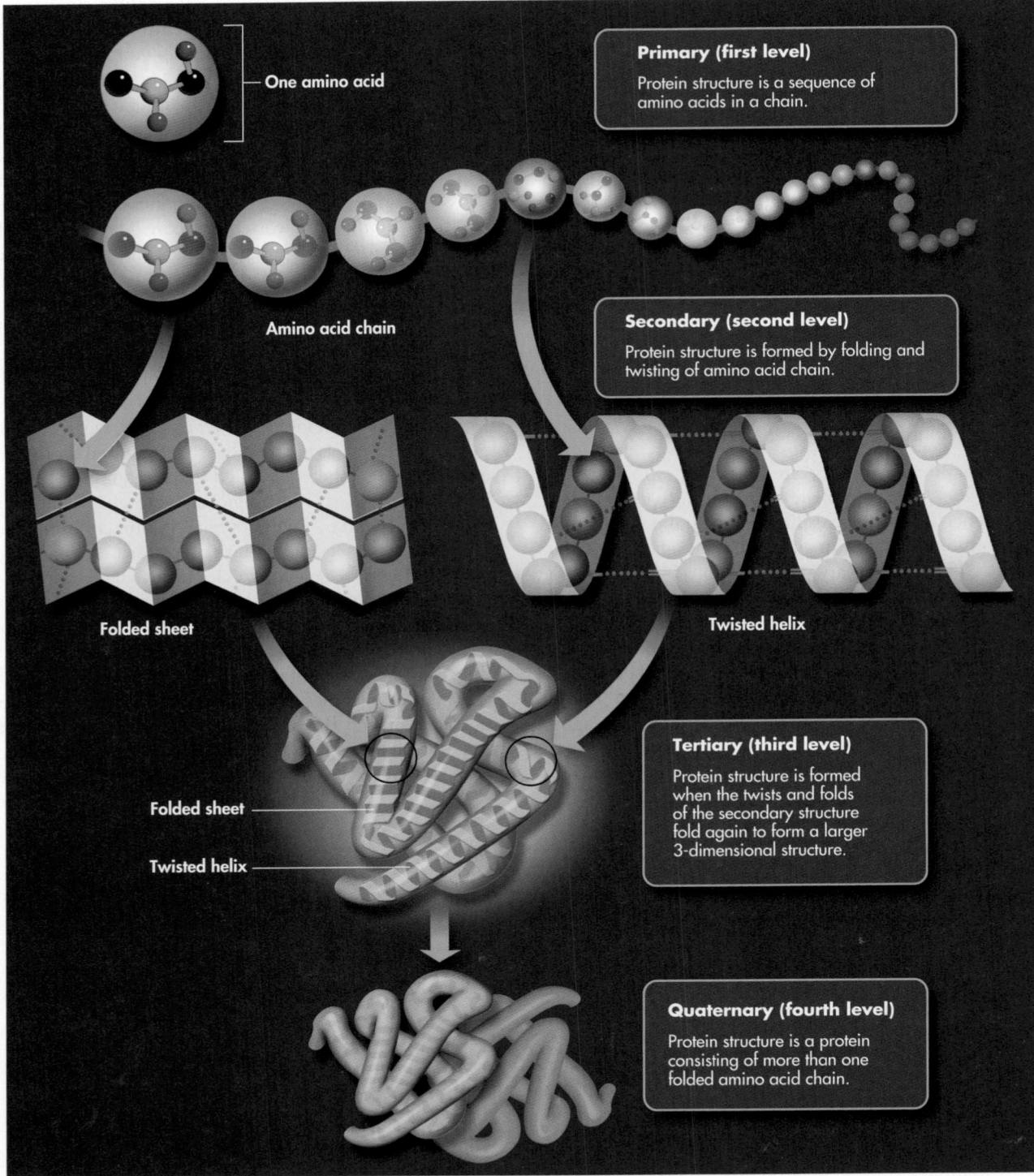

Figure 2-20 *Structural levels of protein. Primary structure:* determined by the number, kind, and sequence of amino acids in the chain. *Secondary structure:* hydrogen bonds stabilize folds or helical spirals. *Tertiary structure:* globular shape maintained by strong (covalent) intramolecular bonding and by stabilizing hydrogen bonds. *Quaternary structure:* results from bonding between more than one polypeptide unit.

BOX 2-5: FYI
Visualizing Proteins

Only a few decades ago, the usual way for a biochemist to demonstrate the three-dimensional structure of a protein molecule was by building wooden ball-and-stick models. These models were less than ideal because they took a long time to build, often fell apart, and were not easy to handle. But now there are many sophisticated, but easy-to-use computer programs that can "build" protein molecules on the monitor screen. These "virtual protein molecules" can be rotated and viewed from nearly any angle. They can also be changed when trying to design a new protein molecule for therapeutic or other purposes. Computer modeling has become the usual way to represent the complex shapes of protein molecules in the twenty-first century. Here, three common types of protein models are shown. The *ribbon model* shows the areas where alpha helices and folded sheets form within the molecule. The *space-filling model* shows each atom as a "cloud" filling up the space occupied by that atom. The *surface-rendering model* shows the three-dimensional boundaries of the whole protein molecule, often also color-coding for charged regions on the surface of the protein.

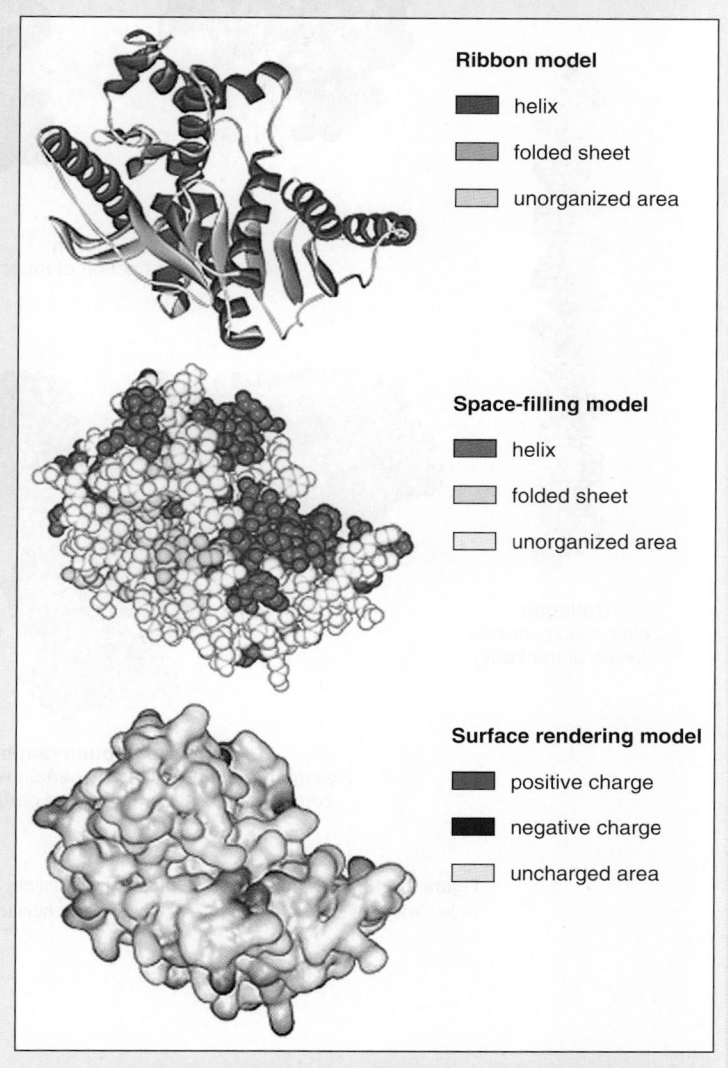

Ribbon model
- helix
- folded sheet
- unorganized area

Space-filling model
- helix
- folded sheet
- unorganized area

Surface rendering model
- positive charge
- negative charge
- uncharged area

Three ways to visualize the same folded protein molecule.

teins such as enzymes (Box 2-6), certain protein hormones, antibodies, albumin, and hemoglobin are globular, often soluble, and chemically active molecules.

Simply stated, proteins perform their roles by having the right shape for their job—whatever that job is. Given the nearly infinite variety of different amino acid sequences and the complexity of how proteins are folded, you can see that the body can make just about any tool or building block it needs for a variety of jobs. Consider also that if one of the body's proteins loses its shape, or **denatures,** it will lose its function (Figure 2-22). If a protein is built incorrectly or it denatures, the whole body may be in peril (Box 2-7). Factors that can cause a protein to denature include changes in temperature, changes in pH, radiation, and the presence of certain hazardous chemicals. Depending on the circumstances, restoring the proper chemical environment of a protein may allow it to *renature* and begin functioning normally once again.

QUICK CHECK

21. What element is present in all proteins but not in carbohydrates?
22. Identify the building blocks of proteins and explain what common chemical features they all share.
23. Explain the four levels of protein structure.

Lipids

Lipids, according to one definition, are water-insoluble organic biomolecules. Lipids ordinarily do not dissolve in water because lipid molecules are generally **nonpolar** (without charged regions) and thus do not cling to the charged areas of the polar water mol-

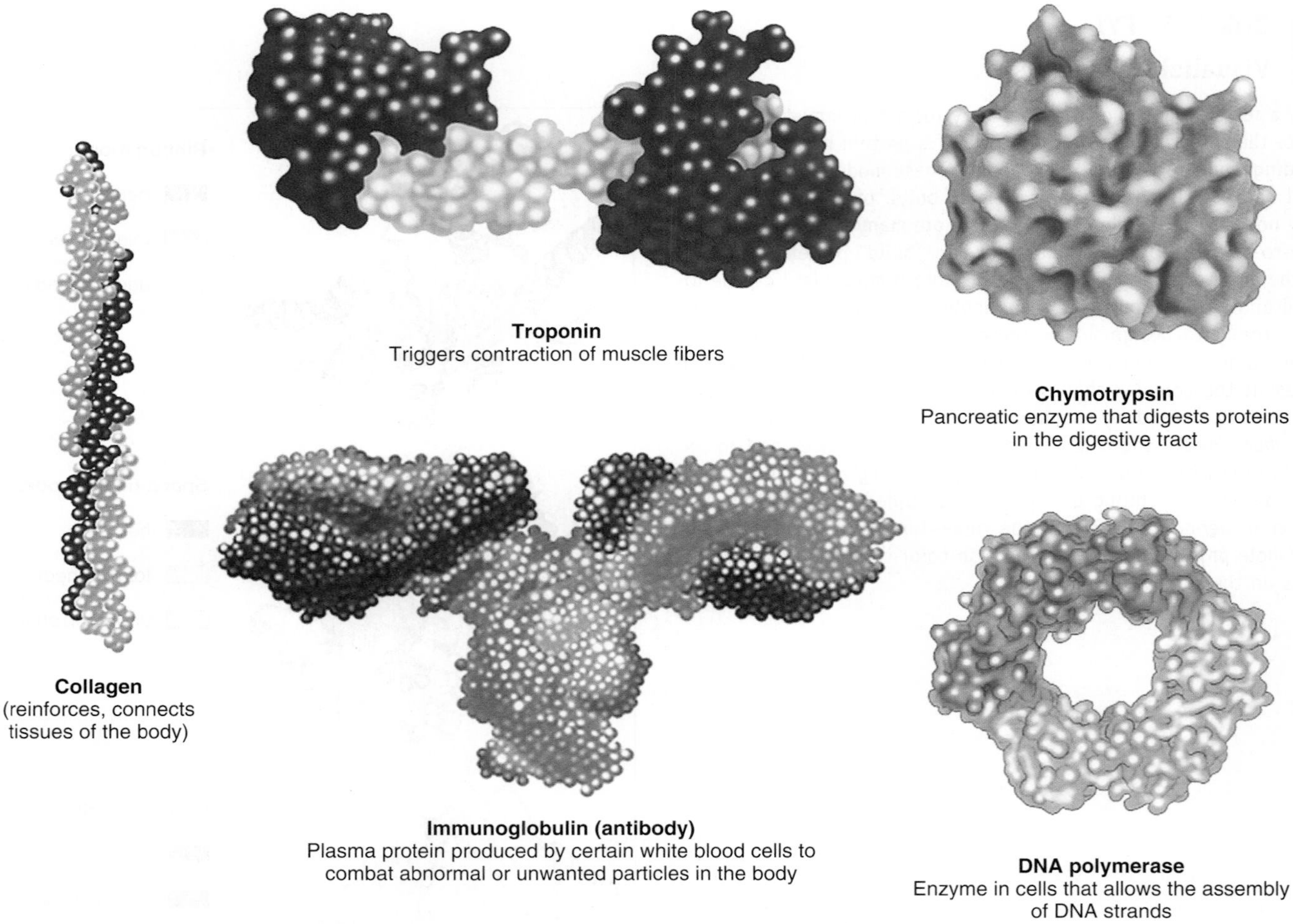

Troponin
Triggers contraction of muscle fibers

Chymotrypsin
Pancreatic enzyme that digests proteins
in the digestive tract

Collagen
(reinforces, connects
tissues of the body)

Immunoglobulin (antibody)
Plasma protein produced by certain white blood cells to
combat abnormal or unwanted particles in the body

DNA polymerase
Enzyme in cells that allows the assembly
of DNA strands

Figure 2-21 *Variety of protein shapes.* These images of folded protein structures provide examples of the wide variety of shapes that proteins have in the human body.

Figure 2-22 *Denatured protein.* When a protein loses its normal folded organization and thus loses its functional shape, it is called a *denatured protein*. Denatured proteins are not able to function normally. However, if the protein shape is restored, the *renatured protein* may resume its normal function.

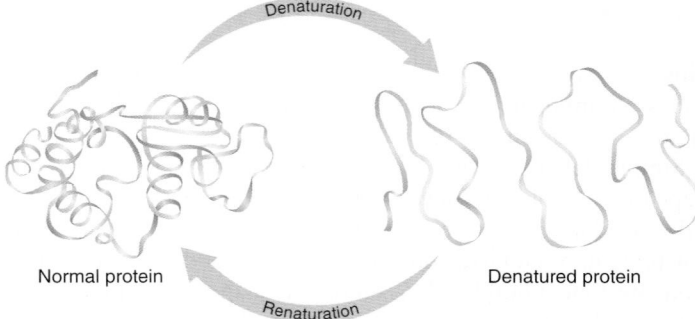

Denaturation

Normal protein

Denatured protein

Renaturation

BOX 2-6: FYI

How Enzymes Function

Enzymes, the largest group of proteins in the body, are *chemical catalysts*. This means that they help a chemical reaction occur but are not reactants or products themselves. They participate in chemical reactions but are not changed by the reactions. Enzymes act to speed up the rate at which metabolic reactions occur. Thousands of enzymes are known, and each is responsible for speeding up the rate of a very particular and unique chemical reaction—sometimes by a factor of 10 and often by a million or more. No reaction in the body occurs fast enough unless the specific enzymes needed for that reaction are present. These important proteins are able to accomplish their function even when present in very small quantities. This is possible because they are not "used up" in a reaction but remain unchanged to be used again and again as needed.

Note how shape is important to the function of enzyme molecules. Each enzyme, by means of its uniquely shaped binding sites, binds to a very specific substance, called a substrate, that it works on—much as a key fits a specific lock. This explanation of enzyme action is sometimes called the lock-and-key model (see figure).

Molecules called *cofactors* may assist enzymes to catalyze a chemical reaction. *Coenzymes* such as NAD (see Figure 2-30) are organic, nonprotein cofactors that often shuttle energy-containing particles from one chemical pathway to another.

More details of enzyme function in the cell are outlined in Chapter 4.

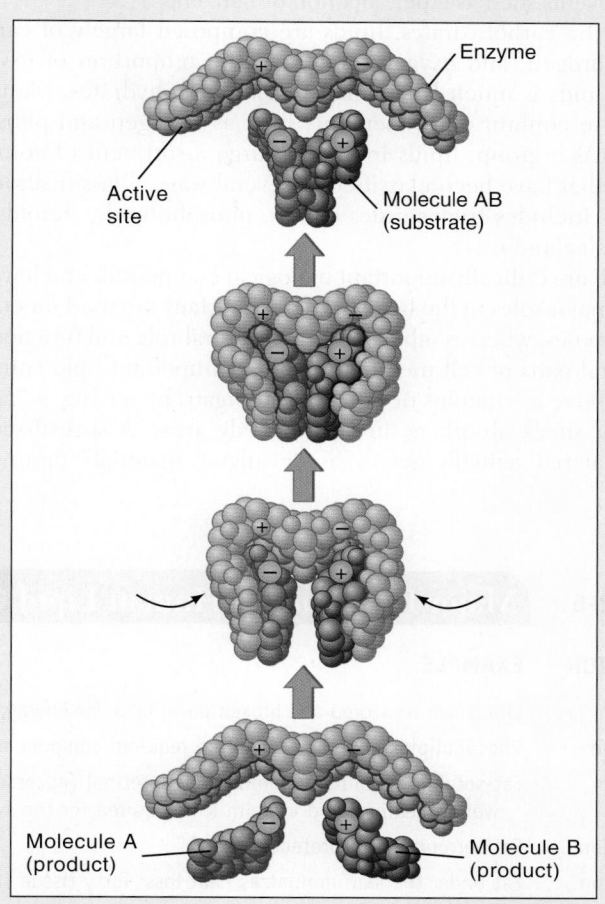

Lock-and-key model of enzyme action. Enzymes are functional proteins whose molecular shape allows them to catalyze chemical reactions. Substrates (molecules *A* and *B*) are brought together by the enzyme to form a larger molecule *(AB)*.

BOX 2-7: HEALTH MATTERS

Phenylketonuria

Phenylketonuria (PKU) is a genetic disease caused by the lack of a single enzyme *(phenylalanine hydroxylase)* required to break down or metabolize the amino acid phenylalanine. PKU is one example of a group of genetic diseases called *inborn errors of metabolism,* which are discussed in Chapter 34. In this instance, phenylalanine metabolism is impaired because the gene required to produce the necessary enzyme for its breakdown is defective. The disease occurs in 1 of every 12,000 births in North America and, if untreated, causes a buildup of phenylalanine in the tissues that results in severe mental retardation and other neurobehavioral symptoms. Traditional treatment of PKU consisted of strict dietary restriction of phenylalanine-containing foods during the first 4 to 8 years of life, followed by some liberalization of diet. Recent data suggest that continuation of the diet into adulthood may be necessary.

Since the mid-1960s every state has mandated testing of newborn infants for PKU before discharge from the nursery. As a result, the number of undiagnosed and untreated PKU cases that formerly resulted in mental retardation has decreased dramatically. A phenylalanine level above 4 mg/dl is considered positive for the disease. However, the screening procedure, called the Guthrie test, is less accurate if blood is drawn before the infant is 48 hours old. Because early discharge of both the mother and baby from the hospital after childbirth is becoming more common, a second, or repeat, screening test at 2 weeks of age may be necessary. The test is performed on a drop of blood, which is obtained from the baby by a "heel stick" and adsorbed onto a piece of filter paper.

ecules. Although insoluble in water, most lipids, many with an oil-like consistency and greasy feel, dissolve readily in some organic solvents such as ether, alcohol, or benzene.

Like the carbohydrates, lipids are composed largely of carbon, hydrogen, and oxygen. However, the proportion of oxygen in lipids is much lower than that in carbohydrates. Many lipids also contain other elements such as nitrogen and phosphorus. As a group, lipids include a large assortment of compounds that have been classified in several ways. Classification of lipids includes triglycerides or fats, phospholipids, steroids, and prostaglandins.

Lipids are critically important biological compounds and have several major roles in the body (Table 2-6). Many are used for energy purposes, whereas others serve a structural role and function as integral parts of cell membranes. Other important lipid compounds serve as vitamins or protect vital organs by serving as "fat pads," or shock absorbers, in certain body areas. A specialized lipid material actually serves as "insulator material" around nerves, thus serving to prevent "short circuits" and speed nervous impulse transmissions.

Triglycerides or Fats

Triglycerides (triacylglycerols), or fats, are the most abundant lipids, and they function as the body's most concentrated source of energy. Two types of building blocks are needed to synthesize or build a fat molecule: *glycerol* and *fatty acids*. Each glycerol unit is joined to three fatty acids, and the glycerol building block is the same in each fat molecule. Therefore, it is the specific type of fatty acid molecule or component that identifies and determines the chemical nature of any fat.

Types of Fatty Acids

Fatty acids vary in the length of their carbon chains (number of carbon atoms) and in the number of hydrogen atoms that are attached to, or "saturate," the available bonds around each carbon in the chain. Figure 2-23 shows a structural formula and three-

Table 2-6	Major Functions of Human Lipid Compounds	
FUNCTION	**EXAMPLE**	
Energy	Lipids can be stored and broken down later for energy; they yield more energy per unit of weight than carbohydrates or proteins do	
Structure	Phospholipids and cholesterol are required components of cell membranes	
Vitamins	Fat-soluble vitamins: vitamin A forms retinal (necessary for night vision); vitamin D increases calcium uptake; vitamin E promotes wound healing; and vitamin K is required for the synthesis of blood-clotting proteins	
Protection	Fat surrounds and protects organs	
Insulation	Fat under the skin minimizes heat loss; fatty tissue (myelin) covers nerve cells and electrically insulates them	
Regulation	Steroid hormones regulate many physiological processes. For example, estrogen and testosterone are responsible for many of the differences between females and males; prostaglandins help regulate inflammation and tissue repair	

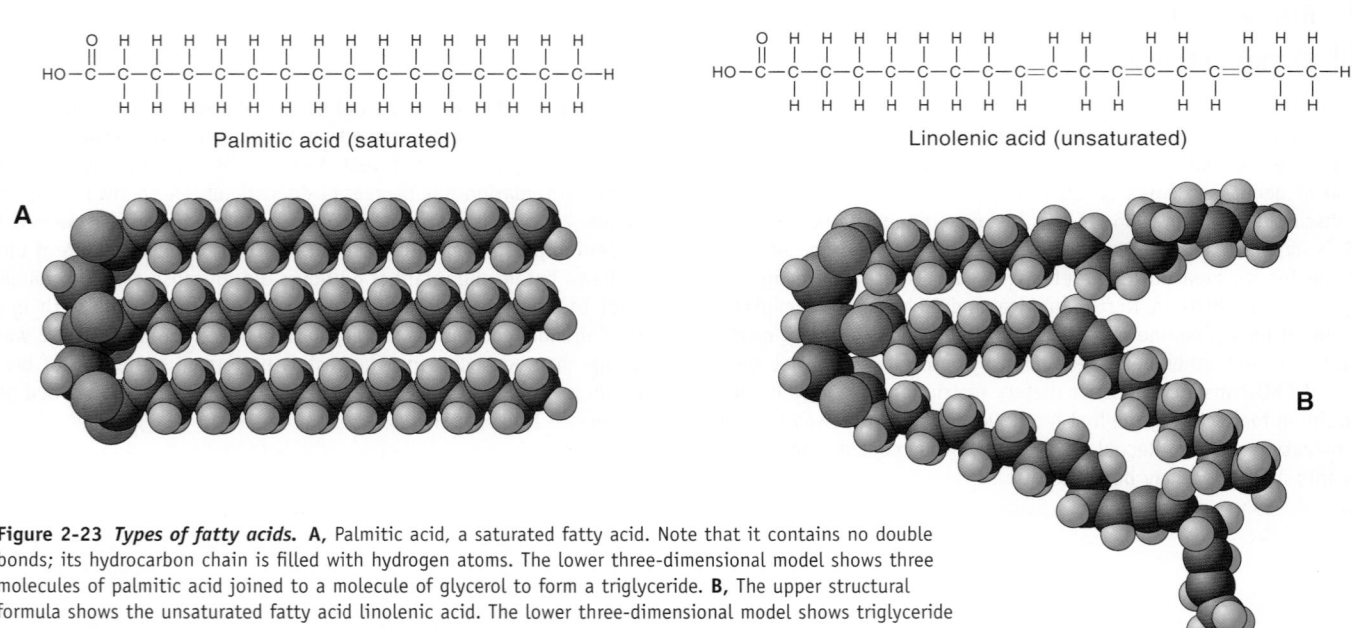

Palmitic acid (saturated)

Linolenic acid (unsaturated)

A

B

Figure 2-23 *Types of fatty acids.* **A,** Palmitic acid, a saturated fatty acid. Note that it contains no double bonds; its hydrocarbon chain is filled with hydrogen atoms. The lower three-dimensional model shows three molecules of palmitic acid joined to a molecule of glycerol to form a triglyceride. **B,** The upper structural formula shows the unsaturated fatty acid linolenic acid. The lower three-dimensional model shows triglyceride exhibiting "kinks" caused by the presence of double bonds in the component fatty acids.

dimensional model for a saturated (palmitic) and unsaturated (linolenic) fatty acid.

By definition, a *saturated fatty acid* is one in which all available bonds of its hydrocarbon chain are filled, that is, saturated, with hydrogen atoms. The chain contains no double bonds (Figure 2-23, A). In contrast, an *unsaturated fatty acid* has one or more double bonds in its hydrocarbon chain because not all the chain's carbon atoms are saturated with hydrogen atoms (Figure 2-23, B). The degree of saturation is the most important factor in determining the physical and chemical properties of fatty acids. For example, animal fats such as tallow and lard are solids at room temperature, whereas vegetable oils are liquids. The difference lies in the extent of unsaturation—animal fats are saturated, whereas vegetable oils are not. Note in Figure 2-23, B, that the presence of double bonds in a fatty acid molecule will cause the chain to "kink" or bend.

Fats become more oily and liquid as the number of unsaturated double bonds increases. The "kinks" and "bends" in the unsaturated molecules keep them from fitting closely together. In contrast, the lack of kinks in saturated fatty acids allows the molecules to fit tightly together to form a solid mass at higher temperatures.

Formation of Triglycerides

Figure 2-24 shows the formation of a triglyceride. Its name, *glycerol tricaproate*, suggests that it contains three molecules of the six-carbon fatty acid *caproic acid* attached to a glycerol molecule. Note that the three caproic acid building blocks attach by their carboxyl groups (COOH) to the three hydroxyl groups (OH) of the glycerol molecule to form the triglyceride and three molecules of water. The process is one you are now familiar with—it is a dehydration synthesis reaction. Keep in mind that although some fats, such as glycerol tricaproate, contain three molecules of the same fatty acid, others may have two or three different fatty acids attached to glycerol. Caproic acid is considered to be a "short-chain" fatty acid; some triglycerides contain fatty acids with a carbon backbone several times longer, thus forming "long-chain" fatty acids.

Phospholipids

Phospholipids are fat compounds similar to triglycerides. They are modified, however, in that one of the three fatty acids attached to glycerol in a triglyceride is replaced in a phospholipid by another type of chemical structure containing phosphorus and nitrogen. The structural formula of a phospholipid is shown in Figure 2-25. Observe that the phospholipid molecule contains glycerol. Joined to the glycerol at one end of the molecule are two fatty acids. Attached to glycerol but extending in the opposite direction is the phosphate group, which is attached to a nitrogen-containing compound.

The head, or end of the molecule containing the phosphorus group, in a phospholipid molecule is polar and is therefore water soluble. The term *hydrophilic*, meaning "water loving," also applies to the phospholipid head. The end formed by the two fatty acids is nonpolar and is therefore fat soluble or *hydrophobic* ("water fearing"). This unique property means that phospholipid molecules can bridge, or join, two different chemical environments—a water environment on one side and a lipid environment on the other. Thus, in water they often form bilayers (double layers) with the fatty tails facing toward one another and the heads forming sheets that face the water on either side of the bilayer. For this reason, phospholipids are a primary component of cell membranes (which are bilayers) and are discussed further in Chapter 3.

Steroids

Steroids are an important and a large group of compounds whose molecules have as their principal component the *steroid nucleus* (Figure 2-26). The steroid nucleus is composed of four attached rings that are structurally similar but may have widely diverse functions related to the differing functional groups that are attached to them. Steroids, also called *sterols*, are widely distributed in the body and are involved in many important structural and functional roles. Cholesterol is a steroid found in the plasma membrane surrounding every body cell (see Chapter 3 and Box 2-8). Its presence helps stabilize this important cellular structure and is required for many reactions that cells must perform to survive. In addition, cholesterol is a constituent part of such im-

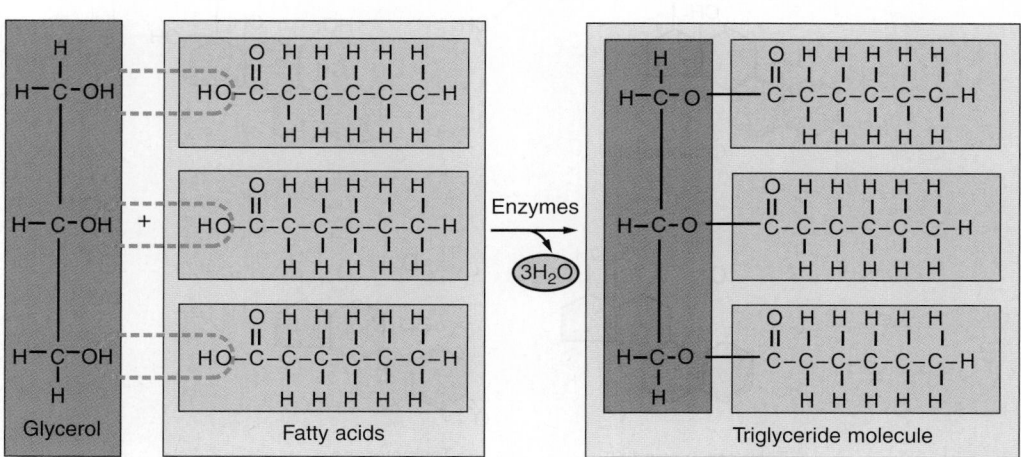

Figure 2-24 *Formation of triglyceride.* Glycerol tricaproate is a composite molecule made up of three molecules of caproic acid (a six-carbon fatty acid) coupled in a dehydration synthesis reaction to a single glycerol backbone. In addition to the triglyceride, this process results in the formation of three molecules of water.

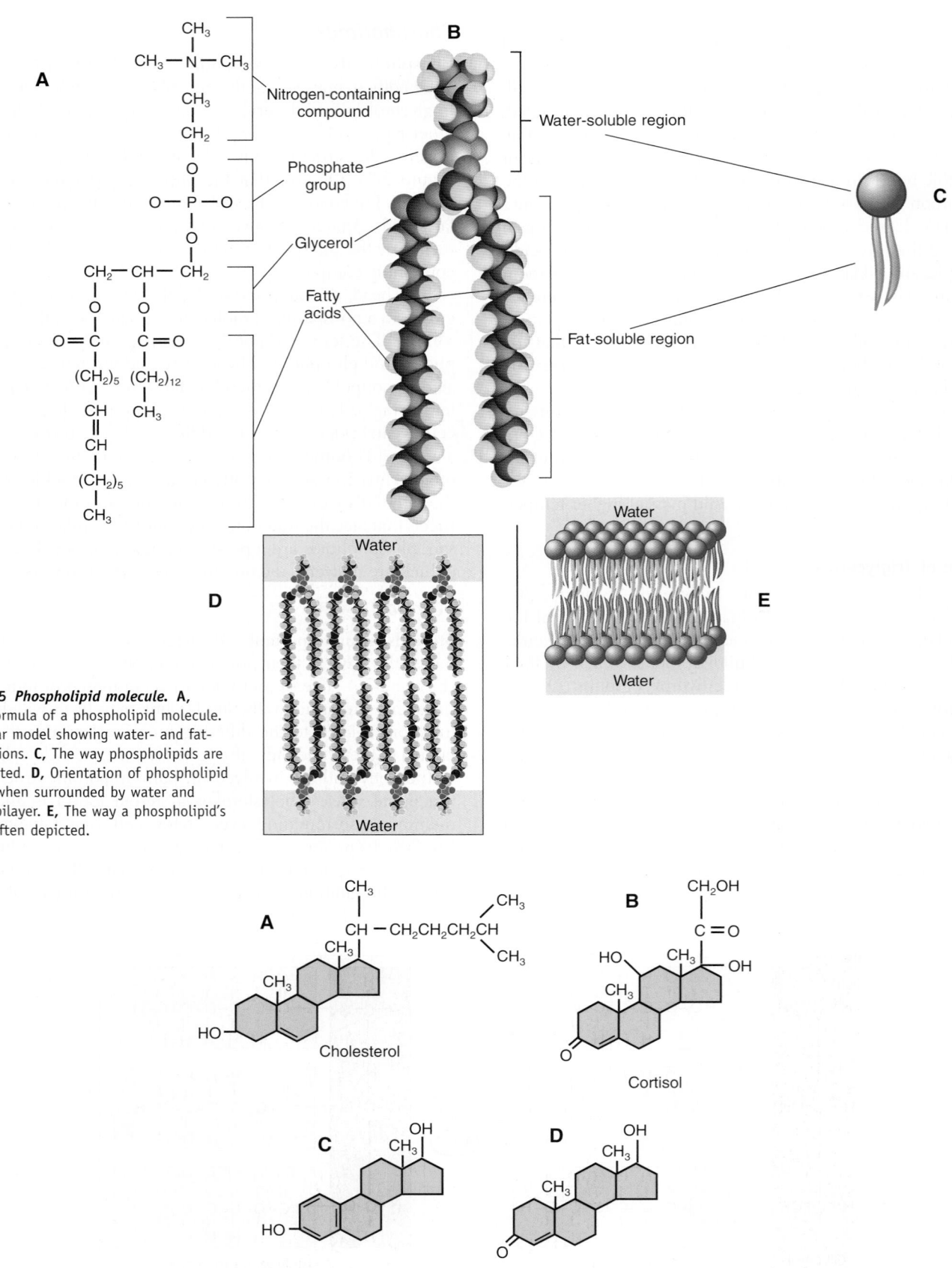

Figure 2-25 *Phospholipid molecule.* **A,** Chemical formula of a phospholipid molecule. **B,** Molecular model showing water- and fat-soluble regions. **C,** The way phospholipids are often depicted. **D,** Orientation of phospholipid molecules when surrounded by water and forming a bilayer. **E,** The way a phospholipid's bilayer is often depicted.

Figure 2-26 *Steroid compounds.* The steroid nucleus—highlighted in *yellow*—found in cholesterol **(A)** forms the basis for many other important compounds such as cortisol **(B)**, estradiol (an estrogen) **(C)**, and testosterone **(D).**

BOX 2-8: HEALTH MATTERS
Blood Lipoproteins

A lipid such as cholesterol can travel in the blood only after it has attached to a protein molecule—forming a lipoprotein. Some of these molecules are called *high-density lipoproteins* (HDLs) because they have a high density of protein (more protein than lipid). Another type of molecule contains less protein (and more lipid), so it is called *low-density lipoprotein* (LDL). The composite nature of a lipoprotein molecule is shown in the figure.

The cholesterol in LDLs is often called *bad* cholesterol because high blood levels of LDL are associated with *atherosclerosis*, a life-threatening blockage of arteries. LDLs carry cholesterol to cells, including the cells that line blood vessels. HDLs, on the other hand, carry so-called good cholesterol *away from cells* and toward the liver for elimination from the body. A high proportion of HDL in the blood is associated with a low risk for atherosclerosis. Factors such as cigarette smoking decrease HDL levels and thus contribute to the risk for atherosclerosis. Factors such as exercise increase HDL levels and thus decrease the risk for atherosclerosis.

The 1985 Nobel Prize in Physiology or Medicine was awarded to Drs. Michael Brown and Joseph Goldstein for their research on specialized receptor sites on LDL molecules, which are elevated in the blood of individuals with certain types of heart disease. Lipid metabolism is discussed in detail in Chapter 27.

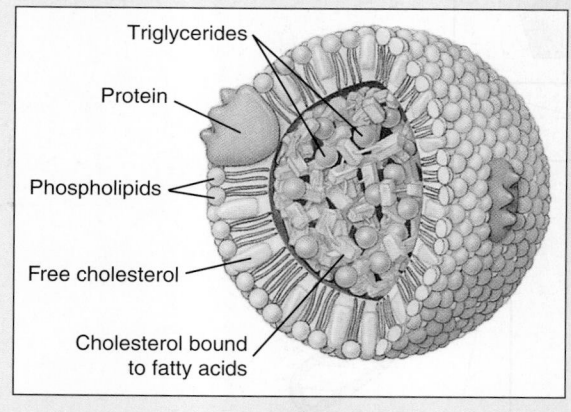

Structure of a lipoprotein.

Figure 2-27 *Prostaglandin.* Prostaglandins such as this example of prostaglandin E (PGE) are 20-carbon unsaturated fatty acids with a 5-carbon ring. Prostaglandins act as local regulators in the body.

shown that these biologically powerful chemical substances are produced by cell membranes located in almost every body tissue. They are formed and then released from cell membranes in response to a particular stimulus. Once released, they have a very local effect and are then inactivated.

The effects of prostaglandins in the body are many and varied. They play a crucial role in regulating the effects of several hormones, influence blood pressure and the secretion of digestive juices, enhance the body's immune system and inflammatory response (Box 2-9), and have an important role in blood clotting and respiration, to name a few. The use of prostaglandins and prostaglandin inhibitors as drugs is an exciting and rapidly growing area in clinical medicine. Treatment of specific disease states,

BOX 2-9: FYI
Aspirin and Prostaglandins

In the presence of an appropriate stimulus such as irritation or injury, fatty acids required for prostaglandin synthesis are released by cell membranes. If a specific type of enzyme, *cyclooxygenase (COX)*, is present to interact with these fatty acids, prostaglandins will be synthesized and released from the cell membrane into the surrounding tissue fluid.

Prostaglandins sometimes serve as inflammatory agents. They cause local dilation of blood vessels with resulting heat (fever), swelling, redness, and pain. Aspirin works to relieve these symptoms by blocking the activity of the COX-1 and COX-2 enzymes. If these enzymes cannot function properly, prostaglandin synthesis will be inhibited, and symptoms will be relieved.

Prostaglandins sometimes serve to regulate blood clotting. Again, aspirin can inhibit prostaglandin synthesis and play a therapeutic role in preventing abnormal blood clots or reducing abnormal clots that have already begun forming. For this reason, some people at risk for a heart attack triggered by abnormal blood clots are advised to take daily low-dose aspirin to reduce the formation of abnormal clots. If a heart attack has already begun, full-dose aspirin taken immediately may stop the clotting and thus increases a person's chances of surviving the episode by about 25%.

The functions of prostaglandins and the actions of other COX enzyme inhibitors are discussed further in Chapter 16.

portant hormones as corticosteroids, estrogen, and testosterone. It is also present in the bile salts needed for digestion. The steroid nucleus is also a part of the active hormone form of vitamin D called *calcitriol.*

Prostaglandins

Prostaglandins, often called "tissue hormones," are lipids composed of a 20-carbon unsaturated fatty acid that contains a 5-carbon ring (Figure 2-27). Many different kinds of prostaglandins exist in the body. We now classify 16 prostaglandin types (PGs) into nine broad categories, called PGA to PGI. Each major grouping of prostaglandins can be further subdivided according to chemical structure and function.

Prostaglandins were first associated with prostate tissue and were named accordingly. Subsequent discoveries, however, have symptoms, or medical conditions with prostaglandins or drugs that inhibit prostaglandin action ranges from their use to relieve menstrual cramps to the treatment of asthma, high blood pressure, and ulcers.

QUICK CHECK

24. What are the building blocks of a triglyceride, or fat?
25. Give an example of a dehydration synthesis reaction.
26. What is a phospholipid, and why is it an important molecule?
27. Identify an important steroid.

Nucleic Acids and Related Molecules

DNA and RNA

Survival of humans as a species—and survival of every other species—depends largely on two kinds of **nucleic acid** molecules. Almost everyone has heard or seen their abbreviated names, DNA and RNA, but their full names are much less familiar. They are deoxyribonucleic and ribonucleic acids. Nucleic acid molecules are polymers of thousands and thousands of smaller molecules called **nucleotides**—deoxyribonucleotides in DNA molecules and ribonucleotides in RNA molecules. A *deoxyribonucleotide* consists of the pentose sugar named *deoxyribose*, a nitrogenous base (either adenine, cytosine, guanine, or thymine) (Figure 2-28). *Ribonucleotides* are similar but contain ribose instead of deoxyribose and uracil instead of thymine (Table 2-7). Two of the bases in a deoxyribonucleotide, specifically adenine and guanine, are called *purine bases* because they derive from *purine*. Purines have a double ring structure. Cytosine and

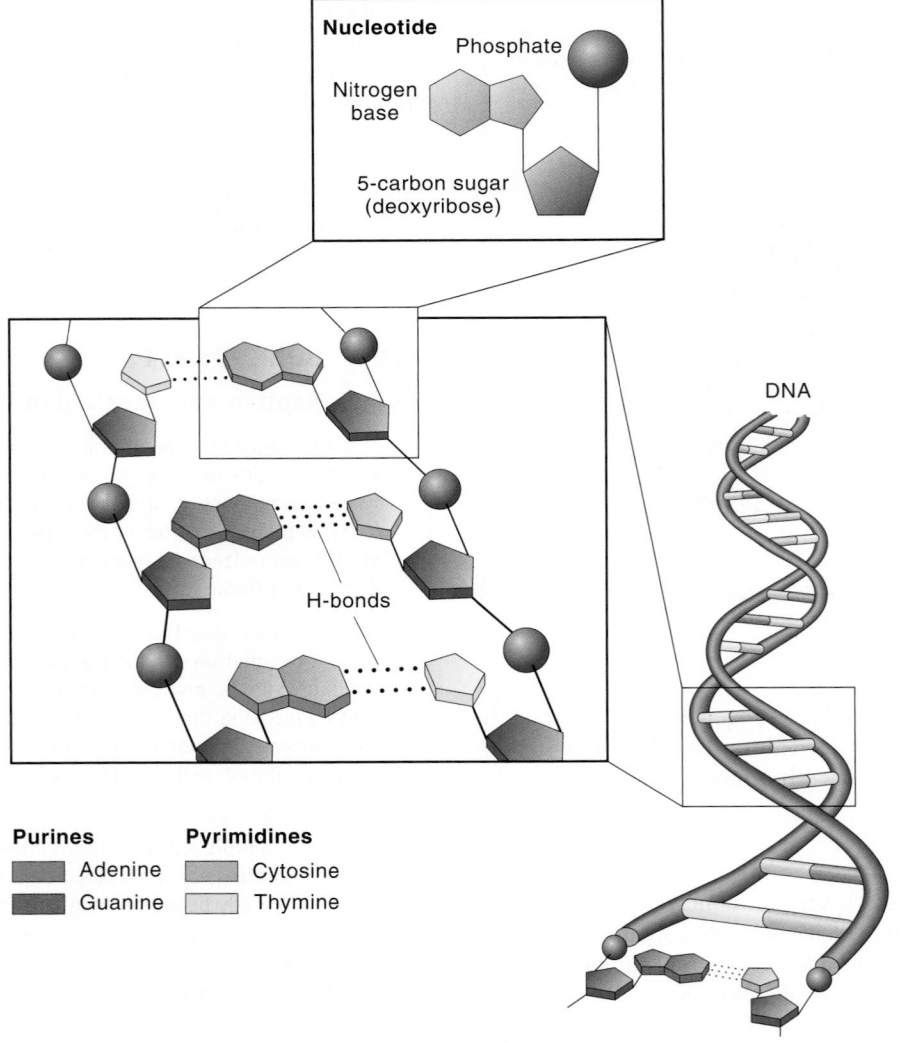

Figure 2-28 *The DNA molecule.* Representation of the DNA double helix showing the general structure of a nucleotide and the two kinds of "base pairs": adenine (A) *(blue)* with thymine (T) *(yellow)* and guanine (G) *(purple)* with cytosine (C) *(red)*. Note that the G–C base pair has three hydrogen bonds and an A–T base pair has two. Hydrogen bonds are extremely important in maintaining the structure of this molecule.

Table 2-7	Comparison of DNA and RNA Structure	
	DNA	**RNA**
Polynucleotide strands	2	1 or 2
Sugar	Deoxyribose	Ribose
Base pairs	Adenine-thymine Guanine-cytosine	Adenine-uracil Guanine-cytosine

thymine derive from *pyrimidine*, so they are known as *pyrimidine bases*. Pyrimidines have a single ring structure. The pyrimidine base uracil replaces thymine in RNA. More about the differences between DNA and RNA is discussed in Chapter 4.

DNA molecules, the largest molecules in the body, are very large polymers composed of many nucleotides. Two long polynucleotide chains compose a single DNA molecule. The chains coil around each other to form a double helix. A helix is a spiral shape similar to the shape of a wire in a spring. Figure 2-28 is a diagram of the double-helix DNA.

Each helical chain in a DNA molecule has its phosphate-sugar backbone toward the outside and its bases pointing inward toward the bases of the other chain. More than that, each base in one chain is joined to a base in the other chain by means of hydrogen bonds to form what is known as a *base pair*. The two polynucleotide chains of a DNA molecule are thus held together by hydrogen bonds between the two members of each base pair (see Figure 2-28). One important principle to remember is that only two kinds of base pairs are present in DNA. What are they? Symbols used to represent them are A-T and G-C. Although a DNA molecule contains only these two kinds of base pairs, it contains

millions of them—more than 100 million pairs estimated in one human DNA molecule! Two other impressive facts are that the millions of base pairs occur in the same sequence in all the millions of DNA molecules in one individual's body but in a slightly different sequence in the DNA of all other individuals. In short, the base pair sequence in DNA is unique to each individual. This fact has momentous significance because DNA functions as the molecule of heredity. It has a weighty responsibility: that of passing the traits of one generation to the next. DNA accomplishes this feat by acting as an "information molecule" that stores the master code of all the recipes (genes) needed to make the various RNA and protein molecules of the body. The details of how the information is stored and retrieved by the cells will be introduced in Chapter 4, and then Chapter 34 will feature even more discussion of the processes of heredity.

Most types of RNA molecules consist of a single strand, but the strand often folds on itself to form a compact folded structure. Each RNA strand is a sequence of ribonucleotides that is essentially copied from a portion of a DNA molecule. Thus, RNA molecules act as "temporary copies" of the master code of hereditary information in the DNA molecules. These RNA "copies" are involved in the process of protein synthesis.

Figure 2-29 shows a type of RNA called *transfer RNA (tRNA)*. tRNA is used by the cell to "grab" a specific amino acid and place it in the correct sequence when building a primary protein strand. The correct location in the sequence is guaranteed by matching tRNA's three-base "anticodon" to the complementary "codon" copied from a gene. In Chapter 4 we will outline the process by which all of this takes place in the cell.

Instead of acting as "information molecules," some RNA molecules regulate cell function. For example, a type of RNA enzyme sometimes called a *ribozyme* is involved in editing the code of RNA strands by removing sections of the code and joining the remaining pieces. A recently discovered type of *double-strand RNA*

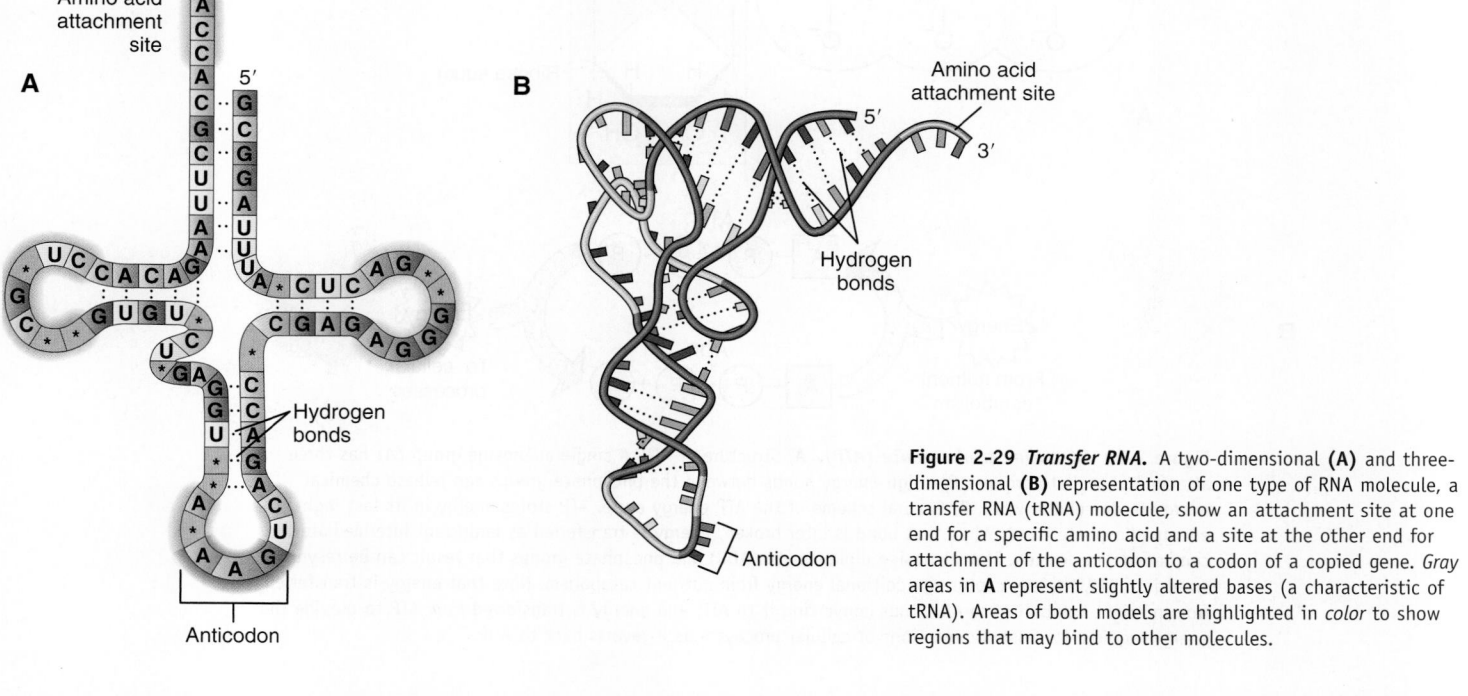

Figure 2-29 *Transfer RNA.* A two-dimensional **(A)** and three-dimensional **(B)** representation of one type of RNA molecule, a transfer RNA (tRNA) molecule, show an attachment site at one end for a specific amino acid and a site at the other end for attachment of the anticodon to a codon of a copied gene. *Gray* areas in **A** represent slightly altered bases (a characteristic of tRNA). Areas of both models are highlighted in *color* to show regions that may bind to other molecules.

(*dsRNA*) is now known to regulate cell function by silencing gene expression in a process called *RNA interference (RNAi)*. These processes will be discussed further in Chapter 4.

Thus, we can say that RNA can act as either an "information molecule" or as a regulatory molecule.

Nucleotides and Related Molecules

Besides joining together to form nucleic acids, nucleotides and related molecules may also play other important roles in the body.

Adenosine triphosphate (ATP is a very important molecule composed of an adenine and ribose sugar (a combination called *adenosine*) to which are attached a string of three phosphate groups (Figure 2-30, *A*). Thus, ATP is really an adenine ribonucleotide with two "extra" phosphate groups attached. The "squiggle" lines indicate covalent bonds that link the phosphate groups. These bonds are called **high-energy bonds** because when they are broken during catabolic chemical reactions, energy is transferred to newly formed compounds. The energy transferred by ATP is used in doing the body's work—the work of muscle contraction and movement, of active transport, and of biosynthesis (Figure 2-30, *B*).

Because ATP is the form of energy that cells generally use, it is an especially important organic molecule. ATP is a molecule that can pick up energy and give it to another chemical process; therefore, it is often called the *energy currency* of cells. Specialized enzyme reaction is required to release the energy that is stored in ATP by splitting it into *adenosine diphosphate (ADP)* and an inorganic phosphate group. It is also possible to split ADP into *adenosine monophosphate (AMP)* and phosphate, with the release of energy. In this case the bond between the second and third phosphate groups is broken.

In prolonged or intense exercise, when ATP is in short supply, muscles turn to *creatine phosphate (CP)* for extra energy. Creatine phosphate is another high-energy molecule made up of an amino acid (not one of those used for proteins) and a phosphate connected with a high-energy bond. When CP releases its phosphate group, the energy can be used to add a phosphate to ATP, thus "recharging" it. In extreme cases, a cell may use ADP for energy.

Chapters 4 and 27 discuss in detail the metabolic pathways that are involved in ATP synthesis and breakdown. A cell at rest has a relatively high ATP concentration, whereas an active cell has less ATP but is constantly rebuilding its stores. An exhausted cell has a high ADP concentration and very low levels of ATP. It must resynthesize needed ATP to sustain its activity over time. Fortunately, cells at rest can recycle ADP and ATP and then reverse the cycle, thus reusing small amounts of ATP on a continuing basis. Exercise physiologists estimate that the body can use up to 0.5 kilograms (1.1 pounds) of ATP per minute during very strenuous physical activity. If reuse was impossible, we would require about 40 kilograms (88 pounds) of ATP per day to remain active.

Other energy-transferring nucleotides such as *nicotinamide adenine dinucleotide (NAD)* and *flavin adenine dinucleotide (FAD)* are also used by cells to transfer energy among molecules (Figure 2-31, *A*). NAD and FAD act as coenzymes (see Box 2-6) to shuttle energy-carrying particles from one metabolic pathway to another during the many complicated steps of transferring energy from food molecules to ATP (Figure 2-31, *B*). The entire process of energy transfer, including the role of NAD and FAD, is discussed in detail in Chapter 27.

Nucleotides are also sometimes used as a signal inside the cell. For example, ATP can be broken down to a one-phosphate mol-

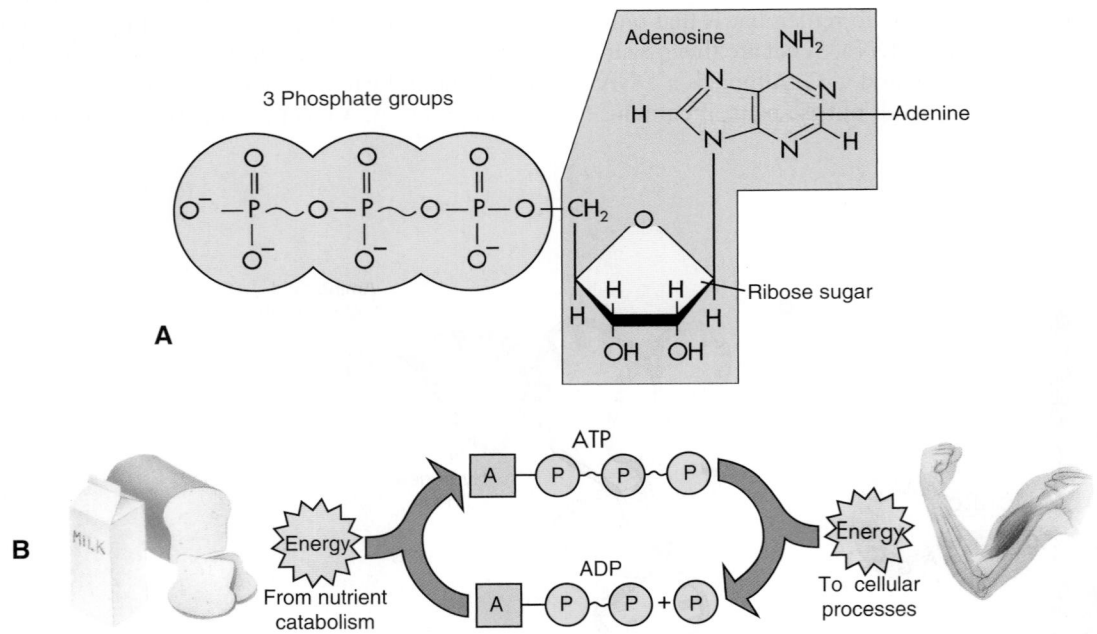

Figure 2-30 *Adenosine triphosphate (ATP)*. A, Structure of ATP. A single adenosine group *(A)* has three attached phosphate groups *(P)*. High-energy bonds between the phosphate groups can release chemical energy to do cellular work. **B,** General scheme of the ATP energy cycle. ATP stores energy in its last high-energy phosphate bond. When that bond is later broken, energy is transferred as important intermediate compounds are formed. The adenosine diphosphate (ADP) and phosphate groups that result can be resynthesized into ATP, thereby capturing additional energy from nutrient catabolism. Note that energy is transferred from nutrient catabolism to ADP, thus converting it to ATP, and energy is transferred *from* ATP to provide the energy required for anabolic reactions or cellular processes as it reverts back to ADP.

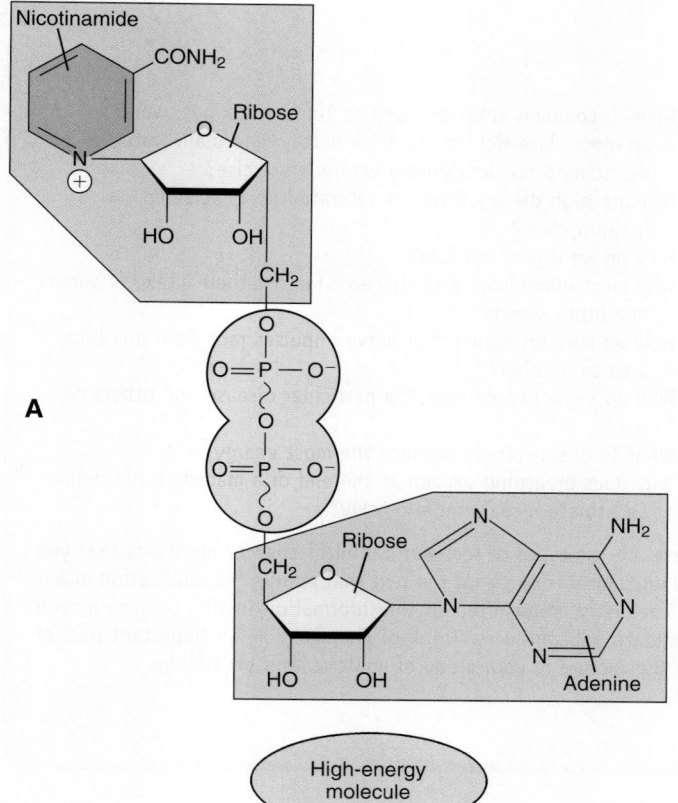

A

Figure 2-31 *Nicotinic adenine dinucleotide (NAD).* NAD is made up of two different ribonucleotides **(A)** and acts as a coenzyme to pick up high-energy particles released from the catabolism of food molecules and shuttle them to another chemical pathway where the energy can be transferred to another molecule **(B).**

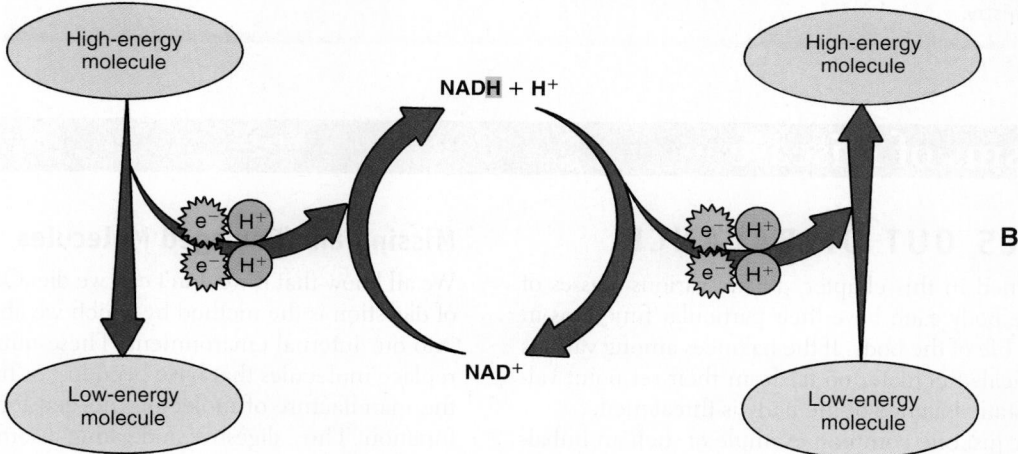

B

ecule called *cAMP (cyclic adenosine monophosphate)* that is used as an intracellular signal in cells. This role of cAMP is outlined in Chapters 12 and 16.

Combined Forms

You have already noticed that large molecules can be joined together to form even larger molecules. Sometimes, only a small addition or alteration is made. For example, in the case of adenosine triphosphate (ATP), two extra phosphate groups are added to an adenine-containing RNA nucleotide. This gives the nucleotide a completely different function. Instead of becoming involved in storing or transmitting genetic information, the ATP molecule transfers energy from one chemical pathway to another. We will learn much more about ATP later. The point now is that macromolecules can be joined to other molecules to make them even larger and to change their functions.

Table 2-4 lists some of the combined or altered macromolecules you will encounter in your study. Notice also the many dif-

ferent important functions performed by these molecules. The names of the combined molecules usually tell you what is in them. *Lipoproteins* contain lipid and protein groups combined into a single molecule. *Glycoproteins* contain carbohydrate (*glyco,* "sweet") and protein. Usually, the base word (*protein* in this case) indicates which component is dominant. The prefix represents the component found in a lesser amount. Thus, glycoproteins have more protein than they do carbohydrate. Review the examples of combined forms in Table 2-4 and their functions in the body.

 QUICK CHECK

28. Name two important *nucleic acids.*
29. What is a *nucleotide?*
30. What is meant by the term *base pair?*
31. What are some roles of *nucleotides* in the body?

THE BIG PICTURE

The Chemical Basis of Life

The importance of the concept of organization at all levels of body structure and function was introduced in Chapter 1 and will be reinforced as you study the individual organ systems of the body in subsequent chapters of the text. Understanding the information in this chapter is a critical first step in connecting the chemistry of life with a real understanding of how the body functions and the relationships that exist between differing functions and body structures.

How the basic chemical building blocks of the body are organized and how they relate to one another are key determinants in understanding normal structure and function, as well as pathological anatomy and disease.

As you learn about the structure and functioning of the various organ systems of the body, the information contained in this chapter will take on new meaning and practical significance. It will help you fully understand and answer many questions that require you to integrate otherwise isolated factual information to make anatomy and physiology emerge as living and dynamic topics of personal interest. Consider the following questions. Each one relates to the study of one or more organ systems covered in subsequent chapters. Your ability to answer these and many other questions correctly will require knowledge of basic chemistry.

How do common antacids, such as Tums or Rolaids, work?

Is an electrolyte-rich sports drink better than plain water in replacing fluids lost during vigorous exercise?

Why are high dietary levels of saturated fat considered inappropriate?

How do we digest our food?

Why must individuals with diabetes restrict their intake of sugars and other sweets?

How do muscles contract or nerve impulses race from one body area to another?

Why do some people inherit a particular disease and others do not?

What food substances produce the most energy?

Why does breathing oxygen at the end of a marathon run help an athlete recover more quickly?

These are the types of real-world, end-of-chapter questions that you will encounter throughout the text that require the application of basic chemistry. Refer often to the information in this chapter as you formulate your answers. Think of chemistry as an important part of the Big Picture in your study of anatomy and physiology.

Mechanisms of Disease

CHEMICALS OUT OF BALANCE

As we have learned in this chapter, all the various classes of chemicals in the body each have their particular functions in maintaining the life of the body. If the balances among various groups of chemicals fluctuate too far from their set point values, the homeostatic balance of the body is threatened.

Let's examine just one common example of such an imbalance. The carbon dioxide (CO_2) concentration in the blood will climb too high, a condition called **hypercapnia**, when the respiratory system fails to remove it from the blood at the normal rate. Thus, we have a CO_2 imbalance. This will have several effects. For one, the high CO_2 levels will inhibit cell metabolism and thus reduce the normal activity of the body. For another, because CO_2 tends to form an acid, the pH of the body's internal environment will drop to below the set point level—a condition called **acidosis**. Acidosis, in turn, may disrupt the shapes of proteins throughout the body and thus interfere with the normal structure and function of the body. Unless CO_2 balance is restored quickly, a person will die.

Various elements of the scenario we just described will be touched on later in appropriate places in the book. However, this is just one of many examples of chemical imbalances that can serve as a mechanism of disease. Nutrient imbalances, ion imbalances, and so on, can all be life-threatening.

Missing and Damaged Molecules

We all know that if we don't eat, we die. Of course, the process of digestion is the method by which we absorb new molecules into our internal environment. These nutrients are needed to replace molecules that have been lost or have been used up in the manufacture of molecules needed for body structure and function. Thus, digestive and eating disorders are in fact disorders of "missing molecules."

We've learned in this chapter that the genes in DNA molecules serve as "recipes" for the structural and functional proteins of the body, which in turn make many other types of molecules in the body such as lipids and carbohydrates. If even one amino acid is missing or out of place, it may not fold correctly and thus the entire protein (and whatever molecule is made by the protein) will be ruined. Even if the protein has a normal sequence of amino acids, it may not fold correctly because of an improper pH or extreme body temperature.

Although some types of molecules can be obtained from our food, some must be made in the body. If certain genes are damaged or missing, we cannot manufacture the chemicals needed for proper body function. For example, in *osteogenesis imperfecta*, the body fails to make normal *collagen*. Collagen is a structural protein needed to hold bones and other tissues together. In osteogenesis imperfecta, the body's bones are brittle

Mechanisms of Disease—cont.

and cannot bear much weight. Most genetic disorders are really disorders of "missing or damaged molecules."

Molecules in the body can also be damaged by physical agents such as radiation or other chemicals. For example, the ultraviolet radiation of the sun can sometimes damage the DNA in skin and cause skin cancer, or radiation from radioactive materials can cause a body-wide breakdown of molecules that results in **radiation sickness.** Depending on the dose of radiation received, a person may experience nausea, vomiting, diarrhea, or headache possibly progressing to loss of hair, cataracts, cancer (especially leukemia), burns, or severe breakdown of body tissues. Figure 2-32 shows a burnlike inflammation and blistering of the skin caused by radiation.

Some chemicals called **toxins** that enter the body can cause damage to our own molecules. Toxins, or poisons, cause their damage by destroying our molecules, combining with our molecules to render them useless, or otherwise disrupting the normal chemical balance and chemical activity of our body. For example, *carbon monoxide (CO)* is a gas that binds to hemoglobin in our blood so tightly that the hemoglobin can no longer carry the oxygen needed for life.

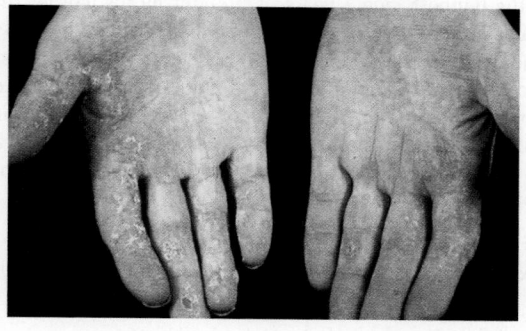

Figure 2-32 *Radiation sickness.* This photo shows burnlike damage to the skin caused by radiation.

LANGUAGE OF SCIENCE *(Cont'd from page 37)*

ionic bond [*ion* to go, *bond* band]

isotopes (EYE-so-topes) [*iso-* equal, *-tope* place]

lipids (LIP-ids) [*lipi-* fat, *-id* form]

metabolism (me-TAB-o-lizm) [*metabol-* change, *-ism* condition]

molecule (MOL-eh-kyool) [*mole-* mass, *-cule* small]

nonpolar (non-PO-lar) [*non-* not, *-pol-* pole, *-ar* pertaining to]

nucleic acid (noo-KLAY-ik AS-id) [*nucle-* nut kernel, *-ic* pertaining to, *acid* sour]

nucleotides (NOO-klee-oh-tides) [*nucleo-* nut or kernel, *-ide* chemical name]

octet rule (ok-TET rool) [*octet* group of eight]

peptide bond (PEP-tyde bond) [*pept-* digest, *-ide* type of chemical]

pH (pee AYCH) [abbreviation for *potenz* power, *hydrogen* hydrogen]

phospholipids (fos-fo-LIP-ids) [*phospho-* phosphorus, *-lip-* fat, *-id* form]

polar (PO-lar) [*pol-* pole, *-ar* pertaining to]

prostaglandins (pross-tah-GLAN-dins) [*pro-* before, *-sta-* stand, *-gland-* acorn, *-in* neutral substance]

proteins (PRO-teens) [*prote-* primary, *-in* neutral substance]

radioactivity (ray-dee-o-ak-TIV-it-ee) [*radio* send out rays]

radioisotope (ray-dee-oh-EYE-so-tope) [*radio-* send out rays, *-iso-* equal, *-tope* place]

reversible reactions (ree-VER-si-bl ree-AK-shuns) [*re-* again, *-vers-* turn, *-ible* able to, *re-* again, *-action* action]

steroids (STAYR-oids) [*ster-* sterol, *-oid* like]

structural proteins (STRUK-cher-al PRO-teens) [*structura-* arrangement, *-al* pertaining to]

synthesis (SIN-the-sis) [*synthes-* put together, *-is* process of]

triglycerides (try-GLISS-er-ydes) [*tri-* three, *glycer-* sweet (glycerine), *-ides* types of chemical]

LANGUAGE OF MEDICINE

acidosis (as-i-DOE-sis) [*acid-* sour, *-osis* condition]

hypercapnia (hye-per-KAP-nee-ah) [*hyper-* above, *-capn-* vapor, *-ia* condition]

radiation sickness (ray-dee-AY-shun SIK-ness) [*radia-* send out rays, *-tion* process]

toxins (TOK-sins) (*tox-* poison, *-in* neutral substance)

CASE STUDY

Johnnie Bennett, age 7 years, is brought by his mother to the clinic because he has vomited and has had diarrheal stools for 3 days. Mrs. Bennett relates that 3 days ago Johnnie started to have a slight fever and cough. Today Johnnie vomited only after eating. His diarrhea is yellow and very liquid. Johnnie has no other symptoms, and his immunizations are up-to-date. Physical examination reflects a cooperative, alert child in no acute distress. There are no signs of dehydration, and his mucous membranes are moist.

1. If pH were tested on the vomitus from Johnnie, would you expect it to be more acid or more alkaline? Why?

   A. Alkaline from loss of stomach hydrochloric acid
   B. Acidic from loss of stomach bicarbonate
   C. Acidic from loss of stomach hydrochloric acid
   D. Alkaline from loss of stomach bicarbonate

2. You are asked to teach Johnnie's mother about appropriate fluid replacement solutions for patients who have nausea and vomiting. Which of the following fluids is BEST to include in this instruction?

   A. Water with sugar added
   B. Water without anything added
   C. Water with electrolytes added
   D. Any of the above would be acceptable

3. Mrs. Bennett tells the clinic staff that Johnnie has always been a poor eater and prefers to eat junk food. She thinks he may be anemic. Which of the following elements is critically needed in the body to prevent anemia?

   A. Aluminum
   B. Iron
   C. Selenium
   D. Zinc

4. While waiting for the results of Johnnie's test for anemia, Mrs. Bennett tells you that her husband has a history of hyperlipidemia and asks why they were told that the high-density cholesterol level should be high. She asks, "Isn't the cholesterol level supposed to be low?" Your best response would include the following information:

   A. "Low-density lipoproteins are associated with a low risk for the development of atherosclerosis."
   B. "Low-density lipoproteins carry 'good' cholesterol away from the cell."
   C. "High-density lipoproteins are associated with a low risk for atherosclerosis."
   D. "High-density lipoproteins carry cholesterol to the cell."

CHAPTER SUMMARY

BASIC CHEMISTRY

A. Elements and compounds (Figure 2-1)
   1. Matter—anything that has mass and occupies space
   2. Element—simple form of matter, a substance that cannot be broken down into two or more different substances
      a. There are 26 elements in the human body
      b. There are 11 "major elements," 4 of which (carbon, oxygen, hydrogen, and nitrogen) make up 96% of the human body
      c. There are 15 "trace elements" that make up less than 2% of body weight
   3. Compound—atoms of two or more elements joined to form chemical combinations
B. Atoms (Figure 2-2)
   1. The concept of an atom was proposed by the English chemist John Dalton
   2. Atomic structure—atoms contain several different kinds of subatomic particles; the most important are
      a. Protons (+ or p)—positively charged subatomic particles found in the nucleus
      b. Neutrons (n)—neutral subatomic particles found in the nucleus
      c. Electrons (− or e)—negatively charged subatomic particles found in the electron cloud

   3. Atomic number and atomic weight
      a. Atomic number (Table 2-1)
         (1) The number of protons in an atom's nucleus
         (2) The atomic number is critically important; it identifies the kind of element
      b. Atomic weight
         (1) The mass of a single atom
         (2) It is equal to the number of protons plus the number of neutrons in the nucleus (p + n)
   4. Energy levels (Figures 2-3 and 2-4)
      a. The total number of electrons in an atom equals the number of protons in the nucleus (in a stable atom)
      b. The electrons form a "cloud" around the nucleus
      c. "Bohr model"—a model resembling planets revolving around the sun, useful in visualizing the structure of atoms
         (1) Exhibits electrons in concentric circles showing relative distances of the electrons from the nucleus
         (2) Each ring or shell represents a specific energy level and can hold only a certain number of electrons
         (3) The number and arrangement of electrons determine whether an atom is chemically stable

(4) An atom with eight, or four pairs of electrons in the outermost energy level is chemically inert

(5) An atom without a full outermost energy level is chemically active

   d. Octet rule—atoms with fewer or more than eight electrons in the outer energy level will attempt to lose, gain, or share electrons with other atoms to achieve stability

5. Isotopes (Figure 2-5)

   a. Isotopes of an element contain the same number of protons but different numbers of neutrons

   b. Isotopes have the same atomic number and therefore the same basic chemical properties as any other atom of the same element, but they have a different atomic weight

   c. Radioactive isotope—an unstable isotope that undergoes nuclear breakdown and emits nuclear particles and radiation

C. Attractions between atoms—chemical bonds

1. Chemical reaction—interaction between two or more atoms that occurs as a result of activity between electrons in their outermost energy levels

2. Molecule—two or more atoms joined together

3. Compound—consists of molecules formed by atoms of two or more elements

4. Chemical bonds—two types unite atoms into molecules

   a. Ionic, or electrovalent, bond (Figure 2-6)—formed by transfer of electrons; strong electrostatic force that binds positively and negatively charged ions together

   b. Covalent bond (Figure 2-7)—formed by sharing of electron pairs between atoms

5. Hydrogen bond (Figures 2-8 and 2-9)

   a. Much weaker than ionic or covalent bonds

   b. Results from unequal charge distribution on molecules

D. Attractions between molecules

1. Hydrogen bonds

   a. Form when electrons are unequally shared

     (1) Example: water molecule

     (2) Polar molecules have regions with partial electrical charges resulting from unequal sharing of electrons among atoms

   b. Areas of different partial charges attract one another and form hydrogen bonds

2. Other weak forces attract molecules to each other through differences in electrical charge

E. Chemical reactions

1. Involve the formation or breaking of chemical bonds

2. There are three basic types of chemical reactions involved in physiology:

   a. Synthesis reaction—combining of two or more substances to form a more complex substance; formation of new chemical bonds: $A + B \rightarrow AB$

   b. Decomposition reaction—breaking down of a substance into two or more simpler substances; breaking of chemical bonds: $AB \rightarrow A + B$

   c. Exchange reaction—decomposition of two substances and, in exchange, synthesis of two new compounds from them: $AB + CD \rightarrow AD + CB$

   d. Reversible reactions—occur in both directions

METABOLISM—ALL OF THE CHEMICAL REACTIONS THAT OCCUR IN BODY CELLS (FIGURE 2-10)

A. Catabolism

1. Chemical reactions that break down complex compounds into simpler ones and release energy; hydrolysis is a common catabolic reaction

2. Ultimately, the end products of catabolism are carbon dioxide, water, and other waste products

3. More than half the energy released is transferred to ATP, which is then used to do cellular work (Figure 2-28)

B. Anabolism

1. Chemical reactions that join simple molecules together to form more complex molecules

2. Chemical reaction responsible for anabolism is dehydration synthesis

ORGANIC AND INORGANIC COMPOUNDS

A. Inorganic compounds—few have carbon atoms and none have C–C or C–H bonds

B. Organic molecules

1. Have at least one carbon atom and at least one C–C or C–H bond in each molecule

2. Often have functional groups attached to the carbon-containing core of the molecule (Figure 2-11)

INORGANIC MOLECULES

A. Water

1. The body's most abundant and important compound

2. Properties of water (Table 2-2)

   a. Polarity—allows water to act as an effective solvent; ionizes substances in solution (Figure 2-8)

   b. The solvent allows transportation of essential materials throughout the body (Figure 2-12)

   c. High specific heat—water can lose and gain large amounts of heat with little change in its own temperature; enables the body to maintain a relatively constant temperature

   d. High heat of vaporization—water requires the absorption of significant amounts of heat to change it from a liquid to a gas; allows the body to dissipate excess heat

B. Oxygen and carbon dioxide—closely related to cellular respiration

1. Oxygen—required to complete decomposition reactions necessary for the release of energy in the body

2. Carbon dioxide—produced as a waste product and also helps maintain the appropriate acid-base balance in the body

C. Electrolytes

1. Large group of inorganic compounds that includes acids, bases, and salts

2. Substances that dissociate in solution to form ions
3. Positively charged ions are cations; negatively charged ions are anions
4. Acids and bases—common and important chemical substances that are chemical opposites
   a. Acids
      (1) Any substance that releases a hydrogen ion (H⁺) when in solution; "proton donor"
      (2) Level of "acidity" depends on the number of hydrogen ions a particular acid will release
   b. Bases
      (1) Electrolytes that dissociate to yield hydroxide ions (OH⁻) or other electrolytes that combine with hydrogen ions (H⁺)
      (2) Described as "proton acceptors"
D. pH scale—measuring acidity and alkalinity (Figure 2-13)
   1. pH indicates the degree of acidity or alkalinity of a solution
   2. pH of 7 indicates neutrality (equal amounts of H⁺ and OH⁻); a pH less than 7 indicates acidity; a pH higher than 7 indicates alkalinity
E. Buffers
   1. Maintain the constancy of pH
   2. Minimize changes in the concentrations of H⁺ and OH⁻ ions
   3. Act as a "reservoir" for hydrogen ions
F. Salts (Table 2-3)
   1. Compound that results from chemical interaction of an acid and a base
   2. Reaction between an acid and a base to form a salt and water is called a *neutralization reaction*

ORGANIC MOLECULES (FIGURE 2-14; TABLE 2-4)

A. "Organic" describes compounds that contain C–C or C–H bonds
B. Carbohydrates—organic compounds containing carbon, hydrogen, and oxygen; commonly called *sugars* and *starches*
   1. Monosaccharides—simple sugars with short carbon chains; those with six carbons are hexoses (e.g., glucose), whereas those with five are pentoses (e.g., ribose, deoxyribose) (Figure 2-15)
   2. Disaccharides and polysaccharides—two (di-) or more (poly-) simple sugars that are bonded together through a synthesis reaction (Figure 2-16)
C. Proteins (Table 2-5)
   1. Most abundant organic compounds
   2. Chainlike polymers
   3. Amino acids—building blocks of proteins (Figures 2-17 to 2-19)
      a. Essential amino acids—eight amino acids that cannot be produced by the human body
      b. Nonessential amino acids—12 amino acids can be produced from molecules available in the human body
      c. Amino acids consist of a carbon atom, an amino group, a carboxyl group, a hydrogen atom, and a side chain

4. Levels of protein structure (Figure 2-20)
   a. Protein molecules are highly organized and show a definite relationship between structure and function
   b. There are four levels of protein organization
      (1) Primary structure—refers to the number, kind, and sequence of amino acids that make up the polypeptide chain
      (2) Secondary structure—polypeptide is coiled or bent into pleated sheets stabilized by hydrogen bonds
      (3) Tertiary structure—a secondary structure can be further twisted and converted to a globular shape; the coils touch in many places and are "welded" by covalent and hydrogen bonds
      (4) Quaternary structure—highest level of organization occurring when protein contains more than one polypeptide chain
5. Two broad categories
   a. Structural proteins form the structures of the body
   b. Functional proteins cause chemical changes in the molecules
6. Shape of protein molecules determines their function (Figure 2-21)
   a. Denatured proteins have lost their shape and therefore their function (Figure 2-22)
   b. Proteins can be denatured by changes in pH, temperature, radiation, and other chemicals
   c. If the chemical environment is restored, proteins may be renatured and function normally
D. Lipids (Table 2-6)
   1. Water-insoluble organic molecules that are critically important biological compounds
   2. Major roles:
      a. Energy source
      b. Structural role
      c. Integral parts of cell membranes
   3. Triglycerides or fats (Figures 2-23 and 2-24)
      a. Most abundant lipids and most concentrated source of energy
      b. The building blocks of triglycerides are glycerol (the same for each fat molecule) and fatty acids (different for each fat and determine the chemical nature)
         (1) Types of fatty acids—saturated fatty acid (all available bonds are filled) and unsaturated fatty acid (has one or more double bonds)
         (2) Triglycerides are formed by a dehydration synthesis
   4. Phospholipids (Figure 2-25)
      a. Fat compounds similar to triglyceride
      b. One end of the phospholipid is water soluble (hydrophilic); the other end is fat soluble (hydrophobic)
      c. Phospholipids can join two different chemical environments
      d. Phospholipids may form double layers called bilayers that make up cell membranes
   5. Steroids (Figure 2-26)
      a. Main component is steroid nucleus
      b. Involved in many structural and functional roles

6. Prostaglandins (Figure 2-27)
   a. Commonly called "tissue hormones"; produced by cell membranes throughout the body
   b. Effects are many and varied; however, they are released in response to a specific stimulus and are then inactivated
E. Nucleic acids and related molecules
1. DNA (deoxyribonucleic acid)
   a. Composed of deoxyribonucleotides, that is, structural units composed of the pentose sugar (deoxyribose), phosphate group, and nitrogenous base (cytosine, thymine, guanine, or adenine)
   b. DNA molecule consists of two long chains of deoxyribonucleotides coiled into a double-helix shape (Figure 2-28)
   c. Alternating deoxyribose and phosphate units form the backbone of the chains
   d. Base pairs hold the two chains of DNA molecule together
   e. Specific sequence of more than 100 million base pairs constitutes one human DNA molecule; all DNA molecules in one individual are identical and different from those of all other individuals
   f. DNA functions as the molecule of heredity
2. RNA (ribonucleic acid) (Figure 2-29, Table 2-7)
   a. Composed of the pentose sugar (ribose), phosphate group, and a nitrogenous base
   b. Nitrogenous bases for RNA are adenine, uracil, guanine, or cytosine (uracil replaces thymine)
   c. Some RNA molecules are temporary copies of segments (genes) of the DNA code and are involved in synthesizing proteins
   d. Some RNA molecules are regulatory and act as enzymes (ribozymes) or silence gene expression (RNA interference).
3. Nucleotides
   a. Nucleotides have other important roles in the body
   b. ATP (Figure 2-30)
      (1) Composed of
         1) Adenosine
            (i) Ribose—a pentose sugar
            (ii) Adenine—a nitrogen-containing molecule
         2) Three phosphate subunits
      (2) High-energy bonds present between phosphate groups
      (3) Cleavage of high-energy bonds releases energy during catabolic reactions
      (4) Energy stored in ATP is used to do the body's work
      (5) ATP often called the *energy currency* of cells
      (6) ATP split into adenosine diphosphate (ADP) and an inorganic phosphate group by a special enzyme
      (7) If ATP is depleted during prolonged exercise, creatine phosphate (CP) or ADP can be used for energy
   c. NAD and FAD (Figure 2-31)
      (1) Used as coenzymes to transfer energy-carrying molecules from one chemical pathway to another

   d. cAMP (cyclic AMP)
      (1) Made from ATP by removing two phosphate groups to form a monophosphate
      (2) Used as an intracellular signal
F. Combined forms—large molecules can be joined together to form even larger molecules
1. Gives the molecules a completely different function
2. Names of combined molecules tell you what is in them
   a. Base word tells which component is dominant
   b. Prefix is the component found in a lesser amount
3. Examples
   a. *Adenosine triphosphate* (ATP)—two extra phosphate groups to a nucleotide
   b. Lipoproteins—lipid and protein groups combined into a single molecule
   c. Glycoproteins—carbohydrate (*glyco*, "sweet") and protein
   d. Examples of combined forms and their functions in the body listed in Table 2-4

REVIEW QUESTIONS

1. Define the following terms: element, compound, atom, molecule.
2. Compare early nineteenth century and present-day concepts of atomic structure.
3. Name and define three kinds of subatomic particles.
4. Are atoms electrically charged particles? Give reason for your answer.
5. What four elements make up approximately 96% of the body's weight?
6. Define and contrast meanings of the terms atomic number and atomic weight.
7. Explain the general rule by which an atom can be listed as chemically inert and unable to react with another atom.
8. Define and give an example of an isotope.
9. Explain what the term radioactivity means.
10. How does radioactivity differ from chemical activity?
11. Define the terms: alpha particles, beta particles, and gamma rays.
12. Explain how radioactive atoms become transformed into atoms of a different element.
13. Explain what the term chemical reaction means.
14. Identify and differentiate between the three basic types of chemical reactions.
15. Define the term inorganic.
16. Explain why water is said to be polar and list at least four functions of water that are crucial to survival.
17. What are electrolytes and how are they formed?
18. What is a cation? an anion? an ion? Give an example of each.
19. Define the terms acid, base, salt, and buffer.
20. Explain how pH indicates the degree of acidity or alkalinity of a solution.

21. What are the structural units, or building blocks, of proteins? of carbohydrates? of triglycerides? of DNA?
22. Explain what a protein molecule's binding site is. What function does it serve in enzymes?
23. Describe some of the functions proteins perform.
24. Proteins, carbohydrates, lipids—which of these are insoluble in water? contain nitrogen? include prostaglandins? include phosphoglycerides?
25. What groups make up a nucleotide?
26. What pentose sugar is present in a deoxyribonucleotide?
27. Describe the size, shape, and chemical structure of the DNA molecule.
28. What base is thymine always paired with in the DNA molecule? What other two bases are always paired?
29. What is the function of DNA?
30. What is catabolism? What function does it serve?
31. Compare catabolism, anabolism, and metabolism.

CRITICAL THINKING QUESTIONS

1. Identify the specific areas of chemistry that would be of interest to a biochemist.

2. In modern blimps, the gas of choice is helium rather than hydrogen. Hydrogen would be lighter, but helium is safer. Compare and contrast the atomic structure of hydrogen and helium. What characteristics of the atomic structure of helium make it so much less reactive than hydrogen?
3. How would you contrast single covalent bonds, double covalent bonds, and ionic bonds?
4. A person has a body weight of 170 pounds. How much of it is due to water?
5. Amylase is an enzyme present in saliva that begins the breakdown of starch. As in all enzymes, it is specific to this chemical reaction. Using the "lock and key model," explain how a change in the shape of this protein might affect this reaction.
6. Amino acids are the building blocks of proteins. Only 20 amino acids make up our proteins. Explain how these 20 amino acids are responsible for the billions of proteins that are used by the body.
7. How does ATP supply the cells with the energy they need to work? Outline the general scheme of the ATP energy cycle.

CHAPTER 3

Anatomy of Cells

LANGUAGE OF SCIENCE

centrioles (SEN-tree-ohls) [*centr-* center, *-ole* small]

centrosome (SEN-troh-sohm) [*centr-* center, *-soma* body]

chromatin (KROH-mah-tin) [*chroma-* color, *-in* substance]

chromosomes (KROH-meh-sohms) [*chroma-* color, *-soma* body]

cilia (SIL-ee-ah) [*cili* eyelid]

composite cell (kahm-PAH-zit sell) [*composite-* to assemble, *cell* storeroom]

cristae [*crista* crest or fold]

cytoplasm (SYE-toh-plaz-em) [*cyto-* cell, *-plasm* to mold]

cytoskeleton (sye-toh-SKEL-e-ton) [*cyto-* cell, *-skeletos* dried body]

desmosomes (DES-mo-sohms) [*desmos-* band, *-soma* body]

electron microscopy (EM) [*electr-* electric, *-on* unit, *micro-* small, *-scop* to see]

endoplasmic reticulum (ER) (en-doh-PLAZ-mik reh-TIK-yoo-lum) [*endo-* inward or within, *-plasm* to mold, *reticulum* little net]

flagella (flah-JEL-ah) [*flagellum* whip]

fluid mosaic model (FLOO-id mo-ZAY-ik MAHD-el)

gap junctions (gap JUNK-shens)

Golgi apparatus (GOL-jee ap-ah-RA-tus) [*Camillo Golgi* Italian histologist]

hydrophilic (hye-dro-FIL-ik) [*hydro-* water, *-philic* to love]

hydrophobic (hye-droh-FOH-bik) [*hydro-* water, *-phobic* fear]

intermediate filaments (in-ter-MEE-dee-it FIL-ah-ments) [*inter-* between, *-mediate* to divide, *fila* threadlike]

light microscope [*micro-* small, *-scop* to see]

lysosomes (LYE-so-sohms) [*lyso-* dissolution, *-soma* body]

microfilaments (my-kroh-FIL-ah-ments) [*micro-* small, *-fila* threadlike]

microtubules (my-kroh-TOOB-yools) [*micro-* small, *-tubule* little tube]

Cont'd on p. 98

In the 1830s, two German scientists, Matthias Schleiden and Theodor Schwann, advanced one of the most important and unifying concepts in biology—the *cell theory*. It states simply that the cell is the fundamental organizational unit of life. Although earlier scientists had seen cells, Schleiden and Schwann were the first to suggest that all living things are composed of cells. Some 100 trillion of them make up the human body. Actually, the study of cells has captivated the interest of countless scientists for over 300 years. However, these small structures have not yet yielded all their secrets, not even to the probing tools of present-day researchers (Box 3-1).

To introduce you to the world of cells, we provide two brief chapters that summarize the essential concepts of cell structure and function. This chapter begins the discussion by describing the functional anatomy of common cell structures. The term *functional anatomy* refers to the study of structures as they relate to function. Chapter 4, Physiology of Cells, continues the discussion by outlining in more detail some important and representative cellular processes.

BOX 3-1 Tools of Microscopic Anatomy

Cell biologists today use very sophisticated tools for studying the microscopic anatomy of cells. Biologists studying macroscopic, or *gross,* anatomy can often get by with only a dissection knife, forceps, and a probe, whereas biologists studying the tiny features of cells and tissues require powerful imaging devices that make the structures appear much larger than they are. In this essay, we briefly outline three tools of microscopic anatomy that are used often in this book to present essential information about the miniature world inside each of our bodies.

Light Microscopy

In the seventeenth century, the amateur Dutch scientist Anton van Leeuwenhoek used a simple (single-lens) optical microscope to observe one-celled organisms in pond water under strong light. The Englishman Robert Hooke used such a **light microscope** (see first figure) to discover that larger organisms had small subunits, which he named "cells" after the tiny rooms of a monastery or prison. Today, in addition to simple light microscopes, we often use compound (multiple-lens) light microscopes to study the anatomy of cells and tissues.

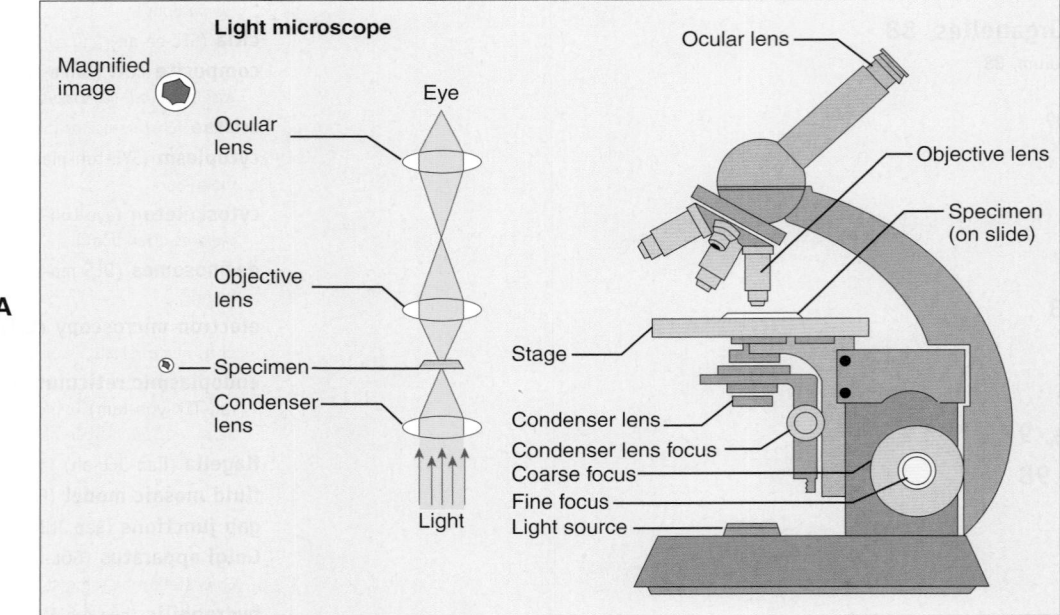

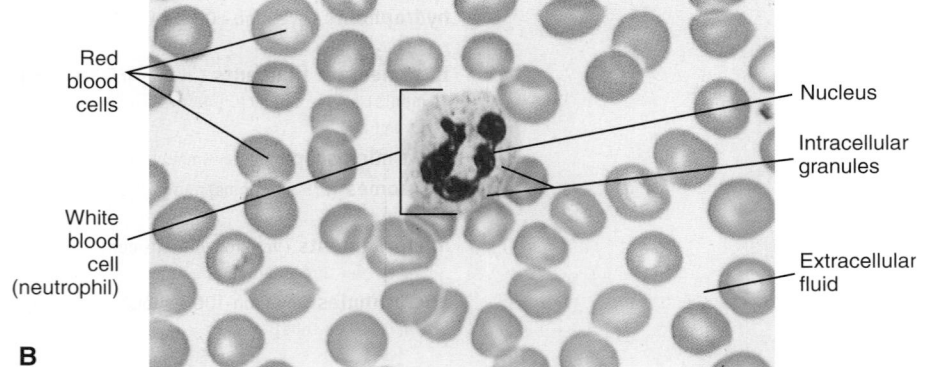

Light microscopy. A, Basic scheme of a compound light microscope. Light shining through the stained specimen is focused by optical lenses to produce a magnified image. **B,** This photograph of the microscopic anatomy of blood cells is used in Chapter 17 to show the structural differences among various types of blood cells. The small reddish cells have their own pigment and hemoglobin, but the larger blood cell is stained to make the transparent nucleus and intracellular granules more visible.

BOX 3-1 Tools of Microscopic Anatomy—cont'd

As the name implies, light microscopy is the observation of tiny structures with visible light. Part *A* of the first figure shows the basic scheme of how a modern compound light microscope works. A thinly sliced specimen is placed on a glass slide and then stained so that the semitransparent structures of cells are more easily seen. The slide with the specimen is then placed on the "stage" of the microscope.

Light shining through the specimen is focused by lenses so a greatly enlarged image can be seen through the eyepiece. Many *light micrographs,* photographs taken with a light microscope, such as that shown in part *B* of the first figure are used throughout this book to illustrate principles of cell anatomy.

The colors seen in light micrographs usually come from the stains used to prepare the specimen, rarely from pigments produced by the cells themselves. Over the last few decades very sophisticated methods of staining, including the use of fluorescent stains, have made light microscopy a very valuable tool for cell anatomists.

Electron Microscopy

Electron microscopy (EM) uses a beam of electrons instead of a beam of light to produce an image of the specimen. An electron beam passes through, or bounces off of, an object just like light does. However, an electron beam can be finely focused by magnets that act as lenses to produce a larger, sharper image than is possible with visible light. Because the human eye cannot see electron beams, an electron detector must be used to produce an image in visible light so that we can see the magnified picture.

Transmission Electron Microscopy

Transmission electron microscopy (TEM) is one of two principal types of electron microscopy used to study the structure of cells. As part *A* of the second figure shows, in TEM an electron gun produces a stream of electrons that is transmitted through the specimen and is focused by a magnetic lens (rather than an optical lens) to produce

A

Transmission electron microscope

Cathode
Electron gun
Anode
Electron beam
Condenser lens magnet
Specimen
Objective lens magnet
Projector lens magnet
Final image on phosphorescent viewing screen

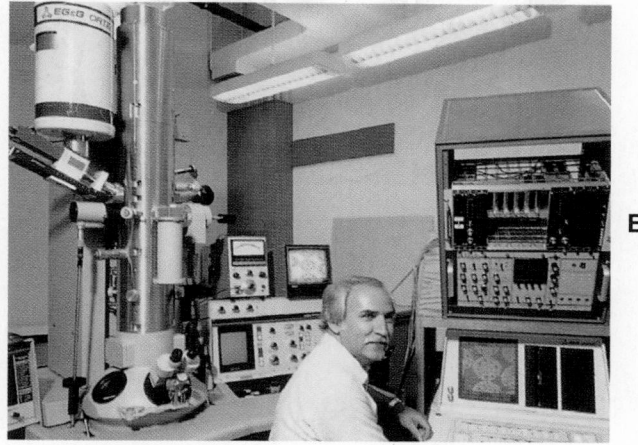

B

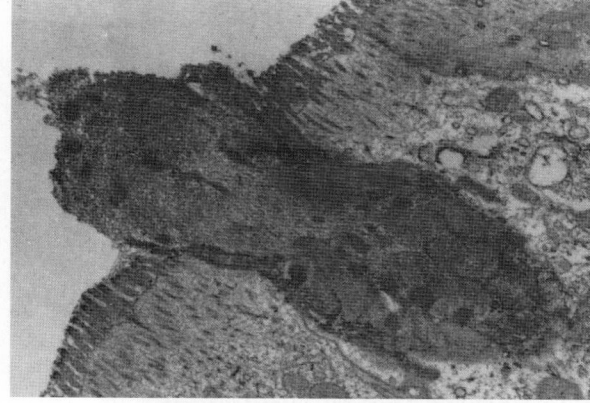

C

D

Electron microscopy. **A**, Diagram of a transmission electron microscope. A stream of electrons passes through the specimen and is focused into a magnified image by a set of magnetic lenses. As electrons hit the phosphors (dots) on the viewing screen, they glow to produce a visible image. **B**, Photograph of a transmission electron microscope. Notice the viewing screen to the right of the operator. The monochrome image transmission electron micrograph (TEM) **(C)** of a type of cell called a *goblet cell* releasing a massive amount of mucus onto the lining of the small intestine was enhanced with color **(D)** to more clearly show the outline of the cell and the mucus.

Continued

BOX 3-1   Tools of Microscopic Anatomy—cont'd

a magnified electron image of the specimen. As electrons hit microscopic dots of phosphorescent material on the viewing screen, the dots glow with visible light. A very highly magnified, highly resolved image then appears on the viewing screen (see part *B* of the second figure).

A photograph of the image produced by a transmission electron microscope is called a **transmission electron micrograph (TEM).** TEMs are ordinarily monochrome (black-and-white) photographs (see part *C* of the second figure) because the electron beam does not have different wavelengths, or colors, like visible white light. However, as part *D* of that figure shows, modern computer graphics techniques can be used to colorize monochrome TEMs to help the viewer recognize details of structure that might otherwise be difficult to see.

Scanning Electron Microscopy

In **scanning electron microscopy (SEM),** a beam of electrons scans the surface of the specimen and electrons reflected from, or knocked off of, the surface are detected by a special sensor (see the third figure in this box). After the signal from the electron detector is amplified, a magnified image of the specimen's surface can be seen on a video monitor. Scanning electron micrographs often gives the illusion of depth in the same way that light and shadow in a regular photograph can give the illusion of depth.

SEMs, like TEMs, are monochrome images. However, SEMS can also be color enhanced by computer graphics techniques to highlight details that may otherwise not be very obvious.

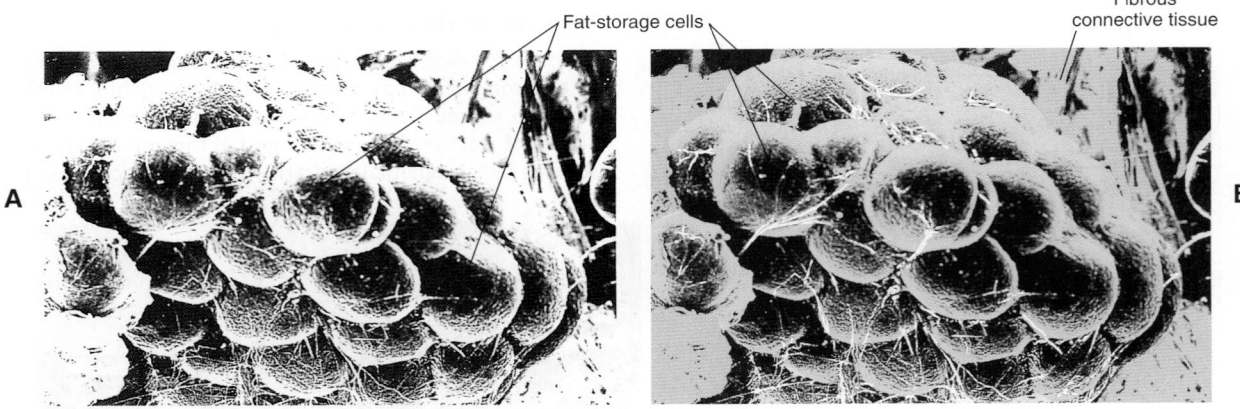

Fat-storage cells

Fibrous connective tissue

A   B

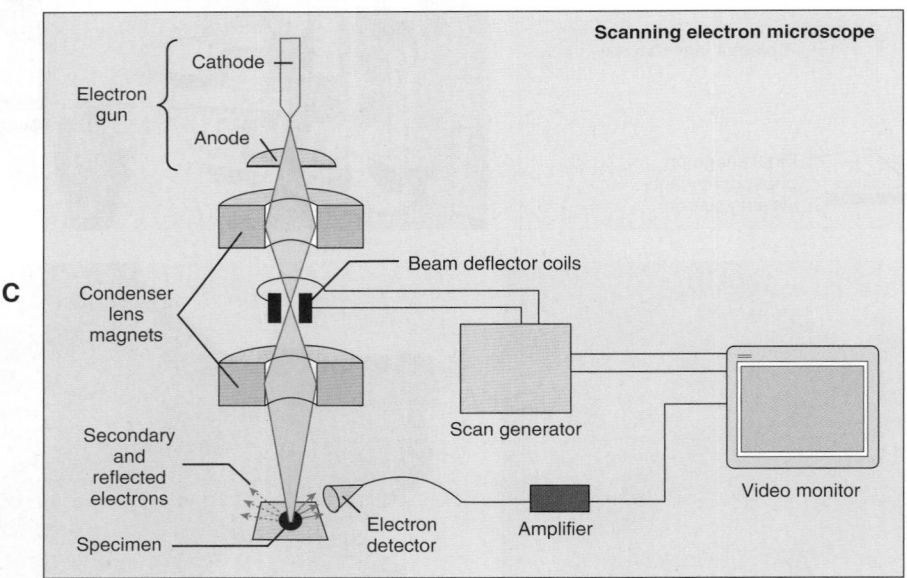

Scanning electron microscope

C

Electron gun {
Cathode
Anode
Condenser lens magnets
Beam deflector coils
Secondary and reflected electrons
Specimen
Electron detector
Scan generator
Amplifier
Video monitor

Scanning electron microscope (SEM). The SEM, used in Chapter 5 to clarify the structure of adipose (fat) tissue **(A)** has been color enhanced **(B)** to highlight the balloonlike fat storage cells. **C,** Diagram of an SEM. An electron gun fires a beam of electrons that is focused by magnetic lenses. Among these lenses is a scan generator that causes the beam to travel back and forth along the surface of the specimen. Electrons from this beam bounce off of the specimen's surface, and secondary electrons are displaced from the surface. These reflected and secondary electrons are detected by a sensor that relays a signal to a video monitor, which displays a magnified image of the specimen.

FUNCTIONAL ANATOMY OF CELLS

The principle of complementarity of structure and function was introduced in Chapter 1 and is evident in the relationships that exist between cell size, shape, and function. Almost all human cells are microscopic in size (Table 3-1). Their diameters range from 7.5 micrometers (μm) (red blood cells) to about 150 μm (female sex cell or ovum). The period at the end of this sentence measures about 100 μm—roughly 13 times as large as our small-est cells and two thirds the size of the human ovum. Like other anatomical structures, cells exhibit a particular size or form be-cause they are intended to perform a specialized activity. A nerve cell, for example, may have threadlike extensions over a meter in length! Such a cell is ideally suited to transmit nervous impulses from one area of the body to another. Muscle cells are specialized to contract or shorten. Other types of cells may serve protective or secretory functions (Table 3-2).

Table 3-1 Units of Size

UNIT	SYMBOL	EQUAL TO	USED TO MEASURE
Centimeter	cm	1/100 meter	Objects visible to the eye
Millimeter	mm	1/10 centimeter	Very large cells; groups of cells
Micrometer (micron)	μm	1/1000 millimeter	Most cells; large organelles
Nanometer	nm	1/1000 micrometer	Small organelles; large biomolecules
Angstrom	Å	1/10 nanometer	Molecules; atoms

Table 3-2 Example of Cell Types

TYPE	EXAMPLE	STRUCTURAL FEATURES	FUNCTIONS
Nerve cells		Surface that is sensitive to stimuli Long extensions	Detect changes in internal or external environment Transmit nerve impulses from one part of the body to another
Muscle cells		Elongated, threadlike Contain tiny fibers that slide together forcefully	Contract (shorten) to allow movement of body parts
Red blood cells		Contain hemoglobin, a red pigment that attracts, then releases, oxygen	Transport oxygen in the bloodstream (from lungs to other parts of the body)
Gland cells		Contain sacs that release a secretion to the outside of the cell	Release substances such as hormones, enzymes, mucus, and sweat
Immune cells		Some have outer membranes able to engulf other cells Some have systems that manufacture anti-bodies Some are able to destroy other cells	Recognize and destroy "nonself" cells such as cancer cells and invading bacteria

The Typical Cell

Despite their distinctive anatomical characteristics and specialized functions, cells have many similarities. There is no cell that truly represents or contains all of the specialized components found in the many types of body cells. As a result, students are often introduced to the anatomy of cells by studying a so-called *typical* or **composite cell**—one that exhibits the most important characteristics of many different cell types. Such a generalized cell is illustrated in Figure 3-1. Keep in mind that no such "typical" cell actually exists in the body; it is a composite structure created for study purposes. Refer to Figure 3-1 and Table 3-3 often as you learn about the principal cell structures described in the paragraphs that follow.

Cell Structures

Ideas about cell structure have changed considerably over the years. Early biologists saw cells as simple, fluid-filled bubbles. Today's biologists know that cells are far more complex than this. Each cell is surrounded by a plasma membrane that separates the cell from its surrounding environment. The inside of the cell is composed largely of a gel-like substance called **cytoplasm** (literally, "cell substance"). The cytoplasm is made of various or-

ganelles and molecules suspended in a watery fluid called *cytosol*, or sometimes *intracellular fluid*. As Figure 3-2 shows, the cytoplasm is crowded with large and small molecules—and various organelles. This crowding of molecules and organelles actually helps improve the efficiency of chemical reactions in the cell.

The nucleus, which is not usually considered to be part of the cytoplasm, is at the center of the cell. Each different cell part is structurally suited to perform a specific function within the cell—much as each of your organs is suited to a specific function within your body. In short, the main cell structures are (1) the plasma membrane; (2) cytoplasm, including the organelles; and (3) the nucleus (see Figure 3-1).

QUICK CHECK

1. What important concept in biology was proposed by Schleiden and Schwann?
2. Give an example of how cell structure relates to its function.
3. List the three main structural components of a typical cell.

Table 3-3	Some Major Cell Structures and Their Functions
CELL STRUCTURE	**FUNCTIONS**
Membranous	
Plasma membrane	Serves as the boundary of the cell, maintains its integrity; protein molecules embedded in plasma membrane perform various functions; for example, they serve as markers that identify cells of each individual, as receptor molecules for certain hormones and other molecules, and as transport mechanisms
Endoplasmic reticulum (ER)	Ribosomes attached to rough ER synthesize proteins that leave cells via the Golgi apparatus; smooth ER synthesizes lipids incorporated in cell membranes, steroid hormones, and certain carbohydrates used to form glycoproteins—also removes and stores Ca^{++} from the cell's interior
Golgi apparatus	Synthesizes carbohydrate, combines it with protein, and packages the product as globules of glycoprotein
Lysosomes	Bags of digestive enzymes break down defective cell parts and ingested particles; a cell's "digestive system"
Proteasomes	Hollow, protein cylinders that break down individual proteins that have been tagged by a chain of ubiquitin molecules
Peroxisomes	Contain enzymes that detoxify harmful substances
Mitochondria	Catabolism; adenosine triphosphate (ATP) synthesis; a cell's "power plants"
Nucleus	Houses the genetic code, which in turn dictates protein synthesis, thereby playing an essential role in other cell activities, namely, cell transport, metabolism, and growth
Nonmembranous	
Ribosomes	Site of protein synthesis; a cell's "protein factories"
Cytoskeleton	Acts as a framework to support the cell and its organelles; functions in cell movement; forms cell extensions (microvilli, cilia, flagella)
Cilia and flagella	Hairlike cell extensions that serve to move substances over the cell surface (cilia) or propel sperm cells (flagella)
Nucleolus	Part of the nucleus; plays an essential role in the formation of ribosomes

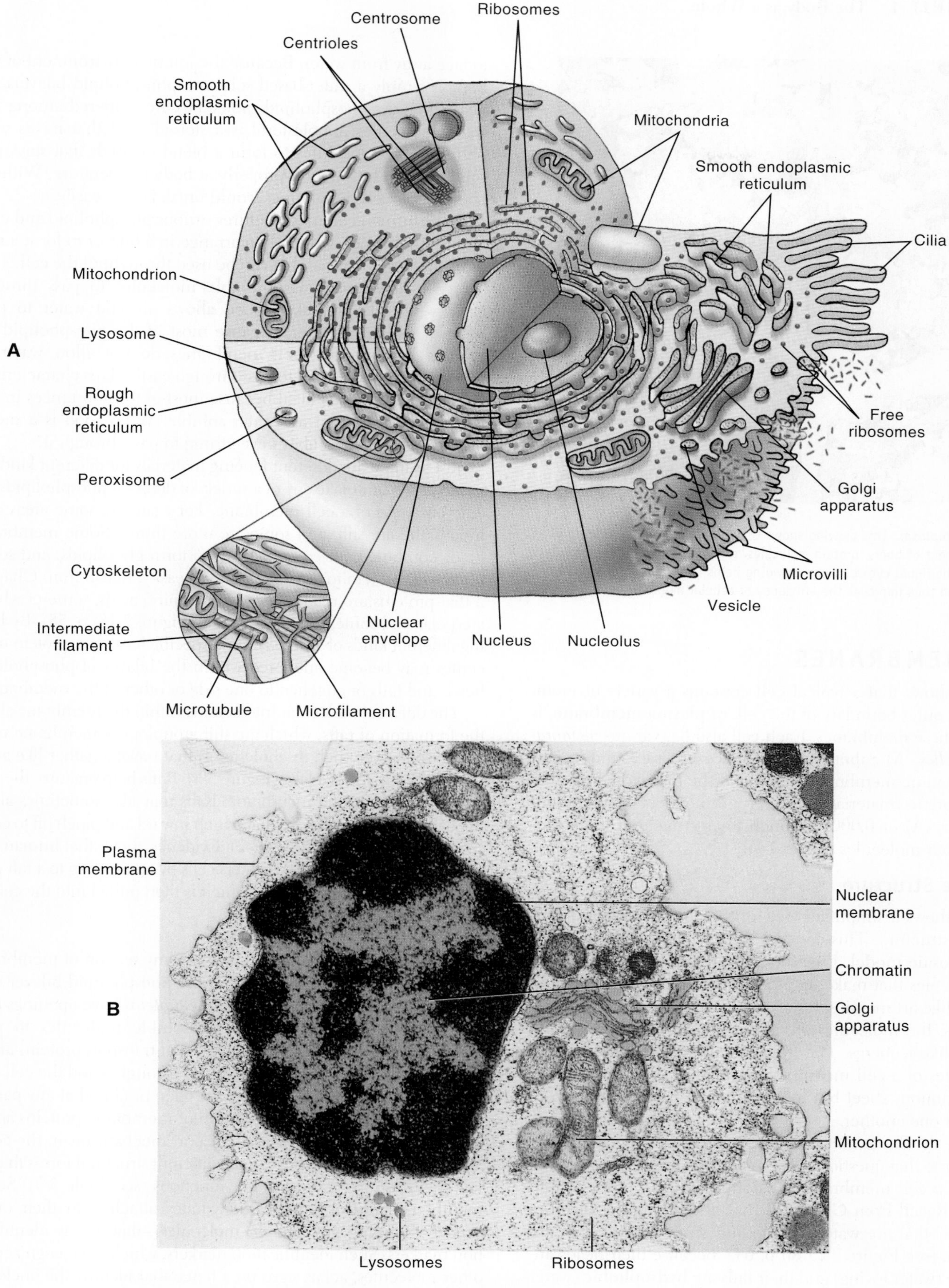

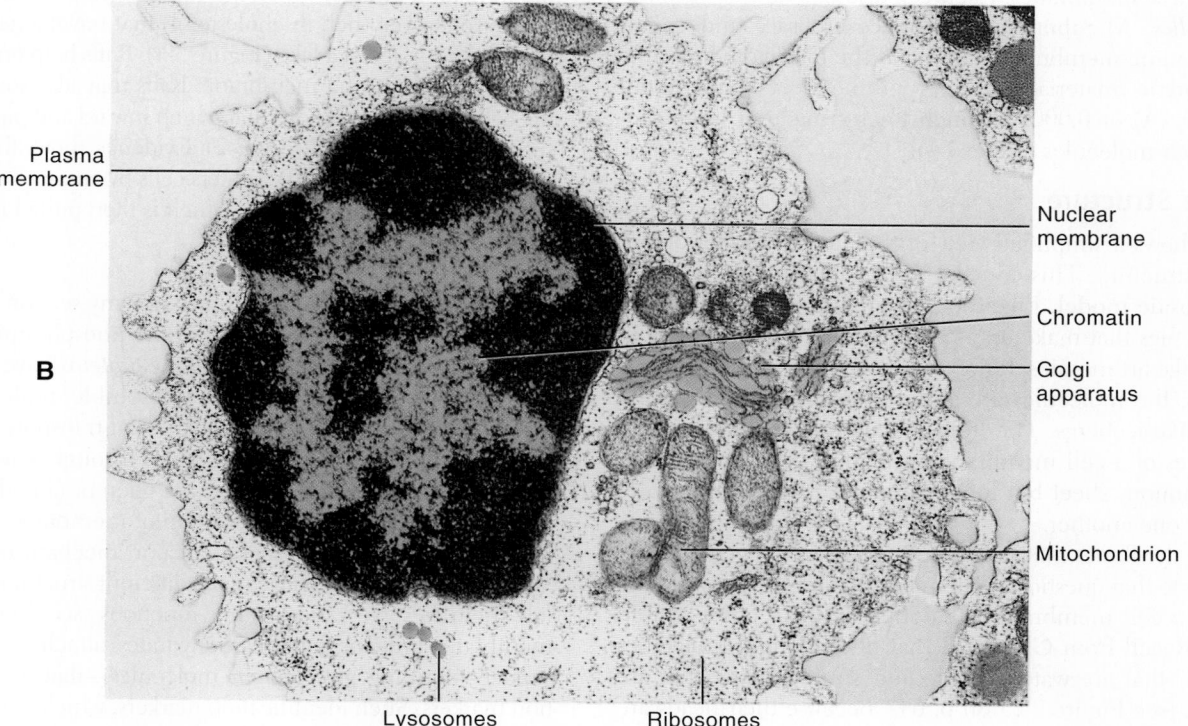

Figure 3-1 *Typical, or composite, cell.* **A,** Artist's interpretation of cell structure. **B,** Color-enhanced electron micrograph of a cell. Both show the many mitochondria, known as the "power plants of the cell." Note, too, the innumerable dots bordering the endoplasmic reticulum. These are ribosomes, the cell's "protein factories."

A

Smooth endoplasmic reticulum
Centrioles
Centrosome
Ribosomes
Mitochondria
Smooth endoplasmic reticulum
Cilia
Mitochondrion
Lysosome
Rough endoplasmic reticulum
Peroxisome
Free ribosomes
Golgi apparatus
Microvilli
Vesicle
Cytoskeleton
Intermediate filament
Nuclear envelope
Nucleus
Nucleolus
Microtubule
Microfilament

B

Plasma membrane
Nuclear membrane
Chromatin
Golgi apparatus
Mitochondrion
Lysosomes
Ribosomes

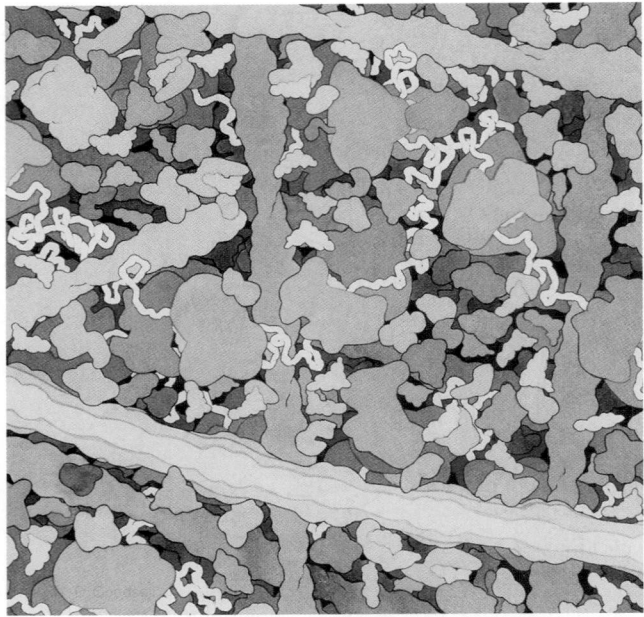

Figure 3-2 *Cytoplasm*. This drawing shows that the cytoplasm is made up of a dense arrangement of fibers, protein molecules, organelles, and other structures, suspended in the liquid cytosol. Such crowding helps molecules interact with one another and thus improves the efficiency of cellular metabolism.

CELL MEMBRANES

Figure 3-1 shows that a typical cell contains a variety of membranes. The outer boundary of the cell, or **plasma membrane,** is just one of these membranes. Each cell also has various *membranous organelles*. Membranous organelles are sacs and canals made of the same membrane material as the plasma membrane. This membrane material is a very thin sheet—only about 75 angstroms (Å) or 0.0000003 inch thick—made of lipid, protein, and other molecules (Table 3-1).

Membrane Structure

Figure 3-3 shows a simplified view of the current model of cell membrane structure. This concept of cell membranes is called the **fluid mosaic model.** Like the tiles in an art mosaic, the different molecules that make up a cell membrane are arranged in a sheet. Unlike art mosaics, however, this mosaic of molecules is *fluid*; that is, the molecules are able to slowly float around the membrane like icebergs. The fluid mosaic model shows us that the molecules of a cell membrane are bound tightly enough to form a continuous sheet but loosely enough that the molecules can slip past one another.

What are the forces that hold a cell membrane together? The short answer to that question is chemical attractions. The primary structure of a cell membrane is a double layer of phospholipid molecules. Recall from Chapter 2 that phospholipid molecules have "heads" that are water soluble and double "tails" that are lipid soluble (see Figure 2-25 on p. 64). Because their heads are **hydrophilic** (water loving) and their tails are **hydrophobic** (water fearing), phospholipid molecules naturally arrange themselves into double layers, or *bilayers*, in water. This allows all the hydrophilic heads to face toward water and all the hydrophobic tails

to face away from water. Because the internal environment of the body is simply a water-based solution, phospholipid bilayers appear wherever phospholipid molecules are scattered among the water molecules. Cholesterol is a steroid lipid that mixes with phospholipid molecules to form a blend of lipids that stays just fluid enough to function properly at body temperature. Without cholesterol, cell membranes would break far too easily.

Each human cell manufactures various phospholipid and cholesterol molecules, which then arrange in a bilayer to form a natural "fencing" material that can be used throughout the cell. This "fence" allows many lipid-soluble molecules to pass through easily—just like a picket fence allows air and water to pass through easily. However, because most of the phospholipid bilayer is hydrophobic, cell membranes do not allow water or water-soluble molecules to pass through easily. This characteristic of cell membranes is ideal because most of the substances in the internal environment are water soluble. What good is a membrane if it allows just about everything to pass through it?

Just as there are different fencing materials for different kinds of fences, cells can make any of a variety of different phospholipids for different areas of a cell membrane. For example, some areas of a membrane are stiff, and some are more flimsy. Some membrane lipids combine with carbohydrates to form glycolipids, and some unite with protein to form lipoproteins easily. Recall from Chapter 2 that proteins are made up of many amino acids, some of which are polar and some nonpolar (see Figure 2-18, p. 55). By having different kinds of amino acids in specific locations, protein molecules may become anchored within the bilayer of phospholipid heads and tails or attached to one side or other of the membrane.

The different molecular interactions within the membrane allow the formation of **rafts,** which are stiff groupings of membrane molecules (often very rich in cholesterol) that travel together like a log raft on the surface of a lake (Figure 3-4). Rafts help organize the various components of a membrane. Rafts may also sometimes allow the cell to form depressions that pouch inward and pinch off to carry substances into the cell (Box 3-2). Evidence shows that human immunodeficiency virus (HIV) enters cells by connecting to a raft protein in the plasma membrane, which is then pulled into the cell.

Membrane Function

A cell can control what moves through any section of membrane by means of proteins embedded in the phospholipid bilayer (see Figure 3-3). Many of these *membrane proteins* have openings that like gates in a fence, allow water-soluble molecules to pass through the membrane. Specific kinds of transport proteins allow only certain kinds of molecules to pass through—and the cell can determine whether these "gates" are open or closed at any particular time. We consider this function of membrane proteins again in Chapter 4 when we discuss transport mechanisms in the cell.

Membrane proteins have many different structural forms that allow them to serve various other functions (see Table 3-3). Some membrane proteins have carbohydrates attached to their outer surface—forming *glycoprotein* molecules—that act as identification markers. Such identification markers, which are recognized by other molecules, act as signs on a fence that identify the enclosed area. Cells and molecules of the immune system can thus distinguish between normal "self" cells and abnormal or "nonself" cells. Not only does this mechanism allow us to attack cancer or bacter-

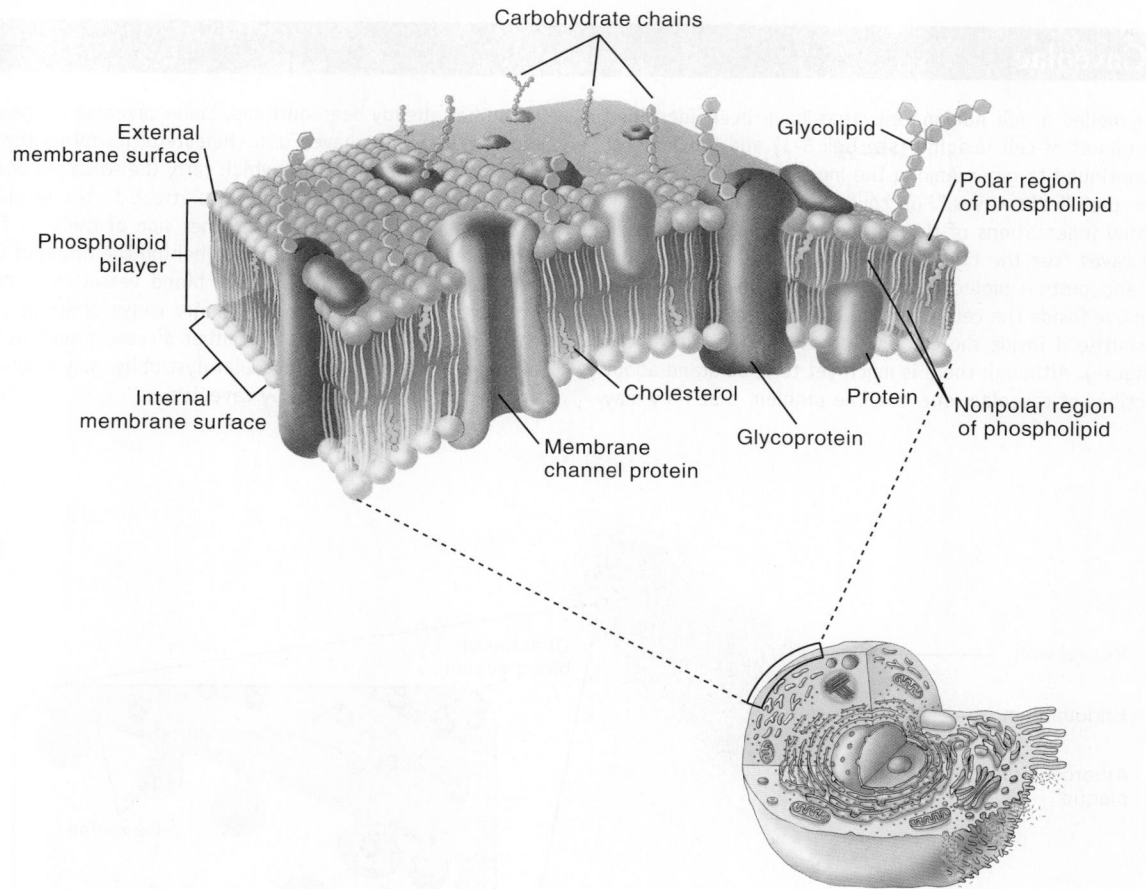

Figure 3-3 *Plasma membrane.* The plasma membrane is made of a bilayer of phospholipid molecules arranged with their nonpolar "tails" pointing toward each other. Cholesterol molecules help stabilize the flexible bilayer structure to prevent breakage. Protein molecules and protein-hybrid molecules may be found on the outer or inner surface of the bilayer—or extending all the way through the membrane.

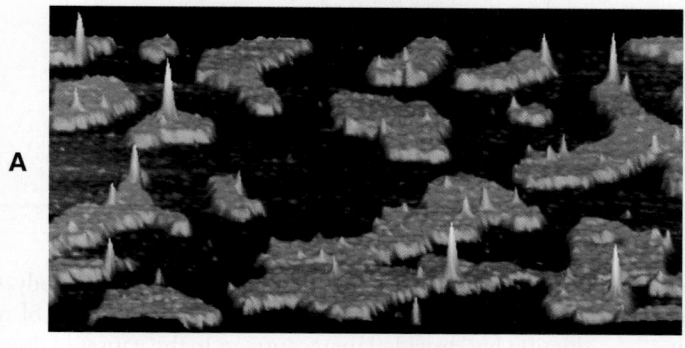

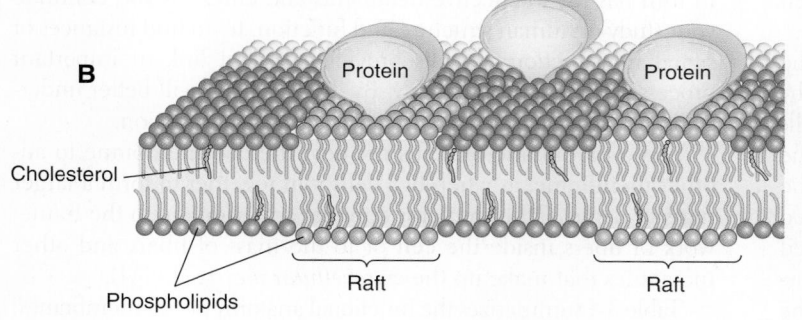

Figure 3-4 *Rafts.* **A,** Atomic force micrograph (AFM) in which an extremely fine-tipped needle drags over the surface of a cell membrane to reveal detailed surface features. Rafts are seen here as raised, orange areas surrounded by black areas of less rigid phospholipid structure. **B,** Diagram showing the basic structure of a membrane with rafts. The raft phospholipids have a richer supply of cholesterol than surrounding regions do and, along with attached proteins, form rather rigid floating platforms in the surface of the membrane. Rafts help organize functions at the surfaces of cells and organelles.

BOX 3-2 Caveolae

The list of organelles inside human cells that have been identified with new techniques of cell imaging (see Box 3-1) and biochemical analysis has continued to grow. Among the more recently discovered organelles are the "little caves," or *caveolae* (singular, *caveola*). Caveolae are tiny indentations of the plasma membrane that indeed resemble tiny caves (see the figure). Caveolae appear to form from rafts of lipid and protein molecules in the plasma membrane that pinch in and move inside the cell. Caveolae can capture extracellular material and shuttle it inside the cell or even all the way across the cell (see the figure). Although there is much yet to understand about the many functions of caveolae, one possible problem that they may

cause has already been outlined. Some caveolae in the cells that line blood vessels may have CD36 cholesterol receptors that attract low-density lipoproteins (LDLs, which carry the so-called *bad cholesterol*). As the figure shows, once the LDLs attach to the receptor, the caveola closes and migrates to the other side of the cell. There, the LDL molecules are released to build up behind the lining of the blood vessel. As the LDLs accumulate, the blood vessel channel narrows and obstructs the flow of blood—a major cause of stroke and heart disease. Researchers believe that other diseases, such as certain forms of diabetes, cancer, and muscular dystrophy, may also result from inappropriate actions taken by caveolae.

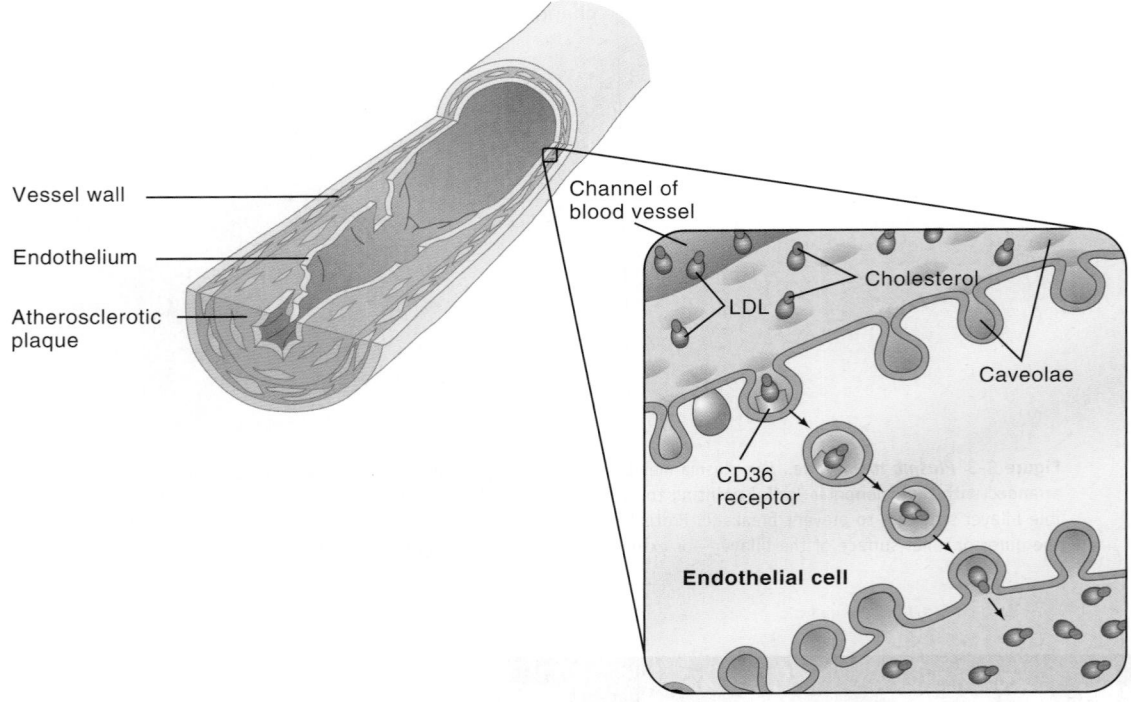

Vessel wall

Endothelium

Atherosclerotic plaque

Channel of blood vessel

Cholesterol

LDL

Caveolae

CD36 receptor

Endothelial cell

Proposed role of caveolae in cholesterol buildup in blood vessel walls. Some caveolae contain CD36 receptors that attract LDL molecules carrying "bad cholesterol" in the bloodstream. In some individuals, caveolae may trap LDLs, shuttle them to the other side of the cell, and release them under the lining of the blood vessel. There they may build up and eventually narrow the vessel's channel enough to obstruct blood flow.

ial cells, but it also prevents us from receiving blood donations from people without similar cell markers. Some membrane proteins are enzymes that catalyze cellular reactions. Some membrane proteins bind to other membrane proteins to form connections between cells or bind to support filaments within the cell to anchor them.

Other membrane proteins are **receptors** that can react to the presence of hormones or other regulatory chemicals and thereby trigger metabolic changes in the cell. The process by which cells translate the signal received by a membrane receptor into a specific chemical change in the cell is called **signal transduction**. The word *transduction* means "carry across," as a message being carried across a membrane. Researchers have only recently discovered some of the details of how signal transduction works. These discoveries have shown the vital importance of signal transduction in the

normal function of cells and therefore the whole body. One of the most active areas of biomedical research, the study of signal transduction has provided many answers to the causes of diseases, which in turn has led to effective treatments and cures. As you continue your study of human structure and function, try to find instances of signal transduction in cells providing a vital link in important processes throughout the body. By doing so, you will better understand the "big picture" of human structure and function.

Some membrane proteins connect the cell membrane to another membrane, as when two cells join together to form a larger mass of tissue. Other proteins connect a membrane to the framework of fibers inside the cell or to the mass of fibers and other molecules that make up the *extracellular matrix (ECM)*.

Table 3-4 summarizes the functional anatomy of cell membranes.

Table 3-4 Functional Anatomy of Cell Membranes

STRUCTURE		FUNCTION
Sheet (bilayer) of phospholipids stabilized by cholesterol		Maintains wholeness (integrity) of a cell or membranous organelle
Membrane protein that act as channels or carriers of molecules		Controlled transport of water-soluble molecules from one compartment to another
Receptor molecules that trigger metabolic changes in membrane (or on other side of membrane)		Sensitivity to hormones and other regulatory chemicals; involved in signal transduction
Enzyme molecules that catalyze specific chemical reactions		Regulation of metabolic reactions
Membrane proteins that bind to molecules outside the cell		Form connections between one cell and another
Membrane proteins that bind to support structures		Support and maintain the shape of a cell or membranous organelle; participate in cell movement; bind to fibers of the extracellular matrix (ECM)
Glycoproteins or proteins in the membrane that act as markers		Recognition of cells or organelles

CYTOPLASM AND ORGANELLES

Cytoplasm is the gel-like internal substance of cells that contains many tiny suspended structures. Early cell scientists believed cytoplasm to be a rather uniform fluid between the plasma membrane and the nucleus. We now know that the cytoplasm of each cell is actually a watery solution called cytosol plus hundreds or even thousands of "little organs," or **organelles,** that thicken the cytoplasm to its gel-like consistency. Table 3-3 lists some of the major types of organelles that we will discuss in this book. Until relatively recently, when newer microscope technology became available, many of these organelles simply could not be seen. If they were able to be seen, they were simply called *inclusions* because their roles as integral parts of the cell were not yet recognized. Undoubtedly, some of the cellular structures we now call inclusions will eventually be recognized as organelles and given specific names.

Many types of organelles have been identified in various cells of the body. To make it easier to study them, they have been classified into two major groups: *membranous organelles* and *nonmembranous organelles.* Membranous organelles are the organelles that are specialized sacs or canals made of cell membrane. Nonmembranous organelles are not made of membrane but are made of microscopic filaments or other particles. As you read through the following sections, be sure to note whether the organelle is membranous or nonmembranous (see Table 3-3).

Endoplasmic Reticulum

Endoplasm means the cytoplasm is located toward the center of a cell. *Reticulum* means network. Therefore, the name **endoplasmic reticulum (ER)** means literally a network located deep inside the cytoplasm. When first seen, it appeared to be just that. Later on, however, more highly magnified views under the electron microscope showed the ER to be distributed throughout the cytoplasm (Figure 3-1). It consists of membranous-walled canals and flat, curving sacs that are arranged in parallel rows.

There are two types of ER: *rough* and *smooth.* Innumerable small granules—ribosomes by name—dot the outer surfaces of the membranous walls of the rough type and give it its "rough" appearance (Figure 3-5). Ribosomes are themselves organelles. The canals of the ER wind tortuously through the cytoplasm and extend all the way from the plasma membrane to the nucleus. The membrane forming the walls of the ER has essentially the same molecular structure as the plasma membrane.

The fact that the structure of the ER is an interconnected system of canals suggests that it might function as a miniature circulatory system for the cell. In fact, proteins do move through the canals. The ribosomes that attach themselves to the rough ER synthesize proteins. The protein structures enter the canals, where they fold and possibly unite with other proteins to form larger molecules. The proteins made in the ER move through the network of canals toward the Golgi apparatus, and some of them eventually leave the cell. Thus, the rough ER functions in both protein synthesis and intracellular transportation of molecules.

No ribosomes border the membranous wall of the smooth ER—hence its smooth appearance and its name. The smooth ER synthesizes certain lipids and carbohydrates. Included among these are the steroid hormones and some of the carbohydrates used to form glycoproteins. Lipids that form cell membranes are

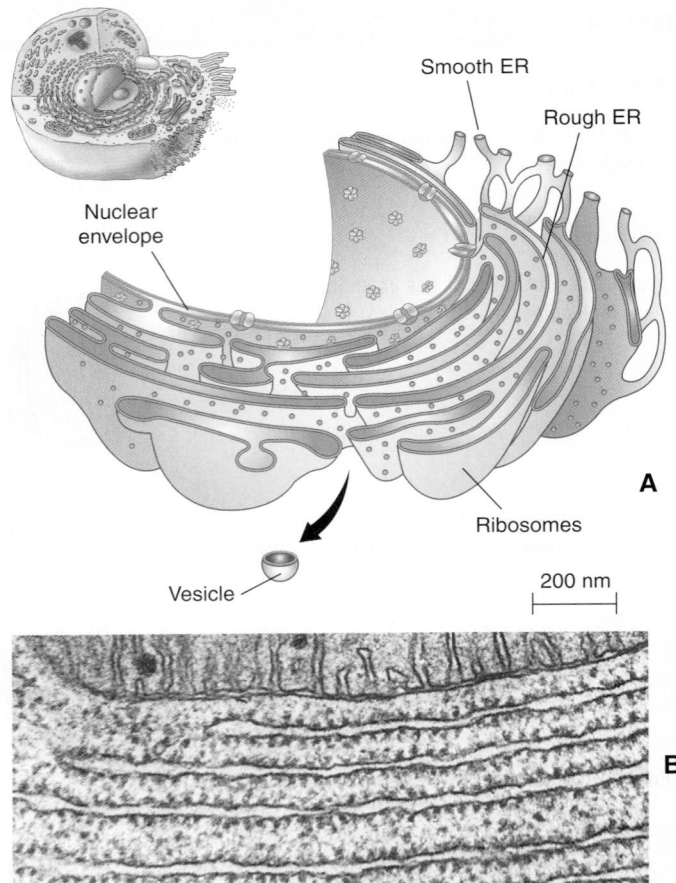

Figure 3-5 *Endoplasmic reticulum (ER).* In both the drawing **(A)** and the transmission electron micrograph **(B)**, the rough ER is distinguished by the presence of tiny ribosomes dotting the boundary of the membrane. The smooth ER lacks ribosomes on its surface. Note also that the ER is continuous with the outer membrane of the nuclear envelope.

also synthesized here. As these membrane lipids are made, they simply become part of the smooth ER's wall. Bits of the ER break off from time to time and travel to other membranous organelles—even the plasma membrane—and add to their membrane. The smooth ER, then, is the organelle that makes membrane for use throughout the cell.

Smooth ER also transports calcium ions (Ca^{++}) from the cytosol into the sacs of the ER, thus helping maintain a low concentration of Ca^{++} in cell's interior. The fact that Ca^{++} is moved into the ER and stored there will prove useful in later chapters, when we discuss several different processes that make use of this function of the ER.

Ribosomes

Every cell contains thousands of **ribosomes.** Many of them are attached to the rough endoplasmic reticulum, and many of them lie free, scattered through the cytoplasm. Find them in both locations in Figure 3-1. Because ribosomes are too small to be seen with a light microscope, no one knew they existed until the electron microscope revealed them. Research has now yielded information about their molecular structure. Each ribosome is a non-

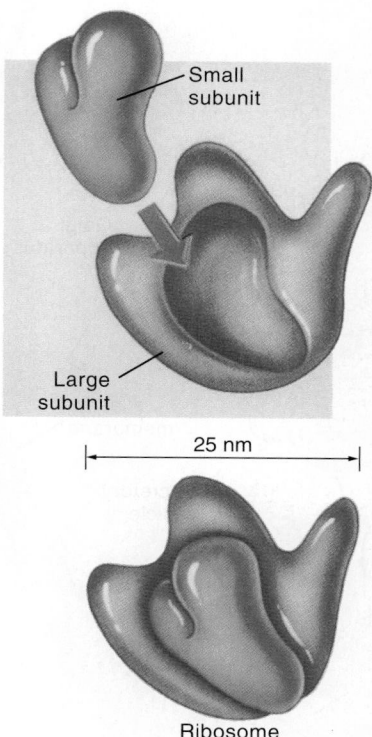

Figure 3-6 *Ribosome.* A ribosome is composed of a small subunit and a large subunit. Before protein synthesis, the subunits come together to form a complete ribosome.

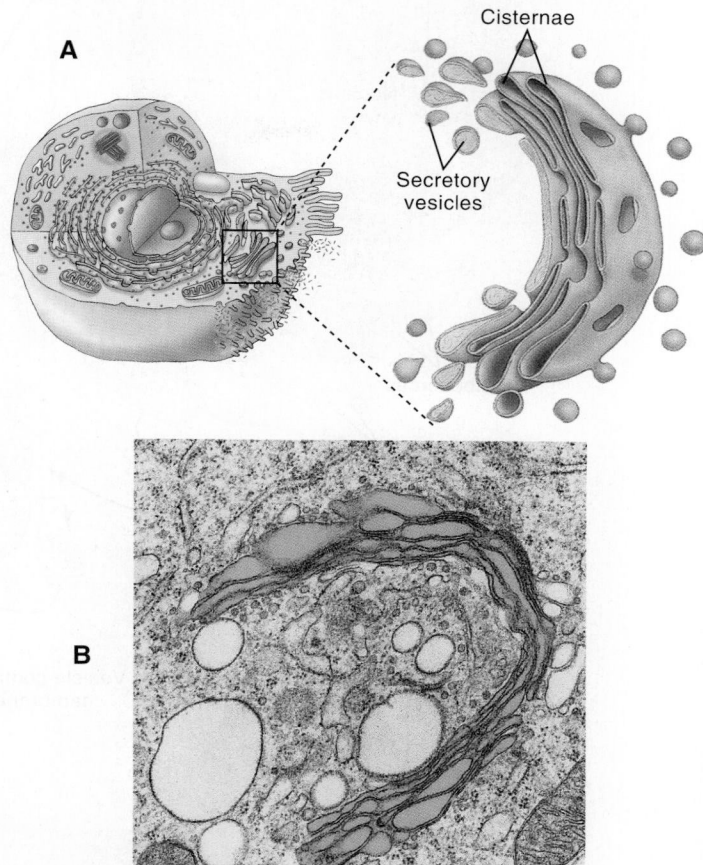

Figure 3-7 *Golgi apparatus.* **A,** Sketch of the structure of the Golgi apparatus showing a stack of flattened sacs, or *cisternae,* and numerous small membranous bubbles, or *secretory vesicles.* **B,** Transmission electron micrograph (TEM) showing the Golgi apparatus highlighted with color.

membranous structure made of two pieces, a large subunit and a small subunit (Figure 3-6). Each subunit is composed of ribonucleic acid (RNA) bonded to protein. Ribosomal RNA is often abbreviated as rRNA. Other types of ribonucleic acid in the cell are messenger RNA (mRNA) and transfer RNA (tRNA). Various types of RNA and their roles in the cell are discussed in more detail in Chapter 4.

The function of ribosomes is protein synthesis. Ribosomes are the molecular machines that make proteins, or to use a popular term, they are the cell's "protein factories." The ribosomes attached to the endoplastic reticulum synthesize proteins mainly for "export" and for use in the cell's plasma membrane. The ribosomes that are free in the cytoplasm make proteins for the cell's own domestic use. They make both its structural and its functional proteins (enzymes). Working ribosomes, those that are actually making proteins, appear to function in groups called *polyribosomes.* Under the electron microscope, polyribosomes look like short strings of beads.

Golgi Apparatus

The **Golgi apparatus** is a membranous organelle consisting of tiny sacs, or *cisternae,* stacked on one another and located near the nucleus (Figure 3-7; see also Figure 3-1). It was first noticed in the nineteenth century by the Italian biologist Camillo Golgi. Like the endoplasmic reticulum, the Golgi apparatus processes molecules within its membranes. As a matter of fact, the Golgi apparatus seems to be part of the same system that prepares protein molecules for export from the cell.

The role of the Golgi apparatus in processing and packaging protein molecules for export from the cell is summarized in Figure 3-8. First, proteins synthesized by ribosomes and transported to the end of an endoplasmic reticulum canal are packaged into tiny membranous bubbles, or vesicles, that break away from the endoplasmic reticulum. The vesicles then move to the Golgi apparatus and fuse with the first cisterna. Protein molecules thus released into the cisterna are then chemically altered by enzymes present there. For example, the enzymes may attach carbohydrate molecules synthesized in the Golgi apparatus to form glycoproteins. The molecules are then "pinched off" in another vesicle, which moves to the next cisterna for further processing. The proteins and glycoproteins eventually end up in the last cisterna, from which vesicles pinch off and move to the plasma membrane. There they release their contents outside the cell in a process called secretion. Some protein and glycoprotein molecules may instead be incorporated into the membrane of a Golgi vesicle. This means that these molecules eventually become part of the plasma membrane, as seen in Figure 3-3 and Table 3-2.

Scientists are using their knowledge of the Golgi apparatus to mimic the cell's function in order to develop treatments—an approach described in Box 3-3.

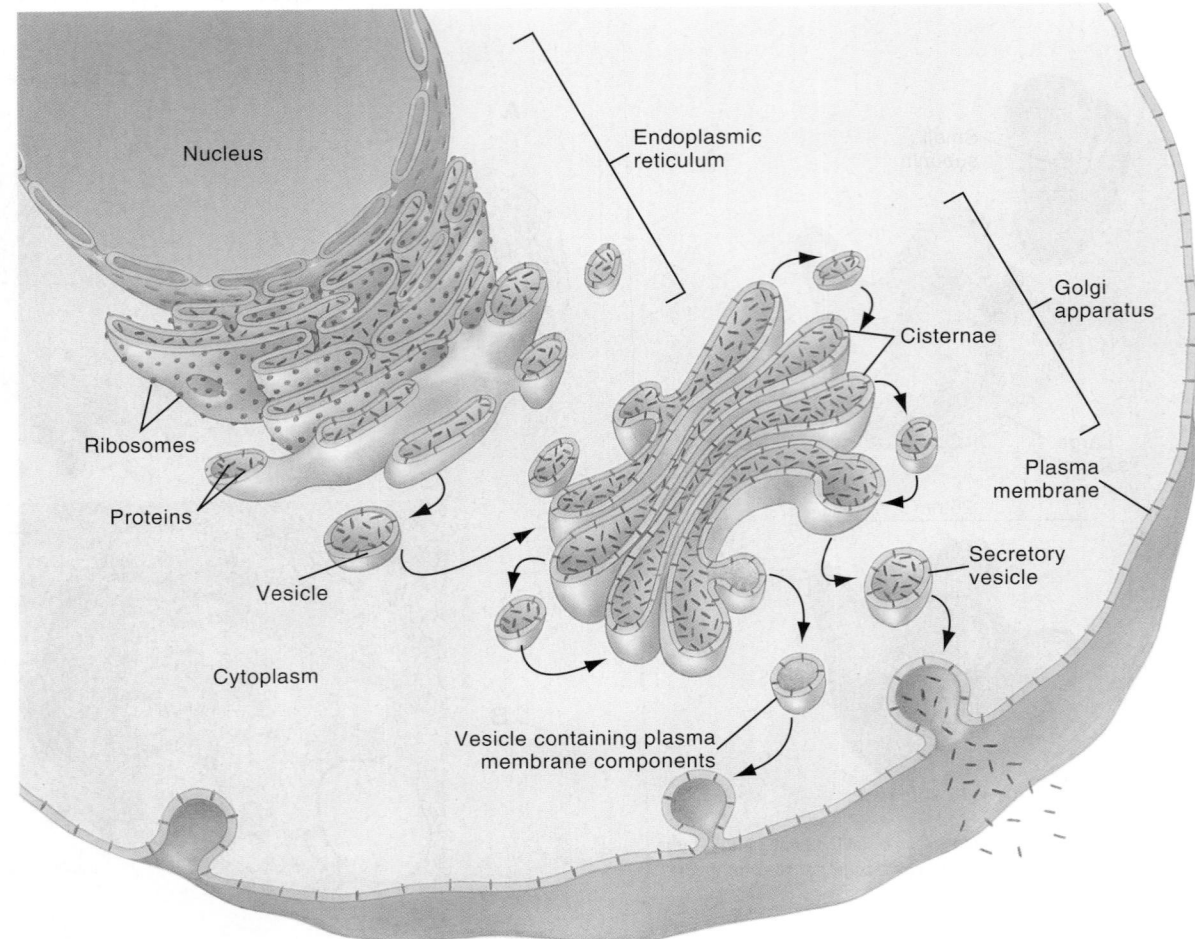

Figure 3-8 *The cell's protein export system.* The Golgi apparatus processes and packages protein molecules delivered from the endoplasmic reticulum by small vesicles. After entering the first cisterna of the Golgi apparatus, a protein molecule undergoes a series of chemical modifications, is sent (by means of a vesicle) to the next cisterna for further modification, and so on, until it is ready to exit the last cisterna. When it is ready to exit, a molecule is packaged in a membranous secretory vesicle that migrates to the surface of the cell and "pops open" to release its contents into the space outside the cell. Some vesicles remain inside the cell for some time and serve as storage vessels for the substance to be secreted.

BOX 3-3 | **Biomimicry**

Researchers are working to mimic, or imitate, the functions of the cell in order to efficiently manufacture synthetic versions of natural substances, drugs, and other complex molecules. This approach is called *biomimicry* because it mimics natural biological mechanisms.

Biomimicry can help reduce waste in making pharmaceuticals, for example. One drug company estimates that with traditional manufacturing methods, making 1 kilogram (kg) of a drug produces between 25 and 100 kg of waste. To overcome this inefficiency, a system of porous polymer vesicles are being developed that carry on the step-by-step manufacturing process in a way similar to the Golgi apparatus in the cell (see Figure 3-8). Each type of porous vesicle contains a different catalyst that promotes a different chemical reaction. Molecules enter a vesicle (through pores in the membrane), and the resulting molecule then leaves the vesicle (through the same pores). The new molecule subsequently meets up with another vesicle, where another step in the process takes place. The remaining steps follow in the same progression through successive vesicles, much like the step-by-step processing in the successive cisternae of the Golgi apparatus.

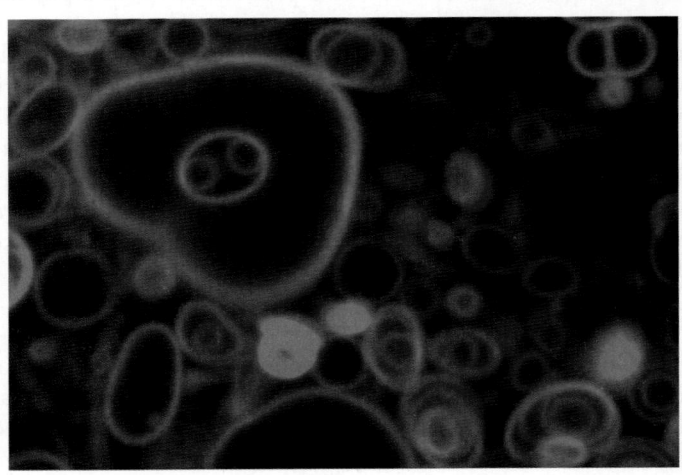

Biomimicry. This micrograph shows fluorescent-stained artificial vesicles that may soon mimic the cell's own machinery to efficiently manufacture pharmaceuticals and other synthetic molecules.

Lysosomes

Like the endoplasmic reticulum and Golgi apparatus, **lysosomes** have membranous walls—indeed, they are vesicles that have pinched off from the Golgi apparatus. The size and shape of lysosomes change with the stage of their activity. In their earliest, inactive stage, they look like mere granules. Later, as they become active, they take on the appearance of small vesicles or sacs (see Figure 3-1). The interior of the lysosome contains various kinds of enzymes capable of breaking down all the protein components of cells. Lysosomal enzymes have several important functions in cells. Chiefly, they help the cell break down proteins that are not needed so that they are not in the way and the amino acids can be reused by the cell. In the process illustrated in Figure 3-9, defective organelles can be thus recycled. Membrane proteins from the plasma membrane can pinch off inside the cell and be recycled in the same manner. Likewise, cells may engulf bacteria or other extracellular particles and destroy them with lysosomal enzymes. Thus, lysosomes deserve their nicknames of "digestive bags" and "cellular garbage disposals."

Proteasomes

The **proteasome** is another protein-destroying organelle in the cell. As Figure 3-10 shows, the proteasome is a hollow, cylindrical "drum" made up of protein subunits. Found throughout the cytoplasm, the proteasome is responsible for breaking down abnormal and misfolded proteins from the ER, as well as destroying normal regulatory proteins in the cytoplasm that are no longer needed. But unlike the lysosome, which destroys large groups of protein molecules all at once, the proteasome destroys protein molecules one at a time.

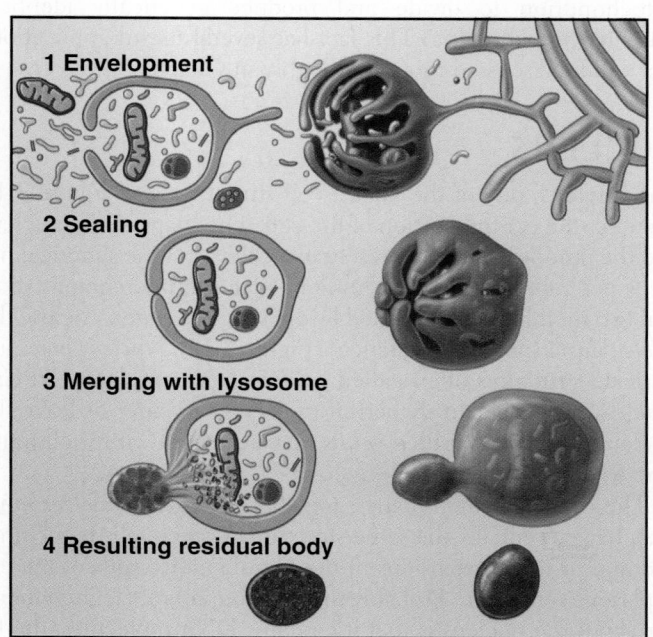

Figure 3-9 *Lysosome.* The process of destroying old cell parts follows these steps: 1. Membrane from the ER or Golgi apparatus encircles the material to be destroyed. 2. A membrane capsule completely traps the material. 3. A lysosome fuses with the membrane capsule, and digestive enzymes from the lysosome enter the capsule and destroy the contents. 4. Undigested material remains in a compact *residual body*.

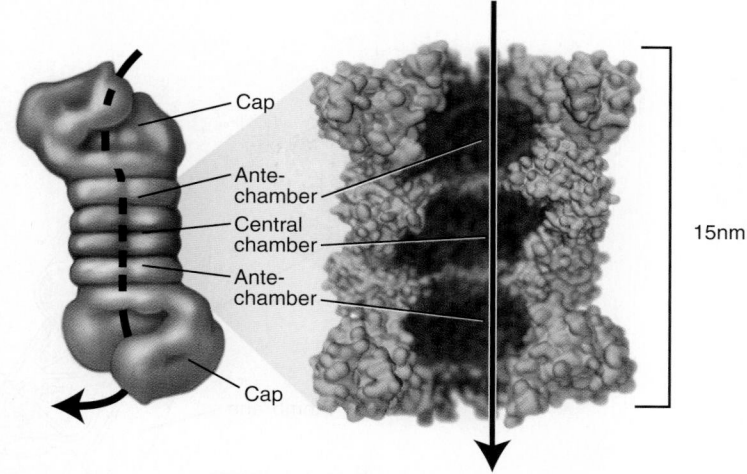

Figure 3-10 *Proteasome.* Made up of protein subunits, the proteasome is a hollow cylinder with regulatory end caps. Following the path shown by *arrows,* an ubiquitin-tagged protein molecule unfolds as it enters the cap and is then broken down into small peptide chains (4 to 25 amino acids), which leave through the other end of the proteasome. The peptides are subsequently broken down into amino acids, which are recycled by the cell. The proteosome is a little over half the size of a ribosome.

Before a protein enters the hollow interior of the proteasome, it must be tagged with a chain of very small proteins called *ubiquitins.* The ubiquitin chain then enters the proteasome, with the rest of the protein following. As it passes through the cap of the proteasome, the protein is unfolded. Peptide bonds are subsequently broken apart by active sites inside the central chamber. Short peptide chains, 4 to 25 amino acids long, then exit the other end of the proteasome. The short peptides are easily broken down into their component amino acids for recycling by the cell.

Proper function of proteasomes is important to prevent abnormal cell function and possibly severe disease. For example, in *Parkinson disease (PD)* the proteasome system fails and improperly folded proteins kill nerve cells in the brain that regulate muscle tension.

Peroxisomes

In addition to lysosomes, another type of smaller, membranous sac-containing enzymes is also present in the cytoplasm of some cells. These organelles, called **peroxisomes,** detoxify harmful substances that may enter cells. They are often seen in kidney and liver cells and serve important detoxification functions in the body. Peroxisomes contain the enzymes *peroxidase* and *catalase,* which are important in metabolic reactions involving hydrogen peroxide—a chemical toxic to cells.

Mitochondria

Note the **mitochondria** shown in Figures 3-1 and 3-11. Magnified thousands of times, as they are there, they look like small, partitioned sausages—if you can imagine sausages only 1.5 μm long and half as wide. (In case you can visualize inches better than micrometers, 1.5 μm equals about 3/50,000 of an inch.) Yet, like all organelles, and tiny as they are, mitochondria have a highly organized molecular structure. Their membranous walls consist of not one but two delicate membranes. They form a sac within a

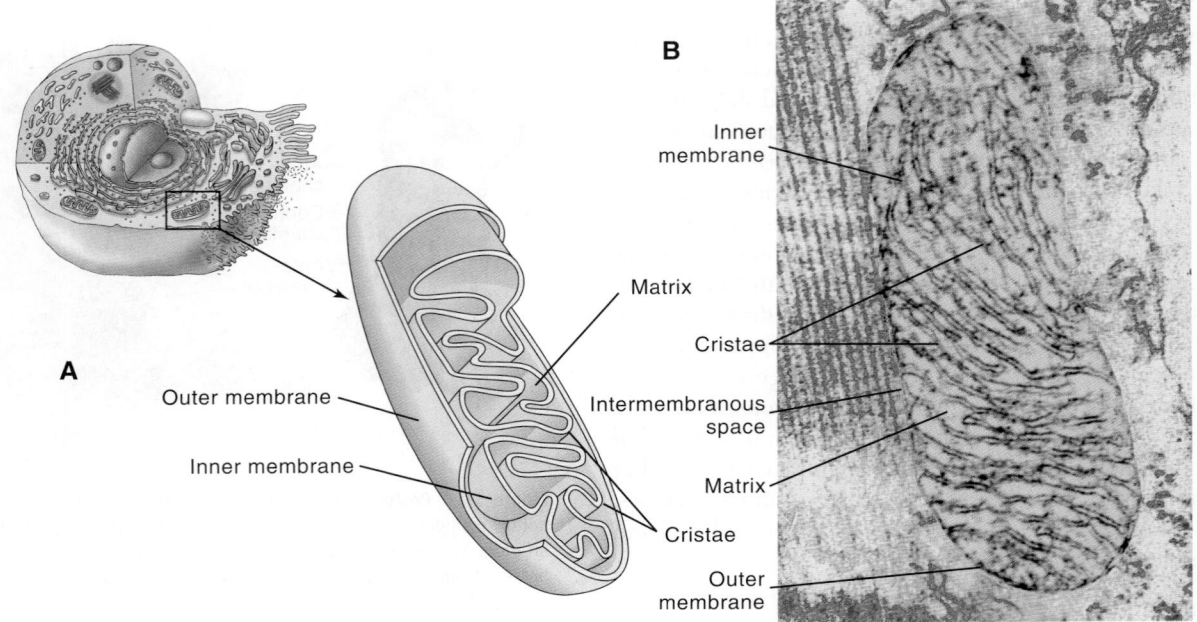

Figure 3-11 *Mitochondrion.* **A,** Cutaway sketch showing outer and inner membranes. Note the many folds (cristae) of the inner membrane. **B,** Transmission electron micrograph of a mitochondrion. Although some mitochondria have the capsule shape shown here, many are round or oval.

sac. The inner membrane is thrown into folds called **cristae.** Embedded in the inner membrane are enzymes that are essential for making one of the most important chemicals in the world. Without this compound, life cannot exist. Its long name, *adenosine triphosphate,* and its abbreviation, *ATP,* were mentioned in Chapter 2. Chapters 4 and 27 present more detailed information about this vital substance and how it is made.

Both the inner and outer membranes of the mitochondrion have essentially the same molecular structure as the cell's plasma membrane. All evidence so far indicates that the proteins in the membranes of the cristae are arranged precisely in the order of their functioning. This is another example, but surely an impressive one, of the principle stressed in Chapter 1 that organization is a foundation stone and a vital characteristic of life.

Enzymes in the mitochondrial inner membrane catalyze oxidation reactions. These are the chemical reactions that provide cells with most of the energy that does all of the many kinds of work that keeps them and the body alive. Thus do mitochondria earn their now familiar title, the "power plants" of cells.

The fact that mitochondria generate most of the power for cellular work suggests that the number of mitochondria in a cell might be directly related to its amount of activity. This principle does seem to hold true. In general, the more work a cell does, the more mitochondria its cytoplasm contains. Liver cells, for example, do more work and have more mitochondria than sperm cells do. A single liver cell contains 1000 or more mitochondria, whereas only about 25 mitochondria are present in a single sperm cell. The mitochondria in some cells multiply when energy consumption increases. For example, frequent aerobic exercise can increase the number of mitochondria inside skeletal muscle cells.

Each mitochondrion has its own DNA molecule, a very surprising discovery indeed! This enables each mitochondrion to make its own enzymes. Having its own DNA also enables each mitochondrion to divide and produce genetically identical daughter mitochondria. This fact has several useful applications that will be discussed more thoroughly in Chapter 34.

NUCLEUS

The **nucleus,** one of the largest cell structures (see Figure 3-1), occupies the central portion of the cell. The shape of the nucleus and the number of nuclei present in a cell vary. One spherical nucleus per cell, however, is common. Electron micrographs show that two membranes perforated by openings, or pores, enclose the *nucleoplasm* (nuclear substance) (Figure 3-12). *Nuclear pores* are intricate structures often called *nuclear pore complexes* (NPCs). Nuclear pores selectively permit molecules to enter or leave the nucleus. Organelles called *vaults* may also assist with such transport, as described in Box 3-4.

These nuclear membranes, together called the *nuclear envelope,* have essentially the same structure as other cell membranes and appear to be extensions of the membranous walls of the endoplasmic reticulum. Probably the most important fact to remember about the nucleus is that it contains DNA molecules, the famous heredity molecules. In nondividing cells, they appear as granules or threads, which are named **chromatin** (from the Greek *chroma,* "color") because they readily take the color of dyes. When the process of cell division begins, DNA molecules become tightly coiled. They then look like short, rodlike structures and are called **chromosomes.** All normal human cells (except mature sex cells)

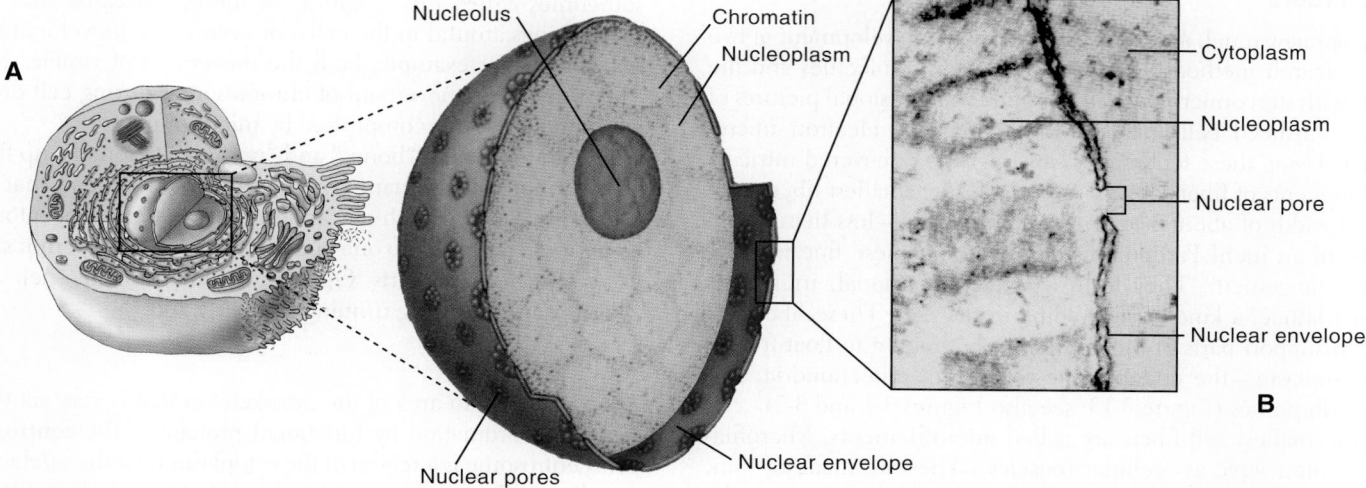

Figure 3-12 *Nucleus*. A, An artist's rendering and, **B,** an electron micrograph show that the nuclear envelope is composed of two separate membranes and is perforated by large openings, or nuclear pores.

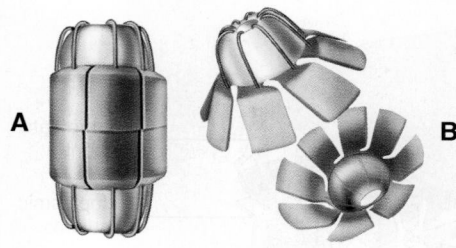

The most prominent structure visible in the nucleus is a small nonmembranous body that stains densely and is called the **nucleolus** (Figure 3-12, A). Like chromosomes, it consists chiefly of a nucleic acid, but the nucleic acid is not DNA. It is RNA, or ribonucleic acid (see pp. 66-68).

The nucleolus functions to synthesize ribosomal RNA (rRNA) and combine it with protein to form the subunits that will later combine to form ribosomes, the protein synthesizers of cells (see Figure 3-6). You might guess, therefore, and correctly so, that the more protein a cell makes, the larger its nucleolus appears. Cells of the pancreas, to cite just one example, make large amounts of protein and have large nucleoli.

QUICK CHECK

4. List at least three functions of the plasma membrane.
5. Define the term *organelle*.
6. Identify three organelles by name and give one function of each.
7. Distinguish between membranous and nonmembranous organelles.

CYTOSKELETON

contain 46 chromosomes, and each chromosome consists of one DNA molecule plus some protein molecules.

The functions of the nucleus are primarily functions of DNA molecules. In brief, DNA molecules contain the master code for making all the RNA plus the many enzymes and other proteins of a cell. Therefore, DNA molecules dictate both the structure and the function of cells. Chapter 4 briefly discusses how a cell transcribes and translates the master code to synthesize specific proteins. DNA molecules are inherited, so DNA plays a pivotal role in the process of heredity—a concept we explore further in Chapter 34.

As its name implies, the **cytoskeleton** is the cell's internal supporting framework. Like the bony skeleton of the body, the cytoskeleton is made up of rather rigid, rodlike pieces that not only provide support but also allow movement. Like the body's musculoskeletal framework, the cytoskeleton has musclelike groups of fibers and other mechanisms that move the cell, or parts of the cell, with great strength and mobility. In this section, we discuss the basic characteristics of the cell's internal skeleton, as well as several organelles that are associated with it.

Cell Fibers

No one knew much about *cell fibers* until the development of two new research methods: one with fluorescent molecules and the other with stereomicroscopy, that is, three-dimensional pictures of whole, unsliced cells made with high-voltage electron microscopes. Using these techniques, investigators discovered intricate arrangements of fibers of varying widths. The smallest fibers seen have a width of about 3 to 6 nanometers (nm)—less than 1 millionth of an inch! Particularly striking about these fine fibers is their arrangement. They form a three-dimensional, irregularly shaped lattice, a kind of scaffolding in the cell. These fibers appear to support parts of the cell formerly thought to float free in the cytoplasm—the endoplasmic reticulum, mitochondria, and "free" ribosomes (Figure 3-13; see also Figures 3-1 and 3-2).

The smallest cell fibers are called **microfilaments.** Microfilaments often serve as "cellular muscles." They are made of thin, twisted strands of protein molecules (Figure 3-14, A) and usually form bundles that lie parallel to the long axis of a cell. In some microfilaments, the proteins can slide past one another and cause shortening of the cell. The most obvious example of such shortening occurs in muscle cells, where many bundles of microfilaments work together to shorten the cells with great force.

Cell fibers called **intermediate filaments** are twisted protein strands that are slightly thicker than microfilaments (Figure 3-14, B). Intermediate filaments are thought to form much of the supporting framework in many types of cells. For example, the protective cells in the outer layer of skin are filled with a dense arrangement of tough intermediate filaments.

The thickest of the cell fibers are tiny, hollow tubes called **microtubules.** As Figure 3-14, C, shows, microtubules are made of protein subunits arranged in a spiral fashion. Microtubules are

sometimes called the "engines" of the cell because they often move things around in the cell—or even cause movement of the entire cell. For example, both the movement of vesicles within the cell and the movement of chromosomes during cell division are thought to be accomplished by microtubules.

In addition to the "bones" and "muscles" that make up the cytoskeleton, there are many types of functional proteins that allow the parts to interact with one another. For example, receptor molecules help parts link to one another and may even permit signals to be sent between parts. Other functional proteins such as enzymes keep everything running smoothly, too.

Centrosomes

An example of an area of the cytoskeleton that is very active and requires coordination by functional proteins is the **centrosome.** The centrosome is a region of the cytoplasm near the nucleus that coordinates the building and breaking of microtubules in the cell. For this reason, this nonmembranous structure is often called the *microtubule-organizing center* (MTOC). Look for the tiny yellow centrosome in Figure 3-14, part C (bottom panel). Radiating out from the centrosome in this micrograph, you can see the red-stained microtubules organized around it. The centrosome also plays an important role during cell division, when a special "spindle" of microtubules is constructed for the purpose of moving chromosomes around the cell.

The boundaries of the centrosome are rather indistinct because it lacks a membranous wall. However, the general location of the centrosome is easy to find because of a pair of cylindrical structures called **centrioles.**

Under the light microscope, centrioles appear as two dots located near the nucleus. The electron microscope, however, re-

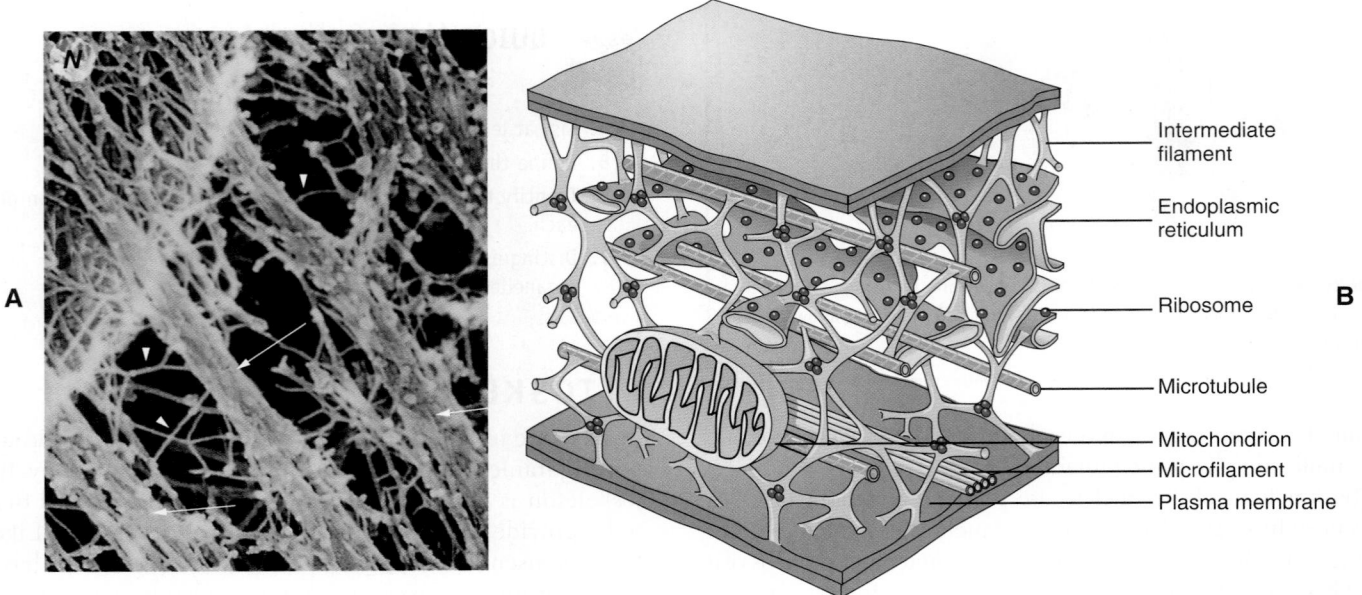

A

B

Intermediate filament

Endoplasmic reticulum

Ribosome

Microtubule

Mitochondrion

Microfilament

Plasma membrane

Figure 3-13 *The cytoskeleton.* **A,** Color-enhanced electron micrograph of a portion of the cell's internal framework. The letter *N* marks the nucleus, the *arrowheads* mark the intermediate filaments, and the *complete arrows* mark the microtubules. **B,** Artist's interpretation of the cell's internal framework. Notice that the "free" ribosomes and other organelles are not really free at all.

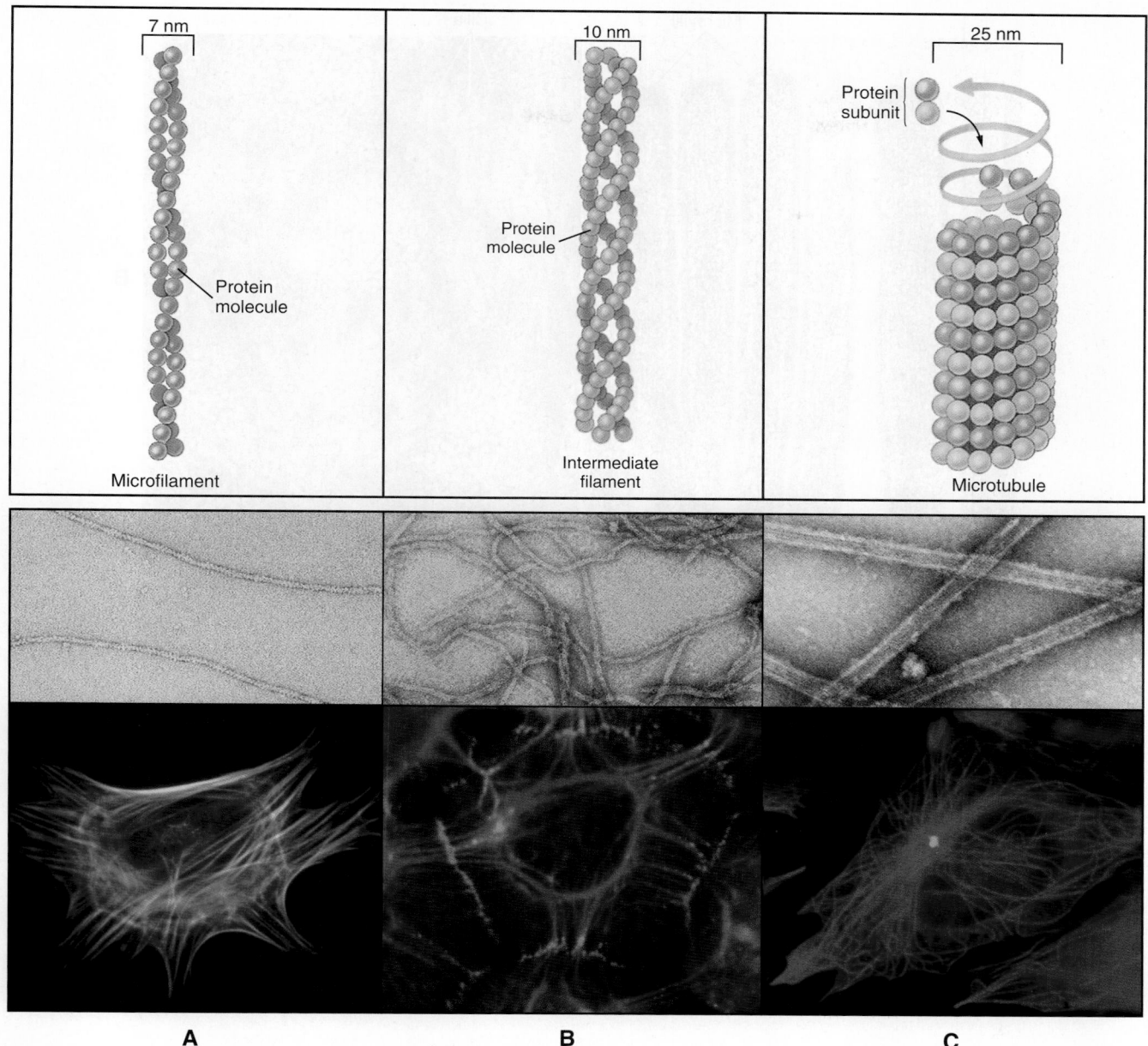

Figure 3-14 *Cell fibers.* The top panel of each part is a sketch, the middle panel is a transmission electron micrograph (TEM), and the bottom panel is a light micrograph using fluorescent stains to highlight specific molecules within each cell. **A,** *Microfilaments* are thin, twisted strands of protein molecules. **B,** *Intermediate filaments* are thicker, twisted protein strands. **C,** *Microtubules* are hollow fibers that consist of a spiral arrangement of protein subunits. Note the bright yellow microtubule-organizing center (centrosome) in the bottom panel.

veals them not as mere dots but as tiny cylinders (see Figure 3-1). The walls of the cylinders consist of nine bundles of microtubules, with three tubules in each bundle. A curious fact about these two tubular-walled cylinders is their position at right angles to each other. This special arrangement occurs when the centrioles separate in preparation for cell division (see p. 132 in Chapter 4). Before separating, a daughter centriole is formed perpendicular to each member of the original pair so that a complete pair may be distributed to each new cell. In addition to their involvement in forming the spindle that appears during cell division, centrioles are involved in the formation of microtubular cell extensions (discussed in the next section).

Cell Extensions

In some cells the cytoskeleton forms projections that extend the plasma membrane outward to form tiny, fingerlike processes. These processes, *microvilli, cilia,* and *flagella,* are present only in certain types of cells—depending, of course, on a cell's specialized functions. **Microvilli,** for example, are found in epithelial cells that line the intestines and other areas where absorption is

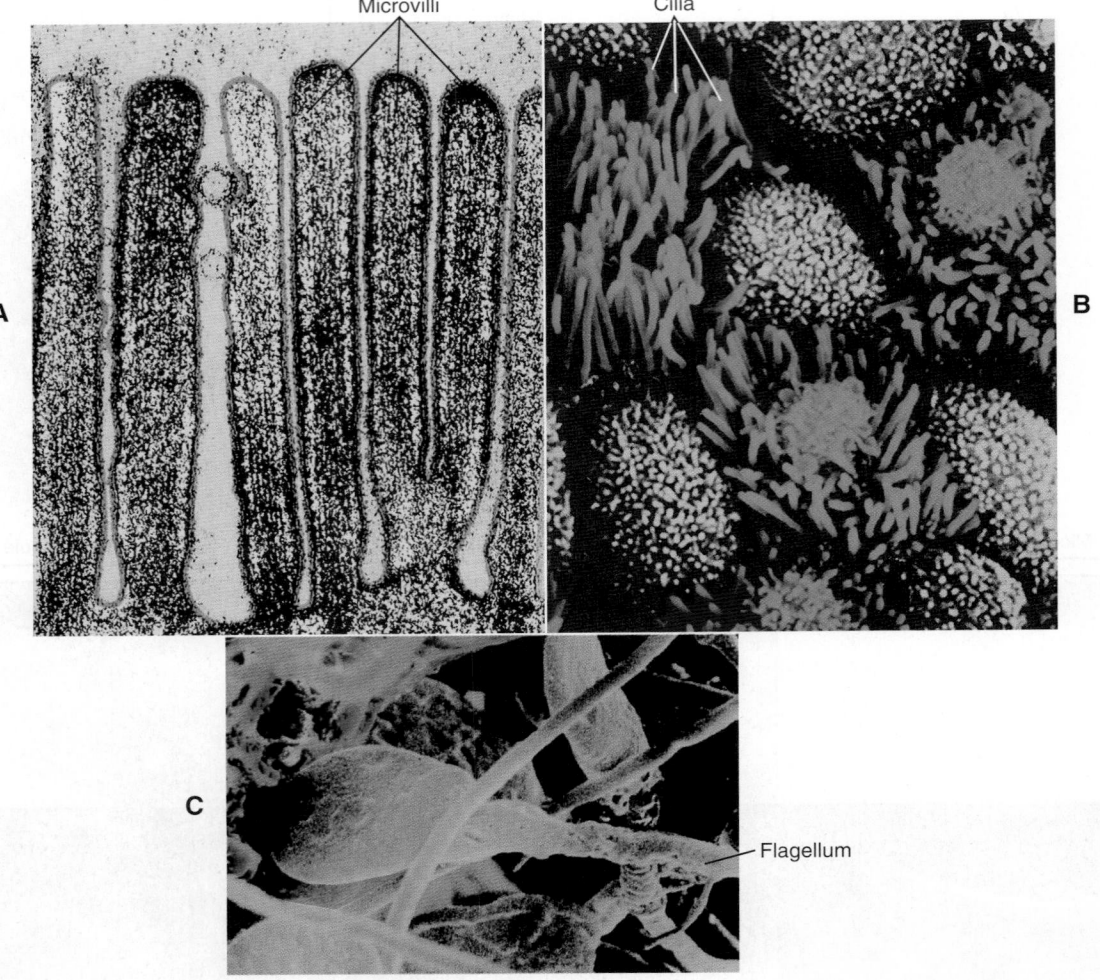

Figure 3-15 *Cell processes.* **A,** *Microvilli* are numerous, fingerlike projections that increase the surface area of absorptive cells. This electron micrograph shows a longitudinal section of microvilli from a cell lining the small intestine. Note the bundles of microfilaments that support the microvilli. **B,** *Cilia* are numerous, fine processes that transport fluid across the surface of the cell. This electron micrograph shows cilia *(long projections)* and microvilli *(small bumps)* on cells lining the lung airways. **C,** In humans, *flagella* are single, elongated processes on sperm cells that enable these cells to "swim."

important (Figure 3-15, A). Like tiny fingers crowded against each other, microvilli cover part of the surface of a cell (see Figure 3-1). A single microvillus measures about 0.5 μm long and only 0.1 μm or less across. Inside each microvillus are microfilament bundles that provide both structural support and the ability to move. Because one cell has hundreds of these projections, the surface area of the cell is increased manyfold—a structural feature that enables the cell to perform its function of absorption at a faster rate.

Cilia and **flagella** are cell processes that have cylinders made of microtubules at their core. Each cylinder is composed of nine double microtubules arranged around two single microtubules in the center—a slightly different arrangement than in centrioles. This particular arrangement is suited to movement. Additional short microtubules at the base of the cylinder react with the longer microtubules and cause them to slip back and forth to produce a "wiggling" movement.

What distinguishes cilia from flagella are their size and number. Cilia are shorter and more numerous than flagella (Figure 3-15, B). Under low magnification, cilia look like tiny hairs. In the lining of the respiratory tract, the movement of cilia keeps contaminated mucus moving toward the throat, where it can be swallowed. In the lining of the female reproductive tract, cilia keep the ovum moving toward the uterus. Flagella are single, long structures in the only type of human cell that has this feature: the human sperm cell (Figure 3-15, C). A sperm cell's flagellum moves like the tail of an eel to allow the cell to "swim" toward the female sex cell (ovum).

Each year brings with it the discovery of new types of cytoskeletal components. As we tease out their functions, we find that the cytoskeleton is an amazingly rich network that is literally the "backbone" of the cell. It provides a variety of many different types of cell movement, both internal and external, depending on the cell and the circumstances. It provides many different kinds

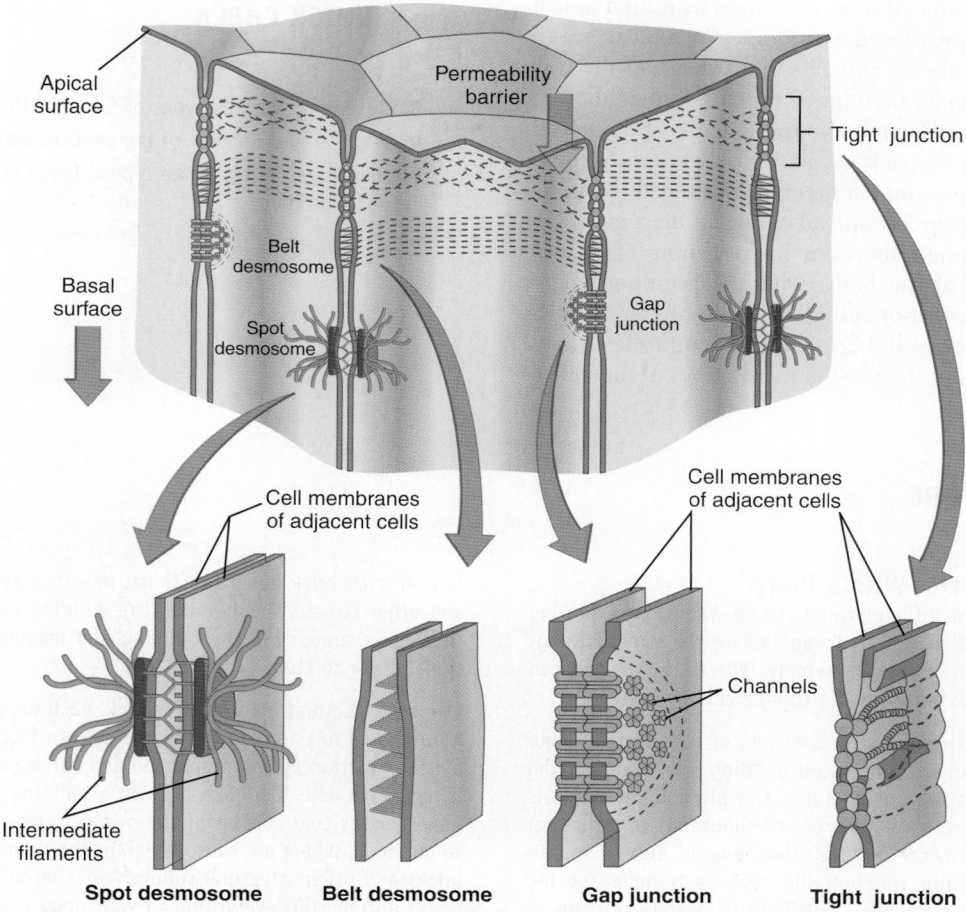

Figure 3-16 *Cell connections.* Spot and belt desmosomes, gap junction, and tight junction.

of structural support for the cell and its parts—and even becomes involved in connections with other cells. Most amazing of all, the cytoskeleton has the ability to organize itself by means of a complex set of signals and reactions so that it can quickly respond to the needs of the cell.

CELL CONNECTIONS

The tissues and organs of the body must be held together, so they must, of course, be connected in some way.

Many cells attach directly to the extracellular material, or *matrix*, that surrounds them. A group of proteins called *integrins* helps hold cells in their place in a tissue. Some integrin molecules span the plasma membrane and often connect the fibers of the cytoskeleton inside the cell to the extracellular fibers of the matrix, thereby anchoring the cell in place. In this manner, also, some cells are held to one another indirectly by fibrous nets that surround groups of cells. Certain muscle cells are held together this way.

Cells may form direct connections with each other. Integrins are sometimes involved in direct cell connections, but other proteins, such as *selectins*, *cadherins*, and *immunoglobulins*, help form most cell-to-cell connections. Not only do such connections hold the cells together, but they sometimes also allow direct com-

munication between the cells. The major types of direct cell connections are summarized in Figure 3-16.

Desmosomes are sometimes like small "spot welds" that hold adjacent cells together. Adjacent skin cells are held together this way. Notice in Figure 3-16 that fibers on the outer surface of each *spot desmosome* interlock with each other. This arrangement resembles Velcro, which holds things together tightly when tiny plastic hooks become interlocked with fabric loops. Notice also that the desmosomes are anchored internally by intermediate filaments of the cytoskeleton. Some cells have a beltlike version of the desmosome structure that completely encircles the cell. This form may be called a *belt desmosome* to distinguish it from a spot desmosome.

Gap junctions are formed when membrane channels of adjacent plasma membranes adhere to each other. As Figure 3-16 shows, such junctions have the following two effects: (1) they form gaps or "tunnels" that join the cytoplasm of two cells, and (2) they fuse the two plasma membranes into a single structure. One advantage of this arrangement is that certain molecules can pass directly from one cell to another. Another advantage is that electrical impulses traveling along a membrane can travel over many cell membranes in a row without stopping because they have "run out of membrane." Heart muscle cells are joined by gap junctions so that a single impulse can travel to, and thus stimulate, many cells at the same time.

Tight junctions occur in cells that are joined near their apical surfaces by "collars" of tightly fused membrane. As you can see in Figure 3-16, rows of membrane proteins that extend all the way around a cell fuse with similar membrane proteins in neighboring cells. An entire sheet of cells can be bound together the way soft drink cans are held in a six-pack by plastic collars—only more tightly. When tight junctions hold a sheet of cells together, molecules cannot easily *permeate*, or spread through, the cracks between the cells. Tight junctions occur in the lining of the intestines and other parts of the body, where it is important to control what gets past a sheet of cells. The only way for a molecule to get past the intestinal lining is through controlled channel, or carrier, molecules in the plasma membranes of the cells.

QUICK CHECK

8. Describe the three types of fibers in the cytoskeleton.
9. What is the function of the centrosome?
10. Name each of the three typical types of cell junctions and describe them.

THE BIG PICTURE

Cell Anatomy and the Whole Body

Probably one of the most difficult things to do when first exploring the microscopic world of the cell is to appreciate the structural significance a single cell has to the whole body. Where are these unseen cells? How does each relate to this big thing I call my body?

One useful way to approach the structural role of cells in the whole body is to think of a large, complicated building. For example, the building in which you take your anatomy and physiology course is made up of thousands, perhaps hundreds of thousands of structural subunits. Bricks, blocks, metal or wood studs, boards, and so on, are individual structures within the building that each have specific parts, or organelles, that somehow contribute to overall function. A brick often has sides of certain dimensions that allow it to fit easily with other bricks or with other building materials. Three or four surfaces are usually textured in an aesthetically pleasing manner, and holes in two of the brick surfaces lighten the weight of the brick, allow other materials such as wires or reinforcing rods to pass through, and permit mortar to form a stronger joint with the brick. The material from which each brick is made has been formulated with certain ratios of sand, clay, pigments, or other materials. Each structural fea-

ture of each brick has a functional role to play. Likewise, each brick and other structural subunit of the building have an important role to play in supporting the building and making it a pleasing, functional place to study.

Like the structural features of a brick, each organelle of each cell has a functional role to play within the cell. In fact, you can often guess a cell's function by the proportion and variety of different organelles it has. Each cell, like each brick in a building, has a role to play in providing its tiny portion of the overall support and function of the whole body. Where are these cells? In the same place you would find bricks and other structural subunits in a building—everywhere! Like bricks and mortar, everything in your body is made of cells and the extracellular material surrounding them. How do cells relate to the whole body? Like bricks that are held together to form a building, cells form the body.

Like a brick building, the human body comprises many structural subunits, each with its own structural features or organelles that contribute to the function of the cell—and therefore to the whole body.

LANGUAGE OF SCIENCE (Cont'd from page 77)

microvillus (my-kroh-VIL-us) [*micro-* small, *-villus* shaggy hair]

mitochondria (my-toh-KON-dree-ah) [*mito-* thread, *-chondro* cartilage]

nucleolus (noo-KLEE-oh-lus) [*nucleo-* nucleus (kernel), *-olus* little]

nucleus (NOO-klee-us) [*nucleus* kernel]

organelles [*organ-* tool or organ, *-elle* small]

peroxisomes (pe-ROKS-i-sohms) [*peroxi-* hydrogen peroxide, *-soma* body]

plasma membrane (PLAZ-mah MEM-brayne) [*plasma-* cell or tissue substance, *membrane* thin skin]

proteasome [*protea-* protein, *-some* body]

rafts

receptors (ree-SEP-tors) [*recept* to receive]

ribosomes (RYE-boh-sohms) [*ribo-* ribose or RNA, *-soma* body]

scanning electron microscopy (SEM)

signal transduction (SIG-nal tranz-DUK-shen)

tight junctions (tite JUNK-shens)

transmission electron micrograph (TEM)

vesicles (VES-i-kuls) [*vesico-* blister, *-cle* little]

CASE STUDY

Ben and Louise Carpenter have come to the clinic to discuss family planning. They decided to stop birth control methods in an attempt to begin a family. Both Ben and Louise are concerned because Louise is older than 35 years. They worry that she will have difficulty conceiving or that something may "go wrong" with the pregnancy or fetus.

1. As part of the routine examination, Mr. and Mrs. Carpenter were checked for sexually transmitted diseases and a Gram stain test was completed. Which one of the following statements is the BEST description of the Gram stain test?
    A. Each cell produces its own pigment that gives color to the specimen.
    B. The specimen is stained to identify structural differences.
    C. The specimen is stained to study the cell as it functions.
    D. The specimen is stained to study the macroscopic anatomy of the cell.

2. Studies are completed as part of the prenatal evaluation to ascertain that Mrs. Carpenter is in good health. Which of the following cells would be the most critical to evaluate?
    A. Nerve cells
    B. Muscle cells
    C. Gland cells
    D. Blood cells

3. Ben and Louise ask about tests to determine genetic problems because there is a previous history within their families of some genetically transmitted diseases. The practitioner's response would include information about which part of the cell?
    A. Golgi apparatus
    B. Lysosome
    C. Mitochondria
    D. Nucleus

4. In discussing how fertilization occurs, which is the BEST description of movement of the sperm and ovum?
    A. Sperm move by means of a flagellum; ova are moved by cilia.
    B. Sperm move by means of a flagellum; ova are moved by microvilli.
    C. Ova move by means of a flagellum; sperm are moved by cilia.
    D. Both sperm and ova have flagella.

CHAPTER SUMMARY

FUNCTIONAL ANATOMY OF CELLS

A. The typical cell (Figure 3-1)
    1. Also called composite cell
    2. Vary in size; all are microscopic (Table 3-1)
    3. Vary in structure and function (Table 3-2)
B. Cell structures
    1. Plasma membrane—separates the cell from its surrounding environment
    2. Cytoplasm—thick gel-like substance inside the cell composed of numerous organelles suspended in watery cytosol; each type of organelle is suited to perform particular functions (Figure 3-2)
    3. Nucleus—large membranous structure near the center of the cell

CELL MEMBRANES

A. Each cell contains a variety of membranes
    1. Plasma membrane (Figure 3-3)
    2. Membranous organelles—sacs and canals made of the same material as the plasma membrane
B. Fluid mosaic model—theory explaining how cell membranes are constructed
    1. Molecules of the cell membrane are arranged in a sheet
    2. The mosaic of molecules is fluid; that is, the molecules are able to float around slowly

3. This model illustrates that the molecules of the cell membrane form a continuous sheet
C. Chemical attractions are the forces that hold membranes together
D. Groupings of membrane molecules form rafts that float as a unit in the membrane (Figure 3-4)
    1. Rafts may pinch inward to bring material into the cell or organelle
    2. Primary structure of a cell membrane is a double layer of phospholipid molecules
    3. Heads are hydrophilic (water loving)
    4. Tails are hydrophobic (water fearing)
    5. Arrange themselves in bilayers in water
    6. Cholesterol molecules are scattered among the phospholipids to allow the membrane to function properly at body temperature
    7. Most of the bilayer is hydrophobic; therefore water or water-soluble molecules do not pass through easily
E. Membrane proteins (Table 3-4)
    1. A cell controls what moves through the membrane by means of membrane proteins embedded in the phospholipid bilayer
    2. Some membrane proteins have carbohydrates attached to them and as a result form glycoproteins that act as identification markers

3. Some membrane proteins are receptors that react to specific chemicals, sometimes permitting a process called signal transduction

CYTOPLASM AND ORGANELLES

A. Cytoplasm—gel-like internal substance of cells that includes many organelles suspended in watery intracellular fluid called *cytosol*
B. Two major groups of organelles (Table 3-3)
  1. Membranous organelles are specialized sacs or canals made of cell membranes
  2. Nonmembranous organelles are made of microscopic filaments or other nonmembranous materials
C. Endoplasmic reticulum (Figure 3-5)
  1. Made of membranous-walled canals and flat, curving sacs arranged in parallel rows throughout the cytoplasm; extend from the plasma membrane to the nucleus
  2. Proteins move through the canals
  3. Two types of endoplasmic reticulum
    a. Rough endoplasmic reticulum
      (1) Ribosomes dot the outer surface of the membranous walls
      (2) Ribosomes synthesize proteins, which move toward the Golgi apparatus and then eventually leave the cell
      (3) Function in protein synthesis and intracellular transportation
    b. Smooth endoplasmic reticulum
      (1) No ribosomes border the membranous wall
      (2) Functions are less well established and probably more varied than for the rough endoplasmic reticulum
      (3) Synthesizes certain lipids and carbohydrates and creates membranes for use throughout the cell
      (4) Removes and stores Ca^{++} from the cell's interior.
D. Ribosomes (Figure 3-6)
  1. Many are attached to the rough endoplasmic reticulum and many lie free, scattered through the cytoplasm
  2. Each ribosome is a nonmembranous structure made of two pieces, a large subunit and a small subunit; each subunit is composed of rRNA
  3. Ribosomes in the endoplasmic reticulum make proteins for "export" or to be embedded in the plasma membrane; free ribosomes make proteins for the cell's domestic use
E. Golgi apparatus
  1. Membranous organelle consisting of cisternae stacked on one another and located near the nucleus (Figure 3-7)
  2. Processes protein molecules from the endoplasmic reticulum (Figure 3-8)
  3. Processed proteins leave the final cisterna in a vesicle; contents may then be secreted to outside the cell
F. Lysosomes (Figure 3-9)
  1. Made of microscopic membranous sacs that have "pinched off" from Golgi apparatus
  2. The cell's own digestive system; enzymes in lysosomes digest the protein structures of defective cell parts, including plasma membrane proteins, and particles that have become trapped in the cell
G. Proteasomes (Figure 3-10)
  1. Hollow, protein cylinders found throughout the cytoplasm
  2. Break down abnormal/misfolded proteins and normal proteins no longer needed by the cell (and which may cause disease)
  3. Break down protein molecules one at a time by tagging each one with a chain of ubiquitin molecules, unfolding it as it enters the proteasome, and then breaking apart peptide bonds
H. Peroxisomes
  1. Small membranous sacs containing enzymes that detoxify harmful substances that enter the cells
  2. Often seen in kidney and liver cells
I. Mitochondria (Figure 3-11)
  1. Made up of microscopic sacs; wall composed of inner and outer membranes separated by fluid; thousands of particles make up enzyme molecules attached to both membranes
  2. The "power plants" of cells; mitochondrial enzymes catalyze series of oxidation reactions that provide about 95% of a cell's energy supply
  3. Each mitochondrion has a DNA molecule, which allows it to produce its own enzymes and replicate copies of itself

NUCLEUS

A. Definition—spherical body in center of cell; enclosed by an envelope with many pores
B. Structure
  1. Consists of a nuclear envelope (composed of two membranes each with essentially the same molecular structure as the plasma membrane) surrounding nucleoplasm; the nuclear envelope has holes called *nuclear pores* (Figure 3-12)
  2. Contains DNA (heredity molecules), which appear as
    a. Chromatin threads or granules in nondividing cells
    b. Chromosomes in early stages of cell division
C. Functions of the nucleus are functions of DNA molecules; DNA determines both the structure and function of cells and heredity

CYTOSKELETON

A. The cell's internal supporting framework made up of rigid, rodlike pieces that provide support and allow movement and mechanisms that can move the cell or its parts (Figure 3-13)
B. Cell fibers
  1. Intricately arranged fibers of varying length that form a three-dimensional, irregularly shaped lattice
  2. Fibers appear to support the endoplasmic reticulum, mitochondria, and "free" ribosomes
  3. Smallest cell fibers are microfilaments (Figure 3-14)
    a. "Cellular muscles"
    b. Made of thin, twisted strands of protein molecules that lie parallel to the long axis of the cell
    c. Microfilaments can slide past each other and cause shortening of the cell
  4. Intermediate filaments are twisted protein strands slightly thicker than microfilaments; form much of the supporting framework in many types of cells

5. Microtubules are tiny, hollow tubes that are the thickest of the cell fibers; they are made of protein subunits arranged in a spiral fashion; their function is to move things around in the cell

C. Centrosome
   1. An area of the cytoplasm near the nucleus that coordinates the building and breaking of microtubules in the cell
   2. Nonmembranous structure also called the *microtubule-organizing center (MTOC)*
   3. Plays an important role during cell division
   4. The general location of the centrosome is identified by the centrioles

D. Cell extensions
   1. Cytoskeleton forms projections that extend the plasma membrane outward to form tiny, fingerlike processes
   2. There are three types of these processes; each has specific functions (Figure 3-15)
      a. Microvilli—found in epithelial cells that line the intestines and other areas where absorption is important; they help increase the surface area manyfold
      b. Cilia and flagella—cell processes that have cylinders made of microtubules at their core; cilia are shorter and more numerous than flagella; flagella are found only on human sperm cells

CELL CONNECTIONS

A. Cells are held together by fibrous nets that surround groups of cells (e.g., muscle cells), or cells have direct connections to each other

B. There are three types of direct cell connections (Figure 3-16)
   1. Desmosome
      a. Fibers on the outer surface of each desmosome interlock with each other; anchored internally by intermediate filaments of the cytoskeleton
      b. Spot desmosomes are like "spot welds" at various points connecting adjacent membranes
      c. Belt desmosomes encircle the entire cell like a collar
   2. Gap junctions—membrane channels of adjacent plasma membranes adhere to each other; have two effects:
      a. Form gaps or "tunnels" that join the cytoplasm of two cells
      b. Fuse two plasma membranes into a single structure

   3. Tight junctions
      a. Occur in cells that are joined by "collars" of tightly fused material
      b. Molecules cannot permeate the cracks of tight junctions
      c. Occur in the lining of the intestines and other parts of the body where it is important to control what gets through a sheet of cells

REVIEW QUESTIONS

1. What is the range in human cell diameters?
2. List the three main cell structures.
3. Describe the location, molecular structure, and width of the plasma membrane.
4. Explain the communication function of the plasma membrane, its transportation function, and its identification function.
5. Briefly describe the structure and function of the following cellular structures/organelles: endoplasmic reticulum, ribosomes, Golgi apparatus, mitochondria, lysosomes, proteasomes, peroxisomes, cytoskeleton, cell fibers, centrosome, centrioles, and cell extensions.
6. Describe the three types of intercellular junctions. What are the special functional advantages of each?
7. Describe briefly the functions of the nucleus and the nucleoli.
8. Name three kinds of micrography used in this book to illustrate cell structures. What perspective does each give that the other two do not?

CRITICAL THINKING QUESTIONS

1. Using the complementarity principle that cell structure is related to its function, discuss how the shapes of the nerve cell and muscle cell are specific to their respective functions.
2. What is the relationship among ribosomes, endoplasmic reticulum, Golgi apparatus, and plasma membrane? How do they work together as a system?

Physiology of Cells

CHAPTER OUTLINE

LANGUAGE OF SCIENCE

active site (AK-tiv site)

actual osmotic pressure (actual os-MOT-ik PRESH-ur) [*osmo-* impulse, *-ic* pertaining to]

aerobic (air-OH-bik) [*aer-* air, *-ic* pertaining to]

allosteric effector (al-o-STEER-ik ee-FECKT-or) [*allo-* another, *-ster-* solid, *-ic* pertaining to, *effect* to accomplish]

anabolism (ah-NAB-o-lizm) [*anabol-* to build up, *-ism* condition]

anaerobic (an-air-OH-bik) [*anaer-* without air, *-ic* pertaining to]

anaphase (AN-ah-fayz) [*ana-* again, *-phase* appearance]

apoptosis (app-o-TOH-sis *or* app-op-TOH-sis) [*apo-* away, *-ptosis* falling]

catabolism (kah-TAB-oh-liz-em) [*cata-* against, *-bol-* to throw, *-ism* condition]

catalyst (KAT-ah-list) [*cata* lower]

cellular respiration (SELL-yoo-lar res-pih-RAY-shun) [*cell* storeroom, *respire* to breathe]

citric acid cycle (SIT-rik ASS-id SYE-kul) [*acid* sour, *cycle* circle]

coenzyme (koh-EN-zyme) [*co-* together, *-en-* in, *-zyme* ferment]

complementary pairing (kom-pleh-MEN-tah-ree PAIR-ing) [*complement* that which competes]

concentration gradient (kahn-sen-TRAY-shun GRAY-dee-ent) [*con-* together, *-centr* center, *gradi* step]

cotransport (koh-TRANZ-port) [*co* together]

countertransport [*contra* against]

cytokinesis (sye-toe-kin-EE-sis) [*cyto-* cell, *-kinesis* activation]

deoxyribonucleic acid (dee-ok-see-rye-boh-noo-klay-ik ASS-id) [*deoxy-* containing a decreased amount of oxygen, *-nucleo,* nucleus (kernel), *acid* sour]

dialysis (dye-AL-i-sis) [*dia-* apart, *-lysis* loosening]

differentiate (dif-er-EN-shee-ayt) [*differ* carry apart]

diffusion (dih-FYOO-shun) [*diffuse* spread out]

diploid (DIP-loid) [*diplo-* double, *-loid* form]

electron transport system (ETS) (eh-LEK-tron TRANZ-port SIS-tem) [*electro* electricity]

Cont'd on p. 139

103

MOVEMENT OF SUBSTANCES THROUGH CELL MEMBRANES

If a cell is to survive, it must be able to move substances to places where they are needed. We already know one way that cells move organelles within the cytoplasm: pushing or pulling by the cytoskeleton. A cell must also be able to move various molecules in and out through the plasma membrane, as well as from one membranous compartment to another within the cell. In this part of the chapter, we briefly describe the basic mechanisms a cell uses to move substances across its membranes. As you read through this section, refer to Table 4-1, which summarizes the various transport mechanisms mentioned in the text.

Before beginning a discussion of individual processes, we must point out that membrane transport processes can be labeled as *passive* or *active*. Passive processes do not require any energy expenditure or "activity" of the cell membrane—the particles move by using energy that they already have. Active processes, on the other hand, do require the expenditure of metabolic energy by the cell. In active processes the transported particles are actively "pulled" across the membrane. Keep this distinction in mind as we explore the basic mechanisms of cell membrane transport.

Passive Transport Processes

Diffusion

Often, molecules simply spread or *diffuse* through the membranes. The term **diffusion** refers to a natural phenomenon caused by the tendency of small particles to spread out evenly within any given space. All molecules in a solution bounce around in short, chaotic paths. As they collide with one another, they tend to spread out, or diffuse. Think of the example of a lump of sugar dissolving in water (Figure 4-1). After the lump is placed in the water, the sugar molecules are very close to one another—there is a very high sugar concentration in the lump. As the sugar molecules dissolve, they begin colliding with one another and thus push each other away. Given enough time, the sugar molecules eventually diffuse evenly through the water.

Notice that during diffusion, molecules move from an area of high concentration to an area of low concentration. It is not surprising, then, that molecules tend to move from the side of the membrane with high concentration to the side of the membrane with a lower concentration of that molecule. Another way of stating this principle is to say that diffusion occurs down a **concentration gradient**. A concentration gradient is simply a measur-

Table 4-1	Some Important Transport Processes		
PROCESS	**DESCRIPTION**		**EXAMPLES**
Passive			
Simple diffusion	Movement of particles through the phospholipid bilayer or through channels from an area of high concentration to an area of low concentration—that is, down the concentration gradient		Movement of carbon dioxide out of all cells; movement of sodium ions into nerve cells as they conduct an impulse
Osmosis	Diffusion of water through a selectively permeable membrane in the presence of at least one impermeant solute (often involves both simple and channel-mediated diffusion)		Diffusion of water molecules into and out of cells to correct imbalances in water concentration
Channel-mediated passive transport (facilitated diffusion)	Diffusion of particles through a membrane by means of channel structures in the membrane (particles move down their concentration gradient)		Diffusion of sodium ions into nerve cells during a nerve impulse
Carrier-mediated passive transport (facilitated diffusion)	Diffusion of particles through a membrane by means of carrier structures in the membrane (particles move down their concentration gradient)		Diffusion of glucose molecules into most cells

Table 4-1	Some Important Transport Processes—cont'd		
PROCESS	**DESCRIPTION**		**EXAMPLES**
Active			
Pumping	Movement of solute particles from an area of low concentration to an area of high concentration (up the concentration gradient) by means of an energy-consuming pump structure in the membrane		In muscle cells, pumping of nearly all calcium ions to special compartments—or out of the cell
Phagocytosis (endocytosis)	Movement of cells or other large particles into cell by trapping it in a section of plasma membrane that pinches off to form an intracellular vesicle; a type of *vesicle-mediated transport*		Trapping of bacterial cells by phagocytic white blood cells
Pinocytosis (endocytosis)	Movement of fluid and dissolved molecules into a cell by trapping them in a section of plasma membrane that pinches off to form an intracellular vesicle; a type of *vesicle-mediated transport*		Trapping of large protein molecules by some body cells
Exocytosis	Movement of proteins or other cell products out of the cell by fusing a secretory vesicle with the plasma membrane; a type of *vesicle-mediated transport*		Secretion of the hormone prolactin by pituitary cells

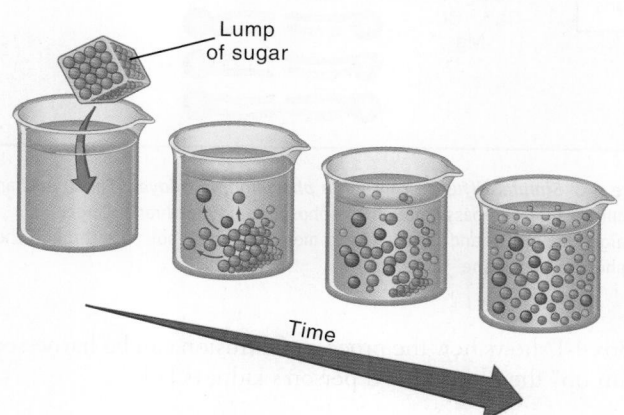

Figure 4-1 *Diffusion.* The molecules of a lump of sugar are very densely packed when they enter the water. As sugar molecules collide frequently in the area of high concentration, they gradually spread away from each other—toward the area of lower concentration. Eventually, the sugar molecules are evenly distributed.

able difference in concentration from one area to another. Because molecules spread from the area of *high* concentration to the area of *low* concentration, they spread down the concentration gradient.

Perhaps the best way to learn the principle of diffusion across a membrane is to look at the example illustrated in Figure 4-2. Suppose a 10% glucose solution is separated from a 20% glucose solution by an artificial membrane. Suppose further that the membrane has pores in it that allow glucose molecules to pass through, as well as pores that allow water molecules to pass. Glucose molecules and water molecules darting about the solution collide with each other and with the membrane. Some inevitably hit the membrane pores from the 20% glucose side, and some hit the membrane pores from the 10% side. Just as inevitably, some pass through the pores in both directions. For a while, more glucose molecules enter the pores from the 20% side simply because they are more numerous there than on the 10% side. More of these particles, therefore, move through the membrane from the 20% glucose solution into the 10% glucose solution than diffuse through it in the opposite direction. In other words, the overall di-

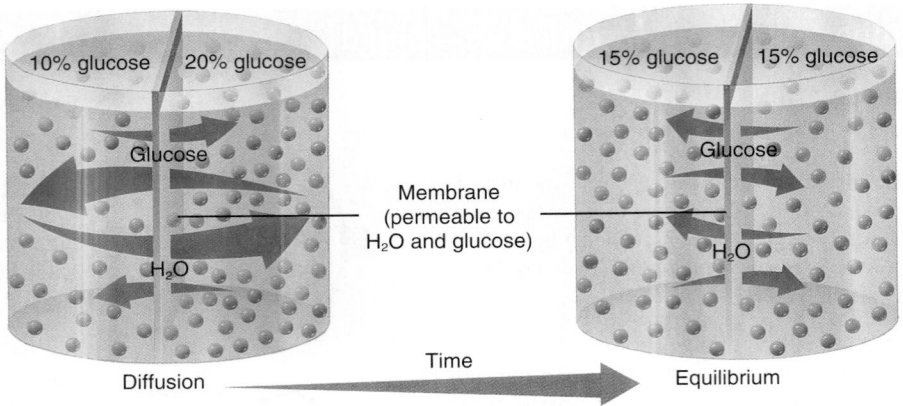

Figure 4-2 *Diffusion through a membrane.* Note that the membrane allows glucose and water to pass and that it separates a 10% glucose solution from a 20% glucose solution. The container on the left shows the two solutions separated by the membrane at the start of diffusion. The container on the right shows the result of diffusion after time.

rection of diffusion is from the side where the concentration is higher (20%) to the side where the concentration is lower (10%).

During the time that diffusion of glucose molecules is taking place, diffusion of water molecules is also going on. Remember, the direction of diffusion of any substance is always down that substance's concentration gradient. Water molecules are more concentrated on the 10% glucose solution side because the solution is more dilute—or watery—on that side. Thus, water molecules move from the 10% glucose solution side to the 20% glucose solution side. As Figure 4-2 shows, diffusion of both kinds of molecules eventually produces a dynamic form of equilibrium in which both solutions have equal concentrations. We say that *equilibration* has occurred.

Dynamic equilibrium is not a static state with no movement of molecules across the membrane. Instead, it is a balanced state in which the number of molecules of a substance bouncing to one side of the membrane exactly equals the number of molecules of that substance that are bouncing to the other side. Once equilibration has occurred, overall diffusion may have stopped, but balanced diffusion of small numbers of molecules may continue.

Simple Diffusion

Now that we know that concentration gradients drive diffusion, we can explore how the molecules actually find a way through a cell membrane. Sometimes molecules diffuse directly through the bilayer of phospholipid molecules that forms most of a cell membrane. As discussed in Chapter 3, lipid-soluble molecules can pass through easily. As Figure 4-3 shows, small hydrophobic molecules such as oxygen (O_2) and carbon dioxide (CO_2) can diffuse directly through the phospholipid bilayer, as can some smaller, uncharged particles such as water (H_2O) and urea. Such molecules simply dissolve in the phospholipid fluid, diffuse through this fluid, and then move into the water solution on the other side of the membrane. When molecules pass directly through the membrane, the process is called **simple diffusion.**

When molecules are allowed to cross a membrane, they are said to *permeate* the membrane. Thus, a particular membrane is *permeable* to a particular molecule only if it can pass through that membrane.

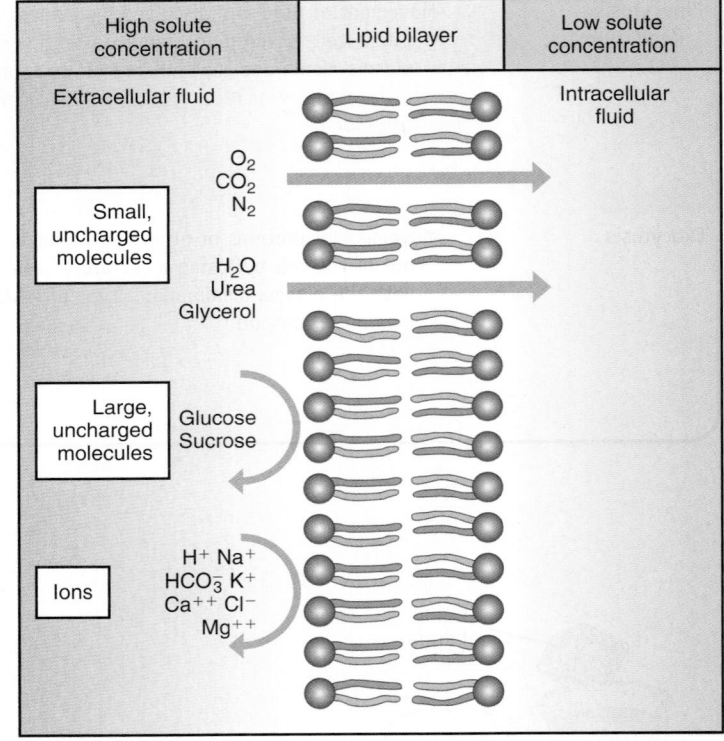

Figure 4-3 *Simple diffusion through a phospholipid bilayer.* Small, uncharged molecules can easily pass through the phospholipid membrane. Larger uncharged molecules and ions (charged molecules) may not pass through the phospholipid membrane, however.

Box 4-1 shows how the process of diffusion can be harnessed to "clean up" the blood after a person's kidneys fail.

Osmosis

A special case of diffusion is called **osmosis.** Osmosis is the diffusion of water through a selectively permeable membrane. Often, water is able to diffuse across a living membrane that does not allow diffusion of one or more other substances (see Figure 4-3). Thus, water can equilibrate its concentration on both

BOX 4-1   Dialysis

Under certain circumstances, a type of diffusion called **dialysis** may occur. Dialysis is a form of diffusion in which the selectively permeable nature of a membrane causes the separation of smaller solute particles from larger solute particles. *Solutes* are the particles dissolved in a *solvent* such as water. Together, the solutes and solvents form a mixture called a *solution*.

Part *A* of the figure illustrates the principle of dialysis. A bag made of dialysis membrane—material with microscopic pores—is filled with a solution containing glucose, water, and albumin (protein) molecules and immersed in a container of pure water. Both water and glucose molecules are small enough to pass through the pores in the dialysis membrane. Albumin molecules, like all protein molecules, are very large and do not pass through the membrane's pores. Because of differences in concentration, glucose molecules diffuse out of the bag as small water molecules diffuse into the bag. Despite a concentra-

tion gradient, the albumin molecules do not diffuse out of the bag. Why not? Because they simply will not fit through the tiny pores in the membrane. After some time has passed, the large solutes are still trapped within the bag, but most of the smaller solutes are outside of it.

The principle of dialysis can be used in medicine to treat patients with kidney failure. In **hemodialysis** (part *B* of the figure), blood pumped from a patient is exposed to a dialysis membrane that separates the blood from a clean, osmotically balanced dialysis fluid. Small solutes such as urea and various ions can diffuse through the membrane to reach an equilibrium, thus removing them from the blood. The larger plasma proteins (including albumin) and blood cells remain in the blood, which is returned to the patient's body. In hemodialysis, the process of dialysis is used to "clean up" the patient's blood because the kidneys have failed to perform this task.

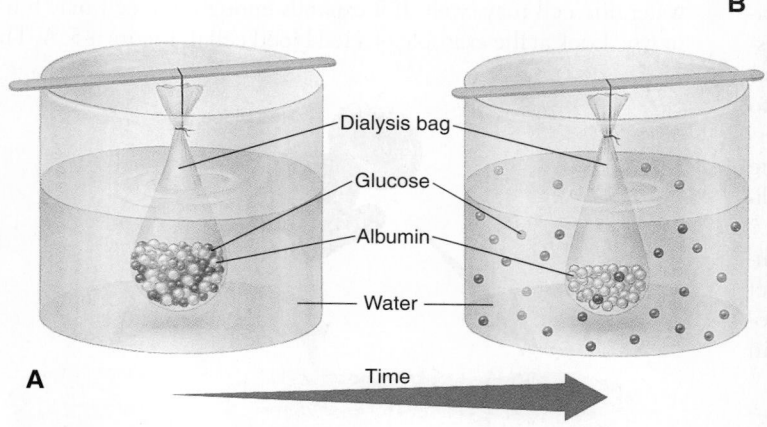

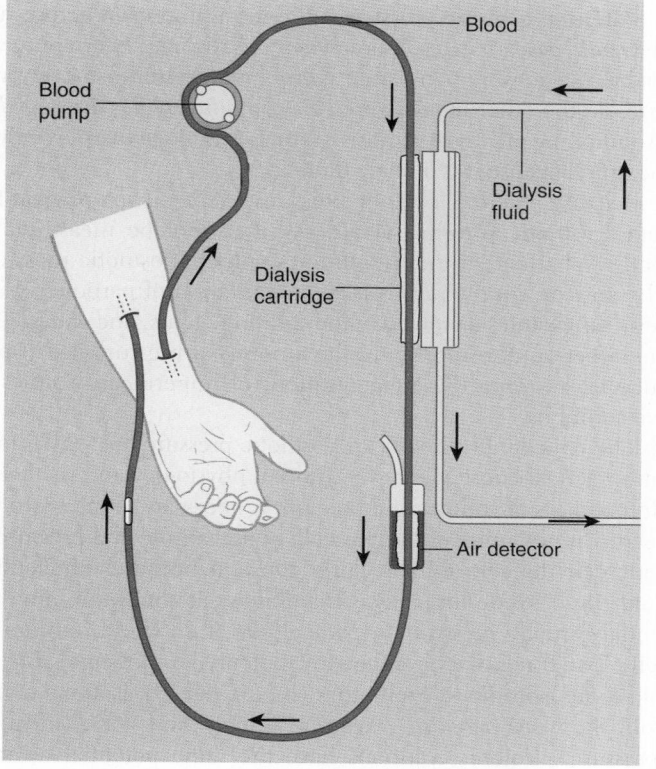

A, Dialysis. A dialysis bag containing glucose, water, and albumin (protein) molecules is suspended in pure water. Over time, the smaller solute molecules (glucose) diffuse out of the bag. The larger solute molecules (albumin) remain trapped in the bag because the bag is impermeable to them. Thus, dialysis is diffusion that results in separation of small and large solute particles. **B, Hemodialysis.** In hemodialysis, the patient's blood is pumped through a dialysis cartridge, which has a semipermeable membrane that separates the blood from the clean dialysis fluid. As dialysis occurs, some of the urea and other small solutes in the blood diffuse into the dialysis fluid whereas the larger solutes (plasma proteins) and blood cells remain in the blood.

sides of the membrane but *impermeant* solutes cannot. This property can have very important consequences in cells, as we shall see.

First, let us look at an example of osmosis. Imagine that you have a 10% albumin solution separated by a membrane from a 5% albumin solution (Figure 4-4). Assume that the membrane is freely permeable to water but impermeable to albumin. The water molecules diffuse or *osmose* through the membrane from the area of high water concentration to the area of low water concentration. That is, the water moves from the more dilute 5% albumin solution to the less dilute 10% albumin solution. Although equilibrium is eventually reached, the albumin does not diffuse

across the membrane. Only the water diffuses. Because of this osmosis, one solution loses volume and the other solution gains volume (see Figure 4-4).

Unlike the open container pictured in Figure 4-4, cells are closed containers. They are enclosed by their plasma membranes. Actually, most of the body is composed of closed compartments such as cells, blood vessels, tubes, and bladders. In closed compartments, such as a toy water balloon, changes in volume also mean changes in pressure. Adding volume to a cell by osmosis increases its pressure, just as adding volume to a water balloon increases its pressure. Water pressure that develops in a solution as a result of osmosis into that solution is called **osmotic pressure**.

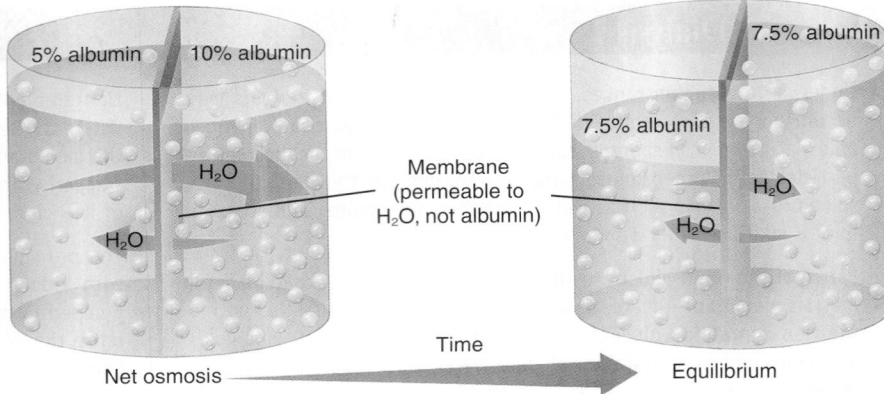

Figure 4-4 *Osmosis.* Osmosis is the diffusion of water through a selectively permeable membrane. The membrane shown in this diagram is permeable to water but not to albumin. Because there are relatively more water molecules in 5% albumin than 10% albumin, more water molecules osmose from the more dilute into the more concentrated solution (as indicated by the larger arrow in the left diagram) than osmose in the opposite direction. The overall direction of osmosis, in other words, is toward the more concentrated solution. Net osmosis produces the following changes in these solutions: their concentrations equilibrate, the volume and pressure of the originally more concentrated solution increase, and the volume and pressure of the other solution decrease proportionately.

Taking this principle a step further, we can state that osmotic pressure develops in the solution that originally has the higher concentration of impermeant solute.

Potential osmotic pressure is the maximum osmotic pressure that *could develop* in a solution when it is separated from pure water by a selectively permeable membrane. **Actual osmotic pressure,** on the other hand, is pressure that *already has developed* in a solution by means of osmosis. Actual osmotic pressure is easy to measure because it is already there.

Because potential osmotic pressure is a *prediction* of what the actual osmotic pressure would be, it cannot be measured directly. What determines a solution's potential osmotic pressure? The answer, simply put, is the concentration of particles of impermeant solutes dissolved in the solution. Thus, one can predict the direction of osmosis and the amount of pressure it will produce by knowing the concentrations of impermeant solutes in two solutions.

The concept of osmosis and osmotic pressure has very important practical consequences in human physiology and medicine. Homeostasis of volume and pressure is necessary to maintain the healthy functioning of human cells. The volume and pressure of body cells tend to remain fairly constant because intracellular fluid (fluid inside the cell) is maintained at about the same potential osmotic pressure as extracellular fluid (fluid outside the cell). Two fluids that have the same potential osmotic pressure are said to be **isotonic** to each other (Figure 4-5, *B*). Isotonic comes from the word parts *iso-*, meaning "same," and *-tonic*, meaning "pressure." Isotonic solutions have the same potential osmotic pressure because they have the same concentration of impermeant solutes.

A human cell placed in a concentrated solution of impermeant solutes will shrivel up. Look at the example of a red blood cell in Figure 4-5, *C*. The pictured cell is in a solution with a higher concentration of impermeant solutes than in the cell and, therefore, a higher potential osmotic pressure. The extracellular solution is said to be **hypertonic** (higher pressure) to the intracellular solution. Cells placed in solutions that are hypertonic to intracellular fluid always shrivel. If cells shrivel too much, they may become permanently damaged—or even die. Obviously, this fact is medically important. Large amounts of solutes cannot be introduced into the body without considering the effect they will have on the concentration of impermeant solutes in the ex-

tracellular fluid. If a treatment or procedure causes extracellular fluid to become hypertonic to the cells of the body, serious damage may occur.

If a human cell is placed in a very dilute solution, such as pure water, the cell may swell. If it expands enough, the cell may burst, or *lyse.* Look at the example of a red blood cell in Figure 4-5, *A.* This

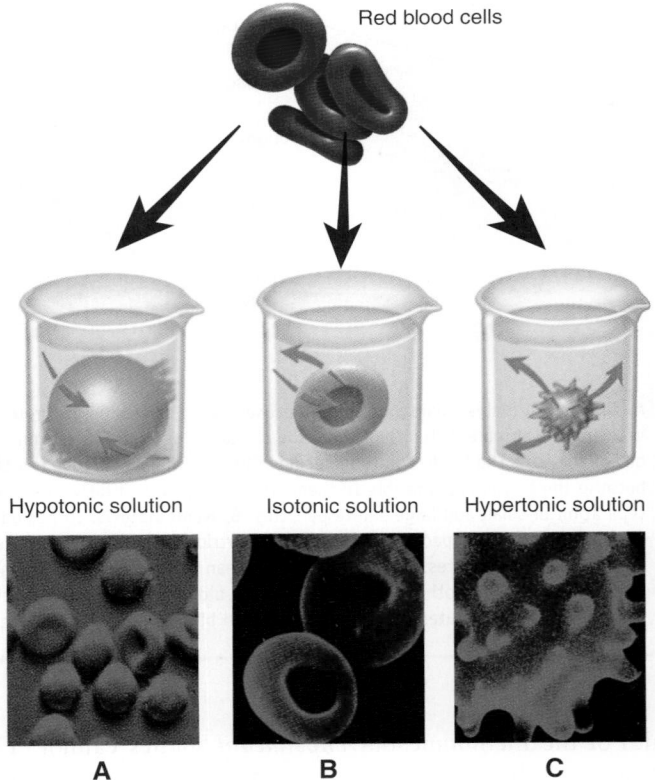

Figure 4-5 *Effects of osmosis on cells.* **A,** Normal red blood cells placed in a hypotonic solution may swell (as the *scanning electron micrograph* shows) or even burst (as the *drawing* shows). This change results from the inward diffusion of water (osmosis). **B,** Cells placed in an isotonic solution maintain constant volume and pressure because the potential osmotic pressure of the intracellular fluid matches that of the extracellular fluid. **C,** Cells placed in a solution that is hypertonic to the intracellular fluid lose volume and pressure as water osmoses out of the cell into the hypertonic solution. The "spikes" seen in the scanning electron micrograph are rigid microtubules of the cytoskeleton. These supports become visible as the cell "deflates."

cell is placed in a solution that is **hypotonic** (lower pressure) to the intracellular fluid. Hypotonic solutions tend to lose pressure because they have a lower concentration of impermeant solutes, and thus a higher water concentration, than the opposite solution. In this example, the extracellular fluid is *hypotonic* to intracellular fluid, which in turn is *hypertonic* to the extracellular fluid. Water always osmoses from the hypotonic solution to the hypertonic solution.

In summary, we can make several generalizations about osmosis. First, osmosis is the diffusion of water across a membrane that limits the diffusion of at least some of the solute molecules. That is, at least one impermeant solute must be present. Second, osmosis results in the gain of volume (and thus pressure) on one side of the membrane and loss of volume (and pressure) on the other side of the membrane. Third, the direction of osmosis and the resulting changes in pressure can be predicted if you know the potential osmotic pressure or *tonicity* of each solution separated by the membrane.

Facilitated Diffusion

For a long time, biologists thought that simple diffusion was the only way that molecules could diffuse through a cell membrane. They found that water-soluble molecules such as sodium ions (Na^+) and glucose molecules could not pass through an artificial phospholipid bilayer easily (Figure 4-3). However, they also found that indeed such small, water-soluble molecules could pass through living cell membranes quickly. Even water molecules, which are small enough to pass through the thin phospholipid membrane with some ease, diffuse even more rapidly through most living cell membranes. It was not until the presence of various transport proteins, such as membrane channels and membrane carriers, was discovered that we understood how these molecules diffuse across cell membranes. These membrane transporters enable a kind of mediated passive transport that is often called **facilitated diffusion.**

Channel-Mediated Passive Transport

As you already know, cell membranes possess protein "tunnels," better known as *membrane channels* (see Table 3-4, p. 87, and Figure 4-6). Membrane channels are pores through which specific ions or other small, water-soluble molecules can pass. For example, sodium ions (Na^+) pass only through *sodium channels* and chloride ions (Cl^-) pass only through *chloride channels*. Membrane channels can exhibit such *specificity* because their molecular structure prevents molecules of the wrong shape and the wrong pattern of charges to pass through the channel. Thus, living membranes can be permeable to some molecules but not to others, depending on the type of channels present.

As Figure 4-6 shows, the permeability of a membrane can also be affected by the opening and closing of membrane channels. Because channels can open or close, they are sometimes called *gated channels*. The active or "open" state can be almost immediately changed to the "closed" state, or changed from closed to open, by various triggering mechanisms. Some gated channels are triggered by electrical changes (voltage), others by light, and still others by mechanical or chemical stimuli. We will be discussing these various types of triggering mechanisms in later chapters. Open gated channels may under certain conditions become inactive, stopping the flow of molecules and becoming insensitive to trigger stimuli, before actually closing and resuming sensitivity to stimuli.

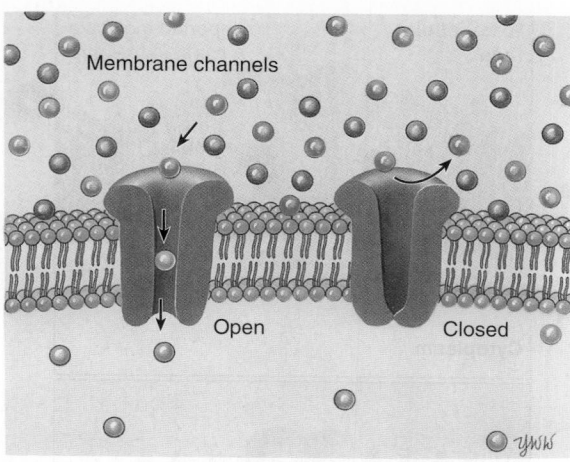

Figure 4-6 *Membrane channels.* Gated channel proteins form tunnels through which only specific molecules may pass—as long as the "gates" are open. Molecules that do not have a specific shape and charge are never permitted to pass through the channel. Notice that the transported molecules move from an area of high concentration to an area of low concentration.

Because a living cell membrane can limit the diffusion of some molecules by opening or closing channels in different situations, we say the membrane is *selectively permeable*. Membrane channel structure often permits diffusion in only one direction. So the cell can also determine whether to allow certain molecules to pass in either direction (depending on the concentration gradient, of course) or in only one direction (when the concentration gradient permits).

Water channels, also known as *aquaporins*, are among the more recently discovered types of membrane channels. As their name suggests, these channels permit water molecules to diffuse through a cell membrane much more rapidly than by simple diffusion. Aquaporins are thought to be responsible for the very rapid changes in blood cell volume during osmosis illustrated in Figure 4-5.

Because ions move down their concentration gradients as they pass through channels, this type of facilitated diffusion is passive and thus called *channel-mediated passive transport*.

Carrier-Mediated Passive Transport

Molecules may move down their concentration gradient by passing through a different type of membrane transporter called a *membrane carrier*. Thus, the carrier may facilitate diffusion in a process called *carrier-mediated passive transport*.

As Figure 4-7 shows, the carrier structure attracts a solute to a binding site, changes its shape, and then releases the solute to the other side of the membrane. This mechanism differs from channel-mediated transport, which does not involve binding the solute molecule and changing shape to release the bound solute. Carrier reactions are reversible and may thus transport molecules in either direction, depending on the concentration gradient.

As in channel-mediated and simple diffusion, carrier-mediated diffusion also transports substances down a concentration gradient (that is, from high to low concentration).

Diffusion, whether simple or facilitated, is a passive process. In other words, the energy for transport through a membrane does not come from the membrane but from the energy of collision already possessed by the moving molecule. The only requirement of

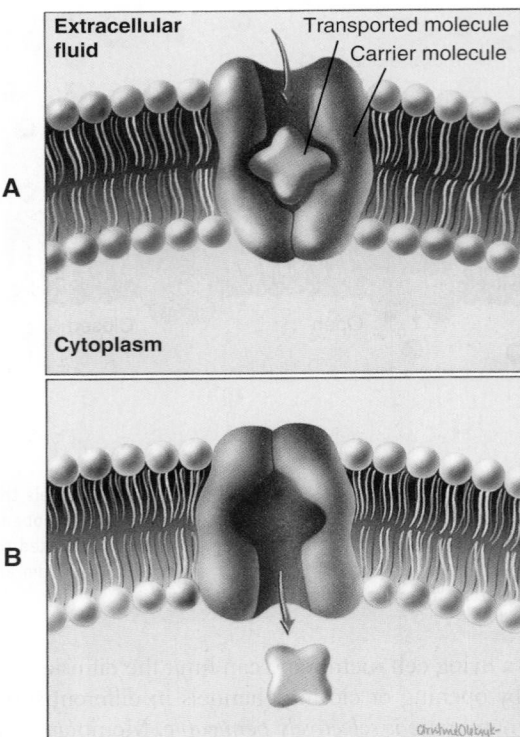

Figure 4-7 *Membrane carrier.* In carrier-mediated transport, a membrane-bound carrier protein attracts a solute molecule to a binding site **(A)** and changes shape in a manner that allows the solute to move to the other side of the membrane **(B).** Passive carriers may transport molecules in either direction, depending on the concentration gradient.

the cell is that it be permeable to the type of molecule in question. Filtration, another type of passive transport process, is discussed in Box 4-2 on p. 110.

Active Transport Processes

All the membrane transport processes that we have discussed so far are passive processes. The force of movement comes from the concentration gradient—that is, from a physical force of nature. The driving force for active transport processes, on the other hand, comes from the cell itself. Energy of metabolism must be used by cells to force particles across a membrane that otherwise would not move across.

Transport by Pumps

Membrane transporters called *membrane pumps* carry out a transport process in which cellular energy is used to move molecules "uphill" through a cell membrane. By "uphill," we mean that the substance moves from an area of low concentration to an area of higher concentration. An actively transported substance moves *against* its concentration gradient. This is exactly the opposite of diffusion, in which a substance is transported *down* its concentration gradient—or "downhill." It is important to remember that molecules will not travel "uphill" on their own, anymore than a ball will roll uphill by itself. Molecules will travel uphill only when they are forced by pump mechanisms powered by cellular energy. Moving solutes in different directions is discussed in Box 4-3 on p. 112.

BOX 4-2   Filtration

Another important passive process for transport in the body is **filtration.** This form of transport involves the passing of water and permeable solutes through a membrane by the force of hydrostatic pressure. *Hydrostatic pressure* is the force, or weight, of a fluid pushing against a surface.

Filtration is movement of molecules through a membrane from an area of high hydrostatic pressure to an area of low hydrostatic pressure—that is, down a hydrostatic pressure gradient. Filtration most often transports substances through a sheet of cells. The force of pressure pushes the molecules through or between the cells that form the sheet. Because the filtration membrane does not allow larger particles through, filtration results in the separation of large and small particles, as you can see in the figure. This is similar to dialysis (Box 4-1, p. 107), except that dialysis is driven by a *concentration gradient*. Filtration is instead driven by a *hydrostatic pressure gradient*.

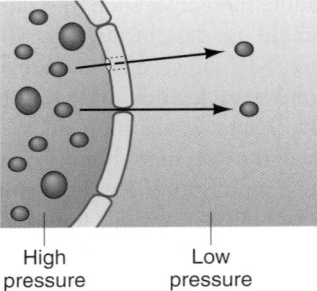

High   Low
pressure pressure

Filtration. Particles small enough to fit through the pores in the filtration membrane move from the area of high hydrostatic pressure to the area of low hydrostatic pressure. This results in separation of small particles from larger particles.

A simple model of filtration is found in many drip-type coffee makers. Ground coffee is placed in a porous paper filter cup in an upper container and boiling water is added. Gravity pulls downward on the mixture in the upper container, generating hydrostatic pressure against the bottom of the filter. The pores in the paper filter are large enough to let water molecules and other small particles pass through to a coffee pot below the filter. Most of the coffee grounds are too large to pass through the filter and thus cannot be filtered. The coffee in the pot below is called the *filtrate*.

How and where does filtration occur in the body? Most often, it occurs in tiny blood vessels called *capillaries,* which are found throughout the body. Hydrostatic pressure of the blood (blood pressure) generated by heart contractions, gravity, and other forces pushes water and small solutes out of the capillaries and into the interstitial spaces of a tissue. Blood cells and large blood proteins are too large to fit through pores in the capillary wall; therefore, they cannot be filtered out of the blood. Capillary filtration allows the blood vessels to supply tissues with water and other essential substances quickly and easily without losing its cells and blood proteins. Capillary filtration is also the first step used by the kidney to form urine.

Active pumping is an extremely important process. It allows cells to move certain ions or other water-soluble particles to specific areas. For example, active *calcium pumps* in the membranes of muscle cells allow the cell to force nearly all of the intracellular calcium ions (Ca^{++}) into special compartments—or out of the cell entirely (Figure 4-8). This is important because a muscle cell cannot operate properly unless the intracellular Ca^{++} concentration is kept low during rest. Other cells use active transport pumps for similar purposes—that is, to create a concentration gradient of a particular solute.

One type of active transport pump, the *sodium-potassium pump*, operates in the plasma membrane of all human cells (Figure 4-9). It is essential for healthy cell survival. As its name suggests, the sodium-potassium pump actively transports sodium ions (Na^+) and potassium ions (K^+)—but in opposite directions. It transports sodium ions *out* of cells and potassium ions *into* cells. By so doing, the sodium-potassium pump maintains a lower sodium concentration in intracellular fluid than in the surrounding extracellular fluid. At the same time, this pump maintains a higher potassium concentration in the intracellular fluid than in the surrounding extracellular fluid. Both ions bind to the same membrane transporter, a molecule known as *sodium-potassium adenosine triphosphatase (Na-K ATPase)*. Figure 4-9 shows that three Na^+ ions bind to sodium-binding sites on the pump's inner face. At the same time, an energy-containing adenosine triphosphate (ATP) molecule produced by the cell's mitochondria binds to the pump. The ATP breaks apart, and its stored energy is transferred to the pump. The pump then changes shape, releases the three Na^+ ions to the outside of the cell, and attracts two K^+ ions

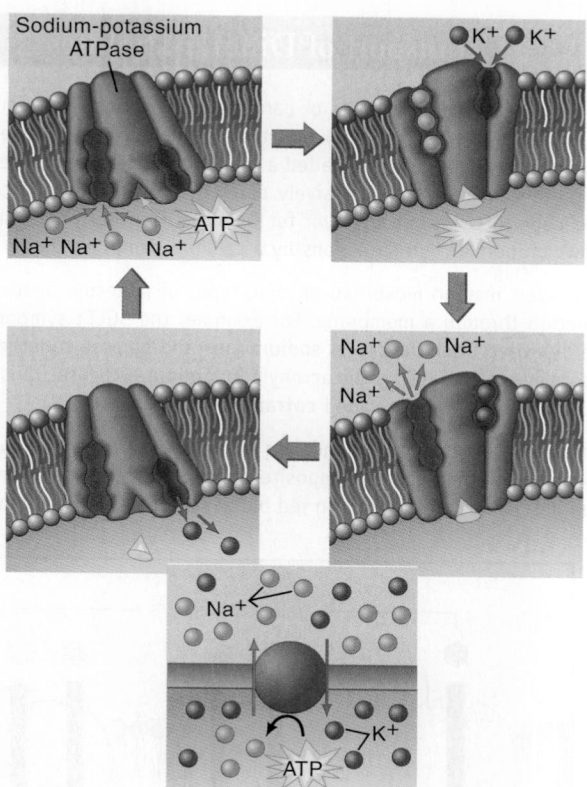

Figure 4-9 *Sodium-potassium pump.* Three sodium ions (Na^+) bind to sodium binding sites on the pump's inner face. At the same time, an energy-containing adenosine triphosphate (ATP) molecule produced by the cell's mitochondria binds to the pump. The ATP breaks apart, and its stored energy is transferred to the pump. The pump then changes shape, releases the three Na^+ ions to the outside of the cell, and attracts two potassium ions (K^+) to its potassium binding sites. The pump then returns to its original shape, and the two K^+ ions and the remnant of the ATP molecule are released to the inside of the cell. The pump is now ready for another pumping cycle. *ATPase,* Adenosine triphosphatase. The small inset is a simplified view of Na^+-K^+ pump activity.

to its potassium binding sites. The pump then returns to its original shape and releases the two K^+ ions and the remnant of the ATP molecule to the inside of the cell.

Transport by Vesicles

Like active transport pumps, mechanisms that carry large groups of molecules into or out of the cell by means of vesicles require the expenditure of metabolic energy by the cell. Such bulk transport mechanisms differ from pump mechanisms in that they allow substances to enter or leave the interior of a cell without actually moving through its plasma membrane (Figure 4-10).

Endocytosis

In **endocytosis** the plasma membrane "traps" some extracellular material and brings it into the cell. The basic mechanism of endocytosis is summarized in Figure 4-10. In endocytosis, the cytoskeleton does all the work by pulling part of the plasma membrane inward, thereby forming a depression, while at the same time pushing the membrane at the edges to form a sort of trap for extracellular material. When the extended edges of membrane

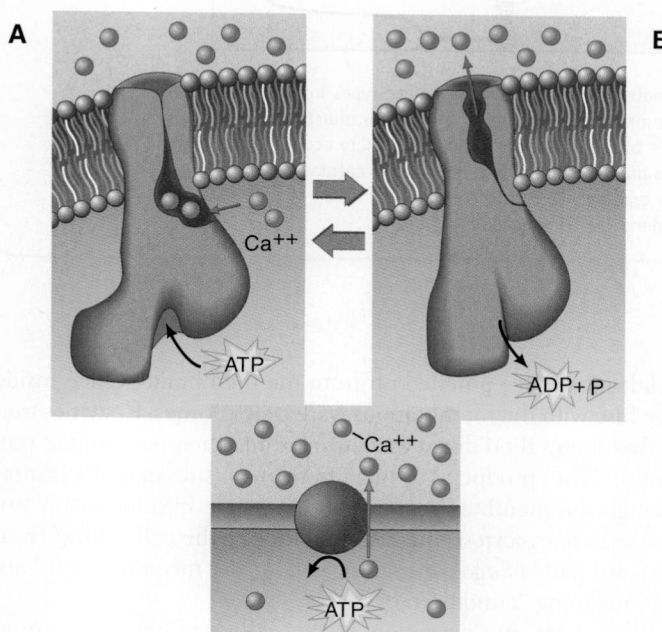

Figure 4-8 *Calcium pump.* **A,** Two calcium ions (Ca^{++}) enter the pump, and then ATP associates with the activating center of the pump. **B,** As the energy released from ATP (forming ADP+P) changes the shape of the pump, the Ca^{++} ions are released on the opposite side of the membrane. The small inset shows a simplified view of a calcium pump's action.

BOX 4-3   Transport of Different Solutes

In looking at different types of carriers and pumps in the body, we see that some transport only one type of molecule at a time. This type of transporter is often called a *uniporter*. For example, the GLUT uniporters in many cells passively move glucose from blood plasma into cells. *GLUT* is an acronym for *GLUcose Transporter*. Figure 4-8 shows uniport of two Ca^{++} ions by a calcium pump.

Symporters instead move two or more types of molecule in the same direction through a membrane. For example, the SGLT1 symporter in the digestive tract transports sodium ions and glucose together into absorptive cells. *SGLT1* is the acronym for *Sodium-GLucose Transporter 1*. Symport can also be called **cotransport.**

Antiporters, on the other hand, are transporters that move two different types of molecule in opposite directions at the same time. For example, *Band 3* antiporters in red blood cells passively exchange bi-

carbonate ions (HCO_3^-) for chloride ions (Cl^-) in opposite directions at the same time. Na^+-K^+-ATPase (sodium-potassium pump) actively antiports sodium ions and potassium ions in all cells of the body. Antiport can also be called **countertransport.**

Sometimes, active transport of one type of solute creates a concentration gradient that drives the passive transport of another solute. For example, in part *B* of the figure the active transport of sodium ions creates a concentration gradient that drives the cotransport of glucose along with sodium by a symport mechanism. In this case, the movement of sodium is an example of *primary active transport*. The movement of glucose, which depends on the sodium concentration gradient created by primary active transport, is an example of *secondary active transport*. This particular example of secondary active transport is often referred to as *sodium cotransport of glucose.*

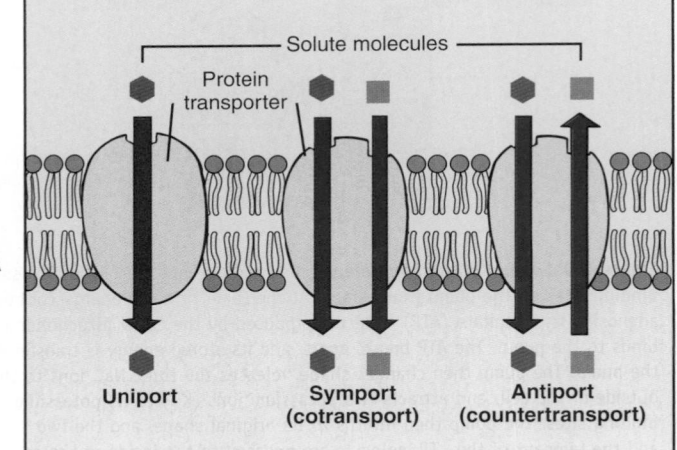

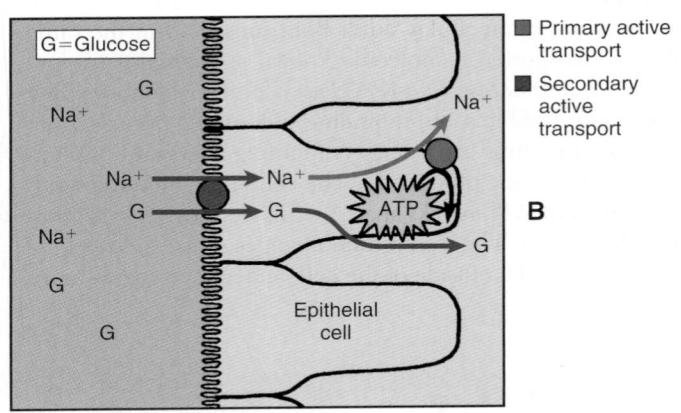

A, Direction of transport. Movement of one solute (uniport), movement of two or more solute types in the same direction (symport or cotransport), and movement of two or more solute types in opposite directions (antiport or countertransport). **B,** Primary and secondary active transport. In this example, primary active transport of sodium by a sodium pump creates a concentration gradient that drives the passive cotransport of glucose along with sodium. Because it depends on the sodium gradient, sodium cotransport of glucose is an example of secondary active transport. *ATP,* Adenosine triphosphate.

fuse, a vesicle is formed. The cytoskeleton then pulls the vesicle containing extracellular material inward.

In a type of endocytosis called *receptor-mediated endocytosis*, receptors in the plasma membrane first bind to specific molecules in the extracellular fluid (Figure 4-11). This causes a portion of the plasma membrane to be pulled inward by the cytoskeleton and form a small pocket around the material to be moved into the cell. The edges of the membranous pocket extend and eventually fuse to form a vesicle. The vesicle is then pulled inward—away from the plasma membrane—by the cytoskeleton. Sometimes, endocytosis picks up various molecules and other particles along with receptor-bound molecules.

There are two basic forms of endocytosis: **phagocytosis** and **pinocytosis.** In phagocytosis, microorganisms or other large particles are engulfed by the plasma membrane and enter the cell in

vesicles that have pinched off from the membrane. Once inside, they fuse with the membranous walls of lysosomes. Enzymes from the lysosomes then digest the particles into their component molecules. The products of digestion may subsequently diffuse through the membranous wall of the vesicle into the cytoplasm. The term *phagocytosis* means "condition of the cell eating" (from the word parts *phago-*, meaning "eat," *-cyto-*, meaning "cell," and *-osis*, meaning "condition").

Pinocytosis, or "condition of the cell drinking," is a similar process in which fluid and the substances dissolved in it enter a cell. Besides providing a way for a cell to bring fluids and solutes into the interior of the cell, pinocytosis also provides a way for the cell to remove material, including membrane receptors and transporters, from the plasma membrane. A cell can thus regulate the function of its plasma membrane (see Table 4-1).

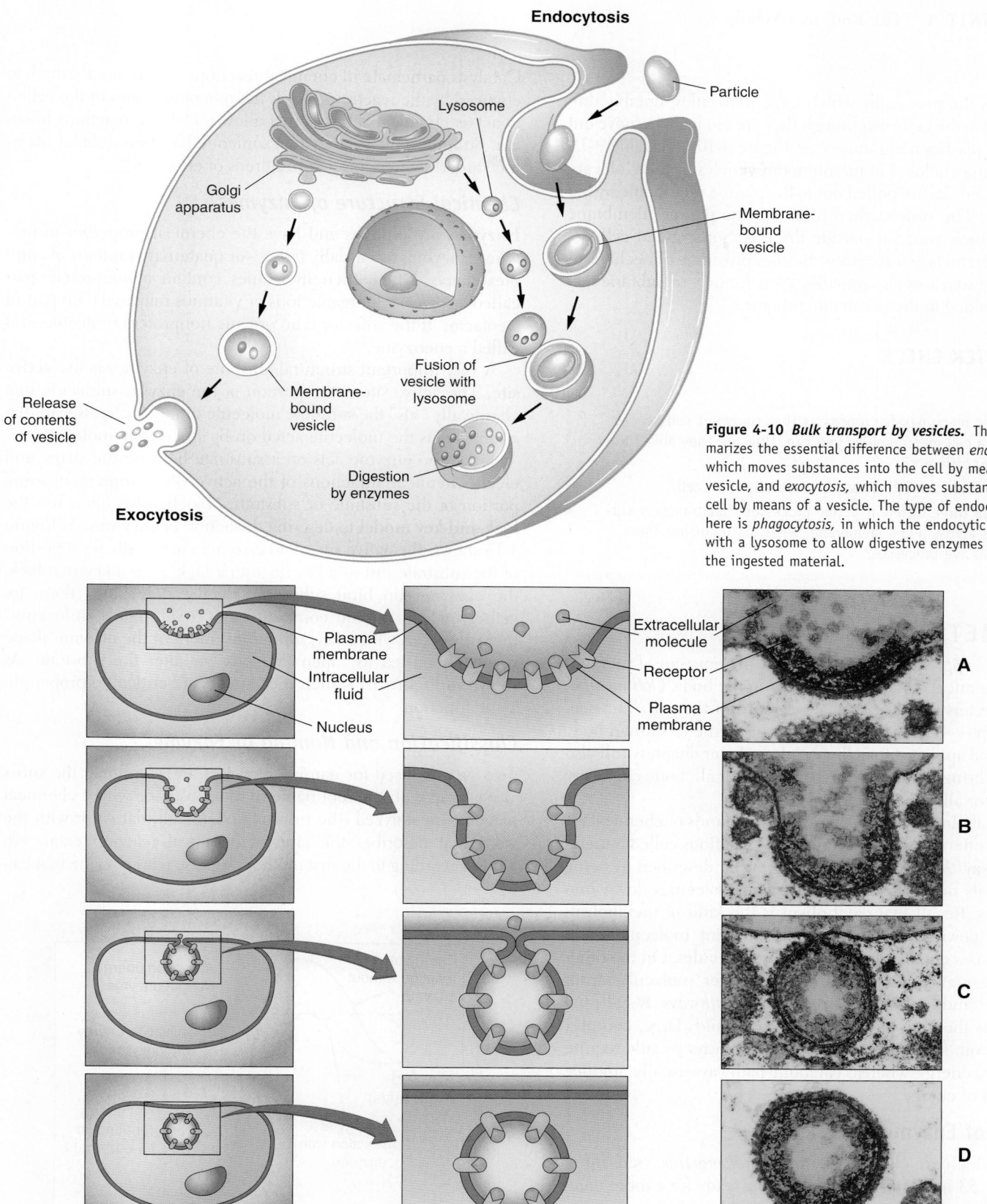

Endocytosis

Particle

Lysosome

Golgi apparatus

Membrane-bound vesicle

Membrane-bound vesicle

Fusion of vesicle with lysosome

Release of contents of vesicle

Digestion by enzymes

Exocytosis

Figure 4-10 *Bulk transport by vesicles.* This sketch summarizes the essential difference between *endocytosis,* which moves substances into the cell by means of a vesicle, and *exocytosis,* which moves substances out of the cell by means of a vesicle. The type of endocytosis shown here is *phagocytosis,* in which the endocytic vesicle fuses with a lysosome to allow digestive enzymes to break down the ingested material.

Plasma membrane

Intracellular fluid

Nucleus

Extracellular molecule

Receptor

Plasma membrane

A

B

C

D

Figure 4-11 *Receptor-mediated endocytosis.* An artist's interpretation *(left* and *center)* and transmission electron micrographs *(right)* show the basic steps of receptor-mediated endocytosis. **A,** Membrane receptors bind to specific molecules in the extracellular fluid. **B,** A portion of the plasma membrane is pulled inward by the cytoskeleton and forms a small pocket around the material to be moved into the cell. **C,** The edges of the pocket eventually fuse and form a vesicle. **D,** The vesicle is then pulled inward—away from the plasma membrane—by the cytoskeleton. In this example, only the receptor-bound molecules enter the cell. In some cases, some free molecules or even entire cells may also be trapped within the vesicle and transported inward.

Exocytosis

Exocytosis is the process by which large molecules, notably proteins, can leave the cell even though they are too large to move out through the plasma membrane (see Figure 4-10 and Table 4-1). After first being enclosed in membranous vesicles by the Golgi apparatus, the vesicles are pulled out to the plasma membrane by the cytoskeleton. The vesicles then fuse with the plasma membrane and release their contents outside the cell. Some gland cells secrete their products by exocytosis. Besides providing a mechanism of transport, exocytosis also provides a way for new membrane material to be added to the plasma membrane.

QUICK CHECK

1. Name as many passive processes that transport substances across a cell membrane as you can. How are they alike? How are they different?
2. What causes osmotic pressure to develop in a cell?
3. Describe three different active processes that transport substances across a cell membrane. What distinguishes them from passive processes?

CELL METABOLISM

In Chapter 2 (pp. 45-46) we introduced the concept of **metabolism**: the chemical reactions that occur in the body. *Cell metabolism*, then, refers to the chemical reactions of the cell. This section of Chapter 4 picks up the important theme of human body chemistry and applies it to cell physiology. Later chapters will also continue to bring up this theme because after all, body chemistry is the basis for all human functions.

Cell metabolism involves many different kinds of chemical reactions that often occur in a sequence of reactions called a **metabolic pathway**. A metabolic pathway can be described as being *catabolic* if its net effect is to break large molecules down into smaller ones. Recall that **catabolism** is the kind of metabolism that breaks down molecules, usually nutrient molecules, and thereby releases energy from the broken molecules. On the other hand, some metabolic pathways build larger molecules from smaller ones and are thus called *anabolic pathways*. Recall that **anabolism** is the kind of metabolism that builds large, complex molecules from smaller ones. Anabolic pathways usually require a net input of energy, whereas catabolic pathways usually produce a net output of energy.

The Role of Enzymes

Enzymes, which are classified as *functional proteins*, were introduced on p. 53 in Chapter 2. We are now ready for a more comprehensive introduction to enzymes.

The series of chemical reactions that make up a metabolic pathway in a cell do not usually just happen on their own. At normal body temperatures, the *activation energy* needed to start a chemical reaction is too great for many molecules to react by themselves. What is needed to make essential chemical reactions happen is a **catalyst**—a chemical that reduces the amount of activation energy needed to start a chemical reaction (Figure 4-12).

Catalysts participate in chemical reactions but are not themselves changed by the reaction. This is the role of enzymes in the cell—to act as chemical catalysts that allow metabolic reactions to occur. So important are they that someone has even defined life as the "orderly functioning of hundreds of enzymes."

Chemical Structure of Enzymes

Enzymes are proteins and have the chemical properties of proteins. Enzymes are usually tertiary or quaternary proteins of complex shape. Often, their molecules contain a nonprotein part called a *cofactor*. Inorganic ions or vitamins may make up part of a cofactor. If the cofactor is an organic nonprotein molecule, it is called a **coenzyme.**

A very important structural attribute of enzymes is the **active site.** The active site is the portion of the enzyme molecule that chemically "fits" the substrate molecule or molecules. Recall that a *substrate* is the molecule acted on by an enzyme molecule.

Since the enzyme acts on a substrate because the shape and electrochemical attractions of the active site complement some portion of the substrate or substrates, biochemists often use the **lock-and-key model** to describe the action of enzymes. As Figure 4-13 shows, the active site of an enzyme chemically fits a portion of the substrate just as a key fits into a lock. Like a key in a lock, the enzyme can bind substrates together ("locking" them together) or can unbind components of a substrate ("unlocking" them). And, as with a key, some movement of the enzyme shape is often required to "open the lock" or alter the substrate. As shown in Figure 4-13, such movements are critical to proper enzyme function.

Classification and Naming of Enzymes

Two systems used for naming enzymes are as follows: the suffix *-ase* is used with the root name of the substance whose chemical reaction is catalyzed (the substrate chemical, that is) or with the word that describes the kind of chemical reaction catalyzed. Thus, according to the first method, sucrase is an enzyme that cat-

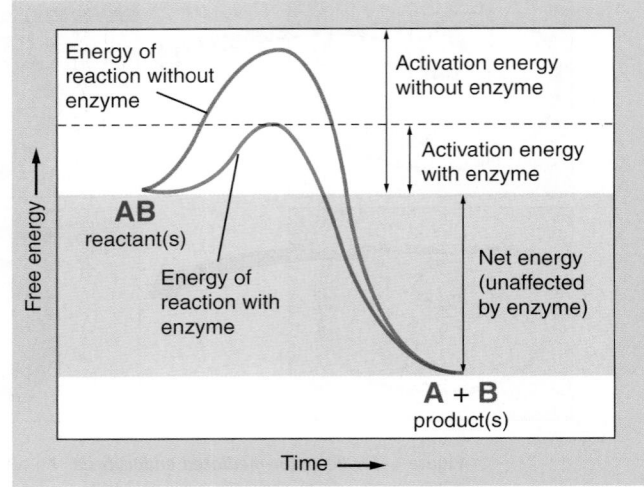

Figure 4-12 *Enzymes as catalysts.* A catalyst is a chemical that reduces the activation energy of a reaction—the energy needed to get a reaction started. Enzymes thus allow reactions to occur at the low level of free energy available at normal human body temperatures.

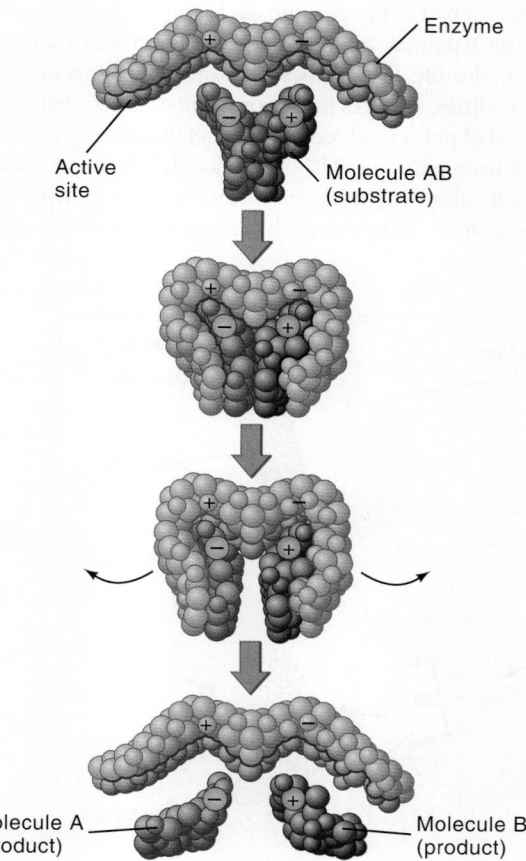

Figure 4-13 *Model of enzyme action.* Enzymes are functional proteins whose molecular shape allows them to catalyze chemical reactions. Substrate molecule *AB* is acted on by a digestive enzyme to yield simpler molecules *A* and *B* as products of the reaction. Notice how the active site of the enzyme chemically fits the substrate—the lock-and-key model of biochemical interaction. Notice also how the enzyme molecule bends its shape in performing its function.

alyzes a chemical reaction in which sucrose takes part. According to the second method, sucrase might also be called a hydrolase because it catalyzes the hydrolysis of sucrose. Enzymes investigated before these methods of nomenclature were adopted are still called by older names, such as *pepsin* and *trypsin*.

Classified according to the kind of chemical reactions catalyzed, enzymes fall into several groups:

Oxidation-reduction enzymes. These are known as oxidases, hydrogenases, and dehydrogenases. Energy release for muscular contraction and all physiological work depends on these enzymes.

Hydrolyzing enzymes, or hydrolases. Digestive enzymes belong to this group. The hydrolyzing enzymes are named after the substrate acted on, for example, lipase, sucrase, and maltase.

Phosphorylating enzymes. These add or remove phosphate groups and are known as phosphorylases or phosphatases.

Enzymes that add or remove carbon dioxide. These are known as carboxylases or decarboxylases.

Enzymes that rearrange atoms within a molecule. These are known as mutases or isomerases.

Hydrases. These add water to a molecule without splitting it, as do hydrolases.

Enzymes are also classified as intracellular or extracellular, depending on whether they act within cells or outside them in the surrounding medium. Most enzymes act intracellularly in the body; an important exception is the digestive enzymes. All digestive enzymes are classified as hydrolases because they catalyze the hydrolysis of food molecules.

General Functions of Enzymes

In general, enzymes regulate cell functions by regulating metabolic pathways (Figure 4-14). As stated earlier, each reaction of a metabolic pathway requires one or more types of enzymes to permit that reaction to occur. An entire metabolic pathway can be turned on or off by the activation or inactivation of any one of the enzymes that catalyze reactions in that particular pathway. A few general principles of enzyme function will help you understand their role in regulating cell metabolism more clearly.

Most enzymes are *specific in their action;* that is, they act only on a specific substrate. This is attributed to their "key-in-a-lock" kind of action, the configuration of the enzyme molecule fitting the configuration of some part of the substrate molecule (see Figure 4-13). This also means that every reaction that occurs in a metabolic pathway requires one or more specific enzymes—or else the reaction will not occur and the entire pathway will be disrupted.

Various physical and chemical agents activate or inhibit enzyme action by changing the shape of enzyme molecules. A molecule or other agent that alters enzyme function by changing its shape is called an

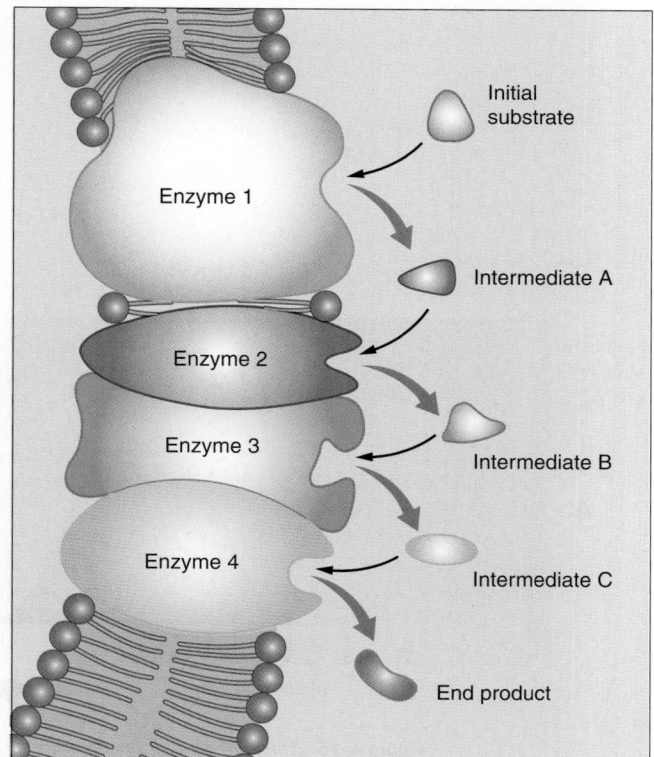

Figure 4-14 *Enzyme regulation of a metabolic pathway.* In a metabolic pathway, the product of one enzyme-regulated reaction becomes the substrate for the next reaction. Thus, a whole series of enzymes is required to keep the pathway functioning. Notice that these enzymes are embedded in a cell membrane whereas other types of enzymes are mobile in the cytosol.

allosteric effector. An "effector" is an agent that accomplishes something, and *allosteric* literally means "pertaining to a change in three-dimensional shape." Thus, an allosteric effector is simply an agent that changes the shape of a molecule. Remember the principle you learned about the shape of proteins in Chapter 2: when the shape changes, so does the function. This certainly applies to enzymes. Some allosteric effectors are molecules that attach to an *allosteric site* on the enzyme molecule and thereby change the shape of the active

site on a different part of the enzyme. As Figure 4-15 shows, allosteric effectors of this type may inhibit or activate enzymes by altering the shape of the active site. Other types of allosteric effectors include certain antibiotic drugs, changes in pH, or changes in temperature. The allosteric effect of pH is produced by the fact that changes in the concentration of hydrogen ions (H^+) influence the chemical attractions that hold molecules—including enzymes—in their complex, multidimensional shapes. Temperature has a similar allosteric, or shape-

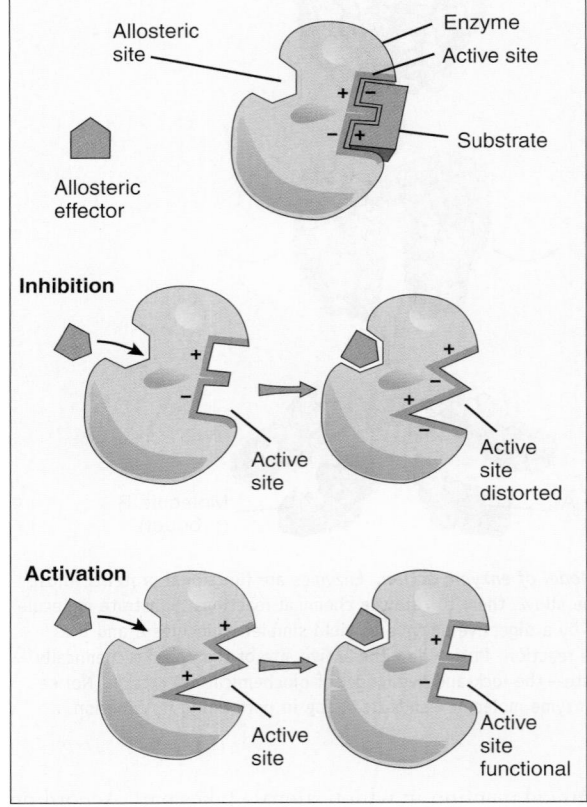

Figure 4-15 *Allosteric effect.* The allosteric effect occurs when some agent, in this case an allosteric effector molecule, binds to the enzyme at an *allosteric site* and thereby changes the shape of the enzyme's active site. Such an allosteric effect may inhibit enzyme action (by distorting the active site) or activate the enzyme (by giving the active site its functional shape).

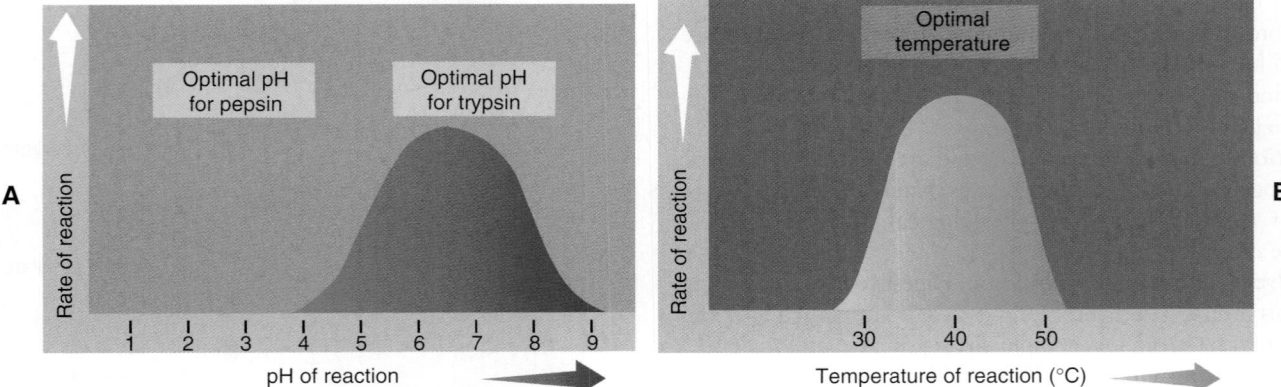

Figure 4-16 *Effects of pH and temperature on enzyme function.* The rate of reactions catalyzed can be affected by the allosteric effects of the chemical or physical properties of the surrounding medium. **A,** Enzymes catalyze chemical reactions with greatest efficiency within a narrow range of pH. For example, *pepsin* (a protein-digesting enzyme in gastric juice) operates within a low pH range, whereas *trypsin* (a protein-digesting enzyme in pancreatic juice) operates within a higher pH range. **B,** Most enzymes in the human body work best within a narrow range of temperatures near 40° C.

changing, effect on enzymes. As Figure 4-16 shows, changing the pH or temperature alters the shape of the active sites of enzyme molecules enough to affect their function. Cofactors, when they are added to or removed from an enzyme molecule, also have an allosteric effect.

In a process known as **end-product inhibition,** a chemical product at the end of a metabolic pathway binds to the allosteric site of one or more enzymes along the pathway that produced it and thereby inhibits the synthesis of more product (Figure 4-17). This is a type of automatic negative feedback mechanism in the cell that prevents the accumulation of an extreme amount of a metabolic product.

Most enzymes catalyze a chemical reaction in both directions, the direction and rate of the reaction being governed by the law of mass action. An accumulation of a product slows the reaction and tends to reverse it.

Enzymes are continually being destroyed and therefore have to be continually synthesized, even though they are not used up in the reactions they catalyze.

Many enzymes are synthesized as inactive **proenzymes.** Substances that convert proenzymes to active enzymes are often called **kinases.** Kinases usually do their job of activating enzymes by means of an allosteric effect (see Figure 4-15). For example, enterokinase changes inactive trypsinogen into active trypsin by changing the shape of the molecule. Within cells, a type of kinase called *kinase* A has been shown to activate enzymes that regulate certain pathways after a hormonal signal is received by the cell.

 QUICK CHECK

4. Describe the structure of an enzyme. How does its structure determine its function?
5. What is an allosteric effector? Give examples.

Catabolism

There are many catabolic pathways that operate inside human cells. Perhaps the most important for a basic understanding of cell catabolism is the pathway known as **cellular respiration.** In this section we briefly outline the basic concepts of the cellular respiratory pathway. Details of this pathway are further outlined in Chapter 27.

Overview of Cellular Respiration

Cellular respiration is the process by which cells break down glucose ($C_6H_{12}O_6$), or a nutrient that has been converted to glucose or one of its simpler products, into carbon dioxide (CO_2) and water (H_2O). As the molecule breaks down, the potential energy that had been stored in its bonds is released. Much of the released energy is converted into heat, but a portion of it is transferred to the high-energy bonds of adenosine triphosphate (ATP). Figure 2-30 on p. 68 shows how ATP is synthesized from adenosine diphosphate (ADP) and inorganic phosphate with energy obtained from cellular respiration.

There are three smaller pathways that are chemically linked together to form the larger catabolic pathway known as cellular respiration:

1. Glycolysis
2. Citric acid cycle
3. Electron transport system

The paragraphs that follow briefly introduce these three basic processes.

Glycolysis

Glycolysis is a catabolic pathway that begins with glucose, which contains six carbon atoms per molecule, and ends with pyruvic acid, which contains only three carbon atoms per molecule. As

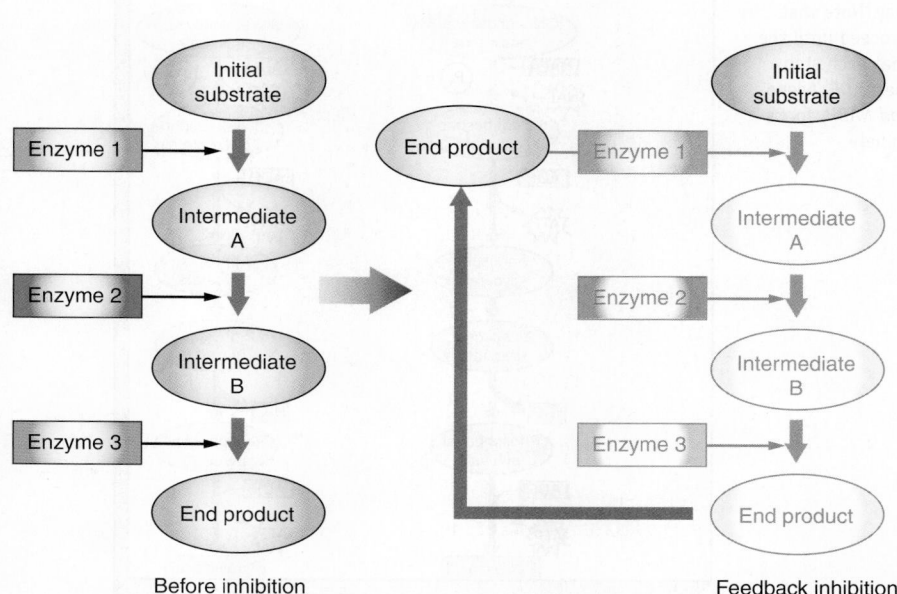

Before inhibition   Feedback inhibition

Figure 4-17 *Feedback inhibition of enzymes.* Formation of an excessive amount of end product can be inhibited by a negative feedback mechanism. In this example, the end product itself inhibits the function of an enzyme needed early in the pathway. Thus, the entire pathway is inhibited—as long as there is an excess of the end product.

Figure 4-18 shows, each glucose molecule that enters this pathway is eventually broken in half. In fact, the name *glycolysis* literally means "breaking glucose."

Glycolysis occurs in the cytosol of cells, outside any particular organelle. The cytosol, then, must contain all the enzymes necessary to catalyze each of the reactions that make up the glycolysis pathway. Because the reactions of glycolysis require no oxygen, glycolysis is said to be **anaerobic.**

Glycolysis releases a small portion of the potential energy stored in the glucose molecule. Some of this energy is transferred to ATP, a molecule that can then transfer the energy to any of a large number of energy-consuming reactions in the cell. Some of the energy is transferred to another energy transfer molecule, a form of nicotinamide adenine dinucleotide, (NADH). NADH may eventually transfer its energy to ATP in the electron transport system, a later step of cellular respiration that is discussed shortly.

Once pyruvic acid is formed by glycolysis, there is a fork in the metabolic pathway. That is, the molecule could enter one of two pathways linked to glycolysis (Figure 4-18). If oxygen is available, the pyruvic acid molecule will follow the *aerobic pathway* and enter the citric acid cycle. This type of respiration is called **aerobic** respiration because oxygen (O_2) is required for this sequence of reactions to occur. If oxygen is unavailable for a particular pyruvic acid molecule to enter the aerobic pathway, it will continue along an anaerobic pathway to form a molecule called *lactic acid*. Lactic acid is later converted back to pyruvic acid or glucose in an energy- and oxygen-requiring pathway.

When the anaerobic pathway is followed, a small amount of the total energy stored in glucose is made available to the cell. However, if enough oxygen is not available to maintain a set point level of ATP by means of the aerobic pathway, the anaerobic pathway can help maintain adequate ATP levels for cellular functions to continue. Because oxygen is later used to process the lactic acid formed by anaerobic respiration, biochemists say that an *oxygen debt* has been incurred. Much of the lactic acid diffuses out of the cell that formed it and is later processed in liver cells, which are specialized to perform this function efficiently.

Citric Acid Cycle

If oxygen is available, the pyruvic acid molecules formed by glycolysis are prepared to enter the next major phase of aerobic cellular respiration—the **citric acid cycle.** This cyclic (repeating) se-

Figure 4-18 *Glycolysis.* This diagram of the reactions involved in glycolysis represents a classic example of a catabolic pathway. Note that each of the chemical reactions in this pathway cannot proceed until the previous step has occurred. Recall from our discussion that each step also requires the presence of one or more specific enzymes. *ADP,* Adenosine diphosphate; *ATP,* adenosine triphosphate; *NAD+ and NADH,* forms of nicotinamide adenine dinucleotide; P_i, inorganic phosphate.

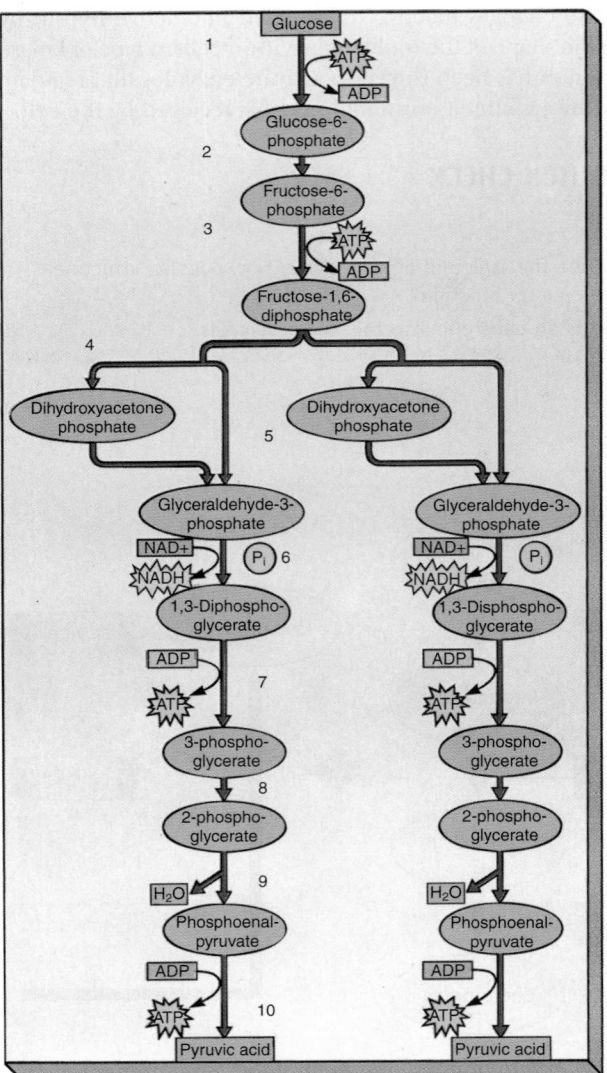

quence of reactions is still sometimes called the *Krebs cycle* after Sir Hans Krebs, who discovered this pathway in the first part of the twentieth century. Figure 4-19 shows that pyruvic acid is converted to acetyl coenzyme A (CoA) and moves into the citric acid cycle. During this transition into the citric acid cycle, the molecule loses one of its carbons, along with some oxygen, producing waste carbon dioxide (CO_2). This cycle also produces some available energy that is transferred to NADH and then to the electron transport system, as explained later.

The citric acid cycle, like glycolysis, is a sequence of many chemical reactions (Figure 4-19). As in glycolysis, each reaction of the citric acid cycle requires one or more specific enzymes. These enzymes are located in the inner chamber of the mitochondrion, so that is where in the cell the citric acid cycle occurs.

In the citric acid cycle, the two-carbon acetyl group that breaks away from its escort, CoA, is further broken down to yield its stored energy. A small amount of this energy is transferred directly to ATP molecules, but most of the available energy is transferred in the form of energized electrons (e^-) and their accompanying protons (H^+) to the coenzyme NAD or flavin adenine dinucleotide (FAD). NAD then becomes NADH and FAD becomes the reduced form of FAD ($FADH_2$). More information on how these coenzymes pick up energy released from a metabolic pathway is found in Chapter 27 on p. 1015. The energized electrons (along with their protons) then move into the next phase of cellular respiration—the electron transport system.

Electron Transport System

As Figure 4-20 shows, NADH and $FADH_2$ transfer the energized electrons to a set of special molecules embedded in the cristae of the inner mitochondrial membrane. These special electron-accepting molecules make up the **electron transport system (ETS)**.

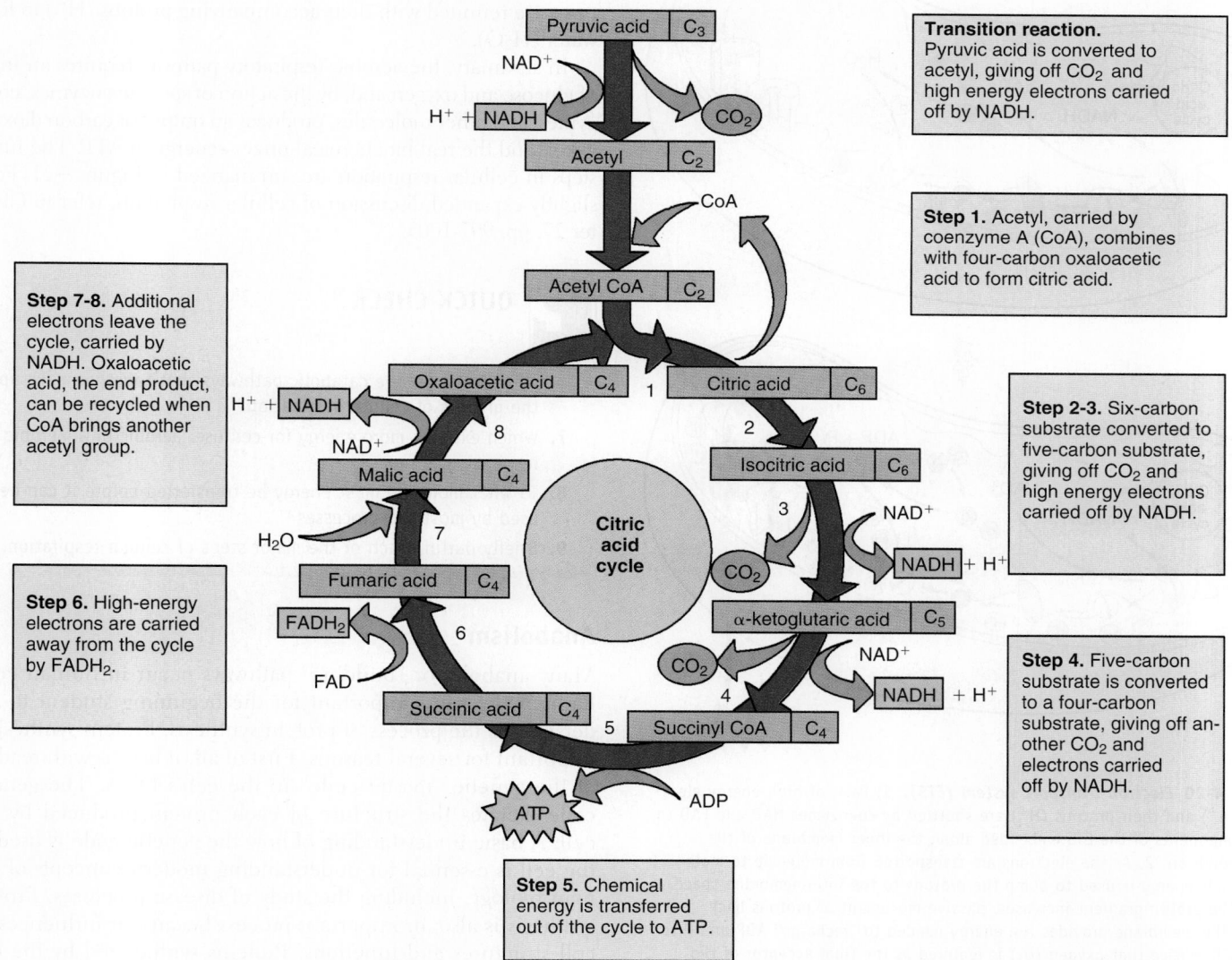

Figure 4-19 *Citric acid cycle.* The citric acid cycle is a circular metabolic pathway that breaks down an acetyl molecule with the release of CO_2 molecules and energized electrons (which along with their protons (H^+), are shuttled away by the coenzymes nicotinamide adenine dinucleotide [NAD] and flavin adenine dinucleotide [FAD]). *ATP,* Adenosine triphosphate.

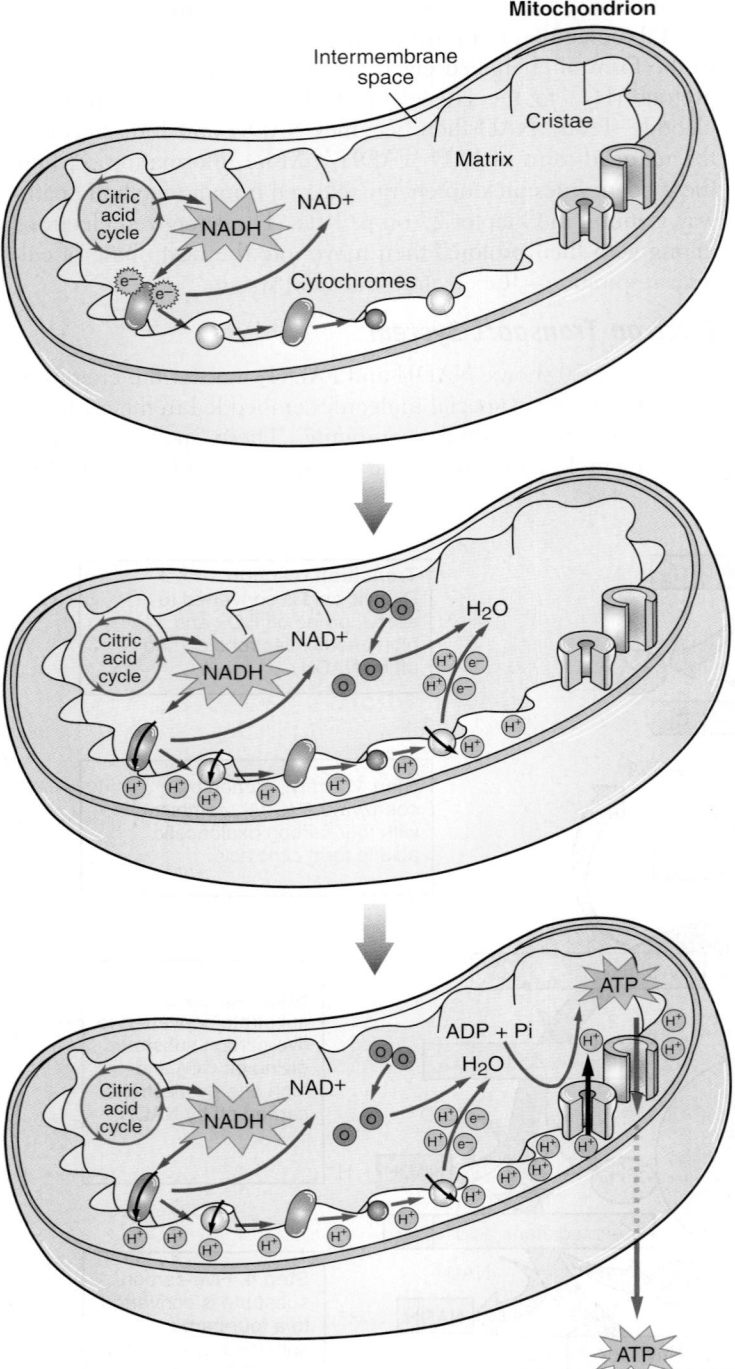

Mitochondrion

Figure 4-20 *Electron transport system (ETS).* **1,** Pairs of high-energy electrons (e⁻) and their protons (H⁺) are shuttled by coenzymes NAD and FAD to the components of the ETS embedded along the inner membrane of the mitochondrion. **2,** As the electrons are transported from molecule to molecule, their energy is used to pump the protons to the intermembrane space. **3,** As the proton gradient increases, passive movement of protons back across the membrane provides the energy needed to "recharge" ADP and P to form ATP. Notice that oxygen (O_2) is required as the final acceptor of the electrons and protons transported through the system, thus forming H_2O as a byproduct.

As energized electrons leap from one of these molecules to the next, their energy is used to pump their accompanying protons (H^+) from the inner chamber of the mitochondrion to the outer chamber. As the protons (H^+) build up in the outer chamber, a concentration gradient of protons develops. Protons then begin movement through the inner mitochondrial membrane into the inner chamber by way of special pump molecules that convert the energy of proton flow into chemical energy, which is transferred to ATP molecules. In short, the energy transferred from the citric acid cycle is used to put protons behind a sort of dam, and then the energy of protons flowing back from behind the dam is transferred to ATP. In most cells, aerobic respiration transfers enough energy from each glucose molecule to form 36 ATP molecules—2 directly from glycolysis and 34 from the rest of the pathway.

The de-energized electrons that come away from the electron transport system are accepted by oxygen molecules. This is why oxygen is required for this metabolic pathway—to act as final electron acceptors. Once they combine with oxygen, the electrons are reunited with their accompanying protons (H^+) to form water (H_2O).

In summary, the aerobic respiratory pathway requires an input of glucose and oxygen and, by the action of specific enzymes, coenzymes, and other molecules, produces an output of carbon dioxide, water, and the real biochemical prize—energy in ATP. The major steps in cellular respiration are summarized in Figure 4-21. For a slightly expanded discussion of cellular respiration, refer to Chapter 27, pp. 997-1005.

 QUICK CHECK

6. What are the three catabolic pathways that together make up the process of cellular respiration?
7. Which extracts more energy for cell use, aerobic or anaerobic respiration?
8. To what molecule must energy be transferred before it can be used by most cell processes?
9. Briefly outline each of the major steps of cellular respiration.

Anabolism

Many anabolic or "building" pathways occur in human cells. Perhaps the most important for the beginning student to understand is the process of protein synthesis. Protein synthesis is important for several reasons. First of all, it begins with reading of the genetic "master code" in the cell's DNA. The genetic code dictates the structure of each protein produced by the cell. A basic understanding of how the genetic code is used by the cell is essential for understanding modern concepts of human biology, including the study of disease processes. Protein synthesis is also an important process because it influences all cell structures and functions. Proteins synthesized by the cell either are structural elements themselves or are enzymes or other functional proteins that direct the synthesis of other structural and functional molecules such as carbohydrates, lipids, and nucleic acids. In short, protein synthesis is the central anabolic pathway for cells.

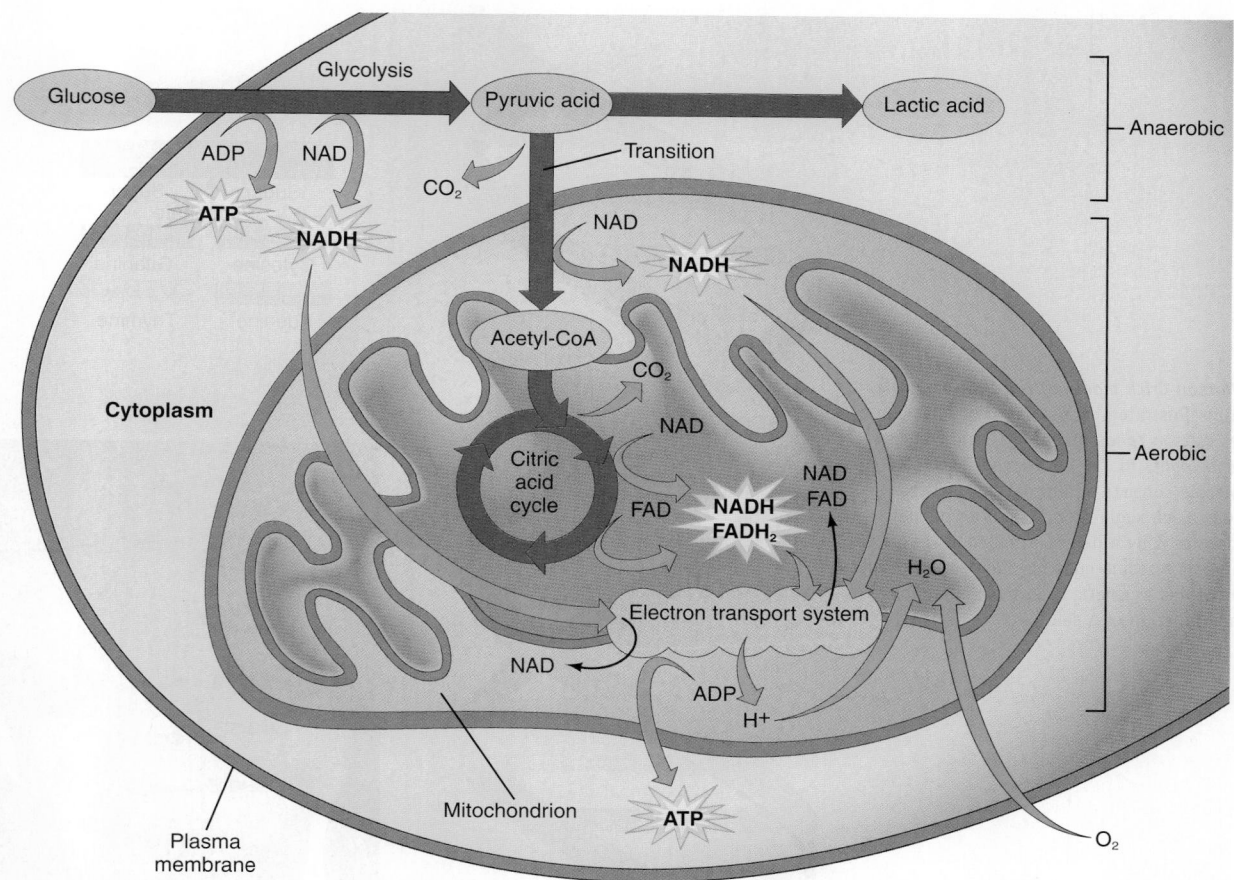

Figure 4-21 *Summary of cellular respiration.* This simplified outline of cellular respiration represents one of the most important catabolic pathways in the cell. Note that one phase (glycolysis) occurs in the cytosol but that the two remaining phases (citric acid cycle and electron transport system) occur within a mitochondrion. Note also the divergence of the anaerobic and aerobic pathways of cellular respiration. *ADP,* Adenosine diphosphate; *ATP,* adenosine triphosphate; *CoA,* coenzyme A; *FADH₂,* form of flavin adenine dinucleotide; *NADH,* form of nicotinamide adenine dinucleotide.

Deoxyribonucleic Acid (DNA)

Over 50 years ago, an American, James Watson, and three British scientists, Francis Crick, Maurice Wilkins, and Rosalind Franklin, won the race to solve the puzzle of DNA's molecular structure (Figure 4-22). Watson and Crick announced their discovery with an understated, one-page report in the journal *Nature* in 1953. A more detailed supportive paper by Wilkins on the chemistry of DNA was included in the same issue. Nine years later, Watson, Crick, and Wilkins received the Nobel Prize for their brilliant and significant work—hailed as the greatest biological discovery of our time. (Franklin died before the Nobel Prize was awarded.) Watson tells his insider's view of how the discovery was made in *The Double Helix* (1969), a book that is now a classic for its science and its intrigue.

Since the original discovery of DNA's structure, we have seen a new branch of biology called *molecular genetics* emerge from our rapidly growing knowledge of DNA and how it works. Indeed, we have seen a revolution in human biology as we witness the continuing application of molecular genetics transform every single aspect of anatomy, physiology, and medicine. We begin our outline of protein synthesis with a discussion of DNA because it truly is, as Watson called it in his book, "the most golden of all molecules."

The **deoxyribonucleic acid** molecule is a giant among molecules. Its size and the complexity of its shape exceed those of most molecules. The importance of its function—in a word, information—surpasses that of any other molecule in the world.

To visualize the shape of the DNA molecule, picture an extremely long, narrow ladder made of a pliable material (see Figure 4-22). Now see it twisting round and round on its axis and taking on the shape of a steep spiral staircase millions of turns long. This is the shape of the DNA molecule—a double spiral or *double helix.*

The DNA molecule is a *polymer,* which means that it is a large molecule made up of many smaller molecules joined together in sequence. DNA is a polymer of millions of pairs of nucleotides. A nucleotide is a compound formed by combining phosphoric acid with a sugar and a nitrogenous base. The DNA molecule has four different kinds of nucleotides. Each consists of a phosphate group that attaches to the sugar deoxyribose, which attaches to one of four bases. Nucleotides differ, therefore, in their nitrogenous base component—containing either adenine or guanine (purine bases) or cytosine or thymine (pyrimidine bases). (Deoxyribose is a sugar that is not sweet and one whose molecules contain only

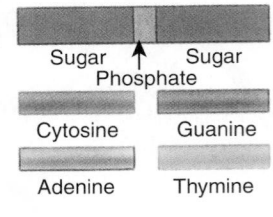

Figure 4-22 *Watson-Crick model of the DNA molecule.* The DNA structure illustrated here is based on that published by James Watson *(photograph, left)* and Francis Crick *(photograph, right)* in 1953. Note that each side of the DNA molecule consists of alternating sugar and phosphate groups. Each sugar group is united to the sugar group opposite it by a pair of nitrogenous bases (adenine-thymine or cytosine-guanine). The sequence of these pairs constitutes a genetic code that determines the structure and function of a cell.

five carbon atoms.) Notice what the name *deoxyribonucleic acid* tells you—that this compound contains deoxyribose, that it occurs in the nucleus, and that it is an acid.

Figure 4-22 reveals additional and highly significant facts about DNA's molecular structure. First, observe which compounds form the sides of the DNA spiral staircase—a long line of phosphate and deoxyribose units joined alternately one after the other. Look next at the stair steps. Notice two facts about them: two bases join (loosely bound by hydrogen bonds) to form each step, and only two combinations of bases occur. The same two bases invariably pair off with each other in a DNA molecule. Adenine always goes with thymine (or vice versa, thymine with adenine), and guanine always goes with cytosine (or vice versa). This fact about DNA molecular structure is called **obligatory base pairing.** Pay particular attention to it, for it is the key to understanding how a DNA molecule is able to duplicate itself. DNA duplication, or replication as it is usually called, is one of the most important of all biological phenomena because it is an essential and crucial part of the mechanism of genetics.

Another fact about DNA's molecular structure that has great functional importance is the sequence of its base pairs. Although the kinds of base pairs possible in all DNA molecules are the same, the *sequence* of these base pairs is not the same in all DNA molecules. For instance, the sequence of the base pairs compos-

ing the seventh, eighth, and ninth steps of one DNA molecule might be cytosine-guanine, adenine-thymine, and thymine-adenine. Such a sequence of three bases forms a code word or "triplet" called a *codon.* In another DNA molecule the coding sequence of the base pairs making up these same steps might be entirely different, perhaps thymine-adenine, guanine-cytosine, and cytosine-guanine. Perhaps these seem to be minor details, but nothing could be further from the truth because it is the sequence of the base pairs in the nucleotides that make up the DNA molecules that identifies each gene. Therefore, it is the sequence of base pairs that determines all hereditary traits.

A human *gene* is a segment of a DNA molecule. One gene consists of a chain of approximately 1000 pairs of nucleotides joined one after the other in a precise sequence. Each gene in DNA is a code. The gene code is first copied to a ribonucleic acid (RNA) molecule, or transcript. Each RNA transcript of a gene may then be translated by the cell and used to build one polypeptide chain. A few RNA transcripts are instead used to support or regulate polypeptide production. One or more polypeptides make up each of a cell's structural proteins and enzymes. Each enzyme catalyzes one or more chemical reactions. Therefore, as Figure 4-23 shows, the nearly 24,000 protein-coding genes constituting a cell's DNA determine the cell's structure and its functions.

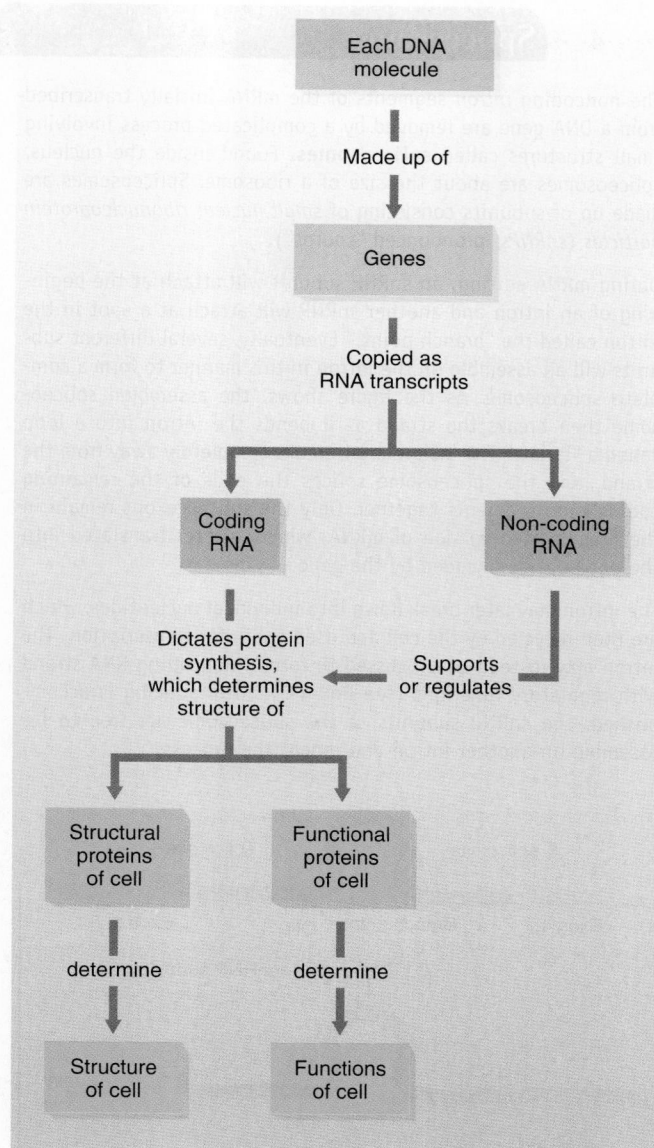

Figure 4-23 *Function of genes.* Genes copied from DNA are copied to RNA molecules, which use the code to determine a cell's structural and functional characteristics. There are approximately 24,000 genes in the human genome (all the DNA molecules together).

Transcription

First, a single strand of RNA (ribonucleic acid) forms along a segment of one strand of a DNA molecule. Figure 4-24 shows how this process happens. RNA differs from DNA in certain respects. Its molecules are smaller than those of DNA, and RNA contains ribose instead of deoxyribose. In addition, one of the four bases in RNA is uracil instead of thymine. As a strand of RNA is forming along a strand of DNA, uracil attaches to adenine, and guanine attaches to cytosine. The process is known as **complementary pairing.** Thus a single-strand molecule of *messenger RNA (mRNA)* is formed. The name "messenger RNA" describes its function. As soon as it is formed, it separates from the DNA strand, is edited, moves out of the nucleus, and carries a "message" to a ribosome in

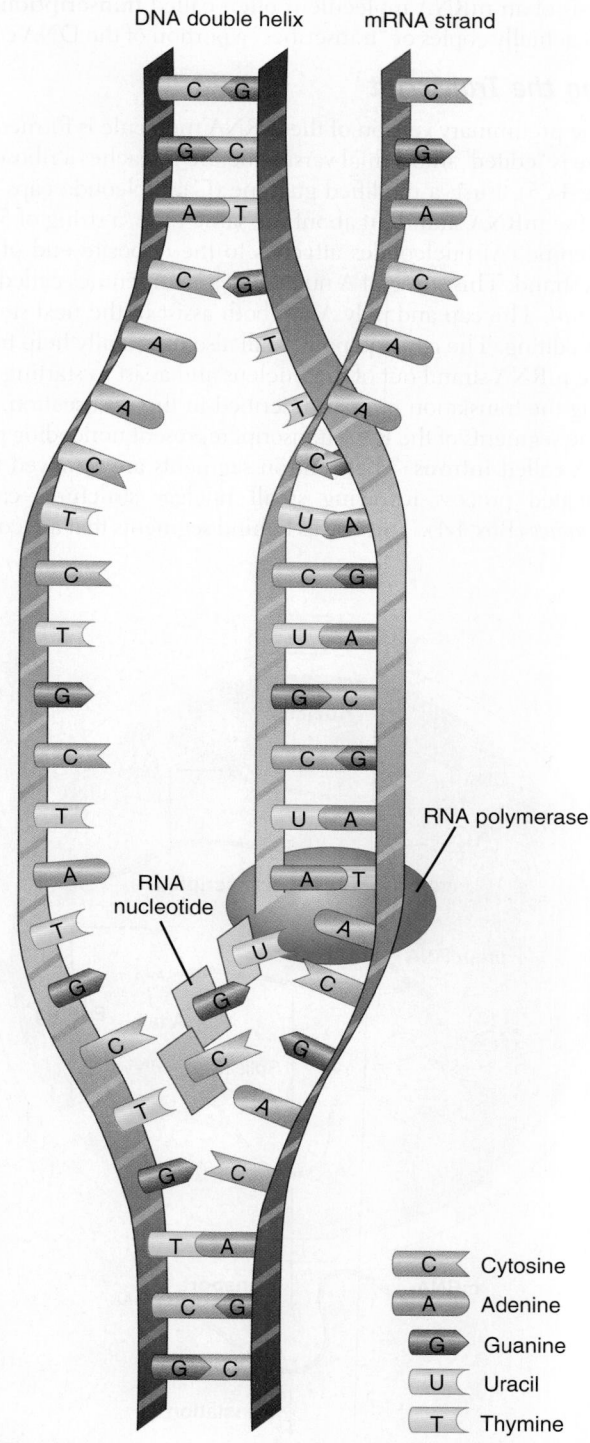

Figure 4-24 *Transcription of messenger RNA (mRNA).* A DNA molecule "unzips" in the region of the gene to be transcribed. RNA nucleotides already present in the nucleus temporarily attach themselves to exposed DNA bases along one strand of the unzipped DNA molecule according to the principle of complementary pairing. As the RNA nucleotides attach to the exposed DNA, they bind to each other and form a chainlike RNA strand called a *messenger RNA (mRNA)* molecule. Notice that the new mRNA strand is an exact copy of the base sequence on the opposite side of the DNA molecule. As in all metabolic processes, the formation of mRNA is controlled by an enzyme—in this case, the enzyme is called *RNA polymerase*.

the cell's cytoplasm to direct the synthesis of a specific polypeptide. Synthesis of an mRNA molecule is often called **transcription** because it actually copies or "transcribes" a portion of the DNA code.

Editing the Transcript

After the preliminary version of the mRNA molecule is formed, its message is "edited" into a final version before it reaches a ribosome (Figure 4-25). First, a modified guanine (G) nucleotide caps one end of the mRNA strand. At about the same time, a string of 50 to 200 adenine (A) nucleotides attaches to the opposite end of the mRNA strand. This string of A nucleotides is sometimes called the *poly A tail*. The cap and poly A tail both assist in the next step of mRNA editing. The cap and poly A tail also eventually help transport the mRNA strand out of the nucleus and assist in starting and stopping the translation process described in the next section.

Some segments of the RNA transcript represent noncoding parts of DNA called **introns**. These intron segments are removed by a complicated process involving small nuclear structures called *spliceosomes* (Box 4-4). This leaves behind segments that are copies

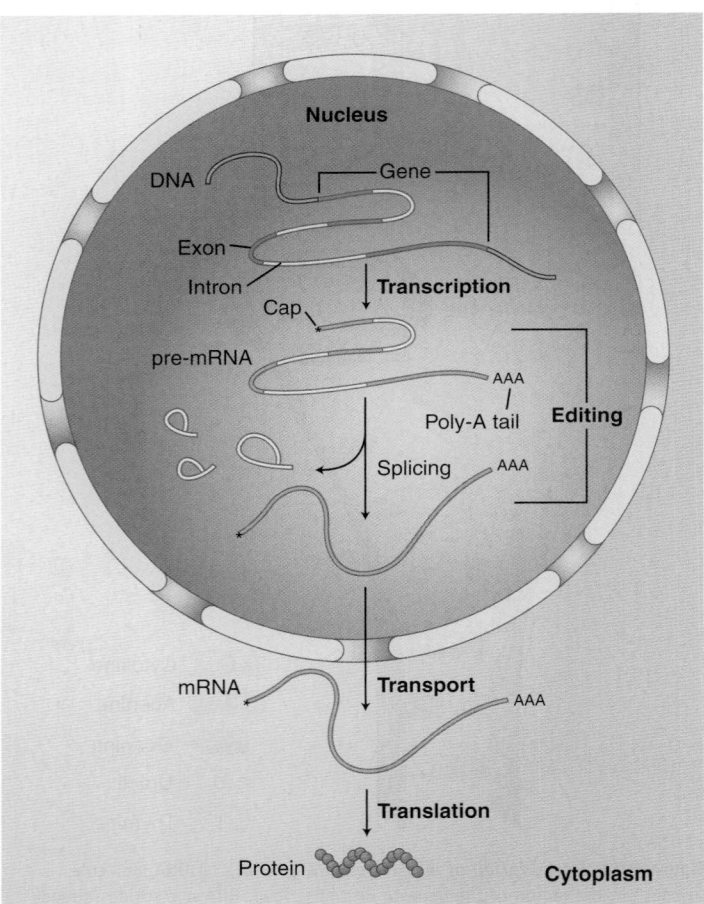

Figure 4-25 *Editing of an mRNA transcript.* After the preliminary mRNA strand is transcribed from a gene in DNA, it is capped with a modified G nucleotide at one end. At the same time, a poly A string is attached at the other end. The intron sections of the transcript are then removed and the remaining exons joined, or *spliced,* together to form the final version of mRNA. The edited mRNA transcript then moves out of the nucleus to participate in translation (see Figure 4-24).

BOX 4-4   Spliceosome

The noncoding *intron* segments of the mRNA initially transcribed from a DNA gene are removed by a complicated process involving small structures called **spliceosomes.** Found inside the nucleus, spliceosomes are about the size of a ribosome. Spliceosomes are made up of subunits consisting of *small nuclear ribonucleoprotein particles* (*snRNPs,* pronounced "snurps").

During mRNA editing, an snRNP subunit will attach at the beginning of an intron and another snRNP will attach at a spot in the intron called the "branch point." Eventually, several different subunits will all assemble on the intron in this manner to form a complete spliceosome. As the figure shows, the assembled spliceosome then breaks the strand as it bends the intron into a loop called a "lariat." The intron lariat breaks completely away from the strand, and the spliceosome splices the ends of the remaining coding *exon* segments together. Only the spliced exons remain in the final, edited version of mRNA, which is later translated into the polypeptide encoded by the gene.

The intron may later break down into individual nucleotides, which are then recycled by the cell and used again for transcription. The intron may instead be processed to form a noncoding RNA strand with regulatory functions (see Box 4-5). After splicing is accomplished, the snRNP subunits of the spliceosome are free to reassemble on another intron and repeat the process.

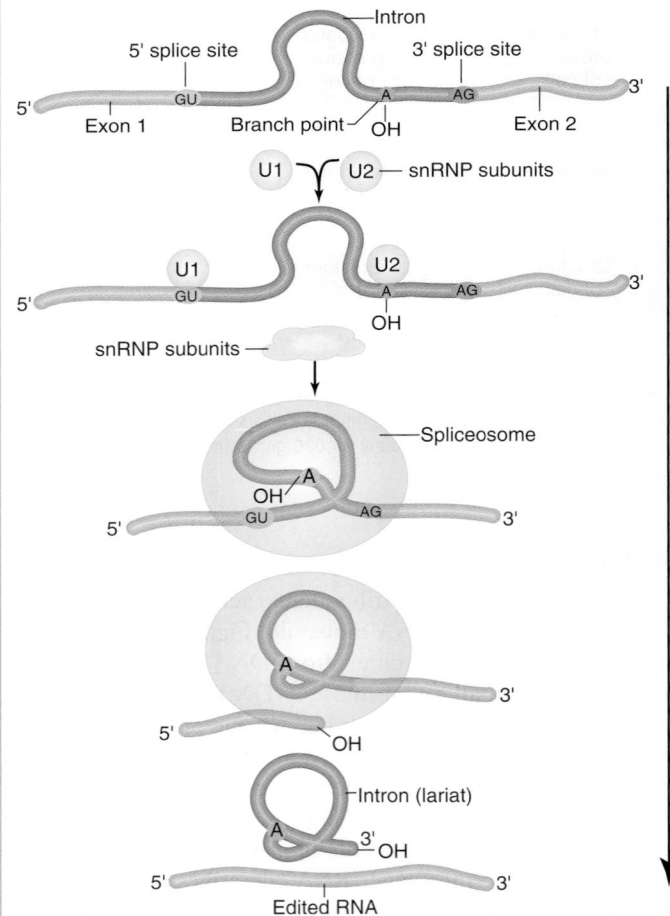

Editing of mRNA by the spliceosome. A simplified view of the complex splicing process.

of the DNA's **exons.** Only these exon copies will be used in the final recipe for the protein. All the RNA segments representing exons are then *spliced* (joined) together by enzymes to form the final edited form of mRNA. It is this edited version of RNA that leaves the nucleus and participates in the next step of protein synthesis.

Translation

In the cytoplasm, the edited mRNA molecule attracts large and small ribosome subunits (see Figure 3-6) that come together around the mRNA molecule to form an "mRNA sandwich." Recall that the two subunits of the now-complete ribosome are composed largely of *ribosomal RNA* (rRNA). The cell is now ready to interpret or "translate" the genetic code and form a specific sequence of amino acids in a process called **translation.**

In translation, yet another type of RNA—*transfer RNA* (tRNA)—becomes involved in protein synthesis (Figure 4-26). As the name implies, tRNA molecules carry or "transfer" amino acids to the ribosome for placement in the prescribed sequence. This function is determined by a unique molecular structure: a binding site for a specific amino acid at one end and a binding site for a specific mRNA codon (base triplet) at the other end (see Figure 2-29 on p. 67). Because tRNA's binding site for mRNA contains the three bases that exactly complement one mRNA codon, this binding site is often called the *anticodon.*

After picking up its amino acid from a pool of 20 different types of amino acids floating free in the cytoplasm, a tRNA molecule moves to the ribosome. There, its anticodon attaches to a complementary mRNA codon—the codon that signifies the specific amino acid carried by that tRNA molecule (Figure 4-27). After the next tRNA brings its amino acid into place, the two amino acids form a peptide bond. The ribosome then moves down the mRNA strand, making the next mRNA codon available for a complementary tRNA molecule bearing the appropriate amino acid. More tRNA molecules, one after the other in rapid sequence, bring more amino acids to the ribosome and fit them into their proper position in the growing chain of amino acids. As Figure 4-26 shows, the ri-

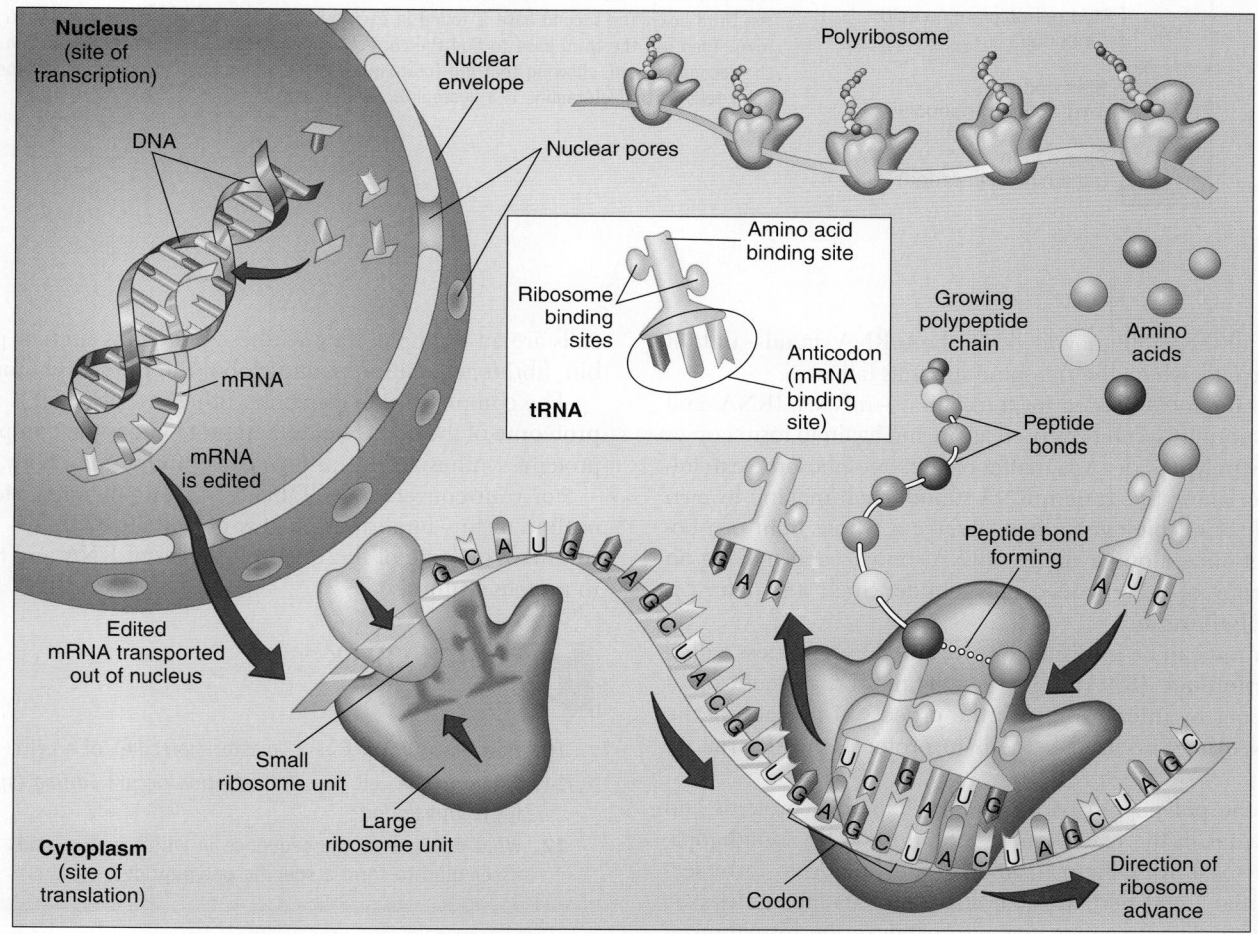

Figure 4-26 *Protein synthesis.* Protein synthesis begins with *transcription,* a process in which an mRNA molecule forms along one gene sequence of a DNA molecule within the cell's nucleus. As it is formed, the mRNA molecule separates from the DNA molecule, is edited, and leaves the nucleus through the large nuclear pores. Outside the nucleus, ribosome subunits attach to the beginning of the mRNA molecule and begin the process of *translation.* In translation, transfer RNA (tRNA) molecules bring specific amino acids—encoded by each mRNA codon—into place at the ribosome site. As the amino acids are brought into the proper sequence, they are joined together by peptide bonds to form long strands called *polypeptides.* Several polypeptide chains may be needed to make a complete protein molecule.

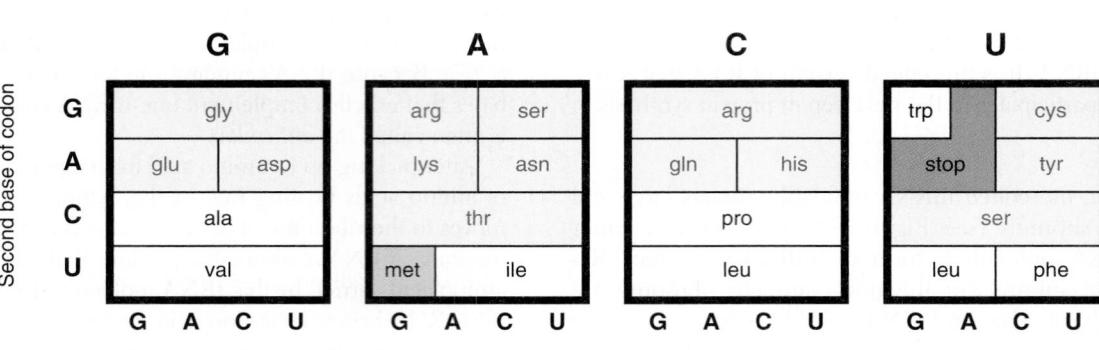

First base of codon

Figure 4-27 *The genetic code.* The first graphic "phrasebook" or "decoder" of the genetic language of the cell was developed in 1966 to summarize the amino acids encoded by various codons (3-base sequences of nucleotides) in RNA. One of the clearest is this version adapted from Ben Fry's decoder, developed when he was a student at MIT. To read the decoder, start with any codon (for example, CGA). Find the first base along the top of the decoder to find the correct box to use (C is the third box). The second base is found in each row of that box, labeled on the left (G is the first row). Then use the third base to find the correct column, labeled at the bottom of each box (A is the second column, showing that gln [glutamine] is the amino acid encoded by CGA; the bluish color tells us that glutamine is a hydrophilic amino acid).

Key

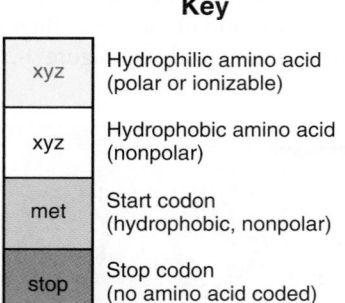

xyz — Hydrophilic amino acid (polar or ionizable)

xyz — Hydrophobic amino acid (nonpolar)

met — Start codon (hydrophobic, nonpolar)

stop — Stop codon (no amino acid coded)

bosome is all the while moving down the mRNA strand—until it reaches the end, where the ribosome subunits fall away.

Each of the tools needed for translation—mRNA, tRNA, and ribosome subunits—can be reused again and again to form copies of the same polypeptide. As a matter of fact, one ribosome can follow another along the same mRNA strand, each making its own polypeptide. Cell biologists often observe a whole train of ribosomes positioned along a single mRNA strand, each making an identical copy of the encoded polypeptide. Such a structure is called a **polyribosome.**

Translation can be inhibited or prevented by a process called **RNA interference (RNAi).** Box 4-5 details the concept of silencing genes by interfering with the process of translating the genes. RNAi may be used by cells to protect against virus infections.

As specific polypeptides are formed, chaperone proteins and other enzymes in the endoplasmic reticulum (ER), Golgi apparatus, or cytosol assist them in folding and linking to form secondary, tertiary, and perhaps quaternary protein molecules (see Figure 3-8, p. 90). Enzymes may also catalyze the formation of hybrid molecules such as lipoproteins or glycoproteins.

Protein anabolism is now complete. It is one of the major kinds of cellular work. One human cell is estimated to synthesize thousands of different enzymes! In addition to this staggering workload, the cell produces many different protein compounds that help form its own structures, and many cells also synthesize special functional proteins for use in other parts of the body. Liver

cells are an example; they synthesize proteins such as prothrombin, fibrinogen, albumin, and globulin for blood plasma.

The complete set of proteins synthesized by a cell is called the **proteome** of the cell. The human proteome is the complete set of proteins synthesized by all the cells of the human body.

For your convenience, Table 4-2 gives a detailed, step-by-step outline of this important process of protein synthesis. Table 4-3 summarizes some of the important types of RNA and their roles in protein synthesis.

 QUICK CHECK

10. How does DNA act as a "master molecule" of a cell?
11. Where in the cell does *transcription* occur? *Editing* (splicing)? *Translation?*
12. What determines the sequence in which amino acids are assembled to form a specific polypeptide?

GROWTH AND REPRODUCTION OF CELLS

Cell growth and reproduction are the most fundamental of all living functions. These two processes together constitute the *cell life cycle* (Figure 4-28). On these processes depend the continued

BOX 4-5
The RNA Revolution

For nearly a half century after the discovery of DNA and its central role in heredity in 1953, scientific understanding of cellular physiology focused primarily on DNA—that "most golden of molecules." RNA was seen as the "go between" molecule, acting mainly as a temporary working copy of a segment of DNA code (gene). However, we soon learned that tRNA (transfer RNA) and rRNA (ribosomal RNA) play a functional role other than direct coding for proteins. This discovery sparked some interest in looking for possible additional regulatory or supportive roles of RNA. Soon, catalytic forms of *noncoding RNA* were shown to be involved in editing mRNA (messenger RNA) transcripts before translation (see Box 4-4). The dawn of the twenty-first century saw an explosion of new discoveries regarding additional roles of RNA in regulating the function of the genome—a scientific revolution that still continues.

One aspect of the RNA revolution involves the surprising treasure trove of noncoding RNA molecules transcribed from the roughly 98% of the DNA genome that does not code for proteins. This includes introns (noncoding segments) that are removed from transcribed (mRNA) protein-coding genes during the editing process before translation. Current scientific databases are chock full of thousands of such regulatory RNA sequences. To cell scientists, it is becoming clear that an extensive RNA regulatory system manages the human genome.

One of the more interesting and useful recent discoveries of the current RNA revolution involves a cellular process of "gene silencing" called **RNA interference (RNAi).** Stated simply, RNAi is a process in which certain genes are silenced and thus synthesis of a particular protein is halted. RNAi occurs naturally in every cell, and scientists are still working to discover the mechanisms that govern the process and the purposes it serves in the cell. In human cells, RNAi is thought to be part of a complex scheme of "fine-tuning" protein synthesis. For example, the cell could use RNAi to inhibit replication of virus components within the cell during a viral infection—thus protecting the cell.

How does RNAi work? A special form of RNA with two strands—double-strand RNA (dsRNA)—seems to be the key, as the figure shows. A small segment of dsRNA, only about 20 or so nucleotides long, called *small interfering RNA (siRNA),* begins the process by combining with protein subunits to form an *RNA-induced silencing complex (RISC).* The siRNA within the RISC unwinds and exposes anticodons that allow the RISC to attach to a particular mRNA transcript. After the RISC attaches to its target mRNA, the mRNA breaks apart and therefore fails to translate its encoded protein. Of course, this effectively silences the gene encoded in the target mRNA. RNAi has also been shown to remodel chromatin, thus changing the "master code" in the DNA molecule!

Scientists are especially excited about using RNAi as a research tool. Researchers can synthesize a particular siRNA double strand and insert it into a cell that they want to study. The siRNA will induce the silencing of a particular gene so that researchers can see what happens to the cell. Knocking out the activity of just one gene at a time allows scientists to test hypotheses regarding how particular genes or particular proteins function in the cell.

The therapeutic possibilities of RNAi are also very exciting. Work is already under way to find an effective means of using RNAi for creat-

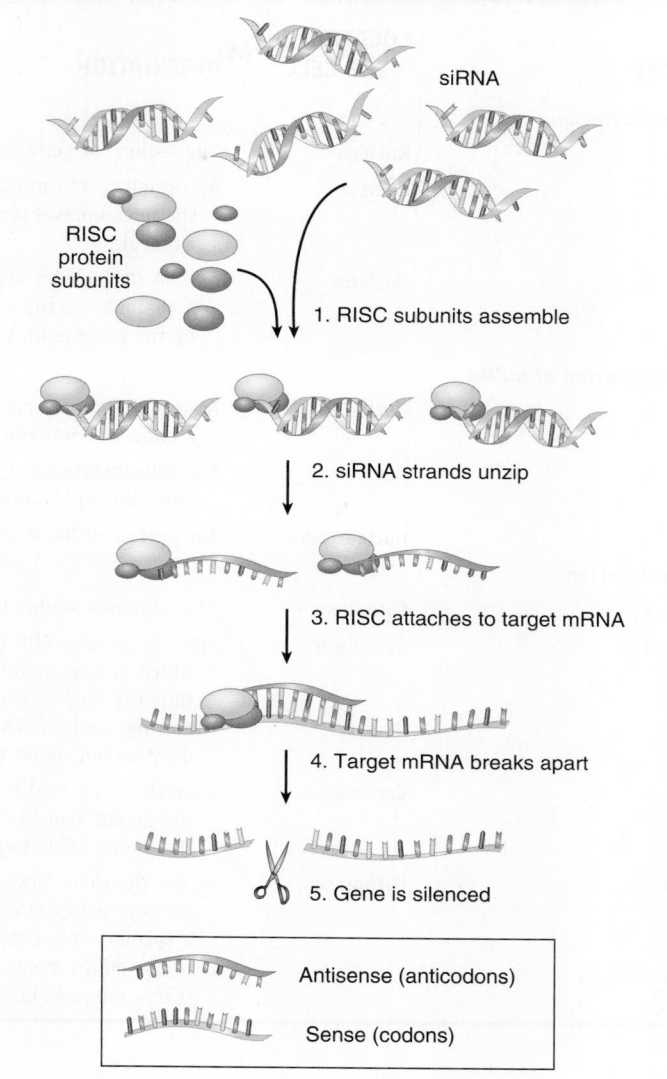

1. RISC subunits assemble

2. siRNA strands unzip

3. RISC attaches to target mRNA

4. Target mRNA breaks apart

5. Gene is silenced

Antisense (anticodons)	
Sense (codons)	

RNA interference. **1,** Short, double-stranded siRNA segments combine with protein subunits to form an RNA-induced silencing complex (RISC). **2,** The siRNA within the RISC unwinds to expose anticodons. **3,** The activated RISC attaches to target mRNA possessing complementary codons. **4,** The RISC causes the mRNA molecule to break apart. **5,** The mRNA can no longer code for a polypeptide, and thus the gene is effectively silenced. Eventually, the mRNA is broken down into its component nucleotides, which can be reused by the cell to build other RNA molecules.

ing antiviral creams containing siRNA to protect against HIV and other viruses. Some researchers are using RNAi to knock out genes that permit permanent tissue damage during heart attacks, kidney failure, and stroke. Researchers are also attempting to harness the power of RNAi to treat cancer and many types of infection.

Table 4-2   Summary of Protein Synthesis

STEP	LOCATION IN THE CELL	DESCRIPTION
Transcription		
1	Nucleus	One region, or gene, of a DNA molecule "unzips" to expose its bases
2	Nucleus	According to the principles of complementary base pairing, RNA nucleotides already present in the nucleoplasm temporarily attach themselves to the exposed bases along one side of the DNA molecule
3	Nucleus	As RNA nucleotides align themselves along the DNA strand, they bind to each other and thus form a chainlike strand called *messenger RNA* (mRNA). This binding of RNA nucleotides is controlled by the enzyme RNA polymerase
Preparation of mRNA		
4	Nucleus	As the preliminary mRNA strand is formed, it peels away from the DNA strand. This mRNA strand is a copy, or *transcript,* of a gene
5	Nucleus	The spliceosome edits the mRNA molecule by removing noncoding portions of the strand (introns) and splicing the remaining pieces (exons)
6	Nuclear pores	The edited mRNA strand is transported out of the nucleus through pores in the nuclear envelope
Translation		
7	Cytoplasm	Two subunits sandwich the end of the mRNA molecule to form a ribosome
8	Cytoplasm	Specific transfer RNA (tRNA) molecules bring specific amino acids into place at the ribosome, which acts as a sort of "holder" for the mRNA strand and tRNA molecules. The kind of tRNA (and thus the kind of amino acid) that moves into position is determined by complementary base pairing: each mRNA codon exposed at the ribosome site will only permit a tRNA with a complementary *anticodon* to attach
9	Cytoplasm	As each amino acid is brought into place at the ribosome, an enzyme in the ribosome binds it to the amino acid that arrived just before it. The chemical bonds formed, called *peptide bonds,* link the amino acids together to form a long chain called a *polypeptide*
10	Cytoplasm	As the ribosome moves along the mRNA strand, more and more amino acids are added to the growing polypeptide chain in the sequence dictated by the mRNA codons. (Each codon represents a specific amino acid to be placed in the polypeptide chain.) When the ribosome reaches the end of the mRNA molecule, it drops off the end and separates into large and small subunits again. Often, enzymes later link two or more polypeptides together to form a whole protein molecule

survival of all organisms already living and the creation of all new organisms. Cell growth depends on using genetic information in DNA to make the structural and functional proteins needed for cell survival. Cell reproduction ensures that the genetic information is passed from one generation of cells to the next and from one generation of organisms to the next. Mistakes in these processes can cause lethal genetic disorders, cancer, and other conditions. Recent advances in our ability to manipulate the genetic code, and thus cell growth and reproduction, now present us with ethical implications more far-reaching than those accompanying the birth of the atomic age. All this makes the cell life cycle a worthy and fascinating topic of study.

Cell Growth

As we have just stated, one of the two major phases of the cell life cycle is the growth phase (see Figure 4-28). It is during this phase that a newly formed cell produces new molecules, from which it

constructs the additional cell membrane, cell fibers, and other structures necessary for growth. As we have just stated, all the structural proteins, plus the enzymes needed to make lipids, carbohydrates, and other substances, are made by the cell with information contained in the genes of DNA molecules. First, we review a simplified account of how these proteins are made and how additional organelles are produced. Later, we discuss how the cell replicates its DNA molecules in anticipation of reproduction.

Production of Cytoplasm

As a cell grows, it must produce additional cytoplasm and the plasma membrane necessary to contain it. One mechanism by which additional cytoplasm is produced is protein synthesis. Recall from the previous section that protein synthesis is an anabolic process, which means that small molecules are joined together to form large molecules. Amino acids are strung together in a specific sequence to form polypeptide chains. Two or more folded

Table 4-3 Types of RNA

ACRONYM	NAME	DESCRIPTION	ROLE IN CELL FUNCTION
RNA Involved in Protein Synthesis			
mRNA	Messenger RNA	Single, unfolded strand of nucleotides	Serves as working copy of one protein-coding gene
rRNA	Ribosomal RNA	Single, folded strand of nucleotides	Component of the ribosome (along with proteins); attaches to mRNA and participates in translation
tRNA	Transfer RNA	Single, folded strand of nucleotides; has an anticodon at one end and an amino acid–binding site at the other end	Carries a specific amino acid to a specific codon of mRNA at the ribosome during translation
snRNP	Small nuclear ribonucleoprotein	Single, folded strand of RNA (combined with polypeptide chains)	Component of the spliceosome (see Box 4-4); attaches to an mRNA transcript to facilitate editing (removal of introns; splicing of exons) into the final version of mRNA
RNA Involved in Gene Silencing (see Box 4-5)			
dsRNA	Double-strand RNA	Double strand of nucleotides (may be up to several hundred nucleotides long)	Involved in RNA interference; see siRNA, which is a type of dsRNA
siRNA	Short interfering RNA	Short segment of double-strand RNA (only 20-25 nucleotides long)	Forms part of the RNA-induced silencing complex (RISC) during RNA interference

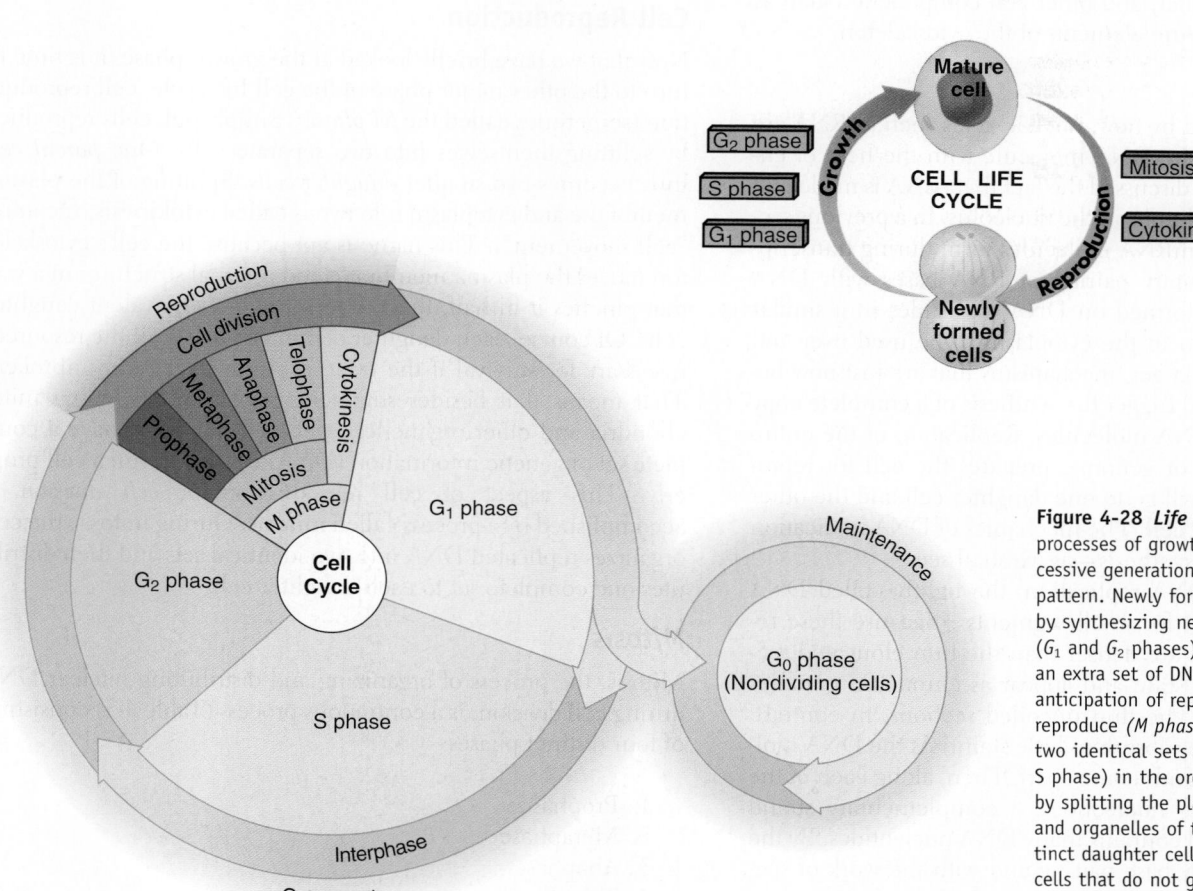

Figure 4-28 *Life cycle of the cell.* The processes of growth and reproduction of successive generations of cells exhibit a cyclic pattern. Newly formed cells grow to maturity by synthesizing new molecules and organelles (G_1 and G_2 phases), including the replication of an extra set of DNA molecules *(S phase)* in anticipation of reproduction. Mature cells reproduce *(M phase)* by first distributing the two identical sets of DNA (produced during the S phase) in the orderly process of *mitosis,* then by splitting the plasma membrane, cytoplasm, and organelles of the parent cell into two distinct daughter cells *(cytokinesis).* Daughter cells that do not go on to reproduce are in a maintenance phase *(G_0).*

polypeptide chains may then be linked to form larger, more complex protein molecules. In the previous paragraphs, we briefly described how the cell "copies" information from genes and interprets it to form specific polypeptides. Refer again to Table 4-2 to review the process of protein synthesis.

Production of additional proteins means not only that more structural proteins are available to contribute to growth of the cytoplasm but also that more enzymes are available to catalyze the production of other organic compounds. Enzymes produced by protein synthesis are used to make additional carbohydrates, lipids, and nucleic acids—as well as more protein molecules. Products of cell anabolism during the growth phase of the cell life cycle may be used in the production of additional organelles and plasma membrane. This is a good opportunity to go back to Chapter 3 and review the process by which the cell membrane—and thus membranous organelles and plasma membrane—is made by the ER and Golgi apparatus.

Most of the membranous organelles and plasma membrane increase in size or number (or both) by the membrane-producing processes referred to in Chapter 3. One notable exception is the mitochondrion. Each mitochondrion is capable of replicating itself in a process similar to that used by some one-celled organisms such as bacteria. Mitochondria often replicate themselves during the cell growth phase so that their total number is very large by the time the cell is ready to reproduce. Nonmembranous organelles grow by anabolic processes associated with the centrosome, or microtubule-organizing center, and other cell components, such as ribosomes, that manufacture elements of the cytoskeleton.

DNA Replication

As you may have noticed by now, nucleic acids such as RNA are synthesized directly on the DNA molecule with the help of enzymes. In Chapter 3, we discussed the fact that rRNA is made and formed into ribosome subunits in the nucleolus. In a previous section, we described how mRNA molecules form during transcription by the complementary pairing of RNA bases with DNA bases. Transfer RNA is formed on DNA molecules in a similar fashion and then remain in the cytoplasm to be used over and over. As a cell becomes larger, mechanisms that are just now beginning to be understood trigger the synthesis of a complete copy of the nucleus' set of DNA molecules. Replication of the entire set of DNA molecules, or genome, prepares the cell for reproduction, when one set will go to one daughter cell and the other set to the other daughter cell. The mechanics of DNA replication resemble those of RNA synthesis—as we shall see.

In the first step of DNA replication, the tightly coiled DNA molecules uncoil except for small segments. (Because these remaining tight little coils are denser than the thin elongated sections, they absorb more stain and appear as chromatin granules under the microscope. The thin uncoiled sections, in contrast, are invisible because they absorb so little stain.) As the DNA molecule uncoils, its two strands come apart. Then, along each of the two separated strands of nucleotides, a complementary strand forms. Intracellular fluid contains many DNA nucleotides. By the mechanism of obligatory base pairing and with the work of specific enzymes, nucleotides become attached at their correct places along each DNA strand (Figure 4-29). This means that new thymine, that is, from the intracellular fluid, attaches to the "old" adenine in the original DNA strand. Conversely, new adenine attaches to old thymine. Also, new guanine joins old cytosine, and new cytosine joins old guanine.

By the end of this part of the growth phase, each of the two DNA strands of the original DNA molecule has a complete new complementary strand attached to it. Each half of the DNA molecule, or strand, in other words, has duplicated itself to create a whole new DNA molecule. Thus, two new chromosomes now replace each original chromosome. However, at this stage (before cell reproduction has actually begun), they are called *chromatids* instead of chromosomes. The two chromatids formed from each original chromosome contain duplicate copies of DNA and, therefore, the same genes as the chromosome from which they were formed. Chromatids are present as attached pairs. *Centromere* is the name of their point of attachment. Table 4-4 gives a detailed, step-by-step account of the process of DNA replication.

Because DNA replication is really the synthesis of new DNA, this part of a cell's growth phase is sometimes called the *synthesis phase* or, simply, the *S phase*. The portions of the growth phase before and after the S phase are simply called the *first growth* (G_1) *phase* and the *second growth* (G_2) *phase*. The first and second growth phases are characterized by a great deal of cytoplasm growth caused by a general increase in the anabolism of protein and other substances, as well as the production of new organelles and plasma membrane.

Cell Reproduction

Now that we have briefly looked at the growth phase, it is time to turn to the other major phase of the cell life cycle: cell reproduction (sometimes called the *M phase*). Simply put, cells reproduce by splitting themselves into two separate cells. One *parent cell* thus becomes two smaller *daughter cells*. Splitting of the plasma membrane and cytoplasm into two is called **cytokinesis** (meaning "cell movement"). This name is apt because the cell's cytoskeleton moves the plasma membrane and internal structures in a way that pinches it in half, thereby forming two equivalent daughter cells. Of course, each daughter cell must possess all the resources necessary for survival if the cycle of life is to remain unbroken. That means that besides sufficient cytoplasm, including mitochondria and other organelles, each cell must also have a complete set of genetic information (DNA) needed to run a cell properly. This aspect of cell reproduction or *cell division* is accomplished by a process called **mitosis**. During mitosis, the cell organizes replicated DNA into two identical sets and then distributes one complete set to each daughter cell.

Mitosis

Mitosis, the process of organizing and distributing nuclear DNA during cell division, is a continuous process (Table 4-5) consisting of four distinct phases:

1. Prophase
2. Metaphase
3. Anaphase
4. Telophase

Figure 4-29 *DNA replication.* When a DNA molecule makes a copy of itself, it "unzips" to expose its nucleotide bases. Through the mechanism of obligatory base pairing, coordinated by the enzyme *DNA polymerase,* new DNA nucleotides bind to the exposed bases. This forms a new "other half" to each half of the original molecule. After all the bases have new nucleotides bound to them, two identical DNA molecules will be ready for distribution to the two daughter cells.

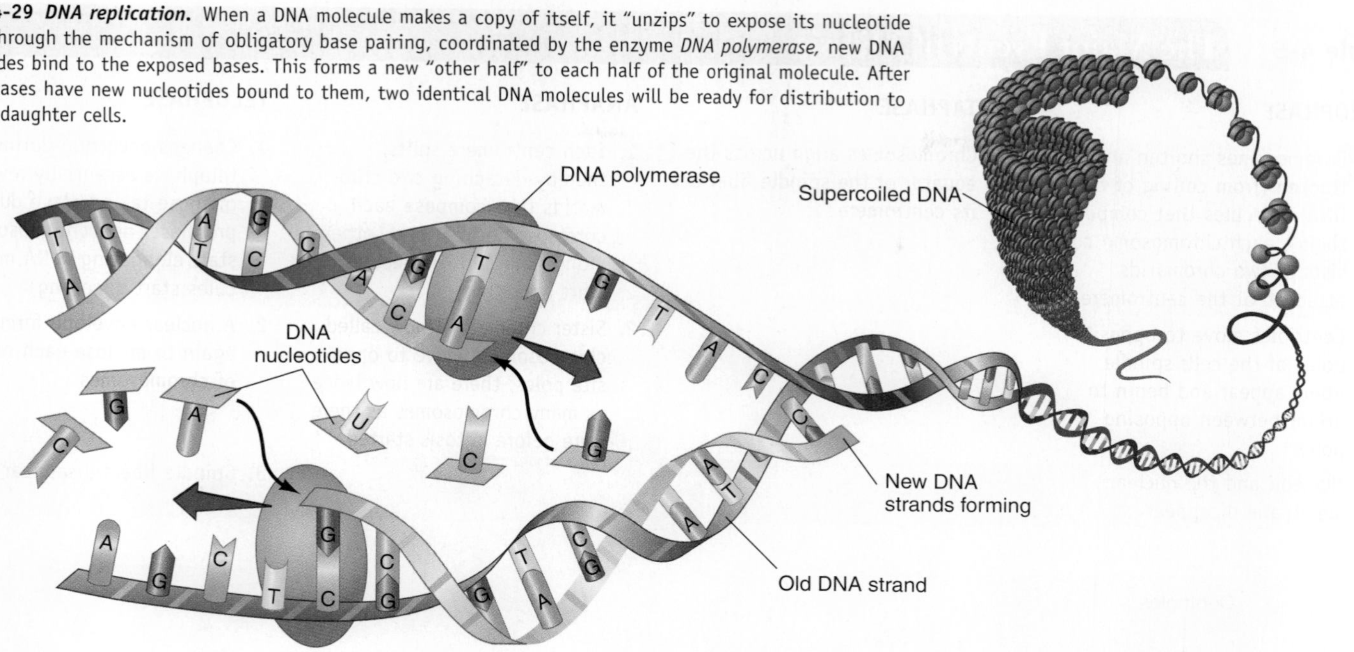

When the cell is not experiencing mitosis—during the growth phase between cell divisions—it is said to be in **interphase** (meaning "between phase"). A cell is not actively reproducing during interphase, but it is actively *preparing for* reproduction. Not only are the DNA molecules being replicated, but the centrioles (the only easily visible part of the centrosome) are also being replicated. Additional cytoplasm, membrane, and other cell structures are likewise being constructed in anticipation of cell division.

The cell enters **prophase** when it begins to divide, usually before cytokinesis, or "pinching in half," becomes apparent. During prophase, which literally means "before phase," the nuclear envelope falls apart as the paired chromatids coil up to form dense, compact *chromosomes* (Figure 4-30). By the end of prophase, each chromosome consists of a pair of short, thick bodies joined together at a centromere. At the same time that chromosomes are forming, the centriole pairs move toward opposite ends, or "poles," of the parent cell. As they move apart, a parallel arrangement of microtubules, or *spindle fibers,* is constructed between them.

The term **metaphase** literally means "position-changing phase." This name is appropriate because during this phase the chromosomes, no longer trapped within a nucleus, are moved by the cytoskeleton into an orderly pattern. The chromosomes are aligned along a plane at the "equator" of the cell about midway between the centriole pairs at opposite poles of the cell (the *equatorial plate*). One chromatid of each chromosome faces one pole of the cell, and its identical sister chromatid faces the opposite pole. Each chromatid then attaches to a spindle fiber.

Anaphase, or the "apart phase," begins as soon as all the chromosomes have aligned along the cell's equator. During this phase of mitosis, the centromere of each chromosome splits to form two chromosomes, each consisting of a single DNA molecule. Each chromosome is pulled toward the nearest pole (that is, toward the

Table 4-4	Summary of DNA Replication
STEP	**DESCRIPTION**
1	DNA molecules uncoil and "unzip" to expose their bases
2	Nucleotides already present in the intracellular fluid of the nucleus attach to the exposed bases according to the principle of obligatory base pairing
3	As nucleotides attach to complementary bases along each DNA strand, the enzyme *DNA polymerase* causes them to bind to each other
4	As new nucleotides fill in the spaces left open on each DNA strand, two identical *daughter molecules* are formed. As the parent DNA molecule completely unzips, the two daughter molecules coil to become distinct, but genetically identical DNA double helices called *chromatids*

nearest centrosome) by a spindle fiber. The effect of this movement is that one set of DNA molecules reaches one end of the cell and another set reaches the other end of the cell—forming two separate, but identical pools of genetic information. By the time a dividing cell has reached anaphase, cytokinesis has usually become apparent. The cell has not completely split yet, but cleavage along the equator may be visible.

Telophase is the "end phase," or "completion phase," of mitosis. It is during this phase that the DNA is returned to its original form and location within the cell. The cell rebuilds the nuclear envelope, and the DNA molecules are trapped within a membranous basket once again. At the same time, the chromosomes elongate back into the chromatin form. Recall that DNA cannot participate in protein synthesis unless at least some of its length is

Table 4-5   The Major Events of Mitosis

PROPHASE	METAPHASE	ANAPHASE	TELOPHASE
1. Chromosomes shorten and thicken (from coiling of the DNA molecules that compose them); each chromosome consists of two chromatids attached at the centromere	1. Chromosomes align across the equator of the spindle fiber at its centromere	1. Each centromere splits, thereby detaching two chromatids that compose each chromosome from each other elongating (DNA molecules start uncoiling)	1. Changes occurring during telophase essentially reverse of those taking place during prophase; new chromosomes start elongating (DNA molecules start uncoiling)
2. Centrioles move to opposite poles of the cell; spindle fibers appear and begin to orient between opposing poles		2. Sister chromatids (now called chromosomes) move to opposite poles; there are now twice as many chromosomes as there were before mitosis started	2. A nuclear envelope forms again to enclose each new set of chromosomes
3. Nucleoli and the nuclear membrane disappear			3. Spindle fibers disappear

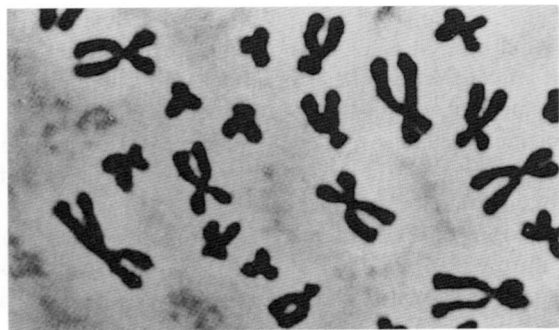

Centrioles

Nucleus

Spindle fibers

Chromatids

Centromere

Sister chromatids

Nuclear envelope

Figure 4-30 *Chromosomes.* Light micrograph showing chromosomes of a normal human cell before cell division is complete. (Additional detailed views of chromosome structure are found in Chapter 34.)

uncoiled. Spindle fibers, which are no longer needed, disappear during telophase. Cytokinesis is usually completed during or just after the telophase portion of mitosis. Each genetically identical daughter cell thus formed is now in interphase. It will grow and develop into a mature cell, perhaps becoming a parent itself.

A series of photographs summarizing the events of mitotic cell division (mitosis and cytokinesis) are shown in Figure 4-31.

Meiosis

Meiosis is the type of cell division that occurs only in primitive sex cells during the process of becoming mature sex cells. As a result of meiosis, the primitive sex cells (*spermatogonia* in the male and *oogonia* in the female) become mature sex cells called **gametes**. Male gametes are named *spermatozoa* but are usually called *sperm*. Female gametes are named *ova* (singular, *ovum*). In humans, all somatic cells contain 46 chromosomes. This total of 46 chromosomes per cell is known as the **diploid** number of chromosomes. *Diploid* comes from the Greek *diploos*, meaning "two" or "pair." In somatic cells the 46 chromosomes are present in 22 homologous pairs—the remaining 2 being the sex chromosomes XY (male) or XX (female). During meiosis, or *reduction division*, the diploid chromosome number (46) of the prim-

itive spermatogonium or oogonium is reduced to the **haploid** number of 23 found in mature sex cells, or gametes. Figure 4-32 shows that meiotic division occurs in two steps: *meiosis I*, during which the number of chromosomes is halved but the chromatid pairs remain together, and *meiosis II*, during which the chromatids finally split apart.

The end result of fertilization is the fusion of two gametes, each containing the haploid number (23) of chromosomes. Fer-

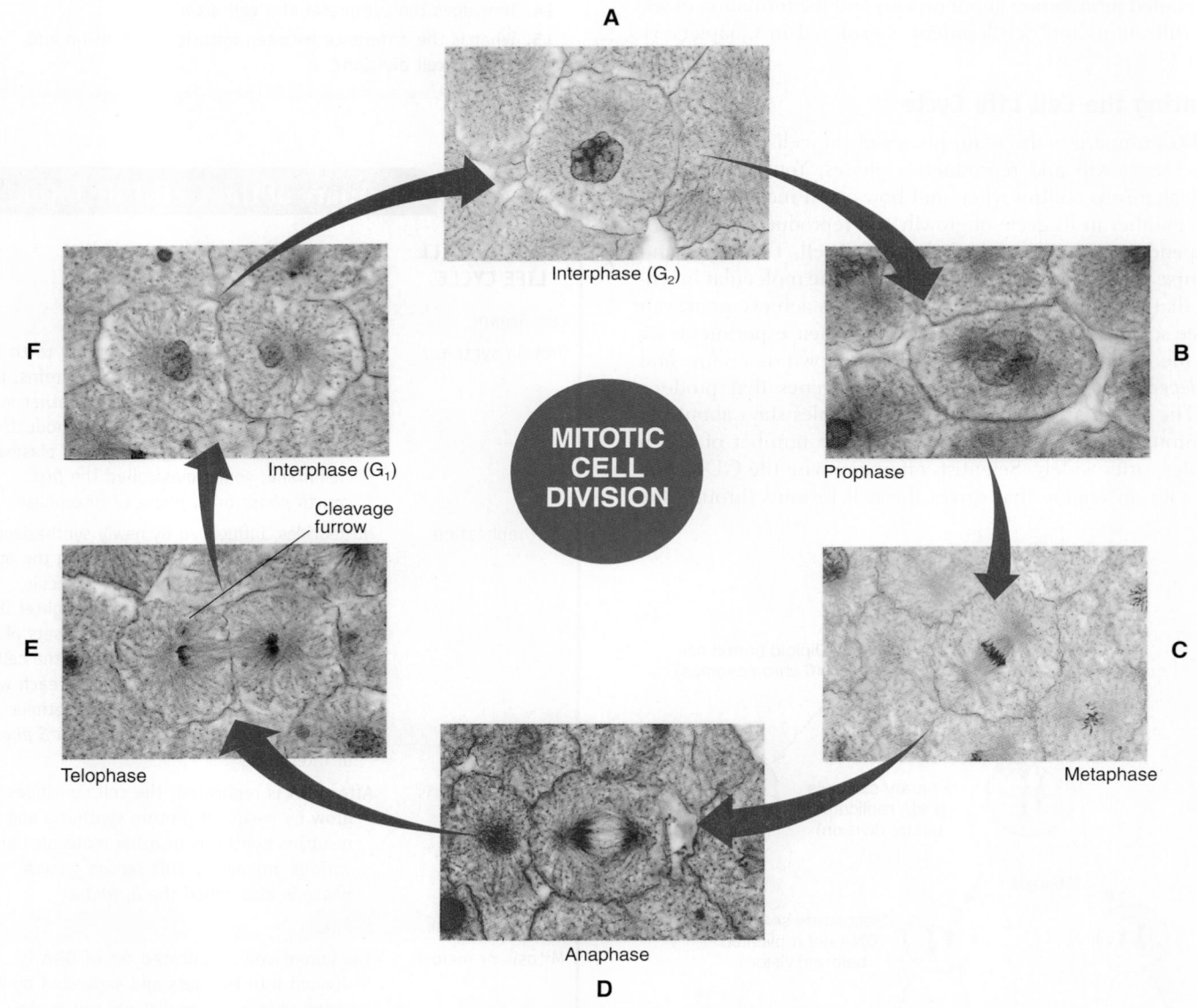

Figure 4-31 *Mitotic cell division.* Series of photomicrographs showing animal cells undergoing a cycle of mitotic cell division. **A,** *Interphase*—division begins at the end of interphase (G_2), after the cell has already replicated the DNA in its nucleus and increased its cytoplasm sufficiently for division. **B,** *Prophase*—chromosomes form from the nuclear chromatin material as the nuclear envelope disappears. **C,** *Metaphase*—chromosomes line up along the cell's equatorial plate, with spindle fibers distinctly visible on either side. **D,** *Anaphase*—spindle fibers pull each of the two chromatids (now called *chromosomes*) toward opposite poles of the cell. **E,** *Telophase*—chromosomes are now at opposite poles, nuclear envelopes begin forming around each group, and the cleavage furrow of cytokinesis becomes apparent. **F,** *Interphase*—after mitosis and cytokinesis are complete, the two daughter cells begin the first growth (G_1) of interphase. When they are ready, each daughter cell may continue the life cycle by undergoing mitotic division.

tilization results in formation of a zygote, which is a diploid cell having 46 chromosomes, 23 chromosomes being contributed by each parent. The zygote is the first cell of the human offspring. It will then undergo mitotic division to form 2 cells, then 4, and so on until 100 trillion cells are eventually formed.

During development, the new daughter cells will specialize, or **differentiate,** to become specific cell types, which in turn will form specific organs. In mature tissues, some cells may opt out of the cycle and remain in a growth and maintenance phase called the G_0 *phase*. We will have a little more to say about the development and maintenance of different tissue types in Chapter 5. More detailed information about meiosis and the formation of sex cells, fertilization, and development is explored in Chapters 31 through 33.

Regulating the Cell Life Cycle

Table 4-6 summarizes the main phases of the cell life cycle, including the growth and reproductive phases. You may wonder what mechanisms control when and how a cell moves from one stage to another in its cycle of growth and reproduction. A series of independent experiments by Leland Hartwell, Tim Hunt, and Paul Nurse paved the way for understanding the molecular mechanisms that regulate the cell's life cycle. This achievement won the three scientists a Nobel Prize in 2001. Their experiments revealed the activity of regulatory proteins known as *cyclins* and *cyclin-dependent kinases* (CDKs) and the genes that produce them. The number of CDK enzyme molecules stays about the same throughout the cell's life cycle, but the number of cyclin molecules varies widely. Scientists often compare the CDK molecules with an engine that drives the cell forward through the

phases of its life cycle. Cyclins act like a gear box that "shifts" the CDK into "drive" and thus moves the cell into the next phase of the cycle. Discoveries about how these molecules and their genes work have already helped us better understand the mechanisms of cancer and other cellular disorders (Box 4-6).

 QUICK CHECK

13. What are the two major phases of the cell life cycle? During which of these phases does mitosis occur?
14. How does the cytoplasm of a cell grow?
15. What is the difference between mitotic cell division and meiotic cell division?

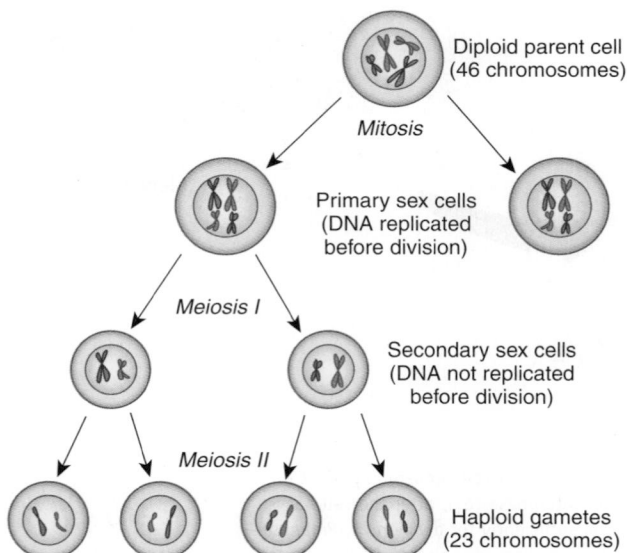

Figure 4-32 *Meiosis.* Meiotic cell division takes place in two steps: *meiosis I* and *meiosis II.* Meiosis is called *reduction division* because the number of chromosomes is reduced by half (from the diploid number to the haploid number). A more detailed diagram of meiosis is presented in Figure 33-1 on p. 1163.

Table 4-6	Summary of the Cell Life Cycle
PHASE OF CELL LIFE CYCLE	**DESCRIPTION**
CELL GROWTH	INTERPHASE
Protein synthesis	Proteins are manufactured according to the cell's genetic code; functional proteins, the enzymes, direct the synthesis of other molecules in the cells and thus the production of more and larger organelles and plasma membrane; sometimes called the *first growth phase* or *G_1 phase* of interphase
DNA replication	Nucleotides, influenced by newly synthesized enzymes, arrange themselves along the open sides of an "unzipped" DNA molecule, thereby creating two identical daughter DNA molecules; produces two identical sets of the cell's genetic code, which enables the cell to later split into two different cells, each with its own complete set of DNA; sometimes called the *[DNA] synthesis* stage or *S phase* of interphase
Protein synthesis	After DNA is replicated, the cell continues to grow by means of protein synthesis and the resulting synthesis of other molecules and various organelles; this *second growth phase* is also called the *G_2 phase*
CELL REPRODUCTION	M PHASE
Mitosis or meiosis	The parent cell's replicated set of DNA is divided into two sets and separated by an orderly process into distinct cell nuclei; mitosis is subdivided into at least four phases: *prophase, metaphase, anaphase,* and *telophase*
Cytokinesis	The plasma membrane of the parent cell "pinches in" and eventually separates the cytoplasm and two daughter nuclei into two genetically identical daughter cells

Cycle of Life

Cells

Different types of cells have highly variable life cycles. The active life span of a single cell may vary from a few minutes to years, depending on its function and level of activity. Some cells may remain dormant or inactive for years. Then, when they are "activated" by some biological need and become functional, their life span may be shortened dramatically.

One example involves cells in the immune system that are programmed to produce antibodies against a specific disease. Other examples include the female sex cells, or ova, which are present from birth. They mature throughout the reproductive life span of the individual so that each month at least one cell will become fully developed and provide an opportunity for fertilization to occur.

Structure follows function at every level of organization in the body. Frequently, function decreases with advancing age, and resulting changes occur in cell numbers and their ability to function effectively. As a result, we lose functional capacity in every body organ system. Our muscles may atrophy, the skin will lose its elasticity, and our respiratory, cardiovascular, and skeletal systems will become affected because of cellular changes that accompany aging.

 THE BIG PICTURE

Cell Physiology and the Whole Body

When exploring the microscopic world of cells, it is easy to get caught up in the intricate mechanisms that operate in each specific organelle. Once you feel comfortable with these mechanisms, try to put them together into a bigger picture of cell function. For example, most of the processes that we discussed in this chapter are going on at about the same time within each and every cell of your body. Each cell is transcribing genes and synthesizing polypeptides, which are then dumped into the ER and transported to the Golgi apparatus for processing and packaging before being sent off to become a lysosome or being secreted by exocytosis. At the same time, energy for this and other cell work is being transferred from food molecules to ATP molecules, which act as energy storage batteries for the cell. The cytoskeleton and the cell membrane are transporting materials into, out of, and around the cytoplasm. Studying cell structure and function is like looking at the score of a symphony for the first time. All the individual parts look unrelated and somewhat confusing, but with a little effort you can combine them to form a coherent whole. The "symphony" of normal cell function results from a coordinated combination of many processes dictated by the cell's "musical score"—the genetic code.

After considering the cell as a whole living unit, take another step back from your mental image of a cell. Picture a huge "society" of trillions of cells—the human body. Each cell contributes to the survival of its society (the body)—and itself—by specializing in functions that help maintain the relative constancy of the internal environment. When we think of that relative constancy, homeostasis, we should appreciate that it is all accomplished by the action of many individual cells. How are individual cells grouped together? How do they function as groups to promote homeostasis? These questions are answered, at least in part, in Chapter 5.

BOX 4-6 Changes in Cell Growth, Reproduction, and Survival

Cells have the ability to adapt to changing conditions. Cells may alter their size, reproductive rate, or other characteristics to adapt to changes in the internal environment. Such adaptations usually allow cells to work more efficiently. However, sometimes cells alter their characteristics abnormally—thereby decreasing their efficiency and threatening the health of the body. Common types of changes in cell growth and reproduction are summarized below.

Cells may respond to changes in function, hormone signals, or the availability of nutrients by increasing or decreasing in size. The term **hypertrophy** (hye-PER-troh-fee) refers to an increase in cell size, and the term **atrophy** (AT-roh-fee) refers to a decrease in cell size. Either type of adaptive change can occur easily in muscle tissue. When a person continually uses muscle cells to pull against heavy resistance, as in weight training, for example, the cells respond by increasing in size. Body builders thus increase the size of their muscles by hypertrophy—increasing the size of muscle cells. Atrophy often occurs in underused muscle cells. For example, when a broken arm is immobilized in a cast for a long period, muscles that move the arm often atrophy. Because the muscles are temporarily out of use, muscle cells decrease in size. Atrophy may also occur in tissues whose nutrient or oxygen supply is diminished. Sometimes cells respond to changes in the internal environment by increasing their rate of reproduction—a process called **hyperplasia** (hye-per-PLAY-zee-ah). The ending -*plasia* comes from a Greek word that means "formation"—referring to the formation of new cells. Because *hyper-* means "excessive," *hyperplasia* means excessive cell reproduction. Like hypertrophy, hyperplasia causes an increase in the size of a tissue or organ. However, hyperplasia is an increase in the *number of cells* rather than an increase in the size of each cell. A common example of hyperplasia occurs in the milk-producing glands of the female breast during pregnancy. In response to hor-

mone signals, the glandular cells reproduce rapidly to prepare the breast for milk production and nursing.

If the body loses its ability to control mitosis normally, abnormal hyperplasia may occur. The new mass of cells thus formed is a tumor or **neoplasm** (NEE-oh-plaz-em). Neoplasms may be relatively harmless growths called **benign** (bee-NYNE) tumors. If tumor cells can break away and travel through the blood or lymphatic vessels to other parts of the body, the neoplasm is a **malignant tumor,** or cancer. Cells in malignant neoplasms often exhibit a characteristic called **anaplasia** (an-ah-PLAY-zee-ah). Anaplasia is a condition in which cells fail to *differentiate* into a specialized cell type. **Dysplasia** (diss-PLAY-zha) is an abnormal change in shape, size, or organization of cells in a tissue and is often associated with neoplasms.

Part *A* of the figure summarizes a few of the changes that can occur in cell growth and reproduction.

Cells also die sometimes. In **necrosis** (neh-KROH-sis), cells die because of an injury or pathological condition, often causing nearby cells to die and triggering an immune response called *inflammation*, which removes the debris if possible. Nonpathological cell death, often called **apoptosis** (ap-o-TOH-sis *or* ap-op-TOH-sis), occurs frequently in the cells of your body. Apoptosis is a type of programmed cell death in which organized biochemical steps within the cell lead to fragmentation of the cell and removal of the pieces by phagocytic cells. Apoptosis occurs when cells are no longer needed or when they have certain malfunctions that could lead to cancer or some other potential problem. Apoptosis is the normal process by which our tissues and other groups of cells remodel themselves throughout the life span. Although apoptosis is a normal cell function, it can be abnormally triggered in some conditions.

BOX 4-6   Changes in Cell Growth, Reproduction, and Survival—cont'd

A

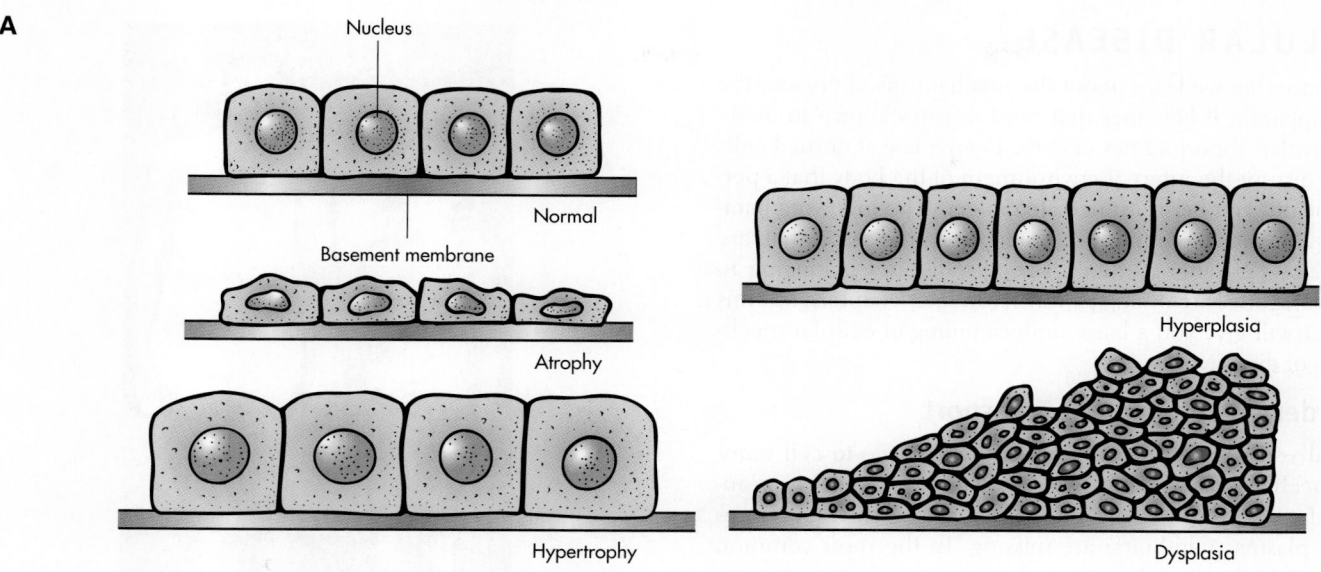

A. Alterations in cell growth and reproduction.

B

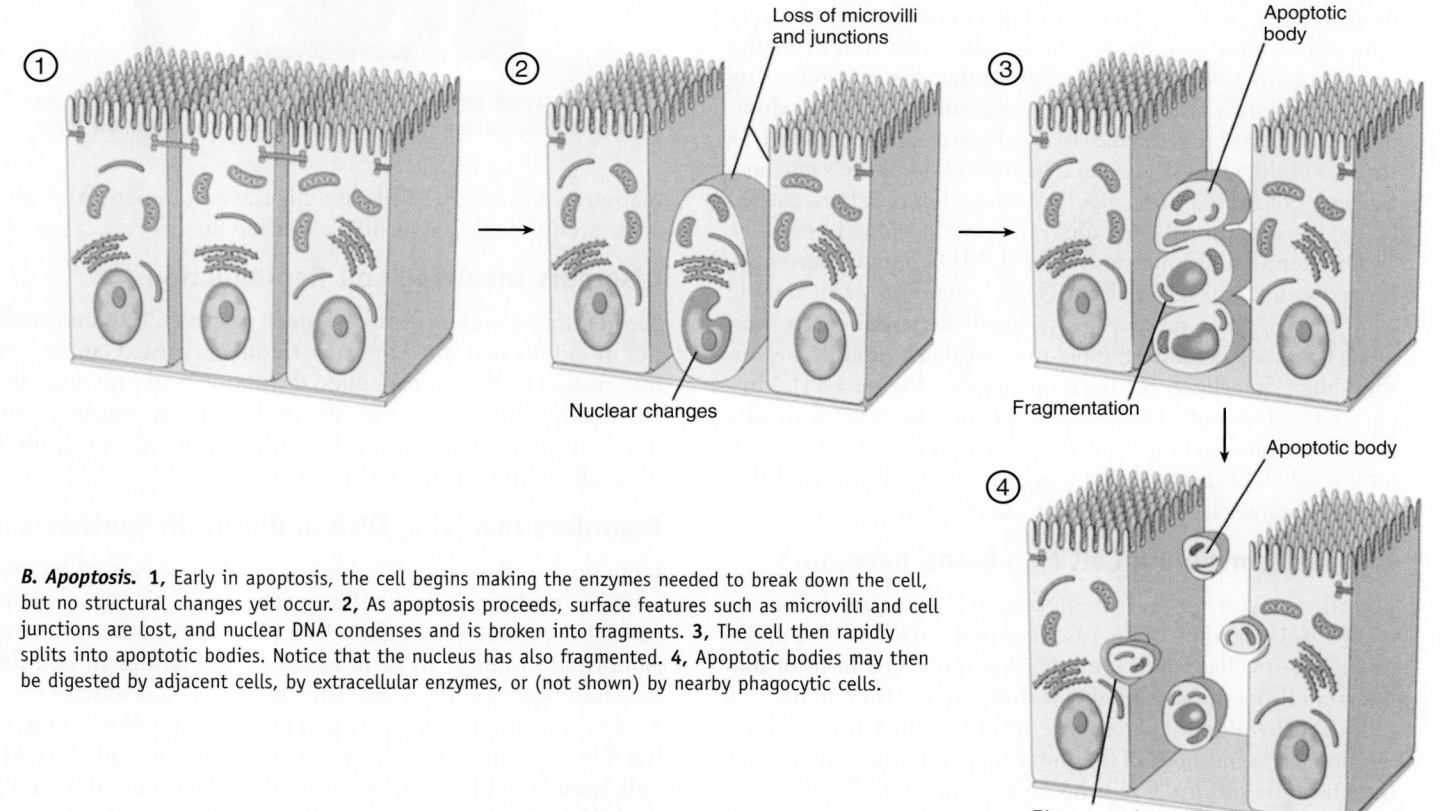

B. Apoptosis. **1,** Early in apoptosis, the cell begins making the enzymes needed to break down the cell, but no structural changes yet occur. **2,** As apoptosis proceeds, surface features such as microvilli and cell junctions are lost, and nuclear DNA condenses and is broken into fragments. **3,** The cell then rapidly splits into apoptotic bodies. Notice that the nucleus has also fragmented. **4,** Apoptotic bodies may then be digested by adjacent cells, by extracellular enzymes, or (not shown) by nearby phagocytic cells.

Mechanisms of Disease

CELLULAR DISEASE

The more that we learn about the mechanisms of disease, the more apparent it becomes that most diseases known to medicine involve abnormalities of cells. Even a few abnormal cells can so disrupt the internal environment of the body that a person's health can be in immediate danger. The following paragraphs list several important categories of disease that are caused by cell abnormalities. Specific diseases belonging to these categories are discussed further in later chapters, but this sampler will give you a basic understanding of cellular mechanisms of disease.

Disorders Involving Cell Transport

Several very severe diseases result from damage to cell transport mechanisms. **Cystic fibrosis (CF),** for example, is an inherited condition in which normal chloride ion (Cl^-) pumps in the plasma membrane are missing. In the most common form of CF, this happens when abnormal Cl^- pump proteins become misfolded in the ER and are thus not sent to the plasma membrane. Because Cl^- transport is altered, secretions such as sweat, mucus, and pancreatic juice are very salty—and often very thick. Abnormally thick mucus in the lungs impairs normal breathing and often leads to recurring lung infections. Thick pancreatic secretions can plug ducts that carry important enzymes to the digestive tract. Figure 4-33 shows a child with CF next to a healthy child of the same age. Because of the breathing, digestive, and other problems caused by the disease, the affected child has not developed normally.

Duchenne muscular dystrophy (DMD), another severe inherited condition, results from "leaky" membranes in muscle cells. *Dystrophin,* a protein that normally helps connect a muscle cell's cytoskeleton to the plasma membrane and to the surrounding extracellular matrix, is missing (see Figure 4-34). Muscle contractions pull at the weakened connections to the plasma membrane and rip holes that allow calcium ions (Ca^{++}) to enter the cell. This flood of Ca^{++} triggers chemical reactions that destroy the muscle, causing life-threatening paralysis.

Disorders Involving Cell Membrane Receptors

Disorders involving cell membrane receptors are being actively investigated by biologists. One particularly common disorder that involves cell membrane receptors is a form of **diabetes mellitus (DM)** called *type 2 diabetes.* This condition is produced by a cellular response to obesity that triggers a reduction in the number of functioning membrane receptors for the hormone *insulin.* Cells throughout the body thus become less sensitive to insulin, the hormone that allows glucose molecules to enter the plasma membrane. Without sufficient stim-

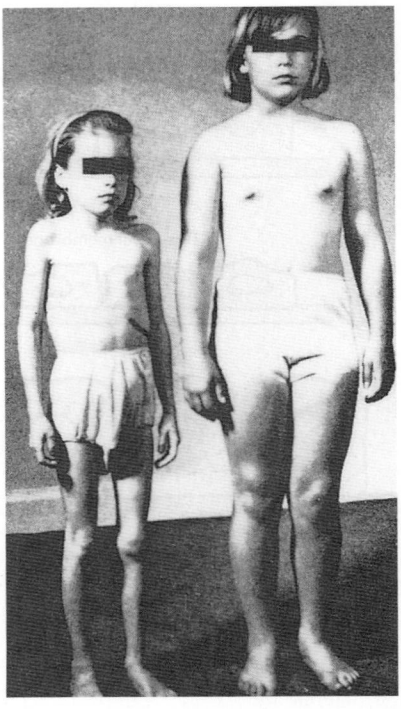

Figure 4-33 *Cystic fibrosis (CF).* The child on the left, who has CF, has failed to develop as normally as the child of the same age on the right.

ulation by insulin, the cells literally starve for energy-rich glucose, even though it is available outside the cell.

Disorders Involving Cell Reproduction

As mentioned earlier in this chapter (see p. 128), abnormalities in mitotic division can cause tumors to arise. **Cancers** are tumors that tend to spread, often disrupting vital functions and eventually killing those with the disease. Even noncancerous tumors can cause significant health impairment—or death—depending on their size and location.

Disorders Involving DNA and Protein Synthesis

Genetic disorders are pathological conditions caused by mistakes, or **mutations,** in a cell's genetic code. Abnormal genes cause the production of abnormal enzymes or other proteins. Abnormal proteins, in turn, cause abnormalities in cellular function—producing a specific disease. Many diseases are known to be caused by this mechanism, including some of the diseases in the other categories. Another example is **sickle cell anemia,** a blood disease caused by the production of abnormal hemoglobin (the protein in red blood cells that carries oxygen). Mistakes in enzyme production can cause a

Mechanisms of Disease—cont.

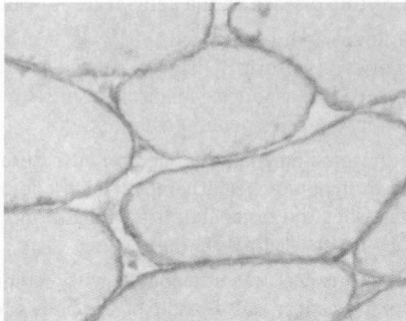

Figure 4-34 *Dystrophin*. This cross section of muscle fibers has been stained for the presence of dystrophin. In Duchenne muscular dystrophy (DMD), this protein would be completely absent, thus causing damage to the membranes of muscle fibers that results in destruction of muscle tissue.

whole group of metabolic disorders called **inborn errors of metabolism.**

Infections

Bacteria and *viruses* can infect cells and thus damage them in ways that produce disease. Bacteria, tiny one-celled organisms, may parasitize cells directly and thus destroy them, or the bacteria may produce toxins that damage cells or elicit violent reactions of the immune system. Viruses are microscopic particles that contain DNA or RNA. Viruses cause disease by taking over the genetic apparatus of a cell and thus using the cell's resources to produce viral products—or even new viruses. If enough cells are damaged during a bacterial or viral infection to disrupt vital functions, death may result.

LANGUAGE OF SCIENCE *(Cont'd from page 103)*

endocytosis (en-doh-sye-TOH-sis) [*endo-* inward or within, *-cyto-* cell, *-osis* condition]

end-product inhibition (end-PROD-ukt in-hib-ISH-un)

enzymes (EN-zymes) [*en-* in, *-zyme* ferment]

exocytosis (eks-o-sye-TOH-sis) [*exo-* outside or outward, *-cyto* cell, *-osis* condition]

exons (EKS-ahns) [*exo-* outside, *-on* unit]

facilitated diffusion (fah-SIL-i-tay-ted di-FYOO-shun) [*facilis* easy, *diffuse* spread out]

filtration (fil-TRAY-shun) [*filter* to strain]

gametes (GAM-eets) [*gamo* sexual union]

haploid (HAP-loid) [*haplo-* single, *-loid* form]

hypertonic (hye-per-TON-ik) [*hyper-* excessive, *-tonic* stretching]

hypotonic (hye-poh-TON-ik) [*hypo-* under, *-tonic* stretching]

interphase (IN-ter-fayz) [*inter-* between, *-phase* appearance]

introns (IN-trahns) [*intra-* within, *-on* unit]

isotonic (eye-soh-TON-ik) [*iso-* equal, *-tonic* stretching]

kinases (KYE-nas-es) [*kinesis-* motion, *-ase* enzyme]

lock-and-key model (lok and kee MAHD-el)

meiosis (my-OH-sis) [*meiosis* becoming smaller]

metabolic pathway (met-ah-BOL-ik PATH-way) [*meta* change]

metabolism (meh-TAB-oh-liz-em) [*meta* change]

metaphase (MET-ah-fayz) [*meta-* change, *-phase* appearance]

mitosis (my-TOH-sis) [*mitos* thread]

obligatory base pairing (oh-BLIG-ah-tor-ee base PAIR-ing)

osmosis (os-MO-sis) [*osmos-* impulse, *-osis* condition]

osmotic pressure (os-MOT-ik PRESH-ur) [*osmos-* impulse, *-ic* pertaining to]

phagocytosis (fag-oh-sye-TOH-sis) [*phago-* eating, *-cyto-* cell, *-osis* condition]

pinocytosis (pin-oh-sye-TOE-sis) [*pino-* to drink, *-cyto-* cell, *-osis* condition]

polyribosome (PAHL-ee-RYE-bo-sohm) [*poly-* many, *-soma* body]

potential osmotic pressure (po-TEN-shal os-MOT-ik PRESH-ur) [*osmos-* impulse, *-ic* pertaining to]

proenzymes (pro-EN-zimes) [*pro-* first, *-en-* in, *-zyme* ferment]

prophase (PRO-fayz) [*pro-* first, *-phase* appearance]

proteome (PRO-tee-ome) [*prote-* protein, *-ome* body (whole set)]

RNA interference (RNAi)

simple diffusion (simple di-FYOO-zhun) [*diffuse* spread out]

spliceosomes (SPLISE-oh-sohms) [*splice-* cut rope and join remaining ends, *-some* body]

telophase (TEL-oh-fayz) [*telo-* end, *-phase* appearance]

transcription (trans-KRIP-shun) [*trans-* across, *-script* to write]

translation (trans-LAY-shun) [*translate* handing over]

LANGUAGE OF MEDICINE

anaplasia (an-ah-PLAY-zha) [*ana-* without, *-plasia* shape]

atrophy (AT-ro-fee) [*atro-* without, *-phy* nourishment]

benign (be-NYNE) [*benign* kind]

cancers (KAN-sers) [*cancer* malignant tumor]

cystic fibrosis (CF) (SIS-tik fye-BRO-sis) [*cyst-* sac, *-ic* pertaining to, *fibro-* fiber, *-osis* condition]

diabetes mellitus (DM) (dye-ah-BEE-teez mell-EYE-tus) [*diabetes* to pass through, *meli* sweet]

Duchenne muscular dystrophy (DMD) (doo-SHEN MUSS-kyoo-lar DISS-troh-fee) [*Duchenne* Guillaume B.A. Duchenne de Boulogne, French neurologist, *muscul-* little mouse (muscle) *–ar* pertaining to, *dys-* bad, *-trophy* nourishment]

dysplasia (diss-PLAY-zha) [*dys-* disordered, *-plasia* to mold]

hemodialysis (he-mo-dye-AL-i-sis) [*hemo-* relating to blood, *-dia-* apart, *-lysis* loosening]

hyperplasia (hye-per-PLAY-zha) [*hyper-* excessive, *-plasia* to mold]

hypertrophy (hye-PER-tro-fee) [*hyper-* excessive, *-trophy* nourishment]

inborn errors of metabolism

malignant tumor (mah-LIG-nant TOO-mer) [*malign* bad, *tumor* swelling]

mutations (myoo-TAY-shuns) [*mutate* to change]

necrosis (ne-KROH-sis) [*necro-* death, *-osis* condition]

neoplasm (NEE-o-plazm) [*neo-* new, *-plasm* tissue]

sickle cell anemia (SIK-ul sell ah-NEE-mee-ah) [*sickle* cresent, *cell* storeroom, *anemia* without blood]

CASE STUDY

Rita Murphy, age 56, was admitted to the hospital to undergo endometrial testing for dysfunctional uterine bleeding. Mrs. Murphy began to pass blood clots larger than a quarter after not having had a menstrual period for 3 years. Her last Pap smear and pelvic examination, done more than 5 years ago, was normal. Blood tests show slight anemia with a hemoglobin level of 10.4. Mrs. Murphy has been taking unopposed estrogen for 2 years because the progesterone prescribed by her gynecologist made her feel bloated. Admission orders include starting an intravenous (IV) solution of dextrose 5% and lactated Ringer's solution.

1. Would you expect the IV solution to be hypertonic, hypotonic, or isotonic? Why?

   A. Hypertonic, to decrease the size of the red blood cell.
   B. Isotonic, because this solution does not change the size of the red blood cell.
   C. Hypotonic, to increase the size of the red blood cell and thus enable it to carry more oxygen.
   D. Hypertonic, to cause fluid shifts from the intracellular space to the extracellular space and thus replace the blood lost by the patient from uterine bleeding.

2. Which types of cell growth and reproduction might you expect to be causing this bleeding?

   A. Hypertrophy
   B. Atrophy
   C. Hyperplasia
   D. Hypoplasia

3. Mrs. Murphy has been diagnosed with cancer. The best description of this disease at the cellular level would be:

   A. Bacteria infect and damage cells.
   B. Abnormalities in mitotic division occur.
   C. Damage to cell transport mechanisms occurs.
   D. Mutations in the cell's genetic code occur.

4. A complete hysterectomy is performed on Mrs. Murphy, and she is scheduled to undergo chemotherapy. Which one of the following describes the rationale for administration of chemotherapy to Mrs. Murphy?

   A. Mrs. Murphy's cancer was a neoplasm with benign growth.
   B. Mrs. Murphy's cancer was a result of the body's loss of its ability to control mitosis.
   C. Mrs. Murphy's tumor cells may have broken away from the primary site and traveled via blood or the lymphatic systems to other parts of the body.
   D. Mrs. Murphy had a neoplasm with malignant growth.

CHAPTER SUMMARY

MOVEMENT OF SUBSTANCES THROUGH CELL MEMBRANES (TABLE 4-1)

A. Passive transport processes—do not require any energy expenditure of the cell membrane
   1. Diffusion—a passive process (Figure 4-1)
      a. Molecules spread through the membranes
      b. Molecules move from an area of high concentration to an area of low concentration, down a concentration gradient (Figure 4-2)
      c. As molecules diffuse, a state of equilibrium will occur
   2. Simple diffusion (Figure 4-3)
      a. Molecules cross through the phospholipid bilayer
      b. Solutes permeate the membrane; therefore we call the membrane permeable
   3. Osmosis (Figure 4-4)
      a. Diffusion of water through a selectively permeable membrane; limits diffusion of at least some of the solute particles
      b. Water pressure that develops as a result of osmosis is called osmotic pressure

   c. Potential osmotic pressure is the maximum pressure that could develop in a solution when it is separated from pure water by a selectively permeable membrane; knowledge of potential osmotic pressure allows prediction of the direction of osmosis and the resulting change in pressure:
      (1) Isotonic—when two fluids have the same potential osmotic pressure (Figure 4-5)
      (2) Hypertonic—"higher pressure"; cells placed in solutions that are hypertonic to intracellular fluid always shrivel as water flows out of them; this has great medical importance: if medical treatment causes the extracellular fluid to become hypertonic to the cells of the body, serious damage may occur
      (3) Hypotonic—"lower pressure"; cells placed in a hypotonic solution may swell as water flows into them; water always osmoses from the hypotonic solution to the hypertonic solution
   d. Osmosis results in gain of volume on one side of the membrane and loss of volume on the other side of the membrane

4. Facilitated diffusion (mediated passive transport)
   a. A special kind of diffusion in which movement of molecules is made more efficient by the action of transporters embedded in a cell membrane
   b. Transports substances down a concentration gradient
   c. Energy required comes from the collision energy of the solute
   d. Channel-mediated passive transport (Figure 4-6)
      (1) Channels are specific—allow only one type of solute to pass through
      (2) Gated channels may be open or closed (or inactive)—may be triggered by any of a variety of stimuli
      (3) Channels allow membranes to be selectively permeable
      (4) Aquaporins are water channels that permit rapid osmosis
   e. Carrier-mediated passive transport (Figure 4-7)
      (1) Carriers attract and bind to the solute, change shape, and release the solute out the other side of the carrier
      (2) Carriers are usually reversible, depending on the direction of the concentration gradient
B. Active transport processes—require the expenditure of metabolic energy by the cell
   1. Transport by pumps
      a. Pumps are membrane transporters that move a substance against its concentration gradient—opposite of diffusion
      b. Examples: calcium pumps (Figure 4-8) and sodium-potassium pumps (Figure 4-9)
   2. Transport by vesicles—allow substances to enter or leave the interior of a cell without actually moving through its plasma membrane
      a. Endocytosis—the plasma membrane "traps" some extracellular material and brings it into the cell in a vesicle
         (1) Two basic types of endocytosis (Figure 4-10)
            (a) Phagocytosis—"condition of cell eating"; large particles are engulfed by the plasma membrane and enter the cell in vesicles; the vesicles fuse with lysosomes, which digest the particles
            (b) Pinocytosis—"condition of cell drinking"; fluid and the substances dissolved in it enter the cell
         (2) Receptor-mediated endocytosis—membrane receptor molecules recognize substances to be brought into the cell (Figure 4-11)
      b. Exocytosis
         (1) Process by which large molecules, notably proteins, can leave the cell even though they are too large to move out through the plasma membrane
         (2) Large molecules are enclosed in membranous vesicles and then pulled to the plasma membrane by the cytoskeleton, where the contents are released
         (3) Exocytosis also provides a way for new material to be added to the plasma membrane

CELL METABOLISM

A. Metabolism is the set of chemical reactions in a cell
   1. Catabolism—breaks large molecules into smaller ones; usually releases energy
   2. Anabolism—builds large molecules from smaller ones; usually consumes energy
B. Role of enzymes
   1. Enzymes are chemical catalysts that reduce the activation energy needed for a reaction (Figure 4-12)
   2. Enzymes regulate cell metabolism
   3. Chemical structure of enzymes
      a. Proteins of a complex shape
      b. The active site is where the enzyme molecule fits the substrate molecule—the lock-and-key model (Figure 4-13)
   4. Classification and naming of enzymes
      a. Enzymes usually have an -ase ending, with the first part of the word signifying the substrate or the type of reaction catalyzed
      b. Oxidation-reduction enzymes—known as oxidases, hydrogenases, and dehydrogenases; energy release depends on these enzymes
      c. Hydrolyzing enzymes—hydrolases; digestive enzymes belong to this group
      d. Phosphorylating enzymes—phosphorylases or phosphatases; add or remove phosphate groups
      e. Enzymes that add or remove carbon dioxide—carboxylases or decarboxylases
      f. Enzymes that rearrange atoms within a molecule—mutases or isomerases
      g. Hydrases add water to a molecule without splitting it
   5. General functions of enzymes
      a. Enzymes regulate cell functions by regulating metabolic pathways (Figure 4-14)
      b. Enzymes are specific in their actions
      c. Various chemical and physical agents known as allosteric effectors affect enzyme action by changing the shape of the enzyme molecule; examples of allosteric effectors include (Figure 4-15)
         (1) Temperature (Figure 4-16, B)
         (2) Hydrogen ion (H^+) concentration (pH) (Figure 4-16, A)
         (3) Ionizing radiation
         (4) Cofactors
         (5) End products of certain metabolic pathways (Figure 4-17)
      d. Most enzymes catalyze a chemical reaction in both directions
      e. Enzymes are continually being destroyed and continually being replaced
      f. Many enzymes are first synthesized as inactive proenzymes

C. Catabolism
  1. Cellular respiration, the pathway by which glucose is broken down to yield its stored energy, is an important example of cell catabolism; cellular respiration has three pathways that are chemically linked (Figure 4-21):
     a. Glycolysis (Figure 4-18)
     b. Citric acid cycle (Figure 4-19)
     c. Electron transport system (ETS) (Figure 4-20)
  2. Glycolysis (Figure 4-18)
     a. Pathway in which glucose is broken apart into two pyruvic acid molecules to yield a small amount of energy (which is transferred to ATP and NADH)
     b. Includes many chemical steps (reactions that follow one another), each regulated by specific enzymes
     c. Is anaerobic (requires no oxygen)
     d. Occurs within cytosol (outside the mitochondria)
  3. Citric acid cycle (Krebs cycle) (Figure 4-19)
     a. Pyruvic acid (from glycolysis) is converted into acetyl CoA and enters the citric acid cycle after losing CO_2 and transferring some energy to NADH
     b. Citric acid cycle is a repeating (cyclic) sequence of reactions that occur inside the inner chamber of a mitochondrion. Acetyl splits from CoA and is broken down to yield waste CO_2 and energy (in the form of energized electrons), which is transferred to ATP, NADH, and $FADH_2$
  4. Electron transport system (ETS) (Figure 4-20)
     a. Energized electrons are carried by NADH and $FADH_2$ from glycolysis and the citric acid cycle to electron acceptors embedded in the cristae of the mitochondrion
     b. As electrons are shuttled along a chain of electron-accepting molecules in the cristae, their energy is used to pump accompanying protons (H^+) into the space between mitochondrial membranes
     c. Protons flow back into the inner chamber through pump molecules in the cristae, and their energy of movement is transferred to ATP
     d. Low-energy electrons coming off the ETS bind to oxygen and rejoin their protons to form water (H_2O)
D. Anabolism
  1. Protein synthesis is a central anabolic pathway in cells (Table 4-2)
  2. Deoxyribonucleic acid (DNA)
     a. A double-helix polymer (composed of nucleotides) that functions to transfer information, encoded in genes, to direct the synthesis of proteins (Figure 4-22)
     b. Gene—a segment of a DNA molecule that consists of approximately 1000 pairs of nucleotides and contains the code for synthesizing one polypeptide (Figure 4-23)
  3. Transcription—mRNA forms along a segment of one strand of DNA (Figure 4-24)
  4. Editing (Figure 4-25)
     a. Noncoding introns are removed and the remaining exons are spliced together to form the final, edited version of the mRNA copy of the DNA segment

     b. Spliceosomes are ribosome-sized structures in the nucleus that splice mRNA transcripts (Box 4-4)
  5. Translation (Figure 4-26)
     a. After leaving the nucleus and being edited, mRNA associates with a ribosome in the cytoplasm
     b. tRNA molecules bring specific amino acids to the mRNA at the ribosome; the type of amino acid is determined by the fit of a specific tRNA's anticodon with mRNA's codon (Figure 4-27)
     c. As amino acids are brought into place, peptide bonds join them—eventually producing an entire polypeptide chain
     d. Translation of genes can be inhibited by RNA interference (RNAi), which protects the cell against viral infection (Box 4-5)
  6. Processing—chaperone molecules and other enzymes in the cytosol, ER, and Golgi apparatus help polypeptides fold and then possibly combine into larger protein molecules or hybrid molecules
  7. Proteome
     a. All the proteins synthesized by a cell make up the cell's proteome
     b. All the proteins synthesized in the whole body is called the human proteome

GROWTH AND REPRODUCTION OF CELLS

A. Cell growth and reproduction of cells are the most fundamental of all living functions and together constitute the cell life cycle (Figure 4-28; Table 4-6)
  1. Cell growth—depends on using genetic information in DNA to make the structural and functional proteins needed for cell survival
  2. Cell reproduction—ensures that genetic information is passed from one generation to the next
B. Cell growth—a newly formed cell produces a variety of molecules and other structures necessary for growth by using the information contained in the genes of DNA molecules; this stage is known as interphase
  1. Production of cytoplasm—more cell material is made, including growth and/or replication of organelles and plasma membrane; a largely anabolic process
  2. DNA replication (Table 4-4)
     a. Replication of the genome prepares the cell for reproduction; the mechanics are similar to RNA synthesis
     b. DNA replication (Figure 4-29)
        (1) The DNA strand uncoils and the strands come apart
        (2) Along each separate strand, a complementary strand forms
        (3) The two new strands are called chromatids instead of chromosomes
        (4) Chromatids are attached pairs, and the centromere is the name of their point of attachment
  3. The growth phase of the cell life cycle can be subdivided into the first growth phase (G_1), the [DNA] synthesis phase (S), and the second growth phase (G_2)

C. Cell reproduction—cells reproduce by splitting themselves into two smaller daughter cells (Table 4-5)
  1. Mitotic cell division—the process of organizing and distributing nuclear DNA during cell division has four distinct phases (Figure 4-31)
    a. Prophase—"before phase"
      (1) After the cell has prepared for reproduction during interphase, the nuclear envelope falls apart as the chromatids coil up to form chromosomes that are joined at the centromere (Figure 4-30)
      (2) As chromosomes are forming, the centriole pairs are seen to move toward the poles of the parent cell, and spindle fibers are constructed between them
    b. Metaphase—"position-changing phase"
      (1) Chromosomes move so that one chromatid of each chromosome faces its respective pole
      (2) Each chromatid attaches to a spindle fiber
    c. Anaphase—"apart phase"
      (1) The centromere of each chromosome has split to form two chromosomes, each consisting of a single DNA molecule
      (2) Each chromosome is pulled toward the nearest pole to form two separate, but identical pools of genetic information
    d. Telophase—"end phase"
      (1) DNA returns to its original form and location within the cell
      (2) After completion of telophase, each daughter cell begins interphase to develop into a mature cell
  2. Meiosis (Figure 4-31; see also Figure 33-1)
D. Regulating the cell life cycle
  1. Cyclin-dependent kinases (CDKs) are activating enzymes that drive the cell through the phases of its life cycle
  2. Cyclins are regulatory proteins that control the CDKs and "shift" them to start the next phase

CYCLE OF LIFE: CELLS

A. Different types of cells have different life cycles
B. Advancing age creates changes in cell numbers and in their ability to function effectively
  1. Examples of decreased functional ability include muscle atrophy, loss of elasticity of the skin, and changes in the cardiovascular, respiratory, and skeletal systems

THE BIG PICTURE: CELLS AND THE WHOLE BODY

A. Most cell processes are occurring at the same time in all of the cells throughout the body
B. The processes of normal cell function result from the coordination dictated by the genetic code

REVIEW QUESTIONS

1. Define the terms diffusion, dialysis, facilitated diffusion, osmosis, and filtration.
2. Explain how a concentration gradient relates to the process of diffusion.
3. Describe and give an example of a membrane channel.
4. Explain the terms isotonic, hypotonic, and hypertonic.
5. State the principle about a solution in which osmotic pressure develops, given appropriate conditions.
6. State the principle about the direction of active transport.
7. Name and describe the active transport pump that operates in the plasma membrane of all human cells.
8. Explain the processes of endocytosis and exocytosis.
9. Describe the classification of enzymes.
10. Discuss three general principles of enzyme function.
11. What is metabolism? Catabolism? Anabolism?
12. Describe briefly each of the three pathways that make up the process of cellular respiration.
13. Describe the size and shape of a DNA molecule.
14. Where is DNA located?
15. Briefly outline the steps of protein synthesis.
16. As a cell grows, how is additional cell material added?
17. What are the steps involved in DNA replication and when does it occur?
18. Define mitosis.
19. Briefly describe the four distinct phases of mitosis.
20. Define meiosis and discuss its four distinct stages.
21. When does the reduction of chromosomes from the diploid to the haploid number take place?
22. Give examples of normal and abnormal hyperplasia.

CRITICAL THINKING QUESTIONS

1. The process of peritoneal dialysis and the process of a white blood cell trapping bacteria by phagocytosis are both transport processes. Compare and contrast these processes and identify them as active or passive.
2. Which type of osmotic pressure can be more easily measured? Explain your answer.
3. Intravenous solutions can be isotonic to blood cells. Therefore, it is very important to know whether sugar, a nonelectrolyte, or salt (NaCl), an electrolyte, is being given in the solution so that the proper amount can be added. What would result if the tonicity of the solution is not isotonic to the blood cells?
4. White blood cells engulf bacteria and solutions that contain dissolved proteins. How would you summarize the processes that allow them to ingest both solids and liquids?
5. Explain how the shape of an enzyme determines its function. What would result if an allosteric effector changed the shape of the enzyme?
6. Compare and contrast aerobic and anaerobic respiration.
7. Why is the mitochondrion such an important organelle for survival of the cell? Explain why some cells, such as skeletal muscle cells, have more mitochondria than others do.
8. Certain antibiotics can damage ribosomes in normal human body cells. People taking these antibiotics need to be carefully monitored. Summarize the result of fewer ribosomes on the process of transcription and translation.
9. DNA is often called the "blueprint of life." However, DNA is composed of only four nitrogen bases as part of its nucleotide

structure. Explain how these four nitrogen bases can influence the genetic makeup of an individual.

10. Watson called DNA "the most golden of all molecules" primarily because of its role in protein synthesis. How would you describe the importance of the process of protein synthesis in the functioning of a cell?

11. A cast is frequently used to immobilize a broken bone. When the cast is removed, the muscles on that limb are usually smaller and weaker than the muscles on the other limb. How would you explain the difference between the two limbs?

12. Compare and contrast hypertrophy and hyperplasia.

13. The nucleus has been called the "brain" or the "control center" of the cell. By applying what you have learned, describe how the nucleus can control growth, development, and day-to-day functions of a cell.

Tissues

LANGUAGE OF SCIENCE

abscess (AB-ses) [*abscess* to go away]

absorption (ab-SORP-shun) [*absorb* to swallow]

adipocytes (AD-i-poh-sytes) [*adipo-* fat, *-cyte* cell]

adipose tissue (AD-i-pohs) [*adipo-* fat, *tissue* fabric]

alveolar (al-VEE-oh-lar) [*alveolus* little hollow]

areolar (ah-REE-oh-lar) [*areola* little space]

avascular (ah-VAS-kyoo-lar) [*a-* without, *-vascular* little vessel]

axon (AK-son) [*axon* axle]

basement membrane [*basement* under base, *membrane* thin skin]

blastocyst (BLASS-toh-sist) [*blasto-* early development state, *-cyst* pouch]

bradykinin (brad-ee-KYE-nin) [*brady-* slow or dull, *-kinin* to move]

canaliculi (kan-ah-LIK-yoo-lye) [*canaliculi* little channel] **sing., cannaliculus**

cancellous bone tissue (KAN-seh-lus) [*cancellous* lattice, *tissue* fabric]

cardiac muscle tissue (KAR-dee-ak) [*cardia* heart, *tissue* fabric]

chemotaxis (kee-moh-TAK-sis) [*chemo-* by chemical reaction, *-taxis* arrangement]

chondrocyte (KON-droh-syte) [*chondro-* cartilage, *-cyte* cell]

collagen (KAHL-ah-jen) [*colla-* glue, *-gen* to produce]

collagenous dense fibrous tissue (kah-LAJ-eh-nus dense FYE-brus TISH-yoo) [*colla-* glue, *-gen* to produce, *dense* thick, *fibro* fiber, *tissue* fabric]

columnar (koh-LUM-nar) [*columnar* column]

compact bone tissue (kom-PAKT) [*compact* to put together, *tissue* fabric]

compound [*compound* to assemble]

connective tissue (koh-NEK-tiv TISH-yoo) [*connective* to bind, *tissue* fabric]

connective tissue membranes (koh-NEK-tiv) [*connective* to bind, *tissue* fabric, *membrane* thin skin]

cuboidal (KYOO-boyd-al) [*cubo* cube]

Cont'd on p. 185

INTRODUCTION TO TISSUES

A **tissue** is a group of similar cells that perform a common function. Tissues can be thought of as the fabric of the body, which is "sewn together" to form the organs of the body and to hold all the organs together as a whole.

Each tissue specializes in performing at least one unique function that helps maintain homeostasis, ensuring the survival of the whole body. The arrangement of cells in a tissue may form a thin sheet only one cell deep, whereas the cells of other tissues form huge masses containing millions of cells. Regardless of the size, shape, or arrangement of cells in a tissue, they all are surrounded by or embedded in a complex extracellular material that often is called simply **matrix.**

The four major types of human tissue that were introduced in Chapter 1 are described in more detail in this chapter. An understanding of the major tissue types will help you understand the next higher levels of organization in the body—organs and organ systems. Eventually, your knowledge of **histology** (the biology of tissues) will give you a better appreciation for the nature of the whole body.

PRINCIPAL TYPES OF TISSUE

Although a number of subtypes are present in the body, all tissues can be classified by their structure and function into four principal types:

1. **Epithelial tissue** covers and protects the body surface, lines body cavities, specializes in moving substances into and out of the body or particular organs (secretion, excretion, and absorption), and forms many glands.
2. **Connective tissue** is specialized to support the body and its parts, connect and hold them together, transport substances through the body, and protect it from foreign invaders. The cells in connective tissue are often relatively far apart and separated by large quantities of matrix.
3. **Muscle tissue** produces movement; it moves the body and its parts. Muscle cells are specialized for contractility and produce movement by shortening the contractile units found in cytoplasm.
4. **Nervous tissue** may be the most complex tissue in the body. It specializes in communication between the various parts of the body and in integration of their activities. This tissue's major function is the generation of complex messages for the coordination of body functions.

The major tissue types are also summarized in Table 5-1.

EXTRACELLULAR MATRIX

Tissues differ in the amount and kind of material between the cells—the **extracellular matrix (ECM).** Figure 5-1 hints at the complex nature of the ECM. The unique and specialized nature of the extracellular matrix in bone and cartilage, for example, contributes to the strength and resiliency of the body. Some tissues have very little ECM. Other tissues are almost entirely extracellular matrix—with only a few cells present. Some types of ECM contain large numbers of structural protein fibers that make them flexible or elastic, some contain many mineral crystals that make them rigid, and others are very fluid.

It is only recently that scientists have discovered the complexities of the ECM and its important role in human biology. Besides water, the ECM is made up mostly of proteins and **proteoglycans** (Table 5-2).

Proteins in the extracellular matrix include various types of structural protein fibers such as *collagen* and *elastin*, both of which will be discussed frequently throughout this book. The ECM also contains many *glycoproteins*, which are mainly protein molecules with attached carbohydrate subunits. For example, *fibronectin* and *laminin* both help connect collagen and proteoglycans to cells by attaching to *integrin* molecules embedded in the plasma membrane of cells (Figure 5-1). This arrangement thus unites the cells and their surroundings into an integral structure. It also provides a communication mechanism between the ECM and the cell that allows coordination of cell and tissue development, as well as guidance of cell movement and shape changes.

Proteoglycans are hybrid molecules made up mostly of carbohydrates attached to a protein backbone, as you can see in Figure 5-1, *B*. Many of the sugars attached to the protein backbone of a proteoglycan molecule are *N-acetylglucosamine* (NAG). Examples of proteoglycans are *chondroitin sulfate*, *heparin*, and *hyaluronate* (Table 5-2).

In some tissues, such as bone, there may also be calcium-containing mineral crystals that make the ECM rigid.

Considering the makeup of ECM, it's no wonder that many dietary supplements that claim to improve the health of bones and joints include calcium, glucosamine, and/or chondroitin sulfate—all are important components of the ECM of cartilage and bone tissues.

In some tissues, it is the ECM that holds the tissue in a single mass. For example, in skeletal muscles, it is mostly a network of structural protein fibers in the ECM that holds skeletal muscle tissue together. In such cases, components of the ECM bind to the integrins in the outer membranes of the cells, and the integrins bind to components of the internal cytoskeleton.

In other tissues, it is primarily the intercellular junctions, such as the desmosomes and tight junctions described in a previous chapter (see Chapter 3, pp. 97-98), that hold groups of cells together to form tissues found in sheets or other continuous masses of cells. The tissue that forms the outer layer of skin is held together this way. In some tissues, the ECM does not bind to tissue cells. For example, the fluid nature of the blood's matrix (plasma) does not hold blood tissue in a solid mass at all.

EMBRYONIC DEVELOPMENT OF TISSUES

The four major tissues of the body appear early in the embryonic period of development (first 2 months after conception). After fertilization has occurred, repeated cell divisions soon convert the single-celled zygote into a hollow ball of cells called a **blastocyst.** The blastocyst implants in the uterus, and within 2 weeks the cells move and regroup in an orderly way into three **primary germ layers** called **endoderm, mesoderm,**

Table 5-1 Major Tissues of the Body

TISSUE TYPE		STRUCTURE	FUNCTION	EXAMPLES IN THE BODY
Epithelial tissue		One or more layers of densely arranged cells with very little extracellular matrix May form either sheets or glands	Covers and protects the body surface Lines body cavities Movement of substances (absorption, secretion, excretion) Glandular activity	Outer layer of skin Lining of the respiratory, digestive, urinary, reproductive tracts Glands of the body
Connective tissue		Sparsely arranged cells surrounded by a large proportion of extracellular matrix often containing structural fibers (and sometimes mineral crystals)	Supports body structures Transports substances throughout the body	Bones Joint cartilage Tendons and ligaments Blood Fat
Muscle tissue		Long fiberlike cells, sometimes branched, capable of pulling loads; extracellular fibers sometimes hold muscle fiber together	Produces body movements Produces movements of organs such as the stomach, heart Produces heat	Heart muscle Muscles of the head/neck, arms, legs, trunk Muscles in the walls of hollow organs such as the stomach, intestines
Nervous tissue		Mixture of many cell types, including several types of neurons (conducting cells) and neuroglia (support cells)	Communication between body parts Integration/regulation of body functions	Tissue of brain and spinal cord Nerves of the body Sensory organs of the body

and **ectoderm** (Figure 5-2). The process by which blastocyst cells move and then differentiate into the three primary germ layers is called **gastrulation**. During this process the cells in each germ layer become increasingly differentiated to form specific tissues. Remodeling themselves by a combination of new growth, differentiation, and apoptosis, these early layers eventually give rise to the structures listed in Figure 5-2, *B* (Box 5-1).

Recent research shows that the connections between the extracellular matrix, plasma membranes, and internal cellular structures such as the cytoskeleton (see Figure 5-1, *A*) are involved in coordinating the early development of tissues. For example, we

know now that migration of tissues cells toward one another, as happens when neurons form connections with each other, are influenced by ECM-cell connections.

In summary, some epithelial tissues develop from each of the primary germ layers, whereas connective and muscle tissues arise from mesoderm and nerve tissue develops from ectoderm. The process of the primary germ layers differentiating into the different kinds of tissues is called **histogenesis**. Histogenesis is a complex process that is assisted by connections between the ECM, cells, and internal cell structures. Chapter 33 provides additional details of human development, including a discussion of differentiation of organs and body systems.

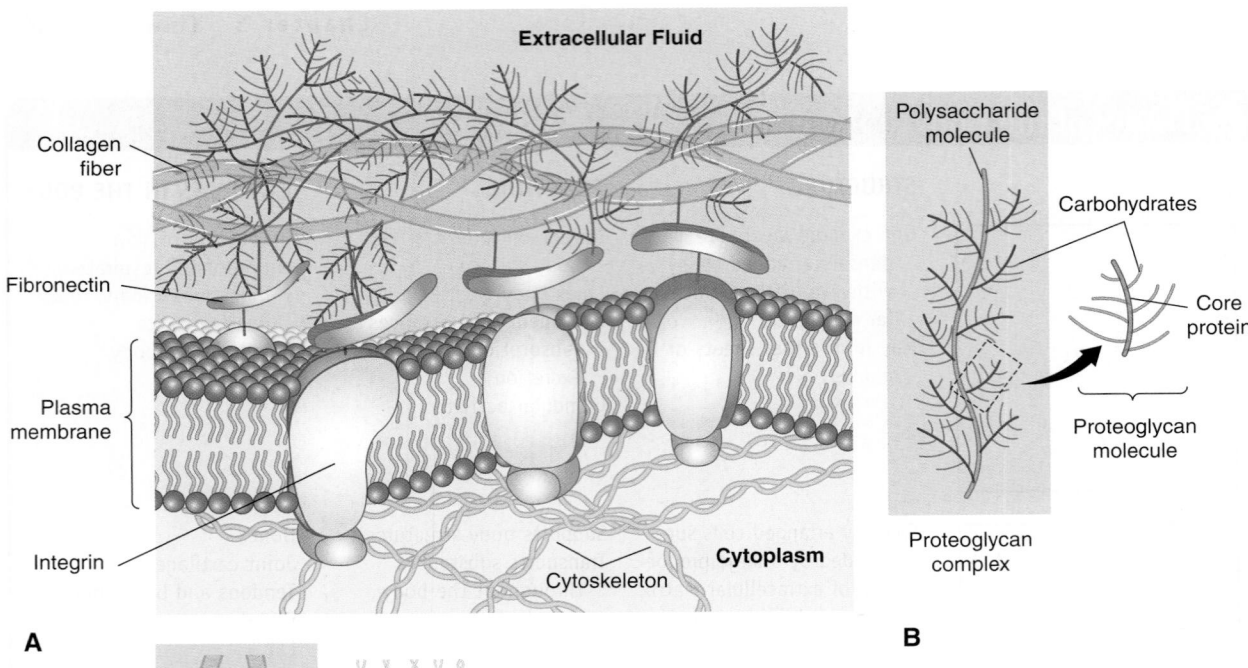

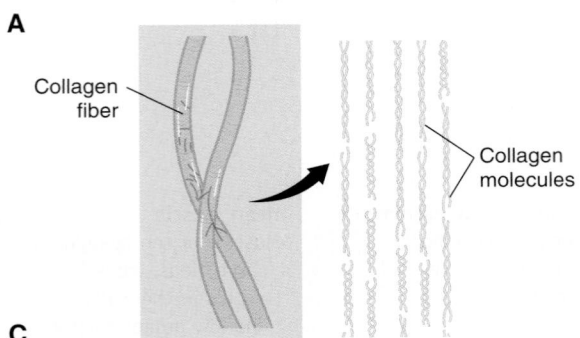

Figure 5-1 *Extracellular matrix (ECM).* **A,** The extracellular matrix is made up of water, proteins/glycoproteins, and proteoglycans that often form large bundles or complexes that bind together and to the cells of the tissue. Although the makeup of ECM varies from tissue to tissue, it usually includes some connections to integrins in the plasma membranes, thereby allowing for structural integrity, as well as communication and coordination within the tissue. **B,** A detailed view of a proteoglycan complex shows many proteoglycans, each with a protein backbone and attached carbohydrate subunits—all held together by a polysaccharide chain. **C,** Detailed view of a collagen bundle showing the individual collagen fibers within it.

Table 5-2 — Components of the Extracellular Matrix (ECM)*

COMPONENT	EXAMPLE	DESCRIPTION	FUNCTION	EXAMPLE OF LOCATION
Water		Water molecules along with a small number of ions (mostly Na⁺ and Cl⁻)	Solvent for dissolved ECM components; provides fluidity of ECM	All tissues of the body
Proteins and glycoproteins	Collagen	Strong, flexible structural protein fiber	Provides flexible strength to tissues	Tendons, ligaments, bones, cartilage, many tissues
	Elastin	Flexible, elastic structural protein fiber	Allows flexibility and elastic recoil of tissues	Skin, cartilage of ear, walls of arteries
	Fibronectin	Rodlike glycoprotein	Binds ECM to cells; communicates with cells through integrins	Many tissues of the body, for example, connective tissues
	Laminin	Glycoproteins arranged as a 3-pronged fork	Binds ECM components together and to cells; communicates with cells through integrins	In basal lamina (basement membrane) of epithelial tissues
Proteoglycans	Various types	Protein backbone with attached chains of various polysaccharides:		
		Chondroitin sulfate	Shock absorber	Cartilage, bone, heart valves
		Heparin	Reduces blood clotting	Lining of some arteries
		Hyaluronate	Thickens fluid; lubricates	Loose connective tissue, joint fluids

*Not all components are present in all ECM; examples of only some of the many major ECM components are provided.

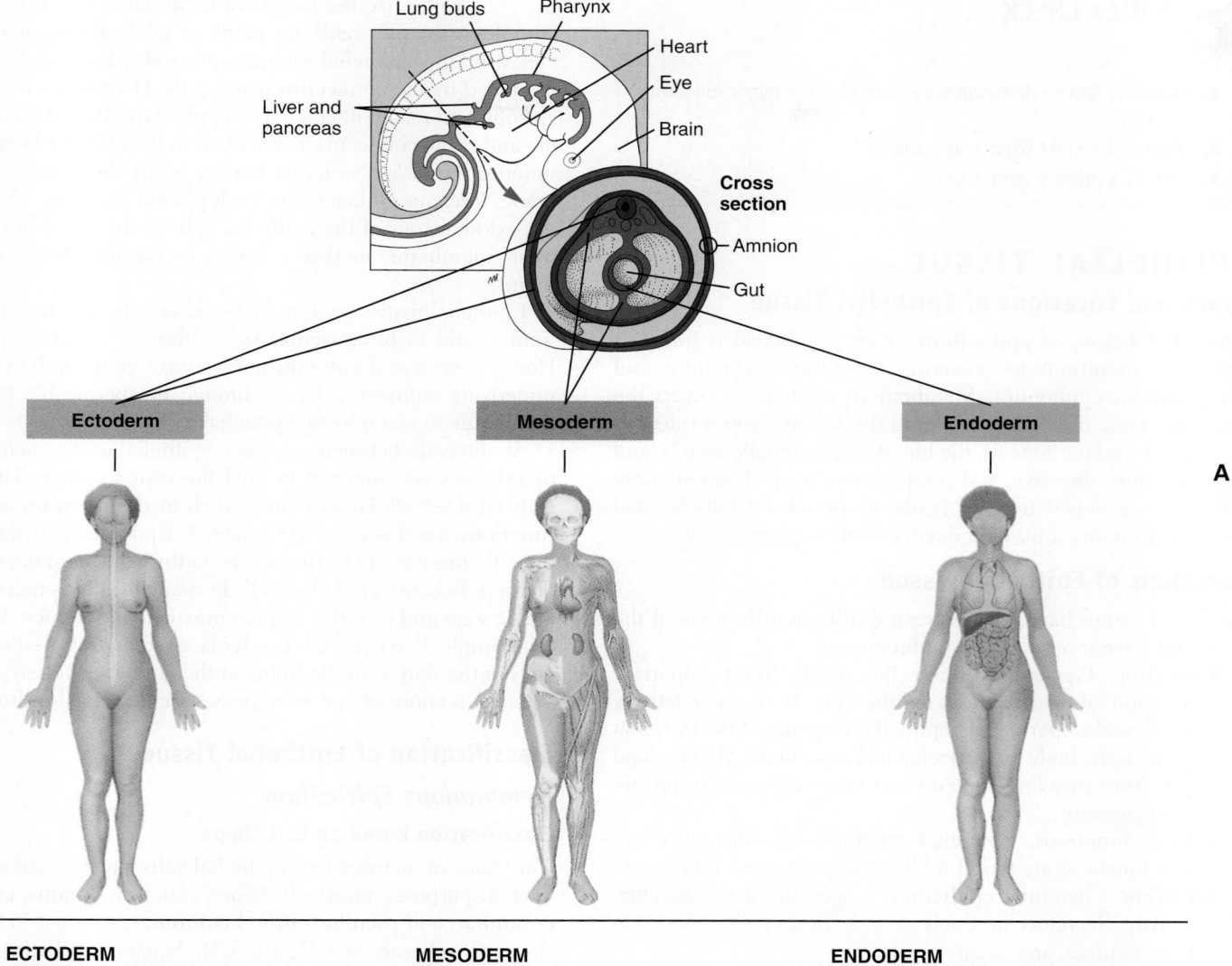

ECTODERM
Epithelium (epidermis) of skin
Lining of mouth, anus, nostrils
Sweat and sebaceous glands
Epidermal derivatives (hair, enamel of teeth)
Nervous system (brain and spinal cord)
Epithelial (sensory) parts of eyes, nose, ear

MESODERM
Muscles
Skeleton (bones and cartilage)
Blood
Epithelial lining of blood vessels
Dermis of skin and dentin of teeth
Organs (except lining) of excretory and reproductive systems
Connective tissue

ENDODERM
Epithelium (lining) of digestive and respiratory systems
Secretory parts of liver and pancreas
Urinary bladder
Epithelial lining of urethra
Thyroid, parathyroid, thymus

Figure 5-2 *Primary germ layers.* A, The illustration shows the primary germ layers and the body systems into which they develop. **B,** Structures derived from the primary germ layers.

BOX 5-1 Stem Cells

During embryonic development, new kinds of cells can be formed from a special kind of undifferentiated cell called a **stem cell.** *Embryonic stem cells* have the potential to reproduce many different kinds of daughter cells, including more stem cells—thus populating the body with all the different cells and tissues needed for body function. Sometimes controversial research is now under way to learn the secrets of embryonic stems cells and develop therapies in which embryonic stem cells might be used to repair or replace damaged tissue in adults.

Adult stem cells are undifferentiated cells found scattered within a differentiated, mature tissue. Many different tissues contain adult

stem cells—we may soon find that all adult tissues have some stem cells. Adult stem cells can usually produce any of the specialized cell types within its particular tissue. However, recent research shows that some adult stem cells can be coaxed into producing a variety of different types of cells. For example, blood cell–producing stem cells from bone marrow have been used to help repair damaged muscle tissue in laboratory animals. Already, therapies for treating degenerative diseases of muscles, the heart, and the brain—even baldness—are being proposed by medical scientists.

EPITHELIAL TISSUE

Types and Locations of Epithelial Tissue

Epithelial tissue, or **epithelium,** often is subdivided into two types: (1) **membranous** (covering or lining) epithelium and (2) **glandular** epithelium. Membranous epithelium covers the body and some of its parts and lines the serous cavities (pleural, pericardial, and peritoneal), the blood and lymphatic vessels, and the respiratory, digestive, and genitourinary tracts. Glandular epithelium is grouped in solid cords or specialized follicles that form the secretory units of endocrine and exocrine glands.

Functions of Epithelial Tissue

Epithelial tissues have a widespread distribution throughout the body and serve several important functions:

Protection. Generalized protection is the most important function of membranous epithelium. It is the relatively tough and impermeable epithelial covering of the skin that protects the body from mechanical and chemical injury and also from invading bacteria and other disease-causing microorganisms.

Sensory functions. Epithelial structures specialized for sensory functions are found in the skin, nose, eye, and ear.

Secretion. Glandular epithelium is specialized for secretory activity. Secretory products include hormones, mucus, digestive juices, and sweat.

Absorption. The lining epithelium of the gut and respiratory tract allows for the absorption of nutrients from the gut and the exchange of respiratory gases between air in the lungs and the blood.

Excretion. The specialized epithelial lining of kidney tubules makes the excretion and concentration of excretory products in the urine possible.

Generalizations About Epithelial Tissue

Most epithelial tissues are characterized by extremely limited amounts of intercellular, or matrix, material. This explains their characteristic appearance, when viewed under a light microscope, of a continuous sheet of cells packed tightly together. With the electron microscope, however, narrow spaces—about 20 nanometers (one millionth of an inch) wide—can be seen around the cells. These spaces, like other intercellular spaces, contain interstitial fluid.

Sheets of epithelial cells compose the surface layer of skin and mucous and serous membranes. The epithelial tissue attaches to an underlying layer of connective tissue by means of a thin noncellular layer of adhesive, permeable material called the **basement membrane** (Figure 5-3). Both epithelial and connective tissue cells synthesize the basement membrane, which is a highly complex structure made up partly of glycoprotein material secreted by the epithelial components and a fine mesh of fibers produced by the connective tissue cells. Histologists refer to the glycoprotein material secreted by epithelial cells as the *basal lamina* and to the connective tissue fibers as the *reticular lamina.* The union of basal and reticular lamina forms the basement membrane. *Integrins* embedded in each plasma membrane help bind the cytoskeletons of the epithelial cells to the fibers of the basement membrane so that a strong connection forms between them.

Epithelial tissues contain no blood vessels. As a result, epithelium is said to be **avascular** (*a,* "without"; *vascular,* "vessels"). Hence oxygen and nutrients must diffuse from capillaries in the underlying connective tissue through the permeable basement membrane to reach living epithelial cells.

At intervals between adjacent epithelial cells, their plasma membranes are modified to hold the cells together. These specialized intercellular structures, such as **desmosomes** and **tight junctions,** are described in Chapter 3. Epithelial cells can reproduce themselves. They frequently go through the process of cell division. Because epithelial cells in many locations meet considerable wear and tear, this fact has practical importance. It means, for example, that new cells can replace old or destroyed epithelial cells in the skin or in the lining of the gut or respiratory tract.

Cross sections of epithelial tissues are described in Box 5-2.

Classification of Epithelial Tissue

Membranous Epithelium

Classification Based on Cell Shape

The shape of membranous epithelial cells may be used for classification purposes. Four cell shapes, called **squamous, cuboidal, columnar,** and **pseudostratified columnar,** are used in this classification scheme (see Figure 5-3). Squamous (Latin, "scaly") cells are flat and platelike. Cuboidal cells, as the name implies, are cube-shaped and have more cytoplasm than the scalelike squamous cells do. Columnar epithelial cells are higher than they are wide and appear narrow and cylindrical. Pseudostratified columnar epithelium has only one layer of oddly shaped columnar cells. Although each cell touches the basement membrane, the tops of some pseudostratified cells do not fully extend to the surface of the membrane. Also, some nuclei are near the "top" of the cell and some near the "bottom" of the cell—rather than all nuclei being near the bottom. The result is a false (*pseudo*) appearance of layering, or stratification, when only a single layer of cells is present.

Classification Based on Layers of Cells

In most cases the location and function of membranous epithelium determine whether its cells will be stacked and layered or arranged in a sheet one cell layer thick. An arrangement of epithelial cells in a single layer is called **simple epithelium.** If epithelial cells are layered one on another, the tissue is called **stratified epithelium. Transitional epithelium** (described later) is a unique arrangement of differing cell shapes in a stratified, or layered, epithelial sheet.

CELL SHAPES   SIMPLE   STRATIFIED

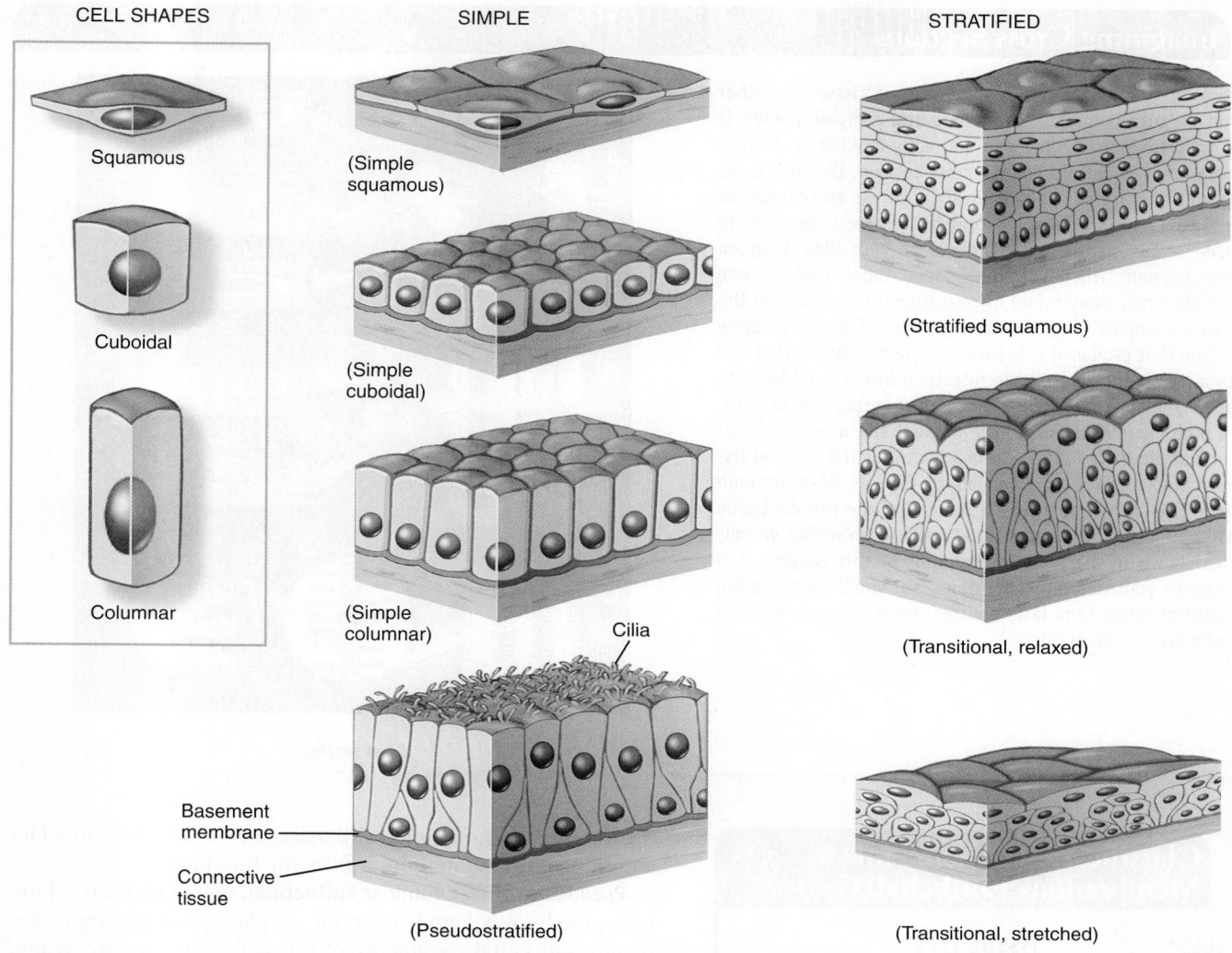

Squamous

Cuboidal

Columnar

(Simple squamous)

(Simple cuboidal)

(Simple columnar)

Cilia

Basement membrane

Connective tissue

(Pseudostratified)

(Stratified squamous)

(Transitional, relaxed)

(Transitional, stretched)

Figure 5-3 *Classification of epithelial tissues.* The tissues are classified according to the shape and arrangement of cells. The color scheme of these drawings is based on a common staining technique used by histologists called *hematoxylin and eosin (H&E)* staining. H&E staining usually renders the cytoplasm pink and the chromatin inside the nucleus a purplish color. The cellular membranes, including the plasma membrane and nuclear envelope, do not generally pick up any stain and are thus transparent. (See Box 5-2 for information on cross sections of membranous tissues.)

If membranous or covering epithelium is classified by the shape and layering of its cells, the specific types listed in Table 5-3 are possible. Notice that stratified tissue types are named for the shape of cells in their top layer only. Each type is described in the paragraphs that follow, and selected examples are illustrated in Figures 5-4 to 5-11.

Simple Epithelium

Simple Squamous Epithelium. Simple squamous epithelium consists of only one layer of flat, scalelike cells (Figure 5-4). Consequently, substances can readily diffuse or filter through this type of tissue. The microscopic air sacs (alveoli) of the lungs, for example, are composed of this kind of tissue, as are the linings of blood and lymphatic vessels and the surfaces of the pleura, pericardium, and peritoneum (Figure 5-5). (Blood and lymphatic ves-

sel linings are called **endothelium,** and the surfaces of the pleura, pericardium, and peritoneum are called **mesothelium.** Some histologists classify these linings as connective tissue because of their embryological origin.)

Simple Cuboidal Epithelium. Simple cuboidal epithelium is composed of one layer of cuboidal cells resting on a basement membrane (Figure 5-6). This type of epithelium is seen in many types of glands and their ducts. It is also found in the ducts and tubules of other organs, such as the kidney.

Simple Columnar Epithelium. Simple columnar epithelium composes the surface of the mucous membrane that lines the stomach, intestine, uterus, uterine tubes, and parts of the respiratory tract (Figure 5-7). It consists of a single layer of cells, many of which have a modified structure. Three common modifications are goblet cells, cilia, and microvilli. Goblet cells have large, se-

BOX 5-2   Imagining Cross Sections

When you look at photomicrographs of epithelial tissues or other structures that are tubelike, saclike, or folded into complex shapes, it is sometimes hard to imagine what you are really looking at. As diagram *A* shows, when you cut a tube on a cross section, the slice looks like a ring (if cut at a right angle) or oval (if cut at an oblique, or slanted, angle). Diagram *A* also shows that if the tube is bent where the cut is made, it can also look like an oval on your slide. Diagram *B* shows what happens when you have many tubes next to one another—your slice has many round or oval rings (depending on the angle of the cut). Compare Diagram *A* to Figure 5-6. Can you imagine the type of tissue that produced this slice? Diagram *C* shows what can happen when a membrane such as the intestinal lining is folded into complex shapes. The slice may look like a sort of zigzag line of cells, or if it is at a high magnification, it may just look like a series of parallel rows. Look at Figure 5-7. Can you imagine what the original tissue must have looked like? Look ahead to Figures 25-9 through 25-21. Can you "see" the context or "situation" of the various tissue slices? Try sketching out the "big picture" of the context of various tissue samples shown in the photomicrographs of this chapter, and then keep them in your notebook. By doing so, you'll be preparing yourself for later chapters (and later courses) by learning to identify the context of a tissue "on sight."

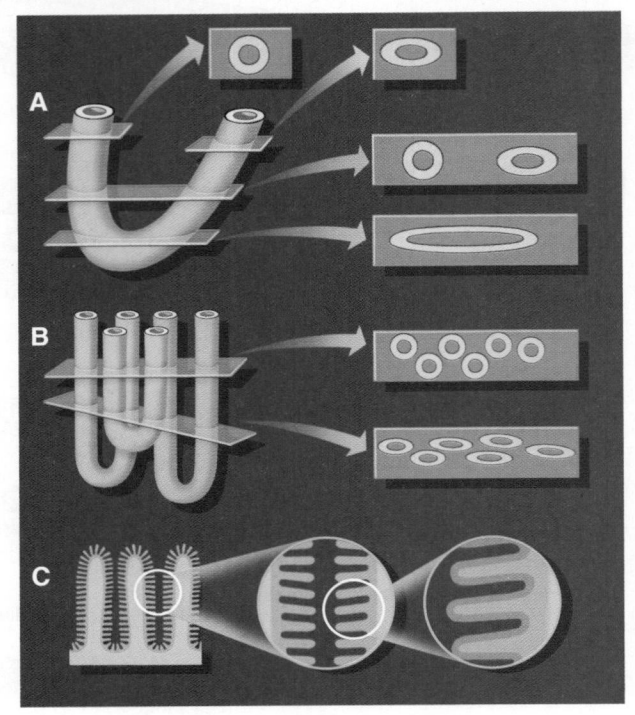

Cross sections.

Table 5-3	Classification Scheme for Membranous Epithelial Tissues

SHAPE OF CELLS*	TISSUE TYPE
One Layer	
Squamous	Simple squamous
Cuboidal	Simple cuboidal
Columnar	Simple columnar
Pseudostratified columnar	Pseudostratified columnar
Several Layers	
Squamous	Stratified squamous
Cuboidal	Stratified cuboidal
Columnar	Stratified columnar
(Varies)	Transitional

*In the top layer (if more than one layer is present in the tissue).

cretory vesicles that give them the appearance of a goblet. The vesicles contain mucus, which goblet cells produce in great quantity and secrete onto the surface of the epithelial membrane. Mucus is a solution of water, electrolytes, and proteoglycans. In the intestine, the plasma membranes of many columnar cells extend out in hundreds and hundreds of microscopic fingerlike projections called **microvilli**. By greatly increasing the surface area of the intestinal mucosa, microvilli make it especially well suited for absorbing nutrients and fluids from the intestine.

Pseudostratified Columnar Epithelium. Pseudostratified columnar epithelium is found lining the air passages of the respiratory system and certain segments of the male reproductive system such as the urethra (Figure 5-8). Although appearing to be stratified, only a single layer of irregularly shaped columnar cells touches the basement membrane. The cells are of differing heights, and many are not tall enough to reach the upper surface of the epithelial sheet. This fact, coupled with placement of cell nuclei at odd and irregular levels in the cells, gives a false (pseudo) impression of stratification. Mucus-secreting goblet cells are numerous and cilia are present. In the respiratory system air passages, uniform motion of the cilia causes a thin layer of tacky mucus to move in one direction only over the free surface of the epithelium. As a result, dust particles and other contaminants in the inspired air are trapped and moved toward the mouth and away from the delicate lung tissues.

Stratified Epithelium

Stratified Squamous (Keratinized) Epithelium. Stratified squamous epithelium is characterized by multiple layers of cells with typical flattened squamous cells at the free, or outer, surface of the epithelial sheet (Figure 5-9). The presence of keratin in these cells contributes to the protective qualities of skin covering the body surface. Details of the histology of this type of epithelium are presented in Chapter 6.

Stratified Squamous (Nonkeratinized) Epithelium. Nonkeratinized stratified squamous epithelium is found lining the

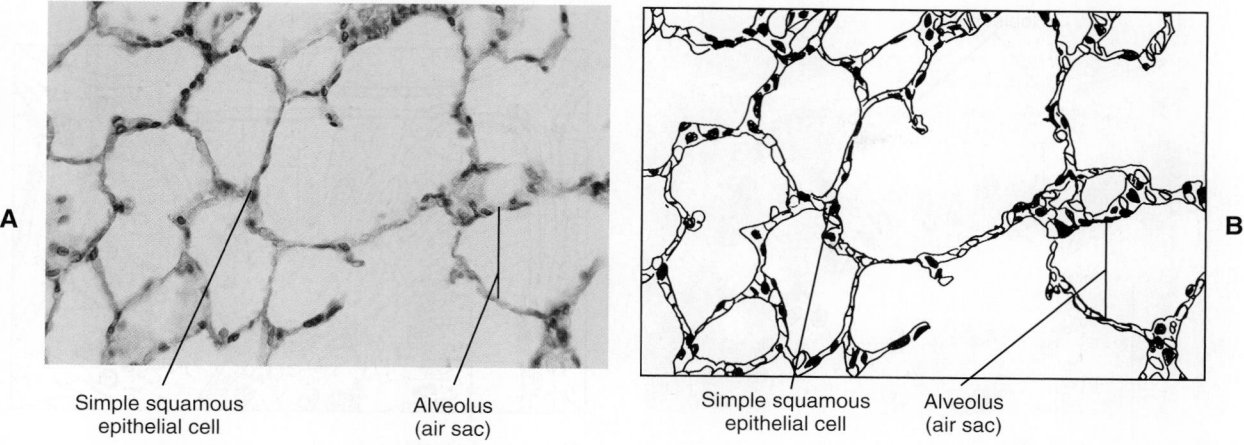

Simple squamous
epithelial cell   Alveolus
                         (air sac)

Simple squamous
epithelial cell   Alveolus
                         (air sac)

Figure 5-4 *Simple squamous epithelium.* **A,** Photomicrograph of lung tissue showing thin simple squamous epithelium lining the tiny air sacs of the lung. Notice how the H&E staining (see Figure 5-3) renders the cytoplasm of each cell pink and each nucleus a purplish color. **B,** Sketch of the micrograph showing the outlines of cellular membranes that are transparent in the photomicrograph.

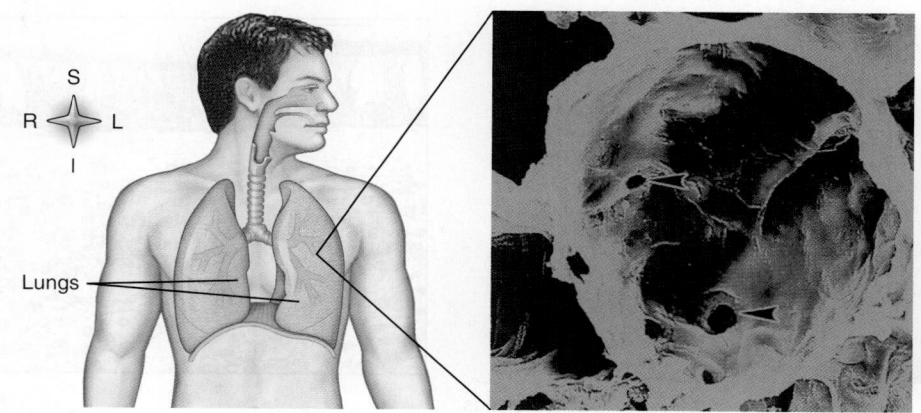

Lungs

Figure 5-5 *Simple squamous lining of the lung.* The tiny air sacs of the lung, which are called *alveoli,* are lined with a thin layer of simple squamous epithelium. This thin epithelium, shown clearly in the scanning electron micrograph in the inset, allows for easy diffusion of oxygen and carbon dioxide across the walls of the alveoli. Arrowheads indicate pores that allow air to pass from one alveolus to another.

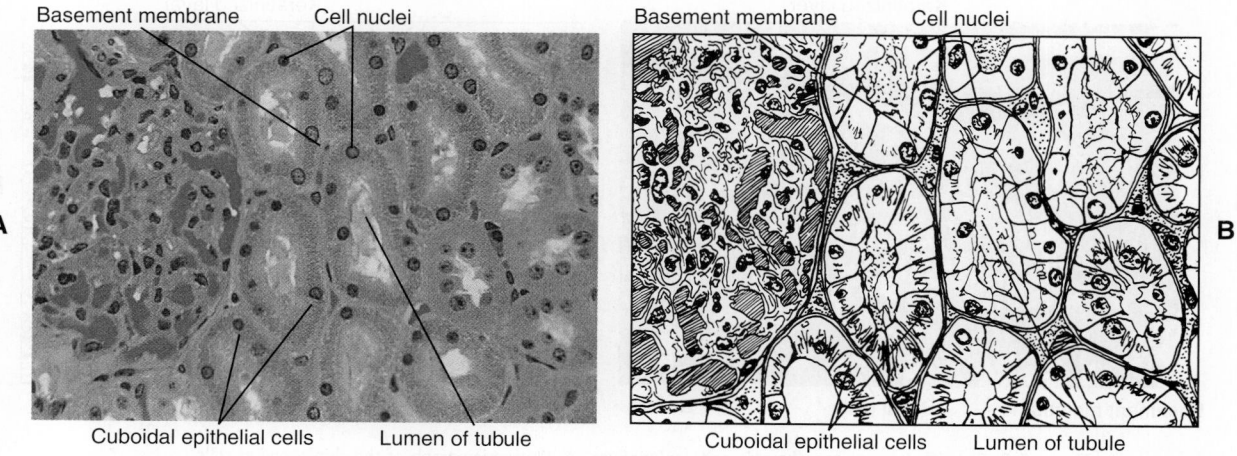

Basement membrane   Cell nuclei

Cuboidal epithelial cells   Lumen of tubule

Basement membrane   Cell nuclei

Cuboidal epithelial cells   Lumen of tubule

Figure 5-6 *Simple cuboidal epithelium.* **A,** Photomicrograph of kidney tubules showing the single layer of cuboidal cells touching a basement membrane. **B,** Sketch of the micrograph. Note the cuboidal cells that enclose the tubule opening (lumen).

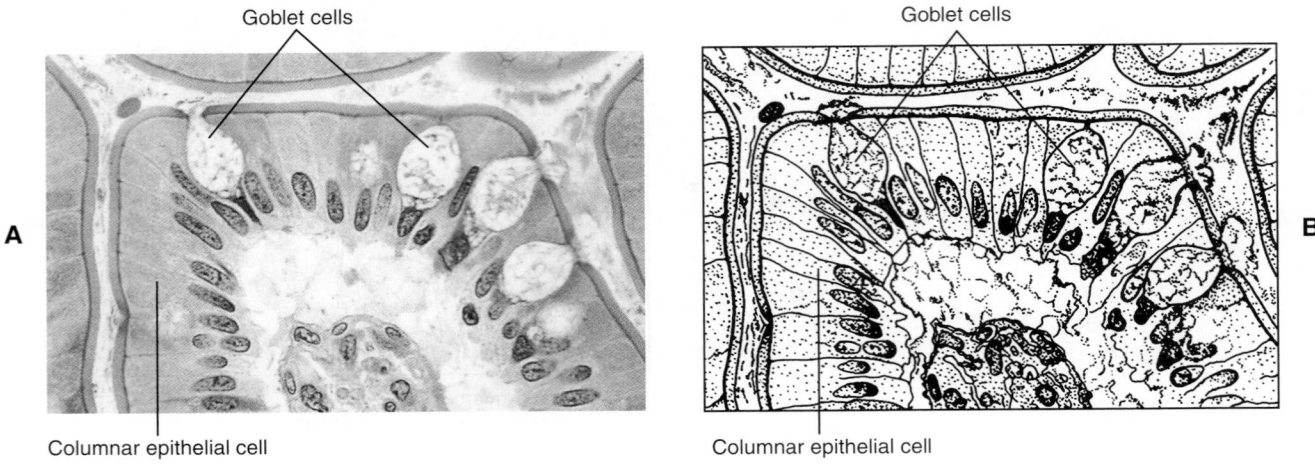

Figure 5-7 *Simple columnar epithelium.* **A,** Photomicrograph of simple columnar epithelium. **B,** Sketch of the photomicrograph. Note the goblet, or mucus-producing, cells present.

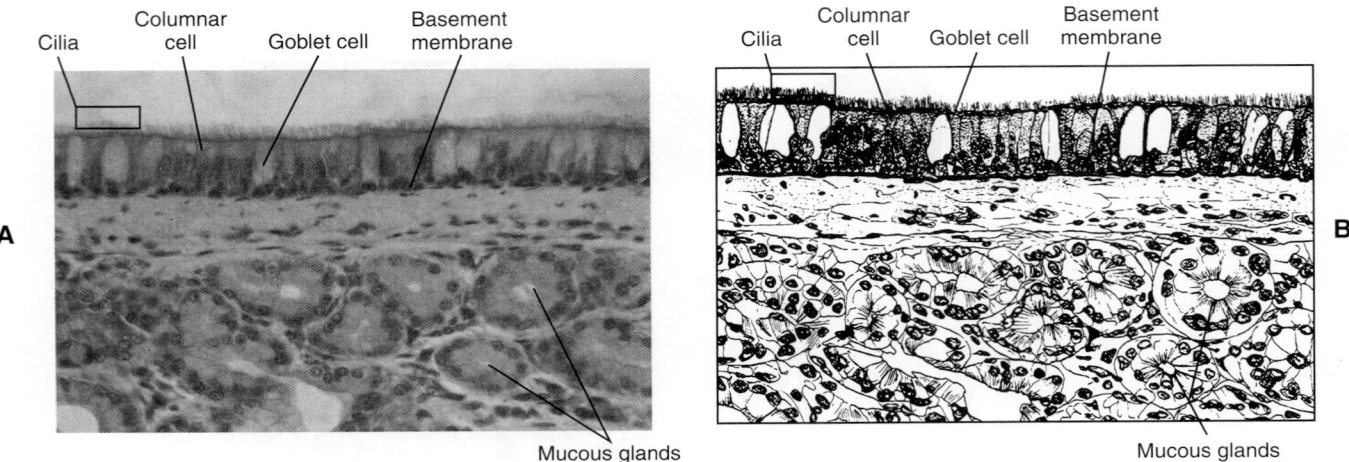

Figure 5-8 *Pseudostratified ciliated epithelium.* **A,** This photomicrograph of the trachea shows that each irregularly shaped columnar cell touches the underlying basement membrane. **B,** Sketch of the photomicrograph. Placement of cell nuclei at irregular levels in the cells gives a false (pseudo) impression of stratification.

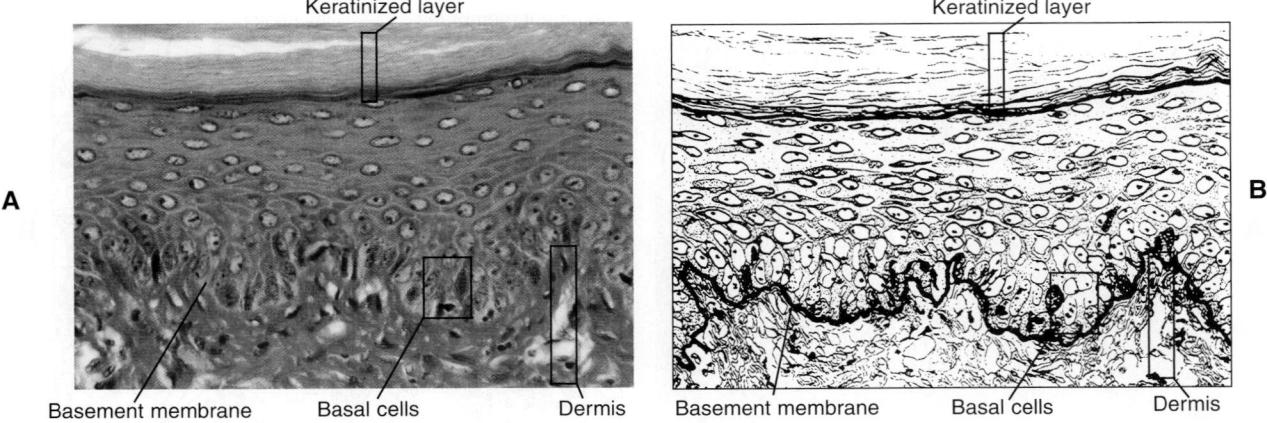

Figure 5-9 *Stratified squamous (keratinized) epithelium.* **A,** Photomicrograph of the skin showing cells becoming progressively flattened and scalelike as they approach the surface and are lost. **B,** Sketch of the photomicrograph. The outer surface of this epithelial sheet contains many flattened cells that have lost their nuclei.

vagina, mouth, and esophagus (Figure 5-10). Its free surface is moist, and the outer epithelial cells, unlike those found in the skin, do not contain keratin. This type of epithelium serves a protective function.

Stratified Cuboidal Epithelium. The cuboidal variety of stratified epithelium also serves a protective function. Typically, two or more rows of low cuboidal cells are arranged randomly over a basement membrane. Stratified cuboidal epithelium can be located in the sweat gland ducts, in the pharynx, and over parts of the epiglottis.

Stratified Columnar Epithelium. Although this protective epithelium has multiple layers of columnar cells, only the most superficial cells are truly columnar in appearance. Epithelium of this type is found in few places in the human body. It is lo-

cated in segments of the male urethra and in the mucous layer near the anus.

Stratified Transitional Epithelium. Transitional epithelium is a stratified tissue typically found in body areas that are subjected to stress and tension changes, such as the wall of the urinary bladder (Figure 5-11). In many instances, 10 or more layers of cuboidal cells of varying shapes are present in the absence of stretching or tension. As tension increases, the epithelial sheet is expanded, the number of observable cell layers decreases, and cell shape changes from cuboidal to squamous in appearance. This ability of transitional epithelium to stretch protects the bladder wall and other distensible structures that it lines from tearing when stretched with great force.

The major types of epithelium are summarized in Table 5-4.

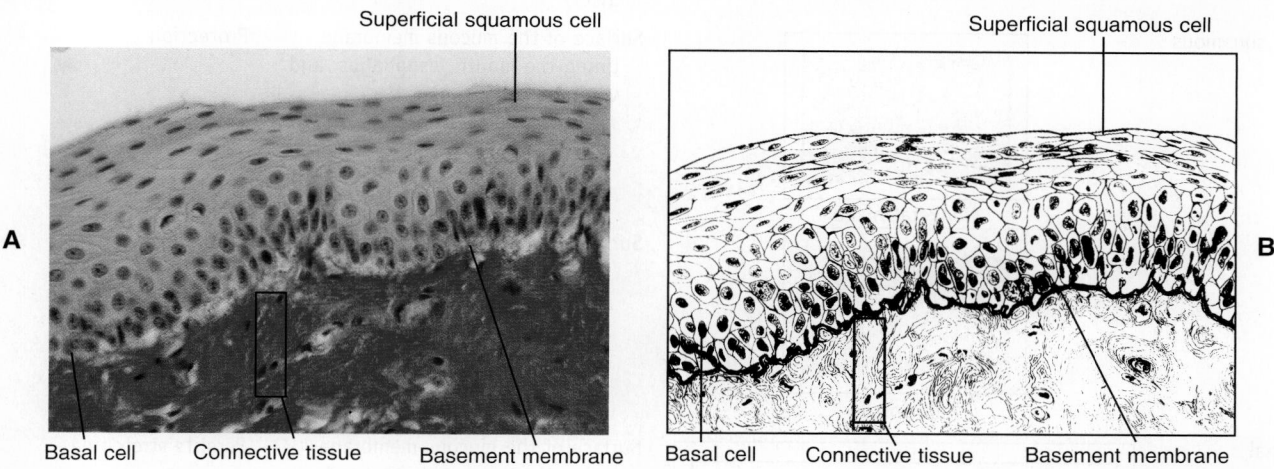

Figure 5-10 *Stratified squamous (nonkeratinized) epithelium.* **A,** Photomicrograph of vaginal tissue. Each cell in the layer is flattened near the surface and attached to the sheet. No flaking of dead cells from the surface occurs. **B,** Sketch of the photomicrograph. All cells have nuclei. Compare with Figure 5-9.

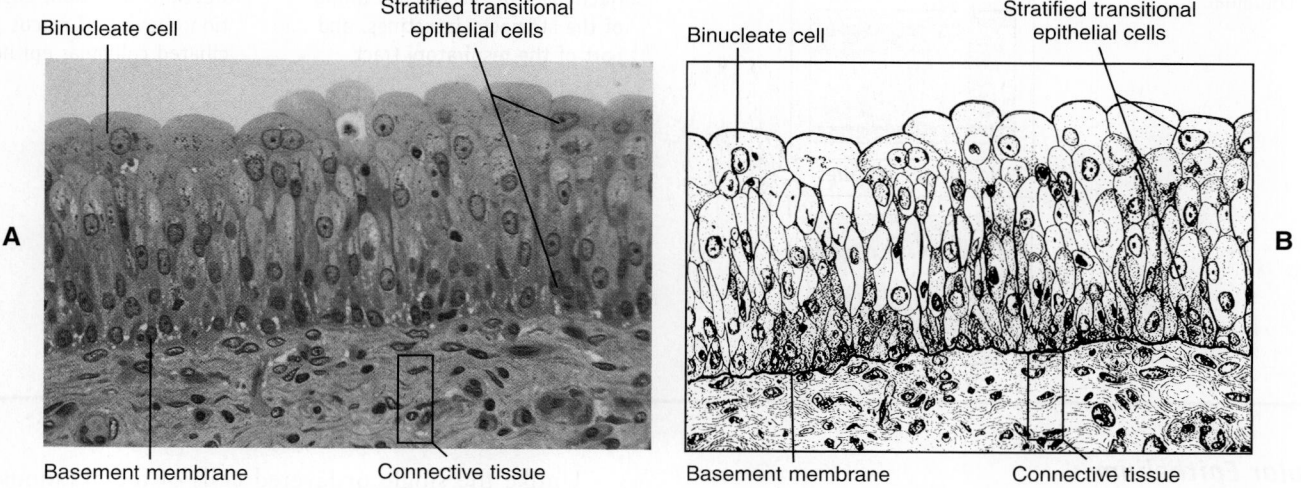

Figure 5-11 *Transitional epithelium.* **A,** Photomicrograph of the urinary bladder showing that its cell shape is variable from cuboidal to squamous. Several layers of cells are present. Intermediate and surface cells do not touch the basement membrane. **B,** Sketch of the photomicrograph.

Table 5-4   Epithelial Tissues

TISSUE		LOCATION	FUNCTION
Membranous			
Simple squamous		Alveoli of lungs	Absorption by diffusion of respiratory gases between alveolar air and blood
		Lining of blood and lymphatic vessels (called endothelium; classified as connective tissue by some histologists)	Absorption by diffusion, filtration, and osmosis
		Surface layer of the pleura, pericardium, and peritoneum (called mesothelium; classified as connective tissue by some histologists)	Absorption by diffusion and osmosis; also, secretion
Stratified squamous	Nonkeratinized	Surface of the mucous membrane lining the mouth, esophagus, and vagina	Protection
	Keratinized	Surface of the skin (epidermis)	Protection
Transitional	Unstretched Stretched	Surface of the mucous membrane lining the urinary bladder and ureters	Permits stretching
Simple columnar	No surface features With microvilli	Surface layer of the mucous lining of the stomach, intestines, and part of the respiratory tract	Protection; secretion; absorption; moving of mucus (by ciliated columnar epithelium)

Glandular Epithelium

Epithelium of the glandular type is specialized for secretory activity. Regardless of the secretory product produced, glandular activity depends on complex and highly regulated cellular activities requiring the expenditure of stored energy.

Unlike the single or layered cells of membranous epithelium typically found in protective coverings or linings, glandular epithelial cells may function singly as **unicellular glands,** or they may function in clusters, solid cords, or specialized follicles as **multicellular glands.** Glandular secre-

Table 5-4   Epithelial Tissues—cont'd

TISSUE		LOCATION	FUNCTION
Simple columnar—continued	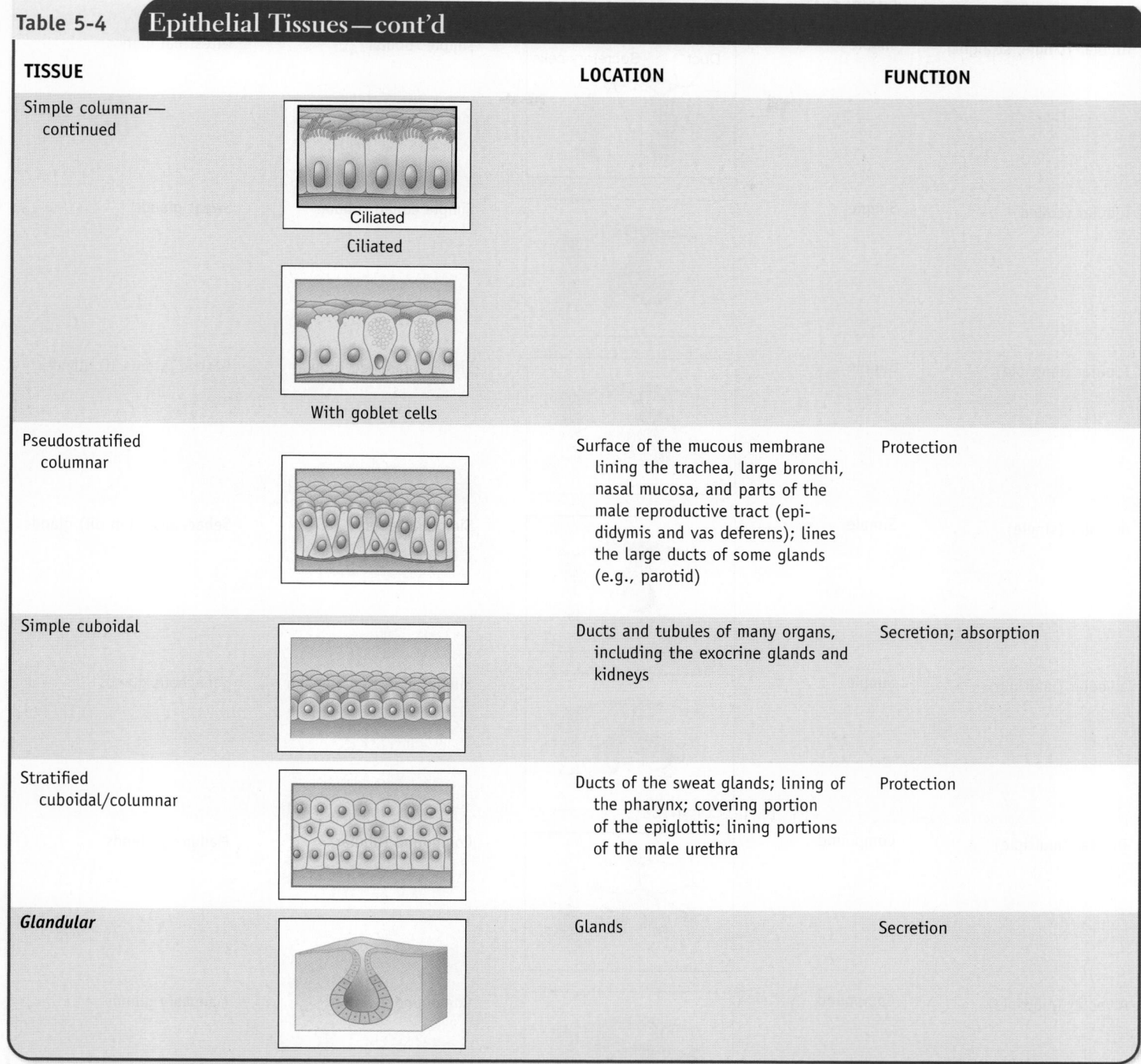 Ciliated Ciliated With goblet cells		
Pseudostratified columnar		Surface of the mucous membrane lining the trachea, large bronchi, nasal mucosa, and parts of the male reproductive tract (epididymis and vas deferens); lines the large ducts of some glands (e.g., parotid)	Protection
Simple cuboidal		Ducts and tubules of many organs, including the exocrine glands and kidneys	Secretion; absorption
Stratified cuboidal/columnar		Ducts of the sweat glands; lining of the pharynx; covering portion of the epiglottis; lining portions of the male urethra	Protection
Glandular		Glands	Secretion

tions may be discharged into ducts, into the lumen of hollow visceral structures, onto the body surface, or directly into the blood.

All **glands** in the body can be classified as either exocrine or endocrine glands. **Exocrine glands,** by definition, discharge their secretion products into ducts. The salivary glands are typical exocrine glands. The secretion product (saliva) is produced in the gland and then discharged into a duct that transports it to the mouth. **Endocrine glands** are often called *ductless glands* because they discharge their secretion products (hormones) directly into blood or interstitial fluid. The

pituitary, thyroid, and adrenal glands are typical endocrine glands.

Structural Classification of Exocrine Glands

Multicellular exocrine glands are most often classified by structure, with the shape of their ducts and the complexity (branching) of their duct systems used as distinguishing characteristics. Shapes include **tubular** and **alveolar** (saclike). **Simple** exocrine glands have only one duct leading to the surface, and **compound** exocrine glands have two or more ducts. Table 5-5 describes some of the major structural types of exocrine glands.

Table 5-5 — Structural Classification of Multicellular Exocrine Glands

SHAPE*	COMPLEXITY†		TYPE	EXAMPLE
Tubular (single, straight)	Simple	Duct / Secretory cells	Simple tubular	Intestinal glands
Tubular (coiled)	Simple		Simple coiled tubular	Sweat glands
Tubular (multiple)	Simple		Simple branched tubular	Gastric (stomach) glands
Alveolar (single)	Simple		Simple alveolar	Sebaceous (skin oil) glands
Alveolar (multiple)	Simple		Simple branched alveolar	Sebaceous glands
Tubular (multiple)	Compound		Compound tubular	Mammary glands
Alveolar (multiple)	Compound		Compound alveolar	Mammary glands
Some tubular; some alveolar	Compound		Compound tubuloalveolar	Salivary glands

*Shape of the distal secreting units of the gland.
†Number of ducts *reaching the surface.*

Figure 5-12 shows examples of exocrine glands in the lining of the stomach.

Functional Classification of Exocrine Glands

In addition to structural differences, exocrine glands also differ in the method by which they discharge their secretion products from the cell. Using these functional criteria, three types of exocrine glands may be identified (Figure 5-13):

1. Apocrine
2. Holocrine
3. Merocrine

Apocrine glands collect their secretory products near the apex, or tip, of the cell and then release them into a duct by pinching off the distended end. This process results in some loss of cytoplasm and damage to the cell. Recovery and repair of cells are rapid, however, and continued secretion occurs. The milk-producing mammary glands are examples of apocrine-type glands.

Holocrine glands—such as the sebaceous glands that produce oil to lubricate the skin—collect their secretory product inside the cell and then rupture completely to release it. These cells literally self-destruct to complete their function.

Merocrine glands discharge their secretion product directly through the cell or plasma membrane. This discharge process is completed without injury to the cell wall and without loss of cytoplasm. Only the secretion product passes from the glandular cell into the duct. Most secretory cells are of this type. The salivary glands are examples of merocrine-type exocrine glands.

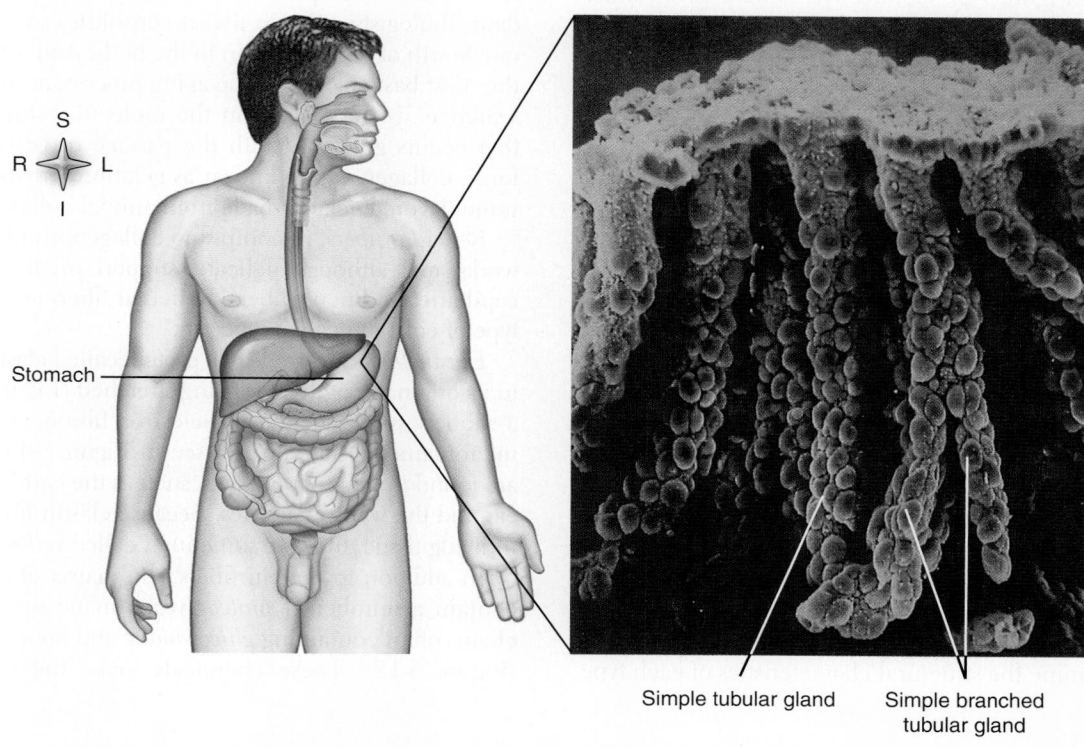

Stomach

Simple tubular gland　　Simple branched tubular gland

Figure 5-12 *Exocrine glands in the stomach.* The inset shows a scanning electron micrograph of exocrine glands, called *gastric glands,* in the lining of the stomach. These glands produce gastric juice—a mixture of water, mucus, enzymes, acid, and other substances.

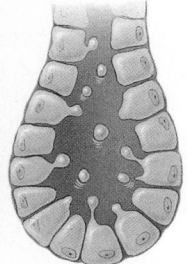
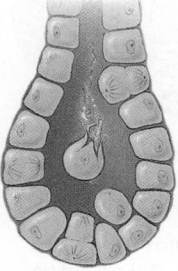
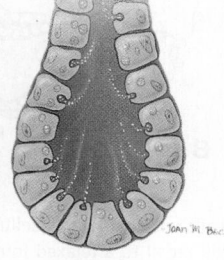

Apocrine gland　　Holocrine gland　　Merocrine gland

Figure 5-13 *Three types of exocrine glands.* Here exocrine glands are classified by the method of secretion.

QUICK CHECK

4. List at least three functions of epithelial tissue.
5. What are the three basic shapes of epithelial cells?
6. Distinguish between a simple epithelial tissue and a stratified epithelial tissue.
7. How do exocrine glands secrete their products?

CONNECTIVE TISSUE

Connective tissue is one of the most widespread tissues in the body and is found in or around nearly every organ of the body. Connective tissue arises during embryonic development from stem cell tissue called *mesenchyme*, most of which originates in the mesoderm (see Figure 5-2). Connective tissue exists in more varied forms than the other three basic tissues do. Among the types that we will discuss are delicate tissue paper webs, tough resilient cords, rubbery elastic sheets, rigid bones, and a fluid (blood).

Functions of Connective Tissue

Connective tissue connects, supports, transports, and defends. It connects tissues to each other, for example. It also connects muscles to muscles, muscles to bones, and bones to bones. It forms a supporting framework for the body as a whole and for its organs individually. One kind of connective tissue—blood—transports a large array of substances between parts of the body. And finally, several kinds of connective tissue cells even defend us against microorganisms and other invaders.

Characteristics of Connective Tissue

Connective tissue consists predominantly of extracellular matrix. Embedded in the matrix are relatively few cells. The ECM of connective tissues is made up of varying numbers and kinds of fibers, fluid, and perhaps other material sometimes called *ground substance*. The qualities of the ECM's fibers and other components largely determine the structural characteristics of each type of connective tissue. The matrix of blood, for example, is a fluid (plasma). It contains numerous blood cells but no fibers, except when it coagulates. Some connective tissues have the consistency of a soft gel, some are firm but flexible, some hard and rigid, some tough, others delicate—and in each case it is their matrix and extracellular fibers that make them so.

A connective tissue's extracellular matrix contains one or more of the following kinds of fibers: collagenous (or white), reticular, or elastic (or yellow). Fibroblasts and some other cells produce these protein fibers. Collagenous fibers are tough and strong, reticular fibers are delicate, and elastic fibers are extensible and elastic.

Collagenous fibers are made of **collagen** (Figure 5-1) and often occur in bundles—an arrangement that provides great tensile strength. Because collagenous fibers look white in living tissue, they are sometimes called *white fibers*. Of all the hundreds of different protein compounds in the body, collagen is the most abundant. Biologists estimate that it constitutes somewhat more than one fourth of all the protein in the body. And interestingly, one of the most basic factors in the aging process, according to some researchers, is the change in the molecular structure of collagen that occurs gradually with the passage of years. In its hydrated form, collagen is also known as gelatin. Perhaps you have eaten some flavored gelatin made from animal collagen.

Reticular fibers, in contrast to collagenous fibers, occur in networks and, although delicate, support small structures such as capillaries and nerve fibers. Reticular fibers are made of a special type of collagen called *reticulin*.

Elastic fibers are made of a protein called **elastin,** which returns to its original length after being stretched (Figure 5-14). Elastin is a rubbery substance that is held in a fibrous shape by long, thin microfilaments—as you can see in Figure 5-14, *B*. Elastic fibers are found in "stretchy" tissues, such as the cartilage of the external ear and the walls of arteries. Because elastin fibers look yellowish in living tissue, they are sometimes called *yellow fibers*.

In addition to protein fibers, the matrix of connective tissues contain a number of *proteoglycans* made up of polysaccharide chains often containing *glucosamine* and bound to a protein core (Figure 5-15). These chemicals make the matrix fluid thick

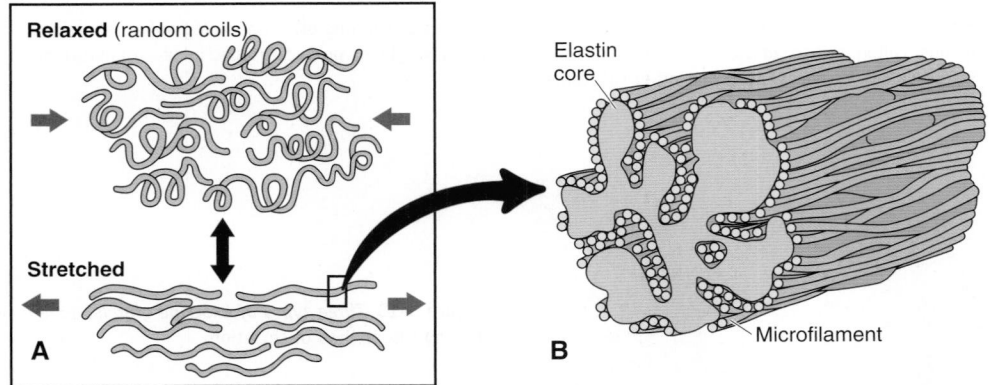

Figure 5-14 *Elastin fibers.* **A,** Elastin fibers form random coils within the extracellular matrix (ECM) that may easily be stretched. When released, the elastin fibers will recoil to a relaxed formation. **B,** Each elastin fiber is made of stretchy, formless elastin that is held in a fibrous shape by an arrangement of microfibrils.

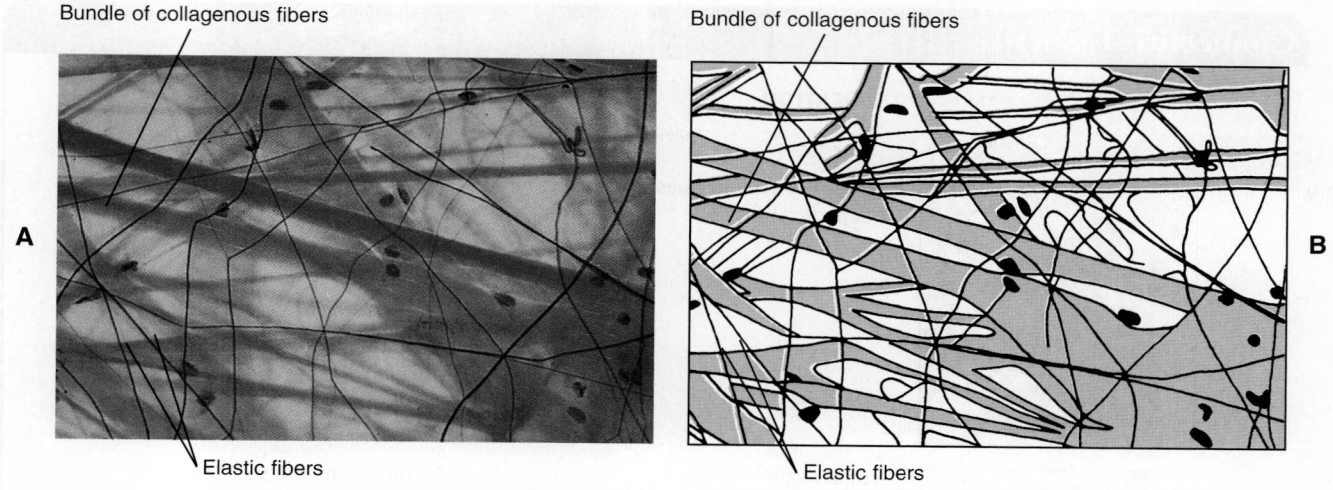

Bundle of collagenous fibers

Bundle of collagenous fibers

A

B

Elastic fibers

Elastic fibers

Figure 5-15 *Loose, ordinary (areolar) connective tissue.* **A,** Photomicrograph. Notice how the H&E staining (see Figure 5-2) renders the bundles of collagen fibers a pinkish color and the elastin fibers and cell nuclei a purplish color. **B,** Sketch of the photomicrograph. Note the loose arrangement of fibers compared with fibers in Figures 5-20, 5-21 and 5-23.

enough to be a barrier to bacteria and other microbes. They also form transparent lubricant and help hold the tissue together. Among the more notable of these compounds are *hyaluronic acid* and *chondroitin sulfate.*

Classification of Connective Tissue

Connective tissues have been classified by histologists in several different ways. Usually they are placed in different categories or types according to the structural characteristics of the intercellular material. The classification scheme we have adopted here is widely used and includes most of the major types:

1. Fibrous
   a. Loose, ordinary (areolar)
   b. Adipose
   c. Reticular
   d. Dense
      (1) Irregular
      (2) Regular
         (a) Collagenous
         (b) Elastic
2. Bone
   a. Compact
   b. Cancellous (spongy)
3. Cartilage
   a. Hyaline
   b. Fibrocartilage
   c. Elastic
4. Blood

Fibrous tissues such as areolar, adipose, reticular, and dense fibrous tissues have many fibers in the ECM as their predominant feature. The type and arrangement of extracellular fibers are what distinguish members of the group from each other. Bone is considered a separate category of connective tissue because it has fibers and a hard mineralized ECM. Cartilage is yet another cat-

egory because besides fibers, it has a specialized ECM that traps water to form a firm gel. Blood, the last category listed, is characterized by the lack of fibers in its matrix. These major types of connective tissues are described further in the following pages and in Table 5-6.

Fibrous Connective Tissue

Loose Connective Tissue (Areolar)

Loose connective tissue, shown in Figure 5-16, is often called *loose, ordinary connective tissue,* or *areolar tissue.* It is loose because it is stretchable and ordinary because it is one of the most widely distributed of all tissues. It is common and ordinary, not special like some kinds of connective tissue (e.g., bone and cartilage) that help form comparatively few structures. *Areolar* was the early name for the loose, ordinary connective tissue that connects many adjacent structures of the body. It acts like a glue spread between them—but an elastic glue that permits movement. The word **areolar** means "like a small space" and refers to the bubbles that appear as areolar tissue is pulled apart during dissection.

The matrix of areolar tissue is a soft, thick gel mainly because it contains hyaluronic acid. An enzyme, hyaluronidase, can change the matrix from its thick gel state to a watery consistency. Physicians have made use of this knowledge for many years. They frequently inject a commercial preparation of hyaluronidase with drugs or fluids. By decreasing the viscosity (thickness) of intercellular material, the enzyme hastens diffusion and absorption of the injected material and lessens tissue tension and pain. Some bacteria, notably *pneumococci* and *streptococci,* spread through connective tissues by secreting hyaluronidase.

The matrix of areolar tissue contains numerous fibers and cells, typically many interwoven collagenous and elastic fibers and about a half dozen kinds of cells. **Fibroblasts** are usually present in the greatest numbers in areolar tissue, and **macrophages** are second. Fibroblasts synthesize the gel-like ground substance

Table 5-6   Connective Tissues

TISSUE		LOCATION	FUNCTION
Fibrous			
Loose, ordinary (areolar)		Between other tissues and organs Superficial fascia	Connection Connection
Adipose (fat)		Under skin Padding at various points	Protection Insulation Support Reserve food
Reticular		Inner framework of spleen, lymph nodes, bone marrow Filtration	Support
Dense Fibrous			
Irregular		Deep fascia Dermis Scars Capsule of kidney, etc.	Connection Support
Regular	 Collagenous	Tendons Ligaments Aponeuroses	Flexible but strong connection

Table 5-6   Connective Tissues—cont'd

TISSUE		LOCATION	FUNCTION
Regular—continued	 Elastic	Walls of some arteries	Flexible, elastic support
Bone			
Compact bone		Skeleton (outer shell of bones)	Support Protection Calcium reservoir
Cancellous (spongy) bone	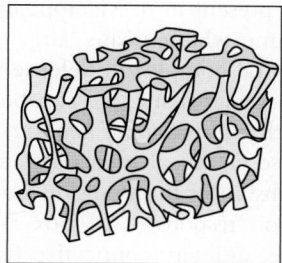	Skeleton (inside bones)	Support Provides framework for blood production
Cartilage			
Hyaline	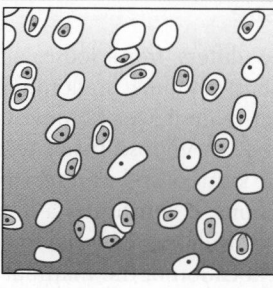	Part of nasal septum Covering articular surfaces of bones Larynx Rings in trachea and bronchi	Firm but flexible support
Fibrocartilage	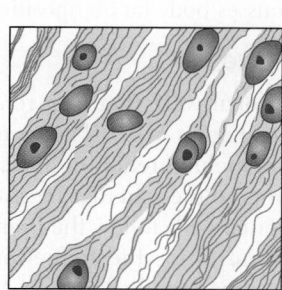	Disks between vertebrae Symphysis pubis	

Continued

Table 5-6	Connective Tissues—cont'd	

TISSUE	LOCATION	FUNCTION
Elastic 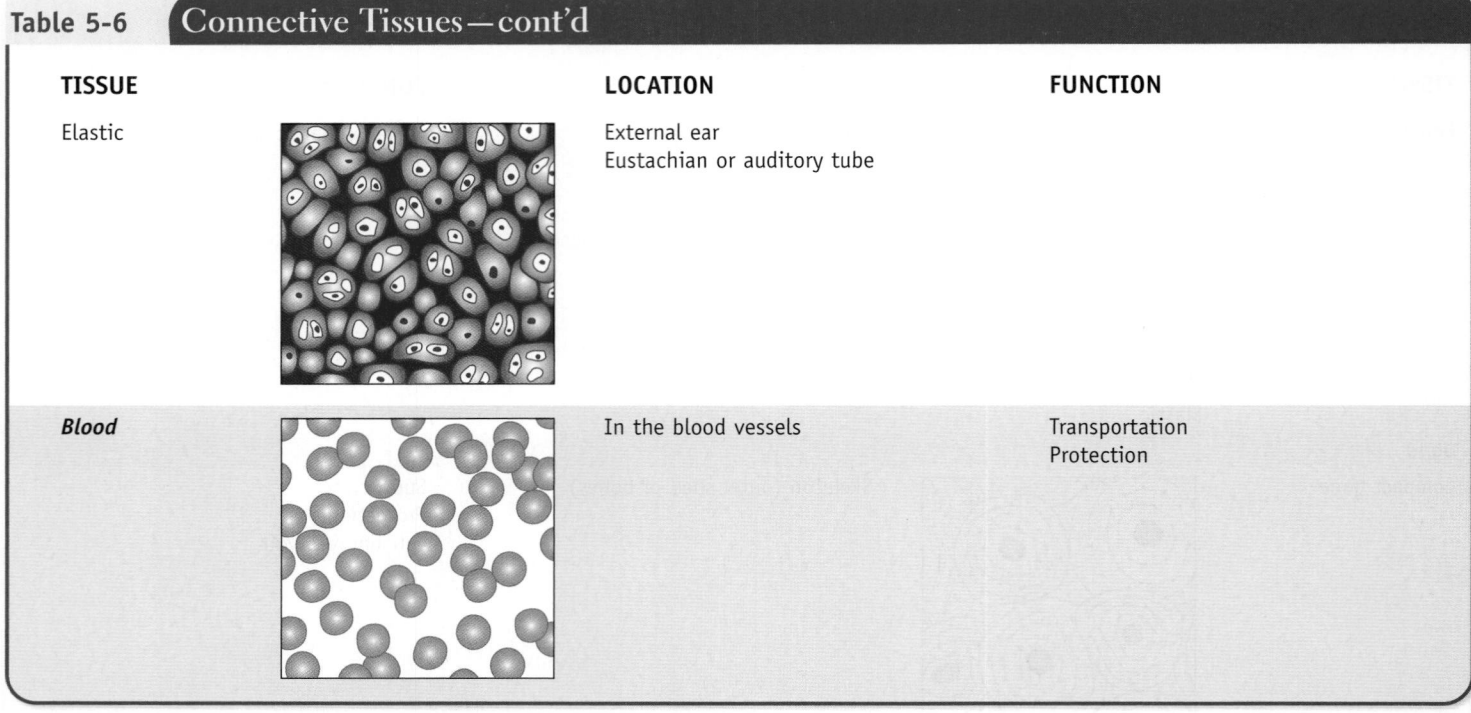	External ear Eustachian or auditory tube	
Blood	In the blood vessels	Transportation Protection

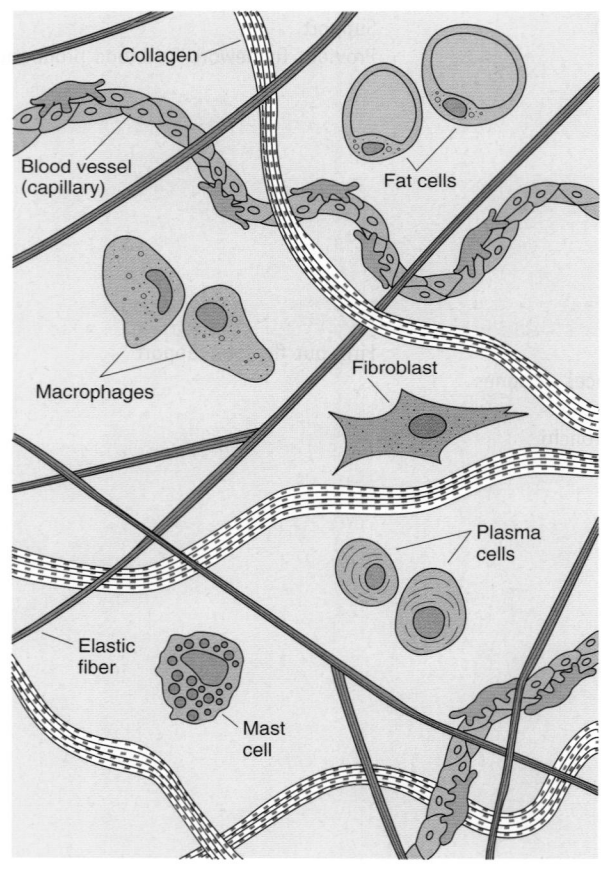

Figure 5-16 *Diagram of loose, ordinary (areolar) connective tissue.* This artist's sketch illustrates the fact that areolar tissue includes a number of different ECM components such as collagenous fibers and elastic fibers, as well as a variety of different cell types.

and the fibers present in it. Macrophages carry on phagocytosis, hence their name, which means "large eater." Phagocytosis is part of the body's vital complex of defense mechanisms. **Mast cells,** also found in areolar tissue, are capable of releasing a variety of molecules such as *histamine, heparin, leukotrienes,* and *prostaglandins.* These chemical mediators are released in response to exposure to substances from outside the body and produce a type of inflammation response (see Box 5-3). Other kinds of cells found in loose, ordinary connective tissue are various forms of white blood cells (leukocytes) and some fat cells.

Adipose Tissue

Adipose tissue differs from loose, ordinary connective tissue mainly in that it contains predominantly fat cells, also called **adipocytes,** and many fewer fibroblasts, macrophages, and mast cells (Figure 5-17). Adipose tissue forms supporting, protective pads around the kidneys and various other structures. It also serves two other functions: it constitutes a storage depot for excess food, and it acts as an insulating material to conserve body heat. Figure 5-18 shows the location of the main fat storage areas.

Box 5-4 discusses body fat composition.

Reticular Tissue

A three-dimensional web, that is, a reticular network, identifies **reticular tissue** (Figure 5-19). The word *reticular* means "like a net." Slender, branching reticulin fibers with reticular cells overlying them compose the reticular meshwork. Branches of the cytoplasm of reticular cells follow the branching reticular fibers.

Reticular tissue forms the framework of the spleen, lymph nodes, and bone marrow. It functions as part of the body's complex mechanism for defending itself against microorganisms and injurious substances (Box 5-5). The reticular meshwork filters in-

BOX 5-3: HEALTH MATTERS
Hay Fever and Asthma

Mast cells in areolar tissues are often involved in *allergy,* or *hypersensitivity,* reactions in local tissues. This is a type of inflammation (see Box 5-5) in response to *allergens,* which are substances that trigger such allergic responses. For example, the inflammation that you experience after a bee sting or when your skin is exposed to poison ivy are all examples of allergic or hypersensitivity reactions. Such reactions are triggered when mast cells encounter an allergen and release any of a group of chemical mediators.

For example, in hay fever (allergies to grasses and other plants), mast cells release the chemical *histamine.* Histamine increases the permeability of blood vessels in the nasal membranes, which in turn causes swelling in the lining of the nose. This gives us a "stuffy feeling" be-

cause the swelling makes it difficult to breathe easily. The stuffiness of hay fever can be relieved by antihistamines such as diphenhydramine (Benadryl) and fexofenadine (Allegra), which block histamine's action.

On the other hand, a different mast cell product is responsible for the breathing difficulties in an asthma attack. In asthma, chemicals called *leukotrienes* trigger muscles in the walls of the respiratory tract to contract—thereby constricting the airways. Thus, leukotriene-blocking drugs such as montelukast (Singulair) and zileuton (Zyflo) are often used along with other drugs that prevent or reduce contraction of airway muscles.

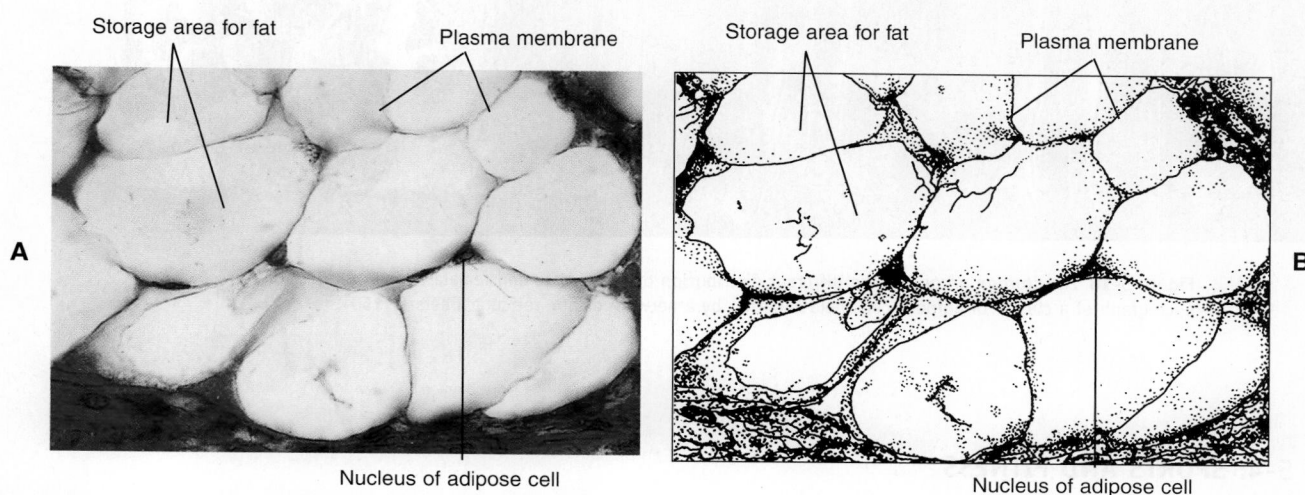

Figure 5-17 *Adipose tissue.* **A,** Photomicrograph. **B,** Sketch of the photomicrograph. Note the large storage spaces for fat inside the adipose tissue cells.

jurious substances out of the blood and lymph, and various types of reticular cells phagocytose (engulf and destroy) them. Another function of some reticular cells is to make reticular fibers.

Dense Fibrous Tissue

Dense fibrous tissue consists mainly of fibers packed densely in the matrix. It contains relatively few fibroblast cells. Some dense fibrous tissues are designated as *regular* and others are designated as *irregular,* depending on the arrangement of fibers.

Dense (Irregular) Fibrous Tissue

In dense (irregular) fibrous tissues, the bundles of collagenous fibers intertwine in irregular, swirling arrangements (Figure 5-20). This irregular pattern forms a thick mat of strong connective tissue that can withstand stresses applied from any direction. Dense (irregular) fibrous tissue forms the strong inner skin layer called the *dermis.* It also forms the outer capsule of such organs as the kidney and the spleen, as well as much of the fascia that surrounds muscles (Box 5-6).

Dense (Regular) Fibrous Tissue

In dense (regular) fibrous tissues, the bundles of fibers are arranged in regular, parallel rows.

One form of dense (regular) fibrous tissue is predominantly bundles of collagenous fibers and may be called **collagenous dense (regular) fibrous tissue** (Figure 5-21). This type of fibrous tissue is flexible but possesses great tensile strength when pulled from either or both ends. These characteristics are desirable in structures that anchor muscle to bone, such as tendons (Figure 5-22). Ligaments (which connect bone to bone) instead have a predominance of elastic fibers. Hence ligaments exhibit some degree of elasticity.

Another form of dense (regular) fibrous tissue contains mostly elastic fibers and may be called **elastic dense (regular) fibrous tissue.** As you can see in Figure 5-23, elastic fibers in this type of tissue are in a parallel arrangement. In the walls of arteries, this arrangement permits the walls to be pushed out by blood pressure without breaking and then recoil to a smaller diameter when the blood pressure decreases.

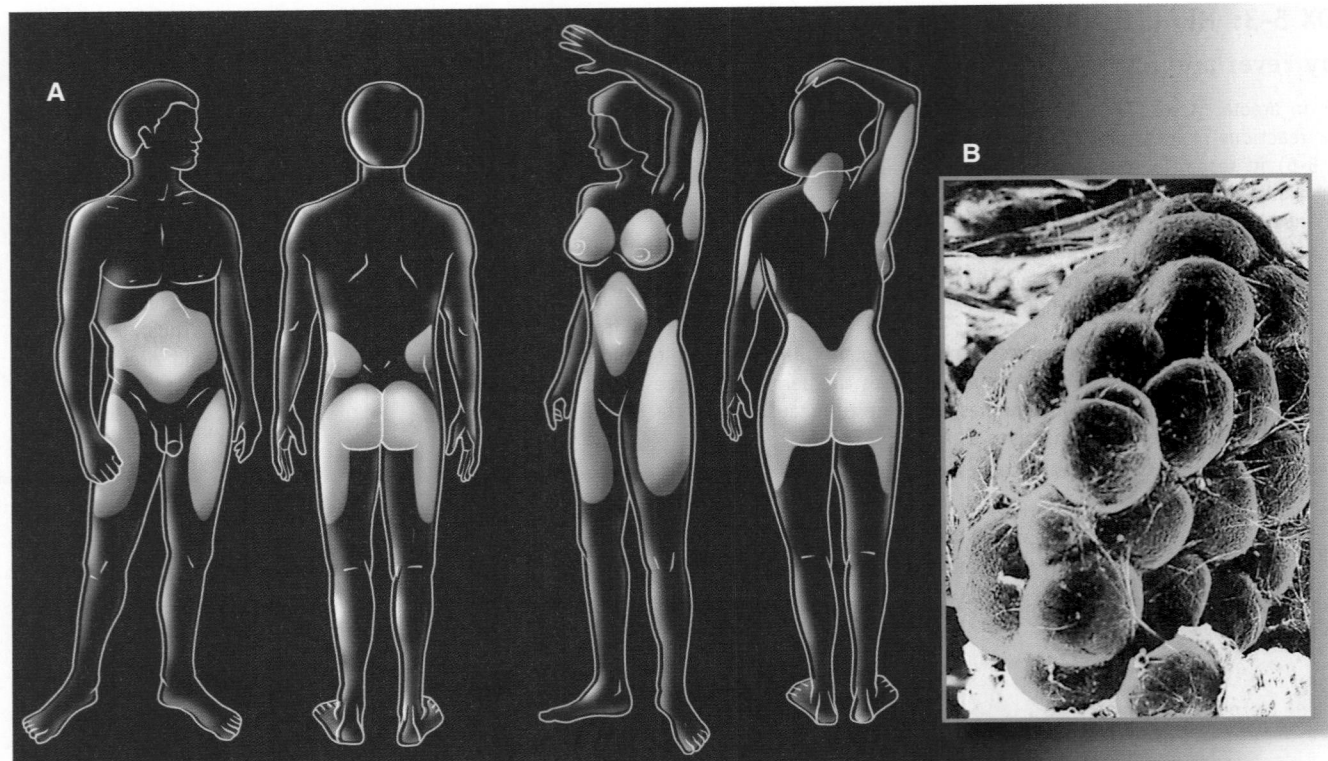

Figure 5-18 *Fat storage areas.* **A,** The different distribution of fat in male and female bodies. **B,** Electron micrograph of a cluster of adipose cells held together by a network of fine reticular fibers (×150).

BOX 5-4: SPORTS AND FITNESS

Tissues and Fitness

Achieving and maintaining ideal body weight is a health-conscious goal. However, a better indicator of health and fitness is **body composition.** Exercise physiologists assess body composition to identify the percentage of the body made of lean tissue and the percentage made of fat. Body fat percentage is often determined by using calipers to measure the thickness of skin folds at certain locations on the body (see figures). The thickness measurements, which reflect the volume of adipose tissue under the skin, are then used to estimate the percentage of fat in the entire body. A much more accurate method is to weigh a subject totally immersed in a tank of water. Fat has very low density and therefore increases the buoyancy of the body. Thus, the lower a person's weight while immersed, the higher the body fat percentage.

A person with low body weight may still have a high ratio of fat to muscle, an unhealthy condition. In this case the individual is "underweight" but "overfat." In other words, fitness depends more on the percentage and ratio of specific tissue types than on the overall amount of tissue present. Therefore, one goal of a good fitness program is a desirable body fat percentage. For men, the ideal is 15% to 18%, and for women, the ideal is 20% to 22%.

Because fat contains stored energy (measured in calories), a low fat percentage means a low energy reserve. High body fat percentages

are associated with several life-threatening conditions, including cardiovascular disease. A balanced diet and an exercise program ensure that the ratio of fat to muscle tissue stays at a level appropriate for maintaining homeostasis.

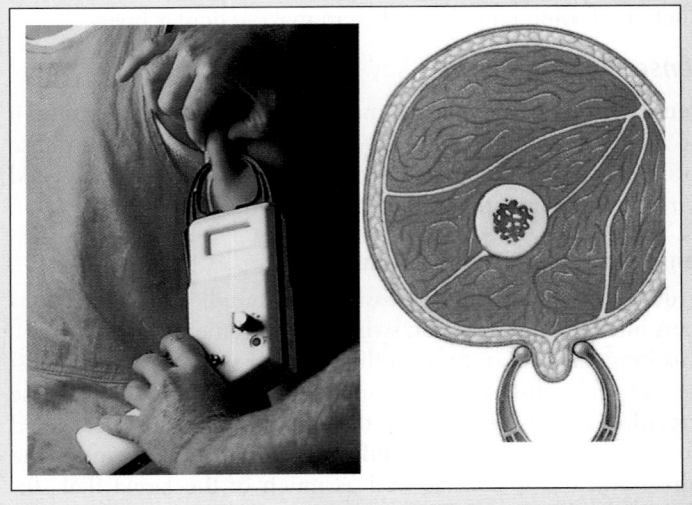

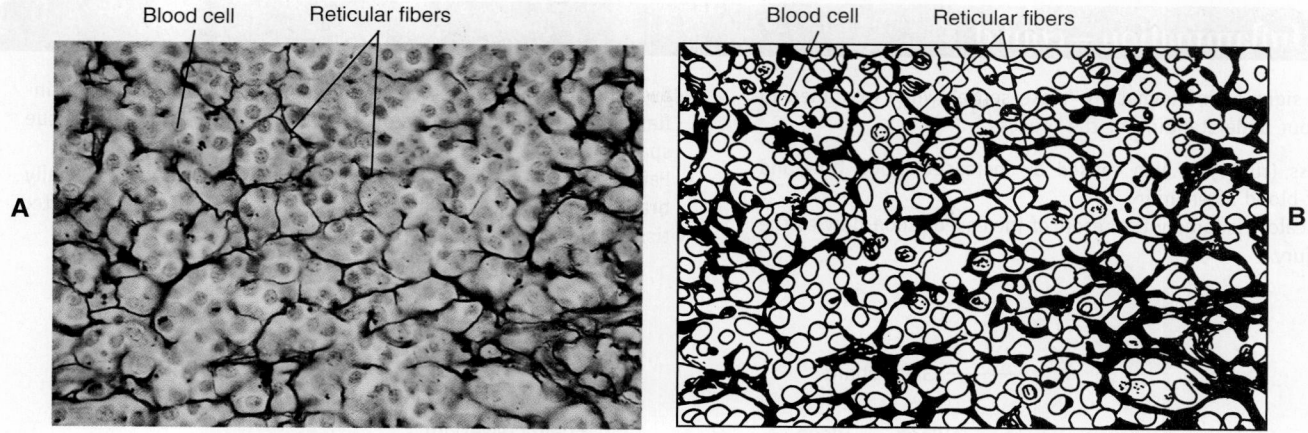

Figure 5-19 *Reticular connective tissue.* **A,** The supporting framework of reticular fibers are stained black in this section of spleen tissue. **B,** Sketch of the photomicrograph.

BOX 5-5   Inflammation

The terms **inflammation** or **inflammatory response** are used to describe the complex way in which cells and tissues react to injury. Many of the events, which are now identified as steps in the inflammatory response, are so dramatic that for centuries they were often thought to be a primary disease. It was in the first century AD that the Roman physician Celsus first established inflammation as an entity by describing its four cardinal signs: *rubor* (redness), *calor* (heat), *tumor* (swelling), and *dolor* (pain). His accurate and detailed description of the visible signs that signal the body's response to injury is considered a classic in the annals of medicine.

The inflammatory response can best be described as a series of sequenced events that occur as a result of an inflammatory stimulus or "insult." Heat, physical pressure, caustic chemicals, toxins released by harmful bacteria, or any other type of noxious stimulus initiate an inflammatory response. Early studies with rabbits, using a device known as a *transparent ear chamber*, permitted scientists to view for prolonged periods changes in living tissue after an injury. Skin over an animal's external ear was subjected to very slight injury and then viewed through the chamber under a microscope.

Immediately after an injury occurs, there is a very brief constriction of surrounding blood vessels that lasts but a moment. Then, almost immediately, blood vessels dilate, or open, and blood flow increases.

Injured tissues release a number of chemicals that affect blood vessels. These chemicals include **histamine, serotonin,** and a group of chemically related compounds called **kinins.** All of these substances result in vasodilation and an increase in the permeability of blood vessels so that components that would normally be retained in the blood are permitted to leak out into the tissue spaces.

In the absence of injury, blood flow through a small vessel is such that the cells tend to pass in large measure within the central two thirds of the lumen, with a thin layer of plasma flowing closest to the outer walls. This is called *axial flow.* After an injury, blood cells no longer pass in a central stream. Microscopic examination of vessels near an injured site shows that white cells begin to accumulate in the vessel near the point of injury and then stick, or *marginate,* to the wall. This *margination of leukocytes* continues until the endothelial surface of the vessel is covered with adherent white cells. Within minutes these cells begin to pass through the endothelial lining and out of the vessels into the interstitial spaces near the injury. One of the important functions of many white blood cells is **phagocytosis**—the process of engulfing and destroying bacteria. Movement of white cells into the area of injury or infection is called **diapedesis.** The term **chemotaxis** describes the attraction of leukocytes, especially neutrophils, into the interstitial spaces. The attractive force is produced by the release of kinins and other chemicals by injured tissue. **Leukocytosis** means an increase in the number of leukocytes in the blood. A substance called **leukocytosis-promoting (LP) factor** is also released by injured tissue. It stimulates the release of white cells from storage areas and increases the number of circulating white blood cells.

The accumulation of dead leukocytes and tissue debris may lead to the formation of **pus** at the focal point of infection. Should this occur, an **abscess,** or cavity, formed by the disintegration of tissues may fill with pus and require surgical drainage.

Increased permeability of blood vessels, increased blood flow, and migration and accumulation of white blood cells all contribute to the formation of **inflammatory exudate,** which accumulates in the interstitial spaces in the area of injury. The result is often swelling, or **edema,** and pain. In addition to white blood cells and tissue debris, inflammatory exudate contains the "leaked" substances normally retained in the blood but allowed to escape into the interstitial spaces because of increased capillary permeability. One such substance is a soluble protein that is soon converted into fibrin in the interstitial spaces. Fibrin formation results in development of a clot, which helps seal off the infected area and decrease the spread of bacteria or other infectious material.

Continued

BOX 5-5   Inflammation—cont'd

The cardinal signs of inflammation "make sense" when examined in the light of our understanding of the process.

- The redness (rubor) is also caused by increased blood flow and pooling of blood after injury.
- The heat (calor) is largely the result of increased blood flow to the area of injury.

- Swelling (tumor) results because of edema and accumulation of inflammatory exudate and clot formation in the affected tissue spaces.
- Pain (dolor) is caused by chemicals such as the kinins (especially **bradykinin**) and other chemical mediators that are released after tissue injury and cellular death.

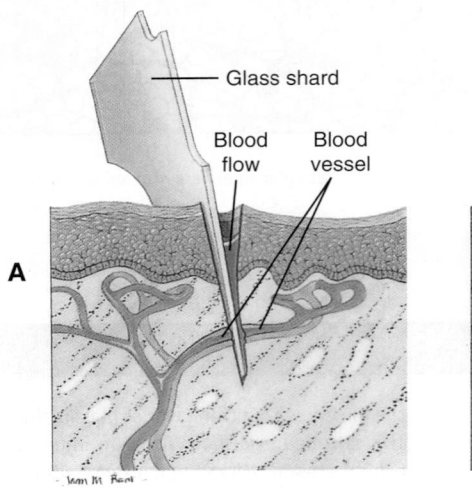
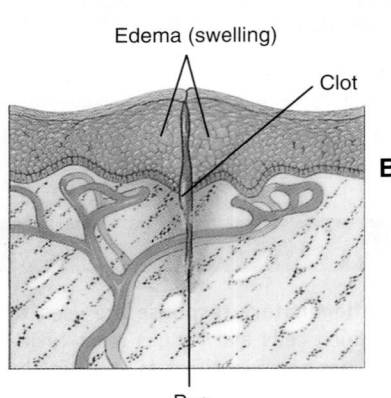

Inflammation in a section of traumatized skin. **A,** Injury or insult to tissue resulting in dilation of blood vessels and an increase in blood flow. **B,** Inflammatory response—the body attempts to seal off an area of insult to limit bacterial invasion.

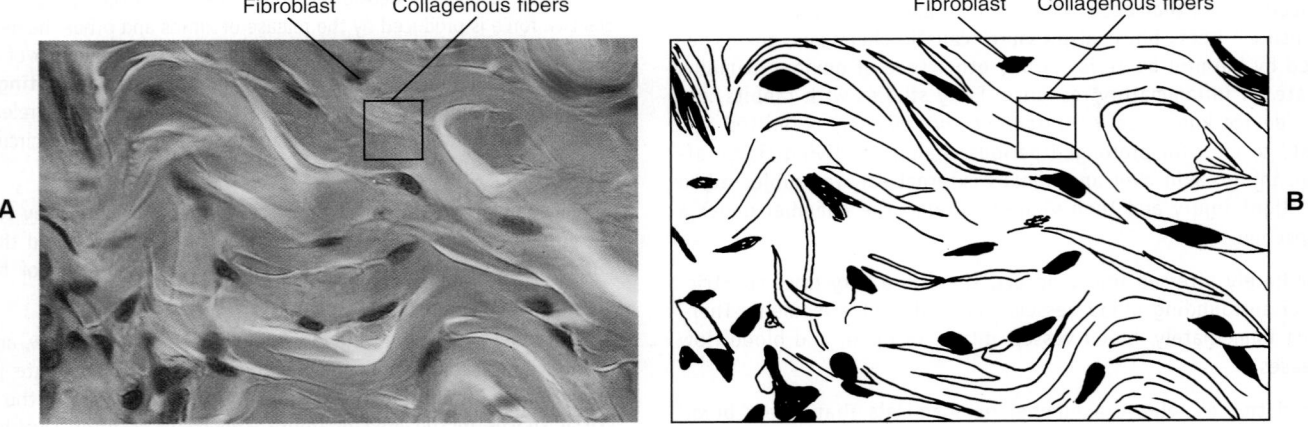

Figure 5-20 *Dense (irregular) fibrous connective tissue.* **A,** Section of skin (dermis) showing arrangements of collagenous fibers (pink) and purple-staining fibroblast cell nuclei. **B,** Sketch of the photomicrograph.

BOX 5-6   Fascia

The term **fascia** (FAH-sha) is a general name for the fibrous connective tissue masses that can be seen by the unaided eye in many locations throughout the body. The word *fascia* is Latin for "band." This literal translation is helpful because it summarizes the general structure and function of fascia: fibrous tissue that binds together the structures of the body.

Fascia is always some form of fibrous connective tissue and almost always features many collagenous fibers that are interwoven in an irregular arrangement. In some areas of the body, for example under the skin, fascia is mostly adipose tissue. In other areas, such as around some of the muscles, it is dense irregular fibrous tissue. Often, some of the fibers of fascia extend into the tissue of nearby organs, thus strongly binding to them.

For convenience, anatomists often distinguish between *superficial fascia,* which is just under the skin, and *deep fascia,* which extends well into the body and surrounds muscles, blood vessels, and other organs (see the figure).

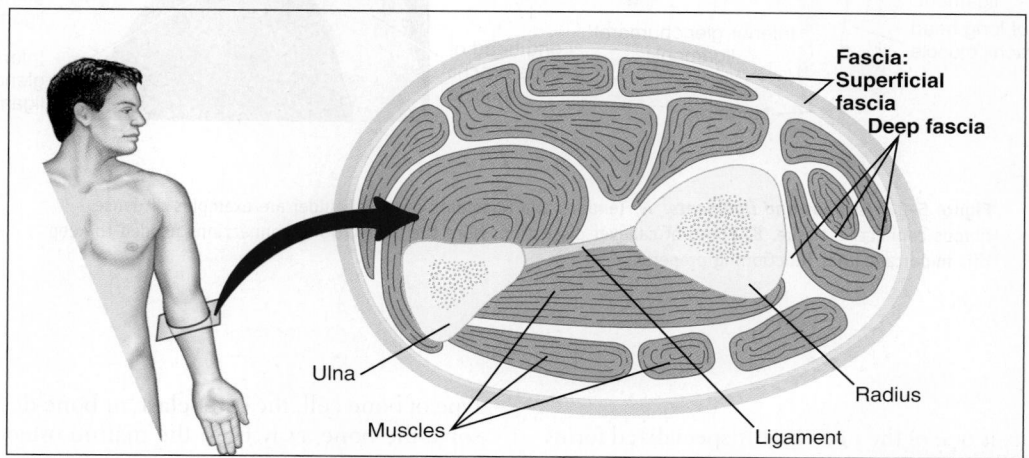

Fascia. Horizontal section of the forearm showing the superficial fascia under the skin and deep fascia below that, extending to surround the individual muscles.

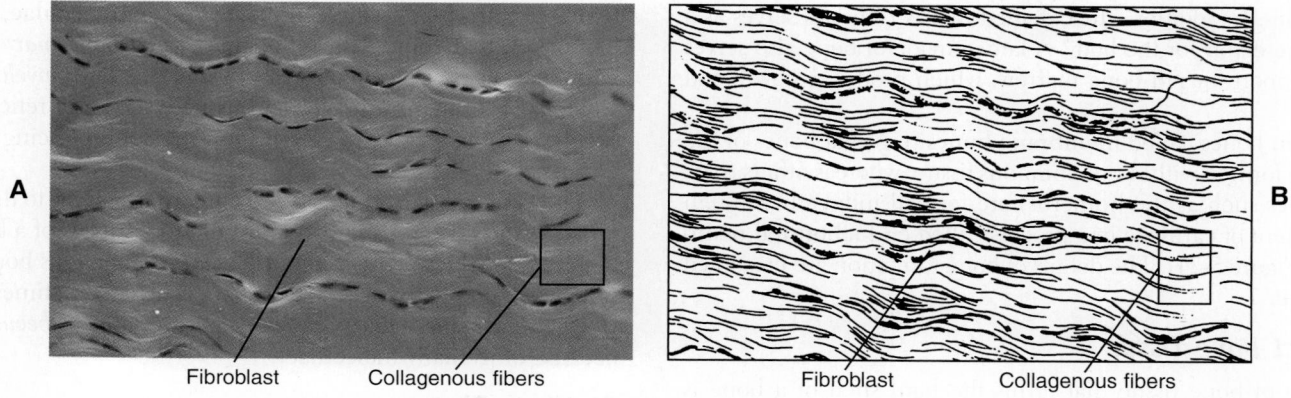

Figure 5-21 *Collagenous dense (regular) fibrous connective tissue.* **A,** Photomicrograph of tissue in a tendon. **B,** Sketch of the photomicrograph. Note the multiple (regular) bundles of collagenous fibers arranged in parallel rows.

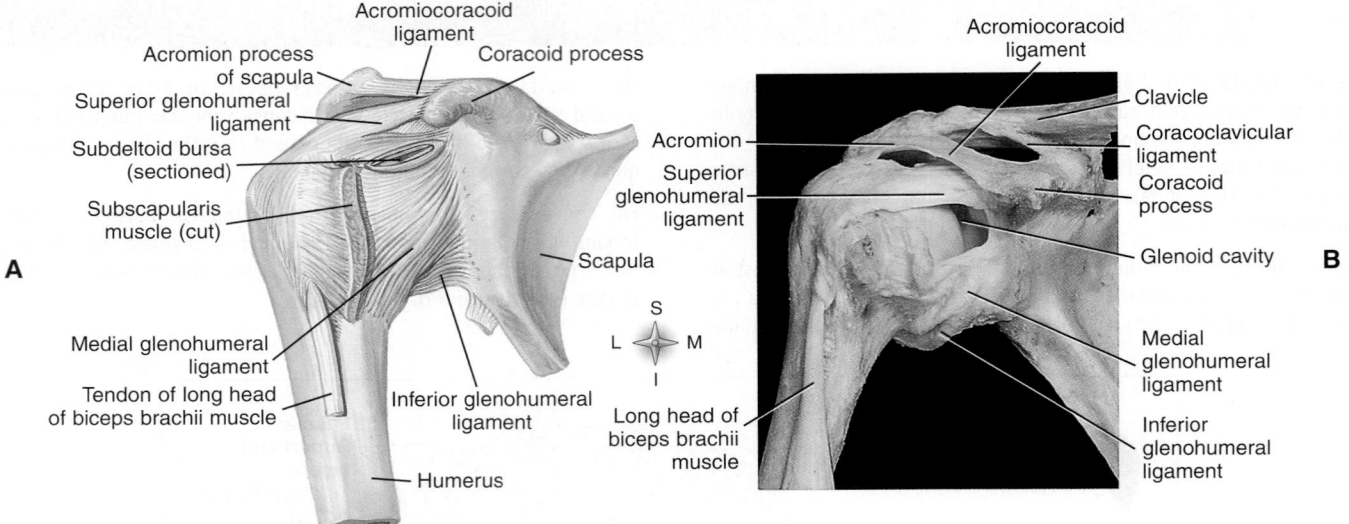

Figure 5-22 *Tendons and ligaments.* A, Tendons and ligaments of the shoulder are examples of dense fibrous connective tissue. **B,** Photo of cadaver dissection. Note the many strong connections needed to keep this important joint functioning properly.

Bone Tissue

Bone, or *osseous tissue,* is one of the most highly specialized forms of connective tissue. The mature cells of bone, **osteocytes,** are embedded in a unique matrix material containing both collagen fibers and mineral salt crystals. The inorganic mineral crystals make up about 66% of the total extracellular matrix. These mineral crystals are responsible for the hardness of bone.

Bones are the organs of the skeletal system. They provide support and protection for the body and serve as points of attachment for muscles. In addition, the calcified matrix of bones serves as a mineral reservoir for the body. A lattice made of bone also serves as the support for red bone marrow, which produces new blood cells.

Certain bones called **membrane bones** (e.g., flat bones of the skull) are formed within membranous tissue, whereas others (e.g., long bones such as the humerus) are formed indirectly through replacement of cartilage in a process called **endochondral ossification** (Figure 5-24). The details of bone formation are presented in Chapter 7.

Compact Bone Tissue

The type of bone tissue that forms the hard shell of a bone is called **compact bone tissue** (Figure 5-25). The basic organizational or structural unit of compact bone is the microscopic **osteon** or *Haversian system* (Figure 5-26). Osteocytes, or bone cells, are located in small spaces, or **lacunae,** which are arranged in concentric layers of bone matrix called **lamellae.** Small canals called **canaliculi** connect each lacuna and osteocyte with nutrient blood vessels found in the central or Haversian canal.

Mature osteocytes are actually trapped in hard bone matrix. At one time they were active, bone-forming cells called **osteoblasts.** However, as they surround themselves with bone, they become trapped and cease making new bone matrix. Another

type of bone cell, the **osteoclast,** or bone-destroying cell, may dissolve the bone away from the mature osteocyte and release it to again become an active osteoblast. Mature bone can thus grow and be reshaped by the simultaneous activity of osteoclasts breaking down and remove existing bone tissue as osteoblasts lay down new bone.

Cancellous (Spongy) Bone Tissue

Inside many bones is a lattice of thin beams of **cancellous bone tissue** (Figure 5-27). These thin beams, or **trabeculae,** form a framework that supports a softer tissue—*red bone marrow.* Red bone marrow is also called *myeloid tissue* (the term *myeloid* literally means "of marrow"). Myeloid tissue is a type of reticular tissue that contains the stem cells responsible for producing the various types of blood cells.

The lattice of trabeculae also give internal support to the bone, much as the crisscrossing pattern of the roof trusses of a building help support the weight of a roof. Because cancellous bone looks somewhat like a sponge at first glance, it is sometimes called *spongy bone tissue.* This type of bone is also called *trabecular bone* because of its many trabeculae.

Cartilage Tissue

Cartilage differs from other connective tissues in that only one cell type, the **chondrocyte,** is present. Chondrocytes produce the fibers and the tough, gristlelike ground substance of cartilage. Chondrocytes, like bone cells, are found in small openings called *lacunae.* Cartilage is avascular (lacking blood vessels), so nutrients must reach the cells by diffusion. Movement is through the matrix from blood vessels located in a specialized connective tissue membrane called the **perichondrium,** which surrounds the cartilage mass. Injuries to cartilage heal slowly, if at all, because of this inefficient method of nutrient delivery.

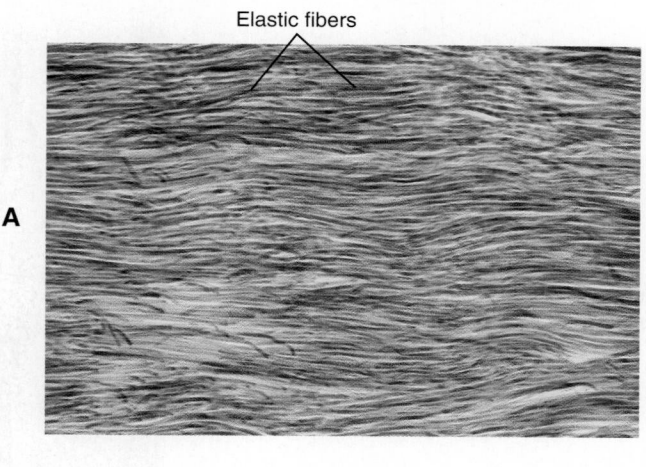

Elastic fibers

A

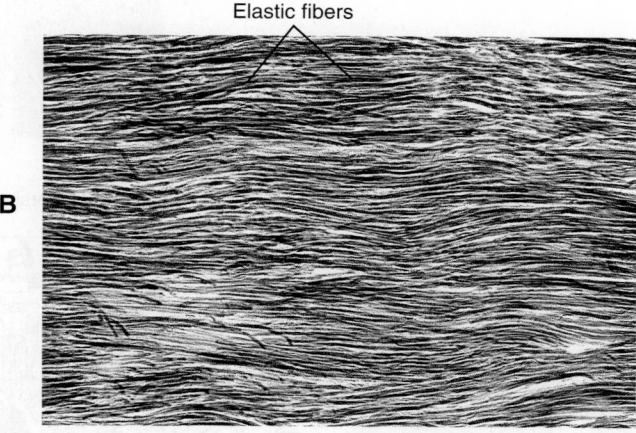

Elastic fibers

B

Figure 5-23 *Elastic dense (regular) fibrous connective tissue.* **A,** Photomicrograph of tissue. **B,** Sketch of the photomicrograph. Note the roughly parallel arrangement of short, darkly stained elastic fibers.

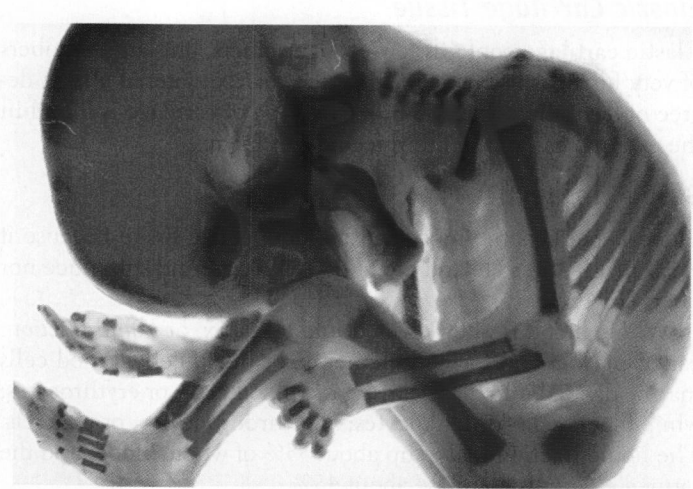

Figure 5-24 *Ossification of the skeleton.* The darkly stained areas of mineralization seen in this 14-week-old fetus show how the cartilage and membrane skeleton is beginning to develop into bone.

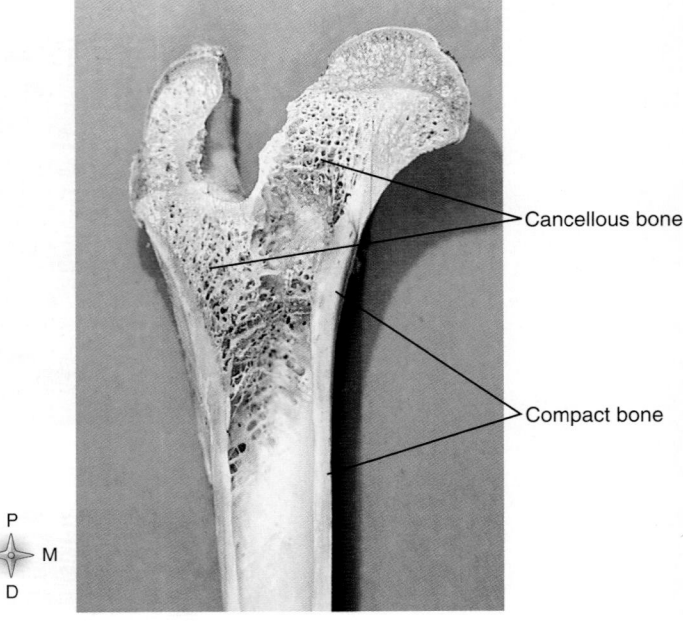

Cancellous bone

Compact bone

Figure 5-25 *Types of bone tissue.* Compact bone is found forming most of the hard shell of a bone. Cancellous (spongy) bone forms a network of hard beams of bone tissue inside many bones.

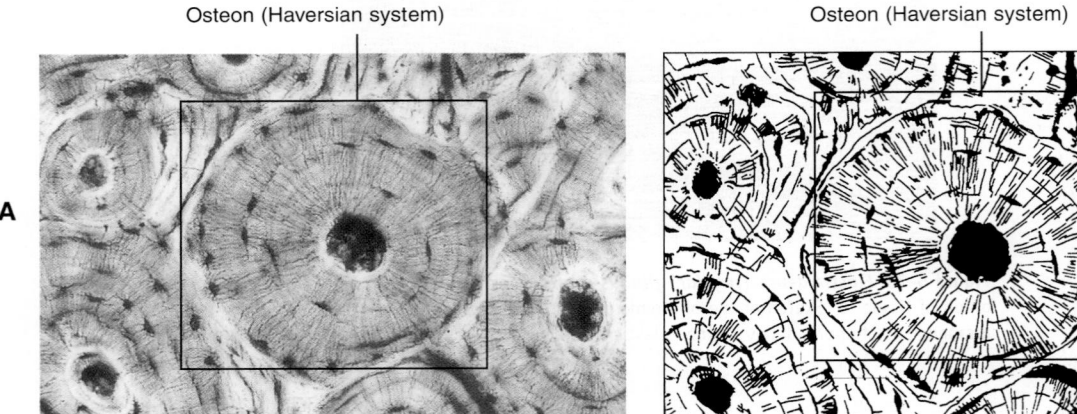

Osteon (Haversian system)

Osteon (Haversian system)

A

B

Figure 5-26 *Compact bone tissue.* **A,** Photomicrograph of dried, ground compact bone. **B,** Sketch of the photomicrograph. Many wheel-like structural units of bone, known as *osteons* or *Haversian systems,* are apparent in this section.

Hyaline Cartilage Tissue

Hyaline cartilage takes its name from the Greek word *hyalos* or "glass." The name is appropriate because the low amount of collagen in the matrix gives hyaline cartilage a shiny and translucent appearance. This is the most prevalent type of cartilage and is found in the support rings of the respiratory tubes and covering the ends of bones that articulate at joints (Figure 5-28).

Fibrocartilage Tissue

Fibrocartilage is the strongest and most durable type of cartilage (Figure 5-29). The matrix is rigid and filled with a dense packing of strong white collagen fibers. Fibrocartilage disks serve as shock absorbers between adjacent vertebrae (intervertebral disks) and in the knee joint. Damage to the fibrocartilage pads or joint menisci (curved pads) in the knee occurs frequently as a result of sports-related injuries.

Elastic Cartilage Tissue

Elastic cartilage contains few collagen fibers, but large numbers of very fine elastic fibers that give the matrix material a high degree of flexibility (Figure 5-30). This type of cartilage is found in the external ear and in the voice box, or larynx.

Blood Tissue

Blood is perhaps the most unusual connective tissue because it exists in a liquid state and contains neither ground substance nor fibers (Figure 5-31).

Whole blood is often divided into a matrix, or *liquid fraction,* called **plasma** and **formed elements,** or blood cells. Blood cells may be divided into three classes: red blood cells, or **erythrocytes;** white blood cells, or **leukocytes;** and **thrombocytes,** or platelets. The liquid fraction makes up about 55% of whole blood, and the formed elements compose about 45%.

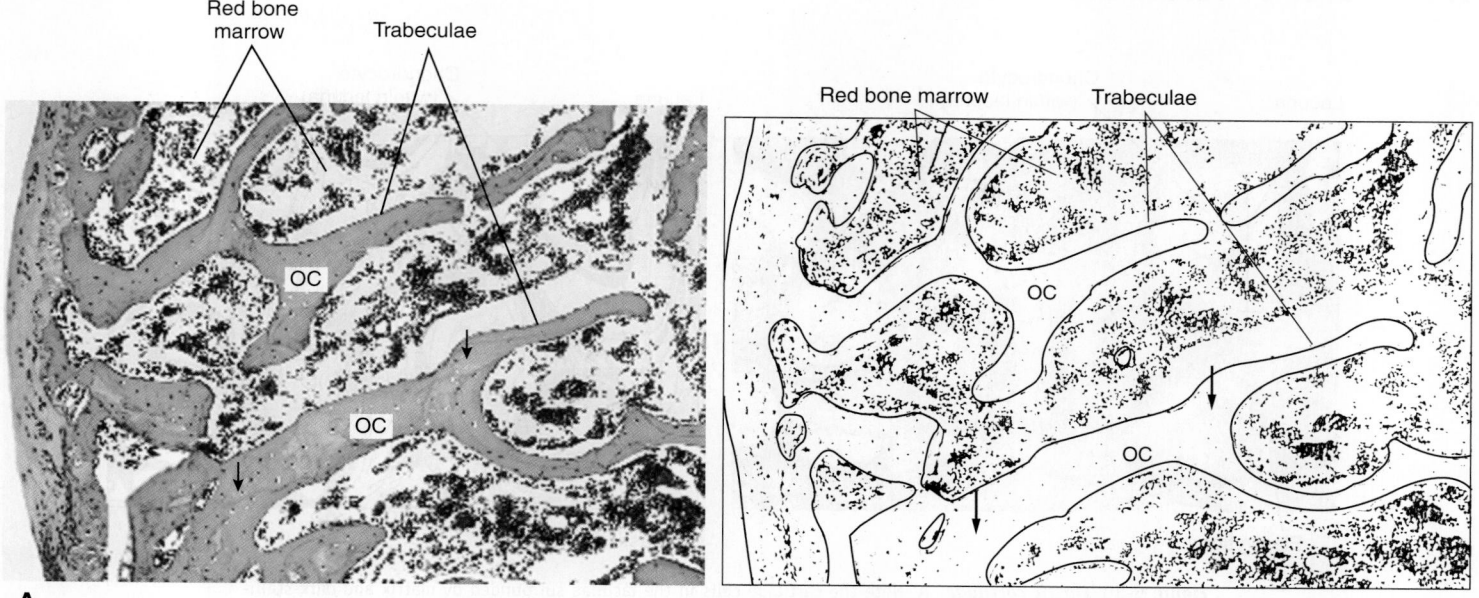

Figure 5-27 *Cancellous bone tissue.* **A,** Photomicrograph of cancellous (or *spongy*) bone. The pink-stained mineralized bone tissue forms a lattice of irregular beams, or *trabeculae,* that support the softer reticular tissue of the bone marrow. The darkly stained nuclei of osteocytes (OC) are visible, as well as the dark boundaries *(arrows)* of mineralized bone layers or lamellae. **B,** Sketch of the photomicrograph.

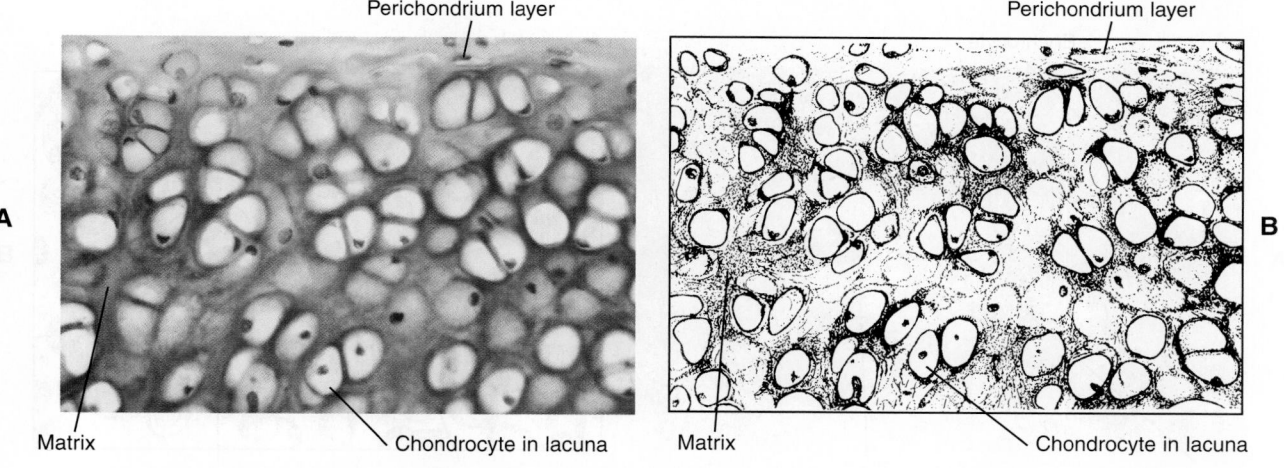

Figure 5-28 *Hyaline cartilage.* **A,** Photomicrograph of the trachea. Note the many spaces, or lacunae, in the gel-like matrix. **B,** Sketch of the photomicrograph.

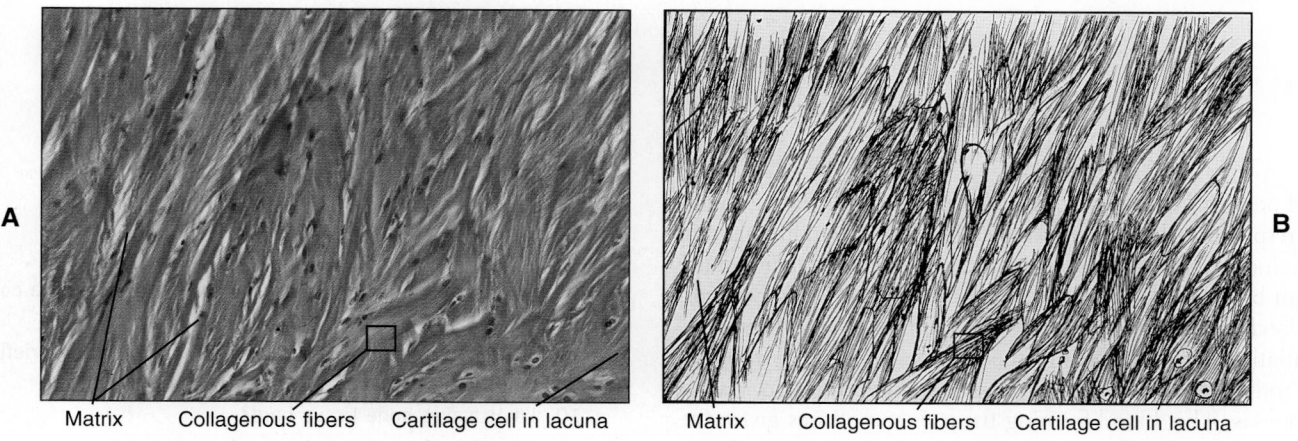

Figure 5-29 *Fibrocartilage.* **A,** Photomicrograph of the pubic symphysis joint. The strong dense fibers that fill the matrix convey shock-absorbing qualities. **B,** Sketch of the photomicrograph.

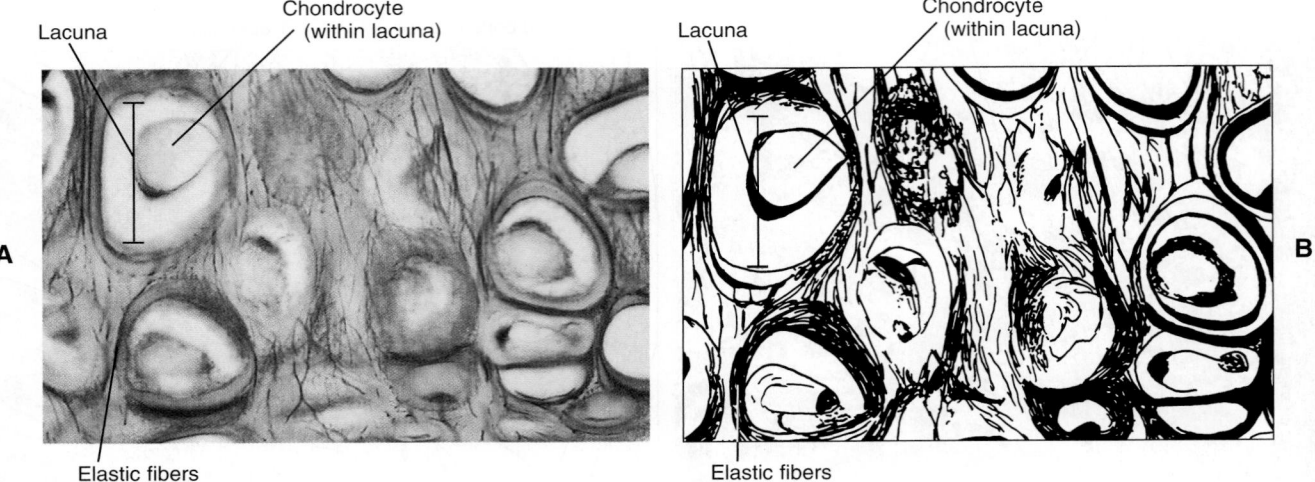

Figure 5-30 *Elastic cartilage.* **A,** Note the cartilage cells in the lacunae surrounded by matrix and dark-staining elastic fibers. **B,** Sketch of the photomicrograph.

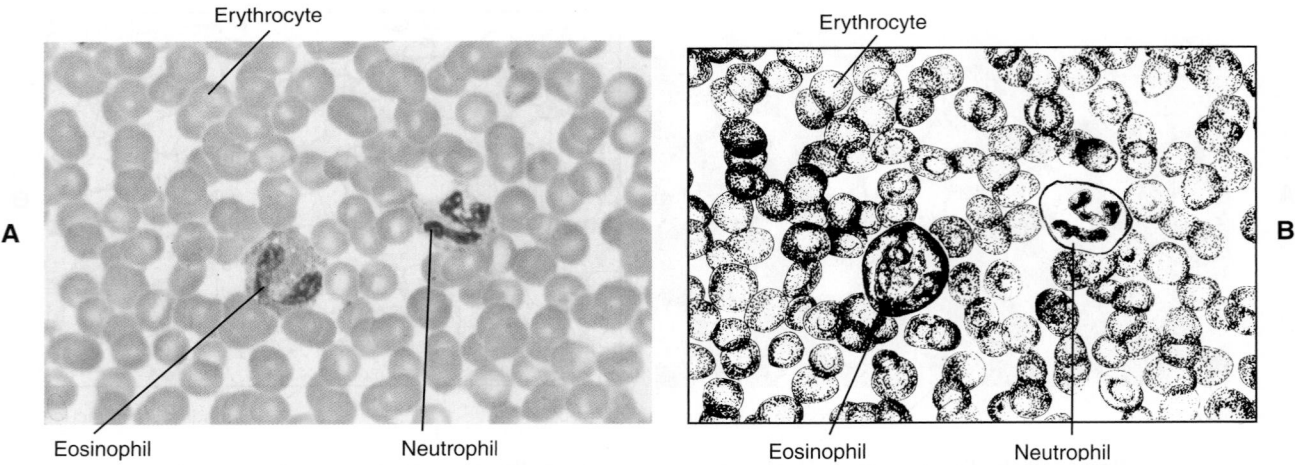

Figure 5-31 *Blood.* **A,** Photomicrograph of a human blood smear showing two white blood cells, or leukocytes, surrounded by numerous smaller red blood cells. **B,** Sketch of the photomicrograph.

Blood performs many body transport functions, including movement of respiratory gases (oxygen and carbon dioxide), nutrients, and waste products. In addition, blood plays a critical role in maintaining a constant body temperature and regulating the pH of body fluids. White blood cells function in destroying harmful microorganisms.

Circulating blood tissue is formed in the *red marrow* of bones and in other tissues by a process of differentiation called *hematopoiesis.* This blood-forming tissue is sometimes given the status of a separate connective tissue type: **hematopoietic tissue.**

Blood and its formation are described in detail in Chapter 17.

QUICK CHECK

8. Name three kinds of fibers that may be present in a connective tissue ECM. Of what are they made?

9. Name four types of fibrous connective tissue and briefly describe each.

10. What makes bone tissue hard?

11. What is unique about the matrix of blood tissue?

MUSCLE TISSUE

Three types of **muscle tissue** are present in the body—skeletal muscle, smooth muscle, and cardiac muscle (Table 5-7). Their names suggest their locations. **Skeletal muscle tissue** (Figure 5-32) makes up most of the muscles attached to bones; these are the organs that we think of as our muscles. **Smooth muscle tissue,** also sometimes called *visceral muscle tissue* (Figure 5-33), is found in the walls of the viscera (hollow internal organs—e.g., the stomach, intestines, and blood vessels; Figure 5-34). **Cardiac muscle tissue** makes up the wall of the heart (Figure 5-35). Another name for skeletal muscle is *striated voluntary* muscle. The term *striated* refers to cross striations (stripes) visible on microscopic slides of the tissue. The term

Table 5-7	Muscle and Nervous Tissues		
TISSUE		**LOCATION**	**FUNCTION**
Muscle Skeletal (striated voluntary)	 Nuclei Muscle fiber Striations	Muscles that attach to bones Extrinsic eyeball muscles Upper third of the esophagus	Movement of bones Eye movements First part of swallowing
Smooth (nonstriated, involuntary, or visceral)	 Smooth muscle cell Nucleus	In the walls of tubular viscera of the digestive, respiratory, and genitourinary tracts In the walls of blood vessels and large lymphatic vessels In the ducts of glands Intrinsic eye muscles (iris and ciliary body) Arrector muscles of hairs	Movement of substances along the respective tracts Change diameter of blood vessels, thereby aiding in regulation of blood pressure Movement of substances along ducts Change diameter of pupils and shape of the lens Erection of hairs (gooseflesh)
Cardiac (striated involuntary)	 Intercalated disc Nucleus Cardiocytes Striations	Wall of the heart	Contraction of the heart
Nervous	 Axon Nuclei of glial cells Cell body Dendrites	Brain Spinal cord	Excitability Conduction Nerves

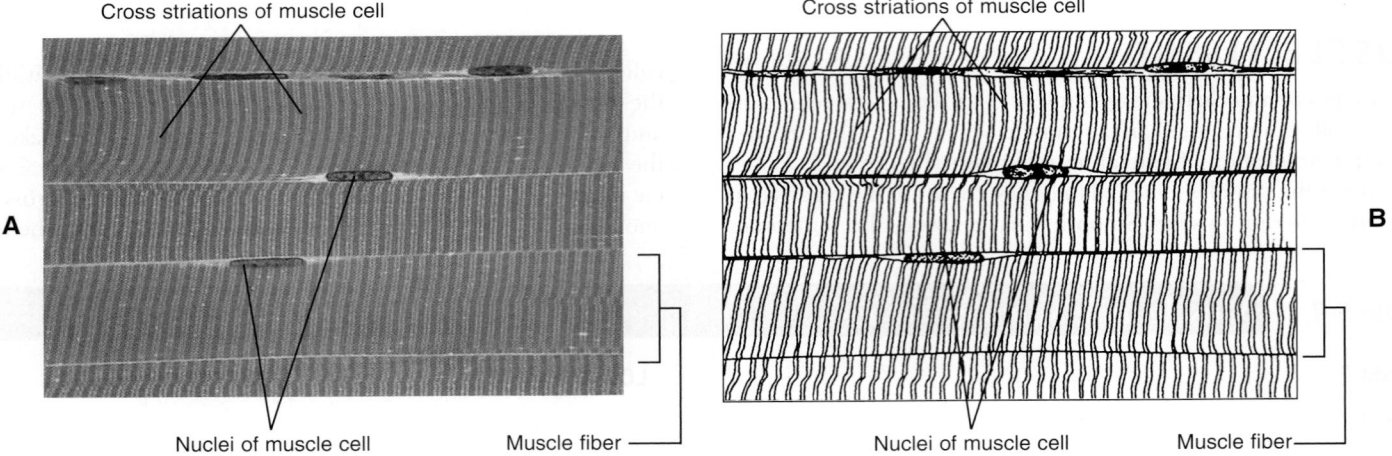

Figure 5-32 *Skeletal muscle.* **A,** Photomicrograph. **B,** Sketch of the photomicrograph. Note the striations of the muscle cell fibers in longitudinal section.

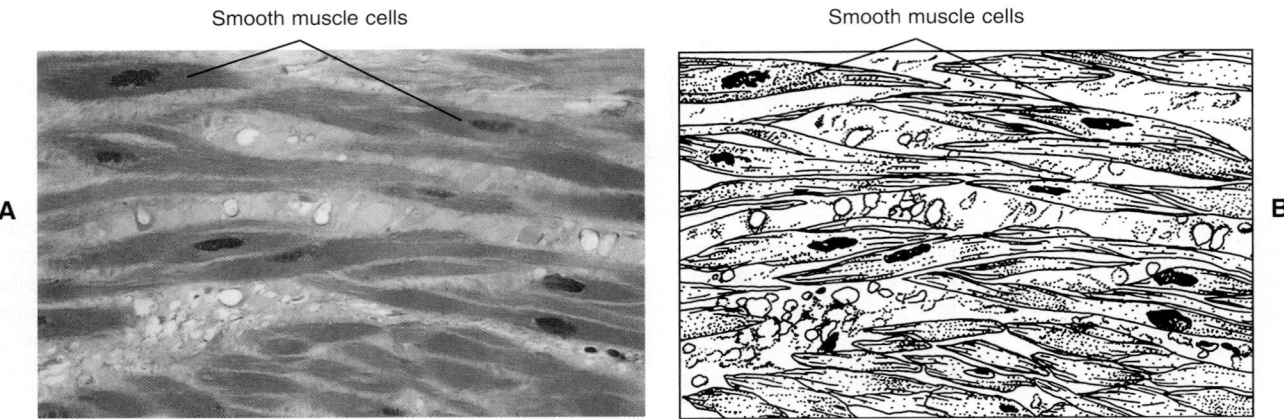

Figure 5-33 *Smooth muscle.* **A,** Photomicrograph, longitudinal section. **B,** Sketch of the photomicrograph. Note the central placement of nuclei in the spindle-shaped smooth muscle fibers.

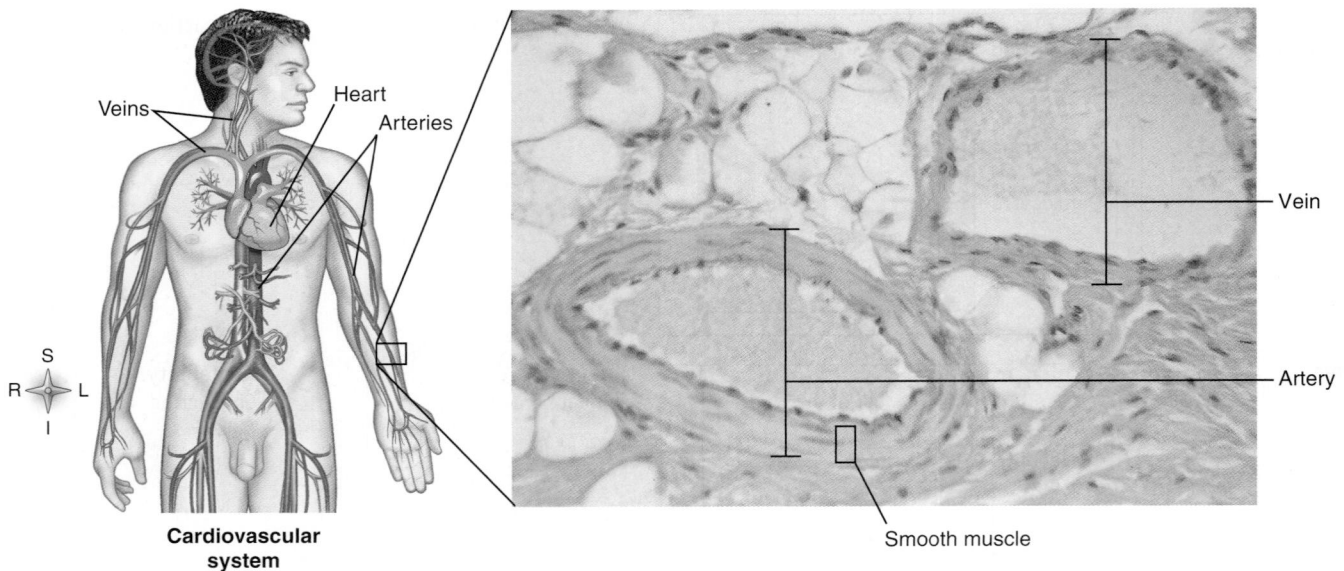

Figure 5-34 *Location of smooth muscle.* Smooth muscle is found in the walls of most internal hollow organs such as the stomach, intestines, and blood vessels. For example, the cross section of an artery shown in the inset reveals that the arterial wall has a thick layer of smooth muscle, which contracts or relaxes to control the diameter of the artery—thus controlling the flow of blood through this vessel.

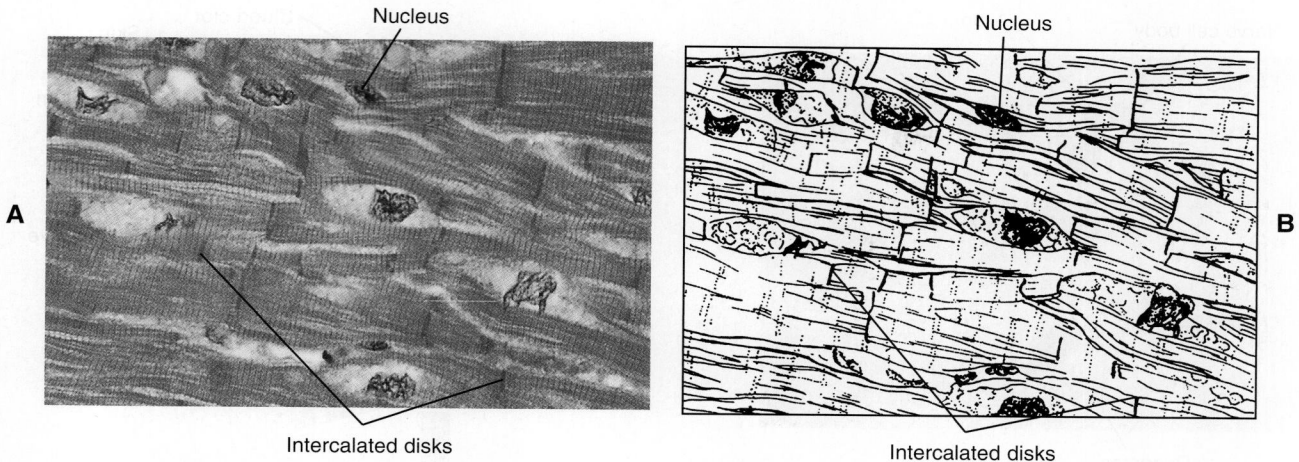

Nucleus

Nucleus

A

B

Intercalated disks

Intercalated disks

Figure 5-35 *Cardiac muscle.* **A,** Photomicrograph. **B,** Sketch of the photomicrograph. The dark bands, called *intercalated disks,* which are characteristic of cardiac muscle, are easily identified in this tissue section.

voluntary indicates that voluntary or willed control of skeletal muscle contractions is possible. Another name for smooth muscle is *nonstriated involuntary.* Smooth muscle has no cross striations and cannot ordinarily be controlled by the will. Another name for cardiac muscle is *striated involuntary* muscle. Like skeletal muscle, cardiac muscle has cross striations, and like smooth muscle, its contractions cannot ordinarily be controlled by will.

Look now at Figure 5-32 and observe the following structural characteristics of skeletal muscle cells: many cross striations, many nuclei per cell, and long, narrow, threadlike shape of the cells. Skeletal muscle cells may have a length of more than 3.75 cm, but they have diameters of only 10 to 100 μm. Because this gives them a threadlike appearance, muscle cells are often called muscle fibers. Chapter 11 gives more detailed information about the structure of skeletal muscle tissue.

Smooth muscle cells are also long, narrow fibers, but not nearly as long as striated fibers. One can see the full length of a smooth muscle fiber in a microscopic field, but only a small part of a striated fiber. According to one estimate, the longest smooth muscle fibers measure about 500 μm and the longest striated fibers about 40,000 μm. As Figure 5-33 shows, smooth muscle fibers have only one nucleus per fiber and are nonstriated or smooth in appearance. Like skeletal muscle fibers, smooth muscle fibers are shaped like a spindle—that is, they are cylinders that taper at both ends.

Under the light microscope, cardiac muscle fibers (see Figure 5-35) have cross striations and unique dark bands (intercalated disks) that join together the untapered ends of fibers. They also seem to be incomplete cells that branch into each other to form a big continuous mass of cytoplasm (a syncytium). The electron microscope, however, reveals that the intercalated disks are actually places where the plasma membranes of two cardiac fibers abut. Cardiac fibers branch in and out, but a complete plasma membrane encloses each cardiac fiber—around its end (at intercalated disks) and its sides.

Muscle cells are the movement specialists of the body. Because their cytoskeletons include bundles of microfilaments spe-

cialized for movement, they have a higher degree of contractility (ability to shorten or contract) than cells of any other tissue.

NERVOUS TISSUE

The basic function of the nervous system is to rapidly regulate, and thereby integrate, the activities of the different parts of the body. Functionally, rapid communication is possible because **nervous tissue** has much more developed excitability and conductivity characteristics than any other type of tissue does.

The organs of the nervous system are the brain, the spinal cord, and the nerves. Actual **nerve tissue** is ectodermal in origin and consists of two basic kinds of cells: nerve cells, or **neurons,** which are the conducting units of the system, and special connecting, protective, and supporting cells called **neuroglia** (Table 5-7, Figure 5-36).

All neurons are characterized by a cell body called the **soma** and, generally, at least two processes: one **axon,** which transmits nerve impulses away from the cell body, and one or more **dendrites,** which carry nerve signals toward the axon. Most neurons are located within the organs of the central nervous system.

There are many types of neuroglia, all with different structures and functions. In addition to physically supporting neurons, neuroglia are known to have important coordinating roles in the nervous system—a concept that we will explore further in Chapter 12.

The anatomy and physiology of the nervous system are presented in Chapters 12 through 14.

QUICK CHECK

12. Name the two types of involuntary muscle. Where is each found in the body?

13. What are the two principal types of cell in nervous tissue? What is the function of each?

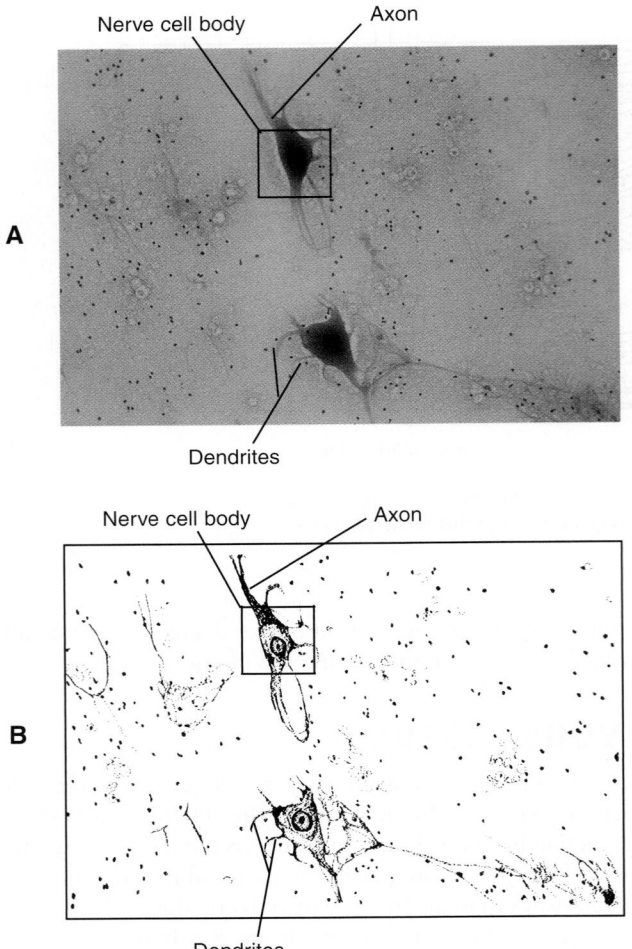

A

B

Figure 5-36 *Nervous tissue.* A, Photomicrograph showing multipolar neurons surrounded by smaller neuroglia in a smear of spinal cord tissue. Both large neurons in this photomicrograph show characteristic soma, or cell bodies, and multiple cell processes. **B,** Sketch of the photomicrograph.

TISSUE REPAIR

When damaged by mechanical or other injuries, tissues have varying capacity to repair themselves. Damaged tissue regenerates or is replaced by tissue we know as scars. Tissues usually repair themselves by allowing phagocytic cells to remove dead or injured cells and then filling in the gaps that are left. This growth of functional new tissue is called **regeneration.**

Epithelial and connective tissues have the greatest capacity to regenerate (Figure 5-37). When a break in an epithelial membrane occurs, as in a cut, cells quickly divide to form daughter cells that fill the wound. In connective tissues, cells that form collagen fibers become active after an injury and fill in a gap with an unusually dense mass of fibrous connective tissue. If this dense mass of fibrous tissue is small, it may be replaced by normal tissue later. If the mass is deep or large or if cell damage was extensive, it may remain as a dense fibrous mass, called a **scar.** An atypical and unusually thick scar that may develop in the lower layer of the skin, such as that shown in Figure 5-38, is called a **keloid.**

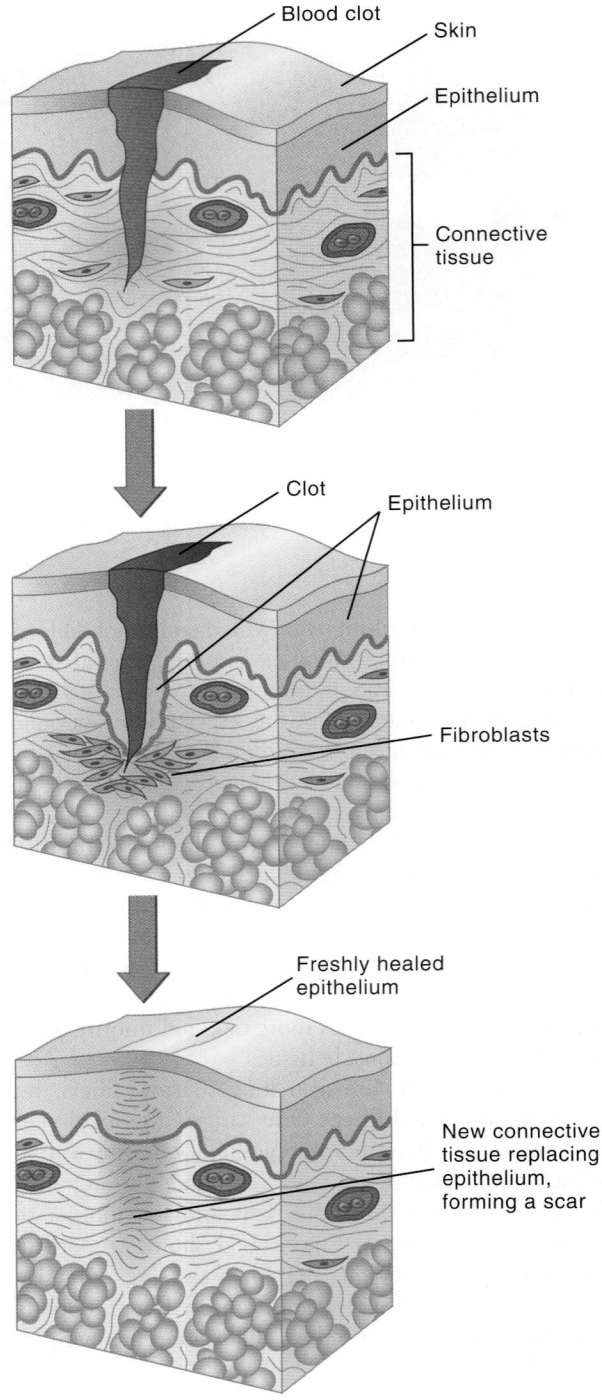

Figure 5-37 *Healing of a minor wound.* When a minor injury damages a layer of epithelium and the underlying connective tissue (as in a minor skin cut), the epithelial tissue and the connective tissue can self-repair.

Muscle tissue, on the other hand, has a limited capacity to regenerate and thus heal itself. Damaged muscle is often replaced with fibrous connective tissue instead of muscle tissue. When this happens, the organ involved loses some or all of its ability to function.

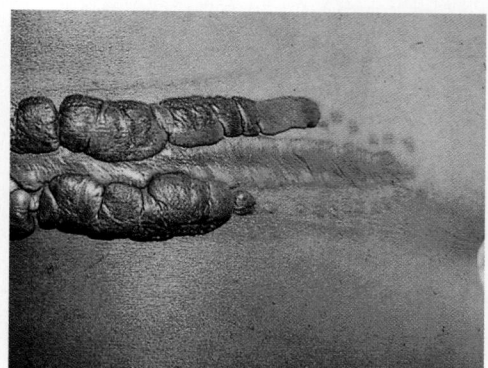

Figure 5-38 *Keloid.* Keloids are thick scars that form in the lower layer of the skin in predisposed individuals. This photograph shows keloids that formed at suture marks after surgery.

Like muscle tissue, nerve tissue also has a limited capacity to regenerate. Neurons outside the brain and spinal cord can sometimes regenerate, but very slowly and only if certain neuroglia are present to "pave the way." In the normal adult brain and spinal cord, neurons do not usually grow back when injured. Thus, brain and spinal cord injuries nearly always result in permanent damage. Fortunately, the discovery of *nerve growth factors* produced by neuroglia offers the promise of treating brain damage by stimulating the release of these factors.

BODY MEMBRANES

The term **membrane** refers to a thin, sheetlike structure that may have many important functions in the body. Membranes cover and protect the body surface, line body cavities, and cover the inner surfaces of hollow organs such as the digestive, reproductive, and respiratory passageways. Some membranes anchor organs to each other or to bones, and others cover the internal organs. In certain areas of the body, membranes secrete lubricating fluids that reduce friction during organ movements such as beating of the heart or lung expansion and contraction. Membrane lubricants also decrease friction between bones and joints. Two major categories, or types, of body membranes exist (Figure 5-39):

1. **Epithelial membranes,** composed of epithelial tissue and an underlying layer of specialized connective tissue
2. **Connective tissue membranes,** composed exclusively of various types of connective tissue; no epithelial cells are present in this type of membrane

Epithelial Membranes

There are three types of epithelial tissue membranes in the body: (1) cutaneous membrane, (2) serous membranes, and (3) mucous membranes (Figure 5-40).

Cutaneous Membranes

The **cutaneous membrane** covers body surfaces that are exposed to the external environment. The cutaneous membrane, or **skin**, is the primary organ of the integumentary system. It is one of the most important and certainly one of the largest and most visible organs of the body. In most individuals the skin composes approximately 16% of body weight. It fulfills the requirements nec-

essary for an epithelial tissue membrane in that it has a superficial layer of epithelial cells and an underlying layer of supportive connective tissue. It also contains many sweat and oil glands that produce a surface film over the skin. The structure of the skin is uniquely suited to its many functions. The skin is discussed in depth in Chapter 6.

Serous Membranes

Serous membrane lines cavities that are not open to the external environment and covers many of the organs inside these cavities. Serous membranes are sometimes called by their Latin name, *serosa.* Like all epithelial membranes, a serous membrane is composed of two distinct layers of tissue. One of the layers, the epithelial sheet, is a thin layer of simple squamous epithelium. The other layer, the connective tissue layer, forms a very thin sheet that holds and supports the epithelial cells.

The serous membrane that lines body cavities and covers the surfaces of organs in these cavities is in reality a single, continuous sheet covering two different surfaces. The *parietal membrane* is the portion that lines the wall of the cavity like wallpaper; the *visceral membrane* covers the surface of the viscera (organs within the cavity). Two important serous membranes are identified in Figure 5-39: the *pleura,* which surrounds a lung and lines the thoracic cavity, and the *peritoneum,* which covers the abdominal viscera and lines the abdominal cavity. Another example is the *pericardium,* which surrounds the heart (Box 5-7).

Serous membranes secrete a thin, watery fluid that lubricates organs as they rub against one another and against the walls of cavities.

Mucous Membranes

Mucous membranes are epithelial membranes that line body surfaces opening directly to the exterior. Mucous membranes are sometimes called by their Latin name, *mucosa.* Examples of mucous membranes include those lining the respiratory, digestive, urinary, and reproductive tracts. The epithelial component of the mucous membrane varies, depending on its location and function. In the esophagus, for example, a tough, abrasion-resistant stratified squamous epithelium is found. A thin layer of simple columnar epithelium covers the walls of the lower segments of the digestive tract. The fibrous connective tissue underlying the epithelium in mucous membranes is called the **lamina propria.**

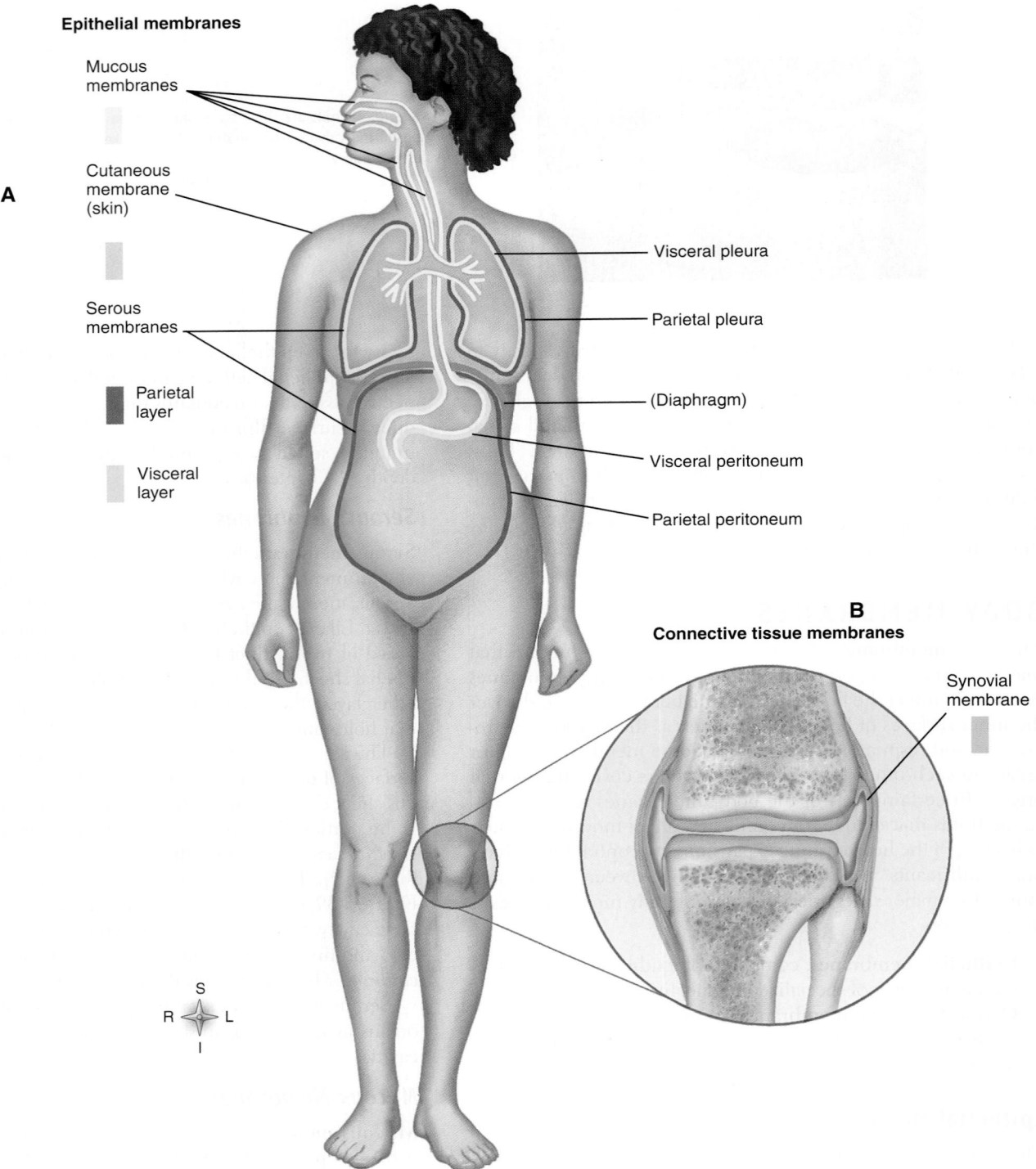

Epithelial membranes

Mucous membranes

Cutaneous membrane (skin)

A

Serous membranes

Parietal layer

Visceral layer

Visceral pleura

Parietal pleura

(Diaphragm)

Visceral peritoneum

Parietal peritoneum

B
Connective tissue membranes

Synovial membrane

S
R L
I

Figure 5-39 *Types of body membranes.* A, Epithelial membranes, including cutaneous membrane (skin), serous membranes (parietal and visceral pleura and peritoneum), and mucous membranes. **B,** Connective tissue membranes, including synovial membranes.

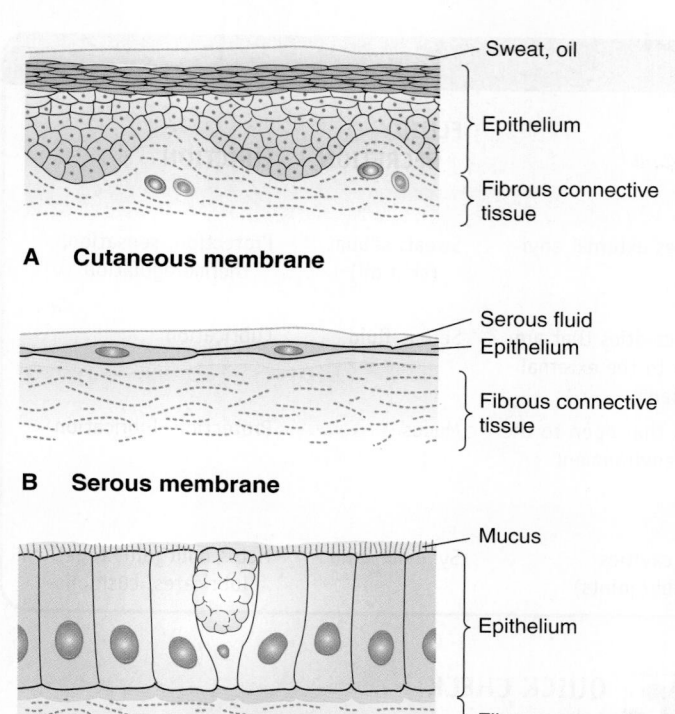

A   **Cutaneous membrane**

- Sweat, oil
- Epithelium
- Fibrous connective tissue

B   **Serous membrane**

- Serous fluid
- Epithelium
- Fibrous connective tissue

C   **Mucous membrane**

- Mucus
- Epithelium
- Fibrous connective tissue (lamina propia)

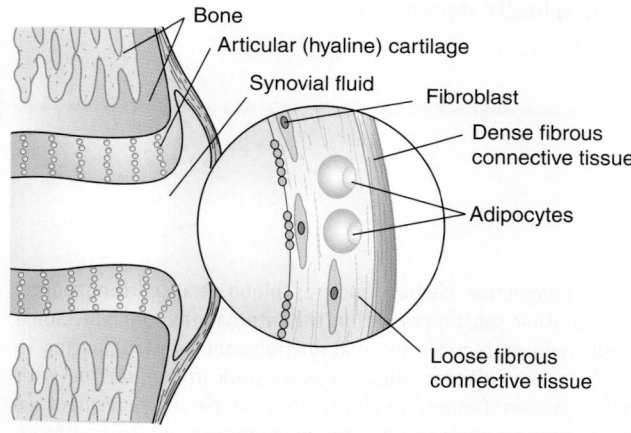

D   **Synovial membrane**

- Bone
- Articular (hyaline) cartilage
- Synovial fluid
- Fibroblast
- Dense fibrous connective tissue
- Adipocytes
- Loose fibrous connective tissue

Figure 5-40 *Structure of body membranes.*

Mucous membranes get their name from the fact that they produce a film of **mucus** that coats and protects the underlying cells. Besides protection, mucus also serves other purposes. For example, mucus acts as a lubricant for food as it moves along the digestive tract. In the respiratory tract, it serves as a sticky trap for contaminants.

Mucus is a watery secretion that contains a mixture of *mucins*, which are a group of about two dozen different proteoglycans. Different mucins are found in different locations in the body. Some mucins are attached directly to the plasma membranes of the epithelial cells to form a protective coat. Other mucins are re-

 BOX 5-7: HEALTH MATTERS
Inflammation of Serous Membranes

Pleurisy (PLOOR-i-see) (also called *pleuritis*) is a very painful pathological condition characterized by inflammation of the serous membranes (pleurae) that line the chest cavity and cover the lungs. Pain is caused by irritation and friction as the lungs rub against the walls of the chest cavity. In severe cases the inflamed surfaces of the pleura fuse, and permanent damage may develop. The term **peritonitis** (pair-i-toe-NYE-tis) is used to describe inflammation of the serous membranes in the abdominal cavity. Peritonitis is sometimes a serious complication of an infected appendix.

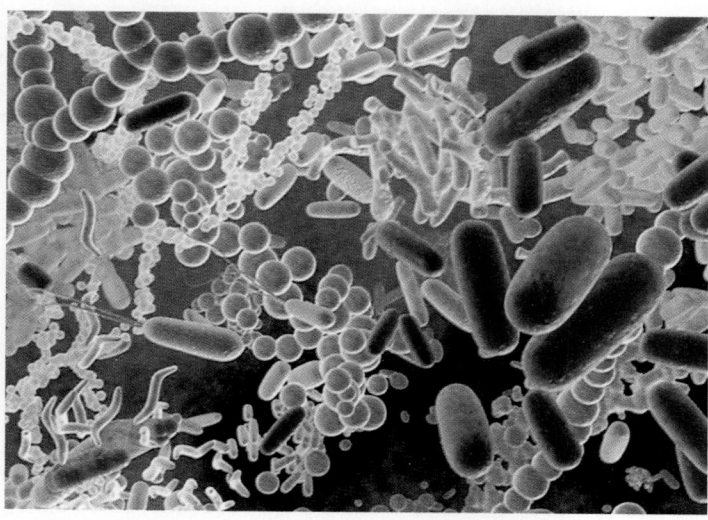

Figure 5-41 *Colonization of mucous membranes by microbes.* This colorized scanning electron micrograph shows many different types of bacteria that colonize the mucous membranes of the digestive tract and urinary tract within a short time after birth (*yellow,* bifidobacteria; *red,* coliform bacteria; *purple,* lactobaccili; *light blue,* streptococci; *green,* other coccal bacteria).

leased by goblet cells and form a protective, sometimes lubricating gel. Thus, mucous membranes have two layers of protective coating.

The mucous membrane is clinically important because it is the place that our body will most likely interact with microorganisms from the external environment. In fact, a very active area of research is attempting to understand the microbial colonization of mucous membranes in the digestive tract, respiratory tract, urinary tract, and reproductive tract (Figure 5-41). As researchers learn more about how mucus defends our internal environment from attack, the better able they will be to develop new strategies to avoid life-threatening infections. Chapter 21 discusses the role of mucous membranes in immunity.

Connective Tissue Membranes

Unlike cutaneous, serous, and mucous membranes, connective tissue membranes do not contain epithelial components. The **synovial membranes** lining the spaces between bones and joints

Table 5-8 Membranes of the Body

TYPE	SUPERFICIAL LAYER	DEEP LAYER	LOCATION	FLUID SECRETION	FUNCTION
Epithelial					
Cutaneous (skin)	Keratinized stratified squamous epithelium (epidermis)	Dense irregular fibrous connective tissue (dermis)	Directly faces external environment	Sweat; sebum (skin oil)	Protection, sensation, thermoregulation
Serous	Simple squamous epithelium	Fibrous connective tissue	Lines body cavities that are not open to the external environment	Serous fluid	Lubrication
Mucous	Various types of epithelium	Fibrous connective tissue (lamina propria)	Lines tracts that open to the external environment	Mucus	Protection, lubrication
Connective					
Synovial	Dense fibrous connective tissue	Loose fibrous connective tissue	Lines joint cavities (in movable joints)	Synovial fluid	Helps hold joint together, lubricates, cushions

that move are classified as connective tissue membranes. These membranes are smooth and slick and secrete a thick and colorless lubricating fluid called **synovial fluid**. The membrane itself, with its specialized fluid, helps reduce friction between the opposing surfaces of bones in movable joints. Synovial membranes also line the small, cushionlike sacs called *bursae* found between some moving body parts.

Table 5-8 summarizes the main characteristics of the four major types of membranes in the body.

 QUICK CHECK

14. Which two of the four major tissue types have the greatest capacity to regenerate after an injury?
15. Name the four principal types of body membranes. Which are epithelial membranes?

 THE BIG PICTURE

Tissues, Membranes, and the Whole Body

Tissues and body membranes are sometimes called "the fabric of the body." Like the pieces of fabric in a garment, tissues and body membranes are portions of a larger integrated structure. Just as each type of fabric in a complex garment has a different functional role determined by its structural characteristics, so does each type of tissue within the body. One of the ultimate functional goals of most tissues and membranes is maintenance of relative constancy in the body: homeostasis.

How do the major tissue types help maintain homeostasis? Epithelial tissues promote constancy of the body's internal environment in several ways. They form membranes that contain and protect the internal fluid environment, they absorb nutrients and other substances needed to maintain an optimum concentration in the body, and they secrete various products that regulate body functions involved in homeostasis. Connective tissues hold organs and systems together to form a whole, connected body. They also form structures that support the body and permit movement, such as the components of the skele-

ton. Some connective tissues, such as blood, transport nutrients, wastes, and other substances within the internal environment. Some blood cells help protect the internal environment by participating in the body's immune system. Muscle tissues work in conjunction with connective tissues (bones, tendons, etc.) to permit movement (a function needed to avoid injury); to communicate; and to find food, shelter, and other requirements. Nervous tissue works with glandular epithelial tissue to regulate various body functions in a way that maintains homeostatic balance.

Now that you have a basic knowledge of the various types of body "fabric"—the tissues and body membranes—you are ready to study the structure and function of specific organs and systems. As you take this next step in your studies, pay close attention to the tissue types that make up each organ. If you do, you will find it easier to understand the characteristics of a particular organ, and you will also improve your understanding of the integrated nature of the whole body.

Mechanisms of Disease

TUMORS AND CANCER

Neoplasms

The term **neoplasm** literally means "new matter" and refers to any abnormal growth of cells. Also called **tumors**, neoplasms can be distinct lumps of abnormal cells or, in blood tissue, can be diffuse. Neoplasms are often classified as **benign** or **malignant**. Benign tumors are called that because they do not spread to other tissues and they usually grow very slowly. Their cells are often well differentiated, unlike the undifferentiated cells typical of malignant tumors. Cells in a benign tumor tend to stay together, and they are often surrounded by a capsule of dense tissue. Benign tumors are not usually life threatening, but can be if they disrupt the normal function of a vital organ. Malignant tumors, or **cancers**, on the other hand, are not encapsulated and tend to spread to other regions of the body. For example, cells from malignant breast tumors usually form new

(secondary) tumors in bone, brain, and lung tissues. The cells migrate by way of lymphatic or blood vessels. This manner of spreading is called **metastasis**. Cells that do not metastasize can spread another way: they grow rapidly and extend the tumor into nearby tissues (Figure 5-42). Malignant tumors may replace part of a vital organ with abnormal, undifferentiated tissue—a life-threatening situation.

Neoplasms are classified further into subgroups, depending on the tissue in which they originate. Both benign and malignant tumors can be divided into three types: epithelial tumors, connective tissue tumors, and miscellaneous tumors. Benign tumor types that arise from epithelial tissues include *papilloma* (a fingerlike projection), *adenoma* (glandular tumor), and *nevus* (small, pigmented skin tumors). Benign tumor types that arise from connective tissues include *lipoma* (adipose tumor),

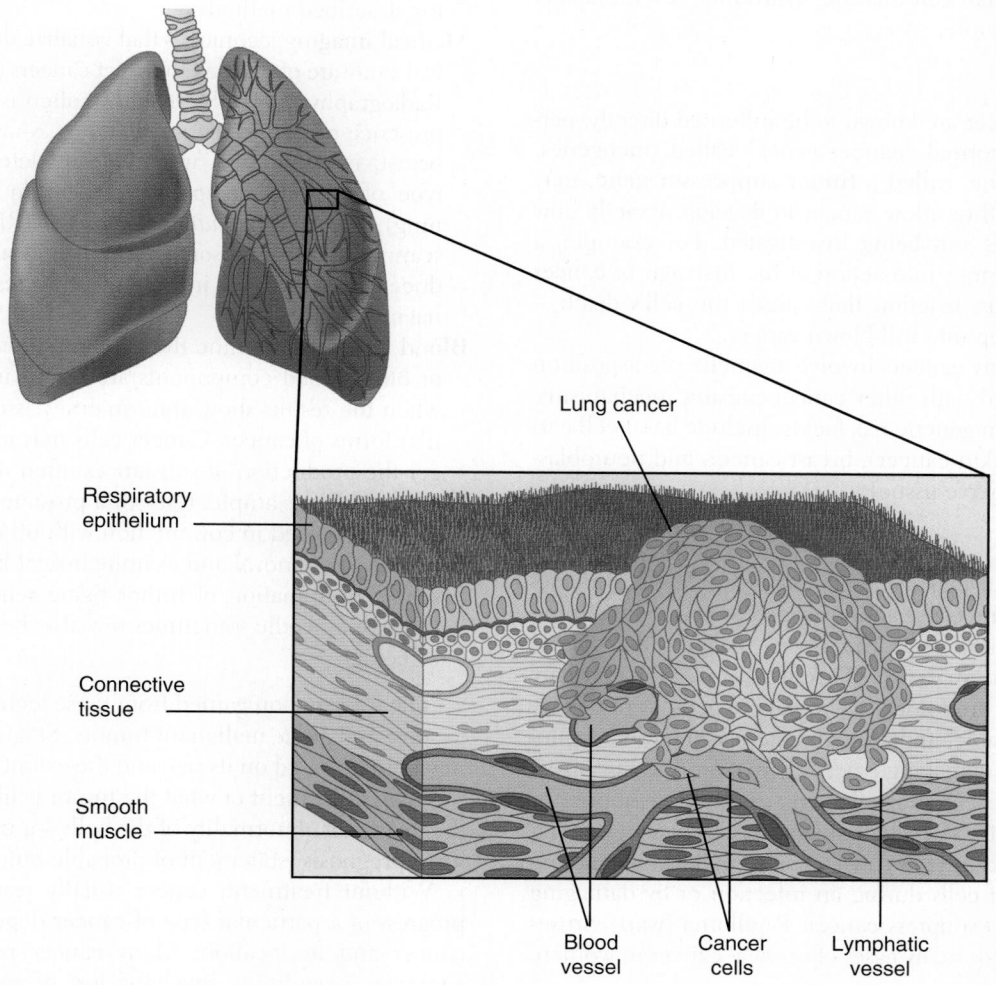

Respiratory epithelium

Connective tissue

Smooth muscle

Lung cancer

Blood vessel

Cancer cells

Lymphatic vessel

Figure 5-42 *Cancer.* This abnormal mass of proliferating cells in the lining of lung airways is a malignant tumor—lung cancer. Notice how some cancer cells are leaving the tumor and entering the blood and lymph vessels.

Mechanisms of Disease—cont.

osteoma (bone tumor), and *chondroma* (cartilage tumor). Malignant tumors that arise from epithelial tissues are generally called **carcinomas.** Examples include *melanoma* (cancer of skin pigment cells) and *adenocarcinoma* (glandular cancer). Malignant tumors that arise from connective tissues are generally called **sarcomas.** Examples include *lymphoma* (lymphatic cancer), *osteosarcoma* (bone cancer), and *fibrosarcoma* (cancer of fibrous connective tissue). Miscellaneous tumors are those that do not fit either of the previous categories. For example, an *adenofibroma* is a benign neoplasm formed by epithelial and connective tissues. Another example is *neuroblastoma*, a malignant tumor that arises from nerve tissue.

The etiologies (origins) of various forms of cancer puzzle medical science no less today than a hundred years ago. We do know that cancer involves uncontrolled cell division: **hyperplasia** (too many cells) and/or **anaplasia** (abnormal, undifferentiated cells). Thus, the mechanism of all cancers is a mistake or problem in cell division. What we are uncertain of is the cause of the abnormal cell division. Currently, several factors are known to play a role.

Genetic Factors

Many forms of cancer are known to be inherited directly, perhaps involving abnormal "cancer genes" called **oncogenes.** Another type of gene, called a **tumor suppressor gene,** may fail to operate and thus allow cancer to develop. Exactly how these genes work is still being investigated. For example, a gene called *p53* springs into action at the first sign of cancer and initiates a chain reaction that causes the cell's death—before it can develop into full-blown cancer.

Presumably, many cancers involve a genetic predisposition (risk factor) coupled with other cancer-causing mechanisms. Cancers with known genetic risk factors include basal cell carcinoma (a type of skin cancer), breast cancer, and neuroblastoma (a cancer of nerve tissue).

Carcinogens

Carcinogens (cancer makers) are agents that affect genetic activity in some way and cause abnormal cell reproduction. Some carcinogens are **mutagens** (mutation makers) that cause changes or mutations in a cell's DNA structure. Although many industrial products are known to be carcinogens, various natural mineral, vegetable, and animal materials are also carcinogenic. Exposure to damaging types of radiation or other physical injuries can be carcinogenic. For example, sunburns or chronic exposure to sunlight can cause skin cancer. Even viruses have been known to cause cancer, perhaps by altering the genetic code of cells during an infection or by damaging the body's ability to suppress cancer. Papilloma (wart) viruses have been blamed for many cases of cervical cancer in women.

Age

Certain cancers are found primarily in young people (e.g., leukemia) and others primarily in older adults (e.g., colon cancer). The age factor may result from changes in the genetic activity of cells over time or from the accumulated effects of cell damage.

DETECTION AND TREATMENT OF CANCER

Signs of cancer are those one would expect of a malignant neoplasm: the appearance of abnormal, rapidly growing tissue. Cancer specialists, or *oncologists*, have stressed that early detection of cancer is important because it is in the early stages of development of primary tumors, before metastasis and the development of secondary tumors have begun, that cancer is most treatable. Some methods currently used to detect the presence of cancer include the following:

Self-examination for the early signs of cancer previously described is one method for detection of cancer. For example, women are encouraged to perform a monthly breast self-examination. Likewise, males are encouraged to perform a monthly testicular self-examination. If an abnormality is found, it can be further investigated with one of the following described methods.

Medical imaging techniques that visualize deep tissues for medical study are often used to detect cancers (see pp. 17 and 18). Radiography (x-ray photography) often is used to detect the presence of tumors. *Mammography*, x-ray photography of a breast, is considered an important detection tool for this type of cancer. *Computed tomography (CT)* (x-ray scanning), *magnetic resonance imaging (MRI)* (electromagnetic scanning), and *ultrasonography* (ultrasound scanning) produce cross-sectional images of body regions suspected of having tumors.

Blood tests to determine the concentration of ions, enzymes, or other blood components are useful in detecting cancer when the results show abnormalities associated with particular forms of cancer. Cancer cells may also produce or trigger the production of substances often referred to as *tumor markers*. For example, tests for a prostate cancer marker are now being used in conjunction with other diagnostic tests.

Biopsy is the removal and examination of living tissue. Microscopic examination of tumor tissue removed surgically or through a needle sometimes reveals whether it is malignant or benign.

The information gained from these techniques can be used to *stage* and *grade* malignant tumors. Staging involves classifying a tumor based on its size and the extent of its spread. Grading is an assessment of what the tumor is likely to do based on the degree of abnormality of the cells—a useful basis for making a prognosis (statement of probable outcome).

Without treatment, cancer usually results in death. The progress of a particular type of cancer depends on the type of cancer and its location. Many cancer patients suffer from *cachexia*, a syndrome involving loss of appetite, weight loss, and general weakness. Various anatomical or functional abnormalities may arise as a result of damage to particular organs. The ultimate causes of death in cancer patients include secondary infection by pathogenic microorganisms, organ fail-

Mechanisms of Disease—cont.

ure, hemorrhage (blood loss), and in some cases, undetermined factors.

Of course, once cancer has been identified, every effort is made to treat it and thus prevent or delay its development. Surgical removal of cancerous tumors is sometimes performed, but the probability that malignant cells have been left behind must be addressed. **Chemotherapy,** or "chemical therapy" with *cytotoxic* (cell-killing) compounds or antineoplastic drugs, can be used after surgery to destroy any remaining malignant cells. **Radiation therapy,** also called *radiotherapy,* involves the use of destructive x-ray or gamma radiation alone or with chemotherapy to destroy any remaining cancer cells. **Laser therapy,** in which an intense beam of light destroys a tumor, is also sometimes coupled with chemotherapy or radiation therapy. **Immunotherapy,** a newer type of cancer treatment, bolsters the body's own defenses against cancer cells. Because viruses cause some types of cancer, oncologists hope that vaccines against certain forms of cancer will be developed.

Perhaps the most promising new research area in the battle against cancer is *molecular oncology.* This rapidly growing medical specialty attempts to use advances in molecular biol-

ogy and genetics to develop so-called *rational drugs* that unlike conventional chemotherapy agents, target only those specific molecules, enzymes, receptors, or other features unique to cancer cells or tumor growth. Ideally, such treatments would affect only the cancer and spare normal cells and body functions, thus increasing efficiency and reducing side effects. Three of the most promising new classes of "rational" drugs used to treat various types of cancer include the following:

Monoclonal antibodies (see p. 821), for example, trastuzumab (Herceptin) for breast cancer and cetuximab (Erbitux) for colorectal cancer

Antiangiogenesis (anti–blood vessel formation) *agents,* for example, vascular endothelial growth factor (VEGF) for various solid tumors

Tyrosine kinase inhibitors (enzyme inhibitors), for example, imatinib mesylate (Gleevec) or ST1571 for chronic myeloid leukemia

Some oncologists believe that future advances in rational drug design could mean for cancer what antibiotics meant for infectious diseases.

LANGUAGE OF SCIENCE   (Cont'd from page 145)

cutaneous membrane (kyoo-TAYN-ee-us) [*cutaneus* pertaining to skin]

dendrites (DEN-drytes) [*dentr* tree or branches]

dense fibrous tissue (dense FYE-brus TISH-yoo) [*dense* thick, *fibr* fiber, *connective,* to bind, *tissue* fabric]

desmosomes (DES-moh-sohms) [*desmo-* band, *-soma* body]

diapedesis (dye-ah-peh-DEE-sis) [*dia-* through, *-pedesis* an oozing]

ectoderm (EK-toh-derm) [*ecto-* outside, *-derm* skin]

elastic cartilage (eh-LAS-tik KAR-ti-lij) [*elastic* to drive, *cartilag* cartilage]

elastic dense fibrous tissue (eh-LAS-tik dense FYE-brus TISH-yoo) [*elastic* to drive, *dense* thick, *fibr* fiber, *tissue* fabric]

elastin (e-LAS-tin) [*elastic* to drive]

endochondral ossification (en-doh-KON-dral os-i-fi-KAY-shun) [*endo-* inward or within, *-chondro* cartilage, *-oss* bone, *-fication* to make]

endocrine glands (EN-doh-krin) [*endo-* inward or within, *-crine* to secrete]

endoderm (EN-doh-derm) [*endo-* inward or within, *-derm* skin]

endothelium (en-doh-THEE-lee-um) [*endo-* inward or within, *-theli* nipple]

epithelial membrane (ep-i-THEE-lee-al) [*epi-* on or upon, *-theli* nipple, *membrane* thin skin]

epithelial tissue (ep-i-THEE-lee-al) [*epi-* on or upon, *-theli* nipple, *tissue* fabric]

epithelium (ep-i-THEE-lee-um) [*epi-* on or upon, *theli* nipple]

erythrocytes (eh-RITH-roh-sytes) [*erythro-* red, *-cyte* cell]

excretion (eks-KREE-shun) [*excrete* to separate]

exocrine gland (EK-soh-krin) [*exo-* outside or outward, *-crine* to secrete]

extracellular matrix (ECM) (eks-trah-SEL-yoo-lar MAY-triks) [*extra-* beyond, *-cell* storeroom, *matrix* womb]

fascia (FAY-shah) [*fascia* band]

fibroblasts (FYE-broh-blasts) [*fibro-* fiber, *-blast* embryonic state of development]

fibrocartilage (fye-broh-KAR-ti-lij) [*fibro-* fiber, *cartilag* cartilage]

formed elements

gastrulation (gas-troo-LAY-shun) [*gastrula* embryonic stage after the blastula]

glands [*gland* acorn]

glandular (GLAN-dyoo-lar) [*glandular* small gland]

hematopoietic tissue (hee-mah-toh-poy-ET-ik) [*hema-* relating to blood, *-poiein-* to make, *-ic* pertaining to, *tissue* fabric]

histamine (HIS-tah-meen) [*hist-* tissue, *-amine* ammonia]

histogenesis (his-toh-JEN-eh-sis) [*histo-* tissue, *-genesis* origin]

histology (his-TOL-oh-jee) [*histo-* tissue, *-ology* study of]

hyaline cartilage (HYE-ah-lin KAR-ti-lij) [*hyaline* glass, *cartilag* cartilage]

inflammation (in-flah-MAY-shun) [*inflame* to set afire]

inflammatory exudate (in-FLAM-ah-toh-ree EKS-oo-dayt) [*inflame* to set afire, *exudate* to sweat out]

kinins (KYE-nins) [*kinin* to move]

lacunae (lah-KOO-nay) [*lacuna* pit] **sing., lacuna**

lamellae (lah-MEL-eye) [*lamella* small plate] **sing., lamella**

lamina propria (LAM-in-ah PROH-pree-ah) (*lamina* thin plate, *propria* proper]

leukocytes (LOO-koh-sytes) [*leuko-* white, *-cyte* cell]

loose connective tissue (LOOS koh-NEK-tiv TISH-yoo)

macrophages (MAK-roh-fayj-ez) [*macro-* large, *-phage* to eat]

mast cell [*mast* fattening, *cell* storeroom]

matrix (MAY-triks) [*matrix* womb]

membrane [*membrane* thin skin]

membrane bones [*membrane* thin skin]

membranous (MEM-brah-nus) [*membrane* thin skin]

mesoderm (MEZ-oh-derm) [*meso-* middle, *-derm* skin]

LANGUAGE OF SCIENCE

(Cont'd from page 185)

mesothelium (mez-oh-THEE-lee-um) [*meso-* middle or median, *-theli* nipple]

microvilli (my-kroh-VIL-ee) [*micro-* small, *-villi* shaggy hair] **sing., microvillus**

mucus (MYOO-kus) [*mucus* slime]

mucous membrane (MYOO-kus) [*mucus* slime, *membrane* thin skin]

multicellular glands [*multi-* many, *-cell* storeroom, *gland* acorn]

muscle tissue (MUSS-el TISH-yoo)

nervous tissue [*nervous* pertaining to nerves, *tissue* fabric]

neuroglia (noo-ROG-lee-ah) [*neuro-* nerve, *-glia* glue]

neurons (NOO-rons) [*neuron* nerve]

osteoblasts (OS-tee-oh-blasts) [*osteo-* bone, *-blast* embryonic state of development]

osteoclast (OS-tee-oh-klast) [*osteo-* bone, *-clast* breaking]

osteocytes (OS-tee-oh-sytes) [*osteo-* bone, *-cyte* cell]

osteon (OS-tee-on) [*osteo* bone]

perichondrium (pair-i-KON-dree-um) [*peri-* around, *-chondrium* cartilage]

phagocytosis (fag-oh-sye-TOH-sis) [*phago-* to eat, *-cyte-* cell, *-osis* condition]

plasma (PLAZ-mah) [*plasma* something formed]

primary germ layers

proteoglycans (PRO-tee-oh-GLYE-kanz) [*proteo-* protein, *-glycan* sweet]

pseudostratified columnar epithelium (SOOD-oh-STRAT-i-fyed KAHL-um-nar ep-i-THEE-lee-um) [*pseudo-* false, *-strati* layer, *columnar* column]

pus [*pus* corrupt matter]

regeneration (ree-jen-er-AY-shun) [*re-* again, *generat* beget or produce]

reticular tissue (reh-TIK-yoo-lar) [*reticular* little net, *tissue* fabric]

secretion (se-KREE-shun) [*secrete* to separate]

serotonin (sair-oh-TOH-nin) [*sero-* whey, *-tonin* tone]

serous membrane (SEE-rus) [*serous* whey, *membrane* thin skin]

simple epithelium (ep-i-THEE-lee-um) [*simple* not mixed, *epi-* on, *-theli* nipple]

skeletal muscle tissue (SKEL-eh-tal) [*skeletal* skeleton, *mus-* mouse, *-cle* small, *tissue* fabric]

skin

smooth muscle tissue [*mus-* mouse, *-cle* small, *tissue* fabric]

soma

squamous (SKWAY-muss) [*squamous* scale]

stratified epithelium (STRAT-i-fyde ep-i-THEE-lee-um) [*strati-* layer, *epi-* on, *-theli* nipple]

synovial fluid (si-NO-vee-all) [*synovia* clear, *fluid* to flow]

synovial membrane (si-NO-vee-all) [*synovia* clear, *membrane* thin skin]

thrombocytes (THROM-boh-sytes) [*thrombo-* clot, *-cyte* cell]

tight junctions

tissue [*tissue* fabric]

trabeculae (trah-BEK-yoo-lee) [*trabecula* little beam]

transitional epithelium (tranz-I-shen-al ep-i-THEE-lee-um) [*transition* to go through, *epi-* on, *-theli* nipple]

tubular (TOOB-yoo-lar) [*tubular* tube]

unicellular glands [*uni-* one, *-cell* storeroom, *gland* acorn]

LANGUAGE OF MEDICINE

anaplasia (an-ah-PLAY-zee-ah) [*ana-* without, *-plasia* to shape]

benign (bee-NYNE) [*benign* kind]

biopsy (BYE-op-see) [*bio-* life, *-opsy* view]

blood test

body composition

cancers [*cancer* crab]

carcinogens (kar-SIN-oh-jens) [*carcino-* cancer, *-gen* to produce]

carcinomas (kar-si-NO-mahs) [*carcin-* cancer, *-oma* tumor].

chemotherapy (kee-moh-THAYR-ah-pee) [*chemo-* by chemical reaction, *-therapy* to treat]

edema (eh-DEE-mah) [*oidema* a swelling]

genetic factors (jeh-NET-ik FAK-tors) [*genesis* origin]

hyperplasia (hye-per-PLAY-zee-ah) [*hyper-* excessive, *-plasia* to mold]

immunotherapy (im-yoo-no-THAYR-ah-pee) [*immuno-* free, *-therapy* to treat]

inflammatory response (in-FLAM-ah-toh-ree) [*inflame* to set afire]

keloid (KEE-loyd) [*kel-* fibrous growth, *-loid* to form]

laser therapy (LAY-zer THAYR-ah-pee)

leukocytosis (loo-koh-sye-TOH-sis) [*leuko-* white, *-cyto-* cell, *-osis* condition]

leukocytosis-promoting (LP) factor (loo-koh-sye-TOH-sis) [*leuko-* white, *-cyto-* cell, *-osis* condition]

malignant (mah-LIG-nant) [*malign* bad]

medical imaging

metastasis (meh-TAS-tah-sis) [*meta-* change, *-stasis* standing]

mutagens (MYOO-tah-jens) [*mutate-* to change, *-gen* to produce]

neoplasm (NEE-oh-plaz-em) [*neo-* new, *-plasma* formation]

oncogenes (ON-koh-jeens) [*onco-* tumor, *-gene* to produce]

peritonitis (pair-i-toh-NYE-tis) [*peri-* around, *-ton-* to stretch, *-itis* inflammation]

pleurisy (PLOOR-i-see) [*pleura-* rib, *-itis* inflammation]

radiation therapy (ray-dee-AY-shun THAYR-ah-pee) [*radiate* to emit rays, *therapy* to treat]

sarcomas (SAR-koh-mahs) [*sarco-* flesh, *-oma* tumor]

scar

self-examination

stem cell

tumors (TOO-mers) [*tumor* swollen state]

tumor suppressor gene (TOO-mer soo-PRESS-or jeen)

CASE STUDY

Antonio Garza, age 64 years, is brought to the rural health clinic by his granddaughter. Mr. Garza has had non–insulin-dependent diabetes for the past 10 years. He is currently taking hypoglycemic medication, glyburide (Micronase), 5 mg twice a day, and following a diabetic diet to help control his disease process. Additionally, Mr. Garza is taking benazepril (Lotensin), 10 mg daily, for hypertension. Forty-eight hours ago, Mr. Garza sustained a burn on his right foot from burning trash. The area is blistered, swollen, hot, and tender to the touch. The area surrounding the burn has an increased redness. Vital signs reflect a low-grade fever of 100.4° F (38° C), with other vital signs and blood sugar level being within normal limits.

1. Based on the structure and function of types of tissue, what type of injury do you suspect Mr. Garza has sustained?
   A. Epithelial
   B. Connective
   C. Muscle tissue
   D. Nervous tissue

2. Which one of the functions of this tissue type is most affected by this injury?
   A. Secretion and absorption
   B. Protection and sensory
   C. Protection and secretion
   D. Digestion and absorption

3. Which one of the following might you expect to be elevated in Mr. Garza's blood cell count?
   A. Erythrocytes
   B. Platelets
   C. Thrombocytes
   D. Leukocytes

4. Because the basement membrane was not completely destroyed in Mr. Garza's injury, what type of tissue repair would you expect to occur?
   A. Nonregeneration of scar tissue
   B. Keloid formation
   C. Replacement by fibrous connective tissue
   D. Regeneration

5. Immediately after Mr. Garza's injury, an inflammatory reaction occurred in response to the injury. Which of the following is the correct description of this expected response?
   A. Constriction of blood vessels surrounding the injury, increase in red blood cell production, vasodilation, and resulting phagocytosis by red blood cells
   B. Short-term immediate vasoconstriction, quickly followed by blood vessel dilation, and release of chemicals by injured tissues to cause vasodilation and increased permeability of blood vessels
   C. Chemical release of histamine, serotonin, and kinins to assist in vasoconstriction and maintenance of intracellular fluids
   D. Axial flow, or increased blood flow, through the small vessels toward the central two thirds of the lumen to maintain equal osmotic pressures

CHAPTER SUMMARY

INTRODUCTION
A. Tissue—group of similar cells that perform a common function
B. Matrix—nonliving intercellular material

PRINCIPAL TYPES OF TISSUE (TABLE 5-1)
A. Epithelial tissue
B. Connective tissue
C. Muscle tissue
D. Nervous tissue

EXTRACELLULAR MATRIX (ECM)
A. Complex, nonliving material between cells in a tissue (Figure 5-1)
   1. Some tissues have a large amount of ECM; other tissues have hardly any ECM
   2. Different kinds of components give ECM in different tissues a variety of characteristics

B. Components (Table 5-2)
   1. Water
   2. Proteins
      a. Structural proteins
         (1) Collagen—strong, flexible protein fiber
         (2) Elastin—elastic fibers
      b. Includes glycoproteins—proteins with a few carbohydrate attachments
         (1) Fibronectin and laminins help connect the ECM components to cells by binding with integrins in plasma membranes
         (2) Glycoprotein attachments also allow local communication within a tissue
   3. Proteoglycans
      a. Hybrid molecules that are mostly carbohydrates attached to a protein backbone
      b. Examples: chondroitin sulfate, heparin, and hyaluronate
      c. Different proteoglycans give different characteristics to ECM, such as thickness, shock absorption (Table 5-2)

C. Functions
  1. Help bind tissues together structurally
    a. ECM components bind to each other and to integrins in plasma membranes of cells
    b. In some tissues, it is primarily intercellular junctions that hold cells together
  2. Allow local communication among ECM and various cells—through connection via integrins in plasma membranes

EMBRYONIC DEVELOPMENT OF TISSUES

A. Primary germ layers (Figure 5-2)
  1. Endoderm
  2. Mesoderm
  3. Ectoderm
B. Gastrulation—process of cell movement and differentiation that results in the development of primary germ layers
C. Histogenesis—the process of the primary germ layers differentiating into different kinds of tissue

EPITHELIAL TISSUE

A. Types and locations
  1. Epithelium is divided into two types:
    a. Membranous (covering or lining) epithelium
    b. Glandular epithelium
  2. Locations
    a. Membranous epithelium—covers the body and some of its parts; lines the serous cavities, blood and lymphatic vessels, and respiratory, digestive, and genitourinary tracts
    b. Glandular epithelium—secretory units of endocrine and exocrine glands
B. Functions
  1. Protection
  2. Sensory functions
  3. Secretion
  4. Absorption
  5. Excretion
C. Generalizations about epithelial tissue
  1. Limited amount of matrix material
  2. Membranous type attached to a basement membrane
  3. Avascular
  4. Cells are in close proximity, with many desmosomes and tight junctions
  5. Capable of reproduction
D. Classification of epithelial tissue
  1. Membranous (covering or lining) epithelium (Table 5-3)
    a. Classification based on cell shape (Figure 5-3)
      (1) Squamous
      (2) Cuboidal
      (3) Columnar
      (4) Pseudostratified columnar
    b. Classifications based on layers of cells (Table 5-4)
      (1) Simple epithelium
        (a) Simple squamous epithelium (Figures 5-4 and 5-5)

          (i) One-cell layer of flat cells
          (ii) Permeable to many substances
          (iii) Examples: endothelium—lines blood vessels; mesothelium—pleura
        (b) Simple cuboidal epithelium (Figure 5-6)
          (i) One-cell layer of cuboidal-shaped cells
          (ii) Found in many glands and ducts
        (c) Simple columnar epithelium (Figure 5-7)
          (i) Single layer of tall, column-shaped cells
          (ii) Cells often modified for specialized functions such as goblet cells (secretion), cilia (movement), microvilli (absorption)
          (iii) Often lines hollow visceral structures
        (d) Pseudostratified columnar epithelium (Figure 5-8)
          (i) Columnar cells of differing heights
          (ii) All cells rest on basement membrane but may not reach the free surface above
          (iii) Cell nuclei at odd and irregular levels
          (iv) Found lining air passages and segments of male reproductive system
          (v) Motile cilia and mucus are important modifications
      (2) Stratified epithelium
        (a) Stratified squamous (keratinized) epithelium
          (i) Multiple layers of flat, squamous cells (Figure 5-9)
          (ii) Cells filled with keratin
          (iii) Covering outer skin on body surface
        (b) Stratified squamous (nonkeratinized) epithelium (Figure 5-10)
          (i) Lining vagina, mouth, and esophagus
          (ii) Free surface is moist
          (iii) Primary function is protection
        (c) Stratified cuboidal epithelium
          (i) Two or more rows of cells are typical
          (ii) Basement membrane is indistinct
          (iii) Located in sweat gland ducts and pharynx
        (d) Stratified columnar epithelium
          (i) Multiple layers of columnar cells
          (ii) Only most superficial cells are typical in shape
          (iii) Rare
          (iv) Located in segments of male urethra and near anus
        (e) Stratified transitional epithelium (Figure 5-11)
          (i) Located in lining of hollow viscera subjected to stress (e.g., urinary bladder)
          (ii) Often 10 or more layers thick
          (iii) Protects organ walls from tearing
  2. Glandular epithelium
    a. Specialized for secretory activity
    b. Exocrine glands—discharge secretions into ducts
    c. Endocrine glands—"ductless" glands; discharge secretions directly into blood or interstitial fluid
    d. Structural classification of exocrine glands (Figure 5-12; Table 5-5)

(1) Multicellular exocrine glands are classified by the shape of their ducts and the complexity of their duct system

(2) Shapes include tubular and alveolar

(3) Simple exocrine glands—only one duct leads to the surface

(4) Compound exocrine glands—have two or more ducts

e. Functional classification of exocrine glands (Figure 5-13)

(1) Apocrine glands

(a) Secretory products collect near apex of cell and are secreted by pinching off the distended end

(b) Secretion process results in some damage to cell wall and some loss of cytoplasm

(c) Mammary glands are good examples of this secretory type

(2) Holocrine glands

(a) Secretion products, when released, cause rupture and death of the cell

(b) Sebaceous glands are holocrine

(3) Merocrine glands

(a) Secrete directly through cell membrane

(b) Secretion proceeds with no damage to cell wall and no loss of cytoplasm

(c) Most numerous gland type

CONNECTIVE TISSUE

A. Functions, characteristics, and types

1. General function—connects, supports, transports, and protects

2. General characteristics—extracellular matrix (ECM) predominates in most connective tissues and determines its physical characteristics; consists of fluid, gel, or solid matrix, with or without extracellular fibers (collagenous, reticular, and elastic) and proteoglycans or other compounds that thicken and hold together the tissue (Figures 5-1 and 5-14)

3. Four main types (Table 5-6)

a. Fibrous

(1) Loose, ordinary (areolar)

(2) Adipose

(3) Reticular

(4) Dense

(a) Irregular

(b) Regular (collagenous and elastic)

b. Bone

(1) Compact bone

(2) Cancellous bone

c. Cartilage

(1) Hyaline

(2) Fibrocartilage

(3) Elastic

d. Blood

B. Fibrous connective tissue

1. Loose, ordinary (areolar) connective tissue (Figure 5-15)

a. One of the most widely distributed of all tissues

b. Intercellular substance is prominent and consists of collagenous and elastic fibers loosely interwoven and embedded in soft viscous ground substance

c. Several kinds of cells present, notably, fibroblasts and macrophages, also mast cells, plasma cells, fat cells, and some white blood cells (Figure 5-16)

d. Function—stretchy, flexible connection

2. Adipose tissue (Figures 5-17 and 5-18)

a. Similar to loose connective tissue but contains mainly fat cells

b. Functions—protection, insulation, support, and food reserve

3. Reticular tissue (Figure 5-19)

a. Forms framework of spleen, lymph nodes, and bone marrow

b. Consists of network of branching reticular fibers with reticular cells overlying them

c. Functions—defense against microorganisms and other injurious substances; reticular meshwork filters out injurious particles and reticular cells phagocytose them

4. Dense fibrous tissue

a. Matrix consists mainly of fibers packed densely and relatively few fibroblast cells

(1) Irregular—fibers intertwine irregularly to form a thick mat (Figure 5-20)

(2) Regular—bundles of fibers are arranged in regular parallel rows

(a) Collagenous—mostly collagenous fibers in ECM (Figures 5-21 and 5-22)

(b) Elastic—mostly elastic fibers in ECM (Figure 5-23)

b. Locations—composes structures that need great tensile strength, such as tendons and ligaments; also dermis and the outer capsule of the kidney and spleen

c. Function—furnishes flexible connections that are strong or stretchy

C. Bone tissue

1. Highly specialized connective tissue type

a. Cells—osteocytes—embedded in a calcified matrix

b. Inorganic component of matrix accounts for 65% of total bone tissue

2. Functions

a. Support

b. Protection

c. Point of attachment for muscles

d. Reservoir for minerals

e. Supports blood-forming tissue

3. Compact bone (Figures 5-25 and 5-26)

a. Osteon (Haversian system)

(1) Structural unity of bone

(2) Spaces for osteocytes called lacunae

(3) Matrix present in concentric rings called lamellae

(4) Canaliculi are canals that join lacunae with the central Haversian canal

b. Cell types
    (1) Osteocyte—mature, inactive bone cell
    (2) Osteoblast—active bone-forming cell
    (3) Osteoclast—bone-destroying cell
c. Formation (ossification) (Figure 5-24)
    (1) In membranes—e.g., flat bones of skull
    (2) From cartilage (endochondral)—e.g., long bones, such as the humerus
4. Cancellous bone (Figures 5-25 and 5-27)
    a. Trabeculae—thin beams of bone
    b. Supports red bone marrow
        (1) Myeloid tissue—a type of reticular tissue
        (2) Produces blood cells
    c. Called spongy bone because of its sponge-like appearance
D. Cartilage
    1. Chondrocyte is the only cell type present
    2. Lacunae house cells as in bone
    3. Avascular—therefore nutrition of cells depends on diffusion of nutrients through matrix
    4. Heals slowly after injury because of slow nutrient transfer to cells
    5. Perichondrium is membrane that surrounds cartilage
    6. Types
        a. Hyaline (Figure 5-28)
            (1) Appearance is shiny and translucent
            (2) Most prevalent type of cartilage
            (3) Located on ends of articulating bones
        b. Fibrocartilage (Figure 5-29)
            (1) Strongest and most durable type of cartilage
            (2) Matrix is semirigid and filled with strong white fibers
            (3) Found in intervertebral disks and pubic symphysis
            (4) Serves as shock-absorbing material between bones at the knee (menisci)
        c. Elastic (Figure 5-30)
            (1) Contains many fine elastic fibers
            (2) Provides strength and flexibility
            (3) Located in external ear and larynx
E. Blood
    1. A liquid tissue (Figure 5-31)
    2. Contains neither ground substance nor fibers
    3. Composition of whole blood
        a. Liquid fraction (plasma) is the matrix—55% of total blood volume
        b. Formed elements contribute 45% of total blood volume
            (1) Red blood cells, erythrocytes
            (2) White blood cells, leukocytes
            (3) Platelets, thrombocytes
    4. Functions
        a. Transportation
        b. Regulation of body temperature
        c. Regulation of body pH
        d. White blood cells destroy bacteria
    5. Circulating blood tissue is formed in the red bone marrow by a process called hematopoiesis; the blood-forming tissue is sometimes called hematopoietic tissue

MUSCLE TISSUE

A. Types (Table 5-7)
    1. Skeletal, or striated voluntary (Figure 5-32)
    2. Smooth, or nonstriated involuntary, or visceral (Figures 5-33 and 5-34)
    3. Cardiac, or striated involuntary (Figure 5-35)
B. Microscopic characteristics
    1. Skeletal muscle—threadlike cells with many cross striations and many nuclei per cell
    2. Smooth muscle—elongated narrow cells, no cross striations, one nucleus per cell
    3. Cardiac muscle—branching cells with intercalated disks (formed by abutment of plasma membranes of two cells)

NERVOUS TISSUE

A. Functions—rapid regulation and integration of body activities
B. Specialized characteristics
    1. Excitability
    2. Conductivity
C. Organs
    1. Brain
    2. Spinal cord
    3. Nerves
D. Cell types (Table 5-7)
    1. Neuron—conducting unit of system (Figure 5-36)
        a. Cell body, or soma
        b. Processes
            (1) Axon (single process)—transmits nerve impulse away from the cell body
            (2) Dendrite (one or more)—transmits nerve impulse toward the cell body and axon
    2. Neuroglia—special connecting, supporting, coordinating cells that surround neurons

TISSUE REPAIR

A. Tissues have a varying capacity to repair themselves; damaged tissue regenerates or is replaced by scar tissue
B. Regeneration—growth of new tissue (Figure 5-37)
C. Scar—dense fibrous mass; unusually thick scar is a keloid (Figure 5-38)
D. Epithelial and connective tissues have the greatest ability to regenerate
E. Muscle and nervous tissues have limited capacity to regenerate

BODY MEMBRANES

A. Thin tissue layers that cover surfaces, line cavities, and divide spaces or organs (Figure 5-39, Table 5-8)
B. Epithelial membranes are most common type (Figure 5-40)
    1. Cutaneous membrane (skin)
        a. Primary organ of integumentary system
        b. One of the most important organs
        c. Composes approximately 16% of body weight
    2. Serous membrane (serosa)
        a. Parietal membranes—line closed body cavities
        b. Visceral membranes—cover visceral organs

   c. Pleura—surrounds a lung and lines the thoracic cavity

   d. Peritoneum—covers the abdominal viscera and lines the abdominal cavity

  3. Mucous membrane (mucosa)

   a. Lines and protects organs that open to the exterior of the body

   b. Found lining ducts and passageways of the respiratory, digestive, and other tracts

   c. Lamina propria—fibrous connective tissue underlying mucous epithelium

   d. Mucus is made up mostly of water and mucins—proteoglycans that form a double-layer protection against environmental microbes (Figure 5-41)

C. Connective tissue membranes

  1. Do not contain epithelial components

  2. Synovial membranes—line the spaces between bone in joints

  3. Have smooth and slick membranes that secrete synovial fluid

  4. Help reduce friction between opposing surfaces in a movable joint

  5. Synovial membranes also line bursae

THE BIG PICTURE: TISSUES, MEMBRANES, AND THE WHOLE BODY

A. Tissues and membranes maintain homeostasis

  1. Epithelial tissues

   a. Form membranes that contain and protect the internal fluid environment

   b. Absorb nutrients

   c. Secrete products that regulate functions involved in homeostasis

  2. Connective tissues

   a. Hold organs and systems together

   b. Form structures that support the body and permit movement

  3. Muscle tissues

   a. Work with connective tissues to permit movement

  4. Nervous tissues

   a. Work with glandular epithelial tissue to regulate body function

REVIEW QUESTIONS

1. Define the term *tissue* and identify the four principal tissue types.

2. List at least three structures derived from each of the primary germ layers.

3. What are the five most important functions of epithelial tissue?

4. Which of the following best describes the number of blood vessels in epithelial tissue: none, very few, very numerous?

5. Explain how the shape of epithelial cells is used for classification purposes. Identify the four types of epithelium described in this classification process.

6. Classify epithelium according to the layers of cells present.

7. List the types of simple and stratified epithelium and give examples of each.

8. What is glandular epithelium? Give examples.

9. Discuss the structural classification of exocrine glands. Give examples of each type.

10. Describe loose connective tissue.

11. How do the types of dense fibrous connective tissue differ from one another?

12. Discuss and compare the microscopic anatomy of bone and cartilage tissue.

13. Compare the structure of the three major types of cartilage tissue. Locate and give an example of each type.

14. List the components of whole blood and discuss the basic function of each fraction or cell type.

15. List the three major types of muscle tissue.

16. Identify the two basic types of cells in nervous tissue.

17. What are the four cardinal signs of inflammation? What causes each?

18. Describe the regenerative capacity of muscle and nerve tissues.

19. Name the two major categories or types of body membranes. Give examples of each.

20. What is a neoplasm?

CRITICAL THINKING QUESTIONS

1. A baby was born with congenital problems in the skeleton and muscle system. From what primary germ layers do these systems arise? What is the earliest possible developmental stage during which a problem could have affected just one primary germ layer?

2. Summarize the structural characteristics of epithelial tissues that enable them to perform their specific functions.

3. Does the production of saliva, milk, or oil cause the most damage to the cell that produces it? Explain.

4. Describe the role of the fiber types in the classification of connective tissue. What examples can you find of these various types?

5. People with arthritis sometimes find relief from their condition by taking a dietary supplement of glucosamine and chondroitin. What do you know about the cartilage and ligaments in joints that might help explain this?

6. Many athletes work to reduce their body fat to the lowest possible percentage. What would happen if too little body fat were present?

7. If a tendon is badly damaged, it may need to be replaced surgically. Based on what you know about the structural and functional differences, explain why a tendon rather than a ligament must replace it.

8. Develop a flow chart or other diagram to describe the process by which tissues respond to injury.

9. If a small, but deep cut involving skin and muscle occurs, predict which tissue will probably heal first and which will heal more completely. Explain your answer.

10. When a joint swells, sometimes it is necessary to remove a thick colorless liquid from the joint. What is it, where did it come from, and what is its normal function?

Career Choices Emergency Clinical Nurse Specialist

As an emergency clinical nurse specialist at Dartmouth Hitchcock Medical Center in Lebanon, New Hampshire, my responsibilities involve the nursing care delivered to emergency patients. I teach classes, orient new nurses, evaluate supplies and equipment, troubleshoot patient care problems, set up systems to enhance care delivery, and help take care of patients.

I've wanted to be a nurse since I was 4 or 5 years old. I was drawn to the emergency care setting because I like action and variety. We see it all, cradle to grave, benign to life threatening. I've delivered babies in the parking lot; I've reassured parents that their child's "blue" hands are related to a new pair of jeans. I've seen teenagers with "lesions" that turned out to be pimples; I've held the hands of the dying as I administered blood to try to save their lives.

Change is a current trend in my field; new medications, new equipment, and new therapies are constantly being introduced. The reassuring thing about change in the twenty-first century is that it is more likely to be evidence based than in the past. In other words, we have scientific data to support doing things differently. We're questioning things we've always done and evaluating them in terms of their impact on the patient's outcome.

When I'm taking care of patients, the rewards are obvious: relief of pain and suffering (physical and emotional), saving

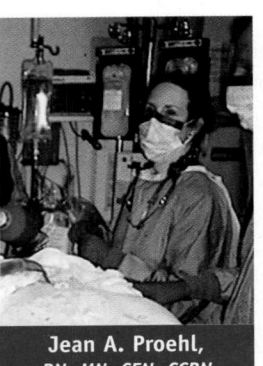

Jean A. Proehl,
RN, MN, CEN, CCRN

lives, and helping people return to an optimum state of health. When I'm teaching other nurses, the reward is in promoting improved knowledge and skills so that the professional practice of the nurse is enhanced and the care that she or he delivers to patients is improved.

I use anatomy specifically for accurate and concise descriptions of my physical examination findings. Physiology is interwoven into the patient's overall care. In assessment, I relate the signs, symptoms, and laboratory results to potential underlying physiological causes. When implementing interventions such as medication administration or positioning, I must consider their intended effect on the patient and potential adverse affects. To evaluate the effectiveness of interventions, I rely on my understanding of physiology to gauge outcome. For example, if I am giving intravenous fluid to a patient in shock, I know that the glomerular filtration rate reflects perfusion to vital organs, so urine output is a valuable parameter to monitor.

One professor I had always emphasized that nurses needed to know anatomy because some day we might be the only one available who could identify something. I found an anatomy coloring book to be a great study aid. Mnemonics were also a lifesaver; I still recite "On Old Olympus' Tiny Tops A Friendly Viking Grew Vines And Hops" when naming cranial nerves.

Support and Movement

The six chapters in Unit 2 describe the outer covering of the body, as well as the bones, muscles, and articulations, or joints, of the body. The skin (and its appendages) is selected in Chapter 6 as the first organ system to be studied. Chapter 7—Skeletal Tissues—provides information on the types of skeletal tissues and how they are formed, grow, are repaired if injured, and function. Skeletal tissues protect and support body structures and function as storage sites for important mineral elements vital to many body functions. Blood cell formation also occurs within the red marrow of bones.

The organs of the skeletal system, bones, are identified by size, shape, and specialized markings and then are organized or grouped into major subdivisions in Chapter 8. Movement between bones occurs at joints, or articulations, which are classified in Chapter 9 according to both structure and potential for movement.

Anatomy of the major muscle groups and muscle physiology are discussed in Chapters 10 and 11. The microscopic and molecular structure of muscle cells and tissues is related to function, as is the gross structure of individual muscles and muscle groups. Movement and heat production constitute the two most important functions of muscle. Discussion of muscle groups in Chapter 10 focuses on how muscles function and on how they attach to bones, how they are named, and how they are integrated functionally with other body organ systems.

SEEING THE BIG PICTURE

Skin and Its Appendages

LANGUAGE OF SCIENCE

albinism (AL-bi-niz-em) [albus- white, -ism condition of]

apocrine sweat glands (AP-oh-krin) [apo- from, -crine to separate]

arrector pili muscle (ah-REK-tor PYE-lye) [arrector raiser, pili of hair]

calluses (KAL-us-ez) [callus hard skin]

cerumen (seh-ROO-men) [cera wax]

ceruminous glands (seh-ROO-mi-nus) [cera- wax, -ous full of]

cleavage lines (KLEE-vij)

cortex (KOHR-teks) [cortex bark]

cutaneous membrane (kyoo-TAYN-ee-us) [cutis- skin, -eous composed of]

cuticle (KYOO-ti-kul) [cuticle little skin]

dermal-epidermal junction (DER-mal EP-i-der-mal) [derma skin, epi- on or upon, -derma skin]

dermal papilla (DER-mal pah-PIL-ah) [derma skin, papilla nipple]

dermis (DER-mis) [derma skin]

desquamation (des-kwah-MAY-shun) [de- to remove, -squama scale]

eccrine sweat glands (EK-rin) [eccrine to secrete]

eleidin (eh-LEE-din) [eleidin olive tree]

epidermis (ep-i-DER-mis) [epi- on or upon, -derma skin]

eumelanin (yoo-MEL-ah-nin) [eu- true, -melan- black, -in substance]

friction ridges

germinal matrix (JER-mi-nal MAY-triks) [germ sprout, matrix womb]

hair follicles (FOHL-i-kulz) [follicle small bag]

hair papilla (pah-PIL-ah) [papilla nipple]

hyperkeratosis (hye-per-ker-ah-TOH-sis) [hyper- excessive, -kera- cornified, -osis condition]

hypodermis (hye-poh-DER-mis) [hypo- under, -dermis skin]

integument (in-TEG-yoo-ment) [integument covering]

Cont'd on p. 224

Vital, diverse, complex, extensive—these adjectives describe the body's largest, thinnest, and one of its most important organs, the skin. It forms a self-repairing and protective boundary between the internal environment of the body and an often hostile external world. The skin surface is as large as the body, an area in average-sized adults of roughly 1.6 to 1.9 m² (17 to 20 square feet). Its thickness varies from slightly less than 0.05 cm (1/50 inch) to slightly more than 0.3 cm (1/8 inch).

As you know, the body is characterized by a "nested," or hierarchical, type of organization. Complexity progresses from cells to tissues and then to organs and organ systems. This chapter discusses the skin and its appendages—hair, nails, and skin glands—as an organ system. Ideally, by studying the skin and its appendages before you proceed to the more complex organ systems in the chapters that follow, you will improve your understanding of how structure is related to function. **Integument** is another name for the skin. **Integumentary system** is a term used to denote the skin and its appendages.

STRUCTURE OF THE SKIN

The skin is a thin, relatively flat organ classified as a membrane—the **cutaneous membrane.** As Figure 6-1 shows, two primary layers compose the skin:

1. Epidermis—superficial, thinner layer
2. Dermis—the deep, thicker layer

The epidermis is an epithelial layer derived from the ectodermal germ layer of the embryo (see Figure 5-2 on p. 149). By the seventeenth week of gestation, the epidermis of the developing baby has all the essential characteristics of the adult. The deeper dermis is derived from the mesoderm. The dermis is a relatively dense and vascular connective tissue layer that may average more than 4 mm in thickness in some body areas. The specialized area where the cells of the epidermis meet the connective tissue cells of the dermis is called the **dermal-epidermal junction** (Figure 6-2). Beneath the dermis lies a loose **hypodermis** rich in fat and areolar tissue.

Thin and Thick Skin

Most of the body surface is covered by skin that is classified as *thin skin.* The hairless skin covering the palms of the hands (and fingertips), soles of the feet, and other body areas subject to friction is classified as *thick skin.* These terms refer only to the epidermal layer and not to overall or total skin thickness, which includes the epidermis and dermis. Total skin thickness varies from 0.5 mm in areas such as the eyelids to more than 5 mm over the back, with most of the difference accounted for by variation in depth of the dermis.

In thick skin, each of the five strata, or layers, of the epidermis described below are present, and each stratum is generally several cell layers thick. The outermost stratum—the stratum corneum—is especially noticeable in thick skin and is generally composed of many cell layers. Furthermore, in thick skin the underlying dermal papillae are raised in curving parallel **friction ridges** to form fingerprints or footprints that are visible on the overlying epidermis. Hair is not found in thick skin. As their name implies, friction ridges increase friction—similar to the function of the ridges of tire treads. These ridges help us pick up and manipulate small objects with the hands and provide slip resistance to the soles of the feet. The surface of thick skin is seen in Figure 6-3, A.

In thin skin, the number of cell layers in each epidermal stratum is less than in thick skin, and one or more strata may be ab-

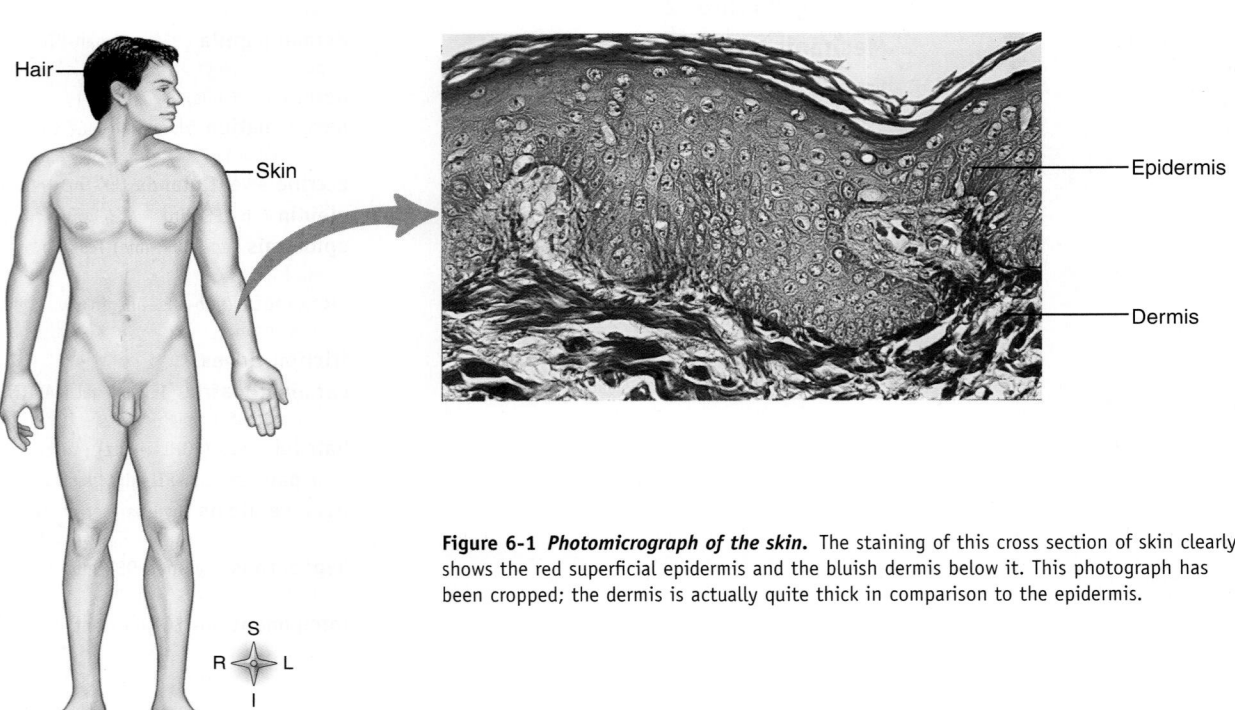

Figure 6-1 *Photomicrograph of the skin.* The staining of this cross section of skin clearly shows the red superficial epidermis and the bluish dermis below it. This photograph has been cropped; the dermis is actually quite thick in comparison to the epidermis.

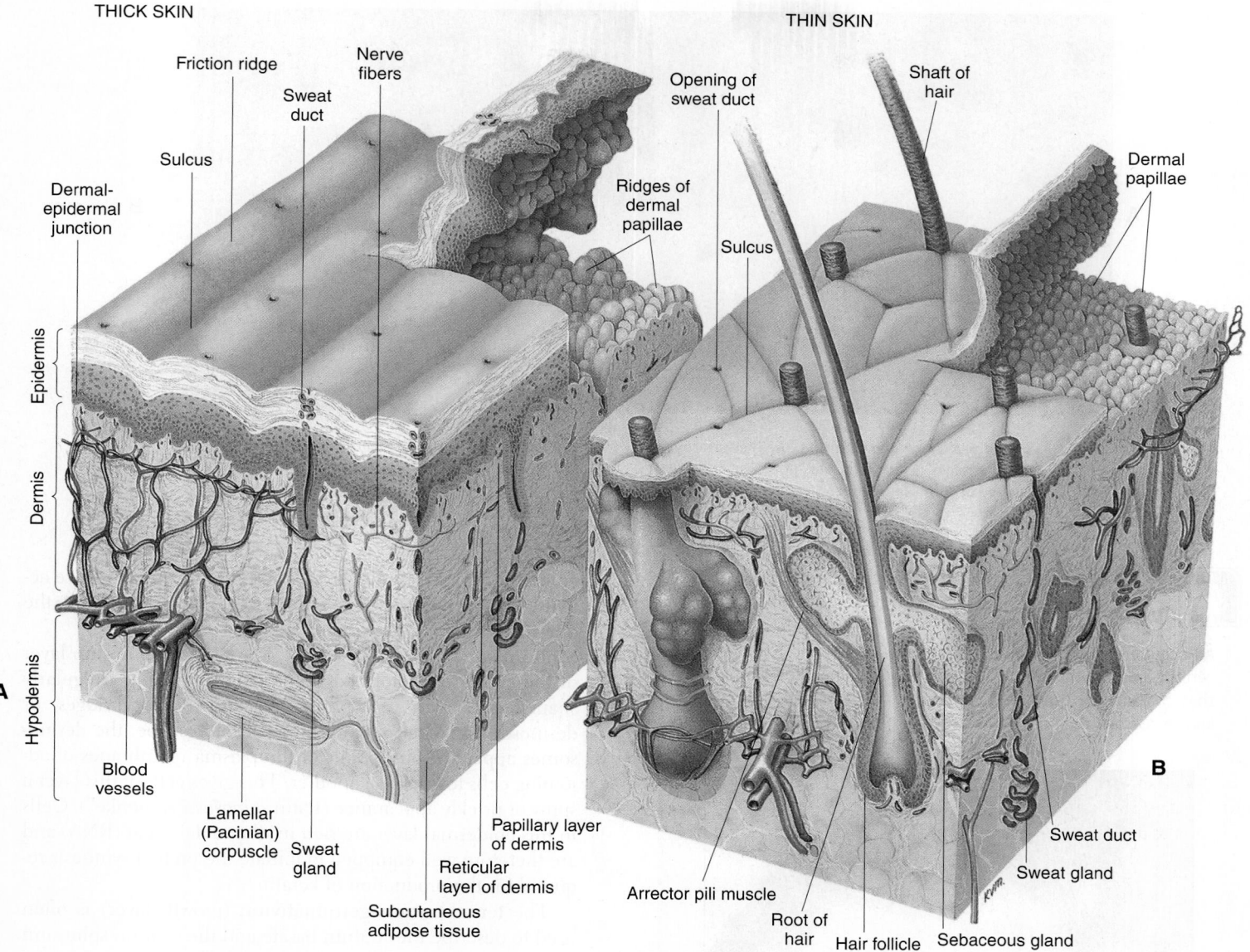

THICK SKIN

Friction ridge
Nerve fibers
Sweat duct
Sulcus
Dermal-epidermal junction
Epidermis
Dermis
Hypodermis
A
Blood vessels
Lamellar (Pacinian) corpuscle
Sweat gland
Reticular layer of dermis
Subcutaneous adipose tissue
Papillary layer of dermis

THIN SKIN

Opening of sweat duct
Shaft of hair
Ridges of dermal papillae
Dermal papillae
Sulcus
Sweat duct
Sweat gland
Sebaceous gland
Hair follicle
Root of hair
Arrector pili muscle
B

Figure 6-2 *Diagram of skin structure.* **A,** Thick skin, found on surfaces of the palms and soles of the feet. **B,** Thin skin, found on most surface areas of the body. In each diagram, the epidermis is raised at one corner to reveal the papillae of the dermis.

sent entirely. Raised parallel ridges are not present in the dermis of thin skin (Figure 6-3, *B*). Instead, the dermal papillae project upward individually and therefore no "fingerprints" are formed on the more superficial epidermis above.

Epidermis

Cell Types

The epidermis is composed of several types of epithelial cells.

Keratinocytes become filled with a tough, fibrous protein called **keratin.** These cells, arranged in distinct *strata,* or layers, are by far the most important cells in the epidermis. They make up more than 90% of the epidermal cells and form the principal structural element of the outer skin.

Melanocytes contribute color to the skin and serve to decrease the amount of ultraviolet (UV) light that can penetrate into the deeper layers of the skin. Although they may compose more than 5% of the epidermal cells, melanocytes may be completely absent from the skin in certain nonlethal conditions (Box 6-1).

Langerhans cells are *dendritic cells (DCs),* branched cells that play a role in immunity. As a type of *antigen-presenting cell (APC),* each Langerhans cell finds markers (antigens) on bacteria and other invaders and presents them to other immune system cells for recognition and destruction—an important defensive function. These cells originate in the bone marrow but migrate to the deep cell layers of the epidermis early in life. APCs such as Langerhans cells and similar DCs are discussed further in Chapter 21.

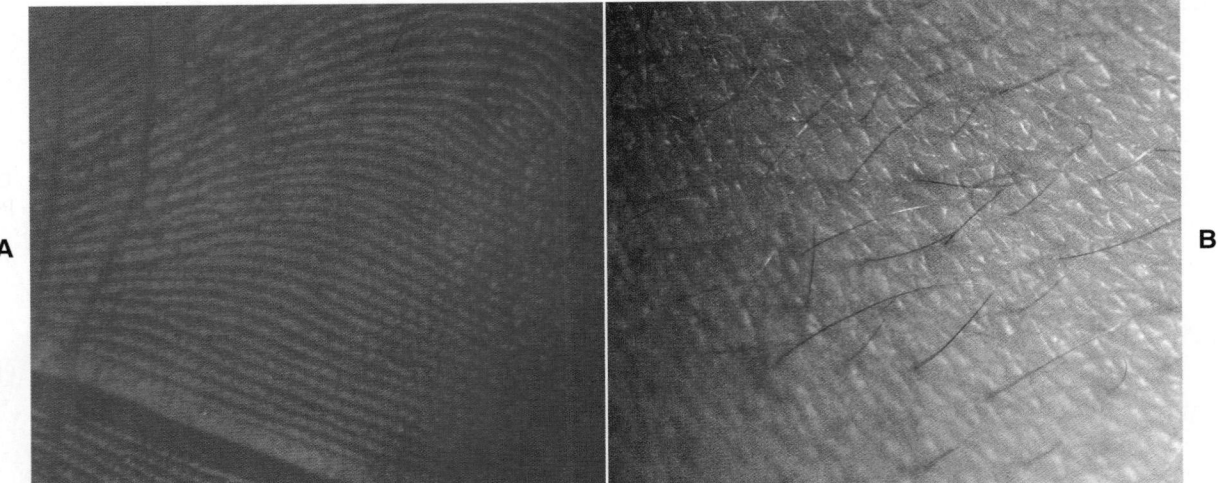

Figure 6-3 *Thick and thin skin.* **A,** The surface of thick skin is hairless and features regular, deep sulci and friction ridges, which form the "prints" of the palmar and plantar surfaces of the hands and feet. **B,** The surface of thin skin features irregular sulci (grooves) and hairs.

 BOX 6-1: HEALTH MATTERS

Vitiligo

An acquired condition called **vitiligo** results in loss of pigment in certain areas of the skin. The patches of depigmented white skin that characterize this condition contain melanocytes, but for unknown reasons they no longer produce pigment.

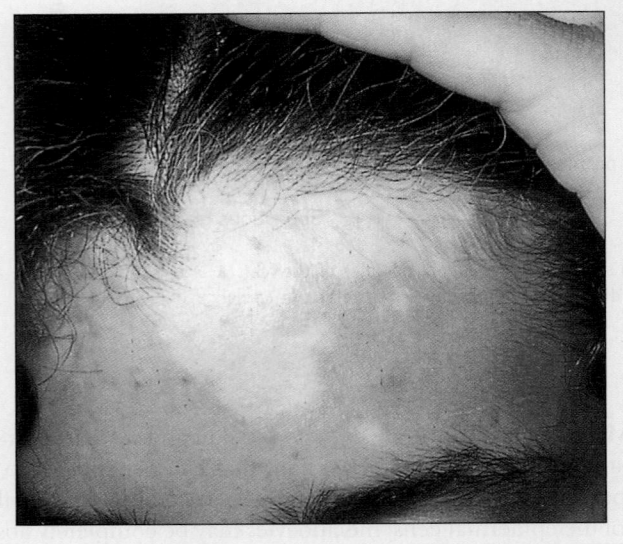

Cell Layers

The cells of the epidermis are found in up to five distinct layers, or **strata.** Each stratum (meaning "layer") is named for its structural or functional characteristics. The strata of the epidermis are listed here in order from deepest to most superficial.

1. **Stratum basale** (base layer). The stratum basale is a single layer of columnar cells. Only the cells in this deepest stratum of the epithelium undergo mitosis. As a result of this regenerative activity, cells transfer or migrate from the basal layer through the other layers until they are shed from the skin surface.

2. **Stratum spinosum** (spiny layer). The stratum spinosum layer of the epidermis is formed from 8 to 10 layers of irregularly shaped cells with very prominent intercellular bridges, or desmosomes. When viewed under a microscope, the desmosomes appear to pull points of the plasma membranes of adjoining cells toward one another. This gives cells of this layer a spiny or prickly appearance (Latin *spinosus,* "spinelike"). Cells in this epidermal layer are rich in ribonucleic acid (RNA) and are therefore well equipped to initiate the protein synthesis required for the production of keratin.

    The term **stratum germinativum** (growth layer) is often used to describe the stratum basale and the stratum spinosum together. Occasionally, anatomists apply this name only to the base layer of cells.

3. **Stratum granulosum** (granular layer). The process of surface keratin formation begins in the stratum granulosum of the epidermis. Cells are arranged in a sheet two to four layers deep and are filled with intensely staining granules called **keratohyalin,** which is required for surface keratin formation. At this stage, the keratinocytes also form small bodies of *glycophospholipids* (part sugar, part phospholipid) built up in multiple layers. Even though there is some important biochemical activity at this stage, cells in the stratum granulosum have started to degenerate. As a result, high levels of lysosomal enzymes are present in the cytoplasm, and the nuclei are in the process of breaking down. In thin skin, not many cells at this stage are present, so this layer of the epidermis may not be visible.

4. **Stratum lucidum** (clear layer). The keratinocytes in the stratum lucidum are very flat, closely packed, and clear. Typically, nuclei are absent, and the cell outlines are now indistinct. These dying cells are filled with a substance called **eleidin,** which will eventually be transformed to keratin. This layer is

absent in thin skin but is apparent in sections of thick skin from the soles of the feet and the palms of the hands.

5. **Stratum corneum** (horny layer). The stratum corneum is the most superficial layer of the epidermis. It is composed of very thin squamous (flat) cells, which at the skin surface are dead and continually being shed and replaced. Much of the cytoplasm in these cells has been replaced by a dense network of keratin fibers. The glycophospholipids from the multilayer bodies cement the keratin fibers into a strong, waterproof barrier. The desmosomes that hold adjacent keratinocytes together strengthen this layer even more and permit it to withstand considerable wear and tear. The process by which cells in this layer are formed from cells in deeper layers of the epidermis and then filled with keratin and moved to the surface is called **keratinization.**

The stratum corneum is sometimes called the *barrier area* of the skin because it functions as a barrier to water loss and to many environmental threats ranging from microorganisms and harmful chemicals to physical trauma. Once this barrier layer is damaged, the effectiveness of the skin as a protective covering is greatly reduced, and most contaminants can easily pass through the lower layers of the cellular epidermis. If the glycophospholipid barrier is washed away by prolonged soaking in water (especially if the water contains lipid-dissolving detergents), the keratin may absorb water and make the skin appear puffy and wrinkled—most easily seen in thick skin because of its high water content. Certain diseases of the skin cause the stratum corneum layer of the epidermis to thicken far beyond normal limits—a condition called **hyperkeratosis.** The result is a thick, dry, scaly skin that is inelastic and subject to painful fissures.

Table 6-1 clearly shows each of these layers of the epidermis.

Epidermal Growth and Repair

The most important function of the integument—protection—largely depends on the special structural features of the epidermis and its ability to create and repair itself after injury or disease. *Turnover* time and *regeneration* time are terms used to describe the period required for a population of cells to mature and reproduce. Obviously, as the surface cells of the stratum corneum are lost, replacement of keratinocytes by mitotic activity must occur. New cells must be formed at the same rate that old keratinized cells flake off from the stratum corneum to maintain a constant thickness of the epidermis. Cells push upward from the stratum basale into each successive layer, die, become keratinized, and eventually desquamate (fall away), as did their predecessors. This fact illustrates a physiological principle: while life continues, the body's work is never done. Even at rest it is producing millions on millions of new cells to replace old ones.

Current research suggests that the regeneration time required for completion of mitosis, differentiation, and movement of new keratinocytes from the stratum basale to the surface of the epidermis is about 35 days. The process can be accelerated by abrasion of the skin surface, which tends to peel off a few of the cell layers of the stratum corneum. The result is an intense stimulation of mitotic activity in the stratum basale and a shortened turnover pe-

riod. If abrasion continues over a prolonged period, the increase in mitotic activity and shortened turnover time will result in an abnormally thick stratum corneum and the development of **calluses** at the point of friction or irritation. Although callus formation is a normal and protective response of the skin to friction, several skin diseases are also characterized by abnormally high mitotic activity in the epidermis. In such conditions, the thickness of the corneum is dramatically increased. As a result, scales accumulate and skin lesions often develop.

Normally, about 10% to 12% of all cells in the stratum basale enter mitosis each day. Cells migrating to the surface proceed upward in vertical columns from discrete groups of 8 to 10 of these basal cells undergoing mitosis. Each group of active basal cells, together with its vertical columns of migrating keratinocytes, is called an *epidermal proliferating unit*, or EPU. Keratinization proceeds as the cells migrate toward the stratum corneum. As mitosis continues and new basal cells enter the column and migrate upward, fully keratinized "dead" cells are sloughed off at the skin surface. Numerous skin diseases are characterized by an abnormally high rate of keratinization.

QUICK CHECK

1. Identify the two main or primary layers of skin. What tissue type dominates each layer?
2. The terms *thin* and *thick* skin refer to which primary layer of skin? How do thin and thick skin differ?
3. Identify the two main cell types found in the epidermis.
4. List the five layers or strata of the epidermis.

Dermal-Epidermal Junction

Electron microscopy and histochemical studies have demonstrated the existence of a specialized area between the epidermis and dermis called the *dermal-epidermal junction.* It is composed chiefly of an easily identified basement membrane. In addition, the junction includes specialized fibrous elements and a unique polysaccharide gel that together with the basement membrane, cement the superficial epidermis to the dermis below. The junction "glues" the two layers together and provides mechanical support for the epidermis, which is attached to its upper surface. In addition, it serves as a partial barrier to the passage of some cells and large molecules. Certain dyes, for example, if injected into the dermis cannot passively diffuse upward into the epidermis unless the junctional barrier is damaged by heat, enzymes, or other chemicals that change its permeability characteristics. Although the junction is remarkably effective in preventing separation of the two skin layers, even when they are subjected to relatively high shear force, this barrier is thought to have only a limited role in preventing passage of harmful chemicals or disease-causing organisms through the skin from the external environment. Any generalized detachment of a large area of epidermis from the dermis is an extremely serious condition that may result in overwhelming infection and death.

Table 6-1 Structure of the Skin

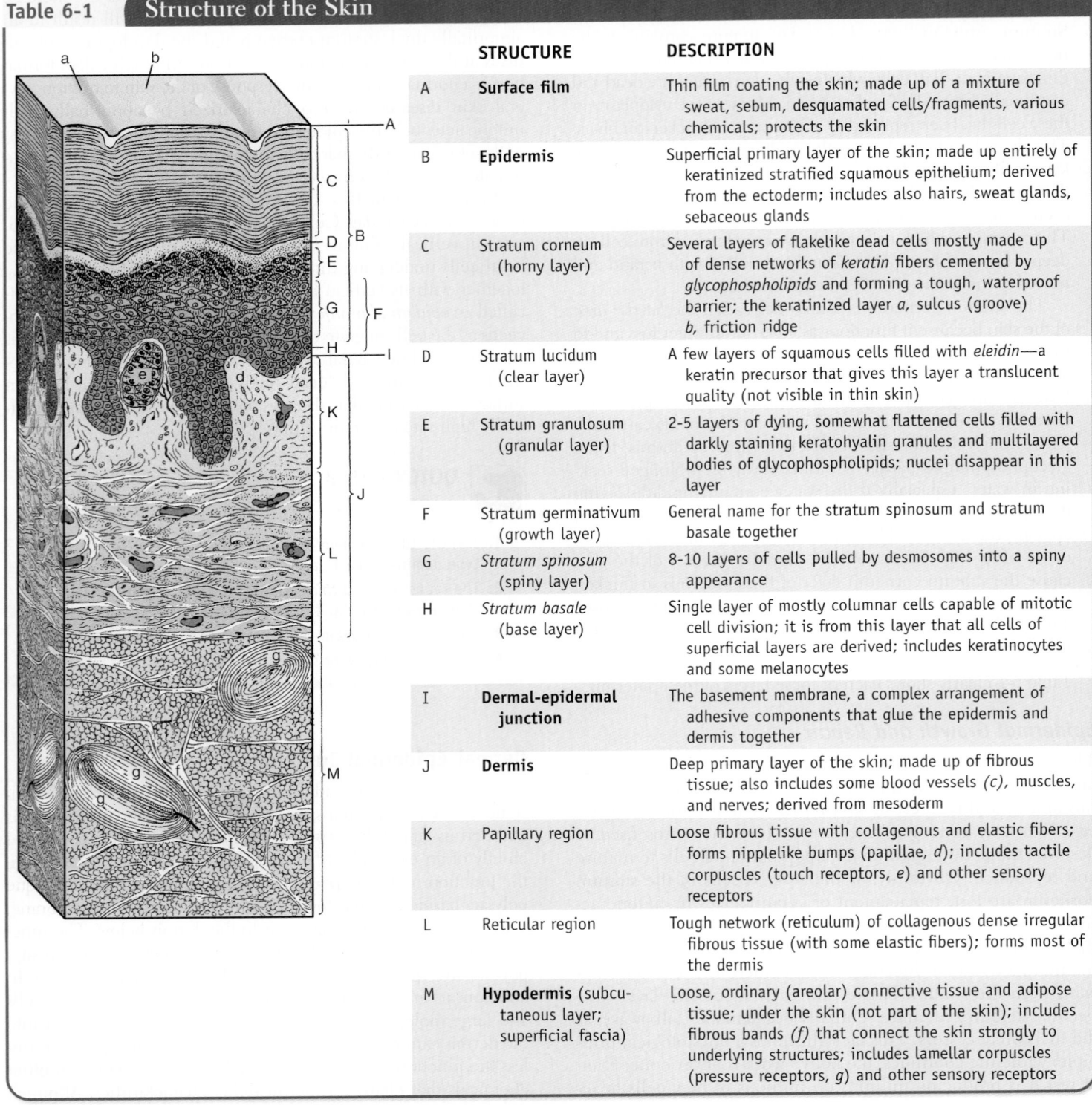

	STRUCTURE	DESCRIPTION
A	**Surface film**	Thin film coating the skin; made up of a mixture of sweat, sebum, desquamated cells/fragments, various chemicals; protects the skin
B	**Epidermis**	Superficial primary layer of the skin; made up entirely of keratinized stratified squamous epithelium; derived from the ectoderm; includes also hairs, sweat glands, sebaceous glands
C	Stratum corneum (horny layer)	Several layers of flakelike dead cells mostly made up of dense networks of *keratin* fibers cemented by *glycophospholipids* and forming a tough, waterproof barrier; the keratinized layer *a*, sulcus (groove) *b*, friction ridge
D	Stratum lucidum (clear layer)	A few layers of squamous cells filled with *eleidin*—a keratin precursor that gives this layer a translucent quality (not visible in thin skin)
E	Stratum granulosum (granular layer)	2-5 layers of dying, somewhat flattened cells filled with darkly staining keratohyalin granules and multilayered bodies of glycophospholipids; nuclei disappear in this layer
F	Stratum germinativum (growth layer)	General name for the stratum spinosum and stratum basale together
G	*Stratum spinosum* (spiny layer)	8-10 layers of cells pulled by desmosomes into a spiny appearance
H	*Stratum basale* (base layer)	Single layer of mostly columnar cells capable of mitotic cell division; it is from this layer that all cells of superficial layers are derived; includes keratinocytes and some melanocytes
I	**Dermal-epidermal junction**	The basement membrane, a complex arrangement of adhesive components that glue the epidermis and dermis together
J	**Dermis**	Deep primary layer of the skin; made up of fibrous tissue; also includes some blood vessels *(c)*, muscles, and nerves; derived from mesoderm
K	Papillary region	Loose fibrous tissue with collagenous and elastic fibers; forms nipplelike bumps (papillae, *d*); includes tactile corpuscles (touch receptors, *e*) and other sensory receptors
L	Reticular region	Tough network (reticulum) of collagenous dense irregular fibrous tissue (with some elastic fibers); forms most of the dermis
M	**Hypodermis** (subcutaneous layer; superficial fascia)	Loose, ordinary (areolar) connective tissue and adipose tissue; under the skin (not part of the skin); includes fibrous bands *(f)* that connect the skin strongly to underlying structures; includes lamellar corpuscles (pressure receptors, *g*) and other sensory receptors

BOX 6-2: HEALTH MATTERS

Subcutaneous and Intradermal Injections

Although the hypodermis, or subcutaneous layer, is not part of the skin itself, it carries the major blood vessels and nerves to the skin above. The rich blood supply and loose spongy texture of this area make it an ideal site for the rapid and relatively pain-free absorption of injected material. Liquid medicines, such as insulin, and pelleted implant materials are often administered by *subcutaneous (SQ) injection* with a *hypodermic needle* into this spongy and porous layer beneath the skin.

Intradermal (ID) injections, on the other hand, place the medication in the skin proper. ID injections, which are ideal for some vaccines, are hard to administer with a typical needle because it is so easy to punch through to the hypodermis. Therefore, this type of injection is sometimes given with special needle-free injectors.

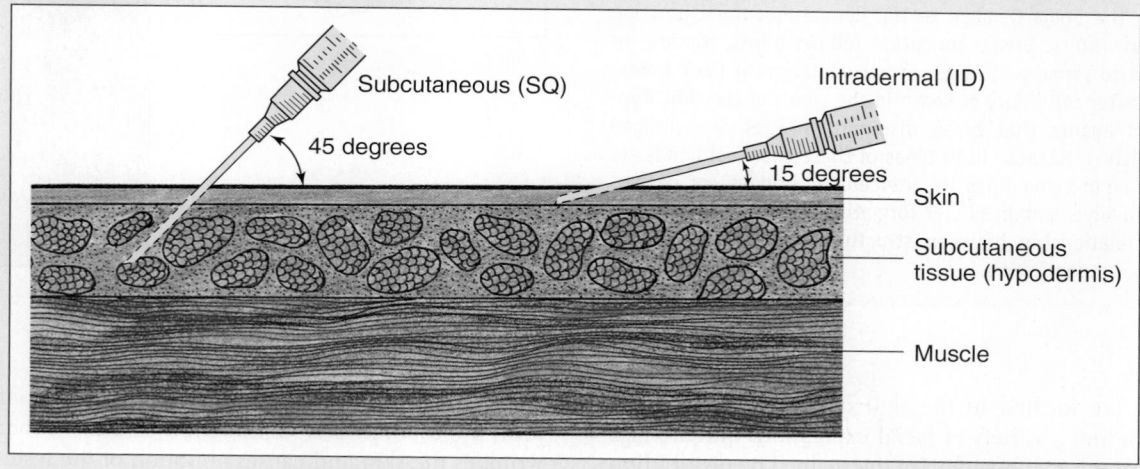

Dermis

The dermis, or *corium,* is sometimes called the "true skin." It is composed of a thin **papillary** and a thicker **reticular** layer. The dermis is much thicker than the epidermis and may exceed 4 mm on the soles and palms. It is thinnest on the eyelids and penis, where it seldom exceeds 0.5 mm. As a rule of thumb, the dermis on the ventral surface of the body and over the appendages is generally thinner than on the dorsal surface. The mechanical strength of the skin is in the dermis. In addition to serving a protective function against mechanical injury and compression, this layer of the skin provides a reservoir storage area for water and important electrolytes. A specialized network of nerves and nerve endings in the dermis called *somatic sensory receptors* also process sensory information such as pain, pressure, touch, and temperature. These specialized receptors are discussed in detail in Chapter 15. At various levels of the dermis extend a variety of muscle fibers, hair follicles, sweat and sebaceous glands, and many blood vessels. It is the rich vascular supply of the dermis that plays a critical role in regulation of body temperature—a function described later in the chapter.

Papillary Layer

Note in Figure 6-2 that the thin superficial layer of the dermis forms bumps, called **dermal papillae,** that project into the epidermis. *Papilla* (plural, *papillae*) is the Latin word for "nipple" and is used often in anatomy to name any small, nipplelike bump. The papillary layer takes its name from the papillae on its surface. Between the sculptured surface of the papillary layer and the stratum basale lies the important dermal-epidermal junction.

The papillary layer and its papillae are composed essentially of loose connective tissue elements and a fine network of thin collagenous and elastic fibers. The thin epidermal layer of the skin conforms tightly to the ridges of dermal papillae. As a result, the epidermis also has characteristic ridges on its surface. Epidermal ridges are especially well defined on the tips of the fingers and toes. In each of us they form a unique pattern—an anatomical fact made famous by the art of fingerprinting. These ridges perform a function that is very important to human survival: they allow us to grip surfaces well enough to walk upright on slippery surfaces and to grasp and use tools. For that reason, they are usually called *friction ridges*.

Reticular Layer

The thick reticular layer of the dermis consists of a much more dense *reticulum,* or network of fibers, than is seen in the papillary layer above it. It is this dense layer of tough and interlacing white collagenous fibers that when commercially processed from animal skin, results in leather. Although most of the fibers in this layer are of the collagenous type, which gives toughness to the skin, elastic fibers are also present. These fibers make the skin stretchable and elastic (able to rebound).

The dermis serves as a point of attachment for numerous skeletal (voluntary) and smooth (involuntary) muscle fibers. Several

BOX 6-3: HEALTH MATTERS

Blisters

Blisters (see figures) may result from injury to cells in the epidermis or from separation of the dermal-epidermal junction. Regardless of cause, they represent a basic reaction of skin to injury. Any irritant that damages the physical or chemical bonds that hold adjacent skin cells or layers together, such as poison ivy, initiates blister formation. The specialized junctions (desmosomes) that hold adjacent cells in the epidermis together are essential for integrity of the skin. If these intercellular bridges, sometimes described as "spot welds" between adjacent cells, are weakened or destroyed, the skin literally falls apart and away from the body. Damage to the dermal-epidermal junction produces similar results. Blister formation follows burns, friction injuries, exposure to primary irritants, or accumulation of toxic breakdown products after cell injury or death in the layers of the skin. Typically, chemical agents that break disulfide linkages or hydrogen bonds cause blisters. Because both types of these chemical bonds are the functional connecting links in intercellular bridges (or desmosomes), their involvement in blister formation serves as a good example of the relationship between structure and function at the chemical level of organization.

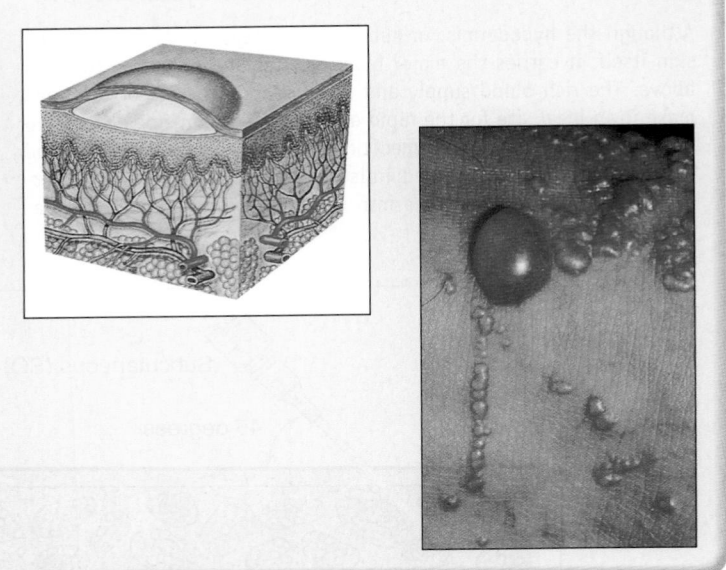

skeletal muscles are located in the skin of the face and scalp. These muscles permit a variety of facial expressions and are also responsible for voluntary movement of the scalp. The distribution of smooth muscle fibers in the dermis is much more extensive than the skeletal variety. Each hair follicle has a small bundle of involuntary muscles attached to it. These are the **arrector pili muscles.** Contraction of these muscles makes the hair "stand on end"—as in extreme fright, for example, or from cold. As the hair is pulled into an upright position, shown in Figure 6-4, it raises the skin around it into what we commonly call "goose bumps." In the dermis of the skin of the scrotum and in the pigmented skin

called the areolae surrounding the nipples, smooth muscle cells form a loose network. Contraction of these smooth muscle cells wrinkles the skin and causes elevation of the testes or erection of the nipples.

Millions of specialized somatic sensory receptors are located in the dermis of all skin areas (Figure 6-5; see also Figure 6-2). They permit the skin to serve as a sense organ transmitting sensations of pain, pressure, touch, and temperature to the brain.

Hair follicles and various skin glands, made up of epithelial tissues that extend from the surface of the epidermis, have most of their structures within the reticular layer of the dermis.

Figure 6-4 *Arrector pili muscle.* When the arrector pili muscle contracts, it pulls the follicle and hair into a more perpendicular position, thus "fluffing up" the hair. Notice how a "goose bump" is raised around the hair shaft.

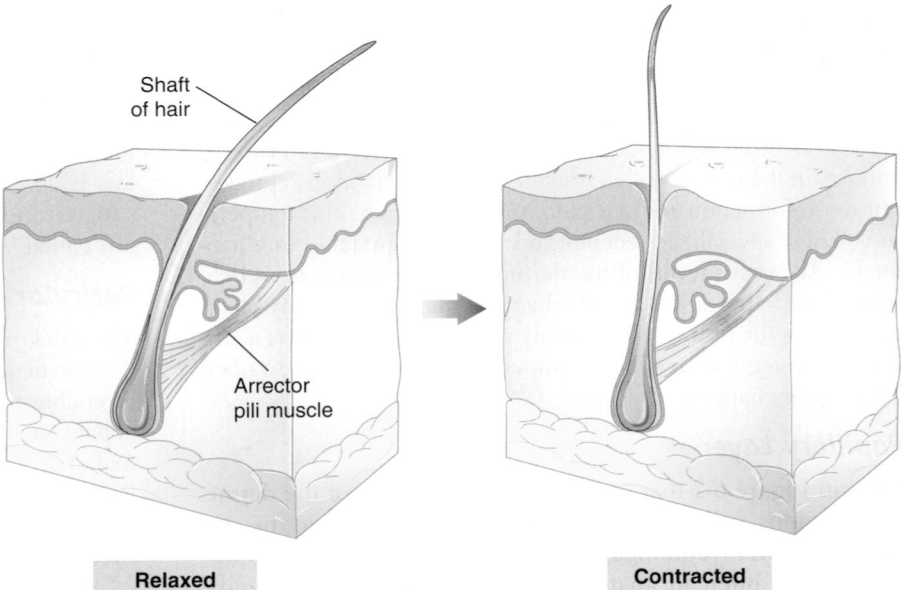

Shaft of hair

Arrector pili muscle

Relaxed   Contracted

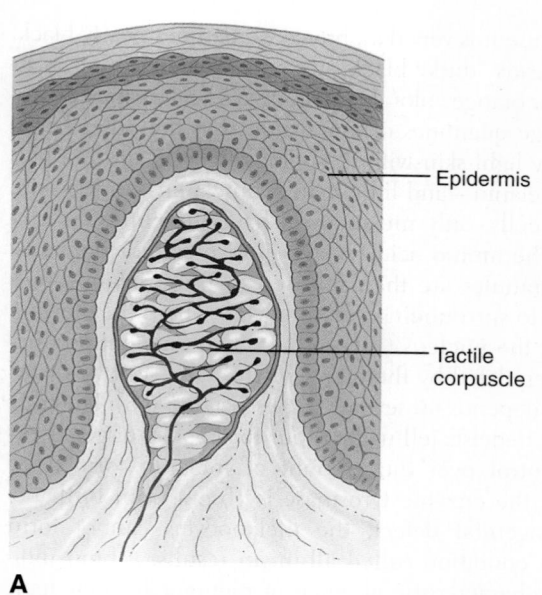

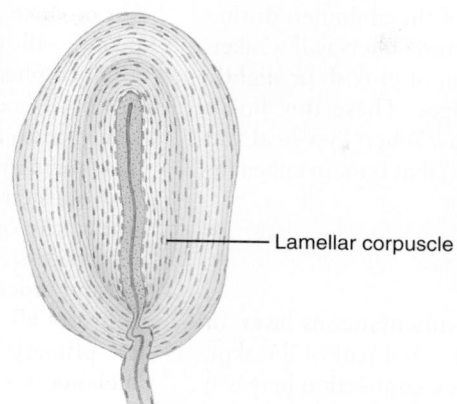

Epidermis

Tactile corpuscle

Lamellar corpuscle

A

B

Figure 6-5 *Skin receptors.* Receptors are specialized nerve endings that make it possible for the skin to act as a sense organ. **A,** This *tactile (Meissner) corpuscle* is capable of detecting light touch (slight pressure). **B,** Another skin receptor is the *lamellar corpuscle,* also called a *pacinian corpuscle,* which detects sensations of deep pressure.

Dermal Growth and Repair

Unlike the epidermis, the dermis does not continually shed and regenerate. It does maintain itself, but rapid regeneration of connective tissue in the dermis occurs only during unusual circumstances, as in the healing of wounds (see Figure 5-37, p. 178). In the healing of a wound such as a surgical incision, fibroblasts in the dermis quickly reproduce and begin forming an unusually dense mass of new connective tissue fibers. If this dense mass is not replaced by normal tissue, it remains as a *scar.*

The dense bundles of white collagenous fibers that characterize the reticular layer of the dermis tend to orient themselves in patterns that differ in appearance from one body area to another. The result is formation of patterns called **cleavage lines** (Figure 6-6). If surgical incisions are made parallel to the cleavage lines, or

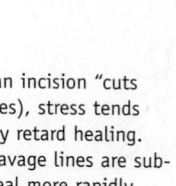

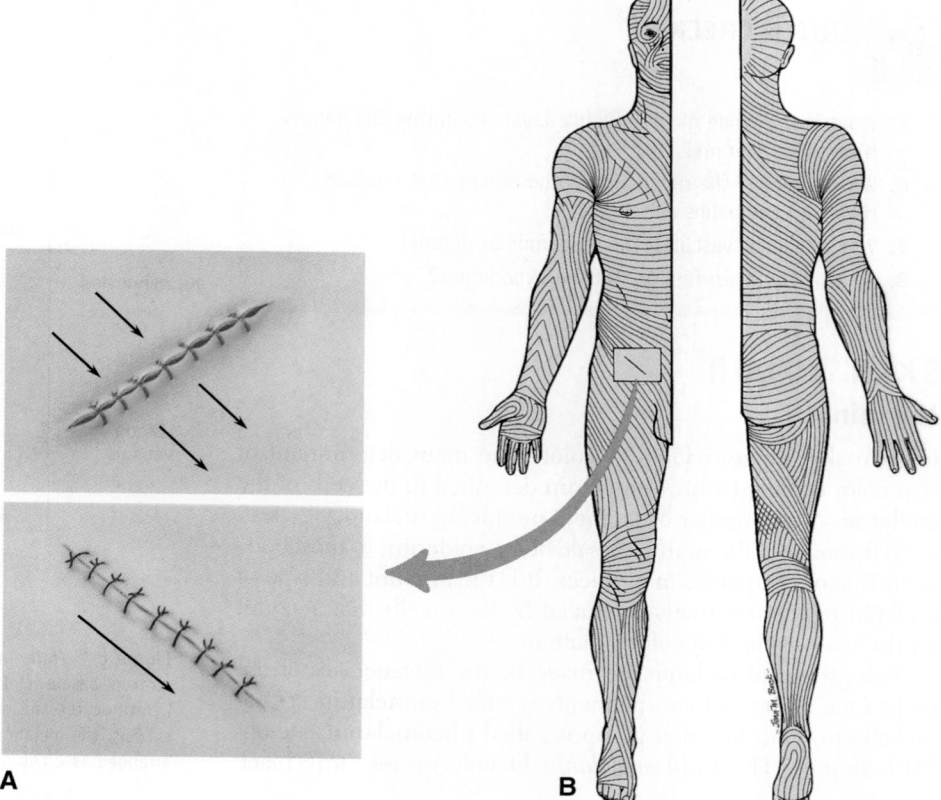

Figure 6-6 *Cleavage lines.* **A,** If an incision "cuts across" cleavage lines (Langer's lines), stress tends to pull the cut edges apart and may retard healing. **B,** Surgical incisions parallel to cleavage lines are subjected to less stress and tend to heal more rapidly.

A

B

Langer's lines, the resulting wound will have less tendency to gape open and will tend to heal with a thin and less noticeable scar.

If the elastic fibers in the dermis are stretched too much—for example, by a rapid increase in the size of the abdomen during pregnancy or as a result of great obesity—these fibers will weaken and tear. The initial result is the formation of pinkish or slightly bluish depressed furrows with jagged edges. These tiny linear markings *(stretch marks)* are really tiny tears. When they heal and lose their color, the *striae* (Latin, "furrows") that remain appear as glistening silver-white scar lines.

HYPODERMIS

The hypodermis is sometimes called the **subcutaneous layer,** or **superficial fascia** (see Box 5-6, p. 169). It is not part of the skin, but lies deep to the dermis and thus forms a connection between the skin and the underlying structures of the body. Thus, the hypodermis is often discussed along with the skin because of the close structural and functional association of these two superficial body structures.

The hypodermis is mostly loose fibrous and adipose tissue. The fat content of the hypodermis varies with the state of nutrition and in obese individuals may exceed 10 cm in thickness in certain areas. The density and arrangement of fat cells and collagen fibers in this area determine the relative mobility of the skin. Bands of fibers running through the hypodermis help hold the skin to underlying structures such as deep fascia and muscles (Table 6-1, *f*). When skin is removed from an animal by blunt dissection, separation occurs in the "cleavage plane" that exists between the superficial fascia and the underlying structures.

QUICK CHECK

5. What is the name of the gluelike layer separating the dermis from the epidermis?
6. Which layer of the dermis forms the bumps that produce ridges on the palms and soles?
7. Which layer is vascular: the epidermis or dermis?
8. What is the main function of the hypodermis?

SKIN COLOR

Melanin

Human skin ranges widely in color. The main determinant of skin color is the quantity of **melanin** deposited in the cells of the epidermis. The number of pigment-producing melanocytes scattered throughout the stratum basale of the epidermis in most body areas is about the same in all races. It is the amount and type of melanin pigment actually produced by these cells that account for the majority of skin color variations.

Two groups of melanin are made by the melanocytes of the body. One group of these pigments is called **eumelanin** (YOO-mel-ah-nin) and the other group is called **pheomelanin** (fee-oh-MEL-ah-nin). The word *eumelanin* literally means "true black

substance" because it is very dark brown, sometimes nearly black. *Pheomelanin* means "dusky black substance"—which hints at its lighter reddish or orange color. Dark-skinned and dark-haired people produce large quantities of eumelanin. On the other hand, people with very light skin with red-orange freckles and red hair produce pheomelanin—and little if any eumelanin—in the skin.

Of all body cells, only melanocytes have the ability to routinely convert the amino acid *tyrosine* into melanin pigments. The pigment granules are then released in tiny **melanosomes** and transferred to surrounding keratinocytes, where they form a sort of cap over the nucleus (Figure 6-7). The pigment-producing process is regulated by the enzyme *tyrosinase*. But this conversion process depends on several factors (Figure 6-8). Heredity is first of all. Geneticists tell us that four to six pairs of genes exert primary control over the amount of melanin formed by melanocytes. If the enzyme tyrosinase is absent from birth because of a congenital defect, the melanocytes cannot form melanin, and a condition called **albinism** results. Albino individuals have a characteristic absence of pigment in their hair, skin, and eyes. Thus, heredity determines how dark or light one's skin color will be (see Chapter 34).

Other factors can modify the expression of the genes for melanin production. Sunlight is an obvious example. Prolonged exposure to the ultraviolet (UV) radiation in sunlight in light-skinned individuals causes melanocytes to increase melanin pro-

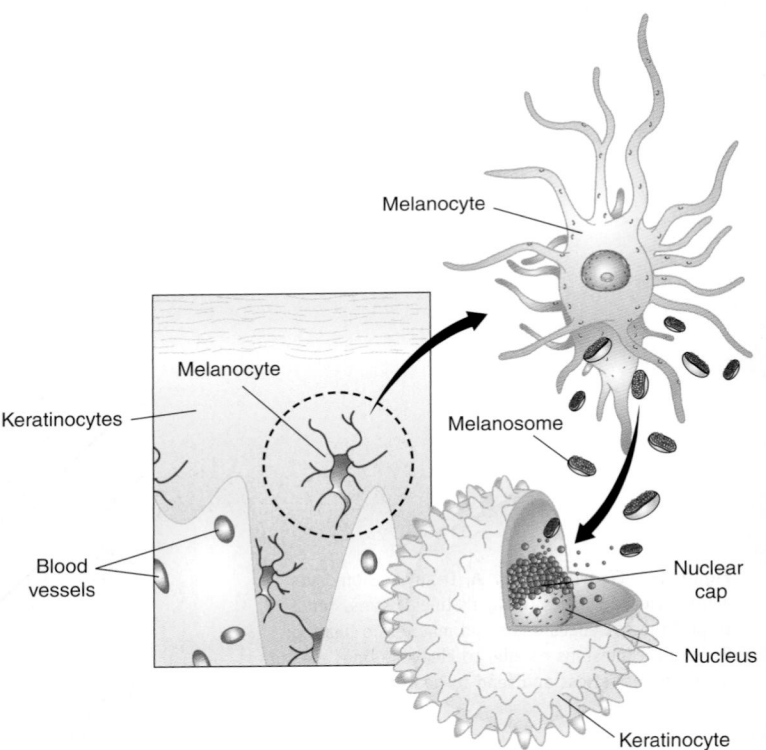

Figure 6-7 *Melanin production.* Melanin is produced by melanocytes in the stratum basale. Melanocytes have long projections that reach between the keratinocytes and release packets of pigment called melanosomes. By endocytosis, the melanosomes are brought into the keratinocytes, where they are arranged as a cap over the nucleus—protecting it from UV radiation from above.

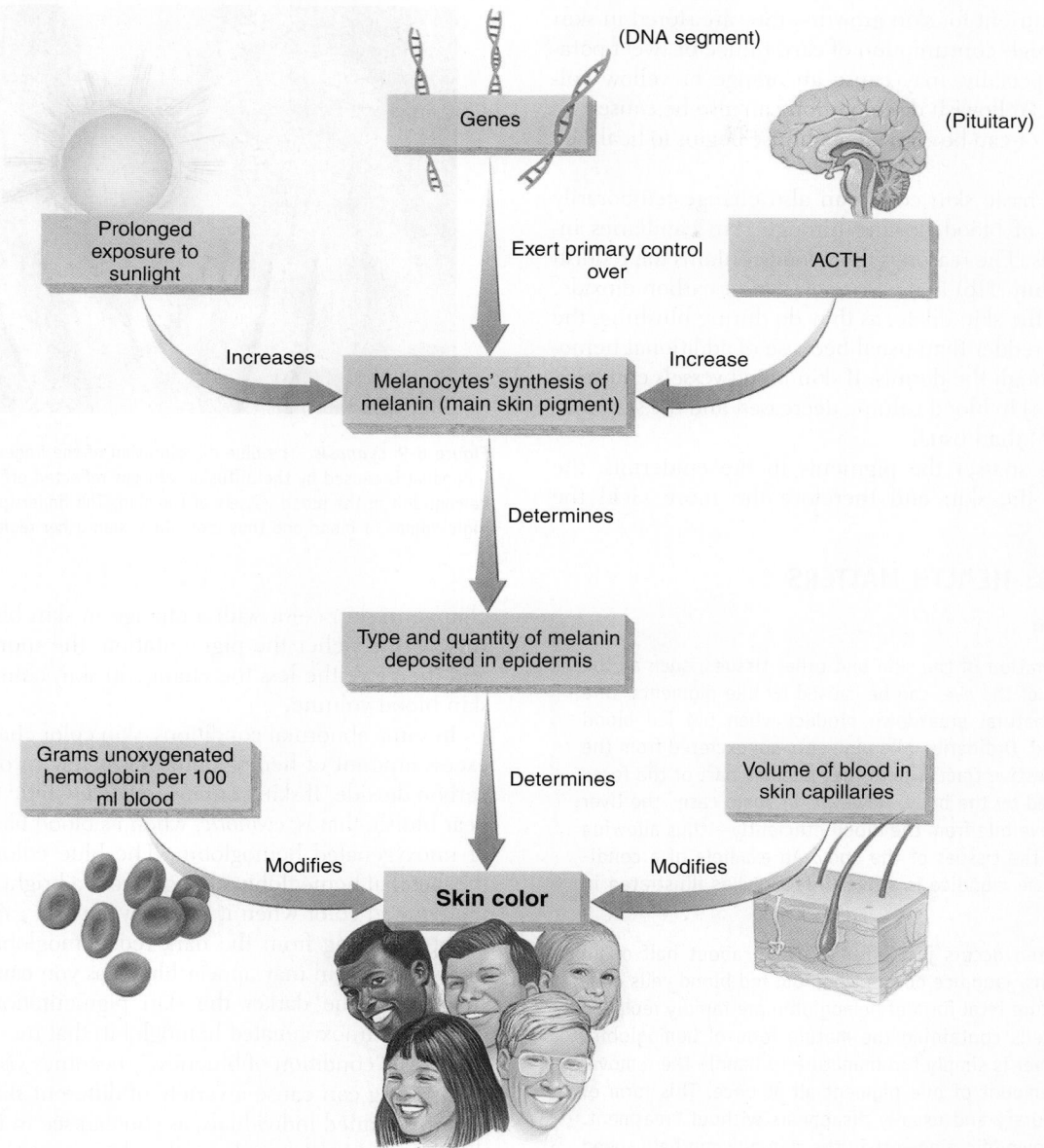

Figure 6-8 *How genes affect skin color.* Genes determine an individual's basic skin color by controlling the amount and type of melanin synthesized and deposited in the epidermis. However, as the diagram shows, other factors may modify the basic skin color.

duction and darken skin color. Notice in Figure 6-7 that the melanosomes form a cap over the top of the nucleus in each keratinocyte. The melanin in this nuclear cap absorbs UV radiation before it can reach the DNA inside the nucleus, where it can cause severe damage that can lead to skin cancer and other problems. Unless protected by melanin, UV radiation can also break down other important molecules, such as the vitamin *folic acid* (B_9). Eumelanin absorbs more UV radiation than pheomelanin does—which explains why very dark-skinned individuals have less risk of developing skin cancer than very light-skinned, reddish freckled individuals do.

Melanin production can also be stimulated by excess secretion of adrenocorticotropic hormone (ACTH) by the anterior pituitary gland. Increasing age may also influence melanocyte activity. In many individuals, apoptosis of melanocyte stem cells in hair follicles produces the graying of the hair often associated with age.

Other Pigments

In addition to melanin, other pigments such as the yellow pigment *beta-carotene (β-carotene)* found in many vegetables and roots (for example, carrots) also contribute to skin color. Because β-carotenes can be converted by the body into vitamin A—a critically important nutrient for skin growth—they are stored in skin tissue. Extremely high consumption of carrot juice or sweet potatoes, in infants especially, may cause an orange or yellow coloration of the skin. Yellowish discoloration can also be caused by jaundice (Box 6-4) or can be seen after a bruise begins to heal (see below).

An individual's basic skin color can also change temporarily when the volume of blood flowing through skin capillaries increases or decreases. The reason is that blood contains the reddish pigment hemoglobin (Hb) that carries oxygen or carbon dioxide. If blood vessels in the skin dilate, as they do during blushing, the skin appears to be redder than usual because of additional hemoglobin flowing through the dermis. If skin blood vessels constrict, on the other hand, skin blood volume decreases, and the skin may turn paler (less red) than usual.

In general, the sparser the pigments in the epidermis, the more transparent the skin and therefore the more vivid the

BOX 6-4: HEALTH MATTERS
Jaundice

Yellowish discoloration of the skin and other tissues, such as the "white" or sclera of the eye, can be caused by bile pigments. Bile pigments are a natural breakdown product when old red blood cells are destroyed. Ordinarily, bile pigments are excreted from the liver into the digestive tract, where they become part of the feces and are eliminated by the body. However, in some cases the liver is unable to remove bile from the blood efficiently—thus allowing the bile to stain the tissues of the body. An example of a condition that can cause jaundice is a liver infection, as illustrated in the photograph.

Jaundice also often occurs just after birth. In about half of all full-term newborns, jaundice occurs when old red blood cells containing an immature fetal form of hemoglobin are rapidly replaced with red blood cells containing the mature form of hemoglobin. Often, a baby's liver is simply too immature to handle the removal of such a large amount of bile pigment all at once. This form of jaundice is temporary and usually disappears without treatment. UV light breaks down bile pigments in the skin and can help speed recovery in infants with moderate to severe cases of jaundice.

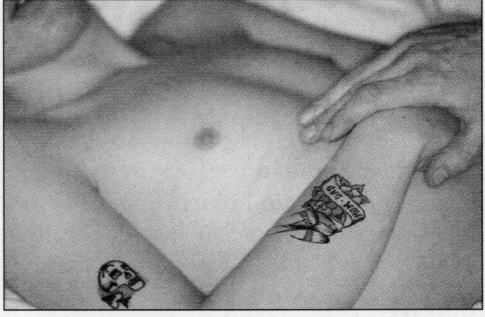

Jaundice. Yellowish discoloration of the skin and other tissues by bile pigments was, in the case seen here, caused by a liver infection. The patient was infected with hepatitis B by a contaminated tattoo needle. The yellow tinge of the skin can be best seen by comparing the patient's skin color with that of the physician's hand.

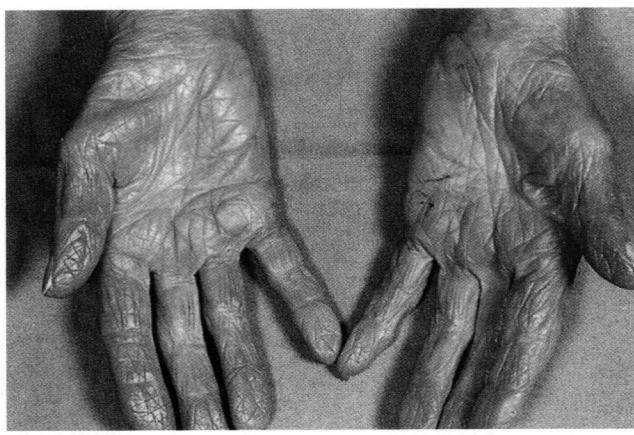

Figure 6-9 *Cyanosis.* The blue discoloration of the fingers of this light-skinned individual is caused by the diffusion of light reflected off dark, unoxygenated hemoglobin in the blood vessels of the skin. The fingertips have an especially high volume of blood and thus look bluer than other regions of the skin.

change in skin color with a change in skin blood volume. Conversely, the richer the pigmentation, the more opaque the skin and therefore the less the change in skin color with a change in skin blood volume.

In some abnormal conditions, skin color changes because of an excess amount of hemoglobin that is low in oxygen and high in carbon dioxide. If skin contains relatively little melanin, it will appear bluish, that is, *cyanotic*, when its blood has a high proportion of unoxygenated hemoglobin. The blue coloration results from the fact that hemoglobin changes from a bright red color to a deep maroon-red color when it loses oxygen and gains carbon dioxide. Light reflecting from the dark red hemoglobin and diffused by fibers in the skin may appear blue—as you can see in Figure 6-9. In general, the darker the skin pigmentation, the greater the amount of unoxygenated hemoglobin that must be present before **cyanosis** ("condition of blueness") becomes visible.

Bruising can cause a variety of different skin colors to appear in light-skinned individuals, as you can see in Figure 6-10. When damage to blood vessels in the skin permits the release of red blood cells, their reddish color begins to darken and produce the bluish colors described previously when hemoglobin loses oxygen and gains carbon dioxide. As the blood clots, it may begin to appear darker blue or even black. Macrophages remove the hemoglobin and break it down into iron-containing *hemosiderin* (a brownish pigment) and several iron-free *bile pigments* that are greenish and yellowish.

Pigments from cosmetics or from tattoos (see the photo in Box 6-3) can also change the coloring of the skin.

FUNCTIONS OF THE SKIN

Skin functions are crucial to maintenance of homeostasis and thus to survival itself. They are also diverse and include such different processes as protection, sensation, growth, synthesis of important chemicals and hormones (such as vitamin D), excretion, temperature regulation, and immunity. Because of its structural flexibility, the skin permits body growth and movement to occur

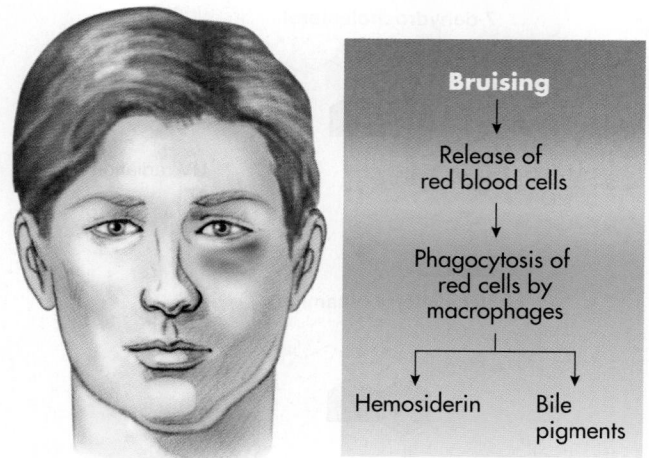

Figure 6-10 *Color changes in a bruise.* In this light-skinned individual, different colors appear in the skin as hemoglobin becomes unoxygenated and turns bluish (see Figure 6-9) and perhaps even black as the blood clots. As macrophages consume the hemoglobin, it is broken down into brownish, greenish, and yellowish pigments—sometimes producing a rainbow of skin colors.

without injury. We also know that certain substances can be absorbed through the skin, including the fat-soluble vitamins (A, D, E, and K), estrogens and other sex hormones, corticoid hormones, and certain drugs such as nicotine and nitroglycerin.

The skin also produces melanin—the pigment that serves as an extremely effective screen to potentially harmful ultraviolet

light—and keratin—one of nature's most flexible, yet enduring protective proteins. Refer to Table 6-2 as you read about the seven functions of the skin described in the paragraphs that follow.

Protection

The keratinized stratified squamous epithelial cells that cover the epidermis makes the skin a formidable barrier. It protects underlying tissues against invasion by hordes of microorganisms, bars entry of most harmful chemicals, and minimizes mechanical injury to underlying structures that might otherwise be harmed by even the relatively minor types of trauma experienced on a regular basis.

In addition to protection from microbiological entry, chemical hazards, and mechanical trauma, the skin also protects us from dehydration caused by loss of internal body fluids and from unwanted entry of fluids from the external environment. The ability of the pigment melanin to protect us from the harmful effects of overexposure to ultraviolet light is yet another protective function of the skin.

Surface Film

The ability of the skin to act as a protective barrier against an array of potentially damaging assaults from the environment begins with the proper functioning of a thin film of emulsified material spread over its surface. The **surface film** is produced by the mixing of residue and secretions from sweat and sebaceous glands with epithelial cells constantly being cast off from the epidermis. The shedding of epithelial elements from the skin surface is

Table 6-2	Functions of the Skin	
FUNCTION	**EXAMPLE**	**MECHANISM**
Protection	From microorganisms From dehydration From ultraviolet radiation From mechanical trauma	Surface film/mechanical barrier Keratin Melanin Tissue strength
Sensation	Pain Heat and cold Pressure Touch	Somatic sensory receptors
Permits movement and growth without injury	Body growth and change in body contours during movement	Elastic and recoil properties of skin and subcutaneous tissue
Endocrine	Vitamin D production	Activation of precursor compound in skin cells by ultraviolet light
Excretion	Water Urea Ammonia Uric acid	Regulation of sweat volume and content
Immunity	Destruction of microorganisms and interaction with immune system cells (helper T cells)	Phagocytic cells and Langerhans cells
Temperature regulation	Heat loss or retention	Regulation of blood flow to the skin and evaporation of sweat

called **desquamation.** Functions of surface film include the following:

Antibacterial and antifungal activity
Lubrication
Hydration of the skin surface
Buffering of caustic irritants
Blockade of many toxic agents

The chemical composition of surface film includes (1) amino acids, sterols, and complex phospholipids from the breakdown of sloughed epithelial cells; (2) fatty acids, triglycerides, and waxes from sebum; and (3) water, ammonia, lactic acid, urea, and uric acid from sweat. The specific chemical composition of surface film is variable, and samples taken from skin covering one body area often have a different "mix" of chemical components than film covering skin in another area does. This difference helps explain the unique and localized distribution patterns of certain skin diseases and also explains why the skin covering one area of the body is sometimes more susceptible to attack by certain bacteria or fungi.

Sensation

The widespread placement of the millions of different somatic sensory receptors found in the skin enables it to function as a sophisticated sense organ covering the entire body surface (see Figures 6-1 and 6-3). The receptors serve as antennas that detect stimuli, which eventually produce the general, or somatic, senses, including pressure, touch, temperature, pain, and vibration. When these receptors are activated by their respective stimuli, they make it possible for the body to respond to changes occurring in both the external and internal environments. A full discussion of the anatomy and physiology of the somatic sense receptors occurs in Chapter 15.

Flexibility

Contraction of our muscles produces the purposeful movement that serves as one of the most easily observed "Characteristics of Life" discussed in Chapter 1. For movement of the body to occur without injury, the skin must be supple and elastic. It grows as we grow and exhibits stretch and recoil characteristics that permit changes in body contours to occur without tearing or laceration.

Excretion

By regulating the volume and chemical content of sweat, the body, through a function of the skin, can influence both its total fluid volume and the amounts of certain waste products, such as uric acid, ammonia, and urea, that are excreted. In most circumstances the skin plays only a minor role in the overall excretion of body wastes. However, it can become a more important function in certain disease states or pathological conditions.

Hormone (Vitamin D) Production

The first step in the production of vitamin D in the body occurs when the skin is exposed to ultraviolet light. When this occurs, molecules of a chemical called *7-dehydrocholesterol*, which is normally found in skin cells, are converted into a precursor substance called *cholecalciferol* (Figure 6-11). This chemical is then transported in the blood to the liver and kidneys where it is converted into an active form of vitamin D—a compound that influ-

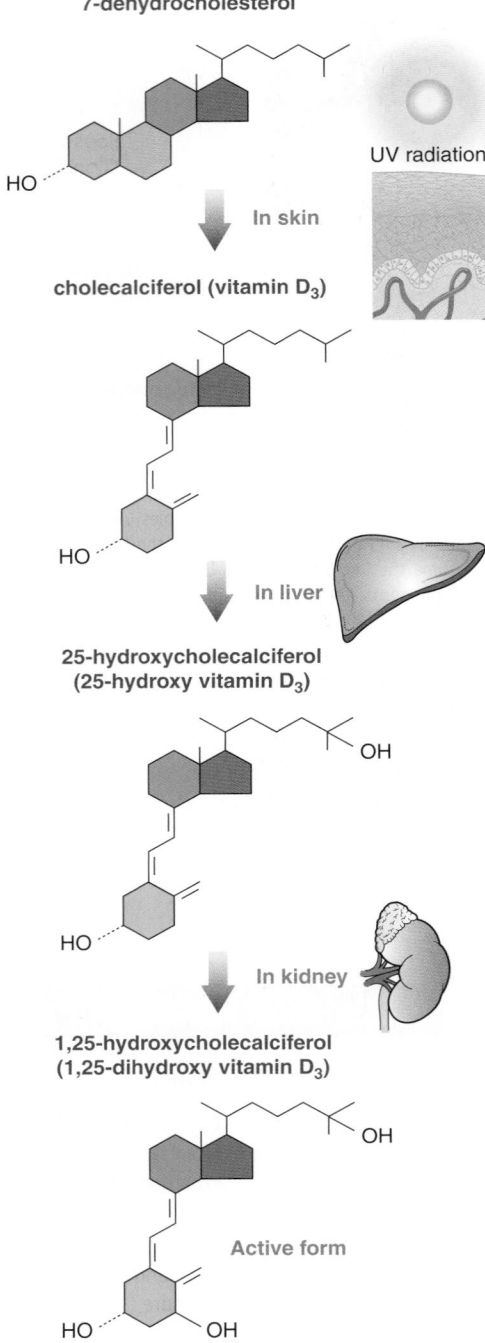

Figure 6-11 *Vitamin D production.* The essential first step in producing the active form of vitamin D in the body occurs in the presence of ultraviolet (UV) light in the skin.

ences several important chemical reactions in the body. In this respect, vitamin D fulfills the requirements required for a substance to be classified as a *hormone.* In general, any chemical substance produced in one body area and then transported in the blood to another location where it has its effect is called a *hormone.* Hormones are critically important regulators of homeostasis and are discussed in detail in Chapter 16.

Melanin in the skin must be dark enough to protect the skin from damage from UV radiation. However, if there is too much melanin, the body cannot synthesize enough vitamin D for normal function. In such a case, vitamin D from outside the body is necessary to compensate for the failure of skin to make enough vitamin D.

Immunity

Specialized cells that attach to and destroy pathogenic microorganisms are found in the skin and play an important role in immunity. In addition, Langerhans cells function with helper T cells to trigger helpful immune reactions in certain diseases.

Homeostasis of Body Temperature

Despite sizable variations in environmental temperature, humans maintain a remarkably constant core body temperature. The functioning of the skin in homeostasis of body temperature is critical to survival and is examined in some detail.

In most people, body temperature moves up and down very little in the course of a day. It hovers close to a set point of about 37° C, perhaps increasing to 37.6° C by late afternoon and decreasing to around 36.2° C by early morning. This homeostasis of body temperature is of the utmost importance. Why? Because healthy survival depends on biochemical reactions taking place at certain rates—and these rates depend on normal enzyme functioning, which depends on body temperature staying within the narrow range of normal.

To maintain an even temperature, the body must balance the amount of heat it produces with the amount it loses. This means that if extra heat is produced in the body, this same amount of heat must be lost from it. Obviously, if this does not occur, if increased heat loss does not closely follow increased heat production, body temperature climbs steadily upward. If body temperature increases above normal for any reason, the skin plays a critical role in heat loss by the physical phenomena of evaporation, radiation, conduction, and convection.

Heat Production

Heat is produced by one means—metabolism of foods. Because the muscles and glands (especially the liver) are the most active tissues, they carry on more metabolism and therefore produce more heat than any of the other tissues. So the chief determinant of how much heat the body produces is the amount of muscular work it does. During exercise and shivering, for example, metabolism and heat production increase greatly. But during sleep, when very little muscular work is being done, metabolism and heat production decrease.

Heat Loss

As already stated, one mechanism the body uses to maintain relative constancy of internal temperature is to regulate the amount of heat loss. Some 80% or more of this transfer of heat occurs through the skin; the remainder takes place in mucous membranes. As Figure 6-12 shows, heat loss can be regulated by altering the flow of blood in the skin. If heat must be conserved to maintain a constant body temperature, dermal blood vessels constrict (vasoconstriction) to keep most of the warm blood circulating deeper in the body. If heat loss must be increased to maintain a constant tem-

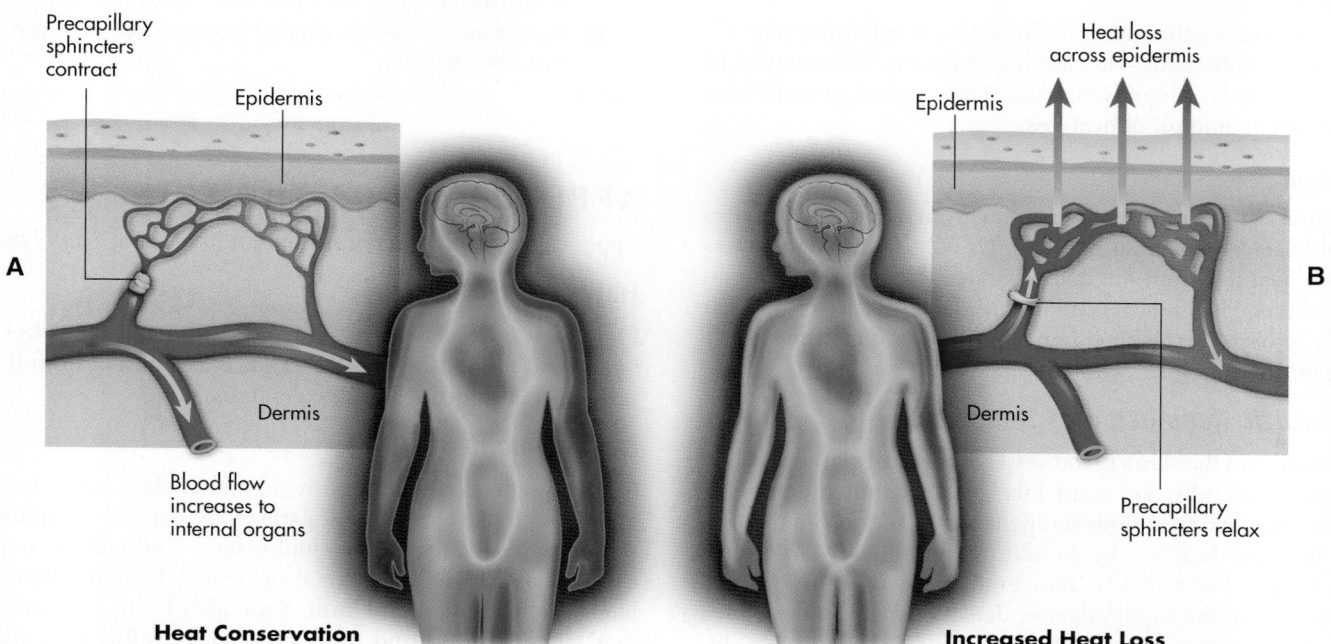

Figure 6-12 *The skin as a thermoregulatory organ.* When homeostasis requires that the body conserve heat, blood flow in the warm organs of the body's core increases. **A,** When heat must be lost to maintain stability of the internal environment, flow of warm blood to the skin increases. **B,** Heat can be lost from the blood and skin by means of radiation, conduction, convection, and evaporation. The head, hands, and feet are major areas of heat loss.

perature, dermal blood vessels widen (vasodilation) to increase the skin's supply of warm blood from deeper tissues. Heat transferred from the warm blood to the epidermis can then be lost to the external environment through the physical processes of evaporation, radiation, conduction, and convection.

Evaporation

Heat energy must be expended to evaporate any fluid. Evaporation of water constitutes one method by which heat is lost from the body, especially from the skin. Evaporation is especially important in high environmental temperatures, when it is the only method by which heat can be lost from the skin. A humid atmosphere necessarily retards evaporation and therefore lessens the cooling effect derived from it—the explanation for the fact that the same degree of temperature seems hotter in humid climates than in dry ones. At moderate temperatures, evaporation accounts for about half as much heat loss as radiation does.

Radiation

Radiation is the transfer of heat from the surface of one object to that of another without actual contact between the two. Heat radiates from the body surface to nearby objects that are cooler than the skin and radiates to the skin from those that are warmer than the skin. This is the principle of heating and cooling systems. In cool environmental temperatures, radiation accounts for a greater percentage of heat loss from the skin than conduction and evaporation combined do. In hot environments, no heat is lost by radiation but may be gained by radiation from warmer surfaces to the skin.

Conduction

Conduction means the transfer of heat to any substance actually in contact with the body—to clothing or jewelry, for example, or even to cold foods or liquids ingested. This process accounts for a relatively small amount of heat loss.

Convection

Convection is the transfer of heat away from a surface by movement of heated air or fluid particles. Usually, convection causes very little heat loss from the body's surface. However, under certain conditions, it can account for considerable heat loss, as you know if you have ever stepped from your shower into even slightly moving air from an open window.

Homeostatic Regulation of Heat Loss

The operation of the skin's blood vessels and sweat glands must be coordinated carefully and must take into account moment-by-moment fluctuations in body temperature. Like most homeostatic mechanisms, heat loss by the skin is controlled by a negative feedback loop (Figure 6-13). Temperature receptors in a part of the brain, called the *hypothalamus*, detect changes in the body's internal temperature. If some disturbance, such as exercise, increases body temperature above the set point value of 37° C, the hypothalamus acts as an integrator and sends a nervous signal to the sweat glands and blood vessels of the skin. The sweat glands and blood vessels (the effectors) respond to the signal by acting in ways that promote heat loss. That is, sweat glands increase their

output of sweat (increasing heat loss by evaporation), and blood vessels increase their diameter (increasing heat loss by radiation and other means). This process continues until the homeostatic set point for normal body temperature is reached (Box 6-5).

QUICK CHECK

9. What is the one means of heat production in the body? In what type of organs does most heat production occur?
10. Name three of the four physical processes by which heat is lost from the body.

APPENDAGES OF THE SKIN

Appendages of the skin consist of *hair, nails,* and *skin glands.*

Hair

Only a few areas of the skin are hairless—notably the palms of the hands and the soles of the feet. Hair is also absent from the lips, nipples, and some areas of the genitalia.

Development of Hair

Many months before birth, tiny tubular pockets called **hair follicles** begin to develop in most parts of the skin. By about the sixth month of pregnancy the developing fetus is all but covered by an extremely fine and soft hair coat, called **lanugo.** Most of the lanugo hair is lost before birth. Soon after birth, any lanugo hair that remains is lost and then replaced by **vellus** hair, which is stronger, fine, and usually less pigmented (see Figure 6-3, A). The replacement hair growth after birth first appears on the scalp, eyelids, and eyebrows, and the coarse pubic and axillary hair that develops at puberty is called **terminal hair.** In the adult male, terminal hair replaces 80% to 90% of the vellus hair on the chest and

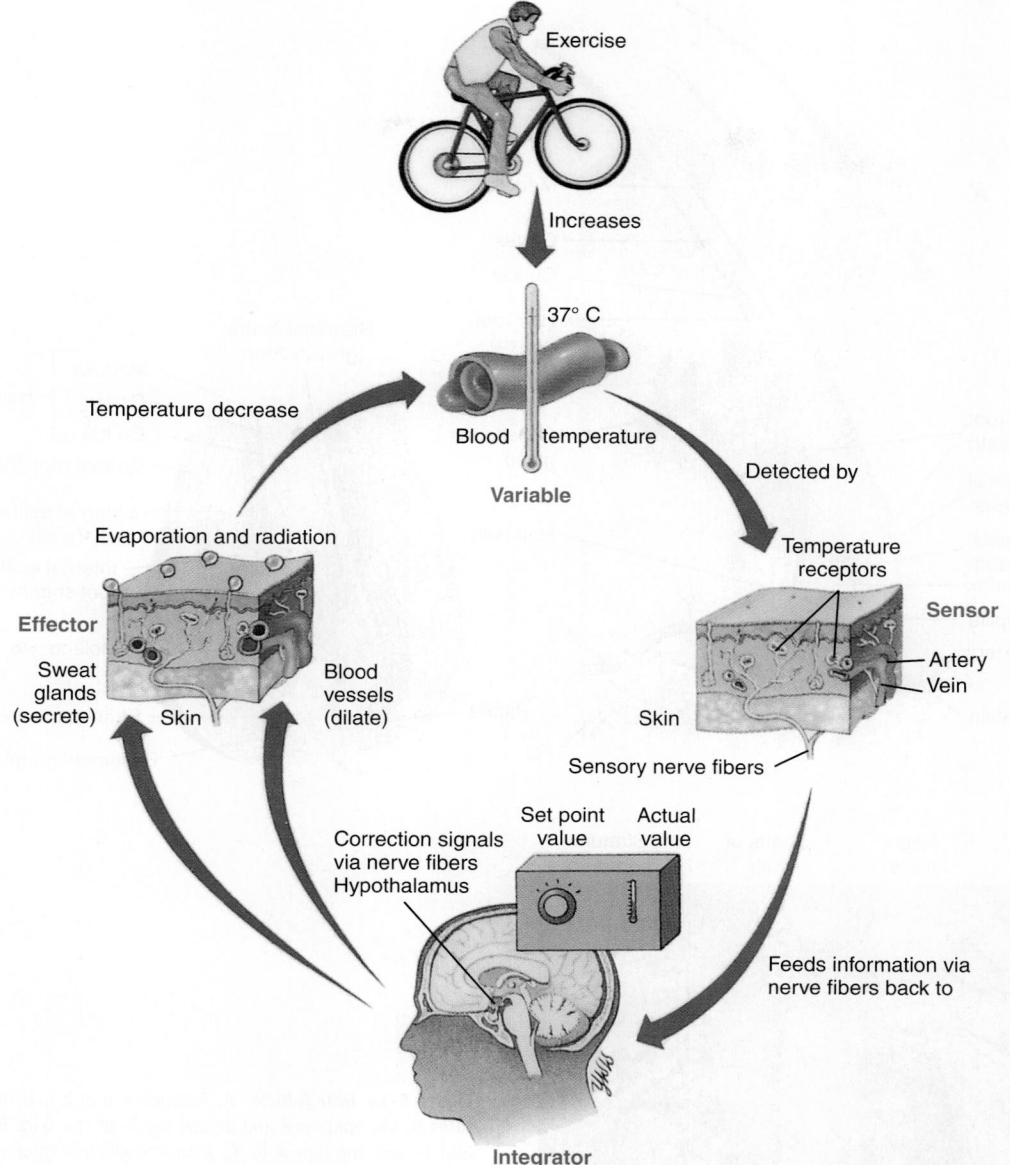

Figure 6-13 *Role of skin in homeostasis of body temperature.* Body temperature is continually monitored by nerve receptors in the skin and other parts of the body. These receptors feed information back to the hypothalamus of the brain, which compares the actual temperature with the set point temperature and then sends out an appropriate correction signal to effectors. If actual body temperature is above the set point temperature, sweat glands in the skin are signaled to increase their secretion and thus promote evaporation and cooling. At the same time, blood vessels in the dermis are signaled to dilate and thus promote radiation of heat away from the skin's surface.

extremities and also makes up the beard. In the female, far less vellus hair is replaced with the more coarse terminal variety except in the pubic and axillary areas.

Hair growth begins when cells of the epidermis spread down into the dermis to form a small tube, the follicle (Figure 6-14). The follicle wall consists of two primary layers: (1) an outer dermal root sheath and (2) an epithelial root sheath, which is subdivided into external and internal layers (Figure 6-14, *B*). The stratum germinativum develops into the follicle's innermost layer and at the bottom

of the follicle forms a cap-shaped cluster of cells known as the **germinal matrix.** Protruding into the germinal matrix is a small mound of the dermis, called the **hair papilla,** which is an important structure because it contains the blood capillaries that nourish the germinal matrix (see Figure 6-14). Cells of the germinal matrix are responsible for forming hairs. They undergo repeated mitosis, push upward in the follicle, and become keratinized to form a hair. As long as cells of the germinal matrix remain alive, hair regenerates even though it is cut, plucked, or otherwise removed.

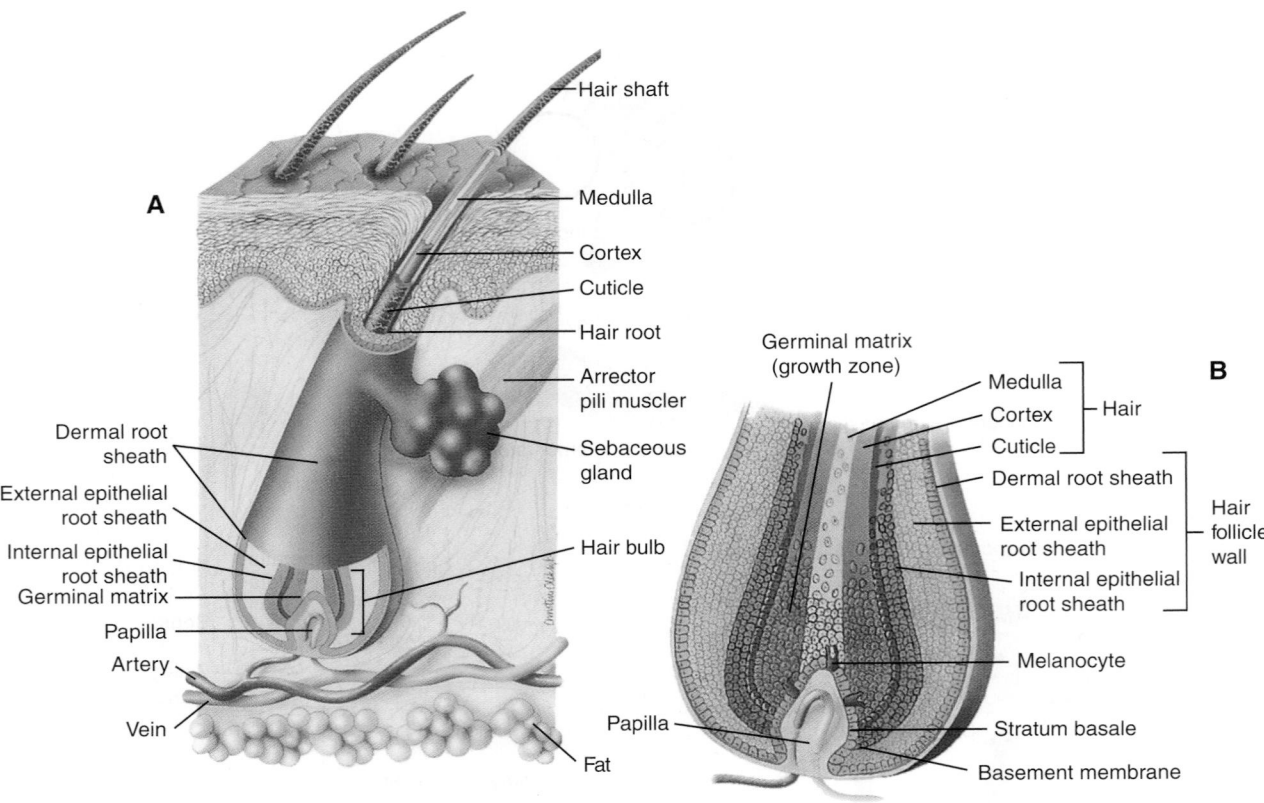

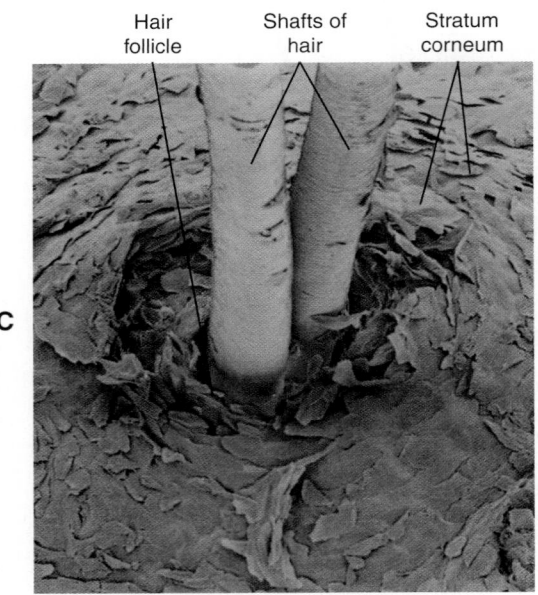

Figure 6-14 *Hair follicle.* **A,** Relationship of a hair follicle and related structures to the epidermal and dermal layers of the skin. **B,** Enlargement of a hair follicle wall and hair bulb. **C,** Scanning electron micrograph showing shafts of hair extending from their follicles.

Part of the hair, namely, the root, lies hidden in the follicle. The visible part of a hair is called the **shaft.** The inner core of a hair is known as the **medulla,** the more superficial portion around it is called the **cortex,** and the covering layer is called the *cuticle.* If hair is plucked from the follicle, the internal epithelial root sheath will be torn free as well and appears as a whitish layer adherent to the hair cuticle. Layers of keratinized cells make up the cortex.

Appearance of Hair

Deposited in the cells of the hair are varying amounts of melanins, the pigments responsible for hair color. Although reflections of light can affect hair color somewhat, it is largely the amount, type, and distribution of melanin that determine the color of hair. Varying amounts of eumelanins in the cortex and/or medulla can produce many shades of blonde and brunette hair. Pheomelanins give hair a reddish tint. Advancing age often produces an increasing smattering of white hairs—the proverbial "crown of glory." White hairs have no pigments at all but appear white because of the diffusion of light through the translucent hair shaft. White hair results from failure to maintain melanocytes in the hair follicle—thus losing the ability to produce pigment in the hair. Gray hair, seen in Figure 6-15, is usually a "salt-and-pepper" mixed arrangement of white and dark hairs. Besides adding color, melanin also imparts strength to the hair shaft.

Whether hair is straight or wavy depends mainly on the shape of the shaft. Straight hair has a round, cylindrical shaft. Wavy hair, in contrast, has a flat shaft that is not as strong. As a result, it is more easily broken or damaged than straight hair is. Two or more small **sebaceous glands** secrete **sebum,** an oily substance, into each hair follicle (see Figure 6-14). The sebaceous gland secretions lubricate and condition the hair and surrounding skin to keep it from becoming dry, brittle, and easily damaged. Lipid-containing hair and skin conditioners can add to this damage-reducing effect.

Hair alternates between periods of growth and rest. On average, hair on the head grows a little less than 12 mm (½ inch) per month, or about 11 cm (5 inches) a year. Body hair grows more slowly. Head hairs reportedly live between 2 and 6 years, then die and are shed. Normally, however, new hairs replace those lost. But baldness can develop. The common type of baldness occurs only when two requirements are met: genes for baldness must be inherited and the male sex hormone, testosterone, must be present. When the right combination of causative factors exist, common baldness or **male pattern baldness** (Figure 6-16) inevitably results. An available treatment that may actually slow or stop the development of male pattern baldness is the drug minoxidil (Rogaine). Unfortunately, the treatment is expensive, must be continued for life if new hair growth is to be retained, and does not always produce as dramatic an effect as hoped.

Contrary to what many people believe, hair growth is not stimulated by frequent cutting or shaving. In addition, stories about hair or beard growth continuing after death are also false. What may appear to be continuing beard growth after death is in reality only a more visible beard in a skin surface dehydrated by environmental conditions or the embalming process.

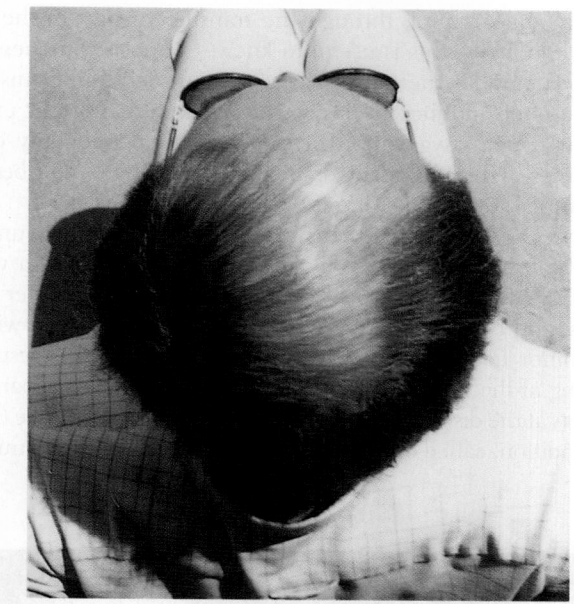

Figure 6-16 *Male pattern baldness.*

Nails

Heavily keratinized epidermal cells compose fingernails and toenails. The visible part of each nail is called the **nail body.** The rest of the nail, namely, the **root,** lies in a flat sinus hidden by a fold of skin bordered by the **cuticle.** The nail body nearest the root has a crescent-shaped white area known as the **lunula,** or "little moon." Under the nail lies a layer of epithelium called the **nail bed** (Figure 6-17). Because it contains abundant blood vessels, it

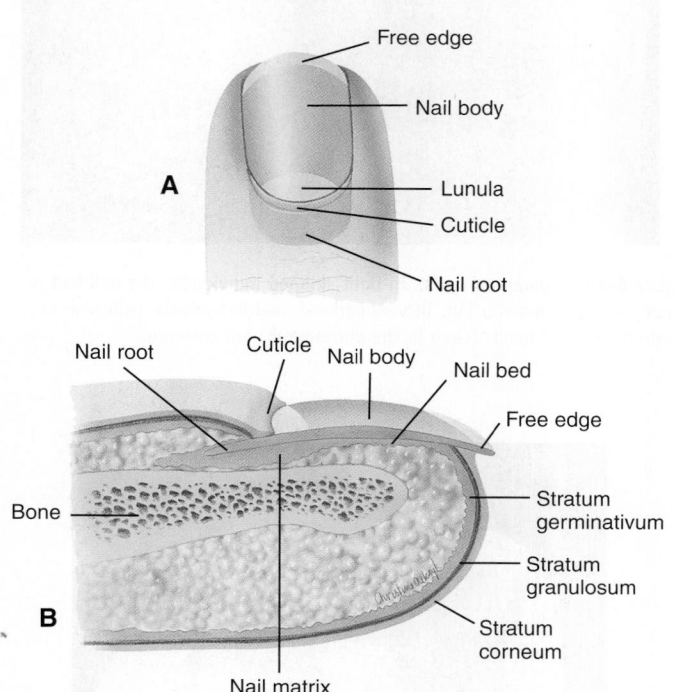

Figure 6-17 *Structure of nails.* **A,** Fingernail viewed from above. **B,** Sagittal section of a fingernail and associated structures.

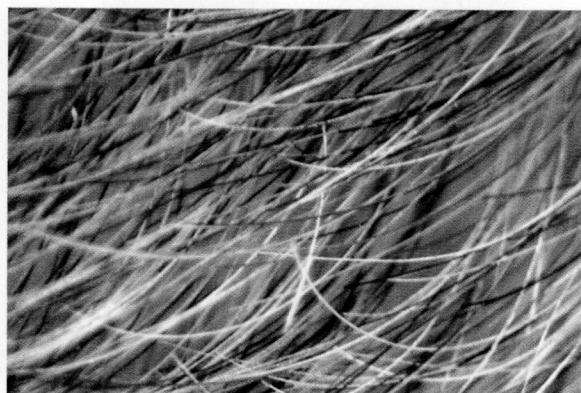

Figure 6-15 *Gray hair.* Gray hair results from a scattered arrangement of both white (pigmentless) hairs and darker (pigmented) hairs.

normally appears pink through the translucent part of the nail bodies. Clinically, this is useful to know. Cyanosis (blueness) often appears under the nails first (see Figure 6-9), and thus nail beds are often monitored closely during surgery or other procedures in which oxygenation of the blood may suddenly drop. Sometimes, lightly pigmented streaks appear in the nail beds of dark-skinned individuals (Figure 6-18).

Nails grow by mitosis of cells in the stratum germinativum beneath the lunula. On average, nails grow about 0.5 mm a week. Fingernails, however, as you may have noticed, grow faster than toenails, and both grow faster in the summer than in the winter. Even minor trauma to long fingernails can sometimes result in loosening of the nail from the nail bed with a resulting separation that starts at the distal or free edge of the affected nail (Figure 6-19). The condition, called **onycholysis** (ahn-ik-oh-LYE-sis), is common.

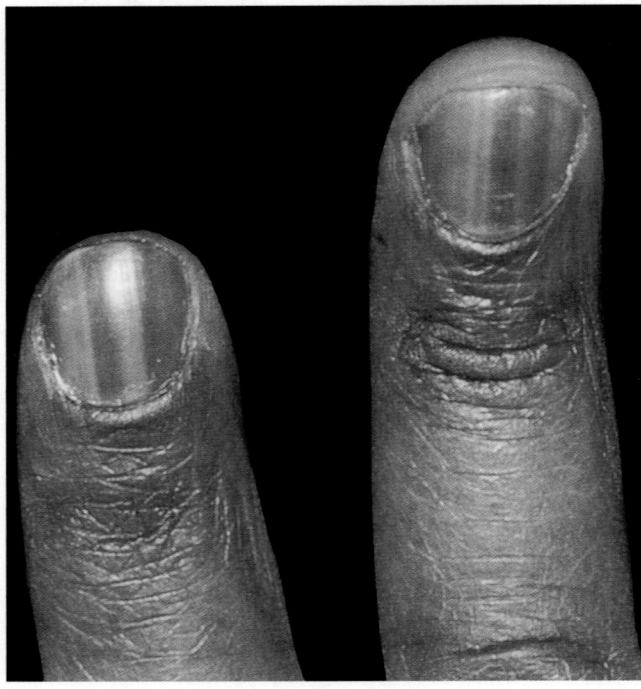

Figure 6-18. *Pigmented nails.* In light-skinned individuals, the nail bed is usually free of pigmentation. In very dark-skinned individuals, yellowish or brown pigmented bands (seen in the photograph) are common.

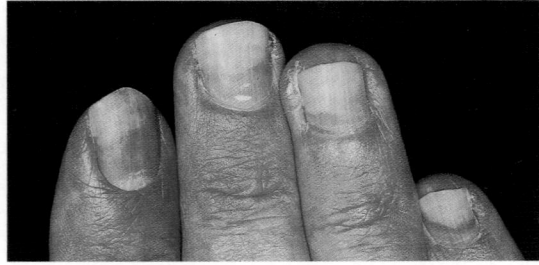

Figure 6-19 *Onycholysis.* Note that separation of this nail from the nail bed begins at the free edge. Minor trauma to long fingernails is the most common cause.

QUICK CHECK

11. Identify the pigment that determines hair color.
12. List seven functions of the skin.
13. How does surface film contribute to the protective function of the skin?
14. List the appendages of the skin.

Skin Glands

The skin glands include three kinds of microscopic glands: sweat, sebaceous, and ceruminous (Figure 6-20).

Sweat Glands

Sweat or **sudoriferous** (soo-doh-RIF-er-us) **glands** are the most numerous of the skin glands. They can be classified into two groups—*eccrine* and *apocrine*—based on the type of secretion, location, and nervous system connections.

Eccrine sweat glands are by far the most numerous, important, and widespread sweat glands in the body. They are small, with a secretory portion less than 0.4 mm in diameter, and are distributed over the total body surface with the exception of the lips, ear canal, glans penis, and nail beds. Eccrine sweat glands are a simple, coiled, tubular type of gland. They function throughout life to produce a transparent watery liquid (*perspiration*, or *sweat*) rich in salts, ammonia, uric acid, urea, and other wastes. In addition to elimination of waste, sweat plays a critical role in helping the body maintain a constant core temperature. Histologists estimate that a single square inch of skin on the palms of the hands contains about 3000 sweat glands. Eccrine sweat glands are also numerous on the soles of the feet, forehead,

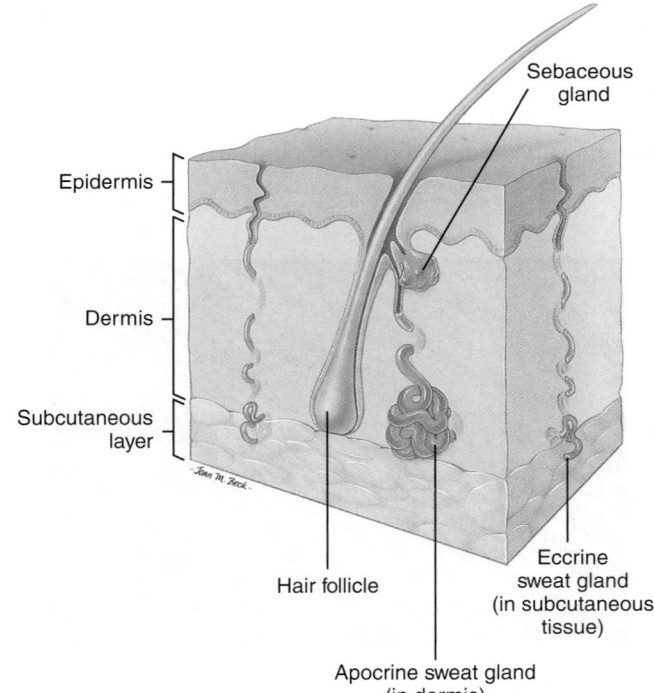

Figure 6-20 *Skin glands.* Several types of exocrine glands occur in the skin.

and upper part of the torso. With a good magnifying glass you can locate the openings of these sweat gland ducts on the skin ridges of the palms and on the skin of the palmar surfaces of the fingers. Although the ducts of eccrine glands travel through both the dermis and epidermis to open on the skin surface, the actual secretory portion is located in the subcutaneous tissue.

Apocrine sweat glands are located deep in the subcutaneous layer of the skin in the armpit (axilla), the areola of the breast, and the pigmented skin areas around the anus. They are much larger than eccrine glands and often have secretory units that reach 5 mm or more in diameter. They are connected with hair follicles and are classified as simple, branched tubular glands. Apocrine glands enlarge and begin to function at puberty; they produce a more viscous and colored secretion than eccrine glands do. In the female, apocrine gland secretions show cyclic changes linked to the menstrual cycle. The odor often associated with apocrine gland secretion is not caused by the secretion itself. Instead, it is caused by contamination and decomposition of the secretion by skin bacteria.

Sebaceous Glands

Sebaceous glands secrete oil for the hair and skin. Wherever hairs grow from the skin, there are sebaceous glands, at least two for each hair. The oil, or **sebum**, keeps the hair supple and the skin soft and pliant. It is nature's own protective skin cream that prevents excessive water loss from the epidermis. Because sebum is rich in chemicals that have an antifungal effect, such as triglycerides, waxes, fatty acids, and cholesterol, it also contributes to reducing fungal activity on the skin surface. This property of sebum increases the effectiveness of the skin's surface film and helps protect the skin from numerous types of fungal infections. Sebaceous glands are simple branched glands of varying size that are found in the dermis, except in the skin of the palms and soles. Although almost always associated with hair follicles, some specialized sebaceous glands do open directly on the skin surface in such areas as the glans penis, lips, and eyelids. Sebum secretion increases during adolescence because it is stimulated by increased blood levels of sex hormones. Frequently, sebum accumulates in and enlarges some of the ducts of the sebaceous glands, thereby forming white pimples. With oxidation, this accumulated sebum darkens and forms a **blackhead**.

Ceruminous Glands

Ceruminous glands are a special variety or modification of apocrine sweat glands. Histologically, they appear as simple coiled tubular glands with excretory ducts that open onto the free surface of the skin in the external ear canal or with sebaceous glands into the necks of hair follicles in this area. The mixed secretions of sebaceous and ceruminous glands form a brown waxy substance called **cerumen**. Although it serves a useful purpose in protecting the skin of the ear canal from dehydration, excess cerumen can harden and cause blockage in the ear, resulting in loss of hearing.

 QUICK CHECK

15. What are the two types of sweat glands? How do they differ?
16. List two functions of sebum.
17. What substances make up the skin's surface film?

Cycle Of Life
Skin

Everyone is aware of the dramatic changes in skin that each person experiences from birth through the mature years. Infants and young children have relatively smooth and unwrinkled skin characterized by the elasticity and flexibility associated with extreme youth. Because the skin tissues are in an active phase of new growth, healing of skin injuries is often rapid and efficient. Young children have fewer sweat glands than adults do, so their bodies rely more on increased blood flow to maintain normal body temperature. This explains why preschoolers often become "red-faced" while playing outdoors on a warm day.

As adulthood begins, at puberty, hormones stimulate the development and activation of sebaceous glands and sweat glands. After the sebaceous glands become active, especially during the initial years, they often overproduce sebum and thus give the skin an unusually oily appearance. Sebaceous ducts may become clogged or infected and form acne pimples or other blemishes on the skin (Box 6-6). Activation of apocrine sweat glands during puberty causes increased sweat production—an ability needed to maintain an adult body properly—and also the possibility of increased "body odor." Body odor is caused by wastes produced by bacteria that feed on the organic compounds found in apocrine sweat and on the surface of the skin.

 BOX 6-6: HEALTH MATTERS
Acne

Common *acne,* or *acne vulgaris,* occurs most frequently in the adolescent years as a result of overactive secretion by the sebaceous glands, with blockage and inflammation of their ducts. The rate of sebum secretion increases more than fivefold between 10 and 19 years of age. As a result, sebaceous gland ducts may become plugged with sloughed skin cells and sebum contaminated with bacteria. The inflamed plug is called a *comedo* and is the most characteristic sign of acne. Pus-filled *pimples* or *pustules* result from secondary infections within or beneath the epidermis, often in a hair follicle or sweat pore.

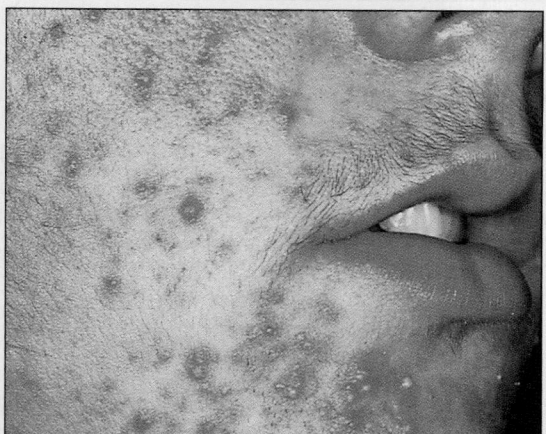

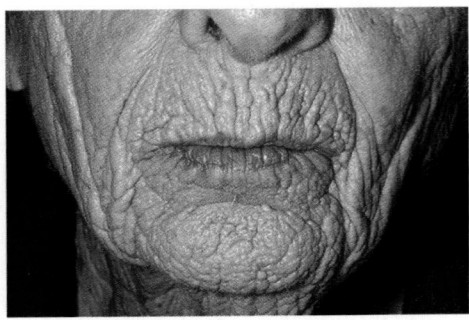

Figure 6-21 *Aged skin.* In late adulthood, wrinkles in the skin often develop—especially in areas of frequent movement such as on the hands, around the mouth, and around the eyelids.

As one continues past early adulthood, the sebaceous and sweat glands become less active. Although this can provide a welcome relief to those who suffer from acne or other problems associated with overactivity of these glands, it can affect normal function of the body. For example, the reduction of sebum production can cause the skin and hair to become less resilient and therefore more likely to wrinkle or crack. Wrinkling can also be caused by an overall degeneration in the skin's ability to maintain itself as efficiently as it did during the early years of development (Figure 6-21). Loss of function in sweat glands as adulthood advances adversely affects the body's ability to cool itself during exercise or when the external temperature is high. Thus, elderly individuals are more likely to suffer severe problems during hot weather than young adults are.

THE BIG PICTURE
The Skin and the Whole Body

The skin is one of the major components of the body's structural framework. It is continuous with the connective tissues that hold the body together, including the fascia, bones, tendons, and ligaments. The integumentary, skeletal, and muscular systems work together to protect and support the whole body.

As stated several times in this chapter, the skin is a barrier that separates the internal environment from the external environment. Put another way, the skin *defines* the internal environment of the body. The barrier formed by the skin is a formidable one indeed; the skin possesses numerous mechanisms for protecting internal structures from the sometimes harsh external environment. Without these protective mechanisms, the internal environment could not maintain a relative constancy that is independent of the external environment.

First, the dermis and epidermis work together to form a tough, waterproof envelope that protects us from drying out and from the dangers of chemical or microbial contamination. The sebaceous secretions of the skin, along with other components of the skin's surface film, enhance the skin's ability to protect the internal environment. Protection against mechanical injuries is provided by hair, calluses, and the layers of the skin itself. Pigmentation in the skin and our

ability to regulate its concentration protect us from the harmful effects of solar radiation.

Although its primary functions are support and protection, the skin has other important roles in maintaining homeostasis. For example, the skin is also an important agent in the regulation of body temperature—it serves as a sort of "radiator" that can be activated or deactivated as needed. It helps maintain a constant level of calcium in the body by producing vitamin D, which is necessary for normal absorption of calcium in the digestive tract. Ridges on the palms and fingers allow us to make and use tools for getting food, building shelters, and conducting other survival tasks. The skin's flexibility and elasticity permit the free movement required to perform such tasks. Sensory nerve receptors in the dermis allow the skin to be a "window on the world." Information about the external environment is relayed from skin receptors to nervous control (integration) centers, where it is used to coordinate the function of other organs.

As you continue your study of the various organ systems of the body, keep in mind that none of them could operate properly without the structural and functional assistance of the integumentary system.

Mechanisms of Disease

SKIN DISORDERS

Any disorder of the skin can be called a **dermatosis,** which means "skin condition." Many dermatoses involve inflammation of the skin, or **dermatitis.** Various disorders involving the skin have already been discussed in this chapter. A few more representative disorders are described here (see Boxes 6-7 and 6-8 on pp. 217-218).

Skin Infections

The skin is the first line of defense against microorganisms that might invade the body's internal environment. It is no wonder that the skin is a common site of infection. In adults, the antimicrobial characteristics of sebum in the skin's surface film often inhibit skin infections. In children, the lack of sebum in surface film makes the skin less resistant to infection.

Many different viruses, bacteria, and fungi cause skin conditions. Here are a few examples of skin infections caused by different types of pathogenic (disease-causing) organisms:

1. **Impetigo.** This highly contagious bacterial condition results from *Staphylococcus* or *Streptococcus* infection and occurs most often in young children. Impetigo starts as a reddish discoloration, or *erythema,* but soon develops into vesicles (blisters) and yellowish crusts. Occasionally, the infection becomes systemic (bodywide) and thus is life threatening.

2. **Tinea.** Tinea is the general name for many different *mycoses* (fungal infections) of the skin. Ringworm, jock itch, and athlete's foot are all classified as tinea. Signs of tinea in-

Mechanisms of Disease—cont.

BOX 6-7: HEALTH MATTERS
Skin Cancer

Each year in the United States more than 800,000 new cases of skin cancer are diagnosed. These *neoplasms,* or abnormal growths, result from cell changes in the epidermis and are the most common form of cancer in humans. They account for nearly 25% of all cancer in men and 14% of reported cases in women. Two forms of the disease, called **basal cell carcinoma** and **squamous cell carcinoma,** account for more than 95% of all reported cases of skin cancer. Both types are very responsive to treatment and seldom *metastasize* (meh-TAS-tah-size), or spread to other body areas. However, if left untreated, these cancers can cause significant damage to adjacent tissues that results in disfigurement and loss of function. A third type of skin cancer called **malignant melanoma** has a tendency to spread to other body areas and is a much more serious form of the disease than either the basal cell or squamous cell varieties. About 35,000 new cases of malignant melanoma are reported each year, and of that number more than 7000 die of the disease.

1. **Basal cell carcinoma.** This is the most common type of skin cancer. The malignancy begins in cells at the base of the epidermis (stratum basale) and most often appears on the nose and face (part *A* of the figure). Although lesions may occur at any age, the incidence increases after the age of 40 years. Basal cell tumors rarely metastasize but may cause widespread destruction of normal tissue if left untreated.

2. **Squamous cell carcinoma.** Like basal cell carcinoma, this skin cancer is slow growing and arises in the epidermis. It occurs most frequently in middle-aged and elderly individuals and is typically found on sun-exposed areas such as the scalp and forehead, backs of the hands, and top of the ears (part *B* of the figure). Some forms of squamous cell carcinoma may metastasize, but as a group, they are far less likely to spread to other body areas than the malignant lesions described later are.

3. **Malignant melanoma.** Malignant melanoma is the most deadly of all skin cancers. During the past 20 years, it has shown a steady increase in incidence at a rate of nearly 4% a year. The highest incidence is found in older individuals (median age at diagnosis of 53 years) with light skin, eyes, and hair who have poor ability to tan and have previously suffered severe sunburns.

    This type of cancer sometimes develops from a pigmented *nevus* (mole) to become a dark, spreading lesion (part *C* of the figure). Benign moles should be checked regularly for warning signs of melanoma because early detection and removal are essential in treating this rapidly spreading cancer. The "ABCD" rule of self-examination of moles is summarized here:

Asymmetry: Benign moles are *symmetrical;* their halves are mirror images of each other. Melanoma lesions are asymmetrical, or lopsided.

Border: Benign moles are outlined by a distinct border, but malignant melanoma lesions are often irregular or indistinct.
Color: Benign moles may be any shade of brown but are relatively evenly colored. Melanoma lesions tend to be unevenly colored and exhibit a mixture of shades or colors.
Diameter: By the time a melanoma lesion exhibits characteristics A, B, and C, it is also probably larger than 6 mm (¼ inch).

We now know that only about 5% of melanoma patients can effectively mount the type of immune response required to find and kill migrating melanoma cells should they escape initial treatment. Fortunately, vaccines already available in Canada and elsewhere are being tested in the United States for both prevention of recurrence of the disease and as a less toxic alternative to chemotherapy for the treatment of advanced melanoma.

Although genetic predisposition also plays a role, many pathophysiologists believe that exposure to the sun's ultraviolet radiation is the most important factor in causing the common skin cancers. UV radiation damages the DNA in skin cells, causing the mistakes in mitosis that produce cancer.

One of the rarer skin cancers, **Kaposi** (KAH-poh-see) **sarcoma** now appears in many cases of AIDS and other immune deficiencies. Kaposi sarcoma, first appearing as purple papules (part *D* of the figure), quickly spreads to the lymph nodes and internal organs. Some researchers believe that a virus or other agent, perhaps transmitted along with the HIV, is a possible cause of this cancer.

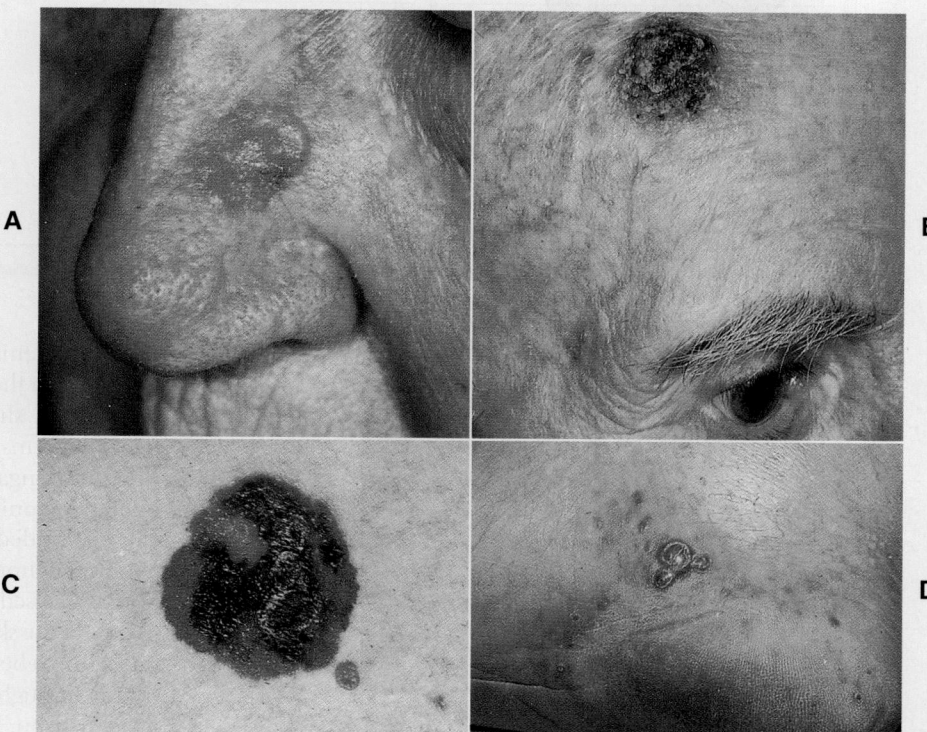

Mechanisms of Disease—cont.

BOX 6-8
Diagnostic Study

Mohs Surgery

More than 60 years ago, Dr. Frederick Mohs, working at the University of Wisconsin, described a microscopically guided method for removing certain skin cancers that have fingerlike projections extending into deeper areas of the skin, such as basal cell carcinomas. These cancerous projections from the tumor must be identified and totally excised or the disease will recur. In some areas of the body, excision by standard surgical techniques may result in incomplete removal of the cancer and considerable loss of normal tissue.

In a Mohs surgery, the tumor and a minimal amount of surrounding tissue are removed in a number of thin layers. Each cut section is divided into quadrants, frozen, stained, and microscopically examined to create a series of "maps" showing areas of remaining tumor that must be excised. Sections continue to be removed until a cancer-free plane is reached. The technique is time consuming and technically difficult, but it preserves maximum amounts of normal tissue and ensures complete removal of the tumor and its fingerlike projections. Cure rates of 95% to 99% are typical after this surgery.

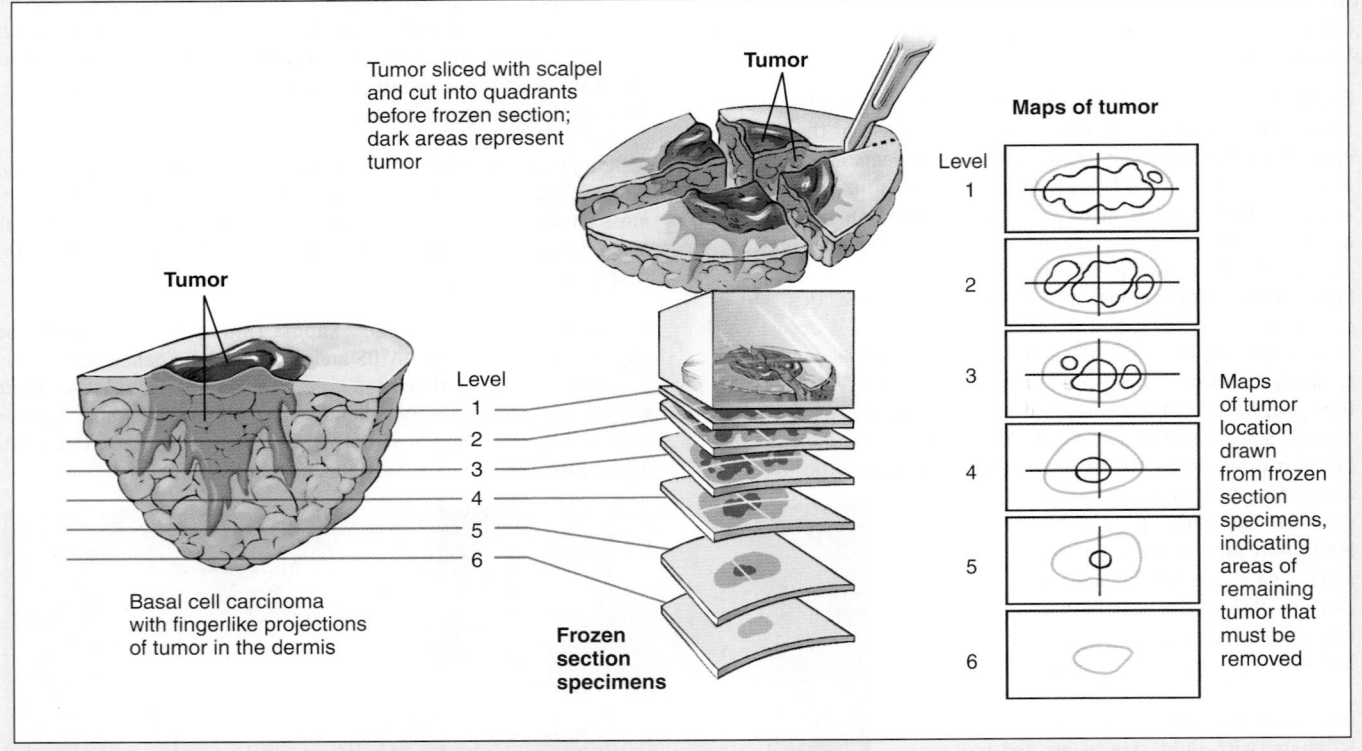

Tumor sliced with scalpel and cut into quadrants before frozen section; dark areas represent tumor

Tumor

Maps of tumor

Tumor

Level 1
2
3
4
5
6

Basal cell carcinoma with fingerlike projections of tumor in the dermis

Frozen section specimens

Maps of tumor location drawn from frozen section specimens, indicating areas of remaining tumor that must be removed

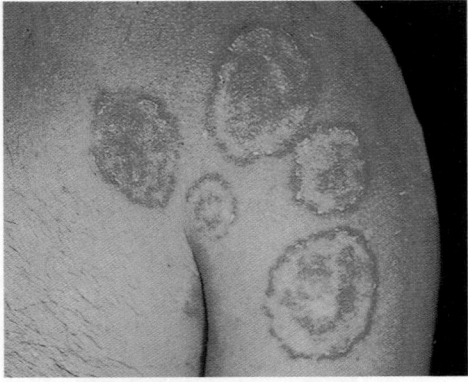

Figure 6-22 Tinea *infection (ringworm).* Note that the lesions heal in the center, with inflamed rings left on the skin's surface.

clude erythema, scaling, and crusting. Occasionally, fissures, or cracks, in the epidermis develop at creases in the epidermis. Figure 6-22 shows a case of ringworm, a tinea infection that typically forms a round rash that heals in the center to form a ring. Antifungal agents usually stop the acute infection but are unable to completely destroy the fungus. Recurrence of tinea can be avoided by keeping the skin dry because fungi require a moist environment to grow.

3. **Warts.** Caused by papillomaviruses, warts are nipplelike neoplasms of the skin. Although they are usually benign, some warts transform to become malignant. Transmission of warts generally occurs through direct contact with warts on the skin of an infected person. Warts can be removed by freezing, drying, laser therapy, or the application of chemicals.

Mechanisms of Disease—cont.

BOX 6-8

Diagnostic Study—cont'd

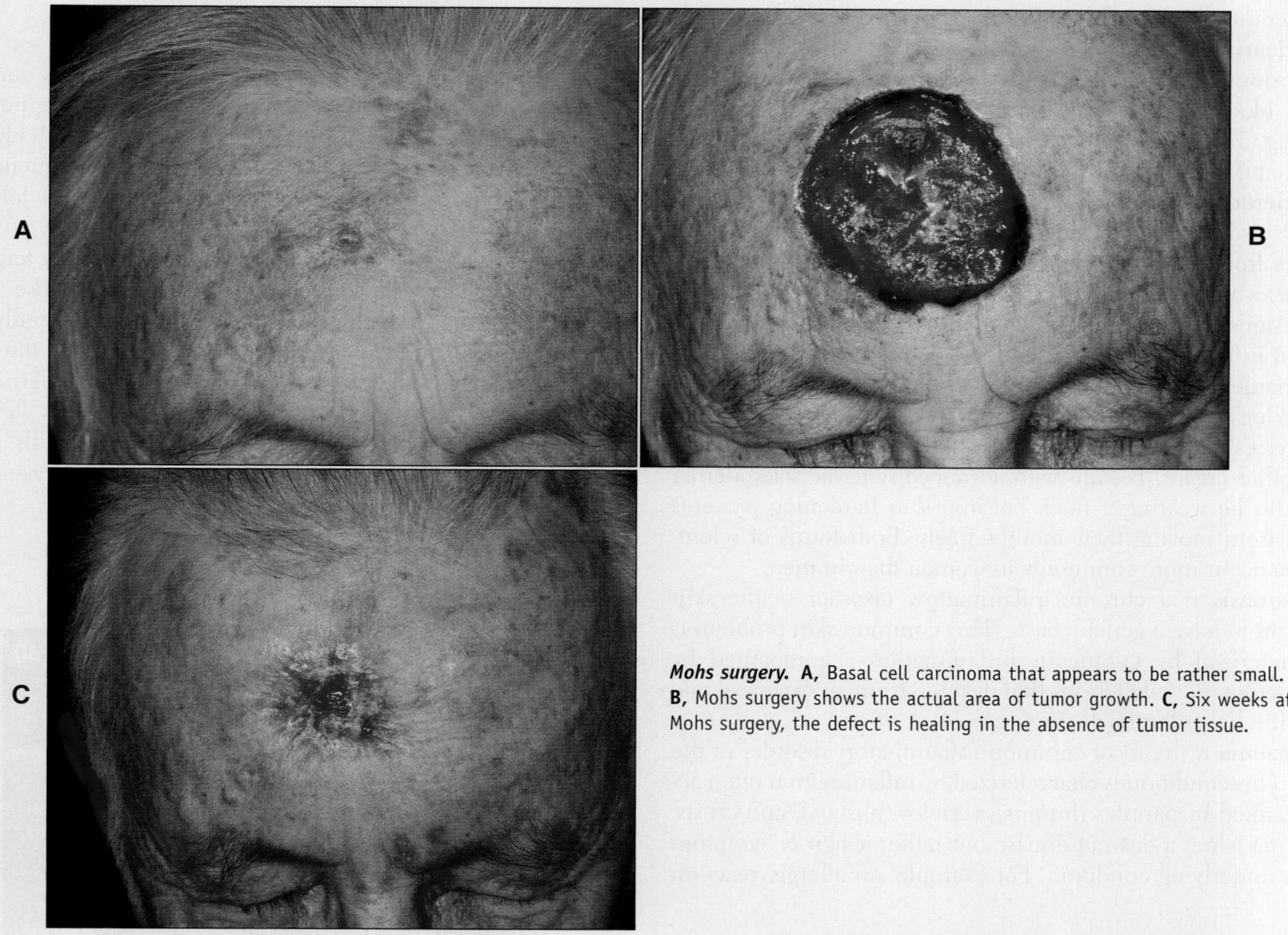

Mohs surgery. **A,** Basal cell carcinoma that appears to be rather small. **B,** Mohs surgery shows the actual area of tumor growth. **C,** Six weeks after Mohs surgery, the defect is healing in the absence of tumor tissue.

4. **Boils.** Also called *furuncles,* boils are local *Staphylococcus* infections of hair follicles characterized by large, inflamed, pus-filled lesions. A group of untreated boils may fuse into even larger lesions called *carbuncles.*

Vascular and Inflammatory Skin Disorders

Everyone who might ever be called on to provide care to a bedridden or otherwise immobilized individual should be aware of the causes and nature of pressure sores, or **decubitus ulcers** (Figure 6-23). *Decubitus* means "lying down," a name that hints at a common cause of pressure sores: lying in one position for long periods. Also called *bedsores,* these lesions appear after blood flow to a local area of skin slows because of pressure on skin covering bony prominences such as the an-

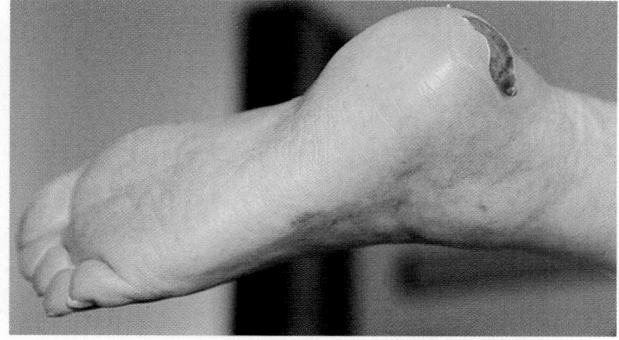

Figure 6-23 *Decubitus ulcer.* The skin forms craterlike lesions when the blood supply is diminished and skin tissue cannot be maintained in the area of reduced blood flow.

Mechanisms of Disease—cont.

kles. Ulcers form and infections develop as lack of blood flow causes tissue damage. Frequent changes in body position and soft support cushions help prevent decubitus ulcers.

A common type of skin disorder that involves blood vessels is **urticaria,** or *hives.* This condition is characterized by raised red lesions, called *wheals,* caused by leakage of fluid from the skin's blood vessels. Urticaria is often associated with severe itching. Hypersensitivity or allergic reactions, physical irritants, and systemic diseases are common causes.

Scleroderma is an autoimmune disease that affects the blood vessels and connective tissues of the skin. The name *scleroderma* comes from the word parts *sclera-,* which means "hard," and *derma,* which means "skin." Hard skin is a good description of the lesions characteristic of scleroderma. Scleroderma begins as a mild inflammation that later develops into a patch of yellowish, hardened skin. Scleroderma most commonly remains a mild, localized condition. Very rarely, localized scleroderma progresses to a systemic form that affects large areas of the skin and other organs. Persons with advanced systemic scleroderma seem to be wearing a mask because skin hardening prevents them from moving their mouths freely. Both forms of scleroderma occur more commonly in women than in men.

Psoriasis is a chronic inflammatory disorder of the skin thought to have a genetic basis. This common skin problem is characterized by cutaneous inflammation accompanied by silver-colored, scaly lesions that develop from an excessive rate of epithelial cell growth (Figure 6-24).

Eczema is the most common inflammatory disorder of the skin. This condition is characterized by inflammation often accompanied by papules (bumps), vesicles (blisters), and crusts. Eczema is not a distinct disease but rather a sign or symptom of an underlying condition. For example, an allergic reaction

called *contact dermatitis* can progress to become eczematous. Poison ivy is a form of contact dermatitis—occurring on *contact* with chemicals produced by the poison ivy plant.

Abnormal Body Temperature

Maintenance of body temperature within a narrow range is necessary for normal functioning of the body. Figure 6-25 shows that straying too far out of the normal range of body temperatures can have very serious physiological consequences. A few important conditions related to body temperature follow:

1. **Fever,** or a *febrile* state, is an unusually high body temperature associated with a systemic inflammatory response. In the case of infections, chemicals called **pyrogens** (literally "fire makers") cause the thermostatic control centers of the hypothalamus to produce a fever. Because the body's "thermostat" is reset to a higher setting, a person feels a need to warm up to this new temperature and often experiences "chills" as the febrile state begins. The high body temperature associated

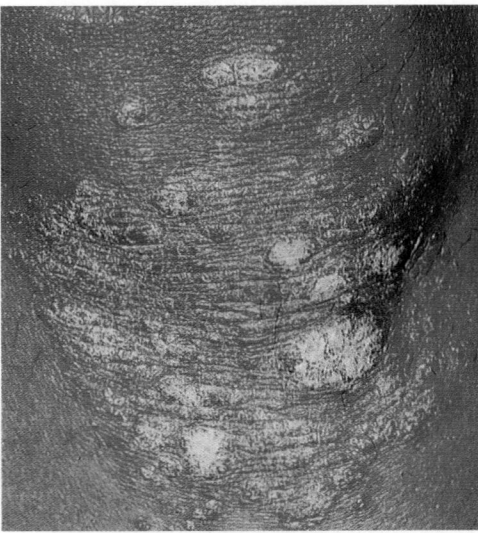

Figure 6-24 *Psoriasis.* Inflammation and scaling are characteristic of this skin disorder.

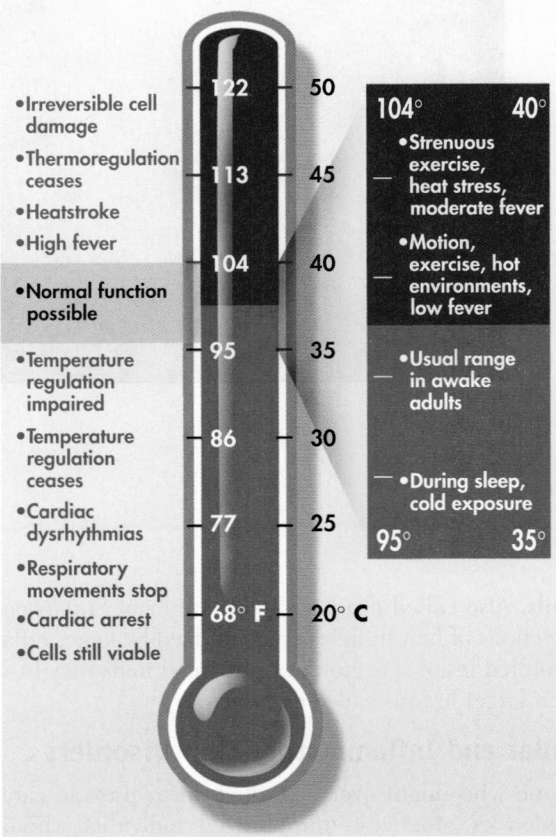

Figure 6-25 *Body temperature.* This diagram, modeled after a thermometer, shows some physiological consequences of abnormal body temperature. The inset shows a range of body temperatures at which normal function is possible. Ideally, body temperature should be near 37° C (98.6° F), but the body can operate normally within the range shown if conditions are not ideal.

Mechanisms of Disease—cont.

with infectious fever is thought to enhance the body's immune responses to eliminate the pathogen. Strategies aimed at reducing the temperature of a febrile person are normally counteracted by the body's heat-generating mechanisms and have the effect of further weakening the infected person. Under ordinary circumstances, it is best to let the fever "break" on its own after the pathogen is destroyed.

2. **Malignant hyperthermia (MH),** an inherited condition, is characterized by abnormally increased body temperature (hyperthermia) and muscle rigidity when exposed to certain anesthetics or muscle relaxants (succinylcholine, for example). The drug *dantrolene (Dantrium)*, which inhibits heat-producing muscle contractions, has been used to prevent or relieve the effects of this condition.

3. **Heat exhaustion** occurs when the body loses a large amount of fluid from heat loss mechanisms. This usually happens when environmental temperatures are high. Although normal body temperature is maintained, the loss of water and electrolytes can cause weakness, dizziness, nausea, and possibly loss of consciousness. Heat exhaustion may also be accompanied by skeletal muscle cramps often called *heat cramps*. Heat exhaustion is treated with rest (in a cool environment) accompanied by fluid replacement.

4. **Heat stroke,** or *sunstroke*, is a severe, sometimes fatal condition resulting from the inability of the body to maintain normal temperature in an extremely warm environment. Such thermoregulatory failure may result from factors such as old age, disease, drugs that impair thermoregulation, or simply overwhelming elevated environmental temperatures. Heat stroke is characterized by body temperatures of 41° C (105° F) or higher, tachycardia (rapid heart rate), headache, and hot, dry skin. Confusion, convulsions, or loss of consciousness may occur. Unless the body is cooled and body fluids replaced immediately, death may result.

5. **Hypothermia** is the inability to maintain normal body temperature in extremely cold environments. Hypothermia is characterized by body temperatures lower than 35° C (95° F), shallow and slow respirations, and a faint, slow pulse. Hypothermia is usually treated by slowly warming the affected person's body.

6. **Frostbite** is local damage to tissues caused by extremely low temperature. Damage to tissues results from the formation of ice crystals accompanied by a reduction in local blood flow. **Necrosis** (tissue death) and even **gangrene** (decay of dead tissue) can result from frostbite.

Burns

Typically we think of a burn as a thermal injury or lesion caused by contact of the skin with some hot object or fire. In addition, overexposure to ultraviolet light (sunburn) (Box 6-9) or contact with an electric current or corrosive chemicals causes injury or death to skin cells. The injuries that result can all be classified as *burns*.

Estimating Body Surface Area

When burns involve large areas of the skin, treatment and the prognosis for recovery depend in large part on the total area involved and the severity of the burn. The severity of a burn is determined by the depth and extent (percentage of body surface burned) of the lesion. There are several ways to estimate the extent of body surface area burned. One method is called the **"rule of palms"** and is based on the assumption that the palm size of a burn victim is about 1% of the total body surface area. Therefore, estimating the number of "palms" that are burned approximates the percentage of body surface area involved.

The **"rule of nines"** (Figure 6-26) is another and more accurate method of determining the extent of a burn injury. In this technique the body is divided into 11 areas of 9%, with the area around the genitals, called the *perineum*, representing the additional 1% of body surface area. As Figure 6-26 shows, 9% of the skin covers the head and each upper extremity, including the front and back surfaces. Twice as much, or 18%, of the total skin area covers the front and back of the trunk and each lower extremity, including the front and back surfaces. The rule of nines works well with adults but does not reflect the differences in body surface area in small children. Special tables called *Lund-Browder charts*, which take the large surface area of certain body areas (such as the head) in a growing child into account, are used by physicians to estimate burn percentages in children.

BOX 6-9: HEALTH MATTERS
Sunburn and Skin

Burns caused by exposure to harmful UV radiation in sunlight are commonly called *sunburns*. As with any burn, serious sunburns can cause tissue damage and lead to secondary infections and fluid loss. Cancer researchers have recently theorized that blistering (second-degree) sunburns during childhood may trigger the development of malignant melanoma later in life. Some epidemiological studies show that adults who had more than two blistering sunburns before the age of 20 years have a greater risk for melanoma than someone who experienced no such burns. If this theory is true, it could explain the dramatic increase in skin cancer rates in the United States in recent years. Those who grew up as sunbathing and "suntans" became popular in the 1950s and 1960s are now, as adults, exhibiting melanoma at a much higher rate than in the previous generation.

Mechanisms of Disease—cont.

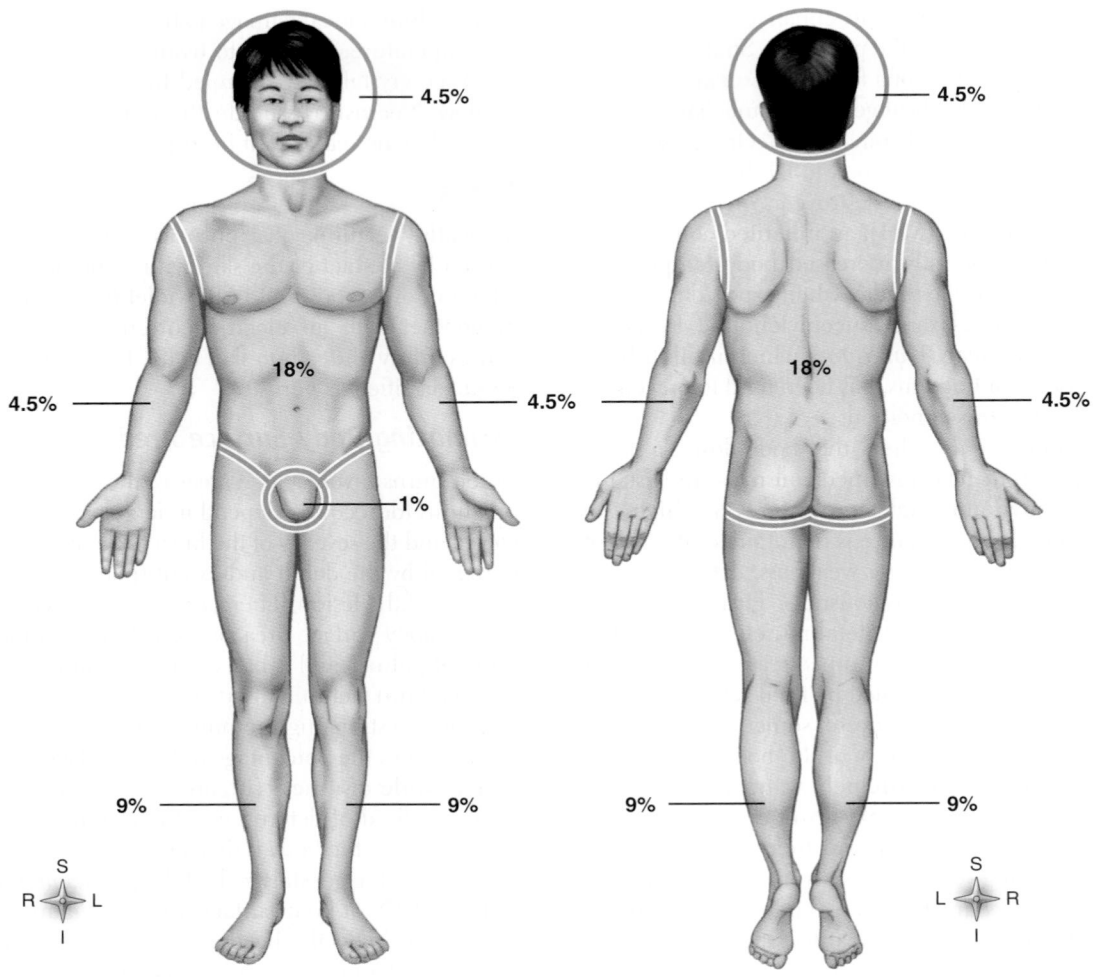

Figure 6-26 *"Rule of nines."* The "rule of nines" is one method used to estimate amount of skin surface burned in an adult.

The depth of a burn injury depends on the tissue layers of the skin that are involved (Figure 6-27). A **first-degree burn** (typical sunburn) causes minor discomfort and some reddening of the skin. Although the surface layers of the burned area may peel in 1 or 2 days, no blistering occurs and the actual tissue destruction is minimal. First- and second-degree burns are called **partial-thickness burns.**

Second-degree burns involve the deep epidermal layers and always cause injury to the upper layers of the dermis. In deep second-degree burns, damage to sweat glands, hair follicles, and sebaceous glands may occur, but tissue death is not complete. Blisters, severe pain, generalized swelling, and edema characterize this type of burn. Scarring is common.

Third-degree, or **full-thickness, burns** are characterized by destruction of both the epidermis and dermis. Tissue death extends below the hair follicles and sweat glands. If burning involves underlying muscles, fasciae, or bone, it may be called a **fourth-degree burn.** A distinction between second- and third- or fourth-degree burning is the fact that a third- or fourth-degree lesion is insensitive to pain immediately after injury because of the destruction of nerve endings. Scarring is a serious problem.

Mechanisms of Disease —cont.

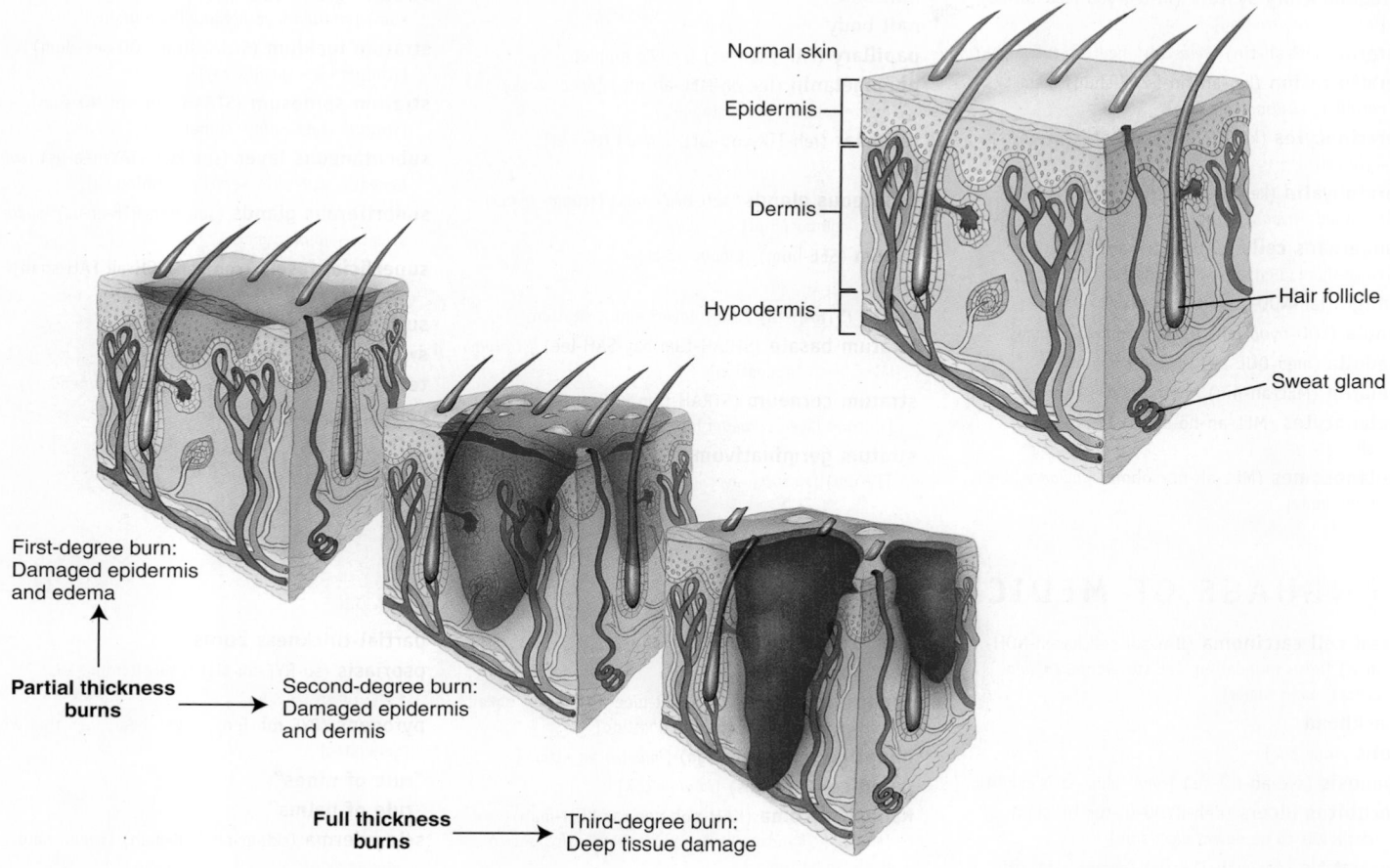

Normal skin

Epidermis

Dermis

Hypodermis

Hair follicle

Sweat gland

First-degree burn:
Damaged epidermis
and edema

**Partial thickness
burns**

Second-degree burn:
Damaged epidermis
and dermis

**Full thickness
burns**

Third-degree burn:
Deep tissue damage

Figure 6-27 *Classification of burns.* Partial-thickness burns include first- and second-degree burns. Full-thickness burns include third-degree burns.

LANGUAGE OF SCIENCE (Cont'd from page 195)

integumentary system (in-teg-yoo-MEN-tar-ee) [*integument* covering]

keratin (KER-ah-tin) [*kera-* cornified, *-in* substance]

keratinization (ker-ah-tin-i-ZAY-shun) [*kera-* cornified, *-ization* to cause]

keratinocytes (keh-RAT-i-no-sytes) [*kera-* cornified, *-cyte* cell]

keratohyalin (ker-ah-toh-HYE-ah-lin) [*kera-* cornified, *-hyaline* glass]

Langerhans cells (LAHNG-er-hanz) [Paul Langerhans, German pathologist]

lanugo (lah-NOO-go) [*lanugo* down]

lunula (LOO-nyoo-lah) [*luna-* moon, *-ula* small]

medulla (meh-DUL-ah) [*medulla* marrow]

melanin (MEL-ah-nin) [*melan-* black, *-in* substance]

melanocytes (MEL-ah-no-sytes) [*melano-* black, *-cyte* cell]

melanosomes (MEL-ah-no-sohms) [*melano-* black, *-soma* body]

nail bed

nail body

papillary (PAP-i-lair-ee) [*papilla* nipple]

pheomelanin (fee-oh-MEL-ah-nin) [*pheo-* dusky, *-melan-* black, *-in* substance]

reticular (reh-TIK-yoo-lar) [*reticu* little net]

root

sebaceous glands (seh-BAY-shus) [*sebum-* sweat, *-eous* composed of]

sebum (SEE-bum) [*sebum* sweat]

shaft

strata (STRAH-ta) [*strata* layer] sing., stratum

stratum basale (STRAH-tum bay-SAH-lee) [*stratum* layer, *basis* foundation]

stratum corneum (STRAH-tum KOR-nee-um) [*stratum* layer, *corneum* horny]

stratum germinativum (STRAH-tum jer-mi-nah-TIV-um) [*stratum* layer, *germ* sprout]

stratum granulosum (STRAH-tum gran-yoo-LOH-sum) [*stratum* layer, *granul* little grain]

stratum lucidum (STRAH-tum LOO-see-dum) [*stratum* layer, *lucid* clear]

stratum spinosum (STRAH-tum spi-NO-sum) [*stratum* layer, *spino-* spine]

subcutaneous layer (sub-kyoo-TAY-nee-us) [*sub-* beneath, *-cut-* skin, *-eous* pertaining to]

sudoriferous glands (soo-doh-RIF-er-us) [*sudo-* sweat, *-ferous* to make]

superficial fascia (soo-per-FISH-all FAH-shah) [*superficial* surface, *fascia* band]

surface film

sweat glands

terminal hair (TUR-mih-nal) [*termin* boundary]

vellus (VEL-us) [*vellus* wool]

LANGUAGE OF MEDICINE

basal cell carcinoma (BAY-sal cell kar-si-NOH-mah) [*basis* foundation, *cell* storeroom, *carcino-* cancer, *-oma* tumor]

blackhead

boils [from *bile*]

cyanosis (sye-ah-NO-sis) [*cyan-* blue, *-osis* condition]

decubitus ulcers (deh-KYOO-bi-tus UL-sers) [*decubitus* to lie down, *ulcer* sore]

dermatitis (der-mah-TYE-tis) [*derma-* skin, *-itis* inflammation]

dermatosis (der-mah-TOH-sis) [*derma-* skin, *-osis* condition]

eczema (EK-zeh-mah) [*eczema* to boil over]

fever (FEE-ver)

first-degree burn

fourth-degree burn

frostbite (FROST-byte)

full-thickness burns

gangrene (GANG-green) [*gangrene* a gnawing sore]

heat exhaustion (ek-ZAWS-chun)

heat stroke

hypothermia (hye-poh-THER-mee-ah) [*hypo-* under, *-therm-* heat, *-ia* abnormal condition]

impetigo (im-peh-TYE-go) [*impetus* an attack]

jaundice (JAWN-dis) [*jaun* yellow]

Kaposi sarcoma (KAH-poh-see sar-KOH-mah) [Moritz K. Kaposi, Hungarian dermatologist, *sarco-* flesh, *-oma* tumor]

male pattern baldness

malignant hyperthermia (MH) (mah-LIG-nant hye-per-THERM-ee-ah) [*malign* bad, *hyper-* excessive, *-therm-* heat, *-ia* abnormal condition]

malignant melanoma (mah-LIG-nant mel-ah-NO-mah) [*malign* bad, *melan-* black, *-oma* tumor]

necrosis (neh-KROH-sis) [*necro-* death, *-osis* condition]

onycholysis (ahn-ik-oh-LYE-sis) [*onycho-* nail, *-lysis* loosen]

partial-thickness burns

psoriasis (so-RYE-ah-sis) [*psor-* itching, *-sis* condition]

pyrogens (PYE-roh-jens) [*pyro-* heat, *-gen* that which generates]

"rule of nines"

"rule of palms"

scleroderma (skleer-oh-DER-mah) [*sclero-* hard, *-derma* skin]

second-degree burns

squamous cell carcinoma (SKWAY-muss sell kar-si-NO-mah) [*squama* scale, *cell* storeroom, *carcin-* cancer, *-oma* tumor]

third-degree burns

tinea (TIN-ee-ah) [*tinea* worm]

urticaria (er-ti-KAIR-ee-ah) [*urtica* nettle]

vitiligo (vit-i-LYE-go) [*vitilgo* blemish]

warts

CASE STUDY

Jane Winston is a 45-year-old woman who has come to see her nurse practitioner because she is concerned about a mole (nevus) on her arm. Jane has always been healthy and active in outdoor activities. In her teens she worked many summers as a lifeguard. Jane never protected her skin from the sun because she considered being tan "attractive" and has always tried to get as deeply tanned as possible. She remembers having blistering sunburns at least twice before the age of 17 years. Jane tells you that the mole on her arm has been present since high school but recently it "looked different." She says it is "not painful but does itch." The nurse practitioner refers Jane to a surgeon who excises the mole and several surrounding lymph nodes. On pathological analysis by the pathologist, the diagnosis is confirmed as malignant melanoma without metastasis.

1. Which of the following is thought to be the strongest causative factor in the development of melanoma?
   A. Heredity
   B. Solar radiation
   C. Presence of moles
   D. Location of moles

2. Which of the following descriptions would you expect to be most likely that of a melanoma?
   A. The mole is hard, painless, and 3 mm in diameter.
   B. The mole is symmetrical with a raised center and crusted crater.
   C. The mole is asymmetrical and has an irregular border with a 6-mm diameter.
   D. The mole is dark brown, has distinct borders, and is painful on palpation.

3. Which one of the following best explains the pathological mechanism in malignant melanoma?
   A. DNA damage leads to mistakes in mitosis that produce cancer.
   B. Loss of melanin from the cells makes the skin more susceptible to cancer.
   C. Hormonal changes of puberty trigger malignant activity in cells.
   D. The skin is unable to repair damage, and this predisposes it to cancer.

CHAPTER SUMMARY

INTRODUCTION

A. Skin (integument) is body's largest organ
B. Approximately 1.6 to 1.9 m² in average-sized adult
C. Integumentary system describes the skin and its appendages—the hair, nails, and skin glands

STRUCTURE OF THE SKIN

A. Skin classified as a cutaneous membrane
B. Two primary layers—epidermis and dermis; joined by dermal-epidermal junction (Figures 6-1 and 6-2)
C. Hypodermis lies beneath dermis
D. Thin and thick skin (Figure 6-3)
   1. "Thin skin"—covers most of body surface (1 to 3 mm thick)
   2. "Thick skin"—soles and palms (4 to 5 mm thick)
E. Epidermis
   1. Cell types
      a. Keratinocytes—constitute over 90% of cells present; principal structural element of the outer skin
      b. Melanocytes—pigment-producing cells (5% of the total); contribute to skin color; filter ultraviolet light
      c. Langerhans cells—dendritic (branched) antigen-presenting cells (APCs), they play a role in immune response

2. Cell layers
   a. Stratum germinativum (growth layer)—describes the stratum spinosum and stratum basale together
      (1) Stratum basale (base layer)—single layer of columnar cells; only these cells undergo mitosis and then migrate through the other layers until they are shed
      (2) Stratum spinosum (spiny layer)—cells arranged in 8 to 10 layers with desmosomes that pull cells into spiny shapes; cells rich in RNA
   b. Stratum granulosum (granular layer)—cells arranged in two to four layers and filled with keratohyalin granules; contain high levels of lysosomal enzymes
   c. Stratum lucidum (clear layer)—cells filled with keratin precursor called eleidin; absent in thin skin
   d. Stratum corneum (horny layer)—most superficial layer; dead cells filled with keratin (barrier area)
3. Epidermal growth and repair
   a. Turnover or regeneration time refers to time required for epidermal cells to form in the stratum basale and migrate to the skin surface—about 35 days
   b. Shortened turnover time will increase the thickness of the stratum corneum and result in callus formation
   c. Normally 10% to 12% of all cells in stratum basale enter mitosis daily
   d. Each group of 8 to 10 basal cells in mitosis with their vertical columns of migrating keratinocytes is called an epidermal proliferating unit, or EPU

F. Dermal-epidermal junction
1. A definite basement membrane, specialized fibrous elements, and a polysaccharide gel serve to "glue" the epidermis to the dermis below
2. The junction serves as a partial barrier to the passage of some cells and large molecules
G. Dermis
1. Sometimes called "true skin"—much thicker than the epidermis and lies beneath it
2. Gives strength to the skin
3. Serves as a reservoir storage area for water and electrolytes
4. Contains various structures
   a. Arrector pili muscles and hair follicles (Figure 6-4)
   b. Sensory receptors (Figure 6-5)
   c. Sweat and sebaceous glands
   d. Blood vessels
5. Rich vascular supply plays a critical role in temperature regulation
6. Layers of dermis
   a. Papillary layer—composed of dermal papillae that project into the epidermis; contains fine collagenous and elastic fibers; contains the dermal-epidermal junction; forms a unique pattern that gives individual fingerprints
   b. Reticular layer—contains dense, interlacing white collagenous fibers and elastic fibers to make the skin tough yet stretchable; when processed from animal skin, produces leather
7. Dermal growth and repair
   a. The dermis does not continually shed and regenerate itself as does the epidermis
   b. During wound healing, fibroblasts begin forming an unusually dense mass of new connective fibers; if not replaced by normal tissue, this mass remains a scar
   c. Cleavage lines (Figure 6-6)—patterns formed by the collagenous fibers of the reticular layer of the dermis; also called Langer's lines
H. Hypodermis
1. Also called the subcutaneous layer or superficial fascia
2. Located deep to the dermis; forms connection between skin and other structures
3. Not part of the skin

SKIN COLOR
A. Melanin
1. Basic determinant is quantity, type, distribution of melanin
2. Types of melanin
   a. Eumelanin—group of dark brown (almost black) melanins
   b. Pheomelanin—group of reddish and orange melanins
3. Melanin formed from tyrosine by melanocytes (Figure 6-7)
   a. Melanocytes release melanin in packets called melanosomes
   b. Melanosomes are ingested by surrounding keratinocytes and form a cap over the nucleus

4. Albinism—congenital absence of melanin
5. Process regulated by tyrosinase, exposure to sunlight (UV radiation), and certain hormones, including ACTH (Figure 6-8)
B. Other pigments
1. Beta-carotene (group of yellowish pigments from food) can also contribute to skin color
2. Hemoglobin—color changes also occur as a result of changes in blood flow
   a. Redder skin color when blood flow to skin increases
   b. Cyanosis—bluish color caused by darkening of hemoglobin when it loses oxygen and gains carbon dioxide (Figure 6-9)
   c. Bruising can cause a rainbow of different colors to appear in the skin (Figure 6-10)
3. Other pigments—from cosmetics, tattoos, and bile pigments in jaundice (Box 6-4)

FUNCTIONS OF THE SKIN (TABLE 6-2)
A. Protection
1. Physical barrier to microorganisms
2. Barrier to chemical hazards
3. Reduces potential for mechanical trauma
4. Prevents dehydration
5. Protects (via melanin) excess UV exposure
6. Surface film
   a. Emulsified protective barrier formed by mixing of residue and secretions of sweat and sebaceous glands with sloughed epithelial cells from skin surface; shedding of epithelial elements is called desquamation
   b. Functions
      (1) Antibacterial, antifungal activity
      (2) Lubrication
      (3) Hydration of skin surface
      (4) Buffer of caustic irritants
      (5) Blockade of toxic agents
   c. Chemical composition
      (1) From epithelial elements—amino acids, sterols, and complex phospholipids
      (2) From sebum—fatty acids, triglycerides, and waxes
      (3) From sweat—water, ammonia, urea, and lactic and uric acid
B. Sensation
1. Skin acts as a sophisticated sense organ
2. Somatic sensory receptors detect stimuli that permit us to detect pressure, touch, temperature, pain, and other general senses
C. Flexibility
1. Skin is supple and elastic, thus permitting change in body contours without injury
D. Excretion
1. Water
2. Urea/ammonia/uric acid

E. Hormone (vitamin D) production (Figure 6-11)
  1. Exposure of skin to UV light converts 7-dehydrocholesterol to cholecalciferol—a precursor to vitamin D
  2. Blood transports precursor to liver and kidneys where vitamin D is produced
  3. Process and end result fulfill the necessary steps required for vitamin D to be classified as a hormone
F. Immunity
  1. Phagocytic cells destroy bacteria
  2. Langerhans cells trigger helpful immune reaction working with "helper T cells"
G. Homeostasis of body temperature
  1. To maintain homeostasis of body temperature, heat production must equal heat loss; skin plays a critical role in this process
  2. Heat production
    a. By metabolism of foods in skeletal muscles and liver
    b. Chief determinant of heat production is the amount of muscular work being performed
  3. Heat loss—approximately 80% of heat loss occurs through the skin; remaining 20% occurs through the mucosa of the respiratory, digestive, and urinary tracts (Figure 6-12)
    a. Evaporation—to evaporate any fluid, heat energy must be expended; this method of heat loss is especially important at high environmental temperatures when it is the only method by which heat can be lost from the skin
    b. Radiation—transfer of heat from one object to another without actual contact; important method of heat loss in cool environmental temperatures
    c. Conduction—transfer of heat to any substance actually in contact with the body; accounts for relatively small amounts of heat loss
    d. Convection—transfer of heat away from a surface by movement of air; usually accounts for a small amount of heat loss
  4. Homeostatic regulation of heat loss (Figure 6-13)
    a. Heat loss by the skin is controlled by a negative-feedback loop
    b. Receptors in the hypothalamus monitor the body's internal temperature
    c. If body temperature is increased, the hypothalamus sends a nervous signal to the sweat glands and blood vessels of the skin
    d. The hypothalamus continues to act until the body's temperature returns to normal

APPENDAGES OF THE SKIN

A. Hair (Figure 6-14)
  1. Development of hair
    a. Distribution—over entire body except palms of hands and soles of feet and a few other small areas
    b. Fine and soft hair coat existing before birth called lanugo
    c. Coarse pubic and axillary hair that develops at puberty called terminal hair
    d. Hair follicles and hair develop from epidermis; stratum germinativum forms innermost layer of follicle and germinal matrix; mitosis of cells of germinal matrix forms hairs
    e. Papilla—cluster of capillaries under germinal matrix
    f. Root—part of hair embedded in follicle in dermis
    g. Shaft—visible part of hair
    h. Medulla—inner core of hair; cortex—outer portion
  2. Appearance of hair
    a. Color—result of different amounts, distribution, types of melanin in cortex of hair (Figure 6-15)
    b. Growth—hair growth and rest periods alternate; hair on head averages 5 inches of growth per year
    c. Sebaceous glands—attach to and secrete sebum (skin oil) into follicle
    d. Male pattern baldness results from combination of genetic tendency and male sex hormones (Figure 6-16)
B. Nails (Figure 6-17)
  1. Consist of epidermal cells converted to hard keratin
  2. Nail body—visible part of each nail
  3. Root—part of nail in groove hidden by fold of skin, the cuticle
  4. Lunula—moon-shaped white area nearest root
  5. Nail bed—layer of epithelium under nail body; contains abundant blood vessels
    a. Appears pink under translucent nails
    b. Nails may have pigmented streaks (Figure 6-18)
    c. Separation of a nail from the nail bed is called *onycholysis* (Figure 6-19)
  6. Growth—nails grow by mitosis of cells in stratum germinativum beneath the lunula; average growth about 0.5 mm per week, or slightly over 1 inch per year
C. Skin glands (Figure 6-20)
  1. Two types of sweat glands
    a. Eccrine glands
      (1) Most numerous sweat glands; quite small
      (2) Distributed over total body surface with exception of a few small areas
      (3) Simple, coiled, tubular glands
      (4) Function throughout life
      (5) Secrete perspiration or sweat; eliminate wastes and help maintain a constant core temperature
    b. Apocrine glands
      (1) Located deep in subcutaneous layer
      (2) Limited distribution—axilla, areola of breast, and around anus
      (3) Large (often more than 5 mm in diameter)
      (4) Simple, branched, tubular glands
      (5) Begin to function at puberty
      (6) Secretion shows cyclic changes in female with menstrual cycle
  2. Sebaceous glands
    a. Secrete sebum—oily substance that keeps hair and skin soft and pliant; prevents excessive water loss from skin
    b. Lipid components have antifungal activity
    c. Simple, branched glands
    d. Found in dermis except in palms and soles
    e. Secretion increases in adolescence; may lead to formation of pimples and blackheads

3. Ceruminous glands
   a. Modified apocrine sweat glands
   b. Simple, coiled, tubular glands
   c. Empty contents into external ear canal alone or with sebaceous glands
   d. Mixed secretions of sebaceous and ceruminous glands called cerumen (wax)
   e. Function of cerumen to protect area from dehydration; excess secretion can cause blockage of ear canal and loss of hearing

CYCLE OF LIFE: SKIN

A. Children
   1. Skin is smooth, unwrinkled, and characterized by elasticity and flexibility
   2. Few sweat glands
   3. Rapid healing
B. Adults
   1. Development and activation of sebaceous and sweat glands
   2. Increased sweat production
      a. Body odor
   3. Increased sebum production
      a. Acne
C. Old age
   1. Decreased sebaceous and sweat gland activity
      a. Wrinkling (Figure 6-21)
      b. Decrease in body's ability to cool itself

THE BIG PICTURE: SKIN AND THE WHOLE BODY

A. Skin is a major component of the body's structural framework
B. Skin defines the internal environment of the body
C. Primary functions are support and protection

REVIEW QUESTIONS

1. List and briefly discuss several of the different functions of the skin.
2. List three cell types found in the epidermis.
3. List and describe the cell layers of the epidermis from superficial to deep.
4. What layer of the epidermis is sometimes called the barrier area?
5. Discuss the process of epidermal growth and repair.
6. What is keratin? Where is it found and how is it formed?

7. What part of the skin contains blood vessels?
8. Why is the process of blister formation a good example of the relationship between the skin's structure and function?
9. List the two layers of the dermis. Which layer helps make the skin stretchable and able to rebound?
10. What are arrector pili muscles?
11. List the appendages of the skin.
12. Identify each of the following: hair papilla, germinal matrix, hair root, hair shaft, follicle.
13. List the three primary types of skin glands.
14. What is the difference between eccrine and apocrine sweat glands?
15. Discuss the importance of the surface film of the skin.
16. How is the "rule of nines" used in determining the extent of a burn injury?
17. What are the differences between first-, second-, and third-degree burns?

CRITICAL THINKING QUESTIONS

1. The stratum corneum, hair follicle, nails, stratum germinativum, sweat glands, oil glands, and stratum basale are all discussed in this chapter. Can you make a distinction regarding whether these structures are part of the integumentary system only or both the integumentary system and the integument?
2. Identify what affects the thickness of the skin (dermis and epidermis). What is the relationship between what affects the thickness of the skin and the thickness of the hypodermis?
3. What is the dermal-epidermal junction? By analyzing its anatomical components, what inference can you make regarding how it functions?
4. In terms of leaving a less noticeable scar, why should an incision on the front of the thigh be made at a different angle than an incision on the back of the thigh?
5. Concern about skin cancer is reducing the amount of time people spend in the sun. If this caution is carried to the extreme, how would you explain the impact on skin function?
6. How would you explain why a light-skinned individual would be more susceptible to malignant melanoma?
7. An individual running a marathon expends a great deal of energy. Much of this energy generates heat, which increases core body temperature. What is the role of sweat glands in balancing body temperature during this strenuous exercise? With this loss of fluid, what suggestions do you have to avoid dehydration?

CHAPTER 7

Skeletal Tissues

CHAPTER OUTLINE

LANGUAGE OF SCIENCE

appositional growth (ap-oh-ZISH-un-al) [*apposit* to put near]

articular cartilage (ar-TIK-yoo-lar KAR-ti-lij) [*articular* to divide into joints]

bone

bone matrix (MAY-triks) [*matrix* womb]

cancellous bone (KAN-seh-lus) [*cancellous* lattice]

cartilage (KAR-ti-lij)

centers of chondrification (kon-dri-fi-KAY-shun)

chondrocytes (KON-droh-sytes) [*chondro-* cartilage, *-cyte* cell]

compact bone

diaphysis (dye-AF-i-sis) [*dia-* through or apart, *-physis* growth]

elastic cartilage (eh-LAS-tik KAR-ti-lij) [*elastic* to drive]

endochondral ossification (en-doh-KON-dral os-i-fi-KAY-shun) [*endo-* within, *-chondro* cartilage, *os-* bone, *-fication* to make]

endosteum (en-DOS-tee-um) [*end-* within, *-osteum* bone]

epiphyseal plate (ep-i-FEEZ-ee-al) [*epi-* on, *-physis* growth, *plate* flat]

epiphyses (eh-PIF-i-sees) [*epi-* on, *-physis* growth] sing., epiphysis

fibrocartilage (fye-broh-KAR-ti-lij) [*fibro* fiber]

flat bones

Haversian canal (hah-VER-shun) [*Clopton Havers*, English physician]

Haversian systems (hah-VER-shun) [*Clopton Havers*, English physician]

hyaline cartilage (HYE-ah-lin KAR-ti-lij) [*hyaline* glass]

interstitial (in-ter-STISH-al) [*inter-* occurring between, *-stitial* to stand]

intramembranous ossification (in-trah-MEM-brah-nus os-i-fi-KAY-shn) [*intra-* formed, *-membrane* thin skin, *os-* bone, *-fication* to make]

irregular bones

long bones

Cont'd on p. 249

This chapter focuses on two highly specialized types of connective tissues—**bone** and **cartilage.** In addition, the functional characteristics of cartilage and a comparison of cartilage and bone are presented later in the chapter.

Other types of tissue in the skeletal system include fibrous and loose connective tissue, blood, nervous tissue, epithelium, lymphatic tissue, myeloid tissue (bone marrow), and adipose, or fat, tissue.

Individual bones, which are considered separate, discrete organs, are discussed in Chapter 8. *Articulations,* or joints, are points of contact between bones that make movement possible and are considered in Chapter 9.

TYPES OF BONES

Structurally, we can name four types of bones. Their names suggest their shapes: long bones, short bones, flat bones, and irregular bones. Figure 7-1 gives an illustration of each type. Bones serve differing needs, and their size, shape, and appearance will vary to meet those needs. Some bones must bear great weight; others serve a protective function or serve as delicate support structures, such as for the fingers and toes. Bones differ in size and shape and also in the amount and proportion of the two different types of bone tissue that compose them. **Compact bone** is dense and "solid" in appearance; **cancellous,** or **spongy, bone** is characterized by open space partially filled by an assemblage of

needlelike structures. Both types are discussed when the microscopic structure of bone is described later in the chapter.

All four types of bone discussed below have varying amounts of cancellous and compact bone in their structure.

Long bones are easily identified by their extended longitudinal axes and by their expanded and often uniquely shaped articular ends. The femur of the thigh and humerus of the arm are examples.

Short bones are often described as cube- or box-shaped structures that are about as broad as they are long. Examples include the wrist (carpals) and ankle (tarsal) bones.

Flat bones are generally broad and thin with a flattened and often curved surface. Certain bones of the skull, the shoulder blades (scapulae), ribs, and breastbone (sternum) are typical flat bones. Red marrow fills the spaces in the cancellous bone inside a few flat bones—the sternum, which contains red marrow even in adulthood, is one example. To help in the diagnosis of leukemia and certain other diseases, a physician may decide to perform a needle puncture of one of these bones. In this type of diagnostic procedure a needle is inserted through the skin and compact bone into the red marrow, and a small amount of the marrow is then aspirated and examined under the microscope for normal or abnormal blood cells. The procedure, called aspiration, biopsy, and cytology (ABC), is discussed on p. 667.

Irregular bones are often clustered in groups and come in various sizes and shapes. The vertebral bones that form the spine and

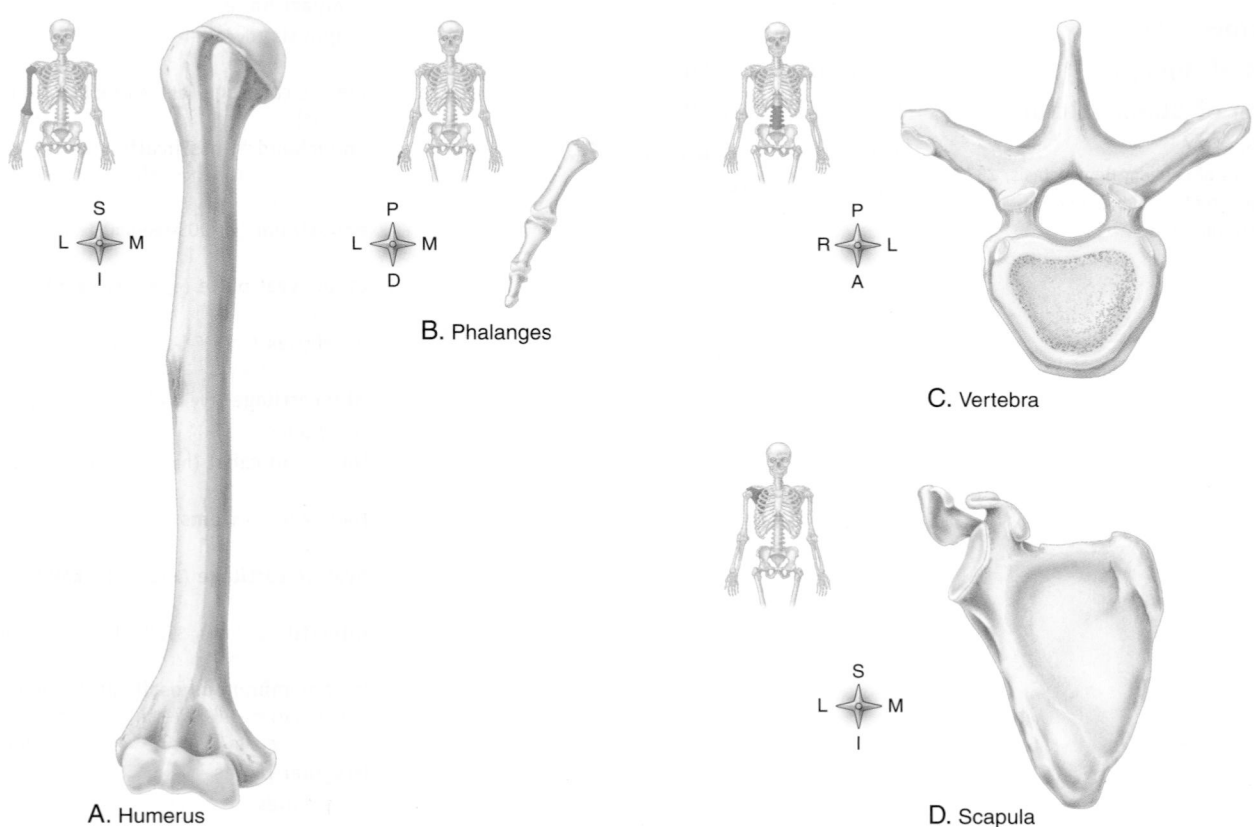

Figure 7-1 *Types of bones.* Examples of bone types include **A,** long bones (humerus); **B,** flat bones (scapula); **C,** short bones (phalanx); and **D,** irregular bones (vertebra).

the facial bones are good examples. Unique irregular bones, which often appear singly rather than in groups, are called *sesamoid bones*. There are only a few sesamoid bones in the body, and they are generally found embedded in the substance of tendons close to the joints. The kneecap, or patella, is a good example.

 QUICK CHECK

1. Name the two major types of connective tissue found in the skeletal system.
2. Name the two different types of bone tissue.

Parts of a Long Bone

A long bone consists of the following structures visible to the naked eye: diaphysis, epiphyses, articular cartilage, periosteum, medullary (marrow) cavity, and endosteum. Identify each of these structures in the tibia shown in Figure 7-2, A. The tibia is the longer, stronger, and more medially located of the two leg bones.

1. **Diaphysis** (dye-AF-i-sis). Main shaftlike portion. Its hollow, cylindrical shape and the thick compact bone that composes it adapt the diaphysis well to its function of providing strong support without cumbersome weight.
2. **Epiphyses** (eh-PIF-i-seez). Both ends of a long bone. Epiphyses have a bulbous shape that provides generous space near

joints for muscle attachments and also gives stability to joints. Look at Figure 7-2 to note the innumerable small spaces in the bone of the epiphysis. They make this kind of bone look a little like a sponge—hence its name spongy, or cancellous, bone. A specialized type of soft connective tissue, called red marrow, fills the spaces within this spongy bone. Early in development, epiphyses are separated from the diaphysis by a layer of cartilage, the *epiphyseal plate*. The region between the epiphyses and diaphysis (in a mature bone) or the epiphyseal plate region (in a growing bone) is called the *metaphysis* (meh-TAF-i-sis).

3. **Articular cartilage.** Thin layer of hyaline cartilage that covers the articular or joint surfaces of epiphyses. The resiliency of this material cushions jolts and blows.
4. **Periosteum** (pair-ee-OS-tee-um). Dense, white fibrous membrane that covers bone except at joint surfaces, where articular cartilage forms the covering. Many of the periosteum's fibers penetrate the underlying bone and weld these two structures to each other. In addition, muscle tendon fibers interlace with periosteal fibers, thereby anchoring muscles firmly to bone. The periosteum is a critically important membrane that depending on its location, also contains bone-forming and bone-destroying cells and blood vessels that become incorporated into bones during their initial growth or subsequent remodeling and repair. This important membrane is essential for bone cell survival and for bone formation, a process that continues throughout life.
5. **Medullary (or marrow) cavity.** A tubelike hollow space in the diaphysis of a long bone. In the adult the medullary cavity is

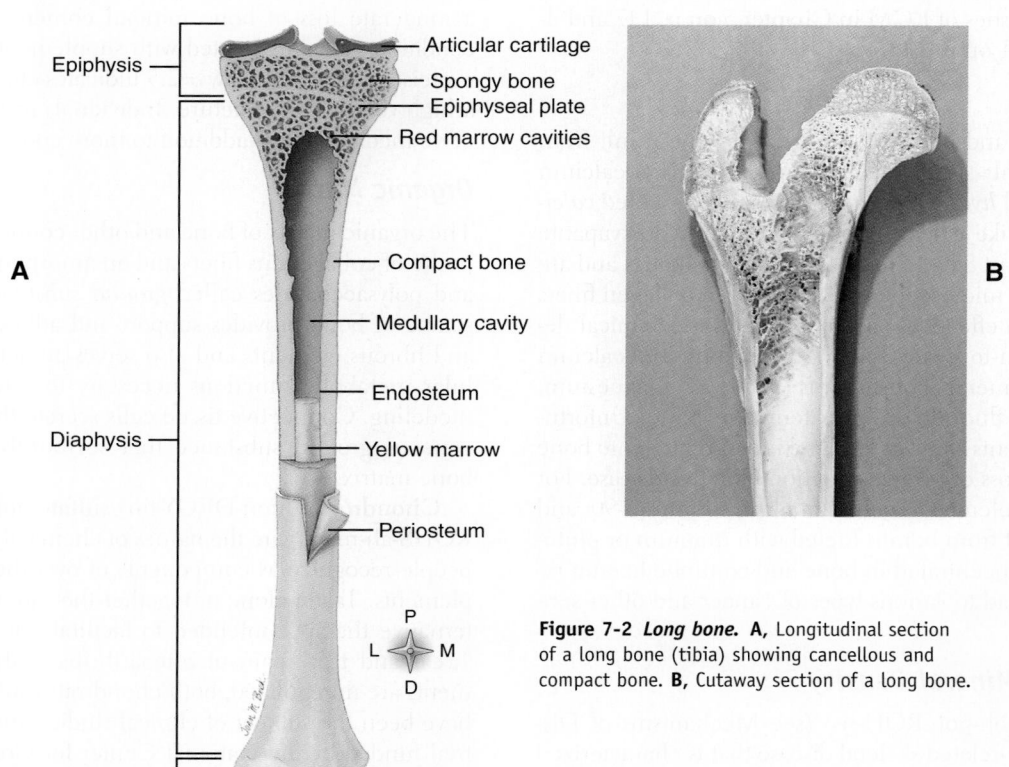

Figure 7-2 *Long bone.* A, Longitudinal section of a long bone (tibia) showing cancellous and compact bone. **B,** Cutaway section of a long bone.

filled with connective tissue rich in fat—a substance called *yellow marrow.*

6. **Endosteum** (en-DOS-tee-um). A thin epithelial membrane that lines the medullary cavity of long bones.

BONE TISSUE

Bone (osseous) tissue is perhaps the most distinctive form of connective tissue in the body. It is typical of other connective tissues in that it consists of cells, fibers, and extracellular material, or *matrix.* However, its extracellular components are hard and *calcified.* In bone the extracellular material, or matrix, predominates. It is much more abundant than the bone cells, and it contains many fibers of collagen (the body's most abundant protein). The rigidity of bone enables it to serve supportive and protective functions.

As a tissue, bone is ideally suited to its functions, and the concept that structure and function are interrelated is apparent in this highly specialized tissue. It has a tensile strength nearly equal to that of cast iron but at less than one third the weight. Bone is organized so that its great strength and minimal weight result from the interrelationships of its structural components. The relationship of structure to function is apparent in its chemical, cellular, tissue, and organ levels of organization.

Composition of Bone Matrix

The extracellular matrix (ECM) of bone, or **bone matrix**, can be subdivided into two principal chemical components: *inorganic salts* and *organic matrix.* About two thirds of the matrix by dry weight analysis consists of inorganic salts and one-third, organic material. You may find it helpful to review the section describing the unique characteristics of ECM in Chapter 5 on p. 147 and illustrated in Figure 5-1 on p. 148.

Inorganic Salts

The calcified nature and thus the hardness of bone result from the deposition of highly specialized chemical crystals of calcium and phosphate, called *hydroxyapatite.* The process is called *calcification.* The needlelike calcium and phosphate hydroxyapatite crystals constitute about 85% of the total inorganic matrix and are found oriented in the microscopic spaces between collagen fibers so that they can most effectively resist stress and mechanical deformation. In addition to hydroxyapatite and about 10% calcium carbonate, other mineral constituents such as magnesium, sodium, sulfate, and fluoride are also found in bone. Unfortunately, harmful elements can also become incorporated into bone matrix and result in loss of normal function or active disease. For example, radioactive elements such as radium, strontium-90, and constituents in fallout from bombs fueled with uranium or plutonium can become concentrated in bone and continue to emit radiation, which can lead to various types of cancer and other serious disease.

Measuring Bone Mineral Density

Osteoporosis (os-tee-oh-poh-ROH-sis) (see Mechanisms of Disease, p. 247) is an age-related skeletal disease that is characterized by declining estrogen levels, loss of bone mineral density, increased bone fragility, and susceptibility to fractures—especially of the spine, forearm, and hip. The disease affects nearly 45% of untreated women during the first 10 years after menopause. During this time they may lose as much as 4% to 8% of their bone density on a yearly basis. Of this group, and white women are particularly susceptible, about 40% will suffer some type of osteoporotic fracture and 15% will break a hip, resulting in costly and devastating medical outcomes.

Treatment of osteoporosis using high doses of vitamin D, short term, low dose estrogen replacement therapy (ERT), bone-building nonhormonal drugs such as Fosamax (alendronate sodium) other so-called "designer estrogens," such as Evista (raloxifene), and Actonel (risedronate), or Miacalcin (calcitonin) is required if nutritional calcium supplements and weight-bearing exercise are ineffective in maintaining bone mineral density.

Unfortunately, regular x-rays do not detect significant reductions in bone mineral content until 30% or more of total bone mass has been lost. Accurate measurement of bone mineral density is a requirement for effective osteoporosis screening and treatment programs. In the past, hip bone biopsy was often employed to either diagnose or judge the effectiveness of osteoporosis therapy and was considered the clinical "gold standard" for accuracy. However, noninvasive diagnostic techniques are now generally preferred over biopsy and are used extensively. Examples include use of a densitometer scanner, and technology called dual energy x-ray absorptiometry (DXA). In addition, advances in an old technique called *radiographic absorptiometry* (RA) allows use of a plain x-ray of the hand, wrist, or heel with a standardized wedge of aluminum, to make an accurate assessment of bone mineral density. Test results are often reported as a "T-score." A zero score is the normal baseline. A T-score between −1 and −2.5 indicates a moderate loss of bone mineral content called *osteopenia*—a condition generally treated with supplements and weight bearing exercise. A T-score below −2.5 indicates clinical osteoporosis with a high risk of bone fracture. Individuals in this category often receive medications in addition to more conservative treatments.

Organic Matrix

The organic matrix of bone and other connective tissues is a composite of collagenous fibers and an amorphous mixture of protein and polysaccharides called *ground substance.* The ground substance of bone provides support and adhesion between cellular and fibrous elements and also serves an active role in many cellular metabolic functions necessary for growth, repair, and remodeling. Connective tissue cells secrete the gel-like and homogeneous ground substance that surrounds the fibers found in bone matrix.

Chondroitin (kon-DROY-tin) **sulfate** and **glucosamine** (gloo-KOHS-ah-meen) are the names of chemical substances that many people recognize as components of over-the-counter dietary supplements. Taken alone or together they are widely used as an "alternative therapy" intended to facilitate healing and reduce the "wear and tear" pain of osteoarthritis. Although dietary supplements are unregulated, both chondroitin sulfate and glucosamine have been the subject of clinical studies, including a multicenter trial funded by the National Center for Complementary and Alternative Medicine at the National Institutes of Health (NIH). At this time, available data suggest that these substances are similar in effectiveness to ibuprofen, aspirin, and other nonsteroidal anti-inflammatory drugs (NSAIDs) in relieving the pain of mild to moderate osteoarthritis.

Chondroitin sulfate and glucosamine are found naturally in the body and are important constituents of the ground substance in both bone and cartilage. Chemically, chondroitin sulfate is a large protein molecule that helps cartilage remain compressible and elastic and may slow its destruction. When sold as a dietary supplement, it is harvested from animal cartilage often obtained from sharks or from cattle tracheas. Glucosamine is an amino sugar important in cartilage formation, maintenance, and repair. Its commercial source is from the shells of shrimp, lobster, and crab.

Components of the organic matrix help cartilage maintain a smooth surface and springy consistency. They add to overall strength and also give bone some degree of plasticlike resilience so that applied stress—within reasonable limits—does not result in frequent crush or fracture injuries.

 QUICK CHECK

3. List the six structural components of a typical long bone that are visible to the naked eye.
4. Identify the two principal chemical components of bone matrix.
5. What disease is characterized by loss of bone mineral density?

MICROSCOPIC STRUCTURE OF BONE

The basic structural components and cell types of bone were described briefly in Chapter 5. In the paragraphs that follow, additional information about bone structure and cell types will serve as a basis for learning the functional characteristics of this important tissue. Understanding how a bone forms and grows, how it repairs itself after injury, and how it interacts with other tissues and organs in maintaining various important homeostatic mechanisms is based on a knowledge of its basic structure—a structure as unique as its chemical composition.

Compact Bone

Compact bone constitutes about 80% of the total bone mass. It contains many cylinder-shaped structural units called **osteons** (AHS-tee-onz), or **Haversian systems** (in honor of Clopton Havers, a seventeenth-century English anatomist who first described them). Note in Figure 7-3 that each osteon surrounds a canal that runs lengthwise through the bone. Living bone cells in these units are literally cemented together to constitute the structural framework of compact bone. The unique structure of the osteon permits delivery of nutrients and removal of waste products from metabolically active, but imprisoned bone cells.

Four types of structures make up each osteon, or Haversian system: lamellae, lacunae, canaliculi, and a Haversian (central) canal. As you read the following definitions, identify each structure in Figure 7-3.

Lamellae (lah-MEL-ay). Concentric, cylinder-shaped layers of calcified matrix. **Interstitial** *lamellae* are islands of calcified matrix between osteons. They are the remnants of older osteons that have been altered by bone growth or remodeling.

Lacunae (lah-KOO-nay) (Latin for "little lakes".) Small spaces containing tissue fluid in which bone cells lie imprisoned between the hard layers of the lamellae.

Canaliculi (kan-ah-LIK-yoo-lye). Ultrasmall canals radiating in all directions from the lacunae and connecting them to each other and into a larger canal, the **Haversian canal.**

BOX 7-1   Bone Scan

As a living dynamic tissue with a rich blood supply, bone readily "takes up" or absorbs radioactive tracer compounds injected intravenously. Radioactive emissions from the tracer compound can then be picked up by specialized scanning equipment and converted to x-ray–like images. Normally, most areas of bone appear light gray because the tracer compound is uniformly distributed. However, if uptake varies because of metabolic differences caused by injury, disease, or periods of altered growth, so-called hot or cold spots appear as dark or very light areas easily noted in the bone scan image. The illustration shows a bone scan of a 3-month-old abused infant. Note the dark "hot spots" in the child's chest, which indicate multiple rib fractures. The appearance of a "hot spot" in an unexpected area of the skeleton in a young infant often permits the diagnosis of so-called hidden, or occult, fractures that might not be visible on a regular radiograph. The prominent hot spots seen in the proximal end of the humerus on both sides are areas of rapid skeletal growth activity and are normal at this age. In some diseases, such as rheumatoid arthritis, tracer uptake is decreased in the affected area, thus causing less radioactive emission and the appearance of a lighter "cold spot" in the bone scan image.

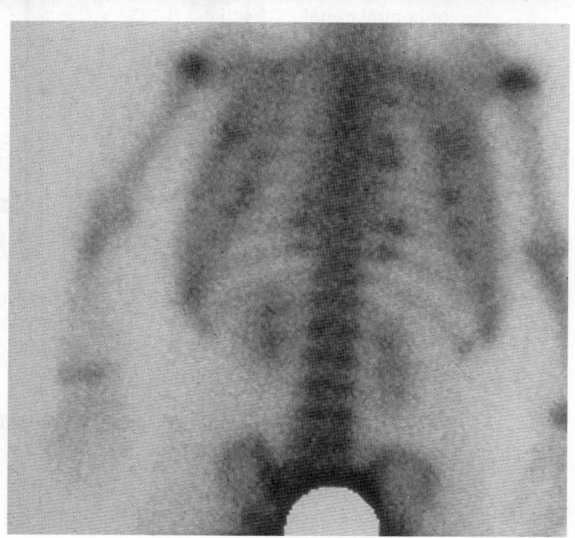

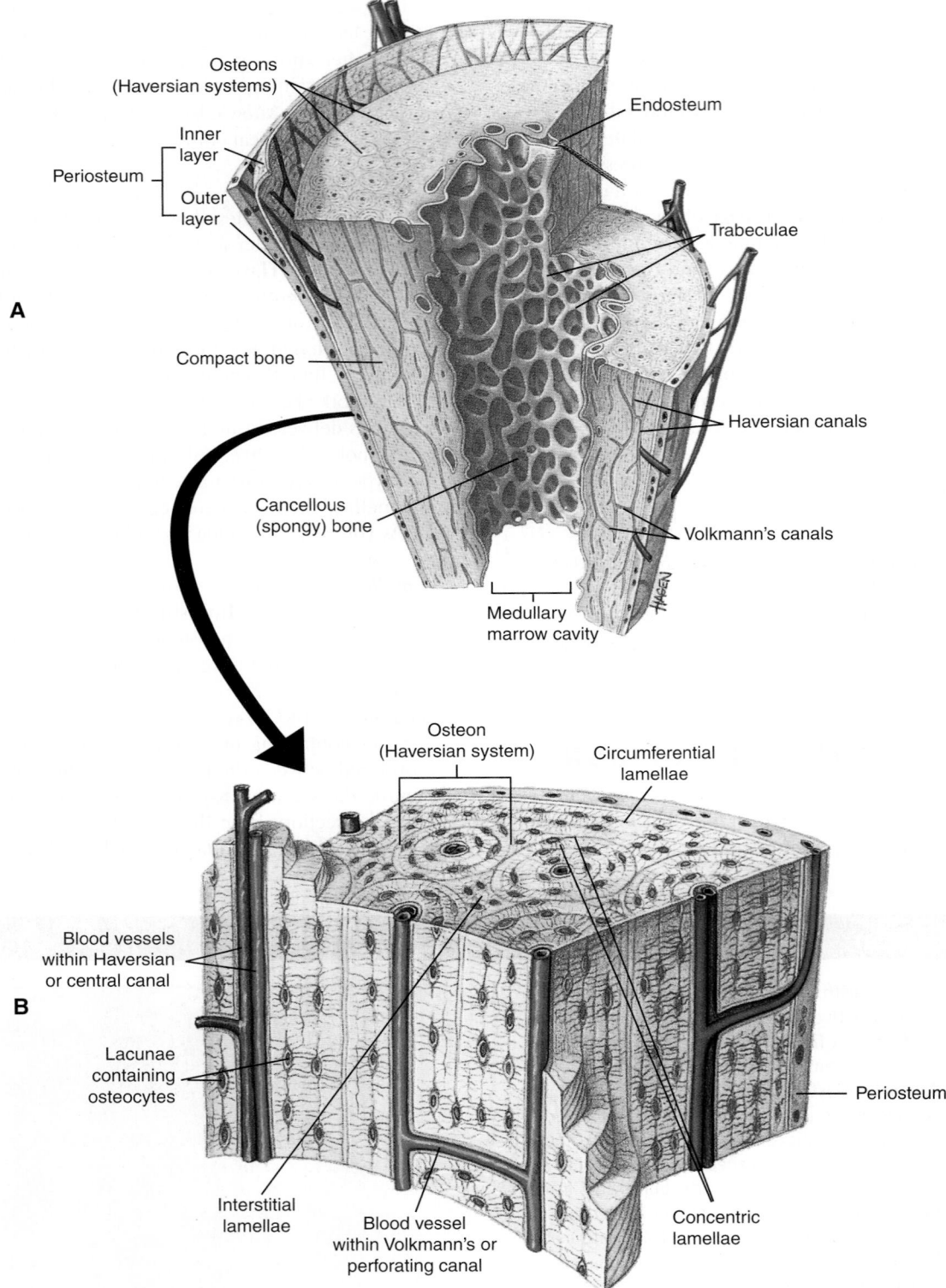

A

Osteons (Haversian systems)

Periosteum — Inner layer / Outer layer

Compact bone

Cancellous (spongy) bone

Medullary marrow cavity

Endosteum

Trabeculae

Haversian canals

Volkmann's canals

B

Osteon (Haversian system)

Circumferential lamellae

Blood vessels within Haversian or central canal

Lacunae containing osteocytes

Interstitial lamellae

Blood vessel within Volkmann's or perforating canal

Concentric lamellae

Periosteum

Figure 7-3 *Structure of compact and cancellous bone.* **A,** Longitudinal section of a long bone showing both cancellous and compact bone. **B,** Magnified view of compact bone.

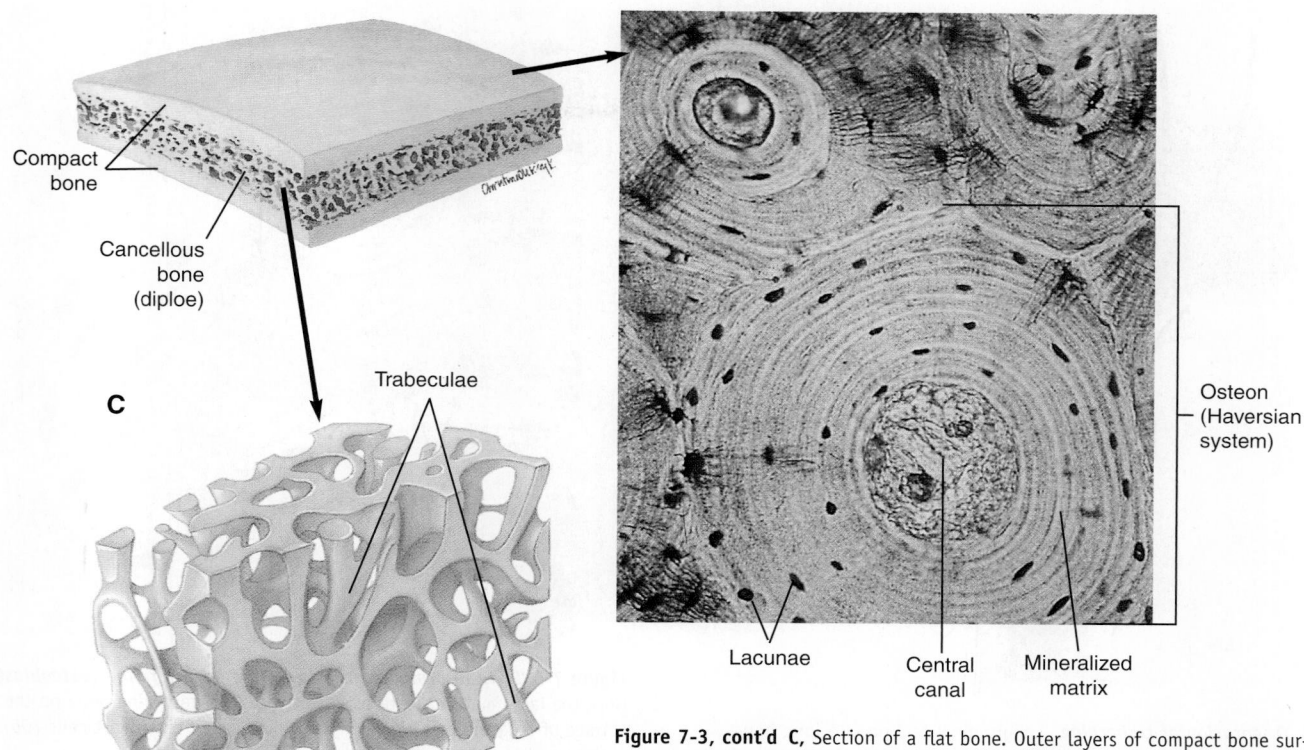

C

Compact bone

Cancellous bone (diploe)

Trabeculae

Osteon (Haversian system)

Lacunae

Central canal

Mineralized matrix

Figure 7-3, cont'd C, Section of a flat bone. Outer layers of compact bone surround cancellous bone. Note the fine structure of compact and cancellous bone.

Haversian (central) canal. Extends lengthwise through the center of each Haversian system; contains blood vessels, lymphatic vessels, and nerves from the Haversian canal; nutrients and oxygen move through canaliculi to the lacunae and their bone cells—a short distance of about 0.1 mm or less.

Lengthwise-running Haversian (central) canals are connected to each other by transverse **Volkmann canals.** These communicating canals contain nerves and vessels that carry blood and lymph from the exterior surface of the bone to the osteons.

Cancellous Bone

Cancellous, or spongy, bone constitutes about 20% of the total bone mass and differs in microscopic structure from compact bone. As you recall, the structural unit of compact bone is the highly organized osteon, or Haversian system. There are no osteons in cancellous bone. Instead, it consists of needlelike bony spicules called **trabeculae.** Bone cells are found within the trabeculae. Nutrients are delivered to the cells and waste products are removed by diffusion through tiny canaliculi that extend to the surface of the very thin spicules.

Note that the cancellous bone shown in Figure 7-3, C, lies between two layers of compact bone, much like the filling in a sandwich. The middle layer of spongy bone is called the *diploe* (DIP-low-ee). This layered organization is typical of flat bones such as those found in the skull. The placement of trabeculae in spongy bone is not as random and unorganized as it might first appear. The bony spicules are actually arranged along lines of stress, and their orientation will therefore differ between individual bones according to the nature and magnitude of the applied load (Figure 7-4). This feature greatly enhances a bone's strength and is yet another example of the relationship between structure and function.

Locked within a seemingly lifeless calcified matrix, bone cells are active metabolically. They must, like all living cells, continually receive food and oxygen and excrete their wastes, so blood supply to bone is both important and abundant. One or more arteries supply the bone marrow in the internal medullary cavity and provide nutrients to areas of cancellous bone. In addition, blood vessels from the periosteum, when they eventually become covered by new bone in the development process, become incorporated into the bone itself and then, by way of the Volkmann canals and by connections with other vessels in adjacent Haversian systems, ultimately serve the nutrient needs of cells that because of their own secretions, have become surrounded by calcified matrix in compact bone. The mechanism by which periosteal blood vessels become "imprisoned" by the deposition of new bone during growth and remodeling is shown in Figure 7-12 and described on p. 242.

QUICK CHECK

6. Identify the four structures that form the osteon, or Haversian system.
7. Name the transverse canals that connect blood vessels between adjacent Haversian systems.
8. Name the needlelike spicules present in cancellous bone.

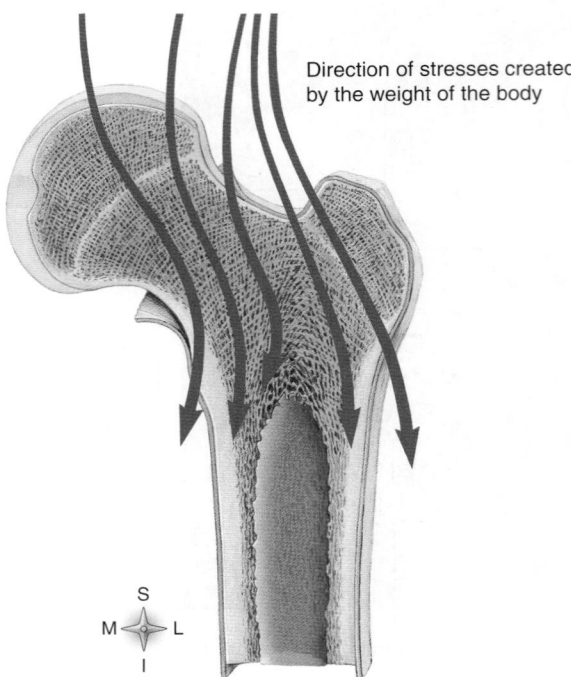

Figure 7-4 *Orientation of trabeculae.* Longitudinal section of a long bone showing trabeculae oriented along lines of stress.

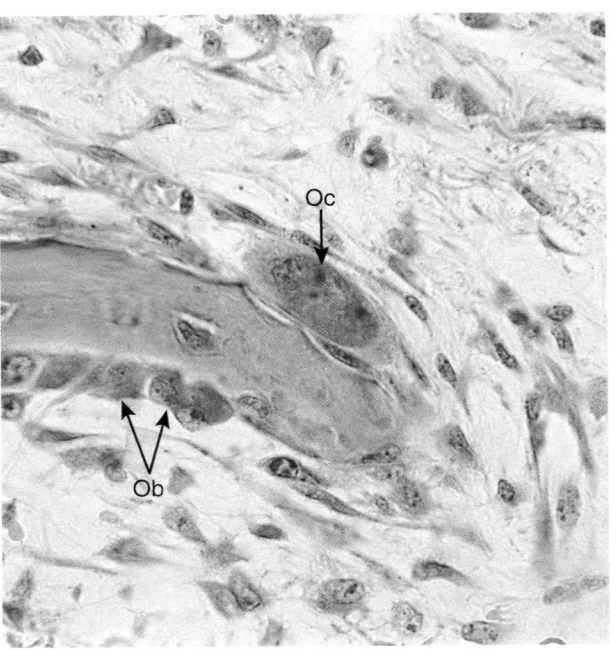

Figure 7-5 *Bone-resorbing (osteoclast) and bone-forming (osteoblast) cells.* Note the large multinucleate osteoclast cell *(Oc)* resorbing bone on the upper surface of a developing bony spicule while smaller osteoblast cells *(Ob)* on the under surface of the spicule are secreting new osteoid.

Types of Bone Cells

Three major types of cells are found in bone: *osteoblasts* (bone-forming cells), *osteoclasts* (bone-reabsorbing cells), and *osteocytes* (mature bone cells). All bone surfaces are covered with a continuous layer of cells that is critical to the survival of bone. This layer is composed of relatively large numbers of osteoblasts interspersed with a much smaller population of osteoclasts.

Osteoblasts (OS-tee-oh-blasts) are small cells that synthesize and secrete a specialized organic matrix, called *osteoid,* that is an important part of the ground substance of bone. Collagen fibrils line up in regular arrays in the osteoid and serve as a framework for the deposition of calcium and phosphate. The process ultimately results in accumulation of mineralized bone. *Osteogenic* (os-tee-oh-JEN-ik) *stem cells,* found in the endosteum and lining the Haversian (central) canals, undergo cell division to form osteoblasts.

Osteoclasts (OS-tee-oh-klasts) are giant multinucleate cells (Figure 7-5) that are responsible for the active erosion of bone minerals. They are formed by the fusion of several precursor cells and contain large numbers of mitochondria and lysosomes. Each 24-hour period sees an alternation of primarily osteoblast, then osteoclast activity. Note in Figure 7-5 how a spicule of bone is being "sculpted" by an osteoclast eroding existing bone mineral from its upper surface while osteoblasts are simultaneously depositing new osteoid on its under surface. This process characterizes bone as a dynamic tissue that undergoes continuous change and remodeling.

Osteocytes (OS-tee-oh-sytes) are mature, nondividing osteoblasts that have become surrounded by matrix and now lie within lacunae. Figure 7-6 is a scanning electron micrograph showing a mature osteocyte within a lacuna. Note that a cytoplasmic process from the cell is extending into a canaliculus below. Numerous collagen fibers are seen in the ground substance and mineralized bone surrounding the osteocyte.

The way in which these cell types work together to produce bone is described in detail when the development of bone is discussed later in this chapter.

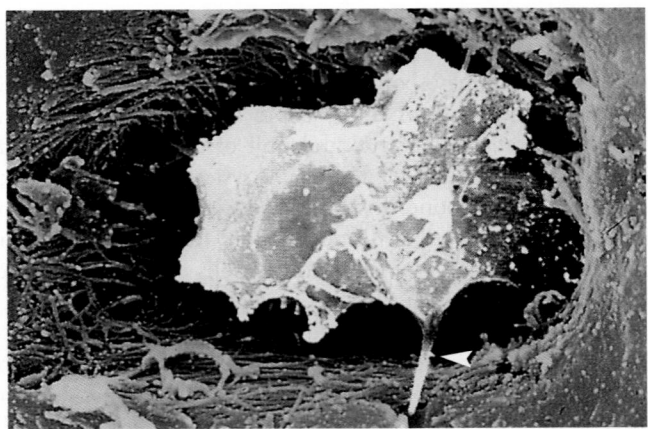

Figure 7-6 *Osteocyte.* Scanning electron micrograph showing an osteocyte within a lacuna. Note the cytoplasmic process *(arrowhead)* extending into a canaliculus below. The cell is surrounded by collagen fibers and mineralized bone.

BONE MARROW

Bone marrow is a specialized type of soft, diffuse connective tissue called **myeloid tissue.** It serves as the site for production of blood cells and is found in the medullary cavities of certain long bones and in the spaces of spongy bone in some areas.

During the lifetime of an individual, two types of marrow exist. In an infant's or child's body, virtually all of the bones contain *red marrow*. This is named for its function in the production of red blood cells. As an individual ages, the red marrow is gradually replaced by *yellow marrow*. In yellow marrow the marrow cells have become saturated with fat and, as a result, are inactive in blood cell production. With advancing age, yellow marrow becomes almost rust-colored, less fatty, and more gelatinous in consistency.

The main bones in an adult that still contain red marrow include the ribs, bodies of the vertebrae, and ends of the humerus in the upper part of the arm, the pelvis, and the femur, or thigh bone. During times of decreased blood supply, yellow marrow in an adult can alter to become red marrow. Such a transition may occur during periods of prolonged anemia caused by chronic blood loss, exposure to radiation or toxic chemicals, and certain diseases.

If the bone marrow is severely damaged, a **bone marrow transplant** can be a lifesaving treatment. In this procedure, red marrow from a compatible donor is introduced into the recipient intravenously. If the recipient's immune system does not reject the new tissue, which is always a danger in tissue transplants, the donor cells may establish a colony of new, healthy tissue in the bone marrow.

FUNCTIONS OF BONE

Bones perform five functions for the body. Each is important for maintaining homeostasis and optimal body function.

1. **Support.** Bones serve as the supporting framework of the body, much as steel girders are the supporting framework of our modern buildings. They contribute to the shape, alignment, and positioning of the body parts.
2. **Protection.** Hard, bony "boxes" serve to protect the delicate structures they enclose. For example, the skull protects the brain, and the rib cage protects the lungs and heart.
3. **Movement.** Bones with their joints constitute levers. Muscles are anchored firmly to bones. As muscles contract and shorten, they pull on bones, thereby producing movement at a joint. This process is discussed further in Chapter 9.
4. **Mineral storage.** Bones serve as the major reservoir for calcium, phosphorus, and certain other minerals. Homeostasis of the blood calcium concentration—essential for healthy survival—depends largely on changes in the rate of calcium movement between blood and bones. If, for example, the blood calcium concentration increases above normal, calcium moves more rapidly out of blood into bones and more slowly in the opposite direction. The result? Blood calcium concentration decreases—usually to its homeostatic level.

5. **Hematopoiesis.** *Hematopoiesis*, or blood cell formation, is a vital process carried on by red bone marrow, or *myeloid tissue*. Myeloid tissue, in the adult, is located primarily in the ends, or epiphyses, of certain long bones, in the flat bones of the skull, in the pelvis, and in the sternum and ribs.

REGULATION OF BLOOD CALCIUM LEVELS

The bones of the skeletal system serve as a storehouse for about 98% of the body calcium reserves. As the major reservoir for this physiologically important body mineral, bones play a key role in maintaining the constancy of blood calcium levels. To maintain homeostasis of blood calcium levels within a very narrow range, calcium is mobilized and moves into and out of blood during the continuous remodeling of bone. It is the balance between deposition of bone by osteoblasts and breakdown and resorption of bone matrix by osteoclasts that helps regulate blood calcium levels. During bone formation, osteoblasts serve to remove calcium from blood, thus lowering its circulating levels. However, when osteoclasts are active and breakdown of bone predominates, calcium is released into blood and circulating levels will increase.

Homeostasis of the calcium ion concentration is essential not only for bone formation, which is described below, but also for normal blood clotting, transmission of nerve impulses, and maintenance of skeletal and cardiac muscle contraction. The primary homeostatic mechanisms involved in the regulation of blood calcium levels involve the secretion of two hormones: (1) *parathyroid hormone* by the parathyroid glands and (2) *calcitonin* by the thyroid gland.

Mechanisms of Calcium Homeostasis

Parathyroid Hormone

The actions of parathyroid hormone are fully discussed in Chapter 16. However, the importance of this hormone as the primary regulator of calcium homeostasis and its effect on bone remodeling warrant a brief description here. When the level of calcium in blood passing through the parathyroid glands decreases below its normal homeostatic "set point" level, osteoclasts are stimulated to initiate increased breakdown of bone matrix, which results in the release of calcium into blood and the return of calcium levels to normal (see Figure 16-21 on p. 612). In addition, parathyroid hormone also increases renal absorption of calcium from urine, thus reducing its loss from the body. Another effect of parathyroid hormone is to stimulate vitamin D synthesis, which increases the efficiency of absorption of calcium from the intestine. If blood passing through the parathyroid glands has an elevated calcium level, osteoclast activity will be suppressed, thus reducing the breakdown of bone matrix and the level of calcium circulating in blood. The multiple effects of this type of hormonal control permit the body to precisely regulate a number of homeostatic mechanisms that have an effect both directly and indirectly on the blood levels of this important mineral.

Parathyroid hormone is the most critical factor in homeostasis of blood calcium levels. As a result of its actions and the ability of the body to regulate its formation and release, bones remain strong and calcium levels are maintained within normal limits during both bone formation and bone resorption.

Calcitonin

Calcitonin, also discussed in Chapter 16, is a protein hormone produced by specialized cells in the thyroid gland. It is produced in response to high blood calcium levels and functions to stimulate bone deposition by osteoblasts and inhibit osteoclast activity. As a result, calcium will move into the bones from the blood and circulating levels will decrease. A specialized nasal spray containing calcitonin (Miacalcin) is available to treat postmenopausal osteoporosis. Although calcitonin does play a role in the homeostasis of blood calcium levels, it is far less important than parathyroid hormone.

DEVELOPMENT OF BONE

When the skeleton begins to form in an infant before birth, it consists not of bones but of cartilage and fibrous structures shaped like bones. Gradually, these cartilage "models" become transformed into real bones when the carftilage is replaced with calcified bone matrix. This process of constantly "remodeling" a growing bone as it changes from a small cartilage model to the characteristic shape and proportion of the adult bone requires continuous activity by the bone-forming osteoblasts and bone-resorbing osteoclasts. The laying down of calcium salts in the gel-like matrix of the forming bones is an ongoing process. This calcification process is what makes bones as "hard as bone." The combined action of osteoblasts and osteoclasts sculpts bones into their adult shape. The term **osteogenesis** is used to describe this process.

"Sculpting" by the bone-forming and bone-resorbing cells allows bones to respond to stress or injury by changing size, shape, and density. The stress placed on certain bones during exercise increases the rate of bone deposition. For this reason, athletes or dancers may have denser, stronger bones than less active people.

Most bones of the body are formed from cartilage models in a process called **endochondral ossification,** meaning "formed in cartilage." A few flat bones are formed within fibrous membrane, rather than cartilage, in the process of **intramembranous ossification.** Figure 7-7 illustrates osseous development of an infant at birth.

BOX 7-2: SPORTS AND FITNESS
Exercise and Bone Density

Walking, jogging, and other forms of exercise subject bones to stress. They respond by laying down more collagen fibers and mineral salts in the bone matrix. This, in turn, makes bones stronger. But inactivity and lack of exercise tend to weaken bones because of decreased collagen formation and excessive calcium withdrawal. To prevent these changes, as well as many others, astronauts regularly perform special exercises in space because of the lack of gravity. Regular weight-bearing exercise is also an important part of the prevention and treatment of osteoporosis.

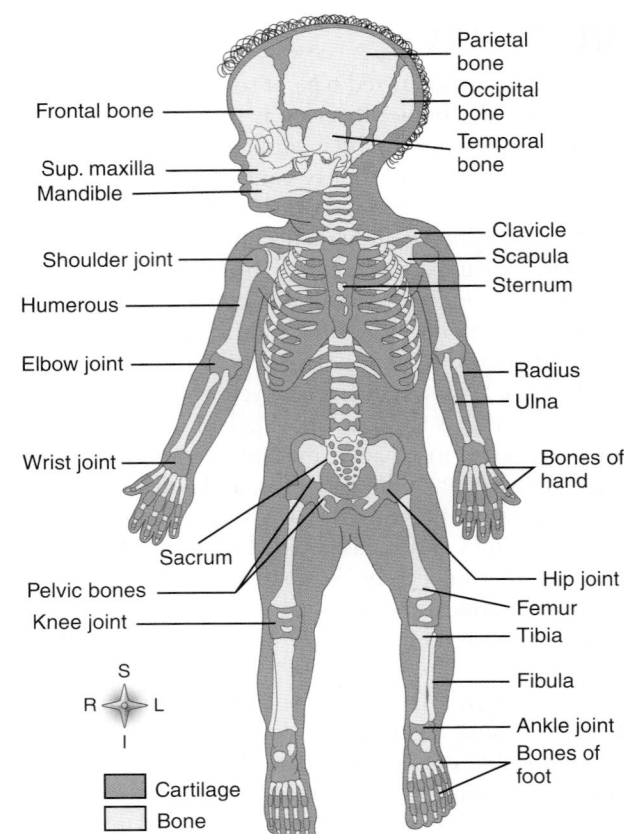

Figure 7-7 *Osseous development.* Diagram showing osseous development at birth.

Intramembranous Ossification

Intramembranous ossification takes place, as its name implies, within a connective tissue membrane. The flat bones of the skull, for example, begin to take shape when groups of osteogenic stem cells within the membrane differentiate into osteoblasts. These clusters of osteoblasts are called *centers of ossification.* They secrete matrix material and collagenous fibrils. The Golgi apparatus in an osteoblast specializes in synthesizing and secreting carbohydrate compounds of the type called *mucopolysaccharides,* and its endoplasmic reticulum makes and secretes collagen, a protein. In time, relatively large amounts of the mucopolysaccharide substance, or *ground substance,* accumulate around each osteoblast. Numerous bundles of collagenous fibers then become embedded in the ground substance. Together, the ground substance and collagenous fibers constitute the organic bone matrix. Calcification of the organic bone matrix occurs when complex calcium salts are deposited in it.

As calcification of bone matrix continues, the trabeculae appear and join in a network to form spongy bone. In time the core layer of spongy bone (diploe) will be covered on each side by plates of compact, or dense, bone. Once formed, a flat bone grows in size by the addition of osseous tissue to its outer surface. The process is called **appositional growth**. Flat bones cannot grow by interior expansion as is the case with endochondral bone growth described in the following section.

Endochondral Ossification

Most bones of the body are formed from cartilage models, with bone formation spreading essentially from the center to the ends. The steps of endochondral ossification are illustrated in Figure 7-8. The cartilage model of a typical long bone, such as the tibia, can be identified early in embryonic life (Figure 7-8, *A*). The cartilage model then develops a periosteum (Figure 7-8, *B*) that soon enlarges and produces a ring, or collar, of bone. Bone is deposited by osteoblasts, which differentiate from cells on the inner surface of the covering periosteum. Soon after appearance of the ring of bone,

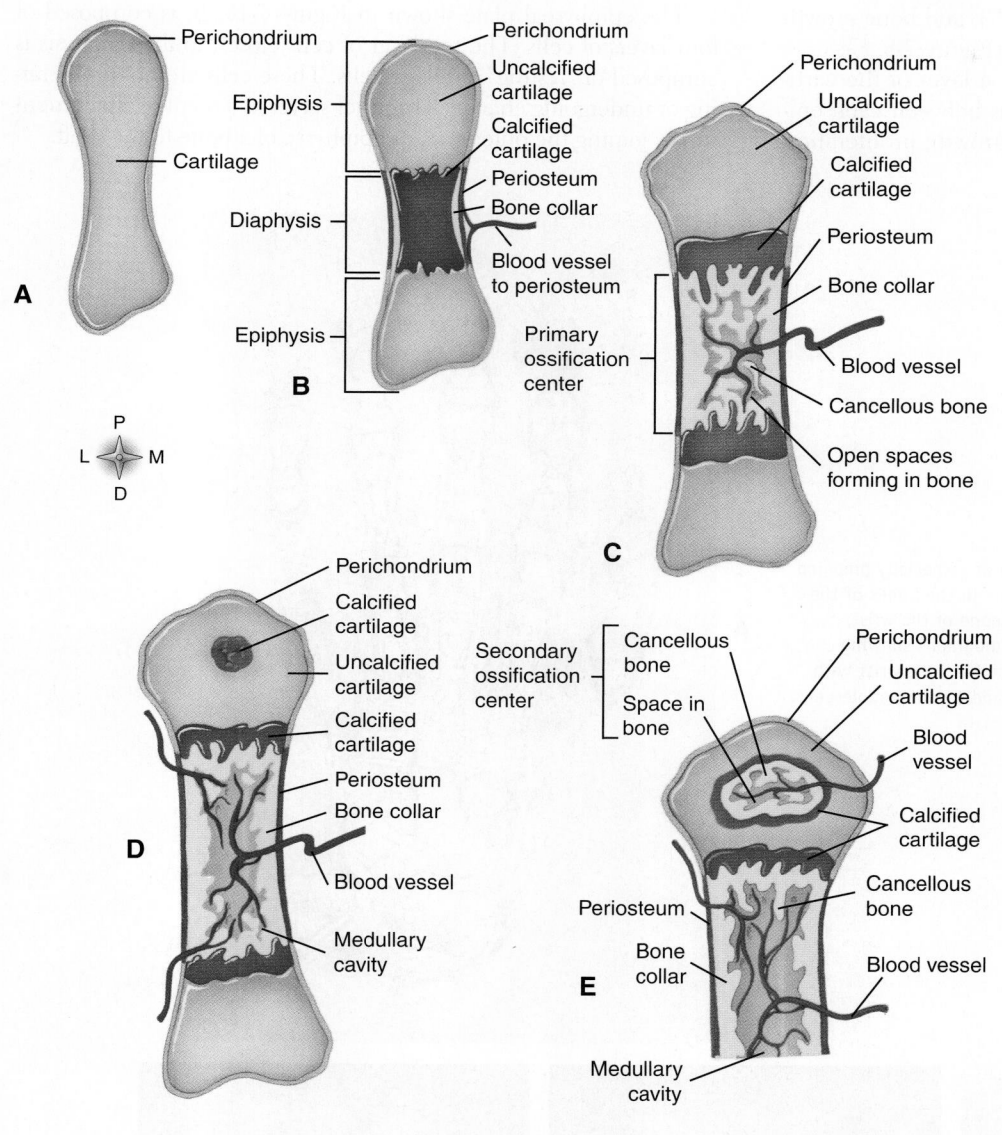

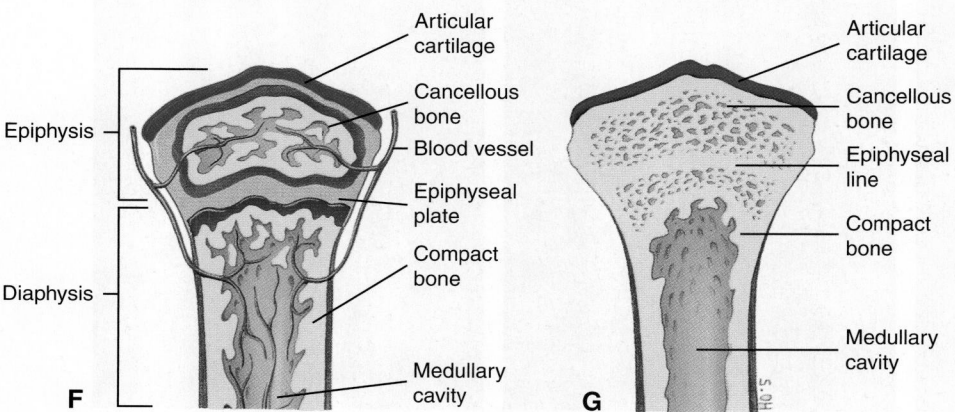

Figure 7-8 *Endochondral bone formation.*
A, Cartilage model. **B,** Subperiosteal bone collar formation. **C,** Development of the primary ossification center and entrance of a blood vessel. **D,** Prominent medullary cavity with thickening and lengthening of the collar. **E,** Development of secondary ossification centers in epiphyseal cartilage. **F,** Enlargement of secondary ossification centers, with bone growth proceeding toward the diaphysis from each end. **G,** With cessation of bone growth, the lower, then upper epiphyseal plates disappear.

the cartilage begins to calcify (Figure 7-8, *C*), and a **primary ossification center** forms when a blood vessel enters the rapidly changing cartilage model at the midpoint of the diaphysis. Endochondral ossification progresses from the diaphysis toward each epiphysis (see Figure 7-8, *D*), and the bone grows in length. The process is called *interstitial growth.* Eventually, **secondary ossification centers** appear in the epiphyses (see Figure 7-8, *E*), and bone growth proceeds toward the diaphysis from each end (Figure 7-8, *F*).

Until bone growth in length is complete, a layer of the cartilage, known as the **epiphyseal plate,** remains between each epiphysis and the diaphysis. During periods of growth, proliferation of epiphyseal cartilage cells brings about a thickening of this layer. **Ossification** of the additional cartilage nearest the diaphysis follows—that is, osteoblasts synthesize organic bone matrix, and the matrix undergoes calcification (see Figure 7-9). As a result, the bone becomes longer. It is the epiphyseal plate that allows the diaphysis of a long bone to increase in length.

The epiphyseal plate shown in Figure 7-10, *B*, is composed of four layers of cells. The top layer of cells closest to the epiphysis is composed of "resting" cartilage cells. These cells are not proliferating or undergoing change. This layer serves as a point of attachment firmly joining the outer end, or epiphysis, of a bone to the shaft.

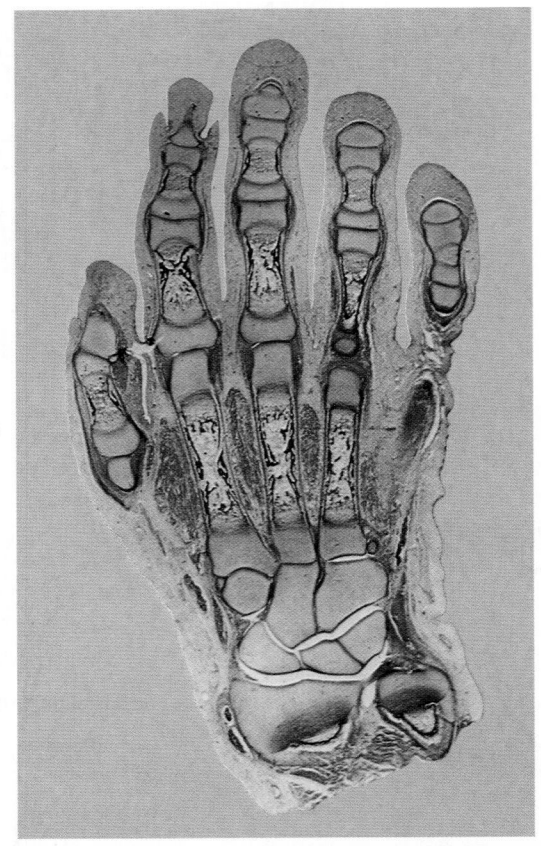

A

Figure 7-9 *Ossification centers.* A, Survey photograph of a specially prepared fetal hand specimen showing primary ossification centers in the bones of the hand (metacarpals) and fingers (phalanges). Note that none of the wrist (carpal) bones show any evidence of ossification. **B,** Radiographs showing increasing numbers of ossification centers becoming visible in the wrist with increasing age. *Left to right,* toddler, school-age child, and a young adolescent.

B

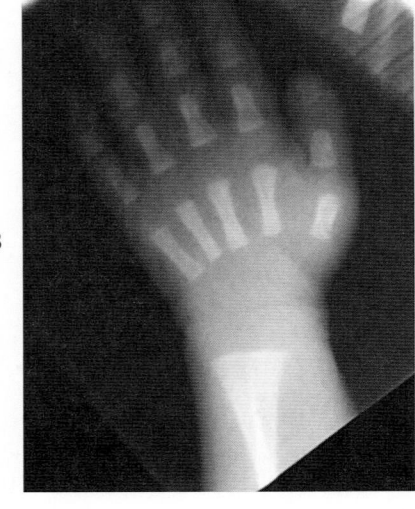
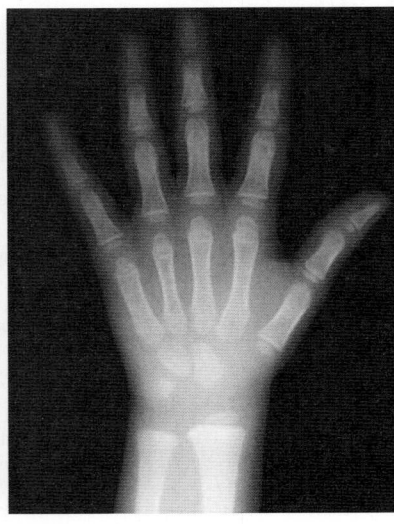
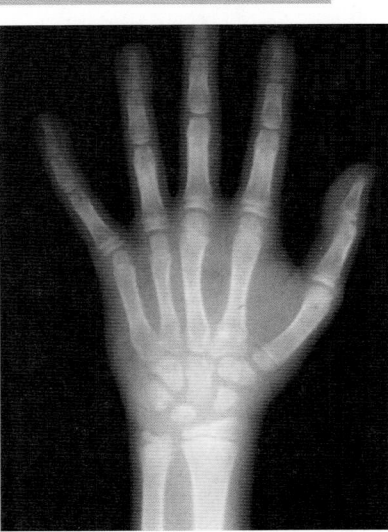

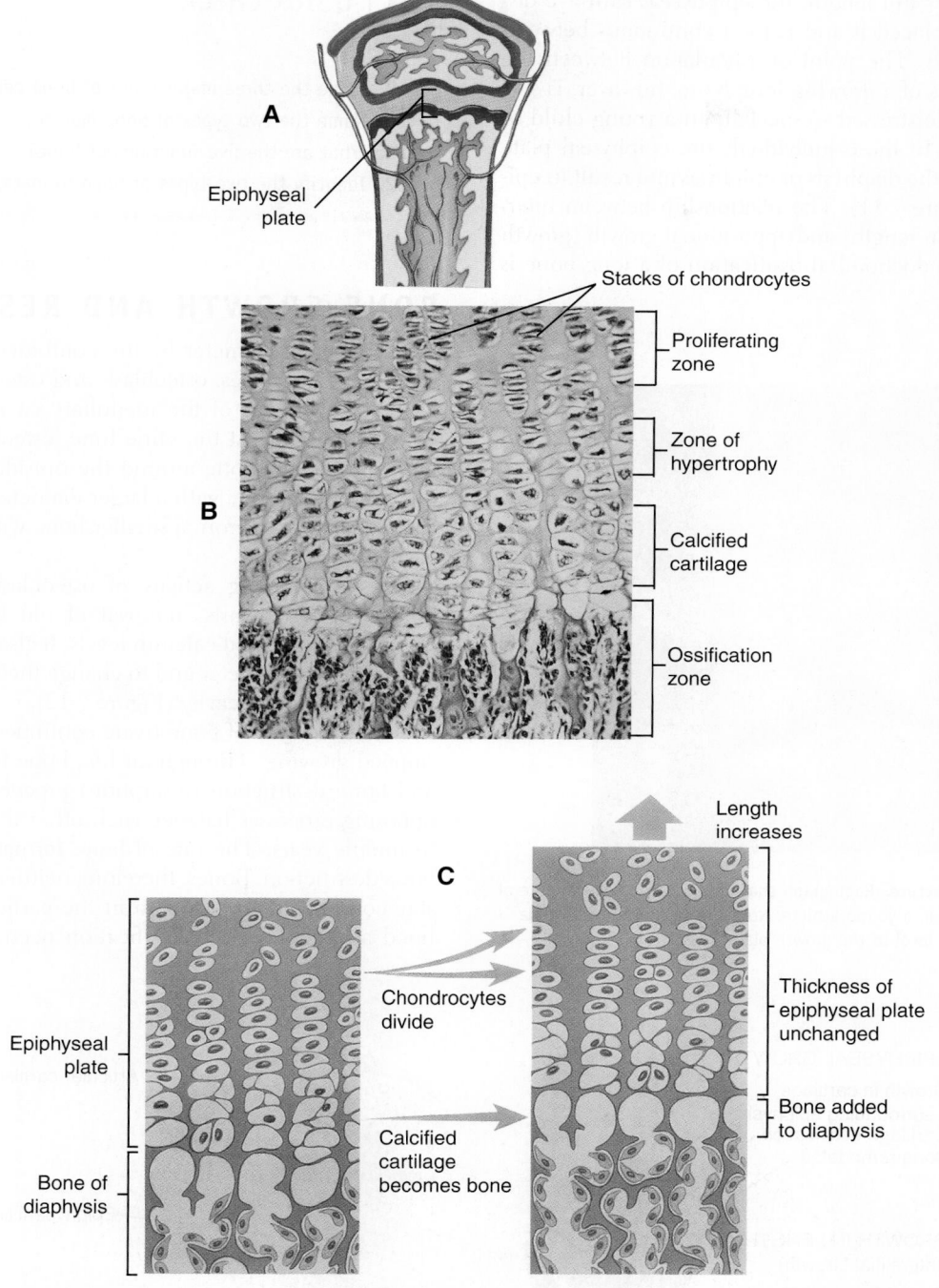

Figure 7-10 *Epiphyseal plate.* **A,** Localization of the epiphyseal plate between the epiphyses and diaphyses of a long bone. **B,** Zones of the epiphyseal plate. **C,** Composite showing steps in ossification on either side of the epiphyseal plate.

The second layer of cells shown in Figure 7-10, *B*, is called the *proliferating zone.* It is composed of cartilage cells that are undergoing active mitosis. As a result of mitotic division and increased cellular activity, the layer thickens and the plate as a whole increases in length.

The third layer of cells, called the *zone of hypertrophy,* is composed of older, enlarged cells that are undergoing degenerative changes associated with calcium deposition.

The layer closest to the diaphysis is a thin layer composed of dead or dying cartilage cells undergoing rapid calcification. As the process of calcification progresses, this layer becomes fragile and disintegrates. The resulting space is soon filled with new bone tissue, and the bone as a whole grows in length.

When epiphyseal cartilage cells stop multiplying and the cartilage has become completely ossified, bone growth ends. Radiographs can reveal any epiphyseal cartilage still present. When bones have grown their full length, the epiphyseal cartilage disappears—bone has replaced it and is then continuous between epiphysis and diaphysis. The point of articulation between the epiphysis and diaphysis of a growing long bone, however, is susceptible to injury if overstressed—especially in a young child or preadolescent athlete. In these individuals the epiphyseal plate can be separated from the diaphysis or epiphysis and result in **epiphyseal fracture** (Figure 7-11). The relationship between interstitial growth (growth in length) and appositional growth (growth in diameter) during endochondral ossification of a long bone is shown in Figure 7-12.

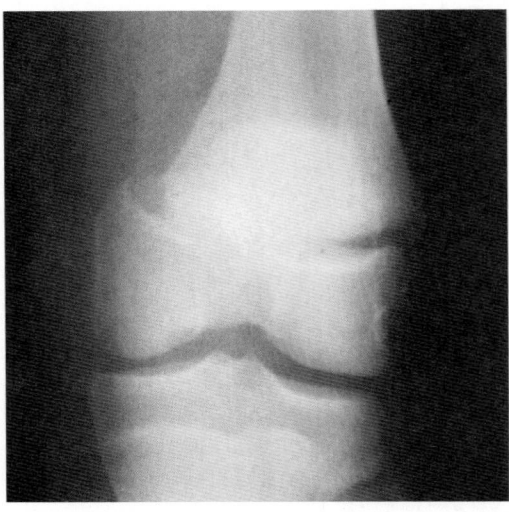

Figure 7-11 *Epiphyseal fracture.* Radiograph showing an epiphyseal fracture of the distal end of the femur in a young athlete. Note the separation of the diaphysis and epiphysis at the level of the growth plate.

BONE GROWTH AND RESORPTION

Bones grow in diameter by the combined action of two of the three bone cell types: osteoblasts and osteoclasts. Osteoclasts enlarge the diameter of the medullary cavity by eating away the bone of its walls. At the same time, osteoblasts from the periosteum build new bone around the outside of the bone. By this dual process, a bone with a larger diameter and larger medullary cavity is produced from a smaller bone with a smaller medullary cavity.

This remodeling activity of osteoblasts (deposition of new bone) and osteoclasts (removal of old bone) is important in homeostasis of blood calcium levels. It also permits bones to grow in length and diameter and to change their overall shape and the size of the marrow cavity (Figure 7-12).

The formation of bone tissue continues long after bones have stopped growing. Throughout life, bone formation (ossification) and bone destruction (resorption) proceed concurrently. These opposing processes balance each other during adulthood's early to middle years. The rate of bone formation equals the rate of bone destruction. Bones, therefore, neither grow nor shrink. They stay constant in size. Not so in the earlier years. During childhood and adolescence, ossification occurs at a faster rate than

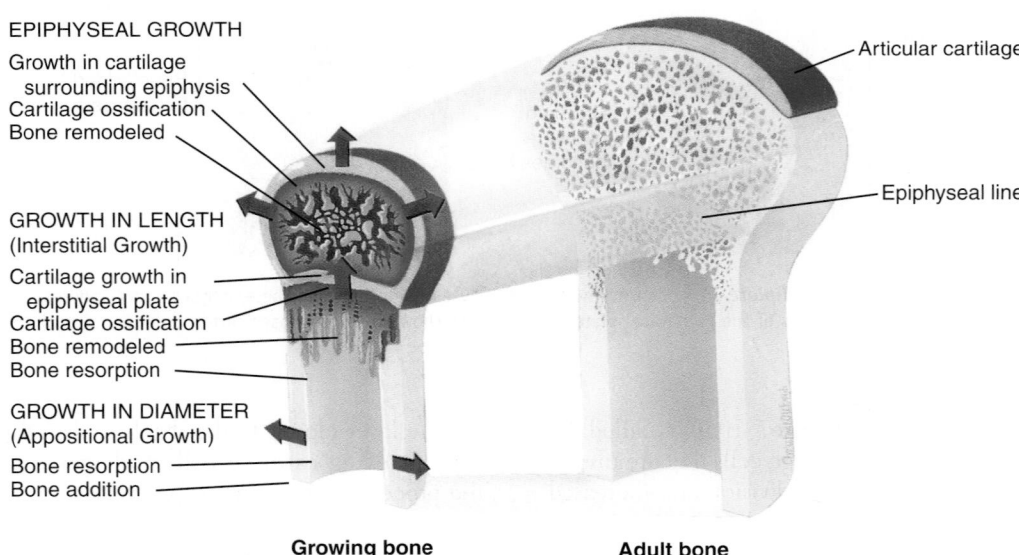

EPIPHYSEAL GROWTH
Growth in cartilage surrounding epiphysis
Cartilage ossification
Bone remodeled

GROWTH IN LENGTH (Interstitial Growth)
Cartilage growth in epiphyseal plate
Cartilage ossification
Bone remodeled
Bone resorption

GROWTH IN DIAMETER (Appositional Growth)
Bone resorption
Bone addition

Articular cartilage
Epiphyseal line

Growing bone   **Adult bone**

Figure 7-12 *Bone remodeling.* Bone formation on the outside of the shaft coupled with bone resorption on the inside increases the bone's diameter. Endochondral growth during bone remodeling increases the length of the diaphyses and causes the epiphysis to enlarge.

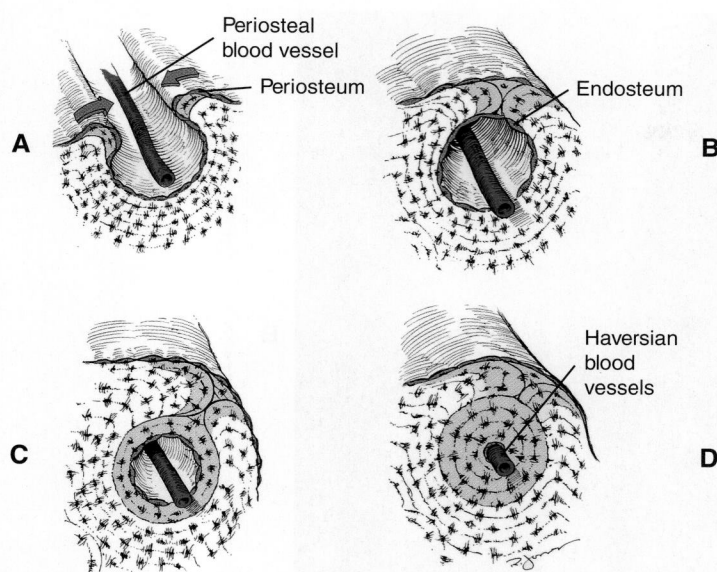

Figure 7-13 *New osteon formation.* **A,** Erosion of the bone surface produces grooves that contain periosteal blood vessels. **B,** New bone added to the groove edges forms a tunnel. **C,** The periosteum of the former groove now becomes the endosteum of the tunnel. **D,** Bone deposition fills the tunnel, thereby forming a new osteon.

bone resorption does. Bone gain outstrips bone loss, and bones grow larger. But between the ages of 35 and 40 years, the process reverses, and from that time on, bone loss exceeds bone gain. Bone gain occurs slowly at the outer, or periosteal, surfaces of bones. Bone loss, on the other hand, occurs at the inner, or endosteal, surfaces and takes place at a somewhat faster pace. More bone is lost on the inside than gained on the outside, and inevitably bones become remodeled as the years go by.

Remodeling in compact bone involves the formation of new Haversian systems (osteons). The process begins when osteoclasts in the covering periosteum are activated and erode the outer surface of the bone, forming grooves. Periosteal blood vessels lie in these grooves, which are eventually surrounded by new bone formed as a result of osteoblast activity. As the grooves are transformed into tunnels, additional layers of bone are deposited by osteoblasts in the lining endosteum. With the passage of time, new lamellae are added and osteon formation occurs (Figure 7-13).

REPAIR OF BONE FRACTURES

The term *fracture* is defined as a break in the continuity of a bone. Types of bone fractures are discussed in Chapter 8, pp. 299-300. *Fracture healing* is considered the prototype of bone repair. The complex bone tissue repair process that follows a fracture is apparently initiated by bone death or by damage to periosteal and Haversian system blood vessels.

A bone fracture invariably tears and destroys blood vessels that carry nutrients to osteocytes. It is this vascular damage that initiates the highly regulated and generally very successful repair process described below. When uncomplicated fractures occur in healthy children and young adults, the healing process often results in a repair that is all but impossible to detect 6 months after

injury. However, especially in older age groups, fracture repair may be complicated by underlying disease or other health problems, such as osteoporosis, diabetes, infection, or diminished blood supply to the injured area. In these cases, delayed healing, instability, deformity, or even nonunion of the fractured bone can result in very serious medical complications.

The process of fracture healing is shown in Figure 7-14, *A* to *D*. Vascular damage occurring immediately after a fracture results in hemorrhage and the pooling of blood at the point of injury. The resulting blood clot is called a *fracture hematoma* (Figure 7-14, *B*). The fracture hematoma quickly becomes "organized," develops a fibrin mesh, and transforms into a soft mass of *granulation tissue* containing specialized inflammatory cells, fibroblasts, bone- and cartilage-forming cells, and new capillaries. Soon, islands of carti-

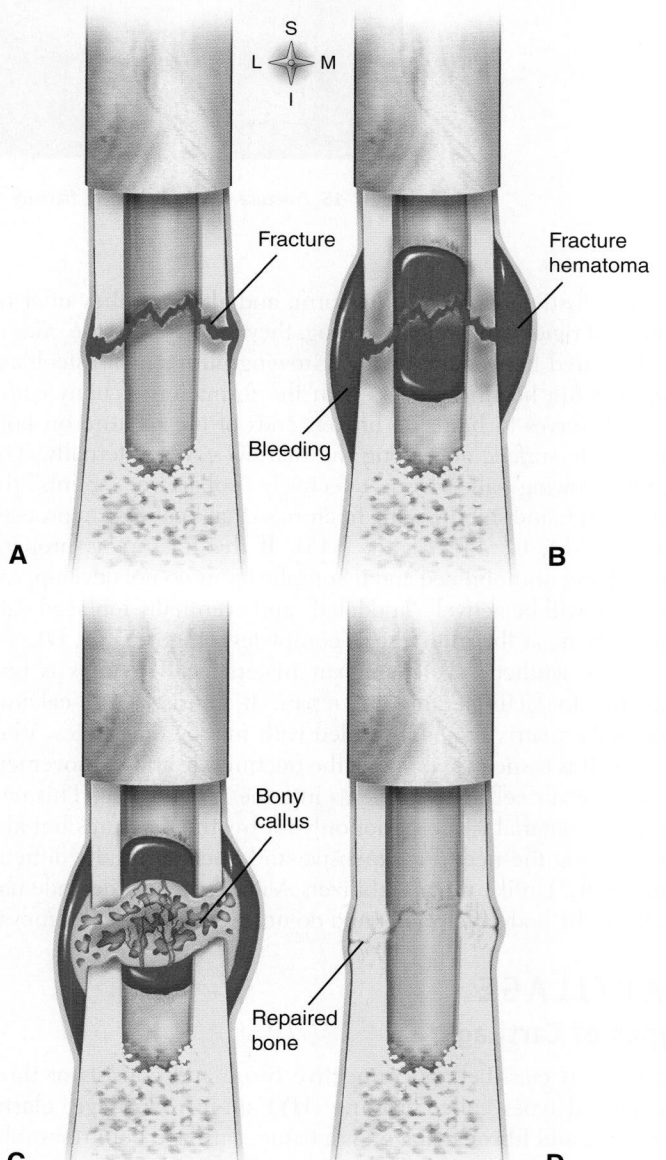

Figure 7-14 *Bone fracture and repair.* **A,** Fracture of the femur. **B,** Formation of a fracture hematoma. **C,** Formation of internal and external bony callus. **D,** Bone remodeling complete.

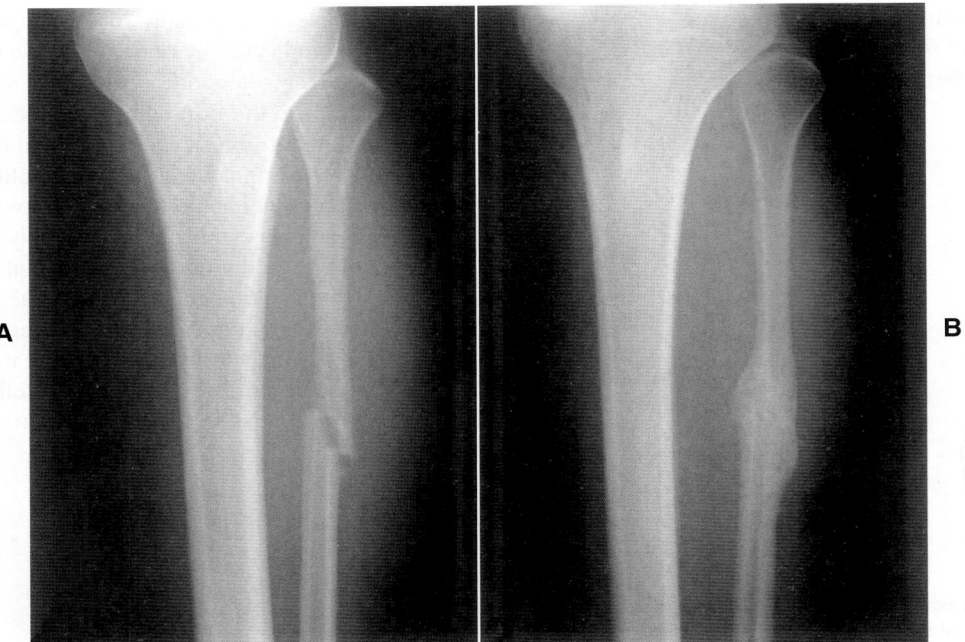

Figure 7-15 *Fracture repair.* **A**, Recent fracture of the fibula. **B**, Bony callus formation 6 weeks after fracture.

laginous tissue called *procallus* form, and although they offer no structural rigidity for weight bearing, they help anchor the ends of the fractured bone more firmly. Growing numbers of osteoblasts continue the healing process with the formation of bony *callus* tissue. It serves to bind the broken ends of the fracture on both the outside surface and along the marrow cavity internally. The rapidly growing callus tissue effectively "collars" or "splints" the broken ends and stabilizes the fracture so that healing can proceed (Figure 7-14, *C*, and Figure 7-15). If the fracture is properly aligned and immobilized and if complications do not develop, callus tissue will be actively "modeled" and eventually replaced with normal bone as the injury heals completely (Figure 7-14, *D*).

A new synthetic skeletal repair material called *Vitos* is now available to facilitate fracture repair. It consists of a calcium spongelike matrix material riddled with microscopic holes. Vitos assists callus tissue in stabilizing the fracture site and in movement of bone repair cells and nutrients into the injured area. This new synthetic material is useful not only for treating fractures but also for reducing the need for expensive and often surgically difficult bone grafts. Unlike metal stabilizers, Vitos "patches" degrade naturally in the body after repair and do not require surgical removal.

CARTILAGE
Types of Cartilage

Cartilage is classified as connective tissue and consists of three specialized types called **hyaline** (HYE-ah-lin) **cartilage, elastic cartilage,** and **fibrocartilage.** As a tissue, cartilage both resembles and differs from bone. Innumerable collagenous fibers reinforce the matrix of both tissues, and as with bone, cartilage consists more of extracellular substance than cells. However, in cartilage the fibers are embedded in a firm gel instead of a calcified cement substance. Hence, cartilage has the flexibility of a firm plastic material, whereas bone has the rigidity of cast iron. Another difference is that no canal system and no blood vessels penetrate the cartilage matrix. Cartilage is avascular and bone is abundantly vascular. Nevertheless, cartilage cells, as with bone cells, lie in lacunae. However, because no canals and blood vessels interlace cartilage matrix, nutrients and oxygen can reach the scattered, isolated **chondrocytes** (cartilage cells) only by diffusion. They diffuse through the matrix gel from capillaries in the fibrous covering of the cartilage—the **perichondrium**—or from synovial fluid in the case of articular cartilage.

The three cartilage types differ from one another largely by the amount of matrix material that is present and also by the relative amounts of elastic and collagenous fibers that are embedded in them. *Hyaline* is the most abundant type, and both elastic cartilage and fibrocartilage are considered modifications of the hyaline type. Collagenous fibers are present in all three types, but are most numerous in *fibrocartilage*. Hence it has the greatest tensile strength. *Elastic* cartilage matrix contains elastic fibers, as well as collagenous fibers, and thus has elasticity, as well as firmness.

Cartilage is an excellent skeletal support tissue in the developing embryo. It forms rapidly and yet retains a significant degree of rigidity, or stiffness. A majority of the bones that eventually form the axial and the appendicular skeleton described in Chapter 8 first appear as cartilage models. Skeletal maturation involves replacement of the cartilage models with bone.

After birth there is a decrease in the total amount of cartilage tissue present in the body. However, it continues to play an important role in the growth of long bones until skeletal maturity and is found throughout life as the material that covers the articular surfaces of bones in joints. The three types of cartilage also serve numerous specialized functions throughout the body.

Hyaline Cartilage

Hyaline, in addition to being the most common type of cartilage, serves many specialized functions. It resembles milk glass in appearance (Figure 7-16, A). In fact, its name is derived from the Greek word meaning "glassy." Sometimes called *gristle*, it is semitransparent and has a bluish, opalescent cast.

In the embryo, hyaline cartilage forms from differentiation of specialized mesenchymal cells that become crowded together in so-called **centers of chondrification.** As the cells enlarge, they secrete matrix material that surrounds the delicate collagen fibrils. Eventually, the continued production of matrix separates and isolates the cells, or chondrocytes, into compartments, which as in bone, are called *lacunae.* Like bone, the organic matrix of hyaline cartilage is a mixture of ground substance and collagenous fibers. The ground substance is rich in both chondroitin sulfate and a unique gel-like polysaccharide. Both substances are secreted from chondrocytes in much the same way protein and carbohydrates are secreted from glandular cells.

In addition to covering the articular surfaces of bones, hyaline cartilage forms the costal cartilages that connect the anterior ends of the ribs with the sternum, or breastbone. It also forms the cartilage rings in the trachea, bronchi of the lungs, and tip of the nose.

Elastic Cartilage

Elastic cartilage gives form to the external ear, the epiglottis that covers the opening of the respiratory tract when swallowing, and the eustachian, or auditory, tubes that connect the middle ear and nasal cavity. The collagenous fibers of hyaline cartilage are also present—but in fewer numbers—in elastic cartilage. Large numbers of easily stained elastic fibers confer the elasticity and resiliency typical of this form of cartilage. In most stained sections, elastic cartilage has a yellowish color and greater opacity than the hyaline variety does (Figure 7-16, B).

Fibrocartilage

Fibrocartilage is characterized by small quantities of matrix and abundant fibrous elements (Figure 7-16, C). It is strong, rigid, and most often associated with regions of dense connective tissue in the body. It occurs in the symphysis pubis, in intervertebral disks, and near the points of attachment of some large tendons to bones.

Histophysiology of Cartilage

The gristlelike nature of cartilage permits it to sustain great weight when covering the articulating surfaces of bones or when serving as a shock-absorbing pad between articulating bones in the spine. In other areas, such as the external ear, nose, or respiratory passages, cartilage provides a strong, yet pliable support structure that resists deformation or collapse of tubular passageways. Cartilage permits growth in the length of long bones and is largely responsible for their adult shape and size.

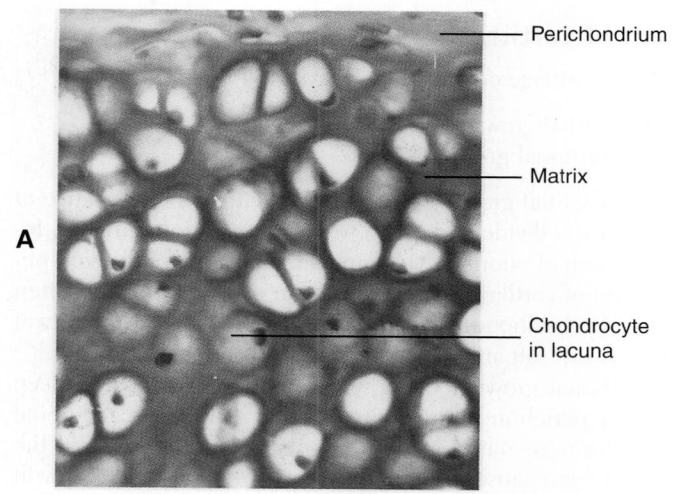
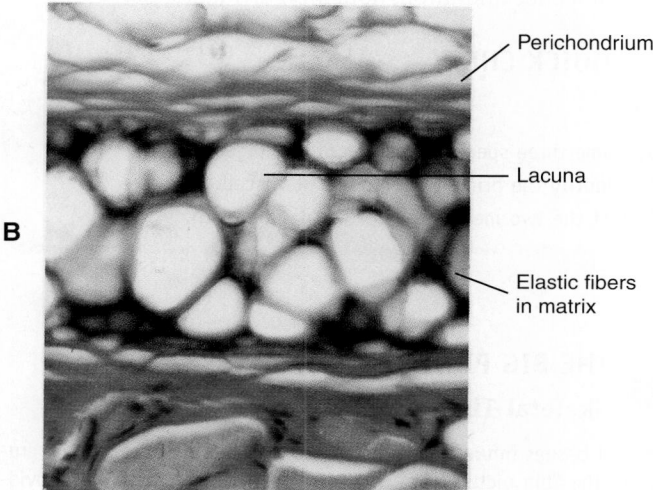
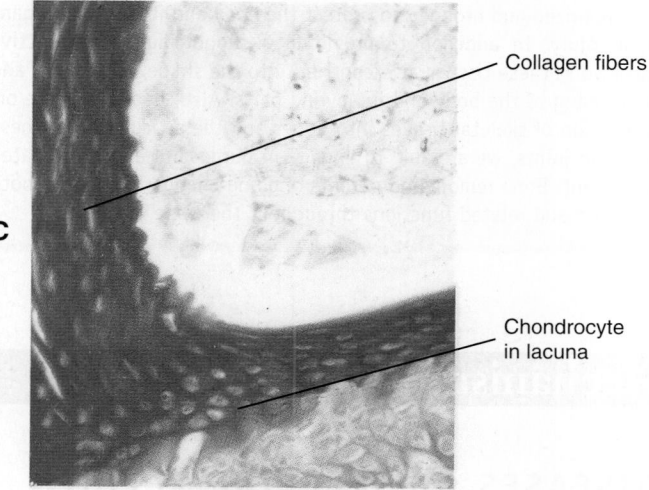

Figure 7-16 *Types of cartilage.* **A**, Hyaline cartilage of the trachea. **B**, Elastic cartilage of the epiglottis. Note the black elastic fibers in the cartilage matrix and the perichondrium layers on both surfaces. **C**, Fibrocartilage of an intervertebral disk.

Growth of Cartilage

Growth of cartilage occurs in two ways:

1. Interstitial growth
2. Appositional growth

During interstitial growth, cartilage cells within the substance of the tissue mass divide and begin to secrete additional matrix. Internal division of chondrocytes is possible because of the soft, pliable nature of cartilage tissue. This form of growth is most often seen during childhood and early adolescence, when a majority of cartilage is still soft and capable of expansion from within.

Appositional growth occurs when chondrocytes in the deep layer of the perichondrium begin to divide and secrete additional matrix. The new matrix is then deposited on the surface of the cartilage, which causes it to increase in size. Appositional growth is unusual in early childhood but, once initiated, continues beyond adolescence and throughout an individual's life.

QUICK CHECK

13. Name three specialized types of cartilage.
14. Identify the primary type of cartilage cell.
15. List the two mechanisms of cartilage growth.

Cycle of Life
Skeletal Tissues

This chapter focused on the changes that occur in bone and cartilage tissue from the time before birth to advanced old age. For instance, we have outlined in some detail the process by which the soft cartilage and membranous skeleton become ossified over a period of years. By the time a person is a young adult in the mid-twenties, the skeleton has become fully ossified. A few areas of soft tissue—the cartilaginous areas of the nose and ears, for example—may continue to grow and ossify very slowly throughout adulthood so that by advanced old age, some structural changes are apparent.

Changes in skeletal tissue that occur during adulthood usually result from specific conditions. For example, the mechanical stress of weight-bearing exercise can trigger dramatic increases in the density and strength of bone tissue. Pregnancy, nutritional deficiencies, and illness can all cause loss of bone density accompanied by loss of structural strength.

In advanced adulthood, degeneration of bone and cartilage tissue becomes apparent. Replacement of hard bone matrix by softer connective tissue results in a loss of strength that increases susceptibility to injury. This is especially true in older women who suffer from osteoporosis. Fortunately, even very light exercise by elderly individuals can counteract some of the skeletal tissue degeneration associated with old age.

THE BIG PICTURE
Skeletal Tissues

Skeletal tissues influence many important body functions that are crucial to the "big picture" of overall health and survival of the individual. The pervasive importance of the protective and support functions of skeletal tissues is immediately apparent. Bone and cartilage tissues are organized and grouped to protect the body and its internal organs from injury. In addition to providing a supporting and protective framework, these tissues also contribute to the shape, alignment, and positioning of the body and its myriad parts. Also, because of the organization of skeletal tissues into bones and the articulation of these bones in joints, we are able to engage in purposeful and coordinated movement. Bone remodeling permits ongoing change to occur in both structure and related functions throughout the cycle of life.

The homeostatic function of skeletal tissues is beautifully illustrated by the role they play in mineral storage and release. For example, regulation of blood calcium levels is important in such diverse areas as nerve transmission, muscle contraction, and normal clotting of blood. Because they contain red marrow, the bones also serve in the important role of hematopoiesis, or blood cell formation. This function ties skeletal tissues to such diverse homeostatic functions as regulation of body pH and the transport of respiratory gases and vital nutrients. By viewing skeletal tissues in such a broad and interrelated functional context, you can more readily sense the "connectedness" that unites these otherwise seemingly isolated body structures to the "big picture" of overall health and survival.

Mechanisms of Disease

DISEASES OF SKELETAL TISSUES
Malignant Tumors of Bone and Cartilage

Osteosarcoma (osteogenic sarcoma) (Figure 7-17, A) is the most common primary malignant tumor of skeletal tissue and is often the most fatal. It appears more frequently in males, with a peak age of incidence between 10 and 25 years. Common sites of involvement are the tibia, femur, and humerus. Roughly 10% of patients experience metastases to the lungs, and if left untreated, the course can involve widespread metastases and death within 1 year. Therapy commonly involves surgery followed by chemotherapy.

Chondrosarcoma is a malignant tumor of hyaline cartilage that arises from chondroblasts (Figure 7-17). It is a large, bulky,

Mechanisms of Disease—cont.

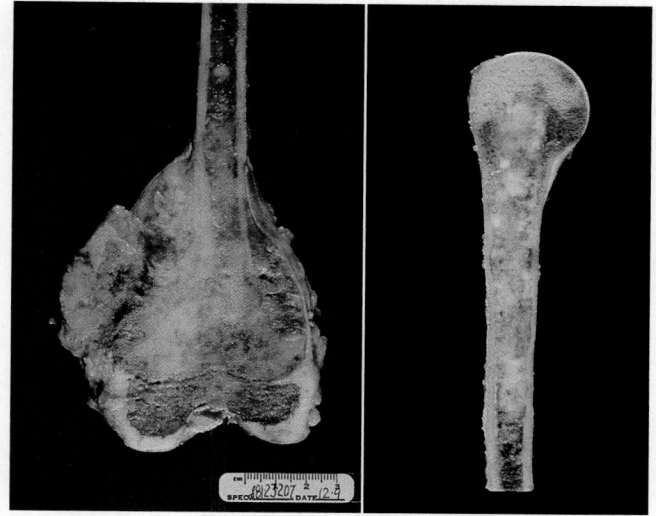

Figure 7-17 *Malignant tumors of bone and cartilage.* **A,** Osteosarcoma of the distal end of the femur. The tumor has broken out of the medullary cavity and is growing on the surface of the bone. **B,** Chondrosarcoma of the proximal end of the humerus. Note the glistening appearance of the hyaline cartilage tumor in the medullary cavity.

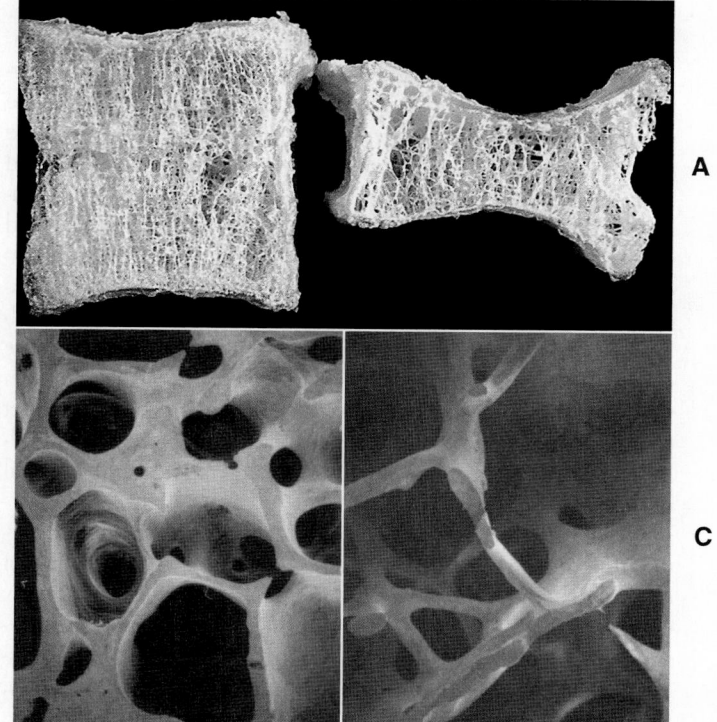

Figure 7-18 *Osteoporosis.* **A,** Compare the normal vertebral body *(left)* with the osteoporotic specimen *(right)*. Note that the osteoporotic vertebral body has been shortened by compression fractures. **B,** Scanning electron micrograph (SEM) of normal bone. **C,** SEM of osteoporotic bone. Note the loss of trabeculae and appearance of enlarged pores caused by osteoporosis.

slow-growing tumor occurring most frequently in middle-aged persons. Common sites of involvement include the humerus, femur, spine, pelvis, ribs, and scapula. Large excisions or amputation of the affected extremity can improve survival rates. Chemotherapy has not been proven to be effective.

Metabolic Bone Diseases

Metabolic bone diseases are disorders of bone remodeling.

You may want to review the overview of osteoporosis that was included in the discussion of bone mineral density on p. 232. As a serious and very common bone disease, especially in older age groups, osteoporosis has been the subject of intense scientific research and widespread public interest and concern. Characterized by increased bone porosity and reduced mineral density and mass, osteoporotic bones fracture easily. Some bones, such as the vertebral bodies in the spinal column, are particularly susceptible to damage. These bones have a large percentage of cancellous bone with trabeculae that become less numerous and weak in osteoporosis. The result is extensive and progressive microfracture of many of the remaining trabeculae and eventual vertebral collapse (Figure 7-18). Over time, increasing numbers of vertebral compression fractures result in a shortened stature and cause an abnormal backward curvature of the spine called "dowager's hump." The condition is, unfortunately, a common sign of long-standing osteoporosis in many untreated older women. Although osteoporosis can occur in both men and women, 1 out of every 2 women who survive 40 years after menopause will suffer an osteoporosis-related fracture, in contrast to 1 in 40 men of similar age.

In most women, a majority of the bone loss characteristic of osteoporosis occurs during the decade after menopause and beyond. However, significant loss can occur much earlier—even during adolescence. Therefore, health care providers are suggesting earlier screening for the disease and lifestyle changes that include weight-bearing exercise and a diet adequate in calcium and vitamin D.

 BOX 7-3: HEALTH MATTERS
Cartilage and Nutritional Deficiencies

Certain nutritional deficiencies and other metabolic disturbances have an immediate and very visible effect on cartilage. It is for this reason that changes in cartilage often serve as indicators of inadequate vitamin, mineral, or protein intake. Vitamin A and protein deficiency, for example, will decrease the thickness of epiphyseal plates in the growing long bones of young children—an effect immediately apparent on x-ray examination. The opposite effect occurs in vitamin D deficiencies. As the epiphyseal cartilage increases in thickness but fails to calcify, the growing bones become deformed and bend under weight bearing. The bent long bones are a sign of rickets (see Mechanisms of Disease p. 248).

Mechanisms of Disease—cont.

Rickets and Osteomalacia

Rickets in young children and **osteomalacia** (os-tee-oh-mah-LAY-shah) in adults are metabolic skeletal diseases that affect significant numbers of individuals worldwide. Both diseases are characterized by demineralization or loss of minerals from bone related to vitamin D deficiency. The loss of minerals is coupled with increased production of unmineralized matrix. Rickets involves demineralization of developing bones in infants and young children before skeletal maturity. In osteomalacia, mineral content is lost from bones that have already matured. In rickets, the lack of rigidity caused by the demineralization of developing bones results in gross skeletal changes, including the classic "bowing of the legs" symptom (Figure 7-19). The demineralization of bones in osteomalacia does not generally affect overall skeletal contours but does result in increased susceptibility to fractures, especially in the vertebral bodies and femoral necks.

Paget's disease, also known as *osteitis deformans*, is a disorder affecting older adults. It is characterized by proliferation of osteoclasts and compensatory increased osteoblastic activity (Figure 7-20). The result is rapid and disorganized bone remodeling. The bones formed are poorly constructed and weakened. It commonly affects the skull, humerus, femur, vertebra, and pelvic bones. Clinical manifestations may include bone pain, tenderness, and fractures. However, the majority of patients experience minimal changes and never know they have the disease. No treatment is recommended in an asymptomatic patient.

Osteomyelitis is a bacterial infection of the bone and marrow tissue. Infections of bone are often more difficult to treat than soft tissue infections because of the decreased blood supply and density of the bone. Bacteria, viruses, fungi, and other pathogens may cause osteomyelitis. *Staphylococcus* bacteria are the most common pathogens. Osteomyelitis is associated with extension of another infection (e.g., bacteremia, urinary

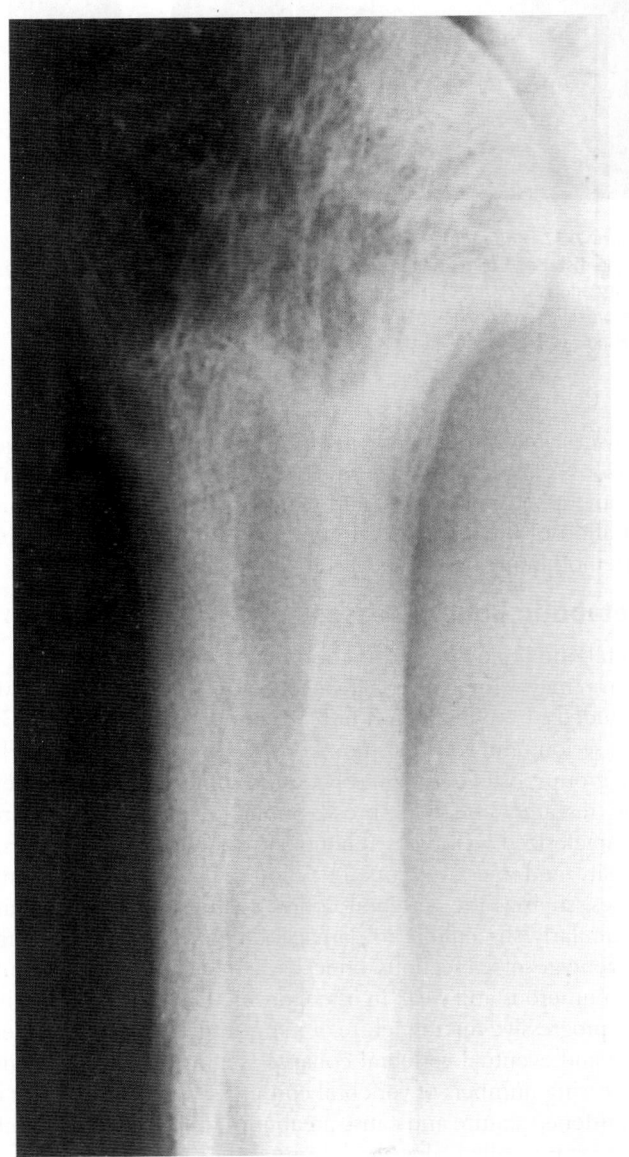

Figure 7-20 *Paget disease of the humerus.* Note the resorption and thinning of bone near the neck and head and the dense thickening of the distal shaft nearly filling the marrow cavity.

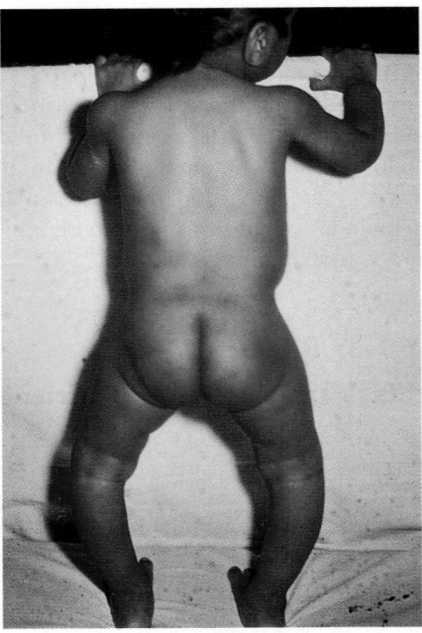

Figure 7-19 *Rickets.* Bowing of legs in this toddler is due to poorly mineralized bones.

Mechanisms of Disease—cont.

tract infection, vascular ulcer) or direct bone contamination (e.g., gunshot wound, open fracture). Patients who are elderly, poorly nourished, or diabetic are also at risk. Thrombosis of blood vessels in osteomyelitis often results in ischemia and bone necrosis (Figure 7-21). As a result, infection can extend under the periosteum and spread to adjacent soft tissues and joints. Signs and symptoms may include an area swollen, warm, tender to touch, and painful. Early recognition of infection and antimicrobial management are required. Sometimes patients may require many weeks of antibiotic therapy.

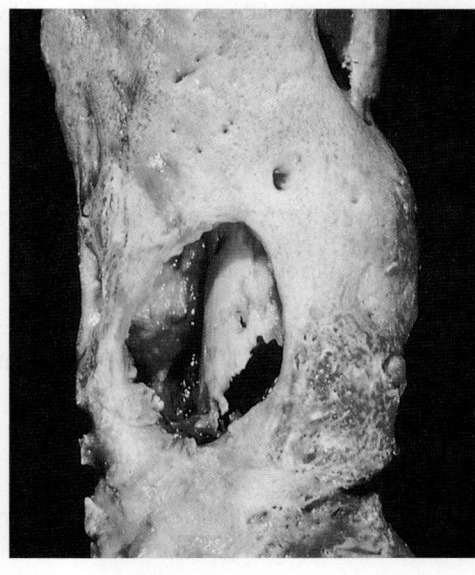

Figure 7-21 *Osteomyelitis.* Segment of femur surgically removed from a patient with chronic osteomyelitis. Necrosis and erosion of bone have occurred at the focal point of infection.

LANGUAGE OF SCIENCE   *(Cont'd from page 229)*

medullary cavity (MED-oo-lair-ee) [*medulla* marrow]

myeloid tissue (MY-eh-loyd) [*myel* bone marrow, *tissue* fabric]

ossification (os-i-fi-KAY-shun) [*os-* bone, *-fication* to make]

osteoblasts (OS-tee-oh-blasts) [*osteo-* bone, *-blast* germ]

osteoclasts (OS-tee-oh-klasts) [*osteo-* bone, *-clast* breaking]

osteocytes (OS-tee-oh-sytes) [*osteo-* bone, *-cyte* cell]

osteons (AHS-tee-onz) [*osteo-* bone, *-on* unit]

perichondrium (pair-i-KON-dree-um) [*peri-* around, *-chondro* cartilage]

periosteum (pair-ee-OS-tee-um) [*peri-* around, *-osteum* bone]

primary ossification center (os-i-fi-KAY-shun) [*primary* first, *os-* bone, *-fication* to make]

secondary ossification center (SEK-on-dair-ee os-i-fih-KAY-shun) [*second* second, *os-* bone, *-fication* to make]

short bones

spongy bones

trabeculae (trah-BEK-yoo-lee) [*trabeculae* little beam] sing., trabecula

Volkmann canal (VOLK-man) [*Richard von Volkmann,* German surgeon]

LANGUAGE OF MEDICINE

bone marrow transplant (MAIR-oh TRANS-plant)

chondroitin sulfate (kon-DROY-tin SUHL-fayt) [*chondroid* resembling cartilage]

epiphyseal fracture (ep-i-FEEZ-ee-al) [*epi-* on, *-physis* growth, *fract* to break]

glucosamine (gloo-KOHS-ah-meen) [*gluco-* sweetness or glucose]

osteogenesis (os-tee-oh-JEN-eh-sis) [*osteo-* bone, *-genesis* origin]

osteomalacia (os-tee-oh-mah-LAY-shah) [*osteo-* bone, *-malacia* softening]

osteoporosis (os-tee-oh-poh-ROH-sis) [*osteo-* bone, *-poro* passage, *-osis* condition]

rickets (RIK-ets) [unknown origin]

CASE STUDY

Johnny, age 5, was riding his tricycle when he fell and injured one of his long bones. He was taken to the emergency department where an x-ray revealed a fracture through the epiphyseal plate.

1. Which one of the following bones did Johnny most likely fracture?
   A. Wrist (carpals)
   B. Tibia
   C. Clavicle
   D. Pelvis

2. Why is there greater significance in this type of fracture in a child Johnny's age versus an adult?
   A. Because in a child there is little cartilage located at the epiphyseal plate.
   B. Because once older than the age of 12 years there is no need for the epiphyseal plate to adequately control bone growth.
   C. Because the epiphyseal plate allows the diaphysis of a long bone to increase in length.
   D. Because the epiphyseal plate controls the zone of hypertrophy and thus bone regeneration in the child.

CHAPTER SUMMARY

TYPES OF BONES

A. Structurally, there are four types of bones (Figure 7-1)
   1. Long bones
   2. Short bones
   3. Flat bones
   4. Irregular bones
B. Bones serve various needs, and their size, shape, and appearance will vary to meet those needs
C. Bones vary in the proportion of compact and cancellous (spongy) bone; compact bone is dense and solid in appearance, whereas cancellous bone is characterized by open space partially filled with needlelike structures
D. Parts of a long bone (Figure 7-2)
   1. Diaphysis
      a. Main shaft of a long bone
      b. Hollow, cylindrical shape and thick compact bone
      c. Function is to provide strong support without cumbersome weight
   2. Epiphyses
      a. Both ends of a long bone, made of cancellous bone filled with marrow
      b. Bulbous shape
      c. Function is to provide attachments for muscles and give stability to joints
   3. Articular cartilage
      a. Layer of hyaline cartilage that covers the articular surface of epiphyses
      b. Function is to cushion jolts and blows
   4. Periosteum
      a. Dense, white fibrous membrane that covers bone
      b. Attaches tendons firmly to bones
      c. Contains cells that form and destroy bone
      d. Contains blood vessels important in growth and repair
      e. Contains blood vessels that send branches into bone
      f. Essential for bone cell survival and bone formation

   5. Medullary (or marrow) cavity
      a. Tubelike, hollow space in the diaphysis
      b. Filled with yellow marrow in adults
   6. Endosteum—thin epithelial membrane that lines the medullary cavity
E. Short, flat, and irregular bones
   1. Inner portion is cancellous bone covered on the outside with compact bone
   2. Spaces inside the cancellous bone of a few irregular and flat bones are filled with red marrow

BONE TISSUE

A. Most distinctive form of connective tissue
B. Extracellular components are hard and calcified
C. Rigidity of bone allows it to serve its supportive and protective functions
D. Tensile strength nearly equal to that of cast iron at less than one third the weight
E. Composition of bone matrix
   1. Inorganic salts
      a. Hydroxyapatite—highly specialized chemical crystals of calcium and phosphate contribute to bone hardness
      b. Slender needlelike crystals are oriented to most effectively resist stress and mechanical deformation
      c. Magnesium and sodium are also found in bone
   2. Measuring bone mineral density
   3. Organic matrix
      a. Composite of collagenous fibers and an amorphous mixture of protein and polysaccharides called ground substance
      b. Ground substance is secreted by connective tissue cells
      c. Adds to overall strength of bone and gives some degree of resilience to bone

MICROSCOPIC STRUCTURE OF BONE (FIGURE 7-3)
A. Compact bone
　1. Contains many cylinder-shaped structural units called osteons, or Haversian systems
　2. Osteons surround canals that run lengthwise through bone and are connected by transverse Volkmann canals
　3. Living bone cells are located in these units, which constitute the structural framework of compact bone
　4. Osteons permit delivery of nutrients and removal of waste products
　5. Four types of structures make up each osteon
　　a. Lamella—concentric, cylinder-shaped layers of calcified matrix
　　b. Lacunae—small spaces containing tissue fluid in which bone cells are located between hard layers of the lamella
　　c. Canaliculi—ultrasmall canals radiating in all directions from the lacunae and connecting them to each other and to the Haversian canal
　　d. Haversian canal—extends lengthwise through the center of each osteon and contains blood vessels and lymphatic vessels
B. Cancellous bones (Figure 7-4)
　1. No osteons in cancellous bone; instead, it has trabeculae
　2. Nutrients are delivered and waste products removed by diffusion through tiny canaliculi
　3. Bony spicules are arranged along lines of stress to enhance the bone's strength
C. Blood supply
　1. Bone cells are metabolically active and need a blood supply, which comes from the bone marrow in the internal medullary cavity of cancellous bone
　2. Compact bone, in addition to bone marrow and blood vessels from the periosteum, penetrate bone and then, by way of Volkmann canals, connect with vessels in the Haversian canals
D. Types of bone cells
　1. Osteoblasts
　　a. Bone-forming cells found in all bone surfaces
　　b. Small cells synthesize and secrete osteoid, an important part of the ground substance
　　c. Collagen fibrils line up in osteoid and serve as a framework for the deposition of calcium and phosphate
　2. Osteoclasts (Figure 7-5)
　　a. Giant multinucleated cells
　　b. Responsible for the active erosion of bone minerals
　　c. Contain large numbers of mitochondria and lysosomes
　3. Osteocytes—mature, nondividing osteoblasts surrounded by matrix and lying within lacunae (Figure 7-6)

BONE MARROW
A. Specialized type of soft, diffuse connective tissue; called myeloid tissue
B. Site for the production of blood cells

C. Found in the medullary cavities of long bones and in the spaces of spongy bone
D. Two types of marrow occur during a person's lifetime
　1. Red marrow
　　a. Found in virtually all bones in an infant's or child's body
　　b. Functions to produce red blood cells
　2. Yellow marrow
　　a. As an individual ages, red marrow is replaced by yellow marrow
　　b. Marrow cells become saturated with fat and are no longer active in blood cell production
E. The main bones in an adult that still contain red marrow include the ribs, bodies of the vertebrae, humerus, pelvis, and femur
F. Yellow marrow can change to red marrow during times of decreased blood supply, such as anemia, exposure to radiation, and certain diseases

FUNCTIONS OF BONE
A. Support—bones form the framework of the body and contribute to the shape, alignment, and positioning of body parts
B. Protection—bony "boxes" protect the delicate structures they enclose
C. Movement—bones with their joints constitute levers that move as muscles contract
D. Mineral storage—bones are the major reservoir for calcium, phosphorus, and other minerals
E. Hematopoiesis—blood cell formation is carried out by myeloid tissue

REGULATION OF BLOOD CALCIUM LEVELS
A. Skeletal system serves as a storehouse for about 98% of body calcium reserves
　1. Helps maintain constancy of blood calcium levels
　　a. Calcium is mobilized and moves into and out of blood during bone remodeling
　　b. During bone formation, osteoblasts remove calcium from blood and lower circulating levels
　　c. During breakdown of bone, osteoclasts release calcium into blood and increase circulating levels
　2. Homeostasis of calcium ion concentration essential for the following:
　　a. Bone formation, remodeling, and repair
　　b. Blood clotting
　　c. Transmission of nerve impulses
　　d. Maintenance of skeletal and cardiac muscle contraction
B. Mechanisms of calcium homeostasis
　1. Parathyroid hormone
　　a. Primary regulator of calcium homeostasis
　　b. Stimulates osteoclasts to initiate breakdown of bone matrix and increase blood calcium levels
　　c. Increases renal absorption of calcium from urine
　　d. Stimulates vitamin D synthesis

2. Calcitonin
   a. Protein hormone produced in the thyroid gland
   b. Produced in response to high blood calcium levels
   c. Stimulates bone deposition by osteoblasts
   d. Inhibits osteoclast activity
   e. Far less important in homeostasis of blood calcium levels than parathyroid hormone is

DEVELOPMENT OF BONE

A. Osteogenesis—development of bone from small cartilage model to adult bone
B. Intramembranous ossification
   1. Occurs within a connective tissue membrane
   2. Flat bones begin when groups of cells differentiate into osteoblasts
   3. Osteoblasts are clustered together in centers of ossification
   4. Osteoblasts secrete matrix material and collagenous fibrils
   5. Large amounts of ground substance accumulate around each osteoblast
   6. Collagenous fibers become embedded in the ground substance and constitute the bone matrix
   7. Bone matrix calcifies when calcium salts are deposited
   8. Trabeculae appear and join in a network to form spongy bone
   9. Appositional growth occurs by adding osseous tissue
C. Endochondral ossification (Figure 7-8)
   1. Most bones begin as a cartilage model with bone formation spreading essentially from the center to the ends
   2. Periosteum develops and enlarges to produce a collar of bone
   3. Primary ossification center forms
   4. Blood vessel enters the cartilage model at the midpoint of the diaphysis
   5. Bone grows in length as endochondral ossification progresses from the diaphysis toward each epiphysis
   6. Secondary ossification centers appear in the epiphysis, and bone growth proceeds toward the diaphysis
   7. Epiphyseal plate remains between the diaphysis and each epiphysis until bone growth in length is complete (Figure 7-10)
   8. Epiphyseal plate is composed of four layers (Figure 7-9)
      a. "Resting" cartilage cells—point of attachment joining the epiphysis to the shaft
      b. Zone of proliferation—cartilage cells undergoing active mitosis, which causes the layer to thicken and the plate to increase in length
      c. Zone of hypertrophy—older, enlarged cells undergoing degenerative changes associated with calcium deposition
      d. Zone of calcification—dead or dying cartilage cells undergoing rapid calcification

BONE GROWTH AND RESORPTION (FIGURES 7-11 AND 7-12)

A. Bones grow in diameter by the combined action of osteoclasts and osteoblasts
B. Osteoclasts enlarge the diameter of the medullary cavity
C. Osteoblasts from the periosteum build new bone around the outside of the bone

REPAIR OF BONE FRACTURES

A. Fracture—break in the continuity of a bone
B. Fracture healing (Figure 7-13)
   1. Fracture tears and destroys blood vessels that carry nutrients to osteocytes
   2. Vascular damage initiates repair sequence
   3. Callus—specialized repair tissue that binds the broken ends of the fracture together
   4. Fracture hematoma—blood clot occurring immediately after the fracture, which is then resorbed and replaced by callus

CARTILAGE

A. Characteristics
   1. Avascular connective tissue
   2. Fibers of cartilage are embedded in a firm gel
   3. Has the flexibility of firm plastic
   4. No canal system or blood vessels
   5. Chondrocytes receive oxygen and nutrients by diffusion
   6. Perichondrium—fibrous covering of the cartilage
   7. Cartilage types differ because of the amount of matrix present and the amounts of elastic and collagenous fibers
B. Types of cartilage (Figure 7-14)
   1. Hyaline cartilage
      a. Most common type
      b. Covers the articular surfaces of bones
      c. Forms the costal cartilages, cartilage rings in the trachea, bronchi of the lungs, and the tip of the nose
      d. Forms from specialized cells in centers of chondrification, which secrete matrix material
      e. Chondrocytes are isolated into lacunae
   2. Elastic cartilage
      a. Forms external ear, epiglottis, and eustachian tubes
      b. Large number of elastic fibers confers elasticity and resiliency
   3. Fibrocartilage
      a. Occurs in symphysis pubis and intervertebral disks
      b. Small quantities of matrix and abundant fibrous elements
      c. Strong and rigid
C. Histophysiology of cartilage
   1. Gristlelike nature permits cartilage to sustain great weight or serve as a shock absorber
   2. Strong yet pliable support structure
   3. Permits growth in length of long bones
D. Growth of cartilage
   1. Interstitial or endogenous growth
      a. Cartilage cells divide and secrete additional matrix
      b. Seen during childhood and early adolescence while cartilage is still soft and capable of expansion from within
   2. Appositional or exogenous growth
      a. Chondrocytes in the deep layer of the perichondrium divide and secrete matrix
      b. New matrix is deposited on the surface, thereby increasing its size
      c. Unusual in early childhood, but once initiated, continues throughout life

CYCLE OF LIFE: SKELETAL TISSUES

A. Skeleton fully ossified by mid-twenties
   1. Soft tissue may continue to grow—ossifies more slowly
B. Adults—changes occur from specific conditions
   1. Increased density and strength from exercise
   2. Decreased density and strength from pregnancy, nutritional deficiencies, and illness
C. Advanced adulthood—apparent degeneration
   1. Hard bone matrix replaced by softer connective tissue
   2. Exercise can counteract degeneration

REVIEW QUESTIONS

1. Describe the microscopic structure of bone and cartilage.
2. Describe the structure of a long bone.
3. Explain the functions of the periosteum.
4. Describe the two principal chemical components of extracellular bone.
5. List and discuss each of the major anatomical components that together constitute an osteon.
6. Compare and contrast the three major types of cells found in bone.
7. Discuss and discriminate between the sequence of steps characteristic of fracture healing.

8. Compare and contrast the basic structural elements of bone and cartilage.
9. Compare the structure and function of the three types of cartilage.

CRITICAL THINKING QUESTIONS

1. Cancer treatment may generate the need for a bone marrow transplant. Osteoporosis is a condition characterized by an excessive loss of calcium in bone. These two conditions are disruptions or failures of two bone functions. Identify these two functions and explain what their normal functioning should be.
2. Compare and contrast bone formation in intramembranous and endochondral ossification.
3. Can you make a distinction between the growth processes of cartilage and bone?
4. Explain why a bone fracture along the epiphyseal plate may have serious implications in children and young adults.
5. During the aging process, adults face the issue of a changing skeletal framework. Describe these changes and explain how these skeletal framework changes affect the health of older adults.

CYCLE OF LIFE: SKELETAL TISSUE

A. Skeleton fully formed by adulthood
B. Infant—bone continues to grow—bones move about
1. Adults—changes occur less rapidly, continuous
2. Increased density and strength (teen years)
3. Decreased density and strength from pregnancy, nutritional deficiencies, and illness
C. Advanced adulthood—support, movement, shape
1. Hard bone matrix replaced by softer connective tissue
2. Exercise can counteract degeneration

REVIEW QUESTIONS

1. Describe the microscopic structure of bone and cartilage.
2. Describe the structure of a long bone.
3. Explain the functions of the periosteum.
4. Describe the two processes by which bone is formed. Name the major chemical compounds that, together, constitute bone matrix.
6. Compare and contrast the structure of spongy bone.
7. Discuss and explain the difference between endochondral and intramembranous ossification.

CRITICAL THINKING QUESTIONS

1. Cancer treatment may require the need for a bone marrow transplant. Osteoporosis is a condition often marked by an excessive loss of calcium in bone. These two conditions are disruptions of two bone functions. Identify these two functions and explain what their normal functions should be.
2. Compare and contrast bone deposition in intramembranous and endochondral ossification.
3. Compare and contrast the extent the growth process of cartilage and bone.
4. Explain why a bone fracture along the epiphyseal plate may have a serious implication in children and not in adults.
5. During the aging process, adults face the onset of skeletal framework. Describe these changes and explain how these skeletal framework changes affect the health of older adults.

Compare and contrast the basic structural element of bone and cartilage.
Compare the structure and function of the three types of cartilage.

Skeletal System

LANGUAGE OF SCIENCE

appendicular skeleton (ah-pen-DIK-yoo-lar SKEL-eh-ton) [*appendicular* pertaining to an appendage]

axial skeleton (AK-see-all SKEL-eh-ton) [*axi* axis]

carpal bones (KAR-pul bohns) [*carp* wrist]

cervical vertebrae (SER-vi-kal VER-teh-bray) [*cervi* neck, *vertebra* joint]

clavicle (KLAV-i-kul) [*clavicle* little key]

coxal bone (KOK-sal) [*coxa* hip]

cribriform plate (KRIB-ri-form)

cranium (KRAY-nee-um) [*cranium* skull]

ethmoid (ETH-moyd) [*ethmos-* sieve, *-oid* having the appearance]

face

false pelvis (PEL-vis) [*false* to deceive, *pelvis* basin]

femurs (FEE-murs) [*femur* thigh]

fibula (FIB-yoo-lah) [*fibula* clasp]

fontanels (FON-tah-nelz) [*fontaine* fountain]

foot

frontal bone (FRON-tal) [*frons* forehead]

humerus (HYOO-mer-us) [*humerus* shoulder]

ilium (IL-ee-um) [*ilium* flank]

inferior nasal conchae (in-FEER-ee-or NAY-zal KONG-kee) [*infer* lower, *nas-* nose, *-al* pertaining to]

ischium (IS-kee-um) [*ischium* hip joint]

lacrimal bone (LAK-ri-mal) [*lacrima* tear]

lateral longitudinal arch (LAT-er-al lon-jih-TOO-di-nal) [*lateral* pertaining to the side, *longitudo* length]

longitudinal arch (lon-ji-TOO-dih-nal) [*longitudo* length]

lumbar vertebrae (LUM-bar VER-teh-bray) [*lumbus* loin, *vertebra* joint]

mandible (MAN-di-bal) [*mandere* to chew]

maxilla (mak-SIH-lah) [*maxilla* upper jaw]

medial longitudinal arch (MEE-dee-al lon-ji-TOO-dih-nal) [*medial* middle, *longitudo* length]

metacarpal bones (met-ah-KAR-pal) [*meta-* beyond, *-carp* wrist]

nasal bones (NAY-zal) [*nas-* nose, *-al* pertaining to]

Cont'd on p. 305

Just as skeletal tissues are organized to form bones, the bones are organized or grouped to form the major subdivisions of the skeletal system described below. The rigid bones lie buried within the muscles and other soft tissues, thus providing support and shape to the body. An understanding of the relationship of bones to each other and to other body structures provides a basis for understanding the function of many other organ systems. Coordinated movement, for example, is possible only because of the way bones are joined in joints and the way muscles are attached to those bones. In addition, knowledge of the placement of bones within the soft tissues assists in locating and identifying other body structures.

The adult skeleton is composed of 206 separate bones. Variations in the total number of bones in the body may occur as a result of certain anomalies, such as extra ribs, or from failure of certain small bones to fuse in the course of development.

In Chapter 7 the basic types of skeletal tissue, including bone and cartilage, were discussed. Comparisons between the structural and functional characteristics of dense (compact) and cancellous (spongy) bone provided the background for study in this chapter of individual bones and their interrelationships in the skeleton. Chapter 9 takes your studies one step farther, by considering articulations—that is, how the bones form joints.

DIVISIONS OF THE SKELETON

The human skeleton consists of two main divisions—the axial skeleton and the appendicular skeleton (Figure 8-1). Eighty bones make up the **axial skeleton.** This includes 74 bones that form the upright axis of the body and 6 tiny middle ear bones. The **appendicular skeleton** consists of 126 bones—more than half again as many as in the axial skeleton. Bones of the appendicular skeleton form the appendages to the axial skeleton: the shoulder girdles, arms, wrists, and hands and the hip girdles, legs, ankles, and feet.

One of the first things you should do in studying the skeleton is to familiarize yourself with the names of the individual bones listed in Table 8-1. Next, look at Table 8-2, which lists some terms often used to name or describe *bone markings*—specific features on an individual bone. After this preparation, begin a step-by-step exploration of the skeletal system by studying the illustrations, text, and tables that constitute the rest of this chapter.

A picture is worth a thousand words. The illustrations and tables contained in this chapter were carefully selected and compiled to assist you in visualizing and organizing the material discussed. If in addition to your textbook, you have access to individual bones or an articulated skeleton in a laboratory setting, frequent reference to chapter illustrations and tabular material will prove immensely helpful in your study efforts.

AXIAL SKELETON
Skull

Twenty-eight irregularly shaped bones form the skull (Figures 8-2 to 8-8). Figures 8-2 through 8-7 show the articulated bones of the skull in full or sectioned views. Figure 8-8 is a multipart series highlighting the individual bones and shows their relationship to the skull as a whole. As you study the skull, refer often to the illustrations and the descriptive information contained in Tables 8-3 to 8-5. The skull consists of two major divisions: the **cranium,** or brain case, and the **face.** The cranium is formed by eight bones, namely, the frontal, two parietal, two temporal, the occipital, the sphenoid, and the ethmoid (Table 8-3). The 14 bones that form the face are the two maxillae, two zygomatic (malar), two nasal, the mandible, two lacrimal, two palatine, two inferior nasal conchae (turbinates), and the vomer (Table 8-4). Note that all the face bones are paired except for the mandible and vomer. All the cranial bones, on the other hand, are single (unpaired) except for the parietal and temporal bones, which are paired. The frontal and ethmoid bones of the skull help shape the face but are not numbered among the facial bones.

Cranial Bones

The **frontal bone** forms the forehead and the anterior part of the top of the cranium (see Figure 8-8, *C*). It contains mucosa-lined, air-filled spaces, or **sinuses**—the *frontal sinuses.* The frontal sinuses, with similar sinuses in the sphenoid, ethmoid, and maxillae, are often called *paranasal sinuses* because they have narrow channels that open into the nasal cavity (Figure 8-9). The paranasal sinuses are also discussed in Chapter 23, pp. 858-859. A portion of the frontal bone forms the upper part of the orbits. It unites with the two parietal bones posteriorly in an immovable joint, or **suture**—the *coronal suture.* Several of the more prominent frontal bone markings are described in Table 8-3.

The two **parietal bones** give shape to the bulging topside of the cranium (see Figure 8-8, *A*). They form immovable joints with several bones: the *lambdoidal suture* with the occipital bone, the *squamous suture* with the temporal bone and part of the sphenoid, and the *coronal suture* with the frontal bone.

The lower sides of the cranium and part of its floor are fashioned from two **temporal bones** (see Figure 8-8, *B*). They house the middle and inner ear structures and contain the *mastoid sinuses,* notable because of the occurrence of *mastoiditis,* an inflammation of the mucous lining of these spaces (see Mechanisms of Disease, p. 303). For a description of several other temporal bone markings, see Table 8-3.

The **occipital bone** creates the framework of the lower, posterior part of the skull (see Figure 8-8, *D*). It forms immovable joints with three other cranial bones—the parietal, temporal, and sphenoid—and a movable joint with the first cervical vertebra. Table 8-3 lists a description of some of its markings.

The shape of the **sphenoid bone** resembles a bat with its wings outstretched and legs extended down and back. Note in Figures 8-4 and 8-8, *E,* the location of the sphenoid bone in the central portion of the cranial floor. Here it serves as the keystone in the architecture of the cranium and anchors the frontal, parietal, occipital, and ethmoid bones. The sphenoid bone also forms part of the lateral wall of the cranium and part of the floor of each orbit (see Figures 8-2 and 8-3). The sphenoid bone contains fairly large mucosa-lined, air-filled spaces—the *sphenoid sinuses* (see Figure 8-6). Several prominent sphenoid markings are described in Table 8-3.

Text continued on page 274

Table 8-1 — Bones of the Skeleton (206 Total)*

AXIAL SKELETON (80 BONES TOTAL)

PART OF BODY	NAME OF BODY
Skull (28 bones total)	
Cranium (8 bones)	Frontal (1)
	Parietal (2)
	Temporal (2)
	Occipital (1)
	Sphenoid (1)
	Ethmoid (1)
Face (14 bones)	Nasal (2)
	Maxillary (2)
	Zygomatic (malar) (2)
	Mandible (1)
	Lacrimal (2)
	Palatine (2)
	Inferior nasal conchae (turbinates) (2)
	Vomer (1)
Ear bones (6 bones)	Malleus (hammer) (2)
	Incus (anvil) (2)
	Stapes (stirrup) (2)
Hyoid bone (1)	
Spinal column (26 bones total)	Cervical vertebrae (7)
	Thoracic vertebrae (12)
	Lumbar vertebrae (5)
	Sacrum (1)
	Coccyx (1)
Sternum and ribs (25 bones total)	Sternum (1)
	True ribs (14)
	False ribs (10)

APPENDICULAR SKELETON (126 BONES TOTAL)

PART OF BODY	NAME OF BODY
Upper extremities (including shoulder girdle) (64 bones total)	Clavicle (2)
	Scapula (2)
	Humerus (2)
	Radius (2)
	Ulna (2)
	Carpals (16)
	Metacarpals (10)
	Phalanges (28)
Lower extremities (including hip girdle) (62 bones total)	Innominate (2)
	Fibula (2)
	Femur (2)
	Patella (2)
	Tibia (2)
	Tarsals (14)
	Metatarsals (10)
	Phalanges (28)

*An inconstant number of small, flat, round bones known as **sesamoid bones** (because of their resemblance to sesame seeds) are found in various tendons in which considerable pressure develops. Because the number of these bones varies greatly between individuals, only two of them, the patellae, have been counted among the 206 bones of the body. Generally, two of them can be found in each thumb (in the flexor tendon near the metacarpophalangeal and interphalangeal joints) and great toe, plus several others in the upper and lower extremities. **Wormian bones,** the small islets of bone frequently found in some of the cranial sutures, have not been counted in this list of 206 bones because of their variable occurrence.

Table 8-2 — Terms Used to Describe Bone Markings

TERM	MEANING
Angle	A corner
Body	The main portion of a bone
Condyle	Rounded bump; usually fits into a fossa on another bone to form a joint
Crest	Moderately raised ridge; generally a site for muscle attachment
Epicondyle	Bump near a condyle; often gives the appearance of a "bump on a bump"; for muscle attachment
Facet	Flat surface that forms a joint with another facet or flat bone
Fissure	Long, cracklike hole for blood vessels and nerves
Foramen	Round hole for vessels and nerves (pl. foramina)
Fossa	Depression; often receives an articulating bone (pl. fossae)
Head	Distinct epiphysis on a long bone, separated from the shaft by a narrowed portion (or neck)
Line	Similar to a crest but not raised as much (is often rather faint)
Margin	Edge of a flat bone or flat portion of the edge of a flat area
Meatus	Tubelike opening or channel (pl. meati)
Neck	A narrowed portion, usually at the base of a head
Notch	A V-like depression in the margin or edge of a flat area
Process	A raised area or projection
Ramus	Curved portion of a bone, like a ram's horn (pl. rami)
Sinus	Cavity within a bone
Spine	Similar to a crest but raised more; a sharp, pointed process; for muscle attachment
Sulcus	Groove or elongated depression (pl. sulci)
Trochanter	Large bump for muscle attachment (larger than a tubercle or tuberosity)
Tuberosity	Oblong, raised bump, usually for muscle attachment; a small tuberosity is called a *tubercle*

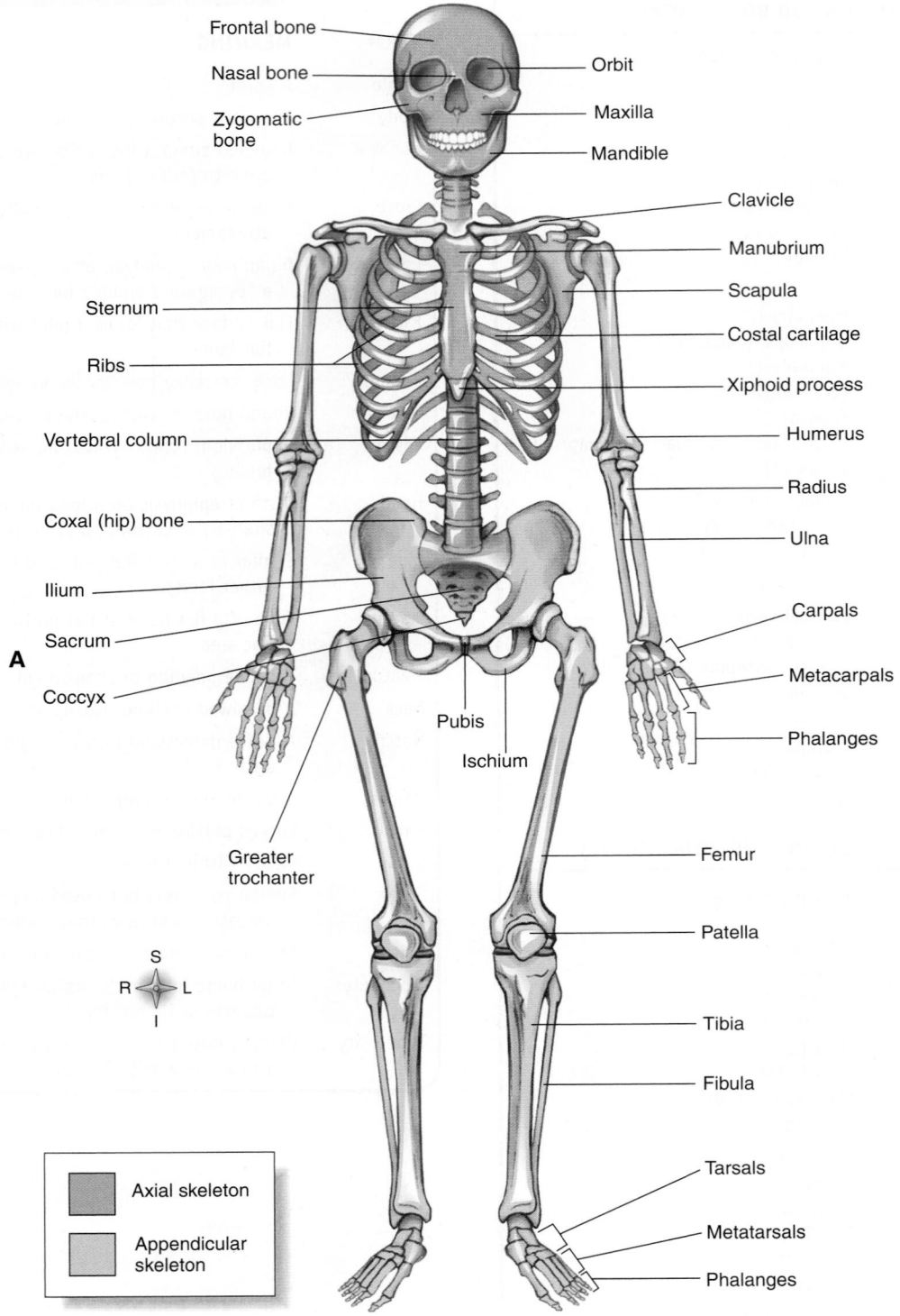

Frontal bone
Nasal bone
Zygomatic bone
Orbit
Maxilla
Mandible
Clavicle
Manubrium
Scapula
Sternum
Costal cartilage
Ribs
Xiphoid process
Vertebral column
Humerus
Radius
Coxal (hip) bone
Ulna
Ilium
Sacrum
Carpals
Metacarpals
Coccyx
Phalanges
Pubis
Ischium
Greater trochanter
Femur
Patella
Tibia
Fibula
Tarsals
Metatarsals
Phalanges

A

S
R — L
I

Axial skeleton

Appendicular skeleton

Figure 8-1 *Skeleton.* **A,** Anterior view.

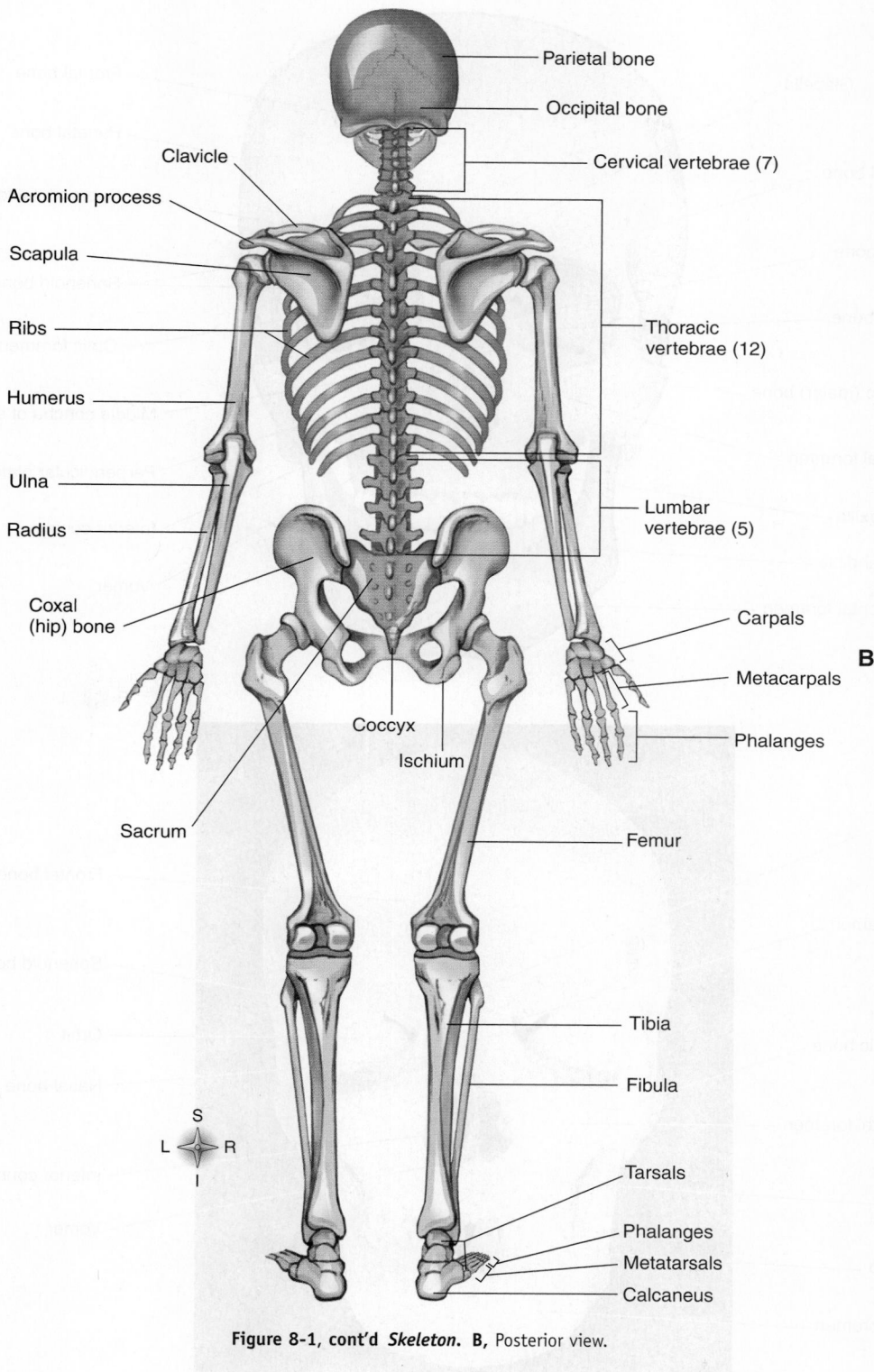

Parietal bone

Occipital bone

Clavicle

Acromion process

Cervical vertebrae (7)

Scapula

Ribs

Thoracic
vertebrae (12)

Humerus

Ulna

Radius

Lumbar
vertebrae (5)

Coxal
(hip) bone

Carpals

B

Metacarpals

Coccyx

Phalanges

Ischium

Sacrum

Femur

S

L ⬥ R

I

Tibia

Fibula

Tarsals

Phalanges

Metatarsals

Calcaneus

Figure 8-1, cont'd *Skeleton*. B, Posterior view.

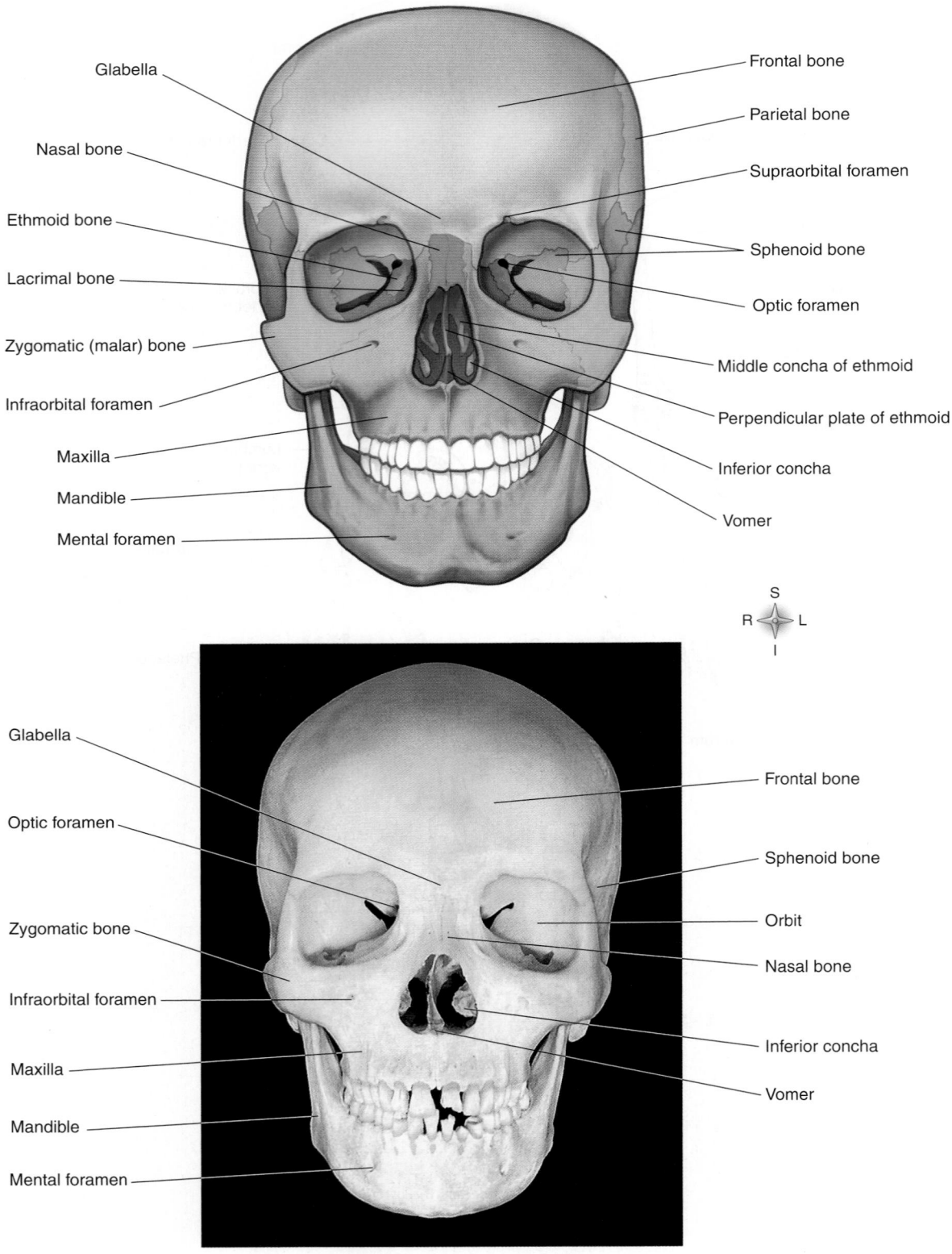

Glabella

Nasal bone

Ethmoid bone

Lacrimal bone

Zygomatic (malar) bone

Infraorbital foramen

Maxilla

Mandible

Mental foramen

Frontal bone

Parietal bone

Supraorbital foramen

Sphenoid bone

Optic foramen

Middle concha of ethmoid

Perpendicular plate of ethmoid

Inferior concha

Vomer

Glabella

Optic foramen

Zygomatic bone

Infraorbital foramen

Maxilla

Mandible

Mental foramen

Frontal bone

Sphenoid bone

Orbit

Nasal bone

Inferior concha

Vomer

Figure 8-2 *Anterior view of the skull.*

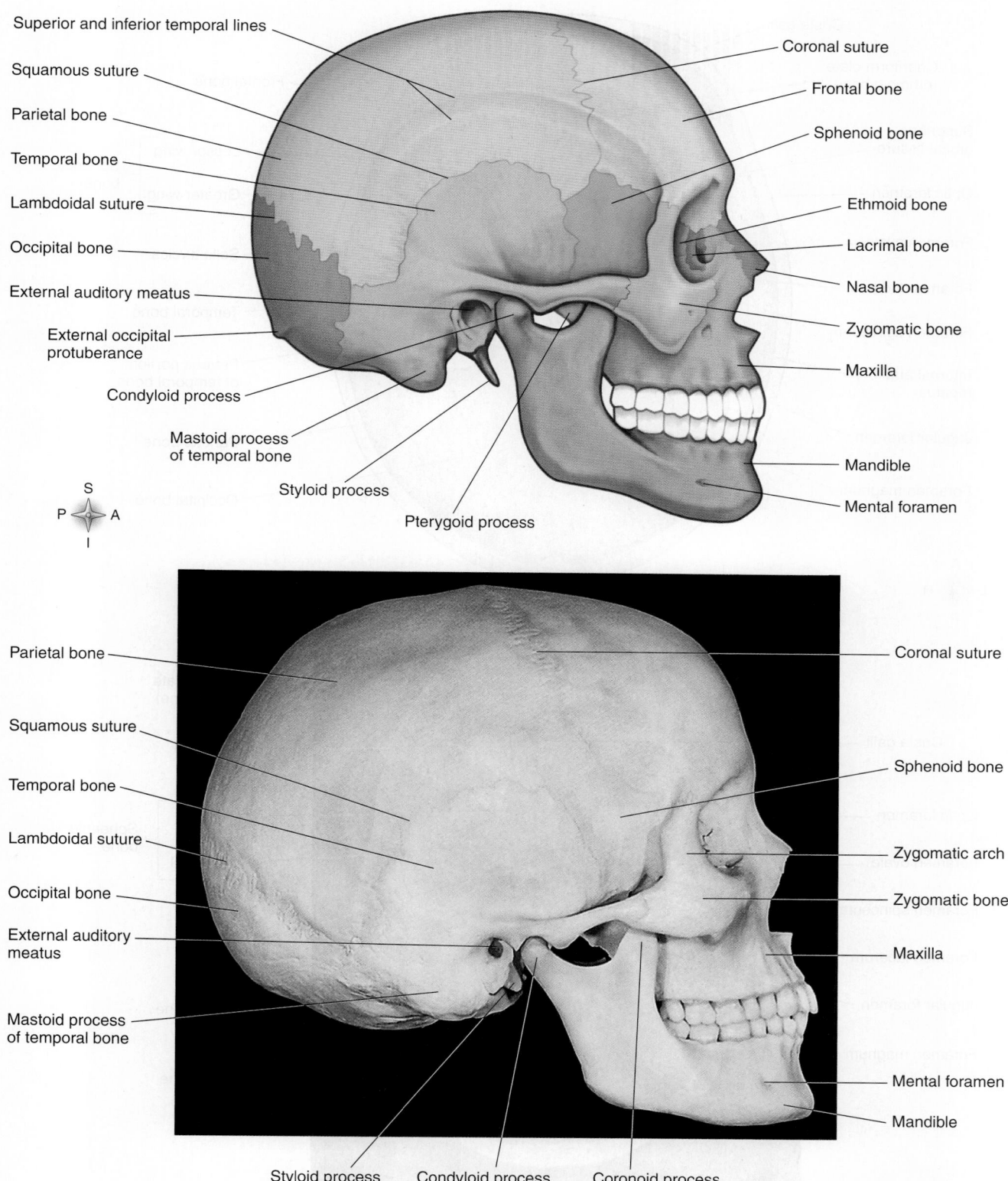

Superior and inferior temporal lines

Squamous suture

Parietal bone

Temporal bone

Lambdoidal suture

Occipital bone

External auditory meatus

External occipital protuberance

Condyloid process

Mastoid process of temporal bone

Styloid process

Pterygoid process

Coronal suture

Frontal bone

Sphenoid bone

Ethmoid bone

Lacrimal bone

Nasal bone

Zygomatic bone

Maxilla

Mandible

Mental foramen

S
P — A
I

Parietal bone

Squamous suture

Temporal bone

Lambdoidal suture

Occipital bone

External auditory meatus

Mastoid process of temporal bone

Coronal suture

Sphenoid bone

Zygomatic arch

Zygomatic bone

Maxilla

Mental foramen

Mandible

Styloid process   Condyloid process   Coronoid process

Figure 8-3 *Skull viewed from the right side.*

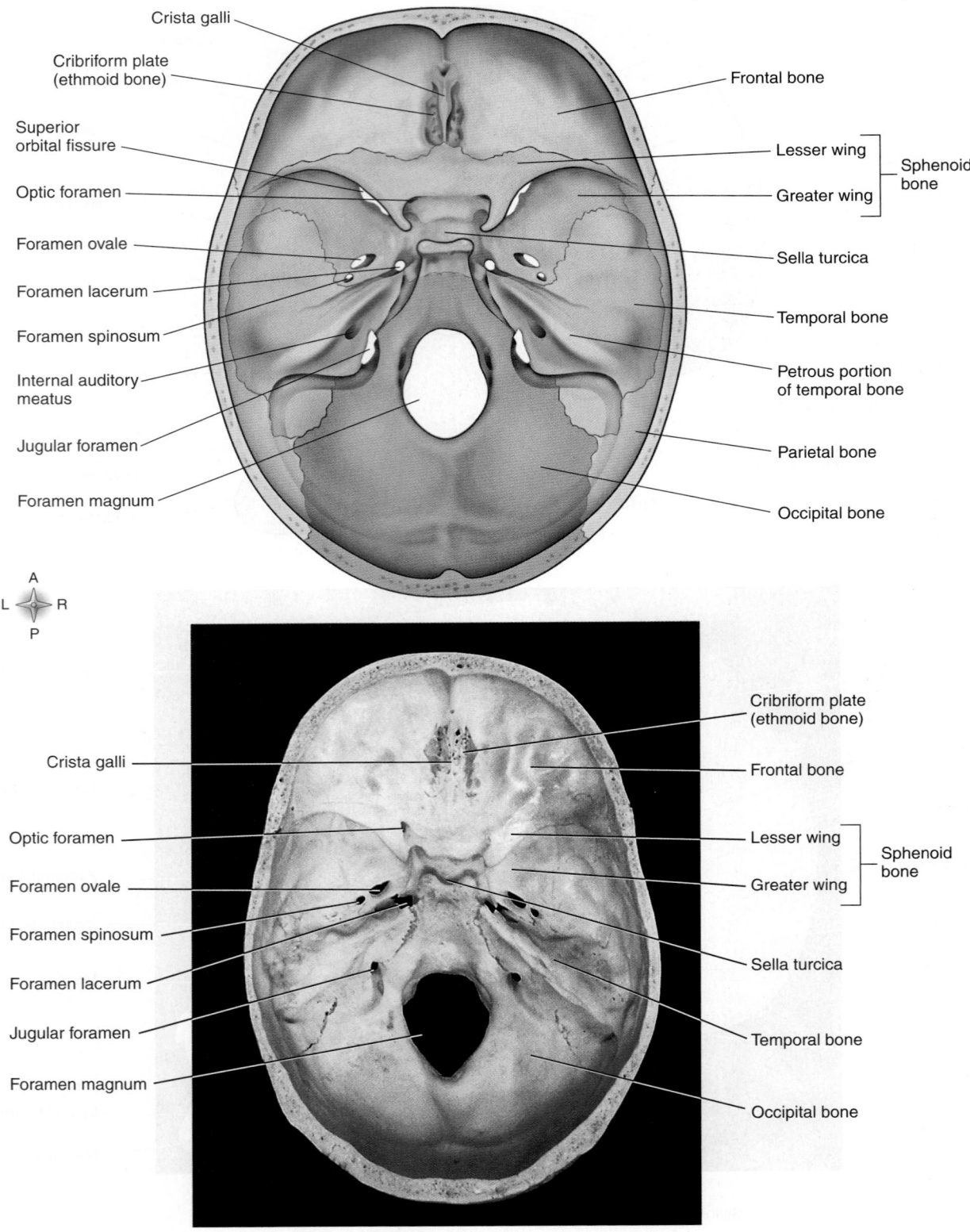

Figure 8-4 *Floor of the cranial cavity viewed from above.*

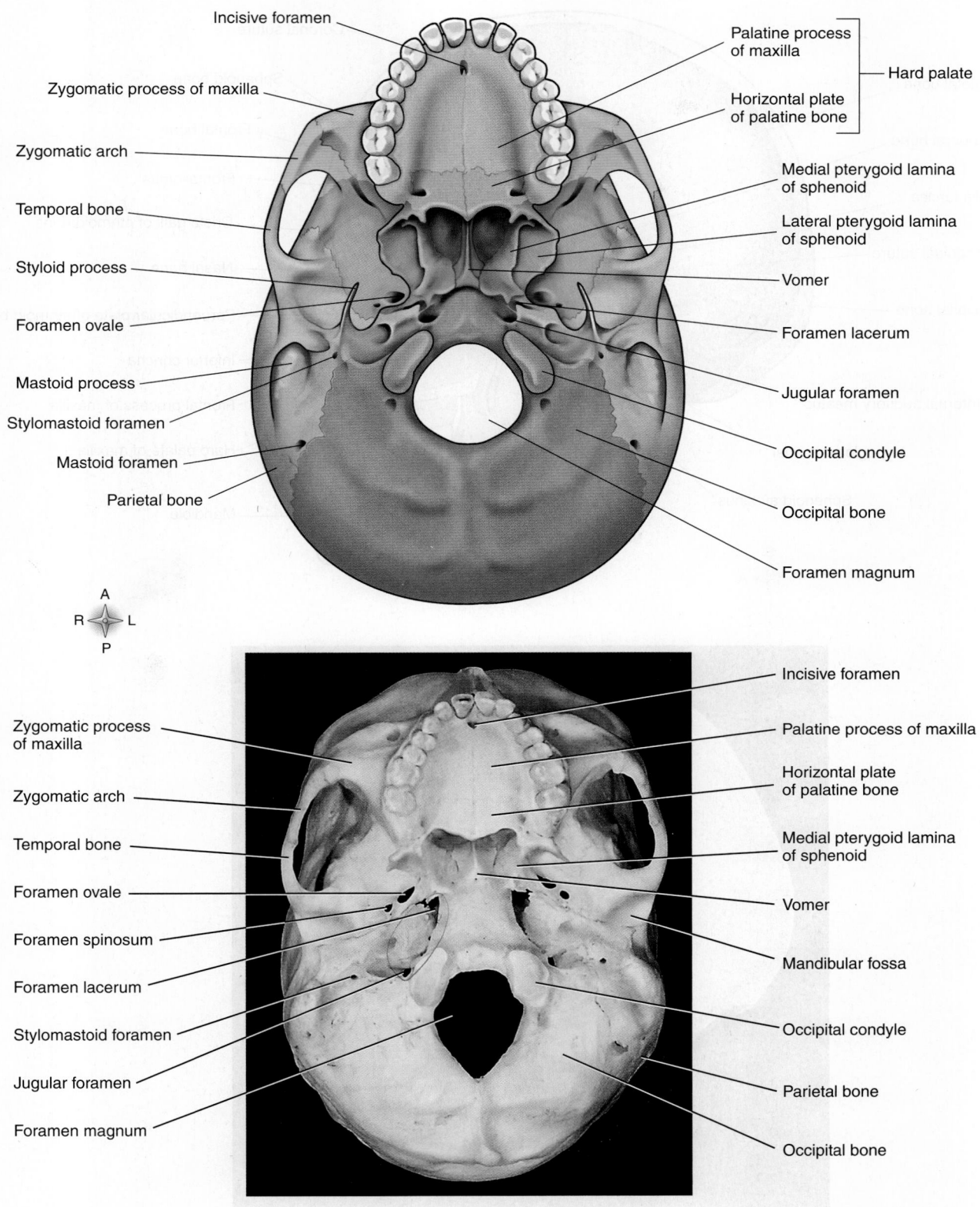

Incisive foramen

Zygomatic process of maxilla

Zygomatic arch

Temporal bone

Styloid process

Foramen ovale

Mastoid process

Stylomastoid foramen

Mastoid foramen

Parietal bone

Palatine process of maxilla

Horizontal plate of palatine bone

Hard palate

Medial pterygoid lamina of sphenoid

Lateral pterygoid lamina of sphenoid

Vomer

Foramen lacerum

Jugular foramen

Occipital condyle

Occipital bone

Foramen magnum

A
R — L
P

Zygomatic process of maxilla

Zygomatic arch

Temporal bone

Foramen ovale

Foramen spinosum

Foramen lacerum

Stylomastoid foramen

Jugular foramen

Foramen magnum

Incisive foramen

Palatine process of maxilla

Horizontal plate of palatine bone

Medial pterygoid lamina of sphenoid

Vomer

Mandibular fossa

Occipital condyle

Parietal bone

Occipital bone

Figure 8-5 *Skull viewed from below.*

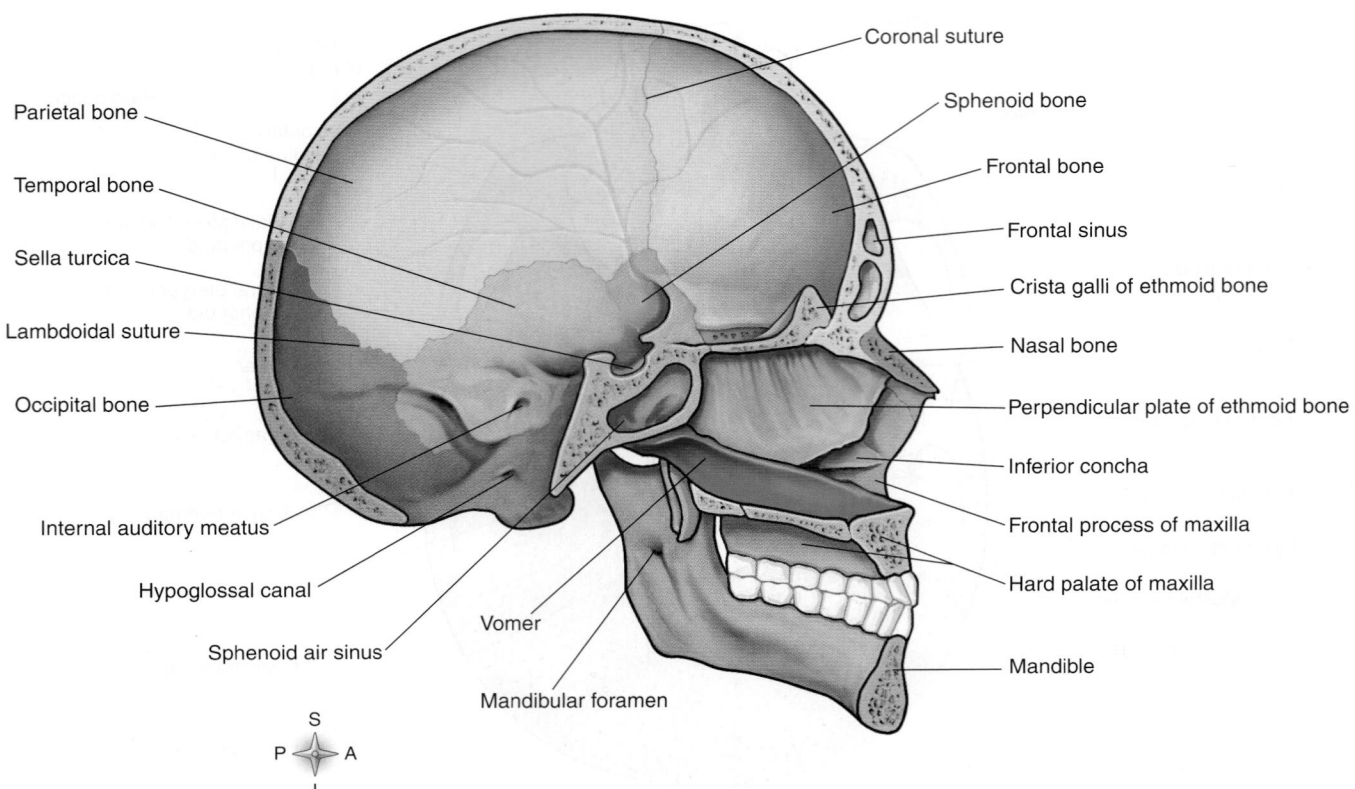

Parietal bone

Temporal bone

Sella turcica

Lambdoidal suture

Occipital bone

Internal auditory meatus

Hypoglossal canal

Sphenoid air sinus

Vomer

Mandibular foramen

Coronal suture

Sphenoid bone

Frontal bone

Frontal sinus

Crista galli of ethmoid bone

Nasal bone

Perpendicular plate of ethmoid bone

Inferior concha

Frontal process of maxilla

Hard palate of maxilla

Mandible

S
P — A
I

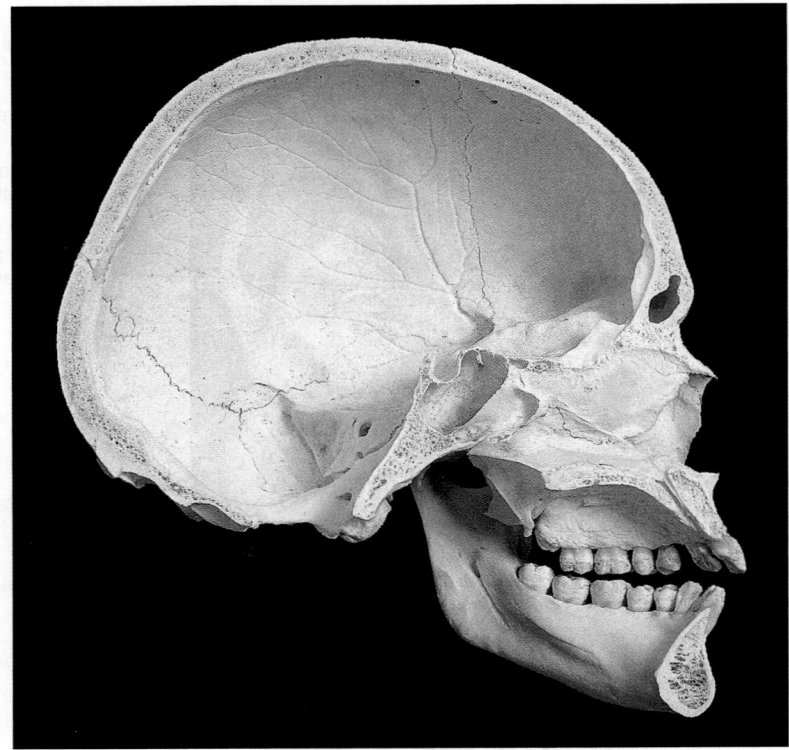

Figure 8-6 *Left half of the skull viewed from within.*

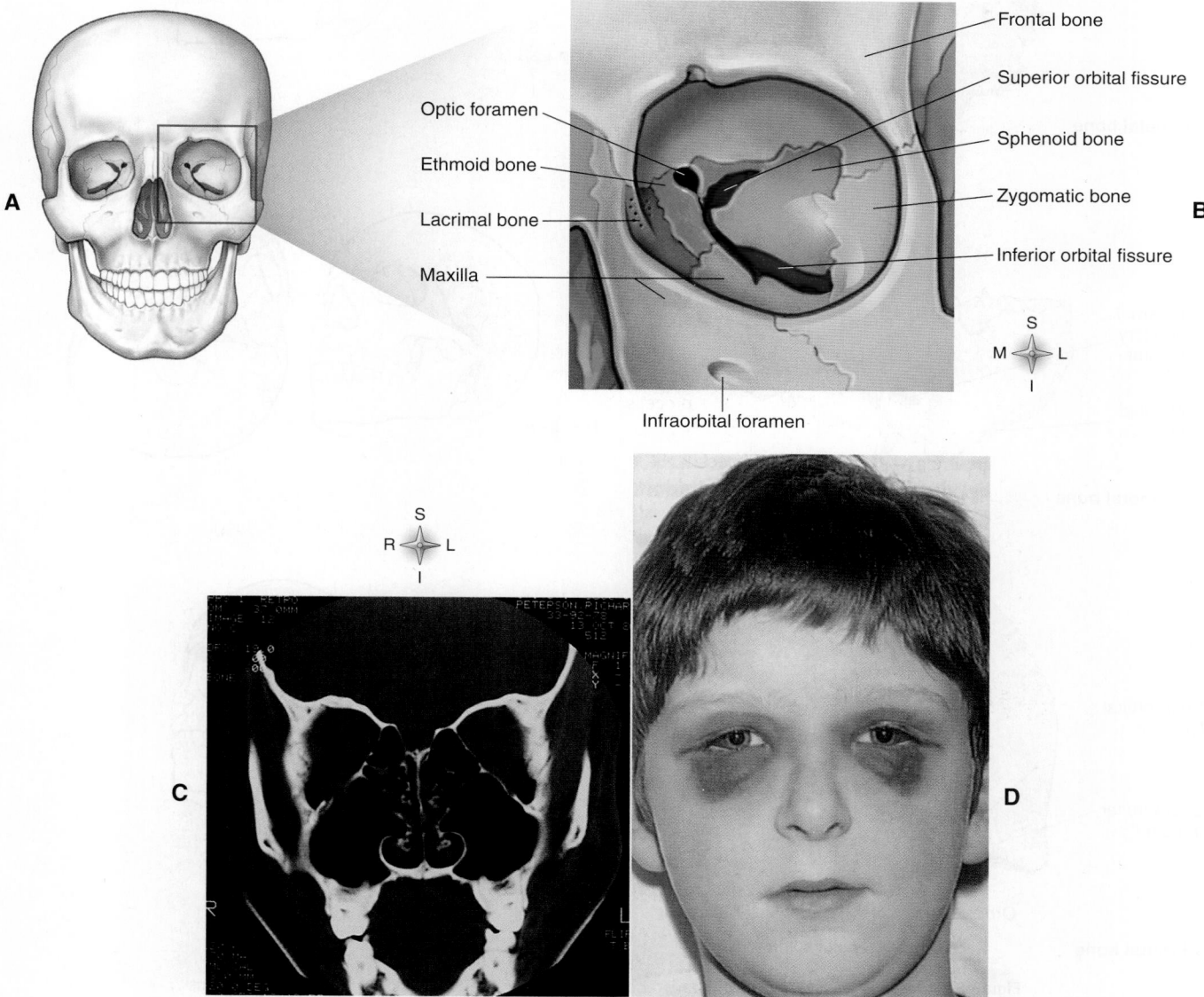

Optic foramen

Ethmoid bone

Lacrimal bone

Maxilla

Frontal bone

Superior orbital fissure

Sphenoid bone

Zygomatic bone

Inferior orbital fissure

Infraorbital foramen

Figure 8-7 *Bones of the left orbit and signs of fracture.* **A**, Skull showing an area of enlargement. **B**, Diagram showing detail of the orbital bones. **C**, CT scan showing orbital blowout fractures—the right orbital fracture is more obvious than the left. **D**, "Raccoon eyes" secondary to orbital floor fracture on both sides.

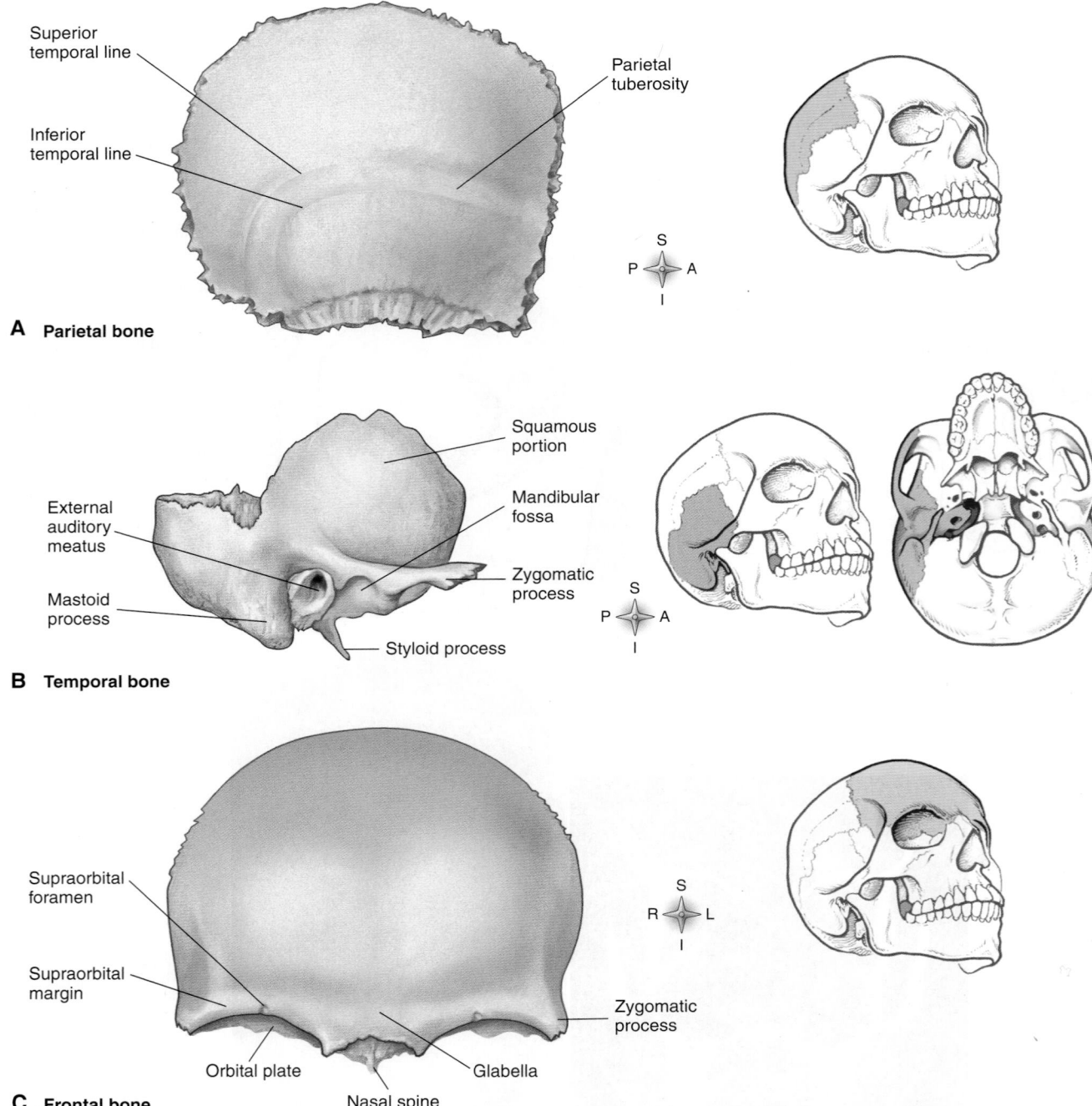

A **Parietal bone**

- Superior temporal line
- Inferior temporal line
- Parietal tuberosity

B **Temporal bone**

- External auditory meatus
- Mastoid process
- Squamous portion
- Mandibular fossa
- Zygomatic process
- Styloid process

C **Frontal bone**

- Supraorbital foramen
- Supraorbital margin
- Orbital plate
- Nasal spine
- Glabella
- Zygomatic process

Figure 8-8 *Bones of the skull.* **A,** Right parietal bone viewed from the lateral side. **B,** Right temporal bone viewed from the lateral side. **C,** Frontal bone viewed from the front and slightly above.

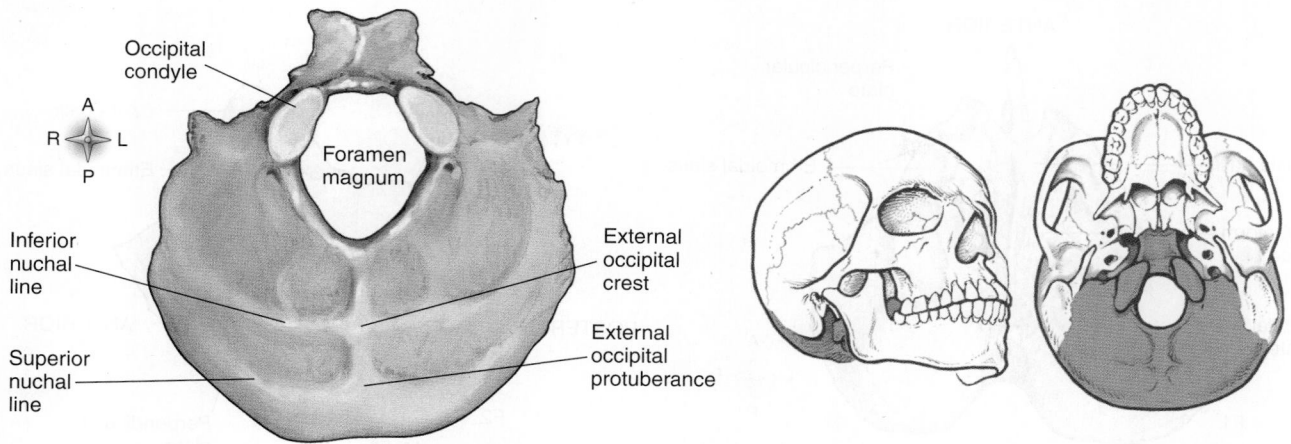

D   Occipital bone

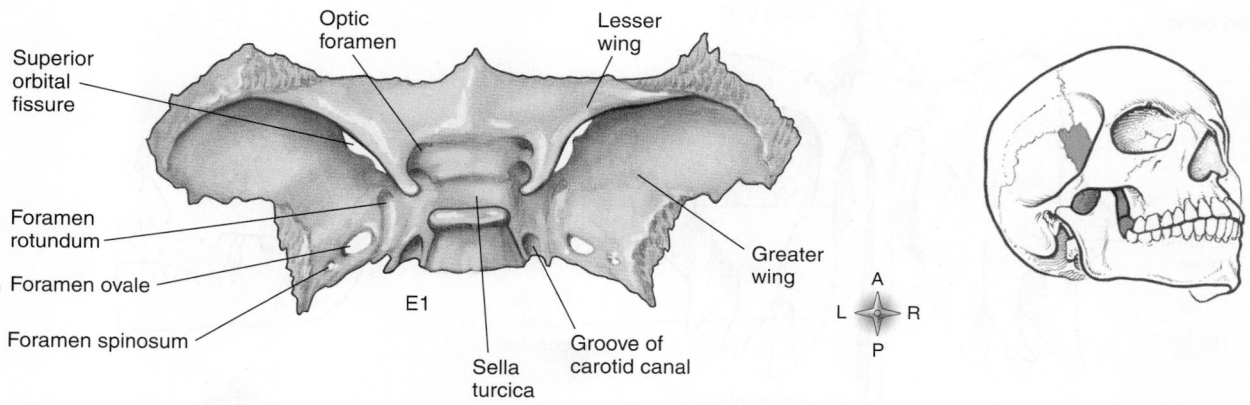

E   Sphenoid bone

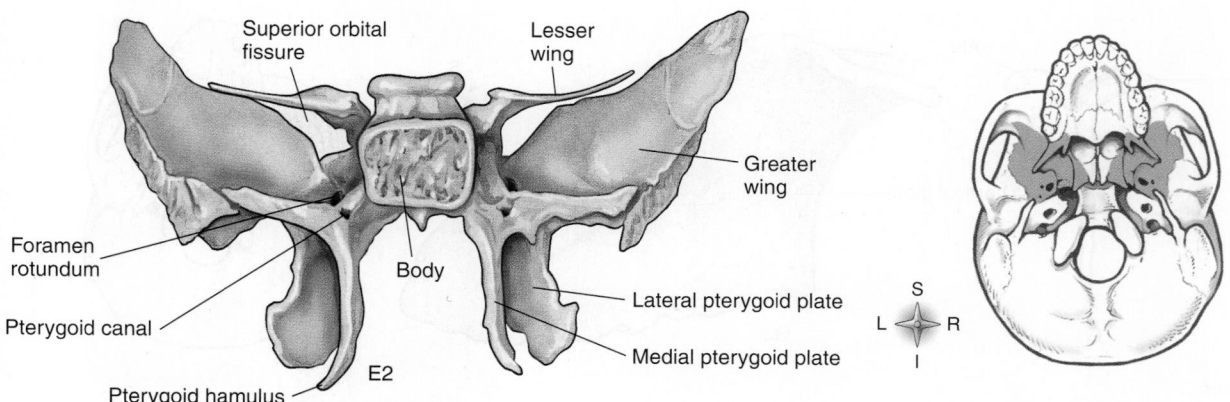

Figure 8-8, cont'd *Bones of the skull.* **D,** Occipital bone viewed from below. **E,** Sphenoid bone. *E1,* Superior view; *E2,* posterior view.   *Continued*

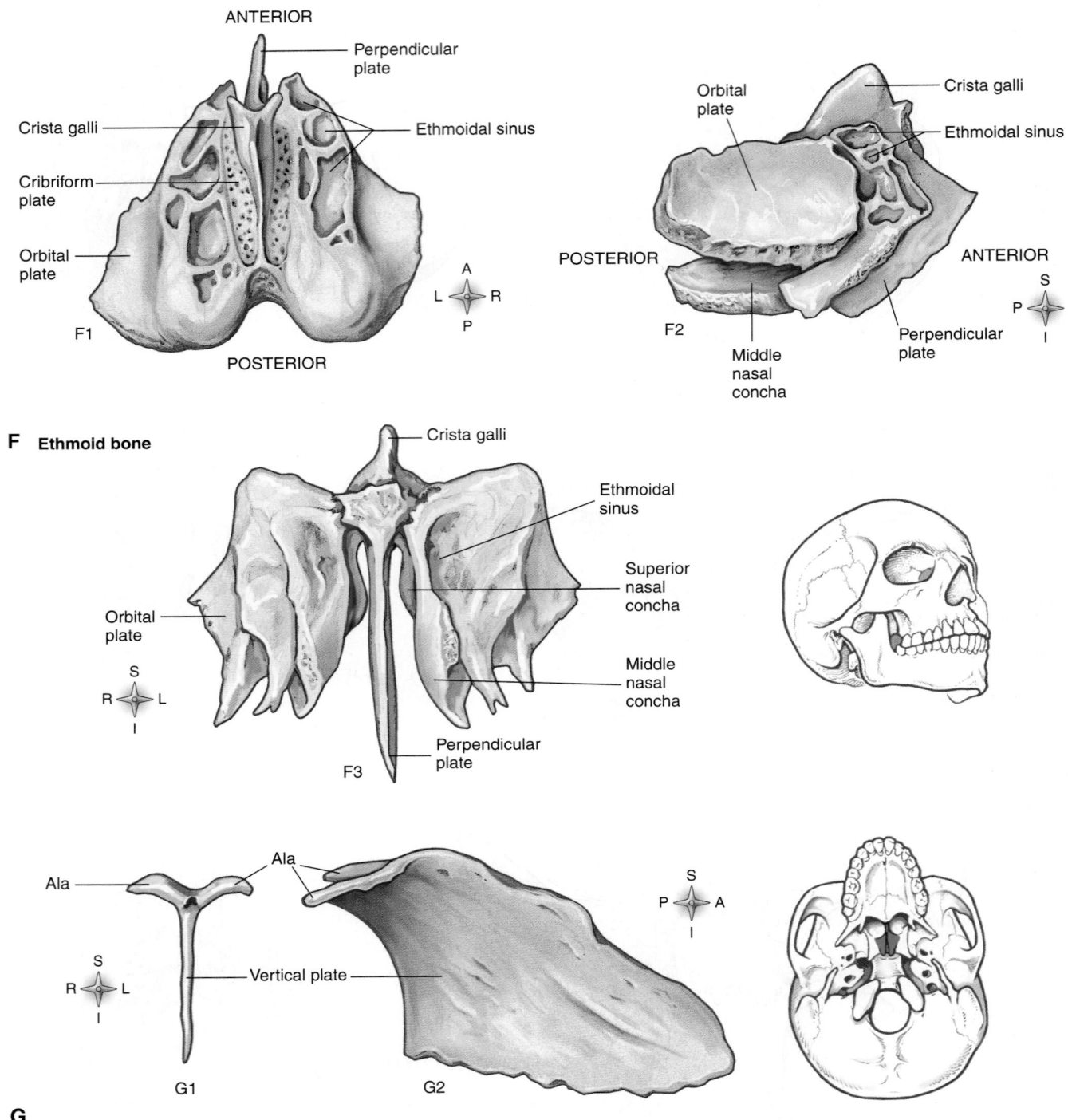

F Ethmoid bone

G Vomer

Figure 8-8, cont'd *Bones of the skull.* **F,** Ethmoid bone. *F1,* Superior view; *F2,* lateral view; *F3,* anterior view. **G,** Vomer. *G1,* Anterior view; *G2,* lateral view.

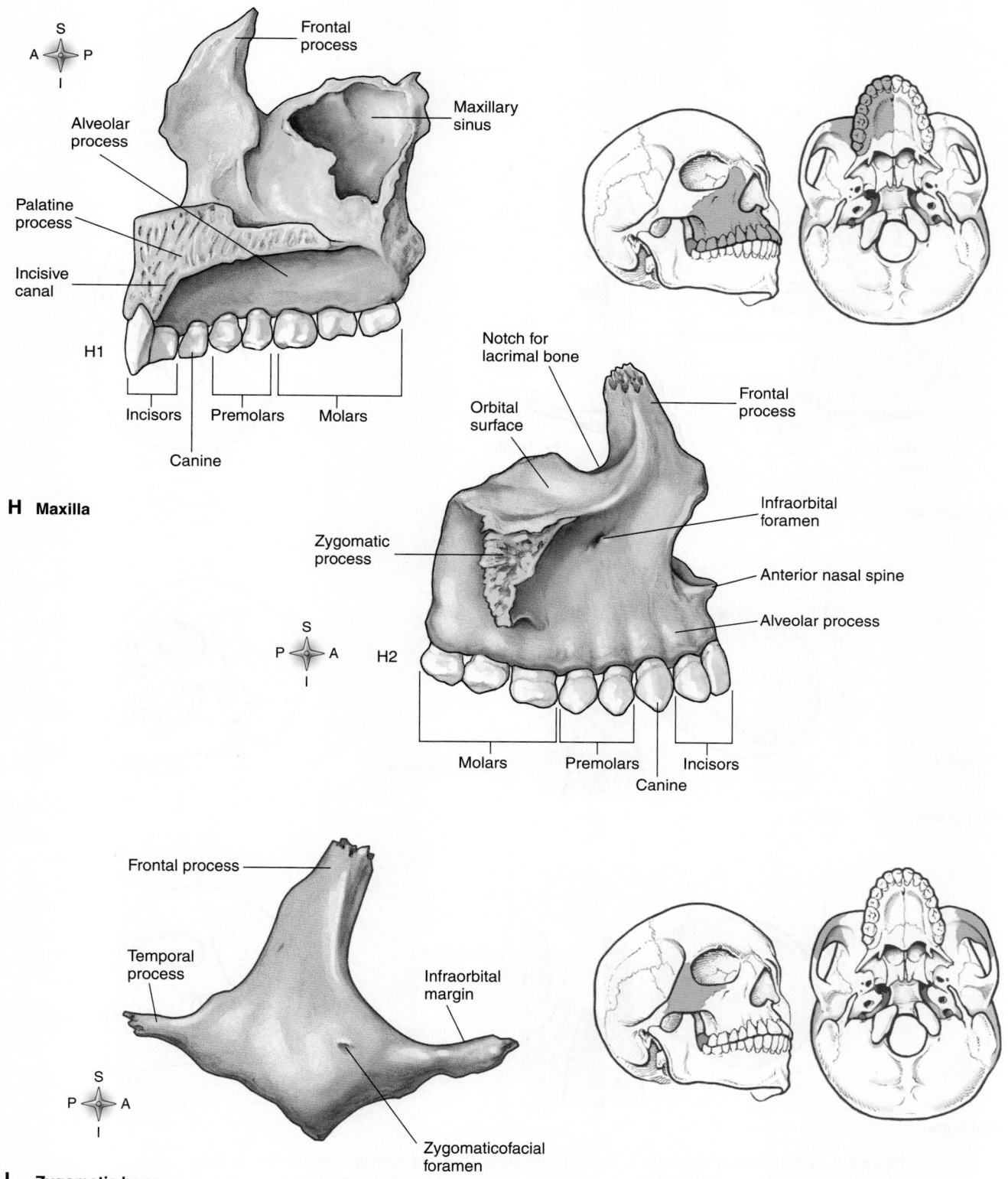

H **Maxilla**

I **Zygomatic bone**

Figure 8-8, cont'd *Bones of the skull.* **H,** Right maxilla. *H1,* Medial view; *H2,* lateral view. **I,** Right zygomatic bone viewed from the lateral side.

Continued

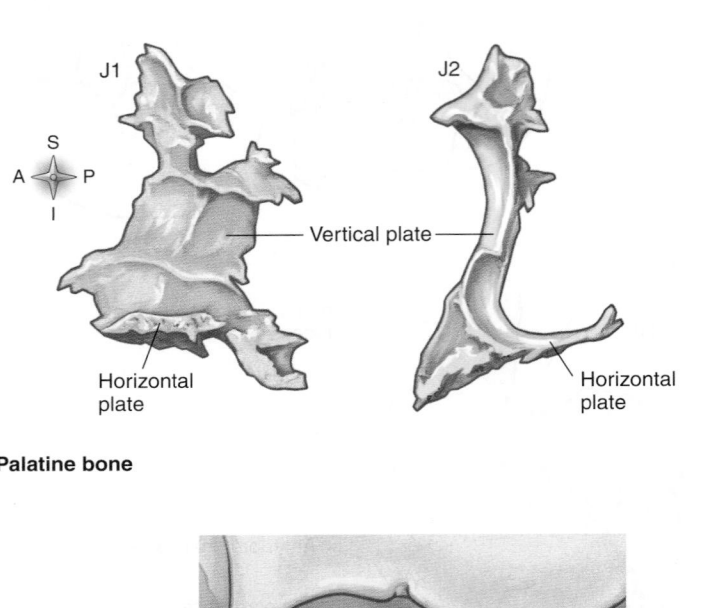

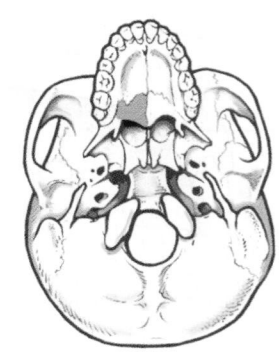

Vertical plate

Horizontal plate

Horizontal plate

J **Palatine bone**

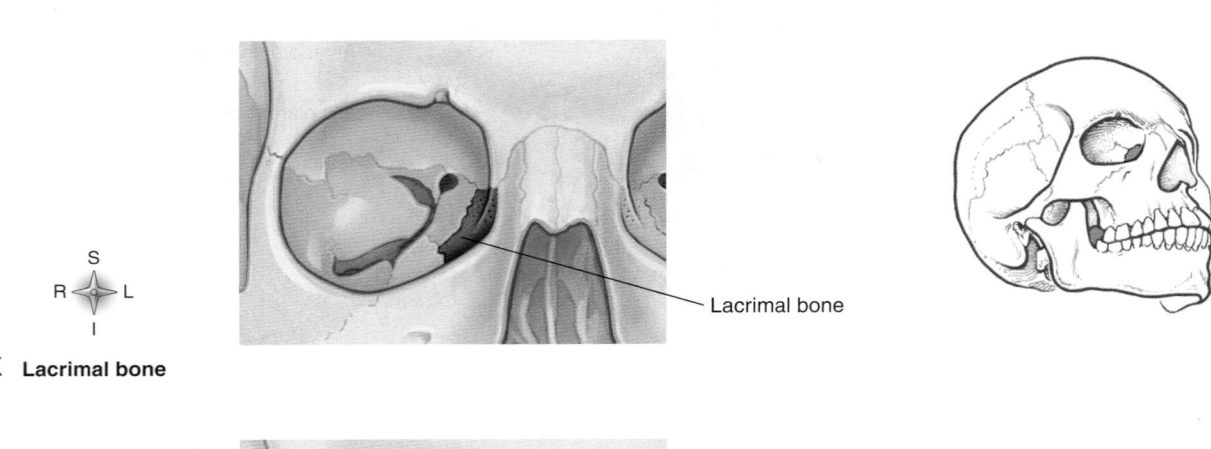

Lacrimal bone

K **Lacrimal bone**

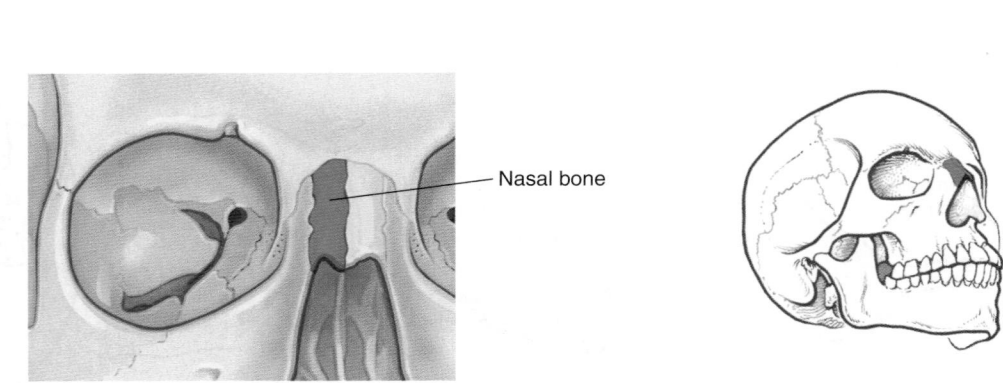

Nasal bone

L **Nasal bone**

Figure 8-8, cont'd *Bones of the skull.* **J,** Right palatine bone. *J1,* Medial view; *J2,* anterior view. **K,** Right lacrimal bone viewed from the lateral side. **L,** Right nasal bone viewed from the lateral side.

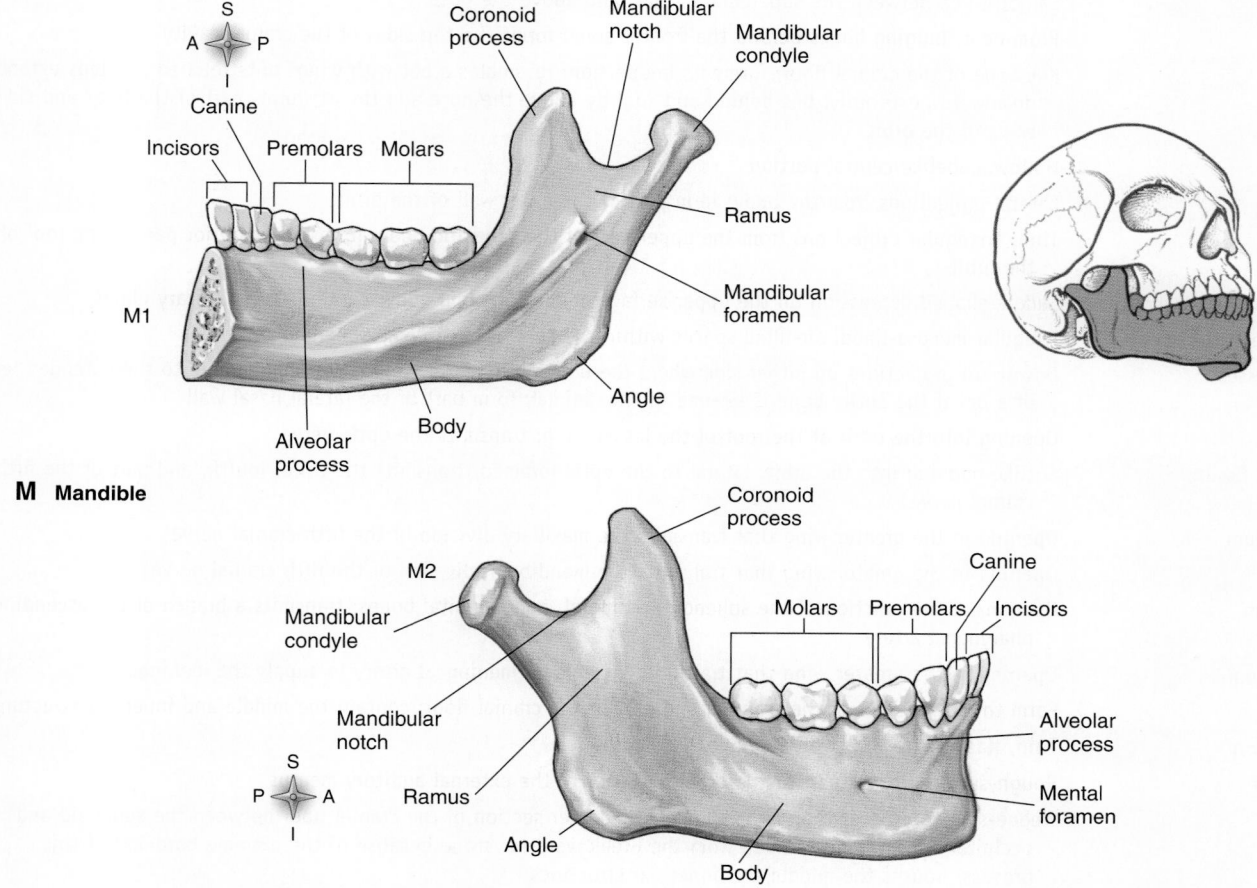

Figure 8-8, cont'd *Bones of the skull.* **M,** Right half of the mandible. *M1,* Medial view; *M2,* lateral view.

Table 8-3	Cranial Bones and Their Markings

BONES AND MARKINGS	DESCRIPTION
Frontal	Forehead bone; also forms most of the roof of the orbits (eye sockets) and the anterior part of the cranial floor
Supraorbital margin	Arched ridge just below eyebrow; forms the upper edge of the orbit
Frontal sinuses	Cavities inside the bone just above supraorbital margin; lined with mucosa; contain air
Frontal tuberosities	Bulge above each orbit; most prominent part of forehead
Superciliary ridges	Ridges caused by projection of the frontal sinuses; eyebrows lie superficial to these ridges
Supraorbital foramen (sometimes notch)	Foramen or notch in the supraorbital margin slightly medial to its midpoint; transmits supraorbital nerve and blood vessels
Glabella	Smooth area between the superciliary ridges and above the nose
Parietal	Prominent, bulging bones behind the frontal bone; forms the top sides of the cranial cavity
Sphenoid	Keystone of the cranial floor; forms its midportion; resembles a bat with wings outstretched and legs extended downward posteriorly; lies behind and slightly above the nose and throat; forms part of the floor and sidewalls of the orbit
Body	Hollow, cubelike central portion
Greater wings	Lateral projections from the body; form part of the outer wall of the orbit
Lesser wings	Thin, triangular projections from the upper part of the sphenoid body; form the posterior part of the roof of the orbit
Sella turcica (or *Turk's saddle*)	Saddle-shaped depression on the upper surface of the sphenoid body; contains the pituitary gland
Sphenoid sinuses	Irregular mucosa-lined, air-filled spaces within the central part of the sphenoid
Pterygoid processes	Downward projections on either side where the body and greater wing unite; comparable to the extended legs of a bat if the entire bone is likened to this animal; form part of the lateral nasal wall
Optic foramen	Opening into the orbit at the root of the lesser wing; transmits the optic nerve
Superior orbital fissure	Slitlike opening into the orbit; lateral to the optic foramen; transmits the third, fourth, and part of the fifth cranial nerves
Foramen rotundum	Opening in the greater wing that transmits the maxillary division of the fifth cranial nerve
Foramen ovale	Opening in the greater wing that transmits the mandibular division of the fifth cranial nerve
Foramen lacerum	Opening at the junction of the sphenoid, temporal, and occipital bones; transmits a branch of the ascending pharyngeal artery
Foramen spinosum	Opening in the greater wing that transmits the middle meningeal artery to supply the meninges
Temporal	Form the lower sides of the cranium and part of the cranial floor; contain the middle and inner ear structures
Squamous portion	Thin, flaring upper part of the bone
Mastoid portion	Rough-surfaced lower part of the bone posterior to the external auditory meatus
Petrous portion	Wedge-shaped process that forms part of the center section of the cranial floor between the sphenoid and occipital bones; name derived from the Greek word for stone because of the extreme hardness of this process; houses the middle and inner ear structures
Mastoid process	Protuberance just behind the ear
Mastoid air cells	Mucosa-lined, air-filled spaces within the mastoid process
External auditory meatus (or canal)	Tube extending into the temporal bone from the external ear opening to the tympanic membrane
Zygomatic process	Projection that articulates with the malar (or zygomatic) bone
Internal auditory meatus	Fairly large opening on the posterior surface of the petrous portion of the bone; transmits the eighth cranial nerve to the inner ear and the seventh cranial nerve on its way to the facial structures
Mandibular fossa	Oval-shaped depression anterior to the external auditory meatus; forms the socket for the condyle of the mandible
Styloid process	Slender spike of bone extending downward and forward from the undersurface of the bone anterior to the mastoid process; often broken off in a dry skull; several neck muscles and ligaments attach to the styloid process
Stylomastoid foramen	Opening between the styloid and mastoid processes where the facial nerve emerges from the cranial cavity

Table 8-3   Cranial Bones and Their Markings—cont'd

BONES AND MARKINGS	DESCRIPTION
Jugular fossa	Depression on the undersurface of the petrous portion; dilated beginning of the internal jugular vein lodged here
Jugular foramen	Opening in the suture between the petrous portion and occipital bone; transmits the lateral sinus and ninth, tenth, and eleventh cranial nerves
Carotid canal (or foramen)	Channel in the petrous portion; best seen from the undersurface of the skull; transmits the internal carotid artery
Occipital	Forms the posterior part of the cranial floor and walls
Foramen magnum	Hole through which the spinal cord enters the cranial cavity
Condyles	Convex, oval processes on either side of the foramen magnum; articulate with depressions on the first cervical vertebra
External occipital protuberance	Prominent projection on the posterior surface in the midline a short distance above the foramen magnum; can be felt as a definite bump
Superior nuchal line	Curved ridge extending laterally from the external occipital protuberance
Inferior nuchal line	Less well defined ridge paralleling the superior nuchal line a short distance below it
Internal occipital protuberance	Projection in the midline on the inner surface of the bone; grooves for the lateral sinuses extend laterally from this process and one for sagittal sinus extends upward from it
Ethmoid	Complicated irregular bone that helps make up the anterior portion of the cranial floor, medial wall of the orbits, upper parts of the nasal septum, and sidewalls and part of the nasal roof; lies anterior to the sphenoid and posterior to the nasal bones
Horizontal (cribriform) plate	Olfactory nerves pass through numerous holes in this plate
Crista galli	Meninges (membranes around the brain) attach to this process
Perpendicular plate	Forms the upper part of the nasal septum
Ethmoid sinuses	Honeycombed, mucosa-lined air spaces within the lateral masses of the bone
Superior and middle conchae (turbinates)	Help form the lateral walls of the nose
Lateral masses	Compose the sides of the bone; contain many air spaces (ethmoid cells or sinuses); the inner surface forms the superior and middle conchae

Table 8-4   Facial Bones and Their Markings

BONES AND MARKINGS	DESCRIPTION
Palatine	Form the posterior part of the hard palate, floor, and part of the sidewalls of the nasal cavity and floor of orbit
Horizontal plate	Joined to the palatine processes of the maxillae to complete part of the hard palate
Mandible	Lower jawbone; largest, strongest bone of the face
Body	Main part of the bone; forms the chin
Ramus	Process, one on either side, that projects upward from the posterior part of the body
Condyle (or head)	Part of each ramus that articulates with the mandibular fossa of the temporal bone
Neck	Constricted part just below the condyles
Alveolar process	Teeth set into this arch
Mandibular foramen	Opening on the inner surface of the ramus; transmits nerves and vessels to the lower teeth
Mental foramen	Opening on the outer surface below the space between the two bicuspids; transmits the terminal branches of the nerves and vessels that enter the bone through the mandibular foramen; dentists inject anesthetics through these foramina
Coronoid process	Projection upward from the anterior part of each ramus; the temporal muscle inserts here
Angle	Juncture of posterior and inferior margins of ramus
Maxilla	Upper jaw bones; form part of the floor of the orbit, anterior part of the roof of the mouth, and floor of the nose and part of the sidewalls of the nose

Table 8-5	Special Features of the Skull
FEATURE	**DESCRIPTION**
Sutures	Immovable joints between skull bones
Squamous	Line of articulation along the top curved edge of the temporal bone
Coronal	Joint between the parietal bones and frontal bone
Lambdoidal	Joint between the parietal bones and occipital bone
Sagittal	Joint between the right and left parietal bones
Fontanels	"Soft spots" where ossification is incomplete at birth; allow some compression of the skull during birth; also important in determining the position of the head before delivery; six such areas located at angles of the parietal bones
Frontal (or anterior)	At the intersection of the sagittal and coronal sutures (juncture of the parietal bones and frontal bone); diamond shaped; largest of the fontanels; usually closed by 1½ years of age
Occipital (or posterior)	At the intersection of the sagittal and lambdoidal sutures (juncture of the parietal bones and occipital bone); triangular; usually closed by the second month
Sphenoid (or anterolateral)	At the juncture of the frontal, parietal, temporal, and sphenoid bones
Mastoid (or posterolateral)	At the juncture of the parietal, occipital, and temporal bones; usually closed by the second year
Air Sinuses	Spaces, or cavities, within bones; those that communicate with the nose are called *paranasal sinuses* (frontal, sphenoidal, ethmoidal, and maxillary); mastoid cells communicate with the middle ear rather than the nose, so not included among the paranasal sinuses
Orbits Formed by	
Frontal	Roof of the orbit
Ethmoid	Medial wall
Lacrimal	Medial wall
Sphenoid	Lateral wall
Zygomatic	Lateral wall
Maxillary	Floor
Palatine	Floor
Nasal Septum Formed by	Partition in the midline of the nasal cavity; separates the cavity into right and left halves
Perpendicular Plate of the Ethmoid Bone	Forms the upper part of the septum
Vomer bone	Forms the lower, posterior part
Cartilage	Forms the anterior part
Wormian Bones	Small islets of bone in sutures
Malleus, Incus, Stapes	Tiny bones, referred to as *auditory ossicles,* in the middle ear cavity in the temporal bones; resemble, respectively, a miniature hammer, anvil, and stirrup

The **ethmoid,** a complicated, irregular bone, lies anterior to the sphenoid but posterior to the nasal bones. It helps fashion the anterior part of the cranial floor (see Figures 8-4 and 8-8, *F*), the medial walls of the orbits (see Figures 8-2 and 8-7), the upper parts of the nasal septum (see Figure 8-2) and the sidewalls of the nasal cavity (see Figure 8-10), and the part of the nasal roof (the cribriform plate) perforated by small foramina through which olfactory nerve branches reach the brain. The lateral masses of the ethmoid bone are honeycombed with sinus spaces (see Figure 8-9). For more ethmoid bone markings, see Table 8-3.

Facial Bones

The two **maxillae** serve as the keystone in the architecture of the face, just as the sphenoid bone acts as the keystone of the cranium. Each maxilla articulates with the other maxilla and also

with a nasal, a zygomatic, an inferior concha, and a palatine bone (see Figure 8-8, *H*). Of all the facial bones, only the mandible does not articulate with the maxillae. The maxillae form part of the floor of the orbits, part of the roof of the mouth, and part of the floor and sidewalls of the nose. Each maxilla contains a mucosa-lined space, the *maxillary sinus* (see Figure 8-9). This sinus is the largest of the paranasal sinuses, that is, sinuses connected by channels to the nasal cavity. For other markings of the maxillae, see Table 8-4.

Unlike the upper jaw, which is formed by the articulation of the two maxillae, the lower jaw, because of fusion of its halves during infancy, consists of a single bone, the **mandible** (see Figure 8-8, *M*). It is the largest, strongest bone of the face. It articulates with the temporal bone in the only movable joint of the skull. Its major markings are identified in Table 8-4.

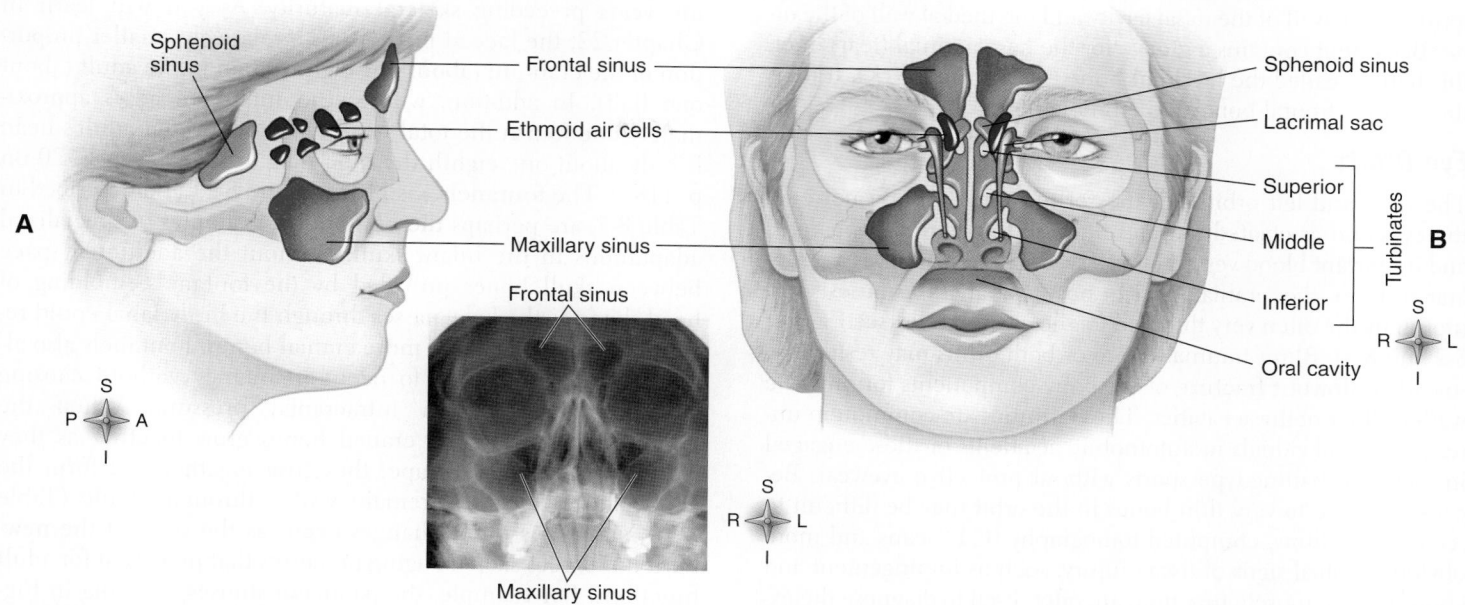

Figure 8-9 *The paranasal sinuses.* **A**, Lateral view. **B**, Frontal view. The photo inset is a "projection" radiograph (from behind) showing the outlines of the facial bones and a number of the paranasal sinuses.

BOX 8-1: HEALTH MATTERS

The Cribriform Plate

Separation of the nasal and cranial cavities by the **cribriform plate** of the ethmoid bone has great clinical significance. The cribriform plate is perforated by many small openings that permit branches of the olfactory nerve responsible for the special sense of smell to enter the cranial cavity and reach the brain. Separation of these two cavities by a thin, perforated plate of bone presents real hazards. If the cribriform plate is damaged as a result of trauma to the nose, it is possible for potentially infectious material to pass directly from the nasal cavity into the cranial fossa. If fragments of a fractured nasal bone are pushed through the cribriform plate, they may tear the coverings of the brain or enter the substance of the brain itself.

The cheek is shaped by the underlying **zygomatic**, or malar, bone (see Figure 8-8, *I*). This bone also forms the outer margin of the orbit and, with the zygomatic process of the temporal bone, makes the zygomatic arch. It articulates with four other facial bones: the maxillary, temporal, frontal, and sphenoid bones.

Shape is given to the nose by the two **nasal bones,** which form the upper part of the bridge of the nose (see Figure 8-8, *L*), and by the *septal cartilage*, which forms the lower part (see Figure 8-10). Although small, the nasal bones enter into several articulations: with the perpendicular plate of the ethmoid bone, the cartilaginous part of the nasal septum, the frontal bone, the maxillae, and each other.

An almost paper-thin bone, shaped and sized about like a fingernail, lies just posterior and lateral to each nasal bone. It helps

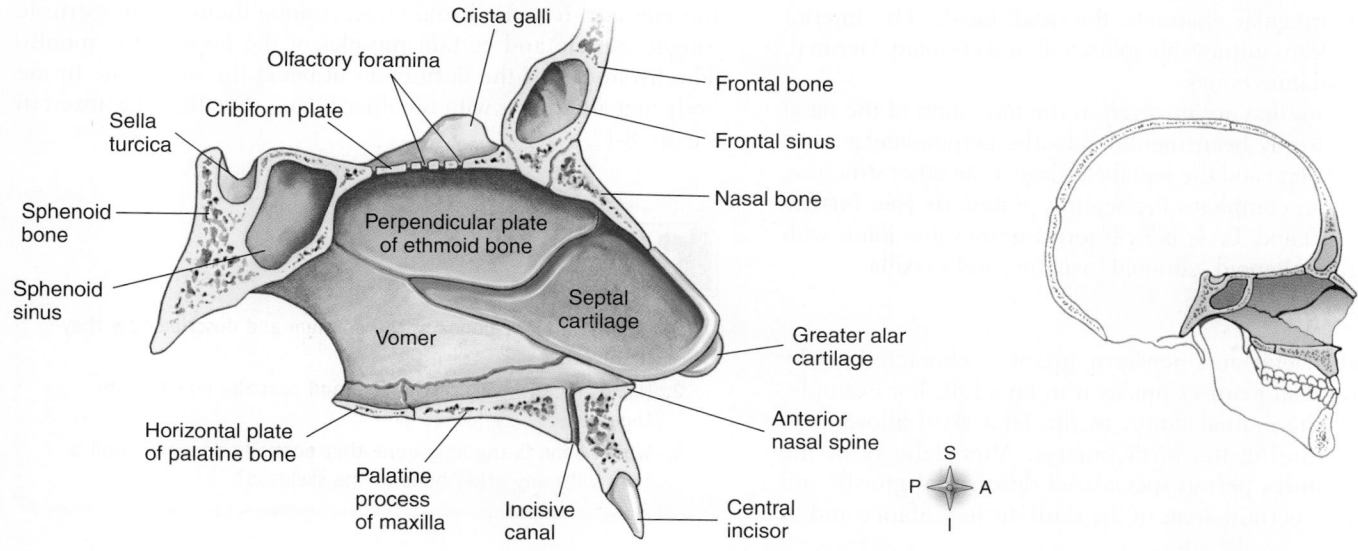

Figure 8-10 *Bones of the nasal cavity.*

form the sidewall of the nasal cavity and the medial wall of the orbit. Because it contains a groove for the nasolacrimal (tear) duct, this bone is called the **lacrimal bone** (see Figure 8-8, *K*). It joins the maxilla, frontal bone, and ethmoid bone.

Eye Orbits

The right and left orbital cavities of the skull contain not only the eyes and associated muscles but also the lacrimal apparatus and important blood vessels and nerves. These structures are separated from the cranial cavity, nose, paranasal sinuses, and mouth by the often very thin and fragile orbital walls (see Figures 8-2 and 8-7). Blunt trauma to one or both orbits may result in a so-called **blowout fracture** of the bony components forming the walls or floor of these cavities. These injuries are common in unrestrained individuals in automobile accidents or those engaged in racket or batting-type sports without protective eyewear. Because damage to very thin bones in the orbit may be difficult to see on x-ray films, computed tomography (CT) scans and more obvious clinical signs of tissue injury, such as impingement and abnormal eye muscle function, are often used to diagnose the extent of the injury. One of the classic signs of orbital fracture is the appearance of **raccoon eyes** (Figure 8-7, *D*) caused by fracture-related hemorrhage into the loose tissues surrounding the eyes. The common "black eye" is generally caused by a much less serious blow to the face and cheek than what is required to fracture an orbit.

The two **palatine bones** join to each other in the midline like two L's facing each other. Their united horizontal portions form the posterior part of the hard palate (see Figure 8-8, *J*). The vertical portion of each palatine bone forms the lateral wall of the posterior part of each nasal cavity. The palatine bones articulate with the maxillae and the sphenoid bone.

There are two **inferior nasal conchae** (turbinates). Each concha is scroll-shaped and forms a kind of ledge projecting into the nasal cavity from its lateral wall. Each nasal cavity has three such ledges. The superior and middle conchae (which are projections of the ethmoid bone) form the upper and middle ledges. The inferior concha (which is a separate bone) forms the lower ledge. They are covered by mucosa and divide each nasal cavity into three narrow, irregular channels, the *nasal meati*. The inferior nasal conchae form immovable joints with the ethmoid, lacrimal, maxilla, and palatine bones.

Two structures that are involved in the formation of the nasal septum have already been mentioned—the perpendicular plate of the ethmoid bone and the septal cartilage. One other structure, the **vomer bone,** completes the septum posteriorly (see Figures 8-8, *G*, and 8-10 and Table 8-5). It forms immovable joints with four bones: the sphenoid, ethmoid, palatine, and maxillae.

The Fetal Skull

The skull of a fetal and newborn infant is characterized by unique anatomical features not seen in an adult. For example, placement of the cranial bones in the fetal skull allows it to change shape during the birth process. After delivery of the baby, other features permit specialized differential growth and development of certain areas of the skull during infancy and in the years preceding skeletal maturity. As you will learn in Chapter 22, the face at birth forms a relatively smaller proportion of the cranium (about one eighth) than in the adult (about one half). In addition, whereas an infant's head is approximately one fourth the total height of the body, an adult's head is only about one eighth the total height (see Figure 33-20 on p. 1181). The **fontanels,** visible in Figure 8-11 and described in Table 8-5, are perhaps the best known examples of specialized adaptations in the infant skull. Without the additional space between skull bones provided by the fontanels, molding of head shape as the baby passes through the birth canal could result in fracture of one or more cranial bones. Fontanels also allow rapid brain growth to occur in infancy without causing damaging increases in intracranial pressure. When the fontanels close and the cranial bones grow together as they reach adult size and shape, they fuse together and form the adult suture lines that remain visible throughout life (Table 8-5). Numerous other changes occur as the skull of the newborn undergoes skeletal aging processes that prepare it for adult functions. For example, the paranasal sinuses, showing in Figure 8-5 and described in detail in Chapter 23, undergo dramatic changes in size and placement during the time between birth and skeletal maturity. They are listed as special features of the skull in Table 8-5. In Figure 8-11, elevations that cover the developing deciduous, or baby, teeth can be seen in the body of the mandible. Over time these teeth will erupt and eventually be replaced by the adult dentition. These and other anatomical changes in the skull are possible because of the orderly sequence of aging processes that occur between fetal life and skeletal maturity.

Other special features of the skull include the orbits, nasal septum, wormian bones, and the auditory ossicles, all of which are described in Table 8-5.

Hyoid Bone

The hyoid bone is a single bone in the neck—a part of the axial skeleton (Table 8-6). Its U shape may be felt just above the larynx (voice box) and below the mandible, where it is suspended from the styloid processes of the temporal bones (Figure 8-12). Several muscles attach to the hyoid bone. Among them are an extrinsic tongue muscle and certain muscles of the floor of the mouth. The hyoid claims the distinction of being the only bone in the body that articulates with no other bones. (See the x-ray insert in Figure 8-12.)

QUICK CHECK

1. Name the eight bones of the cranium and describe how they fit together.
2. Name the 14 bones of the face and describe how they fit together.
3. Which bone is the only bone that normally does not form a joint with any other bone of the skeleton?

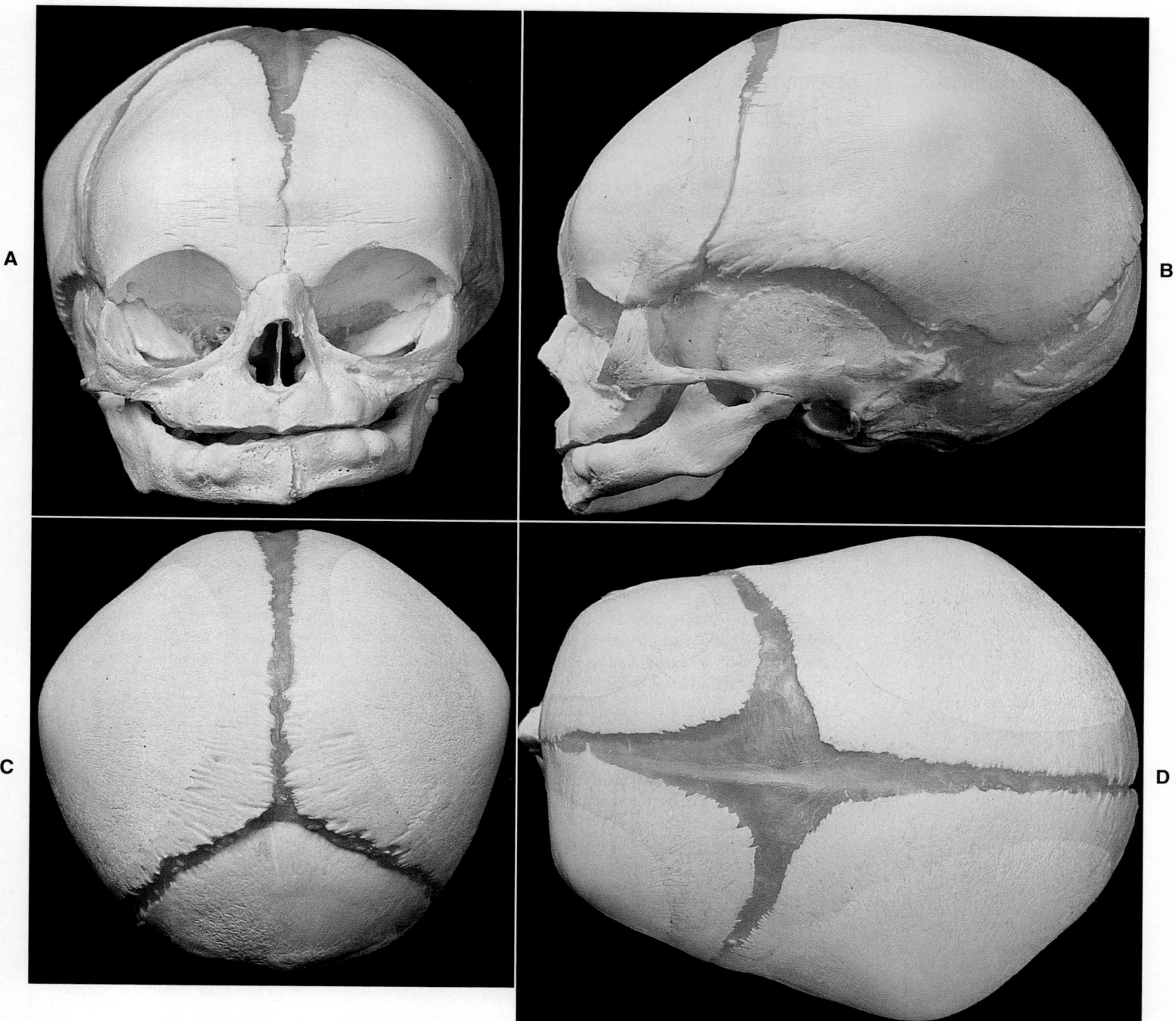

Figure 8-11 *Skull at birth.* **A,** Viewed from the front. **B,** Viewed from the left and slightly below. **C,** Viewed from behind. **D,** Viewed from above.

Vertebral Column

The vertebral, or spinal, column forms the longitudinal axis of the skeleton. It is a flexible rather than a rigid column because it is segmented. As Figure 8-13 shows, the vertebral column consists of 24 **vertebrae** plus the sacrum and coccyx. Joints between the vertebrae permit forward, backward, and sideways movement of the column. Consider too these further facts about the vertebral column. The head is balanced on top, the ribs are suspended in front, the lower extremities are attached below, and the spinal cord is enclosed within. It is indeed the "backbone" of the body.

The seven **cervical vertebrae** constitute the skeletal framework of the neck (see Figure 8-13). The next 12 vertebrae are called **thoracic vertebrae** because of their location in the posterior part of the chest, or the thoracic region. The next five, the **lumbar vertebrae,** support the small of the back. Below the lumbar vertebrae lie the *sacrum* and *coccyx.* In an adult the sacrum is a single bone that has formed from the fusion of five separate vertebrae, and the

Table 8-6	Hyoid, Vertebrae, and Thoracic Bones and Their Markings
BONES AND MARKINGS	**DESCRIPTION**
Hyoid	U-shaped bone in the neck between the mandible and upper part of the larynx; distinctive as the only bone in the body not forming a joint with any other bone; suspended by ligaments from the styloid processes of the temporal bones
Vertebral Column	Not actually a column, but a flexible, segmented curved rod; forms the axis of the body; head balanced above, ribs and viscera suspended in front, and lower extremities attached below; encloses the spinal cord
General features	Anterior part of each vertebra (except the first two cervical) consists of the body; posterior part of the vertebrae consists of the neural arch, which in turn consists of two pedicles, two laminae, and seven processes projecting from the laminae
Body	Main part; flat, round mass located anteriorly; supporting or weight-bearing part of the vertebra
Pedicles	Short projections extending posteriorly from the body
Lamina	Posterior part of the vertebra to which pedicles join and from which processes project
Neural arch	Formed by the pedicles and laminae; protects the spinal cord posteriorly; congenital absence of one or more neural arches is known as *spina bifida* (the cord may protrude right through the skin)
Spinous process	Sharp process projecting inferiorly from laminae in the midline
Transverse processes	Right and left lateral projections from laminae
Superior articulating processes	Project upward from laminae
Inferior articulating processes	Project downward from laminae; articulate with the superior articulating processes of vertebrae below
Spinal foramen	Hole in the center of the vertebra formed by union of the body, pedicles, and laminae; spinal foramina, when vertebrae are superimposed on one another, form the spinal cavity that houses the spinal cord
Intervertebral foramina	Opening between the vertebrae through which the spinal nerves emerge
Cervical Vertebrae	First or upper seven vertebrae; the foramen in each transverse process for transmission of the vertebral artery, vein, and plexus of nerves; short bifurcated spinous processes except on the seventh vertebra, where it is extra long and may be felt as a protrusion when head is bent forward; the bodies of these vertebrae are small, whereas spinal foramina are large and triangular
Atlas	First cervical vertebra; lacks a body and spinous process; superior articulating processes are concave ovals that act as rockerlike cradles for the condyles of the occipital bone; named *atlas* because it supports the head as Atlas supports the world in Greek mythology
Axis (epistropheus)	Second cervical vertebra, so named because the atlas rotates about this bone in rotating movements of the head; the *dens,* or odontoid process, is a peglike projection upward from the body of the axis that forms a pivot for rotation of the atlas
Thoracic Vertebrae	Next 12 vertebrae; 12 pairs of ribs attached to these vertebrae; stronger, with more massive bodies than the cervical vertebrae; no transverse foramina; two sets of facets for articulations with the corresponding rib: one on the body, the second on the transverse process; the upper thoracic vertebrae have elongated spinous processes
Lumbar Vertebrae	Next five vertebrae; strong, massive; superior articulating processes directed medially instead of upward; inferior articulating processes, laterally instead of downward; short, blunt spinous processes
Sacrum	Five separate vertebrae until about 25 years of age; then fused to form one wedge-shaped bone
Sacral promontory	Protuberance from the anterior, upper border of the sacrum into the pelvis; of obstetrical importance because its size limits the anteroposterior diameter of the pelvic inlet
Coccyx	Four or five separate vertebrae in a child but fused into one in an adult
Curves	Curves have great structural importance because they increase the carrying strength of the vertebral column, make balance possible in an upright position (if the column was straight, the weight of the viscera would pull the body forward), absorb jolts from walking (a straight column would transmit jolts straight to the head), and protect the column from fracture
Primary	Column curves at birth from the head to the sacrum with the convexity posteriorly; after the child stands, the convexity persists only in the *thoracic* and *sacral* regions, which are therefore called *primary curves*
Secondary	Concavities in the *cervical* and *lumbar* regions; the cervical concavity results from the infant's attempts to hold the head erect (2 to 4 months); the lumbar concavity, from balancing efforts in learning to walk (10 to 18 months)
Sternum	Breastbone; flat dagger-shaped bone; the sternum, ribs, and thoracic vertebrae together form a bony cage known as the *thorax*

Table 8-6   **Hyoid, Vertebrae, and Thoracic Bones and Their Markings—cont'd**

BONES AND MARKINGS	DESCRIPTION
Body	Main central part of the bone
Manubrium	Flaring, upper part
Xiphoid process	Projection of cartilage at the lower border of the bone
Ribs	
True ribs	Upper seven pairs; fasten to the sternum by costal cartilages
False ribs	False ribs do not attach to the sternum directly; the upper three pairs of false ribs attach by means of the costal cartilage of the seventh ribs; the last two pairs do not attach to the sternum at all and are therefore called "floating" ribs
Head	Projection at the posterior end of a rib; articulates with the corresponding thoracic vertebra and one above, except the last three pairs, which join the corresponding vertebrae only
Neck	Constricted portion just below the head
Tubercle	Small knob just below the neck; articulates with the transverse process of the corresponding thoracic vertebra; missing in the lowest three ribs
Body or shaft	Main part of a rib
Costal cartilage	Cartilage at the sternal end of true ribs; attaches ribs (except floating ribs) to the sternum

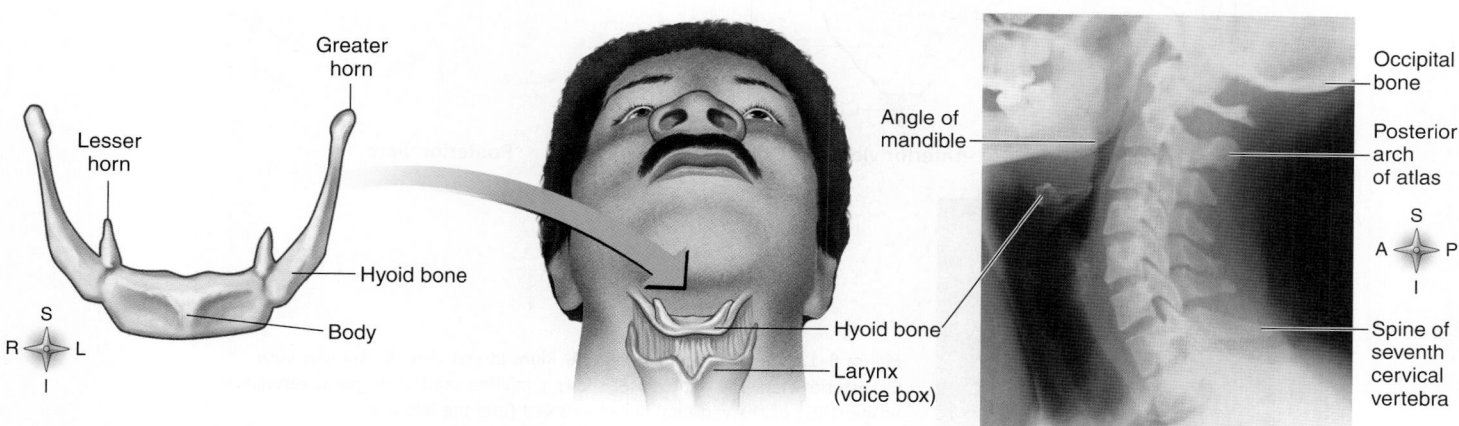

Figure 8-12 *Hyoid bone.* The photo insert (radiograph) shows the relationship of the hyoid bone to the base of the skull and cervical spine. Note that the hyoid does not articulate with any other bony structure.

coccyx is a single bone that has formed from the fusion of four or five vertebrae.

All the vertebrae resemble each other in certain features and differ in others. For example, all except the first cervical vertebra have a flat, rounded body placed anteriorly and centrally, plus a sharp or blunt *spinous process* projecting inferiorly in the posterior midline and two transverse processes projecting laterally (Figure 8-14). All but the sacrum and coccyx have a central opening, the *vertebral foramen*. An upward projection (the *dens*) from the body of the second cervical vertebra furnishes an axis for rotating the head. A long, blunt spinous process, which can be felt at the back of the base of the neck, characterizes the seventh cervical ver-

tebra. Each thoracic vertebra has articular facets for the ribs. More detailed descriptions of separate vertebrae are given in Table 8-6. The vertebral column, as a whole, articulates with the head, ribs, and iliac bones. Individual vertebrae articulate with each other in joints between their bodies and between their articular processes.

To increase the carrying strength of the vertebral column and to make balance possible in the upright position, the vertebral column is curved. At birth there is a continuous posterior convexity from the head to the coccyx. Later, as the child learns to sit and stand, secondary posterior concavities necessary for balance develop in the cervical and lumbar regions (see Figures 33-21 and 33-22, pp. 1182-1183).

Text continued on page 284

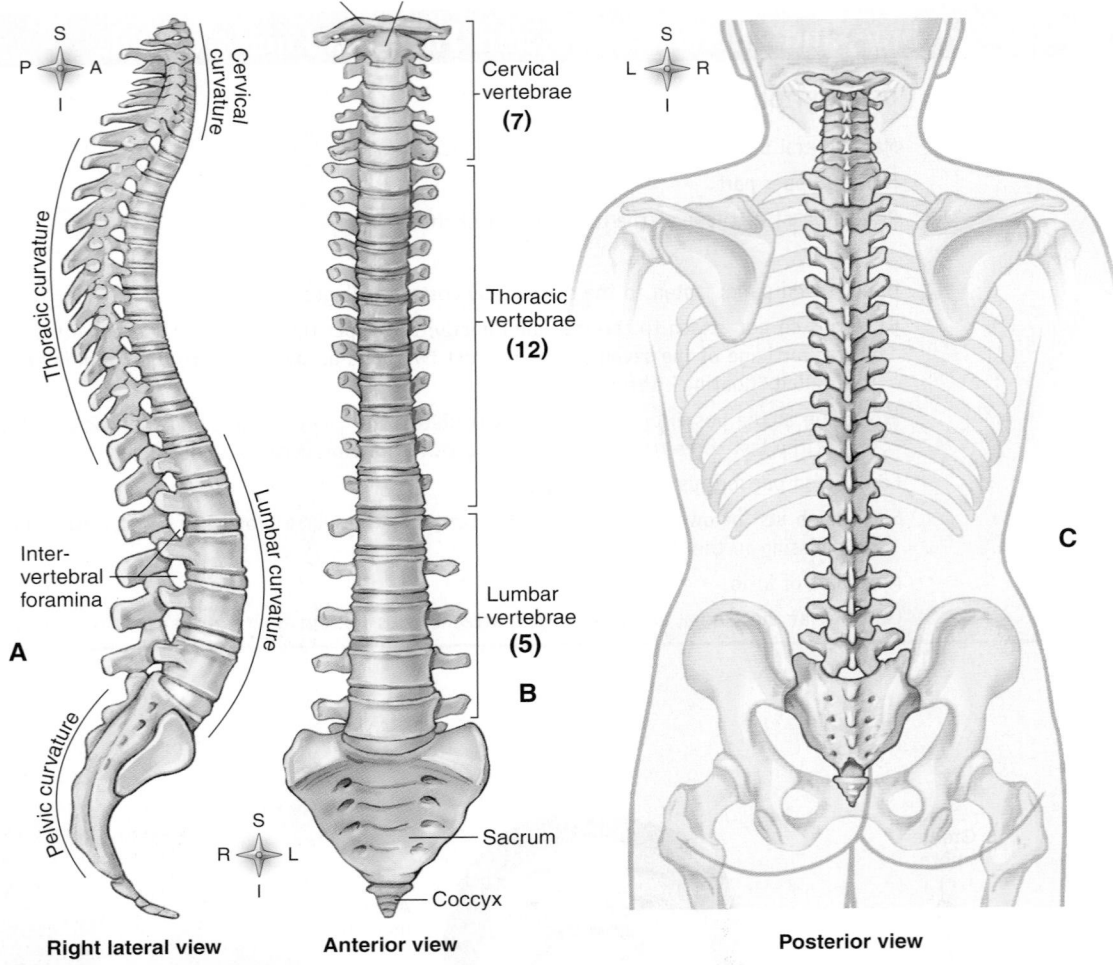

Cervical
curvature

Thoracic curvature

Inter-
vertebral
foramina

A

Lumbar curvature

Pelvic curvature

Cervical
vertebrae
(7)

Thoracic
vertebrae
(12)

Lumbar
vertebrae
(5)

B

Sacrum

Coccyx

Right lateral view   **Anterior view**   **Posterior view**

C

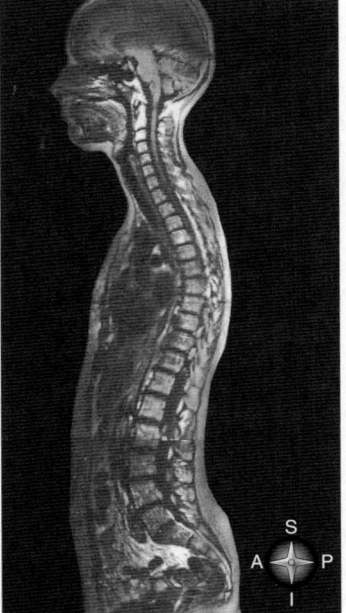

Figure 8-13 *The vertebral column.* **A**, Right lateral view. **B**, Anterior view.
C, Posterior view. The photo inset shows a midline sagittal magnetic resonance
image (MRI) of the vertebral column viewed from the left side.

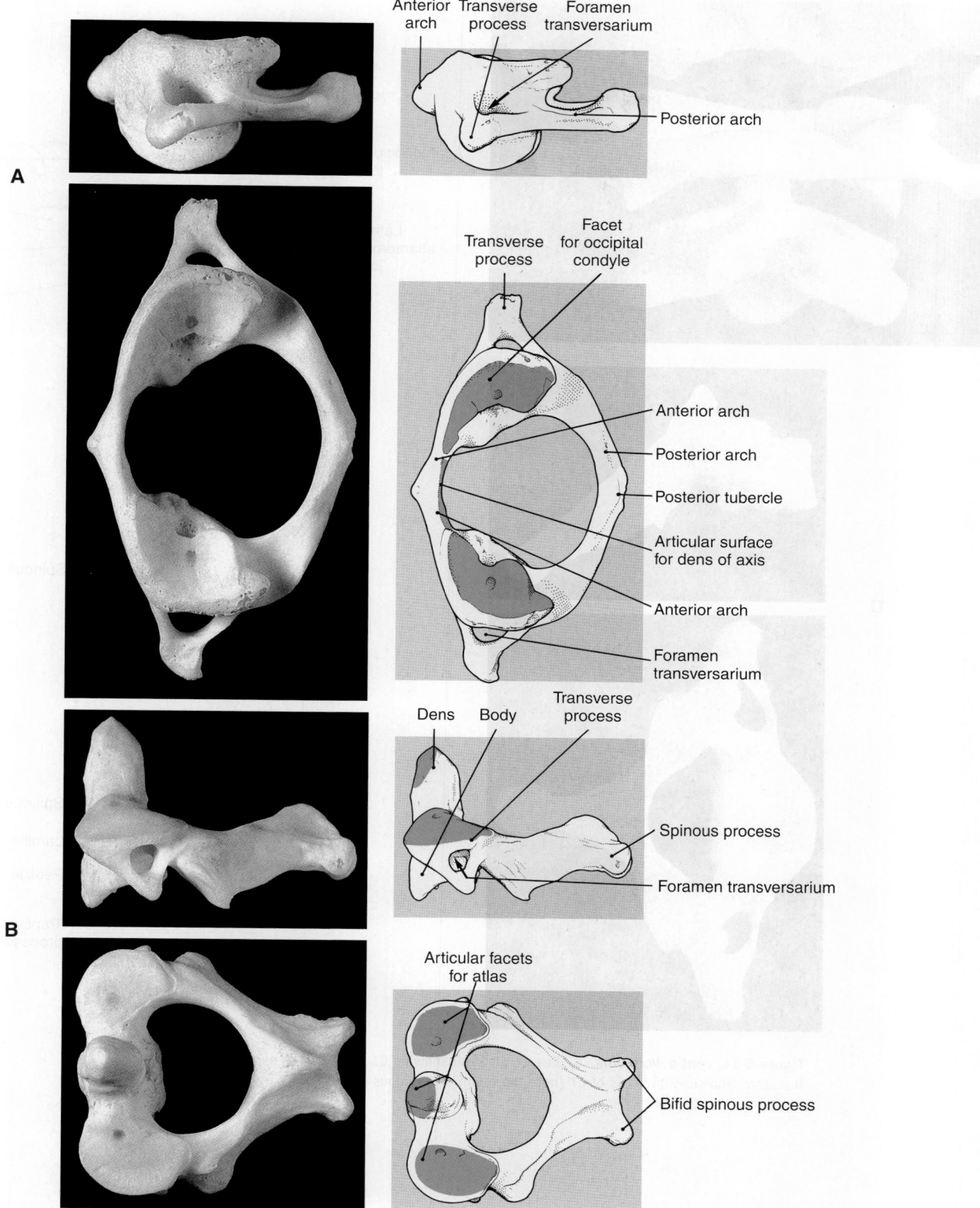

Figure 8-14 *Vertebrae.* **A,** Lateral and superior views of C1, the atlas. **B,** Lateral and superior views of C2, the axis.

Continued

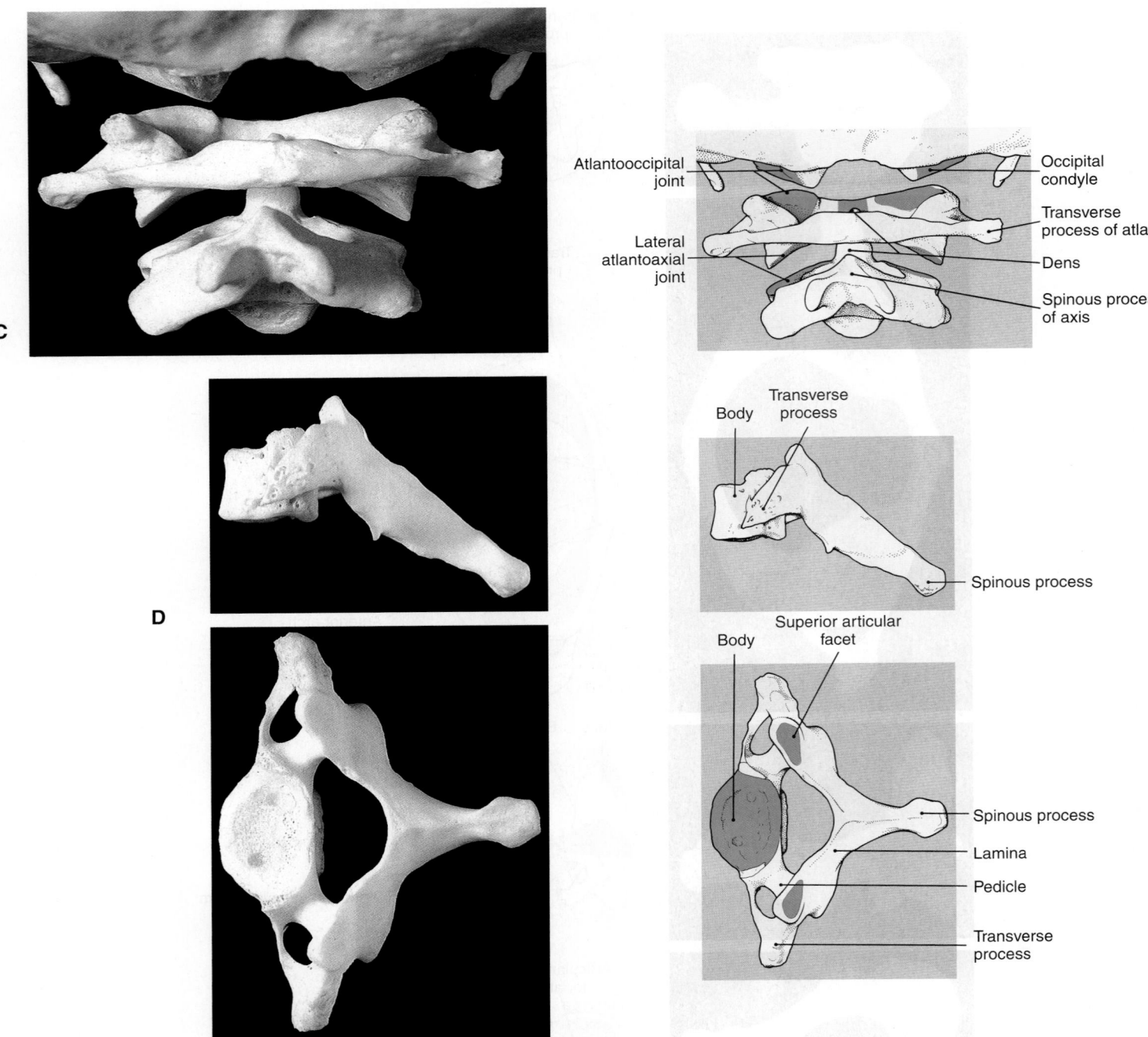

Figure 8-14, cont'd *Vertebrae.* **C,** Base of the skull showing C1 and C2 in an expanded posterior view. **D,** Lateral and superior views of C7 (note the prominent spinous process).

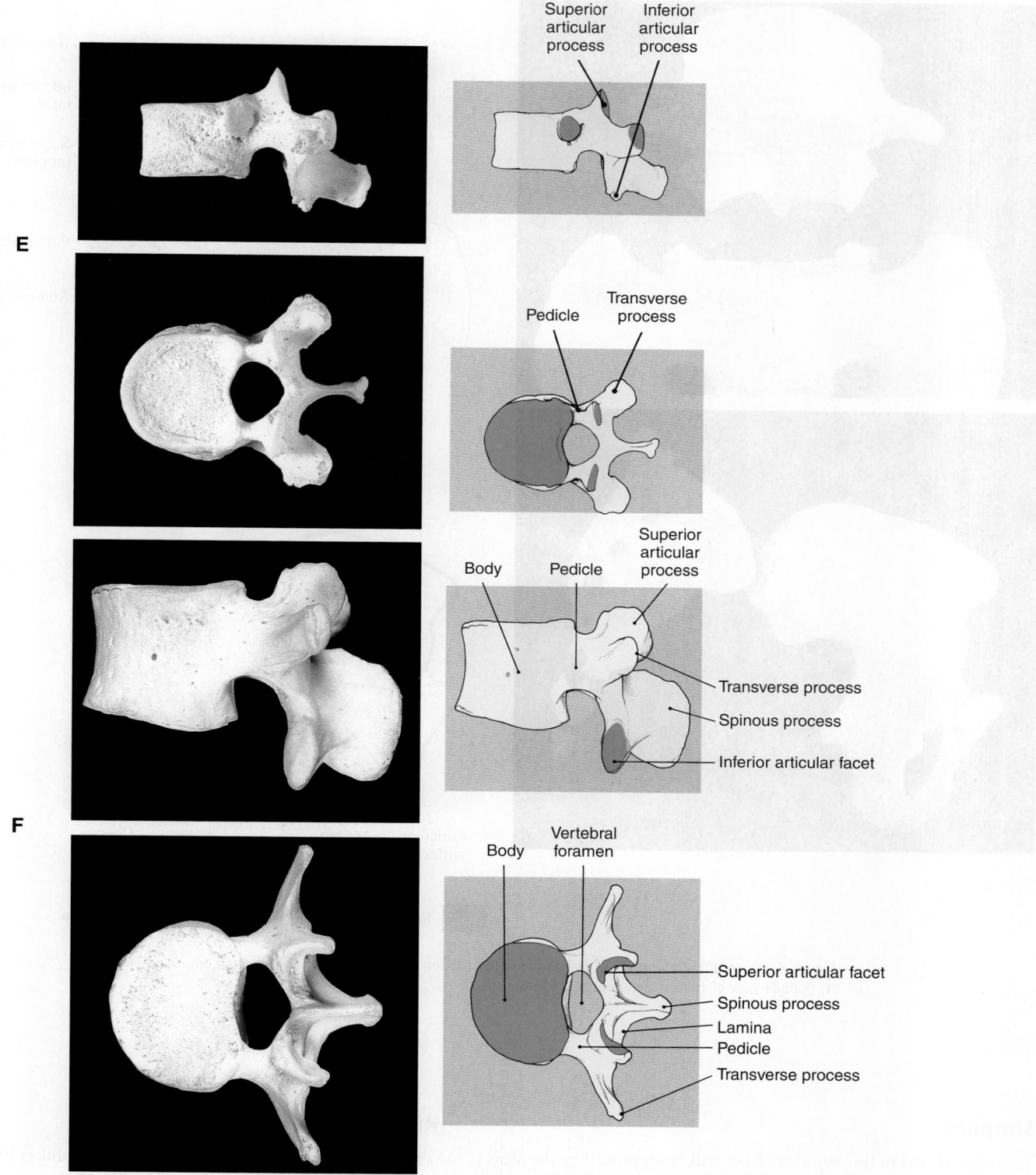

Figure 8-14, cont'd *Vertebrae.* **E,** Lateral and superior views of T10. **F,** Lateral and superior views of L3.

Continued

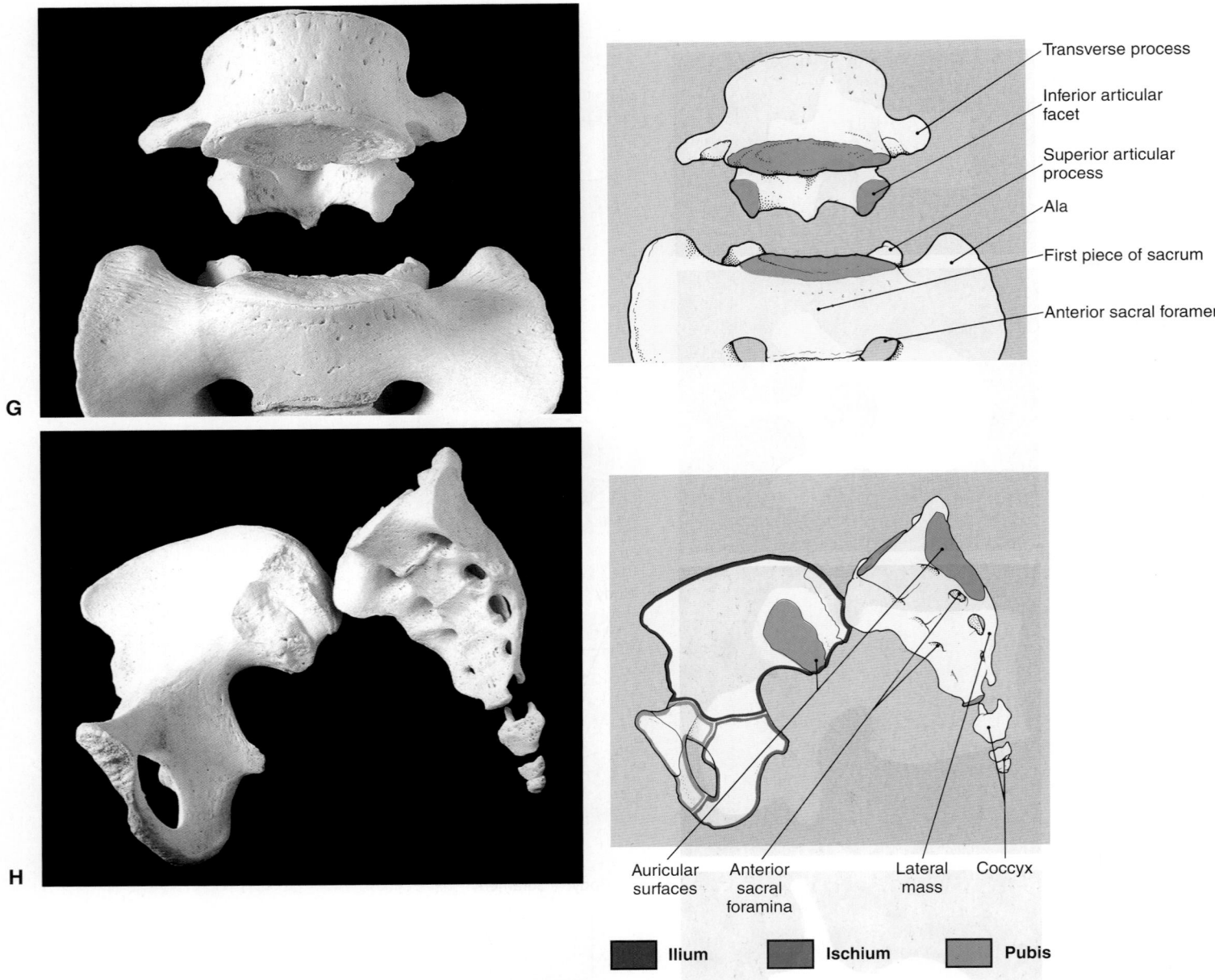

Transverse process

Inferior articular facet

Superior articular process

Ala

First piece of sacrum

Anterior sacral foramen

Auricular surfaces

Anterior sacral foramina

Lateral mass

Coccyx

Ilium   Ischium   Pubis

Figure 8-14, cont'd *Vertebrae.* **G**, L5 and the upper portion of the sacrum in an expanded anterior view. **H**, Oblique view of the sacrum, coccyx, and right hip bone.

Sternum

The medial part of the anterior chest wall is supported by the **sternum,** a somewhat dagger-shaped bone consisting of three parts: the upper handle part, the *manubrium*; the middle blade part, the *body*; and a blunt cartilaginous lower tip, the *xiphoid process.* The last ossifies during adult life. The manubrium articulates with the clavicle and first rib, whereas the next nine ribs join the body of the sternum, either directly or indirectly, by means of the *costal cartilages* (Figure 8-15).

Ribs

Twelve pairs of ribs, together with the vertebral column and sternum, form the bony cage known as the *thoracic cage* or, simply, the **thorax.** Each rib articulates with both the body and the transverse process of its corresponding thoracic vertebra. The head of each rib articulates with the body of the corresponding thoracic vertebra, and the tubercle of each rib articulates with the vertebra's transverse process (Figure 8-16). In addition, the second through ninth ribs articulate with the body of the vertebra above. From its

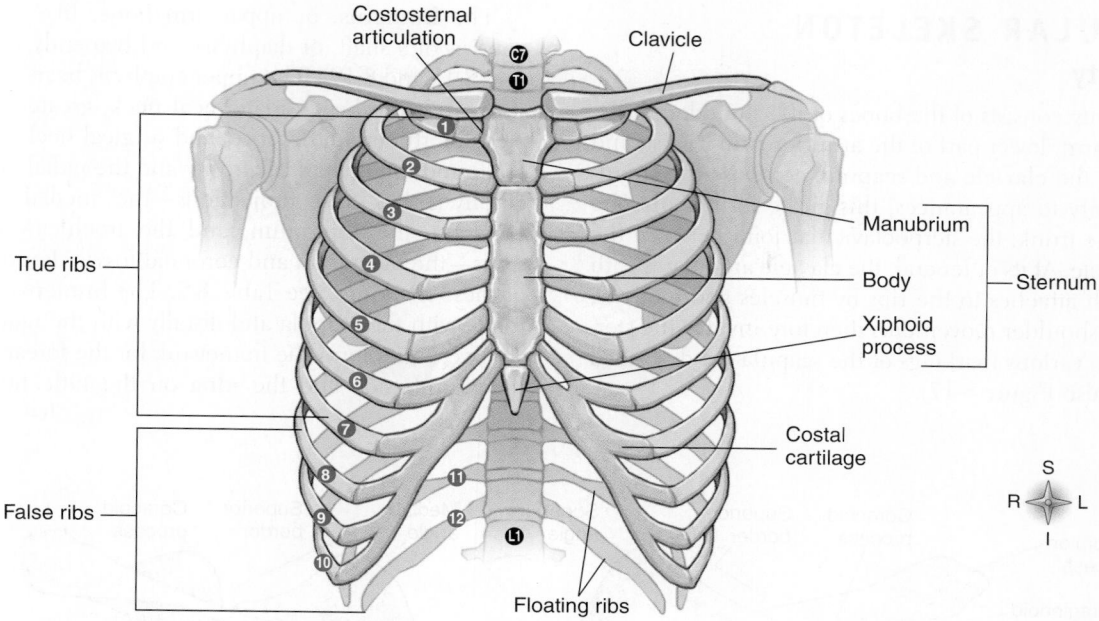

Figure 8-15 *Thoracic cage.* Note the costal cartilages and their articulations with the body of the sternum.

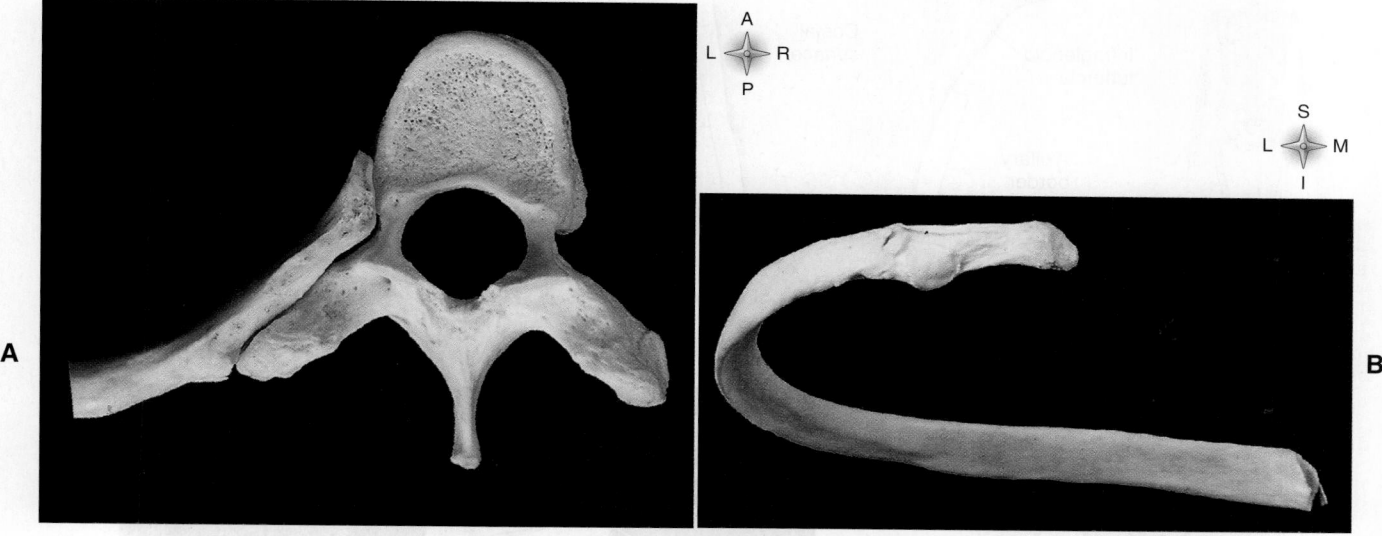

Figure 8-16 *Articulation of a rib and vertebra.* A, Note the head of the rib articulating with the vertebral body and the tubercle of the rib articulating with the transverse process of the vertebra. **B,** Anatomical components of a typical rib (fifth rib) viewed from behind.

vertebral attachment, each rib curves outward, then forward and downward (see Figures 8-1 and 8-16), a mechanical fact important for breathing. Anteriorly, each rib of the first seven pairs joins a costal cartilage that attaches to the sternum. For this reason, these ribs are often called the *true ribs*. Ribs of the remaining five pairs, the *false ribs*, do not attach directly to the sternum. Instead, each costal cartilage of pairs 8, 9, and 10 attaches to the costal cartilage of the rib above it—indirectly attaching it to the sternum. Ribs of the last two pairs of false ribs are designated *floating ribs* because they do not attach even indirectly to the sternum (see Figure 8-15).

 QUICK CHECK

4. Name the three types of vertebrae and how many of each type are found in the vertebral column.
5. What bones make up the bony cage known as the *thorax*? How do these bones fit together to form this structure?
6. What is a floating rib?

APPENDICULAR SKELETON

Upper Extremity

The upper extremity consists of the bones of the shoulder girdle, upper part of the arm, lower part of the arm (forearm), wrist, and hand. Two bones, the **clavicle** and **scapula**, compose the **shoulder girdle.** Contrary to appearances, this girdle forms only one bony joint with the trunk: the sternoclavicular joint between the sternum and clavicle. At its outer end, the clavicle articulates with the scapula, which attaches to the ribs by muscles and tendons, not by a joint. All shoulder movements therefore involve the sternoclavicular joint. Various markings of the scapula are described in Table 8-7 (see also Figure 8-17).

The **humerus,** or upper arm bone, like other long bones, consists of a shaft, or diaphysis, and two ends, or epiphyses (Figures 8-18 and 8-19). The upper epiphysis bears several identifying structures: the head, anatomical neck, greater and lesser tubercles, intertubercular groove, and surgical neck. On the diaphysis are found the deltoid tuberosity and the radial groove. The distal epiphysis has four projections—the medial and lateral epicondyles, the capitulum, and the trochlea—and two depressions—the olecranon and coronoid fossae. For descriptions of all of these markings, see Table 8-7. The humerus articulates proximally with the scapula and distally with the radius and the ulna.

Two bones form the framework for the forearm: the **radius** on the thumb side and the **ulna** on the little finger side. At the

Text continued on page 289

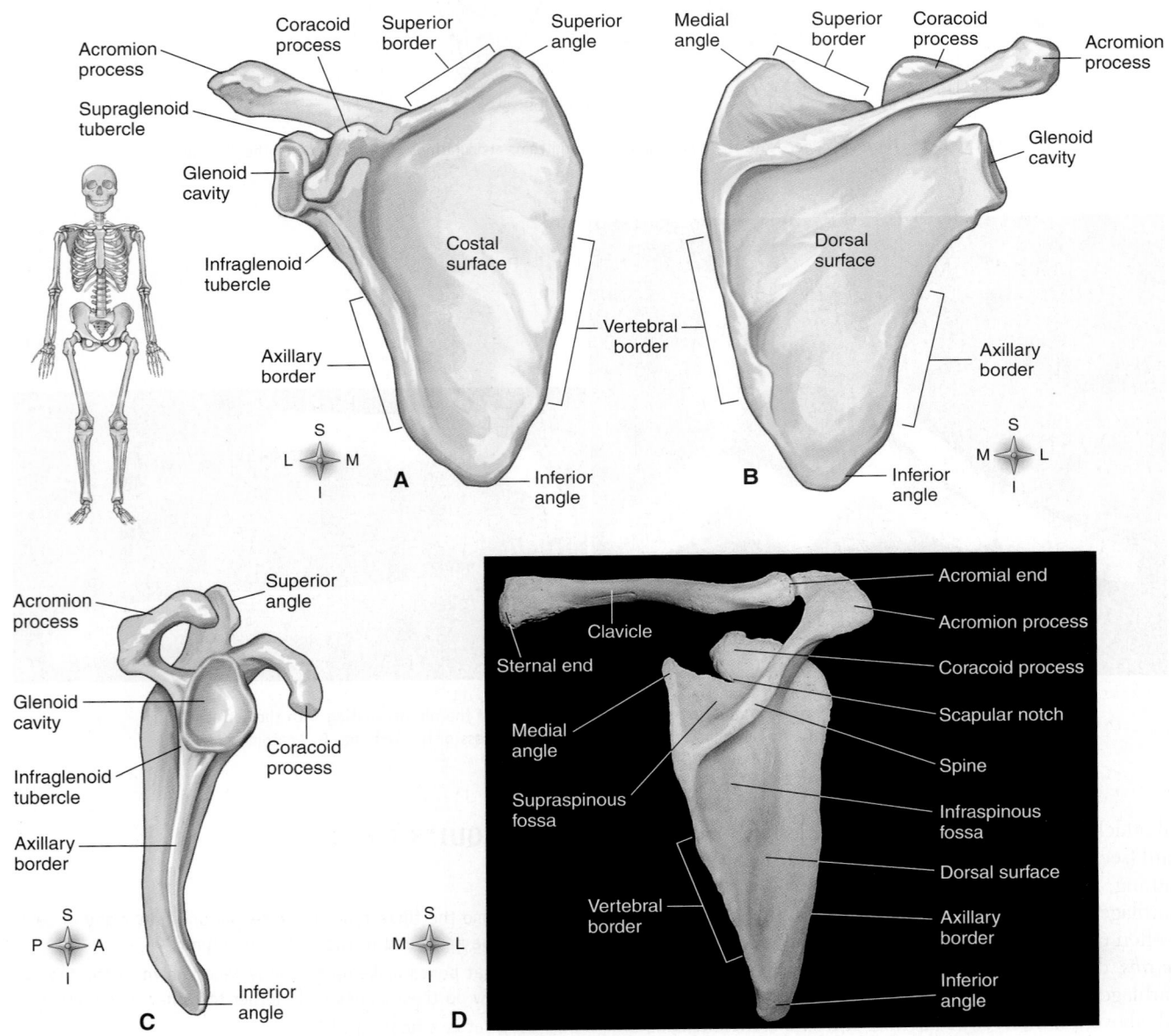

Figure 8-17 *Right scapula.* A, Anterior view. **B,** Posterior view. **C,** Lateral view. **D,** Posterior view showing articulation of the right scapula with the clavicle. (The inset shows the relative position of the right scapula within the entire skeleton.)

Table 8-7 Upper Extremity Bones and Their Markings

BONES AND MARKINGS	DESCRIPTION
Clavicle	Collar bones; the shoulder girdle is joined to the axial skeleton by articulation of the clavicles with the sternum (the scapula does not form a joint with the axial skeleton)
Scapula	Shoulder blades; the scapulae and clavicles together make up the shoulder girdle
Borders	
Superior	Upper margin
Vertebral	Margin toward the vertebral column
Axillary	Lateral margin
Spine	Sharp ridge running diagonally across the posterior surface of the shoulder blade
Acromion process	Slightly flaring projection at the lateral end of the scapular spine; may be felt at the tip of the shoulder; articulates with the clavicle
Coracoid process	Projection on the anterior surface from the upper border of the bone; may be felt in the groove between the deltoid and pectoralis major muscles, about 1 inch below the clavicle
Glenoid cavity	Arm socket
Humerus	Long bone of the upper part of the arm
Head	Smooth, hemispherical enlargement at the proximal end of the humerus
Anatomical neck	Oblique groove just below the head
Greater tubercle	Rounded projection lateral to the head on the anterior surface
Lesser tubercle	Prominent projection on the anterior surface just below the anatomical neck
Intertubercular groove	Deep groove between the greater and lesser tubercles; the long tendon of the biceps muscle lodges here
Surgical neck	Region just below the tubercles; so named because of its liability to fracture
Deltoid tuberosity	V-shaped, rough area about midway down the shaft where the deltoid muscle inserts
Radial groove	Groove running obliquely downward from the deltoid tuberosity; lodges the radial nerve
Epicondyles (medial and lateral)	Rough projections at both sides of the distal end
Capitulum	Rounded knob below the lateral epicondyle; articulates with the radius; sometimes called the *radial* head of the humerus
Trochlea	Projection with a deep depression through the center similar to the shape of a pulley; articulates with the ulna
Olecranon fossa	Depression on the posterior surface just above the trochlea; receives the olecranon process of the ulna when the lower part of the arm extends
Coronoid fossa	Depression on the anterior surface above the trochlea; receives the coronoid process of the ulna in flexion of the lower part of the arm
Radius	Bone of the thumb side of the forearm
Head	Disk-shaped process forming the proximal end of the radius; articulates with the capitulum of the humerus and with the radial notch of the ulna
Radial tuberosity	Roughened projection on the ulnar side, a short distance below the head; the biceps muscle inserts here
Styloid process	Protuberance at the distal end on the lateral surface (with the forearm in the anatomical position)
Ulna	Bone of the little finger side of the forearm; longer than the radius
Olecranon process	Elbow
Coronoid process	Projection on the anterior surface of the proximal end of the ulna; the trochlea of the humerus fits snugly between the olecranon and coronoid processes
Semilunar notch	Curved notch between the olecranon and coronoid process into which the trochlea fits
Radial notch	Curved notch lateral and inferior to the semilunar notch; the head of the radius fits into this concavity
Head	Rounded process at the distal end; does not articulate with the wrist bones but with the fibrocartilaginous disk
Styloid process	Sharp protuberance at the distal end; can be seen from outside on the posterior surface
Carpals	Wrist bones; arranged in two rows at the proximal end of the hand; proximal row (from the little finger toward the thumb)—*pisiform, triquetrum, lunate,* and *scaphoid;* distal row—*hamate, capitate, trapezoid,* and *trapezium*
Metacarpals	Long bones forming the framework of the palm of the hand; numbered I through V
Phalanges	Miniature long bones of the fingers, three (proximal, middle, distal) in each finger, two (proximal, distal) in each thumb

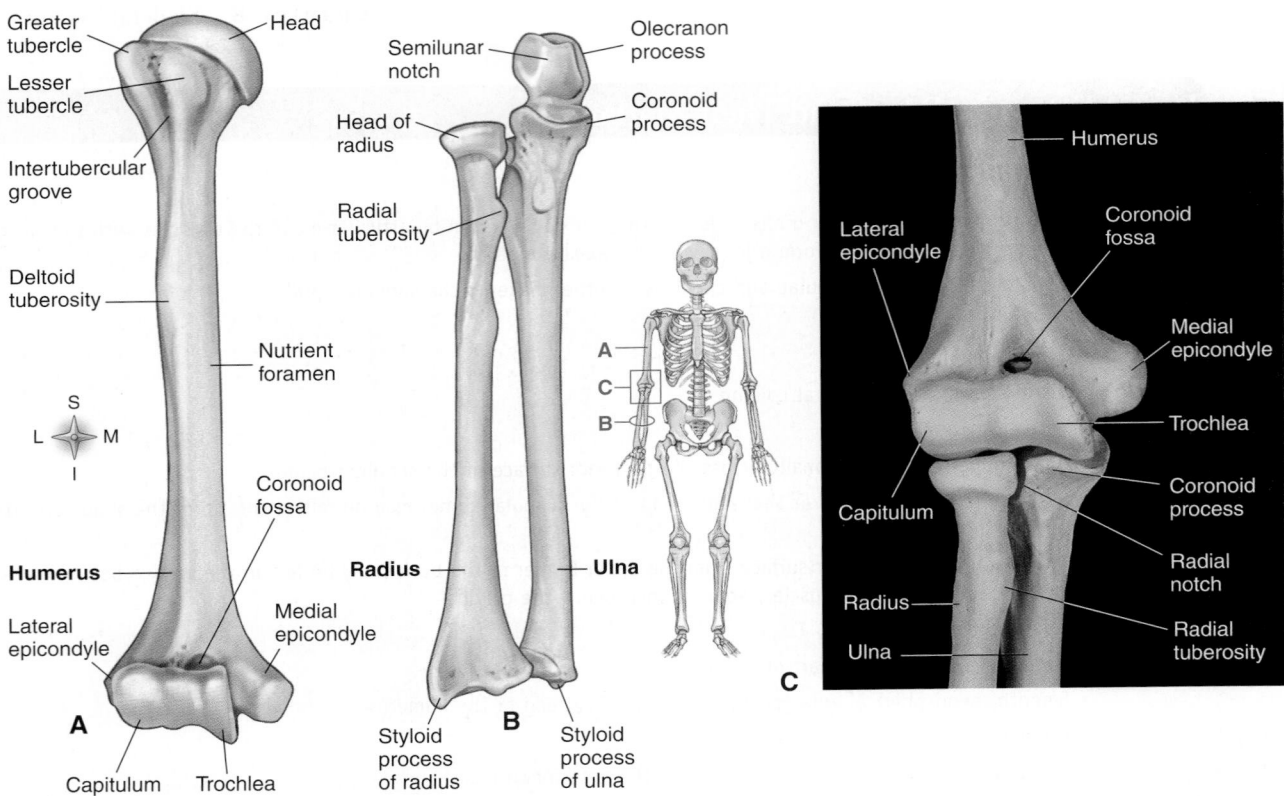

Figure 8-18 *Bones of the arm (right arm, anterior view).* A, Humerus (upper part of the arm). **B,** Radius and ulna (forearm). **C,** Elbow joint, showing how the distal end of the humerus joins the proximal ends of the radius and ulna. (The inset shows the relative position of the right arm bones within the entire skeleton.)

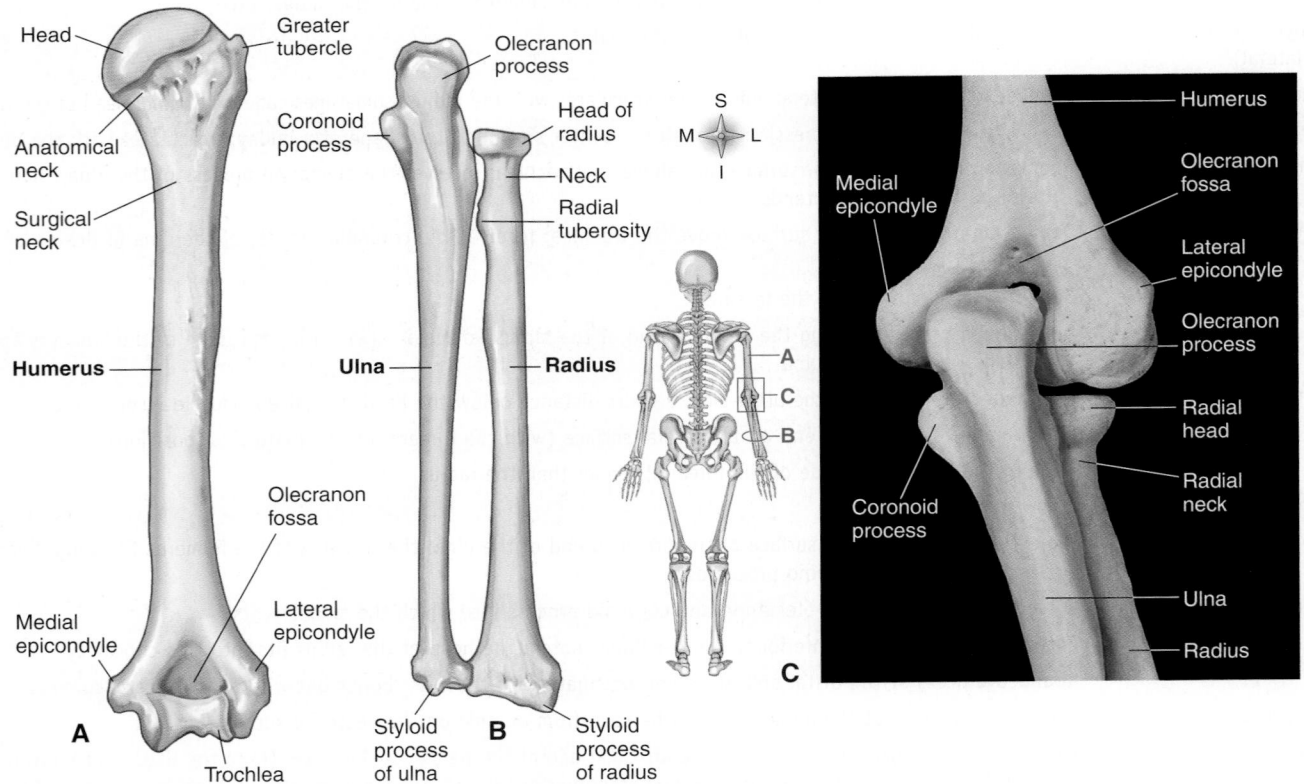

Figure 8-19 *Bones of the arm (right arm, posterior view).* A, Humerus (upper part of the arm). **B,** Radius and ulna (forearm). **C,** Elbow joint, showing how the distal end of the humerus joins the proximal ends of the radius and ulna. (The inset shows the relative position of the right arm bones within the entire skeleton.)

proximal end of the ulna the olecranon process projects posteriorly and the coronoid process projects anteriorly. There are also two depressions: the semilunar notch on the anterior surface and the radial notch on the lateral surface. The distal end has two projections: a rounded head and a sharper styloid process. For more detailed identification of these markings, see Table 8-7. The ulna articulates proximally with the humerus and radius and distally with a fibrocartilaginous disk, but not with any of the carpal bones.

The radius has three projections: two at its proximal end, the head and radial tuberosity, and one at its distal end, the styloid process (see Figures 8-18 and 8-19). There are two proximal articulations: one with the capitulum of the humerus and the other with the radial notch of the ulna. The three distal articulations are with the scaphoid and lunate carpal bones and with the head of the ulna.

The eight **carpal bones** (Figure 8-20) form what most people think of as the upper part of the hand but what, anatomically speaking, is the wrist. Only one of these bones is evident from the outside, the *pisiform bone*, which projects posteriorly on the little finger side as a small rounded elevation. Ligaments bind the carpals closely and firmly together in two rows of four each: proximal row (from the little finger toward the thumb)—pisiform, triquetrum, lunate, and scaphoid bones; distal row—hamate, capitate, trapezoid, and trapezium bones. The joints between the carpals and radius permit wrist and hand movements.

Of the five **metacarpal bones** that form the framework of the hand, the thumb metacarpal forms the most freely movable joint with the carpals. This fact has great significance. Because of the wide range of movement possible between the thumb metacarpal and the trapezium, particularly the ability to oppose the thumb to the fingers, the human hand has much greater dexterity than the forepaw of any animal, which has enabled humans to manipulate even small objects in their environment effectively. The heads of the metacarpals, prominent as the proximal knuckles of the hand, articulate with the phalanges.

 QUICK CHECK

7. What bones make up the shoulder girdle? Where does the shoulder girdle form a joint with the axial skeleton?
8. What are the two bones of the forearm? In the anatomical position, which one is lateral?
9. Name the bones of the hand and wrist.

Lower Extremity

Bones of the hip, thigh, lower part of the leg, ankle, and foot constitute the lower extremity (Table 8-8). Strong ligaments bind each **coxal bone** (*os coxae*, or *innominate bone*) to the sacrum

Table 8-8	Lower Extremity Bones and Their Markings
BONES AND MARKINGS	**DESCRIPTION**
Coxal	Large hip bone; with the sacrum and coccyx, forms the basinlike pelvic cavity; lower extremities attached to the axial skeleton by the coxal bones
Ilium	Upper, flaring portion
Ischium	Lower, posterior portion
Pubic bone (pubis)	Medial, anterior section
Acetabulum	Hip socket; formed by union of the ilium, ischium, and pubis
Iliac crests	Upper, curving boundary of the ilium
Iliac spines	
Anterior superior	Prominent projection at the anterior end of the iliac crest; can be felt externally as the "point" of the hip
Anterior inferior	Less prominent projection short distance below anterior superior spine
Posterior superior	At the posterior end of the iliac crest
Posterior inferior	Just below the posterior superior spine
Greater sciatic notch	Large notch on the posterior surface of the ilium just below the posterior inferior spine
Ischial tuberosity	Large, rough, quadrilateral process forming the inferior part of the ischium; in an erect sitting position the body rests on these tuberosities
Ischial spine	Pointed projection just above the tuberosity
Symphysis pubis	Cartilaginous, amphiarthrotic joint between the pubic bones
Superior ramus of the pubis	Part of the pubis lying between the symphysis and acetabulum; forms the upper part of the obturator foramen
Inferior ramus	Part extending down from the symphysis; unites with the ischium
Pubic arch	Angle formed by the two inferior rami
Pubic crest	Upper margin of the superior ramus
Pubic tubercle	Rounded process at the end of the crest
Obturator foramen	Large hole in the anterior surface of the os coxa; formed by the pubis and ischium; largest foramen in the body

Table 8-8 Lower Extremity Bones and Their Markings—cont'd

BONES AND MARKINGS	DESCRIPTION
Pelvic brim (or inlet)	Boundary of the aperture leading into the true pelvis; formed by the pubic crests, iliopectineal lines, and sacral promontory; the size and shape of this inlet have obstetrical importance because if any of its diameters are too small, the infant's skull cannot enter the true pelvis for natural birth
True pelvis (or pelvis minor)	Space below the pelvic brim; true "basin" with bone and muscle walls and a muscle floor; pelvic organs located in this space
False pelvis (or pelvis major)	Broad, shallow space above the pelvic brim, or the pelvic inlet; name "false pelvis" is misleading because this space is actually part of the abdominal cavity, not the pelvic cavity
Pelvic outlet	Irregular circumference marking the lower limits of the true pelvis; bounded by the tip of the coccyx and two ischial tuberosities
Pelvic girdle (or bony pelvis)	Complete bony ring; composed of two hip bones (ossa coxae), the sacrum, and the coccyx; forms a firm base by which the trunk rests on the thighs and for attachment of the lower extremities to the axial skeleton
Femur	Thigh bone; largest, strongest bone of the body
Head	Rounded upper end of the bone; fits into the acetabulum
Neck	Constricted portion just below the head
Greater trochanter	Protuberance located inferiorly and laterally to the head
Lesser trochanter	Small protuberance located inferiorly and medially to the greater trochanter
Intertrochanteric line	Line extending between the greater and lesser trochanter
Linea aspera	Prominent ridge extending lengthwise along the concave posterior surface
Supracondylar ridges	Two ridges formed by division of the linea aspera at its lower end; the medial supracondylar ridge extends inward to the inner condyle, the lateral ridge to the outer condyle
Condyles	Large, rounded bulges at the distal end of the femur; one medial and one lateral
Epicondyles	Blunt projections from the sides of the condyles; one on the medial aspect and one on the lateral aspect
Adductor tubercle	Small projection just above the medial condyle; marks the termination of the medial supracondylar ridge
Trochlea	Smooth depression between the condyles on the anterior surface; articulates with the patella
Intercondyloid fossa (notch)	Deep depression between the condyles on the posterior surface; the cruciate ligaments, which help bind the femur to the tibia, lodge in this notch
Patella	Kneecap; largest sesamoid bone of the body; embedded in the tendon of the quadriceps femoris muscle
Tibia	Shin bone
Condyles	Bulging prominences at the proximal end of the tibia; upper surfaces concave for articulation with the femur
Intercondylar eminence	Upward projection on the articular surface between the condyles
Crest	Sharp ridge on the anterior surface
Tibial tuberosity	Projection in the midline on the anterior surface
Medial malleolus	Rounded downward projection at the distal end of the tibia; forms the prominence on the medial surface of the ankle
Fibula	Long, slender bone of the lateral side of the lower part of the leg
Lateral malleolus	Rounded prominence at the distal end of the fibula; forms the prominence on the lateral surface of the ankle
Tarsals	Bones that form the heel and proximal or posterior half of the foot
Calcaneus	Heel bone
Talus	Uppermost of the tarsals; articulates with the tibia and fibula; boxed in the medial and lateral malleoli
Longitudinal arches	Tarsals and metatarsals so arranged as to form an arch from the front to the back of the foot
Medial	Formed by the calcaneus, talus, navicular, cuneiforms, and three medial metatarsals
Lateral	Formed by the calcaneus, cuboid, and two lateral metatarsals
Transverse (or metatarsal) arch	Metatarsals and distal row of tarsals (cuneiforms and cuboid) so articulated as to form an arch across the foot; bones kept in two arched positions by means of powerful ligaments in the sole of the foot and by muscles and tendons
Metatarsals	Long bones of the feet
Phalanges	Miniature long bones of the toes; two in each great toe; three in the other toes

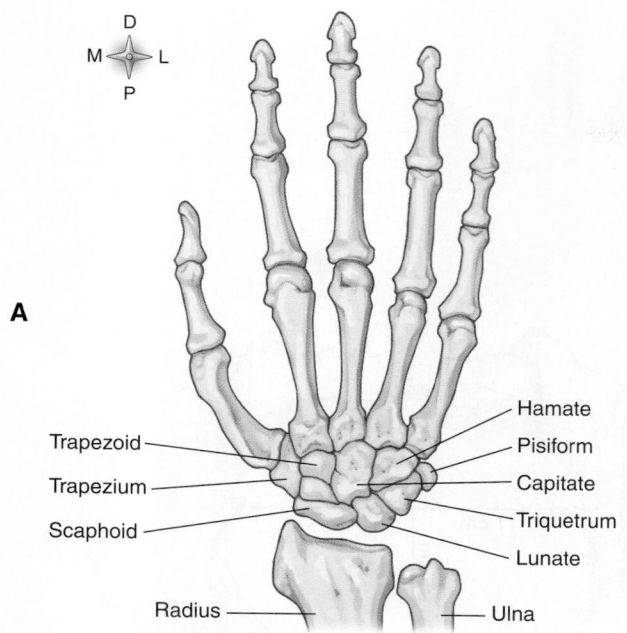

A

Trapezoid
Trapezium
Scaphoid

Hamate
Pisiform
Capitate
Triquetrum
Lunate

Radius — Ulna

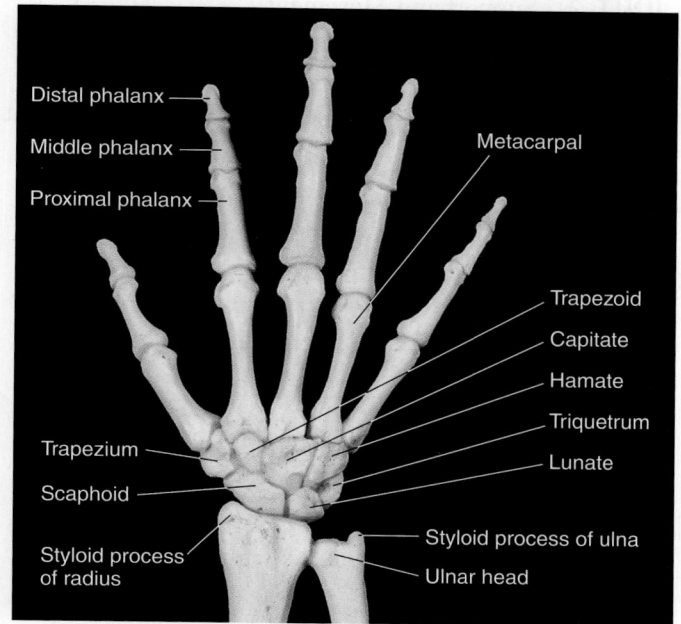

Distal phalanx
Middle phalanx
Proximal phalanx

Metacarpal

Trapezoid
Capitate
Hamate
Triquetrum

Trapezium
Scaphoid

Lunate

Styloid process of radius

Styloid process of ulna
Ulnar head

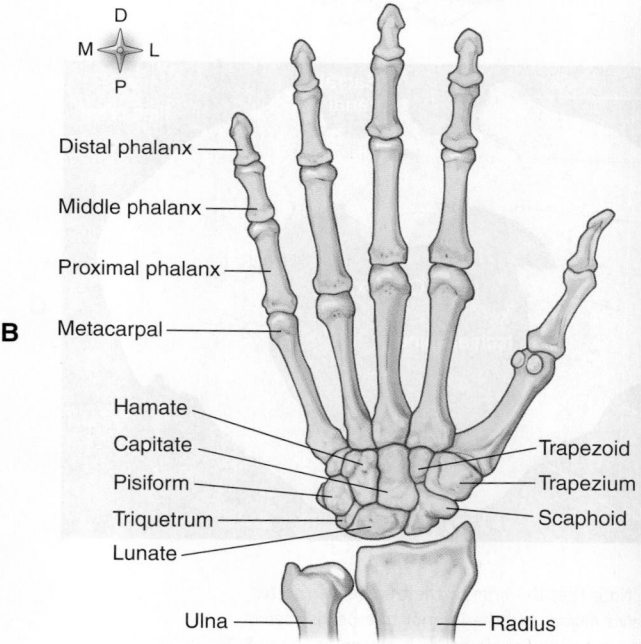

B

Distal phalanx
Middle phalanx
Proximal phalanx
Metacarpal

Hamate
Capitate
Pisiform
Triquetrum
Lunate

Trapezoid
Trapezium
Scaphoid

Ulna — Radius

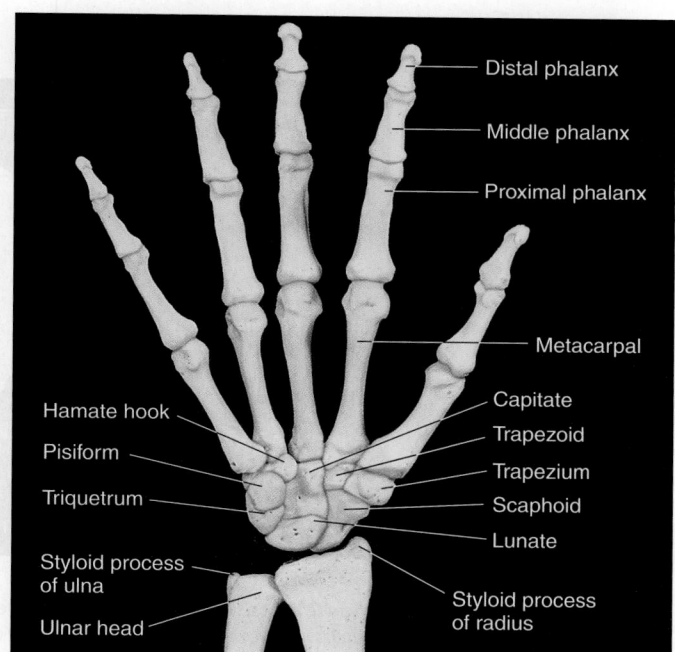

Distal phalanx
Middle phalanx
Proximal phalanx

Metacarpal

Hamate hook
Pisiform
Triquetrum

Capitate
Trapezoid
Trapezium
Scaphoid
Lunate

Styloid process of ulna
Ulnar head

Styloid process of radius

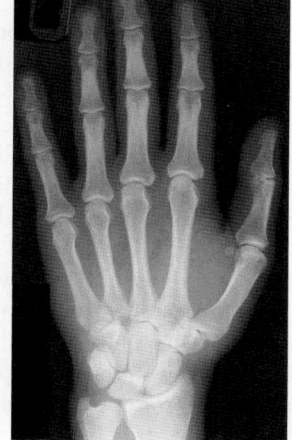

C

Figure 8-20 *Bones of the hand and wrist.* A, Dorsal view of the right hand and wrist. **B,** Palmar view of the right hand and wrist. **C,** How many bones of the hand and wrist can you identify in the x-ray image?

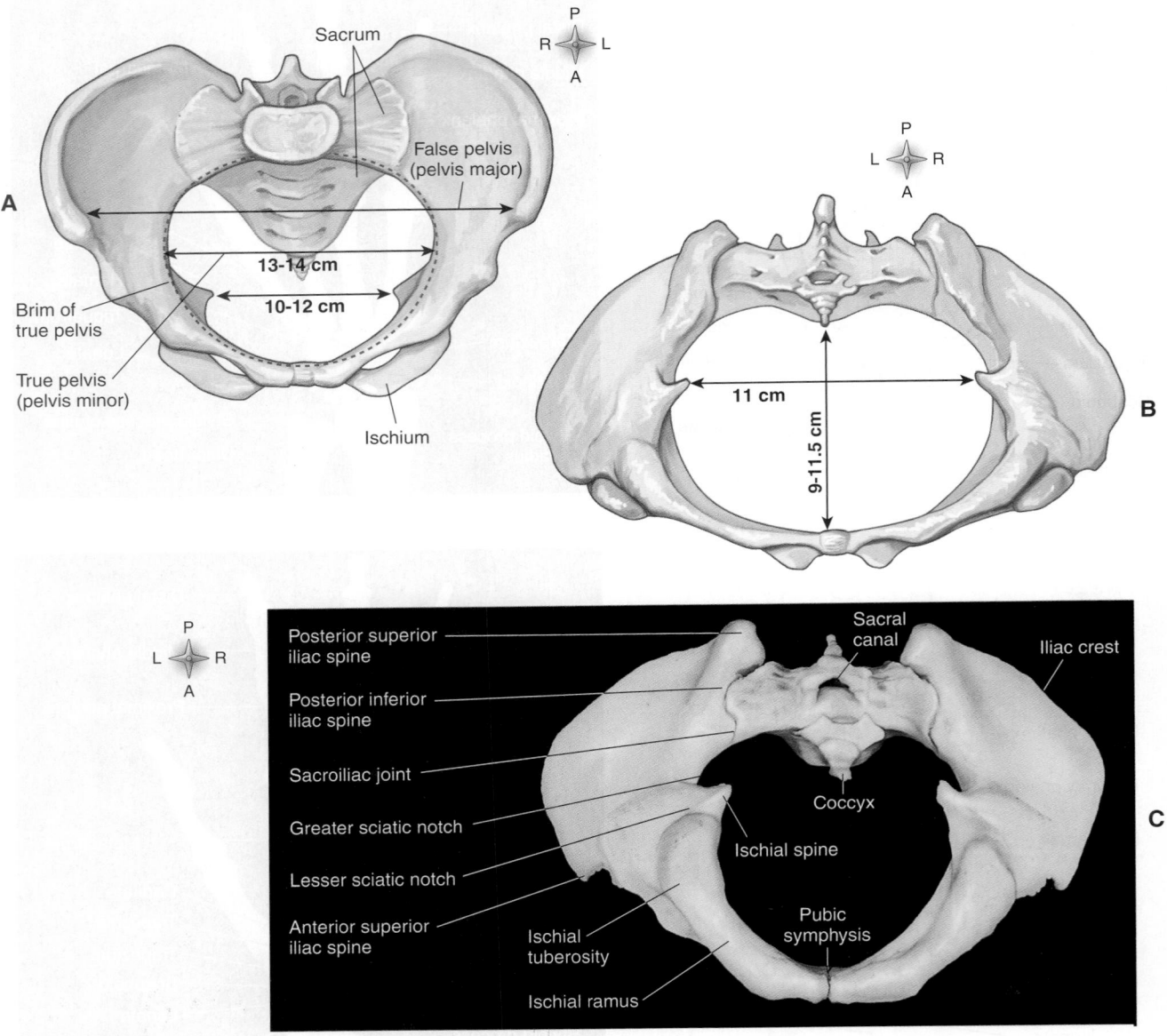

Figure 8-21 *The female pelvis.* **A,** Pelvis viewed from above. Note that the brim of the true pelvis *(dotted line)* marks the boundary between the superior false pelvis *(pelvis major)* and the inferior true pelvis *(pelvis minor)*. **B** and **C,** Pelvis viewed from below. A comparison of the male and female pelvis is shown in Figure 8-26.

posteriorly and to each other anteriorly to form the **pelvic girdle** (see Figures 8-21 and 8-26), a stable, circular base that supports the trunk and attaches the lower extremities to it. In early life, each coxal bone is made up of three separate bones. Later, they fuse into a single, massive irregular bone that is broader than any other bone in the body. The largest and uppermost of the three bones is the **ilium,** the strongest and lowermost is the **ischium,** and the most anteriorly placed is the **pubis.** Numerous markings are present on the three bones (see Table 8-8 and Figures 8-21 and 8-22).

The pelvis can be divided into two parts by an imaginary plane called the *pelvic inlet.* The edge of this plane, outlined in Figure 8-21, is called the *pelvic brim,* or *brim of the true pelvis.*

The structure above the pelvic inlet, termed the **false pelvis,** is bordered by muscle in the front and bone along the sides and back. The structure below the pelvic inlet, the so-called *true pelvis,* creates the boundary of another imaginary plane called the *pelvic outlet.* It is through the pelvic outlet that the digestive tract empties. The female reproductive tract also passes through the pelvic outlet; this is a fact of great importance in childbirth. The *pelvic outlet* is just large enough for the passage of a baby during delivery; however, careful positioning of the baby's head is required. Measurements such as those shown in Figure 8-21 are routinely made by obstetricians to ensure successful delivery. Despite its apparent rigidity, the joint between the pubic portions of each coxal bone, the *symphysis pubis,* softens before delivery.

footnote in Table 8-1.) When the knee joint is extended, the patellar outline may be distinguished through the skin, but as the knee flexes, it sinks into the intercondylar notch of the femur and can no longer be easily distinguished.

The **tibia** is the larger and stronger and more medially and superficially located of the two leg bones. The **fibula** is smaller and more laterally and deeply placed. At its proximal end it articulates with the lateral condyle of the tibia. The proximal end of the tibia, in turn, articulates with the femur to form the knee joint, the largest and one of the most complex and frequently injured joints of the body. Distally, the tibia articulates with the fibula and also with the talus. The latter fits into a boxlike socket (ankle joint) formed by the medial and lateral malleoli, projections of the tibia and fibula, respectively. For other tibial markings, see Table 8-8 and Figure 8-23.

The structure of the **foot** is similar to that of the hand, with certain differences that adapt it for supporting weight (Figure 8-24). One example is the much greater solidity and the more limited mobility of the great toe than the thumb. Then, too, the foot bones are held together in such a way that they form springy lengthwise and crosswise arches (Figure 8-25). This arrangement is architecturally sound because arches furnish more supporting strength per given amount of structural material than any other type of construction does. Hence the two-way arch construction makes a highly stable base. The **longitudinal arch** has an inner, or medial, portion and an outer, or lateral, portion. Both are formed by the placement of the tarsals and metatarsals. Specifically, some of the tarsals (calcaneus, talus, navicular, and cuneiforms) and the first three metatarsals (starting with the great toe) form the **medial longitudinal arch**. The calcaneus and cuboid tarsals plus the fourth and fifth metatarsals shape the **lateral longitudinal arch**. The transverse arch results from the relative placement of the distal row of tarsals and the five metatarsals. (See Table 8-8 for specific bones of different arches.) Strong ligaments and leg muscle tendons normally hold the foot bones firmly in their arched positions. Not infrequently, however, these structures weaken and cause the arches to flatten—a condition aptly called **fallen arches**, or **flatfeet** (see Figure 8-25, *B*). Look at Figure 8-25, *D*, to see what shoes with high heels do to the position of the foot. They give a forward thrust to the body, which forces an undue amount of weight on the heads of the metatarsals. The resulting shift from the normal weight-bearing position of the feet may cause injury and chronic pain. Normally, the tarsals and metatarsals have the major role in functioning of the foot as a supporting structure, with the phalanges being relatively unimportant. The reverse is true for the hand. Here, manipulation is the main function rather than support. Consequently, the phalanges of the fingers are all important, and the carpals and metacarpals are subsidiary.

In Chapter 7 the sesamoid bones were described as unique irregular bones generally found embedded in the substance of tendons close to joints. Although the kneecap, or patella, is the largest of the sesamoid bones, they also appear quite frequently in tendons near the distal end (head) of the first metatarsal bone of the big toe (Figure 8-24, *C*).

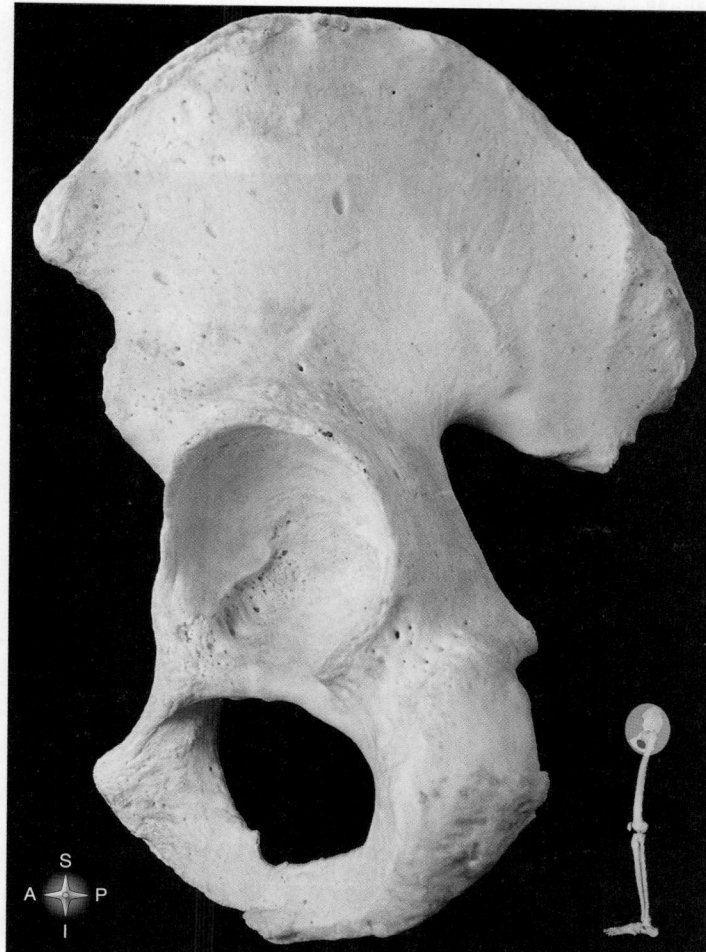

Figure 8-22 *Left coxal (hip) bone.* The left coxal bone is disarticulated from the bony pelvis and viewed from the side.

This softening allows the pelvic outlet to expand to accommodate the newborn's head as it passes out of the birth canal. The tiny coccyx bone, which protrudes into the pelvic outlet, sometimes breaks when the force of labor contractions pushes the newborn's head against it.

The two thigh bones, or **femurs**, have the distinction of being the longest and heaviest bones in the body. Several prominent markings characterize them. For example, three projections are conspicuous at each epiphysis: the head and greater and lesser trochanters proximally and the medial and lateral condyles and adductor tubercle distally (Figure 8-23). Both condyles and the greater trochanter may be felt externally. For a description of the various femur markings, see Table 8-8.

The largest sesamoid bone in the body, and the one that is almost universally present, is the **patella**, or kneecap, located in the tendon of the quadriceps femoris muscle as a projection to the underlying knee joint. Although some individuals have sesamoid bones in the tendons of other muscles, lists of bone names do not usually include them because they are not always present, are not found in any particular tendons, and are less important. (See the

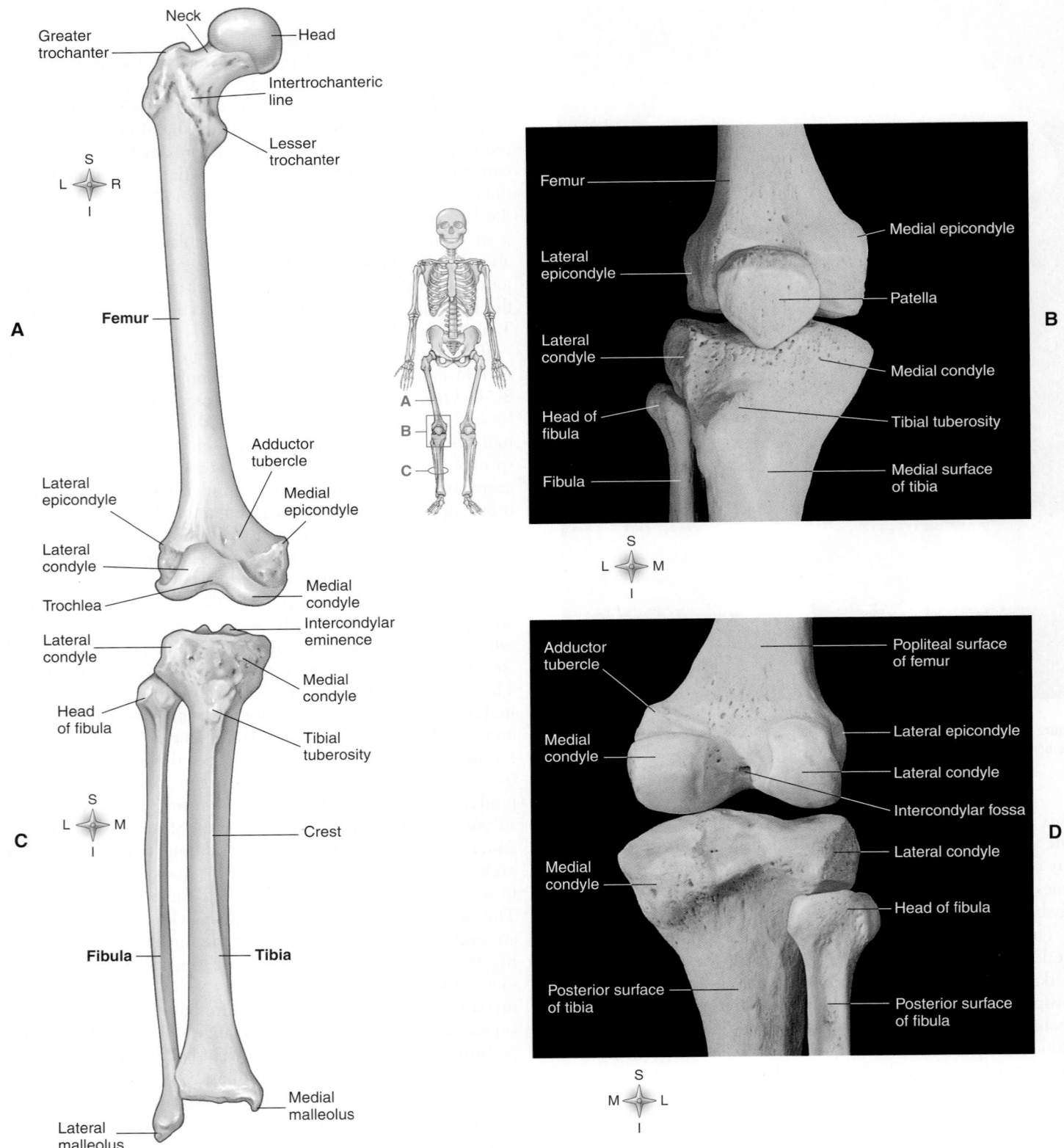

A

Neck
Greater trochanter
Head
Intertrochanteric line
Lesser trochanter

S
L — R
I

Femur

Adductor tubercle

Lateral epicondyle
Medial epicondyle

Lateral condyle
Trochlea
Medial condyle

Intercondylar eminence
Lateral condyle
Medial condyle

Head of fibula

Tibial tuberosity

S
L — M
I

C

Crest

Fibula   **Tibia**

Medial malleolus
Lateral malleolus

B

Femur
Medial epicondyle
Lateral epicondyle
Patella
Lateral condyle
Medial condyle
Head of fibula
Tibial tuberosity
Fibula
Medial surface of tibia

S
L — M
I

D

Adductor tubercle
Popliteal surface of femur
Medial condyle
Lateral epicondyle
Lateral condyle
Intercondylar fossa
Medial condyle
Lateral condyle
Head of fibula
Posterior surface of tibia
Posterior surface of fibula

S
M — L
I

Figure 8-23 *Bones of the thigh and leg.* A, Right femur, anterior surface. **B,** Anterior aspect of the right knee skeleton. **C,** Right tibia and fibula, anterior surface. **D,** Posterior aspect. (The inset shows the relative position of the bones of the thigh and leg within the entire skeleton.)

BOX 8-2: FYI

Palpable Bony Landmarks

Health professionals often identify externally palpable bony landmarks when dealing with the sick and injured. **Palpable** bony landmarks are bones that can be touched and identified through the skin. They serve as reference points in identifying other body structures.

There are externally palpable bony landmarks throughout the body. Many skull bones, such as the zygomatic bone, can be palpated. The medial and lateral epicondyles of the humerus, the olecranon process of the ulna, and the styloid process of the ulna and the radius at the wrist can be palpated on the upper extremity. The highest corner of the shoulder is the acromion process of the scapula.

When you put your hands on your hips, you can feel the superior edge of the ilium, called the *iliac crest*. The anterior end of the crest, called the *anterior superior iliac spine,* is a prominent landmark often used as a clinical reference. The sacral promontory is a prominent anteriorly projecting ridge or border on the superior aspect of the sacrum. It frequently serves as a palpable reference point when measuring the pelvis during obstetrical examinations. The medial malleolus of the tibia and the lateral malleolus of the fibula are prominent at the ankle. The calcaneus, or heel bone, is easily palpated on the posterior aspect of the foot. On the anterior aspect of the lower extremity, examples of palpable bony landmarks include the patella, or kneecap; the anterior border of the tibia, or shin bone; and the metatarsals and phalanges of the toes. Try to identify as many of the externally palpable bones of the skeleton as possible on your own body. Using these as points of reference will make it easier for you to visualize the placement of other bones that cannot be touched or palpated through the skin.

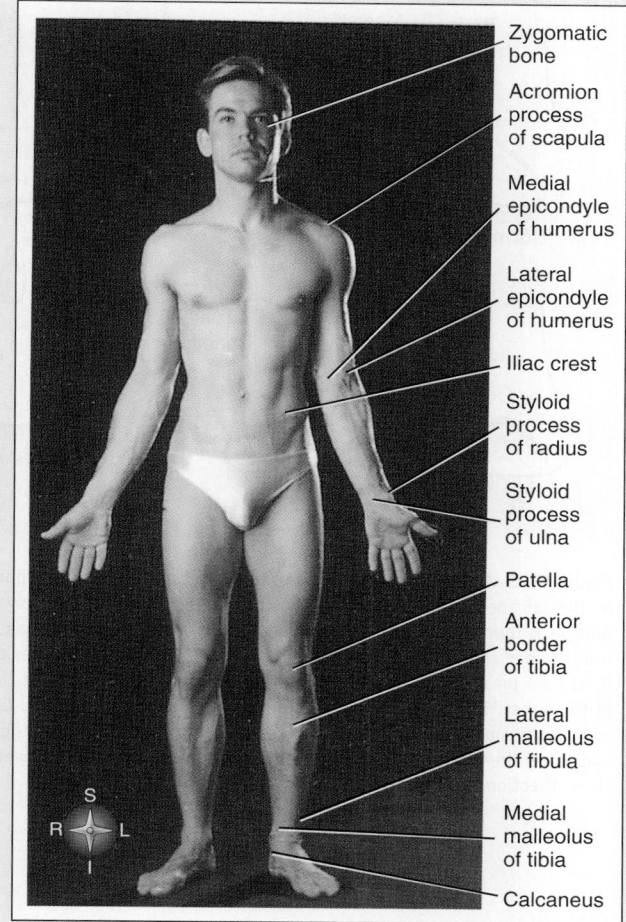

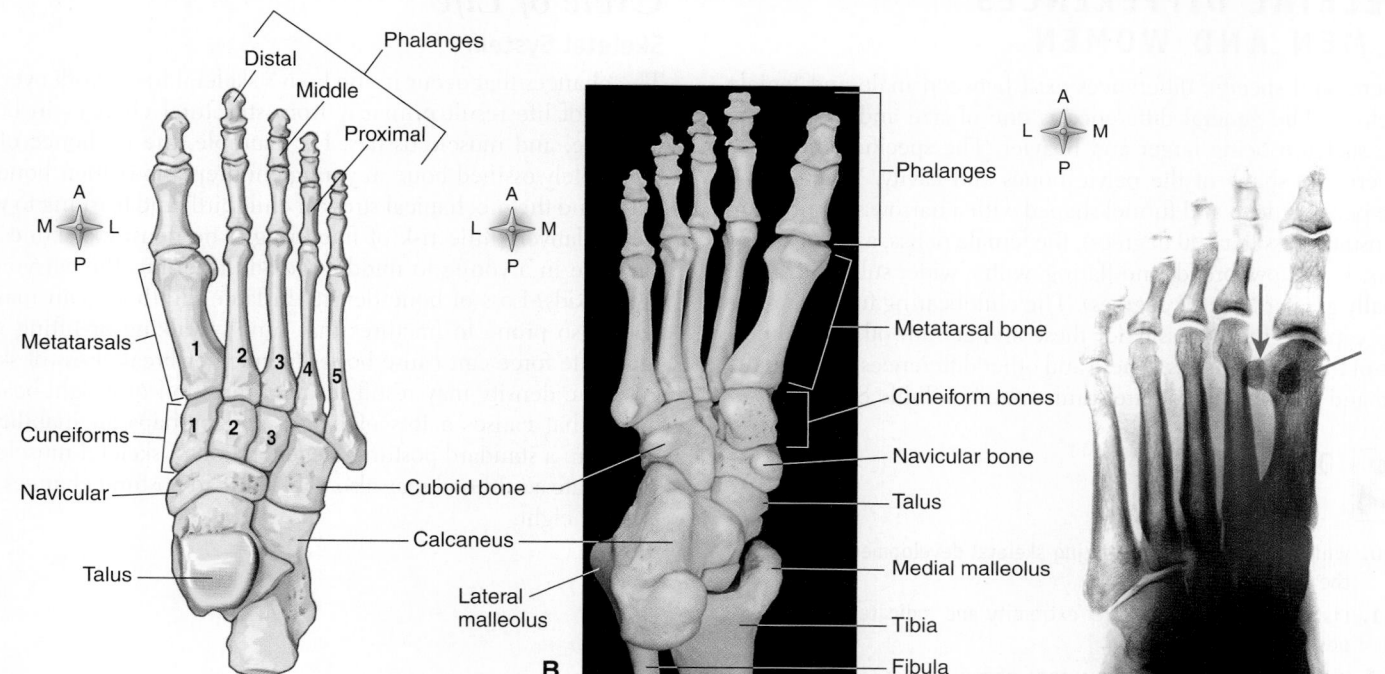

Figure 8-24 *The foot.* **A,** Bones of the right foot viewed from above. The tarsal bones consist of the cuneiforms, navicular, talus, cuboid, and calcaneus. **B,** Posterior aspect of the right ankle skeleton and inferior aspect of the right foot skeleton. **C,** X-ray film of the left foot showing prominent sesamoid bones *(arrows)* near the distal end (head) of the first metatarsal bone of the great toe.

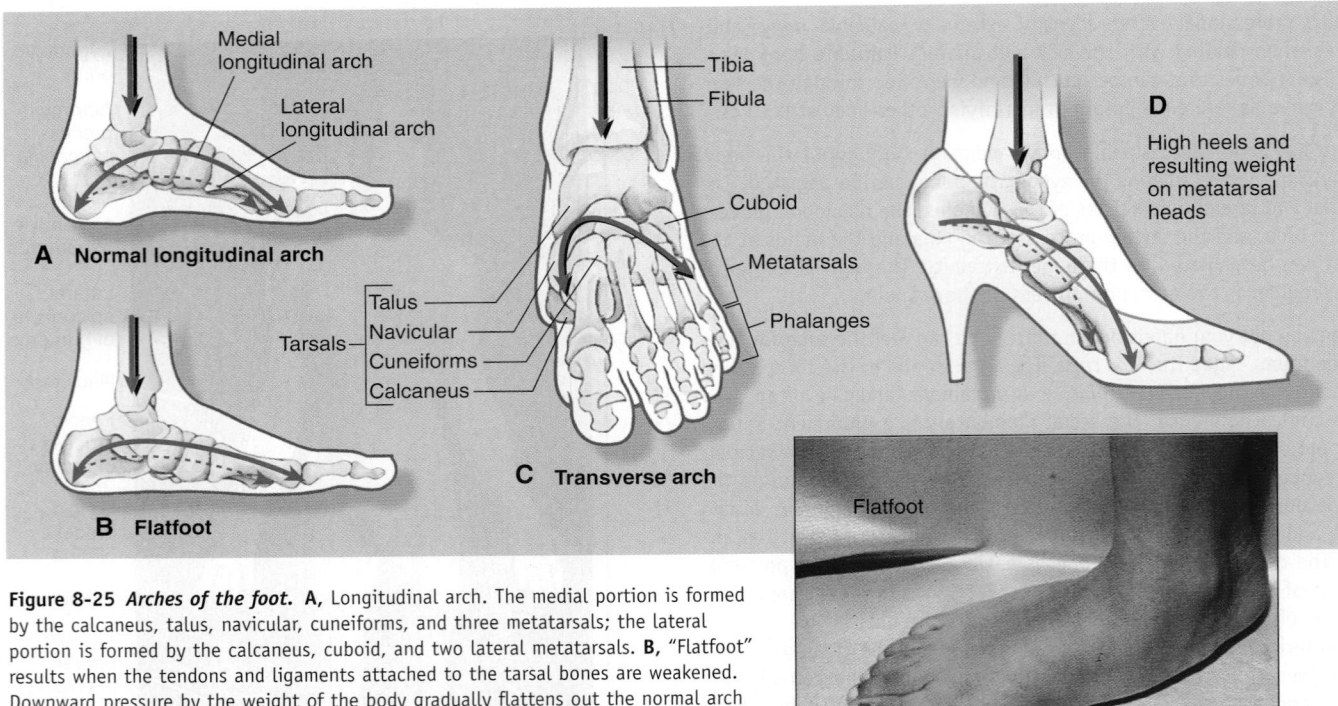

Figure 8-25 *Arches of the foot.* **A,** Longitudinal arch. The medial portion is formed by the calcaneus, talus, navicular, cuneiforms, and three metatarsals; the lateral portion is formed by the calcaneus, cuboid, and two lateral metatarsals. **B,** "Flatfoot" results when the tendons and ligaments attached to the tarsal bones are weakened. Downward pressure by the weight of the body gradually flattens out the normal arch of the bones. The photo shows the clinical appearance of a flatfoot. **C,** Transverse arch in the metatarsal region of the left foot. **D,** High heels throw the weight forward and cause the heads of the metatarsals to bear most of the body's weight. (Arrows show direction of force.)

SKELETAL DIFFERENCES IN MEN AND WOMEN

General and specific differences exist between male and female skeletons. The general difference is one of size and weight, the male skeleton being larger and heavier. The specific differences concern the shape of the pelvic bones and cavity. Whereas the male pelvis is deep and funnel-shaped with a narrow subpubic angle (usually less than 90 degrees), the female pelvis, as Figure 8-26 shows, is shallow, broad, and flaring, with a wider subpubic angle (usually greater than 90 degrees). The childbearing function obviously explains the necessity for these and certain other modifications of the female pelvis. These and other differences between the male and female skeleton are summarized in Table 8-9, p. 297.

QUICK CHECK

10. Which three bones fuse during skeletal development to form the coxal (hip) bone?
11. List the bones of the lower extremity and indicate their positions in the skeleton.
12. What is the functional advantage of foot arches?
13. Name two differences between typical male and female skeletons.

Cycle of Life
Skeletal System

The changes that occur in the body's skeletal framework over the course of life result primarily from structural changes in bone, cartilage, and muscle tissues. For example, the resilience of incompletely ossified bone in young children allows their bones to withstand the mechanical stress of childbirth and learning to walk with relatively little risk of fracturing. The density of bone and cartilage in a young to middle-age adult permits the carrying of great loads. Loss of bone density in later adulthood can make a person so prone to fractures that simply walking or lifting with moderate force can cause bones to crack or break. Loss of skeletal tissue density may result in a compression of weight-bearing bones that causes a loss of height and perhaps an inability to maintain a standard posture. Degeneration of skeletal muscle tissue in late adulthood may also contribute to postural changes and loss of height.

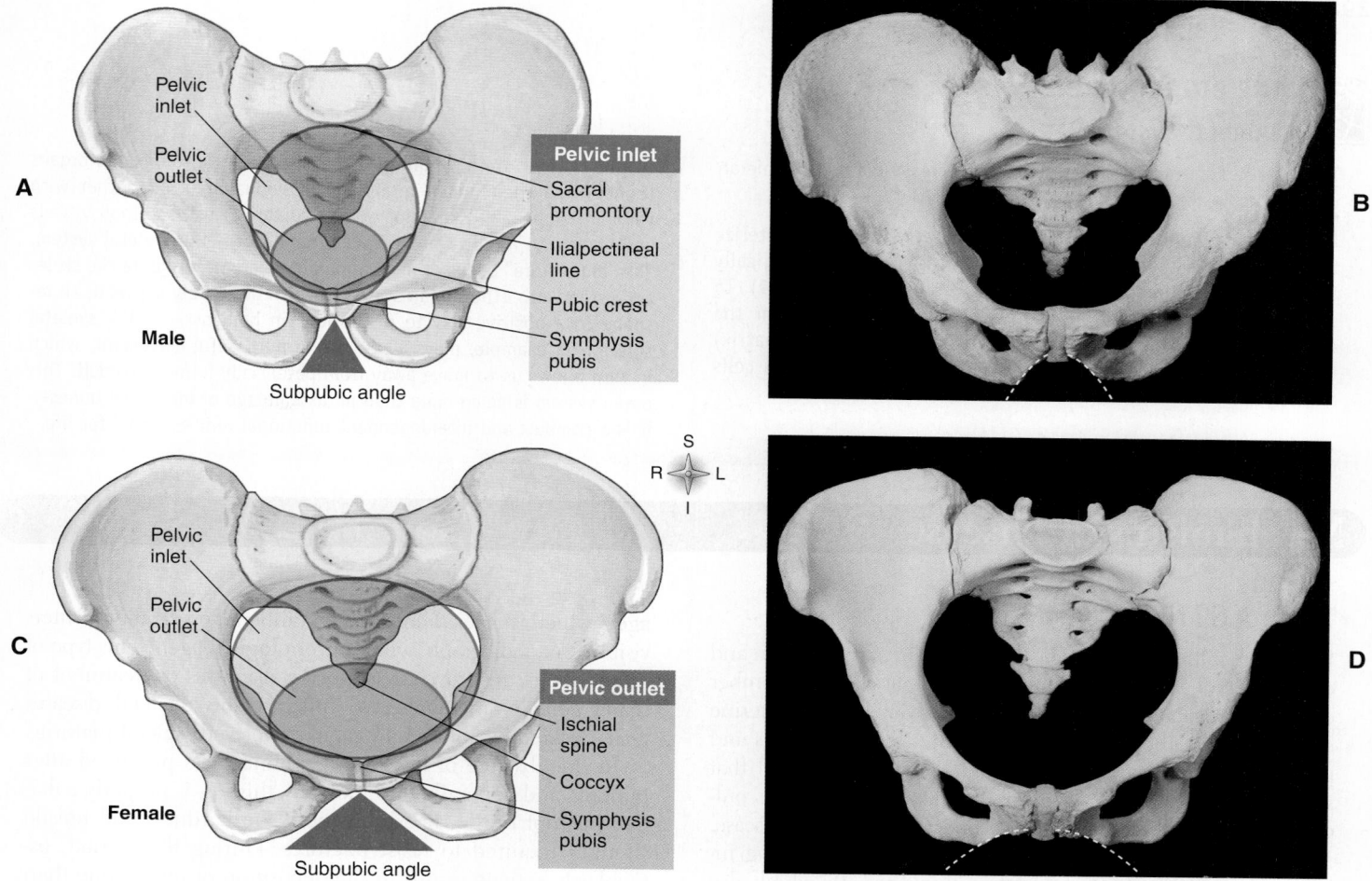

Figure 8-26 *Comparison of the male and female bony pelvis.* **A** and **C**, diagrams showing components of male and female bony pelvis. **B** and **D**, photographs showing the bony pelvis of the male and female from in front and above.

Table 8-9 — Comparison of Male and Female Skeletons

PORTION OF SKELETON	MALE	FEMALE
General form	Bones heavier and thicker	Bones lighter and thinner
	Muscle attachment sites more massive	Muscle attachment sites less distinct
	Joint surfaces relatively large	Joint surfaces relatively small
Skull	Forehead shorter vertically	Forehead more elongated vertically
	Mandible and maxillae relatively larger	Mandible and maxillae relatively smaller
	Facial area more pronounced	Facial area rounder, with less pronounced features
	Processes more prominent	Processes less pronounced
Pelvis		
Pelvic cavity	Narrower in all dimensions	Wider in all dimensions
	Deeper	Shorter and roomier
	Pelvic outlet relatively small	Pelvic outlet relatively large
Sacrum	Long, narrow, with a smooth concavity (sacral curvature); sacral promontory more pronounced	Short, wide, flat concavity more pronounced in a posterior direction; sacral promontory less pronounced
Coccyx	Less movable	More movable and follows the posterior direction of the sacral curvature
Pubic arch	60- to 90-degree angle	90- to 120-degree angle
Symphysis pubis	Relatively deep	Relatively shallow
Ischial spine, ischial tuberosity, and anterior superior iliac spine	Turned more inward	Turned more outward and further apart
Greater sciatic notch	Narrow	Wide

THE BIG PICTURE
Skeletal System

The skeletal system is a good example of increasing structural hierarchy or complexity in the body.

Recall from Chapter 1 that "levels of organization" characterize body structure so that all of our anatomical components logically fit together and function effectively (see Figure 1-3 on p. 9). In studying skeletal tissues in Chapter 7, we proceeded from the chemical level of organization (inorganic salts and organic matrix) to a discussion of the specialized skeletal and cartilaginous cells and tissues.

In this chapter, we have grouped skeletal tissues into discrete organs (bones) and then joined groups of individual bones together with varying numbers and kinds of other structures, such as blood vessels and nerves, to form a complex operational unit—the skeletal system. The "big picture" becomes more apparent when we integrate the skeletal system with other organ systems, which ultimately allows us to respond in a positive way to disruptions in homeostasis. The skeletal system, for example, plays a key role in purposeful movement, which in turn allows us to move away from potentially harmful stimuli. This organ system is much more than an assemblage of individual bones—it is a complex and interdependent functional unit essential for life.

Mechanisms of Disease

THE AGING SKELETON

It is said that the body's aging processes begin at fertilization and continue until death. In the case of the skeletal system, a number of changes related to aging first become obvious early in uterine development. Other changes then occur at differing times and proceed at differing rates throughout life. Keep in mind that some of these changes are normal and produce positive outcomes whereas others are indicators of deterioration and disease.

Recall from Chapter 7 that flat bones enlarge during intramembranous ossification by appositional growth, which also permits the shafts of long bones to widen. It is endochondral ossification that replaces cartilaginous "models" of long bones present in the fetal skeleton with ossified matrix and then allows the new calcified bones to increase in length by interstitial growth until skeletal maturity occurs. The appearance and progression of these highly regulated skeletal aging processes is determined by genetic control factors, hormones, and other homeostatic control mechanisms.

Differing rates of bone deposition or resorption, including the nature and effectiveness of the remodeling process, often characterize aging-associated skeletal changes—both normal and pathological. For example, in the skull, fontanels close, sutures appear, teeth erupt, the paranasal sinuses enlarge and assume their adult placement, and the size of the cranium enlarges to accommodate a growing brain. All these changes occur in a predictable sequence over time and result from normal skeletal aging processes. In long bones, both the age of appearance of ossification centers and, later, the closure of epiphyseal growth plates are age-related skeletal changes important in clinical medicine (Figure 7-9). Ossification centers (see Figure 7-8 and Figure 7-9) and epiphyseal growth plates appear, enlarge, mature, and close in an orderly fashion. Such orderly progression makes it possible to estimate a child's potential for growth and ultimate height from the number and location of ossification centers and the status of epiphyseal plate closures visible on radiographs. Once the epiphyseal plates in long bones ossify, they can no longer lengthen and growth in height ceases. Physicians often compare an individual's "**bone age**," which is determined by the number of ossification centers visible on a radiograph, with "**chronological age**." This type of information can prove useful in the diagnosis or treatment of many genetic, metabolic, or inflammatory skeletal diseases characterized by abnormally rapid or delayed skeletal maturity.

In the absence of disease, new bone mass produced after puberty, and generally into the early thirties, is properly calcified but not brittle. It exhibits both the hardness and tensile strength required to resist fractures. During this period, osteoblastic activity results in the deposition of more bone than is resorbed in the remodeling process. Before puberty this process results in bones that continue to grow until they reach adult size and proportions. Instead of growth in bone size, the skeletal remodeling process in the immediate postpuberal years and, generally, into the second decade of life results in the replacement of younger and weaker bone with bone that is more dense and has higher-quality organic matrix. Thus, the skeletal aging process in early adulthood produces a positive outcome—stronger bones. Unfortunately, the efficiency of the remodeling process during the next decade begins a progressive decline that can be slowed by good nutrition and a healthy lifestyle, but never fully stopped. The negative outcomes of skeletal aging soon appear, and they continue over a lifetime.

Degenerative aging processes affect spongy bone before compact bone and generally begin between 30 and 40 years of age. In many cases, problems arise because of abnormalities in bone remodeling that result from both a decrease in osteoblasts and a gradual increase in osteoclast cell numbers and activity. In addition, the osteoblasts that are formed during late adulthood are less metabolically active and produce lower-quality bone matrix. Furthermore, in compact bone, older osteocytes tend to shrink and coalesce, resulting in the appearance of tiny, honeycomb-like open spaces. The result is bone thinning and loss of density. The skeleton as a whole loses strength, bones become brittle, and the risk for fracture increases.

It is the loss of trabeculae and resulting compression fractures in the spongy bone of vertebrae that cause the progressive decrease in height that is a typical consequence of aging. It's

Mechanisms of Disease—cont.

BOX 8-3 ## Balloon Kyphoplasty

Balloon kyphoplasty is a relatively new orthopedic procedure used to treat the vertebral compression fractures that occur in osteoporosis, as a result of certain tumors, or after prolonged use of steroid drugs. Fractured (collapsed) vertebrae (see Figure 7-18 on p. 247) result in shortened height and spinal deformity and may be the cause of chronic pain. The balloon kyphoplasty procedure involves the insertion of an inflatable balloon–like device called a **"bone tamp"** through a small incision in the skin and then through a channel drilled into the body of the fractured vertebra. The balloon is then deflated and removed, and the surgeon uses the needle to fill the cavity with a type of "super glue" bone cement that quickly hardens and thus stabilizes and seals the fracture. An earlier procedure, called **vertebroplasty,** also involves injecting bone cement, but without using a balloon. Both procedures are cost-effective, have short recovery periods, and in many cases may eliminate the need for difficult and expensive spinal surgery. However, like all surgical procedures, there is a need for both the physician and patient to fully evaluate the risks and rewards. Specialized physician training and experience in the procedure are important, as is the availability of hospital spinal surgery service facilities and personnel if problems should occur.

true—all of us "shrink with age." And the process begins earlier than you might think—about age 35. However, a regular weight-bearing exercise program coupled with good nutrition, especially adequate calcium and vitamin D intake, will help minimize the degenerative effects of skeletal aging.

The severe and repeated vertebral compression fractures common in osteoporosis cause not only a decrease in height but also spinal deformity and loss of mobility. Recall that **osteoporosis** (see p. 247) is characterized by loss of bone mass, especially spongy bone, and demineralization of matrix. No

discussion of skeletal aging is complete without emphasizing the importance of this very common and serious bone disease. You are encouraged to review the discussion of osteoporosis in Chapter 7. Unfortunately, advancing age is often accompanied by the appearance of other metabolic, genetic, or inflammatory disease conditions that affect skeletal tissues, bones, and joints.

Fractures and Abnormal Spinal Curvatures

Bone Fractures

A *bone fracture* is defined as a partial or complete break in the continuity of a bone that most often occurs under mechanical stress. The most common cause of a fracture is traumatic injury. However, bone cancer, cysts, or metabolic bone disorders such as osteoporosis can also cause what are described as **pathological or spontaneous fractures** (see Figure 8-28, A). In these types of fractures, the bone is so weak that it fractures under very little stress and in the absence of any significant trauma. Most fractures can be identified visually during physical examination or on a standard radiograph. One exception is the so-called **stress fracture,** which often occurs in the absence of any clinically visible damage to bone or surrounding tissue. In addition, most x-ray images appear normal with this type of fracture. The microscopic bone damage typical of a stress fracture frequently appears in one or more bones of the lower part of the leg or foot after repetitive trauma. Long-distance (marathon) runners and ballet dancers are notoriously "at risk." Often called hidden or occult fractures because they frequently remain unseen on regular radiographs, they are clearly visible on bone scans (see Box 7-1 on p. 233).

A **displaced or open fracture,** also known as a *compound fracture* (Figure 8-27, A), is one in which broken bone projects through surrounding tissue and skin, thereby inviting the pos-

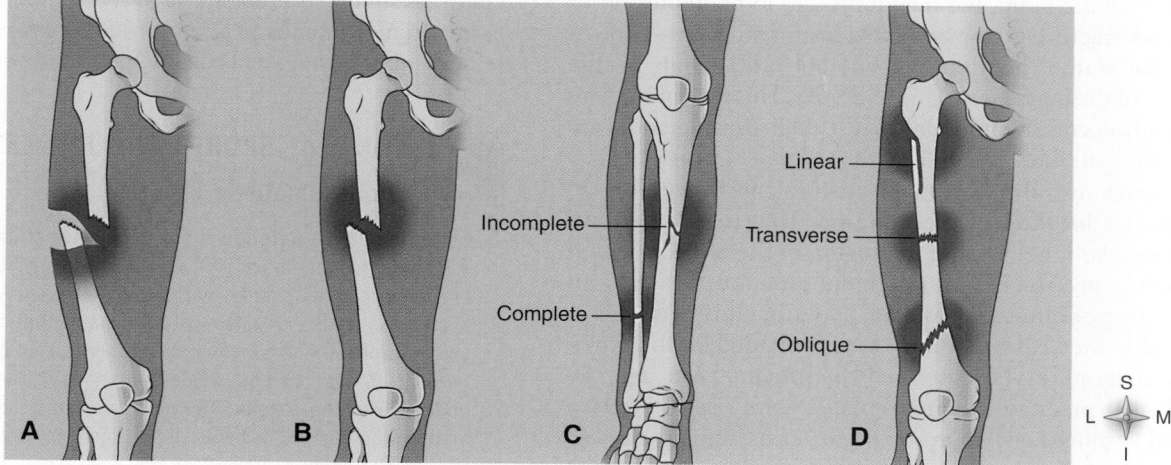

Figure 8-27 *Bone fractures.* **A,** Open. **B,** Closed. **C,** Incomplete and complete. **D,** Linear, transverse, and oblique.

Mechanisms of Disease—cont.

sibility of osteomyelitis (see pp. 248-249) or other types of infection. A **nondisplaced or closed fracture,** also known as a *simple fracture* (Figure 8-27, *B*) does not produce a break in the skin and therefore does not pose an immediate danger of bone infection. A closed fracture that results in one end of the broken bone being driven into the marrow cavity or diaphysis of the other bone segment is called an **impacted fracture.** As Figure 8-27, *C*, shows, fractures also are classified as "complete" or "incomplete." A **complete fracture** involves a break across the entire section of bone, whereas an **incomplete fracture** involves only a partial break, with the bone fragments still being partially joined.

The size and appearance of bone fragments at the fracture site identify two additional types of fracture. A **dentate fracture** results in fragmented ends of the bone being jagged and opposing each other, fitting together like teeth on a gear. In a **comminuted fracture,** the crushed bone fragments that lie between or near the broken ends of the bone are small and appear "crumbled." An **avulsion fracture** occurs when bone fragments are pulled free of the underlying bone surface by some force or trauma (see Osgood-Schlatter disease described below) or when a bone is completely severed from its point of attachment to the body.

Sometimes the angle of the fracture line or crack is used in labeling fracture types (Figure 8-27, *D*). A **linear fracture** involves a fracture line parallel to the bone's long axis. A fracture line at a right angle to the bone's long axis is labeled a **transverse fracture** (see Figure 8-27, *D*). **Oblique fractures** occur at slanted, or diagonal, angles to the longitudinal axis of the bone. As the name implies, in a **spiral fracture** the fracture line spirals around the long axis of the bone. In bones of the skull, especially, if the fracture line is visible but very small and the opposing bone segments remain fully aligned, the term **hairline fracture** is often used. If trauma results in a fractured skull bone being pushed downward into the cranial cavity or substance of the brain, the fracture is said to be **depressed.** A **greenstick fracture** is an incomplete fracture in a long bone in which the bone is bent on one side but broken only on the outer arc of the bend (see Figure 8-28, *B*). This type of fracture commonly occurs in children because their growing bones are less brittle than those of adults.

Fractures are also described anatomically according to the bone or location of the bone involved (e.g., humerus, mandible, skull, pelvis) and the region of the bone in which the fracture occurs (e.g., distal, midportion, or proximal). In the practice of clinical medicine, some fractures, as well as many other medical conditions, are still named for the physician or anatomist who described them many years ago. Although that practice is being replaced with use of modern medical terminology, some of these terms remain in use. **Pott fracture,** for example, is a break of the lower part of the tibia that often results in serious ankle injury in athletes, and **Colles fracture** is a break in the distal end of the radius that

is common in osteoporosis. Both types of fractures are named for surgeons who practiced well over 200 years ago. Another example involves so-called **Le Fort fractures** of the face and base of the skull. These often horrendous injuries are named after the French physician who first described them in the mid-1800s. Many other fracture types are also identified by specific names. Some of these names are helpful in understanding the nature of a specific type of fracture and some are not. For example, **hangman's fracture** is a fracture of posterior elements in the upper cervical spine—especially the axis. Although we do not believe that this particular type of fracture was named after a distinguished anatomist, it seems the name is perhaps more descriptive than some! Figure 8-28 is a grouping of medical images showing a number of different fracture types.

The term **Osgood-Schlatter disease** is used to describe a unique type of avulsion fracture. It involves partial separation of bone fragments from the surface of the tibial tuberosity. The bone fragments are literally "pulled away" from the underlying bone. It is common in adolescent athletes, especially muscular boys, who have recently overused the quadriceps muscles located on the top of the thigh. The common tendon of the four quadriceps muscles (see Chapter 10), called the quadriceps tendon, inserts into the patella. It then extends below the patella and becomes the patellar ligament, which attaches to the tibial tuberosity (see Figure 8-23). In individuals with Osgood-Schlatter disease, the growing bone in the tibial tuberosity is weaker than the patellar ligament attached to it. When the ligament is subjected to powerful contractile forces, the weaker bone tissue will fragment before the ligament tears.

Symptoms, which include gradually increasing swelling and pain just below the knee, often occur in both legs (Figure 8-29, *A*). In most cases, pain-relieving medication, curtailment of activity, or short periods of immobilization will resolve what is generally a self-limited problem. However, bothersome symptoms lasting from 3 months to a year or longer may require surgical removal of bone fragments that do not resorb (Figure 8-29, *B*).

BOX 8-4: SPORTS AND FITNESS
Chondromalacia Patellae

Chondromalacia patellae is a degenerative process that results in softening (degeneration) of the articular surface of the patella. The symptoms associated with chondromalacia of the patella are a common cause of knee pain in many individuals—especially young athletes. The condition is usually caused by irritation of the patellar groove, with subsequent changes in the cartilage on the underside of the patella. The most common complaint is of pain arising from behind or beneath the kneecap, especially during activities that require flexion of the knee, such as climbing stairs, kneeling, jumping, or running.

Mechanisms of Disease—cont.

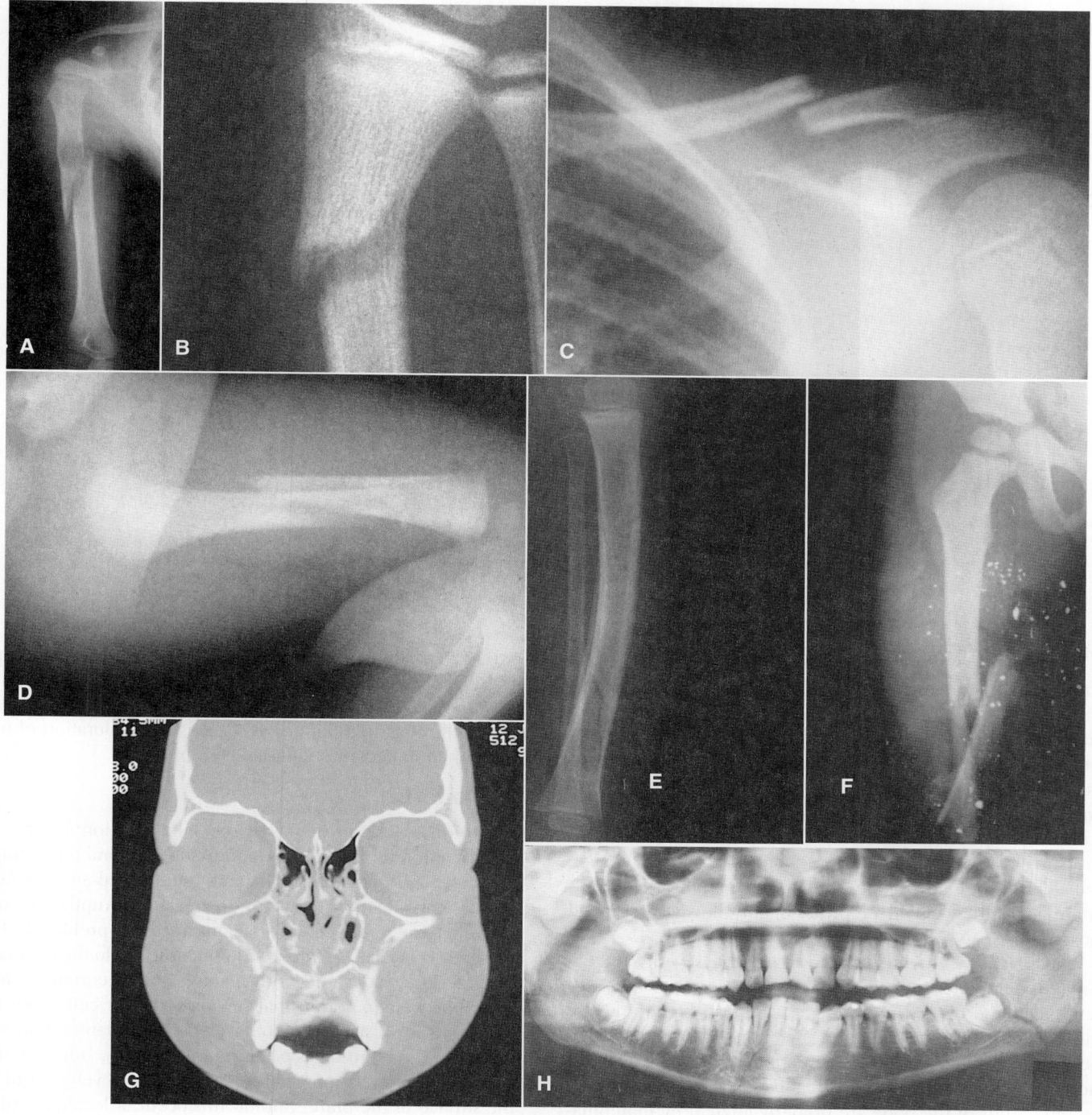

Figure 8-28 *CT scan and x-ray images showing different fracture types.* **A**, Bone cyst resulting in a pathological fracture of the humerus. **B**, Greenstick fracture of the distal end of the radius. **C**, Transverse fracture of the midportion of the clavicle. **D**, Oblique fracture of the midportion of the femur. **E**, Spiral fracture of the distal end of the tibia in a child. **F**, Comminuted fracture of the femur. **G**, CT scan showing a Le Fort fracture separating the maxilla from the facial bones on both sides. **H**, Panoramic radiograph showing a nondisplaced mandibular fracture extending through a wisdom tooth bud. Can you locate the second fracture—a tiny hairline fracture near the midline on the patient's right?

Mechanisms of Disease—cont.

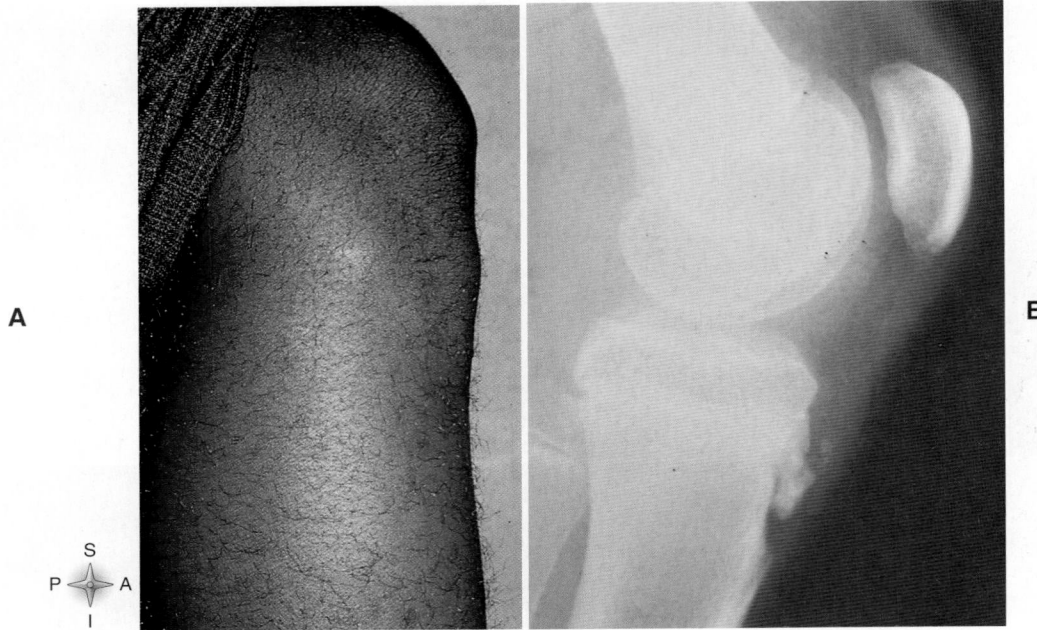

Figure 8-29 *Osgood-Schlatter disease.* **A,** Note the swelling over the tibial tuberosity just below the knee. **B,** Radiograph showing irregular fragments of the tibial tuberosity in a severe case of Osgood-Schlatter disease.

Treatment of Fractures

Orthopedics is the medical specialty dealing with skeletal injury and disease, and it is orthopedic physicians who are often called on to treat especially complex or severe fracture injuries. However, other health care providers often deal with fractures that are not life threatening or do not pose a significant risk of infection, disfigurement, or permanent loss of function. In general, clinical signs and symptoms of a fracture are pain, loss of function or false motion, soft tissue edema, and deformity. Occasionally, the sound of bone fragments rubbing together, called **crepitus,** occurs during attempted movements.

The type of treatment used depends on the severity and nature of the fracture itself, as well as the bone involved, the patient's age, and other circumstances such as systemic disease and the presence or potential for infection. In most cases, initial treatment begins with realignment and immobilization of the bone fragments. In a process called **closed reduction,** the fractured ends of the bone are properly aligned by manipulation without the need for surgery. In **open reduction,** a surgical procedure is required to align the broken ends of the bone, which are then held together by screws, plates, wires, or other specialized orthopedic devices. Occasionally, so-called fixators, composed of rods and pins, are attached to the bone and then surgically extended to the body surface to stabilize a fracture or fracture site. In most cases, once reduction of the fracture has been accomplished, it is immobilized by devices such

as casts, splints, or specialized bandages. In children, especially, traction may be used to immobilize certain fractures (e.g., femur) for relatively short periods. When physical healing of the fractured bone is accomplished, restoration of normal function becomes the treatment priority.

Mastoiditis

Mastoiditis (mas-toyd-EYE-tis) is inflammation of the air spaces within the mastoid portion of the temporal bone. Infectious material may find its way into the mastoid air cells from middle ear infections and, unless treated promptly and successfully, can produce very serious medical problems. The mastoid air cells do not drain into the nose, as do the paranasal sinuses. As a result, infectious material that accumulates may erode the bony partition that separates the air cells from the cranial cavity. If such erosion develops, the inflammation may spread to the brain or its covering membranes. Infection and an accumulation of pus (abscess) may then develop in or on the surface of the brain. Should this occur, a localized infection within the middle ear and mastoid air cells in the substance of the temporal bone escalates into a life-threatening medical emergency involving the central nervous system (Figure 8-30). Externally, clinical signs of mastoiditis include redness and swelling of the external ear and the skin overlaying the mastoid process. The area is very painful when touched. In severe cases, the external ear may be pushed forward and downward by the underlying swelling (Figure 8-31).

Mechanisms of Disease—cont.

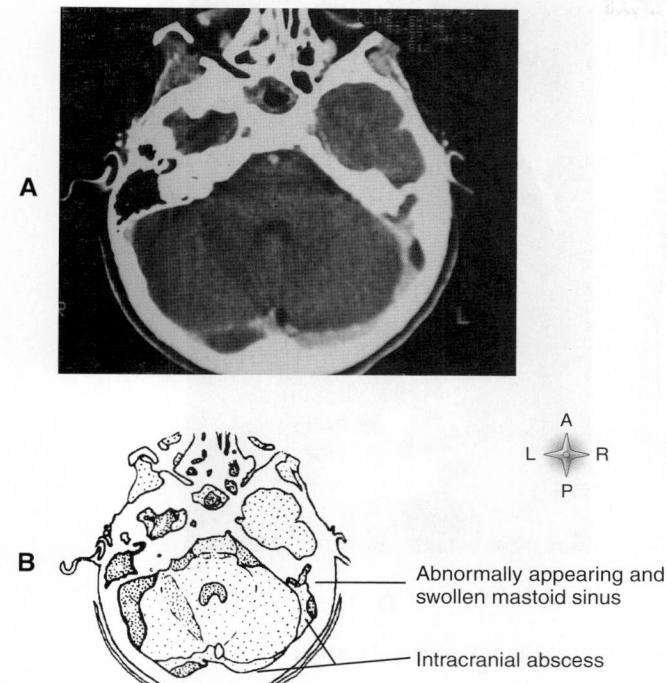

Figure 8-30 *Spread of mastoiditis infection.* **A**, CT image showing a left-sided mastoid infection that has eroded through the sinus wall and created an abscess in the cranial cavity. **B**, Diagram of a CT image showing the location of an abnormally appearing sinus and intracranial abscess.

Prompt treatment of middle ear infection or **otitis media** with antibiotics has made life-threatening cases of mastoiditis rare. Exceptions occur in instances of antibiotic resistance or in cases of inadequate or untreated chronic middle ear infections. If intensive antibiotic therapy fails and the subsequent accumulation of infectious material that fills the middle ear and mastoid air cells threatens extension into the cranial cavity, surgical intervention may be required. A surgical incision of the eardrum is often performed to reduce pressure and release pus from the middle ear. Though rarely performed today, surgical removal of part of the diseased mastoid portion of the temporal bone, a procedure called **mastoidectomy**, may also be necessary to drain trapped infectious material from the air cells. Fortunately, modern antibiotics and prompt medical treatment have dramatically reduced the need for this procedure with its attendant scarring and hearing loss.

Abnormal Spinal Curvatures

The normal curvature of the spine is convex through the thoracic region and concave through the cervical and lumbar regions (see Figure 8-13). This gives the spine strength to support the weight of the rest of the body and balance necessary to stand and walk. A curved structure has more strength than a straight one of the same size and material does. Poor posture or disease may cause the lumbar curve to be abnormally accentuated—a condition known as "sway back," or **lordosis** (Figure 8-32, A). This condition is frequently seen during preg-

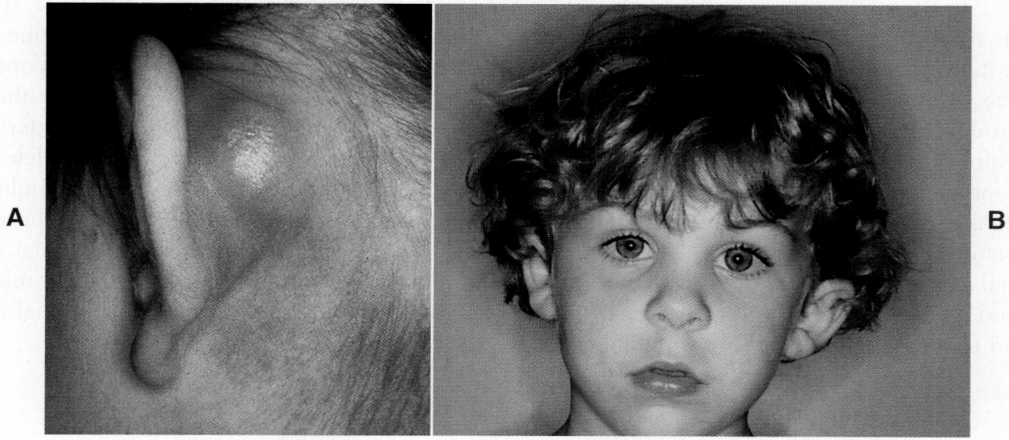

Figure 8-31 *Mastoiditis.* **A**, Note redness and swelling over the mastoid process of the temporal bone. **B**, Swelling has displaced the ear forward and downward.

Mechanisms of Disease — cont.

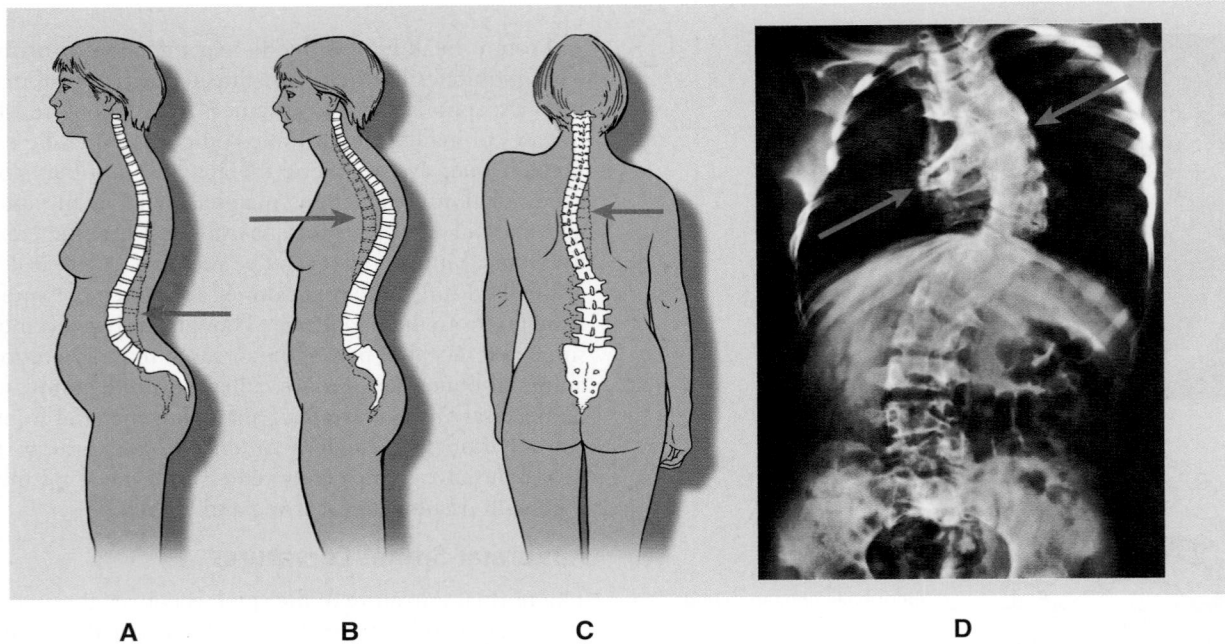

Figure 8-32 *Abnormal spine curvatures.* **A,** Lordosis. **B,** Kyphosis. **C,** Scoliosis. **D,** X-ray film of scoliosis curvature.

nancy as the woman adjusts to changes in her center of gravity. It may also be secondary to traumatic injury or a degenerative process of the vertebral bodies. **Kyphosis,** or "hunchback," is an abnormally increased roundness in the thoracic curvature (Figure 8-32, *B*). It is frequently seen in elderly people with osteoporosis or chronic arthritis, those with neuromuscular diseases, or individuals with compression fractures of the thoracic vertebrae. In a condition called **Scheuermann disease,** kyphosis can develop in children at puberty. Abnormal side-to-side curvature is called **scoliosis** (Figure 8-32, *C*). This too may be idiopathic or a result of damage to the supporting muscles along the spine. It is a relatively common condition that appears before adolescence.

All three abnormal curvatures can interfere with normal breathing, posture, and other vital functions. The degree of abnormal curvature and resulting deformity of the vertebral column determine the various treatments instituted. The traditional treatment of scoliosis is the use of a supportive brace, called the **Milwaukee brace,** that is worn on the upper part of the body 23 hours per day for up to several years. A newer approach to straighten abnormal curvature is **transcutaneous stimulation.** In this method, muscles on one side of the vertebral column are electrically stimulated to contract and pull the vertebrae into a more normal position. If these methods fail, correction may be achieved by surgical intervention in which pieces of bone from elsewhere in the skeleton, or metal rods, are grafted to the deformed vertebrae to hold them in proper alignment. If treated early enough, kyphosis resulting from poor posture can be corrected with special exercises and instructions for appropriate posture. Kyphosis resulting from pathological causes may also require special braces or surgical intervention.

LANGUAGE OF SCIENCE

(Cont'd from page 255)

occipital bone (awk-SIP-it-al) [*occipit-* back of head, *-al* pertaining to]

palatine bones (PAL-ah-tine) [*palatine* palate]

palpable (PAL-pah-bul) [*palpate* to touch gently]

parietal bones (pah-RYE-i-tal) [*paries-* wall, *-al* pertaining to]

patella (pah-TEL-ah) [*patella* small dish]

pelvic girdle (PEL-vic GER-dul) [*pelvis* basin, *-ic* pertaining to]

pubis (PYOO-biss) [short for *os pubis* bone of the groin]

radius (RAY-dee-us) [*radius* ray]

scapula (SKAP-yoo-lah) [*scapula* shoulder blade]

shoulder girdle

sinuses (SYE-nus-ez) [*sinus* hollow]

sphenoid bone (SFEE-noyd) [*spheno-* wedge, *-oid* having the appearance]

sternum (STER-num)

suture (SOO-chur) [*sutura* a seam]

temporal bones (TEM-poh-ral) [*tempora-* the temples, *-al* pertaining to]

thoracic vertebrae (thor-ASS-ik VER-teh-bray) [*thoracic* pertaining to the chest, *vertebra* joint]

thorax (THOR-aks) [*thorax* chest]

tibia (TIB-ee-ah) [*tibia* shin bone]

ulna (UL-nah) [*ulna* elbow]

vertebrae (VER-teh-bray) [*vertebra* joint] sing., vertebra

vomer bone (VOH-mer) [*vomer* plowshare]

zygomatic bones (zye-goh-MAT-ik) [*zygo-* union or fusion, *-ic* pertaining to]

LANGUAGE OF MEDICINE

Medical Terminology

Balloon kyphoplasty (KYE-foh-plas-tee) [*kypho-* hump, *-plasty* surgical repair]

Bone age

Bone tamp

Chondromalacia patellae (kon-droh-mah-LAY-shee-ah pah-TEL-ah) [*chondro-* cartilage, *-malacia* softening, *patella* small dish]

Chronological age

Closed reduction

Crepitus (KREP-i-tus) [*crepitus* crackling]

Fallen arches (flatfeet)

Kyphosis (kye-FOH-sis) [*kyphos-* hunchback, *-osis* condition]

Lordosis (lor-DOH-sis) [*lordos-* bent forward, *-osis* condition]

Mastoidectomy (mass-toyd-EK-toh-mee) [*mastoid-* mastoid process, *-ectomy* surgical removal]

Mastoiditis (mass-toyd-EYE-tis) [*mastoid-* mastoid process, *-itis* inflammation]

Milwaukee brace (mil-WAWK-ee) [*Milwaukee* after the city in Wisconsin, U.S.A.]

Open reduction

Orthopedics (or-thoh-PEE-diks) [*ortho-* straight or normal, *-pedic* feet]

Osgood-Schlatter disease (OZ-good SCHLAYT-er dih-ZEEZ) [*Robert B. Osgood,* American surgeon, *Carl Schlatter,* Swiss surgeon]

Osteoporosis (os-tee-oh-poh-ROH-sis) [*osteo-* bone, *-poro-* passage, *-osis* condition]

Otitis media (oh-TYE-tis MEE-dee-ah) [*otic-* pertaining to the ear, *-itis* inflammation]

Raccoon eyes

Scheuermann disease (SHOY-er-man) [*Holger W. Scheuermann,* Danish surgeon]

Scoliosis (skoh-lee-OH-sis) [*scolio-* twisted or crooked, *-osis* condition]

Transcutaneous stimulation (trans-kyoo-TAYNE-ee-us) [*trans-* across, *-cutis-* skin, *-eous* composed of]

Vertebroplasty (ver-tee-broh-PLAS-tee) [*vertebra* joint, *-plasty* surgical repair]

Listing of Fracture Types

Avulsion fracture (ah-VUL-shun) [*avulsio-* a pulling away, *fract* a breaking]

Blowout fracture [*fract* a breaking]

Colles fracture (KOL-ez) [*Abraham Colles,* Irish surgeon]

Comminuted fracture (Kom-i-NOO-ted) [*comminuere* to break into pieces, *fract* a breaking]

Complete fracture [*complere* to fill up, *fract* a breaking]

Dentate fracture (DEN-tayt)

Depressed fracture [*de-* to do the opposite, *-comprimere* to press together, *fract* a breaking]

Displaced, open, or compound fracture [*compound* to assemble, *fract* a breaking]

Greenstick fracture

Hairline fracture

Hangman's fracture

Impacted fracture [*impingere* to drive against, *fract* a breaking]

Incomplete fracture (in-kom-PLEET FRAK-chur) [*in-* not, *-complere* to fill, *fract* a breaking]

Le Fort fracture (leh-FOHR FRAK-chur) (also called *Guérin fracture*) [*Rene Le Fort* French surgeon, *Alphonse F.M. Guérin,* French surgeon]

Linear fracture (LIN-ee-ar) [*linea* line, *fract* a breaking]

Nondisplaced, closed, or simple fracture [*simplex* not mixed, *fract* a breaking]

Oblique fractures (oh-BLEEK) [*obliquus* slanted, *fract* a breaking]

Pathological (spontaneous) fractures (pah-THOL-lah-jik-al) [*patho-* disease, *-logos-* science, *-al* pertaining to, *fract* a breaking]

Pott fracture [*Percival Pott,* English physician]

Spiral fracture [*spir* coiled, *fract* a breaking]

Stress fracture

Transverse fracture (TRANS-vers) [*trans-* across, *fract* a breaking]

CASE STUDY

Wendy Jones, age 12, during a school physical examination is noted to have uneven shoulder and hip levels and a rotational deformity producing a rib hump on forward flexion.

1. Based on the above information, it would be expected that Wendy has which one of the following?

   A. Kyphosis
   B. Lordosis
   C. Osteoporosis
   D. Scoliosis

2. The best test to confirm the diagnosis would be

   A. Spinal radiograph
   B. Spinal CAT scan
   C. MRI
   D. Ultrasound

3. One of the treatments for this abnormal curvature might be the use of a Milwaukee brace. Which one of the following instructions should be included in the teaching session for Wendy and her parents?

   A. Wendy should remove the brace for only 1 hour a day to bathe.
   B. Wendy will need to wear the brace for approximately 2 months.
   C. The brace is the best available method to treat Wendy's problem.
   D. Wendy will be allowed to take the brace off every night so that she can sleep but must put it back on every morning.

CHAPTER SUMMARY

INTRODUCTION

A. Skeletal tissues form bones—the organs of the skeletal system
B. The relationship of bones to each other and to other body structures provides a basis for understanding the function of other organ systems
C. The adult skeleton is composed of 206 separate bones

DIVISIONS OF SKELETON (FIGURE 8-1; TABLE 8-1)

A. Axial skeleton—the 80 bones of the head, neck, and torso; composed of 74 bones that form the upright axis of the body and six tiny middle ear bones
B. Appendicular skeleton—the 126 bones that form the appendages to the axial skeleton; the upper and lower extremities

AXIAL SKELETON

A. Skull—made up of 28 bones in two major divisions: cranial bones and facial bones (Figures 8-2 to 8-7; Table 8-3)
  1. Cranial bones
    a. Frontal bone (Figure 8-8, C)
      (1) Forms the forehead and anterior part of the top of the cranium
      (2) Contains the frontal sinuses
      (3) Forms the upper portion of the orbits
      (4) Forms the coronal suture with the two parietal bones
    b. Parietal bones (Figure 8-8, A)
      (1) Form the bulging top of the cranium
      (2) Form several sutures: lambdoidal suture with the occipital bone; squamous suture with the temporal bone and part of the sphenoid; and coronal suture with the frontal bone
    c. Temporal bones (Figure 8-8, B)
      (1) Form the lower sides of the cranium and part of the cranial floor
      (2) Contain the inner and middle ears
    d. Occipital bone (Figure 8-8, D)
      (1) Forms the lower, posterior part of the skull
      (2) Forms immovable joints with three other cranial bones and a movable joint with the first cervical vertebra
    e. Sphenoid bone (Figure 8-8, E)
      (1) A bat-shaped bone located in the central portion of the cranial floor
      (2) Anchors the frontal, parietal, occipital, and ethmoid bones and forms part of the lateral wall of the cranium and part of the floor of each orbit (Figure 8-7)
      (3) Contains the sphenoid sinuses
    f. Ethmoid bone (Figure 8-8, F)
      (1) A complicated, irregular bone that lies anterior to the sphenoid and posterior to the nasal bones
      (2) Forms the anterior cranial floor, medial orbit walls, upper parts of the nasal septum, and sidewalls of the nasal cavity
      (3) The cribriform plate is located in the ethmoid
  2. Facial bones (Table 8-4)
    a. Maxilla (upper jaw) (Figure 8-8, H)
      (1) Two maxillae form the keystone of the face
      (2) Maxillae articulate with each other and with the nasal, zygomatic, inferior concha, and palatine bones
      (3) Forms parts of the orbital floors, roof of the mouth, and floor and sidewalls of the nose
      (4) Contains maxillary sinuses

b. Mandible (lower jaw) (Figure 8-8, *M*)
  (1) Largest, strongest bone of the face
  (2) Forms the only movable joint of the skull with the temporal bone
c. Zygomatic bone (Figure 8-8, *I*)
  (1) Shapes the cheek and forms the outer margin of the orbit
  (2) Forms the zygomatic arch with the zygomatic process of the temporal bones
d. Nasal bone (Figures 8-8, *L* and 8-10)
  (1) Both nasal bones form the upper part of the bridge of the nose, whereas cartilage forms the lower part
  (2) Articulates with the ethmoid, nasal septum, frontal bone, maxillae, and the other nasal bone
e. Lacrimal bone (Figure 8-8, *K*)
  (1) Paper-thin bone that lies just posterior and lateral to each nasal bone
  (2) Forms the nasal cavity and medial wall of the orbit
  (3) Contains a groove for the nasolacrimal (tear) duct
  (4) Articulates with the maxilla, frontal, and ethmoid bones
f. Palatine bone (Figure 8-8, *J*)
  (1) Two bones form the posterior part of the hard palate
  (2) Vertical portion forms the lateral wall of the posterior part of each nasal cavity
  (3) Articulates with the maxillae and the sphenoid bone
g. Inferior nasal conchae (turbinates)
  (1) Form the lower edge projecting into the nasal cavity and form the nasal meati
  (2) Articulate with ethmoid, lacrimal, maxillary, and palatine bones
h. Vomer bone (Figure 8-8, *G*)
  (1) Forms the posterior portion of the nasal septum
  (2) Articulates with the sphenoid, ethmoid, palatine, and maxillae
B. Eye orbits (Figure 8-7)
  1. Right and left eye orbits
    a. Contain eyes, associated eye muscles, lacrimal apparatus, blood vessels, and nerves
    b. Thin and fragile orbital walls separate orbital structures from the cranial and nasal cavities and paranasal sinuses
    c. Traumatic injuries may result in "blowout fractures" (Figure 8-7, *C*)
    d. "Raccoon eyes"—clinical sign of blowout fracture (Figure 8-7, *D*)
C. Fetal skull (Figure 8-11)
  1. Characterized by unique anatomical features not seen in adult skull
  2. Fontanels or "soft spots" (4) allow the skull to "mold" during the birth process and for rapid growth of the brain (Table 8-5)
  3. Permits differential growth or appearance of skull components over time
    a. Face—smaller proportion of total cranium at birth ($\frac{1}{8}$) than in adult ($\frac{1}{2}$)
    b. Head at birth is $\frac{1}{4}$ the total height; at maturity about $\frac{1}{8}$ body height

c. Sutures appear with skeletal maturity (Table 8-5)
d. Paranasal sinuses—change in size and placement with skeletal maturity (Figure 8-9)
e. Appearance of deciduous and, later, permanent teeth
D. Hyoid bone (Figure 8-12)
  1. U-shaped bone located just above the larynx and below the mandible
  2. Suspended from the styloid processes of the temporal bone
  3. Only bone in the body that articulates with no other bones
E. Vertebral column (Figure 8-13)
  1. Forms the flexible longitudinal axis of the skeleton
  2. Consists of 24 vertebrae plus the sacrum and coccyx
  3. Segments of the vertebral column:
    a. Cervical vertebrae, 7
    b. Thoracic vertebrae, 12
    c. Lumbar vertebrae, 5
    d. Sacrum—in adults, results from the fusion of five separate vertebrae
    e. Coccyx—in adults, results from the fusion of four or five separate vertebrae
  4. Characteristics of the vertebrae (Figures 8-14; Table 8-6)
    a. All vertebrae, except the first, have a flat, rounded body anteriorly and centrally, a spinous process posteriorly, and two transverse processes laterally
    b. All but the sacrum and coccyx have a vertebral foramen
    c. Second cervical vertebrae has an upward projection, the dens, to allow rotation of the head
    d. Seventh cervical vertebra has a long, blunt spinous process
    e. Each thoracic vertebra has articular facets for the ribs
  5. Vertebral column as a whole articulated with the head, ribs, and iliac bones
  6. Individual vertebrae articulate with each other in joints between their bodies and between their articular processes
F. Sternum (Figure 8-15)
  1. Dagger-shaped bone in the middle of the anterior chest wall made up of three parts:
    a. Manubrium—the upper handle part
    b. Body—middle blade part
    c. Xiphoid process—blunt cartilaginous lower tip, which ossifies during adult life
  2. Manubrium articulates with the clavicle and first rib
  3. Next nine ribs join the body of the sternum, either directly or indirectly, by means of the costal cartilages
G. Ribs (Figures 8-15 and 8-16)
  1. Twelve pairs of ribs, with the vertebral column and sternum, form the thorax
  2. Each rib articulates with the body and transverse process of its corresponding thoracic vertebra
  3. Ribs 2 through 9 articulate with the body of the vertebra above
  4. From its vertebral attachment, each rib curves outward, then forward and downward

5. Rib attachment to the sternum:
  a. Ribs 1 through 8 join a costal cartilage that attaches it to the sternum
  b. Costal cartilage of ribs 8 through 10 joins the cartilage of the rib above to be indirectly attached to the sternum
  c. Ribs 11 and 12 are floating ribs because they do not attach even indirectly to the sternum

APPENDICULAR SKELETON

A. Upper extremity (Table 8-7)
  1. Consists of the bones of the shoulder girdle, upper and lower parts of the arm, wrist, and hand
  2. Shoulder girdle (Figure 8-17)
    a. Made up of the scapula and clavicle
    b. Clavicle forms the only bony joint with the trunk, the sternoclavicular joint
    c. At its distal end, the clavicle articulates with the acromion process of the scapula
  3. Humerus (Figures 8-18 and 8-19)
    a. The long bone of the upper part of the arm
    b. Articulates proximally with the glenoid fossa of the scapula and distally with the radius and ulna
  4. Ulna
    a. The long bone found on the little finger side of the forearm
    b. Articulates proximally with the humerus and radius and distally with a fibrocartilaginous disk
  5. Radius
    a. The long bone found on the thumb side of the forearm
    b. Articulates proximally with the capitulum of the humerus and the radial notch of the ulna; articulates distally with the scaphoid and lunate carpals and with the head of the ulna
  6. Carpal bones (Figure 8-20)
    a. Eight small bones that form the wrist
    b. Carpals are bound closely and firmly by ligaments and form two rows of four carpals each
      (1) Proximal row is made up of the pisiform, triquetrum, lunate, and scaphoid
      (2) Distal row is made up of the hamate, capitate, trapezoid, and trapezium
    c. The joints between the radius and carpals allow wrist and hand movements
  7. Metacarpal bones
    a. Form the framework of the hand
    b. The thumb metacarpal forms the most freely movable joint with the carpals
    c. Heads of the metacarpals (the knuckles) articulate with the phalanges
B. Lower extremity
  1. Consists of the bones of the hip, thigh, lower part of the leg, ankle, and foot (Table 8-8)
  2. Pelvic girdle is made up of the sacrum and the two coxal bones bound tightly by strong ligaments (Figure 8-21)
    a. A stable circular base that supports the trunk and attaches the lower extremities to it

  b. Each coxal bone is made up of three bones that fuse together (Figure 8-22):
    (1) Ilium—largest and uppermost
    (2) Ischium—strongest and lowermost
    (3) Pubis—anteriormost
  3. Femur—longest and heaviest bone in the body (Figure 8-23)
  4. Patella—largest sesamoid bone in the body
  5. Tibia
    a. The larger, stronger, and more medially and superficially located of the two leg bones
    b. Articulates proximally with the femur to form the knee joint
    c. Articulates distally with the fibula and talus
  6. Fibula
    a. The smaller, more laterally and deeply placed of the two leg bones
    b. Articulates with the tibia
  7. Foot (Figures 8-24 and 8-25)
    a. Structure is similar to that of the hand with adaptations for supporting weight
    b. Foot bones are held together to form spring arches
      (1) Medial longitudinal arch is made up of the calcaneus, talus, navicular, cuneiforms, and medial three metatarsals
      (2) Lateral longitudinal arch is made up of the calcaneus, cuboid, and fourth and fifth metatarsals

SKELETAL DIFFERENCES IN MEN AND WOMEN

A. Male skeleton is larger and heavier than female skeleton
B. Pelvic differences (Figure 8-26; Table 8-9)
  1. Male pelvis—deep and funnel-shaped with a narrow pubic arch
  2. Female pelvis—shallow, broad, and flaring with a wider pubic arch

CYCLE OF LIFE: THE AGING SKELETON

A. Aging changes begin at fertilization and continue over a lifetime
  1. Changes can be positive or negative
B. Normal bone development is a skeletal aging process
  1. Intramembranous ossification
  2. Endochondral ossification
  3. Appearance of ossification centers and closure of epiphyseal plates can be used to estimate potential growth and height
C. Characteristics of bone during age
  1. Bone produced early in life is properly calcified but not brittle
    a. Osteoblastic activity during early periods of bone remodeling results in deposition of more bone than is resorbed
      (1) Before puberty results in growth of bones
      (2) After puberty to early thirties replaced bone is stronger

2. Negative outcomes of skeletal aging begin between 30 and 40 years of age
   a. Decrease in osteoblast numbers with production of lower-quality matrix
   b. Increase in osteoclast numbers and activity with increased bone loss
   c. Coalescence and shrinkage of mature osteocytes produces a honeycomb of tiny holes in compact bone
   d. Skeleton as a whole loses strength and fracture risk increases
   e. Decrease in number of trabeculae in spongy bone in vertebral bodies and other bones results in "spontaneous" as well as compression fractures
   f. Overall height decreases beginning about age 35
   g. Osteoporosis is a common and very serious bone disease in old age

THE BIG PICTURE

A. Skeletal system is a good example of increasing structural hierarchy in the body
   1. Skeletal tissues grouped into discrete organs—bones
   2. Skeletal system consists of bones, blood vessels, nerves, and other tissues grouped to form a complex operational unit
   3. Integration of skeletal system with other body organ systems permits homeostasis to occur
   4. Skeletal system more than an assemblage of individual bones—it represents a complex and interdependent functional unit of the body

MECHANISMS OF DISEASE
Bone Fractures

A. Fracture defined as partial or complete break in continuity of a bone
   1. Mechanical stress and traumatic injury are most common causes
   2. Pathological or spontaneous fractures occur in absence of trauma
   3. Stress fractures may not be apparent on clinical examination or standard x-ray images but can be seen on bone scans
      a. Bone damage is microscopic
      b. Caused by repetitive trauma (e.g., marathon runners)
   4. Displaced, open or compound fractures—do not produce a break in the skin and pose less danger of infection
   5. Nondisplaced, closed or simple fractures—do not produce a break in the skin and pose less danger of infection
   6. Fracture types:
      a. Impacted—one end of the fracture driven into the diaphysis of the other fragment
      b. Complete—break extends across the entire section of bone
      c. Incomplete—some fracture components still partially joined
      d. Dentate—fracture components jagged and fit together like teeth on a gear
      e. Comminuted—crushed, small, crumbled-type bone fragments near the fracture
      f. Avulsion—bone fragments pulled away from the underlying bone surface or bone totally torn from the body part
      g. Linear—fracture line parallel to the bone's long axis
      h. Transverse—fracture line at a right angle to the long axis of the bone
      i. Oblique—fracture line slanted or diagonal to the longitudinal axis
      j. Spiral—fracture line spiraling around the long axis
      k. Hairline—common in the skull—fracture components are small and aligned. If the fracture is pushed downward, called a depressed fracture
      l. Greenstick—bone bent but broken only on one side (common in children)
      m. Pott's—fracture of the lower part of the tibia
      n. Colles'—fracture of the distal end of the radius
      o. Le Fort—fracture of the face and/or base of the skull
      p. Hangman's—fracture of the posterior elements in the upper cervical spine, especially the axis
      q. Blowout—fracture of the eye orbit
   7. Osgood-Schlatter disease (Figure 8-29)
      a. Avulsion fracture of bone fragments from the surface of the tibial tuberosity
      b. Caused by powerful contraction of the quadriceps muscle group pulling on the patellar ligament attached to the tibial tuberosity
      c. Common in adolescent athletes when the patellar ligament is stronger than the underlying bone

Treatment of Fractures

   1. Clinical signs of fracture include pain, loss of function, false motion, soft tissue edema, deformity, and crepitus
   2. Initial treatment is realignment and immobilization of bone fragments
      a. Closed reduction—alignment completed without surgery
      b. Open reduction—surgery required to align and internally immobilize bone fragments with screws, wires, plates, or other orthopedic devices
      c. After reduction, immobilization generally accomplished by casts, splints, and bandages
      d. Traction sometimes used—especially in children
      e. Restoration of function is treatment priority after healing

MASTOIDITIS (FIGURES 8-30 AND 8-31)

A. Inflammation of air spaces within the mastoid portion of the temporal bone
   1. Pus may enter mastoid air spaces from middle ear infection or otitis media
   2. Mastoid air cells do not drain into the nose as do the paranasal sinuses
   3. Infectious material may erode the thin bony partition separating air cells from the cranial cavity and cause intracranial infection
   4. Treatment is antibiotic therapy and surgical incision of the eardrum to drain pus from the middle ear
   5. Surgical removal of part of the mastoid process of the temporal bone—mastoidectomy—is rare

Abnormal Spinal Curvatures (FIGURE 8-32)

A. Normal curvature of the spine is convex through the cervical and lumbar regions
   1. Normal curves give the spine strength for support of the body and balance required to stand and walk
B. Abnormal curvatures
   1. Lordosis—abnormally accentuated lumbar curve ("sway back")
      a. Frequently seen during pregnancy
      b. May be secondary to traumatic injury
   2. Kyphosis—abnormally accentuated thoracic curvature ("hunchback")
      a. Frequent consequence of vertebral compression fractures in osteoporosis
      b. Sign of Scheuermann's disease that may develop in children at puberty
   3. Scoliosis—abnormal side-to-side spinal curvature
      a. Often appears before adolescence
      b. Treatments vary with severity of curvature
         (1) Milwaukee brace
         (2) Transcutaneous stimulation
         (3) Surgical grafting of bone from elsewhere in the skeleton or metal rods to the deformed vertebrae

REVIEW QUESTIONS

1. Describe the skeleton as a whole and identify its two major subdivisions.
2. Identify and differentiate the bones in the cranium and face.
3. Name and locate the fontanels and sutures of the skull.
4. Name the five pairs of bony sinuses in the skull.
5. Discuss the clinical (medical) importance of the cribriform plate of the ethmoid bone and the mastoid air cells in the temporal bone.
6. Identify and discuss the normal primary and secondary curves of the spine.
7. Describe and distinguish between the different kinds of bone fractures and discuss symptoms and treatments of broken bones.
8. Identify the bony components of the thorax.
9. Identify the bones of the shoulder and pelvic girdles.
10. Identify, compare, and organize the bones of the arm, forearm, wrist, and hand with those of the thigh, lower part of the leg, ankle, and foot.
11. Discuss the arches of the foot and point out the functional importance of each.
12. Describe the role of the symphysis pubis during childbirth.
13. Describe vertebroplasty.

CRITICAL THINKING QUESTIONS

1. How would you compare and contrast the differences between the skeletal structure of males and females? What are the physiological reasons for these differences?

Articulations

LANGUAGE OF SCIENCE

abduction (ab-DUK-shun) [*abducere* to take away]
adduction (ad-DUK-shun) [*adducere* to bring to]
amphiarthroses (am-fee-ar-THROH-seez) [*amphi-* on both sides or around, *-arthr* joints or articulations]
angular movements
articular cartilage (ar-TIK-yoo-lar KAR-ti-lij) [*artic* joint]
articulation (ar-tik-yoo-LAY-shun) [*artic* joint]
biaxial joints (bye-AK-see-al) [*bi-* two, *-axon* axle, *joint* to join]
bursae (BER-see) [*bursa* wineskin] sing., bursa
carpometacarpal joints (kar-pol-met-ah-KAR-pal) [*carpo-* wrist, *-meta-* beyond, *-carp-* wrist, *-al* pertaining to]
circular movements
circumduction (ser-kum-DUK-shun) [*circum-* around, *-ducere* to lead]
depression [*depress* to press down]
diarthroses (dye-ar-THROH-seez) [*dia-* between, *-arthr* joints] sing., diarthrosis
distal interphalangeal (DIP) (DIS-tal in-ter-fah-LAN-gee-al)
dorsiflexion (dor-si-FLEK-shun) [*dorsi-* back, *-flexion* to bend]
elevation [*elevate* to lift]
eversion (ee-VER-shun) [*evert* to turn out]
extension [*extend* to stretch]
flexion (FLEK-shun)
gliding
gliding movements
gomphoses (gom-FOH-seez) [*gomphos* bolt]
goniometer (gon-ee-OM-eh-ter) [*gonio-* angle, *-meter* measuring instrument]
hyperextension (hye-per-ek-STEN-shun) [*hyper-* excessive, *-extend* to stretch]
intercarpal joints (in-ter-KAR-pal) [*inter-* occurring between, *-carp* wrist]
interphalangeal joints (in-ter-fah-LAN-jee-al) [*inter-* occurring between, *-phalangeal* pertaining to the phalanx]

Cont'd on p. 342

An **articulation**, or joint, is a point of contact between bones. Although most joints in the body allow considerable movement, others are completely immovable or permit only limited motion or motion in only one plane or direction. In the case of immovable joints, such as the sutures of the skull, adjacent bones are bound together into a strong and rigid protective plate. In some joints, movement is possible but highly restricted. For example, joints between the bodies of the spinal vertebrae perform two seemingly contradictory functions. They help firmly bind the components of the spine to each other and yet permit normal, but restricted movement to occur. Most joints in the body allow considerable movement to occur as a result of skeletal muscle contractions. It is the existence of such joints that permits us to execute complex, highly coordinated, and purposeful movements. Functional articulations between bones in the extremities, such as the shoulder, elbow, hip, and knee, contribute to controlled and graceful movement and provide a large measure of our enjoyment of life. This chapter begins by classifying joints and describing their identifying features. Coverage of joint classification and structure is followed by a discussion of body movements and a description of selected major joints. The chapter concludes by describing life cycle changes and some common joint diseases.

CLASSIFICATION OF JOINTS

Joints may be classified into three major categories by using a structural or a functional scheme. If a *structural classification* is used, joints are named according to the type of connective tissue that joins the bones together (fibrous or cartilaginous joints) or by the presence of a fluid-filled joint capsule (synovial joints). If a *functional classification* scheme is used, joints are divided into three classes according to the degree of movement they permit: **synarthroses** (immovable), **amphiarthroses** (slightly movable), and **diarthroses** (freely movable). Table 9-1 classifies joints according to structure, function, and range of movement. Refer often to this table and to the illustrations that follow as you read about each of the major joint types in this chapter.

Fibrous Joints (Synarthroses)

The articulating surfaces of bones that form fibrous joints fit closely together. The different types and amount of connective tissue joining bones in this group may permit very limited movement in some fibrous joints, but most are fixed. There are three subtypes of fibrous joints: *syndesmoses*, *sutures*, and *gomphoses*.

Syndesmoses

Syndesmoses (sin-dez-MO-seez) are joints in which fibrous bands (ligaments) connect two bones. The joint between the distal ends of the radius and ulna is joined by the *radioulnar interosseous ligament* (Figure 9-1). Although this joint is classified as a fibrous joint, some movement is possible because of ligament flexibility.

Sutures

Sutures are found only in the skull. In most sutures, teethlike projections jut out from adjacent bones and interlock with each other with only a thin layer of fibrous tissue between them. Sutures become ossified in older adults and form extremely strong lines of fusion between opposing skull bones (see Figure 9-1).

Gomphoses

Gomphoses (gom-FO-seez) are unique joints that occur between the root of a tooth and the alveolar process of the mandible or maxilla (see Figure 9-1). The fibrous tissue between the tooth's root and the alveolar process is called the *periodontal membrane*.

Cartilaginous Joints (Amphiarthroses)

The bones that articulate to form cartilaginous joints are joined together by either hyaline cartilage or fibrocartilage. Joints characterized by the presence of hyaline cartilage between articulating bones are called *synchondroses*, and those joined by fibrocartilage are called *symphyses*. Cartilaginous joints permit only very limited movement between articulating bones in certain circumstances. During childbirth, for example, slight movement at the symphysis pubis facilitates the baby's passage through the pelvis.

Synchondroses

Note that joints identified as **synchondroses** (sin-kon-DROH-seez) have hyaline cartilage between articulating bones. Examples include the articulation between the first rib and sternum and the joint present during the growth years between the epiphyses of a long bone and its diaphysis (Figure 9-2). The epiphyseal plate between the epiphysis and diaphyses of a long bone is a temporary synchondrosis that does not move. The plate of hyaline cartilage is totally replaced by bone at skeletal maturity. (Most cartilaginous joints are amphiarthroses, or slightly movable joints.)

Symphyses

A **symphysis** (SIM-fi-sis) is a joint in which a pad or disk of fibrocartilage connects two bones. The tough fibrocartilage disks in these joints may permit slight movement when pressure is applied

Table 9-1	Primary Joint Classifications		
FUNCTIONAL NAME	**STRUCTURAL NAME**	**DEGREE OF MOVEMENT PERMITTED**	**EXAMPLE**
Synarthroses	Fibrous	Immovable	Sutures of the skull
Amphiarthroses	Cartilaginous	Slightly movable	Symphysis pubis
Diarthroses	Synovial	Freely movable	Shoulder joint

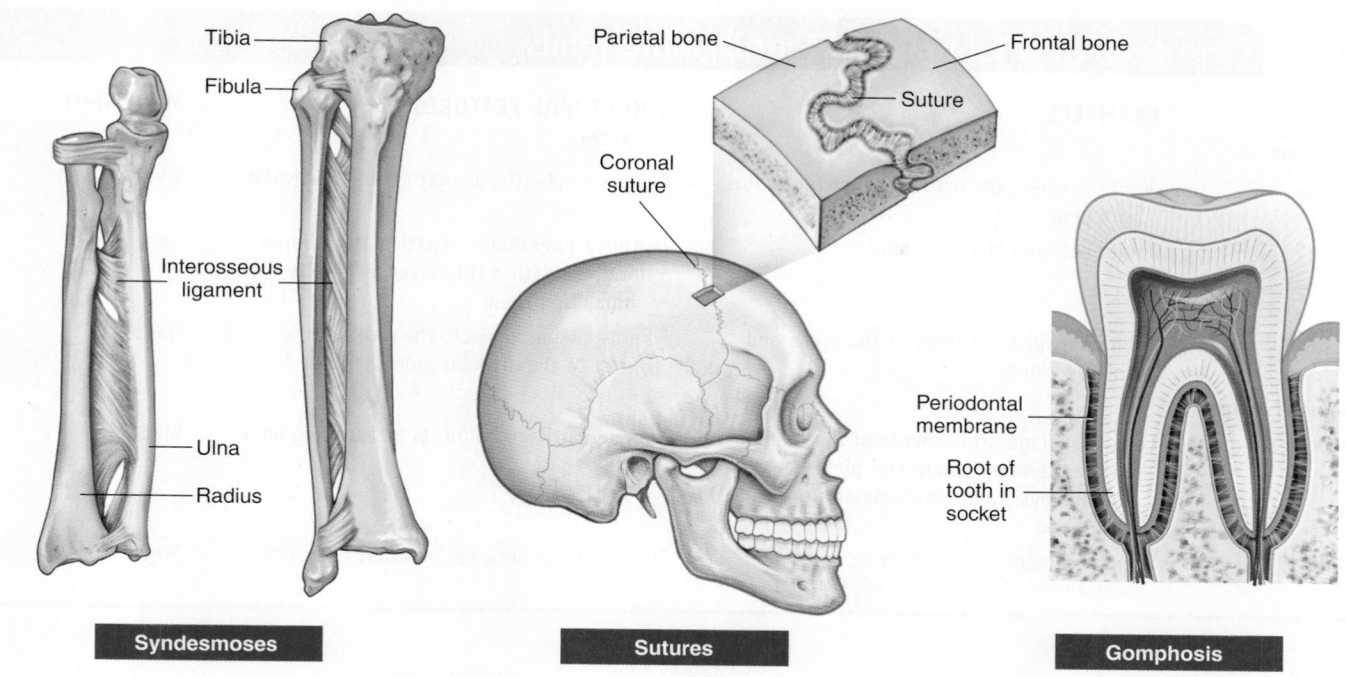

Figure 9-1 *Fibrous joints.* Examples of the types of fibrous joints.

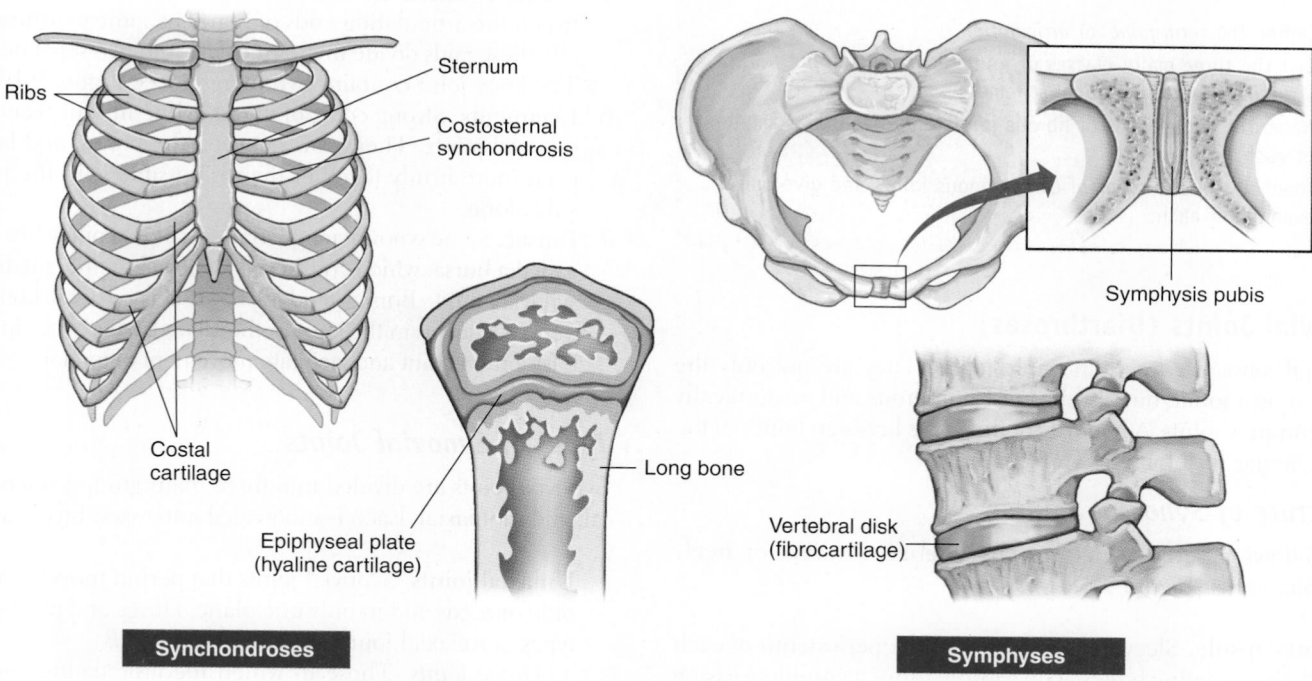

Figure 9-2 *Cartilaginous joints.* Examples of the types of cartilaginous joints.

between the bones. Most symphyses are located in the midline of the body. Examples of symphyses include the symphysis pubis and the articulation between the *bodies* of adjacent vertebrae (see Figure 9-2). The intervertebral disk in these joints is composed of tough and resilient fibrocartilage that absorbs shock and permits limited movement. The bones of the vertebral column have numerous points of articulation between them. Collectively, these

joints permit limited motion of the spine in a very restricted range. The articulation between the *bodies* of adjacent vertebrae is classified as a cartilaginous joint. The points of contact between the *articular facets* of adjacent vertebrae are considered synovial joints and are described later in the chapter.

Table 9-2 summarizes the different kinds of fibrous and cartilaginous joints.

Table 9-2	Classification of Fibrous and Cartilaginous Joints		
TYPES	**EXAMPLES**	**STRUCTURAL FEATURES**	**MOVEMENT**
Fibrous Joints			
Syndesmoses	Joints between the distal ends of the radius and ulna	Fibrous bands (ligaments) connect articulating bones	Slight
Sutures	Joints between the skull bones	Teethlike projections of articulating bones interlock with a thin layer of fibrous tissue connecting them	None
Gomphoses	Joints between the roots of the teeth and the jaw bones	Fibrous tissue connects the roots of the teeth to the alveolar processes	None
Cartilaginous Joints			
Synchondroses	Costal cartilage attachments of the first rib to the sternum; epiphyseal plate between the diaphysis and epiphysis of a growing long bone	Hyaline cartilage connects articulating bones	Slight
Symphyses	Symphysis pubis; joints between *bodies* of vertebrae	Fibrocartilage between articulating bones	Slight

QUICK CHECK

1. Define the term *joint,* or *articulation.*
2. List the three major classes of joints according to both a structural and a functional scheme.
3. Name the three types of fibrous joints and give one example of each.
4. Identify the two types of cartilaginous joints and give one example of each.

Synovial Joints (Diarthroses)

Synovial joints are freely movable joints. They are not only the body's most mobile but also its most numerous and anatomically most complex joints. A majority of the joints between bones in the appendicular skeleton are synovial joints.

Structure of Synovial Joints

The following seven structures characterize synovial, or freely movable, joints (Figure 9-3):

1. **Joint capsule.** Sleevelike extension of the periosteum of each of the articulating bones. The capsule forms a complete casing around the ends of the bones, thereby binding them to each other.
2. **Synovial membrane.** Moist, slippery membrane that lines the inner surface of the joint capsule. It attaches to the margins of the articular cartilage. It also secretes synovial fluid, which lubricates and nourishes the inner joint surfaces.
3. **Articular cartilage.** Thin layer of hyaline cartilage covering and cushioning the articular surfaces of bones.
4. **Joint cavity.** Small space between the articulating surfaces of the two bones of the joint. Absence of tissue between articu-

lating bone surfaces permits extensive movement. Synovial joints are therefore diarthroses, or freely movable joints.

5. **Menisci (articular disks).** Pads of fibrocartilage located between the articulating ends of bones in some diarthroses. Usually these pads divide the joint cavity into two separate cavities. The knee joint contains two menisci (see Figure 9-10).
6. **Ligaments.** Strong cords of dense, white fibrous tissue at most synovial joints. They grow between the bones, and lash them even more firmly together than is possible with the joint capsule alone.
7. **Bursae.** Some synovial joints contain a closed pillowlike structure called a **bursa,** which consists of a synovial membrane filled with synovial fluid. Bursae tend to be associated with bony prominences (such as in the knee or the elbow), where they function to cushion the joint and facilitate movement of tendons.

Types of Synovial Joints

Synovial joints are divided into three main groups: uniaxial, biaxial, and multiaxial. Each is subdivided into two subtypes as follows:

1. **Uniaxial joints.** Synovial joints that permit movement around only one axis and in only one plane. Hinge and pivot joints are types of uniaxial joints (Figure 9-4, A and B).
   a. *Hinge joints.* Those in which the articulating ends of the bones form a hinge-shaped unit. Like a common door hinge, hinge joints permit only back-and-forth movements, namely, flexion and extension. If you have access to an articulated skeleton, examine the articulating end of the humerus (the trochlea) and the ulna (the semilunar notch). Observe their interaction as you flex and extend the forearm. Do you see why you can flex and extend your forearm but cannot move it in any other way at this joint? The shapes of the trochlea and the semilunar notch (see Figures 8-18 and 8-19, p. 288) permit only the uniaxial, horizontal-plane

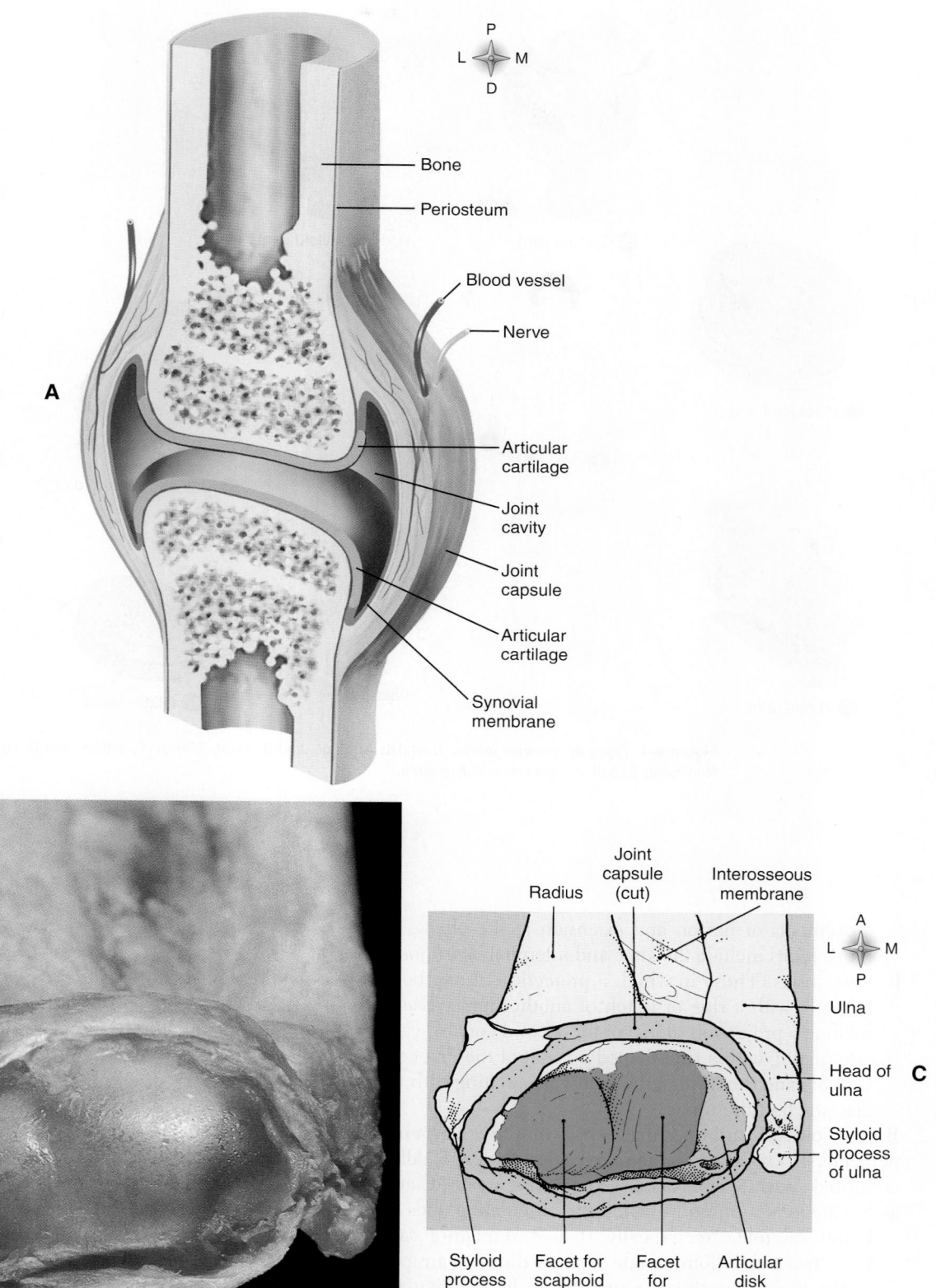

Figure 9-3 *Structure of synovial joints.* A, Artist's interpretation (composite drawing) of a typical synovial joint. **B,** Dissection photo showing the articular surface of a typical synovial joint—the distal radiocarpal joint (the joint capsule has been cut). **C,** Labeled diagram of dissection photo. See text p. 314 for a description of this joint.

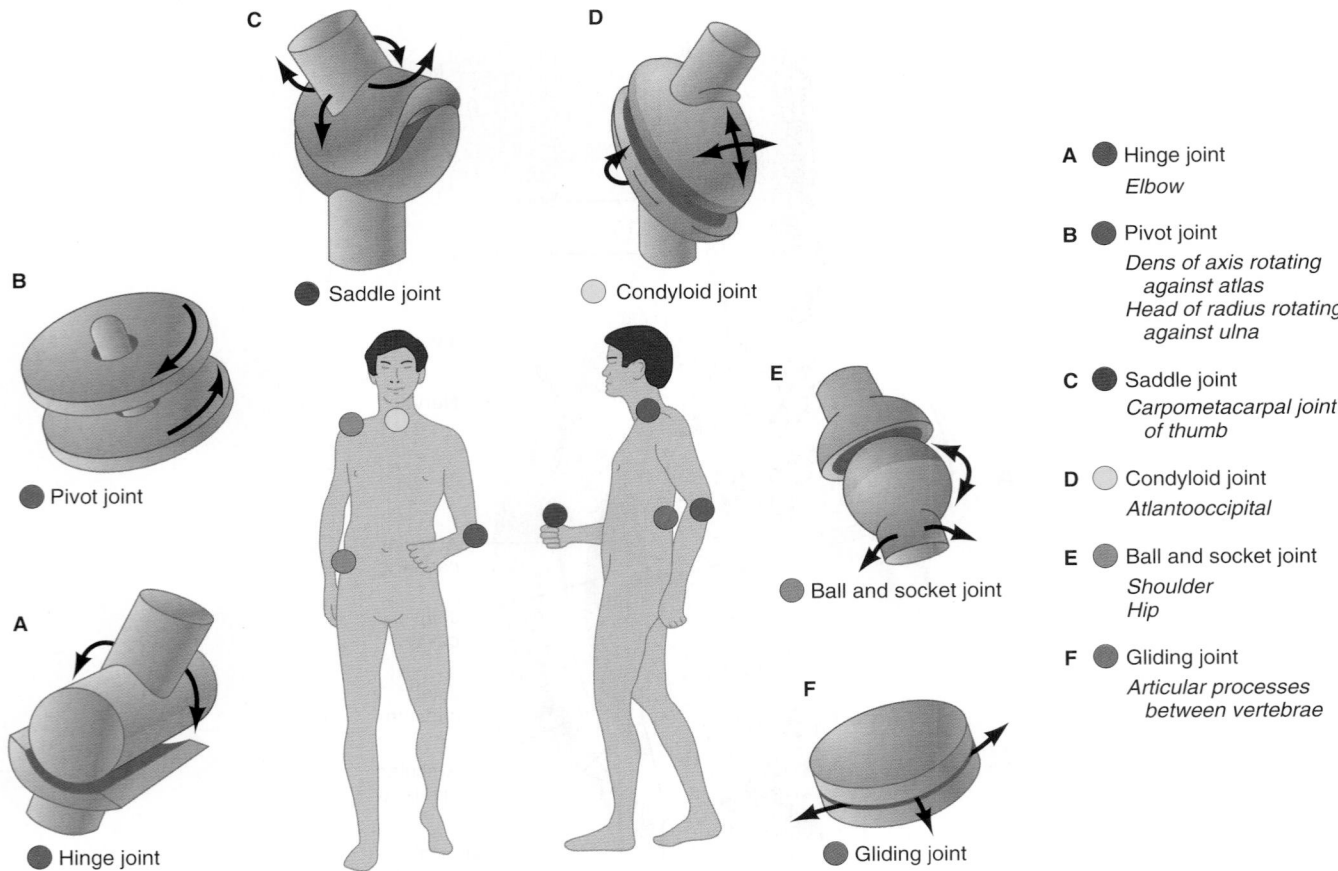

Figure 9-4 *Types of synovial joints.* Uniaxial: **A**, hinge, and **B**, pivot. Biaxial: **C**, saddle, and **D**, condyloid. Multiaxial: **E**, ball and socket, and **F**, gliding.

movements of flexion and extension at the elbow. Other hinge joints include the knee and interphalangeal joints.

b. *Pivot joints.* Those in which a projection of one bone articulates with a ring or notch of another bone. Examples include a projection (dens) of the second cervical vertebra articulating with a ring-shaped portion of the first cervical vertebra and the head of the radius articulating with the radial notch of the ulna.

2. **Biaxial joints.** Diarthroses that permit movement around two perpendicular axes in two perpendicular planes. Saddle and condyloid joints are types of biaxial joints (Figure 9-4, C and D).

a. *Saddle joints.* Those in which the articulating ends of the bones resemble reciprocally shaped miniature saddles. Only two saddle joints—one in each thumb—are present in the body. The thumb's metacarpal bone articulates in the wrist with a carpal bone (trapezium). The saddle-shaped articulating surfaces of these bones make it possible for the thumb to move over to touch the tips of the fingers—that is, to oppose the fingers. How important is this? To answer this for yourself, consider the following. Opposing the thumb to the fingers enables us to grip small objects. Were it not for this movement, we would have much

less manual dexterity. A surgeon could not grasp a scalpel or suture needle effectively, and none of us could easily hold a pen or pencil for writing.

b. *Condyloid (ellipsoidal) joints.* Those in which a condyle fits into an elliptical socket. Examples include condyles of the occipital bone fitting into elliptical depressions of the atlas and the distal end of the radius fitting into depressions of the carpal bones (scaphoid, lunate, and triquetrum).

3. **Multiaxial joints.** Joints that permit movement around three or more axes and in three or more planes (Figure 9-4, E and F).

a. *Ball-and-socket joints (spheroid joints).* Our most movable joints. A ball-shaped head of one bone fits into a concave depression on another, thereby allowing the first bone to move in many directions. Examples include the shoulder and hip joints.

b. *Gliding joints.* Characterized by relatively flat articulating surfaces that allow limited gliding movements along various axes. Examples include the joints between the articular surfaces of successive vertebrae. (Articulations between the bodies of successive vertebrae are symphysis-type cartilaginous joints.) As a group, gliding joints are the least movable of the synovial joints.

Table 9-3 summarizes the classes of synovial joints.

Table 9-3   Classification of Synovial Joints

TYPES	EXAMPLES		TYPE	MOVEMENT
Uniaxial				*Around one axis; in one place*
Hinge	Elbow joint		Spool-shaped process fits a into concave socket	Flexion and extension only
Pivot	Joint between the first and second cervical vertebrae		Arch-shaped process firsts around a peglike process	Rotation
Biaxial				*Around two axes, perpendicular to each other; in two planes*
Saddle	Thumb joint between the first metacarpal and carpal bone		Saddle-shaped bone fits into a socket that is concave-convex-concave	Flexion, extension in one plane; abduction, adduction in the other plane; opposing the thumb to the fingers
Condyloid (ellipsoidal)	Joint between the radius and carpal bones		Oval condyle fits into an elliptical socket	Flexion, extension in one plane; abduction, adduction in the other plane
Multiaxial				*Around many axes*
Ball and socket	Shoulder joint and hip		Ball-shaped process fits into a concave socket	Widest range of movement; flexion, extension, abduction, adduction, rotation, circumduction
Gliding	Joints between the articular facets of adjacent vertebrae; joints between the carpal and tarsal bones		Relatively flat articulating surfaces	Gliding movements without any angular or circular movements

QUICK CHECK

5. List the seven structures that characterize synovial joints.
6. Name the three main categories of synovial joints grouped according to axial movement and list the two subtypes of joints found in each category.
7. Name one specific joint as an example of each of the six types of synovial joints.

REPRESENTATIVE SYNOVIAL JOINTS
Humeroscapular Joint

The joint between the head of the humerus and the glenoid cavity of the scapula is the one we usually refer to as the *shoulder joint* (Figure 9-5). It is our most mobile joint. One anatomical detail, the shallowness of the glenoid cavity, largely accounts for this mobility. The shallowness offers little interference to movement of the head of the humerus. Were it not for the *glenoidal labrum* (Figure 9-5, *C*), a narrow rim of fibrocartilage around the glenoid

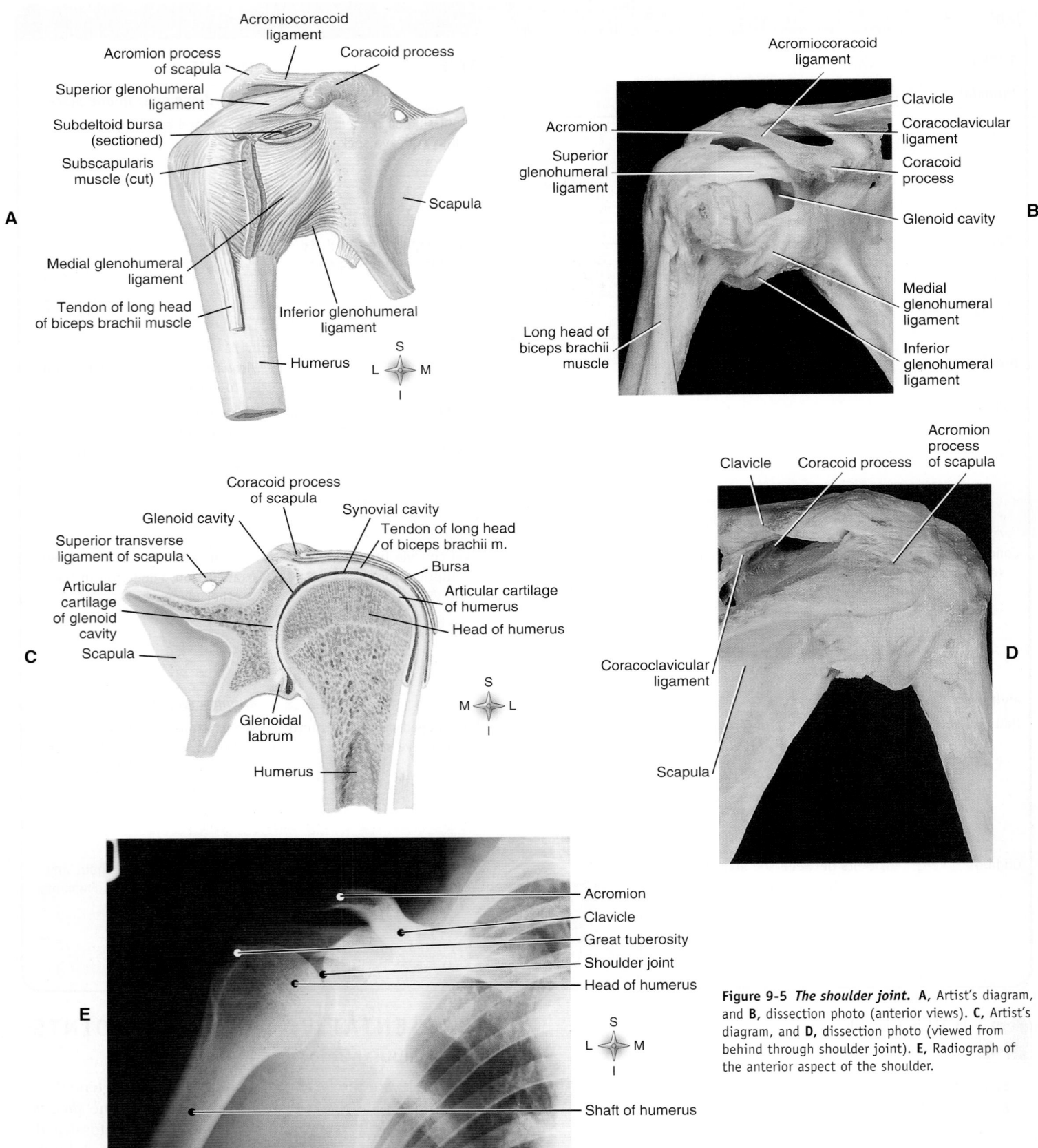

A

Acromiocoracoid ligament
Acromion process of scapula
Coracoid process
Superior glenohumeral ligament
Subdeltoid bursa (sectioned)
Subscapularis muscle (cut)
Medial glenohumeral ligament
Tendon of long head of biceps brachii muscle
Inferior glenohumeral ligament
Humerus
Scapula

S
L — M
I

B

Acromiocoracoid ligament
Acromion
Superior glenohumeral ligament
Long head of biceps brachii muscle
Clavicle
Coracoclavicular ligament
Coracoid process
Glenoid cavity
Medial glenohumeral ligament
Inferior glenohumeral ligament

C

Coracoid process of scapula
Glenoid cavity
Synovial cavity
Tendon of long head of biceps brachii m.
Superior transverse ligament of scapula
Bursa
Articular cartilage of glenoid cavity
Articular cartilage of humerus
Head of humerus
Scapula
Glenoidal labrum
Humerus

S
M — L
I

D

Clavicle
Coracoid process
Acromion process of scapula
Coracoclavicular ligament
Scapula

E

Acromion
Clavicle
Great tuberosity
Shoulder joint
Head of humerus

S
L — M
I

Shaft of humerus

Figure 9-5 *The shoulder joint.* **A,** Artist's diagram, and **B,** dissection photo (anterior views). **C,** Artist's diagram, and **D,** dissection photo (viewed from behind through shoulder joint). **E,** Radiograph of the anterior aspect of the shoulder.

cavity, it would have scarcely any depth at all. Some structures strengthen the shoulder joint and give it a degree of stability, notably several ligaments, muscles, tendons, and bursae. Note in Figure 9-5, *A,* for example, the superior, medial, and inferior glenohumeral ligaments. Each of these ligaments is a thickened portion of the joint capsule. Also in this figure, identify the subscapularis muscle, the tendon of the long head of the biceps brachii muscle, and a bursa. Shoulder muscles and tendons form a cufflike arrangement around the joint. It is called the *rotator cuff.* Baseball pitchers frequently injure the rotator cuff in the shoulder of their pitching arms. The main bursa of the shoulder joint is the *subdeltoid bursa.* It lies wedged between the inferior surface of the deltoid muscle and the superior surface of the joint capsule. Other bursae of the shoulder joint are the subscapular, subacromial, and subcoracoid bursae. All in all, the shoulder joint is more mobile than stable. Dislocations of the head of the humerus from the glenoid cavity occur frequently.

Elbow Joint

The elbow joint (Figure 9-6) is a classic hinge joint formed by two articulations occurring between the distal end of the humerus and the proximal ends of the radius and ulna. Laterally it is the

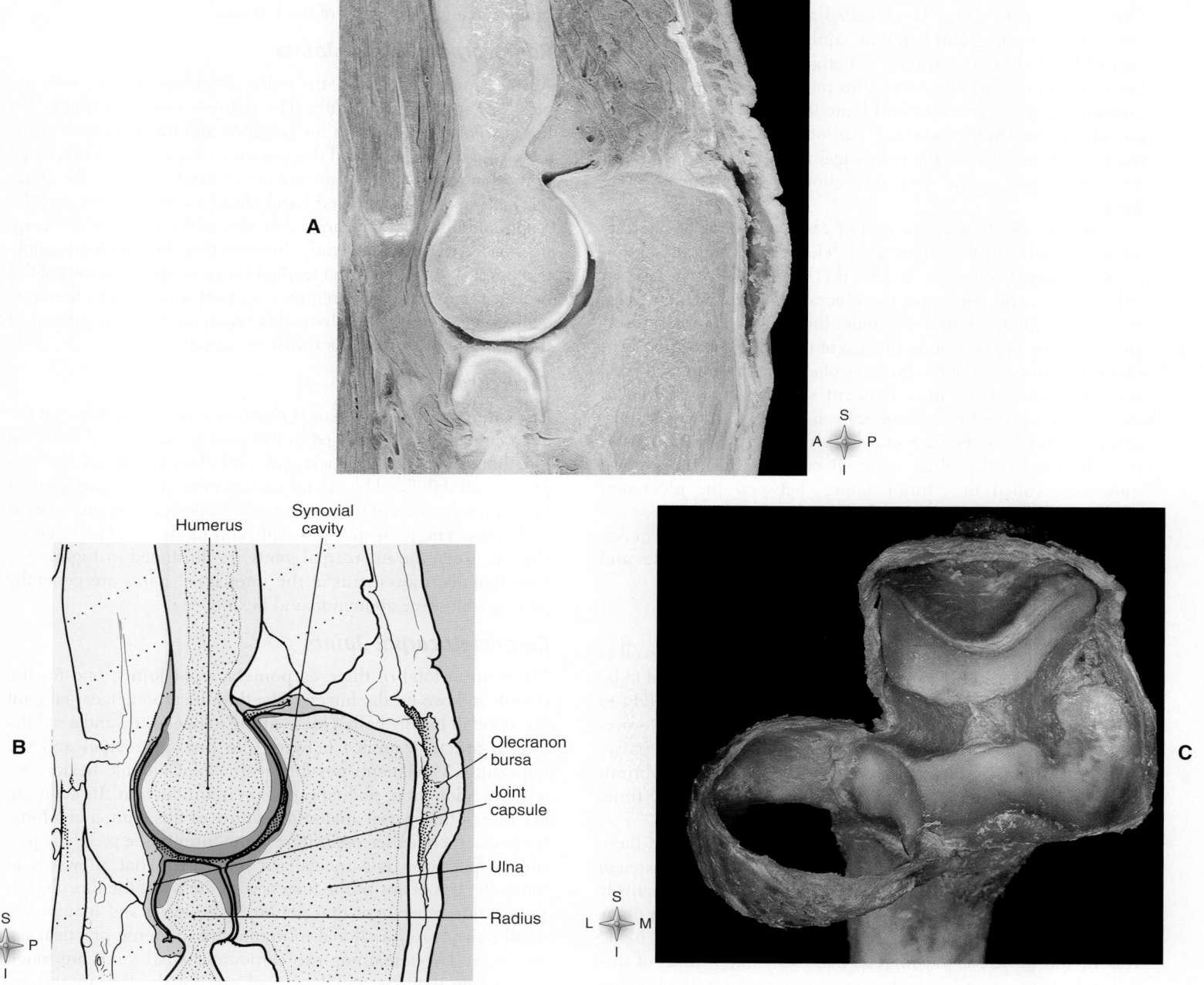

Figure 9-6 *The elbow joint.* **A,** Sagittal section through the elbow joint. **B,** Artist's drawing of a dissection photo. **C,** Inset dissection photo showing the anterior view of the proximal end of the ulna with attached annular ligament.

capitulum of the humerus that articulates with the head of the radius in the *humeroradial joint*. Medially, the trochlea of the humerus articulates with the trochlear notch of the ulna in the *humeroulnar joint*. Both these joint components are surrounded by a single joint capsule and by ligaments on either side, called *collateral ligaments*, that fuse with the capsule to stabilize the joint and help prevent disarticulation. The articulation between the proximal ends of the radius and ulna just below the joint capsule is called the *proximal radioulnar joint*. The point of articulation is between the head of the radius and the radial notch of the ulna and is stabilized by the *annular ligament*. Although it is not a part of the elbow joint involved in hinge movements, it does permit rotation of the forearm as the radial head moves on the ulna. Dislocation of the radial head, called a "pulled elbow," is seen more often in young children than adults. The reason? The disk-shaped head of the radius does not attain its conical shape until late in childhood and therefore slips more easily from under the annular ligament. The inset in Figure 9-6 shows the annular ligament attached to the proximal end of the ulna in an adult. It firmly holds the head of the radius against the articular surface of the radial notch on the ulna, thus allowing the radius to rotate freely.

A number of other anatomical or clinical "points of interest" are associated with the elbow joint. The medial and lateral epicondyles are palpable bony landmarks (see Box 8-2 on p. 295) on either side of the joint, and the **olecranon bursa,** which helps cushion the joint, is found just under the skin on its posterior surface overlying the olecranon process of the ulna. **Olecranon bursitis** is inflammation of the bursa associated with prolonged pressure. And, one of the most frequent sites for venepuncture to obtain blood is the median cubital vein located in the fossa just anterior to the joint. Perhaps one of the most important structures near the joint is the ulnar nerve. It courses along the groove, sometimes called the "funny bone," between the olecranon process and the medial epicondyle. Blows to this area produce unpleasant sensations in the hand and fingers supplied by the nerve. More severe injuries may result in paralysis of hand muscles and produce "clawhand" or a reduction in wrist movements.

Forearm, Wrist, Hand, and Finger Joints

Proper function of the forearm, wrist, hand, and fingers, as well as the joints that permit movement in these areas, is often said to be the functional "reason" for the upper extremity. Our ability to grasp and manipulate small objects, to properly focus the movement and strength of our upper extremities, and the many movements required to coordinate or restore balance and equilibrium are only a few examples of activities that require normal functioning of our forearms, wrists, hands, and fingers.

In clinical medicine, treatment or repair of injuries to these important functional areas constitutes a highly specialized area of surgical practice. In addition to bones and joints, the relatively small area of the wrist and hand, especially, is packed with many other anatomical structures, such as muscles, ligaments, blood vessels, and nerves. Hand surgery, like many other forms of specialized surgery, requires both a mastery of complex techniques and an in-depth understanding of the anatomy involved.

There are seven categories of synovial joints between the bones of the forearm, wrist, hand, and fingers. They are named after the bones that touch or articulate with one another in the joints involved. Some of the movements produced are slight and subtle, whereas others are more apparent. All are important functionally and are illustrated later in the chapter.

Radioulnar Joints

The head of the radius and the radial notch of the ulna articulate just below the elbow joint to form the *proximal radioulnar joint* described earlier. Together with the *distal radioulnar joint*, which is the point of articulation between the ulnar notch of the radius and head of the ulna just above the wrist, the two joints permit pronation and supination of the forearm.

Radiocarpal (Wrist) Joints

As the name implies, only the radius articulates directly with the wrist or carpal bones distally. The point of articulation between the head of the radius and the scaphoid and lunate carpals forms a typical synovial joint and is apparent in Figure 9-7. A full range of motions occur at the joint and are illustrated later in the chapter. Falls on an outstretched hand often result in fractures of the scaphoid bone. It is the carpal that transmits a majority of the applied force to the radial head. Unfortunately, loss of blood supply to a portion of the fractured scaphoid may result in necrosis of the broken fragment, which requires surgical removal or other specialized treatment. Note that a disk separates the terminal part of the ulna from direct contact with the carpals.

Intercarpal Joints

The **intercarpal joints** occur at points of articulation between the eight carpal bones. Arranged in two rows (proximal and distal) of four bones each, the carpals and joints between them are supported and stabilized by numerous ligaments and by passage over or around the wrist of forearm muscle tendons acting on the wrist and digits. The joint spaces usually communicate. The relationships between the intercarpal joints are illustrated in Figure 9-7. Functionally, movements at the intercarpal joints are generally gliding with some abduction and flexion.

Carpometacarpal Joints

There are a total of three **carpometacarpal joints**, one for the thumb and two for the fingers. The thumb carpometacarpal joint is unique in having both a loose-fitting joint capsule and a saddle-shaped articular surface between the first metacarpal and the trapezium. Movements possible at this joint include flexion, extension, adduction, abduction, and circumduction. In addition, because of the special anatomic features of this joint, a combination movement called *opposition* of the thumb (see p. 316) is possible. Opposition is a specialized movement that allows us to touch the tips of any of the fingers with the tip of the thumb. It is this unique movement that permits us to grasp and manipulate small objects—a functional movement of immense practical significance. The remaining two carpometacarpal joints are much less mobile than the saddle-shaped one for the thumb and are limited to essentially gliding-type movements.

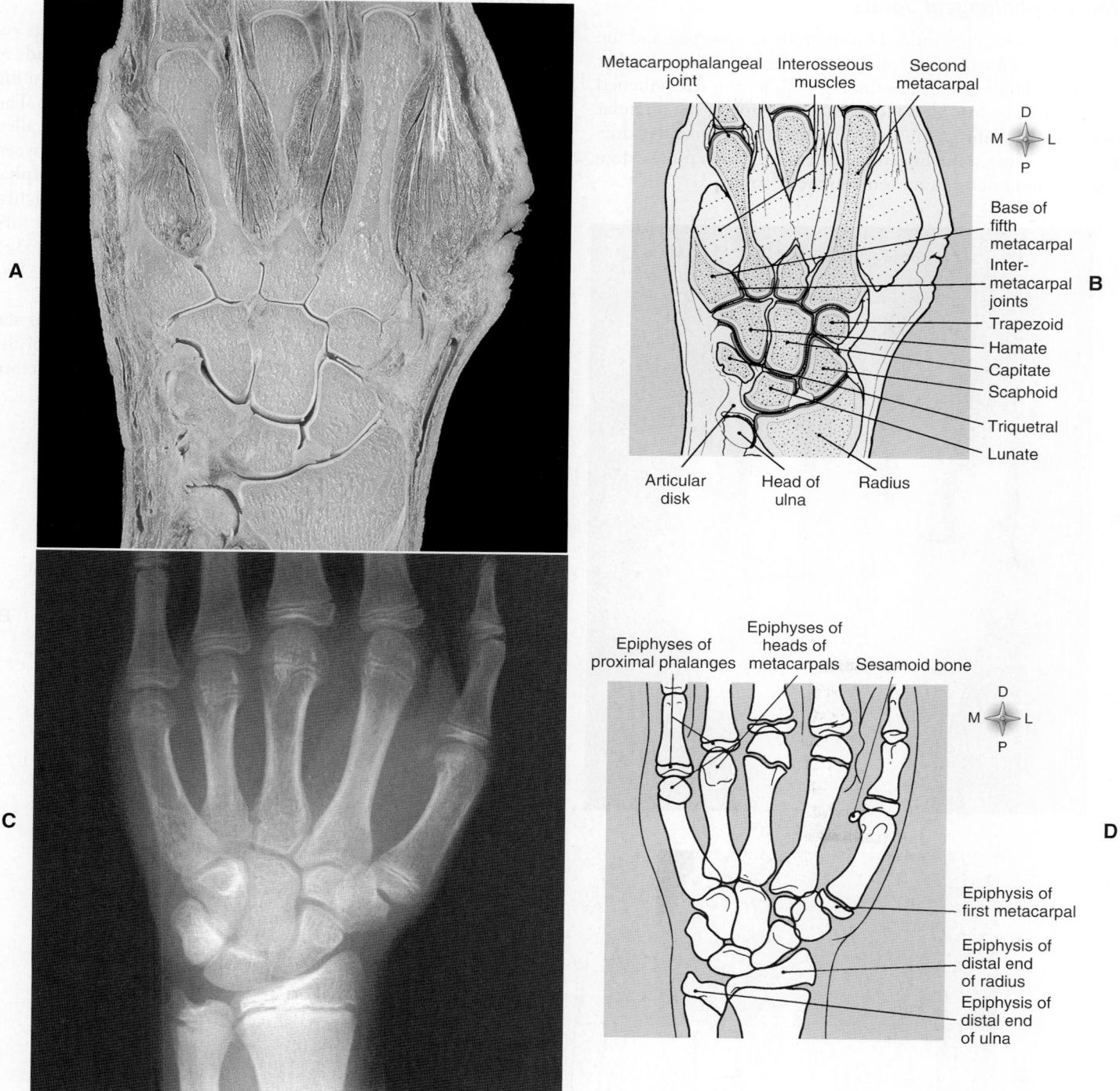

Figure 9-7 *Joints of the wrist*. A, Coronal section of the hand showing the joints of the wrist. **B,** Labeled diagram of the dissection photo. Note that the thumb and little finger are not in the plane of section. **C,** Radiograph of an adolescent hand and wrist. Note that the epiphyseal plates are present. **D,** Labeled diagram of the radiograph.

Metacarpophalangeal Joints

In this type of joint the rounded heads of the metacarpals and the concave bases of the proximal phalanges articulate with each other (Figure 9-8). The capsule surrounding each joint is strengthened by collateral ligaments on both sides. The shape of the articular surfaces in these joints permits only limited adduction and abduction—and only when the fingers are extended. Much more extensive flexion and extension movements are possible.

Interphalangeal Joints

Interphalangeal joints are typical hinge-type synovial joints capable of flexion and extension. They exist between the heads of the phalanges and the bases of the more distal phalanges. In the fingers, two types of interphalangeal joints can be identified. The joints between the proximal and middle phalanges are called **proximal interphalangeal (PIP) joints,** whereas those between the middle and distal phalanges are called the **distal interphalangeal (DIP) joints.** Understanding the relationship of articulations between the phalanges will help in understanding the various finger movements discussed later in the chapter (see p. 334).

Hip Joint

The first characteristic to remember about the hip joint is stability; the second is mobility (Figure 9-9). The stability of the hip joint derives largely from the shapes of the head of the femur

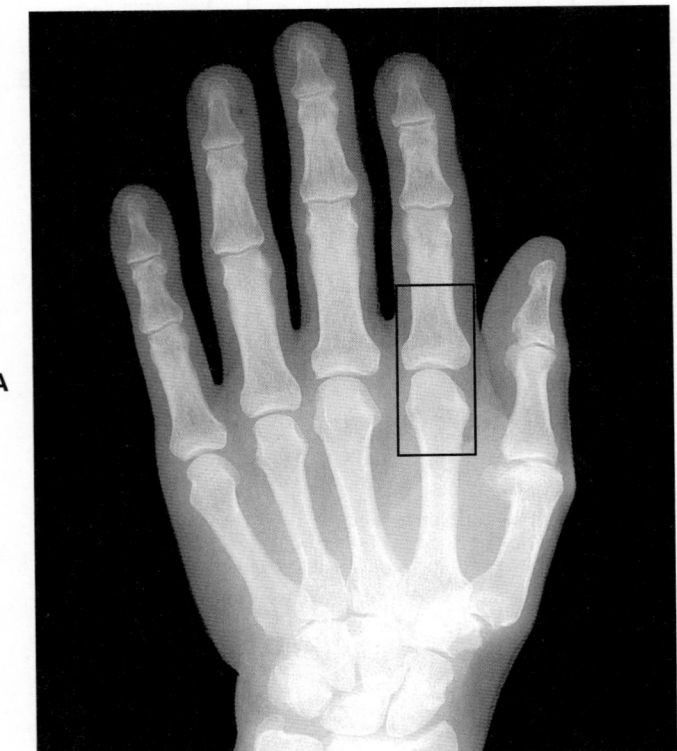

A

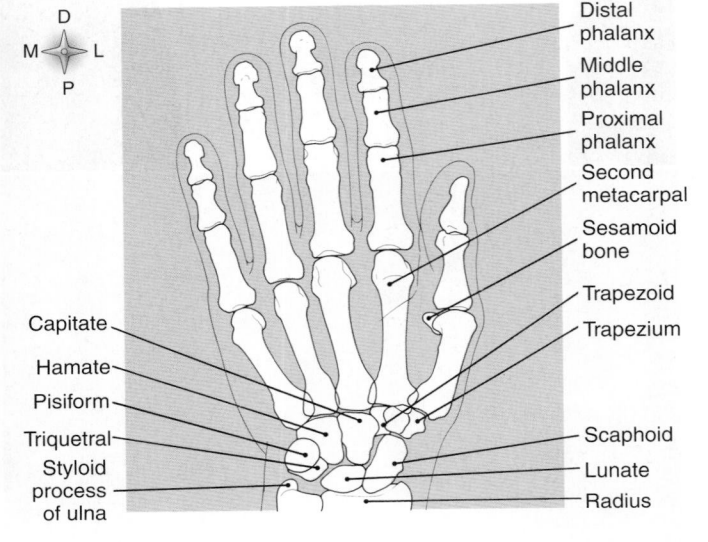

B

Distal phalanx
Middle phalanx
Proximal phalanx
Second metacarpal
Sesamoid bone
Trapezoid
Trapezium
Scaphoid
Lunate
Radius
Capitate
Hamate
Pisiform
Triquetral
Styloid process of ulna

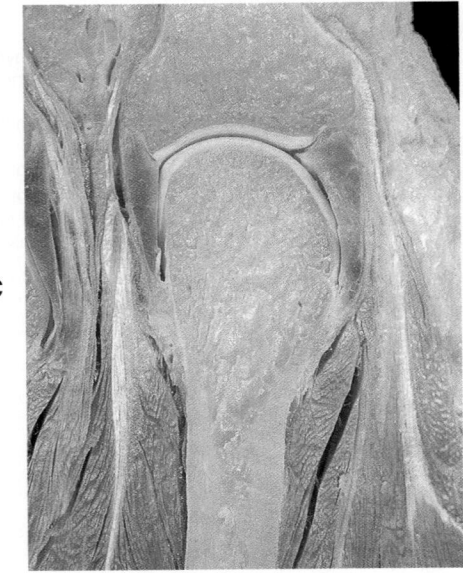

C

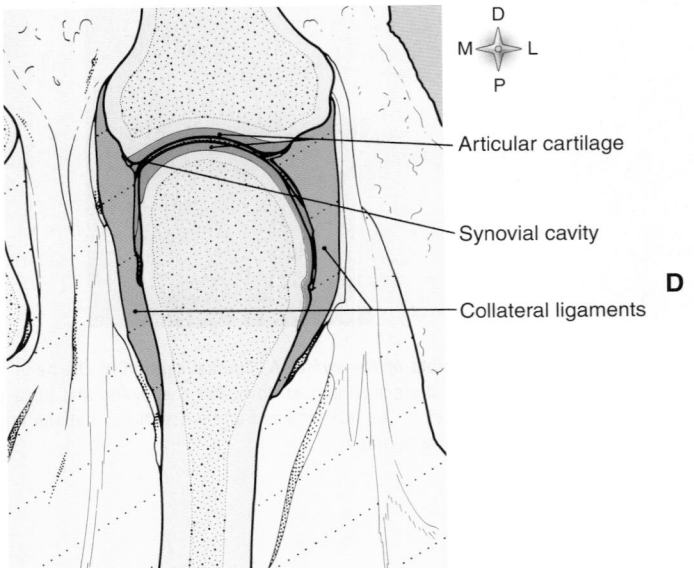

D

Articular cartilage
Synovial cavity
Collateral ligaments

Figure 9-8 *Joints of the hand and fingers.* **A,** Radiograph of an adult hand, and **B,** labeled diagram of the radiograph. **C,** Dissection photo showing a coronal section through the second metacarpophalangeal joint (represents enlargement of the inset box on the radiograph). **D,** Labeled diagram of the dissection photo. Note that the collateral ligaments are thickenings of the joint capsule.

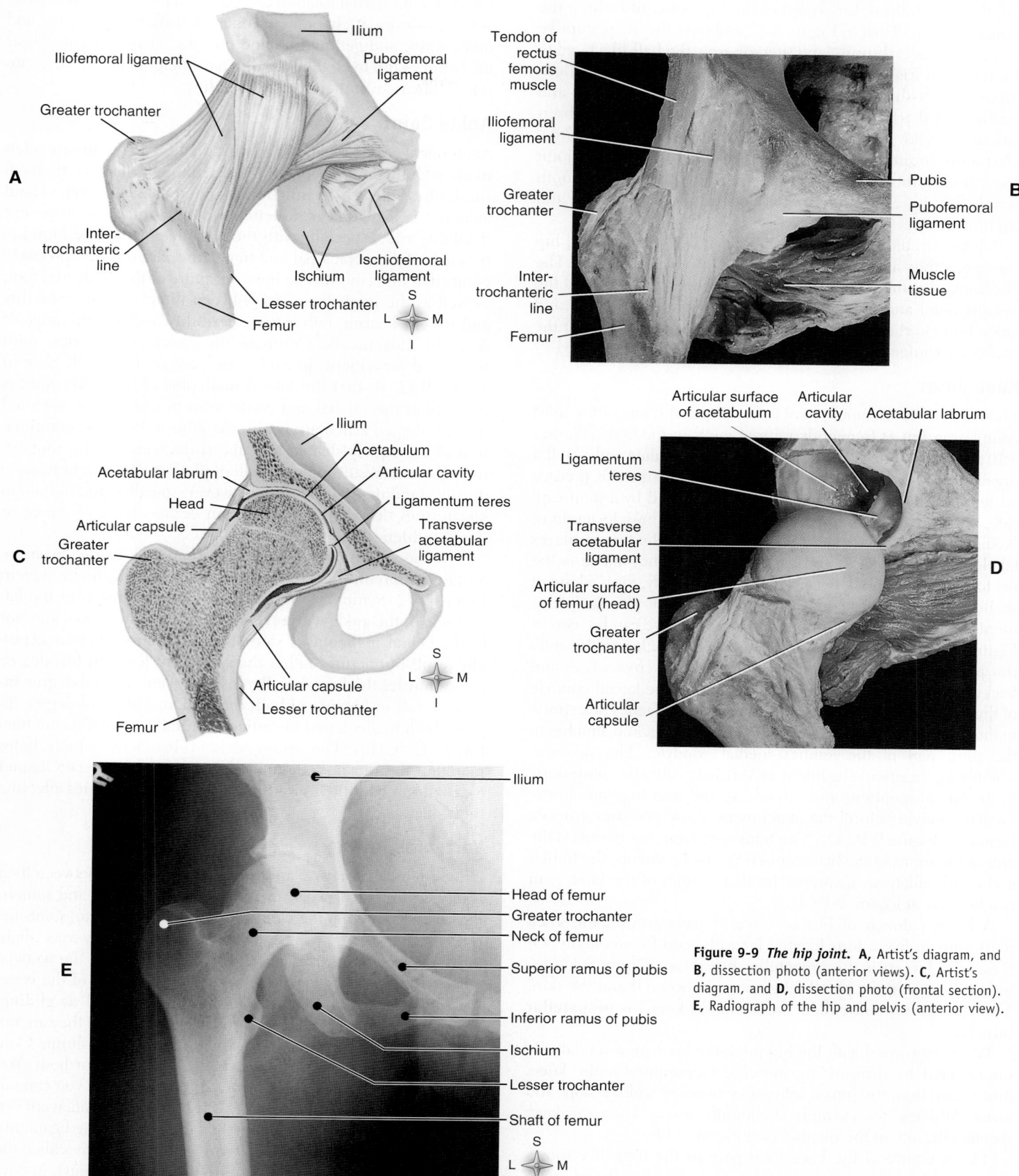

Figure 9-9 *The hip joint.* **A,** Artist's diagram, and **B,** dissection photo (anterior views). **C,** Artist's diagram, and **D,** dissection photo (frontal section). **E,** Radiograph of the hip and pelvis (anterior view).

and the acetabulum, the socket of the hip bone into which the femur head fits. Turn to Figure 8-22 and note the deep, cuplike shape of the acetabulum, and then observe the ball-like head of the femur in Figure 8-23, A. Compare these with the shallow, almost saucer-shaped glenoid cavity (see Figure 8-17, C) and the head of the humerus (see Figure 8-18). From these observations, do you see why the hip joint necessarily has a somewhat more limited range of movement than the shoulder joint does? Both joints, however, allow multiaxial movements. Both permit flexion, extension, abduction, adduction, rotation, and circumduction.

A joint capsule and several ligaments hold the femur and hip bones together and contribute to the hip joint's stability. The iliofemoral ligament connects the ilium with the femur, and the ischiofemoral and pubofemoral ligaments join the ischium and pubic bone to the femur. The iliofemoral ligament is one of the strongest ligaments in the body.

Knee Joint

The knee, or tibiofemoral, joint is the largest and one of the most complex and most frequently injured joints in the body (Figures 9-10 and 9-11). The condyles of the femur articulate with the flat upper surface of the tibia. Although this arrangement is precariously unstable, counteracting forces are supplied by a joint capsule, cartilages, and numerous ligaments and muscle tendons. Note, for example, in Figure 9-10 the shape of the two cartilages labeled *medial meniscus* and *lateral meniscus*. They attach to the flat top of the tibia and, because of their concavity, form a kind of shallow socket for the condyles of the femur. Of the many ligaments that hold the femur bound to the tibia, five can be seen in Figure 9-10. The anterior cruciate ligament attaches to the anterior part of the tibia between its condyles, then crosses over and backward and attaches to the posterior part of the lateral condyle of the femur. The posterior cruciate ligament attaches posteriorly to the tibia and lateral meniscus, then crosses over and attaches to the front part of the femur's medial condyle. The posterior tibiofibular ligament (ligament of Wrisberg) attaches posteriorly to the lateral meniscus and extends up and over to attach to the medial condyle behind the attachment of the posterior cruciate ligament (Figure 9-10, C). The transverse ligament connects the anterior margins of the two menisci. Strong ligaments, the fibular and tibial collateral ligaments, located at sides of the knee joint can be seen in Figure 9-10, C.

A baker's dozen of bursae serve as pads around the knee joint: four in front, four located laterally, and five medially. Of these, the largest is the prepatellar bursa (see Figure 9-11) inserted in front of the patellar ligament, between it and the skin. The painful ailment called "housemaid's knee" is **prepatellar bursitis.**

When compared with the hip joint, the knee joint is relatively unprotected by surrounding muscles. Consequently, the knee, more often than the hip, is injured by blows or sudden stops and turns. Athletes, for example, frequently tear a knee cartilage, specifically, one of the menisci (see Figure 9-11).

The structure of the knee joint permits the hingelike movements of flexion and extension. Also, with the knee flexed, some

internal and external rotation can occur. In most of our day-to-day activities—even such ordinary ones as walking, going up and down stairs, and getting into and out of chairs—our knees bear the brunt of the load; they are the main weight bearers. Knee injury or disease can therefore be crippling.

Ankle Joint

Anatomical comparisons of the wrist, hand, and fingers are often made with the ankle, foot, and toes. Although many anatomical similarities do exist, functional differences abound. Hand anatomy is a marvel of functional design intended to permit flexibility and, especially with the presence of a saddle joint between the first metacarpal and trapezium, discrete and precisely controlled movement. The bony structure of the ankle and foot, as well as the joints that exist between them, enhance stability and weight bearing rather than flexibility and a wide range of different movements. Compare the articulating bones, joint types, and movements in each area listed in Table 9-4. Note in Figure 9-12, B, that the lateral malleolus of the ankle joint is lower than the medial and contributes to a wedge or so-called mortise shaped articulation with the talus below. The combination of a uniquely shaped articular surface and strong supporting ligaments results in an excellent platform for weight bearing during standing and walking. However, so-called rotational ankle injuries do occur—especially in certain types of dance or during athletic activity.

The most common type of **sprained ankle** is caused by an internal rotation injury to the anterior talofibular ligament, seen in Figure 9-12. Symptoms include pain and swelling over the lateral side of the ankle with severe "point tenderness" just anterior to the lateral malleolus. External ankle rotation injuries generally result in fractures rather than ligament tears. In first-degree ankle injuries the lateral malleolus is broken. Second-degree injuries result in fracture of both malleoli, and in third-degree injuries both malleoli and the articular surface of the tibia are fractured (OUCH!). The strong deltoid ligament, which helps maintain the medial longitudinal arch of the foot (see Chapter 8), is also frequently injured in severe twisting injuries affecting the ankle.

Vertebral Joints

One vertebra connects to another by several joints—between their *bodies*, as well as between their articular, transverse, and spinous *processes*. Recall that the cartilaginous (amphiarthrotic) joints between the *bodies* of adjacent vertebrae permit only very slight movement and are classified as symphyses. However, the synovial (diarthrotic) joints between the articulating surfaces of the vertebral *processes* are more movable and are classified as **gliding.** These joints hold the vertebrae firmly together so that they are not easily dislocated, but these joints also form a flexible column. Consider how many ways you can move the trunk of your body. You can flex it forward or laterally, you can extend it, and you can circumduct or rotate it (see Figure 9-17). The bodies of adjacent vertebrae are connected by intervertebral disks and strong ligaments. Fibrous tissue and fibrocartilage form a disk's outer rim (called the *annulus fibrosus*). Its central core (the *nucleus pulposus*), in con-

Text continued on p. 327

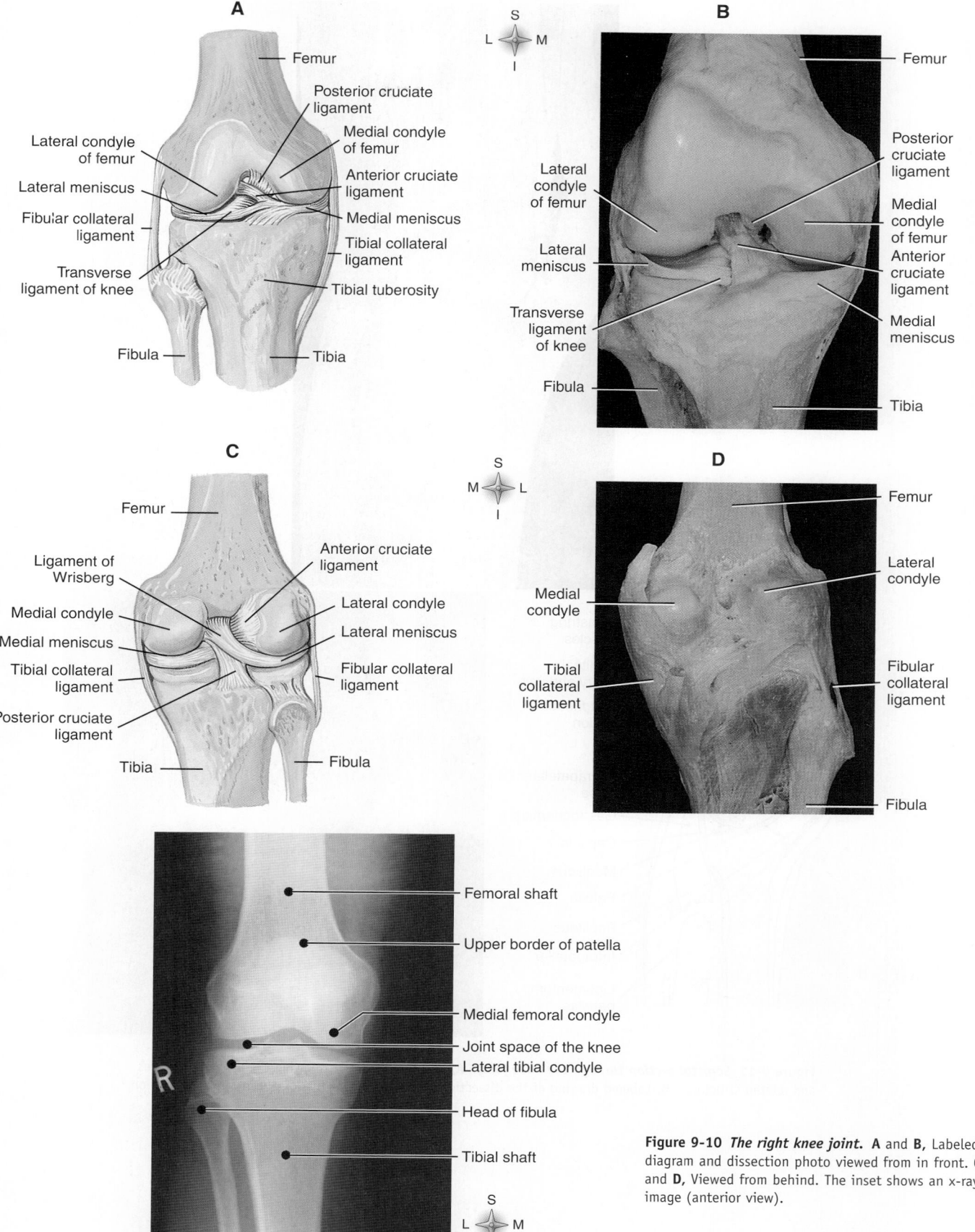

Figure 9-10 *The right knee joint.* A and **B,** Labeled diagram and dissection photo viewed from in front. **C** and **D,** Viewed from behind. The inset shows an x-ray image (anterior view).

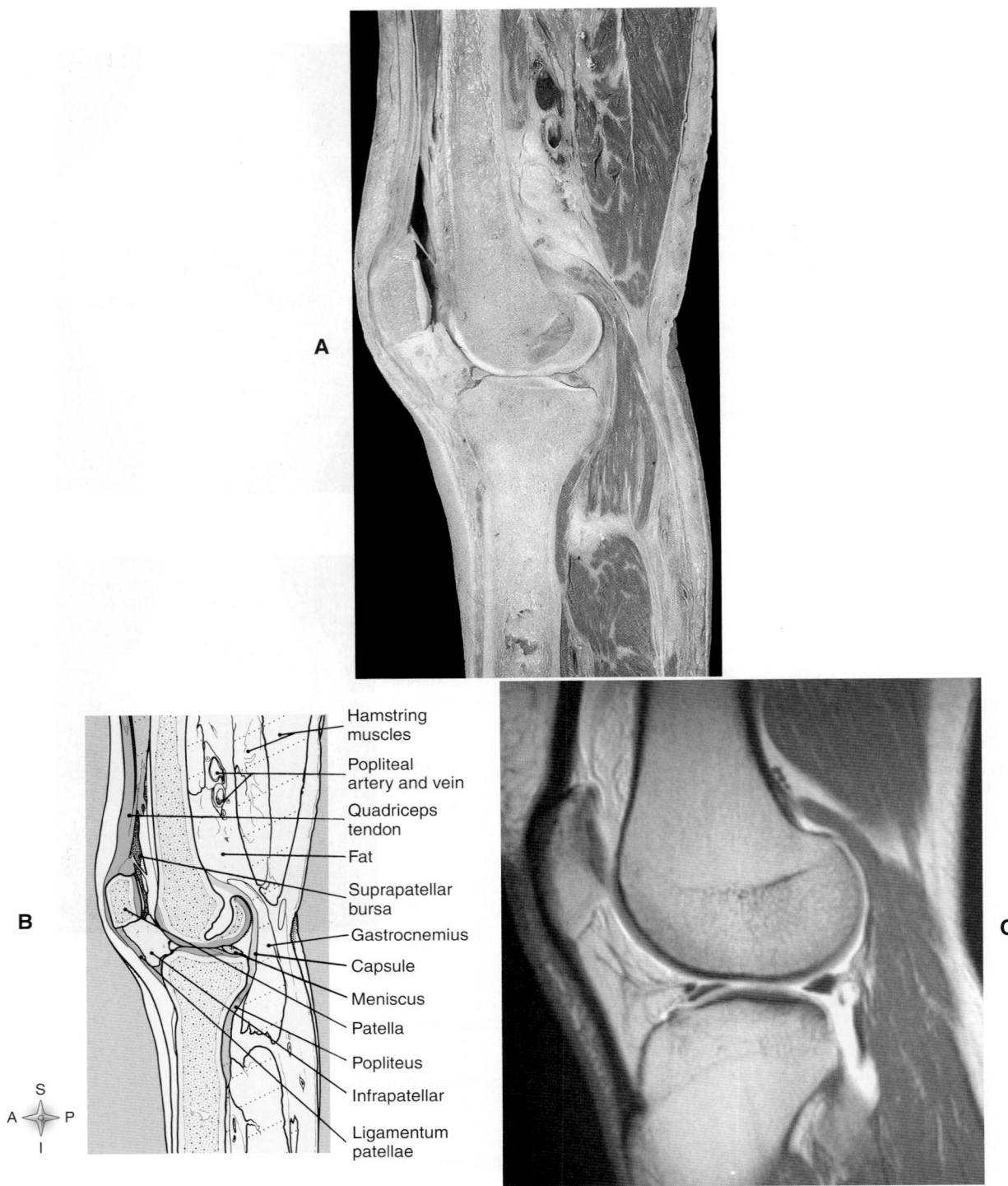

Figure 9-11 *Sagittal section through the knee joint.* A, Cadaver dissection showing the articular surfaces and related structures. **B,** Labeled drawing of the dissection photo. **C,** MRI of the knee joint, sagittal section.

Labels (B):
Hamstring muscles
Popliteal artery and vein
Quadriceps tendon
Fat
Suprapatellar bursa
Gastrocnemius
Capsule
Meniscus
Patella
Popliteus
Infrapatellar
Ligamentum patellae

BOX 9-1: SPORTS AND FITNESS
The Knee Joint

The knee is the largest and most vulnerable joint. Because the knee is often subjected to sudden, strong forces during athletic activity, knee injuries are among the most common type of athletic injury. Sometimes the articular cartilages on the tibia become torn when the knee twists while bearing weight. The ligaments holding the tibia and femur together and the medial and lateral menisci can also be injured in this way. Knee injuries may also occur when a weight-bearing knee is hit by another person, especially from the side.

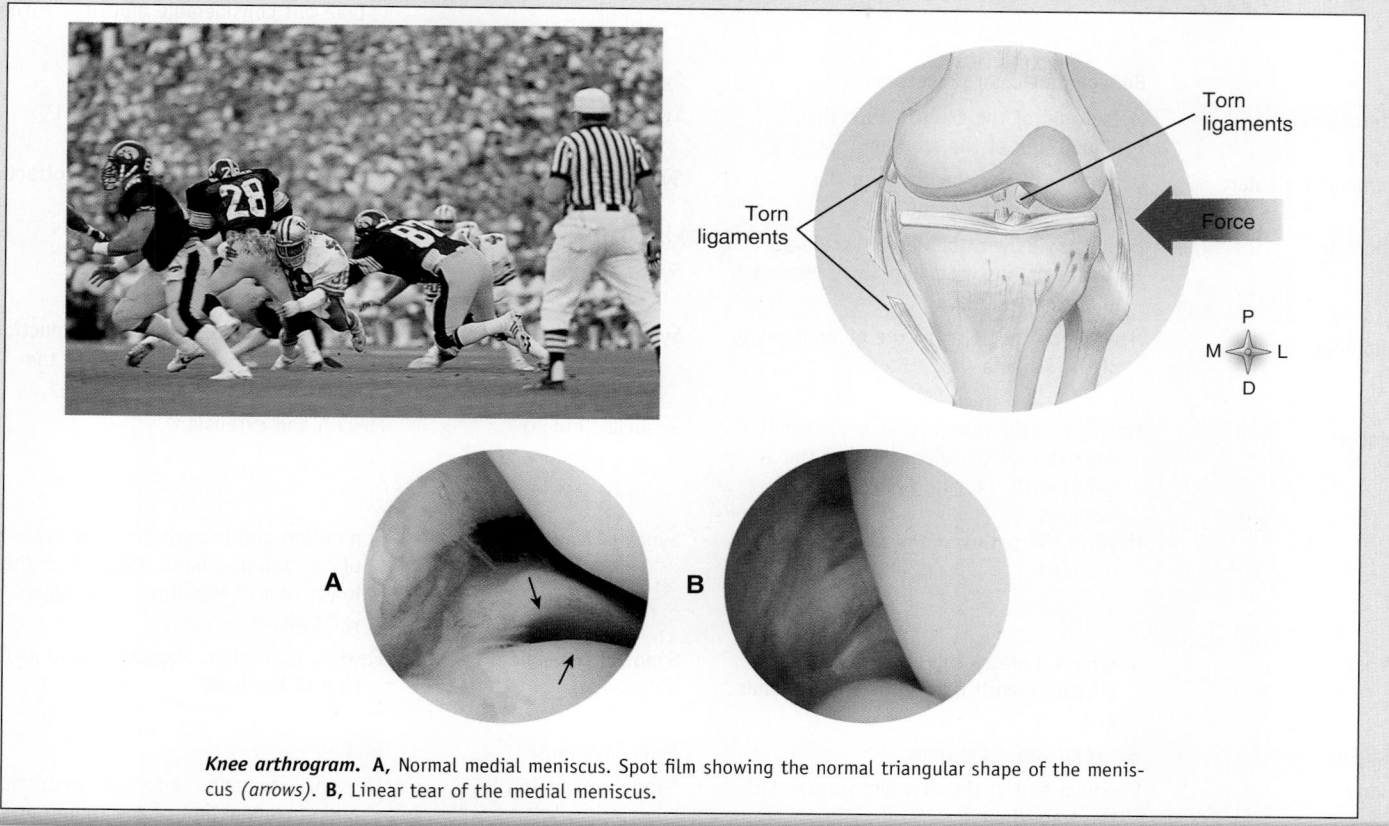

Knee arthrogram. **A,** Normal medial meniscus. Spot film showing the normal triangular shape of the meniscus *(arrows).* **B,** Linear tear of the medial meniscus.

trast, consists of a pulpy, elastic substance (Figure 9-13, *A*). With age, the nucleus loses some of its resiliency. It may then be suddenly compressed by exertion or trauma and pushed through the annulus, with the herniating part protruding into the spinal canal and pressing on spinal nerves or the spinal cord itself. Severe pain results. In medical terminology, this is called a **herniated disk** or *herniated nucleus pulposus* (HNP); in popular language, it is a "slipped disk" (Figure 9-13, *B*).

In Figure 9-14 the following ligaments that bind the vertebrae together can be identified: The *anterior longitudinal ligament*, a strong band of fibrous tissue, connects the anterior surfaces of the vertebral bodies from the atlas down to the sacrum. Connecting the posterior surfaces of the bodies is the *posterior longitudinal ligament*. The *ligamenta flava* bind the laminae of adjacent vertebrae firmly together. Spinous processes are connected by *interspinous ligaments*. In addition, the tips of the

spinous processes of the cervical vertebrae are connected by the *ligamentum nuchae*; its extension, the *supraspinous ligament*, connects the tips of the rest of the vertebrae down to the sacrum. And finally, *intertransverse ligaments* connect the transverse processes of adjacent vertebrae.

Table 9-4 summarizes the entire chapter with descriptions of most of the individual joints of the body.

QUICK CHECK

8. Which joint is the largest, most complex, and most frequently injured in the body?
9. List the two anatomical components of an intervertebral disk.

Table 9-4	Synovial (Diarthrotic) and Two Cartilaginous (Amphiarthrotic) Joints		
NAME	**ARTICULATING BONES**	**TYPE**	**MOVEMENTS**
Atlantoepistropheal	Anterior arch of the atlas rotates about the dens of the axis (epistropheus)	Synovial (pivot)	Pivoting or partial rotation of the head
Vertebral	Between bodies of vertebrae	Cartilaginous (symphyses)	Slight movement between any two vertebrae but considerable motility for the column as a whole
	Between articular processes	Synovial (gliding)	Gliding
Sternoclavicular	Medial end of the clavicle with the manubrium of the sternum	Synovial (gliding)	Gliding
Acromioclavicular	Distal end of the clavicle with the acromion of the scapula	Synovial (gliding)	Gliding; elevation, depression, protraction, and retraction
Thoracic	Heads of ribs with bodies of vertebrae	Synovial (gliding)	Gliding
	Tubercles of ribs with transverse processes of vertebrae	Synovial (gliding)	Gliding
Shoulder	Head of the humerus in the glenoid cavity of the scapula	Synovial (ball and socket)	Flexion, extension, abduction, adduction, rotation, and circumduction of the upper part of the arm
Elbow	Trochlea of the humerus with the semilunar notch of the ulna; head of the radius with the capitulum of the humerus	Synovial (hinge)	Flexion and extension
	Head of the radius in the radial notch of the ulna	Synovial (pivot)	Supination and pronation of the lower part of the arm and hand; rotation of the lower part of the arm on the upper extremity
Wrist	Scaphoid, lunate, and triquetral bones articulate with the radius and articular disk	Synovial (condyloid)	Flexion, extension, abduction, and adduction of the hand
Carpal	Between various carpals	Synovial (gliding)	Gliding
Hand	Proximal end of the first metacarpal with the trapezium	Synovial (saddle)	Flexion, extension, abduction, adduction, and circumduction of the thumb and opposition to the fingers
	Distal end of the metacarpals with the proximal end of the phalanges	Synovial (hinge)	Flexion, extension, limited abduction, and adduction of the fingers
	Between phalanges	Synovial (hinge)	Flexion and extension of finger sections
Sacroiliac	Between the sacrum and two ilia	Synovial (gliding)	None or slight
Symphysis pubis	Between two pubic bones	Cartilaginous (symphysis)	Slight, particularly during pregnancy and delivery
Hip	Head of the femur in the acetabulum of the os coxae	Synovial (ball and socket)	Flexion, extension, abduction, adduction, rotation, and circumduction
Knee	Between the distal end of the femur and proximal end of the tibia	Synovial (hinge)	Flexion and extension; slight rotation of the tibia
Tibiofibular (proximal)	Head of the fibula with the lateral condyle of the tibia	Synovial (gliding)	Gliding
Ankle	Distal end of the tibia and fibula with the talus	Synovial (hinge)	Flexion (dorsiflexion) and extension (plantar flexion)
Foot	Between tarsals	Synovial (gliding)	Gliding; inversion and eversion
	Between metatarsals and phalanges	Synovial (hinge)	Flexion, extension, slight abduction, and adduction
	Between phalanges	Synovial (hinge)	Flexion and extension

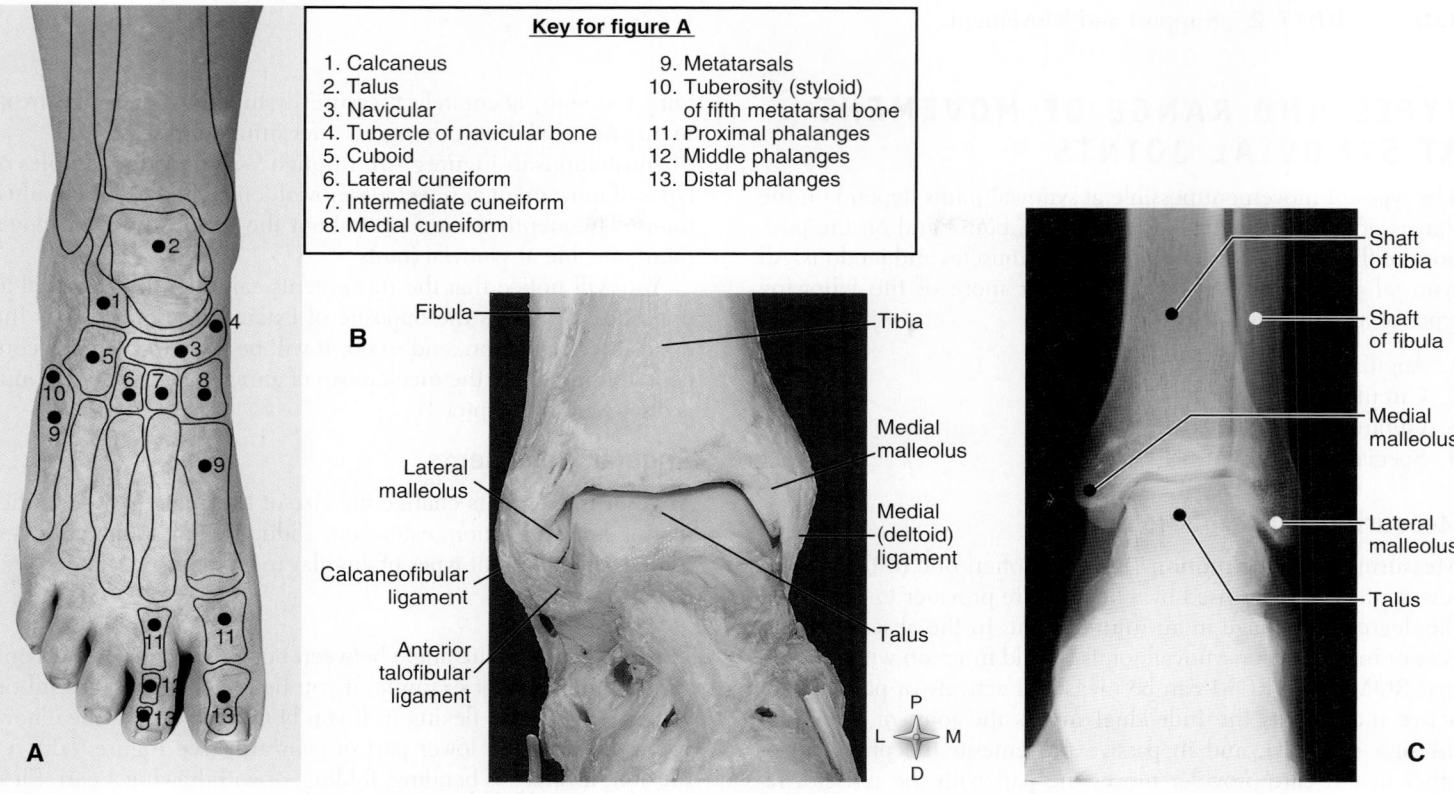

Key for figure A

1. Calcaneus
2. Talus
3. Navicular
4. Tubercle of navicular bone
5. Cuboid
6. Lateral cuneiform
7. Intermediate cuneiform
8. Medial cuneiform
9. Metatarsals
10. Tuberosity (styloid) of fifth metatarsal bone
11. Proximal phalanges
12. Middle phalanges
13. Distal phalanges

B

Fibula
Tibia
Lateral malleolus
Medial malleolus
Calcaneofibular ligament
Medial (deltoid) ligament
Anterior talofibular ligament
Talus

P
L — M
D

C

Shaft of tibia
Shaft of fibula
Medial malleolus
Lateral malleolus
Talus

Figure 9-12 *Ankle joint.* **A,** Dorsum of the ankle and foot showing surface relationships to underlying bones. **B,** Dissection photo of the ankle joint. **C,** Radiograph of the ankle joint, anterior view.

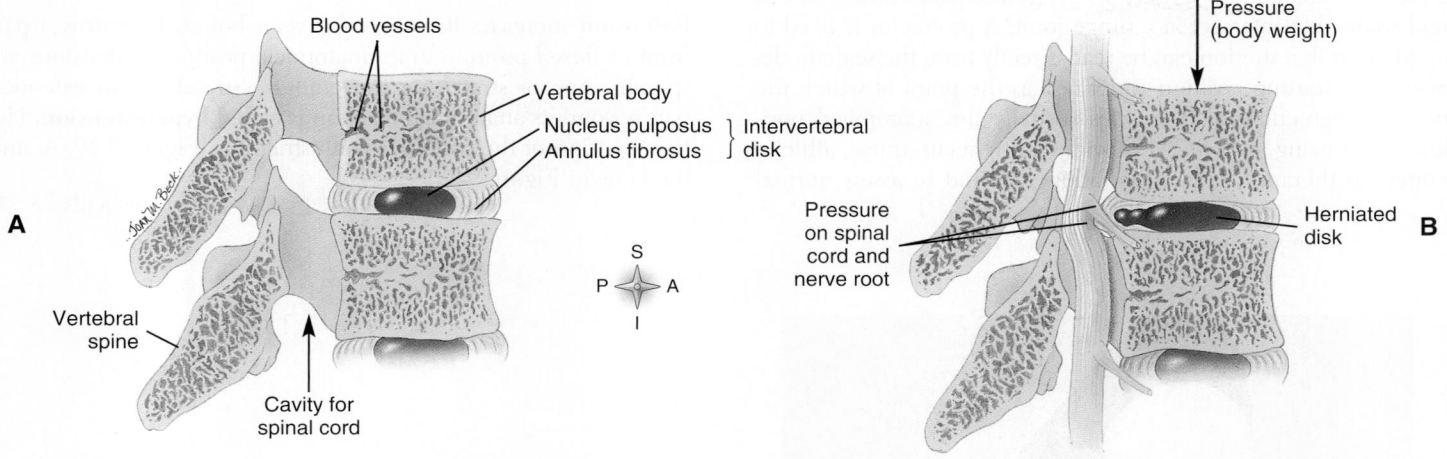

Blood vessels
Vertebral body
Nucleus pulposus
Annulus fibrosus
} Intervertebral disk

A

Vertebral spine
Cavity for spinal cord

S
P — A
I

Pressure (body weight)

Pressure on spinal cord and nerve root
Herniated disk

B

Figure 9-13 *Vertebrae.* Sagittal section of vertebrae showing normal **(A)** and herniated **(B)** disks.

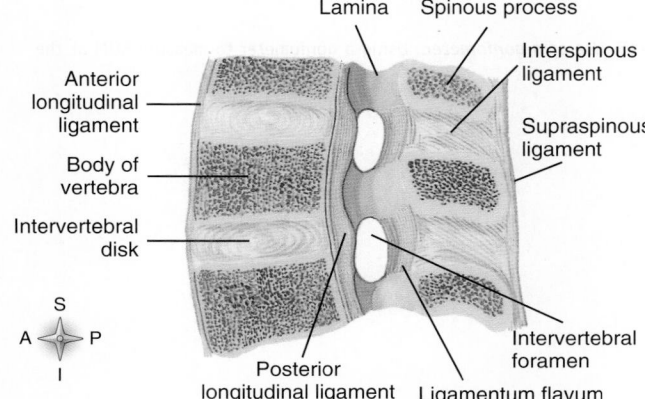

Lamina
Spinous process
Interspinous ligament
Anterior longitudinal ligament
Supraspinous ligament
Body of vertebra
Intervertebral disk
Posterior longitudinal ligament
Ligamentum flavum
Intervertebral foramen

S
A — P
I

Figure 9-14 *Vertebrae and their ligaments.* Sagittal section of two lumbar vertebrae and their ligaments.

TYPES AND RANGE OF MOVEMENT AT SYNOVIAL JOINTS

The types of movement possible at synovial joints depend on the shapes of the articulating surfaces of the bones and on the positions of the joints' ligaments and nearby muscles and tendons. All synovial joints, however, permit one or more of the following types of movements:

1. Angular
2. Circular
3. Gliding
4. Special

Measuring Range of Motion

Measuring **range of motion (ROM)** is often one of the first assessment techniques used by a health care provider to determine the degree of damage in an injured joint. In the absence of disease or injury, major synovial joints should function within a normal ROM. Joint ROM can be measured actively or passively. In active movements the individual moves the joint or body part through its ROM, and in passive movements the physician or other health care provider moves the part with the muscles relaxed. Normally, both active and passive ROM should be about equal. If a joint has an obvious increase or limitation in its range of motion, a specialized instrument, called a **goniometer,** is used to measure the angle (Figure 9-15). A goniometer consists of two rigid shafts that intersect at a hinge joint. A protractor is fixed to one shaft so that motion can be read directly from the scale in degrees. The starting position is defined as the point at which the movable segment is at 0 degrees (usually the anatomical position). Measuring joint ROM provides a physician, nurse, athletic trainer, or therapist with information required to assess normal joint function, accurately measure dysfunction, or gauge treatment and rehabilitative progress after injury or disease.

Illustrations in Figures 9-16 through 9-25 provide examples of types of movement at selected synovial joints. Refer to these illustrations frequently as you read about the various types of movement possible at synovial joints.

You will notice that the movements can often be classified as opposites: flexion is the opposite of extension, protraction is the opposite of retraction, and so on. It will be good to keep this concept in mind when the mechanism of antagonistic muscle groups is discussed in Chapter 10.

Angular Movements

Angular movements change the size of the angle between articulating bones. Flexion, extension, abduction, and adduction are some of the different types of angular movements.

Flexion

Flexion decreases the angle between bones. It bends or folds one part on another. For example, if you bend your head forward on your chest, you are flexing it. If you bend your arm at the elbow, you are flexing the lower part of your arm (see Figure 9-20, A). Flexion, in short, is bending, folding, or withdrawing a part. Flexion can also occur when an extended structure is returned to the anatomical position.

Extension and Hyperextension

Extension increases the angle between bones. It returns a part from its flexed position to its anatomical position. Extensions are straightening or stretching movements. Stretching an extended part beyond its anatomical position is called **hyperextension.** Hyperextension of the shoulder is illustrated in Figure 9-19, A, and the knee in Figure 9-24.

Text continued on p. 336

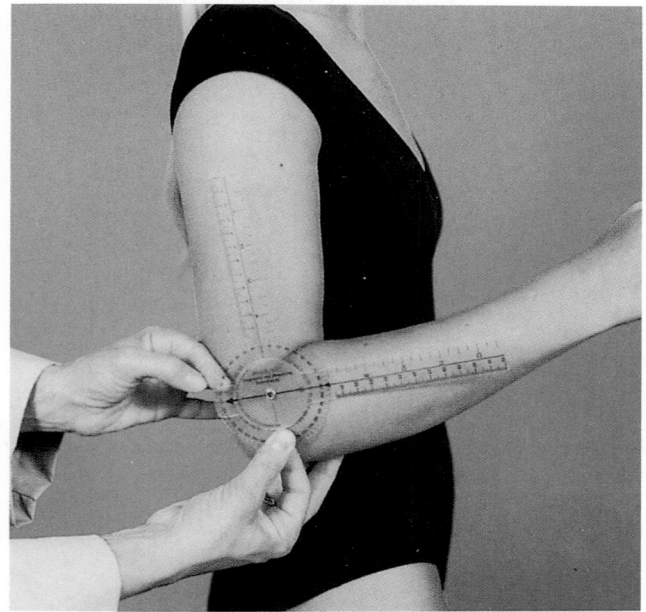

Figure 9-15 *Use of a goniometer.* Using a goniometer to measure ROM at the elbow.

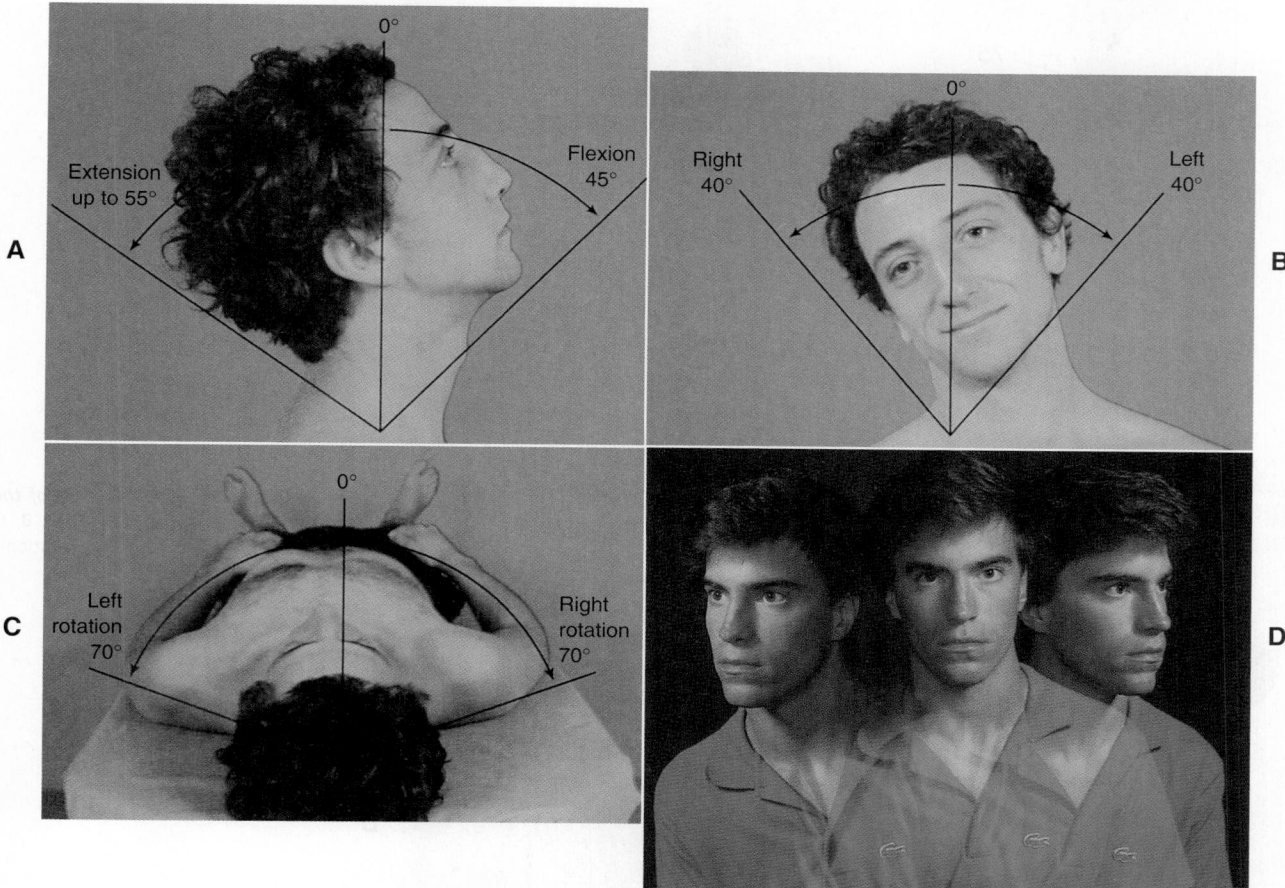

Figure 9-16 *Movements and ROM.* **A,** Flexion and extension. **B,** Lateral bending. **C,** Rotation (supine position). **D,** Rotation (standing position).

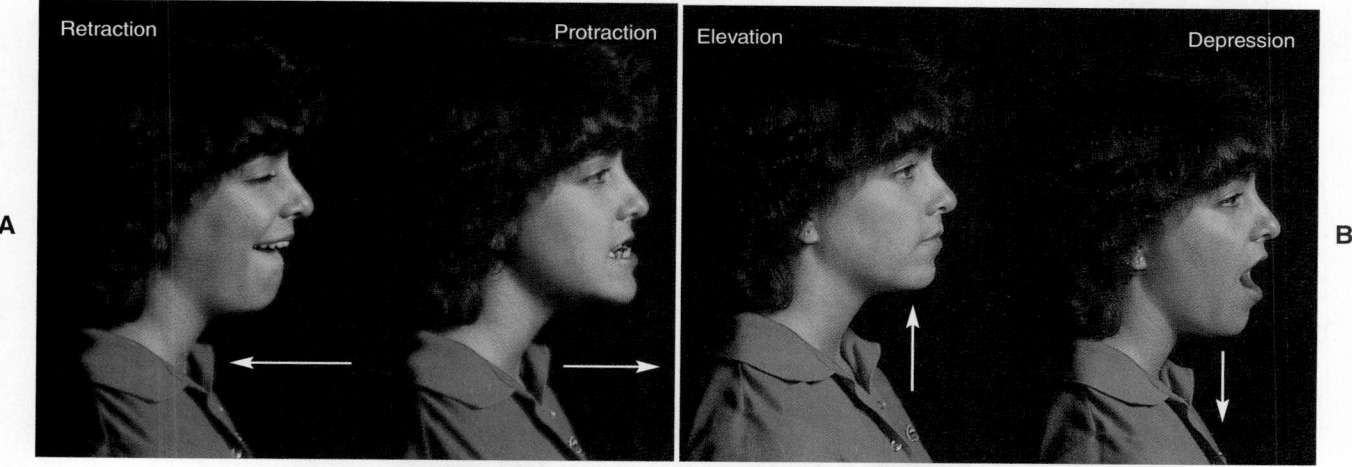

Figure 9-17 *Movements of the jaw.* **A,** Retraction and protraction. **B,** Elevation and depression.

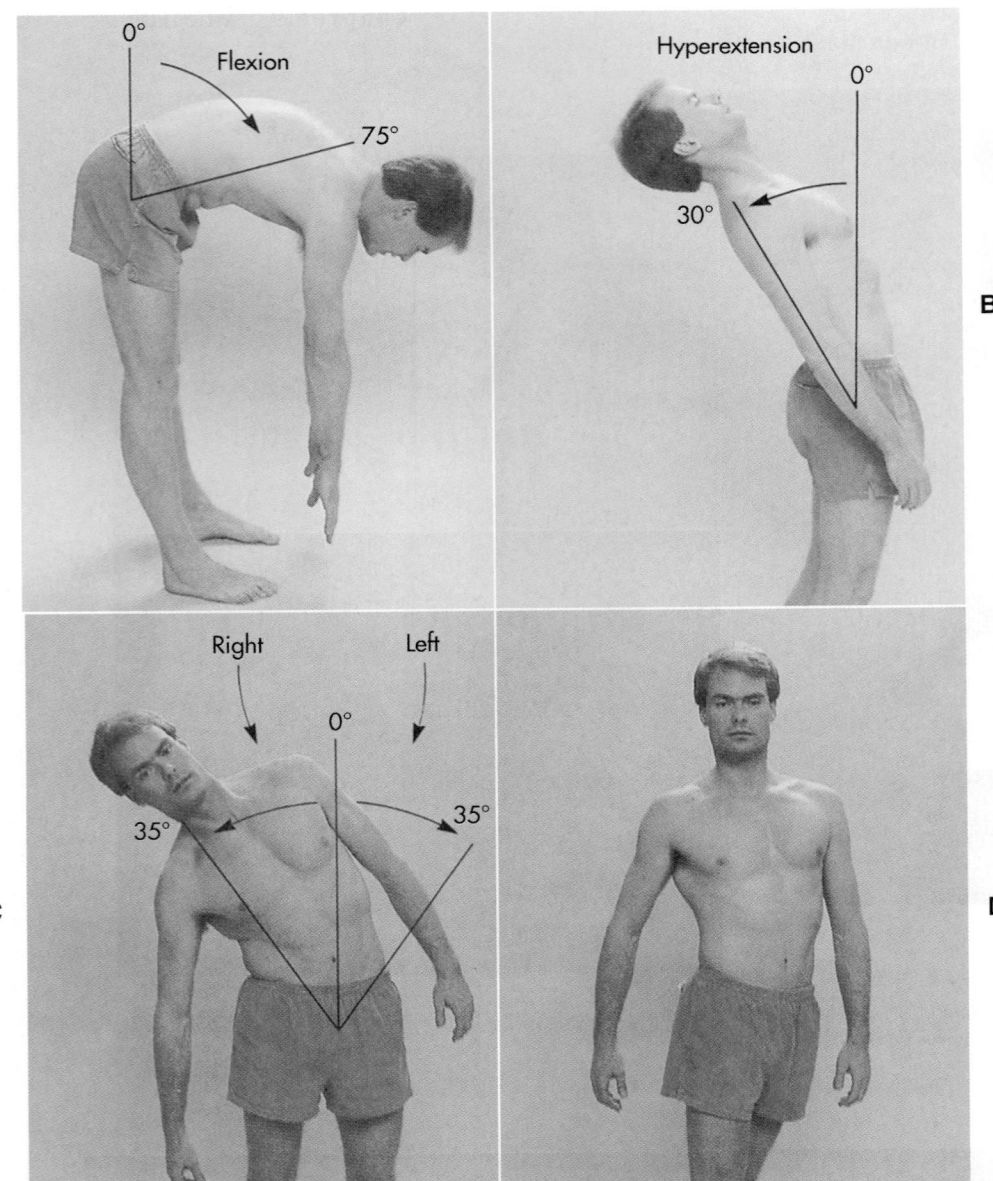

Figure 9-18 *Movements and ROM of the thoracic and lumbar spine.* **A,** Flexion. **B,** Hyperextension. **C,** Lateral bending. **D,** Rotation of the upper part of the trunk.

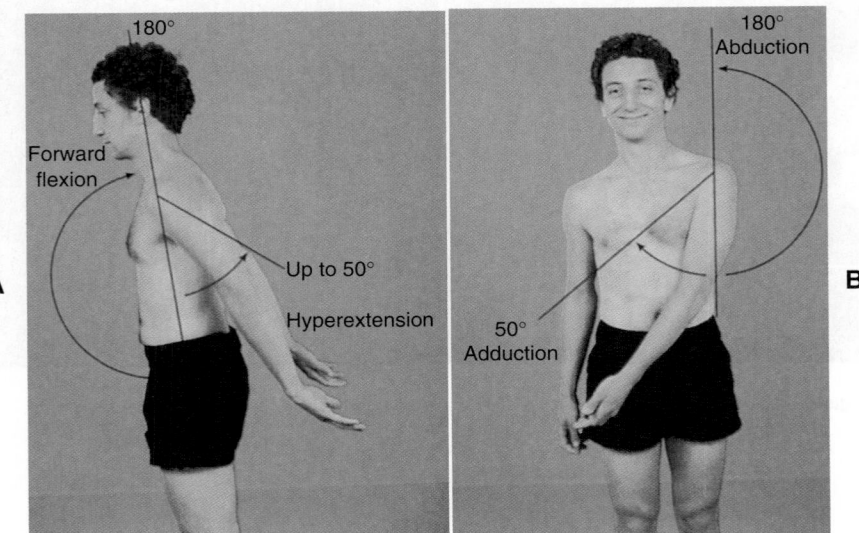

Figure 9-19 *Movements and ROM of the shoulder.* **A,** Forward flexion, extension (back to the anatomical position of 0 degrees), and backward hyperextension up to 50 degrees. **B,** Abduction and adduction.

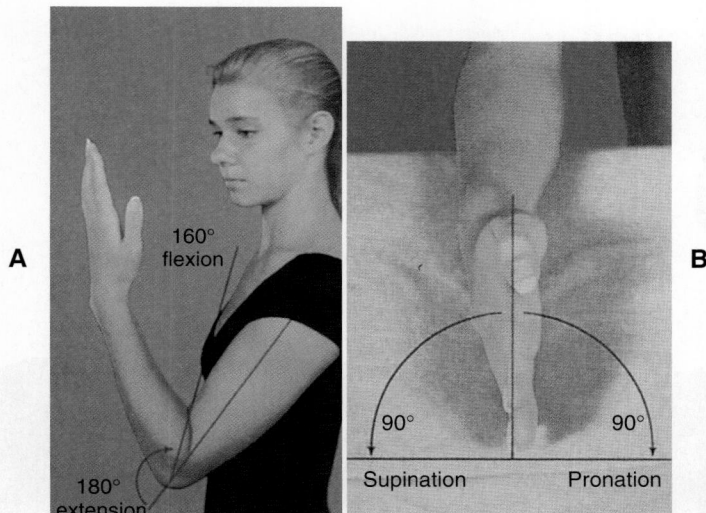

Figure 9-20 *Movements and ROM of the elbow.* **A,** Flexion and extension of the elbow. **B,** Supination and pronation.

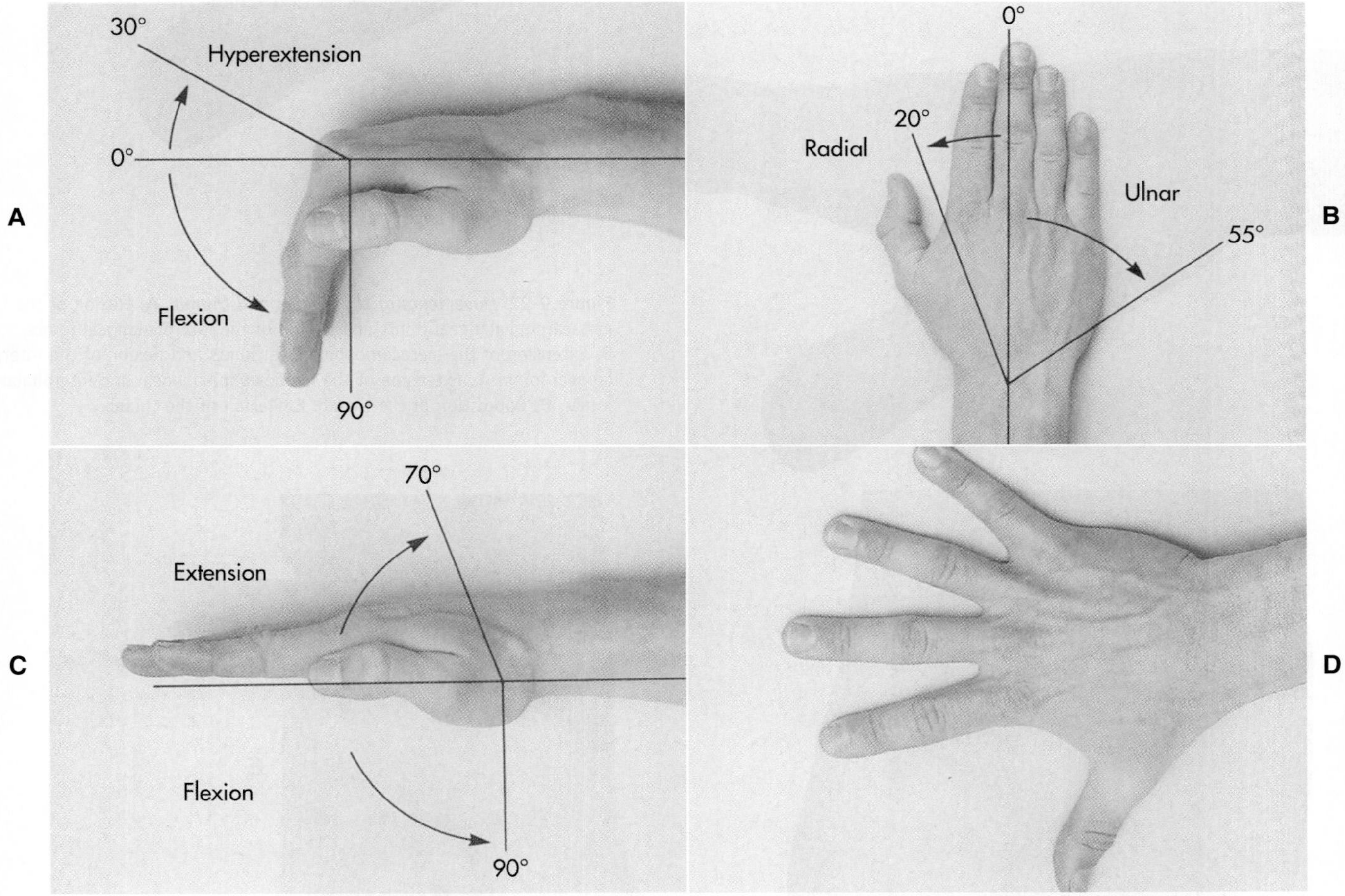

A — 30° Hyperextension — 0° — Flexion — 90°

B — 0° — Radial — 20° — Ulnar — 55°

C — 70° — Extension — Flexion — 90°

D

Figure 9-21 *Movements and ROM of hand and wrist.* **A,** Metacarpophalangeal flexion and hyperextension. **B,** Wrist flexion and hyperextension. **C,** Wrist radial and ulnar movement. **D,** Finger abduction and adduction.

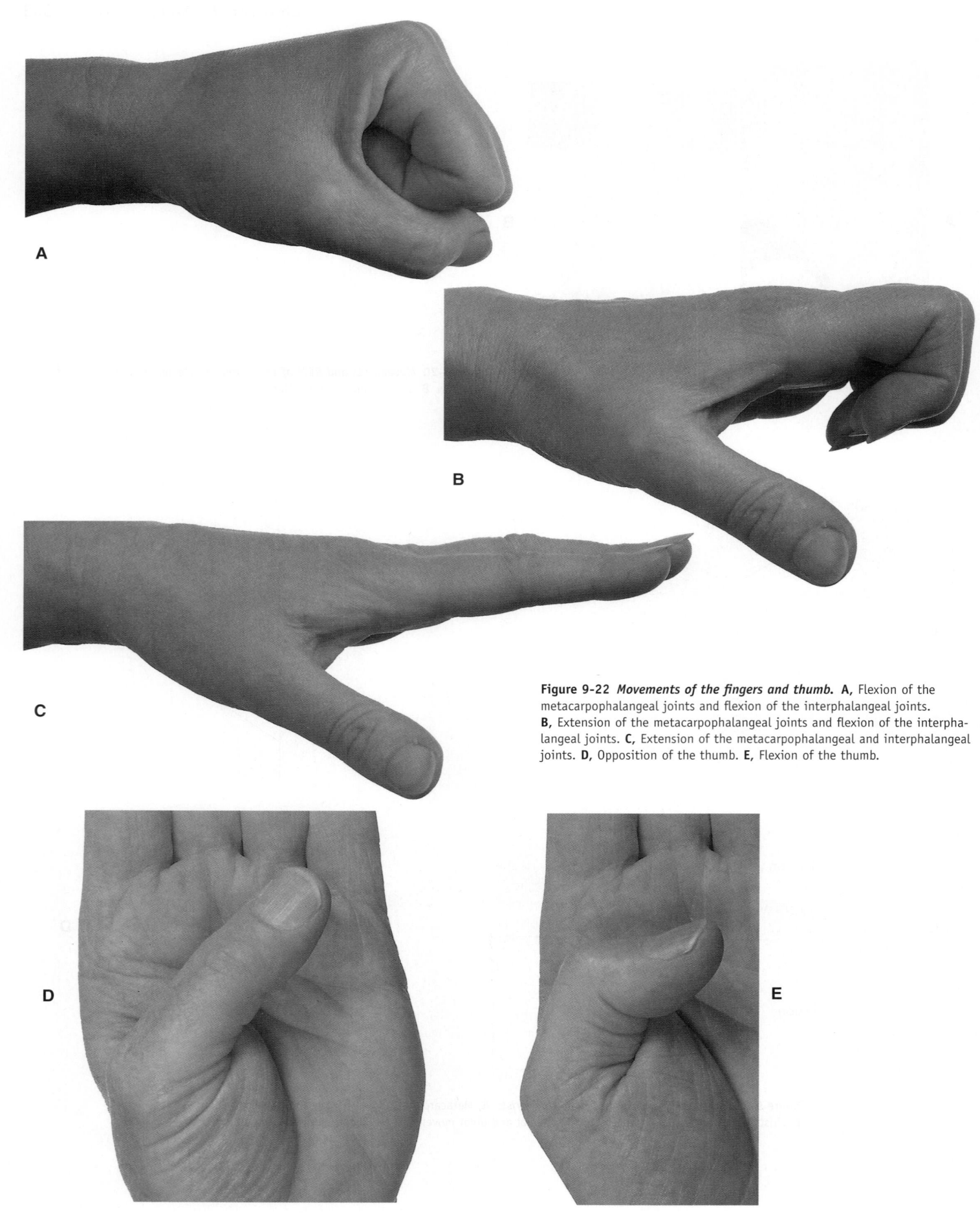

Figure 9-22 *Movements of the fingers and thumb.* **A,** Flexion of the metacarpophalangeal joints and flexion of the interphalangeal joints. **B,** Extension of the metacarpophalangeal joints and flexion of the interphalangeal joints. **C,** Extension of the metacarpophalangeal and interphalangeal joints. **D,** Opposition of the thumb. **E,** Flexion of the thumb.

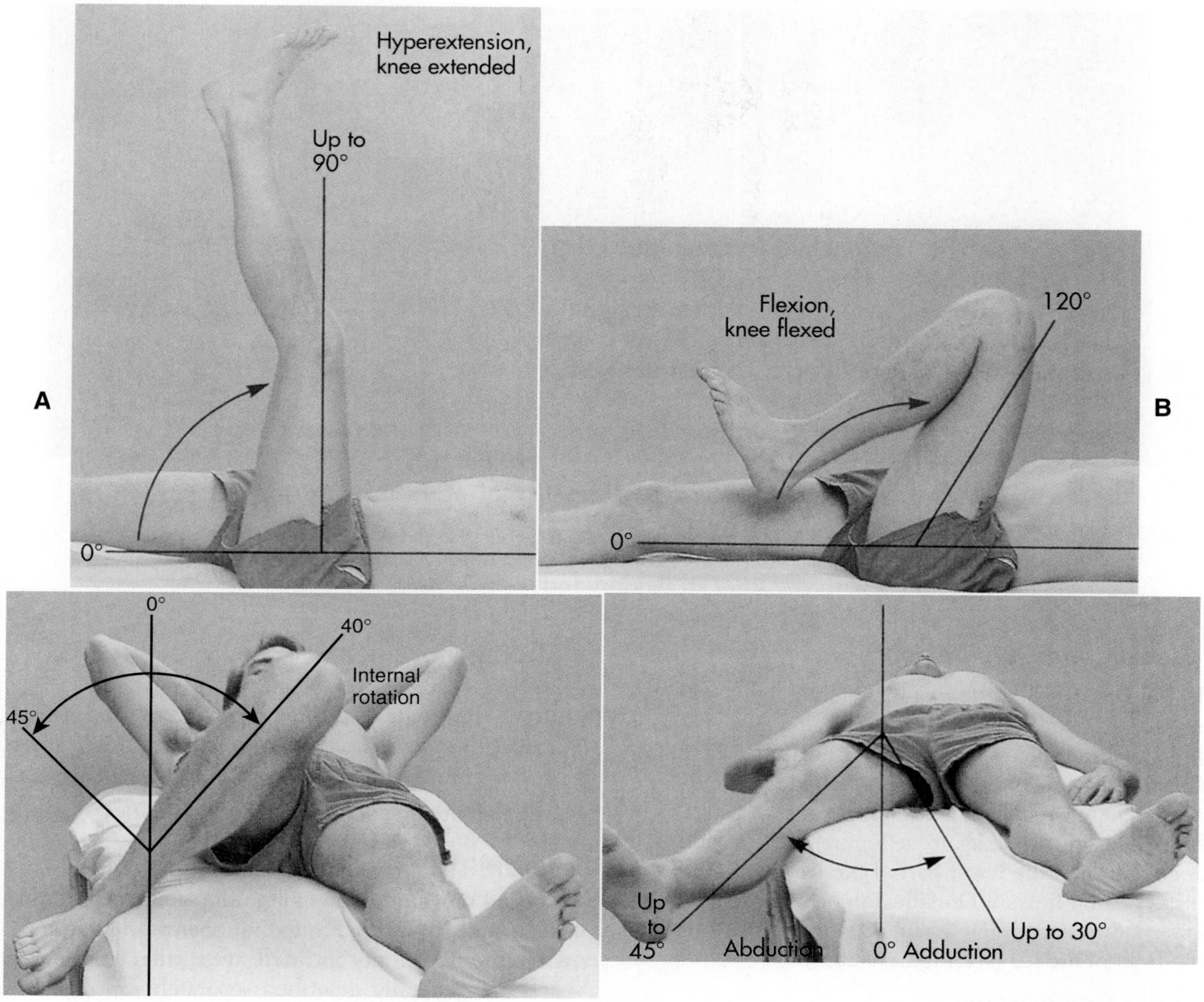

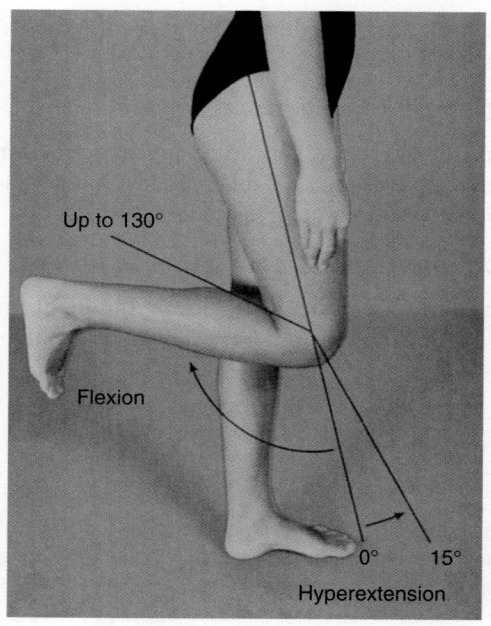

Figure 9-23 *Movements and ROM of the hip.* **A,** Hip hyperextension, knee extended. **B,** Hip flexion, knee flexed. **C,** Internal rotation. **D,** Abduction and adduction.

Figure 9-24 *Movements and ROM of the knee.* Flexion may reach 130 degrees. Extension occurs with movement from a 130-degree flexed position back to the 0-degree position. Hyperextension of 0 degrees to 15 degrees is possible in many individuals.

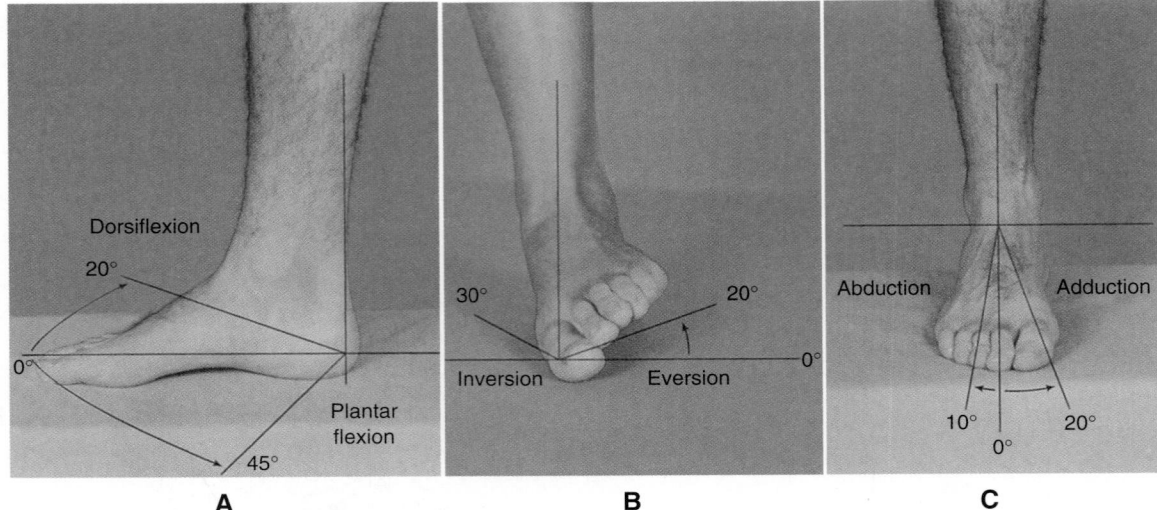

A   **B**   **C**

Figure 9-25 *Movements and ROM of the foot and ankle.* **A,** Dorsiflexion and plantar flexion. **B,** Inversion and eversion. **C,** Abduction and adduction.

Plantar Flexion and Dorsiflexion

Plantar flexion occurs when the foot is stretched down and back (see Figure 9-25, A). This movement increases the angle between the top of the foot and the front of the leg. Recall that the definition of extension refers to an increase in the angle between articulating bones. Therefore, plantar flexion of the foot can also be described as extension. Look again at Figure 9-25, A, and note how the angle between the lower part of the leg and the foot increases during plantar flexion. **Dorsiflexion** occurs when the foot is tilted upward, thus decreasing the angle between the top of the foot and the front of the leg (see Figure 9-25, A).

Abduction and Adduction

Abduction moves a part away from the median plane of the body, as in moving the leg straight out to the side or fingers away from the midline of the hand. **Adduction** moves a part toward the median plane. Examples include bringing the arm back to the side or moving fingers toward the midline of the hand. Opposing movements in the shoulder can be seen in Figure 9-19, B.

Circular Movements

Circular movements result in the arclike rotation of a structure around an axis. The primary circular movements are rotation, circumduction, supination, and pronation (see Figure 9-16, B).

Rotation and Circumduction

Rotation consists of pivoting a bone on its own axis. For example, moving the head from side to side as in indicating "no" (see Figure 9-18, D). **Circumduction** moves a part so that its distal end moves in a circle. When a pitcher winds up to throw a ball, the pitcher circumducts the arm.

Supination and Pronation

Supination and **pronation** of the hand are shown in Figure 9-20, B. Pronation turns the palm of the hand down, whereas supination turns the hand palm side up.

Gliding Movements

Gliding movements are the simplest of all movements. The articular surface of one bone moves over the articular surface of another without any angular or circular movement. Gliding movements occur between the carpal and tarsal bones and between the articular facets of adjoining spinal vertebrae.

Special Movements

Special movements are often unique or unusual movements that occur only in a very limited number of joints. These specialized movements do not fit well into other movement categories and are generally described separately. Special movements include inversion, eversion, protraction, retraction, elevation, and depression.

Inversion and Eversion

Inversion turns the sole of the foot inward, whereas **eversion** turns it outward (see Figure 9-25, B).

Protraction and Retraction

Protraction moves a part forward, whereas **retraction** moves it back. For instance, if you stick out your jaw, you protract it, and if you pull it back, you retract it (see Figure 9-17, A).

Elevation and Depression

Elevation moves a part up, as in closing the mouth (see Figure 9-17, B). **Depression** lowers a part, moving it in the opposite direction from elevation.

 QUICK CHECK

10. Name one example of each of the following types of movement at a synovial joint: angular, circular, gliding, and special.

Cycle of Life
Articulations

This chapter discusses the types of articulations that occur between bones and presents information on the common types of diseases that affect joints during the life cycle.

Changes in the way bones develop and in the sequence of ossification that occurs between birth and skeletal maturity also affect our joints. Fontanels, which exist between bones of the cranium, disappear with increasing age, and the epiphyseal plates, which serve as points of articulation between the epiphyses and diaphyses of growing long bones, ossify at skeletal maturity.

Range of motion at synovial joints is typically greater early in life. With advancing age, joints are often described as being "stiff," range of motion decreases, and changes in gait are common. The difficulty with locomotion that affects many elderly individuals may result from disease conditions that involve the functional unit comprising muscles, bones, and joints.

Some skeletal diseases have profound effects that are manifested as joint problems. Abnormal bone growth, or "lipping," which results in bone spurs, or sharp projections, on the articular surfaces of bones, dramatically influences joint function. Many of these disease conditions are associated with specific developmental periods, including early childhood, adolescence, and adulthood.

THE BIG PICTURE
Articulations

Anatomists sometimes refer to the hand as "the reason for the upper extremity" and the thumb as "the reason for the hand." Statements such as these, when they are used in a functional context, underscore the relationship that exists between structure and function. Furthermore, they serve as examples of how "big picture" thinking is helpful in examining the interaction between body structures and the rationale for how and why they function as they do. Proper functioning of the joints is crucial for meaningful, controlled, and purposeful movement to occur—movement that permits human beings to respond to and control their environment in ways that other animals are not capable of doing. The joints of the upper extremity are good examples.

The great mobility of the upper extremity is possible because of the following: (1) the arrangement of the bones in the shoulder girdle, arm, forearm, wrist, and hand; (2) the location and method of attachment of muscles to these bones; and (3) proper functioning of the joints involved. We give up a great deal of stability in the upper extremity (especially the shoulder joint) to achieve mobility. We need that mobility and extensive range of motion to position the entire upper extremity, and especially the hand, in ways that permit us to interact with objects in our physical environment effectively. It is the proper functioning of the thumb that allows us to grasp and manipulate objects with great dexterity and control. Although loss of mobility in the upper extremity as a whole presents a significant functional problem, loss of thumb function can present equally serious and disabling problems by dramatically decreasing the overall ability to manipulate and grasp objects. By enabling us to effectively interact with our external environment, the joints of the upper extremity contribute in a significant way to maintenance of homeostasis. It is in the "big picture" context that the hand and thumb are truly functional "reasons" for the upper extremity and the hand, respectively.

Mechanisms of Disease

BURSITIS

Bursitis, or inflammation of a bursa, is a relatively common disorder. It is most often caused by prolonged pressure, excessive or repetitive exercise, or sudden trauma that involves one of these fluid-filled sacs. Although bursae are often associated with joints, where they may serve to cushion the joint or facilitate the movement of tendons over bony prominences, such as the olecranon process of the elbow joint, the bursa itself is often completely separate from the joint space. Look again at Figure 9-6, A. It clearly shows the placement of the olecranon bursa just below the skin and independent of the joint space of the elbow. Therefore, inflammation of the olecranon bursa (Figure 9-26) may occur in the absence of any elbow joint inflammation or pathology. Bursitis near the knee joint is also common, especially in carpet layers, roofers, and others who work on their knees.

Joint Disorders

Joint disorders may be classified as **noninflammatory joint disease** or **inflammatory joint disease.** Serving as fulcrums, the joints permit smooth and precise movement to occur when a muscle contracts and pulls on bones. Because joints, bones, and muscles act together as a functional unit, joint disorders, regardless of type or classification, have profound effects on body mobility.

Noninflammatory Joint Disease

Noninflammatory types of joint disease can be distinguished from inflammatory joint conditions because they do not involve inflammation of the synovial membrane. Further, noninflammatory joint diseases do not produce systemic signs or symptoms, such as fever, or damage other body organs such as the heart and lungs.

Mechanisms of Disease—cont.

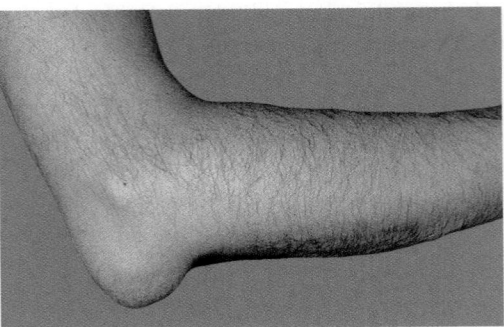

Figure 9-26 *Olecranon bursitis.* Pain, inflammation, and swelling of the bursa may limit motion, although the elbow joint itself may not be affected.

Osteoarthritis, known also as *degenerative joint disease* (DJD), is the most common noninflammatory disorder of movable joints. It is characterized by "wear and tear" degeneration and fracturing of articular cartilage (**see** Box 9-2) and by abnormal formation of new bone ("bone spurs") at joint surfaces. In DJD the articular cartilage no longer acts as a "shock absorber," the joint space narrows, the synovial membrane thickens, and the ligaments calcify.

The articular surface of a bone exhibiting severe osteoarthritis is shown in Figure 9-27. Note that the articular cartilage has degraded and cystlike lesions have formed below the superficial layers of damaged cartilage. Large weight-bearing joints, such as the hips and knees, are often involved first, followed by involvement of joints in the fingers. Swelling deformities of the *distal* interphalangeal joints called **Heberden nodes** and of the *proximal* interphalangeal joints called **Bouchard nodes** are common. Figure 9-28, A, shows the hands of a woman with osteoarthritis. The cause of osteoarthritis is unknown, but obesity, aging, and the "wear and tear" of mechanical stresses on joints over time are known to play a major role.

Unfortunately, no treatment is available to stop the degenerative joint disease process completely. Symptoms, which appear in many people in their fifties and sixties, include morning stiffness, deep "achy" pain on movement that worsens with use, and limited joint motion. In many cases, symptoms can be treated effectively with standard **nonsteroidal anti-inflammatory drugs (NSAIDs),** such as aspirin or ibuprofen, or by nutritional supplements such as glucosamine or chondroitin sulfate (see Chapter 7, p. 233). Injection into the joint space, especially of the knee, of a gelatinous type of lubricating fluid containing *hyaluronic acid* has proved useful in some cases. If the disease is more advanced, various types of prescription medications or partial or total joint replacements may be necessary (Box 9-3). Although the severity of symptoms may plateau or stabilize for varying periods, the disease is generally progressive once established and will continue over a lifetime. Osteoarthritis is the leading cause of long-term disability in aging, but otherwise healthy individuals.

BOX 9-2: HEALTH MATTERS
Joint Mice

The thin layer of cartilage that covers the articulating surfaces of bones in synovial joints may become cracked as a result of traumatic injury—especially to the knee. Such injuries are called **chondral fractures.** They result in fissuring of the articular cartilage, which can be likened to the cracking of an egg shell. If such an injury occurs, small fragments may become detached from the damaged layer of cartilage and enter the joint cavity. These fragments are called "loose bodies" or *joint mice* because they often cause clicking sounds or catching and locking sensations during joint movement. Symptoms often vary over time as cartilage fragments move about the joint cavity in the synovial fluid. Pain and loss of motion will occur if the fragments become lodged between opposing bone surfaces.

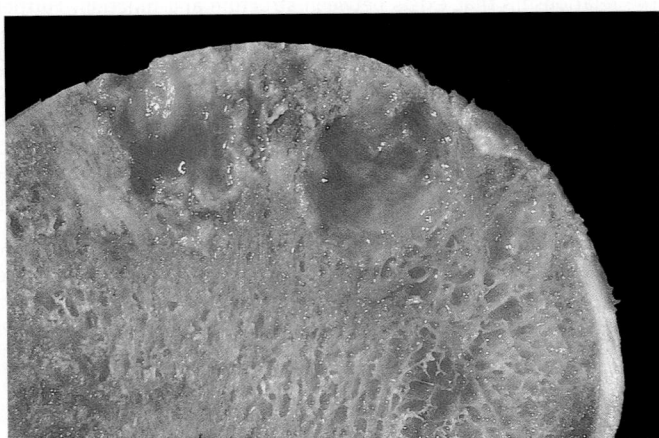

Figure 9-27 *Severe osteoarthritis.* Articular cartilage has eroded from the bone surface, and deeper cystlike lesions have appeared.

Traumatic injuries are a frequent cause of joint injury. **Dislocation** (subluxation) of a joint as a result of trauma can be an emergency because of associated damage to important blood vessels and nerves. In a dislocation, the articular surfaces of bones forming the joint are no longer in proper contact. This displacement of bone alters the normal contour of the joint and can tear surrounding ligaments. Refer once again to Figure 9-6, C. Recall that the annular ligament is intended to hold the head of the radius firmly against the radial notch of the ulna at the proximal radioulnar joint. Functionally, it is this anatomic relationship that allows the radius to rotate freely without dislocation during pronation and supination of the forearm. In children younger than 5 years especially, dislocation of the radial head, called a "pulled elbow," is a common traumatic joint injury. Pain and soft tissue swelling are associated with dislocation injuries. Effective treatment involves early reduction or realignment.

Mechanisms of Disease—cont.

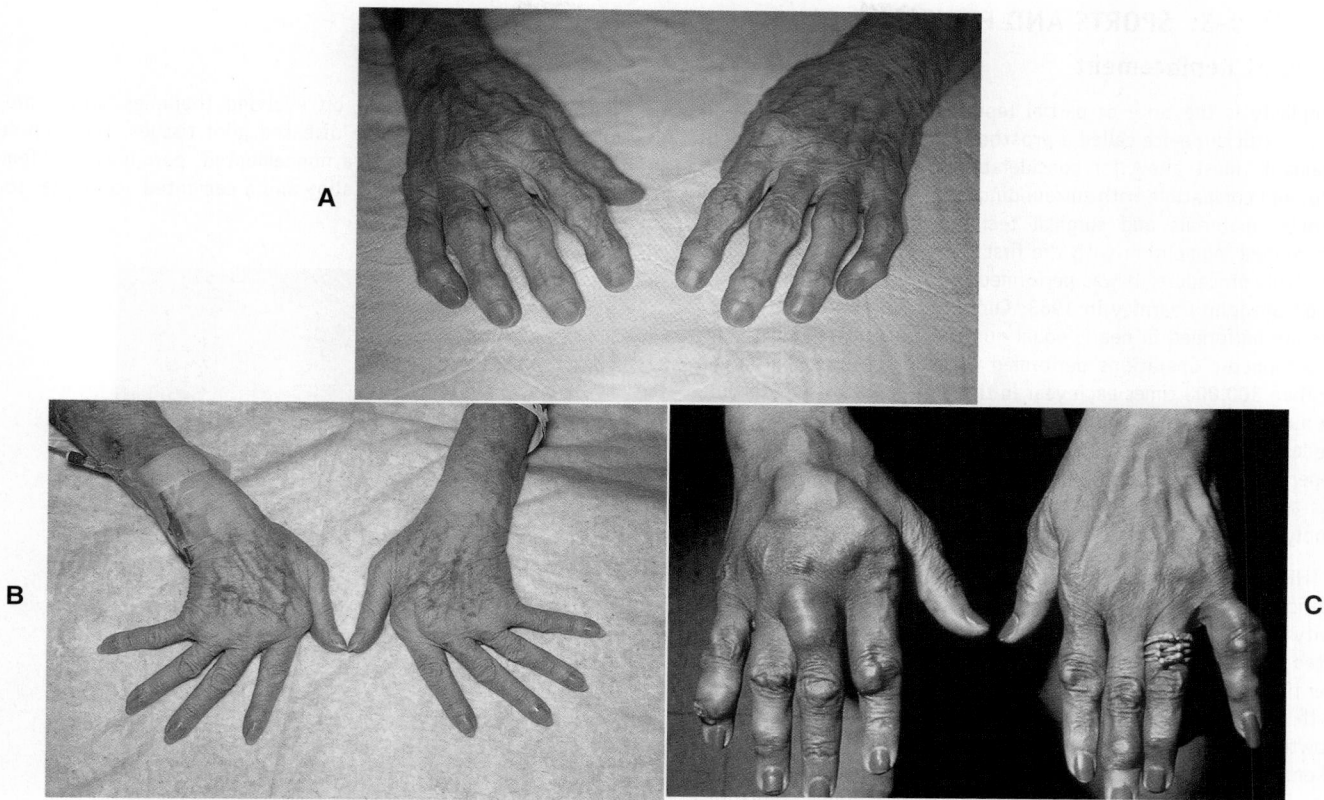

Figure 9-28 *Types of arthritis.* A, Osteoarthritis. Note Heberden's and Bouchard's nodes. **B,** Rheumatoid arthritis. Note the marked ulnar deviation of the fingers. **C,** Gouty arthritis. Note the tophi filled with sodium urate crystals.

One of the most common traumatic joint injuries associated with athletic activity involves damage to the cartilaginous menisci of the knee. Because the menisci act as shock absorbers and stabilize the knee, severe tears may produce edema, pain, instability, and limited motion. Arthroscopic surgical procedures are frequently used in both the diagnosis and treatment of this type of knee injury (Figure 9-29).

Arthroscopy is an imaging technique that allows a physician to examine the internal structure of a joint without the use of extensive surgery (see Figure 9-29, A). A narrow tube with lenses and a fiberoptic light source is inserted into the joint space through a small puncture. Isotonic saline (salt) solution is injected through a needle to expand the volume of the synovial space. This spreads the joint structures and makes viewing easier (see Figure 9-29, B).

Although arthroscopy is often used as a diagnostic procedure, it can be used to perform joint surgery. While the surgeon views the internal structure of the joint through the arthroscope or on an attached video monitor, instruments can be inserted through puncture holes and used to repair or remove damaged tissue. Arthroscopic surgery is much less traumatic than previous methods in which the joint cavity was completely opened.

A **sprain** is an acute musculoskeletal injury to the ligamentous structures surrounding a joint that disrupts the continuity of the synovial membrane. A common cause is a twisting or wrenching movement often associated with "whiplash"-type injuries. Blood vessels may be ruptured, and bruising and swelling may occur. Limitation of joint motion is common in all sprain injuries. External ankle rotation sprain injuries are discussed on p. 324.

Inflammatory Joint Disease

Arthritis is a general name for many different inflammatory joint diseases. Arthritis can be caused by a variety of factors, including infection, injury, genetic factors, and autoimmunity. Here is a brief list of the major types of arthritis:

Rheumatoid arthritis (RA) is a form of systemic autoimmune disease that involves chronic inflammation of many different tissues and organs of the body. Although organs such as the heart, lungs, skin, and muscles may become involved, the disease generally attacks joints first—and affects them most severely. A wide range of pathological joint changes are associated with RA. Initially, the synovial membrane becomes inflamed, thickened, and edematous. Soon,

Mechanisms of Disease—cont.

BOX 9-3: SPORTS AND FITNESS
Joint Replacement

Arthroplasty is the total or partial replacement of a diseased joint with an artificial device called a **prosthesis.** Prosthetic joints or joint components must allow for considerable mobility and be durable, strong, and compatible with surrounding tissues. Development of new prosthetic materials and surgical techniques to replace diseased joints gained momentum with the first successful total hip replacement (THR) procedure. It was performed by noted English orthopedic surgeon Sir John Charnley in 1963. Currently, hip and knee replacements are performed in nearly equal numbers and are the most common orthopedic operations performed on older persons, performed more than 300,000 times each year in the United States. In addition, large numbers of elbow, shoulder, finger, and other joint replacement procedures are performed annually. Osteoarthritis and other types of degenerative bone disease frequently destroy or severely damage joints and cause unremitting pain. Total joint replacement is often the only effective treatment option available.

The THR procedure involves replacement of the femoral head by a metallic alloy prosthesis and the acetabular socket with a high-density polyethylene cup. In the past, both sides were generally cemented into place with methyl methacrylate adhesive. Many of the newer prostheses, however, are "porous coated" to permit natural ingrowth of bone for stability and retention. The advantage of bone ingrowth is prevention of loosening that occurs if the cement bond weakens. With advanced surgical techniques and newer metal prostheses fabricated from alloys of titanium, cobalt/chrome, and surgical steel, 10- to 15-year THR and total knee replacement success rates have risen to approximately 85% in older patients. However, even more durable metal and plastic components are necessary to extend the 10- to 15-year lifespan of most hip and knee prostheses currently in use. At the present time, younger patients who stress artificial joints more than older adults are encouraged to delay re-

placement and rely instead on evolving therapies, which are intended to heal or regenerate diseased joint tissues. The x-ray illustration shows placement of a noncemented, porous-coated femoral prosthesis made of titanium alloy and a cemented acetabular socket made of high-density plastic.

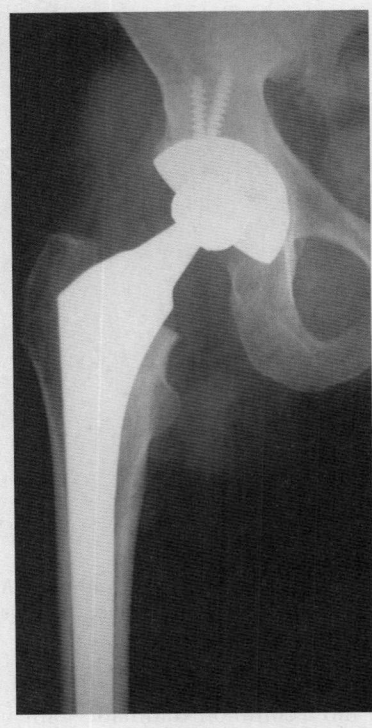

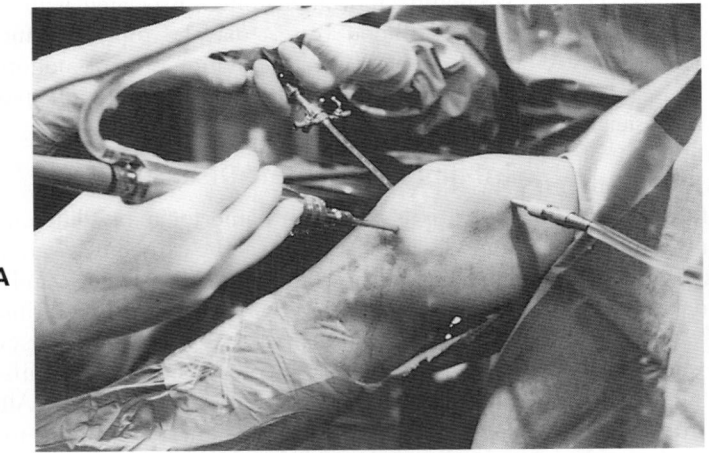

A

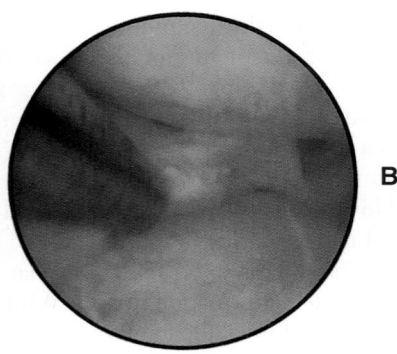

B

Figure 9-29 *Arthroscopy.* **A,** Fiberoptic light source inserted into a joint. **B,** Internal view of the joint.

Mechanisms of Disease—cont.

the diseased synovial membrane produces an aggregation of inflammatory cells, granulation tissue, and fibroblasts called **pannus** that adheres to the articular cartilage. It is the continued growth and spreading of pannus tissue that destroys articular cartilage and bridges the opposing bones. Eventually, the bones fuse together, movement ceases, and permanent crippling occurs. Small joints in the hands and feet are generally affected first. As the disease progresses, the wrists, ankles, elbows, knees, and cervical spine may also be involved. Because the disease is systemic, most patients experience such symptoms as anemia, weight loss, fever, fatigue, and generalized muscle pain, as well as symptoms associated with specific types of joint pathology. A characteristic deformity of the hands in RA is **ulnar deviation** of the fingers (Figure 9-28, B). A number of other common signs and symptoms are related to inflammation and loss of articular surface. They include reduction in joint mobility, pain, nodular swelling, and generalized aching and stiffness. In the past, treatment for many patients was often limited to NSAIDs and other pain or anti-inflammatory–type medication, corticosteroids, and so-called disease-modifying antirheumatic drugs such as methotrexate. Newer drugs influencing the altered immune system response in RA, called **TNF blockers,** are showing promise in many individuals suffering from this disease. *TNF*, or *tumor necrosis factor*, is a substance made in excess by the body's immune system in RA. It is involved in pannus

formation, destruction of articular cartilage, and pain and swelling of joints. By blocking the effects of this substance, people with moderate RA often experience significant relief from symptoms and from continued damage to bone and joints.

Juvenile rheumatoid arthritis (JRA) is often more severe than the adult form but involves similar deterioration and deformity of joints. Onset is generally systemic and associated with rash, high fever, and swelling of the liver and spleen. Targeted joints, which include the wrists, elbows, knees, and ankles, are generally warm and swollen. The joint inflammatory process often destroys the growth of epiphyseal cartilage, and the growth of long bones is arrested. This form begins during childhood and is more common in girls. In addition to articular damage, JRA also affects the heart, lungs, muscles, and kidneys in many young patients.

Gouty arthritis is another type of inflammatory arthritis. It is a metabolic disorder in which excess blood levels of uric acid, a nitrogenous waste, are deposited as sodium urate crystals within the synovial fluid of joints and in other tissues. These deposits, called **tophi,** are often obvious in the soft tissues of patients with gouty arthritis (Figure 9-28, C). They trigger the chronic inflammation and tissue damage seen with the disease. Swelling, tenderness, or pain typically appears in the fingers, wrists, elbows, ankles, and knees. Allopurinol (Zyloprim) is a drug used to treat the disease. It inhibits the synthesis of uric acid.

LANGUAGE OF SCIENCE   (Cont'd from page 311)

inversion (in-VER-shun) [*invert* to turn over]

joint capsule [*joint* to join, *capsula* little box]

joint cavity [*joint* to join, *cavity* hollow]

ligaments (LIG-ah-ments) [*ligare* to bind]

menisci (meh-NIS-kye) [*meniscus* crescent] sing., meniscus

metacarpophalangeal joints (met-ah-KAR-poh-fah-LAN-jee-al) [*meta-* beyond, *-carpo-* wrist, *-phalangeal* pertaining to the phalanx]

multiaxial joints (mul-tee-AK-see-al) [*multi-* many, *-axon* axle]

nonsteroidal anti-inflammatory drugs (NSAIDs) (nahn-STAYR-oyd-al an-ti-in-FLAM-ah-toh-ree [*non-* not, *-steros-* solid, *-al* pertaining to, *anti-* against, *-inflame* to set afire]

olecranon bursa (oh-LEK-rah-non BER-sah) [*olecranon* tip of the elbow, *bursa* wineskin]

plantar flexion (PLAN-tar FLEK-shun) [*planta* sole, *flexion* to bend]

pronation (proh-NAY- shun) [*pronate* to bend forward]

protraction (proh-TRAK- shun) [*protract* to drag forth]

proximal interphalangeal (PIP) (PROK-si-mal in-ter-fah-LAN-gee-al)

radiocarpal joints (RAY-dee-oh-KAR-pal) [*radius-* ray, *-carp* wrist]

radioulnar joints (RAY-dee-oh-UL-nur) [*radius-* ray, *-ulna* elbow or arm]

range of motion (ROM)

retraction [*retractare* to draw back]

rotation (roh-TAY-shun) [*rot* turned or to turn]

special movements

supination (soo-pi-NAY-shun) [*supine* lying on the back]

sutures (SOO-churs) [*sutura* a seam]

symphysis (SIM-fi-sis) [*sym-* union or association, *-physis* growth or growing]

synarthroses (sin-ar-THROH-seez) [*syn-* union or association, *-arthr* joints] sing., synarthrosis

synchondroses (SIN-kon-DROH-seez) [*syn-* union or association, *-chondro-* cartilage] sing., synchodrosis

syndesmoses (SIN-dez-MO-seez) [*syndesmo* connective tissue or ligament] sing., sydesmosis

synovial membrane (si-NO-vee-all) [*syn-* together, *-ovi-* egg, *-al* pertaining to, *membrane* thin skin]

uniaxial joints (yoo-nee-AK-see-al) [*uni-* one, *-axon* axle, *joint* to join]

LANGUAGE OF MEDICINE

arthritis (ar-THRY-tis) [*arthr-* joint, *-itis* inflammation]

Bouchard nodes (boo-SHAR) [*Charles J. Bouchard* French physician]

bursitis (ber-SYE-tiss) [*bursa-* wineskin, *-itis* inflammation]

chondral fractures (KON-dral) [*condr* cartilage, *fract* a breaking]

dislocation [*dis-* opposite of, *-locare* to place]

gouty arthritis (gow-TEE ar-THRY-tis) [*gutta* drop, *arthr-* joint, *-itis* inflammation]

Heberden nodes (HEB-er-den) [*William Heberden* English physician]

herniated disk (HER-nee-ayt-ed) [*hernia* rupture]

inflammatory joint disease (in-FLAM-ah-toh-ree) [*inflame* to set afire, *joint* to join]

juvenile rheumatoid arthritis (JRA) (JOO-veh-neye-al ROO-mah-toyd ar-THRY-tis) [*juvenus* youthful, *rheuma-* flux, *-oid* resembling, *arthr-* joint, *-itis* inflammation]

noninflammatory joint disease (nahn-in-FLAM-ah-toh-ree) [*non-* not, *-inflame* to set afire, *joint* to join]

olecranon bursitis (oh-LEK-rah-nohn ber-SYE-tis) [*olecranon* tip of the elbow, *bursa-* wineskin, *-itis* inflammation]

osteoarthritis (os-tee-oh-ar-THRY-tis) [*osteo-* bone, *-arthr-* joint, *-itis* inflammation]

pannus (PAN-us) [*pannus* cloth]

prepatellar bursitis (pree-pah-TEL-er ber-SYE-tis) [*pre-* in front of, *-patella* small dish, *bursa-* wineskin, *-itis* inflammation]

prosthesis (pros-THEE-sis) [*prosthesis* addition]

rheumatoid arthritis (RA) (ROO-mah-toyd ar-THRY-tis) [*rheuma-* flux, *-oid* resembling, *arthr-* joint, *-itis* inflammation]

sprain

sprained ankle

TNF blockers

tophi (TOH-fee) [*tufa* porus rock]

ulnar deviation (UL-nur dee-vee-AY-shun) [*ulna* elbow]

CASE STUDY

Bubba Brown, age 15, is a running back on his high-school football team. Today during practice, he was running down the field when he was tackled from the side at knee level. Bubba felt his knee "pop" and was unable to walk off the field without assistance. Ice was applied to the injury and Bubba was evaluated by the team athletic trainer, who elevated Bubba's knee and found that the affected area was not swollen. The athletic trainer did note, however, that there was joint line tenderness on the medial side of the knee, limited flexion in the affected knee, and inability to completely extend the leg. When asked to bear weight, Bubba's injured knee "gave way on him." Bubba is referred to the orthopedist for evaluation and follow-up. The orthopedist examines the knee and takes a series of radiographs and an MRI. He then schedules Bubba for arthroscopic surgery. The arthroscopy shows a tear in the medial meniscus. The surgeon repairs the tear, and Bubba is placed in a hinged splint and begins a program of rehabilitation with the sports medicine clinic.

1. Which of the following represents normal movement of the knee joint?

   1. Flexion
   2. Extension
   3. Supination
   4. Hyperextension
      A. 1 and 2
      B. 1, 2, and 3
      C. 2 and 4
      D. 1, 3, and 4

2. Which of the following factors makes the knee joint more vulnerable to injury as just described?

   A. The knee is relatively unprotected by muscle.
   B. The ligaments found here are very fragile.
   C. The knee is a weight-bearing joint.
   D. The knee cannot be pronated.

3. Why is Bubba unable to fully extend his leg on the affected side?

   A. He may have a mechanical obstruction from the displaced meniscus.
   B. Muscle spasm in the thigh may prevent full extension.
   C. Joint effusion may prevent full extension.
   D. All of the above.

CHAPTER SUMMARY

INTRODUCTION
A. Articulation—point of contact between bones
B. Joints are mostly very movable, but others are immovable or allow only limited motion
C. Movable joints allow complex, highly coordinated, and purposeful movements to be executed

CLASSIFICATION OF JOINTS
A. Joints may be classified by using a structural or functional scheme (Table 9-1)
   1. Structural classification—joints are named according to
      a. Type of connective tissue that joins bones together (fibrous or cartilaginous joints)
      b. Presence of a fluid-filled joint capsule (synovial joint)
   2. Functional classification—joints are named according to the degree of movement allowed
      a. Synarthroses—immovable joint
      b. Amphiarthroses—slightly movable
      c. Diarthroses—freely movable
B. Fibrous joints (synarthroses)—bones of joints fit together closely, thereby allowing little or no movement (Figure 9-1)
   1. Syndesmoses—joints in which ligaments connect two bones

   2. Sutures—found only in the skull; teethlike projections from adjacent bones interlock with each other
   3. Gomphoses—between the root of a tooth and the alveolar process of the mandible or maxilla
C. Cartilaginous joints (amphiarthroses)—bones of joints are joined together by hyaline cartilage or fibrocartilage; allow very little motion (Figure 9-2)
   1. Synchondroses—hyaline cartilage present between articulating bones
   2. Symphyses—joints in which a pad or disk of fibrocartilage connects two bones
D. Synovial joints (diarthroses)—freely movable joints (Figure 9-3)
   1. Structures of synovial joints
      a. Joint capsule—sleevelike casing around the ends of the bones that binds them together
      b. Synovial membrane—membrane that lines the joint capsule and also secretes synovial fluid
      c. Articular cartilage—hyaline cartilage covering the articular surfaces of bones
      d. Joint cavity—small space between the articulating surfaces of the two bones of the joint
      e. Menisci (articular disks)—pads of fibrocartilage located between articulating bones

f. Ligaments—strong cords of dense white fibrous tissue that hold the bones of a synovial joint more firmly together

g. Bursae—synovial membranes filled with synovial fluid; cushion joints and facilitate movement of tendons

2. Types of synovial joints (Figure 9-4)

a. Uniaxial joints—synovial joints that permit movement around only one axis and in only one plane

(1) Hinge joints—articulating ends of bones form a hinge-shaped unity that allows only flexion and extension

(2) Pivot joints—a projection of one bone articulates with a ring or notch of another bone

b. Biaxial joints—synovial joints that permit movements around two perpendicular axes in two perpendicular planes

(1) Saddle joints—synovial joints in which the articulating ends of the bones resemble reciprocally shaped miniature saddles; only example in the body is in the thumbs

(2) Condyloid (ellipsoidal) joints—synovial joints in which a condyle fits into an elliptical socket

c. Multiaxial joints—synovial joints that permit movements around three or more axes in three or more planes

(1) Ball-and-socket (spheroid) joints—most movable joints; the ball-shaped head of one bone fits into a concave depression

(2) Gliding joints—relatively flat articulating surfaces that allow limited gliding movements along various axes

REPRESENTATIVE SYNOVIAL JOINTS

A. Humeroscapular joint (Figure 9-5)
1. Shoulder joint
2. Most mobile joint because of the shallowness of the glenoid cavity
3. Glenoid labrum—narrow rim of fibrocartilage around the glenoid cavity that lends depth to the glenoid cavity
4. Structures that strengthen the shoulder joint are ligaments, muscles, tendons, and bursae

B. Elbow joint (Figure 9-6)
1. Humeroradial joint—lateral articulation of the capitulum of the humerus with the head of the radius
2. Humeroulnar joint—medial articulation of the trochlea of the humerus with the trochlear notch of the ulna
3. Both components of the elbow joint surrounded by a single joint capsule and stabilized by collateral ligaments
4. Classic hinge joint
5. Medial and lateral epicondyles are externally palpable bony landmarks
6. Olecranon bursa independent of elbow joint space—inflammation called olecranon bursitis
a. Trauma to nerve results in unpleasant sensations in the fingers and part of the hand supplied by the nerve—severe injury may cause "wristdrop"

C. Proximal radioulnar joint—between the head of the radius and the medial notch of the ulna
1. Stabilized by the annular ligament
2. Permits rotation of the forearm
3. Dislocation of the radial head called a "pulled elbow"

D. Distal radioulnar joint—point of articulation between the ulnar notch of the radius and the head of the ulna
1. Acting with the proximal radioulnar joint permits pronation and supination of the forearm

E. Radiocarpal (wrist) joints (Figure 9-7)
1. Only the radius articulates directly with the carpal bones distally (scaphoid and lunate)
2. Joints are synovial
3. Scaphoid bone is fractured frequently
4. Portion of the fractured scaphoid may become avascular

F. Intercarpal joints (Figure 9-7)
1. Between 8 carpal bones
2. Stabilized by numerous ligaments
3. Joint spaces usually communicate
4. Movements generally gliding with some abduction and flexion

G. Carpometacarpal joints—total of three joints
1. One joint for the thumb—wide range of movements
2. Two joints for the fingers—movements largely gliding type
3. Thumb carpometacarpal joint is unique and important functionally
a. Loose-fitting joint capsule
b. Saddle-shaped articular surface
c. Movements—extension, adduction, abduction, circumduction, and opposition
d. Opposition—ability to touch the tip of the thumb to the tip of other fingers—movement of great functional significance

H. Metacarpophalangeal joints (Figure 9-8)
1. Rounded heads of metacarpals articulate with concave bases of the proximal phalanges
2. Capsule surrounding joints strengthened by collateral ligaments
3. Primary movements are flexion and extension

I. Interphalangeal joints
1. Typical diarthrotic, hinge-type, synovial joints
2. Exist between heads of phalanges and bases of more distal phalanges
3. Two categories:
a. PIP joints—proximal interphalangeal joints (between proximal and middle phalanges)
b. DIP joints—distal interphalangeal joints (between middle and distal phalanges)

J. Hip joint (Figure 9-9)
1. Stable joint because of the shape of the head of the femur and the acetabulum
2. A joint capsule and ligaments contribute to the joint's stability

K. Knee joint (Figures 9-10 and 9-11)
1. Largest and one of the most complex and most frequently injured joints

2. Tibiofemoral joint is supported by a joint capsule, cartilage, and numerous ligaments and muscle tendons
3. Permits flexion, extension, and with the knee flexed, some internal and external rotation
L. Ankle joint (Figure 9-12)
   1. Synovial-type hinge joint
   2. Articulation between the lower ends of the tibia and fibula and the upper part of the talus
   3. Joint is "mortise" or wedge shaped
      a. Lateral malleolus lower than medial malleolus
   4. Internal rotation injury results in common "sprained ankle"
      a. Involves anterior talofibular ligament
   5. Other ankle ligaments may also be involved in sprain injuries—example is deltoid ligament
   6. External ankle rotation injuries generally involve bone fractures rather than ligament tears
      a. First-degree ankle injury—lateral malleolus fractured
      b. Second-degree ankle injury—both malleoli fractured
      c. Third-degree ankle injury—fracture of both malleoli and articular surface of tibia
M. Vertebral joints (Figures 9-13 and 9-14)
   1. Vertebrae are connected to one another by several joints to form a strong flexible column
   2. Bodies of adjacent vertebrae are connected by intervertebral disks and ligaments
   3. Intervertebral disks are made up of two parts
      a. Annulus fibrosus—disk's outer rim, made of fibrous tissue and fibrocartilage
      b. Nucleus pulposus—disk's central core, made of a pulpy, elastic substance

TYPES AND RANGE OF MOVEMENT AT SYNOVIAL JOINTS

A. Measuring range of motion (Figure 9-15)
   1. Range of motion (ROM) assessment used to determine extent of joint injury
   2. ROM can be measured actively or passively; both are generally about equal
   3. ROM measured by instrument called a goniometer
B. Angular movements—change the size of the angle between articulating bones
   1. Flexion—decreases the angle between bones; bends or folds one part on another (Figures 9-16, A; 9-18; and 9-19)
   2. Extension and hyperextension (Figure 9-18)
      a. Extension—increases the angle between bones, returns a part from its flexed position to its anatomical position
      b. Hyperextension—stretching or extending part beyond its anatomical position (Figures 9-19, 9-21, and 9-23)
   3. Plantar flexion and dorsiflexion (Figure 9-25)
      a. Plantar flexion—increases the angle between the top of the foot and the front of the leg
      b. Dorsiflexion—decreases the angle between the top of the foot and the front of the leg

4. Abduction and adduction (Figures 9-19 and 9-23)
   a. Abduction—moves a part away from the median plane of the body
   b. Adduction—moves a part toward the median plane of the body
C. Circular movements
   1. Rotation and circumduction
      a. Rotation—pivoting a bone on its own axis (Figure 9-16, D)
      b. Circumduction—moves a part so that its distal end moves in a circle
   2. Supination and pronation (Figure 9-20, B)
      a. Supination—turns the hand palm side up
      b. Pronation—turns the hand palm side down
D. Gliding movements—simplest of all movements; articular surface of one bone moves over the articular surface of another without any angular or circular movement
E. Special movements
   1. Inversion and eversion (Figure 9-25, B)
      a. Inversion—turning the sole of the foot inward
      b. Eversion—turning the sole of the foot outward
   2. Protraction and retraction (Figure 9-17, A)
      a. Protraction—moves a part forward
      b. Retraction—moves a part backward
   3. Elevation and depression (Figure 9-17, B)
      a. Elevation—moves a part up
      b. Depression—lowers a part

CYCLE OF LIFE: ARTICULATIONS

A. Bone development and the sequence of ossification between birth and skeletal maturity affect joints
   1. Fontanels between cranial bones disappear
   2. Epiphysial plates ossify at maturity
B. Older adults
   1. ROM decreases
   2. Changes in gait
C. Skeletal diseases manifested as joint problems
   1. Abnormal bone growth (lipping)—influences joint motion
   2. Disease conditions can be associated with specific developmental periods

THE BIG PICTURE: ARTICULATIONS

A. Hand—"reason for the upper extremity"; thumb—"reason for the hand"
   1. Examples of "big picture" type of thinking when used in functional context
B. Mobility of the upper extremity is extensive because of the following:
   1. Arrangement of bones in the shoulder girdle, arms, forearm, wrist, and hand
   2. Location and method of attachment of muscles to bones
   3. Proper functioning of joints
C. Mobility and extensive ROM needed to position upper extremity and hand to permit grasping and manipulation of objects, thus enabling effective interaction with objects in the external environment

REVIEW QUESTIONS

1. Classify joints and specify each group according to function and structure.
2. Define the terms *fibrous joints*, *cartilaginous joints*, and *synovial joints*.
3. Name and define three types of fibrous joints. Give an example of each.
4. Name and define two types of cartilaginous joints. Give an example of each.
5. List and describe the different types of synovial joints. Give an example of each.
6. Name and define four kinds of angular movements permitted by some synovial joints.
7. Define and give an example of the following: *rotation, circumduction, pronation, supination*.
8. Describe the function of a goniometer.
9. What joint makes possible much of the dexterity of the human hand? Describe it and the movements it permits.
10. Describe vertebral joints.
11. Describe and differentiate between the following joints: humeroscapular, hip, knee.
12. How do loose bodies, or cartilaginous "joint mice," differ from menisci?
13. Discuss the surgical procedure of total hip replacement (THR).
14. Name the two classifications of joint disorders and give examples of each.

CRITICAL THINKING QUESTIONS

1. The elbow, the joint between the bodies of vertebrae, and the root of a tooth in the jaw are all joints given different functional names. What are these names and what characteristic distinguishes them from one another?
2. A synovial joint is the most freely moving type of joint in the body, but joints require stability and protection also. Evaluate the seven structural characteristics of a synovial joint as primarily necessary for stability, movement, and protection.
3. Compare the range of movement at different joints.
4. Describe the anatomical structure of the knee and explain why the knee joint is the most frequently injured joint in the body.

CHAPTER 10

Anatomy of the Muscular System

LANGUAGE OF SCIENCE

agonist (AG-ah-nist) [*agon* struggle]

anal triangle [*anal* pertaining to the anus]

antagonist (an-TAG-oh-nist) [*ant-* against, *-agon-* struggle]

aponeurosis (ap-oh-nyoo-ROH-sis) [*apo-* from, *-neur-* sinew, *-osis* condition of]

belly

buccinator muscle (BYOOK-si-NAY-tor) [*bucci* cheek]

bulbospongiosus muscle (bul-boh-spun-jee-OH-ses MUSS-el)

calcaneal tendon (kal-KAH-nee-al) [*calcane-* heel, *-al* pertaining to]

circular muscles

coccygeus muscle (kohk-SIJ-ee-us MUSS-el) [*coccyg* coccyx]

convergent muscles (kon-VER-jent)

corrugator supercilii muscle (COR-eh-gay-tor soo-per-SIL-ee-eye) [*corrugare* to wrinkle, *super-* above, *-cillium* eyelash]

deltoid (DEL-toyd) [*delta-* triangle, *-oid* resembling]

diaphragm (DYE-ah-fram) [*dia-* two, *-phragm* septum]

endomysium (en-doh-MEE-see-um) [*endo-* within, *-mys* muscle]

epimysium (ep-i-MIS-ee-um) [*epi-* upon, *-mys* muscle]

erector spinae muscle (eh-REK-tor SPINE-ee) [*erigere* to erect, *spina* spine]

extensor digitorum longus muscle (ek-STEN-ser dij-i-TOH-rum) [*extendere* to stretch, *digit* finger or toe, *longus* long]

external anal sphincter muscle (eks-TER-nal AY-nal SFINGK-ter MUSS-el) [*externa* outward, *anal* pertaining to the anus, *sphingein* to bind]

external intercostal muscle (eks-TER-nal in-ter-KOS-tal) [*externa* outward, *costa* rib]

external oblique muscle (eks-TER-nal oh-BLEEK) [*externa* outward, *obliquus* slanted]

extrinsic foot muscles (eks-TRIN-sik foot) [*extrinsecus* on the outside]

extrinsic muscles (eks-TRIN-sik) [*extrinsecus* on the outside]

Cont'd on p. 390

Survival depends on the ability to maintain a relatively constant internal environment. Such stability often requires movement of the body. Whereas many different systems of the body have *some* role in accomplishing movement, it is the skeletal and muscular systems acting together that actually produce most body movements. We have investigated the architectural plan of the skeleton and have seen how its firm supports and joint structures make movement possible. However, bones and joints cannot move themselves. They must be moved by something. Our subject for now, then, is the large mass of skeletal muscle that moves the framework of the body: the **muscular system.**

Movement is one of the most distinctive and easily observed "characteristics of life." When we walk, talk, run, breathe, or engage in a multitude of other physical activities that are under the "willed" control of the individual, we do so by contraction of skeletal muscle.

There are more than 600 skeletal muscles in the body. Collectively, they constitute 40% to 50% of our body weight. And, together with the scaffolding provided by the skeleton, muscles also determine the form and contours of our body.

Contraction of individual muscle cells is ultimately responsible for purposeful movement. In Chapter 11 the physiology of muscular contraction is discussed. In this preliminary chapter, however, we will learn how contractile units are grouped into unique functioning organs—or muscles. The manner in which muscles are grouped, the relationship of muscles to joints, and how muscles attach to the skeleton determine purposeful body movement. A discussion of muscle shape and how muscles attach to and move bones is followed by information on specific muscles and muscle groups. The chapter ends with a review of the concept of posture.

SKELETAL MUSCLE STRUCTURE
Connective Tissue Components

The highly specialized skeletal muscle cells, or *muscle fibers,* are covered by a delicate connective tissue membrane called the **endomysium** (Figure 10-1). Groups of skeletal muscle fibers, called *fascicles,* are then bound together by a tougher connective tissue envelope called the **perimysium.** The muscle as a whole is covered by a coarse sheath called the **epimysium.** Because all three of these structures are continuous with the fibrous structures that attach muscles to bones or other structures, muscles are firmly harnessed to the structures they pull on during contraction. The epimysium, perimysium, and endomysium of a muscle, for example, may be continuous with fibrous tissue that extends from the muscle as a **tendon,** a strong tough cord continuous at its other end with the fibrous periosteum covering a bone. Alternatively, the fibrous wrapping of a muscle may extend as a broad, flat sheet of connective tissue called an **aponeurosis,** which usually merges with the fibrous wrappings of another muscle. So tough and strong are tendons and aponeuroses that they are not often torn, even by injuries forceful enough to break bones or tear muscles. They are, however, occasionally pulled away from bones. Fibrous connective tissue surrounding the muscle organ and outside the epimysium and tendon is called **fascia.** *Fascia* is a general term for the fibrous connective tissue found under the skin and surrounding many deeper organs, including skeletal muscles and bones. Fascia just under the skin (the hypodermis) is sometimes called *superficial fascia,* and the fascia around muscles and bones is sometimes called *deep fascia.*

Tube-shaped structures of fibrous connective tissue called **tendon sheaths** enclose certain tendons, notably those of the wrist and ankle. Like bursae, tendon sheaths have a lining of synovial membrane. Its moist, smooth surface enables the tendon to move easily, almost without friction, in the tendon sheath.

Size, Shape, and Fiber Arrangement

The structures called *skeletal muscles* are organs. They consist mainly of skeletal muscle tissue plus important connective and nervous tissue components. Skeletal muscles vary considerably in size, shape, and arrangement of fibers. They range from extremely small strands, such as the stapedius muscle of the middle ear, to large masses, such as the muscles of the thigh. Some skeletal muscles are broad in shape and some are narrow. Some are long and tapering and some are short and blunt. Some are triangular, some quadrilateral, and some irregular. Some form flat sheets and others form bulky masses.

The strength and type of movement produced by the shortening of a muscle is related to the orientation of its fibers and overall shape, as well as its attachments to bone and involvement in joints. This is yet another example of the relationship between structure and function. Six muscle shapes are often used to describe and categorize skeletal muscles (Figure 10-2).

1. **Parallel muscles** can vary in length, but long straplike muscles with parallel fascicles are perhaps most typical. The sartorius muscle of the leg is a good example. The rectus abdominis muscles, which run the length of the anterior abdominal wall, have parallel muscle fascicles that are "interrupted" by transverse intersections.
2. **Convergent muscles** have fascicles that radiate out from a small to a wider point of attachment, much like the blades in a fan. The pectoralis major muscle is a good example.
3. **Pennate muscles** are said to be "feather-like" in appearance. Three categories of these muscles have uniquely different types of fascicle attachments that in some ways resemble the feathers in an old-fashioned plume pen. *Unipenate muscles,* such as the soleus, have fascicles that anchor to only one side of the connective tissue shaft. *Bipennate muscles,* such as the rectus femoris in the thigh, have a type of double-feathered attachment of fascicles. In *multipennate muscles,* such as the deltoid, the numerous interconnecting quill-like fascicles converge on a common point of attachment.
4. **Fusiform muscles** have fascicles that may be close to parallel in the center, or "belly," of the muscle but converge to a tendon at one or both ends. The brachioradialis is a good example.
5. **Spiral muscles,** such as the latissimus dorsi, have fibers that twist between their points of attachment.
6. **Circular muscles,** sometimes called *sphincters,* often circle body tubes or openings. The orbicularis oris around the mouth is a good example.

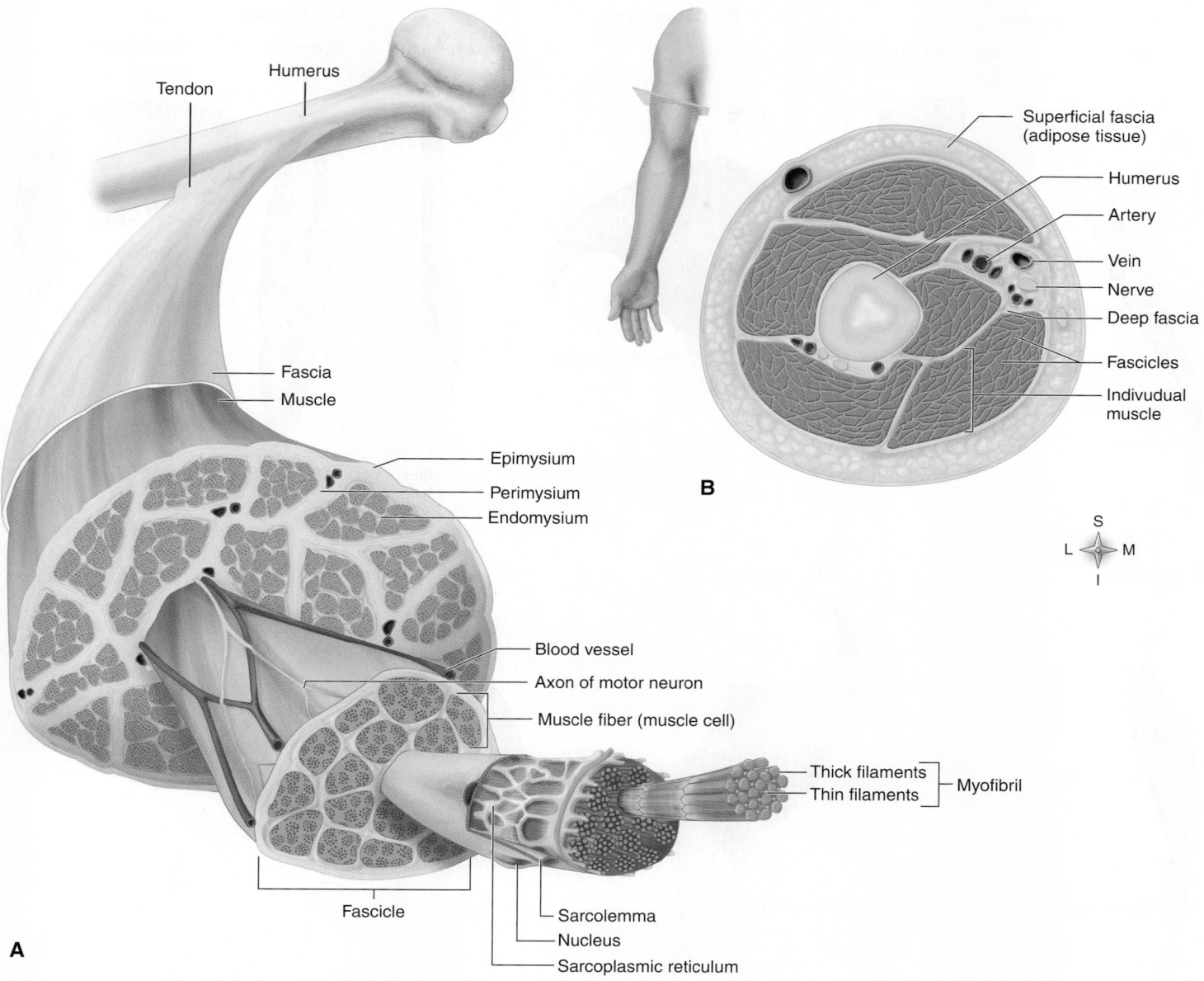

Tendon

Humerus

Fascia

Muscle

Epimysium

Perimysium

Endomysium

Blood vessel

Axon of motor neuron

Muscle fiber (muscle cell)

Fascicle

Sarcolemma

Nucleus

Sarcoplasmic reticulum

Thick filaments

Thin filaments

Myofibril

A

Superficial fascia (adipose tissue)

Humerus

Artery

Vein

Nerve

Deep fascia

Fascicles

Indivudual muscle

B

S
L M
I

Figure 10-1 *Structure of a muscle organ.* **A,** Note that the connective tissue coverings, the epimysium, perimysium, and endomysium, are continuous with each other and with the tendon. Note also that muscle fibers are held together by the perimysium in groups called fascicles. **B,** Diagram showing the arm in cross section. Note the relationships of superficial and deep fascia to individual muscles and other structures in the plane of section.

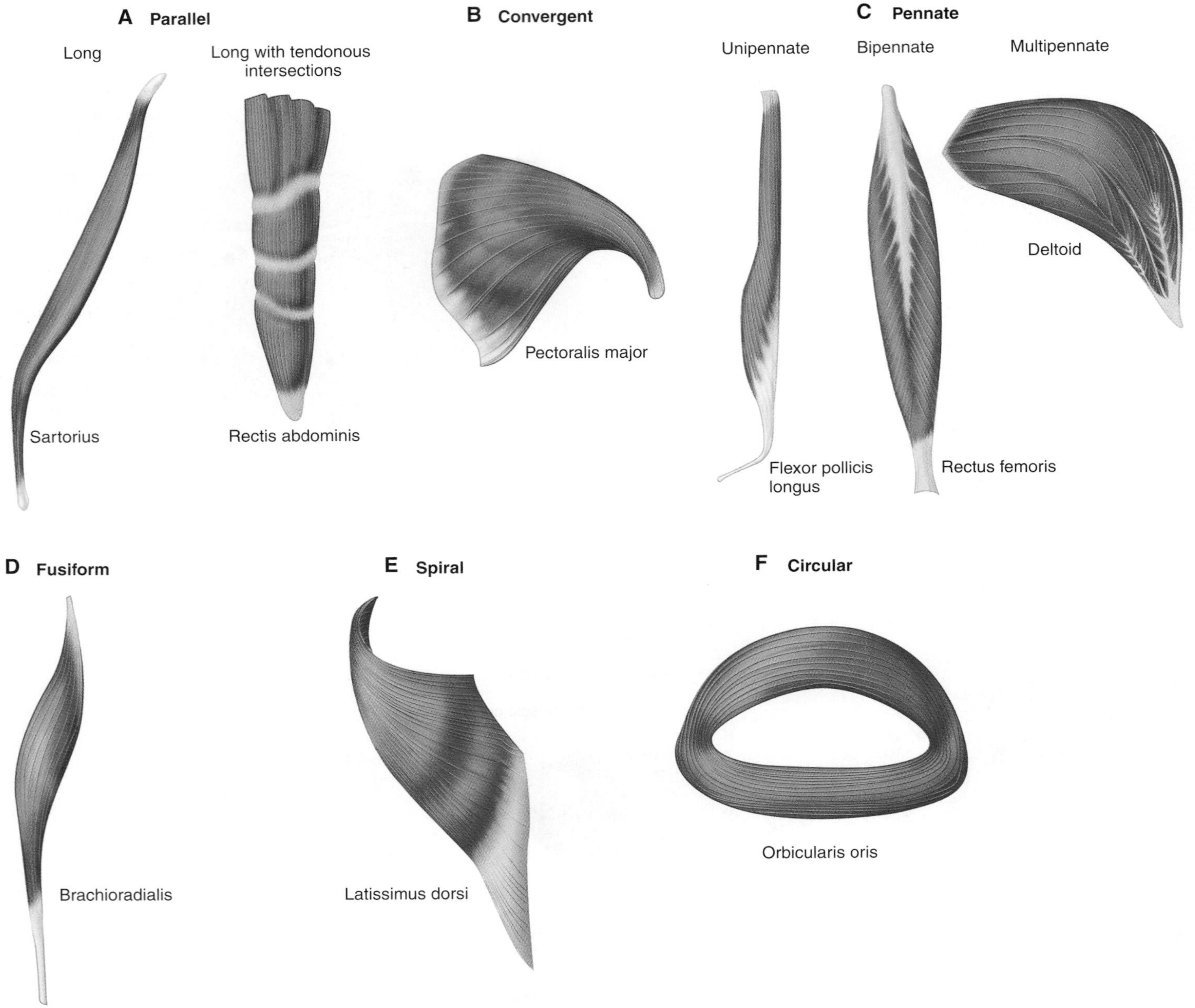

A Parallel

Long

Long with tendonous intersections

Sartorius

Rectis abdominis

B Convergent

Pectoralis major

C Pennate

Unipennate

Bipennate

Multipennate

Deltoid

Flexor pollicis longus

Rectus femoris

D Fusiform

Brachioradialis

E Spiral

Latissimus dorsi

F Circular

Orbicularis oris

Figure 10-2 *Muscle shape and fiber arrangement.* **A,** Parallel. **B,** Convergent. **C,** Pennate: unipennate, bipennate, multipennate. **D,** Fusiform. **E,** Spiral. **F,** Circular.

QUICK CHECK

1. Identify the connective tissue membrane that (a) covers individual muscle fibers, (b) surrounds groups of skeletal muscle fibers (fascicles), and (c) covers the muscle as a whole.
2. Name the tough connective tissue cord that serves to attach a muscle to a bone.
3. Name three types of fiber arrangements seen in skeletal muscle.

Attachment of Muscles

Most of our muscles span at least one joint and attach to both articulating bones. When contraction occurs, one bone usually remains fixed and the other moves. The points of attachment are called the *origin* and *insertion*. The **origin** is the point of attachment that does not move when the muscle contracts. Therefore, the origin bone is the more stationary of the two bones at a joint when contraction occurs. The **insertion** is the point of attachment that moves when the muscle contracts (Figure 10-3). The insertion bone therefore moves toward the origin bone when the

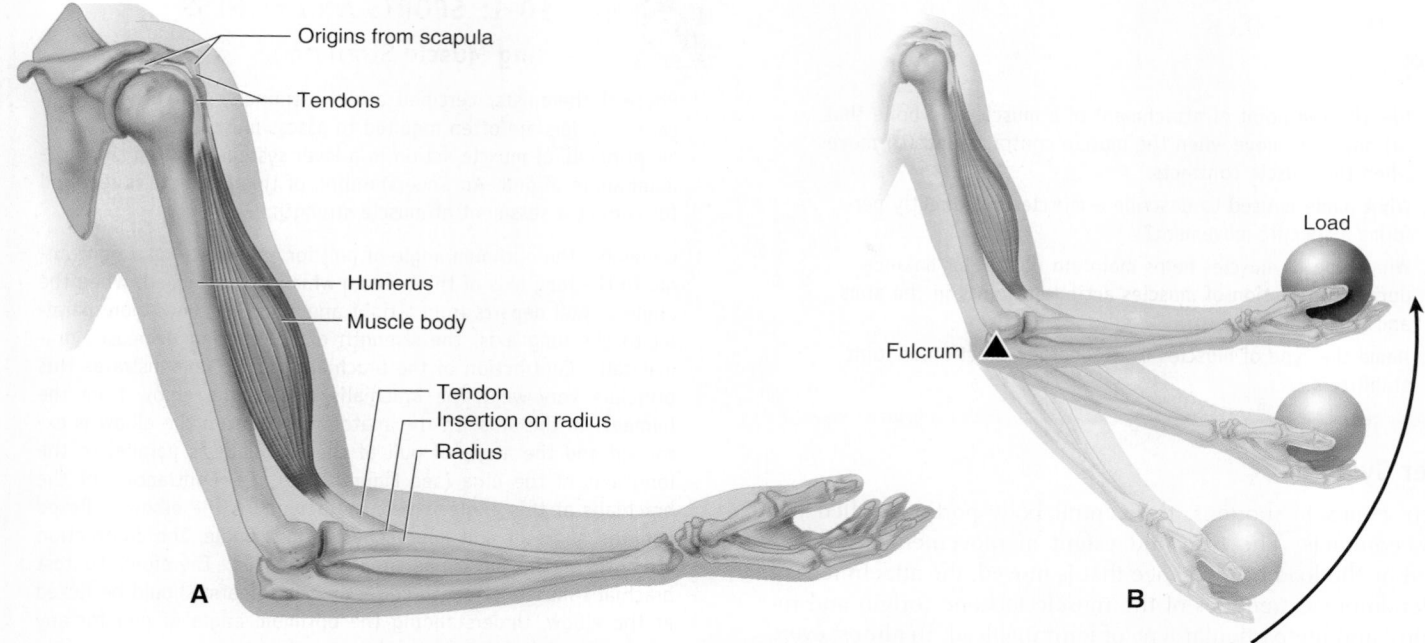

Figure 10-3 *Attachments of a skeletal muscle.* **A**, Origin and insertion of a skeletal muscle. A muscle originates at a relatively stable part of the skeleton (origin) and inserts at the skeletal part that is moved when the muscle contracts (insertion). **B**, Movement of the forearm during weightlifting. Muscle contraction moves bones, which serve as levers, and by acting on joints, which serve as fulcrums for those levers. See the text for discussion and review Figure 10-4, which illustrates types of levers.

muscle shortens. In case you are wondering why both bones do not move because both are pulled on by the contracting muscle, one of them is normally stabilized by isometric contractions of other muscles or by certain features of its own that make it less mobile.

The terms *origin* and *insertion* provide us with useful points of reference. Many muscles have multiple points of origin or insertion. Understanding the functional relationship of these attachment points during muscle contraction helps in deducing muscle actions. The attachment points of the biceps brachii shown in Figure 10-3 help provide functional information. Distal insertion on the radius in the lower part of the arm causes flexion to occur at the elbow when contraction occurs. It should be realized, however, that origin and insertion are points that may change under certain circumstances. For example, not only can you grasp an object above your head and pull it down, but you can also pull yourself up to the object. Although *origin* and *insertion* are convenient terms, they do not always provide the necessary information to understand the full functional potential of muscle action.

Muscle Actions

Skeletal muscles almost always act in groups rather than singly. As a result, most movements are produced by the coordinated action of several muscles. Some of the muscles in the group contract while others relax. The result is a movement pattern that allows for the functional classification of muscles or muscle groups. Several terms are used to describe muscle action during any specialized movement pattern. The terms *prime mover (agonist)*, *antagonist*, *synergist*, and *fixator* are especially important and are

discussed in the following paragraphs. Each term suggests an important concept that is essential to understanding such functional muscle patterns as flexion, extension, abduction, adduction, and other movements discussed in Chapter 9. The term **prime mover** or **agonist** is used to describe a muscle or group of muscles that directly performs a specific movement. The movement produced by a muscle acting as a prime mover is described as the "action" or "function" of that muscle. For example, the biceps brachii shown in Figure 10-3 is acting as a prime mover during flexion of the forearm.

Antagonists are muscles that when contracting, directly oppose prime movers (agonists). They are relaxed while the prime mover is contracting to produce movement. Simultaneous contraction of a prime mover and its antagonist muscle results in rigidity and lack of motion. The term *antagonist* is perhaps unfortunate because muscles cooperate, rather than oppose, in normal movement patterns. Antagonists are important in providing precision and control during contraction of prime movers.

Synergists are muscles that contract at the same time as the prime mover. They facilitate or complement prime mover actions so that the prime mover produces a more effective movement.

Fixator muscles generally function as joint stabilizers. They frequently serve to maintain posture or balance during contraction of prime movers acting on joints in the arms and legs.

Movement patterns are complex, and most muscles function not only as prime movers but also at times as antagonists, synergists, or fixators. A prime mover in a particular movement pattern, such as flexion, may be an antagonist during extension or a synergist or fixator in other types of movement.

QUICK CHECK

4. Identify the point of attachment of a muscle to a bone that (a) does not move when the muscle contracts and (b) moves when the muscle contracts.

5. What name is used to describe a muscle that directly performs a specific movement?

6. What type of muscles helps maintain posture or balance during contraction of muscles acting on joints in the arms and legs?

7. Name the type of muscles that generally function as joint stabilizers.

Lever Systems

When a muscle shortens, the central body portion, called the **belly**, contracts. The type and extent of movement are determined by the load or resistance that is moved, the attachment of the tendinous extremities of the muscle to bone (origin and insertion), and the particular type of joint involved. In almost every instance, muscles that move a part do not lie over that part. Instead, the muscle belly lies proximal to the part moved. Thus, muscles that move the lower part of the arm lie proximal to it, that is, in the upper part of the arm.

Knowledge of **lever systems** is important in understanding muscle action. By definition, a **lever** is any rigid bar free to turn about a fixed point called its *fulcrum*. Bones serve as levers, and joints serve as fulcrums of these levers. A contracting muscle applies a pulling force on a bone lever at the point of the muscle's attachment to the bone. This force causes the insertion bone to move about its joint fulcrum.

A lever system is a simple mechanical device that makes the work of moving a weight or other load easier. Levers are composed of four component parts: (1) a rigid rod or bar (bone); (2) a fixed pivot, or fulcrum (F), around which the rod moves (joint); (3) a load (L), or resistance, that is moved; and (4) a force, or pull (P), which produces movement (muscle contraction). Figure 10-4 shows the three different types of lever arrangements. All three types are found in the human body.

First-Class Levers

As you can see in Figure 10-4, A, the fulcrum in a first-class lever lies between the effort, or pull *(P)*, and the resistance, or load (W), as in a set of scales, a pair of scissors, or a child's seesaw. In the body the head being raised or tipped backward on the atlas is an example of a first-class lever in action. The facial portion of the skull is the load, the joint between the skull and atlas is the fulcrum, and the muscles of the back produce the pull. In the human body first-class levers are not abundant. They generally serve as levers of stability.

Second-Class Levers

In second-class levers the load lies between the fulcrum and the joint at which the pull is exerted. The wheelbarrow is often used as an example. The presence of second-class levers in the human

BOX 10-1: SPORTS AND FITNESS
Assessing Muscle Strength

Physical therapists, certified athletic trainers, and other health care providers are often required to assess muscle strength. A basic principle of muscle action in a lever system is called the optimum angle of pull. An understanding of this principle is required for correct assessment of muscle strength.

Generally, the optimum angle of pull for any muscle is a right angle to the long axis of the bone to which it is attached. When the angle of pull departs from a right angle and becomes more parallel to the long axis, the strength of contraction decreases dramatically. Contraction of the brachialis muscle demonstrates this principle very well. The brachialis crosses the elbow from the humerus to the ulna. In the anatomical position the elbow is extended and the angle of pull of the brachialis is parallel to the long axis of the ulna (see Figure 10-21, *D*). Contraction of the brachialis at this angle is very inefficient. As the elbow is flexed and the angle of pull approaches a right angle, the contraction strength of the muscle is greatly increased. Therefore, to test brachialis muscle strength correctly, the forearm should be flexed at the elbow. Understanding the optimum angle of pull for any given muscle makes a rational approach to correct assessment of functional strength in that muscle possible.

body is a controversial issue. Some authorities interpret the raising of the body on the toes as an example of this type of lever (Figure 10-4, *B*). In this example the point of contact between the toes and the ground is the fulcrum, the load is located at the ankle, and pull is exerted by the gastrocnemius muscle through the Achilles tendon. Opening the mouth against resistance (depression of the mandible) is also considered to be an example of a second-class lever.

Third-Class Levers

In a third-class lever the pull is exerted between the fulcrum and the resistance or load to be moved. Flexing of the forearm at the elbow joint is a frequently used example of this type of lever (Figure 10-4, C). Third-class levers permit rapid and extensive movement and are the most common type found in the body. They allow insertion of a muscle very close to the joint that it moves.

HOW MUSCLES ARE NAMED

The first thing you may notice as you start studying the muscles of the body is that many of the names seem difficult and foreign. The terms are less difficult if you keep in mind that many have Latin roots. Latin-based muscle names are easier to learn if you gain an appreciation for the Latin roots even if a particular name is given in Latin-based English. Although we have strived to use only English names in this edition, some Latin terms still remain in common use. This is especially true in the health-related disciplines. For example, in some texts the deltoid muscle is called the *deltoideus* (Latin) and in others the *deltoid* (Latin-based English). To minimize confusion, if both Latin and English terms are

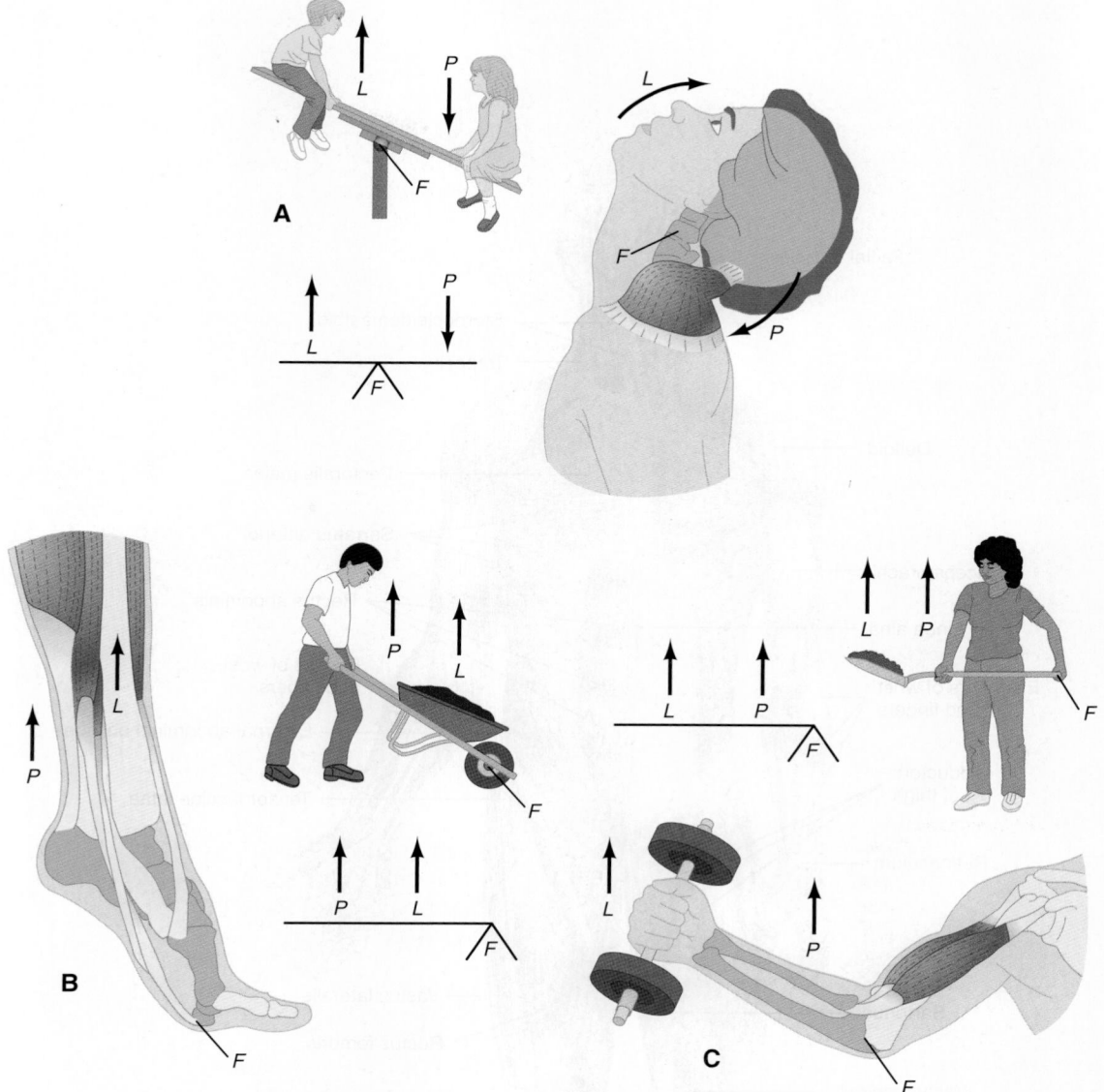

Figure 10-4 *Lever classes.* **A,** Class I: fulcrum *(F)* between the load *(L)* and force or pull *(P);* **B,** Class II: load *(L)* between the fulcrum *(F)* and force or pull *(P);* **C,** Class III: force or pull *(P)* between the fulcrum *(F)* and the load *(L).* The lever rod is yellow in each.

used to identify a particular muscle in the chapter, the terms used will be from the *Terminologia Anatomica* (see Chapter 1).

Regardless of the muscle name used, when one understands the reasons for the term used, it will seem more logical and easier to learn and understand. Many of the muscles of the body shown in Figures 10-5 and 10-6 or listed in Tables 10-1 through 10-5 are named in accordance with one or more of the following features:

- **Location.** Many muscles are named as a result of location. The *brachialis* (arm) muscle and *gluteus* (buttock) muscles are examples. Table 10-1 is a listing of some major muscles grouped by location.

- **Function.** The function of a muscle is frequently a part of its name. The *adductor* muscles of the thigh adduct, or move, the leg toward the midline of the body. Table 10-2 lists selected muscles grouped according to function.

- **Shape.** Shape is a descriptive feature used for naming many muscles. The *deltoid* (triangular) muscle covering the shoulder is deltoid, or triangular, in shape (see Table 10-3).

- **Direction of fibers.** Muscles may be named according to the orientation of their fibers. The term *rectus* means straight. The fibers of the *rectus abdominis* muscle run straight up and down and are parallel to each other (see Table 10-4).

- **Number of heads or divisions.** The number of divisions or heads (points of origin) may be used to name a muscle. The

Text continues on p. 357

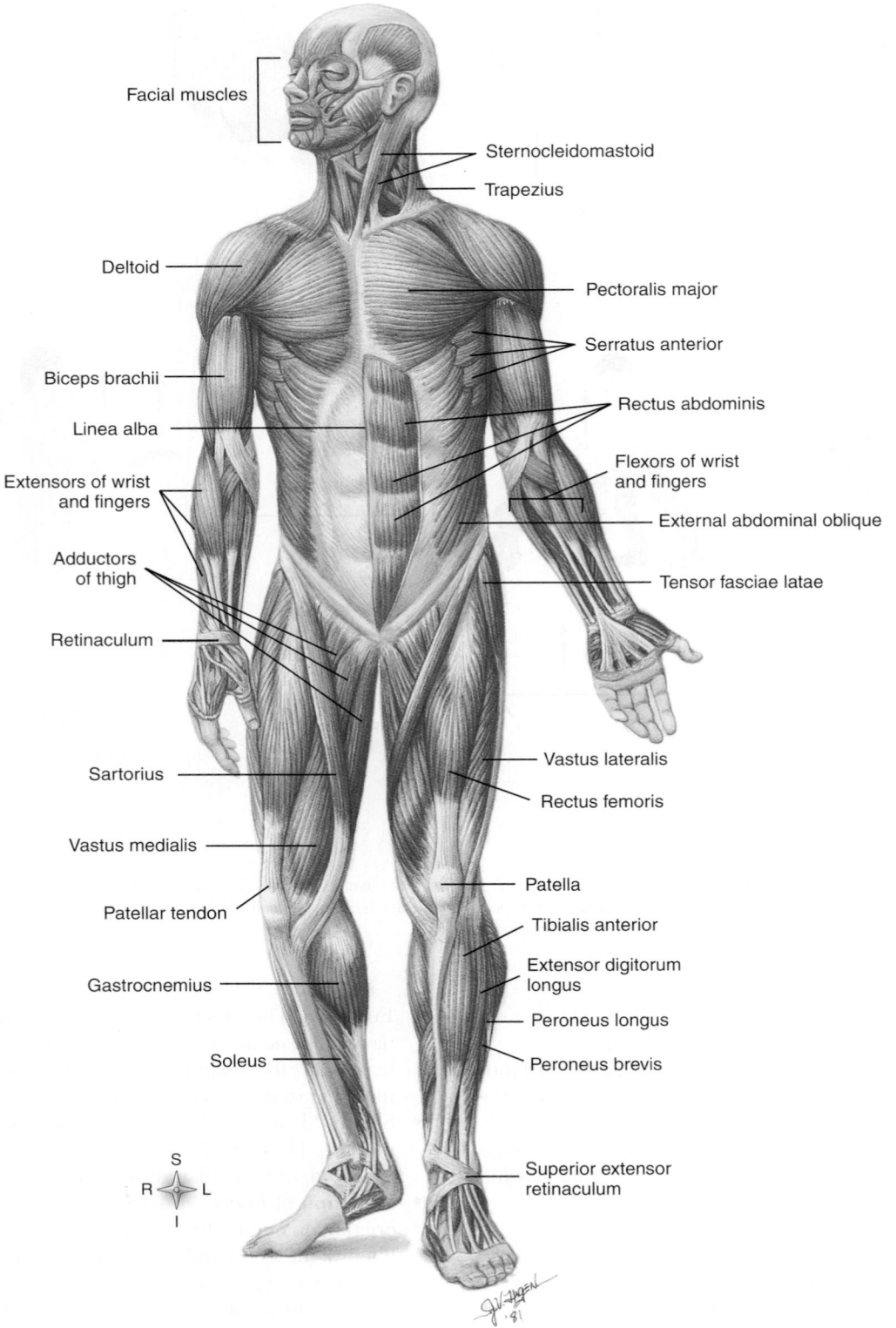

Facial muscles

Sternocleidomastoid

Trapezius

Deltoid

Pectoralis major

Serratus anterior

Biceps brachii

Rectus abdominis

Linea alba

Flexors of wrist and fingers

Extensors of wrist and fingers

External abdominal oblique

Adductors of thigh

Tensor fasciae latae

Retinaculum

Vastus lateralis

Sartorius

Rectus femoris

Vastus medialis

Patella

Patellar tendon

Tibialis anterior

Extensor digitorum longus

Gastrocnemius

Peroneus longus

Soleus

Peroneus brevis

S
R — L
I

Superior extensor retinaculum

Figure 10-5 *General overview of the body's musculature.* Anterior view.

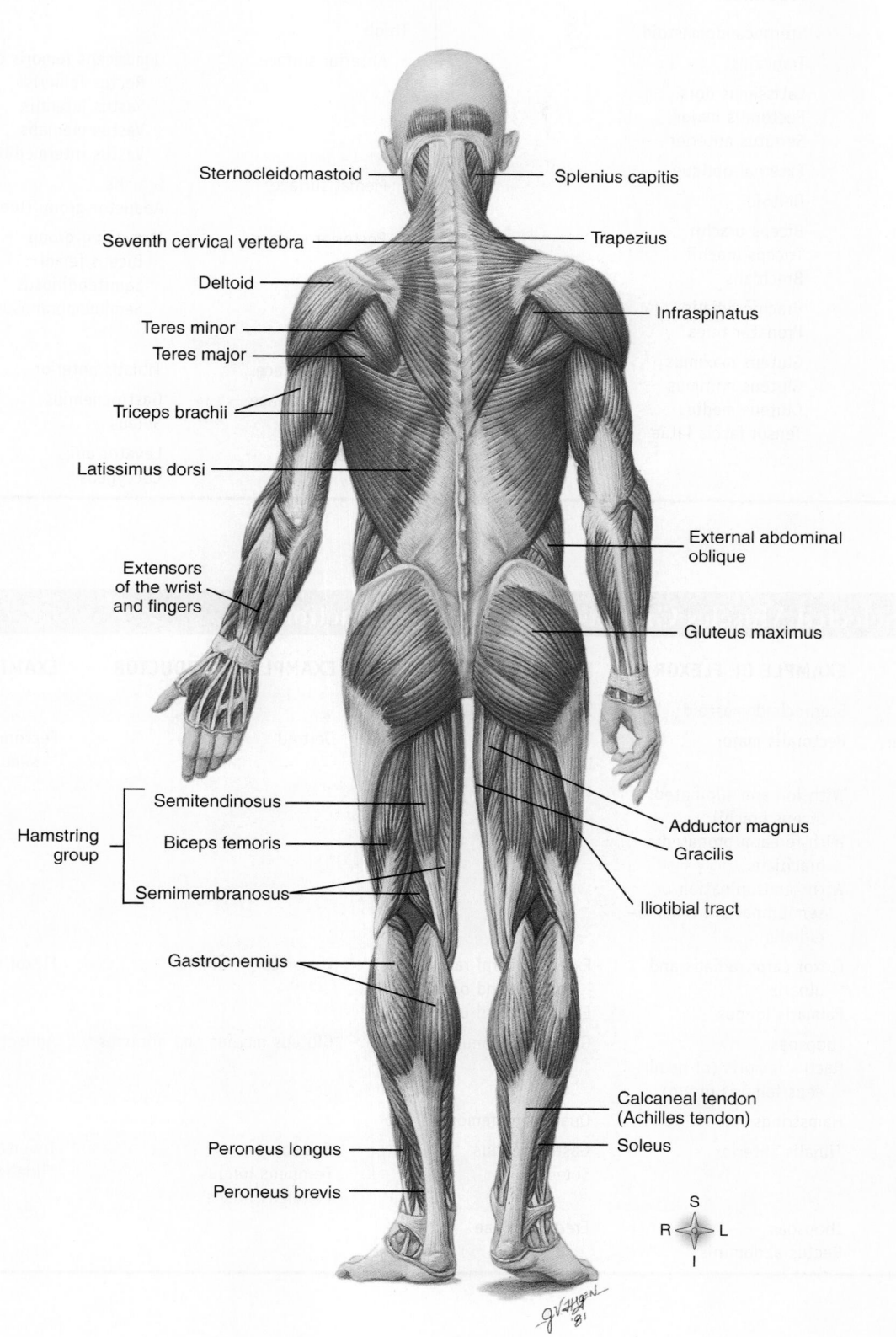

Sternocleidomastoid

Splenius capitis

Seventh cervical vertebra

Trapezius

Deltoid

Infraspinatus

Teres minor

Teres major

Triceps brachii

Latissimus dorsi

External abdominal oblique

Extensors of the wrist and fingers

Gluteus maximus

Semitendinosus

Adductor magnus

Hamstring group

Biceps femoris

Gracilis

Semimembranosus

Iliotibial tract

Gastrocnemius

Calcaneal tendon (Achilles tendon)

Peroneus longus

Soleus

Peroneus brevis

S

R L

I

Figure 10-6 *General overview of the body's musculature.* Posterior view.

Table 10-1	Selected Muscles Grouped According to Location

LOCATION	MUSCLES	LOCATION	MUSCLES
Neck	Sternocleidomastoid	Thigh	
Back	Trapezius	Anterior surface	Quadriceps femoris group
Chest	Latissimus dorsi		Rectus femoris
	Pectoralis major		Vastus lateralis
	Serratus anterior		Vastus medialis
			Vastus intermedius
Abdominal wall	External oblique	Medial surface	Gracilis
Shoulder	Deltoid		Adductor group (brevis, longus, magnus)
Upper part of arm	Biceps brachii	Posterior surface	Hamstring group
	Triceps brachii		Biceps femoris
	Brachialis		Semitendinosus
			Semimembranosus
Forearm	Brachioradialis	Leg	
	Pronator teres	Anterior surface	Tibialis anterior
Buttocks	Gluteus maximus	Posterior surface	Gastrocnemius
	Gluteus minimus		Soleus
	Gluteus medius	Pelvic floor	Levator ani
	Tensor fascia latae		Coccygeus

Table 10-2	Selected Muscles Grouped According to Function

PART MOVED	EXAMPLE OF FLEXOR	EXAMPLE OF EXTENSOR	EXAMPLE OF ABDUCTOR	EXAMPLE OF ADDUCTOR
Head	Sternocleidomastoid	Semispinalis capitis		
Upper part of arm	Pectoralis major	Trapezius Latissimus dorsi	Deltoid	Pectoralis major with latissimus dorsi
Forearm	With forearm supinated: biceps brachii With forearm pronated: brachialis With semisupination or semipronation: brachioradialis	Triceps brachii		
Hand	Flexor carpi radialis and ulnaris Palmaris longus	Extensor carpi radialis, longus, and brevis Extensor carpi ulnaris	Flexor carpi radialis	Flexor carpi ulnaris
Thigh	Iliopsoas Rectus femoris (of quadriceps femoris group)	Gluteus maximus	Gluteus medius and minimus	Adductor group
Leg	Hamstrings	Quadriceps femoris group		
Foot	Tibialis anterior	Gastrocnemius Soleus	Evertors Peroneus longus Peroneus brevis	Invertor Tibialis anterior
Trunk	Iliopsoas Rectus abdominis	Erector spinae		

Table 10-3 Selected Muscles Grouped According to Shape

NAME	MEANING	EXAMPLE
Deltoid	Triangular	Deltoid
Gracilis	Slender	Gracilis
Trapezius	Trapezoid	Trapezius
Serratus	Notched	Serratus anterior
Teres	Round	Pronator teres
Rhomboid	Rhomboidal	Rhomboideus major
Orbicularis	Round or circular	Orbicularis oris
Pectinate	Comblike	Pectineus
Piriformis	Wedge-shaped	Piriformis
Platys	Flat	Platysma
Quadratus	Square	Quadratus femoris
Lumbrical	Wormlike	Lumbricals

Table 10-4 Selected Muscles Grouped According to Number of Heads and Direction of Fiber

NAME	MEANING	EXAMPLE
A. Number of Heads		
Biceps	Two heads	Biceps brachii
Triceps	Three heads	Triceps brachii
Quadriceps	Four heads	Quadriceps
Digastric	Two bellies	Digastric
B. Direction of Fibers		
Oblique	Diagonal	External oblique rectus
Rectus	Straight	Rectus abdominis
Transverse	Transverse	Transversus abdominis
Circular	Around	Orbicularis oris
Spiral	Oblique	Supinator

Table 10-5 Selected Muscles Grouped According to Size

NAME	MEANING	EXAMPLE
Major	Large	Pectoralis major
Maximus	Largest	Gluteus maximus
Minor	Small	Pectoralis minor
Minimus	Smallest	Gluteus minimus
Longus	Long	Adductor longus
Brevis	Short	Extensor pollicis brevis
Latissimus	Very wide	Latissimus dorsi
Longissimus	Very long	Longissimus
Magnus	Very large	Adductor magnus
Vastus	Vast or huge	Vastus medialis

QUICK CHECK

8. Name the four major components of any lever system.
9. Identify the three types of lever systems found in the human body and give one example of each.
10. What type of lever system permits rapid and extensive movement and is the most common type found in the body?
11. List six criteria that may determine a muscle's name, and give an example of a specific muscle named according to each criterion.

Hints on How to Deduce Muscle Actions

To understand muscle actions, you first need to know certain anatomical facts, such as which bones muscles attach to and which joints they pull across. Then, if you relate these structural features to functional principles, you may find your study of muscles more interesting and less difficult than you anticipate. Some specific suggestions for deducing muscle actions follow.

1. Start by making yourself familiar with the names, shapes, and general locations of the larger muscles by using Table 10-1 as a guide.
2. Try to deduce which bones the two ends of a muscle attach to from your knowledge of the shape and general location of the muscle. For example, look carefully at the deltoid muscle as illustrated in Figures 10-5 and 10-17. To which bones does it seem to attach? Check your answer with Table 10-13, p. 372.
3. Next, determine which bone moves when the muscle shortens. (The bone moved by a muscle's contraction is its insertion bone; the bone that remains relatively stationary is its origin bone.) In many cases you can tell which is the insertion bone by trying to move one bone and then another. In some cases either bone may function as the insertion bone. Although not all muscle attachments can be deduced as readily as those of the deltoid, they can all be learned more easily by using this deduction method than by relying on rote memory alone.

word part *-cep* means *head*. *Biceps* (two), *triceps* (three), and *quadriceps* (four) refer to multiple heads, or points of origin. The *biceps brachii* is a muscle having two heads located in the arm (see Table 10-4).

- **Points of attachment.** Origin and insertion points may be used to name a muscle. For example, the *sternocleidomastoid* has its origin on the sternum and clavicle and inserts on the mastoid process of the temporal bone.
- **Size of muscle.** The relative size of a muscle can be used to name a muscle, especially if it is compared to the size of nearby muscles (see Table 10-5). For example, the *gluteus maximus* is the largest muscle of the gluteal (Greek *glautos*, meaning "buttock") region. Nearby, there is a small gluteal muscle, the *gluteus minimus*, and a midsize gluteal muscle, the *gluteus medius*.

4. Deduce a muscle's actions by applying the principle that its insertion moves toward its origin. Check your conclusions with the text. Here, as in steps 2 and 3, the method of deduction is intended merely as a guide and is not adequate by itself for determining muscle actions.

5. To deduce which muscle produces a given action (instead of which action a given muscle produces, as in step 4), start by inferring the insertion bone (bone that moves during the action). The body and origin of the muscle will lie on one or more of the bones toward which the insertion moves—often a bone, or bones, proximal to the insertion bone. Couple these conclusions about origin and insertion with your knowledge of muscle names and locations to deduce the muscle that produces the action.

For example, if you wish to determine the prime mover for the action of raising the upper parts of the arms straight out to the sides, you infer that the muscle inserts on the humerus because this is the bone that moves. It moves toward the shoulder—that is, the clavicle and scapula—so the muscle probably has its origin on these bones. Because you know that the deltoid muscle fulfills these conditions, you conclude, and rightly so, that it is the muscle that raises the upper parts of the arms sideways.

IMPORTANT SKELETAL MUSCLES

The major skeletal muscles of the body are listed, grouped, and illustrated in the tables and figures that follow. Begin your study with an overview of important superficial muscles, shown in Figures 10-5 and 10-6. The remaining figures in this chapter illustrate individual muscles or important muscle groups.

Basic information about many muscles is given in Tables 10-6 to 10-18. Each table has a description of a group of muscles that move one part of the body. The actions listed for each muscle are those for which it is a prime mover. Remember, however, that a single muscle acting alone rarely accomplishes a given action. Instead, muscles act in groups as prime movers, synergists, antagonists, and fixators to bring about movements.

Muscles of Facial Expression

The muscles of facial expression (Table 10-6 and Figures 10-7 and 10-8) are unique in that at least one of their points of attachment is to the deep layers of the skin over the face or neck. Contraction of these muscles produces a variety of facial expressions.

The **occipitofrontalis** (ok-sip-i-toh-fron-TAL-is), or *epicranius*, is in reality two muscles. One portion lies over the forehead (frontal bone); the other covers the occipital bone in back of the head. The two muscular parts, or bellies, are joined by a connective tissue aponeurosis that covers the top of the skull. The frontal portion of the occipitofrontalis raises the eyebrows (surprise) and wrinkles the skin of the forehead horizontally. The **corrugator supercilii** (COR-uh-gay-tor su-per-SIL-ee-eye) draws the eyebrows together and produces vertical wrinkles above the nose (frowning). The **orbicularis oculi** (or-bik-u-LAIR-is OK-yoo-lye) encircles and closes the eye (blinking), whereas the **orbicularis oris** (OR-is) and **buccinator** (BUK-si-NA-tor) pucker the mouth (kissing) and press the lips and cheeks against the teeth. The **zygomaticus** (zye-goh-MAT-ik-us) **major** draws the corner of the mouth upward (laughing).

Muscles of Mastication

The muscles of **mastication** (mass-tih-KAY-shun) shown in Figure 10-9 are responsible for chewing movements. These powerful muscles (see Table 10-6) either elevate and retract the mandible (**masseter**, mass-EET-er, and **temporalis**, tem-poh-RAL-is) or open and protrude it while causing sideways movement (**pterygoids**, TER-i-goids). The pull of gravity helps open the mandible

Table 10-6	Muscles of Facial Expression and Mastication			
MUSCLE	**ORIGIN**	**INSERTION**	**FUNCTION**	**NERVE SUPPLY**
Muscles of Facial Expression				
Occipitofrontalis (epicranius)	Occipital bone	Tissues of eyebrows	Raises eyebrows, wrinkles forehead horizontally	Cranial nerve VII
Corrugator supercilii	Frontal bone (superciliary ridge)	Skin of eyebrow	Wrinkles forehead vertically	Cranial nerve VII
Orbicularis oculi	Encircles eyelid		Closes eye	Cranial nerve VII
Zygomaticus major	Zygomatic bone	Angle of mouth	Laughing (elevates angle of mouth)	Cranial nerve VII
Orbicularis oris	Encircles mouth		Draws lips together	Cranial nerve VII
Buccinator	Maxillae	Skin of sides of mouth	Permits smiling Blowing, as in playing a trumpet	Cranial nerve VII
Muscles of Mastication				
Masseter	Zygomatic arch	Mandible (external surface)	Closes jaw	Cranial nerve V
Temporalis	Temporal bone	Mandible	Closes jaw	Cranial nerve V
Pterygoids (lateral and medial)	Undersurface of skull	Mandible (medial surface)	Grates teeth	Cranial nerve V

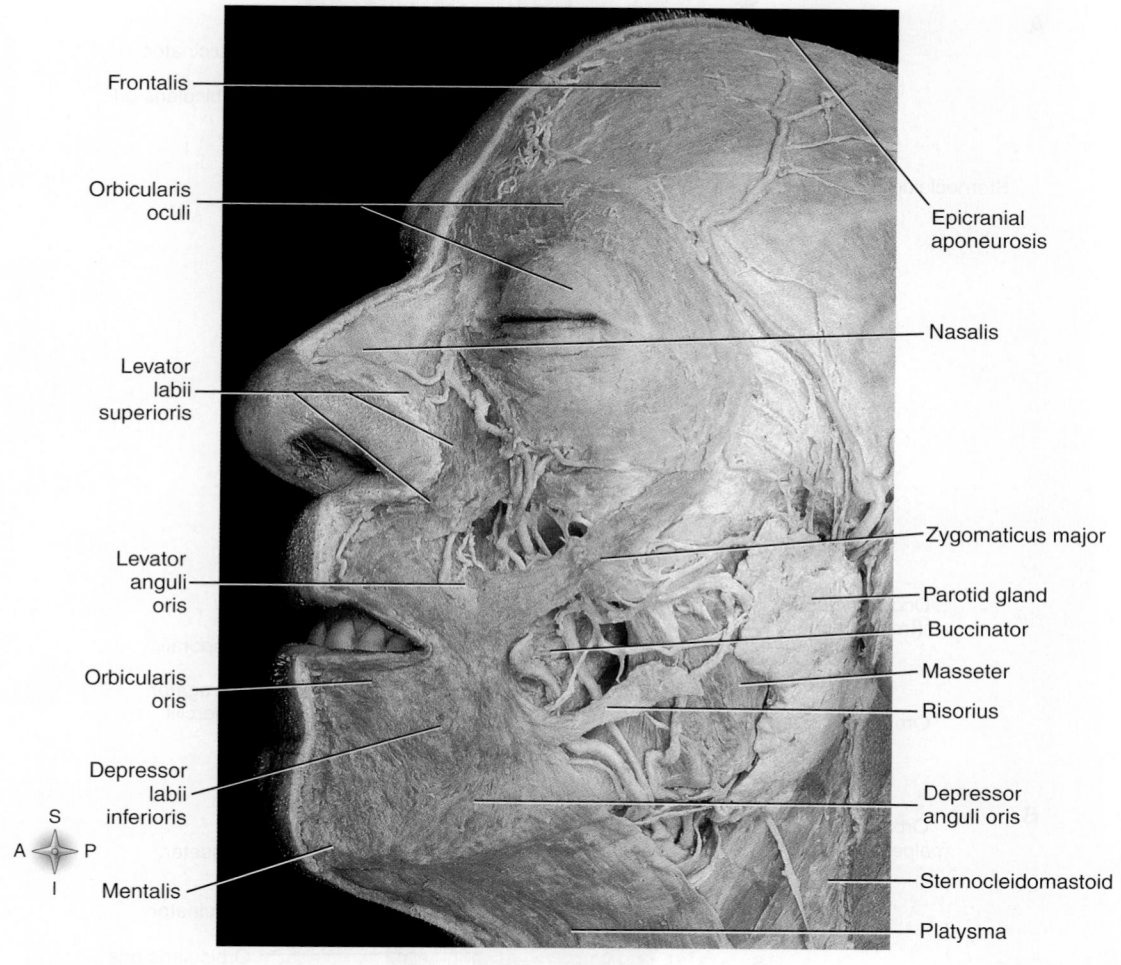

Frontalis

Orbicularis
oculi

Levator
labii
superioris

Levator
anguli
oris

Orbicularis
oris

Depressor
labii
inferioris

Mentalis

Epicranial
aponeurosis

Nasalis

Zygomaticus major

Parotid gland

Buccinator

Masseter

Risorius

Depressor
anguli oris

Sternocleidomastoid

Platysma

S
A ← → P
I

Figure 10-7 *Muscles of facial expression.* The skin and subcutaneous fat have been removed.

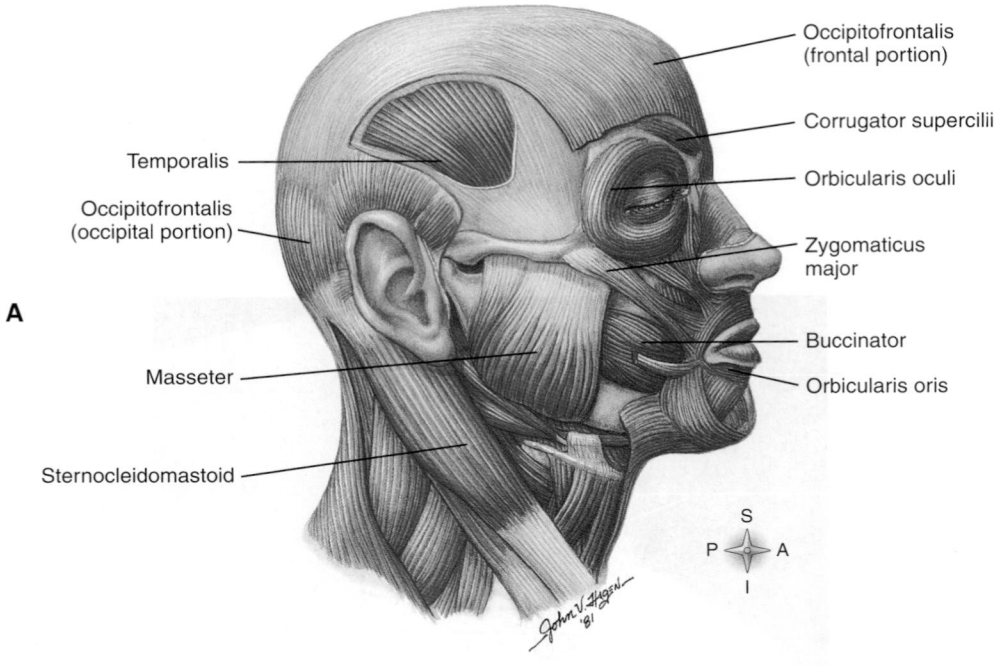

A

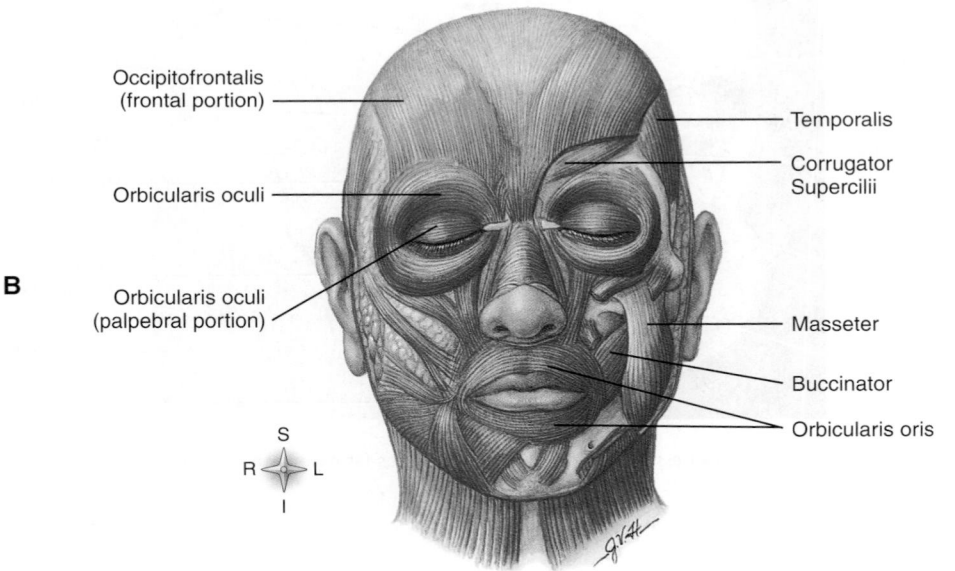

B

Figure 10-8 *Muscles of facial expression and mastication.* **A,** Lateral view. **B,** Anterior view.

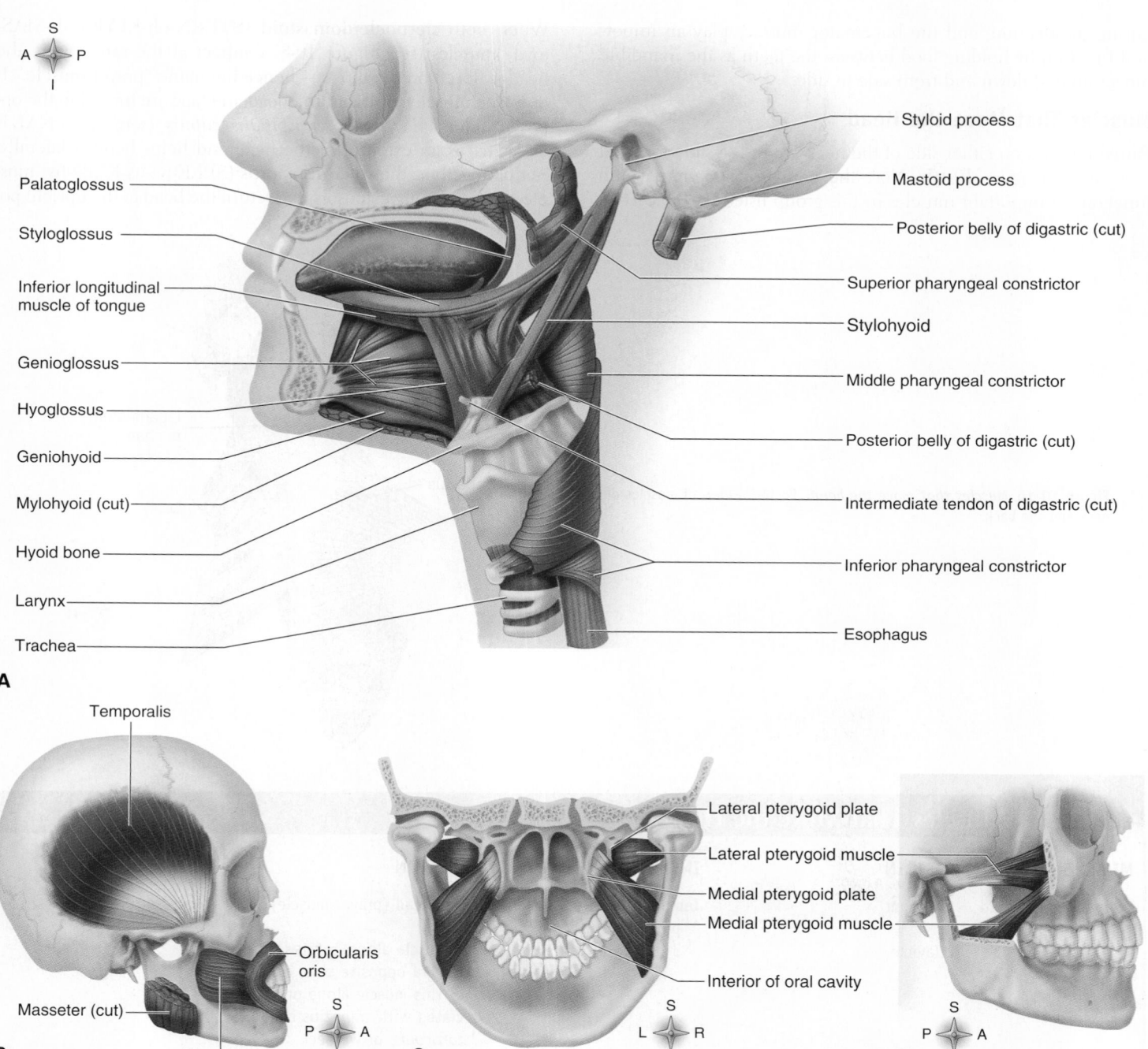

Figure 10-9 *Muscles of mastication.* A, Muscles of the tongue and pharynx. **B,** Right lateral dissection view showing the insertion of the temporalis muscle on the mandible—the masseter muscle is cut and part of the zygomatic arch has been removed. **C,** Lateral and medial pterygoid muscles viewed from the right side after removal of the zygomatic arch. **D,** View of the pterygoids in a posterior dissection view.

during mastication, and the buccinator muscles play an important function by holding food between the teeth as the mandible moves up and down and from side to side.

Muscles That Move the Head

Paired muscles on either side of the neck are responsible for head movements (Figure 10-10). Note the points of attachment and functions of important muscles in this group listed in Table 10-7.

When both **sternocleidomastoid** (STERN-oh-KLYE-doh-MAS-toyd) muscles (see Figure 10-8) contract at the same time, the head is flexed on the thorax—hence the name "prayer muscle." If only one muscle contracts, the head and face are turned to the opposite side. The broad *semispinalis capitis* (sem-ee-spi-NAL-is KAP-i-tis) is an extensor of the head and helps bend it laterally. Acting together, the **splenius capitis** (SPLE-ne-us KAP-i-tis) muscles serve as strong extensors that return the head to the upright po-

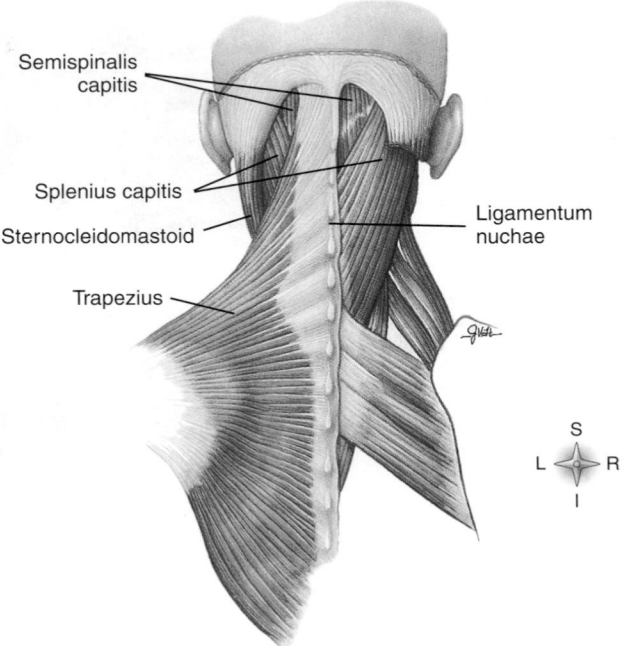

Figure 10-10 *Muscles that move the head.* Posterior view of muscles of the neck and the back.

Table 10-7	Muscles That Move the Head			
MUSCLE	**ORIGIN**	**INSERTION**	**FUNCTION**	**NERVE SUPPLY**
Sternocleidomastoid	Sternum	Temporal bone (mastoid process)	Flexes head (prayer muscle)	Accessory nerve
	Clavicle		One muscle alone, rotates head toward opposite side; spasm of this muscle alone or associated with trapezius called *torticollis,* or wryneck	
Semispinalis capitis	Vertebrae (transverse processes of upper six thoracic, articular processes of lower four cervical)	Occipital bone (between superior and inferior nuchal lines)	Extends head; bends it laterally	First five cervical nerves
Splenius capitis	Ligamentum nuchae	Temporal bone (mastoid process)	Extends head	Second, third, and fourth cervical nerves
	Vertebrae (spinous processes of upper three or four thoracic)	Occipital bone	Bends and rotates head toward same side as contracting muscle	
Longissimus capitis	Vertebrae (transverse processes of upper six thoracic, articular processes of lower four cervical)	Temporal bone (mastoid process)	Extends head Bends and rotates head toward contracting side	Multiple innervation

sition after flexion. When either muscle acts alone, contraction results in rotation and tilting toward that side. The bandlike **longissimus capitis** (lon-JIS-i-mus KAP-i-tis) muscles are covered and not visible in Figure 10-10. They run from the neck vertebrae to the mastoid process of the temporal bone on either side and cause extension of the head when acting together. One contracting muscle will bend and rotate the head toward the contracting side.

QUICK CHECK

12. What is meant by the terms *origin* and *insertion*?
13. Which muscle of facial expression has two parts, one lying over the forehead and the other covering the back of the skull?
14. What group of muscles provides chewing movements?
15. What is the action of the sternocleidomastoid muscle?

TRUNK MUSCLES
Muscles of the Thorax

The muscles of the thorax are of critical importance in respiration (discussed in Chapter 24). Note in Figure 10-11 and Table 10-8 that the **internal** and **external intercostal** (IN-ter-KOS-tal) **muscles** attach to the ribs at different places and their fibers are oriented in different directions. As a result, contraction of the external intercostals elevates and contraction of the internal intercostals depresses the ribs—important in the breathing process. During inspiration the dome-shaped **diaphragm** (DYE-ah-fram) flattens, thus increasing the size and volume of the thoracic cavity. As a result, air enters the lungs.

Muscles of the Abdominal Wall

The muscles of the anterior and lateral abdominal wall (Figures 10-12 and 10-13; Table 10-9) are arranged in three layers, with the fibers in each layer running in different directions much like the layers of wood

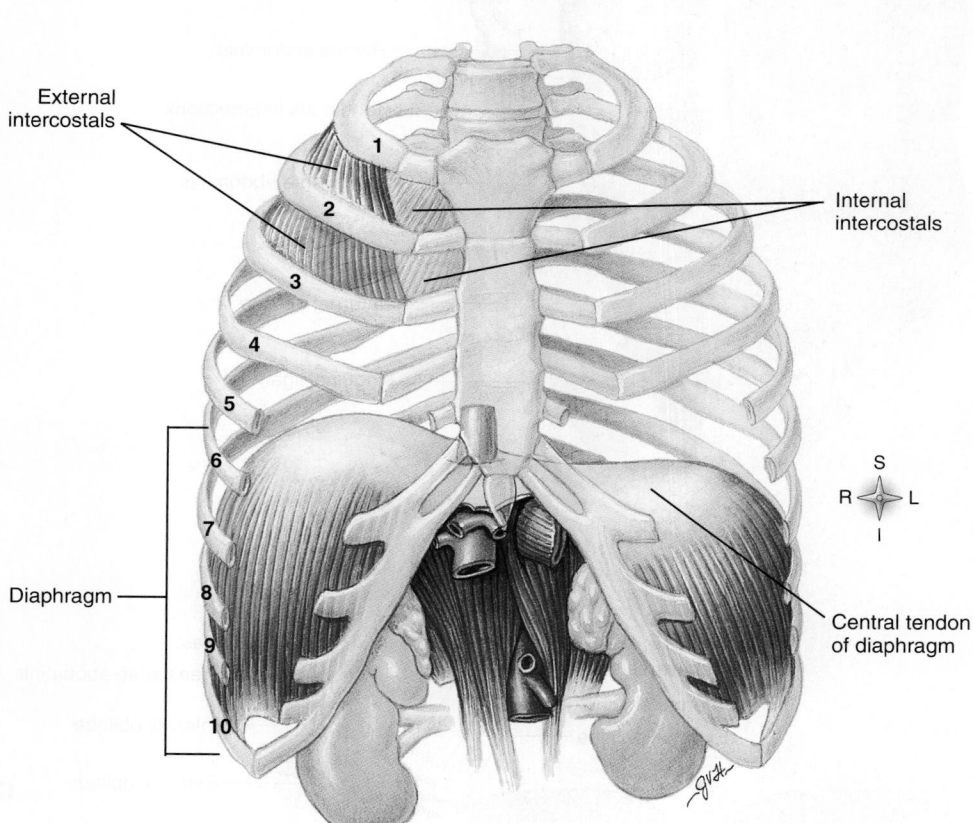

Figure 10-11 *Muscles of the thorax.* Anterior view. Note the relationship of the internal and external intercostal muscles and placement of the diaphragm.

Table 10-8	Muscles of the Thorax			
MUSCLE	**ORIGIN**	**INSERTION**	**FUNCTION**	**NERVE SUPPLY**
External intercostals	Rib (lower border; forward fibers)	Rib (upper border of rib below origin)	Elevate ribs	Intercostal nerves
Internal intercostals	Rib (inner surface, lower border; backward fibers)	Rib (upper border of rib below origin)	Depress ribs	Intercostal nerves
Diaphragm	Lower circumference of thorax (of rib cage)	Central tendon of diaphragm	Enlarges thorax, thereby causing inspiration	Phrenic nerves

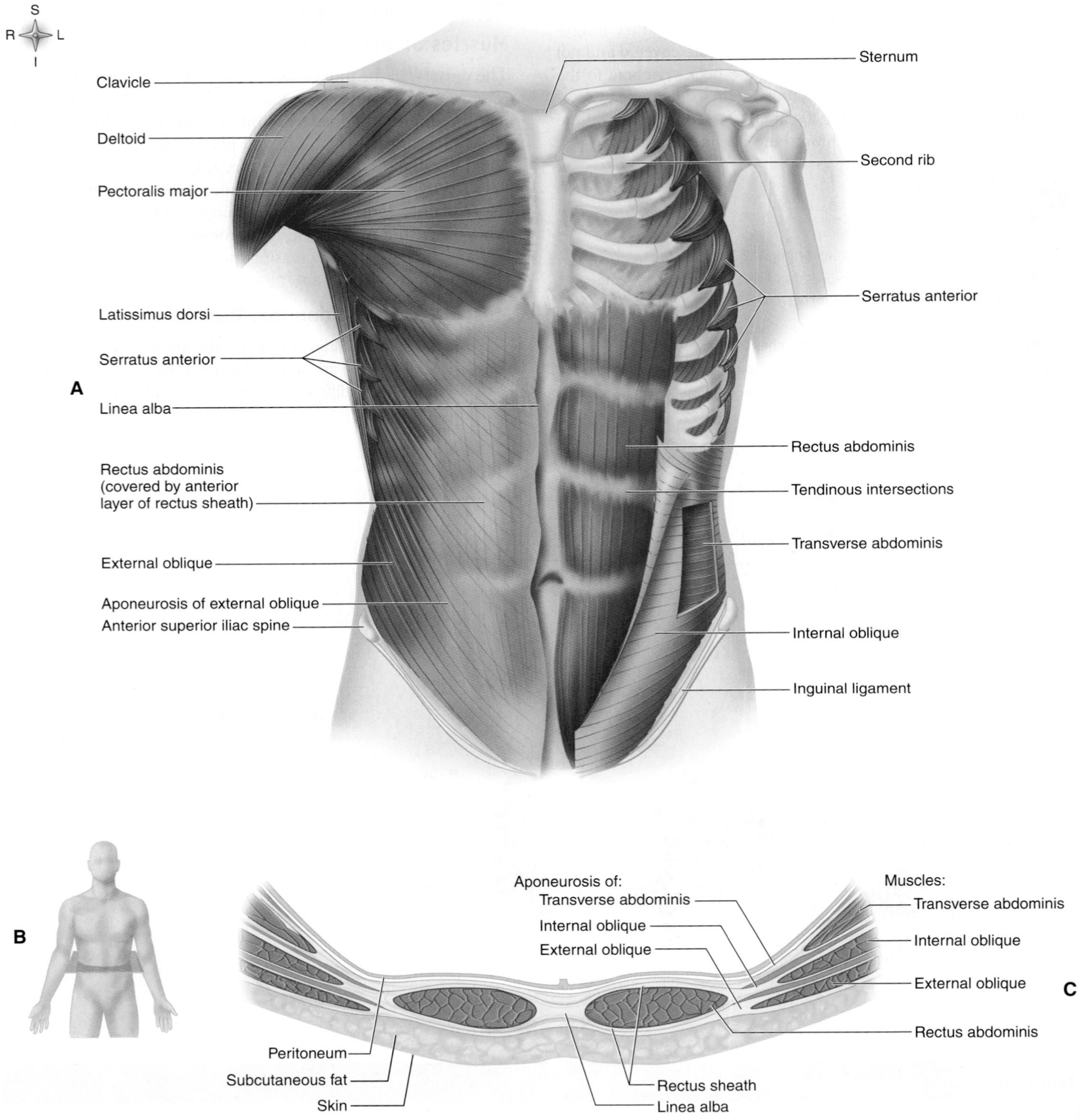

S
R ✦ L
I

Clavicle

Deltoid

Pectoralis major

Latissimus dorsi

Serratus anterior

A

Linea alba

Rectus abdominis
(covered by anterior
layer of rectus sheath)

External oblique

Aponeurosis of external oblique

Anterior superior iliac spine

Sternum

Second rib

Serratus anterior

Rectus abdominis

Tendinous intersections

Transverse abdominis

Internal oblique

Inguinal ligament

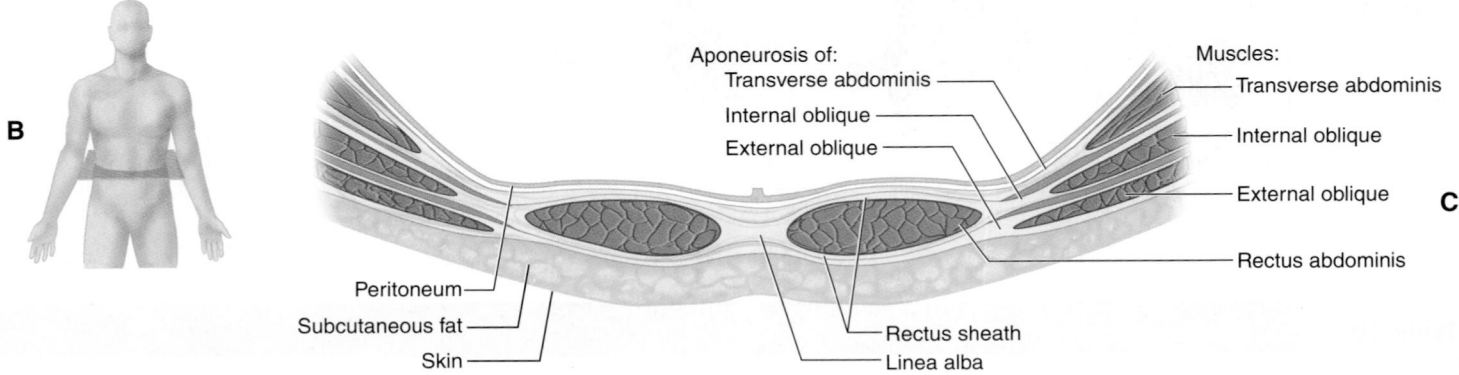

B

Aponeurosis of:
Transverse abdominis
Internal oblique
External oblique

Muscles:
Transverse abdominis
Internal oblique
External oblique

C

Rectus abdominis

Peritoneum

Subcutaneous fat

Skin

Rectus sheath

Linea alba

Figure 10-12 *Muscles of the trunk and abdominal wall.* A, Superficial view. **B,** Deep view. Note that a number of superficial structures have been cut or partially removed. See text for details. **C,** Transverse section of the anterior wall above the umbilicus. See text for details.

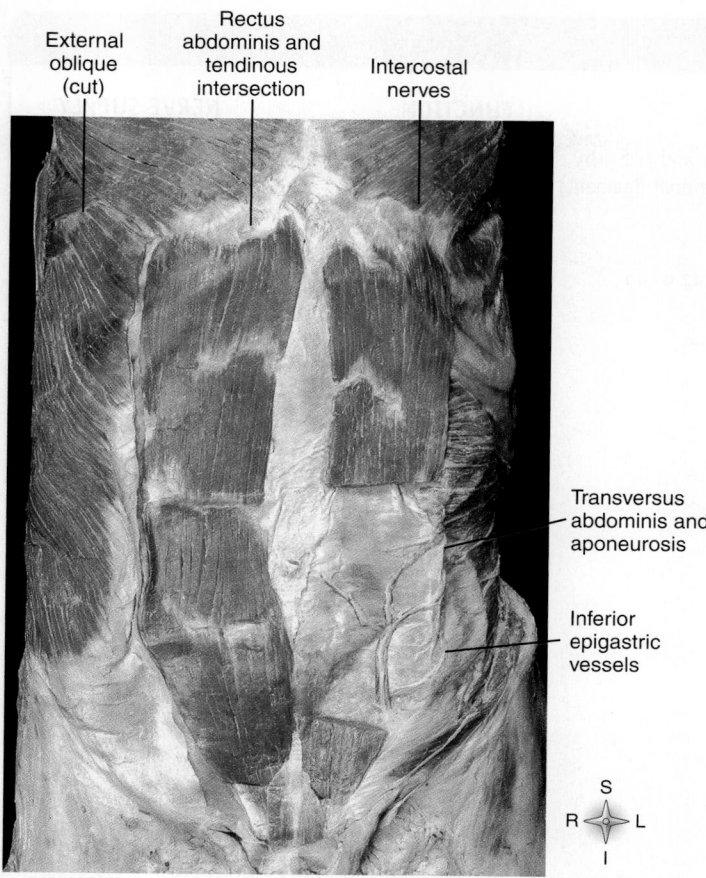

External oblique (cut)

Rectus abdominis and tendinous intersection

Intercostal nerves

Transversus abdominis and aponeurosis

Inferior epigastric vessels

S
R ◆ L
I

Figure 10-13 *Dissection photo of the abdominal wall.* The covering rectus sheaths, the left oblique muscles, and part of the left rectus have been removed.

in a sheet of plywood. The result is a very strong "girdle" of muscle that covers and supports the abdominal cavity and its internal organs.

The fibers in the three layers of muscle in the anterolateral wall are arranged to provide maximum strength. In the **external oblique** the muscle fascicles or fibers extend inferiorly and medially, whereas the fibers of the middle muscle layer, the **internal oblique,** run almost at right angles to those of the external oblique above it. The fibers of the **transversus abdominis,** the innermost muscle layer, are, as the name implies, directed transversely. In addition to these sheetlike muscles, the band- or strap-shaped rectus abdominis muscle runs down the midline of the abdomen from the thorax to the pubis. Note in Figure 10-12 that its parallel running fibers are "interrupted" by three tendinous intersections. When a surgeon "opens" or "closes" an incision through the anterolateral wall, attempts are made to maintain the inherent strength of the wall after surgery by sparing important nerves and blood vessels and by using suturing techniques during closure that will restore the direction of fibers in the cut layers of muscle.

In Figure 10-12, *B,* a number of superficial structures have been cut or partially removed. For example, part of the internal oblique has been removed to expose the underlying transversus abdominis, and the anterior (superficial) layer of the **rectus sheath** has been removed to better visualize the rectus abdominis muscle and its tendinous intersections. Note also in Figure 10-12, *C,* that the aponeuroses of the external oblique, internal oblique, and transversus abdominis muscles form the rectus sheaths that cover the rectus abdominis muscles and then fuse in the midline to form a tough band

of connective tissue called the **linea alba** (white line), that extends from the xiphoid process to the pubis. Occasionally, a surgical procedure will permit an incision through the linea alba rather than through abdominal musculature. Since the linea alba is essentially avascular in some areas, blood loss in such procedures is generally less than what may occur in other approaches.

Working as a group, the abdominal muscles not only protect and hold the abdominal viscera in place, they are responsible for a number of vertebral column movements, including flexion, lateral bending, and some rotation. These important muscles are also involved in respiration and in helping "push" a baby through the birth canal during delivery. They also play a role in assisting in urination, defecation, and vomiting.

Muscles of the Back

Considering the large number of us who suffer from back pain, strain, and injury either occasionally or chronically, you can imagine the importance of the back muscles to health and fitness. Statistics show that 80% of the population in most areas of the world experience backache at some time in their lives. Although symptoms may vary from mild to disabling, "back problems" continue to plague large numbers of individuals and pose significant financial, social, and health care problems. The superficial back muscles play a major role in moving the head and limbs and are illustrated in Figure 10-14, *A.* The deep back muscles (Figure 10-14, *B*) not only allow us to move our vertebral column, thus helping us bend this way and that, but also stabilize our trunk so that

Table 10-9	Muscles of the Abdominal Wall			
MUSCLE	**ORIGIN**	**INSERTION**	**FUNCTION**	**NERVE SUPPLY**
External oblique	Ribs (lower eight)	Pelvis (iliac crest and pubis by way of the inguinal ligament)	Compresses abdomen	Lower seven intercostal nerves and iliohypogastric nerves
		Linea alba by way of an aponeurosis	Rotates trunk laterally	
Internal oblique	Pelvis (iliac crest and inguinal ligament)	Ribs (lower three)		
	Lumbodorsal fascia	Linea alba	Important postural function of all abdominal muscles is to pull the front of the pelvis upward, thereby flattening the lumbar curve of the spine; when these muscles lose their tone, common figure faults of protruding abdomen and lordosis develop	
Transversus abdominis	Ribs (lower six)	Pubic bone	Same as external oblique	Last three intercostal nerves; iliohypogastric and ilioinguinal nerves
	Pelvis (iliac crest, inguinal ligament)	Linea alba		
	Lumbodorsal fascia	Ribs (costal cartilage of fifth, sixth, and seventh ribs)	Same as external oblique	Last five intercostal nerves; iliohypogastric and ilioinguinal nerves
Rectus abdominis	Pelvis (pubic bone and symphysis pubis)	Sternum (xiphoid process)	Same as external oblique; because abdominal muscles compress the abdominal cavity, they aid in straining, defecation, forced expiration, childbirth, etc.; abdominal muscles are antagonists of the diaphragm, relaxing as it contracts and vice versa	Last six intercostal nerves
			Flexes trunk	
Quadratus lumborum	Iliolumbar ligament; iliac crest	Last rib; transverse process of vertebrae (L1-L4)	Flexes vertebral column laterally; depresses last rib	Lumbar

A

Superficial muscles   Intermediate muscles

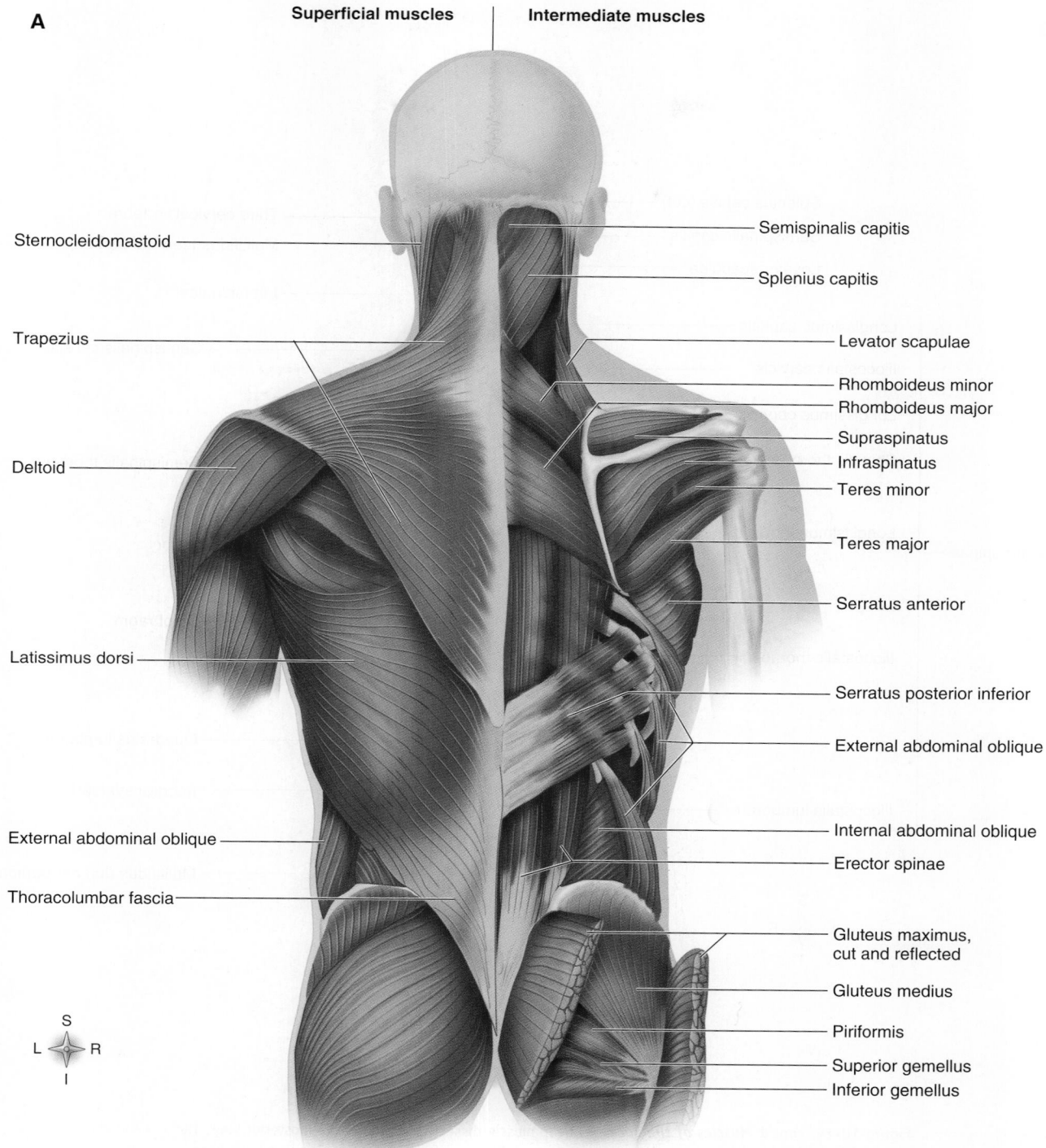

Sternocleidomastoid

Trapezius

Deltoid

Latissimus dorsi

External abdominal oblique

Thoracolumbar fascia

Semispinalis capitis

Splenius capitis

Levator scapulae

Rhomboideus minor
Rhomboideus major

Supraspinatus
Infraspinatus
Teres minor

Teres major

Serratus anterior

Serratus posterior inferior

External abdominal oblique

Internal abdominal oblique

Erector spinae

Gluteus maximus,
cut and reflected

Gluteus medius

Piriformis

Superior gemellus
Inferior gemellus

S
L —✦— R
I

Figure 10-14 *Muscles of the back.* A, Superficial *(left)* and intermediate *(right)* muscle dissection of the back—posterior view. The illustration shows a two-stage dissection. Superficial muscles of the neck and back are shown on the left side and an intermediate-depth dissection is shown on the right.

Continued

B

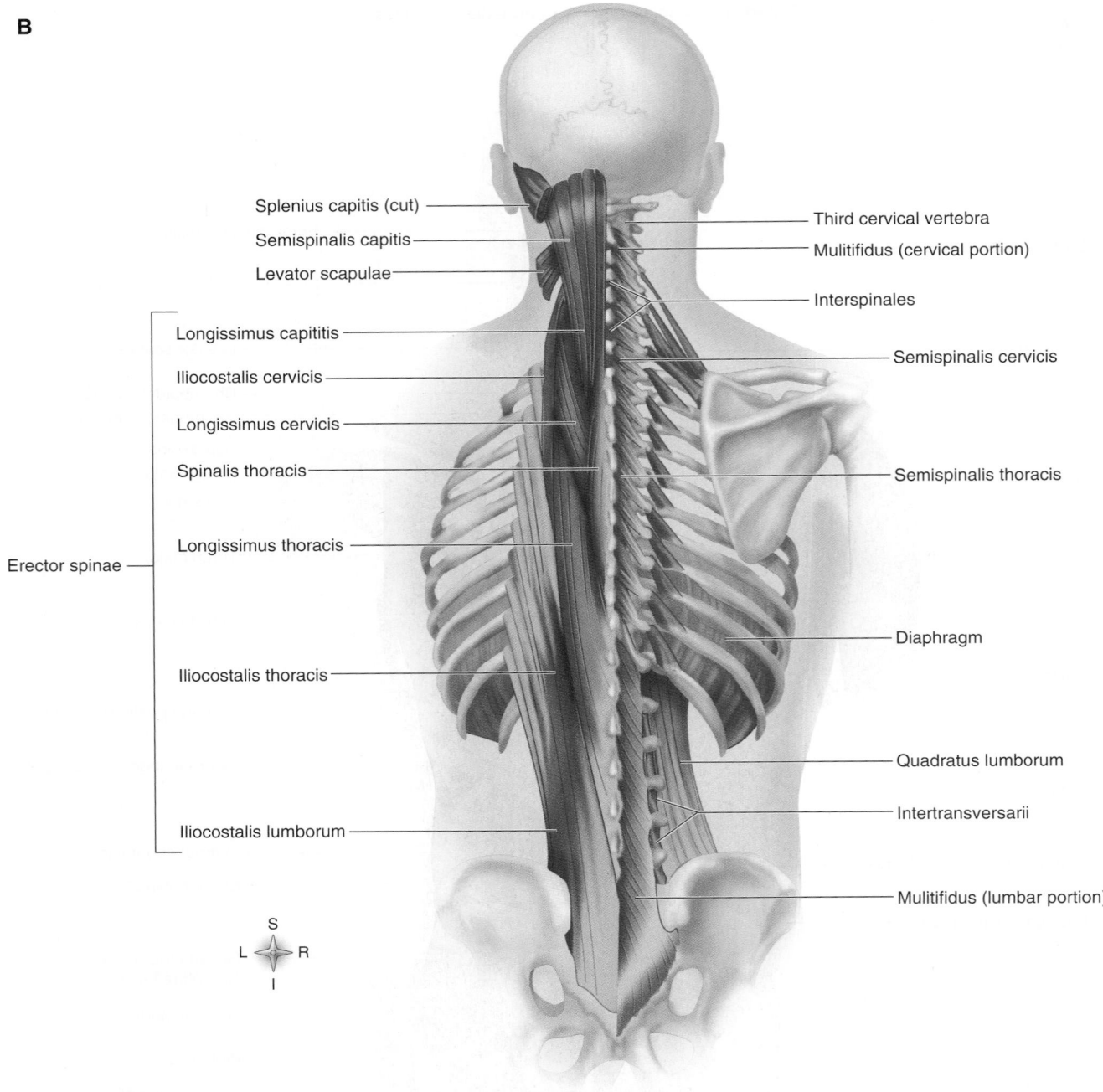

Splenius capitis (cut)

Semispinalis capitis

Levator scapulae

Longissimus capititis

Iliocostalis cervicis

Longissimus cervicis

Spinalis thoracis

Longissimus thoracis

Erector spinae

Iliocostalis thoracis

Iliocostalis lumborum

Third cervical vertebra

Mulitifidus (cervical portion)

Interspinales

Semispinalis cervicis

Semispinalis thoracis

Diaphragm

Quadratus lumborum

Intertransversarii

Mulitifidus (lumbar portion)

S
L ⬥ R
I

Figure 10-14, cont'd *Muscles of the back.* **B,** Deep muscle dissection of the back—posterior view. The superficial and intermediate muscles have been removed. The muscles in the gluteal region have been removed to expose the pelvic insertion of the multifidus.

we can maintain a stable posture. These muscles really get a workout when we lift something heavy because they have to hold the body straight while the load is trying to bend the back.

The **erector spinae** group consists of a number of long, thin muscles that travel all the way down our backs (Figure 10-14). These muscles extend (straighten or pull back) the vertebral column and also flex the back laterally and rotate it a little. Even deeper than the erector spinae muscles are several additional back muscles. The **interspinales** and **multifides groups,** for example, each connect one vertebra to the next; they also help extend the back and neck or flex them to the side. Table 10-10 and Figure 10-14 summarize some of the important deep back muscles.

Muscles of the Pelvic Floor

Structures in the pelvic cavity are supported by a reinforced muscular floor that guards the outlet below. The muscular pelvic floor filling the diamond-shaped outlet is called the **perineum** (pair-ih-NEE-um). Passing through the floor are the anal canal and urethra in both sexes and the vagina in the female.

The two **levator ani** and **coccygeus** muscles form most of the pelvic floor. They stretch across the pelvic cavity like a hammock. This diamond-shaped outlet can be divided into two triangles by a line drawn from side to side between the ischial tuberosities. The **urogenital triangle** is anterior (above) to this line and extends to the symphysis pubis, and the **anal triangle** is posterior (behind it) and ends at the coccyx. Note in Figure 10-15 that structures in the urogenital triangle include the **ischiocavernosus** and **bulbospongiosus** muscles associated with the penis in the

male and the vagina in the female. Constriction of muscles called the **sphincter urethrae,** which encircle the urethra in both sexes, helps control urine flow. The anal triangle allows passage of the anal canal. The terminal portion of the canal is surrounded by the **external anal sphincter,** which regulates defecation. The origin, insertion, function, and innervation of important muscles of the pelvic floor are listed in Table 10-11. The coccygeus muscles lie behind the levator ani and are not visible in Figure 10-15.

QUICK CHECK

16. Name the skeletal muscles that produce respiratory movements.
17. Name two functions of the rectus abdominis muscle.
18. What is the perineum?

UPPER LIMB MUSCLES

The muscles of the upper limb include those acting on the shoulder or pectoral girdle and muscles located in the arm, forearm, and hand.

Muscles Acting on the Shoulder Girdle

Attachment of the upper extremity to the torso is by muscles that have an anterior location (chest) or posterior placement (back and neck). Six muscles (Table 10-12; Figure 10-16) that pass from the axial skeleton to the shoulder or pectoral girdle (scapula and

Table 10-10	Muscles of the Back			
MUSCLE	**ORIGIN**	**INSERTION**	**FUNCTION**	**NERVE SUPPLY**
Erector spinae group				
Iliocostalis group	Various regions of the pelvis and ribs	Ribs and vertebra (superior to the origin)	Extends, laterally flexes the vertebral column	Spinal, thoracic, or lumbar nerves
Longissimus group	Cervical and thoracic vertebrae, ribs	Mastoid process, upper cervical vertebrae, or upper lumbar vertebrae	Extends head, neck, or vertebral column	Cervical or thoracic and lumber nerves
Spinalis group	Lower cervical or lower thoracic/upper lumbar vertebrae	Upper cervical or middle/upper thoracic vertebrae (superior to the origin)	Extends the neck or vertebral column	Cervical or thoracic nerves
Transversospinalis group				
Semispinalis group	Transverse processes of vertebrae (T2-T11)	Spinous processes of vertebrae (C2-T4)	Extends neck or vertebral column	Cervical or thoracic nerves
Multifidus group	Transverse processes of vertebrae; sacrum and ilium	Spinous processes of (next superior) vertebrae	Extends, rotates vertebral column	Spinal nerves
Rotatores group	Transverse processes of vertebrae	Spinous processes of (next superior) vertebrae	Extends, rotates vertebral column	Spinal nerves
Splenius	Spinous processes of vertebrae (C7-T1 or T3-T6)	Lateral occipital/mastoid or transverse processes of vertebrae (C1-C4)	Rotates, extends neck and flexes neck laterally	Cervical nerves
Interspinales group	Spinous processes of vertebrae	Spinous processes (of next superior vertebra)	Extends back and neck	Spinal nerves

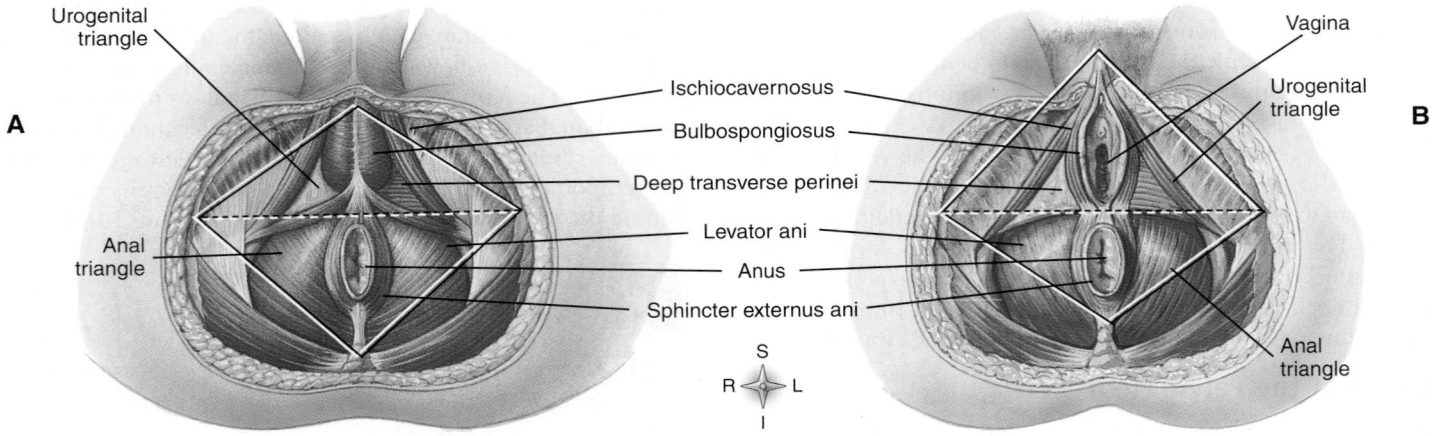

Figure 10-15 *Muscles of the pelvic floor.* **A,** Male, inferior view. **B,** Female, inferior view.

Table 10-11	Muscles of the Pelvic Floor			
MUSCLE	**ORIGIN**	**INSERTION**	**FUNCTION**	**NERVE SUPPLY**
Levator ani	Pubis and spine of the ischium	Coccyx	Together with the coccygeus muscles form the floor of the pelvic cavity and support the pelvic organs	Pudendal nerve
Ischiocavernosus	Ischium	Penis or clitoris	Compress the base of the penis or clitoris	Perineal nerve
Bulbospongiosus				
Male	Bulb of the penis	Perineum and bulb of the penis	Constricts the urethra and erects the penis	Pudendal nerve
Female	Perineum	Base of the clitoris	Erects the clitoris	Pudendal nerve
Deep transverse perinei	Ischium	Central tendon (median raphe)	Support the pelvic floor	Pudendal nerve
Sphincter urethrae	Pubic ramus	Central tendon (median raphe)	Constricts the urethra	Pudendal nerve
Sphincter ani externus	Coccyx	Central tendon (median raphe)	Closes the anal canal	Pudendal and S4

clavicle) serve to not only "attach" the upper extremity to the body but do so in such a way that extensive movement is possible. The clavicle can be elevated and depressed and moved forward and back. The scapula is capable of an even greater variety of movements.

The **pectoralis** (pek-toh-RAL-is) **minor** lies under the larger pectoralis major muscle on the anterior chest wall. It helps "fix" the scapula against the thorax and also raises the ribs during forced inspiration. Another anterior chest wall muscle—the **serratus** (ser-RAY-tus) **anterior**—helps hold the scapula against the thorax to prevent "winging" and is a strong abductor that is useful in pushing or punching movements.

The posterior muscles acting on the shoulder girdle include the **levator scapulae** (leh-VAY-tor SCAP-yoo-lee), which elevates the scapula; the **trapezius** (trah-PEE-zee-us), which is used to "shrug" the shoulders; and the **rhomboideus** (rom-BOYD-ee-us) **major** and **minor** muscles, which serve to adduct and elevate the scapula.

Muscles That Move the Upper arm

The shoulder is a synovial joint of the ball-and-socket type. As a result, extensive movement is possible in every plane of motion. Muscles that move the upper part of the arm can be grouped according to function as flexors, extensors, abductors, adductors, and medial and lateral rotators (Table 10-13; Figure 10-17). The actions listed in Table 10-13 include primary actions and important secondary functions.

The **deltoid** (DEL-toyd) is a good example of a multifunction muscle. It has three groups of fibers and may act as three separate muscles. Contraction of the anterior fibers will flex the arm, whereas the lateral fibers abduct and the posterior fibers serve as extensors. Four other muscles serve as both a structural and functional cap or cuff around the shoulder joint and are referred to as the **rotator cuff muscles** (Figure 10-18). They include the **infraspinatus, supraspinatus, subscapularis,** and **teres minor.**

Table 10-12 | Muscles Acting on the Shoulder Girdle

MUSCLE	ORIGIN	INSERTION	FUNCTION	NERVE SUPPLY
Trapezius	Occipital bone (protuberance)	Clavicle	Raises or lowers the shoulders and shrugs them	Spinal accessory; second, third, and fourth cervical nerves
	Vertebrae (cervical and thoracic)	Scapula (spine and acromion)	Extends the head when the occiput acts as the insertion	
Pectoralis minor	Ribs (second to fifth)	Scapula (coracoid)	Pulls the shoulder down and forward	Medial and lateral anterior thoracic nerves
Serratus anterior	Ribs (upper eight or nine)	Scapula (anterior surface, vertebral border)	Pulls the shoulder down and forward; abducts and rotates it upward	Long thoracic nerve
Levator scapulae	C1-C4 (transverse processes)	Scapula (superior angle)	Elevates and retracts the scapula and abducts the neck	Dorsal scapular nerve
Rhomboid				
Major	T1-T4	Scapula (medial border)	Retracts, rotates, and fixes the scapula	Dorsal scapular nerve
Minor	C6-C7	Scapula (medial border)	Retracts, rotates, elevates, and fixes the scapula	Dorsal scapular nerve

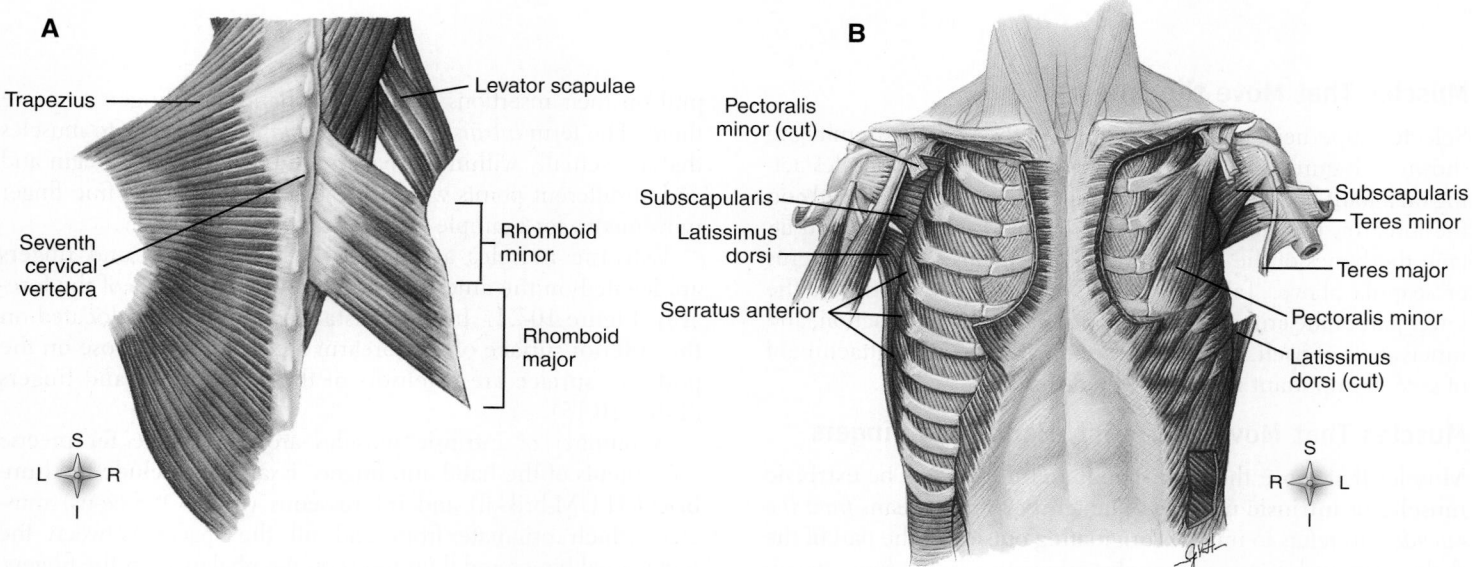

Figure 10-16 *Muscles acting on the shoulder girdle.* **A,** Posterior view. The trapezius has been removed on the right to reveal the deeper muscles. **B,** Anterior view. The pectoralis major has been removed on both sides. The pectoralis minor has also been removed on the right side.

Table 10-13 Muscles That Move the Upper Arm

MUSCLE	ORIGIN	INSERTION	FUNCTION	NERVE SUPPLY
*Axial**				
Pectoralis major	Clavicle (medial half) Sternum Costal cartilages of the true ribs	Humerus (greater tubercle)	Flexes the upper arm Adducts the upper arm anteriorly; draws it across the chest	Medial and lateral anterior thoracic nerves
Latissimus dorsi	Vertebrae (spines of the lower thoracic, lumbar, and sacral) Ilium (crest) Lumbodorsal fascia	Humerus (intertubercular groove)	Extends the upper arm Adducts the upper arm posteriorly	Thoracodorsal nerve
*Scapular**				
Deltoid	Clavicle Scapula (spine and acromion)	Humerus (lateral side about half-way down—deltoid tubercle)	Abducts the upper arm Assists in flexion and extension of the upper arm	Axillary nerve
Coracobrachialis	Scapula (coracoid process)	Humerus (middle third, medial surface)	Adduction; assists in flexion and medial rotation of the arm	Musculocutaneous nerve
Supraspinatus†	Scapula (supraspinous fossa)	Humerus (greater tubercle)	Assists in abducting the arm	Suprascapular nerve
Teres minor†	Scapula (axillary border)	Humerus (greater tubercle)	Rotates the arm outward	Axillary nerve
Teres major	Scapula (lower part, axillary border)	Humerus (upper part, anterior surface)	Assists in extension, adduction, and medial rotation of the arm	Lower subscapular nerve
Infraspinatus†	Scapula (infraspinatus border)	Humerus (greater tubercle)	Rotates the arm outward	Suprascapular nerve
Subscapularis†	Scapula (subscapular fossa)	Humerus (lesser tubercle)	Medial rotation	Suprascapular nerve

*Axial muscles originate on the axial skeleton. Scapular muscles originate on the scapula.
†Muscles of the rotator cuff.

Muscles That Move the Forearm

Selected superficial and deep muscles of the upper extremity are shown in Figures 10-19 and 10-20. Recall that most muscles acting on a joint lie proximal to that joint. Muscles acting directly on the forearm, therefore, are found proximal to the elbow and attach the bones of the forearm (ulna and radius) to the humerus or scapula above. Table 10-14 lists the muscles acting on the lower part of the arm and gives the origin, insertion, function, and innervation of each. Figure 10-21 shows the detail of attachment of several important muscles in this group.

Muscles That Move the Wrist, Hand, and Fingers

Muscles that move the wrist, hand, and fingers can be **extrinsic muscles** or **intrinsic muscles.** The term *extrinsic* means *from the outside* and refers to muscles originating outside of the part of the skeleton moved. Extrinsic muscles originating in the forearm can pull on their insertions in the wrist, hand, and fingers to move them. The term *intrinsic*, meaning *from within*, refers to muscles that are actually within the part moved. Muscles that begin and end at different points within the hand can produce fine finger movements, for example.

Extrinsic muscles acting on the wrist, hand, and fingers are located on the anterior or the posterior surfaces of the forearm (Figure 10-22). In most instances, the muscles located on the anterior surface of the forearm are flexors and those on the posterior surface are extensors of the wrist, hand, and fingers (Table 10-15).

A number of intrinsic muscles are responsible for precise movements of the hand and fingers. Examples include the **lumbrical** (LUM-brik-al) and **interosseous** (in-ter-OSS-ee-us) muscles, which originate from and fill the spaces between the metacarpal bones and then insert on the phalanges of the fingers.

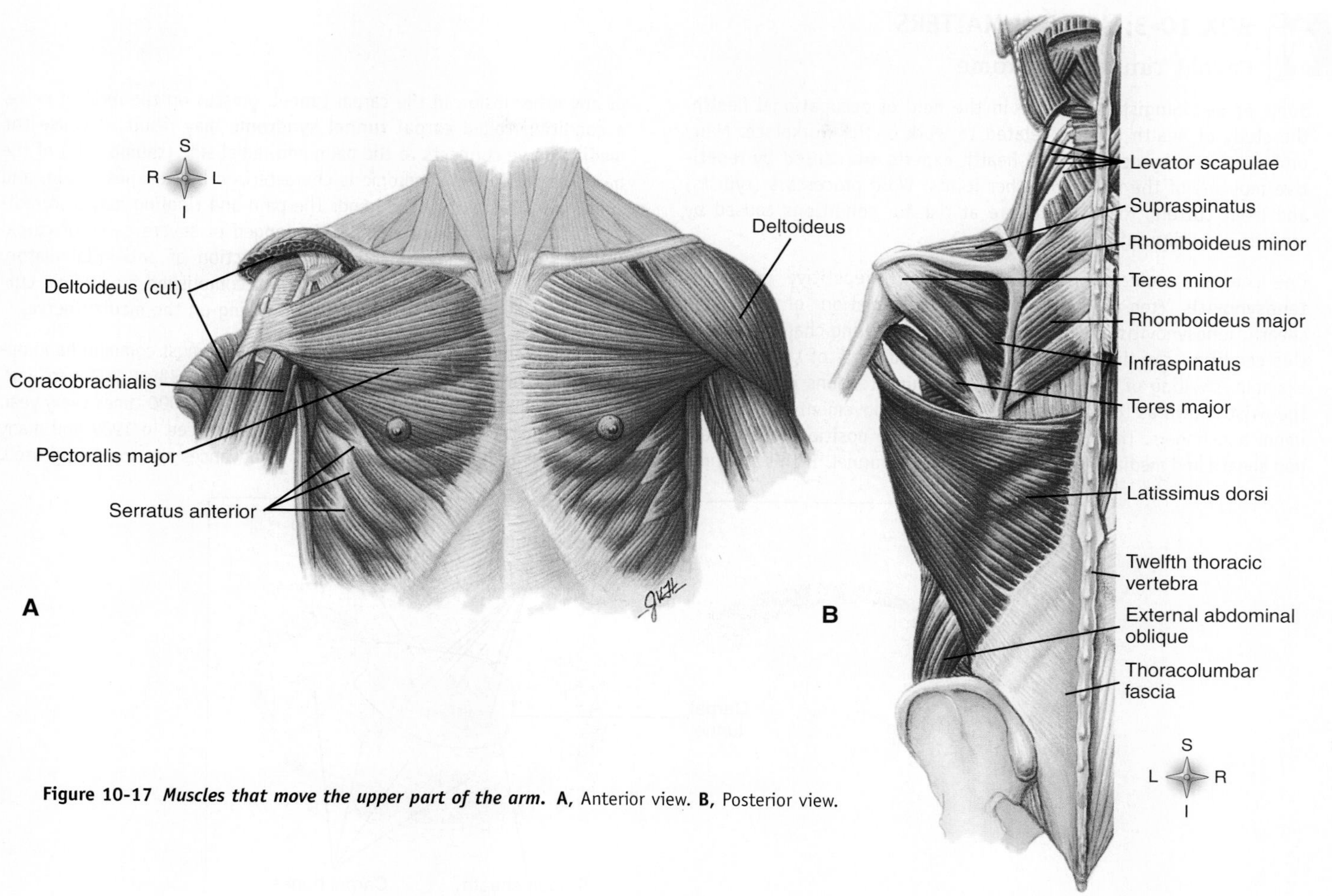

Levator scapulae

Deltoideus

Supraspinatus

Rhomboideus minor

Teres minor

Rhomboideus major

Infraspinatus

Teres major

Latissimus dorsi

Twelfth thoracic vertebra

External abdominal oblique

Thoracolumbar fascia

Deltoideus (cut)

Coracobrachialis

Pectoralis major

Serratus anterior

A

B

Figure 10-17 *Muscles that move the upper part of the arm.* **A,** Anterior view. **B,** Posterior view.

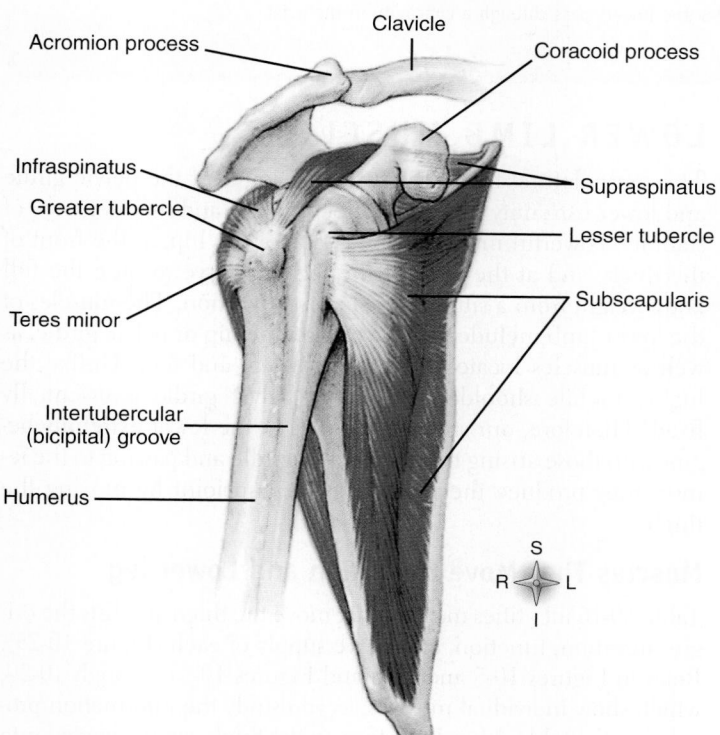

Acromion process

Clavicle

Coracoid process

Infraspinatus

Supraspinatus

Greater tubercle

Lesser tubercle

Teres minor

Subscapularis

Intertubercular (bicipital) groove

Humerus

Figure 10-18 *Rotator cuff muscles.* Note the tendons of the teres minor, infraspinatus, supraspinatus, and subscapularis muscles surrounding the head of the humerus.

BOX 10-3: HEALTH MATTERS
Carpal Tunnel Syndrome

Some epidemiologists specialize in the field of occupational health, the study of health matters related to work or the workplace. Many problems seen by occupational health experts are caused by repetitive motions of the wrists or other joints. Word processors (typists) and meat cutters, for example, are at risk for conditions caused by repetitive motion injuries.

One common problem often caused by such repetitive motion is **tenosynovitis** (ten-oh-sin-oh-VYE-tis)—inflammation of a tendon sheath. Tenosynovitis can be painful, and the swelling characteristic of this condition can limit movement in affected parts of the body. For example, swelling of the tendon sheath around tendons in an area of the wrist known as the *carpal tunnel* can limit movement of the wrist, hand, and fingers. The figure shows the relative positions of the tendon sheath and median nerve within the carpal tunnel. If this swelling,

or any other lesion in the carpal tunnel, presses on the *median nerve*, a condition called **carpal tunnel syndrome** may result. Because the median nerve connects to the palm and radial side (thumb side) of the hand, carpal tunnel syndrome is characterized by weakness, pain, and tingling in this part of the hand. The pain and tingling may also radiate to the forearm and shoulder. Prolonged or severe cases of carpal tunnel syndrome may be relieved by injection of anti-inflammatory agents. A permanent cure is sometimes accomplished by surgical cutting or removal of the swollen tissue pressing on the median nerve.

The procedure called carpal tunnel release is the most common hand operation in the United States. First introduced in 1933 as an "open" surgical procedure, it is now performed more than 200,000 times every year. A less invasive endoscopic approach was introduced in 1989 and many other innovative surgical techniques and advances are now being used.

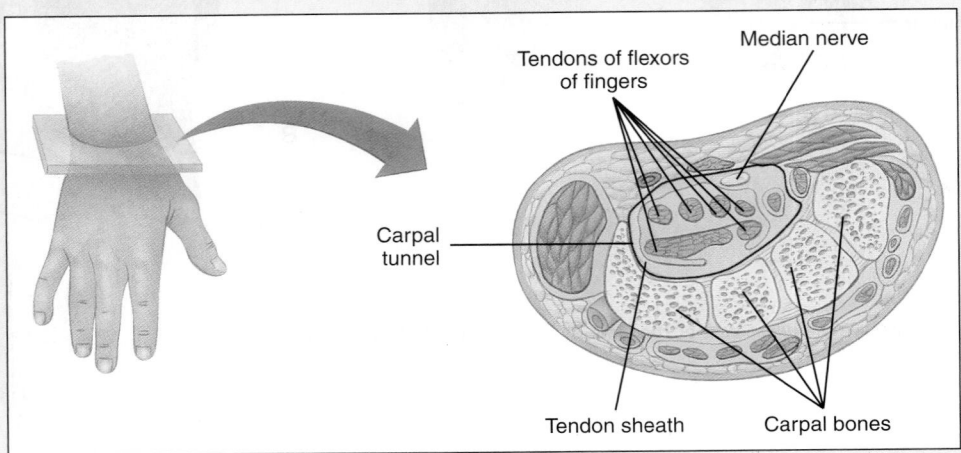

The carpal tunnel. The median nerve and muscles that flex the fingers pass through a concavity in the wrist called the carpal tunnel.

As a group, the intrinsic muscles abduct and adduct the fingers and aid in flexing them. Eight additional muscles serve the thumb and enable it to be placed in opposition to the fingers in tasks requiring grasping and manipulation. The **opponens pollicis** (oh-POH-nenz POL-i-cis) is a particularly important thumb muscle. It allows the thumb to be drawn across the palm to touch the tip of any finger—a critical movement for many manipulative-type activities. Figure 10-23 shows the placement and points of attachment for various individual extrinsic muscles acting on the wrist, hand, and fingers. Figure 10-24 provides a detailed illustration of many of the intrinsic muscles of the hand.

QUICK CHECK

19. What are the functions of the deltoid muscle?
20. What is the function of the biceps brachii muscle?
21. Distinguish the extrinsic from the intrinsic muscles of the hand and wrist.

LOWER LIMB MUSCLES

The musculature, bony structure, and joints of the pelvic girdle and lower extremity function in locomotion and maintenance of stability. Powerful muscles at the back of the hip, at the front of the thigh, and at the back of the leg also serve to raise the full body weight from a sitting to a standing position. The muscles of the lower limb include those acting on the hip or pelvic girdle, as well as muscles located in the thigh, leg, and foot. Unlike the highly mobile shoulder girdle, the pelvic girdle is essentially fixed. Therefore, our study of muscles in the lower extremity begins with those arising from the pelvic girdle and passing to the femur; they produce their effects at the hip joint by moving the thigh.

Muscles That Move the Thigh and Lower leg

Table 10-16 identifies muscles that move the thigh and lists the origin, insertion, function, and nerve supply of each (Figure 10-25). Refer to Figures 10-5 and 10-6 and Figures 10-25 through 10-29, which show individual muscles, as you study the information provided in the table. Muscles acting on the thigh can be divided into

Text continued on p. 383

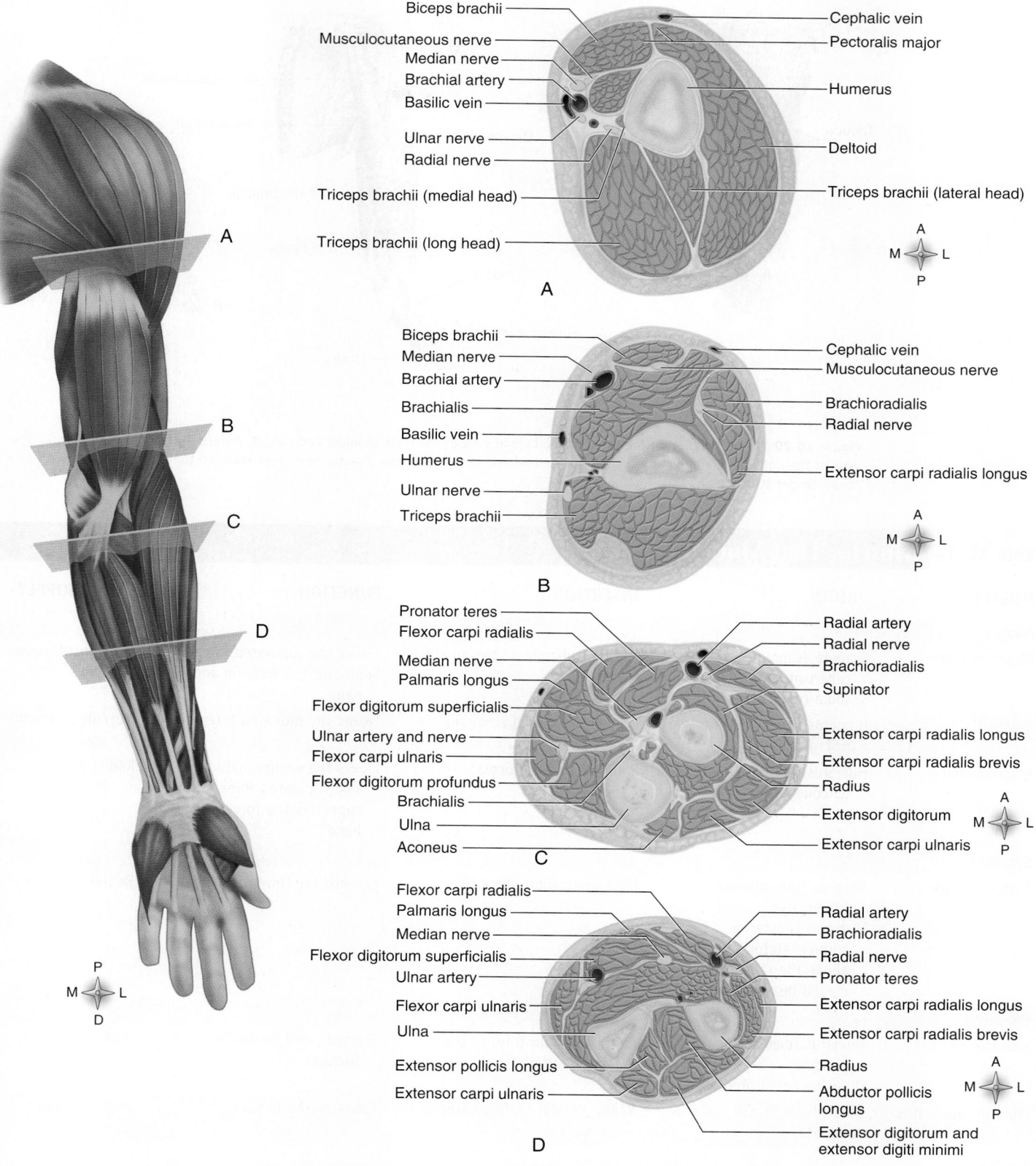

A

Biceps brachii — Cephalic vein
Musculocutaneous nerve — Pectoralis major
Median nerve
Brachial artery
Basilic vein — Humerus
Ulnar nerve
Radial nerve
Triceps brachii (medial head) — Deltoid
Triceps brachii (lateral head)
Triceps brachii (long head)

A
M — L
P

A

B

Biceps brachii — Cephalic vein
Median nerve — Musculocutaneous nerve
Brachial artery
Brachialis — Brachioradialis
Basilic vein — Radial nerve
Humerus — Extensor carpi radialis longus
Ulnar nerve
Triceps brachii

A
M — L
P

B

C

Pronator teres — Radial artery
Flexor carpi radialis — Radial nerve
Median nerve — Brachioradialis
Palmaris longus — Supinator
Flexor digitorum superficialis
Ulnar artery and nerve — Extensor carpi radialis longus
Flexor carpi ulnaris — Extensor carpi radialis brevis
Flexor digitorum profundus — Radius
Brachialis
Ulna — Extensor digitorum
Aconeus — Extensor carpi ulnaris

A
M — L
P

C

D

Flexor carpi radialis
Palmaris longus — Radial artery
Median nerve — Brachioradialis
Flexor digitorum superficialis — Radial nerve
Ulnar artery — Pronator teres
Flexor carpi ulnaris — Extensor carpi radialis longus
Ulna — Extensor carpi radialis brevis
Extensor pollicis longus — Radius
Extensor carpi ulnaris — Abductor pollicis longus
Extensor digitorum and extensor digiti minimi

A
M — L
P

D

P
M — L
D

Figure 10-19 *Cross sections (proximal to distal) through the upper extremity.* **A,** Section at the junction of the proximal and middle thirds of the humerus. **B,** Section just proximal to the medial epicondyle of the humerus. **C,** Section at the level of the radial tuberosity. **D,** Section at the middle of the forearm. In each section you are viewing the superior (proximal) aspect of the specimen.

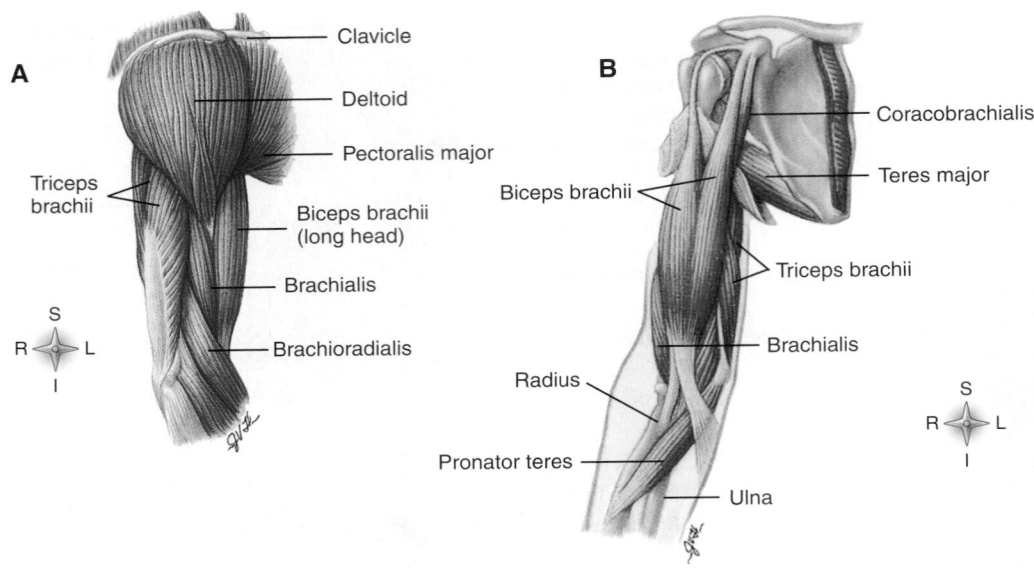

Figure 10-20 *Muscles acting on the forearm.* **A,** Lateral view of the right shoulder and arm. **B,** Anterior view of the right shoulder and arm (deep). The deltoid and pectoralis major muscles have been removed to reveal deeper structures.

Table 10-14 Muscles That Move the Forearm

MUSCLE	ORIGIN	INSERTION	FUNCTION	NERVE SUPPLY
Flexors				
Biceps brachii	Scapula (supraglenoid tuberosity) Scapula (coracoid)	Radius (tubercle at the proximal end)	Flexes the supinated forearm Supinates the forearm and hand	Musculocutaneous nerve
Brachialis	Humerus (distal half, anterior surface)	Ulna (front of the coronoid process)	Flexes the pronated forearm	Musculocutaneous nerve
Brachioradialis	Humerus (above the lateral epicondyle)	Radius (styloid process)	Flexes the semipronated or semisupinated forearm; supinates the forearm and hand	Radial nerve
Extensor				
Triceps brachii	Scapula (infraglenoid tuberosity) Humerus (posterior surface—lateral head above the radial groove; medial head, below)	Ulna (olecranon process)	Extends the lower arm	Radial nerve
Pronators				
Pronator teres	Humerus (medial epicondyle) Ulna (coronoid process)	Radius (middle third of the lateral surface)	Pronates and flexes the forearm	Median nerve
Pronator quadratus	Ulna (distal fourth, anterior surface)	Radius (distal fourth, anterior surface)	Pronates the forearm	Median nerve
Supinator				
Supinator	Humerus (lateral epicondyle) Ulna (proximal fifth)	Radius (proximal third)	Supinates the forearm	Radial nerve

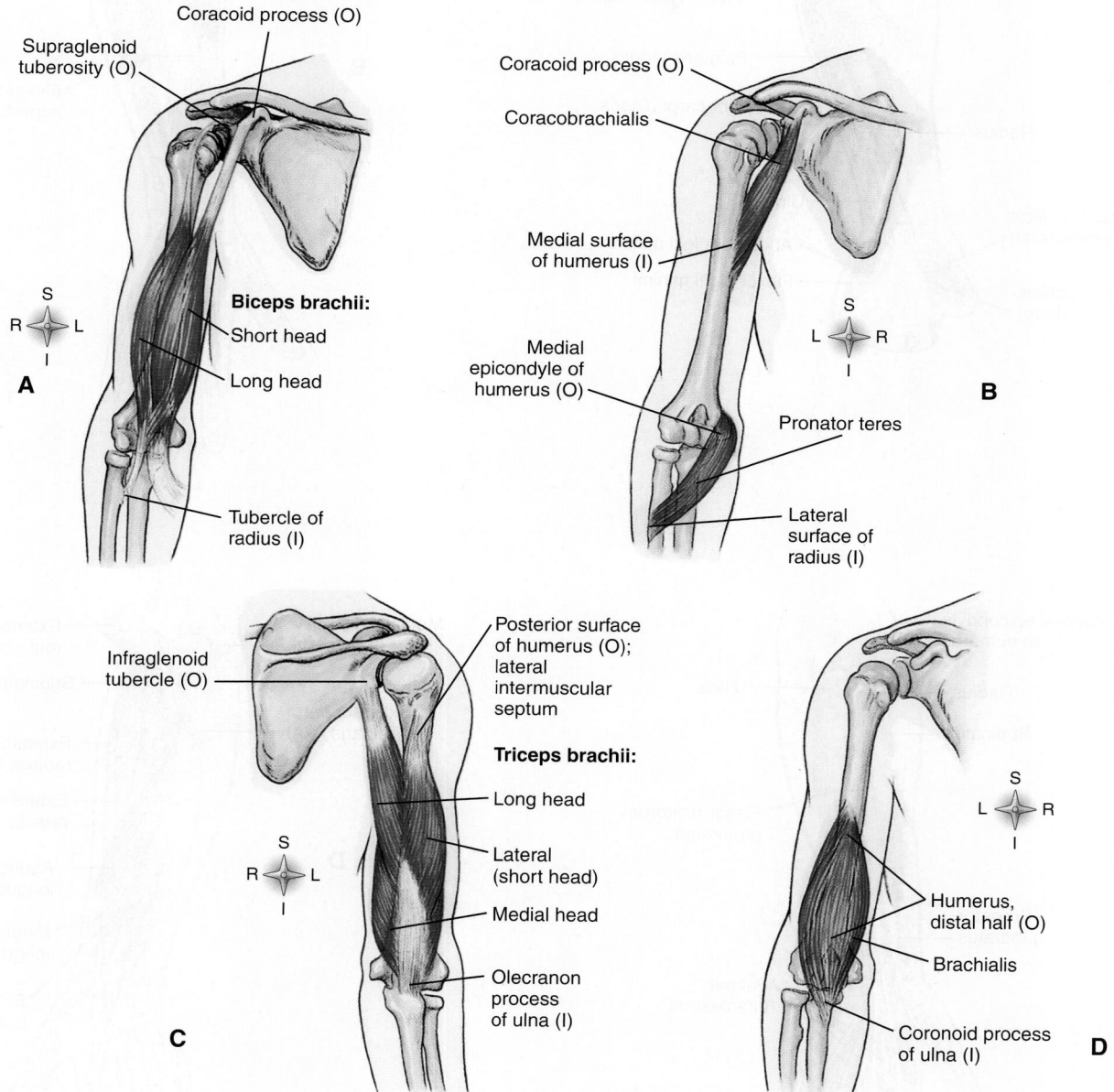

Coracoid process (O)

Supraglenoid
tuberosity (O)

S
R L
I

Biceps brachii:

Short head

Long head

A

Tubercle of
radius (I)

Coracoid process (O)

Coracobrachialis

Medial surface
of humerus (I)

S
L R
I

Medial
epicondyle of
humerus (O)

Pronator teres

B

Lateral
surface of
radius (I)

Infraglenoid
tubercle (O)

Posterior surface
of humerus (O);
lateral
intermuscular
septum

Triceps brachii:

Long head

S
R L
I

Lateral
(short head)

Medial head

Olecranon
process
of ulna (I)

C

S
L R
I

Humerus,
distal half (O)

Brachialis

Coronoid process
of ulna (I)

D

Figure 10-21 *Muscles acting on the forearm.* **A,** Biceps brachii. **B,** Triceps brachii. **C,** Coracobrachialis and pronator teres. **D,** Brachialis. *O,* Origin; *I,* insertion.

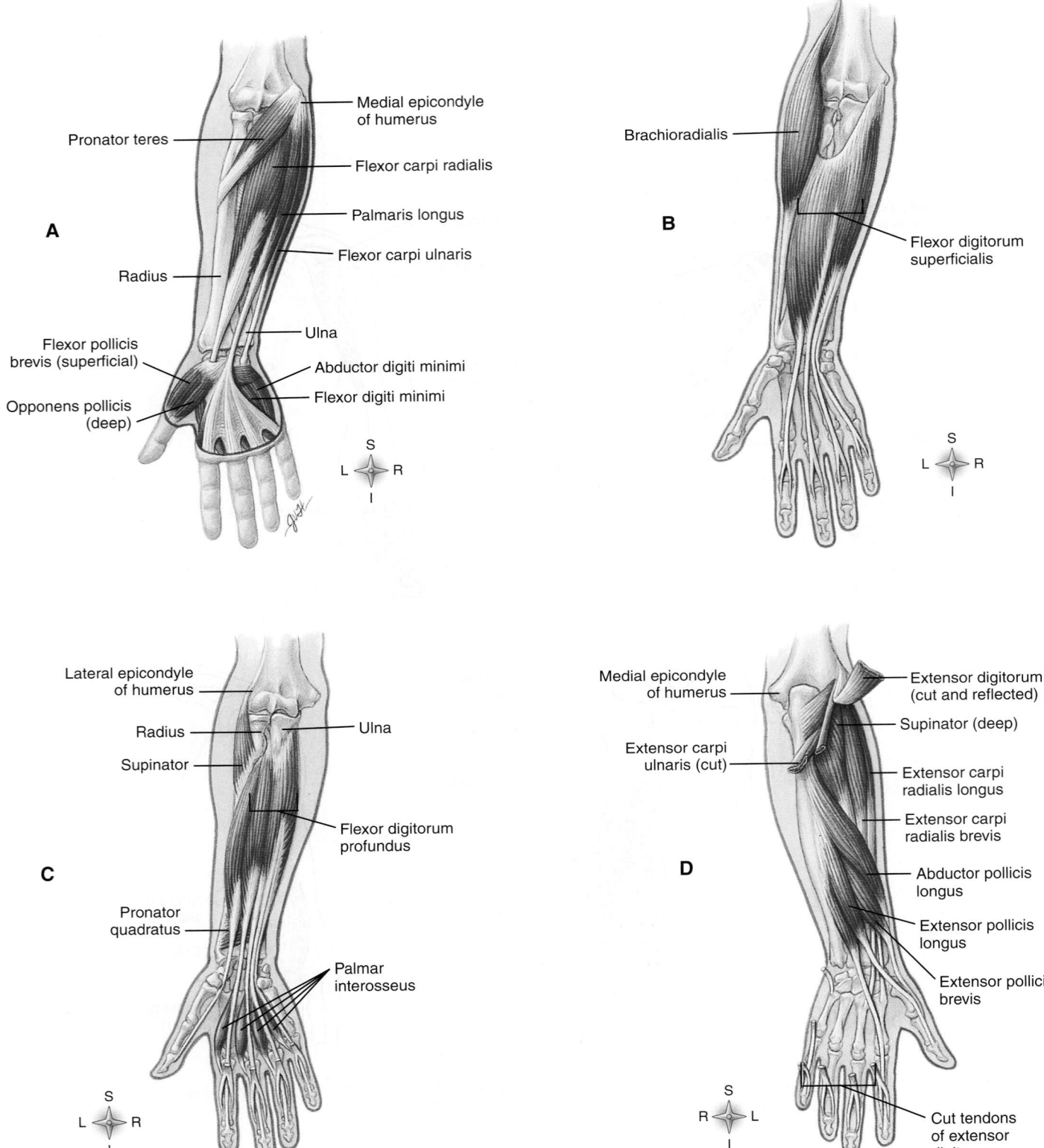

A, Anterior view showing the right forearm (superficial). The brachioradialis muscle has been removed.

Labels in A:
Pronator teres
Medial epicondyle of humerus
Flexor carpi radialis
Palmaris longus
Flexor carpi ulnaris
Radius
Ulna
Flexor pollicis brevis (superficial)
Abductor digiti minimi
Flexor digiti minimi
Opponens pollicis (deep)

Labels in B:
Brachioradialis
Flexor digitorum superficialis

Labels in C:
Lateral epicondyle of humerus
Radius
Ulna
Supinator
Flexor digitorum profundus
Pronator quadratus
Palmar interosseus

Labels in D:
Medial epicondyle of humerus
Extensor digitorum (cut and reflected)
Supinator (deep)
Extensor carpi ulnaris (cut)
Extensor carpi radialis longus
Extensor carpi radialis brevis
Abductor pollicis longus
Extensor pollicis longus
Extensor pollicis brevis
Cut tendons of extensor digitorum

Figure 10-22 *Muscles of the forearm.* A, Anterior view showing the right forearm (superficial). The brachioradialis muscle has been removed. **B,** Anterior view showing the right forearm (deeper than *A*). The pronator teres, flexor carpi radialis and ulnaris, and palmaris longus muscles have been removed. **C,** Anterior view showing the right forearm (deeper than *A* or *B*). The brachioradialis, pronator teres, flexor carpi radialis and ulnaris, palmaris longus, and flexor digitorum superficialis muscles have been removed. **D,** Posterior view showing the deep muscles of the right forearm. The extensor digitorum, extensor digiti minimi, and extensor carpi ulnaris muscles have been cut to reveal deeper muscles.

Table 10-15 Muscles That Move the Wrist, Hand, and Fingers

MUSCLE	ORIGIN	INSERTION	FUNCTION	NERVE SUPPLY
Extrinsic				
Flexor carpi radialis	Humerus (medial epicondyle)	Second metacarpal (base of)	Flexes the hand Flexes the forearm	Median nerve
Palmaris longus	Humerus (medial epicondyle)	Fascia of the palm	Flexes the hand	Median nerve
Flexor carpi ulnaris	Humerus (medial epicondyle) Ulna (proximal two thirds)	Pisiform bone Third, fourth, and fifth metacarpals	Flexes the hand Adducts the hand	Ulnar nerve
Extensor carpi radialis longus	Humerus (ridge above the lateral epicondyle)	Second metacarpal (base of)	Extends the hand Abducts the hand (moves toward the thumb side when the hand is supinated)	Radial nerve
Extensor carpi radialis brevis	Humerus (lateral epicondyle)	Second, third metacarpals (bases of)	Extends the hand	Radial nerve
Extensor carpi ulnaris	Humerus (lateral epicondyle) Ulna (proximal three fourths)	Fifth metacarpal (base of)	Extends the hand Adducts the hand (moves toward the little finger side when the hand is supinated)	Radial nerve
Flexor digitorum profundus	Ulna (anterior surface)	Distal phalanges (fingers 2 to 5)	Flexes the distal interphalangeal joints	Median and ulnar nerves
Flexor digitorum superficialis	Humerus (medial epicondyle) Radius Ulna (coronoid process)	Tendons of the fingers	Flexes the fingers	Median nerve
Extensor digitorum	Humerus (lateral epicondyle)	Phalanges (fingers 2 to 5)	Extends the fingers	Radial nerve
Intrinsic				
Opponens pollicis	Trapezium	Thumb metacarpal	Opposes the thumb to the fingers	Median nerve
Abductor pollicis brevis	Trapezium	Proximal phalanx of the thumb	Abducts the thumb	Median nerve
Adductor pollicis	Second and third metacarpals Trapezoid Capitate	Proximal phalanx of the thumb	Adducts the thumb	Ulnar nerve
Flexor pollicis brevis	Flexor retinaculum	Proximal phalanx of the thumb	Flexes the thumb	Median and ulnar nerves
Abductor digiti minimi	Pisiform	Proximal phalanx of the fifth finger (base of)	Abducts the fifth finger Flexes the fifth finger	Ulnar nerve
Flexor digiti minimi brevis	Hamate	Proximal and middle phalanx of the fifth finger	Flexes the fifth finger	Ulnar nerve
Opponens digiti minimi	Hamate Flexor retinaculum	Fifth metacarpal	Opposes the fifth finger slightly	Ulnar nerve
Interosseous (palmar and dorsal)	Metacarpals	Proximal phalanges	Adducts the second, fourth, and fifth fingers (palmar) Abducts the second, third, and fourth fingers (dorsal)	Ulnar nerve
Lumbricales	Tendons of the flexor digitorum profundus	Phalanges (2 to 5)	Flexes the proximal phalanges (2 to 5) Extends the middle and distal phalanges (2 to 5)	Median nerve (phalanges 2 and 3) Ulnar nerve (phalanges 4 and 5)

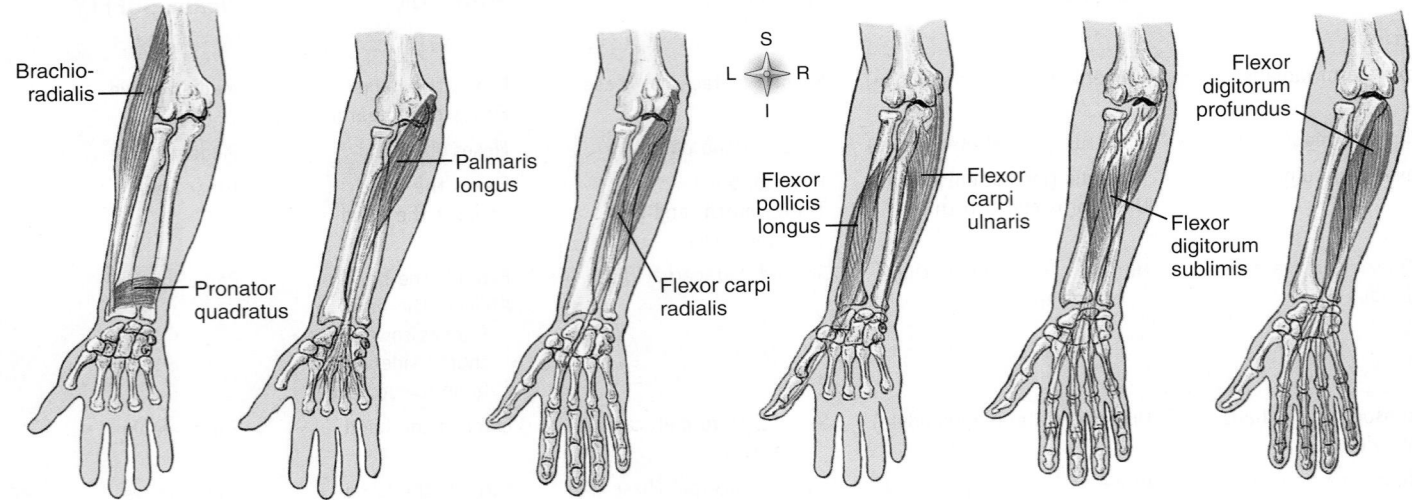

Figure 10-23 *Some muscles of the anterior aspect of the right forearm.*

Brachio-radialis

Pronator quadratus

Palmaris longus

Flexor carpi radialis

Flexor pollicis longus

Flexor carpi ulnaris

Flexor digitorum profundus

Flexor digitorum sublimis

S
L ◆ R
I

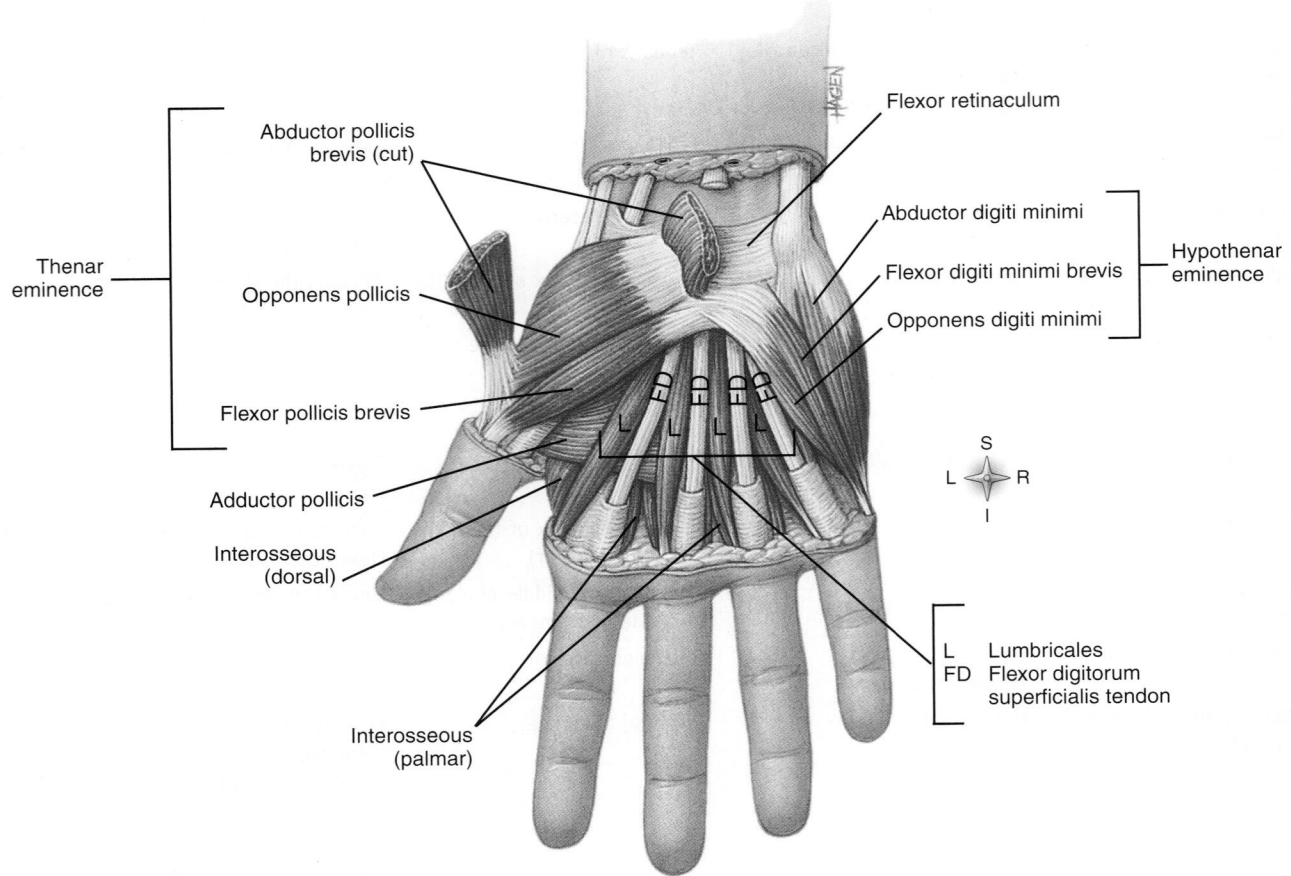

Figure 10-24 *Intrinsic muscles of the hand—anterior (palmar) view.*

Abductor pollicis brevis (cut)

Flexor retinaculum

Abductor digiti minimi

Flexor digiti minimi brevis

Opponens digiti minimi

Hypothenar eminence

Thenar eminence

Opponens pollicis

Flexor pollicis brevis

Adductor pollicis

Interosseous (dorsal)

Interosseous (palmar)

L Lumbricales
FD Flexor digitorum superficialis tendon

S
L ◆ R
I

Table 10-16 Muscles That Move the Thigh

MUSCLE	ORIGIN	INSERTION	FUNCTION	NERVE SUPPLY
Iliopsoas (iliacus, psoas major, and psoas minor)	Ilium (iliac fossa)	Femur (lesser trochanter)	Flexes the thigh	Femoral and second to fourth lumbar nerves
	Vertebrae (bodies of twelfth thoracic to fifth lumbar)		Flexes the trunk (when the femur acts as the origin)	
Rectus femoris	Ilium (anterior, inferior spine)	Tibia (by way of the patellar tendon)	Flexes the thigh Extends the lower leg	Femoral nerve
Gluteal group				
Maximus	Ilium (crest and posterior surface) Sacrum and coccyx (posterior surface) Sacrotuberous ligament	Femur (gluteal tuberosity) Iliotibial tract	Extends the thigh—rotates outward	Inferior gluteal nerve
Medius	Ilium (lateral surface)	Femur (greater trochanter)	Abducts the thigh—rotates outward; stabilizes the pelvis on the femur	Superior gluteal nerve
Minimus	Ilium (lateral surface)	Femur (greater trochanter)	Abducts the thigh; stabilizes the pelvis on the femur Rotates the thigh medially	Superior gluteal nerve
Tensor fasciae latae	Ilium (anterior part of the crest)	Tibia (by way of the iliotibial tract)	Abducts the thigh Tightens the iliotibial tract	Superior gluteal nerve
Adductor group				
Brevis	Pubic bone	Femur (linea aspera)	Adducts the thigh	Obturator nerve
Longus	Pubic bone	Femur (linea aspera)	Adducts the thigh	Obturator nerve
Magnus	Pubic bone	Femur (linea aspera)	Adducts the thigh	Obturator nerve
Gracilis	Pubic bone (just below the symphysis)	Tibia (medial surface behind the sartorius)	Adducts the thigh and flexes and adducts the leg	Obturator nerve

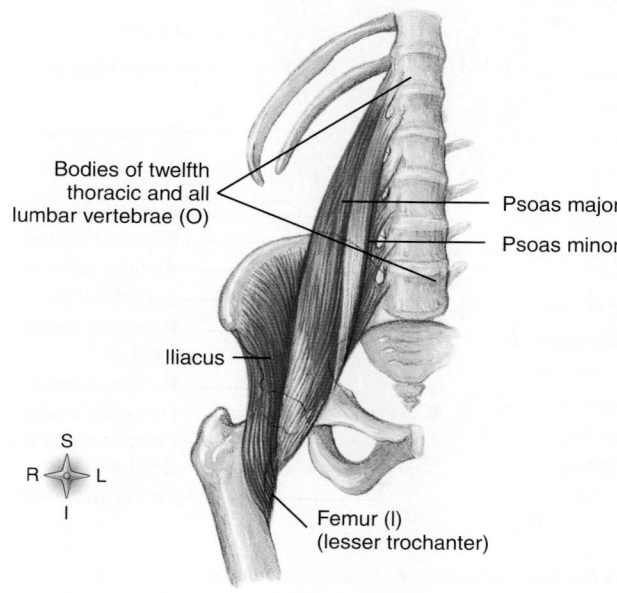

Bodies of twelfth thoracic and all lumbar vertebrae (O)

Psoas major

Psoas minor

Iliacus

Femur (I) (lesser trochanter)

Figure 10-25 *Iliopsoas muscle (iliacus, psoas major, and psoas minor muscles).* O, Origin; I, insertion.

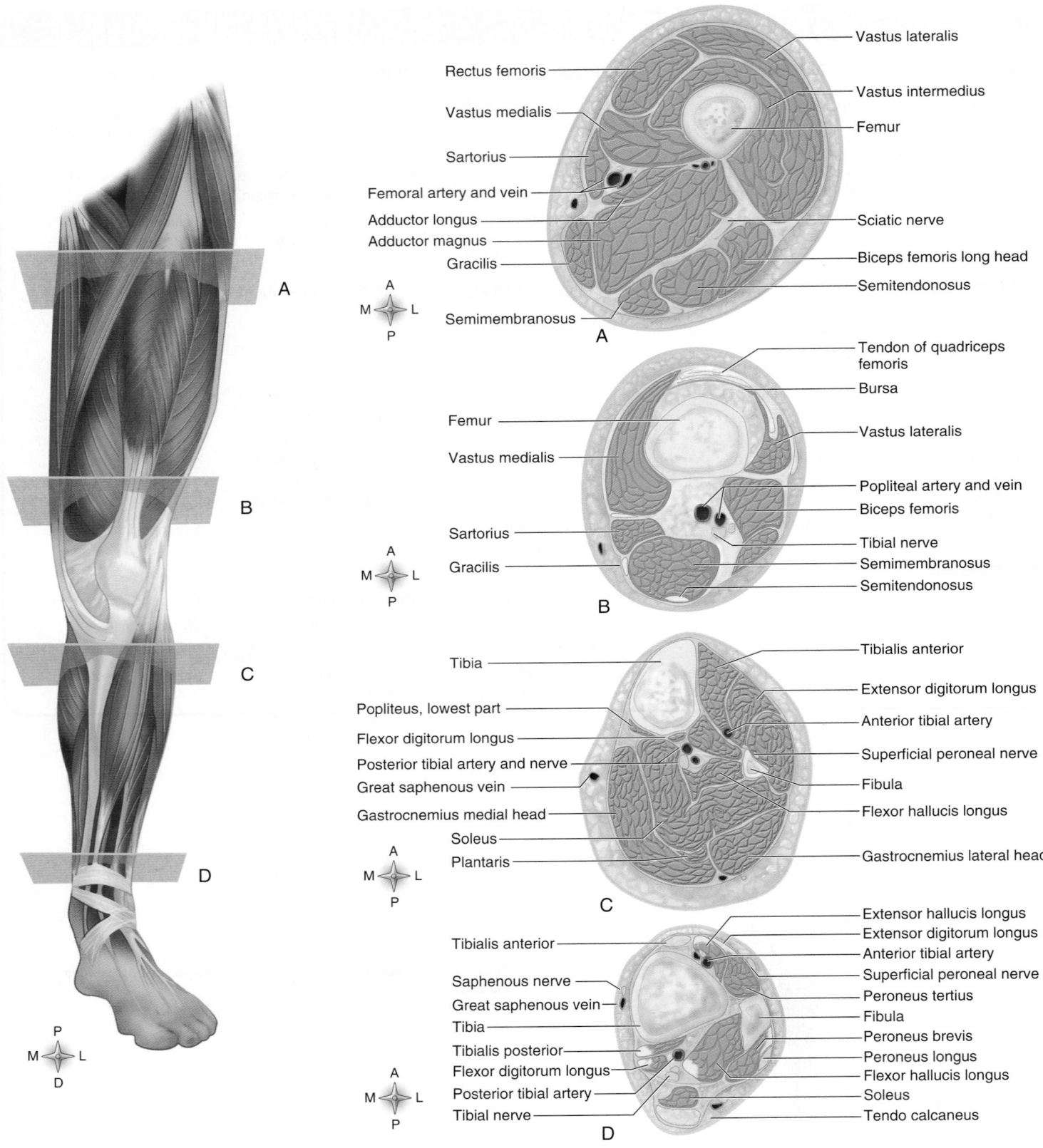

Figure 10-26 *Cross sections (proximal to distal) through the lower extremity.* **A,** Section through the middle of the femur. **B,** Section about 4 cm above the adductor tubercle of the femur. **C,** Section about 10 cm distal to the knee joint. **D,** Section about 6 cm above the medial malleolus. In each section you are viewing the superior (proximal) aspect of the specimen.

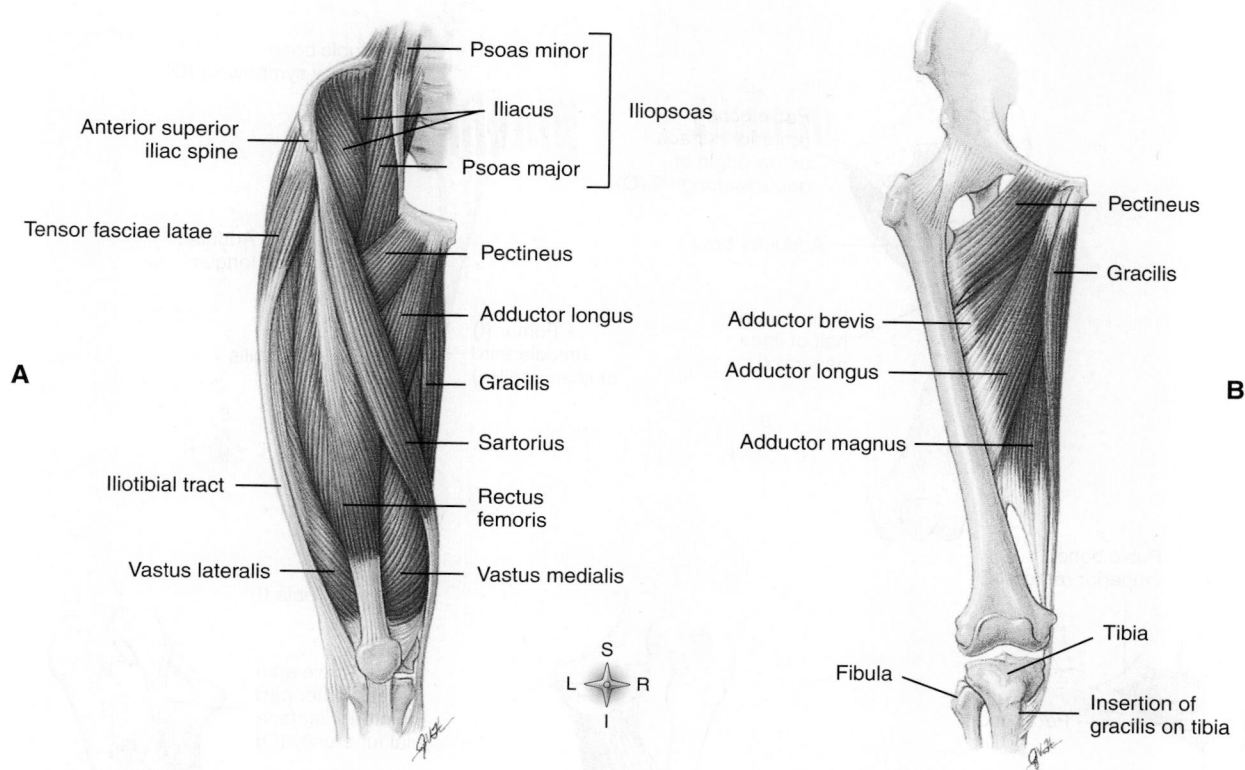

Figure 10-27 *Muscles of the anterior aspect of the thigh.* A, Anterior view of the right thigh. **B,** Adductor region of the right thigh. The tensor fasciae latae, sartorius, and quadriceps muscles have been removed.

three groups: (1) muscles crossing the front of the hip, (2) the three **gluteal** (GLOO-tee-al) muscles, and the **tensor fasciae latae** (TEN-sor FASH-ee LAT-tee), and (3) the thigh adductors.

Table 10-17 identifies muscles that move the lower part of the leg. Again, see Figures 10-5 and 10-6 and refer to Figures 10-30 and 10-31 as you study the table.

Muscles That Move the Ankle and Foot

The muscles listed in Table 10-18 and shown in Figure 10-32 are responsible for movements of the ankle and foot. These muscles, called **extrinsic foot muscles,** are located in the leg but exert their actions by pulling on tendons that insert on bones in the ankle and foot. Extrinsic foot muscles are responsible for such movements as dorsiflexion, plantar flexion, inversion, and eversion of the foot. Muscles located within the foot itself are called **intrinsic foot muscles** (Figure 10-33). They are responsible for flexion, extension, abduction, and adduction of the toes.

The extrinsic muscles listed in Table 10-18 may be divided into four functional groups: (1) dorsal flexors, (2) plantar flexors, (3) invertors, and (4) evertors of the foot.

The superficial muscles located on the posterior surface of the leg form the bulging "calf." The common tendon of the **gastrocnemius** (GAS-trok-NEE-mee-us) and **soleus** is called the **calcaneal,** or *Achilles,* **tendon.** It inserts into the calcaneus, or heel bone. By acting together, these muscles serve as powerful flexors (plantar flexion) of the foot.

Dorsal flexors of the foot, located on the anterior surface of the leg, include the **tibialis** (tib-ee-AL-is) **anterior, peroneus tertius** (per-o-NEE-us TER-shus), and **extensor digitorum longus.** In addition to functioning as a dorsiflexor of the foot the extensor digitorum longus also everts the foot and extends the toes.

 QUICK CHECK

22. Name the three gluteal muscles.
23. What is the function of the gastrocnemius muscle?

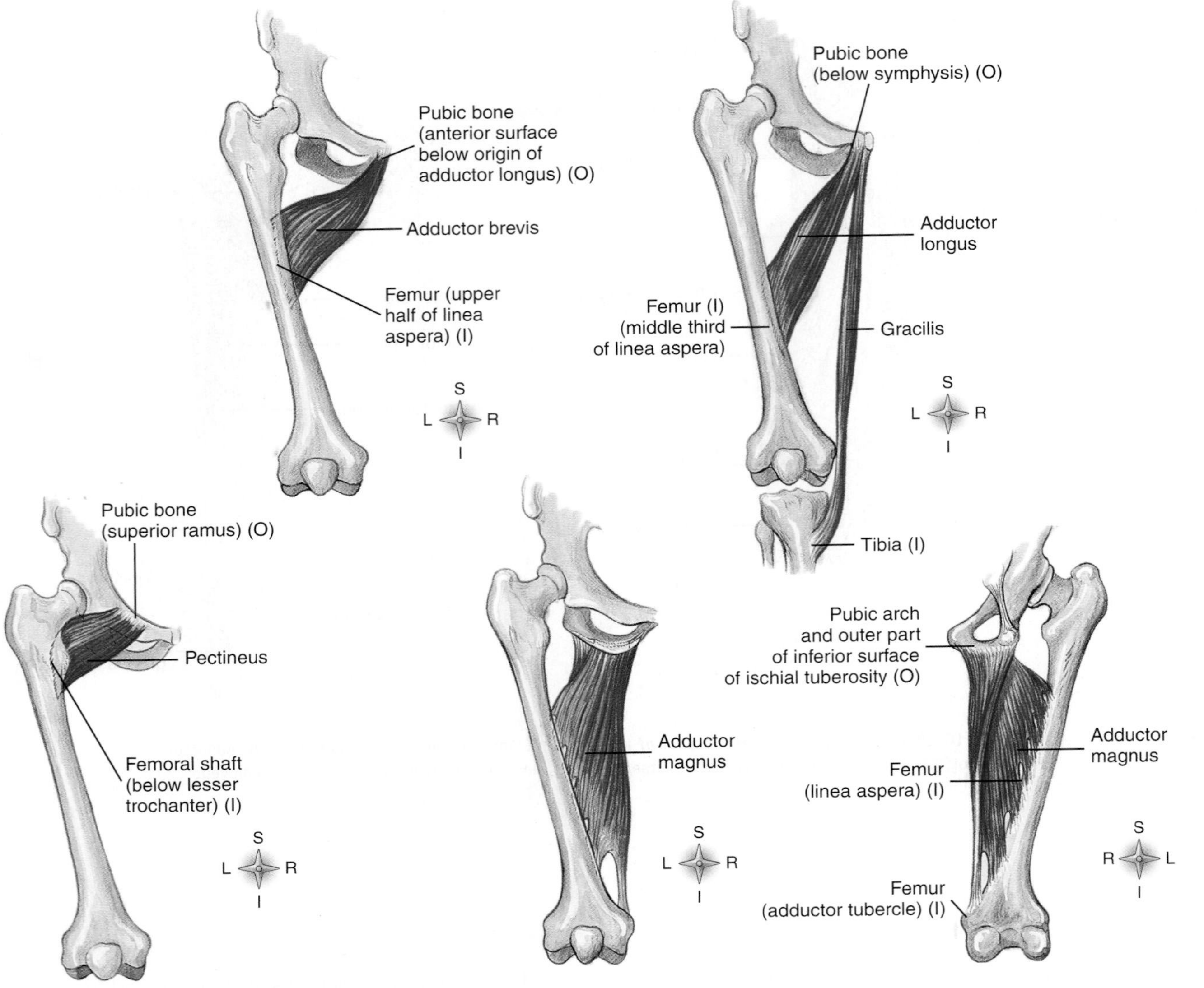

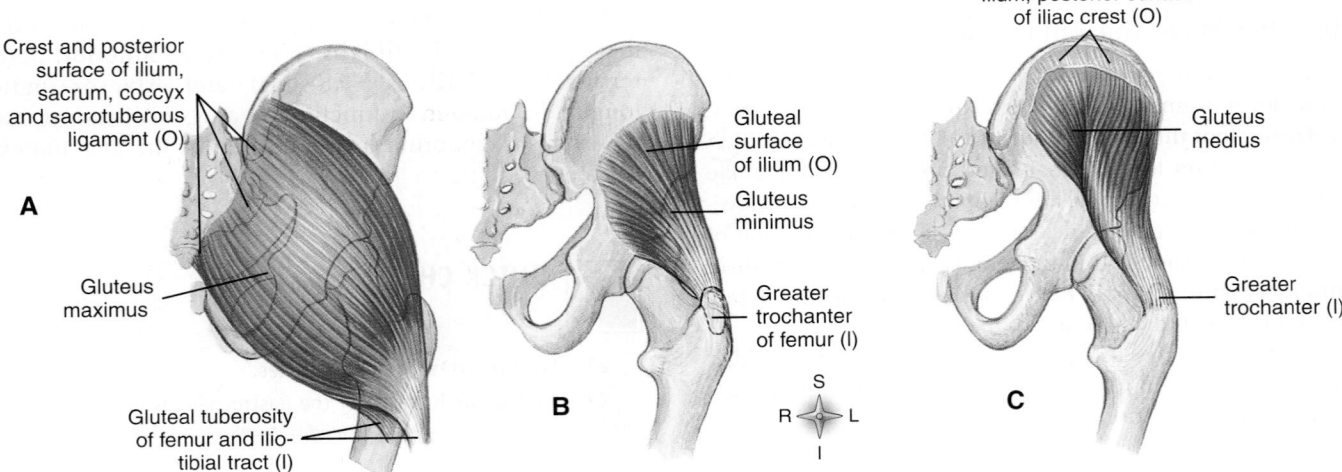

Pubic bone
(anterior surface
below origin of
adductor longus) (O)
Adductor brevis
Femur (upper
half of linea
aspera) (I)
S
L
R
I
Pubic bone
(below symphysis) (O)
Adductor
longus
Femur (I)
(middle third
of linea aspera)
Gracilis
S
L
R
I
Tibia (I)
Pubic bone
(superior ramus) (O)
Pectineus
Femoral shaft
(below lesser
trochanter) (I)
S
L
R
I
Adductor
magnus
S
L
R
I
Pubic arch
and outer part
of inferior surface
of ischial tuberosity (O)
Adductor
magnus
Femur
(linea aspera) (I)
Femur
(adductor tubercle) (I)
S
R
L
I

Figure 10-28 ***Muscles that adduct the thigh.*** *O,* Origin; *I,* insertion.

Crest and posterior
surface of ilium,
sacrum, coccyx
and sacrotuberous
ligament (O)
A
Gluteus
maximus
Gluteal tuberosity
of femur and ilio-
tibial tract (I)
Gluteal
surface
of ilium (O)
Gluteus
minimus
Greater
trochanter
of femur (I)
B
S
R
L
I
Ilium, posterior surface
of iliac crest (O)
Gluteus
medius
Greater
trochanter (I)
C

Figure 10-29 ***Gluteal muscles.*** **A,** Gluteus maximus. **B,** Gluteus minimus. **C,** Gluteus medius.

Table 10-17 Muscles That Move the Lower Leg

MUSCLE	ORIGIN	INSERTION	FUNCTION	NERVE SUPPLY
Quadriceps femoris group				
Rectus femoris	Ilium (anterior inferior spine)	Tibia (by way of patellar tendon)	Flexes the thigh Extends the leg	Femoral nerve
Vastus lateralis	Femur (linea aspera)	Tibia (by way of the patellar tendon)	Extends the leg	Femoral nerve
Vastus medialis	Femur	Tibia (by way of the patellar tendon)	Extends the leg	Femoral nerve
Vastus intermedius	Femur (anterior surface)	Tibia (by way of the patellar tendon)	Extends the leg	Femoral nerve
Sartorius	Coxal (anterior superior iliac spines)	Tibia (medial surface of the upper end of the shaft)	Adducts and flexes the leg Permits crossing of the legs tailor fashion	Femoral nerve
Hamstring group				
Biceps femoris	Ischium (tuberosity)	Fibula (head of)	Extends the thigh	Hamstring nerve (branch of the sciatic nerve)
	Femur (linea aspera)	Tibia (lateral condyle)		Hamstring nerve
Semitendinosus	Ischium (tuberosity)	Tibia (proximal end, medial surface)	Extends the thigh	Hamstring nerve
Semimembranosus	Ischium (tuberosity)	Tibia (medial condyle)	Extends the thigh	Hamstring nerve

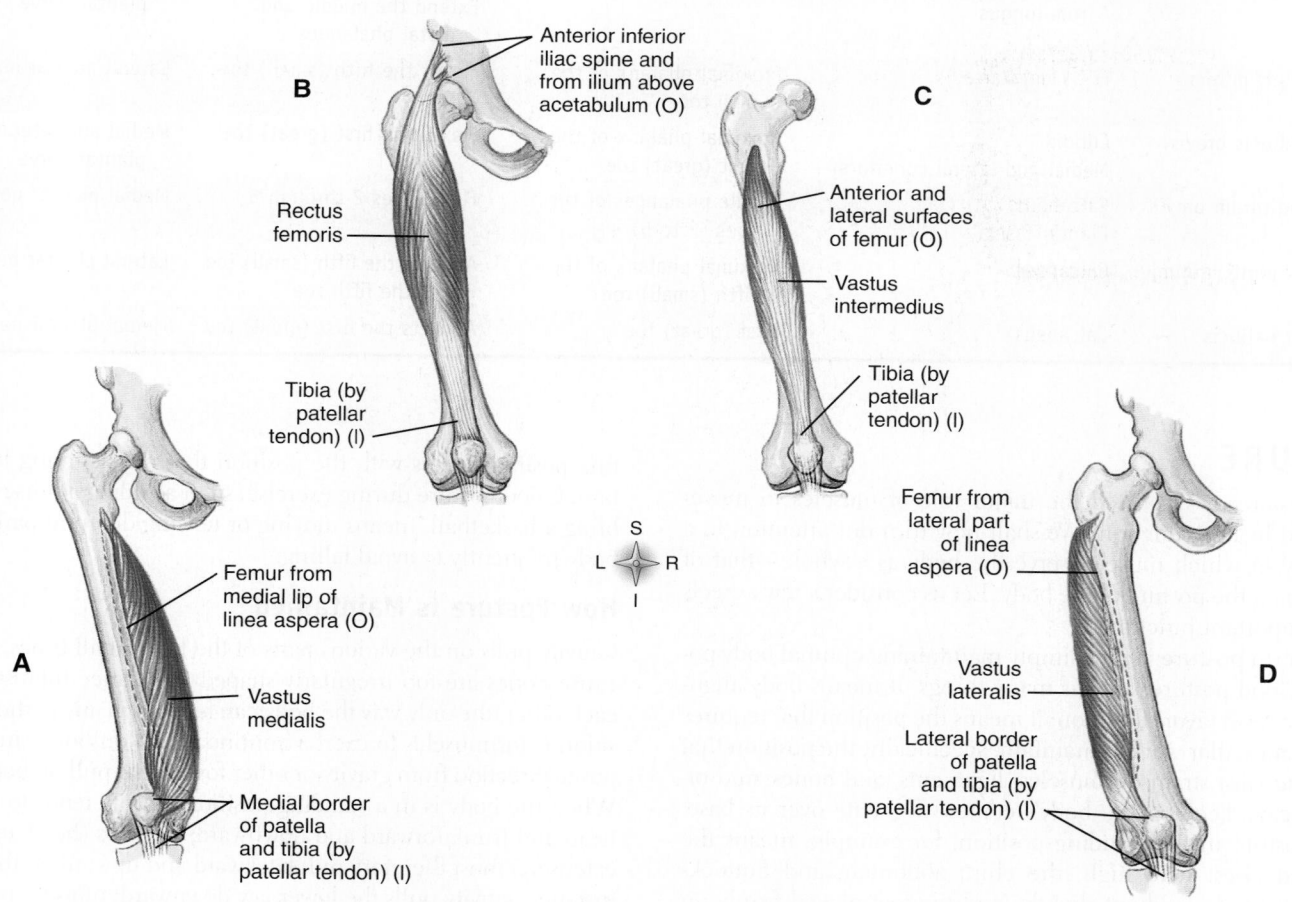

Figure 10-30 *Quadriceps femoris group of thigh muscles.* **A,** Rectus femoris. **B,** Vastus intermedius. **C,** Vastus medialis. **D,** Vastus lateralis. *O,* Origin; *I,* insertion.

Table 10-18	Muscles That Move the Foot			
MUSCLE	**ORIGIN**	**INSERTION**	**FUNCTION**	**NERVE SUPPLY**
Extrinsic				
Tibialis anterior	Tibia (lateral condyle of the upper body)	Tarsal (first cuneiform)	Flexes the foot	Common and deep peroneal nerves
		Metatarsal (base of first)	Inverts the foot	
Gastrocnemius	Femur (condyles)	Tarsal (calcaneus by way of the Achilles tendon)	Extends the foot Flexes lower leg	Tibial nerve (branch of the sciatic nerve)
Soleus	Tibia (underneath the gastrocnemius) Fibula	Tarsal (calcaneus by way of the Achilles tendon)	Extends the foot (plantar flexion)	Tibial nerve
Peroneus longus	Tibia (lateral condyle)	First cuneiform	Extends the foot (plantar flexion)	Common peroneal nerve
	Fibula (head and shaft)	Base of the first metatarsal	Everts the foot	
Peroneus brevis	Fibula (lower two thirds of the lateral surface of the shaft)	Fifth metatarsal (tubercle, dorsal surface)	Everts the foot Flexes the foot	Superficial peroneal nerve
Peroneus tertius	Fibula (distal third)	Fourth and fifth metatarsals (bases of)	Flexes the foot Everts the foot	Deep peroneal nerve
Extensor digitorum longus	Tibia (lateral condyle) Fibula (anterior surface)	Second and third phalanges (four lateral toes)	Dorsiflexion of the foot; extension of the toes	Deep peroneal nerve
Intrinsic				
Lumbricales	Tendons of the flexor digitorum longus	Phalanges (2 to 5)	Flex the proximal phalanges Extend the middle and distal phalanges	Lateral and medial plantar nerve
Flexor digiti minimi brevis	Fifth metatarsal	Proximal phalanx of the fifth toe	Flexes the fifth (small) toe	Lateral plantar nerve
Flexor hallucis brevis	Cuboid Medial and lateral cuneiform	Proximal phalanx of the first (great) toe	Flexes the first (great) toe	Medial and lateral plantar nerve
Flexor digitorum brevis	Calcaneus Plantar fascia	Middle phalanges of the toes (2 to 5)	Flexes toes 2 through 5	Medial plantar nerve
Abductor digiti minimi	Calcaneus	Proximal phalanx of the fifth (small) toe	Abducts the fifth (small) toe Flexes the fifth toe	Lateral plantar nerve
Abductor hallucis	Calcaneus	First (great) toe	Abducts the first (great) toe	Medial plantar nerve

POSTURE

We have already discussed the major role of muscles in movement and heat production. We shall now turn our attention to a third way in which muscles serve the body as a whole—that of maintaining the posture of the body. Let us consider a few aspects of this important function.

The term **posture** means simply maintaining optimal body position. "Good posture" means many things. It means body alignment that most favors function; it means the position that requires the least muscular work to maintain, specifically, the position that places the least strain on muscles, ligaments, and bones; and often it means keeping the body's center of gravity over its base. Good posture in the standing position, for example, means the head and chest held high; the chin, abdomen, and buttocks pulled in; the knees bent slightly; and the feet placed firmly on the ground about 6 inches (15 cm) apart. Good posture in a sitting position varies with the position that one is trying to maintain. Good posture during exercise, such as riding a horse or dribbling a basketball, means moving or tensing different parts of the body frequently to avoid falling.

How Posture is Maintained

Gravity pulls on the various parts of the body at all times, and because bones are too irregularly shaped to balance themselves on each other, the only way the body can be held in any particular position is for muscles to exert a continual pull on bones in the opposite direction from gravity or other forces that pull on body parts. When the body is in a standing position, gravity tends to pull the head and trunk forward and downward; muscles (head and trunk extensors) must therefore pull backward and upward on them. For instance, gravity pulls the lower jaw downward; muscles must pull upward on it. Muscles exert this pull against gravity by virtue of

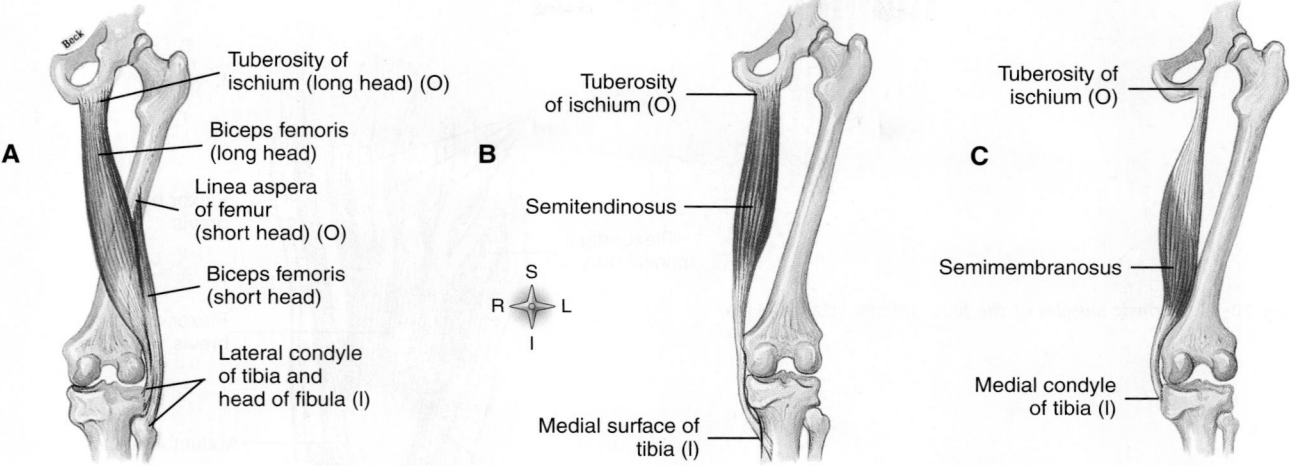

Figure 10-31 *Hamstring group of thigh muscles.* **A,** Biceps femoris. **B,** Semitendinosus. **C,** Semimembranosus. *O,* Origin; *I,* insertion.

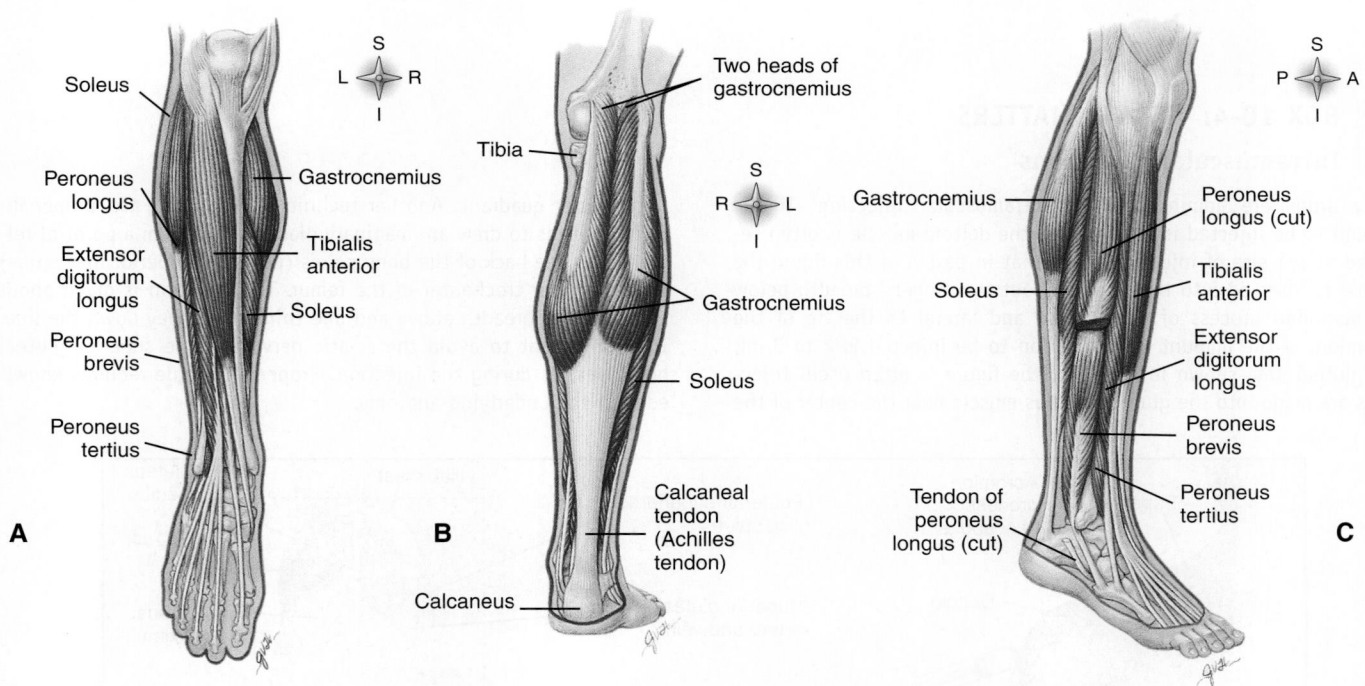

Figure 10-32 *Superficial muscles of the leg.* **A,** Anterior view. **B,** Posterior view. **C,** Lateral view.

their property of **tonicity.** Tonicity, or **muscle tone,** literally means *tension* and refers to the continuous, low level of sustained contraction maintained by all skeletal muscles. Because tonicity is absent during deep sleep, muscle pull does not then counteract the pull of gravity. Hence, we cannot sleep standing up. Interestingly, astronauts in the low-gravity conditions of space station missions can sleep in a standing position, as long as they are secured inside special sleeping bags on the wall of the space station.

Many structures other than muscles and bones play a part in the maintenance of posture. The nervous system is responsible for the existence of muscle tone and also regulates and coordinates the amount of pull exerted by the individual muscles. The respiratory, digestive, circulatory, excretory, and endocrine systems all contribute something toward the ability of muscles to maintain posture. This is one of many examples of the important principle that all body functions are interdependent.

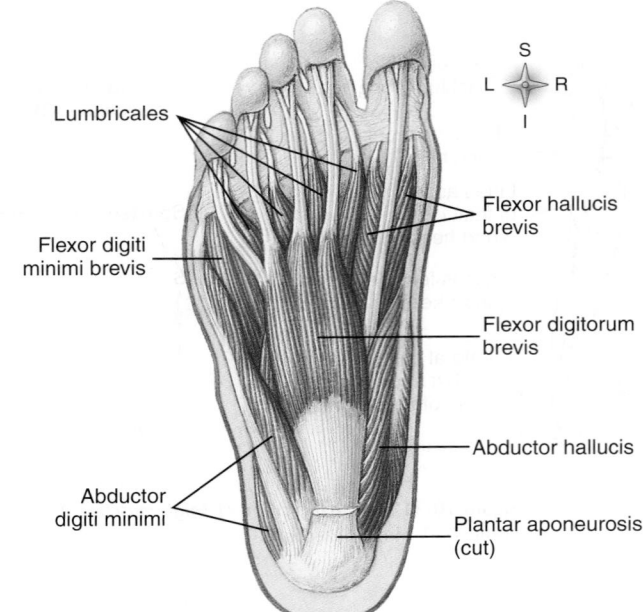

Figure 10-33 *Intrinsic muscles of the foot.* Inferior (plantar) view.

BOX 10-4: HEALTH MATTERS

Intramuscular Injections

Many drugs are administered by intramuscular injection. If the amount to be injected is 2 ml or less, the deltoid muscle is often selected as the site of injection. Note that in part *A* of this figure the needle is inserted into the muscle about two fingers' breadth below the acromion process of the scapula and lateral to the tip of the acromion. If the amount of medication to be injected is 2 to 3 ml, the gluteal area shown in part *B* of the figure is often used. Injections are made into the gluteus medius muscle near the center of the upper outer quadrant. Another technique of locating the proper injection site is to draw an imaginary diagonal line from a point of reference on the back of the bony pelvis (posterior superior iliac spine) to the greater trochanter of the femur. The injection is given about three fingers' breadth above and one third of the way down the line. It is important to avoid the sciatic nerve and the superior gluteal blood vessels during the injection. Proper technique requires knowledge of the underlying anatomy.

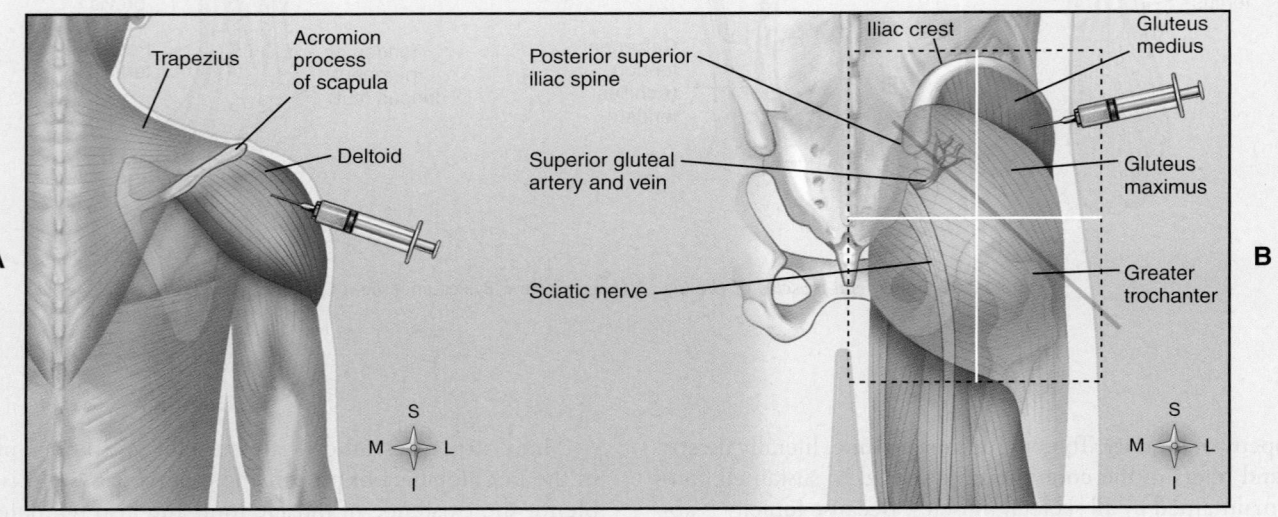

Intramuscular injection sites. **A,** Deltoid injection site for small doses. **B,** Gluteal injection site for larger doses.

The importance of posture can perhaps be best evaluated by considering some of the effects of poor posture. Poor posture throws more work on muscles to counteract the pull of gravity and therefore leads to fatigue more quickly than good posture does. Poor posture puts more strain on ligaments. It puts abnormal strain on bones and may eventually produce deformities. It also interferes with various functions such as respiration, heart action, and digestion.

Cycle of Life
Muscular System

Acting together, the muscular, skeletal, and nervous systems permit us to move in a coordinated and controlled way. However, it is the contraction, or shortening, of muscles that ultimately provides the actual movement necessary for physical activity. Dramatic changes occur in the muscular system throughout the cycle of life. Muscle cells may increase or decrease in size and in their ability to shorten most effectively at different periods in life. In addition to age-related changes,

many pathological conditions occurring at different ages may also affect the muscular system.

Because of the functional interdependence of the musculoskeletal and nervous systems, life cycle changes affecting the muscles are often manifested in other components of the functional unit. During infancy and childhood, the ability to coordinate and control the strength of muscle contraction permits a sequential series of developmental steps to occur. A developing youngster learns to hold the head up, roll over, sit up, stand alone, and then walk and run as developmental changes permit better control and coordination of muscular contraction.

Degenerative changes associated with advancing age often result in replacement of muscle cell volume with nonfunctional connective tissue. As a result, the strength of muscular contraction diminishes. Recent findings show that much, if not all of this age-related decrease in muscle strength actually results from disuse atrophy (see Box 11-6 in Chapter 11) and may thus be avoidable. Pathological conditions associated with specific age ranges can also affect one or more components of the functional unit that permits us to move smoothly and effortlessly.

THE BIG PICTURE
Skeletal Muscles and the Whole Body

As you read through this chapter, what struck you most was probably the large number of individual muscles and their many actions. Although learning the names, locations, origins, insertions, and other details of the major muscles is a worthwhile endeavor, such an activity can cause you to lose sight of the "big picture." Step back from the details a moment to appreciate the fact that the muscles work as coordinated teams of biological engines that move the various components of the flexible skeleton.

As a matter of fact, in this chapter you learned that the fibrous wrappings of each muscle are continuous with its tendons, which in turn are continuous with the fibrous structure of the bone to which they are attached. Thus, we can see that the muscular and skeletal systems are in essence a *single structure*. This fact is very important in seeing a "big picture" comprising many individual muscles and bones. The entire *skeletomuscular system,* as it is often called, is actually a single, continuous structure that provides a coordinated, dynamic framework for the body.

Taking another step back, an even bigger picture comes into focus. Other systems of the body must play a role in the actions of the skeletomuscular system. For example, the nervous system senses changes in body position and degrees of movement—thereby permitting integration of feedback loops that ultimately regulate the muscular contractions that maintain posture and produce movements. The cardiovascular system maintains blood flow in the muscles, and the urinary and respiratory systems rid the body of wastes produced in the muscles. The respiratory and digestive systems bring in the oxygen and nutrients necessary for muscle function.

The picture is still not complete, however. We will see even more when Chapter 11 continues the story of muscle function by delving into the details of how each muscle works as an engine to drive the movement of the body.

LANGUAGE OF SCIENCE

(Cont'd from page 347)

fascia (FAY-shah) [*fasci* band or bundle of fibrous tissue]

fixator muscle (fik-SAY-tor) [*figere* to fasten]

fusiform muscles (FYOO-si-form) [*fusus-* spindle, -*forma* form]

gastrocnemius muscle (GAS-trok-NEE-mee-us) [*gastroknemia* calf of the leg]

gluteal muscles (GLOO-tee-al MUSS-els) [*gloutos* buttocks]

infraspinatus muscle (IN-frah-spy-nah-tus) [*infra-* occurring beneath, -*spina* spine]

insertion [*inserere* to join to]

internal intercostal muscle (in-TER-nal in-ter-KOS-tal) [*internus* inward, *costa* rib]

internal oblique muscle (in-TER-nal oh-BLEEK) [*internus* inward, *obliquus* slanted]

interosseous muscles (in-ter-OSS-ee-us) [*inter-* between, -*os* bone]

interspinales group (in-ter-spy-NAH-leez) [*inter-* between, -*spina* spine]

intrinsic foot muscles (in-TRIN-sik) [*intrinsecus* inside]

intrinsic muscles (in-TRIN-sik) [*instrinsecus* inside]

ischiocavernosus muscle (iss-kee-oh-KAV-er-no-sus) [*ischio-* the ischium or hip, -*caverna* hollow space]

levator ani muscle (leh-VAY-tor) [*levare* to lift up]

levator scapulae (leh-VAY-tor SCAP-yoo-lee) [*levare* to lift up, *scapula* shoulder blade]

lever

lever systems

linea alba (LIN-ee-ah AL-bah) [*linea* line, *albus* white]

longissimus capitis muscle (lon-JIS-i-mus KAP-i-tis) [*longissumus* longest or very long, *capit* head]

lumbrical muscle (LUM-bri-kal) [*lumbrical* resembling a worm]

masseter (mah-SEE-ter) [*masseter* one who chews]

mastication (mass-ti-KAY-shun) [*masticare* to chew]

multifides group (mul-TIF-i-deez) [*multi-* many, -*findere* to split]

muscle tone

muscular system

occipitofrontalis (ok-sip-i-too-fron-TAL-is)

opponens pollicis muscle (oh-POH-nenz POL-i-cis MUSS-el) [*oppenens* opposing, *poli* pole]

orbicularis oculi muscle (or-bik-yoo-LAIR-is OK-yoo-lye MUSS-el) [*orbiculus* little circle, *ocul* pertaining to the eye]

orbicularis oris muscle (or-bik-yoo-LAR-is OR-iss) [*orbiculus* little circle, *oris* mouth]

origin (OR-i-jin)

parallel muscles

pectoralis minor (pek-toh-RAL-is) [*pector-* breast, -*alis* pertaining to]

pennate muscles (PEN-ayt) [*pennate* having feathers]

perimysium (pair-i-MEE-see-um) [*peri-* around, -*mys* muscle]

perineum (pair-i-NEE-um)

peroneus tertius muscle (per-oh-NEE-us TER-tee-us) [*peroneo* related to the fibula, *tertius* third]

posture (POS-chur)

prime mover

pterygoid muscle (TER-i-goid) [*pteryx-* wing, -*oid* resembling]

rectus abdominis muscle (REK-tus ab-DOM-i-nus) [*rectus* straight, *abdominis* abdomen]

rectus sheath (REK-tus sheeth) [*rectus* straight]

rhomboideus major/minor muscle (rom-BOYD-ee-us) [*rhombos* rhombus]

rotator cuff muscles (roh-TAY-tor) [*rot* turned or to turn]

serratus anterior muscle (ser-AYT-us) [*serra* saw teeth, *ante* + *prior* foremost]

soleus muscle (SOH-lee-us) [*solea* sole of foot]

sphincter urethrae muscles (SFINGK-ter yoo-REE-three) [*spingein* to bind]

spiral muscles

splenius capitis (SPLEH-nee-us KAP-i-tis) [*splen* spleen, *capitis* head]

sternocleidomastoid muscle (STERN-oh-KLYE-doh-MAS-toyd) [*sterno-* sternum, -*kleis-* key, -*masto-* breast, -*oid* resembling]

subscapularis muscle (sub-SKAP-yoo-lar-is) [*sub-* beneath, -*scapulae* shoulder blades]

supraspinatus muscle (SOO-prah-spy-nah-tus) [*supra-* above, -*spina* spine]

synergist (SIN-er-jist) [*syn-* together, -*ergein* to work]

temporalis muscle (tem-poh-RAL-is) [*tempo-* temples, -*alis* pertaining to]

tendon [*tendere* to stretch]

tendon sheaths (sheeths) [*tendere* to stretch]

tensor fasciae latae muscle (TEN-sor FASH-ee LAT-tee) [*tendere* to stretch, *fasci* band or bundle]

teres minor (TER-eez) [*teres* rounded]

tibialis anterior muscle (tib-ee-AL-is) [*tibia* the tibia, *ante* + *prior* foremost]

tonicity (toh-NIS-i-tee) [*tonos* stretching]

trapezius muscle (trah-PEE-zee-us) [*trapezion* small table]

urogenital triangle (YOO-roh-JEN-ih-tal) [*uro-* urine, -*genit* birth or reproduction, -*al* pertaining to, *triangulus* three-cornered]

zygomaticus major muscle (zye-goh-MAT-ik-us) [*zygoma* bolt or bar]

LANGUAGE OF MEDICINE

carpal tunnel syndrome (KAR-pul TUN-el SIN-drohm) [*carp* wrist]

intramuscular injections (in-trah-MUSS-kyoo-lar in-JEK-shuns) [*intra* occurring within]

tenosynovitis (ten-oh-sin-oh-VYE-tis) [*teno-* tendon, -*synovi-* joint fluid, -*itis* inflammation]

CASE STUDY

Patricia Rider, age 42, came to the health clinic complaining of pain in her right hand and fingers for the past 2 months, especially at night. She works as a sewing machine operator at a local factory making men's pants. Ms. Rider thinks of herself as being in good health. She walks daily and follows a low-fat diet. Her family history is negative for cancer, arthritis, heart disease, and diabetes. Her physical examination reflects a positive Tinel sign (tingling sensation radiating from the wrist to the hand) with gentle tapping on the inside of her right wrist with a reflex hammer.

1. After a complete physical examination, carpal tunnel syndrome is diagnosed. Which of the following nerves is usually involved with this diagnosis?

   A. Femoral
   B. Median
   C. Ulna
   D. Radial

2. The physician has ordered that Ms. Rider wear a right wrist splint held in place by an Ace bandage. Which one of the following is the BEST rationale for this treatment?

   A. It will stabilize the joint, thus reducing the pain in this area at night, while allowing Ms. Rider to remove the splint and continue her work as a sewing machine operator during the day.
   B. It will decrease the swelling associated with the injury.
   C. It will remind Ms. Rider to avoid heavy lifting with this extremity and thus decrease the likelihood of reinjury to the area.
   D. It will decrease the repetitive motions of the wrist and prevent continuous injury.

3. During the routine examination, Ms. Rider is found to have a slight separation of muscle when she is asked to lift her head off the examination table while in a supine position. Which muscle do you expect to show this separation based on the above description?

   A. External oblique
   B. Internal oblique
   C. Rectus abdominis
   D. Transversus abdominis

4. During the physical examination, the physician tests Ms. Rider's right brachialis muscle strength. Which one of the following correctly describes this procedure?

   A. The forearm should be flexed at the elbow.
   B. The forearm should be extended at the elbow.
   C. Ms. Rider is asked to extend her elbow and push against a hard surface.
   D. Ms. Rider is asked to lift a heavy object above her head.

CHAPTER SUMMARY

INTRODUCTION (FIGURES 10-5 AND 10-6)

A. There are more than 600 skeletal muscles in the body
B. From 40% to 50% of our body weight is skeletal muscle
C. Muscles, along with the skeleton, determine the form and contour of our body

SKELETAL MUSCLE STRUCTURE (FIGURE 10-1)

A. Connective tissue components
  1. Endomysium—delicate connective tissue membrane that covers specialized skeletal muscle fibers
  2. Perimysium—tough connective tissue binding together fascicles
  3. Epimysium—coarse sheath covering the muscle as a whole
  4. These three fibrous components may become a tendon or an aponeurosis
B. Size, shape, and fiber arrangement (Figure 10-2)
  1. Skeletal muscles vary considerably in size, shape, and fiber arrangement
  2. Size—range from extremely small to large masses
  3. Shape—variety of shapes, such as broad, narrow, long, tapering, short, blunt, triangular, quadrilateral, irregular, flat sheets, or bulky masses
  4. Arrangement—variety of arrangements, such as parallel to a long axis, converging to a narrow attachment, oblique, pennate, bipennate, or curved; the direction of fibers is significant because of its relationship to function
C. Attachment of muscles (Figure 10-3)
  1. Origin—point of attachment that does not move when the muscle contracts
  2. Insertion—point of attachment that moves when the muscle contracts
D. Muscle actions
  1. Most movements are produced by the coordinated action of several muscles; some muscles in the group contract while others relax
     a. Prime mover (agonist)—a muscle or group of muscles that directly performs a specific movement
     b. Antagonist—muscles that when contracting, directly oppose prime movers; relax while the prime mover (agonist) is contracting to produce movement; provide precision and control during contraction of prime movers
     c. Synergists—muscles that contract at the same time as the prime movers; they facilitate prime mover actions to produce a more efficient movement
     d. Fixator muscles—joint stabilizers

E. Lever systems
   1. In the human body, bones serve as levers and joints serve as fulcrums; contracting muscle applies a pulling force on a bone lever at the point of the muscle's attachment to the bone, which causes the insertion bone to move about its joint fulcrum
   2. Lever system—composed of four component parts (Figure 10-4)
      a. Rigid bar (bone)
      b. Fulcrum (F) around which the rod moves (joint)
      c. Load (L) that is moved
      d. Pull (P) that produces movement (muscle contraction)
   3. First-class levers
      a. Fulcrum lies between the pull and the load
      b. Not abundant in the human body; serve as levers of stability
   4. Second-class levers
      a. Load lies between the fulcrum and the joint at which the pull is exerted
      b. Presence of these levers in the human body is a controversial issue
   5. Third-class levers
      a. Pull is exerted between the fulcrum and load
      b. Permit rapid and extensive movement
      c. Most common type of lever found in the body

HOW MUSCLES ARE NAMED

A. Muscle names can be in Latin or English (this book uses English)
B. Muscles are named according to one or more of the following features:
   1. Location, function, shape (Tables 10-1; 10-2; 10-3)
   2. Direction of fibers—named according to fiber orientation (Table 10-4)
   3. Number of heads or divisions (Table 10-4)
   4. Points of attachment—origin and insertion points
   5. Relative size—small, medium, or large (Table 10-5)
C. Hints on how to deduce muscle actions

IMPORTANT SKELETAL MUSCLES

A. Muscles of facial expression—unique in that at least one point of attachment is to the deep layers of the skin over the face or neck (Figures 10-7 and 10-8; Table 10-6)
B. Muscles of mastication—responsible for chewing movements (Figure 10-9; Table 10-6)
C. Muscles that move the head—paired muscles on either side of the neck are responsible for head movements (Figure 10-10; Table 10-7)

TRUNK MUSCLES

A. Muscles of the thorax—critical importance in respiration (Figure 10-11; Table 10-8)
B. Muscles of the abdominal wall—arranged in three layers, with fibers in each layer running in different directions to increase strength (Figure 10-12; Table 10-9)
C. Muscles of the back—bend or stabilize the back (Figure 10-14; Table 10-10)

D. Muscles of the pelvic floor—support the structures in the pelvic cavity (Figure 10-15; Table 10-11)

UPPER LIMB MUSCLES

A. Muscles acting on the shoulder girdle—muscles that attach the upper extremity to the torso are located anteriorly (chest) or posteriorly (back and neck); these muscles also allow extensive movement (Figure 10-16; Table 10-12)
B. Muscles that move the upper part of the arm—the shoulder is a synovial joint allowing extensive movement in every plane of motion (Figure 10-17; Table 10-13)
C. Muscles that move the forearm—found proximal to the elbow and attach to the ulna and radius (Figures 10-20 and 10-21; Table 10-14)
D. Muscles that move the wrist, hand, and fingers—these muscles are located on the anterior or posterior surfaces of the forearm (Figures 10-22 through 10-24; Table 10-15)

LOWER LIMB MUSCLES

A. The pelvic girdle and lower extremity function in locomotion and maintenance of stability
B. Muscles that move the thigh and lower part of the leg (Figures 10-5, 10-6, and 10-25 through 10-31; Tables 10-16 and 10-17)
C. Muscles that move the ankle and foot (Figures 10-32 and 10-33; Table 10-18)
   1. Extrinsic foot muscles are located in the leg and exert their actions by pulling on tendons that insert on bones in the ankle and foot; responsible for dorsiflexion, plantar flexion, inversion, and eversion
   2. Intrinsic foot muscles are located within the foot; responsible for flexion, extension, abduction, and adduction of the toes

POSTURE

A. Maintaining the posture of the body is one of the major roles muscles play
B. "Good posture"—body alignment that most favors function; achieved by keeping the body's center of gravity over its base and requires the least muscular work to maintain
C. How posture is maintained
   1. Muscles exert a continual pull on bones in the opposite direction from gravity
   2. Structures other than muscle and bones have a role in maintaining posture
      a. Nervous system—responsible for the existence of muscle tone and also for regulation and coordination of the amount of pull exerted by individual muscles
      b. Respiratory, digestive, excretory, and endocrine systems all contribute to maintain posture

CYCLE OF LIFE: MUSCULAR SYSTEM

A. Muscle cells—increase or decrease in number, size, and ability to shorten at different periods
B. Pathological conditions at different periods may affect the muscular system

C. Life cycle changes—manifested in other components of functional unit:
   1. Infancy and childhood—coordination and control of muscle contraction permit sequential development steps
D. Degenerative changes of advancing age result in replacement of muscle cells with nonfunctional connective tissue
   1. Diminished strength

REVIEW QUESTIONS

1. Define the terms *endomysium*, *perimysium*, and *epimysium*.
2. Identify and describe the most common type of lever system found in the body.
3. Give an example of a muscle named by location, function, shape, fiber direction, number of heads, points of attachment.
4. Name the main muscles of the back, chest, abdomen, neck, shoulder, upper part of the arm, lower part of the arm, thigh, buttocks, leg, and pelvic floor.
5. Name the main muscles that flex, extend, abduct, and adduct the upper part of the arm; that raise and lower the shoulder.
6. Name the main muscles that flex and extend the lower part of the arm; that flex and extend the wrist and hand.
7. Name the main muscles that flex, extend, abduct, and adduct the thigh; that flex and extend the lower part of the leg and thigh; that flex and extend the foot.
8. Name the main muscles that flex, extend, abduct, and adduct the head.
9. Name the main muscles that move the abdominal wall; that move the chest wall.

CRITICAL THINKING QUESTIONS

1. Identify the muscles of facial expression. What muscles permit smiling and frowning?
2. How do the origin and insertion of a muscle relate to each other in regard to actual movement?
3. When the biceps brachii contracts, the elbow flexes. When the triceps brachii contracts, the elbow extends. Explain the role of both muscles in terms of agonist and antagonist in both of these movements.
4. Can you describe how posture is maintained?
5. Describe the clinical significance regarding the difference in size between the large head of the humerus and the small and shallow glenoid cavity of the scapula.
6. If a typist complained of weakness, pain, and tingling in the palm and thumb side of the hand, what type of problem might that typist be experiencing? Explain specifically what was happening to cause this discomfort.
7. Baseball players, particularly pitchers, often incur rotator cuff injuries. List the muscles that make up the rotator cuff and explain the importance of these muscles and their role in joint stability.

Physiology of the Muscular System

LANGUAGE OF SCIENCE

acetylcholine (ACh) (ass-ee-til-KOH-leen) [*acetyl-* vinegar, *-chole-* bile, *-ine* made of]

actin (AK-tin) [*act-* act or do, *-in* substance]

concentric contractions (kon-SEN-trik) [*con-* together, *-centr* center, *con-* together, *-traction* to draw]

contractility (kon-trak-TIL-i-tee) [*con-* together, *-tractil-* to draw, *-ity* quality of]

creatine phosphate (CP) (KREE-ah-tin FOS-fayt) [*creatine* flesh, *phosph-* phosphorus, *-ate* chemical derivative]

cross bridges

eccentric contractions (ek-SENT-rik) [*ec-* out of, *-centr* center, *con-* together, *-traction* to draw]

elastic filaments (eh-LAS-tik FIL-ah-ments) [*elastic* to drive, *fila-* to spin thread, *-ment* result]

excitability (ek-syte-eh-BIL-i-tee) [*excite* to arouse]

extensibility (ek-STEN-si-BIL-i-tee) [*extend* to stretch out]

fast fibers

graded strength principle

hypertrophy (hye-PER-troh-fee) [*hyper-* excessive, *-troph* nourishment]

intermediate fibers (in-ter-MEE-dee-it) [*inter-* occurring between, *-mediate* to divide]

isometric contraction (eye-soh-MET-rik) [*iso-* equal, *-metr-* measure, *-ic* pertaining to]

isotonic contraction (eye-soh-TON-ik) [*iso-* equal, *-tonic* quality of muscle contraction]

lactic acid (LAK-tik) [*lact-* milk, *-ic* pertaining to, *acidus* sour]

M line

motor endplate [*mot* movement]

motor neuron (NOO-ron) [*mot* movement, *neuro* nerve]

motor unit [*mot* movement]

multiunit smooth muscle [*multi* many]

muscle fatigue (fah-TEEG) [*fatigue* to tire]

Cont'd on p. 422

In Chapter 10, we explored the anatomy of skeletal muscle organs and how they work together to accomplish specific body movements. In this chapter, we continue our study of the skeletal muscular system by examining the basic characteristics of skeletal muscle tissue. We uncover the mechanisms that permit skeletal muscle tissue to move the body's framework, as well as perform other functions vital to maintaining a constant internal environment. We also briefly examine smooth and cardiac muscle tissues and contrast them with skeletal muscle tissue.

GENERAL FUNCTIONS

If you have any doubts about the importance of muscle function to normal life, think about what it would be like without it. It is hard to imagine life with this matchless power lost. However, as cardinal as it is, movement is not the only contribution muscles make to healthy survival. They also perform two other essential functions: production of a large proportion of body heat and maintenance of posture.

1. **Movement.** Skeletal muscle contractions produce movement of the body as a whole (locomotion) or movement of its parts.
2. **Heat production.** Muscle cells, like all cells, produce heat by the process known as catabolism (discussed in Chapters 4 and 27). But because skeletal muscle cells are both highly active and numerous, they produce a major share of total body heat. Skeletal muscle contractions therefore constitute one of the most important parts of the mechanism for maintaining homeostasis of temperature.
3. **Posture.** The continued partial contraction of many skeletal muscles makes possible standing, sitting, and maintaining a relatively stable position of the body while walking, running, or performing other movements.

FUNCTION OF SKELETAL MUSCLE TISSUE

Skeletal muscle cells have several characteristics that permit them to function as they do. One such characteristic is the ability to be stimulated, often called **excitability** or *irritability*. Because skeletal muscle cells are excitable, they can respond to regulatory mechanisms such as nerve signals.

Contractility of muscle cells, the ability to contract or shorten, allows muscle tissue to pull on bones and thus produce body movement. Sometimes muscle fibers do work by steadily resisting a load without actually becoming shorter. In such a case, the muscle cell is still said to be contracting. The term contraction, when applied to muscles, is meant in a broad sense of pulling the ends together—regardless of whether the cell actually gets shorter.

Extensibility, the ability to extend or stretch, allows muscles to return to their resting length after having contracted. Muscles may also extend while still exerting force, as when lowering a heavy object in your hand.

All of these characteristics of muscle cells are related to the microscopic structure of skeletal muscle cells. In the following passages, we first discuss the basic structure of a muscle cell. We then explain how a muscle cell's structural components allow it to perform its specialized functions.

Overview of the Muscle Cell

Look at Figure 11-1. As you can see, a skeletal muscle is composed of bundles of skeletal muscle fibers that generally extend the entire length of the muscle. One reason they are called *fibers* instead of *cells* is because of their long, thin, threadlike shape. They are 1 to 40 mm long but have a diameter of only 10 to 100 μm. The gastrocnemius muscle of the calf, for example, has approximately a million threadlike muscle fibers.

Another reason skeletal muscle fibers are not usually called "muscle cells" is because during muscle tissue development, individual precursor cells fuse together to form a new, combined structure with many nuclei—the mature muscle fiber. That is why muscle fibers don't follow the general rule of one nucleus per cell: each fiber is made up of several cells that are now combined into one. Some adult muscle fibers have one of these tiny precursor cells hugging their outer boundary. These "satellite cells" are in fact stem cells that can become active after a muscle injury to produce more muscle fibers. Scientists are now trying to understand this mechanism better in the hope of developing new therapies for replacing lost muscle tissues.

Skeletal muscle fibers have many of the same structural parts as other cells. Several of them, however, bear different names in muscle fibers. For example, **sarcolemma** is a name often used for the plasma membrane of a muscle fiber. **Sarcoplasm** is its cytoplasm. Muscle fibers contain many more mitochondria than usual, and as we have already learned, each fiber has several nuclei (Figure 11-2).

Muscle cells contain networks of tubules and sacs known as the **sarcoplasmic reticulum (SR)**—the muscle fiber's version of smooth endoplasmic reticulum. The function of the SR is to temporarily store calcium ions (Ca^{++}). The membrane of the SR continually pumps Ca^{++} from the sarcoplasm and stores the ions within its sacs.

A structure unique to muscle cells is a system of transverse tubules, or **T tubules.** This name derives from the fact that these tubules extend transversely across the sarcoplasm, at a right angle to the long axis of the cell. As Figures 11-1, *B*, and 11-2 show, T tubules are formed by inward extensions of the sarcolemma. The chief function of T tubules is to allow electrical signals, or *impulses*, traveling along the sarcolemma to move deeper into the cell.

Notice in Figures 11-1, *B*, and 11-2 that a tubular sac of the SR butts up against each side of every T tubule in a muscle fiber. This triplet of tubules (a T tubule sandwiched between sacs of the SR) is called a **triad.** The triad is an important feature of the muscle cell because it allows an electrical impulse traveling along a T tubule to stimulate the membranes of adjacent sacs of the SR. Figure 11-3 shows how this works. At rest, calcium ion pumps in the SR membrane pump Ca^{++} into the terminal sacs of the SR. However, when an electrical impulse travels along the sarcolemma and down the T tubule, gated Ca^{++} channels in the SR open in response to the voltage fluctuation. This floods the sar-

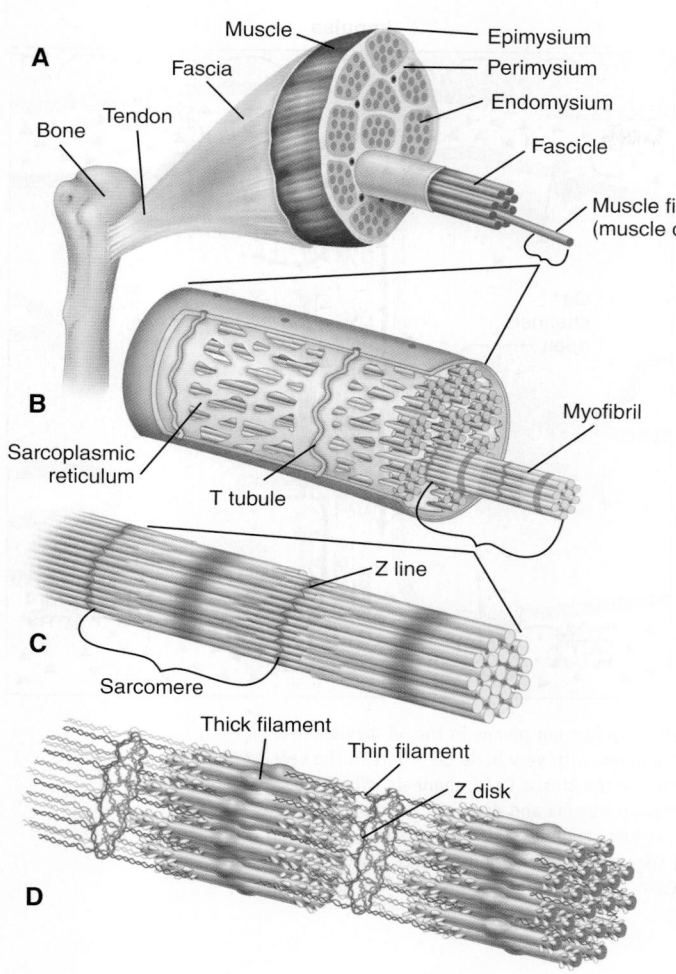

A

Bone
Tendon
Fascia
Muscle
Epimysium
Perimysium
Endomysium
Fascicle
Muscle fiber (muscle cell)

B

Sarcoplasmic reticulum
T tubule
Myofibril

C

Z line
Sarcomere

D

Thick filament
Thin filament
Z disk

Figure 11-1 *Structure of skeletal muscle.* **A,** Skeletal muscle organ composed of bundles of contractile muscle fibers held together by connective tissue. **B,** Greater magnification of a single fiber showing smaller fibers—myofibrils—in the sarcoplasm. Note the sarcoplasmic reticulum and T tubules forming a three-part structure called a *triad*. **C,** Myofibril magnified further to show a sarcomere between successive Z lines (Z disks). Cross striae are visible. **D,** Molecular structure of a myofibril showing thick myofilaments and thin myofilaments.

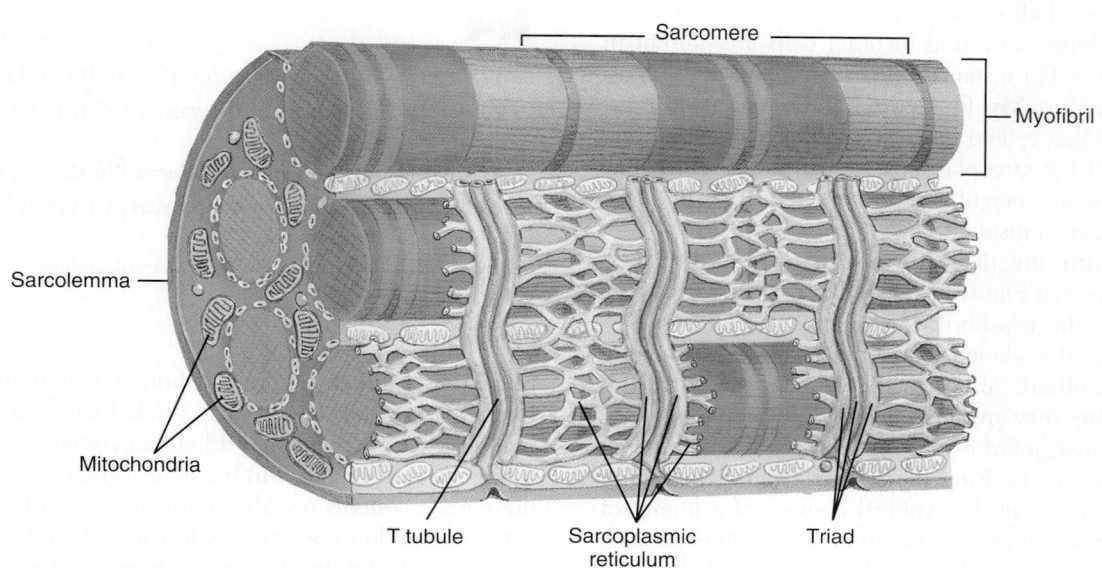

Sarcomere
Myofibril
Sarcolemma
Mitochondria
T tubule
Sarcoplasmic reticulum
Triad

Figure 11-2 *Unique features of the skeletal muscle cell.* Notice especially the T tubules, which are extensions of the plasma membrane, or sarcolemma, and the sarcoplasmic reticulum (SR), a type of smooth endoplasmic reticulum that forms networks of tubular canals and sacs containing stored calcium ions. A triad is a triplet of adjacent tubules: a terminal (end) sac of the SR, a T tubule, and another terminal sac of the SR.

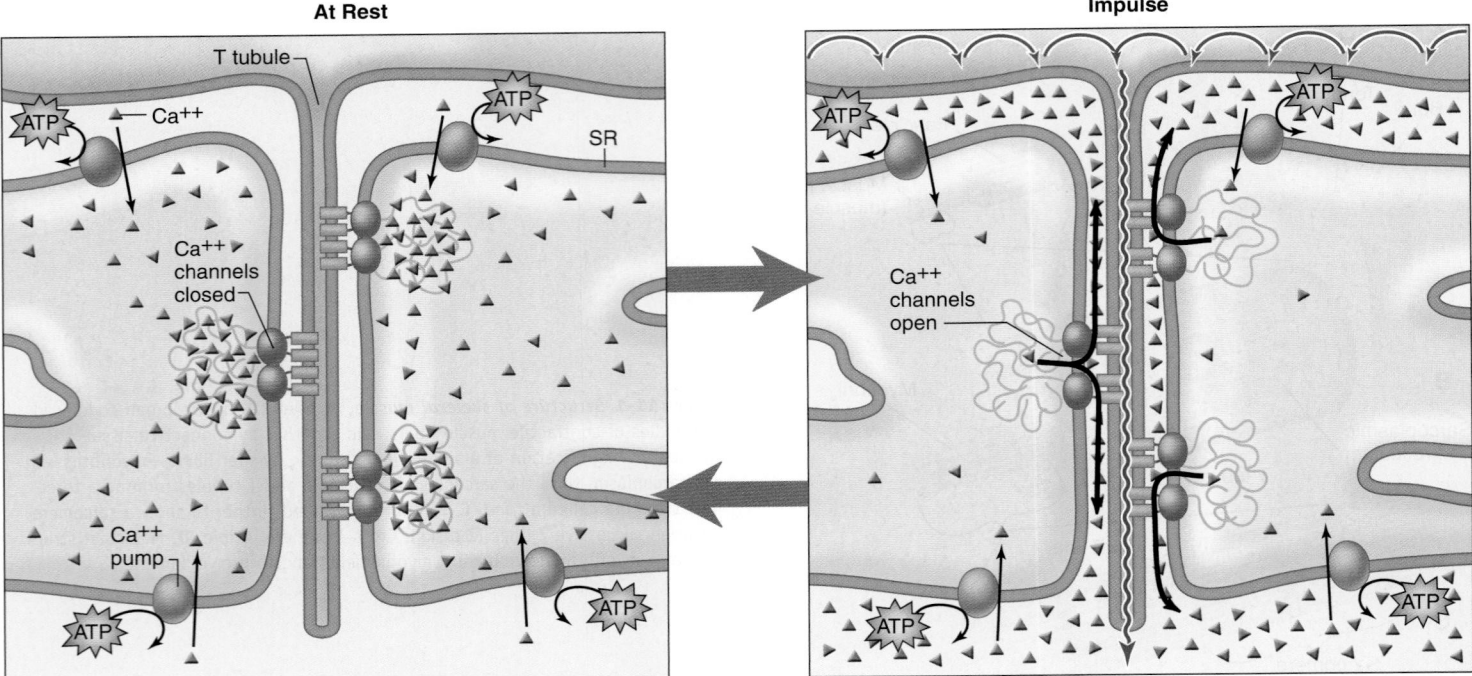

Figure 11-3 *Storage and release of calcium ions.* At rest *(left),* calcium ion pumps in the SR membrane actively pull Ca^{++} into the SR—thus creating a concentration gradient with very little Ca^{++} left in the sarcoplasm. Note that the Ca^{++} is sequestered in "bunches" just inside the closed Ca^{++} channels. When the fiber is stimulated *(right),* an electrical impulse travels from the sarcolemma and down the T tubule, where the voltage fluctuation triggers the opening of voltage-gated calcium channels. This allows passive diffusion of the Ca^{++} inside the SR out to the sarcoplasm, where it will trigger the contraction process. Almost immediately after stimulation, the calcium channels close and the Ca^{++} is once again pumped into the sacs of the SR *(left). ATP,* Adenosine triphosphate.

coplasm with calcium ions—an important step in initiating muscle contraction, as we shall see.

Additional structures not found in other cells are present in skeletal muscle fibers. For instance, the cytoskeleton of the muscle fiber is quite unique. **Myofibrils** are bundles of very fine cytoskeletal filaments that extend lengthwise along skeletal muscle fiber and almost fill the sarcoplasm. A typical muscle fiber may have somewhere in the neighborhood of a thousand or more myofibrils tightly packed inside it.

Myofibrils, in turn, are made up of still finer fibers called **myofilaments.** Notice in Figure 11-1, *D,* that there are two types of myofilaments in the myofibril: *thick filaments* and *thin filaments.* You can see at a glance that some of the filaments are much thicker than others. Although they appear in contrasting colors in the diagram, they are actually colorless.

A **sarcomere** is a segment of the myofibril between two successive **Z lines** (Box 11-1). Find the label *sarcomere* in Figure 11-1, *C.* As you can see, each myofibril consists of a lineup of many sarcomeres. Each sarcomere functions as a contractile unit. The A bands of sarcomeres appear as relatively wide, dark stripes (cross striae) under the microscope, and they alternate with narrower, lighter-colored stripes formed by the I bands (see Figure 11-1, *D*). Because of its cross striations (or *striae*), skeletal muscle is sometimes called *striated muscle.* Electron microscopy of skeletal muscle (Figure 11-4) reveals details that help us understand concepts of its structure and function.

QUICK CHECK

1. What are the three major functions of the skeletal muscles?
2. Name some features of the muscle cell that are not found in other types of cells.
3. What causes the striations observed in skeletal muscle fibers?
4. Why is the triad relationship between T tubules and the SR important?

Myofilaments

Each muscle fiber contains a thousand or more parallel myofibrils that are only about 1 μm thick. Lying side by side in each myofibril are approximately 15,000 sarcomeres, each made up of hundreds of thick and thin filaments. The molecular structure of these myofilaments reveals the mechanism of how muscle fibers contract and do so powerfully. It is wise, therefore, to take a moment to study the structure of myofilaments before discussing the detailed mechanism of muscle contraction.

First of all, four different kinds of protein molecules make up myofilaments: **myosin, actin, tropomyosin,** and **troponin.** The thin filaments are made of a combination of three proteins: actin, tropomyosin, and troponin. Figure 11-5, A, shows that globular actin molecules are strung together like beads to form two fibrous

BOX 11-1   A More Detailed Look at the Sarcomere

The sarcomere is the basic contractile unit of the muscle cell. As you read the explanation of the sarcomere's structure and function, you might wonder what the Z line, M line, and other components really are—and what they do for the muscle cell.

First of all, it is important that you appreciate the three-dimensional nature of the sarcomere. You can then realize that the Z line is actually a dense plate or disk to which the thin filaments directly anchor. As a matter of fact, the Z line is often called the **Z disk.** Besides being an anchor for myofibrils, the Z line is useful as a landmark separating one sarcomere from the next.

Detailed analysis of the sarcomere also shows that the thick (myosin) filaments are held together and stabilized by protein molecules that form the **M line.** Note that the regions of the sarcomere are identified by specific zones or bands:

- **A band**—the segment that runs the entire length of the thick filaments
- **I band**—the segment that includes the Z line and the ends of the thin filaments where they do not overlap the thick filaments
- **H zone**—the middle region of the thick filaments where they do not overlap the thin filaments

Note in Figure 11-2 that the T tubules in human muscle fibers align themselves along the borders between the A band and I band.

Later, as you review the process of contraction, note how the regions listed above change during each step of the process.

In addition to thin and thick filaments, each sarcomere has numerous **elastic filaments.** Elastic filaments, composed of a protein called *titin (connectin),* anchor the ends of the thick filaments to the Z line, as the figure shows. The elastic filaments are believed to give myofibrils, and thus muscle fibers, their characteristic elasticity. *Dystrophin,* not shown here, is a protein that holds the actin filaments to the sarcolemma. Dystrophin and a complex of connected molecules anchors the muscle fiber to surrounding matrix so that the muscle does not break during a contraction. Dystrophin and its role in muscular dystrophy are discussed further on p. 420.

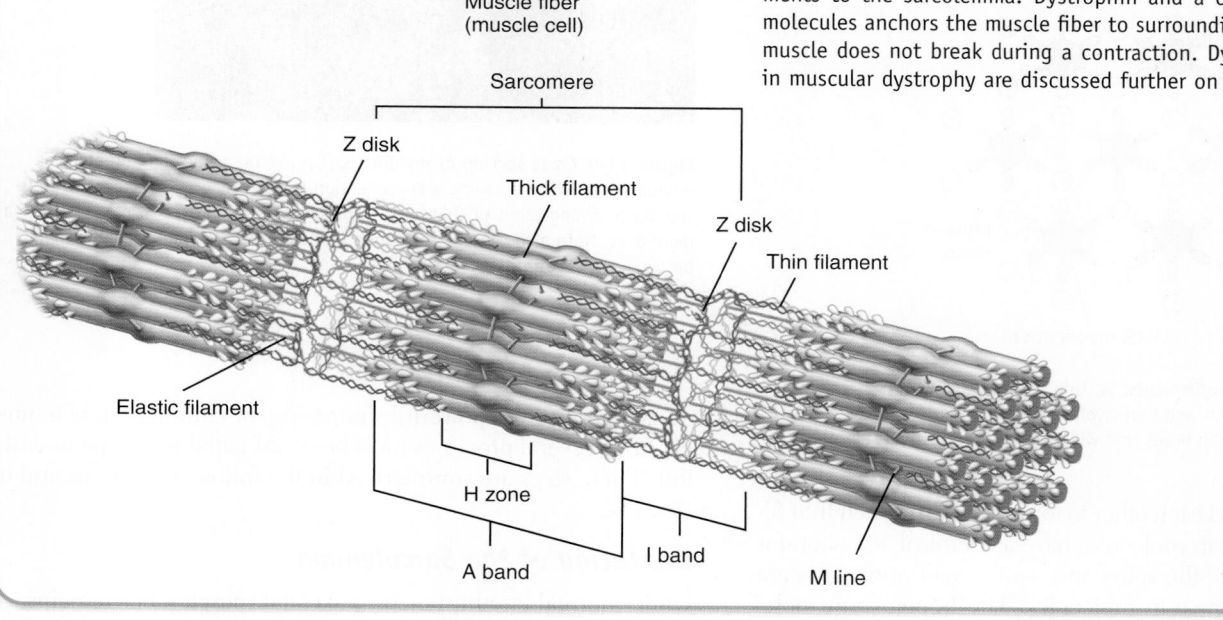

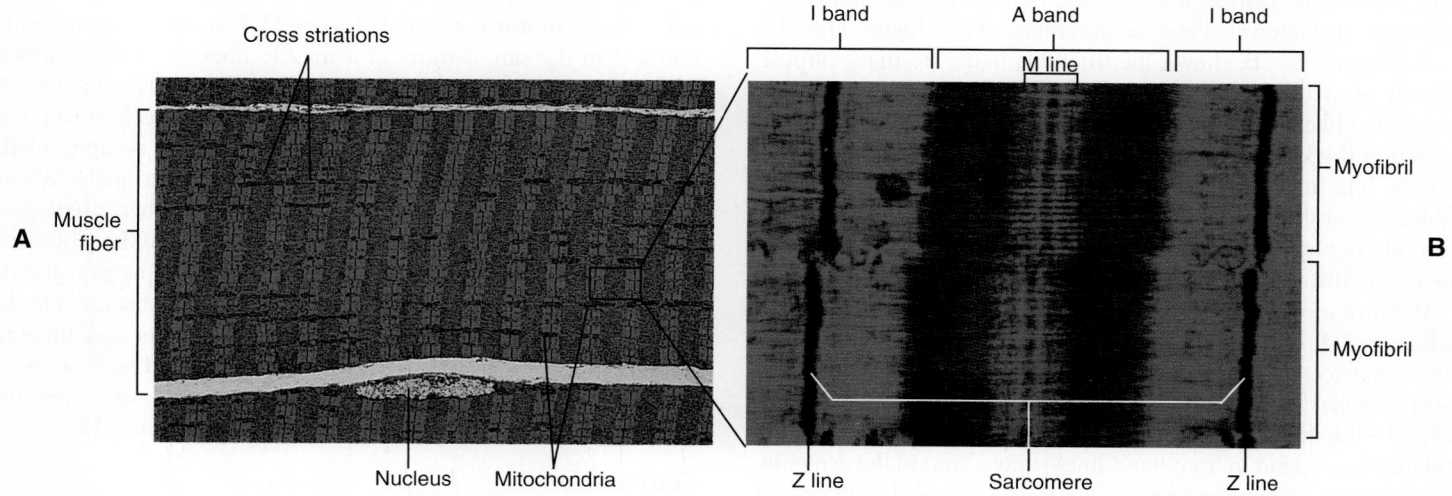

Figure 11-4 *Skeletal muscle striations.* Color-enhanced scanning electron micrographs (SEMs) showing longitudinal views of skeletal muscle fibers. **B** shows detail of **A** at greater magnification. Note that the myofilaments of each myofibril form a pattern that when viewed together, produces the striated (striped) pattern typical of skeletal muscle.

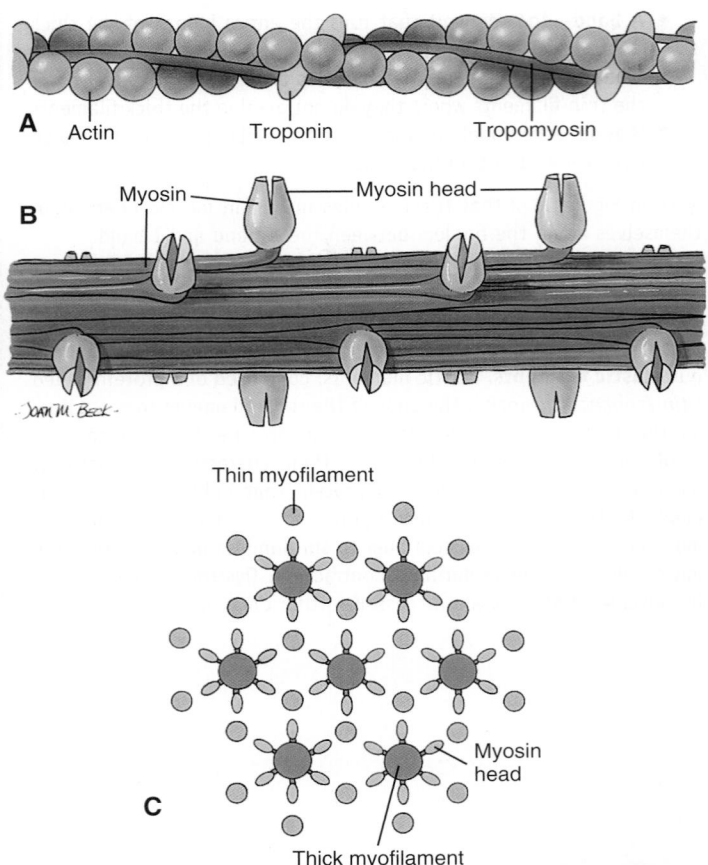

Figure 11-5 *Structure of myofilaments.* **A,** Thin myofilament. **B,** Thick myofilament. **C,** Cross section of several thick and thin myofilaments showing the relative positions of myofilaments and the myosin heads that will form cross bridges between them.

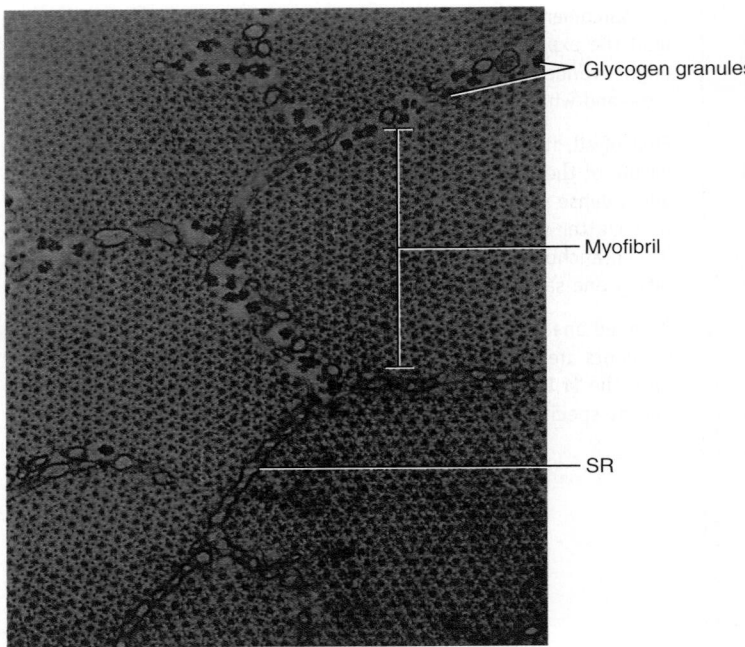

Figure 11-6 *Cross section of myofibrils.* Color-enhanced scanning electron micrographs (SEMs) showing a cross section from a skeletal muscle fiber. Note the dense arrangement of thick and thin filaments, seen here in cross section as mere dots. Note also the dark glycogen granules and SR tubules sandwiched between the myofibrils.

strands that twist around each other to form the bulk of each thin filament. Actin and myosin molecules have a chemical attraction for one another, but at rest, the active sites on the actin molecules are covered by long tropomyosin molecules. The tropomyosin molecules seem to be held in this blocking position by troponin molecules spaced at intervals along the thin filament (see Figure 11-5, A).

As Figure 11-5, B, shows, the thick filaments are made almost entirely of myosin molecules. Notice that the myosin molecules are shaped like golf clubs, with their long shafts bundled together to form a thick filament and their "heads" sticking out from the bundle. The myosin heads are chemically attracted to the actin molecules of the nearby thin filaments, so they angle toward the thin filaments. When they bridge the gap between adjacent myofilaments, the myosin heads are usually called **cross bridges.**

Within a myofibril the thick and thin filaments alternate, as shown in Figure 11-1, D. They are also very close to one another, as you can see in cross section in Figure 11-5, C, and Figure 11-6. This arrangement is crucial for contraction. Another fact important for contraction is that the thin filaments attach to both Z lines (Z disks) of a sarcomere and that they extend in from the Z lines partway toward the center of the sarcomere. When the muscle fiber is relaxed, the thin filaments terminate at the outer edges of the H zones. In contrast, the thick myosin filaments do not attach directly to the Z lines, and they extend only to the length of the A bands of the sarcomeres.

Mechanism of Contraction

To accomplish the powerful shortening, or contraction, of a muscle fiber, several processes must be coordinated in a stepwise fashion. These steps are summarized in the following sections and in Box 11-2.

Excitation of the Sarcolemma

Under normal circumstances, a skeletal muscle fiber remains "at rest" until it is stimulated by a signal from a special type of nerve cell called a **motor neuron.** As Figure 11-7 shows, motor neurons connect to the sarcolemma of a muscle fiber at a folded **motor endplate** to form a junction called a neuromuscular junction. A **neuromuscular junction (NMJ)** is a type of connection called a *synapse* and is characterized by a narrow gap, or synaptic cleft, across which *neurotransmitter* molecules transmit signals. When nerve impulses reach the end of a motor neuron fiber, small vesicles release a neurotransmitter, **acetylcholine (ACh),** into the synaptic cleft. Diffusing swiftly across this microscopic gap, acetylcholine molecules contact the sarcolemma of the adjacent muscle fiber. There they stimulate acetylcholine receptors and thereby initiate an electrical impulse in the sarcolemma. The process of synaptic transmission and induction of an impulse—a process often called *excitation*—is discussed in detail in Chapter 12.

Contraction

The impulse, a temporary electrical voltage imbalance, is conducted over the muscle fiber's sarcolemma and inward along the T tubules (Figure 11-8). The impulse in the T tubules trig-

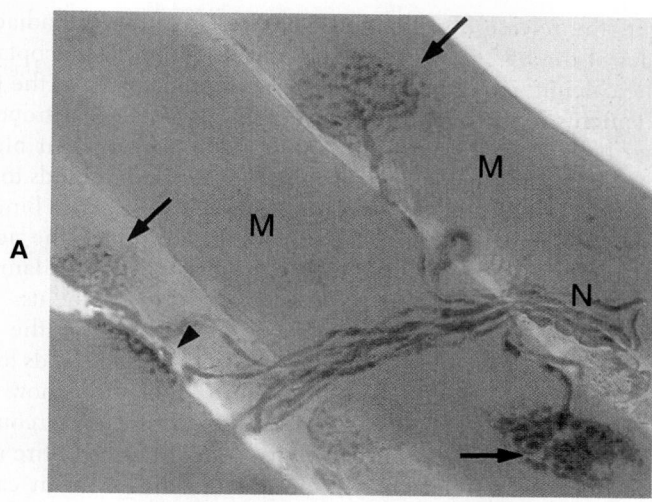

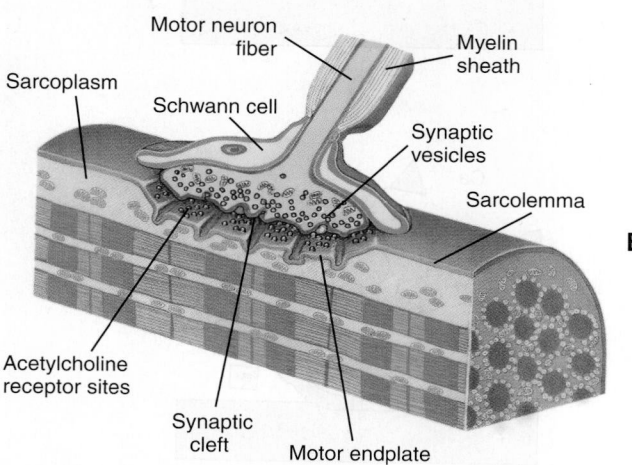

Figure 11-7 *Neuromuscular junction (NMJ).* **A,** Micrograph showing four neuromuscular junctions. Three are surface views *(arrows)* and one is a side view *(arrowhead)*. *N,* Nerve fibers; *M,* muscle fibers. **B,** This sketch shows a side view of the NMJ. Note how the distal end of a motor neuron fiber forms a synapse, or "chemical junction," with an adjacent muscle fiber. Neurotransmitter molecules (specifically, acetylcholine) are released from the neuron's synaptic vesicles and diffuse across the synaptic cleft. There they stimulate receptors in the motor endplate region of the sarcolemma.

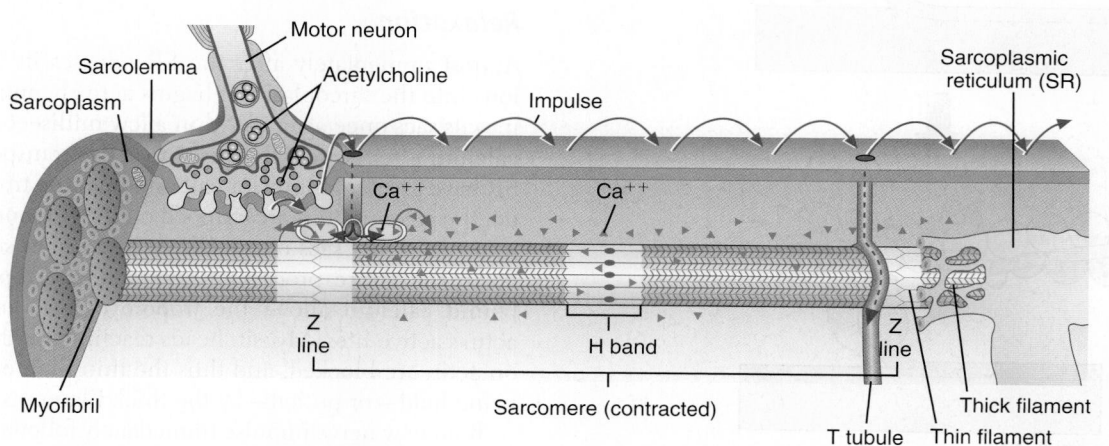

Figure 11-8 *Effects of excitation on a muscle fiber.* Excitation of the sarcolemma by a nerve impulse initiates an impulse in the sarcolemma. The impulse travels across the sarcolemma and through the T tubules, where it triggers adjacent sacs of the SR to release a flood of calcium ions (Ca^{++}) into the sarcoplasm. Ca^{++} is then free to bind to troponin molecules in the thin filaments. This binding, in turn, initiates the chemical reactions that produce a contraction.

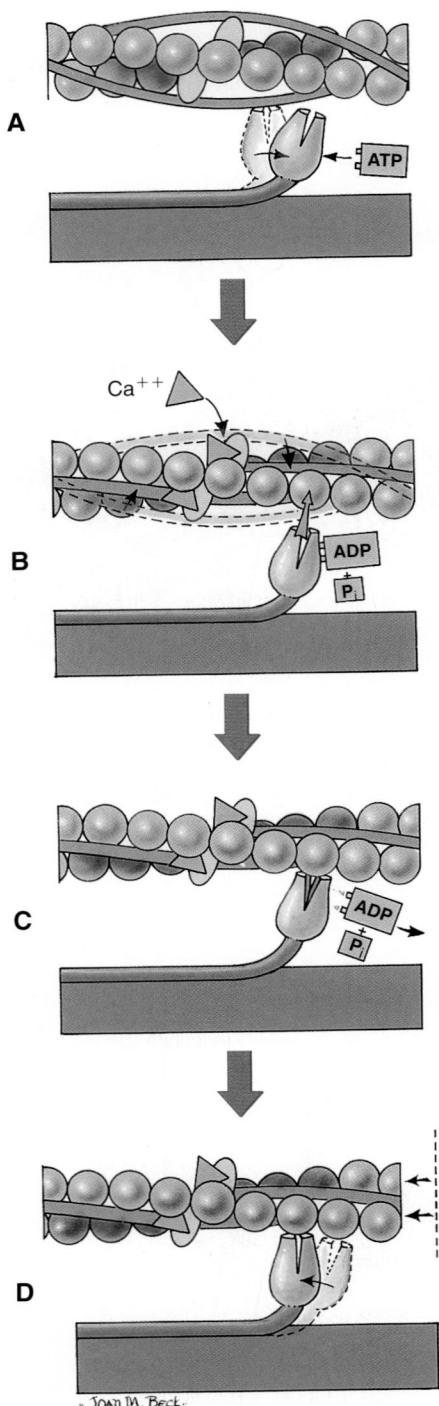

A

Ca++

B

ADP
+
P$_i$

C

ADP
+
P$_i$

D

- Joan M. Beck -

Figure 11-9 *The molecular basis of muscle contraction.* A, Each myosin head in the thick filament moves into a resting position after an ATP molecule binds and transfers its energy. **B,** Calcium ions released from the SR bind to troponin in the thin filament, thereby allowing tropomyosin to shift from its position blocking the active sites of actin molecules. **C,** Each myosin head then binds to an active site on a thin filament and displaces the remnants of ATP hydrolysis—adenosine diphosphate (ADP) and inorganic phosphate (P$_i$). **D,** The release of stored energy from step **A** provides the force needed for each head to move back to its original position and pull actin along with it. Each head will remain bound to actin until another ATP molecule binds to it and pulls it back into its resting position **(A).**

gers the release of a flood of calcium ions from the adjacent sacs of the SR (see Figures 11-3 and 11-8). In the sarcoplasm, the calcium ions combine with troponin molecules in the thin filaments of the myofibrils (Figure 11-9). Recall that troponin normally holds tropomyosin strands in a position that blocks the chemically active sites of actin. When calcium binds to troponin, however, the tropomyosin shifts to expose active binding sites on the actin molecules (Figure 11-10). Once the active sites are exposed, energized myosin heads of the thick filaments bind to actin molecules in the nearby thin filaments. The myosin heads bend with great force, literally pulling the thin filaments past them. Each head then releases itself, binds to the next active site, and pulls again. Figure 11-11 shows how sliding of the thin filaments toward the center of each sarcomere quickly shortens the entire myofibril—and thus the entire muscle fiber. This model of muscle contraction has been called the **sliding-filament model.** Perhaps a better name might be *ratcheting-filament model* because the myosin heads actively ratchet the thin filaments toward the center of the sarcomere with great force.

Muscle fibers usually contract to about 80% of their starting length—rarely to 60% or 70%. Some muscle fibers contract hardly at all because they are pulling on an unmoving load. Even so, such muscle fibers are still said to be "contracting" in a broad sense. Their sarcomeres are still working hard to pull the ends toward one another, as though they are at a steady "draw" in a game of tug-of-war (Figure 11-12). Because the actin-myosin bond is not permanent, the sarcomere cannot "lock up" once it has shortened to hold its position passively. Thus, it takes energy to actively maintain a shortened position—by continually repeating the actin-myosin reaction as long as necessary.

Relaxation

Almost immediately after the SR releases its flood of calcium ions into the sarcoplasm, it begins actively pumping them back into its sacs once again. Within a few milliseconds, much of the calcium is recovered. Because the active transport carriers of the SR have greater affinity for calcium than the troponin molecules do, the calcium ions are stripped off the troponin molecules and returned to the sacs of the SR. As you might suspect, this shuts down the entire process of contraction. Troponin without its bound calcium allows the tropomyosin to once again block actin's active sites. Myosin heads reaching for the next active site on actin are blocked, and thus the thin filaments are no longer being held—or pulled—by the thick filaments.

If no new nerve impulse immediately follows, the muscle fiber now relaxes. The relaxed muscle fiber may remain at its contracted length, but forces outside the muscle fiber are likely to pull it back to its longer resting length.

As we have seen, the contraction process in a skeletal muscle fiber automatically shuts itself off within a small fraction of a second after the initial stimulation. However, a muscle fiber may sustain a contraction for some time if there are many stimuli in rapid succession, thus permitting calcium ions to remain available in the sarcoplasm for a longer period.

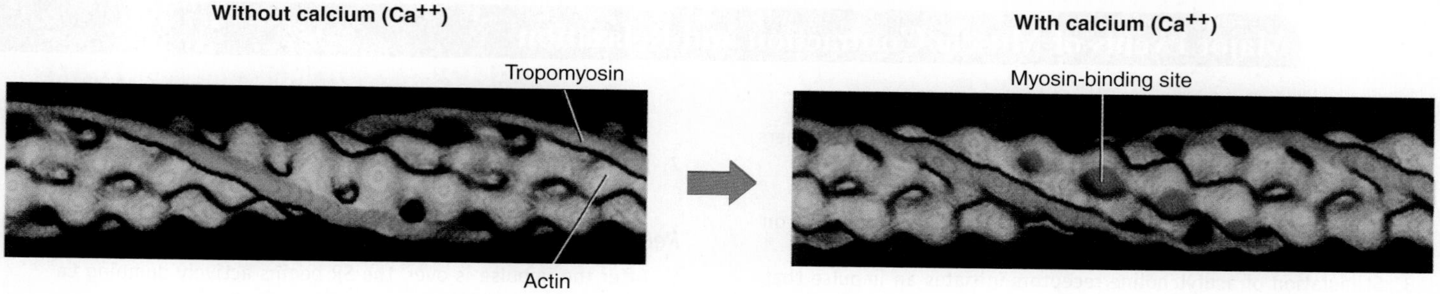

Without calcium (Ca⁺⁺)

Tropomyosin

Actin

With calcium (Ca⁺⁺)

Myosin-binding site

Figure 11-10 *Role of calcium in muscle contraction.* Color-enhanced scanning electron micrograph (SEM) of a thin filament. When calcium is absent, the active myosin binding sites on actin are covered by tropomyosin. However, after calcium becomes available and binds to troponin, the tropomyosin is pulled out of its blocking position and reveals the active binding sites on actin.

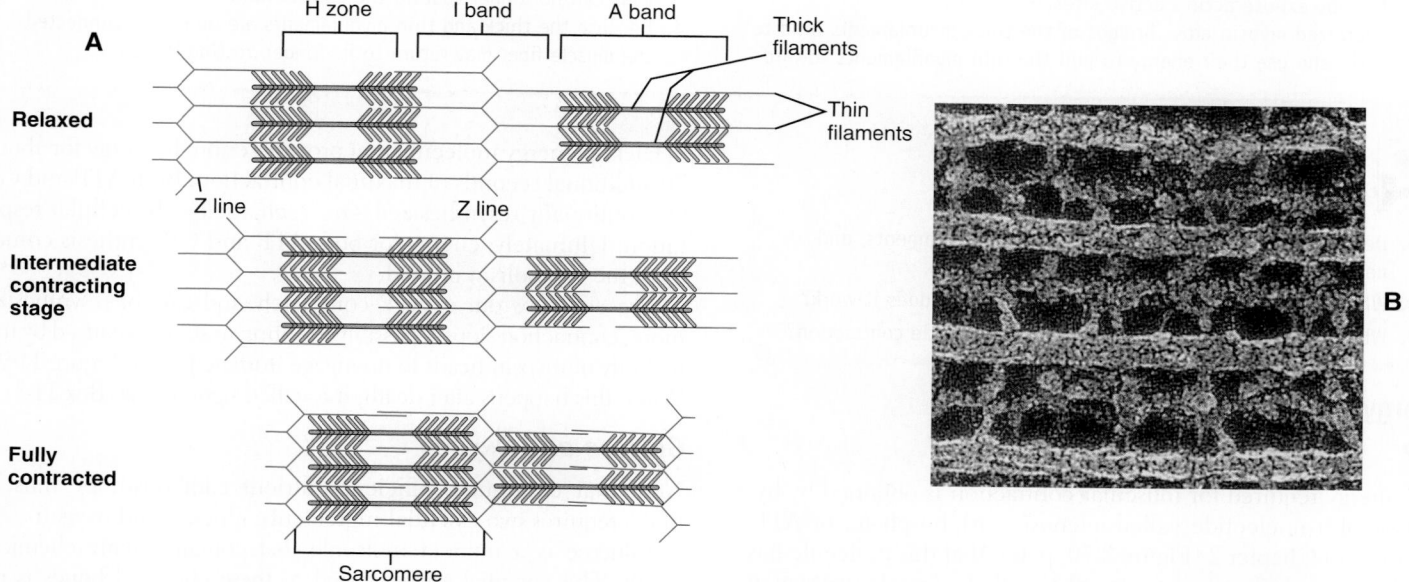

Figure 11-11 *Sliding-filament model.* **A,** During contraction, myosin cross bridges pull the thin filaments toward the center of each sarcomere, thus shortening the myofibril and the entire muscle fiber. **B,** Color-enhanced scanning electron micrograph (SEM) showing the myosin cross bridges that connect the thick filaments to the thin filaments pulling on the thin filaments and causing them to slide.

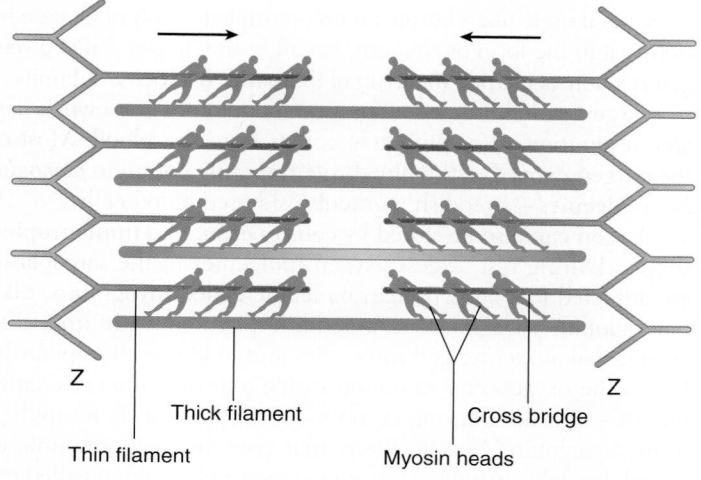

Figure 11-12 *Simplified sarcomere.* This diagram illustrates the concept of muscle contraction as a sort of tug-of-war game in which the myosin heads (shown here as little people) hold onto thin filament "ropes"—thus forming cross bridges. As the myosin heads pull on the thin filaments, the Z lines (Z disks) get closer together—thus shortening the sarcomere. Likewise, the short length of a sarcomere may be held in position by the continued effort of the myosin heads.

BOX 11-2   Major Events of Muscle Contraction and Relaxation

Excitation and Contraction

1. A nerve impulse reaches the end of a motor neuron and triggers release of the neurotransmitter acetylcholine.
2. Acetylcholine diffuses rapidly across the gap of the neuromuscular junction and binds to acetylcholine receptors on the motor endplate of the muscle fiber.
3. Stimulation of acetylcholine receptors initiates an impulse that travels along the sarcolemma, through the T tubules, to the sacs of the SR.
4. Ca^{++} is released from the SR into the sarcoplasm, where it binds to troponin molecules in the thin myofilaments.
5. Tropomyosin molecules in the thin myofilaments shift and thereby expose actin's active sites.
6. Energized myosin cross bridges of the thick myofilaments bind to actin and use their energy to pull the thin myofilaments toward

the center of each sarcomere. This cycle repeats itself many times per second, as long as adenosine triphosphate is available.

7. As the filaments slide past the thick myofilaments, the entire muscle fiber shortens.

Relaxation

1. After the impulse is over, the SR begins actively pumping Ca^{++} back into its sacs.
2. As Ca^{++} is stripped from troponin molecules in the thick myofilaments, tropomyosin returns to its position and blocks actin's active sites.
3. Myosin cross-bridges are prevented from binding to actin and thus can no longer sustain the contraction.
4. Because the thick and thin myofilaments are no longer connected, the muscle fiber may return to its longer, resting length.

QUICK CHECK

5. Describe the structure of thin and thick myofilaments, and name the kinds of proteins that compose them.
6. What is a *neuromuscular junction (NMJ)*? How does it work?
7. What is the role of calcium ions (Ca^{++}) in muscle contraction?

Energy Sources for Muscle Contraction

ATP

The energy required for muscular contraction is obtained by hydrolysis of a nucleotide called adenosine triphosphate, or ATP. Recall from Chapter 2 (Figure 2-30, p. 68) that this molecule has an adenine and ribose group (together called *adenosine*) attached to three phosphate groups. Two of the three phosphate groups in ATP are attached to the molecule by *high-energy bonds*. Breaking of these high-energy bonds provides the energy necessary to pull the thin myofilaments during muscle contraction.

As Figure 11-9, *A*, shows, before contraction occurs, each myosin cross-bridge head moves into a resting position when an ATP molecule binds to it. The ATP molecule breaks its outermost high-energy bond, thereby releasing inorganic phosphate (P_i) and transferring the energy to the myosin head. In a way, this is like pulling back the elastic band of a slingshot—the apparatus is "at rest" but ready to spring. When myosin binds to actin, the stored energy is released, and the myosin head does indeed spring back to its original position. Thus, the energy transferred from ATP is used to do the work of pulling the thin filaments during contraction. Another ATP molecule then binds to the myosin head, which then releases actin and moves into its resting position again—all set for the next "pull." This cycle repeats as long as ATP is available and actin's active sites are unblocked.

Muscle fibers must continually resynthesize ATP because they can store only small amounts of it. There is only enough ATP in a muscle fiber for about 2 to 4 seconds of maximum contraction. However, energy for the resynthesis of ATP can be quickly supplied by the breakdown of another high-energy compound, **creatine phosphate (CP)**, as you can see in Figure 11-13. CP is a sort

of backup energy molecule that provides enough energy for about 20 additional seconds of maximal contraction. Both ATP and CP are continually resynthesized—or "recharged"—by cellular respiration. Ultimately, energy for both ATP and CP synthesis comes from the catabolism of food.

If a cell runs out of ATP completely and cannot resynthesize more, contraction stops—possibly resulting in stiffness caused by the inability of myosin heads to disengage from actin (see Figure 11-9). When this happens after death, it is called *rigor mortis* (Box 11-3).

Glucose and Oxygen

Note that continued, efficient nutrient catabolism by muscle fibers requires two essential ingredients: glucose and oxygen.

Glucose is a nutrient molecule that contains many chemical bonds. The potential energy stored in these chemical bonds is released during catabolic reactions in the sarcoplasm and mitochondria and transferred to ATP or CP molecules. Ultimately, all the glucose needed by muscle fibers comes from the blood. Skeletal muscle fibers are surrounded by a network of blood capillaries, the tiny "exchange" vessels that allow molecules to enter or leave the blood (Figure 11-14).

Some muscle fibers ensure an uninterrupted supply of glucose by storing it in the form of *glycogen*. Recall from Chapter 2 that glycogen is a polysaccharide made up of thousands of glucose subunits.

Oxygen, which is needed for a catabolic process known as *aerobic respiration*, also ultimately comes from the blood. Most of the oxygen carried in the blood is temporarily bound to *hemoglobin* molecules—a reddish pigment inside red blood cells.

Oxygen can also be stored by cells to ensure an uninterrupted supply. During rest, excess oxygen molecules in the sarcoplasm are attracted to a large protein molecule called **myoglobin.** Like hemoglobin, myoglobin is a reddish pigment with iron (Fe) groups that attract oxygen molecules and hold them temporarily. When the oxygen concentration inside a muscle fiber decreases rapidly—as it does during exercise—it can be quickly resupplied from myoglobin. Muscle fibers that contain large amounts of myoglobin take on a deep red appearance and are often called *red fibers*. Muscle fibers with little myoglobin in them are light pink and are often called *white fibers*. Most muscle tissues contain a mixture of red and white fibers (Box 11-4).

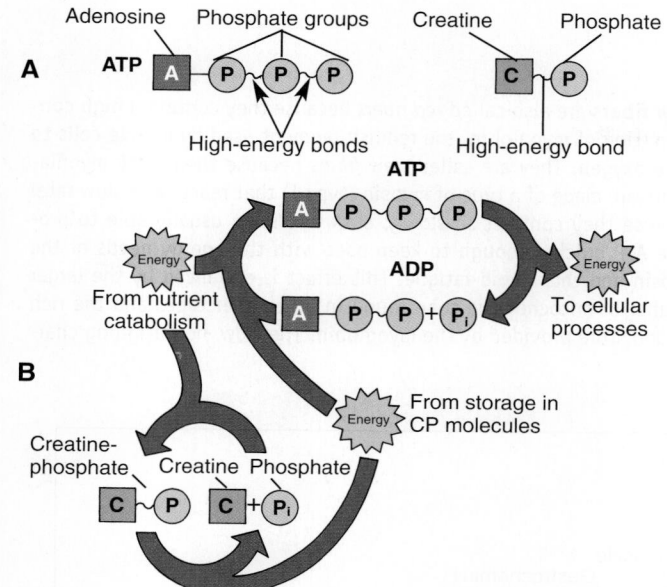

Figure 11-13 *Energy sources for muscle contraction.* **A,** The basic structure of two high-energy molecules in the sarcoplasm: adenosine triphosphate (ATP) and creatine phosphate (CP). **B,** This diagram shows how energy released during the catabolism of nutrients can be transferred to the high-energy bonds of ATP directly or, instead, stored temporarily in the high-energy bond of CP. During contraction, ATP is hydrolyzed and the energy of the broken bond transferred to a myosin head.

 BOX 11-3: FYI
Rigor Mortis

The term **rigor mortis** is a Latin phrase that means "stiffness of death." In a medical context the term *rigor mortis* refers to the stiffness of skeletal muscles sometimes observed shortly after death. What causes rigor mortis? At the time of death, stimulation of muscle cells ceases. However, the muscle fibers of postural muscles may have been in midcontraction at the time of death—when the myosin-actin cross bridges are still intact. In addition, the SR releases much of the Ca^{++} it had been storing, thereby causing even more cross bridges to form. ATP is required to release the cross bridges and "energize" the myosin heads for their next attachment. Because the last of a cell's ATP supply is used up at the time it dies, many cross bridges may be left "stuck" in the contracted position. Thus, muscles in a dead body may be stiff because individual muscle fibers ran out of the ATP required to "turn off" a muscle contraction.

Aerobic Respiration

Aerobic (oxygen-requiring) respiration is a catabolic process that produces the maximum amount of energy available from each glucose molecule. When the oxygen concentration is low, however, muscle fibers can shift toward increased use of another catabolic process: *anaerobic respiration.* As its name implies, anaerobic respiration does not require the immediate use of oxygen. Besides its ability to produce ATP without oxygen, anaerobic respiration has the added advantage of being very rapid. Muscle fibers having difficulty getting oxygen—or fibers that generate a great deal of force very quickly—may rely on anaerobic respiration to resynthesize their ATP molecules.

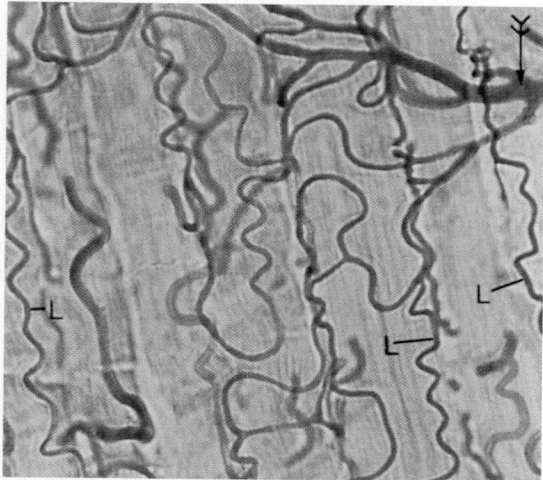

Figure 11-14 *Blood supply of muscle fibers.* Tiny "exchange" vessels called capillaries branch out longitudinally (L) from small arteries *(arrow)* to form a network that supplies muscle fibers with glucose and oxygen and removes carbon dioxide and lactic acid. In this micrograph, the muscle fibers appear translucent and the blood vessels appear reddish.

Anaerobic Respiration

Anaerobic respiration may allow the body to avoid the use of oxygen in the short term, but not in the long term. Anaerobic respiration results in the formation of an incompletely catabolized molecule called **lactic acid.** Lactic acid may accumulate in muscle tissue during exercise and cause a burning sensation. Some of the lactic acid eventually diffuses into the blood and is delivered to the liver, where an oxygen-consuming process later converts it back into glucose. This is one of the reasons that after heavy exercise, when the lack of oxygen in some tissues has caused the production of lactic acid, a person may still continue to breathe heavily. The body is repaying the so-called *oxygen debt* by using the extra oxygen gained by heavy breathing to process the lactic acid that was produced during exercise.

Heat Production

Because the catabolic processes of cells are never 100% efficient, some of the energy released is lost as heat. Because skeletal muscle tissues produce such a massive amount of heat—even when they are doing hardly any work—they have a great effect on body temperature. Recall from Chapter 6 that various heat loss mechanisms of the skin can be used to cool the body when it becomes overheated (see Figure 6-12, p. 209). Skeletal muscle tissues can likewise be used when the body's temperature falls below the set point value determined by the "thermostat" in the hypothalamus of the brain. As Figure 11-15 shows, a low external temperature can reduce body temperature below the set point. Temperature sensors in the skin and other parts of the body feed this information back to the hypothalamus, which compares the actual value with the set point value (usually about 37° C). The hypothalamus responds to a decrease in body temperature by signaling skeletal muscles to contract. The shivering contractions that result produce enough waste heat to warm the body back to the set point temperature—and homeostatic balance is maintained.

The subject of energy metabolism is discussed more thoroughly in Chapter 27.

BOX 11-4: SPORTS AND FITNESS
Types of Muscle Fibers

Skeletal muscle fibers can be classified into three types according to their structural and functional characteristics: (1) *slow* (red) *fibers,* (2) *fast* (white) *fibers,* and (3) *intermediate fibers.* Each type is best suited to a particular type or style of muscular contraction—a fact that is useful in considering how different muscles are used during different athletic activities. Although each muscle organ contains a mix of all three fiber types (part *A* of the figure), different organs have these fibers in different proportions, depending on the types of contraction that they most often perform (Figure B).

Slow fibers are also called *red fibers* because they contain a high concentration of myoglobin, the reddish pigment used by muscle cells to store oxygen. They are called *slow fibers* because their thick myofilaments are made of a type of myosin (type I) that reacts at a slow rate. Because they contract so slowly, slow fibers are usually able to produce ATP quickly enough to keep pace with the energy needs of the myosin and thus avoid fatigue. This effect is enhanced by the larger number of mitochondria than found in other fiber types and the rich oxygen store provided by the myoglobin. The slow, nonfatiguing char-

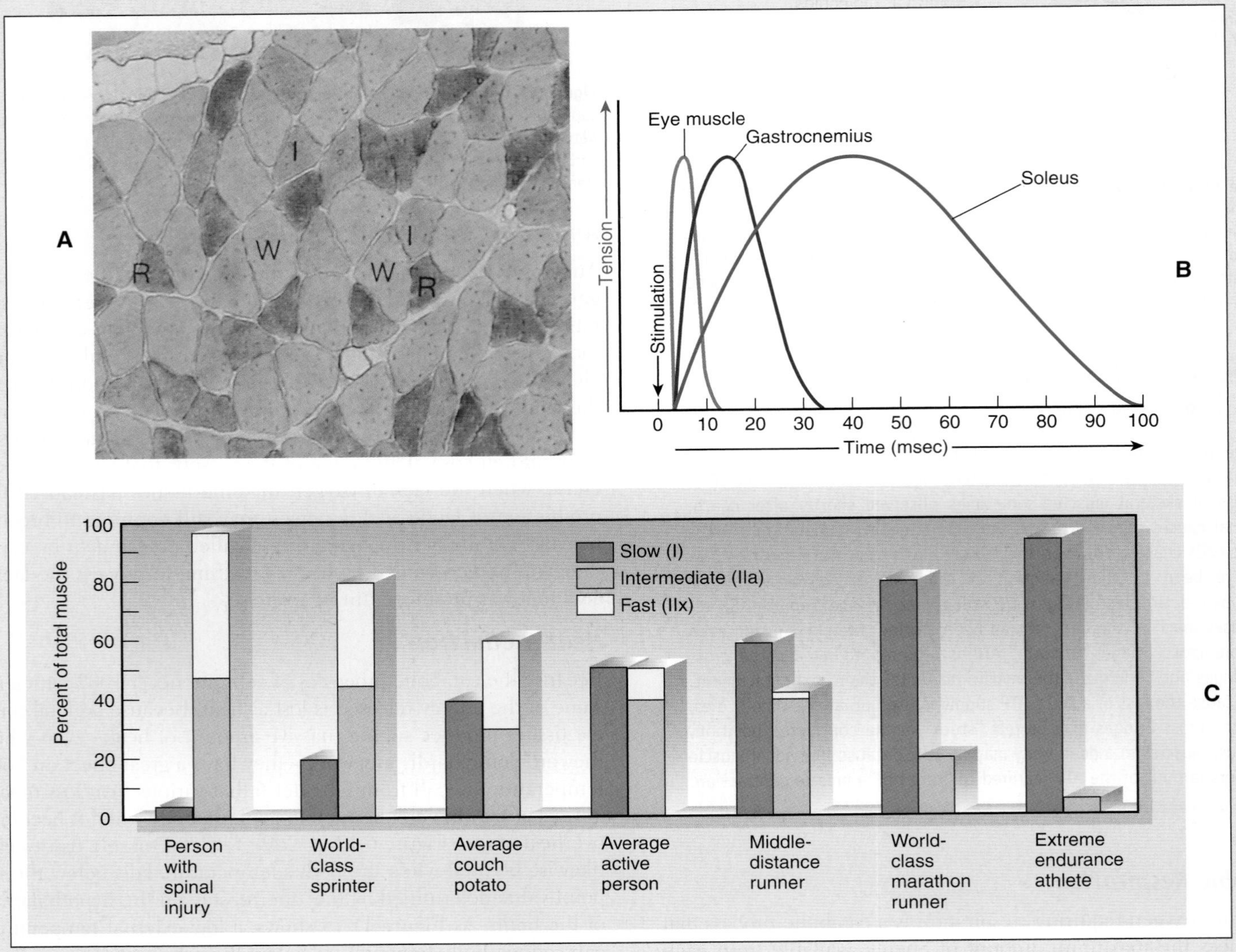

A, *Types of muscle fibers.* Using a special staining technique, this micrograph of a cross section of skeletal muscle tissue shows a mix of fiber types: R, red (slow) fibers; W, white (fast) fibers; and I, intermediate fibers. **B,** *Three styles of twitch contraction.* Different muscle organs have different proportions of slow, fast, and intermediate fibers and thus produces different styles of twitch contractions in a myogram. **C,** *Proportions of fiber types in muscle tissue.* A person with a spinal cord injury eventually loses nearly all of the postural slow fibers while retaining mostly intermediate and fast fibers. In an extreme endurance athlete, on the other hand, the slow fibers develop so much that they greatly dominate the faster fiber types.

BOX 11-4: SPORTS AND FITNESS
Types of Muscle Fibers—cont'd

acteristics of slow fibers make them especially well suited to the sustained contractions exhibited by postural muscles. Postural muscles containing a high proportion of slow fibers can hold the skeleton upright for long periods without fatigue.

Fast fibers are also called *white fibers* because they contain very little myoglobin. Fast fibers can contract much more rapidly than slow fibers because they have a faster type of myosin (type IIx) and because their system of T tubules and SR is more efficient at quickly delivering Ca^{++} to the sarcoplasm. The price of a rapid contraction mechanism is rapid depletion of ATP. Despite the fact that fast fibers typically contain a high concentration of glycogen, they have few mitochondria and thus must rely primarily on anaerobic respiration to regenerate ATP. Because anaerobic respiration produces relatively small amounts of ATP, fast fibers cannot produce enough ATP to sustain a contraction for very long. Because they can generate great force very quickly but not for a long duration, fast fibers

are best suited for muscles that move the fingers and eyes in darting motions.

Intermediate fibers have characteristics somewhere in between the two extremes of fast and slow fibers. They are more fatigue resistant than fast fibers and can generate more force more quickly than slow fibers. This type of muscle fiber predominates in muscles that both provide postural support and are occasionally required to generate rapid, powerful contractions. One example is the *gastrocnemius,* or calf muscle, which helps support the leg but is also used in walking, running, and jumping (part *B* of the figure).

The bar graph in part *C* of the figure shows that the relative proportions of muscle fiber types, body wide, varies with the type of work a person does with his or her muscles. This graph underscores the fact that athletic training, for example, can produce changes in the mix of different fiber types in muscle tissue.

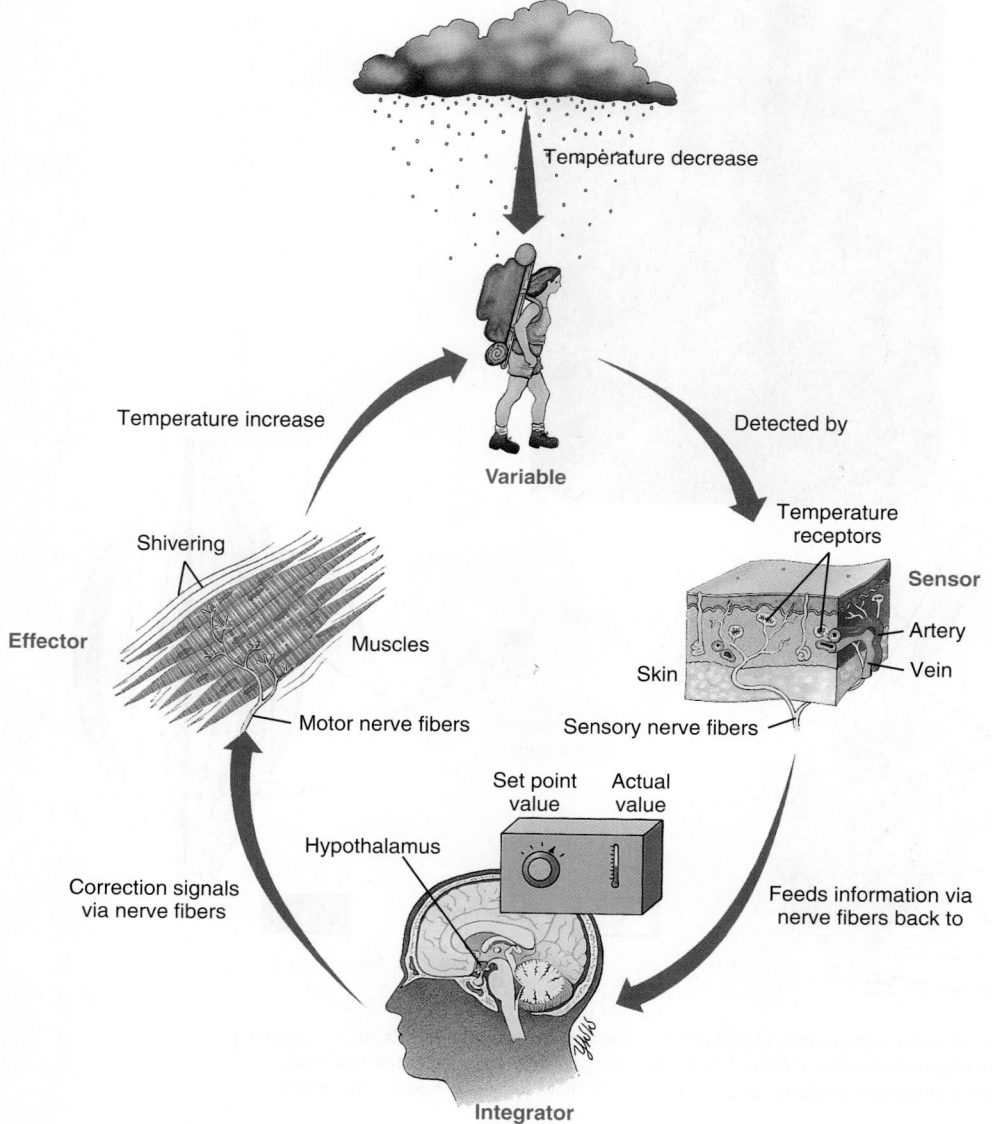

Figure 11-15 *The role of skeletal muscle tissues in maintaining a constant body temperature.* This diagram shows that a drop in body temperature caused by cold weather can be corrected by a negative feedback mechanism that triggers shivering (muscle contraction), which in turn produces enough heat to warm the body.

QUICK CHECK

8. Where does the energy stored in ATP come from?
9. Contrast aerobic respiration and anaerobic respiration in muscle fibers.
10. What is the role of *myoglobin* in muscle fibers?

FUNCTION OF SKELETAL MUSCLE ORGANS

Although each skeletal muscle fiber is distinct from all other fibers, it operates as part of the large group of fibers that form a skeletal muscle organ. Skeletal muscle organs, often simply called *muscles*, are composed of bundle upon bundle of muscle fibers held together by fibrous connective tissues (see Figure 11-1, A). The details of muscle organ anatomy are discussed in Chapter 10. For now, we turn attention to the matter of how skeletal muscle organs function as a single unit.

Motor Unit

Recall that each muscle fiber receives its stimulus from a motor neuron. This neuron, often called a *somatic motor neuron*, is one of several nerve cells that enter a muscle organ together in a bundle called a *motor nerve*. One of these motor neurons, plus the muscle fibers to which it attaches, constitutes a functional unit called a **motor unit** (Figure 11-16). The single fiber of a somatic motor neuron divides into a variable number of branches on entering the skeletal muscle. The neuron branches of some motor

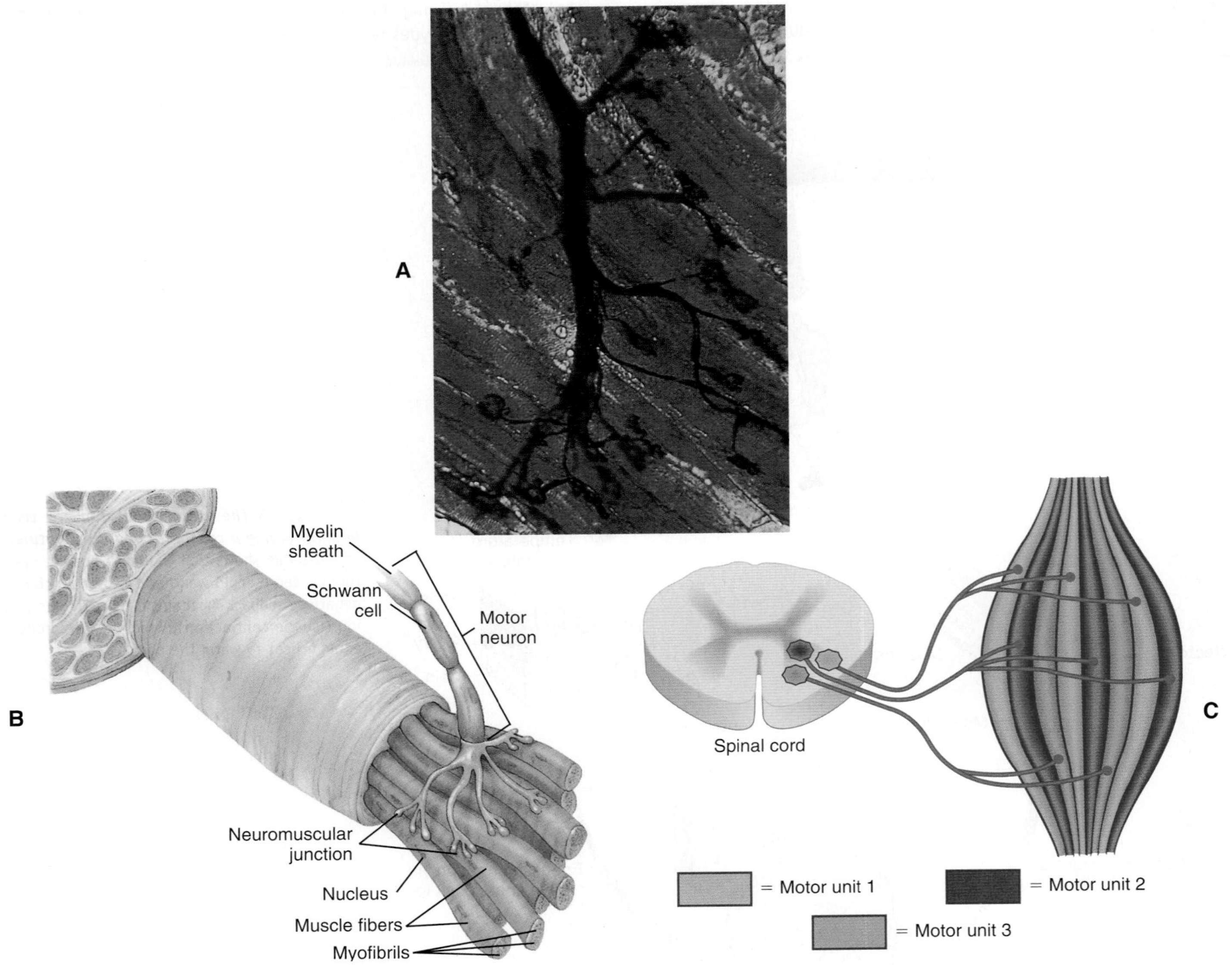

A

B

Myelin sheath

Schwann cell

Motor neuron

Spinal cord

C

Neuromuscular junction

Nucleus

Muscle fibers

Myofibrils

▭ = Motor unit 1   ▬ = Motor unit 2

▭ = Motor unit 3

Figure 11-16 *Motor unit.* A motor unit consists of one somatic motor neuron and the muscle fibers supplied by its branches. **A,** Photomicrograph showing a nerve (black) branching to supply several dozen individual muscle fibers (red). **B,** Sketch showing a single motor unit. **C,** Diagram showing several motor units within the same muscle organ.

units terminate in only a few dozen muscle fibers, whereas others terminate in a few thousand fibers. Consequently, impulse conduction by one motor unit may stimulate only a small part of a muscle organ, whereas conduction by another motor unit may activate a much larger portion of a muscle organ. This fact bears a relationship to the function of the muscle as a whole. As a rule, the fewer the number of fibers supplied by a skeletal muscle's individual motor units, the more precise the movements that muscle can produce. For example, in certain small muscles of the hand, each motor unit includes only a few muscle fibers, and these muscles produce precise finger movements. In contrast, motor units in large abdominal muscles that do not produce precise movements may have thousands of muscle fibers.

Myography

Many experimental methods have been used to study the contractions of skeletal muscle organs. They vary from relatively simple procedures, such as observing or palpating muscles in action, to the more complicated method of *electromyography* (recording electrical impulses from muscles as they contract). One method of studying muscle contraction particularly useful for the purposes of our discussion is called, simply, *myography*. *Myography*, a term that means "muscle graphing," is a procedure in which the force or tension from the contraction of an isolated muscle is recorded as a line that rises and falls as the muscle contracts and relaxes (Figure 11-17). To get the muscle to contract, an electrical stimulus of sufficient intensity (the **threshold stimulus**) is applied to the muscle. A single, brief threshold stimulus produces a quick jerk of the muscle, called a **twitch contraction.**

The Twitch Contraction

The quick, jerky twitch contraction seen in a myogram serves as the fundamental model for how muscles operate. The myogram of a twitch contraction shown in Figure 11-18 shows that the muscle does not begin to contract at the instant of stimulation but rather a fraction of a second later. The muscle then increases its tension (or shortens) until a peak is reached, after which it gradually returns to its resting state. These three phases of the twitch contraction are called, respectively, the *latent period,* the *contraction phase,* and the *relaxation phase.* The entire twitch usually lasts less than one tenth of a second.

During the latent period, the impulse initiated by the stimulation travels through the sarcolemma and T tubules to the SR, where it triggers the release of calcium ions into the sarcoplasm. It is not until the calcium binds to troponin and sliding of the myofilaments begins that contraction is observed. After a few milliseconds, the forceful sliding of the myofilaments ceases and relaxation begins. By the end of the relaxation phase, all of the myosin-actin reactions in all the fibers have ceased.

Twitch contractions of muscle organs rarely happen in the body. Even if we tried to make our muscles twitch voluntarily, they won't. Instead, our nervous system subconsciously "smoothes out" the movements to prevent injury and to make our movements more useful to us. Sustaining a smooth contraction is something that we will discuss a little later. But, to begin our discussion, a look at the twitch contraction gives us important insight about the mechanisms of more typical types of muscle organ contractions.

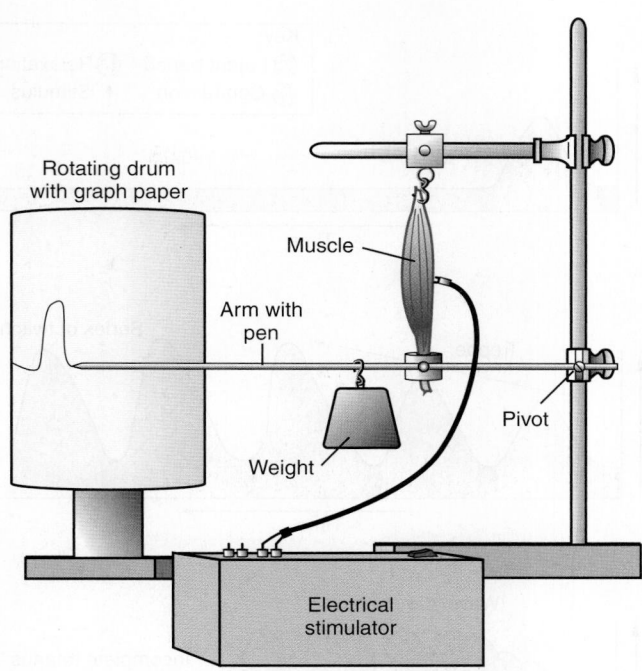

Figure 11-17 *Myography.* This classical type of myography setup, called a *kymograph,* illustrates the principle of recording muscle contractions as graphs showing changes in tension. An isolated muscle moves the pen upward during contraction, and the weight pulls the muscle and pen downward as the muscle relaxes. Electrical voltage simulates a nerve impulse to stimulate the fibers in the muscle. Modern myography often uses computer-based systems that record similar muscle tension graphs.

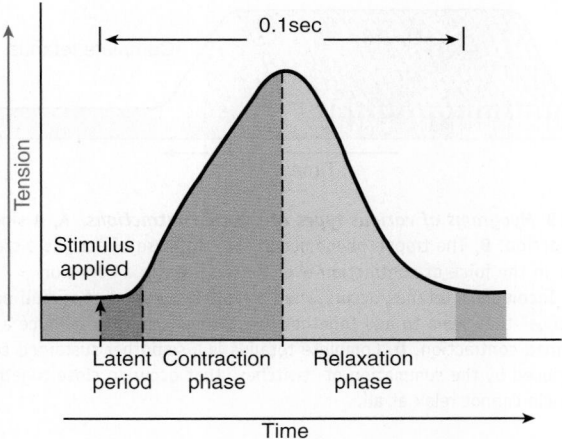

Figure 11-18 *The twitch contraction.* Three distinct phases are apparent: (1) the latent period, (2) the contraction phase, and (3) the relaxation phase.

Treppe: The Staircase Phenomenon

One interesting effect that can be seen in myographic studies of the twitch contraction is called **treppe,** or the *staircase phenomenon.* Treppe is a gradual, steplike increase in the strength of contraction that can be observed in a series of twitch contractions that occur about 1 second apart (Figure 11-19, *B*).

In other words, a muscle contracts more forcefully after it has contracted a few times than when it first contracts—a principle

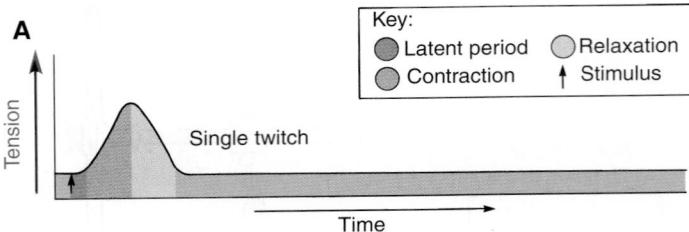

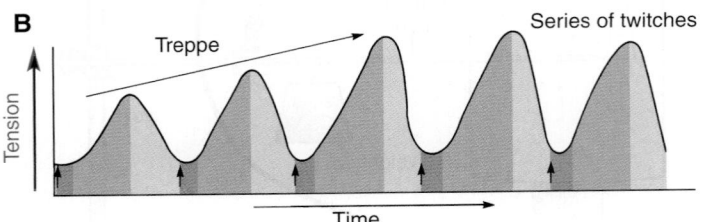

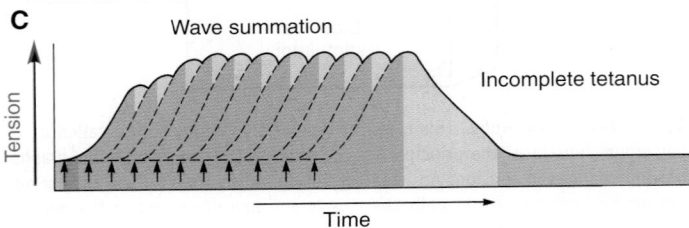

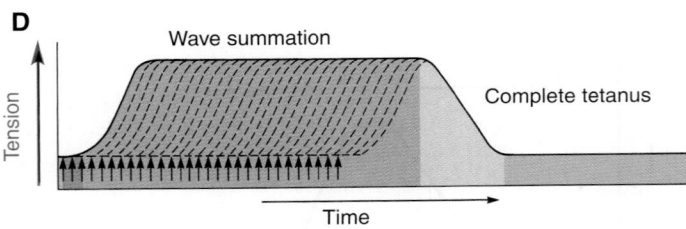

Figure 11-19 *Myograms of various types of muscle contractions.* **A,** A single twitch contraction. **B,** The treppe phenomenon, or "staircase effect," is a step-like increase in the force of contraction over the first few in a series of twitches. **C,** Incomplete tetanus occurs when a rapid succession of stimuli produces "twitches" that seem to add together (wave summation) to produce a rather sustained contraction. **D,** Complete tetanus is a smoother sustained contraction produced by the summation of "twitches" that occur so close together that the muscle cannot relax at all.

used by athletes when they warm up. Several factors contribute to this phenomenon. For example, in warm muscle fibers, calcium ions diffuse through the sarcoplasm more efficiently and more actin-myosin reactions occur. In addition, calcium ions accumulate in the sarcoplasm of muscles that have not had time to relax and pump much of the calcium back into their SR. Thus up to a point, a warm fiber contracts more strongly than a cool fiber does. Thus, after the first few stimuli muscle responds to successive stimuli with maximal contractions. Eventually, it will respond with less and less strong contractions. The relaxation phase becomes shorter and finally disappears entirely. In other words, the muscle stays partially contracted—an abnormal state of prolonged contraction called *contracture.*

Repeated stimulation of muscle in time lessens its excitability and contractility and may result in **muscle fatigue,** a condition in which the muscle does not respond to the strongest stimuli. Complete muscle fatigue can be readily induced in an isolated muscle but very seldom occurs in the body. (See Box 11-5.)

Tetanus

The concept of the simple twitch can help us understand the smooth, sustained types of contraction that are commonly observed in the body. Such smooth, sustained contractions are called *tetanic contractions* or, simply, **tetanus.** Figure 11-19, C, shows that if a series of stimuli come in a rapid enough succession, the muscle does not have time to relax completely before the next contraction phase begins. Muscle physiologists describe this effect as *multiple wave summation*—so named because it seems as though multiple twitch waves have been added together to sustain muscle tension for a longer time. The type of tetanus produced when very short periods of relaxation occur between peaks of tension is called *incomplete tetanus.* It is "incomplete" because the tension is not sustained at a completely constant level. Figure 11-19, D, shows that when the frequency of stimuli increases, the distance between peaks of tension decrease to a point at which they seem to fuse into a single, sustained peak. This produces a very smooth type of tetanic contraction called *complete tetanus.*

In a normal body, tetanus results from two factors working at the same time. One factor is the rapid-fire stimulation of nerve fibers that permits wave summation to occur in each fiber. Another factor is the coordinated contractions of different motor units within the muscle organ. These motor units fire in an asynchronous, overlapping time sequence to produce a "relay team" effect that results in a sustained contraction. Tetanus is the kind of contraction exhibited by normal skeletal muscle organs most of the time.

QUICK CHECK

11. What are the three phases of a twitch contraction? What molecular events occur during each of these phases?
12. What is the difference between a *twitch* contraction and a *tetanic* contraction?
13. How does the *treppe* effect relate to the warm-up exercises of athletes?
14. What is *tetanus*? Is it normal?

Muscle Tone

A **tonic contraction** (*tonus*, "tone") is a continual, partial contraction in a muscle organ. At any one moment a small number of the total fibers in a muscle contract and produce tautness of the muscle rather than a recognizable contraction and movement. Different groups of fibers scattered throughout the muscle contract in relays. Tonic contraction, or **muscle tone**, is the low level of continuous contraction characteristic of the muscles of normal individuals when they are awake. It is particularly important for maintaining posture. A striking illustration of this fact is the following: when a person loses consciousness, muscles lose their tone, and the person collapses in a heap, unable to maintain a sitting or standing posture. Muscles with less tone than normal are described as *flaccid*, and those with more than normal tone are called *spastic*.

Muscle tone is maintained by negative feedback mechanisms centered in the nervous system, specifically in the spinal cord. Stretch sensors in the muscles and tendons detect the degree of stretch in a muscle organ and feed this information back to an integrator mechanism in the spinal cord. When the actual stretch (detected by the stretch receptors) deviates from the set point stretch, signals sent via the somatic motor neurons adjust the strength of tonic contraction. This type of subconscious mechanism is often called a *spinal reflex* (discussed further in Chapters 12 to 15).

THE GRADED STRENGTH PRINCIPLE

Skeletal muscles contract with varying degrees of strength at different times—a fact called the **graded strength principle**. Because muscle organs can generate different grades of strength, we can match the force of a movement to the demands of a specific task (Box 11-6).

Various factors contribute to the phenomenon of graded strength. We have already discussed some of these factors. For example, we stated that the metabolic condition of individual fibers influences their capacity to generate force. Thus, if many fibers of a muscle organ are unable to maintain a high level of ATP and become fatigued, the entire muscle organ suffers some loss in its ability to generate maximum force of contraction. On the other hand, the improved metabolic conditions that produce the treppe effect allow a muscle organ to increase its contraction strength.

Another factor that influences the grade of strength exhibited by a muscle organ is the number of fibers contracting simulta-

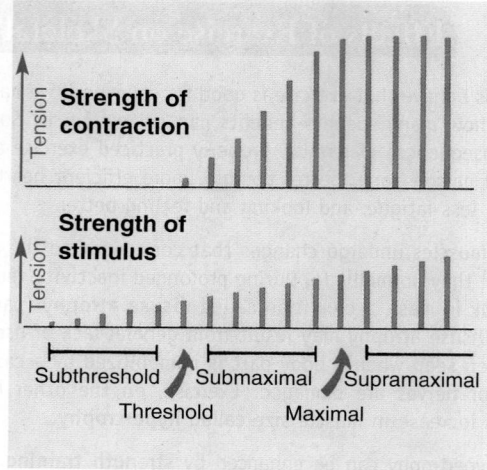

Figure 11-20 *The strength of muscle contraction compared with the strength of the stimulus.* After the threshold stimulus is reached, a continued increase in stimulus strength produces a proportional increase in muscle strength until the maximal level of contraction strength is reached.

neously. Obviously, the more muscle fibers contracting at the same time, the stronger the contraction of the entire muscle organ. How large this number is depends on how many motor units are activated or **recruited.** Recruitment of motor units, in turn, depends on the intensity and frequency of stimulation. In general, the more intense and the more frequent a stimulus, the more motor units that are recruited and the stronger the contraction. Figure 11-20 shows that increasing the strength of the stimulus beyond the threshold level of the most sensitive motor units causes an increase in the strength of contraction. As the threshold level of each additional motor unit is reached, the strength of contraction increases. This process continues as the strength of stimulation increases until the maximal level of contraction is reached. At this point, the limits of the muscle organ to recruit new motor units have been reached. Even if stimulation increases above the maximal level, the muscle cannot contract any more strongly. As long as the supply of ATP holds out, the muscle organ can sustain a tetanic contraction at the maximal level, with motor units contracting and relaxing in overlapping "relays" (see Figure 11-19, *D*).

The maximal strength that a muscle can develop is directly related to the initial length of its fibers—this is the *length-tension relationship* (Figure 11-21). A muscle that begins a contraction from a short initial length cannot develop much tension because its sarcomeres are already compressed. Conversely, a muscle that begins a contraction from an overstretched initial length cannot develop much tension because the thick myofilaments are too far away from the thin myofilaments to effectively pull them and thus compress the sarcomeres. The strongest maximal contraction is possible only when the muscle organ has been stretched to an optimal initial length. To illustrate this point, extend your elbow fully and try to contract the *biceps brachii* muscle on the ventral side of the upper part of your arm. Now flex the elbow just a little and contract the biceps again. Try it a third time with the elbow completely flexed. The greatest tension—seen as the largest "bulge" of the biceps—occurs when the elbow is partly flexed and the biceps only moderately stretched.

BOX 11-6   Effects of Exercise on Skeletal Muscles

Most of us believe that exercise is good for us, even if we have no idea what or how many specific benefits can come from it. Some of the good consequences of regular, properly practiced exercise are greatly improved muscle tone, better posture, more efficient heart and lung function, less fatigue, and looking and feeling better.

Skeletal muscles undergo changes that correspond to the amount of work that they normally do. During prolonged inactivity, muscles usually shrink in mass, a condition called **disuse atrophy** (part *A* of the figure). Disuse atrophy may result from general lack of use, but it is most often seen when a body part is immobilized by a cast or when the motor nerves are damaged. Exercise, on the other hand, may cause an increase in muscle size called **hypertrophy**.

Muscle hypertrophy can be enhanced by **strength training**, which involves contracting muscles against heavy resistance. Isometric exercises and weightlifting are common strength-training activities. This type of training results in increased numbers of myofilaments in each muscle fiber. Although the number of muscle fibers stays the same, the increased number of myofilaments greatly increases the mass of the muscle.

Endurance training, often called **aerobic training** (part *B* of the figure), does not usually result in as much muscle hypertrophy as strength training does. Instead, this type of exercise program increases a muscle's ability to sustain moderate exercise over a long period. Aerobic activities such as running, bicycling, or other primarily isotonic movements increase the number of blood vessels in a muscle (see Figure 11-14). The increased blood flow allows more efficient delivery of oxygen and glucose to muscle fibers during exercise. Aerobic training also causes an increase in the number of mitochondria in muscle fibers. This allows production of more ATP as a rapid energy source.

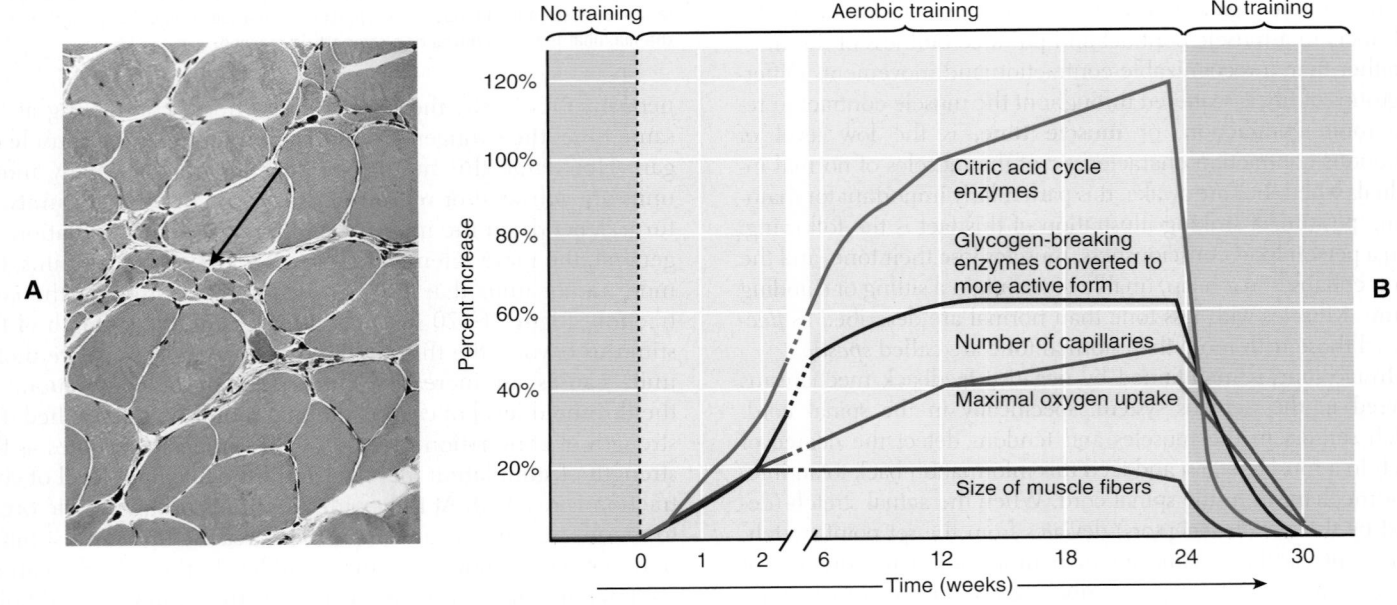

A, *Disuse atrophy.* The arrow points to a group of muscle fibers that have atrophied from disuse (in this case from nerve damage). Notice how much smaller they are than the surrounding, normal fibers.
B, *Effects of aerobic training.* The graph shows that aerobic training increases the metabolic condition of muscles mainly by increasing levels of enzymes and the availability of oxygen. Only moderate increases in muscle fiber size occur.

Another factor that influences the strength of a skeletal muscle contraction is the amount of load imposed on the muscle. Within certain limits, the heavier the load, the stronger the contraction. Lift your hand with palm up in front of you and then put this book in your palm. You can feel your arm muscles contract more strongly as the book is placed in your hand. This occurs because of a *stretch reflex,* a response in which the body tries to maintain constancy of muscle length (Figure 11-22). An increased load threatens to stretch the muscle beyond the set point length that you are trying to maintain. Your body exhibits a negative feedback response when it detects the increased stretch caused by an increased load, feeds the information back to an integrator in the nervous system, and increases its stimulation of the muscle to counteract the stretch. This reflex maintains a relatively constant muscle length as load is increased up to a maximum sustainable level. When the load becomes too heavy and thus threatens to cause injury to the muscle or skeleton, the body abandons this reflex and forces you to relax and drop the load.

The major factors involved in the graded strength principle are summarized in Figure 11-23.

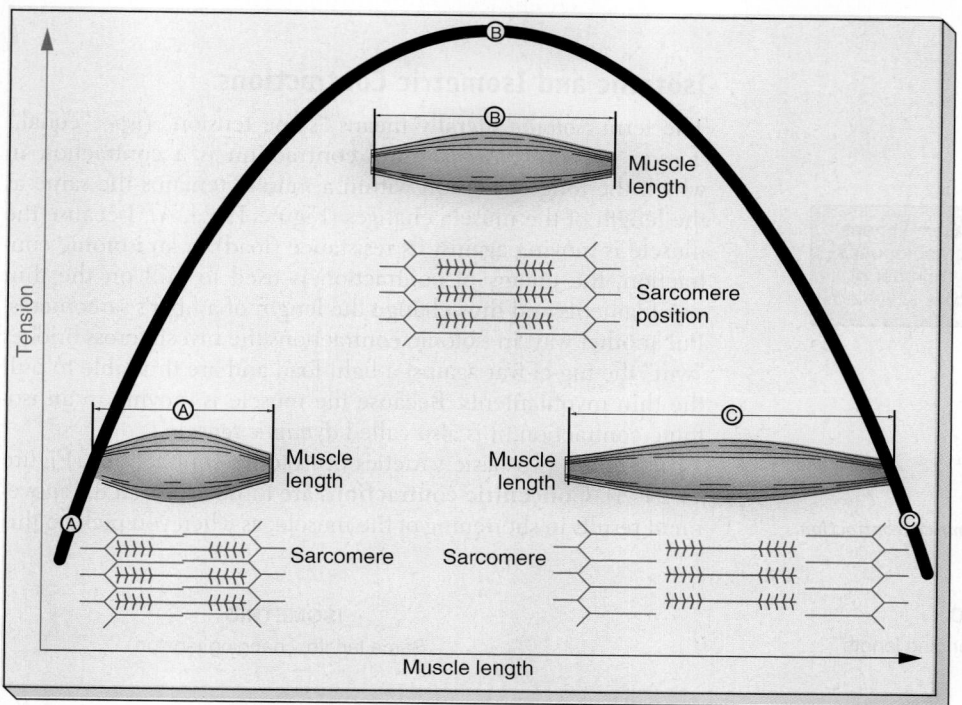

Figure 11-21 *The length-tension relationship.* As this graph of muscle tension shows, the maximum strength that a muscle can develop is directly related to the initial length of its fibers. At a short initial length the sarcomeres are already compressed, and thus the muscle cannot develop much tension (position *A*). Conversely, the thick and thin myofilaments are too far apart in an overstretched muscle to generate much tension (position *B*). Maximum tension can be generated only when the muscle has been stretched to a moderate, optimal length (position *C*).

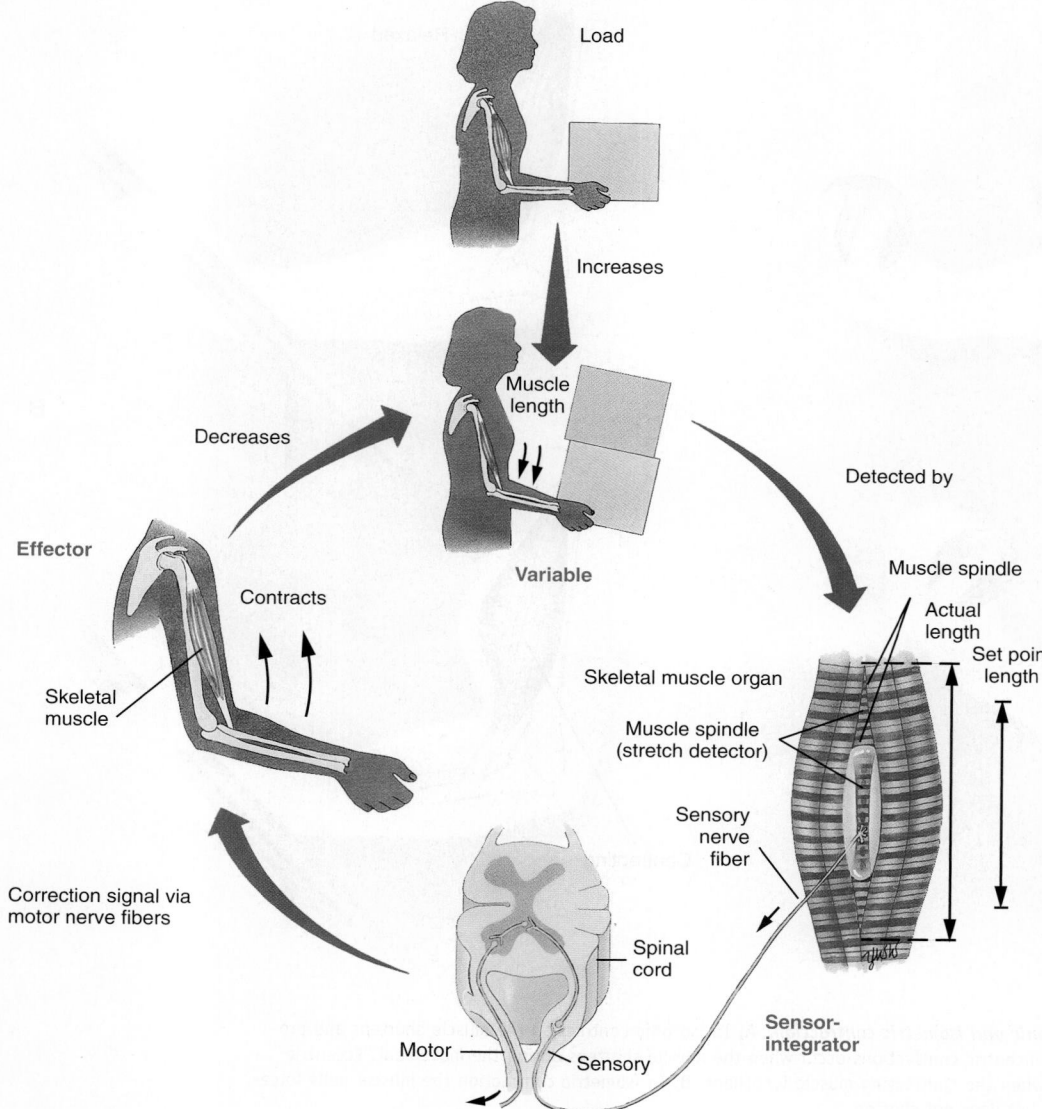

Figure 11-22 *The stretch reflex.* The strength of a muscle organ can be matched to the load imposed on it by a negative feedback response centered in the spinal cord. Increased stretch (caused by increased load) is detected by a sensory nerve fiber attached to a muscle cell (called a muscle spindle) specialized for this purpose. The information is integrated in the spinal cord and a correction signal is relayed through motor neurons back to the same muscle, which increases tension to return to the set point muscle length.

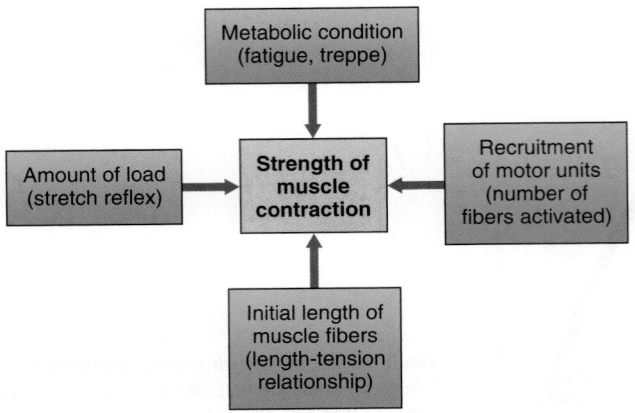

Figure 11-23 *Factors that influence the strength of muscle contraction.*

Isotonic and Isometric Contractions

The term *isotonic* literally means "same tension" (*iso-*, "equal," *-tonic*, "tension"). An **isotonic contraction** is a contraction in which the tone or tension within a muscle remains the same as the length of the muscle changes (Figure 11-24, A). Because the muscle is moving against its resistance (load) in an isotonic contraction, the energy of contraction is used to pull on the thin myofilaments and thus change the length of a fiber's sarcomeres. Put another way, in isotonic contractions the myosin cross bridges "win" the tug-of-war against a light load and are thus able to pull the thin myofilaments. Because the muscle is moving in an isotonic contraction, it is also called *dynamic tension*.

There are two basic varieties of isotonic contractions (Figure 11-24, A). **Concentric contractions** are those in which the movement results in shortening of the muscle, as when you pick up this

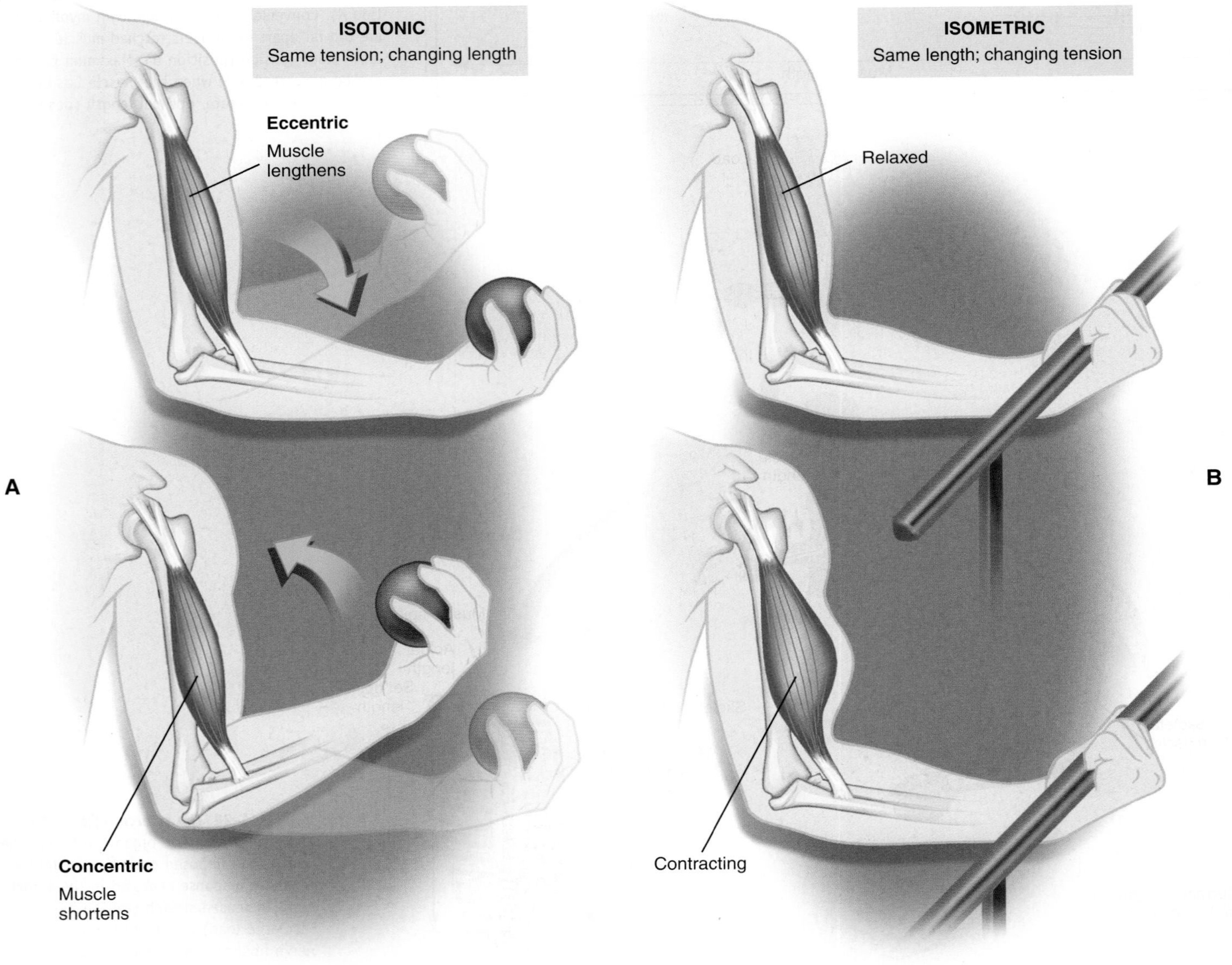

ISOTONIC
Same tension; changing length

Eccentric
Muscle
lengthens

Concentric
Muscle
shortens

ISOMETRIC
Same length; changing tension

Relaxed

Contracting

A

B

Figure 11-24 *Isotonic and isometric contraction.* **A,** In isotonic contraction the muscle shortens and produces movement. Concentric contractions occur when the muscle shortens during the movement. Eccentric contractions occur when the contracting muscle lengthens. **B,** In isometric contraction the muscle pulls forcefully against a load but does not shorten.

book. **Eccentric contractions** are those in which the movement results in lengthening of the muscle being contracted. For example, when you slowly lower the book you have just picked up, you are contracting the same muscle you just used to lift it—but this time you are lengthening the muscle, not shortening it.

An **isometric contraction,** in contrast to an isotonic contraction, is a contraction in which muscle length remains the same while muscle tension increases (Figure 11-18, *B*). The term *isometric* literally means "same length." You can observe isometric contraction by lifting up on a stationary handrail and feeling the tension increase in your arm muscles. Isometric contractions can do work by "tightening" to resist a force, but they do not produce movements. In isometric contractions, the tension produced by the "power stroke" of the myosin cross bridges cannot overcome the load placed on the muscle. Using the tug-of-war analogy, we can say that in isometric contractions the myosin cross bridges reach a "draw"—they hold their own against the load placed on the muscle but do not make any progress in sliding the thin myofilaments. Because muscles remain stable during isotonic contraction, it is also called *static tension.*

QUICK CHECK

15. What is meant by the term *muscle tone?*
16. Name four factors that influence the strength of a skeletal muscle contraction.
17. What is meant by the phrase "recruitment of motor units?"
18. What is the difference between *isotonic* and *isometric* contractions? *Concentric* and *eccentric?*

FUNCTION OF CARDIAC AND SMOOTH MUSCLE TISSUE

Cardiac and smooth muscle tissues operate by mechanisms similar to those in skeletal muscle tissues. Detailed study of cardiac and smooth muscle function will be set aside until we discuss specific smooth and cardiac muscle organs in later chapters. However, it may be helpful to preview some of the basic principles of cardiac and smooth muscle physiology so that we can compare them with those that operate in skeletal muscle tissue. Table 11-1 summarizes the characteristics of the three major types of muscle.

Cardiac Muscle

Cardiac muscle, also known as *striated involuntary muscle,* is found in only one organ of the body: the heart. Forming the bulk of the wall of each heart chamber, cardiac muscle contracts rhythmically and continuously to provide the pumping action necessary to maintain a relative constancy of blood flow through the internal environment. As you shall see, its physiological mechanisms are well adapted to this function.

The functional anatomy of cardiac muscle tissue resembles that of skeletal muscle to a degree, but it exhibits specialized features related to its role in continuously pumping blood. As Figure 11-25 shows, each cardiac muscle fiber contains parallel myofibrils. Each myofibril comprises sarcomeres that give the whole fiber a striated appearance. However, cardiac muscle fiber does not taper like skeletal muscle fiber but, instead, forms strong, electrically coupled junctions (intercalated disks) with other fibers. This feature, along with the branching exhibited by individual cells, allows cardiac fibers to form a continuous, electrically coupled mass called a **syncytium** (meaning "unit of combined cells"). Cardiac muscles thus form a continuous, contractile band around the heart

Table 11-1   Characteristics of Muscle Tissues

	SKELETAL	CARDIAC	SMOOTH
Principal location	Skeletal muscle organs	Wall of heart	Walls of many hollow organs
Principal functions	Movement of bones, heat production, posture	Pumping of blood	Movement in walls of hollow organs (peristalsis, mixing)
Type of control	Voluntary	Involuntary	Involuntary
Structural features			
Striations	Present	Present	Absent
Nucleus	Many near the sarcolemma	Single	Single; near the center of the cell
T tubules	Narrow; form triads with the SR	Large diameter; form diads with the SR, regulate Ca^{++} entry into the sarcoplasm	Absent
Sarcoplasmic reticulum	Extensive; stores and releases Ca^{++}	Less extensive than in skeletal muscle	Very poorly developed
Cell junctions	No gap junctions	Intercalated disks	Visceral: many gap junctions Multiunit: few gap junctions
Contraction style	Rapid twitch contractions of motor units usually summate to produce sustained tetanic contractions; must be stimulated by a neuron	Syncytium of fibers compress the heart chambers in slow, separate contractions (does not exhibit tetanus or fatigue); exhibits autorhythmicity	Visceral: electrically coupled sheets of fibers contract autorhythmically and produce peristalsis or mixing movements Multiunit: individual fibers contract when stimulated by a neuron

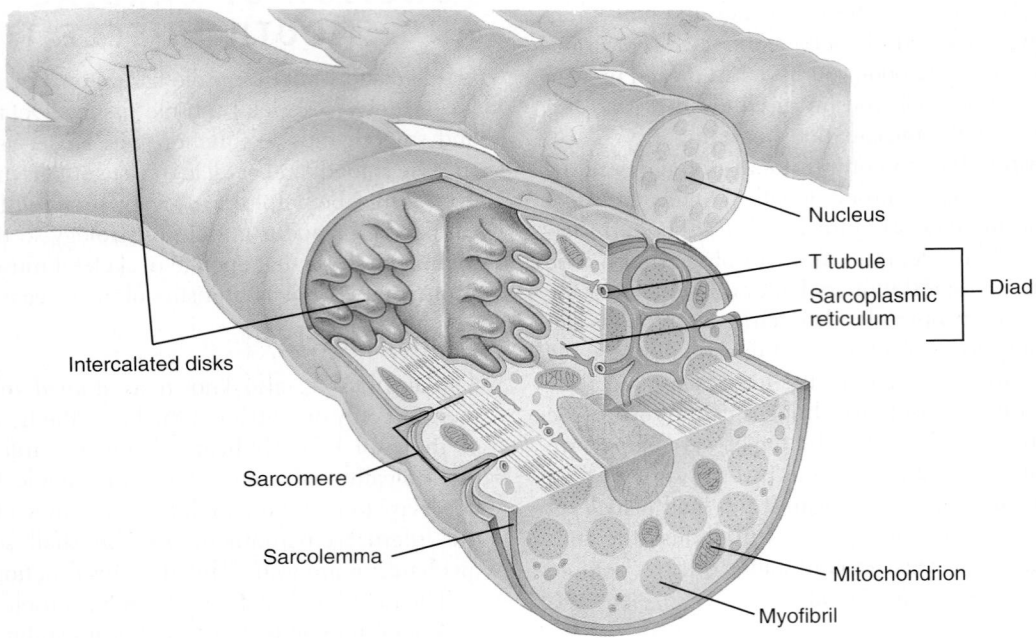

Figure 11-25 *Cardiac muscle fiber.* Unlike other types of muscle fibers, cardiac muscle fiber is typically branched and forms junctions, called *intercalated disks*, with adjacent cardiac muscle fibers. Like skeletal muscle fibers, cardiac muscle fibers contain sarcoplasmic reticula and T tubules—although these structures are not as highly organized as in skeletal muscle fibers.

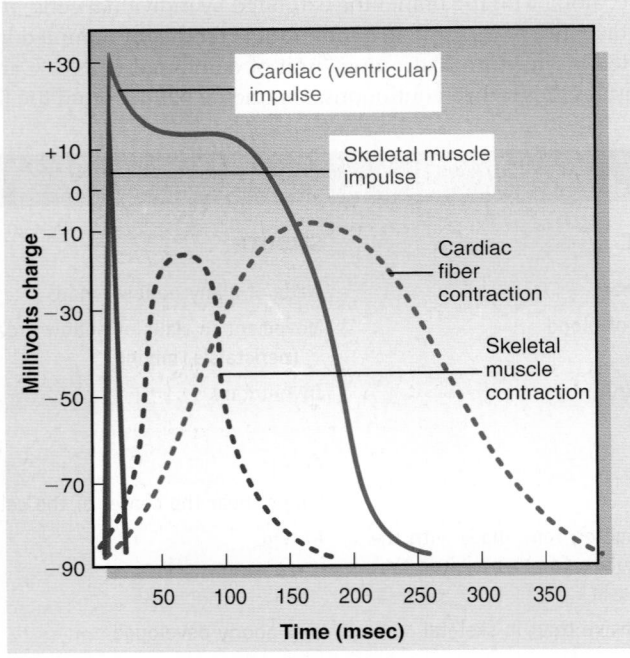

Figure 11-26 *Cardiac and skeletal muscle contractions compared.* A brief nerve impulse triggers a brief twitch contraction in skeletal muscle (blue), but a prolonged impulse in the heart tissue produces a rather slow, drawn out contraction in cardiac muscle (red).

chambers that conducts a single impulse across a virtually continuous sarcolemma—features necessary for an efficient, coordinated pumping action.

Unlike skeletal muscle, in which a nervous impulse excites the sarcolemma to produce its own impulse, cardiac muscle is self-exciting. Cardiac muscle cells thus exhibit a continuing rhythm of excitation and contraction on their own, although the rate of self-induced impulses can be altered by nervous or hormonal input. Figure 11-26 shows that impulses triggering cardiac muscle contractions are much more prolonged than those triggering skeletal muscle contractions. Because the sarcolemma of cardiac muscle sustains each impulse longer than in skeletal muscle,

Ca^{++} remains in the sarcoplasm longer. This means that even though many adjacent cardiac muscle cells contract simultaneously, they exhibit a prolonged contraction rather than a rapid twitch. It also means that impulses cannot come rapidly enough to produce tetanus. Because it cannot sustain long tetanic contractions, cardiac muscle does not normally run low on ATP and thus does not experience fatigue. Obviously, this characteristic of cardiac muscle is vital to keeping the heart continuously pumping.

Although cardiac muscle fiber has T tubules and SR, they are arranged a little differently than in skeletal muscle fibers. The T tubules are larger, and they form diads (double structures) rather than triads (triple structures), with a rather sparse SR. Much of the calcium that enters the sarcoplasm during contraction enters from outside the cells through the T tubules rather than from storage in the SR.

The structure and function of the heart are discussed further in Chapters 18 and 19.

Smooth Muscle

As we mentioned in Chapter 5, smooth muscle is composed of small, tapered cells with single nuclei. Smooth muscle cells do not have T tubules and have only loosely organized sarcoplasmic reticula. The calcium required for contraction comes from outside the cell and binds to a protein called *calmodulin*, rather than to troponin, to trigger a contraction event.

The lack of striations in smooth muscle fibers results from the fact that the thick and thin myofilaments are arranged quite dif-ferently than in skeletal or cardiac muscle fibers. As Figure 11-27 shows, thin arrangements of myofilaments crisscross the cell and attach at their ends to the cell's plasma membrane. When cross bridges pull the thin filaments together, the muscle "balls up" and thus contracts the cell. Because the myofilaments are not organized into sarcomeres, they have more freedom of movement and as a result can contract a smooth muscle fiber to shorter lengths than in skeletal and cardiac muscle.

There are two types of smooth muscle tissue: *single-unit* and *multiunit* (Figure 11-28) smooth muscle.

In *visceral*, or **single-unit**, smooth muscle, gap junctions join individual smooth muscle fibers into large, continuous sheets—much like the syncytium of fibers observed in cardiac muscle. This type of smooth muscle is the most common, and it forms a muscular layer in the walls of many hollow structures such as the digestive, urinary, and reproductive tracts. Like cardiac muscle, this type of smooth muscle commonly exhibits a rhythmic self-excitation, or *autorhythmicity* (meaning "self-rhythm"), that spreads across the entire tissue. When these rhythmic, spreading waves of contraction become strong enough, they can push the contents of a hollow organ progressively along its lumen. This phenomenon, called *peristalsis*, moves food along the digestive tract, assists the flow of urine to the bladder, and pushes a baby out of the womb during labor. Such contractions can also be coordinated to produce mixing movements in the stomach and other organs.

Multiunit smooth muscle tissue does not act as a single unit (as in visceral muscle) but instead is composed of many indepen-

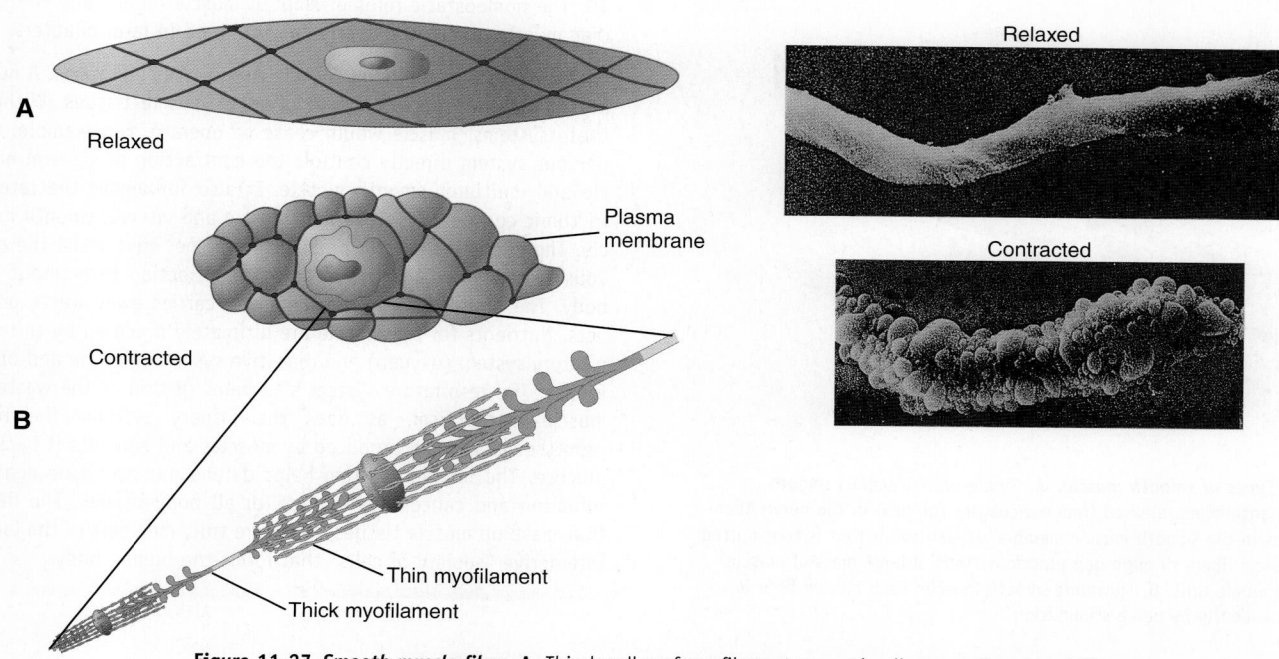

Figure 11-27 *Smooth muscle fiber.* **A,** Thin bundles of myofilaments span the diameter of a relaxed fiber. The scanning electron micrograph *(right)* shows that the surface of the cell is rather flat when the fiber is relaxed. **B,** During contraction, sliding of the myofilaments causes the fiber to shorten by "balling up." The micrograph shows that the fiber becomes shorter and thicker and exhibits "dimples" where the myofilament bundles are pulling on the plasma membrane.

Relaxed

Contracted

Plasma membrane

Thin myofilament

Thick myofilament

dent single-cell units. Each independent fiber does not usually generate its own impulse but rather responds only to nervous input. Although this type of smooth muscle can form thin sheets, as in the walls of large blood vessels, it is more often found in bundles (for example, the *arrector pili* muscles of the skin or muscles that control the lens of the eye) or as single fibers (such as those surrounding small blood vessels).

The structure and function of smooth muscle organs are discussed in later chapters.

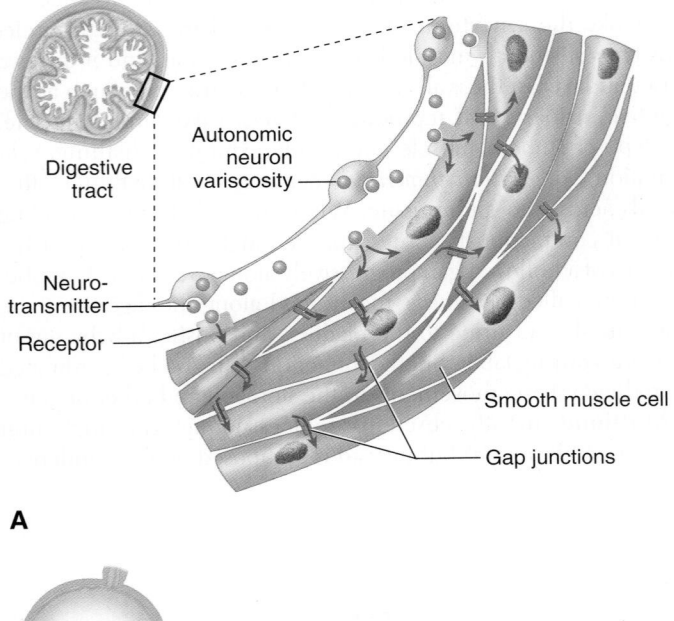

A

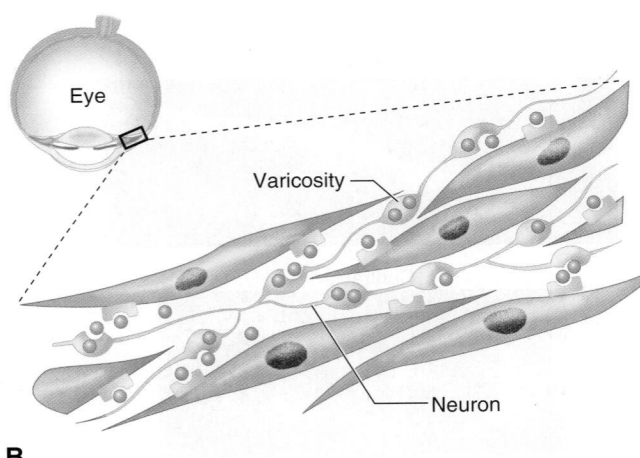

B

Figure 11-28 *Types of smooth muscle.* A, Single-unit (visceral) smooth muscle. Neurotransmitters released from varicosities (bulges) in the nerve fiber trigger impulses in the smooth muscle membranes—an event that is transmitted to adjacent muscle fibers through gap junctions. Thus, a large mass of muscle fibers acts as a single unit. **B,** Multiunit smooth muscle. Each muscle fiber is triggered independently by nerve stimulation.

QUICK CHECK

19. How do slow, separate, autorhythmic contractions of cardiac muscle make it well suited to its role in pumping blood?
20. What produces the striations in cardiac muscle?
21. How are myofilaments arranged in a smooth muscle fiber?
22. What is the difference between *single-unit* and *multiunit* smooth muscle?

THE BIG PICTURE
Muscle Tissue and the Whole Body

The function of all three major types of muscle (skeletal, smooth, and cardiac) is integral to the function of the entire body. What does the function of muscle tissue contribute to homeostasis of the whole body? First, all three types of muscle tissue provide the movement necessary for survival. Skeletal muscle moves the skeleton so that we can seek shelter, gather food, and defend ourselves. All three muscle types produce movements that power vital homeostatic mechanisms such as breathing, blood flow, digestion, and urine flow.

The relative constancy of the body's internal temperature could not be maintained in a cool external environment if not for the "waste" heat generated by muscle tissue—especially the large mass of skeletal muscle found throughout the body. Maintenance of a relatively stable body position—posture—is also a primary function of the skeletal muscular system. Posture, specific body movements, and other contributions of the skeletal muscular system to homeostasis of the whole body were discussed in Chapter 10. The homeostatic roles of smooth muscle organs and the cardiac muscle organ (the heart) are examined in later chapters.

Like all tissues of the body, muscle tissue gives and takes. A number of systems support the function of muscle tissues. Without these systems, muscle would cease to operate. For example, the nervous system directly controls the contraction of skeletal muscle and multiunit smooth muscle. It also influences the rate of rhythmic contractions in cardiac muscle and visceral smooth muscle. The endocrine system produces hormones that assist the nervous system in regulation of muscle contraction throughout the body. The blood delivers nutrients and carries away waste products. Nutrients for the muscle are ultimately procured by the respiratory system (oxygen) and digestive system (glucose and other foods). The respiratory system also helps get rid of the waste of muscle metabolism, as does the urinary system. The liver processes lactic acid produced by muscles and converts it back to glucose. The immune system helps defend muscle tissue against infection and cancer—as it does for all body tissues. The fibers that make up muscle tissues, then, are truly members of the large, interactive "society of cells" that forms the human body.

Mechanisms of Disease

MAJOR MUSCULAR DISORDERS

As you might expect, muscle disorders, or **myopathies**, generally disrupt the normal movement of the body. In mild cases, these disorders vary from inconvenient to slightly troublesome. Severe muscle disorders, however, can impair the muscles used in breathing—a life-threatening situation.

Muscle Injury

Injuries to skeletal muscles caused by overexertion or trauma usually result in a muscle **strain**. Figure 11-29 shows an unusually severe muscle strain that resulted in a massive tear in the entire muscle organ. Muscle strains are characterized by muscle pain, or **myalgia** (my-AL-jee-ah), and involve overstretching or tearing of muscle fibers. If an injury occurs in the area of a joint and a ligament is damaged, the injury may be called a **sprain**. Any muscle inflammation, including that caused by a muscle strain, is termed **myositis** (my-oh-SYE-tis). If tendon inflammation occurs with myositis, as in a charley

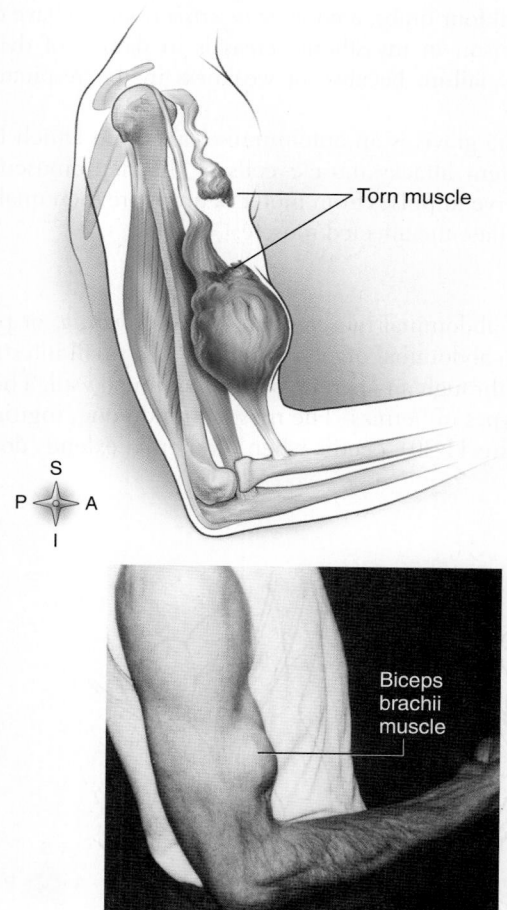

Torn muscle

S
P A
I

Biceps brachii muscle

Figure 11-29 *Muscle strain*. Severe strain of the biceps brachii muscle. In a severe muscle strain, a muscle may break in two pieces and result in a visible gap in muscle tissue under the skin. Notice how the broken ends of the muscle reflexively contract (spasm) to form a knot of tissue.

BOX 11-7: HEALTH MATTERS
Abnormal Muscle Contractions

Cramps are painful muscle spasms (involuntary twitches). Cramps often occur when a muscle organ is fatigued or mildly inflamed, but they can be a symptom of any irritation or ion and water imbalance.

Convulsions are abnormal, uncoordinated tetanic contractions of varying groups of muscles. Convulsions may result from a disturbance in the brain or seizure in which the output along motor nerves increases and becomes disorganized.

Fibrillation is an abnormal type of contraction in which individual fibers contract asynchronously rather than at the same time. This produces a flutter of the muscle but no effective movement. Fibrillation can also occur in cardiac muscle, where it reduces the heart's ability to pump blood.

horse, the condition is termed **fibromyositis** (fye-bro-my-oh-SYE-tis). Although inflammation may subside in a few hours or days, it usually takes weeks for damaged muscle fibers to repair. Some damaged muscle cells may be replaced by fibrous tissue, thereby forming scars. Occasionally, hard calcium is deposited in the scar tissue.

Cramps are painful muscle spasms (involuntary twitches). Cramps often result from mild myositis or fibromyositis, but they can be a symptom of any irritation or an ion and water imbalance.

Minor trauma to the body, especially a limb, may cause a muscle bruise, or **contusion.** Muscle contusions involve local internal bleeding and inflammation. Severe trauma to a skeletal muscle may cause a *crush injury.* Crush injuries greatly damage the affected muscle tissue, and the release of muscle fiber contents into the bloodstream can be life threatening. For example, the reddish muscle pigment myoglobin can accumulate in the blood and cause kidney failure.

Stress-induced muscle tension can result in myalgia and stiffness in the neck and back and is thought to be one cause of "stress headaches." Headache and back pain clinics use various strategies to treat stress-induced muscle tension. These treatments include massage, biofeedback, and relaxation training.

Muscle Infections

Several bacteria, viruses, and parasites may infect muscle tissue—often producing local or widespread myositis. For example, in trichinosis, widespread myositis is common. The muscle pain plus stiffness that sometimes accompanies influenza is another example.

One bacterial infection you have probably heard of is called *tetanus*—a confusing name because that word also refers to normal, sustained muscle contractions (see Figure 11-19). However, tetanus *the infection* is an abnormal condition caused by infection of the central nervous system with the bac-

Mechanisms of Disease—cont.

terium *Clostridium tetani.* This bacterium releases a toxin called *tetanospasmin* that triggers overactivity of the nervous system, often involving painful spasms of the muscles throughout the body. Because the spasms frequently begin in the head and cause the jaw muscles to tense involuntarily, the infection is often called "lockjaw."

Once a tragically common disease, **poliomyelitis** is a viral infection of the nerves that control skeletal muscle movement. Although the disease can be asymptomatic, it often causes paralysis that may progress to death. Virtually eliminated in the United States as a result of a comprehensive vaccination program, it still affects millions in other parts of the world.

Muscular Dystrophy

Muscular dystrophy (DISS-troh-fee) is not a single disorder but a group of genetic diseases characterized by atrophy (wasting) of skeletal muscle tissues. Some, but not all forms of muscular dystrophy can be fatal.

The common form of muscular dystrophy is **Duchenne** (doo-SHEN) **muscular dystrophy (DMD).** This form of the disease is also called *pseudohypertrophy* (meaning "false muscle growth") because the atrophy of muscle is masked by excessive replacement of muscle by fat and fibrous tissue. DMD is characterized by mild leg muscle weakness that progresses rapidly to include the shoulder muscles. The first signs of DMD are apparent at about 3 years of age, and the stricken child is usually severely affected within 5 to 10 years. Death from respiratory or cardiac muscle weakness often occurs by the time the individual is 21 years old.

DMD is caused by a mutation in the X chromosome, although other factors may be involved. DMD occurs primarily in boys. Because girls have two X chromosomes and boys only one, genetic diseases involving X chromosome abnormalities are more likely to occur in boys. This is true because girls with

one damaged X chromosome may not exhibit an "X-linked" disease if their other X chromosome is normal (see Chapter 34). The gene involved in DMD normally codes for the protein **dystrophin** (DIS-trof-in), which forms strands in each skeletal muscle fiber and helps hold the cytoskeleton to the sarcolemma (see Figure 4-34 on p. 139). Dystrophin thus helps keep the muscle fiber from breaking during contractions. Normal dystrophin is missing in DMD because a deletion or mutation of part of the dystrophin gene causes the resulting protein to be nonfunctional (it has the wrong shape to do the job). Therefore, in DMD muscle fibers break apart more easily—causing the symptoms of progressive muscle weakness.

Myasthenia Gravis

Myasthenia gravis (my-es-THEE-nee-ah GRAH-vis) is a chronic disease characterized by muscle weakness, especially in the face and throat. Most forms of this disease begin with mild weakness and chronic muscle fatigue in the face, then progress to wider muscle involvement. When severe muscle weakness causes immobility in all four limbs, a *myasthenic crisis* is said to have occurred. A person in myasthenic crisis is in danger of dying of respiratory failure because of weakness in the respiratory muscles.

Myasthenia gravis is an autoimmune disease in which the immune system attacks muscle cells at the neuromuscular junction. Nerve impulses from motor neurons are then unable to fully stimulate the affected muscle.

Hernias

Weakness of abdominal muscles can lead to a *hernia,* or protrusion, of an abdominal organ (commonly the small intestine or stomach) through an opening in the abdominal wall. There are several types of hernias. The most common one, **inguinal hernia** (Figure 11-30), occurs when the hernia extends down

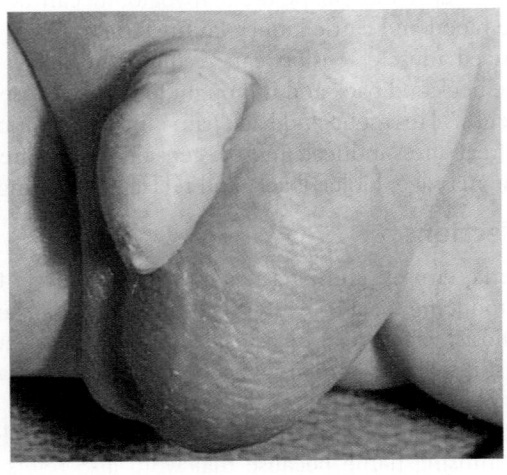

Inguinal hernia.

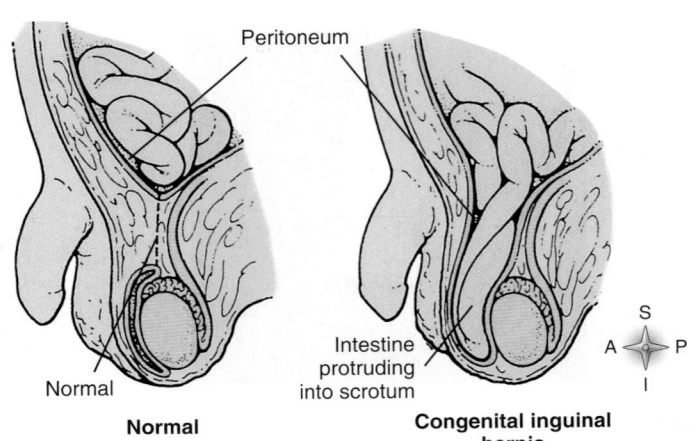

Figure 11-30 *Inguinal hernia in an infant male.*

Mechanisms of Disease—cont.

the inguinal canal, often into the scrotum or labia. Males experience this most often, and it can occur at any age.

An **umbilical hernia** occurs at the navel or umbilicus where the umbilical cord connects to the abdominal wall before birth. In this condition, a loop of intestine bulges through the umbilical opening, as you can see in Figure 11-31. It can happen at any age, but most frequently in newborns.

Another type of hernia occurs in the *femoral ring* where the vessels and nerves exit the abdominopelvic cavity to travel along the femur. A **femoral hernia** occurs when a loop of intestine moves through the femoral ring and into the groin area (Figure 11-32). Women may experience a femoral hernia because of anatomical changes during pregnancy.

When the stomach protrudes through an opening in the diaphragm called the *esophageal hiatus*, the condition is called **hiatal hernia** (Figure 11-33). The word hiatus literally means "opening." Hiatal hernias are very common and often weaken the *lower esophageal sphincter*, which closes the entry to the stomach—thus allowing reflux (backflow) of stomach acids that produce heartburn.

Hernia is referred to as "reducible" when the protruding organ is manipulated back into the abdominal cavity, either naturally by lying down or by manual reduction through a surgical opening in the abdomen. A "strangulated" hernia occurs when the mass is not reducible and blood flow to the affected organ (i.e., intestine) is stopped. Obstruction and gangrene can occur. Pain and vomiting are usually experienced, and emergency surgical intervention is required.

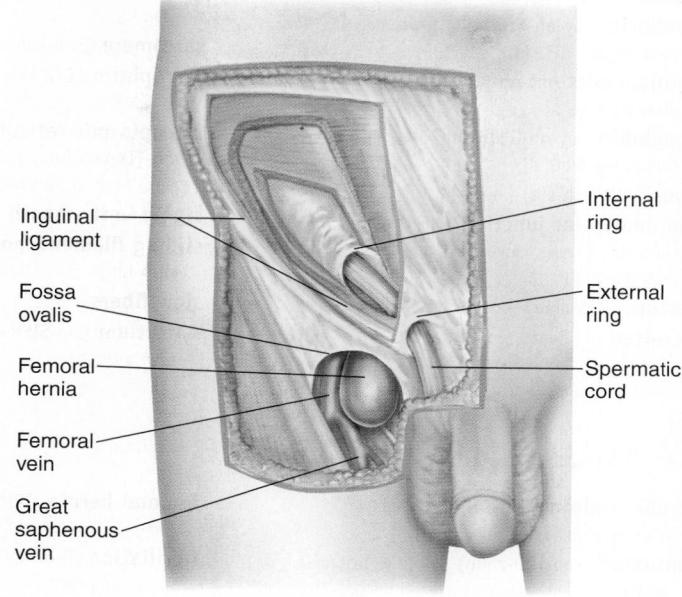

Figure 11-32 *Femoral hernia in an adult male.*

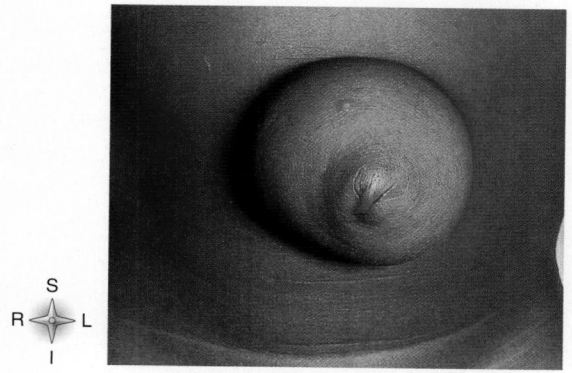

Figure 11-31 *Umbilical hernia in an infant.*

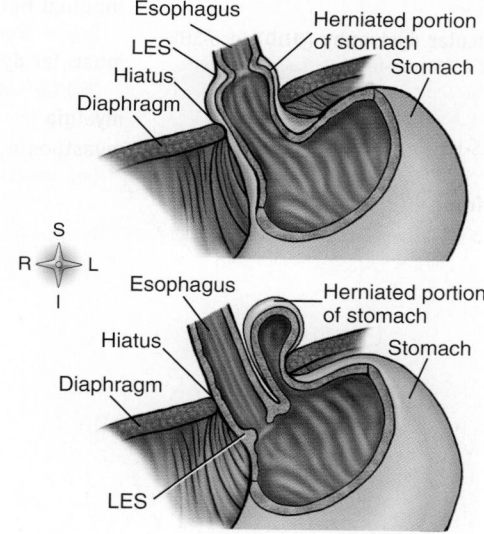

Figure 11-33 *Hiatal hernia.*

LANGUAGE OF SCIENCE

(Cont'd from page 395)

muscle tone [*tone* stretching]

myofibrils (my-oh-FYE-brils) [*myo-* muscle, *-fibril* small fiber]

myofilaments (my-oh-FIL-ah-ments) [*myo-* muscle, *-filament* thread]

myoglobin (my-oh-GLOH-bin) [*myo-* muscle, *-globin* containing protein]

myosin (MY-oh-sin) [*myos-* muscle, *-in* substance]

neuromuscular junction (NMJ) (noo-roh-MUSS-kyoo-lar) [*neuro-* nerve, *-muscul-* muscle, *-ar* pertaining to]

posture (POS-chur) [*postura* position]

recruited [*re-* again, *-cruit* grow]

sarcolemma (sar-koh-LEM-ah) [*sarco-* flesh, *-lemma* sheath]

sarcomere (SAR-koh-meer) [*sarco-* flesh, *-mere* part]

sarcoplasm (SAR-koh-plaz-em) [*sarco-* flesh, *-plasm* to mold]

sarcoplasmic reticulum (SR) (sar-koh-PLAZ-mik reh-TIK-yoo-lum) [*sarco-* flesh, *-plasm-* to mold, *-ic* pertaining to, *reticulum* small net]

single unit smooth muscle

sliding filament model (SLY-ding FILL-ah-ment MOD-uhl)

slow fibers

syncytium (sin-SISH-ee-em) [*syn-* union, *-cyt-* cell, *-ium* a thing]

T tubules (T TOOB-yools)

tetanus (TET-ah-nus) [*tetanus* tension]

threshold stimulus (THRESH-hold STIM-yoo-lus) [*stimul* to excite]

tonic contraction (TAHN-ik) [*ton-* to stretch, *-ic* pertaining to]

treppe (TREP-ee) [*treppe* staircase]

triad (TRY-ad) [*triad* group of three]

tropomyosin (troh-poh-MY-oh-sin) [*tropo-* to turn, *-myo-* muscle, *-in* substance]

troponin (troh-POH-nin) [*tropo-* turn, *-in* substance]

twitch contraction

Z lines (also called *Z disks*)

LANGUAGE OF MEDICINE

aerobic training (air-OH-bik) [*aero-* air, *-ic* pertaining to]

contusion (kon-TOO-zhun) [*contuse-* to bruise, *-ion* result]

convulsions (kon-VUL-shuns) [*convuls-* cramp, *-ion* result]

cramps

disuse atrophy (DIS-yoos AT-roh-fee) [*a-* without, *-troph* nourishment]

Duchenne muscular dystrophy (DMD) (doo-SHEN MUSS-kyoo-lar DISS-troh-fee) [*Duchenne Guillaume B.A. Duchenne de Boulogne* French neurologist, *dys-* bad, *-trophy* nourishment]

dystrophin (DIS-trof-in) [*dys-* bad, *-troph* nourishment]

endurance training

femoral hernia (FEM-or-all HER-nee-ah) [*femor-* femur, *-al* pertaining to, *hernia* rupture]

fibrillation (fi-bri-LAY-shun) [*fibril-* small fiber, *-ation* process]

fibromyositis (fye-broh-my-oh-SYE-tis) [*fibro-* fiber, *-myo-* muscle, *-itis* inflammation]

hiatal hernia (hye-AY-tal HER-nee-ah) [*hiatus* gap, *-al* pertaining to, *hernia* rupture]

inguinal hernia (ING-gwi-nal HER-nee-ah) [*inguin-* groin, *-al* pertaining to, *hernia* rupture]

muscular dystrophy (MUSS-kyoo-lar DISS-troh-fee) [*dys-* bad, *-trophy* nourishment]

myalgia (my-AL-jee-ah) [*my-* muscle, *-algia* pain]

myasthenia gravis (my-es-THEE-nee-ah GRAH-vis) [*my-* muscle, *-asthenia* weakness, *gravis* severe]

myopathies (my-OP-ah-thees) [*myo-* muscle, *-pathy* disease]

myositis (my-oh-SYE-tis) [*myos-* muscle, *-itis* inflammation]

poliomyelitis (pol-ee-oh-my-eh-LYE-tis) [*polio-* gray matter in the nervous system, *-mye-* marrow, *-itis* inflammation]

rigor mortis (RIG-or MOR-tis) [*rigor* stiffness, *mortis* death]

sprain

strain

strength training

umbilical hernia (um-BIL-i-kul HER-nee-ah) [*umbilic-* navel, *-al* pertaining to, *hernia* rupture]

CASE STUDY

Cecelia Pulaski, age 27, noticed changes in her energy level accompanied by muscle weakness. Particularly when she swallowed, she would sometimes feel that food was stuck in her throat. She had difficulty combing her hair, and she noticed that her voice was very weak. The weakness would usually improve when she rested. She was admitted to the hospital for myalgia, paresthesia, and immobility of all extremities. At the time of admission, she was having difficulty breathing. She recently experienced an extremely stressful divorce.

On physical examination, Ms. Pulaski is unable to close her eyes completely. Her pupils respond normally to light and show normal accommodation. She has lost 15 pounds in the last month. Her tongue has several fissures. Her laboratory data are essentially normal except for a positive antibody test, which is indicative of an autoimmune disorder attacking muscle cells at the neuromuscular junction. Electrical testing of the neuromuscular junction shows some blocking of discharges. A pharmacological test using edrophonium chloride is positive. Edrophonium chloride inhibits the breakdown of acetylcholine at the postsynaptic membrane.

1. Based on what is known about myasthenia gravis, which of the following explanations for Cecelia's symptoms would be physiologically correct?

   A. Adenosine triphosphate pulls the thin myofilaments during muscle contraction.
   B. Active sites on the actin molecules are exposed.
   C. A flood of calcium ions combines with troponin molecules in the thin filament myofibrils.
   D. Nerve impulses from motor neuromuscular junctions are unable to fully stimulate the affected muscle.

2. Based on the action of edrophonium chloride, as stated above, how will this drug work in Ms. Pulaski's case? Edrophonium chloride:

   A. Increases the availability of acetylcholine at postsynaptic receptor sites
   B. Decreases the availability of acetylcholine at postsynaptic receptor sites
   C. Increases the attachment of thick myosin filaments to the sarcomere
   D. Decreases electrical impulses in the sarcolemma

3. Based on the action of edrophonium chloride, as stated above, which one of the following physical effects will *most* likely be noted by Ms. Pulaski?

   A. Relaxation of muscle
   B. Decreased muscle excitation and contraction
   C. Increased muscle excitation and contraction
   D. Increased flaccidity of muscle

4. Based on the information presented in the case study, which one of the following disorders does Ms. Pulaski have?

   A. Muscular dystrophy
   B. Poliomyelitis
   C. Fibromyositis
   D. Myasthenia gravis

CHAPTER SUMMARY

INTRODUCTION

A. Muscular system is responsible for moving the framework of the body
B. In addition to movement, muscle tissue performs various other functions

GENERAL FUNCTIONS

A. Movement of the body as a whole or its parts
B. Heat production
C. Posture

FUNCTION OF SKELETAL MUSCLE TISSUE

A. Characteristics of skeletal muscle cells
   1. Excitability (irritability)—ability to be stimulated
   2. Contractility—ability to contract, or shorten, and produce body movement
   3. Extensibility—ability to extend, or stretch, thereby allowing muscles to return to their resting length
B. Overview of the muscle cell (Figures 11-1 and 11-2)
   1. Muscle cells are called fibers because of their threadlike shape
   2. Sarcolemma—plasma membrane of muscle fibers

3. Sarcoplasmic reticulum
   a. Network of tubules and sacs found within muscle fibers
   b. Membrane of the sarcoplasmic reticulum continually pumps calcium ions from the sarcoplasm and stores the ions within its sacs for later release (Figure 11-3)
4. Muscle fibers contain many mitochondria and several nuclei
5. Myofibrils—numerous fine fibers packed close together in sarcoplasm
6. Sarcomere
   a. Segment of myofibril between two successive Z lines
   b. Each myofibril consists of many sarcomeres
   c. Contractile unit of muscle fibers
7. Striated muscle (Figure 11-4)
   a. Dark stripes called A *bands*; light H zone runs across the midsection of each dark A band
   b. Light stripes called I *bands*; dark Z line extends across the center of each light I band
8. T tubules
   a. Transverse tubules extend across the sarcoplasm at right angles to the long axis of the muscle fiber
   b. Formed by inward extensions of the sarcolemma

c. Membrane has ion pumps that continually transport Ca^{++} ions inward from the sarcoplasm

d. Allow electrical impulses traveling along the sarcolemma to move deeper into the cell

9. Triad

a. Triplet of tubules; a T tubule sandwiched between two sacs of sarcoplasmic reticulum. Allows an electrical impulse traveling along a T tubule to stimulate the membranes of adjacent sacs of the sarcoplasmic reticulum

C. Myofilaments (Figures 11-5 and 11-6)

1. Each myofibril contains thousands of thick and thin myofilaments

2. Four different kinds of protein molecules make up myofilaments

  a. Myosin

    (1) Makes up almost all the thick filament

    (2) Myosin "heads" are chemically attracted to actin molecules

    (3) Myosin "heads" are known as *cross bridges* when attached to actin

  b. Actin—globular protein that forms two fibrous strands twisted around each other to form the bulk of the thin filament

  c. Tropomyosin—protein that blocks the active sites on actin molecules

  d. Troponin—protein that holds tropomyosin molecules in place

3. Thin filaments attach to both Z lines (Z disks) of a sarcomere and extend part way toward the center

4. Thick myosin filaments do not attach to the Z lines

D. Mechanism of contraction

1. Excitation and contraction (Figures 11-7 through 11-12; Table 11-1)

  a. A skeletal muscle fiber remains at rest until stimulated by a motor neuron

  b. Neuromuscular junction—motor neurons connect to the sarcolemma at the motor endplate (Figure 11-7)

  c. Neuromuscular junction is a synapse where neurotransmitter molecules transmit signals

  d. Acetylcholine—the neurotransmitter released into the synaptic cleft that diffuses across the gap, stimulates the receptors, and initiates an impulse in the sarcolemma

  e. Nerve impulse travels over the sarcolemma and inward along the T tubules, which triggers the release of calcium ions

  f. Calcium binds to troponin, which causes tropomyosin to shift and expose active sites on actin

  g. Sliding filament model (Figures 11-11 and 11-12)

    (1) When active sites on actin are exposed, myosin heads bind to them

    (2) Myosin heads bend and pull the thin filaments past them

    (3) Each head releases, binds to the next active site, and pulls again

    (4) The entire myofibril shortens

2. Relaxation

  a. Immediately after the Ca^{++} ions are released, the sarcoplasmic reticulum begins actively pumping them back into the sacs (Figure 11-3)

  b. Ca^{++} ions are removed from the troponin molecules, thereby shutting down the contraction

3. Energy sources for muscle contraction (Figure 11-13)

  a. Hydrolysis of ATP yields the energy required for muscular contraction

  b. ATP binds to the myosin head and then transfers its energy to the myosin head to perform the work of pulling the thin filament during contraction

  c. Muscle fibers continually resynthesize ATP from the breakdown of creatine phosphate (CP)

  d. Catabolism by muscle fibers requires glucose and oxygen

  e. At rest, excess O_2 in the sarcoplasm is bound to myoglobin (Box 11-4)

    (1) Red fibers—muscle fibers with high levels of myoglobin

    (2) White fibers—muscle fibers with little myoglobin

  f. Aerobic respiration occurs when adequate O_2 is available

  g. Anaerobic respiration occurs when low levels of O_2 are available and results in the formation of lactic acid

  h. Glucose and oxygen supplied to muscle fibers by blood capillaries (Figure 11-14)

  i. Skeletal muscle contraction produces waste heat that can be used to help maintain the set point body temperature (Figure 11-15)

FUNCTION OF SKELETAL MUSCLE ORGANS

A. Muscles are composed of bundles of muscle fibers held together by fibrous connective tissue

B. Motor unit (Figure 11-16)

1. Motor unit—motor neuron plus the muscle fibers to which it attaches

2. Some motor units consist of only a few muscle fibers, whereas others consist of numerous fibers

3. Generally, the smaller the number of fibers in a motor unit, the more precise the movements available; the larger the number of fibers in a motor unit, the more powerful the contraction available

C. Myography—method of graphing the changing tension of a muscle as it contracts (Figure 11-17)

D. Twitch contraction (Figure 11-18)

1. A quick jerk of a muscle that is produced as a result of a single, brief threshold stimulus (generally occurs only in experimental situations)

2. The twitch contraction has three phases

  a. Latent phase—nerve impulse travels to the sarcoplasmic reticulum to trigger release of Ca^{++}

  b. Contraction phase—Ca^{++} binds to troponin and sliding of filaments occurs

  c. Relaxation phase—sliding of filaments ceases

E. Treppe—the staircase phenomenon (Figure 11-19, *B*)
   1. Gradual, steplike increase in the strength of contraction that is seen in a series of twitch contractions that occur 1 second apart
   2. Eventually, the muscle responds with less forceful contractions, and the relaxation phase becomes shorter
   3. If the relaxation phase disappears completely, a contracture occurs
F. Tetanus—smooth, sustained contractions
   1. Multiple wave summation—multiple twitch waves are added together to sustain muscle tension for a longer time
   2. Incomplete tetanus—very short periods of relaxation occur between peaks of tension (Figure 11-19, *C*)
   3. Complete tetanus—the stimulation is such that twitch waves fuse into a single, sustained peak (Figure 11-19, *D*)
G. Muscle tone
   1. Tonic contraction—continual, partial contraction of a muscle
   2. At any one time, a small number of muscle fibers within a muscle contract and produce a tightness or muscle tone
   3. Muscles with less tone than normal are flaccid
   4. Muscles with more tone than normal are spastic
   5. Muscle tone is maintained by negative feedback mechanisms
H. Graded strength principle
   1. Graded strength principle—skeletal muscles contract with varying degrees of strength at different times
   2. Factors that contribute to the phenomenon of graded strength (Figure 11-23)
      a. Metabolic condition of individual fibers
      b. Number of muscle fibers contracting simultaneously; the greater the number of fibers contracting, the stronger the contraction
      c. Number of motor units recruited
      d. Intensity and frequency of stimulation (Figure 11-20)
      e. Length-tension relationship (Figure 11-21)
         (1) Maximal strength that a muscle can develop bears a direct relationship to the initial length of its fibers
         (2) A shortened muscle's sarcomeres are compressed; therefore, the muscle cannot develop much tension
         (3) An overstretched muscle cannot develop much tension because the thick myofilaments are too far from the thin myofilaments
         (4) Strongest maximal contraction is possible only when the skeletal muscle has been stretched to its optimal length
      f. Stretch reflex (Figure 11-22)
         (1) The load imposed on a muscle influences the strength of a skeletal contraction
         (2) Stretch reflex—the body tries to maintain constancy of muscle length in response to increased load
         (3) Maintains a relatively constant length as load is increased up to a maximum sustainable level

I. Isotonic and isometric contractions (Figure 11-24)
   1. Isotonic contraction
      a. Contraction in which the tone or tension within a muscle remains the same as the length of the muscle changes
         (1) Concentric—muscle shortens as it contracts
         (2) Eccentric—muscle lengthens while contracting
      b. Isotonic—literally means "same tension"
      c. All of the energy of contraction is used to pull on thin myofilaments and thereby change the length of a fiber's sarcomeres
   2. Isometric contraction
      a. Contraction in which muscle length remains the same while muscle tension increases
      b. Isometric—literally means "same length"
   3. Most body movements occur as a result of both types of contractions

FUNCTION OF CARDIAC AND SMOOTH MUSCLE TISSUE

A. Cardiac muscle (Figure 11-25)
   1. Found only in the heart; forms the bulk of the wall of each chamber
   2. Also known as *striated involuntary muscle*
   3. Contracts rhythmically and continuously to provide the pumping action needed to maintain constant blood flow
   4. Cardiac muscle resembles skeletal muscle but has specialized features related to its role in continuously pumping blood
      a. Each cardiac muscle contains parallel myofibrils (Figure 11-25)
      b. Cardiac muscle fibers form strong, electrically coupled junctions (intercalated disks) with other fibers; individual cells also exhibit branching
      c. Syncytium—continuous, electrically coupled mass
      d. Cardiac muscle fibers form a continuous, contractile band around the heart chambers that conducts a single impulse across a virtually continuous sarcolemma
      e. T tubules are larger and form diads with a rather sparse sarcoplasmic reticulum
      f. Cardiac muscle sustains each impulse longer than in skeletal muscle; therefore, impulses cannot come rapidly enough to produce tetanus (Figure 11-26)
      g. Cardiac muscle does not run low on ATP and does not experience fatigue
      h. Cardiac muscle is self-stimulating
B. Smooth muscle
   1. Smooth muscle is composed of small, tapered cells with single nuclei (Figure 11-27)
   2. No T tubules are present, and only a loosely organized sarcoplasmic reticulum is present
   3. Ca^{++} comes from outside the cell and binds to calmodulin instead of troponin to trigger a contraction
   4. No striations because thick and thin myofilaments are arranged differently than in skeletal or cardiac muscle fibers; myofilaments are not organized into sarcomeres

5. Two types of smooth muscle tissue (Figure 11-28)
   a. Single unit (visceral)
      (1) Gap junctions join smooth muscle fibers into large, continuous sheets
      (2) Most common type; forms a muscular layer in the walls of hollow structures such as the digestive, urinary, and reproductive tracts
      (3) Exhibits autorhythmicity and produces peristalsis
   b. Multiunit
      (1) Does not act as a single unit but is composed of many independent cell units
      (2) Each fiber responds only to nervous input

THE BIG PICTURE: MUSCLE TISSUE AND THE WHOLE BODY

A. Function of all three major types of muscle is integral to function of the entire body
B. All three types of muscle tissue provide the movement necessary for survival
C. Relative constancy of the body's internal temperature is maintained by "waste" heat generated by muscle tissue
D. Maintains the body in a relatively stable position

REVIEW QUESTIONS

1. Define the terms *sarcolemma*, *sarcoplasm*, and *sarcoplasmic reticulum*.
2. Describe the function of the sarcoplasmic reticulum.
3. How are acetylcholine, Ca^{++}, and adenosine triphosphate (ATP) involved in the excitation and contraction of skeletal muscle?
4. Describe the general structure of ATP and tell how it relates to its function.
5. How does ATP provide energy for muscle contraction?
6. Describe the anatomical arrangement of a motor unit.

7. List and describe the different types of skeletal muscle contractions.
8. Define the term recruited.
9. Describe rigor mortis.
10. What are the effects of exercise on skeletal muscles?

CRITICAL THINKING QUESTIONS

1. Explain how skeletal muscles provide movement, heat, and posture. Are all of these functions unique to muscles? Explain your answer.
2. The characteristic of excitability is shared by what other system? Relate contractility and extensibility to the concept of agonist and antagonist discussed in Chapter 10.
3. What structures are unique to skeletal muscle fibers? Which of the structures are involved primarily in contractility and which are involved in excitability?
4. Explain how the structure of myofilaments is related to their function.
5. Explain how the sliding filament theory allows for the shortening of a muscle fiber.
6. Compare and contrast the role of Ca^{++} in excitation, contraction, and relaxation of skeletal muscle.
7. People who exercise seriously are sometimes told to work a muscle until they "feel the burn." In terms of how the muscle is able to release energy, explain what is going on in the muscle early in the exercise and when the muscle is "burning."
8. Using fiber types, design a muscle for a marathon runner and a different muscle for a 100-yard–dash sprinter. Explain your choice.
9. Explain the meaning of a "unit of combined cells" as it relates to cardiac muscle. How does this structural arrangement affect its function?
10. Which of the two smooth muscle types would be most affected by damage to the nerves that stimulate them?

Career Choices | Massage Therapist

Massage therapy improves circulation and helps correct imbalances in the soft tissue areas of the body, as in muscles and fascia. As a massage therapist, I am self-employed. My clientele consists of hospital-referred patients for lymph drainage therapy (see Chapter 20) and people who seek traditional massage for various reasons. The majority of my work consists of home visits; however, I also do some work at health facilities and with organized groups.

I have been massaging people since I was 5 years old. I massaged my other classmates whenever they had problems or hurt themselves. It has always been natural and enjoyable to work on others with my hands. Now, as an adult, I have professional certification in many different areas of therapeutic massage and related healing arts.

Some of the current trends in the massage profession are La Stone Therapy, treatment of

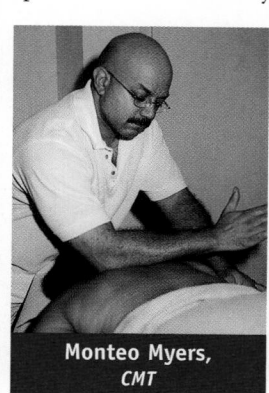

Monteo Myers,
CMT

cancer patients (with a doctor's approval), and stress and pain relief.

My job is extremely rewarding. I rarely encounter a client in a bad mood before a massage, and never after completion! It is very rewarding to see a client or patient's problem improve, or simply see a look of contentment on their face. There is a personal, respectful bond between the therapist and the client. In addition, this profession can be financially rewarding.

Knowledge of anatomy and physiology is definitely needed to effectively treat muscular/skeletal problems because isolation of the involved muscle is necessary. I still review anatomy and physiology of the muscular, skeletal, and lymphatic areas all the time. My advice is to study anatomy and physiology in a quiet room, do not study an entire chapter in one sitting, spread the chapter out, and take notes!

Communication, Control, and Integration

The anatomical structures and functional mechanisms that permit communication, control, and integration of bodily functions are discussed in the chapters of Unit 3. To maintain homeostasis, the body must have the ability to monitor and then respond appropriately to changes that may occur in either the internal or external environment. The nervous and endocrine systems provide this capability. Information originating in sensory nerve endings found in complex special sense organs such as the eye and in simple receptors located in skin or other body tissues provides the body with the necessary input. Nervous signals traveling rapidly from the brain and spinal cord over nerves to muscles and glands initiate immediate coordinating and regulating responses. Slower-acting chemical messengers, hormones produced by endocrine glands, serve to effect more long-term changes in physiological activities to maintain homeostasis.

SEEING THE BIG PICTURE

CHAPTER 12

Nervous System Cells

LANGUAGE OF SCIENCE

absolute refractory period (AB-so-loot ree-FRAK-toh-ree) [*absolutus* set loose, *refringere* to break apart, *periodos* circuit]

acetylcholine (ACh) (ass-ee-til-KOH-leen) [*acetyl-* vinegar, *-chole-* bile, *-ine* made of]

action potential (AK-shun poh-TEN-shal)

afferent division (AF-fer-ent) [*ad-* toward, *-ferre* to carry]

afferent (sensory) neurons (AF-fer-ent NOO-rons) [*ad-* toward, *-ferre* to carry, *neurons* nerves]

amine (AM-een) [*amine* ammonia compound or amino acid]

astrocytes (ASS-troh-sytes) [*astro-* star shaped, *-cyte* cell]

autonomic nervous system (ANS) (aw-toh-NOM-ik) [*auto-* self, *-nom-* rule, *-ic* pertaining to]

axon (AK-son) [*axon* axle]

axon hillock (AK-son HILL-ok) [*axon* axle, *hillock* hill]

axonal transport (AK-soh-nal trans-PORT) [*axon-* axle, *-al* pertaining to, *trans-* across, *-portare* to carry]

bipolar neurons (bye-POH-lar NOO-rons) [*bi-* two, *-polus* pole, *neurons* nerves]

catecholamine (kat-eh-KOHL-ah-meen)

cell body [*cella* storeroom]

central nervous system (CNS) [*kentron-* center, *-al* pertaining to, *nervus* nerves]

convergence (kon-VER-jens) [*convergere* to bend together]

dendrites (DEN-drytes) [*dentr* tree or branches]

depolarization (dee-poh-lar-i-ZAY-shun) [*de-* to do the opposite, *-polus* pole, *-tion* process of]

divergence (dye-VER-jens) [*di-* to separate, *-vergere* to incline]

efferent division (EF-fer-ent) [*effere* to carry out]

efferent (motor) neurons (EF-fer-ent NOO-rons) [*effere* to carry out, *neurons* nerves]

endoneurium (en-doh-NOO-ree-um) [*endo-* inward, *neuron* nerve]

Cont'd on p. 464

The nervous system and the endocrine system together perform a vital function for the body—communication. Homeostasis and therefore survival depend on this function. Why? Because communication provides the means for controlling and integrating the many different functions performed by organs, tissues, and cells. Integrating means unifying. Unifying body functions means controlling them in ways that make them work together like parts of one machine to accomplish homeostasis and thus survival. Communication makes possible control; control makes possible integration; integration makes possible homeostasis; homeostasis makes possible survival.

The **nervous system**—made up of the brain, spinal cord, and nerves—is probably the most intriguing body system (Figure 12-1). Facts, theories, and questions about this system are as fascinating as they are abundant. After all, it is the complex functioning of our nervous system that sets us apart from the rest of creation. We shall begin our study of the nervous system by considering in this chapter the cells of the nervous system and how

they work together to accomplish their function. Then in Chapter 13 we discuss the brain and spinal cord. Chapter 14 presents the nerves of the body, and Chapter 15 continues the discussion by describing the structure and function of the sense organs.

ORGANIZATION OF THE NERVOUS SYSTEM

The nervous system is organized to detect changes (stimuli) in the internal and external environment, evaluate that information, and possibly respond by initiating changes in muscles or glands. To make this complex network of information lines and processing circuits easier to understand, biologists have subdivided the nervous system into the smaller "systems" and "divisions" described in the following paragraphs and illustrated in Figure 12-2. Note as you read through the next several sections that the nervous system can be divided in various ways: according to structure, direction of information flow, or control of effectors.

Central and Peripheral Nervous Systems

The classical manner of subdividing the nervous system is based on the gross dissections of early anatomists. It simply categorizes all nervous system tissues according to their relative positions in the body: central or peripheral.

The **central nervous system (CNS)** is, as its name implies, the structural and functional center of the entire nervous system. Consisting of the brain and spinal cord, the CNS integrates incoming pieces of sensory information, evaluates the information, and initiates an outgoing response. Today, neurobiologists include only those cells that begin and end within the anatomical boundaries of the brain and spinal cord as part of the CNS. Cells that begin in the brain or cord but extend out through a nerve are thus not included in the central nervous system.

The **peripheral nervous system (PNS)** consists of the nerve tissues that lie in the periphery, or "outer regions," of the nervous system. Nerves that originate from the brain are called *cranial nerves*, and nerves that originate from the spinal cord are called *spinal nerves.*

The terms *central* and *peripheral* are often used as directional terms in the nervous system. For example, nerve cell extensions called *nerve fibers* may be called *central fibers* if they extend from the cell body toward the CNS. Likewise, they may be called *peripheral fibers* if they extend from the cell body away from the CNS.

Figure 12-1 represents the anatomical components of the CNS and PNS and their relative positions in the body. Figure 12-2 represents the relationship of the CNS and PNS in diagram form.

Afferent and Efferent Divisions

The tissues of both the central and the peripheral nervous systems include nerve cells that form incoming information pathways and outgoing pathways. For this reason, it is often convenient to categorize the nervous pathways into divisions according to the direction in which they carry information. The **afferent division** of the nervous system consists of all of the incoming *sensory* or *afferent* pathways. The **efferent division** of the nervous system consists of all the outgoing *motor* or *efferent* pathways. The literal meanings of the terms *afferent* (carry toward) and *efferent* (carry away) may

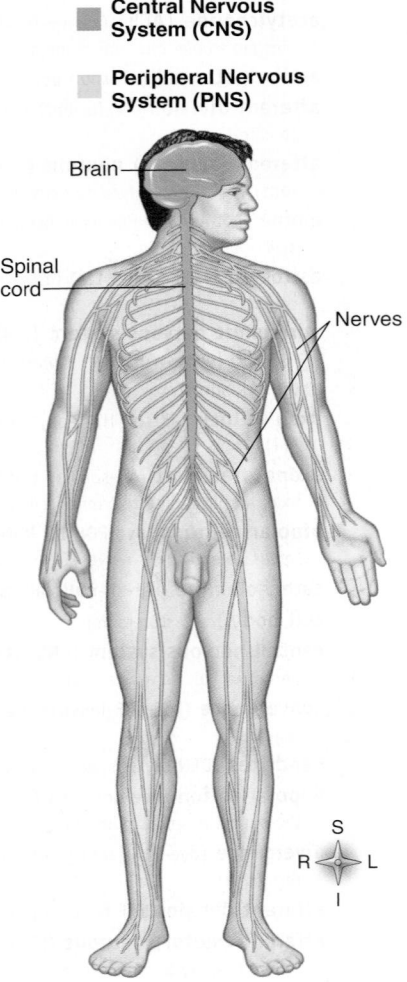

Central Nervous System (CNS)

Peripheral Nervous System (PNS)

Brain

Spinal cord

Nerves

S
R — L
I

Figure 12-1 *The nervous system.* Major anatomical features of the human nervous system include the brain, the spinal cord, and each of the individual nerves. The brain and spinal cord make up the central nervous system (CNS), and all the nerves and their branches make up the peripheral nervous system (PNS). Nerves originating from the brain are classified as *cranial nerves,* and nerves originating from the spinal cord are called *spinal nerves.*

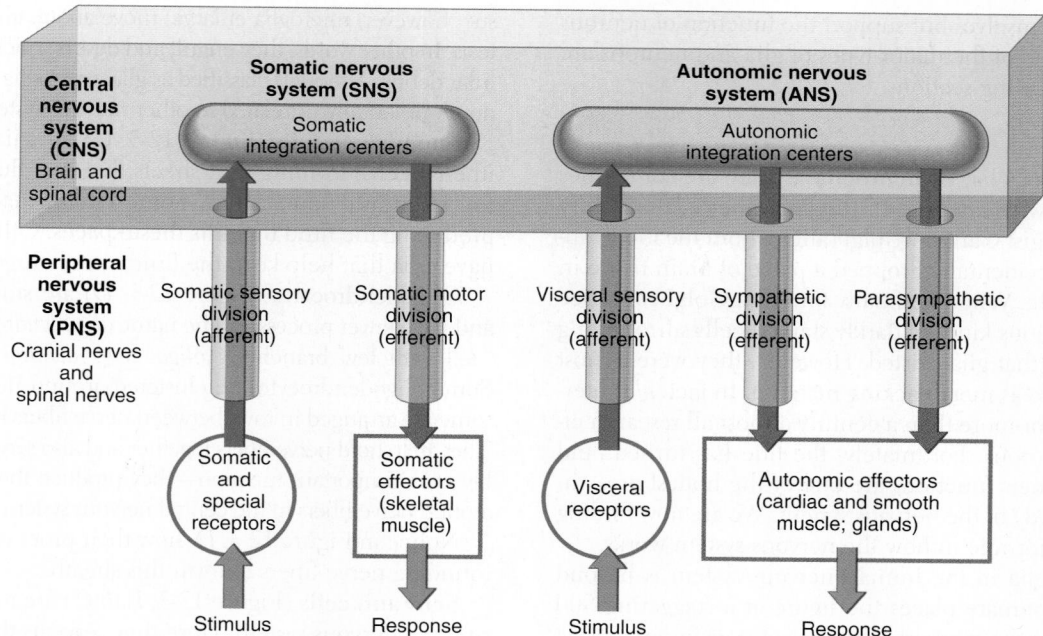

Figure 12-2 *Organizational plan of the nervous system.* Diagram summarizes the scheme used by most neurobiologists in studying the nervous system. Both the somatic nervous system (SNS) and the autonomic nervous system (ANS) include components in the CNS and PNS. Somatic sensory pathways conduct information toward integrators in the CNS, and somatic motor pathways conduct information toward somatic effectors. In the ANS, visceral sensory pathways conduct information toward CNS integrators, whereas the sympathetic and parasympathetic pathways conduct information toward autonomic effectors.

help you distinguish between these two divisions of the nervous system more easily.

Look at Figure 12-2 and try to distinguish the afferent and efferent pathways represented there. Here, and throughout the rest of this book, afferent pathways are typically represented in blue and efferent pathways in red.

Somatic and Autonomic Nervous Systems

Yet another way to organize the components of the nervous system for ease of study is to categorize them according to the type of effectors they regulate. Some pathways of the **somatic nervous system (SNS)** carry information to the *somatic effectors*, which are the skeletal muscles. These motor pathways make up the somatic motor division. As Figure 12-2 shows, the somatic nervous system also includes the afferent pathways, making up the **somatic sensory division,** that provide feedback from the somatic effectors. The SNS also includes the integrating centers that receive the sensory information and generate the efferent response signal.

Pathways of the **autonomic nervous system (ANS)** carry information to the *autonomic,* or *visceral,* effectors, which are the smooth muscles, cardiac muscle, and glands. As its name implies, the autonomic nervous system seems autonomous of voluntary control—it usually appears to govern itself without our conscious knowledge. We now know that the autonomic nervous system is influenced by the conscious mind, but the historical name for this part of the nervous system has remained.

The efferent pathways of the ANS can be divided into the *sympathetic division* and the *parasympathetic division.* The sympathetic division, made up of pathways that exit the middle portions

of the spinal cord, is involved in preparing the body to deal with immediate threats to the internal environment. It produces the "fight-or-flight" response. The parasympathetic pathways exit at the brain or lower portions of the spinal cord and coordinate the body's normal resting activities. The parasympathetic division is thus sometimes called the "rest-and-repair" division.

The afferent pathways of the ANS belong to the **visceral sensory division,** which carries feedback information to the autonomic integrating centers in the central nervous system.

Figure 12-2 summarizes the various ways in which the nervous system is subdivided and combines these approaches into a single "big picture."

 QUICK CHECK

1. List the major subdivisions of the human nervous system.
2. What two organs make up the central nervous system?
3. Contrast the somatic nervous system with the autonomic nervous system.

CELLS OF THE NERVOUS SYSTEM

Two main types of cells compose the nervous system, namely, *neurons* and *glia.* **Neurons** are excitable cells that conduct the impulses that make possible all nervous system functions. In other words, they form the "wiring" of the nervous system's information circuits. **Glia,** or *glial cells,* on the other hand, do not usually conduct information themselves but support the function of neurons in various ways. Some of the major types of glia and neurons are described in the following sections.

Glia

Our understanding of glia, or **neuroglia** as they are sometimes called, has been slow in coming. In the late nineteenth century the Italian cell biologist Camillo Golgi (after whom the Golgi apparatus is named) accidentally dropped a piece of brain tissue in a bath of silver nitrate. When he finally found it, Golgi could see a vast network of various kinds of darkly stained cells surrounding the neurons—proof that glia existed. However, they were almost immediately set aside as mere packing material. In fact, *glia* literally means "glue." For more than a century almost all research efforts focused on neurons. Fortunately, the tide has turned, and studies of glia and their functions are one of the hottest areas in *neurobiology*, the study of the nervous system. We are now finding that they have a major role in how the nervous system works.

The number of glia in the human nervous system is beyond imagination. One estimate places the figure at a staggering 900 billion, or nine times the estimated number of stars in our galaxy! Unlike neurons, glial cells retain their capacity for cell division throughout adulthood. Although this characteristic gives them the ability to replace themselves, it also makes them susceptible to abnormalities of cell division—such as cancer. Most benign and malignant tumors found in the nervous system originate in glial cells.

As stated earlier, glia serve various roles in supporting the function of neurons. To get a sense of this variety of functions, we shall briefly examine five major types of glia (Figure 12-3):

1. Astrocytes
2. Microglia
3. Ependymal cells
4. Oligodendrocytes
5. Schwann cells

The star-shaped glia, **astrocytes** (Figure 12-3, *A*), derive their name from the Greek *astron*, "star." Found only in the central nervous system, they are the largest and most numerous type of glia. Their long, delicate "points" extend through brain tissue, attaching to both neurons and the tiny blood capillaries of the brain. Astrocytes have recently been called "stars of the nervous system" because of the many important functions they perform. Astrocytes actually "feed" the neurons by picking up glucose from the blood, converting it to lactic acid, and passing it along to the neurons to which they are connected. (See Chapter 27 for a fuller discussion of the role of lactic acid in cellular respiration.) Because webs of astrocytes form tight sheaths around the brain's blood capillaries, they help form the *blood-brain barrier* (BBB). The BBB is a double barrier made up of astrocyte "feet" and the endothelial cells that make up the walls of the capillaries. Small molecules (e.g., oxygen, carbon dioxide, water, alcohol) diffuse rapidly through the barrier to reach brain neurons and other glia. Larger molecules penetrate it slowly or not at all (Box 12-1). More recent findings suggest that astrocytes may not only influence the growth of neurons and how the neurons connect to form circuits, but also may transmit information along "astrocyte pathways" themselves.

Microglia (Figure 12-3, *B*) are small, usually stationary cells found in the central nervous system. In inflamed or degenerating brain tis-

sue, however, microglia enlarge, move about, and carry on phagocytosis. In other words, they engulf and destroy microorganisms and cellular debris. Although classified as glia, microglia are functionally and developmentally unrelated to other nervous system cells.

Ependymal cells (Figure 12-3, *C*) are glia that resemble epithelial cells, forming thin sheets that line fluid-filled cavities in the brain and spinal cord. Some ependymal cells take part in producing the fluid that fills these spaces. Other ependymal cells have cilia that help keep the fluid circulating within the cavities.

Oligodendrocytes (Figure 12-3, *D*) are smaller than astrocytes and have fewer processes. The name *oligodendrocytes* literally means "cell with few branches" (*oligo-* few, *-dendro-* branch, *-cyte* cell). Some oligodendrocytes lie clustered around nerve cell bodies; and some are arranged in rows between nerve fibers in the brain and cord. They help hold nerve fibers together and also serve another and probably more important function—they produce the fatty **myelin sheath** around nerve fibers in the central nervous system (Box 12-2).

Notice in Figure 12-3, *D*, how their processes wrap around surrounding nerve fibers to form this sheath.

Schwann cells (Figure 12-3, *E* to *G*) are found only in the peripheral nervous system. Here they serve as the functional equivalent of the oligodendrocytes, supporting nerve fibers and sometimes forming a myelin sheath around them. As Figure 12-3, *F*, shows, many Schwann cells can wrap themselves around a single nerve fiber. The myelin sheath is formed by layers of Schwann cell membrane containing the white, fatty substance **myelin.** Microscopic gaps in the sheath, between adjacent Schwann cells, are called **nodes of Ranvier.** The myelin sheath and its tiny gaps are important in the proper conduction of impulses along nerve fibers in the peripheral nervous system. As you can see in Figure 12-4, the developing Schwann cell wraps around the nerve fiber in a way that forms an inner core of many layers of plasma membrane, made up mostly of the myelin (a type of phospholipid). Notice that the Schwann cell's nucleus and cytoplasm are squeezed to the perimeter to form the **neurilemma.** The neurilemma is essential to the regeneration of injured nerve fibers. Schwann cells are also called *neurilemmocytes*.

Figure 12-3, *E*, shows that some Schwann cells do not wrap around nerve fibers to form a thick myelin sheath but simply hold fibers together in a bundle. Nerve fibers with many Schwann cells forming a thick myelin sheath are called **myelinated fibers**, or *white fibers*. When several nerve fibers are held by a single Schwann cell that does not wrap around them to form a thick myelin sheath, the fibers are called *unmyelinated fibers*, or *gray fibers*.

Figure 12-3, *G*, shows a type of Schwann cell often called a **satellite cell.** Like satellites positioned around a planet, these special Schwann cells surround the cell body of a neuron. Satellite cells support neuronal cell bodies in regions called *ganglia* in the peripheral nervous system.

QUICK CHECK

4. What are the five main types of glia?
5. Describe the myelin sheath found on some nerve fibers.
6. What is a neurilemma?
7. Describe the three different forms of Schwann cells.

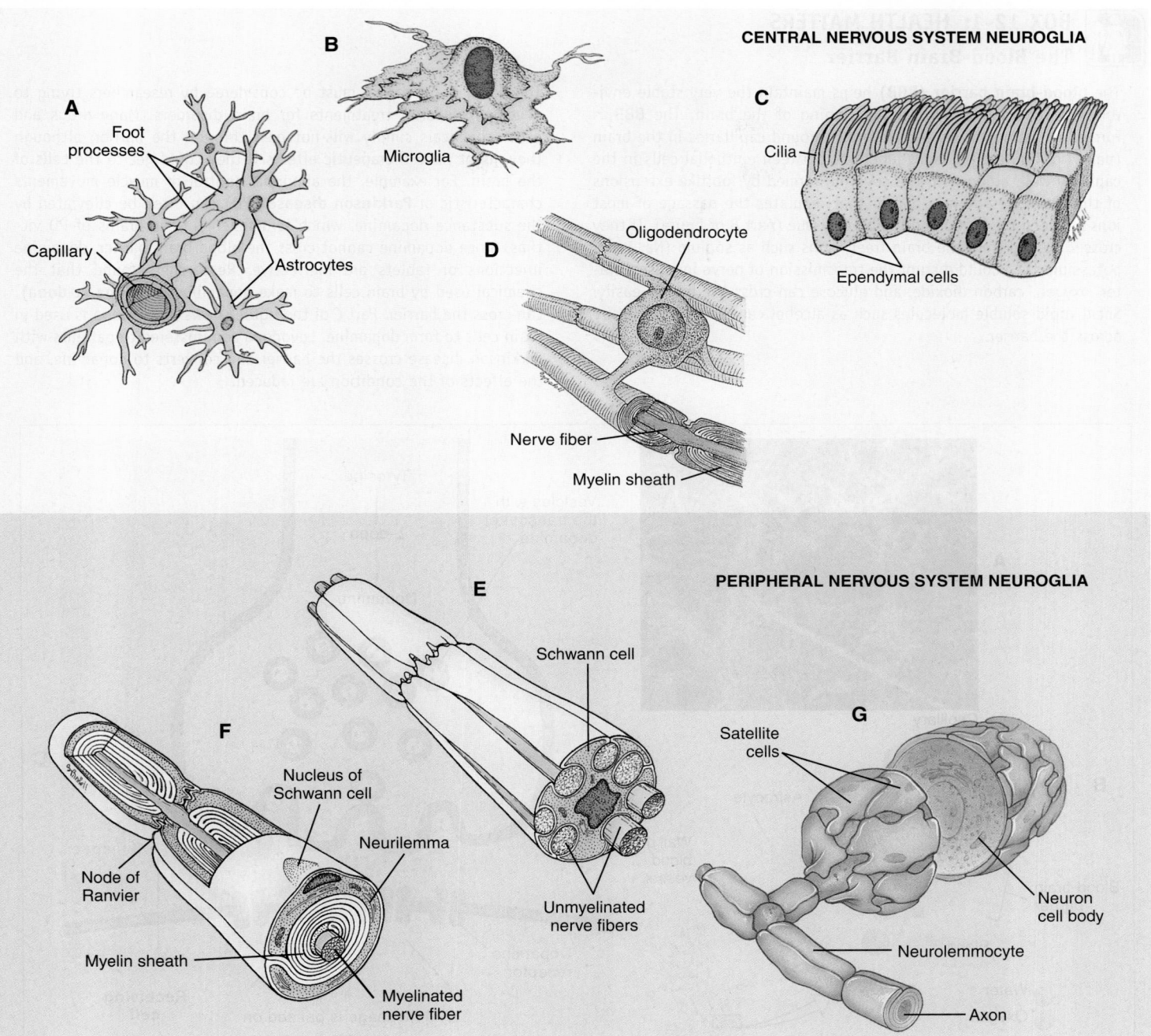

CENTRAL NERVOUS SYSTEM NEUROGLIA

A — Foot processes, Capillary, Astrocytes

B — Microglia

C — Cilia, Ependymal cells

D — Oligodendrocyte, Nerve fiber, Myelin sheath

PERIPHERAL NERVOUS SYSTEM NEUROGLIA

E — Schwann cell, Unmyelinated nerve fibers

F — Nucleus of Schwann cell, Neurilemma, Node of Ranvier, Myelin sheath, Myelinated nerve fiber

G — Satellite cells, Neuron cell body, Neurolemmocyte, Axon

Figure 12-3 *Types of neuroglia.* **Neuroglia of the CNS: A,** Astrocytes attached to the outside of a capillary blood vessel in the brain. **B,** A phagocytic microglial cell. **C,** Ciliated ependymal cells forming a sheet that usually lines fluid cavities in the brain. **D,** An oligodendrocyte with processes that wrap around nerve fibers in the CNS to form myelin sheaths. **Neuroglia of the PNS: E,** A Schwann cell supporting a bundle of nerve fibers in the PNS. **F,** Another type of Schwann cell wrapping around a peripheral nerve fiber to form a thick myelin sheath. **G,** Satellite cells, another type of Schwann cell, surround and support cell bodies of neurons in the PNS.

BOX 12-1: HEALTH MATTERS
The Blood-Brain Barrier

The **blood-brain barrier (BBB)** helps maintain the very stable environment required for normal functioning of the brain. The BBB is formed as astrocytes wrap their "feet" around capillaries in the brain (part *A* of figure). The tight junctions between epithelial cells in the capillary wall, along with the covering formed by footlike extensions of the astrocytes, form a barrier that regulates the passage of most ions between the blood and the brain tissue (part *B* of figure). If they crossed to and from the brain freely, ions such as sodium (Na^+) and potassium (K^+) could disrupt the transmission of nerve impulses. Water, oxygen, carbon dioxide, and glucose can cross the barrier easily. Small, lipid-soluble molecules such as alcohol can also diffuse easily across the barrier.

The blood brain barrier must be considered by researchers trying to develop new drug treatments for brain disorders. Many drugs and other chemicals simply will not pass through the barrier, although they might have therapeutic effects if they could get to the cells of the brain. For example, the abnormal control of muscle movements characteristic of **Parkinson disease (PD)** can often be alleviated by the substance dopamine, which is deficient in the brains of PD victims. Since dopamine cannot cross the blood brain barrier, dopamine injections or tablets are ineffective. Researchers found that the chemical used by brain cells to make dopamine, **levodopa (L-dopa)**, can cross the barrier. Part *C* of the figure shows how L-dopa is used in brain cells to form dopamine. Levodopa administered to patients with Parkinson disease crosses the barrier and converts to dopamine, and the effects of the condition are reduced.

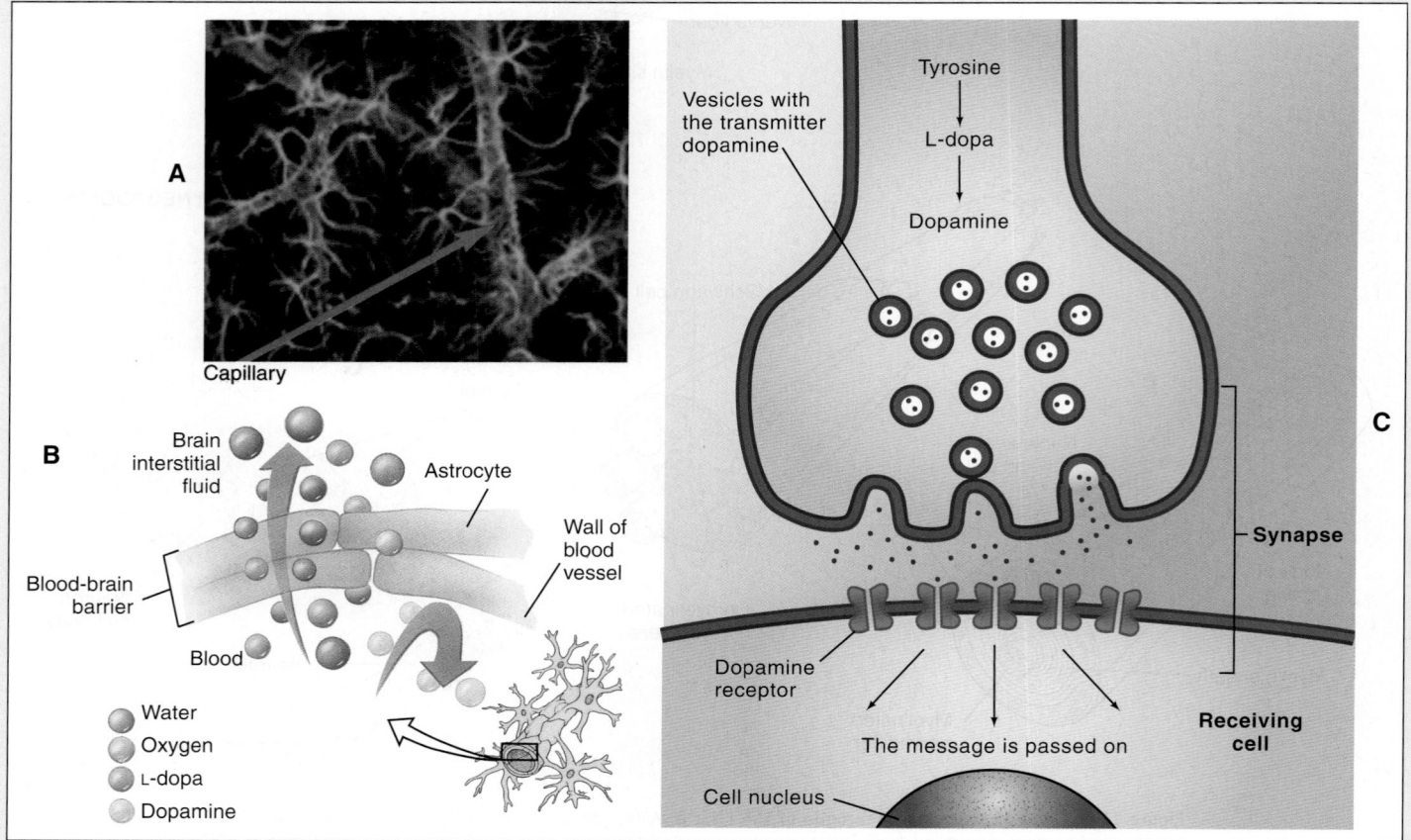

Blood-brain barrier. **A,** Micrograph of fluorescent-stained astrocytes shows how they attach to capillaries, forming a coating around these tiny blood vessels. **B,** Diagram showing a cross section of the blood-brain barrier, made up of the foot processes of astrocytes and the wall of the blood capillary. **C,** Diagram showing how dopamine is formed from L-dopa before being released as a neurotransmitter.

BOX 12-2: HEALTH MATTERS

Multiple Sclerosis (MS)

Several diseases are associated with disorders of the oligodendrocytes. Because these glial cells are involved in myelin formation, the diseases are called **myelin disorders.** The most common primary disease of the central nervous system is a myelin disorder called **multiple sclerosis (MS).** It is characterized by myelin loss and destruction accompanied by varying degrees of oligodendrocyte injury and death. The result is demyelination throughout the white matter of the central nervous system. Hard plaquelike lesions replace the destroyed myelin, and affected areas are invaded by inflammatory cells. As the myelin surrounding nerve fibers is lost, nerve conduction is impaired and weakness, loss of coordination, visual impairment, and speech disturbances occur. Although the disease occurs in both sexes and among all age-groups, it is most common in women between 20 and 40 years of age.

The cause of MS is thought to be related to autoimmunity and to viral infections in some individuals. Susceptibility to MS is inherited in some individuals. MS is characteristically relapsing and chronic in nature, but some cases of acute and unremitting disease have been reported. In most instances the disease is prolonged, with remissions and relapses occurring over a period of many years. There is no known cure.

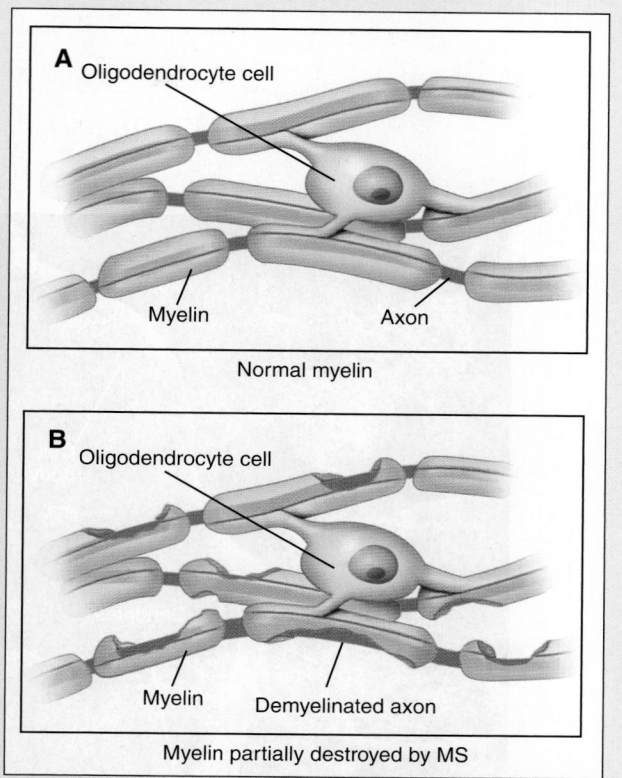

Effects of multiple sclerosis (MS). **A,** A normal myelin sheath allows rapid conduction. **B,** In MS, the myelin sheath is damaged, disrupting nerve conduction.

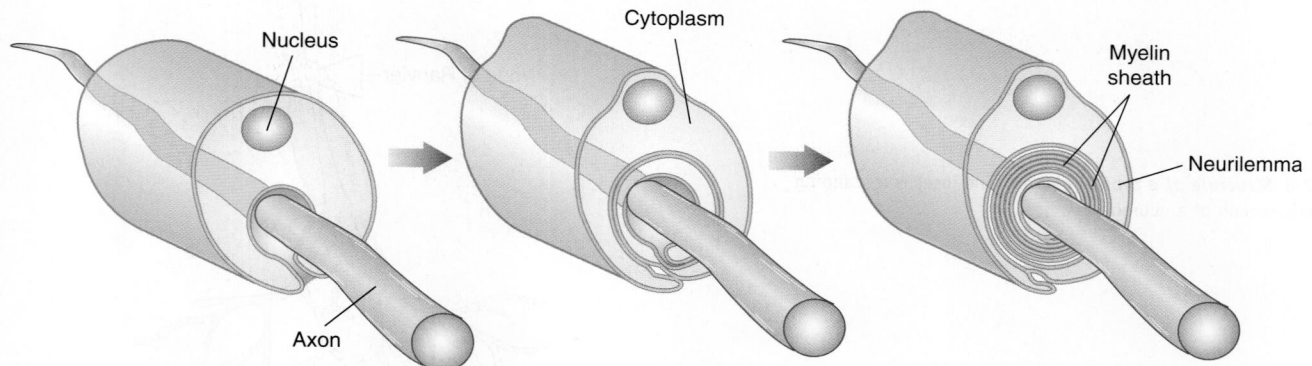

Figure 12-4 *Development of the myelin sheath.* A Schwann cell (neurilemmocyte) migrates to a neuron and wraps around an axon. The Schwann cell's cytoplasm is pushed to the outer layer, leaving a dense multilayered covering of plasma membrane around the axon. Because the plasma membrane of the Schwann cell is mostly the phospholipid myelin, the dense wrapping around the axon is called a *myelin sheath*. The outer layer of cytoplasm is called the *neurilemma*. The extensions of oligodendrocytes also wrap around axons to form a myelin sheath.

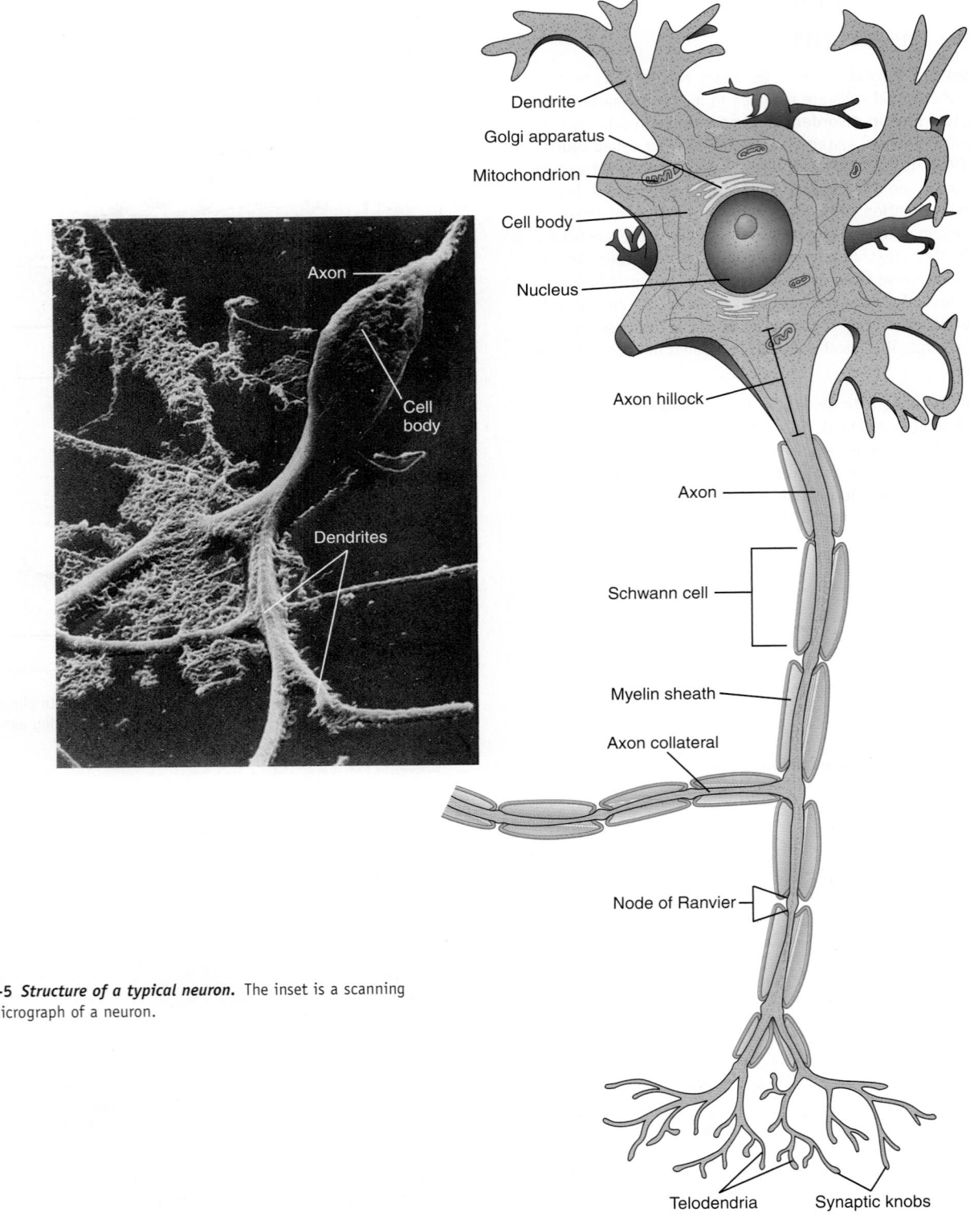

Figure 12-5 *Structure of a typical neuron.* The inset is a scanning electron micrograph of a neuron.

Neurons

The human brain is estimated to contain about 100 billion neurons, or about 10% of the total number of nervous system cells in the brain. All neurons consist of a **cell body** (also called the **perikaryon,** or *soma*) and at least two processes: one **axon** and one

or more **dendrites** (Figure 12-5). Because dendrites and axons are threadlike extensions from a neuron's cell body, they are often called **nerve fibers**.

In many respects the cell body, the largest part of a nerve cell, resembles other cells. It contains a nucleus, cytoplasm, and vari-

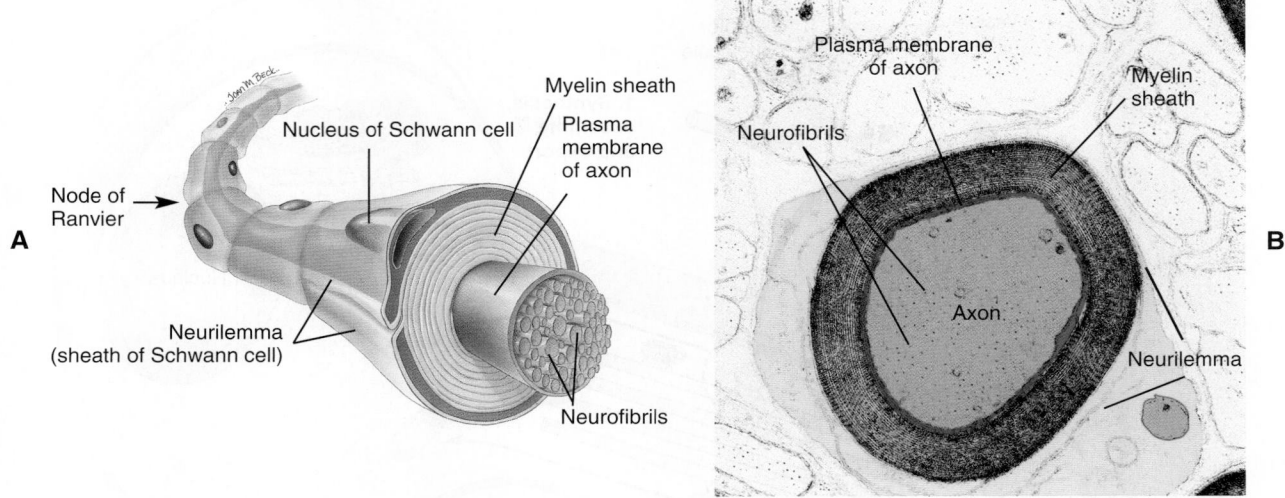

Figure 12-6 *Myelinated axon*. A, The diagram shows a cross section of an axon and its coverings formed by a Schwann cell: the myelin sheath and neurilemma. **B,** Transmission electron micrograph showing how the densely wrapped layers of the Schwann cell's plasma membrane form the fatty myelin sheath.

ous organelles found in other cells, for example, mitochondria and a Golgi apparatus. The location of the nucleus in the cell body not only makes it easy to find in a microscopic specimen, but also provides the name *perikaryon* (literally, "surrounding the nucleus"). A neuron's cytoplasm extends through its cell body and its processes. A plasma membrane encloses the entire neuron.

In the cell body, rough endoplasmic reticulum (ER) and its attached ribosomes provide protein molecules for the neuron. Some of these proteins are then processed and packaged into vesicles by the Golgi apparatus. Some protein molecules in these vesicles are needed for the transmission of nerve signals from one neuron to another. Such proteins are called **neurotransmitters.** Other proteins are used in the maintenance and repair of the neuron.

The cell body also contains many mitochondria, which replicate themselves in the cell body. Some of the resulting mitochondria are transported to the end of the axon to provide energy (ATP) for nerve signaling there.

Dendrites usually branch extensively from the cell body—like tiny trees. In fact, their name derives from the Greek word for *tree*. The distal ends of dendrites of sensory neurons may be called *receptors* because they receive the stimuli that initiate nerve signals. Some dendrites in the brain have small knoblike *dendritic spines*, which serve as connection points for other neurons. Dendrites receive stimuli and conduct electrical signals toward the cell body and/or axon of the neuron.

The axon of a neuron is a single process that usually extends from a tapered portion of the cell body called the **axon hillock.** Axons conduct impulses away from the cell body. Although a neuron has only one axon, that axon often has one or more side branches, called *axon collaterals*. Moreover, the distal tips of axons form branches called **telodendria** that each terminate in a **synaptic knob** (see Figure 12-5). Each synaptic knob contains mitochondria and numerous vesicles.

Axons vary in both length and diameter. Some are a meter long. Some, however, measure only a few millimeters. Axon diameters also vary considerably, from about 20 μm down to about 1 μm—a point of interest because axon diameter relates to velocity of impulse conduction. In general, the larger the diameter, the more rapid the conduction. A neuron's axon conducts impulses away from its cell body.

Whether an axon is myelinated or not also affects the speed of impulse conduction. Figure 12-6 shows a cross section of a typical myelinated axon. Notice in the figure how a series of Schwann cells have grown over the axon in a spiral fashion to form the myelin sheath and neurilemma (see also Figure 12-4). Only axons may have a myelin sheath—dendrites do not. The role of the myelin sheath and nodes of Ranvier in impulse conduction is discussed later.

Extending through the cytoplasm of each neuron are fine strands sometimes called **neurofibrils** (Figure 12-6). Neurofibrils are bundles of intermediate filaments called *neurofilaments.* Microtubules and microfilaments are additional components of the neuron's cytoskeleton. Along with providing structural support, a neuron's cytoskeleton forms a sort of "railway" for the rapid transport of small organelles to and from the far ends of a neuron. Figure 12-7 shows how small "motor molecules" attach to mitochondria and vesicles containing neurotransmitters and carry them to the end of the axon. The "used" vesicles and transmitters are then returned to the cell body by the same process, but in reverse, for recycling. This type of movement is called **axonal transport**.

Figure 12-8 summarizes the different functional regions of the neurons, based on their role in receiving and conducting nerve signals. The dendrites and cell body act primarily as an *input zone*, receiving nerve stimulation and initiating nerve impulses in response. The axon hillock acts as a *summation zone* by adding together all the nerve impulses arriving from the cell body and

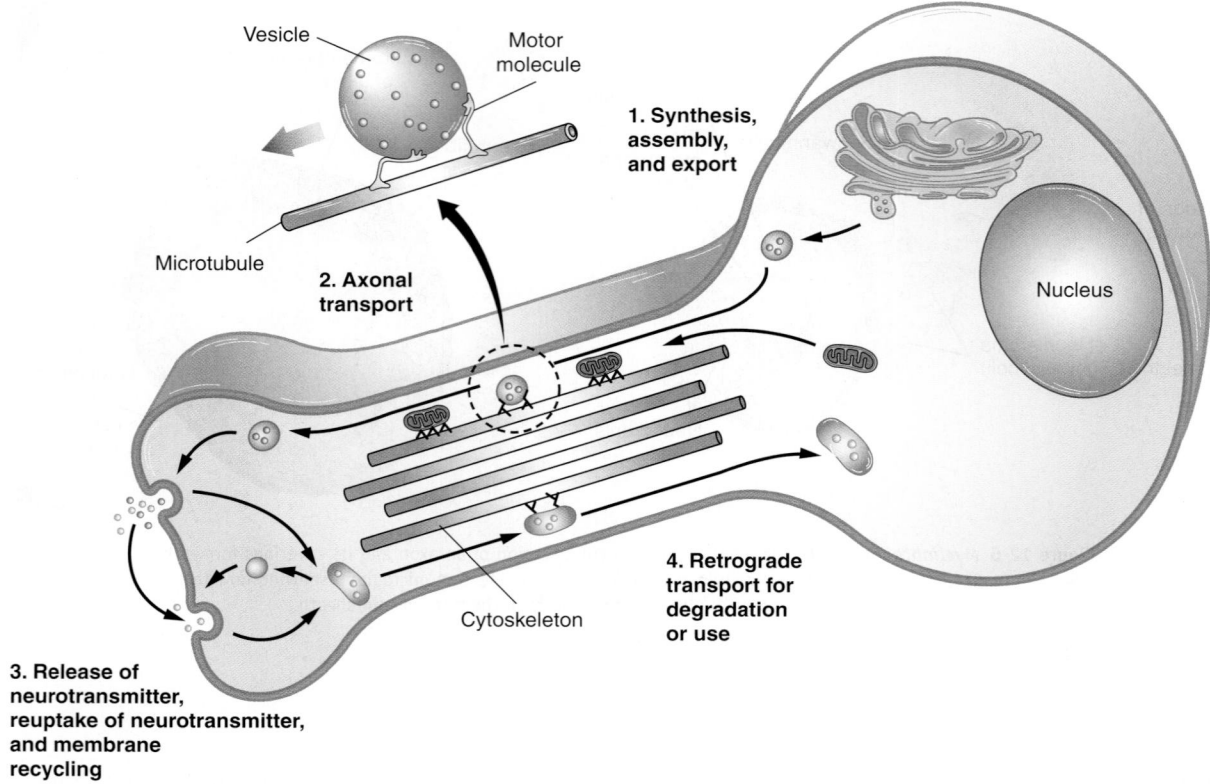

Figure 12-7 *Axonal transport.* Various cellular materials such as mitochondria and vesicles containing neurotransmitter can be shuttled quickly and efficiently from their point of origin in the soma (perikaryon) all the way to the end of the axon by using the neuron's cytoskeleton as a kind of railway system. The inset shows how motor molecules "walk" the material along a microtubule in the axon. The system also permits reverse transport, such as bringing the membranes of spent neurotransmitter vesicles back up to the soma.

dendrites—and deciding whether to send the impulse any farther in the neuron. The axon is the *conduction zone* because its primary job is to conduct the nerve impulse from the axon hillock all the way to the end of the neuron. The telodendria of the axon, along with their synaptic knobs, together act as an *output zone* where vesicles of neurotransmitter are released for possible reception by a nearby neuron or effector cell (muscle or gland cell).

Classification of Neurons

Structural Classification

The three types of neurons classified according to the number of their extensions from the cell body (Figure 12-9) are as follows:

1. Multipolar
2. Bipolar
3. Unipolar

Multipolar neurons have only one axon but several dendrites. Most of the neurons in the brain and spinal cord are multipolar. **Bipolar neurons** have only one axon and also only one highly branched dendrite. Bipolar neurons are the least numerous kind of neuron. They are found in the retina of the eye, in the inner ear, and in the olfactory pathway. **Unipolar (pseudounipolar) neurons** have a single process extending from the cell body. This single process branches to form a central process (toward the

CNS) and a peripheral process (away from the CNS). These two processes together form an axon, conducting impulses away from the dendrites found at the distal end of the peripheral process. Unipolar neurons are always sensory neurons, conducting information toward the central nervous system.

Functional Classification

The three types of neurons classified according to the direction in which they conduct impulses are as follows:

1. Afferent neurons
2. Efferent neurons
3. Interneurons

Figure 12-10 shows one neuron of each of these types. **Afferent (sensory) neurons** transmit nerve impulses to the spinal cord or brain. **Efferent (motor) neurons** transmit nerve impulses away from the brain or spinal cord to or toward muscles or glands. **Interneurons** conduct impulses from afferent neurons to or toward motor neurons. Interneurons lie entirely within the central nervous system (brain and spinal cord).

Reflex Arc

Notice in Figure 12-10 that neurons are often arranged in a pattern called a **reflex arc.** Basically, a reflex arc is a signal conduction route to and from the central nervous system (the brain and

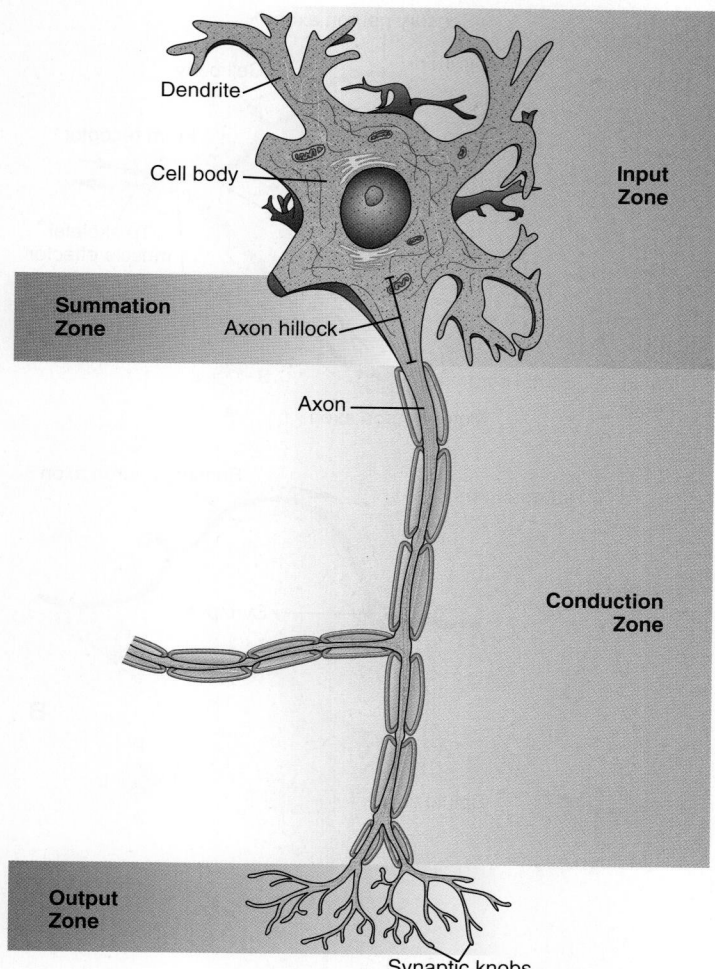

Input Zone

Dendrite

Cell body

Summation Zone

Axon hillock

Axon

Conduction Zone

Output Zone

Synaptic knobs

Figure 12-8 *Functional regions of the neuron's plasma membrane.* The *input zone* (dendrites and soma) receives input from other neurons or from sensory stimuli (stimulus-gated ion channels present). The *summation zone* (axon hillock) serves as the site where the nerve impulses combine and possibly trigger an impulse that will be conducted along the axon—or *conduction zone*. Both the summation (trigger) zone and conduction zone have many voltage-gated Na^+ channels and K^+ channels imbedded in the plasma membrane. The *output zone* (distal end of axon) is where the nerve impulse triggers the release of neurotransmitters. The output zone includes many voltage-gated Ca^{++} channels in the membrane.

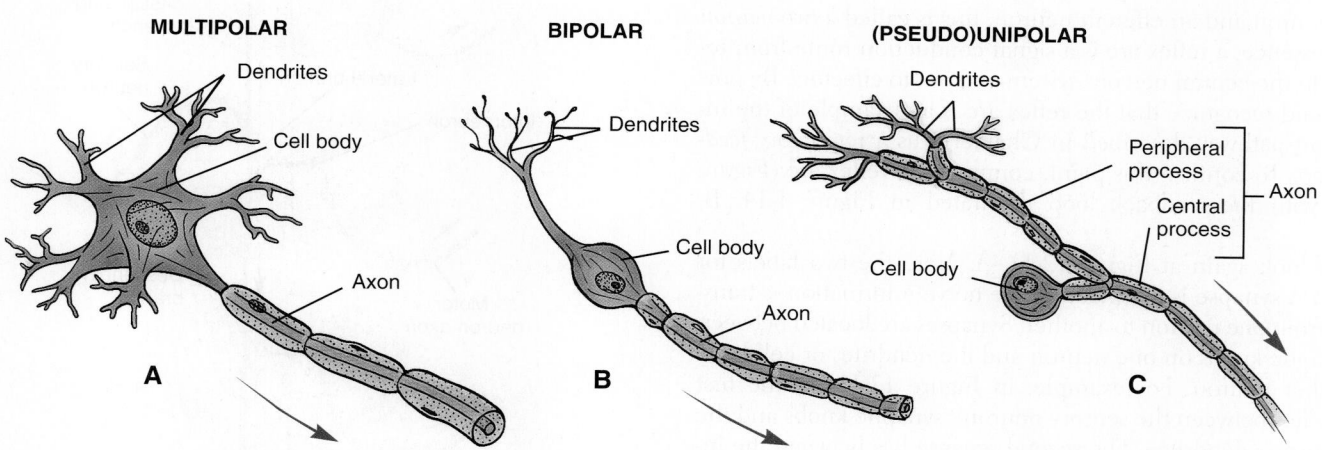

MULTIPOLAR

Dendrites

Cell body

Axon

A

BIPOLAR

Dendrites

Cell body

Axon

B

(PSEUDO)UNIPOLAR

Dendrites

Peripheral process

Central process

Axon

Cell body

C

Figure 12-9 *Structural classification of neurons.* **A,** Multipolar neuron: neuron with multiple extensions from the cell body. **B,** Bipolar neuron: neuron with exactly two extensions from the cell body.
C, (Pseudo) unipolar neuron: neuron with only one extension from the cell body. The central process is an axon; the peripheral process is a modified axon with branched dendrites at its extremity. (The red arrows show the direction of impulse travel.)

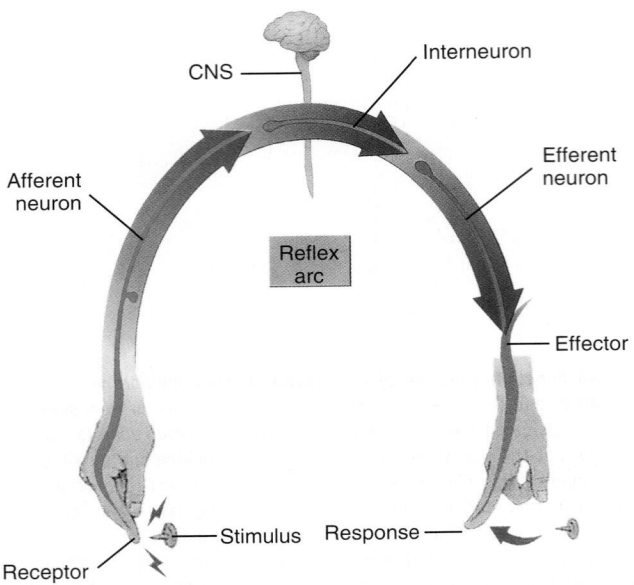

Figure 12-10 *Functional classification of neurons in a reflex arc.* Neurons can be classified according to the direction in which they conduct impulses. Notice that the most basic route of signal conduction follows a pattern called the reflex arc.

spinal cord). The most common form of reflex arc is the three-neuron arc (Figure 12-11, *A*). It consists of an afferent neuron, an interneuron, and an efferent neuron. Afferent, or sensory, neurons conduct signals to the central nervous system from *sensory receptors* in the peripheral nervous system. Efferent neurons, or *motor neurons*, conduct signals from the central nervous system to effectors. An effector is muscle tissue or glandular tissue. Interneurons conduct signals from afferent neurons toward or to motor neurons. In its simplest form, a reflex arc consists of an afferent neuron and an efferent neuron; this is called a *two-neuron arc*. In essence, a reflex arc is a signal conduction route from receptors to the central nervous system and out to effectors. By now you should recognize that the reflex arc is an example of the information pathway described in Chapter 1 as a regulatory *feedback loop*. To confirm this point, compare the reflex arc (Figure 12-10) with the feedback loop illustrated in Figure 1-14, *B*, p. 24.

Now look again at Figure 12-11, *A*. Note the two labels for synapse. A **synapse** is the place where nerve information is transmitted from one neuron to another. Synapses are located between the synaptic knobs on one neuron and the dendrites or cell body of another neuron. For example, in Figure 12-11, *A*, the first synapse lies between the sensory neuron's synaptic knobs and the interneuron's dendrites. The second synapse lies between the interneuron's synaptic knobs and the motor neuron's dendrites. This reflex arc is called an *ipsilateral reflex arc* because the receptors and effectors are located on the same side of the body. Figure 12-11, *B*, shows a *contralateral reflex arc*, one whose receptors and effectors are located on opposite sides of the body.

Besides simple two-neuron and three-neuron arcs, *intersegmental arcs* (Figure 12-11, *C*)—even more complex multineuron, multisynaptic arcs—also exist. An important principle is this: all

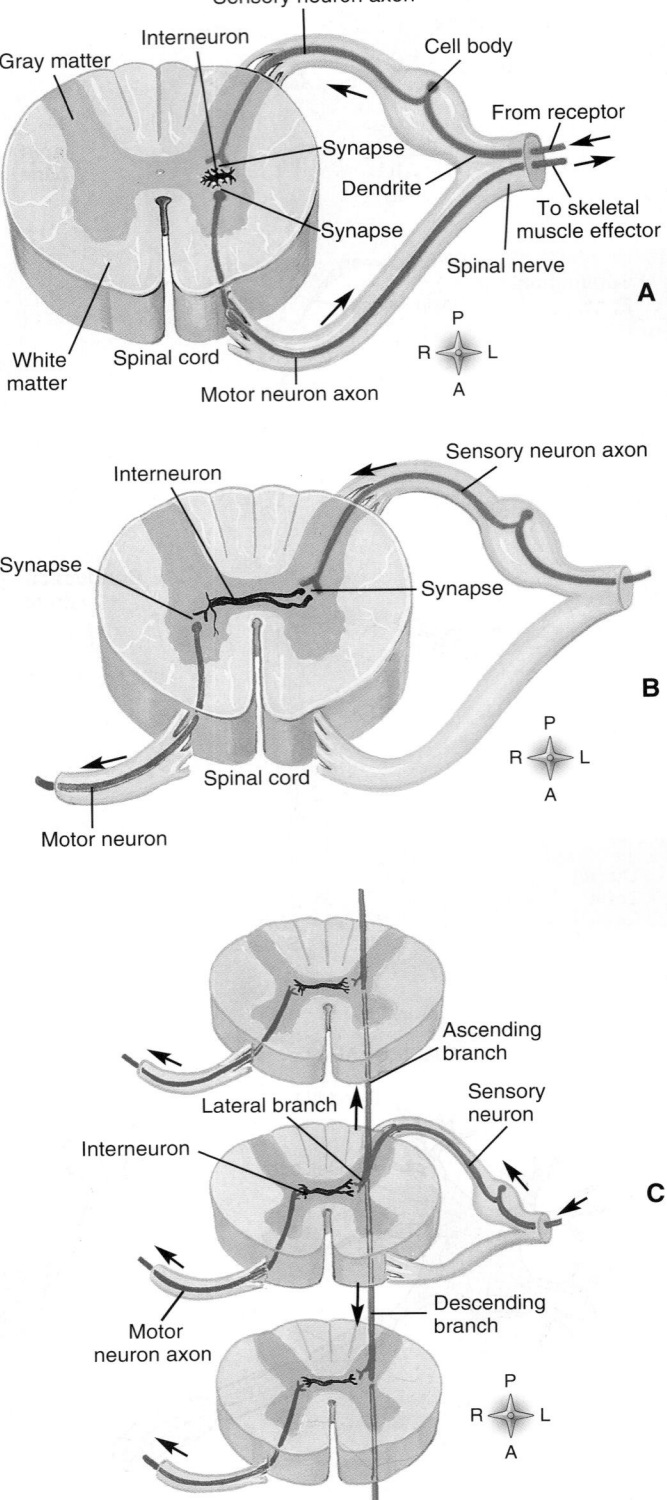

Figure 12-11 *Examples of reflex arcs.* **A,** Three-neuron ipsilateral reflex arc. Sensory information enters on the same side of the CNS as the motor information leaves the CNS. **B,** Three-neuron contralateral reflex arc. Sensory information enters on the opposite side of the CNS from the side that motor information exits the CNS. **C,** Intersegmental contralateral reflex arc. Divergent branches of a sensory neuron bring information to several segments of the CNS at the same time. Motor information leaves each segment on the opposite side of the CNS.

electrical signals that start in receptors do not invariably travel over a complete reflex arc and terminate in effectors. Many signals fail to be conducted across synapses. Moreover, all signals that terminate in effectors do not invariably start in receptors. Many of them, for example, are thought to originate in the brain.

 QUICK CHECK

8. List the characteristics of life in humans.
9. Define the term metabolism as it applies to the characteristics of life.

NERVES AND TRACTS

Nerves are bundles of peripheral nerve fibers held together by several layers of connective tissues (Figure 12-12). Surrounding each nerve fiber is a delicate layer of fibrous connective tissue called the **endoneurium.** Bundles of fibers (each with its own endoneurium), called *fascicles,* are held together by a connective tissue layer called the **perineurium.** Numerous fascicles, along with

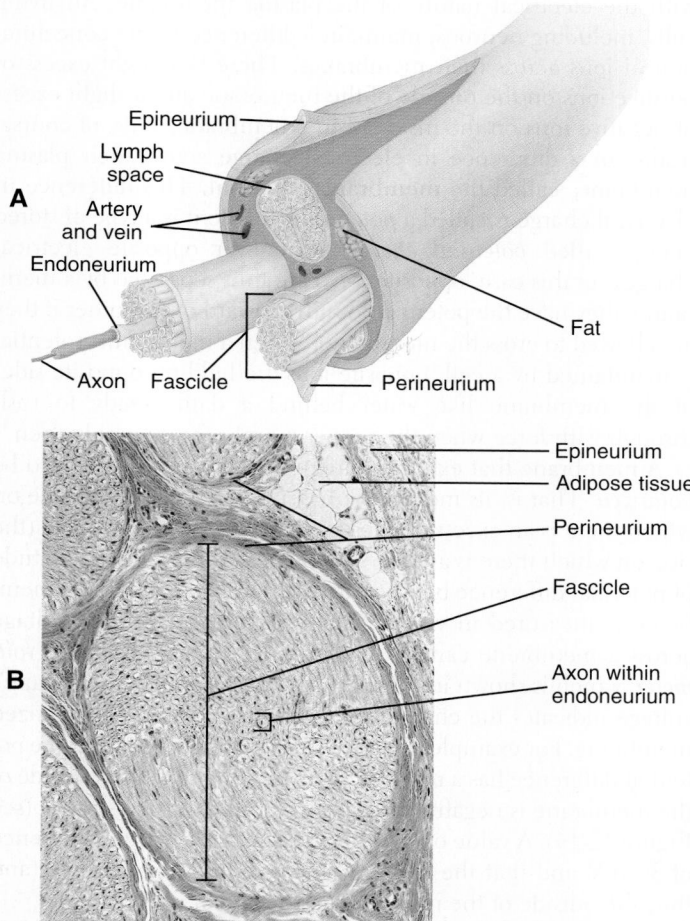

Figure 12-12 *The nerve.* A, Each nerve contains axons bundled into fascicles. A connective tissue epineurium wraps the entire nerve. Perineurium surrounds each fascicle. **B,** A scanning electron micrograph of a cross section of a nerve.

the blood vessels that supply them, are held together to form a complete nerve by a fibrous coat called the **epineurium.** Within the central nervous system, bundles of nerve fibers are called *tracts* rather than nerves.

The creamy white color of myelin distinguishes bundles of myelinated fibers from surrounding unmyelinated tissues, which appear darker in comparison. Bundles of myelinated fibers make up the so-called **white matter,** or *white substance,* of the nervous system. In the peripheral nervous system, white matter consists of myelinated nerves; in the central nervous system, white matter consists of myelinated tracts.

Cell bodies and unmyelinated fibers make up the darker **gray matter,** or *gray substance,* of the nervous system. Small, distinct regions of gray matter within the central nervous system are usually called *nuclei.* In peripheral nerves, similar regions of gray matter are more often called *ganglia.*

Most nerves in the human nervous system are *mixed nerves.* That is, they contain both sensory (afferent) and motor (efferent) fibers. Nerves that contain predominantly afferent fibers are often called *sensory nerves.* Likewise, nerves that contain mostly efferent fibers are called *motor nerves.*

REPAIR OF NERVE FIBERS

Because mature neurons are incapable of cell division, damage to nervous tissue can be permanent. Damaged neurons cannot be replaced; the only option for healing injured or diseased nervous tissue is repairing the neurons that are already present. Unfortunately, neurons have a very limited capacity to repair themselves. Nerve fibers can sometimes be repaired if the damage is not extensive, when the cell body and neurilemma remain intact, and when scarring has not occurred. Figure 12-13 shows the stages of the healing process in the axon of a peripheral motor neuron. Immediately after the injury occurs, the distal portion of the axon degenerates, as does its myelin sheath. Macrophages then move into the area and remove the debris. The remaining neurilemma and endoneurium form a pathway or tunnel from the point of injury to the effector. New Schwann cells grow within this tunnel, maintaining a path for regrowth of the axon.

Meanwhile, the cell body of the damaged neuron has reorganized its ribosomes (sometimes called Nissl bodies, or *Nissl substance*) to provide the proteins necessary to extend the remaining healthy portion of the axon. Several growing axon "sprouts" appear. When one of these growing fibers reaches the tunnel, it increases its growth rate—growing as much as 3 to 5 mm per day. If all goes well, the neuron's connection with the effector is quickly reestablished. Notice in Figure 12-13 that the skeletal muscle cell supplied by the damaged neuron atrophies during the absence of nervous input. Only after the nervous connection is reestablished and stimulation resumed does the muscle cell grow back to its original size. If the damaged axon fails to repair itself, a nearby, undamaged axon from an adjacent neuron may form a sprout that reaches the effector cell and thus reestablishes a connection with the nervous system. If the damaged nerve is connected to another nerve, the receiving nerve may also wither and die. In fact, because the preceding neuron may now have no functioning neuron to send a signal to, it may wither as well. In short, damage to

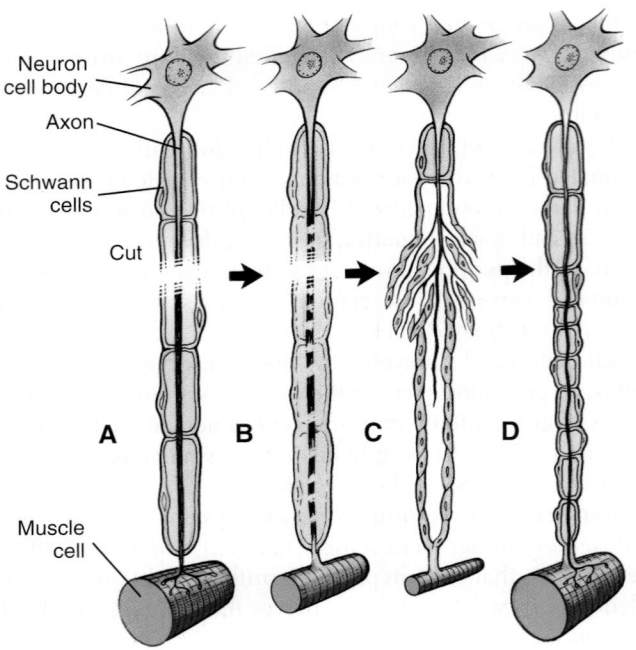

Figure 12-13 *Repair of a peripheral nerve fiber.* **A,** An injury results in a cut nerve. **B,** Immediately after the injury occurs, the distal portion of the axon degenerates, as does its myelin sheath. **C,** The remaining neurilemma tunnels from the point of injury to the effector. New Schwann cells grow within this tunnel, maintaining a path for regrowth of the axon. Meanwhile, several growing axon "sprouts" appear. When one of these growing fibers reaches the tunnel, it increases its growth rate—growing as much as 3 to 5 mm per day. (The other sprouts eventually disappear.) **D,** The neuron's connection with the effector is reestablished.

BOX 12-3   Neuroglobin

Neuroglobin (Ngb) is a protein molecule very similar to the oxygen-binding proteins *hemoglobin (Hb)* in red blood cells (see Chapters 17 and 24) and *myoglobin* in muscle fibers (see Chapter 11). As the myoglobin in muscle fibers, neuroglobin temporarily stores a "backup" supply of oxygen for times when oxygen availability is low. In brain tissues, such a circumstance may occur during a stroke, or *cerebrovascular accident (CVA)*, when the blood supply to the brain is disrupted. It could also occur during respiratory accidents, such as suffocation. Until normal blood flow is restored, neuroglobin can supply oxygen to the cell for a short time.

---

a single axon can shut down an entire nerve pathway if not repaired—and repaired quickly.

In the central nervous system, similar repair of damaged nerve fibers is very unlikely. First of all, neurons in the central nervous system lack the neurilemma needed to form the guiding tunnel from the point of injury to the distal connection. Second, astrocytes quickly fill damaged areas and thus block regrowth of the axon with scar tissue. Despite great strides recently made by researchers looking for ways to stimulate the repair of neurons in the central nervous system, most injuries to the brain and spinal cord cause permanent damage (Box 12-3).

 QUICK CHECK

10. What are the three layers of connective tissues that hold the fibers of a nerve together?
11. What is the difference between a nerve and a tract?
12. How does white matter differ from gray matter?
13. Under what circumstances can a nerve fiber be repaired?

NERVE IMPULSES

Neurons are remarkable among cells because they initiate and conduct signals called *nerve impulses*. Expressed differently, neurons exhibit both *excitability* and *conductivity*. What exactly is a nerve impulse? How is a neuron able to conduct this signal along its entire length—sometimes a full meter? These questions, and more, are answered in the paragraphs that follow.

Membrane Potentials

One way to describe a nerve impulse is as a wave of electrical fluctuation that travels along the plasma membrane. To understand this phenomenon more fully, however, requires some familiarity with the electrical nature of the plasma membrane. All living cells, including neurons, maintain a difference in the concentration of ions across their membranes. There is a slight excess of positive ions on the outside of the membrane and a slight excess of negative ions on the inside of the membrane. This, of course, results in a difference in electrical charge across their plasma membranes called the **membrane potential.** This difference in electrical charge is called a *potential* because it is a type of stored energy called *potential energy.* Whenever opposite electrical charges (in this case, opposite ions) are thus separated by a membrane, they have the potential to move toward one another if they are allowed to cross the membrane. When a membrane potential is maintained by a cell, opposite ions are held on opposite sides of the membrane like water behind a dam—ready to rush through with force when the proper membrane channels open.

A membrane that exhibits a membrane potential is said to be *polarized.* That is, its membrane has a negative pole (the side on which there is an excess of negative ions) and a positive pole (the side on which there is an excess of positive ions). The magnitude of potential difference between the two sides of a polarized membrane is measured in volts (V) or millivolts (mV). The voltage across a membrane can be measured by a device called a *voltmeter,* which is shown in Figure 12-14. The sign of a membrane's voltage indicates the charge on the inside surface of a polarized membrane. For example, the value -70 mV indicates that the potential difference has a magnitude of 70 mV and that the inside of the membrane is negative with respect to the outside surface (see Figure 12-14). A value of $+30$ mV indicates a potential difference of 30 mV and that the inside of the membrane is positive (and thus the outside of the membrane is negative).

To understand the electrical activity of the neuron—which is an essential concept of neurobiology—we will need to understand the different membrane potentials that occur in different circumstances in the plasma membrane of a neuron. Table 12-1

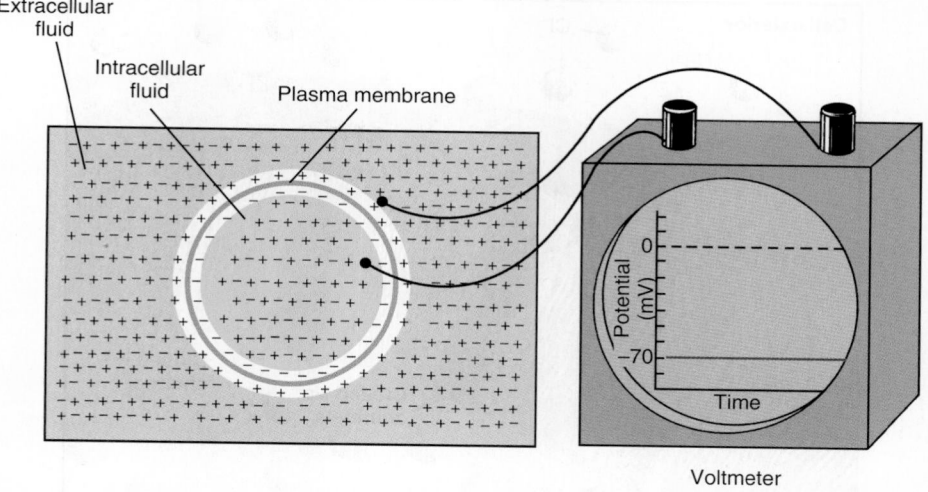

Figure 12-14 *Membrane potential.* The diagram on the left represents a cell maintaining a very slight difference in the concentration of oppositely charged ions across its plasma membrane. The voltmeter records the magnitude of electrical difference over time, which, in this case, does not fluctuate from −70 mV (voltage recorded over time as a red line).

Table 12-1   Types of Membrane Potentials

MEMBRANE POTENTIAL	POLARIZATION	TYPICAL VOLTAGE*	SUMMATION	CONDUCTION	DESCRIPTION
Resting membrane potential (RMP)	Polarized	−70 mV	Not applicable	Not applicable	Membrane voltage when the neuron is not excited and not conducting an impulse
Local potential	Depolarized (excitatory; EPSP)	Graded; varies higher than −70 mV	Yes	Decremental	Temporary fluctuation in a local region of the membrane in response to a sensory or nerve stimulus; may be an upward or downward fluctuation in voltage; loses amplitude as it spreads along membrane
	Hyperpolarized (inhibitory; IPSP)	Graded; varies lower than −70 mV	Yes	Decremental	
Threshold potential	Depolarized	−59 mV	Yes	Triggers action potential	Minimum local depolarization needed to trigger voltage-gated channels that produce the action potential
Action potential	Depolarized	+30 mV	No	Nondecremental	Temporary maximum depolarization of membrane voltage that travels to end of axon without losing amplitude

EPSP, Excitatory postsynaptic potential; *IPSP,* inhibitory postsynaptic potential.
*Example used in this chapter; actual values in body vary depending on many diverse factors.

summarizes characteristics of some important types of membrane potentials that we will be discussing in the sections that follow.

Resting Membrane Potentials

When a neuron is not conducting electrical signals, it is said to be "resting." At rest, a neuron's membrane potential is typically maintained at about −70 mV (see Figure 12-14). The membrane potential maintained by a nonconducting neuron's plasma membrane is called the **resting membrane potential (RMP)**.

The mechanisms that produce and maintain the RMP do so by promoting a slight ionic imbalance across the neuron's plasma membrane. Specifically, these mechanisms produce a slight excess of positive ions on its outer surface. This imbalance of ion concentrations is produced primarily by ion transport mechanisms in the neuron's plasma membrane.

Recall from Chapters 3 and 4 that the permeability characteristics of each cell's plasma membrane are determined in part by the presence of specific membrane transport channels. Many of these channels are *gated channels*, allowing specific molecules to diffuse across the membrane only when the "gate" of each channel is open (see Figure 4-6 on p. 109). In the neuron's plasma membrane, channels for the transport of the major anions (nega-

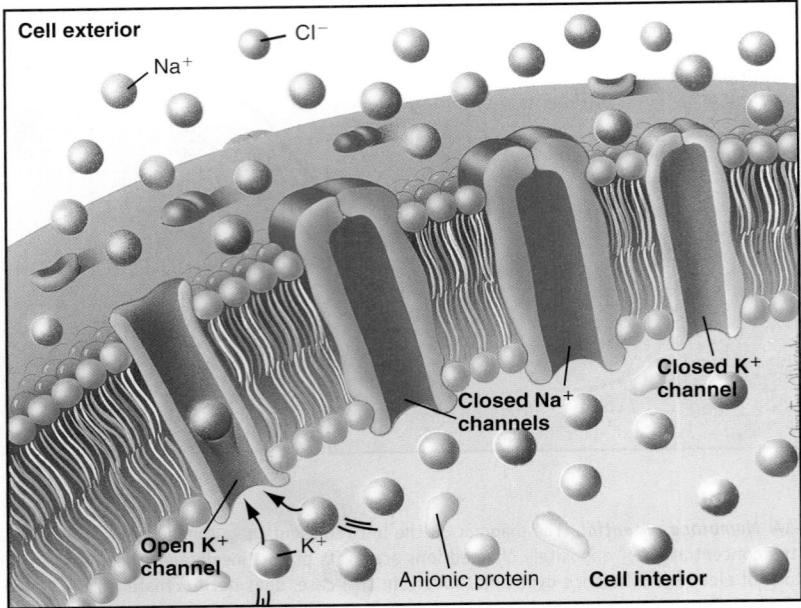

Figure 12-15 *Role of ion channels in maintaining the resting membrane potential (RMP).* Some K⁺ channels are open in a "resting" membrane, allowing K⁺ to diffuse down its concentration gradient (out of the cell) and thus add to the excess of positive ions on the outer surface of the plasma membrane. Diffusion of Na⁺ in the opposite direction would counteract this effect but is prevented from doing so by closed Na⁺ channels. Compare this figure with Figure 12-16.

tive particles) are either nonexistent or closed. For example, there are no channels to allow the exit of the large anionic protein molecules that dominate the intracellular fluid. Chloride ions (Cl⁻), the dominant extracellular anions, are likewise "trapped" on one side of the membrane because chloride ions are repelled by the protein anions inside the cell. This means that the only ions that can move efficiently across a neuron's membrane are the positive ions sodium and potassium. In a resting neuron, many of the potassium channels are open, but most of the sodium channels are closed (Figure 12-15). This means that potassium ions pumped into the neuron can diffuse back out of the cell in an attempt to equalize its concentration gradient, but very little of the sodium pumped out of the cell can diffuse back into the neuron (see Figure 12-16). Thus the membrane's selective permeability characteristics create and maintain a slight excess of positive ions on the outer surface of the membrane.

Another mechanism also operates to maintain the RMP. The sodium-potassium pump is an active transport mechanism in the plasma membrane that transports sodium ions (Na⁺) and potassium ions (K⁺) in opposite directions and at different rates (Figure 12-16). It moves three sodium ions out of a neuron for every two potassium ions it moves into it. If, for instance, the pump transports 100 potassium ions into a neuron from the extracellular fluid, it concurrently transports 150 sodium ions out of the cell. The sodium-potassium pump thus maintains an imbalance in the distribution of positive ions, maintaining a difference in electrical charge across the membrane. As this pump operates, the inside surface of the membrane becomes slightly less positive—that is, slightly negative—with respect to its outer surface.

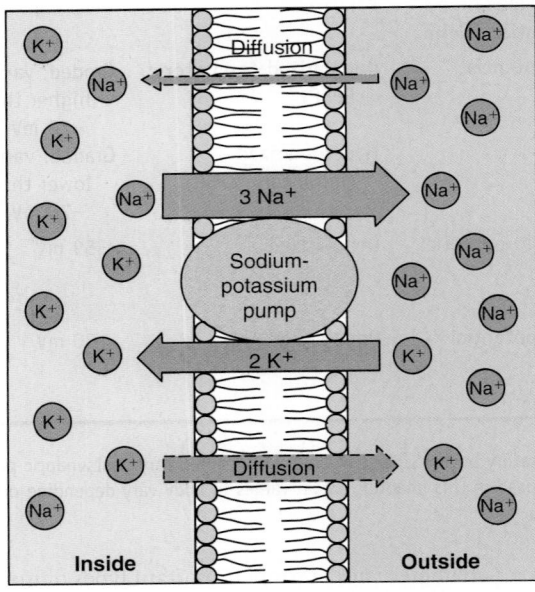

Figure 12-16 *Sodium-potassium pump.* This mechanism in the plasma membrane actively pumps sodium ions (Na⁺) out of the neuron and potassium ions (K⁺) into the neuron—at an unequal (3:2) rate. Because very little sodium reenters the cell via diffusion, this maintains an imbalance in the distribution of ions and thus maintains the resting potential. (See Figure 12-15 for the role of ion channels in the diffusion of ions.)

The RMP can be maintained by a cell as long as its sodium-potassium pumps continue to operate and its permeability characteristics remain stable. If either of these mechanisms is altered, the RMP is altered as well.

Local Potentials

In neurons, membrane potentials can fluctuate above or below the resting membrance potential in response to certain stimuli (Figure 12-17). A slight shift away from the RMP in a specific region of the plasma membrane is often called a **local potential**.

Excitation of a neuron occurs when a stimulus triggers the opening of stimulus-gated Na^+ channels. **Stimulus-gated channels** are ion channels that open in response to a sensory stimulus or a chemical stimulus from another neuron. Many stimulus-gated channels are located in the membrane of the neuron's input zone—the dendrites and soma (see Figure 12-8). The opening of stimulus-gated Na^+ channels in response to a stimulus permits more Na^+ to enter the cell. As the excess of positive ions outside the plasma membrane decreases, the magnitude of the membrane potential is reduced. Such movement of the membrane potential toward zero is called **depolarization**. In *inhibition*, a stimulus triggers the opening of stimulus-gated K^+ channels. As more K^+ diffuses out of the cell, the excess of positive ions outside the plasma membranes increases—increasing the magnitude of the membrane potential. Movement of the membrane potential away from zero (thus below the usual RMP) is called **hyperpolarization**.

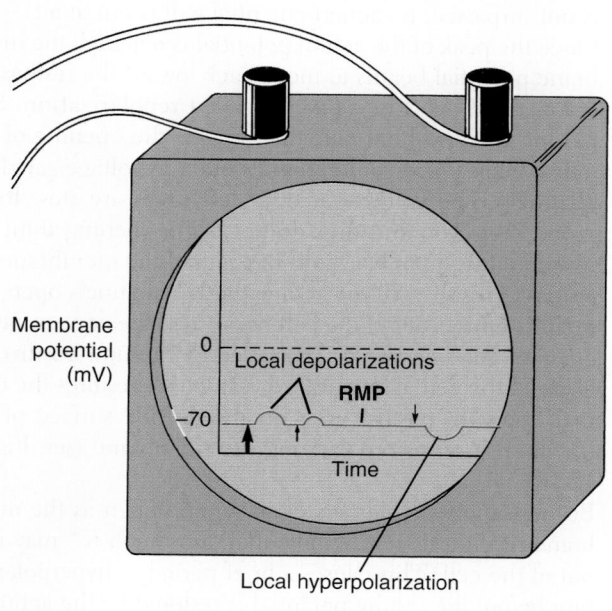

Figure 12-17 *Local potentials.* Recording voltmeter shows that excitatory stimuli (↑) cause depolarizations (movements toward 0 mV) in proportion to the strength of the stimuli. Inhibitory stimuli (↓) cause hyperpolarizations, local deviations away from 0 mV that cause the membrane potential to dip below the level of the RMP. Voltage is recorded as a red line on the screen of the voltmeter. Upward deviations of the red line are depolarizations; downward deviations of the red line are hyperpolarizations.

Local potentials are called *graded potentials* because the magnitude of deviation from the RMP is proportional to the magnitude of the stimulus. In short, local potentials can be large or small—they are not all-or-none events. Local potentials are called "local" nerve signals because they are more or less isolated to a particular region of the plasma membrane. That is, local potentials do not spread all the way to the end of a neuron's axon.

QUICK CHECK

14. What mechanisms are involved in producing the resting membrane potential?
15. In a resting neuron, what positive ion is most abundant outside the plasma membrane? What positive ion is most abundant inside the plasma membrane?
16. How does depolarization of a membrane differ from hyperpolarization?

ACTION POTENTIAL

An **action potential** is, as the term suggests, the membrane potential of an active neuron, that is, one that is conducting an impulse. A synonym commonly used for action potential is *nerve impulse*. The action potential, or nerve impulse, is an electrical fluctuation that travels along the surface of a neuron's plasma membrane. A step-by-step description of the mechanism that produces the action potential is given in the following paragraphs and in Table 12-2. Refer to Figures 12-18 and 12-19 as you read each step.

1. When an adequate stimulus is applied to a neuron, the stimulus-gated Na^+ channels at the point of stimulation open. Na^+ diffuses rapidly into the cell because of the concentration gradient and electrical gradient, producing a local depolarization (Figure 12-18, *B*).
2. If the magnitude of the local depolarization surpasses a limit called the **threshold potential** (typically −59 mV), voltage-gated Na^+ channels are stimulated to open. **Voltage-gated channels** are ion channels open in response to voltage fluctuations. There are many voltage-gated Na^+ channels and K^+ channels in the membrane of the neuron's summation zone (axon hillock) and conduction zone (axon) (Figure 12-8). The threshold potential is the minimum magnitude a voltage fluctuation in the summation or conduction zone must have to trigger the opening of a voltage-gated ion channel.
3. As more Na^+ rushes into the cell, the membrane moves rapidly toward 0 mV and then continues in a positive direction to a peak of +30 mV (see Figure 12-19). The positive value at the peak of the action potential indicates that there is an excess of positive ions *inside* the membrane. If the local depolarization fails to cross the threshold of −59 mV, the voltage-gated Na^+ channels do not open, and the membrane simply recovers back to the resting potential of −70 mV without producing an action potential.
4. Voltage-gated Na^+ channels stay open for only about 1 millisecond (ms) before they automatically close. This means

Table 12-2   Steps of the Mechanism that Produces an Action Potential

STEP	DESCRIPTION
1	A stimulus triggers stimulus-gated Na⁺ channels to open and allow inward Na⁺ diffusion. This causes the membrane to depolarize.
2	As the threshold potential is reached, voltage-gated Na⁺ channels open.
3	As more Na⁺ enters the cell through voltage-gated Na⁺ channels, the membrane depolarizes even further.
4	The magnitude of the action potential peaks (at +30 mV) when voltage-gated Na⁺ channels close.
5	Repolarization begins when voltage-gated K⁺ channels open, allowing outward diffusion of K⁺.
6	After a brief period of hyperpolarization, the resting potential is restored by the sodium-potassium pump and the return of ion channels to their resting state.

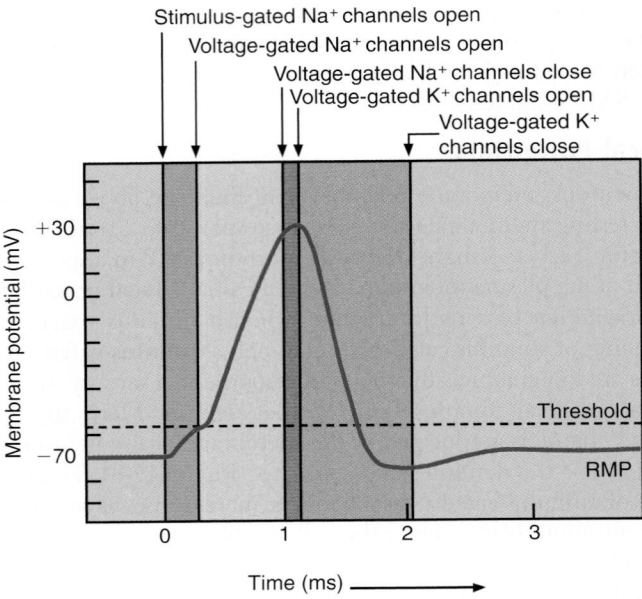

Figure 12-19 *The action potential.* Changes in membrane potential in a local area of a neuron's membrane result from changes in membrane permeability.

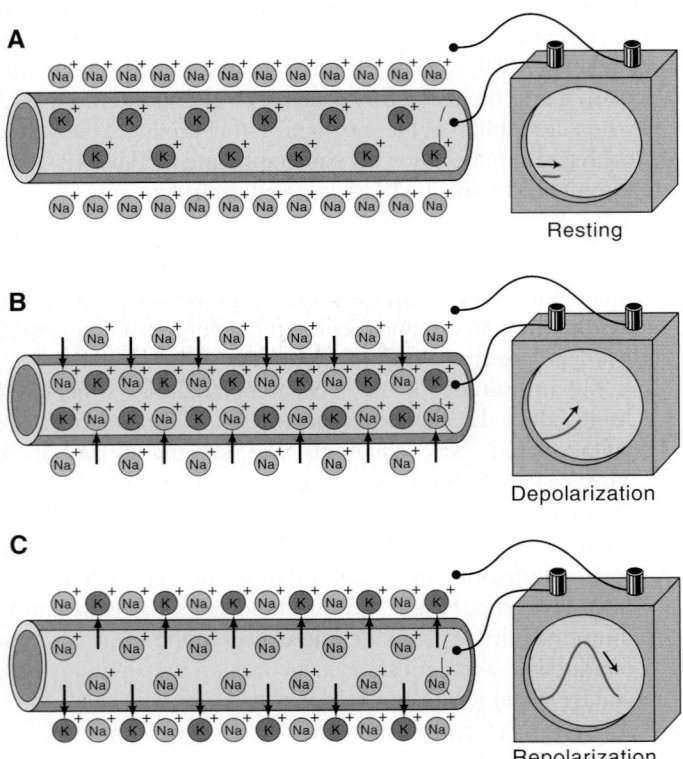

Figure 12-18 *Depolarization and repolarization.* **A,** RMP results from an excess of positive ions on the outer surface of the plasma membrane. More Na⁺ ions are on the outside of the membrane than K⁺ ions are on the inside of the membrane. **B,** Depolarization of a membrane occurs when Na⁺ channels open, allowing Na⁺ to move to an area of lower concentration (and more negative charge) *inside* the cell—reversing the polarity to an inside-positive state. **C,** Repolarization of a membrane occurs when K⁺ channels then open, allowing K⁺ to move to an area of lower concentration (and more negative charge) *outside* the cell—reversing the polarity back to an inside-negative state. Each voltmeter records the changing membrane potential as a red line.

that once they are stimulated, the Na⁺ channels always allow sodium to rush in for the same amount of time, which in turn produces the same magnitude of action potential. In other words, the action potential is an *all-or-none* response. If the threshold potential is surpassed, the full peak of the action potential is always reached; if the threshold potential is not surpassed, no action potential will occur at all.

5. Once the peak of the action potential is reached, the membrane potential begins to move back toward the resting potential (−70 mV) in a process called **repolarization.** Surpassing the threshold potential triggers the opening of not only voltage-gated Na⁺ channels but also voltage-gated K⁺ channels. The voltage-gated K⁺ channels are slow to respond, however, and thus do not begin opening until the inward diffusion of Na⁺ ions has caused the membrane potential to reach +30 mV. Once the K⁺ channels open, K⁺ rapidly diffuses out of the cell because of the concentration gradient and because it is repulsed by the now-positive interior of the cell. The outward rush of K⁺ restores the original excess of positive ions on the outside surface of the membrane—thus repolarizing the membrane (see Figure 12-18).

6. Because the K⁺ channels often remain open as the membrane reaches its resting potential, too much K⁺ may rush out of the cell. This causes a brief period of hyperpolarization before the resting potential is restored by the action of the sodium-potassium pump and the return of ion channels to their resting state.

Refractory Period

The refractory period is a brief period during which a local area of an axon's membrane resists restimulation (Figure 12-20). For about 0.5 ms after the membrane surpasses the threshold poten-

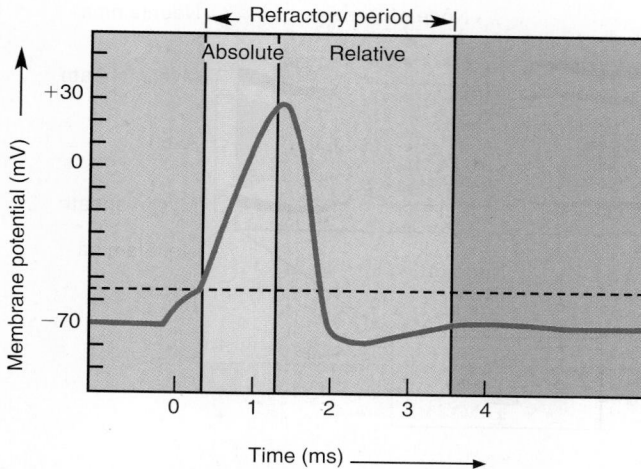

Figure 12-20 *Refractory period.* During the absolute refractory period, the membrane will not respond to any stimulus. During the relative refractory period, however, a very strong stimulus may elicit a response in the membrane.

tial, it will not respond to any stimulus, no matter how strong. This is called the **absolute refractory period**. The **relative refractory period** is the few milliseconds after the absolute refractory period—the time during which the membrane is repolarizing and restoring the resting membrane potential. During the relative refractory period the membrane will respond only to very strong stimuli.

Because only very strong stimuli can produce an action potential during the relative refractory period, a series of closely spaced action potentials can occur only when the magnitude of the stimulus is great. The greater the magnitude of the stimulus, the earlier a new action potential can be produced, and thus the greater the frequency of action potentials. This means that although the magnitude of the stimulus does not affect the magnitude of the action potential, which is an all-or-none response, it does cause a proportional increase in the frequency of impulses. Thus the nervous system uses the frequency of nerve impulses to code for the strength of a stimulus—not changes in the magnitude of the action potential.

Conduction of the Action Potential

At the peak of the action potential, the inside of the axon's plasma membrane is positive relative to the outside. That is, its polarity is now the reverse of that of the resting membrane potential. Such reversal in polarity causes electrical current to flow between the site of the action potential and the adjacent regions of membrane. This local current flow triggers voltage-gated Na$^+$ channels in the next segment of membrane to open. As Na$^+$ rushes inward, this next segment exhibits an action potential. The action potential thus has moved from one point to the next along the axon's membrane (Figure 12-21). This cycle repeats itself because each action potential always causes enough local current flow to surpass the threshold potential for the next region of membrane. Because each action potential is an all-or-none phenomenon, the fluctuation in membrane potential moves along the membrane without any decrement, or decrease, in magnitude.

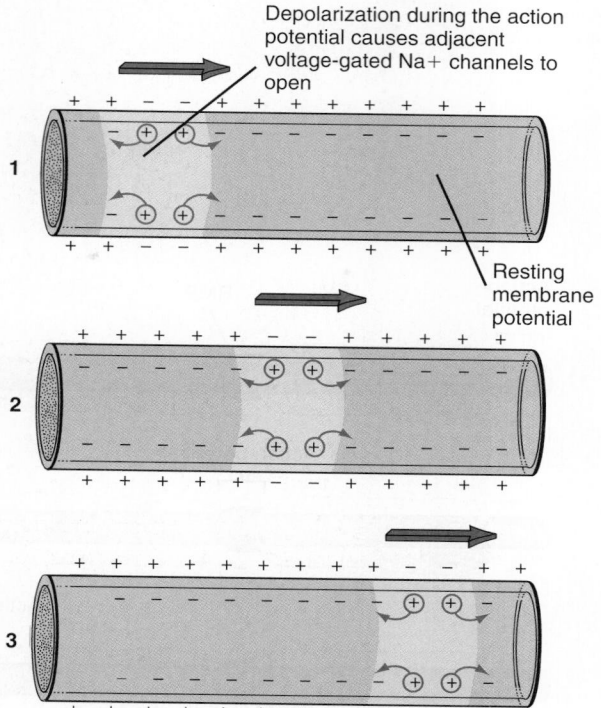

Figure 12-21 *Conduction of the action potential.* The reverse polarity characteristic of the peak of the action potential causes local current flow to adjacent regions of the membrane *(small arrows)*. This stimulates voltage-gated Na$^+$ channels to open and thus create a new action potential. This cycle continues, producing wavelike conduction of the action potential from point to point along a nerve fiber. Adjacent regions of membrane behind the action potential do not depolarize again because they are still in their refractory period.

The action potential never moves backward, restimulating the region from which it just came. It is prevented from doing so because the previous segment of membrane remains in a refractory period too long to allow such restimulation. This is the mechanism responsible for the one-way movement of action potentials along axons.

In myelinated fibers, the insulating properties of the thick myelin sheath resist ion movement and the resulting local flow of current. Electrical changes in the membrane can only occur at gaps in the myelin sheath, that is, at the nodes of Ranvier. Figure 12-22 shows that when an action potential occurs at one node, most of the current flows *under* the insulating myelin sheath to the next node. This stimulates regeneration of an action potential at that node by opening voltage-gated channels, which in turn stimulates the next node. Thus the action potential seems to "leap" from node to node along the myelinated fiber. This type of impulse regeneration is called **saltatory conduction** (from the Latin verb *saltare*, "to leap").

How fast does a nerve fiber conduct impulses? It depends on its diameter and on the presence or absence of a myelin sheath. The speed of conduction of a nerve fiber is proportional to its diameter: the larger the diameter, the faster it conducts impulses. Myelinated fibers conduct impulses more rapidly than unmyelinated fibers. This is because saltatory conduction is more rapid

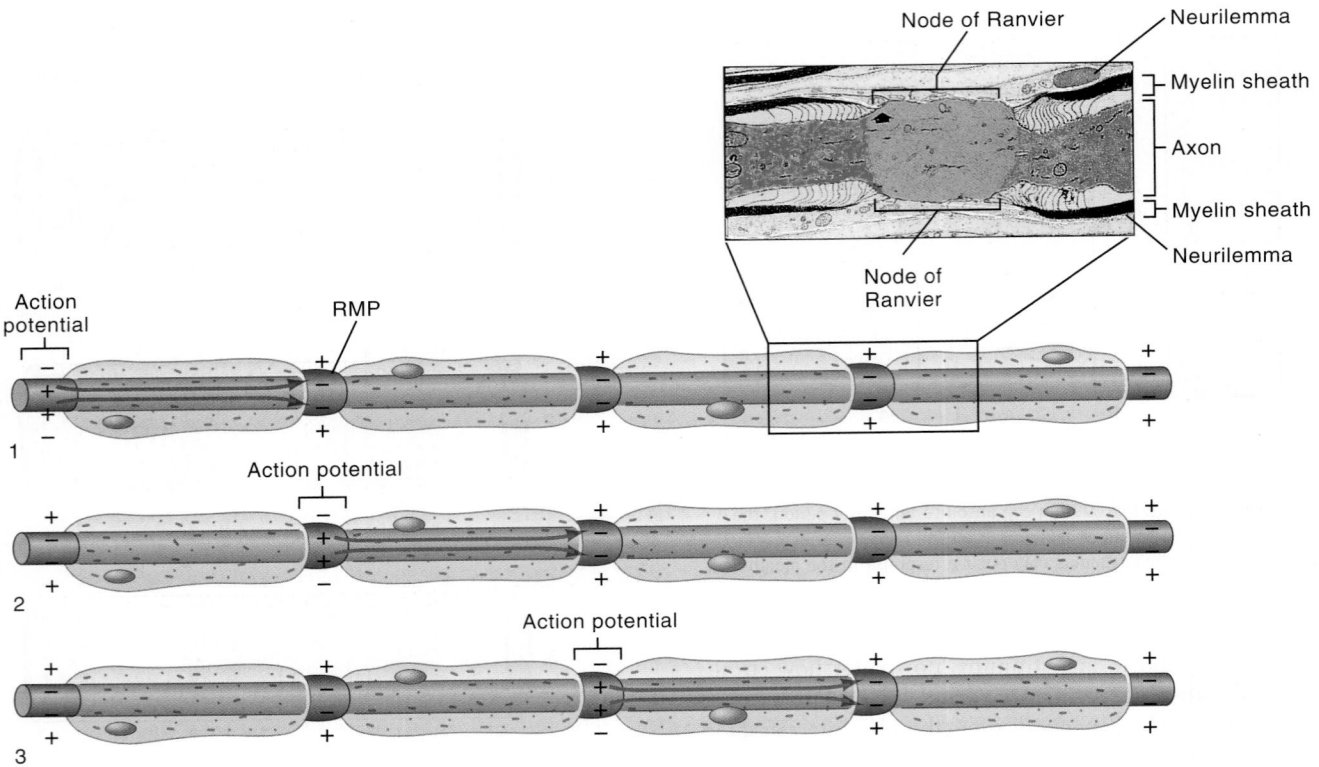

Figure 12-22 *Saltatory conduction.* This series of diagrams shows that the insulating nature of the myelin sheath prevents ion movement everywhere but at the nodes of Ranvier. The action potential at one node triggers current flow *(arrows)* across the myelin sheath to the next node—producing an action potential there. The action potential thus seems to "leap" rapidly from node to node. The inset is a transmission electron micrograph showing a node of Ranvier in a myelinated fiber.

 BOX 12-4: HEALTH MATTERS
Reducing Damage to Nerve Fibers

Crushing and bruising cause most injuries to the spinal cord— often damaging nerve fibers irreparably. This usually results in **paralysis** or loss of function in the muscles normally supplied by the damaged fibers. Unfortunately, the inflammation of the injury site usually damages even more fibers and thus increases the extent of the paralysis. However, early treatment of the injury with the anti-inflammatory drug *methylprednisolone* can reduce the inflammatory response in the damaged tissue and thus limit the severity of a spinal cord injury. Although early studies failed to confirm the effectiveness of standard doses of this steroid drug, later studies showed that very large doses administered within 8 hours of the injury reduced the extent of nerve cell damage dramatically. Since about 95% of the 10,000 Americans suffering spinal cord injuries each year are admitted for treatment well before the 8-hour limit, this drug may prove to be the first effective therapy for spinal cord injuries.

than point-to-point conduction. The fastest fibers, such as those that innervate the skeletal muscles, can conduct impulses up to about 130 meters per second (close to 300 miles per hour). The slowest fibers, such as those from sensory receptors in the skin, may conduct impulses at only about 0.5 meter per second (little more than 1 mile per hour).

Box 12-4 outlines how disrupting conduction of nerve impulses can be used to block pain signals.

 QUICK CHECK

17. List the events that lead to the initiation of an action potential.
18. What is meant by the term *threshold potential?*
19. How does impulse conduction in an unmyelinated fiber differ from impulse conduction in a myelinated fiber?

SYNAPTIC TRANSMISSION
Structure of the Synapse

A **synapse** is the place where signals are transmitted from one neuron, called the *presynaptic neuron,* to another neuron, called the *postsynaptic neuron.* The postsynaptic cell could also be an effector, such as a muscle.

Types of Synapses

There are two types of synapses: electrical synapses and chemical synapses.

Electrical synapses occur where two cells are joined end-to-end by gap junctions (Figure 12-23, A). Because the plasma

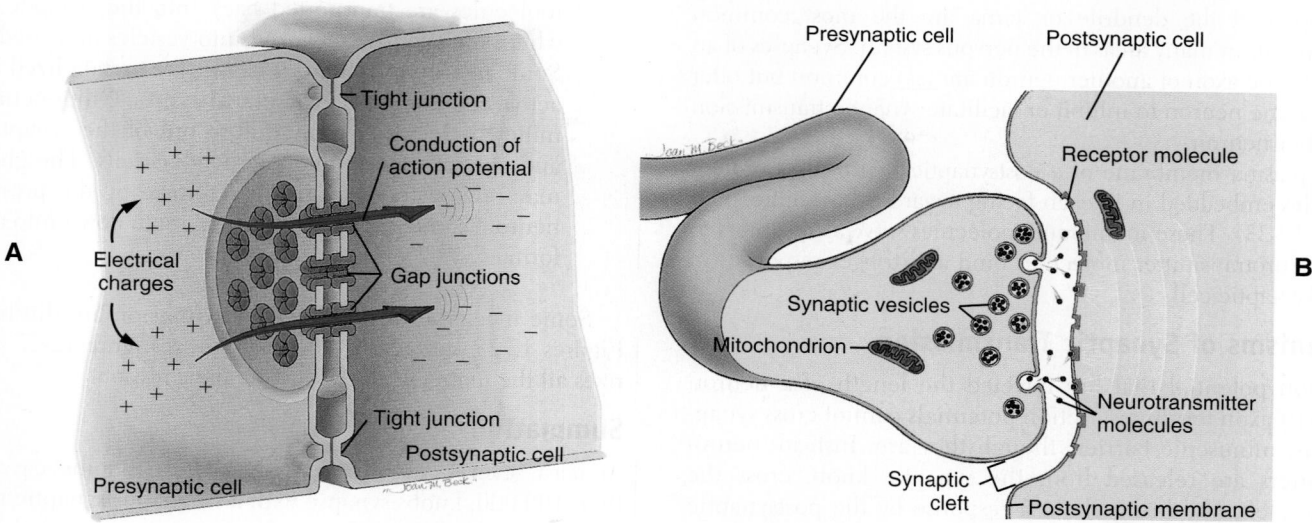

Figure 12-23 *Electrical and chemical synapses.* **A,** Electrical synapses involve gap junctions that allow action potentials to move from cell to cell directly by allowing electrical current to flow between cells. **B,** Chemical synapses involve transmitter chemicals (neurotransmitters) that signal postsynaptic cells, possibly inducing an action potential.

membranes and cytoplasm are functionally continuous in this type of junction, an action potential can simply continue along the postsynaptic plasma membrane as if it belonged to the same cell. Electrical synapses occur between cardiac muscle cells and between some types of smooth muscle cells. Electrical synapses are found in the nervous system early in development but are thought to be replaced by the more complex chemical synapses by the time the nervous system is mature.

Chemical synapses are called that because they use a chemical transmitter called a *neurotransmitter* to send a signal from the **presynaptic** cell to the **postsynaptic** cell (Figure 12-23, *B*). Because of its importance in understanding the function of the adult nervous system, it is the chemical synapse that we will consider carefully in the following paragraphs.

Chemical Synapse

Three structures make up a chemical synapse:

1. A synaptic knob
2. A synaptic cleft
3. The plasma membrane of a postsynaptic neuron

A **synaptic knob** is a tiny bulge at the end of a terminal branch of a presynaptic neuron's axon (Figure 12-23). Each synaptic knob, or synaptic terminal, contains numerous small sacs or vesicles. Each vesicle contains about 10,000 neurotransmitter molecules. Some of these are released from the presynaptic neuron into a **synaptic cleft**—the space between a synaptic knob and the plasma membrane of a postsynaptic neuron.

The synaptic cleft is an incredibly narrow space—only 20 to 30 nanometers (nm), or about one millionth of an inch in width. Identify the synaptic cleft in Figure 12-23. The synaptic cleft is

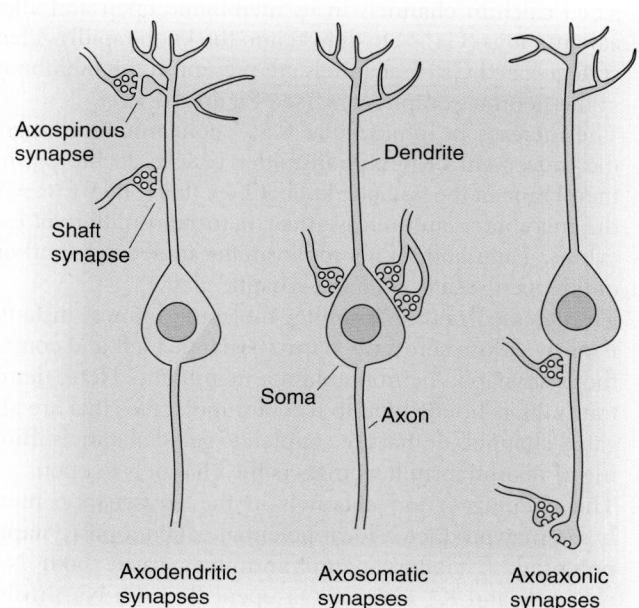

Figure 12-24 *Arrangements of synapses.* The axon of a presynaptic neuron may form a synapse at the dendrite, soma, or axon of another neuron.

not an empty space—it contains fluid, enzymes, adhesion molecules, and other substances.

Figure 12-24 shows that an axon may form a synapse at any of several points along a postsynaptic neuron:

At the dendrite (axodendritic synapses)
At the soma (axosomatic synapses)
At the axon (axoaxonic synapses)

Synapses at the dendrite or soma are the most common arrangements in many areas of the nervous system. Synapses of an axon with the axon of another neuron are less common but offer a way for one neuron to inhibit or facilitate synaptic transmission by another neuron.

The plasma membrane of a postsynaptic neuron has protein molecules embedded in it, each facing toward the synaptic knob (Figure 12-23). These membrane molecules serve as receptors to which neurotransmitter molecules bind and trigger responses in the postsynaptic cell.

Mechanisms of Synaptic Transmission

An action potential that has traveled the length of a neuron stops at its axon terminals. Action potentials cannot cross synaptic clefts, minuscule barriers though they are. Instead, neurotransmitters are released from the synaptic knob, cross the synaptic cleft, and bring about a response by the postsynaptic neuron. *Excitatory* neurotransmitters cause depolarization of the postsynaptic membrane, whereas *inhibitory* neurotransmitters cause hyperpolarization of the postsynaptic membrane (see Figure 12-17).

The mechanism of synaptic transmission, summarized in Figures 12-23 and 12-25, consists of the following sequence of events:

1. When an action potential reaches a synaptic knob, voltage-gated calcium channels in its membrane open and allow calcium ions (Ca^{++}) to diffuse into the knob rapidly. Many voltage-gated Ca^{++} channels are present in the membrane of the neuron's output zone (see Figure 12-8).
2. The increase in intracellular Ca^{++} concentration triggers the movement of neurotransmitter vesicles to the plasma membrane of the synaptic knob. Once there, they fuse with the membrane and release their neurotransmitter via exocytosis. Thousands of neurotransmitter molecules spurt out of the open vesicles into the synaptic cleft.
3. The released neurotransmitter molecules almost instantaneously diffuse across the narrow synaptic cleft and contact the postsynaptic neuron's plasma membrane. Here, neurotransmitters briefly bind to receptor molecules that are also gated channels or that are coupled to gated channels. Binding of neurotransmitters triggers the channels to open.
4. The opening of ion channels in the postsynaptic membrane may produce a local potential called a **postsynaptic potential.** Excitatory neurotransmitters cause both Na^+ channels and K^+ channels to open. Because Na^+ rushes inward faster than K^+ rushes outward, there is a temporary depolarization called an **excitatory postsynaptic potential (EPSP).** Inhibitory neurotransmitters cause K^+ channels and/or Cl^- channels to open. If K^+ channels open, K^+ rushes outward; if Cl^- channels open, Cl^- rushes inward. Either event makes the inside of the membrane even more negative than at the resting potential. This temporary hyperpolarization is called an **inhibitory postsynaptic potential (IPSP).**
5. Once a neurotransmitter binds to its postsynaptic receptors, its action is quickly terminated (Figure 12-26). Several mechanisms bring this about. Some neurotransmitter

molecules are transported back into the synaptic knobs, where they can be repacked into vesicles and used again. Some neurotransmitter molecules are metabolized into inactive compounds by synaptic enzymes. Other neurotransmitter molecules simply diffuse out of the synaptic cleft and are transported into nearby glial cells. The glial cells may release them again for reuptake by the presynaptic neuron, sometimes after breaking them down into another form.

Some mechanisms of synaptic transmission are illustrated in Figures 12-25 and 12-26. The diagram in Figure 12-27 summarizes all the main events of synaptic transmission.

Summation

At least several, usually thousands, and in some cases more than 100,000, knobs synapse with a single postsynaptic neuron. The amount of excitatory neurotransmitter released by one knob is not enough to trigger an action potential. It may, however, *facilitate* initiation of an action potential by producing a local depolarization of the synaptic membrane—an EPSP. When several knobs are activated simultaneously, neurotransmitters stimulate different locations on the postsynaptic membrane. These local potentials may spread far enough to reach the axon hillock, where they may add together, or *summate*. If the sum of the local potentials reaches the threshold potential, voltage-gated channels in the axon membrane open, producing an action potential (Figure 12-28, *A*). This phenomenon is called **spatial summation.** Likewise, when synaptic knobs stimulate a postsynaptic neuron in rapid succession, their effects can add up over a brief period of time to produce an action potential (Figure 12-28, *B*). This type of summation is called **temporal summation.**

Usually both excitatory and inhibitory transmitters are released at the same postsynaptic neuron. The excitatory neurotransmitters produce local EPSPs, and the inhibitory neurotransmitters produce local IPSPs. Summation of these opposing local potentials occurs at the axon hillock, where many voltage-gated ion channels are located. If the EPSPs predominate over the IPSPs enough to depolarize the membrane to the threshold potential, the voltage-gated channels will respond and produce an action potential (Figure 12-28, *C*). The action potential is then conducted without decrement along the axon's membrane. On the other hand, if the IPSPs predominate over the EPSPs, the membrane will not reach the threshold potential. The voltage-gated channels at the axon hillock will not respond, and no action potential will be conducted along the axon.

QUICK CHECK

20. What are the three structural components of a synapse?
21. List the steps of synaptic transmission.
22. What is an EPSP? What is an IPSP?
23. How does temporal summation differ from spatial summation?

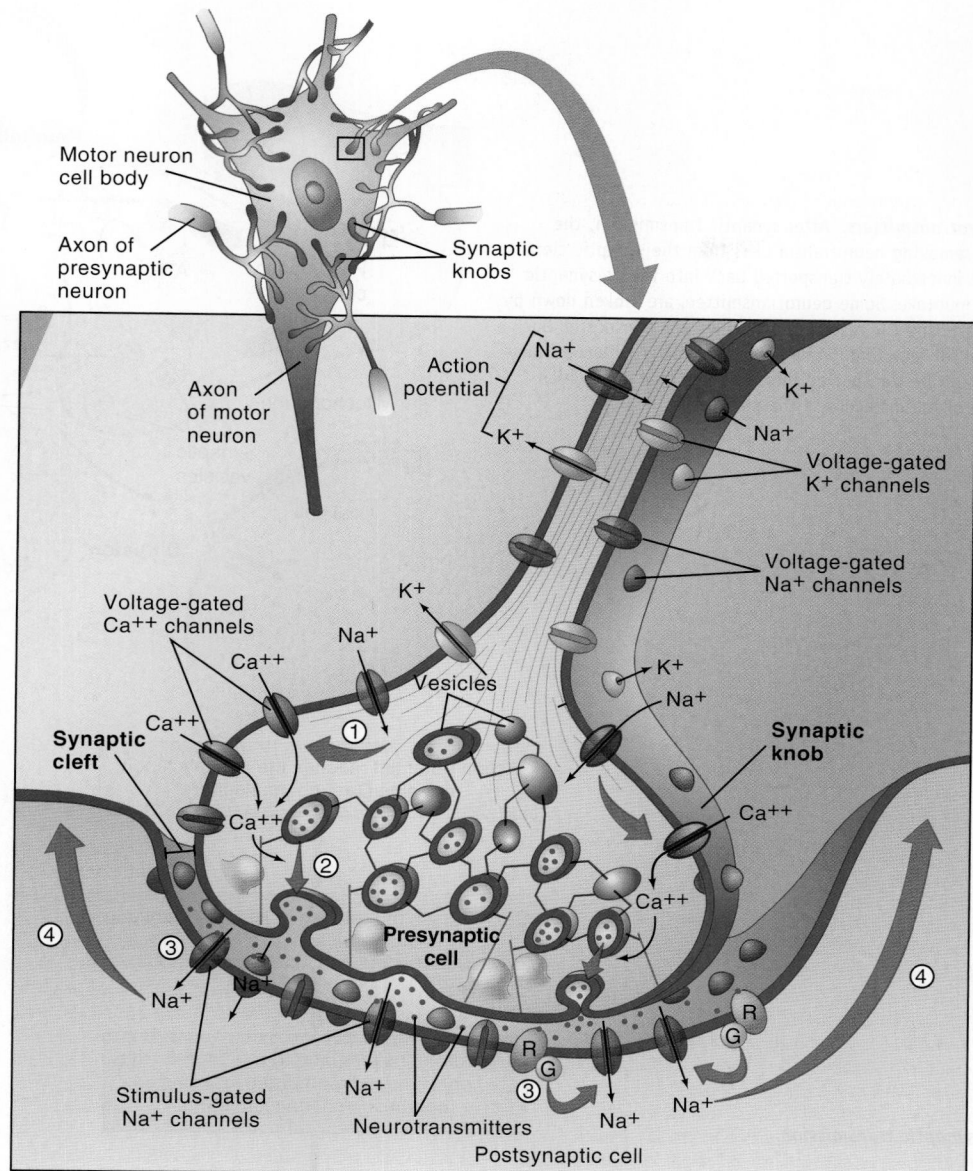

Figure 12-25 *The chemical synapse.* Diagram shows detail of synaptic knob, or axon terminal, of presynaptic neuron, the plasma membrane of a postsynaptic neuron, and a synaptic cleft. On the arrival of an action potential at a synaptic knob, voltage-gated Ca^{++} channels open and allow extracellular Ca^{++} to diffuse into the presynaptic cell *(step 1)*. In step 2, the Ca^{++} triggers the rapid exocytosis of neurotransmitter molecules from vesicles in the knob. In step 3, neurotransmitter diffuses into the synaptic cleft and binds to receptor molecules in the plasma membrane of the postsynaptic neuron. The postsynaptic receptors directly or indirectly trigger the opening of stimulus-gated ion channels, initiating a local potential in the postsynaptic neuron. In step 4, the local potential may move toward the axon, where an action potential may begin.

Reuptake
Enzyme degradation
Presynaptic cell
ATP
Mitochondrion
Synaptic vesicles
Neurotransmitter molecules
Receptor
Glial cell
Diffusion
Postsynaptic cell

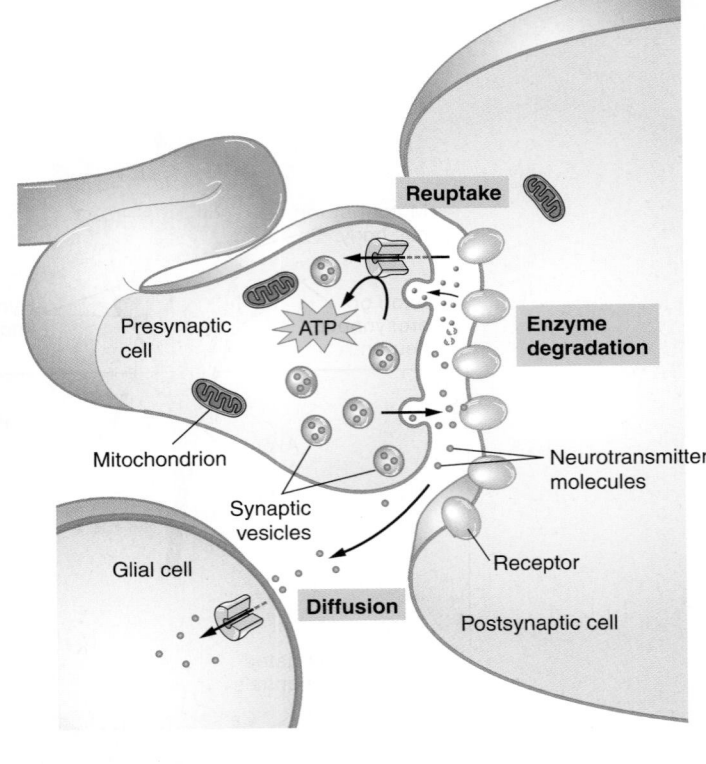

Figure 12-26 ***Fate of neurotransmitters.*** After synaptic transmission, the signal must be stopped by removing neurotransmitters from the synaptic cleft. Many neurotransmitters are immediately transported back into the presynaptic neuron in a process called reuptake. Some neurotransmitters are broken down by enzymes in the synaptic cleft, and the resulting molecules are transported back into the presynaptic neuron for recycling. Some neurotransmitter molecules may diffuse out of the synapse and be transported into a nearby glial cell (which may return an altered form of the molecule to the presynaptic neuron).

Action potential reaches the synaptic knob, opening Ca^{++} channels there
Ca^{++} causes the release of neurotransmitters across the synaptic cleft
Neurotransmitters bind to receptors in the postsynaptic membrane, causing certain ion channels to open
Na^+ and K^+ channels open
Na^+ rushes in faster than K^+ rushes out, causing the inside of the postsynaptic membrane to become more positive
This depolarization is the excitatory postsynaptic potential (EPSP)
If the EPSP reaches the threshold potential, an action potential is initiated
K^+ and/or Cl^- channels open
Outward rush of K^+ and/or inward rush of Cl^- causes postsynaptic membrane to become less positive (more negative)
This hyperpolarization is the inhibitory postsynaptic potential (IPSP)
The postsynaptic membrane is now less likely to reach the threshold potential; initiation of an action potential is thus inhibited

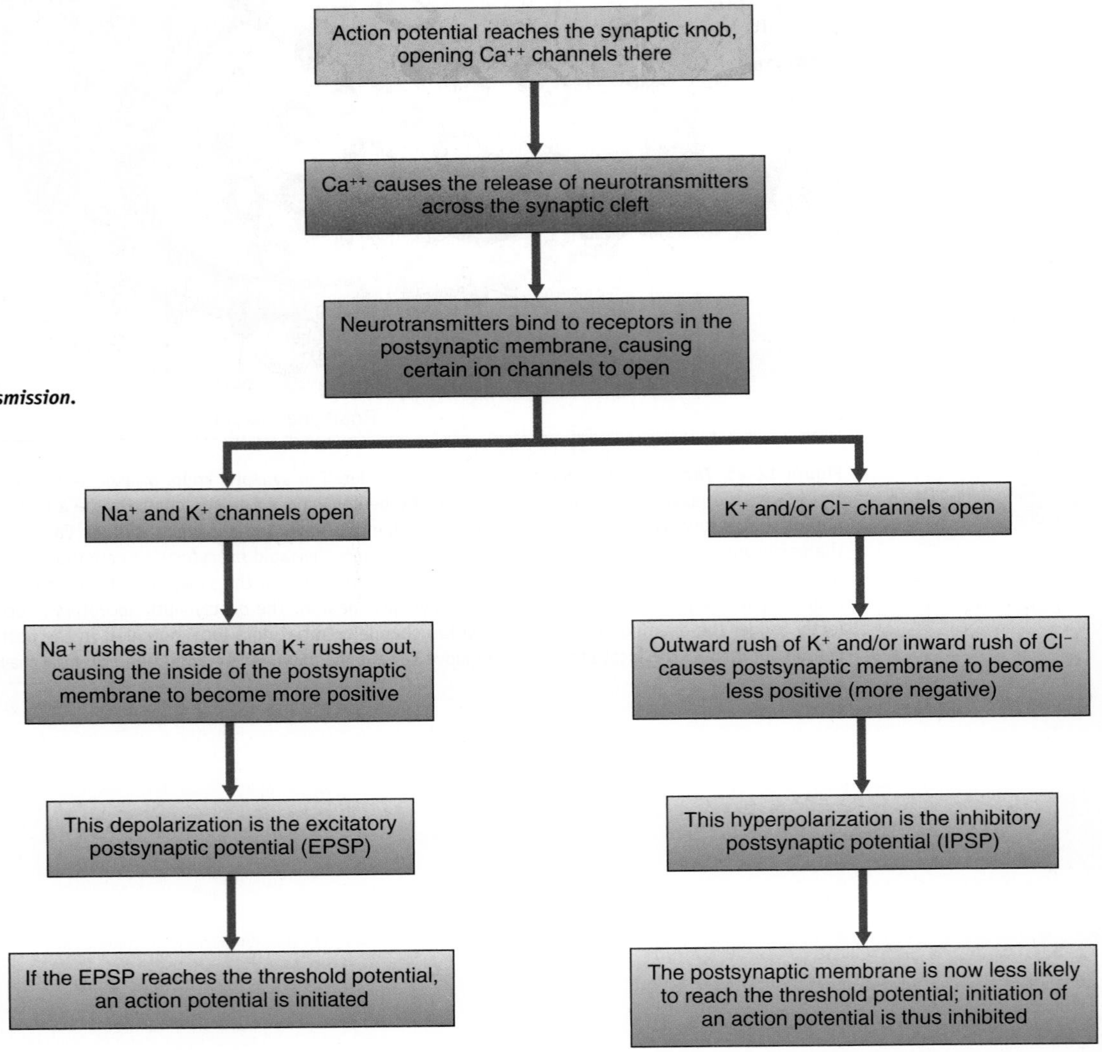

Figure 12-27 ***Summary of synaptic transmission.***

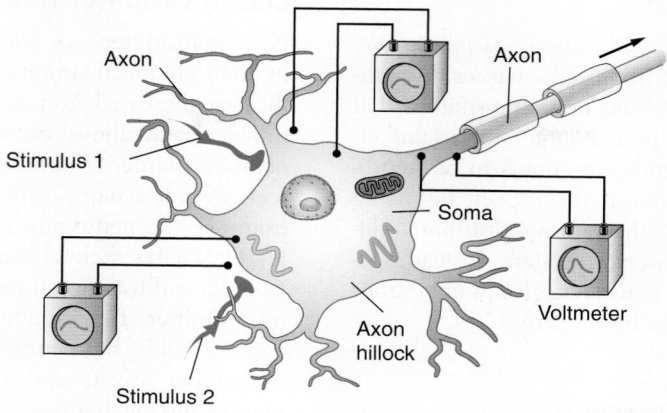

Axon

Axon

Stimulus 1

Soma

Voltmeter

Axon hillock

Stimulus 2

A   Spatial Summation

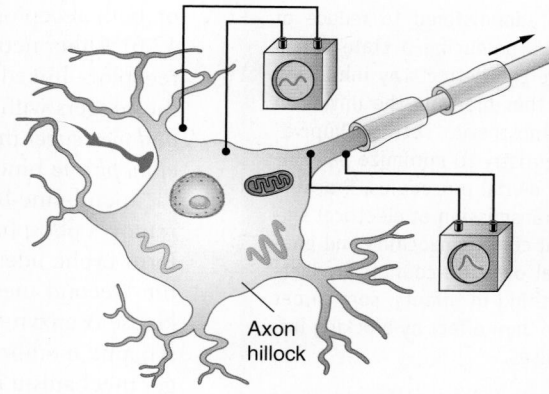

Axon hillock

B   Temporal Summation

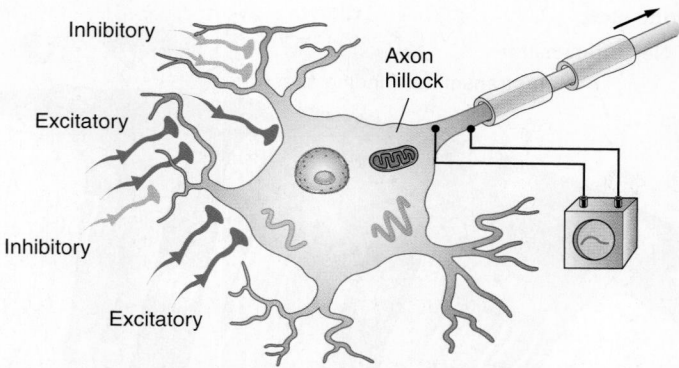

Inhibitory

Axon hillock

Excitatory

Inhibitory

Excitatory

C   Summation of Excitatory and Inhibitory Signals

Figure 12-28 *Summation.* **A,** Spatial summation is the effect produced by simultaneous stimulation by a number of synaptic knobs on the same postsynaptic neuron. Voltmeters placed near the site of stimulation show small depolarizations, and a voltmeter at the axon hillock shows the large depolarization resulting from the combined effect of both smaller depolarizations. **B,** Temporal summation is the effect produced by a rapid succession of stimuli on a single postsynaptic neuron. The figure shows two stimuli in a short burst that produce two small depolarizations that combine at the axon hillock to produce a large depolarization. **C,** Summation of many excitatory and inhibitory effects produces the potential at the axon hillock. Here, depolarizations triggered by excitatory presynaptic neurons *(light blue)*, are offset by hyperpolarizations triggered by inhibitory presynaptic neurons *(violet)* to produce only a small depolarization at the axon hillock only if there is sufficient depolarization to surpass the threshold potential.

NEUROTRANSMITTERS

Neurotransmitters are the means by which neurons talk to one another. At billions, or more likely, trillions, of synapses throughout the body, presynaptic neurons release neurotransmitters that act to facilitate, stimulate, or inhibit postsynaptic neurons and effector cells. More than 50 compounds are known to be neurotransmitters. At least 50 other compounds are suspected of being neurotransmitters. For the most part, they are not distributed diffusely or at random throughout the nervous system. Instead, specific neurotransmitters are localized in discrete groups of neurons and thus released in specific nerve pathways (Box 12-5).

BOX 12-5: HEALTH MATTERS
Anesthetics

Anesthetics are substances that are administered to reduce or eliminate the sensation of pain—thus producing a state called **anesthesia**. Many anesthetics produce their effects by inhibiting the opening of sodium channels and thus blocking the initiation and conduction of nerve impulses. Anesthetics such as bupivacaine (Marcaine) are often used in dentistry to minimize pain involved in tooth extractions and other dental procedures. Procaine has likewise been used to block the transmission of electrical signals in sensory pathways of the spinal cord. Benzocaine and phenol, local anesthetics found in several over-the-counter products that relieve pain associated with teething in infants, sore throat pain, and other ailments, also produce their effect by blocking initiation and conduction of nerve impulses.

Classification of Neurotransmitters

Neurotransmitters are commonly classified by their function or by their chemical structure, depending on the context in which they are discussed. You are already familiar with two major functional classifications: *excitatory neurotransmitters* and *inhibitory neurotransmitters*. Some neurotransmitters can have inhibitory effects at some synapses and excitatory effects at other synapses. For example, the neurotransmitter *acetylcholine*, discussed in Chapter 11, excites skeletal muscle cells but inhibits cardiac muscle cells. This illustrates an important point about the action of neurotransmitters: their function is determined by the postsynaptic receptors, not by the neurotransmitters themselves.

Another way to classify neurotransmitters by function is to identify the mechanism by which they cause a change in the postsynaptic neuron or effector cell. Some neurotransmitters trigger the opening or closing of ion channels directly, by binding to one or both receptor sites on the channel itself (see figure in Box 12-6). Other neurotransmitters produce their effects by binding to receptors linked to G proteins that, in turn, activate chemical messengers within the postsynaptic cell. An example of this *second messenger* mechanism occurs when the neurotransmitter *norepinephrine* binds to its receptors, causing G proteins to activate the membrane-bound enzyme adenyl cyclase. The adenyl cyclase removes phosphate groups from adenosine triphosphate (ATP) to form cyclic adenosine monophosphate (cAMP). Cyclic AMP is the "second messenger," triggering the activation of the protein kinase A enzyme that eventually causes Na channels in the postsynaptic membrane to open (Figure 12-29). The second messenger mechanism is usually slower and longer-lasting than the direct

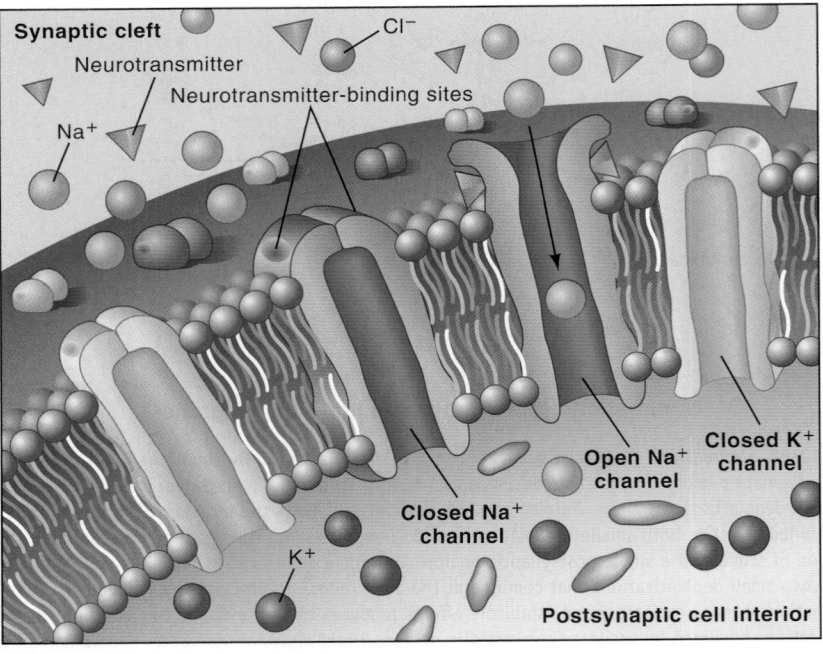

Figure 12-29 *Direct stimulation of postsynaptic receptor.* Some neurotransmitters, such as acetylcholine, initiate nerve signals by binding directly to one or both neurotransmitter-binding sites on the stimulus-gated ion channel. Such binding causes the channel to change its shape to an open position. When the neurotransmitter is removed, the channel again closes.

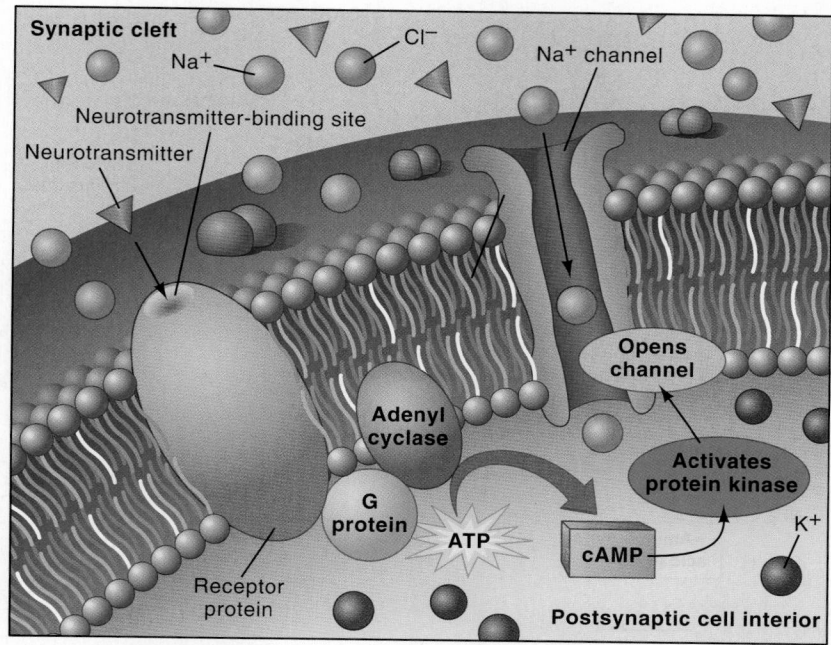

Figure 12-30 *Second messenger stimulation of postsynaptic receptor.* Norepinephrine and many other neurotransmitters initiate nerve signals indirectly by binding to a receptor linked to a G protein that changes shape to activate the enzyme *adenylate cyclase,* which in turn catalyzes the conversion of ATP to cyclic AMP (cAMP). cAMP is a "second messenger" that induces a change in the shape of a stimulus-gated channel. (Compare with Figure 12-29.)

mechanism (Figure 12-30). And because the second messenger mechanism involves intracellular messengers, it can regulate other cellular processes, such as cytoskeleton movement, gene expression, and shuttling of proteins along the axonal transport system.

Because the functions of specific neurotransmitters vary by location, it is often most useful to classify them according to their chemical structure. Neurotransmitters can thus be grouped into two main groupings: *small-molecule transmitters* and *large-molecule transmitters*.

Small-molecule neurotransmitters are, as their name implies, molecules of a smaller size than those in the large-molecule category. Small-molecule neurotransmitters are amino acids or are derived from individual amino acids. Large-molecule neurotransmitters, on the other hand, are made up of more than one amino acid—usually chains of 2 to 40 amino acids.

Small-molecule transmitters are subdivided into four main chemical classes:

1. Acetylcholine
2. Amines
3. Amino acids
4. Other small molecules

Large-molecule neurotransmitters are all neuropeptides—chains of several amino acids strung together by peptide bonds. Examples of neurotransmitters in each of these groupings are given in the following discussion, in Figures 12-31 and 12-32, and in Table 12-3.

Acetylcholine

The neurotransmitter **acetylcholine (ACh)** is in a class of its own because it has a chemical structure unique among neurotransmitters. It is synthesized in neurons by combining an acetate (acetyl-coenzyme-A) with choline (sometimes listed as a B vitamin)—as you can see in Figure 12-31. Postsynaptic membranes contain the enzyme *acetylcholinesterase,* which rapidly inactivates the acetylcholine bound to postsynaptic receptors. Choline molecules released by this reaction are transported back into the presynaptic neuron, where they are combined with acetate to form more acetylcholine.

As Table 12-3 shows, acetylcholine is found in various locations in the nervous system. In many of these locations, it has an excitatory effect (for example, at the neuromuscular junctions of skeletal muscles). In others, it has an inhibitory effect (for example, at the neuromuscular junctions of cardiac muscle tissue).

Amines

Amine neurotransmitters are synthesized from amino acid molecules, such as tyrosine, tryptophan, or histidine. Amines include the neurotransmitters of the **monoamine** subclass: *serotonin* and *histamine.* Also included are neurotransmitters of the **catecholamine** subclass: *dopamine, epinephrine,* and *norepinephrine.* Figure 12-31 shows how these subclasses of neurotransmitters are derived from different amino acids.

The amine neurotransmitters are found in various regions of the brain, where they affect learning, emotions, motor control, and other activities (Box 12-6). Dopamine, for example, is known

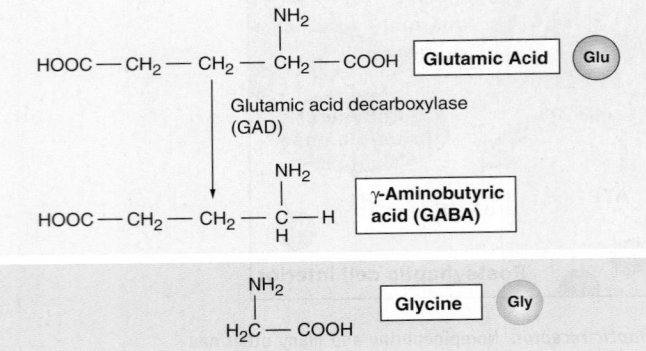

Figure 12-31 *Examples of small-molecule neurotransmitters.* Many of the small-molecule transmitters are amino acid molecules or are derived from the amino acids tryptophan, histidine, or tyrosine.

Table 12-3   Examples of Neurotransmitters

NEUROTRANSMITTER	LOCATION*	FUNCTION*
Small-Molecule Transmitters		
Acetylcholine (Ach)	Junctions with motor effectors (muscles, glands); many parts of brain	Excitatory or inhibitory; involved in memory
AMINES		
Monoamines		
Serotonin (5-HT†)	Several regions of the CNS	Mostly inhibitory; involved in moods and emotions, sleep
Histamine	Brain	Mostly excitatory; involved in emotions and regulation of body temperature and water balance
Catecholamines		
Dopamine (DA)	Brain; autonomic system	Mostly inhibitory; involved in emotions/moods and in regulating motor control
Epinephrine	Several areas of the CNS and in the sympathetic division of the ANS	Excitatory or inhibitory; acts as a hormone when secreted by sympathetic neurosecretory cells of the adrenal gland
Norepinephrine (NE)	Several areas of the CNS and in the sympathetic division of the ANS	Excitatory or inhibitory; regulates sympathetic effectors; in brain, involved in emotional responses
AMINO ACIDS		
Glutamate (glutamic acid, Glu)	CNS	Excitatory; most common excitatory neurotransmitter in CNS
Gamma-aminobutyric acid (GABA)	Brain	Inhibitory; most common inhibitory neurotransmitter in brain
Glycine (Gly)	Spinal cord	Inhibitory; most common inhibitory neurotransmitter in brain
OTHER SMALL MOLECULES		
Nitric oxide (NO)	Several regions of the nervous system	May be a signal from postsynaptic to presynaptic neuron
Large-Molecule Transmitters		
NEUROPEPTIDES		
Vasoactive intestinal peptide (VIP)	Brain; some ANS and sensory fibers; retina; gastrointestinal tract	Function in nervous system uncertain
Cholecystokinin (CCK)	Brain; retina	Function in nervous system uncertain
Substance P	Brain, spinal cord, sensory pain pathways; gastrointestinal tract	Mostly excitatory; transmits pain information
Enkephalins	Several regions of CNS; retina; intestinal tract	Mostly inhibitory; act like opiates to block pain
Endorphins	Several regions of CNS; retina; intestinal tract	Mostly inhibitory; act like opiates to block pain
Neuropeptide Y (NPY)	Brain, some ANS fibers	Variety of functions including enhancing blood vessel constriction by ANS, regulation of energy balance, learning, and memory

*These are examples only; most of these neurotransmitters are also found in other locations, and many have additional functions.
†5-hydroxytryptamine (synonym for serotonin).
CNS, Central nervous system; *ANS,* autonomic nervous system.

to have an inhibitory effect on certain somatic motor pathways. When dopamine is deficient, the tremors and general overstimulation of muscles characteristic of *parkinsonism* occur. Epinephrine and norepinephrine are also involved in motor control, specifically in the sympathetic pathways of the autonomic nervous system. Some autonomic neurons in the adrenal gland do not terminate at a postsynaptic effector cell but instead release their neurotransmitters directly into the bloodstream. When this occurs, epinephrine and norepinephrine are called *hormones* instead of neurotransmitters.

Amino Acids

Many biologists now believe that amino acids are among the most common neurotransmitters in the central nervous system. For example, it is thought that the amino acid *glutamate (glutamic acid)* is responsible for up to 75% of the excitatory signals in the brain. *Gamma-aminobutyric acid (GABA),* which is derived from glutamate, is the most common inhibitory neurotransmitter in the brain. In the spinal cord, the most widely distributed inhibitory neurotransmitter is the simple amino acid *glycine.* These three neurotransmitters are shown in Figure 12-31.

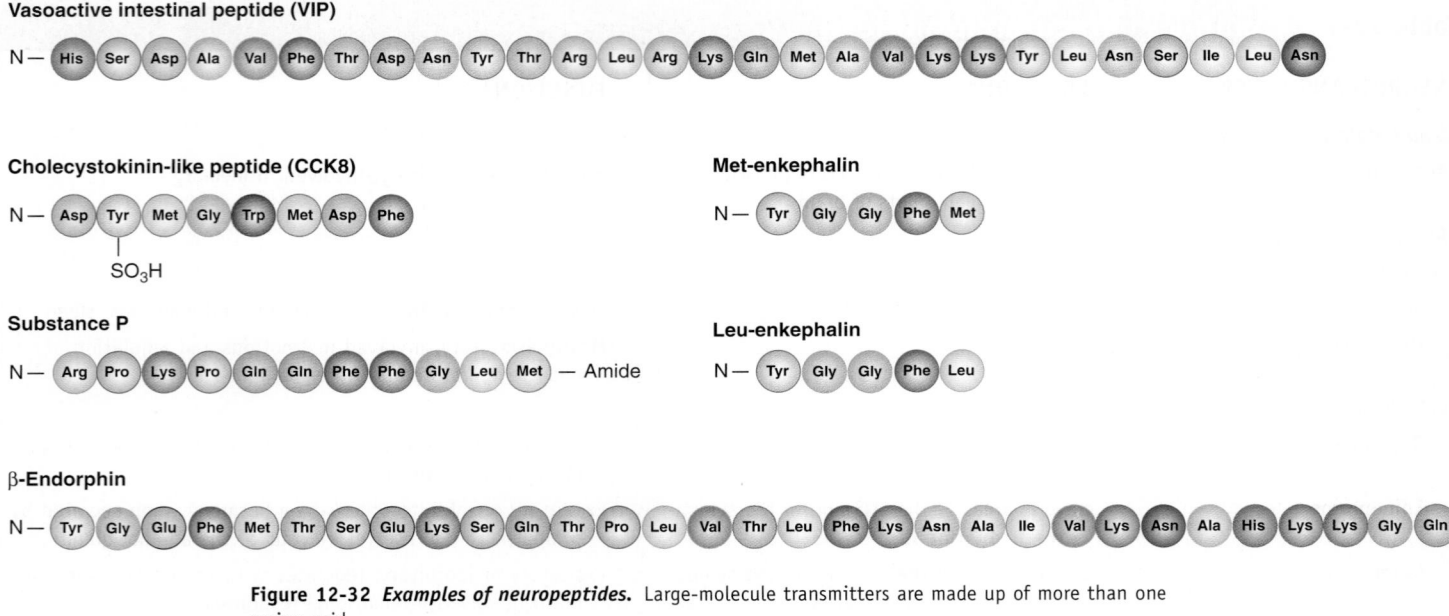

Figure 12-32 *Examples of neuropeptides.* Large-molecule transmitters are made up of more than one amino acid.

Amino acids are found in all cells of the body, where they are used to synthesize various structural and functional proteins. In the nervous system, however, they are also stored in synaptic vesicles and used as neurotransmitters. Specialized receptors in the postsynaptic membrane are sensitive to high quantities of certain amino acids and thus trigger specific responses in the postsynaptic cell. It is believed that an imbalance of certain amino acids in the body could produce similar effects, and thus alter the function of the nervous system.

Other Small-Molecule Transmitters

The term "other" is not a very descriptive term for a category of compounds, but the discovery of a new group of small neurotransmitter molecules is so recent that a standard name for this group has not been put forth. The prime example of this group is **nitric oxide (NO)**. Nitric oxide is a small gas molecule that the cell makes from the amino acid arginine. NO was the first gas to be identified as a transmitter, but carbon monoxide (CO) may be another. Nitric acid is released from postsynaptic neurons and *diffuses backward* toward the presynaptic neuron, where it has its biochemical effects. This gives postsynaptic neurons a mechanism by which they can "talk back" to the presynaptic neuron, establishing the opportunity for feedback.

Neuropeptides

The **neuropeptide** neurotransmitters are short strands of amino acids called polypeptides or, more simply, *peptides* (Figure 12-32). Neuropeptide neurotransmitters are often called *neuroactive peptides*. Peptides were first discovered to have regulatory effects in the digestive tract, where they act as hormones and regulate digestive function. In the 1970s some of these "gut" polypeptides, such as *vasoactive intestinal peptide* (VIP), *cholecystokinin* (CCK), and *substance P*, were also found to be acting as neurotransmitters

in the brain. In addition, researchers found that receptors of many of the gut-brain peptides also bind morphine and other opium derivatives. For example, two subclasses of peptides—*enkephalins* and *endorphins*—that bind to the opiate receptors serve as the body's own supply of opiates. Enkephalins and endorphins have important pain-relieving effects in the body.

Although neuropeptides may be secreted by a synaptic knob by themselves, some may be secreted along with a second, or even third, neurotransmitter. In such cases, the neuropeptide is thought to act as a **neuromodulator**. A neuromodulator is a "cotransmitter" that regulates or modulates the effects of the neurotransmitter(s) released along with it.

An important class of neuropeptides in the nervous system is the **neurotrophins**, or *neurotrophic factors*. When first discovered, researchers found that they stimulate neuron development, hence the name neurotrophin (literally, "nerve growth factor"). Box 12-6, part A of figure, shows how neurotrophins help regulate neuron growth. More recently, we have discovered that neurotrophins also participate in synaptic transmission and neuromodulation. In fact, one type of neurotrophin released by neurons is required to form any memory lasting more than a day.

 QUICK CHECK

24. How do excitatory neurotransmitters differ from inhibitory neurotransmitters?
25. What are the four chemical classes of neurotransmitters?
26. What are neuromodulators?

BOX 12-6 Neural Networks

Nervous pathways, or **neural networks**, conduct information along complex series of neurons joined together by synapses.

Research shows that such networks develop during a person's early life and are influenced by the various sensory learning experiences that we have as our nerve tissue develops. One process that facilitates the formation of connections involves *neurotrophic factors* or *neurotrophins*—nerve growth factors that are released by various cells of the body. As part *A* of the figure shows, the axons of several developing neurons grow toward the cell that releases neurotrophins. However, only enough axons will remain that can be supported by the amount of available neurotrophin—the rest of the neurons will degenerate. Many other factors, such as repeated stimulation of a pathway, influence this process. Once neural networks become mature in adulthood, they are less likely to accept the addition of new neurons—even though neural stem cells do exist in the adult nervous system.

Something that makes the pathways of neural networks complex is that they often *converge* and *diverge*.

Convergence occurs when more than one presynaptic axon synapses with a single postsynaptic neuron (part *B* of the figure). Convergence allows information from several different pathways to be funneled into a single pathway. For example, the pathways that innervate the skeletal muscles may originate in several different areas of the central nervous system. Because pathways from each of these motor control areas converge on a single motor neuron, each area has an opportunity to control the same muscle.

Divergence occurs when a single presynaptic axon synapses with many different postsynaptic neurons (part *C* of the figure). Divergence allows information from one pathway to be "split" or "copied" and sent to different destinations in the nervous system. For example, a single bit of visual information may be sent to many different areas of the brain for processing.

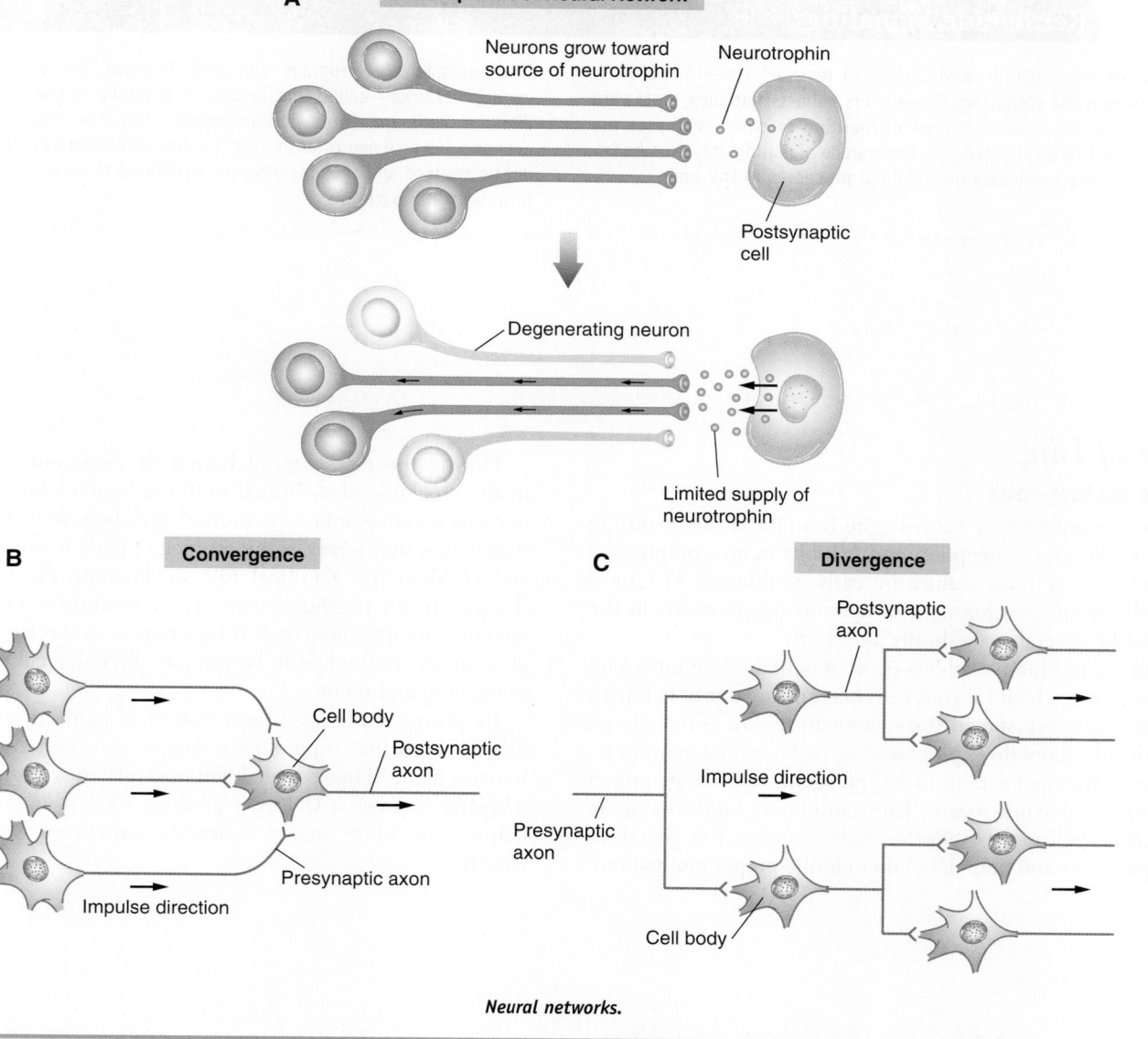

Neural networks.

BOX 12-7: HEALTH MATTERS
Antidepressants

Severe psychic depression occurs when a deficit of norepinephrine, dopamine, serotonin, and other amines exists in certain brain synapses. This fact led to the development of **antidepressant** drugs.

Certain of these antidepressants inhibit *catechol-O-methyl transferase* (COMT), the enzyme that inactivates norepinephrine. When COMT is inhibited by an antidepressant drug, the amount of active norepinephrine in brain synapses increases—relieving the symptoms of depression. Antidepressants such as phenelzine (Nardil) block the action of *monoamine oxidase* (MAO), the enzyme that inactivates dopamine and serotonin. Such drugs are in a class called **MAOIs (monamine oxidase inhibitors)**.

Widely used antidepressants such as imipramine (Tofranil) and amitriptyline (Elavil) increase amine levels at brain synapses by blocking their uptake into the axon terminals. The extremely popular drug fluoxetine (Prozac) and related drugs called **SSRIs (serotonin-specific reuptake inhibitors)** produce antidepressant effects by inhibiting the uptake of serotonin.

Cocaine, which is often used in medical practice as a local anesthetic, produces a temporary feeling of well-being in cocaine abusers by similarly blocking the uptake of dopamine. Unfortunately, cocaine and similar drugs can also adversely affect blood flow and heart function when taken in large amounts—leading to death in some individuals.

BOX 12-8   Retrograde Signaling

Can the nervous system send signals in reverse? Yes—in a process called **retrograde signaling.** Researchers believe that such "backward signaling" is involved in a type of learning in which synapses are strengthened in local pathways. Retrograde signaling may also be involved in other previously unexplained processes in the brain.

One example of retrograde signaling involves the release of an *endocannabinoid*—called that because it is similar to the active ingredient of marijuana, *tetrahydrocannabinol (THC)*. The endocannabinoid acts as a "reverse neurotransmitter" to stop inhibition by certain presynaptic neurons and thus increase the likelihood of an action potential in a particular pathway.

Cycle of Life
Nervous System Cells

The development of nerve tissue begins from the *ectoderm* during the first weeks after conception and exhibits many complicated stages before becoming mature by early adulthood. The most rapid and obvious development of nervous tissue occurs in the womb and for several years shortly after birth.

One of the most remarkable aspects of neural development involves the way in which nervous cells become organized to form a coordinated network spread throughout the body. Although we know very little about these processes, we do know that neurons require the coordinated actions of several agents to promote proper "wiring" of the nervous system. For example, we know that *nerve growth factors* released by effector cells stimulate the growth of neuron processes and help direct them to the proper destination.

During the first years of neural development, synapses are made, broken, and re-formed until the basic organization of the nervous system is intact. Neurobiologists believe that the sensory stimulation that serves as the essence of early learning in infants and children has a critical role in directing the formation of synapses in the nervous system. The formation of new synapses, selective strengthening of existing synapses, and selective removal of synapses are thought to be primary physiological mechanisms of learning and memory.

In advanced old age, degeneration of neurons, glia, and the blood vessels that supply them may destroy certain portions of nervous tissue. This, coupled with age-related syndromes such as *Alzheimer's disease (AD)*, may produce a loss of memory, coordination, and other neural functions sometimes referred to as *senility.*

THE BIG PICTURE

Nervous System Cells and the Whole Body

Neurons, the conducting cells of the nervous system, act as the "wiring" that connects the structures needed to maintain the internal constancy that is so vital to our survival. They also form processing "circuits" that make decisions regarding appropriate responses to stimuli that threaten our internal constancy. Sensory neurons act as sensors or receptors that detect changes in our external and internal environment that may be potentially threatening. Sensory neurons then relay this information to integrator mechanisms in the central nervous system. There, the information is processed—often by one or more interneurons—and an outgoing response signal is relayed to effectors by way of motor neurons. At the effector, a chemical messenger or neurotransmitter triggers a response that tends to restore homeostatic balance. Neurotransmitters released into the bloodstream, where they are called *hormones*, can enhance and prolong such homeostatic responses.

Of course, neurons do much more than simply respond to stimuli in a preprogrammed manner. Circuits of interneurons are capable of remembering, learning new responses, generating rational and creative thought, and implementing other complex processes. The exact mechanisms of many of these complex integrative functions are yet to be discovered, but you will learn some of what we already know in Chapter 13, in our discussion of the central nervous system.

Mechanisms of Disease

DISORDERS OF NERVOUS SYSTEM CELLS

Most disorders of nervous system cells involve glia rather than neurons. **Multiple sclerosis (MS),** one of the **myelin disorders** discussed on p. 437, is a good example of this principle. A few other important disorders involving glia are described in the following paragraphs.

The general name for tumors arising in nervous system structures is **neuroma** (noo-ROH-mah). Tumors do not usually develop directly from neurons but from glia, membrane tissues, and blood vessels. A common type of brain tumor—**glioma** (glee-OH-mah)—occurs in glia. Gliomas are usually benign but may still be life threatening. Because they often develop in deep areas of the brain, they are difficult to treat. Untreated gliomas may grow to a size that disrupts normal brain function—perhaps leading to death. Most malignant tumors of glia and other tissues in the nervous system do not arise there but are secondary tumors resulting from metastasis of cancer cells from the breast, lung, or other organs.

Tumors in the Central Nervous System

Astrocytoma is a type of glioma that originates from astrocytes. It is a slow-growing, infiltrating tumor of the brain that usually appears during the fourth decade of life. Seizures, headaches, or neurological deficits indicative of the area of the brain involved are usual presenting symptoms. **Glioblastoma multi-**forme, a highly malignant form of astrocytic tumor, spreads throughout the white matter of the brain. Because of its invasive nature, surgical removal is difficult and the average survival is less than 1 year. **Ependymoma** is a glial tumor arising from ependymal cells, which line the fluid-filled cavities (*ventricles*) of the brain and spinal cord. This is the most common glioma in children but can occur in adults. Because of its location, fluid pathways are obstructed, causing increased pressure in the brain, which in turn causes neurological damage. Surgical correction is possible, and the average postoperative survival is roughly 5 years. Glioma of oligodendrocytes is called **oligodendroglioma.** This tumor commonly occurs in the anterior portion of the brain and has a peak incidence at 40 years of age. The prognosis is better, with an average survival of 10 years after the onset of symptoms.

Tumors in the Peripheral Nervous System

Glial tumors can also develop in or on the cranial nerves. **Acoustic neuroma** is a lesion of the sheath of Schwann cells surrounding the eighth cranial nerve, responsible for hearing and balance. This tumor may be the size of a pea or walnut, but the person typically experiences difficulty deciphering speech through the affected ear, dizziness, tinnitus (ringing in the ear), and a slow, progressive hearing loss. With the use of microsurgical techniques, the tumor can be removed, but some nerve damage caused by the surgical procedure is common.

Mechanisms of Disease—cont.

Glial tumors can also appear in other regions of the peripheral nervous system. **Multiple neurofibromatosis** (MUL-tih-pul noo-roh-fye-broh-mah-TOH-sis) is an inherited disease characterized by numerous fibrous neuromas throughout the body (Figure 12-33). The tumors are benign, appearing first as small nodules in the Schwann cells of nerve fibers in the skin. In some cases, involvement spreads as large, disfiguring fibrous tumors in many areas of the body, including muscles, bones, and internal organs.

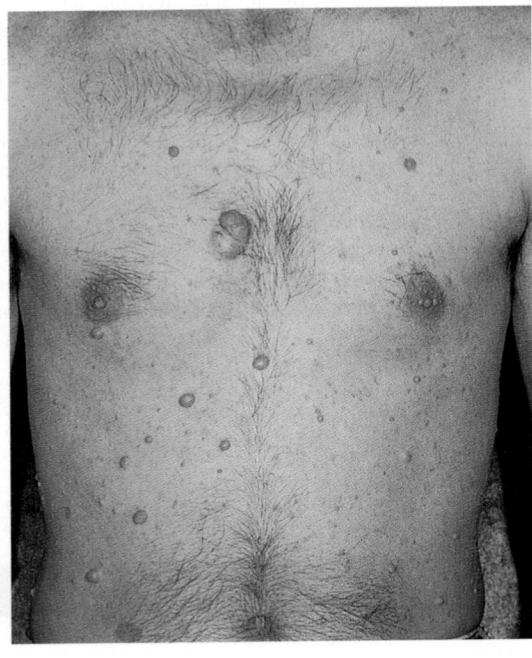

Figure 12-33 *Multiple neurofibromatosis.* Multiple tumors of Schwann cells in nerves of the skin that are characteristic of this inherited condition.

LANGUAGE OF SCIENCE *(Cont'd from page 431)*

ependymal cells (eh-PEN-di-mal) [*ependyma* an upper garment, *cella-* storeroom]

epineurium (ep-i-NOO-ree-um) [*epi-* on, *-neuron* nerve]

excitatory postsynaptic potential (EPSP) (ek-SYE-tah-toh-ree post-si-NAP-tik poh-TEN-shal) [*excitare* to rouse, *post-* after, *-synaptein-* to join, *-ic* pertaining to]

glia (GLEE-ah) [*glia* glue]

gray matter

hyperpolarization (hye-per-pol-lar-i-ZAY-shun) [*hyper-* excessive, *-polus-* pole, *-tion* process of]

inhibitory postsynaptic potential (IPSP) (in-HIB-i-tor-ee post-si-NAP-tik poh-TEN-shal) [*inhibere* to restrain, *post-* after, *-synaptein-* to join, *-ic* pertaining to]

interneurons (in-ter-NOO-rons) [*inter-* occurring between, *-neurons* nerves]

local potential

membrane potential [*membrana* thin skin]

microglia (my-KROG-lee-ah) [*micro-* small *-glia* glue]

monoamine (mon-oh-ah-MEEN) [*mono-* single, *-amine* amino acid]

multipolar neurons (mul-ti-POL-ar NOO-rons) [*multi-* many, *-polus* pole, *neurons* nerves]

myelin (MY-eh-lin) [*myelin* marrow]

myelin sheath (MY-eh-lin sheath) [*myelin* marrow]

myelinated fibers (MY-eh-li-nay-ted) [*myelin* marrow]

nerves

nerve fibers

nervous system [*nervus* nerve]

neurilemma (noo-ri-LEM-mah) [*neuri-* neuron, *-lemma* sheath]

neurofibrils (noo-roh-FYE-brils) [*neuro-* nerves, *-fibrilla* small fiber]

neuroglia (noo-ROG-lee-ah) [*neuro-* nerve, *-glia* glue]

neuromodulator (noo-roh-MOD-yoo-lay-tor) [*neuro* nerves]

neurons (NOO-rons) [*neuron* nerve]

neuropeptide (noo-roh-PEP-tyde) [*neuro-* nerves, *-peptein* to digest]

neurotransmitters (noo-roh-trans-MIT-ters) [*neuro-* nerves, *-transmittere* to transmit]

neurotrophins (noo-roh-TROF-inz) [*neuro-* nerve, *-troph-* nutrition, *-in* substance]

nitric oxide (NO) (NYE-trik AWK-side)

nodes of Ranvier (rahn-vee-AY) [*nodus* knot, *Louis A. Ranvier* French pathologist]

oligodendrocytes (ohl-i-go-DEN-droh-sytes) [*oligo-* few, *-dendro-* branch, *-cyte* cell]

perikaryon (pair-i-KAR-ee-on) [*peri-* around, *-karyon* nut]

perineurium (pair-i-NOO-ree-um) [*peri-* around, *-neuron* nerve]

peripheral nervous system (PNS) (peh-RIF-er-al) [*periphereia* circumference, *nervus* nerves]

postsynaptic (post-si-NAP-tik) [*post-* after, *-synaptein-* to join, *-ic* pertaining to]

postsynaptic potential (post-sih-NAP-tik poh-TEN-shal) [*post-* after, *-synaptein-* to join, *-ic* pertaining to]

presynaptic (pree-sih-NAP-tik) [*pre-* before, *-synaptein-* to join, *-ic* pertaining to]

reflex arc [*reflectere* to bend back]

LANGUAGE OF SCIENCE *(Cont'd)*

relative refractory period (ree-FRAK-tor-ee) [*refringere* to break apart, *periodos* circuit]

repolarization (ree-poh-lah-ri-ZAY-shun) [*re-* back again, *-polus-* pole, *-tion* process of]

resting membrane potential (RMP) [*membrana* thin skin]

retrograde signaling (RET-roh-grayd SIG-nah-ling) [*retro-* backward, *-gradus* step, *signum* mark]

saltatory conduction (SAL-tah-tor-ee) [*saltare* to leap]

satellite cell (SAT-i-lyte) [*satell-* attendant, *-ite* pertaining to, *cell* storeroom]

Schwann cells (shwon or shvon) [*Theodor Schwann* German anatomist]

somatic nervous system (SNS) (so-MAH-tik) [*soma-* body, *-ic* pertaining to, *nervus* nerves]

somatic sensory division (so-MAH-tik) [*soma-* body, *-ic* pertaining to, *sentire* to feel]

spatial summation (SPAY-shal sum-MAY-shun) [*spatium* space, *summa* total]

stimulus-gated channels (STIM-yoo-lus GAY-ted CHAN-nuls) [*stimulare* to incite or gated, *canalis* pipe]

synapse (SIN-aps) [*synapein* to join]

synaptic cleft (si-NAP-tik kleft) [*synaptein-* to join, *-ic* pertaining to]

synaptic knob (si-NAP-tik nob) [*synaptein-* to join, *-ic* pertaining to]

telodendria (tel-oh-DEN-dree-ah) [*telo-* end, *-dendria* branch]

temporal summation (TEM-poh-ral sum-MAY-shun) [*tempo-* temples, *-al* pertaining to]

threshold potential (THRESH-hold poh-TEN-shal)

unipolar (pseudounipolar) neurons (yoo-nee-POH-lar [SOO-doh-yoo-nee-POH-lar] NOO-rons) [*uni-* single, *-polus* pole, *pseudo* false, *neuron-* nerves]

visceral sensory division (VISS-er-al) [*viscus-* internal organs, *-al* pertaining to, *sentire* to feel]

voltage-gated channels (VOL-tij GAY-ted CHAN-nuls) [*Alessandro Volta* Italian physicist, gated, *canalis* pipe]

white matter

LANGUAGE OF MEDICINE

acoustic neuroma (ah-KOOS-tik noo-ROH-mah) [*acous-* to hear, *-ic* pertaining to, *neuro-* nerve, *-oma* tumor]

anesthesia (an-es-THEE-zhah) [*anaisthesia* lack of feeling]

antidepressant (an-tee-deh-PRESS-ant) [*anti-* against, *-deprimere* to press down]

astrocytoma (ass-troh-sye-TOH-mah) [*astro-* star shaped, *-cyt-* cell, *-oma* tumor]

blood-brain barrier (BBB)

cocaine (koh-KAYN) [*coca-* type of shrub, *-ine* made of]

ependymoma (eh-pen-di-MOH-mah) [*ependyma-* an upper garment, *-oma* tumor]

glioblastoma multiforme (glye-oh-blas-TOH-ma mul-ti-FOR-mee) [*glio-* glue, *-blastos* germ, *-oma* tumor, *multi-* many, *-forma* form]

glioma (glee-OH-mah) [*glio-* neuroglia, *-oma* tumor]

levodopa (L-dopa) (LEEV-oh-doh-pah) [*levo-* left (form of molecule), *-dopa* acronym denoting 3,4-dihydroxyphenylalanine]

monoamine oxidase inhibitors (MAOIs) (mon-oh-ah-MEEN OK-si-dase in-HIB-i-tors) [*mono-* single, *-amine* amino acid, *oxid-* oxygen compound, *-ase* enzyme, *inhibare* to prevent]

multiple neurofibromatosis (MUL-ti-pul noo-roh-fye-broh-mah-TOH-sis) [*multi-* many, *-plica* fold, *neuro-* nerves, *-fibra-* fiber, *-oma* tumor, *-osis* condition]

multiple sclerosis (MS) (MUL-ti-pul skleh-ROH-sis) [*multi-* many, *-plica* fold, *sclera-* hard, *-osis* condition]

myelin disorders (MY-eh-lin) [*myelin* marrow]

neural networks (NOOR-al) [*neur-* nerves, *-al* pertaining to]

neuroglobin (Ngb) (NOO-roh-gloh-bin)

neuroma (noo-ROH-mah) [*neur-* nerves, *-oma* tumor]

oligodendroglioma (ohl-i-go-DEN-droh-glye-OH-mah) [*oligo-* few, *-dendro-* branch, *-glio* neurolglia, *-oma* tumor]

paralysis (pah-RAL-i-sis) [*paralyein-* to be palsied, *-sis* condition]

Parkinson disease (PD) (PARK-in-son) [*James Parkinson* English physician]

serotonin-specific reuptake inhibitors (SSRIs) (sair-oh-TOH-nin speh-SIF-ik ree-UP-tayk in-HIB-it-orz) [*sero-* serum (watery fluid), *-ton-* stretch or firmness, *-in* substance, *specificus* kind, *re* again, *inhibere* to restrain]

CASE STUDY

Mickey Cooper is a 32-year-old white male who recently moved to Los Angeles from the Pacific Northwest, where he lived since birth. He is being seen in the physician's office for a routine physical examination for his employment as a cross-country truck driver. He mentions to the nurse that he has had some changes in his vision and has some nonspecific weakness and numbness in his hands and fingers. These symptoms seem to be aggravated when he plays racquetball and during extremely hot weather. He complains of leg spasms that occur at night. During the interview his speech is slow and he has difficulty forming some words.

On further examination, it is noted that he has abnormally increased deep tendon reflexes and slight tremors in both hands. Laboratory findings are all within normal range, but a computerized tomography (CT) scan shows increased demyelination of the white matter of the central nervous system and hard plaquelike lesions.

1. Based on the information presented, Mickey Cooper is *most* likely suffering from:

   A. Glioblastoma multiforme
   B. Multiple sclerosis
   C. Parkinson disease
   D. Myasthenia gravis

2. Because Mickey's disease involves the glia cells, it would be expected that his symptoms result from the lack of:

   A. Functions that support the neuron
   B. Sensory receptors and motor neurons
   C. Molecular transport through the nervous system
   D. Electrical conduction in the nervous system

3. The cause of Mickey's disease is believed to be related to autoimmunity and to viral infections in some individuals. To combat these causes by reducing damage to the nerve fibers, the physician would most likely prescribe which one of the following types of medications?

   A. Monoamine oxidase inhibitors
   B. Dopamine precursors
   C. Anesthetics
   D. Anti-inflammatory drugs

4. When counseling Mickey about the prognosis of his disease, the nurse would include which one of the following statements? This disease is characterized by:

   A. Its adverse effects on blood flow and heart function
   B. Relapses and remissions before it is cured
   C. Myelin loss and destruction
   D. Abnormal control of muscle movements

CHAPTER SUMMARY

INTRODUCTION

A. The function of the nervous system, along with the endocrine system, is to communicate

B. The nervous system is made up of the brain, spinal cord, and nerves (Figure 12-1)

ORGANIZATION OF THE NERVOUS SYSTEM

A. Organized to detect changes in internal and external environments, evaluate the information, and initiate an appropriate response

B. Subdivided into smaller "systems" by location (Figure 12-2)
   1. Central nervous system (CNS)
      a. Structural and functional center of the entire nervous system
      b. Consists of the brain and spinal cord
      c. Integrates sensory information, evaluates it, and initiates an outgoing response
   2. Peripheral nervous system (PNS)
      a. Nerves that lie in the "outer regions" of the nervous system
      b. Cranial nerves—originate from the brain
      c. Spinal nerves—originate from the spinal cord

C. Afferent and efferent divisions
   1. Afferent division—consists of all incoming sensory pathways
   2. Efferent division—consists of all outgoing motor pathways

D. "Systems" according to the types of organs they innervate
   1. Somatic nervous system (SNS)
      a. Somatic motor division carries information to the somatic effectors (skeletal muscles)
      b. Somatic sensory division carries feedback information to somatic integration centers in the CNS
   2. Autonomic nervous system (ANS)
      a. Efferent division of ANS carries information to the autonomic or visceral effectors (smooth and cardiac muscles and glands)
         (1) Sympathetic division—prepares the body to deal with immediate threats to the internal environment; produces "fight-or-flight" response
         (2) Parasympathetic division—coordinates the body's normal resting activities; sometimes called the "rest-and-repair" division
      b. Visceral sensory division carries feedback information to autonomic integrating centers in the CNS

CELLS OF THE NERVOUS SYSTEM

A. Glia
   1. Glial cells support the neurons
   2. Five major types of glia (Figure 12-3)
      a. Astrocytes
         (1) Star-shaped, largest, and most numerous type of glia
         (2) Cell extensions connect to both neurons and capillaries

(3) Astrocytes transfer nutrients from the blood to the neurons

(4) Form tight sheaths around brain capillaries, which, with tight junctions between capillary endothelial cells, constitute the blood-brain barrier (BBB)

b. Microglia

(1) Small, usually stationary cells

(2) In inflamed brain tissue, they enlarge, move about, and carry on phagocytosis

c. Ependymal cells

(1) Resemble epithelial cells and form thin sheets that line fluid-filled cavities in the CNS

(2) Some produce fluid; others aid in circulation of fluid

d. Oligodendrocytes

(1) Smaller than astrocytes with fewer processes

(2) Hold nerve fibers together and produce the myelin sheath

e. Schwann cells

(1) Found only in PNS

(2) Support nerve fibers and form myelin sheaths (Figure 12-4)

(3) Gaps in the myelin sheath are called nodes of Ranvier

(4) Neurilemma is formed by cytoplasm of Schwann cell (neurilemmocyte) wrapped around the myelin sheath; essential for nerve regrowth

(5) Satellite cells are Schwann cells that cover and support cell bodies in the PNS

B. Neurons

1. Excitable cells that initiate and conduct impulses that make possible all nervous system functions

2. Components of neurons (Figure 12-5)

a. Cell body (perikaryon)

(1) Ribosomes, rough endoplasmic reticulum (ER), Golgi apparatus

(a) Provide protein molecules (neurotransmitters) needed for transmission of nerve signals from one neuron to another

(b) Neurotransmitters are packaged into vesicles

(c) Provide proteins for maintaining and regenerating nerve fibers

(2) Mitochondria provide energy (ATP) for neuron; some are transported to end of axon

b. Dendrites

(1) Each neuron has one or more dendrites, which branch from the cell body

(2) Conduct nerve signals to the cell body of the neuron

(3) Distal ends of dendrites of sensory neurons are receptors

(4) Dendritic spines—small knoblike protrusions on dendrites of some brain neurons; serve as connection points for axons of other neurons

c. Axon

(1) A single process extending from the axon hillock, sometimes covered by a fatty layer called a *myelin sheath* (Figure 12-6)

(2) Conducts nerve impulses away from the cell body of the neuron

(3) Distal tips of axons are telodendria, each of which terminates in a synaptic knob

d. Cytoskeleton

(1) Microtubules and microfilaments, as well as neurofibrils (bundles of neurofilaments)

(2) Allow the rapid transport of small organelles (Figure 12-7)

(a) Vesicles (some containing neurotransmitters), mitochondria

(b) Motor molecules shuttle organelles to and from the far ends of a neuron

e. Functional regions of the neuron (Figure 12-8)

(1) Input zone—dendrites and cell body

(2) Summation zone—axon hillock

(3) Conduction zone—axon

(4) Output zone—telodendria and synaptic knobs of axon

C. Classification of neurons

1. Structural classification—classified according to number of processes extending from cell body (Figure 12-9)

a. Multipolar—one axon and several dendrites

b. Bipolar—only one axon and one dendrite; least numerous kind of neuron

c. Unipolar (pseudounipolar)—one process comes off neuron cell body but divides almost immediately into two fibers: central fiber and peripheral fiber

2. Functional classification (Figure 12-10)

a. Afferent (sensory) neurons—conduct impulses to spinal cord or brain

b. Efferent (motor) neurons—conduct impulses away from spinal cord or brain toward muscles or glandular tissue

c. Interneurons

D. Reflex arc

1. A signal conduction route to and from the CNS, with the electrical signal beginning in receptors and ending in effectors

2. Three-neuron arc—most common; consists of afferent neurons, interneurons, and efferent neurons (Figure 12-11)

a. Afferent neurons—conduct impulses to the CNS from the receptors

b. Efferent neurons—conduct impulses from the CNS to effectors (muscle or glandular tissue)

3. Two-neuron arc—simplest form; consists of afferent and efferent neurons

4. Synapse

a. Where nerve signals are transmitted from one neuron to another

b. Two types: electrical and chemical; chemical synapses are typical in the adult

c. Chemical synapses are located at the junction of the synaptic knob of one neuron and the dendrites or cell body of another neuron

NERVES AND TRACTS

A. Nerves—bundles of peripheral nerve fibers held together by several layers of connective tissue (Figure 12-12)
1. Endoneurium—delicate layer of fibrous connective tissue surrounding each nerve fiber
2. Perineurium—connective tissue holding together fascicles (bundles of fibers)
3. Epineurium—fibrous coat surrounding numerous fascicles and blood vessels to form a complete nerve
B. Within the CNS, bundles of nerve fibers are called *tracts* rather than *nerves*
C. White matter
1. PNS—myelinated nerves
2. CNS—myelinated tracts
D. Gray matter
1. Made up of cell bodies and unmyelinated fibers
2. CNS—referred to as *nuclei*
3. PNS—referred to as *ganglia*
E. Mixed nerves
1. Contain sensory and motor neurons
2. Sensory nerves—nerves with predominantly sensory neurons
3. Motor nerves—nerves with predominantly motor neurons

REPAIR OF NERVE FIBERS

A. Mature neurons are incapable of cell division; therefore damage to nervous tissue can be permanent
B. Neurons have limited capacity to repair themselves
C. If the damage is not extensive, the cell body and neurilemma are intact, and scarring has not occurred, nerve fibers can be repaired
D. Stages of repair of an axon in a peripheral motor neuron (Figure 12-13)
1. Following injury, distal portion of axon and myelin sheath degenerates
2. Macrophages remove the debris
3. Remaining neurilemma and endoneurium form a tunnel from the point of injury to the effector
4. New Schwann cells grow in the tunnel to maintain a path for regrowth of the axon
5. Cell body reorganizes its Nissl bodies to provide the needed proteins to extend the remaining healthy portion of the axon
6. Axon "sprouts" appear
7. When "sprout" reaches tunnel, its growth rate increases
8. The skeletal muscle cell atrophies until the nervous connection is reestablished
E. In CNS, similar repair of damaged nerve fibers is unlikely

NERVE IMPULSES

A. Membrane potentials
1. All living cells maintain a difference in the concentration of ions across their membranes
2. Membrane potential—slight excess of positively charged ions on the outside of the membrane and slight deficiency of positively charged ions on the inside of the membrane (Figure 12-14)

3. Difference in electrical charge is called potential because it is a type of stored energy
4. Polarized membrane—a membrane that exhibits a membrane potential
5. Magnitude of potential difference between the two sides of a polarized membrane is measured in volts (V) or millivolts (mV); the sign of a membrane's voltage indicates the charge on the inside surface of a polarized membrane
B. Resting membrane potential (RMP)
1. Membrane potential maintained by a nonconducting neuron's plasma membrane; typically -70 mV
2. The slight excess of positive ions on a membrane's outer surface is produced by ion transport mechanisms and the membrane's permeability characteristics
3. The membrane's selective permeability characteristics help maintain a slight excess of positive ions on the outer surface of the membrane (Figure 12-15)
4. Sodium-potassium pump (Figure 12-16)
   a. Active transport mechanism in plasma membrane that transports Na^+ and K^+ in opposite directions and at different rates
   b. Maintains an imbalance in the distribution of positive ions, resulting in the inside surface becoming slightly negative with respect to its outer surface
C. Local potentials
1. Local potentials—slight shift away from the resting membrane in a specific region of the plasma membrane (Figure 12-17)
2. Excitation—when a stimulus triggers the opening of additional Na^+ channels, allowing the membrane potential to move toward zero (depolarization)
3. Inhibition—when a stimulus triggers the opening of additional K^+ channels, increasing the membrane potential (hyperpolarization)
4. Local potentials are called *graded potentials* because the magnitude of deviation from the resting membrane potential is proportional to the magnitude of the stimulus

ACTION POTENTIAL

A. Action potential—the membrane potential of a neuron that is conducting an impulse; also known as a *nerve impulse*
B. Mechanism that produces the action potential (Figures 12-18 and 12-19)
1. When an adequate stimulus triggers stimulus-gated Na^+ channels to open, allowing Na^+ to diffuse rapidly into the cell, producing a local depolarization
2. As threshold potential is reached, voltage-gated Na^+ channels open and more Na^+ enters the cell, causing further depolarization
3. The action potential is an all-or-none response
4. Voltage-gated Na^+ channels stay open for only about 1 ms before they automatically close
5. After action potential peaks, membrane begins to move back toward the resting membrane potential when K^+ channels open, allowing outward diffusion of K^+; process is known as *repolarization*

6. Brief period of hyperpolarization occurs, and then the resting membrane potential is restored by the sodium-potassium pumps

C. Refractory period (Figure 12-20)
   1. Absolute refractory period—brief period (lasting approximately 0.5 ms) during which a local area of a neuron's membrane resists restimulation and will not respond to a stimulus, no matter how strong
   2. Relative refractory period—time during which the membrane is repolarized and restoring the resting membrane potential; the few milliseconds after the absolute refractory period; will respond only to a very strong stimulus

D. Conduction of the action potential
   1. At the peak of the action potential, the plasma membrane's polarity is now the reverse of the RMP
   2. The reversal in polarity causes electrical current to flow between the site of the action potential and the adjacent regions of membrane and triggers voltage-gated Na^+ channels in the next segment to open; this next segment exhibits an action potential (Figure 12-21)
   3. This cycle continues to repeat
   4. The action potential never moves backward because of the refractory period
   5. In myelinated fibers, action potentials in the membrane only occur at the nodes of Ranvier; this type of impulse conduction is called *saltatory conduction* (Figure 12-22)
   6. Speed of nerve conduction depends on diameter and on the presence or absence of a myelin sheath

SYNAPTIC TRANSMISSION

A. Two types of synapses (junctions) (Figure 12-23)
   1. Electrical synapses occur where cells joined by gap junctions allow an action potential to simply continue along postsynaptic membrane
   2. Chemical synapses occur where presynaptic cells release chemical transmitters (neurotransmitters) across a tiny gap to the postsynaptic cell, possibly inducing an action potential there

B. Structure of the chemical synapse (Figure 12-25)
   1. Synaptic knob—tiny bulge at the end of a terminal branch of a presynaptic neuron's axon that contains vesicles housing neurotransmitters
   2. Synaptic cleft—space between a synaptic knob and the plasma membrane of a postsynaptic neuron
   3. Arrangements of synapses
      a. Axodendritic—axon signals postsynaptic dendrite; common
      b. Axosomatic—axon signals postsynaptic soma; common
      c. Axoaxonic—axon signals postsynaptic axon; may regulate action potential of postsynaptic axon
   4. Plasma membrane of a postsynaptic neuron—has protein molecules that serve as receptors for the neurotransmitters

C. Mechanism of synaptic transmission (Figure 12-25)—sequence of events is:
   1. Action potential reaches a synaptic knob, causing calcium ions to diffuse into the knob rapidly

2. Increased calcium concentration triggers the release of neurotransmitter by way of exocytosis
3. Neurotransmitter molecules diffuse across the synaptic cleft and bind to receptor molecules, causing ion channels to open
4. Opening of ion channels produces a postsynaptic potential, either an excitatory postsynaptic potential (EPSP) or an inhibitory postsynaptic potential (IPSP)
5. The neurotransmitter's action is quickly terminated by neurotransmitter molecules being transported back into the synaptic knob (reuptake) and/or metabolized into inactive compounds by enzymes and/or diffused and taken up by nearby glia (Figure 12-26)

D. Summation (Figure 12-28)
   1. Spatial summation—adding together the effects of several knobs being activated simultaneously and stimulating different locations on the postsynaptic membrane, producing an action potential
   2. Temporal summation—when synaptic knobs stimulate a postsynaptic neuron in rapid succession, their effects can summate over a brief period of time to produce an action potential

NEUROTRANSMITTERS

A. Neurotransmitters—means by which neurons communicate with one another; there are more than 30 compounds known to be neurotransmitters, and dozens of others are suspected

B. Classification of neurotransmitters—commonly classified by:
   1. Function—the function of a neurotransmitter is determined by the postsynaptic receptor; two major functional classifications are excitatory neurotransmitters and inhibitory neurotransmitters; can also be classified according to whether the receptor directly opens a channel or instead uses a second messenger mechanism involving G proteins and intracellular signals (Figures 12-29 and 12-30)
   2. Chemical structure—the mechanism by which neurotransmitters cause a change; there are four main classes; since the functions of specific neurotransmitters vary by location, they are usually classified according to chemical structure

C. Small-molecule neurotransmitters (Figure 12-31)
   1. Acetylcholine
      a. Unique chemical structure; acetate (acetyl- coenzyme-A) with choline
      b. Acetylcholine is deactivated by acetylcholinesterase, with the choline molecules being released and transported back to presynaptic neuron to combine with acetate
      c. Present at various locations, sometimes in an excitatory role, other times, inhibitory
   2. Amines
      a. Synthesized from amino acid molecules
      b. Two categories: monoamines and catecholamines
      c. Found in various regions of the brain, affecting learning, emotions, motor control, and so on

3. Amino acids
    a. Believed to be among the most common neurotransmitters of the CNS
    b. In the PNS, amino acids are stored in synaptic vesicles and used as neurotransmitters
4. Other small transmitters
    a. Nitric oxide (NO) derived from an amino acid
    b. NO from a postsynaptic cell signals the presynaptic neuron, providing feedback in a neural pathway
D. Large-molecule neurotransmitters—neuropeptides (Figure 12-32)
    1. Peptides made up of two or more amino acids
    2. May be secreted by themselves or in conjunction with a second or third neurotransmitter; in this case, neuropeptides act as a neuromodulator, a "co-transmitter" that regulates the effects of the neurotransmitter released along with it
    3. Neurotrophins (neurotrophic [nerve growth] factors) stimulate neuron development but also can act as neurotransmitters or neuromodulators

CYCLE OF LIFE: NERVOUS SYSTEM CELLS

A. Nerve tissue development
    1. Begins in ectoderm
    2. Occurs most rapidly in womb and in first 2 years
B. Nervous cells organize into body network
C. Synapses
    1. Form and re-form until nervous system is intact
    2. Formation of new synapses and strengthening or elimination of old synapses stimulate learning and memory
D. Aging causes degeneration of the nervous system, which may lead to senility

THE BIG PICTURE: NERVOUS SYSTEM CELLS AND THE WHOLE BODY

A. Neurons act as the "wiring" that connects structures needed to maintain homeostasis
B. Sensory neurons—act as receptors to detect changes in the internal and external environment; relay information to integrator mechanisms in the CNS
C. Information is processed, and a response is relayed to the appropriate effectors through the motor neurons
D. At the effector, neurotransmitter triggers a response to restore homeostasis
E. Neurotransmitters released into the bloodstream are called *hormones*
F. Neurons are responsible for more than just responding to stimuli; circuits are capable of remembering or learning new responses, generation of thought, and so on

REVIEW QUESTIONS

1. Briefly explain the general function the nervous system performs for the body.
2. Identify the other body system that performs the same general function.

3. Describe the function of sodium and potassium in the generation of an action potential.
4. Why is an action potential an all-or-none response?
5. How does the myelin sheath affect the speed of an action potential? What about the diameter of the nerve fiber?
6. Describe the structure of a synapse.
7. Describe the series of events that mediate conduction across synapses.
8. List and describe the four main chemical classes of transmitters.
9. Describe the actions of acetylcholine.
10. List and describe disorders of the nervous system cells.

CRITICAL THINKING QUESTIONS

1. Compare and contrast the characteristics of the central nervous system and the peripheral nervous system. Also, explain how the somatic and autonomic nervous systems could be included in the afferent and efferent divisions.
2. There are relatively few neurons in the brain. How would you describe the other cells in the brain, and what do they do?
3. All neurons have axons and dendrites. Explain where these parts of the neuron are located in multipolar, bipolar, and unipolar neurons.
4. Compare and contrast white and gray matter. What would result if there were a loss of myelination?
5. Explain how high doses of methylprednisolone act to reduce the severity of spinal cord injuries.
6. Nerve fibers in the peripheral nervous system are much more successful than nerve fibers in the central nervous system in repairing themselves. How would you explain this difference?
7. Define *potential*. The following is a list of times and charges taken during a nerve impulse. Identify them as either an action potential or not and as occurring either during the absolute or relative refractory period. (Assume the stimulus starts at 0 ms.)
    0 mV at 1 ms
    −50 mV at 0.5 ms
    +25 mV at 1.5 ms
    +25 mV at 3 ms
8. Describe the mechanisms that allow subthreshold stimuli to generate an action potential.
9. Many antibiotics that should kill the causative agents of meningitis are ineffective if given orally. Can you explain the anatomical mechanism that could cause this ineffectiveness?
10. Explain one of the ways an anesthetic may relieve the sensation of pain.
11. Referred pain is a condition in which the pain is not felt where the pain is actually occurring but rather is felt over a wider area or in some other part of the body. What structural aspect of the nervous system might explain referred pain?

Central Nervous System

LANGUAGE OF SCIENCE

amygdaloid nucleus (ah-MIG-dah-loyd NOO-klee-us) [*amygdal-* almond, *-oid* resembling]

anterior corticospinal tracts (an-TEER-ee-or KOR-ti-koh-spy-nal) [*ante* before, *cortic-* cortex, *-spin-* backbone, *-al* pertaining to, *tractus* trail]

anterior spinothalamic tracts (an-TEER-ee-or SPY-no-tha-lam-ik) [*ante* before, *spino-* backbone, *-thalam-* inner chamber, *-ic* pertaining to]

arbor vitae (AR-bor VI-tay) [*arbor* tree, *vitae* of life]

ascending tracts (ah-SEND-ing) [*ascend* climb, *tractus* trail]

basal nuclei (BAY-sal NOO-klee-eye) [*bas-* foundation, *-al* pertaining to, *nucleus-* nut or kernel]

biological clock (bye-oh-LOJ-i-kal) [*bio-* life, *-logos* words (science of)]

brainstem

cauda equina (KAW-da eh-KWINE-ah) [*caud-* tail, *-al* pertaining to, *equina* of a horse]

caudate nucleus (KAW-dayt NOO-klee-us) [*caudate* having a tail, *nucleus* nut or kernel]

central sulcus (fissure of Rolando) (SUL-kus, FISH-ur of roh-LAHN-doh) [*sulcus* trench, *Luigi Rolando* Italian physician]

cerebellum (sair-eh-BELL-um) [*cerebellum* small brain]

cerebral hemispheres (seh-REE-bral HEM-i-sfeerz) [*cerebr-* cerebrum, *-al* pertaining to, *hemi-* half, *-sphere* spherical body]

cerebral peduncles (seh-REE-bral peh-DUNG-kuls) [*cerebr-* cerebrum, *-al* pertaining to, *peduncul* footstalk]

cerebrospinal fluid (CSF) (seh-ree-broh-SPY-nal FLOO-id) [*cerebr-* cerebrum, *-spin-* backbone, *-al* pertaining to]

cerebrum (SAIR-eh-brum) [*cerebrum* brain]

choroid plexuses (KOH-royd PLEK-sus-es) [*chorio-* skin, *-oid* resembling, *plexus* network]

conus medullaris (KOH-nus MED-yoo-lair-is) [*conus* cone, *medulla* marrow]

convolution (kon-voh-LOO-shun) [*convolut* rolled together]

Cont'd on p. 506

Recall from Chapter 12 that the nervous system is said to be composed of two major divisions: the central nervous system (CNS) and the peripheral nervous system (PNS). The reason for designating two distinct divisions is to make the study of the nervous system easier. In this chapter, we discuss the part of the nervous system that lies at the center of the regulatory process: the central nervous system. Comprising both the brain and the spinal cord, the central nervous system is the principal integrator of sensory input and motor output. Thus the central nervous system is capable of evaluating incoming information and formulating responses to changes that threaten our homeostatic balance.

This chapter begins with a description of the protective coverings of the brain and spinal cord. After that, we briefly discuss the watery *cerebrospinal fluid* (CSF) and the spaces in which it is found. We then outline the overall structure and function of the major organs of the central nervous system, beginning at the bottom with the spinal cord; this is the simplest and least complex part of the CNS. Then our focus moves upward to the more complex brain, beginning first with the narrow brainstem (Figure 13-1) and the roughly spherical cerebellum attached to its dorsal surface. Again shifting our attention upward, we describe the structure and function of the diencephalon and then move on to a discussion of the cerebrum. As we move up the central nervous system, the complexity of both structure and function increases. The spinal cord mediates simple reflexes, whereas the brainstem and diencephalon are involved in the regulation of the more complex maintenance functions, such as regulation of heart rate and breathing. The cerebral hemispheres, which together form the largest part of the brain, perform complex integrative functions such as conscious thought, learning, memory, language, and problem solving. We end the chapter with a discussion of the somatic sensory pathways and the somatic motor pathways. This prepares us for Chapter 14, which covers the peripheral nervous system, and Chapter 15, which covers the sense organs.

COVERINGS OF THE BRAIN AND SPINAL CORD

Because the brain and spinal cord are both delicate and vital, nature has provided them with two protective coverings. The outer covering consists of bone: cranial bones encase the brain; vertebrae encase the spinal cord. The inner covering consists of membranes known as **meninges.** Three distinct layers compose the meninges:

1. Dura mater
2. Arachnoid mater
3. Pia mater

Observe their respective locations in Figures 13-2 and 13-3. The dura mater, made of strong white fibrous tissue, serves as the outer layer of the meninges and also as the inner periosteum of the cranial bones. The arachnoid mater, a delicate, cobweblike layer, lies between the dura mater and the pia mater, or innermost layer of the meninges. The transparent pia mater adheres to the outer surface of the brain and spinal cord and contains blood vessels.

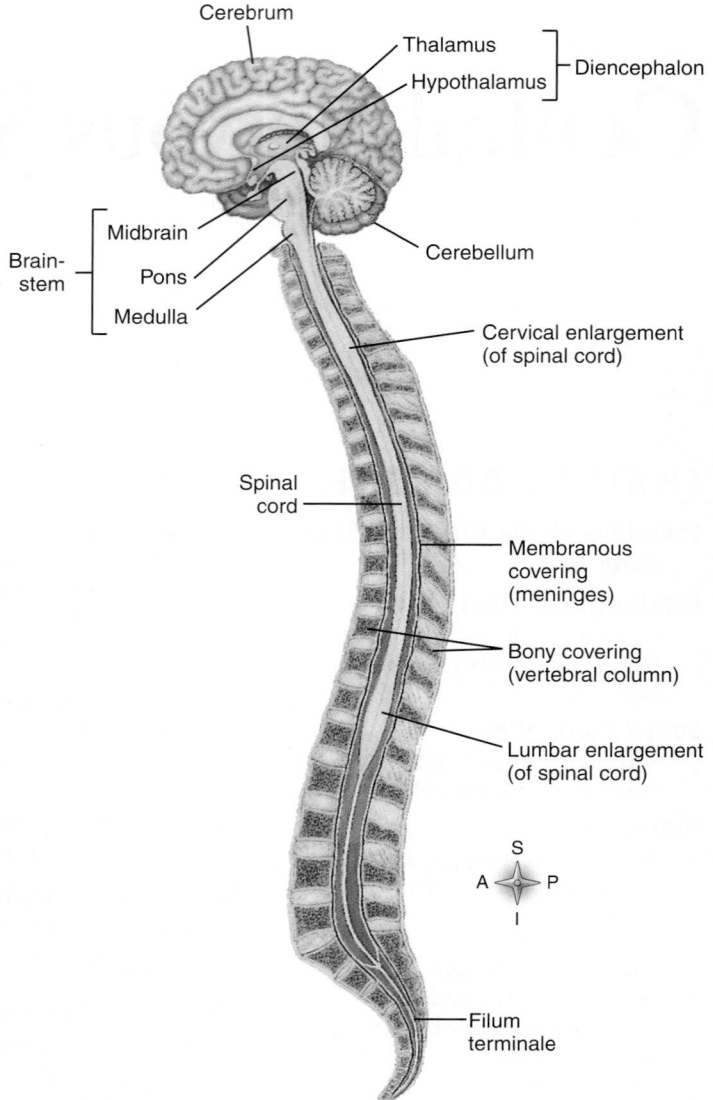

Figure 13-1 *The central nervous system.* Details of both the brain and the spinal cord are easily seen in this figure.

The dura mater has three important inward extensions:

1. **Falx cerebri.** The falx cerebri projects downward into the longitudinal fissure to form a kind of partition between the two cerebral hemispheres. The Latin word *falx* means "sickle" and refers to the curving sickle shape of this partition as it extends from the roof of the cranial cavity (Figure 13-2, *B*).
2. **Falx cerebelli.** The falx cerebelli is a sickle-shaped extension that separates the two halves, or hemispheres, of the cerebellum.
3. **Tentorium cerebelli.** The tentorium cerebelli separates the cerebellum from the cerebrum. It is called a *tentorium* (meaning "tent") because it forms a tentlike covering over the cerebellum.

Figure 13-2 shows a large space within the dura, where the falx cerebri begins to descend between the left and right cerebral

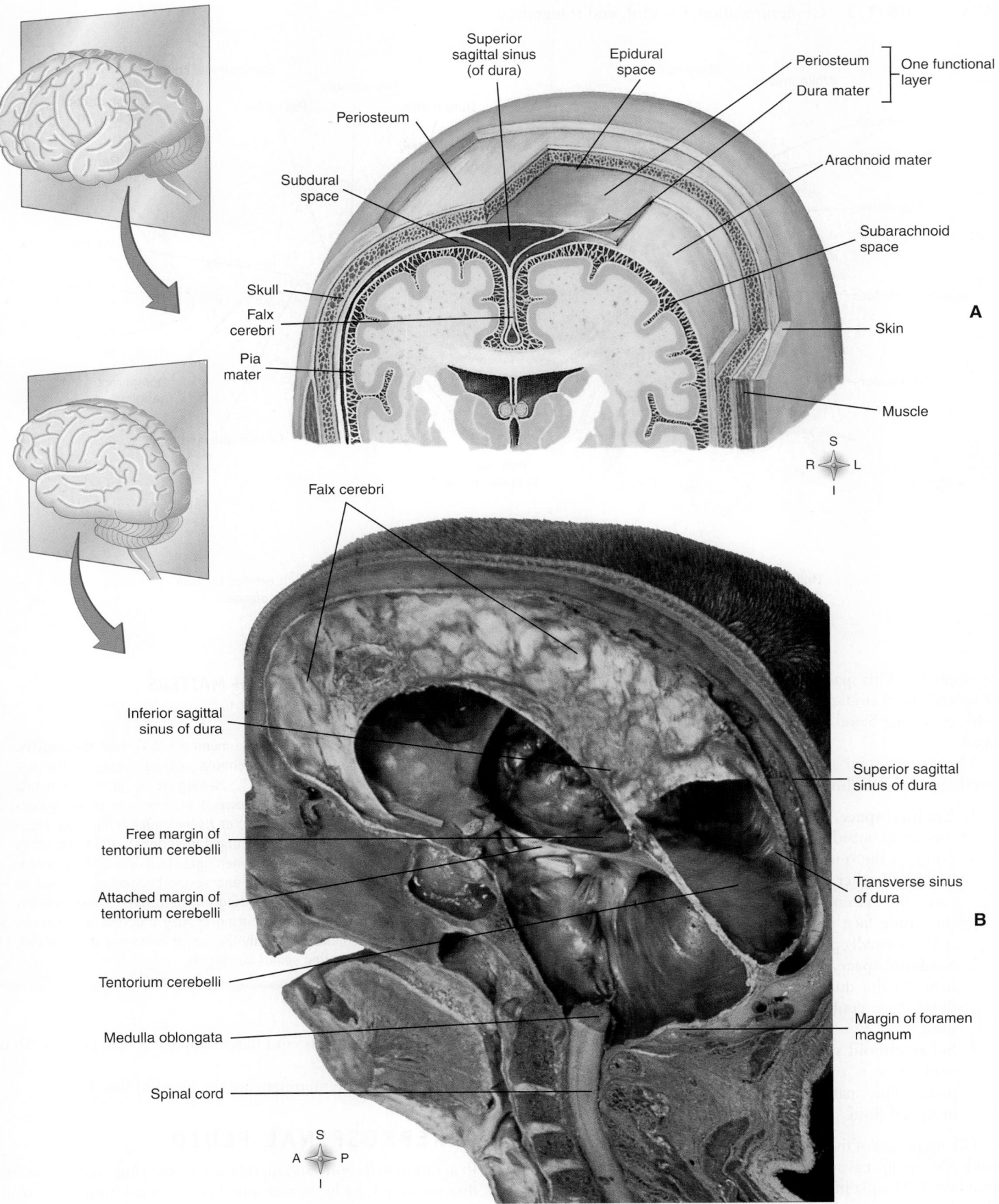

Figure 13-2 *Coverings of the brain*. A, Frontal section of the superior portion of the head, as viewed from the front. Both the bony and the membranous coverings of the brain can be seen. **B,** Transverse section of the skull, viewed from below. The dura mater has been retained in this specimen to show how it lines the inner roof of the cranium and the falx cerebri extending inward.

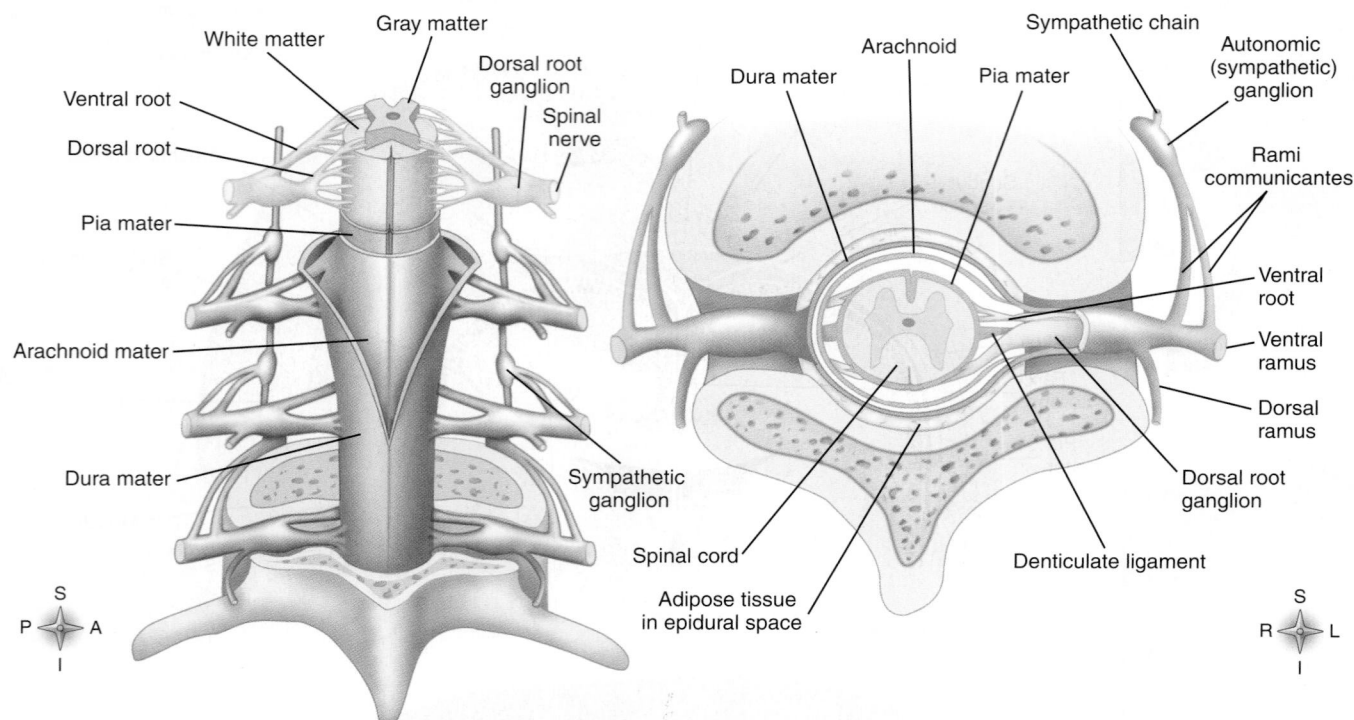

Figure 13-3 *Coverings of the spinal cord.* The dura mater is shown in purple. Note how it extends to cover the spinal nerve roots and nerves. The arachnoid is highlighted in pink and the pia mater in orange.

hemispheres. This space, called the *superior sagittal sinus*, is one of several dural sinuses. Dural sinuses function as venous reservoirs, collecting blood from brain tissues for the return trip to the heart.

There are several spaces between and around the meninges (see Figure 13-2). Three of these spaces are the following:

1. **Epidural space.** The epidural ("on the dura") space is immediately outside the dura mater but inside the bony coverings of the spinal cord. It contains a supporting cushion of fat and other connective tissues. Around the brain, because the dura mater is continuous with the periosteum on the inside face of the cranial bones there is no epidural space normally present.
2. **Subdural space.** The subdural ("under the dura") space is between the dura mater and arachnoid mater. The subdural space contains a small amount of lubricating serous fluid.
3. **Subarachnoid space.** As its name suggests, the *subarachnoid space* is under the arachnoid and outside the pia mater. This space contains a significant amount of cerebrospinal fluid.

The meninges of the cord (see Figure 13-3) continue on down inside the spinal cavity for some distance below the end of the spinal cord. The pia mater forms a slender filament known as the **filum terminale** (see Figure 13-1). At the level of the third segment of the sacrum, the filum terminale blends with the dura

BOX 13-1: HEALTH MATTERS
Meningitis

Infection or inflammation of the meninges is termed **meningitis**. It most often involves the arachnoid and pia mater, or the *leptomeninges* ("thin meninges"). Meningitis is most commonly caused by bacteria such as *Neisseria meningitidis* (meningococcus), *Streptococcus pneumoniae*, or *Haemophilus influenzae*. However, viral infections, mycoses (fungal infections), and tumors may also cause inflammation of the meninges. Individuals with meningitis usually complain of fever and severe headaches, as well as neck stiffness and pain. Depending on the primary cause, meningitis may be mild and self-limiting or may progress to a severe, perhaps fatal, condition. If only the spinal meninges are involved, the condition is called *spinal meningitis*.

mater to form a fibrous cord that disappears in the periosteum of the coccyx.

Infections of the meninges are discussed in Box 13-1.

CEREBROSPINAL FLUID

In addition to its bony and membranous coverings, nature has further protected the brain and spinal cord against injury by providing a cushion of fluid both around the organs and within them. This fluid is the **cerebrospinal fluid (CSF)**. The cerebrospinal

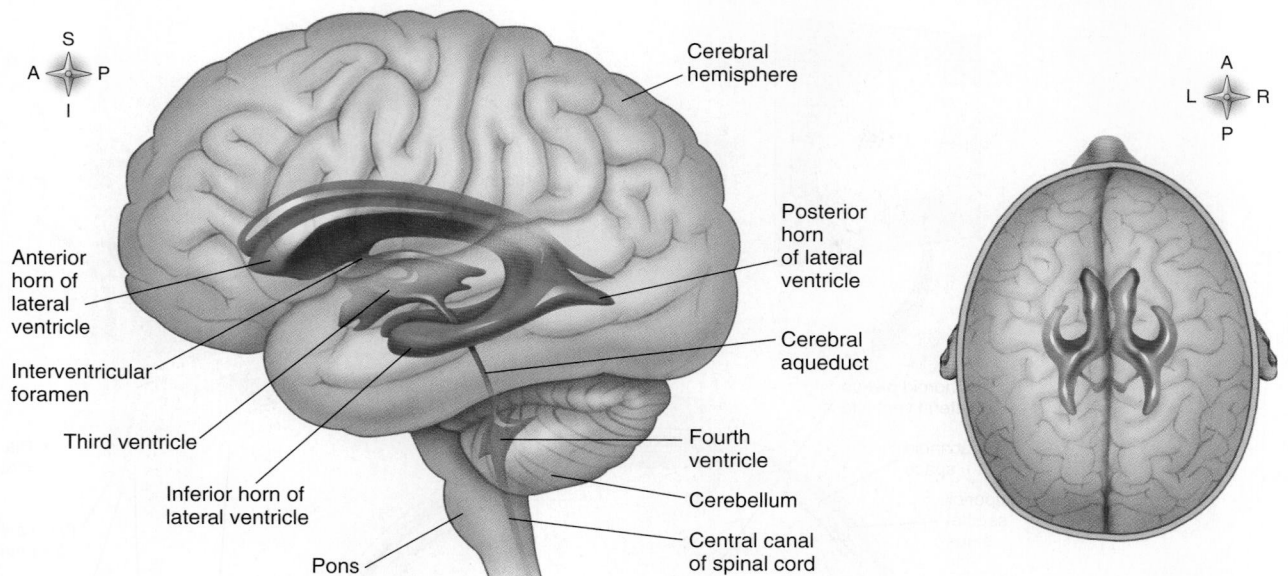

Figure 13-4 *Fluid spaces of the brain.* The large figure shows the ventricles highlighted within the brain in a left lateral view. The small figure shows the ventricles from above.

fluid does more than simply provide a supportive, protective cushion, however. It is also a reservoir of circulating fluid that, along with blood, the brain monitors for changes in the internal environment. For example, changes in the carbon dioxide (CO_2) content of CSF trigger homeostatic responses in the respiratory control centers of the brainstem that help regulate the overall CO_2 content and pH of the body.

Fluid Spaces

Cerebrospinal fluid is found in the subarachnoid space around the brain and spinal cord and within the cavities and canals of the brain and spinal cord.

The four large, fluid-filled spaces within the brain are called **ventricles.** Two of them, the lateral (or first and second) ventricles, are located one in each hemisphere of the cerebrum. As you can see in Figure 13-4, the third ventricle is little more than a thin, vertical pocket of fluid below and medial to the lateral ventricles. The fourth ventricle is a tiny, diamond-shaped space where the cerebellum attaches to the back of the brainstem. Actually, the fourth ventricle is simply a slight expansion of the central canal extending up from the spinal cord.

Formation and Circulation of Cerebrospinal Fluid

Formation of cerebrospinal fluid occurs mainly by separation of fluid from blood in the **choroid plexuses.** Choroid plexuses are networks of capillaries that project from the pia mater into the lateral ventricles and into the roofs of the third and fourth ventricles. Each choroid plexus is covered with a sheet of a special type of ependymal (glial) cell that releases the cerebrospinal fluid into the fluid spaces. From each lateral ventricle the fluid seeps through an opening, the *interventricular foramen (of Monro),* into the third ventricle, then through a narrow chan-

nel, the *cerebral aqueduct (of Sylvius),* into the fourth ventricle (Figure 13-5). Some of the fluid moves from the fourth ventricle directly into the central canal of the cord. Some of it moves out of the fourth ventricle through openings in its roof, two *lateral foramina (of Luschka)* and one *median foramen (of Magendie).* These openings allow cerebrospinal fluid to move into the *cisterna magna,* a space behind the medulla that is continuous with the subarachnoid space around the brain and cord. The fluid circulates in the subarachnoid space and then is absorbed into venous blood through the *arachnoid villi* (finger-like projections of the arachnoid mater into the brain's venous sinuses). Briefly, here is the circulation route of cerebrospinal fluid: it is formed by separation of fluid from blood in the choroid plexuses into the ventricles of the brain, circulates through the ventricles and into the central canal and subarachnoid spaces, and is absorbed back into blood.

The amount of cerebrospinal fluid in the average adult is about 140 ml (about 23 ml in the ventricles and 117 ml in the subarachnoid space of brain and cord). Box 13-2 explains the diagnostic value of testing a patient's cerebrospinal fluid. Box 13-3 shows what happens when the circulation of cerebrospinal fluid is blocked.

 QUICK CHECK

1. Name the three membranous coverings of the central nervous system in order, beginning with the outermost layer.
2. Trace the path of cerebrospinal fluid from its formation by a choroid plexus to its reabsorption into the blood.

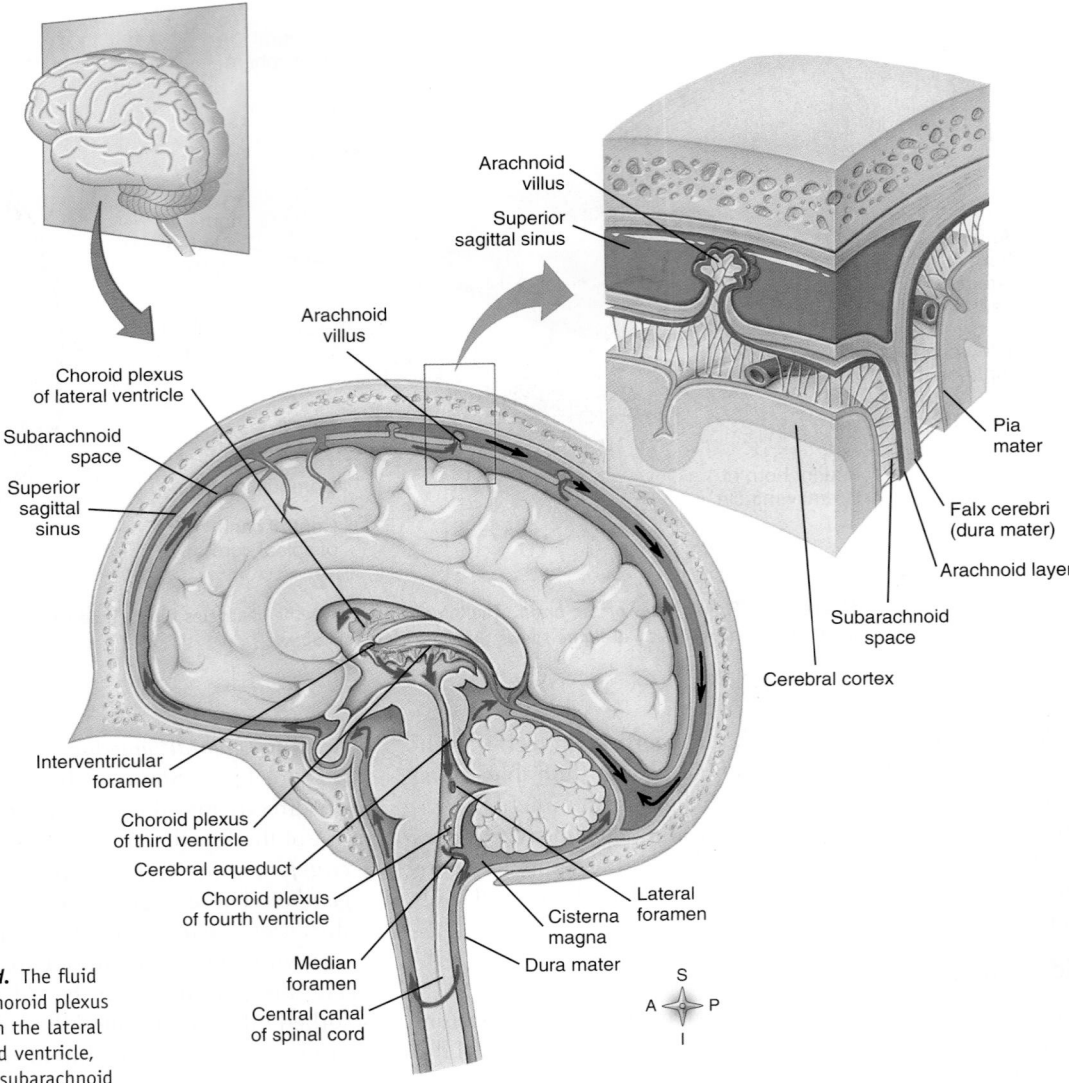

Figure 13-5 *Flow of cerebrospinal fluid.* The fluid produced by filtration of blood by the choroid plexus of each ventricle flows inferiorly through the lateral ventricles, interventricular foramen, third ventricle, cerebral aqueduct, fourth ventricle, and subarachnoid space and to the blood.

BOX 13-2

Diagnostic Study

Lumbar Puncture

The extension of the meninges beyond the cord is convenient for performing lumbar punctures without danger of injuring the spinal cord. A **lumbar puncture** is a withdrawal of some of the cerebrospinal fluid from the subarachnoid space in the lumbar region of the spinal cord. The physician inserts a needle just above or below the fourth lumbar vertebra, knowing that the spinal cord ends 1 inch or more above that level. The fourth lumbar vertebra can be easily located because it lies on a line with the iliac crest. Placing a patient on the side and arching the back by drawing the knees and chest together separate the vertebrae sufficiently to introduce the needle.

Cerebrospinal fluid removed through a lumbar puncture can be tested for the presence of blood cells, bacteria, or other abnormal characteristics that may indicate an injury or infection. A sensor called a *manometer* is sometimes attached to the needle to determine the pressure of the cerebrospinal fluid within the subarachnoid space. The lumbar puncture can also be used to introduce diagnostic agents, such as radiopaque dyes for x-ray photography, into the subarachnoid space.

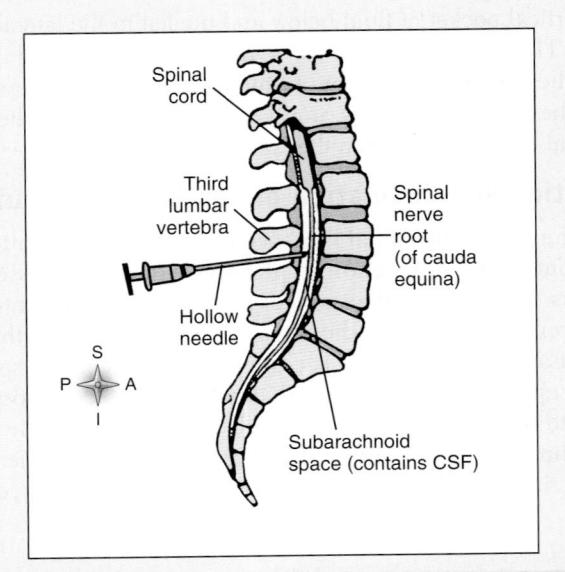

BOX 13-3: HEALTH MATTERS
Hydrocephalus

Occasionally, some condition interferes with the circulation of cerebrospinal fluid. For example, a brain tumor may press against the cerebral aqueduct, shutting off the flow of fluid from the third to the fourth ventricle. In such an event the fluid accumulates within the lateral and third ventricles because it continues to form even though its drainage is blocked. This condition is known as internal **hydrocephalus.** If the fluid accumulates in the subarachnoid space around the brain, external hydrocephalus results. Subarachnoid hemorrhage, for example, may lead to the formation of blood clots that block drainage of the cerebrospinal fluid from the subarachnoid space. With decreased drainage, an increased amount of fluid remains in the space.

When internal hydrocephalus occurs in an infant whose skull has not completely ossified, the increasing fluid pressure in the ventricles causes the cranium to swell (see figure below). This condition can be treated by surgical placement of a shunt, or tube, to drain the excess cerebrospinal fluid (see figure below). When this condition occurs in an older child or adult, the skull will not yield to the increasing fluid pressure. The pressure instead compresses the soft nervous tissue of the brain, potentially leading to coma or even death.

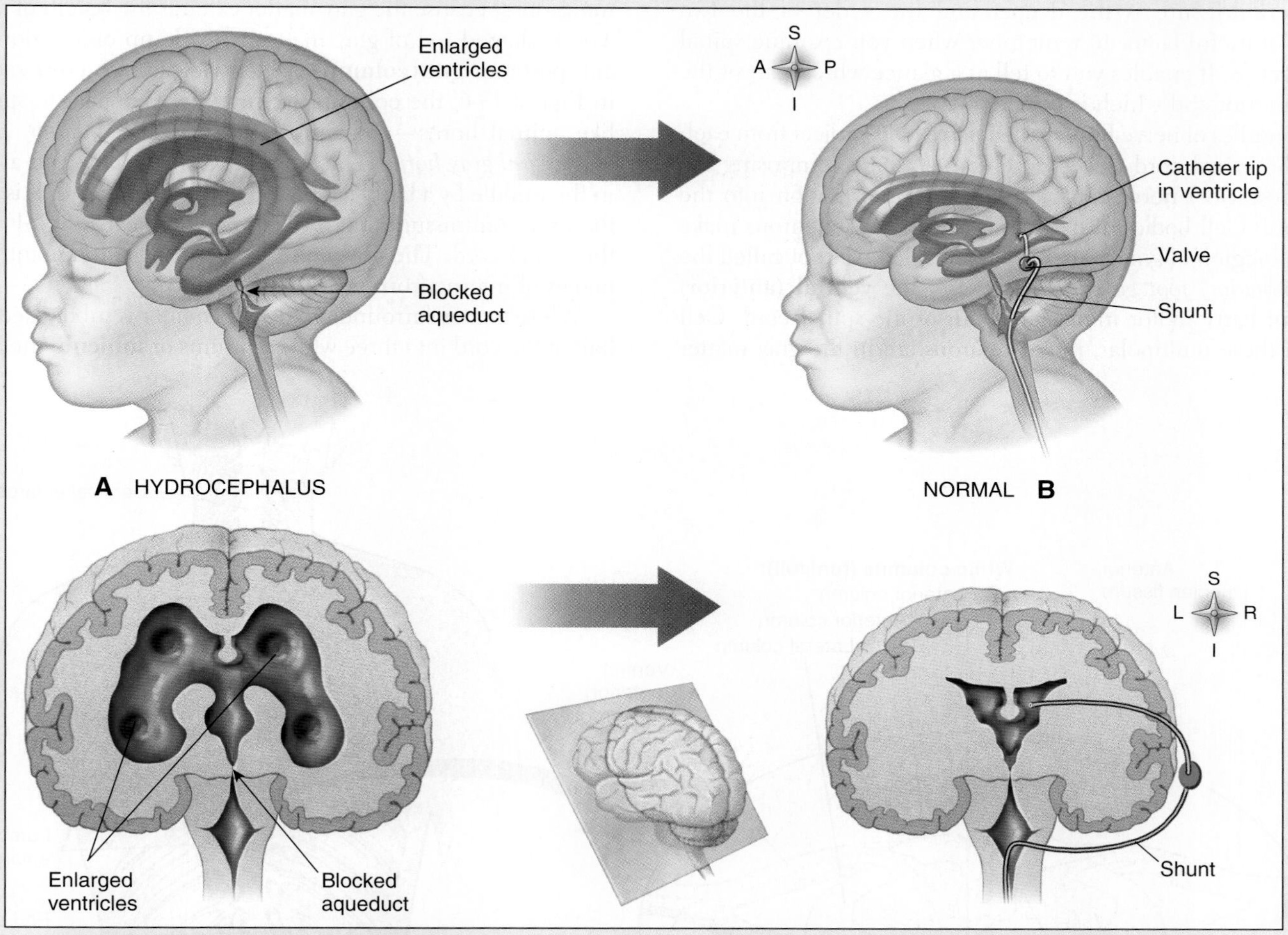

Hydrocephalus. **A,** Ventricles become swollen when the flow of cerebrospinal fluid is blocked. **B,** Surgical placement of a shunt can restore flow of CSF and reduce swelling.

SPINAL CORD
Structure of the Spinal Cord

The spinal cord lies within the spinal cavity, extending from the foramen magnum to the lower border of the first lumbar vertebra (Figure 13-6), a distance of about 45 cm (18 inches) in the average body. The spinal cord does not completely fill the spinal cavity—it also contains the meninges, cerebrospinal fluid, a cushion of adipose tissue, and blood vessels.

The spinal cord is an oval-shaped cylinder that tapers slightly as it descends and has two bulges, one in the cervical region and the other in the lumbar region (see Figure 13-6). Two deep grooves, the *anterior median fissure* and the *posterior median sulcus*, just miss dividing the cord into separate symmetrical halves. The anterior fissure is the deeper and the wider of the two grooves—a useful factor to remember when you examine spinal cord diagrams. It enables you to tell at a glance which part of the cord is anterior and which is posterior.

Two bundles of nerve fibers called *nerve roots* project from each side of the spinal cord (see Figure 13-6). Fibers comprising the **dorsal (posterior) nerve root** carry sensory information into the spinal cord. Cell bodies of these unipolar, sensory neurons make up a small region of gray matter in the dorsal nerve root called the *dorsal (posterior) root ganglion*. Fibers of the **ventral (anterior) nerve root** carry motor information out of the spinal cord. Cell bodies of these multipolar, motor neurons are in the gray matter

that composes the inner core of the spinal cord. Numerous interneurons are also located in the spinal cord's gray matter core. On each side of the spinal cord, the dorsal and ventral nerve roots join together to form a single mixed nerve called, simply, a **spinal nerve.** Spinal nerves, components of the peripheral nervous system, are considered in more detail in the next chapter.

The spinal cord ends at vertebra L1 in a tapered cone called the **conus medullaris.** As you can see in Figure 13-7, many nerve roots extending from the conus medullaris form a sort of "horse tail" of spinal nerve roots called the **cauda equina.** Within the cauda equina the long cordlike *filum terminale* is formed from the spinal meninges.

Although the spinal cord's core of gray matter looks like a flat letter **H** in transverse sections of the cord, it actually has three dimensions, because the gray matter extends the length of the cord. The H-shaped rod of gray matter is made up of anterior, lateral, and posterior **gray columns.** When viewed in a cross section, as in Figure 13-6, the columns forming the **H** appear to spread out like animal horns—and thus are also called *anterior, posterior,* and *lateral gray horns.* The left and right gray columns are joined in the middle by a band called the **gray commissure.** It is through the gray commissure that the central canal carries CSF through the spinal cord. The gray columns consist predominantly of cell bodies of interneurons and motor neurons.

White matter surrounding the gray matter is subdivided in each half of the cord into three white columns or **funiculi:** the anterior,

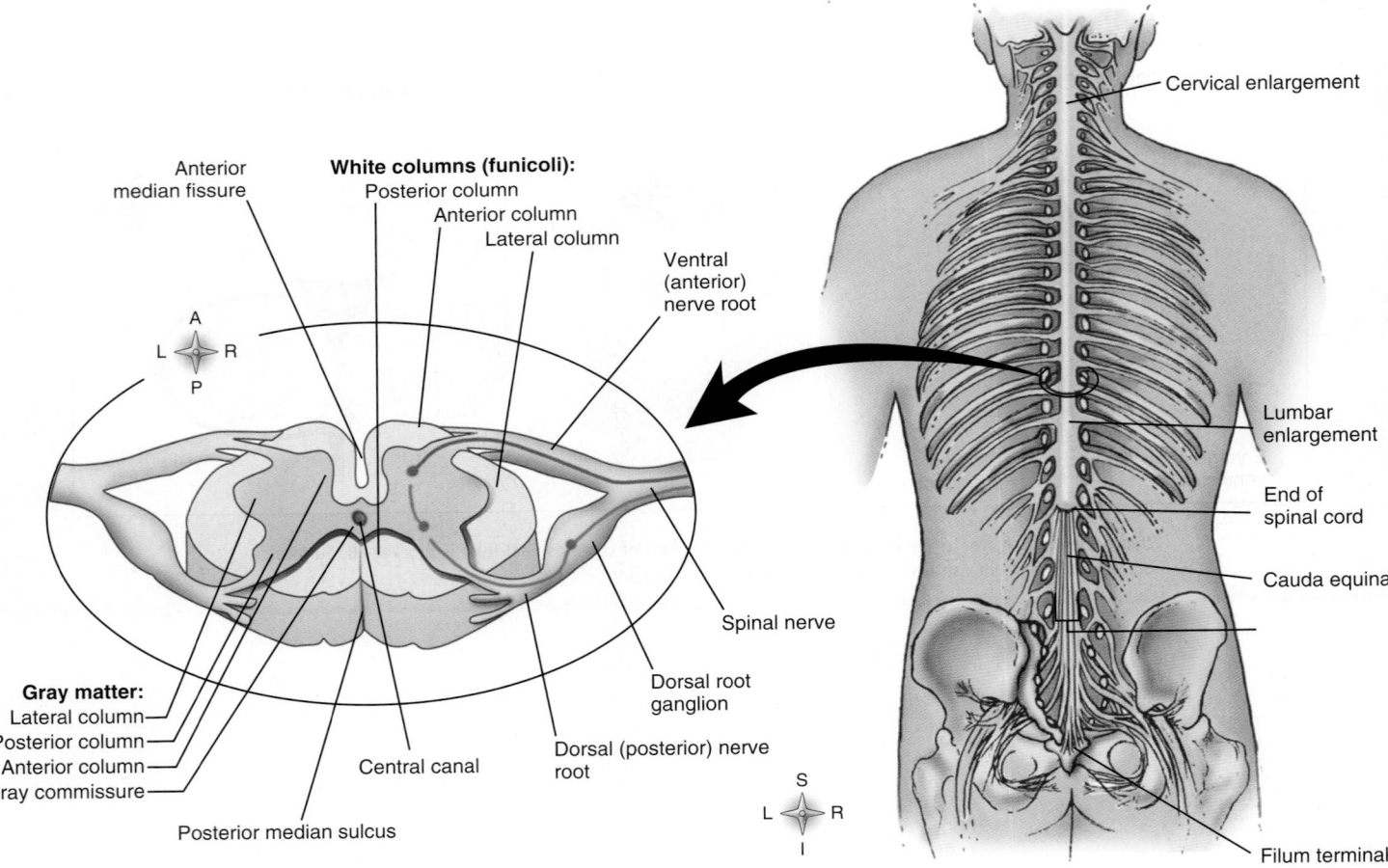

Figure 13-6 *Spinal cord.* The inset illustrates a transverse section of the spinal cord shown in the broader view.

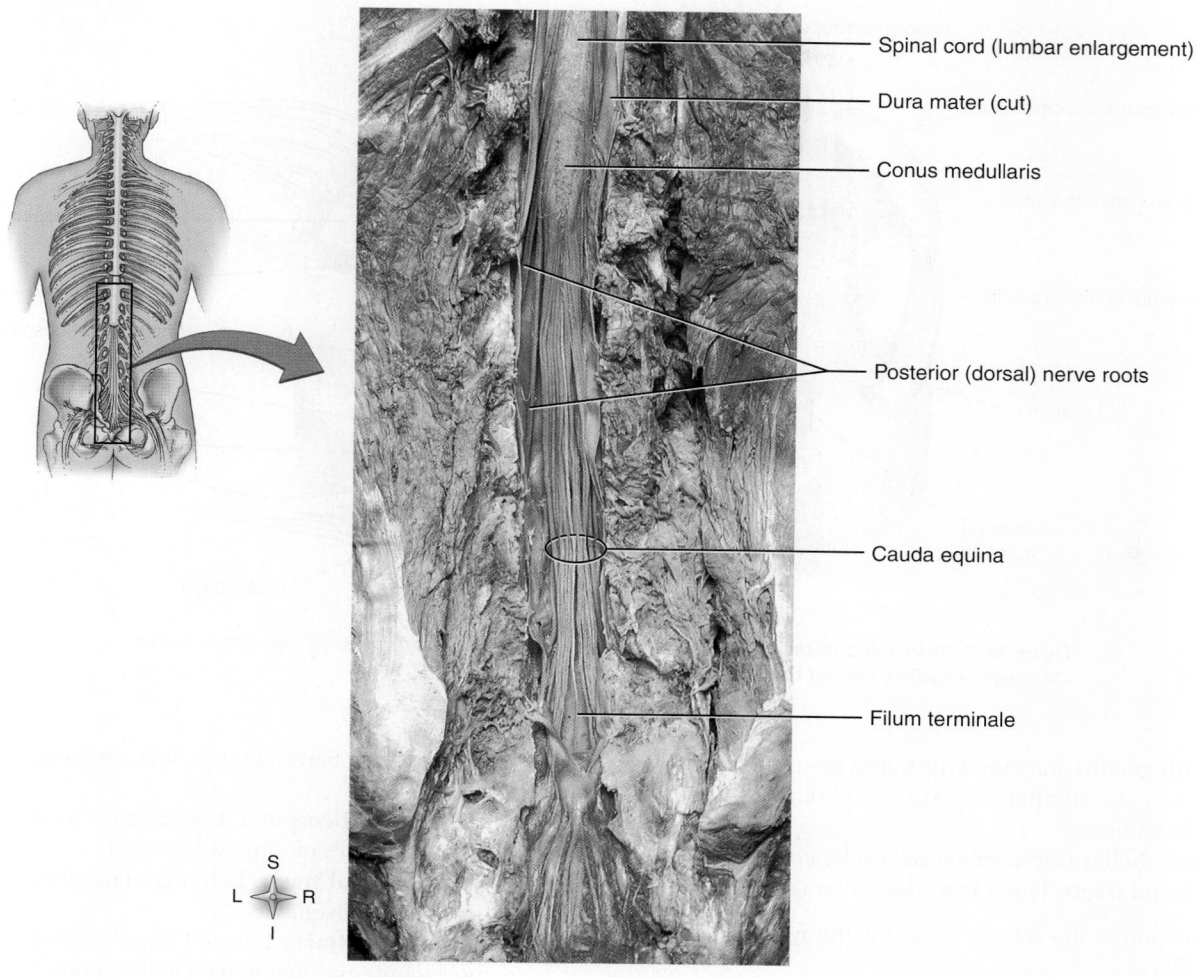

Spinal cord (lumbar enlargement)

Dura mater (cut)

Conus medullaris

Posterior (dorsal) nerve roots

Cauda equina

Filum terminale

Figure 13-7 *Cauda equina.* Photographs shows the inferior portion of the dura mater dissected posteriorly, revealing the cauda equine and nearby structures.

posterior, and lateral white columns. Each white column, or funiculus, consists of a large bundle of nerve fibers (axons) divided into smaller bundles called **spinal tracts**, shown in Figure 13-8. The names of most spinal cord tracts indicate the white column in which the tract is located, the structure in which the axons that make up the tract originate, and the structure in which they terminate. For example, the lateral corticospinal tract is located in the lateral white column of the cord. The axons that compose it originate from neuron cell bodies in the spinal cortex (of the cerebrum) and terminate in the spinal cord. The anterior spinothalamic tract lies in the anterior white column. The axons that compose it originate from neuron cell bodies in the spinal cord and terminate in a portion of the brain called the *thalamus.*

You may wish to refer to the atlas that accompanies this book, where you will find detailed photographs of a human spinal cord. How many structures can you identify in the atlas by sight?

Functions of the Spinal Cord

The spinal cord performs two general functions. Briefly, it provides conduction routes to and from the brain and serves as the integrator, or reflex center, for all spinal reflexes.

Spinal cord tracts provide conduction paths to and from the brain. **Ascending tracts** conduct sensory impulses up the cord to the brain.

Descending tracts conduct motor impulses down the cord from the brain. Bundles of axons compose all tracts. Tracts are both structural and functional organizations of these nerve fibers. They are *structural* organizations: all the axons of any one tract originate from neuron cell bodies located in the same area of the central nervous system, and all the axons terminate in a single structure elsewhere in the central nervous system. For example, all the fibers of the spinothalamic tract are axons originating from neuron cell bodies located in the spinal cord and terminating in the thalamus. Tracts are *functional* organizations: all the axons that compose one tract serve one general function. For instance, fibers of the spinothalamic tracts serve a sensory function. They transmit impulses that produce the sensations of crude touch, pain, and temperature.

Because so many different tracts make up the white columns of the cord, we mention only a few of the more important ones. Locate each tract in Figure 13-8. Consult Tables 13-1 and 13-2 for a brief summary of these tracts.

Five important ascending, or sensory, tracts and their functions, stated very briefly, are as follows:

1. **Lateral spinothalamic tracts:** crude touch, pain, and temperature
2. **Anterior spinothalamic tracts:** crude touch and pressure

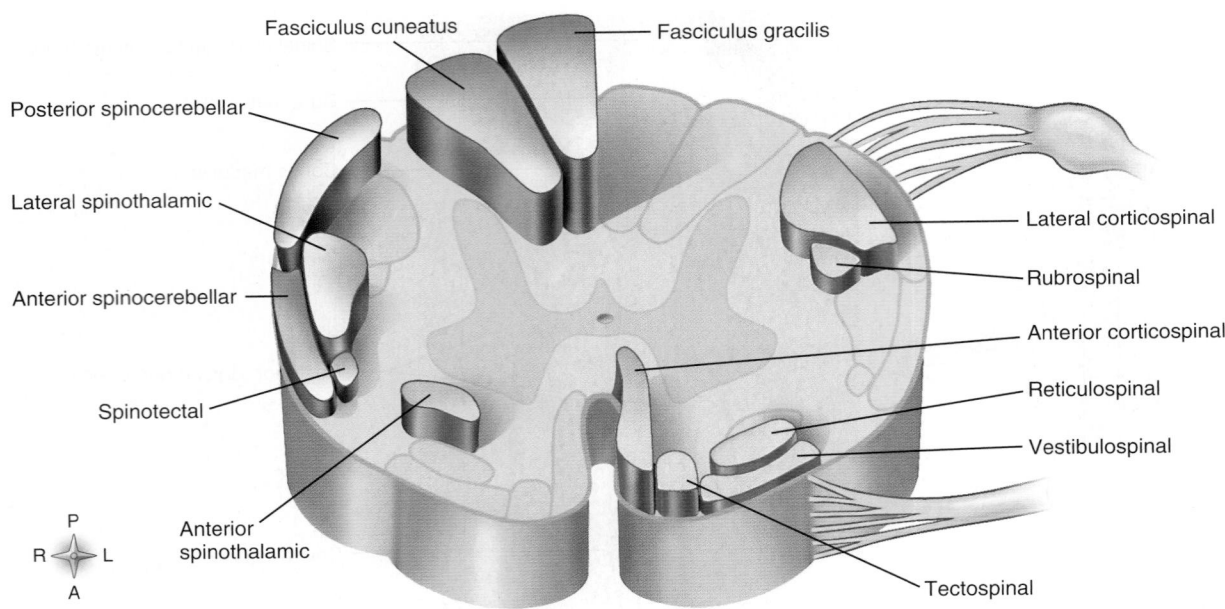

Figure 13-8 *Major tracts of the spinal cord.* The major ascending (sensory) tracts are highlighted in blue. The major descending (motor) tracts are highlighted in red.

3. **Fasciculi gracilis** and **cuneatus tracts:** discriminating touch and conscious sensation of position and movement of body parts (kinesthesia)
4. **Spinocerebellar tracts:** subconscious kinesthesia
5. **Spinotectal tracts:** touch that triggers visual reflexes

Further discussion of the sensory neural pathways may be found on pp. 499-501.

Six important descending, or motor, tracts and their functions in brief are as follows:

1. **Lateral corticospinal tracts:** voluntary movement; contraction of individual or small groups of muscles, particularly those moving hands, fingers, feet, and toes on opposite side of body
2. **Anterior corticospinal tracts:** same as preceding except mainly muscles of same side of body
3. **Reticulospinal tracts:** help maintain posture during skeletal muscle movements
4. **Rubrospinal tracts:** transmit impulses that coordinate body movements and maintenance of posture
5. **Tectospinal tracts:** head and neck movement related to visual reflexes
6. **Vestibulospinal tracts:** coordination of posture and balance

Table 13-1 Major Ascending Tracts of Spinal Cord

NAME	FUNCTION	LOCATION	ORIGIN*	TERMINATION†
Lateral spinothalamic	Pain, temperature, and crude touch on opposite side	Lateral white columns	Posterior gray column on opposite side	Thalamus
Anterior spinothalamic	Crude touch and pressure	Anterior white columns	Posterior gray column on opposite side	Thalamus
Fasciculi gracilis and cuneatus	Discriminating touch and pressure sensations, including vibration, stereognosis, and two-point discrimination; also conscious kinesthesia	Posterior white columns	Spinal ganglia on same side	Medulla
Anterior and posterior spinocerebellar	Unconscious kinesthesia	Lateral white columns	Anterior or posterior gray column	Cerebellum
Spinotectal	Touch related to visual reflexes	Lateral white columns	Posterior gray columns	Superior colliculus (midbrain)

*Location of cell bodies of neurons from which axons of tract arise.
†Structure in which axons of tract terminate.

Table 13-2	Major Descending Tracts of Spinal Cord				
NAME	**FUNCTION**	**LOCATION**	**ORIGIN***	**TERMINATION†**
Lateral corticospinal (or crossed pyramidal)	Voluntary movement, contraction of individual or small groups of muscles, particularly those moving hands, fingers, feet, and toes of opposite side	Lateral white columns	Motor areas or cerebral cortex of opposite side from tract location in cord	Lateral or anterior gray columns
Anterior corticospinal (direct pyramidal)	Same as lateral corticospinal except mainly muscles of same side	Anterior white columns	Motor cortex but on same side as location in cord	Lateral or anterior gray columns
Reticulospinal	Maintain posture during movement	Anterior white columns	Reticular formation (midbrain, pons, medulla)	Anterior gray columns
Rubrospinal	Coordination of body movement and posture	Lateral white columns	Red nucleus (of midbrain)	Anterior gray columns
Tectospinal	Head and neck movement during visual reflexes	Anterior white columns	Superior colliculus (midbrain)	Medulla and anterior gray columns
Vestibulospinal	Coordination of posture/balance	Anterior white columns	Vestibular nucleus (pons, medulla)	Anterior gray columns

*Location of cell bodies of neurons from which axons of tract arise.
†Structure in which axons of tract terminate.

Further discussion of motor neural pathways may be found on pp. 502-504.

The spinal cord also serves as the reflex center for all spinal reflexes. The term *reflex center* means the center of a reflex arc or the place in the arc where incoming sensory impulses become outgoing motor impulses. They are structures that switch impulses from afferent to efferent neurons. In two-neuron arcs, reflex centers are merely synapses between neurons. In all other arcs, reflex centers consist of interneurons interposed between afferent and efferent neurons. Spinal reflex centers are located in the gray matter of the cord. Box 13-4 discusses how reflex centers can act as pain control areas.

BOX 13-4   Pain Control Areas

The term **pain control area** means a place in the pain conduction pathway where impulses from pain receptors can be inhibited. The first pain control area suggested was a segment of the posterior gray horns of the spinal cord. *Substantia gelatinosa* is the name of this segment. Here, axon terminals of neurons that conduct impulses from pain receptors to the spinal cord synapse with neurons that conduct pain impulses up the spinal cord to the brain. Several decades ago, researchers made a surprising discovery about these synapses in the substantia gelatinosa. They found they could inhibit pain conduction across them by stimulating skin touch receptors in a painful area. From this knowledge, a new theory about pain developed—the aptly named *gate-control theory of pain*. According to this theory, the substantia gelatinosa functions as a gate that can close and thus bar the entry of pain impulses into ascending paths to the brain. One way to close the gate is to stimulate skin touch receptors in a painful area. Today this is usually done by a device called the **transcutaneous electrical nerve stimulation (TENS) unit** (see figure). A patient uses the TENS unit to apply a low level of stimulation for a long period. This results in closure of the spinal cord pain gate and relief of pain.

In recent years, pain control areas have been identified in the brain, notably in the gray matter around the cerebral aqueduct and around the third ventricle. Neurons in these areas send their axons down the spinal cord to terminate in the substantia gelatinosa, where they release enkephalins. These chemicals act to prevent pain impulse conduction across synapses in the substantia gelatinosa. In short, enkephalins tend to close the spinal cord pain gate. Brief, intense transcutaneous stimulation at trigger or acupuncture points has been found to relieve pain in distant sites. The intense stimulation is postulated to activate brain pain control areas. They then send impulses down the cord to the substantia gelatinosa, closing the spinal cord pain gate.

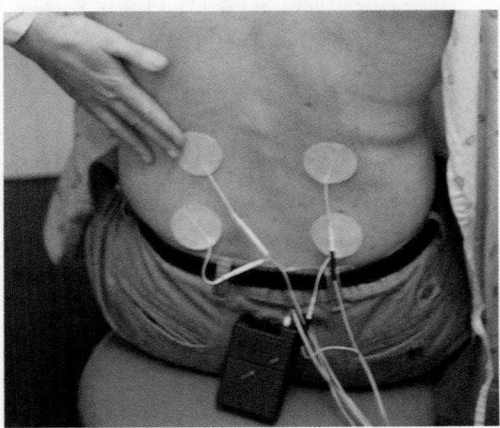

Transcutaneous electrical nerve stimulation (TENS) unit.

QUICK CHECK

3. What are spinal nerve roots? How does the dorsal root differ from the ventral root?
4. Name the regions of the white and gray matter seen in a horizontal section of the spinal cord.
5. Contrast ascending tracts and descending tracts of the spinal cord. Can you give an example of each?

BRAIN

The brain is one of the largest organs in adults (Box 13-5). It consists, in round numbers, of 100 billion neurons and 900 billion glia. In most adults, it weighs about 1.4 kg (3 pounds). Neurons of the brain undergo mitotic cell division only during the prenatal period and the first few months of postnatal life. Although they grow in size after that, they do not increase in number. Malnutrition during the crucial prenatal months of neuron multiplication is reported to hinder the process and result in fewer brain cells. The brain attains full size by about the eighteenth year but grows rapidly only during the first 9 years or so.

Six major divisions of the brain, named from below, upward, are as follows: *medulla oblongata, pons, midbrain, cerebellum, diencephalon,* and *cerebrum.* Very often the medulla oblongata,

BOX 13-5: FYI
Studying Human Brains

As with any part of the body, the best way to learn the anatomy of the brain is to look at actual specimens. Recall from Chapter 1 (Figure 1-1, p. 5) that this is the traditional method of learning human anatomy. When cadavers are not available for this purpose, one useful alternative is using photographs of well-dissected cadavers.

Part *A* of the figure below is an oblique frontal (coronal) section of the human brain, as seen from the front of the subject. This photograph shows details of the internal features of all the major brain divisions. Compare this view of the brain with part *B*, which shows the

brain cut on horizontal (transverse) planes at two slightly different levels, as seen from above. Then compare these photographs with those shown in the small atlas that accompanies this book. How do sections in different planes of the brain benefit your understanding of the three-dimensional aspects of brain anatomy? What benefits are there to studying the brains of human cadavers, rather than relying solely on artists' renderings of the brain? Are there any advantages to using artists' renderings of brain anatomy?

Refer to these photographs—and those in the atlas that accompanies this book—often as you study the details of brain anatomy.

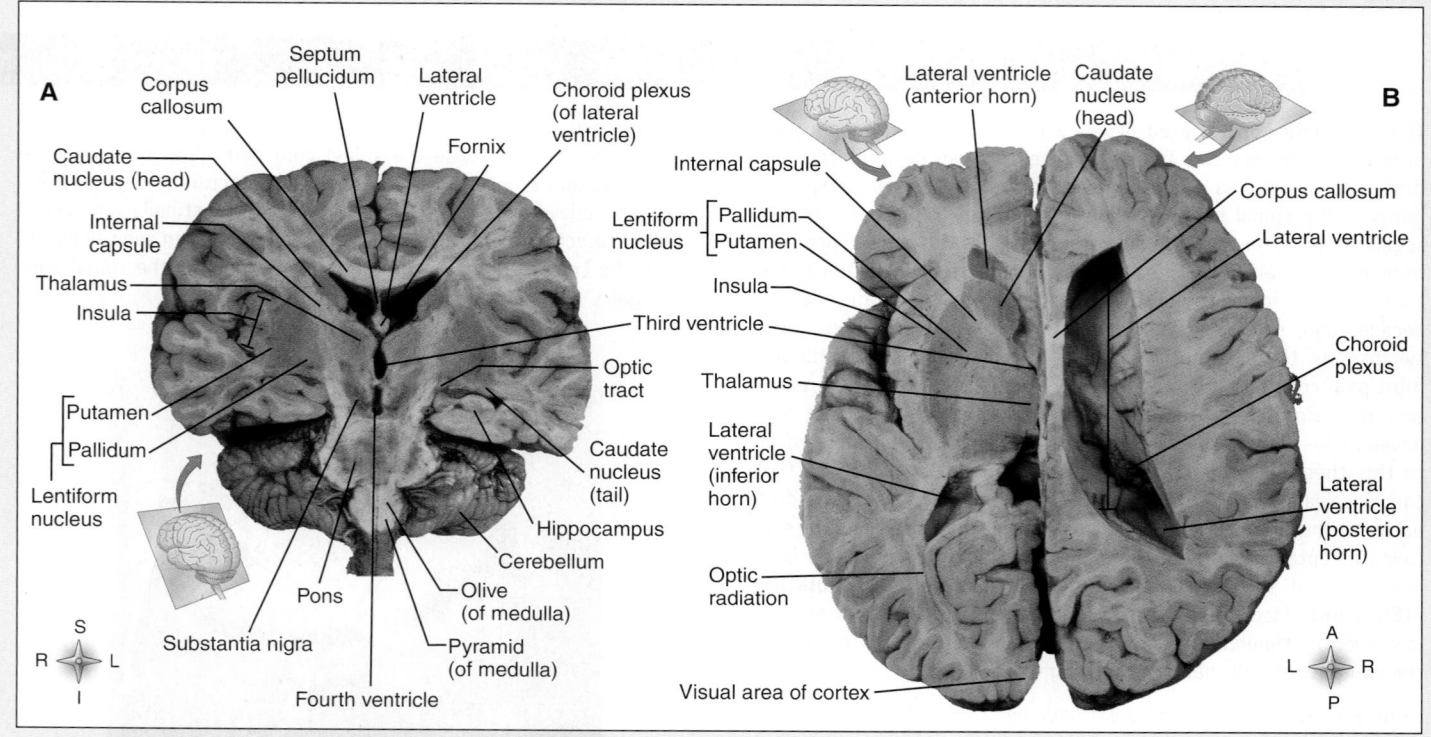

Human brain specimens. **A,** Oblique frontal section. **B,** Horizontal sections (left section is slightly inferior to right section).

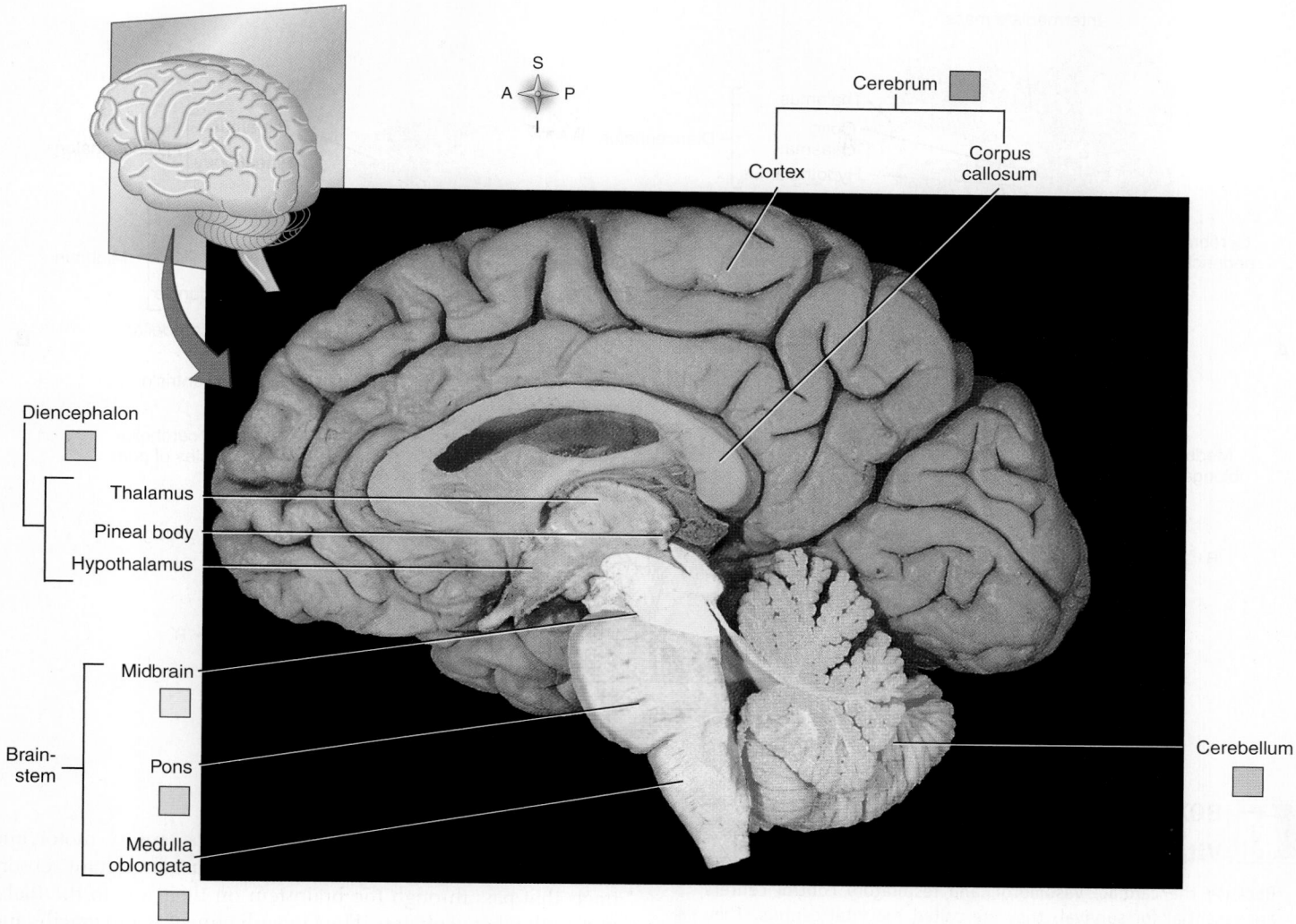

Figure 13-9 *Divisions of the brain.* A midsagittal section of the brain reveals features of its major divisions.

pons, and midbrain are referred to collectively as the *brainstem.* Look at these three structures in Figure 13-9. Do you agree that they seem to form a stem for the rest of the brain?

Structure of the Brainstem

Three divisions of the brain make up the **brainstem.** The **medulla oblongata** forms the lowest part of the brainstem, the **midbrain** forms the uppermost part, and the **pons** lies between them, that is, above the medulla and below the midbrain.

Medulla Oblongata

The medulla oblongata is the part of the brain that attaches to the spinal cord. It is, in fact, an enlarged extension of the spinal cord located just above the foramen magnum. It measures only a few centimeters (about 1 inch) in length and is separated from the pons above by a horizontal groove. It is composed of white matter (projection tracts) and a network of gray and white matter called the **reticular formation** (peek ahead to Figure 13-21).

The **pyramids** (Figure 13-10) are two bulges of white matter located on the ventral surface of the medulla. Fibers of the so-called pyramidal tracts form the pyramids.

The **olive** (see Figure 13-10) of the medulla is an oval projection appearing one on each side of the ventral surface of the medulla, lateral to the pyramids.

Located in the medulla's reticular formation are various *nuclei,* or clusters of neuron cell bodies. Some nuclei are called control centers—for example, the cardiac, respiratory, and vasomotor control centers (Box 13-6).

Pons

Just above the medulla lies the pons, composed, like the medulla, of white matter and reticular formation. Fibers that run transversely across the pons and through the middle cerebellar peduncles into the cerebellum make up the external white matter of the pons and give it its arching, bridgelike appearance.

Midbrain

The midbrain (*mesencephalon*) is appropriately named. It forms the midsection of the brain, because it lies above the pons and below the cerebrum. Both white matter (tracts) and reticular formation compose the midbrain. Extending divergently through it are two ropelike masses of white matter named **cerebral peduncles**

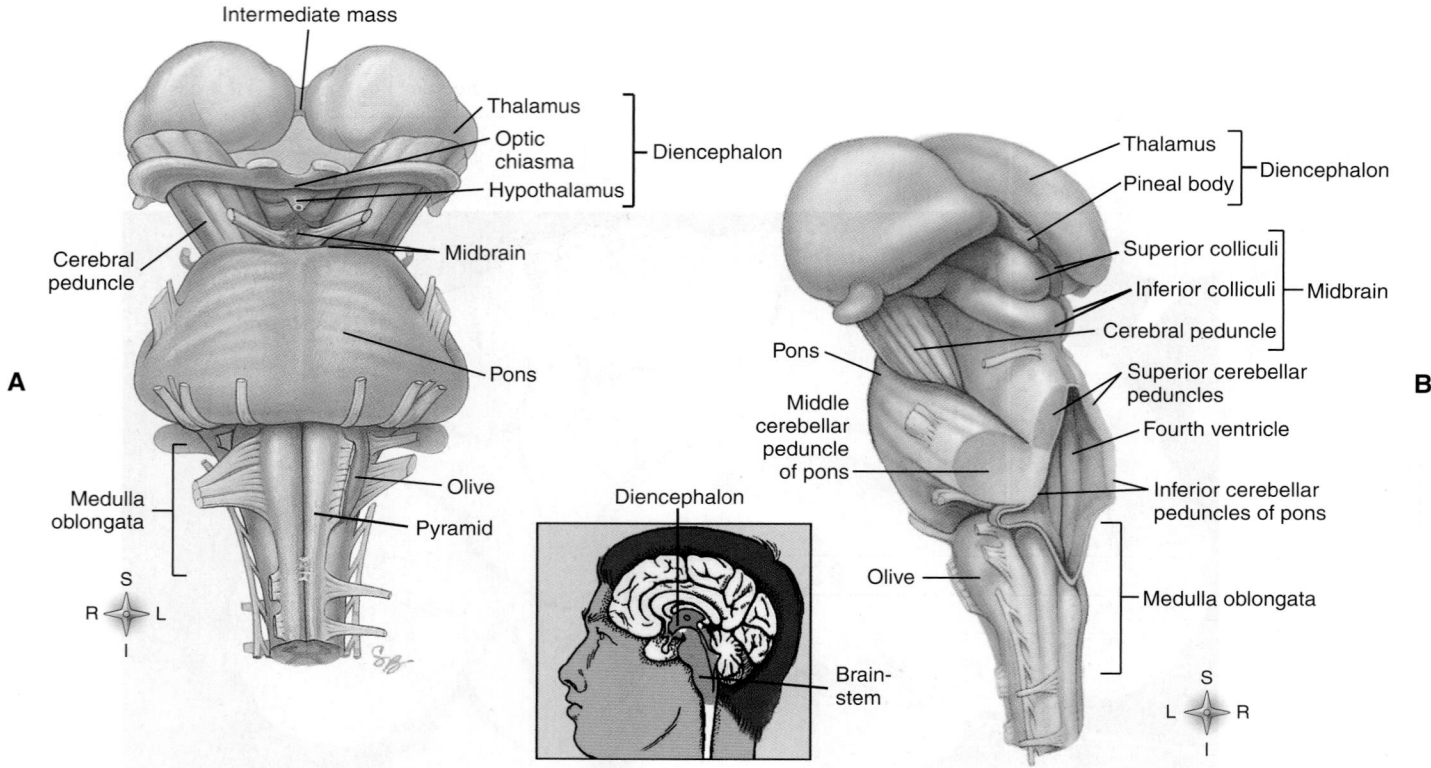

Figure 13-10 *The brainstem and diencephalon.* **A,** Anterior aspect. **B,** Posterior aspect (shifted slightly to lateral).

BOX 13-6: HEALTH MATTERS

Vital Centers

Because the cardiac, vasomotor, and respiratory control centers are essential for survival, they are called the vital centers. They serve as the centers for various reflexes controlling heart action, blood vessel diameter, and respiration. Because the medulla contains these centers, it is the most vital part of the entire brain—so vital that injury or disease of the medulla often proves fatal. Blows at the base of the skull and bulbar poliomyelitis, for example, cause death if they interrupt impulse conduction in the vital respiratory centers.

(see Figure 13-10). Tracts in the peduncles conduct impulses between the midbrain and cerebrum. In addition to the cerebral peduncles, another landmark of the midbrain is the **corpora quadrigemina** (literally, "body of fourfold twins"). The corpora quadrigemina are two **inferior colliculi** and two **superior colliculi**. Note the location of the two sets of twin colliculi, or the corpora quadrigemina, in Figure 13-10, *B.* They form the posterior, upper part of the midbrain, the part that lies just above the cerebellum. Certain auditory centers are located in the inferior colliculus. The superior colliculus contains visual centers. Two other midbrain structures are the *red nucleus* and the *substantia nigra.* Each of these consists of clusters of cell bodies of neurons involved in muscular control. The substantia nigra (literally, "black matter") gets its name from the dark pigment in some of its cells.

Functions of the Brainstem

The brainstem, like the spinal cord, performs sensory, motor, and reflex functions. The spinothalamic tracts are important sensory tracts that pass through the brainstem on their way to the thalamus in the diencephalon. The fasciculi cuneatus and gracilis and the spinoreticular tracts are sensory tracts whose axons terminate in the gray matter of the brainstem. Corticospinal and reticulospinal tracts are two of the major tracts present in the white matter of the brainstem.

Nuclei in the medulla contain a number of reflex centers. Of first importance are the cardiac, vasomotor, and respiratory centers. Other centers present in the medulla are for various nonvital reflexes such as vomiting, coughing, sneezing, hiccupping, and swallowing.

The pons contains centers for reflexes mediated by the fifth, sixth, seventh, and eighth cranial nerves. The locations and functions of these peripheral nerves are discussed in Chapter 14. In addition, the pons contains the pneumotaxic centers that help regulate respiration.

The midbrain, like the pons, contains reflex centers for certain cranial nerve reflexes, for example, pupillary reflexes and eye movements, mediated by the third and fourth cranial nerves, respectively.

Structure of the Cerebellum

The **cerebellum** (literally "little brain") is located just below the posterior portion of the cerebrum and is partially covered by it (Figure 13-11). A transverse fissure separates the cerebellum from the cerebrum. The cerebellum is the second largest part of the brain (after the cerebrum) but has more neurons than all the other

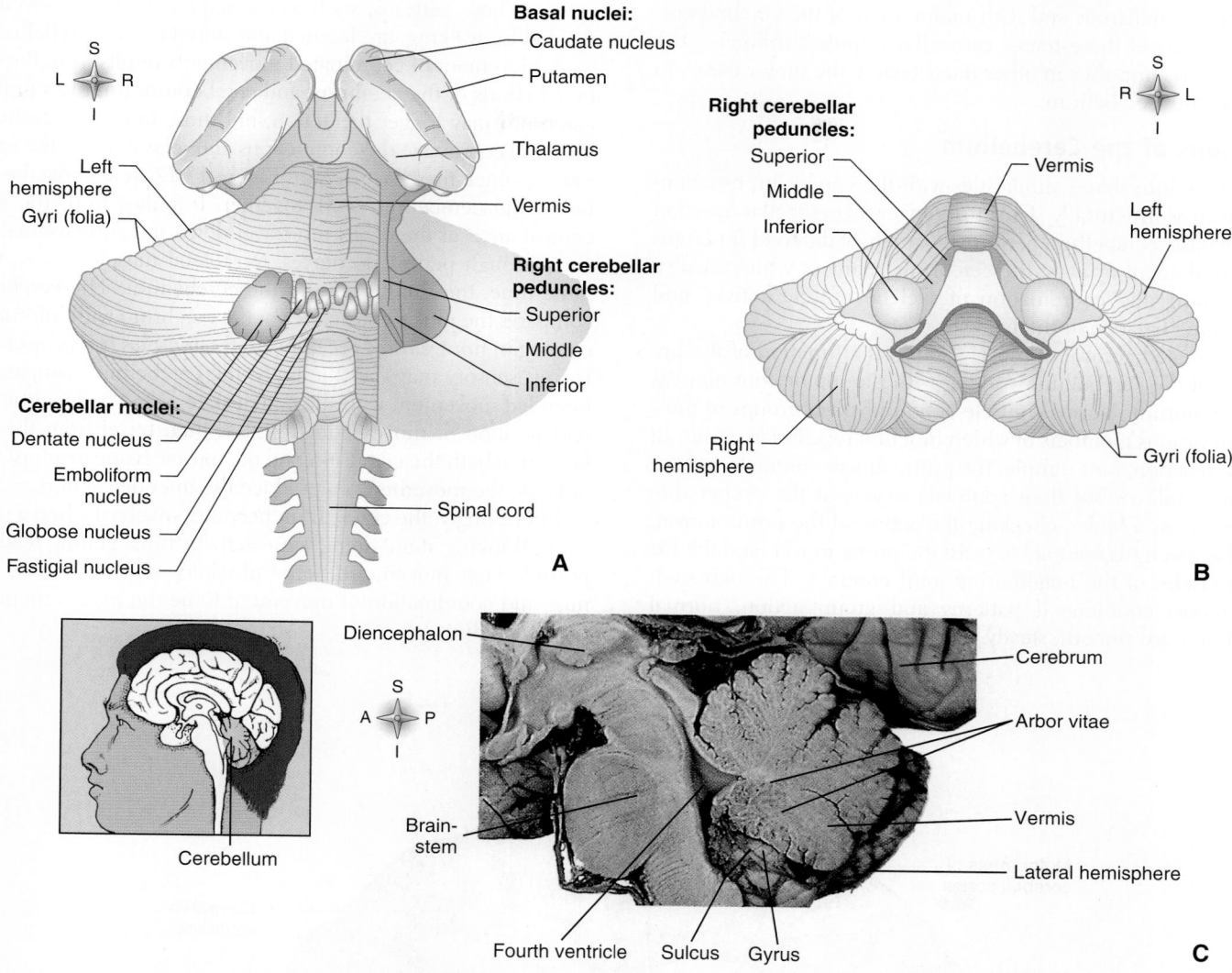

Figure 13-11 *The cerebellum.* A, Posterior view of the surface of the cerebellum. **B,** Anterior view of the cerebellum (with the brainstem removed). **C,** Photograph of midsagittal brain section shows internal features of the cerebellum and surrounding structures of the brain.

parts of the nervous system combined! Thus the cerebellum has a lot of "computing power" compared with other parts of the brain.

The cerebrum and cerebellum have several structural characteristics in common. For instance, gray matter makes up the outer portion, or *cortex*, of each. White matter predominates in the interior of each. Look at Figure 13-11, *C*, and find the internal white matter of the cerebellum called the **arbor vitae** (literally "tree of life"). Note the arbor vitae's distinctive pattern, similar to the branches of a tree. Note, too, that the surfaces of both the cerebellum and the cerebrum have numerous grooves *(sulci)* and raised areas *(gyri)*. The gyri of the cerebellum, however, are much more slender and less prominent than those of the cerebrum. These delicate, roughly parallel gyri are also called **folia** (literally "leaves"). Like the cerebrum, the cerebellum consists of two large lateral masses, the left and right *cerebellar hemispheres*, and a central section called the **vermis.**

The internal white matter of the cerebellum is composed of some short and some long tracts. The shorter tracts conduct impulses from neuron cell bodies located in the cerebellar cortex to

neurons whose dendrites and cell bodies compose nuclei located in the interior of the cerebellum. The longer tracts conduct impulses to and from the cerebellum. Fibers of the longer tracts enter or leave the cerebellum by way of its three pairs of peduncles (see Figure 13-11, *B*), as follows:

1. **Inferior cerebellar peduncles:** composed chiefly of tracts into the cerebellum from the medulla and cord (notably spinocerebellar, vestibulocerebellar, and reticulocerebellar tracts)
2. **Middle cerebellar peduncles:** composed almost entirely of tracts into the cerebellum from the pons, that is, pontocerebellar tracts
3. **Superior cerebellar peduncles:** composed principally of tracts from dentate nuclei in the cerebellum through the red nucleus of the midbrain to the thalamus

An important pair of cerebellar nuclei is the **dentate nuclei,** one of which lies in each hemisphere. Tracts connect these nuclei with the thalamus and with motor areas of the cerebral cortex. By means of these tracts, cerebellar impulses influence the motor cortex. Impulses in other tracts enable the motor cortex to influence the cerebellum.

Functions of the Cerebellum

The cerebellum shares similarities with the cerebrum, functionally as well as structurally. The current view of cerebellar function states that the cerebellum performs a variety of different functions that complement or assist the cerebrum, many of which involve the planning and coordination of skeletal muscle activity and maintaining balance in the body.

Coordinated control of muscle action is a function of the upper part of the cerebellum working with the motor control areas of the cerebrum. Normal muscle action involves groups of muscles, the various members of which function together as a unit. In any given action, for example, the prime mover contracts and the antagonist relaxes but then contracts weakly at the proper moment to act as a brake, checking the action of the prime mover. Also, the synergists contract to assist the prime mover, and the fixation muscles of the neighboring joint contract. Through such harmonious, coordinated patterns and group action, normal movements are smooth, steady, and precise as to force, rate, and

extent. These patterns, such as the sequence of leg movements needed for walking, are learned and stored in the cerebellum.

Achievement of coordinated movements results from the combined efforts of the cerebrum and cerebellum. Impulses from the cerebrum may trigger the action, but those from the cerebellum plan and coordinate the contractions and relaxations of the various muscles once they have begun. Figure 13-12 shows how the cerebrum and cerebellum work together. Impulses from the motor control areas of the cerebrum travel down the corticospinal tract and, through peripheral nerves, to skeletal muscle tissue. At the same time, the impulses go to the cerebellum. The cerebellum compares the motor commands of the cerebrum with information coming in from sensory receptors in the muscles (proprioception). Using "sensory maps" of the body, the cerebellum compares the intended movement with the actual state of the body and its current position or movement. Impulses then travel from the cerebellum to both the cerebrum and the muscle tissue to adjust or coordinate the movements to produce the intended action.

Interestingly, the cerebellum becomes involved when a person is just thinking about doing some activity, thus "getting ready" for possible later movement. Some physiologists consider the planning and coordination of movement to be the main functions of the cerebellum.

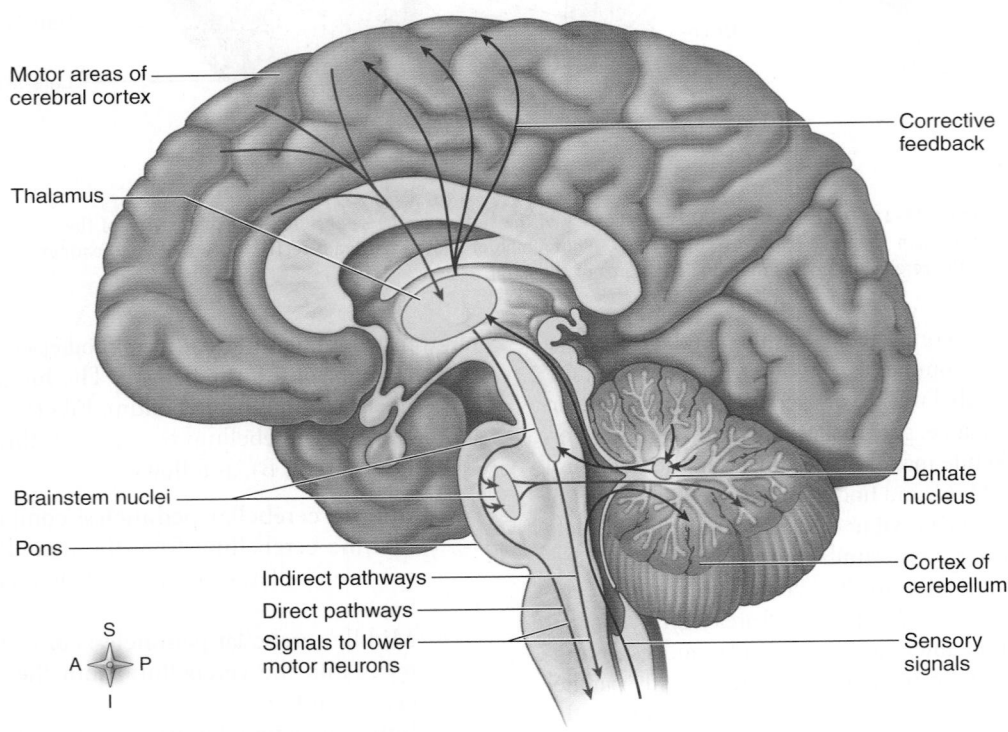

Figure 13-12 *Coordinating function of the cerebellum.* Impulses from the motor control areas of the cerebrum travel down to skeletal muscle tissue and to the cerebellum at the same time. The cerebellum, which also receives and evaluates sensory information, compares the intended movement with the actual movement. It then sends impulses to both the cerebrum and the muscles, thus coordinating and "smoothing" muscle activity.

The cerebellum is also thought to be concerned with both exciting and inhibiting the postural reflexes that help us maintain a stable body position. Sensory impulses from equilibrium (balance) receptors in the ear reach the cerebellum. Using this information, the cerebellum then stimulates or inhibits various muscles to maintain stability of the body.

Not only does the cerebellum work with the cerebrum as a sort of "executive assistant" to coordinate and plan movement and maintain balance—current evidence suggests that the cerebellum is an all-around assistant or planner of a variety of functions normally associated with the cerebrum. In fact, it is becoming clear that the cerebellum coordinates incoming sensory information as much or more than it coordinates outgoing motor information. With the rapid advance of scientific technology, there is no doubt that we will soon discover the full extent of cerebellar function.

Cerebellar disease (e.g., abscess, hemorrhage, tumors, trauma) produces certain characteristic symptoms. Predominant among them are ataxia (muscle incoordination), hypotonia, tremors, and disturbances of gait and balance. One example of ataxia is overshooting a mark or stopping before reaching it when trying to touch a given point on the body (finger-to-nose test). Drawling, scanning, and singsong speech are also examples of ataxia. Tremors are particularly pronounced toward the end of the movements and with the exertion of effort. Disturbances of gait and balance vary, depending on the muscle groups involved. The walk, for instance, is often characterized by staggering or lurching and by a clumsy manner of raising the foot too high and bringing it down with a clap. Paralysis does not result from loss of cerebellar function.

To briefly summarize the general functions of the cerebellum:

1. Acts with the cerebral cortex to produce skilled movements by planning and coordinating the activities of groups of muscles
2. Helps control posture: functions below the level of consciousness to make movements smooth instead of jerky, steady instead of trembling, and efficient and coordinated instead of ineffective, awkward, and uncoordinated
3. Controls skeletal muscles to maintain balance
4. Coordinates incoming sensory information and acts in other ways to complement and assist various functions of the cerebrum

 QUICK CHECK

6. Name the three major divisions of the brainstem, and briefly describe the function of each.
7. What are *gyri* or *folia*? What are *sulci*?
8. How does the cerebellum work with the cerebrum to coordinate muscle activity?

Diencephalon

The **diencephalon** (literally, "between brain") is the part of the brain located between the cerebrum and the midbrain (mesencephalon). Although the diencephalon consists of several structures located around the third ventricle, the main ones are the **thalamus** and **hy-** **pothalamus.** The diencephalon also includes the **optic chiasma**, the **pineal gland**, and several other small but important structures.

Thalamus

The **thalamus** is a dumbbell-shaped mass of gray matter made up of many nuclei. As Figures 13-10 and 13-13 show, each *lateral mass* of the thalamus forms one lateral wall of the third ventricle. Extending through the third ventricle, and thus joining the two lateral masses of the thalamus, is the *intermediate mass*. Two important groups of nuclei that make up the thalamus are the *geniculate bodies*, located in the posterior region of each lateral mass. The geniculate bodies play a role in processing auditory and visual input.

Large numbers of axons conduct impulses into the thalamus from the spinal cord, brainstem, cerebellum, basal nuclei, and various parts of the cerebrum. These axons terminate in thalamic nuclei, where they synapse with neurons whose axons conduct impulses out of the thalamus to virtually all areas of the cerebral cortex. Thus the thalamus serves as the major relay station for sensory impulses on their way to the cerebral cortex.

The thalamus performs the following primary functions:

1. Plays two parts in the mechanism responsible for sensations
   a. Impulses from appropriate receptors, on reaching the thalamus, produce conscious recognition of the crude, less critical sensations of pain, temperature, and touch
   b. Neurons whose dendrites and cell bodies lie in certain nuclei of the thalamus relay all kinds of sensory impulses, except possibly olfactory, to the cerebrum
2. Plays a part in the mechanism responsible for emotions by associating sensory impulses with feelings of pleasantness and unpleasantness
3. Plays a part in the arousal or alerting mechanism
4. Plays a part in mechanisms that produce complex reflex movements

Hypothalamus

The **hypothalamus** consists of several structures that lie beneath the thalamus and form the floor of the third ventricle and the lower part of its lateral walls. Prominent among the structures composing the hypothalamus are the *supraoptic nuclei*, the *paraventricular nuclei*, and the *mamillary bodies*. The supraoptic nuclei consist of gray matter located just above and on either side of the *optic chiasma*. The optic chiasma is the **X**-shaped junction of the optic tracts and optic nerves. The paraventricular nuclei of the hypothalamus are named for their location close to the wall of the third ventricle. The midportion of the hypothalamus gives rise to the **infundibulum**, the stalk leading to the posterior lobe of the *pituitary gland (neurohypophysis)*. The posterior part of the hypothalamus consists mainly of the mamillary bodies (see Figure 13-13, *inset*), which are involved with the olfactory sense (smell).

The hypothalamus is a small but functionally important area of the brain. It weighs little more than 7 g ($\frac{1}{4}$ oz), yet it performs many functions of the greatest importance both for survival and for the enjoyment of life. For instance, it functions as a link between the psyche (mind) and the soma (body). It also links the nervous system to the endocrine system. Certain areas of the hypothalamus function as pleasure centers or reward centers for the

Figure 13-13 Diencephalon. This midsagittal section highlights the largest regions of the diencephalon, the thalamus and hypothalamus, but also shows the smaller optic chiasma and pineal gland. Note the position of the diencephalon between the midbrain and the cerebrum. Compare this view of the diencephalon with that in Figure 13-10.

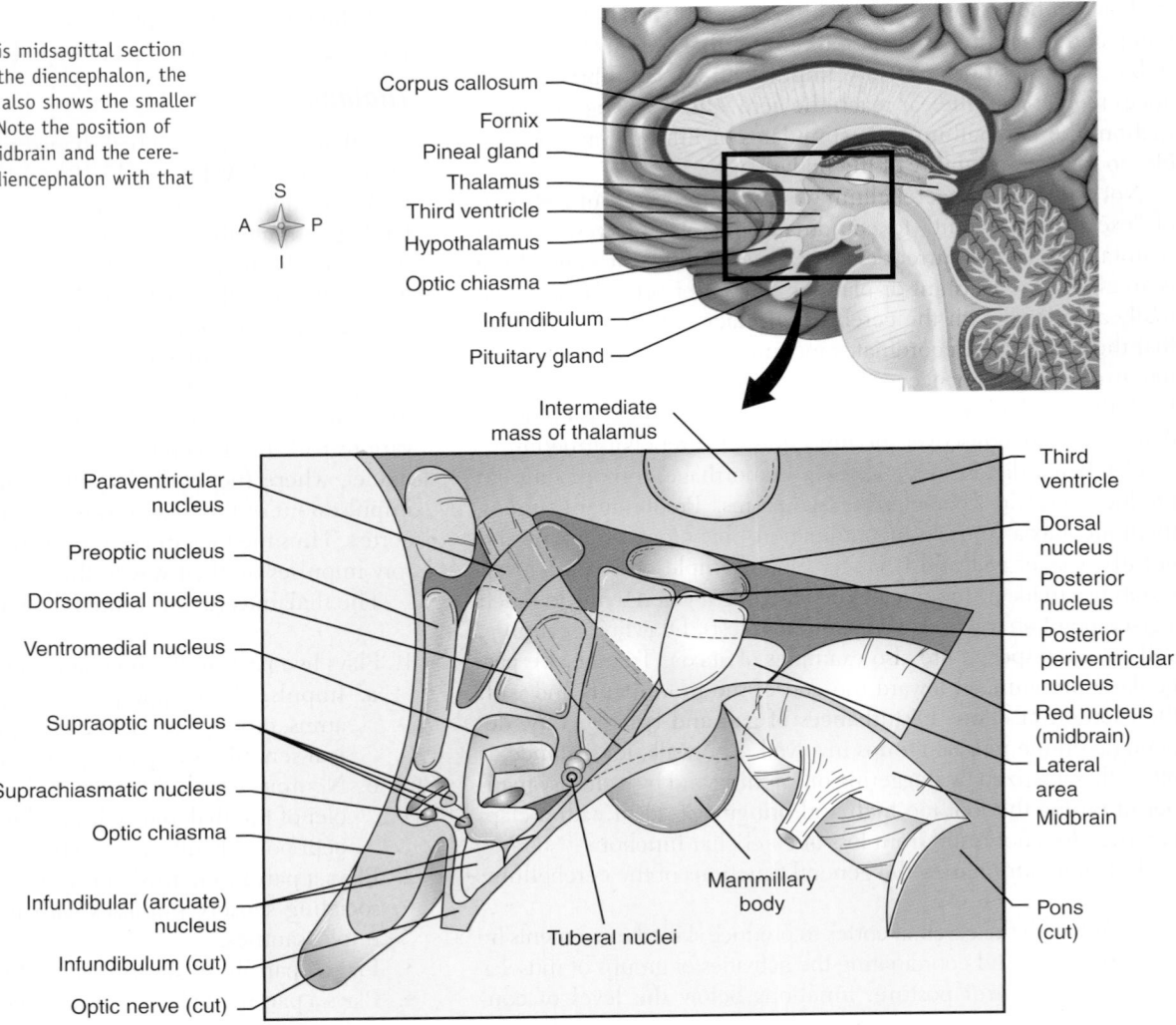

primary drives such as eating, drinking, and sex. The following list briefly summarizes hypothalamic functions.

1. The hypothalamus functions as a higher autonomic center or, rather, as several higher autonomic centers. By this we mean that axons of neurons whose dendrites and cell bodies lie in nuclei of the hypothalamus extend in tracts from the hypothalamus to both parasympathetic and sympathetic centers in the brainstem and cord. Thus impulses from the hypothalamus can simultaneously or successively stimulate or inhibit few or many lower autonomic centers. In other words, the hypothalamus serves as a regulator and coordinator of autonomic activities. It helps control and integrate the responses made by autonomic (visceral) effectors all over the body.

2. The hypothalamus functions as the major relay station between the cerebral cortex and lower autonomic centers. Tracts conduct impulses from various centers in the cortex to the hypothalamus. Then, by way of numerous synapses in the hypothalamus, these impulses are relayed to other tracts that conduct them on down to autonomic centers in the brainstem and cord and also to spinal cord somatic centers (anterior horn motor neurons). Thus the hypothalamus functions as the link between the cerebral cortex and lower centers—hence between the psyche and the soma. It provides a crucial part of the route by which emotions can express themselves in changed bodily functions. It is the all-important relay station in the neural pathways that makes possible the mind's influence over the body—sometimes, unfortunately, even to the profound degree of producing psychosomatic disease. The positive benefits of this mind-body link are the dramatic influences our conscious mind can have in healing the body of various illnesses.

3. Neurons in the supraoptic and paraventricular nuclei of the hypothalamus synthesize the hormones released by the posterior pituitary gland (neurohypophysis). Because one of these hormones affects the volume of urine excreted, the hypothalamus plays an indirect but essential role in maintaining water balance (see Chapters 28 and 29).

4. Some neurons in the hypothalamus have endocrine functions. Their axons secrete chemicals, *releasing hormones*, into blood, which circulate to the anterior pituitary gland. Releasing hormones control the release of certain anterior pituitary hormones—specifically growth hormone and hormones that control hormone secretion by sex glands, thyroid

gland, and the adrenal cortex (discussed in Chapter 16). Thus indirectly the hypothalamus helps control the functioning of every cell in the body.

5. The hypothalamus plays an essential role in maintaining the waking state. Presumably it functions as part of an arousal or alerting mechanism. Clinical evidence of this is that somnolence (sleepiness) characterizes some hypothalamic disorders.

6. The hypothalamus functions as a crucial part of the mechanism for regulating appetite and therefore the amount of food intake. Experimental and clinical findings indicate the presence of an "appetite center" in the lateral part of the hypothalamus and a "satiety center" located medially. For example, an animal with an experimental lesion in the ventromedial nucleus of the hypothalamus consumes tremendous amounts of food. Similarly, a human with a tumor in this region of the hypothalamus may eat insatiably and gain an enormous amount of weight.

7. The hypothalamus functions as a crucial part of the mechanism for maintaining normal body temperature. Hypothalamus neurons whose fibers connect with autonomic centers for vasoconstriction, dilation, and sweating and with somatic centers for shivering constitute heat-regulating centers. Marked elevation of body temperature frequently characterizes injuries or other abnormalities of the hypothalamus.

Pineal Gland

Although the thalamus and hypothalamus account for most of the tissue that makes up the diencephalon, there are several smaller structures of importance. For example, the *optic chiasma* is a region where the right and left *optic nerves* cross each other before entering the brain—exchanging fibers as they do so. The resulting bundles of fibers are called the *optic tracts*. Various small nuclei just outside the

thalamus and hypothalamus, collectively referred to as the **epithalamus**, are also included among the structures of the diencephalon. One of the most intriguing of the epithalamic structures is the *pineal gland* or *pineal body* (formerly known as the *epiphysis*).

As Figures 13-10 and 13-13 show, the pineal gland is located just above the corpora quadrigemina of the midbrain. Its name comes from the fact that it resembles a pine nut.

The functions of the pineal gland are still not completely understood. However, we do know that the tiny pineal gland is an important part of the body's biological clock mechanism. The body's **biological clock** depends partly on the pineal gland varying its secretion of the hormone **melatonin**. Melatonin is a *hormone* because it is a molecule released into the blood to regulate functions elsewhere in the body—a concept we will explore further in Chapter 16. However, melatonin is in fact simply an altered form of the neurotransmitter serotonin.

As Figure 13-14 shows, changing light levels throughout the day and night (circadian) cycle trigger changes in the rate of melatonin secretion. When sunlight levels are high, melatonin secretion decreases (Figure 13-15). When light levels are low, melatonin levels increase proportionally. The changing levels of blood melatonin during the day synchronize various body functions with each other and with external stimuli. This kind of biological clock mechanism is called a daily or *circadian* clock. Melatonin is often called the "timekeeping hormone"—but because high blood levels of melatonin signal the body that it is time to sleep, it is also called the "sleep hormone."

When a person travels to another time zone, or when the seasons change, the altered sunlight patterns cause corresponding time shifts to the melatonin cycle. Likewise, often subtle changes occur in each melatonin cycle depending on how much moonlight (reflected sunlight) is present each evening. Thus the body

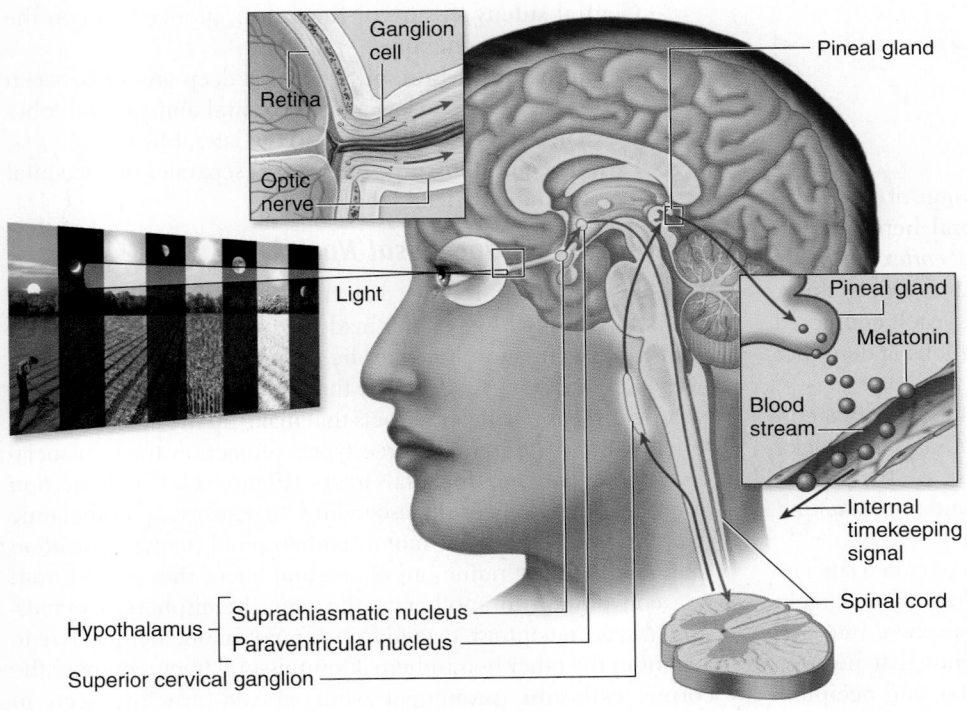

Figure 13-14 *Role of pineal gland in timekeeping.* Changing external levels of light are detected by special receptors in the retina of the eye, and the information is relayed to the suprachiasmatic nucleus (SCN) of the hypothalamus. When light levels decrease, signals from the SCN increase, triggering the paraventricular nucleus and a pathway of nervous system signals that eventually result in the release of increased amounts of melatonin from the pineal gland. The changing levels of melatonin throughout the day serve as an internal timekeeping signal.

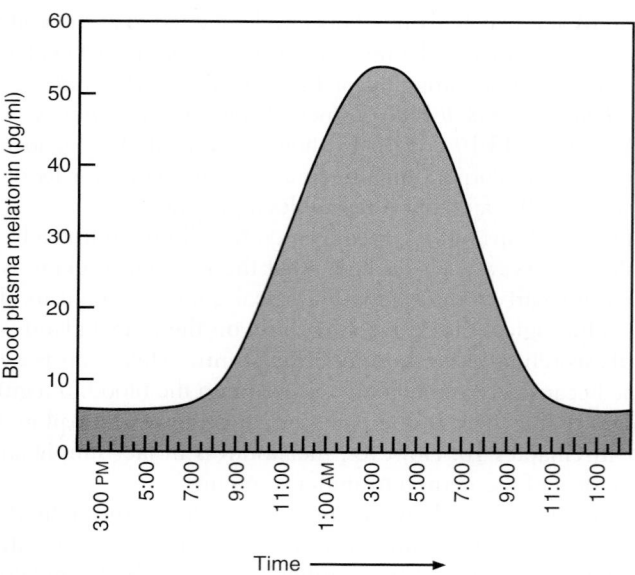

Figure 13-15 *Melatonin.* Graph comparing typical blood melatonin levels throughout the day. Sunlight suppresses melatonin secretion during the day. As the sun goes down, however, melatonin levels begin to rise—dropping again sharply when the sun comes up.

can sometimes tell what time of the (lunar) month it is—a mechanism that may help regulate the female reproductive cycle.

 QUICK CHECK

9. What are the two main components of the diencephalon? Where are they located?
10. Name three general functions of the thalamus.
11. Name three general functions of the hypothalamus.
12. What is the pineal gland's primary function?

Structure of the Cerebrum

Cerebral Cortex

The **cerebrum**, the largest and uppermost division of the brain, consists of two halves, the right and left **cerebral hemispheres.** The surface of the cerebrum—called the *cerebral cortex*—is made up of gray matter only 2 to 4 mm (roughly $\frac{1}{12}$ to $\frac{1}{6}$ inch) thick. But despite its thinness, the cortex has six layers, each composed of millions of axon terminals synapsing with millions of dendrites and cell bodies of other neurons.

If one uses a little imagination, the surface of the cerebral cortex looks like a group of small sausages. Each "sausage" is actually a **convolution**, or gyrus. Names of some of these are the *precentral gyrus, postcentral gyrus, cingulate gyrus,* and *hippocampal gyrus (hippocampus).*

Between adjacent gyri lie either shallow grooves called sulci or deeper grooves called *fissures.* Fissures, as well as a few, largely imaginary boundaries, divide each cerebral hemisphere into five *lobes.* Four of the lobes are named for the bones that lie over them: **frontal lobe, parietal lobe, temporal lobe,** and **occipital lobe** (Figure 13-16). A fifth lobe, the **insula** *(island of Reil),* lies

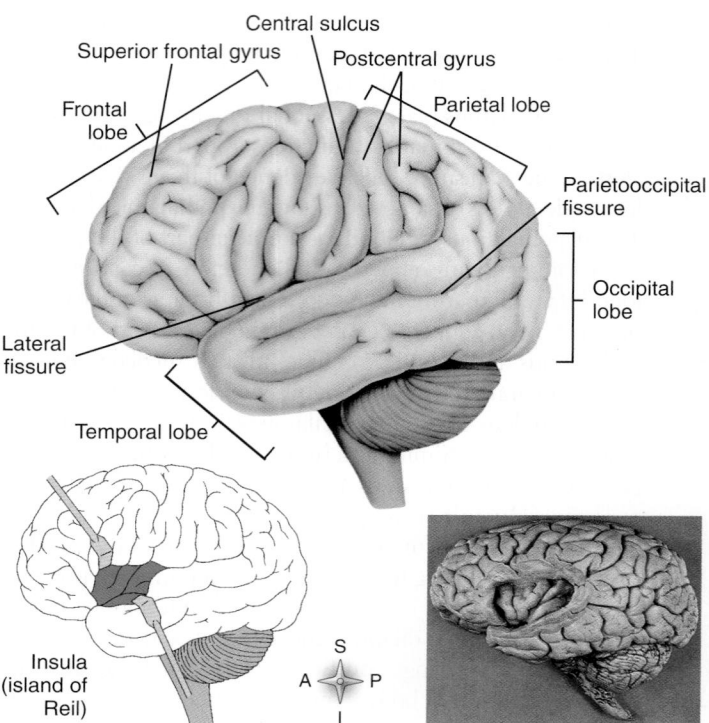

Figure 13-16 *Left hemisphere of cerebrum, lateral surface.* Note the highlighted lobes of the cerebrum.

hidden from view in the lateral fissure. The lobes are highlighted in Figure 13-16. The insula can also be seen in the photographs on p. 482. Names and locations of prominent cerebral fissures are the following (see Figure 13-16):

1. **Longitudinal fissure:** the deepest groove in the cerebrum; divides the cerebrum into two hemispheres
2. **Central sulcus (fissure of Rolando):** groove between the frontal and parietal lobes
3. **Lateral fissure (fissure of Sylvius):** a deep groove between the temporal lobe below and the frontal and parietal lobes above; island of Reil lies deep in the lateral fissure
4. **Parietooccipital fissure:** groove that separates the occipital lobe from the parietal lobe

Cerebral Tracts and Basal Nuclei

Beneath the cerebral cortex lies the large interior of the cerebrum. It is mostly white matter made up of numerous tracts. A few islands of gray matter, however, lie deep inside the white matter of each hemisphere. Collectively these are called **basal nuclei** (or historically, *basal ganglia*). Tracts that make up the cerebrum's internal white matter are of three types: projection tracts, association tracts, and commissural tracts (Figure 13-17). *Projection tracts* are extensions of the ascending, or sensory, spinothalamic tracts and descending, or motor, corticospinal tracts. *Association tracts* are the most numerous of cerebral tracts; they extend from one convolution to another in the same hemisphere. *Commissural tracts,* in contrast, extend from a point in one hemisphere to a point in the other hemisphere. Commissural tracts compose the **corpus callosum** (prominent white curved structure seen in Figure 13-9) and the anterior and posterior commissures.

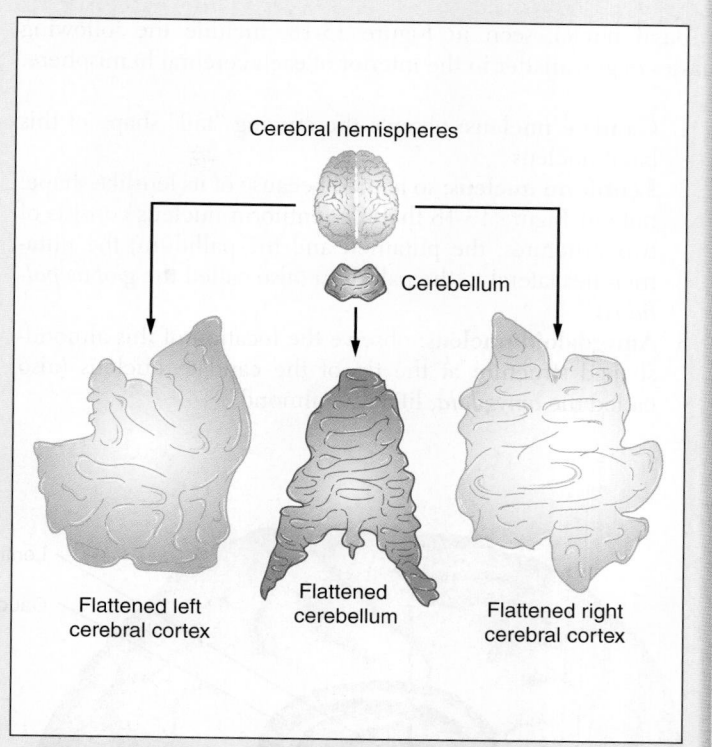
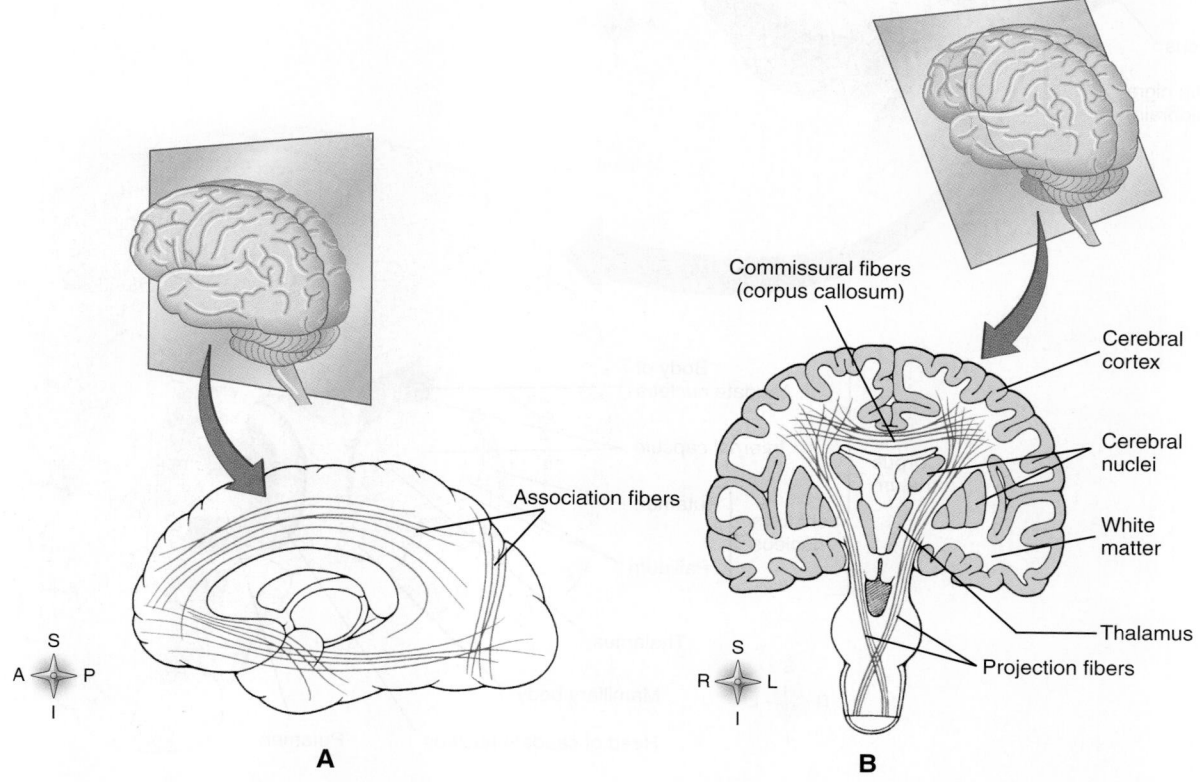

Figure 13-17 *Cerebral tracts.* A, Lateral perspective, showing various association fibers. **B,** Frontal (coronal) perspective, showing commissural fibers that make up the corpus callosum and the projection fibers that communicate with lower regions of the nervous system.

Basal nuclei, seen in Figure 13-18, include the following masses of gray matter in the interior of each cerebral hemisphere:

1. **Caudate nucleus:** observe the curving "tail" shape of this basal nucleus
2. **Lentiform nucleus:** so named because of its lenslike shape; note in Figure 13-18 that the lentiform nucleus consists of two structures, the putamen and the pallidum; the putamen lies lateral to the pallidum (also called the *globus pallidus*)
3. **Amygdaloid nucleus:** observe the location of this almond-shaped structure at the tip of the caudate nucleus (also called the *amygdala,* literally "almond")

A structure associated with the basal nuclei is the **internal capsule.** It is a large mass of white matter located, as Figure 13-18 shows, between the caudate and lentiform nuclei and between the lentiform nucleus and thalamus. The caudate nucleus, internal capsule, and lentiform nucleus constitute the corpus striatum. The term means "striped body."

Researchers are still investigating the exact functions of the basal nuclei, but we already know that this part of the cerebrum plays an important role in regulating voluntary motor functions. For example, most of the muscle contractions involved in maintaining posture, walking, and performing other gross or repetitive movements seem to be initiated or modulated in the basal nuclei (Box 13-8).

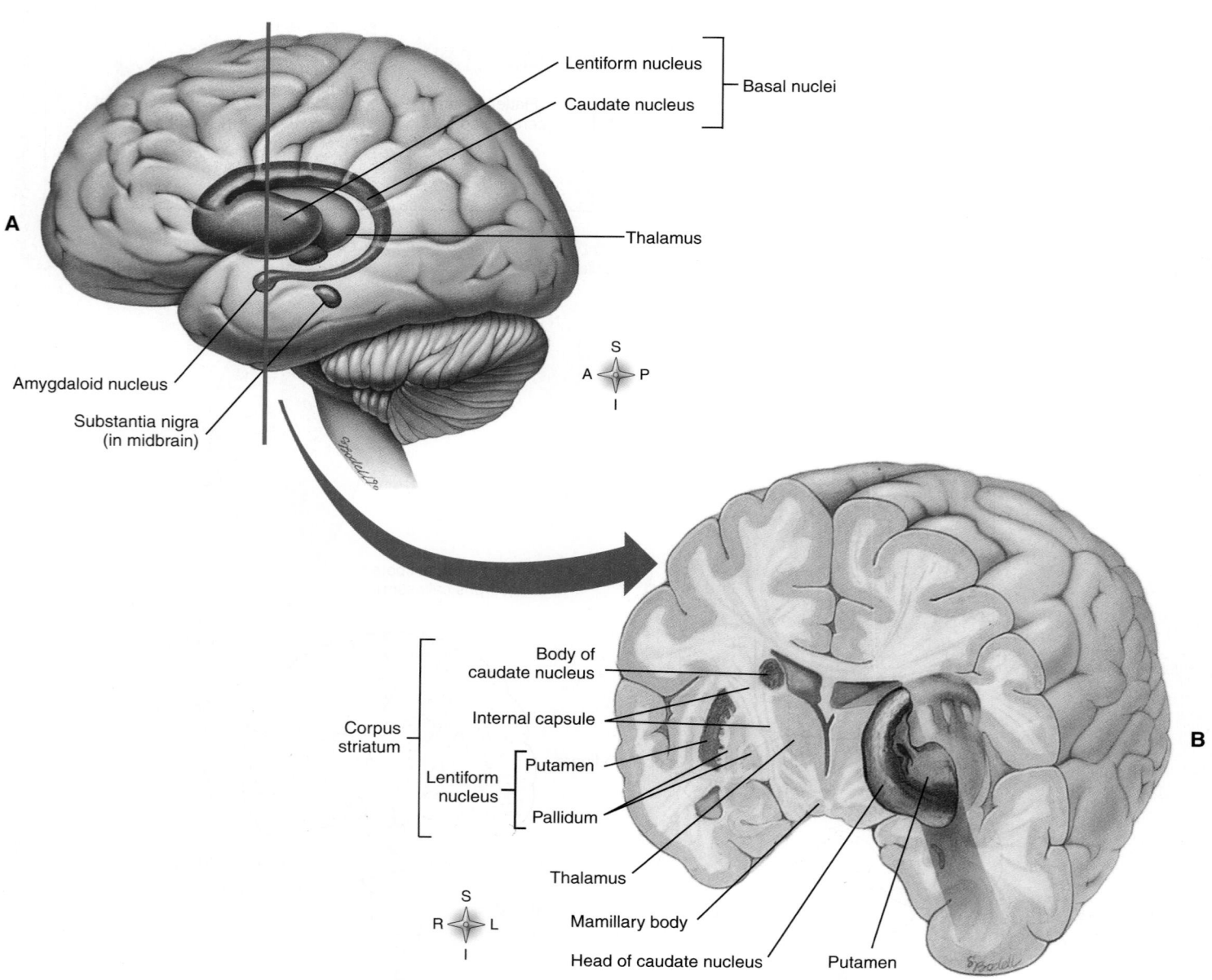

Figure 13-18 *Basal nuclei.* **A,** The basal nuclei seen through the cortex of the left cerebral hemisphere. **B,** The basal nuclei seen in a frontal (coronal) section of the brain.

BOX 13-8: HEALTH MATTERS
Parkinson Disease

The importance of the basal nuclei in regulating voluntary motor functions is made clear in cases of **Parkinson disease (PD).** Normally, neurons that lead from the substantia nigra to the basal nuclei secrete dopamine. Dopamine inhibits the excitatory effects of acetylcholine produced by other neurons in the basal nuclei. Such inhibition by *dopaminergic* (dopamine-producing) neurons produces a balanced, restrained output of muscle-regulating signals from the basal nuclei. In PD, however, neurons leading from the substantia nigra degenerate and thus do not release normal amounts of dopamine. Without dopamine, the excitatory effects of acetylcholine are not restrained, and the basal nuclei produce an excess of signals that affect voluntary muscles in several areas of the body. Overstimulation of postural muscles in the neck, trunk, and upper limbs produces the syndrome of effects that typify this disease: rigidity and tremors of the head and limbs; an abnormal, shuffling gait; absence of relaxed arm-swinging while walking; and a forward tilting of the trunk.

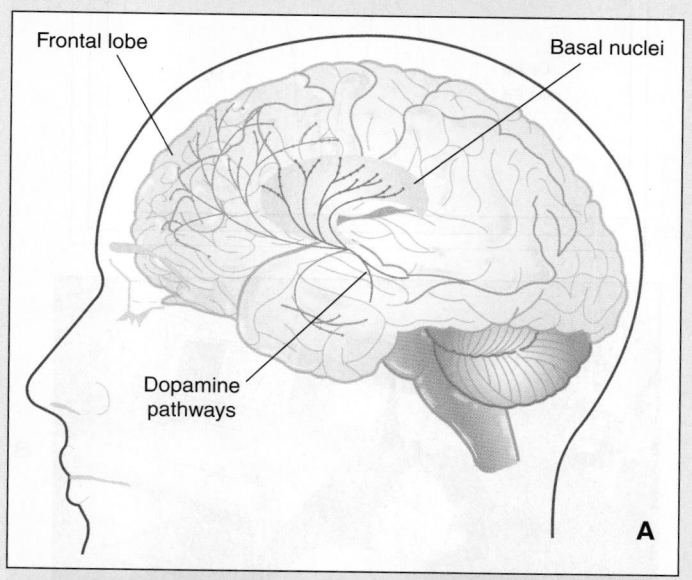
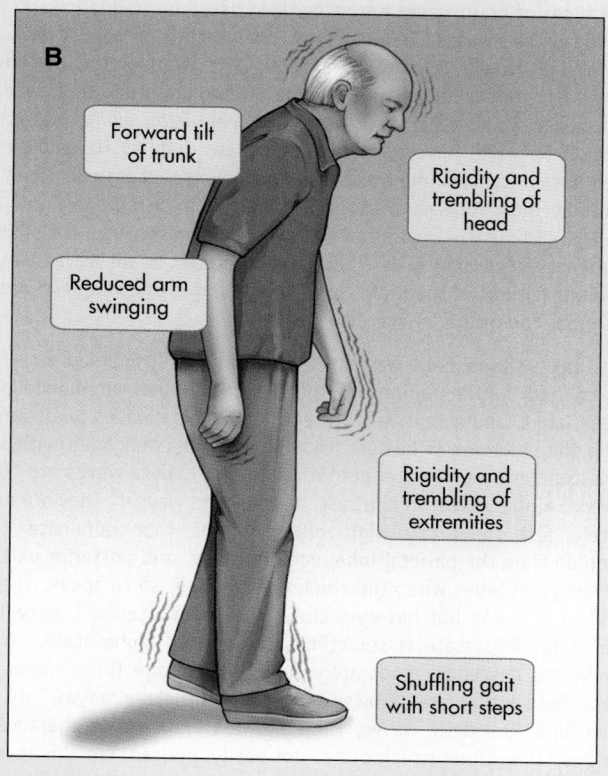

A, Dopaminergic pathways of the brain. **B,** Signs of Parkinson disease (PD).

QUICK CHECK

13. Name the five lobes that make up each cerebral hemisphere. Where is each located?
14. Name the basal nuclei, and describe where they are located within the cerebrum.

Functions of the Cerebral Cortex
Functional Areas of the Cortex

During the past few decades, research scientists in various fields—neurophysiology, neurosurgery, neuropsychiatry, and others—have added mountains of information to our knowledge about the brain. However, questions come faster than answers, and a clear, complete understanding of the brain's mechanisms still eludes us. Perhaps it always will. Perhaps the capacity of the human brain falls short of the ability to fully understand its own complexity.

We do know that certain areas of the cortex in each hemisphere of the cerebrum engage predominantly in one particular function—at least on the average. Differences between genders and among individuals of both genders are not uncommon. The fact that many cerebral functions have a typical location is known as the concept of cerebral localization. The fact that localization of function varies from person to person, and even at different times in an individual when the brain is damaged, is called **cerebral plasticity.**

The function of each region of the cerebral cortex depends on the structures with which it communicates. For example, the

BOX 13-9

Diagnostic Study

The Electroencephalogram (EEG)

Cerebral activity goes on as long as life itself. Only when life ceases (or moments before) does the cerebrum cease its functioning. Only then do all its neurons stop conducting impulses. Proof of this has come from records of brain electrical potentials known as **electroencephalograms**, or **EEGs**. These records are usually made from a number of electrodes placed on different regions of the scalp, and they consist of waves—*brain waves* (parts *A* and *B* of figure).

Four types of brain waves are recognized based on frequency and amplitude of the waves. Frequency, or the number of wave cycles per second, is usually referred to as *hertz* (Hz, from Hertz, a German physicist). Amplitude means voltage. Listed in order of frequency from fastest to slowest, brain wave names are *beta, alpha, theta,* and *delta*. *Beta waves* have a frequency of more than 13 Hz and a relatively low voltage. *Alpha waves* have a frequency of 8 to 13 Hz and a relatively high voltage. *Theta waves* have both a relatively low frequency—4 to 7 Hz—and a low voltage. *Delta waves* have the slowest frequency—less than 4 Hz—but a high voltage. Brain waves vary in different regions of the brain, in different states of awareness, and in abnormal conditions of the cerebrum.

Fast, low-voltage beta waves characterize EEGs recorded from the frontal and central regions of the cerebrum when an individual is awake, alert, and attentive, with eyes open. Beta waves predominate when the cerebrum is busiest, that is, when it is engaged with sensory stimulation or mental activities. In short, beta waves are "busy waves." Alpha waves, in contrast, are "relaxed waves." They are moderately fast, relatively high-voltage waves that dominate EEGs recorded from the parietal lobe, occipital lobe, and posterior parts of the temporal lobes when the cerebrum is idling, so to speak. The individual is awake but has eyes closed and is in a relaxed, nonattentive state. This state is sometimes called the "alpha state." When drowsiness descends, moderately slow, low-voltage theta waves appear. Theta waves are "drowsy waves." "Deep sleep waves," on the other hand, are delta waves. These slowest brain waves characterize the deep sleep from which one is not easily aroused. For this reason, deep sleep is referred to as slow-wave sleep.

Physicians use electroencephalograms (EEGs) to help localize areas of brain dysfunction, to identify altered states of consciousness, and often to establish death. Two flat EEG recordings (no brain waves) taken 24 hours apart in conjunction with no spontaneous respiration and total absence of somatic reflexes are criteria accepted as evidence of brain death.

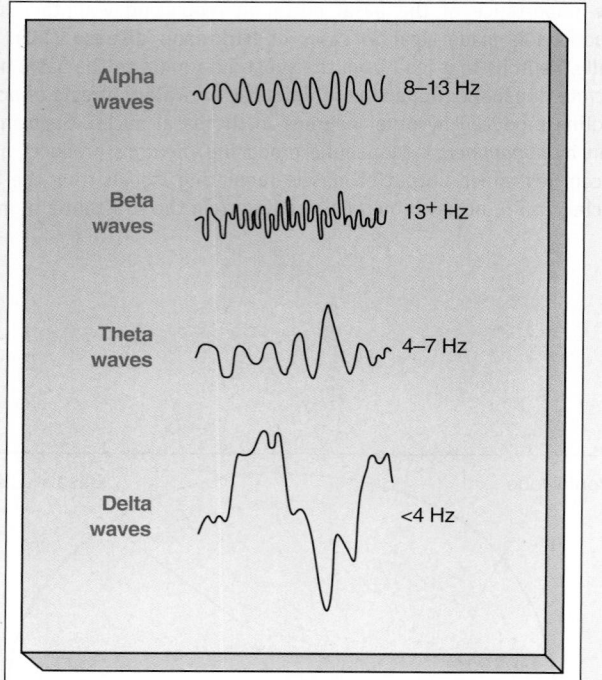

A

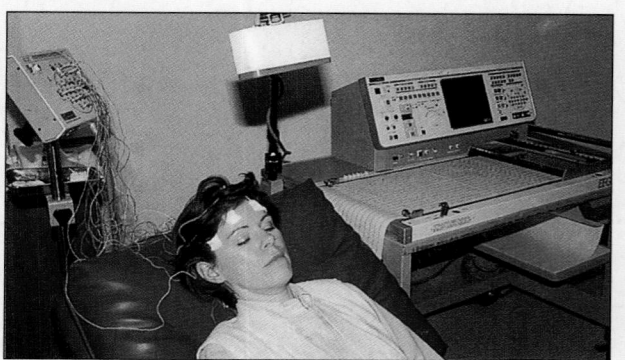

B

The electroencephalogram (EEG). **A,** Examples of alpha, beta, theta, and delta waves seen on an EEG. **B,** Photograph showing a person undergoing an EEG test. Notice the scalp electrodes that detect voltage fluctuations within the cranium.

---

postcentral gyrus (Figures 13-19 and 13-20) functions mainly as a general somatic sensory area. It receives impulses from receptors activated by heat, cold, and touch stimuli. The precentral gyrus, on the other hand, functions chiefly as the somatic motor area (see Figures 13-19 and 13-20). Impulses from neurons in this area descend over motor tracts and eventually stimulate somatic effectors, the skeletal muscles. The transverse gyrus of the temporal lobe serves as the primary auditory area. The primary visual areas are in the occipital lobe. It is important to remember that no part of the brain functions alone. Many structures of the central nervous system must function together for any one part of the brain to function normally.

Sometimes, specific functional areas of the cortex are labeled with numbers (1 through 47) called *Brodmann numbers*. They are named for Korbinian Brodmann, a German neurologist who first proposed his numbered functional map of the cortex in 1908.

Sensory Functions of the Cortex

Various areas of the cerebral cortex are essential for normal functioning of the somatic, or "general," senses, as well as the so-called "special" senses. The somatic senses include sensations of touch, pressure, temperature, body position (proprioception), and similar perceptions that do not require complex sensory organs. The special senses include vision, hearing, and other types of percep-

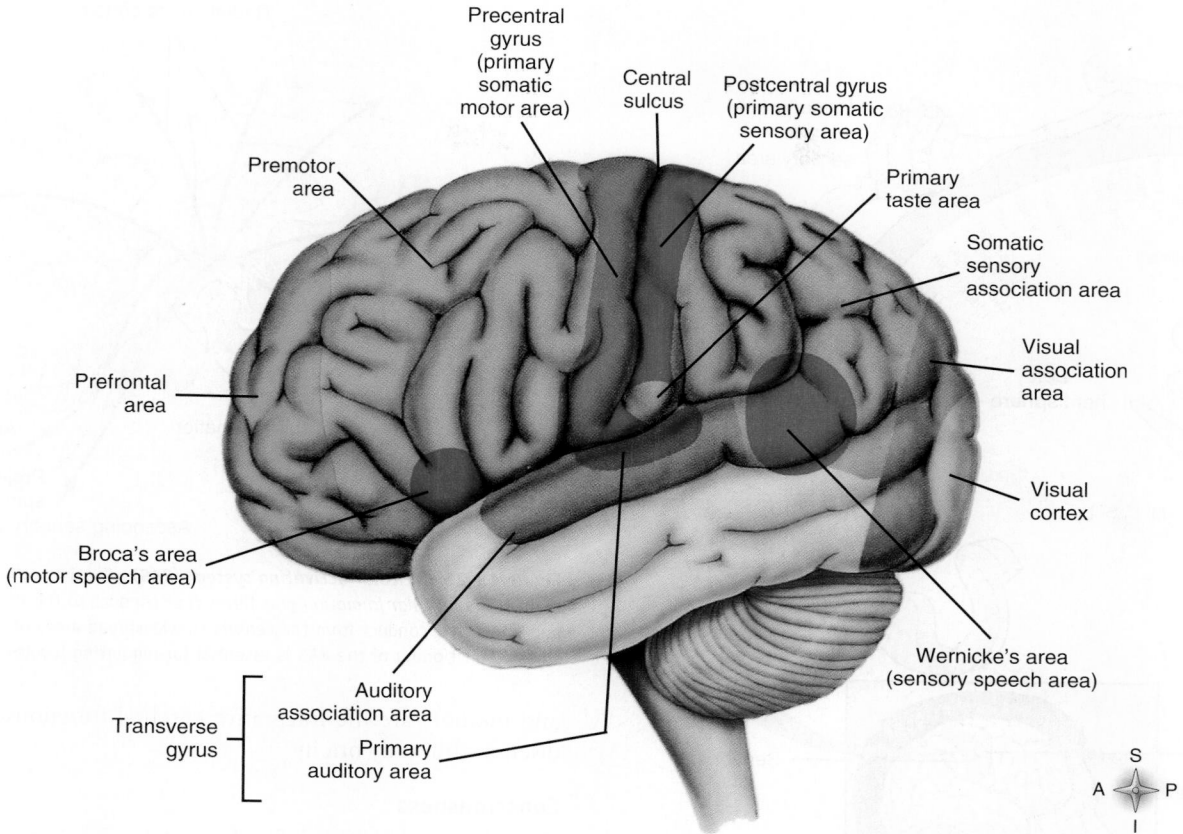

Figure 13-19 *Functional areas of the cerebral cortex.*

tion that require complex sensory organs, for example, the eye and the ear.

As stated earlier, the postcentral gyrus serves as a primary area for the general somatic senses. As Figure 13-20, *A*, shows, sensory fibers carrying information from receptors in specific parts of the body terminate in specific regions of the somatic sensory area. In other words, the cortex contains a sort of "somatic sensory map" of the body. Areas such as the face and hand have a proportionally larger number of sensory receptors, so their part of the somatic sensory map is larger. Likewise, information regarding vision is mapped in the visual cortex, and auditory information is mapped in the primary auditory area (see Figure 13-19).

The cortex does more than just register separate and simple sensations, however. Information sent to the primary sensory areas is in turn relayed to the various sensory association areas, as well as to other parts of the brain. There the sensory information is compared and evaluated. Eventually, the cortex integrates separate bits of information into whole perceptions.

Suppose, for example, that someone put an ice cube in your hand. You would, of course, see it and sense something cold touching your hand. But also you would probably know that it was an ice cube because you would perceive a total impression compounded of many sensations such as temperature, shape, size, color, weight, texture, and movement and position of your hand and arm.

Discussion of somatic sensory pathways begins on p. 499. The special senses are discussed in Chapter 15.

Motor Functions of the Cortex

Mechanisms that control voluntary movements are extremely complex and imperfectly understood. It is known, however, that for normal movements to take place, many parts of the nervous system—including certain areas of the cerebral cortex—must function.

The precentral gyrus, that is, the most posterior gyrus of the frontal lobe, constitutes the primary somatic motor area (see Figures 13-19 and 13-20, *B*). A secondary motor area lies in the gyrus immediately anterior to the precentral gyrus. Neurons in the precentral gyrus are said to control individual muscles, especially those that produce movements of distal joints (wrist, hand, finger, ankle, foot, and toe movements). Notice in Figure 13-20 that the primary somatic motor area is mapped according to the specific areas of the body it controls. Neurons in the premotor area just anterior to the precentral gyrus are thought to activate groups of muscles simultaneously.

Motor pathways descending from the cerebrum through the brainstem and spinal cord are discussed on pp. 502-504. Autonomic motor pathways are discussed in Chapter 14.

Integrative Functions of the Cortex

"Integrative functions" is a nebulous phrase. Even more obscure, however, are the neural processes it designates. They consist of all events that take place in the cerebrum between its reception of sensory impulses and its sending out of motor impulses. Integrative functions of the cerebrum include consciousness and mental activities of all kinds. Consciousness, use of language, emotions,

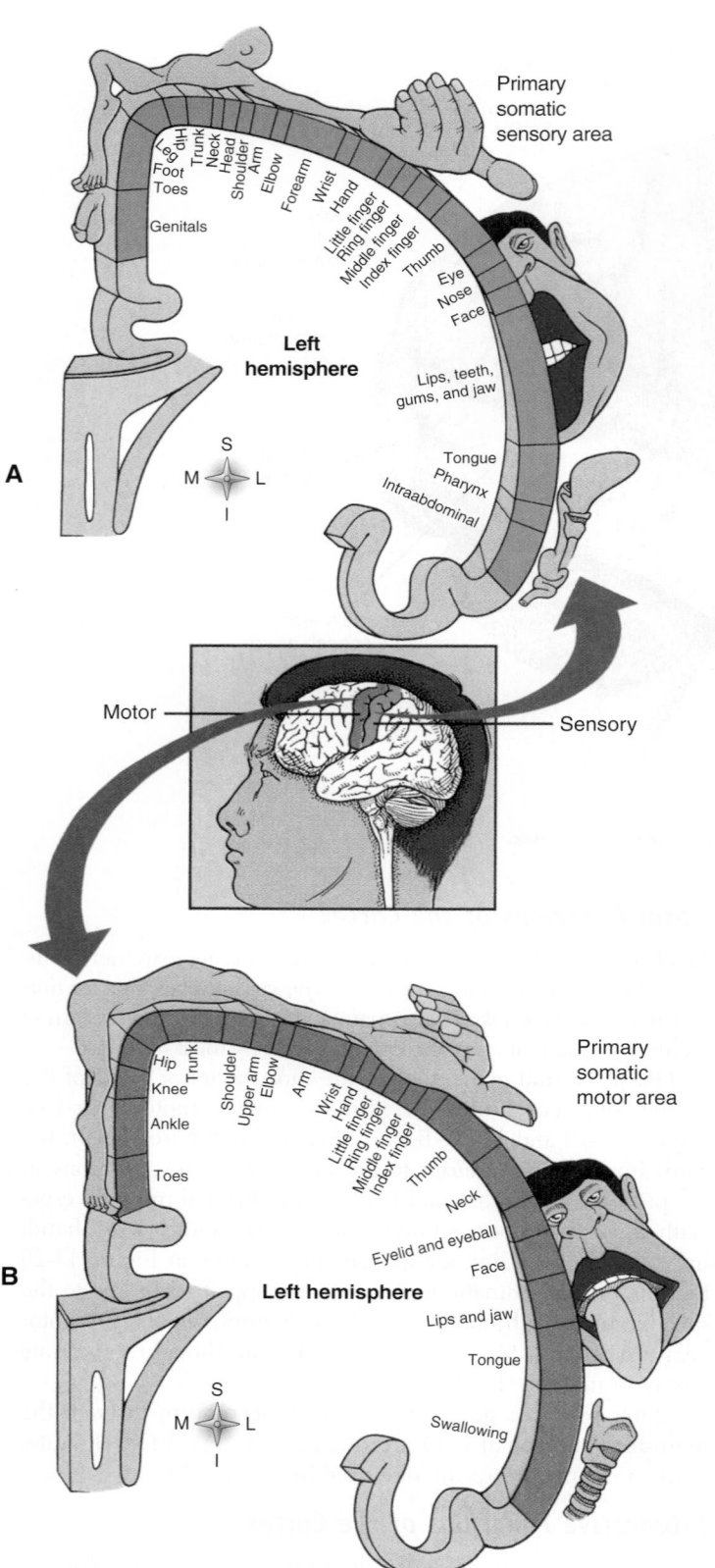

Figure 13-20 *Primary somatic sensory, A, and motor, B, areas of the cortex.* The body parts illustrated here show which parts of the body are "mapped" to specific areas of each cortical area. The exaggerated face indicates that more cortical area is devoted to processing information to and from the many receptors and motor units of the face than for the leg or arm, for example.

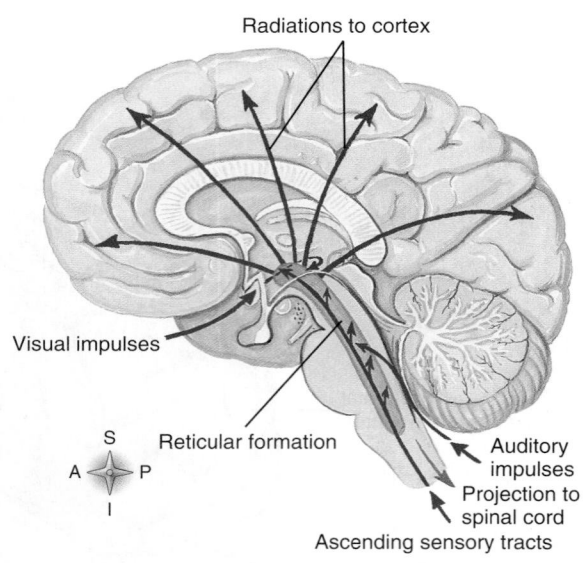

Figure 13-21 *Reticular activating system (RAS).* Consists of centers in the brainstem's *reticular formation* plus fibers that conduct to the centers from below and fibers that conduct from the centers to widespread areas of the cerebral cortex. Functioning of the RAS is essential for regulating levels of consciousness.

and memory are the integrative cerebral functions that we shall discuss—but only briefly.

Consciousness

Consciousness may be defined as a state of awareness of one's self, one's environment, and other humans. Very little is known about the neural mechanisms that produce consciousness. We do know, however, that consciousness depends on excitation of cortical neurons by impulses conducted to them by a network of neurons known as the reticular activating system. The **reticular activating system (RAS)** consists of centers in the brainstem's *reticular formation* that receive impulses from the spinal cord and relay them to the thalamus and from the thalamus to all parts of the cerebral cortex (Figure 13-21). Both direct spinal reticular tracts and collateral fibers from the specialized sensory tracts (spinothalamic, lemniscal, auditory, and visual) relay impulses over the reticular activating system to the cortex. Without continual excitation of cortical neurons by reticular activating impulses, an individual is unconscious and cannot be aroused. Here, then, are two current concepts about the reticular activating system: (1) it functions as the arousal or alerting system for the cerebral cortex, and (2) its functioning is crucial for maintaining consciousness. Drugs known to depress the reticular activating system decrease alertness and induce sleep.

Barbiturates, for example, act this way. On the other hand, amphetamine, a drug known to stimulate the cerebrum and to enhance alertness and produce wakefulness, probably acts by stimulating the reticular activating system.

Certain variations in the levels or state of consciousness are normal. All of us, for example, experience different levels of wakefulness. At times, we are highly alert and attentive. At other times, we are relaxed and nonattentive. All of us also experience different levels of sleep (see Box 13-10).

In addition to the various normal states of consciousness, altered states of consciousness also occur under certain conditions. Anesthetic drugs produce an altered state of consciousness,

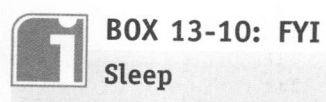

BOX 13-10: FYI
Sleep

Sleep research has been going on full-force since 1953 when Nathaniel Kleitman and his student Eugene Aserinsky finally overthrew the belief that sleeping was merely a shutting down of most brain activities. Even so, there are still many unanswered questions about the process of sleep and its function in the human body.

One thing we do know about sleep is that it occurs in a repeating sequence of stages called the sleep cycle. The figure outlines typical human sleep cycles. Note that brain activity is high during the waking state, usually in the alpha or beta state (see Box 13-9 on p. 494).

During stage 1 of the sleep cycle, which really is more of a drowsy "twilight zone" than true sleep, slower activity in the brain is seen as theta waves on the EEG (part *B* of figure). As a person enters stage 2, a light sleep descends. The brain is still in the theta state but shows spontaneous periods of muscle tension and then relaxation. During stage 2, the heart rate slows down and the body temperature drops.

Stages 3 and 4 are called **slow-wave sleep (SWS).** Slow-wave sleep takes its name from the slow-frequency, high-voltage delta waves seen in the EEG during deep sleep. Notice that the EEG waves become more rhythmic (less chaotic) during deep sleep. SWS is almost entirely a dreamless sleep.

Each of these stages lasts anywhere from about 5 to 20 minutes. After about 90 to 120 minutes, a person may then enter **rapid eye movement (REM) sleep**—sometimes called stage 5 sleep. As its name implies, REM sleep is characterized by rapid movements of the eyes. The REM stage is associated with our most vivid dreaming. Ordinarily during sleep, even dreaming, the reticular activating system (RAS) shuts off motor signals to the skeletal muscles—perhaps keeping us from acting out our dreams. However, motor signals to the muscles that move the eye are not inhibited by the RAS and thus allow our eyes to move while we dream.

The REM stage, or dream stage, of sleep lasts anywhere from 10 to 60 minutes. Usually, the first REM episode is short and the length of the REM stage gets longer and longer with each subsequent sleep cycle. You can see this in part *C* of the figure. Note also in parts *B* and *C* that before REM, the stages reverse themselves back to a higher level of brain activity. After REM, the stages then progress back to a deeper sleep. Part *C* shows that each sleep cycle gets shorter and shorter—and less and less deep. Then a person finally awakes.

Sleep scientists are still unsure about the functions of sleep and dreaming. Certainly, we need both sleep and dreams to feel rested and maintain normal, alert function; but we don't know why this is so. Perhaps non-REM sleep allows time for neurons to repair accumulated damage from free radicals or undergo other maintenance activity. The higher activity during REM may aid in the development of new neural pathways—especially during our early years when REM activity is at its highest. During REM sleep, the release of monoamine neurotransmitters (serotonin, dopamine, histamine) stops. Perhaps their receptors then have a chance to "rest" and thus restore their full sensitivity. This may help the mood pathways and other monoamine pathways function normally when we finally awake.

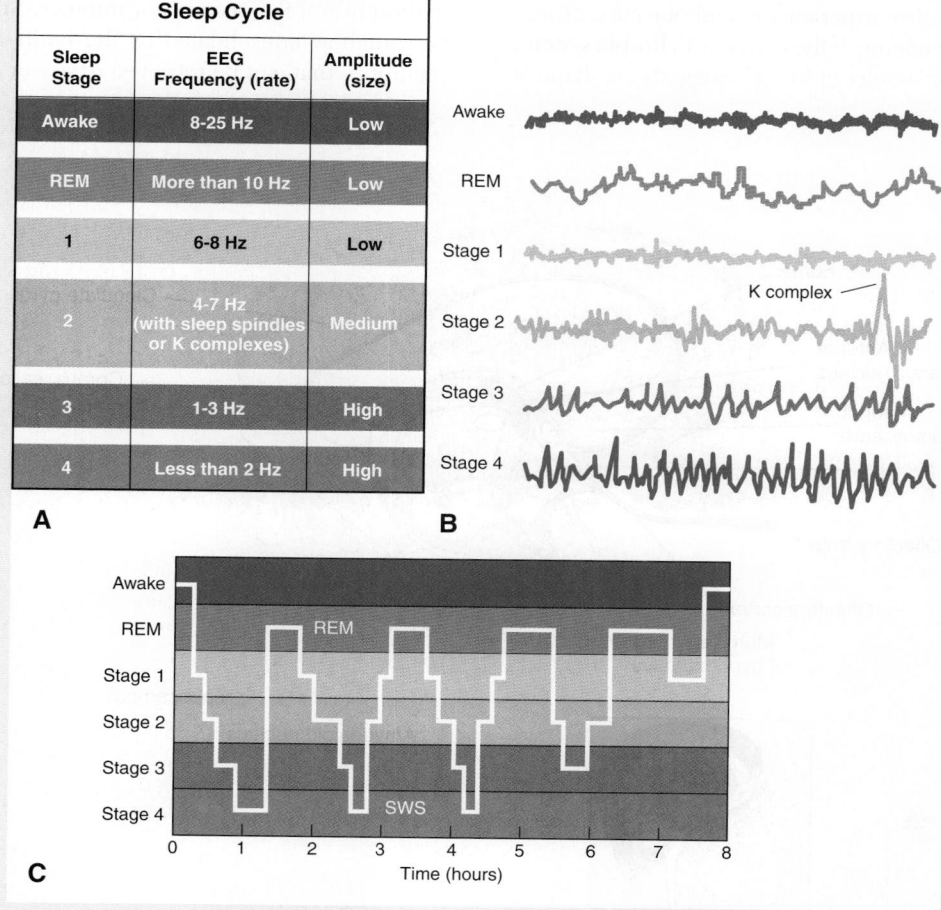

Sleep cycles.

namely, *anesthesia*. Disease or injury of the brain may produce an altered state called **coma**.

Peoples of various cultures have long been familiar with an altered state called *meditation*. Meditation is a waking state but differs markedly in certain respects from the usual waking state. According to some, meditation is a "higher" or "expanded" level of consciousness. This higher consciousness is accompanied, almost paradoxically, by a high degree of both relaxation and alertness. With training in meditation techniques and practice, an individual can enter the meditative state at will and remain in it for an extended period of time.

Language

Language functions consist of the ability to speak and write words and the ability to understand spoken and written words. Certain areas in the frontal, parietal, and temporal lobes serve as speech centers—as crucial areas, that is, for language functions. The left cerebral hemisphere contains these areas in about 90% of the population; in the remaining 10%, either the right hemisphere or both hemispheres contain them. Lesions in speech centers give rise to language defects called *aphasias*. For example, with damage to an area in the inferior gyrus of the frontal lobe (Broca's area, see Figure 13-19), a person becomes unable to articulate words but can still make vocal sounds and understand words heard and read.

Emotions

Emotions—both the subjective experiencing and objective expression of them—involve functioning of the cerebrum's **limbic system.** The name *limbic* (Latin for "border or fringe") suggests the shape of the cortical structures that make up the system. They form a curving border around the corpus callosum, the structure that connects the two cerebral hemispheres. Look now at Figure 13-22. Here on the medial surface of the cerebrum lie most of the structures of the limbic system. They are the cingulate gyrus and the hippocampus (the extension of the hippocampal gyrus that protrudes into the floor of the inferior horn of the lateral ventricle). These limbic system structures have primary connections with various other parts of the brain, notably the thalamus, fornix, septal nucleus, amygdaloid nucleus (the tip of the caudate nucleus, one of the basal nuclei), and the hypothalamus. Some physiologists therefore include these connected structures as parts of the limbic system.

The limbic system (or to use its more descriptive name, the *emotional brain*) functions in some way to make us experience many kinds of emotions—anger, fear, sexual feelings, pleasure, and sorrow, for example. To bring about the normal expression of emotions, parts of the cerebral cortex other than the limbic system must also function. Considerable evidence exists that limbic activity without the modulating influence of the other cortical areas may bring on the attacks of abnormal, uncontrollable rage suffered periodically by some unfortunate individuals.

Memory

Memory is one of our major mental activities. The cortex is capable of storing and retrieving both *short-term memory* and *long-term memory*. Short-term memory involves the storage of information over a few seconds or minutes. Short-term memories can be somehow consolidated by the brain and stored as long-term memories that can be retrieved days—or even years—later.

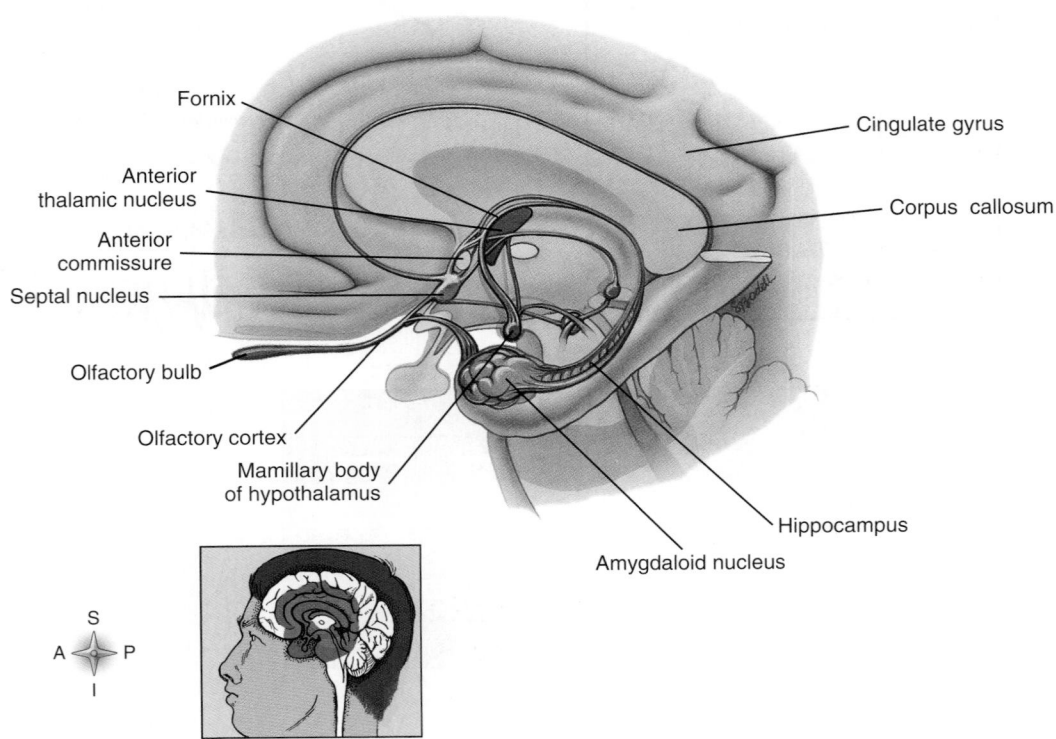

Figure 13-22 *Structures of the limbic system.* Sometimes referred to as "the emotional brain," this network of brain structures functions in our experience of emotions.

Both short-term memory and long-term memory are functions of many parts of the cerebral cortex, especially of the temporal, parietal, and occipital lobes. Findings by Dr. Wilder Penfield, a noted Canadian neurosurgeon, first gave evidence of this almost 100 years ago. He electrically stimulated the temporal lobes of epileptic patients undergoing brain surgery. They responded, much to his surprise, by recalling in the most minute detail songs and events from their past.

Such long-term memories are believed to consist of some kind of structural changes—called *engrams*—in the cerebral cortex. Widely accepted today is the theory that an engram consists of some kind of permanent change in the synapses in a specific circuit of neurons. Repeated impulse conduction over a given neuronal circuit produces the synaptic change. What the change is still is a matter of speculation. Two suggestions are that it represents an increase in the number of presynaptic axon terminals or an increase in the number of receptor proteins in the postsynaptic neuron's membrane. Other suggestions involve changes in the average concentrations of neurotransmitters at certain synapses or changes in the functions of astrocytes. Whatever the changes are, they somehow facilitate impulse transmission at the synapses.

More recent data suggests that different kinds of memories may be stored in different ways. Some memories are perhaps stored as changes at synapses and some by changes in the neurons themselves. For example, some neurons have been shown to react only when a specific person is seen or mentioned—thus somehow acting as a "recognition neuron" for the particular individual who is recognized.

Many research findings indicate that the cerebrum's limbic system—the "emotional brain"—plays a key role in memory. To mention one role, when the hippocampus (part of the limbic system) is removed, the patient loses the ability to recall new information. Personal experience substantiates a relationship between emotion and memory.

Specialization of Cerebral Hemispheres

The right and left hemispheres of the cerebrum specialize in different functions—a concept called **hemisphericity**. For example, as already noted, the left hemisphere specializes in language functions—it does most of the talking, so to speak. The left hemisphere also appears to dominate the control of certain kinds of hand movements, notably skilled and gesturing movements. Most people use their right hands for performing skilled movements, and the left side of the cerebrum controls the muscles on the right side that execute these movements. The next time you are with a group of people who are talking, observe their gestures. The chances are about 9 to 1 that they will gesture mostly with their right hands—indicative of left cerebral control.

Evidence that the right hemisphere of the cerebrum specializes in certain functions has also been reported. It seems that one of the right hemisphere's specialties is the perception of certain kinds of auditory stimuli. For instance, some studies show that the right hemisphere perceives nonspeech sounds such as melodies, coughing, crying, and laughing better than the left hemisphere. The right hemisphere may also function better at tactual perception and for perceiving and visualizing spatial relationships.

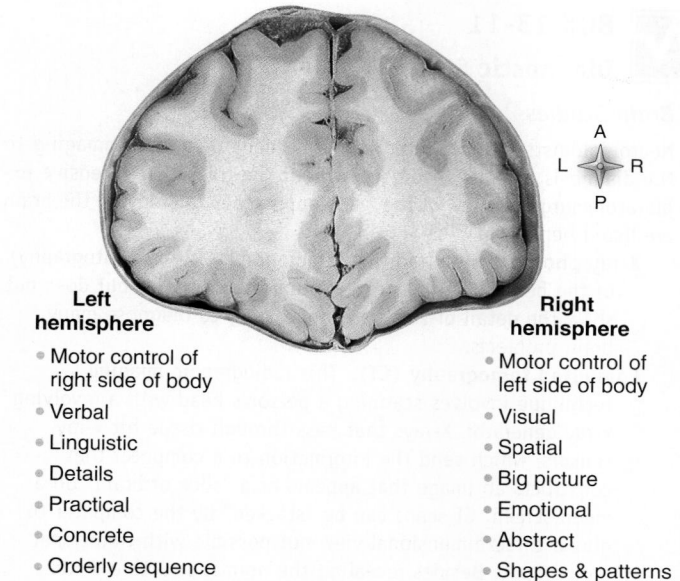

Left hemisphere
- Motor control of right side of body
- Verbal
- Linguistic
- Details
- Practical
- Concrete
- Orderly sequence

Right hemisphere
- Motor control of left side of body
- Visual
- Spatial
- Big picture
- Emotional
- Abstract
- Shapes & patterns

Figure 13-23 *Hemisphericity.* Each hemisphere of the cerebral cortex often specializes in, or becomes dominant in, certain complex functions of the cerebrum—some of which are listed here.

Some of the proposed "special talents" of each hemisphere are shown in Figure 13-23. Despite the specializations of each cerebral hemisphere, both sides of a normal person's brain communicate with each other by way of the corpus callosum to integrate information and accomplish the many complex functions of the brain.

 QUICK CHECK

15. Where is the primary somatic motor area of the cerebral cortex? Where is the primary somatic sensory area?
16. What does the reticular activating system have to do with alertness?
17. What is the function of the limbic system?

SOMATIC SENSORY PATHWAYS IN THE CENTRAL NERVOUS SYSTEM

For the cerebral cortex to perform its *sensory* functions, impulses must first be conducted to its sensory areas by way of relays of neurons referred to as sensory pathways. Most impulses that reach the sensory areas of the cerebral cortex have traveled over at least three pools of sensory neurons. We shall designate these as *primary sensory neurons*, *secondary sensory neurons*, and *tertiary sensory neurons* (Figure 13-24).

Primary sensory neurons of the relay conduct from the periphery to the central nervous system. Secondary sensory neurons conduct from the cord or brainstem up to the thalamus. Their dendrites and cell bodies are located in spinal cord or brainstem gray matter. Their axons ascend in ascending tracts up the cord, through the brainstem, and terminate in the thalamus. Here they synapse with dendrites or cell bodies of tertiary sensory neurons (see Figure 13-24). Tertiary sensory neurons conduct from the thalamus to the postcentral gyrus of the parietal lobe, the somaticosensory area. Bundles of

BOX 13-11
Diagnostic Study

Brain Studies

Neurobiologists have adapted many methods of medical imaging to the diagnosis of brain disorders without the trauma of extensive exploratory surgery. A few of the many approaches to studying the brain are listed here:

X-ray photography. Traditional radiography (x-ray photography) of the head sometimes reveals tumors or injuries but does not show the detail of soft tissue necessary to diagnose many brain problems.

Computed tomography (CT). This radiographic imaging technique involves scanning a person's head with a revolving x-ray generator. X-rays that pass through tissue hit x-ray sensors, which send the information to a computer that constructs an image that appears as a "slice of brain" on a video screen. CT scans can be "stacked" by the computer to give a three-dimensional view not possible with traditional radiography. Besides revealing the normal structure of the brain, CT scanning can often detect hemorrhages, tumors, and other abnormalities.

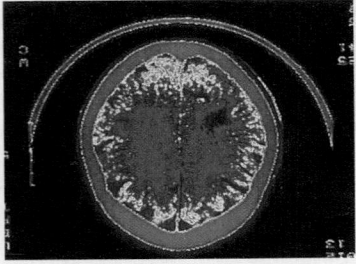

CT scan.

Positron-emission tomography (PET). PET scanning is a variation of CT scanning in which a radioactive substance is introduced into the blood supply of the brain. The radioactive material shows up as a bright spot on the image. Different substances are taken up by the brain in different amounts, depending on the type of tissue and the level of activity, enabling radiologists to determine the functional characteristics of specific parts of the brain.

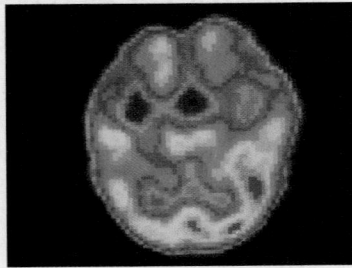

PET scan.

Single-photon emission computed tomography (SPECT). SPECT is a method of scanning similar to PET, but using more stable substances and different detectors. SPECT is used to visualize blood flow patterns in the brain—making it useful in diagnosing cerebrovascular accidents (CVAs or strokes) and brain tumors.

Ultrasonography. In this method, high-frequency sound (ultrasound) waves are reflected off anatomical structures to form images. This is similar to the way radar works. Because it does not use harmful radiation, ultrasonography is often used in diagnosing hydrocephalus or brain tumors in infants. When used on the brain, ultrasonography is often called *echoencephalography*.

Magnetic resonance imaging (MRI). Also called *nuclear magnetic resonance (NMR)* imaging, this scanning method also has the advantage of avoiding the use of harmful radiation. In MRI, a magnetic field surrounding the head induces brain tissues to emit radio waves that can be used by a computer to construct a sectional image. MRI has the added advantage of producing sharper images than CT scanning and ultrasound. This makes it very useful in detecting small brain abnormalities. A newer form of MRI study called the **functional MRI (fMRI)** can detect which areas of the brain are most active by measuring oxygen consumption by neurons. fMRI images can highlight which areas of the brain are most involved in specific tasks or mental processes performed by the subject.

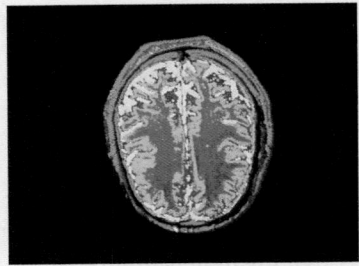

MRI scan.

Evoked potential (EP). As discussed in this chapter, electroencephalography is the measurement of the electrical activity of the brain. The EP test is similar to the electroencephalogram (EEG), but the brain waves observed are caused (evoked) by specific stimuli such as a flash of light or a sudden sound. This information can also be analyzed by a computer that then produces a color-coded graphic image of the brain generated on a video screen, a **brain electrical activity map (BEAM).** Changes in color on the BEAM represent changes in brain activity evoked by each stimulus that is given. This technique is useful in diagnosing abnormalities of the visual or auditory systems because it reveals whether a sensory impulse is reaching the appropriate part of the brain.

Magnetoencephalography (MEG). This method of measuring brain activity uses a sensitivity machine called a *biomagnetometer*, which detects the very small magnetic fields generated by neural activity. It can accurately pinpoint the part of the brain involved in a CVA (stroke), seizure, or other disorder or injury. Based on technology first developed by the military to locate submarines, MEG promises to improve our ability to diagnose epilepsy, Alzheimer disease, and perhaps even certain types of addiction.

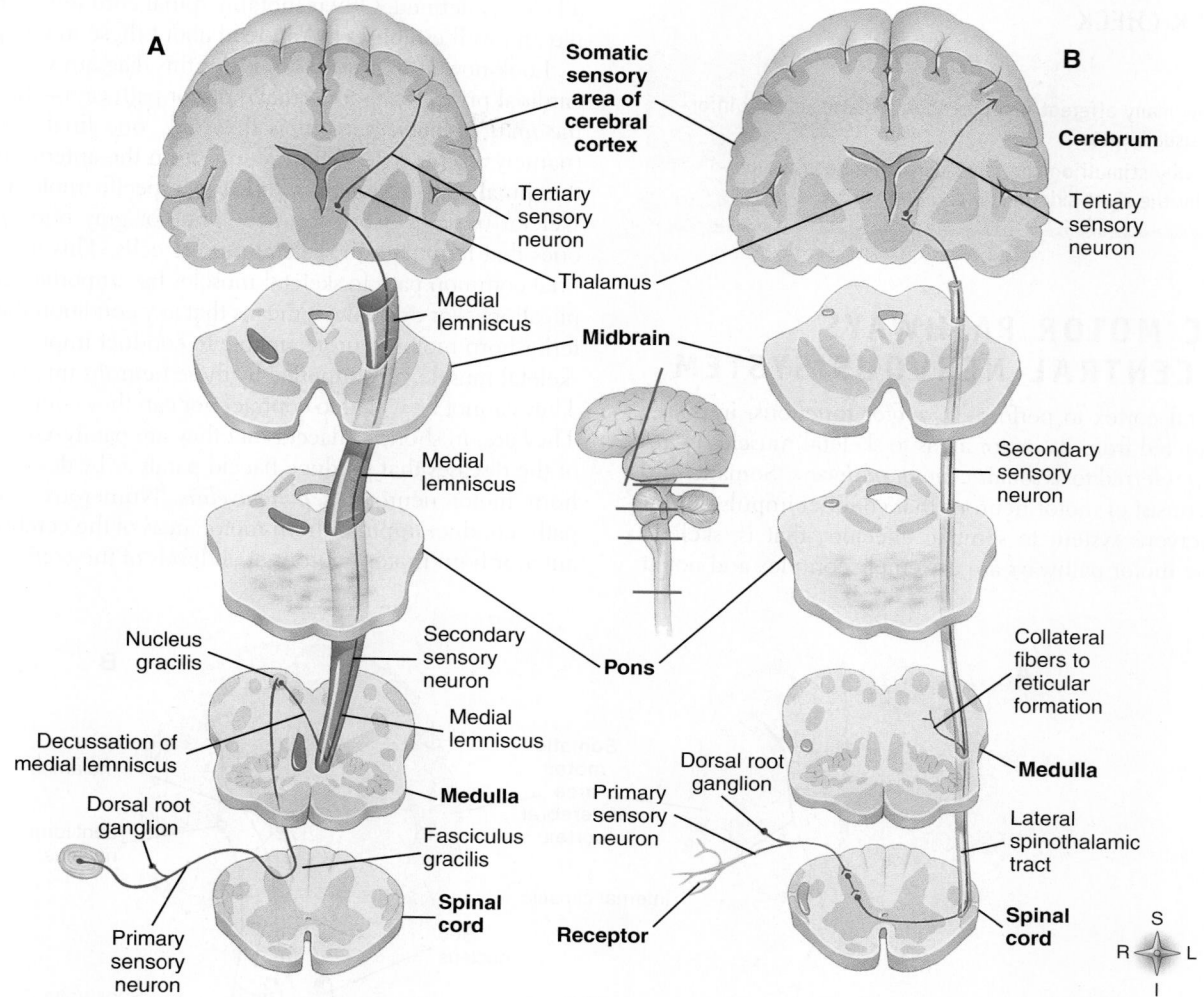

Figure 13-24 *Examples of somatic sensory pathways.* **A,** A pathway of the medial lemniscal system that conducts information about discriminating touch and kinesthesia. **B,** A spinothalamic pathway that conducts information about pain and temperature.

axons of tertiary sensory neurons form thalamocortical tracts. They extend through the portion of cerebral white matter known as the *internal capsule* to the cerebral cortex (see Figure 13-24).

For the most part, sensory pathways to the cerebral cortex are crossed pathways. This means that each side of the brain registers sensations from the opposite side of the body. Look again at Figure 13-24. The axons that *decussate* (cross from one side to the other) in these sensory pathways are which sensory neurons: primary, secondary, or tertiary? Usually it is the axon of a secondary sensory neuron that decussates at some level in its ascent to the thalamus. Thus general sensations of the right side of the body are predominantly experienced by the left somatic sensory area. General sensations of the left side of the body are predominantly experienced by the right somatic sensory area.

Two sensory pathways conduct impulses that produce sensations of touch and pressure, namely, the *medial lemniscal system* and the *spinothalamic pathway* (see Figure 13-24). The medial lemniscal system consists of the tracts that make up the posterior white columns of the cord (the fasciculi cuneatus and gracilis)

plus the *medial lemniscus*, a flat band of white fibers extending through the medulla, pons, and midbrain. (The term *lemniscus* literally means "ribbon," referring to this tract's flattened shape.)

The fibers of the medial lemniscus, like those of the spinothalamic tracts, are axons of secondary sensory neurons. They originate from cell bodies in the medulla, decussate, and then extend upward to terminate in the thalamus on the opposite side. The function of the medial lemniscal system is to transmit impulses that produce our more discriminating touch and pressure sensations, including *stereognosis* (awareness of an object's size, shape, and texture), precise localization, two-point discrimination, weight discrimination, and sense of vibrations. The sensory pathway for kinesthesia (sense of movement and position of body parts) is also part of the medial lemniscal system.

Crude touch and pressure sensations are functions of the **spinothalamic pathway.** Knowing that something touches the skin is a crude touch sensation, whereas knowing its precise location, size, shape, or texture involves the discriminating touch sensations of the medial lemniscal system.

QUICK CHECK

18. Over how many afferent neurons does somatic sensory information usually pass?

19. Explain why stimuli on the left side of the body are perceived by the right side of the cerebral cortex.

SOMATIC MOTOR PATHWAYS IN THE CENTRAL NERVOUS SYSTEM

For the cerebral cortex to perform its *motor* functions, impulses must be conducted from its motor areas to skeletal muscles by relays of neurons referred to as *somatic motor pathways*. Somatic motor pathways consist of motor neurons that conduct impulses from the central nervous system to somatic effectors, that is, skeletal muscles. Some motor pathways are extremely complex and not at all clearly defined. Others, notably spinal cord reflex arcs, are simple and well established. You read about these in Chapter 12.

Look now at Figure 13-25. From this diagram you can derive a cardinal principle about somatic motor pathways—the *principle of the final common path*. It is this: only one final common path (namely, each single motor neuron from the anterior gray horn of the spinal cord) conducts *impulses* to a specific motor unit within a skeletal muscle. Axons from the anterior gray horn are the only ones that terminate in skeletal muscle cells. This principle of the final common path to skeletal muscles has important practical implications. For example, it means that any condition that makes anterior horn motor neurons unable to conduct impulses also makes skeletal muscle cells supplied by these neurons unable to contract. They cannot be willed to contract nor can they contract reflexively. They are, in short, so flaccid that they are paralyzed. Most famous of the diseases that produce flaccid paralysis by destroying anterior horn motor neurons is *poliomyelitis*. Numerous somatic motor paths conduct impulses from motor areas of the cerebrum down to anterior horn motor neurons at all levels of the cord.

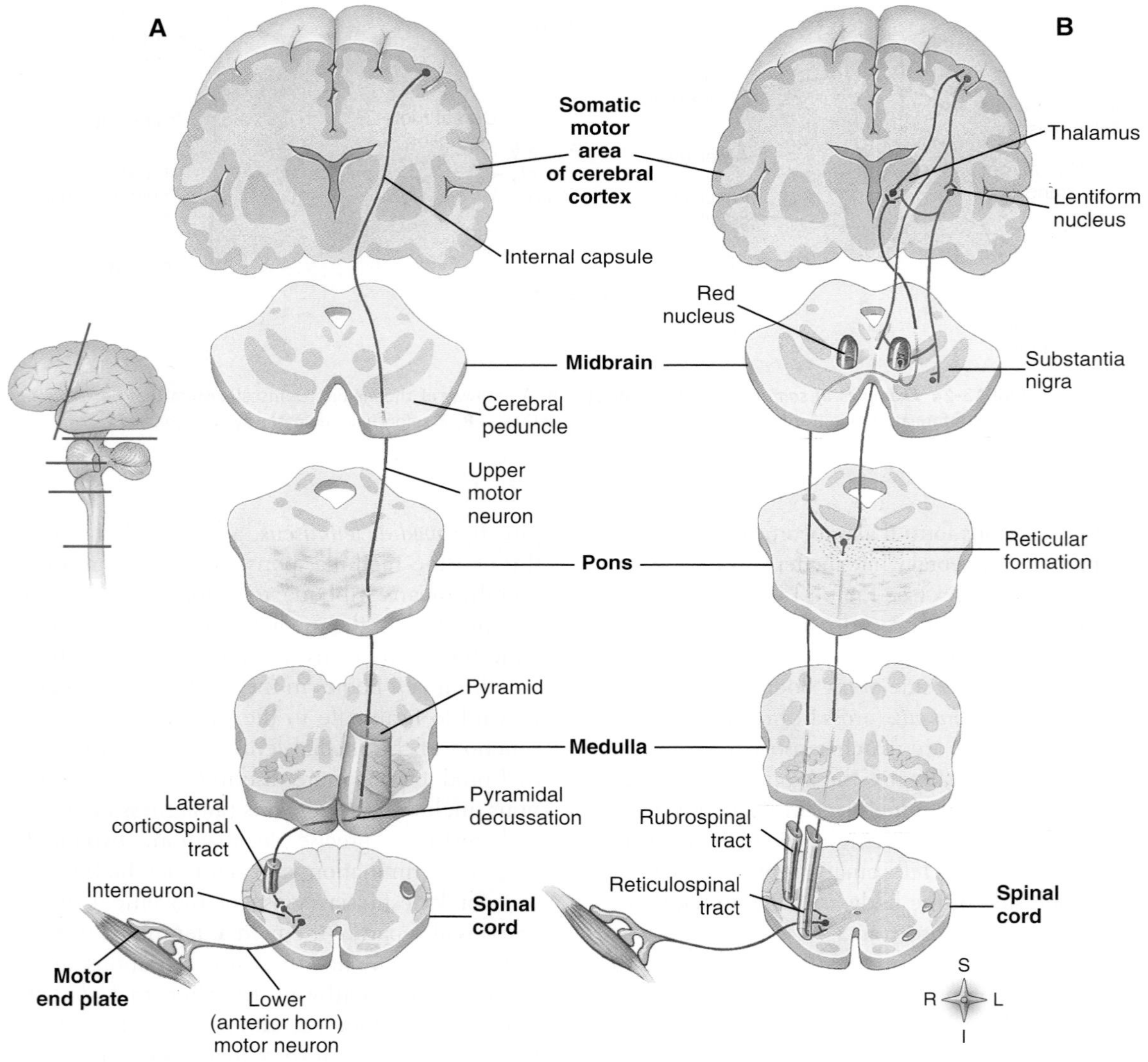

Figure 13-25 *Examples of somatic motor pathways.* **A,** A pyramidal pathway, through the lateral corticospinal tract. **B,** Extrapyramidal pathways, through the rubrospinal and reticulospinal tracts.

Labels for diagram A and B:
Somatic motor area of cerebral cortex — Thalamus — Lentiform nucleus — Internal capsule — Red nucleus — Substantia nigra — Midbrain — Cerebral peduncle — Upper motor neuron — Reticular formation — Pons — Pyramid — Medulla — Lateral corticospinal tract — Pyramidal decussation — Rubrospinal tract — Interneuron — Reticulospinal tract — Spinal cord — Motor end plate — Lower (anterior horn) motor neuron

Two methods are used to classify somatic motor pathways—one based on the location of their fibers in the medulla and the other on their influence on the lower motor neurons. The first method divides them into pyramidal and extrapyramidal tracts. The second classifies them as facilitatory and inhibitory tracts.

Pyramidal tracts are those whose fibers come together in the medulla to form the pyramids, hence their name (Figure 13-25, A). Because axons composing the pyramidal tracts originate from neuron cell bodies located in the cerebral cortex, they also bear another name—*corticospinal tracts*. About three fourths of their fibers decussate (cross over from one side to the other) in the medulla. After decussating, they extend down the spinal cord in the crossed corticospinal tract located on the opposite side of the spinal cord in the lateral white column. About one fourth of the corticospinal fibers do not decussate. Instead, they extend down the same side of the spinal cord as the cerebral area from which they came. One pair of uncrossed tracts lies in the anterior white columns of the cord, namely, the anterior corticospinal tracts. The other uncrossed corticospinal tracts form part of the lateral corticospinal tracts. About 60% of corticospinal fibers are axons that arise from neuron cell bodies in the precentral (frontal lobe) region of the cortex. About 40% of corticospinal fibers originate from neuron cell bodies located in postcentral areas of the cortex, areas classified as sensory; now, more accurately, they are often called *sensorimotor areas*.

Relatively few corticospinal tract fibers synapse directly with anterior horn motor neurons. Most of them synapse with interneurons, which in turn synapse with anterior horn motor neurons. All corticospinal fibers conduct impulses that depolarize resting anterior horn motor neurons. The effects of depolarizations occurring rapidly in any one neuron add up, or summate. Each impulse, in other words, depolarizes an anterior horn motor neuron's resting potential a little bit more. If sufficient numbers of impulses impinge rapidly enough on a neuron, its potential reaches the threshold level. At that moment the neuron starts conducting impulses—it is stimulated. Stimulation of anterior horn motor neurons by corticospinal tract impulses results in stimulation of individual muscle groups (mainly of the hands and feet). Precise control of their contractions is, in short, the function of the corticospinal tracts. Without stimulation of anterior horn motor neurons by impulses over corticospinal fibers, willed movements cannot occur. This means that paralysis results whenever pyramidal corticospinal tract conduction is interrupted. For instance, the paralysis that so often follows cerebrovascular accidents (CVAs, or strokes) comes from pyramidal neuron injury—sometimes of their cell bodies in the motor areas, sometimes of their axons in the internal capsule (see Figure 13-25).

Extrapyramidal tracts are much more complex than pyramidal tracts. They consist of all motor tracts from the brain to the spinal cord anterior horn motor neurons except the corticospinal (pyramidal) tracts. Within the brain, extrapyramidal tracts consist of numerous relays of motor neurons between motor areas of the cortex, basal nuclei, thalamus, cerebellum, and brainstem. In the cord, some of the most important extrapyramidal tracts are the reticulospinal tracts.

Fibers of the *reticulospinal tracts* originate from cell bodies in the reticular formation of the brainstem and terminate in gray matter of the spinal cord, where they synapse with interneurons that synapse with lower (anterior horn) motor neurons. Some reticulospinal tracts function as facilitatory tracts and others as inhibitory tracts. Summation of these opposing influences determines the lower motor neuron's response. It initiates impulse conduction only when facilitatory impulses exceed inhibitory impulses sufficiently to decrease the lower motor neuron's negativity to its threshold level.

Conduction by extrapyramidal tracts plays a crucial part in producing our larger, more automatic movements because extrapyramidal impulses bring about contractions of groups of muscles in sequence or simultaneously. Such muscle action occurs, for example, in swimming and walking and, in fact, in all normal voluntary movements.

Conduction by extrapyramidal tracts plays an important part in our emotional expressions. For instance, most of us smile automatically at things that amuse us and frown at things that irritate us. It is extrapyramidal, not pyramidal, impulses that produce the smiles or frowns.

Axons of many different neurons converge on, that is, synapse with each anterior horn motor neuron (see Figure 13-25). Hence many impulses from diverse sources—some facilitatory and some inhibitory—continually bombard this final common path to skeletal muscles. Together the added, or summated, effect of these opposing influences determines lower motor neuron functioning. Facilitatory impulses reach these cells by way of sensory neurons (whose axons, you will recall, lie in the posterior roots of spinal nerves), pyramidal (corticospinal) tracts, and extrapyramidal facilitatory reticulospinal tracts. Impulses over facilitatory reticulospinal fibers facilitate the lower motor neurons that supply extensor muscles. At the same time, they reciprocally inhibit the lower motor neurons that supply flexor muscles. Hence facilitatory reticulospinal impulses tend to increase the tone of extensor muscles and decrease the tone of flexor muscles.

Inhibitory impulses reach lower motor neurons mainly by way of inhibitory reticulospinal fibers that originate from cell bodies located in the *bulbar inhibitory area* in the medulla. They inhibit the lower motor neurons to extensor muscles (and reciprocally stimulate those to flexor muscles). Hence inhibitory reticulospinal impulses tend to decrease extensor muscle tone and increase flexor muscle tone—note that these are opposite effects from facilitatory reticulospinal impulses.

The set of coordinated commands that control the programmed muscle activity mediated by extrapyramidal pathways is often called a **motor program.** Traditionally, the primary somatic motor areas of the cerebral cortex were thought to be the principal organizer of motor programs sent along the extrapyramidal pathway. That view has been replaced by the concept illustrated in Figure 13-26. This newer concept holds that motor programs result from the interaction of several different centers in the brain. Apparently, many voluntary motor programs are organized in the basal nuclei and cerebellum—perhaps in response to a willed command by the cerebral cortex. Impulses that constitute the motor program are then channeled through the thalamus and back to the cortex, specifically to the primary motor area. From there, the motor program is sent down to the inhibitory and facilitatory regions of the brainstem. Signals from the brainstem then continue on down one or more spinal tracts and out to the muscles by way of the lower (anterior horn) motor neurons. All along the way, neural connections among the various motor control centers allow refinement and adjustment of the motor program. If all this

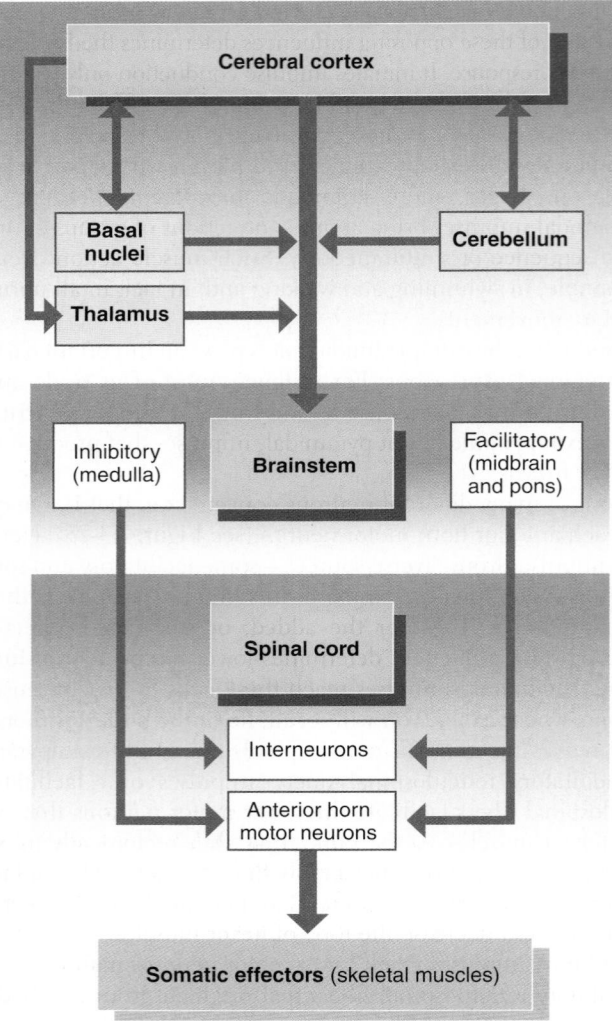

Figure 13-26 *Concept of extrapyramidal motor control.* Programmed movements result from a set of impulses called a motor program. The motor program organized in the basal nuclei and cerebellum in response to a command from the cortex is sent back to the primary motor control area of the cortex. From there, it is sent to the brainstem and then on through the spinal cord to the skeletal muscles. All along the motor pathway, the motor program can be refined by various components of this complex pathway.

sounds complicated and confusing, imagine what it must be like to be a neurobiologist trying to figure out how all this works! In fact, scientists working in this field admit that they have not worked out all the details of the extrapyramidal circuits—or exactly how they control muscle activity. However, the model shown in Figure 13-26 summarizes the current notion that it is a complex, interactive process.

QUICK CHECK

20. What is the "principle of final common path" as it pertains to somatic motor pathways?
21. Distinguish between pyramidal and extrapyramidal pathways.

BOX 13-12: HEALTH MATTERS

Signs of Motor Pathway Injury

Injury of upper motor neurons (those whose axons lie in either pyramidal or extrapyramidal tracts) produces symptoms frequently referred to as "pyramidal signs," notably a spastic type of paralysis, exaggerated deep reflexes, and a positive Babinski reflex (see p. 534). Actually, pyramidal signs result from interruption of both pyramidal and extrapyramidal pathways. The paralysis stems from interruption of pyramidal tracts, whereas the spasticity (rigidity) and exaggerated reflexes come from interruption of inhibitory extrapyramidal pathways.

Injury to lower motor neurons produces symptoms different from those of upper motor neuron injury. Anterior horn cells or lower motor neurons, you will recall, constitute the final common path by which impulses reach skeletal muscles. This means that if they are injured, impulses can no longer reach the skeletal muscles they supply. This, in turn, results in the absence of all reflex and willed movements produced by contraction of the muscles involved. Unused, the muscles soon lose their normal tone and become soft and flabby (flaccid). In short, absence of reflexes and flaccid paralysis are the chief "lower" motor neuron signs.

Cycle of Life

Central Nervous System

If the most obvious *structural* changes over the life span are the overall growth and then degeneration of the skeleton and other body parts, then the most obvious *functional* changes are the development and then degeneration of the complex integrative capacity of the central nervous system.

Although the development of the brain and spinal cord begins in the womb, further development is required by the time a baby is born. The lack of development of the central nervous system in a newborn is evidenced by lack of the more complex integrative functions, such as language, complex memory, comprehension of spatial relationships, and complex motor skills such as walking. As childhood proceeds, one can easily see evidence of the increasing capacity of the central nervous system for complex function. A child learns to use language, to remember both concrete and abstract ideas, to walk, and even to behave in ways that conform to the norms of society. By the time a person reaches adulthood, most, if not all, of these complex functions have become fully developed. We use them throughout adult life to help us maintain internal stability in an unstable external world.

As we enter very late adulthood, the tissues of the brain and spinal cord may degenerate. If they do degenerate—and they may not—the degree of change varies from one individual to the next. In some cases, the degeneration is profound—or it occurs in a critical part of the brain—and an older person becomes unable to communicate, to walk, or to perform some other complex functions. In many cases, however, the degeneration produces milder effects, such as temporary lapses in memory or fumbling with certain very complex motor tasks. As our understanding of this process increases, we are finding ways to avoid entirely such changes associated with aging.

THE BIG PICTURE

The Central Nervous System and the Whole Body

The central nervous system is the ultimate regulator of the entire body. It serves as the anatomical and functional center of the countless feedback loops that maintain the relative constancy of the internal environment. The central nervous system directly or indirectly regulates, or at least influences, nearly every organ in the body.

The intriguing thing about the way in which the central nervous system regulates the whole body is that it is able to integrate, or bring together, literally millions of bits of information from all over the body and make sense of it all. Not only does the central nervous system make sense of all this information, but also it compares it with previously stored memories and makes decisions based on its own conclusions about the data. The complex integrative functions of human language, consciousness, learning, and memory enable us to adapt to situations that less complex organisms could not. Thus our wonderfully complex central nervous system is essential to our survival.

Mechanisms of Disease

DISORDERS OF THE CENTRAL NERVOUS SYSTEM

Destruction of Brain Tissue

Injury or disease can destroy neurons. A common example is the destruction of neurons of the motor area of the cerebrum that results from a **cerebrovascular accident (CVA)**. A CVA, or *stroke*, is a hemorrhage from or cessation of blood flow through cerebral blood vessels. When this happens, the oxygen supply to portions of the brain is disrupted and neurons cease functioning. If the lack of oxygen is prolonged, the neurons die. If the damage occurs in a motor control area of the brain (see Figures 13-19 and 13-20), a person can no longer voluntarily move the parts of the body controlled by the affected area or areas. Because motor neurons cross over from side to side in the brainstem (see Figure 13-25), flaccid paralysis appears on the side of the body opposite the side of the brain on which the CVA occurred. The term **hemiplegia** (hem-ee-PLEE-jee-ah) refers to paralysis (loss of voluntary muscle control) of one whole side of the body.

One of the most common crippling diseases that appears during childhood, **cerebral palsy**, also results from damage to brain tissue. Cerebral palsy involves permanent, nonprogressive damage to motor control areas of the brain. Such damage is present at birth or occurs shortly after birth and remains throughout life. Possible causes of brain damage include prenatal infections or diseases of the mother; mechanical trauma to the head before, during, or after birth; nerve-damaging poisons; reduced oxygen supply to the brain; and other factors. The resulting impairment to voluntary muscle control can manifest in various ways. Many people with cerebral palsy exhibit **spastic paralysis**, a type of paralysis characterized by involuntary contractions of affected muscles. In cerebral palsy, spastic paralysis often affects one entire side of the body (**hemiplegia**), both legs (**paraplegia**), both legs and one arm (**triplegia**), or all four extremities (**quadriplegia**).

Dementia

Various degenerative diseases can result in destruction of neurons in the brain. This degeneration can progress to adversely affect memory, attention span, intellectual capacity, personality, and motor control. The general term for this syndrome is **dementia** (de-MEN-shah).

Alzheimer disease (AD) (AHLZ-hye-mer dih-ZEEZ) is characterized by dementia. Its characteristic lesions develop in the cortex during the middle to late adult years. Exactly what causes dementia-producing lesions to develop in the brains of individuals with Alzheimer disease is not known. There is strong evidence that this disease has a genetic basis—at least in some families. A current theory is that more than one of the handful of different genes associated with AD has to be abnormal before AD occurs. Other evidence indicates that environmental factors may have a role. Because the exact cause of Alzheimer disease is still not known, development of an effective treatment has proven difficult. Currently, people diagnosed with this disease are treated by helping them maintain their remaining mental abilities and looking after their hygiene, nutrition, and other aspects of personal health management. A few drugs such as memantin (Namenda) and donepezil (Aricept) are available to slow down the progression of AD or lessen the severity of some of the symptoms.

Huntington disease (HD) is an inherited disease characterized by chorea (involuntary, purposeless movements) that progresses to severe dementia and death. The initial symptoms of this disease first appear between ages 30 and 40 years, with death generally occurring by age 55 years. The gene responsible for Huntington disease causes the body to make the protein *huntingtin* incorrectly. In brain cells, the abnormal form of huntingtin apparently clings to molecules too tightly and thus prevents normal function.

Acquired immune deficiency syndrome (AIDS), caused by HIV (human immunodeficiency virus) infection, can also cause dementia. The immune deficiency characteristic of AIDS results from HIV infection of white blood cells that are critical to the proper function of the immune system (see Chapter 21). However, HIV also infects neurons and can cause progressive degeneration of the brain—resulting in dementia.

Diseases caused by prions, pathogenic protein molecules, can also cause dementia. For example, *bovine spongiform encephalopathy*, also known as BSE or "mad cow disease," is a degenerative disease of the central nervous system caused by

Mechanisms of Disease—cont.

prions that convert normal proteins of the nervous system into abnormal proteins, causing loss of nervous system function, including dementia. **Creutzfeldt-Jakob** (KROYTS-felt YAH-kobe) **disease (CJD)** is another prion disease that similarly reduces brain function, causing dementia. These diseases have caused controversy recently because animal brains (the tissue that carries prions to other organisms) were fed to other animals in the human food chain, increasing the risk of infecting large numbers of humans. The mechanism of prion disease was outlined in Chapter 1, p. 27.

Seizure Disorders

Some of the most common nervous system abnormalities belong to the group of conditions called *seizure disorders*. These disorders are characterized by **seizures**—sudden bursts of abnormal neuron activity that result in temporary changes in brain function. Seizures may be very mild, causing subtle changes in level of consciousness, motor control, or sensory perception. On the other hand, seizures may be quite severe—resulting in jerky, involuntary muscle contractions called *convulsions* or even unconsciousness.

    Recurring or chronic seizure episodes constitute a condition called **epilepsy.** Although some cases of epilepsy can be traced to specific causes such as tumors or chemical imbalances, most epilepsy is idiopathic (of unknown cause). Epilepsy is often treated with anticonvulsive drugs such as *phenobarbital, phenytoin,* or *valproic acid* that block neurotransmitters in affected areas of the brain. By thus blocking synaptic transmission, such drugs inhibit the explosive bursts of neuron activity associated with seizures. With proper med-

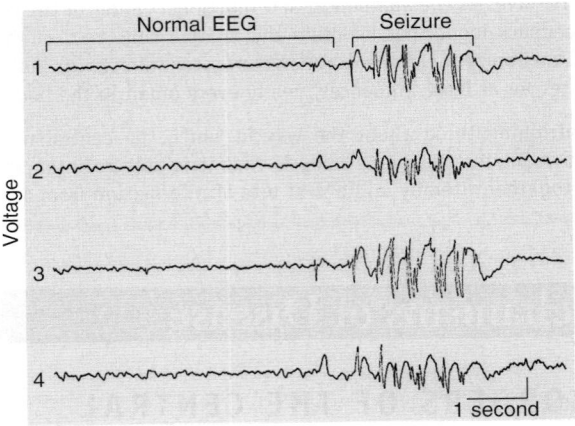

Figure 13-27 *Electroencephalogram (EEG).*

ication, many people with epilepsy lead normal lives without the fear of experiencing uncontrollable seizures. Even those who have not responded well to drug therapies have often gained some relief from surgeries that cut or destroy areas of the brain prone to severe seizures.

    Diagnosis and evaluation of epilepsy or any seizure disorder often rely on **electroencephalography** (see Box 13-9, p. 494). As Figure 13-27 illustrates, a normal EEG shows the moderate rise and fall of voltage in various parts of the brain, but a seizure manifests as an explosive increase in the size and frequency of voltage fluctuations. Different classifications of epilepsy are based on the location or locations and the duration of these changes in brain activity.

LANGUAGE OF SCIENCE   (Cont'd from page 471)

corpora quadrigemina (KOHR-pohr-ah kwod-ri-JEM-i-nah) [*corpora* body, *quadri-* fourfold, *-gemina* twin]

corpus callosum (KOHR-pus kah-LOH-sum) [*corpus* body, *callosum* callous]

cuneatus tracts (KYOO-nee-ay-tus) [*cuneatus* like a wedge, *tractus* trail]

dentate nuclei (DEN-tayt NOO-klee-eye) [*dentate* having teeth, *nucleus* nut or kernel]

descending tracts [*descend* move downward, *tractus* trail]

diencephalon (dye-en-SEF-ah-lon) [*di-* between, *-cephal* head]

dorsal (posterior) nerve root (DOR-sal) [*dors-* back, *-al* pertaining to]

epidural space (ep-i-DOO-ral) [*epi-* on or upon, *-dura-* hard, *-al* pertaining to]

epithalamus (ep-i-THAL-ah-mus) [*epi-* on or upon, *-thalamos* chamber]

extrapyramidal tracts (eks-trah-pi-RAH-mi-dal) [*extra-* outside, *-pyramid-* pyramid, *-al* pertaining to, *tractus* trail]

falx cerebelli (falks ser-eh-BEL-lee) [*falx* sickle, *cerebelli* of the cerebellum]

falx cerebri (falks SER-eh-bree) [*falx* sickle, *cerebri* of the cerebrum]

fasciculi gracilis (fah-SIK-yoo-lee GRAH-sil-iss) [*fasciculi* little bundles, *gracilis* thin]

filum terminale (FYE-lum ter-mi-NAL-ee) [*filum* thread, *termin-* boundary, *-al* pertaining to]

folia (FOH-lee-ah) [*folia* leaves]

frontal lobe (FRON-tal) [*front-* forehead, *-al* pertaining to]

funiculi (fuh-NIK-yoo-lee) [*funiculi* cords]

gray columns

gray commissure (KOHM-is-shoor) [*commissure* a joining together]

hemisphericity (HEM-i-sfeer-is-i-tee) [*hemi-* half, *-spher-* spherical body, *-ic-* pertaining to, *-ity* state of]

hypothalamus (hye-poh-THAL-ah-muss) [*hypo-* beneath, *-thalamos* chamber]

inferior cerebellar peduncles (SAIR-eh-bell-ar peh-DUNG-kuls) [*infer-* lower, *cerebellar* pertaining to the cerebellum, *peduncul* footstalk]

inferior colliculi (koh-LIK-yoo-lee) [*inferior* lower, *colliculus* small elevation]

infundibulum (in-fun-DIB-yoo-lum) [*infundibulum* funnel]

insula (IN-soo-lah) [*insula* island]

internal capsule [*intern-* inward, *-al* pertaining to, *capsula* a small box]

lateral corticospinal tracts (LAT-er-al kohr-ti-koh-SPY-nal) [*later-* side, *-al* pertaining to, *cortico-* cortex, *-spin-* backbone, *-al* pertaining to, *tractus* trail]

LANGUAGE OF SCIENCE *(Cont'd)*

lateral fissure (fissure of Sylvius) (LAT-er-al FISH-ur [FISH-ur of SIL-vi-us]) [*later-* side, *-al* pertaining to, *fissura* split, *Franciscus Sylvius* German medical professor]

lateral spinothalamic tracts (LAT-er-al spy-no-tha-LAM-ik trakts) [*later* side, *spino-* backbone, *-thalam-* inner chamber, *-ic* pertaining to, *tractus* trail]

lentiform nucleus (LEN-ti-form NOO-klee-us) [*lent-* lentil (lens), *-form* shape, *nucleus* nut or kernel]

limbic system (LIM-bik) [*limb-* edge, *-ic* pertaining to]

longitudinal fissure (lon-ji-TOO-dih-nal FISH-ur) [*longitud-* length, *-al* pertaining to, *fissura* split]

medulla oblongata (meh-DUL-ah ob-long-GAH-tah) [*medulla* marrow, *oblongata* oblong]

melatonin (mel-ah-TOH-nin) [*mela-* black, *-ton-* tone, normal state, *-in* substance]

meninges (meh-NIN-jeez) [*mening* membrane]

midbrain

middle cerebellar peduncles (SAIR-eh-bell-ar peh-DUNG-kuls) [*cerebell* pertaining to the cerebellum, *peduncul* footstalk]

motor program

occipital lobe (ok-SIP-it-al) [*occipit-* back of the head, *-al* pertaining to]

olive

optic chiasma (OP-tik kye-AS-mah) [*opti-* vision, *-ic* pertaining to, *chiasma* crossed lines]

parietal lobe (pah-RYE-eh-tal) [*pariet-* wall, *-al* pertaining to]

parietooccipital fissure (pah-RYE-eh-toh-ok-SIP-i-tal FISH-ur) [*parieto-* wall, *-occipit-* back of head, *-al* pertaining to, *fissure* split]

pineal gland (PIN-ee-al) [*pineus* pine cone]

pons (ponz) [*pons* bridge]

pyramidal tracts (pi-RAM-i-dal) [*pyramis-* pyramid, *-al* pertaining to, *tractus* trail]

pyramids (PEER-ah-mids) [*pyrami* pyramid]

reticular activating system (RAS) (reh-TIK-yoo-lar) [*reticul-* small net, *-al* pertaining to]

reticular formation (reh-TIK-yoo-lar) [*reticul-* small net, *-al* pertaining to, *tion* process of]

reticulospinal tracts (reh-TIK-yoo-loh-SPY-nal) [*reticul-* small net, *-spin-* backbone, *-al* pertaining to, *tractus* trail]

rubrospinal tracts (roo-broh-SPY-nal) [*rubro-* red, *-spin-* backbone, *-al* pertaining to, *tractus* trail]

spinal nerve (SPY-nal) [*spin-* backbone, *-al* pertaining to]

spinal tracts (SPY-nal) [*spin-* backbone, *-al* pertaining to, *tractus* trail]

spinocerebellar tracts (SPY-no-sair-eh-BELL-ar) [*spino-* backbone, *-cerebell-* small brain, *-ar* belonging to]

spinotectal tracts (SPY-no-TEK-tal) [*spino-* backbone, *-tectal* roof, *tractus* trail]

spinothalamic pathway (spy-no-thah-LAM-ik) [*spino-* backbone, *-thalam-* inner chamber, *-ic* pertaining to]

subarachnoid space (sub-ah-RAK-noyd) [*sub-* beneath, *-arachn-* spider, *-oid* resembling]

subdural space (sub-DOO-ral) [*sub-* beneath, *-dura-* hard or tough, *-al* pertaining to]

superior cerebellar peduncles (SAIR-eh-bell-ar peh-DUNG-kuls) [*superior* higher, *cerebellar* pertaining to the cerebellum, *peduncul* footstalk]

superior colliculi (koh-LIK-yoo-lee) [*superior* higher, *colliculus* small elevation]

tectospinal tracts (tek-toh-SPY-nal) [*tecto-* roof, *-spin-* backbone, *-al* pertaining to, *tractus* trail]

temporal lobe (TEM-poh-ral) [*tempor-* temples of head, *-al* pertaining to]

tentorium cerebelli (ten-TOR-ee-um sair-eh-BEL-lee) [*tentorium* tent, *cerebelli* of the cerebrum]

thalamus (THAL-ah-muss) [*thalamus* inner chamber]

ventral (anterior) nerve root (VEN-tral) [*ventr-* belly, *-al* pertaining to]

ventricles (VEN-tri-kuls) [*ventr-* belly, *-icle* small]

vermis (VER-mis) [*vermis* worm]

vestibulospinal tracts (ves-TIB-yoo-loh-SPY-nal) [*vestibul-* courtyard, *-spino-* backbone, *-al* pertaining to, *tractus* trail]

LANGUAGE OF MEDICINE

acquired immune deficiency syndrome (AIDS) (ah-KWYERD IM-yoon deh-FISH-en-see SIN-drohm)

Alzheimer disease (AD) (AHLZ-hye-mer) [*Alois Alzheimer* German neurologist]

brain electrical activity map (BEAM)

cerebral palsy (seh-REE-bral PAWL-zee) [*cerebr-* cerebrum, *-al* pertaining to, *para-* beyond, *-lysis* loosening]

cerebral plasticity (seh-REE-bral plas-TIS-i-tee) [*cerebr-* cerebrum, *-al* pertaining to, *plastic-* moldable, *-ity* state of]

cerebrovascular accident (CVA) (SAIR-eh-broh-VAS-kyoo-lar) [*cerebr-* cerebrum, *-vascul-* little vessel, *-ar* pertaining to]

coma (KOH-mah) [*koma* deep sleep]

computed tomography (CT) (kom-PYOO-ted toh-MOG-rah-fee) [*tomo-* section, *-graphy* a kind of recording]

Creutzfeldt-Jakob disease (CJD) (KROYTS-felt YAH-kobe) [*Hans G. Cruetzfeld* German neurologist, *Alfons M. Jakob* German neurologist]

dementia (de-MEN-shah) (deh-MEN-shah) [*de-* off, *-mens-* the mind, *-ia* condition of]

electroencephalograms (EEGs) (eh-lek-troh-en-SEF-ah-loh-grams) [*electro-* electricity, *-en-* inside, *-cephal-* head, *-gram* written record]

electroencephalography (eh-lek-troh-en-SEF-ah-log-rah-fee) [*electro-* electricity, *-en-* inside, *-cephal-* head, *-graphy* kind of recording]

epilepsy (EP-i-lep-see) [*epilepsia* seizure]

evoked potential (EP) (ee-VOHKT poh-TEN-shul)

functional MRI (fMRI) [*MRI* magnetic resonance imaging]

hemiplegia (hem-ee-PLEE-jee-ah) [*hemi-* half, *-plegia* stroke]

Huntington disease (HD) (HUN-ting-ton) [*George S. Huntington* American physician]

hydrocephalus (hye-droh-SEF-ah-lus) [*hydro-* water, *-cephalus* head]

lumbar puncture (LUM-bar) [*lumb-* loin, *-ar* pertaining to]

magnetic resonance imaging (MRI) (mag-NET-ik REZ-ah-nens IM-ah-jing) [*magne-* lodestone, *resona-* to sound again, *imag* image]

magnetoencephalography (MEG) (mag-NET-oh-en-sef-eh-LOG-reh-fee) [*magneto-* lodestone, *-en-* inside, *-cephalo-* head, *-graphy* kind of recording]

meningitis (men-in-JYE-tis) [*mening-* membrane, *-itis* inflammation]

pain control area

paraplegia (pair-ah-PLEE-jee-ah) [*para-* similar or besides, *-plegia* stroke]

Parkinson disease (PD) (PAR-kin-son) [*James Parkinson* English physician]

positron-emission tomography (PET) (POZ-i-tron eh-MISH-un toh-MOG-rah-fee) [*positron* positive electron, *emission* to send out, *tomo-* section, *-graphy* a kind of recording]

quadriplegia (kwod-ri-PLEE-jee-ah) [*quadri-* four, *-plegia* stroke]

rapid eye movement (REM) sleep

seizures (SEE-zhurz) [*saisir* to seize]

single-photon emission computed tomography (SPECT) (single-FOH-ton eh-MISH-un kom-PYOO-ted toh-MOG-rah-fee) [*photo-* light, *emission* to send out, *tomo-* section, *-graphy* a kind of recording]

slow-wave sleep (SWS)

spastic paralysis (SPAS-tik pah-RAL-i-sis) [*spastikos* drawing in, *paralyein* to be palsied]

transcutaneous electrical nerve stimulation (TENS) unit (trans-kyoo-TAY-nee-us) [*trans-* across, *-cutaneo* pertaining to skin]

triplegia (try-PLEE-jee-ah) [*tri-* three, *-plegia* stroke]

ultrasonography (ul-trah-son-OG-rah-fee) [*ultra-* beyond, *-sono-* sound, *-graphy* a kind of recording]

x-ray photography [*x* unknown, *photo-* light, *-graphy* a kind of recording]

CASE STUDY

Emilio Hernandez is a 19-year-old college student who was rock climbing and fell 30 feet to the ground. He is 6 feet 7 inches tall and very active on his college basketball team. His medical history is negative except for the usual childhood illnesses and minor accidents. At the time of his injury, he was unable to move any extremities and complained of neck discomfort. He was awake, alert, and able to answer questions appropriately. He had no feeling in his arms or legs. On physical examination, Emilio's pupils were equal and reactive to light. There were several scrapes on his arms. His blood pressure and heart rate were 110/72 and 86 beats/min, respectively. His respirations were 18/min, unlabored, and regular but shallow. Paramedics applied a cervical collar at the scene, placed him on a backboard, immobilized his head, and transported him to the medical center by helicopter.

In the emergency room, further physical examination revealed no deep tendon reflexes. Emilio did have sensory perception from his head to just above the nipple line of his chest. He was unable to expand his chest wall. His color was poor, but his skin was warm and dry. His blood pressure and heart rate were now 110/60 and 68 beats/min, respectively, with respirations at 24/min.

1. Which one of the following explanations *best* describes the changes in Emilio's vital signs?

   A. Lying on his back during transport
   B. The level and extent of injury

   C. Anxiety related to the injury
   D. Lack of mobility during transport

2. Emilio asks why he can feel a little sensation in some parts of his arms. The nurse's response would be based on the understanding that:

   A. A concussion is causing these transient signs.
   B. His injury affected the autonomic nervous center in the cerebellum.
   C. These findings are consistent with a C5 and C6 injury level.
   D. The brain pathways are too complex to depict the exact etiology.

3. In considering Emilio's respiratory status you would be concerned if:

   A. He continues to take shallow breaths.
   B. His anxiety causes episodes of shallow breathing.
   C. He is breathing only with his diaphragm.
   D. His breath sounds are bronchovesicular.

4. Because of the nature of Emilio's injury, all the following central nervous system functions will be affected *except*:

   A. Control of posture
   B. Conduction route to the brain
   C. Reflex activity for the spine
   D. Conduction route from the brain

CHAPTER SUMMARY

COVERINGS OF THE BRAIN AND SPINAL CORD

A. Two protective coverings (Figure 13-2)
   1. Outer covering is bone; cranial bones encase the brain, and vertebrae encase the spinal cord (Figure 13-1)
   2. Inner covering is the meninges; the meninges of the cord continue inside the spinal cavity beyond the end of the spinal cord
B. Meninges—three membranous layers (Figure 13-3)
   1. Dura mater—strong, white fibrous tissue; outer layer of meninges and inner periosteum of the cranial bones; has three important extensions
      a. Falx cerebri
         (1) Projects downward into the longitudinal fissure between the two cerebral hemispheres
         (2) Dural sinuses—function as veins, collecting blood from brain tissues for return to the heart
         (3) Superior sagittal sinus—one of several dural sinuses
      b. Falx cerebelli—separates the two hemispheres of the cerebellum
      c. Tentorium cerebelli—separates the cerebellum from the cerebrum
   2. Arachnoid mater—delicate, cobweb-like layer between the dura mater and pia mater
   3. Pia mater—innermost, transparent layer; adheres to the outer surface of the brain and spinal cord; contains blood

vessels; beyond the spinal cord, forms a slender filament called *filum terminale*, at level of sacrum, blends with dura mater to form a fibrous cord that disappears into the periosteum of the coccyx
   4. Several spaces exist between and around the meninges
      a. Epidural space—located between the dura mater and inside the bony covering of the spinal cord; contains a supporting cushion of fat and other connective tissues (virtually absent around brain because dura is continuous with periosteum of bone)
      b. Subdural space—located between the dura mater and arachnoid mater; contains lubricating serous fluid
      c. Subarachnoid space—located between the arachnoid and pia mater; contains a significant amount of cerebrospinal fluid

CEREBROSPINAL FLUID

A. Functions
   1. Provides a supportive, protective cushion
   2. Reservoir of circulating fluid, which is monitored by the brain to detect changes in the internal environment
B. Fluid spaces
   1. Cerebrospinal fluid—found within the subarachnoid space around the brain and spinal cord and within the cavities and canals of the brain and spinal cord

2. Ventricles—fluid-filled spaces within the brain; four ventricles within the brain (Figure 13-4)
   a. First and second ventricles (lateral)—one located in each hemisphere of the cerebrum
   b. Third ventricle—thin, vertical pocket of fluid below and medial to the lateral ventricles
   c. Fourth ventricle—tiny, diamond-shaped space where the cerebellum attaches to the back of the brainstem
C. Formation and circulation of cerebrospinal fluid (Figure 13-5)
   1. Occurs by separation of fluid from blood in the choroid plexuses
      a. Fluid from the lateral ventricles seeps through the interventricular foramen (of Monro) into the third ventricle
      b. From the third ventricle goes through the cerebral aqueduct into the fourth ventricle
      c. From the fourth ventricle fluid goes to two different areas
         (1) Some fluid flows directly into the central canal of the spinal cord
         (2) Some fluid leaves the fourth ventricle through openings in its roof into the cisterna magna, a space that is continuous with the subarachnoid space
      d. Fluid circulates in the subarachnoid space and then is absorbed into venous blood through the arachnoid villi

SPINAL CORD

A. Structure of the spinal cord (Figure 13-6)
   1. Lies within the spinal cavity and extends from the foramen magnum to the lower border of the first lumbar vertebra
   2. Oval-shaped cylinder that tapers slightly from above downward
   3. Two bulges, one in the cervical region and one in the lumbar region
   4. Anterior median fissure and posterior median sulcus are two deep grooves; anterior fissure is deeper and wider
   5. Nerve roots
      a. Fibers of dorsal nerve root
         (1) Carry sensory information into the spinal canal
         (2) Dorsal root ganglion—cell bodies of unipolar, sensory neurons make up a small region of gray matter in the dorsal nerve root
      b. Fibers of ventral nerve root
         (1) Carry motor information out of the spinal cord
         (2) Cell bodies of multipolar, motor neurons are in the gray matter of the spinal cord
   6. Interneurons are located in the spinal cord's gray matter core
   7. Spinal nerve—a single mixed nerve on each side of the spinal cord where the dorsal and ventral nerve roots join together
   8. Cauda equina—bundle of nerve roots extending (along with the filum terminale) from the conus medullaris (inferior end of spinal cord) (Figure 13-7)
   9. Gray matter
      a. Columns of gray matter extend the length of the cord
      b. Consists predominantly of cell bodies of interneurons and motor neurons
      c. In transverse section, looks like an **H** with the limbs being called the *anterior, posterior,* and *lateral horns of gray matter*; crossbar of **H** is the *gray commissure*

10. White matter
    a. Surrounds the gray matter and is subdivided in each half on the cord into three funiculi: anterior, posterior, and lateral white columns
    b. Each funiculus consists of a large bundle of axons divided into tracts
    c. Names of spinal tracts indicate the location of the tract, the structure in which the axons originate, and the structure in which they terminate
B. Functions of the spinal cord
    1. Provides conduction routes to and from the brain
       a. Ascending tracts—conduct impulses up the cord to the brain
       b. Descending tracts—conduct impulses down the cord from the brain
       c. Bundles of axons compose all tracts
       d. Tracts are both structural and functional organizations of nerve fibers
          (1) Structural—all axons of any one tract originate in the same structure and terminate in the same structure
          (2) Functional—all axons that compose one tract serve one general function
       e. Important ascending (sensory) tracts (Figure 13-8)
          (1) Lateral spinothalamic tracts—crude touch, pain, and temperature
          (2) Anterior spinothalamic tracts—crude touch, pressure
          (3) Fasciculi gracilis and cuneatus—discriminating touch and conscious kinesthesia
          (4) Spinocerebellar tracts—subconscious kinesthesia
          (5) Spinotectal—touch related to visual reflexes
       f. Important descending (motor) tracts (Figure 13-8)
          (1) Lateral corticospinal tracts—voluntary movements on opposite side of the body
          (2) Anterior corticospinal tracts—voluntary movements on same side of body
          (3) Reticulospinal tracts—maintain posture during movement
          (4) Rubrospinal tracts—transmit impulses that coordinate body movements and maintenance of posture
          (5) Tectospinal tracts—head and neck movements during visual reflexes
          (6) Vestibulospinal tracts—coordination of posture and balance
       g. Spinal cord—reflex center for all spinal reflexes; spinal reflex centers are located in the gray matter of the cord

BRAIN

A. Structures of the brainstem (Figures 13-9 and 13-10)
    1. Medulla oblongata
       a. Lowest part of the brainstem
       b. Part of the brain that attaches to spinal cord, located just above the foramen magnum
       c. A few centimeters in length and separated from the pons above by a horizontal groove
       d. Composed of white matter and a network of gray and white matter called the reticular formation network

e. Pyramids—two bulges of white matter located on the ventral side of the medulla; formed by fibers of the pyramidal tracts

f. Olive—oval projection located lateral to the pyramids

g. Nuclei—clusters of neuron cell bodies located in the reticular formation

2. Pons
   a. Located above the medulla and below the midbrain
   b. Composed of white matter and reticular formation

3. Midbrain
   a. Located above the pons and below the cerebrum; forms the midsection of the brain
   b. Composed of white tracts and reticular formation
   c. Extending divergently through the midbrain are cerebral peduncles; conduct impulses between the midbrain and cerebrum
   d. Corpora quadrigemina—landmark in midbrain
      (1) Made up of two inferior colliculi and two superior colliculi
      (2) Forms the posterior, upper part of the midbrain that lies just above the cerebellum
      (3) Inferior colliculus—contains auditory centers
      (4) Superior colliculus—contains visual centers
   e. Red nucleus and substantia nigra—clusters of cell bodies of neurons involved in muscular control

B. Functions of the brainstem
   1. Performs sensory, motor, and reflex functions
   2. Spinothalamic tracts—important sensory tracts that pass through the brainstem
   3. Fasciculi cuneatus and gracilis and spinoreticular tracts—sensory tracts whose axons terminate in the gray matter of the brainstem
   4. Corticospinal and reticulospinal tracts—two of the major tracts present in the white matter of the brainstem
   5. Nuclei in medulla—contain reflex centers
      a. Of primary importance—cardiac, vasomotor, and respiratory centers
      b. Nonvital reflexes—vomiting, coughing, sneezing, and so on
   6. Pons—contains reflexes mediated by fifth, sixth, seventh, and eighth cranial nerves and pneumotaxic centers that help regulate respiration
   7. Midbrain—contains centers for certain cranial nerve reflexes

C. Structure of the cerebellum (Figure 13-11)
   1. Second largest part of the brain—contains more neurons than the rest of the nervous system
   2. Located just below the posterior portion of the cerebrum; transverse fissure separates these two parts of the brain
   3. Gray matter makes up the cortex, and white matter predominates in the interior
   4. Arbor vitae—internal white matter of the cerebellum; distinctive pattern similar to the veins of a leaf
   5. Cerebellum has numerous sulci and delicate, parallel gyri (folia)
   6. Consists of the cerebellar hemispheres and the vermis

7. Internal white matter—composed of short and long tracts
   a. Shorter tracts—conduct impulses from neuron cell bodies located in the cerebellar cortex to neurons whose dendrites and cell bodies compose nuclei located in the interior of the cerebellum
   b. Longer tracts—conduct impulses to and from the cerebellum; fibers enter or leave by way of three pairs of peduncles
      (1) Inferior cerebellar peduncles—composed chiefly of tracts into the cerebellum from the medulla and cord
      (2) Middle cerebellar peduncles—composed almost entirely of tracts into the cerebellum from the pons
      (3) Superior cerebellar peduncles—composed principally of tracts from dentate nuclei in the cerebellum through the red nucleus of the midbrain to the thalamus

8. Dentate nuclei
   a. Important pair of cerebellar nuclei, one of which is located in each hemisphere
   b. Nuclei connected with thalamus and with motor areas of the cerebral cortex by tracts
   c. By means of the tracts, cerebellar impulses influence the motor cortex, and the motor cortex influences the cerebellum

D. Functions of the cerebellum
   1. Cerebellum compares the motor commands of the cerebrum with the information coming from proprioceptors in the muscle; impulses travel from the cerebellum to both the cerebrum and muscles to coordinate movements to produce the intended action (Figure 13-12)
   2. General functions
      a. Acts with cerebral cortex to produce skilled movements by coordinating the activities of groups of muscles
      b. Controls skeletal muscles to maintain balance
      c. Controls posture; operates at subconscious level to smooth movements and make movements efficient and coordinated
      d. Processes sensory information; complements and assists various functions of the cerebrum

E. The diencephalon (Figure 13-13)
   1. Located between the cerebrum and the midbrain
   2. Consists of several structures located around the third ventricle: thalamus, hypothalamus, optic chiasma, pineal gland, and several others
   3. The thalamus
      a. Dumbbell-shaped mass of gray matter made up of many nuclei
      b. Each lateral mass forms one lateral wall of the third ventricle
      c. Intermediate mass—extends through the third ventricle and joins the two lateral masses
      d. Geniculate bodies—two of the most important groups of nuclei comprising the thalamus; located in posterior region of each lateral mass; play role in processing auditory and visual input

e. Serves as a major relay station for sensory impulses on their way to the cerebral cortex

f. Performs the following primary functions:

(1) Plays two parts in mechanism responsible for sensations

(a) Impulses produce conscious recognition of the crude, less critical sensations of pain, temperature, and touch

(b) Neurons relay all kinds of sensory impulses, except possibly olfactory, to the cerebrum

(2) Plays part in the mechanism responsible for emotions by associating sensory impulses with feeling of pleasantness and unpleasantness

(3) Plays part in arousal mechanism

(4) Plays part in mechanisms that produce complex reflex movements

4. The hypothalamus

a. Consists of several structures that lie beneath the thalamus

b. Forms floor of the third ventricle and lower part of lateral walls

c. Prominent structures found in the hypothalamus

(1) Supraoptic nuclei—gray matter located just above and on either side of the optic chiasma

(2) Paraventricular nuclei—located close to the wall of the third ventricle

(3) Mamillary bodies—posterior part of hypothalamus, involved with olfactory sense

d. Infundibulum—the stalk leading to the posterior lobe of the pituitary gland

e. Small but functionally important area of the brain, performs many functions of greatest importance for survival and enjoyment

f. Links mind and body

g. Links nervous system to endocrine system

h. Summary of hypothalamic functions

(1) Regulator and coordinator of autonomic activities

(2) Major relay station between the cerebral cortex and lower autonomic centers; crucial part of the route by which emotions can express themselves in changed bodily functions

(3) Synthesizes hormones secreted by posterior pituitary and plays an essential role in maintaining water balance

(4) Some neurons function as endocrine glands

(5) Plays crucial role in arousal mechanism

(6) Crucial part of mechanism regulating appetite

(7) Crucial part of mechanism maintaining normal body temperature

5. Pineal gland

a. Located just above the corpora quadrigemina of the midbrain

b. Involved in regulating the body's biological clock (Figure 13-14)

c. Produces melatonin as a "timekeeping hormone"

(1) Melatonin is made from the neurotransmitter serotonin

(2) Melatonin levels increase when sunlight is absent; decrease when sunlight is present, thus regulating the circadian (daily) biological clock (Figure 13-15)

(3) Melatonin is the "sleep hormone"

F. Structure of the cerebrum

1. Cerebral cortex

a. Largest and uppermost division of the brain; consists of right and left cerebral hemispheres; each hemisphere is divided into five lobes (Figure 13-16)

(1) Frontal lobe

(2) Parietal lobe

(3) Temporal lobe

(4) Occipital lobe

(5) Insula (island of Riel)

b. Cerebral cortex—outer surface made up of six layers of gray matter

c. Gyri—convolutions; some are named: precentral gyrus, postcentral gyrus, cingulate gyrus, and hippocampal gyrus

d. Sulci—shallow grooves

e. Fissures—deeper grooves, divide each cerebral hemisphere into lobes; four prominent cerebral fissures

(1) Longitudinal fissure—deepest fissure; divides cerebrum into two hemispheres

(2) Central sulcus (fissure of Rolando)—groove between frontal and parietal lobes

(3) Lateral fissure (fissure of Sylvius)—groove between temporal lobe below and parietal lobes above; island of Reil lies deep in lateral fissure

(4) Parietooccipital fissure—groove that separates occipital lobe from parietal lobes

2. Cerebral tracts and basal nuclei

a. Basal nuclei—islands of gray matter located deep inside the white matter of each hemisphere (Figure 13-18); include the following:

(1) Caudate nucleus

(2) Lentiform nucleus—consists of putamen and pallidum

(3) Amygdaloid nucleus

b. Cerebral tracts make up cerebrum's white matter; there are three types (Figure 13-17)

(1) Projection tracts—extensions of the sensory spinothalamic tracts and motor corticospinal tracts

(2) Association tracts—most numerous cerebral tracts; extend from one convolution to another in the same hemisphere

(3) Commissural tracts—extend from one convolution to a corresponding convolution in the other hemisphere; compose the corpus callosum and anterior and posterior commissures

c. Corpus striatum—composed of caudate nucleus, internal capsule, and lentiform nucleus

G. Functions of the cerebral cortex
1. Functional areas of the cortex—certain areas of the cerebral cortex engage in predominantly one particular function (Figures 13-19 and 13-20)
   a. Postcentral gyrus—mainly general somatic sensory area; receives impulses from receptors activated by heat, cold, and touch stimuli
   b. Precentral gyrus—chiefly somatic motor area; impulses from neurons in this area descend over motor tracts and stimulate skeletal muscles
   c. Transverse gyrus—primary auditory area
   d. Occipital lobe—primary visual areas
2. Sensory functions of the cortex
   a. Somatic senses—sensations of touch, pressure, temperature, proprioception, and similar perceptions that require complex sensory organs
   b. Cortex contains a "somatic sensory map" of the body
   c. Information sent to primary sensory areas is relayed to sensory association areas, as well as to other parts of the brain
   d. The sensory information is compared and evaluated, and the cortex integrates separate bits of information into whole perceptions
3. Motor functions of the cortex
   a. For normal movements to occur, many parts of the nervous system must function
   b. Precentral gyrus—primary somatic motor area; controls individual muscles
   c. Secondary motor area—in the gyrus immediately anterior to the precentral gyrus; activates groups of muscles simultaneously
4. Integrative functions of the cortex
   a. Consciousness (Figure 13-21)
      (1) State of awareness of one's self, one's environment, and other humans
      (2) Depends on excitation of cortical neurons by impulses conducted to them by the reticular activating system
      (3) Two current concepts about the reticular activating system
         (a) Functions as the arousal system for the cerebral cortex
         (b) Its functioning is crucial for maintaining consciousness
   b. Language
      (1) Ability to speak and write words and ability to understand spoken and written words
      (2) Speech centers—areas in the frontal, parietal, and temporal lobes
      (3) Left cerebral hemisphere contains speech centers in approximately 90% of the population; in the remaining 10%, contained in either the right hemisphere or both
      (4) Aphasias—lesions in speech centers
   c. Emotions (Figure 13-22)
      (1) Subjective experiencing and objective expressing of emotions involve functioning of the limbic system

      (2) Limbic system—also known as the "emotional brain"
         (a) Most structures of limbic system lie on the medial surface of the cerebrum; they are the cingulate gyrus and hippocampus
         (b) Have primary connections with other parts of the brain, such as the thalamus, fornix, septal nuclei, amygdaloid nucleus, and hypothalamus
   d. Memory
      (1) One of our major mental activities
      (2) Cortex is capable of storing and retrieving both short- and long-term memory
      (3) Temporal, parietal, and occipital lobes are among the areas responsible for short- and long-term memory
      (4) Engrams—structural traces in the cerebral cortex that make up long-term memories
      (5) Cerebrum's limbic system plays a key role in memory
5. Hemisphericity—specialization of cerebral hemispheres
   a. Right and left hemispheres of the cerebrum specialize in different functions; however, both sides of a normal person's brain communicate with each other to accomplish complex functions
   b. Left hemisphere is responsible for:
      (1) Language functions
      (2) Dominating control of certain hand movements
   c. Right hemisphere is responsible for:
      (1) Perception of certain kinds of auditory material
      (2) Tactual perception
      (3) Perceiving and visualizing spatial relationships

SOMATIC SENSORY PATHWAYS IN THE CENTRAL NERVOUS SYSTEM

A. For the cerebral cortex to perform its sensory functions, impulses must first be conducted to the sensory areas by sensory pathways (Figure 13-24)
B. Three main pools of sensory neurons
1. Primary sensory neurons—conduct impulses from the periphery to the central nervous system
2. Secondary sensory neurons
   a. Conduct impulses from the cord or brainstem to the thalamus
   b. Dendrites and cell bodies are located in the gray matter of the cord and brainstem
   c. Axons ascend in ascending tracts up the cord and through the brainstem, terminating in the thalamus, where they synapse with dendrites or cell bodies of tertiary sensory neurons
3. Tertiary sensory neurons
   a. Conduct impulses from thalamus to the postcentral gyrus of the parietal lobe
   b. Bundle of axons of tertiary sensory neurons form the thalamocortical tracts
   c. Extend through the internal capsule to the cerebral cortex
C. Sensory pathways to the cerebral cortex are crossed

D. Two sensory pathways conduct impulses that produce sensations of touch and pressure
  1. Medial lemniscal system
    a. Consists of tracts that make up the fasciculi cuneatus and gracilis, and the medial lemniscus
    b. Axons of secondary sensory neurons make up medial lemniscus
    c. Functions—transmit impulses that produce discriminating touch and pressure sensations and kinesthesia
  2. Spinothalamic pathway—functions are crude touch and pressure sensations

SOMATIC MOTOR PATHWAYS IN THE CENTRAL NERVOUS SYSTEM

A. For the cerebral cortex to perform its motor functions, impulses are conducted from its motor areas to skeletal muscles by somatic motor pathways
B. Consist of motor neurons that conduct impulses from the central nervous system to skeletal muscles; some motor pathways are extremely complex, and others are very simple
C. Principle of the final common path—cardinal principle about somatic motor pathways; only one final common path, the motor neuron from the anterior gray horn of the spinal cord, conducts impulses to skeletal muscles
D. Two methods used to classify somatic motor pathways
  1. Divides pathways into pyramidal and extrapyramidal tracts (Figure 13-25)
    a. Pyramidal tracts—also known as *corticospinal tracts*
      (1) Approximately three quarters of the fibers decussate in the medulla and extend down the cord in the crossed corticospinal tract located on the opposite side of the spinal cord in the lateral white column
      (2) Approximately one quarter of the fibers do not decussate but extend down the same side of the spinal cord as the cerebral area from which they came
    b. Extrapyramidal tracts—much more complex than pyramidal tracts
      (1) Consist of all motor tracts from the brain to the spinal cord anterior horn motor neurons except the corticospinal tracts
      (2) Within the brain, consist of numerous relays of motor neurons between motor areas of the cortex, basal nuclei, thalamus, cerebellum, and brainstem
      (3) Within the spinal cord, some important tracts are the reticulospinal tracts
      (4) Conduction by extrapyramidal tracts plays a crucial part in producing large, automatic movements
      (5) Conduction by extrapyramidal tracts plays an important part in emotional expressions
      (6) Motor program—set of coordinated commands that control the programmed motor activity mediated by extrapyramidal pathways (Figure 13-26)

CYCLE OF LIFE: CENTRAL NERVOUS SYSTEM

A. The development and degeneration of the central nervous system is the most obvious functional change over the life span
B. Development of brain and spinal cord begins in the womb

C. Lack of development in the newborn is evidenced by lack of complex integrative functions
  1. Language
  2. Complex memory
  3. Comprehension of spatial relationships
  4. Complex motor skills
D. Complex functions develop by adulthood
E. Late adulthood—tissues degenerate
  1. Profound degeneration—unable to perform complex functions
  2. Milder degeneration—temporary memory lapse or difficulty with complex motor tasks

THE BIG PICTURE: THE CENTRAL NERVOUS SYSTEM AND THE WHOLE BODY

A. Central nervous system—ultimate regulator of the body; essential to survival
B. Able to integrate bits of information from all over the body, make sense of it, and make decisions

REVIEW QUESTIONS

1. What term means the membranous covering of the brain and cord? What three layers compose this covering?
2. What are the large fluid-filled spaces within the brain called? How many are there? What do they contain?
3. Describe the formation and circulation of cerebrospinal fluid.
4. Describe the spinal cord's structure and general functions.
5. List the major components of the brainstem, and identify their general functions.
6. Describe the general functions of the cerebellum.
7. Describe the general functions of the thalamus.
8. Describe the general functions of the hypothalamus.
9. Describe the general functions of the cerebrum.
10. What general functions does the cerebral cortex perform?
11. Define consciousness. Name the normal states, or levels, of consciousness.
12. Name some altered states of consciousness.
13. Identify the following kinds of brain waves according to their frequency, voltage, and the level of consciousness in which they predominate: alpha, beta, delta, theta.
14. Locate the dendrite, cell body, and axon of primary, secondary, and tertiary sensory neurons.

CRITICAL THINKING QUESTIONS

1. Explain what the term *reflex center* means. Is an interneuron necessary for a reflex center? Explain your answer.
2. Explain briefly what is meant by the arousal or alerting mechanism.
3. Some people claim meditation has a wide range of benefits. What benefits can be supported by scientific evidence?
4. If a researcher discovered that a substantial reduction in neurotransmitter concentration caused difficulty in forming memory, what theory of memory formation would be refuted?

5. In Chapter 12, an action potential was explained in terms of electrical activity. Explain the process of measuring that activity to differentiate the types of brain waves.

6. A person having an absence of any reflex and a person having exaggerated deep tendon reflexes are showing signs of different motor pathway injuries. Using these symptoms, explain the motor pathway that was damaged and what other symptoms each person might have.

7. Compare pyramidal tract and extrapyramidal tract functions.

8. A patient with a brain infection can be diagnosed by culturing cerebrospinal fluid. The greatest concentration of the disease-causing organism can be drawn from the fluid as soon as it leaves the brain at the level of the third or fourth vertebra. Why would this be an unwise place to take the sample? Where would a better location be? Explain your answer.

Peripheral Nervous System

LANGUAGE OF SCIENCE

abdominal reflex (ab-DOM-i-nal REE-fleks) [*abdomin-* belly, *-al* pertaining to, *re-* again, *-flex* bend]

abducens nerve (ab-DOO-sens) [*abducens* drawing away]

accessory nerve (ak-SES-oh-ree)

acetylcholine (Ach) (ass-ee-til-KOH-leen) [*acetyl-* vinegar, *-chole-* bile, *-ine* made of]

alpha (α) receptors (AL-fah ree-SEP-tors) [*alpha* first letter of the Greek alphabet, *recipere* to receive]

ankle jerk reflex [*re-* again, *-flex* bend]

autonomic nervous system (ANS) (aw-toh-NOM-ik) [*auto-* self, *-nomo-* law, *-ic* pertaining to]

autonomic (visceral) reflexes (aw-toh-NOM-ik REE-fleks-es) [*auto-* self, *-nomo-* law, *-ic* pertaining to, *re-* again, *-flex* bend]

Babinski reflex (bah-BIN-skee REE-fleks) [*Joseph F.F. Babinski* French neurologist, *re-* again, *-flex* bend]

beta (β) receptors (BAY-tah ree-SEP-tors) [*beta* second letter of the Greek alphabet]

brachial plexus (BRAY-kee-all PLEK-sus) [*brachi-* arm, *-al* pertaining to, *plexus* network]

cauda equina (KAW-dah eh-KWINE-ah) [*cauda* tail, *equina* of a horse]

cervical plexus (SER-vih-kal PLEK-sus) [*cervic-* neck, *-al* pertaining to, *plexus* network]

collateral ganglia (koh-LAT-er-al GANG-glee-ah) [*co-* together, *-later-* side, *-al* pertaining to, *ganglion* knot]

corneal reflex (KOR-nee-al REE-fleks) [*corn-* horn, *-al* pertaining to, *re-* again, *-flex* bend]

co-transmission (koh-tranz-MISH-un) [*co-* together]

cranial nerves (KRAY-nee-al) [*crani-* skull, *-al* pertaining to]

cranial reflex (KRAY-nee-al REE-fleks) [*crani-* skull, *-al* pertaining to, *re-* again, *-flex* bend]

dermatome (DER-mah-tohm) [*derma-* skin, *-tome* segment or region]

dorsal ramus (DOR-sal RAY-mus) [*dors-* the back, *-al* pertaining to, *ramus* branch]

Cont'd on p. 546

In Chapter 13, you learned about the structure and function of the central nervous system (CNS). In this chapter we explore the nerve pathways that lead to and from the CNS, which together make up the peripheral nervous system (PNS). The PNS is made up of the 31 pairs of spinal nerves that emerge from the spinal cord, the 12 pairs of cranial nerves that emerge from the brain, and all the smaller nerves that branch from these "main" nerves.

Recall from earlier discussions that *afferent* fibers carry information into the CNS. Afferent fibers, part of the sensory nervous system, help us maintain homeostasis by sensing changes in our internal or external environment—providing the feedback necessary to keep the body functioning normally. Afferent fibers belonging to the somatic sensory or special sensory nervous system feed back information regarding changes detected by receptors in the skin, skeletal muscles, and special sense organs. Afferent fibers belonging to the autonomic nervous system (ANS) feed back information regarding the effects of autonomic control of the viscera. Recall also that *efferent* fibers carry information away from the CNS. Efferent fibers may belong to the somatic nervous system (SNS) that regulates skeletal muscles, allowing us to survive by defending ourselves, getting food, or performing other essential tasks. Efferent fibers may, on the other hand, belong to the ANS. Autonomic regulation controls smooth and cardiac muscle and glands in ways that help us maintain homeostasis of the internal environment.

The first sections of this chapter explore concepts regarding the structure and function of the spinal and cranial nerves. The last portion of the chapter is devoted to discussion of the peripheral elements of the SNS and the ANS, with an emphasis on efferent (motor) pathways. Chapter 15 then picks up the story, emphasizing afferent (sensory) pathways.

SPINAL NERVES

Thirty-one pairs of **spinal nerves** are connected to the spinal cord. They have no special names but are merely numbered according to the level of the vertebral column at which they emerge from the spinal cavity (Figure 14-1). Although there are only seven cervical vertebrae, there are eight cervical nerve pairs (C1 through C8), twelve thoracic nerve pairs (T1 through T12), five lumbar nerve pairs (L1 through L5), five sacral nerve pairs (S1 through S5), and one coccygeal pair of spinal nerves. The first pair of cervical nerves emerges from the cord in the space above the first cervical vertebra, and the eighth cervical nerve emerges between the last cervical vertebra and the first thoracic vertebra. Thus nerve pair C1 passes between the skull and vertebra C1, nerve pair C2 passes between vertebrae C1 and C2, nerve pair C3 passes between vertebrae C2 and C3, and so on. All the thoracic nerves pass out of the spinal cavity horizontally through the intervertebral foramina below their respective vertebrae.

Lumbar, sacral, and coccygeal nerve roots, on the other hand, descend from their point of origin at the lower end of the spinal cord (which terminates at the level of the first lumbar vertebra) before reaching the intervertebral foramina of their respective vertebrae, through which the nerves then emerge. This gives the lower end of the cord, with its attached spinal nerve roots, the appearance of a horse's tail. In fact, it bears the name **cauda equina,** which is the Latin equivalent for "horse's tail" (see Figure 14-1).

Structure of Spinal Nerves

Each spinal nerve attaches to the spinal cord by means of two short roots, a **ventral** (anterior) **root** and a **dorsal** (posterior) **root.** The dorsal root of each spinal nerve is easily recognized by a swelling called the dorsal root **ganglion,** or *spinal ganglion* (Figure 14-2). The roots and dorsal ganglia lie within the spinal cavity, as Figure 14-2 shows.

As described in Chapter 13, the ventral root includes motor neurons that carry information from the CNS and toward effectors (muscles and glands). Recall that in each somatic motor pathway, a single motor fiber stretches from the anterior gray horn of the spinal cord, through the ventral root, and on through the spinal nerve toward a skeletal muscle. Autonomic fibers, which also carry motor information toward effectors, may also pass through the ventral root to become part of a spinal nerve. The dorsal root of each spinal nerve includes sensory fibers that carry information from receptors in the peripheral nerves. The dorsal root ganglion contains the cell bodies of the sensory neurons. Because all spinal nerves contain both motor and sensory fibers, they are designated **mixed nerves.**

Soon after each spinal nerve emerges from the spinal cavity, it forms several large branches, each of which is called a **ramus** (plural, *rami*). As Figure 14-2 shows, each spinal nerve splits into a distinct **dorsal ramus** and **ventral ramus.** The dorsal ramus supplies somatic motor and sensory fibers to several smaller nerves. These smaller nerves, in turn, innervate the muscles and skin of the posterior surface of the head, neck, and trunk. The structure of the ventral ramus is a little more complex. Autonomic motor fibers split away from the ventral ramus, heading toward a ganglion of the *sympathetic chain.* There, some of the autonomic fibers synapse with autonomic neurons that eventually continue on to autonomic effectors by way of *splanchnic nerves* (see Figure 14-2). However,

BOX 14-1: HEALTH MATTERS
Peripheral Neuropathy

The term **peripheral neuropathy** tells you exactly what it signifies: disease of the peripheral nerves. The term designates many different diseases with many different causes—but all involve damage to peripheral nerves. These disorders can also be called *peripheral neuritis.* If many nerves are involved, peripheral neuropathy can also be called *polyneuropathy* or *polyneuritis.* Probably the most common cause of peripheral nerve damage in the United States and Europe is diabetes mellitus (see Box 16-12, p. 627), affecting some 17 million people. But additional causes include other metabolic problems such as kidney disease, injuries such as sports mishaps, poisoning as in drug neurotoxicity and alcohol abuse, infections such as in *leprosy (Hansen disease; Mycobacterium leprae)* or shingles (see Box 14-3, p. 523), and genetic mutations such as *Leber hereditary optic neuropathy* (see Chapter 34, p. 1210).

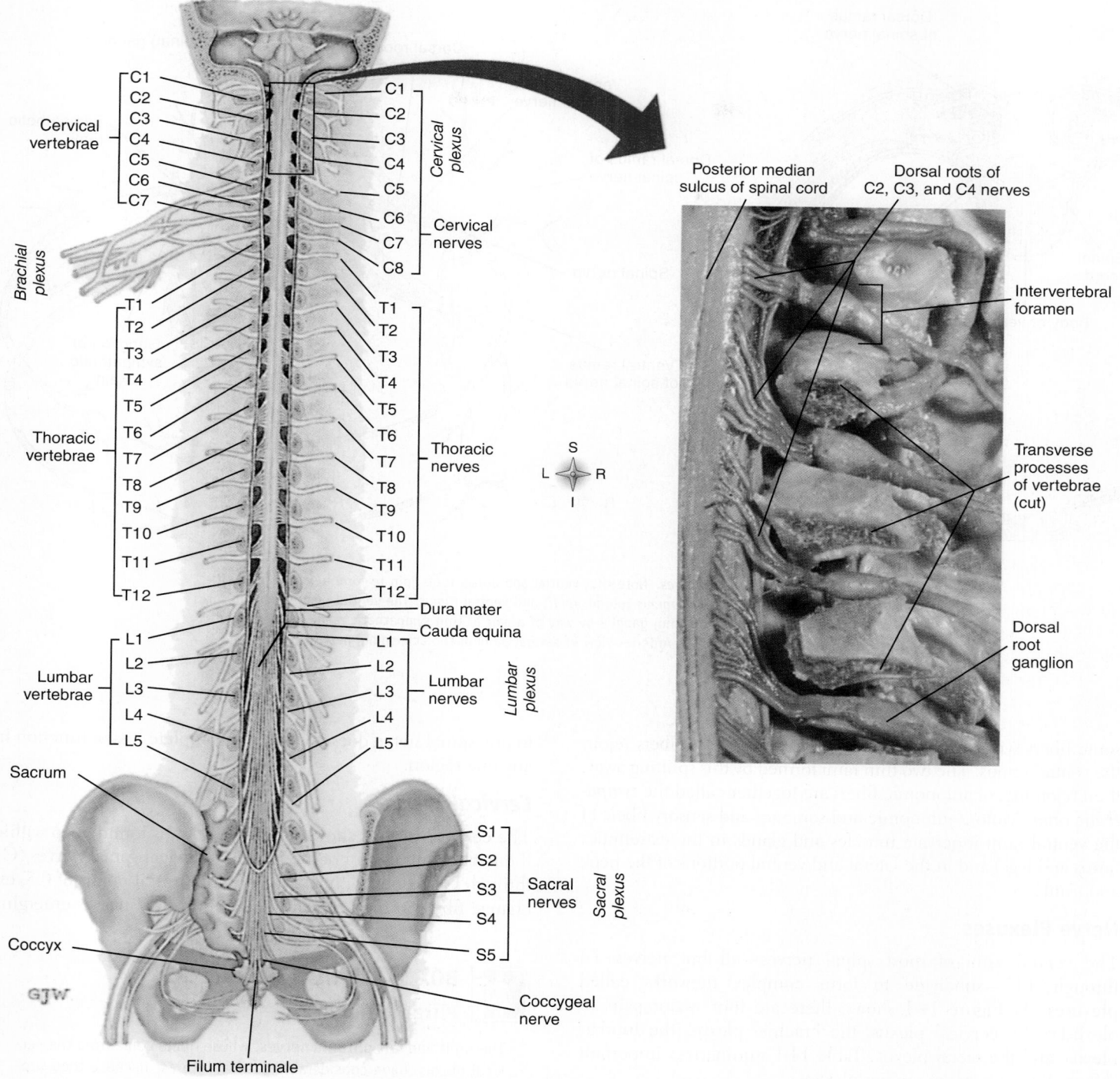

Cervical vertebrae

C1
C2
C3
C4
C5
C6
C7

Brachial plexus

C1
C2
C3
C4 } Cervical plexus
C5
C6 } Cervical nerves
C7
C8

Thoracic vertebrae

T1
T2
T3
T4
T5
T6
T7
T8
T9
T10
T11
T12

T1
T2
T3
T4
T5
T6 } Thoracic nerves
T7
T8
T9
T10
T11
T12

Posterior median sulcus of spinal cord

Dorsal roots of C2, C3, and C4 nerves

Intervertebral foramen

Transverse processes of vertebrae (cut)

Dorsal root ganglion

Dura mater
Cauda equina

Lumbar vertebrae

L1
L2
L3
L4
L5

L1
L2
L3 } Lumbar nerves
L4
L5

Lumbar plexus

Sacrum

Coccyx

S
L ✦ R
I

S1
S2
S3 } Sacral nerves
S4
S5

Sacral plexus

Coccygeal nerve

Filum terminale

GJW

Figure 14-1 *Spinal nerves.* Each of 31 pairs of spinal nerves exits the spinal cavity from the intervertebral foramina. The names of the vertebrae are given on the left and the names of the corresponding spinal nerves on the right. Note that after leaving the spinal cavity, many of the spinal nerves interconnect to form networks called plexuses. The inset shows a dissection of the cervical region, showing a posterior view of cervical spinal nerves exiting intervertebral foramina on the right side.

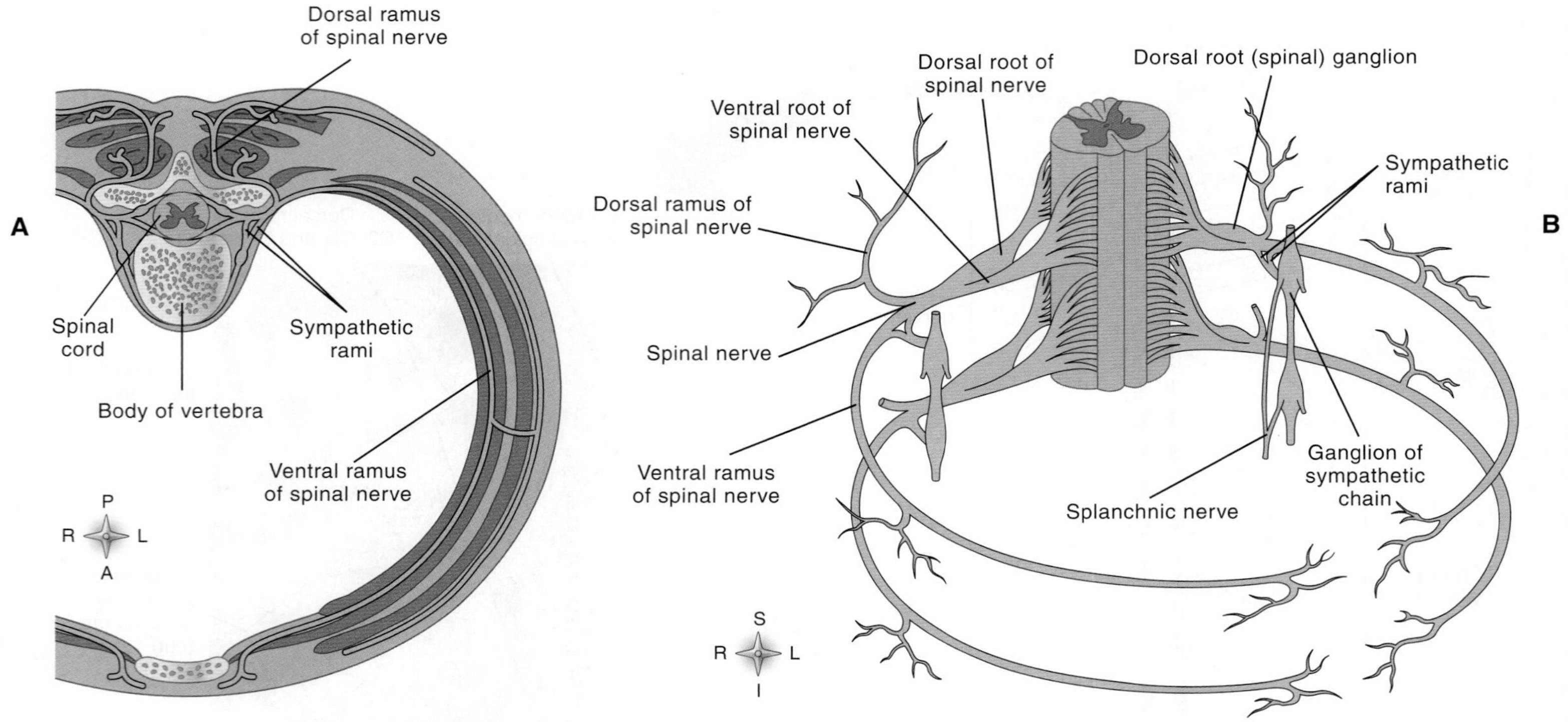

Figure 14-2 *Rami of the spinal nerves.* Note that ventral and dorsal roots join to form a spinal nerve. The spinal nerve then splits into a dorsal *ramus* (plural, *rami*) and ventral *ramus*. The ventral ramus communicates with a chain of sympathetic (autonomic) ganglia by way of a pair of thin sympathetic rami. **A,** Superior view of a pair of thoracic spinal nerves. **B,** Anterior view of several pairs of thoracic spinal nerves.

some fibers synapse with autonomic neurons whose fibers rejoin the ventral ramus. The two thin rami formed by this splitting away, then rejoining. of autonomic fibers are together called the *sympathetic rami.* Motor (autonomic and somatic) and sensory fibers of the ventral rami innervate muscles and glands in the extremities (arms and legs) and in the lateral and ventral portions of the neck and trunk.

Nerve Plexuses

The ventral rami of most spinal nerves—all but nerves T2 through T12—subdivide to form complex networks called **plexuses.** As Figure 14-1 shows, there are four major pairs of plexuses: the cervical plexus, the brachial plexus, the lumbar plexus, and the sacral plexus. Table 14-1 summarizes important information about these major plexuses.

The term *plexus* is the Latin word for "braid." This is an apt name for a structure in which fibers of several different rami join together to form individual nerves. Each individual nerve that emerges from a plexus contains all the fibers that innervate a particular region of the body. In fact, the destination of each nerve serves as a basis for its name (see Table 14-1). Because spinal nerve fibers are thus rearranged according to their ultimate destination, the plexus reduces the number of nerves needed to supply each body part. And because each body region is innervated by fibers that originate from several adjacent spinal nerves, damage

to one spinal nerve does not mean a complete loss of function in any one region.

Cervical Plexus

The **cervical plexus,** shown in Figure 14-3, is found deep within the neck. Ventral rami of the first four cervical spinal nerves (C1 through C4), along with a branch of the ventral ramus of C5, exchange fibers in the cervical plexus. Individual nerves emerging

BOX 14-2: HEALTH MATTERS
Phrenic Nerves

The right and left **phrenic nerves,** whose fibers come from the cervical plexus, have considerable clinical interest because they supply the diaphragm muscle. Contraction of the diaphragm permits inspiration, and relaxation of the diaphragm permits expiration. If the neck is broken in a way that severs or crushes the spinal cord above this level, nerve impulses from the brain can no longer reach the phrenic nerves, and therefore the diaphragm stops contracting. Unless artificial respiration of some kind is provided, the patient dies of respiratory paralysis as a result of the broken neck. Any disease or injury that damages the spinal cord between the third and fifth cervical segments may paralyze the phrenic nerve and, therefore, the diaphragm.

Table 14-1   Spinal Nerves and Peripheral Branches

SPINAL NERVES	PLEXUSES FORMED FROM ANTERIOR RAMI	SPINAL NERVE BRANCHES FROM PLEXUSES	PARTS SUPPLIED
Cervical 1 2 3 4	Cervical plexus	Lesser occipital Greater auricular Cutaneous nerve of neck Supraclavicular nerves Branches to muscles	Sensory to back of head, front of neck, and upper part of shoulder; motor to numerous neck muscles
Cervical 5 6 7 8	Brachial plexus	Phrenic nerve Suprascapular and dorsoscapular Thoracic nerves, medial and lateral branches	Diaphragm Superficial muscles* of scapula Pectoralis major and minor
Thoracic (or Dorsal) 1		Long thoracic nerve Thoracodorsal Subscapular	Serratus anterior Latissimus dorsi Subscapular and teres major muscles
2 3 4		Axillary (circumflex) Musculocutaneous	Deltoid and teres minor muscles and skin over deltoid Muscles of front of arm (biceps brachii, coracobrachialis, brachialis) and skin on outer side of forearm
5 6 7	No plexus formed; branches run directly to intercostal muscles and skin of thorax	Ulnar	Flexor carpi ulnaris and part of flexor digitorum profundus; some muscles of hand; sensory to medial side of hand, little finger, and medial half of fourth finger
8 9		Median	Rest of muscles of front of forearm and hand; sensory to skin of palmar surface of thumb, index, and middle fingers
10 11		Radial	Triceps muscle and muscles of back of forearm; sensory to skin of back of forearm and hand
12		Medial cutaneous	Sensory to inner surface of arm and forearm
Lumbar 1 2		Iliohypogastric ⎤ Sometimes Ilioinguinal ⎦ fused	Sensory to anterior abdominal wall Sensory to anterior abdominal wall and external genitalia; motor to muscles of abdominal wall
3 4		Genitofemoral	Sensory to skin of external genitalia and inguinal region
5	Lumbosacral plexus	Lateral femoral cutaneous Femoral	Sensory to outer side of thigh Motor to quadriceps, sartorius, and iliacus muscles; sensory to front of thigh and medial side of lower leg (saphenous nerve)
Sacral 1 2 3		Obturator	Motor to adductor muscles of thigh
4 5		Tibial† (medial popliteal)	Motor to muscles of calf of leg; sensory to skin of calf of leg and sole of foot
Coccygeal 1	Coccygeal plexus	Common peroneal (lateral popliteal)	Motor to evertors and dorsiflexors of foot; sensory to lateral surface of leg and dorsal surface of foot
		Nerves to hamstring muscles	Motor to muscles of back of thigh
		Gluteal nerves	Motor to buttock muscles and tensor fasciae latae
		Posterior femoral cutaneous	Sensory to skin of buttocks, posterior surface of thigh, and leg
		Pudendal nerve	Motor to perineal muscles; sensory to skin of perineum

*Although nerves to muscles are considered motor, they do contain some sensory fibers that transmit proprioceptive impulses.
†Sensory fibers from the tibial and peroneal nerves unite to form the medial cutaneous (or sural) nerve that supplies the calf of the leg and the lateral surface of the foot. In the thigh the tibial and common peroneal nerves are usually enclosed in a single sheath to form the sciatic nerve, the largest nerve in the body with a width of approximately 3/4 inch (2 cm). About two thirds of the way down the posterior part of the thigh, it divides into its component parts. Branches of the sciatic nerve extend into the hamstring muscles.

Cervical plexus

Brachial plexus

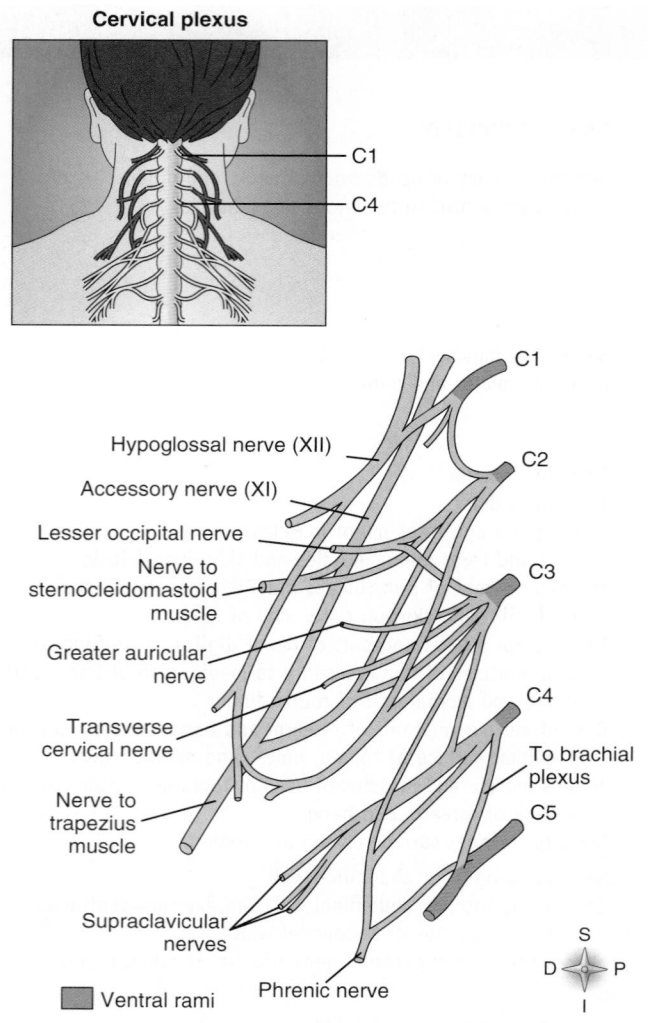

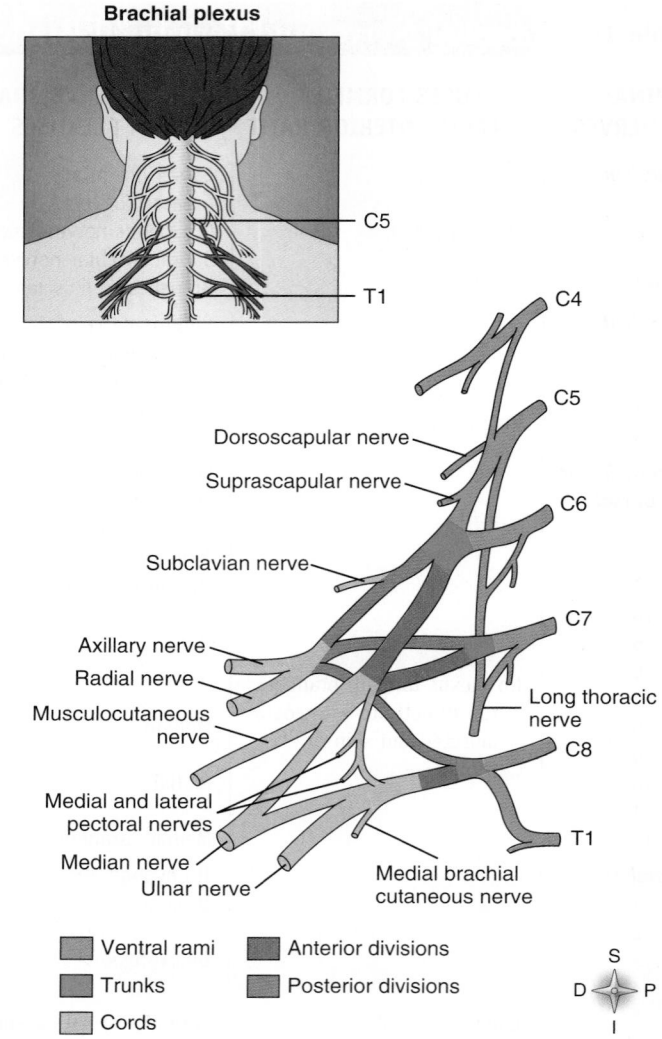

Figure 14-3 *Cervical plexus.* Ventral rami of the first four cervical spinal nerves (C1 through C4) exchange fibers in this plexus found deep within the neck. Notice that some fibers from C5 also enter this plexus to form a portion of the phrenic nerve.

Figure 14-4 *Brachial plexus.* From the five rami, C5 through T1, the plexus forms three "trunks." Each trunk in turn subdivides into an anterior and posterior "division." The divisional branches then reorganize into three "cords." The cords then give rise to the individual nerves that exit this plexus.

from this plexus innervate the muscles and skin of the neck, upper shoulders, and part of the head. Also exiting this plexus is the **phrenic nerve,** which innervates the diaphragm. Two cranial nerves, the accessory nerve (XI) and the hypoglossal nerve (XII), receive small branches that emerge from the cervical plexus.

Brachial Plexus

The **brachial plexus,** shown in Figure 14-4, is found deep within the shoulder. It passes from the ventral rami of spinal nerves C5 through T1, beneath the clavicle (collarbone), and toward the upper arm. Individual nerves that emerge from the brachial plexus innervate the lower part of the shoulder and the entire arm.

Lumbar Plexus

Another spinal nerve plexus is the **lumbar plexus,** which is formed by the intermingling of fibers from the first four lumbar nerves (Figure 14-5). This network of nerves is located in the lumbar region of the back near the psoas muscle. The large

femoral nerve is one of several nerves emerging from the lumbar plexus. It divides into many branches supplying the thigh and leg.

Sacral Plexus and Coccygeal Plexus

Fibers from the fourth and fifth lumbar nerves (L4 and L5) and the first four sacral nerves (S1 through S4) form the **sacral plexus.** It lies in the pelvic cavity on the anterior surface of the piriformis muscle. Because of their close proximity and overlap of fibers, the lumbar and sacral plexuses are often considered together as the *lumbosacral* plexus (see Figure 14-5). Among other nerves that emerge from the sacral plexus are the tibial and common peroneal nerves. In the thigh, they form the largest nerve in the body, the great sciatic nerve. It pierces the buttocks and runs down the back of the thigh. Its many branches supply nearly all the skin of the leg, the posterior thigh muscles, and the leg and foot muscles. *Sciatica,* or neuralgia of the sciatic nerve, is a fairly common and very painful condition.

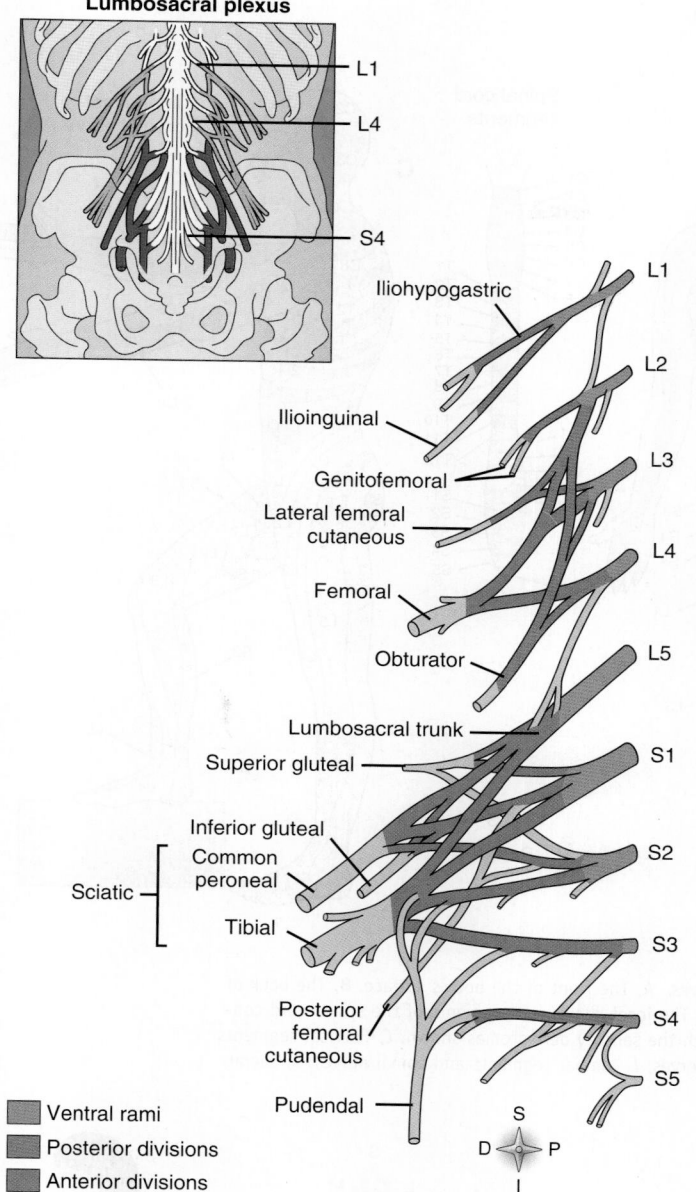

Lumbosacral plexus

L1
L4
S4

Iliohypogastric
Ilioinguinal
Genitofemoral
Lateral femoral cutaneous
Femoral
Obturator
Lumbosacral trunk
Superior gluteal
Inferior gluteal
Common peroneal
Sciatic
Tibial
Posterior femoral cutaneous
Pudendal

L1
L2
L3
L4
L5
S1
S2
S3
S4
S5

Ventral rami
Posterior divisions
Anterior divisions

S
D — P
I

Figure 14-5 *Lumbosacral plexus.* This plexus is formed by the combination of the lumbar plexus with the sacral plexus, as shown in the inset. Notice that the ventral rami split into anterior and posterior "divisions" before reorganizing into the various individual nerves that exit this plexus.

The last sacral spinal nerve (S5), along with a few fibers from S4, joins with the coccygeal nerve to form a small *coccygeal plexus*. Nerves arising from this plexus innervate the floor of the pelvic cavity and some surrounding areas.

Dermatomes and Myotomes

At first glance the distribution of spinal nerves does not appear to follow an ordered arrangement, but detailed mapping of the skin surface has revealed a close relationship between the spinal origin of each spinal nerve and the region of the body it innervates. Knowledge of the segmental arrangement of spinal nerves and the region of the body innervated by each segment has proved useful to health professionals. For instance, one can identify the site of spinal cord or nerve abnormality from the area of the body insensitive to a pinprick. Each skin surface area supplied by sensory fibers of a given spinal nerve is called a **dermatome**, a name that means "skin section" (Figures 14-6 and 14-7). The body maps in Figure 14-7 show sharp boundaries between sensory der-

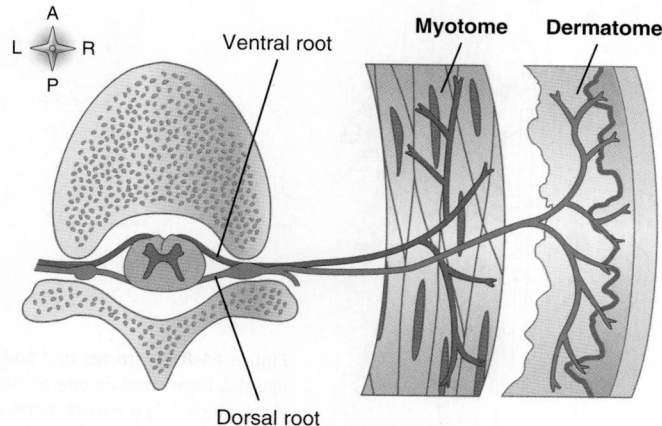

A
L — R
P

Ventral root
Myotome
Dermatome
Dorsal root

Figure 14-6 *Segmental distribution of spinal nerves.* A dermatome is a region of skin supplied by afferent (sensory) fibers of a given spinal nerve. A myotome is a region of skeletal muscle innervated by efferent (motor) fibers of a given spinal nerve. Spinal nerves at different segments of the spinal cord innervate different sets of dermatomes and myotomes.

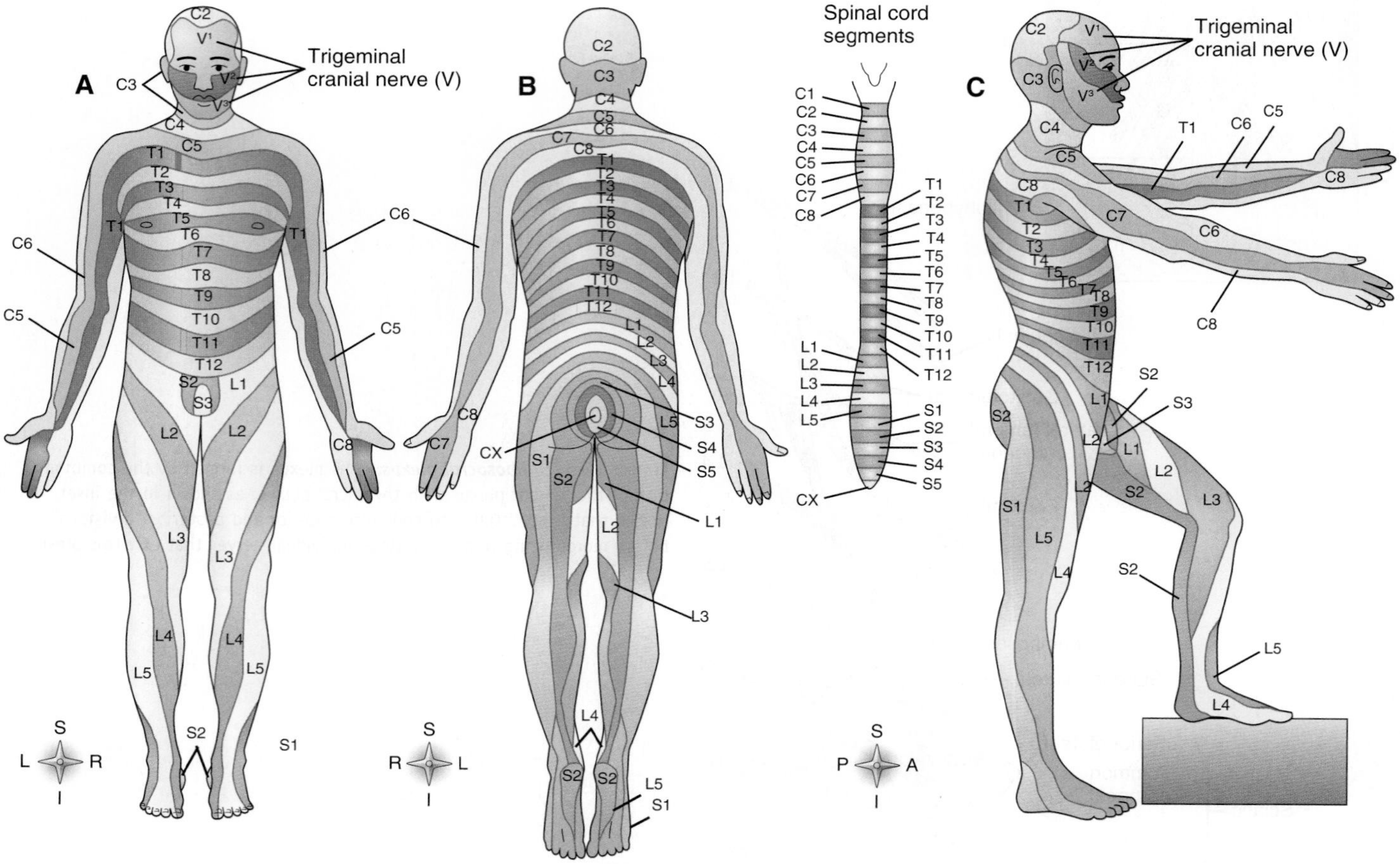

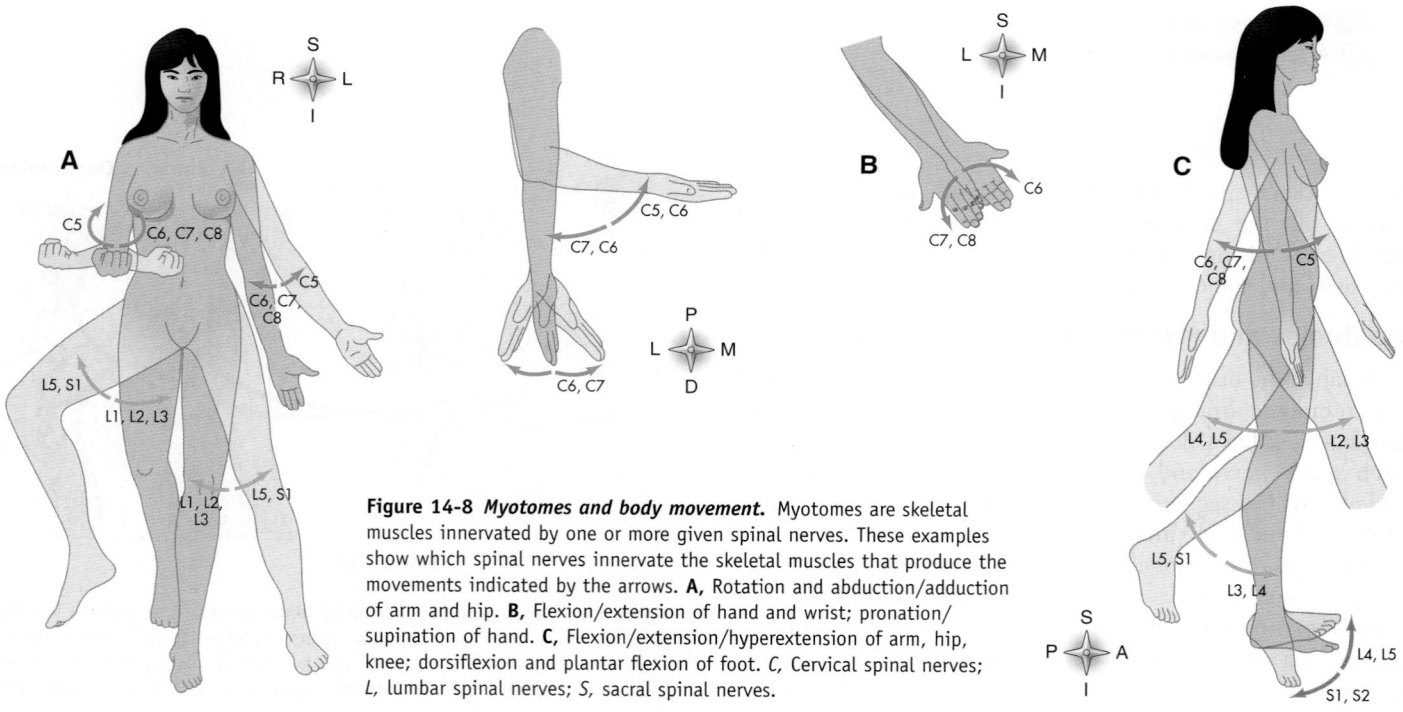

Figure 14-7 *Dermatome distribution of spinal nerves.* A, The front of the body's surface. **B,** The back of the body's surface. **C,** The side of the body's surface. The inset shows the segments of the spinal cord connected with each of the spinal nerves associated with the sensory dermatomes shown. *C,* Cervical segments and spinal nerves; *T,* thoracic segments and spinal nerves; *L,* lumbar segments and spinal nerves; *S,* sacral segments and spinal nerves.

Figure 14-8 *Myotomes and body movement.* Myotomes are skeletal muscles innervated by one or more given spinal nerves. These examples show which spinal nerves innervate the skeletal muscles that produce the movements indicated by the arrows. **A,** Rotation and abduction/adduction of arm and hip. **B,** Flexion/extension of hand and wrist; pronation/supination of hand. **C,** Flexion/extension/hyperextension of arm, hip, knee; dorsiflexion and plantar flexion of foot. *C,* Cervical spinal nerves; *L,* lumbar spinal nerves; *S,* sacral spinal nerves.

BOX 14-3: HEALTH MATTERS
Herpes Zoster

Herpes zoster, or **shingles,** is a unique viral infection that almost always affects the skin of a single dermatome. It is caused by the varicella zoster virus of chickenpox. About 3% of the population will suffer from shingles at some time in their lives. In most cases the disease results from reactivation of the varicella virus. The virus most likely traveled through a cutaneous nerve and remained dormant in a dorsal root ganglion for years after an episode of chickenpox. If the body's immunological protective mechanism becomes diminished in the elderly, or after stress, or in individuals undergoing radiation therapy or taking immunosuppressive drugs, the virus may reactivate. If this occurs, the virus will travel over the sensory nerve to the skin of a single dermatome. The figure shows involvement of dermatome T4 in a 13-year-old boy. The result is a painful eruption of red swollen plaques or vesicles that eventually rupture and crust before clearing in 2 to 3 weeks. In severe cases, extensive inflammation, hemorrhagic blisters, and secondary bacterial infection may lead to permanent scarring. In most cases of shingles, the eruption of vesicles is preceded by 4 to 5 days of preeruptive pain, burning, and itching in the affected dermatome. Unfortunately, an attack of herpes zoster does not confer lasting immunity. Many individuals have three or more episodes in a lifetime.

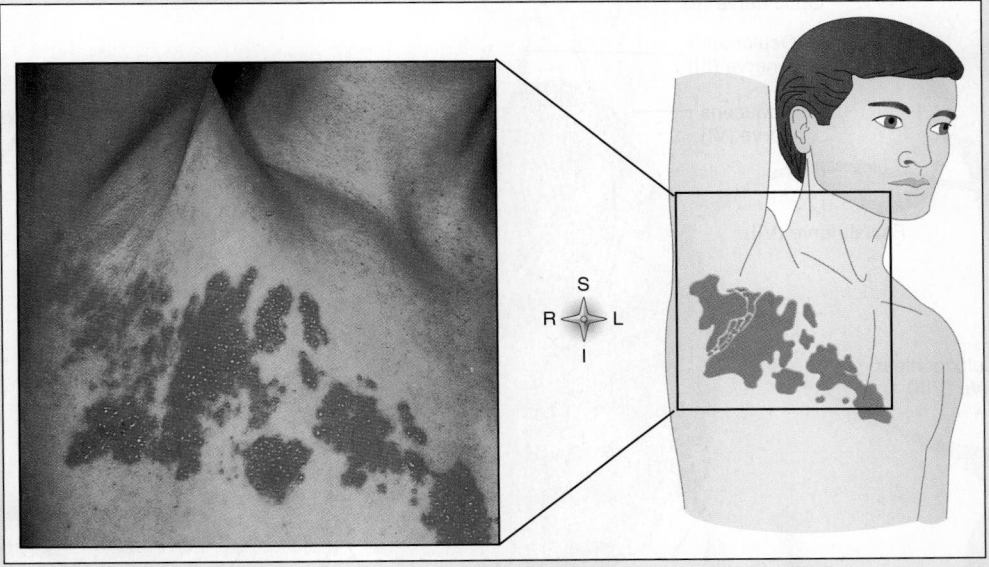

matomes, but there is actually quite a bit of overlap between adjacent dermatomes. Overlap of dermatomes results from differences in the way various types of sensory receptors are distributed in the skin.

A **myotome** is a skeletal muscle or group of muscles that receives motor axons from a given spinal nerve. There is some overlap among myotomes also; thus some skeletal muscle organs may be innervated by motor axons from more than one spinal nerve. Figure 14-8 shows examples of how myotomes relate to specific movements of the body.

QUICK CHECK

1. How many pairs of spinal nerves are there? Can you name them?
2. What is a plexus? Name the four major pairs of plexuses.
3. What is a dermatome? A myotome?

CRANIAL NERVES

Twelve pairs of **cranial nerves** connect to the undersurface of the brain (Figure 14-9), mostly on the brainstem. Cranial nerves pass through small foramina (holes) in the cranial cavity of the skull, allowing them to extend to or from their peripheral destinations. Both names and numbers identify the cranial nerves. Their names suggest either their distribution or their function. Their numbers indicate the order in which they connect to the brain from anterior to posterior. As with all nerves, cranial nerves are made up of bundles of axons. **Mixed cranial nerves** contain axons of sensory and motor neurons. **Sensory cranial nerves** consist of sensory axons only, and **motor cranial nerves** consist mainly of motor axons. Cranial nerves classified as "motor nerves" contain a small number of sensory fibers. These sensory fibers are *proprioceptive* fibers that carry information regarding tension in the muscles controlled by the motor fibers of the same motor nerve. Table 14-2 lists the name, number, and functional classification of each of the 12 pairs of cranial nerves. Details of the structure and function of each cranial nerve are described in Table 14-3 and in the paragraphs that follow.

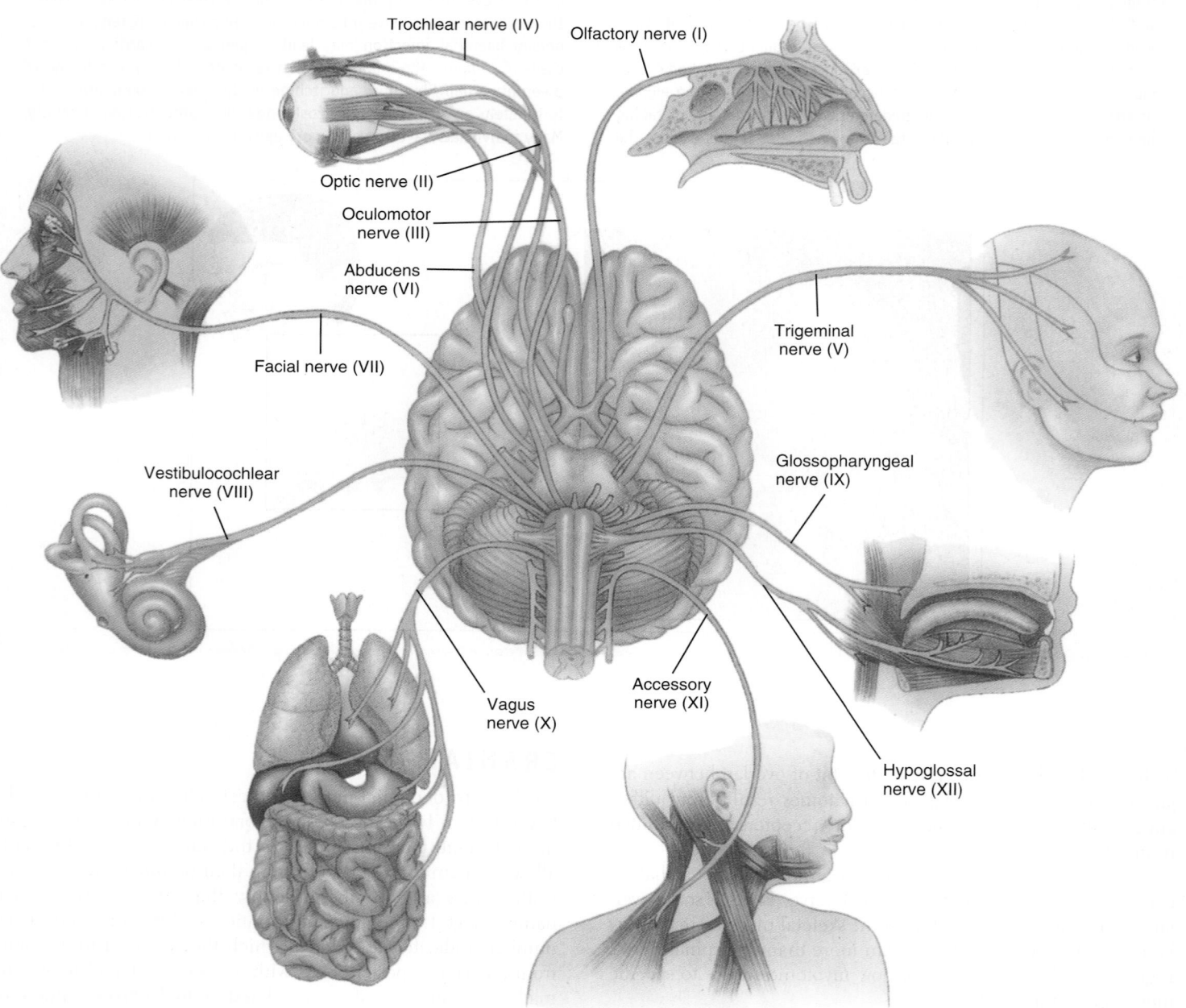

Figure 14-9 *Cranial nerves.* Ventral surface of the brain showing attachment of the cranial nerves.

Table 14-2	Names, Numbers, and Functional Classifications of Cranial Nerves*		
NAME		**NUMBER**	**FUNCTIONAL CLASSIFICATION**
Olfactory		I	Sensory
Optic		II	Sensory
Oculomotor		III	Motor
Trochlear		IV	Motor
Trigeminal		V	Mixed
Abducens		VI	Motor
Facial		VII	Mixed
Vestibulocochlear		VIII	Sensory
Glossopharyngeal		IX	Mixed
Vagus		X	Mixed
Accessory		XI	Motor
Hypoglossal		XII	Motor

*The first letters of the words in the following sentence are the first letters of the names of the cranial nerves, in the correct order. Many anatomy students find that using this sentence, or one like it, helps in memorizing the names and numbers of the cranial nerves. It is "**O**n **O**ld **O**lympus' **T**iny **T**ops, **A F**riendly **V**iking **G**rew **V**ines **A**nd **H**ops." The functional classification of each cranial nerve can be remembered by using this sentence: "**S**ome **S**ay '**M**arry **M**oney,' **B**ut **M**y **B**rothers **S**ay '**B**ad **B**usiness, **M**arry **M**oney.'" In this sentence, **S** indicates *sensory*, **M** indicates *motor*, and **B** indicates *both* sensory and motor (mixed).

Olfactory Nerve (I)

The **olfactory nerves** are composed of axons of neurons whose dendrites and cell bodies lie in the nasal mucosa, high up along the septum and superior conchae (turbinates). Axons of these neurons form about 20 small bundles of fibers that pierce each cribriform plate and terminate in the olfactory bulbs (see Figure 15-4, p. 561). Here they synapse with the second pool of olfactory neurons, whose axons compose the olfactory tracts. The olfactory nerves carry information about the sense of smell.

Optic Nerve (II)

Axons from the innermost layer of sensory neurons of the retina compose the second pair of cranial nerves. They are called **optic nerves** because they carry visual information from the eyes to the brain. After entering the cranial cavity through the optic foramina, the two optic nerves unite to form the *optic chiasma*, in which some of the fibers of each nerve cross to the opposite side and continue in the *optic tract* of that side (see Figure 15-25, p. 579). Thus each optic nerve contains fibers only from the retina of the same side, whereas each optic tract has fibers in it from both retinas, an important fact in interpreting certain visual disorders. Most of the optic tract fibers terminate in the thalamus (in the portion known as the *lateral geniculate nucleus*). From here a new relay of fibers runs to the visual area of the occipital lobe cortex. A few optic tract

fibers terminate in the superior colliculi of the midbrain, where they synapse with motor fibers to the external eye muscles (cranial nerves III, IV, VI).

Oculomotor Nerve (III)

Fibers of each **oculomotor nerve** originate from cells in the oculomotor nucleus in the ventral part of the midbrain and extend to the various external eye muscles, with the exception of the superior oblique and the lateral rectus. Autonomic fibers are also present in the oculomotor nerves. They extend to the intrinsic muscles of the eye, which regulate the amount of light entering the eye and aid in focusing on near objects. Yet a third group of fibers is found in the third cranial nerves; namely, sensory fibers from proprioceptors in the eye muscles.

Trochlear Nerve (IV)

Motor fibers of each **trochlear nerve** have their origin in cells in the midbrain, from which they extend to the superior oblique muscles of the eye. The name *trochlear*, from a Greek word meaning "pulley," refers to the fact that the superior oblique muscles of the eye pass through a pulley-like ligament (see Figure 15-18, p. 575). Afferent fibers from proprioceptors in these muscles are also contained in the trochlear nerves.

Trigeminal Nerve (V)

The fifth pair of cranial nerves is called **trigeminal nerves** because they each split into three large branches (*tri-*, "three"; *-gemina-*, "twins or pair"). The three branches of each pair are the *ophthalmic nerve, maxillary nerve,* and *mandibular nerve* (Figure 14-10). Sensory neurons in all three branches of the trigeminal nerve carry afferent impulses from the skin and mucosa of the head and from the teeth to cell bodies in the trigeminal ganglion (lodged in the petrous portion of the temporal bone). Fibers extend from the ganglion to the main sensory nucleus of the fifth cranial nerve situated in the pons. Motor fibers of the trigeminal nerve originate in the trifacial motor nucleus located in the pons just medial to the sensory nucleus. These motor fibers run to the muscles of mastication by way of the mandibular nerve.

Abducens Nerve (VI)

Each **abducens nerve** is a motor nerve with fibers originating from a nucleus in the pons in the floor of the fourth ventricle and extending to the lateral rectus muscles of the eyes (see Figure 15-18, p. 575). The lateral rectus muscle *abducts* the eye to which it is attached, hence the name *abducens* for this nerve. The sixth cranial nerve also contains some afferent fibers from proprioceptors in the lateral rectus muscles.

Facial Nerve (VII)

The motor fibers of the **facial nerve** arise from a nucleus in the lower part of the pons, from which they extend by way of several branches to the superficial muscles of the face and scalp

Text continued on p. 530

Table 14-3	Structure and Function of the Cranial Nerves					☐ sensory (afferent) ☐ motor (efferent)
	SENSORY FIBERS			**MOTOR FIBERS**		
NERVE	**RECEPTORS**	**CELL BODIES**	**TERMINATION**	**CELL BODIES**	**TERMINATION**	**FUNCTIONS**
I Olfactory 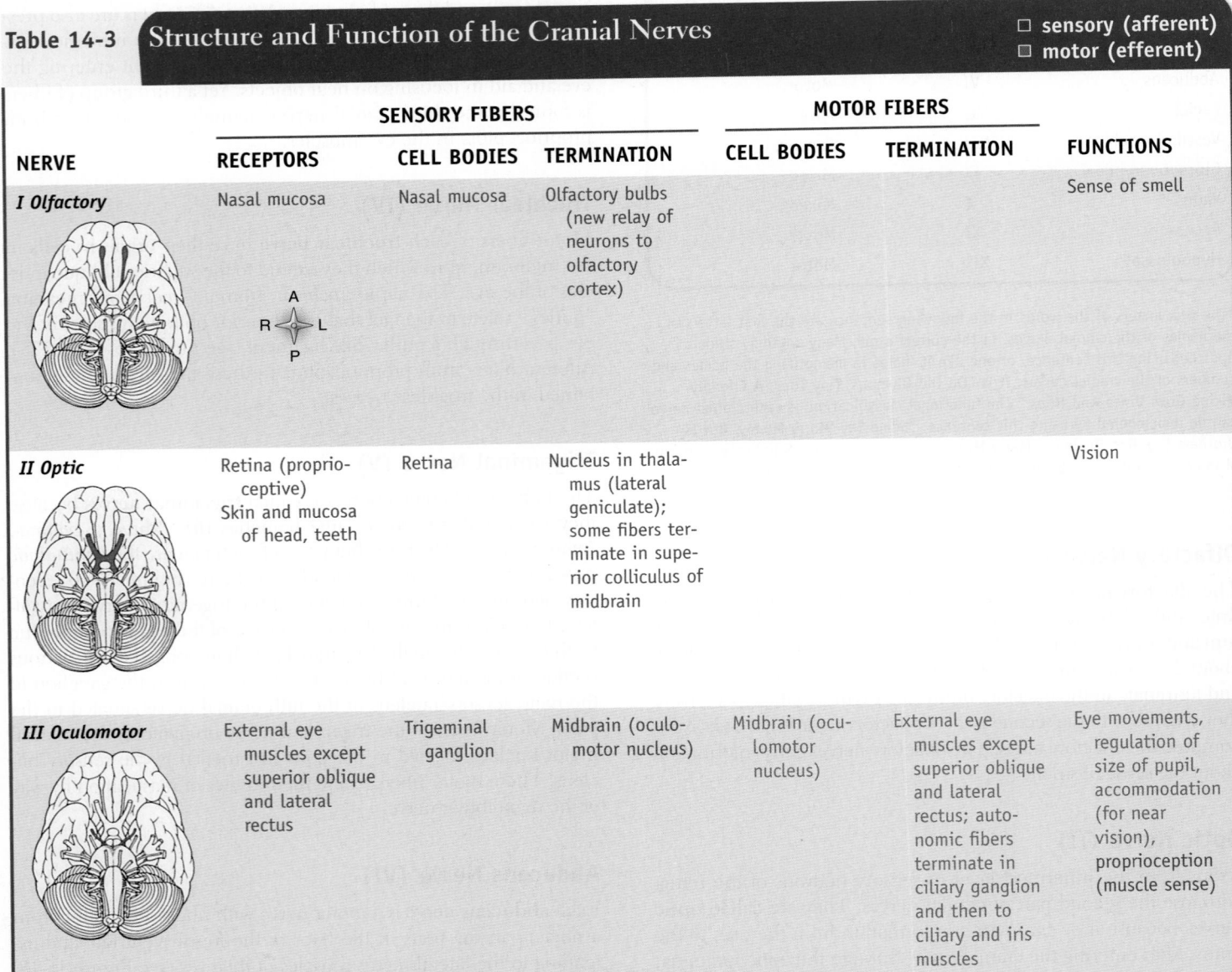	Nasal mucosa	Nasal mucosa	Olfactory bulbs (new relay of neurons to olfactory cortex)			Sense of smell
II Optic	Retina (proprioceptive) Skin and mucosa of head, teeth	Retina	Nucleus in thalamus (lateral geniculate); some fibers terminate in superior colliculus of midbrain			Vision
III Oculomotor	External eye muscles except superior oblique and lateral rectus	Trigeminal ganglion	Midbrain (oculomotor nucleus)	Midbrain (oculomotor nucleus)	External eye muscles except superior oblique and lateral rectus; autonomic fibers terminate in ciliary ganglion and then to ciliary and iris muscles	Eye movements, regulation of size of pupil, accommodation (for near vision), proprioception (muscle sense)

Table 14-3	Structure and Function of the Cranial Nerves—cont'd					☐ sensory (afferent) ☐ motor (efferent)
	SENSORY FIBERS			**MOTOR FIBERS**		
NERVE	**RECEPTORS**	**CELL BODIES**	**TERMINATION**	**CELL BODIES**	**TERMINATION**	**FUNCTIONS**
IV Trochlear	Superior oblique (proprioceptive)	Trigeminal ganglion	Midbrain	Midbrain	Superior oblique muscle of eye	Eye movements, proprioception
V Trigeminal	Skin and mucosa of head, teeth	Trigeminal ganglion	Pons (sensory nucleus)	Pons (motor nucleus)	Muscles of mastication	Sensations of head and face, chewing movements, proprioception
VI Abducens	Lateral rectus (proprioceptive)	Trigeminal ganglion	Pons	Pons	Lateral rectus muscle of eye	Abduction of eye, proprioception

Continued

Table 14-3 — Structure and Function of the Cranial Nerves—cont'd

☐ sensory (afferent)
☐ motor (efferent)

NERVE	SENSORY FIBERS			MOTOR FIBERS		FUNCTIONS
	RECEPTORS	CELL BODIES	TERMINATION	CELL BODIES	TERMINATION	
VII Facial	Taste buds of anterior two thirds of tongue	Geniculate ganglion	Medulla (nucleus solitarius)	Pons	Superficial muscles of face and scalp; autonomic fibers to salivary and lacrimal glands	Facial expressions, secretion of saliva and tears, taste
VIII Vestibulocochlear						
VESTIBULAR BRANCH	Semicircular canals and vestibule (utricle and saccule)	Vestibular ganglion	Pons and medulla (vestibular nuclei)			Balance or equilibrium sense
COCHLEAR OR AUDITORY BRANCH	Organ of Corti in cochlear duct	Spiral ganglion	Pons and medulla (cochlear nuclei)			Hearing
IX Glossopharyngeal	Pharynx; taste buds and other receptors of posterior one third of tongue	Jugular and petrous ganglia	Medulla (nucleus solitarius)	Medulla (nucleus ambiguus)	Muscles of pharynx	Sensations of tongue, swallowing movements, secretion of saliva, aid in reflex control of blood pressure and respiration
	Carotid sinus and carotid body	Jugular and petrous ganglia	Medulla (respiratory and vasomotor centers)	Medulla at junction of pons (nucleus salivatorius)	Otic ganglion and then to parotid salivary gland	

Table 14-3 **Structure and Function of the Cranial Nerves—cont'd**   □ sensory (afferent)
□ motor (efferent)

NERVE	SENSORY FIBERS			MOTOR FIBERS		FUNCTIONS
	RECEPTORS	CELL BODIES	TERMINATION	CELL BODIES	TERMINATION	
X Vagus	Pharynx, larynx, carotid body, thoracic and abdominal viscera	Jugular and nodose ganglia	Medulla (nucleus solitarius), pons (nucleus of fifth cranial nerve)	Medulla (dorsal motor nucleus)	Ganglia of vagal plexus and then to muscles of pharynx, larynx, and autonomic fibers to thoracic and abdominal viscera	Sensations and movements of organs supplied; e.g., slows heart, increases peristalsis, contracts muscles for voice production
XI Accessory	Trapezius and sternocleido-mastoid (proprio-ceptive)	Upper, cervical ganglia	Spinal cord	Medulla (dorsal motor nucleus of vagus and nucleus ambiguus)	Muscles of thoracic and abdominal viscera (autonomic), pharynx, and larynx	Shoulder movements, turning movements of head, movements of viscera, voice production, proprioception
				Anterior gray column of first five or six cervical segments of spinal cord	Trapezius and sternocleido-mastoid muscle	
XII Hypoglossal	Tongue muscles (proprioceptive)	Trigeminal ganglion	Medulla (hypoglossal nucleus)	Medulla (hypoglossal nucleus)	Muscles of tongue and throat	Tongue movements, proprioception

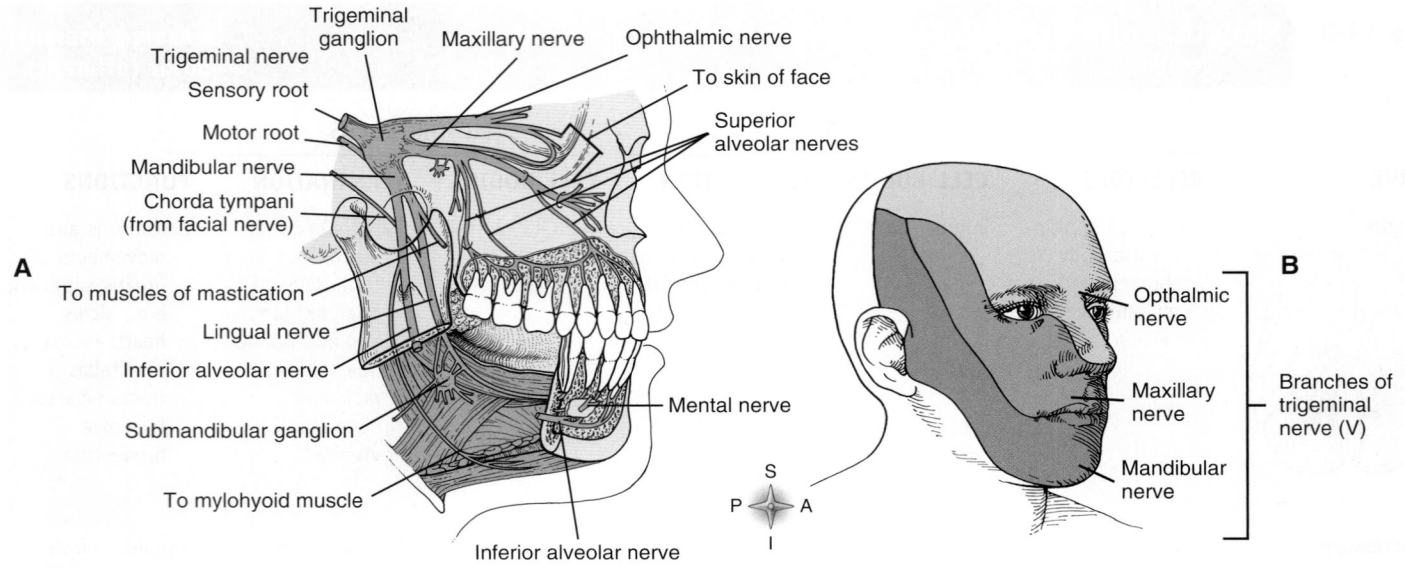

Figure 14-10 *Trigeminal nerve (V).* **A,** The route of the trigeminal nerve and its major branches. **B,** Sensory fibers of the trigeminal nerve form three branch nerves (ophthalmic, maxillary, and mandibular nerves), each of which conducts information from a different region of the face.

 BOX 14-4: HEALTH MATTERS

Trigeminal Neuralgia

Compression, inflammation, or degeneration of the fifth cranial nerve, the trigeminal nerve, may result in a condition called **trigeminal neuralgia,** or *tic douloureux* (tic doo-loo-ROO). This condition is characterized by recurring episodes of intense stabbing pain, or *neuralgia,* radiating from the angle of the jaw along a branch of the trigeminal nerve (see the figure).

One method of doing away with this pain is to remove the trigeminal ganglion on the posterior portion of the nerve (see Figure 14-10). This large ganglion contains the cell bodies of the nerve's afferent fibers. After such an operation the patient's face, scalp, teeth, and conjunctiva on the side treated show anesthesia. Special care, such as wearing protective goggles and irrigating the eye frequently, is therefore prescribed. The patient is instructed also to visit the dentist regularly because he or she can no longer experience a toothache as a warning of diseased teeth.

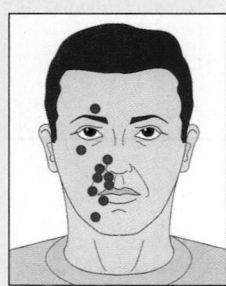

Pain zones in trigeminal neuralgia.

(Figure 14-11). Efferent autonomic fibers of the facial nerve extend to the submaxillary and sublingual salivary glands, as well as to the lacrimal (tear) glands. Sensory fibers from the taste buds of the anterior two thirds of the tongue run in the facial nerve to cell bodies in the geniculate ganglion, a small swelling on the facial nerve, where it passes through a canal in the temporal bone. From the ganglion, fibers extend to a nucleus in the medulla.

Vestibulocochlear Nerve (VIII)

The **vestibulocochlear nerve** has two distinct divisions: the *vestibular nerve* and the *cochlear nerve* (see Figure 15-9, p. 565). Both are sensory. Fibers from the semicircular canals in the inner ear run to the vestibular ganglion (in the internal auditory meatus). Here, the neurons' cell bodies are located, and from them, fibers extend to the vestibular nuclei in the pons and medulla. Together, these fibers constitute the *vestibular nerve.* Some of its fibers run to the cerebellum. The vestibular nerve transmits impulses that result in sensations of equilibrium. The *cochlear nerve* consists of peripheral axon fibers starting in the organ of Corti in the cochlea of the inner ear. They have their cell bodies in the spiral ganglion in the cochlea, and their central axon fibers terminate in the cochlear nuclei located between the medulla and pons. Conduction by the cochlear nerve results in sensations of hearing. Because of its role in hearing, the eighth cranial nerve is sometimes still called the *auditory* or *acoustic* nerve.

Glossopharyngeal Nerve (IX)

Both sensory and motor fibers compose the **glossopharyngeal nerve.** This nerve supplies fibers not only to the tongue and pharynx (throat), as its name implies, but also to other structures

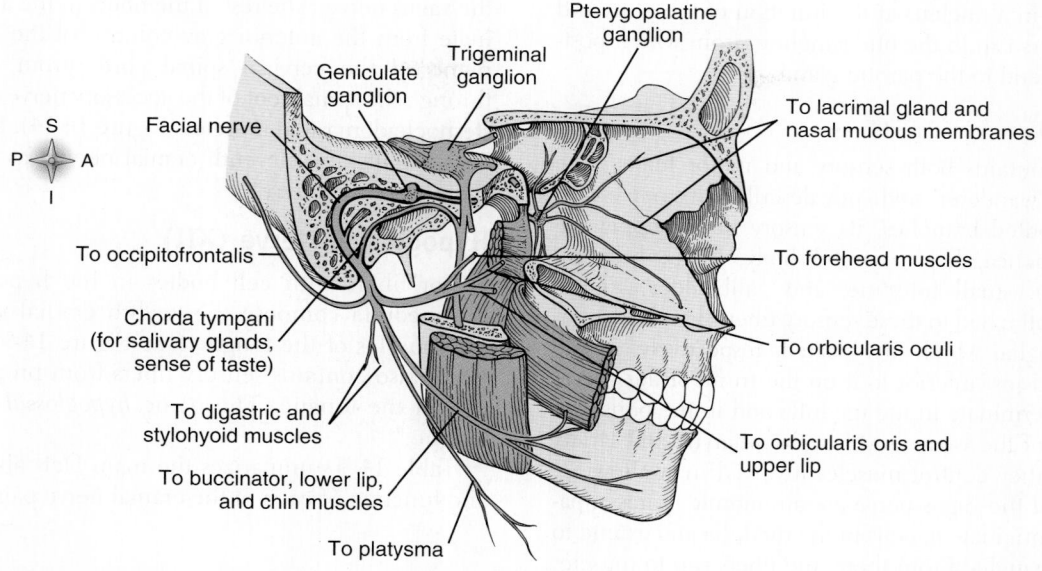

Figure 14-11 *Facial nerve (VII).* Artist's interpretation of the location of the various branches of the facial nerve.

BOX 14-5: HEALTH MATTERS

Cranial Nerve Damage

Severe head injuries often damage one or more of the cranial nerves, producing symptoms that reflect loss of the functions mediated by the affected nerve or nerves. For example, injury of the sixth cranial nerve causes the eye to turn in because of paralysis of the abducting muscle of the eye. Injury of the eighth cranial nerve, on the other hand, produces deafness. Injury to the facial nerve results in a poker-faced expression and a drooping of the corner of the mouth from paralysis of the facial muscles.

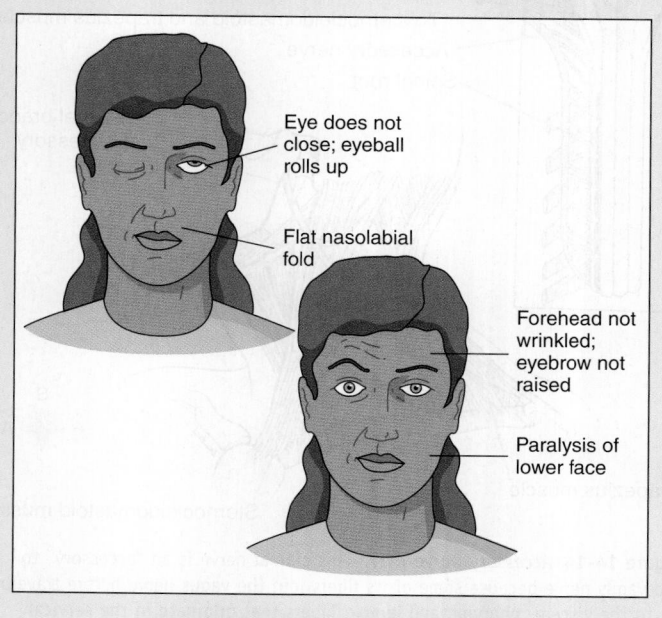

Signs of damage to the facial nerve (VII).

(Figure 14-12). One is the carotid sinus; it plays an important part in the control of blood pressure. Sensory fibers, with their receptors in the pharynx and posterior one third of the tongue, have their cell bodies in the jugular (superior) and petrous (inferior) ganglia. These are located, respectively, in the jugular foramen and the petrous portion of the temporal bone. The ganglia fibers extend to a nucleus in the medulla. The motor fibers of the ninth cranial nerve originate in another medullary nucleus and run to muscles of the pharynx. Also present in this nerve are autonomic

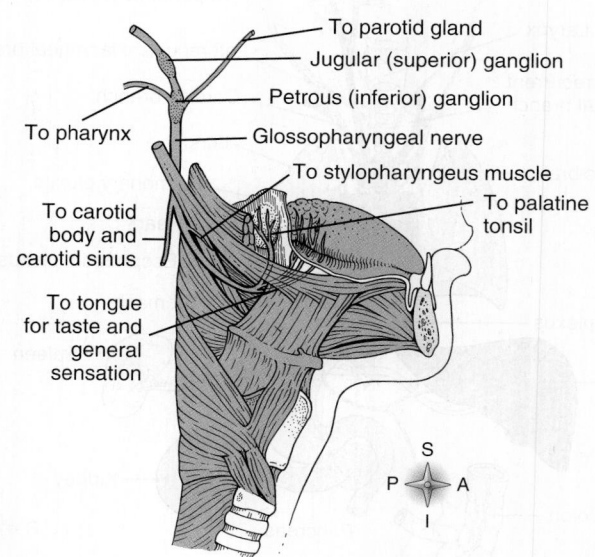

Figure 14-12 *Glossopharyngeal nerve (IX).* As its name implies (*glosso-*, "tongue"; *-pharyng-*, "throat"), the ninth cranial nerve supplies fibers to the tongue and throat. It is a mixed nerve that carries both sensory and motor fibers.

fibers that originate in a nucleus at the junction of the pons and medulla. These fibers run to the otic ganglion, from which postganglionic fibers extend to the parotid gland.

Vagus Nerve (X)

The **vagus nerve** contains both sensory and motor fibers. The name vagus means "wanderer" and aptly describes this nerve with many widely distributed branches. Its sensory fibers supply the pharynx, larynx, trachea, heart, carotid body, lungs, bronchi, esophagus, stomach, small intestine, and gallbladder (Figure 14-13). Cell bodies attached to these sensory fibers lie in the jugular and nodose ganglia, which are located, respectively, in the jugular foramen and just inferior to it on the trunk of the nerve. The sensory axons terminate in the medulla and in the pons. Somatic motor fibers of the vagus travel to the pharynx and larynx (voice box), where they control muscles involved in swallowing. Most motor fibers of the vagus nerve are autonomic (parasympathetic) fibers. They originate in cells in the medulla and extend to various autonomic ganglia. From there, the fibers run to muscles of the pharynx, larynx, and thoracic and abdominal organs, where they control heart rate and other "visceral" activities.

Accessory Nerve (XI)

The **accessory nerve** is a motor nerve. Some of its fibers originate in cells in the medulla and pass by way of branches of the vagus nerve to thoracic and abdominal viscera, as well as the pharynx and larynx. Thus this nerve can be considered an "accessory" to the vagus nerve. The rest of the fibers in the accessory nerve originate from the anterior gray column of the first five or six segments of the cervical spinal cord. From there, they extend through the spinal root of the accessory nerve to the trapezius and sternocleidomastoid muscles (Figure 14-14). Because of its spinal components, the eleventh cranial nerve was formerly called the *spinal accessory nerve.*

Hypoglossal Nerve (XII)

Motor fibers with cell bodies in the hypoglossal nucleus of the medulla compose the twelfth cranial nerve. They supply the muscles of the tongue (see Figure 14-9). The **hypoglossal nerve** also contains sensory fibers from proprioceptors in muscles of the tongue. The name *hypoglossal* means "under the tongue."

Table 14-3 summarizes the main facts about the distribution and function of each of the cranial nerve pairs.

QUICK CHECK

4. List the names and numbers of the 12 pairs of cranial nerves.
5. Identify the primary function of each pair of cranial nerves.
6. Distinguish between a motor nerve, sensory nerve, and mixed nerve.

Left vagus nerve
Pharyngeal branch
Jugular (superior) ganglion
Nodose (inferior) ganglion
Right vagus nerve
Superior laryngeal branch
Larynx
Left recurrent laryngeal branch
Right recurrent laryngeal branch
Cardiac branch
Cardiac branch
Lung
Pulmonary plexus
Heart
Esophageal plexus
Stomach
Celiac plexus
Spleen
Liver
Kidney
Colon
Pancreas
Small intestine

Figure 14-13 *Vagus nerve (X).* The vagus nerve is a mixed cranial nerve with many widely distributed branches—hence the name *vagus,* which is the Latin word for "wanderer."

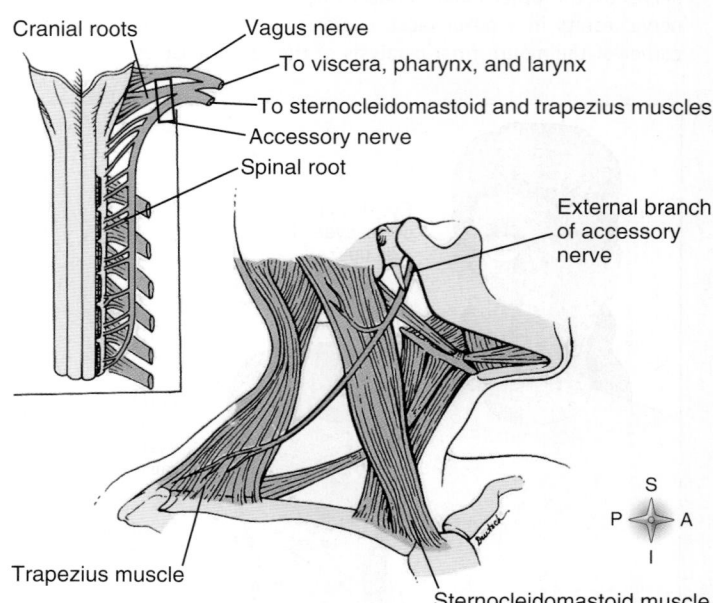

Figure 14-14 *Accessory nerve (XI).* This cranial nerve is an "accessory" to the vagus nerve because some of its fibers join the vagus nerve before traveling on to the viscera, pharynx, and larynx. Fibers that originate in the cervical spinal segments travel through an external nerve branch to the trapezius and sternocleidomastoid muscles of the neck.

DIVISIONS OF THE PERIPHERAL NERVOUS SYSTEM

Recall from Chapter 12 that the PNS includes all the nervous pathways *outside* the brain and spinal cord (see Figure 12-2, p. 433). Thus the entire PNS comprises the fibers present in the cranial nerves, the spinal nerves, and all their individual branches. Although many of these nerves are *mixed nerves*—containing both sensory and motor fibers—it is often convenient to consider the PNS as having two functional divisions: the sensory (afferent) division and the motor (efferent) division. Details of the sensory division of the PNS are discussed in Chapter 15. In this chapter, we concentrate on some essential details of the motor division of the PNS. First, we briefly discuss the *somatic motor nervous system*, and then we move on to discuss the efferent pathways of the *autonomic nervous system*.

Somatic Motor Nervous System

Basic Principles of Somatic Motor Pathways

The somatic motor nervous system includes all the voluntary motor pathways outside the CNS. That is, it involves the peripheral pathways to the skeletal muscles, which are the *somatic effectors.* Recall from Chapter 13 that all these pathways operate according to the principle of *final common path.* This means that all the somatic motor pathways involve a single motor neuron whose axon stretches from the cell body in the CNS all the way to the effector innervated by that neuron. For fibers that originate in the spinal cord, this means that the axon extends from the anterior gray horn, through the ventral nerve root, and out to a skeletal muscle.

Another important principle of somatic motor pathways was mentioned in Chapter 11, when we were discussing the function of skeletal muscles. This principle states that the axon of the last somatic motor neuron—also called the *anterior horn neuron* or *lower motor neuron*—stimulates effector cells by means of the neurotransmitter **acetylcholine.** This fact will have added importance later when we compare the somatic motor nervous system with the ANS.

Because the basic plan of the somatic motor pathways was discussed in Chapter 13, and the effect of acetylcholine on the somatic effectors (skeletal muscles) has already been outlined in Chapter 10, we will end our discussion of the somatic motor division of the PNS with a brief review of the concept of the reflex arc and how it relates to peripheral motor pathways.

Somatic Reflexes

Nature of a Reflex

The action that results from a nerve impulse passing over a reflex arc is called a **reflex** (see Figure 12-10, p. 442). In other words, a reflex is a predictable response to a stimulus. It may or may not be conscious. Usually the term is used to mean only involuntary responses rather than those directly willed; that is, involving cerebral cortex activity. If the center of a reflex arc is in the brain, the response it mediates is called a **cranial reflex.** If the center of a reflex arc is in the spinal cord, the response is called a spinal reflex.

A reflex consists of either muscle contraction or glandular secretion. **Somatic reflexes** are contractions of skeletal muscles. Impulse conduction over somatic reflex arcs—arcs whose motor neurons are somatic motor neurons (i.e., anterior horn neurons or lower motor neurons)—produces somatic reflexes. **Autonomic (visceral) reflexes** consist of contractions of smooth or cardiac muscle or secretion by glands; they are mediated by impulse conduction over autonomic reflex arcs, the motor neurons of which are autonomic neurons (discussed later). The following paragraphs describe only somatic reflexes.

Somatic Reflexes of Clinical Importance

Clinical interest in reflexes stems from the fact that they deviate from normal in certain diseases. Therefore the testing of reflexes is a valuable diagnostic aid. The following reflexes are frequently tested: knee jerk, ankle jerk, Babinski reflex, corneal reflex, and abdominal reflex.

Knee Jerk Reflex

The *knee jerk*, or patellar, reflex is an extension of the lower leg in response to tapping of the patellar tendon (Figure 14-15). The tap stretches the tendon and its muscles, the quadriceps femoris, and thereby stimulates muscle spindles (receptors) in the muscle and initiates conduction over the following two-neuron reflex arc:

1. Sensory neurons
   a. Peripheral axon fibers—in femoral and second, third, and fourth lumbar nerves
   b. Cell bodies—second, third, and fourth lumbar ganglia
   c. Central axon fibers—in posterior roots of second, third, and fourth lumbar nerves; terminate in these segments of the spinal cord; synapse directly with lower motor neurons
2. Reflex center—synapses in anterior gray column between axons of sensory neurons and dendrites and cell bodies of lower motor neurons
3. Motor neurons
   a. Dendrites and cell bodies—in spinal cord anterior gray column
   b. Axons—in anterior roots of second, third, and fourth lumbar spinal nerves and femoral nerves; terminate in quadriceps femoris muscle

The knee jerk can be classified in various ways as follows:

- As a *spinal cord reflex*—because the center of the reflex arc (which transmits the impulses that activate the muscles producing the knee jerk) lies in the spinal cord gray matter
- As a *segmental reflex*—because impulses that mediate it enter and leave the same segment of the cord
- As an *ipsilateral reflex*—because the impulses that mediate it come from and go to the same side of the body
- As a *stretch reflex* or *myotatic reflex* (from the Greek *mys-*, "muscle," *-tasis*, "stretching")—because of the kind of stimulation used to evoke it
- As an *extensor reflex*—because it is produced by extensors of the lower leg (muscles located on the anterior surface of the thigh, which extend the lower leg)

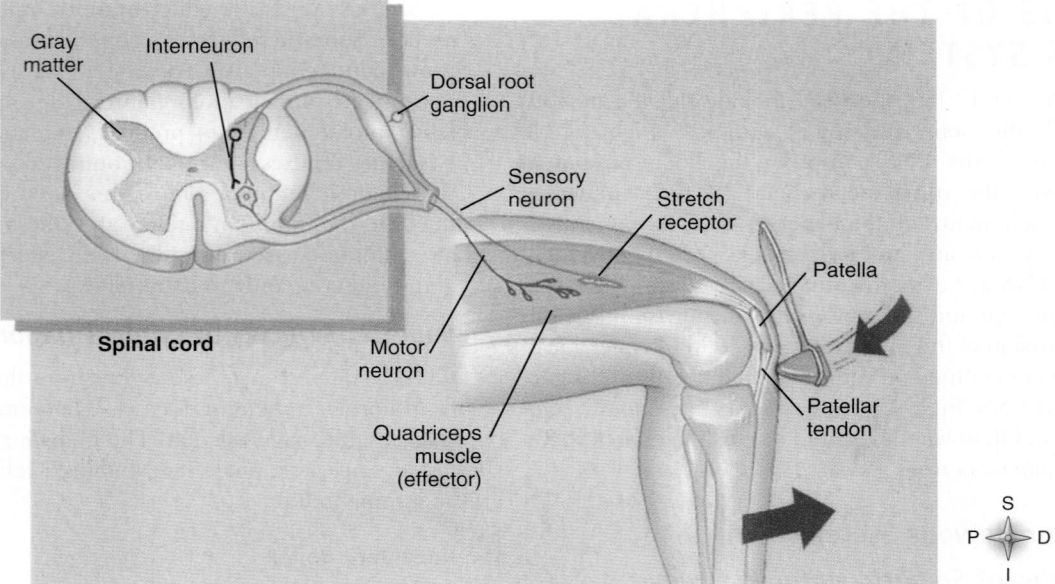

Figure 14-15 *Patellar reflex.* Neural pathway involved in the patellar (knee jerk) reflex.

- As a *tendon reflex*—because tapping of a tendon is the stimulus that elicits it
- As a *deep reflex*—because of the deep location (in tendon and muscle) of the receptors stimulated to produce this reflex (as opposed to *superficial reflexes*—those elicited by stimulation of receptors located in the skin or mucosa)

When testing a patient's reflexes, a physician interprets the test results on the basis of what is known about the reflex arcs that must function to produce normal reflexes.

To illustrate, suppose that a patient has been diagnosed as having poliomyelitis. During the examination the physician finds that she cannot elicit the knee jerk when she taps the patient's patellar tendon. She knows that the poliomyelitis virus attacks anterior horn motor neurons. She also knows the information previously related about which spinal cord segments contain the reflex centers for the knee jerk. On the basis of this knowledge, therefore, she deduces that in this patient the poliomyelitis virus has damaged the second, third, and fourth lumbar segments of the spinal cord. Do you think that this patient's leg would be paralyzed, that he would be unable to move it voluntarily? What neurons would not be able to function that must function to produce voluntary contractions?

Ankle Jerk Reflex

Ankle jerk reflex, or *Achilles reflex*, is an extension (plantar flexion) of the foot in response to tapping of the Achilles tendon. As with the knee jerk, it is a tendon reflex and a deep reflex mediated by two-neuron spinal arcs. The centers for the ankle jerk lie in the first and second sacral segments of the cord.

Babinski Reflex

The **Babinski reflex** is an extension of the great toe, with or without fanning of the other toes, in response to stimulation of the outer margin of the sole of the foot. Normal infants, up until they are about 1½ years old, show this Babinski reflex. By about this time, corticospinal fibers have become fully myelinated and the Babinski reflex becomes suppressed. Just why this is so is not clear; but at any rate it is, and a Babinski reflex after this age is abnormal. From then on the normal response to stimulation of the outer edge of the sole is the *plantar reflex*. It consists of a curling under of all the toes (plantar flexion) plus a slight turning in and flexion of the anterior part of the foot. A positive Babinski reflex is one of the pyramidal signs (see Box 13-12, p. 504) and is interpreted to mean destruction of pyramidal tract (corticospinal) fibers.

Corneal Reflex

The **corneal reflex** is blinking in response to touching the cornea. It is mediated by reflex arcs with sensory fibers in the ophthalmic branch of the fifth cranial nerve, centers in the pons, and motor fibers in the seventh cranial nerve.

Abdominal Reflex

The **abdominal reflex** is drawing in of the abdominal wall in response to stroking the side of the abdomen. It is mediated by arcs with sensory and motor fibers in the ninth to twelfth thoracic spinal nerves and centers in these segments of the cord. It is classified as a superficial reflex. A decrease in this reflex or its absence occurs in lesions involving pyramidal tract upper motor neurons. It can, however, be absent without any pathological condition—in pregnancy, for example.

 QUICK CHECK

7. What are the somatic effectors?
8. State the principle of final common path as it applies to the somatic motor nervous system.
9. Explain how the knee jerk reflex can be both a stretch reflex and a spinal reflex.

Autonomic Nervous System

The **autonomic nervous system (ANS)** is a subdivision of the nervous system that regulates involuntary effectors (see Figure 12-2, p. 433). It is the ANS that carries efferent signals to the autonomic, or *visceral*, effectors—cardiac muscle, smooth muscle, and glandular epithelial tissue (Table 14-4). The major function of the ANS is to regulate heartbeat, smooth muscle contraction, and glandular secretion in ways that maintain homeostasis. Although the ANS also includes sensory (afferent) pathways that provide the feedback necessary to regulate effectors, we will emphasize the motor (efferent) pathways in this chapter.

The ANS has two efferent divisions: the **sympathetic division** and the **parasympathetic division**. The sympathetic division consists of neural pathways that are separate from the parasympathetic pathways. However, many autonomic effectors are *dually innervated*. That is, many autonomic effectors receive input from both sympathetic and parasympathetic pathways.

Usually the effects of the two systems are antagonistic: one inhibits the effector, and the other stimulates the effector. This allows a doubly innervated effector to be controlled with remarkable precision. It also allows a doubly innervated effector to participate in timed events, such as the sexual response, in which effectors must be stimulated and then rapidly inhibited (or vice versa) in a specific timed sequence. Some autonomic effectors are singly innervated, receiving input from only the sympathetic division.

Structure of the Autonomic Nervous System
Basic Plan of Autonomic Pathways

Each efferent autonomic pathway, whether sympathetic or parasympathetic, is made up of autonomic nerves, ganglia, and plexuses. These structures, in turn, are made up of efferent autonomic neurons. They conduct impulses away from the brainstem or spinal cord to autonomic effectors. As with all efferent neurons, efferent autonomic neurons function in reflex arcs. Thus as with somatic motor regulation, efferent autonomic regulation ultimately depends on feedback from the sensory pathways.

A relay of two autonomic neurons conducts information from the CNS to the autonomic effectors. The first is called a **preganglionic neuron**—an awkward name but a descriptive one. Preganglionic neurons conduct impulses from the brainstem or spinal cord to an autonomic ganglion, locating them *before* the ganglion—thus *pre*ganglionic. Within an autonomic ganglion, the preganglionic neuron synapses with a second efferent neuron. Because this second neuron conducts impulses away from the ganglion and to the effector, it is called the **postganglionic neuron**. As you can see in Figure 14-16, A, this plan fundamentally differs from the efferent pathways of the somatic motor nervous system. Conduction to somatic effectors requires only one efferent neuron, the somatic motor neuron that originates in the anterior gray horn of the spinal cord. Conduction to autonomic effectors, however, requires a sequence of two efferent neurons from the CNS to the effector. Essential features of the somatic motor pathways and autonomic efferent pathways are further compared and contrasted in Table 14-5.

Table 14-4	Autonomic Effector Tissues and Organs		
CARDIAC MUSCLE	**SMOOTH MUSCLE**	**GLANDULAR EPITHELIUM**	
Heart	Blood vessels	Sweat glands	
	Bronchial tubes	Lacrimal glands	
	Stomach	Digestive glands (salivary gastric, pancreas, liver)	
	Gallbladder	Adrenal medulla	
	Intestines		
	Urinary bladder		
	Spleen		
	Eye (iris, ciliary muscles)		
	Hair follicles		

Table 14-5	Comparison of Somatic Motor and Autonomic Pathways	
FEATURE	**SOMATIC MOTOR PATHWAYS**	**AUTONOMIC EFFERENT PATHWAYS**
Direction of information flow	Efferent	Efferent
Number of neurons between CNS and effector	One (somatic motor neuron)	Two (preganglionic and postganglionic)
Myelin sheath present	Yes	Preganglionic: yes Postganglionic: no
Location of peripheral fibers	Most cranial nerves and all spinal nerves	Most cranial nerves and all spinal nerves
Effector innervated	Skeletal muscle (voluntary)	Smooth and cardiac muscle, glands (involuntary)
Neurotransmitter	Acetylcholine	Acetylcholine or norepinephrine

CNS, Central nervous system.

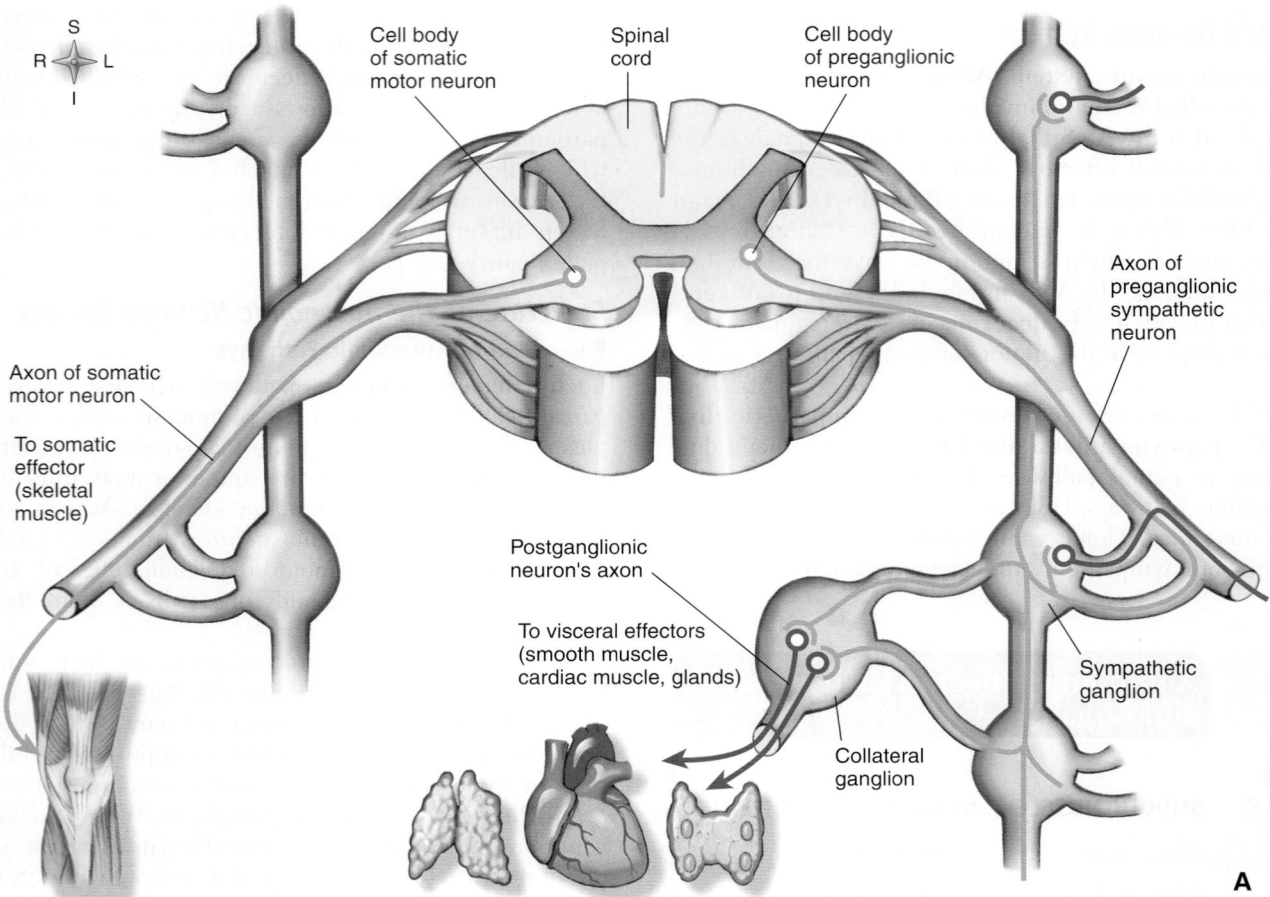

Figure 14-16 *Autonomic conduction paths.* **A,** The left side of the diagram shows that one somatic motor neuron conducts impulses all the way from the spinal cord to a somatic effector. Conduction from the spinal cord to any visceral effector, however, requires a relay of at least two autonomic motor neurons—a preganglionic and a postganglionic neuron, shown on the right side of the diagram.

Structure of the Sympathetic Pathways
Sympathetic Preganglionic Neurons

Sympathetic preganglionic neurons begin within the spinal cord. Specifically, they have their dendrites and cell bodies within the lateral gray horns of the thoracic and lumbar segments of the spinal cord (see Figure 14-16, *A*). For this reason, the sympathetic division has also been called the *thoracolumbar division.*

Most of the ganglia of the sympathetic division lie along either side of the anterior surface of the vertebral column (see Figure 13-3, p. 474). Both gray rami and white rami are sympathetic rami. Sympathetic axon collaterals bridge the gap between adjacent ganglia that lie on the same side of the vertebral column, forming a structure that resembles a chain of beads (see 14-16, *B*). For this reason, the linked ganglia are often referred to as the *sympathetic chain ganglia,* or the **sympathetic trunk.**

Each chain runs from the second cervical vertebra in the neck all the way down to the level of the coccyx. There are usually 22 sympathetic chain ganglia on each side of the vertebral column: three cervical, eleven thoracic, four lumbar, and four

sacral. Axons of sympathetic preganglionic neurons leave the cord by way of the ventral roots of the thoracic and first four lumbar spinal nerves. From there, they split away from other spinal nerve fibers by means of a small branch called the *white ramus.* The white ramus gets its name from the fact that most of the sympathetic preganglionic fibers within it are myelinated axons. Note in Figure 14-16 that the sympathetic preganglionic axon extends through the white ramus to a sympathetic chain ganglion. If we trace the axon inside the sympathetic chain ganglion, we see that the preganglionic fiber may branch along any of three paths:

1. It can synapse with a sympathetic postganglionic neuron.
2. It can send ascending or descending branches through the sympathetic trunk to synapse with postganglionic neurons in other chain ganglia.
3. It can pass through one or more ganglia without synapsing.

Preganglionic neurons that pass through chain ganglia without synapsing continue on through **splanchnic nerves** to other sym-

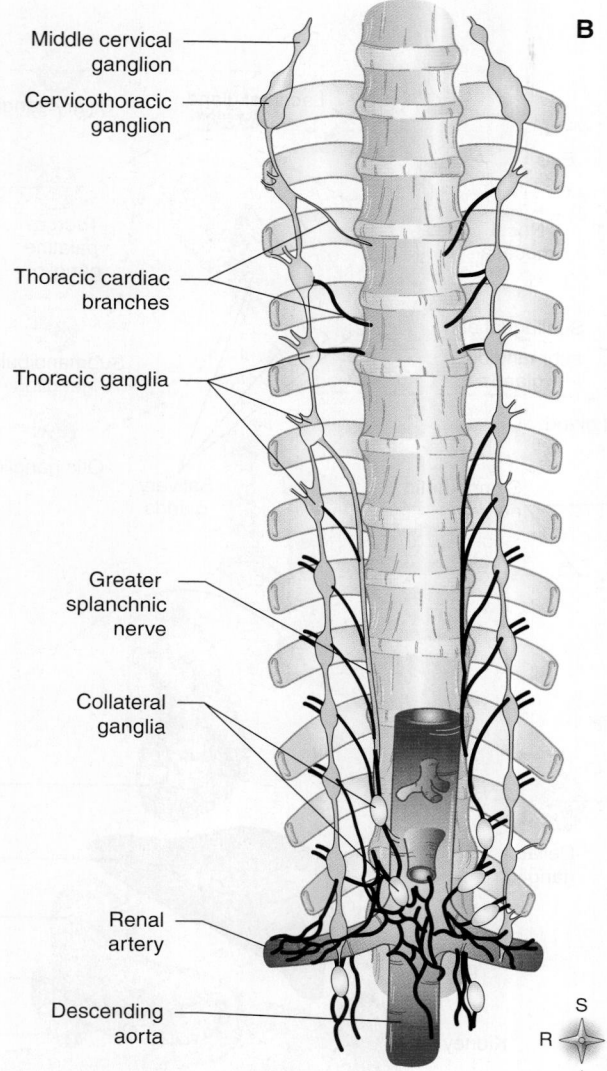

B

Middle cervical ganglion

Cervicothoracic ganglion

Thoracic cardiac branches

Thoracic ganglia

Greater splanchnic nerve

Collateral ganglia

Renal artery

Descending aorta

Figure 14-16, cont'd B, Sketch showing a portion of the left and right sympathetic chain or trunk.

pathetic ganglia (see Figures 14-16 and 14-17). These **collateral ganglia,** or *prevertebral ganglia,* are pairs of sympathetic ganglia located a short distance from the spinal cord. The collateral ganglia are named for nearby blood vessels. For example, the *celiac ganglion* (also called the *solar plexus*) is a large ganglion that lies next to the celiac artery just below the diaphragm. Other examples include the *superior mesenteric ganglion* and the *inferior mesenteric ganglion,* each located close to the beginning of an artery of the same name. Some of the preganglionic fibers that enter the celiac ganglion do not synapse there but continue on to the central portion (medulla) of the *adrenal gland.* Within the adrenal medulla, they synapse with modified postganglionic neurons. These modified postganglionic cells are actually endocrine cells that release hormones (mostly epinephrine) into the bloodstream. These chemical messengers may reach the various sym-

pathetic effectors, where they enhance and prolong the effects of sympathetic stimulation.

Sympathetic Postganglionic Neurons

Most of the sympathetic postganglionic neurons have their dendrites and cell bodies in the sympathetic chain ganglia or collateral ganglia. Some postganglionic axons return to a spinal nerve by way of a short branch called the *gray ramus,* so named because most postganglionic fibers are unmyelinated (see Figure 14-16). Once in the spinal nerve, the postganglionic fibers are distributed with other nerve fibers to the various sympathetic effectors. On the other hand, some postganglionic fibers are distributed to sympathetic effectors by way of separate autonomic nerves. The course of postganglionic fibers through these autonomic nerves is complex, involving the redistribution of fibers

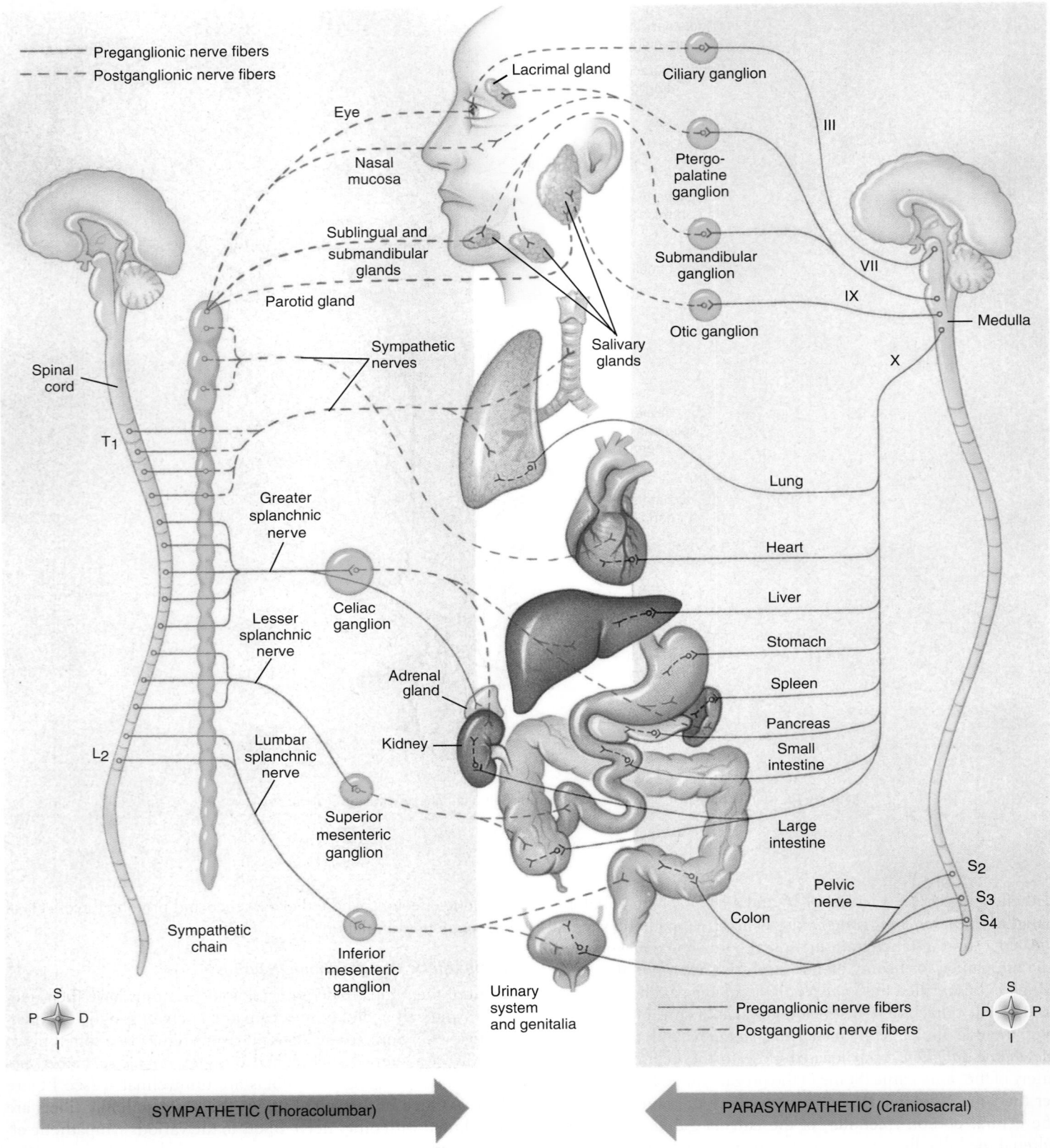

Preganglionic nerve fibers
Postganglionic nerve fibers

Lacrimal gland

Ciliary ganglion

Eye

III

Nasal mucosa

Ptergo-palatine ganglion

Sublingual and submandibular glands

Submandibular ganglion

Parotid gland

VII

Spinal cord

Sympathetic nerves

Salivary glands

IX

Otic ganglion

Medulla

X

T1

Lung

Greater splanchnic nerve

Heart

Liver

Celiac ganglion

Lesser splanchnic nerve

Stomach

Spleen

Adrenal gland

Pancreas

Kidney

Small intestine

Lumbar splanchnic nerve

L2

Superior mesenteric ganglion

Large intestine

S2

Sympathetic chain

Pelvic nerve

S3

Colon

S4

Inferior mesenteric ganglion

Urinary system and genitalia

S
P D
I

Preganglionic nerve fibers
Postganglionic nerve fibers

S
D P
I

SYMPATHETIC (Thoracolumbar)

PARASYMPATHETIC (Craniosacral)

Figure 14-17 *Major autonomic pathways.*

in autonomic plexuses before they reach their respective destinations.

In the sympathetic division, preganglionic neurons are relatively short, and postganglionic neurons are relatively long.

The axon of any one sympathetic preganglionic neuron synapses with many postganglionic neurons, and these frequently terminate in widely separated organs. This anatomical fact partially explains a well-known physiological principle—sympathetic responses are usually widespread, involving many organs and not just one.

QUICK CHECK

10. Do autonomic pathways follow the principle of final common path?
11. Why is the sympathetic division of the ANS also known as the thoracolumbar division?
12. Describe the path generally taken by an impulse along a sympathetic pathway from the CNS to an autonomic effector.

Structure of the Parasympathetic Pathways
Parasympathetic Preganglionic Neurons

Parasympathetic preganglionic neurons have their cell bodies in nuclei in the brainstem or in the lateral gray columns of the sacral cord. For this reason, the parasympathetic division has also been called the *craniosacral division*. Axons of parasympathetic preganglionic neurons are contained in cranial nerves III, VII, IX, and X and in some pelvic nerves. They extend a considerable distance before synapsing with postganglionic neurons. For example, at least 75% of all parasympathetic preganglionic fibers travel in the *vagus nerve (X)* for a distance of 30 cm (1 foot) or more before synapsing with postganglionic fibers in **terminal ganglia** near effectors in the chest and abdomen (see Figure 14-17 and Table 14-6).

Parasympathetic Postganglionic Neurons

Parasympathetic postganglionic neurons have their dendrites and cell bodies in parasympathetic ganglia. Unlike sympathetic ganglia that lie near the spinal column, parasympathetic ganglia lie near or embedded in autonomic effectors. For example, note the ciliary ganglion in Figure 14-17. This and the other ganglia shown near it are parasympathetic ganglia located in the skull. In a parasympathetic ganglion, preganglionic axons synapse with

Table 14-6 Comparison of Structural Features of the Sympathetic and Parasympathetic Pathways

NEURONS	SYMPATHETIC	PARASYMPATHETIC
Preganglionic Neurons		
Dendrites and cell bodies	In lateral gray columns of thoracic and first four lumbar segments of spinal cord	In nuclei of brainstem and in lateral gray columns of sacral segments of cord
Axons	In anterior roots of spinal nerves to spinal nerves (thoracic and first four lumbar), to and through white rami to terminate in sympathetic ganglia at various levels or to extend through sympathetic ganglia, to and through splanchnic nerves to terminate in collateral ganglia	From brainstem nuclei through cranial nerve III to ciliary ganglion From nuclei in pons through cranial nerve VII to sphenopalatine or submaxillary ganglion From nuclei in medulla through cranial nerve IX to otic ganglion or through cranial nerves X and XI to cardiac and celiac ganglia, respectively
Distribution	Short fibers from CNS to ganglion	Long fibers from CNS to ganglion
Neurotransmitter	Acetylcholine	Acetylcholine
Ganglia	Sympathetic chain ganglia (22 pairs); collateral ganglia (celiac, superior, inferior mesenteric)	Terminal ganglia (in or near effector)
Postganglionic Neurons		
Dendrites and cell bodies	In sympathetic and collateral ganglia	In parasympathetic ganglia (e.g., ciliary, sphenopalatine, submaxillary, otic, cardiac, celiac) located in or near visceral effector organs
Receptors	Cholinergic (nicotinic)	Cholinergic (nicotinic)
Axons	In autonomic nerves and plexuses that innervate thoracic and abdominal viscera and blood vessels in these cavities In gray rami to spinal nerves, to smooth muscle of skin blood vessels and hair follicles, and to sweat glands	In short nerves to various visceral effector organs
Distribution	Long fibers from ganglion to widespread effectors	Short fibers from ganglion to single effector
Neurotransmitter	Norepinephrine (many); acetylcholine (few)	Acetylcholine

CNS, Central nervous system.

postganglionic neurons that send their short axons into the nearby autonomic effector. A parasympathetic preganglionic neuron therefore usually synapses with postganglionic neurons to a single effector. For this reason, parasympathetic stimulation frequently involves response by only one organ. Sympathetic stimulation, on the other hand, usually evokes responses by numerous organs.

Autonomic Neurotransmitters and Receptors

Axon terminals of autonomic neurons release either of two neurotransmitters: **norepinephrine (NE)** or **acetylcholine (Ach)**. Axons that release norepinephrine are known as *adrenergic fibers*. Axons that release acetylcholine are called *cholinergic fibers*. Autonomic cholinergic fibers are the axons of preganglionic sympathetic neurons and of both preganglionic and postganglionic parasympathetic neurons. This leaves the axons of postganglionic

sympathetic neurons as the only autonomic adrenergic fibers, and as you can see in Figure 14-18, not all of these are adrenergic. Sympathetic postganglionic axons to sweat glands and some blood vessels are cholinergic fibers.

Norepinephrine and its Receptors. Norepinephrine affects visceral effectors by first binding to *adrenergic receptors* in their plasma membranes. The adrenergic receptors are of two main types, one named **alpha (α) receptors** and the other named **beta (β) receptors** (Figure 14-19, A). Different subtypes of alpha and beta receptors, such as alpha-1 (α_1), alpha-2 (α_2), beta-1 (β_1), and beta-2 (β_2), exist among cells that have adrenergic receptors.

The binding of norepinephrine to alpha receptors in the smooth muscle of blood vessels has a stimulating effect on the muscle that causes the vessel to constrict. The binding of norepinephrine to beta receptors in smooth muscle of a different

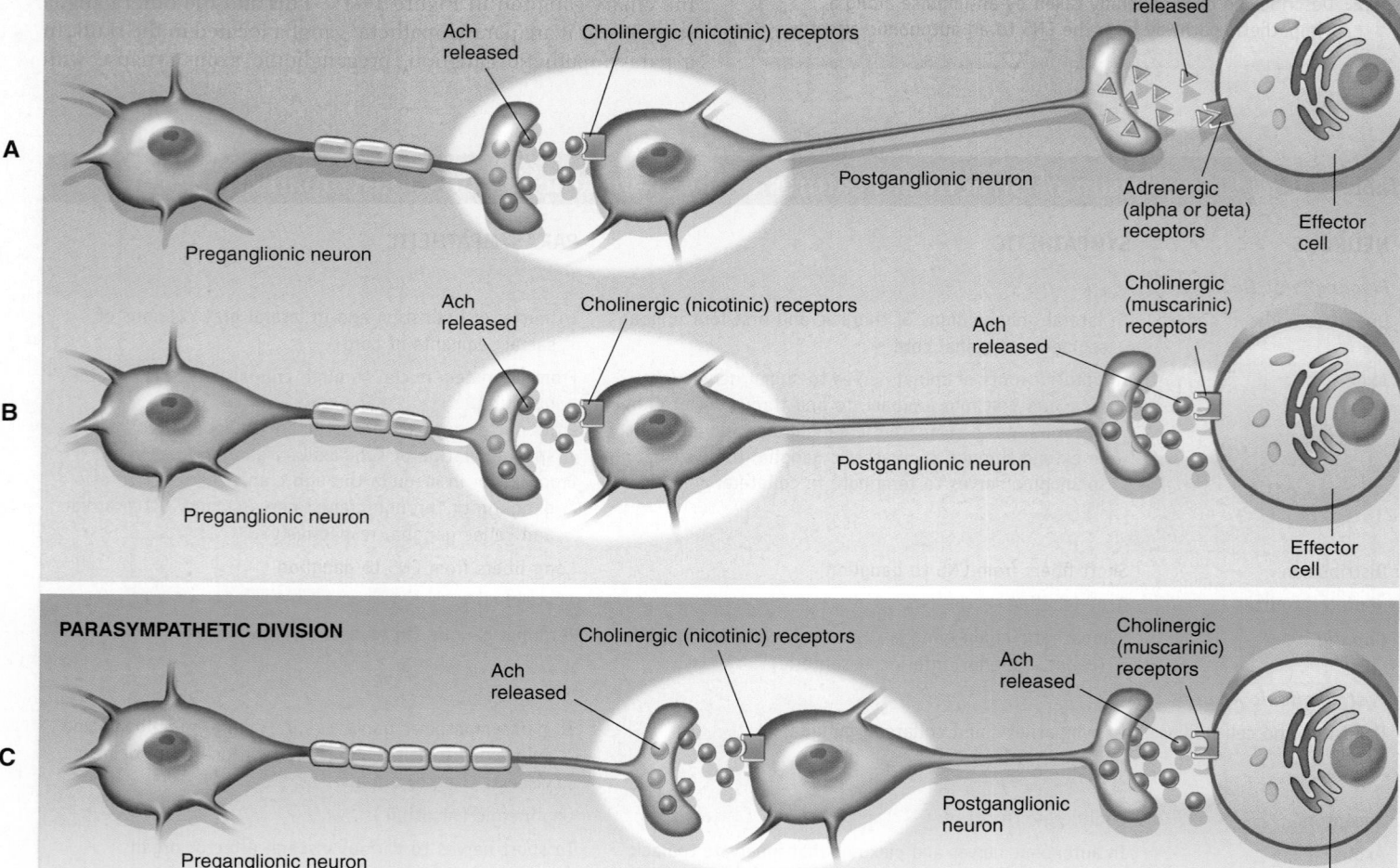

Figure 14-18 *Locations of neurotransmitters and receptors of the autonomic nervous system.* In all pathways, preganglionic fibers are cholinergic, secreting acetylcholine *(Ach)*, which stimulates nicotinic receptors in the postganglionic neuron. Most sympathetic postganglionic fibers are adrenergic **(A)**, secreting norepinephrine *(NE)*, thus stimulating alpha or beta adrenergic receptors. A few sympathetic postganglionic fibers are cholinergic, stimulating muscarinic receptors in effector cells **(B)**. All parasympathetic postganglionic fibers are cholinergic **(C)**, stimulating muscarinic receptors in effector cells.

blood vessel produces opposite effects. It inhibits the muscle, causing the vessel to dilate. But the binding of norepinephrine to beta receptors in cardiac muscle has a stimulating effect that results in a faster and stronger heartbeat. Epinephrine released by the sympathetic postganglionic cells in the adrenal medulla also stimulates the adrenergic receptors, enhancing and prolonging the effects of sympathetic stimulation. Because epinephrine has a greater effect on some beta receptors than norepinephrine, effectors with a large proportion of these beta

receptors are more sensitive to epinephrine. All these facts point to an important principle about nervous regulation: the effect of a neurotransmitter on any postsynaptic cell is determined by the characteristics of the receptor and not by the neurotransmitter itself.

The actions of norepinephrine and epinephrine are terminated in two ways. Most of the neurotransmitter molecules are taken back up by the synaptic knobs of postganglionic neurons, where they are broken down by the enzyme *monoamine oxidase*

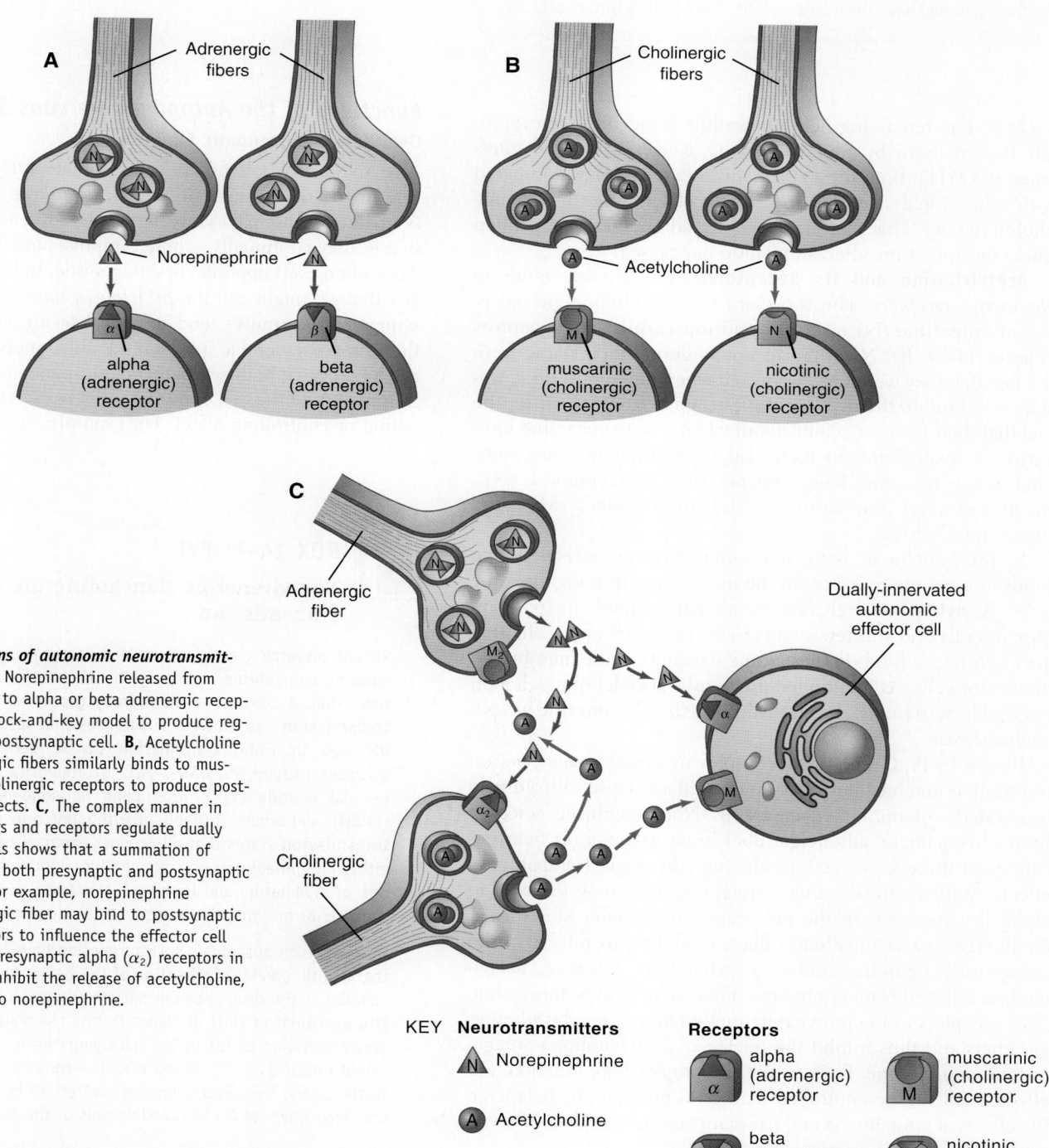

Figure 14-19 *Functions of autonomic neurotransmitters and receptors.* **A,** Norepinephrine released from adrenergic fibers binds to alpha or beta adrenergic receptors according to the lock-and-key model to produce regulatory effects in the postsynaptic cell. **B,** Acetylcholine released from cholinergic fibers similarly binds to muscarinic or nicotinic cholinergic receptors to produce postsynaptic regulatory effects. **C,** The complex manner in which neurotransmitters and receptors regulate dually innervated effector cells shows that a summation of effects on receptors at both presynaptic and postsynaptic locations may occur. For example, norepinephrine released by an adrenergic fiber may bind to postsynaptic alpha (or beta) receptors to influence the effector cell and may also bind to presynaptic alpha (α_2) receptors in a cholinergic fiber to inhibit the release of acetylcholine, a possible antagonist to norepinephrine.

KEY **Neurotransmitters**

N Norepinephrine

A Acetylcholine

Receptors

α alpha (adrenergic) receptor

β beta (adrenergic) receptor

M muscarinic (cholinergic) receptor

N nicotinic (cholinergic) receptor

BOX 14-6: HEALTH MATTERS
Beta Blockers

Drugs that bind to beta receptors, and thus block the binding of norepinephrine and epinephrine, are informally called **beta blockers.** Propranolol (Inderal) is one of nearly a dozen beta blockers used to treat irregular heartbeats and hypertension (high blood pressure). Beta blockers achieve their therapeutic effects by preventing the increased rate and strength of the heart's pumping action triggered by sympathetic stimulation. For patients who also have had a previous heart attack (myocardial infarction), these drugs possibly have the added benefit of preventing further attacks.

QUICK CHECK

13. Describe the pathway taken by an impulse traveling along a parasympathetic pathway.
14. What is the difference between a cholinergic fiber and an adrenergic fiber? Between a cholinergic receptor and an adrenergic receptor?
15. Name the two major types of cholinergic receptors and the two major types of adrenergic receptors.

(MAO). The remaining neurotransmitter molecules are eventually broken down by another enzyme, *catechol-O-methyl transferase (COMT)*. Both these mechanisms are very slow compared with the rapid deactivation of acetylcholine by acetylcholinesterase. This fact explains why adrenergic effects often linger for some time after stimulation has ceased.

Acetylcholine and its Receptors. Acetylcholine binds to *cholinergic receptors*. The two main types of cholinergic receptors are **nicotinic (N) receptors** and **muscarinic (M) receptors** (Figure 14-19, *B*). Nicotinic receptors derive their name from the fact that they were first discovered when nicotine, a drug, was shown to bind to them. Muscarinic receptors are named for the fact that their discovery came about when it was shown that muscarine, a toxin, binds to them. Like the adrenergic receptors, cholinergic receptors have subtypes such as nicotinic-1 (N_1), nicotinic-2 (N_2), muscarinic-1 (M_1), muscarinic-2 (M_2), and muscarinic-3 (M_3).

In the ganglia of both autonomic divisions, acetylcholine binds to nicotinic receptors in the membranes of postganglionic cells. Acetylcholine, released by all parasympathetic postganglionic cells and the few sympathetic postganglionic cells that are cholinergic, binds to muscarinic receptors in the membranes of effector cells. As mentioned previously, acetylcholine's action is quickly terminated by its hydrolysis by the enzyme *acetylcholinesterase*.

Figure 14-19, *C*, shows the complex manner in which neurotransmitters and receptors may function at a synapse with a dually innervated autonomic effector cell. Norepinephrine released from a sympathetic adrenergic fiber binds to alpha (or beta) receptors of the effector cell, producing adrenergic (sympathetic) effects. As the figure also shows, norepinephrine may also bind to alpha (a_2) receptors in the presynaptic membrane of a nearby cholinergic (parasympathetic) fiber, inhibiting its release of the antagonistic neurotransmitter, acetylcholine. Likewise, acetylcholine released from cholinergic fibers may bind to muscarinic (M_2) receptors in the presynaptic membranes of nearby adrenergic fibers and thus inhibit the release of acetylcholine's antagonist, norepinephrine. Because of this complexity of function, the effector cell can be controlled with great precision by balancing the effects of sympathetic and parasympathetic stimulation in various ways.

Functions of the Autonomic Nervous System
Overview of Autonomic Function

The ANS as a whole functions to regulate autonomic effectors in ways that tend to maintain or quickly restore homeostasis. Both sympathetic and parasympathetic divisions are *tonically active*; that is, they continually conduct impulses to autonomic effectors. They often exert opposite, or antagonistic, influences on them—a fact that we might call the *principle of autonomic antagonism*. If sympathetic impulses tend to stimulate an effector, parasympathetic impulses tend to inhibit it. Doubly innervated effectors continually receive both sympathetic and parasympathetic impulses. Summation of the two opposing influences determines the dominating or controlling effect. For example, continual sympathetic

BOX 14-7: FYI
Nonadrenergic-Noncholinergic (NANC) Transmission

Recent research clearly shows that many neurotransmitters besides norepinephrine and acetylcholine bind to autonomic receptors. Thus a concept of **nonadrenergic-noncholinergic** (NANC) **transmission** has evolved. Substances that research shows to be involved in NANC transmission across peripheral autonomic synapses include NO, GABA, ATP, neuropeptide Y, vasoactive intestinal peptide (VIP), luteinizing hormone–releasing hormone (LHRH), and others. A theory of autonomic transmission called **cotransmission** states that all or most postganglionic fibers release either norepinephrine or acetylcholine along with NANC transmitters or modulators and that each substance combines with postsynaptic or presynaptic receptors to produce regulatory effects.

Discoveries in autonomic co-transmission have led to a number of therapeutic breakthroughs. One of the better known of these discoveries is the drug sildenafil citrate (Viagra) used to treat **erectile dysfunction** *(ED)* in males. During the male sexual response, co-transmission of NO in the parasympathetic nerves helps relax smooth muscles in the blood vessels—thus increasing blood flow to the penis. This drug enhances that effect by reducing the natural breakdown of NO in blood vessels of the penis.

impulses to the heart tend to accelerate the heart rate whereas continual parasympathetic impulses tend to slow it. The actual heart rate is determined by whichever influence dominates. To find other examples of autonomic antagonism, examine Table 14-7.

The ANS does not function autonomously as its name suggests. It is continually influenced by impulses from the so-called autonomic centers. These are clusters of neurons located at various levels in the brain whose axons conduct impulses directly or indirectly to autonomic preganglionic neurons. Autonomic centers function as a hierarchy in their control of the

ANS (Figure 14-20). Highest ranking in the hierarchy are the autonomic centers in the cerebral cortex, for example, in the frontal lobe and limbic system (structures near the medial surface of the cerebrum that form a border around the corpus callosum). Neurons in these centers send impulses to other autonomic centers in the brain, notably in the hypothalamus. Then neurons in the hypothalamus send either stimulating or inhibiting impulses to parasympathetic and sympathetic preganglionic neurons located in the lower autonomic centers of the brainstem and cord.

Table 14-7 Autonomic Functions

AUTONOMIC EFFECTOR	EFFECT OF SYMPATHETIC STIMULATION (NEUROTRANSMITTER: NOREPINEPHRINE UNLESS OTHERWISE STATED)	EFFECT OF PARASYMPATHETIC STIMULATION (NEUROTRANSMITTER: ACETYLCHOLINE)
Cardiac Muscle	Increased rate and strength of contraction (beta receptors)	Decreased rate and strength of contraction
Smooth Muscle of Blood Vessels		
Skin blood vessels	Constriction (alpha receptors)	No effect
Skeletal muscle blood vessels	Dilation (beta receptors)	No effect
Coronary blood vessels	Constriction (alpha receptors) Dilation (beta receptors)	Dilation
Abdominal blood vessels	Constriction (alpha receptors)	No effect
Blood vessels of external genitalia	Constriction (alpha receptors)	Dilation of blood vessels causing erection
Smooth Muscle of Hollow Organs and Sphincters		
Bronchioles	Increased peristalsis	Constriction
Digestive tract, except sphincters	Relaxation	Increased peristalsis
Sphincters of digestive tract	Contraction	Relaxation
Urinary bladder	Relaxation	Contraction
Urinary sphincters	Relaxation	Relaxation
Reproductive ducts	Constriction	Relaxation
Eye		
Iiris	Contraction of radial muscle; dilated pupil Relaxation; accommodates for far vision	Contraction of circular muscle; constricted pupil Contraction; accommodates for near vision
Ciliary (pilomotor muscles)	Contraction produces goose pimples, or piloerection (alpha receptors)	No effect
Glands		
Sweat	Increased sweat (neurotransmitter, acetylcholine)	No effect
Lacrimal	No effect	Increased secretion of tears
Digestive (salivary, gastric, etc.)	Decreased secretion of saliva; not known for others	Increased secretion of saliva
Pancreas, including islets	Decreased secretion	Increased secretion of pancreatic juice and insulin
Liver	Increased glycogenolysis (beta receptors); increased blood sugar level	No effect
Adrenal medulla*	Increased epinephrine secretion	No effect

*Sympathetic preganglionic axons terminate in contact with secreting cells of the adrenal medulla. Thus the adrenal medulla functions, to quote someone's descriptive phrase, as a "giant sympathetic postganglionic neuron."

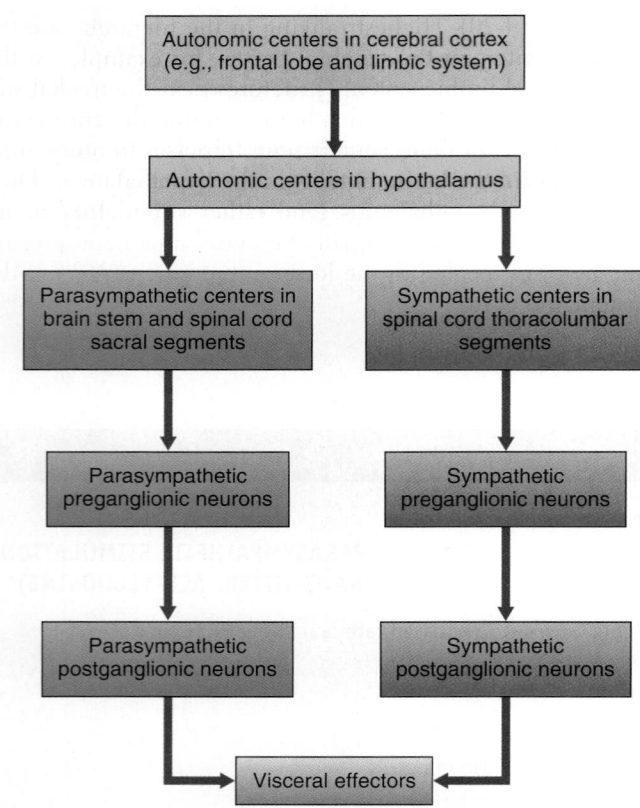

Figure 14-20 *Central nervous system hierarchy that regulates autonomic functions.*

BOX 14-8: FYI

Biofeedback

We now know that individuals can learn to control specific autonomic effectors if two conditions are fulfilled. They must be informed that they are achieving the desired response, and they must be rewarded for it. Various kinds of **biofeedback** instruments have been developed to provide these conditions. For example, a biofeedback instrument that detects slight temperature changes has been used with patients who suffer from migraine headaches. (Migraine headaches initially involve distention of blood vessels in the head.) The instrument is attached to their hands and emits a high sound each time their hand blood vessels dilate. The reward for such patients is a lessening of the migraine pain—presumably because of the shunting of blood away from the head to the hands.

You may be wondering why the name *autonomic system* was ever chosen if the system is really not autonomous. Originally the term seemed appropriate. The autonomic system seemed to be self-regulating and independent of the rest of the nervous system. Common observations furnished abundant evidence of its independence from cerebral control, from direct control by the will, that is; but later even this was found to be not entirely true. Some rare and startling exceptions were discovered; for example, a man in a brightly lighted amphitheater can change the size of his pupils from small, constricted dots (normal response to bright lights) to widely dilated circles. It is also possible to will the smooth muscle of the hairs on the arms to contract, producing gooseflesh.

Functions of the Sympathetic Division

Under ordinary, resting conditions the sympathetic division can act to maintain the normal functioning of doubly innervated autonomic effectors. It does this by opposing the effects of parasympathetic impulses to these structures. For example, by counteracting parasympathetic impulses that tend to slow the heart and weaken its beat, sympathetic impulses function to maintain the heartbeat's normal rate and strength. The sympathetic division also serves another important function under usual conditions. Because only sympathetic fibers innervate the smooth muscle in blood vessel walls, sympathetic impulses function to maintain the normal tone of this muscle. By so doing, the sympathetic system

plays a crucial role in maintaining blood pressure under usual conditions.

The major function of the sympathetic division, however, is that it serves as an "emergency" system. When we perceive that the homeostasis of the body might be threatened—that is, when we are under physical or psychological stress—outgoing sympathetic signals increase greatly. In fact, one of the very first steps in the body's complex defense mechanism against stress is a sudden and marked increase in sympathetic activity. This brings about a group of responses that all go on at the same time. Together they make the body ready to expend maximum energy and thus to engage in the maximum muscular exertion needed to deal with the perceived threat—as, for example, in running or fighting. Walter B. Cannon coined the descriptive and now famous phrase the **"fight-or-flight" reaction** as his name for this group of sympathetic responses. Read Table 14-8 to find many of the fight-or-flight physiological changes. Some particularly important changes for maximum energy expenditure by skeletal muscles are faster, stronger heartbeat, dilated blood vessels in skeletal muscles, dilated bronchi, and increased blood sugar levels from stimulated glycogenolysis (conversion of glycogen to glucose). Also, sympathetic impulses to the medulla of each adrenal gland stimulate its secretion of epinephrine and some norepinephrine. These hormones reinforce and prolong effects of the norepinephrine released by sympathetic postganglionic fibers.

The fight-or-flight reaction is a normal response in times of stress. Without such a response, we might not be able to resist or retreat from something that actually threatens our well-being. However, chronic exposure to stress can lead to dysfunction of sympathetic effectors—and perhaps even to the dysfunction of the ANS itself. Some current concepts regarding the effects of chronic stress are discussed in Chapter 22.

Functions of the Parasympathetic Division

The parasympathetic division is the dominant controller of most autonomic effectors most of the time. Under quiet, nonstressful conditions, more impulses reach autonomic effectors

Table 14-8   Summary of the Sympathetic "Fight-or-Flight" Reaction

RESPONSE	ROLE IN PROMOTING ENERGY USE BY SKELETAL MUSCLES
Increased heart rate	Increased rate of blood flow, thus increased delivery of oxygen and glucose to skeletal muscles
Increased strength of cardiac muscle contraction	Increased rate of blood flow, thus increased delivery of oxygen and glucose to skeletal muscles
Dilation of coronary vessels of the heart	Increased delivery of oxygen and nutrients to cardiac muscle to sustain increased rate and strength of heart contractions
Dilation of blood vessels in skeletal muscles	Increased delivery of oxygen and nutrients to skeletal muscles
Constriction of blood vessels in digestive and other organs	Shunting of blood to skeletal muscles to increase oxygen and glucose delivery
Contraction of spleen and other blood reservoirs	More blood discharged into general circulation, causing increased delivery of oxygen and glucose to skeletal muscles
Dilation of respiratory airways	Increased loading of oxygen into blood
Increased rate and depth of breathing	Increased loading of oxygen into blood
Increased sweating	Increased dissipation of heat generated by skeletal muscle activity
Increased conversion of glycogen into glucose	Increased amount of glucose available to skeletal muscles

by cholinergic parasympathetic fibers than by adrenergic sympathetic fibers. If the sympathetic division dominates during times that require fight-or-flight, then the parasympathetic division dominates during the in-between times of "rest-and-repair." Acetylcholine, the neurotransmitter of the parasympathetic system, tends to slow the heartbeat but acts to promote digestion and elimination. For example, it stimulates digestive gland secretion. It also increases peristalsis by stimulating the smooth muscle of the digestive tract. Can you identify other parasympathetic effects in Table 14-7?

 QUICK CHECK

16. What is the principle of autonomic antagonism? Give an example.
17. Name the responses that occur in the fight-or-flight reaction. How does each of these prepare the body to expend a maximum amount of muscular energy?
18. Which division of the ANS is the dominant controller of autonomic effectors when the body is at rest?

 THE BIG PICTURE
Peripheral Nervous System and the Whole Body

The peripheral nervous system is made up of all the afferent nervous pathways coming into the CNS and all the efferent pathways going out of the CNS. We emphasized the peripheral efferent, or motor, pathways in this chapter. We will emphasize the peripheral sensory pathways in Chapter 15. As we have stated, the main role of the nervous system as a whole is to detect changes in the internal and external environment, to evaluate those changes in terms of their effect on homeostatic balance, and to regulate effectors accordingly. The peripheral motor pathways are simply those nervous pathways that lead from the integrator (CNS) to the effectors.

The somatic motor pathways lead to skeletal muscle effectors, and the autonomic efferent pathways lead to the cardiac muscle effectors, smooth muscle effectors, and glandular effectors. Thus all together the peripheral motor pathways serve as an information-carrying network that allows the CNS to communicate regulatory information to all the nervous effectors in the body. Because some motor pathways communicate with endocrine glands, endocrine effectors throughout the body can also be regulated through nervous mechanisms. Every major organ of the body is thus influenced, directly or indirectly, by nervous output. In essence, the CNS is the ultimate controller of the major homeostatic functions of the body, and peripheral motor pathways are the means to exert that control.

LANGUAGE OF SCIENCE   *(Cont'd from page 515)*

dorsal root (DOR-sal) [*dors-* the back, *-al* pertaining to]

facial nerve (FAY-shal) [*faci-* face, *-al* pertaining to]

"fight-or-flight" reaction

ganglia (GANG-lee-ah) [*ganglion* knot]

glossopharyngeal nerve (glos-oh-fah-RIN-jee-al) [*glosso-* tongue, *-pharyng-* throat, *-al* pertaining to]

hypoglossal nerve (hye-poh-GLOS-al) [*hypo-* below, *-gloss-* tongue, *-al* pertaining to]

knee jerk reflex [*re-* again, *-flex* bend]

lumbar plexus (LUM-bar PLEK-sus) [*lumb-* loin, *-ar* pertaining to, *plexus* network]

mixed cranial nerves (KRAY-nee-al) [*crani-* skull, *-al* pertaining to]

mixed nerves

motor cranial nerves (motor KRAY-nee-al nervs) [*crani-* skull, *-al* pertaining to]

muscarinic (M) receptors (mus-kah-RIN-ik ree-SEP-tors) [*musca-* fly, *-ic* pertaining to]

myotome (MY-oh-tohm) [*myo-* muscle, *-tome* segment or region]

nicotinic (N) receptors (NIK-oh-tin-ik ree-SEP-tors) [*Jean Nicot Villemain* French ambassador to Portugal, *-in-* substance, *-ic* pertaining to]

norepinephrine (NE) (nor-ep-i-NEF-rin) [*nor-* chemical prefix, *-epi-* on, *-nephr-* kidney, *-ine* substance]

oculomotor nerve (awk-yoo-loh-MOH-tor) [*oculo-* eye, *-motor* mover]

olfactory nerves (ol-FAK-tor-ee) [*olfact-* sense of smell]

optic nerves (OP-tik) [*optic* pertaining to the eyes or sight]

parasympathetic division (pair-ah-sim-pah-THET-ik) [*para-* similar, *-sympathe-* to feel with, *-ic* pertaining to]

phrenic nerve (FREN-ik) [*phren-* mind, *-ic* pertaining to]

plexuses (PLEK-sus-es) [*plexus* network]

postganglionic neuron (post-gang-glee-ON-ik NOO-ron) [*post-* after, *-ganglion-* knot, *-ic* pertaining to, *neuron* nerve]

preganglionic neuron (pree-gang-glee-ON-ik NOO-ron) [*pre-* before, *-ganglion-* knot, *-ic* pertaining to, *neuron* nerve]

ramus (RAY-mus) [*ramus* branch]

reflex (REE-fleks) [*re-* again, *-flex* bend]

sacral plexus (SAY-kral PLEK-sus) [*sacr-* sacred, *-al* pertaining to, *plexus* network]

sensory cranial nerves (SEN-sor-ee KRAY-nee-al) [*crani-* skull, *-al* pertaining to]

somatic reflexes (so-MAH-tik REE-fleks-es) [*soma-* body, *-tic* pertaining to, *re-* again, *-flex* bend]

spinal nerves (SPY-nal) [*spine-* backbone, *-al* pertaining to]

splanchnic nerves (SPLANK-nik) [*splanchn-* visceral organs, *-al* pertaining to]

sympathetic division (sim-pah-THET-ik) [*sympathe-* to feel with, *-ic* pertaining to]

sympathetic trunk (sim-pah-THET-ik) [*sympathe-* to feel with, *-ic* pertaining to]

terminal ganglia (TER-mih-nal GANG-glee-ah) [*termin-* boundary, *-al* pertaining to, *ganglion* knot]

trigeminal nerve (try-JEM-i-nal) [*tri-* three, *-gemina-* twin or pair, *-al* pertaining to]

trochlear nerve (TROK-lee-ar) [*trochlea-* pulley, *-al* pertaining to]

vagus nerve (VAY-gus) [*vagus* wanderer]

ventral ramus (VEN-tral RAY-mus) [*ventr-* belly, *-al* pertaining to]

ventral root (VEN-tral root) [*ventr-* belly, *-al* pertaining to]

vestibulocochlear nerve (ves-TIB-yoo-loh-kok-lee-ar) [*vestibulo-* courtyard, *-cochle-* snail shell, *-ar* pertaining to]

LANGUAGE OF MEDICINE

beta blockers (BAY-tah) [*beta* second letter of the Greek alphabet]

biofeedback (bye-oh-FEED-bak) [*bio-* life, *-feedback*]

erectile dysfunction (ED) (eh-REK-tyle dis-FUNK-shun) [*erect-*, *-ile* pertaining to, *dys-* bad or painful, *-function* performance]

herpes zoster (HER-peez ZOS-ter) [*herpe* to creep, *zoster* girdle]

nonadrenergic-noncholinergic transmission (non-AD-ren-er-jik non-KOHL-in-er-jik tranz-MISH-un) [*non-* not, *-ad-* toward, *-ren-* kidney, *-erg-* work, *-ic* pertaining to, *-non-* not, *-chole-* bile, *-erg-* work, *-ic* pertaining to]

peripheral neuropathy (peh-RIF-er-al noo-ROP-ah-thee) [*periph-* circumference, *-al* pertaining to, *neuro-* nerves, *-pathy* disease]

shingles (SHING-guls)

trigeminal neuralgia (try-JEM-i-nal noo-RAL-jee-ah) [*tri-* three, *-gemina-* twins or pair, *-al* pertaining to, *neur-* nerves, *-algia* pain]

CASE STUDY

Charles Smythe, 66 years old, arrived at the clinic complaining of pain behind the left ear for the last few days. This morning he woke up with difficulty moving the entire left side of his face. He has a masklike appearance to his face as he talks about his pain. He has a history of type 2 diabetes mellitus, which is controlled by diet, and he has no known complications from this disease. He takes propranolol but does not know why. His medical history is otherwise unremarkable. He is 6 feet tall with a weight of 190 pounds. He exercises regularly by walking several miles each day. Until this recent facial paralysis, Mr. Smythe felt well and had no complaints.

On physical examination, it is noted that Mr. Smythe feels a drawing sensation on the left side. He is unable to close the eye, wrinkle his forehead, smile, whistle, or grimace. His face sags on the left side. His blood pressure and pulse are 146/88 and 92 beats/min, respectively. His hand grips are strong and equal bilaterally. He has no weakness in either of his legs. Based on the history and physical examination, Mr. Smythe is diagnosed with Bell palsy and started on prednisone, 60 mg/day. He is also given analgesics for the pain.

1. Based on the case presented, which one of Mr. Smythe's cranial nerves is *most* likely affected?

   A. VI   C. VIII
   B. VII   D. IX

2. Prednisone is the drug of choice in this case because of its properties as a(n):

   A. Antibiotic   C. Anti-inflammatory
   B. Immunosuppressant   D. Analgesic

3. Mr. Smythe's propranolol was most likely prescribed because of its ability to:

   A. Block beta stimulation
   B. Block alpha stimulation
   C. Enhance beta stimulation
   D. Enhance alpha stimulation

4. In Mr. Smythe's case, the propranolol was *most* likely started to control:

   A. Diabetes mellitus
   B. Irregular heartbeats
   C. The risk of another heart attack
   D. Hypertension

CHAPTER SUMMARY

SPINAL NERVES

A. Overview
1. Thirty-one pairs of spinal nerves are connected to the spinal cord (Figure 14-1)
2. No special names; are numbered by level of vertebral column at which they emerge from the spinal cavity
   a. Eight cervical nerve pairs (C1 through C8)
   b. Twelve thoracic nerve pairs (T1 through T12)
   c. Five lumbar nerve pairs (L1 through L5)
   d. Five sacral nerve pairs (S1 through S5)
   e. One coccygeal nerve pair
3. Lumbar, sacral, and coccygeal nerve roots descend from point of origin to the lower end of the spinal cord (level of first lumbar vertebra) before reaching the intervertebral foramina of the respective vertebrae, through which the nerves emerge
4. Cauda equina—describes the appearance of the lower end of the spinal cord and its spinal nerves as a horse's tail
B. Structure of spinal nerves
1. Each spinal nerve attaches to spinal cord by a ventral (anterior) root and a dorsal (posterior) root
2. Dorsal root ganglion—swelling in the dorsal root of each spinal nerve
3. All spinal nerves are mixed nerves
4. Ramus
   a. One of several large branches formed after each spinal nerve emerges from the spinal cavity (Figure 14-2)

   b. Dorsal ramus—supplies somatic motor and sensory fibers to smaller nerves that innervate the muscles and skin of the posterior surface of the head, neck, and trunk
   c. Ventral ramus
      (1) Structure is more complex than that of dorsal ramus
      (2) Autonomic motor fibers split from the ventral ramus and head toward a ganglion of the sympathetic chain
      (3) Some autonomic fibers synapse with neurons that continue on to autonomic effectors through splanchnic nerves; others synapse with neurons whose fibers rejoin the ventral ramus
      (4) Sympathetic rami—splitting and rejoining of autonomic fibers
      (5) Motor and sensory fibers innervate muscles and glands in the extremities and lateral and ventral portions of neck and trunk
C. Nerve plexuses
1. Plexuses—complex networks formed by the ventral rami of most spinal nerves (not T2 through T12) subdividing and then joining together to form individual nerves
2. Each individual nerve that emerges contains all the fibers that innervate a particular region of the body
3. In plexuses, spinal nerve fibers are rearranged according to their ultimate destination, reducing the number of nerves needed to supply each body part

4. Four major pairs of plexuses
    a. Cervical plexus (Figure 14-3)
        (1) Located deep within the neck
        (2) Made up of ventral rami of C1 through C4 and a branch of the ventral ramus of C5
        (3) Individual nerves emerging from cervical plexus innervate the muscles and skin of the neck, upper shoulders, and part of the head
        (4) Phrenic nerve exits the cervical plexus and innervates the diaphragm
    b. Brachial plexus (Figure 14-4)
        (1) Located deep within the shoulder
        (2) Made up of ventral rami of C5 through T1
        (3) Individual nerves emerging from brachial plexus innervate the lower part of the shoulder and the entire arm
    c. Lumbar plexus (Figure 14-5)
        (1) Located in the lumbar region of the back in the psoas muscle
        (2) Formed by intermingling fibers of L1 through L4
        (3) Femoral nerve exits the lumbar plexus, divides into many branches, and supplies the thigh and leg
    d. Sacral plexus and coccygeal plexus (Figure 14-5)
        (1) Located in the pelvic cavity in the anterior surface of the piriformis muscle
        (2) Formed by intermingling of fibers from L4 through S4
        (3) Tibial, common peroneal, and sciatic nerves exit the sacral plexus and supply nearly all the skin of the leg, posterior thigh muscles, and leg and foot muscles
D. Dermatomes and myotomes (Figure 14-6)
    1. Dermatome—region of skin surface area supplied by afferent (sensory) fibers of a given spinal nerve (Figure 14-7)
    2. Myotome—skeletal muscle or muscles supplied by efferent (motor) fibers of a given spinal nerve (Figure 14-8)

CRANIAL NERVES (TABLES 14-2 AND 14-3)

A. Overview
    1. Twelve pairs of cranial nerves connect to the brain, mostly the brainstem (Figure 14-9)
    2. Identified by name (determined by either distribution or function) or number (order in which they emerge, anterior to posterior) or both
    3. Made up of bundles of axons
        a. Mixed cranial nerve—axons of sensory and motor neurons
        b. Sensory cranial nerve—axons of sensory neurons only
        c. Motor cranial nerve—mainly axons of motor neurons and a small number of sensory fibers (proprioceptors)
B. Olfactory nerve (I)
    1. Composed of axons of neurons whose dendrites and cell bodies lie in nasal mucosa and terminate in olfactory bulbs
    2. Carries information about sense of smell
C. Optic nerve (II)
    1. Composed of axons from the innermost layer of sensory neurons of the retina
    2. Carries visual information from the eyes to the brain

D. Oculomotor nerve (III)
    1. Fibers originate from cells in the oculomotor nucleus and extend to some of the external eye muscles
    2. Efferent autonomic fibers are also present, which extend to the intrinsic muscles of the eye to regulate amount of light entering eye and aid focusing on near objects
    3. Sensory fibers from proprioceptors in the eye muscles are also present
E. Trochlear nerve (IV)
    1. Motor fibers originate in cells of the midbrain and extend to the superior oblique muscles of the eye
    2. Also contains afferent fibers from proprioceptors in the superior oblique muscles of the eye
F. Trigeminal nerve (V) (Figure 14-10)
    1. Has three branches: ophthalmic nerve, maxillary nerve, and mandibular nerve
    2. Sensory neurons carry afferent impulses from skin and mucosa of head and teeth to cell bodies in the trigeminal ganglion
    3. Motor fibers originate in trifacial motor nucleus and extend to the muscles of mastication through the mandibular nerve
G. Abducens nerve (VI)
    1. Motor nerve with fibers originating from a nucleus in the pons on the floor of the fourth ventricle and extending to the lateral rectus muscles of the eye
    2. Contains afferent fibers from proprioceptors in the lateral rectus muscles
H. Facial nerve (VII)
    1. Motor fibers originate from a nucleus in lower part of pons and extend to superficial muscles of the face and scalp (Figure 14-11)
    2. Autonomic fibers extend to submaxillary and sublingual salivary glands
    3. Also contains sensory fibers from taste buds of anterior two thirds of the tongue
I. Vestibulocochlear nerve (VIII)
    1. Two distinct divisions that are both sensory: vestibular nerve and cochlear nerve
    2. Vestibular nerve fibers originate in the semicircular canals in inner ear and transmit impulses that result in sensations of equilibrium
    3. Cochlear nerve fibers originate in the organ of Corti in the cochlea of the inner ear and transmit impulses that result in sensations of hearing
J. Glossopharyngeal nerve (IX)
    1. Composed of sensory, motor, and autonomic nerve fibers
    2. Supplies fibers to tongue, pharynx, and carotid sinus (Figure 14-12)
K. Vagus nerve (X)
    1. Composed of sensory and motor fibers with many widely distributed branches
    2. Sensory fibers supply pharynx, larynx, trachea, heart, carotid body, lungs, bronchi, esophagus, stomach, small intestine, and gallbladder (Figure 14-13)
    3. Somatic motor fibers innervate the pharynx and larynx and are mostly autonomic fibers

L. Accessory nerve (XI)
1. Motor nerve that is an "accessory" to the vagus nerve
2. Innervates thoracic and abdominal viscera, pharynx, larynx, trapezius, and sternocleidomastoid (Figure 14-14)
M. Hypoglossal nerve (XII)
1. Composed of motor and sensory fibers
2. Motor fibers innervate the muscles of the tongue
3. Contains sensory fibers from proprioceptors in muscles of the tongue

DIVISIONS OF THE PERIPHERAL NERVOUS SYSTEM

A. Two functional divisions of the peripheral nervous system
1. Afferent (sensory) division
2. Efferent (motor) division
B. Efferent division is divided further into the somatic motor nervous system and the efferent portions of the autonomic nervous system

SOMATIC MOTOR NERVOUS SYSTEM

A. Basic principles of somatic motor pathways
1. Somatic nervous system—includes all voluntary motor pathways outside the central nervous system
2. Somatic effectors—skeletal muscles
B. Somatic reflexes
1. Nature of a reflex
a. Reflex—action that results from a nerve impulse passing over a reflex arc; predictable response to a stimulus
(1) Cranial reflex—center of reflex arc is in the brain
(2) Spinal reflex—center of reflex arc is in the spinal cord
b. Reflex consists of either muscle contraction or glandular secretion
(1) Somatic reflex—contraction of skeletal muscles
(2) Autonomic (visceral) reflex—either contraction of smooth or cardiac muscle or secretion by glands
2. Some somatic reflexes of clinical importance—reflexes deviate from normal in certain diseases, and reflex testing is a valuable diagnostic aid
a. Knee jerk (also known as patellar reflex)— extension of the lower leg in response to tapping the patellar tendon; tendon and muscles are stretched, stimulating muscle spindles and initiating conduction over a two-neuron reflex arc (Figure 14-15); may be classified in several different ways
b. Ankle jerk (also known as *Achilles reflex*)— extension of the foot in response to tapping the Achilles tendon; tendon reflex and deep reflex mediated by two-neuron spinal arcs; centers lie in first and second sacral segments of the cord
c. Babinski reflex—extension of great toe, with or without fanning of other toes, in response to stimulation of outer margin of sole of foot; present in normal infants until approximately $1\frac{1}{2}$ years of age and then becomes suppressed when corticospinal fibers become fully myelinated; in humans older than $1\frac{1}{2}$ years of age, a positive Babinski reflex is one of the pyramidal signs indicating destruction of corticospinal (pyramidal tract) fibers

d. Plantar reflex—plantar flexion of all toes and a slight turning in and flexion of the anterior part of the foot in response to stimulation of the outer edge of the sole
e. Corneal reflex—winking in response to touching the cornea; mediated by reflex arcs with sensory fibers in the ophthalmic branch of the fifth cranial nerve, centers in the pons, and motor fibers in the seventh cranial nerve
f. Abdominal reflex—drawing in of the abdominal wall in response to stroking the side of the abdomen; superficial reflex; mediated by arcs with sensory and motor fibers in T9 through T12 and centers in these segments of the cord; decreased or absent reflex may involve lesions of pyramidal tract upper motor neurons
(1) Spinal cord reflex—center of reflex arc located in spinal cord gray matter
(2) Segmental reflex—mediating impulses enter and leave at same cord segment
(3) Ipsilateral reflex—mediating impulses come from and go to the same side of the body
(4) Stretch or myotatic reflex—result of type of stimulation used to evoke reflex
(5) Extensor reflex—produced by extensors of the lower leg
(6) Tendon reflex—tapping tendon is stimulus that elicits reflex
(7) Deep reflex—result of deep location of receptors stimulated to produce reflex

AUTONOMIC NERVOUS SYSTEM

A. Overview
1. Contains afferent (sensory) and efferent (motor) components (the efferent components are emphasized here)
2. Carries fibers to and from the autonomic effectors
3. Major function—to regulate heartbeat, smooth muscle contraction, and glandular secretions to maintain homeostasis
4. Two efferent divisions—sympathetic division and parasympathetic division
5. Sympathetic division consists of neural pathways that are separate from parasympathetic pathways
6. Many autonomic effectors are dually innervated, which allows remarkably precise control of effector
B. Structure of the autonomic nervous system
1. Basic plan of efferent autonomic pathways (Figure 14-16)
a. Each pathway is made up of autonomic nerves, ganglia, and plexuses, which are made of efferent autonomic neurons
b. All autonomic neurons function in reflex arcs
c. Efferent autonomic regulation ultimately depends on feedback from sensory receptors
d. Relay of two efferent autonomic neurons conducts information from central nervous system to autonomic effectors
(1) Preganglionic neuron—conducts impulses from the central nervous system to an autonomic ganglion
(2) Postganglionic neuron—efferent neuron with which a preganglionic neuron synapses within autonomic ganglion

2. Structure of the sympathetic pathways
   a. Sympathetic chain ganglia
      (1) Most ganglia of the sympathetic division lie along either side of the anterior surface of the vertebral column and are joined with the other ganglia located on the same side
      (2) Each chain runs from the second cervical vertebra to the level of the coccyx
      (3) Usually there are 22 sympathetic chain ganglia on each side of vertebral column: three cervical, eleven thoracic, four lumbar, and four sacral
   b. Thoracolumbar division
      (1) Sympathetic preganglionic neurons with dendrite and cell bodies in lateral gray horns of the thoracic and lumbar segments of the spinal cord
      (2) Axons leave the cord by way of the ventral roots of the thoracic and first four lumbar spinal nerves and split away from other spinal nerve fibers by the white ramus to a sympathetic chain ganglion
   c. Preganglionic fiber may take one of three paths once inside the sympathetic chain ganglion
      (1) Synapse with sympathetic postganglionic neuron
      (2) Send ascending or descending branches through the sympathetic trunk to synapse with postganglionic neurons in other chain ganglia
      (3) Pass through one or more chain ganglia without synapsing
   d. Sympathetic postganglionic neurons
      (1) Dendrites and cell bodies are mostly in sympathetic chain ganglia or collateral ganglia
      (2) Gray ramus—short branch by which some postganglionic axons return to a spinal nerve
   e. In the sympathetic division, preganglionic neurons are relatively short, and postganglionic neurons are relatively long
   f. Axon of one sympathetic preganglionic neuron synapses with many postganglionic neurons, terminating in widely spread organs (Figure 14-17)
3. Structure of the parasympathetic pathways
   a. Parasympathetic preganglionic neurons—cell bodies are located in nuclei in the brainstem or lateral gray columns of the sacral cord; extend a considerable distance before synapsing with postganglionic neurons
   b. Parasympathetic postganglionic neurons—dendrites and cell bodies are located in parasympathetic ganglia, which are embedded in or near autonomic effectors
   c. Parasympathetic postganglionic neurons synapse with postganglionic neurons that each lead to a single effector (Figure 14-17)
4. Autonomic neurotransmitters (Figures 14-18 and 14-20)
   a. Axon terminal of autonomic neurons releases either of two neurotransmitters: norepinephrine or acetylcholine
   b. Adrenergic fibers—release norepinephrine; axons of postganglionic sympathetic neurons
   c. Cholinergic fibers—release acetylcholine; axons of preganglionic sympathetic neurons and of preganglionic and postganglionic parasympathetic neurons
   d. Norepinephrine affects visceral effectors by first binding to one of two types of adrenergic receptors in plasma membranes: alpha receptors or beta receptors
      (1) Binding of norepinephrine to alpha receptors in smooth muscle of blood vessels is stimulating, causing the vessels to constrict
      (2) Binding of norepinephrine to beta receptors in smooth muscle of blood vessels is inhibitory, causing blood vessels to dilate; in cardiac muscle, has stimulating effect
   e. Epinephrine also stimulates adrenergic receptors, enhancing and prolonging effects of sympathetic stimulation
   f. Effect of a neurotransmitter on any postsynaptic cell is determined by characteristics of the receptors, not by the neurotransmitter
   g. Termination of actions of norepinephrine and epinephrine
      (1) Monoamine oxidase (MAO)—enzyme that breaks up neurotransmitter molecules taken back up by the synaptic knobs
      (2) Catechol-O-methyl transferase (COMT)—enzyme that breaks down the remaining neurotransmitter
   h. Acetylcholine binds to two types of cholinergic receptors: nicotinic receptors and muscarinic receptors
   i. Termination of action of acetylcholine is by the enzyme acetylcholinesterase
   j. Autonomic neurotransmitters and receptors may influence different types of presynaptic and postsynaptic receptors at synapses with dually innervated effectors—this summation of effects increases precision of control
C. Functions of the autonomic nervous system
   1. Overview of autonomic function
      a. The autonomic nervous system functions to regulate visceral effectors in ways that tend to maintain or quickly restore homeostasis
      b. Sympathetic and parasympathetic divisions are tonically active, often exerting antagonistic influences on visceral effectors
      c. Doubly innervated effectors continually receive both sympathetic and parasympathetic impulses, and the summation of the two determines the controlling effect
   2. Functions of the sympathetic division
      a. Under resting conditions, the sympathetic division can act to maintain the normal functioning of doubly innervated autonomic effectors
      b. Sympathetic impulses function to maintain normal tone of the smooth muscle in blood vessel walls
      c. Major function of sympathetic division is that it serves as an "emergency" system—the "fight-or-flight" reaction (review Table 14-8)
   3. Functions of the parasympathetic division (review Table 14-7)
      a. Dominant controller of most autonomic effectors most of the time
      b. Acetylcholine—slows heartbeat and acts to promote digestion and elimination

THE BIG PICTURE: THE PERIPHERAL NERVOUS SYSTEM AND THE WHOLE BODY

A. The peripheral nervous system is made of all the afferent nervous pathways coming into the CNS and all the efferent pathways going out of the central nervous system

B. Peripheral pathways are pathways that lead from the integrator central nervous system to the effectors

C. Peripheral motor pathways serve as an information-carrying network that allows the central nervous system to communicate regulatory information to all the nervous effectors in the body

D. Every major organ is influenced, directly or indirectly, by peripheral nervous system output

REVIEW QUESTIONS

1. Identify the direction of the information carried by the ventral root of a spinal nerve.
2. Define mixed nerves.
3. Identify the areas innervated by the individual nerves emerging from the cervical plexus.
4. Explain the concept of a dermatome.
5. Explain the correlation between a myotome and a specific movement of the body.
6. Which cranial nerves transmit impulses that result in vision? In eye movement?
7. Which cranial nerves transmit impulses that result in hearing? In taste sensations?
8. What pathways are found in the somatic motor nervous system?
9. Describe an autonomic preganglionic neuron.
10. Identify the paths that a nerve signal may take once it is inside the sympathetic chain ganglion.
11. Differentiate between the sympathetic preganglionic and postganglionic neurons in terms of length.
12. How do parasympathetic ganglia differ from sympathetic ganglia in terms of location?
13. Describe the actions of norepinephrine. How are these actions terminated?
14. Describe the responses caused by acetylcholine release. How are these actions terminated?
15. Name the main types of adrenergic and cholinergic receptors.
16. Which efferent division of the autonomic nervous system is the dominant controller of most autonomic effectors most of the time?
17. Describe what happens when the fifth cranial nerve is compressed.
18. Describe the condition of shingles.

CRITICAL THINKING QUESTIONS

1. Can you distinguish between the reflexes that cause cardiac muscle to react and those that cause skeletal muscles to react?
2. An adult has an extension of the great toe and the fanning of the other toes in response to stimulation of the outer margin of the sole of the foot. Why is this a concern, and where would the problem most likely be?
3. A single reflex can be classified in several ways. What example can you find that would show this? In what ways can this reflex be classified?
4. Using the control of the speed of a car as an example, explain dual innervation in the autonomic nervous system.
5. Can you distinguish between the function of autonomic transmitters and receptors? How do they regulate dually innervated effector cells?
6. Based on what you know, explain why the autonomic nervous system is not an "automatic" nervous system with little control from the higher brain centers.
7. In times of stress, there are reports of people performing almost superhuman feats of strength. Specifically, what is happening in the body that would allow this to happen?
8. Can you predict the respiratory consequences of an injury that damaged the spinal cord between the third and fifth cervical segments? What specific mechanism would cause these consequences?
9. The drug propranolol (Inderal) blocks the effect of a portion of the sympathetic nervous system. To what class of drugs does propranolol (Inderal) belong, and what are its therapeutic effects?
10. Can you identify substances (other than acetylcholine and norepinephrine) that respond to receptors for the autonomic nervous system? What name is given to this group of substances, and what is their most likely role?

CHAPTER 15

Sense Organs

LANGUAGE OF SCIENCE

accommodation (ah-kom-oh-DAY-shun)
[*accommodare* to adjust to]

acute pain (ah-KYOOT) [*acut* sharp]

adaptation (ad-ap-TAY-shun) [*adaptare* to fit]

alpha motor neurons (AL-fah MOH-tor NOO-rons)
[*alpha* (α) first letter of the Greek alphabet]

anterior cavity (an-TEER-ee-or KAV-i-tee) [*ante* +
prior foremost, *cavus* hollow space]

auditory (AW-di-toh-ree) [*audire* to hear]

auditory ossicles (AW-di-toh-ree OS-ik-uls) [*audire*
to hear, *os-* bone, *-icle* small]

basal cells (BAY-sal) [*basis* foundation, *cella*
storeroom]

basilar membrane (BAYS-i-lar) [*basis* foundation,
membrane thin skin]

bleaching

blind spot

chemoreceptors (kee-moh-ree-SEP-tors) [*chemo-* by
chemical reaction, *-ceptor* to receive]

chronic pain [*chronos* time]

ciliary body (SIL-ee-air-ee) [*ciliary* eyelids or
eyelashes]

cones

crista ampullaris (KRIS-tah am-pyoo-LAIR-iss)
[*crista* ridge, *ampulla* flasklike bottle]

cupula (KYOO-pyoo-lah)

dynamic equilibrium (dye-NAM-ik ee-kwi-LIB-ree-
um) [*dynamis* force, *equi-* equal, *-libra* balance]

eardrum

endolymph (EN-doh-limf) [*endo-* inward or within,
-lympha water]

epithelial support cells (ep-i-THEE-lee-al)
[*epi-* on or upon, *-theli* nipple]

equilibrium (e-kwi-LIB-ree-um) [*equi-* equal, *-libra*
balance]

eustachian (yoo-STAY-shun) [*Bartolomeo Eustachio*
Italian anatomist]

external ear

exteroceptors (eks-ter-oh-SEP-tors) [*externus-*
outside, *-ceptor* to receive]

Cont'd on p. 584

The body has millions of sense organs. They fall into two main categories: general sense organs and special sense organs. Of these, by far the most numerous are the general sense organs, or receptors. Receptors function to produce the general, or somatic, senses (e.g., touch, temperature, pain) and to initiate various reflexes necessary for maintaining homeostasis. Special sense organs function to produce the special senses (vision, hearing, balance, taste, smell), and they too initiate reflexes important for homeostasis. In this chapter we begin with a description of receptors and follow with information related to the special senses.

SENSORY RECEPTORS

Sense organs called **sensory receptors** make it possible for the body to respond to stimuli caused by changes occurring in our external or internal environment. This function is crucial to survival. The abilities to see and hear, for example, may provide the necessary warning to help us avoid injury from dangers in our external environment. Internal sensations ranging from pain and pressure to hunger and thirst help us maintain homeostasis of our internal environment.

Receptor Response

The general function of receptors is to respond to stimuli by converting them to nerve impulses. Receptors are often described as the specialized dendritic endings or "end organs" of sensory neurons. As a rule, different types of receptors respond to different types of stimuli. They are specialized in that a particular type of receptor responds, under normal physiological conditions, to a particular type of stimulus and is less able or unable to respond to others. Heat receptors, for example, do not respond to light or stretch stimuli.

When an adequate stimulus acts on a receptor, a potential develops in the receptor's membrane. It is called a **receptor potential.** The receptor potential is a graded response, graded to the strength of the stimulus (see Chapter 12, p. 448). When a receptor potential reaches a certain threshold, it triggers an action potential in the sensory neuron's axon. These impulses then travel over sensory pathways to the brain and spinal cord, where they are interpreted as a particular **sensation,** such as heat or cold, or they initiate some type of reflex action, such as withdrawal of a limb from a painful stimulus.

Specialized sensory impulses terminating in the brainstem may affect so-called "vital sign" reflexes that help regulate heart or respiratory rate. Others may end in the thalamus or cerebral cortex where they trigger imprecise or "crude" sensation awareness (thalamus) or very precise and specific awareness of not only a specific type of sensation but also its exact location and level of intensity (cerebral cortex).

Receptors often exhibit a functional characteristic known as adaptation. **Adaptation** means that the magnitude of the receptor potential decreases over a period in response to a continuous stimulus. As a result, the rate of impulse conduction by the sensory neuron's axon also decreases. So too does the intensity of the resulting sensation. A familiar example of adaptation is feeling the touch of your clothing when you first put it on and soon not sensing it at all. Touch receptors adapt rapidly. In contrast, the proprioceptors in our

muscles, tendons, and joints adapt slowly. As long as stimulation of them continues, they continue sending impulses to the brain.

If we not only remain aware of a particular sensation over time, but also interpret what that sensation means in a larger context, the process is called **perception.** Although we may have no perception of certain sensory inputs, they often play a critical role in maintaining homeostasis. Examples might include our ability to sense and respond to changing levels of blood glucose and carbon dioxide—but not at a conscious level.

Distribution of Receptors

Receptors responsible for the special senses of smell, taste, vision, hearing, and equilibrium are grouped into localized areas (nasal mucosa or tongue) or into such complex organs as the eye and ear. The **general sense organs** consist of microscopic receptors widely distributed throughout the body in the skin, mucosa, connective tissues, muscles, tendons, joints, and viscera. Sensations produced by these receptors are often called the *somatic senses*. Their distribution is not uniform in all areas. In some, it is very dense; in others, it is sparse. The skin covering the fingertips, for instance, contains many more receptors to touch than does the skin on the back. A simple procedure, the **two-point discrimination test**, demonstrates this fact. A subject reports the number of touch points felt when an investigator touches the skin simultaneously with two points of a compass. If the skin on the fingertip is touched with the compass points barely $\frac{1}{8}$ inch apart, the subject senses them as two points. If the skin on the back is touched with the compass points this close together, they will be felt as only one point. Unless they are 1 inch or more apart, they cannot be discriminated as two points. Why this difference? Because touch receptors are so densely distributed in the fingertips that two points very close to each other stimulate two different receptors—they are sensed as two points. The situation is quite different in the skin on the back. There, touch receptors are so widely scattered that two points have to be at least 1 inch apart to stimulate two receptors and be felt as two points.

CLASSIFICATION OF RECEPTORS

Receptors can be classified according to (1) their location in the body, (2) the specialized stimulus that causes them to respond, and (3) their structure.

Classification by Location

Three groups or classes of receptors can be identified by their location:

1. Exteroceptors
2. Visceroceptors (or interoceptors)
3. Proprioceptors

Exteroceptors, as the name implies, are located on or very near the body surface and respond most frequently to stimuli that arise external to the body itself. Receptors in this group are sometimes called *cutaneous receptors* because of their placement in the skin. However, the special sense organs, which are described later in the chapter, are also classified as exteroceptors. Examples of exteroceptors include those that detect pressure, touch, pain, and temperature.

BOX 15-1: FYI
Referred Pain

The stimulation of pain receptors in deep structures may be felt as pain in the skin that lies over the affected organ or in an area of skin on the body surface far removed from the site of disease or injury. **Referred pain** is the term for this phenomenon. The cause is related to a mixing or convergence of sensory nerve impulses from both the diseased organ and the skin in the area of referred pain. For example, pain originating in an organ deep in the abdominal cavity is often interpreted as coming from an area of skin whose sensory fibers enter the same segment of the spinal cord as the sensory fibers from the deep structure. A classic example is the referred pain often associated with a heart attack. Sensory fibers from the skin on the chest over the heart and from the tissue of the heart itself enter the first to the fifth thoracic spinal cord segments and so do sensory fibers from the skin areas over the left shoulder and inner surface of the left arm. Sensory impulses from all these areas travel to the brain over a common pathway. It is a misinterpretation in the brain as to the correct location of sensory neurons being stimulated that causes referred pain. In clinical medicine, an understanding of referred pain is often an important determinant in the correct diagnosis of disease (see figure).

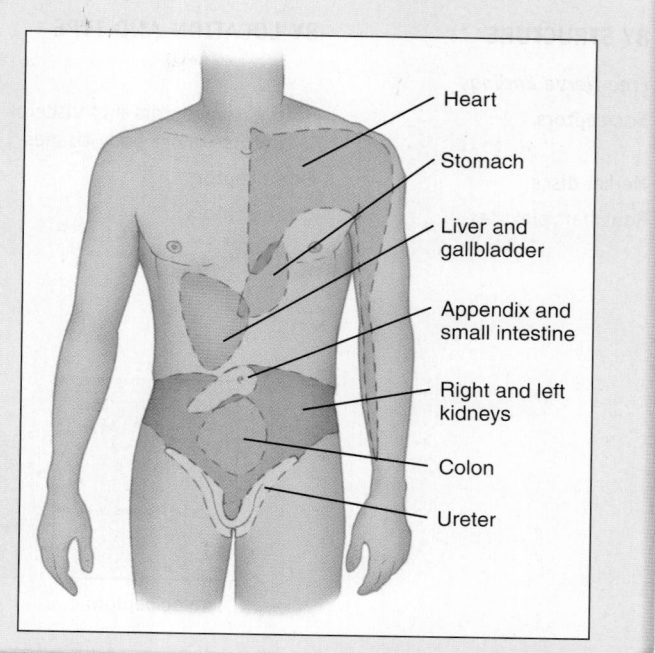

Visceroceptors (interoceptors) are located internally, often within the substance of body organs (viscera), and when stimulated provide information about the internal environment. They are activated by stimuli such as pressure, stretching, and chemical changes that may originate in diverse internal organs, such as the major blood vessels, intestines, and urinary bladder. Visceroceptors are also involved in mediating sensations such as hunger and thirst. **Proprioceptors** are a specialized type of visceroceptor. They are less numerous and generally more specialized than other internally placed receptors, and their location is limited to skeletal muscle, joint capsules, and tendons. These proprioceptors provide us with information about body movement, orientation in space, and muscle stretch. Activation of two types of proprioceptors, called *tonic* and *phasic receptors*, allows us to orient our body in space and provides us with positional information about specific body parts while at rest or during movement. The firing of the nonadapting tonic receptors allows us to locate, for example, our arm, hand, or foot at rest without having to look. Phasic receptors are rapidly adapting receptors, so they are triggered only when there is a change in position. Phasic receptors therefore permit us to feel the changing position of our body parts during continuous movement.

Classification by Stimulus Detected

Receptors are frequently classified into six categories based on the types of stimuli that activate them:

1. **Mechanoreceptors.** Mechanoreceptors are activated by mechanical stimuli that in some way "deform" or change the position of the receptor, resulting in the generation of a receptor potential. Examples include pressure applied to the skin or to blood vessels, or caused by stretch or pressure in muscle, tendon, or lung tissue.
2. **Chemoreceptors.** Receptors of this type are activated by either the amount or the changing concentration of certain chemicals. Our senses of taste and smell depend on chemoreceptors. Specialized chemoreceptors in the body also "sense" the concentration of specific chemicals such as carbon dioxide (CO_2) and blood glucose.
3. **Thermoreceptors.** Thermoreceptors are activated by changes in temperature.
4. **Nociceptors.** Nociceptors are activated by intense stimuli of any type that results in tissue damage. The cause may be a toxic chemical, intense light, sound, pressure, or heat. The sensation produced is one of pain.
5. **Photoreceptors.** Photoreceptors are found only in the eye. Photoreceptors respond to light stimuli if the intensity is great enough to generate a receptor potential.
6. **Osmoreceptors.** Osmoreceptors are specialized receptors concentrated in the hypothalamus that sense levels of osmotic pressure in body fluids. They are important in detecting changes in concentration of electrolytes (osmolarity) in extracellular fluids and stimulating the hypothalamic thirst center.

Classification by Structure

Regardless of their location or how they are activated, the general (somatic) sensory receptors may be classified anatomically as either

1. Free nerve endings, or
2. Encapsulated nerve endings

Refer often to Table 15-1 as you read about the general sensory receptors that are described in the following paragraphs.

Table 15-1   Classification of Somatic Sensory Receptors

BY STRUCTURE	BY LOCATION AND TYPE	BY ACTIVATION STIMULUS	BY SENSATION OR FUNCTION
Free Nerve Endings			
Nociceptors	Both exteroceptors and viscero-ceptors—most body tissues	Almost any noxious stimulus; temperature change; mechanical	Pain; temperature; itch; tickle
Merkel discs	Exteroceptors	Light pressure; mechanical	Discriminative touch
Root hair plexuses	Exteroceptors	Hair movement; mechanical	Sense of "deflection" type of movement of hair

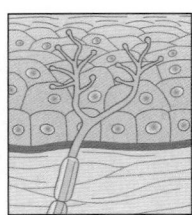 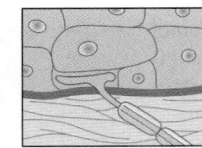 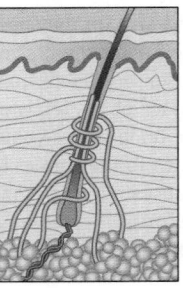

Nociceptors   Merkel discs   Root hair plexuses

BY STRUCTURE	BY LOCATION AND TYPE	BY ACTIVATION STIMULUS	BY SENSATION OR FUNCTION
Encapsulated Nerve Endings			
TOUCH AND PRESSURE RECEPTORS			
Meissner corpuscle	Exteroceptors; epidermis, hairless skin	Light pressure, mechanical	Touch; low-frequency vibration
Krause corpuscle	Mucous membranes	Mechanical	Touch; low-frequency vibration; textural sensation
Ruffini corpuscle	Exteroceptors; dermis of skin	Mechanical	Crude and persistent touch
Pacinian (lamellar) corpuscle	Dermis of skin, joint capsules	Deep pressure, mechanical	Deep pressure; high-frequency vibration; stretch
STRETCH RECEPTORS			
Muscle spindles	Skeletal muscle	Stretch, mechanical	Sense of muscle length
Golgi tendon receptors	Musculotendinous junction	Force of contraction and tendon stretch, mechanical	Sense of muscle tension

Ruffini corpuscle

Meissner corpuscle

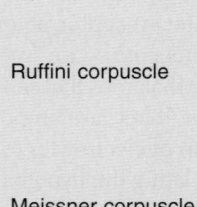

Krause end bulb

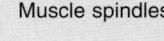

Muscle spindles

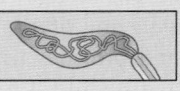

Intrafusal fibers

Pacinian (lamellar) corpuscle

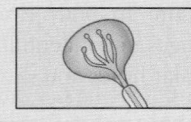

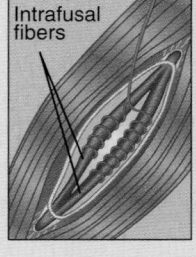

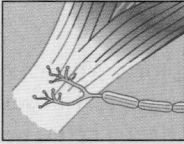

Golgi tendon receptors

Free Nerve Endings

Free nerve endings are the simplest, most common, and most widely distributed sensory receptors. They are located both on the surface of the body (exteroceptors) and in the deep visceral organs (visceroceptors). The term **nociceptor** is used to describe these slender sensory fibers, which, in most cases, terminate in small swellings called *dendritic knobs*. Nociceptors serve as the primary sensory receptors for pain.

Pain Sensations

Nerve fibers that carry pain impulses from nociceptors or free nerve ending receptors to the brain can be divided into two types—acute or **fast (A) pain** fibers and **chronic** or **slow (B) pain** fibers. A fibers are concentrated in the skin, mucous membranes, and other superficial areas. Fast pain is sometimes described as a sharp "take your breath away" type of pain associated with superficial injury or trauma. If you have ever slammed your finger closing a car door you have experienced this type of fast or **somatic pain.** The type of deep or **visceral pain** that develops more slowly over time and travels over B fibers is often described as dull or aching. It originates in deeper body (visceral) structures and can be severe if caused by conditions such as intestinal obstruction or passage of a kidney stone or gallstone. It is important to stress that in responding to powerful stimuli of any kind, including chemical or thermal burns, intense light, sound, or pressure, the generation of a receptor potential in free nerve endings most often results in the sensation of pain—often the first indication of injury or disease.

Brain tissue is unique in that it lacks the type of nociceptors that transmit sensations of pain and is therefore incapable of sensing painful stimuli. Just the opposite is true of many deep visceral organs, in which the presence of free nerve endings makes pain one of the few sensations that can be evoked.

Pain is both a "bad news/good news" type of sensation. The bad news, of course, is its relationship to disease and injury. The good news is the important role pain plays in alerting us to threats in our environment. Individuals who are unable to adequately sense pain on the body surface tend to have other types of sensory loss as well. They are very much at risk for injury because they lack the ability to sense the "warning signs" that a normally functioning sensory system can provide. For example, individuals with uncontrolled or poorly controlled diabetes often lose their ability to sense pain on certain areas of the body surface—especially on the skin of the feet. Other sensations are also frequently diminished or lost over time in diabetics because of nerve damage called **diabetic neuropathy.** For this reason physicians closely monitor the adequacy of cutaneous sensation on the bottom of the foot (Figure 15-1, A). Recognizing early signs of diabetic neuropathy can help prevent later injury. Figure 15-1, B, shows a foreign body protruding from the tip of the third toe in a diabetic patient. The individual lacked sensation in the feet and only after noticing it sought medical attention.

Temperature Sensations

Free nerve endings called **thermoreceptors** mediate sensations of heat and cold. When small cold or warm probes are used to "map" the skin's sensitivity to temperature, small areas called "receptive fields" about 1 mm across can be identified on the skin surface as sensitive to *either* cold or warmth, but not both. Research has shown that these receptive fields represent separate *warm receptors* and *cold receptors* that respond to different thermal sensations and are sensitive to a range of relatively hot or cold temperatures. Thermal maps have also shown that thermoreceptors, like other types of cutaneous receptors, are not spread uniformly across the skin surface. Both types of thermoreceptors undergo rapid adaptation. As a result, the intensity of the warm or cold sensations they initially produce when activated soon fade. We adapt to the heat of a sauna or a cool shower in a short time.

Warm receptors, which are located in the dermis, are activated above about 25° C (77° F) and increase their firing rate un-

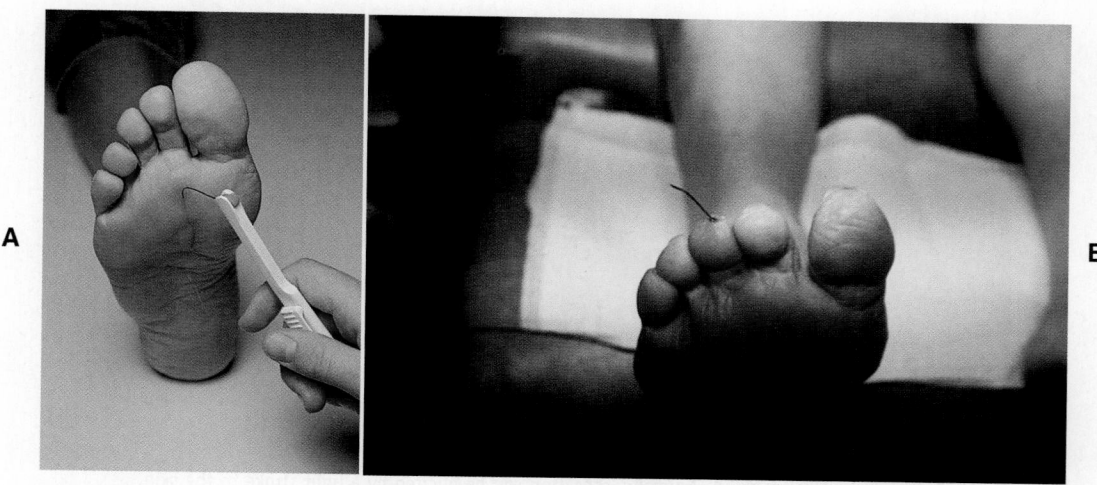

Figure 15-1 *The importance of cutaneous sensation.* **A,** Using a monofilament to check for sensation on the bottom of the foot. **B,** Foreign body in the toe of a diabetic patient. See text for explanation.

til the temperature reaches about 46° C (114° F). Beyond a temperature of about 48° C (118° F) a sensation of burning pain begins. Cold receptors, which are located in the deepest layer of the epidermis, have a broader temperature response than warm receptors. They are active between about 10° C (50° F) and 40° C (104° F). Below 10° C the firing of cold receptors decreases dramatically. The falling temperature first acts as a local anesthetic and then activates nociceptors resulting in a sensation of freezing pain. Between temperature extremes the brain senses a particular temperature sensation by integrating sensory inputs from both receptor types.

Tactile Sensations

In addition to their role in mediation of pain and temperature, free nerve endings and slightly modified free nerve endings are also involved in certain tactile sensations.

Skin Movement

Root hair plexuses are rapidly adapting free nerve endings that are activated when very slight skin movement bends or deforms a hair shaft or follicle surrounded by the receptor. When you feel a mosquito "bite," it may not be caused by the piercing of the skin by the mouth of the insect. Instead, it may be the movement of skin caused by the mosquito's activity that triggers or stimulates a root hair plexus. In any event, it's a good idea to swat the mosquito!

Itch

The term "itch" is used to describe a tactile sensation mediated by free nerve endings. Most people would describe it as a sensation that makes you want to scratch. It can vary in intensity from almost imperceptible to intense and disabling. The cause is generally chemical irritation of free nerve endings by inflammatory chemicals, such as bradykinin or histamine. These types of chemicals are often released by injured tissue following insect bites, or during allergic reactions.

Tickle

Tickle is a unique sensation in that it most often results from tactile stimulation of the skin not by you, but by someone else. It is mediated by free nerve endings and is a good but somewhat baffling example of how perception can alter the conscious interpretation of a nervous impulse when it reaches the brain. We know that neural pathways in tickle involve both the thalamus and cerebellum before the impulses reach the cerebral cortex. However, the circuits involved in the cerebral cortex and how these circuits interact with neurons in other areas of the brain are not well understood. Obviously, knowing consciously that a particular tactile sensation generally occurs only when you are touched by someone else requires complex nervous system involvement. What begins as a simple receptor potential in a free nerve ending ends in a complex, consciously interpreted tactile sensation.

Discriminative Touch

Discriminative touch, or "light touch," refers to an often very subtle sensation that can be located exactly on the skin surface (Figure 15-2). It is mediated by flattened or disc-shaped variations of free nerve endings called tactile (**Merkel**) **discs.** These structures were first described by the German anatomist Friedrich Merkel (1885-1919) when he studied delicate nervous elements near the skin surface. Structurally, this type of tactile disc is a delicate mechanoreceptor that is not encapsulated and has a superficial placement in the epidermis of the skin (Figure 15-2). It is therefore easily "deformed" and capable of generating an action potential when exposed to very minimal stimulation. In addition to discriminative or light touch, this type of tactile (Merkel) disc is also capable of detecting subtle changes in surface form and contours.

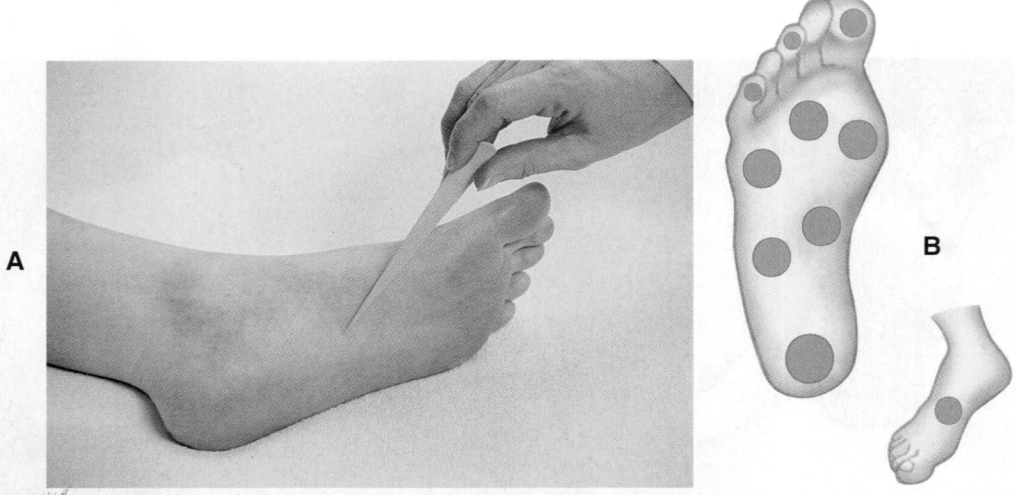

Figure 15-2 *Discriminative touch.* A, Normally, sensation can be elicited by a light stroke to the skin.
B, Examples of specific sites used to verify sensation of discriminative touch.

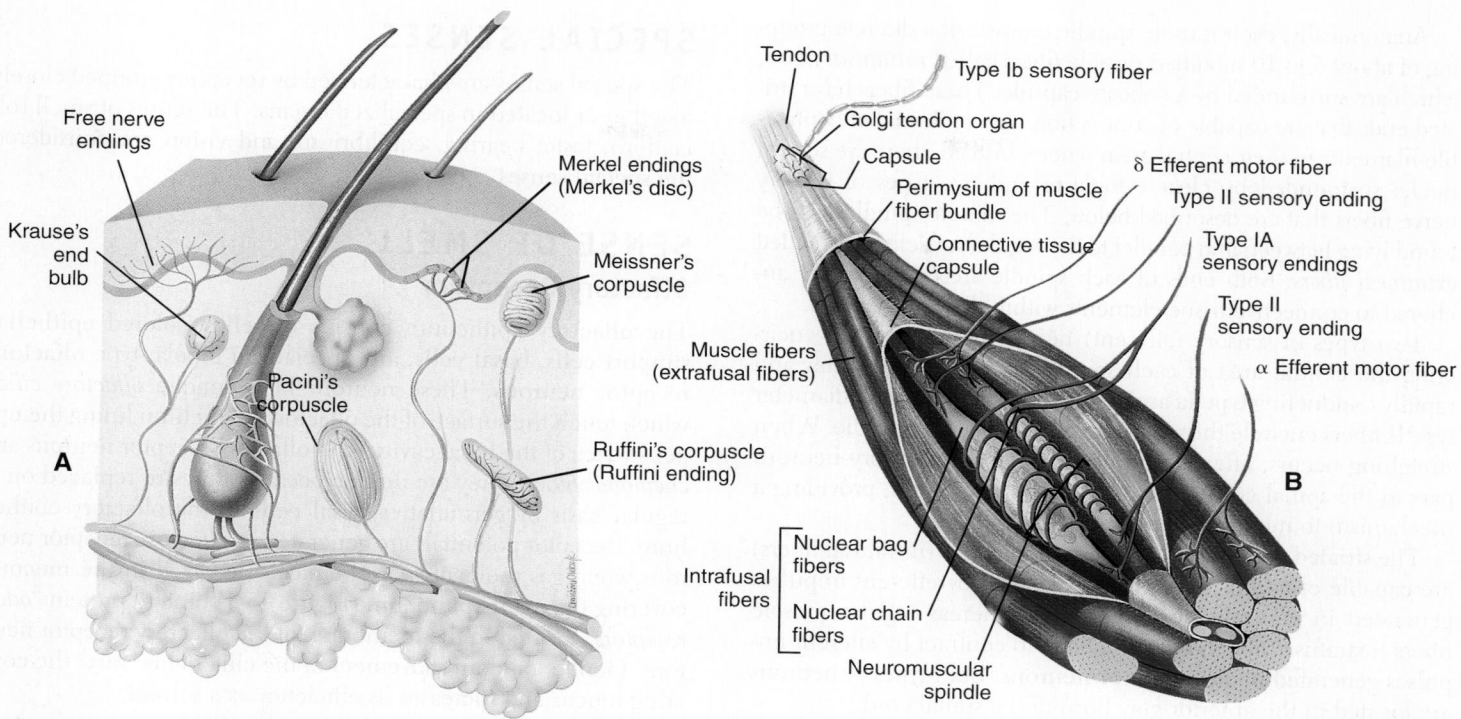

Figure 15-3 *Somatic sensory receptors.* **A,** Exteroceptors. **B,** Proprioceptors. See also Figure 6-1 (p. 196), which shows a microscopic section of the skin.

Encapsulated Nerve Endings

The six types of encapsulated nerve endings have in common some type of connective tissue capsule that surrounds their terminal or dendritic end. In addition, most of these specialized receptors are primary mechanoreceptors; that is, they are most often activated by a mechanical or "deforming" type of stimulus. Encapsulated receptors vary in size and anatomical characteristics, as well as in numbers and distribution throughout the body (see Table 15-1).

Touch and Pressure Receptors

Meissner corpuscles are encapsulated tactile end organs, also called tactile corpuscles, first described by the German physiologist George Meissner (1829-1905). Note in Figure 15-3, A and Table 15-1 that they are superficially placed ovoid- or egg-shaped mechanoreceptors that are larger than tactile discs. The encapsulated receptor fibers are coiled and enmeshed in connective tissue. They are located in or very close to the dermal papillae in hairless skin areas such as the fingertips, lips, nipples, and genitals. Although their covering capsule requires slightly more of a "deforming" stimulus to generate an action potential than a Merkel disc, they do mediate light touch in addition to textural sensations and low-frequency vibration.

Two anatomical variants of Meissner corpuscles called **Krause end bulbs** and **Ruffini corpuscles** are also mechanoreceptors. *Krause end bulbs* are egg shaped, like Meissner corpuscles, but they are somewhat smaller and have fewer and less tightly coiled dendritic endings within their covering capsule. These receptors are more numerous in mucous membranes than in skin and are sometimes called *mucocutaneous corpuscles.* They are involved in touch and low-frequency vibration. Ruffini corpuscles (see Figure 15-2, A) are also considered variants of Meissner corpuscles, but they have a more flattened capsule and are more deeply located in the dermis of the skin. These receptors mediate sensations of crude, heavy, and persistent touch. Because they are slow adapting, they permit the skin of the fingers to remain sensitive to deep pressure for long periods. The ability to grasp an object, such as the steering wheel of a car, for long periods of time and still be able to "sense" its presence between the fingers depends on these receptors. **Pacinian (lamellar) corpuscles** are large mechanoreceptors, which, when sectioned, show thick laminated connective tissue capsules. They are found in the deep dermis of the skin—especially in the hands and feet—and are also numerous in joint capsules throughout the body. Pacinian or lamellar corpuscles respond quickly to sensations of deep pressure, high-frequency vibration, and stretch. Although sensitive and quick to respond, these receptors adapt quickly, and the sensations they evoke seldom last for long.

Stretch Receptors

The most important stretch receptors are associated with muscles and tendons and are classified as proprioceptors. Two types of stretch receptors, called **muscle spindles** and **Golgi tendon receptors**, operate to provide the body with information concerning muscle length and the strength of muscle contraction (Figure 15-3, *B*).

Anatomically, each muscle spindle consists of a discrete grouping of about 5 to 10 modified muscle fibers called *intrafusal fibers,* which are surrounded by a delicate capsule. These fibers have striated ends that are capable of contraction but are devoid of contractile filaments in their central areas where, instead, there are several nuclei surrounded by clear cytoplasm and two types of sensory nerve fibers that are described below. The muscle spindles can be found lying between and parallel to the regular muscle fibers called *extrafusal fibers.* Both ends of each spindle are connected or anchored to connective tissue elements within the muscle mass.

Two types of sensory (afferent) nerve fibers are found encircling the central area of each spindle. Both large-diameter and rapidly conducting type Ia and slower-conducting, small-diameter type II fibers encircle the clear central area of each spindle. When stretching occurs, afferent impulses from these sensory neurons pass to the spinal cord and are relayed to the brain, providing a mechanism to monitor changes in muscle length.

The striated ends of the muscle spindle fibers (intrafusal fibers) are capable of contraction when stimulated by efferent impulses generated in **gamma motor neurons,** whereas regular muscle fibers (extrafusal fibers) are stimulated to contract by efferent impulses generated in **alpha motor neurons.** Both types of neurons are located in the anterior gray horn of the spinal cord.

Muscle spindles are stimulated if the length of a relaxed muscle is stretched and exceeds a certain limit. The result of stimulation is a **stretch reflex** that shortens a muscle or muscle group, thus aiding in the maintenance of posture or the positioning of the body or one of its extremities in a way that may be opposed by the force of gravity (see Figure 11-17, p. 409). We do this unconsciously, because these receptors do not produce a particular sensation, such as pain, heat, or cold.

Golgi tendon organs, like muscle spindles, are proprioceptors. They are located at the point of junction between muscle tissue and tendon. Each Golgi tendon organ consists of dendrites of afferent (sensory) nerves called *type Ib nerve fibers,* which are associated with bundles of collagen fibers from the tendon and surrounded by a capsule. These receptors act in a way opposite that of muscle spindles. They are stimulated by excessive muscle contraction and when activated they cause muscles to relax. This response, called a **Golgi tendon reflex,** protects muscles from tearing internally or pulling away from their tendinous points of attachment to bone because of excessive contractile force.

Table 15-1 summarizes the different types of somatic sense receptors, their locations, and their functions.

QUICK CHECK

1. Classify receptors into four groups based on the type of stimuli that activate them.
2. Distinguish between the special and the general, or somatic, senses.
3. Name the three types of sense receptors classified according to location in the body.
4. Name the receptors associated with pain, touch, pressure, and stretch responses.

SPECIAL SENSES

The special senses are characterized by receptors grouped closely together or located in specialized organs. The senses of smell (olfaction), taste, hearing, equilibrium, and vision are considered the **special senses.**

SENSE OF SMELL
Olfactory Receptors

The **olfactory** epithelium consists of yellow-colored **epithelial support cells, basal cells,** and specialized bipolar-type **olfactory receptor neurons.** These neurons have unique *olfactory cilia,* which touch the surface of the olfactory epithelium lining the upper surface of the nasal cavity. The olfactory receptor neurons are *chemoreceptors.* They are unique because they are replaced on a regular basis by germinative **basal cells** in the olfactory epithelium. Receptor potentials are generated in olfactory receptor neurons when gas molecules or chemicals dissolved in the mucous covering the nasal epithelium (Figure 15-4) bind to protein *"odor receptors"* located in the membrane of the olfactory receptor neurons. Gentle, random movement of the cilia helps "mix" the covering mucus and increases its efficiency as a solvent.

The olfactory epithelium is located in the most superior portion of the nasal cavity (see Figure 15-4). Functionally, this is a poor location because a great deal of inspired air flows around and down the nasal passageways without contacting the protein odorant receptors located in the cell membranes of the olfactory receptor cells. The location of these receptor neurons explains the necessity for sniffing, or drawing air forcefully up into the nose, to smell delicate odors.

The olfactory receptors are extremely sensitive, that is, capable of being stimulated by even very slight odors caused by just a few molecules of a particular chemical. However, rapid adaptation of olfactory sensations in the face of continuous stimulation occurs. It is due to both inhibition of action potentials by specialized **granule cells** in the olfactory bulbs and by fatigue of odorant receptor function caused by ongoing stimulation of the olfactory receptor neurons. Although humans have a sense of smell far less keen than many animals, some individuals can distinguish several thousand different odors, and most of us can easily identify at least several hundred. Examples of well-known primary scents include putrid, floral, peppermint, and musky odors. Many odors are combinations of primary scents.

In 2004 U.S. researchers Drs. Linda Buck and Richard Axel received the Nobel Prize in physiology or medicine for their pioneering work which explained for the first time how the sense of smell or olfaction in humans actually "works." Their research identified about 350 different odorant receptors in the cell membranes of human olfactory receptor neurons. Then, by using sophisticated research methods, they were able to show that by stimulating the entire olfactory epithelium—or specific odorant receptors—or combinations or sequences of these receptor proteins—different odors could be identified. If the receptor proteins are compared to the letters of the alphabet, activation of specific letters, formation of different "words," or word sequences, will cause differing olfactory receptor neurons or groups of neurons to

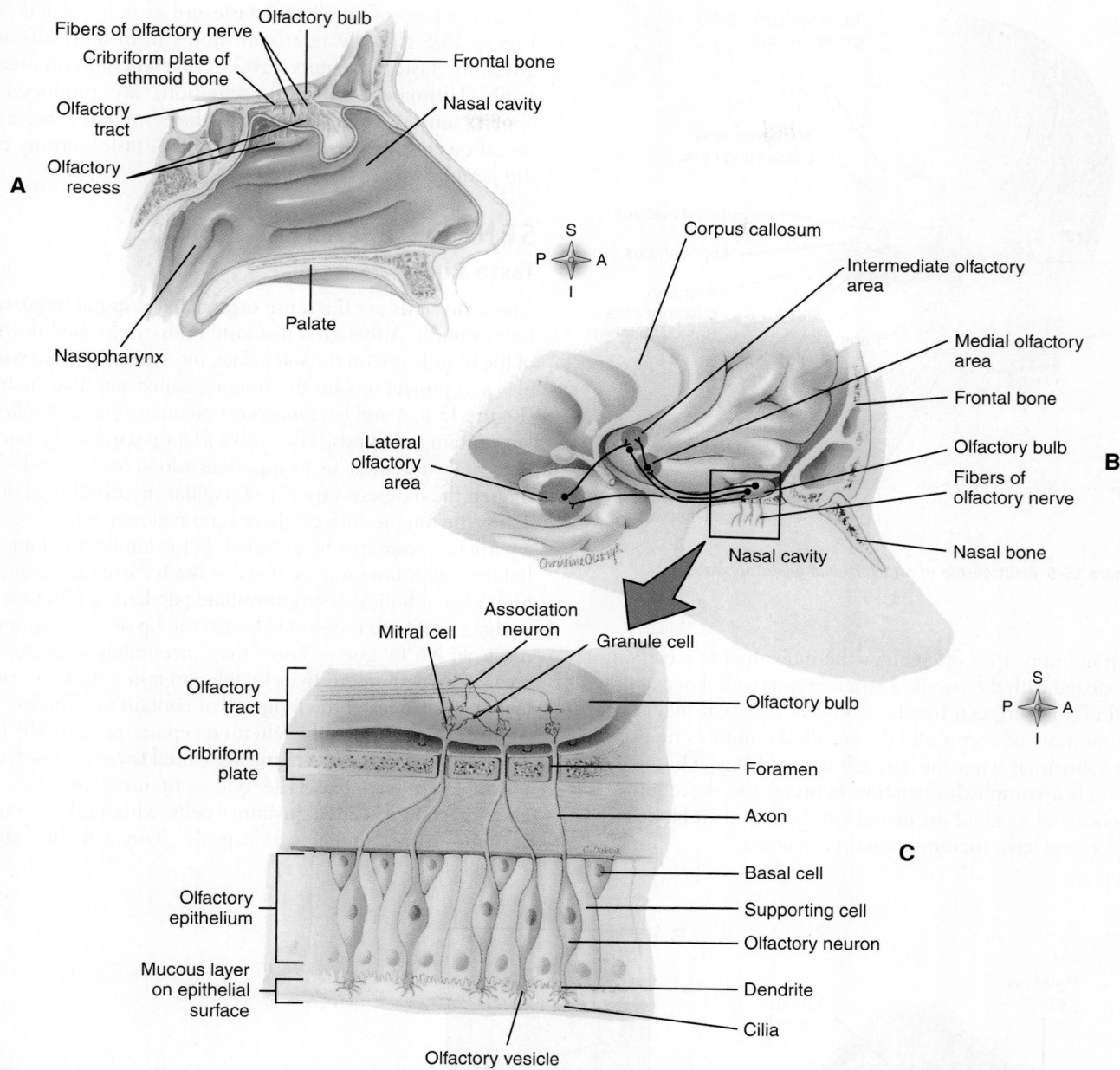

Figure 15-4 *Olfaction.* Location of olfactory epithelium, olfactory bulb, and neuronal pathways involved in olfaction. **A,** Midsagittal section of the nasal area shows the locations of major olfactory sensory structures. **B,** Major olfactory integration centers of the brain. **C,** Details of the olfactory bulb and olfactory epithelium.

transmit impulses to the brain that are perceived as unique odors. Many animals have a keener sense of smell than humans. Mice, for example, have about 1000 different odorant receptor proteins and are capable of detecting a larger number of specific odors than humans.

Olfactory Pathway

If the level of odor-producing chemicals dissolved in the mucus surrounding the olfactory cilia reaches a threshold level, a receptor potential and, then, an action potential will be generated and passed to the olfactory nerves in the olfactory bulb. From there, the impulse passes through the olfactory tract and into the tha-

lamic and olfactory centers of the brain for interpretation, integration, and memory storage.

The sense of smell can create powerful and long-lasting memories. Memories coupled with unique sensory inputs, especially distinctive odors, often persist from early childhood to death. Dental office smells, baby smells, kitchen smells, and new car smells are examples of olfactory "triggers" that often bring back memories of events that occurred years earlier. In addition to the olfactory cortex and thalamic areas of the brain, components of the limbic system, including the angulate gyrus and hippocampus, play a key role in coupling olfactory sense inputs to both short- and long-term memory. Not only can specific odors trigger

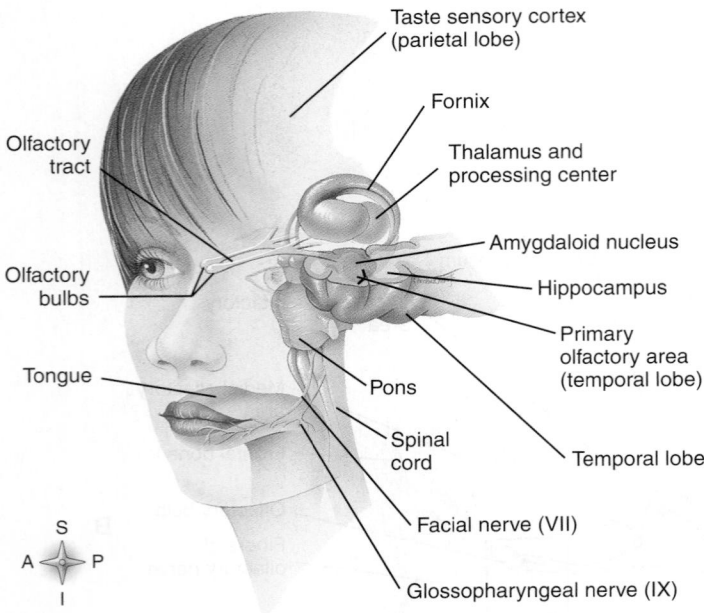

Figure 15-5 *Relationship of olfactory and gustatory pathways.*

long term memory, they often allow the individual to recall emotions associated with the recalled experience as well. For example, the smell of an evergreen tree by a 90 year old man may trigger both the memory of events and the recall of emotions he experienced at Christmas when he was a 9 year old boy. The mechanism of such a complex association between the detection of a specific odor and recall of the actual emotions and ambiance surrounding a long term memory remain unknown.

Our senses of smell and taste are closely related. Note in Figure 15-5 that the neuronal inputs from both olfactory and gustatory (taste) receptors travel in several common areas of the brain. Ultimately, olfactory sensations are produced in the (smell) sensory cortex located in the temporal lobes and taste sensations result from stimulation of the (taste) sensory cortex in the parietal lobes.

SENSE OF TASTE

Taste Buds

The **taste buds** are the sense organs that respond to **gustatory**, or taste, stimuli. Although a few taste buds are located in the lining of the mouth and on the soft palate, most are associated with small, elevated projections on the tongue, called **papillae** (pah-PIL-ee) (Figure 15-6, A and B). *Fungiform, circumvallate,* and *foliate* papillae contain taste buds. Threadlike *filiform* papillae do not contain taste buds but allow us to experience food texture and "feel." Although the different varieties of papillae are distributed differently across the tongue surface, there is no regional difference in where a particular taste can be detected. For example, the long-held belief that bitter taste was localized at the back of the tongue with its high concentration of circumvallate papillae (see Figure 15-6, A) or that sweet taste is detected best at the tip of the tongue is simply not true. No tongue or taste "map" accurately indicates regional areas of particular sensitivity to different tastes. All tastes can be detected in all areas of the tongue that contain taste buds.

Taste buds house the chemoreceptors responsible for taste. They are stimulated by chemicals, called *testants,* dissolved in the saliva. Each grapelike taste bud contains 50 to 125 of these chemoreceptors, called **gustatory cells,** which are surrounded by a supportive epithelial cell capsule. Tiny cilia-like structures,

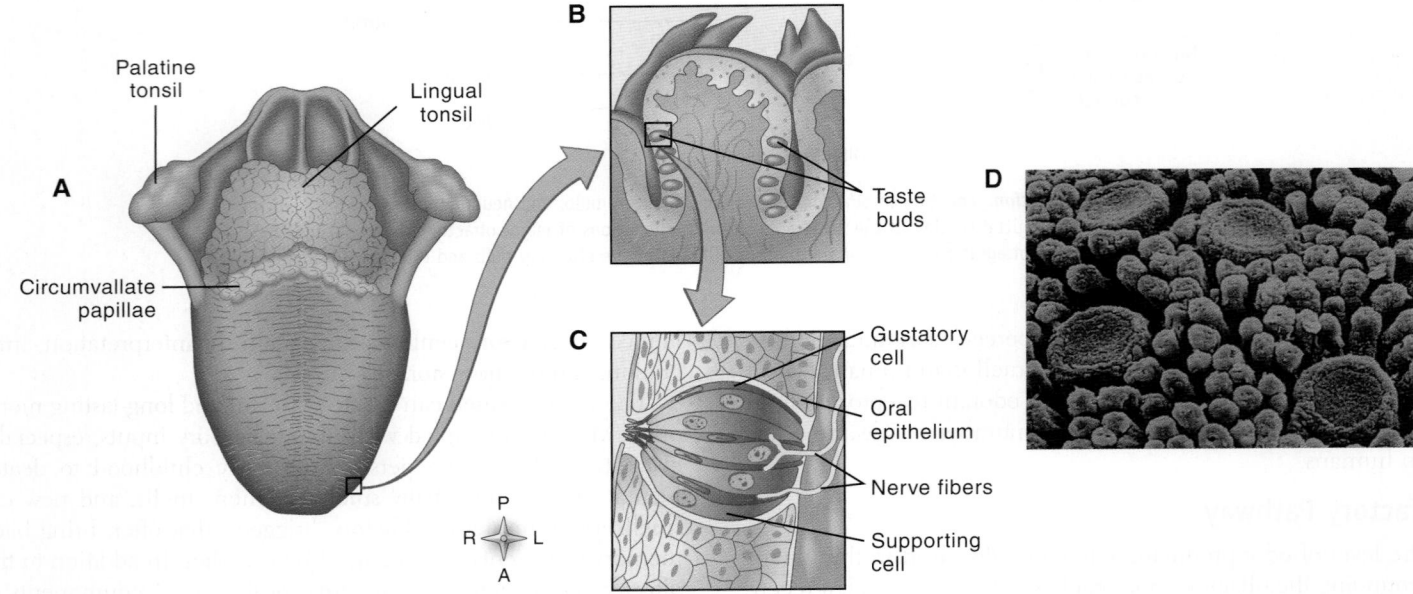

Figure 15-6 *The tongue.* **A,** Dorsal surface of tongue and adjacent structures. **B,** Section through a papilla with taste buds on the side. **C,** Enlarged view of a section through a taste bud. **D,** Scanning electron micrograph of the tongue surface showing the papillae in detail.

called *gustatory hairs*, extend from each of the gustatory cells and project into an opening called the *taste pore*, which is bathed in saliva (Figure 15-6, *C*).

The sense of taste depends on the creation of a receptor potential in gustatory cells. Only then can an action potential be generated and a nerve impulse relayed to the brain for interpretation. Generation of a receptor potential begins when specialized areas called **G protein receptor sites** on the cell membranes or porelike **ion channels** in the covering gustatory hairs bind to taste-producing chemicals (testants) in the saliva. The nature and concentration of the chemicals that bind to either the G protein receptor sites or ion channels determine how fast the receptor potential is generated.

Taste cells appear similar structurally, and all of them can respond at least in some degree to most taste-producing chemicals. Functionally, however, each taste bud responds most effectively to only one of four "primary" taste sensations: sour, sweet, bitter, and salty. Our ability to detect many different flavors, or tastes, is due to combinations of the four primary sensations and to interaction with the sense of smell. The number of "primary" taste sensations (e.g., the pure, or primary, olfactory scents) is likely to increase with expanded research efforts. Metallic taste, for example, may be a fifth primary taste sensation rather than a mixture of other flavors. Researchers have also suggested that yet another primary taste, called *umami*, results from stimulation of G protein receptor sites by amino acids, such as glutamate, found in protein-rich foods such as meat and fish.

The exact mechanism by which a chemical testant binds to a particular G protein receptor site or ion channel on a gustatory hair is unknown. Chemical structure plays a part but is not the only factor involved, because substances that are very different chemically, such as artificial sweeteners, chloroform, and table sugar, produce a sweet taste. Some chemical compounds and specific ions, however, are definitely associated with specific tastes. Sour (H^+) and salty (Na^+ and Cl^-) tastes activate ion channels, whereas sweet and bitter tastes result from stimulation of G protein receptor sites. As chemoreceptors, the taste buds, like olfactory receptors, tend to be quite sensitive but fatigue easily. Very low levels of taste-producing chemicals are required to generate a receptor potential. However, adaptation often begins within a few seconds after a taste sensation is first noticed and is generally complete in a few minutes.

Neuronal Pathway for Taste

The taste sensation begins with creation of a receptor potential in the gustatory cells of a taste bud. The generation and propagation of an action potential, or nerve impulse, then transmit the sensory input to the brain. We generally think of taste in terms of the "primary" sensations of sour, sweet, bitter, and salty (and perhaps metallic and umami). However, touch, texture, and temperature are also involved in taste and whether we sense it as pleasant, neutral, or unpleasant.

Nervous impulses generated in the anterior two thirds of the tongue travel over the facial (VII) nerve, whereas those generated from the posterior one third are conducted by fibers of the glossopharyngeal (IX) nerve. A third cranial nerve, the vagus (X) nerve, plays a minor role in taste. It contains a few fibers that carry taste sensation from a limited number of taste buds located in the walls of the pharynx and on the epiglottis.

All three cranial nerves carry impulses into the medulla oblongata. Relays then carry the impulses into the thalamus and then into the taste, or gustatory, area of the cerebral cortex in the parietal lobe of the brain (see Figure 15-5).

QUICK CHECK

5. List the special senses.
6. Discuss how a receptor potential is generated in olfactory cells.
7. Discuss how a receptor potential is generated in gustatory cells.
8. Trace the path of a nervous impulse carrying (1) olfactory and (2) gustatory sense information from its point of origin to that area of the brain where interpretation occurs.
9. List the four "primary" taste sensations.
10. Locate on the tongue where taste buds are detected.

SENSE OF HEARING AND BALANCE: THE EAR

The ear has dual sensory functions. In addition to its role in hearing, it also functions as the sense organ of balance, or equilibrium. The stimulation, or "trigger," responsible for hearing and balance involves activation of specialized mechanoreceptors called hair cells. Sound waves and fluid movement are the physical forces that act on hair cells to generate receptor potentials and then nerve impulses, which are eventually perceived in the brain as sound or balance. The ear is divided into three anatomical parts: **external ear, middle ear,** and **inner ear** (Figure 15-8).

The structures illustrated in Figure 15-8 are not drawn to scale. Instead, the smaller components of the middle and inner ear are enlarged so that they can be more easily identified. In addition, this type of artistic rendering makes it easier to see the anatomical relationships of these tiny elements to one another and to adjacent structures.

External Ear

The external ear has two divisions: the *auricle* or *pinna*, and the *external auditory meatus* (ear canal). The auricle is the visible appendage on the side of the head surrounding the opening of the external auditory meatus. A number of anatomical components of the auricle are identified in Figure 15-7. Because it lies exposed against the bony surface of the skull, the auricle is frequently injured by blunt trauma. Bruising may then cause an accumulation of blood and tissue fluid between the skin and underlying cartilage. If bruising is left untreated, the classic swelling of "cauliflower ear" may develop and become permanent. The sun-exposed auricle, especially the tops of the helix, is also a frequent site for development of skin cancer, often squamous cell carcinoma (see Chapter 6).

Note in Figure 15-7 that the *tragus* of the auricle lies just in front of the opening to the ear canal. In adults, this canal is about

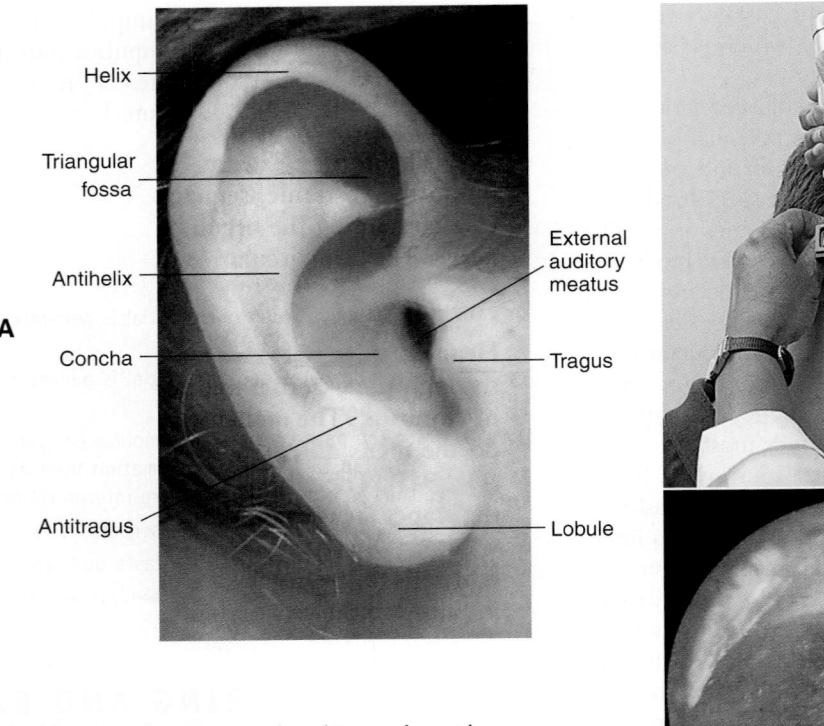

Figure 15-7 *Examining the external ear, ear canal, and tympanic membrane.*
A, Anatomical structures of the auricle. **B,** Using a lighted otoscope to view the
external ear canal and tympanic membrane. **C,** Note the translucent, pearly-grey
appearance of a normal tympanic membrane. The "handle" of the malleus can be
seen attaching near the center of the inner surface of the membrane.

3 cm long. Initially it slants upward a bit and then curves downward before traveling into the temporal bone in an inward and forward direction. It ends at the **tympanic membrane,** or **eardrum,** which stretches across the inner end of the canal, separating it from the middle ear.

A lighted instrument called an **otoscope** is used to examine the external ear canal and outer surface of the tympanic membrane (Figure 15-7, *B*). Changes in the appearance of the ear canal and tympanic membrane can provide a skilled observer with a great deal of information. For example, middle ear infection, called **otitis media,** will cause the eardrum to become red and inflamed and to bulge outward into the ear canal as pus and other fluids accumulate in the middle ear (see Mechanisms of Disease, p. 580). Accumulation of the waxlike substance *cerumen,* which is secreted by modified sweat glands in the auditory canal, may cause pain and temporary deafness. It is easily identified during otoscopic examination of the ear canal.

Middle Ear

The middle ear (tympanic cavity), a tiny epithelial-lined cavity hollowed out of the temporal bone, contains the three **auditory ossicles:** the *malleus, incus,* and *stapes* (see Figure 15-8). The names of these very small bones describe their shapes (hammer, anvil, stirrup). The "handle" of the malleus is attached to the inner surface of the tympanic membrane, whereas the "head" at-

taches to the incus, which in turn attaches to the stapes. There are several openings into the middle ear cavity: one from the external auditory meatus, covered with the tympanic membrane; two into the internal ear, the **oval window** (into which the stapes fits) and the **round window,** which is covered by a membrane; and one into the auditory (eustachian) tube.

Posteriorly the middle ear cavity is continuous with numerous mastoid air spaces in the temporal bone. The clinical importance of these middle ear openings is that they provide routes for infection to travel. Head colds, for example, especially in children, may lead to middle ear or mastoid infections by way of the nasopharynx-auditory tube, middle ear–mastoid path.

The **auditory,** or **eustachian,** tube is composed partly of bone and partly of cartilage and fibrous tissue and is lined with mucosa. It extends downward, forward, and inward from the middle ear cavity to the nasopharynx, or pharyngotympanic tube (the part of the throat behind the nose). The auditory tube serves a useful function: it makes possible equalization of pressure against inner and outer surfaces of the tympanic membrane and therefore prevents membrane rupture and the discomfort that marked pressure differences produce. The way the auditory tube equalizes tympanic membrane pressure is this: when you swallow or yawn, air spreads rapidly through the open tube. Atmospheric pressure then presses against the inner surface of the tympanic membrane. Because atmospheric pressure is continually exerted against its outer surface, the pressures are equal.

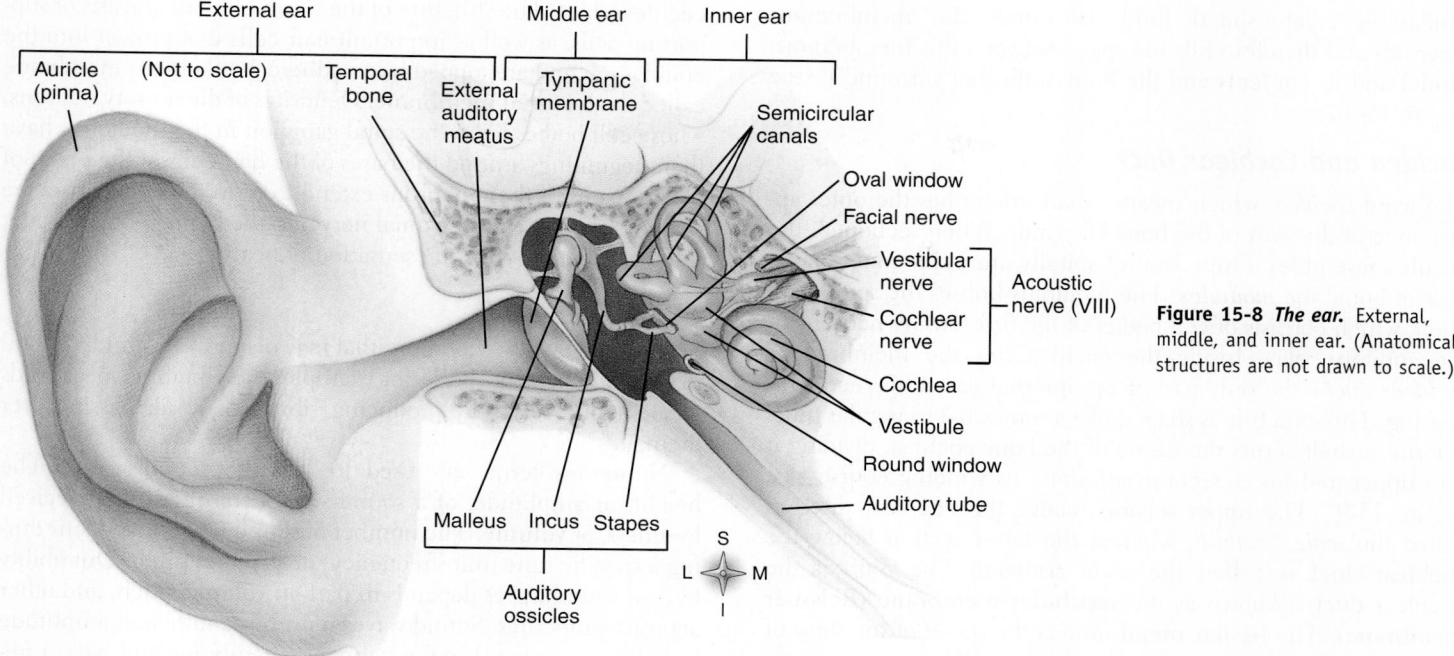

Figure 15-8 *The ear.* External, middle, and inner ear. (Anatomical structures are not drawn to scale.)

Inner Ear

The inner ear is also called the **labyrinth** because of its complicated shape. It consists of two main parts, a bony labyrinth and, inside this, a membranous labyrinth. The bony labyrinth consists of three parts: the *vestibule, cochlea,* and *semicircular canals* (Figure 15-9). The membranous labyrinth consists of the *utricle* and *saccule* inside the

vestibule, the *cochlear duct* inside the cochlea, and the *membranous semicircular canals* inside the bony ones. The vestibule (containing the utricle and saccule) and the semicircular canals are involved in balance; the cochlea is involved in hearing.

The term **endolymph** is used to describe the clear and potassium-rich fluid that fills the membranous labyrinth. **Perilymph,** a fluid

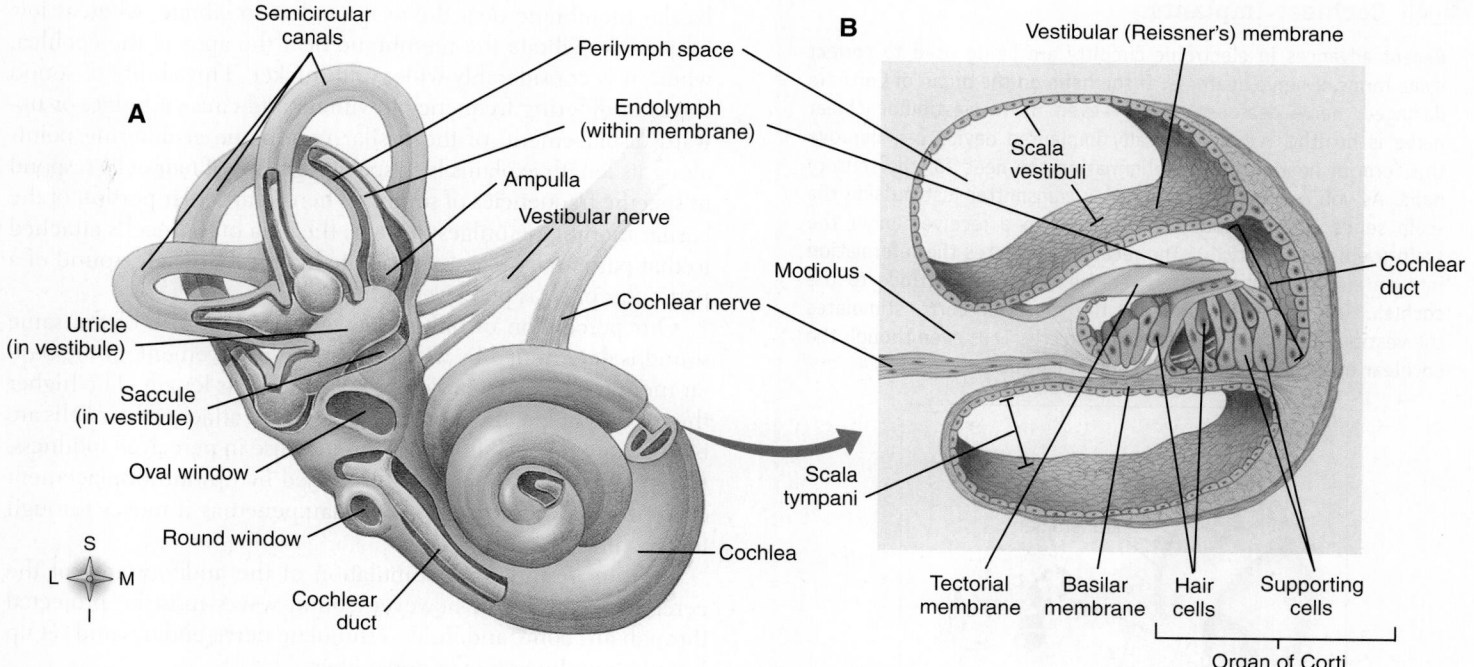

Figure 15-9 *The inner ear.* **A,** The bony labyrinth (*orange*) is the hard outer wall of the entire inner ear and includes semicircular canals, vestibule, and cochlea. Within the bony labyrinth is the membranous labyrinth (*purple*), which is surrounded by perilymph and filled with endolymph. Each ampulla in the vestibule contains a crista ampullaris that detects changes in head position and sends sensory impulses through the vestibular nerve to the brain. **B,** The inset shows a section of the membranous cochlea. Hair cells in the organ of Corti detect sound and send the information through the cochlear nerve. The vestibular and cochlear nerves join to form the eighth cranial nerve.

similar to cerebrospinal fluid, surrounds the membranous labyrinth and therefore fills the space between this membranous tunnel and its contents and the bony walls that surround it (see Figure 15-9).

Cochlea and Cochlear Duct

The word *cochlea*, which means "snail," describes the outer appearance of this part of the bony labyrinth. When sectioned, the cochlea resembles a tube wound spirally around a cone-shaped core of bone, the *modiolus*. The modiolus houses the spiral ganglion, which consists of cell bodies of the first sensory neurons in the auditory relay. Inside the cochlea lies the membranous *cochlear duct*—the only part of the internal ear concerned with hearing. This structure is shaped like a somewhat triangular tube. It forms a shelf across the inside of the bony cochlea, dividing it into upper and lower sections all along its winding course (see Figure 15-9). The upper section (above the cochlear duct) is called the *scala vestibuli*, whereas the lower section below the cochlear duct is called the *scala tympani*. The roof of the cochlear duct is known as the **vestibular membrane (Reissner membrane)**. The **basilar membrane** is the name of the floor of the cochlear duct. It is supported by bony and fibrous projections from the wall of the cochlea. Perilymph fills the scala vestibuli and scala tympani, and endolymph fills the cochlear duct.

The hearing sense organ, named the **organ of Corti**, rests on the basilar membrane throughout the entire length of the

BOX 15-2: HEALTH MATTERS
Cochlear Implants

Recent advances in electronic circuitry are being used to correct some forms of nerve deafness. If the hairs on the organ of Corti are damaged, nerve deafness results—even if the vestibulocochlear nerve is healthy. A new surgically implanted device can improve this form of hearing loss by eliminating the need for the sensory hairs. As you can see in the figure, a transmitter just outside the scalp sends external sound information to a receiver under the scalp (behind the auricle). The receiver translates the information into an electrical code that is relayed down an electrode to the cochlea. The electrode, wired to the organ of Corti, stimulates the vestibulocochlear nerve endings directly. Thus even though the cochlear hair cells are damaged, sound can be perceived.

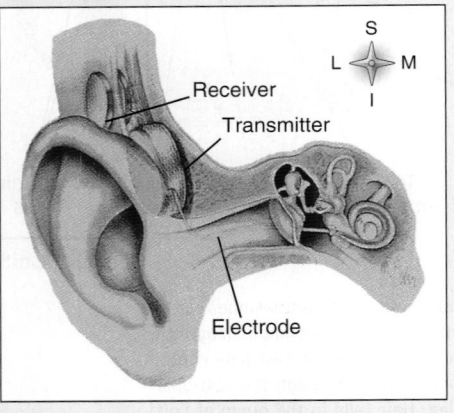

cochlear duct. The structure of the organ of Corti consists of supporting cells, as well as important hair cells that project into the endolymph and are topped by an adherent gelatinous membrane called the **tectorial membrane.** Dendrites of the sensory neurons, whose cell bodies lie in the spiral ganglion in the modiolus, have their beginnings around the bases of the hair cells of the organ of Corti. Axons of these neurons extend to form the cochlear nerve (a branch of the eighth cranial nerve) to the brain. They conduct impulses that produce the sensation of hearing.

Sense of Hearing

Sound is created by vibrations that may occur in air, fluid, or solid material. When we speak, for example, the vibrating vocal cords create sound waves by producing vibrations in air passing over them.

Numerous terms are used to describe sound waves. The height, or amplitude, of a sound wave determines its perceived loudness, or **volume.** The number of sound waves that occur during a specific time unit (frequency) determines **pitch.** Our ability to hear sound waves depends in part on volume, pitch, and other acoustic properties. Sound waves must be of sufficient amplitude to initiate movement of the tympanic membrane and have a frequency that is capable of stimulating the hair cells in the organ of Corti at some point along the basilar membrane.

The basilar membrane is not the same width and thickness throughout its length. Because of this structural fact, different frequencies of sound will cause the basilar membrane to vibrate and bulge upward at different places along its length. Two bulges are shown in Figure 15-10.

High-frequency sound waves cause the narrow portion of the basilar membrane near the oval window to vibrate, whereas low frequencies vibrate the membrane near the apex of the cochlea, where it is considerably wider and thicker. This ability of sound waves of differing frequency to vibrate and cause a bulge, or upward displacement, of the basilar membrane at differing points along its length explains how specific groups of hair cells respond to specific frequencies of sound. When a particular portion of the basilar membrane bulges upward, the cilia on hair cells attached to that particular area are stimulated, and, ultimately, sound of a particular pitch is perceived.

Our perception of different degrees of loudness of the same sound is determined by the amplitude, or movement, of the basilar membrane at any particular point along its length. The higher the upward bulge, the more the cilia on the attached hair cells are bent or stimulated. This causes an increase in perceived loudness. The moving wave of perilymph caused by upward displacement of the basilar membrane is soon dampened as it moves through the cochlea.

Hearing results from stimulation of the auditory area of the cerebral cortex. First, however, sound waves must be projected through air, bone, and fluid to stimulate nerve endings and set up impulse conduction over nerve fibers.

Pathway of Sound Waves

Sound waves in the air enter the external auditory canal with aid from the pinna. At the inner end of the canal, they strike against the tympanic membrane, setting it in vibration. Vibrations of the

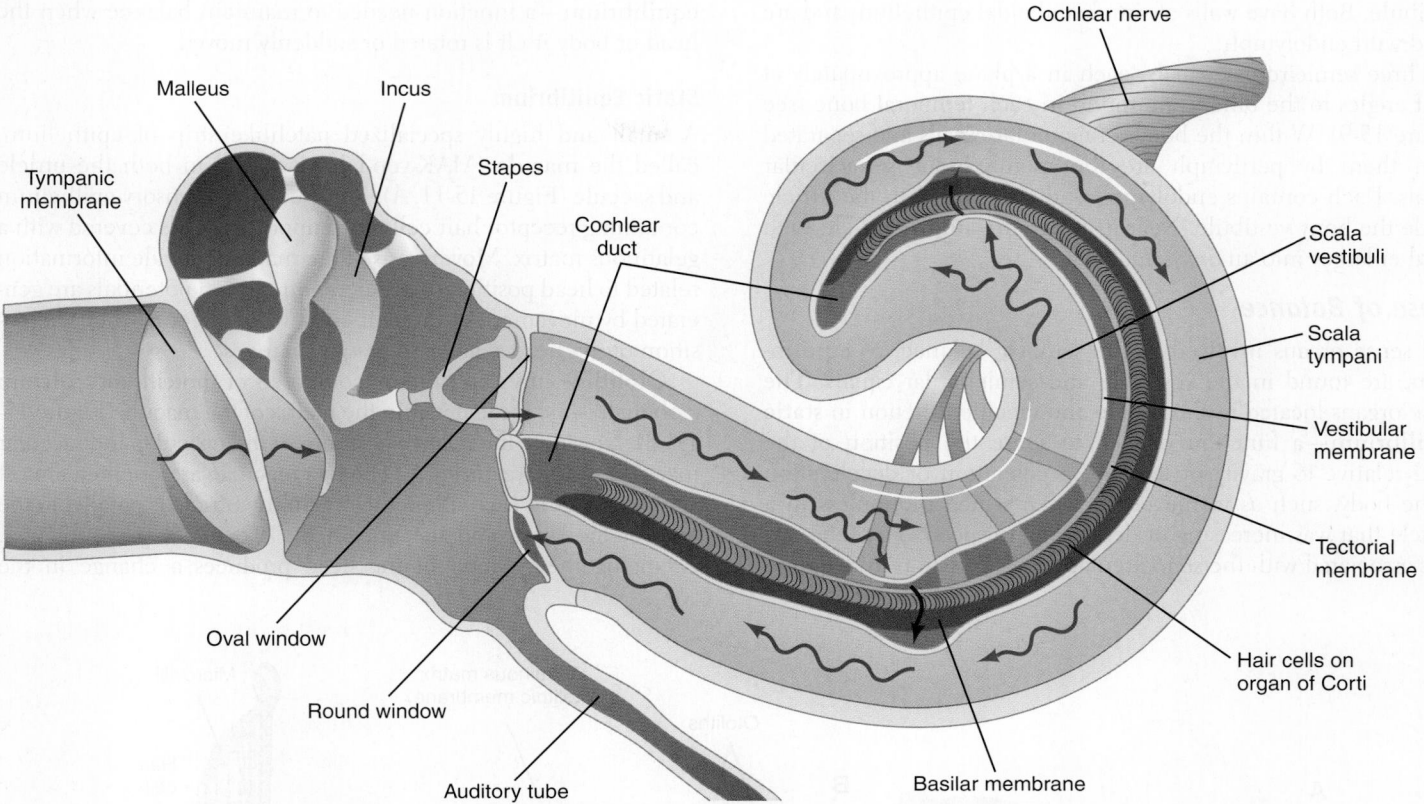

Figure 15-10 *Effect of sound waves on cochlear structures.* Sound waves strike the tympanic membrane and cause it to vibrate. This causes the membrane of the oval window to vibrate, which causes the perilymph in the bony labyrinth of the cochlea and the endolymph in the membranous labyrinth of the cochlea, or cochlear duct, to move. This movement of endolymph causes the basilar membrane to vibrate, which in turn stimulates hair cells on the organ of Corti to transmit nerve impulses along the cranial nerve. Eventually, nerve impulses reach the auditory cortex and are interpreted as sound.

tympanic membrane move the malleus, whose handle attaches to the membrane. The head of the malleus attaches to the incus, and the incus attaches to the stapes. So when the malleus vibrates, it moves the incus, which moves the stapes against the oval window into which it fits so precisely. At this point, fluid conduction of sound waves begins. When the stapes moves against the oval window, pressure is exerted inward into the perilymph in the scala vestibuli of the cochlea. This starts a "ripple" in the perilymph that is transmitted through the vestibular membrane (the roof of the cochlear duct) to endolymph inside the duct and then to the organ of Corti and to the basilar membrane that supports the organ of Corti and forms the floor of the cochlear duct. From the basilar membrane the ripple is next transmitted to and through the perilymph in the scala tympani and finally expends itself against the round window—like an ocean wave as it breaks against the shore but on a much reduced scale.

Figure 15-10 shows the steps involved in hearing.

Neuronal Pathway of Hearing

Dendrites of neurons whose cell bodies lie in the spiral ganglion and whose axons make up the cochlear nerve terminate around the bases of the hair cells of the organ of Corti, and the tectorial membrane adheres to their upper surfaces. The movement of the hair cells against the adherent tectorial membrane somehow stimulates these dendrites and initiates impulse conduction by the cochlear nerve to the brainstem. Before reaching the auditory area of the temporal lobe, impulses pass through "relay stations" in nuclei in the medulla, pons, midbrain, and thalamus.

 QUICK CHECK

11. List the three anatomical divisions of the ear.
12. Identify the three auditory ossicles.
13. Name the divisions of both the membranous and bony labyrinths.
14. Name the sense organ responsible for hearing.

Vestibule and Semicircular Canals

The vestibule constitutes the central section of the bony labyrinth. Look again at Figure 15-9. Note that the bony labyrinth opens into the oval and round windows from the middle ear, as well as the three semicircular canals of the internal ear. The utricle and saccule are the membranous structures within the vestibule. Both have walls of simple cuboidal epithelium and are filled with endolymph.

Three semicircular canals, each in a plane approximately at right angles to the others, are found in each temporal bone (see Figure 15-9). Within the bony semicircular canals and separated from them by perilymph are the membranous semicircular canals. Each contains endolymph and connects with the utricle inside the bony vestibule. Near its junction with the utricle each canal enlarges into an *ampulla*.

Sense of Balance

The sense organs involved in the sense of balance, or **equilibrium,** are found in the vestibule and semicircular canals. The sense organs located in the utricle and saccule function in **static equilibrium**—a function needed to sense the position of the head relative to gravity or to sense acceleration or deceleration of the body, such as would occur when seated motionless in a vehicle that was increasing or decreasing in speed. The sense organs associated with the semicircular canals function in **dynamic**

equilibrium—a function needed to maintain balance when the head or body itself is rotated or suddenly moved.

Static Equilibrium

A small and highly specialized patchlike strip of epithelium, called the **macula** (MAK-yoo-lah), is found in both the utricle and saccule (Figure 15-11, *A*). It is specialized sensory epithelium containing receptor hair cells and supporting cells covered with a gelatinous matrix. Movements of the macula provide information related to head position or acceleration. Action potentials are generated by movement of the hair cells, which occurs when the position of the head relative to gravity changes.

Otoliths—tiny "ear stones" composed of protein and calcium carbonate—are located within the matrix of the macula (Figure 15-11, *B*). Now note the relative positions of the utricular and saccular maculae in Figure 15-11, *A*. The two maculae are oriented almost at right angles to each other: the one in the utricle is parallel to the base of the skull, and the one in the saccule is perpendicular. Changing the position of the head produces a change in the

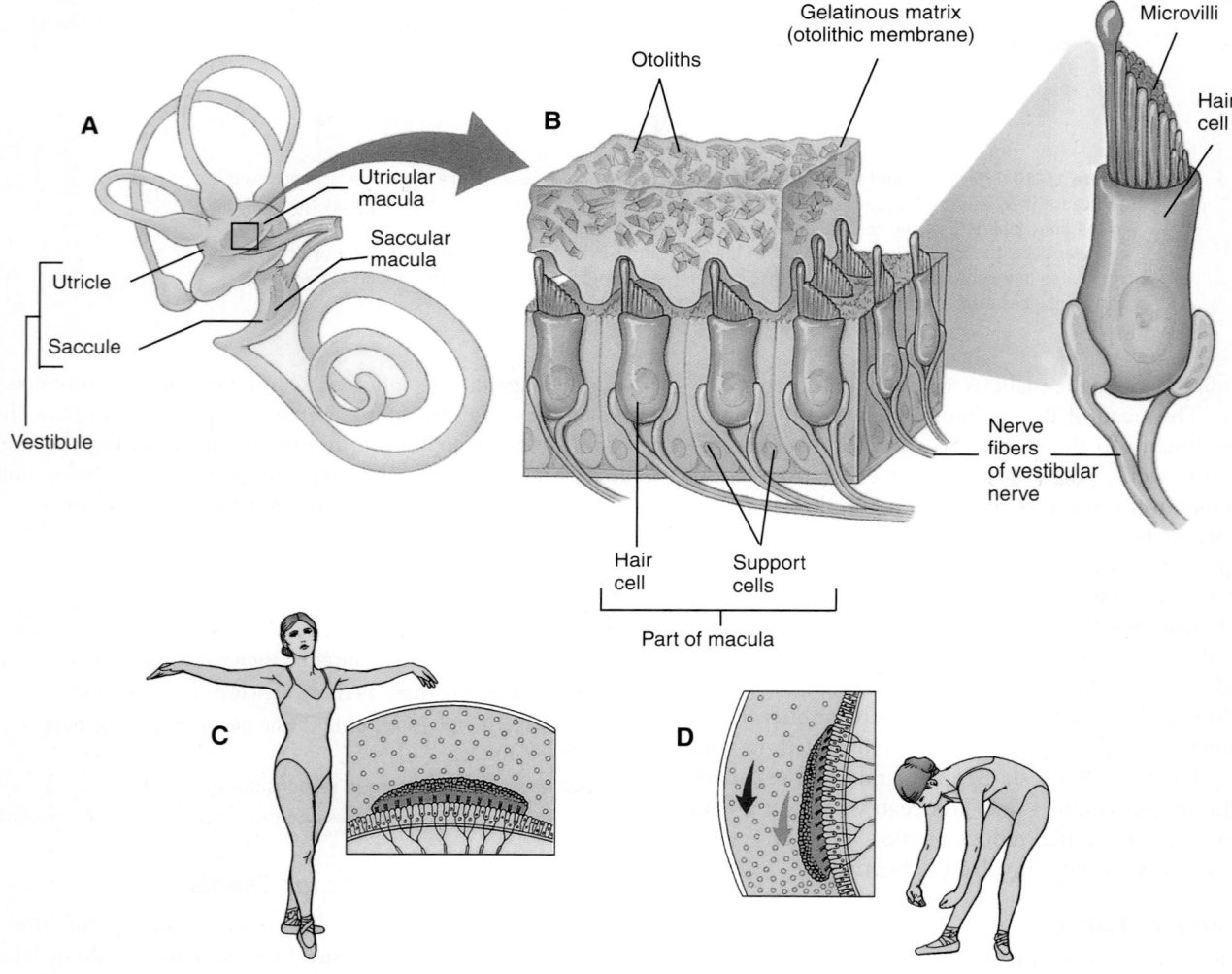

Figure 15-11 *The macula.* **A,** Structure of vestibule showing placement of utricular and saccular maculae. **B,** Section of macula showing otoliths. **C,** Macula stationary in upright position. **D,** Macula displaced by gravity as person bends over.

amount of pressure on the otolith-weighted matrix, which, in turn, stimulates the hair cells (Figure 15-11, *C* and *D*). This stimulates the adjacent receptors of the vestibular nerve. Its fibers conduct impulses to the brain that produce a sense of the position of the head and also a sensation of a change in the pull of gravity, for example, a sensation of acceleration. In addition, stimulation of the macula evokes *righting reflexes*, muscular responses to restore the body and its parts to their normal position when they have been displaced. Impulses from proprioceptors and from the eyes also activate righting reflexes. Interruption of the vestibular, visual, or proprioceptive impulses that initiate these reflexes may cause disturbances of equilibrium, nausea, vomiting, and other symptoms.

Dynamic Equilibrium

Dynamic equilibrium depends on the functioning of the **crista ampullaris,** located in the ampulla of each semicircular canal. This specialized structure is a form of sensory epithelium that is similar in many ways to the maculae. Each cone-shaped crista is composed of many hair cells, with their processes embedded in a gelatinous cap, called the **cupula** (Figure 15-12, *A* and *B*).

The cupula is not weighted with otoliths and does not respond to the pull of gravity. It serves, instead, much like a float that moves with the flow of endolymph in the semicircular canals. Like the maculae, the semicircular canals are placed nearly at right angles to each other. This arrangement enables detection of movement in all directions. As the cupula moves, it bends the hairs embedded in it, producing first a receptor and then an action potential that passes through the vestibular portion of the eighth cranial nerve to the medulla oblongata and, from there, to other areas of the brain and spinal cord for interpretation, integration, and response.

When a person spins (Figure 15-12, *C* to *E*), the semicircular canals move with the body, but inertia keeps the endolymph in them from moving at the same rate. The cupula therefore moves in a direction opposite to head movement until after the initial movement stops. Dynamic equilibrium is thus able to detect changes in both the direction and the rate at which movement occurs.

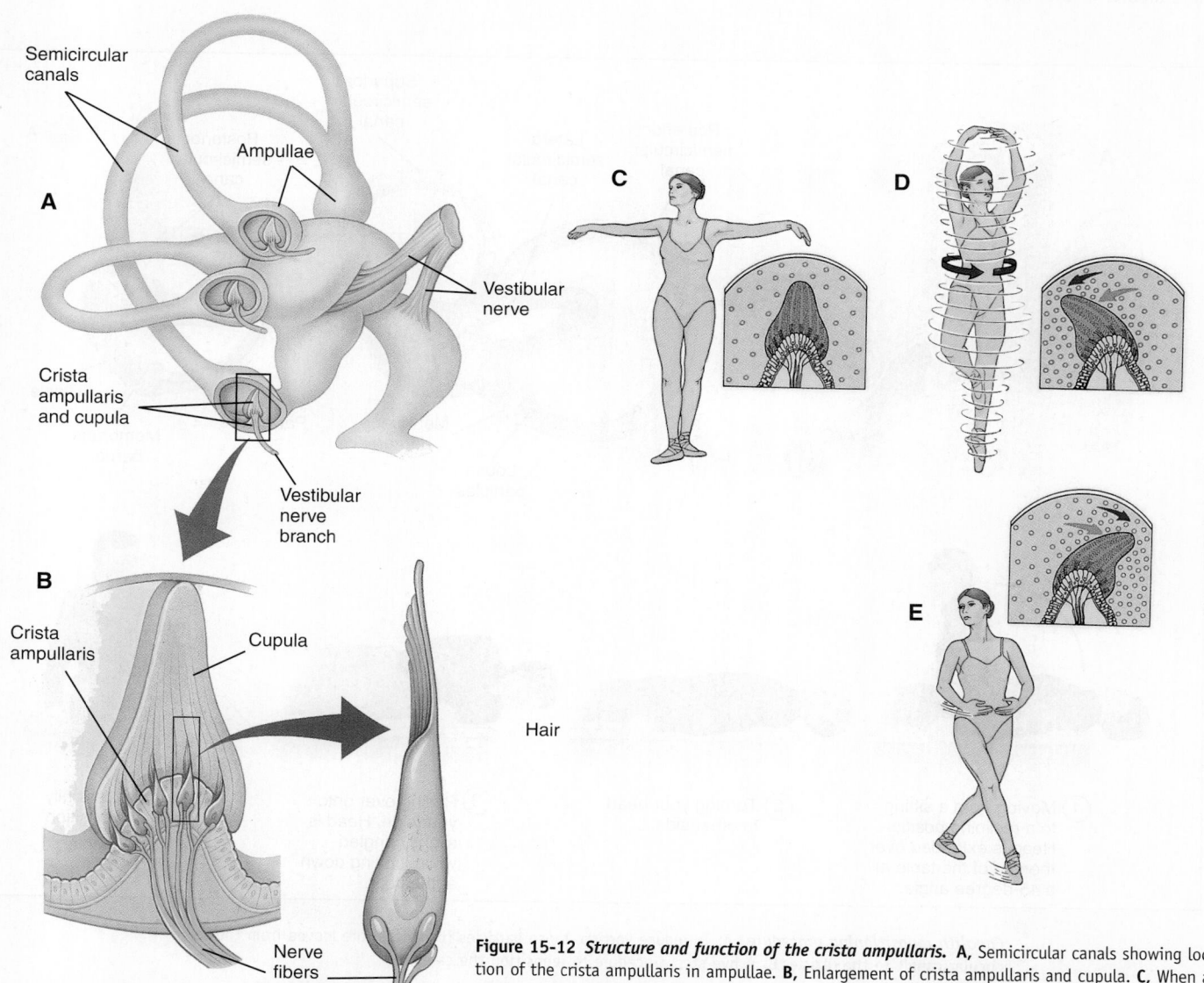

Figure 15-12 *Structure and function of the crista ampullaris.* **A,** Semicircular canals showing location of the crista ampullaris in ampullae. **B,** Enlargement of crista ampullaris and cupula. **C,** When a person is at rest, the crista ampullaris does not move. **D** and **E,** As a person begins to spin, the crista ampullaris is displaced by the endolymph in a direction opposite to the direction of spin.

BOX 15-3: HEALTH MATTERS

Canalith Repositioning Procedure

The term *dizziness* is often used to describe a wide array of symptoms that are associated with unsteadiness and a feeling of movement within the head. A somewhat different and more specialized condition called *vertigo* involves a sensation of one's own body spinning in space or of the external world spinning around the individual. Vertigo can be a frightening and recurring problem often precipitated in affected individuals by sudden changes in body position that may occur when rolling over in bed, bending to pick an object up from the floor, or simply sitting up quickly. Symptoms include nausea, vomiting, and an inability to maintain balance and equilibrium.

A common type of vertigo called **benign paroxysmal** (par-ock-SIZ-mul) **positional vertigo,** or **BPPV** for short, is frequently associated with inner ear problems and is most common in adults over 50 years of age. Although BPPV sometimes follows some type of head trauma or infection, it is most often associated with the aging process. Whatever the cause, release of otolith particles occurs, which then float around in the fluid-filled semicircular canals. These particles—

now called **canaliths**—settle in the posterior semicircular canal and are subject to movement with changes in body position (see part *A* of the figure).

A 5- to 10-minute treatment called the **canalith repositioning procedure** has been developed to reposition or move these particles into the utricle (see part *B* of the figure). Once in the utricle, they tend to attach to the membranous walls and are no longer able to move about and cause the stimulation that results in vertigo. The five-step procedure (see the figure) is suitable only for BPPV—a condition that can be accurately diagnosed by a physician and separated from other types of vertigo or dizziness. If BPPV is diagnosed as the cause of symptoms, this simple office procedure may eliminate the need for more expensive and less well-tolerated exercise treatments or medications that have been used in the past. The treatment is often successful in relieving symptoms on the first attempt but may have to be repeated if the amount of floating debris is substantial. After the repositioning sequence, the individual must keep the head upright for 48 hours.

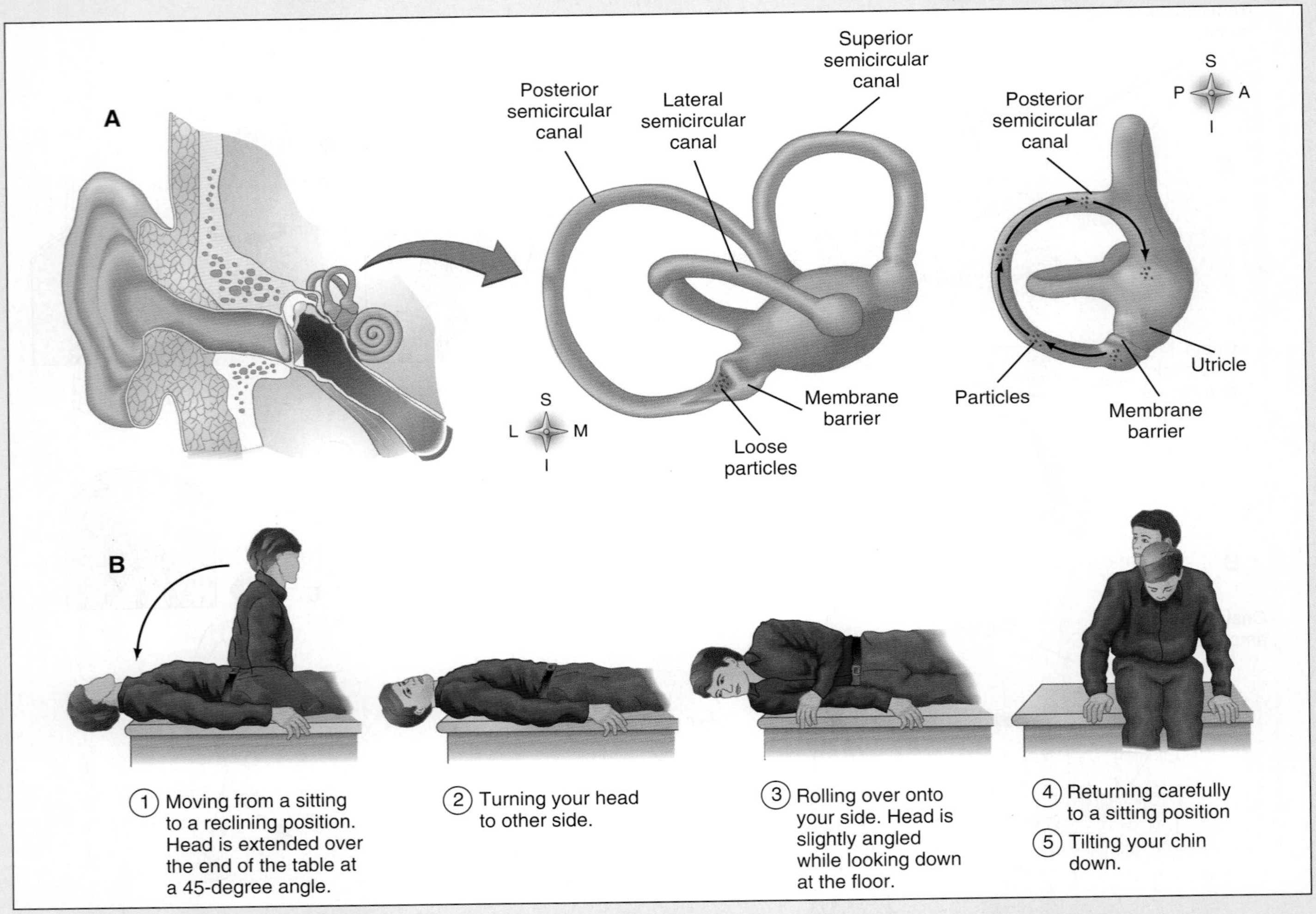

A

Posterior
semicircular
canal

Lateral
semicircular
canal

Superior
semicircular
canal

Posterior
semicircular
canal

S
P — A
I

Loose
particles

Membrane
barrier

S
L — M
I

Particles

Membrane
barrier

Utricle

B

① Moving from a sitting to a reclining position. Head is extended over the end of the table at a 45-degree angle.

② Turning your head to other side.

③ Rolling over onto your side. Head is slightly angled while looking down at the floor.

④ Returning carefully to a sitting position

⑤ Tilting your chin down.

Canalith repositioning procedure. **A,** To reduce vertigo, loose particles (canaliths) are moved from the semicircular canals to the utricle. **B,** A five-step procedure to reposition the canaliths.

VISION: THE EYE

One of the most important sensations involved in maintaining homeostasis is vision, and the eye is the body's sense organ for this important function. Vision is a truly remarkable sensory ability and is used to guide almost all that we do. It allows us to activate and respond to a multitude of warning systems, and provides us with almost constant feedback on various types of form and movement in an ever-changing environment. It is the eye that uses as a stimulus the pervasive nature of light to convert stored photochemical energy into nervous impulses that are ultimately interpreted by the brain as sight.

The study of vision and the visual apparatus is an important area of ongoing research and clinical interest. Clinical **ophthalmology** is a medical practice specialty concerned with pathologic conditions of the eye and the diagnosis and treatment of eye disorders. A number of allied health professionals, including ophthalmic medical technologists, laboratory technicians, and opticians, work closely with ophthalmologists in administrative, research, and clinical environments.

We will first discuss the various structures of the eye and then move on to the ways in which these structures enable the eye to control the amount of light entering it and how the conversion to electrical stimuli actually occurs.

Structure of the Eye

External Eye

A number of the anatomical structures of the external eye are visible in Figure 15-13, *A*, and discussed in the paragraphs that follow. Just as an otoscope is used to view the external auditory canal and surface of the tympanic membrane, a lighted **ophthalmoscope**

(off-THAL-mah-skohp) can be used to examine the retinal surface, called the *fundus*, and internal eye structures by viewing them through the pupil (Figure 15-13, *B* and *C*). This noninvasive technique is a safe and effective aid in the diagnosis of many types of disease (see Mechanisms of Disease, p. 583).

Layers of the Eyeball

Approximately five sixths of the eyeball, a sphere-like globe about 22 to 25 mm in diameter, lies recessed in and is protected by the bony orbit or eye socket. Only the small anterior surface of the eyeball is exposed. Three layers of tissues compose the eyeball. From the outside in, the following three layers of tissues and their component parts or regions form the eyeball.

1. Fibrous layer
   a. Sclera
   b. Cornea
2. Vascular layer
   a. Choroid
   b. Ciliary body
   c. Iris
3. Inner layer
   a. Retina
   b. Optic nerve
   c. Retinal blood vessels

Refer to Figure 15-14 as you read the following paragraphs.

The anterior portion of the sclera is called the *cornea* and lies over the colored part of the eye, the *iris* (Figures 15-13 and 15-14). The cornea is transparent, whereas the rest of the sclera is white and opaque, a fact that explains why the visible anterior surface of the sclera is usually spoken of as the "white" of the eye. No blood vessels are found in the cornea or in the lens. Deep within the an-

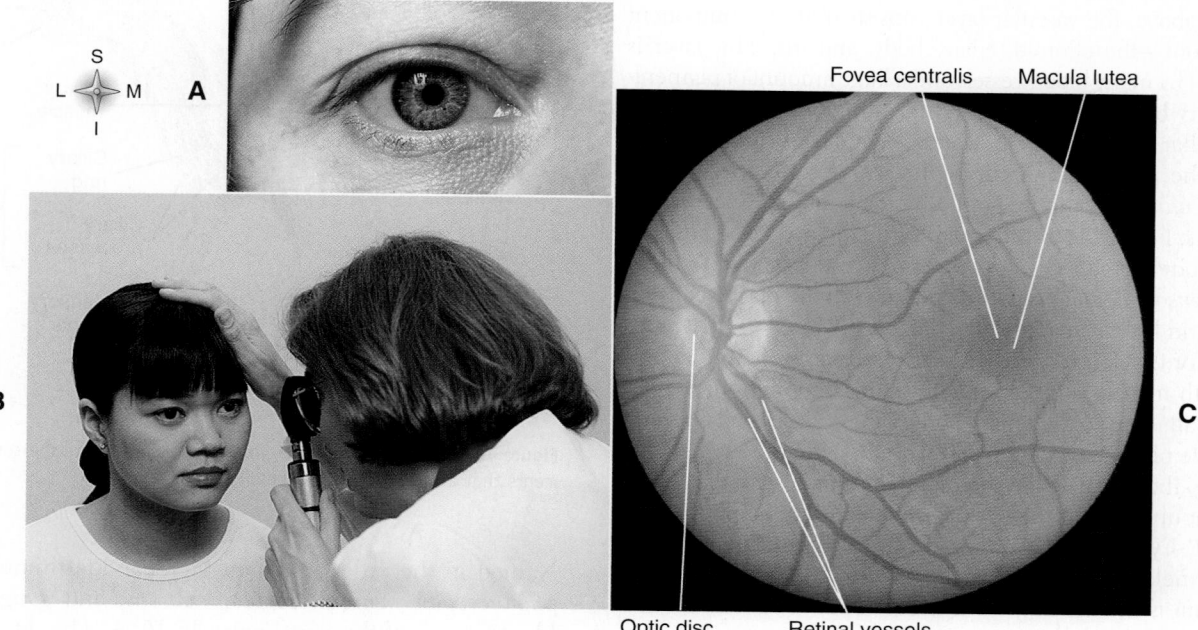

Figure 15-13 *Examining the eye.* **A,** Anatomical structures of the external eye. See text for discussion. **B,** Using the ophthalmoscope to view the retina. **C,** Ophthalmoscopic view of the retina as seen through the pupil.

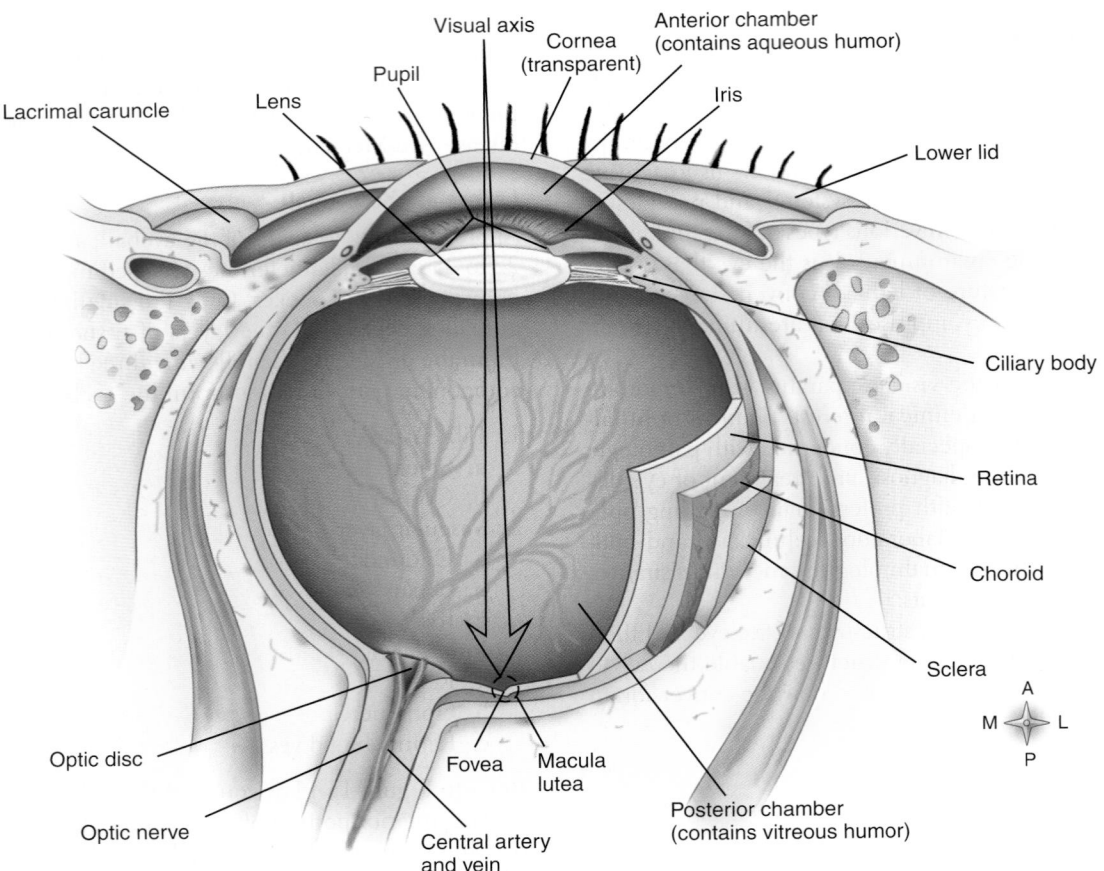

Figure 15-14 *Horizontal section through the eyeball.* The eye is viewed from above.

terior part of the sclera at its junction with the cornea lies a ring-shaped venous sinus, the *canal of Schlemm.*

As noted above, the vascular layer consists of three component parts or regions—the choroid, ciliary body, and iris. This layer is characterized by many blood vessels and a large amount of pigment.

The **ciliary body** is formed by a thickening of the choroid and fits like a collar into the area between the anterior margin of the retina and the posterior margin of the iris (Figure 15-15). The small *ciliary muscle,* composed of both radial and circular smooth muscle fibers, lies in the anterior part of the ciliary body. Folds in the ciliary body are called *ciliary processes,* and attached to these are the *suspensory ligaments,* which blend with the elastic capsule of the *lens* and hold it suspended in place.

The **iris,** or the colored part of the eye, consists of circular and radial smooth muscle fibers arranged to form a doughnut-shaped structure. It attaches to the ciliary body. The hole-shaped opening in the middle of the iris is the **pupil.** By adjusting the diameter of the opening, the iris acts like the diaphragm of a camera. It controls the amount of light that enters the eye by adjusting the size of the pupil. Eye color is a function of the amount, placement, and type of melanin in the iris.

The **retina** is the incomplete innermost coat of the eyeball—incomplete in that it has no anterior portion. Pigmented epithelial cells form the layer of the retina next to the choroid coat. Three layers of neurons make up the major portion of the retina.

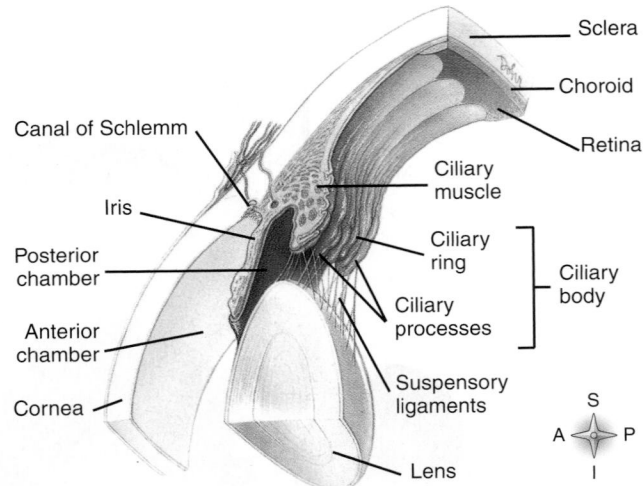

Figure 15-15 *Lens, cornea, iris, and ciliary body.* Note the suspensory ligaments that attach the lens to the ciliary body.

Named in the order in which they conduct impulses, they are *photoreceptor neurons, bipolar neurons,* and *ganglion neurons.* Identify each of these in Figure 15-16, *A.* The distal ends of the dendrites of the photoreceptor neurons have names that describe their shapes. Because some look like tiny rods and others look like

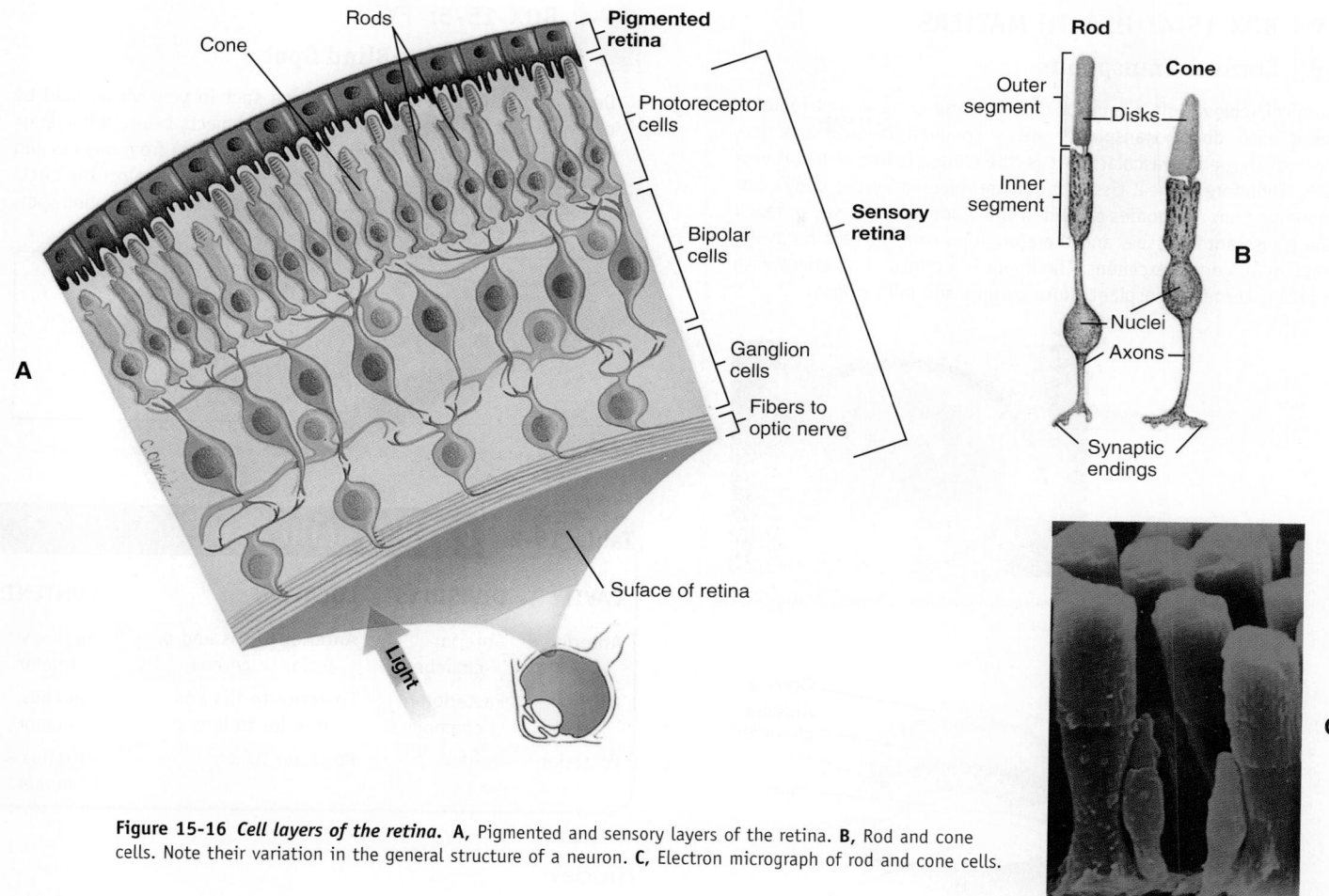

Figure 15-16 *Cell layers of the retina.* **A,** Pigmented and sensory layers of the retina. **B,** Rod and cone cells. Note their variation in the general structure of a neuron. **C,** Electron micrograph of rod and cone cells.

cones, they are called **rods** and **cones,** respectively (Figure 15-16, *B* and *C*). They constitute our visual receptors, structures highly specialized for stimulation by light rays. They differ in numbers, distribution, and function. Cones are less numerous than rods and are most densely concentrated in the **fovea centralis,** a small depression in the center of a yellowish area, the **macula lutea,** found near the center of the retina (see Figures 15-13 and 15-14). They become less and less dense from the fovea outward. Rods, on the other hand, are absent entirely from the fovea and macula and increase in density toward the periphery of the retina. How these anatomical facts relate to rod and cone functions is discussed on p. 578.

All the axons of ganglion neurons extend back to a small circular area in the posterior part of the eyeball known as the *optic disc.* This part of the sclera contains perforations through which the fibers emerge from the eyeball as the **optic nerve** (second cranial nerve). The optic disc is also called the **blind spot** because light rays striking this area cannot be seen. Why? Because it contains no rods or cones, only nerve fibers.

Cavities and Humors

The eyeball is not a solid sphere but contains a large interior space that is divided into two cavities: anterior and posterior.

The **anterior cavity** has two subdivisions, known as the *anterior* and *posterior chambers.* As Figure 15-14 shows, the entire anterior cavity lies in front of the lens. The posterior chamber of the anterior cavity consists of the small space directly posterior to the iris but anterior to the lens. And the anterior chamber of the anterior cavity is the space anterior to the iris but posterior to the cornea. *Aqueous humor* fills both chambers of the anterior cavity. This substance is clear and watery and often leaks out when the eye is injured.

The **posterior cavity** of the eyeball is considerably larger than the anterior, because it occupies all the space posterior to the lens, suspensory ligament, and ciliary body (see Figure 15-14). It contains *vitreous humor,* a substance with a consistency comparable to soft gelatin. This semisolid material, along with the aqueous humor, helps maintain sufficient intraocular pressure to prevent the eyeball from collapsing.

Aqueous humor forms from blood in capillaries (located mainly in the ciliary body). The ciliary body actively secretes aqueous humor into the posterior chamber, but passive filtration from capillary blood contributes also to aqueous humor formation. From the posterior chamber, aqueous humor moves from the area between the iris and the lens through the pupil into the anterior chamber. From here, it drains into the canal of Schlemm and moves into small veins (Figure 15-17). Normally, aqueous

BOX 15-4: HEALTH MATTERS
Corneal Transplants

Surgical removal of opaque or deteriorating corneas and replacement with donor transplants are a common medical practice. Corneal tissue is avascular; that is, the cornea is free of blood vessels. Therefore corneal tissue is seldom rejected by the body's immune system. Antibodies carried in the blood have no way to reach the transplanted tissue, and therefore long-term success following implant surgery is excellent. The figure is a photo of a patient with a recent corneal transplant. Note sutures and mild edema.

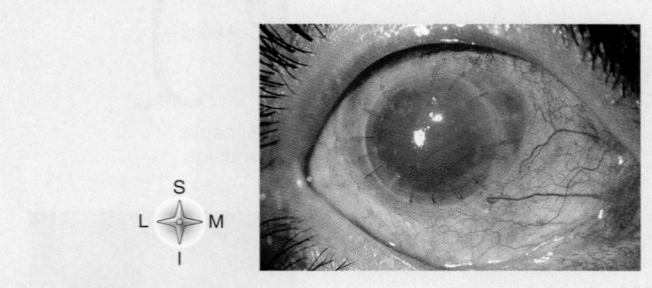

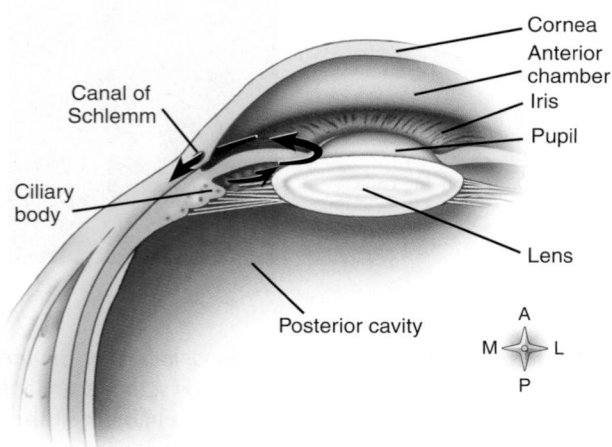

Figure 15-17 *Formation of aqueous humor.* Aqueous humor *(heavy arrows)* is believed to be formed mainly by secretion by the ciliary body into the posterior chamber. It passes into the anterior chamber through the pupil, from which it is drained away by the ring-shaped canal of Schlemm, and finally into the anterior ciliary veins.

humor drains out of the anterior chamber at the same rate at which it enters the posterior chamber, so the amount of aqueous humor in the eye remains relatively constant—and so, too, does intraocular pressure. But sometimes something happens to upset this balance, and intraocular pressure increases above the normal level of about 20 to 25 mm Hg. The individual then has the eye disease known as *glaucoma*, which, if untreated, can lead to retinal damage and blindness. Either excess formation or, more often, decreased drainage is seen as an immediate cause of this condition, but underlying causes are unknown (see Mechanisms of Disease, p. 583).

An outline summary of the cavities of the eye appears in Table 15-2.

BOX 15-5: FYI
Finding Your Blind Spot

Demonstrate the location of the blind spot in your visual field by covering your left eye and looking at the objects below. While staring at the square, begin about 30 cm (12 inches) from objects and slowly bring the figures closer to your eye. At one point, the circle will seem to disappear because its image has fallen on the blind spot.

Table 15-2	Cavities of the Eye		
CAVITY	**DIVISIONS**	**LOCATION**	**CONTENTS**
Anterior	Anterior chamber	Anterior to iris and posterior to cornea	Aqueous humor
	Posterior chamber	Posterior to iris and anterior to lens	Aqueous humor
Posterior	None	Posterior to lens	Vitreous humor

Muscles

Eye muscles are of two types: *extrinsic* and *intrinsic*. **Extrinsic eye muscles** are skeletal muscles that attach to the outside of the eyeball and to the bones of the orbit. They move the eyeball in any desired direction and are, of course, voluntary muscles. Four of them are straight muscles, and two are oblique. Their names describe their positions on the eyeball. They are the superior, inferior, medial, and lateral rectus muscles and superior and inferior oblique muscles (Figure 15-18).

Intrinsic eye muscles are smooth, or involuntary, muscles located within the eye. Their names, as we have discussed, are the *iris* and the *ciliary muscles*. Incidentally, the eye is the only organ in the body in which both voluntary and involuntary muscles are found. The iris regulates the size of the pupil. The ciliary muscle controls the shape of the lens. As the ciliary muscle contracts, it releases the suspensory ligament from the backward pull usually exerted on it, and this allows the elastic lens, suspended in the ligament, to bulge, or become more convex. The role of both these muscles in vision is discussed later in this chapter.

Accessory Structures

Accessory structures of the eye include the eyebrows, eyelashes, eyelids, and lacrimal apparatus.

Eyebrows and Eyelashes

The eyebrows and eyelashes serve a cosmetic purpose and give some protection against the entrance of foreign objects into the eyes. They also help shade the eyes and provide at least minimal

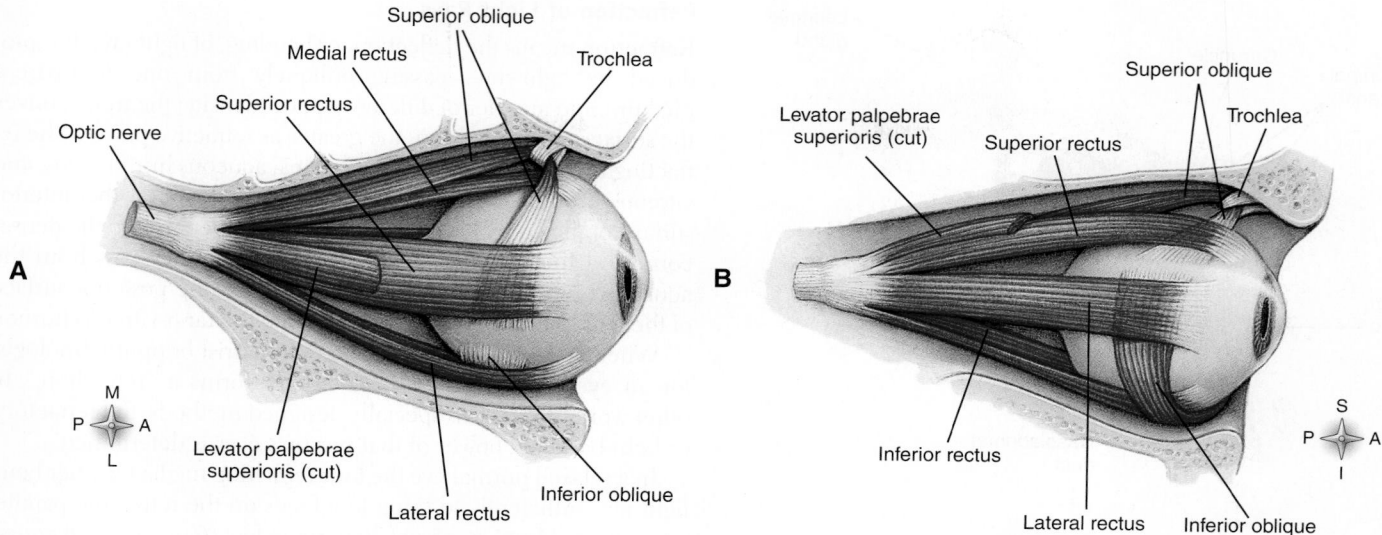

Figure 15-18 *Extrinsic muscles of the right eye.* A, Superior view. **B,** Lateral view.

protection from direct light. Small glands located at the base of the lashes secrete a lubricating fluid. They frequently become infected, forming a *sty.*

Eyelids

The eyelids, or *palpebrae,* consist mainly of voluntary muscle and skin, with a border of thick connective tissue at the free edge of each lid, known as the *tarsal plate.* One can feel the tarsal plate as a ridge

when turning back the eyelid to remove a foreign object. Mucous membrane, called *conjunctiva,* lines each lid (Figure 15-19). It continues over the surface of the eyeball, where it is modified to give transparency. Inflammation of the conjunctiva (conjunctivitis) is a fairly common infection. Because it produces a pinkish discoloration of the eye's surface, it is called *pinkeye* (Figure 15-20).

The opening between the eyelids bears the technical name of *palpebral fissure.* The height of the fissure determines the apparent size of the eyes. If the eyelids are habitually held wide open, the eyes appear large, although there is little difference in size between eyeballs of different adults. Eyes appear small if the upper eyelids droop. Plastic surgeons correct this common aging change with an operation called **blepharoplasty.** The upper and lower eyelids join, forming an angle or corner known as a *canthus.* The inner canthus is the medial corner of the eye. The outer canthus is the lateral corner.

Lacrimal Apparatus

The *lacrimal apparatus* consists of the structures that secrete tears and drain them from the surface of the eyeball. They are the lacrimal glands, lacrimal ducts, lacrimal sacs, and nasolacrimal ducts (Figure 15-21).

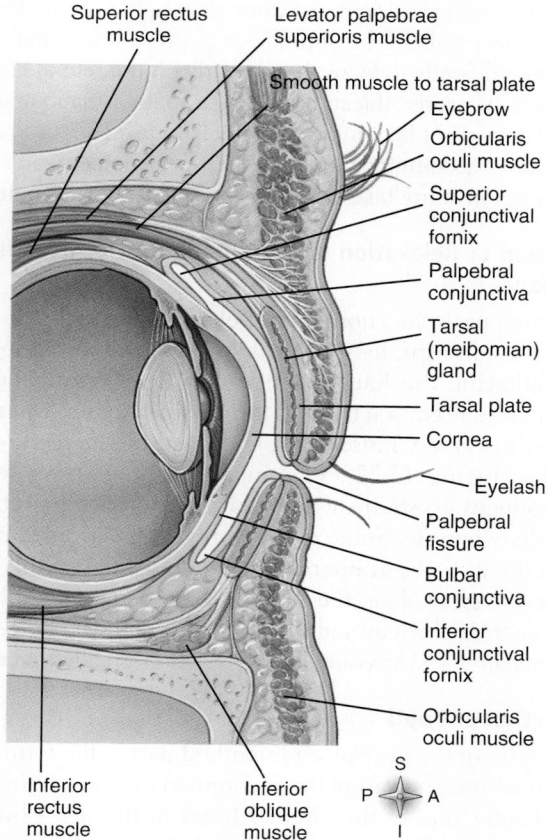

Figure 15-19 *Accessory structures of the eye.* Lateral view with eyelids closed.

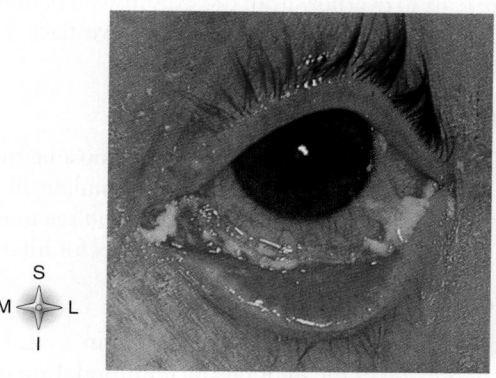

Figure 15-20 *Acute bacterial conjunctivitis.* Note the discharge of pus characteristic of this highly contagious infection of the conjunctiva.

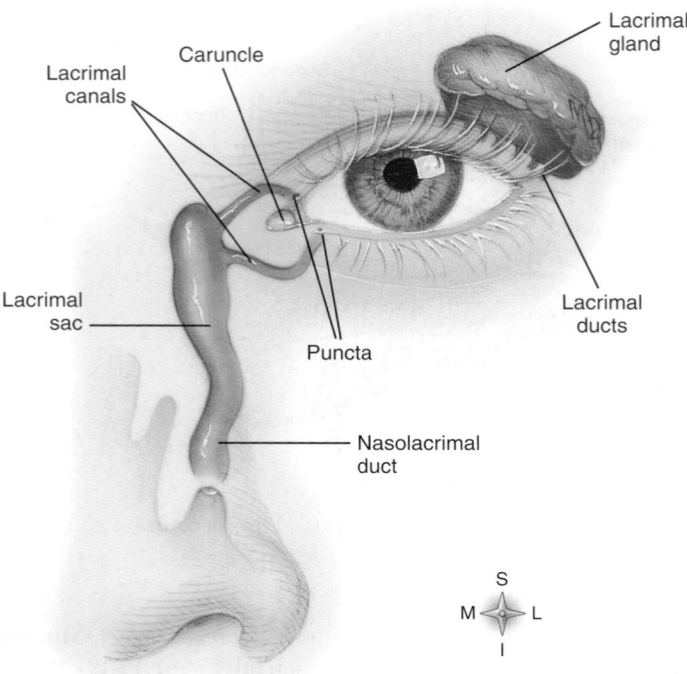

Figure 15-21 *Lacrimal apparatus.* Fluid produced by lacrimal glands (tears) streams across the eye surface, enters the canals, and then passes through the lacrimal sac and nasolacrimal duct to enter the nose.

The *lacrimal glands*, comparable in size and shape to a small almond, are located in a depression of the frontal bone at the upper outer margin of each orbit. Approximately a dozen small ducts lead from each gland, draining the tears onto the conjunctiva at the upper outer corner of the eye.

The *lacrimal canals* are small channels, one above and the other below each *caruncle* (small red body at inner canthus). They empty into the lacrimal sacs. The openings into the canals are called *puncta* and can be seen as two small dots at the inner canthus of the eye. The *lacrimal sacs* are located in a groove in the lacrimal bone. The *nasolacrimal ducts* are small tubes that extend from the lacrimal sac into the inferior meatus of the nose. All the tear ducts are lined with mucous membrane, an extension of the mucosa that lines the nose. When this membrane becomes inflamed and swollen, the nasolacrimal ducts become plugged, causing the tears to overflow from the eyes instead of draining into the nose as they do normally. Hence when we have a common cold, "watering" eyes add to our discomfort.

The Process of Seeing

For vision to occur, the following conditions must be fulfilled: an image must be formed on the retina to stimulate its receptors (rods and cones), and the resulting nerve impulses must be conducted to the visual areas of the cerebral cortex for interpretation.

Formation of Retinal Image

Four processes focus light rays so that they form a clear image on the retina: *refraction* of the light rays, **accommodation** of the lens, *constriction* of the pupil, and *convergence* of the eyes.

Refraction of Light Rays

Refraction means the deflection, or bending, of light rays. It is produced by light rays passing obliquely from one transparent medium into another of different optical density; the more convex the surface of the medium, the greater its refractive power. The refracting media of the eye are the cornea, aqueous humor, lens, and vitreous humor. Light rays are bent, or refracted, at the anterior surface of the cornea as they pass from the rarer air into the denser cornea, at the anterior surface of the lens as they pass from the aqueous humor into the denser lens, and at the posterior surface of the lens as they pass from the lens into the rarer vitreous humor.

When an individual goes to an optometrist or ophthalmologist for an eye examination, the doctor performs a "refraction." In other words, by various specially designed methods, the refractory, or light-bending, power of that person's eyes is determined.

In a relaxed normal eye the four refracting media together bend light rays sufficiently to bring to a focus on the retina the parallel rays reflected from an object 20 or more feet (6 m) away. Of course a normal eye can also focus on objects located much nearer than 20 feet (6 m) from the eye. This is accomplished by a mechanism known as *accommodation* (discussed next). Many eyes, however, show errors of refraction; that is, they are not able to focus the rays on the retina under the stated conditions. Some common errors of refraction that are discussed later in the chapter are nearsightedness (myopia), farsightedness (hyperopia), and astigmatism.

Accommodation of Lens

Accommodation for near vision necessitates three changes: increase in the curvature of the lens, constriction of the pupils, and convergence of the two eyes. Light rays from objects 20 or more feet away are practically parallel. The normal eye, as previously noted, refracts such rays sufficiently to focus them clearly on the retina. However, light rays from nearer objects are divergent rather than parallel. So obviously they must be bent more acutely to bring them to a focus on the retina. Accommodation of the lens or, in other words, an increase in its curvature takes place to achieve this greater refraction.

Contraction or Relaxation of the Ciliary Muscle Affecting Lens Shape

Contraction pulls the choroid layer closer to the lens (see Figure 15-15). This, in turn, loosens the tension of the suspensory ligaments, allowing the lens to bulge. For near vision, then, the ciliary muscle is contracted and the lens is bulging, whereas for far vision the ciliary muscle is relaxed and the lens is comparatively flat (Figure 15-22). Continual use of the eyes for near work produces eyestrain because of the prolonged contraction of the ciliary muscle. Some of the strain can be avoided by looking into the distance at intervals while doing close work.

As people grow older, they tend to become farsighted because lenses lose their elasticity and therefore their ability to bulge and to accommodate for near vision. This condition is called *presbyopia*.

Constriction of Pupil

The muscles of the iris play an important part in the formation of clear retinal images. Part of the accommodation mechanism consists of contraction of the circular fibers of the iris, which constricts the pupil. This prevents divergent rays from the object from

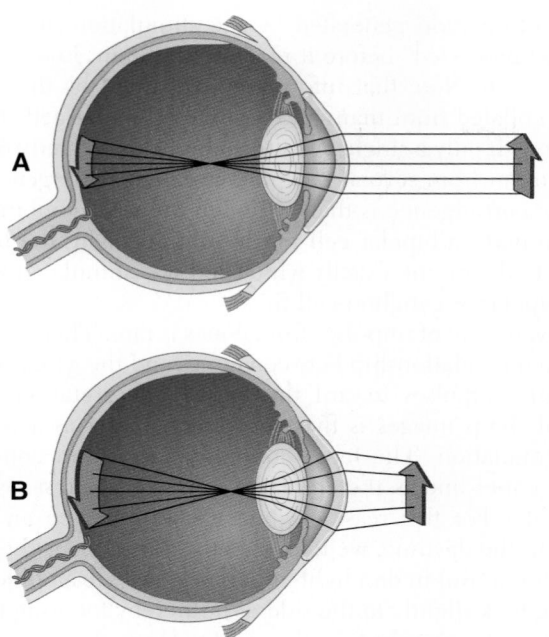

Figure 15-22 *Accommodation of lens.* **A,** Distant image: the lens is flattened (ciliary muscle relaxed), and the image is focused on the retina. **B,** Close image: the lens is rounded (ciliary muscle contracted), and the image is focused on the retina.

BOX 15-6: FYI
Visual Acuity

Visual *acuity* is the clearness or sharpness of visual perception. Acuity is affected by our focusing ability, the efficiency of the retina, and the proper function of the visual pathway and processing centers in the brain.

One common way to measure visual acuity is to use the familiar test chart on which letters or other objects of various sizes and shapes are printed. The subject is asked to identify the smallest object that he or she can see from a distance of 20 feet (6.1 m). The resulting determination of visual acuity is expressed as a double number such as "20/20." The first number represents the distance (in feet) between the subject and the test chart; the standard is 20. The second number represents the number of feet a person with normal acuity would have to stand to see the same objects clearly. Thus a finding of 20/20 is normal because the subject can see at 20 feet what a person with normal acuity can see at 20 feet. A person with 20/100 vision can see objects at 20 feet that a person with normal vision can see at 100 feet.

People whose acuity is worse than 20/200 after correction are considered to be legally blind. Legal blindness is a designation that is used to identify the severity of a wide variety of visual disorders so that laws that involve visual acuity can be enforced. For example, laws that govern the awarding of driving licenses require that drivers have a minimum level of visual acuity.

---

entering the eye through the periphery of the cornea and lens. Such peripheral rays could not be refracted sufficiently to be brought to a focus on the retina and therefore would cause a blurred image. Constriction of the pupil for near vision is called the *near reflex* of the pupil and occurs simultaneously with accommodation of the lens in near vision. The pupil constricts also in bright light (*photopupil reflex* or *pupillary light reflex*) to protect the retina from stimulation that is too intense or too sudden.

Convergence of Eyes

Single binocular vision (seeing only one object instead of two when both eyes are used) occurs when light rays from an object fall on corresponding points of the two retinas. The foveae and all points lying equidistant and in the same direction from the foveae are corresponding points. Whenever the eyeballs move in unison, either with the visual axes parallel (for far objects) or converging on a common point (for near objects), light rays strike corresponding points of the two retinas. *Convergence* is the movement of the two eyeballs inward so that their visual axes come together, or converge, at the object viewed. The nearer the object is, the greater the degree of convergence necessary to maintain single vision. A simple procedure demonstrates the fact that single binocular vision results from stimulation of corresponding points on two retinas. Gently press one eyeball out of line while viewing an object. Instead of one object, you will see two. To achieve unified movement of the two eyeballs, a functional balance between the antagonistic extrinsic muscles must exist. For clear distant vision, the muscles must hold the visual axes of the two eyes parallel. For clear near vision, they must converge them. These conditions cannot be met if, for exam-

ple, the medial rectus muscle of one eye contracts more forcefully than its antagonist, the lateral rectus muscle. That eye is then pulled in toward the nose. The movement of its visual axis does not coordinate with that of the other eye. Light rays from an object then fall on noncorresponding points of the two retinas, and the object is seen double (**diplopia**). **Strabismus** (cross-eye or squint) is an exaggerated condition that cannot be overcome by neuromuscular effort (Figure 15-23). An individual with strabismus usually does not have double vision, as you might expect, because he or she learns to suppress one of the images.

The Role of Photopigments

Both rods and cones contain *photopigments*, or light-sensitive pigmented compounds that are found in the outer (distal) area of both types of photoreceptors near the pigmented retina (see

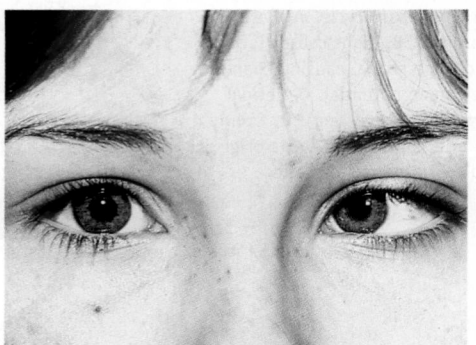

Figure 15-23 *Strabismus.* This child exhibits convergent left eye strabismus.

Figure 15-16). Chemically, all photopigments can be broken down into a glycoprotein called **opsin** and a vitamin A derivative called **retinal,** which acts as the light-absorbing portion of all photopigments.

Rods

The single photopigment found in rods is named **rhodopsin.** Rhodopsin is so highly light sensitive that even dim light causes a rapid breakdown of the photopigment into its opsin and retinal components. Light causes retinal to change its shape and the opsin molecule to expand, or open. When opsin and retinal open and separate in the presence of light (a process called **bleaching**), active sites are exposed and an action potential is created in the rod cell (Figure 15-24). This signal then travels to the brain for interpretation. Objects are seen in shades of gray but not in colors. Energy is required to bring opsin back to its original shape and reattach retinal to it. Until this occurs, the photopigment is unable to respond to light.

Cones

Three types of cones are present in the retina. Each contains a photopigment different from the rhodopsin found in rod cells. Each of the three primary colors (red, green, blue) reflects light rays of a different wavelength. Each wavelength acts primarily on one type of cone, causing its particular photopigment to break down and initiate impulse conduction by the cone. Our perception of a range of colors results from the combined neural input from varying numbers of the three different cone types.

Because cone photopigments are less sensitive to light than rhodopsin, brighter light is necessary for their breakdown. Cones therefore function to produce vision in bright light. In addition, cones contribute more than rods to the perception of sharp images. The reason for this difference involves the way in which information generated by the stimulation of rods and cones is "processed" before it reaches the brain. Look again at Figure 15-16. Note that information obtained by the bipolar cells is collated from many rods, whereas bipolar cells tend to synapse with only a single cone receptor. This property of combining input from several receptors is called *convergence*. The result of convergence is that, although the rods combine their input to make a bipolar cell fire in dimmer light, the brain is unable to determine exactly which rod was stimulated when a given bipolar or ganglion cell fires.

Convergence of impulses from cones is rare. There is almost a one-to-one relationship between cones and the ganglion cells that carry impulses toward the brain. Interpretation by the brain of sharp images is therefore much better as a result of cone stimulation. The fovea contains the greatest concentration of cones and is therefore the point of clearest vision in good light. For this reason, when we want to see an object clearly in the daytime, we look directly at it to focus the image on the fovea. But in dim light or darkness, we see an object better if we look slightly to the side of it, thereby focusing the image nearer the periphery of the retina, where the more plentiful rods can collate the lesser amount of light information and generate an image.

Neuronal Pathway of Vision

Fibers that conduct impulses from the rods and cones reach the visual cortex in the occipital lobes by way of the optic nerves, optic chiasma, optic tracts, and optic radiations. "Relay stations" along the way include the superior colliculi and the lateral geniculate nuclei of the thalamus. Look closely at Figure 15-25. Note that each optic nerve contains fibers from only one retina but that the optic chiasma contains fibers from the nasal portions of both retinas. Each optic tract also contains fibers from both retinas. These anatomical facts explain certain peculiar visual abnormal-

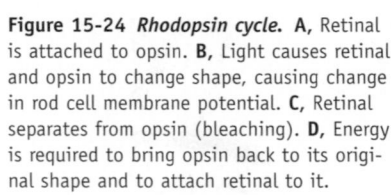

Figure 15-24 *Rhodopsin cycle.* A, Retinal is attached to opsin. **B,** Light causes retinal and opsin to change shape, causing change in rod cell membrane potential. **C,** Retinal separates from opsin (bleaching). **D,** Energy is required to bring opsin back to its original shape and to attach retinal to it.

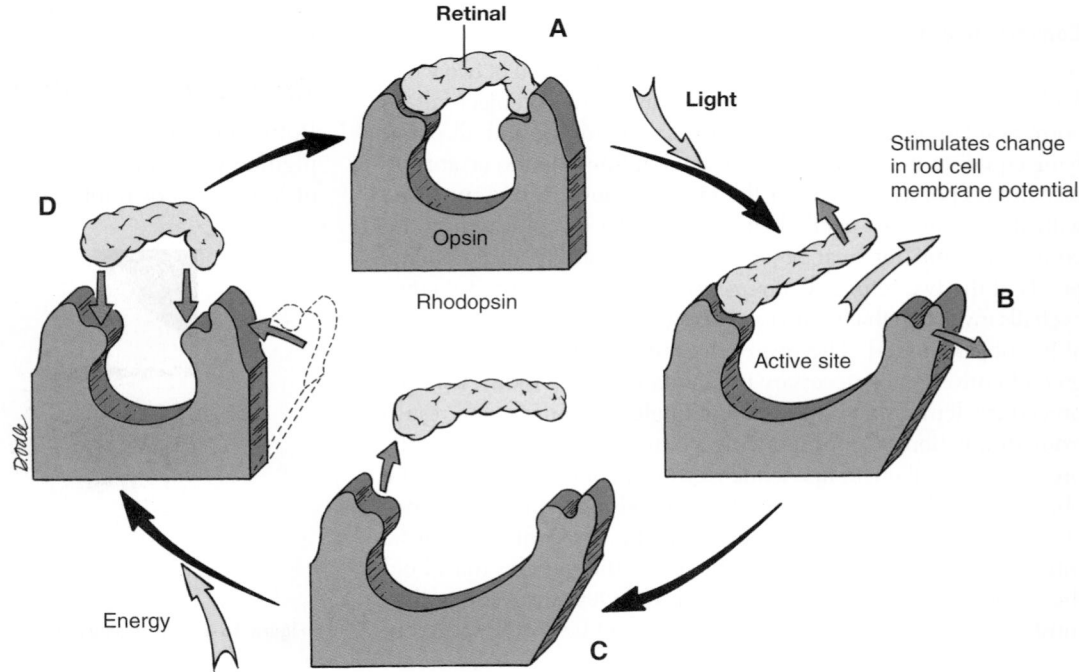

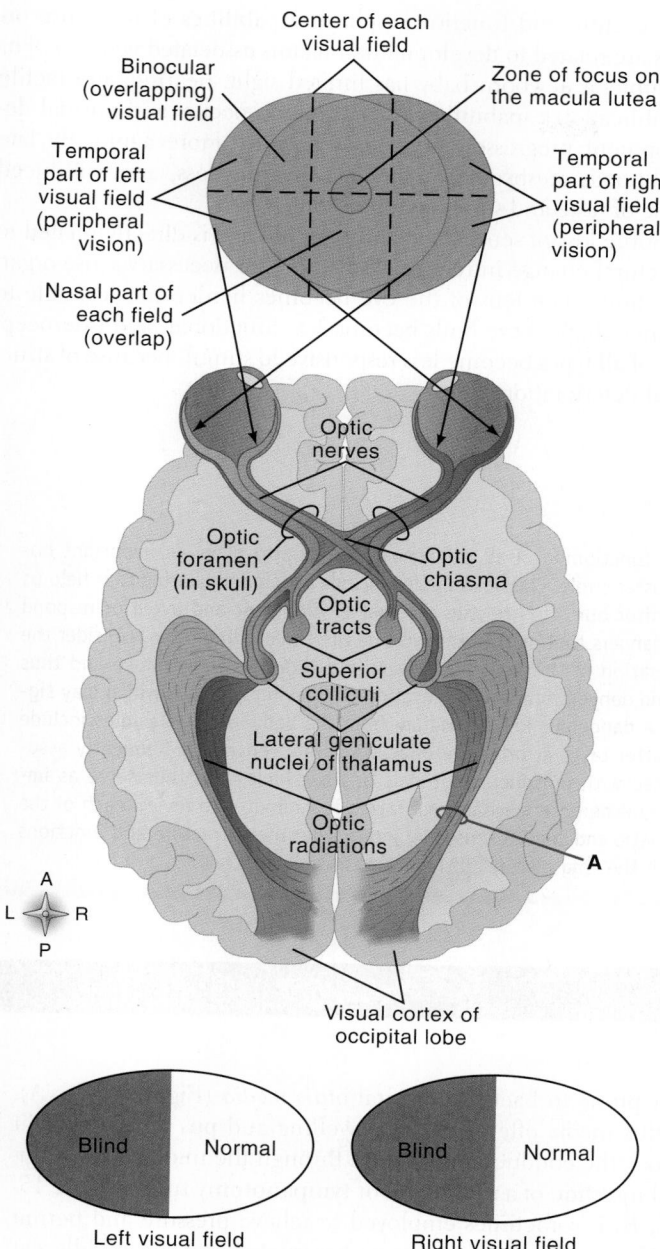

Figure 15-25 *Visual fields and neuronal pathways of the eye.* Note the structures that make up each pathway: optic nerve, optic chiasma, lateral geniculate body of thalamus, optic radiations, and visual cortex of occipital lobe. Fibers from the nasal portion of each retina cross over to opposite side at the optic chiasma and terminate in the lateral geniculate nuclei. Location of a lesion in the visual pathway determines the resulting visual defect. Damage at point *A,* for example, would cause blindness in the right nasal and left temporal visual fields, as the ovals beneath indicate. (Trace the visual pathway from point *A* back to the visual field map to see why this is so.) What would be the effect of pressure on the optic chiasma—by a pituitary tumor, for instance? (Answer: It would produce blindness in both temporal visual fields. Why? Because it destroys fibers from the nasal side of both retinas.)

ities that sometimes occur. Suppose a person's right optic tract were injured so that it could not conduct impulses—say, at point A in Figure 15-25. This person would be totally blind in neither eye but partially blind in both eyes. Specifically, this person would be blind in the right nasal and left temporal visual fields.

BOX 15-7: FYI
Color Blindness

Color blindness, usually an inherited condition, is caused by mistakes in producing three chemicals, called *photopigments,* in the cones. Each photopigment is sensitive to one of the three primary colors of light: green, blue, and red. In many cases, the green-sensitive photopigment is missing or deficient; other times, the red-sensitive photopigment is abnormal. (Deficiency of the blue-sensitive photopigment is rare.) Color-blind individuals see colors, but they cannot distinguish between them normally.

Figures such as parts *A* and *B* shown here are often used to screen individuals for color blindness. A person with red-green color blindness cannot see the *74* in part *A* of the figure, whereas a person with normal vision can. To determine which photopigment is deficient, a color-blind person may try a figure similar to part *B.* Persons with a deficiency of red-sensitive photopigment can distinguish only the *2;* those deficient in green-sensitive photopigment can only see the *4.*

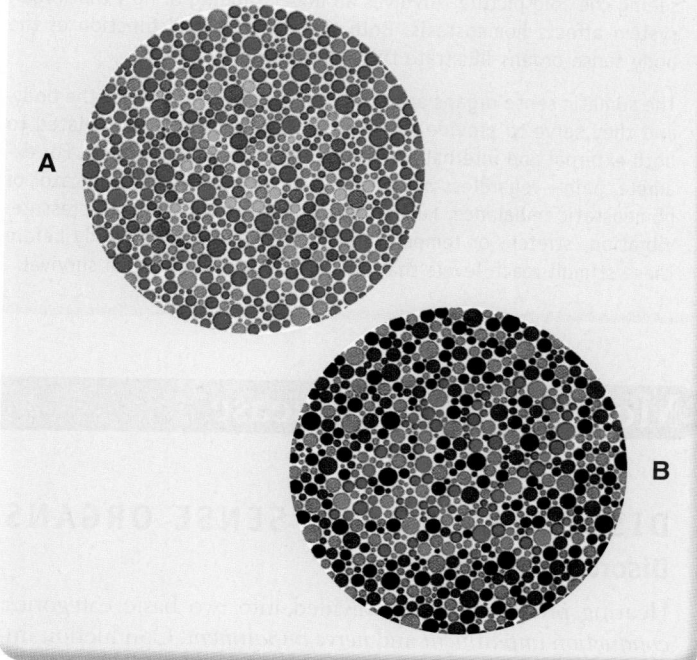

Here are the reasons: the right optic tract contains fibers from the right retina's temporal area, the area that sees the right nasal visual field. In addition, the right optic tract contains fibers from the left retina's nasal area, the area that sees the left temporal visual field.

QUICK CHECK

15. Name the layers, or coats, of the eyeball.
16. Identify the layers of the retina.
17. Name the four processes that function to focus a clear image on the retina.
18. Outline the steps of the rhodopsin cycle.

Cycle of Life
Sense Organs

The ability of the sense organs to respond to stimuli caused by changes in the body's internal or external environment varies during life. Ultimately, all sensory information is acquired through depolarization of sensory nerve endings. Anything that interferes with the generation of a receptor potential or its transmission to and interpretation by areas of the central nervous system influences sensory acuity. Age, disease, structural defects, and lack of maturation all affect our ability to identify and respond to sensory input.

Structure and function response capabilities of the sense organs are related to developmental factors associated with age. For example, a newborn baby has limited sight, hearing, and tactile identification capabilities. As maturation occurs and normal development progresses, the senses become more acute. By late adulthood, presbyopia, progressive hearing loss, and a reduced sense of taste and smell are common.

Some loss of sensory capability in old age is directly related to structural change in receptor cells or other necessary sense organ structures. The lens of the eye becomes harder and less able to change shape, taste buds become less functional, and exteroceptors of all types become less responsive to stimuli because of structural deterioration.

THE BIG PICTURE
Sense Organs

Almost invariably, as you study the various organ systems of the body, seeing the "big picture" involves an understanding of how that organ system affects homeostasis. Both the structure and function of the body sense organs illustrate this relationship.

The somatic sense organs are widely distributed throughout the body, and they serve to provide the body with vital information related to both external and internal conditions that affect homeostasis. For example, pain—regardless of cause—is very often the first indicator of homeostatic imbalance. Further, the ability to sense touch, pressure, vibration, stretch, or temperature changes on or in the body before these stimuli reach levels that may cause injury is vital to survival.

The functioning of all the special senses also plays an important homeostatic role. Classic examples include vision and hearing that help us monitor our often hostile external environment and avoid or respond to dangers that might otherwise be life threatening. Also, consider the sensation of thirst. It helps us to regulate our water intake and thus avoid dehydration or the sensation or "craving" for salt, which may signal a dangerous loss of sodium from the body. Other examples include a bitter taste or offensive odor, which are sensations frequently associated with poisonous materials or toxic fumes and thus serve as important defense mechanisms. Take a few minutes to review each of the somatic and special senses by integrating their structure and functions with the "big picture" of homeostasis and survival.

Mechanisms of Disease

DISORDERS OF THE SENSE ORGANS
Disorders of the Ear

Hearing problems can be divided into two basic categories: *conduction impairment* and *nerve impairment*. Conduction impairment refers to the blocking of sound waves as they are conducted through the external and middle ear to the sensory receptors of the inner ear (the conduction pathway). Nerve impairment results in insensitivity to sound because of inherited or acquired nerve damage.

The most obvious cause of conduction impairment is blockage of the external auditory canal. Waxy buildup of cerumen (Figure 15-26, *C*) commonly blocks conduction of sound toward the tympanic membrane. Foreign objects (Figure 15-26, *D*), tumors, and other matter can block conduction in the external or middle ear. An inherited bone disorder called **otosclerosis** (oh-toh-skleh-ROH-sis) impairs conduction by causing structural irregularities in the stapes. Otosclerosis usually first appears during childhood or early adulthood as **tinnitus** (tih-NYE-tus), or "ringing in the ear."

Temporary conduction impairment often results from ear infection, or **otitis**. The structure of the auditory tube, especially its connection with the nasopharynx, makes the middle ear prone to bacterial or viral *otitis media* (Figure 15-26, *A*). Otitis media often produces swelling and pus formation that block the conduction of sound through the middle ear. Surgical insertion of a *ventilation* or **tympanotomy tube** (Figure 15-26, *B*) is sometimes employed to relieve pressure and permit drainage. Permanent damage to structures of the middle ear occasionally occurs in severe cases.

Hearing loss because of nerve impairment is common in the elderly. Called **presbycusis** (pres-bih-KYOO-sis), this progressive hearing loss associated with aging results from degeneration of nerve tissue in the ear and the vestibulocochlear nerve. A similar type of hearing loss occurs after chronic exposure to loud noises that damages receptors in the organ of Corti. Because different sound *frequencies* (tones) stimulate different regions of the organ of Corti, hearing impairment is limited to only those frequencies associated with the portion of the organ of Corti that is damaged. For example, the portion of the organ of Corti that degenerates first in presbycusis is normally stimulated by high-frequency sounds. Thus the inability to hear high-pitched sounds is common among older adults.

Two small muscles, the *tensor tympani* and *stapedius*, help prevent damage to hearing caused by prolonged loud noise. The tensor tympani attaches to and limits movement of the eardrum,

Mechanisms of Disease—cont.

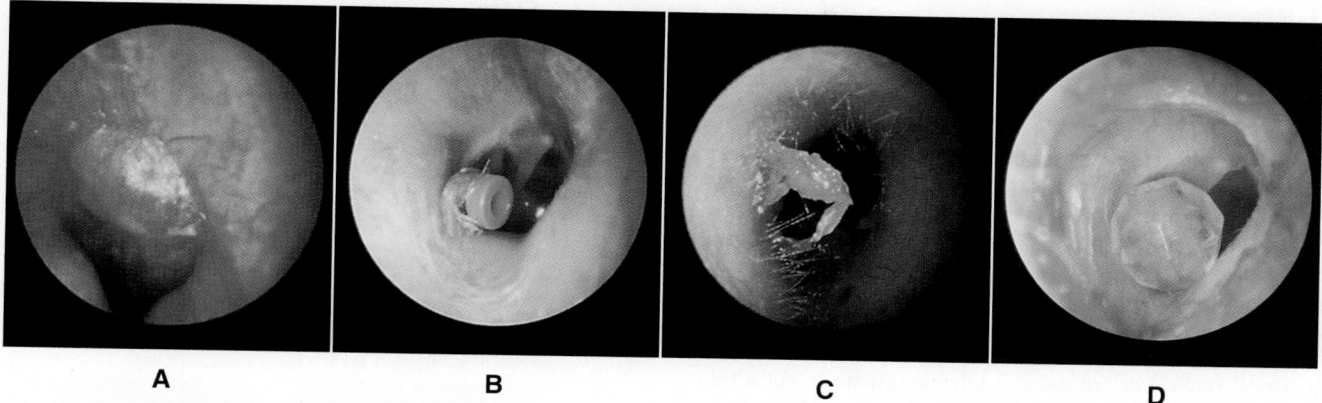

Figure 15-26 *Disorders of the ear.* **A,** Acute otitis media. Note the red, thickened, and bulging tympanic membrane. **B,** Ventilation or tympanotomy tube. Inserted to relieve pressure and permit drainage in otitis media. **C,** Cerumen in ear canal. **D,** Foreign object (plastic bead) in ear canal of a small child.

thus preventing excess displacement caused by prolonged loud sounds. In addition, the smallest of all body muscles, the stapedius, limits excess movement of the stapes and thus protects the oval window from prolonged noise-related damage. Only protective devices, such as earplugs, can protect hearing from damage caused by very sudden loud noises, such as a gunshot.

Nerve damage can also occur in **Ménière** (men-ee-AIR) **disease,** a chronic inner ear disease of unknown cause. Ménière disease is characterized by tinnitus, progressive nerve deafness, and **vertigo** (sensation of spinning).

Disorders of the Eye

Healthy vision requires three basic processes: formation of an image on the retina (refraction), stimulation of rods and cones, and conduction of nerve impulses to the brain. Malfunction of any of these processes can disrupt normal vision.

Refraction Disorders

Focusing a clear image on the retina is essential for good vision. In the normal eye, light rays enter the eye and are focused into a clear, upside-down image on the retina (Figure 15-27, A).

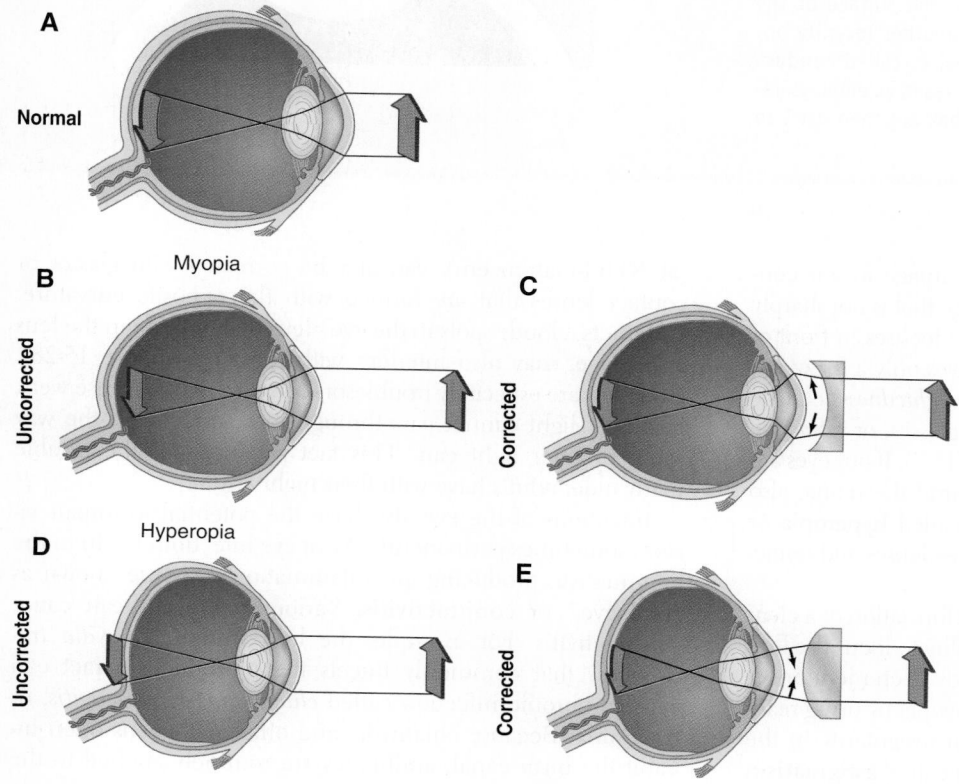

Figure 15-27 *Correcting refraction disorders.* **A,** Normal eye structure results in a well-focused retinal image. **B,** In myopia, an elongated eyeball causes the image to focus in front of the retina. **C,** An external lens can refocus the image on the retina, correcting the effects of myopia. **D,** In hyperopia, a flattened eyeball causes the image to focus behind the retina. **E,** An external lens can refocus the image on the retina.

Mechanisms of Disease—cont.

BOX 15-8: HEALTH MATTERS
Refractive Eye Surgery

A surgery technique to treat myopia (nearsightedness) without the use of eyeglasses or contact lenses became available almost 25 years ago. The procedure, called **radial keratotomy** (RK), involves surgical placement of six or more radial slits (incisions) in a spokelike pattern around the cornea. As a result, the cornea flattens and the ability to focus improves. Other incisional types of refractory eye surgery included astigmatic keratotomy (AK), which involves treatment of astigmatism by placement of transverse cuts across the corneal surface, and automated lamellar keratoplasty (ALK). The ALK technique uses a special surgical device called a **microkeratome** to cut a thin cap off the corneal surface and then shave and reshape the underlying tissue. At the end of the procedure the corneal cap is replaced and heals without the need for sutures. ALK is used to treat both myopia and hyperopia (farsightedness).

More recent advances in refractory eye surgery involve the use of surgical lasers. Excimer laser surgery, also called **photorefractive keratectomy** (PRK), uses a "cool" excimer laser beam to vaporize corneal tissue. It is used to flatten the cornea to correct mild to moderate nearsightedness. A recent refractive eye surgery procedure to correct myopia is called **laser-assisted in situ keratomileusis** *(LASIK)*. This procedure employs both PRK and ALK techniques. First a microkeratome is used to create a hinged cap of tissue, which is lifted off the corneal surface (see part *A* of the figure). An excimer laser is then used to vaporize and reshape the underlying tissue *(B)*. At the end of the procedure, the cap is replaced *(C)*. Another laser surgery recently approved by the U.S. FDA for treating hyperopia (farsightedness) is laser thermal keratoplasty (LTK). Ultra-short bursts of laser energy (lasting 3 seconds) are used to reshape the surface of the cornea, with no surgical cutting involved. Yet another recently approved treatment for correction of farsightedness is called **conductive keratoplasty (CK).** Instead of a scalpel or laser, it employs radiofrequency energy to heat hair-thin probes that are then used to change the shape of the cornea.

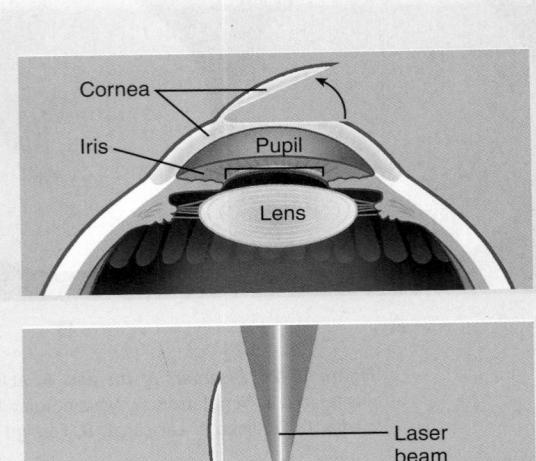

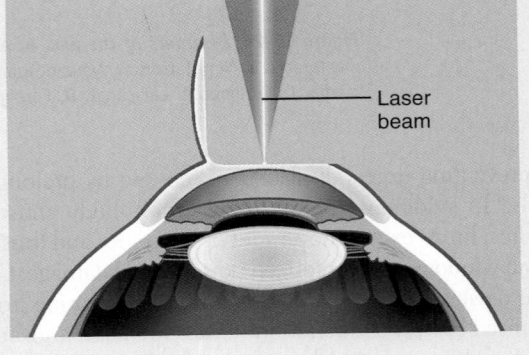

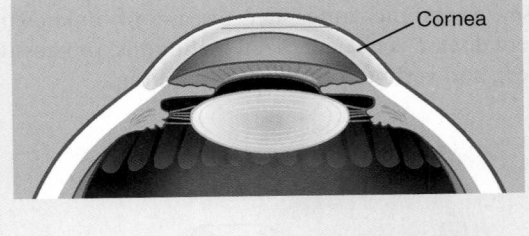

The brain can easily right the upside-down image in our conscious perception but cannot correct an image that is not sharply focused. If our eyes are elongated, the image focuses in front of the retina rather than on it. The retina receives only a fuzzy image. This condition, called **myopia** or *nearsightedness*, can be corrected by using concave contact lenses, glasses, or refractive eye surgery (Figure 15-27, *B* and *C*, and Box 15-8). If our eyes are shorter than normal, the image focuses behind the retina, also producing a fuzzy image. This condition, called **hyperopia** or *farsightedness*, can also be corrected by convex lenses and refractive eye surgery (Figure 15-27, *D* and *E*).

Various other conditions can prevent the formation of a clear image on the retina. For example, the inability to focus the lens properly as we age, or **presbyopia,** has already been mentioned. Older individuals can compensate for presbyopia by using reading glasses when near vision is needed. An irregularity in the curvature of the cornea or lens, a condition called **astigmatism** (ah-STIG-mah-tiz-em), can also be corrected with glasses or contact lenses that are formed with the opposite curvature. **Cataracts**, cloudy spots in the eye's lens that develop in the lens as we age, may also interfere with focusing (Figure 15-28). Cataracts are especially troublesome in dim light because weak beams of light cannot pass through the cloudy spots the way some brighter light can. This fact accounts for the trouble many older adults have with their night vision.

Infections of the eye also have the potential to impair vision, sometimes permanently. Most eye infections begin in the conjunctiva, producing an inflammation response known as "pink-eye," or **conjunctivitis.** Various pathogens can cause conjunctivitis. For example, the bacterium *Chlamydia trachomatis* that commonly infects the reproductive tract can cause a chronic infection called *chlamydial conjunctivitis*, or **trachoma.** Because chlamydia and other pathogens often inhabit the birth canal, antibiotics are routinely applied to the

Mechanisms of Disease—cont.

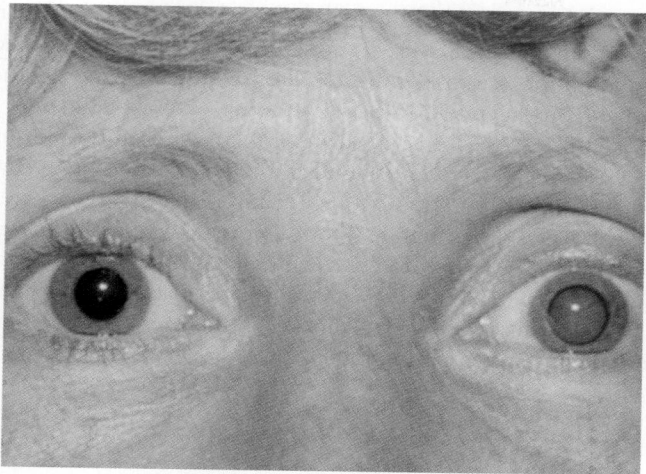

Figure 15-28 *Cataract.* Note the prominent cataract of the left eye.

eyes of newborns to prevent conjunctivitis. Highly contagious *acute bacterial conjunctivitis,* characterized by drainage of a mucous pus, is most commonly caused by bacteria such as *Staphylococcus* and *Haemophilus.* Conjunctivitis may produce lesions on the inside of the eyelid that can damage the cornea and thus impair vision. Occasionally infections of the conjunctiva spread to the tissues of the eye proper and cause permanent injury—even total blindness. Besides infection, conjunctivitis may also be caused by allergies. The red, itchy, watery eyes commonly associated with allergic reactions to pollen and other substances result from an allergic inflammatory response of the conjunctiva.

Disorders of the Retina

Damage to the retina impairs vision because even a well-focused image cannot be perceived if some or all of the light receptors do not function properly. For example, in a condition called **retinal detachment,** part of the retina falls away from the tissue supporting it (Figure 15-29, *A*). This condition may result from aging, eye tumors, or blows to the head—as in

a sporting injury. Common warning signs include the sudden appearance of floating spots that may decrease over a period of weeks and odd "flashes of light" that appear when the eye moves. If left untreated, the retina may detach completely and cause total blindness in the affected eye.

Diabetes mellitus, a disorder involving the hormone insulin, may cause a condition known as **diabetic retinopathy** (ret-in-AH-path-ee). In this disorder the diabetes causes small hemorrhages in retinal blood vessels that disrupt the oxygen supply to the photoreceptors (Figure 15-29, *B*). The eye responds by building new, but abnormal, vessels that block vision and may cause detachment of the retina. Diabetic retinopathy is one of the leading causes of blindness in the United States. Diabetic retinopathy and other complications of diabetes are often associated with high blood pressure or *hypertension,* which also causes retinal hemorrhages (Figure 15-29, *C*).

Another condition that can damage the retina is **glaucoma.** Recall that glaucoma is excessive *intraocular* (in-trah-AHK-yoo-lar) *pressure* caused by abnormal accumulation of aqueous humor. As fluid pressure against the retina increases above normal, blood flow through the retina slows. Reduced blood flow causes degeneration of the retina and thus leads to loss of vision. Although acute forms of glaucoma can occur, most cases of glaucoma develop slowly over a period of years. This chronic form may not produce any symptoms, especially in its early stages. For this reason, routine eye examinations typically include a screening test for glaucoma. As chronic glaucoma progresses, damage first appears at the edges of the retina—causing a gradual loss of peripheral vision. Blurred vision and headaches may also occur. As the damage becomes more extensive, "halos" are seen around bright lights. If untreated, glaucoma eventually produces total, permanent blindness. One of the most common characteristic early retinal changes associated with glaucoma is swelling or "cupping" of the optic disc (Figure 15-29, *D*).

Degeneration of the retina can cause difficulty seeing at night or in dim light. This condition, called **nyctalopia** (nik-tah-LOH-pee-ah), or "night blindness," can also be caused by a deficiency of vitamin A. Recall that vitamin A is needed to

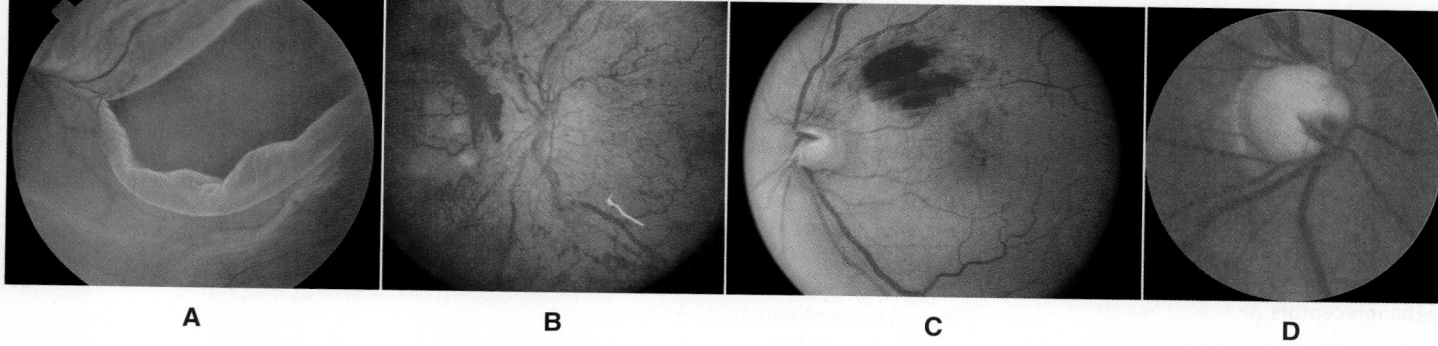

A   B   C   D

Figure 15-29 *Retinal pathological conditions.* **A,** Retina tear and detachment. **B,** Diabetic retinopathy. Note the abnormal blood vessels and hemorrhages in the retina caused by diabetes. **C,** Hypertensive retinopathy. Large "flame hemorrhages" in the retina are associated with high blood pressure. **D,** Glaucoma. Note "cupping" of the optic disc caused by increased intraocular pressure.

Mechanisms of Disease—cont.

make retinal, a component of rhodopsin. A deficiency of rhodopsin impairs the function of rod cells, which are needed for dim light vision.

Disorders of the Visual Pathway

Damage or degeneration in the optic nerve, the brain, or any part of the visual pathway between them, can impair vision. For example, the pressure associated with glaucoma can also damage the optic nerve. Diabetes, already cited as a cause of retina damage, can also cause degeneration of the optic nerve.

Damage to the visual pathway does not always result in total loss of sight. Depending on where the damage occurs, only a part of the visual field may be affected. For example, a certain form of neuritis (nerve inflammation), often associated with multiple sclerosis, can cause loss of only the center of the visual field—a condition called **scotoma** (skoh-TOH-mah).

A stroke can cause vision impairment when the resulting tissue damage occurs in one of the regions of the brain that processes visual information. For example, damage to an area that processes information about colors may result in a rare condition called **acquired cortical color blindness.** This condition is characterized by difficulty in distinguishing any color—not just one or two colors as in the more common inherited forms of color blindness.

LANGUAGE OF SCIENCE (Cont'd from page 553)

extrinsic eye muscles (eks-TRIN-sik) [extr- outside or beyond, -insic beside]

fast (A) pain

fovea centralis (FOH-vee-ah sen-TRAL-is) [fovea pit, centralis center]

G protein receptor sites

gamma motor neurons (GAM-mah MOH-tor NOO-roons)

general sense organs

Golgi tendon organs (GOL-jee) [Camillo Golgi Italian histologist, tendon to stretch]

Golgi tendon receptors (GOL-jee) [Camillo Golgi Italian histologist, tendon to stretch, ceptor to receive]

Golgi tendon reflex (GOL-jee) [Camillo Golgi Italian histologist, tendon to stretch, reflectere to bend back]

granule cells (GRAN-yool) [granulum little grain]

gustatory (GUS-tah-tor-ee) [gustare to taste]

gustatory cells (GUS-tah-tor-ee sells) [gustare to taste, cella storeroom]

inner ear

interoceptors (in-ter-oh-SEP-tors) [internus- inward, -capere to take]

intrinsic eye muscles (in-TRIN-sik) [intr- inside or within, -insic beside]

ion channels (EYE-on)

iris [iris rainbow]

Krause end bulbs (KROWS) [Wilhelm J. F. Krause German anatomist]

labyrinth (LAB-i-rinth) [labyrinth maze]

macula (MAK-yoo-lah) [macula spot]

macula lutea (MAK-yoo-lah LOO-tee-ah) [macula spot, lute yellow]

mechanoreceptors (mek-an-oh-ree-SEP-tors) [mechano- mechanical, -ceptor to receive]

Merkel discs (MER-kuhl) [Friedrich Sigmund Merkel German anatomist]

middle ear

muscle spindles

nociceptors (noh-see-SEP-tors) [noci- to cause harm, -ceptor to receive]

olfactory (ohl-FAK-tor-ee) [olfacere to smell]

olfactory receptor neuron (ol-FAK-tor-ee ree-SEP-tor NOO-ron)

opsin (OP-sin)

optic nerve (OP-tik) [optic sight]

organ of Corti (OR-gan of KOR-tee) [Alfonso Corti Italian anatomist]

osmoreceptors (os-moh-ree-SEP-tors) [osmo-impulse or osmosis, -ceptor to receive]

otoliths (OH-toh-liths) [oto- ear, -lithos stone]

oval window

pacinian (lamellar) corpuscles (pah-SIN-ee-an KOHR-pus-uls) [Filippo Pacini Italian anatomist]

papillae (pah-PIL-ee) [papillae nipple]

perception

perilymph (PAIR-i-limf) [peri- around, -lympha water]

photoreceptors (FOH-toh-ree-sep-tors) [photo-light, -ceptor to receive]

pitch

posterior cavity (pohs-TEER-ee-or KAV-i-tee) [posterior behind, cavus hollow space]

proprioceptors (proh-pree-oh-SEP-tors) [proprius-one's own, -ceptor to receive]

pupil (PYOO-pill) [pupa doll]

receptor potential (ree-SEP-tor poh-TEN-shal) [ceptor to receive]

refraction (ree-FRAK-shun) [refringere to break apart]

Reissner membrane (RYZ-ner) [Ernst Reissner German anatomist]

retina (RET-i-nah) [rete net]

retinal (RET-i-nal) [retin- retina, -al pertaining to]

rhodopsin (roh-DOP-sin) [rhodo red]

rods

round window

Ruffini corpuscles (roo-FEE-nee KOR-pus-uls) [Angelo Ruffini Italian anatomist, corpusculum little body]

sensation [sentire to feel]

sensory receptors (SEN-soh-ree ree-SEP-tors) [sentire to feel, ceptor to receive]

slow (B) pain

somatic pain (so-MAH-tik) [soma body]

special senses

static equilibrium (STAT-ik ee-kwi-LIB-ree-um) [statikos causing to stand, equi- equal, -libra balance]

stretch reflex

taste buds

tectorial membrane (tek-TOH-ree-al) [tecto-rooflike, -al pertaining to, membrane thin skin]

thermoreceptors (ther-moh-ree-SEP-tors) [thermo-heat, -ceptor to receive]

tympanic membrane (tim-PAN-ik) [tympanum drum, membrane thin skin]

vestibular membrane (ves-TIB-yoo-lar) [vestibulum courtyard, membrane thin skin]

visceral pain (VISS-er-al) [viscus internal organs]

visceroceptors (viss-er-oh-SEP-tors) [viscero- organs of the body, -ceptor to receive]

volume

LANGUAGE OF MEDICINE

acquired cortical color blindness (ah-KWYERD KOHR-tih-kahl) [*cortic* cortex or bark]

astigmatism (ah-STIG-mah-tiz-em) [*a-* not, *-stigma-* without point, *-ism* condition of]

benign paroxysmal positional vertigo (BPPV) (be-NYNE par-ock-SIZ-mul poh-ZISH-i-nal VER-ti-go) [*benign* kind, *paroxysmos* irritation, *vertere* to turn]

blepharoplasty (blef-ar-oh-PLAS-tee) [*blepharo-* eyelid or eyelash, *-plasty* surgical repair]

canalith repositioning procedure (KAN-ih-lith) [*canal-* channel, *-lith* stone]

canaliths (KAN-ih-liths) [*canal-* channel, *-lith* stone]

cataracts (KAT-ah-rakts) [*katarrhakies* waterfall]

conductive keratoplasty (CK) (kon-DUK-tiv ker-ah-toh-PLAS-tee) [*kera-* horn, *-plasty* surgical repair]

conjunctivitis (kon-junk-ti-VYE-tis) [*conjunctivus-* connecting, *-itis* inflammation]

diabetic neuropathy (dye-ah-BET-ik nyoo-ROP-ah-thee) [*diabetes-* a syphon, *-ic* pertaining to, *neuro-* neuron, *-pathy* disease]

diabetic retinopathy (dye-ah-BET-ik ret-in-AH-path-ee) [*diabetes-* a syphon, *-ic* pertaining to, *retin-* retina, *-pathy* disease]

diplopia (di-PLOH-pee-ah) [*di-* double, *-opia* vision condition]

glaucoma (glaw-KOH-mah) [*glauco* gray or silver]

hyperopia (hye-per-OH-pee-ah) [*hyper-* excessive or above, *-opia* vision condition]

laser-assisted in situ keratomileusis (LASIK) (LAY-zer ah-SIS-ted in-SYE-too kair-at-oh-mill-YOO-sis)

Ménière disease (men-ee-AIR) [*Prosper Ménière* French physician]

microkeratome (my-kroh-KAR-ah-tohm) [*micro-* small, *-kera-* horn, *-tome* cutting instrument]

myopia (my-OH-pee-ah) [*myops-* nearsighted, *-opia* vision condition]

nyctalopia (nik-tah-LOH-pee-ah) [*nycto-* night or darkness, *-opia* vision condition]

ophthalmology (off-thal-MOL-eh-jee) [*ophthalmo-* eye, *-ology* the study or science of]

ophthalmoscope (off-THAL-mah-skohp) [*ophthalmo-* eye, *-scope* visual examination]

otitis (oh-TYE-tis) [*ot-* ear, *-itis* inflammation]

otitis media (oh-TYE-tis MEE-dee-ah) [*ot-* ear, *-itis* inflammation, *medi* middle]

otosclerosis (oh-toh-skleh-ROH-sis) [*oto-* ear, *-sclero-* hard, *-sis* condition]

otoscope (OH-toh-skohp) [*oto-* ear, *-scope* visual examination]

photorefractive keratectomy (FOH-toh-ree-frak-tiv kair-ah-TEK-toh-mee) [*photo-* light, *-refringere* to break apart, *kera-* hard, *-ectomy* surgical removal]

presbycusis (pres-bih-KYOO-sis) [*presby-* aging or being elderly, *-akousis* hearing]

presbyopia (pres-bee-OH-pee-ah) [*presby-* aging or being elderly, *-opia* vision condition]

radial keratotomy (RAY-dee-al KAR-ah-tah-toh-mee) [*radius* ray, *kera-* horn, *-tomy* surgical incision]

referred pain

retinal detachment (RET-ih-nal) [*retin-* retina, *-al* pertaining to]

scotoma (skoh-TOH-mah) [*scoto-* darkness, *-oma* tumor]

strabismus (strah-BIS-mus) [*strab-* squinting, *-ismus* condition]

tinnitus (tih-NYE-tus) [*tinnitis* a ringing or tinkling]

trachoma (trah-KOH-mah) [*trachoma* roughness]

two-point discrimination test

tympanotomy tube (tim-peh-NOT-eh-mee) [*tympan-* the tympanic membrane, *-tomy* surgical incision]

vertigo (VER-ti-go) [*vertigo* a turning around]

CASE STUDY

Adam Bassett, 50 years old, visits the health clinic for a complete physical examination. Mr. Bassett is very concerned because he has had periods of vertigo on and off for the past 2 days. He describes these episodes as occurring with sudden motion and then "everything just seems to spin around" for several minutes. He admits that these episodes are becoming more frequent.

Past medical history includes a recent ear infection that was treated with a 10-day course of the antibiotic amoxicillin while he was on vacation in Florida. He denies any current ear pain, headaches, fever, or vomiting but says his ear still feels "full." He has noticed that he has blurred vision, especially when reading the newspaper. He does not wear corrective lenses, and his last eye examination was more than 10 years ago. Mrs. Bassett accompanied her husband to the examination and says she feels her husband does not hear as well as he once did because she finds she often has to repeat things to him. Mr. Bassett denies any difference in his hearing or tinnitus.

Family history is negative for diabetes, heart disease, and cancer.

1. Identify one of the following as an *incorrect* method of physical examination of the ear for Mr. Bassett:

   A. Pull the auricle down and back to straighten the canal and visualize the tympanic membrane.
   B. Inspect the external ear by looking at the auricle or pinna and the ear canal.
   C. Attempt to visualize blocking cerumen of the canal by use of an otoscope.
   D. Move the auricle up and down to identify pain in the external ear.

2. Physical examination reveals that the external canal is free of cerumen, that the tympanic membrane is dull and inflamed, and that the landmarks of the middle ear are not clearly visible. It is decided that Mr. Bassett has had otitis media that has failed to completely resolve. Which one of the following hearing problems would you expect Mr. Bassett to currently have based on this information?

   A. Nerve impairment
   B. Conductive impairment
   C. Presbycusis
   D. Ménière disease

3. Based on the information provided, what is the *most likely* diagnosis for Mr. Bassett's vertigo?

   A. Benign paroxysmal positional vertigo
   B. Presbycusis
   C. Otitis media
   D. Myopia

4. The vision screening has determined that Mr. Bassett has become farsighted. What is the best explanation for this new development?

   A. As people get older, muscles have difficulty holding the visual axes parallel.
   B. As people get older, the ciliary muscle is more contracted, thus causing the lens to bulge.
   C. As people get older, the lenses lose their elasticity and their ability to bulge.
   D. As people get older, the ability to constrict the pupil when focusing on near objects is lost.

5. In addition to his farsightedness, Mr. Bassett scores 20/40 on the Snellen vision chart. Which one of the following is the best description of this finding?

   A. Mr. Bassett is legally blind.
   B. Mr. Bassett sees at 20 feet what others see at 40 feet.
   C. Mr. Bassett sees at 40 feet what others see at 20 feet.
   D. Mr. Bassett has developed hyperopia.

CHAPTER SUMMARY

SENSORY RECEPTORS

A. Sensory receptors make it possible for the body to respond to stimuli caused by changes occurring in our internal or external environment
B. Receptor response
   1. General function—responds to stimuli by converting them to nerve impulses
   2. Different types of receptors respond to different stimuli
   3. Receptor potential
      a. The potential that develops when an adequate stimulus acts on a receptor; it is a graded response
      b. When a threshold is reached, an action potential in the sensory neuron's axon is triggered
      c. Impulses travel over sensory pathways to the brain and spinal cord where either they are interpreted as a particular sensation or they initiate a reflex action
   4. Adaptation—a functional characteristic of receptors; receptor potential decreases over time in response to a continuous stimulus, which leads to a decreased rate of impulse conduction and a decreased intensity of sensation
C. Distributions of receptors
   1. Receptors for special senses of smell, taste, vision, hearing, and equilibrium are grouped into localized areas or into complex organs
   2. General sense organs of somatic senses are microscopic receptors widely distributed throughout the body in the skin, mucosa, connective tissue, muscles, tendons, joints, and viscera

CLASSIFICATION OF RECEPTORS BY: (TABLE 15-1)

   1. Location
   2. Type of stimulus that causes response

A. Classification by location
1. Exteroceptors
a. On or near body surface
b. Often called *cutaneous receptors*; examples: pressure, touch, pain, temperature
2. Visceroceptors (interoceptors)
a. Located internally—often within body organs, or viscera
b. Provide body with information about internal environment; examples: pressure, stretch, chemical changes, hunger, thirst
3. Proprioceptors: specialized type of visceroceptor
a. Location limited to skeletal muscle, joint capsules, and tendons
b. Provide information on body movement, orientation in space, and muscle stretch
c. Two types: tonic and phasic receptors provide positional information on body or body parts while at rest or during movement
B. Classification by stimulus detected
1. Mechanoreceptors—activated when "deformed" to generate receptor potential
2. Chemoreceptors—activated by amount or changing concentration of certain chemicals, e.g., taste and smell
3. Thermoreceptors—activated by changes in temperature
4. Nociceptors—activated by intense stimuli that may damage tissue; sensation produced in pain
5. Photoreceptors—found only in the eye; respond to light stimuli if the intensity is great enough to generate a receptor potential
6. Osmoreceptors—concentrated in the hypothalamus; activated by changes in concentration of electrolytes (osmolarity) in extracellular fluids.
C. Classification by structure (Figure 15-2): divides sensory receptors into either those with free nerve endings or those with encapsulated nerve endings
1. Free nerve endings
a. Most widely distributed sensory receptor
b. Include both exteroceptors and visceroceptors
c. Called nociceptors—primary receptors for pain
d. Other sensations mediated: itching, tickling, touch, movement, mechanical stretching
e. Primary receptors for heat and cold
f. Two types of nerve fibers carry pain impulses from nociceptors to the brain
(1) Acute (A) fibers—mediate sharp, intense, localized pain
(2) Chronic (B) fibers—mediate less intense but more persistent dull or aching pain
2. Other free nerve ending receptors
a. Root hair plexuses
(1) Weblike arrangements of free nerve endings around hair follicles
b. Merkel discs
(1) Mediate sensations of discriminative touch

3. Encapsulated nerve endings—six types; all have connective tissue capsules and are mechanoreceptors
a. Touch and pressure receptors
(1) Meissner corpuscle (tactile corpuscle), relatively large and superficial in placement; mediates touch and low-frequency vibration; large numbers in hairless skin areas, such as nipples, fingertips, and lips
(a) Two anatomical variations of Meissner corpuscle
(i) Krause end bulbs—small, with less tightly coiled dendritic endings within their capsule; involved in touch, low-frequency vibrations
(ii) Ruffini corpuscles, have a flattened capsule and are deeply located in the dermis; mediate crude and persistent touch; may be secondary temperature receptors for heat (85° to 120° F [29° to 49° C])
(2) Pacinian or lamellar corpuscles—large mechanoreceptors that respond quickly to sensations of deep pressure, high-frequency vibration, and stretch; found in deep dermis and in joint capsules—they adapt quickly, and sensations they evoke seldom last for long periods
b. Stretch receptors—two types (muscle spindles and Golgi tendon receptors) operate to provide body with information concerning muscle length and strength of muscle contraction
(1) Muscle spindle—composed of 5 to 10 intrafusal fibers lying between and parallel to regular (extrafusal) muscle fibers
(a) Large-diameter and rapid-conducting type Ia and smaller-diameter and slower-conducting type II afferent fibers carry messages to brain concerning changes in muscle length
(b) If the length of a muscle exceeds a certain limit, a stretch reflex is initiated to shorten the muscle, thus helping to maintain posture
(2) Golgi tendon organs—located at junction between muscle tissue and tendon (Figure 15-2)
(a) Type Ib sensory neurons are stimulated by excessive contraction—when stimulated, they cause muscle to *relax*
(b) Golgi tendon reflex protects muscle from tearing internally because of excessive contractile force

SPECIAL SENSES

A. Characterized by receptors grouped closely together or grouped in specialized organs; sense of smell, taste, hearing, equilibrium, and vision

SENSE OF SMELL

A. Olfactory receptors
1. Olfactory sense organs consist of epithelial support cells and specialized olfactory receptor neurons (Figure 15-4)
a. Olfactory cilia—located on olfactory receptor neurons that touch the olfactory epithelium lining the upper surface of the nasal cavity

b. Olfactory cells—chemoreceptors; gas molecules or chemicals dissolved in the mucus covering the nasal epithelium stimulate the olfactory cells

c. Olfactory epithelium—located in most superior portion of the nasal cavity

d. Olfactory receptors—extremely sensitive and easily fatigued

B. Olfactory pathway—when the level of odor-producing chemicals reaches a threshold level, the following occurs (Figure 15-5):

1. Receptor potential and then action potential are generated and passed to the olfactory nerves in the olfactory bulb

2. The impulse then passes through the olfactory tract and into the thalamic and olfactory centers of the brain for interpretation, integration, and memory storage

SENSE OF TASTE

A. Taste buds—sense organs that respond to gustatory, or taste, stimuli; associated with papillae

1. Chemoreceptors that are stimulated by chemicals dissolved in the saliva

2. Gustatory cells—specialized cells in taste buds; gustatory hairs extend from each gustatory cell into the taste pore

3. Sense of taste depends on the creation of a receptor potential in gustatory cells because of taste-producing chemicals in the saliva

4. Taste buds are similar structurally; functionally, each taste bud responds most effectively to one of four primary taste sensations: sour, sweet, bitter, and salty (and perhaps metallic and umami) (Figure 15-6)

5. Adaptation and sensitivity thresholds differ for each of the primary taste sensations

B. Neuronal pathway for taste

1. Taste sensation begins with a receptor potential in the gustatory cells of a taste bud; generation and propagation of an action potential then transmits the sensory input to the brain

2. Nerve impulses from the anterior two thirds of the tongue travel over the facial nerve; those from the posterior one third of the tongue travel over the glossopharyngeal nerve; vagus nerve plays a minor role in taste

3. Nerve impulses are carried to the medulla oblongata, relayed into the thalamus, and then relayed into the gustatory area of the cerebral cortex in the parietal lobe of the brain

SENSE OF HEARING AND BALANCE: THE EAR

A. External ear—two divisions (Figures 15-7 and 15-8)

1. Auricle, or pinna—the visible portion of the ear

2. External auditory meatus—tube leading from the auricle into the temporal bone and ending at the tympanic membrane

B. Middle ear (Figure 15-8)

1. Tiny, epithelium-lined cavity hollowed out of the temporal bone

2. Contains three auditory ossicles

a. Malleus (hammer)—attached to the inner surface of the tympanic membrane

b. Incus (anvil)—attached to the malleus and stapes

c. Stapes (stirrup)—attached to the incus

3. Openings into the middle ear cavity

a. Opening from the external auditory meatus covered with tympanic membrane

b. Oval window—opening into inner ear; stapes fits here

c. Round window—opening into inner ear; covered by a membrane

d. Opening into the auditory (eustachian) tube

C. Inner ear (Figure 15-9, A)

1. Structure of the inner ear

a. Bony labyrinth—made up of the vestibule, cochlea, and semicircular canals

b. Membranous labyrinth—made up of utricle and saccule inside the vestibule, cochlear duct inside the cochlea, and membranous semicircular canals inside the bony ones

c. Vestibule and semicircular canals are involved with balance

d. Cochlea—involved with hearing

e. Endolymph—clear, potassium-rich fluid filling the membranous labyrinth

f. Perilymph—similar to cerebrospinal fluid, surrounds the membranous labyrinth, filling the space between the membranous tunnel and its contents and the bony walls that surround it

2. Cochlea and cochlear duct (Figure 15-9, B)

a. Cochlea—bony labyrinth

b. Modiolus—cone-shaped core of bone that houses the spiral ganglion, which consists of cell bodies of the first sensory neurons in the auditory relay

c. Cochlear duct

(1) Lies inside the cochlea; only part of the internal ear concerned with hearing; contains endolymph

(2) Shaped like a triangular tube

(3) Divides the cochlea into the scala vestibuli, the upper section, and the scala tympani, the lower section; both sections filled with perilymph

(4) Vestibular membrane—the roof of the cochlear duct

(5) Basilar membrane—floor of the cochlear duct

(6) Organ of Corti—rests on the basilar membrane; consists of supporting cells and hair cells

(7) Axons of the neurons that begin around the organ of Corti, extend in the cochlear nerve to the brain to produce the sensation of hearing

3. Sense of hearing

a. Sound is created by vibrations

b. Ability to hear sound waves depends on volume, pitch, and other acoustic properties

c. Sound waves must be of sufficient amplitude to move the tympanic membrane and have a frequency capable of stimulating the hair cells in the organ of Corti

d. Basilar membrane is not the same width and thickness throughout its length; high-frequency sound waves vibrate the narrow portion near the oval window, whereas low frequencies vibrate the wider, thicker portion near the apex of the cochlea; this fact allows different hair cells to be stimulated and different pitches of sound to be perceived

e. Perception of loudness is determined by the amplitude of movement of the basilar membrane; the greater the movement, the louder the perceived sound

f. Hearing—results from stimulation of the auditory area of the cerebral cortex

g. Pathway of sound waves (Figure 15-10)

(1) Enter external auditory canal

(2) Strike tympanic membrane, causing vibrations

(3) Tympanic vibrations move the malleus, which in turn moves the incus and then the stapes

(4) The stapes moves against the oval window, which begins the fluid conduction of sound waves

(5) The perilymph in the scala vestibuli of the cochlea begins a "ripple" that is transmitted through the vestibular membrane to the endolymph inside the duct, to the basilar membrane, and then to the organ of Corti

(6) From the basilar membrane, the ripple is transmitted through the perilymph in the scala tympani and then expends itself against the round window

h. Neuronal pathway of hearing

(1) A movement of hair cells against the tectorial membrane stimulates the dendrites that terminate around the base of the hair cells and initiates impulse conduction by the cochlear nerve to the brainstem

(2) Impulses pass through "relay stations" in the nuclei in the medulla, pons, midbrain, and thalamus before reaching the auditory area of the temporal lobe

4. Vestibule and semicircular canals (Figure 15-9, A)

a. Vestibule—the central section of the bony labyrinth; the utricle and saccule are the membranous structures within the vestibule

b. Semicircular canals—three, each at right angles to the others, are found in each temporal bone; within bony semicircular canals are membranous semicircular canals, each containing endolymph and connecting with the utricle; near this junction, each canal enlarges into an ampulla

c. Sense of balance

(1) Static equilibrium—ability to sense the position of the head relative to gravity or to sense acceleration or deceleration (Figure 15-11)

(a) Movements of the maculae, located in both the utricle and saccule almost at right angles to each other, provide information related to head position or acceleration

(b) Otoliths are located within the matrix of the macula

(c) Changing head position produces a change of pressure on the otolith-weighted matrix, which stimulates the hair cells that in turn stimulate the receptors of the vestibular nerve

(d) Vestibular nerve fibers conduct impulses to the brain and produce a sensation of the position of the head and also a sensation of a change in the pull of gravity

(e) Righting reflexes—muscular responses to restore the body and its parts to their normal position when they have been displaced; caused by stimuli of the macula and impulses from proprioceptors and from the eyes

(2) Dynamic equilibrium—needed to maintain balance when the head or body is rotated or suddenly moved; able to detect changes both in direction and rate at which movement occurs (Figure 15-12)

(a) Depends on the functioning of the cristae ampullaris, which are located in the ampulla of each semicircular canal

(b) Cupula—gelatinous cap in which the hair cells of each crista are embedded; does not respond to gravity; it moves with the flow of endolymph in the semicircular canals

(c) Semicircular canals are almost placed at right angles to each other to detect movement in all directions

(d) When the cupula moves, hair cells are bent, producing a receptor potential followed by an action potential; the action potential passes through the vestibular portion of the eighth cranial nerve to the medulla oblongata, where it is sent to other areas of the brain and spinal cord for interpretation, integration, and response

VISION: THE EYE

A. Structure of the eye (Figures 15-13 and 15-14)

1. Coats of the eyeball—three layers of tissues compose the eyeball

a. Sclera—outer coat

(1) Tough, white, fibrous tissue

(2) Cornea—the transparent anterior portion that lies over the iris; no blood vessels found in the cornea or in the lens

(3) Canal of Schlemm—ring-shaped venous sinus found deep within the anterior portion of the sclera at its junction with the cornea

b. Choroid—middle coat

(1) Contains many blood vessels and a large amount of pigment

(2) Anterior portion has three different structures (Figure 15-14)

(a) Ciliary body—thickening of choroid, fits between anterior margin of retina and posterior margin of iris; ciliary muscle lies in anterior part of ciliary body; ciliary processes—fold in the ciliary body

(b) Suspensory ligament—attached to the ciliary processes and blends with the elastic capsule of the lens, to hold it in place

(c) Iris—colored part of the eye; consists of circular and radial smooth muscle fibers that form a doughnut-shaped structure; attaches to the ciliary body

c. Retina—incomplete innermost coat of the eyeball
 (1) Three layers of neurons make up the sensory retina (Figure 15-16)
  (a) Photoreceptor neurons—visual receptors, highly specialized for stimulation by light rays
   (i) Rods—absent from the fovea and macula; increased in density toward the periphery of the retina
   (ii) Cones—less numerous than rods; most densely concentrated in the fovea centralis in the macula lutea
  (b) Bipolar neurons
  (c) Ganglionic neurons—all axons of these neurons extend back to the optic disc; part of the sclera, which contains perforations through which the fibers emerge from the eyeball as the optic nerve
2. Cavities and humors
 a. Cavities—eyeball has a large interior space divided into two cavities
  (1) Anterior cavity—lies in front of the lens; has two subdivisions
   (a) Anterior chamber—space anterior to the iris and posterior to the cornea
   (b) Posterior chamber—small space posterior to the iris and anterior to the lens
  (2) Posterior cavity—larger than the anterior cavity; occupies all the space posterior to the lens, suspensory ligament, and ciliary body
 b. Humors
  (1) Aqueous humor—fills both chambers of the anterior cavity; clear, watery fluid that often leaks out when the eye is injured; formed from blood in capillaries located in the ciliary body (Figure 15-17)
  (2) Vitreous humor—fills the posterior cavity; semisolid material; helps to maintain sufficient intraocular pressure, with aqueous humor, to give the eyeball its shape
3. Muscles—two types of eye muscles
 a. Extrinsic eye muscles (Figure 15-18)—skeletal muscles that attach to the outside of the eyeball and to the bones of the orbit; named according to their position on the eyeball; the muscles are superior, inferior, medial, and lateral rectus muscles and superior and inferior oblique muscles
 b. Intrinsic eye muscles—smooth muscles located within the eye: iris—regulates size of pupil; ciliary muscle—controls shape of lens
4. Accessory structures (Figure 15-19)
 a. Eyebrows and eyelashes—give some protection against foreign objects entering the eye; cosmetic purposes
 b. Eyelids—consist of voluntary muscle and skin with a tarsal plate; lined with conjunctiva, a mucous membrane; palpebral fissure—opening between the eyelids; canthus—where the upper and lower eyelids join

 c. Lacrimal apparatus—structures that secrete tears and drain them from the surface of the eyeball (Figure 15-21)
  (1) Lacrimal glands—size and shape of a small almond; located at the upper, outer margin of each orbit; approximately a dozen small ducts lead from each gland; drain tears onto the conjunctiva
  (2) Lacrimal canals—small channels that empty into lacrimal sacs
  (3) Lacrimal sacs—located in a groove in the lacrimal bone
  (4) Nasolacrimal ducts—small tubes that extend from the lacrimal sac into the inferior meatus of the nose
B. The process of seeing
 1. Formation of retinal image
  a. Refraction of light rays—deflection, or bending, of light rays produced by light rays passing obliquely from one transparent medium into another of different optical density; cornea, aqueous humor, lens, and vitreous humor are the refracting media of the eye
  b. Accommodation of lens—increase in curvature of the lens to achieve the greater refraction needed for near vision (Figure 15-22)
  c. Constriction of pupil—muscles of iris are important to formation of a clear retinal image; pupil constriction prevents divergent rays from object from entering eye through periphery of the cornea and lens; near reflex—constriction of pupil that occurs with accommodation of the lens in near vision; photopupil reflex—pupil constricts in bright light
  d. Convergence of eyes—movement of the two eyeballs inward so that their visual axes come together at the object viewed; the closer the object, the greater the degree of convergence necessary to maintain single vision; for convergence to occur, a functional balance between antagonistic extrinsic muscles must exist
 2. The role of photopigments—light-sensitive pigmented compounds undergo structural changes that result in generation of nerve impulses, which are interpreted by the brain as sight
  a. Rods—photopigment in rods is rhodopsin; highly light sensitive; breaks down into opsin and retinal; separation of opsin and retinal in the presence of light causes an action potential in rod cells; energy is needed to re-form rhodopsin (Figure 15-24)
  b. Cones—three types of cones are present in the retina, with each having a different photopigment; cone pigments are less light sensitive than rhodopsin and need brighter light to break down
 3. Neuronal pathway of vision (Figure 15-25)
  a. Fibers that conduct impulses from the rods and cones reach the visual cortex in the occipital lobes by way of the optic nerves, optic chiasma, optic tracts, and optic radiations
  b. Optic nerve contains fibers from only one retina, but optic chiasma contains fibers from the nasal portion of both retinas; these anatomical facts explain peculiar visual abnormalities that sometimes occur

CYCLE OF LIFE: SENSE ORGANS

A. Sensory information is acquired through depolarization of sensory nerve endings
   1. Age, disease, structural defects, or lack of maturation affects ability to identify and respond
B. Structure and function response capabilities are related to developmental factors associated with age
C. Senses become more acute with maturation
D. Late adulthood—loss of sensory capability
   1. Structural change in receptor cells or other sense organ structures

REVIEW QUESTIONS

1. Define adaptation.
2. Describe each of the following: mechanoreceptors, chemoreceptors, thermoreceptors, nociceptors, photoreceptors.
3. Identify the pathway involved for the production of the sense of smell.
4. What are G protein receptor sites?
5. Describe the main features of the middle ear.
6. Name the parts of the bony and membranous labyrinths and describe the relationship of those parts.
7. In what ear structure or structures is the hearing sense organ located? The equilibrium sense organs?
8. What is the name of the hearing sense organ? Of the equilibrium sense organs?
9. Describe the path of sound waves as they enter the ear.
10. Define vertigo. How is vertigo treated?
11. Describe the role of the basilar membrane in hearing.
12. What is the difference between static and dynamic equilibrium? Describe the general mechanisms by which each is maintained.
13. Name two involuntary muscles in the eye. Explain their functions.
14. Define the term *refraction*. Name the refractory media of the eye.
15. Explain briefly the mechanism for accommodation for near vision.
16. Name the photopigments present in rods and in cones. Explain their role in vision.
17. Explain why "night blindness" may occur in marked vitamin A deficiency.
18. Describe the phenomenon of referred pain.

CRITICAL THINKING QUESTIONS

1. With regard to location, can you explain the difference between an above-threshold receptor potential and the sensation of that stimulus?
2. Which free nerve ending is able to respond to light touch? Name the free nerve ending that is the most common receptor in the brain.
3. Describe the neuronal pathways for taste and smell. What would be most likely to stimulate a memory: the taste of apple pie or the smell of apple pie? How can you explain your answer?
4. Can you identify the three coats of the eye? Integrate their specific functions within your answer.
5. How would you compare and contrast the receptors for vision in dim light and those for vision in bright light?
6. How is sensory response related to age?

Endocrine System

LANGUAGE OF SCIENCE

adenohypophysis (ad-eh-no-hye-POF-i-sis) [*adeno-* gland, *-hypo-* beneath, *-physis* growth]

adrenal cortex (ah-DREE-nal KOR-teks) [*ad-* toward, *-ren-* kidney, *-al* pertaining to, *cortex* bark]

adrenal glands (ah-DREE-nal) [*ad-* toward, *-ren-* kidney, *-al* pertaining to]

adrenal medulla (ah-DREE-nal meh-DUL-eh) [*ad-* toward, *-ren-* kidney, *-al* pertaining to, *medulla* marrow or middle]

adrenocorticotropic hormone (ACTH) (ah-dree-no-kor-teh-koh-TROH-pic HOR-mohn) [*adreno-* gland, *-cortico-* cortex, *-trop-* affecting, *-ic* pertaining to, *hormaien* spur on]

aldosterone (AL-doh-steh-rohn *or* al-DAH-stair-ohn) [*aldo-* aldehyde, *-stero-* solid or steroid derivative, *-one* chemical derivative]

alpha cell (AL-fah) [*alpha* first letter of the Greek alphabet, *cella* storeroom]

amino acid derivative hormones (ah-MEE-no ASS-id deh-RIV-i-tiv HOR-mohns)

anabolic hormone (an-ah-BOL-ik HOR-mohn) [*anabol-* build up, *-ic* pertaining to, *hormaien* spur on]

anabolic steroids (an-ah-BOL-ik STAYR-oyds) [*anabol-* build up, *-ic* pertaining to, *stero-* solid, *-oid* resembling]

angiotensin I (an-jee-oh-TEN-sin) [*angio-* blood vessel, *-tens-* pressure or stretch, *-in* substance, *I* Roman numeral one]

angiotensin II (an-jee-oh-TEN-sin) [*angio-* blood vessel, *-tens-* pressure or stretch, *-in* substance, *II* Roman numeral two]

angiotensinogen (an-jee-oh-TEN-sin-oh-jen) [*angio-* blood vessel, *-tens-* pressure or stretch, *-in-* substance, *-gen* that which generates]

antagonism (an-TAG-oh-niz-em) [*antagon-* struggle, *-ism* condition of]

antidiuretic hormone (ADH) (an-tee-dye-yoo-RET-ik HOR-mohn) [*anti-* against, *-dia-* through, *-uret-* urination, *-ic* pertaining to, *hormaien* spur on]

arginine vasopressin (AVP) (AHR-jih-neen vas-oh-PRES-in) [*arginine* type of amino acid, *vaso-* vessel or duct, *-press-* pressure, *-in* substance]

Cont'd on p. 636

ORGANIZATION OF THE ENDOCRINE SYSTEM

The endocrine system and nervous system both function to achieve and maintain stability of the internal environment. Each system may work alone or in concert with each other as a single **neuroendocrine system,** performing the same general functions within the body: communication, integration, and control.

Both the endocrine system and the nervous system perform their regulatory functions by means of chemical messengers sent to specific cells. In the nervous system, neurons secrete neurotransmitter molecules to signal nearby cells that have the appropriate receptor molecules. In the endocrine system, secreting cells send **hormone** (from the Greek *hormaein,* "to excite") molecules by way of the bloodstream to signal specific **target cells** throughout the body. Tissues and organs that contain endocrine target cells are called *target tissues* and *target organs,* respectively. As with postsynaptic cells, endocrine target cells must have the appropriate receptor to be influenced by the signaling chemical—a process called *signal transduction.* Many cells have receptors for neurotransmitters and hormones, so they can be influenced by both types of chemicals.

Whereas neurotransmitters are sent over very short distances across a synapse, hormones diffuse into the blood to be carried to nearly every point in the body. The nervous system can directly control only muscles and glands that are innervated with efferent fibers, whereas the endocrine system can regulate most cells in the body. The effects of neurotransmitters are rapid and short lived compared with the effects of hormones, which appear more slowly and last longer. Table 16-1 compares endocrine structure and function with nervous structure and function (Figure 16-1).

Endocrine glands secrete their products, hormones, directly into the blood. Because they do not have ducts, they are often called "ductless glands." This characteristic distinguishes *endocrine glands* from *exocrine glands,* which secrete their products into ducts (see Chapter 5, p. 157). Many endocrine glands are made of glandular epithelium, whose cells manufacture and secrete hormones. However, a few endocrine glands are made of **neurosecretory tissue.** Neurosecretory cells are simply modified neurons that secrete chemical messengers that diffuse into the bloodstream rather than across a synapse. In such cases, the chemical messenger is called a *hormone* rather than a *neurotransmitter.* For example, when norepinephrine is released by neurons, diffuses across a synapse, and binds to an adrenergic receptor in a postsynaptic neuron, we call norepinephrine a *neurotransmitter.* On the other hand, we call norepinephrine a *hormone* when it diffuses into the blood (because there is no postsynaptic cell present) and then binds to an adrenergic receptor in a distant target cell.

Table 16-1	Comparison of Features of the Endocrine System and Nervous System	
FEATURE	**ENDOCRINE SYSTEM**	**NERVOUS SYSTEM**
Overall Function	Regulation of effectors to maintain homeostasis	Regulation of effectors to maintain homeostasis
Control by regulatory feedback loops	Yes (endocrine reflexes)	Yes (nervous reflexes)
Effector tissues	Endocrine effectors: virtually all tissues	Nervous effectors: muscle and glandular tissue only
Effector cells	Target cells (throughout the body)	Postsynaptic cells (in muscle and glandular tissue only)
Chemical Messenger	Hormone	Neurotransmitter
Cells that secrete the chemical messenger	Glandular epithelial cells or neurosecretory cells (modified neurons)	Neurons
Distance traveled (and method of travel) by chemical messenger	Long (by way of circulating blood)	Short (across a microscopic synapse)
Location of receptor in effector cell	On the plasma membrane or within the cell	On the plasma membrane
Characteristics of regulatory effects	Slow to appear, long lasting	Appear rapidly, short lived

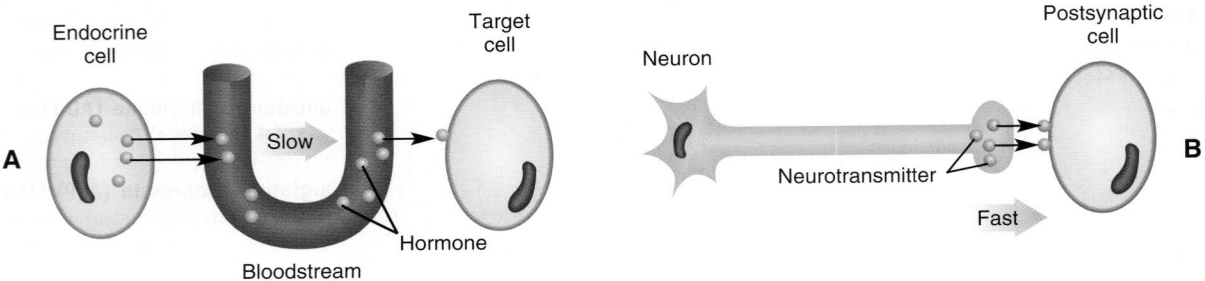

Figure 16-1 *Mechanisms of endocrine* (A) *and nervous* (B) *signals.*

Glands of the endocrine system are widely scattered throughout the body. New discoveries in *endocrinology* continue to add to the long list of hormone-secreting tissues. However, even the most newly discovered endocrine tissues and their hormones operate according to some basic physiological principles.

In this chapter, we will focus our discussion primarily on a few major endocrine glands and their principal hormones. Figure 16-2

Table 16-2	Names and Locations of Some Major Endocrine Glands	
NAME	**LOCATION**
Hypothalamus	Cranial cavity (brain)
Pituitary gland	Cranial cavity
Pineal gland	Cranial cavity (brain)
Thyroid gland	Neck
Parathyroid glands	Neck
Thymus	Mediastinum
Adrenal glands	Abdominal cavity (retroperitoneal)
Pancreatic islets	Abdominal cavity (pancreas)
Ovaries	Pelvic cavity
Testes	Scrotum
Placenta	Pregnant uterus

and Table 16-2 summarize the names and locations of these representative endocrine glands. After you are familiar with the basic principles of endocrinology and the major examples of glands and their hormones, you will be prepared for additional examples that you will encounter as you continue your study of the human body.

QUICK CHECK

1. What is meant by the term *target cell*?
2. Describe how the nervous system and the endocrine system differ in the way they control effectors.

HORMONES
Classification of Hormones

Hormone molecules can be classified in various useful ways. For example, when classified by general function, hormones can be identified as **tropic hormones** (hormones that target other endocrine glands and stimulate their growth and secretion), **sex hormones** (hormones that target reproductive tissues), **anabolic hormones** (hormones that stimulate anabolism in their target cells), and many other functional names. Another useful way to classify hormones is by their chemical structure. Because this method of classifying hormones is so widely used, we will briefly describe it in the following paragraphs.

Steroid Hormones

All of the many hormones secreted by endocrine tissues can be classified simply as **steroid** or **nonsteroid** (Figure 16-3). Steroid hormone molecules are manufactured by endocrine cells from cholesterol, an important type of lipid in the human body (see Chapter 2, p. 59). As Figure 16-4 shows, because all steroid hor-

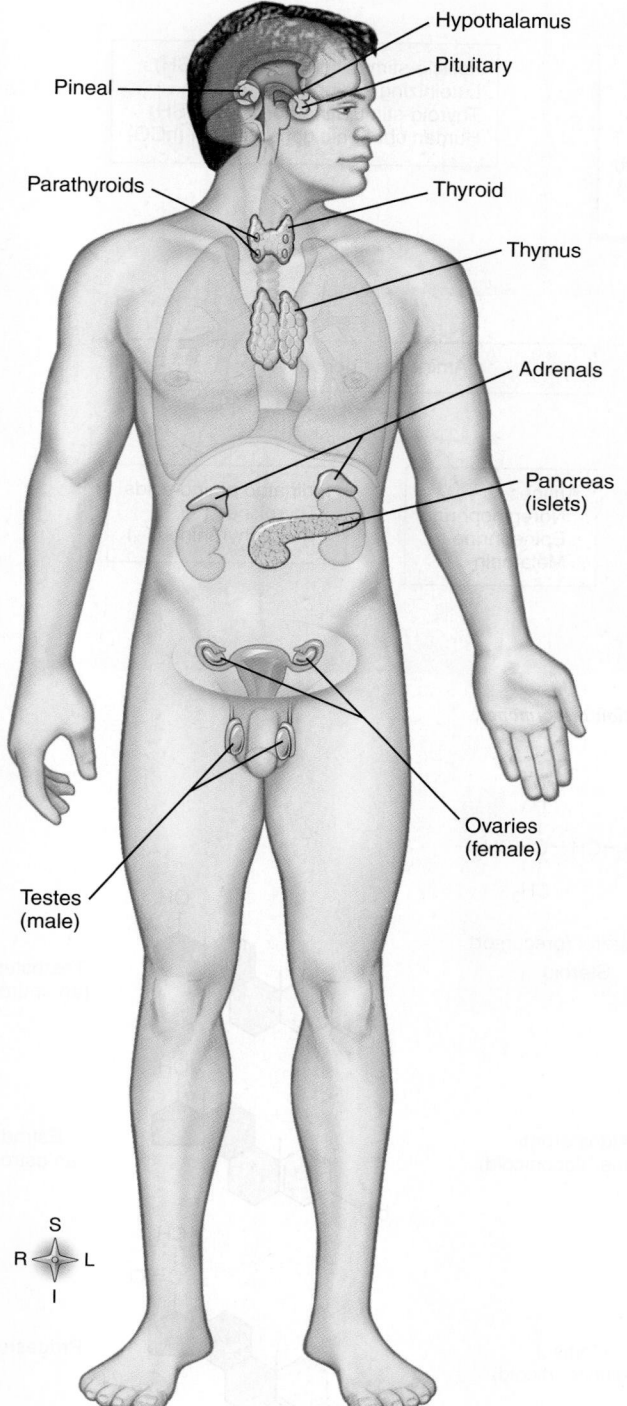

Figure 16-2 *Locations of some major endocrine glands.*

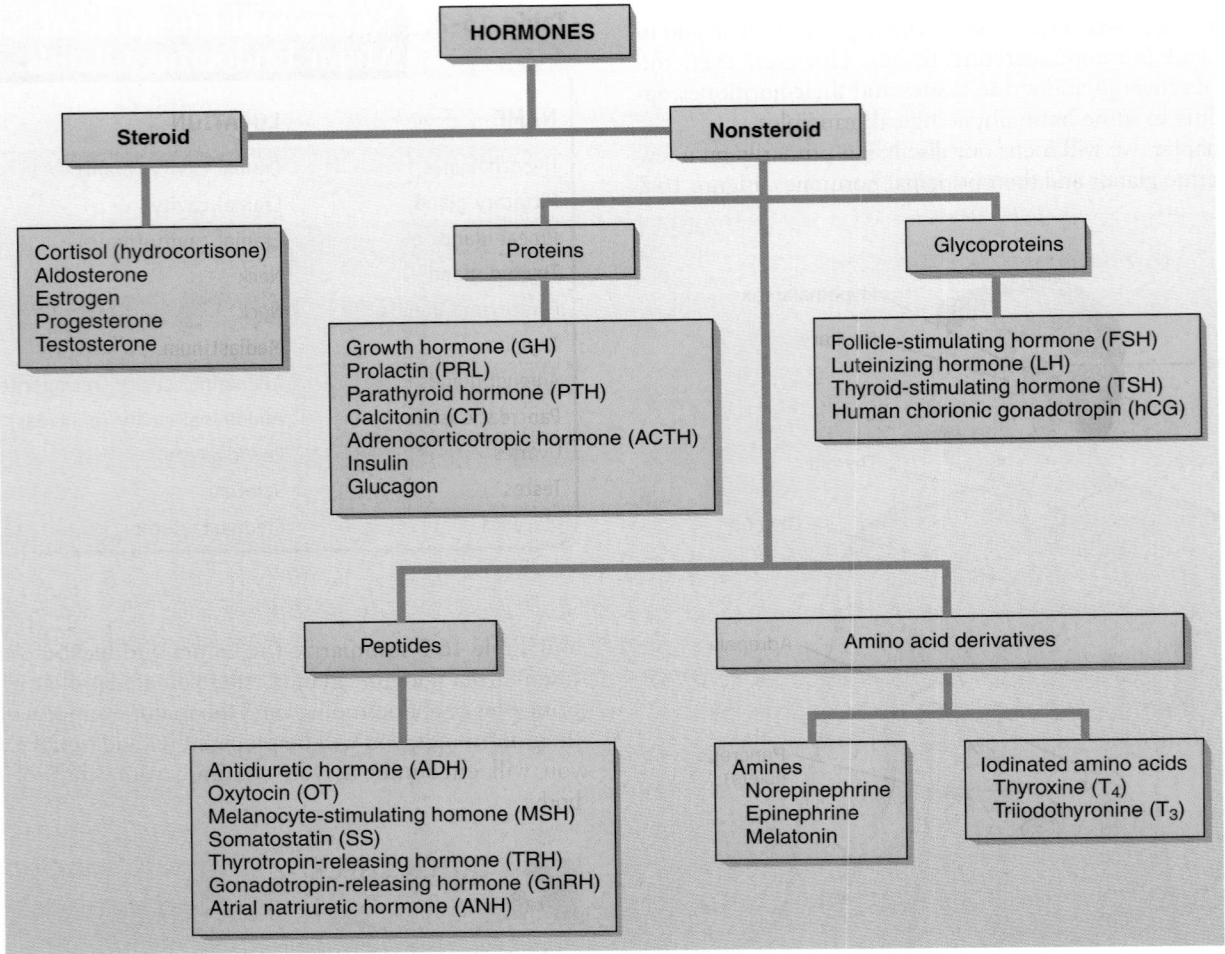

Figure 16-3 *Chemical classification of hormones.*

Figure 16-4 *Steroid hormone structure.*

As these examples show, steroid hormone molecules are very similar in structure to cholesterol *(top left)*, particularly in having a 4-ring steroid nucleus at their core. Cholesterol is the precursor molecule from which the steroid hormones are all derived.

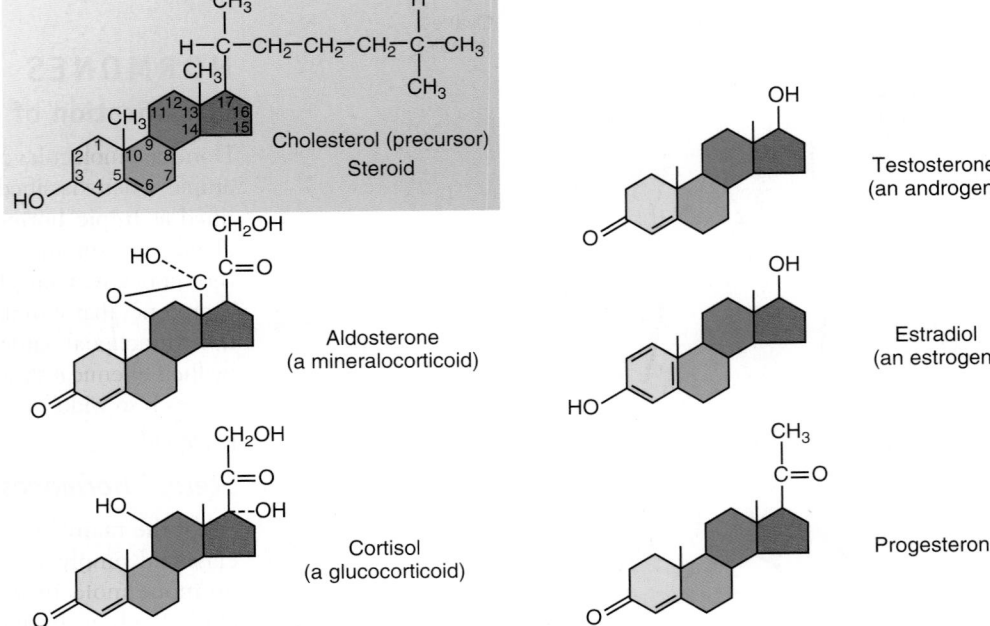

mones are derived from a common molecule, cholesterol, they have a characteristic chemical group at the core of each molecule. Because steroids are lipid soluble, they can easily pass through the phospholipid plasma membrane of target cells. Figure 16-5 outlines the metabolic pathways used by endocrine cells to convert cholesterol into steroid hormones. Examples of important steroid hormones include cortisol, aldosterone, estrogen, progesterone, and testosterone.

Nonsteroid Hormones

Most nonsteroid hormones are synthesized primarily from amino acids rather than from cholesterol (Figure 16-6). Some nonsteroid hormones are **protein hormones.** These hormones are long, folded chains of amino acids, a structure typical of protein molecules of any sort (see Chapter 2, p. 53). Included among the protein hormones are insulin, parathyroid hormone, and others listed in Figure 16-3. Protein hormones that have carbohydrate groups attached to their amino acid chains are often classified separately as **glycoprotein hormones** (Figure 16-6, D).

Another major category of nonsteroid hormones consists of the peptide hormones. **Peptide hormones** such as oxytocin and antidiuretic hormone are smaller than the protein hormones. They are each made of a short chain of amino acids, as Figure 16-6, B, shows. Figure 16-3 lists examples of peptide hormones.

Yet another category of nonsteroid hormones consists of the **amino acid derivative hormones.** Each of these hormones is derived from only a single amino acid molecule. There are two major subgroups within this category. One subgroup, the *amine hormones*, is synthesized by modifying a single molecule of the amino acid, tyrosine. Amine hormones such as epinephrine and norepinephrine are produced by neurosecretory cells (where they are secreted as hormones) and by neurons (where they are secreted as neurotransmitters). Another subgroup of amino acid derivatives produced by the thyroid gland are all synthesized by adding iodine (I) atoms to a tyrosine molecule (Figure 16-6, C). Figure 16-3 lists examples of hormones derived from single amino acids.

QUICK CHECK

3. How are steroid hormones able to pass through a cell's plasma membrane easily?
4. Name some of the different general types of nonsteroid hormones. Can you give an example of each?

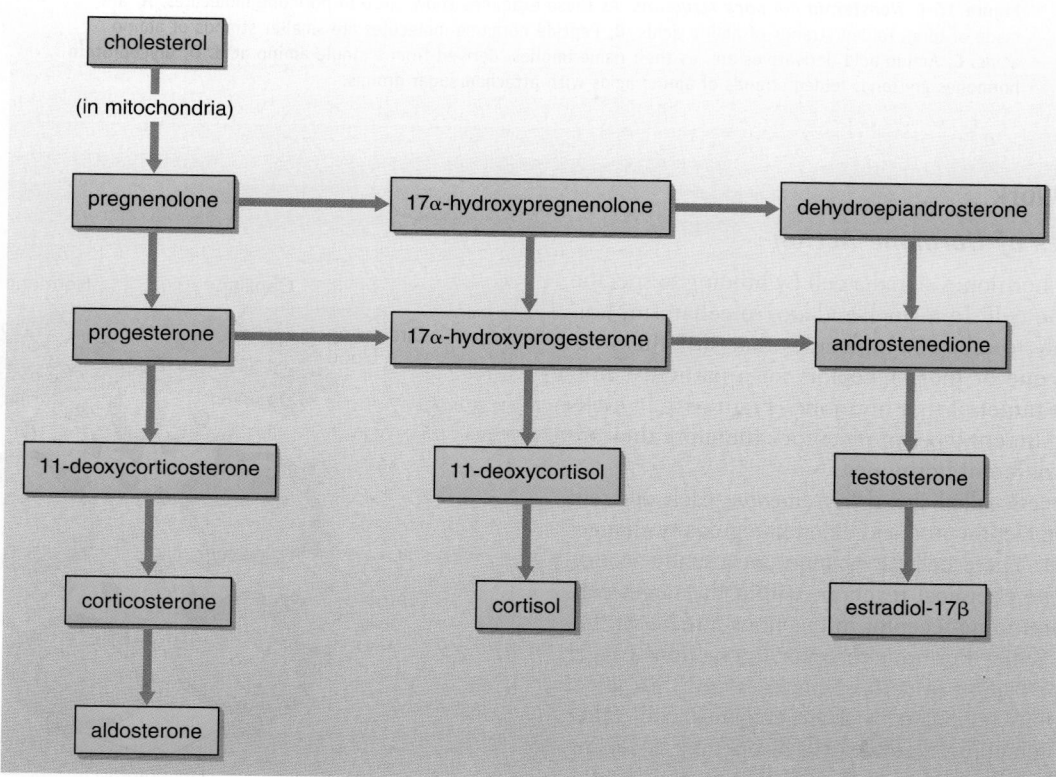

Figure 16-5 *Synthesis of steroid hormones.* Steroid hormones are synthesized in various endocrine tissues using cholesterol as a precursor (see Figure 16-4). Notice that many steroid hormones are converted from one form to another in human cells.

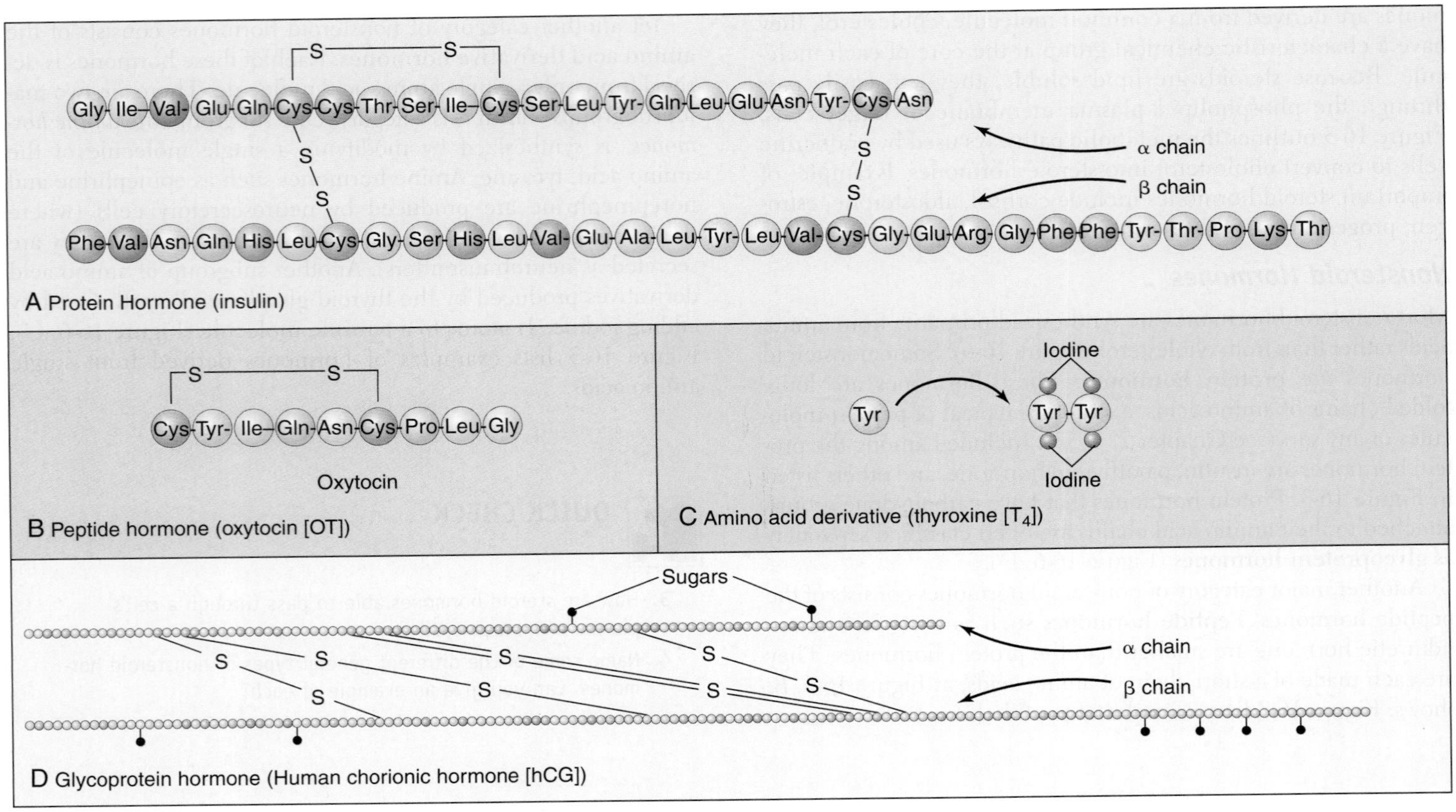

S
S
Gly Ile Val Glu Gln Cys Cys Thr Ser Ile Cys Ser Leu Tyr Gln Leu Glu Asn Tyr Cys Asn
S
S
S
S
α chain
β chain
Phe Val Asn Gln His Leu Cys Gly Ser His Leu Val Glu Ala Leu Tyr Leu Val Cys Gly Glu Arg Gly Phe Phe Tyr Thr Pro Lys Thr
A Protein Hormone (insulin)
S
S
Cys Tyr Ile Gln Asn Cys Pro Leu Gly
Oxytocin
B Peptide hormone (oxytocin [OT])
Iodine
Tyr
Tyr Tyr
Iodine
C Amino acid derivative (thyroxine [T4])
Sugars
S S S S
S S S S
α chain
β chain
D Glycoprotein hormone (Human chorionic hormone [hCG])

Figure 16-6 ***Nonsteroid hormone structure.*** As these examples show, protein hormone molecules, **A,** are made of long, folded strands of amino acids. **B,** Peptide hormone molecules are smaller strands of amino acids. **C,** Amino acid derivatives are, as their name implies, derived from a single amino acid. **D,** Glycoprotein hormones are long, folded strands of amino acids with attached sugar groups.

How Hormones Work

General Principles of Hormone Action

As previously stated, hormones signal a cell by binding to specific receptors on or in the cell. In a "lock-and-key" mechanism, hormones will bind only to receptor molecules that "fit" them exactly. Any cell with one or more receptors for a particular hormone is said to be a **target** of that hormone (Figure 16-7). Cells usually have many different types of receptors; therefore they are target cells of many different hormones.

In a complex process called *signal transduction,* each different hormone-receptor interaction produces different regulatory changes within the target cell. These cellular changes are usually accomplished by altering the chemical reactions within the target cell. For example, some hormone-receptor interactions initiate synthesis of new proteins. Other hormone-receptor interactions trigger the activation or inactivation of certain enzymes and thus affect the metabolic reactions regulated by those enzymes. Still other hormone-receptor interactions regulate cells by opening or closing specific ion channels in the plasma membrane. Specific mechanisms of signal transduction and endocrine effects in target cells are outlined in the next section.

Different hormones may work together to enhance each other's influence on a target cell. In a phenomenon called **synergism,**

Capillary
Hormone
Circulating blood
Target cells
Nontarget cells
Receptors

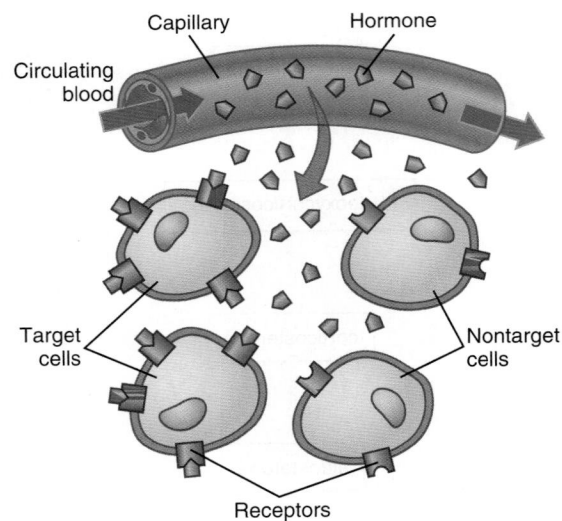

Figure 16-7 ***The target cell concept.*** A hormone acts only on cells that have receptors specific to that hormone because the shape of the receptor determines which hormone can react with it. This is an example of the lock-and-key model of biochemical reactions.

combinations of hormones have a greater effect on a target cell than the sum of the effects that each would have if acting alone. Combined hormone actions may exhibit instead the phenomenon of **permissiveness.** Permissiveness occurs when a small amount of one hormone allows a second hormone to have its full effect on a target cell; the first hormone "permits" the full action of the second hormone. A common type of combined action of hormones is seen in the phenomenon of **antagonism.** In antagonism, one hormone produces the opposite effect of another hormone. Antagonism between hormones can be used to "fine tune" the activity of target cells with great accuracy, signaling the cell exactly when (and by how much) to increase or decrease a certain cellular process.

Although our discussion is focused on the primary actions of a few selected hormones, recent studies make it clear that hormones usually have many diverse secondary functions in the body. For example, prolactin (PRL) is a hormone with primary actions that regulate milk production (lactation) and reproduction, but it also has about 300 secondary actions in the body. Most secondary effects of hormones modulate, or influence, the activity of other regulatory mechanisms. The primary effect of a hormone, on the other hand, is a more direct regulatory mechanism.

As previously stated, hormones travel to their target cells by way of the circulating bloodstream. This means that all hormones travel throughout the body. Because they only affect their target cells, however, the effects of a particular hormone may be limited to specific tissues in the body. Some hormone molecules are attached to plasma proteins while they are carried along the bloodstream. Such hormones must free themselves from the plasma protein to leave the blood and combine with their receptors. Because blood carries hormones nearly everywhere in the body, even where there are no target cells, endocrine glands produce more hormone molecules than actually hit their target. Unused hormones usually are quickly excreted by the kidneys or broken down by metabolic processes.

Mechanism of Steroid Hormone Action

Steroid hormones are lipids and thus are not very soluble in blood plasma, which is mostly water. Instead of traveling in the plasma as free molecules, they attach to soluble plasma proteins. As you can see in Figure 16-8, a steroid hormone molecule dissociates from its carrier before approaching the target cell. Because steroid hormones are lipid soluble and thus can pass into cells easily, it is not surpris-

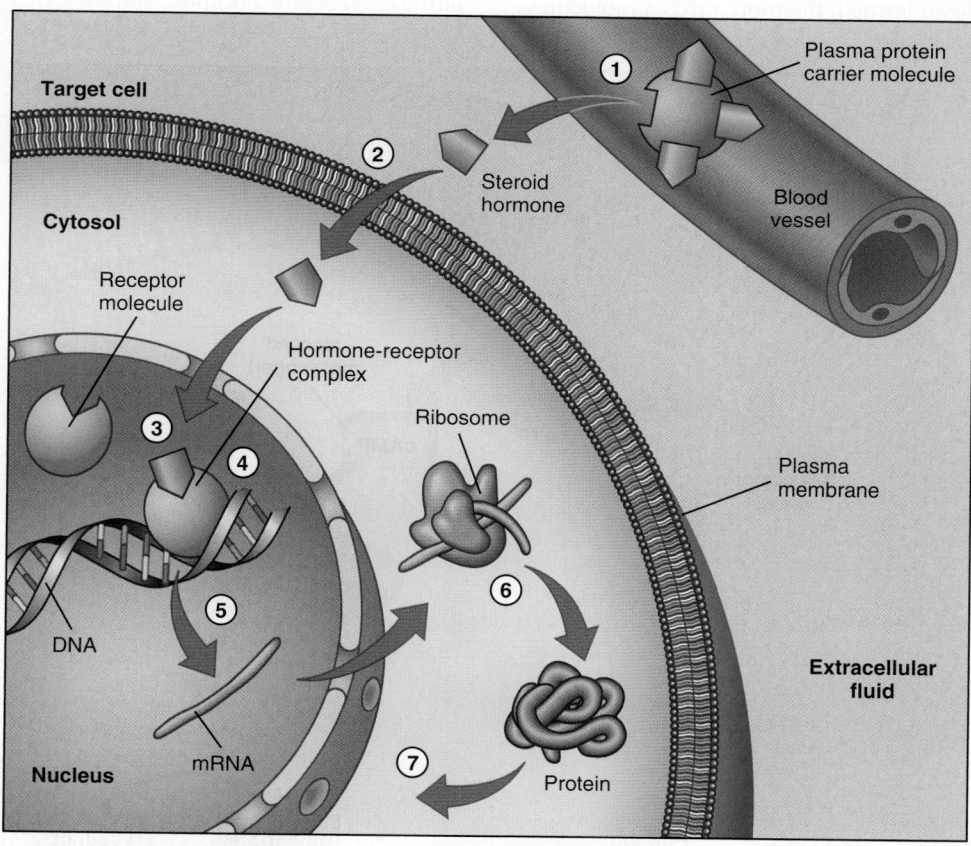

Figure 16-8 *Steroid hormone mechanism.* According to the mobile-receptor model, lipid-soluble steroid hormone molecules detach from a carrier protein *(1)* and pass through the plasma membrane *(2)*. The hormone molecules then pass into the nucleus where they bind with a mobile receptor to form a hormone-receptor complex *(3)*. This complex then binds to a specific site on a DNA molecule *(4)*, triggering transcription of the genetic information encoded there *(5)*. The resulting mRNA molecule moves to the cytosol, where it associates with a ribosome, initiating synthesis of a new protein *(6)*. This new protein—usually an enzyme or channel protein—produces specific effects in the target cell *(7)*. Some steroid hormones also have additional secondary effects such as influencing signal transduction pathways at the plasma membrane.

ing that many of their receptors are found inside the cell rather than on the surface of the plasma membrane. After a steroid hormone molecule has diffused into its target cell, it passes into the nucleus, where it binds to a mobile-receptor molecule to form a *hormone-receptor complex*. Some hormones must be activated by enzymes before they can bind to their receptors. Because steroid hormone receptors are not attached to the plasma membrane, but seem to move freely in the nucleoplasm, this model of hormone action has been called the **mobile-receptor model** or the *nuclear-receptor model*.

Once formed, the hormone-receptor complex activates a certain gene sequence to begin transcription of messenger RNA (mRNA) molecules. The newly formed mRNA molecules then move out of the nucleus into the cytosol, where they associate with ribosomes and begin synthesizing protein molecules.

The new protein molecules synthesized by the target cell would not have been made if not for the arrival of the steroid hormone molecule. Steroid hormones regulate cells by regulating their production of certain critical proteins, such as enzymes that control intracellular reactions or membrane proteins that alter the permeability of a cell.

This mechanism of steroid hormone action implies several things about the effects of these hormones. For one thing, the more hormone-receptor complexes formed, the more mRNA molecules

are transcribed, the more new protein molecules are formed, and thus the greater the magnitude of the regulatory effect. In short, the *amount* of steroid hormone present determines the magnitude of a target cell's response. Also, because transcription and protein synthesis take some time, responses to steroid hormones can be slow—from 45 minutes to several days before the full effect is seen.

Mechanisms of Nonsteroid Hormone Action

The Second Messenger Mechanism

Nonsteroid hormones typically operate according to a mechanism originally called the **second messenger model.** This concept of hormone action—first proposed in the last century by Dr. Earl W. Sutherland—was a milestone in endocrinology for which he received the 1971 Nobel Prize in Medicine or Physiology. According to this concept of signal transduction, a nonsteroid hormone molecule acts as a "first messenger," delivering its chemical message to fixed receptors in the target cell's plasma membrane. The "message" is then passed into the cell where a "second messenger" triggers the appropriate cellular changes. This concept of nonsteroid hormone action is also called the *fixed-membrane-receptor model*.

In the example illustrated in Figure 16-9, formation of the hormone-receptor complex causes a membrane protein, called

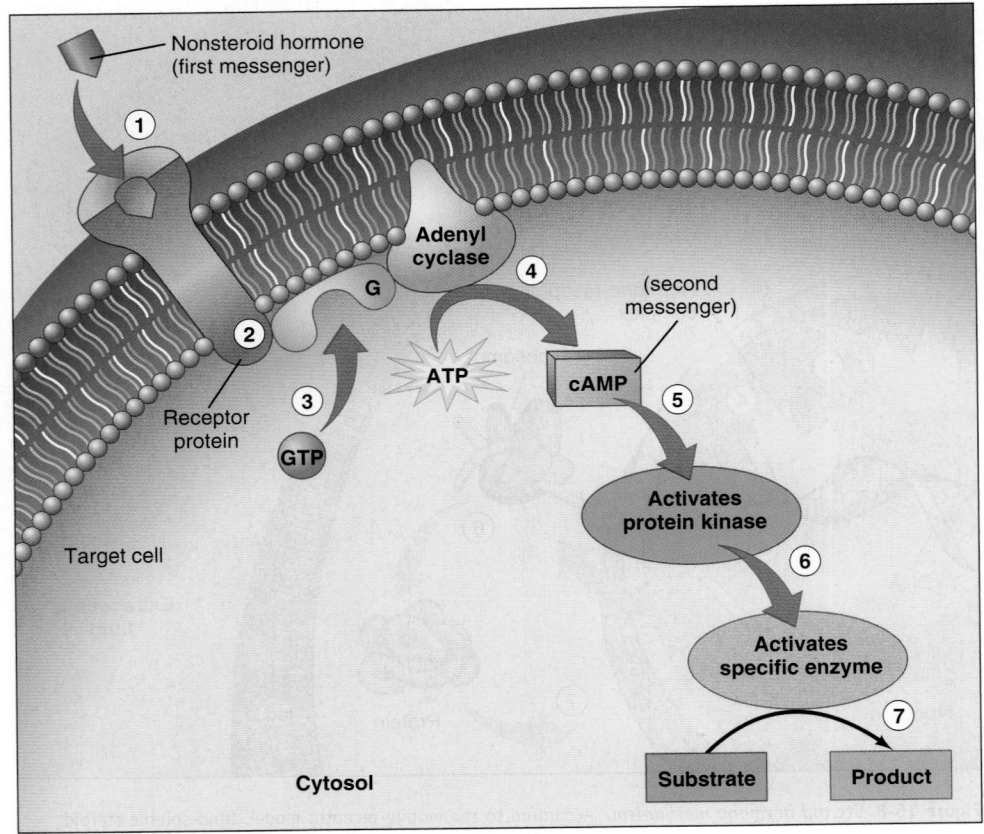

Figure 16-9 *Example of a second messenger mechanism.* A nonsteroid hormone (first messenger) binds to a fixed receptor in the plasma membrane of the target cell *(1)*. The hormone-receptor complex activates the G protein *(2)*. The activated G protein reacts with GTP, which in turn activates the membrane-bound enzyme adenyl cyclase *(3)*. Adenyl cyclase removes phosphates from ATP, converting it to cAMP (second messenger) *(4)*. cAMP activates or inactivates protein kinases *(5)*. Protein kinases activate specific intracellular enzymes *(6)*. These activated enzymes then influence specific cellular reactions, thus producing the target cell's response to the hormone *(7)*.

the **G protein**, to bind to a nucleotide called *guanosine triphosphate (GTP)*. This in turn activates another membrane protein, *adenyl cyclase*. Adenyl cyclase is an enzyme that promotes the removal of two phosphate groups from adenosine triphosphate (ATP) molecules in the cytosol. The product thus formed is cyclic adenosine monophosphate *(cAMP)*. The cAMP molecule acts as a second messenger within the cell. cAMP activates *protein kinases*, a set of enzymes that activate other types of enzymes. It is this final set of specific enzymes, which are now activated, that catalyze the cellular reactions that characterize the target cell's response. In short, the hormone first messenger binds to a membrane receptor, triggering formation of an intracellular second messenger, which activates a cascade of chemical reactions that produces the target cell's response.

Since the time Sutherland first began his pioneering work, other second messenger systems for signal transduction have been discovered. As a matter of fact, the study of second messenger mechanisms is currently a central area of physiology research. Although many nonsteroid hormones seem to use cAMP as the second messenger, we now know that some hormones use nucleotides such as inositol triphosphate (IP$_3$) and cyclic guanosine monophosphate (cGMP) as the second messenger.

Still other hormones produce their effects by triggering the opening of calcium (Ca^{++}) channels in the target cell's membranes, as you can see in Figure 16-10. Binding of a hormone to a fixed membrane receptor activates a chain of membrane proteins (G protein and *phosphodiesterase*, PIP$_2$) that in turn trigger the opening of calcium channels in the plasma membrane. Ca^{++} ions that enter the cytosol when the channels open bind to an intracellular molecule called *calmodulin*. The Ca^{++}-calmodulin complex thus formed acts as a second messenger, influencing the enzymes that produce the target cell's response.

Research findings also show that in second messenger systems, the hormone-receptor complexes may be taken into the cell by means of endocytosis. Although the purpose of this may be primarily to break down the complexes and recycle the receptors, the hormone-receptor complex may continue to have physiological effects after it is taken into the cell.

The second messenger mechanism of signal transduction produces target cell effects that differ from steroid hormone effects in several important ways (Table 16-3). First, the cascade of reactions produced in the second messenger mechanism greatly amplifies the effects of the hormone. Thus the effects of many nonsteroid hormones are disproportionately great when compared with the amount of hormone present. Recall that steroid hormones produce effects in proportion to the amount of hormone present. Also, the second messenger mechanism operates much more quickly than the steroid mechanism. Many nonsteroid hor-

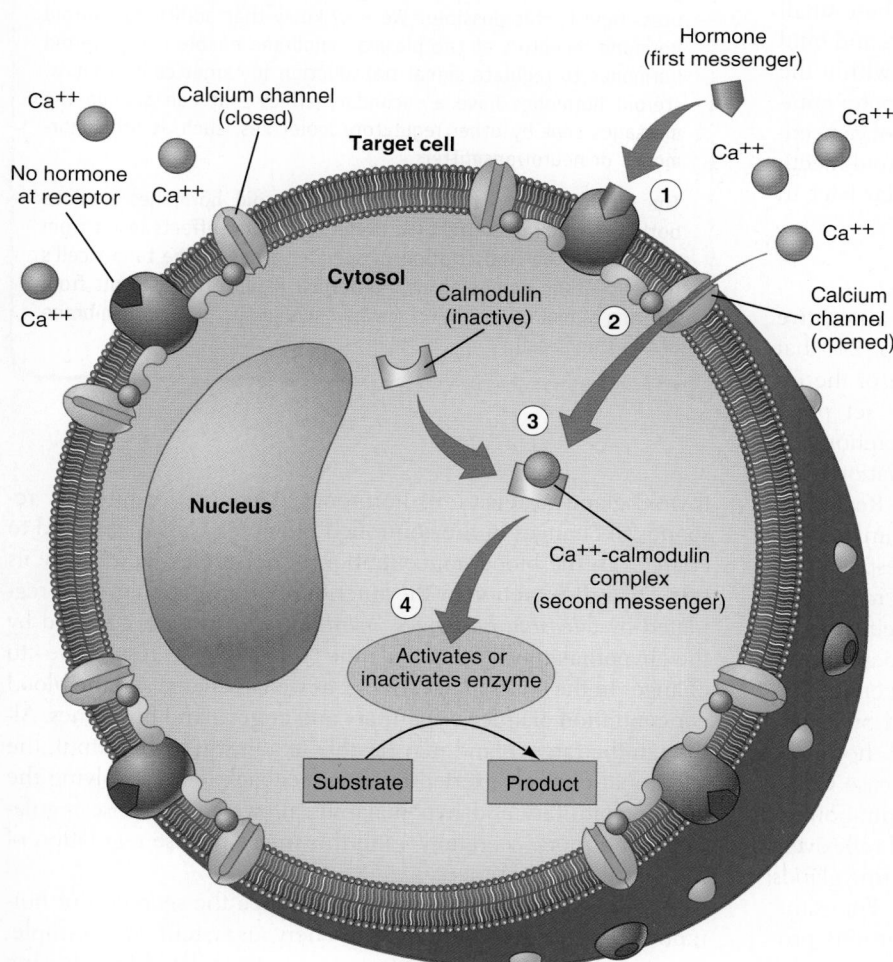

Figure 16-10 *Calcium-calmodulin as a second messenger.* In this example of a second messenger mechanism, a nonsteroid hormone (first messenger) first binds to a fixed receptor in the plasma membrane *(1)*, which activates membrane-bound proteins (G protein and PIP$_2$) that trigger the opening of calcium channels *(2)*. Calcium ions, which are normally at a higher concentration in the extracellular fluid, diffuse into the cell and bind to a calmodulin molecule *(3)*. The Ca^{++}-calmodulin complex thus formed is a second messenger that binds to an enzyme to produce an allosteric effect that promotes or inhibits the enzyme's regulatory effect in the target cell *(4)*.

Table 16-3 **Comparison of Steroid and Nonsteroid Hormones**

CHARACTERISTIC	STEROID HORMONES*	NONSTEROID HORMONES†
Chemical structure	Lipid	One or more amino acids, sometimes with added sugar groups
Stored in secretory cell	No	Yes; stored in secretory vesicles before release
Interaction with plasma membrane	No; simple diffusion through plasma membrane and into target cell	Yes; binds to specific plasma membrane receptor
Receptor	Mobile receptor in cytoplasm or nucleus	Embedded in plasma membrane
Action	Regulates gene activity (transcription of new proteins that eventually produce effects in the cell)	Triggers signal transduction cascade, producing internal "second messengers" that trigger rapid effects in the target cell
Response time	One hour to several days	Several seconds to a few minutes

*See Box 16-1 on this page for additional emerging concepts of steroid hormone characteristics.
†Some nonsteroid hormones derived from amino acids (e.g., thyroid hormones T_3 and T_4) have gene-activating actions similar to steroid hormones.

mones produce their full effects within seconds or minutes of initial binding to the target cell receptors—not the hours or days sometimes seen with steroid hormones.

The Nuclear-Receptor Mechanism

Not all nonsteroid hormones operate according to the second messenger model. The notable exception is the pair of thyroid hormones, thyroxine (T_4) and triiodothyronine (T_3). These small iodinated amino acids apparently enter their target cells and bind to receptors already associated with a DNA molecule within the nucleus of the target cell. Formation of a hormone-receptor complex triggers transcription of mRNA and the synthesis of new enzymes in a signal transduction system similar to the steroid mechanism. More information on thyroid hormones is given later in this chapter.

Regulation of Hormone Secretion

The control of hormonal secretion is usually part of a negative feedback loop. Recall from Chapter 1 (Figure 1-14, p. 24) that negative feedback loops tend to reverse any deviation of the internal environment away from its stable point (the set point value). Rarely, a positive feedback loop controls the secretion of a hormone. Recall that in positive feedback control, deviation from the stable point is exaggerated rather than reversed. Responses that result from the operation of feedback loops within the endocrine system are called *endocrine reflexes*, just as responses to nervous feedback loops (reflex arcs) are called *nervous reflexes*.

For a moment, let us focus our attention on the specific mechanisms that regulate the release of hormones from endocrine cells (Box 16-1). The simplest mechanism operates when an endocrine cell is sensitive to the physiological changes produced by its target cells (Figure 16-11). For example, parathyroid hormone (PTH) produces responses in its target cells that increase Ca^{++} concentration in the blood. When blood Ca^{++} concentration exceeds the set point value, parathyroid cells sense it and reflexively reduce their output of PTH. Secretion by many endocrine glands is regulated by a hormone produced by another gland. For example, the pituitary gland (specifically, the anterior portion) produces thyroid-stimulating hormone (TSH), which stimulates the

BOX 16-1 **Rapid Responses to Steroid Hormones**

The conventional view of steroid hormone regulation of gene activity states that one or more hours pass before the effects of the hormone reach their peak. Steroid hormones, however, can also produce some rapid effects in target cells—in just seconds or minutes. How is this possible? We now know that additional steroid hormone receptors at the plasma membrane enable many steroid hormones to regulate signal transduction in target cells. That is, steroid hormones have a secondary effect that can change the messages sent by other regulatory molecules, such as other hormones or neurotransmitters.

The emerging view therefore is that steroid hormones produce both slow and rapid effects in target cells. Slow effects result from stimulating the transcription of specific genes in the target cell's nucleus—thus producing new proteins. Rapid effects result from altering signal transduction mechanisms in the plasma membrane of the target cell.

thyroid gland to release its hormones. The anterior pituitary responds to changes in the controlled physiological variable and to changes in the blood concentration of hormones secreted by its target gland. Secretion by the anterior pituitary can in turn be regulated by *releasing hormones* or *inhibiting hormones* secreted by the hypothalamus. Hypothalamic secretion is responsive to changes in the controlled variable, as well as changes in the blood concentration of anterior pituitary and target gland hormones. Although the target gland may be able to adjust its own output, the additional controls exerted by long feedback loops involving the anterior pituitary and hypothalamus allow more precise regulation of hormone secretion—and thus more precise regulation of the internal environment.

Another mechanism that may influence the secretion of hormones by a gland is input from the nervous system. For example, secretion by the posterior pituitary is not regulated by releasing

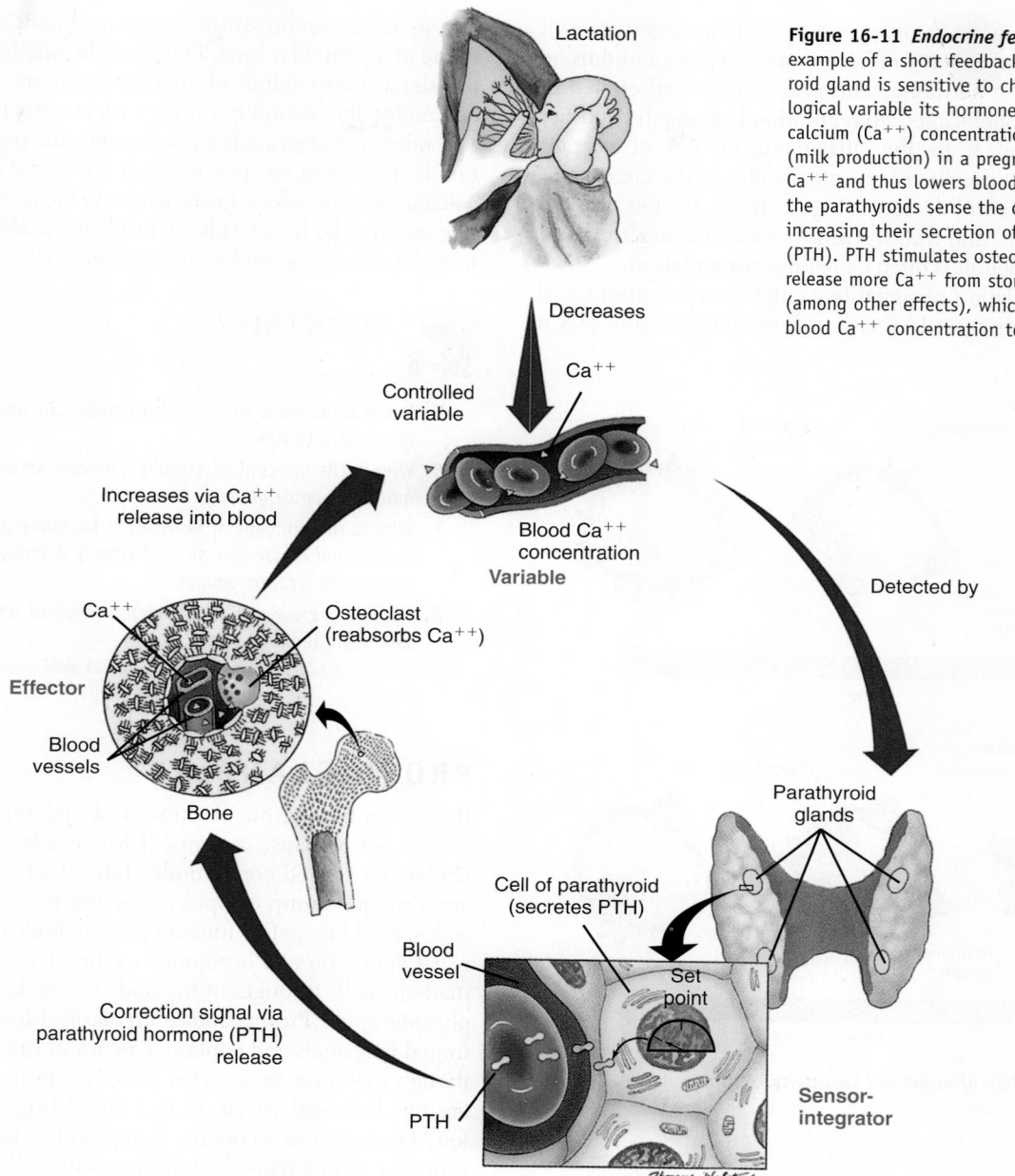

Lactation

Decreases

Ca^{++}

Controlled
variable

Blood Ca^{++}
concentration
Variable

Detected by

Increases via Ca^{++}
release into blood

Ca^{++}

Osteoclast
(reabsorbs Ca^{++})

Effector

Blood
vessels

Bone

Parathyroid
glands

Cell of parathyroid
(secretes PTH)

Blood
vessel

Set
point

Correction signal via
parathyroid hormone (PTH)
release

PTH

**Sensor-
integrator**

Figure 16-11 *Endocrine feedback loop.* In this example of a short feedback loop, each parathyroid gland is sensitive to changes in the physiological variable its hormone (PTH) controls—blood calcium (Ca^{++}) concentration. When lactation (milk production) in a pregnant woman consumes Ca^{++} and thus lowers blood Ca^{++} concentration, the parathyroids sense the change and respond by increasing their secretion of parathyroid hormone (PTH). PTH stimulates osteoclasts in bone to release more Ca^{++} from storage in bone tissue (among other effects), which increases maternal blood Ca^{++} concentration to the set point level.

hormones but by direct nervous input from the hypothalamus. Likewise, sympathetic nerve impulses that reach the medulla of the adrenal glands trigger the secretion of epinephrine and norepinephrine. Many other glands, including the pancreas, are also influenced to some degree by nervous input. That the nervous system operates with hormonal mechanisms to produce endocrine reflexes emphasizes the close functional relationship between these two systems.

Although the operation of long feedback loops tends to minimize wide fluctuations in secretion rates, the output of several hormones typically rises and falls dramatically within a short period. For example, the concentration of insulin—a hormone that can correct a rise in blood glucose concentration—increases to a high level just after a meal high in carbohydrates. The level of insulin decreases only after the blood glucose concentration returns

to its set point value. Likewise, threatening stimuli can cause a sudden, dramatic increase in the secretion of epinephrine from the adrenal medulla as part of the fight-or-flight response.

Specific examples of feedback control of hormone secretion are given later in this chapter.

Regulation of Target Cell Sensitivity

The sensitivity of a target cell to any particular hormone partly depends on how many receptors for that hormone it has. The more receptors, the more sensitive the target cell. Hormone receptors, as with other cell components, are constantly broken down by the cell and replaced with newly synthesized receptors. This mechanism not only ensures that all cell parts are "new" and working properly, but also provides a method by which the number of receptors can be changed from time to time.

If synthesis of new receptors occurs faster than degradation of old receptors, the target cell will have more receptors and thus be more sensitive to the hormone. This phenomenon, illustrated in Figure 16-12, *A*, is often called **up-regulation** because the number of receptors goes up. If, on the other hand, the rate of receptor degradation exceeds the rate of receptor synthesis, the target cell's number of receptors will decrease (Figure 16-12, *B*). Because the number of receptors, and thus the sensitivity of the target cell, go down, this phenomenon is often called **down-regulation.**

Of course, regulation of signal transduction mechanisms and gene transcription triggered by various hormones can also play a role in adjusting the sensitivity of target cells to a particular hormone at a particular time. For example, one hormone may affect the signal transduction of another hormone, thus inhibiting or enhancing the second hormone's effects (Box 16-2).

Endocrinologists are just now learning the mechanisms that control the process of receptor turnover and signal transduction in the cell and how this affects the functions of the target cell. Information uncovered so far has already led to a better understanding of important and widespread endocrine disorders such as diabetes mellitus.

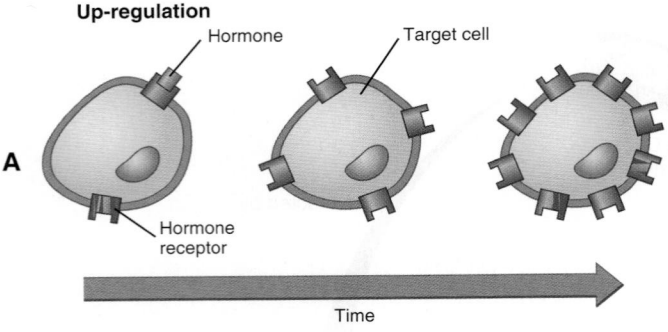

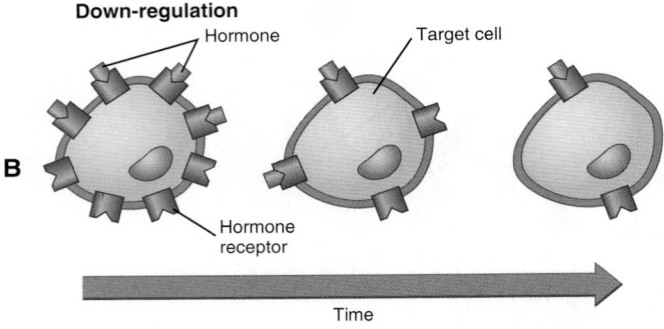

Figure 16-12 *Regulation of target cell sensitivity.* **A,** Up-regulation. **B,** Down-regulation.

 BOX 16-2: HEALTH MATTERS
"Too Much or Too Little"

Diseases of the endocrine system are numerous, varied, and sometimes spectacular. Tumors or other abnormalities frequently cause the glands to secrete too much or too little of their hormones. Production of too much hormone by a diseased gland is called **hypersecretion.** If too little hormone is produced, the condition is called **hyposecretion.**

Various endocrine disorders that appear to result from hyposecretion are actually caused by a problem in the target cells. If the usual target cells of a particular hormone have damaged receptors, too few receptors, or some other abnormality, they will not respond to that hormone properly. In other words, lack of target cell response could be a sign of hyposecretion or a sign of target cell insensitivity. *Diabetes mellitus,* for example, can result from insulin hyposecretion or from the target cells' insensitivity to insulin.

 QUICK CHECK

5. Name some ways in which hormones can work together to regulate a tissue.
6. Why is the concept of steroid hormone action called the *mobile-receptor model?*
7. Why is the concept of nonsteroid hormone action often called the *second messenger model?* Why is it known as the *fixed-membrane-receptor model?*
8. Name some ways that the secretion of an endocrine cell can be controlled.

PROSTAGLANDINS

Before continuing our discussion of endocrine glands and hormones, let us pause a moment to consider the **prostaglandins (PGs)** and related compounds (Table 16-4). The prostaglandins are a unique group of lipid molecules that serve important and widespread integrative functions in the body but do not meet the usual definition of a hormone (see Box 16-3). Prostaglandins are made by cells throughout the body by breaking apart membrane phospholipids. Prostaglandins are derived from 20-carbon unsaturated fatty acids and contain a 5-carbon ring (Figure 16-13). Although they may be secreted directly into the bloodstream, they are rapidly metabolized, so that circulating levels are extremely low. The term *tissue hormone* is appropriate because the secretion is produced in a tissue and diffuses only a short distance to other cells within the same tissue. Whereas typical hormones integrate activities of widely separated organs, prostaglandins tend to integrate activities of neighboring cells.

There are at least 16 different prostaglandins, falling into nine structural classes—prostaglandin A through prostaglandin I. Prostaglandins have been isolated and identified from a variety of

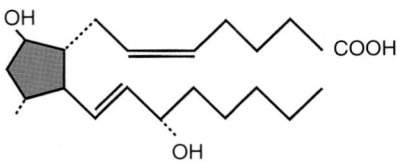

Figure 16-13 *Structure of the prostaglandin molecule.* Structure of prostaglandin $F_{2\alpha}$ ($PGF_{2\alpha}$), showing the typical 20-carbon unsaturated fatty acid structure with a characteristic 5-carbon ring (highlighted in color). See also Figure 2-27 on p. 65.

Table 16-4   Prostaglandins and Related Hormones

HORMONE	SOURCE	TARGET	PRINCIPAL ACTION
Prostaglandins (PGs)	Many diverse tissues of the body	Local cells within the source tissue	Diverse local (paracrine/autocrine) effects such as regulation of inflammation, muscle contraction in blood vessels
Thromboxanes (TXs)	Platelets	Other platelets; muscles in blood vessel walls	Increase stickiness of platelets; promote blood clotting; cause constriction of blood vessels
Leukotrienes	Several types of white blood cells (leukocytes)	Local cells of various types	Produce local inflammation triggered by allergens, including constriction of airways (as in asthma) and other inflammatory responses

BOX 16-3   Local Hormones

The classical view of hormones is that they are secreted from a tissue into the bloodstream and have their effects in target cells at some distance from their source. This is the standard definition of an **endocrine** hormone (part *A* of figure). However, some hormones have primarily local effects—that is, effects *within* the source tissue. To distinguish classical endocrine hormones from local regulators such as prostaglandins and related compounds, scientists often use these more precise terms:

Paracrine hormones—hormones that regulate activity in nearby cells within the same tissue as their source (part *B* of figure)

Autocrine hormones—hormones that regulate activity in the secreting cell itself (part *C* of figure)

To avoid confusion, scientists most often refer only to endocrine hormones simply as "hormones." For local regulators, scientists use general terms such as "tissue hormone" or "local regulator"—or more specific terms such as "paracrine" or "autocrine" factor.

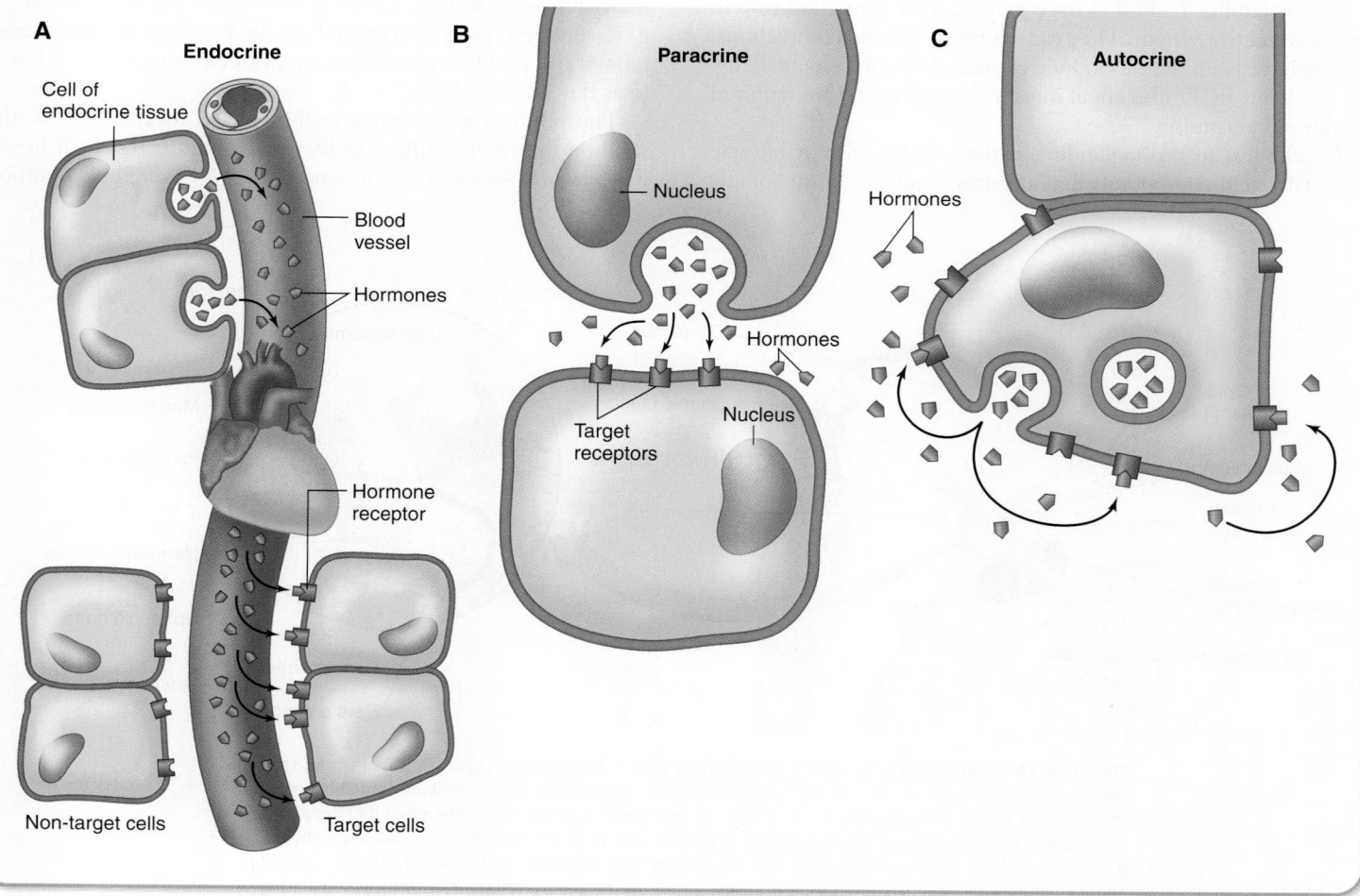

tissues. The first prostaglandin was discovered in semen, so it was attributed to the prostate gland (hence the name, prostaglandin). Later, researchers found that the seminal vesicles, not the prostate, secreted the prostaglandin that they had found. Many other tissues are now known to secrete prostaglandins. As a group, the prostaglandins have diverse physiological effects and are among the most varied and potent of any naturally occurring biological compounds. They are intimately involved in overall endocrine regulation by influencing adenyl cyclase–cAMP interaction within the cell's plasma membrane (see Figure 16-9). Specific biological effects depend on the class of prostaglandin.

Intraarterial infusion of prostaglandins A (PGAs) results in an immediate fall in blood pressure accompanied by an increase in regional blood flow to several areas, including the coronary and renal systems. PGAs apparently produce this effect by causing relaxation of smooth muscle fibers in the walls of certain arteries and arterioles.

Prostaglandins E (PGEs) have an important role in various vascular, metabolic, and gastrointestinal functions. Vascular effects include regulation of red blood cell deformability and platelet aggregation (see Chapter 17). PGEs also have a role in systemic inflammations, such as fever. Common anti-inflammatory agents such as aspirin and ibuprofen produce some of their effects by inhibiting PGE synthesis. These drugs act by blocking one or more of the *cyclooxygenase* enzymes such as *COX-1* and *COX-2*, as you may recall from Box 2-9 on p. 65. PGE also regulates hydrochloric acid secretion in the stomach, helping to prevent gastric ulcers.

Prostaglandins F (PGFs) have an especially important role in the reproductive system. They cause uterine muscle contractions, so they have been used to induce labor and thus accelerate delivery of a baby. PGFs also affect intestinal motility and are required for normal peristalsis.

In addition to prostaglandins, various tissues also synthesize other fatty acid compounds that are structurally and functionally similar to prostaglandins. Examples include the **thromboxanes** and **leukotrienes.** As with prostaglandins, these compounds may also be referred to as "tissue hormones" because of their local, yet potent, regulatory effects.

The potential therapeutic use of prostaglandins and related compounds, which are found in almost every body tissue and are capable of regulating hormone activity on the cellular level, has been described as the most revolutionary development in medicine since the advent of antibiotics. They are likely to play increasingly important roles in the treatment of diverse conditions, such as hypertension, coronary thrombosis, asthma, and ulcers.

QUICK CHECK

9. Why are prostaglandins sometimes called *tissue hormones?*
10. Why are prostaglandins considered to be important in clinical applications?

PITUITARY GLAND
Structure of the Pituitary Gland

The **pituitary gland** (formerly called the *hypophysis*) is a small but mighty structure. It measures only 1.2 to 1.5 cm (about ½ inch) across. By weight, it is even less impressive—only about 0.5 g (¹⁄₆₀ ounce)! And yet so crucial are the functions of the anterior lobe of the pituitary gland that, in past centuries, it was referred to as the "master gland."

The pituitary gland has a well-protected location within the skull on the ventral surface of the brain (Figure 16-14). It lies in the *pituitary fossa* of the sella turcica and is covered by a portion

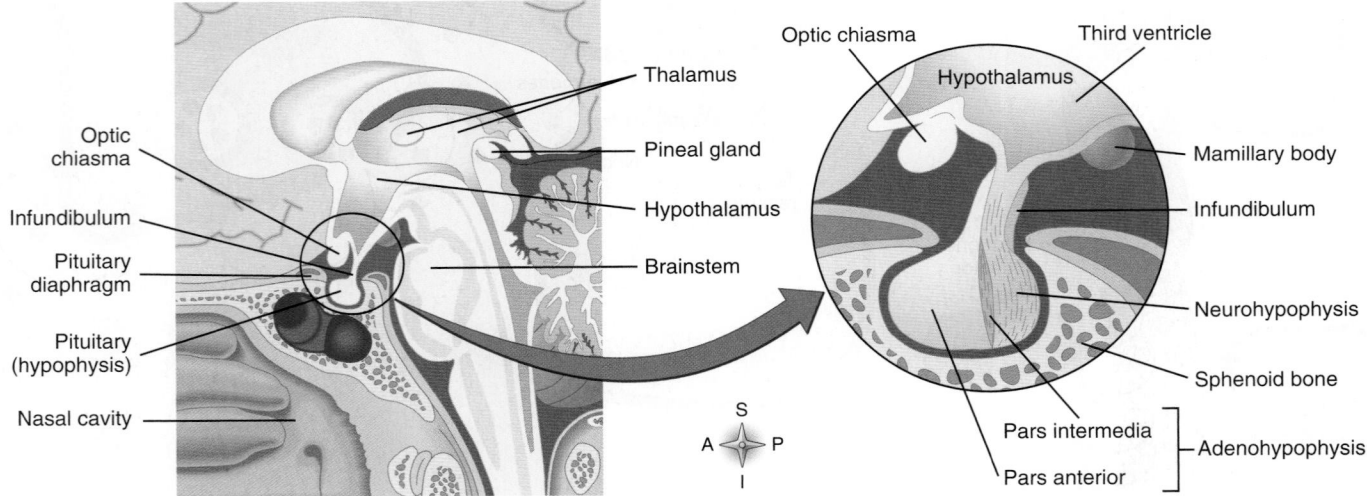

Figure 16-14 *Location and structure of the pituitary gland.* The pituitary gland is located within the sella turcica of the skull's sphenoid bone and is connected to the hypothalamus by a stalklike infundibulum. The infundibulum passes through a gap in the portion of the dura mater that covers the pituitary (the pituitary diaphragm). The inset shows that the pituitary is divided into an anterior portion, the adenohypophysis, and a posterior portion, the neurohypophysis. The adenohypophysis is further subdivided into the pars anterior and pars intermedia. The pars intermedia is almost absent in the adult pituitary. *FSH,* Follicle-stimulating hormone; *LH,* luteinizing hormone.

of the dura mater called the *pituitary diaphragm.* The gland has a stemlike stalk, the **infundibulum,** which connects it to the hypothalamus of the brain.

Although the pituitary looks like one gland, it actually consists of two separate glands—the **adenohypophysis,** or anterior pituitary gland, and the **neurohypophysis,** or posterior pituitary gland. In the embryo, the adenohypophysis develops from an upward projection of the pharynx and is composed of regular endocrine tissue. The neurohypophysis, on the other hand, develops from a downward projection of the brain and is composed of neurosecretory tissue. These histological differences are incorporated into their names—*adeno* means "gland," and *neuro* means "nervous." As you may suspect, the hormones secreted by the adenohypophysis serve very different functions from those released by the neurohypophysis.

Adenohypophysis (Anterior Lobe of Pituitary)

The **adenohypophysis,** the anterior portion of the pituitary gland, is divided into two parts—the *pars anterior* and the *pars intermedia.* The pars anterior forms the major portion of the adenohypophysis and is divided from the tiny pars intermedia by a narrow cleft and some connective tissue (Figure 16-14).

The tissue of the adenohypophysis is composed of irregular clumps of secretory cells supported by fine connective tissue fibers and surrounded by a rich vascular network.

Traditionally, histologists have identified three types of cells according to their affinity for certain types of stains: *chromophobes* (literally "afraid of color"), *acidophils* ("acid [stain] lovers"), and *basophils* ("base [stain] lovers"). All three types are visible in the photomicrograph shown in Figure 16-15. Currently, however, cells of the adenohypophysis are more often classified by their secretions into five types:

1. **Somatotrophs**—secrete growth hormone (GH)
2. **Corticotrophs**—secrete adrenocorticotropic hormone (ACTH)
3. **Thyrotrophs**—secrete thyroid-stimulating hormone (TSH)
4. **Lactotrophs**—secrete prolactin (PRL)
5. **Gonadotrophs**—secrete luteinizing hormone (LH) and follicle-stimulating hormone (FSH)

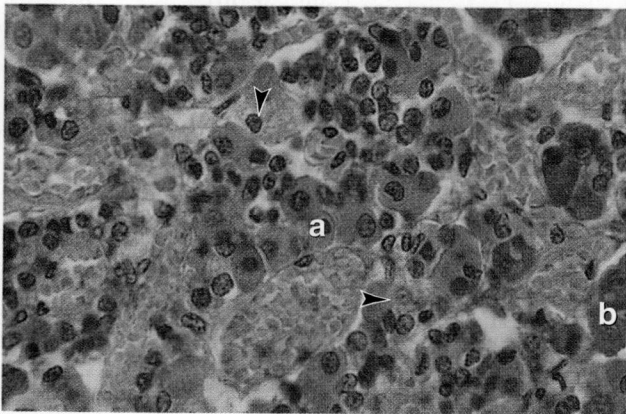

Figure 16-15 *Histology of the adenohypophysis.* In this light micrograph, nonstaining chromophobes are indicated by arrowheads. Examples of hormone-secreting cells are labeled *a* (acidophil) and *b* (basophil).

Figure 16-16 summarizes the hormones of the adenohypophysis and shows the primary locations of their target cells.

Growth Hormone

Growth hormone (GH), or **somatotropin (STH),** is thought to promote bodily growth indirectly by stimulating the liver to produce certain growth factors, which, in turn, accelerate amino acid transport into cells. Rapid entrance of amino acids from the blood into the cells allows protein anabolism within the cells to accelerate. Increased protein anabolism allows an increased rate of growth. GH promotes the growth of bone, muscle, and other tissues (Box 16-4).

BOX 16-4: HEALTH MATTERS
Growth Hormone Abnormalities

Hypersecretion of GH during the growth years (before ossification of the epiphyseal plates) causes an abnormally rapid rate of skeletal growth. This condition is known as **gigantism** (see figure, *left*). Hypersecretion after skeletal fusion has occurred can result in **acromegaly,** a condition in which cartilage still left in the skeleton continues to form new bone. This abnormal growth may result in a distorted appearance because of the enlargement of the hands, feet, face, jaw (causing separation of the teeth), and other body parts. Overlying soft tissue may also be affected—for instance, the skin often thickens and the pores become more pronounced.

Hyposecretion of GH during growth years may result in stunted body growth, known as **pituitary dwarfism** (see figure, *right*). Formerly, patients were treated only with GH extracted from human tissues. Since the 1980s, the availability of human GH produced by genetically engineered bacteria has made the treatment obtainable for many more patients. However, concerns have been raised about possible adverse side effects associated with human GH from bacterial sources.

Growth hormone abnormalities. The man on the left exhibits gigantism, and the man on the right exhibits pituitary dwarfism. The two men in the middle are of average height.

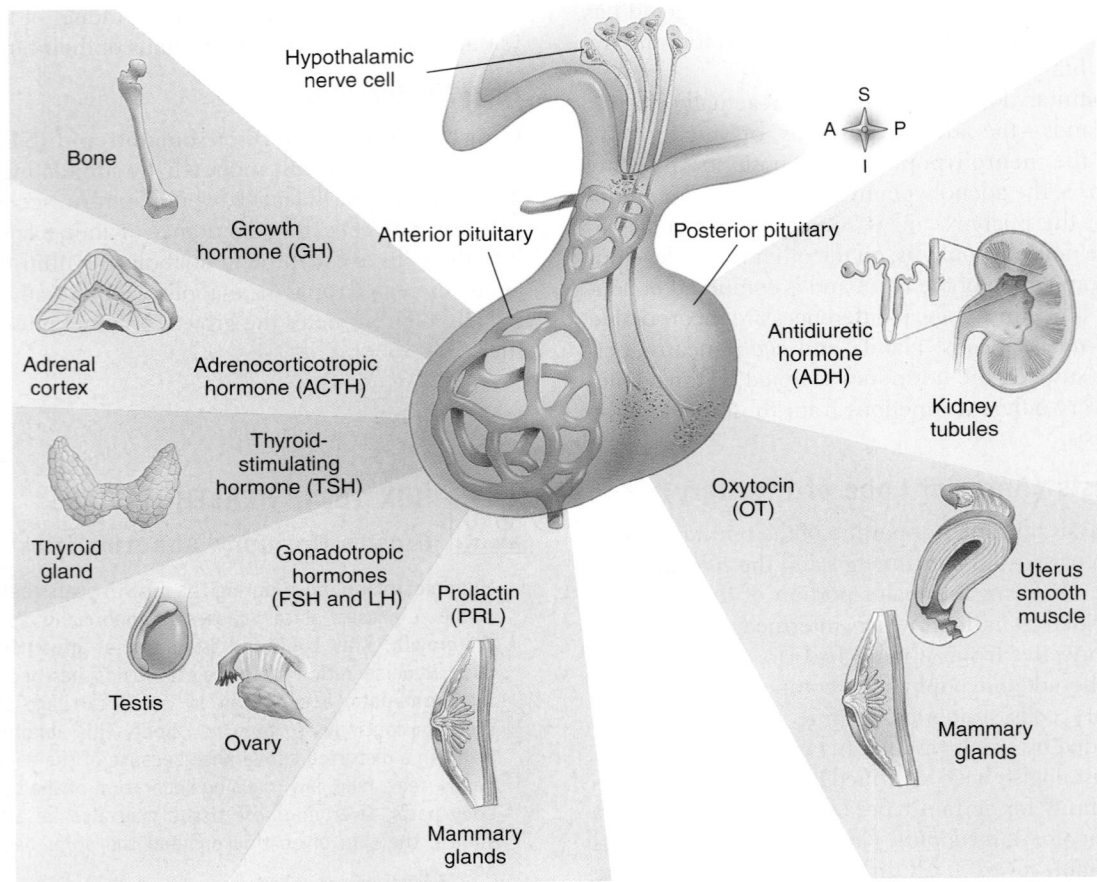

Figure 16-16 *Pituitary hormones*. Some of the major hormones of the adenohypophysis and neurohypophysis and their principal target organs.

In addition to stimulating protein anabolism, GH also stimulates fat metabolism. GH accelerates mobilization of lipids from storage in adipose cells and also speeds up the catabolism of those lipids after they have entered another cell. In this way, GH tends to shift a cell's use of nutrients away from carbohydrate (glucose) catabolism and toward lipid catabolism as an energy source. Because less glucose is then removed from the blood by cells, the blood glucose levels tend to rise. Thus GH is said to have a *hyperglycemic effect*. Insulin (from the pancreas) has the opposite effect—it promotes glucose entry into cells, producing a *hypoglycemic effect*. Therefore GH and insulin function as antagonists. The balance between these two hormones is vital to maintaining a homeostasis of blood glucose levels.

GH affects metabolism in these ways:

- Promotes protein anabolism (growth, tissue repair)
- Promotes lipid mobilization and catabolism
- Indirectly inhibits glucose metabolism
- Indirectly increases blood glucose levels

Prolactin

Prolactin (PRL), produced by acidophils in the pars anterior, is also called *lactogenic hormone*. Both names of this hormone suggest its function in "generating" or initiating milk secretion (lac-

tation). During pregnancy, a high level of PRL promotes the development of the breasts in anticipation of milk secretion. At the birth of an infant, PRL in the mother stimulates the mammary glands to begin milk secretion.

Hypersecretion of PRL may cause lactation in nonnursing women, disruption of the menstrual cycle, and impotence in men. **Hyposecretion** of PRL is usually insignificant except in women who want to nurse their children. Milk production cannot be initiated or maintained without PRL.

Tropic Hormones

Tropic hormones are hormones that have a stimulating effect on other endocrine glands. These hormones stimulate the development of their target glands and tend to stimulate synthesis and secretion of the target hormone (Box 16-5). Four principal tropic hormones are produced and secreted by the basophils of the pars anterior:

1. **Thyroid-stimulating hormone (TSH),** or *thyrotropin*, promotes and maintains the growth and development of its target gland—the thyroid. TSH also causes the thyroid gland to secrete its hormones.
2. **Adrenocorticotropic hormone (ACTH),** or *adrenocorticotropin*, promotes and maintains normal growth and de-

BOX 16-5: HEALTH MATTERS
Tropic Hormone Abnormalities

Hypersecretion of the tropic hormones may result from a pituitary tumor. The tropic hormones, produced at higher than normal levels, cause hypersecretion in their target glands. This may result in various effects throughout the body. Early hypersecretion of gonadotropins may lead to an abnormally early onset of puberty.

Hyposecretion of tropic hormones often causes their target glands to secrete less than a normal amount of their hormones. This may disrupt reproduction, kidney function, overall metabolism, and other processes.

velopment of the cortex of the adrenal gland. ACTH also stimulates the adrenal cortex to synthesize and secrete some of its hormones.

3. **Follicle-stimulating hormone (FSH)** stimulates structures within the ovaries, primary follicles, to grow toward maturity. Each follicle contains a developing egg cell (ovum), which is released from the ovary during ovulation. FSH also stimulates the follicle cells to synthesize and secrete estrogens (female sex hormones). In the male, FSH stimulates the development of the seminiferous tubules of the testes and maintains spermatogenesis (sperm production) by them.

4. **Luteinizing hormone (LH)** stimulates the formation and activity of the corpus luteum of the ovary. The corpus luteum (meaning "yellow body") is the tissue left behind when a follicle ruptures to release its egg during ovulation. The corpus luteum secretes progesterone and estrogens when stimulated by LH. LH also supports FSH in stimulating the maturation of follicles. In males, LH stimulates interstitial cells in the testes to develop and then synthesize and secrete testosterone (the male sex hormone).

FSH and LH are called **gonadotropins** because they stimulate the growth and maintenance of the gonads (ovaries and testes). During childhood the adenohypophysis secretes insignificant amounts of the gonadotropins. A few years before puberty, gonadotropin secretion is gradually increased. Then, suddenly, their secretion spurts, and the gonads are stimulated to develop and begin their normal functions.

Control of Secretion in the Adenohypophysis

The cell bodies of neurons in certain parts of the hypothalamus synthesize chemicals that their axons secrete into the blood. These chemicals, generally called **releasing hormones,** travel through a complex of small blood vessels called the **hypophyseal portal system** (Figure 16-17). A *portal system* is an arrangement of blood vessels in which blood exiting one tissue is immediately carried to a second tissue before being returned to the heart and lungs for oxygenation and redistribution. The hypophyseal portal system carries blood from the hypothalamus directly to the adenohypophysis, where the target cells of the releasing hormones are located. The advantage of a portal system in the hypophysis is

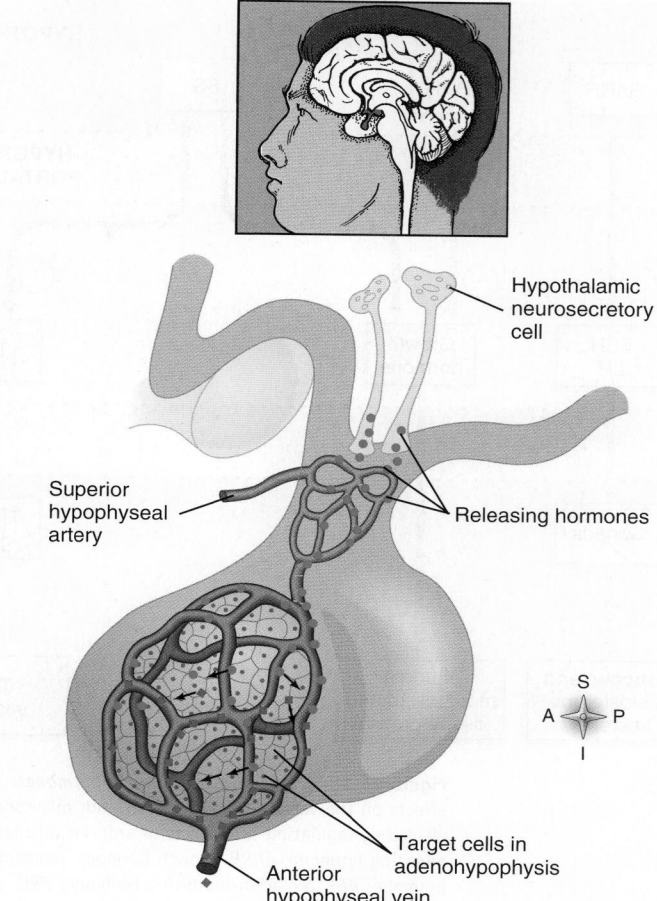

Figure 16-17 *Hypophyseal portal system.* Neurons in the hypothalamus secrete releasing hormones into veins that carry the releasing hormones directly to the vessels of the adenohypophysis, thus bypassing the normal circulatory route.

that a small amount of hormone can be delivered directly to its target tissue without the great dilution that would occur in the general circulation. The releasing hormones that arrive in the adenohypophysis by means of this portal system influence the secretion of hormones by acidophils and basophils. In this manner, the hypothalamus directly regulates the secretion of the adenohypophysis. You can see that the supposed "master gland" really has a master of its own—the hypothalamus.

The following is a list of some of the important hormones secreted by the hypothalamus into the hypophyseal portal system:

- Growth hormone–releasing hormone (GHRH)
- Growth hormone–inhibiting hormone (GHIH) (also called somatostatin [SS])
- Corticotropin-releasing hormone (CRH)
- Thyrotropin-releasing hormone (TRH)
- Gonadotropin-releasing hormone (GnRH)
- Prolactin-releasing hormone (PRH)
- Prolactin-inhibiting hormone (PIH)

Figure 16-18 and Table 16-5 list functions of each releasing hormone. Before consulting the figure or table, try to deduce their functions from their names.

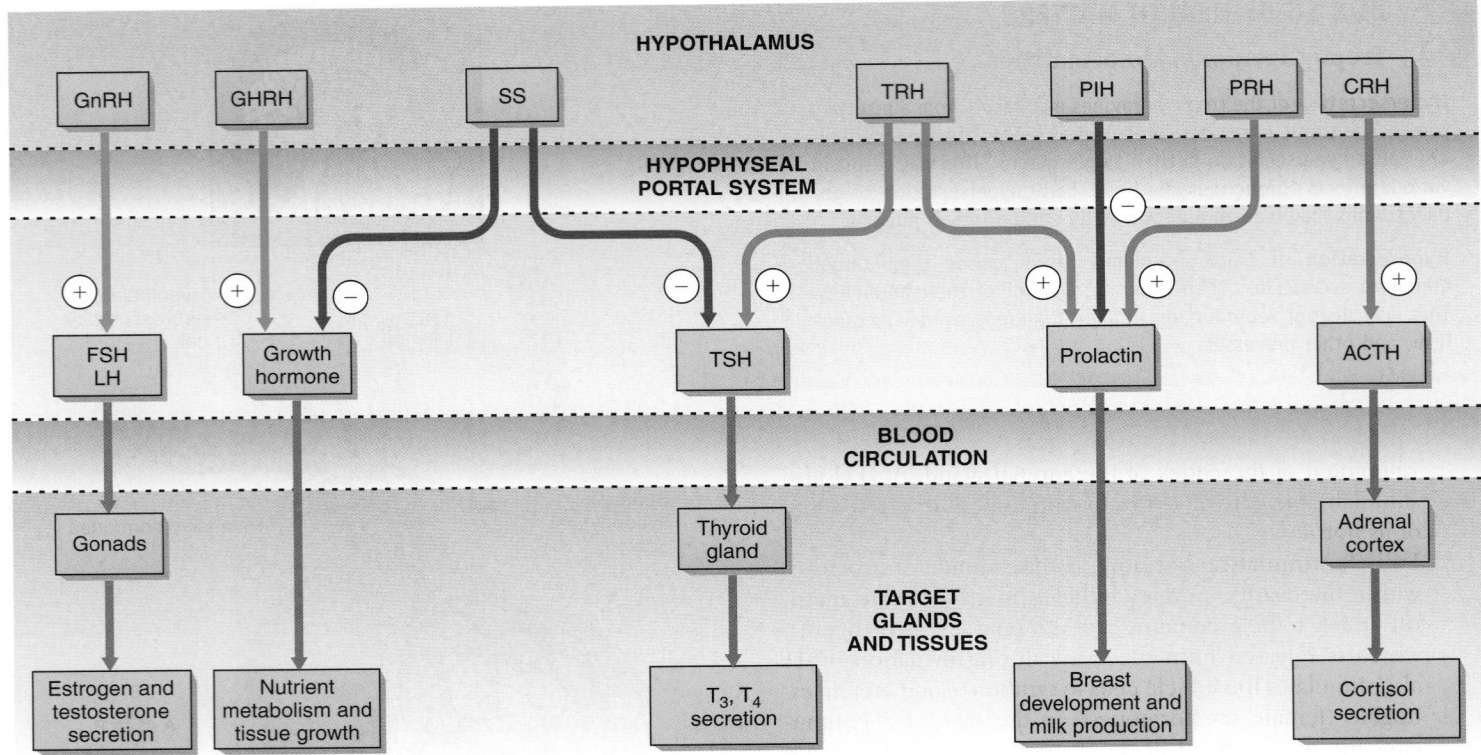

Figure 16-18 *Action of hypothalamic hormones.* Hypothalamic hormones have releasing or inhibiting effects on the various cells of the anterior pituitary, thus regulating anterior pituitary secretion—and thus ultimately regulating the effects of anterior pituitary hormones throughout the body. *GnRH,* Gonadotropin-releasing hormone; *GHRH,* growth hormone–releasing hormone; *SS,* somatostatin; *TRH,* thyroid-releasing hormone; *PIH,* prolactin-inhibiting hormone; *PRH,* prolactin-releasing hormone; *CRH,* corticotropin-releasing hormone; *FSH,* follicle-stimulating hormone; *LH,* luteinizing hormone; *TSH,* thyroid-stimulating hormone; *ACTH,* adrenocorticotropic hormone; *T₃,* triiodothyronine; *T₄,* thyroxine.

Table 16-5 Hormones of the Hypothalamus

HORMONE	SOURCE	TARGET	PRINCIPAL ACTION
Growth hormone–releasing hormone (GHRH)	Hypothalamus	Adenohypophysis (somatotrophs)	Stimulates secretion (release) of growth hormone
Growth hormone–inhibiting hormone (GHIH), or somatostatin	Hypothalamus	Adenohypophysis (somatotrophs)	Inhibits secretion of growth hormone
Corticotropin-releasing hormone (CRH)	Hypothalamus	Adenohypophysis (corticotrophs)	Stimulates release of adrenocorticotropic hormone (ACTH)
Thyrotropin-releasing hormone (TRH)	Hypothalamus	Adenohypophysis (thyrotrophs)	Stimulates release of thyroid-stimulating hormone (TSH)
Gonadotropin-releasing hormone (GnRH)	Hypothalamus	Adenohypophysis (gonadotrophs)	Stimulates release of gonadotropins (FSH and LH)
Prolactin-releasing hormone (PRH)	Hypothalamus	Adenohypophysis (corticotrophs)	Stimulates secretion of prolactin
Prolactin-inhibiting hormones (PIH)	Hypothalamus	Adenohypophysis (corticotrophs)	Inhibits secretion of prolactin

Through negative feedback mechanisms, the hypothalamus adjusts the secretions of the adenohypophysis, and the adenohypophysis adjusts the secretions of its target glands, which in turn adjust the activity of their target tissues (Box 16-6). For example, Figure 16-19 shows the negative feedback control of the secretion of TSH and thyroid hormone (T₃ and T₄).

Hormone secretion can occur in pulses or peaks, as we see in a graph of minute by minute variations in GH secretion (Figure 16-20). The peaks result from many somatotroph cells in the pituitary collectively increasing their rate of secretion of GH. Such increases result from pulses in GHRH secretion by the hypothalamus (see Figure 16-18), which are especially large during sleep. Exercise,

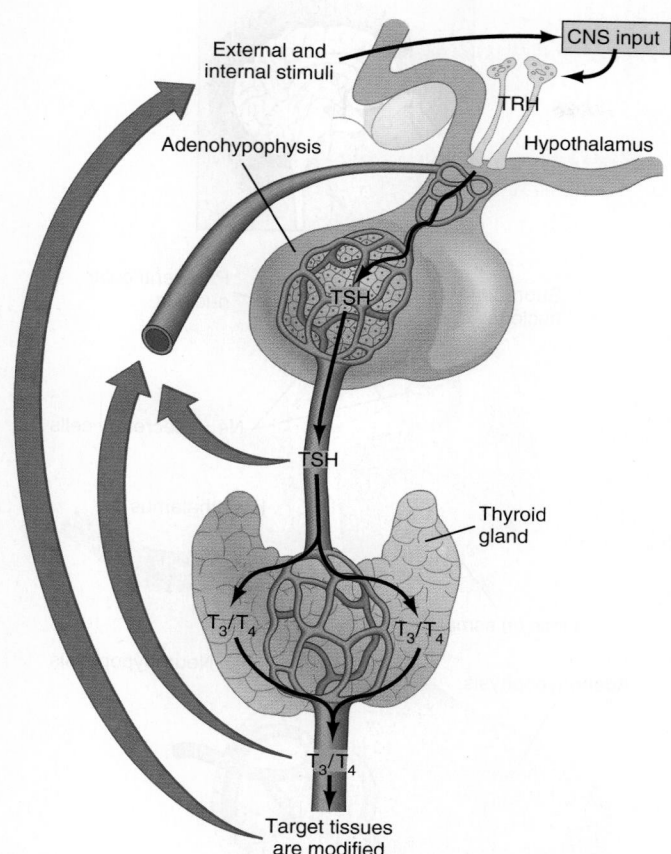

BOX 16-6: HEALTH MATTERS
Clinical Evidence of Feedback Control

Clinical facts furnish interesting evidence about feedback control of hormone secretion by the anterior pituitary gland and its target glands. For instance, patients who have the pituitary gland removed (**hypophysectomy**) surgically or by radiation must be given hormone replacement therapy for the rest of their lives. If not, they will develop thyroid, adrenal cortical, and gonadotropic deficiencies—deficiencies of the anterior pituitary's target gland hormones. Another well-known clinical fact is that estrogen deficiency develops in women between 40 and 50 years of age. By then the ovaries seem to have tired of producing hormones and ovulating each month. They no longer respond to FSH stimulation, so estrogen deficiency develops, brings about menopause, and persists after menopause. What therefore would you deduce is true of the blood concentration of FSH after menopause? Apply the principle that a low concentration of a target gland hormone stimulates tropic hormone secretion by the anterior pituitary gland and you will understand that FSH levels rise, indeed, after menopause.

Figure 16-19 *Negative feedback control by the hypothalamus.* In this example, the secretion of thyroid hormone (T_3 and T_4) is regulated by a number of negative feedback loops. A long negative feedback loop *(thin line)* allows the CNS to influence hypothalamic secretion of thyrotropin-releasing hormone (TRH) by nervous feedback from the targets of T_3/T_4 (and from other nerve inputs). The secretion of TRH by the hypothalamus and thyroid-stimulating hormone (TSH) by the adenohypophysis is also influenced by shorter feedback loops *(thicker lines)*, allowing great precision in the control of this system.

stress, and high-protein meals can cause an increase in the frequency of these peaks.

Before leaving the subject of control of pituitary secretion, we want to call attention to another concept about the hypothalamus. It functions as an important part of the body's complex machinery for responding to stress situations. For example, in severe pain or intense emotions, the cerebral cortex—especially the limbic area—sends impulses to the hypothalamus. The impulses stimulate the hypothalamus to secrete its releasing hormones into the hypophyseal portal veins. Circulating quickly to the adenohypophysis, they stimulate it to secrete more of its hormones. These in turn stimulate increased activity by the pituitary's target struc-

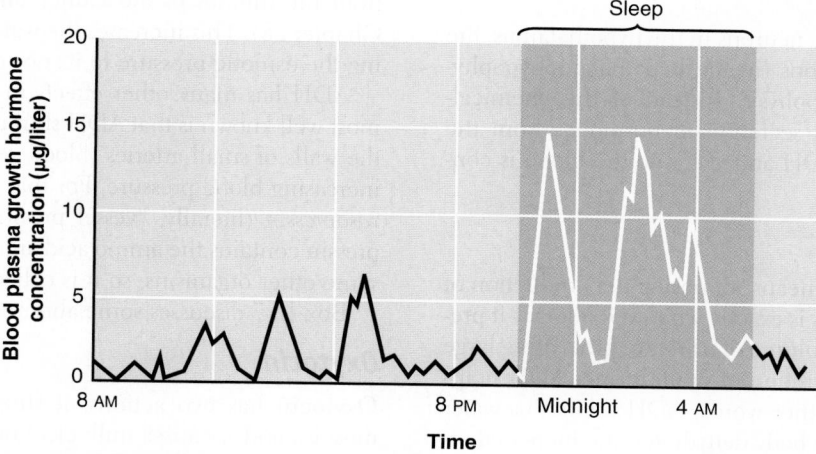

Figure 16-20 *Secretion of growth hormone (GH).* The amount of GH in the blood plasma of a 23-year-old woman was plotted on a graph every 5 minutes over the course of 24 hours. Note that GH fluctuates chaotically minute by minute, with occasional peaks that get larger after she falls asleep. The surges of GH result in part from pulses of GHRH (GH-releasing hormone) from the hypothalamus.

tures. In essence, what the hypothalamus does through its releasing hormones is to translate nerve impulses into hormone secretion by endocrine glands. Thus the hypothalamus links the nervous system to the endocrine system. It integrates the activities of these two great integrating systems—particularly, it seems, in times of stress. When survival is threatened, the hypothalamus can take over the adenohypophysis and thus gain control of literally every cell in the body. We will discuss stress in more detail in Chapter 22.

The mind-body link provided by the hypothalamus has tremendous implications. It means that the cerebrum can do more than just receive sensory impulses and send out impulses to muscles and glands. It means that our thoughts and emotions—our minds—can, by way of the hypothalamus, influence the functions of all of our billions of cells. In short, the brain has two-way contact with every tissue of the body. Thus the state of the body can influence mental processes, and the state of the mind can affect the functioning of the body. Therefore both psychosomatic (mind influencing the body) and somatopsychic (body influencing the mind) relationships exist between human body systems and the brain.

QUICK CHECK

11. What are the two main divisions of the pituitary called? How are they distinguished by location and histology?
12. Name three hormones produced by the adenohypophysis, and give their main functions.
13. What is a tropic hormone? A releasing hormone?

Neurohypophysis (Posterior Lobe of Pituitary)

The **neurohypophysis** serves as a storage and release site for two hormones: **antidiuretic hormone (ADH)** and **oxytocin (OT)**. The cells of the neurohypophysis do not themselves make these hormones. Instead, neurons whose bodies are in either the *supraoptic* or the *paraventricular nuclei* of the hypothalamus synthesize them (Figure 16-21).

From the cell bodies of these neurons in the hypothalamus, the hormones pass down along axons (in the hypothalamohypophyseal tract) into the neurohypophysis. Instead of the chemical-releasing factors that triggered secretion of hormones from the adenohypophysis, release of ADH and OT into the blood is controlled by nervous stimulation.

Antidiuretic Hormone

The term *antidiuresis* literally means "opposing the production of a large urine volume." And this is exactly what ADH does—it prevents the formation of a large volume of urine. In preventing large losses of fluid through the excretion of dilute urine, ADH helps the body conserve water. In other words, ADH maintains water balance in the body. When the body dehydrates, the increased osmotic pressure of the blood is detected by special *osmoreceptors* near the supraoptic nucleus. This triggers the release of ADH from the neurohypophysis. ADH causes water to be reabsorbed

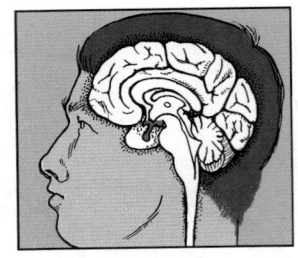

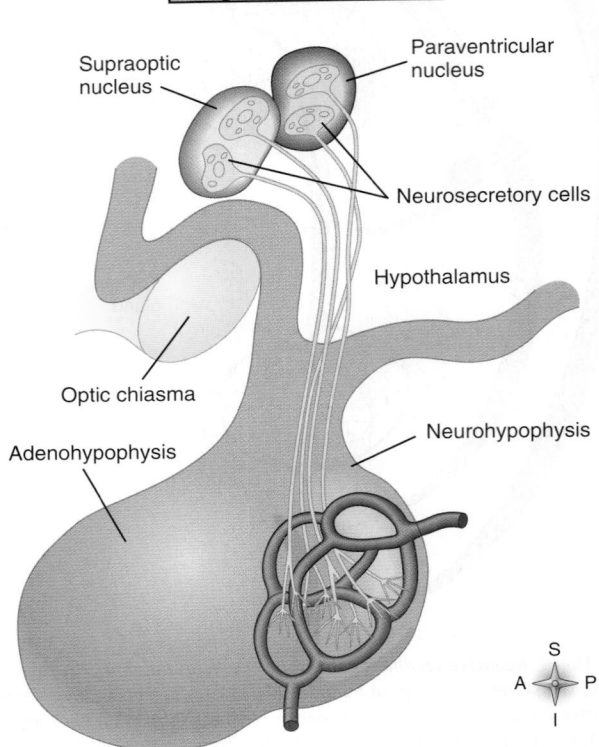

Figure 16-21 *Relationship of the hypothalamus and neurohypophysis.* Neurosecretory cells have their cell bodies in the hypothalamus and their axon terminals in the neurohypophysis. Thus hormones synthesized in the hypothalamus are actually released from the neurohypophysis.

from the tubules of the kidney and returned to the blood (see Chapter 28). This increases the water content of the blood, restoring the osmotic pressure to its normal lower level.

ADH has many other effects in the body as well. One of the most well known is that ADH stimulates contraction of muscles in the walls of small arteries (blood vessels that supply tissues), thus increasing blood pressure. For that reason, ADH is also known as *vasopressin* (literally, "vessel pressure substance"). Human vasopressin contains the amino acid arginine, unlike the vasopressin of some other organisms, so it is called **arginine vasopressin (AVP)**.

Box 16-7 discusses some abnormalities associated with ADH.

Oxytocin

Oxytocin has two actions: it stimulates contraction of uterine muscles and it causes milk ejection from the breasts of lactating women. Under the influence of OT, milk-producing *alveolar* cells release their secretion into the ducts of the breast. This is very important because milk cannot be removed by suckling un-

Table 16-6   Hormones of the Pituitary Gland

HORMONE	SOURCE	TARGET	PRINCIPAL ACTION
Growth hormone (GH) (somatotropin [STH])	Adenohypophysis (somatotrophs)	General	Promotes growth by stimulating protein anabolism and fat mobilization
Prolactin (PRL) (lactogenic hormone)	Adenohypophysis (lactotrophs)	Mammary glands (alveolar secretory cells)	Promotes milk secretion
Thyroid-stimulating hormone (TSH)*	Adenohypophysis (thyrotrophs)	Thyroid gland	Stimulates development and secretion in the thyroid gland
Adrenocorticotropic hormone (ACTH)*	Adenohypophysis (corticotrophs)	Adrenal cortex	Promotes development and secretion in the adrenal cortex
Follicle-stimulating hormone (FSH)*	Adenohypophysis (gonadotrophs)	Gonads (primary sex organs)	Female: promotes development of ovarian follicle; stimulates estrogen secretion Male: promotes development of testes; stimulates sperm production
Luteinizing hormone (LH)*	Adenohypophysis (gonadotrophs)	Gonads	Female: triggers ovulation; promotes development of corpus luteum Male: stimulates production of testosterone
Antidiuretic hormone (ADH), or arginine vaso-pressin (AVP)	Neurohypophysis	Kidney	Promotes water retention by kidney tubules; raises blood pressure by stimulating muscles in walls of small arteries
Oxytocin (OT)	Neurohypophysis	Uterus and mammary glands	Stimulates uterine contractions; stimulates ejection of milk into ducts of mammary glands

*Tropic hormones.

BOX 16-7: HEALTH MATTERS

Antidiuretic Hormone (ADH) Abnormalities

Hyposecretion of ADH can lead to **diabetes insipidus,** a condition in which the patient produces abnormally large amounts of urine. ADH, administered under the name vasopressin (Pitressin), can alleviate this symptom. Studies have shown that ADH may be involved in learning and memory, so investigators are looking into the possibility of administering ADH to reverse the memory loss associated with senility.

less it has first been ejected into the ducts. Throughout nursing, the mechanical and psychological stimulation of the baby's suckling action triggers the release of more OT. In other words, OT secretion is regulated by a *positive feedback* mechanism: the baby suckles, which increases OT levels, which provides more milk, so the baby continues to suckle, which increases OT levels, and so on. OT, together with prolactin, ensures successful nursing. Prolactin prepares the breast for milk production and stimulates cells to produce milk. The milk is not released, however, until OT permits it to do so.

It is OT's other action—its stimulation of uterine contractions—that gives it its name: oxytocin (literally "swift childbirth"). OT stimulates the uterus to strengthen the strong, muscular labor contractions that occur during childbirth. OT secretion is regulated here again by means of a positive feedback mechanism. After they have begun, uterine contractions push on receptors in the pelvis, which triggers the release of more OT, which again pushes on the pelvic receptors, and so on. The wavelike contractions

continue to some degree after childbirth, which helps the uterus expel the placenta and then return to its unstretched shape. Commercial preparations of OT have been given to stimulate contractions after childbirth to lessen the danger of uterine hemorrhage. Important characteristics of the hormones secreted by the pituitary—both the adenohypophysis and the neurohypophysis—are summarized in Table 16-6.

PINEAL GLAND

The **pineal gland,** or *pineal body,* is a tiny (1 cm [about ⅜ inch]) structure resembling a pine nut located on the dorsal aspect of the brain's diencephalon region (see Figure 16-14). It is a member of two body systems because it acts as a part of the nervous system (because it receives and processes nerve stimuli relayed from other parts of the nervous system) and as a part of the endocrine system (because it secretes hormones).

The pineal gland produces small amounts of many different hormones, but the principal hormone is melatonin. As you recall from Chapter 13, melatonin is an altered form of serotonin that acts as a hormone because it is released by neurosecretory cells of the pineal into the blood to regulate functions throughout the body. Recall also that melatonin levels rise and fall in a cycle related to the changing levels of sunlight throughout the day—melatonin levels rise when sunlight is absent, triggering sleepiness. Thus the pineal gland and melatonin act as important parts of a person's *biological clock*—the timekeeping mechanism of the body.

Melatonin, whose secretion is inhibited by the presence of sunlight, may also affect a person's mood. This is not surprising, given melatonin's similarity to another mood-altering molecule—serotonin. A mental disorder called *seasonal affective disorder (SAD)*, in which a patient suffers severe depression only in winter (when day length is shorter), has been linked to the pineal gland. Patients suffering from this "winter depression" are often advised to expose themselves to special high-intensity lights for several hours each evening during the winter months. Apparently, light stimulates the pineal gland for a longer period, which reduces the blood levels of mood-altering melatonin. The symptoms of depression are thus reduced or eliminated.

QUICK CHECK

14. Where are the hormones of the neurohypophysis manufactured? From which location in the body are they released into the bloodstream?
15. Name the two hormones of the neurohypophysis.
16. How does the pineal gland adjust the body's biological clock?

THYROID GLAND
Structure of the Thyroid Gland

Two large **lateral lobes** and a narrow connecting isthmus make up the **thyroid gland** (Figure 16-22). There is often a thin wormlike piece of thyroid tissue, called the *pyramidal lobe*, extending upward from the isthmus. The weight of the gland in the adult is variable, but it is around 30 g (1 ounce). The thyroid is located in the neck, on the anterior and lateral surfaces of the trachea, just below the larynx.

Thyroid tissue is composed of tiny structural units called **follicles**, the site of thyroid hormone synthesis. Each follicle is a small hollow sphere with a wall of simple cuboidal glandular epithelium (Figure 16-23). The interior is filled with a thick fluid called **thyroid colloid.** The colloid is produced by the cuboidal cells of the follicle wall (**follicular cells**) and contains protein-iodine complexes known as **thyroglobulins**—the precursors of thyroid hormones.

Thyroid Hormone

The substance that is often called *thyroid hormone* (TH) is actually two different hormones. The most abundant TH is **tetraiodothyronine (T_4)**, or **thyroxine.** The other is called **triiodothyronine**

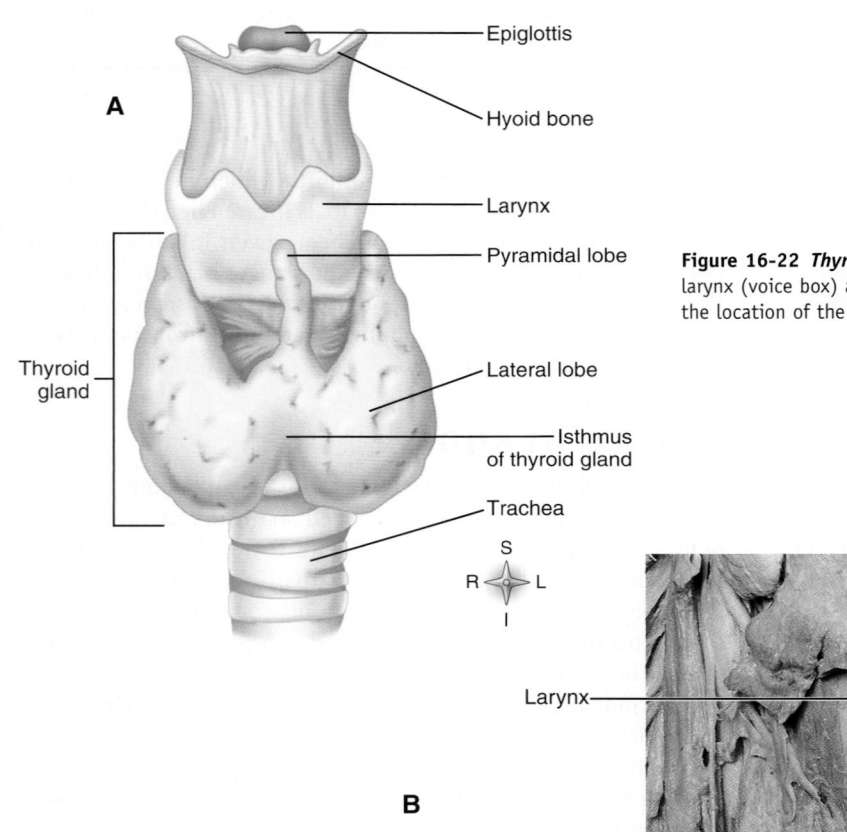

Figure 16-22 *Thyroid gland.* A, In this drawing, the relationship of the thyroid to the larynx (voice box) and to the trachea is easily seen. **B,** In this photo of a dissected cadaver, the location of the thyroid relative to the carotid arteries and jugular veins is seen.

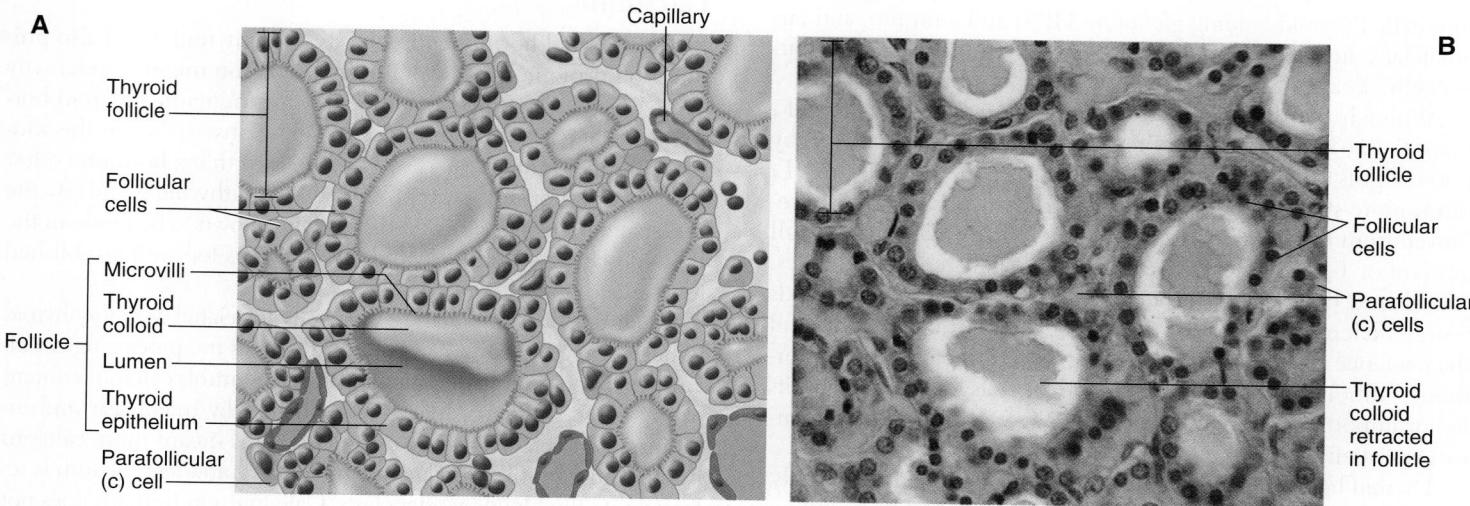

Figure 16-23 *Thyroid gland tissue.* In the drawing, **A,** and the photomicrograph, **B,** note that each of the thyroid follicles is filled with colloid. In the micrograph (×140), the thyroid colloid has separated from the follicular cells during preparation of the specimen.

(T₃). One molecule of T₄ contains four iodine atoms, and one molecule of T₃ contains three iodine atoms. After synthesizing a preliminary form of its hormones, the thyroid gland stores considerable amounts of them before secreting them (Figure 16-24). This is unusual because none of the other endocrine glands stores its hormones in another form for later release. T₃ and T₄ form in the colloid of the follicles on globulin molecules, forming thyroglobulin complexes. When they are to be released, T₃ and T₄ detach from the globulin and enter the blood. Once in the bloodstream, however, they attach to plasma proteins, principally a glob-

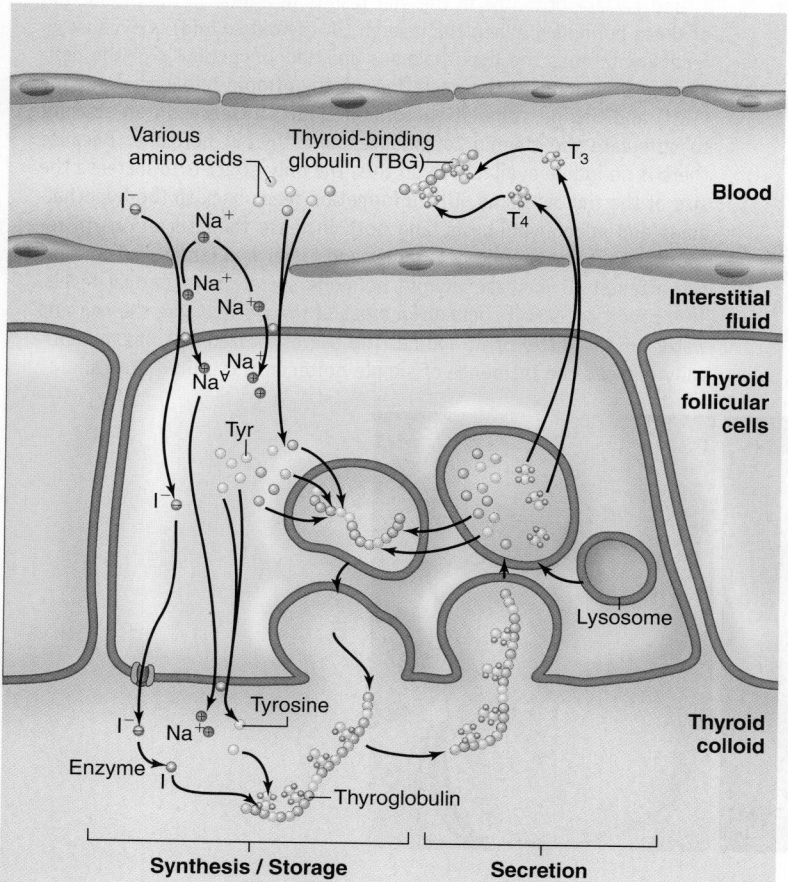

Figure 16-24 *Synthesis, storage, and release of thyroid hormone (T₃ and T₄).* *1,* Iodide ions *(I)* present in the blood enter follicular cells in the thyroid by sodium co-transport (see Box 4-3, p. 112). I⁻ then moves into the thyroid follicle through an ion channel, after which is converted to iodine *(I).* *2,* At the same time, various amino acids (including tyrosine) enter follicular cells by sodium co-transport. *3,* Tyrosine amino acids move into the follicle. *4,* Some of the amino acids are used to make the polypeptide "backbone" for thyroglobulin, which is released into the follicle. *5,* The iodine *(I)* and tyrosine *(tyr)* molecules are added to the polypeptide backbone to form thyroglobulin. *6,* When needed, thyrotropin molecules move into follicular cells by endocytosis, where they are digested and thus release "free" T₃ and T₄ molecules. *7,* Thyroid hormone *(T₃ and T₄)* are secreted into the bloodstream, where they bind to plasma proteins called *thyroid-binding globulins (TBGs)* and travel to other parts of the body.

ulin called *thyroid-binding globulin* (TBG) and albumin, and circulate as a hormone-globulin complex. When they near their target cells, T_3 and T_4 detach from the plasma globulin.

Although the thyroid gland releases about 20 times more T_4 than T_3, T_3 is much more potent than T_4 and is considered by physiologists to be the principal thyroid hormone. Why is this? T_4 binds more strongly to plasma globulins than T_3, so T_4 is not removed from the blood by target cells as quickly as T_3. The small amount of T_4 that enters target tissues is usually converted to T_3. Add this to the fact that experiments have shown that T_3 binds more efficiently than T_4 to nuclear receptors in target cells, and the evidence is overwhelming that T_3 is the principal thyroid hormone. Although T_4 may influence target cells to some extent, its major importance is as a precursor to T_3. Such hormone precursors are often called *prohormones*.

Thyroid hormone helps regulate the metabolic rate of all cells, as well as the processes of cell growth and tissue differentiation (Box 16-8). Because thyroid hormone can potentially interact with any cell in the body, it is said to have a "general" target.

Calcitonin

Besides thyroid hormone (T_3 and T_4), the thyroid gland also produces a hormone called **calcitonin (CT)**. You might wonder why some hormones of the thyroid qualify for the name "thyroid hormone," whereas calcitonin does not. The answer lies in the simple fact that for many years, we had no idea that a hormone other than thyroid hormone was produced by the thyroid gland. By the time calcitonin was discovered, and later shown to be made in the thyroid gland, the term *thyroid hormone* was too well established to change it easily.

Produced by *parafollicular cells* (cells associated with the thyroid follicles) called **C cells**, calcitonin influences the processing of calcium by bone cells. Calcitonin apparently controls calcium content of the blood by increasing bone formation by osteoblasts and inhibiting bone breakdown by osteoclasts. This means more calcium is removed from the blood by the osteoblasts, and less calcium is released into the blood by osteoclasts. Calcitonin in humans does not seem to have a great effect—but it may slightly decrease blood calcium levels and promote conservation of hard bone matrix. Parathy-

BOX 16-8: HEALTH MATTERS
Thyroid Hormone Abnormalities

Hypersecretion of thyroid hormone occurs in **Graves disease,** which is thought to be an autoimmune condition. Graves disease patients may suffer from unexplained weight loss, nervousness, increased heart rate, and **exophthalmos** (protrusion of the eyeballs resulting, in part, from edema of tissue at the back of the eye socket; see part A of the figure).

Hyposecretion of thyroid hormone during growth years may lead to **cretinism.** Cretinism is a condition characterized by a low metabolic rate, retarded growth and sexual development, and possibly mental retardation. People with profound manifestations of this condition are said to have deformed *dwarfism* (as opposed to the proportional dwarfism caused by hyposecretion of growth hormone). Hyposecretion later in life produces a condition characterized by decreased metabolic rate, loss of mental and physical vigor, gain in weight, loss of hair, yellow dullness of the skin, and myxedema. **Myxedema** is a swelling (edema) and firmness of the skin caused by accumulation of mucopolysaccharides in the skin.

In a condition called **simple goiter,** the thyroid enlarges when there is a lack of iodine in the diet (part B of figure). This condition is an interesting example of how the feedback control mechanisms illustrated in Figure 16-19 operate. Because iodine is required for the synthesis of T_3 and T_4, lack of iodine in the diet results in a drop in the production of these hormones. When the reserve (in thyroid colloid) is exhausted, feedback informs the hypothalamus and adenohypophysis of the deficiency. In response, the secretion of thyrotropin-releasing hormone (TRH) and thyroid-stimulating hormone (TSH) increases in an attempt to stimulate the thyroid to produce more thyroid hormone. Because there is no iodine available to do this, the only effect is to increase the size of the thyroid gland. This information feeds back to the hypothalamus and adenohypophysis, and both increase their secretions in response. Thus the thyroid gets larger and larger and larger—all in a futile attempt to increase thyroid hormone secretion to normal levels. This condition is still common in areas of the world where the soil and water contain little or no iodine. The use of iodized salt has dramatically reduced the incidence of simple goiter in the United States.

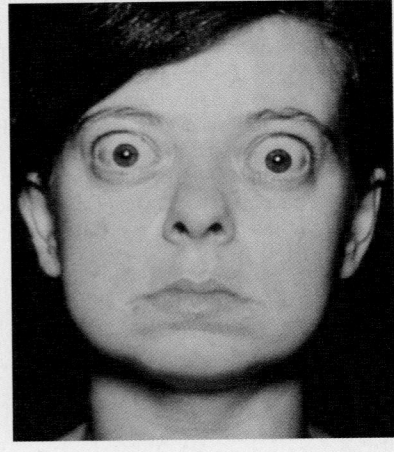
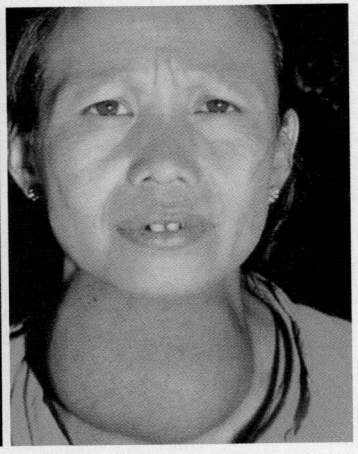

Thyroid hormone abnormalities. **A,** Exophthalmos goiter. **B,** Simple goiter.

Table 16-7 Hormones of the Thyroid and Parathyroid Glands

HORMONE	SOURCE	TARGET	PRINCIPAL ACTION
Triiodothyronine (T_3)	Thyroid gland (follicular cells)	General	Increases rate of metabolism
Tetraiodothyronine (T_4), or thyroxine	Thyroid gland (follicular cells)	General	Increases rate of metabolism (usually converted to T_3 first)
Calcitonin (CT)	Thyroid gland (parafollicular cells)	Bone tissue	Increases calcium storage in bone, lowering blood Ca^{++} levels
Parathyroid hormone (PTH), or parathormone	Parathyroid glands	Bone tissue and kidney	Increases calcium removal from storage in bone and produces the active form of vitamin D in the kidneys, increasing absorption of calcium by intestines and increasing blood Ca^{++} levels

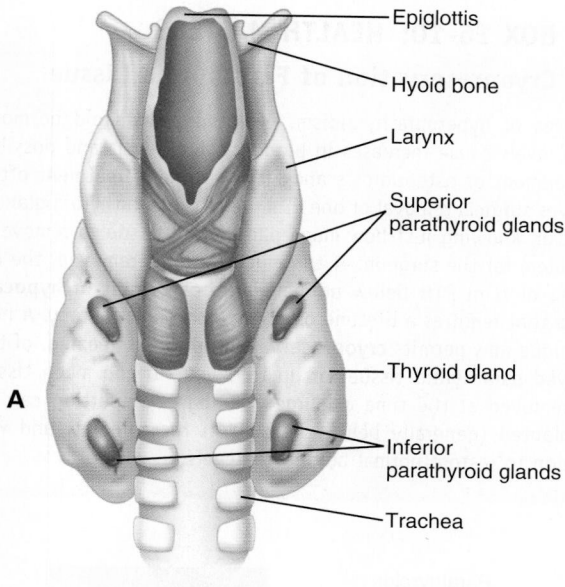

A

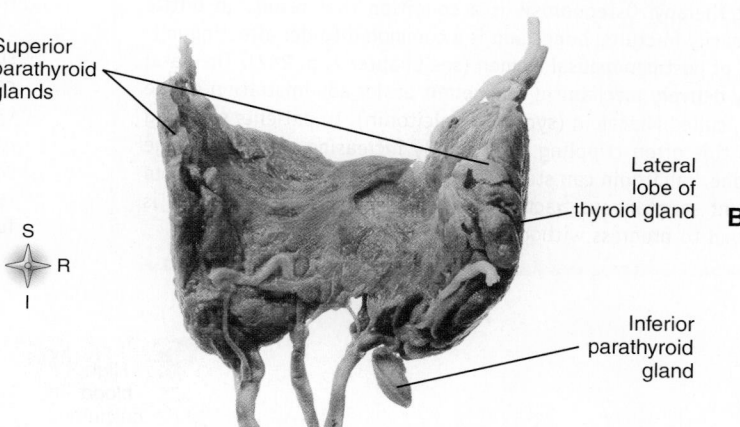

B

Figure 16-25 *Parathyroid gland.* **A,** In this drawing from a posterior view, note the relationship of the parathyroid glands to each other, to the thyroid gland, to the larynx (voice box), and to the trachea. **B,** Photo of a cadaver dissection (also from a posterior view) showing several parathyroid glands on the posterior surface of the lateral lobes of an isolated thyroid gland.

roid hormone, discussed later, is an antagonist to calcitonin, because it has the opposite effect. Together, calcitonin and parathyroid hormone help maintain calcium homeostasis (see Figure 16-27).

Table 16-7 summarizes hormones of the thyroid gland.

PARATHYROID GLANDS
Structure of the Parathyroid Glands

There are usually four or five **parathyroid glands** embedded in the posterior surface of the thyroid's lateral lobes (see Figure 16-25). They appear as tiny rounded bodies within thyroid tissue formed by compact, irregular rows of cells (Figure 16-26).

Parathyroid Hormone

The parathyroid glands secrete **parathyroid hormone (PTH)**, or *parathormone* (see Table 16-7). PTH is the main hormone the body uses to maintain calcium homeostasis. PTH acts on bone and kidney cells by increasing the release of calcium into the blood. The bone cells are especially affected, causing less new bone to be

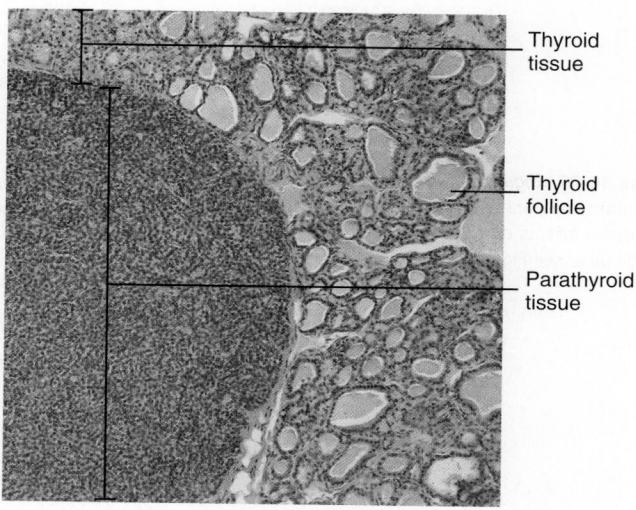

Figure 16-26 *Parathyroid tissue.* This microscopic specimen shows a portion of a parathyroid gland bordered by the surrounding thyroid tissue ($\times35$).

formed and more old bone to be dissolved, yielding calcium and phosphate. These minerals are then free to move into the blood, elevating blood levels of calcium and phosphate. In the kidney, however, only calcium is reabsorbed from urine into the blood. Under the influence of PTH, phosphate is secreted by kidney cells *out of* the blood and into the urine to be excreted. PTH also increases the body's absorption of calcium from food by activating **vitamin D** (*cholecalciferol*) in the kidney, which then permits Ca^{++} to be transported through intestinal cells and into the blood.

The maintenance of calcium homeostasis, achieved through the interaction of PTH and calcitonin, is very important for healthy survival (Figure 16-27). Normal neuromuscular excitability, blood clotting, cell membrane permeability, and normal functioning of certain enzymes all depend on the maintenance of normal levels of calcium in the blood. For example, hyposecretion of PTH can lead to hypocalcemia (see also Box 16-9). Hypocal-

cemia increases neuromuscular irritability—sometimes so much that it produces muscle spasms and convulsions. Conversely, high blood calcium levels decrease the irritability of muscle and nerve tissue so that constipation, lethargy, and even coma can result.

 QUICK CHECK

17. Where is the thyroid located? What does it look like?
18. Thyroid hormone is really two distinct compounds—what are they? Which of the two is considered more physiologically active?
19. How do calcitonin and parathyroid hormone act together to regulate homeostasis of blood calcium concentration?

 BOX 16-9: FYI

Osteoporosis

Calcitonin contained in nasal spray can be used for the treatment of **osteoporosis.** This treatment is designed especially for patients who are unable to tolerate the more common estrogen replacement therapy. Osteoporosis is a condition that results in brittle and easily fractured bones and is a common disorder affecting millions of postmenopausal women (see Chapter 7, p. 247). The nasal spray delivery mechanism will permit easier administration of the drug, called Miacalcin (synthetic calcitonin), to patients suffering from this often crippling disorder. By increasing calcium storage in bone, calcitonin can strengthen weakened bone tissue and help prevent spontaneous fractures, which can occur if the disease is allowed to progress without treatment.

 BOX 16-10: HEALTH MATTERS

Cryopreservation of Parathyroid Tissue

In cases of hyperparathyroidism, elevated parathyroid hormone (PTH) levels cause increases in blood calcium levels and possible development of osteoporosis and kidney stones. Treatment often involves surgical removal of one or more of the parathyroid glands. However, knowing just how much parathyroid tissue to remove is a problem for the surgeon. If too much tissue is removed, the resulting drop in PTH below normal limits can result in **hypocalcemia** that requires a lifetime of PTH replacement therapy. A new technique now permits **cryopreservation,** or deep freezing, of the removed parathyroid tissue for up to 1 year. If too much tissue was removed at the time of surgery, a "banked" portion can be reimplanted (generally below the skin in the forearm) and will function to restore normal blood PTH levels.

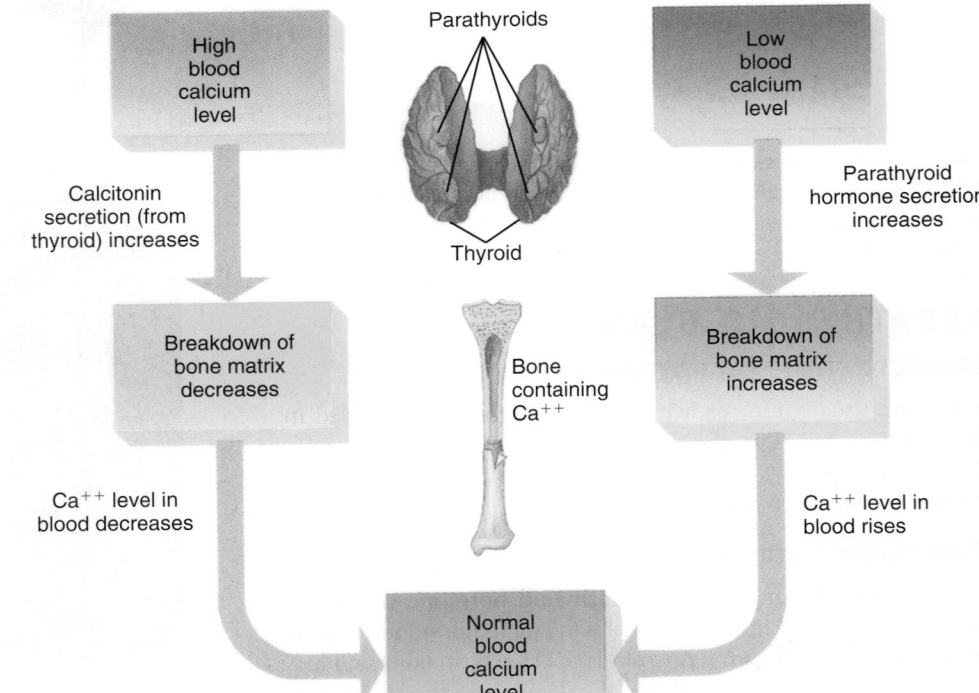

Figure 16-27 *Regulation of blood calcium levels.* Calcitonin and parathyroid hormones have antagonistic (opposite) effects on calcium concentration in the blood. (Also see Figure 16-11.)

ADRENAL GLANDS
Structure of the Adrenal Glands

The **adrenal glands**, or *suprarenal glands*, are located atop the kidneys, fitting like a cap over these organs (see Figure 16-28). The outer portion of the gland is called the **adrenal cortex**, and the inner portion of the gland is called the **adrenal medulla** (Figure 16-29). Even though the adrenal cortex and adrenal medulla are part of the same organ, they have different embryological origins and are structurally and functionally so different that they are often spoken of as if they were separate glands. The adrenal cortex is composed of regular endocrine tissue, but the adrenal medulla is made of neurosecretory tissue (Figure 16-30). As you might guess, each of these tissues synthesizes and secretes a different set of hormones (Figure 16-31 and Table 16-8).

Adrenal Cortex

The adrenal cortex is composed of three distinct layers, or zones, of secreting cells (see Figures 16-29 to 16-31). Starting with the zone directly under the outer connective tissue capsule of the gland, they are the **zona glomerulosa, zona fasciculata,** and **zona reticularis.** Cells of the outer zone secrete a class of hormones called *mineralocorticoids.* Cells of the middle zone secrete *glucocorticoids.* The inner zone secretes small amounts of *glucocorticoids* and *gonadocorticoids* (sex hormones). All these cortical hormones are steroids so together they are known as corticosteroids.

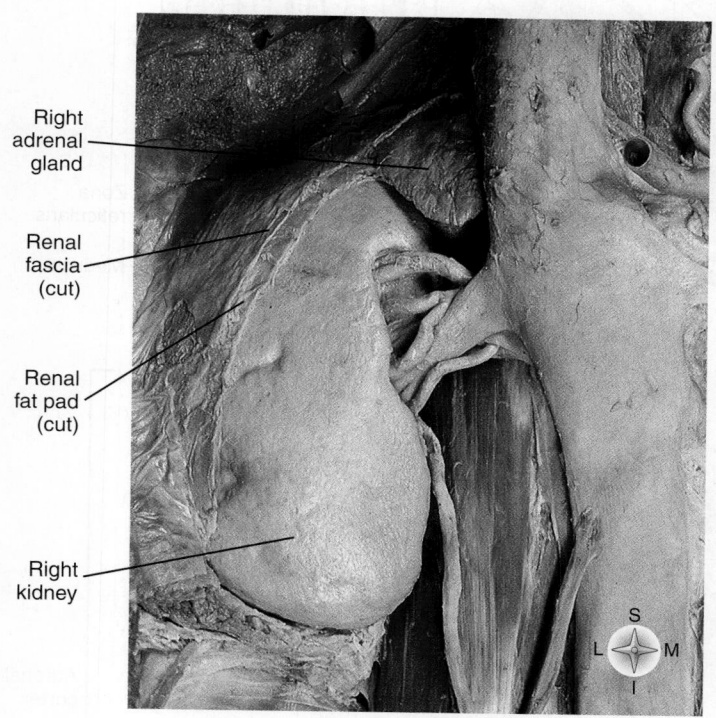

Right adrenal gland

Renal fascia (cut)

Renal fat pad (cut)

Right kidney

Figure 16-28 *Location of the adrenal gland.* Photograph of a cadaver dissection showing the location of the adrenal gland just superior to the kidney. Note that the adrenal glands are covered by the renal fascia but not the renal fat pad, both of which cover the kidney.

Table 16-8 Hormones of the Adrenal Glands

HORMONE	SOURCE	TARGET	PRINCIPAL ACTION
Aldosterone	Adrenal cortex (zona glomerulosa)	Kidney	Stimulates kidney tubules to conserve sodium, which in turn triggers the release of ADH and the resulting conservation of water by the kidney
Cortisol (hydrocortisone)	Adrenal cortex (zona fasciculata)	General	Influences metabolism of food molecules; in large amounts, it has an anti-inflammatory effect
Adrenal androgens	Adrenal cortex (zona reticularis)	Sex organs, other effectors	Exact role uncertain but may support sexual function
Adrenal estrogens	Adrenal cortex (zona reticularis)	Sex organs	Thought to be physiologically insignificant
Epinephrine (adrenaline)	Adrenal medulla	Sympathetic effectors	Enhances and prolongs the effects of the sympathetic division of the autonomic nervous system
Norepinephrine	Adrenal medulla	Sympathetic effectors	Enhances and prolongs the effects of the sympathetic division of the autonomic nervous system

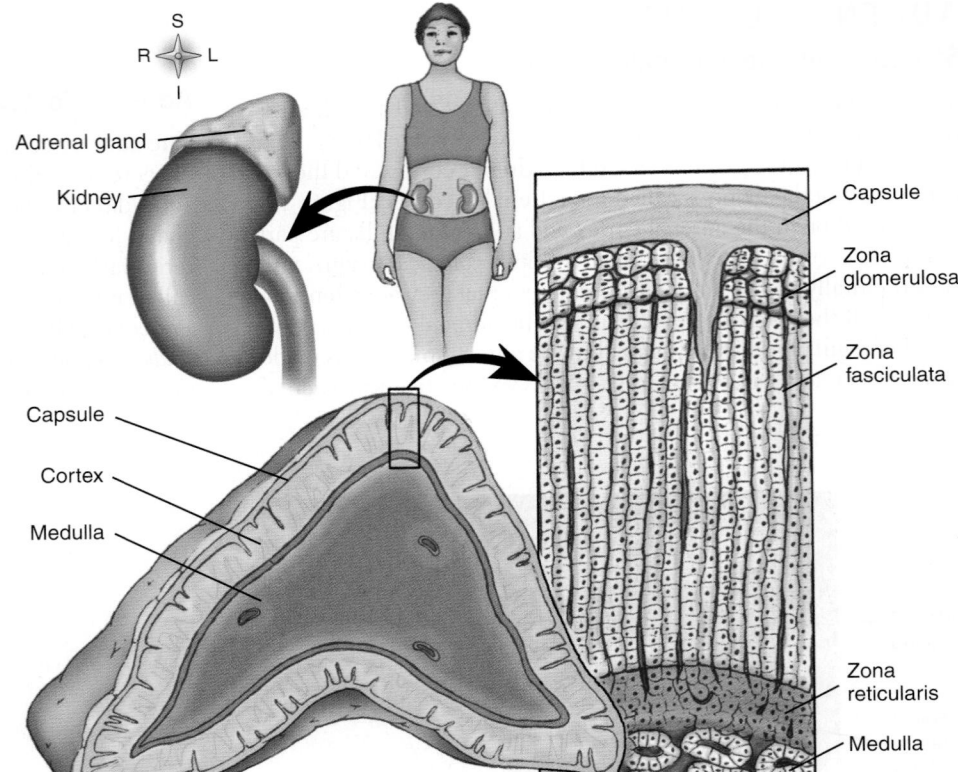

Figure 16-29 *Structure of the adrenal gland.* The zona glomerulosa of the cortex secretes aldosterone. The zona fasciculata secretes abundant amounts of glucocorticoids, chiefly cortisol. The zona reticularis secretes minute amounts of sex hormones and glucocorticoids. A portion of the medulla is visible at lower right at the bottom of the drawing.

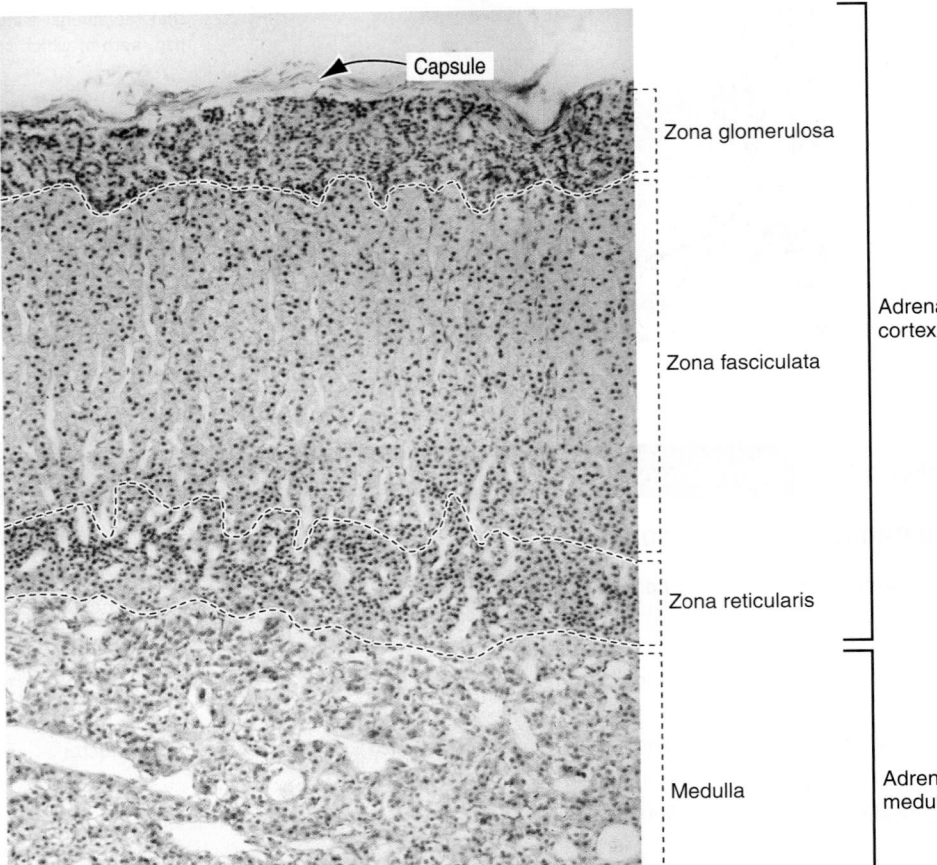

Figure 16-30 *Adrenal tissue.* The major regions of the adrenal gland are shown in a light micrograph of a stained specimen. The cortex is made up of epithelial endocrine tissue, and the medulla is instead made up of neurosecretory tissue. Compare with drawing in Figure 16-29.

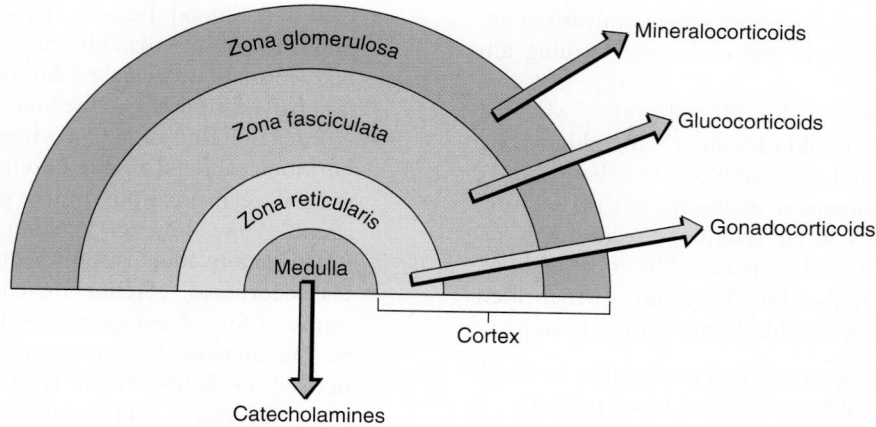

Figure 16-31 *Adrenal hormones.*

Mineralocorticoids

Mineralocorticoids, as their name suggests, have an important role in regulating how mineral salts (electrolytes) are processed in the body. In the human, **aldosterone** is the only physiologically important mineralocorticoid. Its primary function is the maintenance of sodium homeostasis in the blood. Aldosterone accomplishes this by increasing sodium reabsorption in the kidneys. Sodium ions are reabsorbed from the urine back into the blood in exchange for potassium or hydrogen ions. In this way, aldosterone not only adjusts blood sodium levels but also can influence potassium and pH levels in the blood.

Because the reabsorption of sodium ions causes water to also be reabsorbed (partly by triggering the secretion of ADH), aldo-

sterone promotes water retention by the body. Altogether, aldosterone can increase sodium and water retention and promote the loss of potassium and hydrogen ions.

Aldosterone secretion is controlled mainly by the **renin-angiotensin-aldosterone system (RAAS)** and by blood potassium concentration. The RAAS (Figure 16-32) operates as indicated in this sequence of steps:

1. When the incoming blood pressure in the kidneys drops below a certain level, a piece of tissue near the vessels (the *juxtaglomerular apparatus*) secretes **renin** into the blood.
2. Renin, an enzyme, causes **angiotensinogen** (a normal constituent of blood) to be converted to **angiotensin I.**

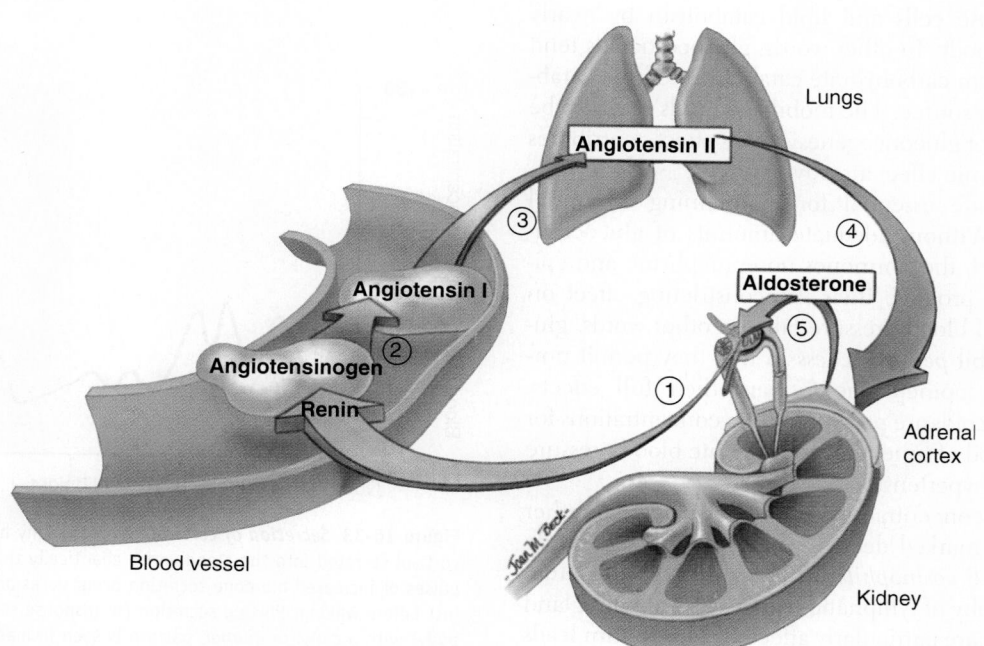

Figure 16-32 *Renin-angiotensin-aldosterone system (RAAS) for regulating aldosterone secretion.* The numbers correspond to the steps outlined in the text.

3. Angiotensin I circulates to the lungs, where converting enzymes in the capillaries split the molecule, forming **angiotensin II.**

4. Angiotensin II circulates to the adrenal cortex, where it stimulates the secretion of aldosterone. (Some aldosterone is also synthesized in the heart and blood vessels.)

5. Aldosterone causes increased reabsorption of sodium, which causes increased water retention. As water is retained, the volume of blood increases. The increased volume of blood creates higher blood pressure—which then causes the renin-angiotensin-aldosterone system to stop.

The renin-angiotensin-aldosterone system is a negative feedback mechanism that helps maintain homeostasis of blood pressure.

Glucocorticoids

The chief glucocorticoids secreted by the zona fasciculata of the adrenal cortex are **cortisol** (also called *hydrocortisone*), *cortisone,* and *corticosterone.* Of these, only cortisol is secreted in significant quantities in the human. Glucocorticoids affect every cell in the body. Although much remains to be discovered about their precise mechanisms of action, we do know enough to make some generalizations:

- Glucocorticoids accelerate the breakdown of proteins into amino acids (except in liver cells). These "mobilized" amino acids move out of the tissue cells and into the blood. From there, they circulate to the liver cells, where they are changed to glucose in a process called *gluconeogenesis.* A prolonged high blood concentration of glucocorticoids in the blood therefore results in a net loss of tissue proteins ("tissue wasting") and hyperglycemia (high blood glucose). Glucocorticoids are protein mobilizing, gluconeogenic, and hyperglycemic.

- Glucocorticoids tend to accelerate mobilization of both lipids from adipose cells and lipid catabolism by nearly every cell in the body. In other words, glucocorticoids tend to cause a shift from carbohydrate catabolism to lipid catabolism as an energy source. The mobilized lipids may also be used in the liver for gluconeogenesis. This effect contributes to the hyperglycemic effect already observed.

- Glucocorticoids are essential for maintaining a normal blood pressure. Without adequate amounts of glucocorticoids in the blood, the hormones norepinephrine and epinephrine cannot produce their vasoconstricting effect on blood vessels, and blood pressure falls. In other words, glucocorticoids exhibit *permissiveness* in that they permit norepinephrine and epinephrine to have their full effects. When glucocorticoids are present in high concentrations for a prolonged period of time, they may elevate blood pressure beyond normal (hypertension).

- A high blood concentration of glucocorticoids rather quickly causes a marked decrease in the number of white blood cells, called *eosinophils,* in the blood (eosinopenia) and marked atrophy of lymphatic tissues. The thymus gland and lymph nodes are particularly affected. This in turn leads to a decrease in the number of lymphocytes and plasma cells in the blood. Because of the decreased number of lymphocytes and plasma cells (antibody-processing cells), antibody formation decreases. Antibody formation is an important part of immunity—the body's defense against infection.

- Normal amounts of glucocorticoids act with epinephrine, a hormone secreted by the adrenal medulla, to bring about normal recovery from injury produced by inflammatory agents. How they act together to bring about this anti-inflammatory effect is still uncertain.

- Glucocorticoid secretion increases as part of the stress response. One advantage gained by increased secretion may be the increase in glucose available for skeletal muscles needed in fight-or-flight responses. However, prolonged stress can lead to immune dysfunction, probably as a result of prolonged exposure to high levels of glucocorticoids (see Chapter 22).

- Except during the stress response, glucocorticoid secretion is controlled mainly by means of a negative feedback mechanism that involves ACTH from the adenohypophysis.

- As with many hormones, including all the adrenal cortical hormones, glucocorticoid secretion occurs in pulses and also shows a daily pattern of pulses of different amounts of hormone secretion (Figure 16-33).

Gonadocorticoids

The term *gonadocorticoid* refers to sex hormones that are released from the zona fasciculata and zona reticularis of the adrenal cortex rather than the gonads. The normal adrenal cortex secretes small amounts of male hormones (androgens). Normally, not enough androgen is produced to give women masculine characteristics, but it is sufficient to influence the appearance of pubic and axillary hair in both boys and girls.

Box 16-11 discusses some adrenal cortical hormone disorders.

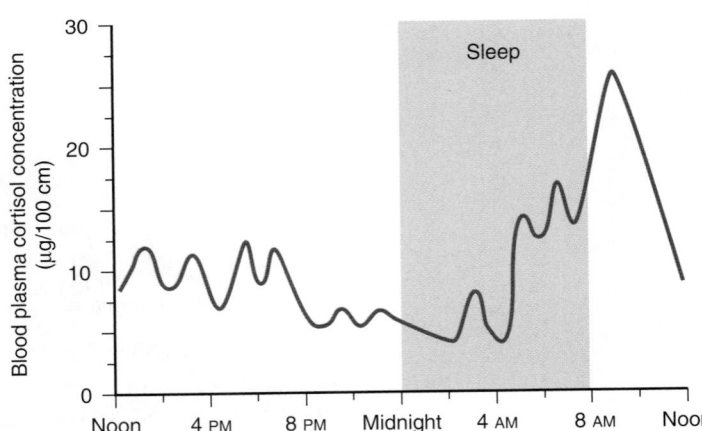

Figure 16-33 *Secretion of cortisol.* As with many hormones, the amount of cortisol secreted into the blood varies chaotically throughout the day. However, pulses of increased hormone secretion occur occasionally with the highest peaks just before waking. Pulsing secretion (in response to specific conditions in the body) with a daily, or *diurnal,* pattern is seen in many hormones—including all of the adrenal cortical hormones.

BOX 16-11: HEALTH MATTERS
Adrenal Cortical Hormone Abnormalities

Hypersecretion of cortisol from the adrenal cortex often produces a collection of symptoms called *Cushing syndrome*. **Hypersecretion** of glucocorticoids results in a redistribution of body fat. The fatty "moon face" and thin, reddened skin characteristic of Cushing syndrome are shown in part *A* of the figure. Part *B* shows the face of the same boy 4 months after treatment. Hypersecretion of aldosterone, *aldosteronism*, leads to increased water retention and muscle weakness resulting from potassium loss. Hypersecretion of androgens can result from tumors of the adrenal cortex, called *virilizing tumors*. They are so called because the increased blood level of male hormones in women can cause them to acquire male characteristics. People with Cushing syndrome may suffer from all the symptoms described above.

Hyposecretion of mineralocorticoids and glucocorticoids, as in *Addison disease*, may lead to a drop in blood sodium and blood glucose, an increase in blood potassium levels, dehydration, and weight loss.

Pharmacological preparations of glucocorticoids have been used for many years to temporarily relieve the symptoms of severe inflammatory conditions such as rheumatoid arthritis. More recently, over-the-counter creams and ointments containing hydrocortisone have become available for use in treating the pain, itching, swelling, and redness of skin rashes.

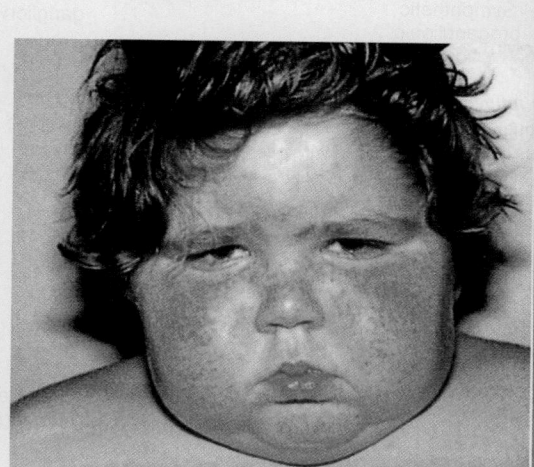
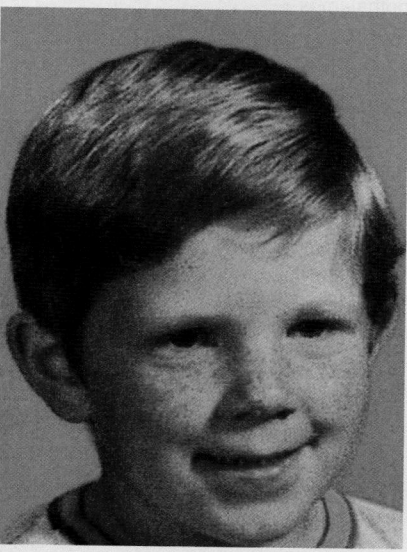

Cushing syndrome. **A,** Fatty "moon face" in a body with Cushing syndrome. **B,** The face of the same boy 4 months after treatment.

Adrenal Medulla

The adrenal medulla is composed of neurosecretory tissue, that is, tissue composed of neurons specialized to secrete their products into the blood rather than across a synapse. Actually, the medullary cells are modified versions of sympathetic postganglionic fibers of the autonomic nervous system. They are innervated by sympathetic preganglionic fibers, so that when the sympathetic nervous system is activated (as in the stress response), the medullary cells secrete their hormones directly into the blood.

The adrenal medulla secretes two important hormones, both of which are in the class of nonsteroid hormones called catecholamines. **Epinephrine,** or *adrenaline*, accounts for about 80% of the medulla's secretion. The other 20% is **norepinephrine** (**NE** or **NR**). You may recall that norepinephrine is also the neurotransmitter produced by postganglionic sympathetic fibers. Sympathetic effectors such as the heart, smooth muscle, and

glands have receptors for norepinephrine. Both epinephrine and norepinephrine produced by the adrenal medulla can bind to the receptors of sympathetic effectors to prolong and enhance the effects of sympathetic stimulation by the autonomic nervous system (Figure 16-34).

QUICK CHECK

20. Distinguish between the histology of the adrenal cortex and the adrenal medulla.
21. Name some effects of cortisol in the body.
22. How does the function of the adrenal medulla overlap with the function of the autonomic nervous system?

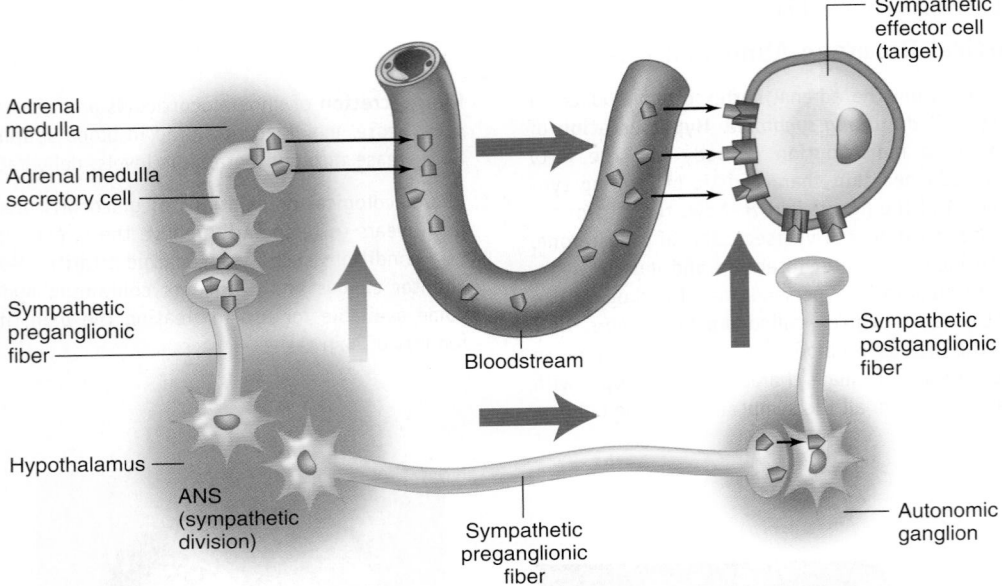

Figure 16-34 *Combined nervous and endocrine influence on sympathetic effectors.* A sympathetic center in the hypothalamus sends efferent impulses through preganglionic fibers. Some preganglionic fibers synapse with postganglionic fibers that deliver norepinephrine across a synapse with the effector cell. Other preganglionic fibers synapse with postganglionic neurosecretory cells in the adrenal medulla. These neurosecretory cells secrete epinephrine and norepinephrine into the bloodstream, where they travel to the target cells (sympathetic effectors). Compare this figure with Figure 16-1. *ANS,* Autonomic nervous system.

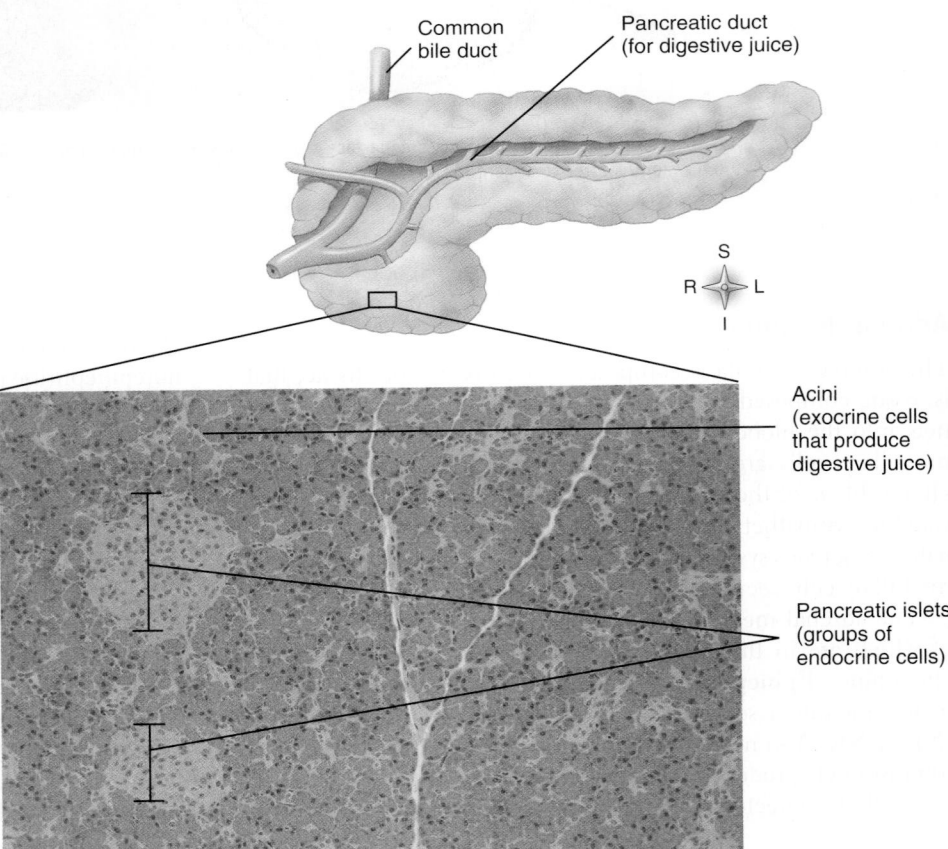

Figure 16-35 *Pancreas.* Two pancreatic islets, or hormone-producing areas, are evident among the pancreatic cells that produce the pancreatic digestive juice. The pancreatic islets are more abundant in the tail of the pancreas than in the body or head.

PANCREATIC ISLETS

Structure of the Pancreatic Islets

The pancreas is an elongated gland (12 to 15 cm [or about 5 to 6 inches] long) weighing up to 100 g (3.5 ounces) (Figure 16-35). The "head" of the gland lies in the C-shaped beginning of the small intestine (duodenum), with its body extending horizontally behind the stomach and its tail touching the spleen.

The tissue of the pancreas is composed of both endocrine and exocrine tissues. The endocrine portion is made up of scattered, tiny islands of cells, called **pancreatic islets** (*islets of Langerhans*) that account for only about 2% or 3% of the total mass of the pancreas. These hormone-producing islets are surrounded by cells called *acini*, which secrete a serous fluid containing digestive enzymes into ducts that drain into the small intestine (see Figure 16-35). The digestive roles of the pancreas are discussed in Chapters 25 and 26. For the moment, we will concentrate on the endocrine part of this gland, the pancreatic islets.

Each of the 1 to 2 million pancreatic islets in the pancreas contains a combination of four primary types of endocrine cells, all joined to each other by gap junctions. Each type of cell secretes a different hormone, but the gap junctions may allow for some coordination of these functions as a single secretory unit. One type of pancreatic islet cell is the **alpha cell** (also called the *A cell*), which secretes the hormone glucagon. **Beta cells** (*B cells*) secrete the hormone insulin; **delta cells** (*D cells*) secrete the hormone somatostatin; and **pancreatic polypeptide cells** (*F, or PP, cells*) secrete pancreatic polypeptide. Beta cells, which account for about three fourths of all the pancreatic islet cells, are usually found near the center of each islet, whereas cells of the other three types are more often found in the outer portion. Figure 16-36 shows how the different cell types can be distinguished by the microscope with special staining techniques.

Pancreatic Hormones

The pancreatic islets produce several hormones, the most important of which are described in Table 16-9 and in the following list:

- **Glucagon,** produced by alpha cells, tends to increase blood glucose levels by stimulating the conversion of glycogen to

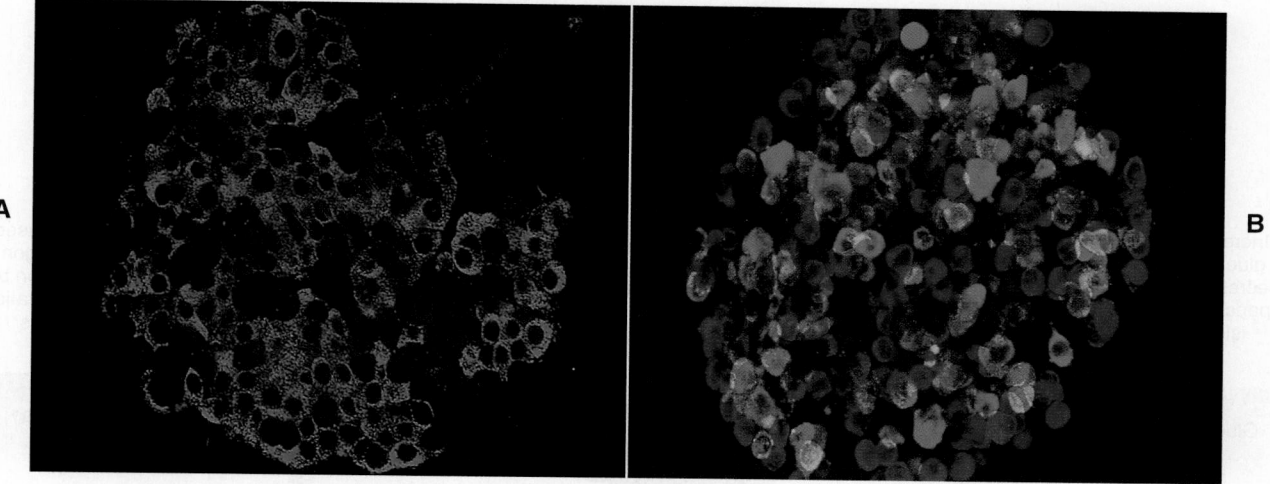

Figure 16-36 *Cells of the pancreatic islet.* A special technique of microscopy, called immunofluorescent staining, allows cells to be distinguished by the different molecules that they contain. **A,** Micrograph shows a central cluster of beta, or B, cells outlined by green dots. The green dots are insulin molecules stained by a special type of antibody coupled to a fluorescent dye. **B,** In this micrograph, peripheral alpha, or A, cells are stained green and delta, or D, cells are stained red. The central beta cells are not stained in this specimen, so they appear only as black spaces in the micrograph.

Table 16-9 Hormones of the Pancreatic Islets

HORMONE	SOURCE	TARGET	PRINCIPAL ACTION
Glucagon	Pancreatic islets (alpha [α] cells, or A cells)	General	Promotes movement of glucose from storage and into the blood
Insulin	Pancreatic islets (beta [β] cells, or B cells)	General	Promotes movement of glucose out of the blood and into cells
Somatostatin	Pancreatic islets (delta [δ] cells, or D cells)	Pancreatic cells and other effectors	Can have general effects in the body, but primary role seems to be regulation of secretion of other pancreatic hormones
Pancreatic polypeptide	Pancreatic islets (pancreatic polypeptide [PP] or F cells)	Intestinal cells and other effectors	Exact function uncertain but seems to influence absorption in the digestive tract

glucose in liver cells. It also stimulates gluconeogenesis (transformation of fatty acids and amino acids into glucose) in liver cells. The glucose produced by way of the breakdown of glycogen and by gluconeogenesis is released into the bloodstream, producing a hyperglycemic effect.

- **Insulin,** produced by beta cells, tends to promote the movement of glucose, amino acids, and fatty acids out of the blood and into tissue cells. Hence insulin tends to lower the blood concentrations of these food molecules and to promote their metabolism by tissue cells. The antagonistic effects that glucagon and insulin have on blood glucose levels are summarized in Figure 16-37 and discussed further in Chapter 27. (See Box 16-12 for a discussion of diabetes mellitus.)
- **Somatostatin,** produced by delta cells, may affect many different tissues in the body, but its primary role seems to be in regulating the other endocrine cells of the pancreatic islets. Somatostatin inhibits the secretion of glucagon, insulin, and pancreatic polypeptide. It also inhibits the secretion of growth hormone (somatotropin) from the anterior pituitary.

- *Pancreatic polypeptide* is produced by PP (or F) cells in the periphery of pancreatic islets. Although much is yet to be learned about pancreatic polypeptide, we do know that it influences the digestion and distribution of food molecules to some degree.

All four of these pancreatic hormones probably work together as a team to maintain a homeostasis of food molecules (glucose, fatty acids, and amino acids). More about their respective roles in overall nutrient metabolism is discussed in Chapter 27.

QUICK CHECK

23. Name two of the four principal hormones secreted by the pancreatic islets.
24. In what way do insulin and glucagon exert antagonistic influences on the concentration of glucose in the blood?

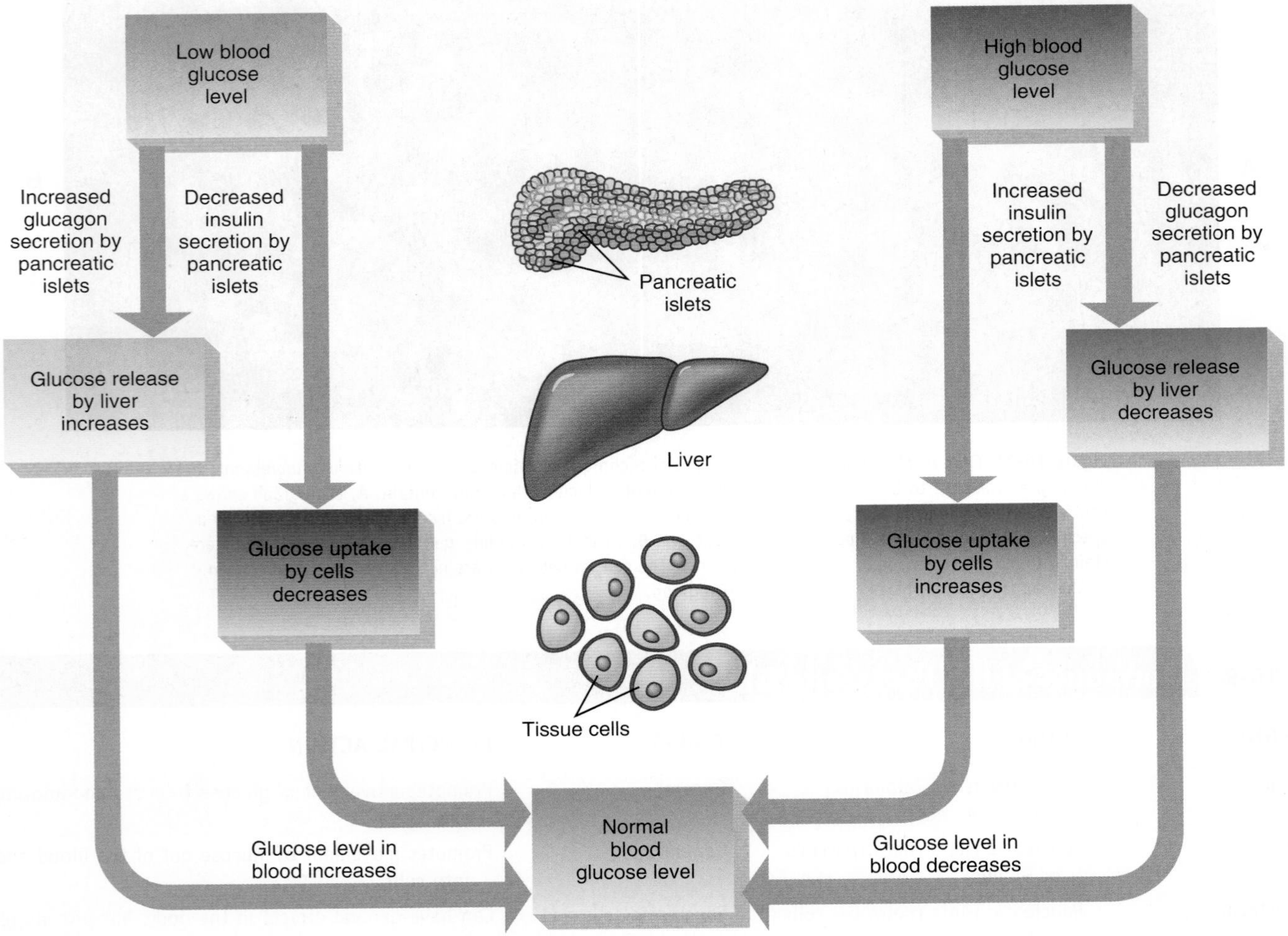

Figure 16-37 *Regulation of blood glucose levels.* Insulin and glucagon, two of the major pancreatic hormones, have antagonistic (opposite) effects on glucose concentration in the blood. Of course, many other hormones, such as GH, cortisol, and others, also influence blood glucose levels.

BOX 16-12   **Diabetes Mellitus**

Diabetes mellitus is one of the most common endocrine disorders. It affects more than 14 million Americans and will have a significant impact on the health of about 7% of the population at some point in their lives. Diabetes is best described as a syndrome—a collection of features that characterizes the disease. Although the features of diabetes tend to vary between individuals and depend on the type, severity, and length of the illness, each sign or symptom is related in some way to abnormal metabolism of nutrients and the consequences that follow (see the figure on p. 629).

In diabetics, an inadequate amount or abnormal type of insulin may be produced. In other affected individuals, diminished numbers of normal insulin receptors or second messenger systems in target cells impair glucose entry into cells even if normal insulin in adequate amounts is present. A newly discovered hormone secreted by fat cells, called **resistin,** may also interfere with insulin action. Research on resistin may help explain the relationship between obesity and Type 2 diabetes (see following discussion). The possible causes of all these insulin-related problems are still under investigation. Probably one or a combination of factors, such as genetic predisposition, hormone problems, viral infection, nutrition and diet, obesity, autoimmune disorders, and exposure to damaging agents, is involved in most cases of diabetes mellitus.

The term *insulin resistance* refers to an inability of fat, muscle, and other cells to respond normally to insulin and permit the entry of blood sugar that the hormone is trying to accomplish. *Insulin insensitivity* is the opposite of insulin resistance. It involves the rapid response of cells to insulin action and subsequent entry of sugar.

The presence of adequate amounts of normal insulin, and sufficient sensitivity to insulin, are key to the entry of glucose into cells. If there are no insulin problems and if the major target tissues such as skeletal muscle, fat, and liver have adequate, normal insulin receptors and signaling mechanisms, glucose will transfer from blood into cells. In diabetes, glucose cannot enter cells normally. The result is one of the most universal symptoms of the disease—chronic elevation in blood glucose levels, a condition called **hyperglycemia.**

Glucose is normally filtered out of the blood and then resorbed from the kidney tubules. However, as blood sugar levels rise in diabetes, the amount of glucose filtered out of the blood exceeds the ability of the kidney tubules to resorb it. The result is a "spilling over" of sugar into the urine. This condition, called **glycosuria,** causes increased urine production **(polyuria),** since additional water is required to carry the sugar load. In effect, the excess glucose acts like an osmotic diuretic. As large quantities of water are lost in the urine, the body dehydrates. The resulting sense of excessive and ongoing thirst **(polydipsia)** and the tendency to drink large quantities of liquid are also classic symptoms of the disease. In addition, since cells are deprived of glucose to burn as energy, people with diabetes often suffer from intense and continuous hunger **(polyphagia).** Their blood sugar level is high, but the cells are literally starving to death, and the body craves food. In screening for diabetes, health care professionals look for glycosuria and the "three polys": polyuria, polydipsia, and polyphagia.

The person with untreated diabetes is unable to utilize glucose for energy, and the body is forced to burn protein and fat. This abnormal metabolic shift results in fatigue and weight loss. If glucose metab-olism is severely restricted, the large quantities of fat that must be burned produce toxic quantities of acetoacetic acid and other acidic metabolites called **ketone bodies.** Buildup of these organic acids lowers blood pH, causing acidosis, and disturbs the normal acid-base balance of the body. (Mechanisms that attempt to restore homeostasis are described in Chapter 30.) Accumulation of ketone bodies in the blood results in **diabetic ketoacidosis.** Signs and symptoms include abdominal pain, nausea, vomiting, fruity odor of the breath, possible alterations in level of consciousness, coma, and even death if left untreated.

Types of Diabetes Mellitus

There are two major types of diabetes mellitus: **type 1** and **type 2.** Hereditary factors play an important role in both types.

Type 1

Type 1 diabetes mellitus was formerly referred to as *juvenile-onset diabetes* because it usually strikes before 30 years of age. In this form of the disease the beta cells of the pancreatic islets are destroyed and there is an absolute deficiency of insulin production. Individuals with type 1 diabetes are required to take insulin injections daily to prevent ketosis and to control hyperglycemia. As a result, type 1 diabetes was sometimes called *insulin-dependent diabetes mellitus (IDDM).* This form of the disease accounts for only about 10% of the total number of diabetic individuals.

The cause of beta cell destruction in type 1 diabetes is still uncertain. Current research, however, suggests that type 1 diabetes is an autoimmune disease that is probably triggered by some type of viral infection in genetically susceptible individuals. Anyone who has a parent, brother, or sister with type 1 diabetes has about a 5% to 7% chance of getting the disease. If an identical twin has type 1 diabetes, the risk increases to about 50%.

Type 2

Type 2 diabetes mellitus, previously called *non–insulin-dependent diabetes mellitus (NIDDM),* is the most common form of the disease, accounting for about 90% of all cases. Because it most often occurs after age 40 years, it was formerly also called *maturity-onset diabetes.* In susceptible individuals who are overweight, the incidence increases with age. In this form of diabetes, insulin is still produced by the beta cells but generally in reduced amounts. In addition, loss of insulin receptors on the surface membranes of target cells leading to insulin resistance also reduces effectiveness of glucose uptake from the blood.

The older name NIDDM was inappropriate because insulin injections, often combined with oral diabetic medicines, may be required to control the disease. However, in many type 2 diabetics, hyperglycemia will frequently respond to changes in lifestyle that result in eating a balanced diet high in fiber, achieving an ideal body weight, getting adequate exercise, and maintaining body weight within normal limits.

Heredity and ethnic background are important determinants in type 2 diabetes. Those who have a family history of the disease are particularly susceptible if overweight and sedentary. Native Americans are at increased risk for type 2 diabetes. In addition, Hispanics and African Americans are more likely than Caucasians to develop this type of diabetes. A specific gene defect is responsible for at least some forms of type 2 diabetes. The gene in question is involved in regulating insulin secretion.

Continued

BOX 16-12   Diabetes Mellitus—cont'd

The symptoms of type 1 diabetes are dramatic and, as a result, most individuals seek medical care soon after the disease occurs. The symptoms of type 2 diabetes, although of the same type as seen in type 1 diabetes, can be much more subtle and hard to recognize. The American Diabetes Association estimates that more than 7 million Americans have diabetes and are unaware of it. Unfortunately, if diabetes is left untreated, over time the hyperglycemia and other effects result in many complications affecting almost every area of the body. Reduced blood flow caused by a buildup of fatty materials in blood vessels (atherosclerosis) is one of the most serious complications. It causes diverse problems, such as heart attack, stroke, and reduced circulation to the extremities, resulting in tingling or numbness in the feet and, in severe cases, gangrene. Retinal changes (diabetic retinopathy) may cause blindness in some people with diabetes who have battled the disease for decades. As you read in Chapter 12, nerve damage, or *neuropathy,* can also result from diabetes. Kidney disease is another common diabetic complication. Most authorities agree that careful regulation of blood sugar levels and reducing cardiovascular risk factors such as hypertension and "bad" (LDL) cholesterol levels are the most important measures that people with diabetes can take to reduce the number of long-term complications of the disease.

Treatment of Type 1 Diabetes

The discovery of insulin in 1921 was one of the most important advances in medicine of the last century. Problems associated with the injection of insulin harvested from animal pancreatic tissue were solved when synthetic human insulin became available.

Insulin can be introduced into the body by regular injection using a needle and syringe or by implanting miniaturized pumps to deliver the insulin as needed. Oral (noninjectable) insulin in pill form that can withstand the destructive action of digestive enzymes may become a reality in the not-too-distant future. In addition, development of alternate ways to administer new types of insulin also shows promise. Special inhalers are being tested to deliver a powder form of insulin into the lungs. A liquid aerosol form of insulin is also under study. It is sprayed into the mouth and absorbed through the epithelial lining of the oral cavity.

Pancreas transplants eliminate the need for insulin injection or administration altogether, and the number of successful pancreatic transplants is increasing with the advent of new surgical techniques and more effective anti-rejection drugs (see Chapter 21). However, transplantation of islet cells rather than the entire pancreas shows even greater promise for treatment of the most serious cases of type 1 diabetes.

One of the newest and most promising of the experimental islet cell transplant techniques is called the **Edmonton Protocol.** It was developed at the University of Alberta in Edmonton, Canada. The procedure begins with the separation and harvesting of islets from a donor pancreas; the islets are then injected into a diabetic patient's portal vein that goes to the liver. After injection, the islets pass into the liver, become lodged there, and produce insulin. The protocol differs from other islet cell transplant procedures by increasing the number of functional islet cells that can be recovered from a donor pancreas and by elimination of steroids and other substances that decrease rejection rates. The procedure, although still experimental, shows great promise in treating severe type 1 diabetics. Successful islet cell transplants from nonliving donors began in 2000. The first successful islet cell transplant from a living donor occurred in Kyoto, Japan in 2005.

Treatment of Type 2 Diabetes

If the lifestyle changes described earlier in this discussion are not totally effective in lowering elevated blood glucose levels in individuals with type 2 diabetes, medicines called oral hypoglycemic agents may be prescribed. One group of **oral hypoglycemic agents,** called *sulfonylureas,* acts by stimulating beta cells to increase insulin supplies (e.g., glyburide [Micronase] and glipizide [Glucotrol]). Other short-acting "insulin stimulators" act like sulfonylureas but much more rapidly (e.g., repaglinide [Prandin] and nateglinide [Starlix]). Another group of oral agents, called *biguanides,* act by inhibiting the release of glucose from the liver (e.g., metformin [Glucophage]). Still others, such as the drugs rosiglitazone (Avandia) and pioglitazone (Actos) (*thiazolidinediones* or *TZDs*), reduce blood levels of resistin, thus decreasing insulin resistance. Newer oral treatments for type 2 diabetes often involve administration of lower doses of more than one drug. A popular combination called Glucovance combines metformin (Glucophage) and the sulfonylurea glyburide (Micronase). In addition, oral agents are given in combination with insulin injections in many patients with type 2 diabetes.

BOX 16-12   Diabetes Mellitus—cont'd

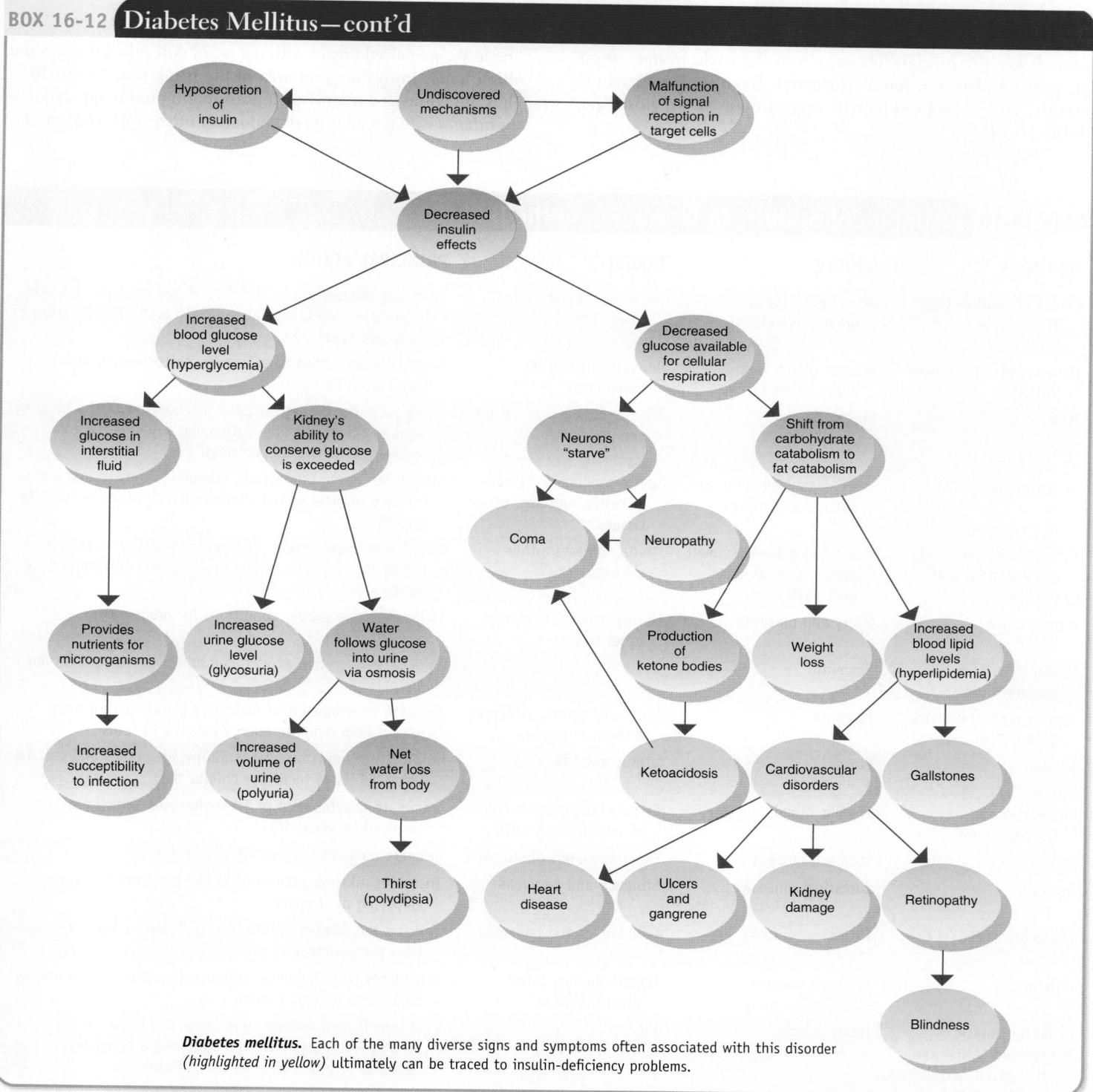

Diabetes mellitus. Each of the many diverse signs and symptoms often associated with this disorder *(highlighted in yellow)* ultimately can be traced to insulin-deficiency problems.

GONADS

Gonads are the primary sex organs in the male (*testes*; singular, *testis*) and in the female (*ovaries*). Each is structured differently, and each produces its own unique set of hormones (Table 16-10).

Testes

The testes are paired organs within a sac of skin called the *scrotum*, which hangs from the groin area of the trunk (see Figure 16-2). They are composed mainly of coils of sperm-producing *seminiferous tubules* with a scattering of endocrine *interstitial cells* found in

Table 16-10	Examples of Additional Hormones of the Body		
HORMONE	**SOURCE**	**TARGET**	**PRINCIPAL ACTION**
Cholecalciferol (vitamin D_3)	Skin, liver, kidney (in progressive steps)	Intestines, bones, most other tissues	Promotes calcium absorption from food, regulates mineral balance in bones, regulates growth and differentiation of many cell types
Dehydroepiandrosterone (DHEA)	Adrenal gland, testis, ovary, other tissues	Converted to other hormones	Eventually converted to estrogens, testosterone, or both (see Figure 16-5)
Melatonin	Pineal gland	Timekeeping tissues of the nervous system	Helps "set" the biological clock mechanisms of the body by signaling light changes during the day, month, and seasons; may help induce sleep
Testosterone	Testis (small amounts in adrenal and ovary)	Sperm-producing tissues of testis, muscles, other tissues	Stimulates sperm production, stimulates growth and maintenance of male sexual characteristics, promotes muscle growth
Estrogen including estradiol (E_2) and estrone	Ovary and placenta (small amounts in adrenal and testis)	Uterus, breasts, other tissues	Stimulates development of female sexual characteristics, breast development, bone and nervous system maintenance
Progesterone	Ovary and placenta	Uterus, mammary glands, other tissues	Helps maintain proper conditions for pregnancy
Human chorionic gonadotropin (hCG)	Placenta	Ovary	Stimulates secretion of estrogen and progesterone during pregnancy
Human placental lactogens (hPLs)	Placenta	Mammary glands; pancreas and other tissues	Promote development of mammary glands during pregnancy; help regulate energy balance in fetus
Relaxin	Placenta	Uterus and joints	Inhibits uterine contractions during pregnancy and softens pelvic joints to facilitate childbirth
Thymosins and thymopoietins	Thymus gland	Certain lymphocytes (type of white blood cell)	Stimulate development of T lymphocytes, which are involved in immunity
Gastrin	Stomach mucosa	Exocrine glands of stomach	Triggers increased gastric juice secretion
Secretin	Intestinal mucosa	Stomach and pancreas	Increases alkaline secretions of the pancreas and slows emptying of stomach
Cholecystokinin (CCK)	Intestinal mucosa	Gallbladder and pancreas	Triggers the release of bile from gallbladder and enzymes from the pancreas
Ghrelin	Stomach mucosa	Hypothalamus; other diverse tissues	Stimulates hypothalamus to boost appetite; affects energy balance in various tissues
Atrial natriuretic hormone (ANH) and other atrial natriuretic peptides (ANPs)	Heart muscle	Kidney	Promotes loss of sodium from body into urine, thus promoting water loss from the body and a resulting decrease in blood volume and pressure
Inhibins	Ovary	Adenohypophysis (anterior pituitary)	Inhibit secretion of FSH by the anterior pituitary, thus helping to regulate the female reproductive cycle
Leptin	Adipose tissue	Hypothalamus; other diverse tissues	Affects energy balance, perhaps as a signal of how much fat is stored; affects various immune, neuroendocrine, and developmental functions throughout body
Resistin	Adipose tissue and macrophages	Liver and other tissues	Reduces sensitivity to insulin (a pancreatic islet hormone), thus increasing blood glucose levels

BOX 16-13: SPORTS AND FITNESS
Steroid Abuse

Some steroid hormones are called **anabolic steroids** because they stimulate the building of large molecules (anabolism). Specifically, they stimulate the building of proteins in muscle and bone. Steroids such as testosterone and its synthetic derivatives may be abused by athletes and others who want to increase their performance. The anabolic effects of the hormones increase the mass and strength of skeletal muscles.

Unfortunately, steroid abuse has other consequences. Prolonged use of testosterone will cause a negative feedback response of the adenohypophysis: gonadotropin levels will drop in response to high blood levels of testosterone. This may lead to atrophy of the testes and, possibly, permanent sterility. Many other adverse effects, including behavioral abnormalities, are known to result from steroid abuse.

areas between the tubules. These interstitial cells produce androgens (male sex hormones); the principal androgen is **testosterone.** Testosterone is responsible for the growth and maintenance of male sexual characteristics and for sperm production (Box 16-13). Testosterone secretion is regulated principally by gonadotropin (especially LH) levels in the blood.

Ovaries

Ovaries are a set of paired glands in the pelvis (see Figure 16-2) that produce several types of sex hormones, including those described briefly in the following list:

- **Estrogens,** including *estradiol* and *estrone,* are steroid hormones secreted by the cells of the ovarian follicles that promote the development and maintenance of female sexual characteristics. With other hormones, they are responsible for breast development and the proper sequence of events in the female reproductive cycle (menstrual cycle). More details of their function are discussed in Chapter 32.
- **Progesterone** is a hormone whose name, which means "pregnancy-promoting steroid," indicates its chief function. Secreted by the corpus luteum (the tissue left behind after the rupture of a follicle during ovulation), progesterone (along with estrogen) maintains the lining of the uterus necessary for successful pregnancy (gestation). This hormone, along with others, is discussed in detail in Chapter 32.

Regulation of ovarian hormone secretion is complex, to say the least, but basically depends on the changing levels of follicle-stimulating hormone (FSH) and LH (gonadotropins) from the adenohypophysis.

PLACENTA

Another reproductive tissue that functions as an important endocrine gland is the placenta. The **placenta,** the tissue that forms on the lining of the uterus as an interface between the circulatory systems of the mother and developing child, serves as a temporary endocrine gland.

The placenta produces **human chorionic gonadotropin (hCG).** This hormone is called "chorionic" because it is secreted by the chorion, a fetal tissue component of the placenta. It is called "gonadotropin" because, as with the gonadotropins of the adenohypophysis, it stimulates development and hormone secretion by maternal ovarian tissues. Chorionic gonadotropin secretion is high during the early part of pregnancy and serves as a signal to the mother's gonads to maintain the uterine lining rather than allow it to degenerate and fall away (as in menstruation).

The discovery of hCG many years ago led to the development of early pregnancy tests. The high levels of hCG in the urine of women who are in the early part of their pregnancies can be detected through several means. The most familiar test involves the use of an over-the-counter kit that tests for hCG in urine by means of an antigen-antibody reaction that can be easily interpreted.

As the placenta develops past the first trimester (3 months) of pregnancy, its production of hCG drops as its production of estrogens and progesterone increases. The placenta therefore more or less takes over the job of ovaries in producing these hormones necessary for a successful pregnancy.

The placenta also produces additional estrogen and progesterone during pregnancy, as well as several other hormones (Table 16-10). These include **human placental lactogen (hPL)** and **relaxin.**

More about how placental hormones work is discussed in Chapter 33.

THYMUS

The **thymus** is a gland in the mediastinum, just beneath the sternum (see Figure 16-2). It is large in children until puberty, when it begins to atrophy. It continues to atrophy throughout adulthood, so that by the time an individual reaches old age, the gland is but a vestige of fat and fibrous tissue.

The anatomy of the thymus is described in Chapter 20.

Although it is considered to be primarily a lymphatic organ (see Chapter 20), the hormones **thymosin** and **thymopoietin** have been isolated from thymus tissue and are considered to be largely responsible for its endocrine activity (Table 16-10). Thymosin and thymopoietin actually refer to two entire families of peptides that together have a critical role in the development of the immune system. Specifically, thymosin and thymopoietin are thought to stimulate the production of specialized lymphocytes involved in the immune response called *T cells.* The role of T cells in the immune system is discussed in Chapter 21.

GASTRIC AND INTESTINAL MUCOSA

The mucous lining of the gastrointestinal (GI) tract, like the pancreas, contains cells that produce both endocrine and exocrine secretions (Table 16-10). GI hormones such as **gastrin, secretin,** and **cholecystokinin (CCK)** have important regulatory roles in coordinating the secretory and motor activities involved in the digestive process. For example, secretin is released when acids make contact with the intestinal mucosa. Secretin carried by the blood triggers its target cells in the stomach to reduce acid secretion. Secretin also triggers its target cells in the pancreas to release an alkaline fluid, and it acts with CCK to trigger the pancreas to release digestive enzymes. CCK triggers the gallbladder to release more bile, which helps break up fat droplets. In effect, secretin and CCK are signals from the intestine to other parts of the digestive system that promote an effective coordination of GI functions.

The hormone **ghrelin** is secreted by endocrine cells in the gastric mucosa. It acts by stimulating the hypothalamus to boost appetite. Since it also acts on other body tissues to slow metabolism and reduce fat burning, it may play an important role in contributing to obesity.

Chapter 26 describes the hormonal control of digestion in the stomach and small intestine in more detail (see Table 26-5, p. 978).

HEART

The heart is another organ with a secondary endocrine role. Although the heart's main function is to pump blood, a specific area in its wall contains some hormone-producing cells. These cells produce several peptide hormones (Table 16-10). This group of hormones is collectively called *atrial natriuretic peptide (ANP)*, and the principal hormone of the group is called **atrial natriuretic hormone (ANH)**. The name of this hormone reveals much about its role in the body. The term *atrial* refers to the fact that ANH is secreted by cells in an upper chamber of the heart called an *atrium*. Atrial cells increase their secretion of ANH in response to an increase in the stretch of the atrial wall caused by abnormally high blood volume or blood pressure. The term *natriuretic* refers to the fact that its principal effect is to promote the loss of sodium (Latin, *natrium*) from the body by means of the urine. When sodium is thus lost from the internal environment, water follows. Water loss results in a decrease in blood volume (and thus a decrease in blood pressure). We can then state that the primary effect of ANH is to oppose increases in blood volume or blood pressure. We can also state that ANH is an antagonist to ADH and aldosterone. ANH is also known by several other names, including *atrial natriuretic factor (ANF)*, *atrial natriuretic peptide*, and, simply, *atrial peptide*.

OTHER ENDOCRINE GLANDS AND HORMONES

In this chapter, we have outlined the structure and function of only a few of the more central endocrine glands. And we have discussed only a few of their principal hormones. Many of the glands we have discussed in this chapter produce many more hormones, all of which are important to normal body function. For example,

the ovaries produce the hormone **inhibin**—a glycoprotein hormone that helps to regulate FSH levels in women. And many hormones have additional effects besides the principal effects listed in this chapter (see Box 16-1, p. 602).

Many other tissues throughout the body produce hormones—perhaps all tissues in the body produce hormones. For example, adipose tissue has been shown to secrete **leptin,** a protein hormone that plays a role in energy balance, regulation of immunity and neuroendocrine function, and development. Box 16-12 on p. 627 describes another hormone secreted by adipose tissue and macrophages—*resistin*. Table 16-10 summarizes a few additional examples of hormones that you are likely to encounter in your studies.

It is neither within the scope of this chapter—nor within the scope of this book—to discuss every known human hormone. Such a discussion would be longer than the whole book is now! Having had this brief preview, however, you will now be prepared for additional examples that you will encounter as you continue your study of the human body.

QUICK CHECK

25. What are the major hormones secreted by reproductive tissues (gonads and the placenta)?
26. Which gland produces a hormone that regulates the development of cells important to the immune system?
27. Secretin was the first substance in the body to be identified as a hormone. What structure produces secretin?
28. Which type of body tissue produces the hormone *leptin?*

Cycle of Life
Endocrine System

Endocrine regulation of body processes first begins during early development in the womb. By the time a baby is born, many of the hormones are already at work influencing the activity of target cells throughout the body. As a matter of fact, new evidence suggests that it is a hormonal signal from the fetus to the mother that signals the onset of labor and delivery. Many of the basic hormones are active from birth, but most of the hormones related to reproductive functions are not produced or secreted until puberty. Secretion of male reproductive hormones follows the same pattern as most nonreproductive hormones: continuous secretion from puberty until there is a slight tapering off in late adulthood. The secretion of female reproductive hormones such as estrogens also declines late in life, but more suddenly and completely—often during or just at the end of middle adulthood.

THE BIG PICTURE

The Endocrine System and the Whole Body

It is important to appreciate the precision of control afforded by the partnership of the two major regulatory systems: the endocrine system and the nervous system. The neuroendocrine system is able to finely adjust the availability and processing of nutrients through a diverse array of mechanisms: growth hormone, thyroid hormone, cortisol, epinephrine, somatostatin, autonomic nervous regulation, and so on. The absorption, storage, and transport of calcium ions are kept in balance by the antagonistic actions of calcitonin and parathyroid hormone (and its effects on vitamin D). Reproductive ability is triggered, developed, maintained, and timed by the complex interaction of the nervous system with follicle-stimulating hormone, luteinizing hormone, estrogen, progesterone, testosterone, chorionic gonadotropin, prolactin, oxytocin, and melatonin. Nearly every process in the human organism is kept in balance by the incredibly complex, but precise, interaction of all these different nervous and endocrine regulatory chemicals.

In this chapter, we have seen the many different structures and regulatory mechanisms that make up the endocrine system. Some of the more important hormones and their characteristics are summarized in tables throughout the chapter. We have seen how the hormones interact with each other, as well as how their functions complement those of the nervous system. Although there is still much to be learned, we can see that a basic understanding of hormonal regulatory mechanisms is required to fully appreciate the nature of homeostasis in the human organism. As we continue our study of human anatomy and physiology, we will often encounter the critical integrative role played by the endocrine system.

Mechanisms of Disease

ENDOCRINE DISORDERS

As we have stated throughout this chapter, endocrine disorders typically result from either elevated or depressed hormone levels (hypersecretion or hyposecretion). At first thought, this may seem very simple and straightforward. In reality, however, nothing could be further from the truth. A variety of specific mechanisms may produce hypersecretion or hyposecretion of hormones. A few of the more well-known mechanisms are briefly explained here. Table 16-11 lists examples of a few major endocrine disorders and their mechanisms.

Mechanisms of Hypersecretion

Excessively high blood concentration of a hormone—or any condition that mimics high hormone levels—is called *hypersecretion*. Specific types of hypersecretion are usually named by placing the prefix *hyper-* in front of the name of the source gland and the suffix *-ism* at the end. For example, hypersecretion of thyroid hormone—no matter what the specific cause—is called **hyperthyroidism**. Hyperthyroidism is not a disease itself but a condition that characterizes several different diseases (e.g., Graves disease, toxic nodular goiter).

Table 16-11	Examples of Endocrine Conditions	
CONDITION	**MECHANISM**	**DESCRIPTION**
Acromegaly	Hypersecretion of growth hormone (GH) during adulthood	Chronic metabolic disorder characterized by gradual enlargement or elongation of facial bones and extremities
Addison disease	Hyposecretion of adrenal cortical hormones (adrenal cortical insufficiency)	Caused by tuberculosis, autoimmunity, or other factors, this life-threatening condition is characterized by weakness, anorexia, weight loss, nausea, irritability, decreased cold tolerance, dehydration, increased skin pigmentation, and emotional disturbance; it may lead to an acute phase (adrenal crisis) characterized by circulatory shock
Aldosteronism	Hypersecretion of aldosterone	Often caused by adrenal hyperplasia, this condition is characterized by sodium retention and potassium loss—producing Conn syndrome: severe muscle weakness, hypertension (high blood pressure), kidney dysfunction, cardiac problems
Cretinism	Hyposecretion of thyroid hormone during early development	Congenital condition characterized by dwarfism, retarded mental development, facial puffiness, dry skin, umbilical hernia, lack of muscle coordination
Cushing disease	Hypersecretion of adrenocorticotropic hormone (ACTH)	Caused by adenoma of the anterior pituitary, increased ACTH causes hypersecretion of adrenal cortical hormones, producing *Cushing syndrome*
Cushing syndrome	Hypersecretion (or injection) of glucocorticoids	Metabolic disorder characterized by fat deposits on upper back, striated pad of fat on chest and abdomen, rounded "moon" face, muscular atrophy, edema, hypokalemia (low blood potassium level), possible abnormal skin pigmentation; occurs in *Cushing disease*

Continued

Mechanisms of Disease—cont.

Table 16-11 Examples of Endocrine Conditions—cont'd

CONDITION	MECHANISM	DESCRIPTION
Diabetes insipidus	Hyposecretion of (or insensitivity to) antidiuretic hormone (ADH)	Metabolic disorder characterized by extreme polyuria (excessive urination) and polydipsia (excessive thirst) because of a decrease in the kidney's retention of water
Gestational diabetes mellitus (GDM)	Temporary decrease in blood levels of insulin during pregnancy	Carbohydrate-metabolism disorder occurring in some pregnant women; characterized by polydipsia, polyuria, overeating, weight loss, fatigue, irritability
Gigantism	Hypersecretion of GH before age 25 years	Condition characterized by extreme skeletal size caused by excess protein anabolism during skeletal development
Graves disease	Hypersecretion of thyroid hormone	Inherited, possibly autoimmune disease characterized by hyperthyroidism, exophthalmos (protruding eyes)
Hashimoto disease	Autoimmune damage to thyroid causing hyposecretion of thyroid hormone	Enlargement of thyroid (goiter) is sometimes accompanied by hypothyroidism, typically occurring between ages 30 and 50 years; 20 times more common in females than males
Hyperparathyroidism	Hypersecretion of parathyroid hormone (PTH)	Condition characterized by increased resorption of calcium from bone tissue and kidneys and increased absorption by the gastrointestinal tract; produces hypercalcemia, resulting in confusion, anorexia, abdominal pain, muscle pain, and fatigue, possibly progressing to circulatory shock, kidney failure, death
Hyperthyroidism (adult)	Hypersecretion of thyroid hormone	Condition characterized by nervousness, tremor, weight loss, excessive hunger, fatigue, heat intolerance, heart arrhythmia, and diarrhea; caused by a general acceleration of body function
Hypothyroidism (adult)	Hyposecretion of thyroid hormone	Condition characterized by sluggishness, weight gain, skin dryness, constipation, arthritis, and general slowing of body function; may lead to myxedema, coma, or death if untreated
Insulin shock	Hypersecretion (or overdose injection) of insulin, decreased food intake, excessive exercise	Hypoglycemic (low blood glucose) shock characterized by nervousness, sweating and chills, irritability, hunger, and pallor—progressing to convulsion, coma, and death if untreated
Myxedema	Extreme hyposecretion of thyroid hormone during adulthood	Severe form of adult hypothyroidism characterized by edema of the face and extremities; often progressing to coma and death
Osteoporosis	Hyposecretion of estrogen in postmenopausal women	Bone disorder characterized by loss of minerals and collagen from bone matrix, producing holes or porosities that weaken the skeleton
Pituitary dwarfism	Hyposecretion of GH before age 25 years	Condition characterized by reduced skeletal size caused by decreased protein anabolism during skeletal development
Simple goiter	Lack of iodine in diet	Enlargement of thyroid tissue results from the inability of the thyroid to make thyroid hormone because of a lack of iodine; a positive feedback situation develops in which low thyroid hormone levels trigger hypersecretion of thyroid-stimulating hormone (TSH) by pituitary—which stimulates thyroid growth
Sterility	Hyposecretion of sex hormones	Loss of reproductive function
Type 1 diabetes mellitus	Hyposecretion of insulin	Inherited condition with sudden childhood onset characterized by polydipsia, polyuria, overeating, weight loss, fatigue, and irritability, resulting from the inability of cells to secure and metabolize carbohydrates
Type 2 diabetes mellitus	Insensitivity of target cells to insulin	Carbohydrate-metabolism disorder with slow adulthood onset thought to be caused by a combination of genetic and environmental factors and characterized by polydipsia, polyuria, overeating, weight loss, fatigue, irritability
Winter (seasonal) depression	Hypersecretion of (or hypersensitivity to) melatonin	Abnormal emotional state characterized by sadness and melancholy resulting from exaggerated melatonin effects; melatonin levels are inhibited by sunlight so they increase when day length decreases during winter

Mechanisms of Disease—cont.

Any of several different mechanisms may be responsible for a particular case of hypersecretion. For example, tumors are often responsible for an abnormal proliferation of endocrine cells and the resulting increase in hormone secretion. Pituitary adenomas, for example, are benign tumors that may cause **hyperpituitarism.** As many as one in five people may have pituitary adenomas, but the majority of tumors are microscopic and asymptomatic. Larger tumors may, however, cause hyperpituitarism with a possible outcome of gigantism, or **acromegaly.**

Another cause of hypersecretion is a phenomenon called **autoimmunity.** In autoimmunity, the immune system functions abnormally. In Graves disease, for example, autoimmune antibodies against the TSH receptor actually stimulate the receptor and mimic the activity of TSH. As a result, the thyroid gland hypertrophies and excess thyroid hormone is produced.

Another possible cause of hypersecretion of a hormone is a failure of the feedback mechanisms that regulate secretion of a particular hormone. For example, a condition called **primary hyperparathyroidism** is characterized by a failure of the parathyroid gland to adjust its output to compensate for changes in blood calcium levels. Instead, the parathyroid gland seems to operate independently of the normal feedback loop and thus overproduces parathyroid hormone.

Mechanisms of Hyposecretion

Depressed blood hormone levels—or any condition that mimics low hormone levels—is termed *hyposecretion.* Specific types of hyposecretion are named in a manner similar to that in which hypersecretion disorders are named: by the addition of the *hypo-* prefix and the *-ism* suffix. For example, hyposecretion of thyroid hormone is called *hypothyroidism.*

Various different mechanisms have been shown to cause hyposecretion of hormones. For example, although most tumors cause oversecretion of a hormone, they may instead cause a gland to undersecrete its hormone or hormones. Tissue death, perhaps caused by a blockage or other failure of the blood sup-ply, can also cause a gland to reduce its hormonal output. Hypopituitarism (hyposecretion by the anterior pituitary) can occur this way. Still another way in which a gland may reduce its secretion below normal levels is through abnormal operation of regulatory feedback loops. An example of this is in the case of hyposecretion of testosterone and gonadotropic hormones in males who abuse anabolic steroids. Men who take testosterone steroids increase their blood concentration of this hormone above set point levels. The body responds to this overabundance by reducing its own output of testosterone (i.e., gonadotropins). This may lead to sterility and other complications.

Abnormalities of immune function may also cause hyposecretion. For example, an autoimmune attack on glandular tissue sometimes has the effect of reducing hormone output. Some endocrinologists theorize that autoimmune destruction of pancreatic islet cells, perhaps in combination with viral and genetic mechanisms, is a culprit in many cases of type 1 (insulin-dependent) diabetes mellitus (Box 16-13).

Many types of hyposecretion disorders have recently been shown to be caused by insensitivity of the target cells to tropic hormones rather than from actual hyposecretion. A few major types of abnormal responses in target cells are as follows:

- An abnormal decrease in the number of hormone receptors
- Abnormal function of hormone receptors, resulting in failure to bind to hormones properly
- Antibodies bind to hormone receptors, thus blocking binding of hormone molecules
- Abnormal metabolic response to the hormone-receptor complex by the target cell
- Failure of the target cell to produce enough second messenger molecules

Type 2 (non–insulin-dependent) diabetes mellitus is thought to be caused by target cell abnormalities that render the cells insensitive to insulin.

LANGUAGE OF SCIENCE *(Cont'd from page 593)*

atrial natriuretic hormone (ANH) (AY-tree-al nay-tree-yoo-RET-ik HOR-mohn) [*atria-* hall (atrium of heart), *-al* pertaining to, *natri-* natrium (sodium), *-uret-* urination, *-ic* pertaining to, *hormaien* spur on]

autocrine (AW-toh-krin) [*auto-* self, *-crine* secrete]

beta cells (BAY-tah) [*beta* second letter of Greek alphabet, *cella* storeroom]

C cells [*C* for calcitonin, *cella* storeroom]

calcitonin (CT) (kal-sih-TOH-nin) [*calci-* lime (calcium), *-toni-* tone, *-in* substance]

cholecystokinin (CCK) (koh-lee-sis-toh-KYE-nin) [*chole-* bile, *-cysto-* bag, *-kin-* movement, *-in* substance]

corticotrophs (kohr-tih-koh-TROHFS) [*cortico-* cortex, *-troph* affecting]

cortisol (KOHR-tih-sol) [*cortis-* cortex (cortisone), *-ol* alcohol]

delta cells [*delta* fourth letter of Greek alphabet, *cella* storeroom]

down-regulation

endocrine (EN-doh-krin) [*endo-* within, *-crine* secrete]

epinephrine (ep-i-NEF-rin) [*epi-* on or upon; *-nephr-* kidney, *-ine* substance]

estrogens (ES-troh-jens) [*estro-* frenzy, *-gen* generate]

follicles (FOL-lih-kuls) [*folliculus* small bag]

follicle-stimulating hormone (FSH) (FOL-lih-kul-STIM-yoo-lay-ting HOR-mohn) [*folliculus* small bag, *hormaien* spur on]

follicular cells (foh-LIK-yoo-lar) [*follicul-* small bag, *-ar* pertaining to, *cella* storeroom]

G protein (PRO-teen) [*G* for guanine nucleotide binding, *prote-* first rank, *-in* substance]

gastrin (GAS-trin) [*gastr-* stomach, *-in* substance]

ghrelin (GRAY-lin) [*ghrel-* acronym for growth hormone releasing peptide, *-in* substance]

glucagon (GLOO-kah-gon) [*gluc-* glucose, *-agon* drive]

glycoprotein hormones (glye-koh-PRO-teen HOR-mohns) [*glyco-* sweetness (glucose), *-prote-* first rank, *-in* substance, *hormaien* spur on]

gonadotrophs (go-NAD-oh-trohfs) [*gonado-* generate, *-troph* affecting]

gonadotropins (go-nah-doh-TROH-pins) [*gonado-* generate, *-trop-* affecting, *-in* substance]

gonads (GO-nads) [*gonad* generate]

growth hormone (GH) (HOR-mohn) [*hormaien* spur on]

hormone (HOR-mohn) [*hormaien* spur on]

human chorionic gonadotropin (hCG) (KOH-ri-on-ik go-na-doh-TROH-pin) [*chorion-* skin, *-ic* pertaining to, *gonado-* generate, *-trop-* affecting, *-in* substance]

human placental lactogen (hPL) (plah-SEN-tal lak-TOH-jen) [*placenta* flat cake, *-al* pertaining to, *lacto-* milk, *-gen* generate]

hypophyseal portal system (hye-poh-FIZ-ee-al POR-tal) [*hypo-* beneath, *-physis-* growth, *-al* pertaining to, *portal* doorway]

infundibulum (in-fun-DIB-yoo-lum) [*infundibulum* funnel]

inhibin (in-HIB-in) [*inhib-* inhibit, *-in* substance]

insulin (IN-suh-lin) [*insul-* island, *-in* substance]

ketone bodies (KEE-tohn) [*keto-* acetone, *-one* chemical derivative]

lactotrophs (lak-toh-TROHFS) [*lacto-* milk, *-troph* affecting]

lateral lobes (LAT-er-all)

leptin (LEP-tin) [*lept-* thin, *-in* substance]

leukotrienes (loo-koh-TRY-eens) [*leuko-* white, *-tri-* three, *-ene* chemical derivative]

luteinizing hormone (LH) (loo-tee-in-EYE-zing HOR-mohn) [*lute-* yellow, *-iz* to cause, *hormaein* to spur on]

mobile-receptor model (MO-bil-ree-SEP-tor)

neuroendocrine system (noo-roh-EN-doh-krin) [*neuro-* nerves, *-endo-* within, *-crine* secrete]

neurohypophysis (noo-roh-hye-POF-i-sis) [*neuro-* nerves, *-hypo-* under, *-physis* growth]

neurosecretory tissue (noo-roh-SEK-reh-tor-ee TISH-yoo) [*neuro-* nerves, *-secretory* pertaining to secretion, *tissue* fabric]

nonsteroid (nahn-STAYR-oyd) [*non-* not, *-stero-* solid, *-oid* similar to]

norepinephrine (NE, NR) (nor-ep-i-NEF-rin) [*nor-* chemical prefix (unbranched C chain), *-epi-* on or upon, *-nephr-* kidney, *-ine* substance]

oxytocin (OT) (ahk-see-TOH-sin) [*oxy-* oxygen, *-toc-* birth, *-in* substance]

pancreatic islets (pan-kree-AT-ik eye-lets) [*pan-* all, *-creat-* flesh, *-ic* pertaining to, *isl-* island, *-et* little]

pancreatic polypeptide cells (pan-kree-AT-ik pol-ee-PEP-tyde) [*pan-* all, *-creat-* flesh, *-ic* pertaining to, *poly-* many, *-pept-* to digest, *-ide* chemical derivative, *cella* storeroom]

paracrine (PAIR-ah-krin) [*para-* beside, *-crine* secrete]

parathyroid glands (pair-ah-THYE-royd) [*para-* beside, *-thyr-* shield, *-oid* similar to, *gland* acorn]

parathyroid hormone (PTH) (pair-ah-THYE-royd HOR-mohn) [*para-* besides, *-thyr-* shield, *-oid* similar to, *hormaien* to spur on]

peptide hormones (PEP-tyde HOR-mohns) [*pept-* to digest, *-ide* chemical derivative, *hormaien* to spur on]

permissiveness (per-MISS-iv-ness)

pineal gland (PIN-ee-al) [*pine-* pine, *-al* pertaining to, *gland* acorn]

pituitary gland (pih-TOO-i-tair-ee) [*pituit-* phlegm, *-ary* pertaining to]

placenta (plah-SEN-tah) [*placenta* flat cake]

progesterone (proh-JES-ter-ohn) [*pro-* provide for, *-gester-* bearing (pregnancy), *-stero-* solid or steroid derivative, *-one* chemical derivative]

prolactin (PRL) (proh-LAK-tin) [*pro-* provide for, *-lact-* milk, *-in* substance]

prostaglandins (PG) (pross-tah-GLAN-dins) [*prosta-* stand before (prostate), *-gland-* acorn (gland), *-in* substance]

protein hormones (PRO-teen HOR-mohns) [*prote-* first rank, *-in* substance, *hormaien* to spur on]

relaxin (reh-LAK-sin) [*relax-* relaxation, *-in* substance]

releasing hormone (ree-LEE-sing HOR-mohn) [*hormaien* to spur on]

renin (REH-nin) [*ren-* kidney, *-in* substance]

renin-angiotensin-aldosterone system (RAAS) (REH-nin-an-jee-oh-TEN-sin-al-DAH-stair-ohn) [*ren-* kidney, *-in* substance, *angio-* blood vessel, *-tens-* pressure or stretch, *-in* substance, *aldo-* aldehyde, *-stero-* solid or steroid derivative, *-one* chemical derivative]

resistin (reh-SIS-tin) [*resist-* resistance, *-in* substance]

second messenger model

secretin (seh-KREE-tin) [*secret-* secretion, *-in* substance]

sex hormones (HOR-mohns) [*hormaien* to spur on]

somatostatin (soh-mah-toh-STAT-in) [*somato-* body, *-stat-* halt, *-in* substance]

somatotrophs (soh-mah-toh-TROHFS) [*somato-* body, *-troph* affecting]

somatotropin (STH) (soh-mah-toh-TROH-pin) [*somato-* body, *-trop-* affecting, *-in* substance]

LANGUAGE OF SCIENCE *(Cont'd)*

steroid (STAYR-oyd) [*stero-* solid, *-oid* resembling]

synergism (SIN-er-jiz-em) [*syn-* together, *-erg-* work, *-ism* condition of]

target

target cells [*cella* storeroom]

testosterone (tes-TOS-teh-rohn) [*testo-* witness (testis), *-stero-* solid or steroid derivative, *-one* chemical derivative]

tetraiodothyronine (T₄) (tet-rah-eye-oh-doh-THY-roh-neen) [*tetra-* four, *-iodo-* violet (iodine), *-thyro-* shield (thyroid gland), *-nine* chemical derivative]

thromboxanes (throm-BOKS-aynes) [*thrombo-* clot, *-oxa-* oxygen, *-ane* chemical derivative]

thymopoietin (thy-moh-POY-eh-tin) [*thymo-* thymus gland, *-poiet-* make, *-in* substance]

thymosin (THY-moh-sin) [*thymos-* thymus gland, *-in* substance]

thymus (THY-mus) [*thymo* thyme flower]

thyroglobulins (thy-roh-GLOB-yoo-linz) [*thyro-* shield (thyroid gland), *-globul-* small ball, *-in* substance]

thyroid colloid (THY-royd KOL-oyd) [*thyro-* shield (thyroid gland), *-oid* similar to, *coll-* glue, *-oid* resembling]

thyroid gland (THY-royd) [*thyro-* shield, *-oid* similar to, *gland* acorn]

thyroid-stimulating hormone (TSH) (THY-royd STIM-yoo-lay-ting HOR-mohn) [*thyreos* small shield, *hormaien* to spur on]

thyrotrophs (thy-roh-TROHFS) [*thyro-* pertaining to thyroid gland, *-troph* affecting]

thyroxine (thy-ROK-sin) [*thyro-* pertaining to thyroid gland, *-ox-* oxygen, *-ine* chemical derivative]

triiodothyronine (T₃) (try-eye-oh-doh-THY-roh-neen) [*tri-* three, *-iodo-* violet (iodine), *-thyro-* shield (thyroid gland), *-nine* chemical derivative]

tropic hormones (TROH-pik HOR-mohns) [*trop-* affecting, *-ic* pertaining to, *hormaien* to spur on]

up-regulation

vitamin D (VYE-tah-min D) [*vit-* life, *-amine* chemical group]

zona fasciculata (ZOH-nah fas-sic-yoo-LAY-tah) [*zona* belt, *fasiculata* having little bundles]

zona glomerulosa (ZOH-nah gloh-mair-yoo-LOH-sah) [*zona* belt, *glomerulosa* having small balls]

zona reticularis (ZOH-nah reh-tik-yoo-LAIR-is) [*zona* belt, *reticularis* having little nets]

LANGUAGE OF MEDICINE

acromegaly (ak-roh-MEG-ah-lee) [*acro-* extremities, *-mega-* great, *-aly* condition]

Addison disease (AD-i-son) [*Thomas Addison* English physician]

autoimmunity (aw-toh-i-MYOO-ni-tee) [*auto-* self, *-immun-* free, *-ity* state of]

chiropractic (kye-roh-PRAK-tik) [*chiro-* hand, *-practic* practical]

cretinism (KREE-tin-iz-em) [*cretin-* idiot, *-ism* condition of]

cryopreservation (krye-oh-prez-er-VAY-shun) [*cryo-* cold, *-preserv-* to keep, *-tion* process of]

Cushing syndrome (KOOSH-ing SIN-drohm) [*Harvey W. Cushing* American neurosurgeon]

diabetes insipidus (dye-ah-BEE-teez in-SIP-i-dus) [*diabetes* to pass through, *insipidus* without zest]

diabetes mellitus (dye-ah-BEE-teez MELL-i-tus) [*diabetes* to pass through, *mellitus* honey sweet]

diabetic ketoacidosis (dye-ah-BET-ik kee-toh-ass-i-DOH-sis) [*diabet-* to pass through (diabetes mellitus), *-ic* pertaining to, *keto-* acetone, *-acid-* sour, *-osis* condition of]

Edmonton Protocol (ED-mon-ton PRO-toh-kol) [*Edmonton* city in Canada]

exophthalmos (ek-soff-THAL-mus) [*exo-* outside, *-ophthalm-* eye, *-ic* pertaining to]

gigantism (jye-GAN-tiz-em) [*gigant-* great, *-ism* condition of]

glycosuria (glye-koh-SOO-ree-ah) [*glyco-* sweetness (glucose), *-ure-* urine, *-ia* condition]

Graves disease (gravz dih-ZEEZ) [*Robert J. Graves* Irish physician]

hyperglycemia (hye-per-glye-SEE-mee-ah) [*hyper-* excessive, *-glyc-* sweet (glucose), *-emia* blood condition]

hyperpituitarism (hye-per-pih-TYOO-i-tar-iz-em) [*hyper-* excessive, *-pituitar-* phlegm (pituitary gland), *-ism* condition of]

hypersecretion (hyp-per-seh-KREE-shun) [*hyper* excessive]

hyperthyroidism (hye-per-THYE-royd-iz-em) [*hyper-* excessive, *-thyr-* shield (thyroid gland), *-oid* similar to, *-ism* condition of]

hypocalcemia (hye-poh-kal-SEE-mee-ah) [*hypo-* deficient, *-calc-* lime (calcium), *-emia* blood condition]

hypophysectomy (hye-poff-i-SEK-toh-mee) [*hypo-* deficient, *-phys-* growth, *-ectomy* cut away (surgical removal)]

hyposecretion (hye-poh-seh-KREE-shun) [*hypo* deficient]

kinesiopathology (kih-nee-see-oh-PATH-ol-oh-jee) [*kinesio-* movement, *-path-* disease, *-ology* science of]

myopathology (mye-oh-PATH-ol-oh-jee) [*myo-* muscle, *-path-* disease, *-ology* science of]

myxedema (mik-seh-DEE-mah) [*myx-* mucus, *-edema* swelling]

oral hypoglycemic agents (OR-al hye-poh-glye-SEE-mik) [*oral* pertaining to the mouth, *hypo-* excessive, *-glyc-* sweet, *-emia* blood condition]

osteoporosis (os-tee-oh-poh-ROH-sis) [*osteo-* bone, *-poros-* passage, *-osis* condition of]

pituitary dwarfism (pih-TOO-i-tair-ee DWARF-iz-em) [*pituita-* phlegm, *-ism* condition of]

polydipsia (pol-ee-DIP-see-ah) [*poly-* many. *-dipsa-* thirst, *-ia* condition]

polyphagia (pol-ee-FAY-jee-ah) [*poly-* many, *-phag-* eat, *-ia* condition]

polyuria (pol-ee-YOO-ree-ah) [*poly-* many, *-uri-* urine, *-ia* condition]

primary hyperparathyroidism (hye-per-pair-ah-THY-royd-iz-em) [*prim-* first, *-ary* pertaining to, *hyper-* excessive, *-para-* beside, *-thyr-* shield, *-oid* similar to, *-ism* condition of]

simple goiter (GOY-ter) [*guttur* throat]

type 1 diabetes mellitus (dye-ah-BEE-teez MELL-i-tus) [*diabetes* to pass through, *mellitus* honey sweet]

type 2 diabetes mellitus (dye-ah-BEE-teez MELL-i-tus) [*diabetes* to pass through, *mellitus* honey sweet]

CASE STUDY

Sonia Dawson, a 30-year-old female, has been diagnosed with Graves disease. She has recently lost 25 pounds despite an increased appetite. Her vision is blurry; her heart races and pounds. She has been irritable but attributes the irritability to insomnia. She has had bouts of diarrhea. She feels hot, even if others are comfortable. The collars on her clothing are becoming too tight. Her last menstrual period was 2 months ago. Before that time her menses were regular.

On physical examination, the nurse notes that Ms. Dawson is thin, pale, and anxious. She moves restlessly around the room. Her eyes are bulging and have a staring appearance. Her eyelids and hands have fine tremors. Her skin is smooth, warm, and moist. She is sweating profusely, although the room temperature is ambient. Her hair is very fine and soft. She weighs 112 pounds. Her oral temperature is 99° F (37.2° C), heart rate is 120 beats/min at rest, respirations are 20/min, and blood pressure is 110/50. Her thyroid gland is enlarged with auscultation of a bruit. Laboratory results reveal a serum T_3 level of 200 mcg/dl (high), a T_4 level of 20 mcg/dl (low), and a TSH level that is 2.5 μU/ml, which is normal. A thyroid antibody test shows a high titer of thyroglobulin antibodies. Radioactive iodine uptake test shows a 40% uptake in 6 hours, which is elevated.

The decision to treat nonsurgically is made. Ms. Dawson is started on methimazole, 10 mg by mouth every day, to suppress her thyroid gland and propranolol, 10 mg twice each day, to decrease her heart rate and control her cardiac symptoms.

1. Which of the following reasons is the *most* likely cause of Ms. Dawson's increased thyroid function?
   A. Hyperplasia of the thyroid
   B. Anterior pituitary tumor
   C. Thyroid carcinoma
   D. Autoimmune response

2. After 1 month of treatment Ms. Dawson returns to the clinic for a checkup. She has not been feeling well and is very tired. However, her previous symptoms have resolved. A TSH level from a few days ago is now 7.5, which is greatly elevated. Which one of the following explanations for this recent elevation would be considered reasonable in Ms. Dawson's case?
   A. Pituitary failure has occurred.
   B. The thyroid gland has been suppressed too much.
   C. The hypothalamus has been suppressed.
   D. The thyroid gland has been overstimulated.

3. Based on the elevated TSH level and the change in Ms. Dawson's condition, what symptoms would you expect her to exhibit?
   A. Dyspnea, increased heart rate, and constipation
   B. Slower heart rate, lethargy, and nervousness
   C. Dyspnea, slight fever, and increased heart rate
   D. Slower heart rate, lethargy, and hair loss

CHAPTER SUMMARY

INTRODUCTION

A. The endocrine and nervous systems function to achieve and maintain homeostasis (Table 16-1)

B. When the two systems work together, referred to as the *neuroendocrine system*, they perform the same general functions: communication, integration, and control

C. In the endocrine system, secreting cells send hormone molecules by way of the blood to specific target cells contained in target tissues or target organs

D. Hormones—carried to almost every point in the body; can regulate most cells; effects work more slowly and last longer than those of neurotransmitters

E. Endocrine glands are "ductless glands"; many are made of glandular epithelium whose cells manufacture and secrete hormones; a few endocrine glands are made of neurosecretory tissue

F. Glands of the endocrine system are widely scattered throughout the body (Figure 16-2; Table 16-2)

HORMONES

A. Classification of hormones
   1. Classification by general function
      a. Tropic hormones—hormones that target other endocrine glands and stimulate their growth and secretion
      b. Sex hormones—hormones that target reproductive tissues
      c. Anabolic hormones—hormones that stimulate anabolism in target cells
   2. Classification by chemical structure (Figure 16-3; Table 16-3)
      a. Steroid hormones
      b. Nonsteroid hormones
   3. Steroid hormones (Figure 16-4)
      a. Synthesized from cholesterol (Figure 16-5)
      b. Lipid soluble and can easily pass through the phospholipid plasma membrane of target cells
      c. Examples of steroid hormones: cortisol, aldosterone, estrogen, progesterone, and testosterone
   4. Nonsteroid hormones (Figure 16-6)
      a. Synthesized primarily from amino acids
      b. Protein hormones—long, folded chains of amino acids; e.g., insulin, parathyroid hormone
      c. Glycoprotein hormones—protein hormones with carbohydrate groups attached to the amino acid chain
      d. Peptide hormones—smaller than protein hormones; short chain of amino acids; e.g., oxytocin, antidiuretic hormone (ADH)

e. Amino acid derivative hormones—each is derived from a single amino acid molecule
   (1) Amine hormones—synthesized by modifying a single molecule of tyrosine; produced by neurosecretory cells and by neurons; e.g., epinephrine, norepinephrine
   (2) Amino acid derivatives produced by the thyroid gland; synthesized by adding iodine to tyrosine

B. How hormones work
   1. General principles of hormone action
      a. Hormones signal a cell by binding to the target cell's specific receptors in a "lock-and-key" mechanism (Figure 16-7)
      b. Different hormone-receptor interactions produce different regulatory changes within the target cell through chemical reactions
      c. Combined hormone actions
         (1) Synergism—combinations of hormones acting together have a greater effect on a target cell than the sum of the effects that each would have if acting alone
         (2) Permissiveness—when a small amount of one hormone allows a second one to have its full effects on a target cell
         (3) Antagonism—one hormone produces the opposite effects of another hormone; used to "fine tune" the activity of target cells with great accuracy
      d. Most hormones have primary effects that directly regulate target cells and many secondary effects that influence or modulate other regulatory mechanisms in target cells
      e. Endocrine glands produce more hormone molecules than actually are needed; the unused hormones are quickly excreted by the kidneys or broken down by metabolic processes
   2. Mechanism of steroid hormone action (Figure 16-8)
      a. Steroid hormones are lipid soluble, and their receptors are normally found in the target cell's cytosol
      b. After a steroid hormone molecule has diffused into the target cell, it binds to a receptor molecule to form a hormone-receptor complex
      c. Mobile-receptor model—the hormone passes into the nucleus, where it binds to a mobile receptor and activates a certain gene sequence to begin transcription of mRNA; newly formed mRNA molecules move into the cytosol, associate with ribosomes, and begin synthesizing protein molecules that produce the effects of the hormone
      d. Steroid hormones regulate cells by regulating production of certain critical proteins
      e. The amount of steroid hormone present determines the magnitude of a target cell's response
      f. Because transcription and protein synthesis take time, responses to steroid hormones are often slow

   3. Mechanisms of nonsteroid hormone action
      a. The second messenger mechanism—also known as the *fixed-membrane-receptor model* (Figure 16-9)
         (1) A nonsteroid hormone molecule acts as a "first messenger" and delivers its chemical message to fixed receptors in the target cell's plasma membrane
         (2) The "message" is then passed by way of a G protein into the cell where a "second messenger" triggers the appropriate cellular changes
         (3) Second messenger mechanism—produces target cell effects that differ from steroid hormone effects in several important ways
            (a) The effects of the hormone are amplified by the cascade of reactions
            (b) There are a variety of second messenger mechanisms—e.g., IP_3, GMP, calcium-calmodulin mechanisms (Figure 16-10)
            (c) The second messenger mechanism operates much more quickly than the steroid mechanism
      b. The nuclear-receptor mechanism—small iodinated amino acids (T_4 and T_3) enter the target cell and bind to receptors associated with a DNA molecule in the nucleus; this binding triggers transcription of mRNA and synthesis of new enzymes

C. Regulation of hormone secretion
   1. Control of hormonal secretion is usually part of a negative feedback loop and is called *endocrine reflexes* (Figure 16-11)
   2. Simplest mechanism—when an endocrine gland is sensitive to the physiological changes produced by its target cells
   3. Endocrine gland secretion may also be regulated by a hormone produced by another gland
   4. Endocrine gland secretions may be influenced by nervous system input; this fact emphasizes the close functional relationship between the two systems

D. Regulation of target cell sensitivity
   1. Sensitivity of target cell depends in part on number of receptors (Figure 16-12)
      a. Up-regulation—increased number of hormone receptors increases sensitivity
      b. Down-regulation—decreased number of hormone receptors decreases sensitivity
   2. Sensitivity of target cell may also be regulated by factors that affect signal transcription or gene transcription

PROSTAGLANDINS

A. Unique group of lipid hormones (20-carbon fatty acid with 5-carbon ring) that serve important and widespread integrative functions in the body but do not meet the usual definition of a hormone (Figure 16-13; Table 16-4)

B. Called *tissue hormones* because the secretion is produced in a tissue and diffuses only a short distance to other cells within the same tissue; PGs tend to integrate activities of neighboring cells

C. Many structural classes of prostaglandins have been isolated and identified
  1. Prostaglandin A (PGA)—intraarterial infusion resulting in an immediate fall in blood pressure accompanied by an increase in regional blood flow to several areas
  2. Prostaglandin E (PGE)—vascular effects: regulation of red blood cell deformability and platelet aggregation; inflammation (which can be blocked with drugs that inhibit PG-producing enzymes such as COX-1 and COX-2); gastrointestinal effects: regulates hydrochloric acid secretion
  3. Prostaglandin F (PGF)—especially important in reproductive system, causing uterine contractions; also affects intestinal motility and is required for normal peristalsis
D. Many tissues are known to secrete PGs
E. PGs have diverse physiological effects

PITUITARY GLAND

A. Structure of the pituitary gland
  1. Formerly known as *hypophysis*
  2. Size: 1.2 to 1.5 cm (about 1/2 inch) across; weight: 0.5 g (1/60 ounce)
  3. Located on the ventral surface of the brain within the skull (Figure 16-14)
  4. Infundibulum—stemlike stalk that connects pituitary to the hypothalamus
  5. Made up of two separate glands, the adenohypophysis (anterior pituitary gland) and the neurohypophysis (posterior pituitary gland)
B. Adenohypophysis (anterior pituitary)
  1. Divided into two parts
    a. Pars anterior—forms the major portion of the adenohypophysis
    b. Pars intermedia
  2. Tissue is composed of irregular clumps of secretory cells supported by fine connective tissue fibers and surrounded by a rich vascular network
  3. Three types of cells can be identified according to their affinity for certain stains (Figure 16-15)
    a. Chromophobes—do not stain
    b. Acidophils—stain with acid stains
    c. Basophils—stain with basic stains
  4. Five functional types of secretory cells exist
    a. Somatotrophs—secrete GH
    b. Corticotrophs—secrete ACTH
    c. Thyrotrophs—secrete TSH
    d. Lactotrophs—secrete prolactin (PRL)
    e. Gonadotrophs—secrete LH and FSH
  5. Growth hormone (GH) (Figure 16-16; Table 16-6)
    a. Also known as *somatotropin* (STH)
    b. Promotes growth of bone, muscle, and other tissues by accelerating amino acid transport into the cells
    c. Stimulates fat metabolism by mobilizing lipids from storage in adipose cells and speeding up catabolism of the lipids after they have entered another cell
    d. GH tends to shift cell chemistry away from glucose catabolism and toward lipid catabolism as an energy source; this leads to increased blood glucose levels

e. GH functions as an insulin antagonist and is vital to maintaining homeostasis of blood glucose levels
6. Prolactin (PRL; Table 16-6)
  a. Produced by acidophils in the pars anterior
  b. Also known as *lactogenic hormone*
  c. During pregnancy, PRL promotes development of the breasts, anticipating milk secretion; after the baby is born, PRL stimulates the mother's mammary glands to produce milk
7. Tropic hormones—hormones that have a stimulating effect on other endocrine glands; four principal tropic hormones are produced and secreted by the basophils of the pars anterior (Table 16-6)
  a. Thyroid-stimulating hormone (TSH), or thyrotropin—promotes and maintains the growth and development of the thyroid; also causes the thyroid to secrete its hormones
  b. Adrenocorticotropic hormone (ACTH), or adrenocorticotropin—promotes and maintains normal growth and development of the cortex of the adrenal gland; also stimulates the adrenal cortex to secrete some of its hormones
  c. Follicle-stimulating hormone (FSH)—in the female, stimulates primary graafian follicles to grow toward maturity; also stimulates the follicle cells to secrete estrogens; in the male, FSH stimulates the development of the seminiferous tubules of the testes and maintains spermatogenesis
  d. Luteinizing hormone (LH)—in the female, stimulates the formation and activity of the corpus luteum of the ovary; corpus luteum secretes progesterone and estrogens when stimulated by LH; LH also supports FSH in stimulating maturation of follicles; in the male, LH stimulates interstitial cells in the testes to develop and secrete testosterone; FSH and LH are called *gonadotropins* because they stimulate the growth and maintenance of the gonads
8. Control of secretion in the adenohypophysis
  a. Hypothalamus secretes releasing hormones into the blood, which are then carried to the hypophyseal portal system (Figure 16-17; Table 16-5)
  b. Hypophyseal portal system carries blood from the hypothalamus directly to the adenohypophysis, where the target cells of the releasing hormones are located (Figure 16-18)
  c. Releasing hormones influence the secretion of hormones by acidophils and basophils
  d. Through negative feedback, the hypothalamus adjusts the secretions of the adenohypophysis, which then adjusts the secretions of the target glands that in turn adjust the activity of their target tissues (Figure 16-19)
  e. Minute-by-minute variations in hormone secretion can exhibit occasional large peaks, caused by pulse in releasing hormone secretion by the hypothalamus (Figure 16-20)
  f. In stress, the hypothalamus translates nerve impulses into hormone secretions by endocrine glands, basically creating a mind-body link

C. Neurohypophysis (posterior pituitary)
 1. Serves as storage and release site for antidiuretic hormone (ADH) and oxytocin (OT), which are synthesized in the hypothalamus (Figure 16-21; Table 16-6)
 2. Release of ADH and OT into the blood is controlled by nervous stimulation
 3. Antidiuretic hormone (ADH)
  a. Prevents the formation of a large volume of urine, thereby helping the body conserve water
  b. Causes a portion of each tubule in the kidney to resorb water from the urine it is forming
  c. Dehydration triggers the release of ADH
  d. Also called arginine vasopressin (AVP) because it stimulates a rise in blood pressure, partly by increasing contraction in small arteries
 4. Oxytocin (OT)—has two actions
  a. Causes milk ejection from the lactating breast; regulated by positive feedback mechanism; PRL cooperates with oxytocin
  b. Stimulates contraction of uterine muscles that occurs during childbirth; regulated by positive feedback mechanism

PINEAL GLAND

A. Tiny, pine cone–shaped structure located on the dorsal aspect of the brain's diencephalon
B. Member of the nervous system because it receives visual stimuli and also a member of the endocrine system because it secretes hormones
C. Pineal gland supports the body's biological clock
D. Principal pineal secretion is melatonin (Table 16-10)

THYROID GLAND

A. Structure of the thyroid gland
 1. Made up of two large lateral lobes and a narrow connecting isthmus (Figure 16-22)
 2. A thin, wormlike projection of thyroid tissue often extends upward from the isthmus
 3. Weight of the thyroid in an adult is approximately 30 g (1 ounce)
 4. Located in the neck, on the anterior and lateral surfaces of the trachea, just below the larynx
 5. Composed of follicles (Figure 16-23)
  a. Small, hollow spheres
  b. Filled with thyroid colloid that contains thyroglobulins
B. Thyroid hormone (Figure 16-24; Table 16-7)
 1. Actually two different hormones
  a. Tetraiodothyronine (T_4), or thyroxine—contains four iodine atoms; approximately 20 times more abundant than T_3; major importance is as a precursor to T_3
  b. Triiodothyronine (T_3)—contains three iodine atoms; considered to be the principal thyroid hormone; T_3 binds efficiently to nuclear receptors in target cells
 2. Thyroid gland stores considerable amounts of a preliminary form of its hormones *before* secreting them
 3. Before being stored in the colloid of follicles, T_3 and T_4 are attached to globulin molecules, forming thyroglobin complexes

4. Before release, T_3 and T_4 detach from globulin and enter the bloodstream
 5. Once in the blood, T_3 and T_4 attach to a plasma protein called *thyroid-binding globulins (TBGs)* and travel as a hormone-globulin complex
 6. T_3 and, to a lesser extent, T_4 detach from plasma globulin as they near the target cells
 7. Thyroid hormone—helps regulate the metabolic rate of all cells and cell growth and tissue differentiation; it is said to have a "general" target
C. Calcitonin (CT) (Table 16-7)
 1. Produced by thyroid gland in the parafollicular cells
 2. In humans, CT may subtly influence the processing of calcium by bone cells by decreasing blood calcium levels and promoting conservation of hard bone matrix
 3. Parathyroid hormone acts as antagonist to calcitonin to maintain calcium homeostasis

PARATHYROID GLANDS

A. Structure of the parathyroid glands
 1. Four or five parathyroid glands embedded in the posterior surface of the thyroid's lateral lobes (Figure 16-25)
 2. Tiny, rounded bodies within thyroid tissue formed by compact, irregular rows of cells (Figure 16-26)
B. Parathyroid hormone (PTH) (Table 16-7)
 1. PTH is an antagonist to calcitonin and is the primary hormone to maintain calcium homeostasis (Figure 16-27)
 2. PTH acts on bone and kidney
  a. Causes more bone to be dissolved, yielding calcium and phosphate, which enters the bloodstream
  b. Causes phosphate to be secreted by the kidney cells into the urine to be excreted
  c. Causes increased intestinal absorption of calcium by stimulating the kidney to produce active vitamin D, which increases calcium absorption in gut

ADRENAL GLANDS

A. Structure of the adrenal glands
 1. Located on top of the kidneys, fitting like caps (Figure 16-28)
 2. Made up of two portions (Figure 16-29; Table 16-8)
  a. Adrenal cortex—composed of endocrine tissue (Figure 16-30)
  b. Adrenal medulla—composed of neurosecretory tissue
B. Adrenal cortex—all cortical hormones are steroids and known as *corticosteroids* (Figure 16-31)
 1. Composed of three distinct layers of secreting cells
  a. Zona glomerulosa—outermost layer, directly under the outer connective tissue capsule of the adrenal gland; secretes mineralocorticoids
  b. Zona fasciculata—middle layer; secretes glucocorticoids
  c. Zona reticularis—inner layer; secretes small amounts of glucocorticoids and gonadocorticoids
 2. Mineralocorticoids
  a. Have an important role in the regulatory process of sodium in the body

b. Aldosterone
  (1) Only physiologically important mineralocorticoid in the human; primary function is maintenance of sodium homeostasis in the blood by increasing sodium resorption in the kidneys
  (2) Aldosterone also increases water retention and promotes the loss of potassium and hydrogen ions
  (3) Aldosterone secretion is controlled by the renin-angiotensin-aldosterone system (RAAS) and by blood potassium concentration (Figure 16-32)

3. Glucocorticoids
  a. Main glucocorticoids secreted by the zona fasciculata are cortisol, cortisone, and corticosterone, with cortisol the only one secreted in significant quantities
  b. Affect every cell in the body
  c. Are protein mobilizing, gluconeogenic, and hyperglycemic
  d. Tend to cause a shift from carbohydrate catabolism to lipid catabolism as an energy source
  e. Essential for maintaining normal blood pressure by aiding norepinephrine and epinephrine to have their full effect, causing vasoconstriction
  f. High blood concentration causes eosinopenia and marked atrophy of lymphatic tissues
  g. Act with epinephrine to bring about normal recovery from injury produced by inflammatory agents
  h. Secretion increases in response to stress
  i. Except during stress response, secretion is mainly controlled by a negative feedback mechanism involving ACTH from the adenohypophysis
  j. Secretion is characterized by several large pulses of increased hormone levels throughout the day—the largest occurring just before waking (Figure 16-33)

4. Gonadocorticoids—sex hormones (androgens) that are released from the adrenal cortex

C. Adrenal medulla
  1. Neurosecretory tissue—tissue composed of neurons specialized to secrete their products into the blood
  2. Adrenal medulla secretes two important hormones—epinephrine and norepinephrine; they are part of the class of nonsteroid hormones called catecholamines
  3. Both hormones bind to the receptors of sympathetic effectors to prolong and enhance the effects of sympathetic stimulation by the ANS (Figure 16-34)

PANCREATIC ISLETS

A. Structure of the pancreatic islets (Figure 16-35)
  1. Elongated gland, weighing approximately 100 g (3.5 ounce); its head lies in the duodenum, extends horizontally behind the stomach, and, then, touches the spleen
  2. Composed of endocrine and exocrine tissues
    a. Pancreatic islets (islets of Langerhans)—endocrine portion
    b. Acini—exocrine portion—secretes a serous fluid containing digestive enzymes into ducts draining into the small intestine

3. Pancreatic islets—each islet contains four primary types of endocrine glands joined by gap junctions
  a. Alpha cells (A cells)—secrete glucagon (Figure 16-36)
  b. Beta cells (B cells)—secrete insulin; account for up to 75% of all pancreatic islet cells
  c. Delta cells (D cells)—secrete somatostatin
  d. Pancreatic polypeptide cells (F, or PP, cells)—secrete pancreatic polypeptides

B. Pancreatic hormones (Table 16-9)—work as a team to maintain homeostasis of food molecules (Figure 16-37)
  1. Glucagon—produced by alpha cells; tends to increase blood glucose levels; stimulates gluconeogenesis in liver cells
  2. Insulin—produced by beta cells; lowers blood concentration of glucose, amino acids, and fatty acids and promotes their metabolism by tissue cells
  3. Somatostatin—produced by delta cells; primary role is regulating the other endocrine cells of the pancreatic islets
  4. Pancreatic polypeptide—produced by F (PP) cells; influences the digestion and distribution of food molecules to some degree

GONADS

A. Testes (Figure 16-2; Table 16-10)
  1. Paired organs within the scrotum in the male
  2. Composed of seminiferous tubules and a scattering of interstitial cells
  3. Testosterone is produced by the interstitial cells and responsible for the growth and maintenance of male sexual characteristics
  4. Testosterone secretion is mainly regulated by gonadotropin levels in the blood

B. Ovaries (Figure 16-2; Table 16-10)
  1. Primary sex organs in the female
  2. Set of paired glands in the pelvis that produce several types of sex hormones
    a. Estrogens—steroid hormones secreted by ovarian follicles; promote development and maintenance of female sexual characteristics
    b. Progesterone—secreted by corpus luteum; maintains the lining of the uterus necessary for successful pregnancy
    c. Ovarian hormone secretion depends on the changing levels of FSH and LH from the adenohypophysis

PLACENTA

A. Tissues that form on the lining of the uterus as a connection between the circulatory systems of the mother and developing child
B. Serves as a temporary endocrine gland that produces human chorionic gonadotropin, estrogens, and progesterone (Table 16-10)

THYMUS (FIGURE 16-2)

A. Gland located in the mediastinum just beneath the sternum
B. Thymus is large in children, begins to atrophy at puberty, and, by old age, is a vestige of fat and fibrous tissue
C. Considered to be primarily a lymphatic organ, but the hormone thymosin has been isolated from thymus tissue (Table 16-10)
D. Thymosin—stimulates development of T cells

GASTRIC AND INTESTINAL MUCOSA

A. The mucous lining of the GI tract contains cells that produce both endocrine and exocrine secretions (Table 16-10)

B. GI hormones, such as gastrin, secretin, and cholecystokinin (CCK), play regulatory roles in coordinating the secretory and motor activities involved in the digestive process

C. Ghrelin—hormone secreted by endocrine cells in gastric mucosa; stimulates hypothalamus to boost appetite; slows metabolism and fat burning; may contribute to obesity

HEART

A. The heart has a secondary endocrine role

B. Hormone-producing cells produce several atrial natriuretic peptides (ANPs), including atrial natriuretic hormone (ANH) (Table 16-10)

C. ANH's primary effect is to oppose increases in blood volume or blood pressure; also an antagonist to ADH and aldosterone

OTHER ENDOCRINE GLANDS AND ORGANS (TABLE 16-10)

A. Major endocrine glands produce more hormones that are outlined in this book (e.g., inhibin secreted by the ovaries)

B. Many tissues (perhaps all tissues) produce hormones, most of which are beyond the scope of this book (e.g., leptin and resistin secreted by adipose tissue).

CYCLE OF LIFE: ENDOCRINE SYSTEM

A. Endocrine regulation begins in the womb

B. Many active hormones are active from birth
   1. Evidence that a hormonal signal from fetus to mother signals the onset of labor

C. Hormones related to reproduction begin at puberty

D. Secretion of male reproductive hormones—continuous production from puberty, slight decline in late adulthood

E. Secretion of female reproductive hormones declines suddenly and completely in middle adulthood

THE BIG PICTURE: THE ENDOCRINE SYSTEM AND THE WHOLE BODY

A. Nearly every process in the human organism is kept in balance by the intricate interaction of different nervous and endocrine regulatory chemicals

B. The endocrine system operates with the nervous system to finely adjust the many processes they regulate

C. Neuroendocrine system adjusts nutrient supply

D. Calcitonin, parathyroid hormone, and vitamin D balance calcium ion use

E. The nervous system and hormones regulate reproduction

REVIEW QUESTIONS

1. Define the terms *hormone* and *target organ*.
2. Describe the characteristic chemical group found at the core of each steroid hormone.
3. Identify the major categories of nonsteroid hormones.
4. Identify the sequence of events involved in a second messenger mechanism.
5. What is the function of calmodulin?
6. List the classes of prostaglandins. Identify the functions of three of these classes.
7. Name the two subdivisions of the adenohypophysis.
8. Discuss and identify, by staining tendency and relative percentages, the cell types present in the anterior pituitary gland.
9. Discuss the functions of growth hormone.
10. What effect does growth hormone have on blood glucose concentration? Fat mobilization and catabolism? Protein anabolism?
11. List the four tropic hormones secreted by the basophils of the anterior pituitary gland. Which of the tropic hormones are also called gonadotropins?
12. How does antidiuretic hormone act to alter urine volume?
13. Describe the positive feedback associated with oxytocin.
14. Discuss the synthesis and storage of thyroxine and triiodothyronine. How are they transported in the blood?
15. Discuss the functions of parathyroid hormone.
16. List the hormones produced by each "zone" of the adrenal cortex, and describe the actions of these hormones.
17. Discuss the normal function of hormones produced by the adrenal medulla.
18. Identify the hormones produced by each of the cell types in the pancreatic islets.
19. Identify the "pregnancy-promoting" hormone.
20. Where is human chorionic gonadotropin produced? What does it do?
21. Describe the role of atrial natriuretic hormone.
22. Identify the conditions resulting from both hypersecretion and hyposecretion of growth hormone during growth years.
23. Describe the face of a patient with Cushing syndrome.
24. How does exercise affect diabetes mellitus?

CRITICAL THINKING QUESTIONS

1. Driving a car requires rapid response of selected muscles. The regulation of blood sugar level requires regulating almost every cell in the body. Based on the characteristics of each system, explain why driving would be a nervous system function and blood sugar regulation would be an endocrine function.
2. How would you explain the ways in which one hormone interacts with another and its impact on the cell?
3. Compare and contrast the action mechanisms of steroid and nonsteroid hormones.
4. Why are thyroid hormones exceptions to the usual mode of nonsteroid hormone functioning?
5. What examples can you find that apply the concept of a negative feedback loop to the regulation of hormone secretion?
6. A hyposecretion of which hormone would make it difficult for a mother to nurse her child? How would you summarize the effects of this hormone?
7. Why do you think the hypothalamus can be called the "mind-body link"?

8. How would you explain the hormonal interaction that helps maintain the set point value for the glucose in the blood?
9. Explain how the cell is able to become more or less sensitive to a specific hormone.
10. What is the relationship between increased blood level concentrations of FSH and menopause?
11. A lack of iodine in the diet will cause a simple goiter. Describe the feedback loop that will cause its formation.
12. A friend of yours is considering the use of anabolic steroids. Based on what you know of their effects, how would you explain not using these steroids?

Career Choices Chiropractor

Chiropractic is a holistic approach to health and illness (the absence of health), recognizing the body's inherent ability to heal itself during times of physical, mental, and environmental stress. I chose to be a Doctor of Chiropractic because I like the mind-body connection concept. I believe health comes from within, not from outside the body with potions and pills. Many chiropractic and medical studies have proven the effectiveness of chiropractic care. I was an athlete with many biomechanical problems, and traditional approaches did not help. Eventually I tried chiropractic, and it worked! Great thinkers such as Hippocrates wrote, "Look well to the spine for the cause of disease." Thomas Edison said, "The doctor of the future will give no medicines, but will interest his patients in the care of the human frame, in diet, and the cause and prevention of disease." I chose to practice in a small-town setting, in a suburb of Philadelphia called Media, Pennsylvania.

Mark Alderman, BS, DC

D.D. Palmer founded chiropractic back in 1885 in Davenport, Iowa. Chiropractic now enjoys worldwide acceptance, even by the medical profession. Chiropractic is the second largest health care profession in the world. Currently, there are more than 65,000 chiropractors worldwide. I feel chiropractic will continue to grow and be even further accepted as long as people understand that health does not come from a bottle or a pill but from within.

The reward of chiropractic is quite clear—it helps patients return to their optimum potential for health. Chiropractic is not a treatment for disease. The purpose of the chiropractic adjustment is to allow the inborn intelligence within the body to be more fully expressed. When the innate (inborn) intelligence within the body is fully expressed, the body functions at its maximum potential for health and healing. To see this occur is the biggest reward for me!

Chiropractors use anatomy and physiology in everyday practice. We use the five components of subluxation that affect the physiology of the body. **Kinesiopathology** deals with the biomechanics, neuropathology deals with the nerves, **myopathology** deals with the muscles, histopathology deals with the cells, and finally, we use the biochemical system. Chiropractic is concerned with the relationship of the spinal column and the musculoskeletal structures of the body to the nervous system. Proper alignment of the spinal column is essential for optimum health because the spine acts as a "switchboard" for the nervous system. When there is nerve interference caused by misalignment in the spine, known as subluxation, pain can occur, and the body's natural defenses are diminished. The chiropractor adjusts the spinal joints to remove subluxation and restore normal nerve function.

The education of the chiropractor is long and challenging. First you must go to undergraduate school to obtain a B.S. with a concentration in the sciences. Then you attend chiropractic school for another 4 years (8 years is a big commitment!). To receive a Doctor of Chiropractic, you have to take four national boards and a state board in the state in which you wish to practice.

Transportation and Defense

The chapters in Unit 4 deal with transportation, how the body defends itself, and stress. Blood (Chapter 17) is discussed as a complex fluid tissue that serves to transport respiratory gases and key nutrients to cells and carry away wastes. It fills the cardiovascular system (Chapters 18 and 19) and is moved through a closed pathway, or circuit, of vessels by the pumping action of the heart.

The elements of the lymphatic system (Chapter 20) provide an open pathway for return of fluid and proteins from the interstitial spaces and for fats absorbed from the intestine into the general circulation. This system is also involved in immunity or resistance to disease and in the removal and destruction of dead red blood cells. The immune system is more fully discussed in Chapter 21. Elements of this system provide a multilayered defense mechanism, involving both phagocytic cells and specialized proteins called **antibodies.** Stress—and the body's frequently maladaptive response to it—are discussed in Chapter 22.

SEEING THE BIG PICTURE

Blood

LANGUAGE OF SCIENCE

adult stem cell

agranulocytes (ah-GRAN-yoo-loh-sytes) [*a-* without, *-granul-* little grains or granules, *-cytes* cells]

antibodies (AN-ti-bod-ees) [*anti* against]

antigens (AN-ti-jens) [*anti-* against, *-gen* to produce]

basophils (BAY-so-fils) [*basis-* foundation, *-phil* to love]

basophilic erythroblasts (BAY-so-fil-ik eh-rith-roh-BLASTS) [*basis-* foundation, *-philic* to love, *erythro-* red, *-blast* embryonic state of development]

bicarbonate ions (bye-KAR-boh-nayte EYE-ons) [*bi-* twice, *-carbo-* coal, *-ate* chemical compound derived from a source]

blood clotting

blood serum (SEER-um) [*sera* watery fluid]

buffy coat

carbonic anhydrase (kar-BON-ik an-HYE-drays) [*carbo-* coal, *-ic* pertaining to, *a-* without, *-hydor-* water, *-ase* enzyme]

diapedesis (dye-ah-peh-DEE-sis) [*dia-* apart or through, *-pedesis* an oozing]

electrolytes (eh-LEK-troh-lytes) [*electro-* electricity, *-lytes* a substance capable or resulting from decomposition]

embryonic stem cells (em-bree-ON-ik) [*embryo-* fetus, *-ic* pertaining to]

eosinophils (ee-oh-SIN-oh-fils) [*eosin-* a rose, red, or dawn color, *-phil* to love]

erythrocyte (eh-RITH-roh-syte) [*erythro-* red, *-cyte* cell]

erythropoiesis (eh-rith-roh-poy-EE-sis) [*erythro-* red, *-poiesis* formation or production of]

erythropoietin (eh-RITH-roh-POY-eh-tin) [*erythro-* red, *-poietin* to make]

extrinsic clotting pathways (eks-TRIN-sik) [*extr-* outside or beyond, *-insic* beside]

extrinsic factor (eks-TRIN-sik) [*extr-* outside or beyond, *-insic* beside]

fibrinolysis (fye-brin-OL-i-sis) [*fibr-* fiber, *-lysis* breaking down]

formed elements

Cont'd on p. 671

Four chapters in this unit deal with transportation, one of the body's vital functions. Homeostasis of the internal environment—and therefore survival itself—depend on continual transportation to and from body cells. This chapter discusses the major transportation fluid, blood. Chapters 18 and 19 consider the major transportation system, the cardiovascular system, and Chapter 20 treats a supplementary drainage system, the lymphatic system.

COMPOSITION OF BLOOD

Blood is much more than the simple liquid it seems to be. It consists of not only a fluid but also cells, including *thrombocytes (platelets)*. The fluid portion of blood, that is, the **plasma,** is one of the three major body fluids (interstitial and intracellular fluids are the other two). As explained below, the quantity of whole blood found in the body (blood volume) is often expressed as a percent of total body weight. However, the measurement of the plasma and formed elements is generally expressed as a percent of the *whole blood volume.* Using these measurement criteria, whole blood constitutes about 8% of total body weight, plasma accounts for 55%, and the formed elements account for about 45% of the total blood volume (Figure 17-1). The term **formed elements** is used to designate the various kinds of blood cells that are normally present in blood.

Blood is a complex transport medium that performs vital pickup and delivery services for the body. It picks up food and oxygen from the digestive and respiratory systems and delivers them to cells while picking up wastes from cells for delivery to excretory organs. Blood also transports hormones, enzymes, buffers, and various other biochemical substances that serve important functions. Blood serves another critical function. It is the keystone of the body's heat-regulating mechanism. Certain physical properties of blood make it especially effective in this role. Its high specific heat and conductivity enable this unique fluid to absorb large quantities of heat without an appreciable increase in its own temperature and to transfer this absorbed heat from the core of the body to its surface, where it can be more readily dissipated (see Figure 6-6, p. 203).

Blood Volume

How much blood does the body contain? The answer is about 8% of total body weight in average-sized adults. In a healthy young female, that amounts to about 4 to 5 liters and in a male about 5 to 6 liters. In addition to gender, blood volume varies with age, body composition, and method of measurement. A *unit* of blood is the amount (about 0.5 liter or just under 1 pint) that is collected from a blood donor for transfusion purposes. It constitutes about 10% of total blood volume in many adults.

Blood volume can be determined using **direct methods** or **indirect methods** of measurement. *Direct* measurement of total blood volume can be accomplished only by complete removal of all blood from an experimental animal. In humans, *indirect* methods of mea-

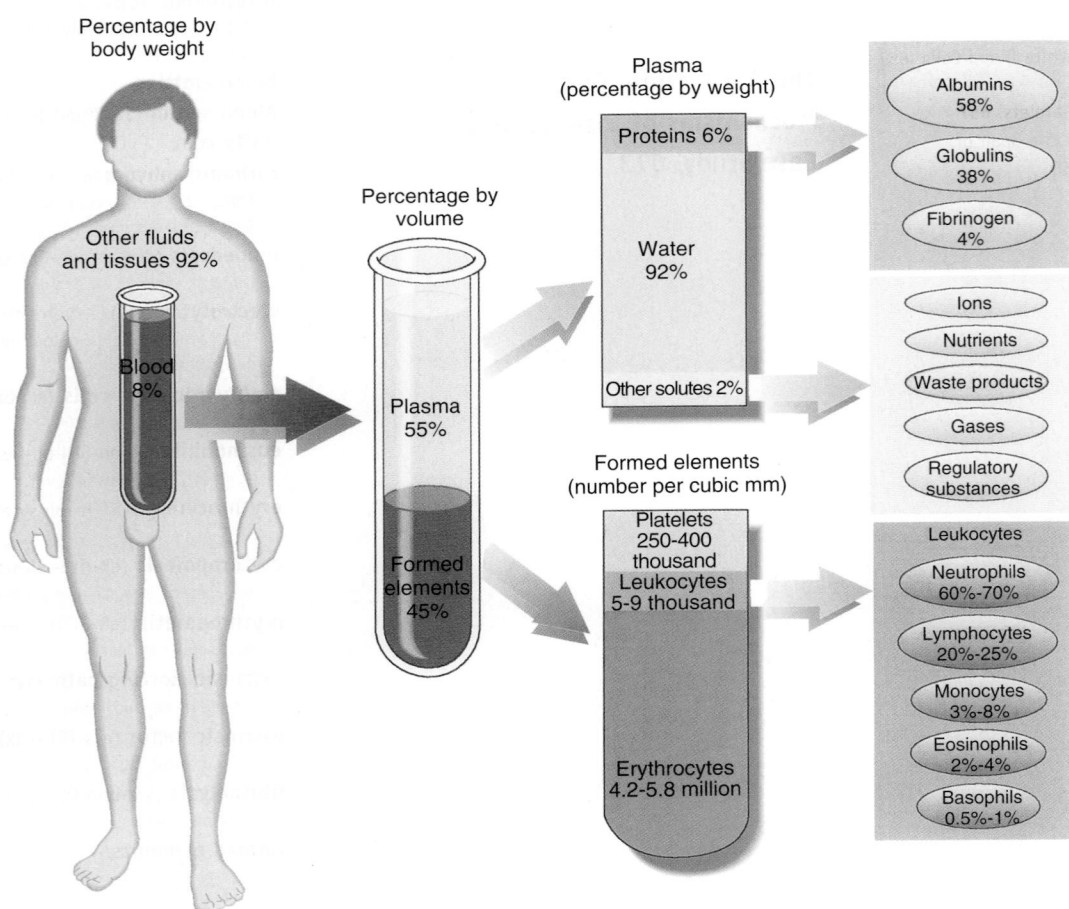

Figure 17-1 *Composition of whole blood.* Approximate values for the components of blood in a normal adult.

surement that employ "tagging" of red blood cells or plasma components with radioisotopes are used. The principle is simply to introduce a known amount of radioisotope into the circulation, allow the material to distribute itself uniformly in the blood, and then analyze its concentration in a representative blood sample. Accurate measurement is important in replacing blood lost because of hemorrhage or in treating other conditions, such as shock.

One of the chief variables influencing normal blood volume is the amount of body fat. Blood volume per kilogram of body weight varies inversely with the amount of excess body fat. This means that the less fat there is in your body, the more blood you have per kilogram of your body weight. Because females normally have a higher percent of body fat than males, they have less blood per kilogram of body weight and therefore a lower blood volume.

FORMED ELEMENTS OF BLOOD

Figure 17-2 illustrates the formed elements of blood. They are as follows:

- Red blood cells (RBCs) (erythrocytes)
- White blood cells (WBCs) (leukocytes)
- Platelets (thrombocytes)

Plasma when separated from "whole blood" is a clear straw-colored fluid that consists of about 90% water and 10% solutes. *Blood transfusions* most often involve three major blood components—plasma, platelets, and red blood cells. In the United

States, red blood cells constitute about 13.5 million transfusions compared with 1.75 million transfusions of platelets and 3.5 million transfusions of plasma each year.

If a tube of whole blood, that is, plasma and formed elements, is allowed to stand or is spun in a centrifuge, separation will occur. The term **packed cell volume (PCV)**, or **hematocrit** (hee-MAT-oh-krit), is used to describe the volume percent of red blood cells in whole blood. In Figure 17-3, *A*, a sample of normal whole blood has been separated by centrifuging so that the formed elements are forced to the bottom. The percentage of plasma is about 55% of the total sample, whereas the packed cell volume, or hematocrit, is 45%. A hematocrit of 45% means that in every 100 ml of whole blood there are 45 ml of red blood cells and 55 ml of fluid plasma.

Normally the average hematocrit for a man is about 45% (40% to 54%, normal range) and for a woman about 42% (38% to 47%, normal range). Conditions that result in decreased RBC numbers (Figure 17-3, *B*) are called **anemias** and are characterized by a reduced hematocrit value. Healthy individuals who live and work in high altitudes often have elevated RBC numbers and hematocrit values (Figure 17-3, *C*). The condition is called **physiological polycythemia** (from the Greek *polys*, "many," *kytos*, "cell," and *haima*, "blood").

White blood cells, or leukocytes, and platelets make up less than 1% of blood volume. Note in Figure 17-3 that a thin white layer of leukocytes and platelets, called the **buffy coat**, is present at the interface between the packed red cells and plasma.

QUICK CHECK

1. Name the fluid portion of whole blood.
2. What constitutes the formed elements of blood?
3. What factors influence blood volume?
4. Identify the component percentages of the normal hematocrit.

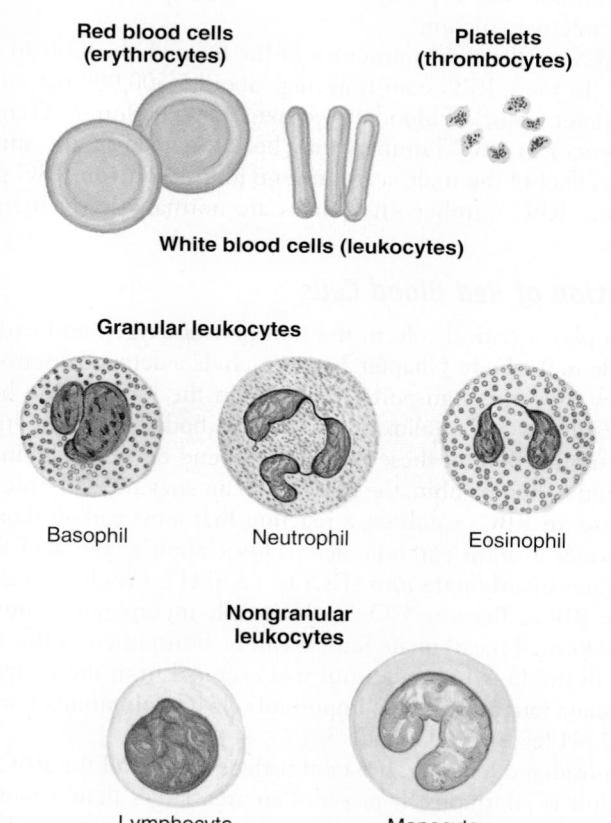

Granular leukocytes

Basophil   Neutrophil   Eosinophil

Nongranular leukocytes

Lymphocyte   Monocyte

Figure 17-2 *The formed elements of blood.* Red blood cells (erythrocytes), white blood cells (leukocytes), and platelets (thrombocytes) constitute the formed elements of blood.

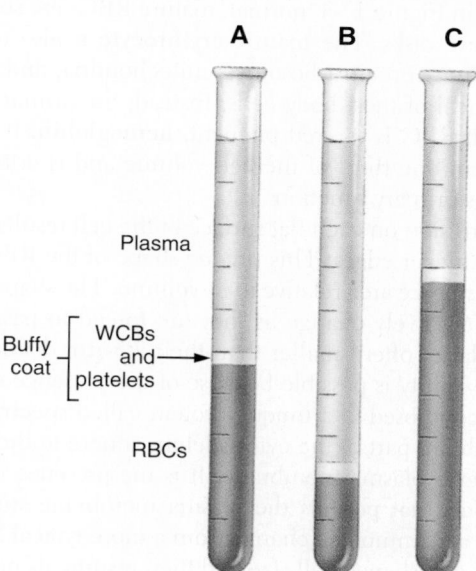

Figure 17-3 *Hematocrit tubes showing normal blood, anemia, and poly-cythemia.* Note the buffy coat located between the packed RBCs and the plasma. **A,** A normal percent of red blood cells. **B,** Anemia (a low percent of red blood cells). **C,** Polycythemia (a high percent of red blood cells).

BOX 17-1

Diagnostic Study

Erythrocyte Sedimentation Rate (ESR)

If a tube of anticoagulated blood from a healthy individual is allowed to stand upright for an hour, gravity will cause sedimentation of the formed elements leaving a layer of clear plasma about 15 mm thick on top of the tube. The rate of fall in one hour is called the **erythrocyte sedimentation rate** or **ESR**.

In individuals suffering from some type of inflammatory process, the clear layer on top of the ESR tube at the end of an hour is often over 40 mm thick. The increased ESR is due to the liver secreting into the circulation a number of large proteins that assist the body in responding to the inflammation. In addition to their large size and high molecular weight, these proteins also cause RBCs to "cluster," thus effectively increasing blood density. The result is an elevated ESR. The ESR is a "nonspecific" medical test because it will increase in many different inflammatory conditions. Although the test will not identify a specific disease, it is widely used by clinicians because it is a safe, easily performed, and cost-effective procedure that provides useful information early in the diagnostic process. For example, it is often used to make an initial determination concerning the possibility of inflammation as the cause of nonspecific complaints such as generalized fatigue, or weakness. If the presence of an inflammatory process is confirmed by an elevated ESR, more specific diagnostic tests can be used to identify the cause.

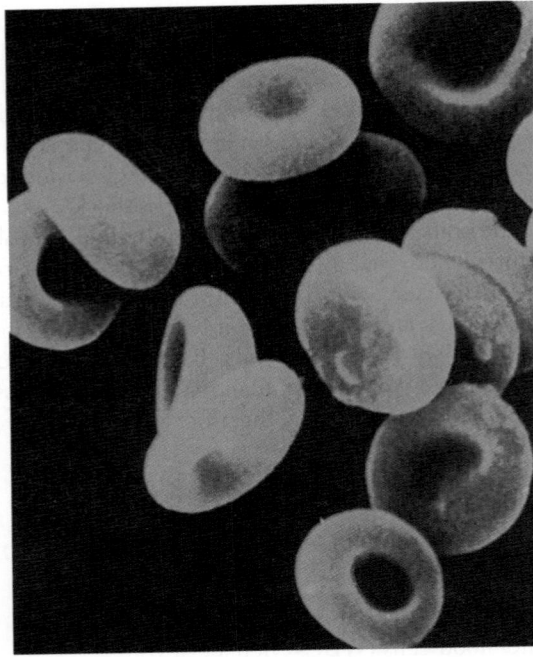

Figure 17-4 *Erythrocytes.* Color-enhanced scanning electron micrograph shows normal erythrocytes.

Red Blood Cells (Erythrocytes)

A normal, mature RBC has no nucleus and is only about 7.5 μm in diameter. More than 1500 of them could be placed side by side in a 1 cm space. Before the cell reaches maturity and enters the bloodstream from the bone marrow, the nucleus is extruded. As you can see in Figure 17-4, normal, mature RBCs are shaped like tiny biconcave disks. The mature **erythrocyte** is also unique in that it does not contain ribosomes, mitochondria, and other organelles typical of most body cells. Instead, the primary component of each RBC is the red pigment, **hemoglobin.** It accounts for more than one third of the cell volume and is critically important to its primary function.

The depression on each flat surface of the cell results in a thin center and thicker edges. This unique shape of the RBC gives it a very large surface area relative to its volume. The shape of erythrocytes can passively change as they are forced to pass through blood capillaries often smaller than their 7.5 μm diameter. This inherent flexibility is possible because of the presence of stretchable fibers composed of a unique protein called **spectrin.** These fibers, which are part of the cytoskeleton, adhere to the inside of the erythrocyte plasma membrane. It is the presence of flexible spectrin fibers that permits the plasma membrane surrounding the RBC to accommodate change from a more typical biconcave to a smaller cup-shaped cell size and then resume its normal size and appearance when deforming pressures are no longer being applied to the surface of the plasma membrane. This ability to change shape is necessary for the survival of RBCs, which are under almost constant mechanical shearing and bursting strains as

they pass through the capillary system. In addition, the degree of cell deformity that is possible influences the speed of blood flow in the microcirculation.

RBCs are the most numerous of the formed elements in the blood. In men, RBC counts average about 5,500,000 per cubic millimeter (mm^3) of blood; in women, 4,800,000/mm^3. Gender differences in RBC numbers may be influenced by the stimulating effect of the male sex hormone testosterone on RBC production. RBC numbers in females are normally lower than in males.

Function of Red Blood Cells

RBCs play a critical role in the transport of oxygen and carbon dioxide in the body. Chapter 24 will include a detailed discussion of how oxygen is transported from air in the lungs to the body cells and how carbon dioxide moves from body cells to the lungs for removal. Both of these functions depend on hemoglobin. In addition to hemoglobin, the presence of an enzyme, **carbonic anhydrase,** in RBCs catalyzes a reaction that joins carbon dioxide and water to form carbonic acid. Dissociation of the acid then generates **bicarbonate ions** (HCO_3^-) and H^+, which diffuse out of the RBCs. Because CO_2 is chemically incorporated into the newly formed bicarbonate ions, it can be transported in this new form in the blood plasma until it is excreted from the body. Bicarbonate ions also have an important role in maintaining normal blood pH levels (see Chapter 30).

Considered together, the total surface area of all the RBCs in an adult is enormous. It provides an area larger than a football field for the exchange of respiratory gases between hemoglobin found in circulating erythrocytes and interstitial fluid that bathes the body cells. This is an excellent example of the relationship between form and function.

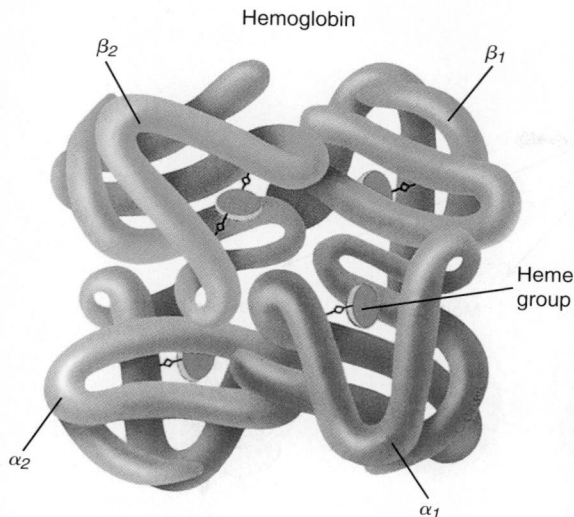

Hemoglobin

β_2 β_1

Heme group

α_2 α_1

Figure 17-5 *Hemoglobin.* Four protein chains (globins), each with a heme group, form a hemoglobin molecule. Each heme contains one iron atom.

Hemoglobin

Packed within each red blood cell are an estimated 200 to 300 million molecules of hemoglobin, which make up about 95% of the dry weight of each cell. Each hemoglobin molecule is composed of four protein chains. Each chain, called a **globin**, is bound to a red pigment, identified in Figure 17-5, as a **heme** molecule. Each heme molecule contains one iron atom. Therefore one hemoglobin molecule contains four iron atoms. This structural fact enables one hemoglobin molecule to unite with four oxygen molecules to form *oxyhemoglobin* (a reversible reaction). Hemoglobin can also combine with carbon dioxide to form *carbaminohemoglobin* (also reversible). But in this reaction the structure of the globin part of the hemoglobin molecule, rather than of its heme part, makes the combining possible.

A man's blood usually contains more hemoglobin than a woman's. In most normal men, 100 ml of blood contains 14 to 16 g of hemoglobin. The normal hemoglobin content of a woman's blood is a little less—specifically, in the range of 12 to 14 g/100 ml. Recall that testosterone in the male tends to stimulate erythrocyte production and cause an increase in RBC numbers. Increased levels of hemoglobin in males are directly related to increased erythrocyte numbers. An adult who has a hemoglobin content of less than 10 g/100 ml of blood is diagnosed as having **anemia** (from the Greek *a-*, "not," and *-haima*, "blood"). In addition, the term may be used to describe a reduction in the number or volume of functional RBCs in a given unit of whole blood. Anemias are classified according to the size and hemoglobin content of RBCs.

Formation of Red Blood Cells

The entire process of RBC formation is called **erythropoiesis.** In the adult, erythrocytes begin their maturation sequence in the red bone marrow from nucleated cells known as **hematopoietic stem cells,** or adult blood-forming stem cells. *Adult stem cells* are cells that have the ability to maintain a constant population of newly differentiating cells of a specific type. These adult blood-forming stem cells divide by mitosis; some of the daughter cells remain as undifferentiated adult stem cells, whereas others go through several stages of devel-

BOX 17-2: HEALTH MATTERS
Artificial Blood

One of the primary benefits of blood transfusion is an increase in oxygen-carrying capacity caused by increases in hemoglobin levels. However, blood is a complex liquid tissue, and transfusion always carries certain risks. For years medical researchers have worked to develop an artificial blood or blood substitute that could duplicate one or more blood functions, such as sustaining circulatory volume and delivering oxygen to the body's tissue, without subjecting a recipient to transfusion reactions, infections, or other health dangers.

These intravenous solutions, called *oxygen therapeutics,* began with a fluorochemical oxygen-carrying compound called *Fluosol.* It has been replaced with newer blood substitutes now in clinical trials. One, called **PolyHeme,** is produced from chemically treated hemoglobin obtained from outdated human blood. Another experimental oxygen therapeutic is called **Hemopure.** It is produced from bovine hemoglobin.

The ideal blood substitute will be one that can be transfused into a recipient with any blood type without "typing" and is free of blood-borne viruses such as HIV and hepatitis C. Other second-generation hemoglobin-based artificial blood products are also under development. If efforts are successful, they will be produced using recombinant hemoglobin technology. By using genetic engineering techniques, production of these products would not require the use of existing human or animal blood products. When available, artificial blood substitutes will be viewed as exciting and welcome developments in transfusion medicine.

opment to become erythrocytes. Figure 17-6 shows the sequential cell types or stages that can be identified as transformation from the immature forms to the mature red blood cell occurs. The entire maturation process requires about 4 days and proceeds from step to step by gradual transition. Just what influences direct **embryonic stem cells** (found in the embryo and in umbilical cord blood) to develop into a specific type of **adult stem cell** (e.g., hematopoietic stem cells) and then eventually differentiate into a specific cell type, such as an RBC, is a topic of intense research.

As you can see in Figure 17-6, all blood cells are derived from hematopoietic stem cells. In RBCs, differentiation begins with the appearance of **proerythroblasts.** Mitotic divisions then produce **basophilic erythroblasts.** The next maturation division produces **polychromatic erythroblasts,** which produce hemoglobin. These cells subsequently lose their nuclei and become **reticulocytes.** Once released into the circulating blood, reticulocytes lose their delicate reticulum and become mature erythrocytes in about 24 to 36 hours. Note in Figure 17-6 that overall cell size decreases as the maturation sequence progresses.

RBCs are formed and destroyed at a breathtaking rate. Normally, every minute of every day of our adult lives, more than 200 billion RBCs are formed to replace an equal number destroyed during that brief time. Because in health the number of RBCs remains relatively constant, efficient homeostatic mechanisms must operate to balance the number of cells formed against the number destroyed. The rate of RBC production soon speeds up if blood oxygen levels reaching the tissues decrease. Oxygen deficiency in-

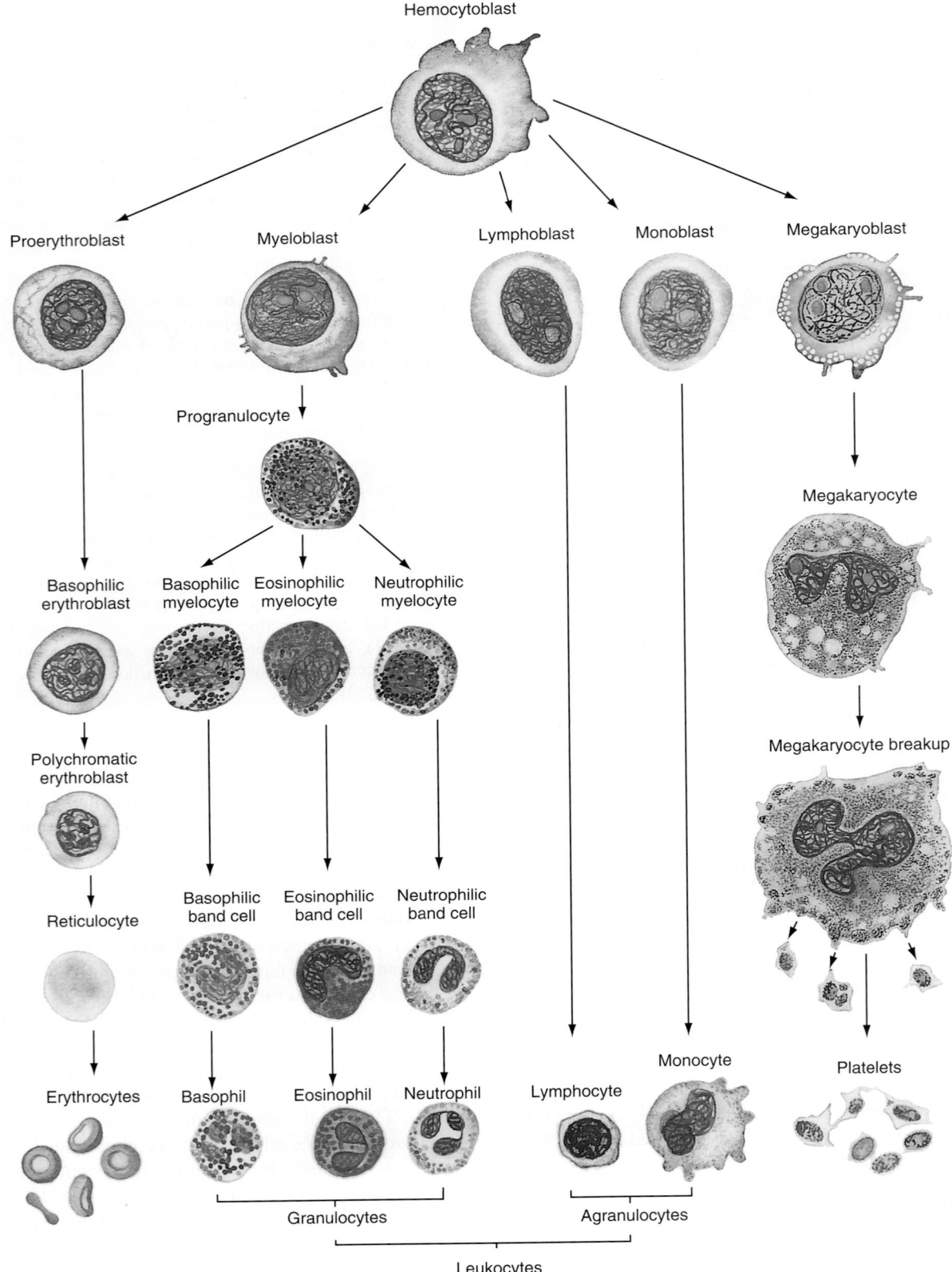

Figure 17-6 *Formation of blood cells.* The hematopoietic stem cell serves as the original stem cell from which all formed elements of the blood are derived. Note that all five precursor cells, which ultimately produce the different components of the formed elements, are derived from the hematopoietic stem cell called a hemocytoblast.

BOX 17-3: FYI

Sickle Cell Anemia

Sickle cell anemia is a severe, sometimes fatal, hereditary disease that is characterized by an abnormal type of hemoglobin. A person who inherits only one defective gene develops a form of the disease called *sickle cell trait*. In sickle cell trait, red blood cells contain a small proportion of a type of hemoglobin that is less soluble than normal. It forms solid crystals when the blood oxygen level is low, causing distortion and fragility of the red blood cell. If two defective genes are inherited (one from each parent), more of the defective hemoglobin is produced, and the distortion of red blood cells becomes severe. In the United States, about 1 in every 500 African-American and 1 in every 1000 Hispanic newborns are affected each year. In these individuals, the distorted red blood cell walls can be damaged by drastic changes in shape. Red blood cells damaged in this way tend to stick to vessel walls, and, if a blood vessel in the brain is affected, a stroke may occur because of the decrease in velocity or blockage of blood flow.

Stroke is one of the most devastating problems associated with sickle cell anemia in children and will affect about 10% of the 2500 youngsters who have the disease in the United States. Studies have shown that frequent blood transfusions in addition to standard care can dramatically reduce the risk of stroke in many children suffering from sickle cell anemia. The illustration shows the characteristic shape of a red cell containing the abnormal hemoglobin.

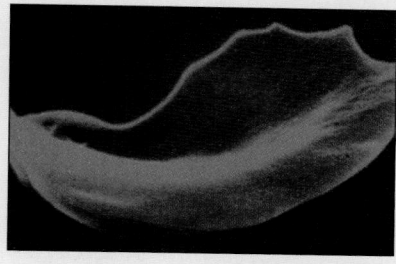

Sickle cell anemia.

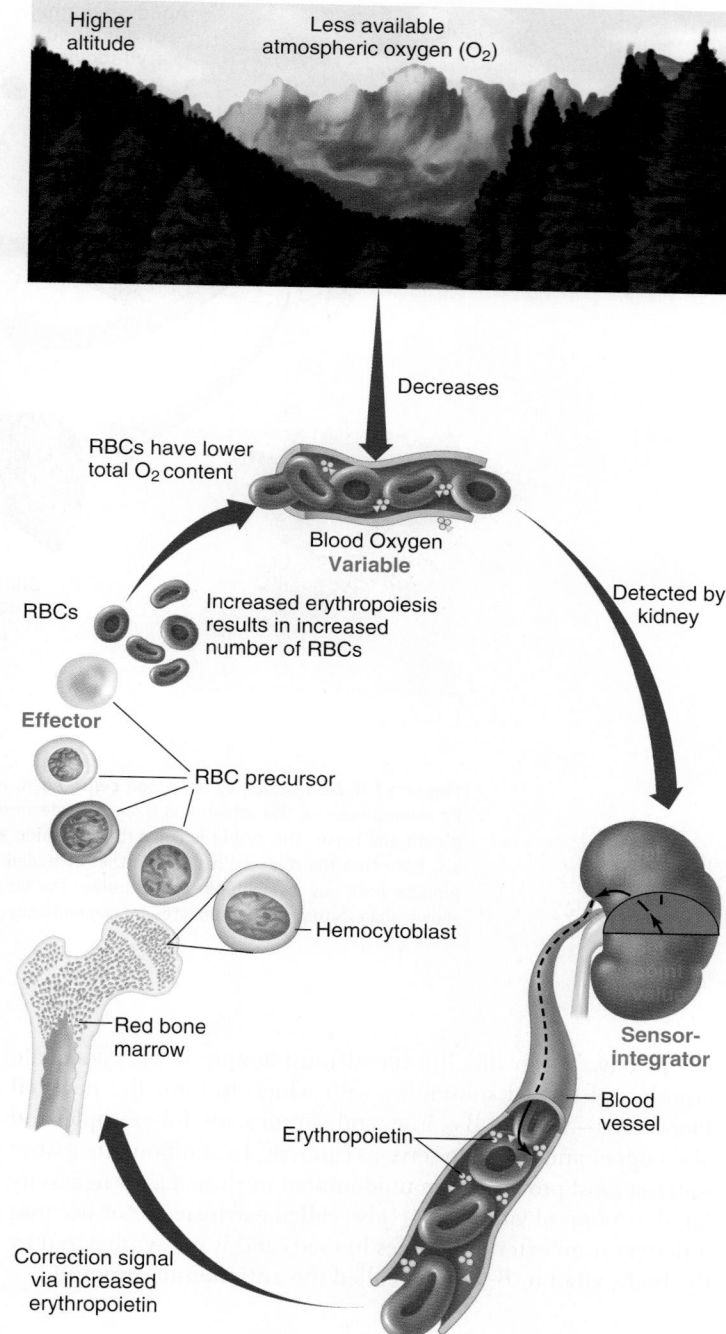

Figure 17-7 *Erythropoiesis.* In response to decreased blood oxygen, the kidneys release erythropoietin, which stimulates erythrocyte production in the red bone marrow.

creases RBC numbers by increasing the secretion of a glycoprotein hormone named **erythropoietin** (eh-RITH-roh-POY-eh-tin). An inactive form of this hormone, called an *erythropoietinogen*, is released into the blood primarily from the liver on an ongoing basis. If oxygen levels decrease, the kidneys release increasing amounts of erythropoietin, which in turn stimulates bone marrow to accelerate its production of red blood cells. With increasing numbers of RBCs, oxygen delivery to tissues increases, and less erythropoietin is produced and is available to stimulate RBC production in the red bone marrow. Figure 17-7 shows how erythropoietin production is controlled by a negative feedback loop that is activated by decreasing oxygen concentration in the tissues.

Destruction of Red Blood Cells

The life span of RBCs circulating in the bloodstream averages about 105 to 120 days. They often break apart, or fragment, in the capillaries as they age. Macrophage cells in the lining of blood vessels, particularly in the liver and spleen, phagocytose (ingest and destroy) the aged, abnormal, or fragmented red blood cells (Figure 17-8). The process results in breakdown of hemoglobin,

with the release of amino acids, iron, and the pigment bilirubin. Iron is returned to the bone marrow for use in synthesis of new hemoglobin, and bilirubin is transported to the liver, where it is excreted into the intestine as part of bile. Amino acids, released from the globin portion of the degraded hemoglobin molecule, are used by the body for energy or for synthesis of new proteins.

For the RBC homeostatic mechanism to succeed in maintaining a normal number of RBCs, the bone marrow must function

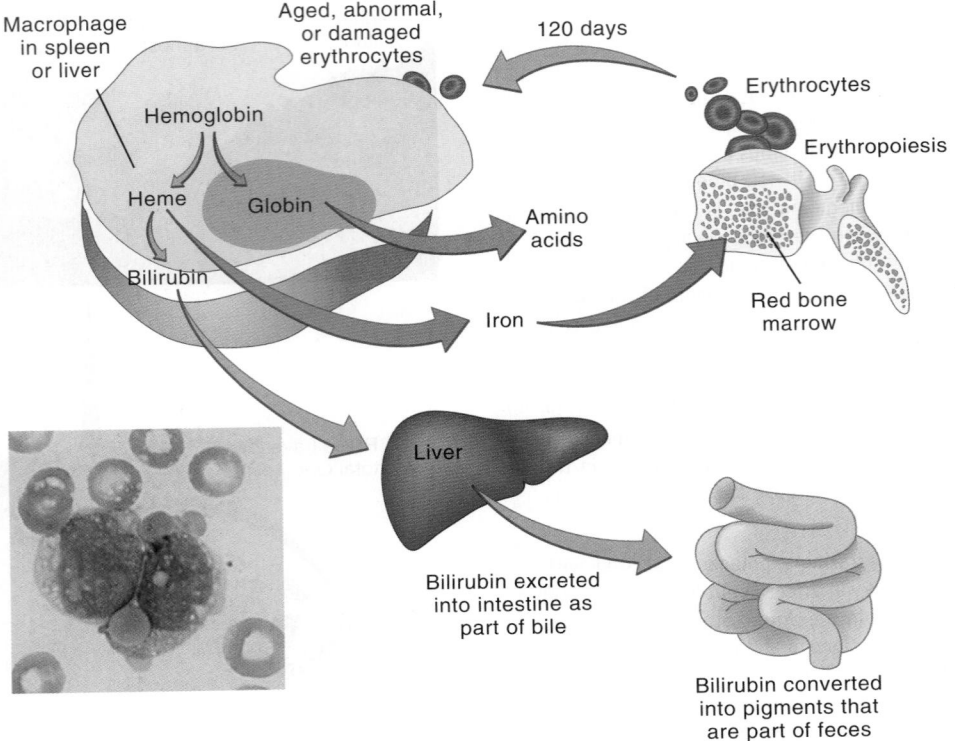

Figure 17-8 *Destruction of red blood cells.* Aged, abnormal, or damaged red blood cells are phagocytosed by macrophages in the spleen and liver. Breakdown of hemoglobin released from the red blood cells yields globin and heme. The globin is converted to amino acids and used as an energy source or for protein synthesis. Note that the released heme is further degraded into iron, which may be stored or used immediately to produce both new hemoglobin and bilirubin; the bilirubin is ultimately excreted in the bile. Photo inset shows phagocytosis of erythrocytes by macrophages in the liver.

adequately. To do this the blood must supply it with adequate amounts of several substances with which to form the new red blood cells—vitamin B_{12}, iron, and amino acids, for example, and also copper and cobalt to serve as catalysts. In addition, the gastric mucosa must provide some unidentified intrinsic factor necessary for absorption of vitamin B_{12} (also called **extrinsic factor** because it derives from external sources in foods and is not synthesized by the body; vitamin B_{12} is also called the **antianemic principle**).

 BOX 17-4

Diagnostic Study

Complete Blood Cell Count

One of the most useful and frequently performed clinical blood tests is called the **complete blood cell count** or simply the **CBC.** The CBC is a collection of tests whose results, when interpreted as a whole, can yield an enormous amount of information regarding a person's health. Standard red blood cell, white blood cell, and thrombocyte counts, the differential white blood cell count, hematocrit, hemoglobin content, and other characteristics of the formed elements are usually included in this battery of tests.

 BOX 17-5: SPORTS AND FITNESS

Blood Doping

Reports that some Olympic and other elite athletes employ transfusions of their own blood to improve performance have surfaced repeatedly in the past 20 years. The practice—called **blood doping** or **blood boosting**—is intended to increase oxygen delivery to muscles. A few weeks before competition, blood is drawn from the athlete and the red blood cells (RBCs) are separated and frozen. Just before competition, the RBCs are thawed and injected. Theoretically, infused RBCs and elevation of hemoglobin levels after transfusion should increase oxygen consumption and muscle performance during exercise. In practice, however, the advantage appears to be minimal. All blood transfusions carry some risk, and unnecessary or questionably indicated transfusions are medically unacceptable.

In addition to blood transfusions, injection of substances that increase RBC levels in an attempt to improve athletic performance has also been condemned by leading authorities in the area of sports medicine and by organized athletic organizations around the world. "Doping" with either the naturally occurring hormone erythropoietin or with synthetic drugs that have similar biological effects—such as Epogen and Procrit—can result in devastating medical outcomes.

QUICK CHECK

5. Name the red pigment found in RBCs, and list the normal range (in grams per 100 ml of blood) for women and men.
6. Trace the formation of an erythrocyte from stem cell precursor to a mature and circulating RBC.
7. Explain the negative feedback loop that controls erythropoiesis.
8. Discuss the destruction of RBCs.

White Blood Cells (Leukocytes)

There are five types of white blood cells, or **leukocytes**, classified according to the presence or absence of granules and the staining characteristics of their cytoplasm. **Granulocytes** include the three WBCs that have large granules in their cytoplasm. They are named according to their cytoplasmic staining properties:

1. Neutrophils
2. Eosinophils
3. Basophils

There are two types of **agranulocytes** (WBCs without cytoplasmic granules):

1. Lymphocytes
2. Monocytes

As a group, the leukocytes appear brightly colored in stained preparations. In addition, they all have nuclei and are generally larger in size than RBCs. Table 17-1 illustrates the formed elements, provides a brief description of each cell type, and lists life span and primary functions.

Granulocytes
Neutrophils

Neutrophils (Figure 17-9) take their name from the fact that their cytoplasmic granules stain a very light purple with neutral dyes. The granules in these cells are small and numerous and tend to give the cytoplasm a coarse appearance. Because their nuclei have two, three, or more lobes, neutrophils are also called **polymorphonuclear leukocytes** or, to avoid that tongue twister, simply *polys*.

Neutrophil numbers average about 65% of the total WBC count in a normal blood sample. These leukocytes are highly mobile and very active phagocytic cells that can migrate out of blood vessels and enter the tissue spaces. The process is called **diapedesis.** The cytoplasmic granules in neutrophils contain powerful lysosomes, the organelles with digestive-like enzymes that are capable of destroying bacterial cells.

Bacterial infections that produce an inflammatory response cause the release of chemicals from damaged cells that attract neutrophils and other phagocytic WBCs to the infection site. The process, called **positive chemotaxis,** helps the body concentrate phagocytic cells at focal points of infection.

Eosinophils

Eosinophils (Figure 17-10) contain cytoplasmic granules that are large, are numerous, and stain orange with acid dyes such as eosin. Their nuclei generally have two lobes. Normally, eosinophils account for about 2% to 5% of circulating WBCs. They are numerous in body areas such as the lining of the respiratory and digestive tracts. Although eosinophils are weak phagocytes, they are capable of ingesting inflammatory chemicals and proteins associated with antigen-antibody reaction complexes. Perhaps their most important functions involve protection against infections caused by parasitic worms and involvement in allergic reactions.

Basophils

Basophils (Figure 17-11) have relatively large, but sparse, cytoplasmic granules that stain a dark purple with basic dyes. They are the least numerous of the WBCs, numbering only 0.5% to 1% of the total leukocyte count. Basophils are both motile and capable of diapedesis. They exhibit **S**-shaped, but indistinct, nuclei. The cytoplasmic granules of these WBCs contain *histamine* (an inflammatory chemical) and *heparin* (an anticoagulant).

Agranulocytes
Lymphocytes

Lymphocytes (Figure 17-12) found in the blood are the smallest of the leukocytes, averaging about 6 to 8 μm in diameter. They have large, spherical nuclei surrounded by a very limited amount of pale blue–staining cytoplasm. Next to neutrophils, lymphocytes are the most numerous WBCs. They account for about 25% of all the leukocyte population. Two types of lymphocytes, called *T lymphocytes* and *B lymphocytes*, have important roles in immunity. T lymphocytes function by directly attacking an infected or

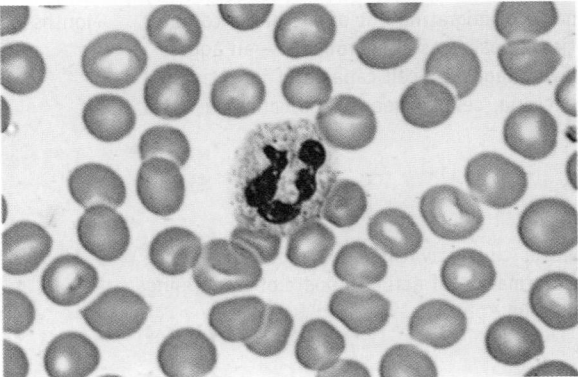

Figure 17-9 *Neutrophil.*

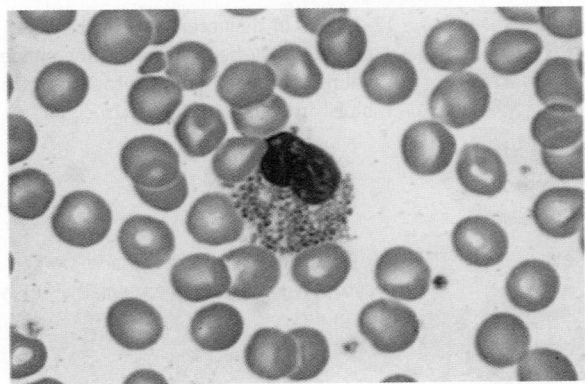

Figure 17-10 *Eosinophil.*

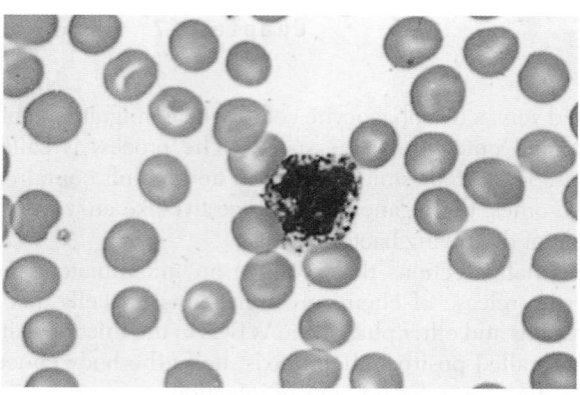

Figure 17-11 *Basophil.*

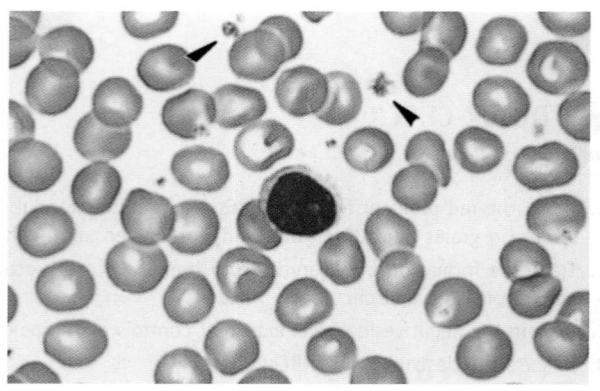

Figure 17-12 *Lymphocyte.* Arrowheads are pointing to platelets.

Table 17-1 ‖ Classes of Blood Cells

CELL TYPE	DESCRIPTION	FUNCTION	LIFE SPAN
Erythrocyte	7 μm in diameter; concave disk shape; entire cell stains pale pink; no nucleus	Transportation of respiratory gases (O_2 and CO_2)	105-120 days
Neutrophil	12-15 μm in diameter; spherical shape; multilobed nucleus; small, pink-purple–staining cytoplasmic granules	Cellular defense—phagocytosis of small pathogenic microorganisms	Hours to 3 days
Basophil	11-14 μm in diameter; spherical shape; generally two-lobed nucleus; large purple-staining cytoplasmic granules	Secretes heparin (anticoagulant) and histamine (important in inflammatory response)	Hours to 3 days
Eosinophil	10-12 μm in diameter; spherical shape; generally two-lobed nucleus; large, orange-red–staining cytoplasmic granules	Cellular defense—phagocytosis of large pathogenic microorganisms, such as protozoa and parasitic worms; releases anti-inflammatory substances in allergic reactions	10-12 days
Lymphocyte	6-9 μm in diameter; spherical shape; round (single-lobed) nucleus; small lymphocytes have scant cytoplasm	Humoral defense—secretes antibodies; involved in immune system response and regulation	Days to years
Monocyte	12-17 μm in diameter; spherical shape; nucleus generally kidney-bean or horseshoe shaped with convoluted surface; ample cytoplasm often "steel blue" in color	Capable of migrating out of the blood to enter tissue spaces as a macrophage—an aggressive phagocytic cell capable of ingesting bacteria, cellular debris, and cancerous cells	Months
Platelet	2-5 μm in diameter; irregularly shaped fragments; cytoplasm contains very small, pink-staining granules	Releases clot-activating substances and helps in formation of actual blood clot by forming platelet "plugs"	7-10 days

Table 17-2	Differential Count of White Blood Cells	
	DIFFERENTIAL COUNT*	
CLASS	**NORMAL RANGE (%)**	**TYPICAL VALUE (%)**
Neutrophils	65-75	65
Eosinophils	2-5	3
Basophils	½-1	1
Lymphocytes (large and small)	20-25	25
Monocytes	3-8	6
TOTAL	100	100

*In any differential count the sum of the percentages of the different kinds of WBCs must, of course, total 100%.

cancerous cell, whereas B lymphocytes produce antibodies against specific antigens. The functions of both types of lymphocytes will be fully discussed in Chapter 21.

Monocytes

Monocytes (Figure 17-13) are the largest of the leukocytes. They have dark, kidney bean–shaped nuclei surrounded by large quantities of distinctive blue-gray cytoplasm. Monocytes are motile and highly phagocytic cells capable of engulfing large bacterial organisms and viral-infected cells.

White Blood Cell Numbers

One cubic millimeter of normal blood usually contains about 5000 to 9000 leukocytes, with different percentages of each type. Because these numbers change in certain abnormal conditions, they have clinical significance. In acute appendicitis, for example, the percentage of neutrophils increases and so, too, does the total WBC count. In fact, these characteristic changes may be the deciding points for surgery.

The procedure in which the different types of leukocytes are counted and their percentage of total WBC count is computed is

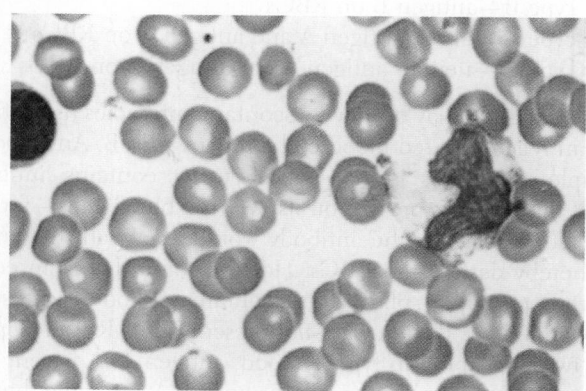

Figure 17-13 *Monocyte.* Compare the large monocyte *(right)* with the smaller lymphocyte *(left)*.

known as a **differential white blood cell count.** In other words, a differential count is a percentage count of WBCs. The different kinds of WBCs and a normal differential count are listed in Table 17-2. A decrease in the number of WBCs is **leukopenia.** An increase in the number of WBCs is **leukocytosis.**

Formation of White Blood Cells

The hematopoietic stem cell serves as the precursor of not only the erythrocytes but also the leukocytes and platelets in blood. Figure 17-6, p. 652, shows the maturation sequence that results in formation of the granular and the agranular leukocytes from the undifferentiated hematopoietic stem cell (hemocytoblast).

Neutrophils, eosinophils, basophils, and a few lymphocytes and monocytes originate, as do erythrocytes, in red bone marrow (myeloid tissue). Most lymphocytes and monocytes derive from hematopoietic adult stem cells in lymphatic tissue. Although many lymphocytes are found in bone marrow, presumably most are formed in lymphatic tissues and carried to the bone marrow by the bloodstream.

Myeloid tissue (bone marrow) and lymphatic tissue together constitute the hematopoietic, or blood cell–forming, tissues of the body. Red bone marrow is myeloid tissue that actually produces blood cells. Its red color comes from the red blood cells it contains. Yellow marrow, on the other hand, is yellow because it stores considerable fat. It is not active in the business of blood cell formation as long as it remains yellow. Sometimes, however, it becomes active and red in color when an extreme and prolonged need for red blood cell production occurs.

Platelets

To compare **platelets** with other blood cells in terms of appearance and size, see Figure 17-12. In circulating blood, platelets are small, nearly colorless bodies that usually appear as irregular spindles or oval disks about 2 to 4 μm in diameter.

Three important physical properties of platelets—namely, agglutination, adhesiveness, and aggregation—make attempts at classification on the basis of size or shape in dry blood smears all but impossible. As soon as blood is removed from a vessel, the platelets adhere to each other and to every surface they contact; in so doing, they assume various shapes and irregular forms.

Platelet counts in adults average about 250,000/mm³ of blood. A range of 150,000 to 400,000/mm³ is considered normal. Newborn infants often show reduced counts, but these rise gradually to reach normal adult values at about 3 months of age. There are no differences between the sexes in platelet count.

Functions of Platelets

Platelets play an important role in both **hemostasis** (from the Greek *stasis*, "a standing") and **blood clotting**, or **coagulation.** The two functions, although interrelated, are separate and distinct. Hemostasis refers to the stoppage of blood *flow* and may occur as an end result of any one of several body defense mechanisms. The role that platelets play in the operation of the blood-clotting mechanism is discussed on pp. 662 to 665.

Within 1 to 5 seconds after injury to a blood capillary, platelets will adhere to the damaged lining of the vessel and to each other to form a hemostatic **platelet plug** that helps to stop the flow of blood into the tissues.

The formation of a temporary platelet plug is an important step in hemostasis. It generally follows vascular spasm, which is caused by constriction of smooth muscle fibers in the wall of damaged blood vessels. Vascular spasm can cause temporary closure of a damaged vessel and lessen blood loss until the platelet plug and subsequent coagulation effectively stop the hemorrhage.

The formation of a platelet plug results when platelets undergo a change caused by an encounter with a damaged capillary wall, or with underlying connective tissue fibers. The transformation results in the creation of *sticky platelets*, which bind to underlying tissues and each other. In addition to forming a physical plug at the site of injury, sticky platelets secrete several chemicals, including adenosine diphosphate (ADP), thromboxane, and a fatty acid (arachidonic acid), that are involved in the coagulation process. When released, these substances affect both local blood flow (by vasoconstriction) and platelet aggregation at the site of injury. If the injury is extensive, the blood-clotting mechanism is activated to assist in hemostasis.

Platelet plugs are extremely important in controlling so-called microhemorrhages, which may involve a break in a single capillary. Failure to arrest hemorrhage from these very minor but numerous and widespread capillary breaks can result in life-threatening blood loss.

Platelet plugs are also believed to be involved in causing intermittent arterial microvascular occlusion in certain types of peripheral vascular disease (see Box 14-1, p. 516).

Formation and Life Span of Platelets

Formation of platelets, sometimes called *thrombocytes*, is referred to as **thrombopoiesis.** It begins with stimulation of precursor cells called megakaryoblasts (Figure 17-6), and is controlled by the hormone *thrombopoietin*. Mature megakaryocytes are huge cells (20 to 100 μm) with large nuclei containing as many as 20 lobes

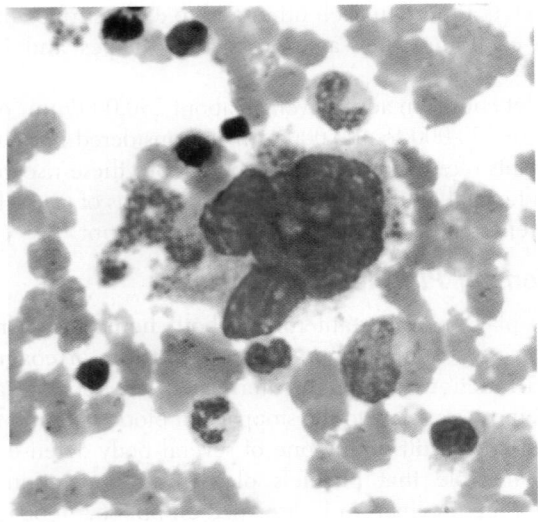

Figure 17-14 *Megakaryocyte.* Note the large size of the cell and multinucleated nucleus. This bone marrow smear also shows a number of WBCs scattered among more numerous RBCs.

(Figure 17-14). They often have a bizarre shape. The cytoplasm is blue to pink in color, is abundant, and contains a variable number of very fine granules. Mature megakaryocytes are largely confined to red bone marrow although some are located in the lungs and, to a lesser extent, the spleen. Between 2000 and more than 3000 platelets are released when the irregular cytoplasmic membrane surrounding the mature megakaryocyte ruptures. The resulting platelets have a limiting plasma membrane but, like RBCs, no nucleus. Platelets have a short life span, an average of about 7 days.

BLOOD TYPES (BLOOD GROUPS)

The term *blood type* refers to the type of antigens, (called *agglutinogens*), present on RBC membranes. (The concept of **antigens** and **antibodies** is discussed in detail in Chapter 21.) Antigens A, B, and Rh are the most important blood antigens as far as transfusions and newborn survival are concerned. Many other blood antigens have also been identified, but they are less important clinically and are seldom discussed.

The term *agglutinins* is often used to describe the antibodies dissolved in plasma that react with specific blood group antigens or agglutinogens. It is the specific agglutinogens or antigens on red cell membranes that characterize the different ABO blood groups, which are described in the following paragraphs.

When a blood transfusion occurs, great care must be taken to prevent a mixture of agglutinogens (antigens) and agglutinins (antibodies) that would result in the agglutination of the donor and recipient blood—a potentially fatal event known as a **transfusion reaction.** Clinical laboratory tests, called *blood typing* and *cross-matching*, ensure the proper identification of blood group antigens and antibodies in both donor and recipient blood and demonstrate the lack of an agglutination reaction when they are mixed together.

The ABO System

Every person's blood belongs to one of the four ABO blood groups. Blood types are named according to the antigens present (agglutinogens) on RBC membranes. Here, then, are the four ABO blood types:

1. Type A—antigen A on RBCs
2. Type B—antigen B on RBCs
3. Type AB—both antigen A and antigen B on RBCs
4. Type O—neither antigen A nor antigen B on RBCs

Blood plasma may or may not contain antibodies (agglutinins) that can react with red blood cell antigens A or B. An important principle about this is that plasma never contains antibodies against the antigens present on its own red blood cells—for obvious reasons. If it did, the antibody would react with the antigen and thereby destroy the RBCs. However (and this is an equally important principle), plasma does contain antibodies against antigen A or antigen B if they are *not* present on its RBCs. Applying these two principles: in type A blood, antigen A is present on its RBCs; therefore its plasma contains no anti-A antibodies but does contain anti-B antibodies. In type B blood, antigen B is present on its RBCs; therefore its plasma contains no anti-B antibodies but does contain anti-A antibodies (Figure 17-15).

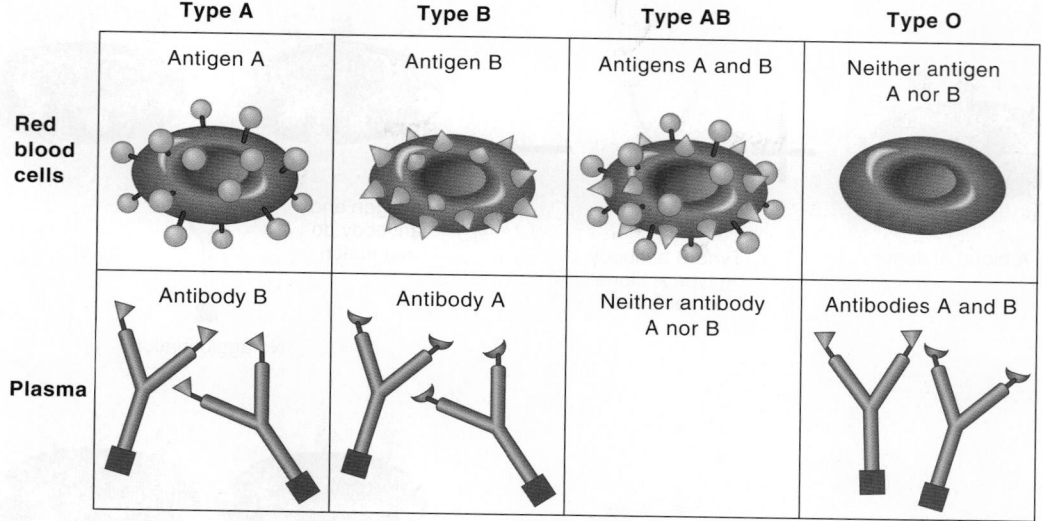

Figure 17-15 *ABO blood types.* Note that antigens characteristic of each blood type are bound to the surface of RBCs. The antibodies of each blood type are found in the plasma and exhibit unique structural features that permit agglutination to occur if exposure to the appropriate antigen occurs.

BOX 17-6: HEALTH MATTERS

Blood Transfusions

If you are a blood donor, your "gift of life," when transfused into the body of another individual, is called a **homologous** (ho-MOL-oh-gus) **transfusion**. If you donate blood and store it for possible reinfusion into your own body, the process is called an **autologous** (aw-TAHL-eh-gus) **transfusion**. People often store their own blood for reinfusion in this way if they are scheduled to undergo elective surgery in which blood loss and the need for transfusion are anticipated.

The need for transfusion can often be reduced during surgery by salvaging lost blood and then immediately reinfusing it back into the patient. The salvaged blood is suctioned from the surgical field or body cavity and then washed or filtered using special equipment to remove debris such as tissue fragments or bone chips.

Recently, the U.S. Food and Drug Administration (FDA) has recommended that all blood used for (homologous) transfusions be filtered to remove leukocytes—a process called **leukoreduction.** Leukocytes are responsible for many posttransfusion reactions, including fever and chills. In addition, white blood cells often contain viruses that may harm premature infants or cancer or acquired immunodeficiency syndrome patients who need multiple transfusions. However, mandating that all donated blood be leukoreduced increases the cost of transfusion, reduces available red cells, and may not benefit the majority of patients. Currently the FDA recommendation to mandate leukoreduction of all donated blood is under review.

Efforts to increase the safety of blood used in transfusions are ongoing. One of the new safety screening procedures is called the **nucleic acid test (NAT).** The test, called the INTERCEPT System by the manufacturer, detects minute particles of genetic material (DNA and RNA) found in infectious agents such as the hepatitis C, hepatitis B, and AIDS viruses even before a newly infected donor develops antibodies to the disease. As a result, new technology makes it possible to detect and inactivate a broad spectrum of disease-causing organisms before they can be recognized by conventional testing.

Note in Figure 17-16, *A*, that type A blood donated to a type A recipient does not cause an agglutination transfusion reaction because the type B antibodies in the recipient do not combine with the type A antigens in the donated blood. However, type A blood donated to a type B recipient causes an agglutination reaction because the type A antibodies in the recipient combine with the type A antigens in the donated blood (Figure 17-16, *B*). Figure 17-17 shows the results of different combinations of donor and recipient blood.

Because type O blood does not contain either antigen A or B, it has been referred to as *universal donor* blood, a term that implies that it can safely be given to any recipient. This, however, is not true because the recipient's plasma may contain agglutinins other than anti-A or anti-B antibodies. For this reason the recipient's and the donor's blood—even if it is type O—should be cross-matched; that is, mixed and observed for agglutination of the donor's red blood cells.

Universal recipient (type AB) blood contains neither anti-A nor anti-B antibodies, so it cannot agglutinate type A or type B donor red blood cells. This does not mean, however, that any type of donor blood may be safely given to an individual who has type AB blood without first cross-matching. Other agglutinins may be present in the so-called universal recipient blood and clump unidentified antigens (agglutinogens) in the donor's blood.

The Rh System

The term *Rh-positive blood* means that Rh antigen is present on its RBCs. *Rh-negative blood*, on the other hand, is blood whose red cells have no Rh antigen present on them.

Blood does not normally contain anti-Rh antibodies. However, anti-Rh antibodies can appear in the blood of an Rh-negative person, provided Rh-positive RBCs have at some time entered the bloodstream. One way this can happen is by giving an Rh-negative person a transfusion of Rh-positive blood. In a short time, the person's body makes anti-Rh antibodies, and these remain in the blood. The other way in which Rh-positive RBCs can enter

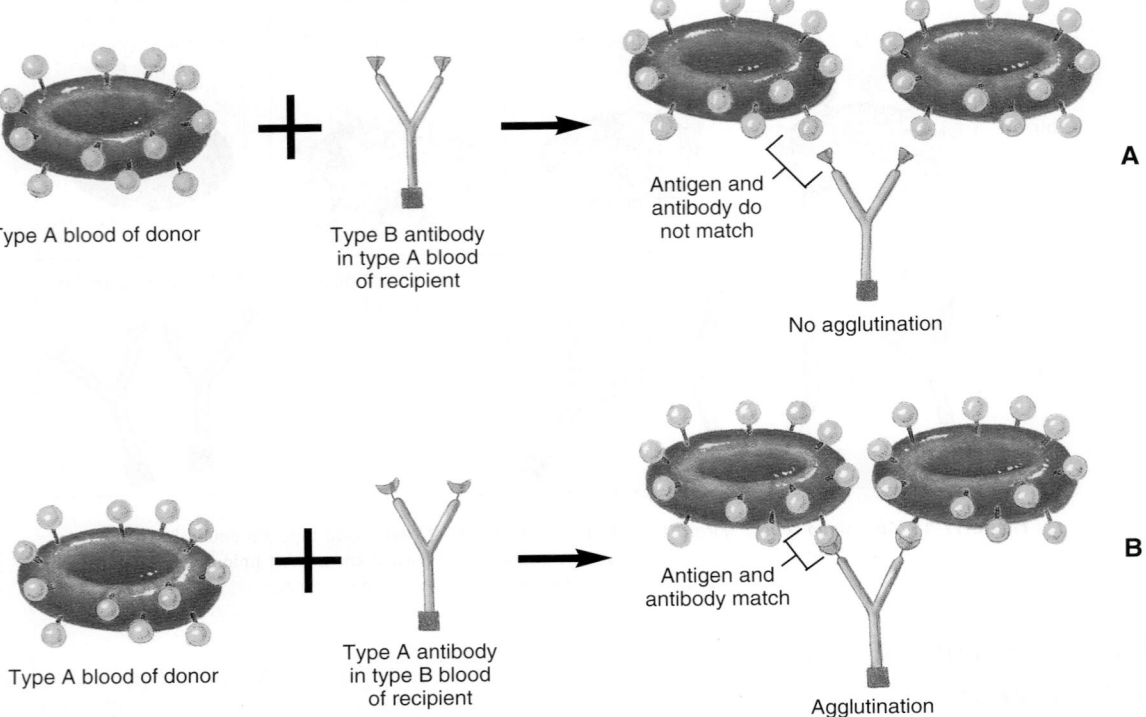

Type A blood of donor

Type B antibody
in type A blood
of recipient

Antigen and
antibody do
not match

No agglutination

A

Type A blood of donor

Type A antibody
in type B blood
of recipient

Antigen and
antibody match

Agglutination

B

Figure 17-16 *Agglutination.* **A,** When mixing of donor and recipient blood of the same type *(A)* occurs, there is no agglutination because only type B antibodies are present. **B,** If type A donor blood is mixed with type B recipient blood, agglutination will occur because of the presence of type A antibodies in the type B recipient blood.

the bloodstream of an Rh-negative individual can happen only to a woman during pregnancy. In this fact lies the danger for a baby born to an Rh-negative mother and an Rh-positive father. If the baby inherits the Rh-positive trait from the father, the Rh factor on the RBCs may stimulate the mother's body to form anti-Rh antibodies. Then, if she later carries another Rh-positive fetus, the fetus may develop a disease called **erythroblastosis fetalis,** caused by the mother's Rh antibodies reacting with the baby's Rh-positive cells (Figure 17-18).

All Rh-negative mothers who carry an Rh-positive baby should be treated with a protein marketed as RhoGAM. RhoGAM stops the mother's body from forming anti-Rh antibodies and thus prevents the possibility of harm to the next Rh-positive baby.

Briefly, the only people who can ever have anti-Rh antibodies in their plasma are Rh-negative men or women who have been transfused with Rh-positive blood or Rh-negative women who have carried an Rh-positive fetus.

BLOOD PLASMA

Plasma is the liquid part of blood—whole blood minus formed elements (Figure 17-19). In the laboratory, whole blood that has not clotted is centrifuged to form plasma. This consists of a rapid whirling process that hurls the blood cells to the bottom of the centrifuge tube. A clear, straw-colored fluid—blood plasma—lies above the cells. Plasma consists of 90% water and 10% solutes. By far the largest quantity of these solutes is pro-

teins; normally, they constitute about 6% to 8% of the plasma. Proteins such as factor VIII regulate blood clotting; and others such as gamma globulins, which are important in treating weakened immune systems, and albumin, a blood volume expander, are common transfusion products. Other solutes present in much smaller amounts in plasma are food substances (principally glucose, amino acids, and lipids), compounds formed by metabolism (e.g., urea, uric acid, creatinine, and lactic acid), respiratory gases (oxygen and carbon dioxide), and regulatory substances (hormones, enzymes, and certain other substances).

Some solutes present in blood plasma are true solutes, or crystalloids. Others are colloids. *Crystalloids* are solute particles less than 1 nm in diameter (ions, glucose, and other small molecules). *Colloids* are solute particles from 1 to about 100 nm in diameter (e.g., proteins of all types). Blood solutes also may be classified as **electrolytes** (molecules that ionize in solution) or **nonelectrolytes**— examples: proteins and inorganic salts are electrolytes; glucose and lipids are nonelectrolytes.

The proteins in blood plasma consist of three main kinds of compounds: *albumins,* *globulins,* and clotting proteins, principally *fibrinogen.* Measuring the amounts of these compounds reveals that 100 ml of plasma contains a total of approximately 6 to 8 g of protein. Albumins constitute about 55% of this total, globulins about 38%, and fibrinogen about 7%.

Plasma proteins are crucially important substances. Fibrinogen, for instance, and a clotting protein named prothrombin have

Recipient's blood		Reactions with donor's blood			
RBC antigens	Plasma antibodies	Donor type O	Donor type A	Donor type B	Donor type AB
None (Type O)	Anti-A Anti-B	(no agglutination)	(agglutination)	(agglutination)	(agglutination)
A (Type A)	Anti-B	(no agglutination)	(no agglutination)	(agglutination)	(agglutination)
B (Type B)	Anti-A	(no agglutination)	(agglutination)	(no agglutination)	(agglutination)
AB (Type AB)	(none)	(no agglutination)	(no agglutination)	(no agglutination)	(no agglutination)

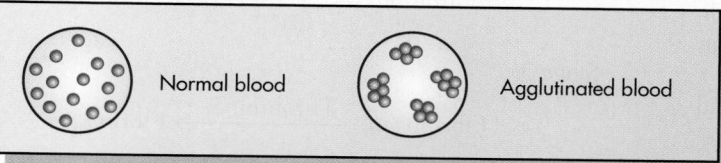

Normal blood   Agglutinated blood

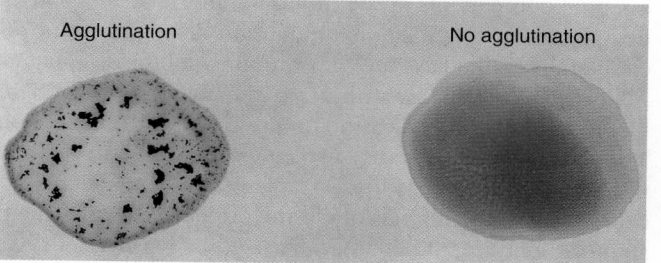

Agglutination   No agglutination

Figure 17-17 *Results of (cross-matching) different combinations (types) of donor and recipient blood.* The left columns show the antigen and antibody characteristics that define the recipient's blood type, and the top row shows the donor's blood type. Cross-matching identifies either a compatible combination of donor-recipient blood (no agglutination) or an incompatible combination (agglutinated blood). Photo inset shows drops of blood showing appearance of agglutinated and non-agglutinated red blood cells.

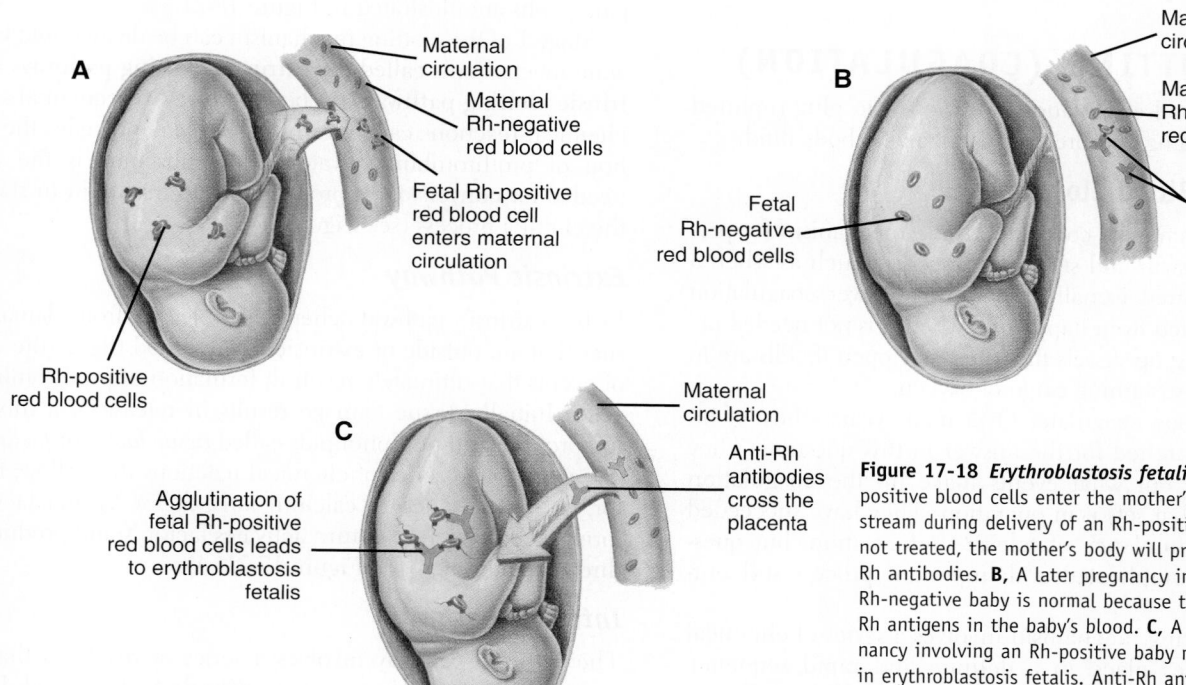

Figure 17-18 *Erythroblastosis fetalis.* **A,** Rh-positive blood cells enter the mother's bloodstream during delivery of an Rh-positive baby. If not treated, the mother's body will produce anti-Rh antibodies. **B,** A later pregnancy involving an Rh-negative baby is normal because there are no Rh antigens in the baby's blood. **C,** A later pregnancy involving an Rh-positive baby may result in erythroblastosis fetalis. Anti-Rh antibodies enter the baby's blood supply and cause agglutination of RBCs with the Rh antigen.

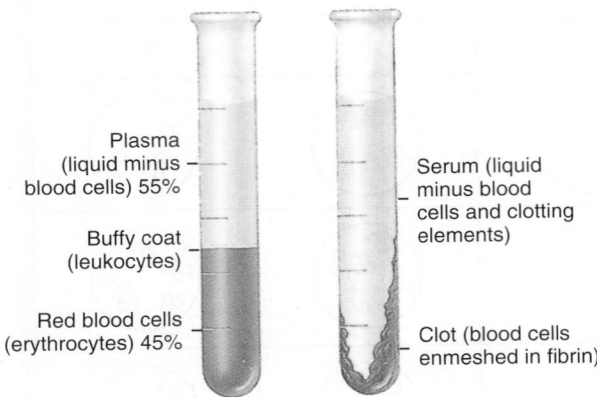

Figure 17-19 *Difference between blood plasma and blood serum.* Plasma is whole blood minus cells. Serum is whole blood minus the clotting elements. Plasma is prepared by centrifuging anticoagulated blood. Serum is prepared by allowing blood to clot.

key roles in the blood-clotting mechanism. Globulins function as essential components of the immunity mechanism. Many modified globulins, called gamma globulins, serve important roles as circulating antibodies (see also immunoglobulins, Chapter 21). All plasma proteins contribute to the maintenance of normal blood viscosity, blood osmotic pressure, and blood volume. Therefore plasma proteins have an essential part in maintaining normal circulation. Synthesis of plasma proteins occurs in liver cells. They form all kinds of plasma proteins, except some of the gamma globulin antibodies synthesized by plasma cells. Cancer of plasma cells, called *multiple myeloma*, results in production of an abnormal myeloma antibody—a gamma globulin—that results in numerous and very serious disease symptoms (see Mechanisms of Disease, p. 667).

BLOOD CLOTTING (COAGULATION)

The purpose of blood coagulation is obvious—to plug ruptured vessels to stop bleeding and prevent loss of a vital body fluid.

Mechanism of Blood Clotting

Because of the function of coagulation, the mechanism for producing it must be swift and sure when needed, such as when a vessel is cut or ruptured. Equally important, however, coagulation needs to be prevented from happening when it is not needed because clots can plug up vessels that must stay open if cells are to receive blood's life-sustaining cargo of oxygen.

What makes blood coagulate? Over many years a host of investigators have searched for the answer to this question. They have tried to find out what events make up the coagulation mechanism and what sets it in operation. They have succeeded in gathering an abundance of relevant information, but questions about this complicated and important process still outnumber answers.

The blood-clotting mechanism involves a series of chemical reactions that takes place in a definite and rapid sequence resulting in a net of fibers that traps red blood cells (see Figure 17-20, *B*).

The so-called classic theory of coagulation was advanced in 1905 and dominated research efforts in this complex area for almost 50 years. It continues as the basis of our current understanding of coagulation. This theory has now been expanded and assumes (1) the interaction of numerous coagulation factors in the presence of calcium ions and (2) that the interaction between the components occurs in three stages.

Scientists working more than 90 years ago discovered that four components were critical to coagulation:

1. Prothrombin
2. Thrombin
3. Fibrinogen
4. Fibrin

These early studies in coagulation research suggested that interactions between these components occurred in what we now designate as stage 2 and stage 3 of the blood-clotting process, which are simplified and illustrated in Figure 17-20:

Stage 2:

$$\text{Prothrombin} \xrightarrow[\text{Ca}^{++}]{\substack{\text{Prothrombin} \\ \text{activator}}} \text{Thrombin}$$

Stage 3:

$$\text{Fibrinogen} \xrightarrow[\text{Ca}^{++}]{\text{Thrombin}} \text{Fibrin}$$

It is interesting that basic reaction stages 2 and 3 have been modified only by the action of a host of additional coagulation factors discovered in recent years (Table 17-3). In addition, the source of prothrombin activator at the end of stage 1 of the current coagulation theory mechanism is now used to divide this stage into *intrinsic* and *extrinsic* systems.

The three stages of coagulation described in the following paragraphs are illustrated in Figure 17-21.

Stage 1 of the clotting mechanism can be divided into two separate mechanisms called the **extrinsic clotting pathways** and **intrinsic clotting pathways.** In both pathways a sequential series of chemical reactions called a *clotting cascade* precedes the formation of prothrombin activator. This substance is the catalyst needed for conversion of prothrombin to thrombin in stage 2 of the clotting process (see Figure 17-21).

Extrinsic Pathway

In the extrinsic pathway, chemicals released from damaged tissues that are outside or extrinsic to the blood trigger the cascade of events that ultimately result in formation of prothrombin activator. Initially, tissue damage results in release of a mixture of lipoproteins and phospholipids called *tissue factor* or *factor III*. In the coordinated series of chemical reactions that follow, this factor, in the presence of calcium ions, factor V, and factor VII, forms a complex that in turn activates factor X and produces prothrombin activator (see Figure 17-21, *A*).

Intrinsic Pathway

The intrinsic pathway involves a series of reactions that begin with factors normally present, or intrinsic to, the blood. Damage to the endothelial lining of blood vessels exposes collagen fibers,

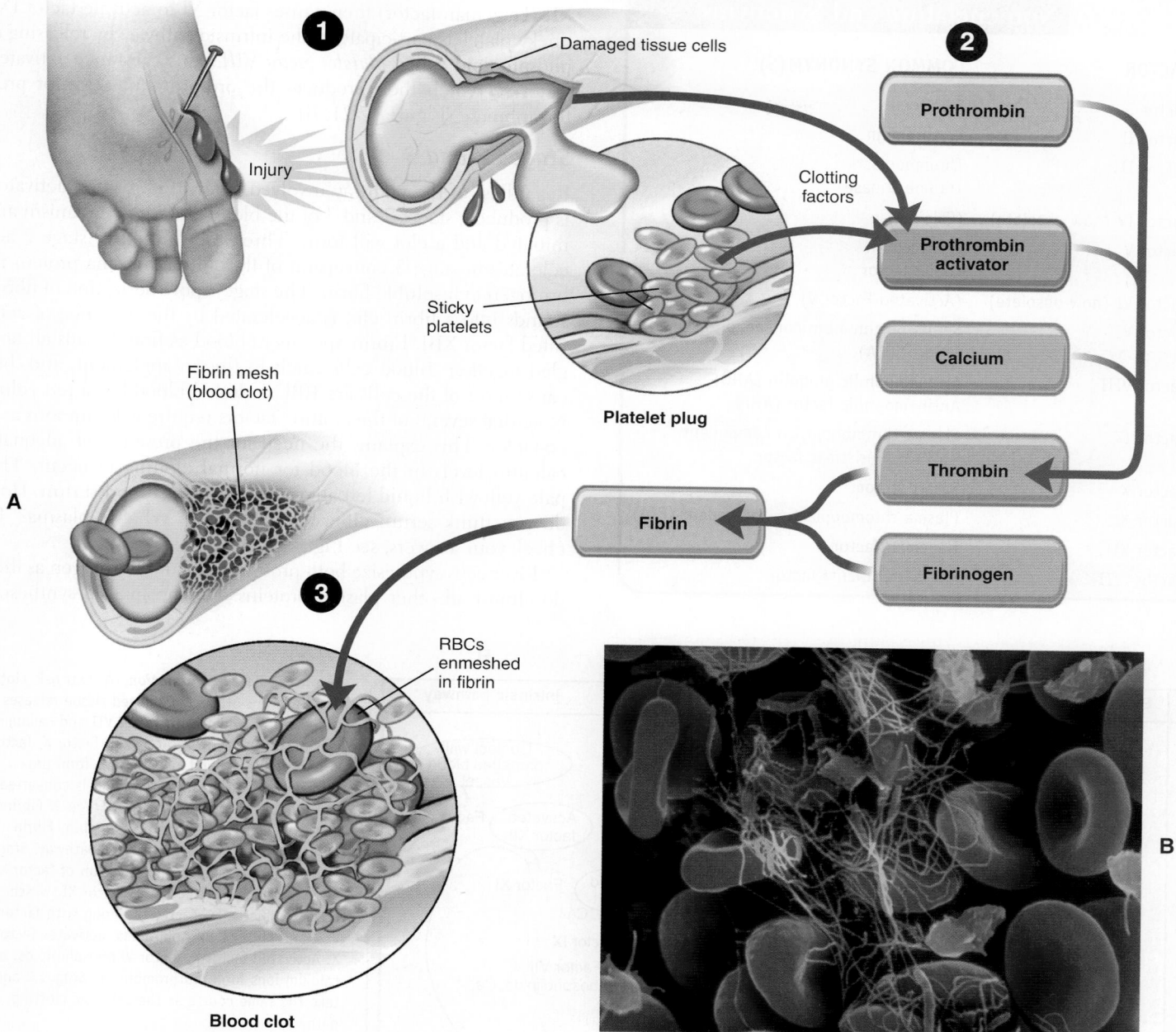

Figure 17-20 *Blood-clotting mechanism.* A, The complex clotting mechanism can be distilled into three basic steps: (*1*) release of clotting factors from both injured tissue cells and sticky platelets at the injury site (which form temporary platelet plug); (*2*) series of chemical reactions that eventually result in the formation of thrombin; and (*3*) formation of fibrin and trapping of blood cells to form a clot. **B,** Photo inset is a colorized electron micrograph showing RBCs and WBCs *(blue)* entrapped in a fibrin *(yellow)* mesh during clot formation.

Table 17-3 — Coagulation Factors—Standard Nomenclature and Synonyms

FACTOR	COMMON SYNONYM(S)
Factor I	Fibrinogen
Factor II	Prothrombin
Factor III	Thromboplastin Thrombokinase
Factor IV (now obsolete)	(Calcium)
Factor V	Proaccelerin Labile factor
Factor VI (now obsolete)	(Activated Factor V)
Factor VII	Serum prothrombin conversion accelerator (SPCA)
Factor VIII	Antihemophilic globulin (AHG) Antihemophilic factor (AHF)
Factor IX	Plasma thromboplastin component (PTC), Christmas factor
Factor X	Stuart factor
Factor XI	Plasma thromboplastin antecedent (PTA)
Factor XII	Hageman factor
Factor XIII	Fibrin-stabilizing factor

which in turn causes the activation of a number of coagulation factors present in plasma. When activated in this manner, factor XII (Hageman factor) then causes factor XI to activate factor IX. Sticky platelets participate in the intrinsic pathway by releasing a phospholipid called *platelet factor VIII*. This substance activates factor X, which then produces the prothrombin activator prothrombinase (Figure 17-21, *B*).

Stages 2 and 3

Regardless of the pathway involved, after prothrombin activator is produced, stages 2 and 3 of the blood-clotting mechanism are initiated and a clot will form. Thrombin formed in stage 2 accelerates in stage 3 conversion of the soluble plasma protein fibrinogen to insoluble fibrin. The stage 3 polymerization of fibrin strands into a fibrin clot is accelerated by the presence of activated factor XIII. Fibrin appears in blood as fine threads all tangled together. Blood cells catch in the entanglement, and, because most of the cells are RBCs, clotted blood has a red color. Note that several of the clotting factors require calcium ions as a co-factor. This explains the need for the presence of adequate calcium levels in the blood for normal clotting to occur. The pale yellowish liquid left after a clot forms is **blood serum.** How do you think serum differs from plasma? What is plasma? To check your answers, see Figure 17-19.

Liver cells synthesize both prothrombin and fibrinogen as they do almost all other plasma proteins. For the liver to synthesize

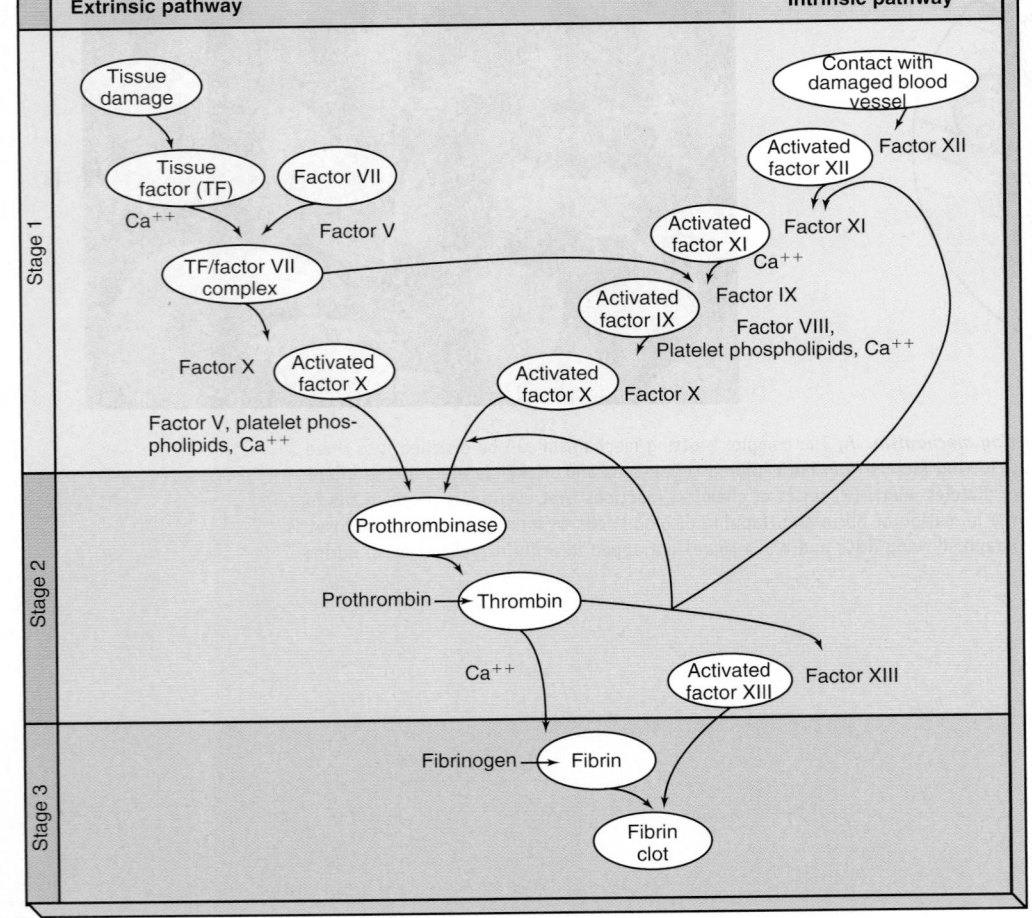

Figure 17-21 *Clot formation.* **A,** Extrinsic clotting pathway. *Stage 1:* Damaged tissue releases tissue factor, which with factor VII and calcium ions activates factor X. Activated factor X, factor V, phospholipids, and calcium ions form prothrombinase. *Stage 2:* Prothrombin is converted to thrombin by prothrombinase. *Stage 3:* Fibrinogen is converted to fibrin by thrombin. Fibrin forms a clot. **B,** Intrinsic clotting pathway. *Stage 1:* Damaged vessels cause activation of factor XII. Activated factor XII activates factor XI, which activates factor IX. Factor IX, along with factor VIII and platelet phospholipids, activates factor X. Activated factor X, factor V, phospholipids, and calcium ions form prothrombinase. Stages 2 and 3 take the same course as the extrinsic clotting pathway.

BOX 17-7: HEALTH MATTERS
Clinical Methods of Hastening Clotting

One way of treating excessive bleeding is to speed up the blood-clotting mechanism. This can be accomplished by increasing any of the substances essential for clotting, for example:

- By applying a rough surface such as gauze or by gently squeezing the tissues around a cut vessel. Each procedure causes more platelets to activate and release more platelet factors. This in turn accelerates the first of the clotting reactions.
- By applying purified thrombin (in the form of sprays or impregnated gelatin sponges that can be left in a wound). Which stage of the clotting mechanism does this accelerate?
- By applying fibrin foam, films, and so on.
- Some superficial bleeding can be stopped by the application of cold, which causes vasoconstriction and slows blood flow.

BOX 17-8: HEALTH MATTERS
Anticoagulant Versus Antiplatelet Drug Treatment

If an individual is at risk for thrombus formation, selecting a so-called targeted or rational drug treatment may depend on the location in the vascular system in which the clots may form. Research has shown that venous thrombi consist mainly of fibrin and red blood cells whereas arterial thrombi consist mainly of platelet aggregates. This information provides a theoretic basis for selecting different types of drug treatment for conditions caused by either venous or arterial thrombi. Namely, anticoagulant drugs such as heparin and warfarin (Coumadin) should be more effective in prevention of venous thrombi and drugs that decrease the tendency for platelets to become sticky and form aggregates (antiplatelet drugs) should be more effective in preventing arterial thrombi. A number of antiplatelet drugs, such as cilostazol (Pletal) and ticlopidine (Ticlid), exert effects by inhibiting an enzyme called phosphodiesterase, which is involved in platelet aggregation activity. This mechanism of action may explain why a condition such as **intermittent claudication,** which causes cramplike pain in the calves after walking and is caused by intermittent arterial microvascular occlusion by platelet plugs, often responds well to antiplatelet but not to anticoagulant drug therapy.

prothrombin at a normal rate, blood must contain an adequate amount of vitamin K. Vitamin K is absorbed into the blood from the intestine. Some foods contain this vitamin, but it is also synthesized in the intestine by certain bacteria (not present for a time in newborn infants). Because vitamin K is fat soluble, its absorption requires bile. Therefore if the bile ducts become obstructed and bile cannot enter the intestine, a vitamin K deficiency develops. The liver cannot then produce prothrombin at its normal rate, and the blood's prothrombin concentration soon falls below normal. A prothrombin deficiency gives rise to a bleeding tendency. As a preoperative safeguard, therefore, patients with obstructive jaundice are generally given some kind of vitamin K preparation.

Conditions that Oppose Clotting

Although blood clotting probably goes on continuously and concurrently with clot dissolution (fibrinolysis), several conditions operate to oppose clot formation in intact vessels. Most important by far is the perfectly smooth surface of the normal endothelial lining of blood vessels. Platelets do not adhere to healthy endothelium; consequently, they do not activate and release platelet factors into the blood. Therefore the blood-clotting mechanism does not begin in normal vessels. As an additional deterrent to clotting, blood contains certain substances called *antithrombins*. The name suggests their function—they oppose (inactivate) thrombin. Thus antithrombins prevent thrombin from converting fibrinogen to fibrin. **Heparin,** a natural constituent of blood, acts as an antithrombin. It was first prepared from liver (hence its name), but other organs also contain heparin. Injections of heparin are used to prevent clots from forming in vessels. *Coumarin* compounds impair the liver's use of vitamin K and thereby slow its synthesis of prothrombin and factors VII, IX, and X. Indirectly, therefore, coumarin compounds retard coagulation. Citrates keep donor blood from clotting before transfusion. Aspirin and other drugs, such as clopidogrel (Plavix) or cilostazol (Pletal), that inhibit platelet aggregation also inhibit coagulation (see Box 17-8 on this page).

Conditions that Hasten Clotting

Two conditions particularly favor thrombus formation: a rough spot in the endothelium (blood vessel lining) and abnormally slow blood flow. Atherosclerosis, for example, is associated with an increased tendency toward thrombosis because of endothelial rough spots in the form of plaques of accumulated cholesterol-lipid material. Body immobility, on the other hand, may lead to thrombosis because blood flow slows down as movements decrease. Incidentally, this fact is one of the major reasons why physicians insist that bed patients must either move or be moved frequently.

Once started, a clot tends to grow. Platelets enmeshed in the fibrin threads activate, releasing more thromboplastin, which, in turn, causes more clotting, which enmeshes more platelets, and so on, in a vicious circle. Clot-retarding substances, available in recent years, have proved valuable for retarding this process.

Clot Dissolution

The physiological mechanism that dissolves clots is known as **fibrinolysis** (Figure 17-22). Evidence indicates that the two opposing processes of clot formation and fibrinolysis go on continuously. Normal blood contains an inactive plasma protein called plasminogen that can be activated by several substances released from damaged cells. These converting substances include thrombin, factor XII, tissue plasminogen activator (t-PA), and lysosomal enzymes. Plasmin hydrolyzes fibrin strands and dissolves the clot (see Figure 17-22).

In today's clinical practice, several different kinds of proteins are used to dissolve blood clots that are causing an acute medical crisis. These are enzymes that generate plasmin when in-

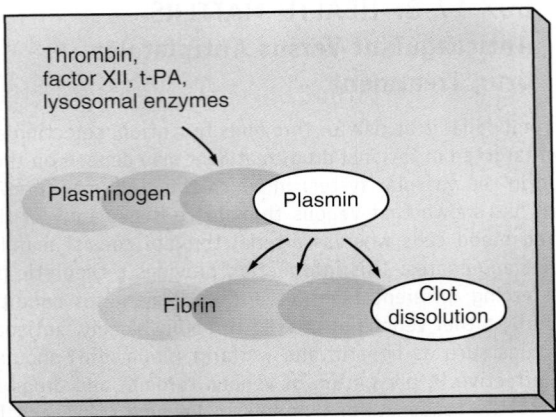

Figure 17-22 *Fibrinolysis*. Plasminogen is converted into the active enzyme plasmin by thrombin, factor XII, tissue plasminogen activator (t-PA), and lysosomal enzymes. Plasmin hydrolyzes fibrin strands and dissolves the clot.

jected into patients. Streptokinase is a plasminogen-activating factor made by certain streptococci bacteria. It and recombinant t-PA can be used to dissolve clots in the large arteries of the heart, which, when blocked, can result in myocardial infarction (heart attack). In addition, t-PA has been recognized as a promising drug for the early treatment of strokes. If given within the first 6 hours after a clot forms in a cerebral vessel, it can often improve blood flow and greatly reduce the serious aftereffects of a stroke.

 QUICK CHECK

9. List the granulocytic and agranulocytic leukocytes.
10. List the normal percentages of the different WBCs in a differential count.
11. List the four ABO blood groups, and identify the antigens and antibodies (if any) associated with each.
12. Identify the two basic coagulation steps.

 THE BIG PICTURE
Blood and the Whole Body

In every chapter of this book we have referred to the notion that the whole body's function is geared toward maintaining stability of the internal fluid environment—that is, *homeostasis*. The fluid that makes up the internal environment in which cells are bathed—the fluid that must be kept stable—includes the plasma of the blood. As a matter of fact, it is the blood plasma that transports substances, and even heat, around the internal environment so that all body tissues are linked together. The various tissues of the body are linked by the plasma, which flows back and forth between any two points served by blood vessels. This, of course, means that substances such as nutrients, wastes, dissolved gases, water, antibodies, and hormones can be transported between almost any two points in the body.

Blood tissue is not just plasma, however. It contains the **formed elements**—the blood cells and platelets. The RBCs participate in the mechanisms that permit the efficient transport of the gases oxygen and carbon dioxide. WBCs are important in the defense mechanisms of the whole body. Their presence in blood ensures that they are available to all parts of the body, all of the time, to fight cancer, resist infectious agents, and clean up injured tissues. Platelets provide mechanisms for preventing loss of the fluid that constitutes our internal environment.

All other organs and systems of the body rely on blood to perform its many functions. No organ or system can maintain proper levels of nutrients, dissolved gases, or water without direct or indirect help from the blood. On the other hand, many other systems help blood do its job. For example, the respiratory system excretes carbon dioxide from the blood and picks up oxygen. Organs of the digestive system pick up nutrients, remove some toxins, and take care of old blood cells. The endocrine system regulates the production of blood cells and the water content of the plasma. Besides removing toxic wastes such as urea, the urinary system has a vital role in maintaining homeostasis of plasma water concentration and pH.

Of course, blood is useless unless it continually and rapidly flows around the whole body—and continues to transport, defend, and maintain balance. The next several chapters outline the structures and functions that make this possible. Chapters 18 and 19 discuss the plan of the blood circulation and how adequate blood flow is maintained. Chapter 20 discusses the role of the lymphatic system in maintaining the fluid balance of the blood. Chapters 21 and 22 deal with the defensive mechanisms of blood and other tissues. As a matter of fact, most of the remaining chapters feature the role of blood in maintaining stability of the whole body.

Mechanisms of Disease

BLOOD DISORDERS

Most blood diseases are disorders of the formed elements. Thus it is not surprising that the basic mechanism of many blood diseases is the failure of the blood-producing myeloid and lymphatic tissues to properly form blood cells. In many cases, this failure is the result of damage by drugs, toxic chemicals, or radiation. In other cases, it results from an inherited defect or even cancer.

If bone marrow failure is the suspected cause of a particular blood disorder, a sample of myeloid tissue may be drawn into a syringe. The bone marrow is obtained from inside the pelvic bone (iliac crest) or the sternum. This procedure, called **aspiration biopsy cytology** (ABC), allows examination of the tissue that may help confirm or reject a tentative diagnosis. If the bone marrow is severely damaged, the choice of a **bone marrow transplant** may be offered to the patient. In this procedure, myeloid tissue from a compatible donor is introduced into the recipient intravenously. If the recipient's immune system does not reject the new tissue, which is always a danger in this type of tissue transplant, the donor cells may establish a colony of new, healthy tissue in the bone marrow. In some cases, infusion of healthy marrow follows total body irradiation. This treatment destroys the diseased marrow, permitting the new tissue to grow.

Red Blood Cell Disorders
Anemia

The term **anemia** is used to describe different disease conditions caused by an inability of the blood to carry sufficient oxygen to the body cells. Anemias can result from inadequate numbers of RBCs or a deficiency of oxygen-carrying hemoglobin. Thus anemia can occur if the hemoglobin in RBCs is inadequate, even if normal numbers of RBCs are present.

Anemia Resulting from Changes in RBC Number

Anemias caused by an actual change in the number of RBCs can occur if blood is lost by hemorrhage, as with accidents or bleeding ulcers or if the blood-forming tissues cannot maintain normal numbers of blood cells. Such failures occur because of cancer, chemotherapy treatment, radiation (x-ray) damage, and certain types of infections. If bone marrow produces an excess of RBCs, the result is a condition called **polycythemia** (pahl-ee-sye-THEE-mee-ah). The blood in individuals suffering from this condition may contain so many RBCs that it may become too thick to flow properly.

One type of anemia characterized by an abnormally low number of red blood cells is **aplastic** (a-PLAS-tik) **anemia.** Although idiopathic forms of this disease occur, most cases result from destruction of bone marrow by drugs, toxic chemicals, or radiation. Less commonly, aplastic anemia results from bone marrow destruction by cancer. Because tissues that produce other formed elements are also affected, aplastic anemia is usually accompanied by a decreased number of WBCs and platelets. Bone marrow transplants have been successful in treating some cases of aplastic anemia.

Pernicious (per-NISH-us) **anemia** is another disorder characterized by a low number of RBCs. Pernicious anemia sometimes results from a dietary deficiency of vitamin B_{12}. Vitamin B_{12} is used in the formation of new RBCs in the bone marrow. In many cases, pernicious anemia results from the failure of the stomach lining to produce *intrinsic factor*—the substance that allows vitamin B_{12} to be absorbed. Pernicious anemia can be fatal if not successfully treated. One method of treatment involves intramuscular injections of vitamin B_{12}.

Folate deficiency anemia is similar to pernicious anemia because it also causes a decrease in the RBC count resulting from a vitamin deficiency. In this condition, it is *folic acid* (vitamin B_9) that is deficient. Folic acid deficiencies are common among individuals with alcoholism and other malnourished individuals. Treatment for folate deficiency **acute anemia** involves taking vitamin supplements until a balanced diet can be restored.

Of course, a significant reduction in the number of RBCs can occur as a result of blood loss. **Blood loss anemia** often occurs after hemorrhages associated with trauma, extensive surgeries, or other situations involving a sudden loss of blood. **Anemia of chronic disease** can be a serious complication of chronic inflammatory diseases and cancer; the cause is often unknown.

Changes in Hemoglobin

The amount and quality of hemoglobin within RBCs are just as important as the number of RBCs. In hemoglobin disorders, RBCs are sometimes classified as **hyperchromic** (abnormally high hemoglobin content) or hypochromic (abnormally low hemoglobin content).

Iron (Fe) is a critical component of the hemoglobin molecule, forming the central core of each heme group (see Figure 17-5). Without adequate iron in the diet, the body cannot manufacture enough hemoglobin. The result is **iron deficiency anemia**—a worldwide medical problem. Although the body carefully protects its iron reserves, they may be depleted through hemorrhage, increased requirements such as wound healing or pregnancy, or low intake. Unfortunately, iron deficiency is the most common nutritional deficiency in the world. The tragic result is that an estimated 10% of the population in some developed countries and up to 50% in developing countries suffers from iron deficiency anemia. Oral administration of iron-containing compounds, such as ferrous sulfate or ferrous gluconate, is very effective in treating the basic iron deficiency seen in the disease. Figure 17-23 shows a peripheral blood smear from an individual suffering from iron deficiency anemia. The numbers of RBCs are only slightly below normal. Note, however, that the cells are small and appear pale (hypochromic) because of the reduction in hemoglobin content.

The term **hemolytic** (hee-moh-LIT-ik) **anemia** applies to any of a variety of inherited blood disorders characterized by

Mechanisms of Disease—cont.

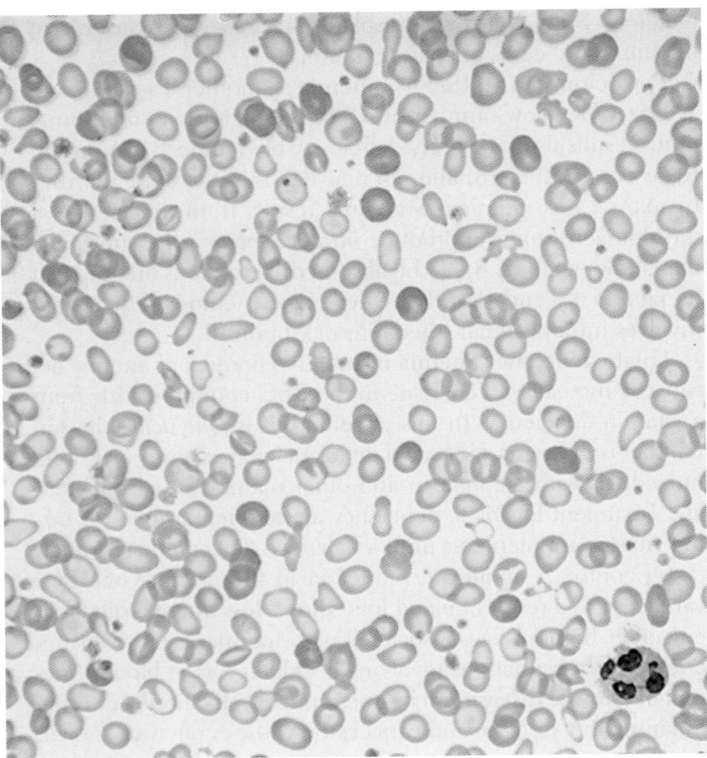

Figure 17-23 *Iron deficiency anemia.* Note the small, pale (hypochromic) RBCs. Lack of adequate color in the RBCs is due to reduced hemoglobin content.

abnormal types of hemoglobin. The term *hemolytic* means "pertaining to blood breakage" and emphasizes the fact that abnormal hemoglobin often causes red blood cells to become distorted and easily broken. An example of a hemolytic anemia is **sickle cell anemia.** Another type of hemolytic anemia is **thalassemia** (thal-ah-SEE-mee-ah). As with sickle cell anemia, thalassemia is an inherited disorder, with both a mild and a severe form (*thalassemia minor* and *thalassemia major*).

White Blood Cell Disorders

The term **leukopenia** refers to an abnormally low WBC count (less than 5000 cells/mm³ of blood). Various disease conditions may affect the immune system and decrease the amount of circulating WBCs. Acquired immunodeficiency syndrome or AIDS results in marked leukopenia. **Leukocytosis** refers to an abnormally high WBC count. It is a much more common problem than leukopenia, is seen in most types of leukemia, and almost always accompanies bacterial infections.

Two major groups of disease conditions constitute a majority of WBC and blood-related cancers, or malignant neoplasms. **Lymphoid neoplasms** arise from lymphoid precursor cells that normally produce B lymphocytes, T lymphocytes, or their descendent cell types. **Myeloid neoplasms** appear as a result of malignant transformation of myeloid stem or precursor

BOX 17-9: HEALTH MATTERS
Treating Chronic Myeloid Leukemia

Scientists have known for several years that a pathologic transformation of the gene that controls the production of an intracellular protein called *tyrosine kinase* is associated with development of **chronic myeloid leukemia (CML)**—a common and deadly form of blood cancer.

One of the most promising treatments for CML is a tyrosine kinase **signal transduction inhibitor** drug marketed as imatinib **(Gleevec).** It acts by seeking out and breaking up the signal transduction between the abnormal tyrosine kinase being produced and the specific receptor sites for this enzyme on the CML cancer cells that send the signal causing runaway proliferation. It is a "rational" drug that acts to treat CML at the molecular level. By doing so, it blocks some of the most serious functional effects of the genetic flaw that causes the disease—not just the symptoms resulting from expression of the genetic defect.

cells that normally produce granulocytic WBCs, monocytes, RBCs, and platelets.

Multiple Myeloma

Multiple myeloma (my-eh-LOH-mah) is cancer of antibody-secreting B lymphocytes called *plasma cells* (Figure 17-24, *B*). It is one of the most common and one of the most deadly forms of blood-related cancers in people older than 65 years of age. The cancerous transformation of plasma cells results in impairment of bone marrow function, production of defective antibodies, recurrent infections (from neutropenia), anemia, and the painful destruction and fracture of bones in the skull and throughout the skeletal system. The x-ray photo in Figure 17-24, *A*, shows the typical "honeycomb" or "punched-out" defects in skull bones caused by the defective myeloma antibody.

Leukemia

Leukemia (loo-KEE-mee-ah) is the term used to describe a number of blood cancers affecting the WBCs. In almost every form of leukemia, marked leukocytosis occurs. Leukocyte counts in excess of 100,000/mm³ in circulating blood are common. The different types of leukemia are identified as either *acute* or *chronic*, based on how quickly symptoms appear after the disease begins, and as *lymphocytic* or *myeloid*, depending on the cell type involved. Four of the most common leukemias are briefly described here.

Chronic Lymphocytic Leukemia (CLL)

Chronic lymphocytic leukemia (CLL) most often affects older adults. Average age of onset is about 65 years, and it appears more often in men than women. In those with CLL, ma-

Mechanisms of Disease—cont.

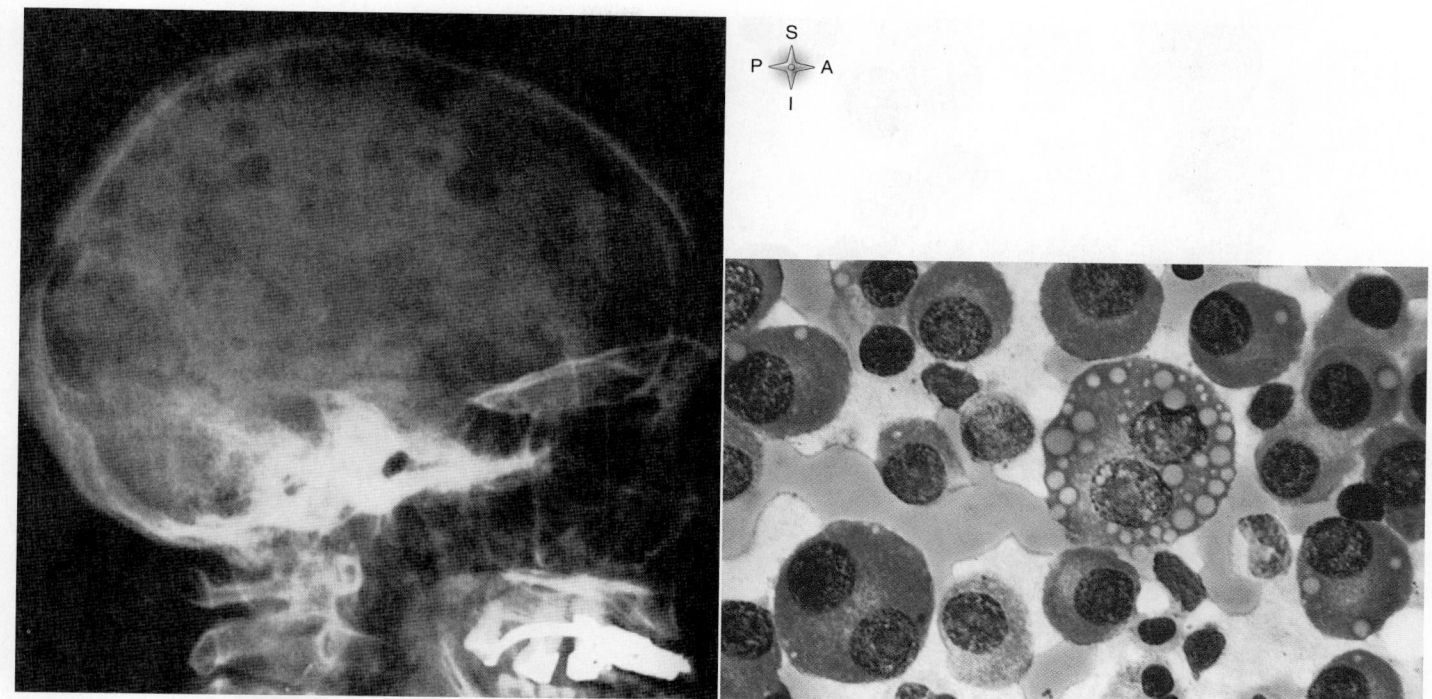

Figure 17-24 *Multiple myeloma.* **A,** X-ray of skull showing "honeycomb" or "punched-out" bone defects caused by diseased antibody from plasma cells. **B,** Malignant plasma cells. Vacuoles contain defective antibodies.

lignant precursor B lymphocytes are produced in great numbers (Figure 17-25, *A*). Early in the disease few symptoms are apparent, and many patients are diagnosed inadvertently as part of a routine physical examination when results of blood tests become available. When symptoms such as lymph node enlargement and fatigue do appear, they are often quite mild. Many patients with CLL live many years after diagnosis with little or no treatment.

Acute Lymphocytic Leukemia (ALL)

Acute lymphocytic leukemia (ALL) is primarily a disease of children and constitutes the most common form of "blood cancer" in children between 3 and 7 years of age (Figure 17-25, *B*). Fully 80% of all children who develop leukemia have this form of the disease. Although always a serious condition, it is highly curable in children but less so when it occurs in adults. As the name implies, onset of the disease is sudden. Symptoms include fever, bone pain, and increased rates of infection. Cancerous cells crowd out other bone marrow cells and decrease the production of RBCs and platelets as well as other nonmalignant lymphocyte cells. Anemia and swelling of lymph nodes, spleen, and liver are common symptoms. Treatment may involve chemotherapy, irradiation, and bone marrow or stem cell transplants.

Chronic Myeloid Leukemia (CML)

Chronic myeloid leukemia (CML) accounts for about 20% of all cases of leukemia and occurs most often in adults between 25 and 60 years of age. CML results from cancerous transformation of granulocytic (neutrophil, eosinophil, and basophil) precursor cells in the bone marrow. Onset is slow, and once the disease is established, it progresses slowly. Diagnosis is often made by discovery of marked elevations of granulocytic WBCs in peripheral blood (Figure 17-25, *C*) and by extreme spleen enlargement. The drug imatinib (Gleevec) constitutes a major advance in treatment of CML. It specifically seeks out and blocks the flawed signals in CML cancer cells that cause runaway proliferation (see Box 17-9).

Acute Myeloid Leukemia (AML)

Acute myeloid leukemia (AML) is caused by pathologic transformation of myeloid stem cells (Figure 17-25, *D*). It accounts for 80% of all cases of acute leukemia in adults and 20% of acute leukemia cases in children. Onset is sudden, and once symptoms appear the disease progresses rapidly. The most common symptoms include anemia and fatigue, recurrent infections, bone and joint pain, and spongy bleeding gums. The prognosis of AML is poor, with only about 50% of children and 30% of adults achieving long-term survival. However, advances

Mechanisms of Disease—cont.

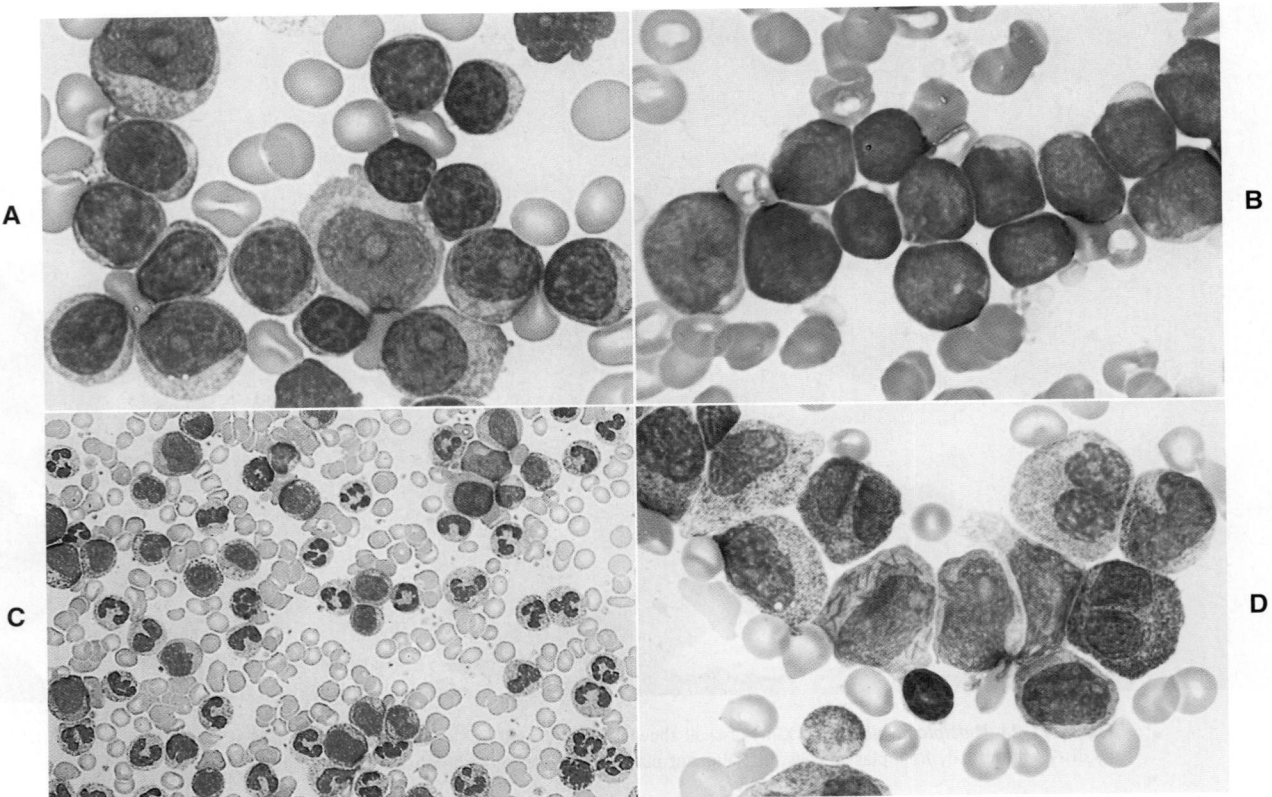

Figure 17-25 *Types of leukemia.* **A,** Chronic lymphocytic leukemia (CLL)—peripheral blood smear showing large numbers of diseased B lymphocytes. **B,** Acute lymphocytic leukemia (ALL)—appearance of B lymphocytes in ALL. **C,** Chronic myeloid leukemia (CML)—note severe granulocytic leukocytosis. **D,** Acute myeloid leukemia (AML)—note large numbers of myelocytic precursor cells.

in bone marrow and stem cell transplantation have increased cure rates in selected patients.

Infectious Mononucleosis

Infectious mononucleosis is a common noncancerous WBC disorder appearing most often in adolescents and young adults between 15 and 25 years of age. It is caused by a virus found in the saliva of infected individuals. Leukocytosis is common early in the disease with total WBC counts averaging between 12,000 and 18,000/mm³. More than 60% of the leukocytes can be identified in a differential WBC count as large, *atypical* (abnormal) lymphocytes that have abundant cytoplasm and a large nucleus (Figure 17-26). Symptoms vary greatly, but, in addition to the leukocytosis and atypical lymphocytes seen in peripheral blood, fever, sore throat, rash, severe fatigue, and enlargement of lymph nodes and the spleen are common findings. Infectious "mono" is generally self-limited and resolves without complications in about 4 to 6 weeks, although fatigue may last for longer periods.

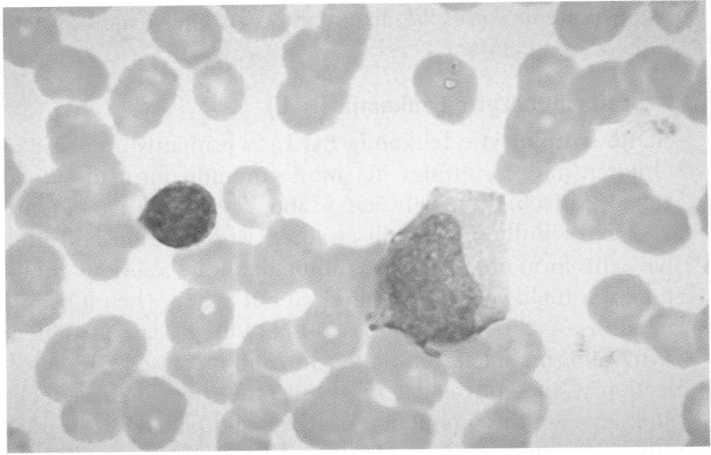

Figure 17-26 *Infectious mononucleosis.* The cell on the left is a typical small lymphocyte with the nucleus almost filling the cell. The larger atypical lymphocyte on the right has much more cytoplasm and a larger nucleus.

Mechanisms of Disease—cont.

Clotting Disorders

Unfortunately, clots sometimes form in unbroken blood vessels of the heart, brain, lungs, or other organs—a dreaded thing because clots may produce sudden death by shutting off the blood supply to a vital organ. When a clot stays in the place where it formed, it is called a **thrombus** (THROM-bus), and the condition is spoken of as **thrombosis** (throm-BOH-sis). If all or part of the clot dislodges and circulates through the bloodstream, it is called an **embolus** (EM-boh-lus), and the condition is called an **embolism** (EM-boh-liz-em). Physicians now have drugs that help prevent thrombosis and embolism. Heparin, for example, can be used to prevent excessive clotting. Heparin inhibits the conversion of prothrombin to thrombin, preventing the formation of a thrombus. Warfarin (Coumadin), an oral anticoagulant, is also frequently used to prevent excessive clotting. It blocks the stimulating effect of vitamin K on the liver, and consequently the liver cells make less prothrombin. The prothrombin content soon falls low enough to prevent abnormal clotting.

Hemophilia is an X-linked inherited disorder that affects 1 in every 10,000 males worldwide. It results from a failure to produce one or more plasma proteins responsible for blood clotting—a process illustrated in Figure 17-20. Thus hemophilia is characterized by a relative inability to form blood clots. Because minor blood vessel injuries are common in ordinary life, hemophilia can be a life-threatening condition.

The most common form is called *hemophilia* A. It is caused by absence of factor VIII protein and affects more than 300,000 people around the world.

Historically, factor VIII was obtained by plasma fractionation. This method cannot meet demand given the shortage of available donated blood. Currently, recombinant methods are used to produce enough recombinant antihemophilic factor VIII (rAHF) to meet the therapeutic needs of the world's hemophiliac population. Increasing amounts are needed to meet the larger quantities required as adolescent patients mature and as physicians prescribe more factor VIII to prevent as well as to treat bleeding episodes.

A more common type of clotting disorder results from a decrease in the platelet count—a condition called **thrombocytopenia** (throm-boh-sye-toh-PEE-nee-ah). This condition is characterized by bleeding from many small blood vessels throughout the body, most visibly in the skin and mucous membranes. If the number of thrombocytes falls to $20,000/mm^3$ or less (normal range is 150,000 to $400,000/mm^3$), catastrophic bleeding may occur. Although a number of different mechanisms can result in thrombocytopenia, the usual cause is bone marrow destruction by drugs or an immune system disease, chemicals, radiation, or cancer. Drugs may cause thrombocytopenia as a side effect. In such cases, stopping the drug usually solves the problem.

LANGUAGE OF SCIENCE   *(Cont'd from page 647)*

globin (GLOH-bin) [*globus* ball]

granulocytes (GRAN-yoo-loh-sytes) [*granul-* little grains or granules, *-cytes* cells]

hematopoietic stem cells (hee-mah-toh-poy-ET-ik) [*hemato-* relating to blood, *-poiesis* formation of]

heme (heem) [*haima* blood]

hemoglobin (hee-moh-GLOH-bin) [*hemo-* blood or blood vessels, *-globus* ball]

heparin (HEP-ah-rin) [*hepar* liver]

intrinsic clotting pathways (in-TRIN-sik) [*intr-* inside or within, *-insic* beside]

leukocytes (LOO-koh-sytes) [*leuko-* white, *-cyte* cell]

lymphocytes (LIM-foh-sytes) [*lympho-* the lymph, *-cyte* cell]

monocytes (MON-oh-sytes) [*mono-* one, *-cyte* cell]

neutrophils (NOO-troh-fils) [*neuter-* neither, *-phil* to love]

nonelectrolytes (non-ee-LEK-troh-lytes) [*non-* not, *-electro-* electricity, *-lyte* a substance capable or resulting in decomposition]

plasma (PLAZ-mah) [*plasma* something formed]

platelets (PLAYT-lets) [*platelet* small dish]

polychromatic erythroblasts (pahl-ee-kroh-MAT-ik ee-RITH-roh-blasts) [*poly-* many, *-chroma-* color, *-ic* pertaining to, *erythro-* red, *-blast* embryonic state of development]

polymorphonuclear leukocytes (pahl-ee-mohr-foh-NYOO-klee-er LOO-koh-sytes) [*poly-* many, *-morph-* form or shape, *-nucleus* nut or kernel, *leuko-* white, *-cytes* cells]

proerythroblasts (proh-eh-rith-roh-BLASTS) [*pro-* first, *-erythro-* red, *-blast* embryonic state of development]

reticulocytes (reh-TIK-yoo-loh-sytes) [*reticul-* netlike, *-cyte* cell]

spectrin (SPEK-trin)

thrombopoiesis (throm-boh-poy-EE-sis) [*thrombo-* clot, *-poiesis* formation of]

thrombus (THROM-bus) [*thrombo* clot]

transportation

LANGUAGE OF MEDICINE

acute anemia (ah-KYOOT ah-NEE-mee-ah) [*acu* sharp, *an-* without, *-emia* blood condition]

acute lymphocytic leukemia (ALL) (ah-KYOOT LIM-foh-sit-ik loo-KEE-mee-ah) [*acu* sharp, *lympho-* without water, *-cyte* cell, *leuko-* white corpuscle, *-emia* blood condition]

acute myeloid leukemia (AML) (ah-KYOOT MY-eh-loyd loo-KEE-mee-ah) [*acu* sharp, *myel-* bone marrow, *-oid* resembling, *leuko-* white corpuscle, *-emia* blood condition]

anemias (ah-NEE-me-ahs) [*an-* without, *-emia* blood condition]

anemia of chronic disease (ah-NEE-mee-ah KRON-ik) [*an-* without, *-emia* blood condition, *chronos* time, *dis-* opposite of, *-aise* ease]

antianemic principle (an-tee-an-NEE-mik) [*anti-* against, *-an-* without, *-emia* blood condition, *-ic* pertaining to]

LANGUAGE OF MEDICINE—cont'd

anticoagulant drug treatment (an-tee-koh-AG-yoo-lant) [*anti-* against, *-coagulare* to curdle]

antiplatelet drug treatment (an-tee-PLAYT-let) [*anti-* against, *-platelet* little dish]

aplastic anemia (a-PLAS-tik ah-NEE-mee-ah) [*a-* without, *-plassein* form, *a-* without, *-nemia* condition of red blood cell deficiency]

aspiration biopsy cytology (ass-pih-RAY-shen BYE-op-see SYE-TOL-oh-jee) [*bio-* life, *-ops* view]

autologous transfusion (aw-TAHL-eh-gus) [*auto* self]

blood boosting

blood doping

blood loss anemia (ah-NEE-mee-ah) [*a-* without, *-nemia* condition of red blood cell deficiency]

bone marrow transplant

chronic lymphocytic leukemia (CLL) (KRON-ik LIM-foh-sit-ik loo-KEE-mee-ah) [*chronos-* time, *-ic* pertaining to, *lympho-* without water, *-cyte* cell, *leuko-* white corpuscle, *-emia* blood condition]

chronic myeloid leukemia (CML) (KRON-ik MY-eh-loyd loo-KEE-mee-ah) [*chronos-* time, *-ic* pertaining to, *myel* bone marrow, *leuko-* white corpuscle, *-emia* blood condition]

coagulation (koh-ag-yoo-LAY-shen) [*coagulare* to curdle]

complete blood cell count (CBC)

differential white blood cell count (dif-er-EN-shal)

direct methods

embolism (EM-boh-liz-em) [*embol-* embolus or plug, *-ism* condition of]

embolus (EM-boh-lus) [*embolus* plug]

erythroblastosis fetalis (eh-rith-roh-blas-TOH-sis feh-TAL-is) [*erythro-* red, *-blast-* embryonic state of development, *-osis* result]

erythrocyte sedimentation rate (ESR) (eh-RITH-roh-syte sed-i-men-TAY-shen)

folate deficiency anemia (FOH-layt deh-FISH-en-see ah-NEE-mee-ah) [*de-* down, *-facere* to make, *a-* without, *-nemia* condition of red blood cell deficiency]

Gleevec

hematocrit (hee-MAT-oh-krit) [*hemato-* relating to blood, *-krinein* to separate]

hemolytic anemia (hee-moh-LIT-ik ah-NEE-mee-ah) [*hemo-* blood, *-lytic* to produce decomposition, *a-* without, *-nemia* condition of red blood cell deficiency]

hemophilia (hee-moh-FIL-ee-ah) [*hemo-* blood, *-philia* to love]

Hemopure

hemostasis (hee-moh-STAY-sis) [*hemo-* blood, *-stasis* stoppage or inhibition]

homologous transfusion (ho-MOL-oh-gus) [*homo* the same]

hyperchromic (hye-per-KROH-mik) [*hyper-* excessive, *-chroma-* color, *-ic* pertaining to]

indirect methods

infectious mononucleosis (in-FEK-shuss mahn-oh-noo-klee-OH-sis) [*inficere* to stain, *mono-* single, *-nucleus-* nut, *-osis* result]

intermittent claudication (in-ter-MIT-tent klaw-dih-KAY-shen) [*inter-* occurring between, *-mittere* to send, *claudicatio* a limping]

iron deficiency anemia (deh-FISH-en-see ah-NEE-mee-ah) [*de-* down, *-facere* to make, *a-* without, *-nemia* condition of red blood cell deficiency]

leukemia (loo-KEE-mee-ah) [*leuko-* white corpuscle, *-emia* blood condition]

leukocytosis (loo-koh-SYE-toh-sis) [*leuko-* white, *-cyt-* cell, *-osis* a result]

leukopenia (loo-koh-PEE-nee-ah) [*leuko-* white corpuscle, *-penia* a specified deficiency]

leukoreduction (loo-koh-ree-DUK-shen) [*leuko-* white corpuscle, *-reducere* to lead back]

lymphoid neoplasms (LIM-foyd NEE-oh-plaz-ems) [*lympho-* the lymph, *-oid* resembling, *neo-* new, *-plasm* cell or tissue substance]

multiple myeloma (my-eh-LOH-mah) [*myel-* bone marrow, *-oma* tumor]

myeloid neoplasms (MY-eh-loyd NEE-oh-plaz-ems) [*myel-* bone marrow, *-oid* resembling, *neo-* new, *-plasm* cell or tissue substance]

nucleic acid test (NAT) (noo-KLAY-ik) [*nucle* nucleus, *acidus-* sour]

packed cell volume (PCV)

pernicious anemia (per-NISH-us ah-NEE-mee-ah) [*perniciosus* destructive, *a-* without, *-nemia* condition of red blood cell deficiency]

physiological polycythemia (fiz-ee-oh-LOJ-i-kal pol-ee-sye-THEE-mee-ah) [*physis-* nature, *-ology* study or science of, *poly-* many, *-kytos-* cell, *-emia* blood condition]

platelet plug (PLAYT-let) [*platelet-* little dish]

polycythemia (pahl-ee-sye-THEE-mee-ah) [*poly-* many, *-cyt-* cell, *-emia* blood condition]

PolyHeme (pahl-ee-HEEM)

positive chemotaxis (kee-moh-TAK-sis) [*chemo-* a chemical, *-taxis* movement of an organism in response to a stimulus]

sickle cell anemia (SIK-ul sell ah-NEE-mee-ah) [*sicol* crescent, *cella* storeroom, *a-* without, *-nemia* condition of red blood cell deficiency]

signal transduction inhibitor (SIG-nal tranz-DUK-shen in-HIB-ih-tor) [*signum* mark, *trans-* across or through, *-ducere* to lead, *inhibere* to restrain]

thalassemia (thal-ah-SEE-mee-ah) [*thalassa-* sea, *-emia* blood condition]

thrombocytopenia (throm-boh-sye-toh-PEE-nee-ah) [*thrombo-* clot, *-cyto-* cell, *-penia* a specified deficiency]

thrombosis (throm-BOH-sis) [*thrombo-* clot, *-osis* a result]

transfusion reaction

CASE STUDY

Cindy Alaniz, age 25 years, has come to the health center to have her routine annual pap and pelvic examination. During her interview, she tells you that her menstrual periods are heavier than normal for the past 6 months and that she has been passing clots about the size of a nickel on the first day of each cycle. She denies any cramping with her menses. She is thinking about trying to become pregnant. Her medical history is negative for serious illness or surgery. Ms. Alaniz is currently taking no medications. She exercises but says that lately she has stopped this practice because she feels "tired all the time." Her diet is high in pasta foods, some fruits, and vegetables, and she only eats meat once each month in an attempt to limit her fat content. She has episodes of dizziness, which she describes as "feeling very weak." She also says she has been having some heart palpitations. The health care provider completes a physical examination and orders a complete blood count and differential.

1. Ms. Alaniz is diagnosed with iron deficiency anemia. Which of the following would most likely reflect her ordered laboratory results?

   A. Low hemoglobin, low hematocrit, high RBC count
   B. Low hemoglobin, low hematocrit, decreased number of WBCs and platelets
   C. Low hemoglobin, low hematocrit, low RBC count with red cells classified as hyperchromic
   D. Low hemoglobin, low hematocrit, low RBC count with red cells classified as hypochromic

2. Along with decreased hematocrit and hemoglobin levels, Ms. Alaniz's laboratory results reflect a normal WBC with a very elevated eosinophil count. Which one of the following might the health care provider suspect?

   A. A bacterial infection
   B. A viral infection
   C. An immunity problem
   D. A parasitic infection

3. Ms. Alaniz is given iron and folic acid supplements and told to come back in 1 month for a reticulocyte count. Which of the following would be the expected results?

   A. A low reticulocyte count, less than 0.5%
   B. A high reticulocyte count, more than 1.5%
   C. A normal reticulocyte count, about 1.0%

4. After 6 months, Ms. Alaniz has improved her diet and has normal blood levels. She and her husband have decided to try to become parents but are concerned because Ms. Alaniz is Rh positive whereas her husband is Rh negative. Which one of the following is your *best* response?

   A. She and her husband should be very concerned and not have children because the child will have a disease called erythroblastosis fetalis.
   B. She and her husband should be concerned, but because this is their first child it will not be affected.
   C. She and her husband do not have to worry. RhoGAM administered to the mother can prevent the complication.
   D. She and her husband do not need to worry about Rh problems because she is not Rh negative.

CHAPTER SUMMARY

COMPOSITION OF BLOOD

A. Introduction (Figure 17-1)
   1. Blood—made up of plasma and formed elements
   2. Blood—complex transport medium that performs vital pickup and delivery services for the body
   3. Blood—keystone of body's heat-regulating mechanism
B. Blood volume
   1. Young adult male has approximately 5 liters of blood
   2. Blood volume varies according to age, body type, sex, and method of measurement

FORMED ELEMENTS OF BLOOD

A. Red blood cells (RBCs; erythrocytes)
   1. Description of mature RBCs (Figure 17-4)
      a. Have no nucleus and shaped like tiny biconcave disks
      b. Do not contain ribosomes, mitochondria, and other organelles typical of most body cells
      c. Primary component is hemoglobin
      d. Most numerous of the formed elements
   2. Function of RBCs
      a. RBCs' critical role in the transport of oxygen and carbon dioxide depends on hemoglobin

b. Carbonic anhydrase—enzyme in RBCs that catalyzes a reaction that joins carbon dioxide and water to form carbonic acid
c. Carbonic acid—dissociates and generates bicarbonate ions, which diffuse out of the RBC and serve to transport carbon dioxide in the blood plasma
   3. Hemoglobin (Figure 17-5)
      a. Within each RBC are approximately 200 to 300 million molecules of hemoglobin
      b. Hemoglobin is made up of four globin chains, with each attached to a heme molecule
      c. Hemoglobin is able to unite with four oxygen molecules to form oxyhemoglobin to allow RBCs to transport oxygen where it is needed
      d. A male has a greater amount of hemoglobin than a female
      e. Anemia—a decrease in number or volume of functional RBCs in a given unit of whole blood
   4. Formation of red blood cells (review Figures 17-6 and 17-7)
      a. Erythropoiesis—entire process of RBC formation
      b. RBC formation begins in the red bone marrow as hematopoietic stem cells and goes through several

stages of development to become erythrocytes; entire maturation process requires approximately 4 days

   c. RBCs are created and destroyed at approximately 100 million per minute in an adult; homeostatic mechanisms operate to balance the number of cells formed against the number of cells destroyed

5. Destruction of RBCs (Figure 17-8)
   a. Life span of a circulating RBC averages 105 to 120 days
   b. Macrophage cells phagocytose the aged, abnormal, or fragmented RBCs
   c. Hemoglobin is broken down, and amino acids, iron, and bilirubin are released

B. White blood cells (leukocytes, WBCs) (review Table 17-1)
1. Granulocytes
   a. Neutrophils (review Figure 17-9)—make up approximately 65% of total WBC count in a normal blood sample; highly mobile and very active phagocytic cells; capable of diapedesis; cytoplasmic granules contain lysosomes
   b. Eosinophils (review Figure 17-10)—account for 2% to 5% of circulating WBCs; numerous in lining of respiratory and digestive tracts; weak phagocytes; capable of ingesting inflammatory chemicals and proteins associated with antigen-antibody reaction complexes; provide protection against infections caused by parasitic worms and allergic reactions
   c. Basophils (review Figure 17-11)—account for only 0.5% to 1% of circulating WBCs; motile and capable of diapedesis; cytoplasmic granules contain histamine and heparin
2. Agranulocytes (Figures 17-12 and 17-13)
   a. Lymphocytes—smallest of the WBCs; second most numerous WBC; account for approximately 25% of circulating WBCs; T lymphocytes and B lymphocytes have an important role in immunity—T lymphocytes directly attack an infected or cancerous cell, and B lymphocytes produce antibodies against specific antigens
   b. Monocytes—largest leukocytes; mobile and highly phagocytic cells
3. WBC numbers—1 mm^3 of normal blood usually contains 5000 to 9000 leukocytes, with different percentages for each type; WBC numbers have clinical significance because they change with certain abnormal conditions
4. Formation of WBCs (review Figure 17-6)
   a. Granular and agranular leukocytes mature from the undifferentiated hematopoietic stem cell
   b. Neutrophils, eosinophils, basophils, and a few lymphocytes and monocytes originate in red bone marrow; most lymphocytes and monocytes develop from hematopoietic stem cells in lymphatic tissue

C. Platelets (review Figure 17-12)
1. Structure
   a. In circulating blood, platelets are small, pale bodies that appear as irregular spindles or oval disks
   b. Three important properties are agglutination, adhesiveness, and aggregation
   c. Platelet counts in adults average 250,000/mm^3 of blood; normal range is 150,000 to 400,000/mm^3

2. Functions of platelets
   a. Important role in hemostasis and blood coagulation
   b. Hemostasis—refers to stoppage of blood flow; however, if injury is extensive, the blood-clotting mechanism is activated to assist
3. Platelet plug formation
   a. One to five seconds after injury to vessel wall, platelets adhere to damaged endothelial lining and to each other, forming a platelet plug
   b. Temporary platelet plug is an important step in hemostasis
   c. Normal platelets (positive charge) adhere to damaged capillary wall and underlying collagen fibers, which both have a negative charge
   d. "Sticky platelets" form physical plug and secrete several chemicals involved in the coagulation process
4. Formation and life span of platelets (7 to 10 days)—formed in red bone marrow, lungs, and spleen by fragmentation of megakaryocytes

BLOOD TYPES (BLOOD GROUPS)

A. The ABO system (Figures 17-15 to 17-17)
1. Every person's blood belongs to one of four ABO blood groups
2. Named according to antigens present on RBC membranes
   a. Type A—antigen A on RBC
   b. Type B—antigen B on RBC
   c. Type AB—both antigen A and antigen B on RBC; known as universal recipient
   d. Type O—neither antigen A nor antigen B on RBC; known as universal donor
B. The Rh system (Figure 17-18)
1. Rh-positive blood—Rh antigen is present on the RBCs
2. Rh-negative—RBCs have no Rh antigen present
3. Anti-Rh antibodies are not normally present in blood; anti-Rh antibodies can appear in Rh-negative blood if it has come in contact with Rh-positive RBCs

BLOOD PLASMA

A. Plasma—liquid part of blood; clear, straw-colored fluid; made up of 90% water and 10% solutes (Figure 17-19)
B. Solutes—6% to 8% of plasma solutes are proteins, consisting of three main compounds
1. Albumins—help maintain osmotic balance of the blood
2. Globulins—essential component of the immunity mechanism
3. Fibrinogen—key role in blood clotting
C. Plasma proteins have an essential role in maintaining normal blood circulation

BLOOD CLOTTING (COAGULATION)

A. Mechanism of blood clotting—goal of coagulation is to stop bleeding and prevent loss of vital body fluid in a swift and sure method; the classic theory (Figure 17-20) is as follows:
1. Classic theory of coagulation advanced in 1905
   a. Identified four components critical to coagulation
     (1) Prothrombin
     (2) Thrombin
     (3) Fibrinogen
     (4) Fibrin

2. Current explanation of coagulation involves three stages (Figure 17-21)
   a. Stage 1—production of thromboplastin activator by either:
      (1) Chemicals released from damaged tissues (extrinsic pathway) or
      (2) Chemicals present in the blood (intrinsic pathway)
   b. Stage II—conversion of prothrombin to thrombin
   c. Stage III—conversion of fibrinogen to fibrin and production of fibrin clot

B. Conditions that oppose clotting
   1. Clot formation in intact vessels is opposed
   2. Several factors oppose clotting
      a. Perfectly smooth surface of the normal endothelial lining of blood vessels does not allow platelets to adhere
      b. Antithrombins—substances in the blood that oppose or inactivate thrombin; prevent thrombin from converting fibrinogen to fibrin; e.g., heparin

C. Conditions that hasten clotting
   1. Rough spot in the endothelium
   2. Abnormally slow blood flow

D. Clot dissolution (Figure 17-22)
   1. Fibrinolysis—physiological mechanism that dissolves
   2. Fibrinolysin—enzyme in the blood that catalyzes the hydrolysis of fibrin, causing it to dissolve
   3. Additional factors are presumed to aid clot dissolution; e.g., substances that activate profibrinolysin

THE BIG PICTURE: BLOOD AND THE WHOLE BODY

A. Blood plasma transports substances, including heat, around the body, linking all body tissues together
   1. Substances can be transported between almost any two points in the body

B. Blood tissue contains formed elements—blood cells and platelets
   1. RBCs assist in the transport of oxygen and carbon dioxide
   2. WBCs assist in the defense mechanisms of the whole body
   3. Platelets prevent loss of the fluid that constitutes the internal environment

C. No organ or system of the body can maintain proper levels of nutrients, gases, or water without direct or indirect help from blood
   1. Other systems assist the blood

D. Blood is useless unless it continues to transport, defend, and maintain balance

REVIEW QUESTIONS

1. What are the formed elements of blood?
2. What is the function of carbonic anhydrase?
3. Describe the structure of hemoglobin.
4. How does the structure of hemoglobin allow it to combine with oxygen?
5. Discuss the steps involved in erythropoiesis.
6. What is the average life span of a circulating red blood cell?
7. How are granulocytes similar to agranulocytes? How do they differ?
8. Define the term *positive chemotaxis*.
9. List the important physical properties of platelets.
10. Explain what is meant by *type AB blood*. Explain *Rh-negative blood*.
11. Which organ is responsible for the synthesis of most plasma proteins?
12. What is the normal plasma protein concentration?
13. What are some functions served by plasma proteins?
14. Describe the role of platelets in hemostasis and blood clotting.
15. What triggers blood clotting?
16. Identify factors that oppose blood clotting. Do the same for factors that hasten blood clotting.
17. Describe the physiological mechanism that dissolves clots.
18. Describe the hemoglobin in a person with sickle cell trait.
19. List and describe three blood disorders.

CRITICAL THINKING QUESTIONS

1. A modern hospital laboratory has machines that can automatically do blood cell counts based on the size of cells. If the lab technician did not want to count erythrocytes, for what size range would the machine be set? What white blood cells would be missed? What would be the classes, functions, and life spans of the white blood cells that were counted?
2. A friend received a report on his physical examination that his hematocrit was below normal. Because he knows you are taking anatomy and physiology, he has come to you for an explanation. Based on what you know, explain to him what a hematocrit value is, how it is determined, and what value would put him below normal.
3. You are a medical examiner, and a body is brought to you to determine the cause of death. You find a very large agglutination (not a clot) in a major vein. What judgment would you make regarding the cause of death?
4. Suppose a person had a thyroid disorder that caused the production of calcitonin to be many times higher than it should be. Can you elaborate on why a possible side effect of this condition might be a very slow blood-clotting time?
5. A patient comes into the emergency room with severe bleeding. What can be done to speed the formation of a clot?
6. Some athletes, seeking a competitive edge, may resort to blood doping. How would you explain blood doping? What information would you use to discourage athletes from participating in the practice of blood doping?
7. How would you summarize the condition *erythroblastosis fetalis*?

Anatomy of the Cardiovascular System

LANGUAGE OF SCIENCE

abdominal aorta (ab-DOM-i-nal ay-OR-tah)
[abdomen- belly, -al pertaining to]

accessory hemiazygos vein (ak-SES-oh-ree hem-EE-ah-ZYE-gos) *[hemi- half, -a- without, -zygon yoke]*

anastomosis (ah-nas-toh-MOH-sis) *[anastomoien- to provide a mouth, -osis increase in a pathologic condition]*

anterior tibial vein (an-TEER-ee-or TIB-ee-al)
[ante- before, tibia shin bone]

aorta (ay-OR-tah) *[aerein to raise]*

aortic arch (ay-OR-tik) *[aerein to raise]*

arterial anastomosis (ar-TEER-ee-al ah-nas-toh-MOH-sis) *[arteria air pipe, anastomoien- to provide a mouth, -osis increase in a pathologic condition]*

arterioles (ar-TEER-ee-ohls) *[arteriola little artery]*

arteriovenous anastomoses (ar-teer-ee-oh-VEE-nus ah-nas-teh-MOH-seez) *[atrio- atrium of the heart, -vena vein, anastomoein- to provide a mouth, -osis increase a pathologic condition]*

artery (AR-ter-ee) *[arteria airpipe]*

ascending aorta (ah-SEND-ing ay-OR-tah)
[ascendere to climb, aerein to raise]

atria (AY-tree-ah) *[atria hall]*

atrioventricular bundle (ay-tree-oh-ven-TRIK-yoo-lar) *[atrio- atrium of the heart, -ventri- belly, -ular pertaining to]*

atrioventricular node (ay-tree-oh-ven-TRIK-yoo-lar) *[atrio- atrium of the heart, -ventri- belly, -ular pertaining to]*

atrioventricular (AV) valves (ay-tree-oh-ven-TRIK-yoo-lar) *[atrio- atrium of the heart, -ventri- belly, -ular pertaining to]*

axillary vein (AK-si-lair-ee) *[axilla wing]*

azygos vein (AZ-i-gohs) *[a- without, -zygon yoke]*

basilic vein (bah-SIL-ik) *[basis- foundation, -ic pertaining to]*

bicuspid (bye-KUSS-pid) *[bi- two, -cuspis point]*

brachial vein (BRAY-kee-al) *[brachi- arm, -al pertaining to]*

brachiocephalic artery (brayk-ee-oh-seh-FAL-ik AR-ter-ee) *[brachi- arm, -cephalic pertaining to the head, arteria air pipe]*

Cont'd on p. 725

The cardiovascular system is sometimes called simply the *circulatory system*. It consists of the heart, which is a muscular pumping device, and a closed system of vessels called *arteries, veins,* and *capillaries*. As the name implies, blood contained in the circulatory system is pumped by the heart around a closed circle or circuit of vessels as it passes again and again through the various circulations of the body (see p. 696).

As in the adult, survival of the developing embryo depends on the circulation of blood to maintain homeostasis and a favorable cellular environment. In response to this need, the cardiovascular system makes its appearance early in development and reaches a functional state long before any other major organ system. Incredible as it seems, the heart begins to beat regularly early in the fourth week after fertilization.

HEART
Location of the Heart

The human heart is a four-chambered muscular organ, shaped and sized roughly like a person's closed fist. It lies in the mediastinum, or middle region of the thorax, just behind the body of the sternum between the points of attachment of the second through the sixth ribs. Approximately two thirds of the heart's mass is to the left of the midline of the body, and one third is to the right (Figure 18-1, *A*).

Posteriorly the heart rests against the bodies of the fifth to the eighth thoracic vertebrae. Because of its placement between the sternum in front and the bodies of the thoracic vertebrae behind, it can be compressed by application of pressure to the lower portion of the body of the sternum using the heel of the hand (Figure 18-1, *B*). Rhythmic compression of the heart in this way can maintain blood flow in cases of cardiac arrest and, if combined with effective artificial respiration, the resulting procedure, called *cardiopulmonary resuscitation* (CPR), can be life saving.

The anatomical position of the heart in the thoracic cavity is shown in Figure 18-1. The lower border of the heart, which forms a blunt point known as the *apex*, lies on the diaphragm, pointing toward the left. To count the apical beat, one must place a stethoscope directly over the apex, that is, in the space between the fifth and sixth ribs (fifth intercostal space) on a line with the midpoint of the left clavicle.

The upper border of the heart, that is, its base, lies just below the second rib. The boundaries, which, of course, indicate its size, have considerable clinical importance, because a marked increase in heart size accompanies certain types of heart disease. Therefore when diagnosing heart disorders, the physician charts the boundaries of the heart. The "normal" boundaries of the heart are, however, influenced by factors such as age, body build, and state of contraction.

Size and Shape of the Heart

At birth the heart is said to be transverse (wide) in type and appears large in proportion to the diameter of the chest cavity. In the infant, it is $\frac{1}{130}$ of the total body weight compared with about $\frac{1}{300}$ in the adult. Between puberty and 25 years of age the heart attains its adult shape and weight—about 310 g is average for the male and 225 g for the female.

In the adult the shape of the heart tends to resemble that of the chest. In tall, thin individuals the heart is frequently described as elongated, whereas in short, stocky individuals it has greater width and is described as transverse. In individuals of average height and weight it is neither long nor transverse but somewhat intermediate between the two (Figure 18-2). Its approximate dimensions are 12 cm ($4\frac{3}{4}$ inches) long, 9 cm ($3\frac{1}{2}$ inches) wide, and 6 cm ($2\frac{1}{2}$ inches) deep. Figure 18-3 shows details of the heart and great vessels in a posterior view and in an anterior view.

 QUICK CHECK

1. In anatomical terms, where is the heart located?
2. Describe the shape of the heart.
3. When does the heart attain its adult shape and weight?

Coverings of the Heart
Structure of the Heart Coverings

The heart has its own special covering, a loose-fitting inextensible sac called the **pericardium.** The pericardial sac, with the heart removed, can be seen in Figure 18-3. The **pericardium** consists of two parts: a fibrous portion and a serous portion (Figure 18-4). The sac itself is made of tough white fibrous tissue but is lined with smooth, moist serous membrane—the parietal layer of the serous pericardium. The same kind of membrane covers the entire outer surface of the heart. This covering layer is known as the *visceral layer* of the serous pericardium or the **epicardium.** The fibrous sac attaches to the large blood vessels emerging from the top of the heart but not to the heart itself (see Figure 18-3). Therefore it fits loosely around the heart, with a slight space between the visceral layer adhering to the heart and the parietal layer adhering to the inside of the fibrous sac. This space is called the **pericardial space.** It contains 10 to 15 ml of lubricating fluid secreted by the serous membrane and called **pericardial fluid.**

The structure of the pericardium can be summarized as follows:

- **Fibrous pericardium**—tough, loose-fitting, and inelastic sac around the heart
- **Serous pericardium**—consisting of two layers
  - *Parietal layer*—lining inside the fibrous pericardium
  - *Visceral layer (epicardium)*—adhering to the outside of the heart; between visceral and parietal layers is a space, the pericardial space, that contains a few drops of pericardial fluid

Function of the Heart Coverings

The fibrous pericardial sac with its smooth, well-lubricated lining provides protection against friction. The heart moves easily in this loose-fitting jacket with no danger of irritation from friction between the two surfaces, as long as the serous pericardium remains normal and continues to produce lubricating serous fluid.

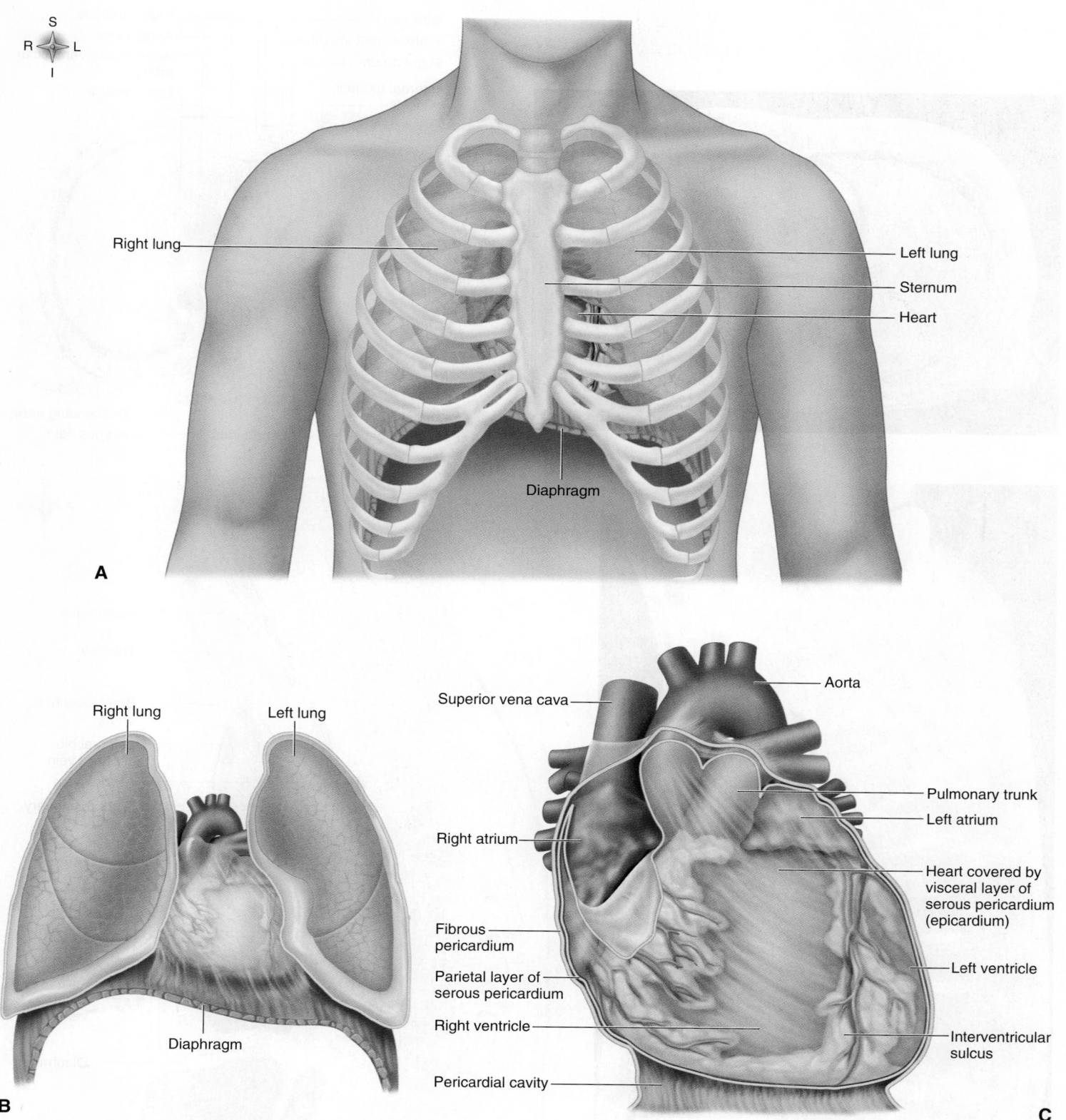

S
R ✦ L
I

A

Right lung

Left lung

Sternum

Heart

Diaphragm

B

Right lung

Left lung

Diaphragm

C

Superior vena cava

Aorta

Right atrium

Pulmonary trunk

Left atrium

Heart covered by
visceral layer of
serous pericardium
(epicardium)

Fibrous
pericardium

Parietal layer of
serous pericardium

Left ventricle

Right ventricle

Interventricular
sulcus

Pericardial cavity

Figure 18-1, A *Location of the heart.* *A,* Heart in mediastinum showing relationship to lungs and other anterior thoracic structures. *B,* Anterior view of isolated heart and lungs. Portions of the parietal pleura and pericardium have been removed. *C,* Detail of heart resting on diaphragm with pericardial sac opened.

Continued

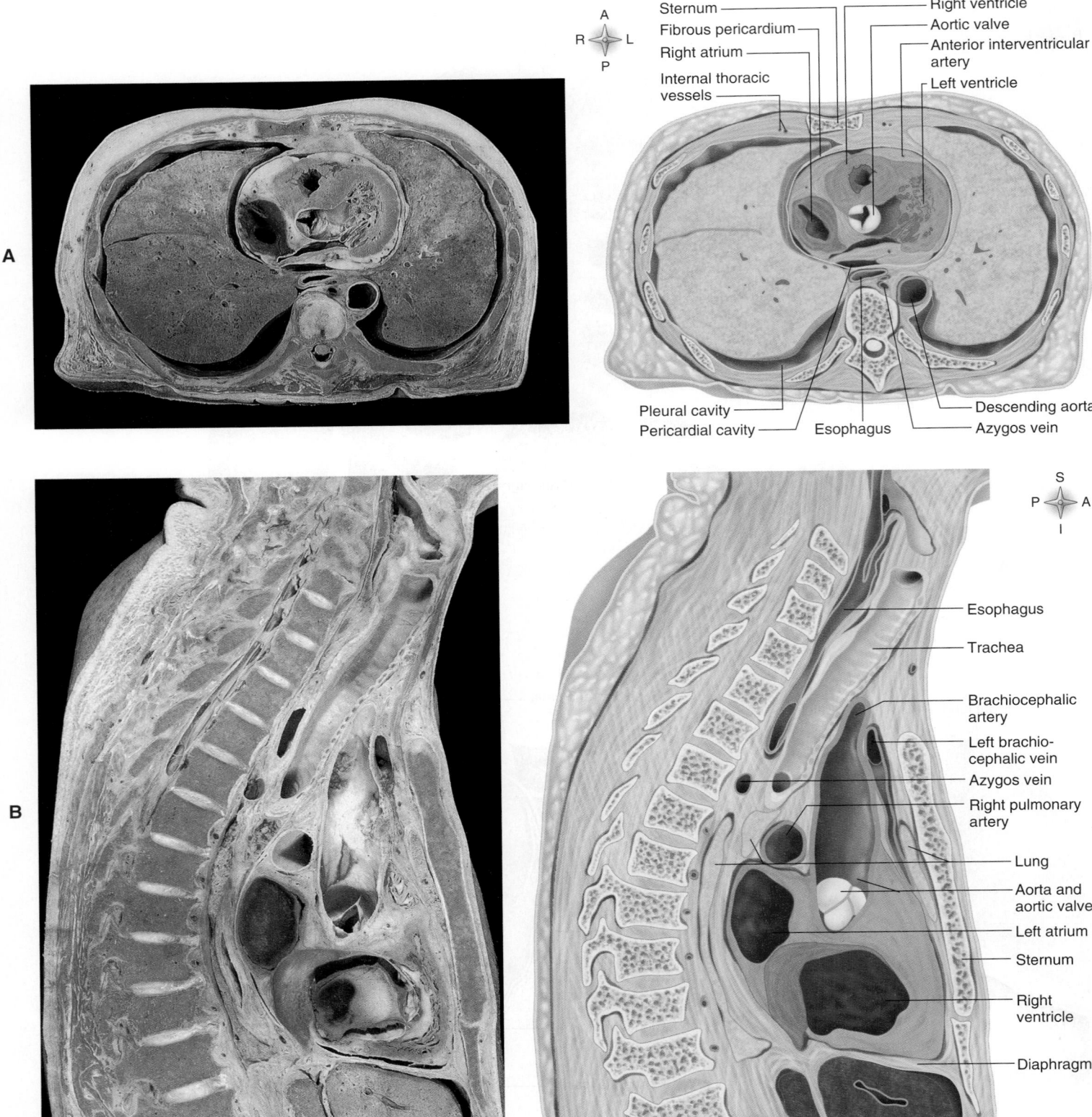

Figure 18-1, cont'd B *Location of the heart.* *A,* Transverse section of cadaver specimen and color drawing of thoracic structures at the level of the sixth thoracic vertebra. Inferior aspect. *B,* Midline sagittal section of cadaver specimen and color drawing showing thorax structures. Note the placement of the heart between the sternum in front and thoracic vertebrae behind.

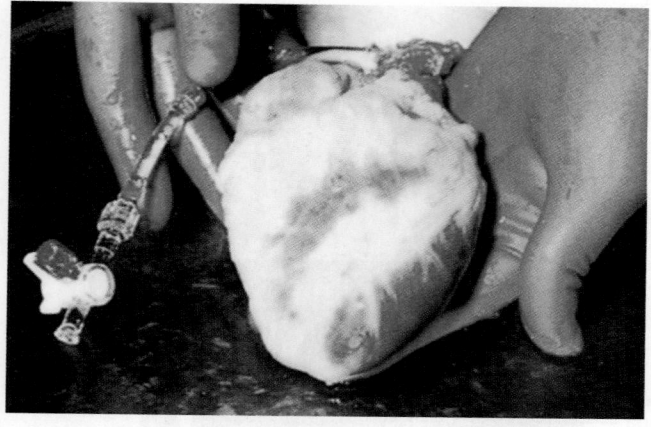

Figure 18-2 *Appearance of the heart.* This photograph shows a living human heart prepared for transplantation into a patient. Note its size relative to the hands that are holding it.

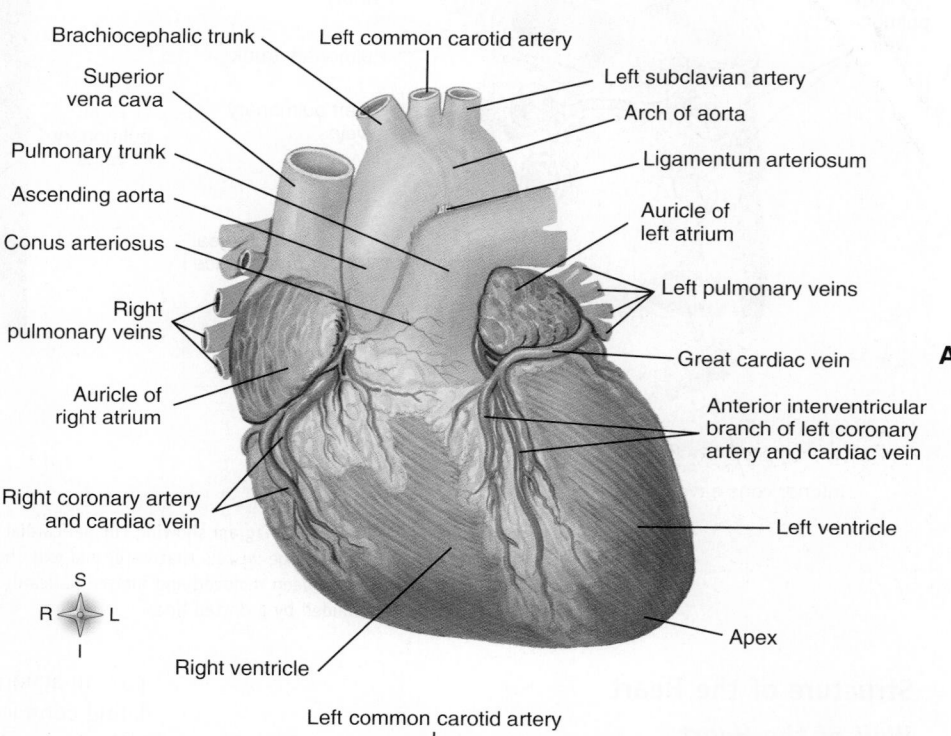

Brachiocephalic trunk
Superior vena cava
Pulmonary trunk
Ascending aorta
Conus arteriosus
Right pulmonary veins
Auricle of right atrium
Right coronary artery and cardiac vein

Left common carotid artery
Left subclavian artery
Arch of aorta
Ligamentum arteriosum
Auricle of left atrium
Left pulmonary veins
Great cardiac vein
Anterior interventricular branch of left coronary artery and cardiac vein
Left ventricle
Apex
Right ventricle

S
R L
I

A

Figure 18-3 *The heart and great vessels.* **A,** Anterior view of the heart and great vessels. **B,** Posterior view of the heart and great vessels.

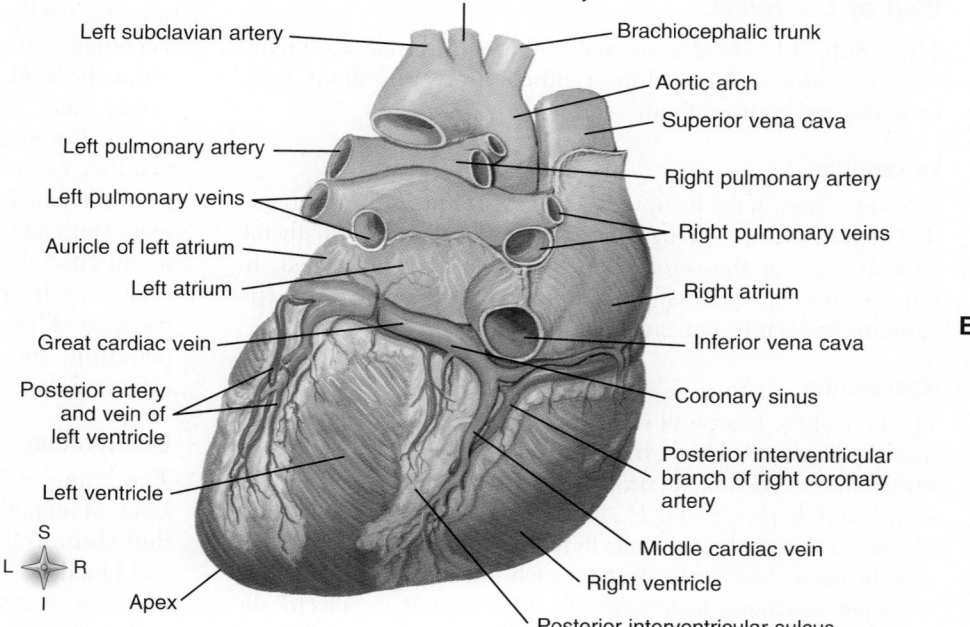

Left common carotid artery
Left subclavian artery
Brachiocephalic trunk
Aortic arch
Superior vena cava
Left pulmonary artery
Right pulmonary artery
Left pulmonary veins
Right pulmonary veins
Auricle of left atrium
Left atrium
Right atrium
Great cardiac vein
Inferior vena cava
Posterior artery and vein of left ventricle
Coronary sinus
Posterior interventricular branch of right coronary artery
Left ventricle
Middle cardiac vein
Apex
Right ventricle
Posterior interventricular sulcus

S
L R
I

B

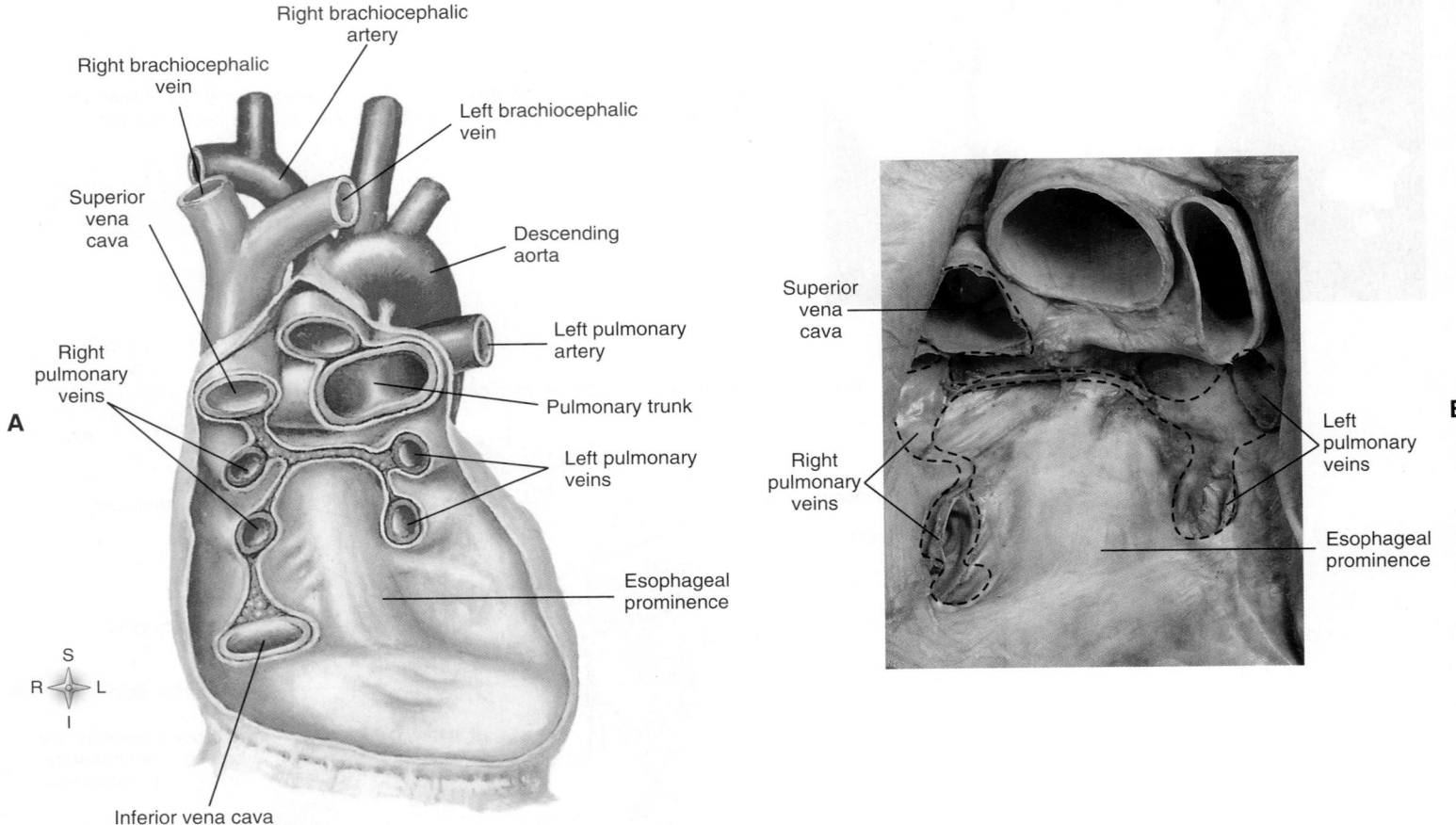

Figure 18-4 *Pericardium.* **A,** Frontal view diagram showing cut pericardial sac with heart removed. Notice that the pericardial sac attaches to the large vessels that enter and exit the heart, not to the heart itself. **B,** Cadaver dissection photo. Heart has been removed and incised pericardium turned back. The orifices of the six vessels identified in **A** are surrounded by a dotted line.

Structure of the Heart

Wall of the Heart

Three distinct layers of tissue make up the heart wall (see Figure 18-5) in both the atria and the ventricles: the epicardium, myocardium, and endocardium.

Epicardium

The outer layer of the heart wall is called the **epicardium,** a name that literally means "on the heart." The epicardium is actually the visceral layer of the serous pericardium already described. In other words, the same structure has two different names: epicardium and serous pericardium.

Myocardium

The bulk of the heart wall is the thick, contractile, middle layer of specially constructed and arranged cardiac muscle cells called the **myocardium.** The minute structure of cardiac muscle has been described in Chapters 5 and 11. Recall that cardiac muscle tissue is composed of many branching cells that are joined into a continuous mass by *intercalated disks.* Because each intercalated disk includes many gap junctions, large areas of cardiac muscle are electrically coupled into a single functional unit called a *syncytium* (meaning "joined cells"). Because they form a syncytium, muscle cells can

pass an action potential along a large area of the heart wall, stimulating contraction in each muscle fiber of the syncytium. Another advantage of the syncytium structure is that the cardiac fibers form a continuous sheet of muscle that wraps entirely around the cavities within the heart. Thus the encircling myocardium can compress the heart cavities, and the blood within them, with great force.

Recall also that cardiac muscles are *autorhythmic,* meaning that they can contract on their own in a slow, steady rhythm. As we explained in Chapter 11, cardiac muscle cells cannot summate contractions to produce tetanus and thus do not fatigue—a useful characteristic for muscle tissue that must maintain a continuous cycle of alternating contraction and relaxation for the entire span of life. Because the muscular myocardium can contract powerfully and rhythmically, without fatigue, the heart is an efficient and dependable pump for blood.

Endocardium

The lining of the interior of the myocardial wall is a delicate layer of endothelial tissue known as the **endocardium. Endothelium** is the type of membranous tissue that lines the heart and blood vessels. Endothelium resembles simple squamous epithelium, except for the fact that during embryonic development endothelium arises from different tissue than does epithelium. Note in Figure 18-5 that the endocardium covers beamlike pro-

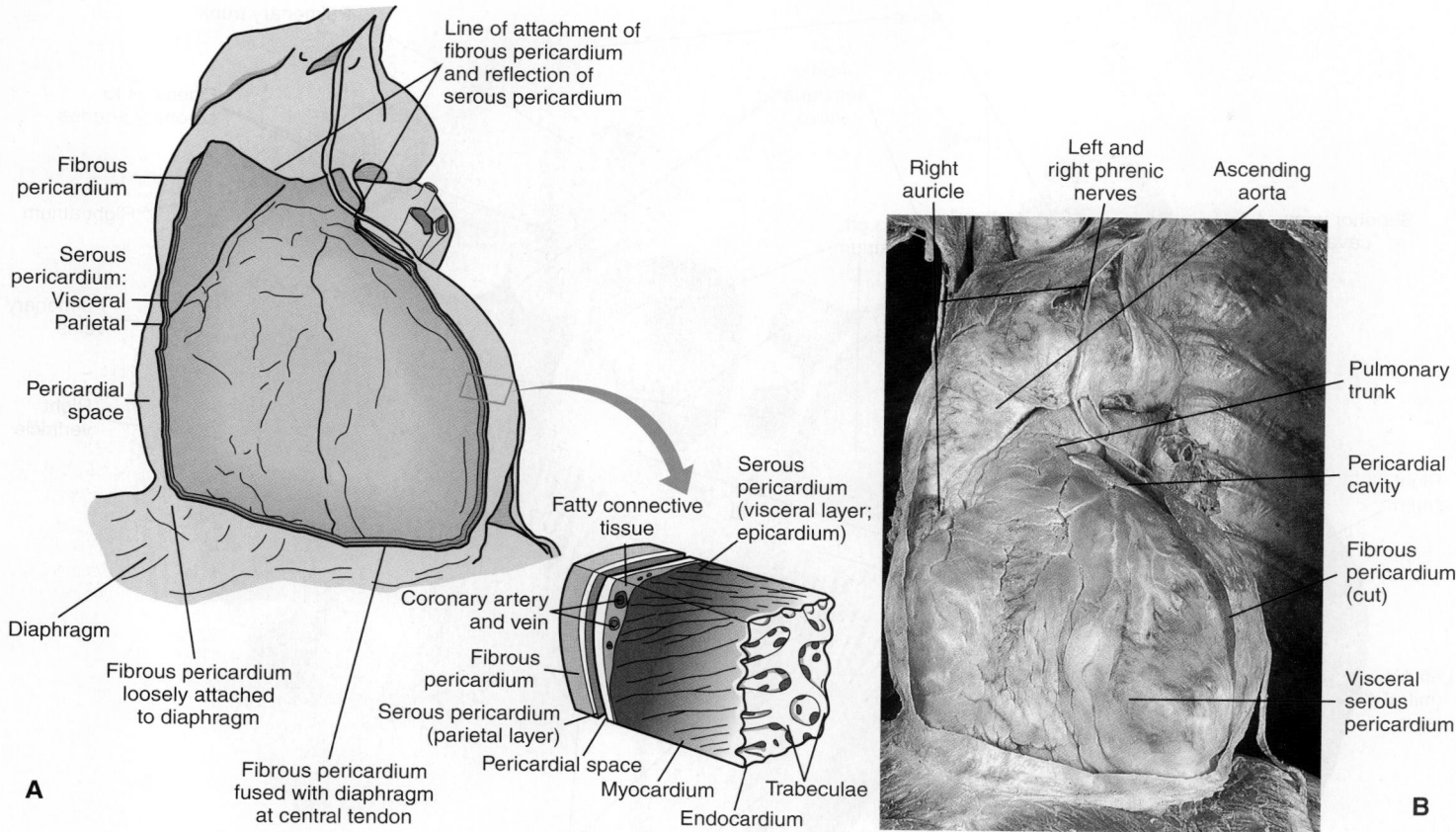

Figure 18-5 *Pericardial sac and wall of the heart.* **A,** Diagram showing the pericardial sac cut and opened to expose the anterior surface of the heart. Both the diagram and cutout section of the heart wall show the outer fibrous pericardium and the parietal and visceral layers of the serous pericardium (with the pericardial space between them). Note in the cutout of the heart wall that a layer of fatty connective tissue is located between the visceral layer of the serous pericardium (epicardium) and the myocardium. Note also that the endocardium covers beamlike projections of myocardial muscle tissue, called trabeculae. **B,** Cadaver dissection photo showing many of the structures identified in the diagram. Note the fusion of the fibrous pericardium with the diaphragm.

jections of myocardial tissue. These muscular projections are called *trabeculae.* Specialized folds or pockets formed by the endocardium make up the functional components of the major valves that regulate the flow of blood through the chambers of the heart.

Chambers of the Heart

The interior of the heart is divided into four cavities, or heart chambers (Figure 18-6). The two upper chambers are called *atria* (singular, *atrium*), and the two lower chambers are called **ventricles.** The left chambers are separated from the right chambers by an extension of the heart wall called the *septum.*

Atria

The two superior chambers of the heart—the **atria**—are often called the "receiving chambers" because they receive blood from vessels called **veins.** Veins are the large blood vessels that return blood from various tissues to the heart so that the blood can be pumped out to tissues again. Figure 18-7 shows how the atria alternately relax and contract to receive blood, then push it into the lower chambers. Because the atria need not generate great pres-

sure to move blood such a small distance, the myocardial wall of each atrium is not very thick.

If you look at Figure 18-3, *A,* you will notice that part of each atrium is labeled as an *auricle.* The term *auricle* (meaning "little ear") refers to the earlike flap protruding from each atrium. Thus the auricles are *part of* the atria. The terms *auricles* and *atria* should not be used synonymously.

Ventricles

The **ventricles** are the two lower chambers of the heart. Because the ventricles receive blood from the atria and pump blood out of the heart into arteries, the ventricles are considered to be the primary "pumping chambers" of the heart. Because more force is needed to pump blood such a distance, the myocardium of each ventricle is thicker than the myocardium of either atrium. The myocardium of the left ventricle is thicker than that of the right ventricle because the left ventricle pushes blood through most vessels of the body, whereas the right ventricle pushes blood only through the vessels that serve the gas exchange tissues of the lungs.

The pumping action of the heart chambers is summarized in Figure 18-7 and described further in Chapter 19.

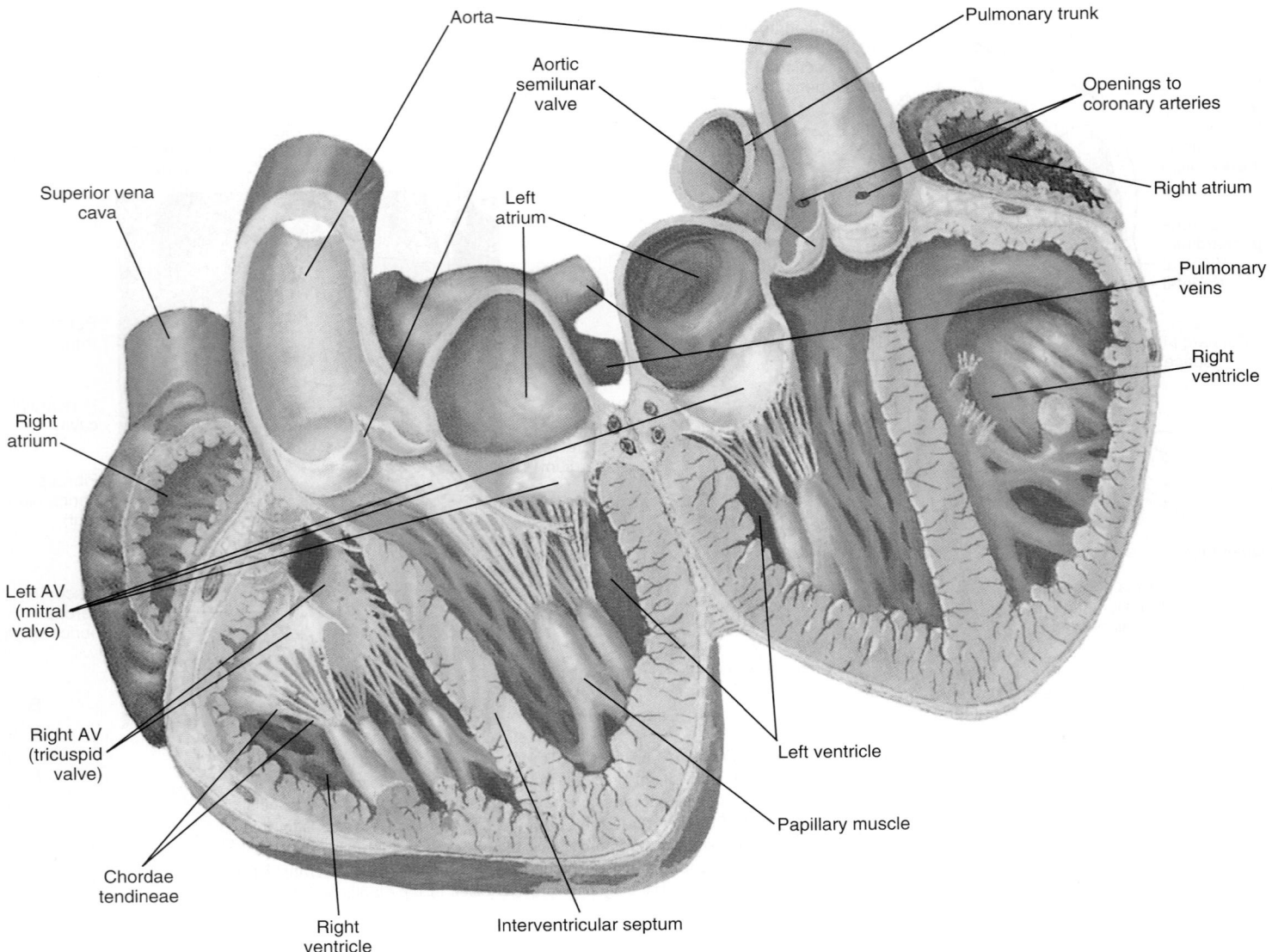

Figure 18-6 *Interior of the heart.* This illustration shows the heart as it would appear if it were cut along a frontal plane and opened like a book. The front portion of the heart lies to the reader's right; the back portion of the heart lies to the reader's left. The four chambers of the heart—two atria and two ventricles—are easily seen. *AV,* Atrioventricular.

Valves of the Heart

The heart valves are mechanical devices that permit the flow of blood in one direction only. Four sets of valves are of importance to the normal functioning of the heart (Figure 18-8; see Figure 18-7). Two of these, the **atrioventricular (AV) valves,** guard the openings between the atria and the ventricles (atrioventricular orifices). The atrioventricular valves are also called **cuspid valves.** The other two heart valves, the **semilunar (SL) valves,** are located where the pulmonary artery and the aorta arise from the right and left ventricles, respectively.

Atrioventricular Valves

The atrioventricular valve guarding the right atrioventricular orifice consists of three flaps (cusps) of endocardium. The free edge of each flap is anchored to the *papillary muscles* of the right ventricle by several cordlike structures called **chordae tendineae.** Be-

cause the right atrioventricular valve has three flaps, it is also called the **tricuspid valve.** The valve that guards the left atrioventricular orifice is similar in structure to the right atrioventricular valve, except that it has only two flaps and is therefore also called the **bicuspid** or, more commonly, the **mitral valve.**

The construction of both atrioventricular valves allows blood to flow from the atria into the ventricles but prevents it from flowing back up into the atria from the ventricles. Ventricular contraction forces the blood in the ventricles hard against the valve flaps, closing the valves and thereby ensuring the movement of the blood upward into the pulmonary artery and aorta as the ventricles contract (see Figure 18-7).

Semilunar Valves

The semilunar valves consist of half-moon–shaped flaps growing out from the lining of the pulmonary artery and aorta. The semi-

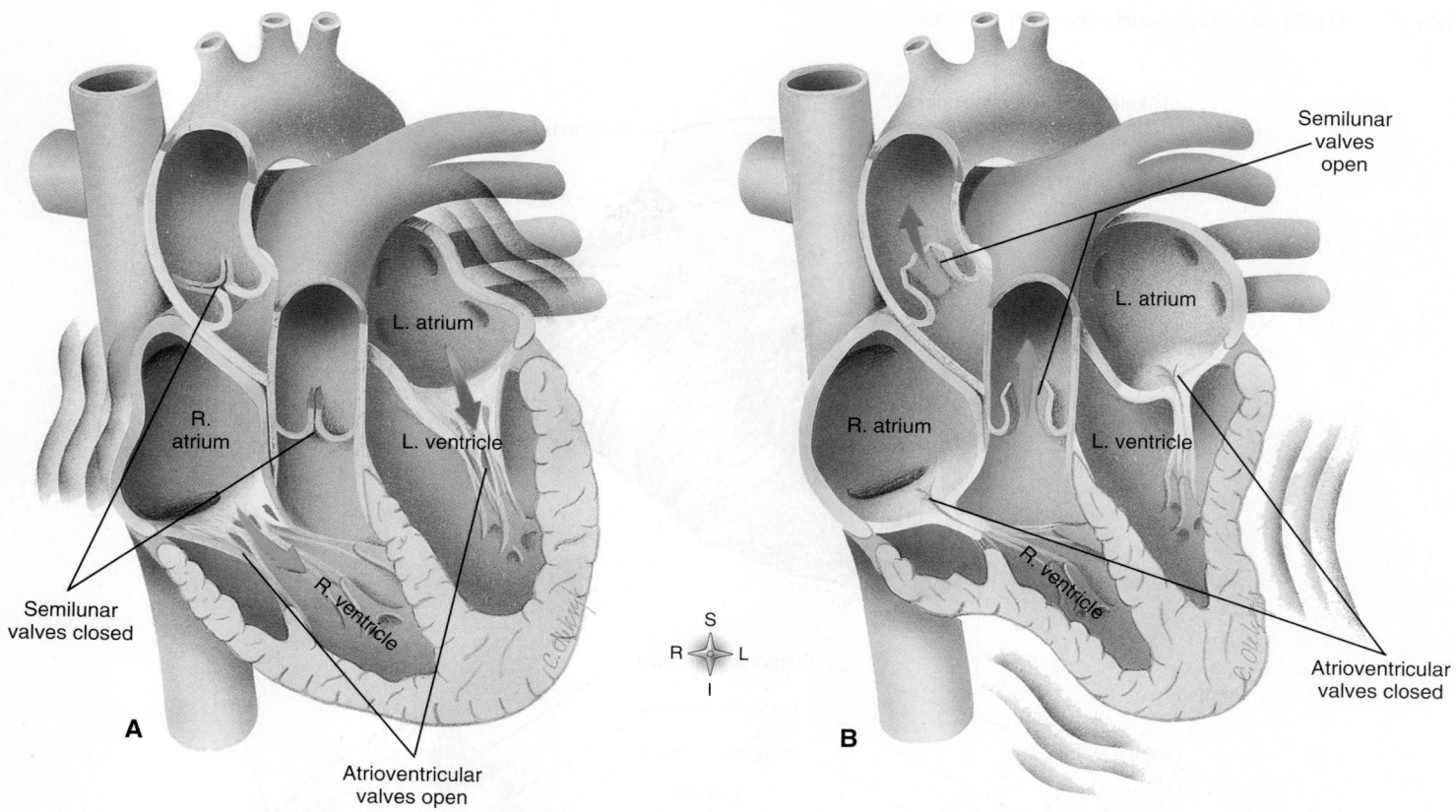

L. atrium

R. atrium

R. ventricle

L. ventricle

Semilunar valves open

Atrioventricular valves closed

R. atrium

L. atrium

R. ventricle

L. ventricle

Semilunar valves closed

Atrioventricular valves open

S

R · L

I

A

B

Figure 18-7 *Chambers and valves of the heart.* These illustrations depict the action of the heart chambers and valves when the atria contract, **A,** and when the ventricles contract, **B.**

lunar valve at the entrance of the pulmonary artery (pulmonary trunk) is called the *pulmonary semilunar valve.* The semilunar valve at the entrance of the aorta is called the *aortic semilunar valve.* When these valves are closed, as in Figures 18-7, *A,* and 18-8, *B,* blood fills the spaces between the flaps and the vessel wall. Each flap then looks like a tiny, filled bucket. Inflowing blood smoothes the flaps against the blood vessel walls, collapsing the buckets and thereby opening the valves (see Figures 18-7, *B,* and 18-8, *C*). Closure of the semilunar valves, as of the atrioventricular valves, simultaneously prevents backflow and ensures forward flow of blood in places where there would otherwise be considerable backflow. Whereas the atrioventricular valves prevent blood from flowing back up into the atria from the ventricles, the semilunar valves prevent it from flowing back down into the ventricles from the aorta and pulmonary artery.

Skeleton of the Heart

Figure 18-8 shows the fibrous structure that is often called the *skeleton of the heart.* It is a set of connected rings that serve as a semirigid support for the heart valves (on the inside of the rings) and for the attachment of cardiac muscle of the myocardium (on the outside of the rings). The skeleton of the heart also serves as an electrical barrier between the myocardium of the atria and the myocardium of the ventricles.

Surface Projection

When listening to the sounds of the heart on the body surface, as with a stethoscope, one must have an idea of the relationship between the valves of the heart and the surface of the thorax. Figure

18-9 indicates the surface relationship of the four heart valves and other features of the heart. It is important to remember, however, that considerable variation within the normal range makes a precise "surface projection" outline of the heart's structure on the chest wall difficult.

Flow of Blood through the Heart

To understand the functional anatomy of the heart and the rest of the cardiovascular system, one should be able to trace the flow of blood through the heart. As we take you through one complete pass through the right side of the heart, then the left side of the heart, trace the path of blood flow with your finger, using Figure 18-7. We can trace the path of blood flow through the right side of the heart by beginning in the right atrium. From the right atrium, blood flows through the right atrioventricular (tricuspid) valve into the right ventricle. From the right ventricle, blood flows through the pulmonary semilunar valve into the first portion of the pulmonary artery, the pulmonary trunk. The pulmonary trunk branches to form the left and right pulmonary arteries, which conduct blood to the gas exchange tissues of the lung. From there, blood flows through pulmonary veins into the left atrium.

We can begin to trace the path of blood flow through the left side of the heart from the left atrium. From the left atrium, blood flows through the left atrioventricular (mitral) valve into the left ventricle. From the left ventricle, blood flows through the aortic semilunar valve into the aorta. Branches of the aorta supply all the tissues of the body except the gas-exchange tissues of the lungs. Blood leaving the head and neck tissues empties into the superior vena cava, and blood leaving the lower body empties into

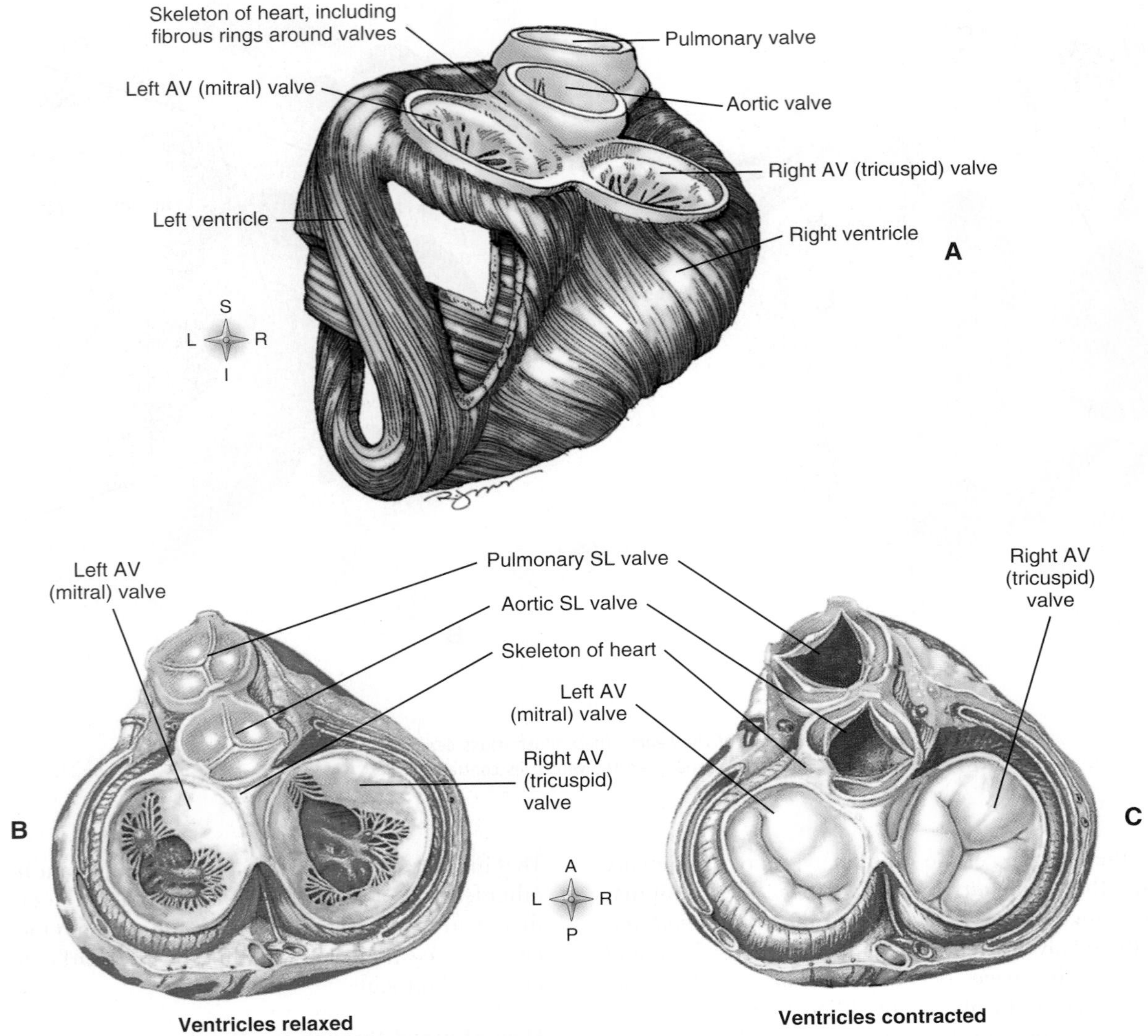

Skeleton of heart, including
fibrous rings around valves

Pulmonary valve

Left AV (mitral) valve

Aortic valve

Right AV (tricuspid) valve

Left ventricle

Right ventricle

A

Left AV
(mitral) valve

Pulmonary SL valve

Right AV
(tricuspid)
valve

Aortic SL valve

Skeleton of heart

Left AV
(mitral) valve

Right AV
(tricuspid)
valve

B

C

Ventricles relaxed

Ventricles contracted

Figure 18-8 *Structure of the heart valves.* A, This posterior view shows part of the ventricular myocardium
with the heart valves still attached. The rim of each heart valve is supported by a fibrous structure, called
the skeleton of the heart, that encircles all four valves. **B,** This figure shows the heart valves as if the figure
in **A** is viewed from above. Note that the semilunar (SL) valves are closed and the atrioventricular (AV) valves
are open, as when the atria are contracting (compare with Figure 18-7, *A*). **C,** This figure is similar to
B, except that the semilunar valves are open and the atrioventricular valves are closed, as when the ventricles are contracting (compare with Figure 18-7, *B*).

the inferior vena cava. Both large vessels conduct blood into the
right atrium, bringing us back to the point where we began.

QUICK CHECK

4. Name the layers of tissues that make up the pericardium.
5. What is the function of the pericardium?
6. Name the three layers of tissue that make up the wall of the
   heart. What is the function of each layer?
7. Name the four chambers of the heart and the valves associated with them.
8. How do atrioventricular valves differ from semilunar valves?

Blood Supply of Heart Tissue
Coronary Arteries

Myocardial cells receive blood by way of two small vessels, the
right and left **coronary arteries**. Because the openings into these
vitally important vessels lie behind flaps of the aortic semilunar
valve, they come off the aorta at its very beginning and are its first
branches. Both right and left coronary arteries have two main
branches, as shown in Figure 18-10.

More than one-half million Americans die every year from
coronary disease, and another 3.5 million or more are estimated
to suffer some degree of incapacitation. Knowledge about the distribution of coronary artery branches therefore has the utmost
practical importance. Here, then, are some principles about the
heart's own blood supply that are worth noting:

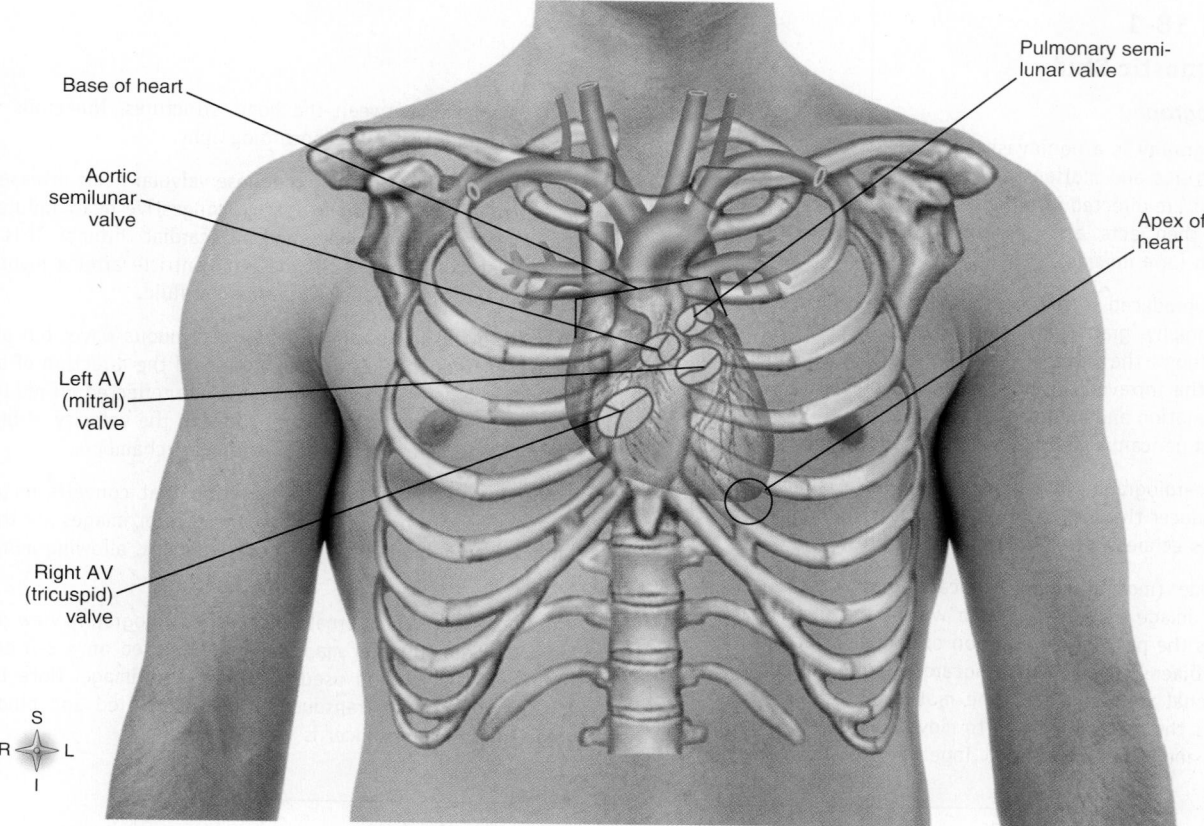

Figure 18-9 *Relation of the heart to the anterior wall of the thorax.* Valves of the heart are projected on the anterior thoracic wall. *AV,* Atrioventricular.

- Both ventricles receive their blood supply from branches of the right and left coronary arteries.
- Each atrium, in contrast, receives blood only from a small branch of the corresponding coronary artery.
- The most abundant blood supply goes to the myocardium of the left ventricle—an appropriate amount, because the left ventricle does the most work and so needs the most oxygen and nutrients delivered to it.
- The right coronary artery is dominant in about 50% of all hearts; the left coronary artery is dominant in about 20%; and in about 30%, neither right nor left coronary artery dominates.

Another fact about the heart's own blood supply—one of life-and-death importance—is that only a few connections, or anastomoses, exist between the larger branches of the coronary arteries. An **anastomosis** consists of one or more branches from the proximal part of an artery to a more distal part of itself or of another artery. Thus anastomoses provide detours in which arterial blood can travel if the main route becomes obstructed. In short, they provide collateral circulation to a part. This explains why the scarcity of anastomoses between larger coronary arteries looms so large as a threat to life. If, for example, a blood clot plugs one of the larger coronary artery branches, as it frequently does in coronary thrombosis or embolism, too little or no blood can reach some of the heart muscle cells. They become ischemic, in other words. Deprived of oxygen, metabolic function is impaired and cell survival is threatened. **Myocardial infarction** (death of ischemic heart

muscle cells) soon results. There is another anatomical fact, however, that brightens the picture somewhat: many anastomoses exist between very small arterial vessels in the heart, and, given time, new ones develop and provide collateral circulation to ischemic areas. In recent years, several surgical procedures have been devised to aid this process (see Mechanisms of Disease, p. 718).

Cardiac Veins

After blood has passed through **capillary** beds in the myocardium, it enters a series of cardiac veins before draining into the right atrium through a common venous channel called the *coronary sinus.* Several veins that collect blood from a small area of the right ventricle do not end in the coronary sinus but instead drain directly into the right atrium. As a rule, the cardiac veins (Figure 18-11) follow a course that closely parallels that of the coronary arteries.

Conduction System of the Heart

Four structures—the **sinoatrial node, atrioventricular node, atrioventricular bundle,** and **Purkinje fibers** *(subendocardial fibers)*—compose the conduction system of the heart (Figure 18-12). Each of these structures consists of cardiac muscle modified enough in structure to differ in function from ordinary cardiac muscle. The main specialty of ordinary cardiac muscle is contraction. In this, it is like all muscle, and like all muscle, ordinary cardiac muscle can also conduct impulses. However, conduction alone is the specialty of the modified cardiac muscle that composes the conduction system structures.

BOX 18-1
Diagnostic Study

Echocardiography

Echocardiography is a noninvasive technique for evaluating the internal structures and motions of the heart and great vessels. Ultrasound beams are directed into the patient's chest by a transducer. The transducer then acts as a receiver of the ultrasonic waves, or "echoes," to form images.

The images produced from the echoes are transmitted to a monitor. Echocardiography graphically demonstrates overall cardiac performance. It shows the internal dimensions of the chambers, size and motion of the intraventricular septum and posterior left ventricular wall, valve motion and anatomy, direction of blood flow, and presence of increased pericardial fluid, blood clots, and cardiac tumors.

Three echocardiographic techniques are used in clinical practice. All use a transducer that emits ultrasonic pulses through the chest wall and receives echoes from the cardiac structures.

- M-mode (motion-mode) echocardiography produces an "ice-pick" image of a narrow area within the ultrasonic beam. It shows the position and motion of cardiac structures.
- Two-dimensional (2-D) echocardiography produces a cross-sectional view and real time motion of cardiac structures. It allows the ultrasonic beam to move quickly, showing the structures and lateral movement. Together this shows the spatial re-

lationship between the heart structures. Numerous views are possible with 2-D echocardiography.

Echocardiography is used to diagnose valvular heart disease, congenital heart disease, cardiomyopathy, congestive heart failure, pericardial disease, cardiac tumors, and intracardiac thrombi. It is also used to evaluate the function of the left ventricle after a myocardial infarction and the presence of pericardial fluid.

Doppler ultrasonography provides continuous waves but also uses a sound, or frequency ultrasound, to record the direction of blood flow through the heart. These sound waves are reflected off red blood cells as they pass through the heart, allowing the velocity of blood to be calculated as it travels through the heart chambers.

Color Doppler mapping is a variation that converts recorded flow frequencies into different colors. These color images are then superimposed on M-mode or 2-D echocardiograms, allowing more detailed evaluation of disorders.

The figure is a suprasternal notch echocardiography view of the aortic arch. Color Doppler mapping superimposed on a 2-D echocardiogram was the method used to obtain this image. Note that blood moving toward the transducer is viewed as red and blood moving away from the transducer is viewed as blue.

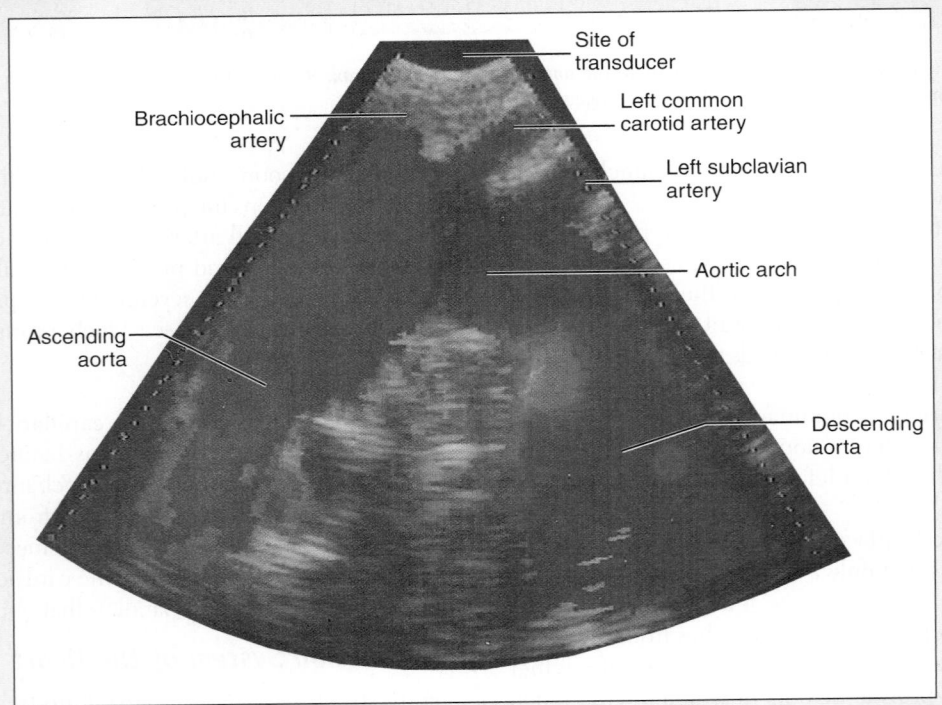

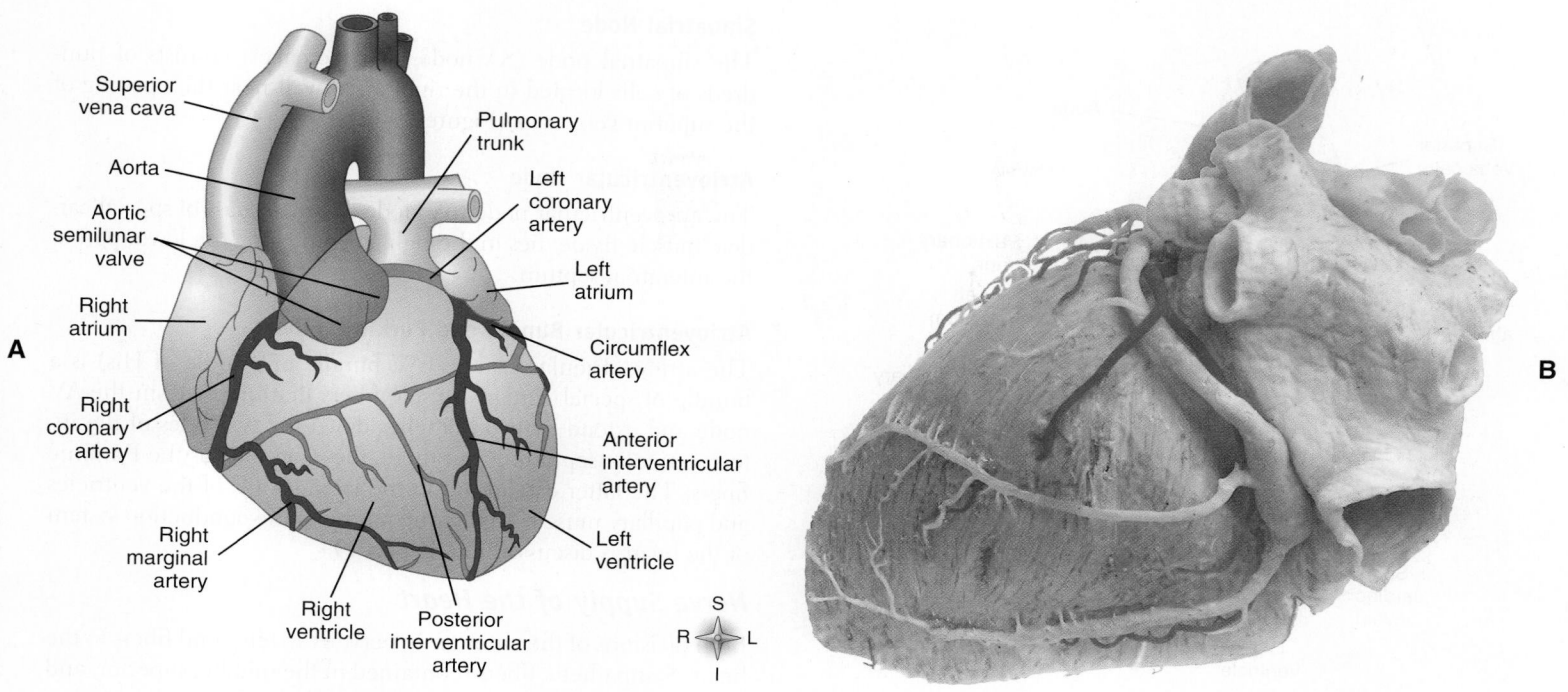

Figure 18-10 *Coronary arteries (anterior view).* **A,** Diagram showing the major coronary arteries. **B,** Heart with coronary arteries injected with red latex (the veins are white). In the heart diagram, vessels near the anterior surface are more darkly colored than vessels of the posterior surface seen through the heart. (Clinicians often refer to the interventricular arteries as *descending* arteries. Thus a cardiologist would refer to the anterior descending artery and an anatomist would refer to the same vessel as the anterior interventricular artery.)

BOX 18-2

Diagnostic Study

Cardiac Enzyme Studies

When the heart muscle is damaged, enzymes contained within the muscle cells are released into the bloodstream, causing increased serum levels. Studies of a number of cardiac enzymes, including creatine kinase (CK), lactic dehydrogenase (LD), and serum glutamic-oxaloacetic transaminase (SGOT), are useful in confirming a myocardial infarction (MI). In addition to absolute levels, the rate of increase of specific cardiac enzymes also varies over time after an infarction. For example, LD blood levels increase steadily 12 to 18 hours after an MI. It is this pattern of enzyme elevation that is of diagnostic importance. Cardiac enzyme studies are especially useful when they are viewed in relation to the complete medical examination, specific tests of heart function such as the EKG, and other tests for inflammation. For example, **C-reactive protein** is a well-known blood marker for inflammation. Research has now confirmed that elevated levels of this blood protein are highly suggestive of either active cardiac disease or early stages of progressive risk factor development. Many cardiologists are now measuring C-reactive protein levels in their heart patients as often as they check blood lipid levels or follow other known risk factors. The intent is to actively treat the disease before it becomes symptomatic or a heart attack actually occurs (also see Box 17-1 on erythrocyte sedimentation rate, p. 650). Enzyme determinations are also useful in observing the course of a myocardial infarction once it has occurred and in detecting an extension of it. Although not an enzymatic determination, the **troponins test** is another very valuable diagnostic aid. It is used to identify a specific biochemical marker present in cardiac disease. The presence of cardiac troponins is particularly useful in differentiating cardiac from noncardiac chest pain and in evaluating angina and suspected coronary ischemic syndrome.

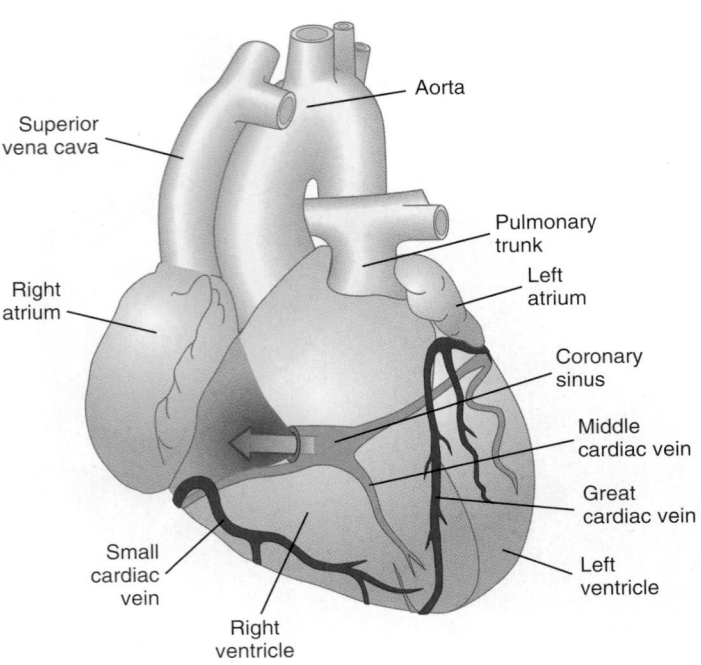

Figure 18-11 *Coronary veins (anterior view).* Vessels near the anterior surface are more darkly colored than vessels of the posterior surface seen through the heart.

Sinoatrial Node

The sinoatrial node (SA node, or pacemaker) consists of hundreds of cells located in the right atrial wall near the opening of the superior vena cava (Figure 18-12).

Atrioventricular Node

The atrioventricular node (AV node), a small mass of special cardiac muscle tissue, lies in the right atrium along the lower part of the interatrial septum.

Atrioventricular Bundle and Purkinje Fibers

The atrioventricular bundle (AV bundle or bundle of His) is a bundle of special cardiac muscle fibers that originate in the AV node and extend by two branches down the two sides of the interventricular septum. From there, they continue as the Purkinje fibers. The latter extend out to the lateral walls of the ventricles and papillary muscles. The functioning of the conduction system of the heart is discussed in Chapter 19.

Nerve Supply of the Heart

Both divisions of the autonomic nervous system send fibers to the heart. Sympathetic fibers (contained in the middle, superior, and inferior cardiac nerves) and parasympathetic fibers (in branches of the vagus nerve) combine to form cardiac plexuses located

Figure 18-12 *Conduction system of the heart.* Specialized cardiac muscle cells in the wall of the heart rapidly conduct an electrical impulse throughout the myocardium. The signal is initiated by the SA node (pacemaker) and spreads to the rest of the atrial myocardium and to the AV node. The AV node then initiates a signal that is conducted through the ventricular myocardium by way of the AV bundle (of His) and Purkinje fibers.

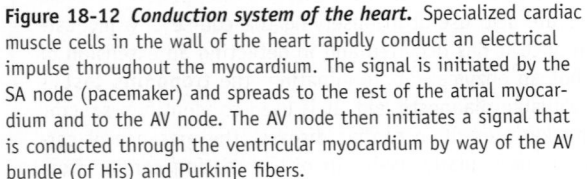
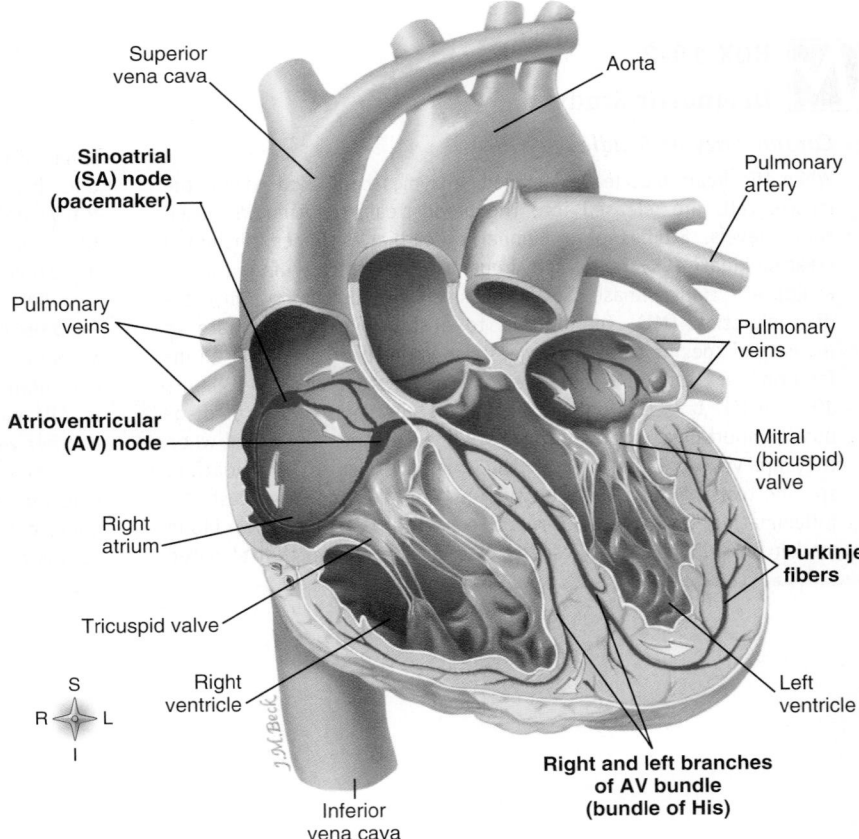

BOX 18-3
Diagnostic Study

Cardiac Nuclear Scanning

Cardiac nuclear scanning is a safe method of recognizing cardiac diseases. Basically, it is used to evaluate blood flow in coronary arteries or to evaluate ventricular function. The patient is given an intravenous (IV) injection of an appropriate radioactive substance. Shortly thereafter, a gamma-ray detector is placed over the heart while the patient is lying down. The detector records the image of the heart, and a photograph of that image is taken. The only discomfort for the patient is the initial injection.

Cardiac scanning is used for various clinical situations, particularly the following: (1) screening, evaluation, and surveillance of adults for old and new myocardial infarctions; (2) evaluating chest pain or unusual results from other heart tests; and (3) evaluation of bypass surgery or therapy.

These studies are nearly free of complications. As a result, noninvasive cardiac nuclear scanning procedures are both widely used and currently viewed as a vital part of the evaluation and diagnosis of many patients with cardiovascular disease.

close to the arch of the aorta. From the cardiac plexuses, fibers accompany the right and left coronary arteries to enter the heart. Here most of the fibers terminate in the sinoatrial node, but some end in the atrioventricular node and in the atrial myocardium. Sympathetic nerves to the heart are also called *accelerator nerves*. Vagus fibers to the heart serve as *inhibitory* or *depressor nerves*.

QUICK CHECK

9. Briefly describe the general structure and the function of the coronary circulation. Why is an understanding of the coronary circulation so critical to understanding major types of heart disease?
10. What is meant by the term *conduction system of the heart*?
11. What is a myocardial infarction?

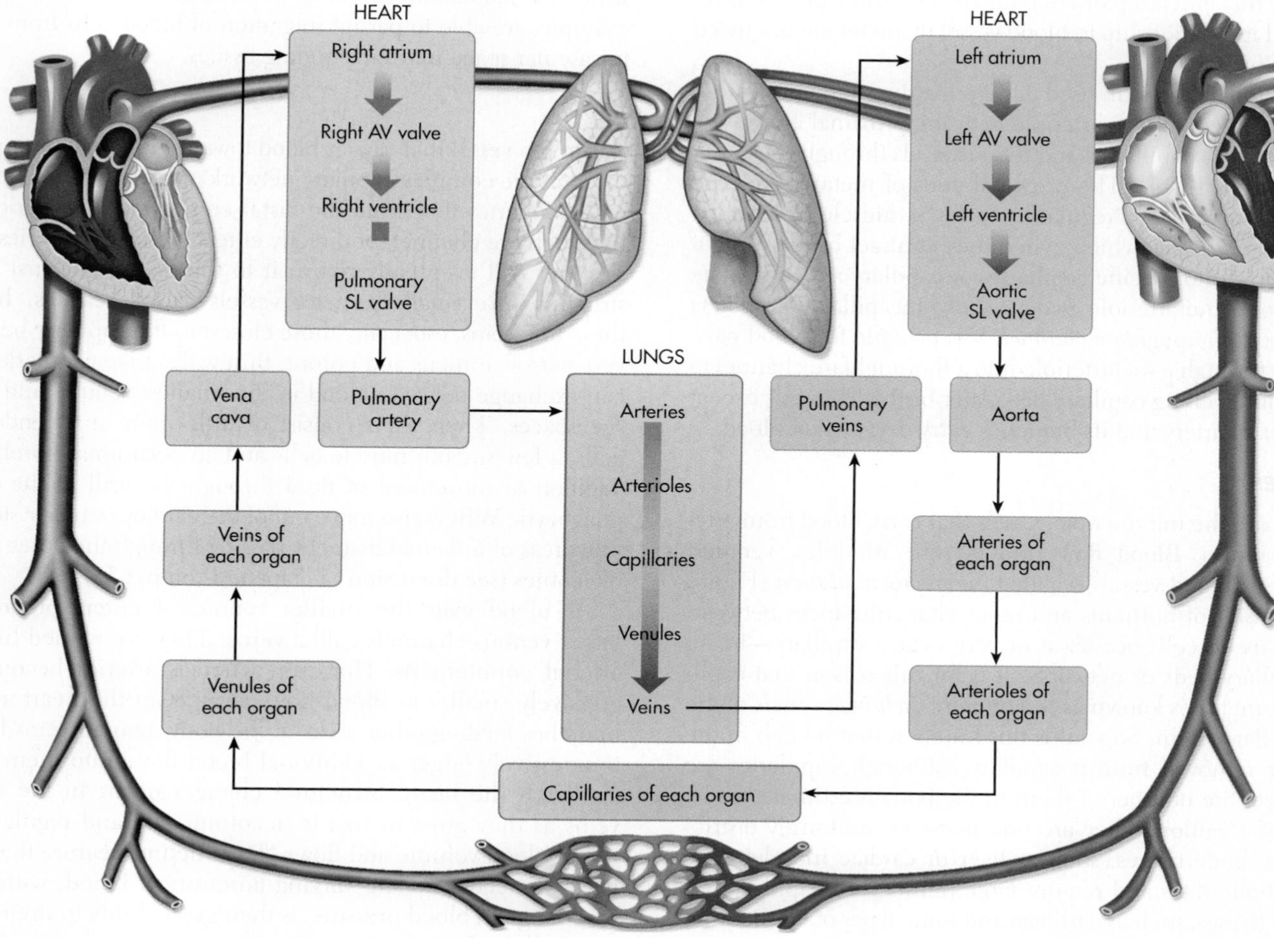

Figure 18-13 *Diagram of blood flow in the circulatory system.* In this illustration blood passes through a single capillary bed in the systemic circulation—the most typical route. After leaving the heart and entering the major arteries of the body, blood travels through arterioles, to a capillary bed, and then to the venules and veins before returning to the opposite side of the heart. Blood flow through portal systems, anastomoses, and arteriovenous shunts is described in the text.

BLOOD VESSELS
Types of Blood Vessels
Arteries

An artery is a vessel that carries blood away from the heart (Figure 18-14). **Elastic arteries** are the largest in the body and include the aorta and some of its major branches. As the name implies, elastic arteries can stretch without injury to accommodate the surge of blood that is forced into them when the heart contracts and then recoil when the ventricles relax.

Muscular or *distributing* **arteries** carry blood farther away from the heart to specific organs and areas of the body. They are smaller in diameter than elastic arteries. However, the muscular layer in the wall of these vessels is proportionately thicker. Like elastic arteries, muscular arteries are given names. Examples include the brachial, gastric, and superior mesenteric arteries.

Arterioles, also called *resistance vessels*, are the smallest arteries. They are not named individually, but, as a group, they are critically important in regulating blood flow throughout the body. They function in this way by variable contraction of smooth muscle in their walls, which in turn increases resistance to blood flow and helps regulate blood pressure as well as determine the quantity of blood that enters a particular organ. The concept of flow resistance and its relationship to blood vessel diameter are discussed further in Chapter 19.

Metarteriole is the term used to describe the short connecting vessel that connects a true arteriole with the proximal end of between 20 and 100 capillaries and then extends through the capillary bed (Figure 18-15). The proximal ends of metarterioles are encircled by specialized "regulatory valves"—muscle cells called *precapillary sphincters*, which, when they contract or relax, regulate blood flow into specific capillaries or capillary networks. The distal end of a metarteriole is devoid of precapillary sphincters and is called a *thoroughfare channel*. It is possible for blood passing directly through a metarteriole into a thoroughfare channel to bypass the intervening capillary bed. After birth all arteries except the pulmonary artery and its branches carry oxygenated blood.

Capillaries

Capillaries are the microscopic vessels that carry blood from arterioles to venules. Blood flow through the arterioles, venules, and these tiny blood vessels is called the *microcirculation* (Figure 18-15). Transfer of nutrients and other vital substances between blood and tissue cells occurs at or very near a capillary—in so-called capillary beds or networks. It is for this reason that capillaries are sometimes known as the *primary exchange vessels* of the cardiovascular system. So vital is this function that no cell in the body is far removed from a capillary. Although capillaries are small in size, the number of them in the body is estimated to be in excess of 1 billion. They are not, however, uniformly distributed. Some body tissues, such as liver or cardiac muscle, have high metabolic rates and require large numbers of capillary vessels. Other tissues, such as cartilage and some types of epithelium, are avascular and lack capillary networks altogether.

True capillaries receive blood flowing out of metarterioles or other small arterioles. Precapillary sphincters regulate the volume of inflow of blood and its rate of passage through a true capillary. If the sphincter is "open" blood flows into the capillary; and, if the

sphincter is closed or partially closed, blood flow into the capillary bed decreases (Figure 18-15). Capillaries are often categorized into three groups by the ease of passage of substances through their walls or by structural differences that affect their permeability.

Continuous capillaries have a continuous lining of endothelial cells with only small openings called intercellular clefts between them (Figure 18-16). This type of capillary is typically found in skeletal muscle, lung, and many types of connective tissue. **Fenestrated capillaries** also have intercellular clefts between their lining endothelial cells. But, in addition, they also have small "holes" or fenestrations through the plasma membrane of the endothelial cell itself. This unique structural adaptation allows for special function. Understanding the anatomy of these tiny vessels will be helpful in understanding the function of a number of specialized organs, such as the kidneys and small intestines, discussed later in the text. **Sinusoid** is the term used to describe a type of capillary that has a much larger lumen and more winding or tortuous course than other capillary vessels. The basement membrane that completely covers other capillaries is either absent or incomplete in sinusoids. In addition, the fenestrations present both between and within the endothelial lining cells are much larger than in other capillary types. The result is great porosity. Because of this structural adaptation, sinusoids in the bone marrow and liver, for example, are able to permit migration of blood cells from the intravascular space into surrounding tissues.

Veins

A vein is a vessel that carries blood toward the heart. After passing through the complex capillary network of vessels, blood from several capillaries flows into the distal end of the metarteriole, the thoroughfare channel, or directly enters the first of a series of vessels that will eventually return it to the heart. The first venous structures are small diameter vessels called **venules.** Initially, these tiny veins, especially those closest to the capillary bed, have very narrow lumens and porous, thin walls. Like capillaries, fluid can exchange between blood in the smallest venules and the tissue spaces. Their walls consist of little more than endothelial cells, a few smooth muscle cells and an occasional fibroblast. In addition to movement of fluid through the wall of the venule, phagocytic WBCs also move out of the cardiovascular system and into areas of inflamed tissue by passing through the pores in walls of venules (see discussion of diapedesis on p. 655).

As blood exits the smaller venules it enters progressively larger venous channels called **veins.** They are named like their arterial counterparts. However, whereas arteries become progressively smaller as blood flows away from the heart and into branches feeding other areas of the body, named veins become progressively larger as additional blood flows into them as they approach the heart. Structural changes occur in the walls of veins as they grow in size to accommodate and regulate additional blood volume and flow. The structural feature that allows them to accommodate varying amounts of blood, with almost no change in blood pressure, is their great ability to stretch. Ease of stretch is also called *capacitance*, and for this reason the veins are often referred to as the *capacitance vessels* of the cardiovascular system. This feature permits veins to serve as reservoirs for blood as well as conduits for its passage back to the heart. Another structural adaptation in veins that has an important func-

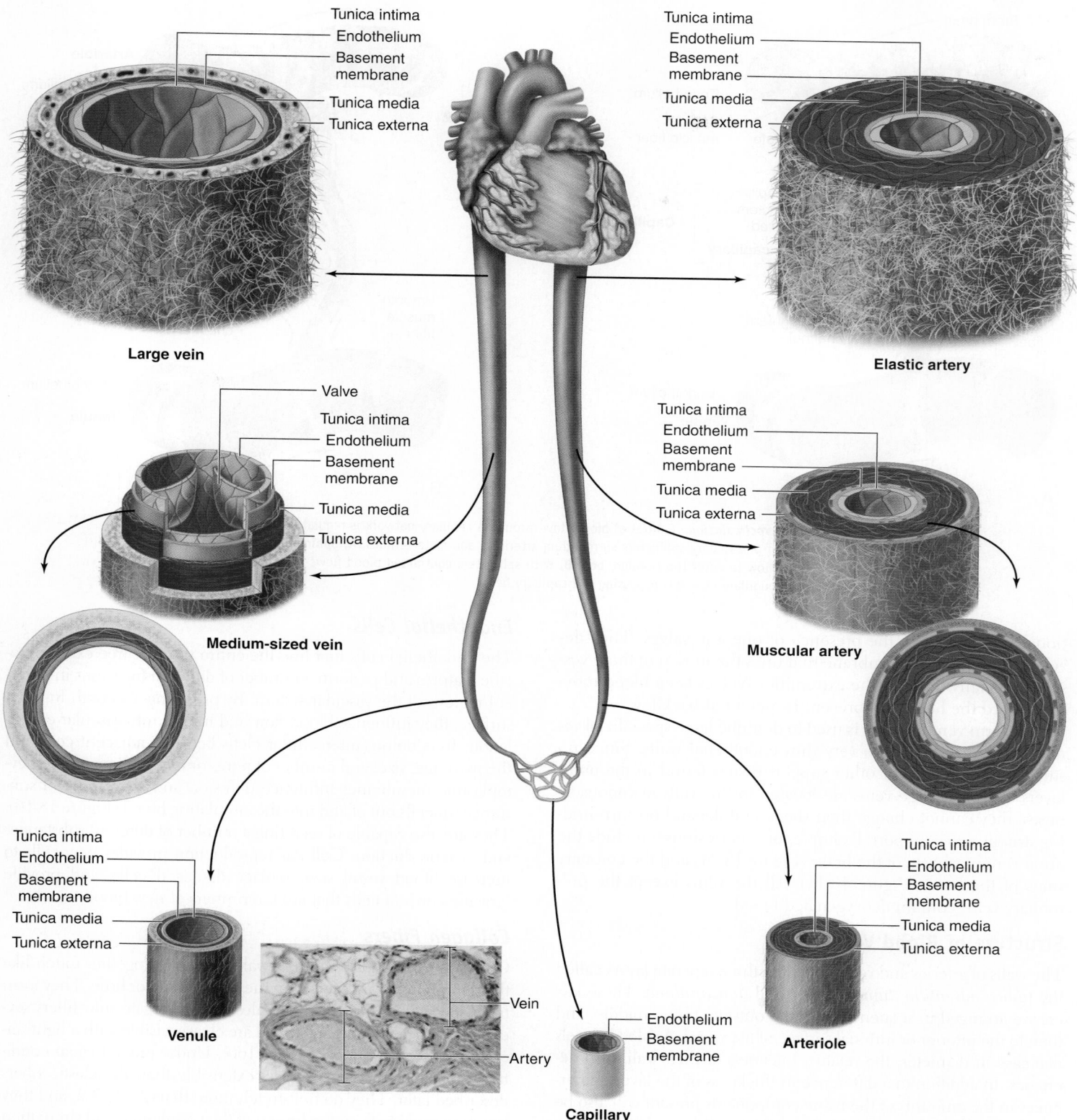

Figure 18-14 *Structure of blood vessels.*

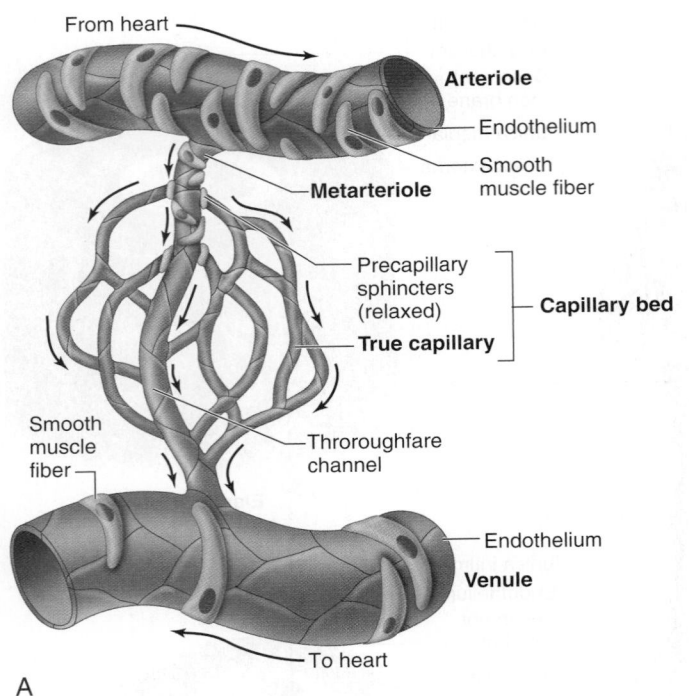

A

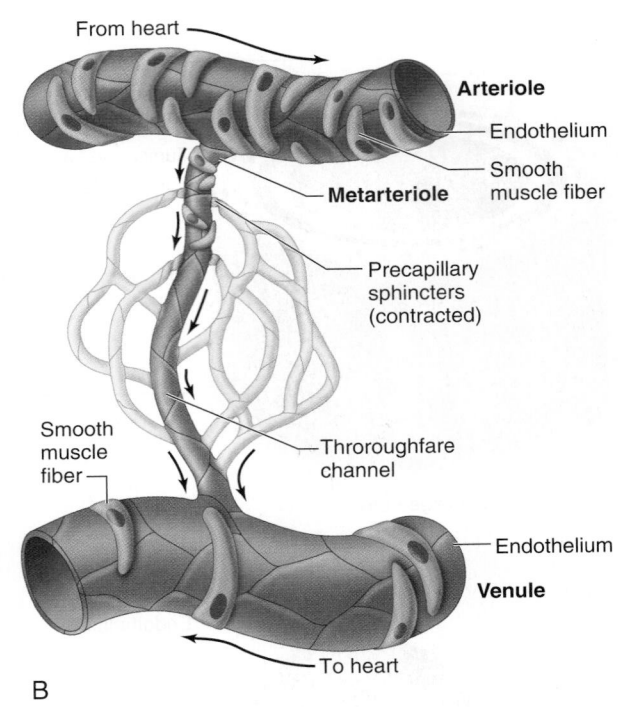

B

Figure 18-15 *Microcirculation.* Control of blood flow through a capillary network is regulated by the relative contraction of precapillary sphincters surrounding arterioles and metarterioles. **A,** Sphincters are relaxed, permitting blood flow to enter the capillary bed. **B,** With sphincters contracted blood flows from metarteriole directly into thoroughfare channel, bypassing the capillary bed.

tional significance is the presence of one-way valves. They develop from the thin membrane that lines the lumen of these vessels, especially those in the extremities. Valves keep blood moving toward the heart and prevent its potential backflow.

The term **venous sinus** is used to describe large specialized venous structures that have very thin endothelial walls. Since no smooth muscle cells or other support tissues found in the outer layers of typical large veins are located in the walls of venous sinuses, they cannot change their shape and depend on surrounding structures for support. Examples of venous sinuses include the dural venous sinuses of the brain (Figure 18-24) and the coronary sinus of the heart (Figure 18-11). All the veins except the pulmonary veins contain deoxygenated blood.

Structure of Blood Vessels

The walls of arteries and veins consist of three separate layers called the *tunica adventitia, tunica media,* and *tunica intima.* These layers are arranged in sequence from the outside, to the middle, and then to the interior or luminal surface of the vessel. As blood vessels decrease in diameter, the relative thickness of their walls also decreases. In addition to a difference in thickness of the layers, differences in the amounts of the tissue components present will also be present in different types of arteries and veins (Figure 18-14). However, regardless of how thick any of the various layers might be, four types of tissue "building blocks" are commonly present: (1) lining *endothelial cells,* (2) *collagen fibers,* (3) *elastic fibers,* and (4) *smooth muscle cells.* When considering the generalized structure of the walls of vessels in the circulatory system, capillaries are the exception. They consist of a single layer of endothelial cells (tunica intima) surrounded by a basement membrane (Figure 18-16).

Endothelial Cells

The endothelial cells that line the entire vascular tree exhibit specific features and perform a number of different functions in different regions of the vascular system. By providing a smooth luminal surface, they influence blood flow and inhibit intravascular coagulation. In addition, intercellular clefts between adjacent cells and the presence, size, and number of pores or fenestrations in their cytoplasmic membranes influence diffusion and movement of substances or cells out of and into the circulating blood (Figure 18-16). They are also capable of secreting a number of different substances and of reproduction. Cellular reproduction provides new cells to increase blood vessel size, replace damaged cells, and provide growing cords of cells that are forerunners of new blood vessels.

Collagen Fibers

Collagen fibers in the vascular wall are woven together much like the reinforcing strands found in the wall of a tire or hose. They form from specialized protein molecules that aggregate into fibers several micrometers in diameter and are clearly visible with a light microscope (see Figure 5-14 on p. 160). Under physiological conditions collagen fibers are far less extensible than the elastic fibers described later. They do not stretch more than 2% to 3%, and they function more to keep the lumen of the vessel open and strengthen the wall than to contribute to overall tension or recoil ability.

Elastic Fibers

Individual elastic fibers are composed largely of an insoluble protein polymer called *elastin* and are quite small (0.1 to 1 μm in diameter). Once secreted into the extracellular matrix, they form into a rubber-like network that is highly elastic and capable of

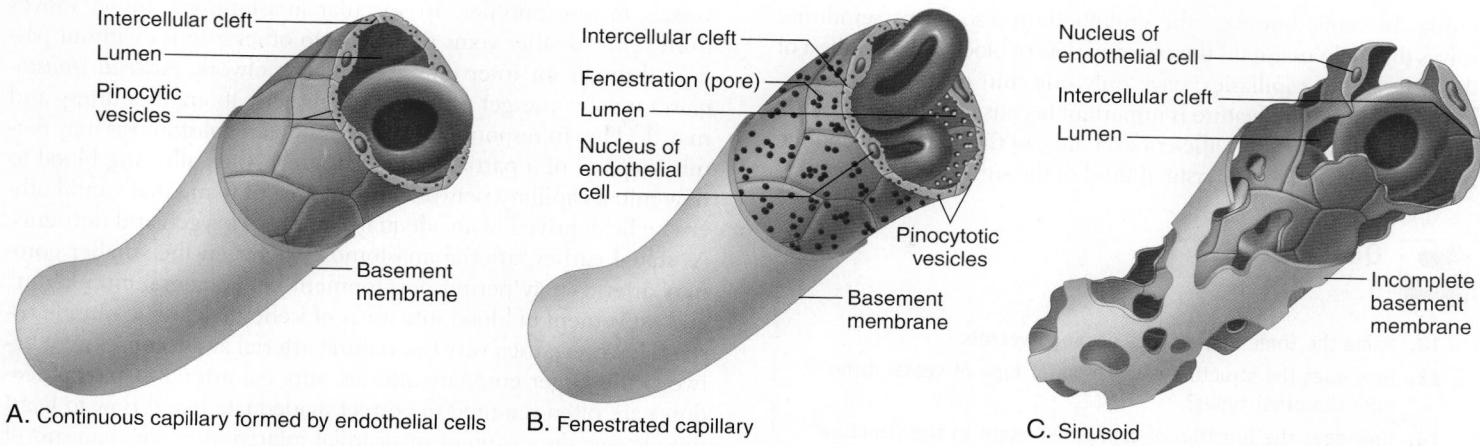

A. Continuous capillary formed by endothelial cells   B. Fenestrated capillary   C. Sinusoid

Figure 18-16 *Types of capillaries.* **A,** Continuous capillary. Note the presence of clefts only between adjacent endothelial cells. **B,** Fenestrated capillary. In addition to intercellular clefts, fenestrations (pores) exist in the plasma membranes of endothelial cells. **C,** Sinusoid. In addition to large intercellular clefts and cellular fenestrations, the basement membrane is incomplete or absent.

stretching more than 100% under physiological conditions (see Figure 5-1 on p. 148). In large elastic arteries, especially, elastic fibers are organized or arranged in concentric, almost circular patterns. Elastic fibers allow for recoil after distention. This property of elastic fibers plays an important role in maintaining *passive tension* in the vessels of the cardiovascular system. This type of tension is required to maintain normal blood pressure levels throughout the cardiac cycle—a process that will be discussed in Chapter 19.

Smooth Muscle Fibers

Smooth muscle cells are found in the wall of all segments of the vascular system except capillaries. They are most numerous in elastic and muscular arteries and exert *active tension* in these vessels when contracting.

Outer Layer

The walls of the larger blood vessels, the arteries and veins, have three layers (Figure 18-14 and Table 18-1). The outermost layer is called the **tunica adventitia** (externa). This Latin name literally means "coat that comes first," referring to how it is found during the dissection of a vessel. The tunica adventitia, also called *tunica externa*, is made of strong, flexible fibrous connective tissue. This

layer helps hold vessels open and prevents tearing of the vessel walls during body movements. In veins, the tunica adventitia is the thickest of the three layers of the venous wall. In arteries, it is usually a little thinner than the middle layer of the arterial wall.

Middle Layer

The middle layer, or **tunica media** (Latin for "middle coat"), is made of a layer of smooth muscle tissue sandwiched together with a layer of elastic connective tissue. Some anatomists consider the elastic portion of the tunica media to be distinct enough to call it a separate *elastic layer* of the wall. The encircling smooth muscles of the tunica media permit changes in blood vessel diameter. The smooth muscle tissue of the tunica media is innervated by autonomic nerves and supplied with blood by tiny *vasa vasorum* ("vessels of vessels"). As a rule, arteries have a thicker layer of smooth muscle than do veins.

Inner Layer

The innermost layer of a blood vessel is called the *tunica intima*—Latin for "innermost coat." The tunica intima is made up of endothelium that is continuous with the endothelium that lines the heart. In arteries, the endothelium provides a completely smooth

Table 18-1	Structure of Blood Vessels		
TYPE OF VESSEL	**TUNICA INTIMA (ENDOTHELIUM)**	**TUNICA MEDIA (SMOOTH MUSCLE; ELASTIC CONNECTIVE TISSUE)**	**TUNICA ADVENTITIA (FIBROUS TYPE OF VESSEL CONNECTIVE TISSUE)**
Arteries	Smooth lining	Allows constriction and dilation of vessels; thicker than in veins; muscle innervated by autonomic fibers	Provides flexible support that resists collapse or injury; thicker than in veins; thinner than tunica media
Veins	Smooth lining with semilunar valves to ensure one-way flow	Allows constriction and dilation of vessels; thinner than in arteries; muscle innervated by autonomic fibers	Provides flexible support that resists collapse or injury; thinner than in arteries; thicker than tunica media
Capillaries	Makes up entire wall of capillary; thinness permits ease of transport across vessel wall	(Absent)	(Absent)

lining. In veins, however, the endothelium also forms semilunar valves that help maintain the one-way flow of blood. The smallest of the vessels, the capillaries, have only one thin coat: the endothelium. This structure feature is important because the thinness of the capillary wall allows for efficient exchange of materials between the blood plasma and the interstitial fluid of the surrounding tissues.

QUICK CHECK

12. Name the three major types of blood vessels.
13. How does the structure of each major type of vessel differ from the other types?
14. How does the function of capillaries relate to the structure of their walls?

MAJOR BLOOD VESSELS

Circulatory Routes

The term *circulation of blood* suggests its meaning, namely, blood flow through vessels arranged to form a circuit or circular pattern. Blood flow from the heart (left ventricle) through blood vessels to all parts of the body (except the lungs) and back to the heart (to the right atrium) is spoken of as **systemic circulation** (Figure 18-13). The left ventricle pumps blood into the ascending aorta. From here it flows into arteries that carry it into the various tissues and organs of the body. Within each structure, blood moves, as indicated in Figure 18-13, from arteries to arterioles to capillaries. Here the vital two-way exchange of substances occurs between the blood and cells. Blood flows next out of each organ by way of its venules and then its veins to drain eventually into the inferior or superior vena cava. These two great veins of the body return venous blood to the heart (to the right atrium) to complete the systemic circulation. But the blood does not quite come full circle back to its starting point, the left ventricle. To do this and start on its way again, it must first flow through another circuit, the **pulmonary circulation.** Observe in Figure 18-13 that venous blood moves from the right atrium to the right ventricle to the pulmonary artery to lung arterioles and capillaries. Here, exchange of gases between blood and air takes place, converting deoxygenated blood to oxygenated blood. This oxygenated blood then flows on through lung venules into four pulmonary veins and returns to the left atrium of the heart. From the left atrium it enters the left ventricle to be pumped again through the systemic circulation.

Movement of blood as shown in Figure 18-13 follows a general rule of thumb often used when studying the circulatory system. Namely, that blood passes through only one capillary network in the systemic circulation from the time it leaves the heart until it returns. Although this is certainly true in most instances, two important exceptions to the rule do occur. In a **portal system,** blood flowing through the systemic circulation passes through two consecutive capillary beds rather than one. The details of the **hepatic portal system** involve movement of blood through the liver. This important circulatory route will be discussed later in the chapter. The term **vascular anastomosis** is used to describe a second type of exception. It involves the direct connection or merger of blood vessels to one another. In vascular anastomoses, blood moves from veins to other veins or arteries to other arteries without passage through an intervening capillary network. *Arterial anastomoses* involve merger of one artery directly to another artery and may develop in response to disease. Such an anastomosis may permit "bypass" of a partially blocked artery, thus allowing blood to flow into a capillary network and an area of tissue that would otherwise be deprived of an adequate supply of oxygen and nutrients. As stated earlier, arterial anastomoses between the smaller coronary arteries may permit development of "collateral circulation" and movement of blood into areas of ischemic cardiac muscle tissue. However, since very few natural arterial anastomoses exist between the larger coronary arteries, surgical arterial bypass procedures are often required to correct inadequate blood flow to heart muscle (see discussion of myocardial infarction in Mechanisms of Disease, p. 719). *Venous anastomoses*, which are much more common, involve direct linkage between different veins. Multiple venous drainage routes from an organ or body area provide a safety mechanism if occlusion of one venous return route should occur. This is especially true of the deep veins. Catastrophic consequences from a so-called *deep venous thrombosis* (DVT) may be prevented or lessened because of these anastomoses. **Arteriovenous anastomoses** or **shunts** occur when blood flows from an artery directly into a vein without passing through a capillary bed. Heat loss occurs when blood passes through capillary beds in the skin. In cases of hypothermia, heat loss can be avoided by shunting of blood directly from skin arteries to veins without permitting it to pass through capillary beds near the skin surface (see Figure 6-12 on p. 209).

Systemic Circulation

The systemic circulatory route includes most of the vessels of the body. The following paragraphs, tables, and illustrations outline some of the major vessels of the systemic circulation as well as tips for understanding this circulatory route.

Systemic Arteries

Locate the arteries listed in Table 18-2 (see also Figure 18-13 and Figures 18-17 to 18-22). You may find it easier to learn the names of blood vessels and the relation of the vessels to each other from diagrams and tables than from narrative descriptions.

General Principles about Arteries

As you learn the names of the main arteries, keep in mind that these are only the major pipelines distributing blood from the heart to the various organs and that, in each organ, the main artery resembles a tree trunk in that it gives off numerous branches that continue to branch and rebranch, forming ever smaller vessels (arterioles), which also branch, forming microscopic vessels, the capillaries. In other words, most arteries eventually diverge into capillaries. Arteries of this type are called **end-arteries.** Important organs or areas of the body supplied by end-arteries are subject to serious damage or death in occlusive arterial disease. As an example, permanent blindness results when the central artery of the retina, an end-artery, is occluded. Therefore occlusive arterial disease such as atherosclerosis is of great concern in clinical medicine when it affects important organs having an end-arterial blood supply.

Table 18-2   Major Systemic Arteries

ARTERY*	REGION SUPPLIED	ARTERY*	REGION SUPPLIED
Ascending Aorta		**Descending Abdominal Aorta**	
Coronary arteries	Myocardium	VISCERAL BRANCHES	Abdominal viscera
Arch of Aorta		Celiac artery (trunk)	Abdominal viscera
BRACHIOCEPHALIC (INNOMINATE)	Head, upper extremity	Left gastric	Stomach, esophagus
Right subclavian	Head, upper extremity	Common hepatic	Liver
Right vertebral†	Spinal cord, brain	Splenic	Spleen, pancreas, stomach
Right axillary (continuation of subclavian)	Shoulder, chest, axillary region	Superior mesenteric	Pancreas, small intestine, colon
Right brachial (continuation of axillary)	Arm, hand	Inferior mesenteric	Descending colon, rectum
		Suprarenal	Adrenal (suprarenal) gland
Right radial	Lower arm, hand (lateral)	Renal	Kidney
Right ulnar	Lower arm, hand (medial)	Ovarian	Ovary, uterine tube, ureter
Superficial and deep palmar arches (formed by anastomosis of branches of radial and ulnar)	Hand, fingers	Testicular	Testis, ureter
		PARIETAL BRANCHES	Walls of abdomen
		Inferior phrenic	Inferior surface of diaphragm, adrenal gland
Digital	Fingers	Lumbar	Lumbar vertebrae, muscles of back
Right common carotid	Head, neck	Median sacral	Lower vertebrae
Right internal carotid†	Brain, eye, forehead, nose	COMMON ILIAC (FORMED BY TERMINAL BRANCHES OF AORTA)	Pelvis, lower extremity
Right external carotid†	Thyroid, tongue, tonsils, ear, etc.	EXTERNAL ILIAC	Thigh, leg, foot
LEFT SUBCLAVIAN	Head, upper extremity	Femoral (continuation of external iliac)	Thigh, leg, foot
Left vertebral†	Spinal cord, brain	Popliteal (continuation of femoral)	Leg, foot
Left axillary (continuation of subclavian)	Shoulder, chest, axillary region		
Left brachial (continuation of axillary)	Arm, hand	Anterior tibial	Leg, foot
		Posterior tibial	Leg, foot
Left radial	Lower arm, hand (lateral)	Plantar arch (formed by anastomosis of branches of anterior and posterior tibial arteries)	Foot, toes
Left ulnar	Lower arm, hand (medial)		
Superficial and deep palmar arches (formed by anastomosis of branches of radial and ulnar)	Hand, fingers		
		Digital	Toes
Digital	Fingers	INTERNAL ILIAC	Pelvis
LEFT COMMON CAROTID	Head, neck	Visceral branches	Pelvic viscera
Left internal carotid†	Brain, eye, forehead, nose	Middle rectal	Rectum
Left external carotid†	Thyroid, tongue, tonsils, ear, etc.	Vaginal	Vagina, uterus
Descending Thoracic Aorta		Uterine	Uterus, vagina, uterine tube, ovary
VISCERAL BRANCHES	Thoracic viscera	Parietal branches	Pelvic wall, external regions
Bronchial	Lungs, bronchi	Lateral sacral	Sacrum
Esophageal	Esophagus	Superior gluteal	Gluteal muscles
PARIETAL BRANCHES	Thoracic walls	Obturator	Pubic region, hip joint, groin
Intercostal	Lateral thoracic walls (rib cage)	Internal pudendal	Rectum, external genitalia, floor of pelvis
Superior phrenic	Superior surface of diaphragm	Inferior gluteal	Lower gluteal region, coccyx, upper thigh

*Branches of each artery are indented below its name.
†See text and/or figures for branches of the artery.

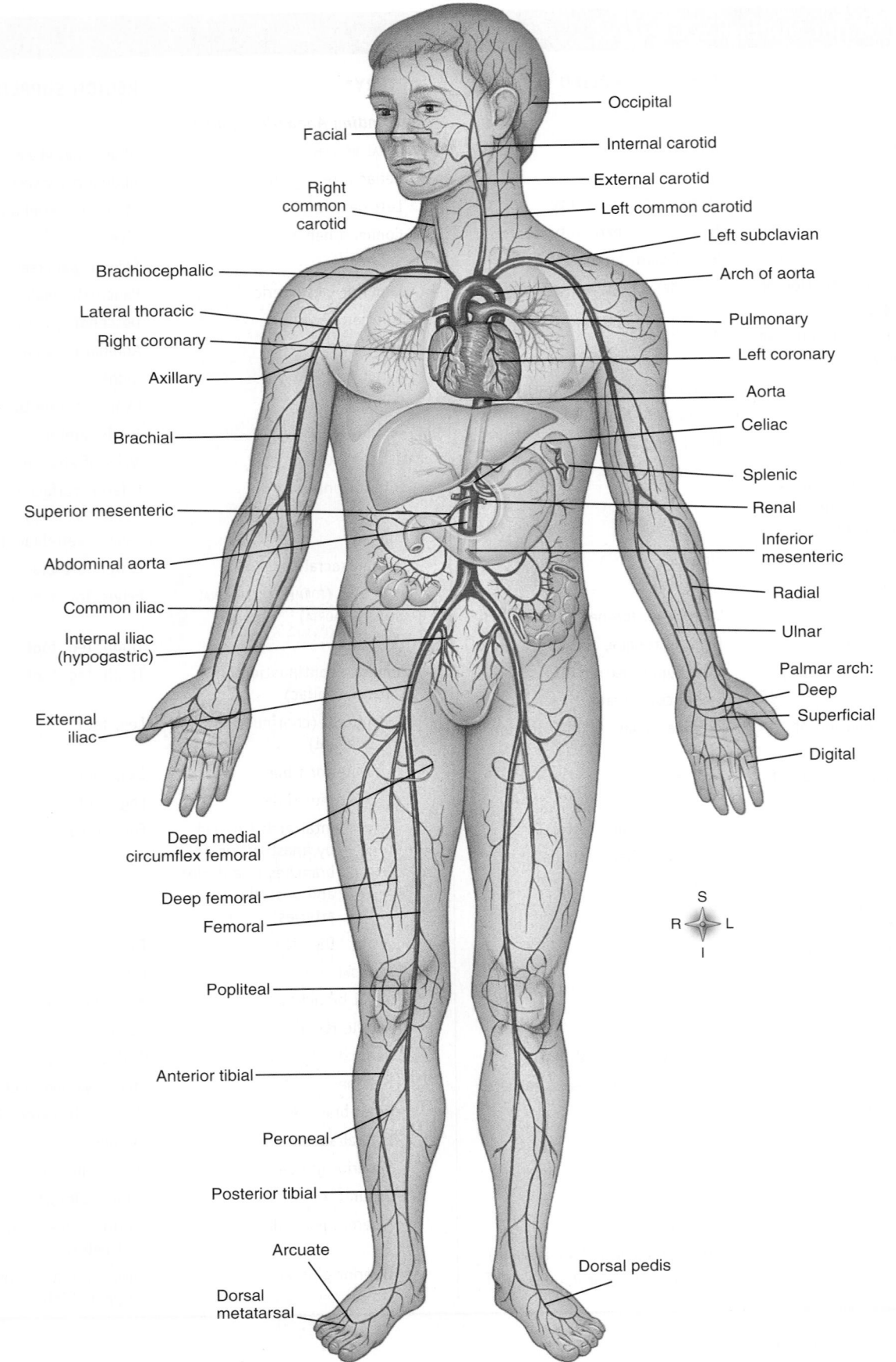

Figure 18-17 *Principal arteries of the body.*

A few arteries open into other branches of the same or other arteries. Such a communication was described earlier as an **arterial anastomosis.** Anastomoses, we have already noted, fulfill an important protective function in that they provide detour routes for blood to travel in the event of obstruction of a main artery. The incidence of arterial anastomoses increases as distance from the heart increases, and smaller arterial branches tend to anastomose more often than larger vessels. Examples of arterial anastomoses

are the palmar and plantar arches and the arterial circle of Willis at the base of the brain (Figure 18-20). Other examples are found around several joints, as well as in other locations.

Another general principle to remember as you study the systemic arteries is that the **aorta** is the major artery that serves as the main trunk of the entire systemic arterial system. Notice in Figures 18-17 and 18-18 that different segments of the aorta are known by different names. Because the first few centimeters of

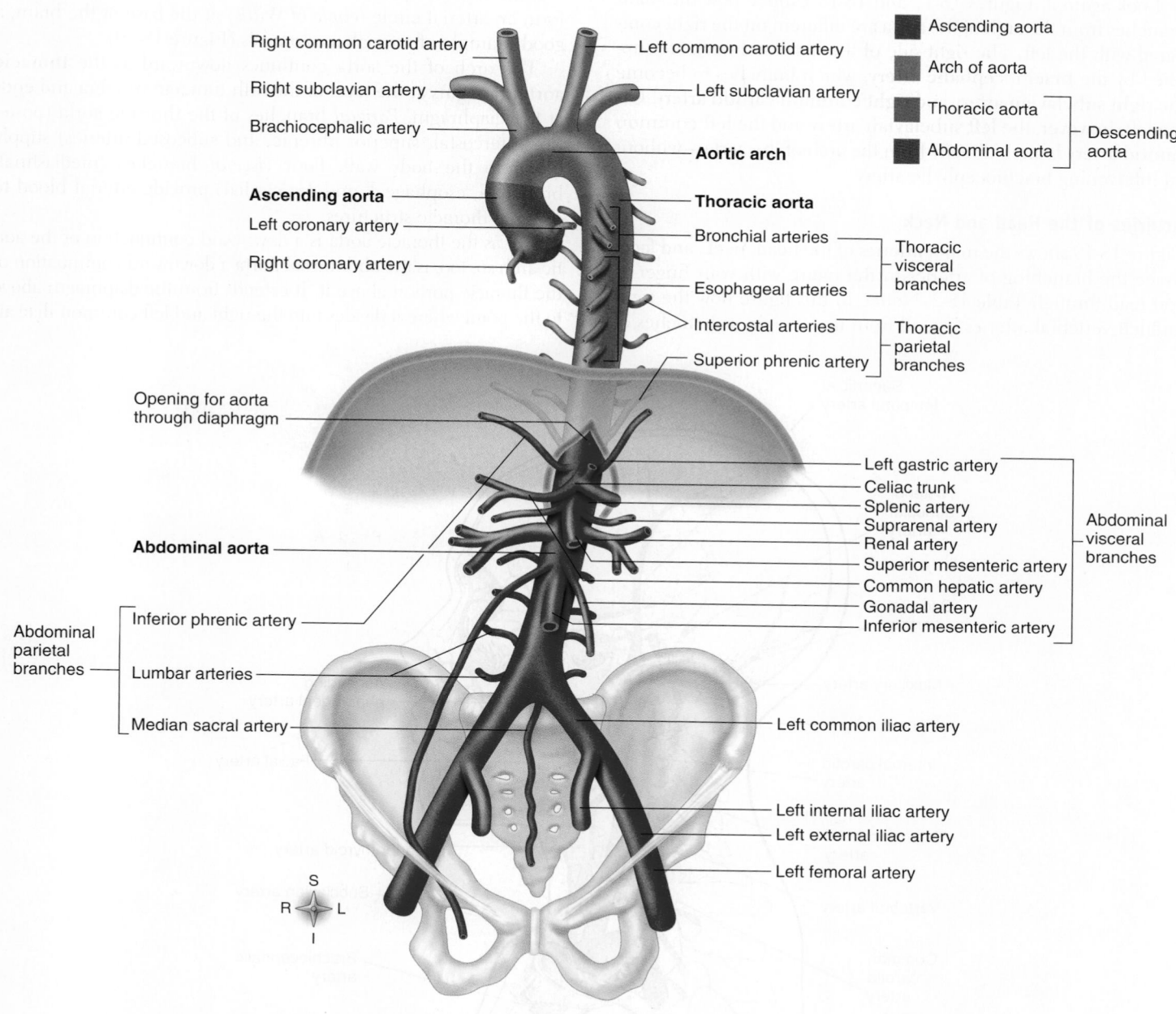

Figure 18-18 *Divisions and primary branches of the aorta.* Anterior view. The aorta is the main systemic artery, serving as a trunk from which other arteries branch. Blood is conducted from the heart first through the ascending aorta, then through the arch of the aorta, and then through the thoracic and abdominal segments of the descending aorta. Note the designation of visceral and parietal branches in the thoracic and abdominal aortic divisions. Tables 18-2 and 18-4 to 18-6 showing branches of the aortic divisions are intended to assist you in interpreting chapter illustrations of arterial vessels.

the aorta conduct blood upward out of the left ventricle, this region is known as the **ascending aorta.** The coronary arteries are branches of the ascending aorta (Figure 18-10 and Table 18-2). The aorta then turns 180 degrees, forming a curved segment called the *arch of the aorta* or simply **aortic arch.** Arterial blood is conducted downward from the arch of the aorta through the **descending aorta.** The descending aorta passes through the thoracic cavity, where it is known as the **thoracic aorta,** to the abdominal cavity, where it is known as the **abdominal aorta.** If you check Table 18-2 or Figure 18-17 you will notice that all systemic arteries branch from the aorta or one of its branches.

Look again at Figures 18-17 and 18-18. Notice how the main branches from the arch of the aorta are different on the right compared with the left. The right side of the head and neck are supplied by the **brachiocephalic artery,** which branches to become the **right subclavian artery** and **right common carotid artery.** On the left, however, the **left subclavian artery** and **left common carotid artery** branch directly from the arch of the aorta—without an intervening brachiocephalic artery.

Arteries of the Head and Neck

Figure 18-19 shows the major arteries of the head, neck, and face. Trace the branching of arteries in the figure with your finger as you read through Table 18-2. Notice in this figure how the right and left vertebral arteries extend from their origin as branches of the subclavian arteries up the neck, through foramina in the transverse processes of the cervical vertebrae, through the foramen magnum, and into the cranial cavity. Next, take a look at Figure 18-20, which shows the arteries at the base of the brain. Note how the vertebral arteries unite on the undersurface of the brainstem to form the *basilar artery,* which shortly branches into the right and left *posterior cerebral arteries* (Figure 18-20). The basilar artery also braches to the pons and cerebellum (Table 18-2). The internal carotid arteries enter the cranial cavity in the mid part of the cranial floor, where they become known as the *anterior cerebral arteries.* Small vessels, the *communicating arteries,* join the anterior and posterior cerebral arteries in such a way as to form an arterial circle (*circle of Willis*) at the base of the brain, a good example of arterial anastomosis (Figure 18-20).

The arch of the aorta continues downward as the **thoracic aorta.** It begins at the level of the fifth thoracic vertebra and ends at the diaphragm. *Parietal* branches of the thoracic aorta (posterior intercostal, superior phrenic, and subcostal arteries) supply blood to the body wall. Four *visceral* branches (mediastinal, bronchial, esophageal, and pericardial) provide arterial blood to internal thoracic structures.

Just as the thoracic aorta is a downward continuation of the aortic arch so, too, is the **abdominal aorta** a downward continuation of the thoracic portion above it. It extends from the diaphragm above to the point where it divides into the right and left common iliac ar-

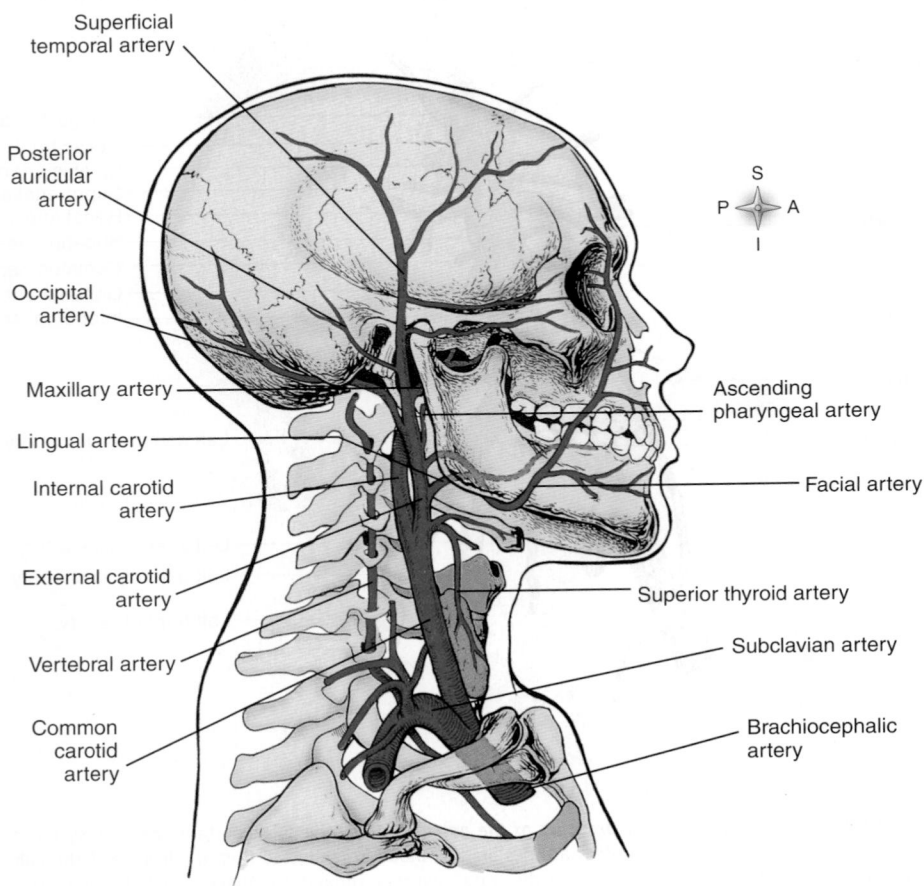

Figure 18-19 *Major arteries of the head and neck.* See Figure 18-20 for arteries at the base of the brain.

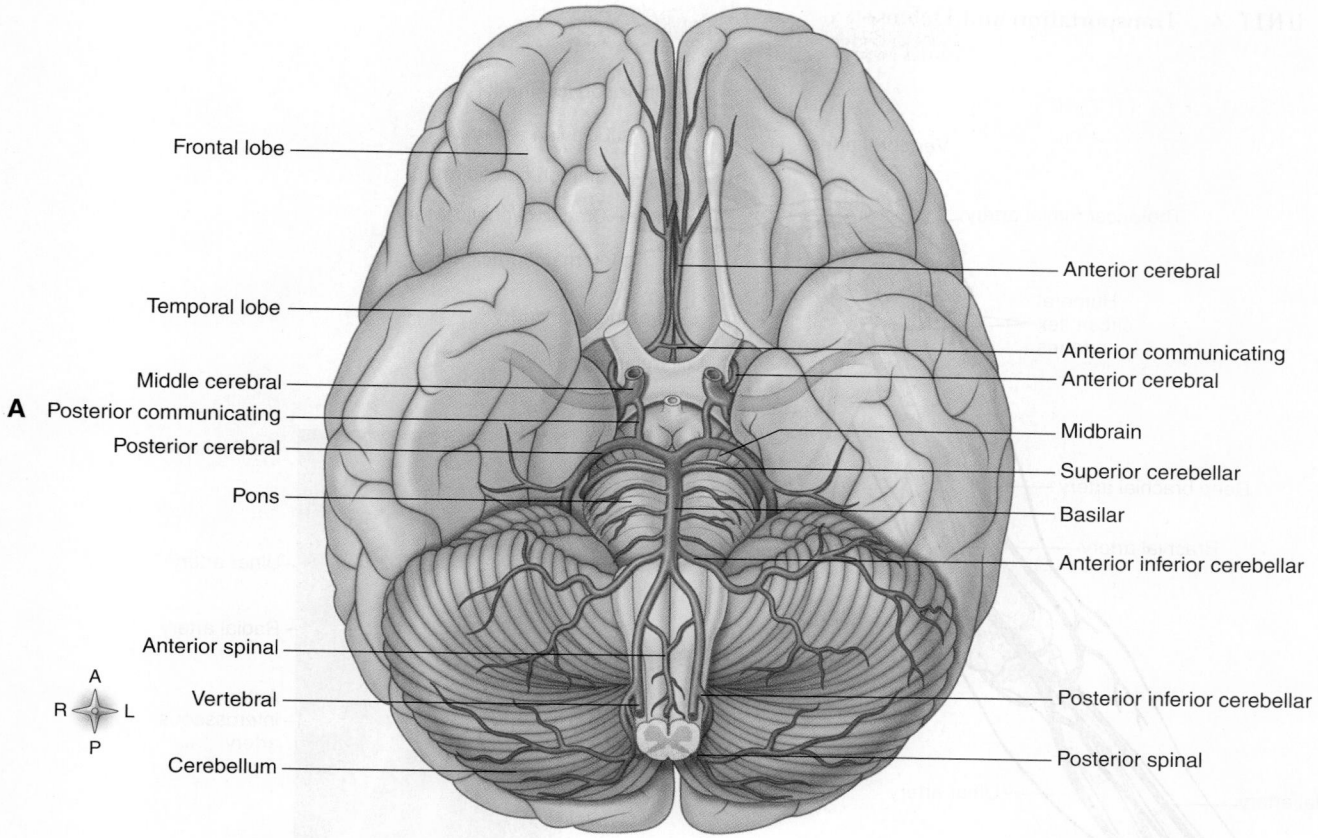

A

Frontal lobe

Temporal lobe

Middle cerebral
Posterior communicating
Posterior cerebral

Pons

Anterior spinal

Vertebral

Cerebellum

Anterior cerebral

Anterior communicating
Anterior cerebral

Midbrain

Superior cerebellar

Basilar

Anterior inferior cerebellar

Posterior inferior cerebellar

Posterior spinal

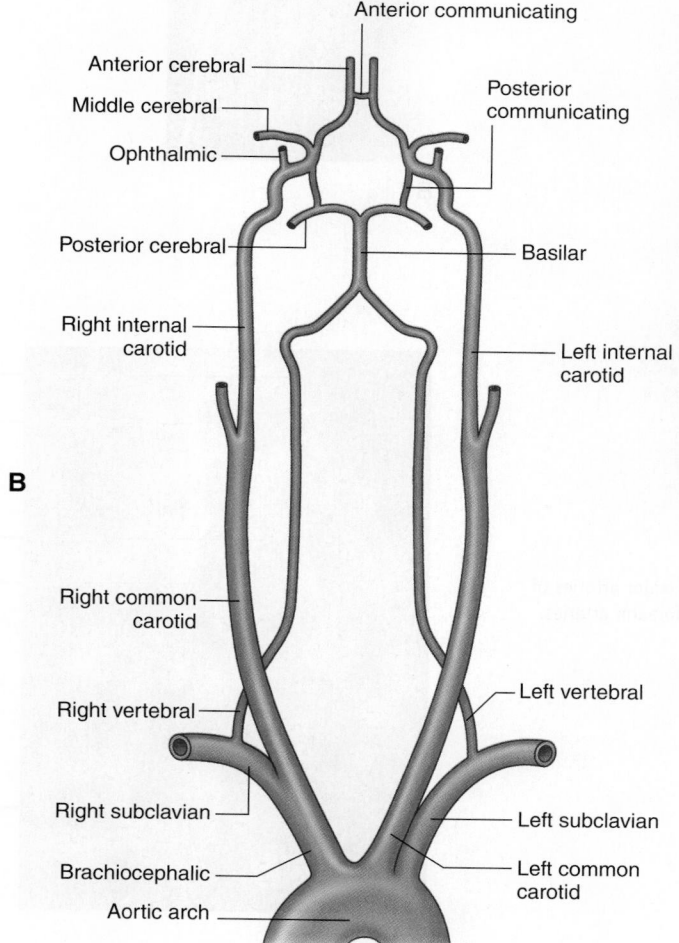

B

Anterior communicating

Anterior cerebral

Middle cerebral

Ophthalmic

Posterior
communicating

Posterior cerebral

Basilar

Right internal
carotid

Left internal
carotid

Right common
carotid

Right vertebral

Left vertebral

Right subclavian

Left subclavian

Brachiocephalic

Left common
carotid

Aortic arch

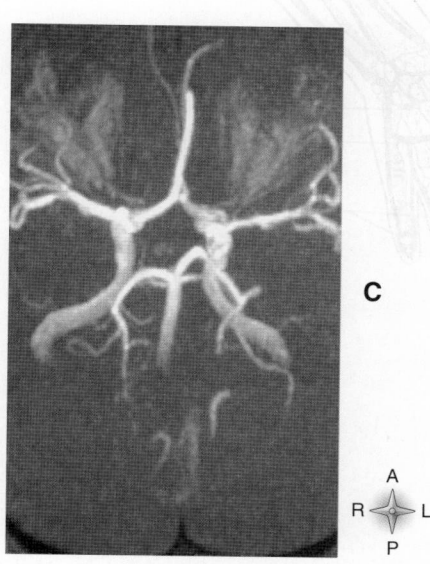

C

Figure 18-20 *Arteries at the base of the brain.* A, Diagram shows the arterial circle of Willis and related structures on the base of the brain. Note the arterial anastomoses. **B,** Origins of blood vessels that form the arterial circle. **C,** Magnetic resonance image of the arterial circle.

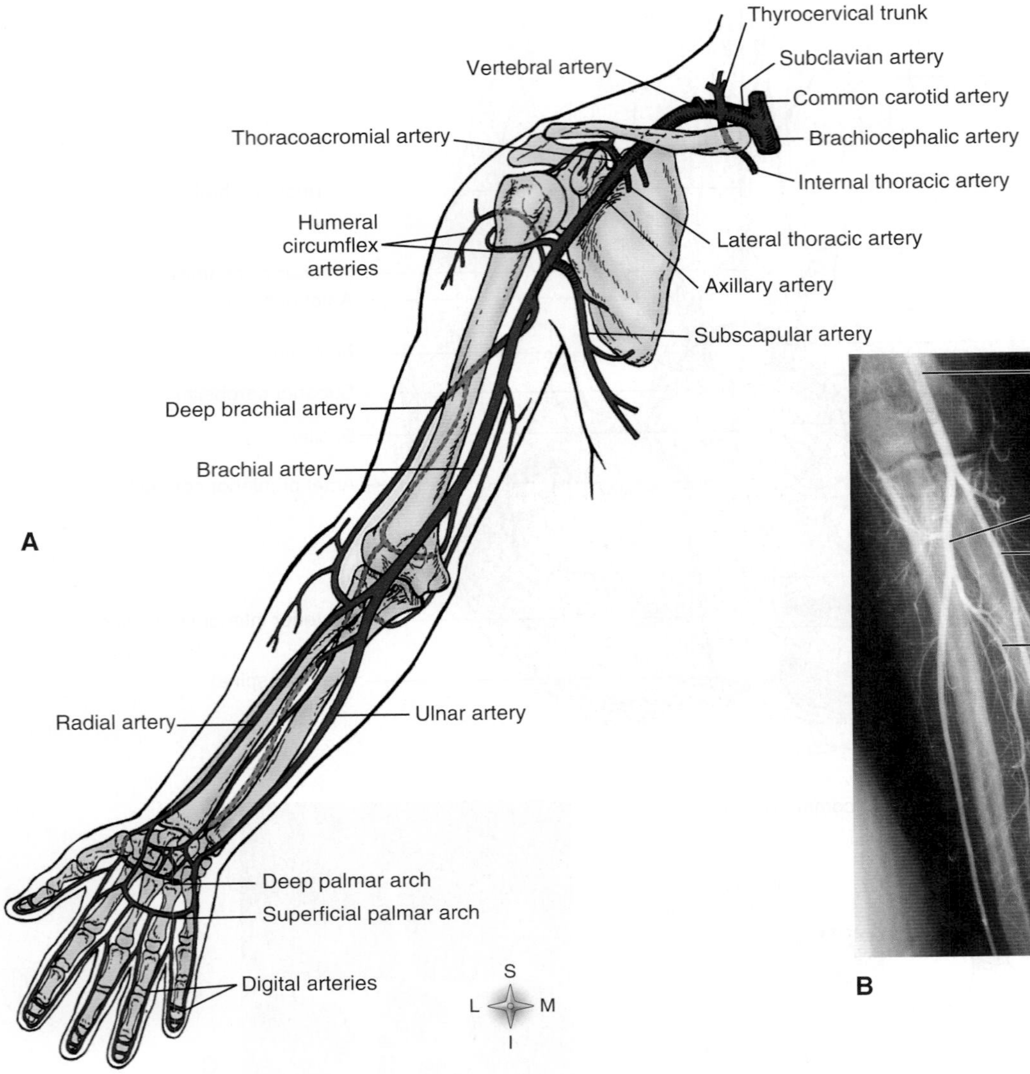

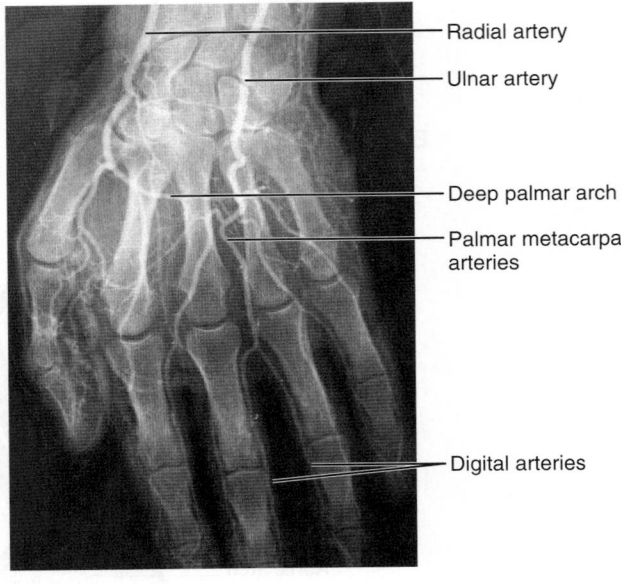

A, Diagram of major arteries of upper extremity. Anterior view.

Thyrocervical trunk
Vertebral artery
Subclavian artery
Common carotid artery
Brachiocephalic artery
Thoracoacromial artery
Internal thoracic artery
Humeral circumflex arteries
Lateral thoracic artery
Axillary artery
Subscapular artery
Deep brachial artery
Brachial artery
Radial artery
Ulnar artery
Deep palmar arch
Superficial palmar arch
Digital arteries

S
L — M
I

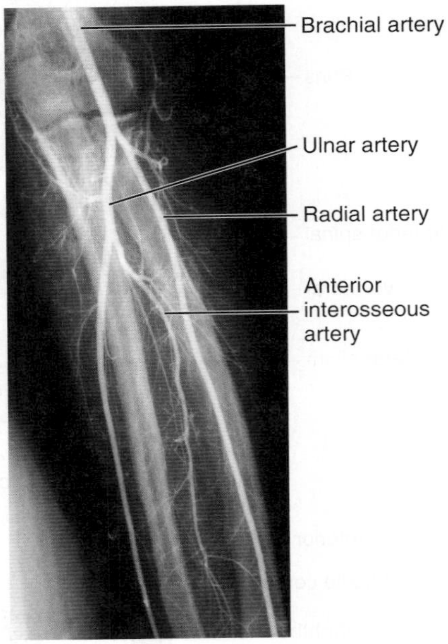

B
Brachial artery
Ulnar artery
Radial artery
Anterior interosseous artery

C
Radial artery
Ulnar artery
Deep palmar arch
Palmar metacarpal arteries
Digital arteries

Figure 18-21 *Major arteries of the upper extremity.* A, Diagram of major arteries of upper extremity. Anterior view. **B,** Arteriogram of brachial and upper forearm arteries. **C,** Arteriogram of hand arteries.

teries at the level of the fourth lumbar vertebra. Note in Figure 18-18 that this segment of the aorta lies just anterior to the vertebral bodies. As a result, a physician can feel the aortic pulse during deep palpation of the abdomen when the vessel is compressed against the underlying vertebrae. The presence of a pulsating swelling along the aorta—an aortic aneurysm—is often diagnosed in this way.

Major abdominal and pelvic branches of the abdominal aorta may also be described as parietal or visceral depending on the location of the end organ or structures they supply with blood. Branches

of this segment of the aorta are illustrated in Figure 18-18 and listed or presented in schematic form in Tables 18-4 to 18-6.

Arteries of the Extremities

Next, take a look at Figure 18-21, which outlines the arteries of the upper extremity, and Figure 18-22, which outlines the arteries of the lower extremity. Trace these arteries with your finger as you read through Tables 18-2 and 18-5. Because of differences in orientation, not every artery listed in the tables appears in every illustration.

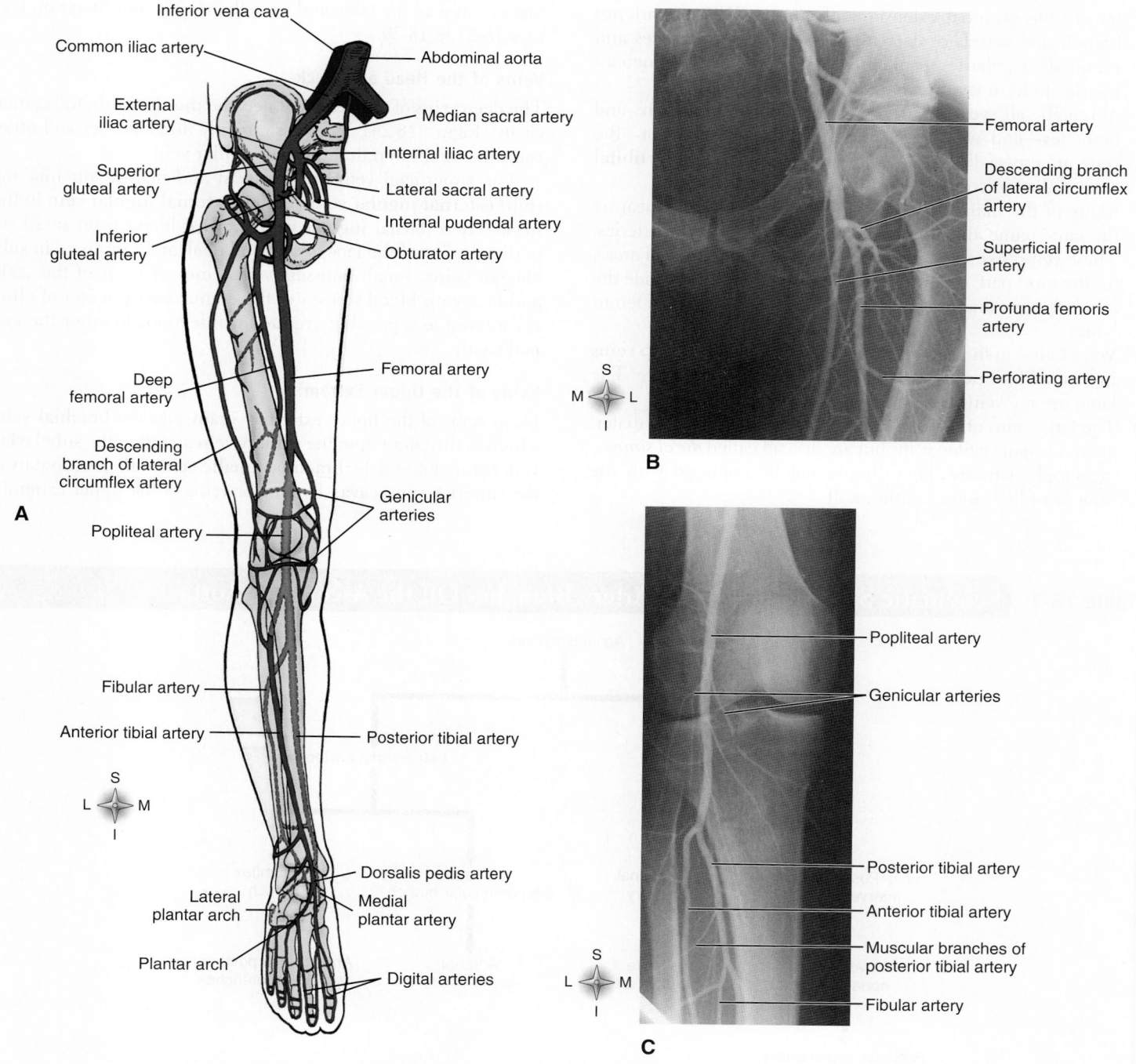

Figure 18-22 *Major arteries of the lower extremity.* **A,** Diagram of major arteries of lower extremity. Anterior view of the right hip and leg. **B,** Femoral arteriogram. Contrast material passing though the external iliac artery in the abdomen has entered the femoral artery and its branches in the thigh. **C,** Popliteal arteriogram.

Systemic Veins

Locate the veins listed in Table 18-3 (see also Figures 18-22 to 18-29). As with the arteries, you may find it easier to learn the names of veins and their anatomical relation to each other from diagrams and tables than from narrative descriptions.

General Principles about Veins

The following facts should be borne in mind while learning the names and locations of veins:

- Veins are the ultimate extensions of capillaries, just as capillaries are the eventual extensions of arteries. Whereas arteries branch into vessels of decreasing size to form arterioles and eventually capillaries, capillaries unite into vessels of increasing size to form venules and, eventually, veins.
- Although all vessels vary considerably in location and branches—and whether they are even present or not—the veins are especially variable. For example, the **median cubital vein** in the forearm is absent in many individuals.
- Many of the main arteries have corresponding veins bearing the same name and are located alongside or near the arteries. These veins, like the arteries, lie in deep, well-protected areas, for the most part, close along the bones. Examples include the femoral artery and femoral vein, both located along the femur bone.
- Veins found in the deep parts of the body are called **deep veins** in contrast to **superficial veins,** which lie near the surface. The latter are the veins that can be seen through the skin.
- The large veins of the cranial cavity, formed by the dura mater, are not usually called veins but are instead called *dural sinuses,* or, simply, **sinuses.** They should not be confused with the bony, air-filled sinuses of the skull.

- Veins communicate (anastomose) with each other in the same way as arteries. In fact, the venous portion of the systemic circulation has even more anastomoses than the arterial portion. Such venous anastomoses provide for collateral return blood flow in cases of venous obstruction.
- Venous blood from the head, neck, upper extremities, and thoracic cavity, with the exception of the lungs, drains into the superior vena cava. Blood from the lower extremities and abdomen enters the inferior vena cava.

Table 18-7 identifies the major systemic veins. Locate each one as you read the table and trace them with your finger on Figures 18-23 to 18-29.

Veins of the Head and Neck

The deep veins of the head and neck lie mostly within the cranial cavity (Figure 18-24). These are mainly dural sinuses and other veins that drain into the **internal jugular vein.**

The superficial veins of the head and neck drain into the **right external jugular vein** and **left external jugular vein** in the neck. The external jugular veins receive blood from small superficial veins of the face, scalp, and neck and terminate in **subclavian veins.** Small emissary veins connect veins of the scalp and face with blood sinuses of the cranial cavity, a fact of clinical interest as a possible avenue for infections to enter the cranial cavity.

Veins of the Upper Extremity

Deep veins of the upper extremity drain into the **brachial vein,** which in turn drain into the **axillary vein** and then the **subclavian vein** before joining the **brachiocephalic vein,** a major tributary of the **superior vena cava.** The major veins of the upper extremity

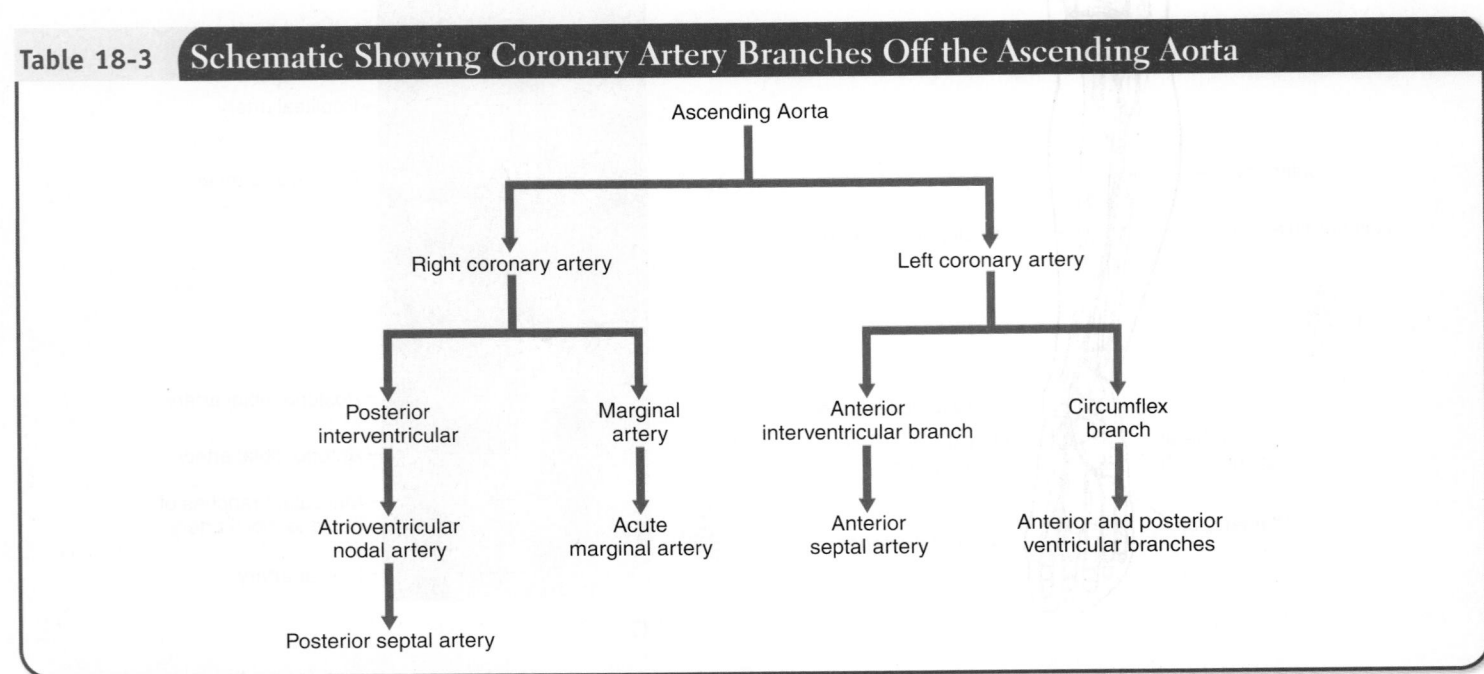

Table 18-3	Schematic Showing Coronary Artery Branches Off the Ascending Aorta

Ascending Aorta

Right coronary artery

Left coronary artery

Posterior interventricular

Marginal artery

Anterior interventricular branch

Circumflex branch

Atrioventricular nodal artery

Acute marginal artery

Anterior septal artery

Anterior and posterior ventricular branches

Posterior septal artery

Table 18-4 Schematic Showing Branches Off the Arch of the Aorta and their Terminal Branches

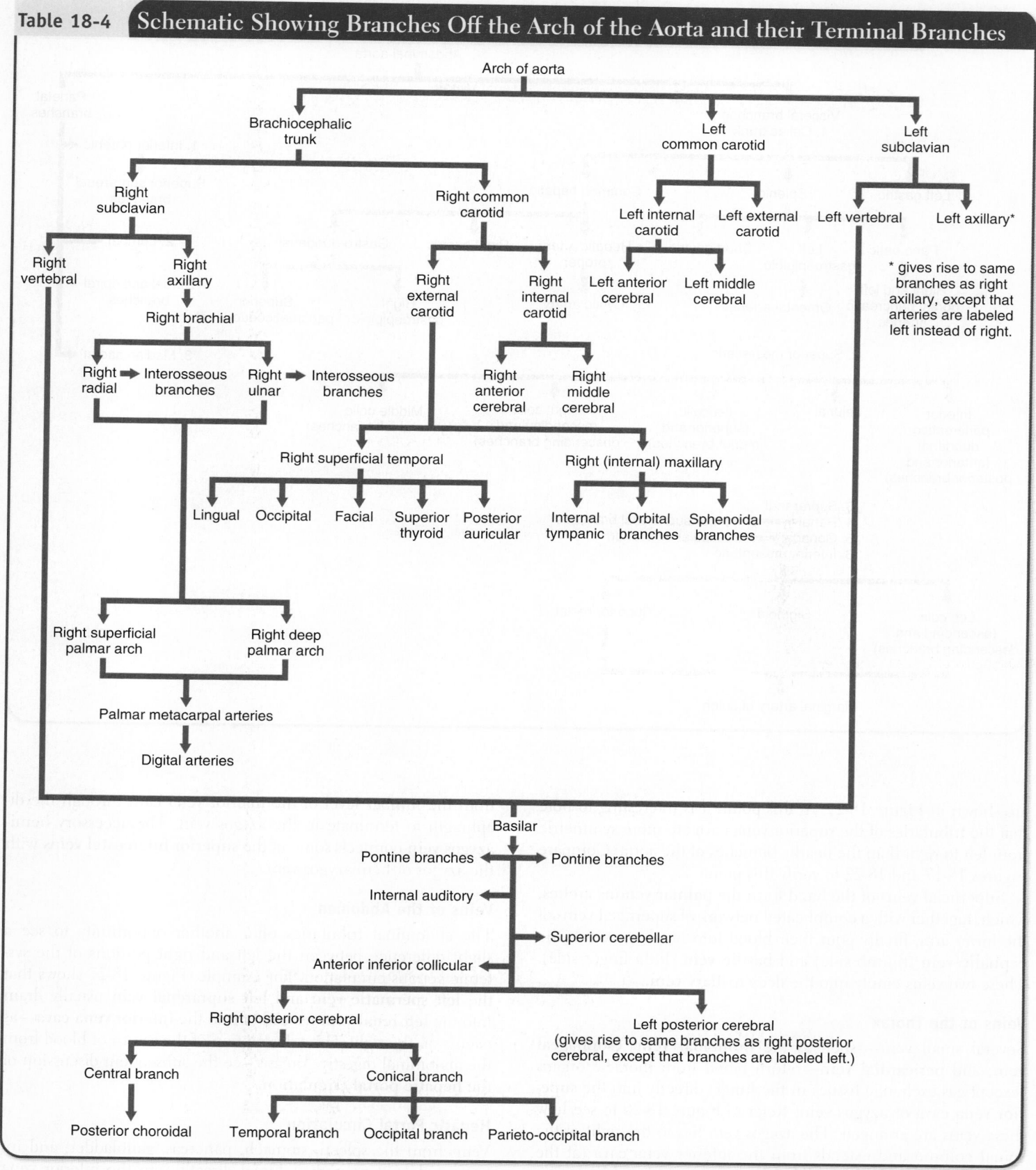

Arch of aorta

Brachiocephalic trunk — Left common carotid — Left subclavian

Right subclavian — Right common carotid

Right vertebral — Right axillary

Right brachial

Right radial → Interosseous branches — Right ulnar → Interosseous branches

Right external carotid — Right internal carotid

Left internal carotid — Left external carotid

Left anterior cerebral — Left middle cerebral

Left vertebral — Left axillary*

* gives rise to same branches as right axillary, except that arteries are labeled left instead of right.

Right anterior cerebral — Right middle cerebral

Right superficial temporal — Right (internal) maxillary

Lingual   Occipital   Facial   Superior thyroid   Posterior auricular

Internal tympanic   Orbital branches   Sphenoidal branches

Right superficial palmar arch — Right deep palmar arch

Palmar metacarpal arteries

Digital arteries

Basilar

Pontine branches ← → Pontine branches

Internal auditory ←

→ Superior cerebellar

Anterior inferior collicular ←

Right posterior cerebral

Left posterior cerebral (gives rise to same branches as right posterior cerebral, except that branches are labeled left.)

Central branch — Conical branch

Posterior choroidal   Temporal branch   Occipital branch   Parieto-occipital branch

Table 18-5	Schematic Showing Branches of the Abdominal Aorta

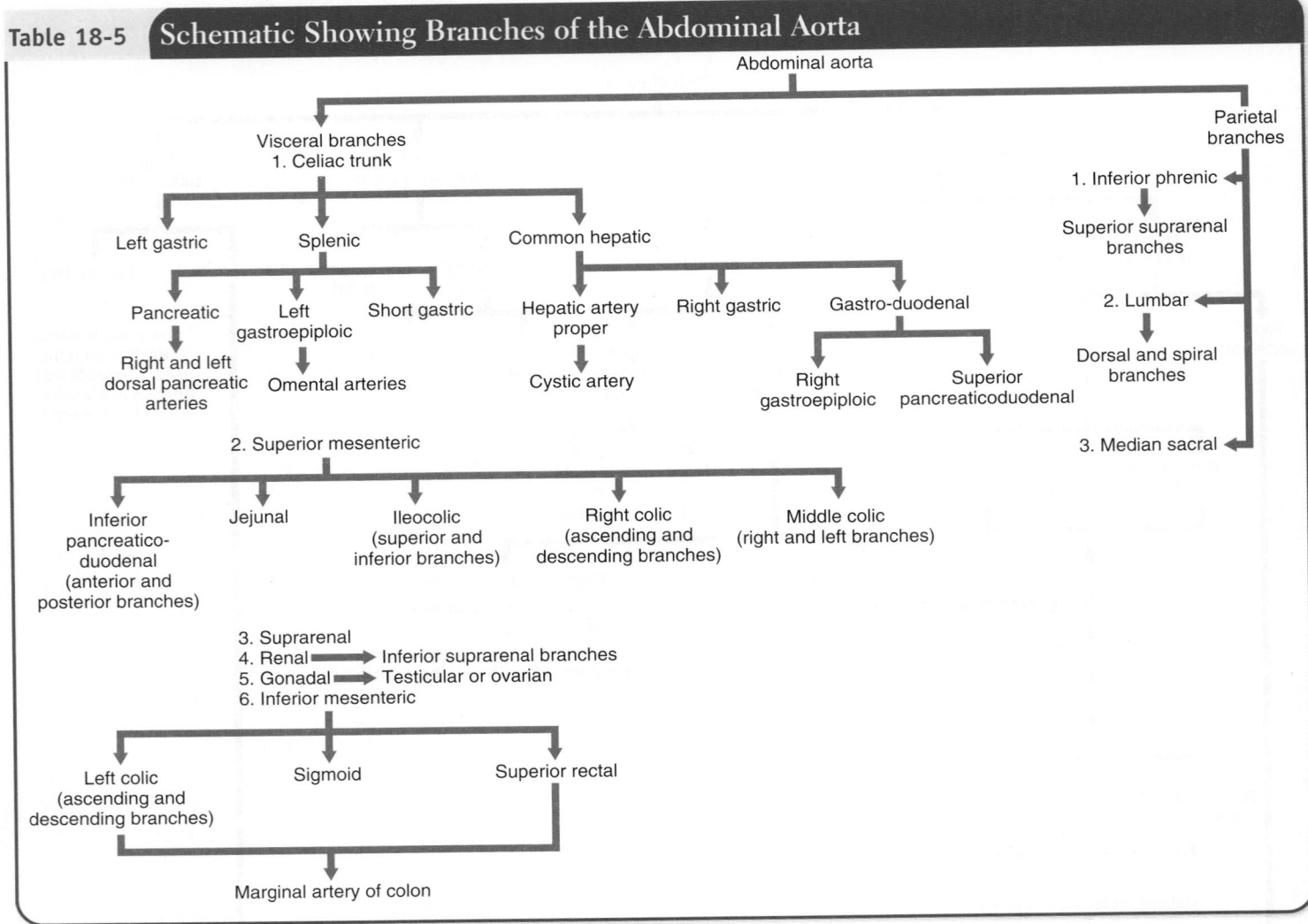

are shown in Figure 18-25. At this point, it is interesting to note that the tributaries of the superior vena cava are more symmetric from left to right than the nearby branches of the aorta. Compare Figures 18-17 and 18-22 to verify this point.

Superficial veins of the hand form the **palmar venous arches,** which, together with a complicated network of superficial veins of the lower arm, finally pour their blood into two large veins: the **cephalic vein** (thumb side) and **basilic vein** (little finger side). These two veins empty into the deep **axillary vein.**

Veins of the Thorax

Several small veins—such as the **bronchial vein, esophageal vein,** and **pericardial vein**—return blood from thoracic organs (except gas exchange tissues in the lungs) directly into the **superior vena cava** or **azygos vein.** Refer to Figure 18-26 to see how these veins are arranged. The azygos vein lies to the right of the spinal column and extends from the inferior vena cava (at the level of the first or second lumbar vertebra) through the diaphragm to the terminal part of the superior vena cava. The **hemiazygos vein** lies to the left of the spinal column, extending

from the lumbar level of the inferior vena cava through the diaphragm to terminate in the azygos vein. The **accessory hemiazygos vein** connects some of the **superior intercostal veins** with the azygos or hemiazygos vein.

Veins of the Abdomen

The abdominal tributaries offer another opportunity to see a slight difference between the left and right portions of the systemic venous circulation. For example, Figure 18-27 shows that the **left spermatic vein** and **left suprarenal vein** usually drain into the **left renal vein** instead of into the **inferior vena cava**—as occurs on the right. For a description of the return of blood from the abdominal digestive organs, see the subsequent discussion of the **hepatic portal circulation.**

Hepatic Portal Circulation

Veins from the spleen, stomach, pancreas, gallbladder, and intestines do not pour their blood directly into the inferior vena cava, as do the veins from other abdominal organs. They send their blood to the liver by means of the hepatic portal vein. Here

Table 18-6   Schematic Showing Arterial Vessels of the Lower Extremity

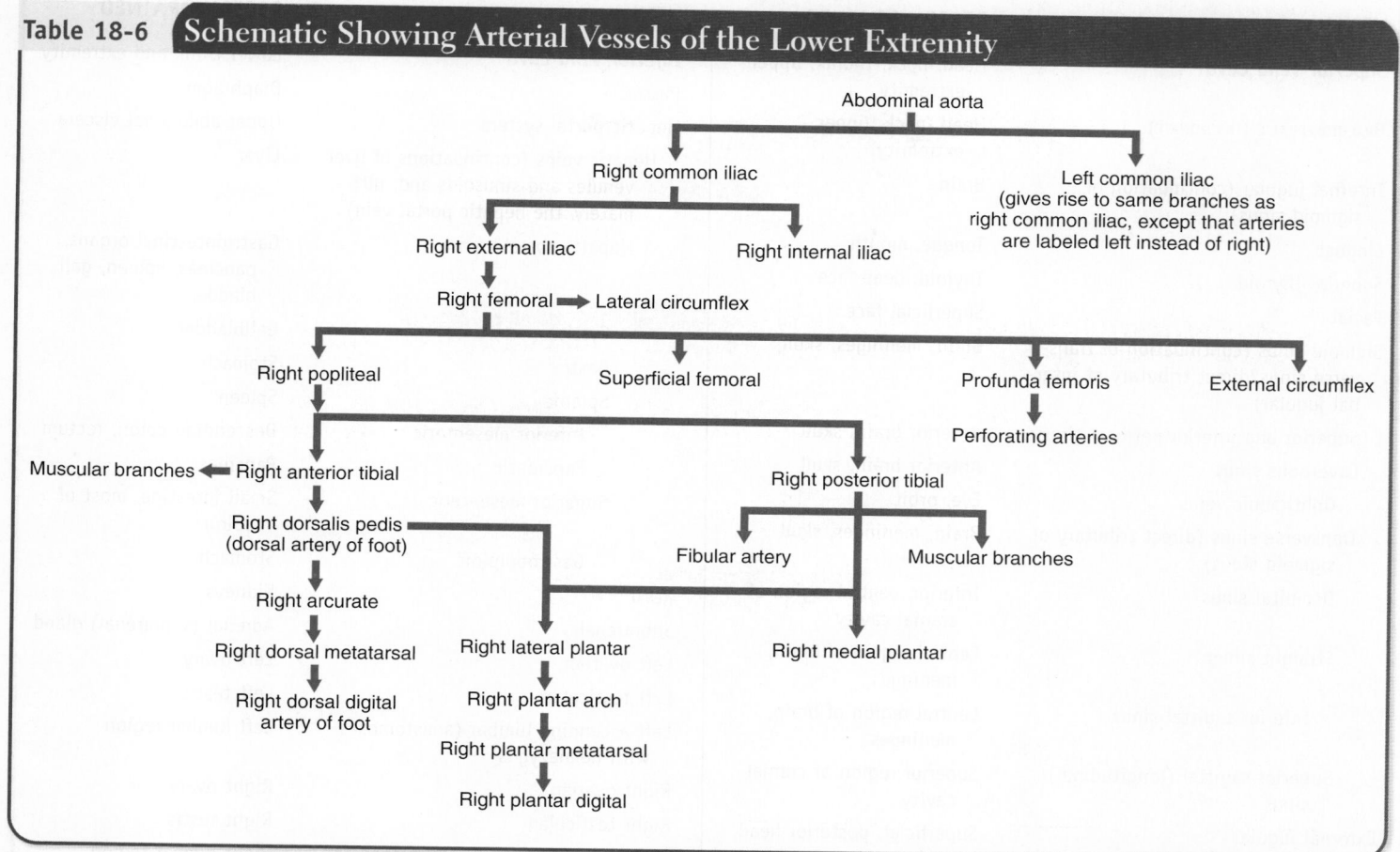

the blood mingles with the arterial blood in the capillaries and is eventually drained from the liver by the hepatic veins that join the inferior vena cava. Any arrangement in which venous blood flows through a second capillary network before returning to the heart is called a *portal* circulatory route. Portal comes from the Latin *porta*, meaning "gateway," and is used here because the liver is a gateway through which blood returning from the digestive tract must pass before it returns to the heart.

There are several advantages to detouring blood from the digestive tract through the liver before it returns to the heart. Shortly after a meal, blood flowing through digestive organs begins absorbing glucose and other simple nutrients. The result is a tremendous increase in the blood glucose level. As the blood travels through the liver, however, excess glucose is removed from the blood and stored in liver cells as glycogen. Thus blood returned to the heart carries only a moderate level of glucose. Many hours after food has yielded its nutrients, low-glucose blood coming from the digestive organs can pick up glucose released from the glycogen stores held in the liver cells before returning to the heart. Another advantage of the hepatic portal scheme is that toxic molecules such as alcohol can be partially removed or detoxified before the blood is distributed to the rest of the body. Additional information regarding the role of the liver, and the advantages of the portal circulation through the liver, is discussed in Chapters 25 and 26.

Figure 18-28 shows the plan of the hepatic portal system. In most individuals the hepatic portal vein is formed by the union of the splenic and superior mesenteric veins, but blood from the gastric, pancreatic, and inferior mesenteric veins drains into the splenic vein before it merges with the superior mesenteric vein.

If either hepatic portal circulation or venous return from the liver is interfered with (as often occurs in certain types of liver disease or heart disease), venous drainage from most of the other abdominal organs is necessarily obstructed also. The accompanying increased capillary pressure accounts, at least in part, for the occurrence of abdominal bloating, or *ascites* (ah-SITE-eez), under these conditions.

Veins of the Lower Extremity

As Figure 18-29 shows, deep veins of the lower leg drain from the **anterior tibial vein,** which continues as the **fibular (peroneal) vein,** and the **posterior tibial vein.** The fibular and posterior tibial veins join to form the **popliteal vein,** which runs behind the knee joint and continues up along the femur as the deep femoral vein. The **femoral vein** continues as the **external iliac vein,** draining into the common iliac vein and from there into the **inferior vena cava.**

Superficial veins of the lower extremity include the **small saphenous vein,** a tributary of the popliteal vein, and the **great**

Table 18-7 Major Systemic Veins

VEIN*	REGION DRAINED	VEIN*	REGION DRAINED
Superior Vena Cava	Head, neck, thorax, upper extremity	**Inferior Vena Cava**	Lower trunk and extremity
BRACHIOCEPHALIC (INNOMINATE)	Head, neck, upper extremity	PHRENIC	Diaphragm
Internal jugular (continuation of sigmoid sinus)	Brain	Hepatic portal system	Upper abdominal viscera
Lingual	Tongue, mouth	Hepatic veins (continuations of liver venules and sinusoids and, ultimately, the hepatic portal vein)	Liver
Superior thyroid	Thyroid, deep face		
Facial	Superficial face	Hepatic portal vein	Gastrointestinal organs, pancreas, spleen, gallbladder
Sigmoid sinus (continuation of transverse sinus/direct tributary of internal jugular)	Brain, meninges, skull		
Superior and inferior petrosal sinuses	Anterior brain, skull	Cystic	Gallbladder
Cavernous sinus	Anterior brain, skull	Gastric	Stomach
Ophthalmic veins	Eye, orbit	Splenic	Spleen
Transverse sinus (direct tributary of sigmoid sinus)	Brain, meninges, skull	Inferior mesenteric	Descending colon, rectum
		Pancreatic	Pancreas
Occipital sinus	Inferior, central region of cranial cavity	Superior mesenteric	Small intestine, most of colon
Straight sinus	Central region of brain, meninges	Gastroepiploic	Stomach
		RENAL	Kidneys
Inferior sagittal sinus	Central region of brain, meninges	Suprarenal	Adrenal (suprarenal) gland
		Left ovarian	Left ovary
Superior sagittal (longitudinal) sinus	Superior region of cranial cavity	Left testicular	Left testis
		Left ascending lumbar (anastomoses with hemiazygos)	Left lumbar region
External jugular	Superficial, posterior head, neck	Right ovarian	Right ovary
		Right testicular	Right testis
SUBCLAVIAN (CONTINUATION OF AXILLARY/DIRECT TRIBUTARY OF BRACHIOCEPHALIC)	Axilla, lower extremity	Right ascending lumbar (anastomoses with azygos)	Right lumbar region
Axillary (continuation of basilic/direct tributary of subclavian)	Axilla, lower extremity	Common iliac (continuation of external iliac; common iliacs unite to form inferior vena cava)	Lower extremity
Cephalic	Lateral and lower arm, hand	External iliac (continuation of femoral/direct tributary of common iliac)	Thigh, leg, foot
Brachial	Deep arm	Femoral (continuation of popliteal/direct tributary of external iliac)	Thigh, leg, foot
Radial	Deep lateral forearm		
Ulnar	Deep medial forearm	Popliteal	Leg, foot
Basilic (direct tributary of axillary)	Medial and lower arm, hand	Posterior tibial	Deep posterior leg
Median cubital (basilic) (formed by anastomosis of cephalic and basilic)	Arm, hand	Medial and lateral plantar	Sole of foot
		Fibular (peroneal) (continuation of anterior tibial)	Lateral and anterior leg, foot
Deep and superficial palmar venous arches (formed by anastomosis of cephalic and basilic)	Hand	Anterior tibial	Anterior leg, foot
		Dorsal veins of foot	Anterior (dorsal) foot, toes
Digital	Fingers	Small (external, short) saphenous	Superficial posterior leg, lateral foot
AZYGOS (ANASTOMOSES WITH RIGHT ASCENDING LUMBAR)	Right posterior wall of thorax and abdomen, esophagus, bronchi, pericardium, mediastinum	Great (internal, long) saphenous	Superficial medial and anterior thigh, leg, foot
		Dorsal veins of foot	Anterior (dorsal) foot, toes
Hemiazygos (anastomoses with left renal)	Left inferior posterior wall of thorax and abdomen, esophagus, mediastinum	Dorsal venous arch	Anterior (dorsal) foot, toes
		Digital	Toes
Accessory hemiazygos	Left superior posterior wall of thorax	INTERNAL ILIAC	Pelvic region

*Tributaries of each vein are indented below its name; deep veins are printed in dark blue, and superficial veins are printed in light blue.

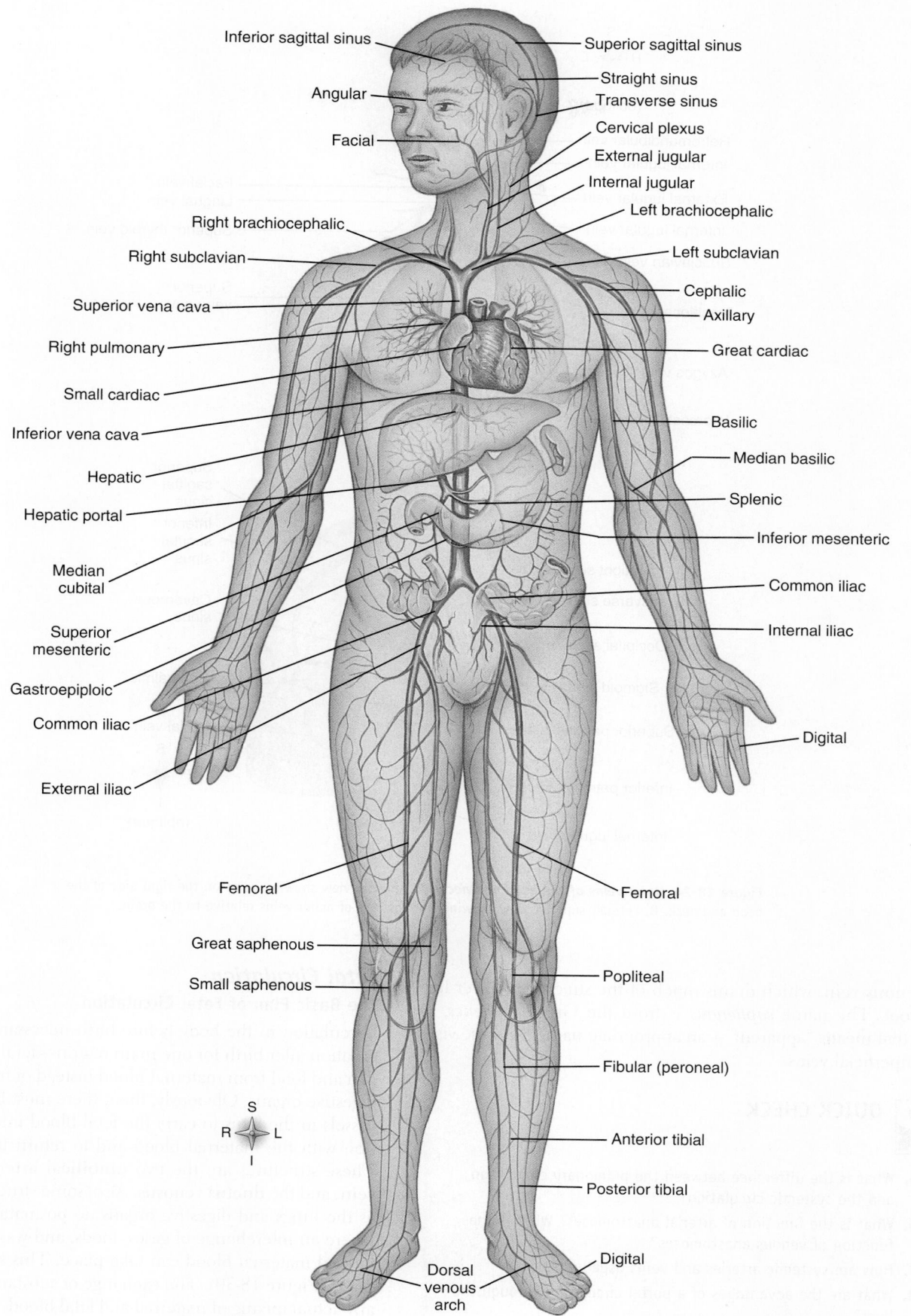

Figure 18-23 *Principal veins of the body.*

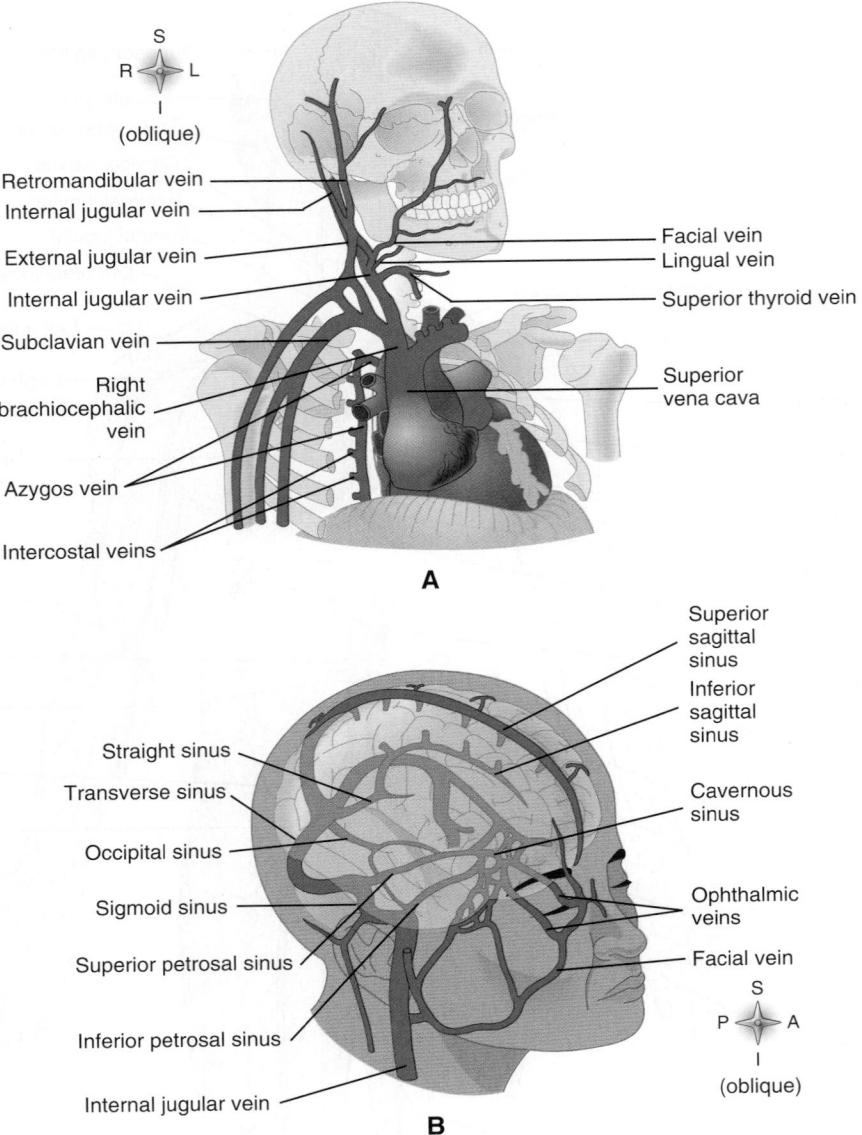

Figure 18-24 *Major veins of the head and neck.* A, Anterior view showing veins on the right side of the head and neck. **B,** Lateral, superior view showing the position of major veins relative to the brain.

saphenous vein, which drains much of the superficial lower leg and foot. The name *saphenous* is from the Greek *saphenes*, a word that means "apparent"—an appropriate name for these visible, superficial veins.

QUICK CHECK

15. What is the difference between the pulmonary circulation and the systemic circulation?
16. What is the function of arterial anastomoses? What is the function of venous anastomoses?
17. How are systemic arteries and veins usually named?
18. What are the advantages of a portal circulation through the liver?

Fetal Circulation

The Basic Plan of Fetal Circulation

Circulation in the body before birth necessarily differs from circulation after birth for one main reason—fetal blood secures oxygen and food from maternal blood instead of from fetal lungs and digestive organs. Obviously, then, there must be additional blood vessels in the fetus to carry the fetal blood into close approximation with the maternal blood and to return it to the fetal body. These structures are the two **umbilical arteries,** the **umbilical vein,** and the **ductus venosus.** Also, some structure must function as the lungs and digestive organs do postnatally, that is, a place where an interchange of gases, foods, and wastes between the fetal and maternal blood can take place. This structure is the *placenta* (Figure 18-30). The exchange of substances occurs without any actual mixing of maternal and fetal blood, because each flows in its own capillaries.

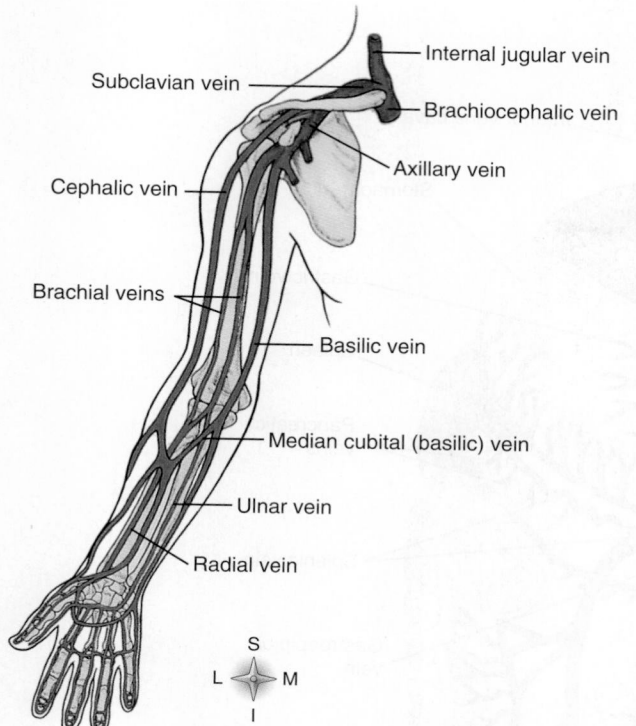

Internal jugular vein
Subclavian vein
Brachiocephalic vein
Cephalic vein
Axillary vein
Brachial veins
Basilic vein
Median cubital (basilic) vein
Ulnar vein
Radial vein

S
L — M
I

Figure 18-25 *Major veins of the upper extremity.* The median cubital (basilic) vein is commonly used for removing blood or giving intravenous infusions (anterior view).

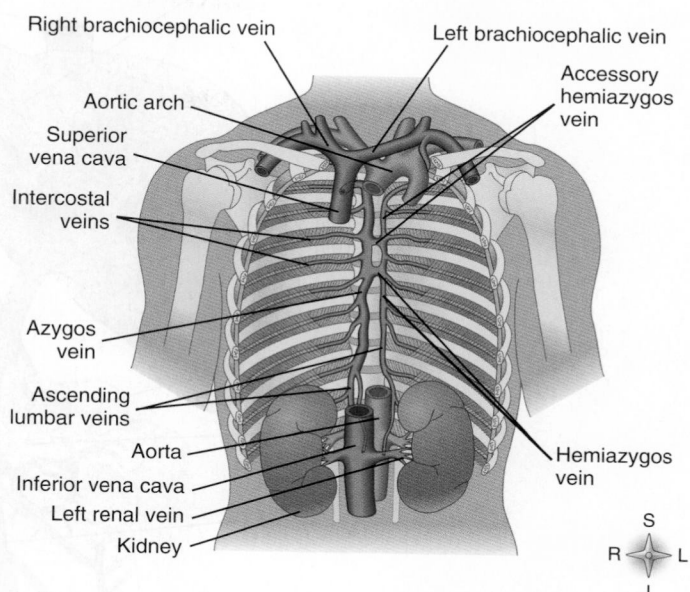

Right brachiocephalic vein
Left brachiocephalic vein
Accessory hemiazygos vein
Aortic arch
Superior vena cava
Intercostal veins
Azygos vein
Ascending lumbar veins
Aorta
Inferior vena cava
Left renal vein
Kidney
Hemiazygos vein

S
R — L
I

Figure 18-26 *Principal veins of the thorax.* Smaller veins of the thorax drain blood into the inferior vena cava or into the azygos vein—both are shown here. The hemiazygos vein and accessory hemiazygos vein on the left drain into the azygos vein on the right.

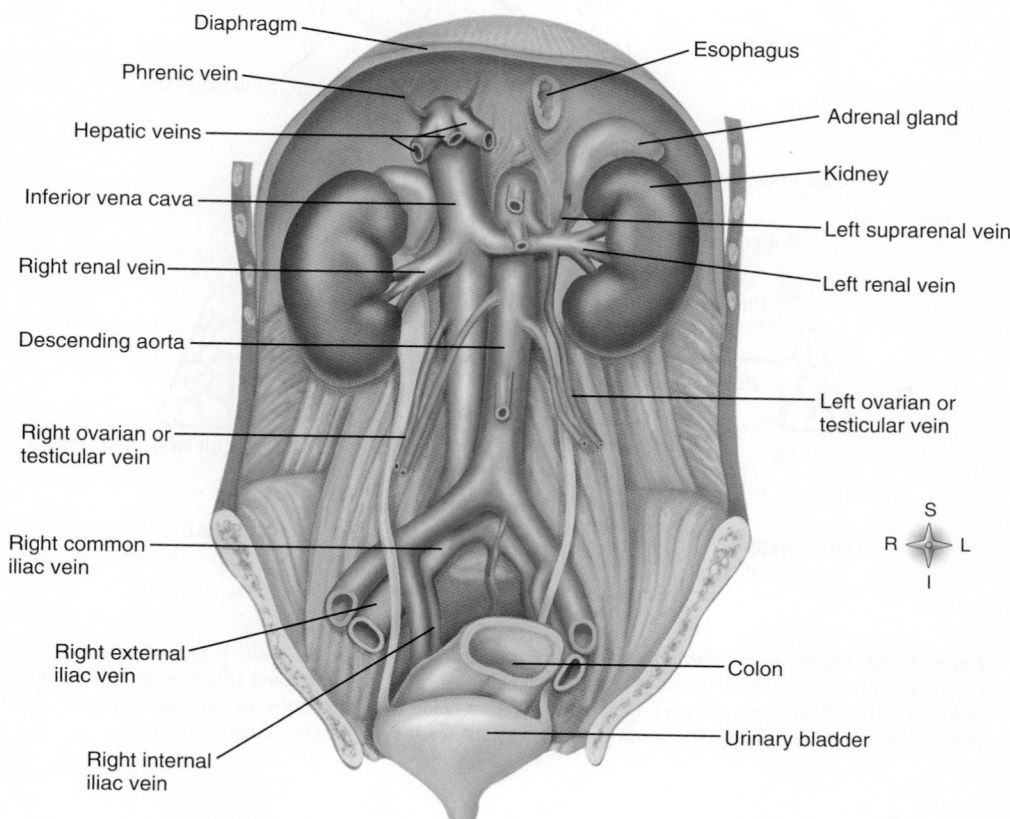

Diaphragm
Phrenic vein
Esophagus
Hepatic veins
Adrenal gland
Inferior vena cava
Kidney
Right renal vein
Left suprarenal vein
Left renal vein
Descending aorta
Right ovarian or testicular vein
Left ovarian or testicular vein
Right common iliac vein
Right external iliac vein
Colon
Right internal iliac vein
Urinary bladder

S
R — L
I

Figure 18-27 *Inferior vena cava and its abdominopelvic tributaries.* Anterior view of abdominopelvic cavity with many of the viscera removed. Note the close anatomical relationship between the inferior vena cava and the descending aorta.

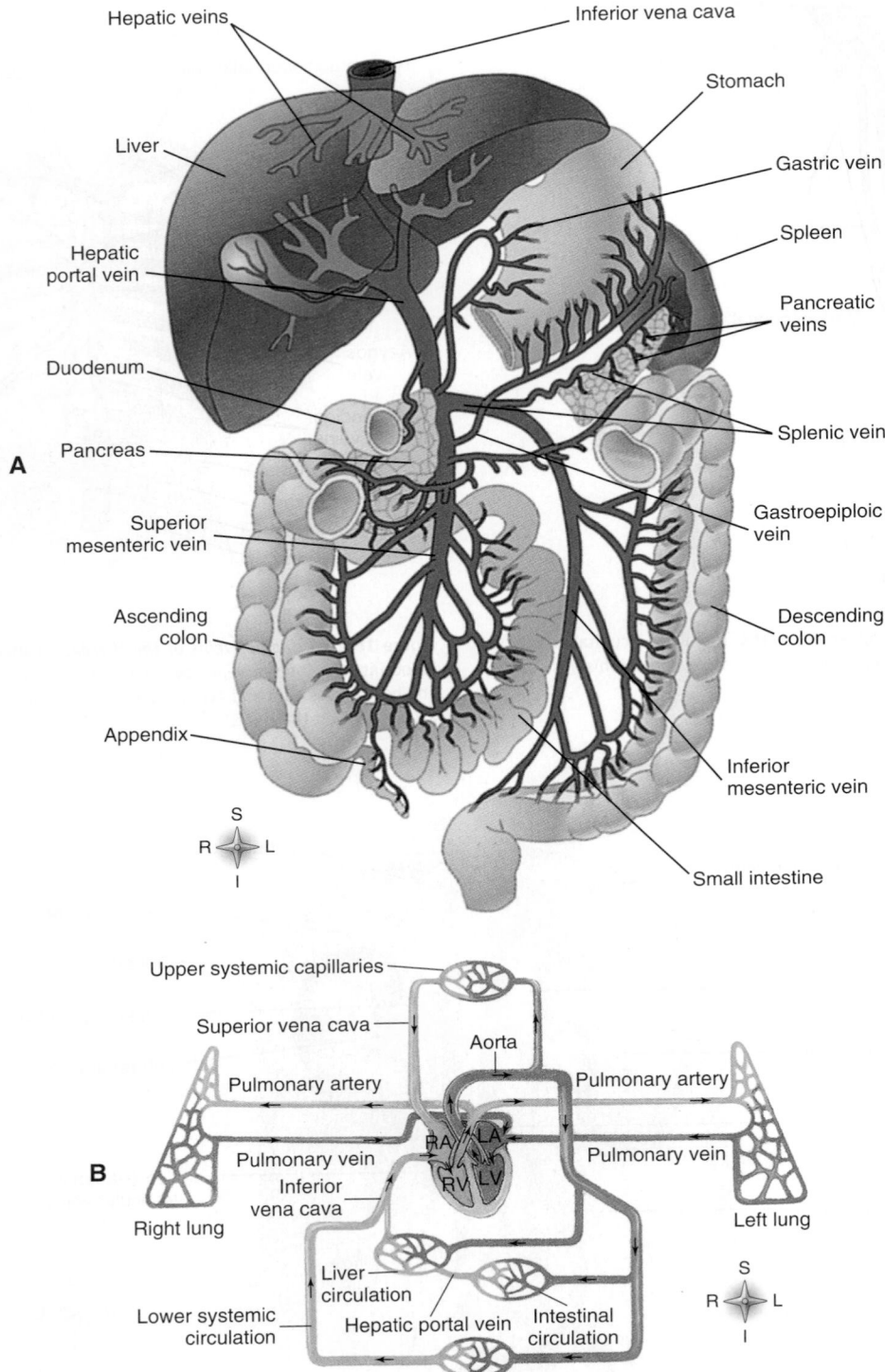

Figure 18-28 *Hepatic portal circulation.* A, In this unusual circulatory route, a vein is located between two capillary beds. The hepatic portal vein collects blood from capillaries in visceral structures located in the abdomen and empties it into the liver. Hepatic veins return blood to the inferior vena cava. **B,** Schematic showing relationship of the hepatic portal circulation to the lower systemic circulation.

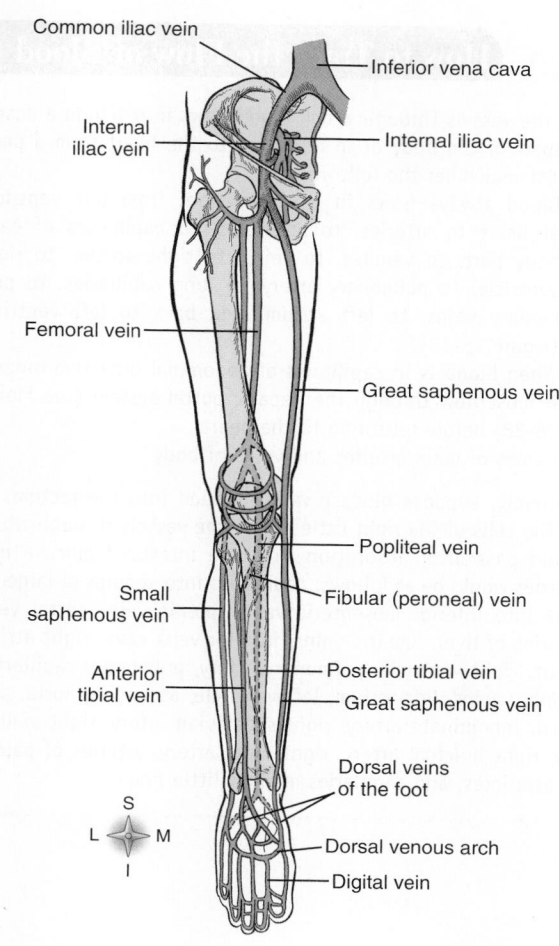

Figure 18-29 *Major veins of the lower extremity (anterior view).*

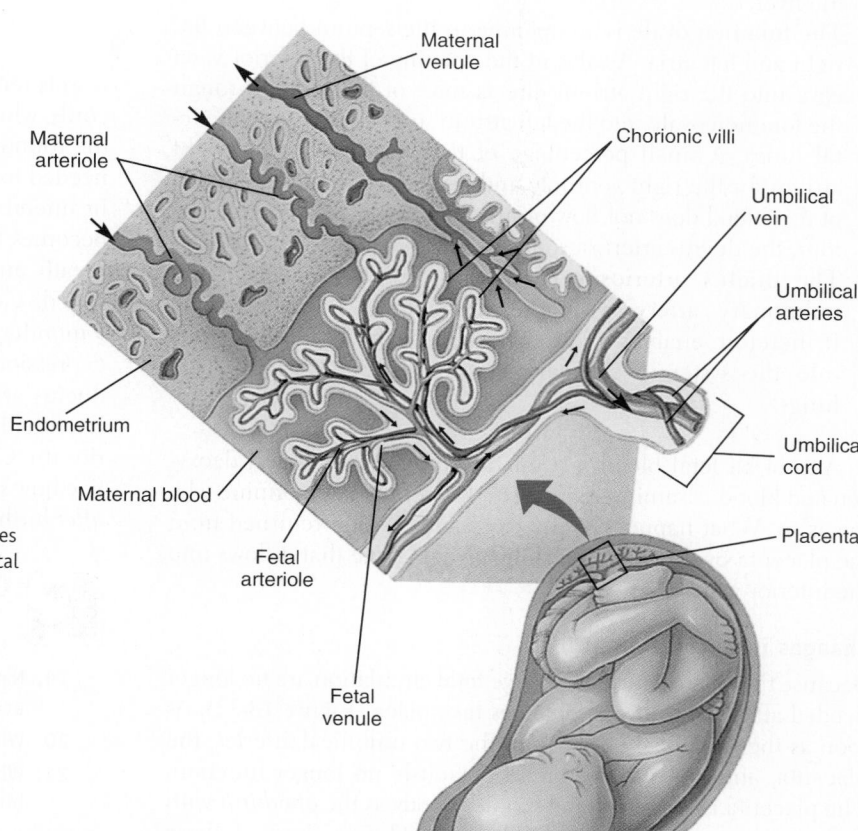

Figure 18-30 *Placental circulation.* The placenta is an organ that permits interchange of blood gases (O_2 and CO_2), nutrients, and wastes between fetal blood and maternal blood. Other special features of fetal circulation are shown in Figures 18-31 and 18-32.

In addition to the placenta and umbilical vessels, three structures located within the fetus's own body play an important part in fetal circulation. One of them (the ductus venosus) serves as a detour by which most of the blood returning from the placenta bypasses the fetal liver. The other two (the foramen ovale and ductus arteriosus) provide detours by which blood bypasses the lungs. A brief description of each of the six structures necessary for fetal circulation follows (Figure 18-31):

1. The two **umbilical arteries** are extensions of the internal iliac (hypogastric) arteries and carry fetal blood to the placenta.
2. The **placenta** is a structure attached to the uterine wall. Exchange of oxygen and other substances between maternal and fetal blood takes place in the placenta, although no mixing of maternal and fetal blood occurs.
3. The **umbilical vein** returns oxygenated blood from the placenta, enters the fetal body through the umbilicus, extends up to the undersurface of the liver where it gives off two or three branches to the liver, and then continues on as the ductus venosus. Two umbilical arteries and the umbilical vein together constitute the umbilical cord and are shed at birth along with the placenta.
4. The **ductus venosus** is a continuation of the umbilical vein along the undersurface of the liver and drains into the inferior vena cava. Most of the blood returning from the placenta bypasses the liver. Only a relatively small amount of blood enters the liver by way of the branches from the umbilical vein into the liver.
5. The **foramen ovale** is an opening in the septum between the right and left atria. A valve at the opening of the inferior vena cava into the right atrium directs most of the blood through the foramen ovale into the left atrium so that it bypasses the fetal lungs. A small percentage of the blood leaves the right atrium for the right ventricle and pulmonary artery. But most of this blood does not flow on into the lungs. Still another detour, the ductus arteriosus, diverts it.
6. The **ductus arteriosus** is a small vessel connecting the pulmonary artery with the descending thoracic aorta. It therefore enables another portion of the blood to detour into the systemic circulation without going through the lungs.

Almost all fetal blood is a mixture of oxygenated and deoxygenated blood. Examine Figure 18-31 carefully to determine why this is so. What happens to the oxygenated blood returned from the placenta by way of the umbilical vein? Note that it flows into the inferior vena cava.

Changes in Circulation at Birth

Because the six structures that serve fetal circulation are no longer needed after birth, several changes take place (Figure 18-32). As soon as the umbilical cord is cut, the two umbilical arteries, the placenta, and the umbilical vein obviously no longer function. The placenta is shed from the mother's body as the *afterbirth* with part of the umbilical vessels attached. The sections of these

vessels remaining in the infant's body eventually become fibrous cords, which remain throughout life (the umbilical vein becomes the round ligament of the liver). The ductus venosus, no longer needed to bypass blood around the liver, eventually becomes the ligamentum venosum of the liver. The foramen ovale normally becomes functionally closed soon after a newborn takes the first breath and full circulation through the lungs becomes established. Complete structural closure, however, usually requires 9 months or more. Eventually, the foramen ovale becomes a mere depression (fossa ovalis) in the wall of the right atrial septum. The ductus arteriosus contracts as soon as respiration is established. Eventually, it also turns into a fibrous cord, the ligamentum arteriosum. Compare the blood flow in Figures 18-31 and 18-32. Notice how separation of oxygenated and deoxygenated blood occurs *after* birth.

QUICK CHECK

19. Name some structures of the fetal circulation that do not occur in the adult.
20. What is the function of the placenta and umbilical vessels?
21. What changes in the circulatory system occur at the time of birth?

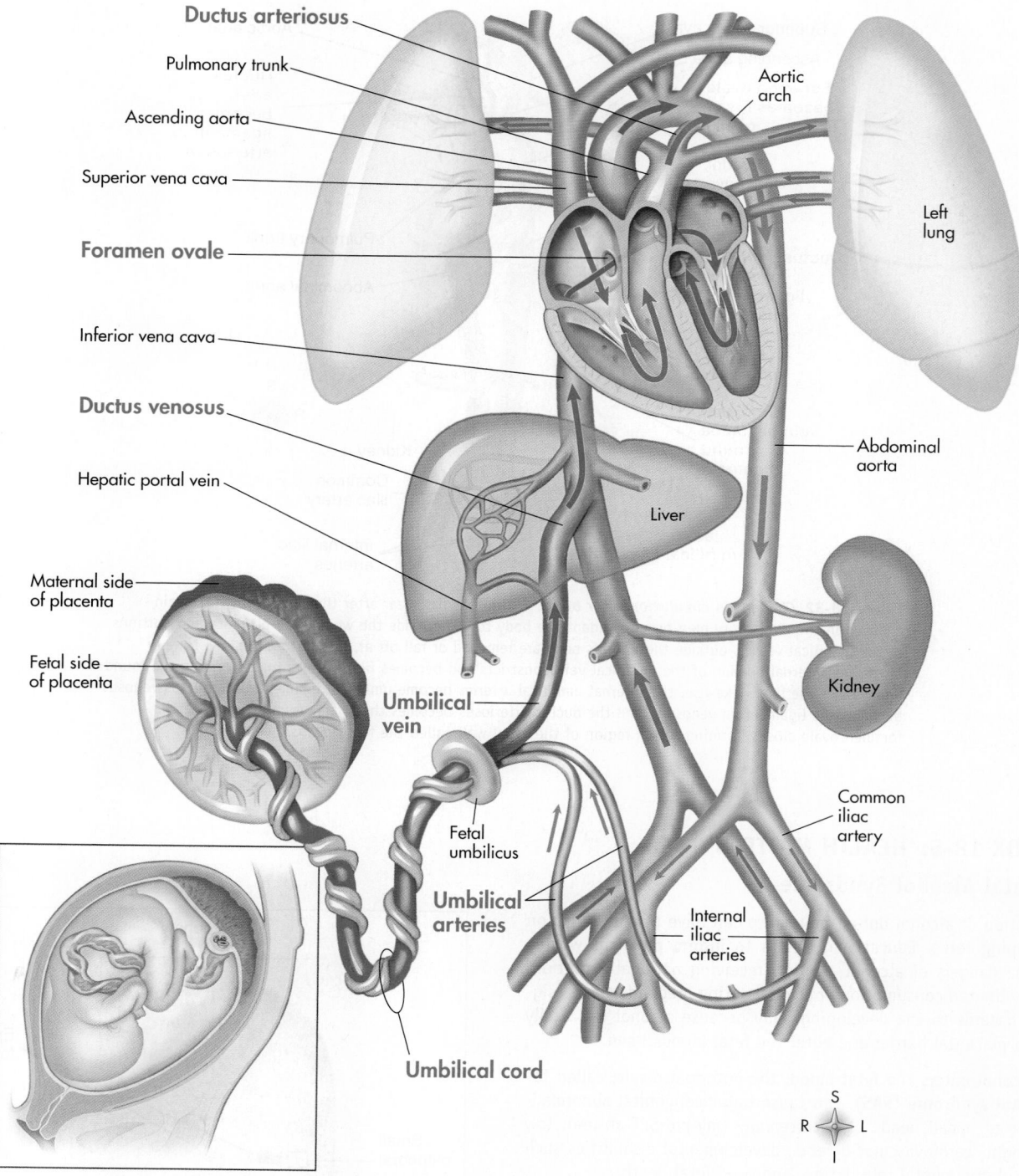

Ductus arteriosus

Pulmonary trunk

Ascending aorta

Superior vena cava

Foramen ovale

Inferior vena cava

Ductus venosus

Hepatic portal vein

Maternal side
of placenta

Fetal side
of placenta

**Umbilical
vein**

**Umbilical
arteries**

Umbilical cord

Aortic
arch

Left
lung

Abdominal
aorta

Liver

Kidney

Fetal
umbilicus

Common
iliac
artery

Internal
iliac
arteries

Figure 18-31 *Plan of fetal circulation.* Before birth, the human circulatory system has several special features that adapt the body to life in the womb. These features *(labeled in red type)* include two umbilical arteries, one umbilical vein, ductus venosus, foramen ovale, ductus arteriosus, and umbilical cord. The placenta, another essential feature of the fetal circulatory plan, is shown in Figure 18-30.

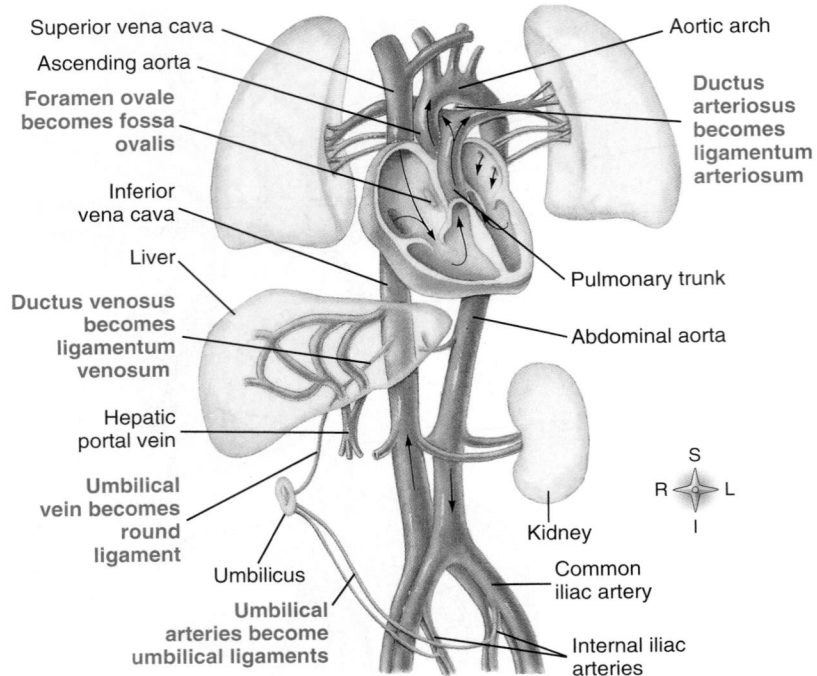

Figure 18-32 *Changes in circulation after birth.* Within the first year after the time of birth, certain changes in the circulatory plan occur to adapt the body to life outside the womb. The placenta and portions of the umbilical vessels outside the infant's body are removed or fall off at, or shortly after, the time of birth. The internal portion of the umbilical vein constricts and becomes fibrous, eventually forming the round ligament of the liver. Likewise, the internal umbilical arteries become umbilical ligaments, the ductus venosus becomes the ligamentum venosum, and the ductus arteriosus becomes the ligamentum arteriosum. The foramen ovale closes, forming a thin region of the atrial wall called the fossa ovalis.

BOX 18-5: HEALTH MATTERS
Fetal Alcohol Syndrome

Consumption of alcohol during pregnancy can have tragic effects on a developing fetus. Educational efforts to inform pregnant women about the dangers of alcohol are now receiving national attention. Even very limited consumption of alcohol during pregnancy poses significant hazards to the developing baby because alcohol can easily cross the placental barrier and enter the fetal bloodstream.

When alcohol enters the fetal blood, the potential result, called **fetal alcohol syndrome (FAS),** can cause tragic congenital abnormalities such as "small head" or *microcephaly* (my-kro-SEF-ah-lee), low birth weight, cardiovascular defects, developmental disabilities such as physical and mental retardation, and even fetal death.

The photograph shows the small head, thinned upper lip, small eye openings (palpebral fissures), epicanthal folds, and receded upper jaw (retrognathia) typical of fetal alcohol syndrome.

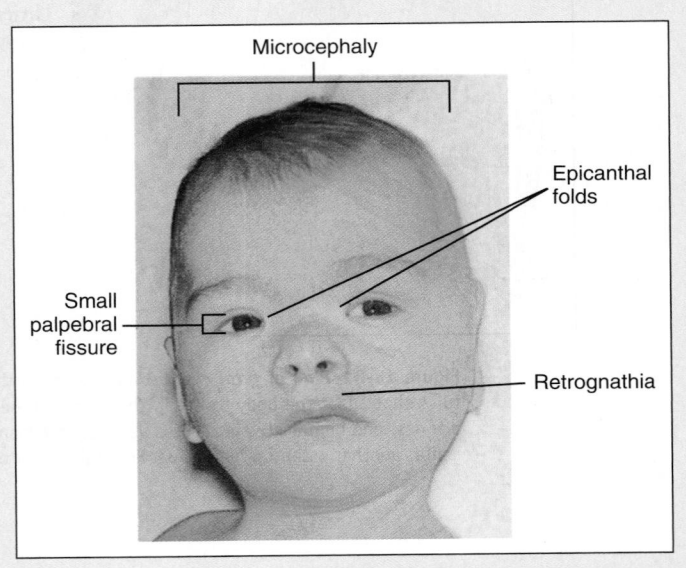

Fetal alcohol syndrome.

Cycle of Life

Cardiovascular Anatomy

As with all body structures, the heart and blood vessels undergo profound anatomical changes during early development in the womb. At birth, the switch from a placenta-dependent system causes another set of profound anatomical changes. Throughout childhood, adolescence, and adulthood, the heart and blood vessels normally maintain their basic structure and function—permitting continued survival of the individual. Perhaps the only apparent normal changes in these structures occur as a result of regular exercise. The myocardium thickens and the supply of blood vessels in skeletal muscle tissues increases in response to increased oxygen and glucose use during prolonged exercise.

As we pass through adulthood, especially later adulthood, various degenerative changes can occur in the heart and blood vessels. For example, a type of "hardening of the arteries," called **atherosclerosis,** can result in blockage or weakening of critical arteries—perhaps causing a myocardial infarction or stroke. The heart valves and myocardial tissues often degenerate with age, becoming hardened or fibrotic and less able to perform their functions properly. This reduces the heart's pumping efficiency and therefore threatens homeostasis of the entire internal environment.

THE BIG PICTURE

Cardiovascular Anatomy and the Whole Body

In this chapter, we have briefly outlined the functional anatomy of many different structures: the heart, major arteries and veins, and even the tiny arterioles, venules, and capillaries.

In essence, we have described a complex, interconnected network of transport pipes for blood. Part of what you learned from this description is how each component of this blood-transporting network is constructed and just a little bit about how it works as an individual unit. Step back from this collection of facts for a moment, and think about how each component of the heart works with the other components to keep blood flowing continuously through the heart. Broaden the picture now in your mind's eye to include each of the other components of the network: the arteries, the veins, and the capillaries. As you continue to study this chapter, keep this "big picture" in mind and add to it your deepening understanding of where each vessel and chamber gets its blood and where it delivers the blood. Before long, you will have a useful understanding of how the entire network functions as a system.

In the next chapter, we will build on your understanding of the functional anatomy of the cardiovascular system by discussing mechanisms of blood flow. In other words, all of the next chapter will deal with issues relating to the "big picture" function of cardiovascular function and its importance to survival.

Mechanisms of Disease

DISORDERS OF THE CARDIOVASCULAR SYSTEM

Disorders of Heart Structure
Disorders Involving the Pericardium

If the **pericardium** becomes inflamed, a condition called **pericarditis** results (Figure 18-33). Pericarditis may be caused by various factors: trauma, viral or bacterial infection, tumors, and other factors. The pericardial edema that characterizes this condition often causes the visceral and parietal layers of the serous pericardium to rub together—causing severe chest pain. Pericardial fluid, pus, or blood (in the case of an injury) may accumulate in the space between the two pericardial layers and impair the pumping action of the heart. This is termed **pericardial effusion** and may develop into a serious compression of the heart called **cardiac tamponade.**

Pericarditis may be acute or chronic, depending on the rate, severity, and duration of symptoms. Clinical manifestations include pericardial pain that increases with respirations or coughing, a "friction rub" (a grating, scratching sound heard over the left sternal border and upper ribs) resulting from movement together of the swollen pericardial layers, difficulty breathing, restlessness, and an accumulation of pericardial fluid. Cardiac tamponade requires immediate pericardial drainage (**pericardiocentesis**). Antibiotics are usually pre-

scribed to treat the causative organism, and nonsteroidal anti-inflammatory agents such as aspirin are prescribed to reduce the inflammation and thus control the symptoms.

Disorders Involving Heart Valves

Disorders of the cardiac valves can have several effects. For example, a congenital defect in valve structure can result in mild to severe pumping inefficiency. Incompetent valves leak, allowing some blood to flow back into the chamber from which it came. **Stenosed valves** are valves that are narrower than normal, slowing blood flow from a heart chamber (Figure 18-34).

Rheumatic heart disease results from a delayed inflammatory response to streptococcal infection that occurs most often in children. A few weeks after an improperly treated streptococcal infection, the cardiac valves and other tissues in the body may become inflamed—a condition called **rheumatic fever.** If severe, the inflammation can result in stenosis or other deformities of the valves, chordae tendineae, or myocardium.

Mitral valve prolapse (MVP), a condition affecting the bicuspid or mitral valve, has a genetic basis in some cases but can result from rheumatic fever or other factors. A prolapsed mitral valve is one whose flaps extend back into the left atrium, causing incompetence (leaking) of the valve (Figure 18-35). Although this condition is common, occurring in up to 1 in every 20 people, most cases are asymptomatic. In severe cases, patients suffer chest pain and fatigue.

Aortic regurgitation is a condition in which blood not only ejects forward into the aorta but also regurgitates back into the left ventricle because of a leaky aortic semilunar valve. This causes a volume overload on the left ventricle, with subse-

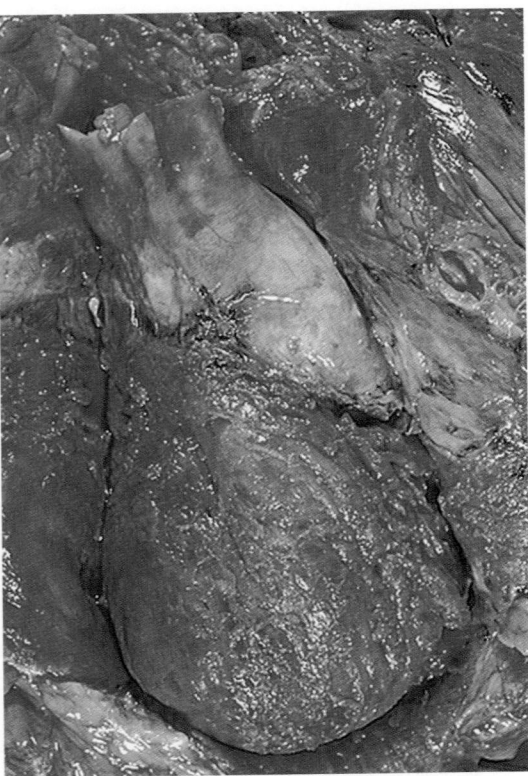

Figure 18-33 *Pericarditis.* Note the acute inflammation of the pericardium caused by extension of infective organisms from lobar pneumonia.

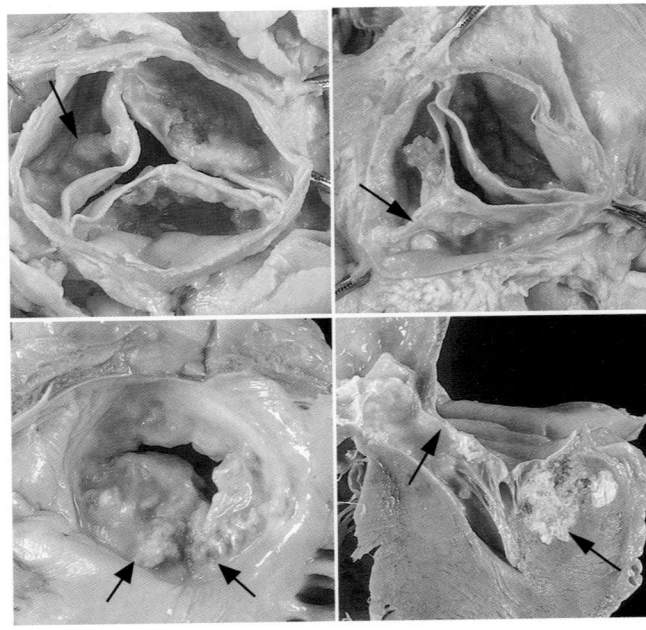

Figure 18-34 *Stenosed mitral valve.* Note the calcific nodules *(arrows)* attached to the cusps, thus narrowing the opening and slowing blood flow.

Mechanisms of Disease—cont.

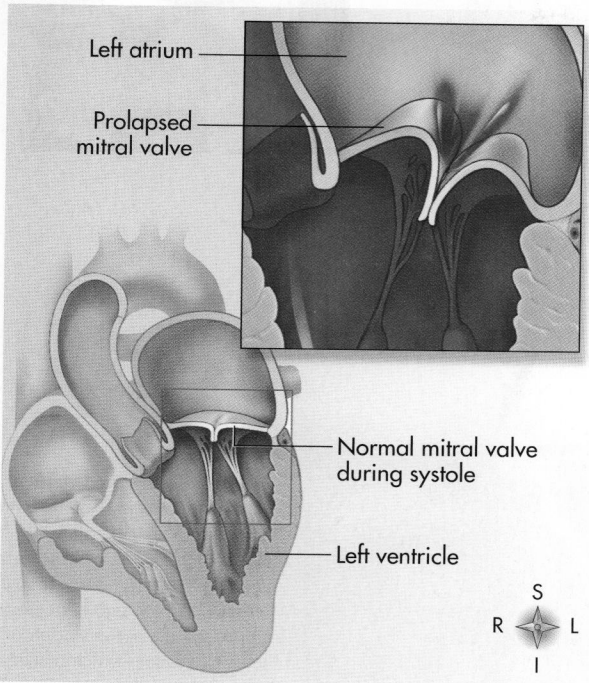

Figure 18-35 *Mitral valve prolapse.* The normal mitral valve *(left)* prevents backflow of blood from the left ventricle into the left atrium during ventricular systole (contraction). The prolapsed mitral valve *(inset)* permits leakage because the valve flaps billow backward, parting slightly.

quent hypertrophy and dilation of the left ventricle. The left ventricle attempts to compensate for the increased load by increasing its strength of contraction—which may eventually stress the heart to the point of causing myocardial ischemia.

Damaged or defective cardiac valves can often be replaced surgically—a procedure called **valvuloplasty.** Artificial valves made from synthetic materials, as well as valves taken from other mammals such as swine, are frequently used in these valve replacement procedures.

Disorders Involving the Myocardium

One of the leading causes of death in the United States is **coronary artery disease (CAD).** This condition can result from many causes, all of which somehow reduce the flow of blood to the vital myocardial tissue. For example, in both coronary thrombosis and coronary embolism, a blood clot occludes, or plugs, some part of a coronary artery. Blood cannot pass through the occluded vessel and so cannot reach the heart muscle cells it normally supplies (Figure 18-36). Deprived of oxygen, these cells soon die or are damaged. In medical terms, a **myocardial infarction (MI),** or tissue death, occurs. A myocardial infarction (heart attack) is a common cause of death during middle and late adulthood. Recovery from a myocardial infarction is possible if the amount of heart tissue damaged was small enough that the remaining undamaged heart mus-

cle can pump blood effectively enough to supply the needs of the rest of the heart, as well as the body.

Coronary arteries may also become blocked as a result of **atherosclerosis,** a type of "hardening of the arteries" in which lipids and other substances build up on the inside wall of blood vessels and eventually calcify, making the vessel wall hard and brittle. Mechanisms of atherosclerosis are discussed elsewhere in this chapter. Coronary atherosclerosis has increased dramatically over the last half century to become the leading cause of death in western countries. Many pathophysiologists believe this increase results from a change in lifestyle. They cite several important risk factors associated with coronary atherosclerosis: cigarette smoking, high-fat and high-cholesterol diets, diabetes, and hypertension (high blood pressure).

The term **angina pectoris** is used to describe the severe chest pain that occurs when the myocardium is deprived of adequate oxygen. It is often a warning that the coronary arteries are no longer able to supply enough blood and oxygen to the heart muscle. **Coronary bypass surgery** is a frequent treatment for those who suffer from severely restricted coronary artery blood flow. In this procedure, veins and sometimes arteries are "harvested" from other areas of the body and used to bypass partial blockages in coronary arteries (Figure 18-37).

Congestive heart failure (CHF), or, simply, *left-side heart failure,* is the inability of the left ventricle to pump blood effectively. Most often, such failure results from chronic systemic hypertension (high blood pressure) or from myocardial infarction caused by coronary artery disease (Figure 18-38, *A*). It is called *congestive heart failure* because it decreases pumping pressure in the systemic circulation, which in turn causes the body to retain fluids. Portions of the systemic circulation thus become congested with extra fluid. As stated previously, left-side heart failure also causes congestion of blood in the pulmonary circulation, termed **pulmonary edema**—possibly leading to right-side heart failure and pulmonary hypertension (Figure 18-38, *B*).

The term **cardiomyopathy** is used to describe a number of different types of heart diseases that result in abnormal enlargement. In addition to left and right ventricular hypertrophy, another form of enlargement is called **hypertrophic cardiomyopathy** and is caused by a genetic defect. Figure 18-39 shows the appearance of the heart in this condition. It is one of the most common causes of sudden unexplained death in young athletes.

Patients in danger of death because of heart failure may be candidates for heart *transplants* or heart *implants.* Heart transplants are surgical procedures in which healthy hearts from recently deceased donors replace the hearts of patients with heart disease. Unfortunately, a continuing problem with this procedure is the tendency of the body's immune system to reject the new heart as a foreign tissue. More details about the rejection of transplanted tissues are found in Chapter 21. Heart implants are artificial hearts that are made of biologically inert synthetic materials. On July 3, 2001, the first artificial heart was successfully implanted into Robert Tools by University of Louisville researchers. The 1-kg (2-pound) **AbioCor Implantable Replace-**

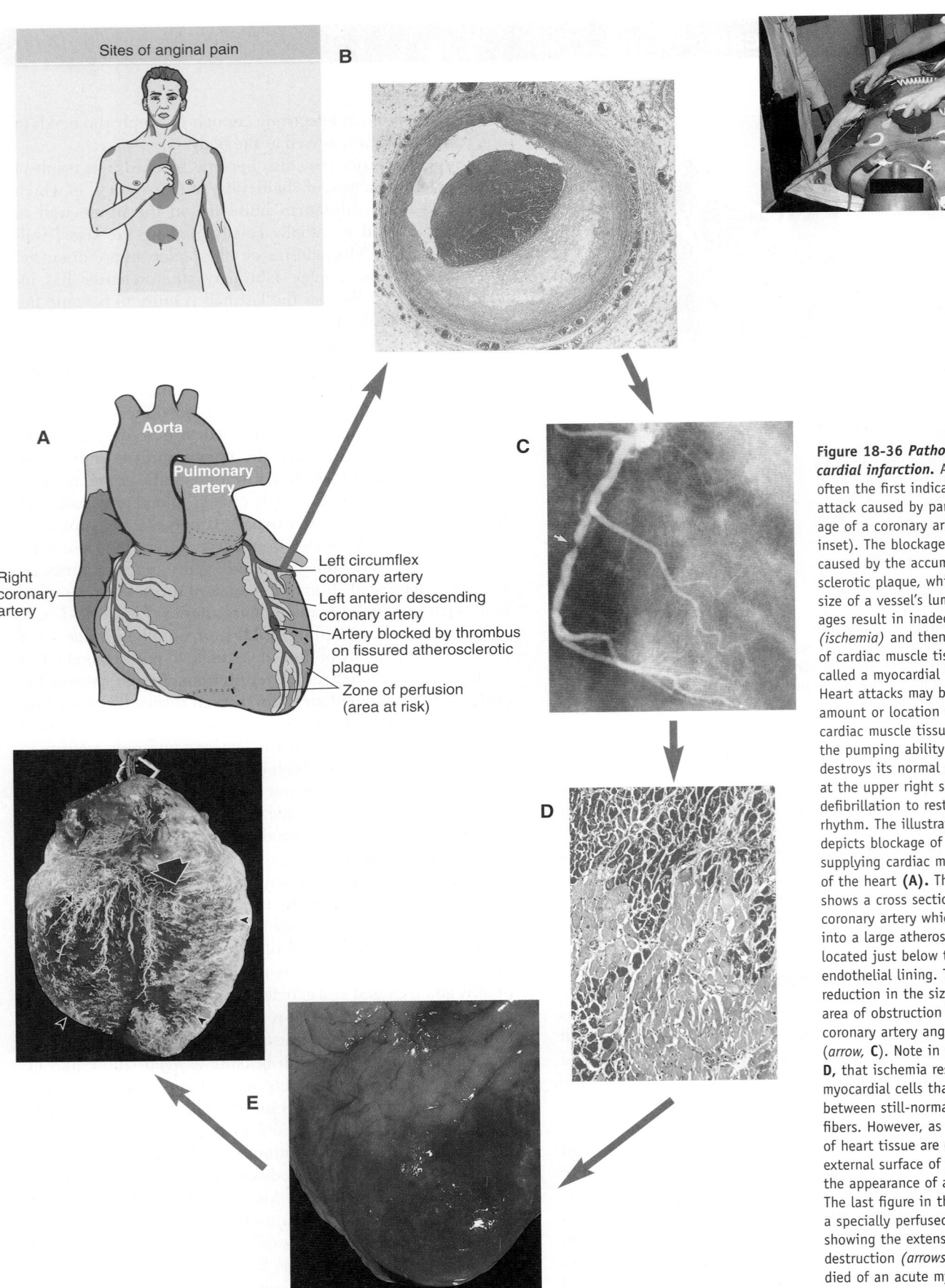

Sites of anginal pain

B

A

Aorta

Pulmonary artery

Right coronary artery

Left circumflex coronary artery

Left anterior descending coronary artery

Artery blocked by thrombus on fissured atherosclerotic plaque

Zone of perfusion (area at risk)

C

D

E

Figure 18-36 *Pathogenesis of a myocardial infarction.* Angina pain is often the first indication of a heart attack caused by partial or total blockage of a coronary artery (upper left inset). The blockage is frequently caused by the accumulation of atherosclerotic plaque, which reduces the size of a vessel's lumen. Such blockages result in inadequate blood flow *(ischemia)* and then death *(infarction)* of cardiac muscle tissue. The result is called a myocardial infarction or MI. Heart attacks may be fatal if the amount or location of the destroyed cardiac muscle tissue seriously limits the pumping ability of the heart or destroys its normal rhythms. The inset at the upper right shows the use of defibrillation to restore normal heart rhythm. The illustration sequence first depicts blockage of a coronary artery supplying cardiac muscle in the apex of the heart **(A).** The next figure **(B)** shows a cross section of the diseased coronary artery which is hemorrhaging into a large atherosclerotic plaque located just below the vessel's endothelial lining. The result is a reduction in the size of the lumen. The area of obstruction is apparent in the coronary artery angiography x-ray *(arrow,* **C).** Note in the tissue section, **D,** that ischemia results in infarcted myocardial cells that can be seen between still-normal cardiac muscle fibers. However, as increasing amounts of heart tissue are destroyed, the external surface of the apex takes on the appearance of an acute infarct **(E).** The last figure in the sequence, **F,** is of a specially perfused postmortem heart showing the extensive areas of tissue destruction *(arrows)* in a patient who died of an acute myocardial infarction.

Mechanisms of Disease—cont.

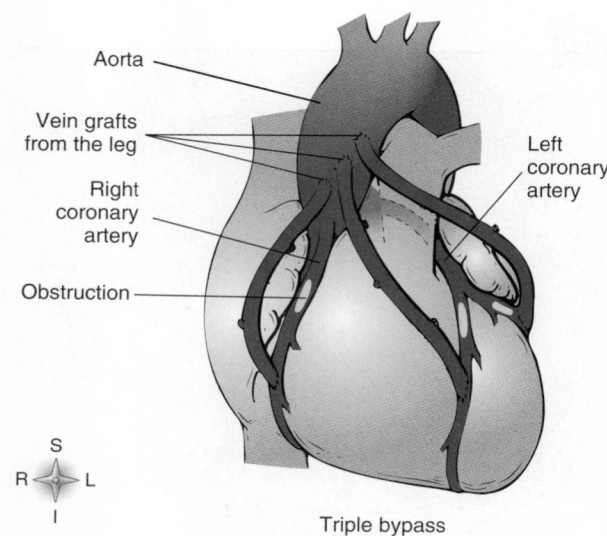

Figure 18-37 *Coronary bypass.* In coronary bypass surgery, blood vessels are "harvested" from other parts of the body and used to construct detours around blocked coronary arteries. Artificial vessels can also be used.

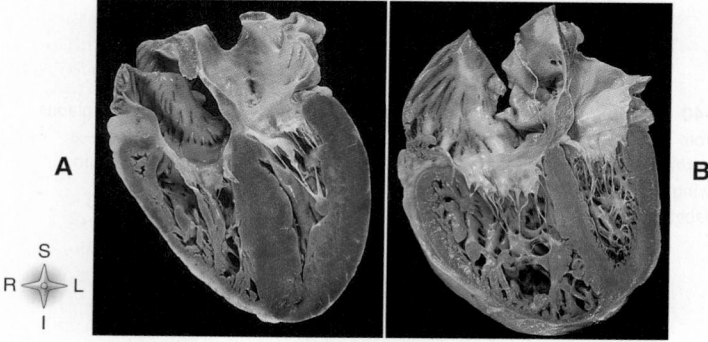

Figure 18-38 *Congestive heart failure.* **A,** Left ventrical hypertrophy (left-sided heart failure). Note the hypertrophy of the left ventricle often caused by chronic systemic hypertension. **B,** Right ventricular hypertrophy (right-sided heart failure or *cor pulmonale*). Note the right ventricular hypertrophy and dilation caused by pulmonary hypertension.

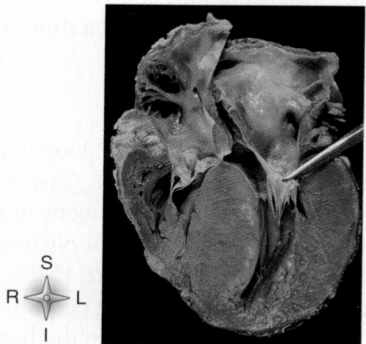

Figure 18-39 *Hypertrophic cardiomyopathy.* Note the dramatic increase of myocardial tissue. This genetic condition may cause sudden heart failure in the apparent absence of warning symptoms. It is a cause of sudden, unexplained death in young athletes.

ment Heart ("artificial heart") allowed the patient to move about freely without any external pumps. Portable external battery packs are used to recharge the small internal battery that powers the internal pumping unit. The success of the AbioCor unit and other artificial heart models gives hope that artificial hearts will one day be a commonplace and effective treatment for patients with severe heart disease.

Disorders of Blood Vessels
Disorders of Arteries

As mentioned earlier in this chapter, arteries contain blood that is maintained at a relatively high pressure. This means the arterial walls must be able to withstand a great deal of force or they will burst. The arteries must also stay free of obstruction; otherwise they cannot deliver their blood to the capillary beds (and thus the tissues they serve).

A common type of vascular disease that occludes (blocks) arteries and weakens arterial walls is called **arteriosclerosis,** or *hardening of the arteries.* Arteriosclerosis is characterized by thickening of arterial walls that progresses to hardening as calcium deposits form. The thickening and calcification reduce the flow of blood to the tissues. If blood slows in peripheral tissues such as the hands, lower legs, or feet, the condition is often referred to as **peripheral vascular disease** and the result is **ischemia.** Ischemia, or decreased blood supply to a tissue, involves the gradual death of cells and may lead to complete tissue death—a condition called **necrosis.** If a large section of tissue becomes necrotic, it may begin to decay because of bacterial action. Necrosis that has progressed this far is called **gangrene.** Gangrene is a very serious consequence of ischemia or blocked blood flow in peripheral vascular disease and is, unfortunately, a common problem in uncontrolled diabetes (Figure 18-40).

Because of the tissue damage involved, arteriosclerosis is not only painful—it is life threatening. As we have previously stated, ischemia of heart muscle can lead to *myocardial infarction* and death (see Figure 18-36).

There are several types of arteriosclerosis, but perhaps the most well known is *atherosclerosis*—described earlier as the blockage of arteries by lipids and other matter (Figure 18-40). Eventually, the fatty deposits in the arterial walls become fibrous and perhaps calcified—resulting in sclerosis (hardening). High blood levels of triglycerides and cholesterol, which may be caused by a high-fat and high-cholesterol diet, smoking, and a genetic predisposition, are associated with atherosclerosis.

In general, arteriosclerosis develops with advanced age, diabetes, high-fat and high-cholesterol diets, hypertension (high blood pressure), and smoking. Arteriosclerosis can be treated by drugs, called **vasodilators,** that trigger the smooth muscles of the arterial walls to relax, thus causing the arteries to dilate (widen). Some cases of atherosclerosis are treated by mechanically opening the affected area of an artery, a type of procedure called **angioplasty.** In one procedure, a deflated balloon attached to a long tube, called a **catheter,** is inserted into a partially blocked artery and inflated (Figure 18-41). As

Mechanisms of Disease—cont.

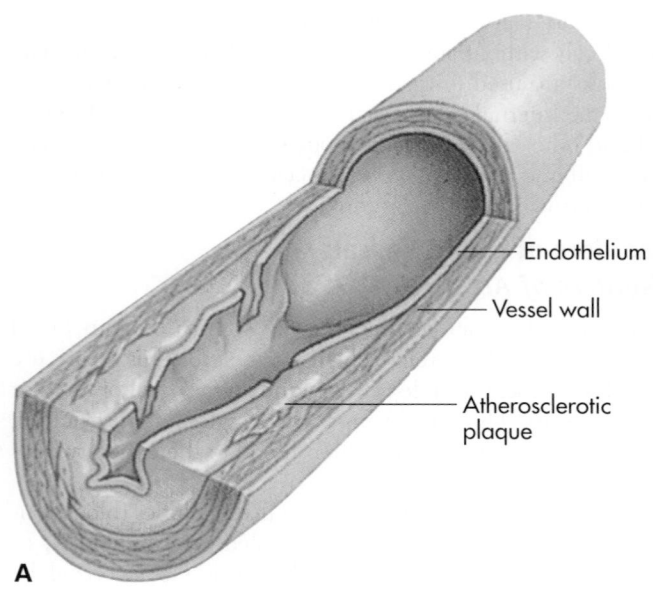

A

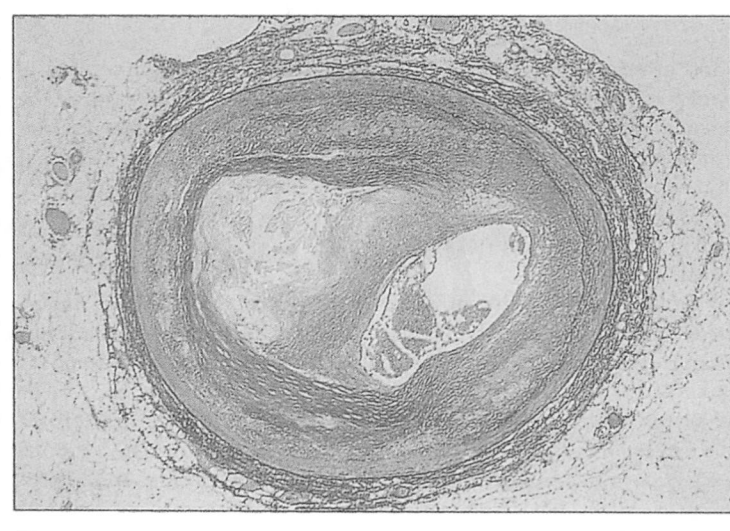

B

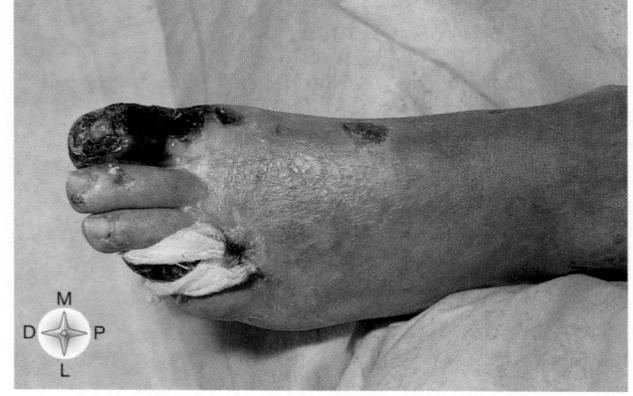

C

Figure 18-40 *Partial blockage of an artery in atherosclerosis.* **A,** Atherosclerotic plaque develops from the deposition of fats and other substances in the wall of the artery—a classic symptom of peripheral vascular disease. **B,** Photograph of a cross section of an artery showing partial blockage of lumen by atherosclerotic plaque. **C,** Photograph showing diabetic gangrene that results from decreased peripheral blood flow.

the balloon inflates, the *plaque* (fatty deposits and tissue) is pushed outward, and the artery widens to allow near-normal blood flow. In a similar procedure, metal springs or mesh tubes, called **stents,** are inserted in affected arteries to hold them open. Other types of angioplasty use lasers, drills, or spinning loops of wire to clear the way for normal blood flow. Severely affected arteries can also be surgically bypassed or replaced.

Damage to arterial walls caused by arteriosclerosis or other factors may lead to the formation of an **aneurysm.** An aneurysm is a section of an artery that has become abnormally widened because of a weakening of the arterial wall. Aneurysms sometimes form a saclike extension of the arterial wall. One reason aneurysms are dangerous is because they, like atherosclerotic plaques, promote the formation of thrombi (abnormal clots). A thrombus may cause an embolism (blockage) in the heart or some other vital tissue. Another reason aneurysms are dangerous is their tendency to burst, causing severe hemorrhaging that

may result in death. A brain aneurysm may lead to a **stroke,** or **cerebrovascular accident (CVA).** A stroke results from ischemia of brain tissue caused by an embolism or ruptured aneurysm. Depending on the amount of tissue affected and the place in the brain the CVA occurs, effects of a stroke may range from hardly noticeable to crippling to fatal.

Disorders of Veins

Varicose veins are enlarged veins in which blood tends to pool rather than continue on toward the heart. Varicosities, also called **varices** (singular, *varix*), commonly occur in *superficial veins* near the surface of the body. The *great saphenous vein,* the largest superficial vein of the leg (see Figure 18-29), often becomes varicose in people who stand for long periods. The force of gravity slows the return of venous blood to the heart in such cases, causing blood-engorged veins to dilate. As the veins dilate, the distance between the flaps of venous valves widens—eventually making them incompetent (leaky) (Figure 18-42).

Mechanisms of Disease—cont.

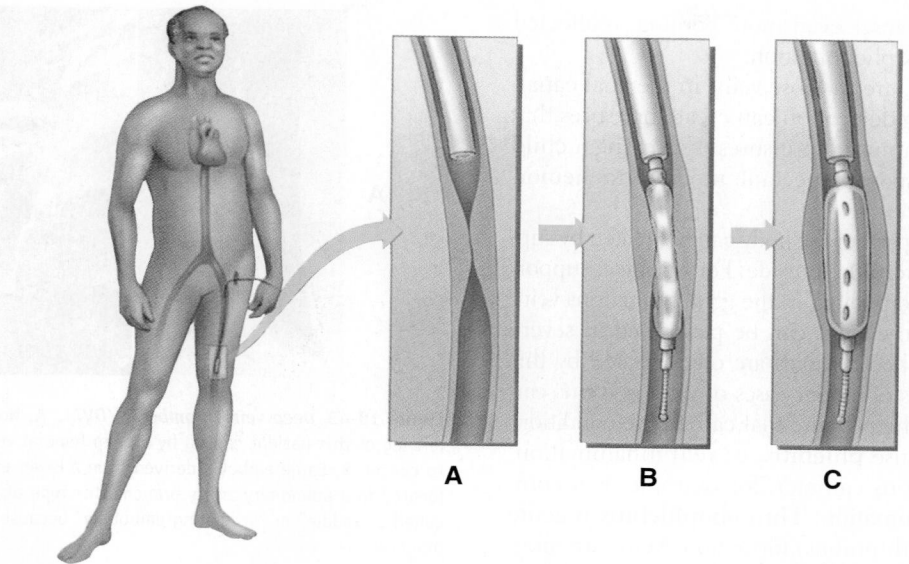

Figure 18-41 *Balloon angioplasty.* **A,** A catheter is inserted into the vessel until it reaches the affected region. **B,** A probe with a metal tip is pushed out the end of the catheter into the blocked region of the vessel. **C,** The balloon is inflated, pushing the walls of the vessel outward. Sometimes metal coils or tubes, called "stents," are inserted to keep the vessel open. Newer drug-coated stents further reduce delayed scarring and reclogging of arteries following angioplasty.

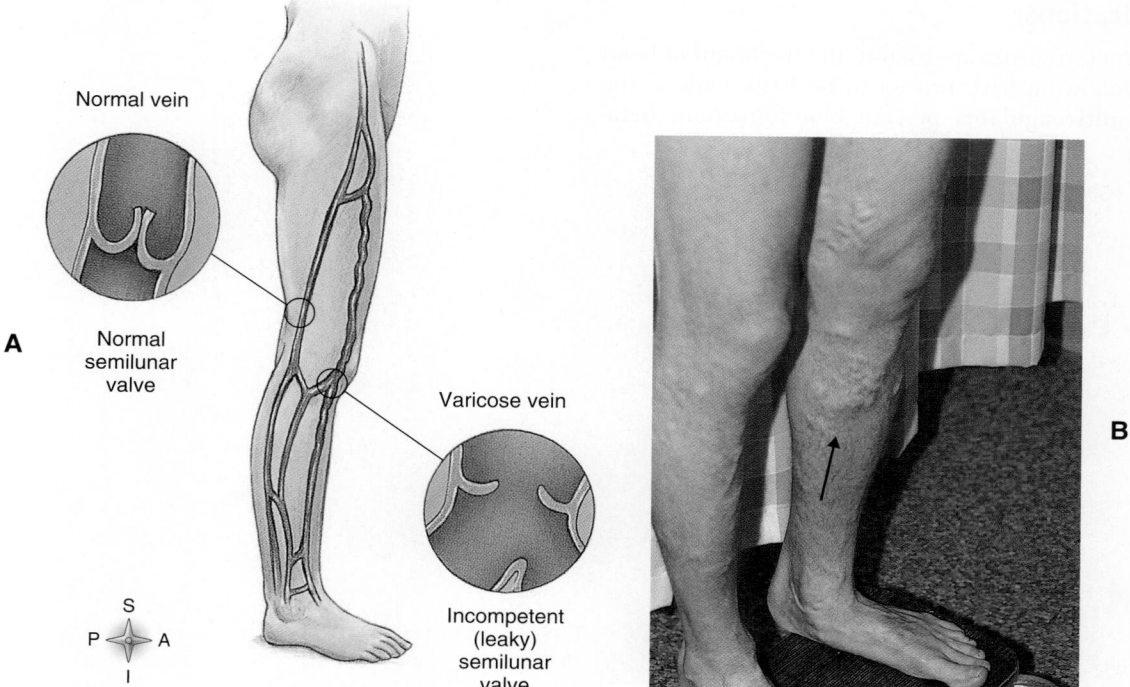

Figure 18-42 *Varicose veins.* **A,** Veins near the surface of the body—especially in the legs—may bulge and cause venous valves to leak. **B,** Photograph showing varicose veins on the surface of the leg.

Mechanisms of Disease — cont.

Incompetence of valves causes even more pooling in affected veins — a positive feedback phenomenon.

Hemorrhoids, or *piles,* are varicose veins in the anal canal. Excessive straining during defecation can create pressures that cause hemorrhoids. The unusual pressures of carrying a child during pregnancy predispose expectant mothers to hemorrhoids and other varicosities.

Varicose veins in some parts of the body can be treated by supporting the dilated veins from the outside. For instance, support stockings can reduce blood pooling in the great saphenous vein. Surgical removal of varicose veins can be performed in severe cases. Advanced cases of hemorrhoids are often treated by this type of surgery. Symptoms of milder cases of varicose veins can be relieved by removing the pressure that caused the condition.

Several factors can cause **phlebitis,** or vein inflammation. Irritation by an intravenous catheter, for example, is a common cause of vein inflammation. **Thrombophlebitis** is acute phlebitis caused by clot (thrombus) formation. Veins are more likely sites of thrombus formation than arteries because venous blood moves more slowly and is under less pressure. A *deep vein thrombosis (DVT)* is a clot that has formed in a deep vein, especially in the legs, and is particularly dangerous. If a piece of a clot breaks free from a DVT, it may cause an embolism when it blocks a blood vessel. **Pulmonary embolism,** for example, could result when an embolus lodges in the circulation of the lung (Figure 18-43). Pulmonary embolism can lead to death quickly if too much blood flow is blocked.

Heart Medications

Although numerous drugs are used in the treatment of heart disease, the following have proven to be basic tools of the cardiologist: **anticoagulants** prevent clot formation; **beta-**

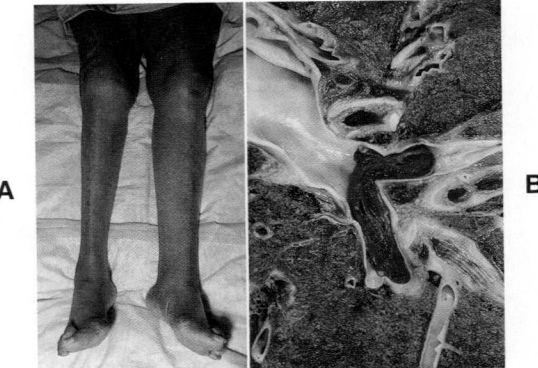

A   **B**

Figure 18-43 *Deep vein thrombosis (DVT).* **A,** Note the swelling in the left leg of this patient caused by a deep femoral vein thrombosis secondary to cancer. **B,** Large embolus derived from a lower extremity DVT and now located in a pulmonary artery branch. This type of blood clot is sometimes called a "saddle" or "straddling embolism" because it straddles a dividing blood vessel.

adrenergic blockers block norepinephrine receptors and thus reduce the strength and rate of heart beats; **calcium channel blockers** reduce heart contractions by preventing the flow of Ca^{++} into cardiac muscle cells; **digitalis** slows and increases the strength of cardiac contractions; **nitroglycerin** dilates coronary blood vessels and thus improves O_2 supply to myocardium; and **tissue plasminogen activator (t-PA)** helps dissolve clots.

LANGUAGE OF SCIENCE   *(Cont'd from page 677)*

brachiocephalic vein (brayk-ee-oh-seh-FAL-ik)
[*brachi-* arm, *-cephalic* pertaining to the head]

bronchial vein (BRONK-kee-al) [*bronchi-* windpipe, *-al* pertaining to]

capillary (KAP-i-lair-ee) [*capillaris* hairlike]

cephalic vein (seh-FAL-ik) [*cephalic* pertaining to the head]

chordae tendineae (KOR-dee ten-DIN-ee) [*chord* string or cord, *tendere* to stretch]

continuous capillaries (KAP-i-lair-eez) [*capillaris* hairlike]

coronary arteries (KOHR-oh-nair-ee AR-ter-eez) [*corona* crown, *arteria* air pipe]

cuspid valves (KUS-pid) [*cuspis* point, *valva* folding door]

deep veins

descending aorta (ay-OR-tah) [*descendere* to descend, *aerein* to raise]

ductus arteriosus (DUK-tus ar-teer-ee-OH-sus) [*ductus* duct, *arterio* artery]

ductus venosus (DUK-tus veh-NO-sus) [*ductus* duct, *veno* vein]

elastic arteries

end-arteries

endocardium (en-doh-KAR-dee-um) [*endo-* inward or within, *-cardi* heart]

endothelium (en-doh-THEE-lee-um) [*endo-* inward or within, *-thele* nipple]

epicardium (ep-i-KAR-dee-um [*epi-* on or upon, *-cardi* heart]

esophageal vein (eh-sof-ah-JEE-al) [*oisophagos* gullet]

external iliac vein (eks-TER-nal IL-ee-ak) [*externus* outward, *ilium* flank]

femoral vein (FEM-or-al) [*femur* thigh]

fenestrated capillaries (fen-es-TRAY-tid KAP-i-lair-eez) [*fenestrae* window, *capillaris* hairlike]

fibrous pericardium (FYE-brus pair-i-KAR-dee-um) [*fibra* fiber, *peri-* around, *-cardi* heart]

fibular (peroneal) vein (FIB-yoo-lar [per-oh-NEE-al]) [*fibula* clasp, *perone* brooch]

foramen ovale (foh-RAY-men oh-VAL-ee) [*for* an opening, *ovalis* egg shaped]

great saphenous vein (sah-FEE-nus) [*saphenes* manifest]

hemiazygos vein (hem-ee-AZ-i-gohs) [*hemi-* half, *-a-* without, *-zygon* yolk]

hepatic portal circulation (heh-PAT-ik POR-tal) [*hepar-* liver, *-ic* pertaining to, *porta* gateway, *circulatio* to go around]

hepatic portal system (heh-PAT-ik POR-tal) [*hepar-* liver, *-ic* pertaining to, *porta* gateway]

inferior vena cava (in-FEER-ee-or VEE-nah KAY-vah) [*inferus* lower, *vena* vein, *cavum* cavity]

internal jugular vein (JUG-yoo-lar) [*internus* inward, *jugulum* neck]

left common carotid artery (kah-ROT-id AR-ter-ee) [*karos* heavy sleep, *arteria* airpipe]

left external jugular vein (eks-TER-nal JUG-yoo-lar) [*externus* outward, *jugulum* neck]

left renal vein (REE-nal) [*ren-* kidney, *-al* pertaining to]

left spermatic vein (left sper-MAT-ik) [*sperm-* seed, *-ic* pertaining to]

left subclavian artery (sub-KLAY-vee-an) [*sub-* below, *-clavicula* little key]

left suprarenal vein (soo-prah-REE-nal) [*supra-* above or over, *-ren-* kidney, *-al* pertaining to]

median cubital vein (KYOO-bi-tall) [*medius* middle, *cubital* pertaining to the elbow]

metarteriole (met-ar-TEER-ee-ohl) [*meta-* change or exchange, *-arteriola* little artery]

mitral valve (MY-tral) [*mitra* turban, *valva* folding door]

muscular arteries (MUSS-kyoo-lar AR-ter-eez)

myocardium (my-oh-KAR-dee-um) [*myo-* muscle, *-cardia* heart]

palmar venous arches (PAHL-mar VEE-nus) [*palmar* pertaining to the palm, *veno* veins]

pericardial fluid (pair-i-KAR-dee-al) [*peri-* around, *-cardi-* heart, *-al* pertaining to]

pericardial space (pair-i-KAR-dee-al) [*peri-* around, *-cardi-* heart, *-al* pertaining to]

pericardial vein (pair-i-KAR-dee-al) [*peri-* around, *-cardi-* heart, *-al* pertaining to]

pericardium (pair-i-KAR-dee-um) [*peri-* around, *-cardi* heart]

placenta (plah-SEN-tah) [*placenta* flat cake]

popliteal vein (pop-lih-TEE-al vayn) [*poples* ham of the knee]

portal system (POR-tal) [*porta* gateway]

posterior tibial vein (pohs-TEER-ee-or TIB-ee-al vayn) [*posterior* behind, *tibia* shin bone]

pulmonary circulation (PUL-moh-nair-ee) [*pulmon* lungs, *circulatio* to go around]

Purkinje fibers (pur-KIN-jee) [*Johannes E. Purkinje* Czech physiologist]

right common carotid artery (kah-ROT-id AR-ter-ee) [*karos* heavy sleep, *arteria* air pipe]

right external jugular vein (eks-TER-nal JUG-yoo-lar) [*externus* outward, *jugulum* neck]

right subclavian artery (sub-KLAY-vee-an AR-ter-ee) [*sub-* below, *-clavicula* little key]

semilunar (SL) valves (sem-i-LOO-nar) [*semi-* one half, *-luna* moon, *valva* folding door]

serous pericardium (SEER-us pair-i-KAR-dee-um) [*serum* whey, *peri-* around, *-cardi* heart]

sinoatrial node (sye-no-AY-tree-al) [*sinus-* hollow, *-atria* hall]

sinuses (SYE-nus-us) [*sinus* hollow]

sinusoid (SYE-nah-soyd) [*sinus-* hollow, *-oid* resembling]

small saphenous vein (sah-FEE-nus) [*saphenes* manifest]

subclavian veins (sub-KLAY-vee-an) [*sub-* below, *-clavicula* little key]

superficial veins (soo-per-FISH-al) [*superficies* surface]

superior intercostal veins (soo-PEER-ee-or in-ter-KOS-tal) [*superior* higher, *inter-* occurring between, *-costa* ribs]

superior vena cava (soo-PER-ee-or VEE-nah KAY-vah) [*superior* higher, *vena* vein, *cavum* cavity]

systemic circulation (sis-TEM-ik ser-kyoo-LAY-shun) [*systema-* an organized whole, *-ic* pertaining to, *circulatio* to go around]

thoracic aorta (tho-RASS-ik ay-OR-tah) [*thorax* chest, *aerein* to raise]

tricuspid valve (try-KUS-pid) [*tri-* three, *-cuspis* point, *valva* folding door]

true capillaries

tunica adventitia (TOO-nih-kah ad-ven-TISH-ah) [*tunica* tunic, *adventitius* coming from abroad]

tunica media (TOO-nih-kah MEE-dee-ah) [*tunica* tunic, *media* middle]

umbilical arteries (um-BIL-i-kul AR-ter-eez) [*umbilicus-* navel, *-al* pertaining to, *arteria* air pipe]

umbilical vein (um-BIL-i-kul) [*umbilicus-* navel, *-al* pertaining to]

vascular anastomosis (VAS-kyoo-lar ah-nas-toh-MOH-sis) [*vas-* vessel, *-ular* pertaining to, *anastomoien-* to provide a mouth, *-osis* increase in a pathologic condition]

veins

venous sinus (VEE-nus SYE-nus) [*veno* vein, *sinus* hollow]

ventricles (VEN-tri-kuls) [*ventri* belly]

venules (VEN-yools) [*venula* small vein]

LANGUAGE OF MEDICINE

AbioCor Implantable Replacement Heart

aneurysm (AN-yoo-riz-em) [*aneurysma* widening]

angina pectoris (an-JYE-nah PEK-tor-iss) [*angor* strangling, *pector-* breast, *-alis* pertaining to]

angioplasty (AN-jee-oh-plass-tee) [*angio-* a blood vessel, *-plasty* surgical repair]

anticoagulants (an-tee-koh-AG-yoo-lants) [*anti-* against, *-coagulare* curdle]

aortic regurgitation (ay-OR-tik ree-gur-jih-TAY-shun) [*aerein* to raise, *re-* again, *-guritare* to flow back]

arteriosclerosis (ar-teer-ee-oh-skleh-ROH-sis) [*arterio-* artery, *-scler-* hardening, *-osis* result]

atherosclerosis (ath-er-oh-skleh-ROH-sis) [*athero-* gruel, *-scler-* hardening, *-osis* result]

beta-adrenergic blockers (BAY-tah-ad-ren-ER-jik) [*ad-* toward, *-ren-* kidney, *-ergon* work]

calcium channel blockers (KAL-see-um CHAN-al)

cardiac nuclear scanning (KAR-dee-ak NOO-klee-ar) [*cardia* heart, *nucleus* nut or kernel]

cardiac tamponade (KAR-dee-ak tam-pon-odd) [*cardia* heart, *tamponner* to plug up]

cardiomyopathy (KAR-dee-oh-my-OP-ah-thee) [*cardio-* heart, *-myo-* muscle, *-pathy* disease]

catheter (KATH-eh-ter) [*katheter* something lowered]

cerebrovascular accident (CVA) (SAIR-eh-broh-VAS-kyoo-lar) [*cerebr-* cerebrum, *-vas-* vessel, *-ular* pertaining to]

color Doppler mapping [*Christian Doppler* Australian physicist and mathematician]

congestive heart failure (CHF) (kon-JES-tive) [*congerere* to accumulate]

coronary artery disease (CAD) (KOR-oh-nair-ee AR-ter-ee) [*corona* crown, *arteria* air pipe, *di-* reversal, *-aise* ease]

coronary bypass surgery (KOR-oh-nair-ee) [*corona* crown]

C-reactive protein

digitalis (dij-i-TAL-is) [*digitalis* of the fingers]

Doppler ultrasonography (ul-trah-son-OG-rah-fee) [*Christian Doppler* Australian physicist and mathematician, *ultra-* beyond, *-sonus-* sound, *-graph* product of drawing or writing]

echocardiography (ek-oh-kar-dee-OG-rah-fee) [*echo-* sound, *-cardi-* heart, *-graph* product of drawing or writing]

fetal alcohol syndrome (FAS) (FEE-tal AL-koh-hol SIN-drohm) [*fetus* fruitful *alkohl* subtle essence *syn-* together, *-dromos* course]

gangrene (GANG-green) [*gangraina* a gnawing sore]

hemorrhoids (HEM-eh-royds) [*haima-* blood, *-rhoia* flow]

hypertrophic cardiomyopathy (hye-PER-troh-fik KAR-dee-oh-my-OP-ah-thee) [*hyper-* excessive, *-trophe* nourishment, *cardi-* heart, *-myo-* muscle, *-pathy* disease]

ischemia (is-KEE-mee-ah) [*ischein-* to hold back, *-emia* blood condition]

mitral valve prolapse (MVP) (MY-tral valv PROH-laps) [*mitra-* turban, *-al* pertaining to, *valva* folding door, *prolapsus* falling]

myocardial infarction (MI) (my-oh-KAR-dee-al in-FARK-shun) [*myo-* muscle, *-cardi-* heart, *-al* pertaining to, *infarcire* to stuff]

necrosis (neh-KROH-sis) [*necro-* death, *-sis* condition]

nitroglycerin (nye-troh-GLIS-eh-rin)

pericardial effusion (pair-i-KAR-dee-all ef-FYOO-shen) [*peri-* around, *-cardi* heart, *effundere* to pour out]

pericardiocentesis (pair-ee-KAR-dee-oh-sen-TEE-sis) [*peri-* around, *-cardi-* heart, *-centesis* pricking]

pericarditis (pair-i-kar-DYE-tis) [*peri-* around, *-cardi-* heart, *-itis* inflammation]

peripheral vascular disease (peh-RIF-er-al VAS-kyoo-lar dih-ZEEZ) [*peri-* around, *-phereia* boundary]

phlebitis (fleh-BYE-tis) [*phleb-* vein, *-itis* inflammation]

pulmonary edema (PUL-moh-nair-ee eh-DEE-mah) [*pulmon* lungs, *oidema* swelling]

pulmonary embolism (PUL-moh-nair-ee EM-boh-liz-em) [*pulmon* lungs, *embolus* plug]

rheumatic fever (roo-MAT-ik FEE-ver) [*rheuma-* flux, *-ic* pertaining to]

rheumatic heart disease (roo-MAT-ik) [*rheuma-* flux, *-ic* pertaining to]

shunts

stenosed valves (steh-NOSD) [*stenos-* narrow, *-osis* condition, *valva* folding door]

stents

stroke

thrombophlebitis (throm-boh-fleh-BYE-tis) [*thrombo-* clot, *-phleb-* vein, *-itis* inflammation]

tissue plasminogen activator (t-PA) (TISH-yoo plaz-MIN-oh-jen AK-ti-vay-tor)

troponins test (TROH-poh-nins)

valvuloplasty (VAL-vyoo-loh-plas-tee) [*valva* folding door, *-plasty* surgical repair]

varices (VAIR-i-seez) [*varices* varicose veins]

varicose veins (VAIR-i-kohse)

vasodilators (vay-so-DYE-lay-tors) [*vaso-* vessel or duct, *-dilatare* to widen]

CASE STUDY

Clyde Banks, a 57-year-old African-American man who works in New York City as a taxi driver, is admitted to the critical care unit with chest pain. He has smoked two packs of cigarettes per day for 25 years. He has high blood pressure and type 2 diabetes mellitus. His father died of a heart attack at 51 years of age. Other than bowling once each week, he leads a sedentary lifestyle. His typical food intake is high in fat and cholesterol.

Mr. Banks was bowling with friends when he felt pain in his chest. At first he ignored the pain, but it became a crushing sensation in his sternal area that spread down his left arm. The pain continued even after he sat down for a while. His friends became concerned and called an ambulance.

On admission to the hospital, his physical examination reveals a height of 5 feet 11 inches and a weight of 230 pounds. His blood pressure is 110/60. He has a pulse of 110 beats/min. His heart sounds irregular with an extra heart sound. He has crackling sounds at the bases of both lungs. His distal peripheral pulses are palpable but weak. His skin temperature is cool, and he is diaphoretic. The cardiac monitor shows irregular electrical activity. A 12-lead electrocardiogram indicates evidence of acute injury to the anterior myocardium. Cardiac enzymes are elevated in a pattern typical of myocardial infarction. High-density lipoprotein (HDL) levels are low, and low-density lipoprotein (LDL) levels are high.

1. Which of Mr. Banks' risk factors for atherosclerosis are considered modifiable?
   1. Age
   2. Hypertension
   3. Smoking
   4. Family history
   5. Male sex
   6. African-American race
   7. Cholesterol, HDL, and LDL levels
   8. Sedentary lifestyle
   9. High fat intake
   10. Body weight
      A. 1, 2, 3, 8, 9, and 10
      B. 2, 3, 7, 8, 9, and 10
      C. 1, 2, 4, 5, 6, and 7
      D. 2, 3, 5, 6, 7, and 10

2. Which of the following factors is the earliest pathologic event leading to Mr. Banks' blocked coronary arteries?
   A. Lipid deposition in the tunica intima
   B. Ulceration of the tunica intima
   C. Cellular proliferation of fatty streaks
   D. Calcification of fibrous plaques

3. Based on the location of the infarcted tissues, which of the following coronary vessels is MOST likely to have atherosclerotic lesions?
   A. Left coronary vein
   B. Left coronary artery
   C. Left subclavian artery
   D. Right coronary vein

4. What was the primary cause of Mr. Banks' chest pain?
   A. Myocardial ischemia
   B. Pulmonary insufficiency
   C. High cardiac output
   D. Oliguria

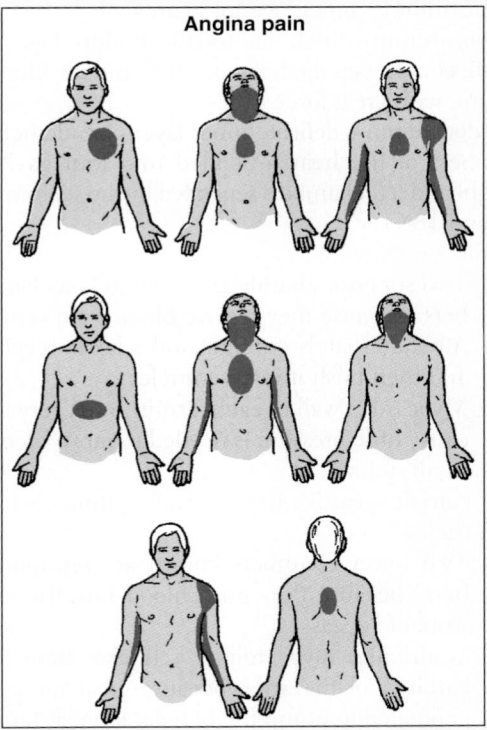

Angina pain

CHAPTER SUMMARY

HEART
A. Location of the heart (Figure 18-1)
   1. Lies in the mediastinum, behind the body of the sternum between the points of attachment of ribs two through six; approximately two thirds of its mass is to the left of the midline of the body, and one third is to the right
   2. Posteriorly the heart rests on the bodies of thoracic vertebrae five through eight
   3. Apex lies on the diaphragm, pointing to the left
   4. Base lies just below the second rib
   5. Boundaries of the heart are clinically important as an aid in diagnosing heart disorders

B. Size and shape of the heart (Figures 18-1 and 18-2)
   1. At birth, transverse and appears large in proportion to the diameter of the chest cavity
   2. Between puberty and 25 years of age the heart attains its adult shape and weight
   3. In adult, the shape of the heart tends to resemble that of the chest
C. Coverings of the heart
   1. Structure of the heart coverings
      a. Pericardium (Figure 18-4)
         (1) Fibrous pericardium—tough, loose-fitting inextensible sac
         (2) Serous pericardium—parietal layer lies inside the fibrous pericardium, and visceral layer (epicardium) adheres to the outside of the heart; pericardial space with pericardial fluid separates the two layers
   2. Function of the heart coverings—provides protection against friction
D. Structure of the heart
   1. Wall of the heart—made up of three distinct layers (Figure 18-5)
      a. Epicardium—outer layer of heart wall
      b. Myocardium—thick, contractile middle layer of heart wall; compresses the heart cavities, and the blood within them, with great force
      c. Endocardium—delicate inner layer of endothelial tissue
   2. Chambers of the heart—divided into four cavities with the right and left chambers separated by the septum (Figures 18-6 and 18-7)
      a. Atria
         (1) Two superior chambers known as "receiving chambers" because they receive blood from veins
         (2) Atria alternately contract and relax to receive blood and then push it into ventricles
         (3) Myocardial wall of each atrium is not very thick, because little pressure is needed to move blood such a small distance
         (4) Auricle—earlike flap protruding from each atrium
      b. Ventricles
         (1) Two lower chambers known as "pumping chambers" because they push blood into the large network of vessels
         (2) Ventricular myocardium is thicker than the myocardium of the atria, because great force must be generated to pump the blood a large distance; myocardium of left ventricle is thicker than the right, because it must push blood much further
   3. Valves of the heart—mechanical devices that permit the flow of blood in one direction only (Figure 18-8)
      a. Atrioventricular (AV) valves—prevent blood from flowing back into the atria from the ventricles when the ventricles contract
         (1) Tricuspid valve (right AV valve)—guards the right atrioventricular orifice; free edges of three flaps of endocardium are attached to papillary muscles by chordae tendineae

   (2) Bicuspid, or mitral, valve (left AV valve)—similar in structure to tricuspid valve except only two flaps
      b. Semilunar (SL) valves—half-moon–shaped flaps growing out from the lining of the pulmonary artery and aorta; prevent blood from flowing back into the ventricles from the aorta and pulmonary artery
         (1) Pulmonary semilunar valve—valve at entrance of the pulmonary artery
         (2) Aortic semilunar valve—valve at entrance of the aorta
      c. Skeleton of the heart
         (1) Set of connected rings that serve as a semirigid support for the heart valves and for the attachment of cardiac muscle of the myocardium
         (2) Serves as an electrical barrier between the myocardium of the atria and that of the ventricles
      d. Surface projection (review Figure 18-9)
      e. Flow of blood through heart (review Figure 18-7)
   4. Blood supply of heart tissue (Figures 18-10 and 18-11)
      a. Coronary arteries—myocardial cells receive blood from the right and left coronary arteries
         (1) First branches to come off the aorta
         (2) Ventricles receive blood from branches of both right and left coronary arteries
         (3) Each ventricle receives blood only from a small branch of the corresponding coronary artery
         (4) Most abundant blood supply goes to the myocardium of the left ventricle
         (5) Right coronary artery is dominant in approximately 50% of all hearts and the left in about 20%; in approximately 30%, neither coronary artery is dominant
         (6) Few anastomoses exist between the larger branches of the coronary arteries
      b. Veins of the coronary circulation
         (1) As a rule, veins follow a course that closely parallels that of coronary arteries
         (2) After going through cardiac veins, blood enters the coronary sinus to drain into the right atrium
         (3) Several veins drain directly into the right atrium
   5. Conduction system of the heart—comprising the sinoatrial (SA) node, atrioventricular (AV) node, AV bundle, and Purkinje fibers; made up of modified cardiac muscle (Figure 18-12)
      a. SA node (pacemaker)—hundreds of cells in the right atrial wall near the opening of the superior vena cava
      b. AV node—small mass of special cardiac muscle in right atrium along the lower part of interatrial septum
      c. AV bundle (bundle of His) and Purkinje fibers
         (1) AV bundle originates in AV node, extends by two branches down the two sides of the interventricular septum, and continues as Purkinje fibers
         (2) Purkinje fibers extend out to the papillary muscles and lateral walls of ventricles

6. Nerve supply of the heart
   a. Cardiac plexuses—located near the arch of the aorta, made up of the combination of sympathetic and parasympathetic fibers
   b. Fibers from the cardiac plexus accompany the right and left coronary arteries to enter the heart
   c. Most fibers end in the SA node, but some end in the AV node and in the atrial myocardium
   d. Sympathetic nerves—accelerator nerves
   e. Vagus fibers—inhibitory, or depressor, nerves

BLOOD VESSELS

A. Types of blood vessels
   1. Arteries
      a. Carry blood away from heart—all arteries except pulmonary artery carry oxygenated blood
      b. Elastic arteries—largest in body
         (1) Examples: aorta and its major branches
         (2) Able to stretch without injury
         (3) Accommodate surge of blood when heart contracts and able to recoil when ventricles relax
      c. Muscular (distributing) arteries
         (1) Smaller in diameter than elastic arteries
         (2) Muscular layer is thick
         (3) Examples: brachial, gastric, superior mesenteric
      d. Arterioles (resistance vessels)
         (1) Smallest arteries
         (2) Important in regulating blood flow to end organs
      e. Metarterioles
         (1) Short connecting vessel between true arteriole and 20 to 100 capillaries
         (2) Encircled by precapillary sphincters
         (3) Distal end called thoroughfare channel, which is free of precapillary sphincters
   2. Capillaries—primary exchange vessels
      a. Microscopic vessels
      b. Carry blood from arterioles to venules—together, arterioles, capillaries, and venules constitute the microcirculation (Figure 18-15)
      c. Not evenly distributed—highest numbers in tissues with high metabolic rate; may be absent in some "avascular" tissues, such as cartilage
      d. Types of capillaries (Figure 18-16)
         (1) True capillaries—receive blood flowing from metarteriole with input regulated by precapillary sphincters
         (2) Continuous capillaries
            (a) Continuous lining of endothelial cells
            (b) Openings called intercellular clefts exist between adjacent endothelial cells
         (3) Fenestrated capillaries
            (a) Have both intercellular clefts and "holes" or fenestrations through plasma membrane to facilitate exchange functions

         (4) Sinusoids
            (a) Large lumen and tortuous course
            (b) Absent or incomplete basement membrane
            (c) Very porous—permit migration of cells into or out of vessel lumen
   3. Veins
      a. Carry blood toward the heart
      b. Act as collectors and as reservoir vessels; called capacitance vessels
B. Structure of blood vessels (Figure 18-14)
   1. Layers
      a. Tunica adventitia—found in arteries and veins
      b. Tunica media—found in arteries and veins
      c. Tunica intima—found in all blood vessels; only layer present in capillaries
   2. "Building blocks" commonly present
      a. Lining endothelial cells
         (1) Only lining found in capillary
         (2) Line entire vascular tree
         (3) Provide a smooth luminal surface—protects against intravascular coagulation
         (4) Intercellular clefts, cytoplasmic pores, and fenestrations allow exchange to occur between blood and tissue fluid
         (5) Capable of secreting a number of substances
         (6) Capable of reproduction
      b. Collagen fibers
         (1) Exhibit woven appearance
         (2) Formed from protein molecules that aggregate into fibers
         (3) Visible with light microscope
         (4) Have only a limited ability to stretch (2% to 3%) under physiological conditions
         (5) Function to strengthen and keep lumen of vessel open
      c. Elastic fibers
         (1) Composed of insoluble protein called elastin
         (2) Form highly elastic networks
         (3) Fibers can stretch more than 100% under physiological conditions
         (4) Play important role in creating passive tension to help regulate blood pressure throughout the cardiac cycle
      d. Smooth muscle fibers
         (1) Present in all segments of vascular system except capillaries
         (2) Most numerous in elastic and muscular arteries
         (3) Exert active tension in vessels when contracting

MAJOR BLOOD VESSELS

A. Circulatory routes (Figure 18-13)
   1. Systemic circulation—blood flows from the left ventricle of the heart through blood vessels to all parts of the body (except gas exchange tissues of lungs) and back to the right atrium

2. Pulmonary circulation—venous blood moves from right atrium to right ventricle to pulmonary artery to lung arterioles and capillaries, where gases are exchanged; oxygenated blood returns to left atrium by way of pulmonary veins; from left atrium, blood enters the left ventricle

B. Systemic circulation
1. Systemic arteries (review Tables 18-2 to 18-6 and Figures 18-17 to 18-22)
   a. Main arteries give off branches, which continue to re-branch, forming arterioles and then capillaries
   b. End-arteries—arteries that eventually diverge into capillaries
   c. Arterial anastomosis—arteries that open into other branches of the same or other arteries; incidence of arterial anastomoses increases as distance from the heart increases
   d. Arteriovenous anastomoses or shunts occur when blood flows from an artery directly into a vein
2. Systemic veins (review Figures 18-23 to 18-29)
   a. Veins are the ultimate extensions of capillaries; unite into vessels of increasing size to form venules and then veins
   b. Large veins of the cranial cavity are called dural sinuses
   c. Veins anastomose the same as arteries
   d. Venous blood from the head, neck, upper extremities, and thoracic cavity (except lungs) drains into superior vena cava
   e. Venous blood from thoracic organs drains directly into superior vena cava or azygos vein
   f. Hepatic portal circulation (Figure 18-28)
      (1) Veins from the spleen, stomach, pancreas, gallbladder, and intestines send their blood to the liver by way of the hepatic portal vein
      (2) In the liver the venous blood mingles with arterial blood in the capillaries and is eventually drained from the liver by hepatic veins that join the inferior vena cava
   g. Venous blood from the lower extremities and abdomen drains into the inferior vena cava

C. Fetal circulation
1. The basic plan of fetal circulation—additional vessels needed to allow fetal blood to secure oxygen and nutrients from maternal blood at the placenta (Figure 18-31)
   a. Two umbilical arteries—extensions of the internal iliac arteries; carry fetal blood to the placenta
   b. Placenta—attached to uterine wall; where exchange of oxygen and other substances between the separated maternal and fetal blood occurs (Figure 18-30)
   c. Umbilical vein—returns oxygenated blood from the placenta to the fetus; enters body through the umbilicus and goes to the undersurface of the liver where it gives off two or three branches and then continues as the ductus venosus
   d. Ductus venosus—continuation of the umbilical vein and drains into inferior vena cava
   e. Foramen ovale—opening in septum between the right and left atria

   f. Ductus arteriosus—small vessel connecting the pulmonary artery with the descending thoracic aorta
2. Changes in circulation at birth (compare Figures 18-31 and 18-32)
   a. When umbilical cord is cut, the two umbilical arteries, the placenta and umbilical vein, no longer function
   b. Umbilical vein within the baby's body becomes the round ligament of the liver
   c. Ductus venosus becomes the ligamentum venosum of the liver
   d. Foramen ovale—functionally closed shortly after a newborn's first breath and pulmonary circulation is established; structural closure takes approximately 9 months
   e. Ductus arteriosus—contracts with establishment of respiration, becomes ligamentum arteriosum

CYCLE OF LIFE: CARDIOVASCULAR ANATOMY

A. Birth—change from placenta-dependent system
B. Heart and blood vessels maintain basic structure and function from childhood through adulthood
   1. Exercise thickens myocardium and increases the supply of blood vessels in skeletal muscle tissue
C. Adulthood through later adulthood—degenerative changes
   1. Atherosclerosis—blockage or weakening of critical arteries
   2. Heart valves and myocardial tissue degenerate—reduces pumping efficiency

REVIEW QUESTIONS

1. Discuss the size, position, and location of the heart in the thoracic cavity.
2. Describe the pericardium, differentiating between the fibrous and serous portions.
3. Exactly where is pericardial fluid found? Explain its function.
4. Define the following terms: *intercalated disks*, *syncytium*, *autorhythmic*.
5. Name and locate the chambers and valves of the heart.
6. Trace the flow of blood through the heart.
7. Identify, locate, and describe the functions of each of the following structures: SA node, AV node, AV bundle.
8. Identify the vessels that join to form the hepatic portal vein.
9. Describe the six unique structures necessary for fetal circulation.
10. Explain how the separation of oxygenated and deoxygenated blood occurs after birth.
11. Explain the purpose of echocardiography.
12. Identify the clinical significance of cardiac enzymes.
13. Why is the internal jugular vein sometimes ligated as a result of middle ear infection?
14. Briefly define the following terms: *aneurysm*, *atherosclerosis*, *phlebitis*.
15. Identify possible causes of coronary artery disease.
16. Define congestive heart failure.
17. Explain how hemorrhoids develop.

CRITICAL THINKING QUESTIONS

1. How is CPR accomplished? What is the significance of the placement of the heart in the thoracic cavity and successful CPR?

2. What would result if there were a lack of anastomosis in the arteries of the heart?

3. The general public thinks the most important structure in the cardiovascular system is the heart. Anatomists know it is the capillary. What information would you use to support this view?

4. Compare and contrast arterial blood in systemic circulation and arterial blood in pulmonary circulation.

5. Can you make the distinction between an occlusion of an end-artery and an occlusion of other small arteries?

6. Determine which of the following veins drain into the superior vena cava and which drain into the inferior vena cava: longitudinal sinus, great saphenous, basilic, internal jugular, azygos, popliteal, and hepatic portal.

7. How can you describe the functional advantage of a portal system?

8. Can you make use of the vessels and organs to trace the path taken by a single RBC? Begin at the right atrium, proceed to the left great toe, and return to the right atrium.

9. Can you explain what happens to the cardiovascular system during the normal cycle of one's life?

CHAPTER 19

Physiology of the Cardiovascular System

CHAPTER OUTLINE

LANGUAGE OF SCIENCE

aortic reflex (ay-OR-tik REE-fleks) [*aort-* raise, *-ic* pertaining to, *reflex* bend back]

atrioventricular (AV) node (ay-tree-oh-ven-TRIK-yoo-lar) [*atrio-* hall, *-ventricul-* little belly, *-ar* pertaining to, *nodus* knot]

atrioventricular (AV) bundle (ay-tree-oh-ven-TRIK-yoo-lar BUN-del) [*atrio-* hall, *-ventricul-* little belly, *-ar* pertaining to]

baroreceptors (bar-oh-ree-SEP-tors) [*baro-* pressure, *-recept-* to receive, *-or* agent]

capillary exchange (KAP-i-lair-ee) [*capillar-* hairlike, *-ary* pertaining to]

cardiac cycle (KAR-dee-ak) [*cardia-* heart, *-ac* pertaining to, *cycle* circle]

cardiac output (CO) (KAR-dee-ak) [*cardia-* heart, *-ac* pertaining to]

cardiac pressoreflexes (KAR-dee-ak pres-oh-REE-fleks-ez) [*cardia-* heart, *-ac* pertaining to, *presso-* pressure, *-reflex* bend back]

carotid sinus reflex (kah-ROT-id SYE-nus REE-fleks) [*caros-* heavy sleep, *-id* pertaining to, *sinus* hollow, *reflex* bend back]

chemoreceptor reflex (kee-moh-ree-SEP-tor REE-fleks) [*chemo-* by chemical reaction, *-recipt-* receive, *-or* agent, *reflex* bend back]

diastole (dye-ASS-toh-lee) [*dia-* through, *-stole* contraction]

end-diastolic volume (EDV) (end-dye-ASS-toh-lik) [*dia-* through, *-stol-* contraction, *-ic* pertaining to]

hemodynamics (hee-moh-dye-NAM-iks) [*hemo-* blood, *-dynami-* force, *-ic* pertaining to]

hypertension (HTN) (hye-per-TEN-shun) [*hyper-* excessive, *-tens-* stretch or pressure, *-ion* state of]

interatrial bundle (in-ter-AT-tree-al) [*inter-* between, *-atri-* hall, *-al* pertaining to]

internodal bundles (in-ter-NOH-dal) [*inter-* between, *-nod-* knot, *-al* pertaining to]

ischemic (is-KEE-mik) [*ischem-* hold back, *-ic* pertaining to]

Cont'd on p. 773

The vital role of the cardiovascular system in maintaining homeostasis depends on the continuous and controlled movement of blood through the thousands of miles of capillaries that permeate every tissue and reach every cell in the body. It is in the microscopic capillaries that blood performs its ultimate transport function. Nutrients and other essential materials pass from capillary blood into fluids surrounding the cells as waste products are removed. Blood must not only be kept moving through its closed circuit of vessels by the pumping activity of the heart, but also must be directed and delivered to those capillary beds surrounding cells that need it most. Blood flow to cells at rest is minimal. In contrast, blood is shunted to the digestive tract after a meal or to skeletal muscles during exercise. The thousands of miles of capillaries could hold far more than the body's total blood volume if it were evenly distributed. Regulation of blood pressure and flow must therefore change in response to cellular activity.

Numerous control mechanisms help to regulate and integrate the diverse functions and component parts of the cardiovascular system to supply blood to specific body areas according to need. These mechanisms ensure a constant *milieu intérieur*, that is, a constant internal environment surrounding each body cell regardless of differing demands for nutrients or production of waste products. This chapter presents information about several of the control mechanisms that regulate the pumping activity of the heart and the smooth and directed flow of blood through the complex channels of the circulation.

HEMODYNAMICS

Hemodynamics is a term used to describe a collection of mechanisms that influence the active and changing—or dynamic—circulation of blood (Figure 19-1). Circulation is, of course, a vital function. It constitutes the only means by which cells can receive materials needed for their survival and can have their wastes removed. Circulation is necessary, and circulation of different volumes of blood per minute is also essential for healthy survival. More active cells need more blood per minute than less active cells. The reason underlying this principle is obvious. The more work cells do, the more energy they use, and the more oxygen and nutrients they remove from the blood. Because blood circulates, it can continually bring in more oxygen and nutrients to replace what is consumed. The greater the activity of any part of the body, the greater the volume of blood circulating through it. This requires that circulation control mechanisms accomplish two functions: maintain circulation (keep blood flowing) and vary the volume and distribution of the blood circulated. Therefore as any structure increases its activity, an increased volume of blood must be distributed to it—must be shifted from the less active to the more active tissues.

To achieve these two ends, a great many factors must operate together as one smooth-running, although complex, machine. Incidentally, this is an important physiological principle that you have no doubt observed by now—that every body function depends on many other functions. A constellation of separate processes or mechanisms acts as a single integrated mechanism. Together, these separate mechanisms perform one large function.

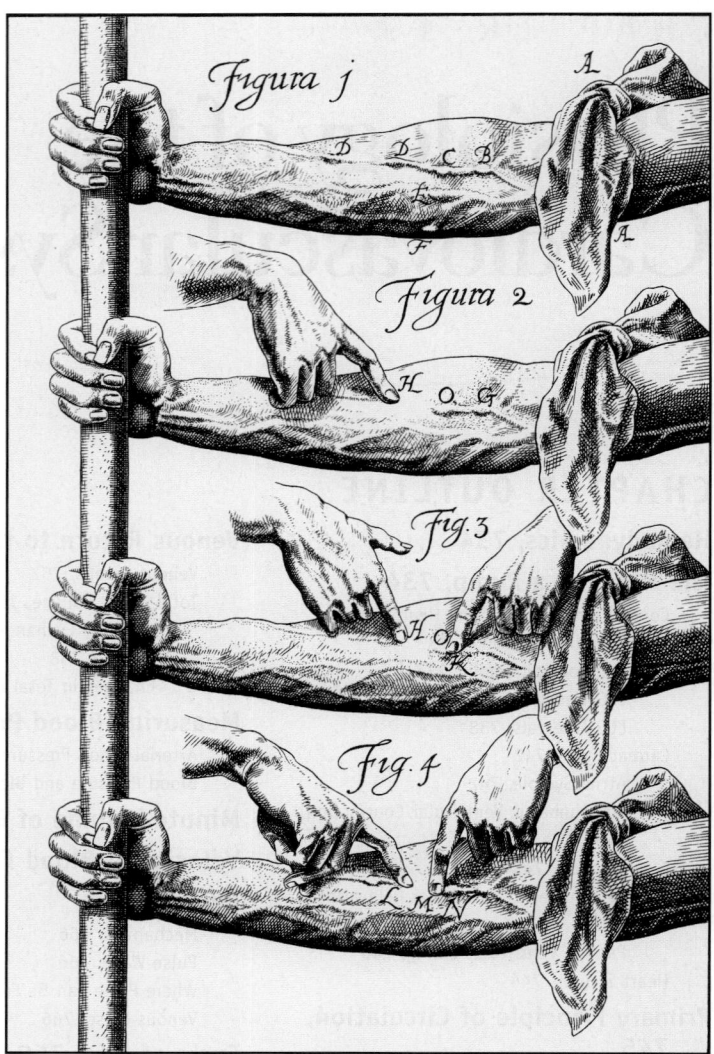

Figure 19-1 *Hemodynamics.* An illustration of a famous experiment in hemodynamics conducted by English scientist William Harvey in the seventeenth century. Through a series of such experiments, Harvey finally proved that blood flows in a circuit out through the arteries and back through the veins—a concept on which most other hemodynamic concepts are based.

For example, many mechanisms together accomplish the large function we call *circulation.*

This chapter is about hemodynamics—the mechanisms that keep blood flowing properly. We begin with a discussion of the heart as a pump and then move on to the even bigger picture of blood flow through the entire cardiovascular system.

THE HEART AS A PUMP

In Chapter 18 we discussed the functional anatomy of the heart. Its four chambers and their valves make up two pumps: a left pump and a right pump. The left pump (left side of the heart) helps move blood through the systemic circulation, and the right pump (right side of the heart) helps move blood through the pulmonary circulation. We will now step back from our previous discussion of the valves and chambers of the heart to look at the bigger picture and see how these two linked pumps function together as a single unit. First, we will discuss the role of the electrical conduction system of the heart in coordinating heart contractions. Then we will discuss how these coordinated contractions produce the pumping cycle of the heart.

Conduction System of the Heart

The anatomy of four structures that compose the conduction system of the heart—**sinoatrial (SA) node**, **atrioventricular (AV) node**, **AV bundle**, and **Purkinje (subendocardial) system**—was discussed briefly in Chapter 18. Each of these structures consists of cardiac muscle modified enough in structure to differ in function from ordinary cardiac muscle. The specialty of ordinary cardiac muscle is contraction. In this, it is like all muscle, and like all muscle, ordinary cardiac muscle can also conduct impulses; but the conduction system structures are more highly specialized, both structurally and functionally, than ordinary cardiac muscle tissue. They are not contractile. Instead, they permit only generation or rapid conduction of an action potential through the heart.

The normal cardiac impulse that initiates mechanical contraction of the heart arises in the SA node (or **pacemaker**), located just below the atrial epicardium at its junction with the superior vena cava (Figure 19-2). Specialized pacemaker cells in the node possess an *intrinsic rhythm*. This means that without any stimulation by nerve impulses from the brain and cord, they themselves initiate impulses at regular intervals. Even if pacemaker cells are removed from the body and placed in a nutrient solution, completely separated from all nervous and hormonal control, they will continue to beat! In an intact living heart, of course, nervous and hormonal regulation does occur and the SA node generates a pace accordingly.

Each impulse generated at the SA node travels swiftly throughout the muscle fibers of both atria. An **interatrial bundle** of conducting fibers facilitates rapid conduction to the left atrium. Thus stimulated, the atria begin to contract. As the action potential enters the AV node by way of three **internodal bundles** of conducting fibers, its conduction slows markedly, thus allowing for complete contraction of both atrial chambers before the impulse reaches the ventricles. After passing slowly through the AV node, conduction velocity increases as the impulse is relayed through the AV bundle (bundle of His) into the ventricles. Here, right and left *bundle branches* and the Purkinje fibers in which they terminate conduct the impulses throughout the muscle of both ventricles, stimulating them to contract almost simultaneously.

Thus the SA node initiates each heartbeat and sets its pace—it is the heart's own natural *pacemaker* (Box 19-1). Under the in-

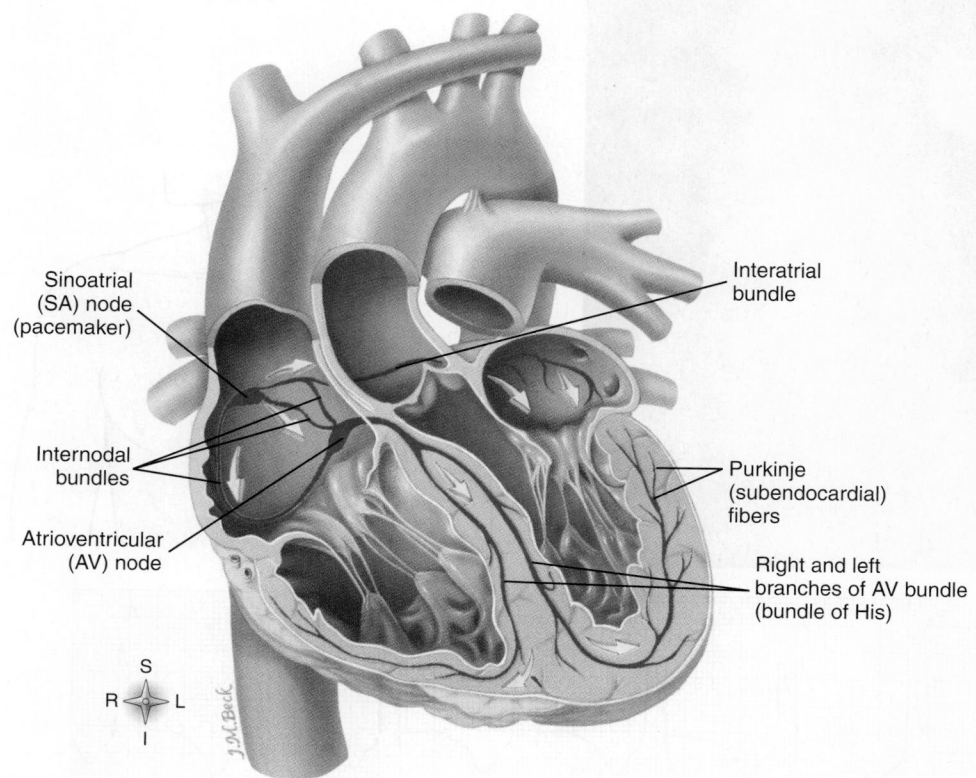

Sinoatrial
(SA) node
(pacemaker)

Internodal
bundles

Atrioventricular
(AV) node

Interatrial
bundle

Purkinje
(subendocardial)
fibers

Right and left
branches of AV bundle
(bundle of His)

Figure 19-2 *Conduction system of the heart.* Specialized cardiac muscle cells in the wall of the heart rapidly initiate or conduct an electrical impulse throughout the myocardium. The signal is initiated by the SA node (pacemaker) and spreads to the rest of the right atrial myocardium directly, to the left atrial myocardium by way of a bundle of interatrial conducting fibers, and to the AV node by way of three internodal bundles. The AV node then initiates a signal that is conducted through the ventricular myocardium by way of the AV (bundle of His) and Purkinje fibers.

BOX 19-1   Artificial Cardiac Pacemakers

Everyone has heard about **artificial pacemakers,** devices that electrically stimulate the heart at a set rhythm (continuously discharging pacemakers) or those that fire only when the heart rate decreases below a preset minimum (demand pacemakers). They do an excellent job of maintaining a steady heart rate and of keeping many individuals with damaged hearts alive for many years. Hundreds of thousands of people currently have permanently implanted cardiac pacemakers.

Several types of artificial pacemakers have been designed to deliver an electrical stimulus to the heart muscle. The stimulus passes through electrodes that are sewn directly to the epicardium on the outer surface of the heart or are inserted by a catheter into a heart chamber, such as the right ventricle, and placed in contact with the endocardium. Modern pacemakers generate a stimulus that lasts from 0.08 to 2 msec and produces a very low current output.

One common method of inserting a permanent pacemaker is by the **transvenous approach.** In this procedure, a small incision is made just above the right clavicle and the electrode is threaded into the jugular vein and then advanced to the apex of the right ventricle. Part *A* of the figure shows the battery-powered stimulus generator, which is placed in a pocket beneath the skin on the right side of the chest just below the clavicle. The proximal end of the electrical lead, or catheter, is then directed through the subcutaneous tissues and attached to the power pack. Part *B* of the figure shows the tip of the electrical lead in the apex of the right ventricle. Part *C* of the figure shows the ECG of an artificially paced heart. Note the uniform, rhythmic "pacemaker spikes" that trigger each heartbeat.

Although life saving, these devices must be judged inferior to the heart's own natural pacemaker. Why? Because they cannot speed up the heartbeat when necessary (e.g., to make strenuous physical activity possible), nor can they slow it down again when the need has passed. The normal SA node, influenced as it is by autonomic impulses and hormones, can produce these changes. Discharging an average of 75 times each minute, this truly remarkable bit of specialized tissue will generate more than 2 billion action potentials in an average lifetime of some 70 years.

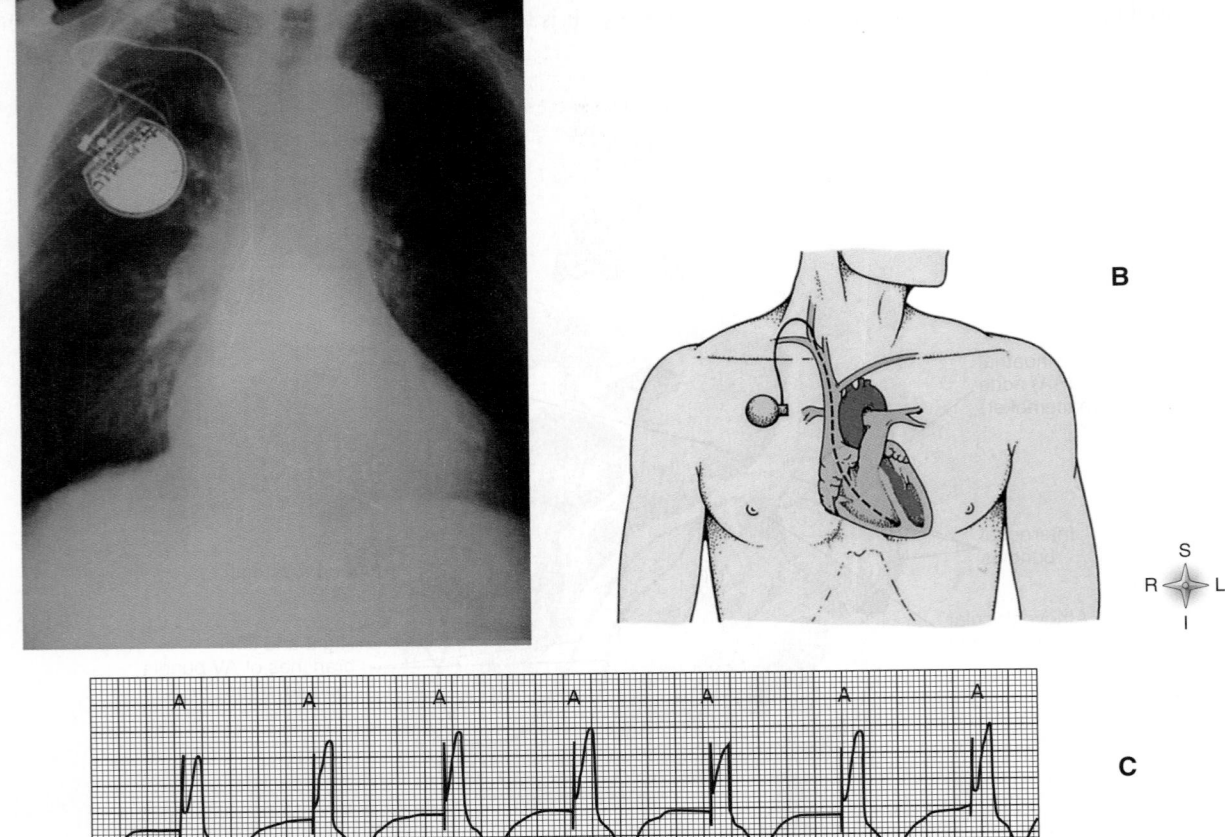

Artificial cardiac pacemaker. **A,** Battery-powered stimulus generator placed below the skin of the chest.
B, The electrical lead extends from the stimulus generator into the right ventricle. **C,** Pacemaker spikes
(A) characterize the ECG of an artificially paced heart.

fluence of autonomic and endocrine control, the SA node will normally "discharge," or "fire," at an intrinsic rhythmical rate of 70 to 75 beats/min under resting conditions. However, if for any reason the SA node loses its ability to generate an impulse, pacemaker activity will shift to another excitable component of the conduction system, such as the AV node or the Purkinje fibers. Pacemakers other than the SA node are called *abnormal*, or **ectopic, pacemakers.** Although ectopic pacemakers fire rhythmically, their rate of discharge is generally much slower than that of the SA node. For example, a pulse of 40 to 60 beats/min would result if the AV node were forced to assume pacemaker activity.

Electrocardiogram (ECG)

Electrocardiography

Impulse conduction generates tiny electrical currents in the heart that spread through surrounding tissues to the surface of the body. This fact has great clinical importance. Why? Because from the skin, visible records of the heart's electrical activity can be made with an instrument called an *electrocardiograph*. Skilled interpretation of these records may sometimes make the difference between life and death.

The **electrocardiogram** (**ECG** when written or **EKG** when spoken) is a graphic record of the heart's electrical activity, its conduction of impulses. It is not a record of the heart's contractions but of the electrical events that precede them. To produce an electrocardiogram, electrodes of a recording voltmeter (electrocardiograph) are attached to the limbs and/or chest of the subject (Figure 19-3, A). Changes in voltage, which represent changes in the heart's electrical activity, are observed as deflections of a line drawn on paper or traced on a video monitor.

Figure 19-4 explains the basic theory behind **electrocardiography.** To keep things simple, a single cardiac muscle fiber is shown with the two electrodes of a recording voltmeter nearby. Before the action potential reaches either electrode, there is no difference in charge between the electrodes, and thus no change

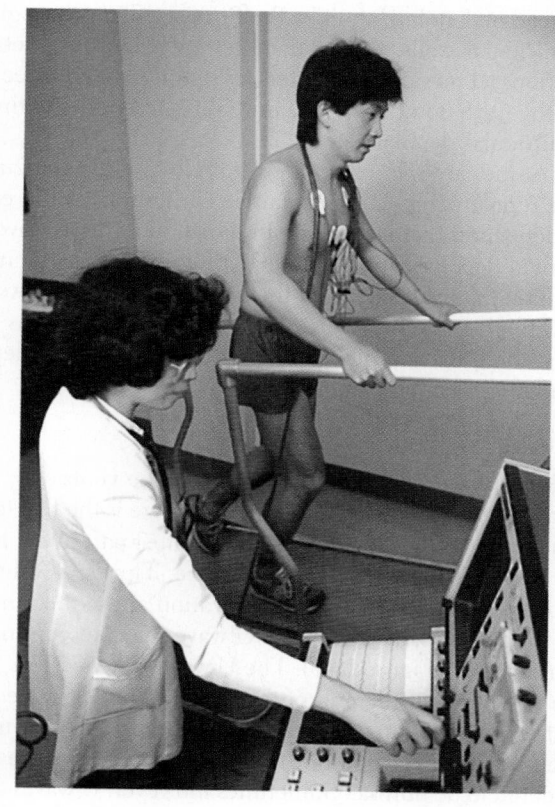

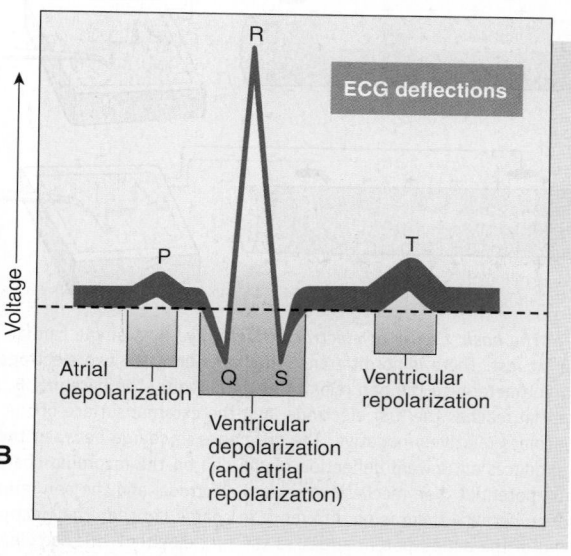

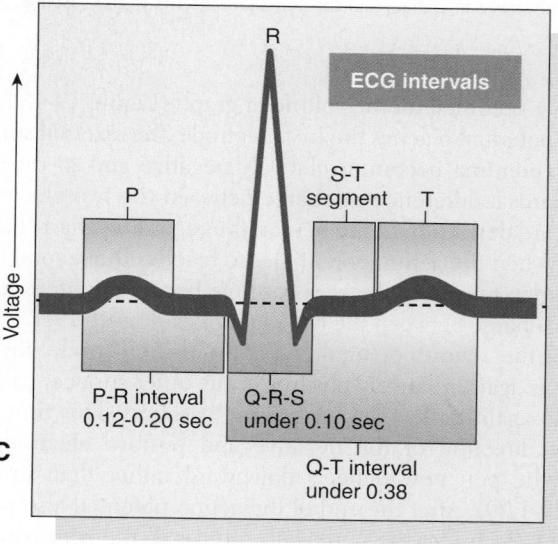

Figure 19-3 *Electrocardiogram.* **A,** A nurse monitors a patient's ECG as he exercises on a treadmill. **B,** Idealized ECG deflections represent depolarization and repolarization of cardiac muscle tissue. **C,** Principal ECG intervals between P, QRS, and T waves. Note that the P-R interval is measured from the start of the P wave to the start of the Q wave.

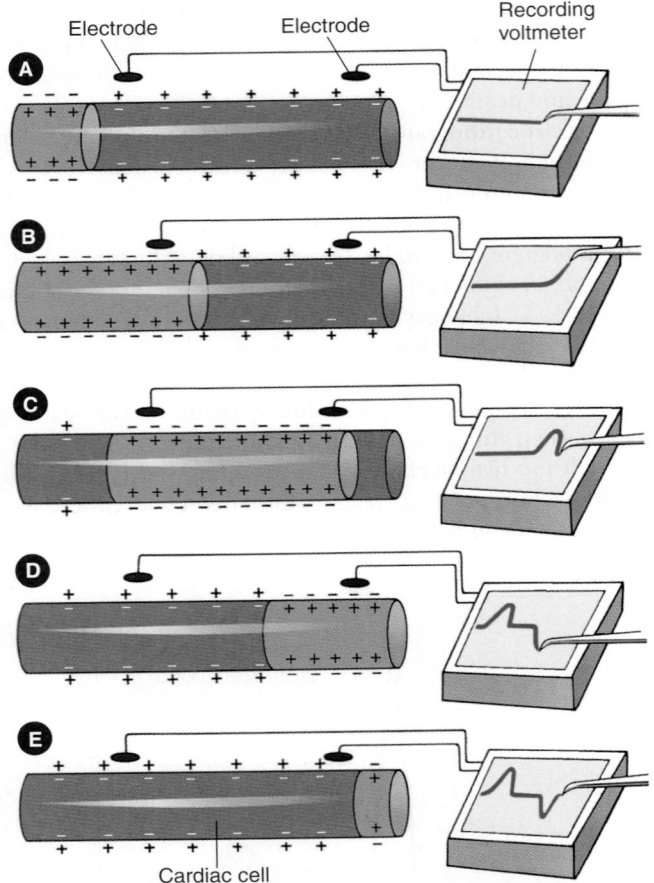

Figure 19-4 *The basic theory of electrocardiography.* **A,** A single cardiac muscle fiber at rest. There is no difference in charge between two electrodes of a recording voltmeter—so the pen remains at 0 millivolts, the baseline. **B,** An action potential reaches the first electrode, and the external surface of the sarcolemma becomes relatively negative. The difference in charge between the two electrodes produces an upward deflection of the pen on the recording chart. **C,** The action potential then reaches the second electrode, and the pen returns to the baseline because there is no difference in charge between the electrodes. **D,** As the end of the action potential passes the first electrode, the sarcolemma is again relatively positive on its outer surface, causing the pen to deflect downward. **E,** After the end of the action potential also passes the second electrode, there is no difference in charge, and the pen again returns to the baseline.

in voltage is recorded on the voltmeter graph (Figure 19-4, *A*). As an action potential reaches the first electrode, the external surface of the sarcolemma becomes relatively negative and so the voltmeter records a difference in charge between the two electrodes as an upward deflection of the pen on the recording chart (Figure 19-4, *B*). When the action potential also reaches the second electrode, the pen returns to the zero baseline because there is no difference in charge between the two electrodes (Figure 19-4, *C*). As the end of the action potential passes the first electrode, the sarcolemma is again relatively positive on its outer surface, causing the pen to again deflect away from the baseline. This time, because the direction of the negative and positive electrodes is reversed, the pen now deflects downward rather than upward (Figure 19-4, *D*). After the end of the action potential also passes the second electrode, the pen again returns to the zero baseline (Figure 19-4, *E*). In short, depolarization of cardiac muscle causes

a deflection of the graphed line; repolarization causes a deflection in the opposite direction.

Electrocardiography electrodes are normally quite some distance from myocardial tissue, but, given the massive size of the myocardial syncytium, it should not be surprising that even cutaneous electrodes can detect changes in the heart's polarity (Box 19-2).

ECG Waves

Because electrocardiography is far too complex a subject to explain fully here, normal ECG *deflection waves* and the ECG *intervals* between them shall be only briefly discussed. As shown in Figures 19-3, *B,* and 19-5, the normal ECG is composed of deflection waves called the *P wave, QRS complex,* and *T wave.* (The letters do not stand for any words but were chosen as an arbitrary sequence of the alphabet.)

P Wave

Briefly, the **P wave** represents depolarization of the atria. That is, the P wave is the deflection caused by the passage of an electrical impulse from the SA node through the musculature of both atria.

QRS Complex

The **QRS complex** represents depolarization of the ventricles. Depolarization of the ventricles is a complex process, involving depolarization of the interventricular septum and the subsequent spread of depolarization by the Purkinje fibers through the lateral ventricular walls. Rather than getting mired in a detailed explanation, let us simplify matters by stating that all three deflections of the QRS complex (Q, R, and S) represent the entire process of ventricular depolarization.

At the same time that the ventricles are depolarizing, the atria are repolarizing. As we explained earlier, we should expect to see a deflection that is opposite in direction to the P wave that represented depolarization. However, the massive ventricular depolarization that is occurring at the same time overshadows the voltage fluctuation produced by atrial repolarization. Thus we can say that the QRS complex represents both ventricular depolarization and atrial repolarization.

T Wave

The **T wave** reflects repolarization of the ventricles. In atria, the first part of the myocardium to depolarize is the first to repolarize. In ventricles, on the other hand, the first part of the myocardium to depolarize is the last to repolarize. Thus ECG deflections for both depolarization and repolarization are in the same direction.

Sometimes, an additional **U wave** may be seen in the electrocardiogram (Figure 19-6). The U wave, when visible, appears as a tiny "hump" at the end of the T wave. The U wave results from late repolarization of Purkinje fibers in the papillary muscle of the ventricular myocardium. If not too big, U waves are usually considered to be normal. Sometimes, however, U waves can be a sign of **hypokalemia** (low blood potassium) or too much **digoxin** (a heart medication).

ECG Intervals

The principal ECG intervals between P, QRS, and T waves are shown in Figure 19-3, *C.* Measurement of these intervals can provide valuable information concerning the rate of conduc-

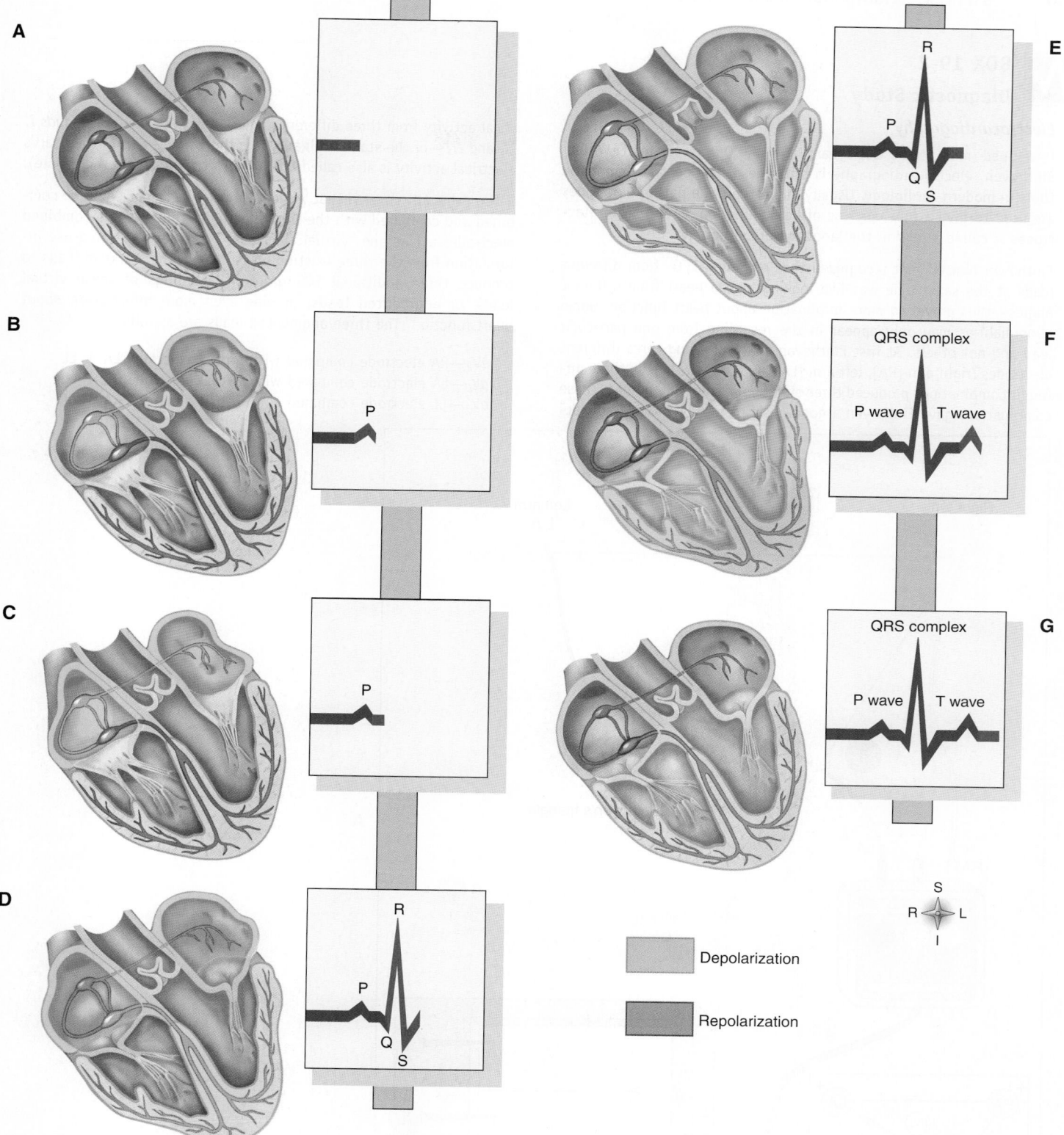

Figure 19-5 *Events represented by the electrocardiogram (ECG).* It is impossible to illustrate the invisible, dynamic events of heart conduction in a few cartoon panels or "snapshots," but the sketches here give you an idea of what is happening in the heart as an ECG is recorded. **A,** The heart wall is completely relaxed, with no change in electrical activity, so the ECG remains constant. **B,** P wave occurs when the AV node and atrial walls depolarize. **C,** Atrial walls are completely depolarized, and thus no change is recorded in the ECG. **D,** The QRS complex occurs as the atria repolarize and the ventricular walls depolarize. **E,** The atrial walls are now completely repolarized, the ventricular walls are now completely depolarized, and thus no change is seen in the ECG. **F,** The T wave appears on the ECG when the ventricular walls repolarize. **G,** Once the ventricles are completely repolarized, we are back at the baseline of the ECG—essentially back where we began in **A.** Note that depolarization triggers contraction in the affected muscle tissue. Thus cardiac muscle contraction occurs *after* depolarization begins.

BOX 19-2
Diagnostic Study

Electrocardiography

Developed more than 100 years ago by Dutch scientist Willem Einthoven, electrocardiography is still one of the central diagnostic tools in modern cardiology. Usually, more than one pair of electrodes are used to give a fuller picture of heart function. Each pair of electrodes is called a *lead* in the language of cardiology.

Einthoven himself first recognized that recording ECGs from different leads at the same time was like looking at the heart from different angles—thus providing more information about heart function. Some abnormalities may only appear in the recording from one particular lead and not others. At first Einthoven saw that using three different electrodes (right arm [RA], left arm [LA], and left leg [LL]) in three different combinations produced three slightly different ECG graphs. These combinations gave Einthoven a good basic picture of the heart's elec-

trical activity from three different angles. Today, we call these *leads I, II,* and *III*—or the **standard leads.** This three-angle view of the heart's electrical activity is also called **Einthoven's triangle** (part *A* of figure).

Later, others discovered that wires from two electrodes could be combined and compared with the third electrode. Thus the two combined electrodes act as one "virtual electrode." This way, we could use information from the three existing electrodes in the standard leads to produce three additional ECG graphs. Recordings of these virtual leads, or **augmented leads,** provide even more information about heart function. The three augmented leads are as follows:

aV_R—RA electrode compared with virtual electrode LA + LL
aV_L—LA electrode compared with virtual electrode RA + LL
aV_F—LL electrode compared with virtual electrode RA + LA

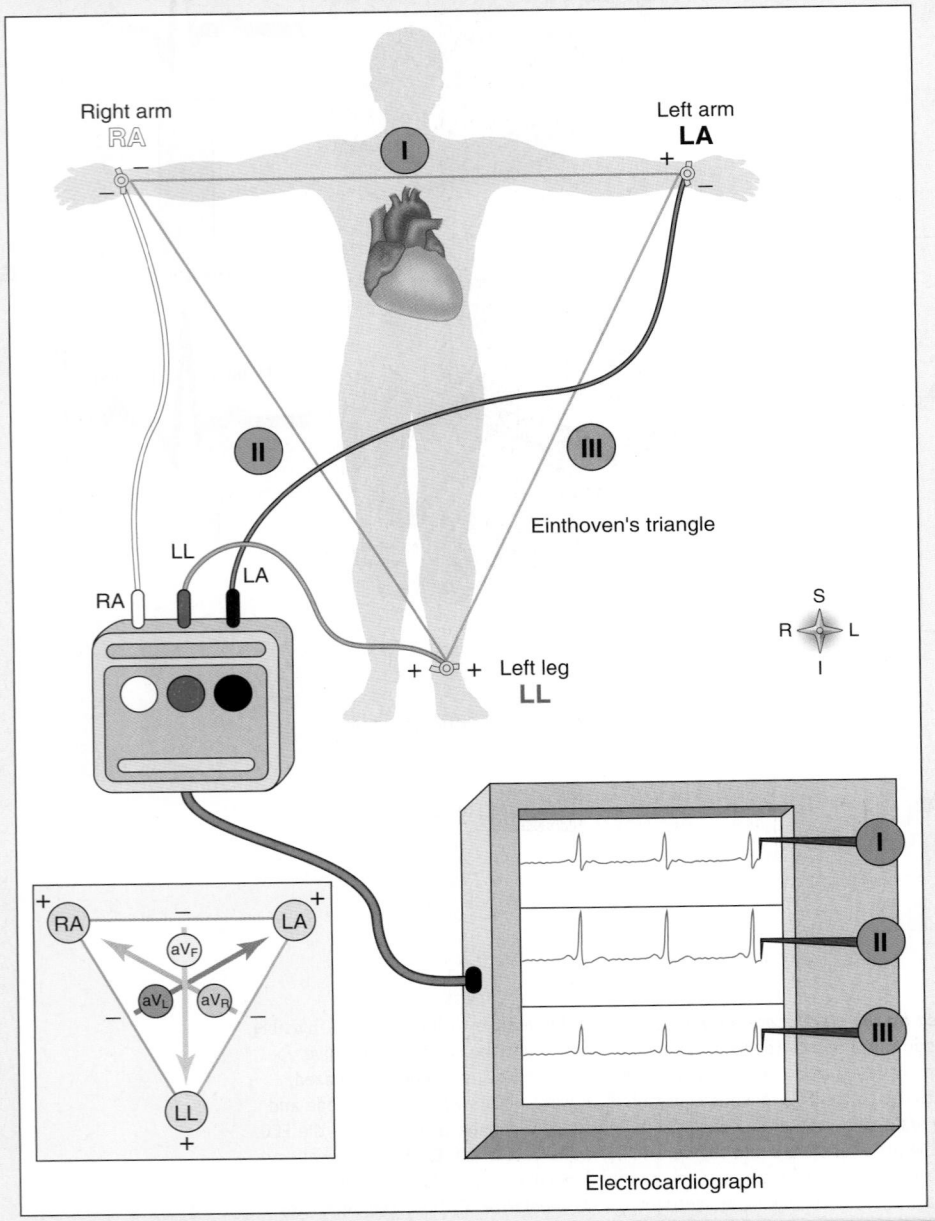

Einthoven's triangle

A

A, ECG limb leads. Electrodes are placed on the right arm (RA) or wrist, on the left arm (LA) or wrist, and on the left leg (LL) or ankle. Einthoven's triangle shows the "electrical angle" of each of the three standard limb leads: I, II, and III. The inset shows the electrical angle of the augmented limb leads (aV_R, aV_L, aV_F), which combine limb leads to form virtual leads given different angles of electrical voltage measurement.

Electrocardiograph

BOX 19-2

Diagnostic Study—cont'd

In the symbols for the augmented leads, "a" is for "augmented," "V" is for "voltage," and the "R, L, or F" is for "right, left, or frontal." All together, the standard leads and the augmented leads can be called the **limb leads** because they use electrodes placed on the limbs.

All three limb electrodes (RA, LA, and LL) can be combined together to form a single virtual electrode that is compared with any or all of the six chest electrodes shown in part B of the figure. Thus standard **chest leads** V_1 through V_6 are also available to record even more in-

formation about heart function. The chest leads are also called **precordial leads.** Additional supplemental chest leads can also be placed on the right side of the chest, especially when damage to the right ventricular myocardium is suspected.

Today, 12-lead ECG recordings are fairly common. A 12-lead ECG is a simultaneous recording of all limb and chest leads: I, II, III, aV_R, aV_L, aV_F, V_1, V_2, V_3, V_4, V_5, and V_6. Part C of the figure shows a typical 12-lead ECG recording.

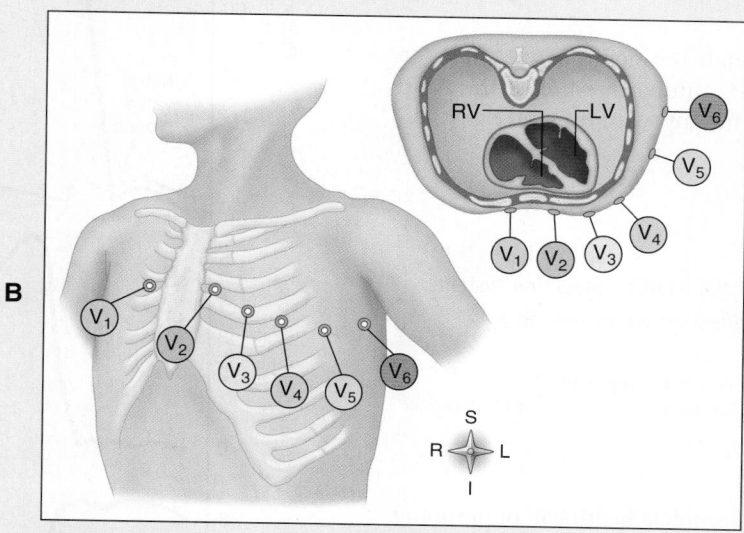

B, *ECG chest leads.* Electrodes are placed at specific locations across the chest. Each of these is compared with a single virtual lead formed by the combination of all three limb electrodes and electrically positioned approximately over the spine.

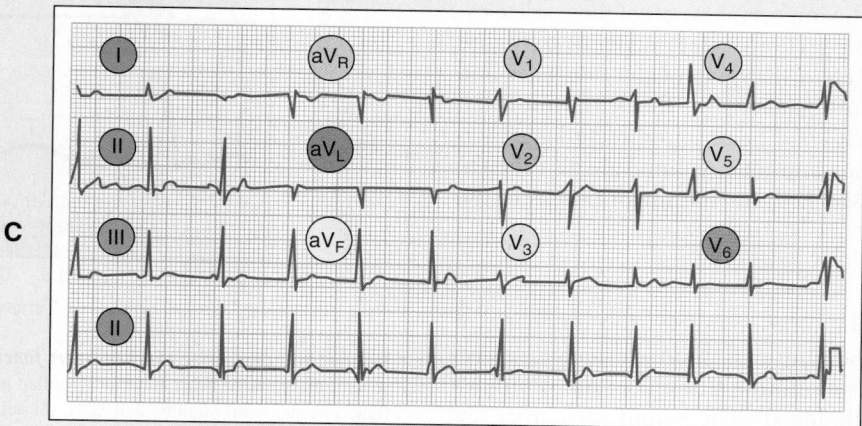

C, *12-lead ECG recording.* This brief recording of a typical 12-lead ECG shows information recorded from all six limb leads and all six chest leads. Note that they are all printed above a continuous recording of lead II, which becomes a point of reference. Health professionals can usually get more information from 12 different ECG leads than from any one ECG lead.

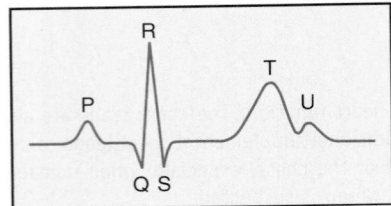

Figure 19-6 *The U wave.* This ECG recording shows the presence of a U wave, which follows the T wave—or appears as a "hump" on the back of the T wave. The U wave represents a sudden, late depolarization of Purkinje fibers in the papillary muscle.

tion of an action potential through the heart, as we shall see later in this chapter. Figure 19-5 summarizes the relationship between the electrical events of the myocardium and the ECG recordings.

 QUICK CHECK

1. List the principal structures of the heart's conduction system.
2. What are the three types of deflection waves seen in a typical ECG?
3. What event does each type of ECG wave represent?

Cardiac Cycle

The term **cardiac cycle** means a complete heartbeat, or pumping cycle, consisting of contraction (**systole**) and relaxation (**diastole**) of both atria and both ventricles. The two atria contract simultaneously. Then, as the atria relax, the two ventricles contract and relax, instead of the entire heart contracting as a unit. This gives a kind of pumping action to the movements of the heart. The atria remain relaxed during part of the ventricular relaxation and then start the cycle over again. The cycle as a whole is often divided into time intervals for discussion and study. The following sections describe several of the important events of the cardiac cycle. As you read through these sections, refer frequently to Figure 19-7, which is a composite chart that graphically illustrates and integrates changes in pressure gradients in the left atrium, left ventricle, and aorta with ECG and heart sound recordings. Aortic blood flow and changes in ventricular volume are also shown. Refer also to Figure 19-8, which shows the major phases of the cardiac cycle.

Atrial Systole

The contracting force of the atria completes the emptying of blood out of the atria into the ventricles. Atrioventricular (or cuspid) valves are necessarily open during this phase; the ventricles are relaxed and filling with blood. The semilunar valves are closed so that blood does not reenter from the pulmonary artery or aorta. This period of the cycle begins with the P wave of the ECG. Passage of the electrical wave of depolarization is then followed almost immediately by actual contraction of the atrial musculature.

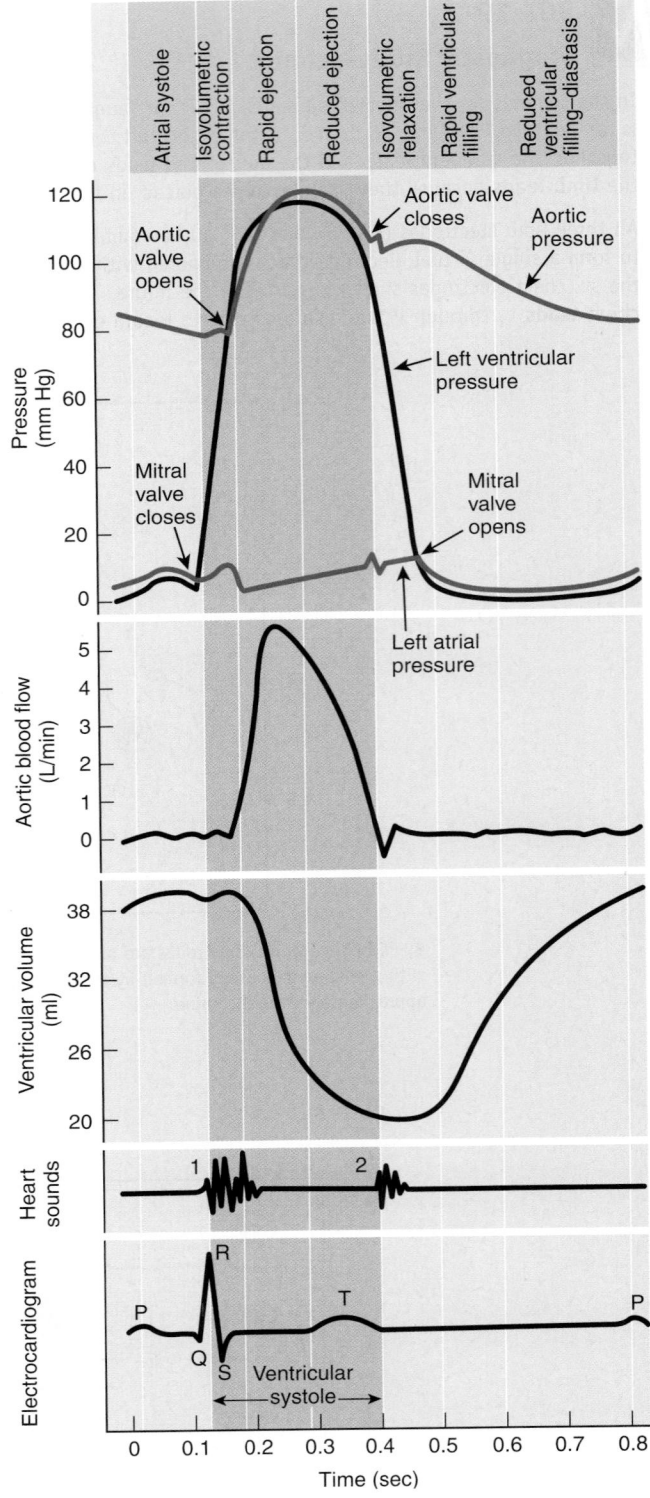

Figure 19-7 *Composite chart of heart function.* This chart is a composite of several diagrams of heart function (cardiac pumping cycle, blood pressure, blood flow, volume, heart sounds, and ECG), all adjusted to the same time scale. Although it appears daunting at first glance, you will find it a valuable reference tool as you proceed through this chapter.

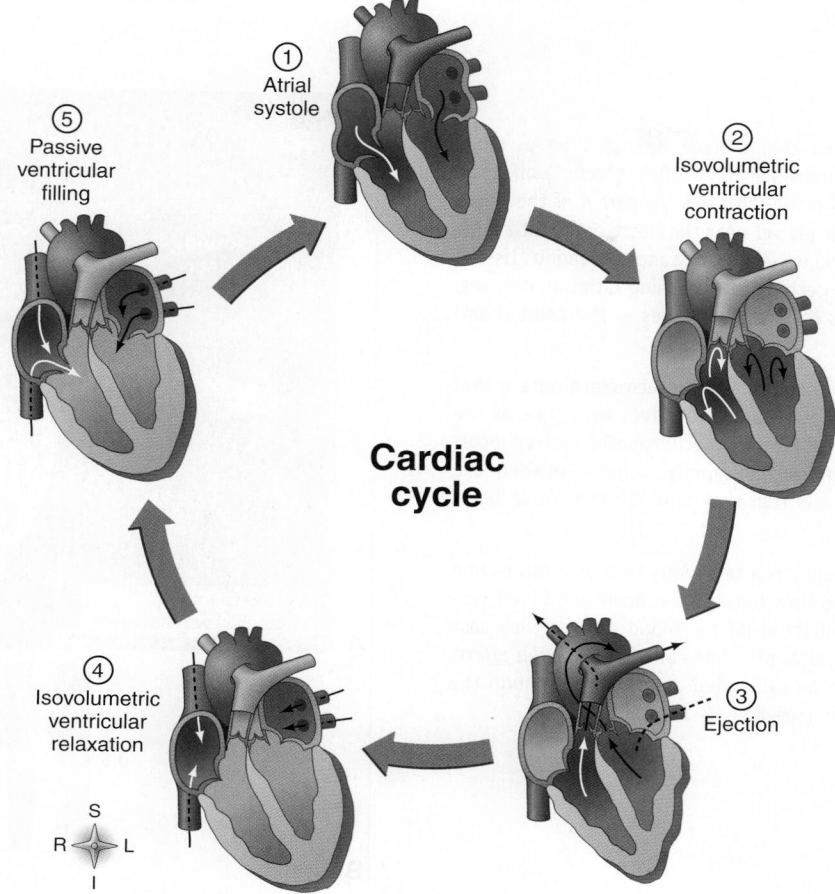

Cardiac cycle

① Atrial systole

② Isovolumetric ventricular contraction

③ Ejection

④ Isovolumetric ventricular relaxation

⑤ Passive ventricular filling

Figure 19-8 *The cardiac cycle.* The five steps of the heart's pumping cycle described in the text are shown as a series of changes in the heart wall and valves.

Isovolumetric Ventricular Contraction

Iso- is a combining form denoting equality or uniformity, and *volumetric* denotes measurement of volume. Thus *isovolumetric* is a term that means "having the same measured volume." During the brief period of isovolumetric ventricular contraction, that is, between the start of ventricular systole and the opening of the semilunar valves, ventricular volume remains constant, or uniform, as the pressure increases rapidly. The onset of ventricular systole coincides with the R wave of the ECG and the appearance of the first heart sound.

Ejection

The semilunar valves open and blood is ejected from the heart when the pressure gradient in the ventricles exceeds the pressure in the pulmonary artery and aorta. An initial, shorter phase, called *rapid ejection,* is characterized by a marked increase in ventricular and aortic pressure and in aortic blood flow (Box 19-3). The T wave of the ECG appears during the later, longer phase of *reduced ejection* (characterized by a less abrupt decrease in ventricular volume). A considerable quantity of blood, called the **residual volume,** normally remains in the ventricles at the end of the ejection period. In heart failure the residual volume remaining in the ventricles may greatly exceed that ejected during systole.

Isovolumetric Ventricular Relaxation

Ventricular diastole, or relaxation, begins with this period of the cardiac cycle. It is the period between closure of the semilunar valves and opening of the atrioventricular valves. At the end of ventricular ejection the semilunar valves close so that blood cannot reenter the ventricular chambers from the great vessels. The atrioventricular valves do not open until the pressure in the atrial chambers increases above that in the relaxing ventricles. The result is a dramatic fall in intraventricular pressure but no change in volume. Both sets of valves are closed, and the ventricles are relaxing. The second heart sound is heard during this period.

Passive Ventricular Filling

Return of venous blood increases intraatrial pressure until the atrioventricular valves are forced open and blood rushes into the relaxing ventricles. The rapid influx lasts about 0.1 second and results in a dramatic increase in ventricular volume. The term *diastasis* is often used to describe a later, longer period of slow ventricular filling at the end of ventricular diastole. The abrupt inflow of blood that occurred immediately after opening of the atrioventricular valves is followed by a slow but continuous flow of venous blood into the atria and then through the open atrioventricular valves into the ventricles. Diastasis lasts about 0.2 second

BOX 19-3
Diagnostic Study

Echocardiography

As we discussed in the previous chapter (see Box 18-1 on p. 688), **echocardiography** uses ultrasound waves that reflect ("echo") off heart structures to produce images of heart functions. As part *A* of the figure shows, an ultrasound transducer placed near the heart emits ultrasonic waves (very high frequency sound waves) that bounce off various tissues differently. After the transducer records the returning ultrasound waves, a computerized image produces distinct boundaries at the heart chamber walls and valves.

In part *B* of the figure, you can see an **M-mode echocardiogram** that shows the motion (M) of the heart walls and valves over time as the transducer is held in a stable position. Another method of echocardiography called *2D echocardiography* rapidly swings between the dashed lines in *A* to produce a virtual slice through the whole heart at the same instant.

Recall that *Doppler ultrasonography* is a technique that uses ultrasound to record the direction of blood flow through the heart and blood vessels. These sound waves are reflected off red blood cells as they pass through the heart. Using a physical principle called the *Doppler effect*, the velocity of blood can then be calculated as it travels through the heart chambers or large vessels (see Box 18-1 on p. 688).

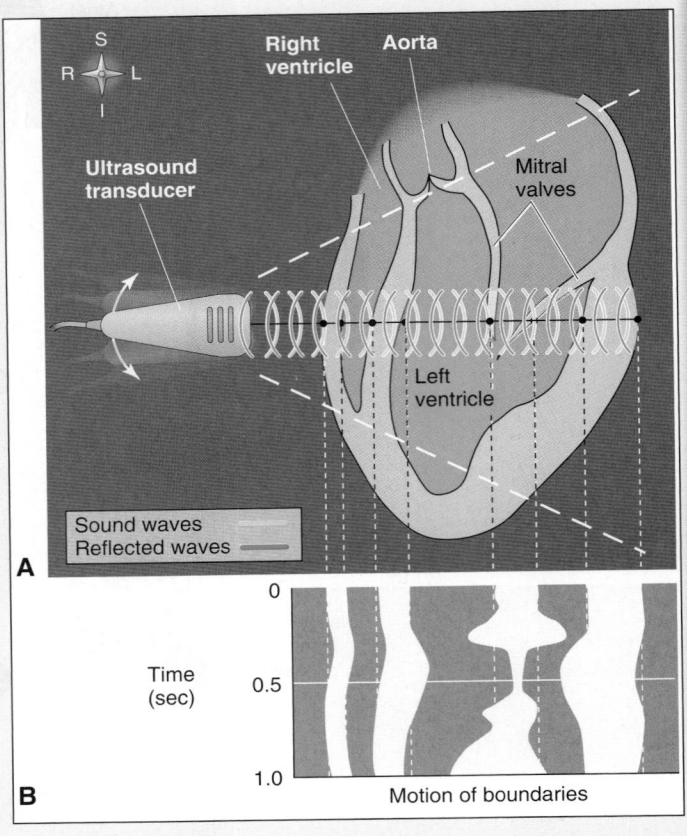

A, *Principle of echocardiography.* A probe, containing an ultrasound transducer is placed near the heart. Very high frequency sound waves (ultrasound) are emitted from the transducer and travel through the tissue of the heart. At boundaries, some of the ultrasound waves are reflected back to the transducer—which sends the data to a computer. The computer then constructs a visual image of the boundaries of the heart structures based on the reflected ultrasound information. **B, *M-mode echocardiogram.*** When holding the transducer stationary, the changes in the boundaries of the heart walls and valves are recorded over time, producing a graphic image of motion (M) in heart structures.

and is characterized by a gradual increase in ventricular pressure and volume.

Heart Sounds

The heart makes certain typical sounds during each cardiac cycle that are described as sounding like "lubb-dupp" through a stethoscope.

The first, or systolic, sound is caused primarily by the contraction of the ventricles and also by vibrations of the closing atrioventricular, or cuspid, valves. It is longer and lower than the second, or diastolic, sound, which is short and sharp and is caused by vibrations of the closing semilunar valves (see Figure 19-7).

Heart sounds have clinical significance, since they give information about the valves of the heart. Any variation from normal in the sounds indicates imperfect functioning of the valves. **Heart** murmur is one type of abnormal sound heard frequently. It is sometimes a "swishing" sound that may signify incomplete closing of the valves (valvular insufficiency) or stenosis (constriction, or narrowing) of them.

QUICK CHECK

4. Using Figure 19-8, describe the major events of the cardiac cycle.
5. As the ventricles contract, their volume remains constant for a period of time. Can you explain why the volume does not begin to decrease immediately?

PRIMARY PRINCIPLE OF CIRCULATION

Blood circulates for the same reason that any fluid flows—whether it is water in a river, water in a garden hose, fluid in hospital tubing, or blood in vessels. A fluid flows because a pressure gradient exists between different parts of its volume (Figure 19-9).

This primary fluid flow principle derives from Newton's first and second laws of motion. In essence, these laws state the following principles:

1. A fluid does not flow when the pressure is the same in all parts of it.
2. A fluid flows only when its pressure is higher in one area than in another, and it flows always from its higher pressure area toward its lower pressure area.

Thus the primary principle about circulation is this: blood flows because of a pressure gradient. Blood circulates from the left ventricle and returns to the right atrium of the heart because a blood pressure gradient exists between these two structures. By blood pressure gradient, we mean the difference between the blood pressure in one structure and the blood pressure in another. For example, a typical normal blood pressure in the aorta, as the left ventricle contracts pumping blood into it, is 120 mm Hg; as the left ventricle relaxes, it decreases to 80 mm Hg. The mean, or average, blood pressure therefore in the aorta in this instance is 100 mm Hg.

Figure 19-9 shows the systolic and diastolic pressures in the arterial system and illustrates the progressive fall in pressure to 0 mm Hg by the time blood reaches the venae cavae and right atrium. The progressive fall in pressure as blood passes through the circulatory system is directly related to resistance. Resistance to blood flow in the aorta is almost zero. Although the pumping action of the heart causes fluctuations in aortic blood pressure (systolic 120 mm Hg; diastolic 80 mm Hg), the mean pressure remains almost constant, dropping perhaps only 1 or 2 mm Hg. The greatest drop in pressure (about 50 mm Hg) occurs across the arterioles because they present the greatest resistance to blood flow.

P_1–P_2 is often used to stand for a pressure gradient, with P_1 the symbol for the higher pressure and P_2 the symbol for the lower pressure. For example, blood enters the arterioles at 85 mm Hg and leaves at 35 mm Hg. Which is P_1? P_2? What is the blood pressure gradient? It would cause blood to flow from the arterioles into capillaries.

Of course, this principle applies to local blood flow as well as an entire circulatory loop. For example, pressure in the arteries and arterioles of the kidney must be higher than the blood pressure in the capillaries and veins of the kidney in order for blood to flow through the tissues of the kidney. This local pressure gradient needed to maintain blood flow in a tissue is called **perfusion pressure** (*perfusion* means "flow through").

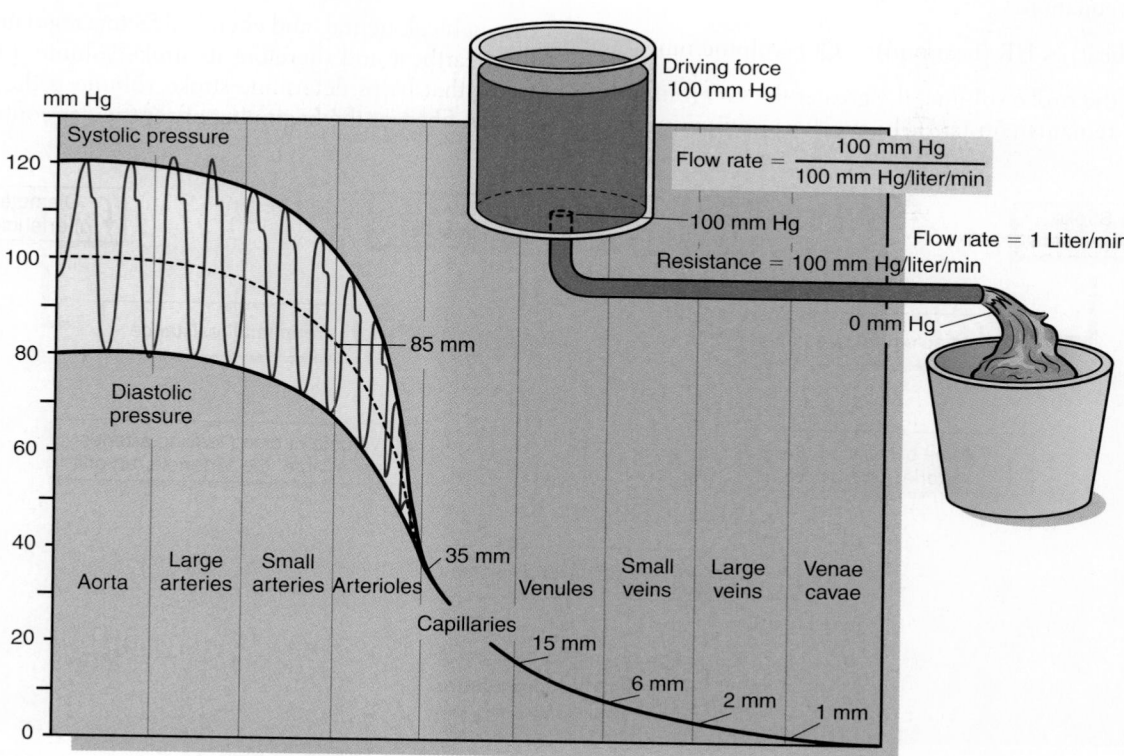

Figure 19-9 *The primary principle of circulation.* Fluid always travels from an area of high pressure to an area of low pressure. Water flows from an area of high pressure in the tank (100 mm Hg) toward the area of low pressure above the bucket (0 mm Hg). Blood tends to move from an area of high average pressure at the beginning of the aorta (100 mm Hg) toward the area of lowest pressure at the end of the venae cavae (0 mm Hg). Blood flow between any two points in the circulatory system can always be predicted by the pressure gradient.

ARTERIAL BLOOD PRESSURE

According to the primary principle of circulation, high pressure in the arteries must be maintained to keep blood flowing through the cardiovascular system. The chief determinant of arterial blood pressure is the volume of blood in the arteries. Arterial blood volume is directly proportional to arterial pressure. This means that an increase in arterial blood volume tends to increase arterial pressure, and, conversely, a decrease in arterial volume tends to decrease arterial pressure.

Many factors determine arterial pressure through their influence on arterial volume. Two of the most important—**cardiac output** and **peripheral resistance**—are directly proportional to blood volume (Figure 19-10).

Cardiac Output

Cardiac output (CO) is the amount of blood that flows out of the heart per unit of time—5000 ml/min, for example. As Figure 19-11 shows, the cardiac output influences the flow rate to the various organs of the body.

Cardiac output is determined by the volume of blood pumped out of the ventricles by each beat (**stroke volume** or **SV**) and by heart rate (HR). Because contraction of the heart is called *systole*, the volume of blood pumped by one contraction is known as *systolic discharge*. Stroke volume means the same thing, the amount of blood pumped by one stroke (contraction) of the ventricle.

Stroke volume, or volume pumped per heartbeat, is one of two major factors that determine CO. CO can be computed by the following simple equation:

$$SV \text{ (volume/beat)} \times HR \text{ (beat/min)} = CO \text{ (volume/min)}$$

Thus the greater the stroke volume, the greater the CO (but only if the heart rate remains constant). In practice, computing the CO is far from simple. It requires introducing a catheter into the right side of the heart (cardiac catheterization) and solving a computation known as *Fick's formula.*

Because the heart's rate and stroke volume determine its output, anything that changes the rate of the heartbeat or its stroke volume tends to change CO, arterial blood volume, and blood pressure in the same direction. In other words, anything that makes the heart beat faster or anything that makes it beat stronger (increases its stroke volume) *tends* to increase CO and therefore arterial blood volume and pressure. Conversely, anything that causes the heart to beat more slowly or more weakly tends to decrease CO, arterial volume, and blood pressure. But do not overlook the word *tends* in the preceding sentences. A change in heart rate or stroke volume does not always change the heart's output, or the amount of blood in the arteries, or the blood pressure. To see whether this is true, do the following simple arithmetic, using the simple formula for computing CO. Assume a normal rate of 72 beats/min and a normal stroke volume of 70 ml. Next, suppose the rate drops to 60 and the stroke volume increases to 100. Does the decrease in heart rate actually cause a decrease in CO in this case? Clearly not—the CO increases. Do you think it is valid, however, to say that a slower rate tends to decrease the heart's output? By itself, without any change in any other factor, would not a slowing of the heartbeat cause CO volume, arterial volume, and blood pressure to fall?

Factors That Affect Stroke Volume

Mechanical, neural, and chemical factors regulate the strength of the heartbeat and therefore its stroke volume. One mechanical factor that helps determine stroke volume is the length of myocardial fibers at the beginning of ventricular contraction.

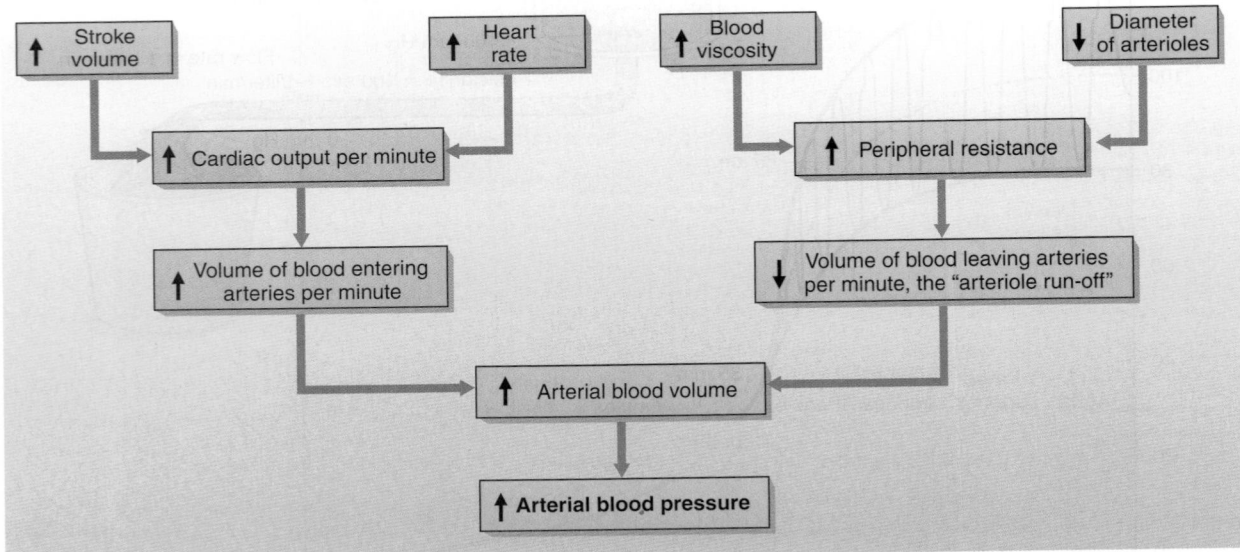

Figure 19-10 *Relationship between arterial blood volume and blood pressure.* Arterial blood pressure is directly proportional to arterial blood volume. Cardiac output (CO) and peripheral resistance (PR) are directly proportional to arterial blood volume, but for opposite reasons: CO affects blood *entering* the arteries, and PR affects blood *leaving* the arteries. If cardiac output increases, the amount of blood entering the arteries increases and tends to increase the volume of blood in the arteries. If peripheral resistance increases, it decreases the amount of blood leaving the arteries, which tends to increase the amount of blood left in them. Thus an increase in either CO or PR results in an increase in arterial blood volume, which increases arterial blood pressure.

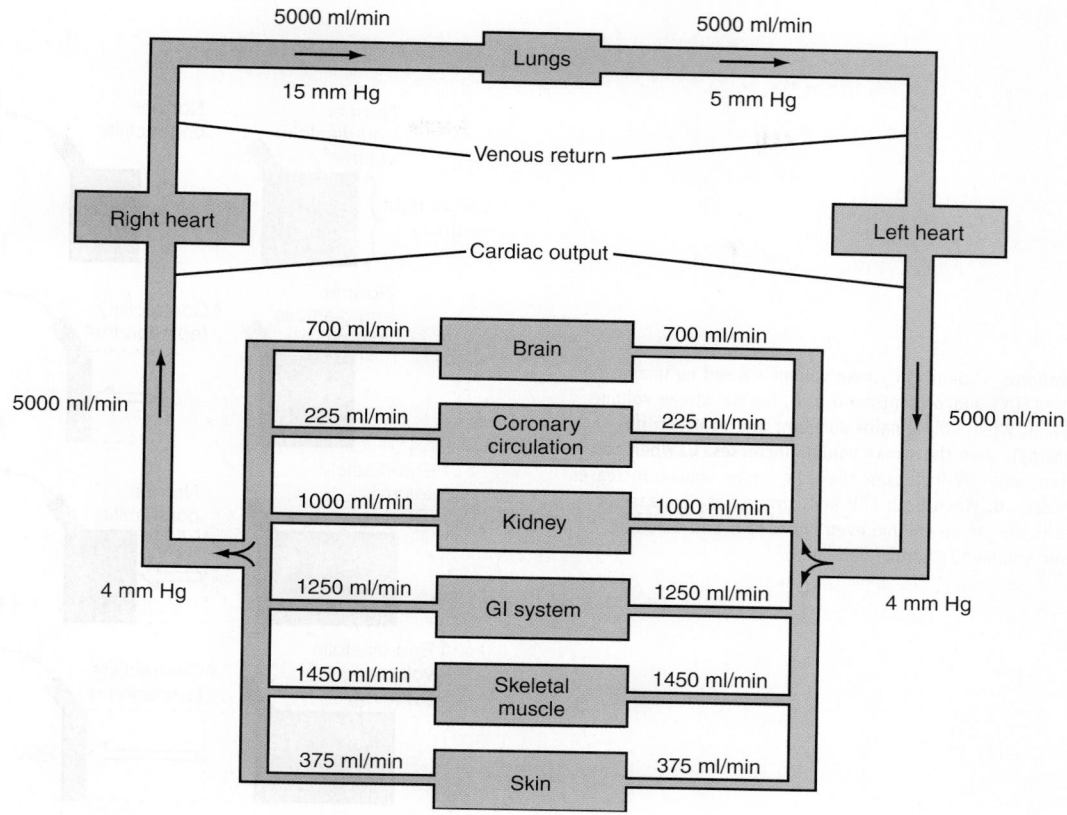

Figure 19-11 *Cardiac output.* This diagram shows that a typical resting cardiac output (CO) of 5000 ml/min (or 5 L/min) is distributed among the various systems and organs of the body. *GI,* Gastrointestinal.

Many years ago, an English physiologist named Ernest Starling described a principle that later became known as **Starling's law of the heart.** Because the principle was partly based on the earlier work of Otto Frank, it is sometimes called the *Frank-Starling mechanism*. In this principle, Starling stated the factor he had observed as the main regulator of heartbeat strength in experiments performed on denervated animal hearts. Starling's law of the heart is this: within limits, the longer, or more stretched, the heart fibers at the beginning of contraction, the stronger is their contraction. Compare this concept with the length-tension relationship in skeletal muscle described in Chapter 11 (p. 396).

The factor determining how stretched the animal hearts were at the beginning of contractions was, as you might deduce, the amount of blood in the hearts at the end of diastole—the **end-diastolic volume (EDV).** The more blood returned to the hearts per minute, the more stretched were their fibers, the stronger were their contractions, and the larger was the volume of blood they ejected with each contraction. If, however, too much blood stretched the hearts beyond a certain critical point, they seemed to lose their elasticity. They then contracted less vigorously, similar to how a band of elastic, stretched too much, rebounds with less force (Figure 19-12).

Thus, according to Starling's law of the heart, the heart pumps out what it receives. That is, within certain limits, the strength of myocardial contraction matches the pumping load—unlike mechanical pumps that do not adjust themselves to their input with every stroke.

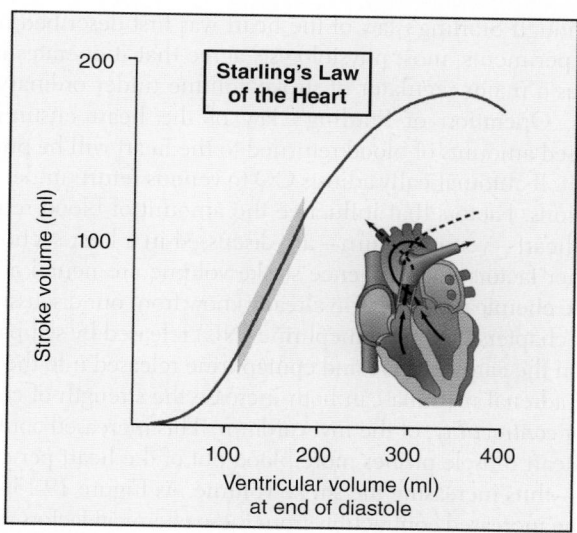

Figure 19-12 *Starling's law of the heart.* This curve represents the relationship between the stroke volume and the ventricular volume at the end of diastole. The range of values observed in a typical heart is shaded. Note that if the ventricle has an abnormally large volume at the end of diastole *(far right portion of the curve),* the stroke volume cannot compensate.

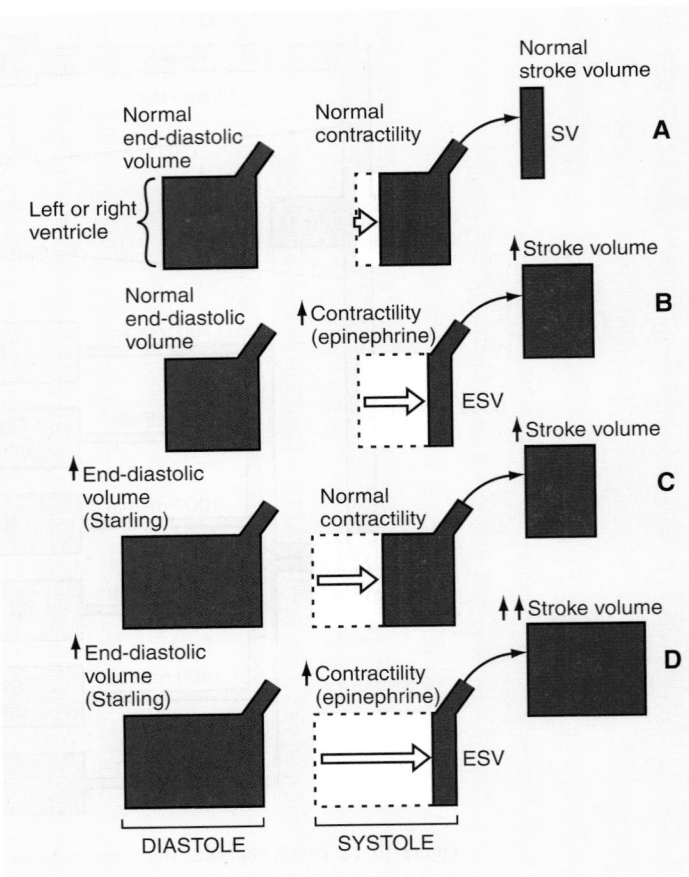

Figure 19-13 *Stroke volume.* Changes in stroke volume caused by increasing the end-diastolic volume (EDV) and/or contractility. **A,** Normal stroke volume (no external influences). **B,** When EDV remains constant and contractility increases (from epinephrine), then the stroke volume increases. **C,** When contractility remains constant and EDV increases, then the stroke volume increases (Starling's law of the heart). **D,** When both EDV and contractility increase, a combined effect increases the stroke volume even more. *EDV,* End-diastolic volume; *ESV,* end-systolic volume; *SV,* stroke volume.

Although Starling's law of the heart was first described in animal experiments, most physiologists agree that it operates in humans as a major regulator of stroke volume under ordinary conditions. Operation of Starling's law of the heart ensures that increased amounts of blood returned to the heart will be pumped out of it. It automatically adjusts CO to venous return under usual conditions. Factors that influence the amount of blood returned to the heart—venous return—are discussed in a later section.

Other factors that influence stroke volume are neural and endocrine chemical factors. You already know from our discussions in earlier chapters that norepinephrine (NE) released by sympathetic fibers in the cardiac nerve and epinephrine released into the blood by the adrenal medulla can both increase the strength of contraction, or *contractility*, of the myocardium. This increased contractility of heart muscle pushes more blood out of the heart per cardiac stroke—thus increasing the stroke volume. As Figure 19-13 shows, both the increased contractility from these chemical factors and the effects of Starling's law of the heart can change the stroke volume—and therefore also cardiac output. Factors such as the stress of exercise can trigger these neural and endocrine responses.

Factors That Affect Heart Rate

Although the sinoatrial node normally initiates each heartbeat, the rate it sets is not an unalterable one. Various factors can and do change the rate of the heartbeat. One major modifier of sinoatrial node activity—and therefore of the heart rate—is the ratio of sympathetic and parasympathetic impulses conducted to the node per minute. Autonomic control of heart rate is the result of opposing influences between parasympathetic (chiefly vagus) and sympathetic (cardiac nerve) stimulation. The results of parasympathetic stimulation on the heart are inhibitory and are mediated by vagal release of acetylcholine, whereas sympathetic (stimulatory) effects result from the release of norepinephrine at the distal end of the cardiac nerve.

Cardiac Pressoreflexes

Receptors sensitive to changes in pressure (**baroreceptors**) are located in two places near the heart (Figure 19-14). Called the *aortic baroreceptors* and *carotid baroreceptors*, they send afferent nerve fibers to cardiac control centers in the medulla oblongata. These stretch receptors, located in the aorta and carotid sinus, constitute a very important heart rate control mechanism because of their effect on the autonomic cardiac control centers—and therefore on parasympathetic and sympathetic outflow. Baroreceptors operate with integrators in the cardiac control centers in negative feedback loops called **pressoreflexes** or *baroreflexes* that oppose changes in pressure by adjusting heart rate.

Carotid Sinus Reflex

The carotid sinus is a small dilation at the beginning of the internal carotid artery just above the branching of the common carotid artery to form the internal and external carotid arteries (see

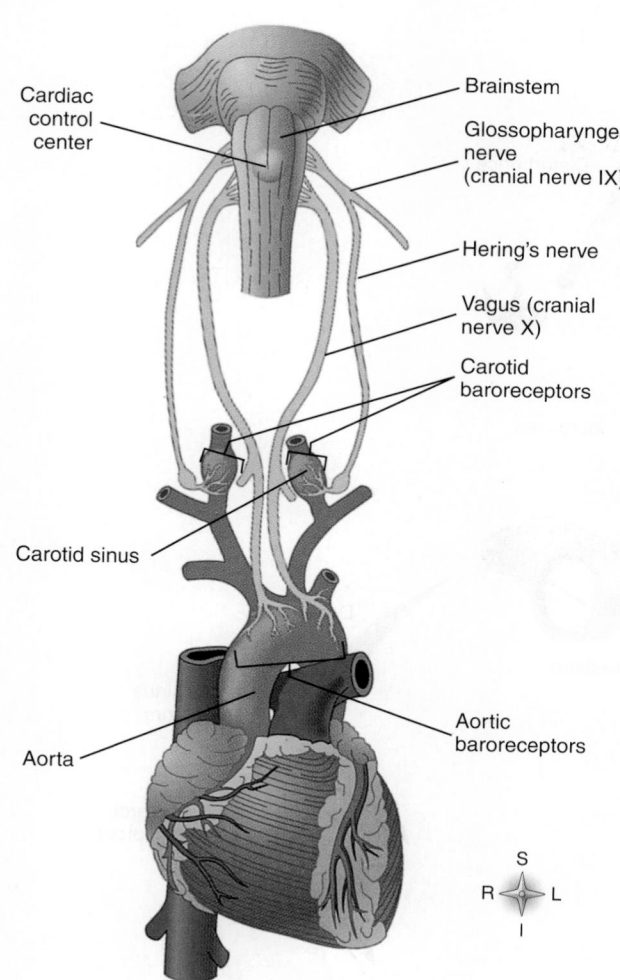

Figure 19-14 *Cardiac baroreceptors.* Location of aortic and carotid baroreceptors and the sensory nerves that carry feedback information about blood pressure back to the central nervous system.

Figure 19-14). The sinus lies just under the sternocleidomastoid muscle at the level of the upper margin of the thyroid cartilage. Sensory (afferent) fibers from carotid sinus baroreceptors (pressure sensors) run through the carotid sinus nerve (of Hering) and on through the glossopharyngeal (or ninth cranial) nerve. These nerves relay feedback information to an integrator area of the medulla called the *cardiac control center.* If the integrators in the cardiac control center detect an increase in blood pressure above the set point, then a correction signal is sent to the SA node by way of efferent parasympathetic fibers in the vagus (tenth cranial) nerve. Acetylcholine released by vagal fibers decreases the rate of SA node firing, thus decreasing the heart rate back toward the set point. The vagus is said to act as a "brake" on the heart—a situation called *vagal inhibition.* Figure 19-15 summarizes this negative feedback loop.

Aortic Reflex

Sensory (afferent) nerve fibers also extend from baroreceptors located in the wall of the arch of the aorta through the aortic nerve and then through the vagus (tenth cranial) nerve to terminate in the cardiac control center of the medulla (see Figure 19-14).

If blood pressure within the aorta or carotid sinus suddenly increases beyond the set point, it stimulates the aortic or carotid baroreceptors, as shown in Figure 19-15. Stimulation of these stretch receptors causes the cardiac control center to increase vagal inhibition, thus slowing the heart and returning blood pressure back toward the normal set point. A decrease in aortic or carotid blood pressure usually allows some acceleration of the heart by way of correction signals through the cardiac nerve. More details of pressoreflex activity are included later in the chapter as part of a mechanism that tends to maintain or restore homeostasis of arterial blood pressure.

Other Reflexes That Influence Heart Rate

Reflexes involving important factors such as emotions, exercise, hormones, blood temperature, pain, and stimulation of various exteroceptors also influence heart rate. Anxiety, fear, and anger often make the heart beat faster. Grief, in contrast, tends to slow it. Emotions produce changes in the heart rate through the influence of impulses from the "higher centers" in the cerebrum by way of the hypothalamus. Such impulses can influence activity of the cardiac control centers.

In exercise the heart normally accelerates. The mechanism is not definitely known, but it is thought to include impulses from the cerebrum through the hypothalamus to the cardiac center. Epinephrine is the hormone most noted as a cardiac accelerator.

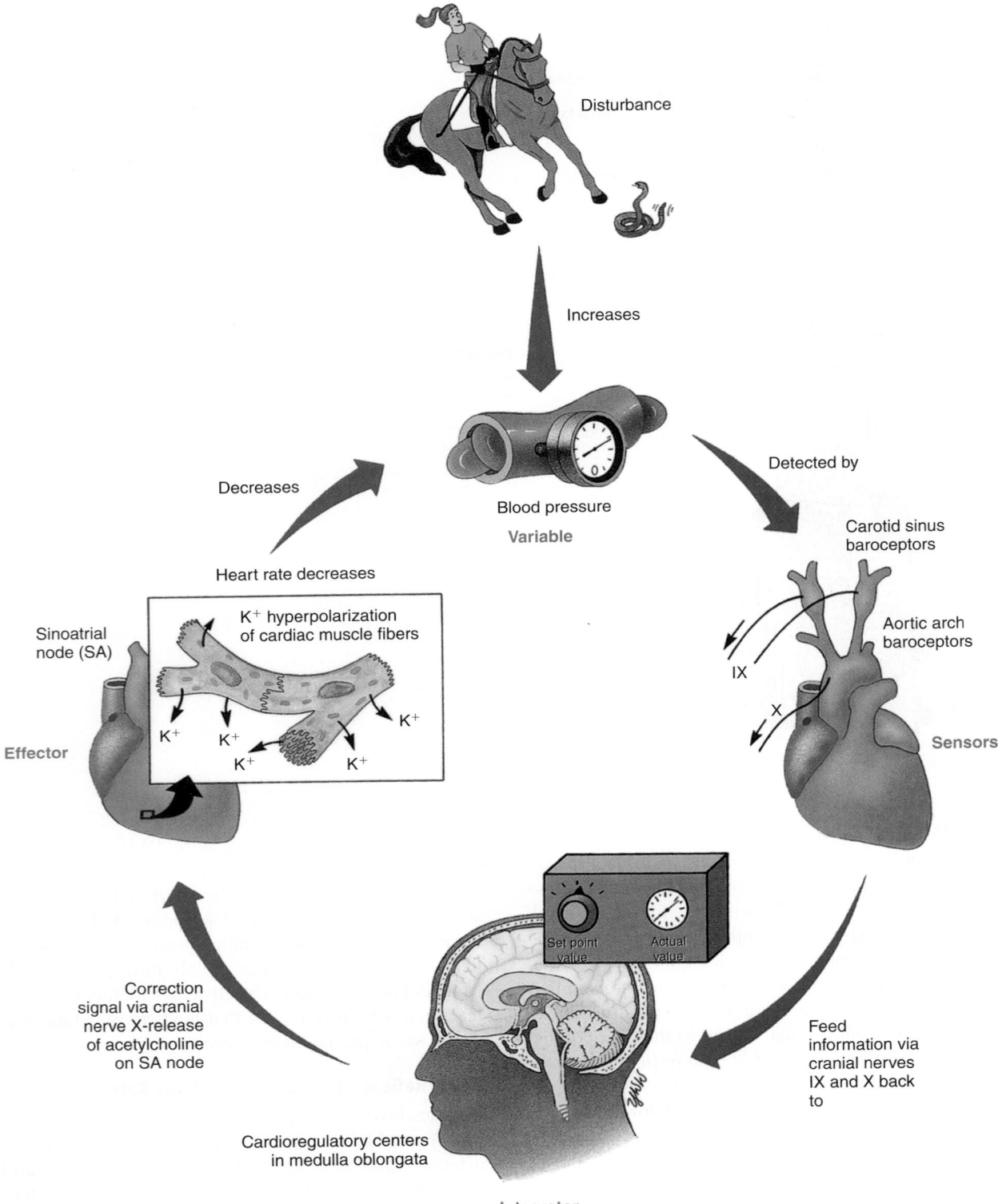

Figure 19-15 *Aortic and carotid sinus pressoreflexes.* These pressoreflexes operate in a feedback loop that maintains the homeostasis of blood pressure by decreasing the heart rate when the blood pressure surpasses the set point.

Increased blood temperature or stimulation of skin heat receptors tends to increase the heart rate, and decreased blood temperature or stimulation of skin cold receptors tends to slow it. Sudden, intense stimulation of pain receptors in visceral structures such as the gallbladder, ureters, or intestines can result in such slowing of the heart that fainting may result.

Reflexive increases in heart rate often result from an increase in sympathetic stimulation of the heart. Sympathetic impulses originate in the cardiac control center of the medulla and reach the heart by way of sympathetic fibers (contained in the middle, superior, and inferior cardiac nerves). Norepinephrine released as a result of sympathetic stimulation increases heart rate and strength of cardiac muscle contraction.

 QUICK CHECK

6. State the primary principle of circulation.
7. Relate stroke volume and heart rate to cardiac output.
8. If the amount of blood returned to the heart increases, what happens to the stroke volume? What principle explains this?
9. How does the body use a pressoreflex to counteract an abnormal increase in heart rate? An abnormal decrease in heart rate?

Peripheral Resistance

How Resistance Influences Blood Pressure

Peripheral resistance helps determine arterial blood pressure. Specifically, arterial blood pressure tends to vary directly with peripheral resistance. Peripheral resistance means the resistance to blood flow imposed by the force of friction between blood and the walls of its vessels. Friction develops partly because of a characteristic of blood—its **viscosity**, or stickiness—and partly from the small diameter of arterioles and capillaries. The resistance offered by arterioles, in particular, accounts for almost one half of the total resistance in systemic circulation.

Blood viscosity stems mainly from the proportion of red blood cells (hematocrit) but also partly from the protein molecules present in blood. An increase in either blood protein concentration or hematocrit tends to increase viscosity, and a decrease in either tends to decrease it (Figure 19-16). Under normal circumstances, blood viscosity changes very little. But under certain abnormal conditions, such as marked anemia or hemorrhage, a decrease in blood viscosity may be the crucial factor lowering peripheral resistance and arterial pressure, even to the point of circulatory failure.

The muscular coat of the arterioles allows them to constrict or dilate and thus change the amount of resistance to blood flow. This muscular mechanism in the vessels is called the **vasomotor mechanism.** Reduction in vessel diameter caused by increased contraction of the muscular coat, or **vasoconstriction,** increases resistance to blood flow, and thus blood flow into the tissue decreases. **Vasodilation,** the increase in vessel diameter caused by relaxation of vascular muscles, decreases resistance to blood flow, and thus blood flow into the tissue increases. As Figure 19-17

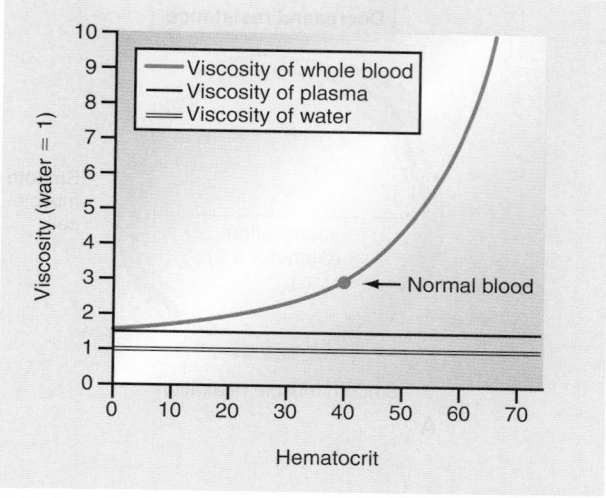

Figure 19-16 *Blood viscosity.* The effect of a changing hematocrit (% RBCs in blood) on blood viscosity is shown here. As hematocrit increases (horizontal axis of graph), the viscosity increases. Water is the reference value at viscosity = 1. Plasma is slightly more viscous than water—at approximately 1.5. As the total viscosity of blood increases, the resistance to blood flow increases.

shows, small changes in diameter can cause proportionally large changes in resistance—and therefore large changes in local blood flow. This makes the vasomotor mechanism well suited for quickly and dramatically changing blood flow under varying conditions in the body, as we shall see.

Peripheral resistance helps determine arterial pressure by controlling the rate of "arteriole runoff," the amount of blood that runs out of the arteries into the arterioles (Figure 19-18). The greater the resistance, the less the arteriole runoff, or outflow, tends to be—and therefore the more blood left in the arteries, the higher the arterial pressure tends to be. This can occur locally, within a particular tissue or organ, or it can occur throughout the systemic loop when enough arterioles constrict and increase the **total peripheral resistance (TPR).**

Vasomotor Control Mechanism

Blood distribution patterns, as well as blood pressure, can be influenced by factors that control changes in the diameter of arterioles. Such factors might be said to constitute the vasomotor control mechanism. Like most physiological control mechanisms, it consists of many parts. An area in the medulla called the *vasomotor center,* or *vasoconstrictor center,* will, when stimulated, initiate an impulse outflow by way of sympathetic fibers that ends in the smooth muscle surrounding resistance vessels, arterioles, venules, and veins of the "blood reservoirs," causing their constriction. Thus the vasomotor control mechanism plays a role both in the maintenance of the general blood pressure and in the distribution of blood to areas of special need.

The main blood reservoirs are the venous plexuses and sinuses in the skin and abdominal organs (especially in the liver and spleen). In other words, blood reservoirs are the venous networks in most parts of the body—all but those in the skeletal muscles, heart, and brain. Figure 19-19 shows that the volume of blood in

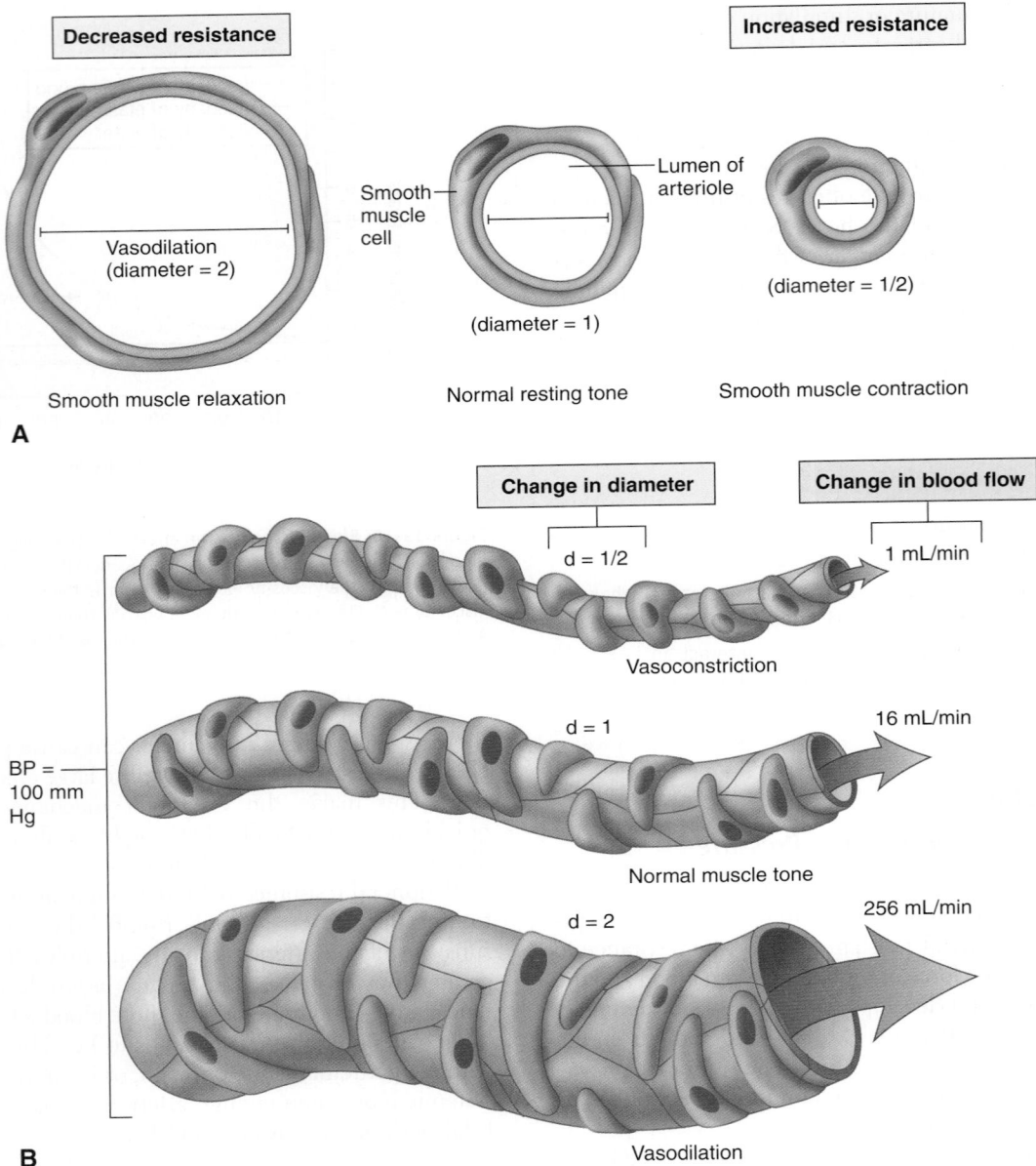

Figure 19-17 *Vessel diameter.* The effect of changing diameter of arterioles on peripheral resistance and blood flow. **A,** Cross sections of an arteriole showing vasodilation *(left),* normal diameter *(center),* and vasoconstriction *(right)* as tension in smooth muscle fibers changes. *D,* diameter **B,** Diagram showing that relatively small changes from the normal diameter of an arteriole (normal = 1) cause very large changes in peripheral resistance to blood flow. For example, reducing diameter to one half of normal reduces blood flow to $1/16$ of normal. Likewise, doubling the vessel diameter does not double the blood flow—it increases blood flow 16 times the normal flow! *P,* Pressure.

the systemic veins and venules in a resting adult is extremely large compared with the volume in other vessels of the body. The term *reservoir* is apt, because the systemic veins and venules serve as a kind of slowly moving stockpile or reserve of blood. Blood can quickly be moved out of blood reservoirs and "shifted" to arteries that supply heart and skeletal muscles when increased activity demands (Figure 19-20). A change in either arterial blood's oxygen or carbon dioxide content sets a chemical vasomotor control

mechanism in operation. A change in arterial blood pressure initiates a **vasomotor pressoreflex.**

Vasomotor Pressoreflexes

A sudden increase in arterial blood pressure stimulates aortic and carotid baroreceptors—the same ones that initiate cardiac reflexes. Not only does this stimulate the cardiac control center to reduce heart rate (see Figure 19-15), but also it inhibits the vaso-

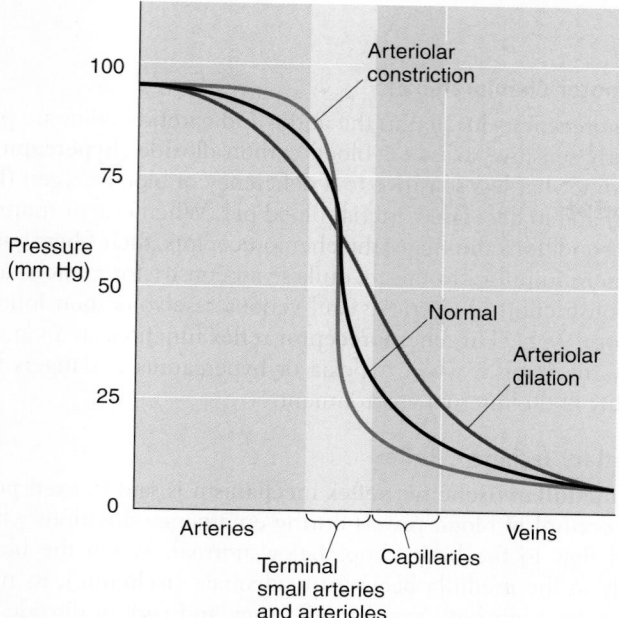

Figure 19-18 *Vasomotor effects on blood pressure.* The normal blood pressures in various vessels are shown by the black line. The green line shows a shift in blood pressures during vasoconstriction of the arterioles. Note that because there is more resistance in the arterioles, blood flow in the arteries backs up and thus increases arterial blood pressure. The purple line shows that vasodilation of the arterioles increases blood flow from the arteries and therefore reduces arterial blood pressure.

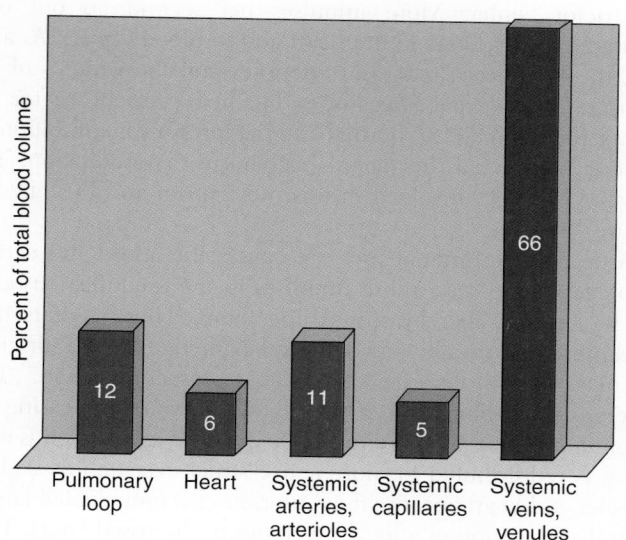

Figure 19-19 *Relative blood volumes.* The relative volumes of blood at rest in different parts of the adult cardiovascular system expressed as percentages of total blood volume. Note that at rest most of the body's blood supply is in the systemic veins and venules.

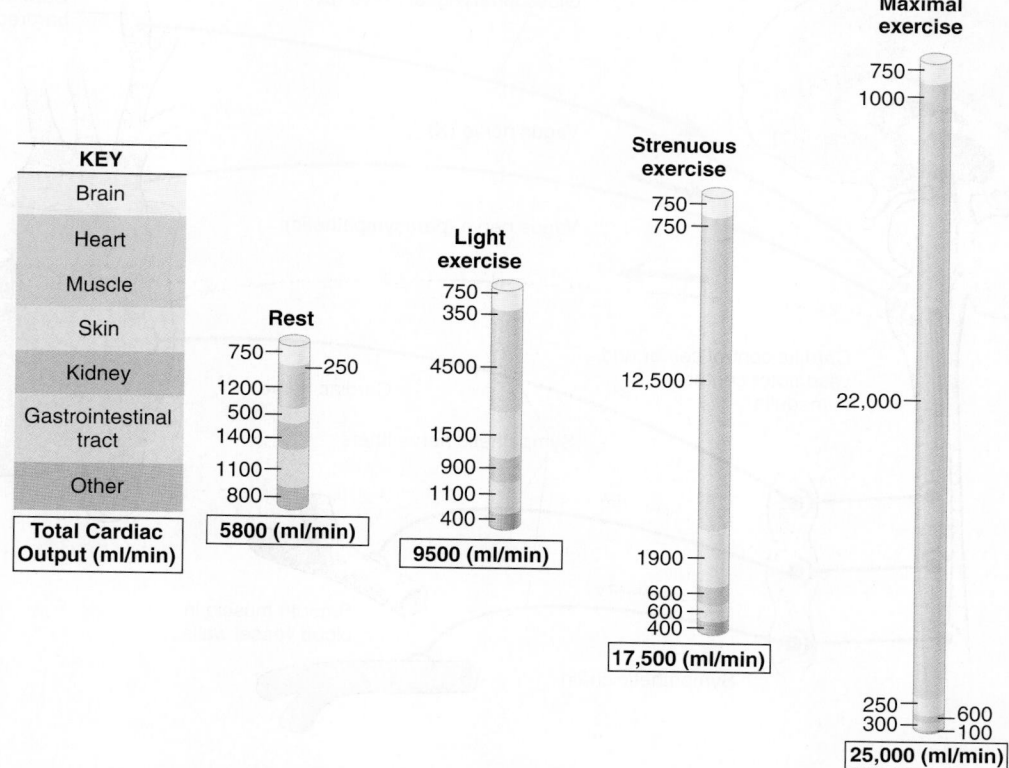

Figure 19-20 *Changes in local blood flow during exercise.* Graphic representation of blood flow (ml/min) in various circumstances: at rest and during light, strenuous, and maximal exercise. During exercise, the vasomotor center of the medulla sends sympathetic signals to certain blood vessels to change diameter and thus shunt blood away from "maintenance" organs, such as the digestive organs in the abdomen, and toward the skeletal muscles. Note that blood flow in the brain is held constant. Note also that the total blood flow (cardiac output) increases as the intensity of exercise increases.

constrictor center. More impulses per second go out over parasympathetic fibers to the heart and to blood vessels. As a result, the heartbeat slows, and arterioles and the venules of the blood reservoirs dilate. Because sympathetic vasoconstrictor impulses predominate at normal arterial pressures, inhibition of these is considered the major mechanism of vasodilation. The nervous pathways involved in this mechanism are illustrated in Figure 19-21.

A decrease in arterial pressure causes the aortic and carotid baroreceptors to send more impulses to the medulla's vasoconstrictor centers, thereby stimulating them. These centers then send more impulses by way of the sympathetic fibers to stimulate vascular smooth muscle and cause vasoconstriction. This squeezes more blood out of the blood reservoirs, increasing the amount of venous blood return to the heart. Eventually, this extra blood is redistributed to more active structures such as skeletal muscles and heart because their arterioles become dilated largely from the operation of a local mechanism (discussed later). Thus the vasoconstrictor pressoreflex and the local vasodilating mechanism together serve as an important device for shifting blood from reservoirs to structures that need it more. It is an especially valuable mechanism during exercise (see Figure 19-20 and Box 19-4).

Vasomotor Chemoreflexes

Chemoreceptors located in the aortic and carotid bodies are particularly sensitive to excess blood carbon dioxide (**hypercapnia**) and somewhat less sensitive to a deficiency of blood oxygen (**hypoxia**) and to decreased arterial blood pH. When one or more of these conditions stimulates the chemoreceptors, their fibers transmit more impulses to the medulla's vasoconstrictor centers, and vasoconstriction of arterioles and venous reservoirs soon follows (Figure 19-22). This **chemoreceptor reflex** functions as an emergency mechanism when hypoxia or hypercapnia endangers the stability of the internal environment.

Medullary Ischemic Reflex

The **medullary ischemic reflex** mechanism is said to exert powerful control of blood vessels during emergency situations when blood flow to the brain drops below normal. When the blood supply to the medulla becomes inadequate (**ischemic**), its neurons suffer from both oxygen deficiency and carbon dioxide excess. But, presumably, it is hypercapnia that intensely and directly stimulates the vasoconstrictor centers to bring about marked arteriole and venous constriction (see Figure 19-22). If the oxygen supply to the medulla decreases below a certain level, its neurons,

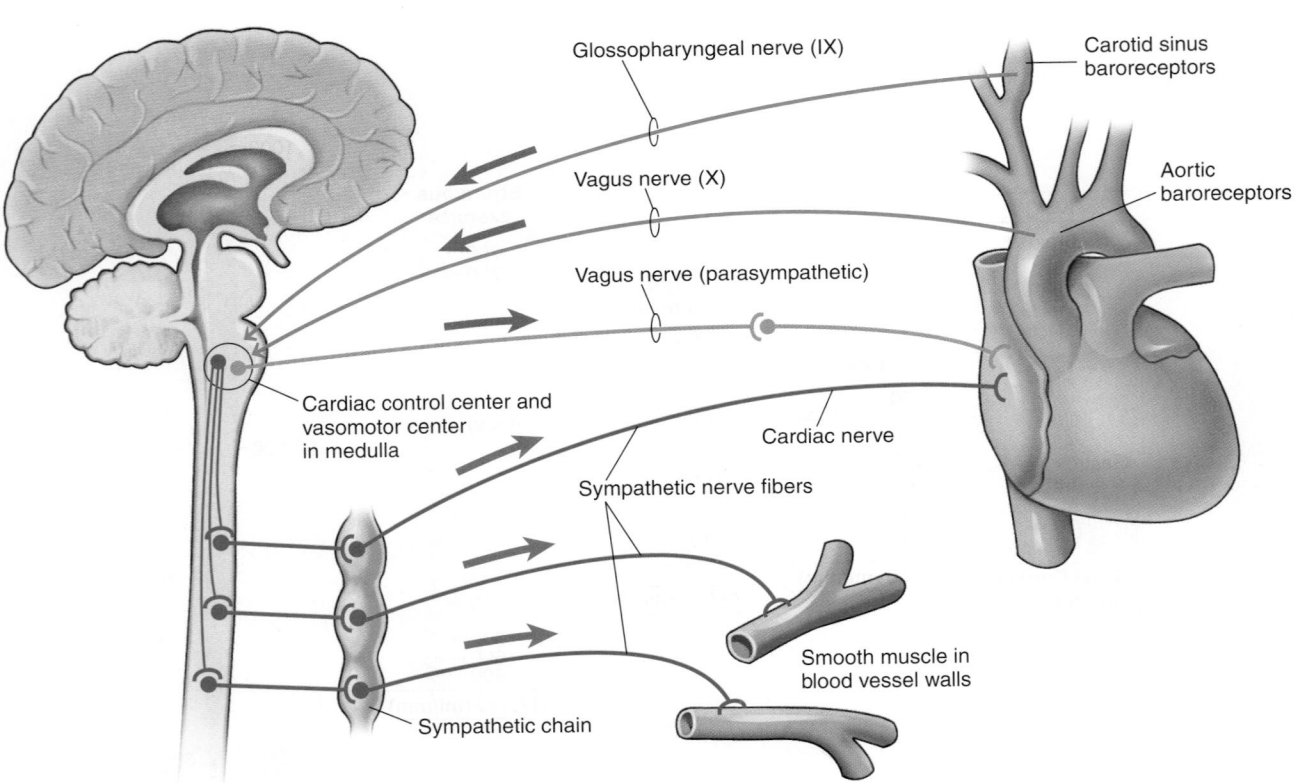

Figure 19-21 *Vasomotor pressoreflexes.* Carotid sinus and aortic baroreceptors detect changes in blood pressure and feed the information back to the cardiac control center and the vasomotor center in the medulla. In response, these control centers alter the ratio between sympathetic and parasympathetic output. If the pressure is too high, a dominance of parasympathetic impulses will reduce it by slowing heart rate, reducing stroke volume, and dilating blood "reservoir" vessels. If the pressure is too low, a dominance of sympathetic impulses will increase it by increasing heart rate and stroke volume and constricting reservoir vessels.

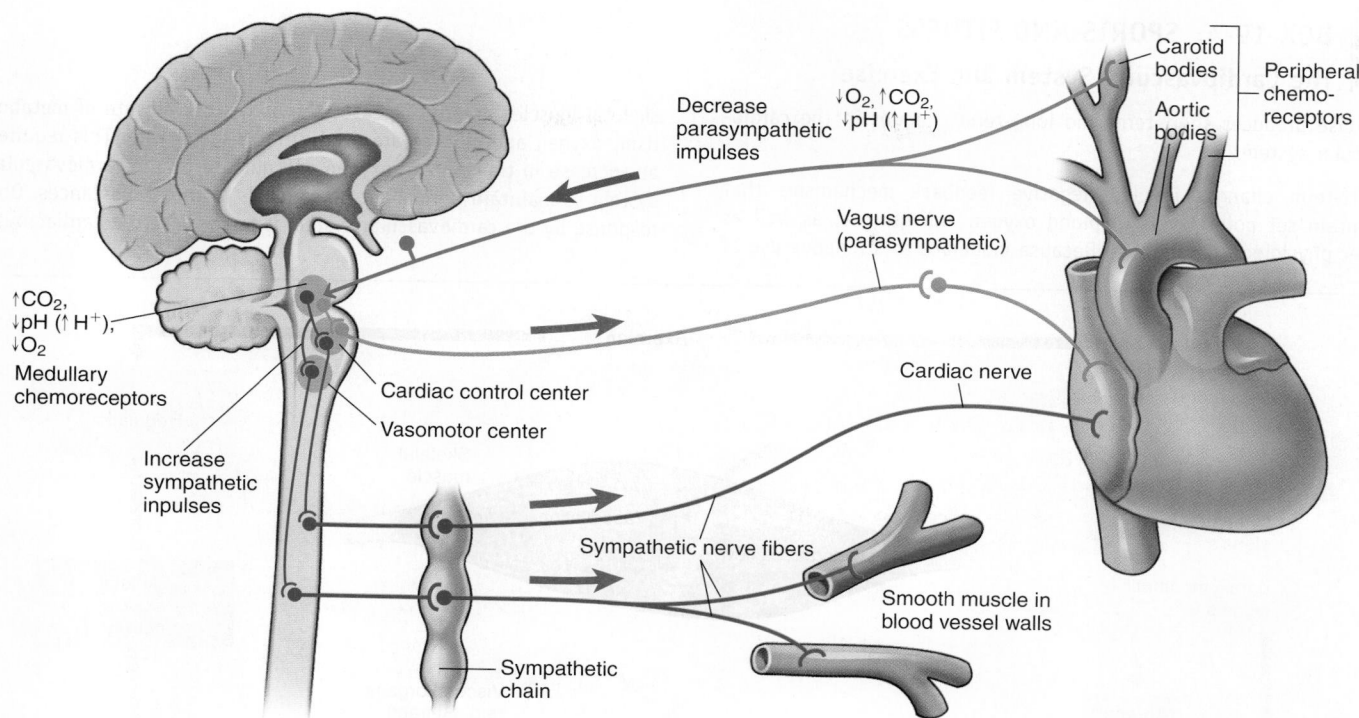

Figure 19-22 *Vasomotor chemoreflexes.* Chemoreceptors in the carotid and aortic bodies, as well as chemoreceptive neurons in the vasomotor center of the medulla itself, detect increases in carbon dioxide (CO_2), decreases in blood oxygen (O_2), and decreases in pH (which is really an increase in H^+). This information feeds back to the cardiac control center and the vasomotor control center of the medulla, which in turn alter the ratio of parasympathetic and sympathetic output. When O_2 drops, CO_2 increases, and/or pH drops, a dominance of sympathetic impulses increases heart rate and stroke volume and constricts reservoir vessels in response.

of course, cannot function, and the medullary ischemic reflex cannot operate.

Vasomotor Control by Higher Brain Centers

Impulses from centers in the cerebral cortex and in the hypothalamus are believed to be transmitted to the vasomotor centers in the medulla and to thereby help control vasoconstriction and dilation. Evidence supporting this view is that vasoconstriction and a rise in arterial blood pressure characteristically accompany emotions of intense fear or anger. Also, laboratory experiments on animals in which stimulation of the posterior or lateral parts of the hypothalamus leads to vasoconstriction support the belief that higher brain centers influence the vasomotor centers in the medulla.

Local Control of Arterioles

Several kinds of local mechanisms operate to produce vasodilation in localized areas. Although not all these mechanisms are clearly understood, they are known to function in times of increased tissue activity. For example, they probably account for the increased blood flow into skeletal muscles during exercise. They also operate in ischemic tissues, serving as a homeostatic mechanism that tends to restore normal blood flow. Some locally produced substances, such as nitric oxide, activate the local vasodilator mechanism, whereas others, such as *endothelin*, constrict the arterioles. Local vasodilation is also referred to as *reactive hyperemia*.

 QUICK CHECK

10. Peripheral resistance is affected by two major factors: blood viscosity and what else?
11. If the diameter of the arteries decreases, what effect does it have on peripheral resistance?
12. In general, how do vasomotor pressoreflexes affect the flow of blood?
13. What is a chemoreflex? How do chemoreflexes affect the flow of blood?

VENOUS RETURN TO THE HEART

Venous return refers to the amount of blood that is returned to the heart by way of the veins. Various factors influence venous return, including the reservoir function of veins, which occurs whenever blood pressure drops and the elasticity of the venous walls adapts the diameter of veins to the lower pressure, thus maintaining blood flow and venous return to the heart. Likewise, when overall blood pressure rises, the elastic nature of blood vessels allows them to expand and adapt to the higher pressure to maintain normal blood flow. This effect, which occurs in all blood vessels to some degree (with certain limitations to its adaptability), is often called the **stress-relaxation effect**.

BOX 19-4: SPORTS AND FITNESS

The Cardiovascular System and Exercise

Exercise produces short-term and long-term changes in the cardiovascular system.

Short-term changes involve negative feedback mechanisms that maintain set point levels of blood oxygen and glucose, as well as other physiological variables. Because moderate to strenuous use of skeletal muscles greatly increases the body's overall rate of metabolism, oxygen and glucose are used up at a faster rate. This requires an increase in transport of oxygen and glucose by the cardiovascular system to maintain normal set point levels of these substances. One response by the cardiovascular system is to increase the cardiac out-

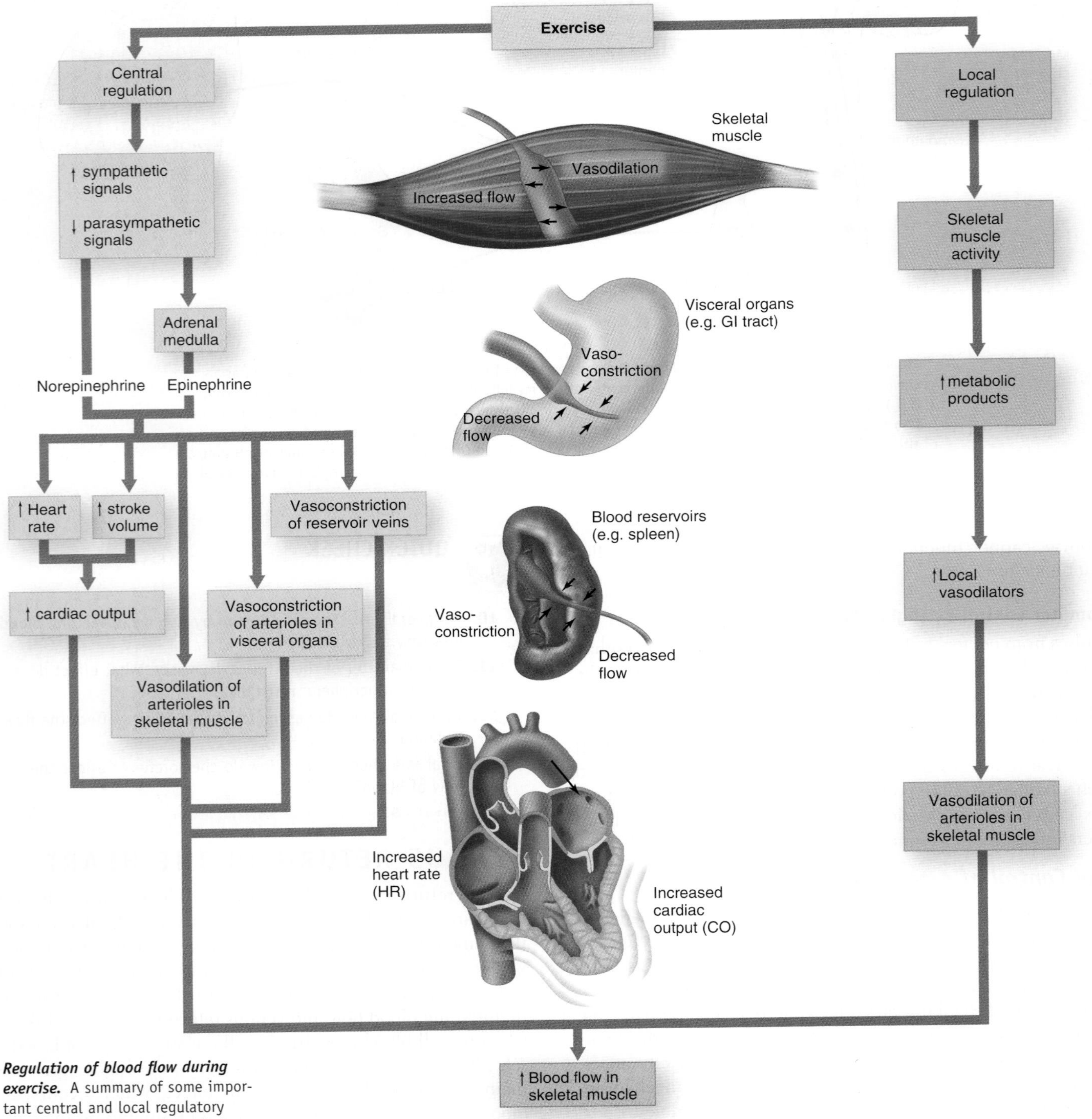

Exercise

Central regulation

↑ sympathetic signals

↓ parasympathetic signals

Adrenal medulla

Norepinephrine   Epinephrine

↑ Heart rate   ↑ stroke volume

Vasoconstriction of reservoir veins

↑ cardiac output

Vasoconstriction of arterioles in visceral organs

Vasodilation of arterioles in skeletal muscle

Skeletal muscle

Vasodilation

Increased flow

Visceral organs (e.g. GI tract)

Vasoconstriction

Decreased flow

Blood reservoirs (e.g. spleen)

Vasoconstriction

Decreased flow

Increased heart rate (HR)

Increased cardiac output (CO)

Local regulation

Skeletal muscle activity

↑ metabolic products

↑ Local vasodilators

Vasodilation of arterioles in skeletal muscle

↑ Blood flow in skeletal muscle

Regulation of blood flow during exercise. A summary of some important central and local regulatory mechanisms. *GI,* Gastrointestinal.

BOX 19-4: SPORTS AND FITNESS
The Cardiovascular System and Exercise—cont'd

put (CO) from 5 to 6 L/min at rest to up to 30 to 40 L/min during strenuous exercise. This represents a fivefold to eightfold increase in the blood output of the heart! Such an increase is accomplished by a reflexive increase in heart rate (see Figure 19-15) coupled with an increase in stroke volume (see Figures 19-12 and 19-13). Exercise can also trigger a reflexive change in local distribution of blood flow to various tissues, shown in Figure 19-20, that results in a larger share of blood flow going to the skeletal muscles than to some other tissues. A number of central and local regulatory effects that operate during exercise are summarized in the figure.

Long-term changes in the cardiovascular system come only when moderate to strenuous exercise occurs regularly over a long period.

Evidence indicates that 20 to 30 minutes of moderate aerobic exercise such as cycling or running three times per week produces profound, health-promoting changes in the cardiovascular system. Among these long-term cardiovascular changes are an increase in the mass and contractility of the myocardial tissue, an increase in the number of capillaries in the myocardium, a lower resting heart rate, and decreased peripheral resistance during rest. The lower resting heart rate coupled with the increase in stroke volume possible with increased myocardial mass and contractility produce a greater range of CO, and, coupled with the other listed effects, a greater maximum CO. Exercise is also known to decrease the risk of various cardiovascular disorders, including arteriosclerosis, hypertension, and heart failure.

Another factor that influences venous return is gravity. Figure 19-23 shows that when a person is reclining, the force of gravity is not pulling blood downward toward the legs. However, when a person is sitting or standing the blood is pulled by gravity toward the legs. Because the venous blood is already at a low blood pressure and because the venous walls are compliant (easily stretched), it is easy for the force of gravity to work against venous return back to the heart and cause some blood to remain in the veins of the limbs. The shift of the blood reservoir to the veins in the legs when standing is often called the **orthostatic effect** because *orthostasis* means "standing upright."

A factor that can help to overcome the influence of gravity is the operation of **venous pumps** that maintain the pressure gradients necessary to keep blood moving into the central veins (e.g., venae cavae) and from there into the atria of the heart. Changes in the total volume of blood in the vessels can also alter venous return. Venous return and total blood volume are discussed in the paragraphs that follow.

Venous Pumps

One important factor that promotes the return of venous blood to the heart is the blood-pumping action of respirations and skeletal muscle contractions. Both actions produce their facilitating effect on venous return by increasing the pressure gradient between the peripheral veins and the venae cavae (central veins).

The process of inspiration increases the pressure gradient between peripheral and central veins by decreasing central venous pressure and also by increasing peripheral venous pressure. Each time the diaphragm contracts, the thoracic cavity necessarily becomes larger and the abdominal cavity smaller. Therefore the pressures in the thoracic cavity, in the thoracic portion of the vena cava, and in the atria decrease, and those in the abdominal cavity and the abdominal veins increase. As Figure 19-24, A, shows, this change in pressure between expiration and inspiration acts as a "respiratory pump" that moves blood along the venous route.

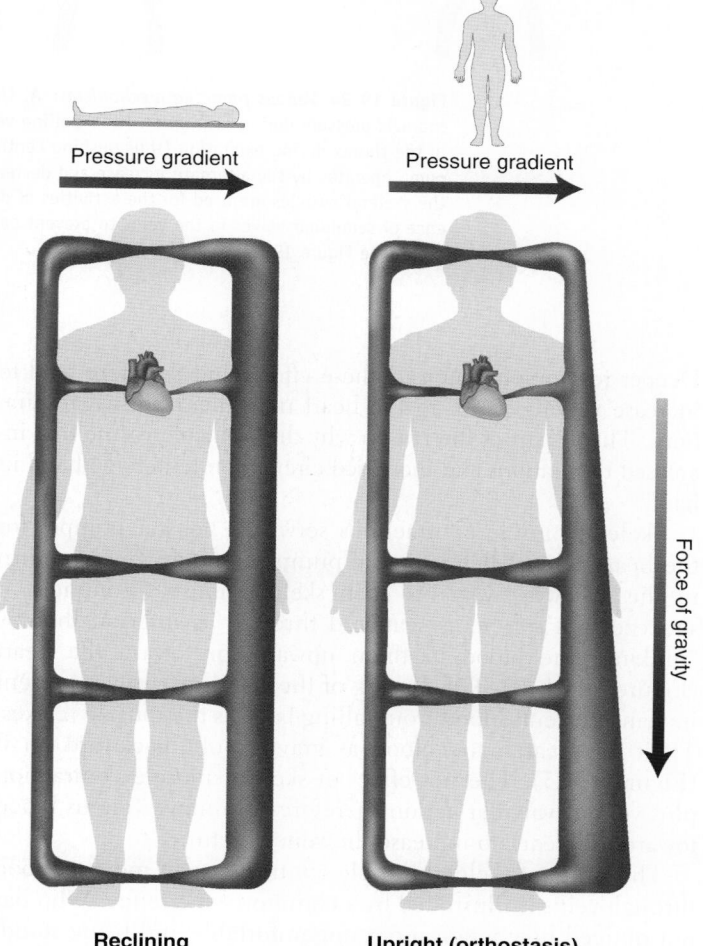

Figure 19-23 *Influence of gravity on blood distribution in veins.* A, When a person is lying flat, the pull of gravity is equal above and below the heart. **B,** When a person stands upright (orthostasis), the pull of gravity combined with the compliance (ease of stretch) of the veins causes a redistribution of venous blood to the lower limbs. This *orthostatic effect* can reduce venous return to the heart if not counteracted by other forces.

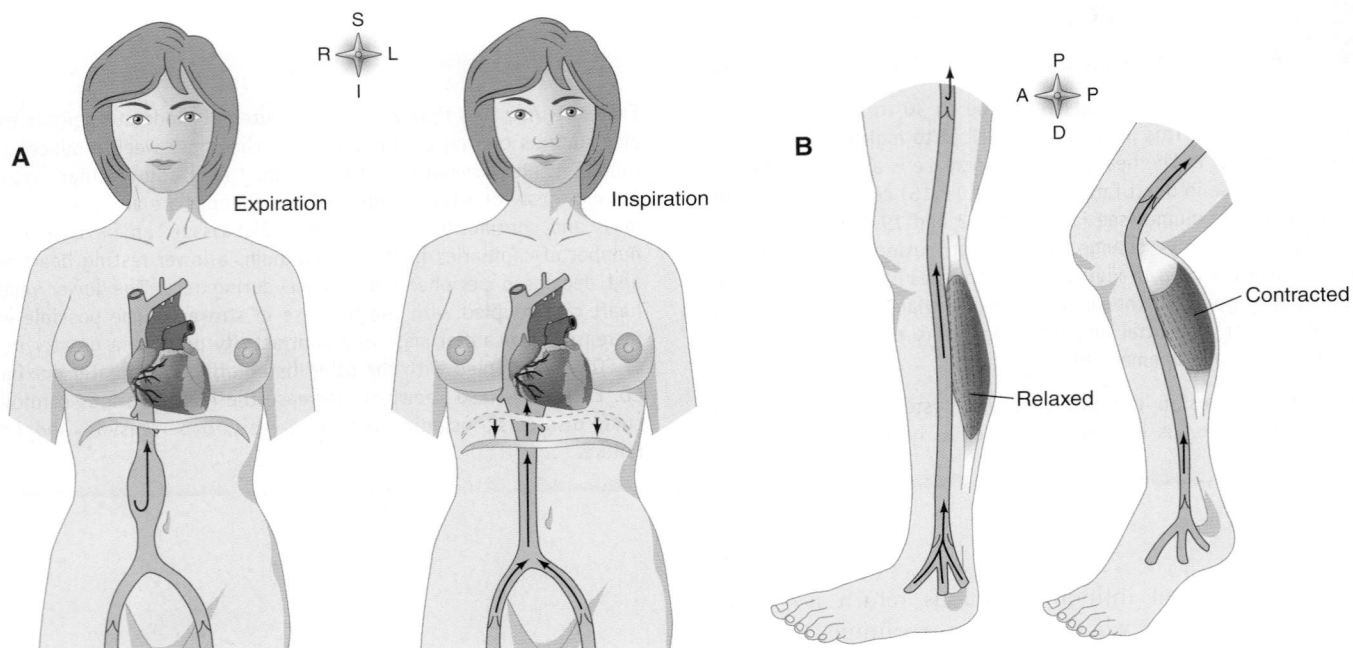

Figure 19-24 *Venous pumping mechanisms.* **A,** The respiratory pump operates by alternately decreasing thoracic pressure during inspiration (thus pulling venous blood into the central veins) and increasing pressure in the thorax during expiration (thus pushing central venous blood into the heart). **B,** The skeletal muscle pump operates by the alternate increase and decrease in peripheral venous pressure that normally occur when the skeletal muscles are used for the activities of daily living. Both pumping mechanisms rely on the presence of semilunar valves in the veins to prevent backflow during the low-pressure points in the pumping cycle (see Figure 19-25).

Deeper respirations intensify these effects and therefore tend to increase venous return to the heart more than normal respirations. This is part of the reason why the principle is true that increased respirations and increased circulation tend to go hand in hand.

Skeletal muscle contractions serve as "booster pumps" for the heart. The skeletal muscle pump promotes venous return in the following way. As each skeletal muscle contracts, it squeezes the soft veins scattered through its interior, thereby "milking" the blood in them upward, or toward the heart (Figure 19-24, *B*). The closing of the semilunar valves present in veins prevents blood from falling back as the muscle relaxes. Their flaps catch the blood as gravity pulls backward on it (Figure 19-25). The net effect of skeletal muscle contraction plus venous valvular action therefore is to move venous blood toward the heart, to increase the venous return.

The value of skeletal muscle contractions in moving blood through veins is illustrated by a common experience. Who has not noticed how much more uncomfortable and tiring standing still is than walking? After several minutes of standing quietly, the feet and legs feel "full" and swollen. Blood has accumulated in the veins because the skeletal muscles are not contracting and squeezing it upward. The repeated contractions of the muscles when walking, on the other hand, keep the blood moving in the veins and prevent the discomfort of distended veins.

Total Blood Volume

The return of venous blood to the heart can be influenced by factors that change the total volume of blood in the closed circulatory pathway. Stated simply, the more the total volume of blood, the greater the volume of blood returned to the heart. What mechanisms can increase or decrease the total volume of blood? The mechanisms that change total blood volume most quickly, making them most useful in maintaining constancy of blood flow, are those that cause water to quickly move into the plasma (increasing total blood volume) or out of the plasma (decreasing total blood volume). Most mechanisms that accomplish such changes in plasma volume operate by altering the body's retention of water.

Capillary Exchange and Total Blood Volume

We begin our discussion of fluid movement into and out of the blood plasma with a brief overview of **capillary exchange**—the exchange of materials between plasma in the capillaries and the surrounding interstitial fluid of the systemic tissues.

According to a principle first proposed by Ernest Starling, several factors govern the movement of fluid (and solutes contained in the fluid) back and forth across a capillary wall—a principle now known as **Starling's law of the capillaries.** These factors, illustrated in Figure 19-26, include inwardly directed forces and outwardly directed forces. It is the balance between these forces that determines whether fluids will move into or out of the plasma at any particular point.

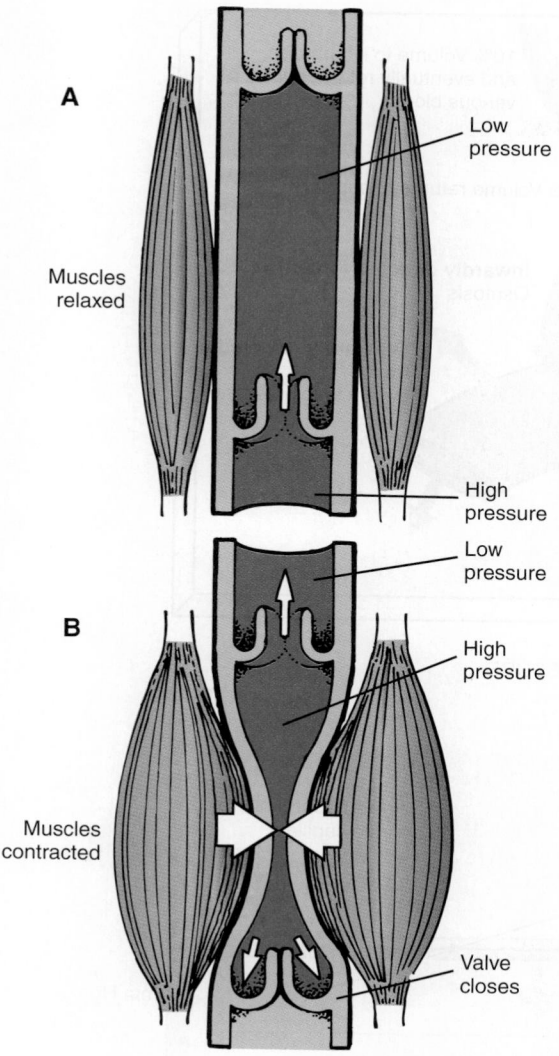

A

Low pressure

Muscles relaxed

High pressure

B

Low pressure

High pressure

Muscles contracted

Valve closes

Figure 19-25 *Semilunar valves.* In veins, semilunar valves aid circulation by preventing backflow of venous blood when pressure in a local area is low. **A,** Local high blood pressure pushes the flaps of the valve to the side of the vessel, allowing easy flow. **B,** When pressure below the valve drops, blood begins to flow backward but fills the "pockets" formed by the valve flaps, pushing the flaps together and thus blocking further backward flow.

One type of force, osmotic pressure, tends to promote diffusion of fluid into the plasma. Osmotic pressure generated by blood colloids (large solute particles such as plasma proteins) in the plasma that cannot cross the vessel wall tends to draw water osmotically into the plasma. At the arterial end of a capillary (and in some thin-walled arterioles) the potential osmotic pressure is small and thus generates only a small, inwardly directed force. However, a much larger, outwardly directed force is operating at the arterial end of a capillary, namely, a hydrostatic pressure gradient. Recall from Chapter 4 that hydrostatic pressure gradients promote filtration across a barrier with filtration pores, such as the capillary wall. At the arterial end of a capillary the blood pressure in the vessel is much greater than the hydrostatic pressure of the interstitial fluid (IF), thus generating a very large, outwardly directed force. In short, the stronger, outwardly directed forces at the arterial end of a capillary drive fluids out of the blood vessel and into the surrounding IF—producing a net loss of blood volume.

At the venous end of the capillary, however, the loss of water has increased the blood colloid osmotic pressure—promoting

osmosis of water back into the plasma. This inwardly directed force is much larger than the hydrostatic pressure gradient, which has dissipated somewhat with the loss of water at the arterial end of the vessel. In short, the capillary recovers much of the fluid it lost—recovering some of the previously lost blood volume.

If you look carefully at Figure 19-26, A, you will notice that about 10% of the fluid lost at the arterial end of a capillary is not recovered by the forces operating in Starling's law of the capillaries. Does that mean that there is a constant loss of blood volume? No. Notice that the 10% fluid loss is recovered by the lymphatic system and returned to the venous blood before it reaches the heart. Details of how the lymphatic system accomplishes this fluid recovery are discussed in the next Chapter. For now, we will simply state that if the lymphatic system operates normally and the osmotic and hydrostatic pressure gradients remain relatively constant, there is no net loss of blood volume resulting from capillary exchange. If any of these factors changes, however, fluid retention by the blood will be affected.

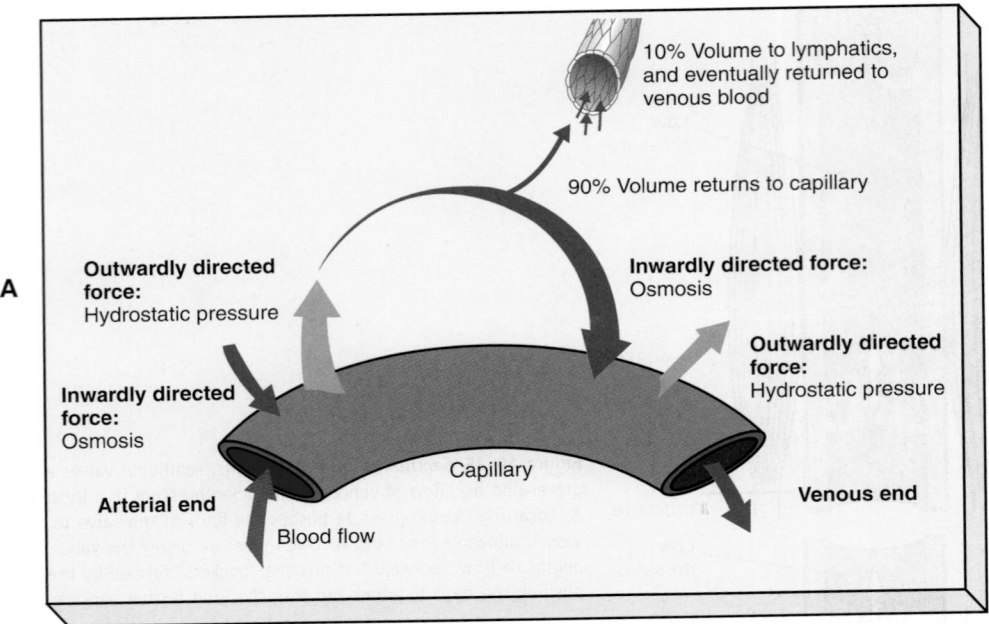

NET FILTRATION PRESSURE

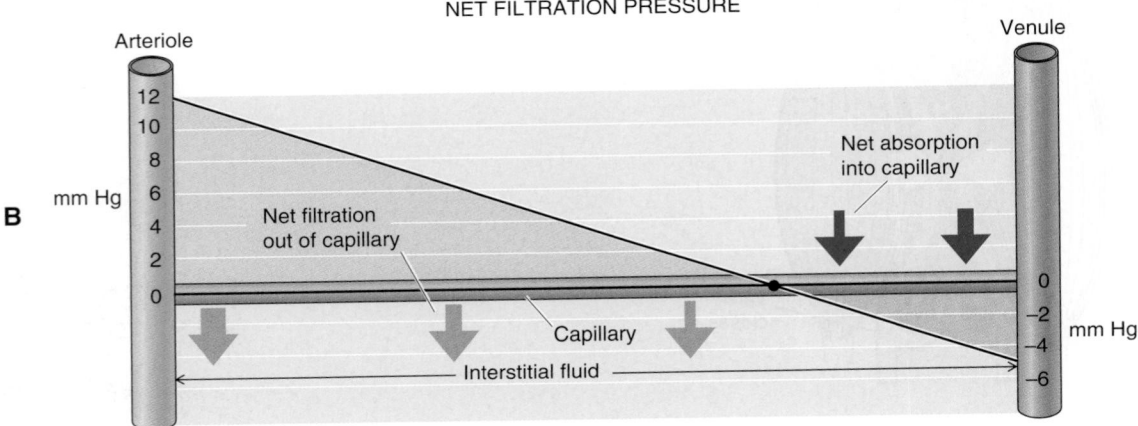

Figure 19-26 *Starling's law of the capillaries.* A, At the arterial end of a capillary the outward driving force of blood pressure is larger than the inwardly directed force of osmosis—thus fluid moves out of the vessel. At the venous end of a capillary the inward driving force of osmosis is greater than the outwardly directed force of hydrostatic pressure—thus fluid enters the vessel. About 90% of the fluid leaving the capillary at the arterial end is recovered by the blood before it leaves the venous end. The remaining 10% is recovered by the venous blood eventually, by way of the lymphatic vessels (see Chapter 20). **B,** Graph showing the shift in net filtration force along the length of a capillary, according to Starling's law of the capillaries. At the arterial end, net filtration is *out of* the capillary; at the venous end, net filtration shifts *into* the capillary.

Changes in Total Blood Volume

You have already studied the primary mechanisms for altering water retention in the body—they are the endocrine reflexes outlined in Chapter 16. One is the *ADH mechanism.* Recall that ADH (antidiuretic hormone) is released by the neurohypophysis (posterior pituitary) and acts on the kidneys in a way that reduces the amount of water lost by the body. ADH does this by increasing the amount of water that the kidneys resorb from urine before the urine is excreted from the body. The more ADH is secreted, the more water will be resorbed into the blood, and the greater the blood plasma volume will become. The ADH mechanism can be triggered by various factors, such as input from baroreceptors and input from osmoreceptors (which detect the balance between water and solutes in the internal environment).

Another mechanism that changes blood plasma volume is the *renin-angiotensin mechanism* of aldosterone secretion (renin-angiotensin-aldosterone system, RAAS). You may want to turn back to Figure 16-32 (p. 621) to see that the enzyme renin is released when blood pressure in the kidney is low. Renin triggers a series of events that leads to the secretion of aldosterone, a hormone of the adrenal cortex. Aldosterone promotes sodium retention by the kidney, which in turn stimulates the osmotic flow of water from kidney

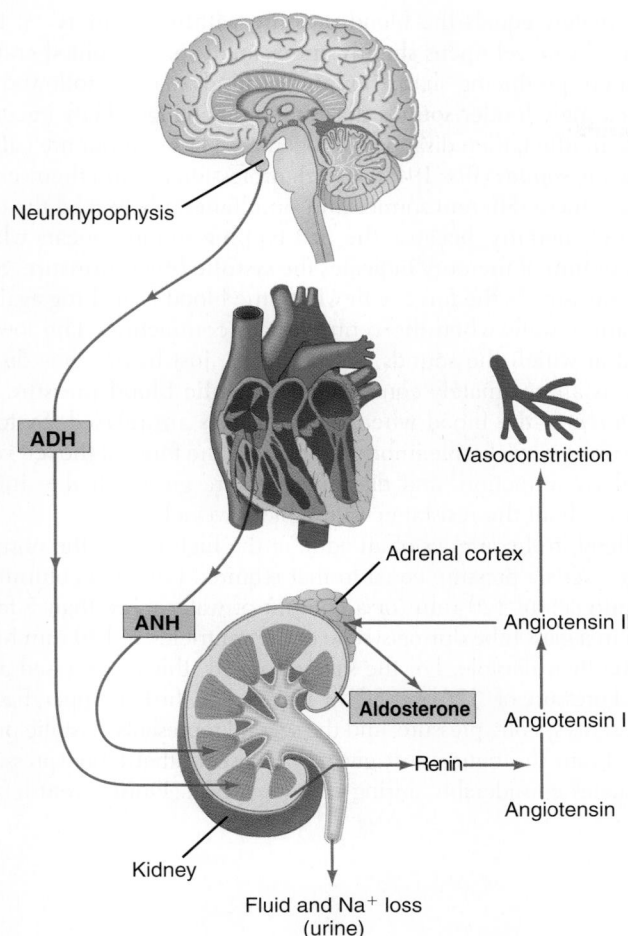

Figure 19-27 *Three mechanisms that influence total plasma volume.* The antidiuretic hormone (ADH) mechanism and renin-angiotensin-aldosterone system (RAAS) tend to increase water retention and thus increase total plasma volume. The atrial natriuretic hormone (ANH) mechanism antagonizes these mechanisms by promoting water loss and thus promoting a decrease in total plasma volume.

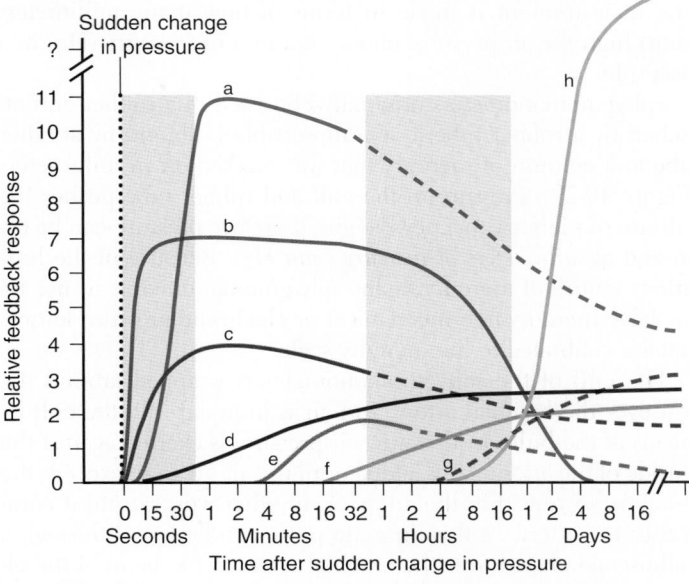

Fast-acting	Intermediate	Long-term
a) CNS ischemic	d) Stress-relaxation	g) Aldosterone
b) Barorecptors	e) Capillary fluid shift	h) Renal-blood volume pressure control
c) Chemoreceptors	f) Renin-angiotensin vasoconstriction	

Figure 19-28 *Feedback responses of various arterial pressure mechanisms.* This composite graph shows the relative regulatory shifts that can be carried out by different regulatory mechanisms in response to a sudden shift in blood pressure—all with the outcome of moving arterial blood pressure back to its set point value. Note that some responses have their maximum effect within a few seconds or minutes. Other responses take longer to reach their maximum effect. *CNS,* Central nervous system.

tubules back into the blood plasma—but only when ADH is present to permit the movement of water. Thus low blood pressure increases the secretion of aldosterone, which in turn stimulates retention of water and thus an increase in blood volume. Another effect of the renin-angiotensin mechanism is the vasoconstriction of blood vessels caused by an intermediate compound called *angiotensin II*. This complements the volume-increasing effects of the mechanism and thus also promotes an increase in overall blood flow.

Yet another mechanism that can change blood plasma volume and thus venous return of blood to the heart is the *ANH mechanism*. Recall that ANH (atrial natriuretic hormone) is secreted by specialized cells in the atrial wall in response to overstretching. Overstretching of the atrial wall, of course, occurs when venous return to the heart is abnormally high. ANH adjusts venous return back down to its set point value by promoting the loss of water from the plasma and the resulting decrease in blood volume. ANH accomplishes this feat by increasing urine sodium loss, which causes water to follow osmotically. Sodium loss also inhibits the secretion of ADH. ANH may also have other complementary effects, such as promoting vasodilation of blood reservoirs.

Thus various mechanisms influence blood volume and therefore venous return. These primary mechanisms are summarized in Figure 19-27. The ANH mechanism opposes ADH, renin-angiotensin, and aldosterone mechanisms to produce a balanced, precise control of blood volume. Precision of blood volume control contributes to precision in controlling venous return, which in turn contributes to precision in the overall control of blood circulation.

In summary, many different factors help regulate blood pressure and therefore regulate blood flow. Figure 19-28 summarizes some of the rapidly acting, intermediate, and long-term responses that can return blood pressure and blood flow back to normal after a sudden change. Most of these responses have been discussed in this chapter and Chapter 18, but some will be explored further in later chapters.

 QUICK CHECK

14. What is meant by the term *venous return?*

15. Briefly describe how the *respiratory pump* and *skeletal muscle pump* work.

16. How does Starling's law of the capillaries explain capillary exchange?

17. What three hormonal mechanisms work together to regulate blood volume?

MEASURING BLOOD PRESSURE
Arterial Blood Pressure

Blood pressure is measured with the aid of an apparatus known as a **sphygmomanometer,** which makes it possible to measure the amount of air pressure equal to the blood pressure in an artery. The measurement is made in terms of how many millimeters (mm) high the air pressure raises a column of mercury (Hg) in a glass tube.

Sphygmomanometers originally consisted of a rubber cuff attached by a rubber tube to a compressible bulb and by another tube to a column of mercury that was marked off in millimeters (Figure 19-29). Pressure in the cuff and rubber tube pushes the column of mercury to a new height; therefore pressure can be expressed as *millimeters of mercury (mm Hg).* Because of the hazardous nature of mercury, many sphygmomanometers in use today have mercury-free mechanical or electronic pressure sensors that are calibrated to the mercury scale.

The cuff of the sphygmomanometer is wrapped around the arm over the brachial artery, and air is pumped into the cuff by means of the bulb. In this way, air pressure is exerted against the outside of the artery. Air is added until the air pressure exceeds the blood pressure within the artery or, in other words, until it compresses the artery. At this time, no pulse can be heard through a stethoscope placed over the brachial artery at the bend of the elbow along the inner margin of the biceps muscle. By slowly releasing the air in the cuff the air pressure is decreased until it ap-

proximately equals the blood pressure within the artery. At this point, the vessel opens slightly and a small spurt of blood comes through, producing sharp "tapping" sounds. This is followed by increasingly louder sounds that suddenly change. They become more muffled, then disappear altogether. These sounds are called *Korotkoff sounds* (Box 19-5). Health professionals train themselves to hear these different sounds and simultaneously to read the column of mercury, because the first tapping sound appears when the column of mercury indicates the **systolic blood pressure.** Systolic pressure is the force with which the blood is pushing against the artery walls when the ventricles are contracting. The lowest point at which the sounds can be heard, just before they disappear, is approximately equal to the **diastolic blood pressure,** or the force of the blood when the ventricles are relaxed. Systolic pressure gives valuable information about the force of the left ventricular contraction, and diastolic pressure gives valuable information about the resistance of the blood vessels.

Blood in the arteries of an adult at the high end of the normal range exerts a pressure equal to that required to raise a column of mercury about 120 mm (or a column of water more than 5 feet) high in a glass tube during systole of the ventricles and 80 mm high during their diastole. For the sake of brevity, this is expressed as a blood pressure of "120 over 80" or 120/80. The first, or upper, figure represents systolic pressure, and the second represents diastolic pressure. From the figures just given, we observe that blood pressure fluctuates considerably during each heartbeat. During ventricular

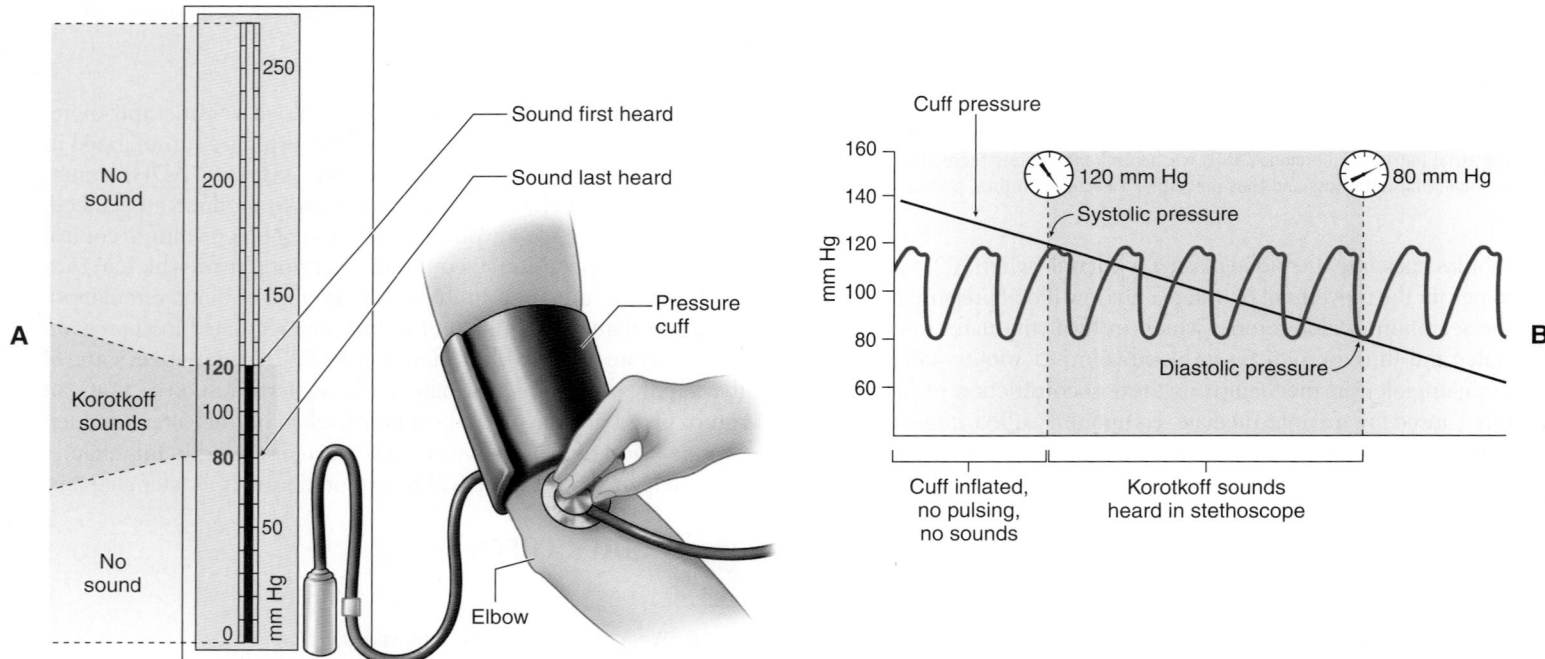

Figure 19-29 *Sphygmomanometer.* This mercury-filled pressure sensor is used in clinical and research settings to quickly and accurately measure arterial blood pressure. **A,** The pressure cuff is pumped with air until the pressure inside the cuff exceeds the expected systolic pressure of the large arteries of the arm. No sound caused by pulsing of blood in the arteries can then be heard with a stethoscope. As the pressure inside the cuff is slowly released from a valve, the air pressure equals the maximum pressure of the pulse waves in the artery—thus the pulsing sounds can now be heard. **B,** The sounds of pulsing (Korotkoff sounds) continue as long as the cuff pressure is equal to pressures of the pulse wave. The sounds disappear at the point that cuff pressure drops below the minimum pulse pressure in the arteries.

BOX 19-5
Diagnostic Study

Focus on Turbulent Blood Flow

Understanding the mechanics of fluid flow through blood vessels helps to be able to see the bigger picture of hemodynamics.

Normally, blood flows through vessels with smooth walls that taper or enlarge only slightly, if at all. The manner in which a fluid, in this case blood, flows through a smooth vessel is termed **laminar flow.** The Latin word *lamina* means layer, an apt description of the way fluids flow through tubes in concentric, cylindrical layers—as you can see in part *A* of the figure. Because of friction against the inner face of the vessel wall, the outer layers of blood flow more slowly than the inner layers. Laminar flow is the normal pattern of blood flow in most healthy vessels.

When smooth, laminar flow is disrupted by branching or narrowing of a vessel, a sharp turn, or a sudden constriction or obstruction, the flow of blood becomes turbulent. **Turbulent flow,** shown in part *B* of the figure, occurs normally at heart valves and contributes to the first and second heart sounds described on p. 744.

Significant turbulent flow in most vessels is not normal. If the sounds of turbulence are detected in a peripheral vessel, an abnormal or unusual constriction may exist. Such constrictions could be temporary and intentional, as in the case of an inflated blood pressure cuff pressing on an artery and creating turbulence. The sounds of turbulence generated by such blood pressure measurements are called **Korotkoff sounds** (see Figure 19-29). These sounds are named for Nicolai Korotkoff, the Russian surgeon who developed the method for using these sounds to indirectly measure arterial blood pressure in 1905.

However, some sounds of turbulence are caused by constrictions that are a serious threat to one's health. For example, in Chapter 18 we discussed the fact that arteriosclerosis can cause such a narrowing of a vessel's channel and can ultimately lead to death by ischemia, thrombosis, or other mechanisms (see Figure 18-40). Part *C* of the figure shows a stethoscope being used to listen for low-pitched blowing sounds called **bruits** (BROO-eez) that can occur in the carotid arteries. Bruits may result from obstruction caused by arteriosclerosis or abnormally increased pulse pressure. A usually benign, but abnormal, condition known as venous hum produces a humming noise in the internal jugular vein. Venous hum is common in children and probably results from vigorous myocardial contraction.

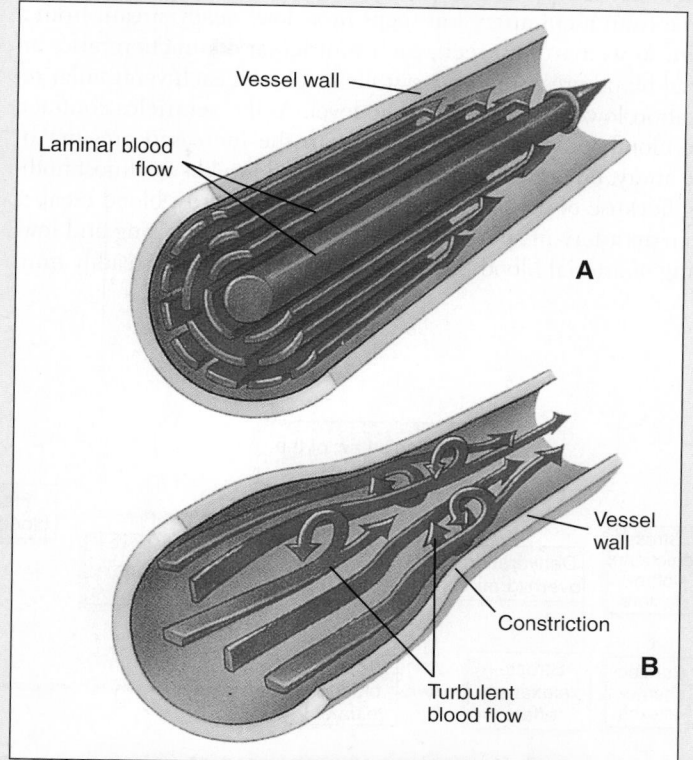

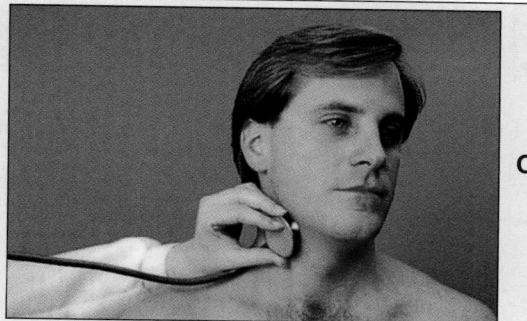

Turbulent blood flow. **A,** Laminar blood flow (no turbulence). **B,** Turbulent blood flow. **C,** Listening for bruits.

systole, the force is great enough to raise the mercury column 40 mm higher than during ventricular diastole. This difference between systolic and diastolic pressure is called **pulse pressure.** It characteristically increases in arteriosclerosis, mainly because systolic pressure increases more than diastolic pressure. Pulse pressure increases even more markedly in aortic valve insufficiency because of both a rise in systolic pressure and a fall in diastolic pressure.

Be aware that although the classic method of indirect arterial blood pressure assessment involves a column of mercury, and the units in which blood pressure is expressed are based on the height of a mercury column, very few modern sphygmomanometers actually use mercury. The reason is twofold. First, mercury is a very hazardous substance and should the mercury column break, an

environmental health hazard would exist. Second, it is much more cost effective to use cheap and durable electronic pressure sensors that are calibrated to the mercury scale than to build fragile columns of mercury.

Blood pressure can also be measured directly in various ways. For example, a small tube called a cannula with a removable pointed insert can be pushed directly into a vessel, the insert withdrawn, and the canula connected to a manometer or electronic pressure sensor. A long, flexible tube called a catheter can likewise be placed in blood vessels or even into a chamber of the heart. These techniques can be used in critical care, but the more indirect methods described previously are more practical for routine screening and clinical evaluation.

Blood Pressure and Bleeding

Because blood exerts a comparatively high pressure in arteries and a very low pressure in veins, it gushes forth with considerable force from a cut artery but seeps in a slow, steady stream from a vein. As we have just seen, each ventricular contraction raises arterial blood pressure to the systolic level, and each ventricular relaxation lowers it to the diastolic level. As the ventricles contract, the blood spurts forth forcefully from the increased pressure in the artery, but as the ventricles relax, the flow ebbs to almost nothing because of the fall in pressure. In other words, blood escapes from an artery in spurts because of the alternate raising and lowering of arterial blood pressure but flows slowly and steadily from a vein because of the low, practically constant pressure. A uniform, instead of a pulsating, pressure exists in the capillaries and veins. Why? Because the arterial walls, being elastic, continue to squeeze the blood forward while the ventricles are in diastole. Therefore blood enters capillaries and veins under a relatively steady pressure (see Figure 19-9).

MINUTE VOLUME OF BLOOD

The volume of blood circulating through the body per minute (**minute volume**) is determined by the magnitude of the blood pressure gradient and the peripheral resistance (Figure 19-30).

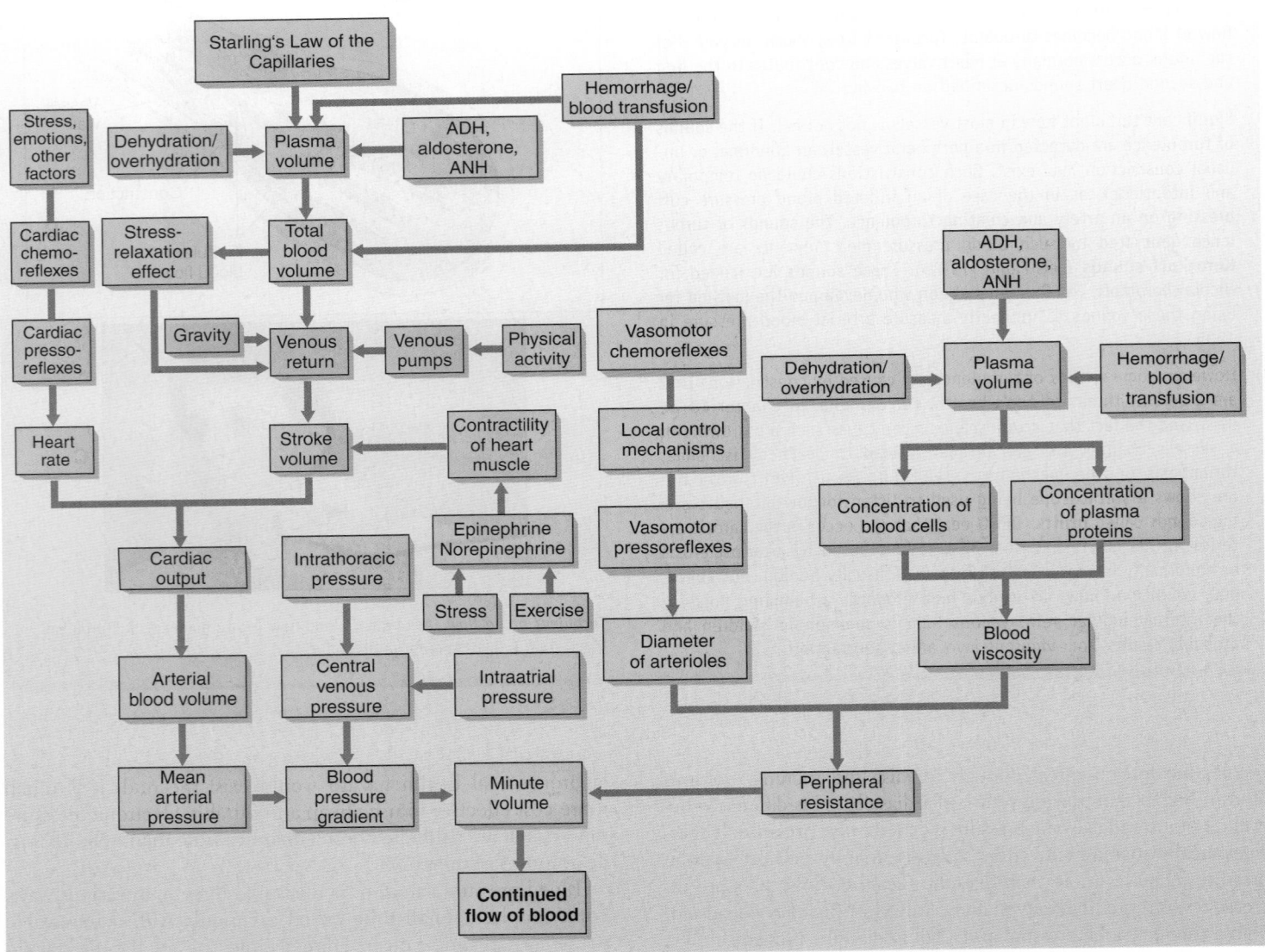

Figure 19-30 *Factors that influence the flow of blood.* The flow of blood, expressed as volume of blood flowing per minute (or *minute volume*), is determined by various factors. This chart shows only some of the major factors that influence blood flow. Note that some factors appear more than once in the chart, indicating that they can influence blood flow in several ways. *ADH,* Antidiuretic hormone; *ANH,* atrial natriuretic hormone.

A nineteenth century physiologist and physicist, Poiseuille (pwah-soo-EE), described the relation between these three factors—pressure gradient, resistance, and minute volume—with a mathematical equation known as **Poiseuille's law**. In general, but with certain modifications, it applies to blood circulation. We can state it in a simplified form as follows: the volume of blood circulated per minute is directly related to mean arterial pressure minus central venous pressure and inversely related to resistance:

Volume of blood circulated per minute =
$$\frac{\text{Mean arterial pressure} - \text{Central venous pressure}}{\text{Resistance}}$$

$$\text{Minute volume} = \frac{\text{Pressure gradient}}{\text{Resistance}}$$

This mathematical relationship needs qualifying with regard to the influence of peripheral resistance on circulation. For instance, according to the equation, an increase in peripheral resistance tends to decrease blood flow. (Why? Increasing peripheral resistance increases the denominator of the fraction in the preceding equation. Increasing the denominator of any fraction necessarily does what to its value? It decreases the value of the fraction.)

Increased peripheral resistance, however, has a secondary action that opposes its primary tendency to decrease blood flow. An increase in peripheral resistance hinders or decreases arteriole runoff. This, of course, tends to increase the volume of blood left in the arteries and so tends to increase arterial pressure. Note also that increasing arterial pressure tends to increase the value of the fraction in Poiseuille's equation. Therefore it tends to increase circulation. In short, to say unequivocally what the effect of an increased peripheral resistance will be on circulation is impossible. It depends also on arterial blood pressure—whether it increases, decreases, or stays the same when peripheral resistance increases. The clinical condition *arteriosclerosis with hypertension* (high blood pressure) illustrates this point. Both peripheral resistance and arterial pressure are increased in this condition. If resistance were to increase more than arterial pressure, circulation (i.e., volume of blood flow per minute) would decrease. But if arterial pressure increases proportionately to resistance, circulation remains normal.

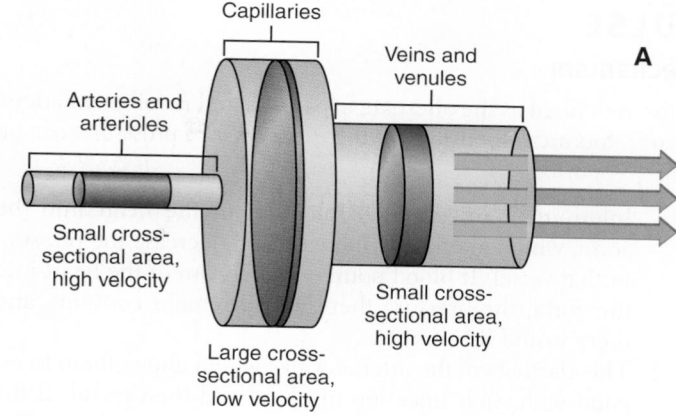

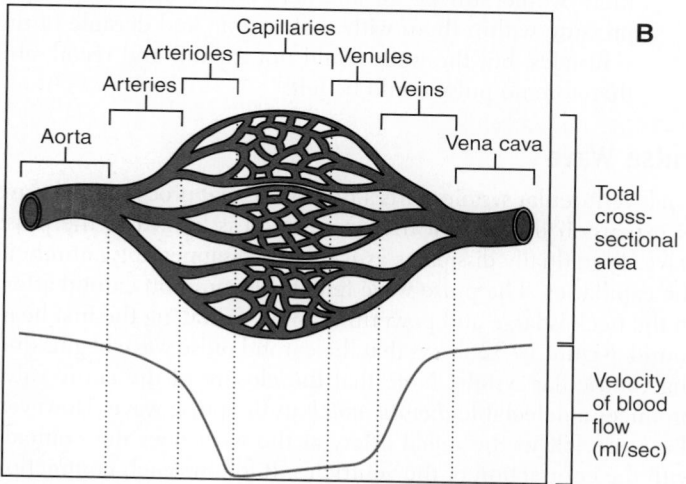

Figure 19-31 *Relationship between cross-sectional area and velocity of blood flow.* As you can see in the simple diagram, **A,** and the blood vessel chart, **B,** blood flows with great speed in the large arteries. However, branching of arterial vessels increases the total cross-sectional area of the arterioles and capillaries, reducing the flow rate. When capillaries merge into venules and venules merge into veins, the total cross-sectional area decreases, causing the flow rate to increase.

 QUICK CHECK

18. What is meant by the term *minute volume?*
19. How is minute volume related to peripheral resistance?

VELOCITY OF BLOOD FLOW

The speed with which the blood flows, that is, distance per minute, through its vessels is governed in part by the physical principle that when a liquid flows from an area of one cross-sectional size to an area of larger size, its velocity slows in the area with the larger cross section (Figure 19-31). For example, a narrow river whose bed widens flows more slowly through the wide section than through the narrow section. In terms of the blood vascular system, the total cross-sectional area of all arterioles together is greater than that of the arteries. Therefore blood flows more slowly through arterioles than through arteries. Likewise, the total cross-sectional area of all capillaries together is greater than that of all arterioles, and therefore capillary flow is slower than arteriole flow. The venule cross-sectional area, on the other hand, is smaller than the capillary cross-sectional area. Therefore blood velocity increases in venules and again in veins, which have a still smaller cross-sectional area. In short, the most rapid blood flow takes place in arteries and the slowest in capillaries. Can you think of a valuable effect stemming from the fact that blood flows most slowly through the capillaries?

PULSE
Mechanism

Pulse is defined as the alternate expansion and recoil of an artery. Two factors are responsible for the existence of a pulse that can be felt:

1. Intermittent injections of blood from the heart into the aorta, which alternately increase and decrease the pressure in that vessel. If blood poured steadily out of the heart into the aorta, the pressure there would remain constant, and there would be no pulse.
2. The elasticity of the arterial walls, which allows them to expand with each injection of blood and then recoil. If the vessels were fashioned from rigid material such as glass, there would still be an alternate raising and lowering of pressure within them with each systole and diastole of the ventricles, but the walls could not expand and recoil, and therefore no pulse could be felt.

Pulse Wave

Each ventricular systole starts a new pulse that proceeds as a wave of expansion throughout the arteries and is known as the pulse wave. It gradually dissipates as it travels, disappearing entirely in the capillaries. The pulse wave felt at the common carotid artery in the neck is large and powerful, rapidly following the first heart sound. Figure 19-32 shows that the carotid pulse wave begins during ventricular systole. Note that the closure of the aortic valve produces a detectable *dicrotic notch* in the pulse wave. However, the pulse felt in the radial artery at the wrist does not coincide with the contraction of the ventricles. It follows each contraction by an appreciable interval (the length of time required for the pulse wave to travel from an aorta to the radial artery). The farther from the heart the pulse is taken, therefore, the longer that interval is.

Almost everyone is aware of the diagnostic importance of the pulse. It reveals important information about the cardiovascular system, heart action, blood vessels, and circulation.

Not everyone is aware of the basic functional role of the pulse wave, however. The pulse wave actually conserves energy produced by the pumping action of the heart (Figure 19-33). The great force of pressure with which blood is ejected from the heart during ventricular systole expands the wall of the aorta. The stretched aortic wall now has potential energy stored in it—just as a stretched rubber band has potential energy. During ventricular diastole, the elastic nature of the aortic wall allows it to recoil. This recoil exerts pressure on the blood and thus keeps it moving. If the wall of the aorta was inelastic, it would not alternately expand and recoil—and would thus not keep blood moving continuously. Instead, arterial blood would simply spurt, then stop, then spurt, then stop, and so on.

Where Pulse Can Be Felt

The pulse can be felt wherever an artery lies near the surface and over a bone or other firm background. Some specific locations where the pulse point is most easily felt are listed here and shown in Figure 19-34:

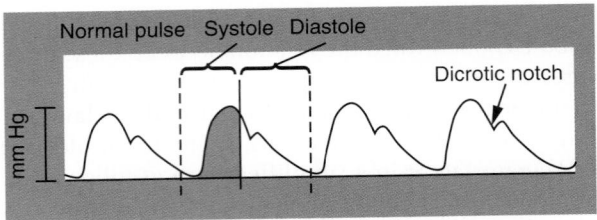

Figure 19-32 *Normal carotid pulse wave.* This series of pulse waves shows rhythmic increases and decreases in pressure as measured at the common carotid artery in the neck. The dicrotic notch represents the pressure fluctuation generated by closure of the aortic valve.

- Radial artery—at wrist
- Temporal artery—in front of ear or above and to outer side of eye
- Common carotid artery—along anterior edge of sternocleidomastoid muscle at level of lower margin of thyroid cartilage
- Facial artery—at lower margin of lower jawbone on a line with corners of mouth and in groove in mandible about one third of way forward from angle
- Brachial artery—at bend of elbow along inner margin of biceps muscle
- Popliteal artery—behind the knee
- Posterior tibial artery—behind the medial malleolus (inner "ankle bone")
- Dorsalis pedis artery—on dorsum (upper surface) of foot

Six important pressure points can be used to stop arterial bleeding:

1. Temporal artery—in front of ear
2. Facial artery—same place as pulse is taken
3. Common carotid artery—point where pulse is taken, with pressure back against spinal column
4. Subclavian artery—behind medial third of clavicle, pressing against first rib
5. Brachial artery—few inches above elbow on inside of arm, pressing against humerus
6. Femoral artery—in middle of groin, where artery passes over pelvic bone; pulse can also be felt here

In trying to stop arterial bleeding by pressure, one must always remember to apply the pressure at the pulse point, or *pressure point*, that lies between the bleeding part and the heart. Why? Blood flows from the heart through the arteries to the part. Pressure between the heart and bleeding point therefore cuts off the source of the blood flow to that point.

Venous Pulse

A detectable pulse exists in the large veins only. It is most prominent in the veins near the heart because of changes in venous blood pressure brought about by alternate contraction and relaxation of the atria of the heart. The clinical significance of venous pulse is not as great as that of arterial pulse, and thus it is less often measured.

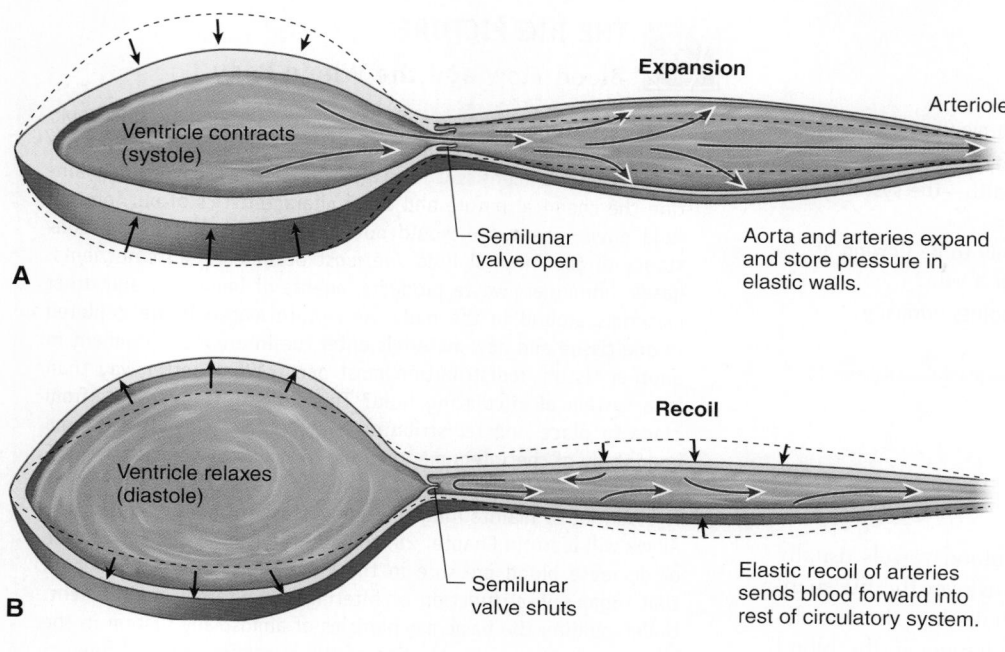

Expansion

Arterioles

Ventricle contracts
(systole)

Semilunar
valve open

Aorta and arteries expand
and store pressure in
elastic walls.

A

Recoil

Ventricle relaxes
(diastole)

Semilunar
valve shuts

Elastic recoil of arteries
sends blood forward into
rest of circulatory system.

B

Figure 19-33 *Functional role of the pulse wave.* The arterial pulse conserves energy by absorbing and storing force from the ventricular contraction by causing elastic expansion of the arterial wall. The energy is used to maintain continued blood flow during ventricular relaxation by elastic recoil of the arterial wall—thus producing enough arterial pressure to keep blood flowing.

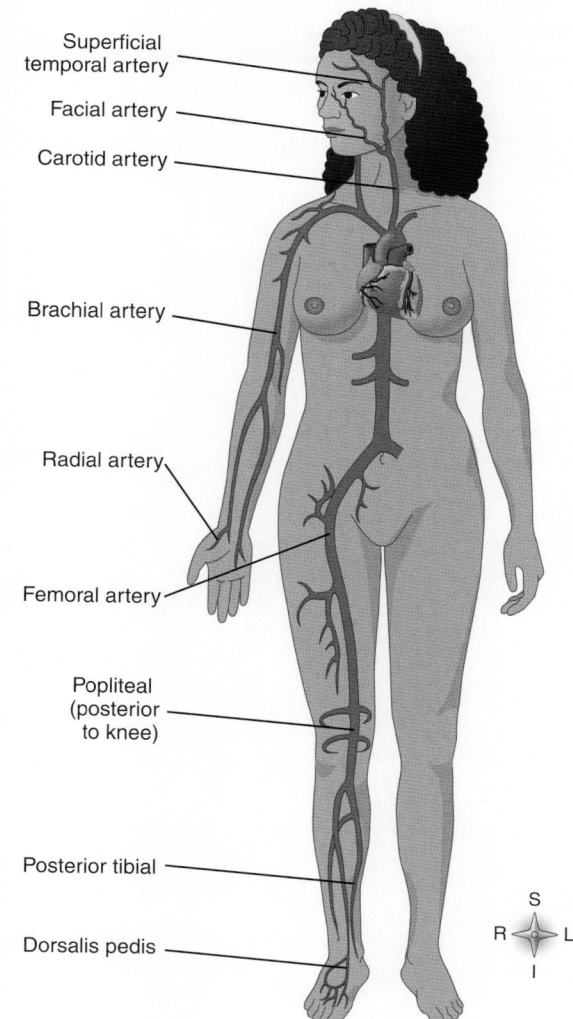

Superficial
temporal artery

Facial artery

Carotid artery

Brachial artery

Radial artery

Femoral artery

Popliteal
(posterior
to knee)

Posterior tibial

Dorsalis pedis

Figure 19-34 *Pulse points.* Each pulse point is named after the artery with which it is associated.

Cycle of Life

Cardiovascular Physiology

Changes in the function of the heart and blood vessels usually parallel the structural changes in these organs over the life span. For example, changes at the time of birth that adapt the circulatory system to life outside the womb cause changes in the blood pressure gradients that alter the flow of blood in many parts of the body. Likewise, the degenerative changes associated with aging reduce the heart's ability to maintain cardiac output and the ability of arteries to withstand high pressure.

Changes in arterial blood pressure are among the most apparent changes in the function of the cardiovascular system associated with the progression through the life cycle. In a newborn, normal arterial blood pressure is only about 90/55 mm Hg—much lower than the arterial pressure of just under 120/80 mm Hg in most healthy young adults. In older adults, arterial blood pressures commonly reach 150/90 mm Hg.

Another commonly observed change in cardiovascular function relates to heart rate. The heart rates of infants and children typically vary more than those in adults. Compared with adults, children often exhibit very large increases in heart rate in response to stressors such as illness, pain, tension, and exercise. Whereas a typical resting heart rate for adults is about 72 beats/min, the resting heart rate of a newborn can range from 120 to 170 beats/min, and the resting heart rate of a preschooler can range from 80 to 160 beats/min. In older adults, resting heart rates range from lows of around 40 beats/min to 100 beats/min.

THE BIG PICTURE
Blood Flow and the Whole Body

As stated in this chapter and many times throughout this book, one of the essential concepts of homeostasis is the fact that our internal environment is a renewable fluid. If we could not maintain the chemical nature and other characteristics of our internal fluid environment, we would not survive. To maintain the constancy of the internal fluid, we must be able to shift nutrients, gases, hormones, waste products, agents of immunity, and other materials around in the body. As certain materials are depleted in one tissue and new materials enter the internal environment in another tissue, redistribution must occur. What better way than in a system of circulating fluid? This fluid shifts materials from place to place and redistributes heat and pressure. Recall from your study of the integumentary and muscular systems that shifting the flow of blood to or from warm tissues at the proper time is essential to maintaining the homeostasis of body temperature. As we will learn in Chapter 28, the ability of our blood to increase or decrease blood pressure in the kidney has a great impact on that organ's vital function of filtering the internal environment. Understanding the basic mechanisms of almost any system in the body requires an understanding of the dynamics of blood flow.

What we have seen in this chapter is a wonderfully complex array of mechanisms that work together in concert with the actions of other systems to maintain the constancy of the *milieu intérieur*—the internal environment.

Mechanisms of Disease

DISORDERS OF CARDIOVASCULAR PHYSIOLOGY

Hypertension

The largest number of office visits to physicians is due to a condition called **hypertension (HTN),** or high blood pressure. More than 60 million cases of HTN have been diagnosed in the United States. This condition occurs when the force of blood exerted by the arterial blood vessel exceeds a blood pressure of 140/90 mm Hg. Ninety percent of HTN cases are classified as *primary-essential,* or idiopathic, with no single known causative etiology. Another classification, secondary (HTN), is caused by kidney disease or hormonal problems or induced by oral contraceptives, pregnancy, or other causes.

Another way of classifying hypertension is illustrated in the accompanying chart adapted from the National High Blood Pressure Education Program (Figure 19-35). This system uses systolic and diastolic blood pressure values to classify hypertension into stages according to severity. The guidelines that accompany this scheme emphasize the belief that there is no precise distinction between normal and abnormal values—thus even those in the high-normal range or the *prehypertension* range may be treated as having HTN.

Many risk factors have been identified in the development of HTN. Genetic factors play a large role. There is an increased susceptibility or predisposition with a family history of HTN. Men experience higher rates of HTN at an earlier age than women, and HTN in African Americans far exceeds that of Caucasians in the United States. There is also a direct relationship between age and high blood pressure. This is because as age advances, the blood vessels become less compliant and there is a higher incidence of atherosclerotic plaque buildup. HTN can also be fatal if undetected in women taking oral contraceptives. Risk factors include high stress levels, obesity, calcium deficiencies, high levels of alcohol and caffeine intake, smoking, and lack of exercise.

Untreated HTN has many potential complications. Ischemic heart disease and heart failure, kidney failure, and stroke are some examples. As many as 400,000 people per year experience a stroke. Because HTN manifests minimal or no overt signs, it is known as the "silent killer." Headaches, dizziness, and fainting have been reported but are not always symptomatic of HTN. Regular screenings at the worksite and screening booths in malls and in hospitals often help to identify asymptomatic HTN.

Heart Failure

Heart failure is the inability of the heart to pump enough blood to sustain life. Heart failure can result from many different heart diseases. Valve disorders can reduce the pumping efficiency of the heart enough to cause heart failure. **Cardiomyopathy** (kar-dee-oh-my-OP-ah-thee), or disease of the myocardial tissue, may reduce pumping effectiveness. A specific event such as myocardial infarction can result in myocardial damage that causes heart failure. Dysrhythmias such as complete heart block or ventricular fibrillation can also impair the pumping effectiveness of the heart and thus cause heart failure (Box 19-6).

Congestive heart failure (CHF) or, simply, *left-side heart failure,* is the inability of the left ventricle to pump blood effectively. Most often, such failure results from myocardial infarction caused by coronary artery disease. It is called *congestive heart failure* because it decreases pumping pressure in the systemic circulation, which in turn causes the body to retain fluids. Portions of the systemic circulation thus become congested with extra fluid. As previously stated, left-side heart failure also causes congestion of blood in the pulmonary circulation (termed *pulmonary edema*)—possibly leading to right-side heart failure.

Failure of the right side of the heart, or *right-side heart failure,* accounts for about one fourth of all cases of heart failure. Right-side heart failure often results from the progression of disease that begins in the left side of the heart. Failure of the left side of the heart results in reduced pumping of blood returning from the lungs. Blood backs up into the pulmonary circulation and then

Blood Pressure (BP) Classification

Classification	Systolic BP		Diastolic BP
Normal	less than 120	and	less than 80
Prehypertension	120-139	or	80-89
High blood pressure			
Stage 1 hypertension	140-159	or	90-99
Stage 2 hypertension	greater than or equal to 160	or	greater than or equal to 100

Figure 19-35 *Classification of hypertension.* This chart is adapted from the National High Blood Pressure Education Program.

Mechanisms of Disease—cont.

BOX 19-6 Cardiac Dysrhythmia

Various conditions such as inflammation of the endocardium (endocarditis) or myocardial infarction (heart attack) can damage the heart's conduction system and thereby disturb the normal rhythmical beating of the heart (part *A* of figure). The term **dysrhythmia** refers to an abnormality of heart rhythm.

One kind of dysrhythmia is called a **heart block.** In AV node block, impulses are blocked from getting through to the ventricular myocardium, resulting in the ventricles contracting at a much slower rate than normal. On an ECG, there may be a large interval between the P wave and the R peak of the QRS complex (part *B* of figure). *Complete heart block* occurs when the P waves do not match up at all with the QRS complexes—as in an ECG that shows two or more P waves for every QRS complex. A physician may treat heart block by implanting in the heart an *artificial pacemaker* (see Box 19-1).

Bradycardia is a slow heart rhythm—below 60 beats/min (part *C* of figure). Slight bradycardia is normal during sleep and in conditioned athletes while they are awake (but at rest). Abnormal bradycardia can result from improper autonomic nervous control of the heart or from a damaged SA node. If the problem is severe, artificial pacemakers can be used to increase the heart rate by taking the place of the SA node.

Tachycardia is a very rapid heart rhythm—more than 100 beats/min (part *D* of figure). Tachycardia is normal during and after exercise and during the stress response. Abnormal tachycardia can result from improper autonomic control of the heart, blood loss or shock, the action of drugs and toxins, fever, and other factors.

Sinus dysrhythmia is a variation in heart rate during the breathing cycle. Typically, the rate increases during inspiration and decreases during expiration. The causes of sinus dysrhythmia are not clear. This phenomenon is common in young people and usually does not require treatment.

Premature contractions, or *extrasystoles,* are contractions that occur before the next expected contraction in a series of cardiac cycles. For example, *premature atrial contractions (PACs)* may occur shortly after the ventricles contract—seen as early P waves on the ECG (part *E* of figure). Premature atrial contractions often occur with lack of sleep, too much caffeine or nicotine, alcoholism, or heart damage. Ventricular depolarizations that appear earlier than expected are called *premature ventricular contractions* (PVCs). PVCs appear on the ECG as early, wide QRS complexes without a preceding related P wave. PVCs are caused by a variety of circumstances that include stress, electrolyte imbalance, acidosis, hypoxemia, ventricular enlargement, or drug reactions. Occasional PVCs are not clinically significant in otherwise healthy individuals. However, in people with heart disease they may reduce cardiac output.

Frequent premature contractions can lead to fibrillation, a condition in which cardiac muscle fibers contract out of step with each other. This event can be seen in an ECG as the absence of regular P waves or abnormal QRS and T waves. In fibrillation, the affected heart chambers do not effectively pump blood. **Atrial fibrillation** occurs commonly in mitral stenosis, rheumatic heart disease, and infarction of the atrial myocardium (part *F* of figure). This condition can be treated with drugs such as digoxin (a **digitalis** preparation) or by *defibrillation*—application of electrical shock to force cardiac muscle fibers to contract in unison. **Ventricular fibrillation** is an immediately life-threatening condition in which the lack of ventricular pumping suddenly stops the flow of blood to vital tissues (part *G* of figure). Unless ventricular fibrillation is corrected immediately by defibrillation or some other method, death may occur within minutes.

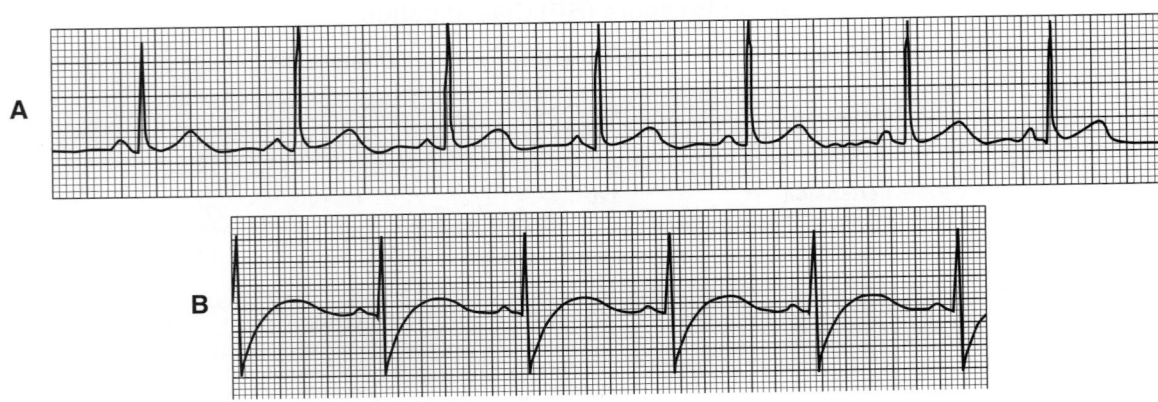

ECG strip chart recording. A, Normal ECG. **B,** AV node block. Very slow ventricular contraction (25 to 45 beats/min at rest); P waves widely separated from peaks of QRS complexes.

BOX 19-6 **Cardiac Dysrhythmia—cont'd**

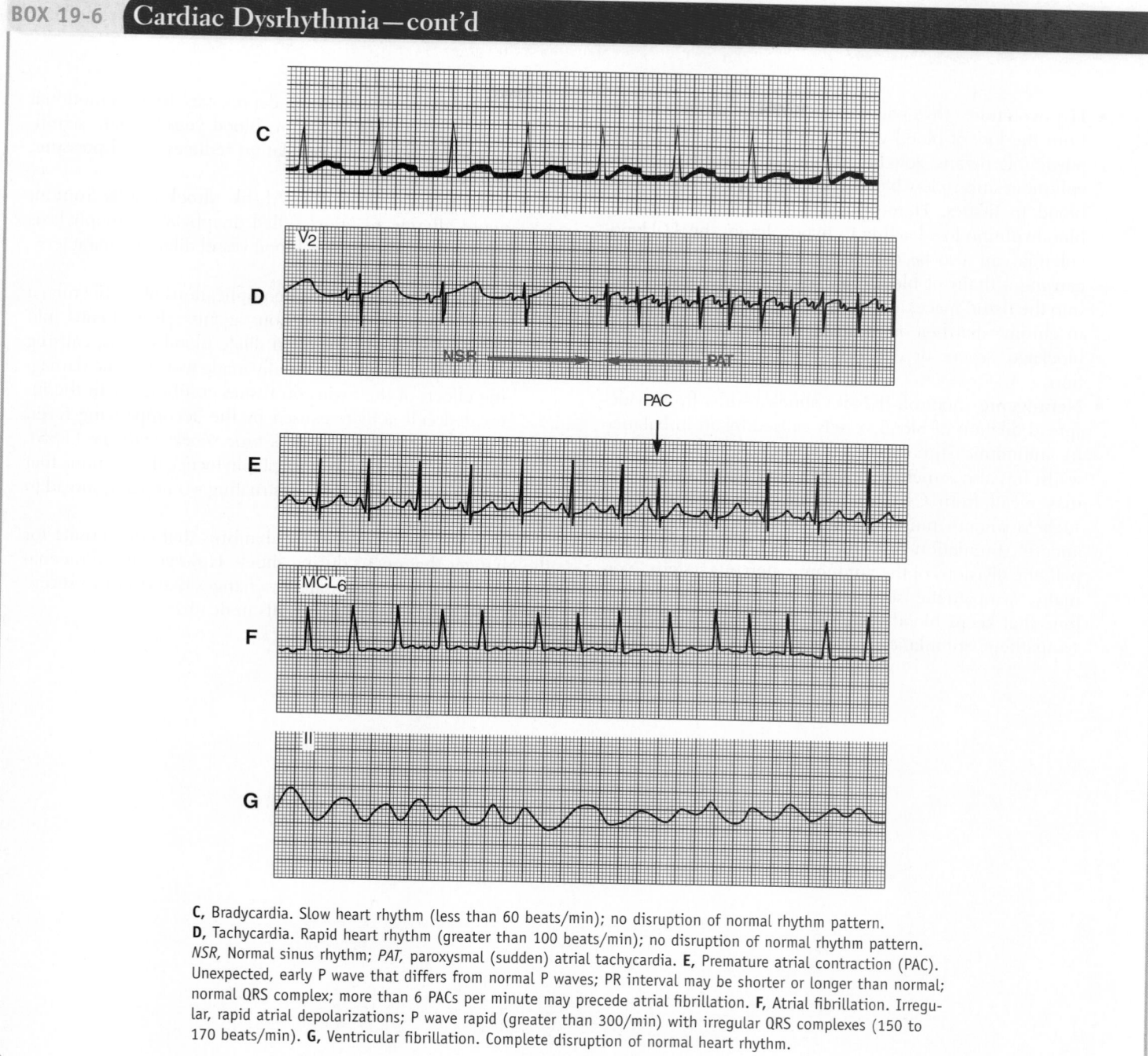

C, Bradycardia. Slow heart rhythm (less than 60 beats/min); no disruption of normal rhythm pattern.
D, Tachycardia. Rapid heart rhythm (greater than 100 beats/min); no disruption of normal rhythm pattern. *NSR,* Normal sinus rhythm; *PAT,* paroxysmal (sudden) atrial tachycardia. **E,** Premature atrial contraction (PAC). Unexpected, early P wave that differs from normal P waves; PR interval may be shorter or longer than normal; normal QRS complex; more than 6 PACs per minute may precede atrial fibrillation. **F,** Atrial fibrillation. Irregular, rapid atrial depolarizations; P wave rapid (greater than 300/min) with irregular QRS complexes (150 to 170 beats/min). **G,** Ventricular fibrillation. Complete disruption of normal heart rhythm.

into the right side of the heart—causing an increase in pressure that the right side of the heart simply cannot overcome. Right-side heart failure can also be caused by lung disorders that obstruct normal pulmonary blood flow and thus overload the right side of the heart—a condition called **cor pulmonale** (kohr pul-mah-nal-ee). (See also Mechanisms of Disease section in Chapter 18.)

Circulatory Shock

The term **circulatory shock** refers to the failure of the circulatory system to adequately deliver oxygen to the tissues, resulting in the impairment of cell function throughout the body. If left untreated, circulatory shock may lead to death. Circulatory failure has many causes, all of which somehow reduce the flow of blood through the blood vessels of the body. Because of the variety of causes, circulatory shock is often classified into the following types:

- **Cardiogenic** (kar-dee-oh-JEN-ik) **shock** results from any type of heart failure, such as that after severe myocardial infarction (heart attack), heart infections, and other heart conditions. Because the heart can no longer pump blood effectively, blood flow to the tissues of the body decreases or stops.

Mechanisms of Disease—cont.

- **Hypovolemic** (hye-poh-voh-LEE-mik) **shock** results from the loss of blood volume in the blood vessels (hypovolemia means "low blood volume"). Reduced blood volume results in low blood pressure and reduced flow of blood to tissues. Hemorrhage is a common cause of blood volume loss leading to hypovolemic shock. Hypovolemia can also be caused by loss of interstitial fluid, causing a drain of blood plasma out of the vessels and into the tissue spaces. Loss of interstitial fluid is common in chronic diarrhea or vomiting, dehydration, intestinal blockage, severe or extensive burns, and other conditions.
- **Neurogenic** (noo-roh-JEN-ik) **shock** results from widespread dilation of blood vessels caused by an imbalance in autonomic stimulation of smooth muscle in vessel walls. It is also sometimes called **vasodilatory shock.** You may recall from Chapter 14 that autonomic effectors such as smooth muscle tissues are controlled by a balance of stimulation from the sympathetic and parasympathetic divisions of the autonomic nervous system. Normally, sympathetic stimulation maintains the muscle tone that keeps blood vessels at their usual diameter. If sympathetic stimulation is disrupted by an injury to the spinal cord or medulla, depressive drugs, emotional stress, or some other factor, blood vessels dilate significantly. Widespread vasodilation reduces blood pressure, thus reducing blood flow.
- **Anaphylactic** (an-ah-fih-LAK-tik) **shock** results from an acute allergic reaction called anaphylaxis. Anaphylaxis causes the same kind of blood vessel dilation characteristic of neurogenic shock.
- **Septic shock** results from complications of septicemia, a condition in which infectious agents release toxins into the blood. The toxins often dilate blood vessels, causing shock. The situation is usually made worse by the damaging effects of the toxins on tissues combined with the increased cell activity caused by the accompanying fever. One type of septic shock is *toxic shock syndrome* (TSS), which usually results from staphylococcal infections that begin in the vagina of menstruating women and spread to the blood.

The body has numerous mechanisms that compensate for the changes that occur during shock. However, these mechanisms may fail to compensate for changes that occur in severe cases, and this failure often results in death.

LANGUAGE OF SCIENCE

(Cont'd from page 733)

medullary ischemic reflex (MED-uh-lair-ee is-KEE-mik REE-fleks) [*medulla-* marrow or middle, *-ary* pertaining to, *ischem-* hold back, *-ic* pertaining to, *reflex* bend back]

minute volume

orthostatic effect (or-thoh-STAT-ik) [*orthostat-* standing upright, *-ic* pertaining to]

P wave

pacemaker (PAYS-may-ker)

perfusion pressure (per-FYOO-shun) [*perfus-* pour over, *-ion* state of]

peripheral resistance (peh-RIF-er-al) [*periphera-* circumference, *-al* pertaining to]

Poiseuille's law (pwah-SWEEZ law) [*Jean L.M. Poiseuille* French physiologist]

pressoreflexes (pres-oh-REE-fleks-es) [*presso-* pressure, *-reflex* bend back]

Purkinje system (pur-KIN-jee) [*Johannes E. Purkinje* Czech physiologist]

QRS complex (Q R S KOM-pleks)

residual volume (ree-ZID-yoo-al) [*residu-* remainder, *-al* pertaining to]

sinoatrial (SA) node (sye-no-AY-tree-al) [*sino-* hollow (sinus), *atri-* hall, *-al* pertaining to]

Starling's law of the capillaries (STAR-lings) [*Ernest H. Starling* English physiologist]

Starling's law of the heart (STAR-lings) [*Ernest H. Starling* English physiologist]

stress-relaxation effect

stroke volume (SV)

subendocardial fibers (sub-en-doh-KAR-dee-al) [*sub-* under, *-endo-* within, *-cardia-* heart, *-al* pertaining to]

systole (SIS-toh-lee) [*systole* contraction]

T wave

total peripheral resistance (TPR) (peh-RIF-er-al) [*perphera-* circumference, *-al* pertaining to, *resist-* withstand, *-ance* act of]

U wave

vasoconstriction (vay-soh-kon-STRIK-shun) [*vaso-* vessel, *-constrict-* draw tight, *-ion* state of]

vasodilation (vay-soh-DYE-lay-shun) [*vaso-* vessel, *-dilat-* to widen, *-ion* state of]

vasomotor chemoreflexes (vay-so-MOH-tor kee-moh-REE-fleks-ez) [*vaso-* vessel, *-motor* move, *chemo-* chemical, *-reflex* bend back]

vasomotor mechanism (vay-so-MOH-tor) [*vaso-* vessel, *-motor* move, *mechan-* machine, *-ism* action of]

vasomotor pressoreflex (vay-so-MOH-tor press-oh-REE-fleks) [*vaso-* vessel, *-motor* move, *presso-* pressure, *-reflex* bend back]

venous pumps (VEE-nus) [*ven-* vein, *-ous* pertaining to]

venous return (VEE-nus) [*ven-* vein, *-ous* pertaining to]

viscosity (vis-KOS-i-tee) [*viscos-* sticky, *-ity* state of]

LANGUAGE OF MEDICINE

anaphylactic shock (an-ah-fih-LAK-tik) [*ana-* without, *-phylact-* protection, *-ic* pertaining to]

artificial pacemakers (ar-ti-FISH-al PAYS-may-kers) [*artifici-* not natural, *-al* pertaining to, *passus-* step, *-macian* to make]

atrial fibrillation (AY-tree-al fi-bril-LAY-shun) [*atri-* hall, *-al* pertaining to, *fibrilla-* small fiber, *-ation* process of]

augmented leads (AWG-men-ted)

bradycardia (bray-dee-KAR-dee-ah) [*brady-* slow, *-cardi-* heart, *-ia* condition]

bruits (BROO-eez) [*bruits* noise]

cardiogenic shock (kar-dee-oh-JEN-ik) [*cardio-* heart, *-gen-* generate, *-ic* pertaining to]

cardiomyopathy (kar-dee-oh-my-OP-ah-thee) [*cardio-* heart, *-myo-* muscle, *-pathy* disease]

chest leads

circulatory shock (SER-kyoo-lah-tor-ee) [*circulat-* go around, *-ory* pertaining to]

congestive heart failure (kon-JES-tiv) [*congest-* accumulate, *-ive* pertaining to]

cor pulmonale (kohr pul-mah-NAL-ee) [*cor* heart, *pulmon-* lungs, *-ale* pertaining to]

diastolic blood pressure (dye-ah-STOL-ik) [*dia-* apart or through, *-stol-* contraction, *-ic* pertaining to]

digitalis (dij-i-TAL-is) [*digit-* finger, *-alis* pertaining to (from finger-shaped flowers of foxglove plant)]

digoxin (di-JOK-sin) [*dig-* finger (from digitalis or foxglove), *-oxin* poison or toxin]

dysrhythmia (dis-RITH-mee-ah) [*dys-* disordered, *-rhythm-* rhythm, *-ia* condition]

echocardiography (ek-oh-kar-dee-OG-rah-fee) [*echo-* sound, *-cardio-* heart, *-graphy* process of recording]

ectopic pacemakers (ek-TOP-ik PAYS-may-kers) [*ec-* out of, *-top-* place, *-ic* pertaining to]

Einthoven's triangle (EYN-to-venz) [*Willem Einthoven* Dutch physiologist]

electrocardiogram (ECG or EKG) (eh-lek-troh-KAR-dee-oh-gram) [*electro-* electricity, *-cardio-* heart, *-gram* written record]

electrocardiography (eh-lek-troh-kar-dee-OG-rah-fee) [*electro-* electricity, *-cardio-* heart, *-graphy* process of recording]

heart block

heart failure

heart murmur [*murmur* hum]

hypercapnia (hye-per-KAP-nee-ah) [*hyper-* excessive, *-capn-* vapor (CO_2), *-ia* condition]

hypokalemia (hye-poh-kal-EE-mee-ah) [*hyper-* excessive, *-kal-* potassium, *-emia* blood condition]

hypovolemic shock (hye-poh-voh-LEE-mik) [*hypo-* under or below, *-vol-* whirl, *-emi(a)* blood condition, *-ic* pertaining to]

hypoxia (hye-POK-see-ah) [*hypo-* under or below, *-ox-* oxygen, *-ia* condition]

Korotkoff sounds (kor-ROT-koff) [*Nicolai Korotkoff* Russian physician]

laminar flow (LAM-i-nar) [*lamina-* plate, *-ar* pertaining to]

limb leads

M-mode echocardiogram (ek-oh-KAR-dee-oh-gram) [*M-* motion, *-mode* manner, *echo-* sound, *-cardio-* heart, *-gram* written record]

neurogenic shock (noo-roh-JEN-ik) [*neuro-* nerve, *-gen-* generate, *-ic* pertaining to]

precordial leads (pree-KOR-dee-al) [*pre-* before, *-cordi-* heart, *-al* pertaining to]

premature contractions

pulse pressure [*pulse* beat]

septic shock (SEP-tik shok) [*septi-* putrid, *-ic* pertaining to]

sinus dysrhythmia (SYE-nus dis-RITH-mee-ah) [*sinus* hollow, *dys-* disordered, *-rhythm-* rhythm, *-ia* condition]

sphygmomanometer (sfig-moh-mah-NOM-eh-ter) [*sphygmo-* pulse, *-mano-* thin, *-meter* measure]

standard leads

systolic blood pressure (sis-TOL-ik) [*systol-* contraction, *-ic* pertaining to]

tachycardia (tak-i-KAR-dee-ah) [*tachy-* swift or rapid, *-cardi-* heart, *-ia* condition]

transvenous approach (tranz-VEE-nus) [*trans-* across, *-ven-* vein, *-ous* pertaining to]

turbulent flow (TER-byoo-lent) [*turbulent* restless or disturbed]

vasodilatory shock (vay-soh-DYE-lah-tor-ee) [*vaso-* vessel, *-dilat-* to widen, *-ory* pertaining to]

ventricular fibrillation (ven-TRIK-yoo-lar fib-ril-LAY-shun) [*ventricul-* little belly, *-ar*, pertaining to, *fibrilla-* little fiber, *-ation* process]

CASE STUDY

Mr. Charles Simpson, 20 years old, is injured in an industrial accident, resulting in a crushed pelvis, ruptured spleen, and associated blood loss. His past medical history is negative for previous trauma or chronic illness and positive for usual childhood diseases. Before the accident Mr. Simpson was in good health and exercised daily. His height is 6 feet and weight 165 pounds.

At midnight a 10-ton forklift fell and pinned Mr. Simpson at the pelvis for about 20 minutes. Paramedics at the scene began intravenous lactated Ringer's solution at 150 ml/hr. Vital signs were heart rate 120 beats/min, blood pressure 90/70, and respirations 46/min. Mr. Simpson was awake and complained of pelvic, back, and abdominal pain. His toes were mottled, pedal pulses were absent, radial pulses weak, and brachial and carotid pulses palpable. The electrocardiogram monitor reflected tachycardia, and he became short of breath with conversation. Skin was cold and clammy, with numerous pinpoint hemorrhages present over his upper thorax, face, and neck.

Mr. Simpson was transported to the hospital, where his injuries were diagnosed and immediate treatment for shock initiated. After stabilization, his ruptured spleen was surgically removed, he was placed in pelvic traction to stabilize his fractures, and he was admitted to the intensive care unit.

1. Mr. Simpson's extracellular fluid volume deficit occurred as a result of which primary mechanism?
   A. Decreased intake of fluids and electrolytes
   B. Excessive loss of blood and fluids
   C. Shifts of fluids and electrolytes into nonaccessible areas

2. Which of Mr. Simpson's signs are the result of compensatory mechanisms directed at maintaining cardiac output?
   A. Increased heart rate and oliguria
   B. Decreased blood pressure and sodium loss
   C. Respiratory acidosis and decreased heart rate
   D. All of the above

3. The mechanism MOST responsible for Mr. Simpson's tachycardia is:
   A. Hypoxemia caused by atelectasis
   B. Anxiety as a result of traumatic injuries and pain
   C. A physiological change in response to decreased blood pressure
   D. Reaction to blood transfusion

4. Based on the information provided, what type of shock did Mr. Simpson experience?
   A. Cardiogenic
   B. Hypovolemic
   C. Neurogenic
   D. Anaphylactic

CHAPTER SUMMARY

A. Vital role of the cardiovascular system in maintaining homeostasis depends on the continuous and controlled movement of blood through the capillaries
B. Numerous control mechanisms help to regulate and integrate the diverse functions and component parts of the cardiovascular system to supply blood in response to specific body area needs

HEMODYNAMICS

A. Hemodynamics—collection of mechanisms that influence the dynamic (active and changing) circulation of blood (Figure 19-1)
B. Circulation of different volumes of blood per minute is essential for healthy survival
C. Circulation control mechanisms must accomplish two functions
   1. Maintain circulation
   2. Vary volume and distribution of the blood circulated

THE HEART AS A PUMP

A. Conduction system (Figure 19-2)
   1. Four major structures that compose the conduction system of the heart
      a. Sinoatrial node (SA node)
      b. Atrioventricular node (AV node)
      c. AV bundle (bundle of His)
      d. Purkinje system
   2. Conduction system structures are more highly specialized than ordinary cardiac muscle tissue and permit only rapid conduction of an action potential through the heart
   3. SA node (pacemaker)
      a. Initiates each heartbeat and sets its pace
      b. Specialized pacemaker cells in the node possess an intrinsic rhythm
   4. Sequence of cardiac stimulation
      a. After being generated by the SA node, each impulse travels throughout the muscle fibers of both atria and the atria begin to contract
      b. As the action potential enters the AV node from the right atrium, its conduction slows to allow complete contraction of both atrial chambers before the impulse reaches the ventricles
      c. After the AV node, conduction velocity increases as the impulse is relayed through the AV bundle into the ventricles
      d. Right and left branches of the bundle fibers and Purkinje fibers conduct the impulses throughout the muscles of both ventricles, stimulating them to contract almost simultaneously

B. Electrocardiogram (ECG or EKG)
1. Graphic record of the heart's electrical activity, its conduction of impulses; a record of the electrical events that precede the contractions of the heart
2. To produce an ECG (Figure 19-3)
   a. Electrodes of an electrocardiograph are attached to the subject
   b. Changes in voltage are recorded that represent changes in the heart's electrical activity (Figure 19-4)
3. Normal ECG (Figures 19-3 and 19-5) is composed of:
   a. P wave—represents depolarization of the atria
   b. QRS complex—represents depolarization of the ventricles and repolarization of the atria
   c. T wave—represents repolarization of the ventricles; may also have a U wave that represents repolarization of the papillary muscle (Figure 19-6)
   d. Measurement of the intervals between P, QRS, and T waves can provide information about the rate of conduction of an action potential through the heart
C. Cardiac cycle—a complete heartbeat consisting of contraction (systole) and relaxation (diastole) of both atria and both ventricles; the cycle is often divided into time intervals (Figures 19-7 and 19-8)
1. Atrial systole
   a. Contraction of atria completes emptying blood out of the atria into the ventricles
   b. AV valves are open; SL valves are closed
   c. Ventricles are relaxed and filling with blood
   d. This cycle begins with the P wave of the ECG
2. Isovolumetric ventricular contraction
   a. Occurs between the start of ventricular systole and the opening of the SL valves
   b. Ventricular volume remains constant as the pressure increases rapidly
   c. Onset of ventricular systole coincides with the R wave of the ECG and the appearance of the first heart sound
3. Ejection
   a. SL valves open and blood is ejected from the heart when the pressure gradient in the ventricles exceeds the pressure in the pulmonary artery and aorta
   b. Rapid ejection—initial, short phase characterized by a marked increase in ventricular and aortic pressure and in aortic blood flow
   c. Reduced ejection—characterized by a less abrupt decrease in ventricular volume, coincides with the T wave of the ECG
4. Isovolumetric ventricular relaxation
   a. Ventricular diastole begins with this phase
   b. Occurs between closure of the SL valves and opening of the AV valves
   c. A dramatic fall in intraventricular pressure but no change in volume
   d. The second heart sound is heard during this period
5. Passive ventricular filling
   a. Returning venous blood increases intraatrial pressure until the AV valves are forced open and blood rushes into the relaxing ventricles

   b. Influx lasts approximately 0.1 second and results in a dramatic increase in ventricular volume
   c. Diastasis—later, longer period of slow ventricular filling at the end of ventricular diastole lasting approximately 0.2 second; characterized by a gradual increase in ventricular pressure and volume
D. Heart sounds
1. Systolic sound—first sound, believed to be caused primarily by the contraction of the ventricles and by vibrations of the closing AV valves
2. Diastolic sound—short, sharp sound; thought to be caused by vibrations of the closing of SL valves
3. Heart sounds have clinical significance because they give information about the functioning of the valves of the heart

PRIMARY PRINCIPLE OF CIRCULATION

A. Blood flows because a pressure gradient exists between different parts of its bed; this is based on Newton's first and second laws of motion (Figure 19-9)
B. Blood circulates from the left ventricle to the right atrium of the heart because a blood pressure gradient exists between these two structures
C. P_1-P_2 is the symbol used to stand for a pressure gradient, with P_1 representing the higher pressure and P_2 the lower pressure
D. Perfusion pressure—the pressure gradient needed to maintain blood flow through a local tissue

ARTERIAL BLOOD PRESSURE

A. Primary determinant of arterial blood pressure is the volume of blood in the arteries; a direct relationship exists between arterial blood volume and arterial pressure (Figure 19-10)
B. Cardiac output (CO)—volume of blood pumped out of the heart per unit of time (ml/min or L/min) (Figure 19-11)
1. General principles and definitions
   a. Cardiac output (CO)—determined by stroke volume and heart rate
   b. Stroke volume (SV)—volume pumped per heartbeat
   c. CO (volume/min)—SV (volume/beat) times HR (beats/min)
   d. In practice, CO is computed by Fick's formula
   e. Heart rate and stroke volume determine CO, so anything that changes either also tends to change CO, arterial blood volume, and blood pressure in the same direction
2. Stroke volume
   a. Starling's law of the heart (Frank-Starling mechanism) (Figure 19-12)
      (1) Within limits, the longer, or more stretched, the heart fibers at the beginning of contraction, the stronger the contraction
      (2) The amount of blood in the heart at the end of diastole determines the amount of stretch placed on the heart fibers
      (3) The myocardium contracts with enough strength to match its pumping load (within certain limits) with each stroke—unlike mechanical pumps

b. Contractility (strength of contraction) can also be influenced by chemical factors (Figure 19-13)
   (1) Neural—norepinephrine; endocrine—epinephrine
   (2) Triggered by stress, exercise
3. Factors that affect heart rate—SA node normally initiates each heartbeat; however, various factors can and do change the rate of the heartbeat
   a. Cardiac pressoreflexes—aortic baroreceptors and carotid baroreceptors, located in the aorta and carotid sinus, are extremely important because they affect the autonomic cardiac control center, and therefore parasympathetic and sympathetic outflow, to aid in control of blood pressure (Figures 19-14 and 19-15)
      (1) Carotid sinus reflex
         (a) Carotid sinus is located at the beginning of the internal carotid artery
         (b) Sensory fibers from carotid sinus baroreceptors run through the carotid sinus nerve and the glossopharyngeal nerve to the cardiac control center
         (c) Parasympathetic impulses leave the cardiac control center, travel through the vagus nerve to reach the SA node
      (2) Aortic reflex—sensory fibers extend from baroreceptors located in the wall of the arch of the aorta through the aortic nerve and through the vagus nerve to terminate in the cardiac control center
   b. Other reflexes that influence heart rate—various important factors influence the heart rate; reflexive increases in heart rate often result from increased sympathetic stimulation of the heart
      (1) Anxiety, fear, and anger often increase heart rate
      (2) Grief tends to decrease heart rate
      (3) Emotions produce changes in heart rate through the influence of impulses from the cerebrum by way of the hypothalamus
      (4) Exercise—heart rate normally increases
      (5) Increased blood temperature or stimulation of skin heat receptors increases heart rate
      (6) Decreased blood temperature or stimulation of skin cold receptors decreases heart rate
C. Peripheral resistance—resistance to blood flow imposed by the force of friction between blood and the walls of its vessels
   1. Factors that influence peripheral resistance
      a. Blood viscosity—the thickness of blood as a fluid (Figure 19-16)
         (1) High plasma protein concentration can slightly increase blood viscosity
         (2) High hematocrit (% RBC) can increase blood viscosity
         (3) Anemia, hemorrhage, or other abnormal conditions may also affect blood viscosity
      b. Diameter of arterioles (Figure 19-17)
         (1) Vasomotor mechanism—muscles in walls of arteriole may constrict (vasoconstriction) or dilate (vasodilation), thus changing diameter of arteriole

   (2) Small changes in blood vessel diameter cause large changes in resistance, making the vasomotor mechanism ideal for regulating blood pressure and blood flow.
2. How resistance influences blood pressure
   a. Arterial blood pressure tends to vary directly with peripheral resistance
   b. Friction caused by viscosity and small diameter of arterioles and capillaries
   c. Muscular coat of arterioles allows them to constrict or dilate and change the amount of resistance to blood flow
   d. Peripheral resistance helps determine arterial pressure by controlling the amount of blood that runs from the arteries to the arterioles (Figure 19-18)
      (1) Increased resistance and decreased arteriole runoff lead to higher arterial pressure
      (2) Can occur locally (in one organ), or the total peripheral resistance (TPR) may increase thus generally raising systemic arterial pressure
3. Vasomotor control mechanism—controls changes in the diameter of arterioles; plays role in maintenance of the general blood pressure and in distribution of blood to areas of special need (Figures 19-19 and 19-20)
   a. Vasomotor pressoreflexes (Figure 19-21)
      (1) Sudden increase in arterial blood pressure stimulates aortic and carotid baroreceptors; results in arterioles and venules of the blood reservoirs dilating
      (2) Decrease in arterial blood pressure results in stimulation of vasoconstrictor centers, causing vascular smooth muscle to constrict
   b. Vasomotor chemoreflexes (Figure 19-22)—chemoreceptors located in aortic and carotid bodies are sensitive to hypercapnia, hypoxia, and decreased arterial blood pH
   c. Medullary ischemic reflex—acts during emergency situation when there is decreased blood flow to the medulla; causes marked arteriole and venous constriction
   d. Vasomotor control by higher brain centers—impulses from centers in cerebral cortex and hypothalamus are transmitted to vasomotor centers in medulla to help control vasoconstriction and dilation
4. Local control of arterioles—several local mechanisms produce vasodilation in localized areas; referred to as reactive hyperemia

VENOUS RETURN TO THE HEART

A. Venous return—amount of blood returned to the heart by the veins
B. Stress-relaxation effect—occurs when a change in blood pressure causes a change in vessel diameter (because of elasticity) and thus adapts to the new pressure to keep blood flowing (works only within certain limits)
C. Gravity—the pull of gravity on venous blood while sitting or standing tends to cause a decrease in venous return (orthostatic effect) (Figure 19-23)

D. Venous pumps—blood-pumping action of respirations and skeletal muscle contractions facilitate venous return by increasing pressure gradient between peripheral veins and venae cavae (Figure 19-24)
   1. Respirations—inspiration increases the pressure gradient between peripheral and central veins by decreasing central venous pressure and also by increasing peripheral venous pressure
   2. Skeletal muscle contractions—promote venous return by squeezing veins through a contracting muscle and milking the blood toward the heart
   3. Semilunar valves in veins prevent backflow (Figure 19-25)
E. Total blood volume—changes in total blood volume change the amount of blood returned to the heart
   1. Capillary exchange—governed by Starling's law of the capillaries (Figure 19-26)
      a. At arterial end of capillary, outward hydrostatic pressure is strongest force; moves fluid out of plasma and into IF
      b. At venous end of capillary, inward osmotic pressure is strongest force; moves fluid into plasma from IF; 90% of fluid lost by plasma at arterial end is recovered
      c. Lymphatic system recovers fluid not recovered by capillary and returns it to the venous blood before it is returned to the heart
   2. Changes in total blood volume—mechanisms that change total blood volume most quickly are those that cause water to quickly move into or out of the plasma (Figure 19-27)
      a. ADH mechanism—decreases the amount of water lost by the body by increasing the amount of water that kidneys resorb from urine before the urine is excreted from the body; triggered by input from baroreceptors and osmoreceptors
      b. Renin-angiotensin-aldosterone system (RAAS)
         (1) Renin—released when blood pressure in kidney is low; leads to increased secretion of aldosterone, which stimulates retention of sodium, causing increased retention of water and an increase in blood volume
         (2) Angiotensin II—intermediate compound that causes vasoconstriction, which complements the volume-increasing effects of renin and promotes an increase in overall blood flow
      c. ANH mechanism—adjusts venous return from an abnormally high level by promoting the loss of water from plasma, causing a decrease in blood volume; increases urine sodium loss, which causes water to follow osmotically
F. A variety of feedback responses restore normal blood pressure after a sudden change in pressure (Figure 19-28)

MEASURING BLOOD PRESSURE

A. Arterial blood pressure
   1. Measured with the aid of a sphygmomanometer and stethoscope; listen for Korotkoff sounds as the pressure in the cuff is gradually decreased (Figure 19-29)
   2. Systolic blood pressure—force of the blood pushing against the artery walls while ventricles are contracting
   3. Diastolic blood pressure—force of the blood pushing against the artery walls when ventricles are relaxed
   4. Pulse pressure—difference between systolic and diastolic blood pressure
B. Relation to arterial and venous bleeding
   1. Arterial bleeding—blood escapes from artery in spurts because of alternating increase and decrease of arterial blood pressure
   2. Venous bleeding—blood flows slowly and steadily because of low, practically constant pressure

MINUTE VOLUME OF BLOOD (FIGURE 19-30)

A. Minute volume—determined by the magnitude of the blood pressure gradient and peripheral resistance
B. Poiseuille's law—Minute volume = Pressure gradient ÷ Resistance

VELOCITY OF BLOOD FLOW

A. Velocity of blood is governed by the physical principle that when a liquid flows from an area of one cross-sectional size to an area of larger size, its velocity decreases in the area with the larger cross section (Figure 19-31)
B. Blood flows more slowly through arterioles than arteries because total cross-sectional area of arterioles is greater than that of arteries and capillary blood flow is slower than arteriole blood flow
C. Venule cross-sectional area is smaller than capillary cross-sectional area, causing blood velocity to increase in venules and then veins with a still smaller cross-sectional area

PULSE

A. Mechanism
   1. Pulse—alternate expansion and recoil of an artery (Figure 19-32)
      a. Clinical significance: reveals important information regarding the cardiovascular system, blood vessels, and circulation
      b. Physiological significance: expansion stores energy released during recoil, conserving energy generated by the heart and maintaining relatively constant blood flow (Figure 19-33)
   2. Existence of pulse is due to two factors
      a. Alternating increase and decrease of pressure in the vessel
      b. Elasticity of arterial walls allows walls to expand with increased pressure and recoil with decreased pressure
B. Pulse wave
   1. Each pulse starts with ventricular contraction and proceeds as a wave of expansion throughout the arteries
   2. Gradually dissipates as it travels, disappearing in the capillaries
C. Where pulse can be felt—wherever an artery lies near the surface and over a bone or other firm background (Figure 19-34)
D. Venous pulse—detectable pulse exists only in large veins; most prominent near the heart; not of clinical importance

THE BIG PICTURE: BLOOD FLOW AND THE WHOLE BODY

A. Blood flow shifts materials from place to place and redistributes heat and pressure

B. Vital to maintaining homeostasis of internal environment

REVIEW QUESTIONS

1. Identify, locate, and describe the function of each of the following structures: SA node, AV node, AV bundle, and Purkinje fibers.
2. What does an electrocardiogram measure and record? List the normal ECG deflection waves and intervals. What do the various ECG waves represent?
3. What is meant by the term *cardiac cycle*?
4. List the "phases" of the cardiac cycle and briefly describe the events that occur in each.
5. What is meant by the term *residual volume* as it applies to the heart?
6. Describe and explain the origin of the heart sounds.
7. What blood vessels present the greatest resistance to blood flow?
8. What is the primary determinant of arterial blood pressure?
9. List the two most important factors that indirectly determine arterial pressure by their influence on arterial volume.
10. How is cardiac output determined?
11. List and give the effect of several factors, such as grief or pain, on heart rate.
12. What mechanisms control peripheral resistance? Cite an example of the operation of one or more parts of this mechanism to increase resistance and to decrease it.
13. What are the components of the vasomotor control mechanism?
14. Explain how antidiuretic hormone can change the total blood volume.
15. What is the effect of low blood pressure in relation to aldosterone and antidiuretic hormone secretion?
16. Describe the measurement of arterial blood pressure.
17. Identify the eight locations where the pulse point is most easily felt. List the six pressure points at which pressure can be applied to stop arterial bleeding distal to that point.
18. Describe the various types of cardiac dysrhythmias.

CRITICAL THINKING QUESTIONS

1. What is an ectopic pacemaker? What would be the effect on the heart rate if an ectopic pacemaker took over for the SA node?
2. State in your own words the primary principle of circulation. How does it govern the various mechanisms involved in blood flow?
3. What is Starling's law of the heart? What role does *venous return* play in Starling's law?
4. Explain how the heart rate is a good example of dual innervation in the autonomic nervous system. Include the nerves and neurotransmitters involved.
5. Explain how blood pressure in the brain is kept from getting too high.
6. What is arterial runoff? What is the relationship of arteriole runoff and peripheral resistance?
7. Explain what occurs when the carbon dioxide level in the medulla rises above set point.
8. By the time blood gets to the veins, almost all the pressure from the contracting ventricles has been lost. What mechanisms does the body use to assist in returning blood to the heart?
9. Explain the forces that act on capillary exchange on both the arterial and venous ends of the capillary. Is the recovery of fluid at the venous end 100% effective?
10. State in your own words Poiseuille's law. What equation expresses this law? According to the equation, an increase in resistance would lower the minute volume. Give an example of where this would not be the case.
11. What two factors determine blood viscosity? What does viscosity mean? Give an example of a condition in which blood viscosity decreases. Explain its effects on resistance, arterial pressure, and circulation.

CHAPTER 20

Lymphatic System

LANGUAGE OF SCIENCE

aggregated lymphoid nodules (ag-rah-GAYT-ed LIM-foyd NOD-yools) [*a[d]-* to, *-grega-* collect, *lymph-* water, *-oid* resembling, *nodu-* knot, *-ule* small]

anastomoses (ah-nas-toh-MOH-seez) [*ana-* again, *-stomo-* mouth or outlet, *-oses* conditions of]

axillary lymph nodes (AK-sil-lair-ee limf nohds) [*axilla-* wing, *-ary* pertaining to, *lymph* water, *nodus* knot]

blood reservoir (REZ-eh-vwahr)

chyle (kile) [*chylos* juice]

cisterna chyli (sis-TER-nah KYE-lye) [*cisterna* vessel, *chyli* of juice]

cortical nodules (KOHR-ti-kal NOD-yools) [*cortic-* cortex, *-al* pertaining to, *nodu* knot, *-ule* small]

defense

glands

hematopoiesis (hee-mah-toh-poy-EE-sis) [*hemato-* blood, *-poiesis* formation of]

iliac lymph nodes (ILL-ee-ak limf nohds) [*ilia-* loin or gut [ileum], *-ac* pertaining to, *lymph* water, *nodus* knot]

inguinal lymph nodes (ING-gwi-nal limf nohds) [*inguin-* groin, *-al* pertaining to, *lympha* water, *-atic* of the kind, *nodus* knot]

interstitial fluid (IF) (in-ter-STISH-al) [*interstiti-* space between, *-al* pertaining to]

involution (in-voh-LOO-shun) [*involu-* wrap up, *-tion* state of]

lacteals (LAK-tee-als) [*lacte-* milk, *-al* pertaining to]

lingual tonsils (LING-gwal TAHN-sils) [*lingua-* tongue, *-al* pertaining to]

lymph (limf) [*lymph* water]

lymph nodes (limf nohds) [*lymph* water, *nodus* knot]

lymphatic capillaries (lim-FAT-ik KAP-i-lair-eez) [*lymph-* water, *-atic* of the kind, *capillary* hairlike]

lymphatic vessels (lim-FAT-ik) [*lymph-* water, *-atic* of the kind, *vess-* vase, *-el* small]

lymphoid tissue (LIM-foyd) [*lymph-* water, *-oid* resembling, *tissu* fabric]

lymphokinesis (lim-foh-kih-NEE-sis) [*lymph-* water, *-kinesis* activation]

Cont'd on p. 796

OVERVIEW OF THE LYMPHATIC SYSTEM

Importance of the Lymphatic System

The lymphatic system serves various functions in the body. The two most important functions of this system are maintenance of fluid balance in the internal environment and immunity. Although both of these important functions are discussed in this chapter, details of immunity are discussed more fully in Chapter 21. The importance of the lymphatic system in maintaining a balance of fluid in the internal environment is best explained by the diagram in Figure 20-1. As this figure shows, plasma filters into interstitial spaces from blood flowing through capillaries. Most of this interstitial fluid (IF) is absorbed by tissue cells or reabsorbed by the blood before it flows out of the tissue. However, a small percentage of the interstitial fluid remains behind. If this continued for even a brief period, the increased interstitial fluid would cause massive edema (swelling) of the tissue. This edema could cause tissue destruction or perhaps even death. This problem is avoided by the presence of lymphatic vessels that act as "drains" to collect the excess tissue fluid and return it to the venous blood just before it reaches the heart.

The lymphatic system is a component of the circulatory system because it consists of a moving fluid (lymph) derived from the blood and tissue fluid and a group of vessels (lymphatics) that return the lymph to the blood. In general, the lymphatic vessels that drain the peripheral areas of the body parallel the venous return. In addition to lymph and the lymphatic vessels, the system includes various structures that contain **lymphoid tissue**—a type of reticular tissue (see Chapter 5) that contains lymphocytes and other specialized cells. For example, lymph nodes are located along the paths of the collecting lymphatic vessels. Isolated nodules of lymphatic tissue, such as the **aggregated lymphoid nodules** called *Peyer's patches* in the intestinal wall, are another example. Additional lymphoid structures include the tonsils, thymus, spleen, and bone marrow (Figure 20-2).

Although it serves a unique transport function by returning tissue fluid, proteins, fats, and other substances to the general circulation, lymph flow differs from the true "circulation" of blood seen in the cardiovascular system. The lymphatic vessels do not, like vessels in the blood vascular system, form a closed ring, or circuit, but instead begin blindly in the intercellular spaces of the soft tissues of the body (see Figure 20-1).

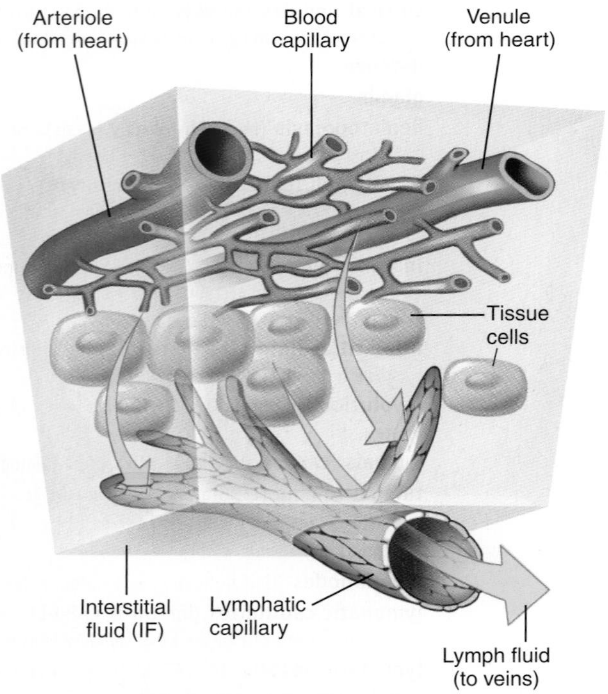

Figure 20-1 *Role of the lymphatic system in fluid balance.* Fluid from plasma flowing through the capillaries moves into interstitial spaces. Although *most* of this interstitial fluid is either absorbed by tissue cells or resorbed by capillaries, *some* of the fluid tends to accumulate in the interstitial spaces. As this fluid builds up, it tends to drain into lymphatic vessels that eventually return the fluid to the venous blood.

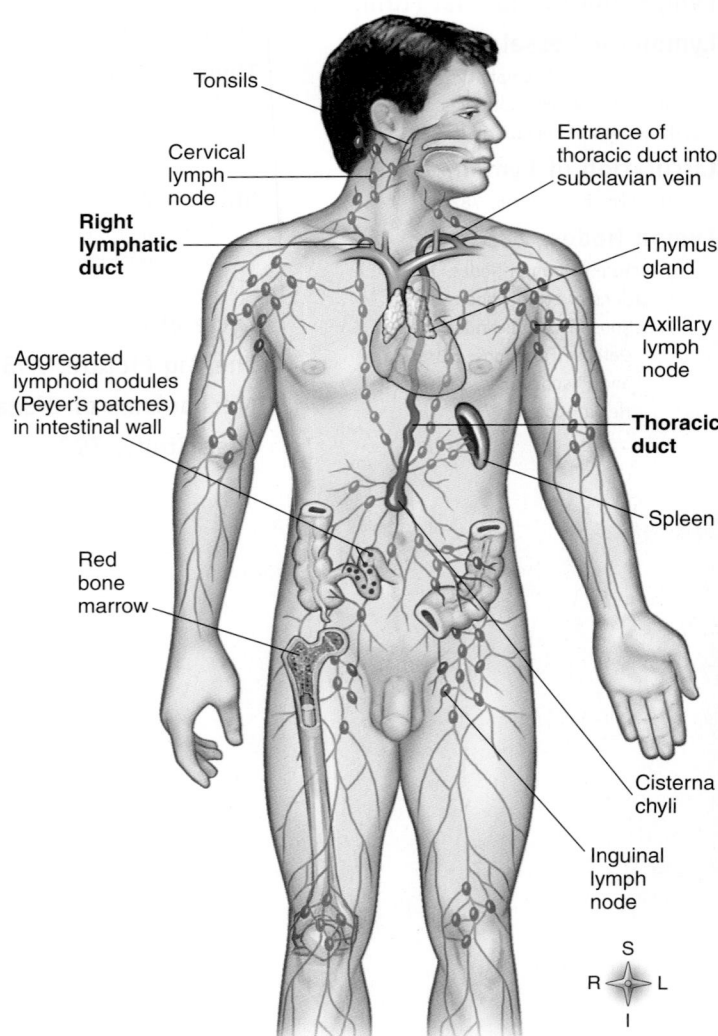

Figure 20-2 *Principal organs of the lymphatic system.*

BOX 20-1: HEALTH MATTERS
Loss of Lymphatic Fluid

Lymph does not clot. Therefore if damage to the main lymphatic trunks in the thorax occurs as a result of penetrating injury, the flow of lymph must be stopped surgically or death ensues. It is impossible to maintain adequate serum protein concentration by dietary means if significant loss of lymph continues over time. As lymph is lost, rapid emaciation occurs, with a progressive and eventually fatal decrease in total blood fat and protein concentration.

LYMPH AND INTERSTITIAL FLUID

Lymph (or *lymphatic fluid*) is the clear, watery-appearing fluid found in the lymphatic vessels. **Interstitial fluid (IF)**, which fills the spaces between the cells, is not the simple fluid it seems to be. Studies show that it is a complex and organized material. It is an important part of the *ECM (extracellular matrix)*. Interstitial fluid and blood plasma together constitute the extracellular fluid compartment of the body, or in the words of Claude Bernard, the "internal environment of the body"—the fluid environment of cells in contrast to the atmosphere, or external environment, of the body.

Both lymph and interstitial fluid closely resemble blood plasma in composition. The main difference is that they contain a lower percentage of proteins than does plasma. Lymph is isotonic and almost identical in chemical composition to interstitial fluid when comparisons are made between the two fluids taken from the same area of the body. However, the average concentration of protein (4 g/100 ml) in lymph taken from the thoracic duct (see Figure 20-2) is about twice that found in most interstitial fluid samples. The elevated protein level of thoracic duct lymph (a mixture of lymph from all areas of the body) results from protein-rich lymph flowing into the duct from the liver and small intestine. A little more than one half of the 2800 to 3000 ml total daily lymph flowing through the thoracic duct is derived from these two organs. Box 20-1 discusses the effects of abnormal loss of lymphatic fluid.

LYMPHATIC VESSELS
Distribution of Lymphatic Vessels

Lymphatic vessels—often simply called *lymphatics*—originate as microscopic blind-end vessels called **lymphatic capillaries.** (Those originating in the villi of the small intestine are called **lacteals**; see Chapter 26.) The wall of each lymphatic capillary consists of a single layer of flattened endothelial cells. Each blindly ending capillary is attached, or fixed, to surrounding cells by tiny connective tissue filaments. Networks of lymphatic capillaries branch and anastomose extensively. Lymphatic networks are located in the intercellular (interstitial) spaces and are widely distributed throughout the body. As a rule, lymphatic and blood capillary networks lie side by side but are always independent of each other.

As twigs of a tree join to form branches, branches join to form larger branches, and larger branches join to form the tree trunk,

so do lymphatic capillaries merge, forming slightly larger lymphatics that join other lymphatics to form still larger vessels, which merge to form the main lymphatic trunks: the **right lymphatic ducts** and the **thoracic duct** (see Figure 20-2). Lymph from the entire body, except the upper right quadrant, drains eventually into the thoracic duct, which drains into the left subclavian vein at the point where it joins the left internal jugular vein (Figure 20-3). Lymph from the upper right quadrant of the body empties into the right lymphatic duct (or, more commonly, into three collecting ducts) and then into the right subclavian vein. Because most of the lymph of the body returns to the bloodstream by way of the thoracic duct, this vessel is considerably larger than the other main lymph channels, the right lymphatic ducts, but is much smaller than the large veins, which it resembles in structure. It has an average diameter of about 5 mm (⅕ inch) and a length of about 40 cm (16 inches). It originates as a dilated structure, the **cisterna chyli** or *chyle cistern*, in the lumbar region of the abdominal cavity—where fatty chyle from the intestinal tract collects. The thoracic duct then ascends a curving pathway to the root of the neck, where it joins the subclavian vein as just described (see Figures 20-2 and 20-3).

Structure of Lymphatic Vessels

Lymphatics resemble veins in structure with these exceptions:

- Lymphatics have thinner walls.
- Lymphatics contain more valves.
- Lymphatics contain lymph nodes located at certain intervals along their course.

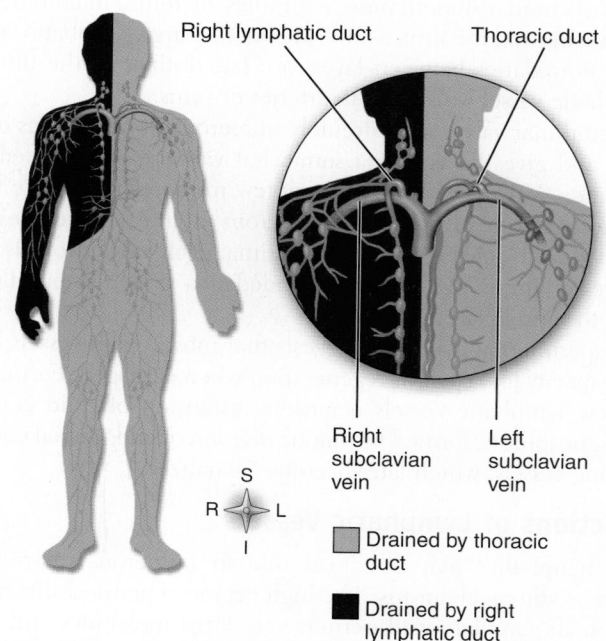

Figure 20-3 *Lymphatic drainage.* The right lymphatic duct drains lymph from the upper right quadrant *(dark blue)* of the body into the right subclavian vein. The thoracic duct drains lymph from the rest of the body *(green)* into the left subclavian vein. The lymphatic fluid is thus returned to the systemic blood just before entering the heart.

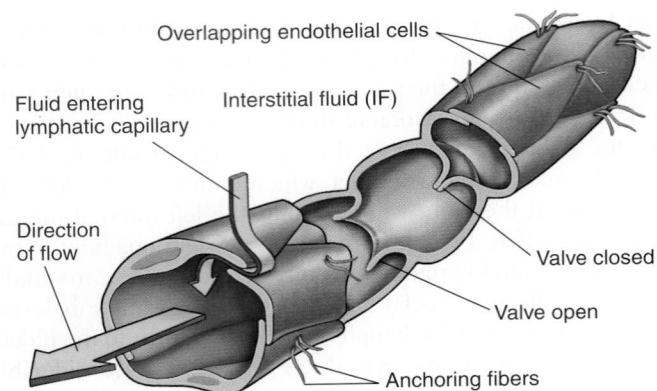

Figure 20-4 *Structure of a typical lymphatic capillary.* Notice that interstitial fluid enters through clefts between overlapping endothelial cells that form the wall of the vessel. Semilunar valves ensure one-way flow of lymph out of the tissue. Small fibers anchor the wall of the lymphatic capillary to the surrounding ECM (extracellular matrix) and cells, thus holding it open to allow entry of fluids and small particles.

The lymphatic capillary wall is formed by a single layer of large but very thin and flat endothelial cells (Figure 20-4). Although the openings (clefts) between endothelial cells of lymphatic capillary walls are small, they are larger than those found in blood capillaries—a fact that explains the remarkable permeability of this system.

As lymph flows from the thin-walled capillaries into vessels with a larger diameter (0.2 to 0.3 mm), the walls become thicker and exhibit the three coats, or layers, typical of arteries and veins (see Table 18-1, p. 695). Interlacing elastic fibers and several strata of circular smooth muscle bundles are found in both the tunica media and the tunica adventitia of the large lymphatic vessel wall. Boundaries between layers are less distinct in the thinner lymphatic vessel walls than in arteries or veins.

Semilunar valves are extremely numerous in lymphatics of all sizes and give the vessels a somewhat varicose and beaded appearance. Valves are present every few millimeters in large lymphatics and are even more numerous in the smaller vessels. Formed from folds of the tunica intima, each valve projects into the vessel lumen in a slightly expanded area circled by bundles of smooth muscle fibers.

Experimental evidence suggests that most lymph vessels have the capacity for repair or regeneration when damaged. Formation of new lymphatic vessels occurs by extension of solid cellular cores, or sprouts, formed by mitotic division of endothelial cells in existing vessels, which later become "canalized."

Functions of Lymphatic Vessels

The lymphatics play a critical role in numerous interrelated homeostatic mechanisms. The high degree of permeability of the lymphatic capillary wall permits very large molecules and even particulate matter, which cannot be absorbed into a blood capillary, to be removed from the interstitial spaces. Proteins that accumulate in the tissue spaces can return to blood only by way of lymphatics. This fact has great clinical importance. For instance, if anything blocks lymphatic return, blood protein concentration and blood osmotic pressure soon fall below normal, and fluid imbalance and death will result (discussed in Chapter 29).

Lacteals (lymphatics in the villi of the small intestine) serve an important function in the absorption of fats and other nutrients. The milky lymph found in lacteals after digestion contains 1% to 2% fat and is called **chyle.** Interstitial fluid has a much lower lipid content than chyle (see Chapter 26).

CIRCULATION OF LYMPH

Water and solutes continually filter out of capillary blood into the interstitial fluid (see Figure 20-1). To balance this outflow, fluid continually reenters blood from the interstitial fluid. Newer evidence has disproved the old idea that healthy capillaries do not "leak" proteins. In truth, each day about 50% of the total blood proteins leak out of the capillaries into the tissue fluid and return to the blood by way of the lymphatic vessels. For more details about fluid exchange between blood and interstitial fluid, see Chapter 29. From lymphatic capillaries, lymph flows through progressively larger lymphatic vessels to eventually reenter blood at the junction of the internal jugular and subclavian veins (see Figure 20-5).

The Lymphatic Pump

Although there is no muscular pumping organ connected with the lymphatic vessels to force lymph onward as the heart forces blood, still lymph moves slowly and steadily along in its vessels. Lymph flows through the thoracic duct and reenters the general circulation at the rate of about 3 liters per day. This occurs despite the fact that most of the flow is against gravity, or "uphill." It moves through the system in the right direction because of the large number of valves that permit fluid flow only in the central direction. What mechanisms establish the pressure gradient required by the basic law of fluid flow? Two of the same mechanisms that contribute to the blood pressure gradient in veins also establish a lymph pressure gradient. These are breathing movements and skeletal muscle contractions (see Figure 19-24, p. 758). Box 20-2 explains one of many reasons a working knowledge of lymphatic flow is important.

Activities that result in central movement, or flow, of lymph are called *lymphokinetic* actions (from the Greek *kinetos,* "movable"). Thus, the flow of lymph may be called **lymphokinesis.** X-ray films taken after radiopaque material is injected into the lymphatics show that lymph pours into the central veins most rapidly at the peak of inspiration. This method of visualizing lymphatic vessels is called **lymphangiography** (Figure 20-6).

> ### BOX 20-2: HEALTH MATTERS
> ### Lymphatic Drainage and Artificial Limbs
>
> An understanding of the anatomy of lymphatic drainage of the skin is critically important in surgical amputation of an extremity. A majority of the lymphatics draining the skin are located in plexus-like networks lying on the deep fascia. To prevent stasis of lymph in the stump following amputation, the surgeon retains the deep fascia and its lymphatic vessels with the skin flaps that are used to cover the cut end of the extremity. This procedure results in minimal edema and swelling—a matter of obvious importance to the patient and artificial-limb fitter.

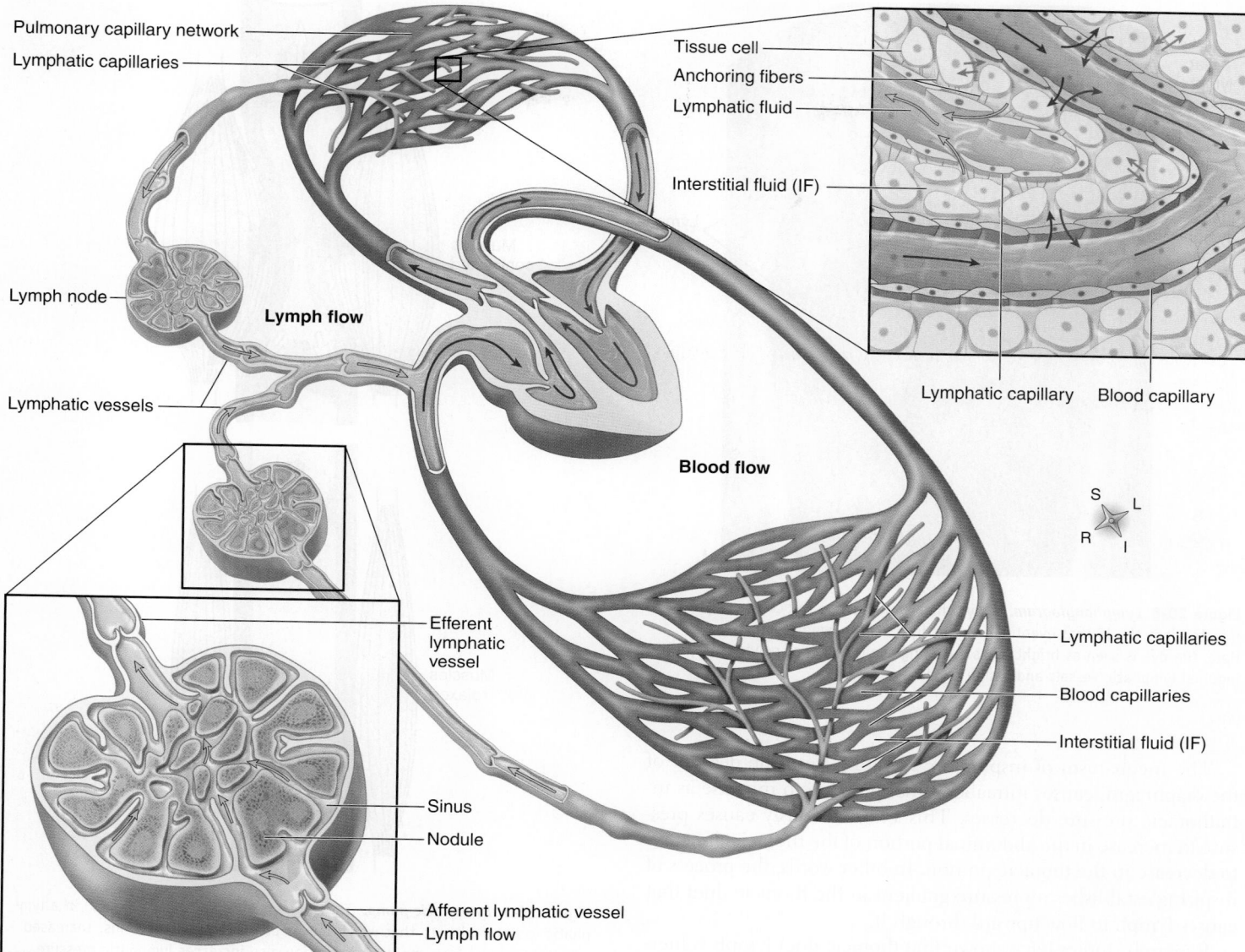

Pulmonary capillary network
Lymphatic capillaries

Tissue cell
Anchoring fibers
Lymphatic fluid

Interstitial fluid (IF)

Lymph node

Lymph flow

Lymphatic vessels

Blood flow

Lymphatic capillary   Blood capillary

Efferent lymphatic vessel

Lymphatic capillaries

Blood capillaries

Interstitial fluid (IF)

Sinus

Nodule

Afferent lymphatic vessel
Lymph flow

Figure 20-5 *Circulation plan of lymphatic fluid.* This diagram outlines the general scheme for lymphatic circulation. Fluids from the systemic and pulmonary capillaries leave the bloodstream and enter the interstitial space, thus becoming part of the IF (interstitial fluid). The IF also exchanges materials with the surrounding tissues. Often, because less fluid is returned to the capillary than moved into it, IF pressure increases—causing IF to flow into the lymphatic capillary. The fluid is now called lymph (lymphatic fluid) and is carried through one or more lymph nodes and finally to large lymphatic ducts. The lymph enters a subclavian vein, where it is returned to the systemic blood plasma. Thus fluid circulates through blood vessels, tissues, and lymphatic vessels in a sort of "open circulation."

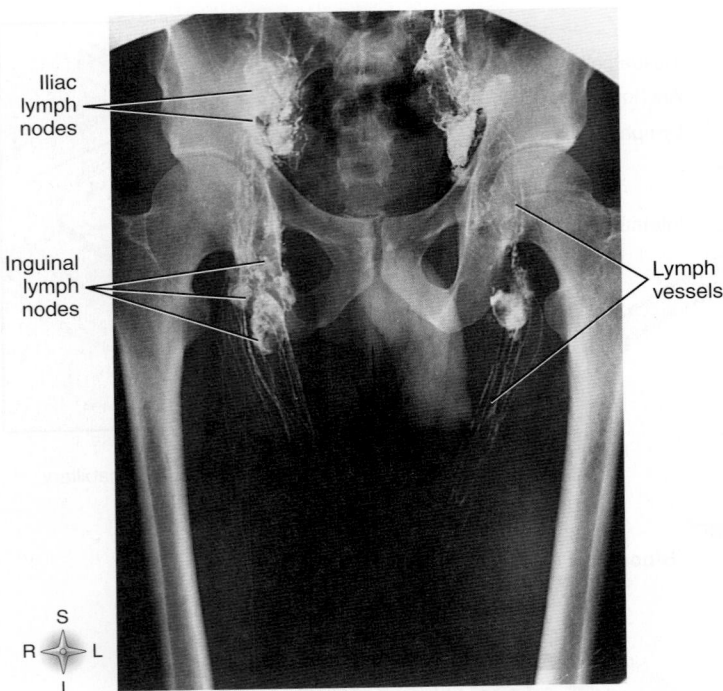

Figure 20-6 *Lymphangiogram.* A dye that is radiopaque (blocks x-rays) is injected into the IF that eventually drains into nearby lymphatic pathways. Here, the dye is seen as bright areas outlining the location of pelvic (iliac) and inguinal lymphatic vessels and lymph nodes.

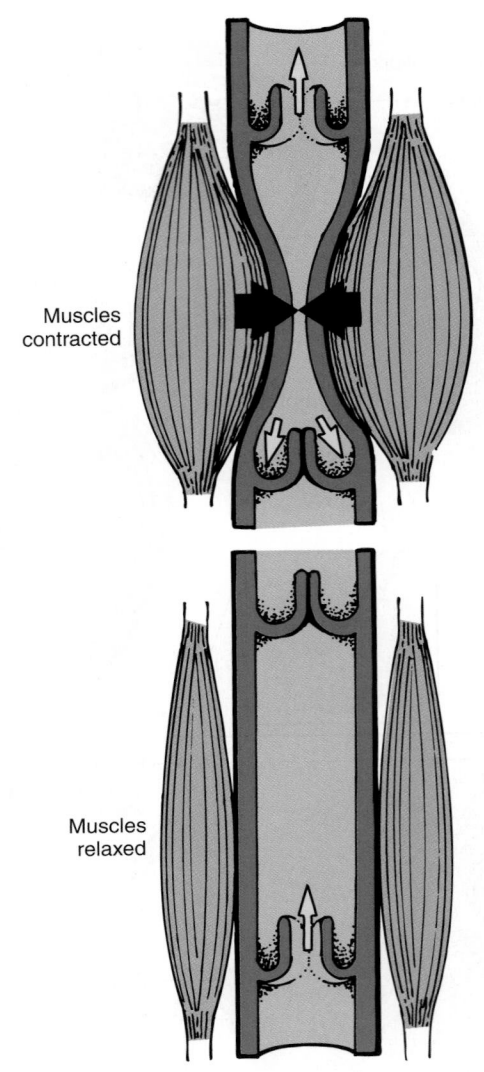

Figure 20-7 *Lymphatic pump.* The diagram shows a "muscle pump" in a lymphatic vessel similar to that which moves blood through the veins. Increased external pressure from muscle contraction also increases lymphatic pressure, pushing it past semilunar (SL) valves. Because the SL valves prevent backflow, the system becomes a pump that keeps lymph moving in one direction (toward a subclavian vein).

The mechanism of inspiration, resulting from the descent of the diaphragm, causes intraabdominal pressure to increase as intrathoracic pressure decreases. This simultaneously causes pressure to increase in the abdominal portion of the thoracic duct and to decrease in the thoracic portion. In other words, the process of inspiring establishes a pressure gradient in the thoracic duct that causes lymph to flow upward through it.

Research studies have shown that thoracic duct lymph is literally "pumped" into the venous system during the inspiration phase of pulmonary ventilation. The rate of flow, or ejection, of lymph into the venous circulation is proportional to the depth of inspiration. The total volume of lymph that enters the central veins during a given period depends on the depth of the inspiration phase and the overall breathing rate.

Most lymph flow in the body is the result of contracting skeletal muscles. As muscles contract, they "milk" the lymphatic vessels and push the lymph forward (Figure 20-7). The actual pressure generated in the system remains very low, and movement of lymph proceeds quite slowly when compared with the circulation of blood. During exercise, lymph flow may increase as much as tenfold to fifteenfold. In addition to lymph flow caused by skeletal muscle contractions, a very limited amount of smooth muscle exists in the walls of the large lymphatic trunks. Contraction of the smooth muscle in the thoracic vessel walls permits lymphatic vessels to pulse rhythmically and thus help move lymph from one valved segment to the next.

Other pressure-generating factors that can compress the lymphatics also contribute to the effectiveness of the "lymphatic pump." Examples of such lymphokinetic factors include interstitial fluid pressure (Figure 20-8), arterial pulsations, postural changes, and passive compression (massage) of the body soft tissues.

Although the volume of lymph that enters the bloodstream during each 24-hour period averages about 3 L, it may enter the system at different rates during the day. The rate of return varies, depending on the level of generalized physical activity and other factors, including changes in the interstitial fluid pressure and the rate and depth of respiration. As physical activity increases, so does the outflow of fluid from blood capillaries into the tissue spaces of the body. The increased flow of lymph that occurs with increased physical activity helps to return this fluid to the cardiovascular system and thus serves as an important balancing or homeostatic mechanism.

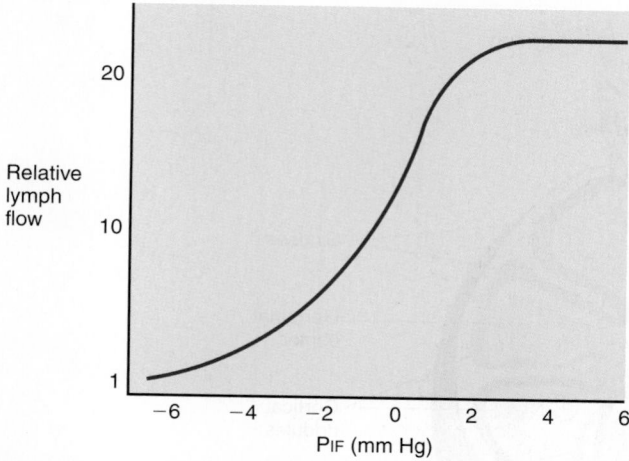

Figure 20-8 *Lymph flow and interstitial fluid (IF) pressure.* As the pressure of IF increases, the flow of lymph increases. Muscle pumps can increase IF pressure, as can other factors, such as accumulation of additional interstitial fluid volume within a tissue.

QUICK CHECK

1. What is the overall function of the lymphatic system?
2. What is the origin of lymph?
3. Compare lymphatic vessels with blood vessels.

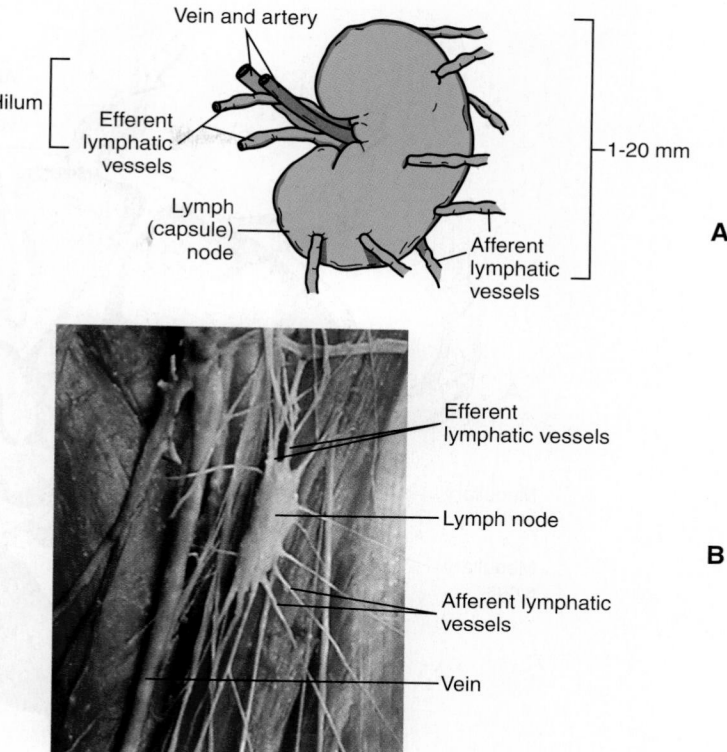

Figure 20-9 *External structure of a lymph node.* **A,** A lymph node is typically a small structure into which afferent lymphatic ducts empty their lymph and out of which efferent lymphatic ducts drain the lymph. An outer fibrous capsule maintains the structural integrity of the node. **B,** Photograph of a dissected cadaver shows a lymph node and its associated lymphatic vessels, along with nearby muscles, nerves, and blood vessels.

LYMPH NODES

Structure of Lymph Nodes

Lymph nodes, or **glands,** as some people call them, are oval-shaped or bean-shaped structures (Figure 20-9). Some are as small as a pinhead, and others are as large as a lima bean. Each lymph node (from 1 mm to more than 20 mm in diameter) is enclosed by a fibrous capsule. Note in Figure 20-10 that lymph moves into a node by way of several afferent lymphatic vessels and emerges by one efferent vessel. Think of a lymph node as a biological filter placed in the channel of several afferent lymph vessels (as you saw in Figure 20-5). Once lymph enters the node, it "percolates" slowly through the spaces known as sinuses before draining into the single efferent exit vessel. One-way valves in both the afferent and efferent vessels keep lymph flowing in one direction.

Fibrous septa, or *trabeculae*, extend from the covering capsule toward the center of the node. **Cortical nodules** within sinuses along the periphery, or cortex, of the node are separated from each other by these connective tissue trabeculae. Each cortical nodule is composed of packed lymphocytes that surround a less dense area called a *germinal center* (see Figure 20-10).

Figure 20-10, *B*, is a low-power (×35) light micrograph of a portion of a typical lymph node. When an infection is present, germinal centers form and the node begins to release lymphocytes. B lymphocytes (B cells) begin their final stages of maturation within

the less dense germinal center of the nodule and then are pushed to the more densely packed outer layers as they mature to become antibody-producing plasma cells. The center, or medulla, of a lymph node is composed of sinuses and medullary cords (see Figure 20-10). Both the cortical and medullary sinuses are lined with specialized reticuloendothelial cells (macrophages) capable of phagocytosis.

Locations of Lymph Nodes

With the exception of comparatively few single nodes, most lymph nodes occur in groups, or clusters, in certain areas. The group locations of greatest clinical importance are as follows:

- **Preauricular lymph nodes** are located just in front of the ear; these nodes drain the superficial tissues and skin on the lateral side of the head and face (Figure 20-11).
- **Submental group** and **submaxillary group** in the floor of the mouth—lymph from the nose, lips, and teeth drains through these nodes (Figure 20-11).
- **Superficial cervical lymph nodes** in the neck along the sternocleidomastoid muscle—these nodes drain lymph (which has already passed through other nodes) from the head and neck (see Figures 20-11 and 20-12).

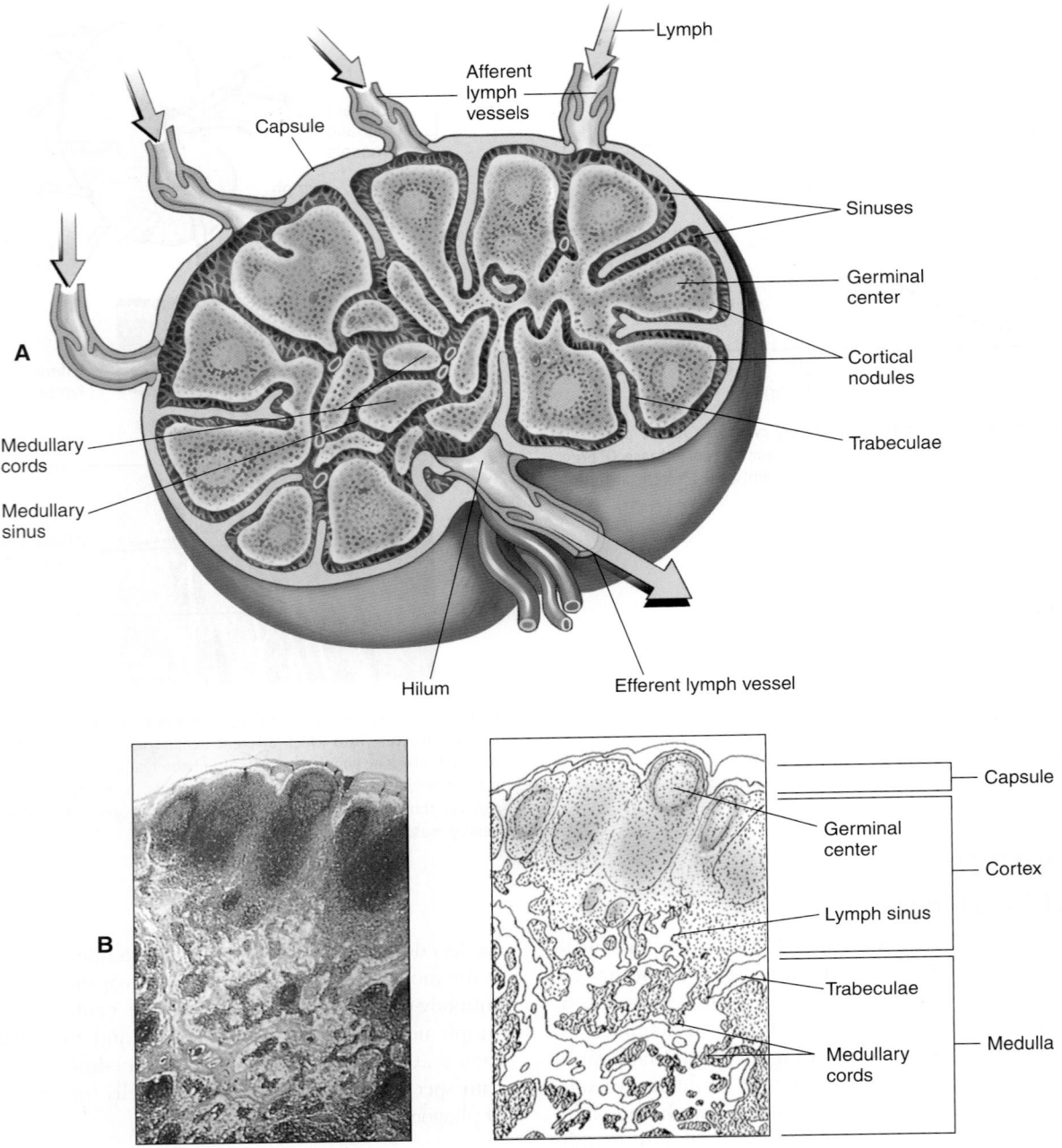

Figure 20-10 *Internal structure of a lymph node.* **A,** Several afferent valved lymphatics bring lymph to the node. In this example, a single efferent lymphatic leaves the node at a concave area called the *hilum*. Note that the artery and vein also enter and leave at the hilum. Arrows show direction of lymph. **B,** Photomicrograph and drawing showing a portion of the cortex and medulla of a lymph node.

- **Superficial cubital lymph nodes (supratrochlear lymph nodes)** located just above the bend of the elbow—lymph from the forearm passes through these nodes (see Figure 20-2).
- **Axillary lymph nodes** (20 to 30 large nodes clustered deep within the underarm and upper chest regions)—lymph from the arm and upper part of the thoracic wall, including the breast, drains through these nodes (see Figure 20-2).

- **Iliac lymph nodes** and **inguinal lymph nodes** in the pelvis and groin—lymph from the pelvic organs, legs, and external genitalia drains through these nodes (Figures 20-6 and 20-13).

Functions of Lymph Nodes

Lymph nodes perform at least two distinct functions: defense and hematopoiesis.

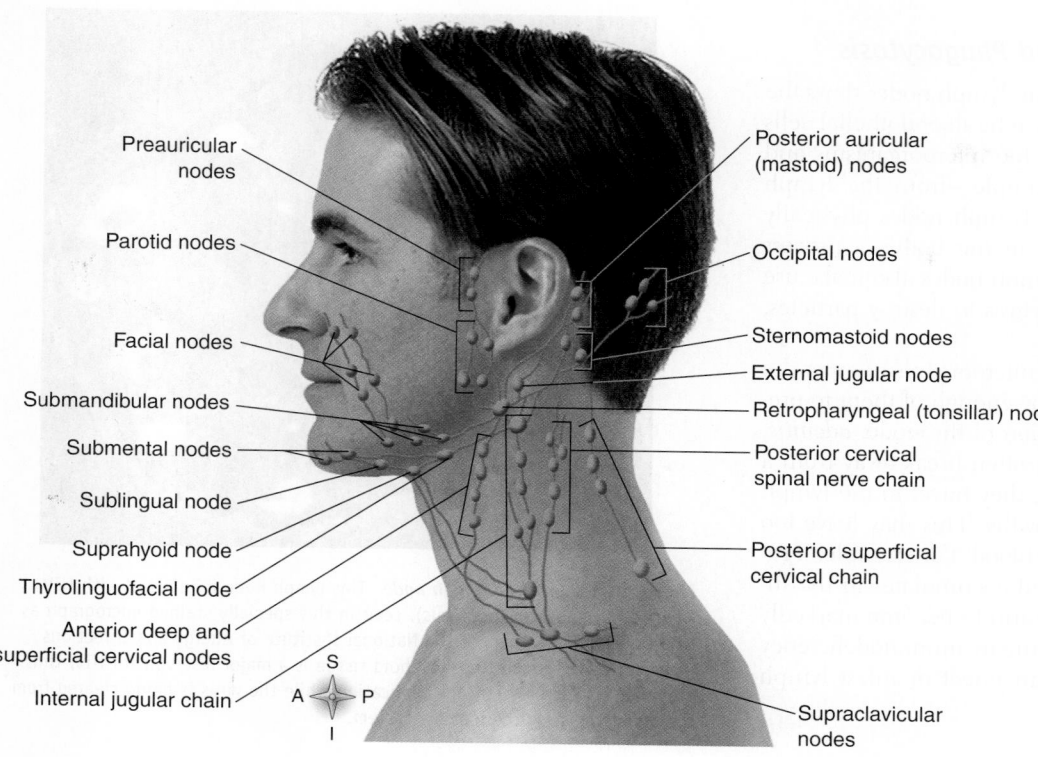

Preauricular nodes

Parotid nodes

Facial nodes

Submandibular nodes

Submental nodes

Sublingual node

Suprahyoid node

Thyrolinguofacial node

Anterior deep and superficial cervical nodes

Internal jugular chain

Posterior auricular (mastoid) nodes

Occipital nodes

Sternomastoid nodes

External jugular node

Retropharyngeal (tonsillar) node

Posterior cervical spinal nerve chain

Posterior superficial cervical chain

Supraclavicular nodes

Figure 20-11 *Lymphatic drainage of the head and neck.* The head and neck contain many lymph nodes (and associated lymphatic vessels) that are often of clinical significance in certain infections and cancers.

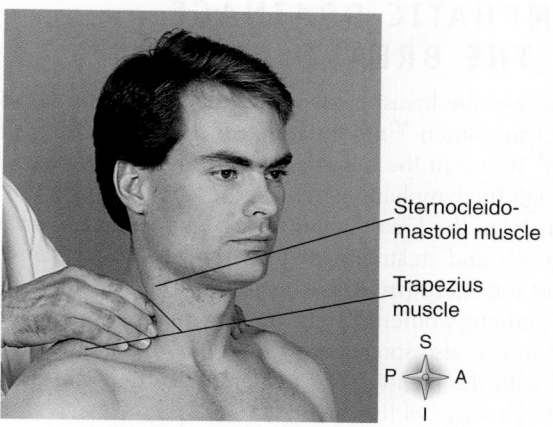

Sternocleido-mastoid muscle

Trapezius muscle

Figure 20-12 *Palpation of the posterior cervical lymph nodes.* The pads of the fingertips are used to palpate (feel) along the anterior surface of the trapezius muscle and then moved slowly in a circular motion to the posterior surface of the sternocleidomastoid muscle. Swollen lymph nodes may be a sign of infection, cancer, blocked lymphatic drainage, or some other health condition.

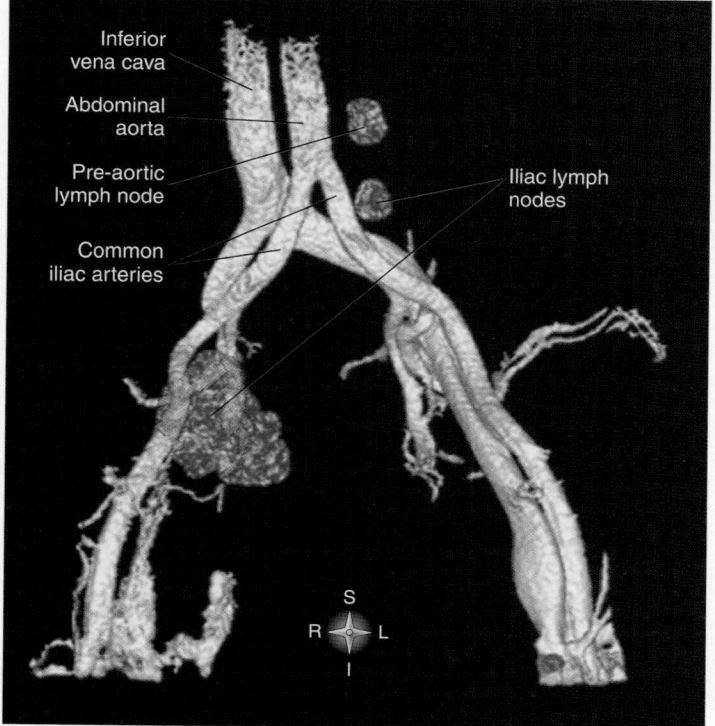

Inferior vena cava

Abdominal aorta

Pre-aortic lymph node

Common iliac arteries

Iliac lymph nodes

Figure 20-13 *MRI of lymph nodes.* This three-dimensional MRI (magnetic resonance image) has been enhanced by the use of specially engineered nanoparticles that are injected into the blood and are later ingested by macrophages in the lymphoid tissue of the lymph nodes. In lymphoid tissue containing metastasized cancer cells, there is a recognizable, abnormal pattern calculated in the MRI software (color coded red in here). Thus cancerous lymph nodes can be identified and accurately located for surgical removal or other therapy.

Defense Functions: Filtration and Phagocytosis

The structure of the sinus channels within lymph nodes slows the lymph flow through them. This gives the reticuloendothelial cells that line the channels time to remove the microorganisms and other injurious particles—soot, for example—from the lymph and phagocytose them (Figure 20-14). Lymph nodes physically stop particles from progressing farther in the body—a process called *mechanical filtration*. Because lymph nodes also make use of biological processes such as phagocytosis to destroy particles, *biological filtration* also occurs here.

Sometimes, however, such hordes of microorganisms enter the nodes that the phagocytes cannot destroy enough of them to prevent their injuring the node. An infection of the node, *adenitis*, then results. Also, because cancer cells often break away from a malignant tumor and enter lymphatics, they travel to the lymph nodes, where they may set up new growths. This may leave too few channels for lymph to return to the blood. For example, if tumors block axillary node channels, fluid accumulates in the interstitial spaces of the arm, causing the arm to become markedly swollen. Even viruses such as HIV (human immunodeficiency virus) and other types of pathogens can infect or infest lymph nodes, as seen in Figure 20-15.

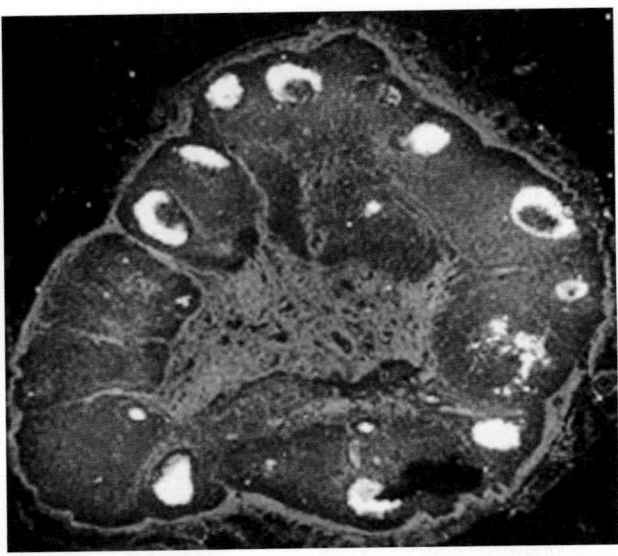

Figure 20-15 *Infected lymph node.* This lymph node is infected with HIV (human immunodeficiency virus), seen in this specially stained micrograph as white areas. Researchers at the National Institute of Allergy and Infectious Disease (NIAID) found that lymphoid tissue is a major reservoir for HIV, thus providing a sanctuary for viral replication while the virus is being cleared from the bloodstream by the immune system.

Hematopoiesis

The lymphoid tissue of lymph nodes serves as the site of the final stages of maturation for some types of lymphocytes and monocytes that have migrated from the bone marrow. In addition to lymph nodes and the specialized lymphatic organs described later in the chapter, small aggregates of diffuse lymphoid tissue and other lymphatic cell types are found throughout the body—especially in connective tissues and under mucous membranes.

LYMPHATIC DRAINAGE OF THE BREAST

Cancer of the breast is one of the most common forms of malignancy in women. Unfortunately, cancerous cells from a single "primary" tumor in the breast often spread to other areas of the body through the lymphatic system. An understanding of the lymphatic drainage of the breast is therefore of particular importance in the diagnosis and treatment of this type of malignancy (Box 20-3). Breast infections (**mastitis**) are also a serious health concern, especially among women who are nursing infants. Breast infections, like cancer, can also spread easily through lymphatic pathways associated with the breast. Refer to Figure 20-16 as you study the lymphatic drainage of the breast.

Distribution of Lymphatics in the Breast

The breast—mammary gland and surrounding tissues—is drained by the following two sets of lymphatic vessels:

1. Lymphatics that originate in and drain the skin over the breast with the exception of the areola and nipple
2. Lymphatics that originate in and drain the substance of the breast itself, as well as the skin of the areola and nipple

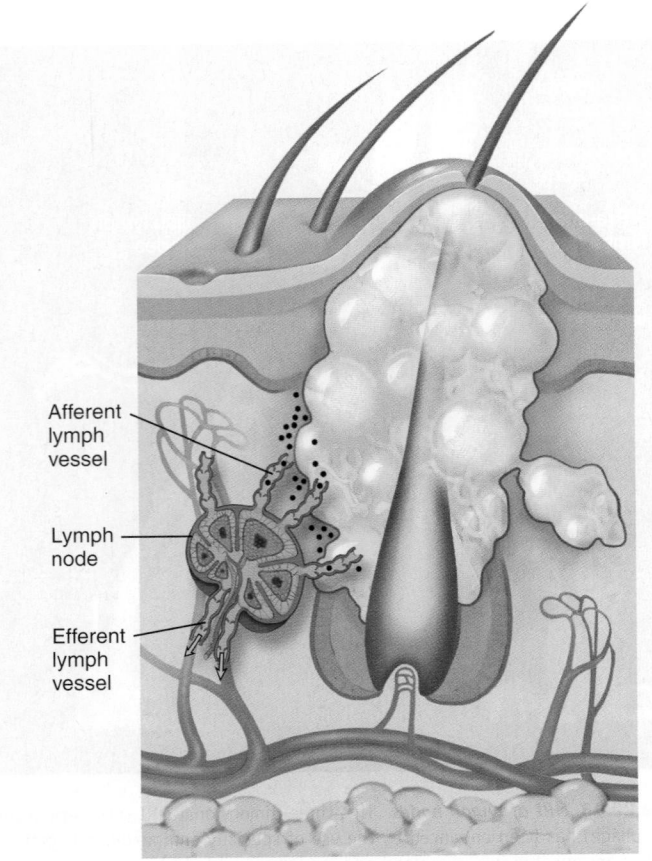

Figure 20-14 *Role of a lymph node in a skin infection.* Yellow areas represent dead and dying cells (pus). Black dots around the yellow areas represent bacteria. Leukocytes phagocytose many bacteria in tissue spaces. Others may enter the lymph nodes by way of afferent lymphatics. The nodes filter out these bacteria; here reticuloendothelial cells usually destroy them all by phagocytosis. (Lymph node is shown smaller than actual size.)

Afferent lymph vessel

Lymph node

Efferent lymph vessel

BOX 20-3: HEALTH MATTERS
Lymphedema after Breast Surgery

Surgical procedures called *mastectomies,* in which some or all of the breast tissues are removed, are sometimes performed to treat breast cancer. Because cancer cells can spread so easily through the extensive network of lymphatic vessels associated with the breast (see Figure 20-16), the lymphatic vessels and their nodes are sometimes also removed. Occasionally, such procedures interfere with the normal flow of lymph fluid from the arm. When this happens, tissue fluid may accumulate in the arm—resulting in swelling, or **lymphedema.** Fortunately, adequate lymph drainage is almost always restored by the reestablishment of new lymphatic vessels, which grow back into the area.

BOX 20-4: HEALTH MATTERS
Lymphatic Anastomoses and Breast Cancer

Anastomoses (connections) occur between superficial lymphatics from both breasts across the middle line. Such communication can result in the spread of cancerous cells in one breast to previously healthy tissue in the other breast.

Both superficial and deep lymphatic vessels also communicate with lymphatics in the fascia of the pectoralis major muscle. Removal of a wide area of deep fascia is therefore required in surgical treatment of advanced or diffuse breast malignancy (radical mastectomy). In addition, cancer cells from a breast tumor sometimes reach the abdominal cavity because of lymphatic communication through the upper part of the linea alba.

Superficial vessels that drain lymph from the skin and surface areas of the breast converge to form a diffuse *cutaneous lymphatic plexus.* Communication between the cutaneous plexus and large lymphatics that drain the secretory tissue and ducts of the breast occurs in the *subareolar plexus (plexus of Sappey)* located under the areola surrounding the nipple.

Box 20-4 discusses the numerous connections with the breast lymphatic pathways.

Lymph Nodes Associated with the Breast

More than 85% of the lymph from the breast enters the lymph nodes of the axillary region (see Figure 20-16). Most of the remainder enters lymph nodes along the lateral edges of the sternum.

Several very large nodes in the axillary region are in actual physical contact with an extension of breast tissue called the *axillary tail (of Spence).* Because of the physical contact between

these nodes and breast tissue, cancerous and infectious cells may spread by both lymphatic extension and contiguity of tissue. Other nodes in the axilla will enlarge and swell after being "seeded" with malignant cells or bacteria as lymph from a cancerous or infected breast flows through them.

QUICK CHECK

4. Describe the overall structure of a typical lymph node.
5. Where are lymph nodes usually found?
6. What functions are carried out by lymph nodes?
7. How do the lymphatic structures of the breast relate to breast cancer?

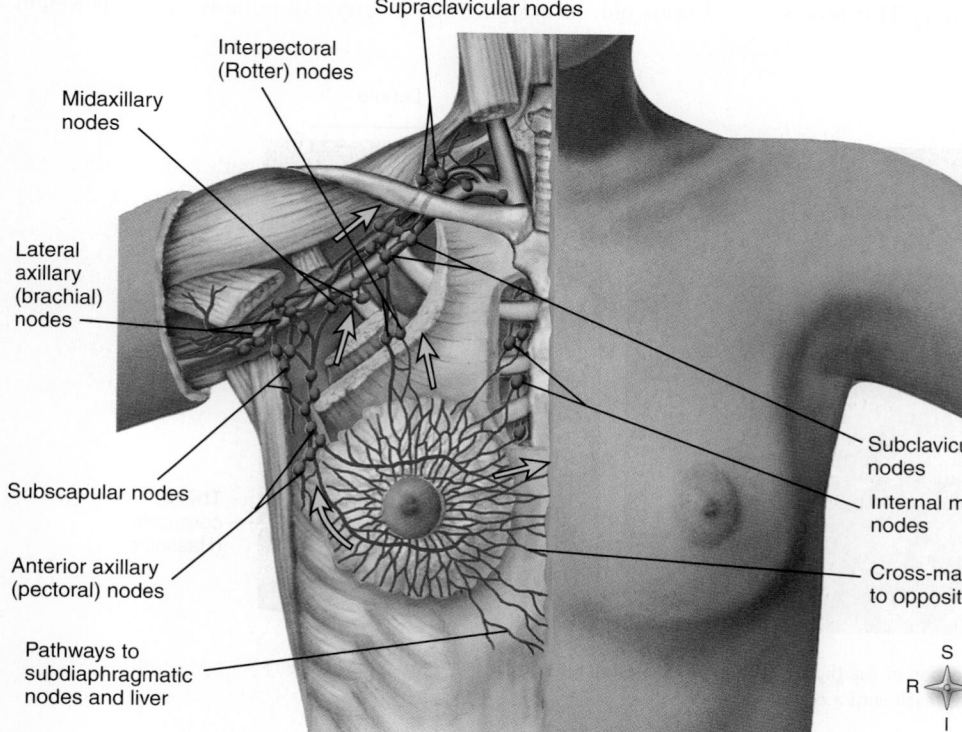

Figure 20-16 *Lymphatic drainage of the breast.* Note the extensive network of lymphatic vessels and nodes that receive lymph from the breast.

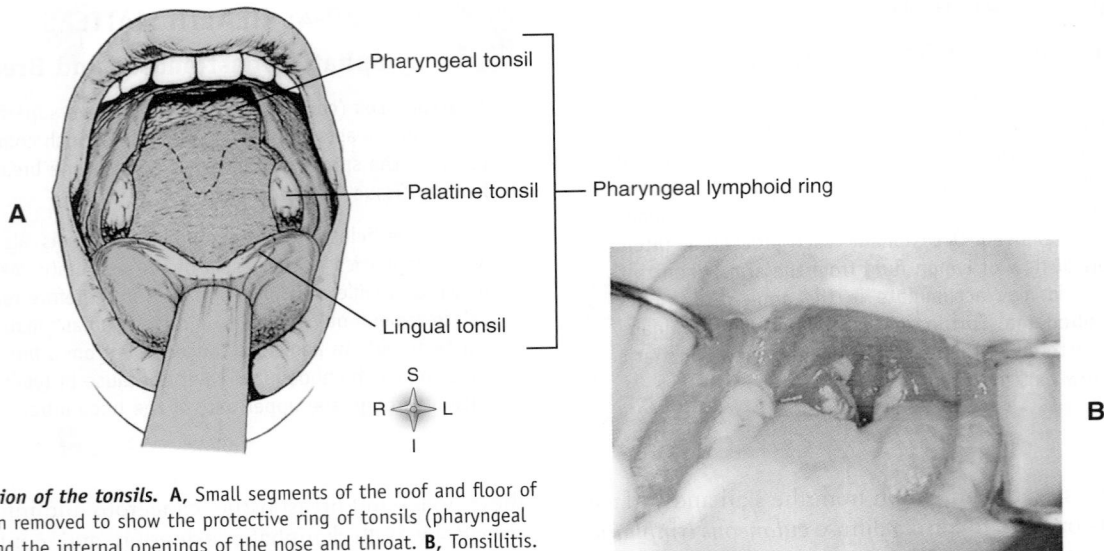

Figure 20-17 *Location of the tonsils.* **A,** Small segments of the roof and floor of the mouth have been removed to show the protective ring of tonsils (pharyngeal lymphoid ring) around the internal openings of the nose and throat. **B,** Tonsillitis. Note the swelling and presence of a white coating (exudate) indicating an infection.

TONSILS

Masses of lymphoid tissue, called **tonsils,** are located in a protective ring under the mucous membranes in the mouth and back of the throat (Figure 20-17, A). This ring is called the *pharyngeal lymphoid ring.* The ring of tonsils helps protect against bacteria that may invade tissues in the area around the openings between the nasal and oral cavities. The **palatine tonsils** are located on each side of the throat. The **pharyngeal tonsils,** known as *adenoids* when they become swollen, are near the posterior opening of the nasal cavity. A third type of tonsil, the **lingual tonsils,** is near the base of the tongue. The tonsils serve as the first line of defense from the exterior and as such are subject to chronic infection, or tonsillitis (Figure 20-17, B). They are sometimes removed surgically if antibiotic therapy is not successful or if swelling impairs breathing. This procedure, called **tonsillectomy,** has become controversial because of the critical immunological role played by the lymphoid tissue.

THYMUS
Location and Appearance of the Thymus

Intensive study and experimentation have identified the **thymus** as a primary organ of the lymphatic system. It is an unpaired organ consisting of two pyramidal lobes with delicate and finely lobulated surfaces. The thymus is located in the mediastinum, extending up into the neck as far as the lower edge of the thyroid gland and inferiorly as far as the fourth costal cartilage (Figure 20-18, A). Its size relative to the rest of the body is largest in a child about 2 years old. Its absolute size is largest at puberty, when its weight

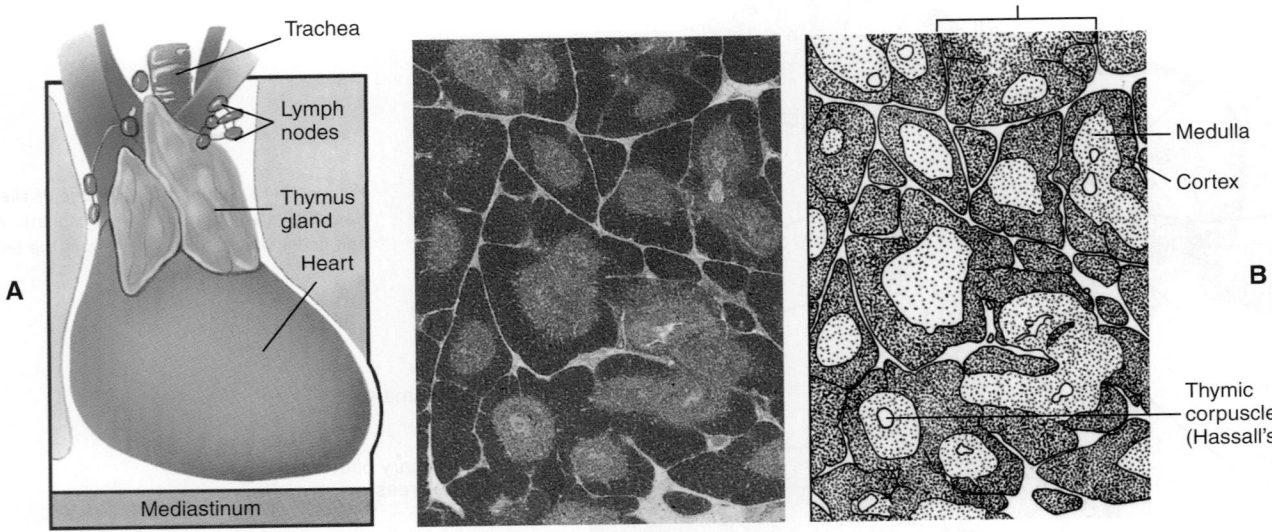

Figure 20-18 *Thymus.* **A,** Location of the thymus within the mediastinum. **B,** Microscopic structure of the thymus showing several lobules, each with a cortex and a medulla.

ranges between 35 and 40 g. From then on, it gradually atrophies until, in advanced old age, it may be largely replaced by fat, weigh less than 10 g, and be barely recognizable. The process of shrinkage of an organ in this manner is called **involution.** The thymus is pinkish gray in color early in childhood but, with advancing age, becomes yellowish as lymphoid tissue is replaced by fat.

Structure of the Thymus

The lobes of the thymus are subdivided into small (1- to 2-mm) lobules by connective tissue septa that extend inward from a fibrous covering capsule. Each lobule is composed of a dense cellular cortex and an inner, less dense medulla (Figure 20-18, *B*). Both cortex and medulla are composed of lymphocytes in an epithelial framework quite different from the supporting connective tissue seen in other lymphoid organs.

In stained histological sections of thymus, medullary tissue can be identified by the presence of rather large (30- to 150-μm) laminated spherical structures called **thymic corpuscles,** or *Hassall's corpuscles.* Composed of concentric layers of keratinized epithelial cells, thymic corpuscles have a unique onion-like appearance. These corpuscles may serve as a place to break down dead, keratinized epithelial cells migrating inward from the outer parts of each lobule.

Function of the Thymus

One of the body's best-kept secrets has been the function of the thymus. Before 1961 there were no significant clues as to its role. Then a young Briton, Dr. Jacques F.A.P. Miller, removed the thymus glands from newborn mice. His findings proved startling and crucial. Almost like a chain reaction, further investigations followed, leading to the gradual uncovering of the thymus's longheld secrets. It is now clear that this small structure plays a critical part in the body's defenses against infections—in its vital immunity mechanism (see Chapter 21).

The thymus performs at least two important functions. First, it serves as the final site of lymphocyte development before birth. (The fetal bone marrow forms immature lymphocytes, which then move to the thymus.) Many lymphocytes leave the thymus and circulate to the spleen, lymph nodes, and other lymphoid tissues. Second, soon after birth the thymus begins secreting a group of hormones (collectively called *thymosin*) and other regulators that enable lymphocytes to develop into mature T cells. Only T cells (T lymphocytes) that pass immunological testing by lymphoid cells such as *macrophages* and *dendritic cells*—only about 5% of the cells that mature each day—are released into the bloodstream. Figure 20-19 outlines some essential steps in T cell development within the thymus.

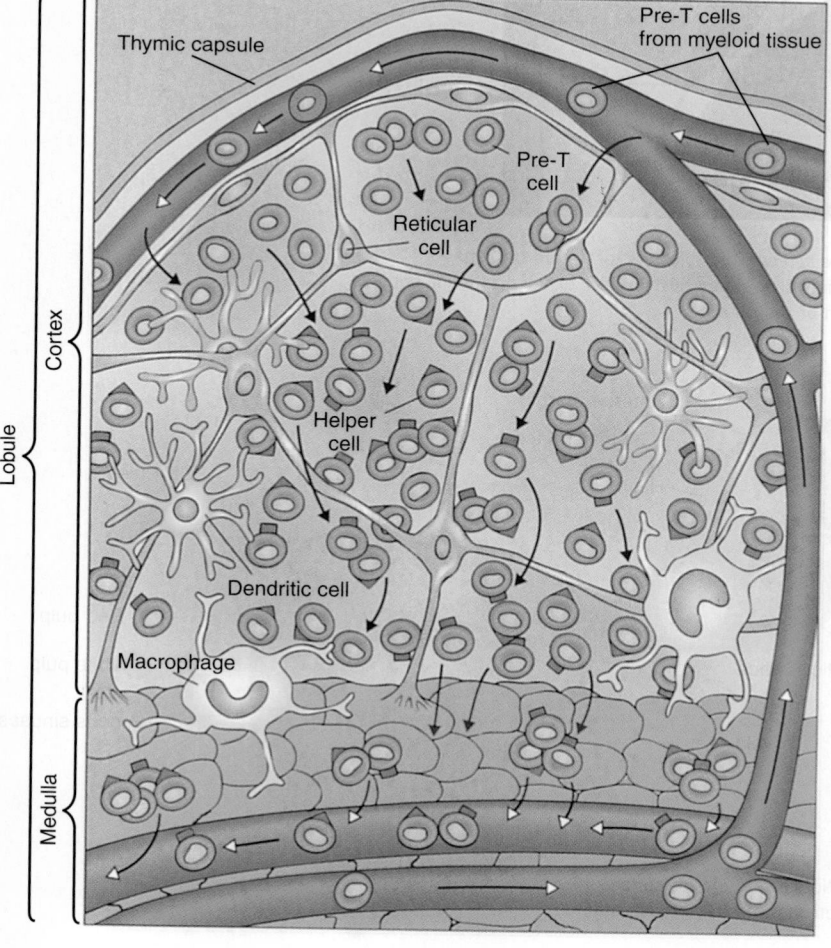

Figure 20-19 *T-cell development in the thymus.* This diagram shows part of a lobule, with the cortex at the top and the medulla at the bottom. Pre-T cells derived from hemopoietic stem cells in red bone marrow (myeloid tissue) travel through the bloodstream to the cortex of the thymus. Under the influence of thymosin and other regulators, the pre-T cells divide and mature as they migrate toward the medulla. Along the way, the developing T cells (T lymphocytes) are tested for immune capability against various lymphoid immune cells, such as dendritic cells and macrophages. Only about 5% of the cells pass these tests and are released into the bloodstream to defend the body.

Because T cells attack foreign or abnormal cells and also serve as regulators of immune function, the thymus functions as an important part of the immune mechanism. It probably completes much of its essential work early in childhood.

SPLEEN

Location of the Spleen

The **spleen** is located in the left hypochondrium of the abdominopelvic cavity, directly below the diaphragm. The spleen is just above most of the left kidney and the descending colon and behind the fundus of the stomach (Figures 20-2 and 20-20). In addition, it is common to find small *accessory spleens* embedded in the double fold of serous membrane that connects the spleen and stomach.

Structure of the Spleen

As Figures 20-20 and 20-21 show, the spleen is roughly ovoid in shape. Its size varies in different individuals and in the same individual at different times (Box 20-5). For example, it hypertrophies during infectious diseases and atrophies in old age.

Like other lymphoid organs, the spleen is surrounded by a fibrous capsule with inward extensions that roughly divide the organ into compartments. One such compartment is shown in Figure 20-21, *B*. Arteries leading into each compartment are surrounded by dense masses (nodules) of developing lymphocytes. Because of its whitish appearance, this tissue is called *white pulp*. Near the outer regions of each compartment is tissue called *red pulp*, made up of a network of fine reticular fibers submerged in blood that

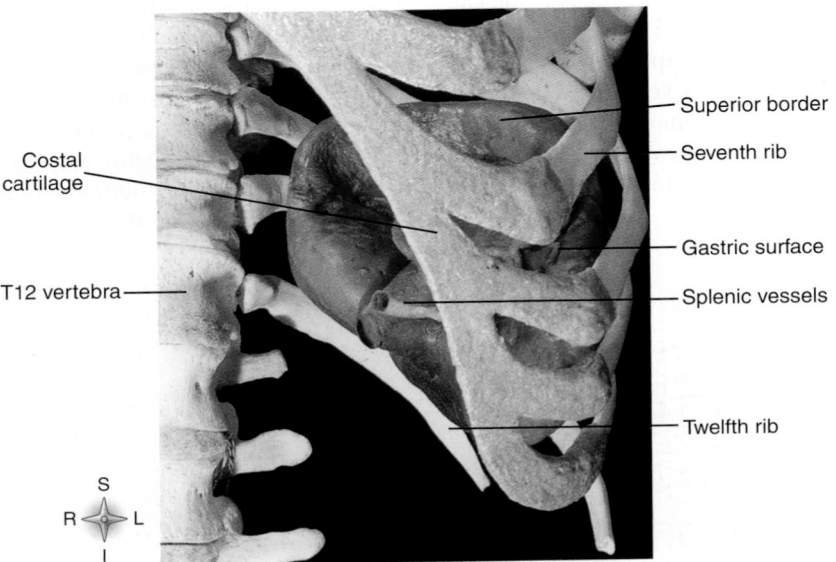

Figure 20-20 *Location of the spleen.* The spleen is located in the left hypochondrium of the abdominopelvic cavity, just inferior to the diaphragm and just deep to the lower portion of the rib cage.

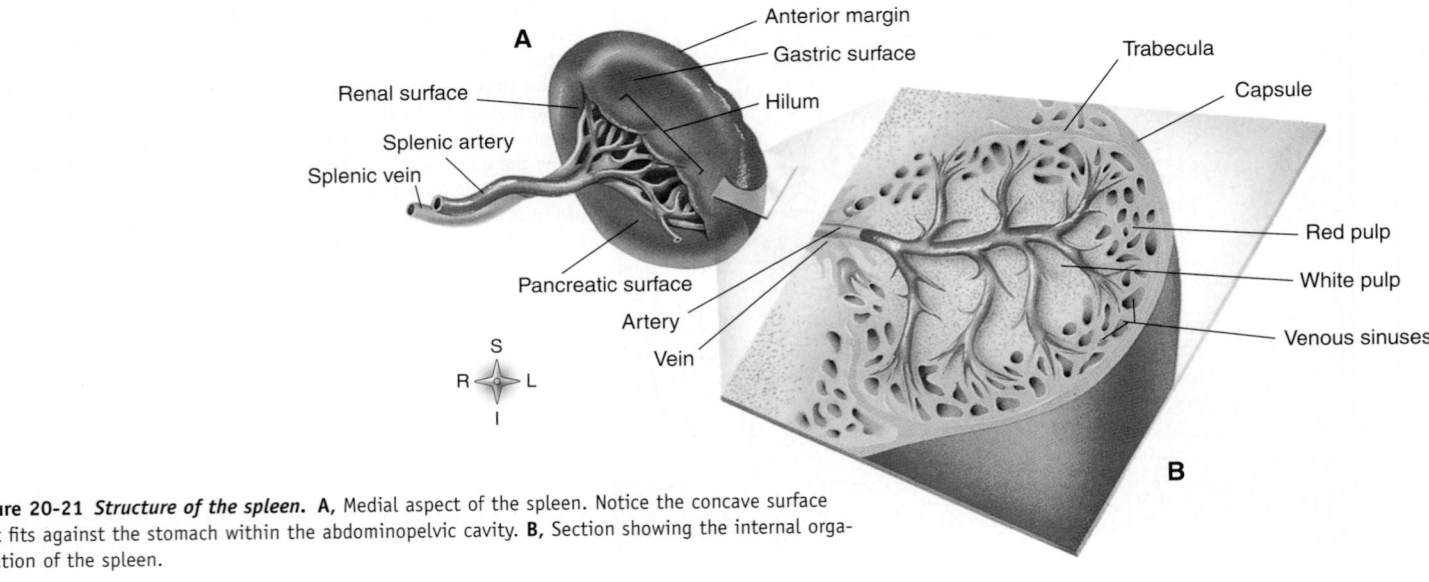

Figure 20-21 *Structure of the spleen.* **A,** Medial aspect of the spleen. Notice the concave surface that fits against the stomach within the abdominopelvic cavity. **B,** Section showing the internal organization of the spleen.

comes from the nearby arterioles. After passing through the reticular meshwork, blood collects in venous sinuses and then returns to the heart through veins.

Functions of the Spleen

The spleen has many and sundry functions, including defense, hematopoiesis, and red blood cell and platelet destruction; it also serves as a reservoir for blood.

- **Defense.** As blood passes through the sinusoids of the spleen, reticuloendothelial cells (macrophages) lining these venous spaces remove microorganisms from the blood and destroy them by phagocytosis. Therefore the spleen plays a part in the body's defense against microorganisms.
- **Hematopoiesis.** Nongranular leukocytes, that is, monocytes and lymphocytes, complete their development to become activated in the spleen. Before birth, red blood cells are also formed in the spleen, but, after birth, the spleen forms red blood cells only in extreme hemolytic anemia.
- **Red blood cell destruction** and **platelet destruction.** Macrophages lining the spleen's sinusoids remove worn-out red blood cells and imperfect platelets from the blood and destroy them by phagocytosis. They also break apart the hemoglobin molecules from the destroyed red blood cells and salvage their iron and globin content by returning them to the bloodstream for storage in bone marrow and liver.

- **Blood reservoir.** At any given point in time the pulp of the spleen and its venous sinuses contain a considerable amount of blood. Although continually moving slowly through the spleen, blood can rapidly be added back into the circulatory system from this functional reservoir if needed. Its normal volume of about 350 ml is said to decrease about 200 ml in less than 1 minute's time following sympathetic stimulation that produces marked constriction of its smooth-muscle capsule. This "self-transfusion" occurs, for example, as a response to the stress imposed by hemorrhage.

Although the spleen's functions make it a most useful organ, it is not a vital one. Dr. Charles Austin Doan in 1933 took the daring step of performing the first **splenectomy.** He removed the spleen from a 4-year-old girl who was dying of hemolytic anemia. Presumably, he justified his radical treatment on the basis of what was then merely conjecture, that is, that the spleen destroys red blood cells. The child recovered, and Dr. Doan's operation proved to be a landmark. It created a great upsurge of interest in the spleen and led to many investigations of this organ.

Cycle of Life
Lymphatic System

Many of the structural features of the lymphatic system exhibit dramatic changes as a person progresses through a life span. Most of the organs containing masses of developing lymphocytes appear before birth and continue growing through most of childhood until just before puberty. After puberty, these lymphoid organs typically begin to slowly atrophy until they reach much smaller size by late adulthood. These organs—including the thymus, lymph nodes, tonsils, and other lymphoid structures—shrink in size and become fatty or fibrous. The notable exception to this principle is the spleen, which develops early in life and remains intact until very late adulthood.

Despite the fact that lymphocyte-producing lymphoid tissues decline after puberty, the overall function of the immune system is maintained until late adulthood. During the late adult years, deficiency of the immune system permits a greater risk of infections and cancer, and hypersensitivity of the immune system may make autoimmune conditions more likely to occur.

THE BIG PICTURE
Lymphatic System and the Whole Body

One way to imagine the role of the lymphatic system in the "society of cells" that makes up the human body is as a sort of waste-water system. Like waste-water systems used in the cities of human society, the lymphatic system drains away excess, or "runoff," water from large areas. After collecting the body's runoff, or lymph, the lymphatic system conducts it through a network of lymphatic vessels, or "drain pipes," to miniature "treatment facilities" called lymph nodes. Contaminants are there removed from lymph, just as contaminants are removed in a waste-water treatment plant. The "clean" fluid is then returned to the bloodstream much as clean waste water is returned to a nearby river or lake. Like municipal waste-water systems, the lymphatic system not only prevents dangerous fluid buildups, or "floods," but also prevents the spread of disease.

All systems of the body benefit from the fluid-balancing and immune functions of the lymphatic system. Some parts of the body such as the digestive and respiratory tracts make special use of the defensive capacities of lymphatic organs such as aggregated lymph nodules (Peyer's patches) and tonsils. Likewise, body structures such as the breasts and limbs make more use of the fluid-draining capacities of the lymphatic system than do other regions of the body. Overall, however, the entire body benefits from the fluid balance and freedom from disease conferred by the proper functioning of this important body system.

Mechanisms of Disease

DISORDERS OF THE LYMPHATIC SYSTEM

Disorders Associated with Lymphatic Vessels

Lymphedema is an abnormal condition in which swelling of tissues in the extremities occurs because of an obstruction of the lymphatics and accumulation of lymph (Figure 20-22). The most common type of lymphedema is *congenital lymphedema* (lymphedema praecox), more often seen in women between the ages of 15 and 25 years. The obstruction in lymphedema can be in both the lymphatic vessels and lymph nodes themselves. Initially, the swelling, or edema, in the extremity will be soft, but as the condition progresses, it becomes firm, painful, and unresponsive to treatment. Frequent infections, involving high fever and chills, may occur with chronic lymphedema. Diuretics (agents that cause water loss) to reduce the swelling have been shown to be effective, along with strict bed rest, massage, and elevation of the involved extremities. If the edema is severe and unresponsive to these measures, or infection has occurred, or the person's mobility is severely compromised, surgical removal of the involved subcutaneous tissue and fascia may be required. Other procedures involving surgically "shunting" of superficial lymphatic drainage into the deep lymphatic system have been tried.

Lymphedema may be caused by small parasitic worms called **filaria** that infest the lymph vessels. This condition is rare in most of North America and is more often seen in the tropics. The flow of lymph is blocked, causing edema in the affected extremities that, in severe cases, become so swollen that they resemble an elephant's limbs (Figure 20-23). For this reason, the condition is referred to as **elephantiasis**—literally "condition of being an elephant." Chronic swelling, thicken-

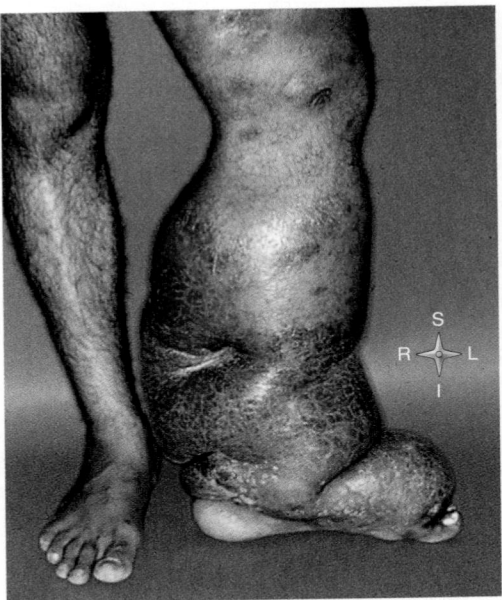

Figure 20-23 *Elephantiasis.* Prolonged infestation of the lymphatic system by *Filaria* worms produces so much swelling (lymphedema) that the affected limbs begin to resemble those of an elephant!

ing of the subcutaneous tissue, and frequent bouts of infections are common in this condition.

Lymphangitis, an acute inflammation of the lymphatic vessels, stems from invasion of an infectious organism. This condition is characterized by thin, red streaks extending from an infected region up the arm or leg (Figure 20-24). The lymph nodes also become enlarged, tender, and reddened. Necrosis, or tissue death, along with development of an abscess (collection of fluid and pus) can occur, leading to a condition known as *suppurative lymphadenitis*. The lymph nodes commonly involved are in the groin, axilla, and cervical regions. The infectious agents that cause lymphangitis may eventually spread into the bloodstream, causing *septicemia* (blood poisoning) and possible death from septic shock, but this is rare if the proper antibiotic therapy is initiated.

Disorders Associated with Lymph Nodes and Other Lymphatic Organs

Lymphoma is a term that refers to a tumor of the cells of lymphoid tissue. Lymphomas are often malignant but, in rare cases, can be benign. They usually originate in isolated lymph nodes but can involve lymphoid tissue in the liver, spleen, and gastrointestinal tract. Widespread involvement is common be-

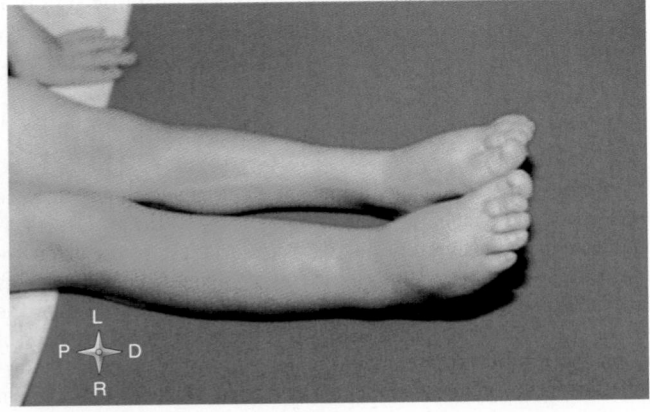

Figure 20-22 *Lymphedema.* Notice the significant swelling in the subject's right leg and foot.

Mechanisms of Disease—cont.

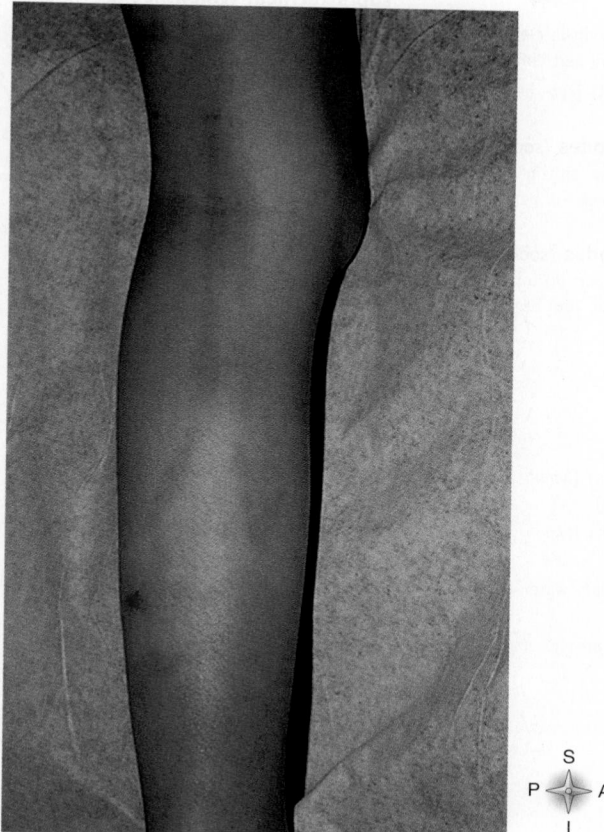

Figure 20-24 *Lymphangitis.* Red streaks mark the location of inflamed lymphatic vessels in this leg.

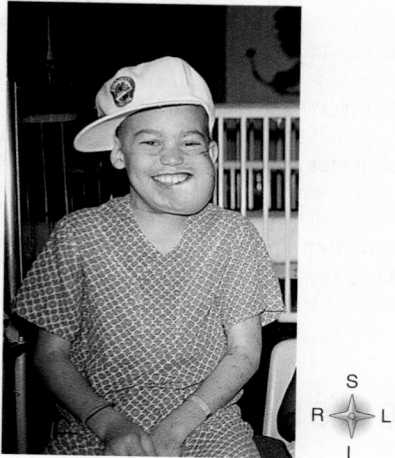

Figure 20-25 *Hodgkin lymphoma.* Enlarged lymph nodes in the neck or axilla characterize this condition.

cause the disease spreads from node to node through the many anastomoses of the lymphatic vessels throughout the body. The exact cause of these neoplasms is unknown.

Two principal categories of lymphomas are *Hodgkin lymphoma (HL)* and *non-Hodgkin lymphoma (NHL)*. Hodgkin lymphoma (or **Hodgkin disease**) is a malignancy with an uncertain etiology. Some pathophysiologists believe that it originates as a pathogen-induced tumor of T cells, although there is currently no evidence to support this conclusively. Other factors such as exposure to chemicals or other environmental hazards may also be involved. This condition usually begins as painless, nontender, enlarged lymph nodes in the neck or axilla (Figure 20-25). Soon, lymph nodes in other regions enlarge in the same manner. If they involve the trachea or esophagus, pressure results in difficulty breathing or swallowing. HL is considered to be one of the most curable forms of cancer if detected early.

Lymphedema caused by blockage of lymph nodes may cause enlargement of the extremities. Occasionally the disease may obstruct flow into or out of the liver, leading to liver enlargement and failure. Anemia, leukocytosis, fever, and weight loss occur as the condition progresses. Hodgkin lymphoma is potentially curable with radiation therapy, provided it has not spread beyond the lymphatic system. Chemotherapy is used in addition to radiation therapy in more advanced cases. Infection, from both the disease and the treatments, is a common complication.

Non-Hodgkin lymphoma is the name given to a malignancy of lymphoid tissue other than Hodgkin lymphoma. Again, the etiology is uncertain but has been hypothesized to be caused by a virus. Patients with immunodeficiencies such as AIDS often develop this condition. Manifestations are similar to Hodgkin lymphoma, but there is usually a more generalized involvement of lymph nodes. The central nervous system is also often involved. Radiation and chemotherapy are treatments of choice.

The tonsils, composed of lymphoid tissue, serve as the first line of defense from the exterior and also are subject to acute or chronic infection, known as **tonsillitis.** Fever, sore throat, and difficulty swallowing are common signs and symptoms. Enlarged pharyngeal tonsils *(adenoids)* may cause nasal obstruction. The infection may extend to the middle ear by way of the auditory (eustachian) tubes, causing *acute otitis media* (middle ear infection) and possible deafness if left untreated. Antibiotics are usually initiated after diagnosis of tonsillitis. If these are unsuccessful, and swelling has endangered the airway and breathing, a *tonsillectomy,* or surgical removal of the tonsils, may be performed.

LANGUAGE OF SCIENCE   *(Cont'd from page 779)*

palatine tonsils (PAL-ah-tine TAHN-sils) [*palat-* palate, *-ine* pertaining to]

pharyngeal tonsils (fair-IN-jee-al TAHN-sils) [*pharyng-* throat, *-al* pertaining to]

platelet destruction (PLAYT-let) [*platelet* small plate]

preauricular lymph nodes (pree-ah-RIK-yoo-lar limf nohds) [*pre-* before, *-auric-* ear, *-ula-* little, *-ar* pertaining to, *lymph* water, *nodus* knot]

red blood cell destruction

right lymphatic ducts (lim-FAT-ik) [*lymph-* water, *-atic* of the kind, *duct* lead or conduct]

spleen

submaxillary group (sub-MAK-sih-lair-ee) [*sub-* beneath, *-maxilla-* upper jaw, *-ary* pertiaining to]

submental group (sub-MEN-tal) [*sub-* beneath, *-ment-* chin, *-al* pertaining to]

superficial cervical lymph nodes (soo-per-FISH-al SER-vi-kal limf nohds) [*super-* on top of, *-fic-* face, *-al* pertaining to, *cervic-* neck, *-al* pertaining to, *lymph* water, *nodus* knot]

superficial cubital lymph nodes (soo-per-FISH-al KYOO-bi-tal limf nohds) [*super-* on top of, *-fic-* face, *-al* pertaining to, *cub-* cube, *-ital* pertaining to, *lymph* water, *nodus* knot]

supratrochlear lymph nodes (soo-prah-TROHK-lee-ar limf nohds) [*supra-* above, *-trochlea-* pulley, *-ar* pertiaining to, *lymph* water, *nodus* knot]

thoracic duct (thoh-RAS-ik) [*thorac-* chest (thorax), *-ic* pertaining to, *duct* lead or conduct]

thymic corpuscles (THYE-mik KOR-pus-uls) [*thym-* thyme or flowers (thymus gland), *-ic* pertaining to, *corpus-* body, *-cle* little]

thymus (THY-muss) [*thymus* thyme or flowers]

tonsils (TAHN-sils)

LANGUAGE OF MEDICINE

elephantiasis (el-eh-fan-TYE-ah-sis) [*elephant-* elephant, *-iasis* condition]

filaria (fi-LAR-ee-a) [*fila-* thread, *-ar-* resembling, *-ia* things]

Hodgkin disease (HOJ-kin) [*Thomas Hodgkin* English physician]

lymphangiography (lim-fan-jee-OG-reh-fee) [*lymph-* water, *-angio-* vessel, *-graphy* process of recording]

lymphangitis (lim-fan-JYE-i-tis) [*lymph-* water, *-angi-* vessel, *-itis* inflammation]

lymphedema (lim-fah-DEE-mah) [*lymph-* water, *-edema* swelling]

lymphoma (lim-FOH-mah) [*lymph-* water, *-oma* tumor]

mastitis (mass-TYE-tis) [*mast-* breast, *-itis* inflammation]

splenectomy (spleh-NEK-toh-mee) [*splen-* spleen, *-ectomy* surgical removal]

tonsillectomy (tahn-sih-LEK-toh-mee) [*ectomy* surgical removal]

tonsillitis (tahn-sih-LYE-tis) [*itis* inflammation]

CASE STUDY

Marc Tyler is a 46-year-old man who has recently been told he has Hodgkin lymphoma. He saw his family physician after he noticed an enlarged lymph node in his right axilla. He has been told that his prognosis is good. His treatment will include 6 months of intensive cancer chemotherapy followed by 1 month of radiation treatments.

1. When Mr. Tyler first discovered the enlarged lymph nodes, he probably also noticed that:

   A. The lymph nodes in his right axilla were painless.
   B. The lymph nodes in the right axilla were tender.
   C. A red streak extended down his right arm.
   D. His right arm was swollen from the axilla to the fingers.

2. Mr. Tyler has been told that his prognosis is good, because:

   A. The lymph system is not involved yet.
   B. Most lymphomas are benign.
   C. The disease has not spread beyond the lymph nodes.
   D. It is hypothesized that it is caused by a virus.

3. During Mr. Tyler's complete physical examination, his physician palpated many lymph nodes, then sent Mr. Tyler to the radiology department for a lymphogram. What information is the physician gathering with these examinations?

   A. The size of Mr. Tyler's lymph nodes
   B. The circulation of Mr. Tyler's lymphatic system
   C. The location of Mr. Tyler's lymph nodes
   D. All of the above

4. Mr. Tyler develops a lung infection following a common cold during his treatment. Which one of the following mechanisms is *most* likely responsible for his increased risk of infection?

   A. Fatigue resulting in poor hygiene
   B. Inadequate hematopoiesis
   C. Inability of the tonsils to keep the common cold from entering
   D. Inadequate functioning of the spleen

CHAPTER SUMMARY

OVERVIEW OF THE LYMPHATIC SYSTEM

A. Importance of the lymphatic system (Figure 20-1)
   1. Two most important functions—maintain fluid balance in the internal environment and immunity
   2. Lymph vessels act as "drains" to collect excess tissue fluid and return it to the venous blood just before it returns to the heart
   3. Lymphatic system—specialized component of the circulatory system; made up of lymph, lymphatic vessels, and isolated structures containing lymphoid tissue: lymph nodes, aggregated lymphoid nodules, tonsils, thymus, spleen, and bone marrow (Figure 20-2)
   4. Transports tissue fluid, proteins, fats, and other substances to the general circulation
   5. Lymphatic vessels begin blindly in the intercellular spaces of the soft tissues; do not form a closed circuit

LYMPH AND INTERSTITIAL FLUID

A. Lymph (lymphatic fluid)
   1. Clear, watery-appearing fluid found in the lymphatic vessels
   2. Closely resembles blood plasma in composition but has a lower percentage of protein; isotonic
   3. Elevated protein concentration in thoracic duct lymph because of protein-rich lymph from the liver and small intestine
B. Interstitial fluid (IF)
   1. Complex, organized fluid that fills the spaces between the cells and is part of the ECM (extracellular matrix)
   2. Resembles blood plasma in composition with a lower percentage of protein
   3. Along with blood plasma, constitutes the extracellular fluid

LYMPHATIC VESSELS

A. Distribution of lymphatic vessels (lymphatics) (Figures 20-2 and 20-3)
   1. Lymphatic capillaries—microscopic blind-end vessels where lymphatic vessels originate; wall consists of a single layer of flattened endothelial cells; networks branch and anastomose freely
   2. Lymphatic capillaries merge to form larger lymphatics and eventually form the main lymphatic trunks, the right lymphatic ducts, and the thoracic duct
   3. Lymph from upper right quadrant empties into right lymphatic duct and then into right subclavian vein
   4. Lymph from rest of the body empties into the thoracic duct, which then drains into the left subclavian vein; thoracic duct originates as the cisterna chyli (chyle cistern)
B. Structure of lymphatic vessels (Figure 20-4)
   1. Similar to veins except lymphatic vessels have thinner walls, have more valves, and contain lymph nodes
   2. Lymphatic capillary wall is formed by a single layer of thin, flat endothelial cells
   3. As the diameter of lymphatic vessels increases from capillary size, the walls become thicker and have three layers
   4. Semilunar valves are present every few millimeters in large lymphatics and even more frequently in smaller lymphatics
C. Functions of the lymphatic vessels
   1. Remove high–molecular-weight substances and even particulate matter from interstitial spaces
   2. Lacteals absorb fats and other nutrients from the small intestine

CIRCULATION OF LYMPH

A. From lymphatic capillaries, lymph flows through progressively larger lymphatic vessels to eventually reenter blood at the junction of the internal jugular and subclavian veins (Figure 20-5).
B. The lymphatic pump
  1. Lymphokinesis—the movement (flow) of lymph; can be visualized in a lymphangiogram (Figure 20-6)
  2. Lymph moves through the system in the right direction because of the large number of valves
  3. Breathing movements and skeletal muscle contractions (Figure 20-7) establish a fluid pressure gradient, as they do with venous blood
  4. Other factors, such as IF pressure, also drive lymphokinesis (Figure 20-8)
  5. Lymphokinetic actions—activities that result in a central flow of lymph

LYMPH NODES

A. Structure of lymph nodes
  1. Lymph nodes are oval-shaped structures enclosed by a fibrous capsule (Figure 20-9)
  2. Nodes are a type of biological filter
  3. Once lymph enters a node, it moves slowly through sinuses to drain into the efferent exit vessel (Figure 20-10)
  4. Trabeculae extend from the covering capsule toward the center of the node
  5. Cortical and medullary sinuses are lined with specialized reticuloendothelial cells capable of phagocytosis
B. Locations of lymph nodes
  1. Most lymph nodes occur in groups
  2. Groups with greatest clinical importance are submental and submaxillary groups and superficial cervical, superficial cubital, axillary, iliac, and inguinal lymph nodes (Figures 20-11 through 20-13)
  3. Preauricular lymph nodes located in front of the ear drain superficial tissues and skin on the lateral side of the head and face
C. Functions of lymph nodes—perform two distinct functions
  1. Defense functions
    a. Filtration
      (1) Mechanical filtration—physically stopping particles from progressing further in the body
      (2) Biological filtration—biological activity of cells destroys and removes particles
    b. Phagocytosis—reticuloendothelial cells remove microorganisms and other injurious particles from lymph and phagocytose them (biological filtration)
    c. If overwhelmed, lymph nodes can become infected or damaged (Figures 20-14 and 20-15)
  2. Hematopoiesis—lymphoid tissue is the site for the final stages of maturation of some lymphocytes and monocytes

LYMPHATIC DRAINAGE OF THE BREAST

A. Clinically important because cancer cells and infections can spread along lymphatic pathways to lymph nodes and other organs of the body
B. Distribution of lymphatics in the breast (Figure 20-16)
  1. Drained by two sets of lymphatic vessels
    a. Lymphatics that drain the skin over the breast with the exception of the areola and nipple
    b. Lymphatics that drain the substance of the breast, as well as the skin of the areola and nipple
  2. Superficial vessels converge to form a diffuse, cutaneous lymphatic plexus
  3. Subareolar plexus—located under the areola surrounding the nipple; where communication between the cutaneous plexus and large lymphatics that drain the secretory tissue and ducts of the breast occurs
C. Lymph nodes associated with the breast
  1. More than 85% of the lymph from the breast enters the lymph nodes of the axillary region
  2. Remainder of lymph enters lymph nodes along the lateral edges of the sternum

TONSILS

A. Form a broken ring under the mucous membranes in the mouth and back of the throat—the pharyngeal lymphoid ring (Figure 20-17)
  1. Palatine tonsils—located on each side of the throat
  2. Pharyngeal tonsils—located near the posterior opening of the nasal cavity
  3. Lingual tonsils—located near the base of the tongue
B. Protect against bacteria that may invade tissues around the openings between the nasal and oral cavities

THYMUS

A. Location and appearance of the thymus (Figure 20-18)
  1. Primary central organ of lymphatic system
  2. Single, unpaired organ located in the mediastinum, extending upward to the lower edge of the thyroid and inferiorly as far as the fourth costal cartilage
  3. Thymus is pinkish gray in childhood; with advancing age, becomes yellowish as lymphoid tissue is replaced by fat
B. Structure of the thymus
  1. Pyramid-shaped lobes are subdivided into small lobules
  2. Each lobule is composed of a dense cellular cortex and an inner, less dense medulla
  3. Medullary tissue can be identified by presence of thymic corpuscles
C. Function of the thymus
  1. Plays vital role in immunity mechanism
  2. Source of lymphocytes before birth
  3. Shortly after birth, thymus secretes thymosin and other regulators, which enables lymphocytes to develop into T cells (Figure 20-19)

SPLEEN

A. Location of the spleen—in the left hypochondrium, directly below the diaphragm, above the left kidney and descending colon, and behind the fundus of the stomach (Figures 20-2 and 20-20)

B. Structure of the spleen (Figure 20-21)
  1. Ovoid in shape
  2. Surrounded by fibrous capsule with inward extensions that divide the organ into compartments
  3. White pulp—dense masses of developing lymphocytes
  4. Red pulp—near outer regions, made up of a network of fine reticular fibers submerged in blood that comes from nearby arterioles
C. Functions of the spleen
  1. Defense—macrophages lining the sinusoids of the spleen remove microorganisms from the blood and phagocytose them
  2. Hematopoiesis—monocytes and lymphocytes complete their development in the spleen
  3. Red blood cell and platelet destruction—macrophages remove worn-out RBCs and imperfect platelets and destroy them by phagocytosis; also salvage iron and globin from destroyed RBCs
  4. Blood reservoir—pulp of spleen and its sinuses store blood

CYCLE OF LIFE: LYMPHATIC SYSTEM

A. Dramatic changes throughout life
B. Organs with lymphocytes appear before birth and grow until puberty
C. Postpuberty
  1. Organs atrophy through late adulthood
    a. Shrink in size
    b. Become fatty or fibrous
  2. Spleen—develops early, remains intact
D. Overall function maintained until late adulthood
  1. Later adulthood
    a. Deficiency permits risk of infection and cancer
    b. Hypersensitivity—likelihood of autoimmune conditions

THE BIG PICTURE: THE LYMPHATIC SYSTEM AND THE WHOLE BODY

A. Lymphatic system drains away excess water from large areas
B. Lymph is conducted through lymphatic vessels to nodes, where contaminants are removed
C. Lymphatic system benefits the whole body by maintaining fluid balance and freedom from disease

REVIEW QUESTIONS

1. List the anatomical components of the lymphatic system.
2. How do interstitial fluid and lymph differ from blood plasma?
3. How do lymphatic vessels originate?
4. Briefly describe the anatomy of the lymphatic capillary wall.
5. Lymph from what body areas enters the general circulation by way of the thoracic duct? The right lymphatic ducts?
6. What is the cisterna chyli?
7. Where does lymph enter the blood vascular system?
8. In general, lymphatics resemble veins in structure. List three exceptions to this general rule.
9. What are the specialized lymphatics that originate in the villi of the small intestine called?
10. What is chyle? Where is it formed?
11. Discuss the "lymphatic pump."
12. List several important groups, or clusters, of lymph nodes.
13. Discuss how lymph nodes function in body defense and hematopoiesis.
14. If cancer cells from breast cancer enter the lymphatics of the breast, where do you think they might lodge and start new growths? Explain, using your knowledge of the anatomy of the lymphatic and circulatory systems.
15. Locate the thymus, and describe its appearance and size at birth, at maturity, and in old age.
16. Discuss the function of the thymus.
17. Describe the location and functions of the spleen.
18. What happens when there is a loss of lymphatic fluid?
19. Explain why lymphedema may occur after breast surgery.

CRITICAL THINKING QUESTIONS

1. Even though the lymphatic system is a component of the circulatory system, why is the term *circulation* not the most appropriate term to describe the flow of lymph?
2. Explain how lymph is formed. What would be the impact on lymph formation if the osmotic force at the venous end of the capillary was more successful at recovering fluid lost at the arterial end?
3. Discuss the importance of valves in the lymphatic system.
4. Explain the role of the lymphatic system in the spread of breast cancer and its surgical treatment.

CHAPTER 21

Immune System

LANGUAGE OF SCIENCE

adaptive immunity (ah-DAP-tiv i-MYOO-ni-tee)
[*adapt-* process of fitting, *-ive* pertaining to, *immun-* free, *-ity* state of]

antibodies (AN-ti-bod-ees) [*anti-* against]

antibody-mediated immunity (AN-ti-bod-ee-MEE-dee-ayt-ed i-MYOO-nih-tee) [*anti* against, *medi-* middle, *-ate* process of, *immun-* free, *-ity* state of]

antigens (AN-ti-jens) [*anti-* against, *-gen* generate]

antigenic determinants (AN-ti-jen-ik dih-TUR-mah-nants) [*anti-* against, *-gen* generate, *-ic* pertaining to, *determin-* limit, *-ant* agent of]

antigen-antibody complex (AN-ti-jen-AN-ti-bod-ee KOM-pleks) [*anti-* against, *-gen* generate, *anti* against, body, *complexus* an embrace]

antigen-presenting cells (APCs) (AN-ti-jen) [*anti-* against, *-gen* generate, presenting, *cella* storeroom]

autoimmunity (aw-toh-i-MYOO-ni-tee) [*auto-* self, *-immun-* free, *-ity* state of]

B cells [*B* bursa-equivalent tissue, *cella* storeroom]

cell-mediated immunity (sell-MEE-dee-ayt-ed i-MYOO-nih-tee) [*cella* storeroom, *medi-* middle, *-ate* process of, *immun-* free, *-ity* state of]

cellular immunity (SELL-yoo-lar i-MYOO-ni-tee) [*cella* storeroom, *-ular* pertaining to, *immun-* free, *-ity* state of]

chemotactic factor (kee-moh-TAK-tik) [*chemo-* chemical, *-tact-* movement, *-ic* pertaining to]

chemotaxis (kee-moh-TAK-sis) [*chemo-* chemical, *-taxis* movement]

clone (klohn) [*klon* a plant cutting]

combining site

complement [*complementum* that which completes]

cytokines (SYE-toh-kynes) [*cyto-* cell, *-kine* movement]

cytolysis (sye-TOL-i-sis) [*cyto-* cell, *-lysis* loosening]

cytotoxic T cells (sye-toh-TOK-sik) [*cyto-* cell, *-toxic* poison, *T* thymus gland, *cella* storeroom]

Cont'd on p. 833

nemies of many kinds and in great numbers assault the body during a lifetime. Among the most threatening are hordes of microorganisms. We live our lives in a virtual sea of protozoa, fungi, bacteria, and viruses. So ever-present and potentially lethal are these small but formidable foes that no newborn could live through infancy, much less survive to adulthood or old age, without effective defenses against them. However, there are also enemies within. Inside the body, abnormal cells appear on an irregular but continual basis. If allowed to survive, these abnormal cells reproduce to form a tumor. A tumor by itself can be life threatening, and there is always the possibility that it could become cancerous and spread (metastasize) to many other locations within the body. Without an internal "security force" to deal with such abnormal cells when they first appear, we would live very short lives. This chapter presents a brief overview of the system that provides defenses against both external and internal enemies—the immune system.

ORGANIZATION OF THE IMMUNE SYSTEM

Like any security force, the components and mechanisms of the immune system are organized in an efficient—almost military—manner. They are not just ready on a moment's notice; they are *continually* patrolling the body for foreign or internal enemies and shoring up the various lines of defense for a possible attack. Before we begin studying the specifics of immunity, we should spend a moment mapping out the overall defensive strategy of the immune system.

First, it is important to recognize that cells, viruses, and other particles have unique molecules and groups of molecules on their surfaces that can be used to identify them. This is similar to military operations in which enemy aircraft, vehicles, or soldiers can be identified by their distinctive insignia that are different from the insignia seen on "our side." Likewise, our own cells have unique cell markers embedded in our plasma membranes that identify each of our cells as **self**—that is, belonging to us as an individual. And foreign cells or particles have **nonself** molecules that serve as recognition markers for our immune system. The ability of our immune system to attack abnormal or foreign cells but spare our own normal cells is called **self-tolerance.**

In human society, any good security force employs numerous different strategies to guard its territory and take action if necessary. It is not surprising, then, to discover that the body's "society of cells" also employs a system that uses many different kinds of mechanisms to ensure the integrity and survival of the internal environment. All of these defense mechanisms can be categorized into one of two major categories of immune mechanisms: **innate immunity** and **adaptive immunity.**

Innate immunity is called that because it is in place before a person is exposed to a particular harmful particle or condition. The word *innate* refers to something that is already present naturally at birth. Because it includes mechanisms that resist a wide variety of threatening agents or conditions, innate immunity is also called **nonspecific immunity.** The term *nonspecific* implies that these immune mechanisms do not act on only one or two specific invaders but rather provide a more general defense by simply acting against a wide variety of particles recognized as *nonself.*

Adaptive immunity, on the other hand, involves mechanisms that recognize *specific* threatening agents and then *adapt*, or respond, by targeting their activity against these agents—and these agents only. Because it targets only specific harmful particles, adaptive immunity is also called **specific immunity.** Adaptive im-

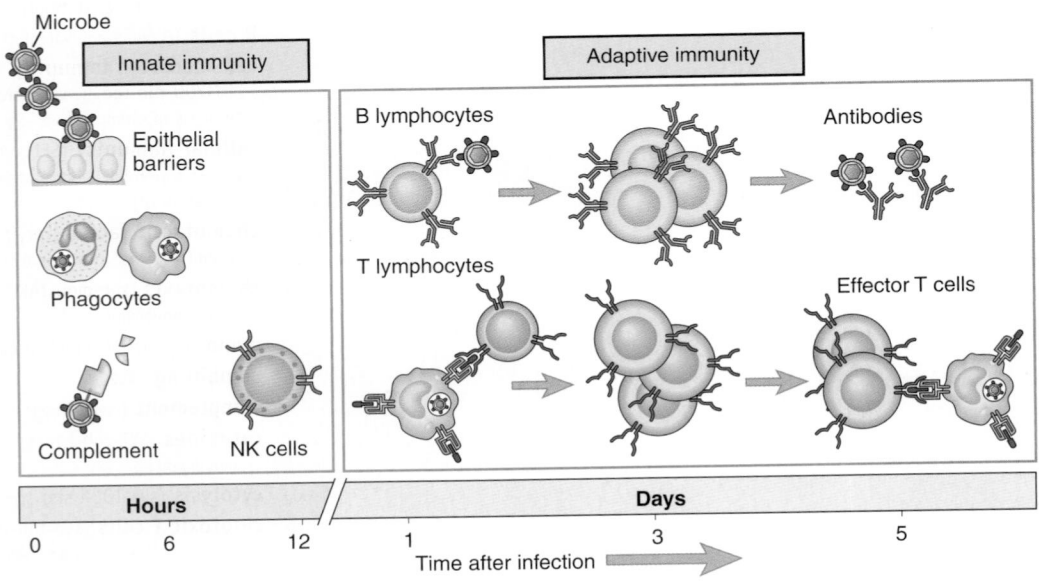

Figure 21-1 *Innate and adaptive immunity.* Innate (nonspecific) immune mechanisms are "built in" and ready for action—thus providing the initial defense against infections and other assaults on the body. Adaptive (specific) immune mechanisms develop later, as lymphocytes are activated to work against specific foreign or abnormal cells and particles. The time frames are generalizations.

Table 21-1 **Innate and Adaptive Immunity**

		INNATE IMMUNITY	ADAPTIVE IMMUNITY
Synonyms	Frequently used alternate terminology	Nonspecific immunity, native immunity, genetic immunity	Specific immunity, acquired immunity
Characteristics			
Specificity	Unique antigens produce unique responses of the immune system	Not specific—recognizes variety of different groups of foreign cells or particles	Specific—recognizes specific antigens on specific cells or particles
Speed of reaction	Reaction time of the immune responses	Rapid: immediate up to several hours	Slower: several hours to several days
Memory	Enhanced responses to repeated exposures to the same antigen	None	Yes
Does not react to self	Prevents injury to the individual's own cells	Yes	Yes
Components			
Barriers	Prevent entry of harmful particles	Skin, mucosa, antimicrobial chemicals	Lymphocytes in epithelia; antibodies released at epithelial surfaces
Blood proteins	Circulate throughout body, providing wide area of protection	Complement, interferon (IFN), others	Antibodies
Cells	Types of leukocytes involved in immunity	Phagocytes (macrophages, neutrophils), natural killer (NK) cells	Lymphocytes (B cells and T cells)

mune mechanisms often take some time to recognize their targets and react with sufficient force to overcome the threat, at least on their first exposure to a specific kind of threatening agent. Innate mechanisms, because they are already in place, have the advantage of being able to meet an enemy as soon as it presents itself. As we discuss examples of each major type of immunity, you will come to understand how each type works and appreciate the distinction between them. You will also come to appreciate the value in having two complementary strategies for defending the body.

As in any body system, the work of the immune system is done by cells or substances made by cells. The primary types of cells involved in innate immunity are these: epithelial barrier cells, phagocytic cells (neutrophils, macrophages), and so-called *natural killer (NK) cells.* The primary types of cells involved in adaptive immunity are two types of lymphocytes called *T cells* and *B cells.*

Cytokines, which are chemicals released from cells to trigger or regulate innate and adaptive immune responses, also participate in innate immunity. Examples of cytokines include *interleukins (ILs), leukotrienes,* and *interferons (IFNs)*—all of which will be described later in this chapter. In addition to cytokines, other chemicals play a regulatory role in immunity, such as *complements,* other enzymes, and the amine *histamine.*

Awesome indeed is the army of cells and molecules that make up the immune system. Over one trillion lymphocytes, for example, and 100 million trillion (10^{20}) plasma protein molecules (antibodies) are a few of the many agents that help your body resist damage and disease. Figure 21-1 and Table 21-1 summarize some of the essential characteristics of innate and adaptive immunity that we will discuss in this chapter.

QUICK CHECK

1. What is the difference between *self* and *nonself?*
2. What is the difference between *adaptive* and *innate* immunity?
3. What is a *cytokine?* What are some examples of cytokines?

INNATE IMMUNITY

The general, innate defensive mechanisms of the body are many and varied (Table 21-2). Only the major types of innate immune mechanisms are listed here; many other examples appear in other chapters throughout this book. As a matter of fact, you will probably recognize examples here that you have encountered already in previous chapters. Macrophages are a good example. They are often referred to by different names when identified in specific body areas. For example, macrophages in the skin were identified as Langerhans cells (Chapter 6, p. 197) and those in brain tissue were called microglia (Chapter 12, p. 434).

Species Resistance

Species resistance refers to a phenomenon in which the genetic characteristics common to a particular kind of organism, or *species,* provide defense against certain *pathogens* (disease-causing agents). The human species *(Homo sapiens),* for example, is resistant to many life-threatening infections and infestations that often spread easily among plants and other animals. For example,

Table 21-2	Mechanisms of Innate Defense

MECHANISM	DESCRIPTION
Species Resistance	Genetic characteristics of the human species protect the body from certain pathogens
Mechanical and Chemical Barriers	Physical impediments to the entry of foreign cells or substances
Skin and mucosa	Forms a continuous wall that separates the internal environment from the external environment, preventing the entry of pathogens
Secretions	Secretions such as sebum, mucus, and enzymes chemically inhibit the activity of pathogens
Inflammation	The inflammatory response isolates the pathogens and stimulates the speedy arrival of large numbers of immune cells
Phagocytosis	Ingestion and destruction of pathogens by phagocytic cells
Neutrophils	Granular leukocytes that are usually the first phagocytic cell to arrive at the scene of an inflammatory response
Macrophages	Monocytes that have enlarged to become giant phagocytic cells capable of consuming many pathogens; often called by other, more specific names when found in specific tissues of the body
Natural Killer (NK) Cells	Group of lymphocytes that kill many different types of cancer cells and virus-infected cells
Interferon	Protein produced by cells after they become infected by a virus; inhibits the spread or further development of a viral infection
Complement	Group of plasma proteins (inactive enzymes) that produce a cascade of chemical reactions that ultimately causes lysis (rupture) of a foreign cell; the complement cascade can be triggered by adaptive or innate immune mechanisms

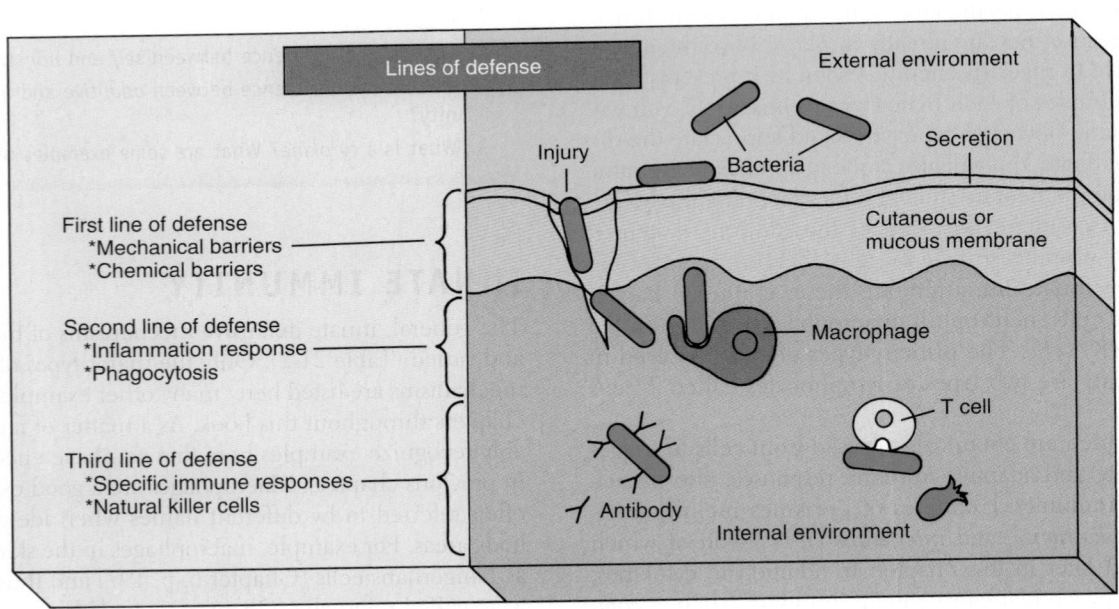

Figure 21-2 *Lines of defense.* Immune function, that is, defense of the internal environment against foreign cells, proteins, and viruses, includes three layers of protection. The first line of defense is a set of barriers between the internal and external environment, the second line of defense involves the innate inflammatory response (including phagocytosis), and the third line of defense includes the adaptive immune responses and the innate defense offered by NK cells. Of course, tumor cells that arise in the body are already past the first two lines of defense and must be attacked by the third line of defense. This diagram is a simplification of the complex function of the immune system; in reality, there is a great deal of crossover of mechanisms between these "lines of defense."

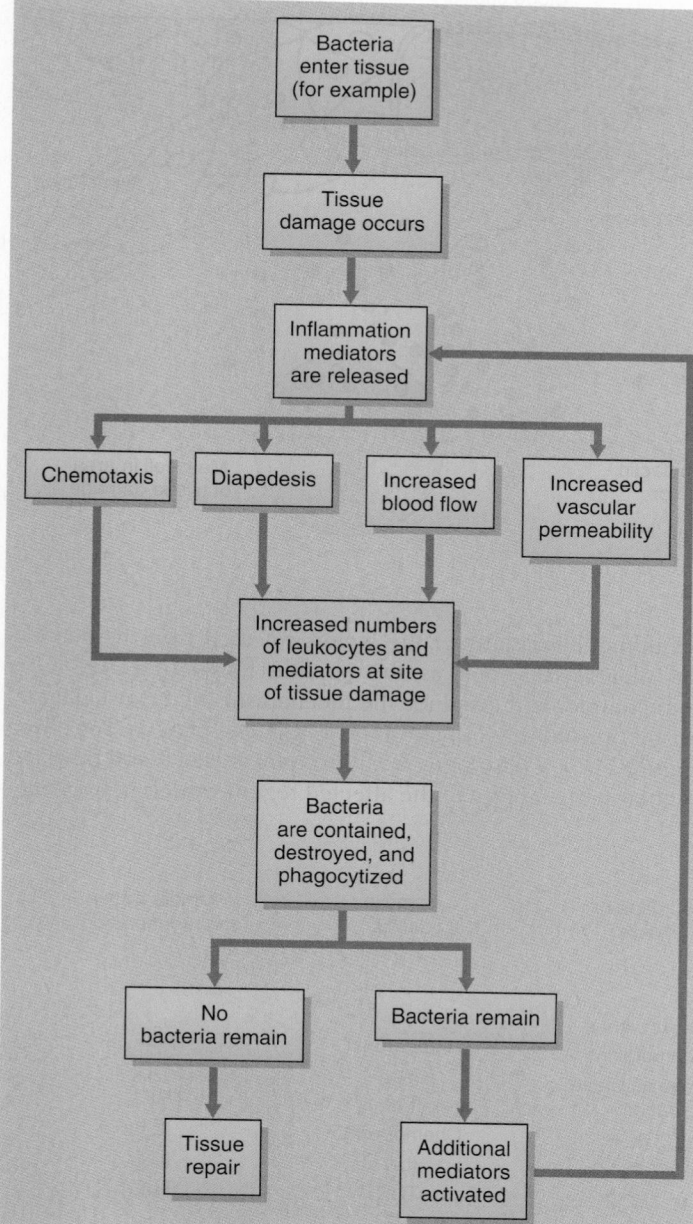

Figure 21-3 *Example of the inflammatory response.* Tissue damage caused by bacteria triggers a series of events that produces the inflammatory response and promotes phagocytosis at the site of injury. These responses tend to inhibit or destroy the bacteria. Similar reactions will occur in the presence of other abnormal or injurious particles or conditions.

humans do not have to worry about getting Dutch elm disease or becoming infected with canine viral distemper to which young dogs are susceptible. Usually, species resistance in humans results from the fact that our internal environment is not suitable for certain pathogens. We may also have resistance because a particular microbe may not be biochemically compatible with the various molecules on our cell membranes that the microbes need to gain entry into a host cell.

Mechanical and Chemical Barriers

The internal environment of the human body is protected by a continuous mechanical barrier formed by the cutaneous membrane (skin) and mucous membranes (see Figure 5-39, p. 180). Often called the *first line of defense*, these membranes provide several layers of densely packed cells and other materials—forming a sort of "castle wall" that protects the internal environment from invasion by foreign cells (Figure 21-2).

Besides forming a protective wall, the skin and mucous membranes operate various additional immune mechanisms. For example, substances such as *sebum* (contains pathogen-inhibiting agents), *mucus* (pathogens may stick and be swept away), enzymes (may hydrolyze pathogens), and hydrochloric acid in gastric mucosa (may destroy pathogens) may also be present to act as innate defense mechanisms. These chemical barriers act as a sort of moat around the castle wall formed by the membranes.

The epithelial barriers of the body are essentially innate, nonspecific defenses. However, the protective epithelial membranes also have adaptive (specific) defenses that reinforce them. The combined innate and adaptive immune functions of protective mucous membranes are discussed later in the chapter in a separate boxed essay (Box 21-8 on p. 828).

Inflammation

The Inflammatory Response

If bacteria or other invaders break through the chemical and mechanical barriers formed by the membranes and their secretions, the body has a *second line of defense*: the inflammatory response (see Figure 21-2). The **inflammatory response** has already been discussed in some detail in Chapter 5 (see p. 167). For the purpose of a quick review recall that tissue damage elicits a host of responses that counteract the injury and promote a return to normal. An example of how local inflammation works is illustrated in Figure 21-3. In the example, bacteria cause tissue damage that, in turn, triggers the release of various inflammation mediators from cells such as the *mast cell* found in connective tissues (Figure 21-4). These inflammation mediators include *histamine, kinins, prostaglandins,*

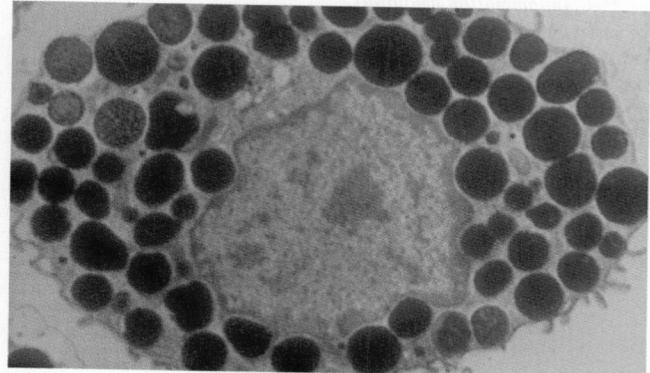

Figure 21-4 *Mast cell.* This colorized micrograph shows a mast cell filled with dense red granules of the inflammation mediator called *histamine.*

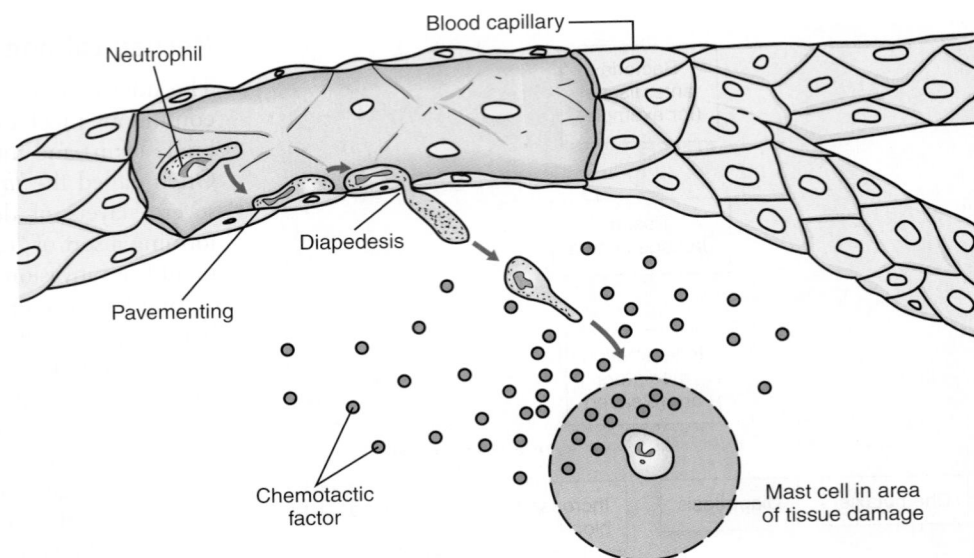

Figure 21-5 *Chemotaxis and diapedesis.* In this example, a neutrophil is attracted by chemotactic agents released by a mast cell in a damaged or infected tissue. After adhering to the inside of the blood capillary (pavementing), the neutrophil exits the capillary by the process of diapedesis. Through chemotaxis, movement directed by chemical attraction, the neutrophil migrates toward the highest concentration of chemotactic factor—the site of the injury—where it can then begin its immune functions.

leukotrienes, interleukins (ILs), and related compounds. Many of these mediators are chemotactic factors, that is, substances that attract white blood cells to the area in a process called **chemotaxis**. As Figure 21-5 shows, chemotaxis is the process by which a cell navigates toward the source of the **chemotactic factor** *(chemotaxin)* by detecting and moving toward higher concentrations of the factor.

Additionally, many of the factors released from tissue cells and phagocytes, such as the peptide fragment called C5α from complement, produce the mechanisms that cause characteristic signs of inflammation: heat, redness, pain, and swelling (Figure 21-6). These signs result from increased blood flow and vascular permeability in the affected region, which help phago-

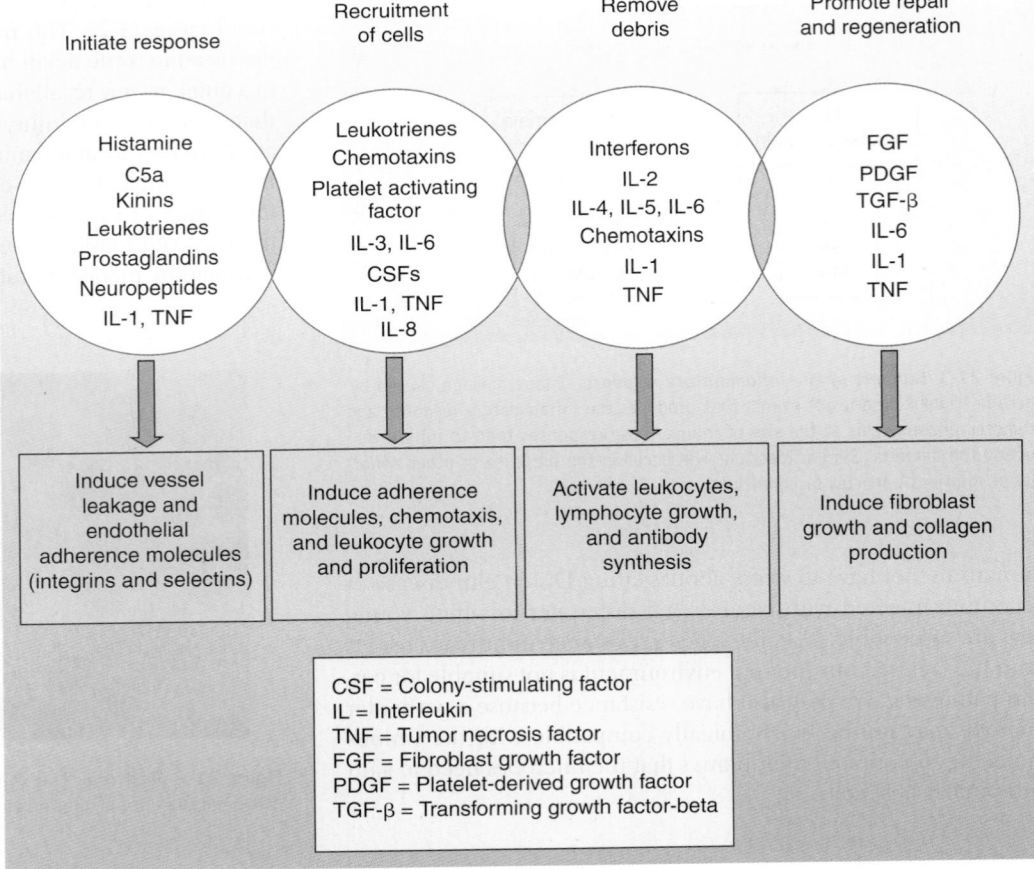

Figure 21-6 *Inflammation mediators.* A wide variety of chemical mediators help regulate the immune response, as this generalized chart shows. Some of the processes shown overlap in time and are thus sometimes concurrent.

cytic white blood cells reach the general area and then enter the affected tissue. Also, some inflammation mediators trigger fibroblasts to grow and produce more collagen fibers to promote repair and regeneration. Besides local inflammation, systemic inflammation may occur when the inflammatory response occurs on a body-wide basis.

Phagocytosis

A major component of the body's second line of defense is the mechanism of **phagocytosis**—the ingestion and destruction of microorganisms or other small particles. There are many types of *phagocytes*, that is, cells capable of phagocytosis, in the body. As Figure 21-7 shows, when phagocytes approach a microorganism, they extend footlike projections (pseudopods) toward it. Soon the pseudopods encircle the organism and form a complete sac, called a *phagosome*, around it. The phagosome then moves into the interior of the cell, where a lysosome fuses with it. The contents of the lysosome, chiefly digestive enzymes and hydrogen peroxide, drain into the phagosome and destroy the microorganisms in it.

Because phagocytosis defends us against various kinds of agents, it is classified as an *innate defense*. However, phagocytes

also "cross over" to play an important role in adaptive immunity as well. After digesting the offending particle, a phagocyte will often process the proteins and display bits of the protein—peptides—on the surface of the phagocyte. These peptides are then recognized by cells of the adaptive immune system as antigens, thus possibly triggering an adaptive immune response. Cells that perform this function are called *antigen-presenting cells (APCs)*.

The most numerous type of phagocyte is the *neutrophil*, a granular, neutral-staining type of white blood cell (WBC). After being released at a site of inflammation or tissue damage, chemotactic factors diffuse into adjoining capillaries. Once in the bloodstream, they cause neutrophils and other phagocytes to adhere to the vessel's endothelial lining in a process called **pavementing** (Figure 21-5). Numbers of these adherent phagocytes pass between the endothelial cells that form the capillary wall, dissolve the underlying basement membrane, and then exit through the vessel wall in the inflamed area. The movement of phagocytes from blood vessel to inflammation site is called **diapedesis**. Phagocytes have a very short life span, and thus dead cells tend to "pile up" at the inflammation site—forming most of the white substance called **pus**.

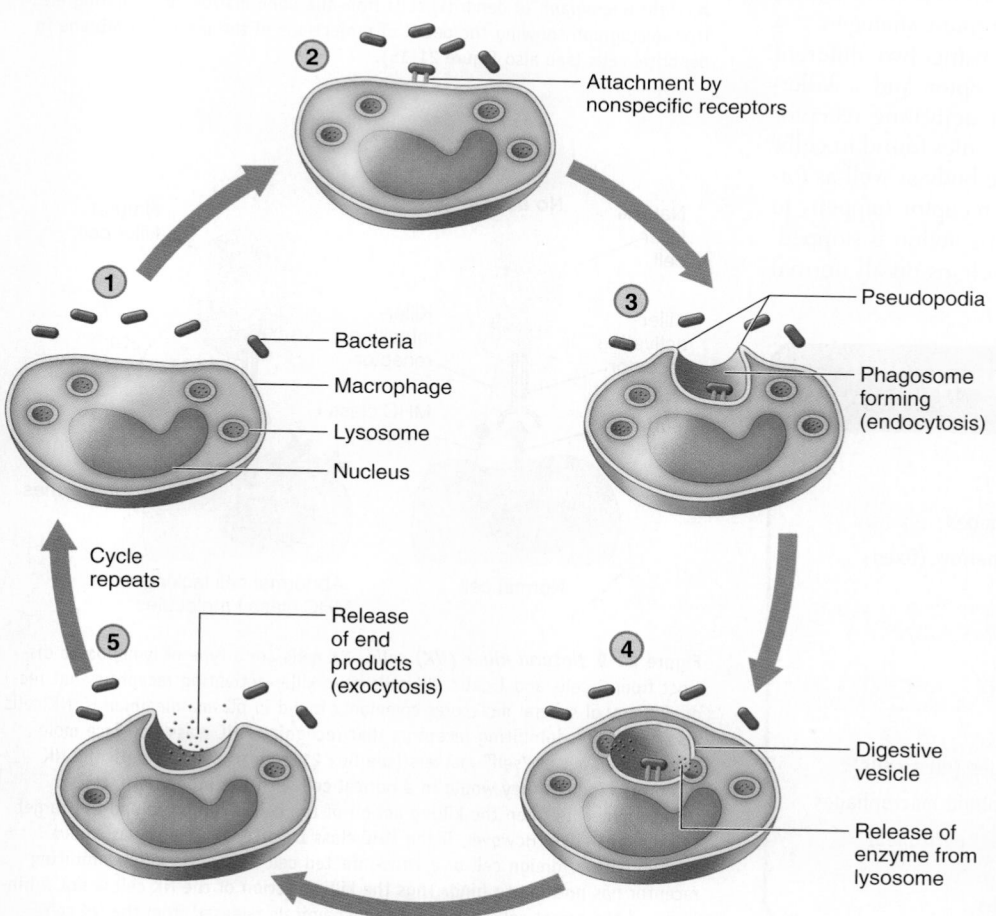

Figure 21-7 *Phagocytosis of bacteria.* Drawing of sequence showing steps in phagocytosis of bacteria. The plasma membrane extends (as a pseudopod) toward the bacterial cells, then envelops them. Once trapped, they are engulfed by the cell and destroyed by lysosomal enzymes.

Another common type of phagocyte is the **macrophage** (meaning "large eater"). Macrophages are phagocytic monocytes (a nongranular WBC) that have grown to several times their original size after migrating out of the bloodstream. Macrophages of various subtypes are present in many areas of the body, even on the outside surface of some mucous membranes (for example, in the respiratory tract). Macrophage subtypes are often known by specific names that designate their location (Table 21-3). One general category of macrophage is the **dendritic cell (DC)** found in many tissues of the body that are in contact with the external environment, such as the skin and mucous membranes (Figure 21-8). This structural type of macrophage is called dendritic because of its many branches (dendr- "branch"). Dendritic cells are also sometimes called *stellate* ("star shaped") *cells*. The importance of macrophages to our overall defense of the body is made clear by the fact that 10% to 15% of all cells in any organ of the body are macrophages!

Natural Killer Cells

Besides phagocytes, the body has another important set of cells that provides innate defense of the body. These are the **natural killer (NK) cells.** NK cells are a group of lymphocytes that kill many types of tumor cells and cells infected by different kinds of viruses. As a group they are produced in the red bone marrow and constitute about 15% of the total lymphocyte cell numbers. NK cells are neither T cells nor B cells, as described below under the topic of adaptive immunity. Because they have such a broad action and do not have to be activated by a specific foreign antigen to become active, they are usually included among the innate immune strategies.

NK cells recognize abnormal cells by using two different recognition receptors: a killer-activating receptor and a killer-inhibiting receptor (Figure 21-9). The killer-activating receptor binds to any of several common surface molecules found in cells. Thus the NK cell can bind to any cell of the body as well as foreign cells. However, if the killer-inhibiting receptor happens to bind to an MHC protein also, then the killing action is stopped. Box 21-1 explains that MHCs are surface proteins on all normal

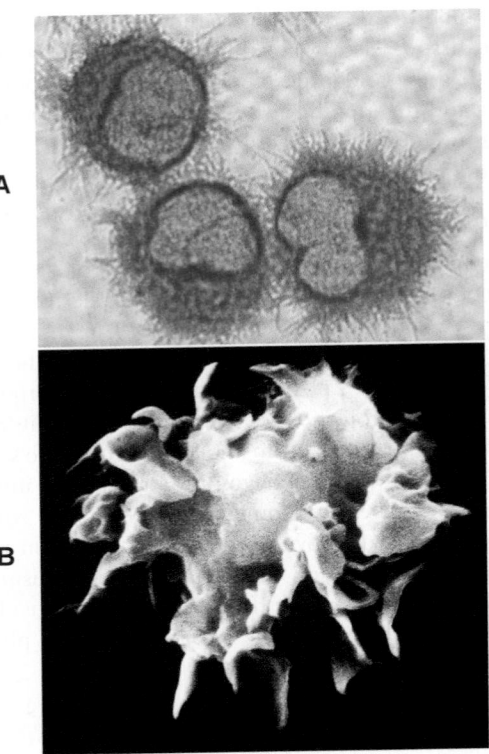

Figure 21-8 *Dendritic cells (DCs).* Dendritic cells, also called stellate (star-shaped) cells, are macrophages that are found in many areas of the body. **A,** Light micrograph of dendritic cells from the bone marrow. **B,** Scanning electron micrograph showing the detail of projections of the plasma membrane in dendritic cells (see also Figure 21-15).

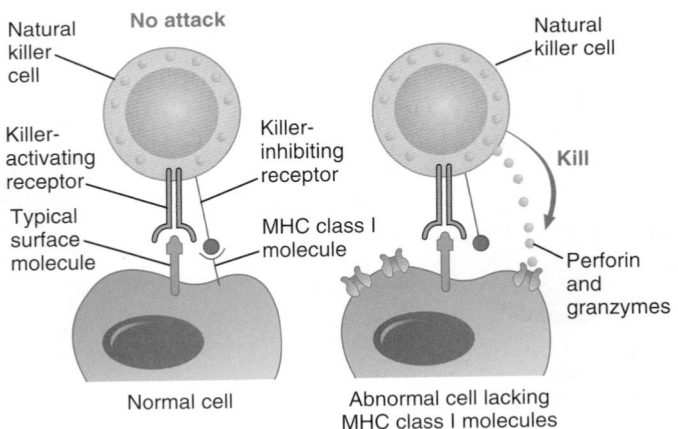

Figure 21-9 *Natural killer (NK) cells.* NK cells are a type of lymphocyte distinct from B cells and T cells. NK cells have killer-activating receptors that recognize any of several molecules commonly found in plasma membranes. NK cells also have killer-inhibiting receptors that recognize MHC class I surface molecules, which act as "self" markers (see Box 21-1 on p. 809). If both the NK receptors bind, as they would in a normal cell with the proper MHC class I surface molecule, then the killing action of the cell is inhibited and the target cell remains alive. However, if the MHC class I marker is absent—as would happen with a foreign cell or a virus-infected cell—then the killer-inhibiting receptor has nothing to bind. Thus the killing action of the NK cell is not inhibited and the target cell is destroyed by chemicals released from the NK cell.

Table 21-3	Examples of Macrophage Locations
LOCATION	**MACROPHAGES**
Bloodstream	Circulating macrophages
Bone and bone marrow	Osteoclasts; bone marrow (fixed) macrophages
Central nervous system	Microglia
Connective tissues	Histiocytes
Epidermis	Langerhans cells
Liver	Kupffer cells
Lung	Alveolar macrophages (dust cells)
Lymph nodes	Fixed and free lymphoid macrophages
Serous fluids	Pleural macrophages; peritoneal macrophages
Spleen	Splenic macrophages

BOX 21-1: FYI

Major Histocompatibility Complex (MHC)

The **major histocompatibility complex (MHC)** is a set of genes in chromosome 6 that all code for antigen-presenting proteins and other immune system proteins (part *A* of figure). Antigens are proteins that potentially trigger a specific immune response. The MHC proteins produced by MHC genes in class I and class II also are called *human leukocyte antigens (HLAs)*. Their function is to present different protein fragments (peptides) at the surface of the cell for possible recognition as either self or nonself antigens by immune system cells.

The MHC class I proteins, or HLAs, are present in every nucleated cell of the body. Their function is to present protein fragments from within the cell at the surface as antigens. An immune cell will then recognize the presented antigen as a *self-antigen* or as a *nonself-antigen* (part *B* of figure). Self-antigens are normally ignored by the immune cell. Nonself-antigens are instead recognized as abnormal and attacked by the mechanisms described later in this chapter. If a normal cell becomes infected with a virus or become cancerous, it may present some abnormal antigens on the surface and thus be identified by the immune system.

MHC class I proteins are also involved in the mechanism by which natural killer (NK) cells recognize abnormal cells. As Figure 21-9 shows, the absence of MHC class I proteins fails to inhibit the killing action of the NK cell. Cells from outside the body are likely to have no MHC class I proteins or have a different version of the MCH class I proteins. Infected or damaged cells are also likely to have missing or damaged MHC class I surface proteins.

MHC class II proteins are expressed in immune cells that specialize in presenting antigens. These "professional" **antigen-presenting cells (APCs)** include macrophages (especially the dendritic cells, or DCs) and B cells, for example. The APCs use their MHC class II proteins to present fragments of proteins that they've brought in from outside the cell—perhaps from a bacterial cell. Thus, they alert the immune system to the presence of these invaders and trigger certain adaptive (specific) immune responses.

MHC class III proteins include a wide variety of different immune-related proteins such as complement components, some cytokines, and a number of immune and nonimmune proteins.

The major histocompatibility complex (MHC) first came to the attention of researchers trying to find out why transplants and tissue grafts were often rejected by the recipient. They found that individuals with different MHC genes rejected tissues from each other. Thus, they coined the term "histocompatibility" for this set of genes because the genes seemed to regulate the compatibility of transplants and grafts.

There are hundreds of different versions or alleles of the principle MHC genes—far more genetic variability than in any other group of genes in the human genome! Scientists are still trying to find a satisfactory explanation for this tremendous variation.

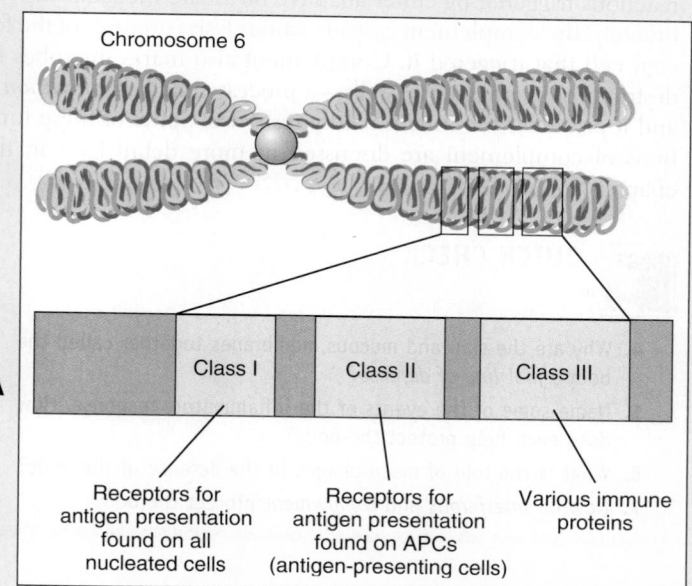

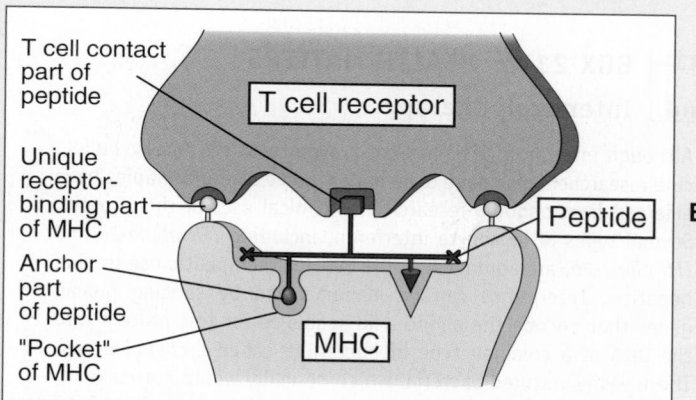

A, Major histocompatibility complex (MHC). The MHC is a region of chromosome 6 that codes for proteins important in immune system function. MHC class I proteins present antigens on the surfaces of all cells of the body that possess a nucleus—which is most cells. MHC class II proteins present antigens on the surfaces of cells that specialize in presenting antigens—the antigen-presenting cells (APCs). MHC class III proteins include complement, cytokines, and other proteins of the immune system. **B, MHC function.** This simplified diagram shows that the MCH protein displays an antigen (protein fragment or peptide) on the surface of the cell. A receptor on the surface of a T cell may then bind to the unique receptor-binding part of the MHC and to a complementary part of the T cell. Antigens (peptides) presented this way can then be recognized by immune cells as being "self" or "nonself."

cells and are unique to each individual person. Thus, only abnormal and foreign cells fail to bind to the killer-inhibitor centers—and therefore are killed by the NK cell.

The NK cells use several different methods for killing cells, most of which involve chemically triggering apoptosis (programmed cell death) that progresses to lysis (breaking apart). NK cells must engage their target cells by direct contact (binding of receptors) to cause cell destruction.

Interferon

Several types of cells, if invaded by viruses, respond rapidly by synthesizing the protein **interferon (IFN)** and releasing some of it into the circulation. As the name suggests, interferon proteins interfere with the ability of viruses to cause disease. One way that they do this is by preventing viruses from multiplying in cells. IFN production is triggered by viral infection in a cell, probably by the presence of viral dsRNA (double-strand RNA). The IFN is then released to nearby cells, where it triggers signal transduction that activates antiviral genes in the neighboring cells. These genes produce an "antiviral state" by producing several enzymes that block viral replication should the cell become infected. Thus, IFN acts as a paracrine (local) hormone that allows virus-infected cells to send an "alarm" to nearby cells that protects the uninfected cells.

Some interferons also promote synthesis of more MHC proteins, thus allowing them to present viral antigens and promote immune destruction of infected cells (Box 21-1). By promoting the destruction of cells that are already infected, the chance of the virus spreading to other cells is reduced.

Interferon comes in several varieties, each with somewhat different antiviral actions. Leukocyte interferon (α), fibroblast interferon (β), and immune interferon (γ) are the three major types of interferon proteins. All three have now been produced by using gene-splicing techniques. Studies exploring antiviral and anticancer activities of interferons are currently underway (Box 21-2).

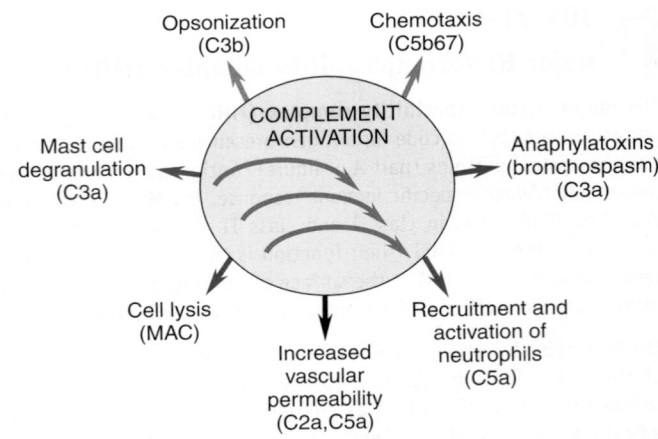

Figure 21-10 *Summary of complement function.* Complement (C), when activated in a cascade of reactions, may participate in a variety of immune functions. *MAC,* Membrane attack complex (see Figure 21-21).

Complement

Complement is the name given to each of a group of about 20 inactive enzymes in the plasma and on cell surfaces. Individual complement proteins are often designated by C (for complement) followed by a number, such as C1, C2, C3, and so on. Subtypes of each complement are usually identified by a lower-case letter, such as C4b or C5a—or by Greek letters, as in C5bα or C8β.

Complement molecules are activated in a cascade of chemical reactions triggered by either adaptive or innate mechanisms. Ultimately, the complement cascade causes lysis (rupture) of the foreign cell that triggered it. Complement also marks microbes for destruction by phagocytic cells—a process called *opsonization*—and it promotes the inflammatory response. Some of these functions of complement are discussed in more detail later in this chapter (Figure 21-10).

BOX 21-2: HEALTH MATTERS

Interferon Therapy

Although interferon (IFN) has not proven to be the "magic bullet" cure researchers may have once hoped, some useful therapies have emerged from ongoing research into clinical uses of this protein. Several types of leukocyte interferon, including *IFN alpha-2a* and *IFN alpha-2b*, are approved by the FDA for therapeutic use in viral hepatitis. Interferons can be manufactured by splicing human genes that encode the amino acid sequence for this protein into the DNA of a common type of bacterium called *Escherichia coli*. These easily cultured bacteria, which normally would not make interferon, are thus converted into miniature "interferon factories." Interferon is often combined with polyethylene glycol to make it last longer in the body—and thus reduce the number of injections needed from three per week to just one per week.

Having both anticancer and antiviral properties, interferons are sometimes used with other therapies in the treatment of AIDS-related Kaposi sarcoma (see p. 217), hairy-cell leukemia, other cancers, and genital warts. A type of interferon called *IFN beta* has also been used to treat multiple sclerosis (MS).

QUICK CHECK

4. Why are the skin and mucous membranes together called the body's *first line of defense*?
5. Name some of the events of the inflammatory response. How does each help protect the body?
6. What is the role of macrophages in the defense of the body?
7. How do *interferons* and *complement* protect the body?

OVERVIEW OF ADAPTIVE IMMUNITY

Unlike the innate, nonspecific mechanisms of immunity, the various types of adaptive immune mechanisms attack *specific* agents that the body recognizes as abnormal or nonself. Adaptive immunity, part of the body's *third line of defense*, is orchestrated by two different classes of a type of white blood cell called the *lymphocyte* (Figure 21-11).

Figure 21-11 *Lymphocytes.* Color-enhanced scanning electron micrograph showing lymphocytes in yellow, red blood cells in red, and platelets in green.

Originally, lymphocytes are formed in the red bone marrow of the fetus. They, like all blood cells, derive from primitive cells known as *hematopoietic stem cells* (see Chapter 17). The stem cells destined to become lymphocytes of the adaptive immune system follow two developmental paths and differentiate into two major classes of lymphocytes—*B lymphocytes* and *T lymphocytes,* or simply **B cells** and **T cells** (Figure 21-12).

B cells do not attack pathogens themselves but instead produce molecules called **antibodies** that attack the pathogens or direct other cells, such as phagocytes, to attack them. B cell mechanisms are therefore often classified as **antibody-mediated**

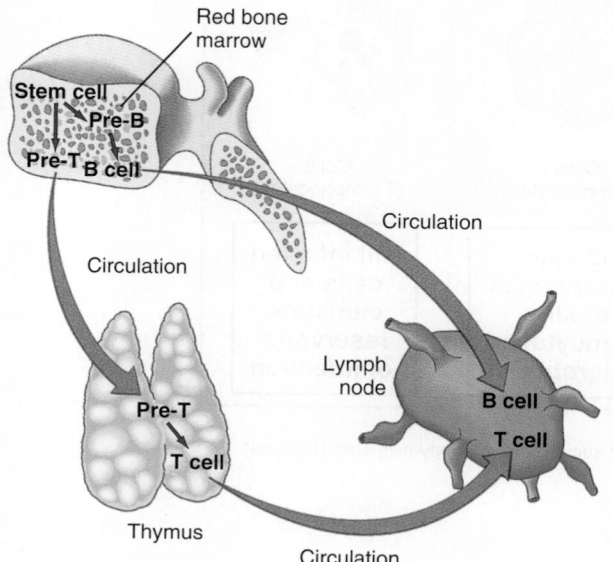

Figure 21-12 *Development of B cells and T cells.* Both types of lymphocytes originate from stem cells in the red bone marrow. Pre-B cells that are formed by dividing stem cells develop in the "bursa-equivalent" tissues in the yolk sac, fetal liver, and bone marrow. Likewise, pre-T cells migrate to the thymus, where they continue developing. Once they are formed, B cells and T cells circulate to the lymph nodes and spleen.

immunity. Because antibodies disperse freely in the blood plasma, where they accomplish their immune functions, this type of immunity is also sometimes called **humoral immunity.** The word *humoral* refers to body fluids, especially blood plasma.

Because T cells attack pathogens more directly, T cell immune mechanisms are classified as **cell-mediated immunity** or, more simply, **cellular immunity** (Figure 21-13).

Lymphocytes express proteins on their surfaces known as *surface markers.* Some of these proteins are unique to lymphocytes; some are shared by other types of cells. B cells and T cells each have some unique surface markers that not only distinguish B cells from T cells but also subdivide these categories into *subsets.* Lymphocyte subsets are clinically meaningful. For example, the condition of a patient's T-cell subsets is very important in understanding AIDS. The international system for naming surface markers on blood cells is the **CD system** (CD stands for cluster of differentiation; the number after "CD" refers to a single, defined surface marker protein). For example, the T-cell subsets that are clinically important in diagnosing AIDS are the CD4 and CD8 T-cell subsets.

Adaptive immunity requires activation of lymphocyte populations, which then begin their immune attack of specific antigens (or cells or viruses bearing those antigens). Such activation requires two activating signals: a specific antigen and a chemical signal (Figure 21-14). Each lymphocyte has receptors both for antigens and for signaling chemicals. Both receptors must be activated for the lymphocyte to begin its active immune function. Chemicals required to stimulate immune function may come from injured/infected cells or from microbes themselves.

The densest populations of lymphocytes occur in the bone marrow, thymus gland, lymph nodes, and spleen (Figure 21-15). From these structures, lymphocytes pour into the blood and then distribute themselves throughout the tissues of the body. After wandering through the tissue spaces, they eventually find their way into lymphatic capillaries. Lymph flow transports the lymphocytes through a succession of lymph nodes and lymphatic vessels and empties them by way of the thoracic and right lymphatic ducts into the subclavian veins. Thus returned to the blood, the lymphocytes embark on still another long journey—through blood, tissue spaces, and lymph and then back to blood. The survival value of the continued recirculation of lymphocytes and of their widespread distribution throughout body tissues seems apparent. It provides these major cells of the immune system ample opportunity to perform their functions of searching out, recognizing, and destroying foreign invaders.

Before continuing our discussion of adaptive immunity, please review the basic terminology in Box 21-3.

 QUICK CHECK

8. What is an antigen? What is the difference between a *self-antigen* and a *nonself-antigen*?
9. What is meant by the term **clone?**

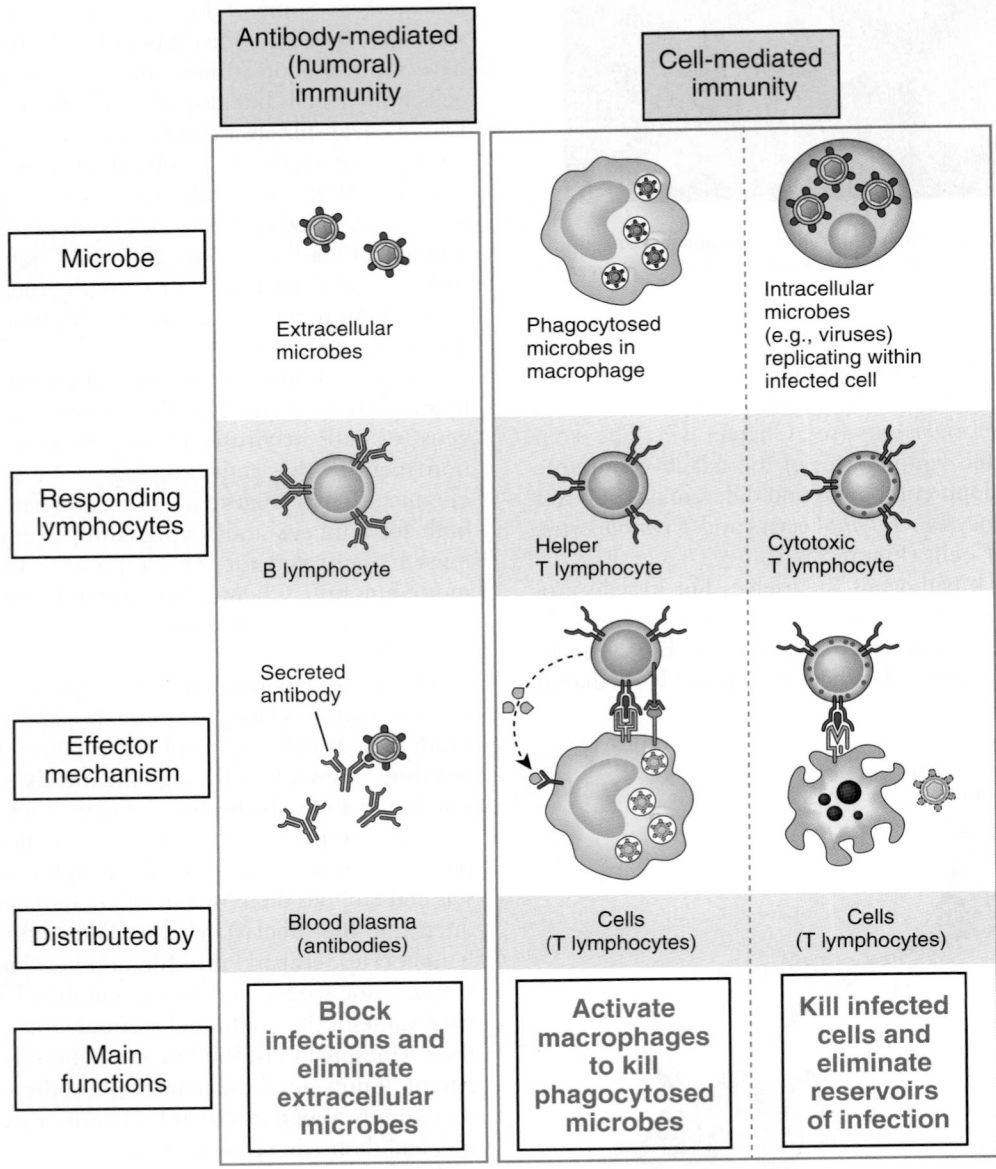

Figure 21-13 *Two strategies of adaptive immunity.* Simplified summary of antibody-mediated (humoral) immunity and cell-mediated (cellular) immunity.

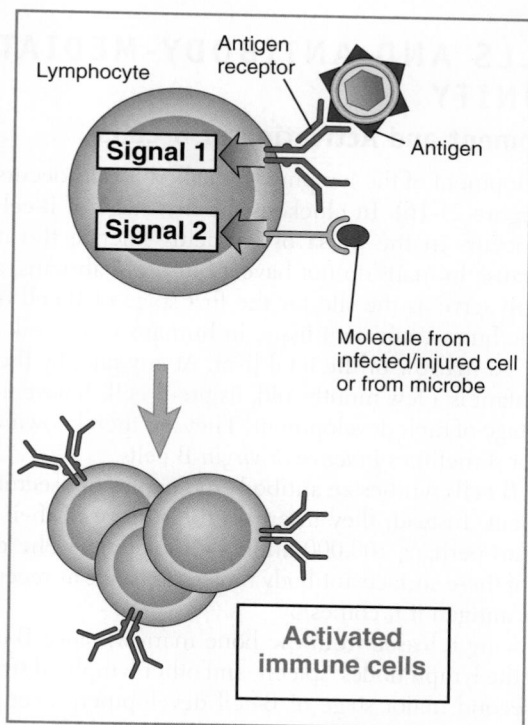

Figure 21-14 *Activation of lymphocytes.* B cells and T cells require two stimuli to activate them for immune function: a specific antigen and a chemical released from damaged or infected cells or from attacking microbes (microbial pathogens).

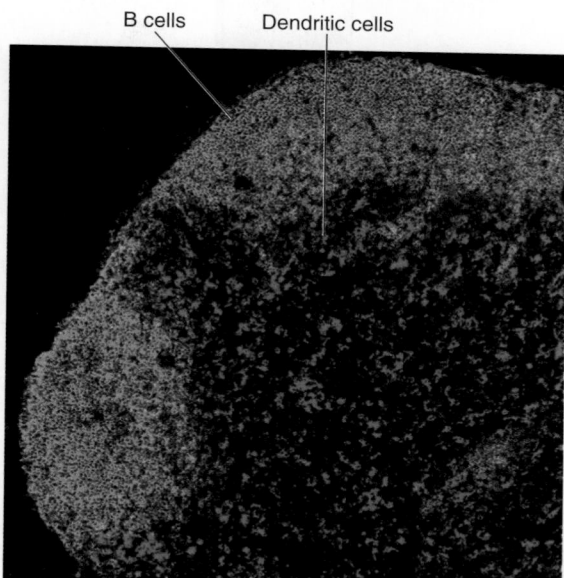

Figure 21-15 *B cells in a lymph node.* This micrograph of a lymph node shows fluorescent green stain in B cells, which are densely packed in the nodules of the lymph node. The cells marked by the red stain are a type of macrophage called dendritic cells (see Figure 21-8).

BOX 21-3: FYI

The Language of Adaptive Immunity

Before learning the mechanisms of adaptive (specific) immunity, you should be familiar with these terms:

Antigens—macromolecules (large molecules) that induce the immune system to make certain responses. Most antigens are foreign proteins. Some, however, are polysaccharides, and some are nucleic acids. Many antigens that enter the body are macromolecules located in the walls or outer membranes of microorganisms or the outer coats of viruses. Of course, antigens on the surfaces of some tumor cells (tumor markers) are not really from outside the body but are "foreign" in the sense that they are recognized as not belonging. The membrane molecules that identify all the normal cells of the body are called *self-antigens* or *major histocompatibility complex (MHC) antigens* (see Box 21-1 on p. 809). Foreign and tumor cell antigens can be called *nonself-antigens.*

Antigenic determinants—variously shaped, small regions on the surface of an antigen molecule; a less cumbersome name is *epitopes.* In a protein molecule, for instance, an epitope consists of a sequence of only about 10 amino acids that are part of a much longer, folded chain of amino acids. The sequence of the amino acids in an epitope determines its shape. Because the sequence differs in different kinds of antigens, each kind of antigen usually has specific and uniquely shaped epitopes.

Antibodies—plasma proteins of the class called *immunoglobulins.* Unlike most antigens, all antibodies are native molecules, that is, they are normally present in the body.

Combining sites—two small concave regions on the surface of an antibody molecule. Like epitopes, combining sites have specific and unique shapes. An antibody's combining sites are shaped so that an antigen's epitope that has a complementary shape can fit into the combining site and thereby bind the antigen to the antibody to form an **antigen-antibody complex.** Because combining sites receive and bind antigens, they are also called *antigen receptors* and *antigen-binding sites.*

Clone—family of cells, all of which have descended from one cell.

Complement—a group of proteins that, when activated, work together to destroy foreign cells.

Effector cell—a B cell or T cell that is actively producing an immune response, such as secreting antibodies (effector B cells) or directly attacking other cells (effector T cells); effector cells usually die during or just after their immune response; effector B cells are also called *plasma cells.*

Memory cell—a B or T cell that has been activated (no longer naïve) but is not an effector cell producing an active response; rather, a memory cell survives for a long period in the lymph nodes and if later exposed to the same specific antigen, forms a clone of cells that rapidly produce a specific immune response.

Naïve—refers to a B or T cell that is inactive, not yet have been exposed to (or having an opportunity to react with) a specific antigen; synonymous with "inactive" or "virgin."

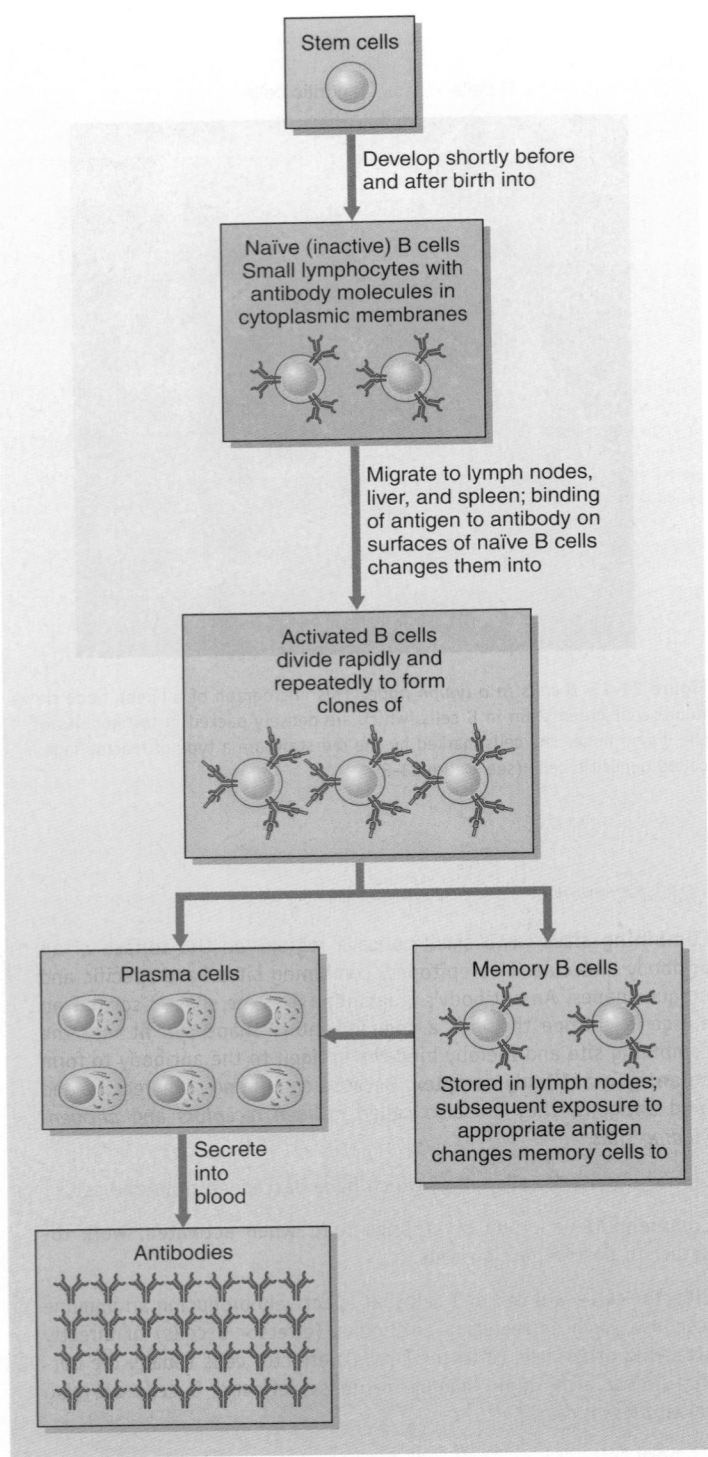

Figure 21-16 B cell development. B cell development takes place in two stages. *First stage:* shortly before and after birth, stem cells develop into naïve B cells. *Second stage* (occurs only if naïve B cell contacts its specific antigen): naïve B cell develops into activated B cell, which divides rapidly and repeatedly to form a clone of plasma cells and a clone of memory cells. Plasma cells secrete antibodies capable of combining with specific antigens that cause naïve B cell to develop into active B cell. Stem cells maintain a constant population of newly differentiating cells.

B CELLS AND ANTIBODY-MEDIATED IMMUNITY

Development and Activation of B Cells

The development of the lymphocytes called *B cells* occurs in two stages (Figure 21-16). In chickens the first stage of B-cell development occurs in the bursa of Fabricius—hence the name *B cells*. Because humans do not have a bursa of Fabricius, another organ must serve as the site for the first stage of B-cell development. The bursa-equivalent tissue in humans is the yolk sac and then the red marrow or the fetal liver. At any rate, by the time a human infant is a few months old, its pre-B cells have completed the first stage of their development. They are then known as **naïve B cells,** or sometimes *inactive* or *virgin* B cells.

Naïve B cells synthesize antibody molecules but secrete few if any of them. Instead, they insert on the surface of their plasma membranes perhaps 100,000 antibody molecules. The combining sites of these surface antibody molecules serve as receptors for a specific antigen if it comes by.

After being released from the bone marrow, naïve B cells circulate to the lymph nodes, spleen, and other lymphoid structures.

The second major stage of B-cell development occurs when the naïve B cells become activated. Activation of a B cell must be initiated by an encounter between a naïve B cell and its specific antigen, that is, one whose epitopes fit the combining sites of the B cell's surface antibodies (Figure 21-16).

The antigen binds to these antibodies on the B cell's surface. Antigen-antibody binding activates the B cell, triggering a rapid series of mitotic divisions. By dividing repeatedly, a single B cell produces a clone, or family, of identical B cells. Some of them differentiate to form **effector B cells** or **plasma cells.** Others do not differentiate completely but remain in the lymphatic tissue as the so-called **memory B cells.**

Plasma cells synthesize and secrete huge amounts of antibody. A single plasma cell, according to one estimate, secretes 2000 antibody molecules per second during the few days that it lives. All the cells in a clone of plasma cells secrete identical antibodies because they have all descended from the same B cell.

Memory B cells do not themselves secrete antibodies, but if they are later exposed to the antigen that triggered their formation, memory B cells then rapidly divide to produce more plasma cells and memory cells. The newly formed plasma cells then quickly secrete antibodies that can combine with the initiating antigen (see Figure 21-16)—thus quickly combating the antigen. Thus the ultimate function of B cells is to serve as ancestors of antibody-secreting plasma cells (effector B cells).

Antibodies (Immunoglobulins)

Structure of Antibody Molecules

Antibodies are proteins of the family called **immunoglobulin (Ig).** Like all proteins, antibodies are very large molecules and are composed of long chains of amino acids (polypeptides). Each immunoglobulin molecule consists of four polypeptide chains—two heavy chains and two light chains. Each polypeptide chain is intricately folded to form globular regions that are joined together in such a way that the immunoglobulin molecule as a whole is Y shaped. Look now at Figure 21-17, A. The twisted strands of red spheres (amino acids) in the diagram represent the light chains, and the two twisted strands of blue spheres represent the heavy chains. Each heavy chain consists of 446 amino acids. Heavy chains therefore are about twice as long and weigh about twice as much as light chains.

The regions with colored bars seen in Figure 21-17, *B*, represent *variable regions*, that is, regions in which the sequence of amino acids varies in different antibody molecules. Note the relative positions of the variable regions of the light and heavy chains; they lie directly opposite each other. Because the amino acid sequence determines conformation or shape, and because different sequences of amino acids occur in the variable regions of different antibodies, the shapes of the sites between the variable regions also differ. At the end of each "arm" of the Y-shaped antibody molecule, the unique shapes of the variable regions form a cleft that serves as the antibody's combining sites, or antigen-binding sites. It is this structural feature that enables antibodies to recognize and combine with specific antigens, both of which are crucial first steps in the body's defense against invading microorganisms and other foreign cells.

In addition to its variable region, each light chain in an antibody molecule also has a constant region. The constant region consists of 106 amino acids whose sequence is identical in all antibody molecules. Each heavy chain of an antibody molecule consists of three constant regions in addition to its one variable region. Identify the constant and variable regions of the light and heavy chains in Figure 21-17. Note the location of two *complement-binding sites* on the antibody molecule (one on each heavy chain).

In summary, an immunoglobulin, or antibody molecule, consists of two heavy and two light polypeptide chains. Each light chain consists of one variable region and one constant region. Each heavy chain consists of one variable region and three constant regions. Disulfide bonds join the two heavy chains to each other; they also bind each heavy chain to its adjacent light chain. An antibody has two antigen-binding sites—one at the top of each pair of variable regions—and two complement-binding sites located as shown in Figure 21-17.

Diversity of Antibodies

Every normal baby is born with an enormous number of different clones of B cells populating his or her bone marrow, lymph nodes, and spleen. All the cells of each clone are committed to

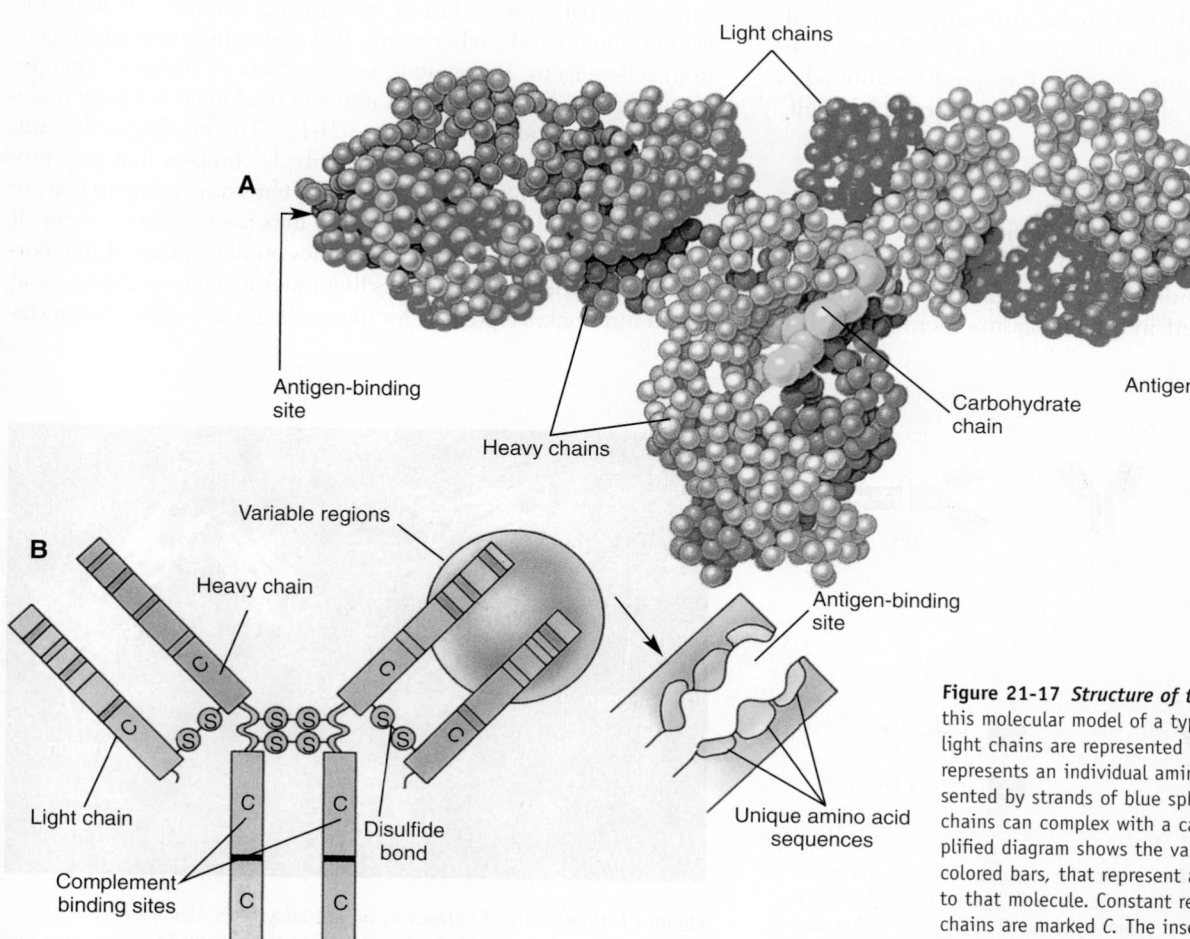

Figure 21-17 *Structure of the antibody molecule.* **A,** In this molecular model of a typical antibody molecule, the light chains are represented by strands of red spheres (each represents an individual amino acid). Heavy chains are represented by strands of blue spheres. Notice that the heavy chains can complex with a carbohydrate chain. **B,** This simplified diagram shows the variable regions, highlighted by colored bars, that represent amino acid sequences unique to that molecule. Constant regions of the heavy and light chains are marked *C.* The inset shows that the variable regions at the end of each arm of the molecule form a cleft that serves as an antigen-binding site.

synthesizing a specific antibody with a sequence of amino acids in its variable regions that is different from the sequence synthesized by any other of the innumerable clones of B cells.

How does this astounding diversity originate? One suggested answer is called the *somatic recombination hypothesis.* According to this explanation, our chromosomes do not contain whole genes for producing the heavy and light polypeptide chains that make up each antibody molecule. Instead, the genetic code is a set of separate sequences that are assembled into whole genes as a B cell develops. Because the separate sequences can be assembled in an astounding number of different combinations to form each whole gene, and since several different polypeptides are needed to make one antibody, any particular B cell is not likely to synthesize *exactly* the same antibody as any other B cell. Thus somatic recombination is a sort of "genetic lottery" that produces millions of unique genes by combining different gene segments and millions of unique antibodies by combining different polypeptides.

Antibody diversity may also be influenced by occasional mutations in the gene segments used to form the genes needed to produce antibodies. Evidence from several different studies shows that random genetic mutations—slight changes in the master DNA code—result in slight differences in the variable regions of antibodies.

If these hypotheses about antibody diversity are correct, it is possible to produce B cells that make antibodies against self-antigens. It is thought that although most such B cells are eliminated early in their development, before they produce antibodies that attack a person's own cells, all humans have some "antiself" B cells in their bodies.

Classes of Antibodies

There are five classes of antibodies, identified by letter names as immunoglobulins M, G, A, E, and D (Figure 21-18). *IgM* (abbreviation for immunoglobulin M) is the antibody that immature B cells synthesize and insert into their plasma membranes. It is also the predominant class of antibody produced after initial contact with an antigen. The most abundant circulating antibody, the one that normally makes up about 75% of all the antibodies in the blood, is *IgG.* It is the predominant antibody of the secondary antibody response—that is, following subsequent contacts with a given antigen. The IgG antibodies are those that cross the placental barrier during pregnancy to impart natural passive immunity to the offspring (see Table 21-4 and Box 21-7). *IgA* is the major class of antibody present in the mucous membranes of the body, in saliva, and in tears (covered later in Box 21-8). *IgE,* although minor in amount, can produce major harmful effects, such as those associated with allergies. *IgD* is present in the blood in very small amounts, and its precise function is as yet unknown. As Figure 21-18 shows, some immunoglobulin molecules are formed by the joining of several basic antibody units.

Functions of Antibodies

The function of antibody molecules—some 100 million trillion of them—is to produce **antibody-mediated immunity.** As we stated earlier, this type of immunity is also called *humoral immunity* because it occurs within plasma, which is one of the humors, or fluids, of the body.

Antigen-Antibody Reactions

Antibodies fight disease first by recognizing substances that are foreign or abnormal. In other words, they distinguish nonself-antigens from self-antigens. Recognition occurs when an antigen's epitopes (small regions on its surface) fit into and bind to an antibody molecule's antigen-binding sites (Figure 21-19). The binding of the antigen to antibody forms an antigen-antibody complex that may produce one or more effects. For example, it transforms antigens that are toxins (chemicals poisonous to cells) into harmless substances. It agglutinates antigens that are molecules on the surface of microorganisms. In other words, it makes them stick together in clumps, and this in turn makes it possible for macrophages and other phagocytes

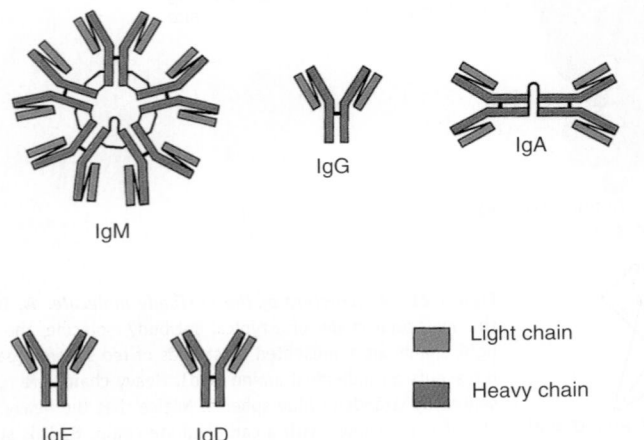

Figure 21-18 *Classes of antibodies.* Antibodies are classified into five major groups: immunoglobulin M (IgM), immunoglobulin G (IgG), immunoglobulin A (IgA), immunoglobulin E (IgE), and immunoglobulin D (IgD). Notice that each IgM molecule has five Y-shaped basic antibody units, IgA has two basic antibody units, and the others have a single basic antibody unit.

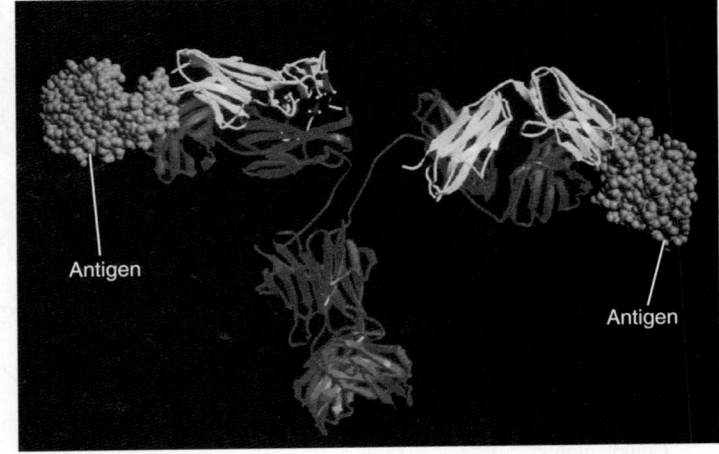

Figure 21-19 *Binding of antigen by an antibody.* This ribbon model of an antibody shows the heavy chains in red and the light chains in yellow. Note the blue antigen molecules bound to each antigen-binding site.

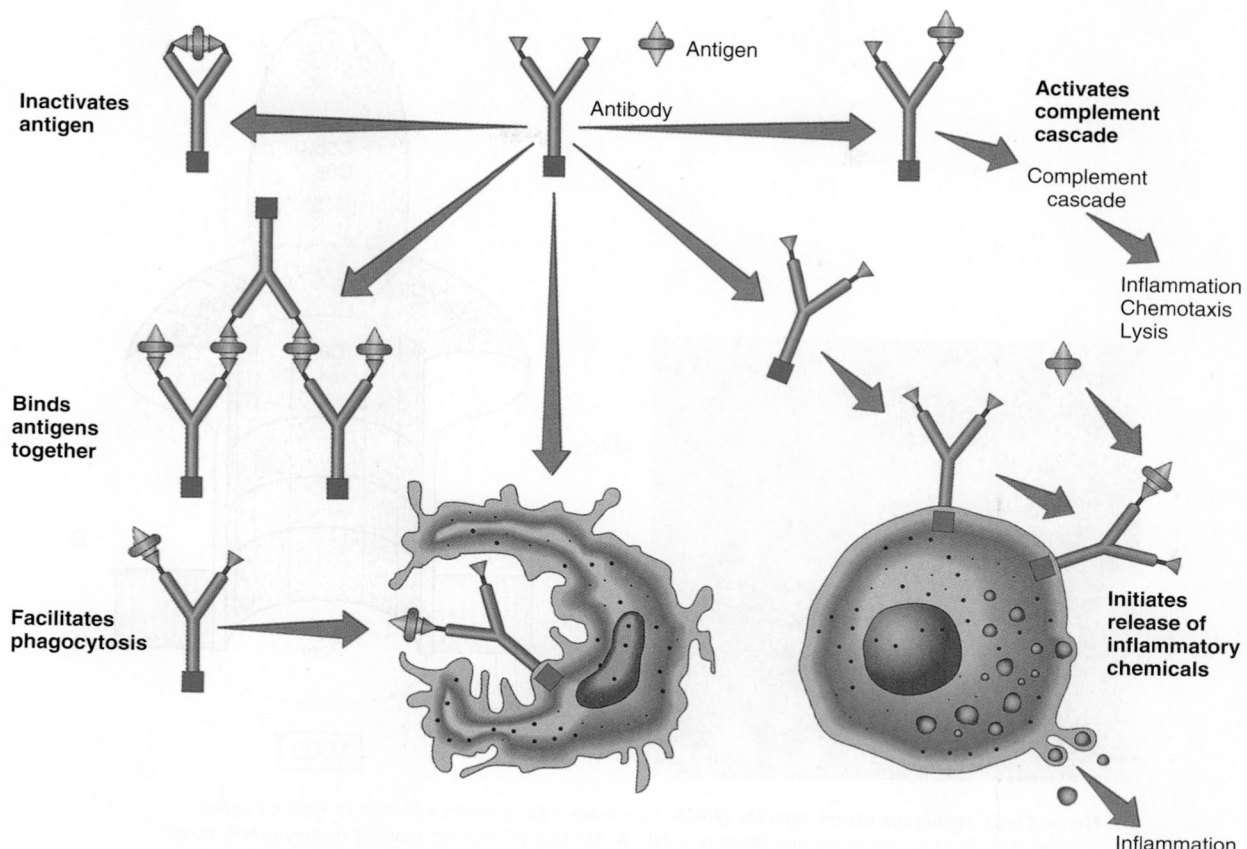

Figure 21-20 *Actions of antibodies.* Antibodies act on antigens by inactivating and bending them together to facilitate phagocytosis and by initiating inflammation and activating the complement cascade.

to dispose of them more rapidly by ingesting and digesting large numbers of them at one time. The binding of antigens to antibodies frequently produces still another effect—it alters the shape of the antibody molecule, not very much, but enough to expose the molecule's previously hidden complement-binding sites. This seems a trivial enough change, but it is not so. It initiates an astonishing series of reactions that culminate in the destruction of microorganisms and other foreign cells (Figure 21-20).

Complement

Complement is a component of blood plasma that consists of about 20 protein compounds. They are inactive enzymes that become activated in a definite sequence to catalyze a series of intricately linked reactions. The binding of an antibody to an antigen located on the surface of a cell alters the shape of the antibody molecule in a way that exposes its complement-binding sites. By binding to these sites, complement protein 1 becomes activated and touches off the catalytic activity of the next complement protein in the series. A rapid sequence, or cascade, of activity by the next protein, then the next, and the next, follows until the entire series of enzymes has functioned. The end result of this rapid-fire activity challenges the imagination. Some of the resulting reactions were summarized in Figure 21-10 (p. 810).

One of the more spectacular results of the complement cascade is the formation of **membrane attack complexes (MACs).** Molecules formed by the reactions of the complement cascade assemble themselves on the enemy cell's surface in such a way as to form a doughnut-shaped structure—complete with a hole in the middle (Figure 21-21). In effect, the complement has drilled a hole through the foreign cell's surface membrane. Ions and water rush into the cell through the MAC; it swells and bursts (Figure 21-22). **Cytolysis** is the technical name for this process. Nucleated cells usually resist cytolysis, but the influx of ions triggers apoptosis and thus kills the target cell another way.

Briefly, then, complement functions to kill foreign cells by cytolysis. In addition, various complement proteins serve other functions. Some, for example, cause vasodilation in the invaded area, and some attract neutrophils to the site and enhance phagocytosis.

The complement cascade can also be initiated by innate immune mechanisms. Complement protein 3 (C3) can become activated without any stimulation by an antigen. C3 is normally inactivated by enzymes, but it can produce the full complement effect if it binds to bacteria or viruses in the presence of a protein called *properdin.* Thus lysis of various foreign cells and viruses by complement can occur even when antibodies are not involved. This method of activating the complement cascade is often called

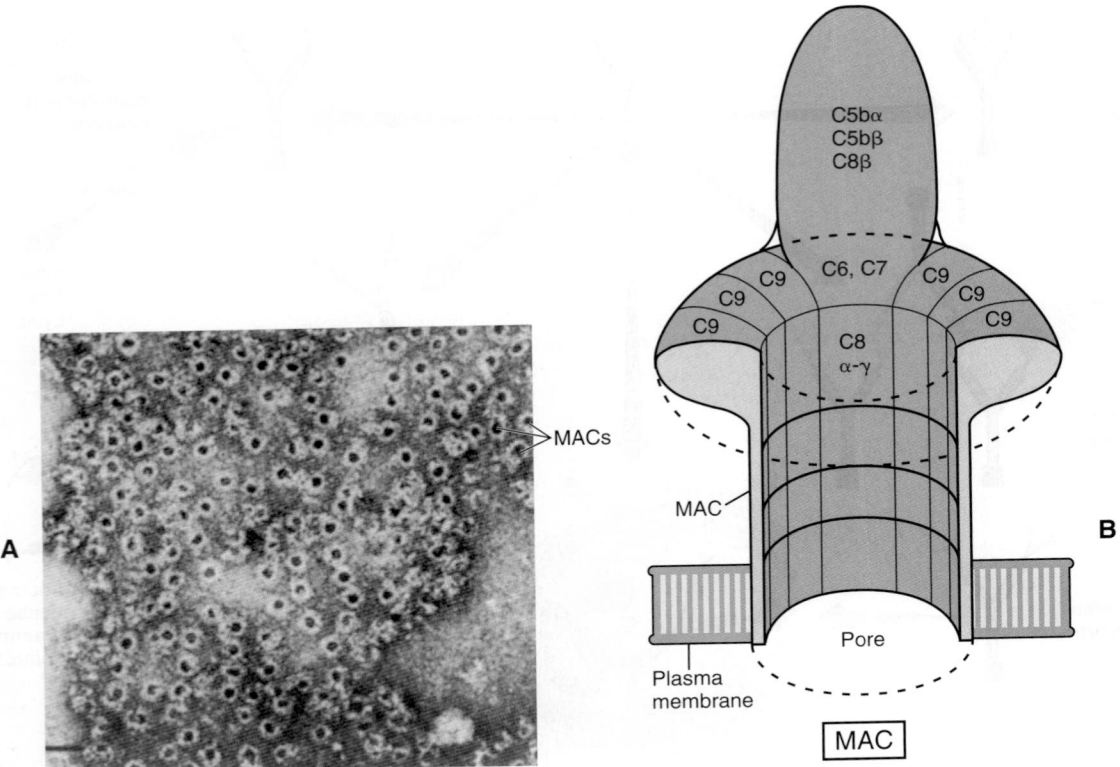

Figure 21-21 *Membrane attack complex (MAC).* Complement components assemble to form a ringlike complex that forms a pore in the membrane of a cell. **A,** Electron micrograph showing numerous MAC pores, each about 100 Å in diameter. **B,** Diagram of the structure of a MAC embedded in a plasma membrane.

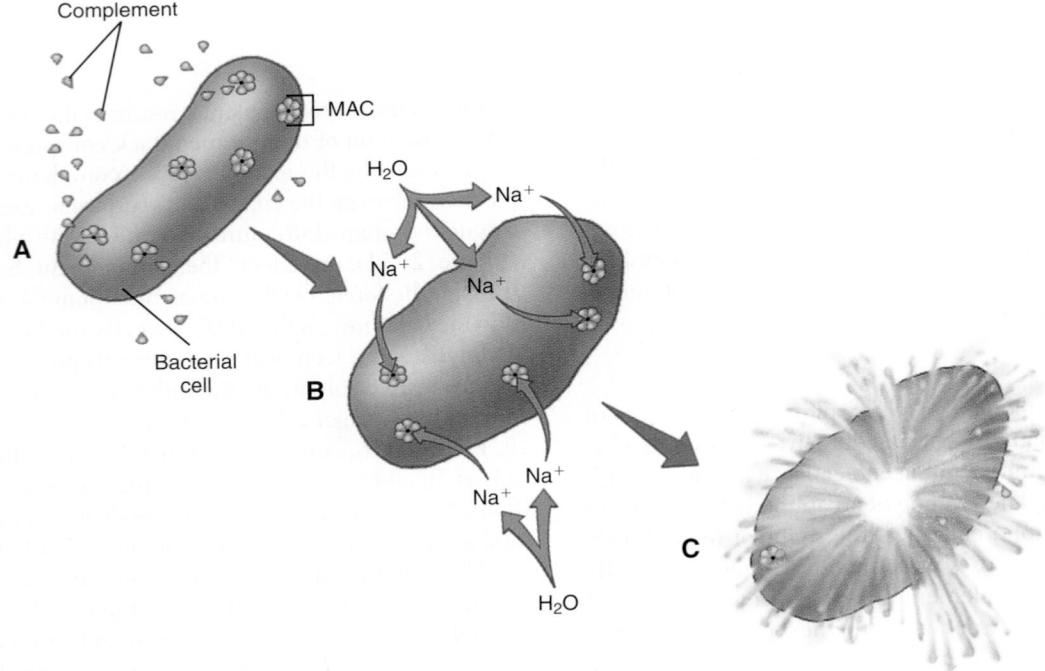

Figure 21-22 *Cytolysis of a bacterial cell.* **A,** Complement molecules activated by antibodies form doughnut-shaped membrane attack complexes (MACs) in a bacterium's plasma membrane. **B,** Holes in the complement complex allow sodium (Na^+) and then water (H_2O) to diffuse into the bacterium. **C,** After enough water has entered, the swollen bacterium bursts. In nucleated cells that resist cytolysis, the influx of calcium ions triggers apoptosis and thus kills the target cell another way.

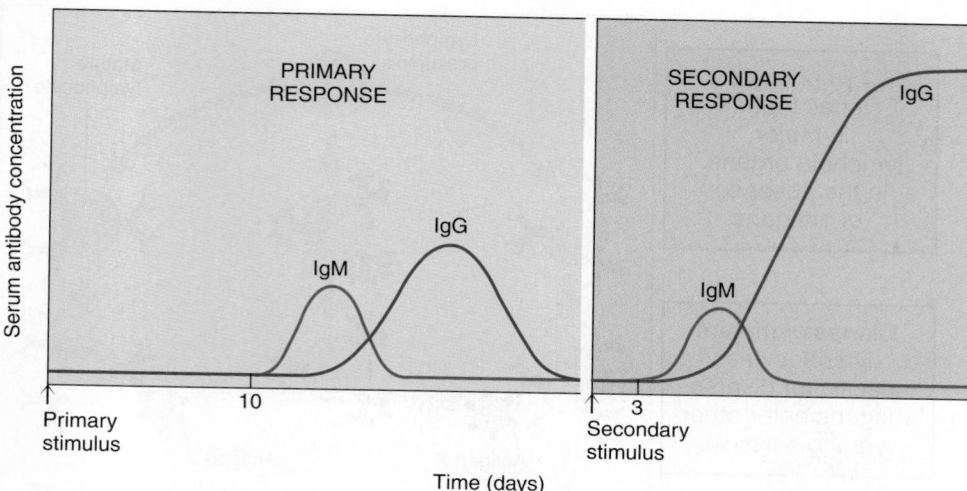

Figure 21-23 *Antibody response times.* The initial encounter with a specific antigen (primary stimulus) produces a primary response (increased production of IgM and IgG) in a few days. A later encounter (secondary stimulus) produces a secondary response in much less time. Note that both IgM production and IgG production occur more quickly in the secondary response—and IgG production also increases in the total amount of antibody produced.

the "alternate pathway" to distinguish it from the "classical pathway" involving antibodies.

Primary and Secondary Responses

As Figure 21-23 shows, an initial encounter with a specific antigen produces a *primary response* of increased antibody production in a few days. As the antigen is dealt with, the antibody levels decrease to their normal baseline levels. However, memory B cells that can respond to the triggering antigen have also been produced and wait for another encounter with the antigen. A later encounter with the same antigen triggers the waiting memory B cells and thus produces a secondary response in much less time. The memory B cells quickly divide to form more memory cells and a large number of plasma cells that produce antibodies against the known antigen. Thus the secondary response can be quicker and thus more effective.

BOX 21-4: HEALTH MATTERS

Immunization

Active immunity can be established artificially by using a technique called **vaccination.** The original vaccine was a live cowpox virus that was injected into healthy people to cause a mild cowpox infection. The term *vaccine* literally means "cow substance." Because the cowpox virus is similar to the deadly *smallpox virus,* vaccinated individuals developed antibodies that imparted immunity against both cowpox and smallpox viruses.

Modern vaccines work on a similar principle; substances that trigger the formation of antibodies against specific pathogens are introduced orally or by injection. Some of these vaccines are killed pathogens or live, *attenuated* (weakened) pathogens. Such pathogens still have their specific antigens intact, so they can trigger formation of the proper antibodies, but they are no longer *virulent* (able to cause disease). Although rare, these vaccines sometimes backfire and actually cause an infection. Many of the newer vaccines avoid this potential problem by using only the part of the pathogen that contains antigens. Because the disease-causing portion is missing, such vaccines cannot cause infection.

The amount of antibodies in a person's blood produced in response to vaccination or an actual infection is called the **antibody titer.** As you can see in the figure, the initial injection of vaccine triggers a rise in the antibody titer that gradually diminishes. Often, a **booster shot,** or second injection, is given to keep the antibody titer high or to raise it to a level that is more likely to prevent infection. The sec-

ondary response is more intense than the primary response because memory B cells are ready to produce a large number of antibodies at a moment's notice. A later accidental exposure to the pathogen will trigger an even more intense response—thus preventing infection.

Toxoids are similar to vaccines but use an altered form of a bacterial toxin to stimulate production of antibodies. Injection of toxoids imparts protection against toxins, whereas administration of vaccines imparts protection against pathogenic organisms and viruses.

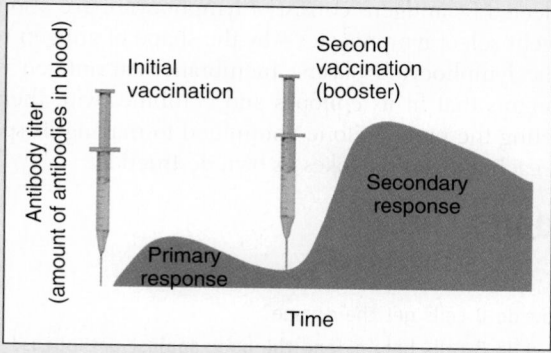

Changes in blood antibody titers following primary and secondary (booster) vaccinations.

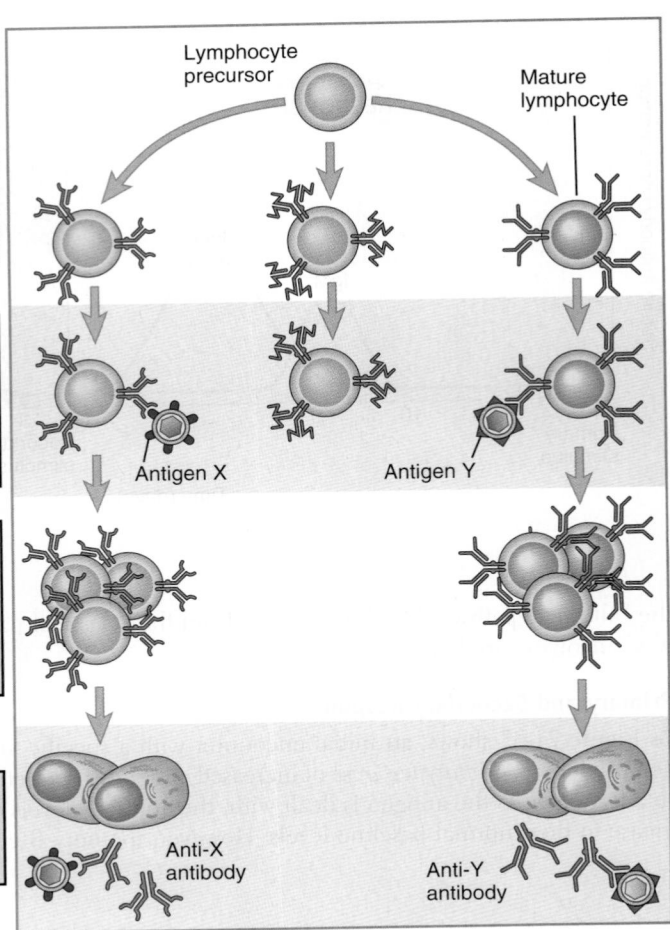

Figure 21-24 *The clonal selection theory.* This theory of immunity states that each specific antigen (here shown as X and Y) activates—selects—a previously produced clone of lymphocytes. The clone is "selected" because it is specifically targeted at the selecting antigen. The clone, when thus activated, produces effector cells that attack the antigen. B cells are shown here, but the same principle also applies to T cells.

Lymphocyte clones mature in major lymphoid organs, in the absence of antigens

Clones of mature lymphocytes specific for diverse antigens enter other lymphoid tissues

Antigen-specific clones are activated ("selected") by antigens

Antigen-specific immune responses occur

Lymphocyte precursor

Mature lymphocyte

Antigen X

Antigen Y

Anti-X antibody

Anti-Y antibody

Clonal Selection Theory

The *clonal selection theory*, which deals with antigen destruction, was first proposed in 1959 by Sir Macfarlane Burnet (Figure 21-24). It has two basic tenets. First, it holds that the body contains an enormous number of diverse clones of cells, each committed by certain of its genes to synthesize a different antibody. Second, the clonal selection theory postulates that when an antigen enters the body, it selects the clone whose cells are committed to synthesizing its specific antibody and stimulates these cells to proliferate and to thereby produce more antibodies. We now know that the clones selected by antigens consist of lymphocytes. We also know how antigens select lymphocytes—by the shape of antigen receptors on the lymphocyte's plasma membrane. An antigen recognizes receptors that fit its epitopes and combines with them. By thus selecting the precise clone committed to making its specific antibody, each antigen provokes its own destruction.

QUICK CHECK

10. How do B cells get their name?
11. How do B cells help defend the body against pathogens?
12. How does the structure of an antibody relate to its function?
13. Describe the mechanism by which complement destroys foreign cells.

T CELLS AND CELL-MEDIATED IMMUNITY

Development of T Cells

T cells, by definition, are lymphocytes that have made a detour through the thymus gland before migrating to the lymph nodes and spleen (see Figure 21-12). During their residence in the thymus, pre-T cells develop into **thymocytes,** cells that proliferate as rapidly as any in the body. Thymocytes divide up to three times each day, and, as a result, their numbers increase enormously in a relatively short time. They stream out of the thymus into the blood and find their way to a new home in areas of the lymph nodes and spleen called *T-dependent zones.* From this time on, they are known as T cells.

Activation and Functions of T Cells

Each T cell, like each B cell, displays antigen receptors on its surface membrane. They are not immunoglobulins as are B-cell receptor molecules but are proteins similar to them. When an antigen (preprocessed and presented by macrophages) encounters a naïve T cell whose surface receptors fit the antigen's epitopes, the antigen binds to the T cell's receptors. Here is where we see one of several differences between antibody-mediated immunity and cell-mediated immunity: antibodies can react to soluble antigens dissolved in the plasma, but T cells can only react to protein fragments presented on the

BOX 21-5: HEALTH MATTERS
Monoclonal Antibodies

Techniques that permit biologists to produce large quantities of pure and very specific antibodies have resulted in dramatic advances in medicine. As a new medical technology, the development of **monoclonal antibodies (MAbs)** has been compared in importance with advances in recombinant DNA, or genetic engineering.

Monoclonal antibodies are specific antibodies produced or derived from a population or culture of identical, or **monoclonal**, cells. In the past, antibodies produced by the immune system against a specific antigen had to be "harvested" from serum containing a large number of other antibodies. The total amount of a specific antibody that could be recovered was very limited, so the cost of recovery was high.

Monoclonal antibody techniques are based on the ability of immune system cells to produce individual antibodies that bind to and react with very specific antigens. We know, for example, that if the body is exposed to the varicella virus of chickenpox, B cells will produce an antibody that will react specifically with that virus and no other. With monoclonal antibody techniques, B lymphocytes that are activated after the injection of a specific antigen are "harvested" and then "fused" with other cells that have been transformed to grow and divide indefinitely in a tissue culture medium. These fused (hybrid) cells, called **hybridomas**, continue to produce the same antibody produced by the original lymphocyte (see the figure). The result is a rapidly growing population of identical, or monoclonal, cells that produce large quantities of a very specific antibody. Monoclonal antibodies have now been produced against an array of different antigens, including disease-producing organisms and various types of cancer cells.

The availability of very pure antibodies against specific disease-producing agents has been used in the commercial preparation of diagnostic tests that can be used to identify viruses, bacteria, and even specific cancer cells in the blood or other body fluids. The use of monoclonal antibodies serves as the basis for specific treatment of many human diseases—and their use in medicine is growing rapidly. For example, the antibody drug infliximab (Remicade) is used to treat rheumatoid arthritis (RA), and trastuzumab (Herceptin) is an antibody therapy aimed at breast cancer.

More recently researchers have been trying a similar approach with antibody fragments and tinier version of an antibody called a **nanobody.** Nanobodies are only one-tenth the size of an antibody, are easier and cheaper to produce, and can be taken orally for mucosal applications. Nanobodies are derived from a smaller, heavy-chain–only version of antibodies found only in llamas and camels. After immunizing llamas with the target antigen, the smaller antibodies are harvested, screened, and genetically engineered to be produced by yeasts or other common microorganisms. Although yet to enter the mainstream of therapy, such fragment-based therapies hold great promise for the treatment of many cancers, autoimmune conditions, and other disorders.

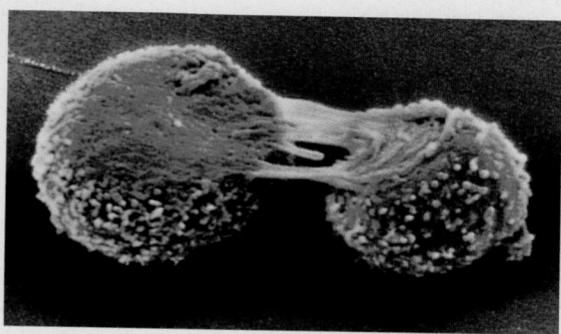

Hybridoma dividing.

surfaces of APCs (antigen-presenting cells). Thus T cells react to cells that are already infected—or have otherwise engulfed the antigen. B cells, on the other hand, react mainly to antigens that are in the plasma.

The presentation of an antigen by an antigen-presenting cell activates or sensitizes the T cell. The T cell then divides repeatedly to form a clone of identical *sensitized T cells* that form **effector T cells** and **memory T cells.** Effector T cells include **cytotoxic T cells,** which cause contact killing of a target cell. Cytotoxic T cells are also called *cytolytic T lymphocytes (CDLs)* or *killer T cells.* T memory cells ultimately produce additional active T cells. The process of T-cell development and activation is summarized in Figure 21-25.

The effector T cells then travel to the site where the antigens originally entered the body. There, in the inflamed tissue, the sensitized T cells bind to antigens of the same kind that led to their formation. However, T cells bind to their specific antigen only if the antigen is presented by an APC such as a macrophage. The antigen-bound sensitized T cells then release chemical messengers into the inflamed tissues.

The chemical messengers released by T cells, as we have stated earlier in this chapter, are called **cytokines.** Because some cytokines are secreted mainly by lymphocytes, such cytokines are sometimes called *lymphokines.* Names of a few individual cytokines are chemotactic factor, migration inhibition factor, macrophage activating factor, interleukin, and lymphotoxin.

Chemotactic factors attract macrophages, causing hundreds of them to migrate into the vicinity of the antigen-bound, sensitized T cell. *Migration inhibition factor* halts macrophage migration. *Macrophage activating factor* prods the assembled macrophages to destroy antigens by phagocytosing them at a rapid rate. **Interleukins (ILs)** are a class of about a dozen different cytokines that are involved in regulating a wide variety of immune functions in different cell types. **Lymphotoxin** is a powerful poison that acts more directly, quickly killing any cell it attacks.

Effector T cells that release lymphotoxin are the *cytotoxic T cells.* Figure 21-15 shows how lymphotoxins work in killing a cell—a cancer cell in this case. After having been activated by the presentation of tumor cell antigens by an APC such as a dendritic cell, the cytotoxic T cell then becomes active and finds a tumor cell bearing that antigen. The cytotoxic T cell binds directly to the surface of the tumor cell and releases two kinds of molecules: perforin and granzymes. The lymphotoxins called *perforin* produce a ringlike hole in the plasma membrane of the

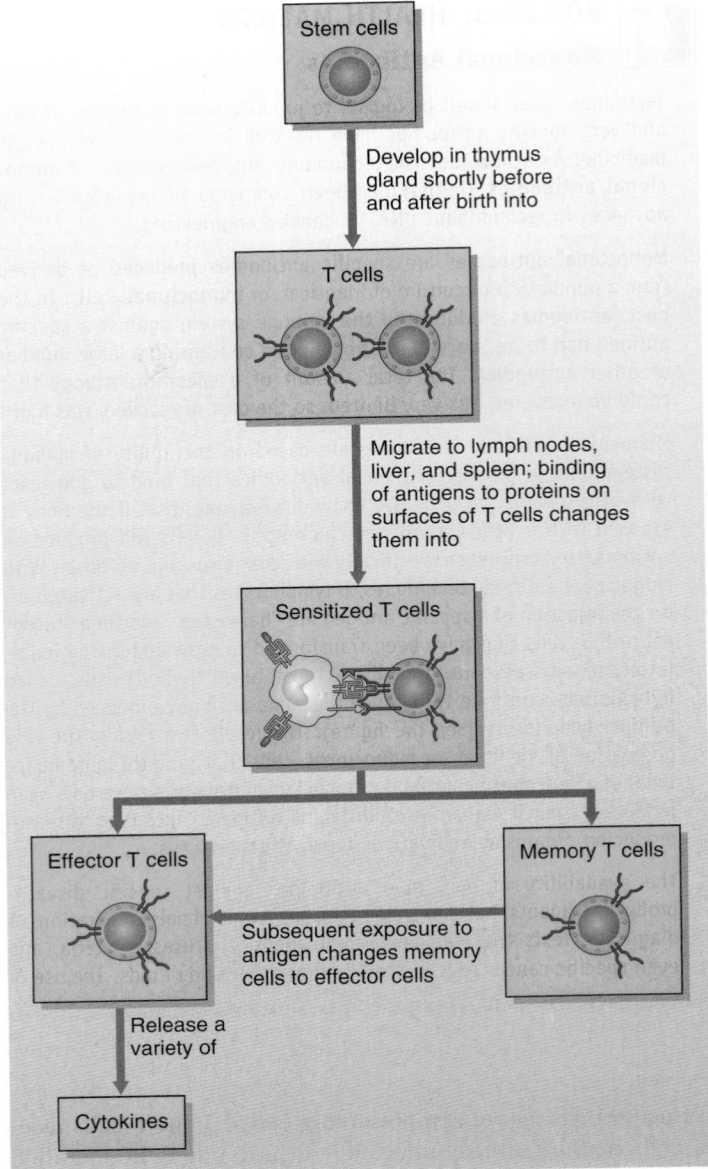

Figure 21-25 *T cell development.* The *first stage* occurs in the thymus gland shortly before and after birth. Stem cells maintain a constant population of newly differentiating cells as they are needed. The *second stage* occurs only if a T cell is presented an antigen, which combines with certain proteins on the T cell's surface.

target cell, similar to the one produced by the MHC formed by complement (see Figure 21-26). The *granzymes* enter the target cell through the perforin-ring hole and trigger apoptosis of the cell—thus killing it.

Besides cytotoxic T cells, at least two other populations of effector T cells are found in the body: **helper T cells (T_H cells)** and **suppressor T cells.** Both types of cells help regulate adaptive immune function by regulating B-cell and T-cell function.

Helper T cells help other lymphocytes by secreting cytokines that stimulate B cells and cytotoxic T cells. T_H cytokines also stimulate phagocytes and other leukocytes (Figure 21-27). These cytokines include *interleukin-2 (IL-2)* and *interleukin-4 (IL-4 or B-cell differentiating factor)*. Like other effector T cells, naïve T_H cells are activated by antigens presented on the surfaces of APCs and form a clone that differentiates into *effector T_H cells* and *memory T_H cells*.

Suppressor T cells, often called *regulator T cells*, act to suppress B-cell differentiation into plasma cells. The antagonistic

action allows the immune system to finely tune its antibody-mediated response. Suppressor T cells also regulate other T cells, helping "turn off" an immune response to restore homeostasis, for example. Suppressor T cells help maintain self-tolerance by reducing T-cell reactions to self-antigens. For this reason, researchers are trying to find ways to enhance suppressor T-cell function to treat autoimmunity (see p. 829) or in organ transplants to prevent rejection of donor tissue.

Summarizing briefly, the function of T cells is to produce cell-mediated immunity. They search out, recognize, and bind to appropriate antigens located on the surfaces of cells. This kills the cells—the ultimate function of killer T cells. Usually these are not the body's own normal cells but are cells that have been invaded by viruses, that have become malignant, or that have been transplanted into the body. Killer T cells therefore function to defend us from viral diseases and cancer, but they also bring about rejection of transplanted tissues or organs. T cells also serve as overall regulators of adaptive immune mechanisms.

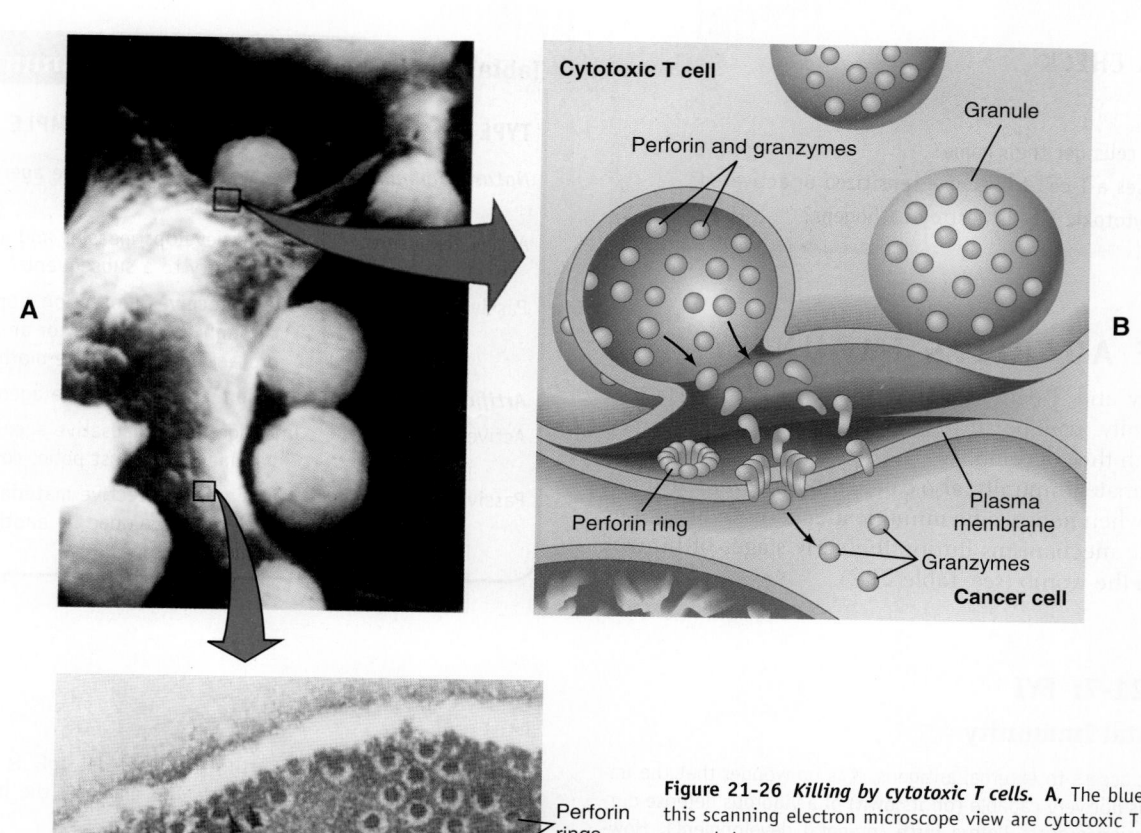

Perforin rings

Figure 21-26 *Killing by cytotoxic T cells.* **A,** The blue spheres seen in this scanning electron microscope view are cytotoxic T cells attacking a much larger cancer cell. T cells are a significant part of our defense against cancer and other abnormal or foreign cells. **B,** After adhering to the tumor cell, the cytotoxic T cell releases perforin, which forms ring-like holes in the tumor cell's membrane, and granzymes, which pass through the perforin rings to trigger apoptosis (programmed cell death) in the tumor cell. **C,** Electron micrograph showing perforin rings with an average diameter of about 160 Å, which is much larger than the MHC rings formed by complement (compare with Figure 21-21).

BOX 21-6: HEALTH MATTERS
Immunity and Cancer

One of the many functions of the immune system is to constantly guard against the development of cancer. Cell mutations occur frequently in the normal body, and many of the mutated cells formed are cancer cells. Cancer cells, you may recall, are cells capable of forming tumors in many different parts of the body—unless they are destroyed before this can happen. Abnormal antigens presented on cancer cells, called **tumor-specific antigens,** are present in the plasma membranes of some cancer cells in addition to self-antigens. Lymphocytes, in their continual wanderings through body tissues, are almost sure to come in contact with newly formed cancer cells. Hopefully, the lymphocytes recognize these cells by their abnormal antigens and quickly initiate reactions that kill the cancer cells. Ab-

normal antigens on cancer cells, called tumor-specific antigens or **tumor markers,** are present in the plasma membranes of some cancer cells in addition to self-antigens or MHC antigens. Examples of cancer markers include (1) **carcinoembryonic antigen (CEA)**—found normally in the fetus and elevated in colorectal and other adult cancers; (2) **alpha-fetoprotein (AFP)**—normal fetal protein, whose presence in the adult strongly suggests liver or germ cell cancer; (3) **CA-125**—tumor antigen associated with ovarian cancer; and (4) **prostate-specific antigen (PSA),** which is elevated in both benign and malignant prostate disease. The relationship between cancer and the immune system continues to be an area of intense research by scientists looking for effective cancer treatments.

14. How do T cells get their name?
15. What causes a T cell to become sensitized or activated?
16. How do cytotoxic T cells destroy pathogens?

TYPES OF ADAPTIVE IMMUNITY

B-cell immunity and T-cell immunity, the two major types of adaptive immunity, can be further classified according to the manner in which they develop.

Recall that innate immunity, also called *inborn* or *inherited immunity*, occurs when nonspecific immune mechanisms are put in place by genetic mechanisms during the early stages of human development in the womb (see Table 21-2).

Table 21-4 Types of Adaptive Immunity

TYPE	DESCRIPTION OR EXAMPLE
Natural Immunity	Exposure to the causative agent is not deliberate
Active (exposure)	A child develops measles and acquires an immunity to a subsequent infection
Passive (exposure)	A fetus receives protection from the mother through the placenta, or an infant receives protection through the mother's milk
Artificial Immunity	Exposure to the causative agent is deliberate
Active (exposure)	Injection of the causative agent, such as a vaccination against polio, confers immunity
Passive (exposure)	Injection of protective material (antibodies) that was developed by another individual's immune system

BOX 21-7: FYI
Prenatal Immunity

Without direct access to external antigens, it is no wonder that the immune system is not very capable (on its own) of a vigorous defense during its maturation process before birth (prenatal development). However, as part A of the figure shows, certain antibodies from the mother (maternal antibodies) can be actively transported across the maternal-fetal blood barrier (trophoblast). Only IgG antibodies in the mother's blood can bind to the receptors, which then trigger endocytosis and transport each IgG antibody across to the fetal bloodstream. This mechanism provides passive natural immunity before and shortly after birth.

Part B of the figure shows that at birth, the newborn has adult levels of IgG—but nearly all of it came from the mother (maternal IgG,

black line). Shortly after birth, the maternal IgG is broken down *(black line)* and replaced with new IgG made by the newborn's own immune system *(purple line)*.

Note also in part B of the figure that the concentration of IgM *(broken black line)* is only about 20% of the adult level at birth but steadily increases after birth. IgM reaches adult levels in about 2 years. IgA, an important component of the mucosal immune system (see Box 21-8), also begins to rise at birth. IgA reaches adult levels in just a few months. All three types of antibody are also found in breast milk, providing another avenue of passive immunity after birth.

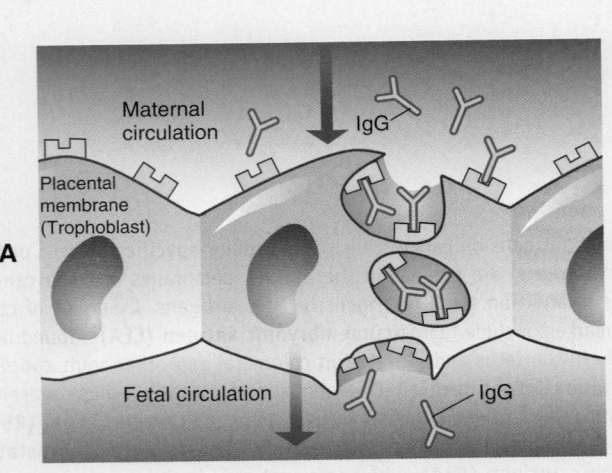
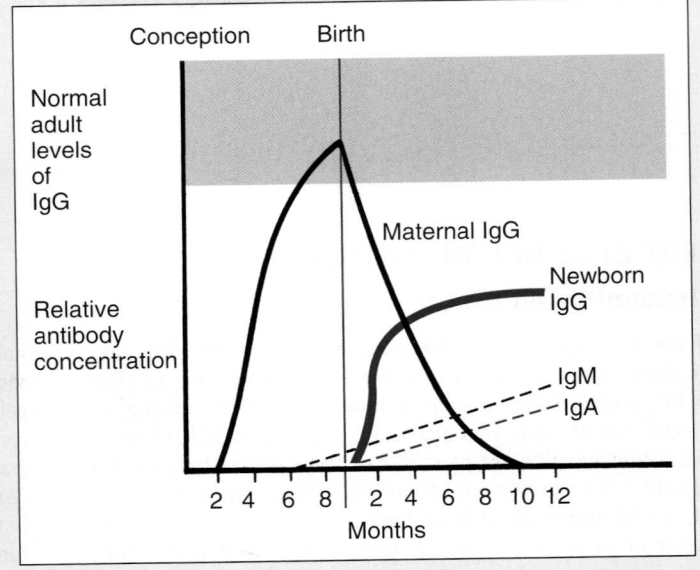

A, *Transport of antibodies across the placenta.* **B,** *Antibody concentrations before and after birth.*

Adaptive immunity, our focus here, is instead a specific kind of resistance that develops after we are born. Acquired immunity may be further classified as either *natural immunity* or *artificial immunity*, depending on how the body is exposed to the antigen. Natural exposure is not deliberate and occurs in the course of everyday living. We are naturally exposed to many disease-causing agents on a regular basis. Artificial, or deliberate, exposure to potentially harmful antigens is called *immunization.*

Natural and artificial immunity may be "active" or "passive." Active immunity occurs when an individual's own immune system responds to a harmful agent, regardless of whether that agent was naturally or artificially encountered. Passive immunity results when immunity to a disease that has developed in

another individual or animal is transferred to an individual who was not previously immune. For example, antibodies in a mother's milk confer passive immunity to her nursing infant (see also Box 21-7). Active immunity generally lasts longer than passive immunity. Passive immunity, although temporary, provides immediate protection. Table 21-4 lists the various forms of adaptive immunity.

SUMMARY OF ADAPTIVE IMMUNITY

Adaptive immunity is specific immunity—that is, it targets specific antigens. Two special types of lymphocytes play a major role in immunity: B cells and T cells. As Figure 21-27 shows, B cells

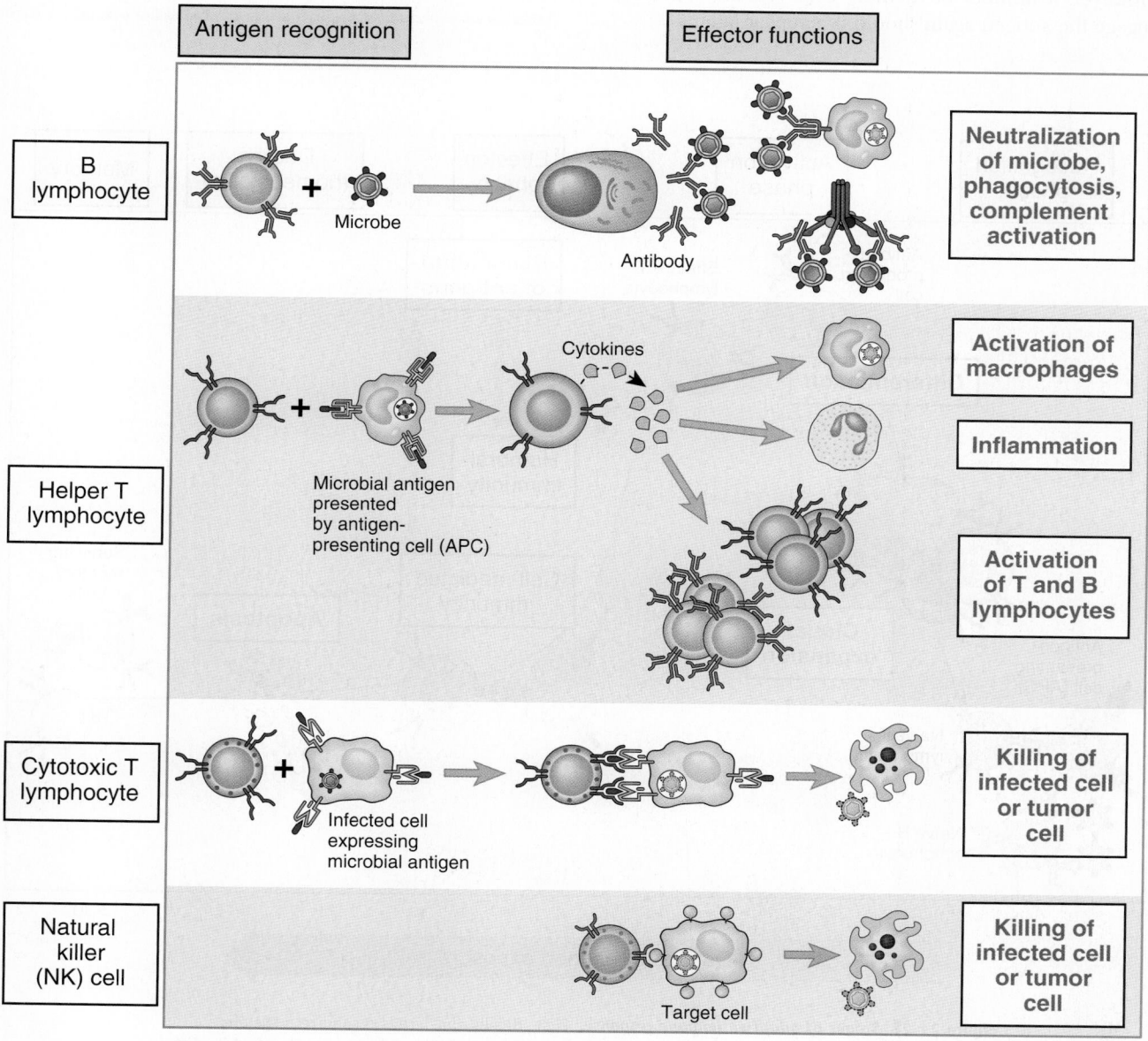

Figure 21-27 *Lymphocyte functions.* This diagram summarizes, in a simplified way, the main functions of the major classes of lymphocytes. B and T lymphocytes participate in adaptive immunity and NK cells participate in innate immunity.

recognize specific antigens and produce specific antibodies (immunoglobulins) to destroy the antigen—antibody-mediated or humoral immunity. T cells recognize antigens presented on cell surfaces to attack infected and abnormal cells in several ways—cell-mediated or cellular immunity.

Adaptive immunity progresses along a pathway of stages outlined in Figure 21-28. First, B cells and T cells recognize a specific antigen. Next, the B and T cells are activated—expanding their population (a clone) and thus producing effector cells and memory cells. Then, the effector cells get to work attacking the source or sources of the antigen. When there are no longer enough antigens to continue stimulating these immune responses, the effector B and effector T cells die off through the process of apoptosis. This represents a return to a homeostatic balance after the immune response. However, a number of memory cells remain—ready to quickly engage the antigen again should it reappear later.

Figure 21-29 summarizes the activity of the adaptive immune system in a different way. This figure shows the interaction of different cells and cytokines (in this case, interleukins) in a simplified model of the cooperative nature of the adaptive immune responses.

QUICK CHECK

17. What is the difference between inherited and acquired immunity?
18. What is the difference between natural and artificial immunity?
19. What is the difference between active and passive immunity?

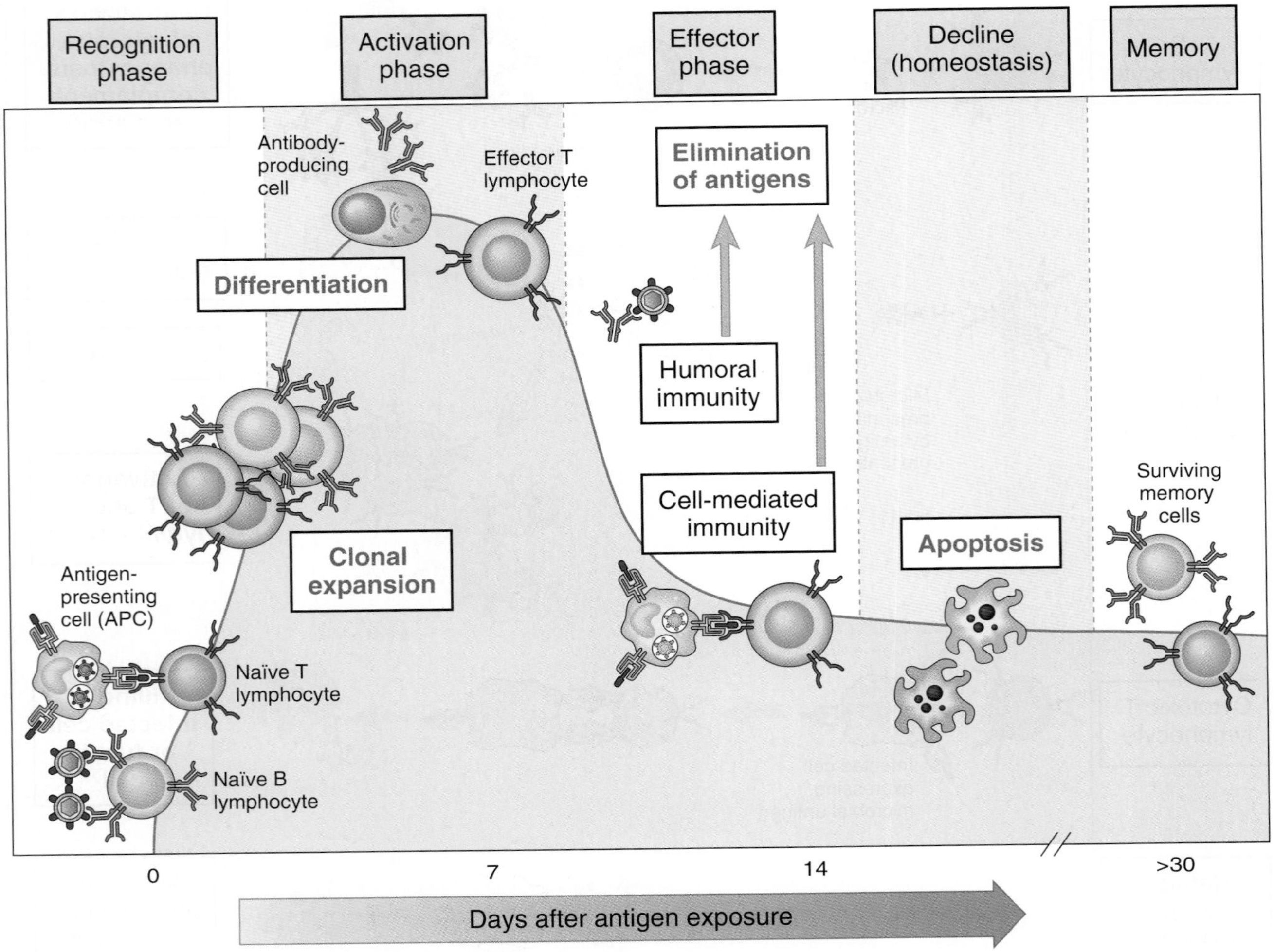

Figure 21-28 *Stages of adaptive immune response.* First, B cells and T cells recognize a specific antigen. Next, the B and T cells are activated—expanding their population (clonal expansion) while differentiating into effector cells and memory cells. Then the effector cells get to work attacking the source or sources of the antigen—humoral (antibody-mediated) immunity and cell-mediated immunity. As antigen levels decline, effector cells die off (apoptosis). Memory cells then remain—ready to quickly engage the antigen again later.

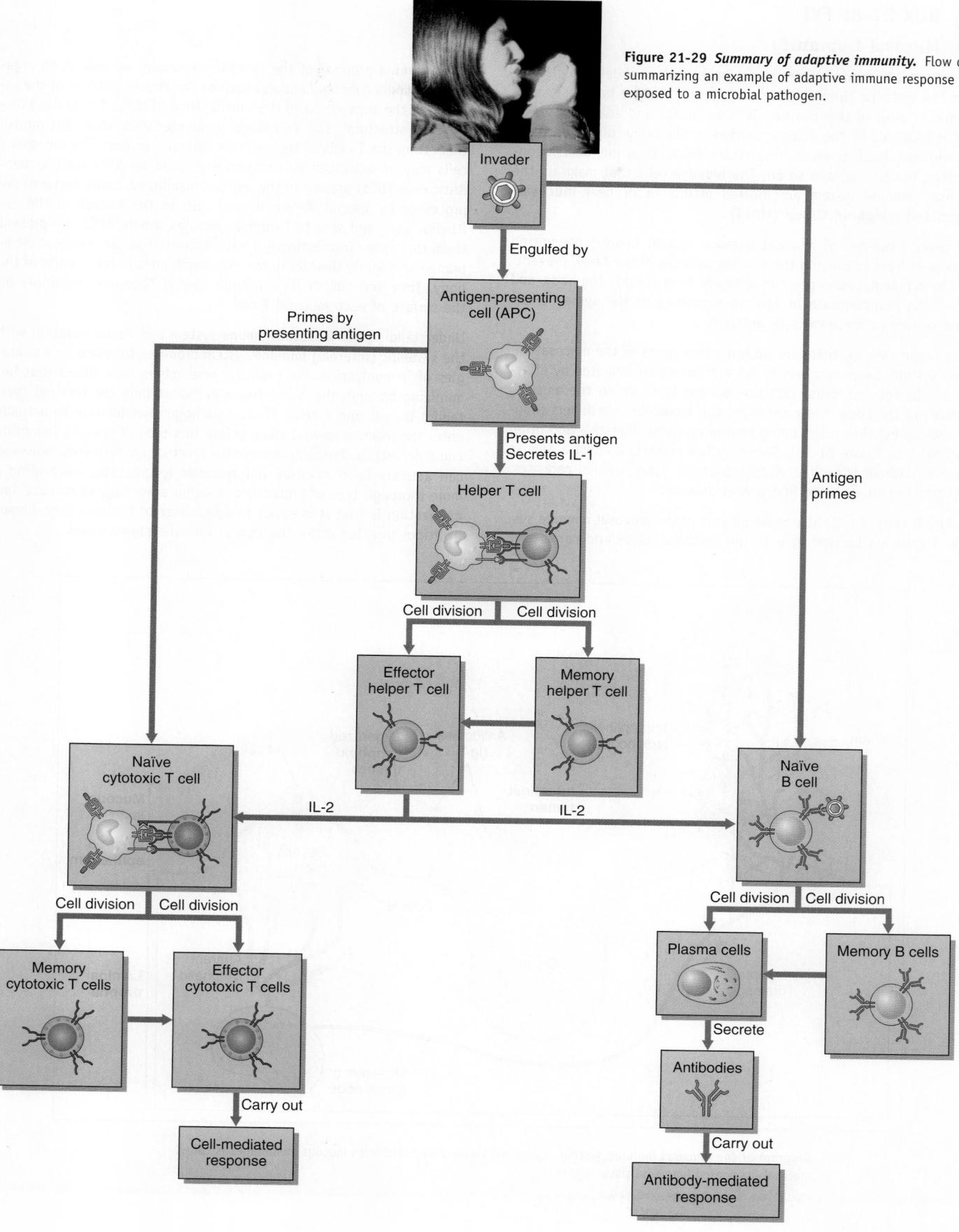

Figure 21-29 *Summary of adaptive immunity.* Flow chart summarizing an example of adaptive immune response when exposed to a microbial pathogen.

BOX 21-8: FYI
Mucosal Immunity

The **mucosal immune system** is a complex system of defense distinct from the systemic (internal) immune system that we have been discussing in most of this chapter. It is an innate and adaptive system that is localized to the mucous barriers of the body: digestive tract, urinary/reproductive tracts, respiratory tract, exocrine ducts, conjunctiva, middle ear, and so on. The immune cells that make up the mucosal immune system are located mainly in or near **mucosal-associated lymphoid tissue (MALT).**

The main functions of mucosal immune system involve preventing pathogens from colonizing the mucous surfaces of the body, preventing the accidental absorption of antigens from outside the body, and preventing inappropriate or intense responses of the systemic immune system to these external antigens.

As the figure shows, there are several components of the mucosal immune system. Large numbers of IgA antibodies are secreted by effector B cells (plasma cells) into the mucous layer lining the mucosal surfaces of the body. These secretory IgA molecules are dimers (double molecules) that resist being broken down by digestive and other enzymes (see Figure 21-18). Secretory IgA protects against a diverse group of pathogens such as viruses, bacteria, fungi, animal parasites, thus forming an effective first line of defense.

Besides B cells, T cells also make up part of the mucosal immune system. T cells are located in both the epithelial layer and connective layer (lamina propria) of the mucous membrane, as well as in organized *regional lymphoid nodules* such as the Peyer's patches of the intestines, the appendix, and the tonsils. Most of these T cells have distinctive structural and functional characteristics that distinguish them from the T cells of the systemic immune system. The mucosal T cells may be activated by antigens presented by APCs such as dendritic cells (DCs) present in the mucous membrane. Some antigens are processed by special *M (membrane) cells* in the surface of the epithelial layer and sent to lymphoid nodules, where APCs can present them to T cells. Interestingly, T cells activated in one mucosal membrane can migrate directly to mucous membranes in other parts of the body. They accomplish this through special "homing" receptors on the surface of each mucosal T cell.

Understanding the mucosal immune system and its cooperation with the systemic (internal) immune system promises to reveal new strategies of immunization. For example, researchers have found that immunizing through the bloodstream activates only the internal (systemic) B cells and T cells. Thus, a pathogen would have to actually enter the internal environment before this type of specific immunity could protect us. Immunization of the mucosal lymphocytes, however, can activate both mucosal and systemic lymphocytes—providing a more thorough type of protection. Another advantage of mucosal immunization is that it is easier to administer to patients than immunizations injected under the skin or into the bloodstream.

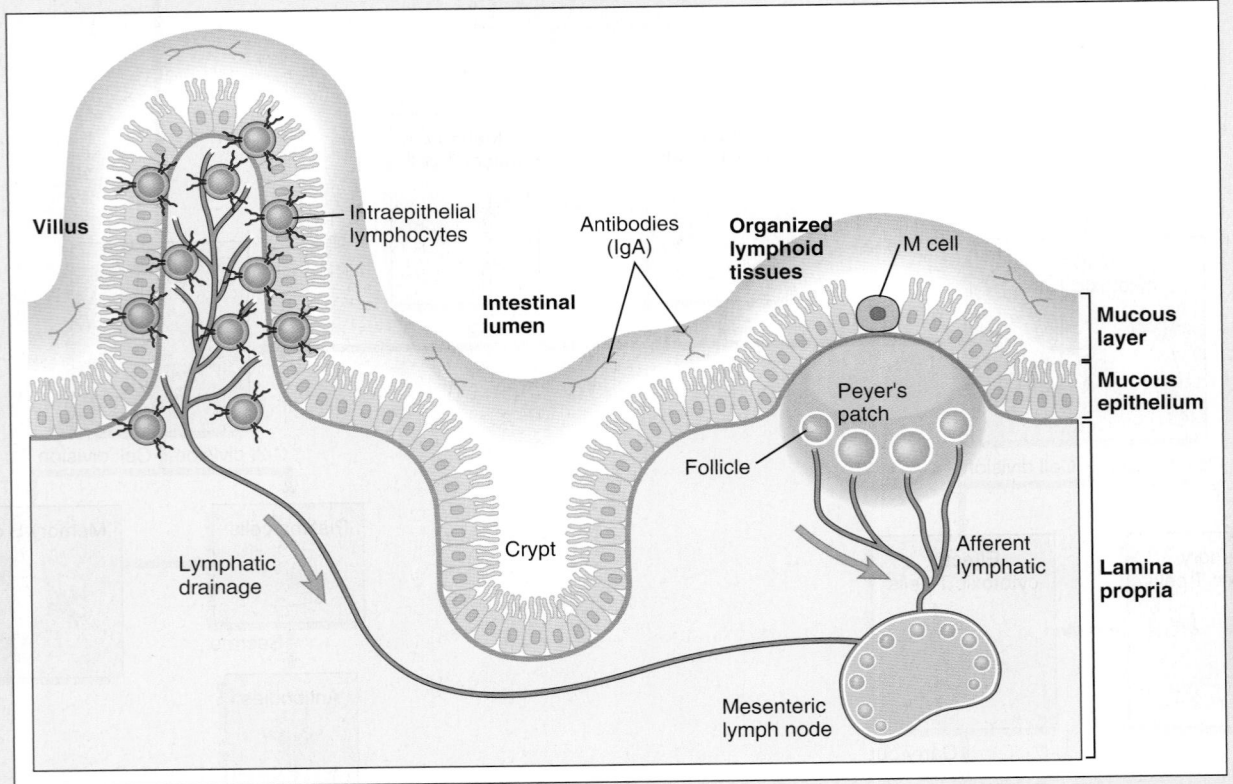

Diagram of the mucosal immune system. Lymphoid tissue associated with mucous membranes is called mucosal-associated lymphoid tissue (MALT).

THE BIG PICTURE
Immune System and the Whole Body

The "big picture" of the immune system's role in maintaining the relative constancy of the internal environment is probably easier to "see" than any other system. After all, its agents—antibodies, lymphocytes, and other substances and cells—are everywhere in the body. They even stand guard on the outside surface of the body. Without the defensive activity of the immune system, our internal constancy would be ruined by cancer, infections, and even minor injuries.

In describing the various mechanisms of the immune system, we have used the analogy of a militaristic-style security force. As useful as this analogy might be, it may mislead us into believing that the immune system is a completely independent group of defensive agents. Nothing could be further from the truth.

First of all, recent evidence has shown scientists that the immune system is regulated to some degree by the nervous and endocrine systems—recently triggering the development of the new field of *neuroimmunology*. These systems, as you already know, are in turn influenced by feedback from all parts of the body.

Second, the agents of the immune system are not a separate, distinct group of cells and substances. They include blood cells, skin cells, mucosal cells, brain cells, liver cells, and many other types of cells and their secretions. Thus the immune system is more like a self-defense force made up of ordinary citizens who work shoulder-to-shoulder with military specialists.

Mechanisms of Disease

DISORDERS OF THE IMMUNE SYSTEM

There are two basic mechanisms for disorders of immunity. The immune defenses can either overreact to antigens or fail to react to an antigen to produce disease. We will briefly describe a few examples of each of these mechanisms.

Hypersensitivity of the Immune System

Hypersensitivity is a type of inappropriate or excessive response of the immune system. The three major types of immune hypersensitivity discussed in the following sections are allergy, autoimmunity, and isoimmunity.

Allergy

The term **allergy** is used to describe hypersensitivity of the immune system to relatively harmless environmental antigens. Antigens that trigger an allergic response are often called **allergens** (AL-er-jens). One in six Americans has a genetic predisposition to an allergy of some kind.

Immediate allergic responses involve antigen-antibody reactions, mainly IgE. Before such a reaction occurs, a susceptible person must be exposed repeatedly to an allergen—triggering the production of antibodies. After a person is thus *sensitized*, exposure to an allergen causes antigen-antibody reactions that trigger the release of histamine, kinins, and other inflammatory substances. These responses usually cause typical allergy symptoms such as runny nose, conjunctivitis, and *urticaria* (hives). In some cases, however, these substances may cause constriction of the airways, relaxation of blood vessels, and irregular heart rhythms that can progress to a life-threatening condition called **anaphylactic shock** (see Chapter 19, p. 772). Drugs called **antihistamines** are sometimes used to relieve the symptoms of this type of allergy.

Delayed allergic responses, on the other hand, involve cell-mediated immunity. In **contact dermatitis,** for example, T cells trigger events that lead to local skin inflammation a few hours or days after initial exposure to an antigen. Exposure to poison ivy, soaps, and certain cosmetics may cause contact dermatitis in this manner. Hypersensitive individuals may use **hypoallergenic** products (products without common allergens) to avoid such allergic reactions.

Autoimmunity

Autoimmunity is an inappropriate and excessive response to self-antigens. Disorders that result from autoimmune responses are called **autoimmune diseases.** Table 21-5 gives examples of autoimmune diseases. Self-antigens are molecules that are native to a person's body and that are used by the immune system to identify components of "self." In autoimmunity, the immune system inappropriately attacks these antigens.

A common autoimmune disease is **systemic lupus erythematosus (SLE),** or simply *lupus.* Lupus is a chronic inflammatory disease that affects many tissues in the body: joints, blood vessels, kidney, nervous system, and skin. The name *lupus erythematosus* refers to the red rash that often develops on the faces of those afflicted with SLE. The "systemic" part of the name comes from the fact that the disease affects many systems throughout the body. The systemic nature of SLE results from the production of IgG antibodies against a person's own DNA.

Isoimmunity

Isoimmunity is a normal but often undesirable reaction of the immune system to antigens from a different individual of the same species. Isoimmunity is important in two situations: pregnancy and tissue transplants.

During pregnancy, antigens from the fetus may enter the mother's blood supply and sensitize her immune system. An-

Mechanisms of Disease—cont.

Table 21-5	Examples of Autoimmune Diseases	
DISEASE	**POSSIBLE SELF-ANTIGEN**	**DESCRIPTION**
Addison disease	Surface antigens on adrenal cells	Hyposecretion of adrenal hormones, resulting in weakness, reduced blood sugar, nausea, loss of appetite, and weight loss
Cardiomyopathy	Cardiac muscle	Disease of cardiac muscle (i.e., the myocardium), resulting in loss of pumping efficiency (heart failure)
Diabetes mellitus (type 1)	Pancreatic islet cells, insulin, insulin receptors	Hyposecretion of insulin by the pancreas, resulting in extremely elevated blood glucose levels (in turn causing a host of metabolic problems, even death if untreated)
Glomerulonephritis	Blood antigens that form immune complexes that deposit in kidney	Disease of the filtration apparatus of the kidney (renal corpuscle), resulting in fluid and electrolyte imbalance and possibly total kidney failure and death
Graves disease (type of hyperthyroidism)	TSH receptors on thyroid cells	Hypersecretion of thyroid hormone and resulting increase in metabolic rate
Hemolytic anemia	Surface antigens on RBCs	Condition of low RBC count in the blood resulting from excessive destruction of mature RBCs (hemolysis)
Hypothyroidism	Antigens in thyroid cells	Hyposecretion of thyroid hormone in adulthood, causing decreased metabolic rate and characterized by reduced mental and physical vigor, weight gain, hair loss, and edema
Multiple sclerosis (MS)	Antigens in myelin sheaths of nervous tissue	Progressive degeneration of myelin sheaths, resulting in widespread impairment of nerve function (especially muscle control)
Myasthenia gravis	Antigens at neuromuscular junction	Muscle disorder characterized by progressive weakness and chronic fatigue
Pernicious anemia	Antigens on parietal cells, intrinsic factor	Abnormally low RBC count resulting from the inability to absorb vitamin B_{12}, a substance critical to RBC production
Reproductive infertility	Antigens on sperm or tissue surrounding ovum (egg)	Inability to produce offspring (in this case, resulting from destruction of gametes)
Rheumatic fever	Cardiac cell membranes (cross reaction with group A streptococcal antigen)	Rheumatic heart disease; inflammatory cardiac damage (especially to the endocardium/valves)
Rheumatoid arthritis (RA)	Collagen	Inflammatory joint disease characterized by synovial inflammation that spreads to other fibrous tissues
Systemic lupus erythematosus (SLE)	Numerous	Chronic inflammatory disease with widespread effects and characterized by arthritis, a red rash on the face, and other signs
Ulcerative colitis	Mucous cells of colon	Chronic inflammatory disease of the colon characterized by watery diarrhea containing blood, mucus, and pus

tibodies that are formed as a result of this sensitization may enter the fetal circulation and cause an inappropriate immune reaction. One example, erythroblastosis fetalis, was discussed in Chapter 17. Other pathological conditions may also be caused by damage to developing fetal tissues resulting from attack by the mother's immune system. Examples include congenital heart defects, Graves disease, and myasthenia gravis.

Tissue or organ **transplants** are medical procedures in which tissue from a donor is surgically *grafted* into the body. For example, skin grafts are often performed to repair dam-

age caused by burns. Donated whole blood tissue is often transfused into a recipient after massive hemorrhaging. A kidney is sometimes removed from a living donor and grafted into a person suffering from kidney failure. Unfortunately, the immune system sometimes reacts against foreign antigens in the grafted tissue, causing what is often called a **rejection syndrome.** The antigens commonly involved in transplant rejection are called MHC proteins—or **human leukocyte antigens (HLAs).**

Rejection of grafted tissues can occur in two ways: (1) **host-versus-graft rejection**—the recipient's immune system recog-

Mechanisms of Disease—cont.

nizes foreign HLAs and attacks them, destroying the donated tissue, and (2) **graft-versus-host rejection**—the donated tissue (e.g., bone marrow) attacks the recipient's HLAs, destroying tissue throughout the recipient's body. Graft-versus-host rejection may lead to death.

There are two ways to prevent rejection syndrome. One strategy is called *tissue typing* in which HLAs and other antigens of a potential donor and recipient are identified. If they match, tissue rejection is unlikely to occur. Another strategy is the use of **immunosuppressive drugs** in the recipient. Immunosuppressive drugs such as *cyclosporine* and *prednisone* suppress the immune system's ability to attack the foreign antigens in the donated tissue.

Deficiency of the Immune System

Immune deficiency, or *immunodeficiency,* is the failure of immune system mechanisms to defend against pathogens. Immune system failure usually results from disruption of lymphocyte (B cell or T cell) function. The chief characteristic of immune deficiency is the development of unusual or recurring severe infections or cancer. Although immune deficiency by itself does not cause death, the resulting infections or cancer can.

The two broad categories of immune deficiencies, based on the mechanism of lymphocyte dysfunction, are *congenital* and *acquired.* Each of these types is outlined in the following discussion.

Congenital Immune Deficiency

Congenital immune deficiency, which is rare, results from improper lymphocyte development before birth. Depending on which stage of the development of stem cells, B cells, or T cells the defect occurs, different diseases can result. For example, improper B-cell development can cause insufficiency or absence of antibodies in the blood. If stem cells are missing or are unable to grow properly, a condition called **severe combined immune deficiency (SCID)** occurs. In most forms of SCID, both humoral immunity and cell-mediated immunity are defective. Temporary immunity can be imparted to children with SCID by injecting them with a preparation of antibodies (gamma globulin). Bone marrow transplants, which replace the defective stem cells with healthy donor cells, have proven effective in treating some cases of SCID.

Acquired Immune Deficiency

Acquired immune deficiency develops after birth (and is not related to genetic defects). Many factors can contribute to acquired immune deficiency: nutritional deficiencies, immunosuppressive drugs or other medical treatments, trauma, stress, and viral infection.

One of the best known examples of acquired immune deficiency is **acquired immune deficiency syndrome (AIDS).**

AIDS affects millions of people worldwide. This syndrome is caused by the **human immunodeficiency virus,** or **HIV.** HIV, a retrovirus, contains RNA that produces its own DNA inside infected cells. The viral DNA often becomes part of the cell's DNA (Figure 21-30). When the viral DNA is activated by cytokines, it directs the cell to synthesize viral RNA and viral proteins—producing new retroviruses. HIV thus "steals" raw materials from the cell. When this occurs in the CD4 subset of T cells (helper T cells), the cell is destroyed and immunity is seriously impaired.

As the T cell dies, it releases new retroviruses that can spread the HIV infection. As the HIV infection progresses, more and more CD4 lymphocytes are lost. This change in CD4 lymphocyte number is one of the principal clinical methods for monitoring AIDS (Figure 21-31, A).

HIV can invade several types of human cells, including brain cells. However, when CD4 T-cell (helper T-cell) function is impaired, infectious organisms and cancer cells can grow and spread much more easily than normal. Infections and tumors that rarely occur in healthy people, such as *Pneumocystis carinii pneumonia* (a protozoal infection) and *Kaposi sarcoma* (a type of skin cancer), frequently are seen in AIDS patients. Because their immune system is deficient, AIDS patients usually die from one of these infections or cancers.

After they are infected with HIV, T cells may not show signs of AIDS for years. This is because the immune system can hold the infection at bay for a long time before finally succumbing to it. Figure 21-31, B, shows the progression of HIV infection to AIDS.

There are several strategies for controlling AIDS and related conditions. Many agencies are trying to slow the spread of AIDS by educating people about how to avoid contact with the HIV retrovirus. HIV is spread by direct contact with body fluids, so preventing such contact reduces HIV transmission. Sexual relations, blood transfusions, breastfeeding, and intravenous use of contaminated needles are the usual modes of HIV transmission.

Most patients with AIDS have an abundance of antibodies against the HIV in their blood. This is another important clinical test for diagnosing and monitoring AIDS patients (frequently referred to as a *Western blot*). Unfortunately, for the majority of patients, the antibody response to HIV is not sufficient to suppress the disease. However, a few patients who demonstrated a strong antibody response did lose the virus completely. These few are fueling an intensive, worldwide research effort to develop a vaccine for treating AIDS. The fast rate at which these viruses mutate (change their protein structure), however, is making vaccine development an extremely difficult challenge.

A way to inhibit symptoms of the disease is by means of chemicals such as zidovudine (Azidothymidine [AZT]) and ri-

Mechanisms of Disease—cont.

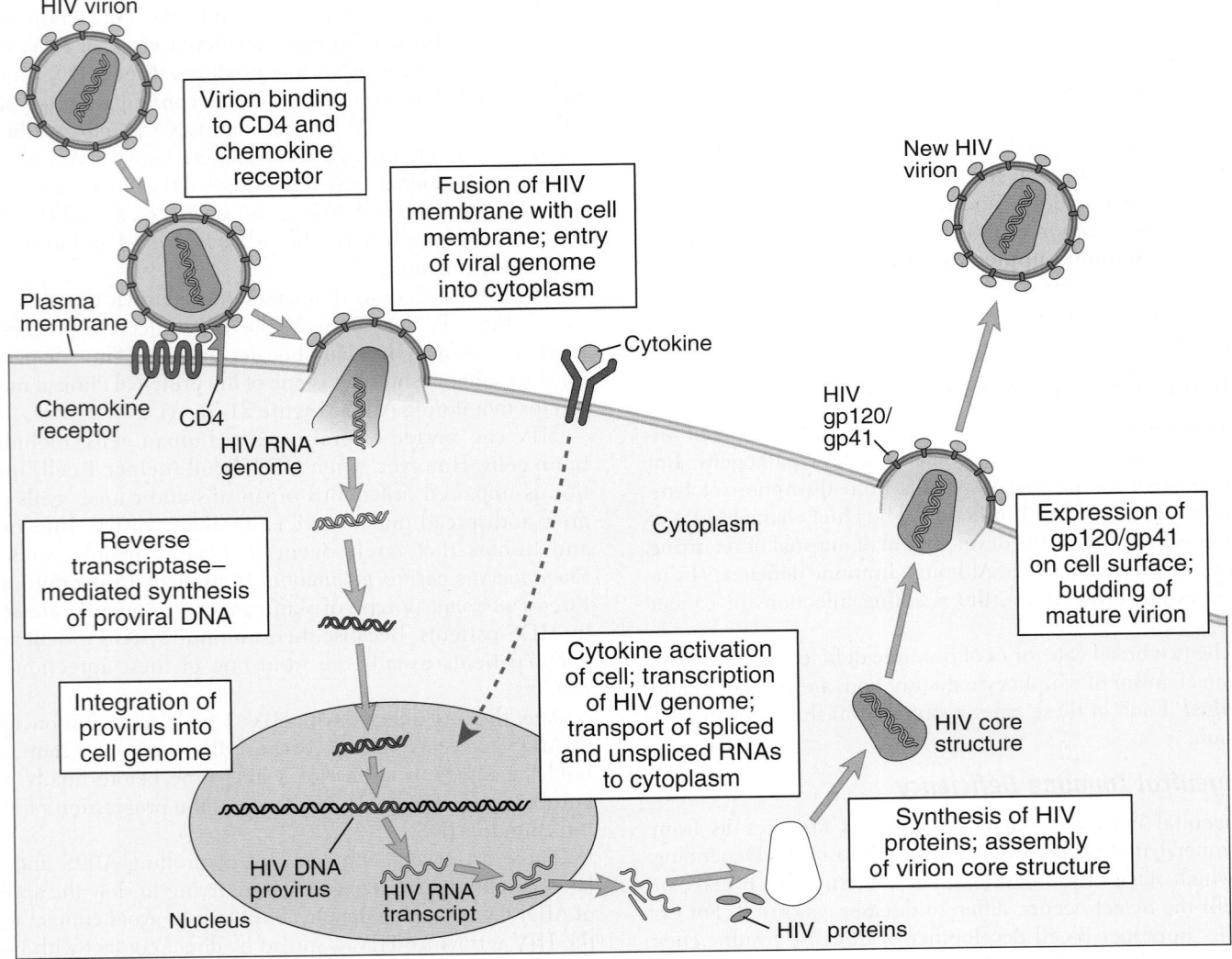

Figure 21-30 *Mechanisms of HIV infection.* HIV is an RNA containing virus *(retrovirus)* that appears to infect T cells by way of the mechanism described in the diagram and simplified in the following steps:
1. HIV virion (complete viral particle) attaches to specific receptors on the surface of a C4 helper T cell.
2. Viral RNA is released into cytoplasm.
3. Viral RNA is used to synthesize a DNA molecule ("reverse transcription").
4. The new "viral" DNA is spliced into the cell's chromosomal DNA.
5. Cytokines trigger the viral DNA to initiate transcription of RNA molecules.
6. The new RNA molecules direct synthesis of HIV, RNA, and HIV proteins.
7. New HIV components are released from the cell by budding of the plasma membrane (exocytosis).
8. Whole HIV particles (HIV virions) are thus released by the infected cell. They are now free to infect more cells.

tonavir (Norvir) that block HIV's ability to reproduce within infected cells. A breakthrough in the treatment of HIV came a few years ago when it was discovered that a "cocktail" of several antiviral drugs working together greatly reduces the number of virus particles in a patient's blood. More than 100 such compounds in various combinations are being evaluated for use in halting the progress of HIV infections. Currently, the recommended treatment for HIV infection is a combination of at least three medications in an individually tailored regimen called *highly active antiretroviral therapy (HAART)*.

Mechanisms of Disease—cont.

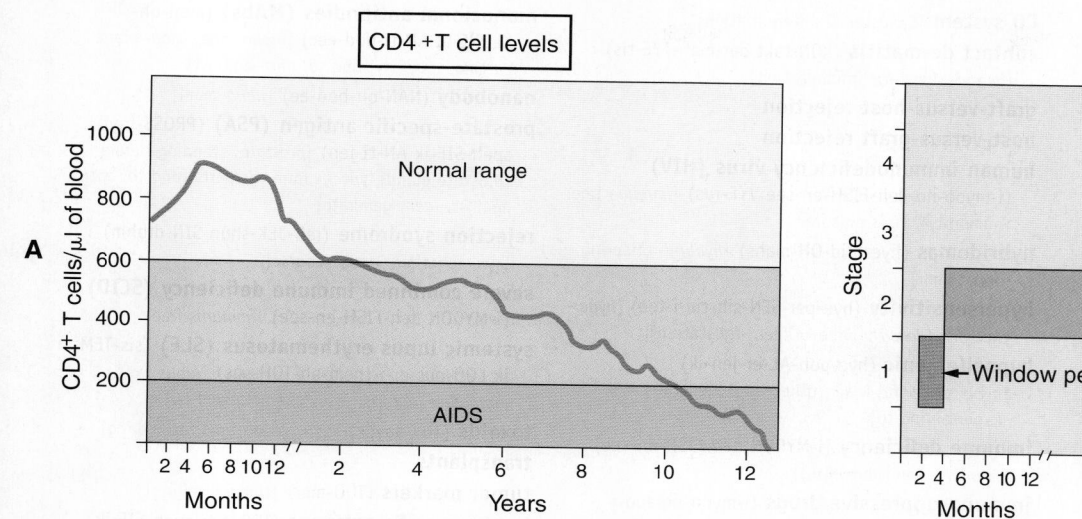

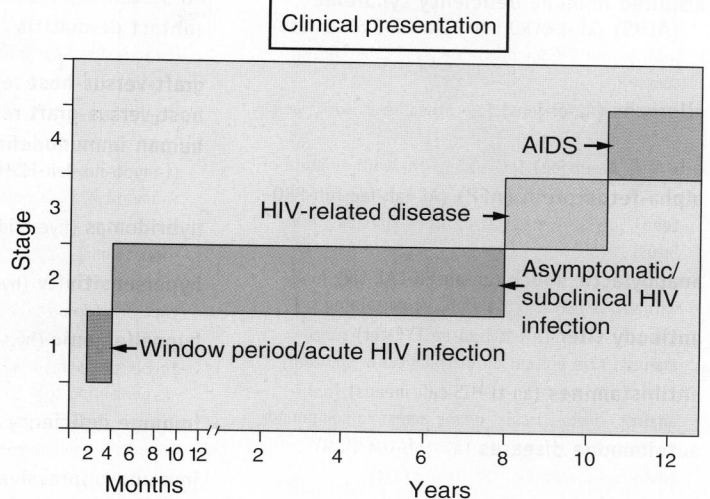

Figure 21-31 *Clinical progression of HIV/AIDS.* **A,** Changing numbers of CD4 T cells as an HIV infection progresses to AIDS. **B,** Progression from initial HIV infection to full-blown AIDS is usually described in four stages: (1) Acute viral infection with common viral symptoms; called "window period" because anti-HIV antibodies are not yet detectable by laboratory tests. (2) Subclinical stage in which there are often no (or minor) symptoms—but the virus is replicating. (3) HIV-related disease, with symptoms of acute viral infection and high levels of anti-HIV antibodies found in laboratory tests. (4) AIDS, including opportunistic infections and cancers.

LANGUAGE OF SCIENCE   *(Cont'd from page 801)*

dendritic cell (DC) (DEN-dri-tik) [*dendrit-* tree branch, *-ic* pertaining to, *cella* storeroom]

diapedesis (dye-ah-peh-DEE-sis) [*dia-* through, *-pedesis* an oozing]

effector cell

effector B cells [*effect-* accomplish, *-or* agent, *B* bursa-equivalent tissue, *cella* storeroom]

effector T cells [*effect-* accomplish, *-or* agent, *T* thymus gland, *cella* storeroom]

helper T cells (T$_H$ cells) [helper, *T* thymus gland, *cella* storeroom]

human leukocyte antigens (HLAs) (LOO-koh-syte AN-ti-jens) [*leuko-* white, *-cyte* cell, *anti-* against, *-gen* generate]

humoral immunity (HYOO-mor-al i-MYOO-ni-tee) [*humor* liquid, *-al* pertaining to, *immun-* free, *-ity* state of]

immunoglobulin (Ig) (i-myoo-noh-GLOB-yoo-lin) [*immuno-* free (immunity), *-globul-* small sphere, *-in* substance]

inflammatory response (in-FLAM-ah-toh-ree) [*inflammare* to set afire, *responsum* reply]

innate immunity (IN-ayt i-MYOO-ni-tee) [*innatus* inborn, *immun-* free, *-ity* state of]

interferon (IFN) (in-ter-FEER-on) [*inter-* between, *-fer-* to strike, *-on* substance]

interleukins (ILs) (in-ter-LOO-kins) [*inter-* between, *-leuko-* white (blood cell), *-in* substance]

lymphotoxin (lim-foh-TAWK-sin) [*lympho-* water (lymphocyte), *-tox-* poison, *-in* substance]

macrophage (MAK-roh-fayj) [*macro-* large, *-phage* to eat]

major histocompatibility complex (MHC) [*histo-* tissue, *-compatibil-* agreeable, *-ity* state of, *complexus* an embrace]

membrane attack complexes (MACs) [*complexus* an embrace]

memory cell

memory B cells [*B* bursa-equivalent tissue, *cella* storeroom]

memory T cells [*T* thymus gland, *cella* storeroom]

mucosal immune system (myoo-KOH-sal i-MYOON) [*mucos-* slime or mucus, *-al* pertaining to, *immunis* free]

mucosal-associated lymphoid tissue (MALT) (myoo-KOH-sal-ah-soh-she-AYT-ed LIM-foyd) [*mucos-* slime or mucus, *-al* pertaining to, *associa-* unite, *-ate* process of, *lymph-* water, *-oid* resembling, *tissu* fabric]

naïve (nye-EVE)

naïve B cells (nye-EVE) [*naïve* natural, *B* bursa-equivalent tissue, *cella* storeroom]

natural killer (NK) cells

nonself

nonspecific immunity (non-speh-SIF-ik i-MYOO-ni-tee) [*non-* not, *-specif-* form or kind, *-ic* pertaining to, *immun-* free, *-ity* state of]

pavementing [*pavimentare* cover with pavement]

phagocytosis (fag-oh-sye-TOH-sis) [*phago-* eating, *-cyt-* cell, *-osis* condition]

plasma cells (PLAZ-mah) [*plasma* something molded (blood plasma), *cella* storeroom]

pus

self

self-tolerance

species resistance (SPEE-sheez ree-SIS-tens) [*species* form or kind resistance]

specific immunity (i-MYOO-ni-tee) [*specif* form or kind, *-ic* pertaining to, *immun-* free, *-ity* state of]

suppressor T cells [*suppress-* press down, *-or* agent, *T* thymus gland, *cella* storeroom]

T cells [*T* thymus gland, *cella* storeroom]

thymocytes (THY-moh-sytes) [*thymo-* thymus gland, *-cyte* cell]

LANGUAGE OF MEDICINE

acquired immune deficiency syndrome (AIDS) (ah-KWYRD i-MYOON deh-FISH-en-see) [*acquire* obtain, *immunis* free, *syn-* together, *-drome* course]

allergens (AL-er-jens) [*all-* other, *-erg-* work, *-gen* generate]

allergy (AL-er-jee) [*all-* other, *-erg* work, *-y* state of]

alpha-fetoprotein (AFP) (AL-fah-fee-toh-PRO-teen) [*alpha-* first letter of Greek alphabet, *feto-* fetus]

anaphylactic shock (an-ah-fih-LAK-tik) [*ana-* without, *-phylact-* protection, *-ic* pertaining to]

antibody titer (AN-ti-bod-ee TYE-ter) [*anti-* against, *titre* proportion of chemical in solution]

antihistamines (an-ti-HIS-tah-meens) [*anti-* against, *-histo-* tissue, *-amine* ammonia compound]

autoimmune diseases (aw-toh-i-MYOON) [*auto-* self, *-immune* free (immunity)]

booster shot

CA-125 [*C* cancer, *A* antigen]

carcinoembryonic antigen (CEA) (kar-sin-oh-em-bree-AHN-ik AN-ti-jen) [*carcino-* cancer, *-em-* in, *-bryo-* to grow, *-ic* pertaining to, *anti-* against, *-gen* generate]

CD system [*C* cluster, *D* differentiation]

contact dermatitis (KON-takt der-mah-TYE-tis) [*derma-* skin, *-itis* inflammation]

graft-versus-host rejection

host-versus-graft rejection

human immunodeficiency virus (HIV) (i-myoo-no-deh-FISH-en-see VYE-rus) [*immuno* free (immunity), *virus* poison]

hybridomas (hye-brid-OH-mahs) [*hydbrid-* offspring, *-oma* tumor]

hypersensitivity (hye-per-SEN-sih-tiv-i-tee) [*hyper-* excessive, *sensitiv-* able to feel, *-ity* state of]

hypoallergenic (hye-poh-AL-er-jen-ik) [*hypo-* deficient, *-aller-* other, *-gen-* generate, *-ic* pertaining to]

immune deficiency (i-MYOON deh-FISH-en-see) [*immunis* free (immunity)]

immunosuppressive drugs (i-myoo-no-soo-PRES-iv) [*immuno* free (immunity), *-suppress-* press down, *-ive* pertaining to]

isoimmunity (eye-soh-i-MYOO-ni-tee) [*iso-* equal, *-immun-* free, *-ity* state of]

monoclonal (mon-oh-KLONE-al) [*mono-* one, *-clon-* plant cutting, *-al* pertaining to]

monoclonal antibodies (MAbs) (mon-oh-KLONE-al AN-ti-bod-eez) [*mono-* one, *-clon-* plant cutting, *-al* pertaining to, *anti-* against]

nanobody (NAN-oh-bod-ee) [*nano* small]

prostate-specific antigen (PSA) (PROSS-tayt speh-SIF-ik AN-ti-jen) [*prostates* standing before (prostate gland), *specif-* form, *-ic* pertaining to, *anti-* against, *-gen* generate]

rejection syndrome (reh-JEK-shun SIN-drohm) [*syn-* together, *-drome* course]

severe combined immune deficiency (SCID) (i-MYOON deh-FISH-en-see) [*immunis* free]

systemic lupus erythematosus (SLE) (sis-TEM-ik LOO-pus er-i-them-ah-TOH-sus) [*lupus* wolf, *erythema-* redness, *-osus* characterized by]

toxoids (TOK-soyds) [*tox-* poison, *-oid* resembling]

transplants

tumor markers (TOO-mer) [*tumor* swelling]

tumor-specific antigens (TOO-mer-speh-SIF-ik AN-ti-jens) [*tumor* swelling, *specif-* form, *-ic* pertaining to, *anti-* against, *-gen* generate]

vaccination (vak-si-NAY-shun) [*vaccin-* cow (cowpox), *-ation* process of]

CASE STUDY

Chen Yau, a 32-year-old Asian-American man, was admitted with symptoms of a depressed immune system, including oral and groin candidiasis (thrush) and a herpesvirus infection. This is his fourth admission in less than 2 years. Mr. Yau has engaged in anal and oral homosexual intercourse since 20 years of age. At 30 years old, he tested positive for HIV (human immunodeficiency virus). Chen has had four inpatient and three outpatient admissions between 30 and 32 years of age. His previous admissions were for *Pneumocystis carinii* pneumonia (PCP), anemia, gastrointestinal problems, and Kaposi sarcoma of the skin. Mr. Yau takes several medications to control his symptoms, including zidovudine (Azidothymidine [AZT]).

Mr. Yau's most recent admission was preceded with symptoms of diarrhea, loss of appetite, a 10-pound weight loss over a 2-week period, a nonproductive cough, some shortness of breath, night sweats, fatigue, and a perianal lesion. On admission, physical examination reveals painful enlargement of multiple lymph nodes, white exudate, and inflammation of the mouth and back of the throat, which is consistent with candidiasis, and purplish brown palpable lesions on his arms, legs, and trunk, which are consistent with Kaposi sarcoma. A comprehensive baseline mental status examination reveals several subtle changes in personality and some mild confusion. Mr. Yau demonstrates limits in his visual field, and an ophthalmoscopic examination of the eye shows evidence of inflammation and hemorrhage of the retina that is consistent with cytomegalovirus. Laboratory studies reveal a low white cell count with a few lymphocytes, a reduced ratio of T4 helper cells to T8 suppressor cells, a positive stool for the *Cryptosporidium* parasite, and an x-ray with diffuse infiltrates compatible with *Pneumocystis carinii* pneumonia.

During this most recent admission, Mr. Yau became extremely confused and also became unable to fully control the muscles of his feet and legs. His condition continued to deteriorate until his death in the hospital a few weeks after his admission.

1. Acquired immune deficiency syndrome (AIDS) manifests as reduced resistance to opportunistic infections and malignancies because of:

    A. An acquired deficiency of isoimmunity
    B. An exaggerated immune response to counteract a T-cell deficiency
    C. Impaired functioning of one of the immune inflammatory responses
    D. Deficiency of antigen (HIV) recognition

2. The T4 helper cell to T8 suppressor cell ratio is normally about 2:1. This ratio in Mr. Yau was severely reduced because:

    A. Bone marrow depression reduces hemoglobin and hematocrit levels.
    B. HIV infects and kills the host T4 cells, decreasing their numbers.
    C. The T8 suppressor cells are stimulated by HIV and their numbers increase.
    D. γ-Interferon is reduced, thereby reducing cytotoxic T8 suppressor cell activity.

3. *Pneumocystis carinii*, a protozoan that is usually not pathogenic, caused severe respiratory problems for Mr. Yau because of his:

    A. T-cell function impairment
    B. Lack of knowledge about his disease
    C. Recent exposure to human immunodeficiency virus
    D. Immunosuppression from the AZT

4. Mr. Yau's confusion and distal peripheral neuropathy are not supported with any serological findings for neurological or infectious disease. His peripheral neuropathy was most likely caused by:

    A. Primary or secondary central nervous system lymphoma
    B. *Toxoplasma,* a protozoan infection of the central nervous system
    C. Disturbance in normal brain patterns from drug therapy
    D. Presence of HIV in peripheral nerves and the central nervous system

CHAPTER SUMMARY

INTRODUCTION

A. The immune system protects against assaults on the body
   1. External assaults include microorganisms—protozoans, bacteria, and viruses
   2. Internal assaults—abnormal cells reproduce and form tumors that may become cancerous and spread

ORGANIZATION OF THE IMMUNE SYSTEM

A. The immune system is continually at work patrolling and protecting the body
B. Identification of cells and other particles
   1. Self markers—molecules on the surface of our cells that are unique to an individual, thus identifying the cell as "self" to the immune system
   2. Nonself markers—molecules on the surface of foreign or abnormal cells or particles that identify the particle as "nonself" to the immune system

   3. Self-tolerance—the ability of our immune system to attack abnormal or foreign cells but spare our own normal cells
C. Two major categories of immune mechanisms—innate immunity and adaptive immunity (Figure 21-1; Table 21-1)
D. Innate immunity provides a general, nonspecific defense against anything that is not "self"
E. Adaptive immunity acts as a specific defense against specific threatening agents
F. Primary cells for innate immunity—epithelial barrier cells, phagocytes (neutrophils, macrophages), and natural killer cells; chemicals used in innate immunity—complement and interferon
G. Primary types of cells for adaptive immunity—lymphocytes called T cells and B cells
H. Cytokines—any of several kinds of chemicals released by cells to promote innate and adaptive immune responses (e.g., interleukin, interferon, leukotriene)

I. Other chemicals (e.g., complement, other enzymes, histamine) also play regulatory roles in immunity

INNATE IMMUNITY (TABLE 21-2)

A. Species resistance—genetic characteristics of an organism or species defend against pathogens
B. Mechanical and chemical barriers—first line of defense (Figure 21-2)
  1. The internal environment of the body is protected by a barrier formed by the skin and the mucous membranes
  2. Skin and mucous membranes provide additional immune mechanisms—sebum, mucus, enzymes, and hydrochloric acid in the stomach
C. Inflammation—second line of defense (Figure 21-3)
  1. Inflammatory response—tissue damage elicits responses to counteract injury and promote normalcy
    a. Inflammation mediators include histamine, kinins, prostaglandins, and related compounds (Figure 21-4)
    b. Chemotactic factors—substances that attract white blood cells to the area in a process called chemotaxis
    c. Characteristic signs of inflammation—heat, redness, pain, and swelling
    d. Systemic inflammation—occurs from a body-wide inflammatory response
  2. Phagocytosis—ingestion and destruction of microorganisms or other small particles by phagocytes (Figure 21-7)
    a. Antigen-presenting cells (APCs)—phagocytes that ingest foreign particles, isolate protein segments (peptides), and display them as antigens on their surface to trigger an immune response when recognized by a specific (adaptive) immune cell
    b. Chemotaxis—chemical attraction of cells to the source of the chemical attractant (Figure 21-6)
    c. Diapedesis—process by which immune cells squeeze through the wall of a blood vessel to get to the site of injury or infection (Figure 21-5)
    d. Neutrophil—most numerous phagocyte; usually first to arrive at site of injury, migrates out of bloodstream during diapedesis, forms pus
    e. Macrophages (Table 21-3)
      (1) Phagocytic monocytes grow larger after migrating from bloodstream
      (2) Dendritic cell (DC)—type of macrophage with long branches or extensions (Figure 21-8)
      (3) Examples are histiocytes in connective tissue, microglia in nervous system, and Kupffer cells in liver
D. Natural killer cells—lymphocytes that kill tumor cells and cells infected by viruses (Figure 21-9)
  1. Method of recognizing abnormal or nonself cells—target cell is killed if killer-inhibiting receptor on NK cell does not bind to a proper MHC surface protein
  2. Method of killing cells—lysing cells by damaging plasma membranes
E. Interferon (IFN)—protein synthesized and released into circulation by certain cells if invaded by viruses to signal other nearby cells to enter a protective antiviral state

F. Complement—group of enzymes that produce a cascade of reactions resulting in a variety of immune responses (Figure 21-10)
  1. Lyse cells when activated by either adaptive or innate mechanisms
  2. Opsonization—mark cells for destruction by phagocytes
  3. Variety of other immune responses

OVERVIEW OF ADAPTIVE IMMUNITY

A. Adaptive immunity is part of the third line of defense consisting of lymphocytes—two different classes of a type of white blood cell (Figure 21-11)
B. Two classes of lymphocytes (Figure 21-12)—B lymphocytes (B cells) and T lymphocytes (T cells)
C. Subsets of lymphocytes are defined by the CD surface markers that the cells carry, for example, CD4 and CD8 cells
D. Lymphocytes flow through the bloodstream, become distributed in tissues, and return to the bloodstream in a continuous recirculation
E. B-cell mechanisms—antibody-mediated immunity (humoral immunity); produce antibodies that attack pathogens (Figure 21-13)
F. T cells attack pathogens more directly—classified as cell-mediated immunity (cellular immunity)
G. Lymphocytes have protein markers on their surfaces that are named using the CD system
H. Activation of lymphocytes require two stimuli: a specific antigen and activating chemicals (Figure 21-14)
I. Lymphocytes are densest where they develop—in bone marrow, thymus gland, lymph nodes, and spleen (Figure 21-15)

B CELLS AND ANTIBODY-MEDIATED IMMUNITY

A. B cells develop in two stages
  1. Pre-B cells develop by a few months of age
  2. The second stage occurs in lymph nodes and spleen—activation of a naïve B cell after it binds to a specific antigen
  3. B cells serve as ancestors to antibody-secreting plasma cells
B. Antibodies—proteins (immunoglobulins) secreted by activated B cell (Figure 21-16)
C. An antibody molecule consists of two heavy and two light polypeptide chains; each molecule has two antigen-binding sites and two complement-binding sites (Figure 21-17)
D. Babies are born with different clones of B cells in bone marrow, lymph nodes, and spleen; cells of the clone synthesize a specific antibody with a sequence of amino acids in its variable region that differs from the sequence synthesized by other clones
E. Five classes of antibodies (Figure 21-18)—immunoglobulins M, G, A, E, and D
  1. IgM—antibody that naïve B cells synthesize and insert into their own plasma membranes; it is the predominant class produced after initial contact with an antigen
  2. IgG—makes up 75% of antibodies in the blood; predominant antibody of the secondary antibody response
  3. IgA—major class of antibody in the mucous membranes, in saliva and tears
  4. IgE—small amount; produces harmful effects such as allergies
  5. IgD—small amount in blood; precise function unknown

F. Antibody molecules produce antibody-mediated immunity (humoral immunity)—within plasma
G. Antibodies resist disease first by recognizing foreign or abnormal substances (Figure 21-19)
  1. Epitopes bind to an antibody molecule's antigen-binding sites, which forms an antigen-antibody complex that may produce several effects (Figure 21-20)
H. Complement—a component of blood plasma consisting of several protein compounds
  1. Complement kills foreign cells by cytolysis or apoptosis (Figures 21-21 and 21-22)
  2. Complement causes vasodilation, enhances phagocytosis, and has other functions
  3. Complement activity can also be initiated by innate immune mechanisms
    a. Formation by innate immunity is called the alternate pathway
    b. Complement protein 3—activated without antigen stimulation—produces full complement effect by binding to bacteria or viruses in presence of properdin
I. Primary and secondary responses (Figure 21-23)
  1. Primary response—initial encounter with a specific antigen triggers the formation and release of specific antibodies that reaches its peak in a few days.
  2. Secondary response—a later encounter with the same antigen triggers a much quicker response; B memory cells rapidly divide, producing more plasma cells and thus more antibodies
J. Clonal selection theory (Figure 21-24)
  1. The body contains many diverse clones of cells, each committed by its genes to synthesize a different antibody
  2. When an antigen enters the body, it selects the clone whose cells are synthesizing its antibody and stimulates them to proliferate and create more antibody
  3. The clones selected by antigens consist of lymphocytes and are selected by the shape of antigen receptors on the lymphocyte's plasma membrane

T CELLS AND CELL-MEDIATED IMMUNITY

A. T cells—lymphocytes that go through the thymus gland before migrating to the lymph nodes and spleen
  1. Pre-T cells develop into thymocytes while in the thymus
  2. Thymocytes stream into the blood and are carried to the T-dependent zones in the spleen and the lymph nodes
B. Activation of T cells
  1. T cells display antigen receptors on their surface membranes that are similar to antibodies
  2. A T cell is activated when an antigen (presented by an APC) binds to its receptors, causing it to divide repeatedly to form a clone of identical T cells (Figure 21-25)
    a. Cells of the clone differentiate into effector T cells and memory T cells
    b. Effector T cells go to the site where the antigen entered, bind to antigens, and begin their attack

C. Cytotoxic T cells—T cells release lymphotoxin to kill cells (Figure 21-26)
D. Helper T cells (T_H cells)—regulate the function of B cells, T cells, phagocytes, and other leukocytes (Figure 21-27)
E. Suppressor T cells—regulatory T cells that suppress lymphocyte function, thus regulating immunity and promoting self-tolerance
F. T cells function to produce cell-mediated immunity and help to regulate adaptive immunity in general

TYPES OF ADAPTIVE IMMUNITY (TABLE 21-4)

A. Innate immunity (inborn or inherited immunity)—genetic mechanisms put innate immune mechanisms in place during development in the womb
B. Adaptive or acquired immunity; resistance developed after birth; two types:
  1. Natural immunity results from nondeliberate exposure to antigens
  2. Artificial immunity results from deliberate exposure to antigens, called *immunization*
C. Natural and artificial immunity may be active or passive
  1. Active immunity—when the immune system responds to a harmful agent regardless of whether it was natural or artificial; lasts longer than passive
  2. Passive immunity—immunity developed in another individual is transferred to an individual who was not previously immune; it is temporary but provides immediate protection

SUMMARY OF ADAPTIVE IMMUNITY

A. Adaptive immunity is specific immunity—targeting specific antigens
B. Adaptive immunity involves two classes of lymphocyte: B cells and T cells (Figure 21-27)
  1. B cells—antibody-mediated (humoral) immunity
  2. T cells—cell-mediated (cellular) immunity
C. Adaptive immunity occurs in a series of stages (Figure 21-28)
  1. Recognition of antigen
  2. Activation of lymphocytes
  3. Effector phase (immune attack)
  4. Decline of antigen causes lymphocyte death (homeostatic balance)
  5. Memory cells remain for later response if needed
D. B cells and T cells work together in a coordinated system of adaptive immunity (Figure 21-29)

THE BIG PICTURE: IMMUNE SYSTEM AND THE WHOLE BODY

A. Immune system regulated to some degree by nervous and endocrine systems
B. Agents of the immune system include blood cells, skin cells, mucosal cells, brain cells, liver cells, and other types of cells and their secretions

REVIEW QUESTIONS

1. Define the term *innate immunity*.
2. List several mechanisms of innate defense, and give a brief description of each one.
3. Identify the body's first line of defense.
4. Describe chemotactic factors released from the mast cell.
5. What is the function of interferon?
6. Define the following terms: *antigens, antibodies, antigenic determinants, combining sites, clone.*
7. What two terms are synonyms for combining sites?
8. When an antigen-antibody complex is formed, what region on the antigen molecule fits into what region on the antibody molecule?
9. Activated B cells develop into clones of what two kinds of cells?
10. What cells synthesize and secrete copious amounts of antibodies?
11. Explain the function of memory cells.
12. Antibodies belong to what class of compounds? Diagram and describe the structure of an antibody.
13. Explain the two basic tenets of Burnet's clonal selection theory.
14. What are cytokines? Lymphotoxins?
15. Differentiate between the classifications of natural and artificial immunity.

16. Describe the process behind the functioning of modern vaccines.
17. What are monoclonal antibodies, and how do they function?

CRITICAL THINKING QUESTIONS

1. The causative organism of tuberculosis has a coat around it that makes it much more resistant than other bacteria to digestive enzymes and hydrogen peroxide. What can you say about this characteristic, and why is it more difficult for the body to fight off these bacteria?
2. If a person had a mutation that prevented the formation of the complement proteins, what capabilities would be lessened in the immune system?
3. How would you explain how the various types of T cells can fine-tune the immune system?
4. Why do you think the development of cancer can be seen as a failure of the immune system?
5. Can you make the distinction between inherited immunity and acquired immunity?
6. One of the best-known examples of acquired immune deficiency is acquired immune deficiency syndrome (AIDS). The human immune deficiency virus (HIV) causes this syndrome. How would you summarize the mechanism of HIV infection?

CHAPTER 22

Stress

CHAPTER OUTLINE

LANGUAGE OF SCIENCE

adaptation (ad-ap-TAY-shun) [*adaptatio* process of adjusting]

alarm reaction

corticoids (KOHR-tih-koyds) [*cortic-* cortex or bark, *-oid* resembling]

fetal programming (FEE-tal) [*fetus* fruitful]

fight-or-flight reaction

hypothalamic-pituitary-adrenal "axis" (HPA) (hye-poh-THAL-ah-mik pi-TOO-i-tair-ee ah-DREE-nal)

neuropeptide Y (NOOR-oh-pep-tyde) [*neuro-* nerves, *-peptein* to digest]

psychological stressor (sye-koh-LOJ-i-kal STRESS-or) [*psycho-* the mind, *-logos-* science, *-al* pertaining to, *estrecier* to tighten]

stage of exhaustion [*exhaurire* to drain away]

stage of resistance [*resistere* to withstand]

stress

stress response

stress triad (TRYE-ad) [*trias-* three]

stressors (STRESS-ors) [*estrecier* to tighten]

LANGUAGE OF MEDICINE

general adaptation syndrome (GAS) [*adaptatio* process of adjusting, *syn-* together, *-dromos* course]

stress syndrome (SIN-drohm) [*estrecier-* to tighten, *syn-* together, *-dromos* course]

stress-age syndrome [*estrecier* to tighten, *syn-* together, *-dromos* course]

Stress affects people of all ages and in all walks of life. Children at play, students preparing for an examination, workers on the job, and even a baby before birth are all subject to stress. Although some stress can result in beneficial outcomes, excessive and long-term stress often has disastrous consequences for the health and quality of life of many people.

People experiencing severe stress are often overwhelmed by tension, anger, fear, and frustration. As a result, adrenaline levels rise, blood pressure and heart rate increase, and breathing patterns change. Blood levels of nutrients such as glucose and fatty acids deviate from their normal set points and the immune system becomes less effective. People who have difficulty in dealing with stress over time suffer from a number of stress-related illnesses. Every individual responds somewhat differently to stress, making diagnosis and treatment difficult. There is often a decrease in efficiency of study habits or job-related problem-solving skills; susceptibility to infections increases; and complaints of stomach pains, heart palpitations, fatigue, and muscle aches are common. Sleep disorders and depression frequently accompany stress, and affected individuals often have difficulty in staying with tasks.

Contemporary medicine now recognizes excessive and long-term stress as a critically important and widespread cause of disease. It causes disruption in the homeostasis of numerous physiological control systems in the body and must be controlled to ensure good health.

Our current concept of stress extends beyond the nonspecific physiological responses suggested by researchers nearly 70 years ago. However, the pioneering and classic research on stress by Hans Selye (described in the next section) continues to provide a scientific basis for ongoing work in this complex and important area.

A growing understanding of stress has resulted from insights and research data from diverse disciplines such as molecular genetics, neurobiology, psychology, endocrinology, immunology, and sociology. The result is a holistic model of stress—one that affirms the importance of mind-body interactions in both health and disease.

SELYE'S CONCEPT OF STRESS

In 1935, Hans Selye of McGill University in Montreal made an accidental discovery that launched him on a lifelong career and led him to conceive the idea of stress. This chapter tells briefly the story of how Selye developed his stress concept and also describes the mechanism of stress that he postulated. It then presents some current ideas about stress.

Development of the Stress Concept

Selye made his accidental discovery in 1935 when he was trying to learn whether there might be another sex hormone besides those already known. He had injected rats with various extracts made from ovaries and placenta, expecting to find that different changes had occurred in animals injected with different hormonal preparations. But to his surprise and puzzlement, he found the same three changes in all the animals. The cortices of their adrenal glands were enlarged, but their lymphatic organs—thymus glands, spleens, and lymph nodes—were atrophied, and bleeding ulcers of the stomach and duodenum had developed in every animal. Then he tried injecting many other substances, for

example, extracts from pituitary glands, kidneys, and spleens, and even a poison, formaldehyde. Every time he found the same three changes: enlarged adrenals, shrunken lymphatic organs, and bleeding gastrointestinal ulcers. Selye felt that these symptoms were a specific syndrome. A syndrome, according to the classical definition, is a set of signs and symptoms that occur together and that characterize one particular disease. The three changes, or "stress triad," Selye had observed occurred together, but they seemed to characterize not any one particular kind of injury but all kinds of harmful stimuli. More experiments using various chemicals and injurious agents confirmed for him that the three changes truly were a syndrome of injury. His first publication on the subject was a short paper entitled "A Syndrome Produced by Diverse Nocuous Agents"; it appeared in the July 1936 issue of the British journal *Nature*. Years later, in 1956, he published his monumental technical treatise, *The Stress of Life*.

Although a majority of the current literature credits Hans Selye with the first published reports on stress, Walter B. Cannon used the term "emotional stress" much earlier (1914) when discussing his theory of homeostasis. It seems Cannon was also convinced stress had both a psychological (emotional) and a physiological origin. It was Selye, however, who brought our knowledge of stress and its importance in health and disease into the forefront of modern medicine.

Definitions

Stress, according to Selye's use of the word, is a state, or condition, of the body produced by "diverse nocuous agents" and manifested by a syndrome of changes. We know as well that stress and its negative effects can also be caused by a wide variety of mental, emotional, and other psychological events an individual may perceive as threatening or undesirable. Selye named the agents that produce stress **stressors** and coined a name—**general adaptation syndrome (GAS)**—for the syndrome or group of changes that make the presence of stress in the body known.

Stressors

A **stressor** is any agent or stimulus that produces stress. However, just how an individual interacts with or relates to a particular type of physiological or cognitive situation may well determine if that particular event is stressful or not. Even the same intensity of a particular physiological stressor, such as temperature change, may be perceived as a threat and activate a stress response in one individual and not another. In addition to variances in ability to handle physiological stress, some people have the ability to cope better than others when faced with difficult events in life, such as divorce or bereavement. As a result, they suffer less from negative stress-related symptoms or illness. Therefore a precise classification of stimuli as stressors or nonstressors is not possible. We can, however, make the following generalizations about the character of stressors.

1. Stressors are extreme stimuli—too much or too little of almost anything. The *perception* of the individual is critical. A particular event or circumstance that is viewed as undesirable or that is beyond the individual's ability to cope may be viewed as "extreme" and cause one or more of the psychological or physiological responses associated with stress. In contrast, almost any-

thing in moderation or mild stimuli are nonstressors. Thus coolness, warmth, and soft sounds are nonstressors, whereas extreme cold, extreme heat, and extremely loud sounds almost always act as stressors. Not only extreme excesses but also extreme deficiencies may act as stressors. One example of this kind of stressor is an extreme lack of social contact stimuli. Solitary confinement in a prison, space travel, social isolation because of blindness or deafness, and, in some cases, old age have all been identified as stressors. But the opposite extreme, an excess of social contact stimuli (e.g., caused by overcrowding), also acts as a stressor (Figure 22-1). We now know that various types of stress in young children and even prenatal infants may occur as a result of negative stimuli that are experienced by a parent. For example, severe psychological or physical trauma, environmental hazards, and nutritional deficiencies or abuses, when encountered by a pregnant or nursing woman, can all have a stress-related impact on her child that may result in immediate or delayed behavioral, physiological, or anatomical anomalies. Intrauterine stress is discussed in detail later in the chapter. Poverty and illness, typically associated with stress, are often coupled with alcoholism, drug abuse, and malnutrition. These conditions often result either directly or indirectly in premature births, developmental anomalies, increased susceptibility to many types of mental and physical diseases, and a myriad of other physical defects that may occur immediately in the children involved or develop later in life.

2. Stressors very often are injurious, unpleasant, or painful stimuli—but not always. "A painful blow and a passionate kiss," Selye wrote, "can be equally stressful."

3. Anything that an individual perceives as a threat, whether real or imagined, arouses fear or anxiety. These emotions act as stressors. So, too, does the emotion of grief.

4. Stressors differ in different individuals and in one individual at different times. A stimulus that is a stressor for you may not be a stressor for me. A stimulus that is a stressor for you today may not be a stressor for you tomorrow and might not have been a stressor for you yesterday. Many factors—including one's physical and mental health, heredity, past experiences, coping habits, and even diet—determine which stimuli are stressors for each individual. A significant stress response that manifests in a different way can occur even in a developing infant and is called *prenatal stress*. In many instances prenatal stress will result from a physical or nutritional stressor experienced by the mother. For example, in the United States alone, fetal alcohol syndrome (FAS) affects between 4000 and 7500 infants, and about 35,000 are born with fetal alcohol effects ranging from poor prenatal and infant growth to mental retardation. In this context, alcohol consumed during pregnancy is a dangerous stressor. Even in small amounts it can produce prenatal brain damage that may permanently affect the ability of the child to concentrate, think abstractly, use appropriate judgment, or learn effectively.

Figure 22-1 *Causes of stress.* **A,** Overcrowding (Asia). **B,** Food shortages (refugee camp, New Delhi, India). **C,** Air pollution (Los Angeles). **D,** Environmental dangers (Antarctic ozone hole).

The decision by a woman to consume alcohol during pregnancy is often related to real or perceived stress. Providing education and coping strategies that will help a pregnant woman manage her stress must be a part of social programs intended to eliminate the consumption of alcohol during pregnancy and the incidence of FAS.

QUICK CHECK

1. Define the terms *stress syndrome* and *stress*.
2. Identify the three changes Hans Selye called the "stress triad."
3. List four characteristics of stressors.

General Adaptation Syndrome

Manifestations

Stress, like health or any other state or condition, is an intangible phenomenon. It cannot be seen, heard, tasted, smelled, felt, or measured directly. How, then, can we know that stress exists? It can be inferred to exist when certain visible, tangible, and measurable responses occur. Selye, for example, inferred that the animals on which he experimented were in a state of stress when he found the syndrome of the three changes previously noted—hypertrophied adrenals, atrophied lymphatic organs, and bleeding gastrointestinal ulcers. Because this syndrome indicated the presence of stress and consisted of three changes, he called them the **"stress triad."** Eventually he found that many other changes also took place as a result of stress. The entire group of changes or responses he named the **general adaptation syndrome (GAS).** For each word in this name he gave his reasons. By the word "general" he wanted to suggest that the syndrome was "produced only by agents that have a general effect on large portions of the body." The word adaptation was meant to imply that the syndrome of changes made it possible for the body to adapt, to cope successfully with stress. Selye thought that these responses seemed to protect the animals from serious damage by extreme stimuli and to promote their healthy survival. He looked on the general adaptation syndrome as a crucial part of the body's complex defense mechanism.

Stages

Changes that make up the general adaptation syndrome do not all take place simultaneously but over a period of time in three stages. Selye named these stages: (1) the alarm reaction, (2) the stage of resistance or adaptation, and (3) the stage of exhaustion. A different syndrome of changes, he noted, characterized each stage. Among the responses characteristic of the **alarm reaction,** for example, were the stress triad already described—hypertrophied adrenal cortex, atrophied lymphatic organs (thymus, spleen, lymph nodes), and bleeding gastric and duodenal ulcers. In addition, the adrenal cortex increased its secretion of glucocorticoids, the number of lymphocytes decreased markedly, and so, too, did the number of eosinophils. Also, the sympathetic nervous system and the adrenal medulla greatly increased

their activity (Figure 22-2). Each of these changes, in turn, produced other widespread changes. Figure 22-3 indicates some of the changes stemming from adrenal cortical hypertrophy. Figure 22-4 shows responses produced by increased sympathetic activity and increased secretion by the adrenal medulla of its hormone, epinephrine (adrenaline).

Quite different responses characterize the **stage of resistance** of the general adaptation syndrome. For instance, the adrenal cortex and medulla return to their normal rates of secreting hormones. The changes that had taken place in the alarm stage as a result of increased corticoid secretion disappear during the stage of resistance. All of us go through the first and second stages of the stress syndrome many times in our lifetimes. Stressors of one kind or another act on most of us every day. They may upset or alarm us, but we soon resist them successfully. In short, we adapt; we cope.

The **stage of exhaustion** develops only when stress is extremely severe or when it continues over long periods. Otherwise, when stress is mild and of short duration, it ends with a successful stage of resistance and adaptation to the stressor. If stress continues to the stage of exhaustion, corticoid secretion and adaptation eventually decrease markedly. The body can no longer cope successfully with the stressor and death may ensue. For a brief summary of the changes characteristic of the three stages of the general adaptation syndrome, see Table 22-1.

Mechanism of Stress

Stressors produce a state of stress. A state of stress in turn inaugurates a series of responses that Selye called the general adaptation syndrome. More simply, a state of stress turns on the stress response mechanism. It activates the organs that produce the responses that make up the general adaptation syndrome; but just how stress—a state of the body—does this is not clear. Selye could

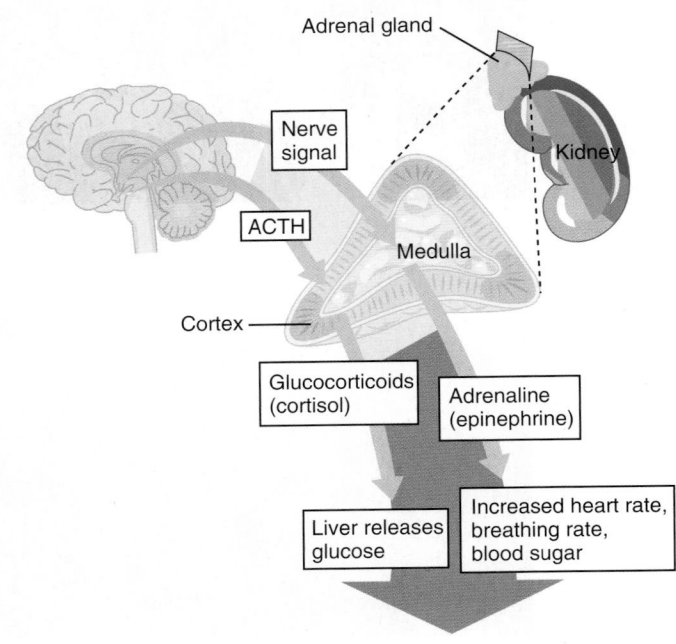

Figure 22-2 *The alarm reaction.* Note the interaction of nervous and hormonal responses. *ACTH,* Adrenocorticotropic hormone.

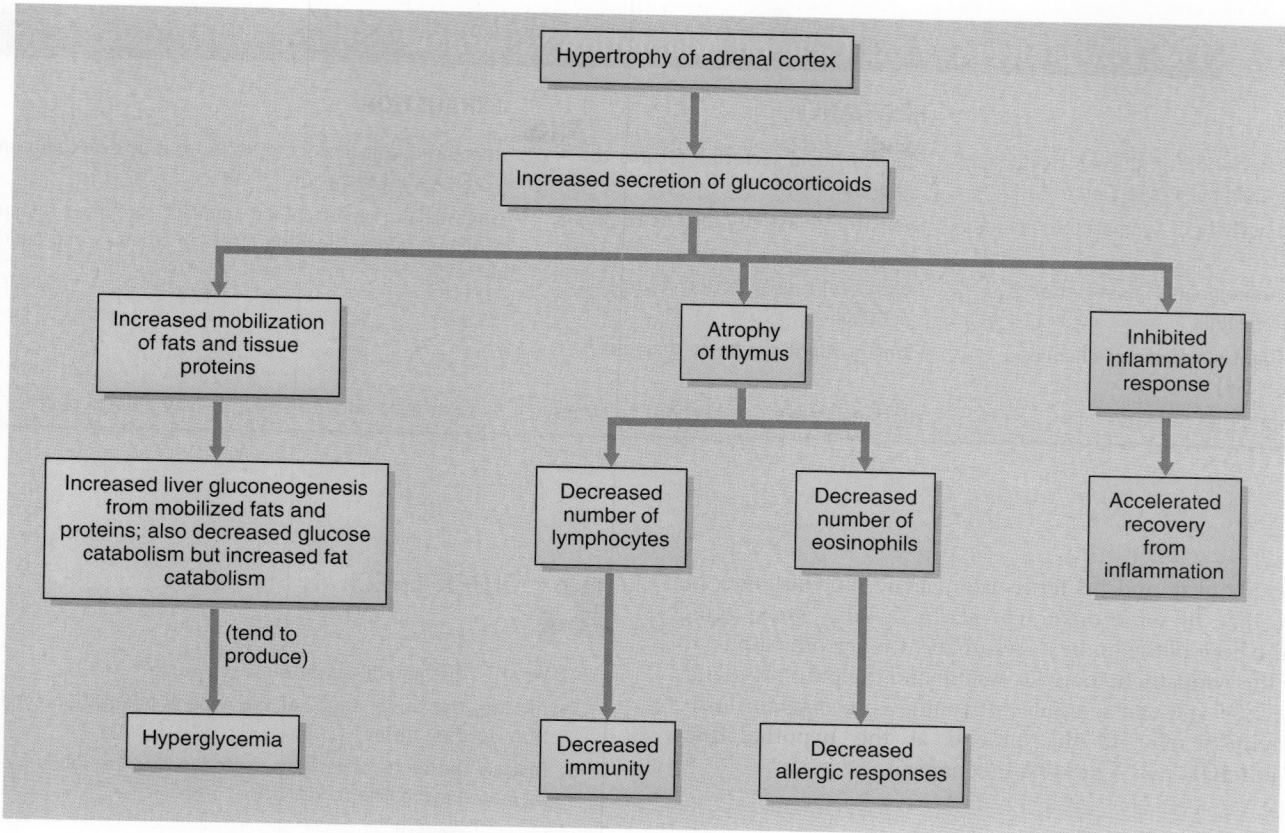

Figure 22-3 *Alarm reaction responses resulting from hypertrophy of adrenal cortex.*

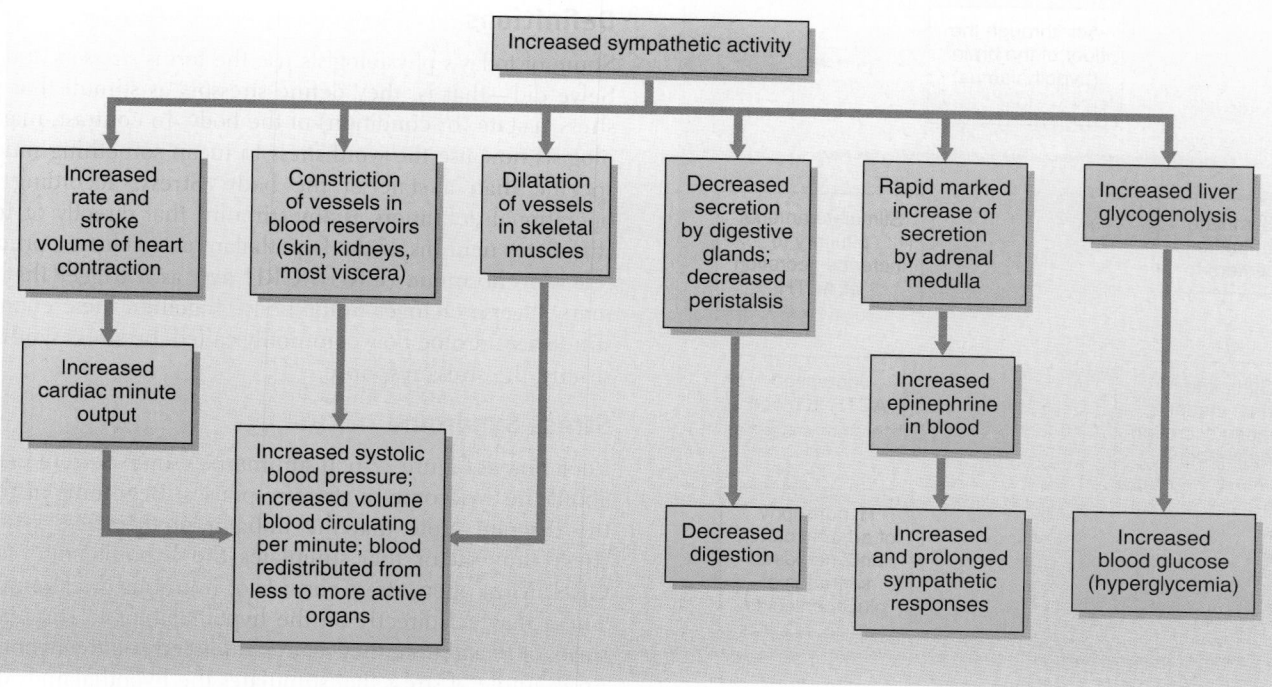

Figure 22-4 *Alarm reaction responses resulting from increased sympathetic activity.* Note that these are the responses commonly referred to as the "fight-or-flight" reaction.

Table 22-1	The Three Stages of the General Adaptation Syndrome	
ALARM	**RESISTANCE**	**EXHAUSTION**
Increased secretion of glucocorticoids and resultant changes (see Figure 22-2)	Glucocorticoid secretion returns to normal	Increased glucocorticoid secretion but eventually marked decreased secretion
Increased activity of sympathetic nervous system	Sympathetic activity returns to normal	Stress triad (hypertrophied adrenals, atrophied thymus and lymph nodes, bleeding ulcers in stomach and duodenum)
Increased norepinephrine secretion by adrenal medulla	Norepinephrine secretion returns to normal	
Fight-or-flight syndrome of changes (Figure 22-4)	Fight-or-flight syndrome disappears	
Low resistance to stressors	High resistance (adaptation) to stressor	Loss of resistance to stressor; may lead to death

only guess about it, and his terms were vague. For instance, he postulated that, by some unknown "alarm signals," stress "acted through the floor of the brain" (presumably the hypothalamus) to stimulate the sympathetic nervous system and the pituitary gland. In Figure 22-5 you can see Selye's hypothesis in diagram form. Many scientists refer to this process as the **hypothalamic-pituitary-adrenal "axis,"** or **HPA** mechanism.

QUICK CHECK

4. What is the general adaptation syndrome?
5. Identify the three stages of the general adaptation syndrome. How do they differ?
6. Discuss the types of responses produced in the body by increased sympathetic activity.

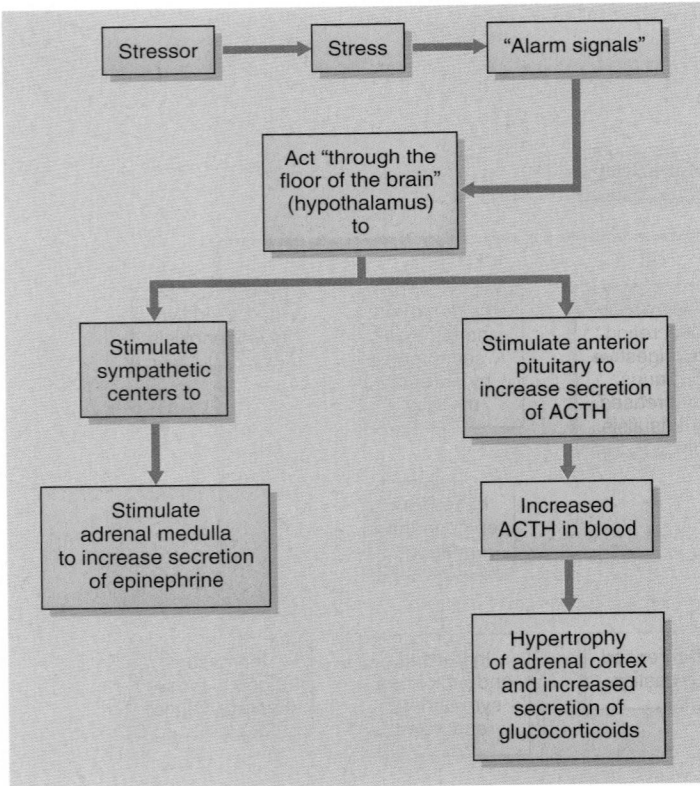

Figure 22-5 *Selye's hypothesis about activation of the stress mechanism often called the hypothalamic-pituitary-adrenal "axis," or HPA mechanism.* *ACTH,* Adrenocorticotropic hormone.

SOME CURRENT CONCEPTS ABOUT STRESS
Definitions

Some of today's physiologists use the terms *stressors* and *stress* as Selye did—that is, they define stressors as stimuli that produce stress, a state (or condition) of the body. In contrast, many physiologists now use the word *stress* to mean something much more specific than a state of the body. **Stress,** according to their operational definition, is any stimulus that directly or indirectly stimulates neurons of the hypothalamus to release corticotropin-releasing hormone (CRH). CRH acts as a trigger that initiates many diverse changes in the body. Together, these changes constitute a syndrome now commonly called the **stress syndrome,** or, simply, the **stress response.**

Stress Syndrome

Look now at Figure 22-6. It summarizes some current major ideas about the syndrome of stress responses. Beginning at the top of the diagram, note that the initiator of the stress syndrome is stress—any factor that stimulates the hypothalamus to release CRH. Most often, stress consists of injurious or extreme stimuli. These may act directly on the hypothalamus to stimulate it. Instead, or in addition, they may act indirectly on the hypothalamus. An example of stress that stimulates the hypothalamus directly is hypoglycemia. A lower than normal concentration of glucose in the blood circulating to the hypothalamus stimulates it to release

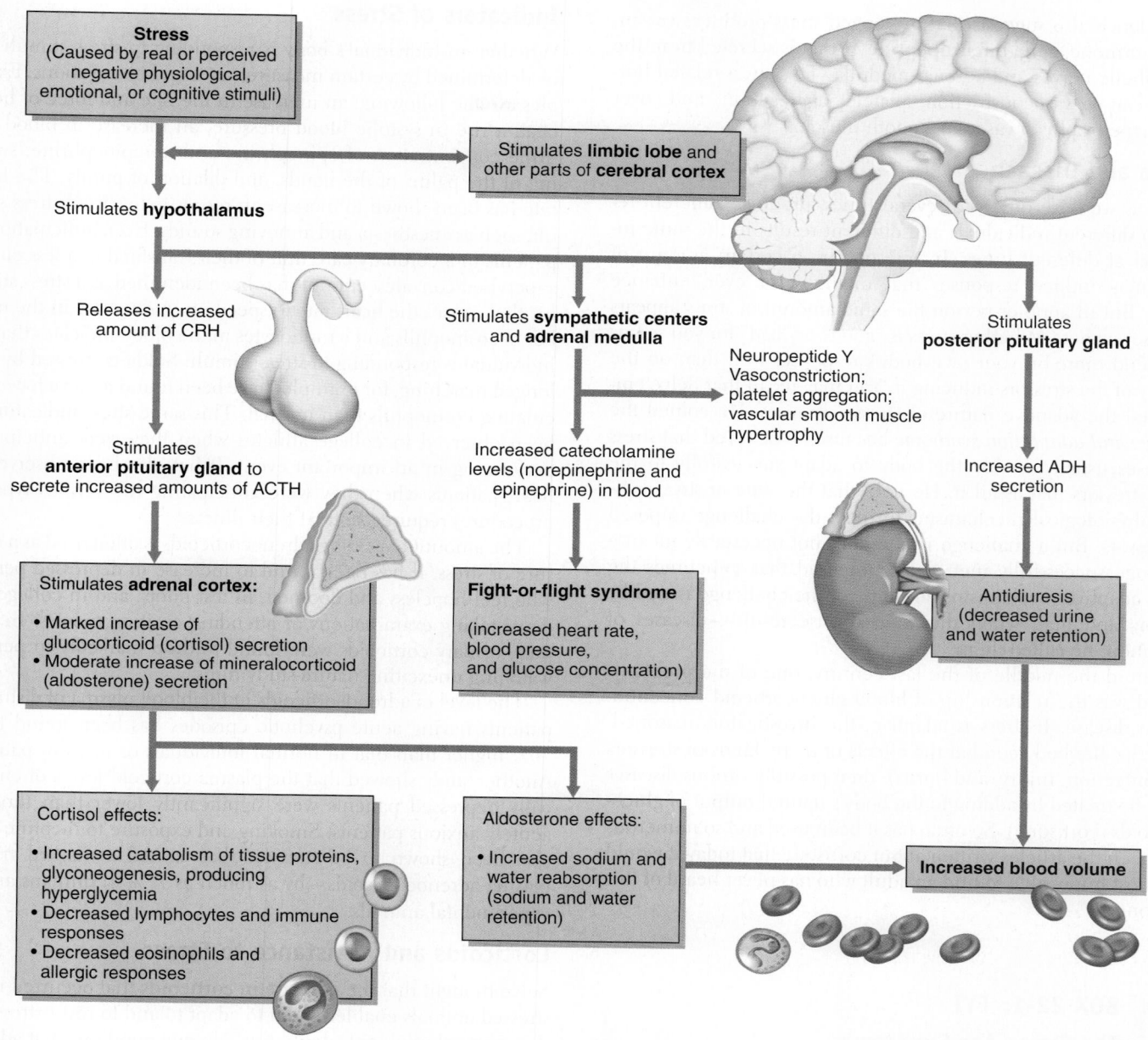

Figure 22-6 *Current concepts of the stress syndrome. CRH,* Corticotropin-releasing hormone; *ACTH,* adrenocorticotropic hormone; *ADH,* antidiuretic hormone.

CRH. Indirect stimulation of the hypothalamus occurs in this way: stress stimulates the limbic lobe—the so-called *emotional brain*—and other parts of the cerebral cortex, and these regions then send stimulating impulses to the hypothalamus, causing it to release CRH. In addition to releasing CRH, note in Figure 22-6 that the stress-stimulated hypothalamus sends stimulating impulses to sympathetic centers and to the posterior pituitary gland.

CRH stimulates the anterior pituitary gland to secrete increased amounts of adrenocorticotropic hormone (ACTH). ACTH stimulates the adrenal cortex to secrete greatly increased amounts of cortisol and more moderately increased amounts of aldosterone. These two hormones induce various stress responses. Some im-

portant ones worth remembering are listed in Figure 22-6 under cortisol effects and aldosterone effects.

Stimulation of sympathetic centers by impulses from the stress-stimulated hypothalamus leads to many stress responses, known collectively as the **fight-or-flight reaction.** They include important changes such as an increase in the rate and strength of the heartbeat, a rise in blood pressure, and hyperglycemia. Other sympathetic stress responses are pallor and coolness of the skin, sweaty palms, and dry mouth.

Water retention and an increase in blood volume are common stress responses. They stem, as you can see in Figure 22-6, from increased ADH and increased aldosterone secretion.

Evidence also suggests that prolonged stress produces yet another hormone, called **neuropeptide Y,** that is secreted from the sympathetic nerves and adrenal medulla. This stress-related hormone causes vasoconstriction, platelet aggregation, and, over time, hypertrophy of vascular smooth muscle.

Stress and Disease

Stress, as we have observed several times, produces different results in different individuals and different results in the same individual at different times. In one person, a certain amount of stress may induce responses that maintain or even enhance health. But in another person the same amount of stress appears to cause sickness. Whether stress is "good" or "bad" for you seems to depend more on your own body's responses to it than on the severity of the stressors inducing it. You may recall that Selye emphasized the adaptive nature of stress responses. He coined the term *general adaptation syndrome* because he believed that stress responses usually enable the body to adapt successfully to the many stressors that assail it. He held that the state of stress activates physiological mechanisms to meet the challenge imposed by stressors. But a challenge issued does not necessarily mean a challenge successfully met. Selye proposed that sometimes the body's adaptive mechanisms fail to meet the challenge issued by stressors and that, when they fail, disease results—diseases of adaptation, he called them.

Around the middle of the last century, one of the problems studied was the relationship of blood glucocorticoid concentration to disease. If stress is adaptive, the investigators reasoned, and helps the body combat the effects of many kinds of stressors (e.g., infection, injury, and burns), then possibly various diseases might be treated by adding to the body's natural output of glucocorticoids (cortisone). So often has it been used and so numerous have been the articles written about cortisone that today it would be almost impossible to find an adult who has never heard of this hormone.

BOX 22-1: FYI
The Stress-Age Syndrome

The **stress-age syndrome** refers to a group of anatomical, neurohormonal, and immune system changes related to aging that influence both physiological and psychological stress responses. The syndrome includes a wide array of changes, including a frequent decrease in coping skills; alteration of limbic lobe and hypothalamus excitability; increases in catecholamines, ACTH, and cortisol; decreases in testosterone, estrogen, thyroxine, and other hormones; immunodepression; decreases in neuromuscular transmitter chemicals resulting in chronic fatigue and reduced physical strength; changes in blood lipid profile; increased potential for hypercoagulation of the blood; and disturbances of the sleep/wake cycle.

Although not all age-related changes connected with stress are damaging, a majority result in a lower potential for positive adaptation to change.

Indicators of Stress

Whether an individual's body is responding to stress stimuli can be determined by certain measurements and observations. Examples are the following: an increase in the rate and force of heartbeat, a rise in systolic blood pressure, an increase in blood and urine concentration of epinephrine and norepinephrine, sweating of the palms of the hands, and dilation of pupils. The heart rate has been shown to increase in response to varied stress stimuli, such as anesthesia and annoying sounds. Even anticipation by patients in a coronary care unit of their transferal to a less closely supervised convalescent unit has been identified as a stress stimulus that causes the heart rate to speed up. A decrease in the number of eosinophils and lymphocytes in the blood indicates that the individual is responding to stress stimuli. Soldiers stressed by prolonged marching, for example, have been found to have fewer circulating eosinophils than normal. This same stress indicator has been observed in college athletes when they were anticipating performing in an important event. It has also been observed in heart patients when they were anticipating the various types of procedures required to treat their illness.

The amount of urinary adrenocorticoids is often used as a measure of stress. It has been found to increase in depressed persons who feel hopeless and doomed, in test pilots, and in college students taking examinations or attending exciting movies. In contrast, urinary corticoids were found to drop markedly in persons watching unexciting nature-study films.

The level of adrenocorticoids in the blood plasma of disturbed patients having acute psychotic episodes has been found to be 70% higher than that in normal individuals or in calm patients. Another study showed that the plasma corticoid levels of chronically depressed patients were significantly lower than those of acutely anxious patients. Smoking and exposure to nicotine have also been shown to be stressors that caused a marked rise in plasma adrenocorticoids—by as much as 77% in humans and in experimental animals.

Corticoids and Resistance to Stress

Selye thought that the increase in **corticoids** that occurred in his stressed animals enabled them to adapt to and to resist stress. Today many physiologists doubt this. No one questions that adrenal cortical hormones increase during stress. That fact has been

BOX 22-2: FYI
Type A Behavior and Stress

High-stress, hard-driving, and competitive type A individuals have been found to be at greater risk of elevated systolic blood pressure and coronary disease than individuals with the more relaxed and less impatient type B personality. Studies show that type A behavior patterns can be reduced by psychological guidance. Counseling of type A individuals helps them reduce stress and may cut in half their chances of suffering a heart attack because of stress.

clearly established. But what many question is how essential this increase is for *resisting stress*. No one has proved by an unequivocal experiment that a higher than normal blood level of corticoids increases an animal's ability to adapt to stress, and increasing corticoid levels may in themselves be problematic—especially in the developing fetus. Some clinical evidence, however, seems to indicate that it does increase a human's coping ability. For instance, patients who have been taking cortisol for some time are known to require increased doses of this hormone to successfully resist stresses such as surgery or severe injury.

Psychological Stress

Stress as defined by Selye is physiological stress, that is, a state of the body. Psychological stress, in contrast, might be defined as a state of the mind. It is caused by psychological stressors and manifested by a syndrome. A **psychological stressor** is anything that an individual perceives as a threat—a threat to survival or to self-image. Moreover, the threat does not need to be real—it needs only to be real to the individual, who must see it as a threat, although in truth, it may not be. Psychological stressors produce a syndrome of subjective and objective responses. Dominant among the subjective reactions is a feeling of anxiety. Other emotional reactions, such as anger, hate, depression, fear, and guilt, are also common subjective responses to psychological stressors. Some characteristic objective responses are restlessness, fidgeting, criticizing, quarreling, lying, and crying. Another objective indicator of psychological stress is that the concentration of lactate in the blood increases.

Does psychological stress relate to physiological stress? The answer is clearly "yes." Physiological stress usually is accompanied by some degree of psychological stress. And, conversely, in most people, psychological stress produces some physiological stress responses. Ancient peoples intuitively recognized this fact. For example, it is said that in ancient times when the Chinese suspected a person of lying, they made that person chew rice powder and then spit it out. If the powder came out dry, not moistened by saliva, they judged the suspect guilty. They seemed to know that lying makes a person nervous—and that nervousness makes a person's mouth dry. We "moderns" also know these facts, but we describe them with more technical language. Lying, we might say, induces psychological stress, and psychological stress acts in some way to cause the physiological stress response of decreased salivation.

Within recent years, a scientific discipline called *psychophysiology* has come into being. Psychophysiologists, using accepted research methods and sophisticated instruments—including polygraphs designed especially for this type of research—have investigated various physiological responses made by individuals subjected to psychological stressors. Their findings amply confirm the principle that psychological stressors often produce physiological stress responses in all age-groups. Especially apparent are the links between the immune, nervous, and endocrine systems. Table 22-2 lists a number of stress-related diseases and conditions. Psychophysiologists have found, however, that identical psychological stressors do not necessarily induce identical physiological responses in different individuals. For example, stress may result in heart rate and blood pressure changes in one individual and changes in breathing patterns in another. Another of their inter-

esting discoveries is that some organ systems become less responsive after they have been stimulated a number of times. We now know that infancy and early childhood, once overlooked as potentially stressful periods in life, do indeed present youngsters with challenges that result in stress. The result is often manifested by physical or behavioral adaptations intended to assist in coping. As young children mature, they continue to face multiple periods of often stressful transition to new roles in social relationships, self-concept, personal identity, and, later, sexual identity and behavior. These challenges are highly individualistic and often temporary—existing for only short periods during childhood and adolescence.

In the adult years, the types of stressors characteristic of earlier periods of development, such as the transition from childhood to sexual maturity, will often change and maladaptive responses will vary. Adults often face higher levels of ongoing stress than children because of the many complex social interactions they must

Table 22-2	Stress-Related Diseases and Conditions
TARGET ORGAN OR SYSTEM	**DISEASE OR CONDITION**
Cardiovascular system	Coronary artery disease Hypertension Stroke Disturbances of heart rhythm
Muscles	Tension headaches Muscle contraction backache
Connective tissues	Rheumatoid arthritis (autoimmune disease) Related inflammatory diseases of connective tissue
Pulmonary system	Asthma (hypersensitivity reaction) Hay fever (hypersensitivity reaction)
Immune system	Immunosuppression or immune deficiency Autoimmune diseases
Gastrointestinal system	Ulcer Irritable bowel syndrome Diarrhea Nausea and vomiting Ulcerative colitis
Genitourinary system	Diuresis Impotence (erectile dysfunction) Frigidity
Skin	Eczema Neurodermatitis Acne
Endocrine system	Diabetes mellitus Amenorrhea
Central nervous system	Fatigue and lethargy Type A behavior Overeating Depression Insomnia

deal with on an ongoing basis, as well as the need to make ethical judgments and decisions related to standards of conduct.

Maladaptive responses to chronic stress in adults often affect not only their physical health but also the emotional, social, and intellectual aspects of their lives. Living with chronic stress as an adult often leads to a decreased quality of life. Ulcers, hypertension, chemical dependency, impaired relationships, and even loss of contact with reality may occur. Adults struggling with unhealthy levels of stress over time often develop multiple problems that limit their ability to function in society as their general level of both physical and psychological health deteriorates.

Managing stress in these individuals is an important component in holistic treatment programs designed to improve both their physical and psychological health. In addition to specific drugs and counseling, such programs may include use of relaxation techniques, such as massage, meditation, and positive imagery, as well as educational components focusing on development of coping strategies.

Elderly people are at high risk for stress-related illness (see Box 22-1). It is estimated that the U.S. population of very elderly individuals (80 years or older) will increase from a current number of about 31 million to more than 65 million by 2030. These individuals must deal with unique and very significant stressors during their later years. Unfortunately, loss of health and general well-being, fear of death or dying, a sense of abandonment and social isolation, poverty, and often the death of a spouse result in frequent maladaptive responses to stress in this age-group. Family support and societal support are particularly important as elderly individuals struggle with stress.

In summary, here are some principles to remember about psychological stress:

- Physiological stress almost always is accompanied by some degree of psychological stress.

- In most people, psychological stress leads to some physiological stress responses. Many of these are measurable autonomic responses, for example, accelerated heart rate and increased systolic blood pressure.

- Identical psychological stressors do not always induce identical physiological responses in different individuals.

- In any one individual, certain autonomic responses are better indicators of psychological stress than others.

Effects of Intrauterine Stress

We now know that a fetus develops both short-term and long-term responses to stress experienced during intrauterine development. Maternal malnutrition is one of the most common "stressors" experienced by a pregnant woman and her unborn child. In malnourished women who smoke, oxygen delivery to the fetus will also decrease. The result is hypoxic as well as nutritional stress. Physicians have known for years that when a fetus is stressed by toxins such as alcohol (see p. 841) or by a lack of necessary nutrients or oxygen during development, the immediate outcome for the fetus is often preterm delivery and low birth weight. Both of these outcomes are associated with potential problems, including developmental delays, anatomical (congenital) anomalies, functional deficits, and overt disease.

We now know that the fetus is not simply a passive recipient of the negative results of maternal stress during pregnancy. And, early delivery, low birth weight, or both are not the only outcomes related to fetal stress. Stress-related changes in the intrauterine environment may also trigger numerous development problems that become apparent during gestation or at the time of delivery. Many are related to oversecretion of endocrine secretions, especially cortisol, by both mother and developing baby. Intrauterine levels of cortisol are influenced by the maternal-fetoplacental unit and by the shared maternal-fetal hypothalamic-pituitary-adrenal or HPA mechanism (Figure 22-5).

Not only do abnormal increases in the intrauterine level of cortisol have an immediate impact on the fetus or newborn infant, but also they produce intermediate and long-term effects on the individual over a lifetime. The process is now called **fetal programming.** It refers to the relationship between events occurring during the course of fetal development and the appearance of specific anatomical, physiological, or disease states that develop later in life. Early research in the area of fetal programming focused on trying to explain the relationship known to exist between the stress of inadequate fetal nutrition, low birth weight, and adult rates of cardiovascular disease and its precursors—including elevated cholesterol levels, high blood pressure, and diabetes.

We now know that fetal programming is strongly influenced by elevated intrauterine levels of cortisol and other stress hormones and that it affects not only the cardiovascular system but also many other body systems and physiological variables over a lifetime. Cortisol-induced fetal programming changes are known to influence body composition; growth rates; age at maturity; the functioning of the adult immune, endocrine, renal, and reproductive systems; and even life expectancy. Although sophisticated experimental work has shown that fetal programming does indeed occur and that it is often influenced by stress before birth, the mechanism to explain exactly how it works is yet to be discovered. Fully understanding the role of cortisol in regulating the "partitioning" or allocation of energy (nutrients) available for use by different organ systems during intrauterine development will certainly be important in solving the puzzle.

Many scientists believe the answer will be related to what are described as *biological tradeoffs*. For example, for a developing fetus, the "tradeoff" of simply surviving a dangerous pregnancy caused by the stress of maternal malnutrition might be to "accept" the dangers associated with preterm delivery and low birth weight. Or, diverting very limited nutrients to support central nervous system development required to sustain life at birth rather than using the same amount of energy resource to protect future reproductive system function might be another "acceptable" biological tradeoff. In a biological sense, in the cycle of life, the "big picture" focuses on survival. It is therefore appropriate to state that stress plays an important role in the health of an individual in every stage of the cycle of life from conception to death.

New research data related to fetal programming and its relationship to stress experienced by the fetus during development have dramatically increased interest in this area in both the scientific and medical communities. For example, knowing that a particular disease or condition is caused by prenatal stress related to a nutritional deficiency in the maternal diet, and not by genetic susceptibility, will drastically change public health and disease prevention strategies intended to reduce the incidence of that disease or condition in the population. A good example is prevention of certain congenital defects in the central nervous system by addition of an important micronutrient, folic acid, to the diet of pregnant women. Expanding our knowledge about prenatal stress has resulted in new commitments being made to improve nutrition for women during pregnancy and to increase access to educational programs intended to decrease smoking and alcohol consumption.

QUICK CHECK

7. What is meant by the phrase "diseases of adaptation"?
8. What is the fight-or-flight reaction?
9. List four indicators of stress.
10. What is the difference between physiological and psychological stress?

THE BIG PICTURE
Stress and the Whole Body

Although our understanding of physiological and psychological stress is still evolving, we are certain of this: stress affects the entire body. Stress responses involve numerous physiological mechanisms, many of which are suspected to occur but are as yet unproven.

So far, we understand that stress responses involve nearly every system of the body. The nervous system detects and integrates the factors, or stressors, that trigger the stress responses. Physiological stress responses result from signals sent from the nervous system directly—or by way of the endocrine system. Many of these "stress signals" have only recently been discovered. Many different kinds of neurotransmitters, hormones, and perhaps other regulatory chemicals influence the function of the skeletal muscles, the digestive system, the urinary system, the reproductive system, the respiratory system, the cardiovascular system, the integumentary system—perhaps every system, organ, and tissue in the body.

Because the regulatory agents associated with stress influence the function of blood cells, stress can have a profound effect on the function of the immune system. Although the fact that stress inhibits immune function has been known for some time, many of the exact mechanisms that accomplish this have only recently been discovered. Biologists now better appreciate the link between the mind and the immune system and thus are better able to explain how stress causes disease—and even death. In fact, an emerging field within human biology is devoted to studying the mind-immunity link—the field of neuroimmunology. Some researchers in this new field have found themselves overlapping diverse fields such as endocrinology, psychology, and hematology.

Considering the effect that stress can have on the entire internal environment—the whole body—and the high levels of stress that characterize the modern cultures in which many of us live and work, advancements in stress research hold the promise of improving the length and quality of our lives.

CASE STUDY

Anthony Draecus is a 72-year-old male with type 2 diabetes mellitus, hypertension, and atherosclerosis. He lives with his oldest daughter and her family. His daughter has six children ranging in age from 7 to 18 years old. Mr. Draecus's wife of 50 years died a few months ago. Until her death he lived with her in their own home about 100 miles away. The daughter insisted that he move in with her in spite of the fact that her husband felt that her father would be better off in his own home. Anthony's daughter has brought him to the rural health clinic since the death of his wife, where he sees a family nurse practitioner. He tells the nurse practitioner that he is always tired and short of breath, his strength has decreased, and he is hungry all the time but when he eats it upsets his stomach. His weight has not changed recently.

On physical examination, Mr. Draecus is 5 feet 10 inches tall, and he weighs 210 pounds. Today his blood pressure and pulse are 170/92 and 92 beats/min, respectively. He demonstrates atherosclerotic and hypertensive changes in the eye. His lungs are clear, and his bowel sounds are normal. He has dependent edema with strong peripheral pulses. His blood glucose level is high at 210 mg/dl. Cholesterol and triglyceride levels are also high.

He takes oral hypoglycemic medication, a diuretic, and an antihypertensive medication. These medications have not been changed in more than 2 years. Until the death of his wife, his diabetes and hypertension were in better control.

1. The nurse practitioner would most likely consider which of the following as the cause of Mr. Draecus's symptoms?
   A. Lack of control of his diabetes mellitus
   B. Age-related body changes
   C. Stress
   D. Lack of control of his hypertension

2. With Selye's general adaptation syndrome as a foundation, Mr. Draecus's symptoms would characterize responses of the:
   A. Alarm reaction
   B. Stage of resistance
   C. Stage of exhaustion
   D. Stress triad

3. Which one of the following actions should the nurse practitioner consider first?
   A. Change the hypoglycemic medication to insulin.
   B. Help Mr. Draecus identify his stressors.
   C. Control his hypertension.
   D. Teach Mr. Draecus's daughter how to manage her father's diseases.

4. Which information from Mr. Draecus's history is characteristic of the fight-or-flight reaction?
   A. Diabetes mellitus, hypertension, and atherosclerosis
   B. Chronic fatigue, blood pressure at 170/92, and pulse at 92 beats/min
   C. Decreased strength, dependent edema, and clear lungs
   D. Hunger, normal bowel sounds, and dependent edema

CHAPTER SUMMARY

SELYE'S CONCEPT OF STRESS

A. Development of the concept
   1. Through many experiments, Selye exposed animals to noxious agents and found that they all responded with the same syndrome of changes, or "stress triad"

B. Definitions
   1. Stress—a state, or condition, of the body produced by "diverse nocuous agents" and manifested by a syndrome of changes
   2. Stressors—agents that produce stress
   3. General adaptation syndrome—group of changes that manifest the presence of stress

C. Stressors
   1. Stressors are extreme stimuli—too much or too little of almost anything
   2. Stressors are very often injurious or painful stimuli
   3. Anything an individual perceives as a threat is a stressor for that individual
   4. Stressors are different for different individuals and different for one individual at different times

D. The general adaptation syndrome
   1. Manifestations—stress triad (hypertrophied adrenals, atrophied thymus and lymph nodes, and bleeding ulcers); many other changes
   2. Stages—three successive phases, namely, alarm reaction, stage of adaptation or resistance, and stage of exhaustion, each characterized by a different syndrome of changes (Table 22-1)

E. Mechanism of stress
   1. Consists of group (syndrome) of responses to internal condition of stress; stress responses nonspecific in that same syndrome of responses occur regardless of kind of extreme change that produced stress
   2. Stimulus that produces stress and thereby activates stress mechanism is nonspecific in that it can be any kind of extreme change in environment
   3. Stress responses, Selye thought, were adaptive, tend to enable body to adapt to and survive extreme change; Selye referred to this syndrome of stress responses as general adaptation syndrome

4. Numerous factors influence stress responses—individual's physical and mental condition, age, sex, socioeconomic status, heredity, and previous experience with similar stressors
5. Stress most often is coped with successfully—namely, results in adaptation, healthy survival, and increased resistance; sometimes, however, stress produces exhaustion and death
6. See Figure 22-4, then Figures 22-2 and 22-3, for a summary of Selye's stress mechanism
F. Stress and disease
1. Selye held that stress could result in disease instead of adaptation

SOME CURRENT CONCEPTS ABOUT STRESS

A. Definitions
1. Stress—any stimulus that directly or indirectly stimulates the hypothalamus to release CRH
2. Stress syndrome—also called the stress response; many diverse changes initiated by stress
B. The stress syndrome—see Figure 22-6
C. Indicators of stress
1. Changes caused by increased sympathetic activity; for example, faster, stronger heartbeat; higher blood pressure; sweaty palms; dilated pupils
2. Changes resulting from increased corticoids—eosinopenia, lymphocytopenia, increased adrenocorticoids in blood and urine
D. Corticoids and resistance to stress
1. Still controversial; not proved that increased corticoids increase an animal's or a person's ability to resist stress but is proved that corticoid levels in blood increase in stress
E. Psychological stress
1. Psychological stressors—anything that an individual perceives as a threat to survival or self-image
2. Psychological stress—a mental state characterized by a syndrome of subjective and objective responses; dominant subjective response is anxiety; some characteristic objective responses are restlessness, irritability, lying, crying
3. Relation to physiological stress (see summary of principles on p. 848)
F. Intrauterine stress
1. Fetus develops short- and long-term responses to stress experienced during intrauterine development
   a. Frequent stressors include maternal malnutrition, hypoxia, and exposure to toxins such as alcohol
   b. Low birth weight and preterm delivery are common immediate responses to intrauterine stress
2. Maternal-fetal blood levels of cortisol are important mediators of stress
   a. Influenced by endocrine secretory activity of maternal-fetoplacental unit and the shared maternal-fetal hypothalamic-pituitary-adrenal (HPA) mechanism
3. Fetal programming responses to stress often result in negative outcomes later in life
   a. Affect many organ systems and even life expectancy
4. Concept of "biological tradeoffs" may be involved in fetal programming outcomes; for example, low birth weight and preterm delivery may be tradeoffs by fetus to survive the stress of maternal malnutrition in a problem pregnancy

THE BIG PICTURE: STRESS AND THE WHOLE BODY

A. Stress affects the entire body
B. Nervous system detects and integrates stressors that trigger stress responses
C. Stress affects immune system
1. Neuroimmunology studies mind-immunity link

REVIEW QUESTIONS

1. Define the terms stress, stressor, and general adaptation syndrome.
2. Make a few generalizations about the kinds of stimuli that constitute stressors.
3. What three stages make up the general adaptation syndrome? What changes characterize each stage?
4. What changes constitute the "stress triad"?
5. An increase in what three hormones brought about the changes that Selye named the general adaptation syndrome?
6. What function, according to Selye, does the general adaptation syndrome serve?
7. Stress, according to Selye, is a state, or condition, of the body. According to a current operational definition, what is stress?
8. What part of the brain, according to current ideas, plays the key role in initiating stress syndrome responses?
9. What parts of the nervous system other than that named in question 8 are involved in inducing the stress syndrome?
10. Briefly, what role, if any, does each of the following hormones play when the body is subjected to stress: ACTH, ADH, aldosterone, cortisol, CRH, epinephrine, and norepinephrine?
11. What has been proved about corticoids in relation to stress?
12. What is the controversial issue about corticoids in relation to stress?
13. Cite several examples of psychological stressors.
14. Are psychological stressors the same for all individuals?
15. Give some examples of subjective indicators of psychological stress.
16. Give some examples of objective responses that are part of psychological stress.

CRITICAL THINKING QUESTIONS

1. Based on what you know regarding Selye's experimental results, how would you explain why stress could be described as a nonspecific response?
2. What is the relationship between stress and disease? If stress levels were high, how might age and personality type affect the risk for specific diseases?
3. What observations and clinical test results would indicate an individual was under stress?
4. How would you use psychophysiology to explain the Chinese rice powder test?
5. Compare and contrast psychological stress and physiological stress. What is the relationship between physiological stress and psychological stress?

Career Choices   Paramedic

As a paramedic my primary function is to maintain life in critically ill or injured people, often during transport.

My involvement with the profession came about while serving as a safety and security specialist at a mental health facility. I found that, in order to do my job effectively, I needed to know more about medical matters. I enrolled in an Emergency Medical Technician (EMT) course, and became a full-time EMT. I've had a lifelong interest in science and technology, so the next natural step was to further my education and become an EMT-Paramedic (EMT-P), eventually obtaining National Registry status (NREMT-P).

Paramedics are trained as generalists, and our scope of practice is expanding yearly. I have medic friends who have chosen to leave "the street" to work in hospital emergency departments, in hyperbaric chambers, and in flight medicine. Continuing education is mandatory to maintain state and national certification, and many paramedics choose to further their careers in other medical fields. I know of one paramedic who is currently completing her residency to become a physician's assistant, and many of us are taking courses to become registered nurses. Credentialing, with the goal of national certification, is becoming key. This will allow lateral (i.e., from state to state) mobility for para-

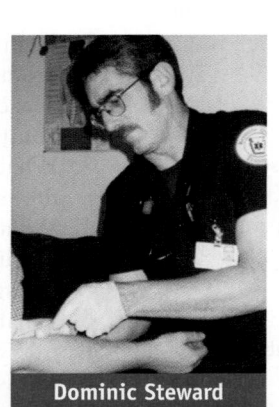

Dominic Steward
NREMT-P

medics. Pay and benefits are increasing, in part because of shortages. There is also a trend for colleges to offer associates' or bachelors' degrees in conjunction with paramedic training. This will allow the profession to grow and mature.

My profession is rewarding in several tangible ways. It's important to me to have the opportunity to help people in need. And, especially after recent national events, paramedics receive public appreciation and recognition. It is an honorable way to make a living, although you won't get rich!

A paramedic's effectiveness hinges on knowledge of the anatomy and physiology of the human body. One of my responsibilities is to precept new paramedics, and those who do not have a broad foundation in A&P do not go on the street. The lab component of the course was key to my comprehension of the text content. If you have the chance, try to attend a dissection. I was once in a lab where we learned about the components of the heart by dissecting a cow's heart in a carcass obtained from a local meat market. The organ was four times the size of a human heart, so it was easy to see the various features. Pharmacology was integrated into the class, and I had the opportunity to inject multiple doses of epinephrine into the tissue. This caused the long-dead organ to begin to beat. To this day I remember all of my heart A&P!

Respiration, Nutrition, and Excretion

The chapters in Unit 5 deal with respiration, digestion, processing of nutrients, and excretion of wastes by the urinary system. The unit ends with chapters discussing both fluid and electrolyte and acid-base balance.

Ultimately, all homeostatic mechanisms function to maintain what can best be described as a "dynamic constancy" at the cellular level. For example, although under normal conditions the O_2 and CO_2 content of blood does not change much over time, the quantity of O_2 and CO_2 that enters and exits the blood can vary widely with exercise. Delivery of oxygen and elimination of carbon dioxide and other wastes resulting from the metabolism of nutrients must be regulated within narrow limits so that cellular function remains normal. Maintaining the dynamic constancy of fluid and electrolyte and acid-base balance at the cellular level is also required for survival. The anatomical structures and functional control mechanisms discussed in this unit all relate, in the last analysis, to cellular homeostasis.

SEEING THE BIG PICTURE

CHAPTER 23

Anatomy of the Respiratory System

LANGUAGE OF SCIENCE

alveolar ducts (al-VEE-oh-lar) [*alveolus* little hollow, *ducere* to draw or lead]

alveoli (al-VEE-oh-lye) [*alveolus* little hollow]

apex (AY-peks) [*apex* tip]

arytenoid (ar-eh-TEH-noyd) [*arytaina-* ladle, *-oid* resembling]

base

bronchial tree (BRONG-kee-al) [*bronchi-* windpipe, *-al* pertaining to]

bronchioles (BRONG-kee-ohls) [*bronchiolus* little windpipe]

bronchopulmonary segments (brong-koh-PUL-moh-nair-ee) [*broncho-* windpipe, *-pulmonis* lung]

conchae (KONG-kee) [*konche* shell]

costal surface (KOS-tal) [*costa-* rib, *-al* pertaining to]

cribriform plate (KRIB-ri-form) [*cribrum* sieve]

epiglottis (ep-i-GLOT-iss) [*epi-* upon, *-glossa* tongue]

false vocal folds

glottis (GLOT-iss)

hilum (HYE-lum) [*hilum* trifle]

horizontal fissure (hor-i-ZON-tal FISH-ur) [*horizein* to encircle, *fissura* cleft]

laryngopharynx (lah-ring-go-FAIR-inks) [*laryngo-* larynx, *-pharynx* throat]

larynx (LAIR-inks)

lingual tonsils (LING-gwal TAHN-sils) [*lingu-* tongue, *-al* pertaining to]

lower respiratory tract (RES-pih-rah-tor-ee) [*respirare* to breathe, *tractus* trail]

nasal mucosa (NAY-zal myoo-KOH-sah) [*nas-* nose, *-al* pertaining to, *mucus* slime]

nasopharynx (nay-zoh-FAIR-inks) [*naso-* nose, *-pharynx* throat]

oblique fissure (oh-BLEEK FISH-ur) [*obliquus* slanted, *fissura* cleft]

Cont'd on p. 879

OVERVIEW OF THE RESPIRATORY SYSTEM

Functions of the Respiratory System

The respiratory system functions as an air distributor and a gas exchanger so that oxygen may be supplied to and carbon dioxide removed from the body's cells. Because most of our billions of cells lie too far from air to exchange gases directly with it, air must first exchange gases with blood, blood must circulate, and finally, blood and cells must exchange gases. These events require the functioning of two systems, namely, the respiratory system and the circulatory system. All parts of the respiratory system—except its microscopic-sized sacs called *alveoli*—function as air distributors. Only the alveoli and the tiny alveolar ducts that open into them serve as gas exchangers.

In addition to air distribution and gas exchange, the respiratory system effectively filters, warms, and humidifies the air we breathe. Respiratory organs also influence sound production, including speech used in communicating oral language. Specialized epithelium in the respiratory tract makes the sense of smell (olfaction) possible. The respiratory system also plays an important role in the regulation, or homeostasis, of pH in the body.

Structural Plan of the Respiratory System

For purposes of study, the respiratory system may be divided into upper and lower tracts, or divisions. The organs of the upper respiratory tract are located outside the thorax, or chest cavity, whereas those in the lower tract, or division, are located almost entirely within it (Figure 23-1).

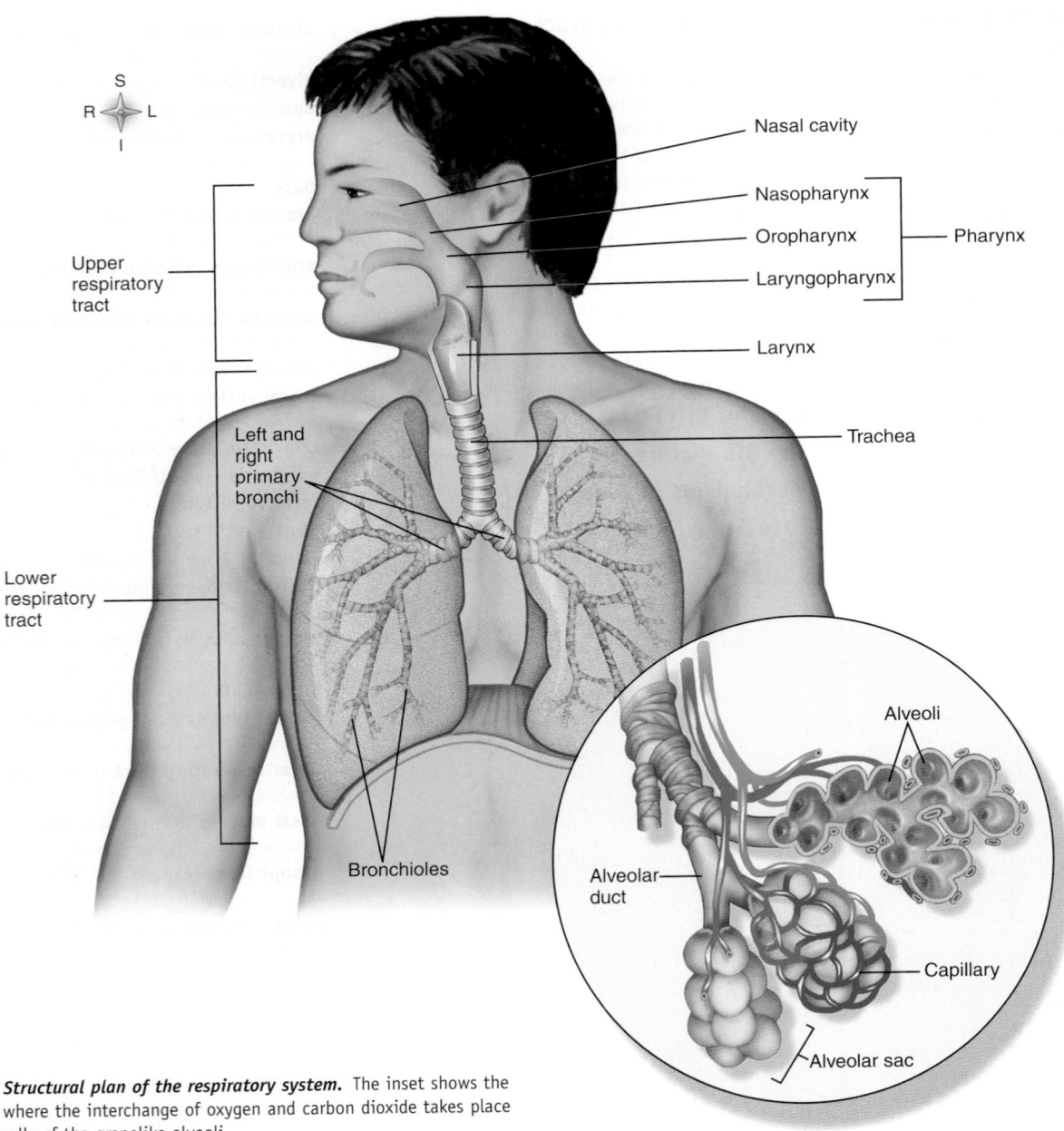

Figure 23-1 *Structural plan of the respiratory system.* The inset shows the alveolar sacs where the interchange of oxygen and carbon dioxide takes place through the walls of the grapelike alveoli.

The **upper respiratory tract** is composed of the nose, nasopharynx, oropharynx, laryngopharynx, and larynx. The **lower respiratory tract**, or division, consists of the trachea, all segments of the bronchial tree, and the lungs. Functionally, the respiratory system also includes several accessory structures, such as the oral cavity, rib cage, and respiratory muscles including the diaphragm. Together, these structures constitute the lifeline, the air supply line of the body. This chapter describes the functional anatomy of these organs. The physiology of the respiratory system as a whole is discussed in Chapter 24. Cells require a constant supply of oxygen for the vital energy conversion process carried out within each cell's mitochondria—a process called *cellular respiration* (Chapter 27). Cellular respiration produces carbon dioxide (CO_2) as a waste product, which must be removed before it accumulates to dangerously high levels.

UPPER RESPIRATORY TRACT

Nose

Structure of the Nose

The nose consists of an external and an internal portion. The external portion, that is, the part that protrudes from the face, consists of a bony and cartilaginous framework overlaid by skin containing

BOX 23-1: FYI

Danger Area of the Face

The fact that the skin of the external nose contains many sebaceous glands has great clinical significance. If these glands become blocked, it is possible for infectious material to enter and pass from facial veins near the nose to one of the intracranial venous sinuses (see Chapter 18). For this reason, the triangle-shaped zone surrounding the external nose is often known as the "danger area of the face."

many sebaceous glands. The two nasal bones meet above where they are surrounded by the frontal bone to form the root of the nose. The nose is surrounded by the maxilla laterally and inferiorly at its base. The flaring cartilaginous expansion forming and supporting the outer side of each oval nostril opening is called the *ala*.

The internal nose, or nasal cavity, lies over the roof of the mouth where the palatine bones, which form the floor of the nose and the roof of the mouth, separate the nasal cavities from the mouth cavity. Sometimes the palatine bones fail to unite completely and produce a condition known as **cleft palate**. When this abnormality exists, the mouth is only partially separated from the nasal cavity, and difficulties arise in swallowing and speaking.

The roof of the nose is separated from the cranial cavity by a portion of the ethmoid bone called the **cribriform plate** (Figures 23-2 and 23-3). The cribriform plate is perforated by many small openings that permit branches of the olfactory nerve responsible for the special sense of smell to enter the cranial cavity and reach the brain.

Separation of the nasal and cranial cavities by a thin, perforated plate of bone presents real hazards. If the cribriform plate is

BOX 23-2   Vomeronasal Organ

Scientists have known for years that a small remnant of unique fetal ectodermal (nervous) tissue is located in the nasal mucosa close to the olfactory nerves and sensory cells. Until recently, this structure, called the **vomeronasal organ (VNO),** was considered vestigial and without physiological significance. However, interest in the tiny organ and its connections to the brain is growing. New research evidence now suggests that it may play a role in detection of complex body chemicals called *pheromones*. These compounds serve as carriers of information between individuals. Once detected, pheromones may influence the production of certain hormones, regulate levels of various brain chemicals, and influence behavior.

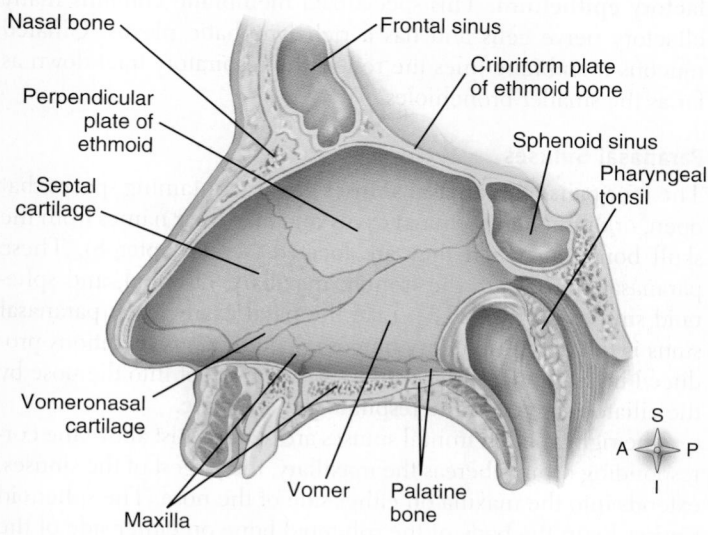

Figure 23-2 *Nasal septum.* The nasal septum consists of the perpendicular plate of the ethmoid bone, the vomer, and the septal and vomeronasal cartilages.

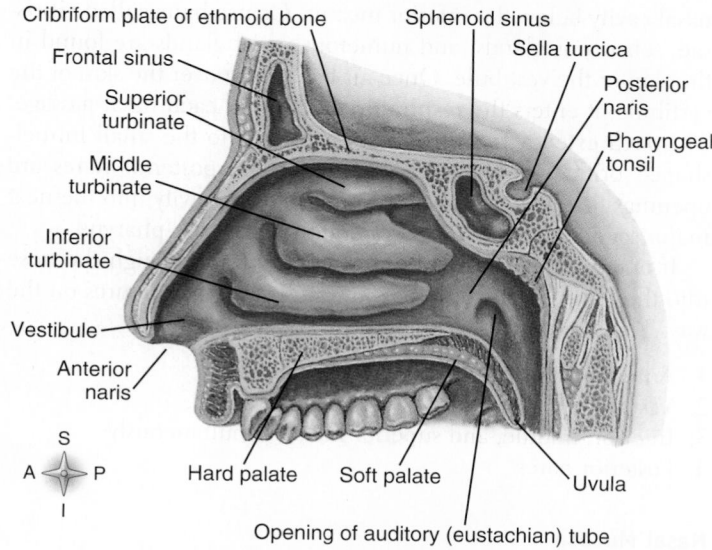

Figure 23-3 *Nasal cavity.* In this midsagittal section through the nose and nasal cavity, the nasal septum has been removed to reveal the turbinates (nasal conchae) of the lateral wall of the nasal cavity.

damaged as a result of trauma to the nose, it is possible for potentially infectious material to pass directly from the nasal cavity into the cranial fossa and infect the brain and its covering membranes.

The hollow nasal cavity is separated by a midline partition, the **septum** (see Figure 23-2), into a right and a left cavity. Note in Figure 23-2 that the nasal septum is made up of four main structures: the perpendicular plate of the ethmoid bone above, the vomer bone, and the septal nasal and vomeronasal cartilages below. In the adult the nasal septum is frequently deviated to one side or the other, interfering with respiration and with drainage of the nose and sinuses. The nasal septum has a rich blood supply. Nosebleeds, or *epistaxis*, often occur as a result of septal contusions caused by a direct blow to the nose. Epistaxis may also result from weak blood vessels combined with high blood pressure. Though dramatic, nosebleeds are seldom serious.

Each nasal cavity is divided into three passageways (named the superior, middle, and inferior meati) by the projection of the **turbinates,** or **conchae,** from the lateral walls of the internal portion of the nose (see Figure 23-3). The superior and middle turbinates are processes of the ethmoid bone, whereas the inferior turbinates are separate bones.

The external openings into the nasal cavities (nostrils) have the technical name of *anterior nares* (singular, naris). They open into an area covered by skin that is reflected from the wings (ala) of the nose. This area, called the **vestibule,** is located just inside the nasal cavity below the inferior meatus. Coarse hairs called **vibrissae,** sebaceous glands, and numerous sweat glands are found in the skin of the vestibule. Once air has passed over the skin of the vestibule, it enters the **respiratory portion** of each nasal passage. This area extends from the inferior meatus to the small funnel-shaped orifices of the *posterior nares.* The posterior nares are openings that allow air to pass from the nasal cavity into the next major segment of the upper respiratory tract—the pharynx.

If one were to "trace" the movement of air through the nose into the pharynx, it would pass through several structures on the way. The sequence is as follows:

1. Anterior nares (nostrils)
2. Vestibule
3. Inferior, middle, and superior meati, simultaneously
4. Posterior nares

Nasal Mucosa

Once air has passed over the skin of the vestibule and enters the respiratory portion of the nasal passage, it passes over the highly specialized **respiratory mucosa.** This mucous membrane has a

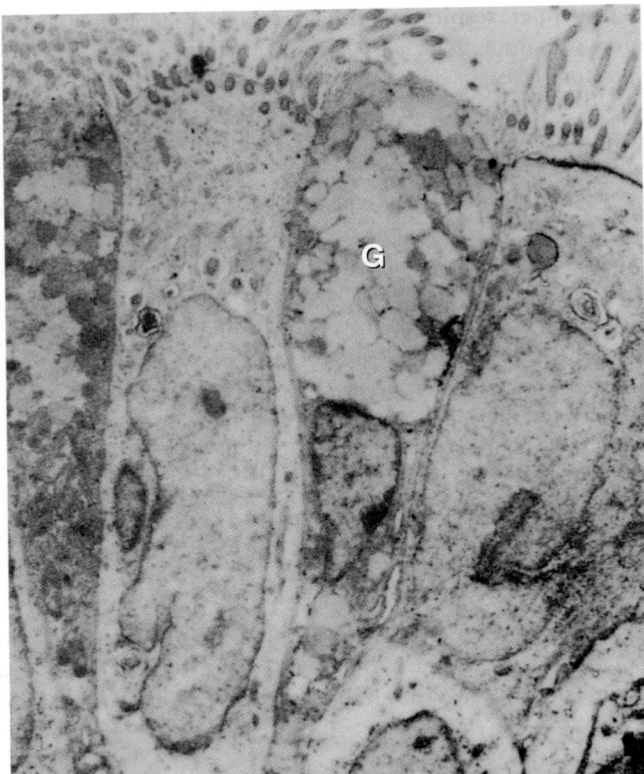

Figure 23-4 *Electron micrograph of respiratory mucosa.* This epithelium is typically ciliated and exhibits numerous goblet cells, *G*, filled with mucus. (×7600.)

pseudostratified ciliated columnar epithelium rich in goblet cells (Figure 23-4). The respiratory mucosa possesses a rich blood supply, especially over the inferior turbinate, and is bright pink or red. Near the roof of the nasal cavity and over the superior turbinate and opposing portion of the septum, the mucosa turns pale and has a yellowish tint. In this area it is referred to as the **olfactory epithelium.** This specialized membrane contains many olfactory nerve cells and has a rich lymphatic plexus. Ciliated mucous membrane lines the rest of the respiratory tract down as far as the smaller bronchioles.

Paranasal Sinuses

The four pairs of **paranasal sinuses** are air-containing spaces that open, or drain, into the nasal cavity and take their names from the skull bones in which they are located (see Chapter 8). These paranasal sinuses are the frontal, maxillary, ethmoid, and sphenoid sinuses (Figure 23-5). Like the nasal cavity, each paranasal sinus is lined by respiratory mucosa. The mucous secretions produced in the sinuses are continually being swept into the nose by the ciliated surface of the respiratory membrane.

The right and left frontal sinuses are located just above the corresponding orbit, whereas the maxillary, the largest of the sinuses, extends into the maxilla on either side of the nose. The sphenoid sinuses lie in the body of the sphenoid bone on either side of the midline in close proximity to the optic nerves and pituitary gland.

Note in Figure 23-5 that the ethmoid sinuses are not single large cavities but a collection of small air cells divided into ante-

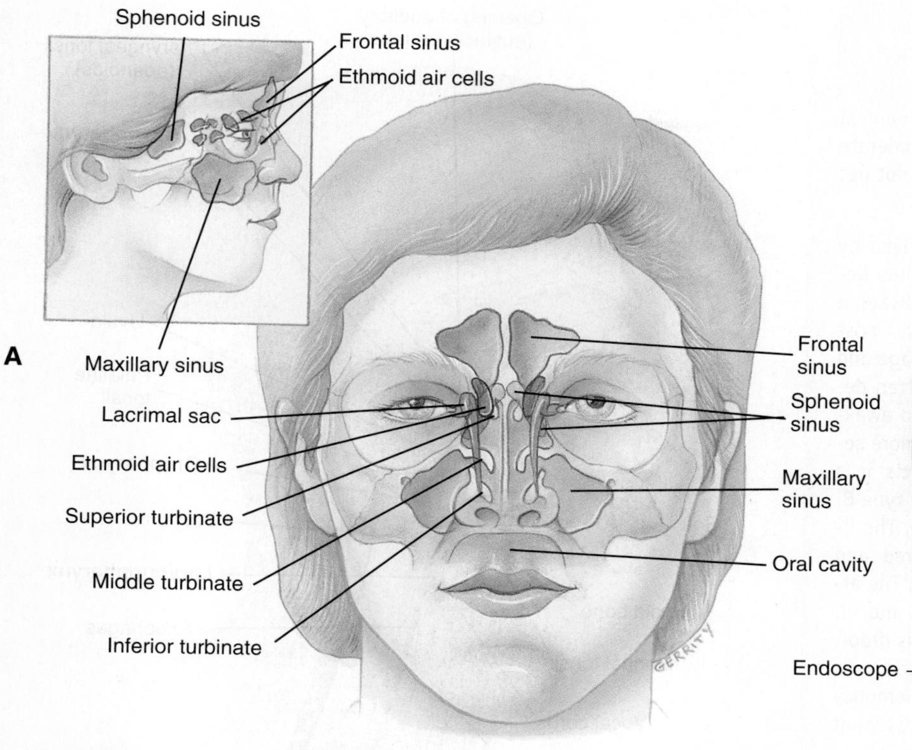

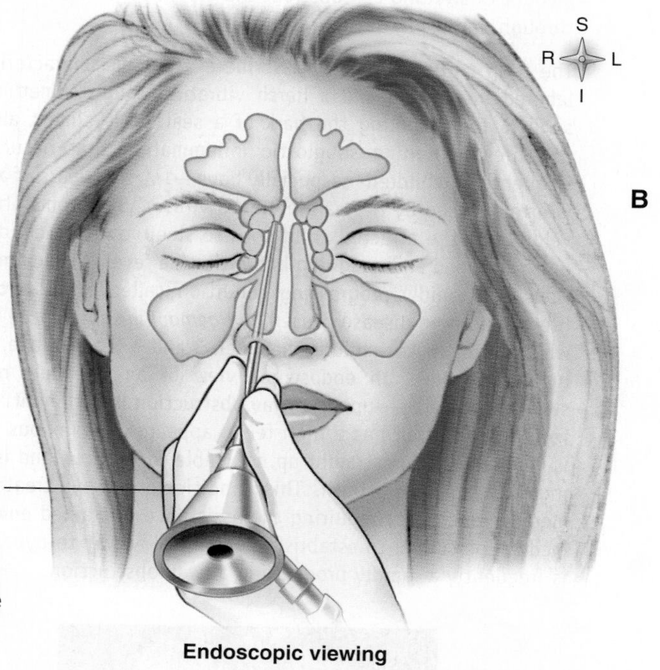

Sphenoid sinus
Frontal sinus
Ethmoid air cells

A

Maxillary sinus

Lacrimal sac

Ethmoid air cells

Superior turbinate

Middle turbinate

Inferior turbinate

Frontal sinus

Sphenoid sinus

Maxillary sinus

Oral cavity

B

Endoscope

Figure 23-5 *The paranasal sinuses.* A, The anterior view shows the anatomical relationship of the paranasal sinuses to each other and to the nasal cavity. The *inset* is a lateral view of the position of the sinuses. **B,** Viewing the sinus cavity with an endoscope (see Figure 8-9 on p. 275).

Endoscopic viewing

rior, middle, and posterior groups that open independently into the upper part of the nasal cavity. The paranasal sinuses drain as follows:

- Into the middle meatus (passageway below the middle turbinate)—frontal, maxillary, anterior, and middle ethmoidal sinuses
- Into the superior meatus—posterior ethmoidal sinuses
- Into the space above the superior turbinates (sphenoethmoidal recess)—sphenoidal sinuses

Functions of the Nose

The nose serves as a passageway for air going to and from the lungs. However, if the nasal passages are obstructed, it is possible for air to bypass the nose and enter the respiratory tract directly through the mouth. Air that enters the system through the nasal cavity is filtered of impurities, warmed, moistened, and chemically examined for substances that might prove irritating to the delicate lining of the respiratory tract. The vibrissae, or hairs, in the vestibule serve as an initial "filter" to screen particulate matter from air that is entering the system. The turbinates, or conchae, then serve as baffles to provide a large mucus-covered surface area over which air must pass before reaching the pharynx. The respiratory membrane produces copious quantities of mucus and possesses a rich blood supply, especially over the inferior turbinates, which permits rapid warming and moistening of the dry inspired air. Mucous secretions provide the final "trap" for removal of any remaining particulate matter from air as it moves

through the nasal passages. Fluid from the lacrimal glands (see Figure 15-18 on p. 575) and additional mucus produced in the paranasal sinuses also help trap particulate matter and moisten air passing through the nose. In addition, the hollow sinuses act to lighten the bones of the skull and serve as resonating chambers for speech. Deflection of air by the middle and superior conchae over the olfactory epithelium makes the special sense of smell possible.

 QUICK CHECK

1. What are the overall functions of the respiratory system? Which other body system is involved in accomplishing these functions?
2. What are the principal organs of the upper respiratory tract? What are the principal organs of the lower respiratory tract?
3. What are the paranasal sinuses? What is their anatomical relationship to the nose?

Pharynx

Structure of the Pharynx

Another name for the **pharynx** is the throat. It is a tubelike structure about 12.5 cm (5 inches) long that extends from the base of the skull to the esophagus and lies just anterior to the cervical verte-

BOX 23-4: HEALTH MATTERS
Laryngeal Edema

Edema of the mucosa covering the vocal cords and other laryngeal tissues can be a potentially lethal condition. Even a moderate amount of swelling can obstruct the glottis so that air cannot get through and asphyxiation results.

The term **croup** is used to describe a syndrome characterized by labored inspiration and a harsh vibrating cough sometimes described as resembling the bark of a seal. It is almost always a non–life-threatening subglottic inflammation that occurs most frequently in children—especially boys—1½ to 3 years of age and is often caused by the parainfluenza viruses. Symptoms often develop after the affected child goes to sleep, and the child awakens frightened and coughing but without a fever. A much more serious and rapidly progressing condition called **epiglottitis** is a life-threatening disease caused by *Haemophilus influenzae* type B, which often strikes children between 3 and 7 years of age. The illustration shows an endoscopic view of an intensely red and swollen epiglottis—upper airway obstruction is imminent. The affected child often has a high fever, appears very anxious and ill, is generally sitting straight up, is unable to swallow, and is drooling from an open mouth. This condition is always treated as a medical emergency requiring the assistance of trained emergency medical personnel to establish an artificial airway to bypass what is frequently a rapidly progressing airway obstruction.

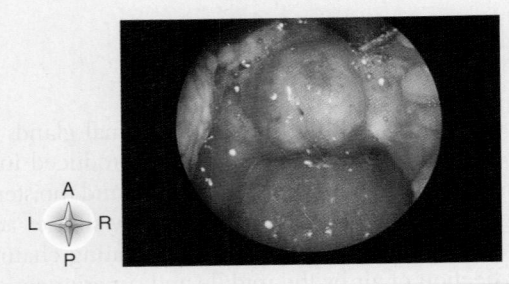

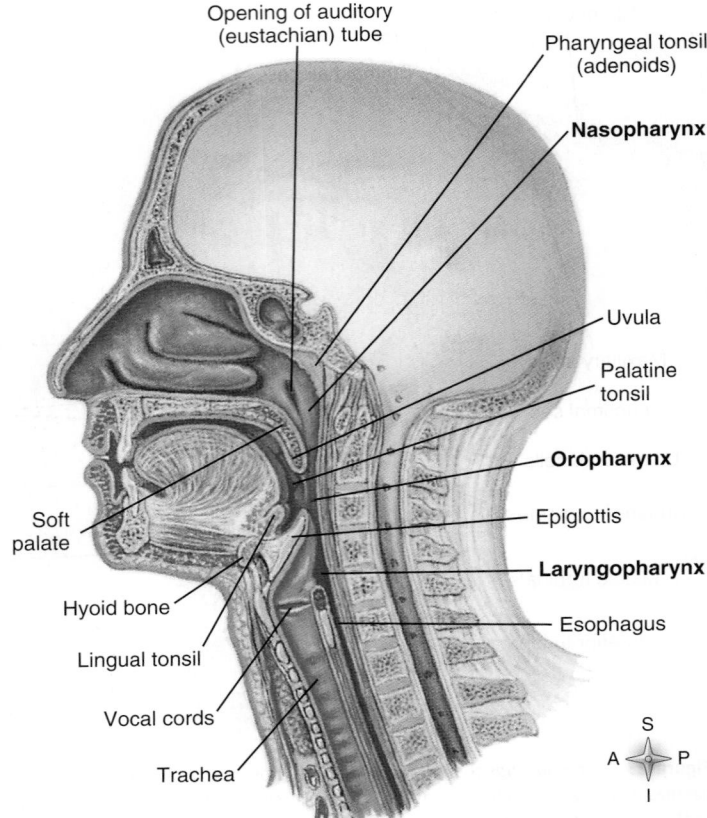

Figure 23-6 *Pharynx.* This midsagittal section shows the three divisions of the pharynx (nasopharynx, oropharynx, and laryngopharynx) and nearby structures.

brae. It is made of muscle, is lined with mucous membrane, and has three anatomical divisions: the **nasopharynx**, located behind the nose and extending from the posterior nares to the level of the soft palate; the **oropharynx**, located behind the mouth from the soft palate above to the level of the hyoid bone below; and the **laryngopharynx**, which extends from the hyoid bone to its termination in the esophagus. Figure 23-6 shows the divisions of the pharynx.

Seven openings are found in the pharynx (see Figures 23-3 and 23-6):

- Right and left auditory (eustachian) tubes opening into the nasopharynx
- Two posterior nares opening into the nasopharynx
- The opening from the mouth, known as the fauces, into the oropharynx
- The opening into the larynx from the laryngopharynx
- The opening into the esophagus from the laryngopharynx

The **pharyngeal tonsils** are located in the nasopharynx on its posterior wall opposite the posterior nares. The pharyngeal tonsils are called *adenoids* when they are enlarged. Although the cavity

of the nasopharynx differs from the oral and laryngeal divisions in that it does not collapse, it may become obstructed. If these tonsils enlarge to become adenoids, they may fill the space behind the posterior nares and make it difficult or impossible for air to travel from the nose into the throat.

Two pairs of organs are found in the oropharynx: the **palatine tonsils**, located behind and below the pillars of the fauces, and the **lingual tonsils**, located at the base of the tongue. The palatine tonsils are the ones most commonly removed by a tonsillectomy. Only rarely are the lingual tonsils also removed.

Functions of the Pharynx

The pharynx serves as a common pathway for the respiratory and digestive tracts, since both air and food must pass through this structure before reaching the appropriate tubes. It also affects phonation (speech production). For example, only by the pharynx changing its shape can the different vowel sounds be formed.

Larynx

Location of the Larynx

The **larynx**, or voice box, lies between the root of the tongue and the upper end of the trachea just below and in front of the lowest part of the pharynx (see Figure 23-1). It might be described as a vestibule opening into the trachea from the pharynx. It normally

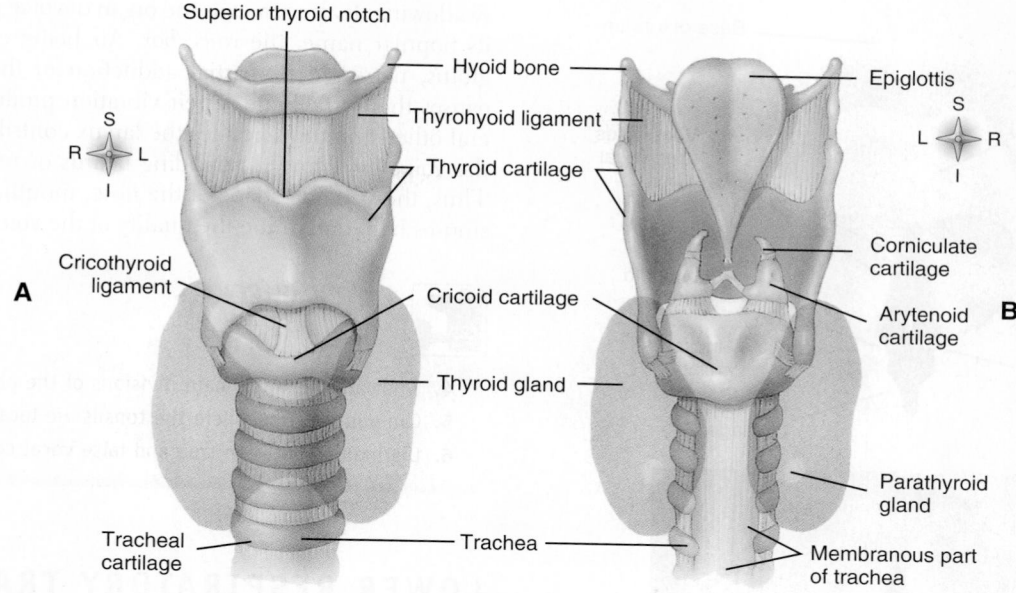

Figure 23-7 *Anatomy of the larynx.* **A,** Anterior view. **B,** Posterior view.

extends between the third, fourth, fifth, and sixth cervical verte-brae but is often somewhat higher in females and during child-hood. The lateral lobes of the thyroid gland and the carotid artery in its covering sheath touch the sides of the larynx.

Structure of the Larynx

The triangle-shaped larynx consists largely of cartilages that are at-tached to one another and to surrounding structures by muscles or by fibrous and elastic tissue components (Figure 23-7). It is lined by a ciliated mucous membrane. The cavity of the larynx extends from its triangle-shaped inlet at the epiglottis to the circular outlet at the lower border of the cricoid cartilage, where it is continuous with the lumen of the trachea (Figure 23-8). The mucous membrane lin-ing the larynx forms two pairs of folds that jut inward into its cavity

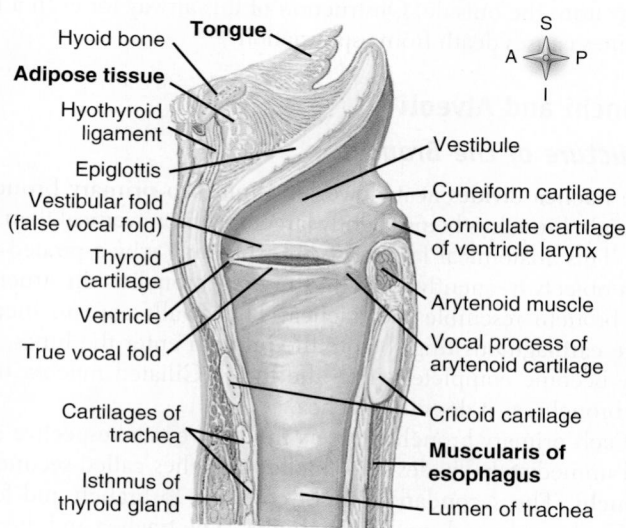

Figure 23-8 *Larynx (sagittal section).*

and divide it into three compartments, or divisions. The upper pair is called the **vestibular,** or **false, vocal folds** for the rather obvious reason that they play no part in vocalization. The lower pair serves as the **true vocal cords.** The slitlike space between the true vocal cords, called the *rima glottidis,* is the narrowest part of the larynx. The true vocal cords and the space between them (rima glottidis) are together designated as the **glottis.** An endoscopic view of the vo-cal cords and related structures is shown in Figure 23-9, *B.*

The division, or compartment, of the laryngeal cavity above the false, or vestibular, vocal folds is called the **vestibule.** The very short middle portion of the cavity between the false and true vo-cal cords is the ventricular division, or **ventricle.**

Cartilages of the Larynx

Nine cartilages form the framework of the larynx. The three largest—the thyroid cartilage, the epiglottis, and the cricoid cartilage—are single structures. There are three pairs of smaller accessory cartilages, namely, the arytenoid, corniculate, and cuneiform cartilages.

- The **thyroid cartilage** (Adam's apple) is the largest cartilage of the larynx and is the one that gives the characteristic triangu-lar shape to its anterior wall. It is usually larger in men than in women and has less of a fat pad lying over it—two reasons why a man's thyroid cartilage protrudes more than a woman's.
- The **epiglottis** is a small leaf-shaped cartilage that projects up-ward behind the tongue and hyoid bone. It is attached below to the thyroid cartilage, but its free superior border can move up and down during swallowing to prevent food or liquids from entering the trachea (see Figures 23-7 and 23-8).
- The pyramid-shaped **arytenoid** cartilages are the most impor-tant of the paired laryngeal cartilages. The base of each carti-lage articulates with the superior border of the cricoid cartilage (see Figure 23-7). The anterior angles of these cartilages serve as points of attachment for the vocal cords.

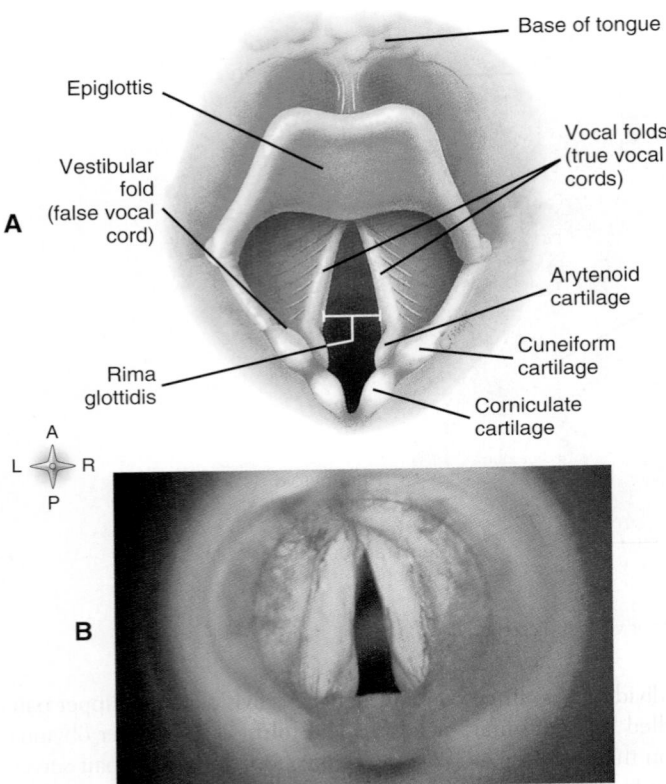

Figure 23-9 **Vocal cords. A,** Vocal cords viewed from above. **B,** Photograph taken with an endoscope showing the vocal cords in the open position.

Muscles of the Larynx

The muscles of the larynx are often divided into intrinsic and extrinsic groups. Intrinsic muscles have both their origin and insertion on the larynx. They are important in controlling vocal cord length and tension and in regulating the shape of the laryngeal inlet. Extrinsic muscles insert on the larynx but have their origin on some other structure—such as the hyoid bone. Therefore, contraction of the extrinsic muscles actually moves or displaces the larynx as a whole. Muscles in both groups play important roles in respiration, vocalization, and swallowing. During swallowing, for example, contraction of the intrinsic aryepiglottic muscles (those that connect the arytenoid cartilages with the epiglottis) prevents entry of food or fluid into the trachea by squeezing the laryngeal inlet shut.

Two other pairs of intrinsic laryngeal muscles function to open and close the glottis by adducting or abducting the true vocal cords. These events are crucial to both respiration and voice production. Certain other intrinsic muscles of the larynx function to influence the pitch of the voice by either lengthening and tensing or shortening and relaxing the vocal cords.

Functions of the Larynx

The larynx functions in respiration because it constitutes part of the vital airway to the lungs. This unique passageway, like the other components of the upper respiratory tract, is lined with a ciliated mucous membrane that helps in removal of dust particles and in warming and humidification of inspired air. In addition, it protects the airway against the entrance of solids or liquids during

swallowing. It also serves as the organ of voice production—hence its popular name, the *voice box.* Air being expired through the glottis, narrowed by partial adduction of the true vocal cords, causes them to vibrate. Their vibration produces the voice. Several other structures besides the larynx contribute to the sound of the voice by acting as sounding boards or resonating chambers. Thus, the size and shape of the nose, mouth, pharynx, and bony sinuses help determine the quality of the voice.

QUICK CHECK

4. What are the three main divisions of the pharynx?
5. Can you describe where the tonsils are located?
6. Distinguish between true and false vocal cords.

LOWER RESPIRATORY TRACT

Trachea

Structure of the Trachea

The **trachea,** or windpipe, is a tube about 11 cm (4.5 inches) long that extends from the larynx in the neck to the primary bronchi in the thoracic cavity (Figure 23-10). Its diameter measures about 2.5 cm (1 inch). Smooth muscle, in which are embedded **C**-shaped rings of cartilage at regular intervals, fashions the wall of the trachea (Figure 23-11). The cartilaginous rings are incomplete on the posterior surface (Figure 23-7). They give firmness to the wall and tend to prevent it from collapsing and shutting off the vital airway.

Note in Figure 23-12 that the trachea is lined with the type of pseudostratified ciliated columnar epithelium typical of the respiratory tract as a whole.

Function of the Trachea

The trachea performs a simple, but vital function—it furnishes part of the open passageway through which air can reach the lungs from the outside. Obstruction of this airway for even a few minutes causes death from asphyxiation.

Bronchi and Alveoli

Structure of the Bronchi

The trachea divides at its lower end into two **primary bronchi,** the right bronchus being slightly larger and more vertical than the left. This anatomical fact assists in explaining why aspirated foreign objects frequently lodge in the right bronchus. In structure the bronchi resemble the trachea. Their walls contain incomplete cartilaginous rings before the bronchi enter the lungs, but they become complete within the lungs. Ciliated mucosa lines the bronchi, as it does the trachea.

Each primary bronchus enters the lung on its respective side and immediately divides into smaller branches called **secondary bronchi.** The secondary bronchi continue to branch and form tertiary bronchi and small **bronchioles.** The trachea and the two primary bronchi and their many branches resemble an inverted tree trunk with its branches and are therefore spoken of as the

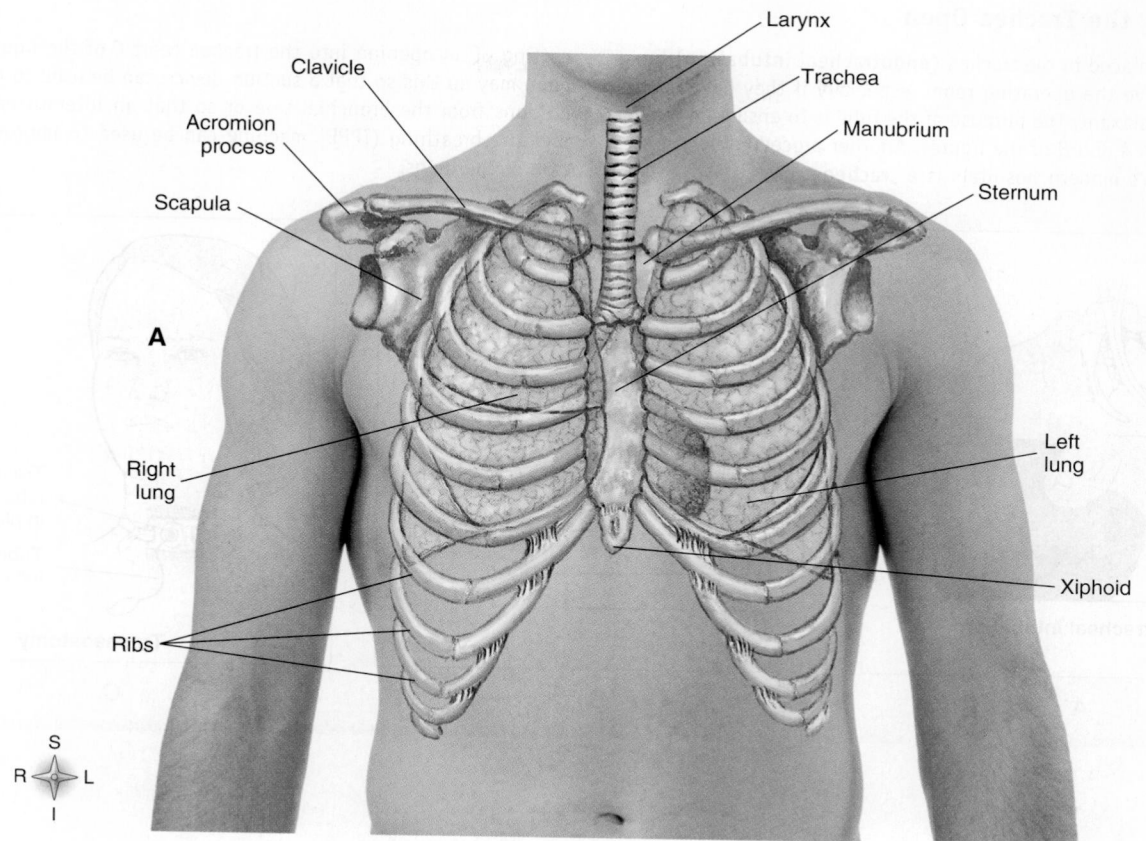

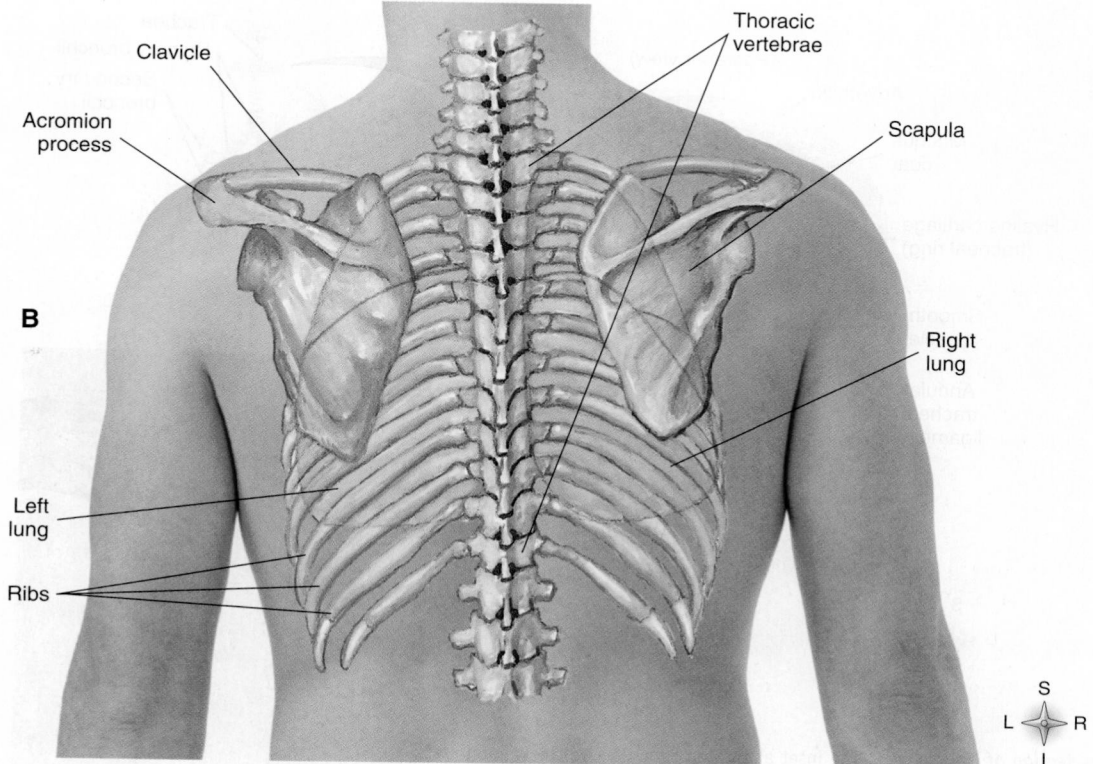

Figure 23-10 *Bony structures of the chest.* These structures form a protective and expandable cage around the lungs and heart. **A,** Anterior view. **B,** Posterior view.

BOX 23-5: HEALTH MATTERS
Keeping the Trachea Open

Often, a tube is placed in the trachea **(endotracheal intubation)** before patients leave the operating room, especially if they have been given a muscle relaxant. The purpose of the tube is to ensure an open airway (see parts *A* and *B* of the figure). Another procedure done frequently in today's modern hospitals is a **tracheostomy,** that is, the

cutting of an opening into the trachea (part *C* of the figure). A surgeon may do this so that a suction device can be used to remove secretions from the bronchial tree or so that an intermittent positive pressure breathing (IPPB) machine can be used to improve ventilation of the lungs.

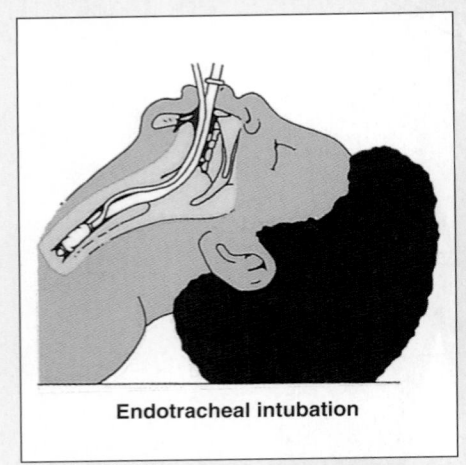
Endotracheal intubation
A

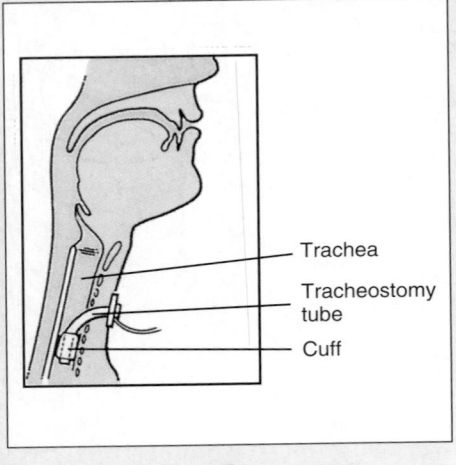
Trachea
Tracheostomy tube
Cuff
B

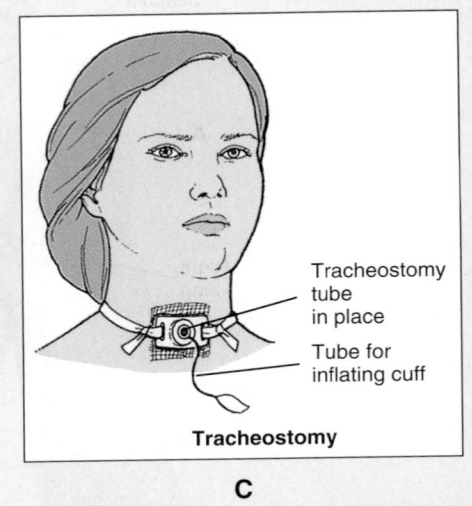
Tracheostomy tube in place
Tube for inflating cuff
Tracheostomy
C

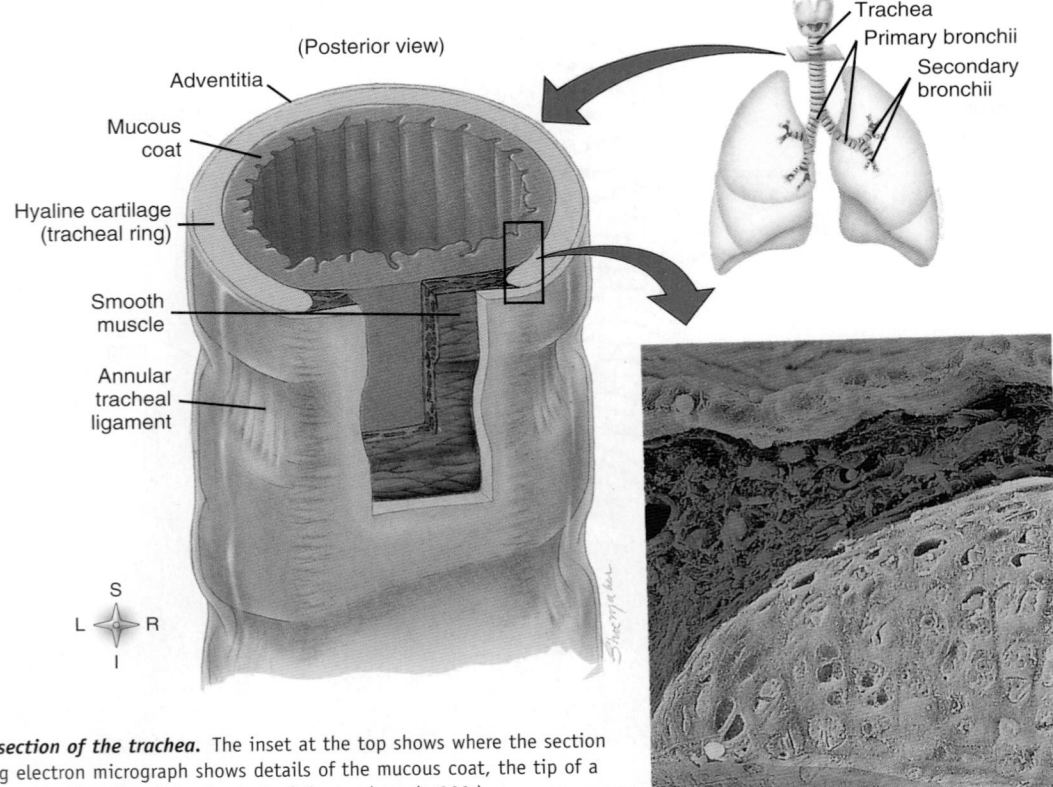

Figure 23-11 *Cross section of the trachea.* The inset at the top shows where the section was cut. The scanning electron micrograph shows details of the mucous coat, the tip of a cartilage ring, and the adventitia that form the wall of the trachea. (×300.)

Cilia

Mucus

Pseudostratified epithelium

Submucosa

Mucous gland

Figure 23-12 *Transverse section of the trachea.* Note the mucosa of ciliated epithelium. Hyaline cartilage occurs below the glandular submucosa and is not visible in this section. (×70.)

bronchial tree (Figure 23-13). The bronchioles subdivide into smaller and smaller tubes, eventually terminating in microscopic branches that divide into **alveolar ducts,** which terminate in several alveolar sacs, the walls of which consist of numerous **alveoli** (Figure 23-14; see also Figure 23-1). The structure of an alveolar duct with its branching alveolar sacs can be likened to a bunch of grapes—the stem represents the alveolar duct, each cluster of grapes represents an alveolar sac, and each grape represents an alveolus. Some 300 million alveoli are estimated to be present in our two lungs.

The structure of the secondary and tertiary bronchi and bronchioles shows some modification of the primary bronchial structure. The cartilaginous rings become irregular and disappear entirely in the smaller bronchioles. By the time the branches of the bronchial tree dwindle to form the alveolar ducts and sacs and the alveoli, only the internal surface layer of cells remains. In other words, the walls of these microscopic structures consist of a single layer of simple, squamous epithelial tissue (Figures 23-14 and 23-15). As we shall see, this structural fact makes it possible for them to perform their functions.

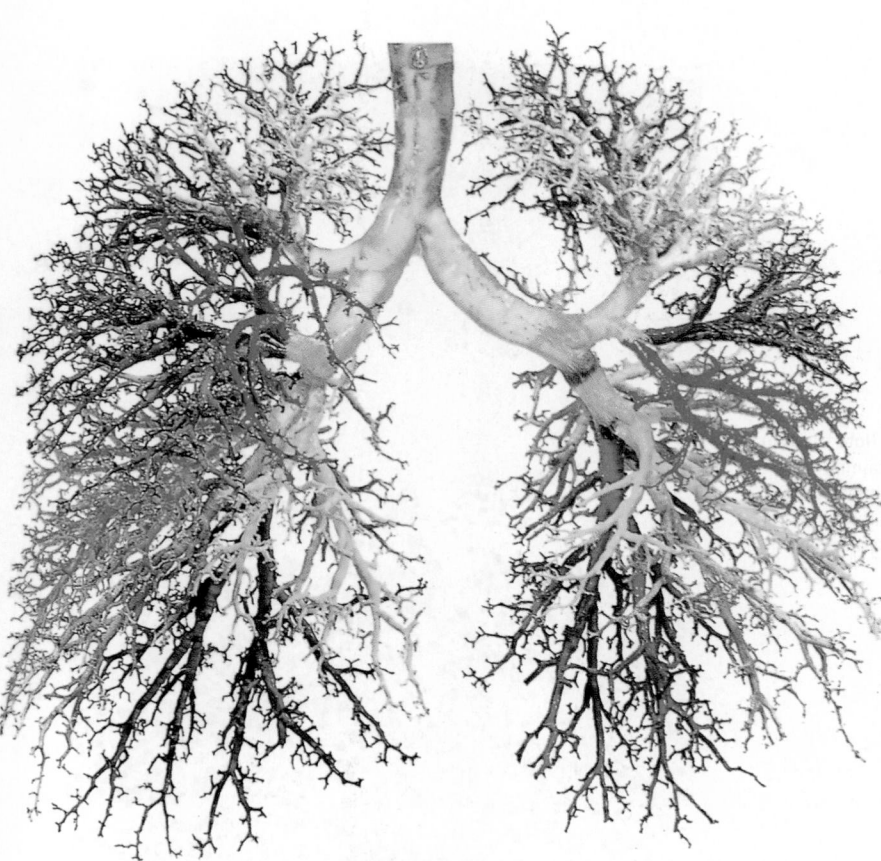

Figure 23-13 *Plastic cast of air spaces of the lungs.* The cast was prepared by pouring liquid plastic into the airways of a human lung—a different color for each bronchopulmonary segment supplied by its own tertiary bronchus. After the plastic hardened, the soft tissue was removed, leaving the branched form of the lower respiratory tract that is pictured here (compare with Figure 23-18).

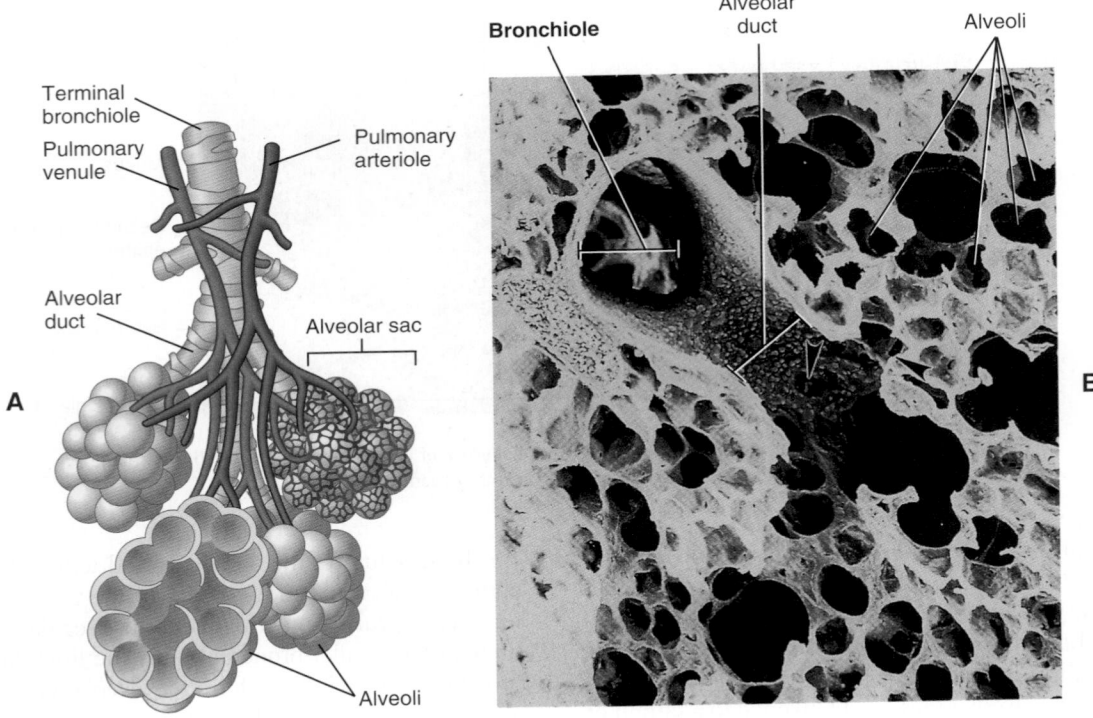

Figure 23-14 *Alveoli.* **A,** Bronchioles subdivide to form tiny tubes called alveolar ducts, which end in clusters of alveoli called alveolar sacs. **B,** Scanning electron micrograph of a bronchiole, alveolar ducts, and surrounding alveoli. The arrowhead indicates the opening of alveoli into the alveolar duct.

Figure 23-15 *Histologic appearance of alveoli.* **A,** Normal alveoli. Note the thin alveolar walls and open alveolar spaces. An occasional pigment containing macrophage can be seen in several of the alveolar spaces and is a normal finding. **B,** Appearance of alveoli in acute pneumonia. The alveolar walls are thickened and the spaces filled with a fibrin exudate and neutrophils (pus)—a classic finding in acute pneumonia (see Mechanisms of Disease p. 873).

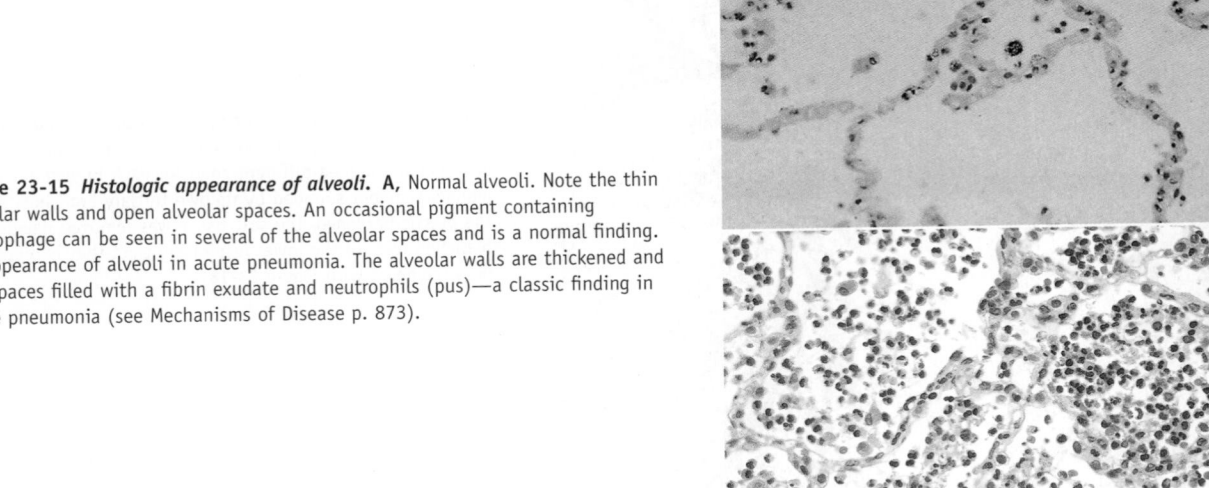

Structure of the Alveoli

The alveoli are the primary gas exchange structures of the respiratory tract (Figures 23-15, A, and 23-16). Alveoli are very effective in the exchange of carbon dioxide (CO_2) and oxygen (O_2) because each alveolus is extremely thin walled, each alveolus lies in contact with blood capillaries, and there are millions of alveoli in each lung. The barrier across which gases are exchanged between alveolar air and blood is called the **respiratory membrane** (see Figure 23-16, *inset*). The respiratory membrane consists of the alveolar epithelium, the capillary endothelium, and their joined basement membranes. The surface of the respiratory membrane inside each alveolus is coated with a fluid containing **surfactant**. Surfactant helps reduce surface tension—the force of attraction between water molecules—of the fluid. Thus it helps prevent each alveolus from collapsing and "sticking shut" as air moves in and out during respiration.

Functions of the Bronchi and Alveoli

The tubes composing the bronchial tree perform the same function as the trachea—that of distributing air to the lung's interior. The alveoli, enveloped as they are by networks of capillaries, ac-complish the lung's main and vital function, that of gas exchange between air and blood. Someone has observed that "the lung passages all serve the alveoli" just as "the circulatory system serves the capillaries."

Recall that in addition to serving as air distribution passageways or gas exchange surfaces, the anatomical components of the respiratory tract and lungs cleanse, warm, and humidify inspired air. Air entering the nose is generally contaminated with one or more common irritants; examples include insects, dust, pollen, and bacterial organisms. A remarkably effective air purification mechanism removes almost every form of contaminant before inspired air reaches the alveoli or terminal air sacs in the lungs.

The layer of protective mucus that covers a large portion of the membrane that lines the respiratory tree serves as the most important air purification mechanism. More than 125 ml of respiratory mucus is produced daily. It forms a continuous sheet, called a *mucus blanket*, that covers the lining of the air distribution tubes in the respiratory tree. This layer of cleansing mucus moves upward to the pharynx from the lower portions of the bronchial tree

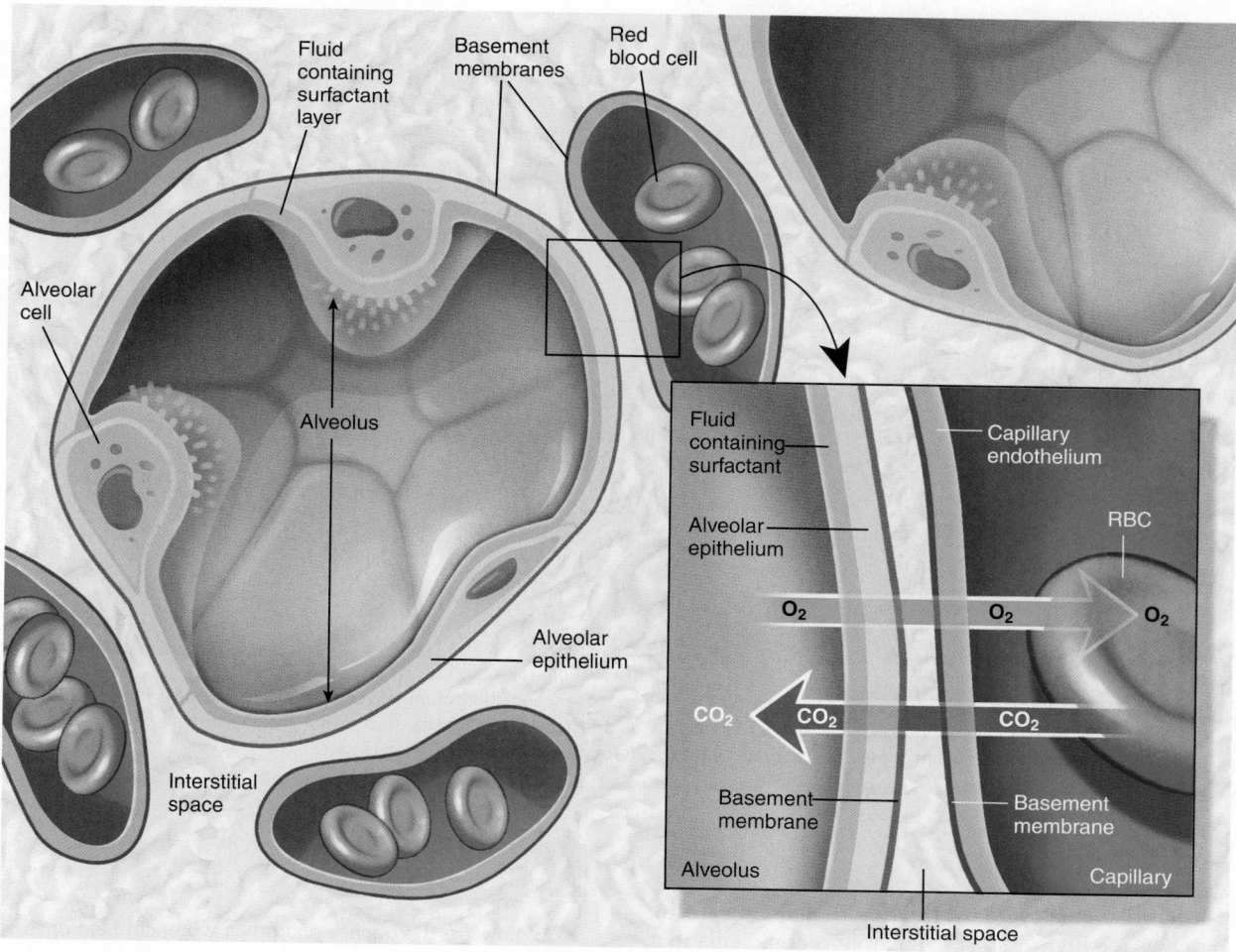

Figure 23-16 *Gas exchange structures of the lung.* Each alveolus is continually ventilated with fresh air. The inset shows a magnified view of the respiratory membrane composed of the alveolar wall (fluid coating, epithelial cells, and basement membrane), interstitial fluid, and the wall of a pulmonary capillary (basement membrane and endothelial cells). The gases—CO_2 (carbon dioxide) and O_2 (oxygen)—diffuse across the respiratory membrane.

BOX 23-6: HEALTH MATTERS
Lung Volume Reduction Surgery

Lung volume reduction surgery (LVRS) is a "treatment of last resort" for severe cases of emphysema. It involves the removal of 20% to 30% of each lung. Diseased tissue is generally removed from the upper or apical areas of the superior lobes. Evidence from a number of large clinical trials has now shown that the LVRS procedure may benefit or at least help stabilize selected emphysema patients whose lung function continues to decline despite aggressive pulmonary rehabilitation efforts and other more conservative forms of treatment.

More than 2 million Americans, most older than age 50 and current or former smokers, have emphysema—a major cause of disability and death in the United States. Emphysema is one of a number of conditions discussed in Chapter 24 and classified as a **chronic obstructive pulmonary disease** or COPD. Although lung damage caused by emphysema is irreversible, in some cases the disease may be halted or its progression slowed by LVRS. In the end stages of this chronic disease, breathing becomes labored as the lungs fill with large irregular spaces resulting from the enlargement and rupture of many alveoli (see illustration). The LVRS procedure removes part of the diseased lung tissue and increases available space in the pleural cavities. As a result, the diaphragm and other respiratory muscles can more effectively move air into and out of the remaining lung tissue, thereby improving pulmonary function and making breathing easier.

LVRS may reduce the need for lung transplantation procedures and augment the effectiveness of such supporting medical treatments as nutritional supplementation and exercise training in the treatment of selected late-stage emphysema patients. Newer and less invasive techniques involving smaller incisions and specialized video equipment inserted into the thoracic cavity (video-assisted thoracic surgery) are now being used for many LVRS procedures. As a result, the relatively long hospital stays and home recovery periods previously required after more traditional open-chest surgery have been shortened.

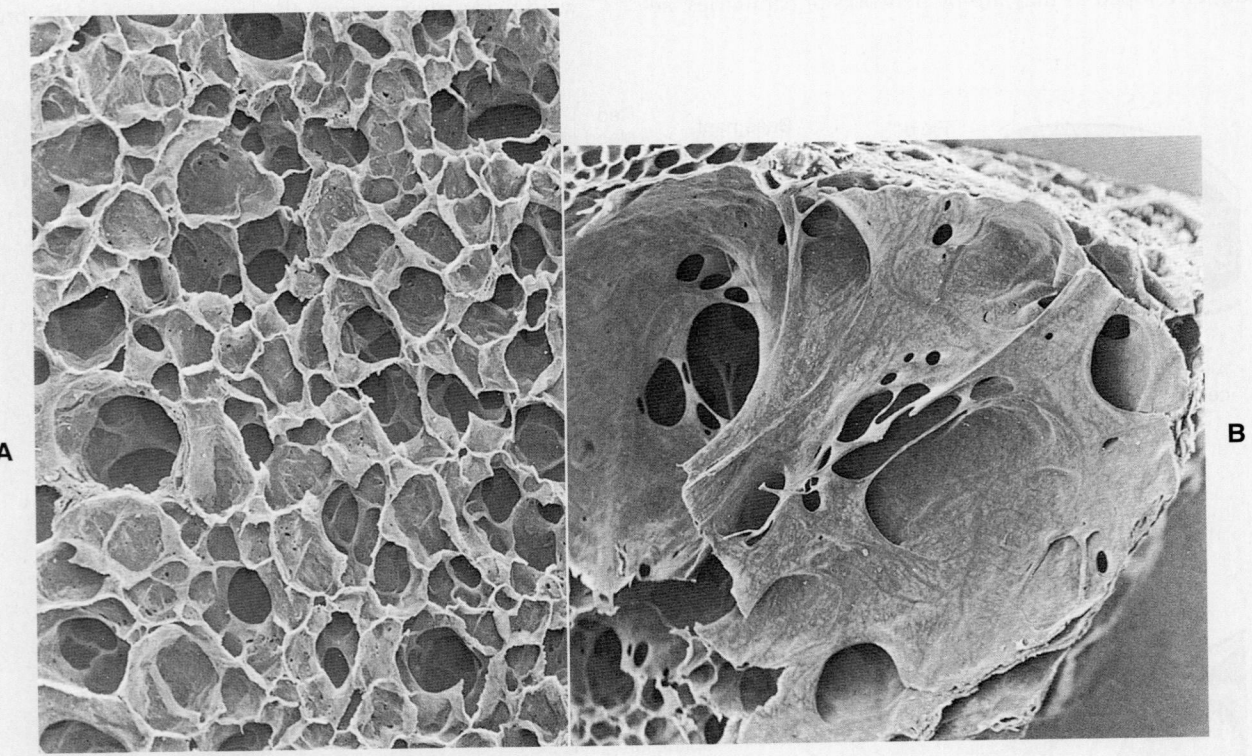

Emphysema. The effects of emphysema can be seen in these scanning electron micrographs of lung tissue. **A,** Normal lung with many small alveoli. **B,** Lung tissue affected by emphysema. Notice that the alveoli have merged into large air spaces, thereby reducing the surface area available for gas exchange.

on millions of hairlike cilia that cover the epithelial cells in the respiratory mucosa (see Figure 23-12). The microscopic cilia that cover epithelial cells in the respiratory mucosa beat or move in only one direction. The result is movement of mucus toward the pharynx. Cigarette smoke paralyzes these cilia and results in accumulations of mucus and the typical smoker's cough, an effort to clear the secretions.

QUICK CHECK

7. How are the trachea and primary bronchi held open so that they do not collapse during inspiration?
8. What is meant by the term *bronchial tree?*
9. What characteristics of alveoli enable them to efficiently exchange gases with blood?

Lungs

Structure of the Lungs

The lungs are cone-shaped organs, large enough to fill the pleural portion of the thoracic cavity completely (see Figure 23-10). They extend from the diaphragm to a point slightly above the clavicles and lie against the ribs both anteriorly and posteriorly. The medial surface of each lung is roughly concave to allow room for the mediastinal structures and for the heart, but the concavity is greater on the left than on the right because of the position of the heart. The primary bronchi and pulmonary blood vessels (bound together by connective tissue to form what is known as the **root** of the lung) enter each lung through a slit on its medial surface called the **hilum.**

The broad inferior surface of the lung, which rests on the diaphragm, constitutes the **base,** whereas the pointed upper margin is the **apex** (Figure 23-17). Each apex projects above a clavicle. The **costal surface** of each lung lies against the ribs and is rounded to match the contours of the thoracic cavity.

Each lung is divided into lobes by fissures. The left lung is partially divided into two lobes (superior and inferior) and the right lung into three lobes (superior, middle, and inferior). Note in Figure 23-18 that an **oblique fissure** is present in both lungs. In the right lung a **horizontal fissure** is also present that separates the superior from the middle lobe. After the primary bronchi enter the lungs, they branch into *secondary,* or *lobar,* bronchi that enter each lobe. Thus, in the right lung, three secondary bronchi are formed that enter the superior, middle, and inferior lobes. Each secondary bronchus is named for the lung lobe that it enters; for example, the superior secondary bronchus enters the superior lobe. The left primary bronchus divides into two secondary bronchi entering the superior and inferior lobes of that lung.

The lobes of the lung can be further subdivided into functional units called **bronchopulmonary segments** (see Figure 23-18). Each bronchopulmonary segment is served by a tertiary bronchus. The interior of each bronchopulmonary segment consists of almost innumerable tubes of dwindling diameter that make up the bronchial tree and serve as air distributors. The smallest tubes terminate in the smallest, but functionally most important structures of the lung—the alveoli, or "gas exchangers."

Visceral pleura covers the outer surfaces of the lungs and adheres to them much as the skin of an apple adheres to the apple (Figure 23-19).

Functions of the Lungs

The lungs perform two functions—air distribution and gas exchange. Air distribution to the alveoli is the function of the tubes of the bronchial tree. Gas exchange between air and blood is the joint function of the alveoli and the networks of blood capillaries that envelop them. These two structures—one part of the respiratory system and the other part of the circulatory system—together serve as highly efficient gas exchangers. Why? Because they provide an enormous surface area, the **respiratory membrane,** where the very thin-walled alveoli and equally thin-walled pulmonary capillaries come in contact (see Figures 23-15 and 23-16). This makes possible extremely rapid diffusion of gases between alveolar air and pulmonary capillary blood. Someone has estimated that if the lungs' 300 million or so alveoli could be opened up flat, they would form a surface about the size of a tennis court, that is, about 85 square meters, or more than 40 times the surface area of the entire body! No wonder such large amounts of oxygen can be so quickly loaded into the blood while large amounts of carbon dioxide are rapidly unloaded from it.

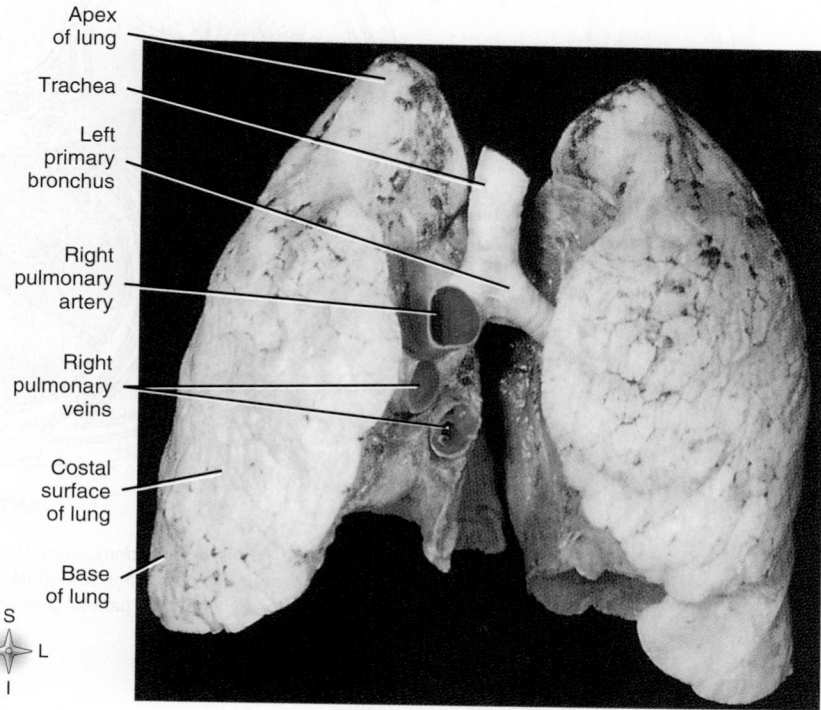

Figure 23-17 *Trachea, bronchi, and lungs.* The lower respiratory tract has been dissected from a cadaver and its organs separated to show them clearly.

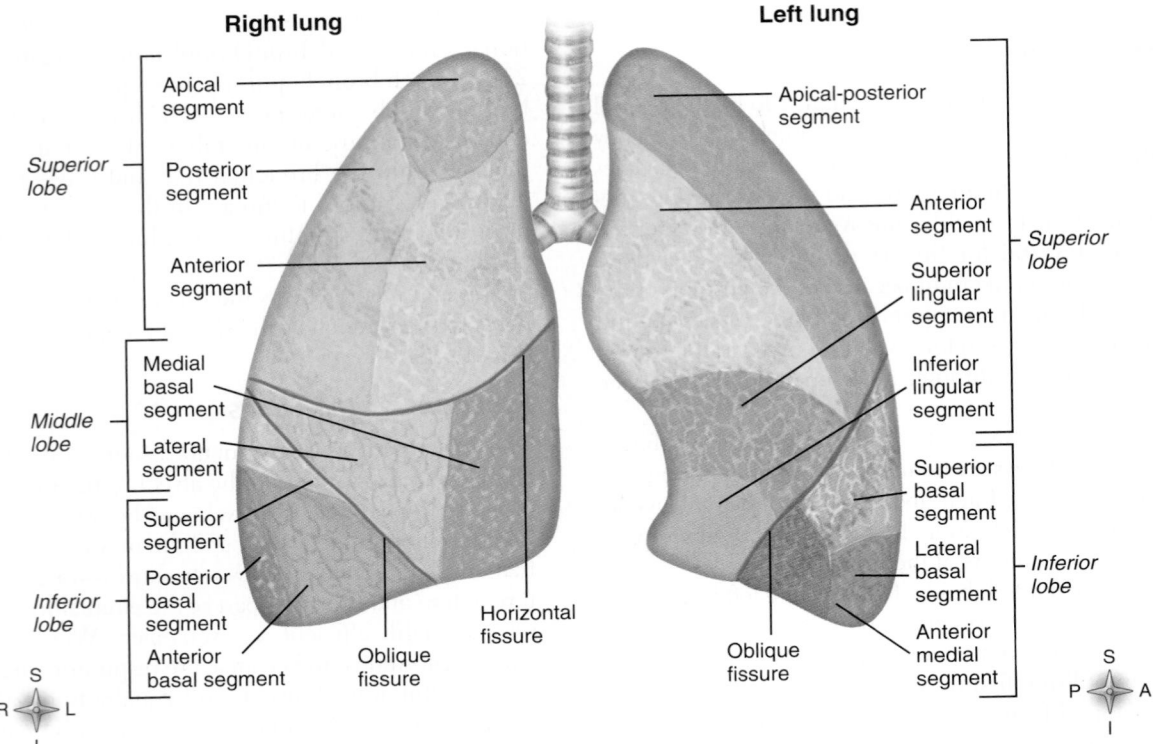

Figure 23-18 *Lobes and bronchopulmonary segments of the lungs.* Anterior view of the left and right lungs, bronchi, and trachea.

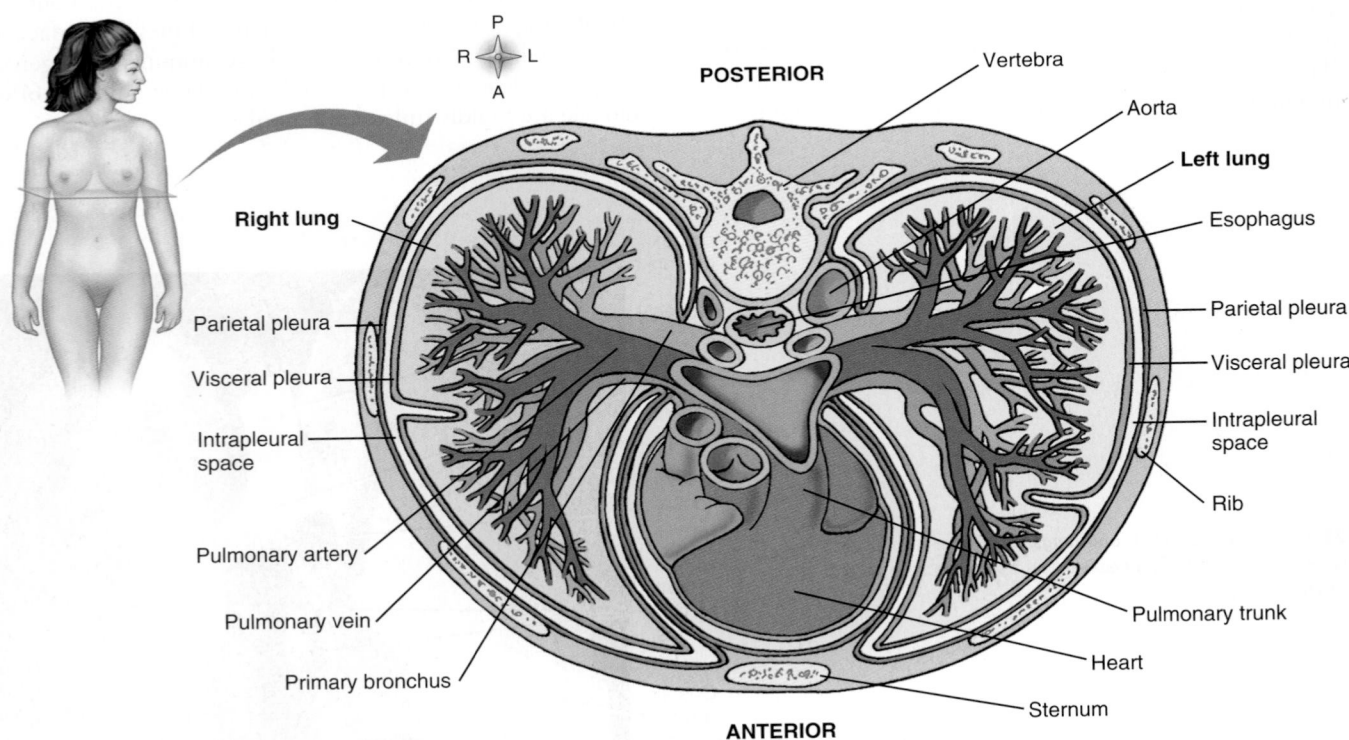

Figure 23-19 *Lungs and pleura (transverse section).* Note the parietal pleura lining the right and left pleural divisions of the thoracic cavity before folding inward near the bronchi to cover the lungs as the visceral pleura. The intrapleural space separates the parietal and visceral pleura. The heart, esophagus, and aorta are shown in the central mediastinum.

Thorax

Structure of the Thoracic Cavity

As described in Chapter 1, the thoracic cavity has three divisions, separated from each other by partitions of **pleura.** The parts of the cavity occupied by the lungs are the *pleural divisions.* The space between the lungs occupied mainly by the esophagus, trachea, large blood vessels, and heart is the *mediastinum* (see Figure 23-19).

The **parietal pleura** lines the entire thoracic cavity. It adheres to the internal surface of the ribs and the superior surface of the diaphragm, and it partitions off the mediastinum. A separate pleural sac thus encases each lung. Because the outer surface of each lung is covered by the visceral layer of the pleura, the **visceral pleura** lies against the parietal pleura, separated only by a potential space (pleural space) that contains just enough pleural fluid for lubrication (see Figure 23-19). Thus, when the lungs inflate with air, the smooth, moist visceral pleura coheres to the smooth, moist parietal pleura. Friction is thereby avoided, and respirations are painless. In *pleurisy* (pleuritis), on the other hand, the pleura is inflamed and respirations become painful.

Functions of the Thoracic Cavity

The thorax plays a major role in respiration. Because of the elliptical shape of the ribs and the angle of their attachment to the spine, the thorax becomes larger when the chest is raised and smaller when it is lowered. Lifting up the chest raises the ribs so that they no longer slant downward from the spine, and because of their elliptical shape, both the depth (from front to back) and the width of the thorax are enlarged. (If this does not sound convincing to you, examine a skeleton to see why it is so.) An even greater change in thoracic volume occurs when the diaphragm contracts and relaxes. When the diaphragm contracts, it flattens out and thus pulls the floor of the thoracic cavity downward—thereby enlarging the volume of the thorax. When the diaphragm relaxes, it returns to its resting, domelike shape—thus reducing the volume of the thoracic cavity. It is these changes in thorax size that bring about inspiration and expiration (discussed on pp. 884-892 in Chapter 24).

QUICK CHECK

10. What is meant by the term *lobe of the lung?* What is a *bronchopulmonary segment?*
11. How does the structure of the diaphragm enable it to participate in breathing movements?

Cycle of Life

Respiratory System

Respiration involves the exchange of O_2 and CO_2 between the organism and its environment. The exchange must occur between air in the lungs and blood and, then, between blood and every body cell. In addition to the structural components of the body through which the respiratory gases must pass, hemoglobin plays a vital role in the respiration process. Each component of the system may be affected by developmental defects, by age-related structural changes, or by loss of function during the life cycle.

Premature birth can cause potentially fatal respiratory problems. A very low–birth weight baby may have inadequate blood flow to the lungs, an inability to ventilate properly, and inadequate quantities of surfactant. Other diseases that cause serious respiratory problems are also associated with specific age groups. Examples include cystic fibrosis and asthma in children and certain types of obstructive pulmonary disease and emphysema in older adults. Pneumothorax occurs more frequently in young adult females.

Numerous age-related changes affect vital capacity, make ventilation difficult, or reduce the oxygen- or carbon dioxide–carrying capacity of blood. For example, in older adulthood the ribs and sternum tend to become more fixed and less able to expand during inspiration, the respiratory muscles are less effective, and hemoglobin levels are often reduced. The result is a general reduction in respiratory efficiency in old age.

THE BIG PICTURE

Anatomy of the Respiratory System

Understanding the relationship of structure to function is critical to an understanding of homeostasis in all of the body organ systems. The anatomy of the respiratory system components permits the distribution of air and the exchange of respiratory gases. This dual function ultimately allows for both exchange of gases between environmental air and blood in the lungs and, finally, gas exchange between blood and individual body cells. In addition to delivery of air to the tiny terminal air passageways and alveoli for gas exchange with blood, components of the upper respiratory tract effectively filter, warm, and humidify the air we breathe.

Respiratory functions are dependent on the structural organization of the system parts and on the interrelationship of those components with other body systems, including the nervous, circulatory, muscular, and immune systems. Understanding the proper functioning and regulation of the physiology of the respiratory system, discussed in Chapter 24, depends on your understanding of its structural components and their relationships to one another and to other body organ systems—the "big picture" of total-body homeostasis.

Mechanisms of Disease

DISORDERS ASSOCIATED WITH RESPIRATORY ANATOMY

Disorders of the Upper Respiratory Tract

Inflammation and Infection

Any infection localized in the mucosa of the upper respiratory tract (nose, pharynx, and larynx) can be called an **upper respiratory infection (URI)** and is often named for the specific structure involved. **Rhinitis** (from the Greek *rhinos*, "nose") is an inflammation of the mucosa of the nasal cavity. It is commonly caused by a viral infection, as in the common cold (caused by rhinoviruses) or flu (caused by influenza viruses). Rhinitis can also be caused by nasal irritants or an allergic reaction to airborne allergens. Allergic rhinitis, or "hay fever," occurs in sensitive people in a seasonal pattern, depending on the allergens involved (for example, pollen). The excessive mucus production that results from the inflammatory response involved in rhinitis can cause fluid to drip down the pharynx and into the esophagus and lower respiratory tract. This dripping may cause sore throat, coughing, and upset stomach. Irritation of the **nasal mucosa** itself often triggers the sneeze reflex. Elimination of the causative factor, rest, and the use of antihistamines and decongestants usually relieve these symptoms.

Pharyngitis is inflammation or infection of the pharynx. Commonly referred to as a "sore throat," it is often due to viral invasion. Bacterial infection by *Streptococcus* bacteria is termed "strep throat." The common complaint is a sore throat, but redness and difficulty swallowing (dysphagia) often accompany it. Throat lozenges, rest, and fluid intake are encouraged, and antibiotics are prescribed for severe infections.

Laryngitis, or inflammation of the mucous lining of the larynx, is characterized by edema of the vocal cords, resulting in hoarseness (dysphonia) or loss of voice. Besides infections, inhalation of toxic or irritating fumes (i.e., smoking), endotracheal intubation, vocal abuse (i.e., public speaking), and alcohol ingestion can precipitate laryngitis. In children younger than 5 years, it may cause difficulty breathing, a condition often called *croup*. Conservative treatment, including limiting speech, is usually effective. A much more severe and rapidly progressing form of laryngeal edema called **epiglottitis** is always treated as a medical emergency because of the potential for airway obstruction (see Box 23-4, p. 860).

Tonsillitis is inflammation of one or more of the masses of lymphatic tissue embedded in the mucous membrane of the pharynx (see p. 860). Most cases of tonsillitis result in inflammation and swelling of the palatine tonsils in the oropharynx and the pharyngeal tonsils or adenoids in the nasopharynx. Repeated episodes of infection and chronic swelling of these tonsillar tissues require aggressive antibiotic therapy and, in severe cases, even surgical removal. However, **tonsillectomy,** the surgical procedure used to remove inflamed and/or enlarged tonsillar tissue, is no longer considered a "first-choice" treatment option in routine cases of tonsillitis. As a result, the number of tonsillectomies performed each year continues to decrease. Physicians now recognize the value of lymphatic tissue in the body's defense mechanism and delay tonsillectomy with its rare, but potentially serious complications—including severe hemorrhage—until more conservative treatment options have proved ineffective. However, surgical removal may be required, especially in children, if adenoidal and tonsillar infection and hypertrophy do not respond to antibiotic therapy. In these cases, sleep disturbance, fatigue, and other potentially serious complications involving the heart and lungs can result from progressive upper airway obstruction over time. Figure 23-20, *A,* shows the fatigued facial appearance of a child with severely enlarged tonsils who must keep his mouth open to breathe. In Figure 23-20, *B,* note how the enlarged tonsils in this child have all but filled the pharynx and are nearly meeting in the midline, thus seriously limiting air flow.

Because the upper respiratory mucosa is continuous with the mucous lining of the sinuses, auditory or eustachian tube, middle ear, and lower respiratory tract, URIs have the unfortunate tendency to spread. It is not unusual to see a common cold progress to **sinusitis** (sinus infection) or **otitis media** (middle ear infection).

Anatomical Disorders

Nasal obstruction can be caused by displacement of the nasal septum from the midline of the nasal cavity, called a **deviated septum.** Most individuals have a small amount of septal cartilage protruding into one nasal passage; however, some are born with a congenital defect that results in various degrees of blockage on one or both sides of the nasal cavity. Damage from injury or infection may also cause a deviated septum. If breathing is impaired, surgical intervention is required to correct the deformity.

A frequent problem associated with a deviated septum is snoring. Pronounced snoring may be a symptom of **sleep apnea.** In these individuals there is a transition during sleep from loud snoring to variable periods of complete cessation of breathing. These periods of apnea are characterized by restlessness and often end in a loud "snort" before a normal breathing pattern resumes. Sleep apnea may be repeated many times each night and cause excessive daytime sleepiness and other symptoms related to chronic lack of oxygen.

Trauma to the nose can occur because the nose projects some distance from the front of the head. Usually, however, common bumps and other injuries cause little, if any serious damage. **Epistaxis,** or nosebleed, can be caused by violent sneezing or nose blowing, chronic infection or inflammation

Mechanisms of Disease—cont.

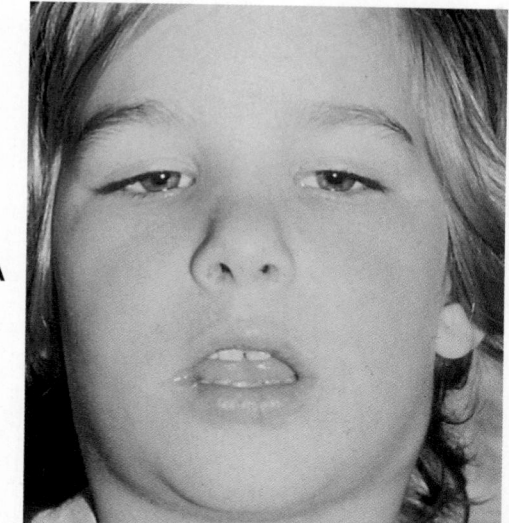

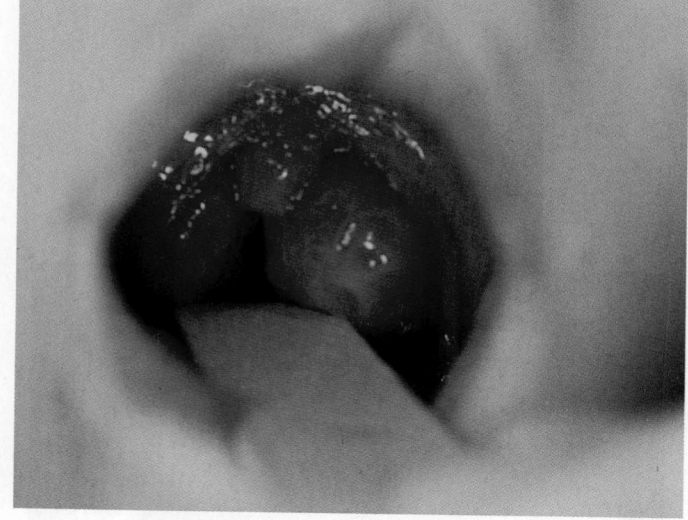

Figure 23-20 *Tonsillitis.* **A,** Facial appearance of a child with marked enlargement of the tonsils and adenoids. He must keep his mouth open to breathe and shows signs of fatigue. **B,** Enlarged tonsils can be seen meeting in the midline of the pharynx.

(as in rhinitis), hypertension, or a strong bump or blow to the nose. Immediate direct pressure with an ice pack will often slow or stop the bleeding.

Disorders of the Lower Respiratory Tract

A range of conditions can interfere with the lower respiratory tract functions of gas exchange and ventilation. Some of these disorders, such as restrictive and obstructive conditions, are discussed in Chapter 24. For now, we will concentrate on infections and lung cancer.

Lower Respiratory Infection

Acute bronchitis is a common condition characterized by acute inflammation of the tracheobronchial tree, most commonly caused by infection. Part of or preceded by an acute URI, it is most prevalent in winter. Predisposing factors include chilling, fatigue, malnutrition, and exposure to air pollutants. The protective functions of the bronchial epithelium are disturbed and excessive fluid accumulates in the bronchi. Acute bronchitis often begins with a nonproductive cough, but malaise, slight fever, back and muscle pain, and a sore throat occur if a URI is present. Rest is indicated until the fever subsides, and cough suppressants may be used if the cough is troublesome.

Pneumonia is a common condition characterized by acute inflammation of the lungs in which the alveoli and bronchi become plugged with a thick fibrin and neutrophil (pus)-containing exudate (see Figure 23-15, *B*). The vast majority of pneumonia cases result from infection by *Streptococcus pneumoniae* bacteria (see Figure 23-21), but pneumonia can also be caused by several other bacteria, viruses, and fungi. For example, **legionnaires' disease** is a form of bacterial pneumonia caused by infection with the *Legionella pneumophila* organism. Contaminated air conditioning cooling towers and whirlpool spas are sources of infection. The term *aspiration pneumonia* is used to describe lung infections caused by the inhalation of vomit or other infective material. It is common in acute alcohol intoxication and as a complication of anesthesia. Pneumonia is characterized by a high fever, chills, headache, cough, and chest pain. Increases in WBC numbers (leukocytosis) and depressed blood oxygen levels (hypoxia) are common findings. The fact that each day over 10,000 liters of potentially contaminated air enters the respiratory system helps explain why pneumonia is such a common illness—especially in individuals with lowered resistance or impaired immune systems. Types include *lobar pneumonia*, which typically affects an entire lobe of the lung, and *bronchopneumonia*, in which patches of infection are scattered along portions of the bronchial tree and generally involve more than one lobe (Figure 23-21). Treatment involves antimicrobial drugs to control the infection and supportive therapy, including the administration of supplemental oxygen. Removal of tracheobronchial secretions may be necessary to maintain airway integrity.

Tuberculosis (TB) is a chronic bacillus infection caused by *Mycobacterium tuberculosis*. It is a highly contagious disease, transmitted by airborne mechanisms (that is, inhalation of infectious droplets). Inflammatory lesions called "tubercles" form around colonies of TB bacilli in the lung (Figure 23-22, *A*) and produce the characteristic symptoms of nonproductive cough, fatigue, chest pain, weight loss, and fever. As TB progresses, lung hemorrhage and dyspnea (labored breathing) may develop. If large areas of the lung are infected

Mechanisms of Disease—cont.

Figure 23-21 *Types of pneumonia.* **A,** Bronchopneumonia. Note the patchy distribution of inflammation involving more than one lobe. **B,** Lobar pneumonia. Exudate fills a single lobe. (See Figure 23-15, *B.*)

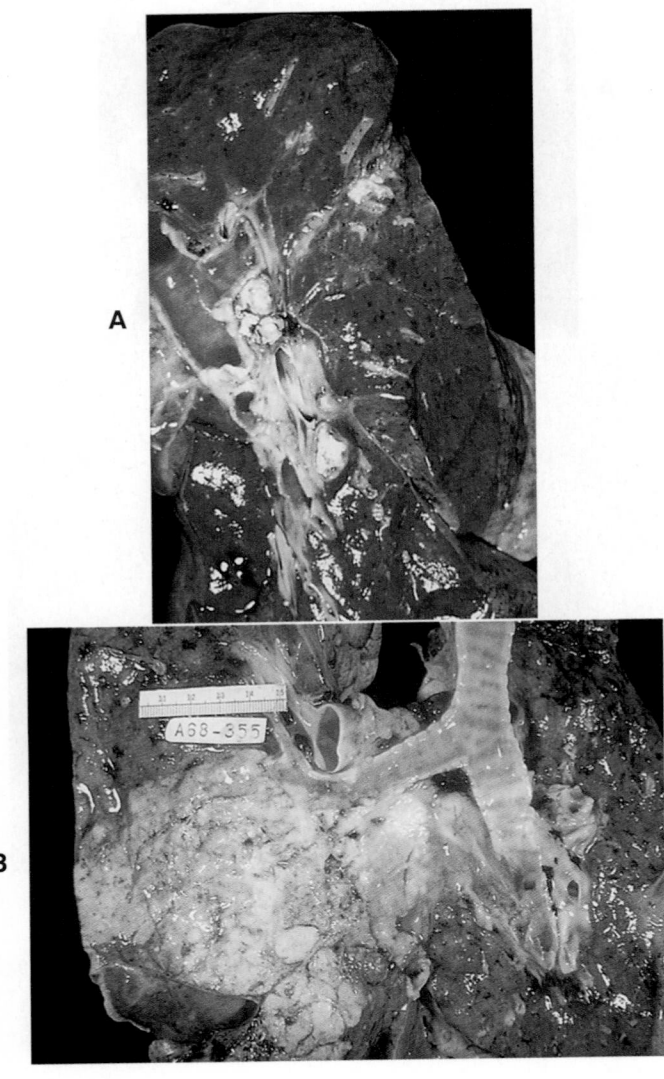

Figure 23-22 *Lung pathology.* **A,** Pulmonary tuberculosis. Gray-white TB lesions may harbor living bacilli for years. **B,** Lung cancer. A large cancerous lesion has extended outward to the lung surface and visceral pleura.

Mechanisms of Disease—cont.

and tissue is destroyed, scar tissue may develop and cause reduced lung volume and restrictive lung disease. TB can invade other tissues or organs such as the lymphatic system, genitourinary system, and bone tissue. Because of the advancement of modern antimicrobial agents, the incidence of TB in the United States dropped dramatically in the last half century. However, various factors have allowed some highly resistant strains of TB to emerge once again as a major health threat in the United States. Although the incidence of TB has not yet reached epidemic proportions, health authorities in many large cities are working hard to prevent a major health crisis.

Lung Cancer

Lung cancer is a malignancy of pulmonary tissue that not only destroys the vital gas exchange tissues of the lungs but like other cancers may also invade other parts of the body (metastasis). Lung cancer most often develops in damaged or diseased lungs (Figure 23-22, *B*). The most common predisposing condition associated with lung cancer is cigarette smoking (accounting for about 75% of lung cancer cases). In 1950, British epidemiologist Sir Richard Doll published the results of a groundbreaking scientific study that established for the first time the deadly link between smoking and lung cancer. His pioneering research was uniquely important in medical history. At the time of his death in 2005 at age 92, Doll was regarded as one of the most eminent scientists of his generation and one whose work will ultimately prevent tens of millions of premature tobacco-related deaths around the world. Other factors thought to cause lung cancer include exposure to "second-hand" cigarette smoke, asbestos, chromium, coal products, petroleum products, rust, and ionizing radiation (as in radon gas).

Lung cancer may be arrested if detected early on routine chest x-ray films or other diagnostic procedures such as bronchoscopy. Depending on the size, location, and exact type of malignancy involved, several strategies are available for treatment. Surgery is perhaps the most effective single treatment for most localized lung cancers. In a lobectomy, only the affected lobe of a lung is removed. **Pneumonectomy** is the surgical removal of an entire lung. Chemotherapy can also cause a cure or remission in selected cases, as can radiation therapy or concurrent (combination) chemotherapy and radiation treatment. Unfortunately, in about 40% of patients diagnosed with the most common type of lung cancer, called "non-small cell" lung cancer, the disease has already spread, or metastasized, to lymph nodes and other organs by the time a diagnosis is made. Non-small cell lung cancer accounts for over 85% of cancer diagnoses in the United States. Results of a recent National Cancer Institute study has shown that lobectomy combined with concurrent chemotherapy and radiation treatment can significantly increase survival time in non-small cell lung cancer patients. Patients in the study who received either surgery or standard concurrent therapy—but not both—had less favorable survival rates 3 and 5 years post diagnosis. So called "small cell" lung cancers usually appear first in the lining of the respiratory passageways. Curable if detected very early, this type of cancerous lesion is often treated by surgical removal, with photodynamic therapy (see Box 23-7), or both. This deadly form of the disease spreads very rapidly and is usually not effectively treated in its later stages with either standard concurrent radiation plus chemotherapy or by the standard treatment options followed by operation.

Pulmonary Radiology

Chest x-ray examinations account for more than half of all radiographs taken in the United States each year. Most are rapid and cost-effective diagnostic procedures that require a relatively low radiation dose. They provide an excellent "survey" examination of the lungs, mediastinal contents, and bony thorax. The most common type of chest x-ray examination uses air in the lungs to provide a natural contrast between questionable thoracic lesions and normal body tissues. In addition to their use for the diagnosis of pathological conditions, x-ray studies are useful in the study of normal respiratory anatomy and physiology.

Terminology

Most chest radiographs are taken during full inspiration while the individual is standing erect with the anterior surface of the chest against the film holder. The x-ray machine is placed about 2 meters (7 feet) behind the subject so that the x-ray beam enters the body from behind (posterior) and exits through the front (anterior). This type of exposure is appropriately called a PA (posterior to anterior) radiograph and is the most common type of chest x-ray examination (Figure 23-23). If the subject is asked to turn completely around so that the x-ray beam enters the chest from in front (anterior) and exits from the back (posterior), the resulting radiograph is termed an AP (anterior to posterior) film.

The x-ray image in Figure 23-24 is a frontal view of a PA chest film in a 42-year-old woman who had no symptoms of lung disease. This type of x-ray is often ordered as part of a routine or preemployment physical examination. The arrow indicates a 2-cm mass in the right superior lobe of the lung that is projecting behind the right clavicle. The mass, called a *granuloma*, was removed and found to be nonmalignant. Before surgery, the growth could not be identified radiographically as cancerous or benign. This illustrates a limitation of many x-ray procedures. Namely, radiographs are very sensitive for detection of abnormal nodules or lesions but are often not specific in identifying the cause or nature of the growths that are revealed.

Mechanisms of Disease—cont.

BOX 23-7: HEALTH MATTERS
Photodynamic Therapy for Lung Cancer

Photodynamic therapy (PDT), is now being used for the treatment of certain types of lung cancer that often begin in the lining of the trachea or bronchial tubes. PDT is a less "invasive" and more cost-effective outpatient procedure that does not require the extended hospitalization and weeks of recovery typical of major lung surgery. In the past, use of this specialized treatment technique was limited to certain eye disorders and some cancers of the esophagus and skin. When used to treat cancers that originate in the lining of the trachea or bronchi, PDT begins with IV injection of a photosensitizing drug (porfimer sodium) that quickly enters and remains in cancerous cells but

not normal cells. Then, a powerful, but "cool" red-light laser beam is delivered through a bronchoscope to illuminate the tumor and trigger a chemical reaction with the photosensitizing drug that destroys the cancerous cells (see the illustration). The dead cells are then sloughed off a few days after treatment. Because the type of laser light used in PDT can penetrate only a few millimeters of cancerous tissue, this type of therapy can be used only on superficial tumors. Traditional or laser photo resection surgery remains the most effective treatment for removing larger lung cancers (see photo insert/diagram). Chemotherapy and radiation may also be appropriate treatment options.

Specialized Radiographs

X-ray examinations that provide specialized information such as *bronchograms* and *arteriograms* are particularly useful for the diagnosis of disease.

A **bronchogram** is an x-ray image of the lung made after the bronchi are coated with a radiopaque substance to increase contrast and enhance detail. Figure 23-25 compares a normal and abnormal bronchogram. In the normal bronchogram,

note how divisions of the bronchial tree enter the bronchopulmonary segments without crowding, and are continuously open and uniformly tapered. In contrast, the abnormal bronchogram shows multiple bead-like local constrictions narrowing the lumina of air passages throughout the entire bronchial tree. They indicate partial occlusion of airways with "bronchial plugs" composed of fluid and exudate. Crowding of bronchi, obvious in the abnormal bronchogram, indicates collapse and compression of lung tissue.

An **arteriogram** is a specialized x-ray of an artery taken after the introduction of contrast material, usually via a catheter. Figure 23-26 is an arteriogram of the right lung showing the presence of a pulmonary embolism, or moving blood clot (black arrow), at the root of the right pulmonary artery. The white arrowheads outline the normal lumen of the artery distal to the clot. The term *deep vein thrombosis*, or DVT, was introduced in Chapter 18 to describe a blood clot that forms in a deep vein, especially in the legs. One of the most serious outcomes of a DVT "breaking free" is the formation of a pulmonary embolism. Look again at Figure 18-43 on p. 724 and compare it with the x-ray image in Figure 23-26.

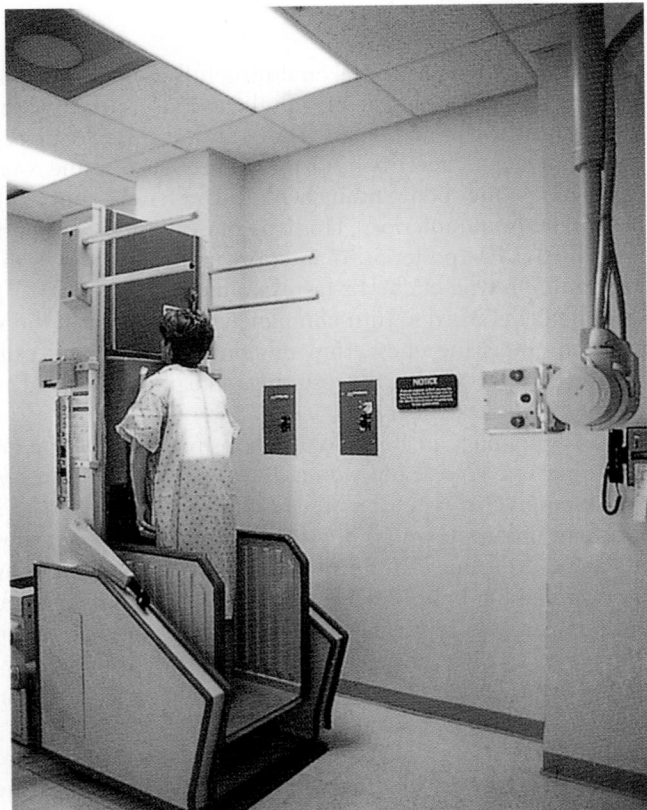

Figure 23-23 *Patient positioned for a PA (posterior-to-anterior) chest radiograph.*

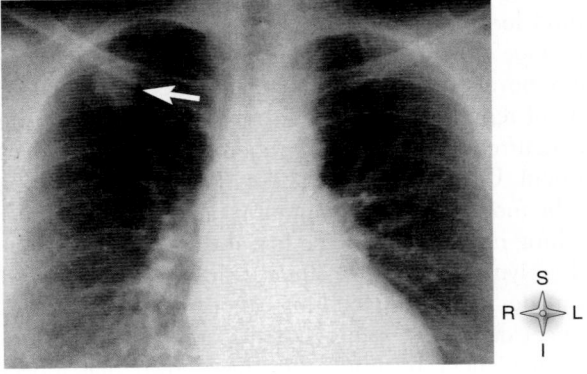

Figure 23-24 *Chest radiograph—PA projection (frontal view).* The arrow points to a 2-cm mass in the superior lobe of the right lung projecting behind the right clavicle.

Mechanisms of Disease—cont.

BOX 23-7: HEALTH MATTERS

Photodynamic Therapy for Lung Cancer—cont.

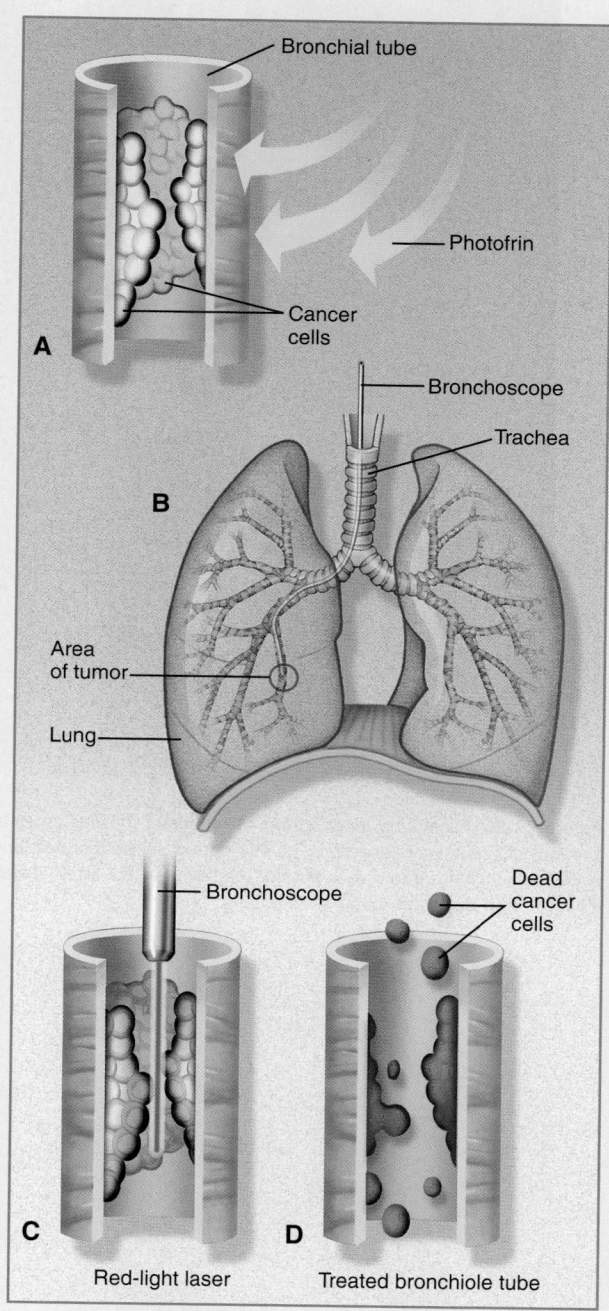

A — Bronchial tube / Photofrin / Cancer cells

B — Bronchoscope / Trachea / Area of tumor / Lung

C — Bronchoscope / Red-light laser

D — Dead cancer cells / Treated bronchiole tube

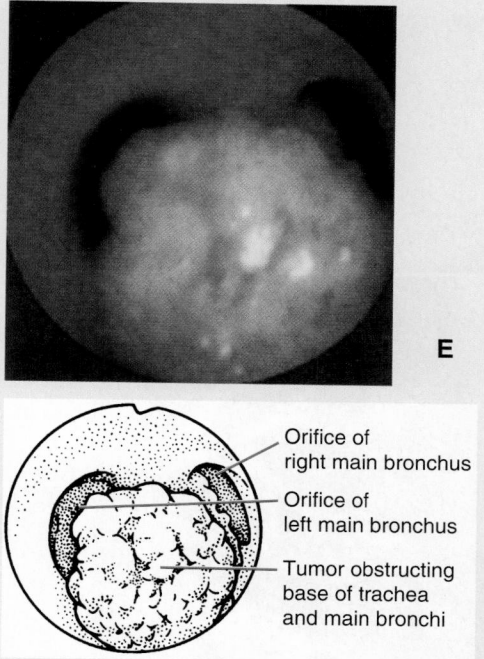

E

Orifice of right main bronchus

Orifice of left main bronchus

Tumor obstructing base of trachea and main bronchi

The bronchoscopy photo shows a cancerous tumor located at the lower end of the trachea. The tumor is partially blocking the opening into both the right and left primary bronchi.

Mechanisms of Disease—cont.

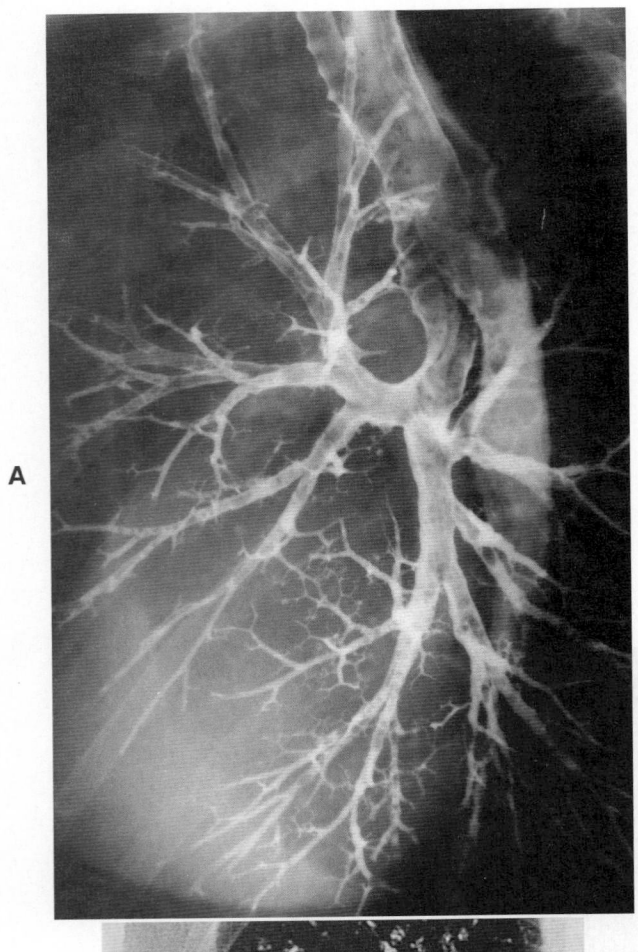

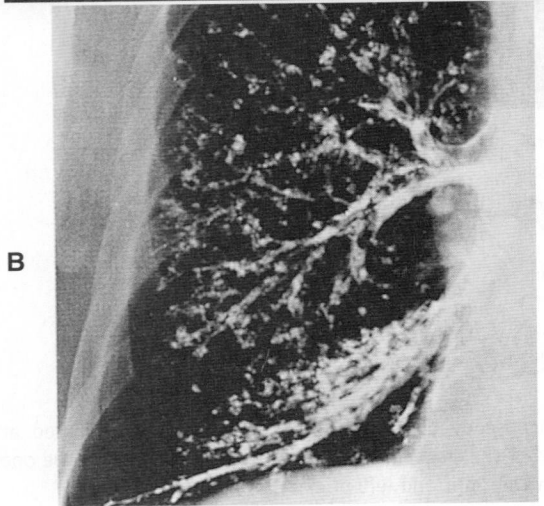

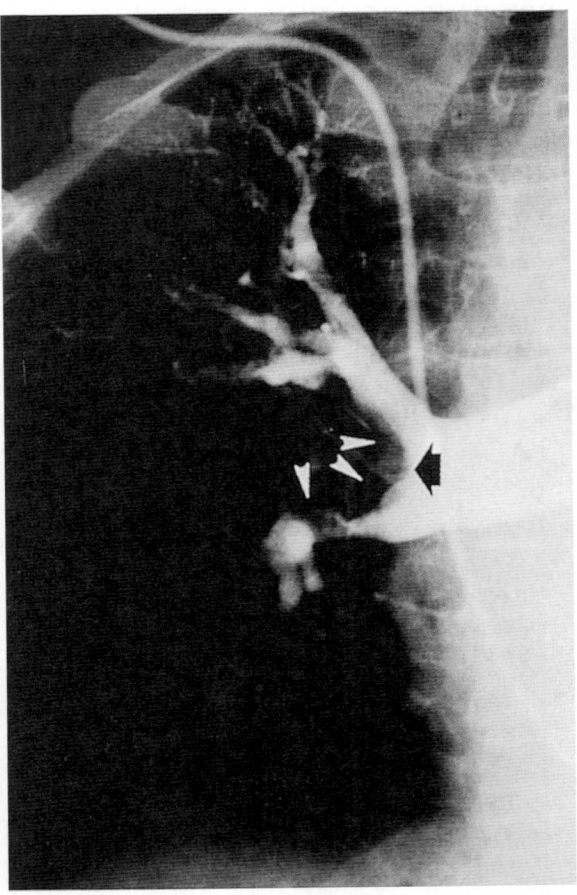

Figure 23-26 *Pulmonary arteriogram.* Arteriogram showing an embolus at the root of the pulmonary artery. The black arrow shows the location of the embolus, and white arrowheads outline the lumen of the artery distal to the clot (see also Figure 18-43 on p. 724).

Figure 23-25 *Comparison of normal and abnormal bronchograms.*
A, Normal bronchogram. Note that the terminal bronchial segments are open with the branches extending freely into the bronchopulmonary segments without the appearance of crowding or compression. **B,** Abnormal bronchogram. The terminal segments of the bronchi appear "spotty" because of multiple points of partial luminal blockage, and the lower bronchi show "crowding" indicative of collapse and compression of lung tissue.

LANGUAGE OF SCIENCE

(Cont'd from page 855)

olfactory epithelium (ohl-FAK-tor-ee ep-ih-THEE-lee-um) [*olfactus* sense of smell, *epi-* on, *-thele* nipple]

oropharynx (or-oh-FAIR-inks) [*oro-* mouth, *-pharynx* throat]

palatine tonsils (PAL-ah-tine TAHN-sils) [*palatum* plate]

paranasal sinuses (pair-ah-NAY-zal SYE-nus-ehz) [*para-* beside, *-nas-* nose, *-al* pertaining to, *sinus* hollow]

parietal pleura (pah-RYE-i-tal PLOO-rah) [*paries-* wall, *-al* pertaining to, *pleura* rib]

pharyngeal tonsils (fah-RIN-jee-al TAHN-sils) [*pharynx-* throat, *-al* pertaining to]

pharynx (FAIR-inks) [*pharynx* throat]

pleura (PLOO-rah) [*pleura* rib]

primary bronchi (BRONG-kye) [*primus* first, *bronchi* windpipe]

respiratory membrane (RES-pih-rah-tor-ee) [*respirare* to breathe, *membrana* thin skin]

respiratory mucosa (RES-pih-rah-tor-ee myoo-KOH-sah) [*respirare* to breathe, *mucus* slime]

respiratory portion (RES-pih-rah-tor-ee POR-shun) [*respirare* to breathe]

root

secondary bronchi (BRONG-kye) [*secundus* second, *bronchi* windpipe]

septum (SEP-tum)

surfactant (sur-FAK-tant) [surf(ace) act(ive) a(ge)nt]

thyroid cartilage (THY-royd KAR-ti-lij) [*thyreos-* shield, *-oid* resembling]

trachea (TRAY-kee-ah) [*tracheia* rough artery]

true vocal cords [*treue* faith, *vocalis* of the voice, *chorde* string]

turbinates (TUR-bih-nayts) [*turbinum* top-shaped]

upper respiratory tract (RES-pih-rah-tor-ee) [*respirare* to breathe, *tractus* trail]

ventricle (VEN-tri-kul) [*ventriculus* little belly]

vestibular vocal folds (ves-TIB-yoo-lar) [*vestibulum* courtyard, *vocalis* of the voice]

vestibule (VES-tih-byool) [*vestibulum* courtyard]

vibrissae (vih-BRISS-ee) [*vibrissa* nostril hair]

visceral pleura (VISS-er-al PLOO-rah) [*viscus-* internal organs, *-al* pertaining to, *pleura* rib]

vomeronasal organ (VNO) (voh-mer-oh-NAY-sal)

LANGUAGE OF MEDICINE

acute bronchitis (ah-KYOOT brong-KYE-tiss) [*acutus* sharp, *bronchos-* windpipe, *-itis* inflammation]

arteriogram (ar-TEER-ee-oh-gram) [*arterio-* artery, *-gram* drawing]

bronchogram (BRONG-koh-gram) [*bronchus-* windpipe, *-gram* drawing]

chronic obstructive pulmonary disease (KRON-ik ob-STRUK-tiv PUL-moh-nair-ee) [*chronos* time, *obstruere* to build against, *pulmonis* lung, *dis-* reversal, *-aise* ease]

cleft palate (kleft PAL-ett)

croup (kroop) [*croup* to croak]

deviated septum (DEE-vee-ay-ted SEP-tum) [*deviare* to turn aside, *septum* partition]

endotracheal intubation (en-doh-TRAY-kee-al in-too-BAY-shun) [*endo-* inward or within, *-tracheia-* rough artery, *-al* pertaining to, *in-* within, *-tubus-* tube, *-ation* process]

epiglottitis (EPP-ih-glaw-tye-tiss) [*epi-* upon, *-glossa-* tongue, *-itis-* inflammation]

epistaxis (ep-i-STAK-sis) [*epistaxis* a dropping]

laryngitis (lar-in-JYE-tis) [*laryngitis* inflammation of larynx]

legionnaires' disease (LEE-jen-airs) [American Legion]

lung cancer [*cancer* crab]

otitis media (oh-TYE-tis MEE-dee-ah) [*otic-* pertaining to the ear, *-itis* inflammation, *media* middle]

pharyngitis (fair-in-JYE-tis) [*pharynx-* throat, *-itis* inflammation]

photodynamic therapy (PDT) (foh-toh-dye-NAM-ik) [*photo-* light, *-dynamis* force, *therapeia* treatment]

pneumonectomy (noo-moh-NEK-toh-mee) [*pneumo-* lungs, *-ectomy* surgical removal]

pneumonia (noo-MOH-nee-ah) [*pneumo* lungs]

pulmonary radiology (PUL-moh-nair-ee RAY-dee-ohl-oh-gee) [*pulmonis* lung, *radio-* radiation, *-ology* the study or science of]

rhinitis (rye-NYE-tis) [*rhino-* nose, *-itis* inflammation]

sinusitis (sye-nyoo-SYE-tis) [*sinus-* hollow, *-itis* inflammation]

sleep apnea (APP-nee-ah) [*a-* not, *-pnein* to breathe]

tonsillectomy (tahn-sih-LEK-toh-mee) [*tonsillectomy* removal of the tonsils]

tonsillitis (tahn-sih-LYE-tis) [*tonsillitis* inflammation of the tonsils]

tracheostomy (tray-kee-OS-toh-mee) [*tracheia-* rough artery, *-ostomy* form a new opening]

tuberculosis (TB) (too-ber-kyoo-LOH-sis) [*tuber-* swelling, *-osis* condition]

upper respiratory infection (RES-pih-rah-tor-ee) [*respirare* to breathe, *inficere* to stain]

CASE STUDY

Baby Linda was born at 35 weeks' gestation and weighed 3 lb, 12 oz. At birth she was suctioned well and given oxygen, and she responded with spontaneous respirations. On admission to the special care nursery, Baby Linda was noted to have nasal flaring, a respiratory rate of 66, and an apical pulse of 150. Her axillary temperature was 96.8° F (36° C). Linda was placed in a radiant warmer and oxygen continued by oxygen hood.

1. Baby Linda is at risk for respiratory distress syndrome. Why?

   A. Linda is not 37 weeks' gestation.
   B. Linda does not weigh 2.2 kg.
   C. Linda will not have the ability to manufacture adequate amounts of surfactant, which helps decrease surface tension within the alveoli.
   D. All of the above.

2. Which of the following statements about respiratory distress syndrome in infants is false?

   A. Treatment includes delivering oxygen under pressure while applying surfactant directly into the baby's airways via an endotracheal tube.

   B. Surfactant helps prevent each alveolus from collapsing and "sticking shut" as air moves in and out.
   C. Infants lacking surfactant are required to expend more energy to breathe than infants with normal surfactant.
   D. Treatment is limited to keeping the alveoli open so that delivery and exchange of oxygen and carbon dioxide can occur.

3. Baby Linda improves and is discharged from the hospital. She returns to the health clinic as scheduled and does well except for minor childhood illnesses until the age of 4. At this time Linda is noted to have a temperature of 101° F (38.3° C) (axillary), no cough, and some nasal congestion. Her parents have noticed that her appetite and activity level have decreased. At the health center, Linda is sitting straight up and seems to have difficulty breathing. Which one of the following would be the expected diagnosis based on the description of Linda's condition?

   A. Croup
   B. Epiglottitis
   C. Acute bronchitis
   D. Laryngitis

CHAPTER SUMMARY

OVERVIEW OF THE RESPIRATORY SYSTEM

A. The respiratory system functions as an air distributor and gas exchanger—supplying oxygen and removing carbon dioxide from cells (Figure 23-1)
   1. Alveoli—sacs that serve as gas exchangers; all other parts of respiratory system serve as air distributors
   2. The respiratory system also warms, filters, and humidifies air
   3. Respiratory organs influence speech, homeostasis of body pH, and olfaction

B. The respiratory system is divided into two divisions
   1. Upper respiratory tract—the organs are located outside the thorax and consist of the nose, nasopharynx, oropharynx, laryngopharynx, and larynx
   2. Lower respiratory tract—the organs are located within the thorax and consist of the trachea, the bronchial tree, and the lungs
   3. Accessory structures include the oral cavity, rib cage, and diaphragm

UPPER RESPIRATORY TRACT

A. The nose
   1. The external portion of the nose consists of a bony and cartilaginous frame covered by skin containing sebaceous glands—the two nasal bones meet and are surrounded by the frontal bone to form the root; the nose is surrounded by the maxilla (Figure 23-2)

   2. The internal nose (nasal cavity) lies over the roof of the mouth, separated by the palatine bones
      a. Cleft palate—the palatine bones fail to unite completely and only partially separate the nose and the mouth, thereby producing difficulty swallowing
      b. Cribriform plate—separates the roof of the nose from the cranial cavity
      c. Septum—separates the nasal cavity into a right and left cavity; it consists of four structures: the perpendicular plate of the ethmoid bone, the vomer bone, the vomeronasal cartilages, and the septal nasal cartilage
   3. Each nasal cavity is divided into three passageways: superior, middle, and inferior meati (Figure 23-3)
   4. Anterior nares—external openings to the nasal cavities; open into the vestibule
   5. Sequence of air through the nose into the pharynx—anterior nares to the vestibule to all three meati simultaneously to the posterior nares
   6. Nasal mucosa—a mucous membrane that air passes over; it contains a rich blood supply (Figure 23-4)
      a. Olfactory epithelium—specialized membrane containing many olfactory nerve cells and a rich lymphatic plexus
   7. Paranasal sinuses—four pairs of air-containing spaces that open or drain into the nasal cavity; each is lined with respiratory mucosa (Figure 23-5)
   8. The nose is a passageway for air traveling to and from the lungs—it filters the air, aids speech, and makes possible the sense of smell

B. Pharynx (throat)
1. The pharynx is a tubelike structure extending from the base of the skull to the esophagus; it is made of muscle and divided into three parts (Figure 23-6)
   a. Nasopharynx
   b. Oropharynx
   c. Laryngopharynx
2. The pharyngeal tonsils—located in the nasopharynx, called adenoids when they become enlarged
3. The oropharynx contains two pair of organs—the palatine tonsils (most commonly removed) and the lingual tonsils (rarely removed)
4. The pharynx is the pathway for the respiratory and digestive tracts
C. Larynx (Figures 23-7 and 23-8)
1. The larynx is located between the root of the tongue and the upper end of the trachea
2. The larynx consists of cartilages attached to each other by muscle and is lined by a ciliated mucous membrane, which forms two pairs of folds (Figure 23-9)
   a. The vestibular (false) vocal folds
   b. The true vocal cords
3. The framework of the larynx is formed by nine cartilages
   a. Single laryngeal cartilages—the three largest cartilages: the thyroid cartilage, the epiglottis, and the cricoid cartilages
   b. Paired laryngeal cartilages—three pairs of smaller cartilages: the arytenoid, the corniculate, and the cuneiform cartilages
4. Muscles of the larynx
   a. Intrinsic muscles both insert and originate within the larynx
   b. Extrinsic muscles insert in the larynx but originate on some other structure
5. The larynx functions as part of the airway to the lungs and produces the voice

LOWER RESPIRATORY TRACT

A. The trachea (windpipe) extends from the larynx to the primary bronchi (Figure 23-11)
1. The trachea furnishes part of the open airway to the lungs—obstruction causes death
B. Bronchi and alveoli
1. The lower end of the trachea divides into two primary bronchi, one on the right and one on the left; the primary bronchi enter the lung and divide into secondary bronchi, which branch into bronchioles and eventually divide into alveolar ducts (Figure 23-13)
2. The alveoli are the primary gas exchange structures
   a. Respiratory membrane—the barrier between which gases are exchanged by alveolar air and blood (Figure 23-16)
   b. The respiratory membrane consists of the alveolar epithelium, the capillary endothelium, and their joined basement membranes
   c. Surfactant—a component of the fluid coating the respiratory membrane that reduces surface tension
3. The bronchi and alveoli distribute air to the lung's interior

C. Lungs
1. The lungs are cone-shaped organs extending from the diaphragm to above the clavicles (Figure 23-18)
   a. Hilum—slit on the lung's medial surface where the primary bronchi and pulmonary blood vessels enter
   b. Base—the inferior surface of the lung that rests on the diaphragm
   c. Costal surface—lies against the ribs
   d. The left lung is divided into two lobes—superior and inferior
   e. The right lung is divided into three lobes—superior, middle, and inferior
   f. The lobes are further divided into functional units—bronchopulmonary segments
      (1) Ten segments in the right lung
      (2) Eight segments in the left lung
2. The lungs have two functions—air distribution and gas exchange
D. The thorax (Figure 23-19)
1. The thoracic cavity has three divisions divided by the pleura
   a. Pleural divisions—the part occupied by the lungs
   b. Mediastinum—part occupied by the esophagus, trachea, large blood vessels, and heart
2. The thorax functions to bring about inspiration and expiration

CYCLE OF LIFE: RESPIRATORY SYSTEM

A. Respiration may be affected by developmental defects, age-related structural changes, or loss of function throughout the life cycle
B. Age-related changes affect vital capacity, make ventilation difficult, or reduce the oxygen- or carbon dioxide–carrying capacity of blood
C. Respiratory efficiency is reduced in old age as a result of changes in ribs, respiratory muscles, and hemoglobin levels

REVIEW QUESTIONS

1. Identify the major anatomical structures of the nose.
2. How are the turbinates arranged in the nose? What are they?
3. Describe the draining of the paranasal sinuses.
4. What organs are found in the nasopharynx?
5. What tubes open into the nasopharynx?
6. The pharynx is common to what two systems?
7. List the divisions of the larynx.
8. What is the voice box? Of what is it composed? What is the Adam's apple?
9. What is the epiglottis? What is its function?
10. What are the true vocal cords? What name is given to the opening between the cords?
11. Describe the structure and function of the trachea.
12. Discuss the component parts of the bronchial tree.
13. Make a diagram showing the termination of a bronchiole in an alveolar duct with alveoli.
14. How many lobes are in the right lung? The left? What are the bronchopulmonary segments?

15. Describe the changes in thorax size during respiration.
16. List the organs that are included in the upper respiratory tract. Do the same for the lower respiratory tract.
17. What is the role of the vomeronasal organ?

CRITICAL THINKING QUESTIONS

1. How would you describe the structure and function of the respiratory mucosa? Include the types of cells it contains and where these cells are located in the respiratory system.
2. Why do you think mucus production is especially important in the olfactory epithelium?
3. What are the features of the pharynx? Include in your answer the location of the openings in the pharynx, location of the tonsils, and the role of the pharynx in phonation.

4. Which of the paired laryngeal cartilages are the most important? What evidence can you find to support your answer?
5. Can you identify which single cartilage is associated with a life-threatening condition in children? What are the symptoms of this condition?
6. In an earlier chapter the characteristic of water called *polarity* was described as the attraction water molecules have for each other. Why is this a problem in the respiratory system, and how is it solved?
7. Can you make the distinction between air distribution and gas exchange in the respiratory system? Identify the organs that serve as air distributors and gas exchangers.

CHAPTER 24

Physiology of the Respiratory System

LANGUAGE OF SCIENCE

alveolar ventilation (al-VEE-oh-lar ven-ti-LAY-shun) [*alveol-* little hollow, *-ar* pertaining to, *ventila-* fan or create wind, *-tion* process of]

anatomical dead space (an-ah-TOM-i-kal) [*ana-* up, *-tom-* to cut, *-ical* pertaining to]

apneustic center (ap-NYOO-stik) [*a-* not, *-pneust-* breathing, *-ic* pertaining to]

arterial blood Po$_2$ (ar-TEER-ee-al) [*arteria-* airpipe, *-al* pertaining to]

arterial blood pressure (ar-TEER-ee-al) [*arteria-* airpipe, *-al* pertaining to]

bicarbonate (bye-KAR-boh-nayt) [*bi-* two, *-carbon-* coal (carbon), *-ate* oxygen compound]

Bohr effect (BOR) [*Christian Bohr* Danish physiologist]

Boyle's law (boils law) [*Robert Boyle* English scientist]

carbaminohemoglobin (kahr-bam-ih-no-hee-moh-GLOH-bin) [*carb-* carbon, *-amino-* ammonia compound, *-hemo-* blood, *-glob-* ball, *-in* substance]

cerebral cortex (seh-REE-bral KOR-teks) [*cerebr-* brain, *-al* pertaining to, *cortex* bark]

Charles' law (Gay-Lussac's law) (charlz law [gay lus-SAKS law]) [*Jacques Alexandre César Charles* French physicist, *Joseph L. Gay-Lussac* French physical scientist]

chloride shift (KLOR-ide) [*chlor-* green, *-ide* chemical ending]

compliance [*compli-* complete, *-ance* act of]

Dalton's law (DAL-tenz law) [*John Dalton* English chemist and physicist]

elastic recoil (eh-LAS-tik REE-koyl) [*elast-* drive or propel, *-ic* pertaining to]

expiration (eks-pih-RAY-shun) [*expirare* to breath out]

fetal hemoglobin (FEE-tal hee-moh-GLOH-bin) [*fet-* offspring, *-al* pertaining to *hemo-* blood, *-glob-* ball, *-in* substance]

flow-volume loop

Haldane effect (HAWL-dayne) [*John Scott Haldane* Scots physiologist]

Cont'd on p. 919

RESPIRATORY PHYSIOLOGY

In Chapter 23 the anatomy of the respiratory system was presented as a basis for understanding the physiological principles that regulate air distribution and gas exchange. This chapter deals with **respiratory physiology**—a complex series of interacting and coordinated processes that have a critical role in maintaining the stability, or constancy, of our internal environment. The proper functioning of the respiratory system ensures the tissues of an adequate oxygen supply and prompt removal of carbon dioxide. This process is complicated by the fact that control mechanisms must permit maintenance of homeostasis throughout a wide range of ever-changing environmental conditions and body demands. Adequate and efficient regulation of gas exchange between body cells and circulating blood under changing conditions is the essence of respiratory physiology. This complex function would not be possible without integration between numerous physiological control systems, including acid-base, water, and electrolyte balance, circulation, and metabolism.

Functionally, the respiratory system is composed of an integrated set of regulated processes that include the following:

1. External respiration: pulmonary ventilation (breathing) and gas exchange in the pulmonary capillaries of the lungs
2. Transport of gases by the blood
3. Internal respiration: gas exchange in the systemic blood capillaries and cellular respiration
4. Overall regulation of respiration

Figure 24-1 summarizes the essential processes of pulmonary function. We will use this set of processes as a general framework for this chapter. Cellular respiration has already been covered in Chapter 4 and will be reviewed again in greater detail in Chapter 27.

PULMONARY VENTILATION

Pulmonary ventilation is a technical term for what most of us call breathing. One phase of it, inspiration, moves air into the lungs and the other phase, expiration, moves air out of the lungs.

Mechanism of Pulmonary Ventilation

Air moves in and out of the lungs for the same basic reason that any fluid (a liquid or a gas) moves from one place to another—briefly, because its pressure in one place is different from that in the other place. Or stated differently, the existence of a pressure gradient (a pressure difference) causes fluids to move. A fluid always moves down its pressure gradient. This means that a fluid moves from the area where its pressure is higher to the area where its pressure is lower. When applied to the flow of air in the pulmonary airways, we can call this central idea the **primary principle of ventilation.**

Under standard conditions, air in the atmosphere exerts a pressure of 760 mm Hg. Air in the alveoli at the end of one **expiration** and before the beginning of another **inspiration** also exerts a pressure of 760 mm Hg. This explains why, at that moment, air is neither entering nor leaving the lungs. The mechanism that produces pulmonary ventilation is one that establishes a gas pressure gradient between the atmosphere and the alveolar air.

When atmospheric pressure is greater than pressure within the lung, air flows down this gas pressure gradient. Then air moves from the atmosphere into the lungs. In other words, inspiration occurs. When pressure in the lungs becomes greater than atmospheric pressure, air again moves down a gas pressure gradient. But now, this means that air moves in the opposite direction. This time, air moves out of the lungs into the atmosphere. The pulmonary ventilation mechanism, therefore, must somehow establish these two gas pressure gradients—one in which alveolar pressure (P_A, pressure within the alveoli of the lungs) is lower than atmospheric pressure (or barometric pressure, P_B) to produce inspiration and one in which it is higher than atmospheric pressure to produce expiration. See Figures 24-2 and 24-3.

These pressure gradients are established by changes in the size of the thoracic cavity, which in turn are produced by contraction and relaxation of respiratory muscles. An understanding of *Boyle's law* is important for understanding the pressure changes that occur in the lungs and thorax during the breathing cycle. It is a familiar principle, stating that the volume of a gas varies inversely with pressure at a constant temperature (Box 24-1). An application is the following: expansion of the thorax (increase in volume) results in a decreased intrapleural (intrathoracic) pressure. This leads to a decreased intraalveolar pressure that causes air to move from the outside into the lungs.

The mechanics of ventilation are often modeled with a balloon in a jar, as you can see in Figure 24-4. The bell-shaped jar represents the rib cage (thoracic cavity), and a rubber sheet across the open bottom of the bell jar represents the diaphragm. A balloon represents the lungs. The space between the balloon and the jar represents the intrapleural space. Expanding the thorax by pulling the diaphragm downward increases thoracic volume—thus decreasing intrapleural pressure (P_{IP}). Because the balloon is compliant (stretchable) the decrease in P_{IP} causes a similar decrease in the balloon pressure (alveolar pressure, P_A). This creates a pressure gradient that results in flow of air into the balloon. The opposite occurs when the elastic diaphragm recoils, decreasing internal air volumes (thus increasing internal air pressure) and forcing air out of the balloon.

Figure 24-5 applies the same principles of the balloon model to the human airways to demonstrate the mechanics of ventilation. The constant alternation between inspiration and expiration is called the **respiratory cycle.** The specific mechanics of the respiratory cycle are outlined in the following sections and in Table 24-1.

Inspiration

Contraction of the diaphragm alone, or contraction of both the diaphragm and the external intercostal muscles, produces quiet inspiration. As the diaphragm contracts, it descends, and this makes the thoracic cavity longer. Contraction of the external intercostal muscles pulls the anterior end of each rib up and out (Figure 24-6, *A*). This also elevates the attached sternum and enlarges the thorax from front to back and from side to side (Figure 24-6, *B*). In addition, contraction of the sternocleidomastoid, pectoralis minor, and serratus anterior muscles can aid in elevation of the sternum and rib cage during forceful inspiration.

As the size of the thorax increases, the intrapleural (intrathoracic) and alveolar pressure decreases (Boyle's law) and inspiration occurs.

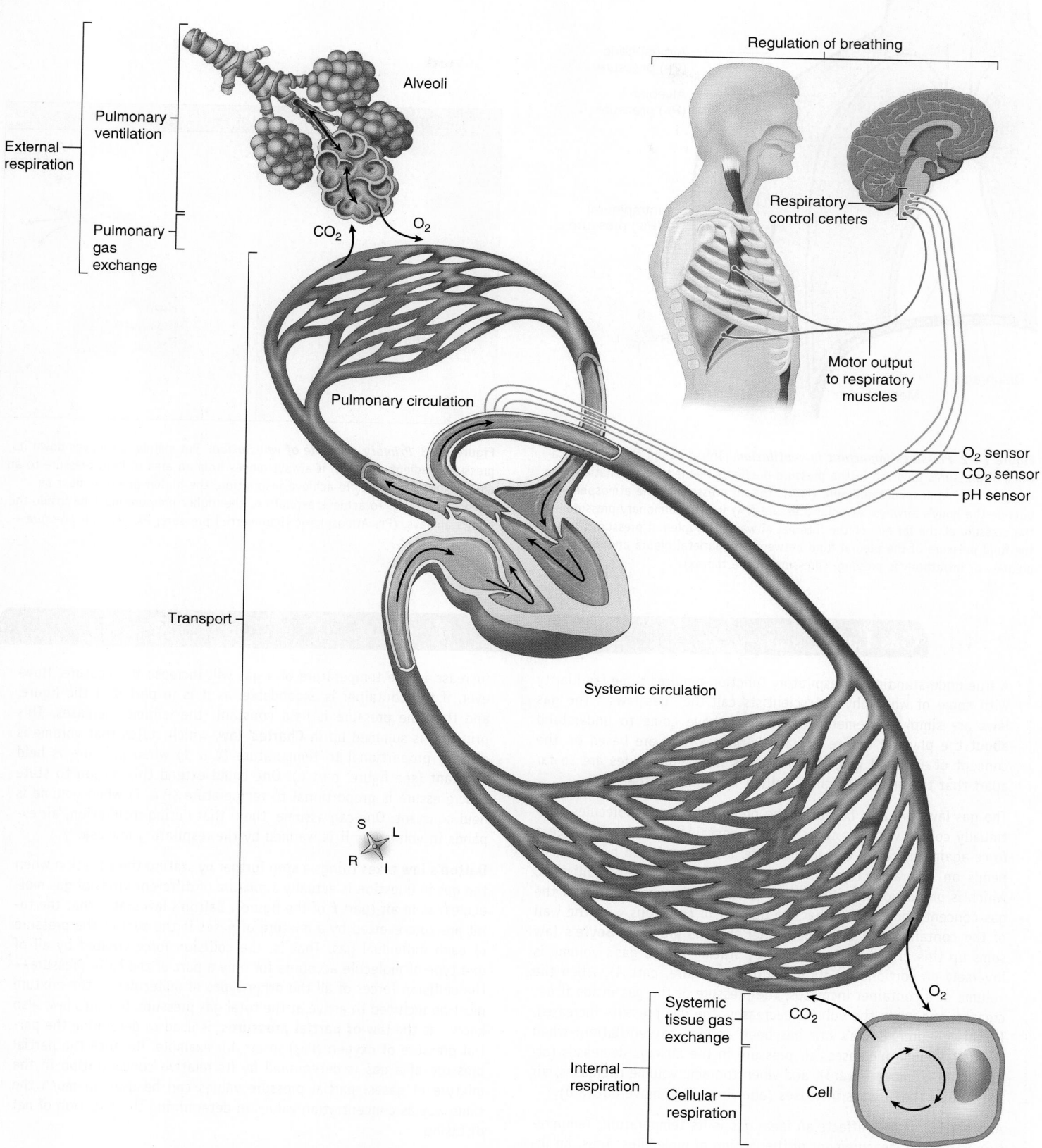

Figure 24-1 *Overview of respiratory physiology.* This chapter is organized around the principle that respiratory function includes external respiration (ventilation and pulmonary gas exchange), transport of gases by blood, and internal respiration (systemic tissue gas exchange and cellular respiration). Cellular respiration is discussed separately (see Chapters 4 and 27). Regulatory mechanisms centered in the brainstem use feedback from blood gas sensors to regulate ventilation.

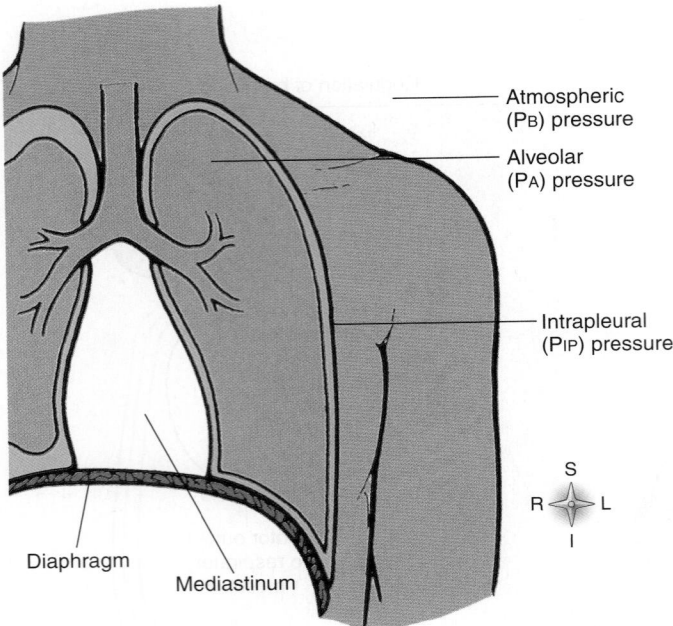

Figure 24-2 *Pressures important in ventilation.* This diagram shows the locations of pressures involved in the pressure gradients needed for ventilation (see Figure 24-3). Atmospheric pressure (P_B) is the air pressure of the atmosphere outside the body's airways. Alveolar pressure (P_A) is intrapulmonary pressure—the pressure at the far end of the internal airways. Intrapleural pressure (P_{IP}) is the fluid pressure of the pleural fluid between the parietal pleura and visceral pleura—or intrathoracic pressure (pressure in the thorax).

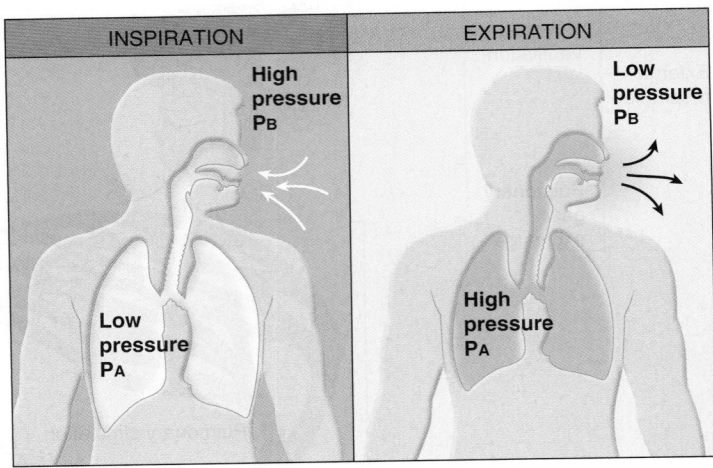

Figure 24-3 *Primary principle of ventilation.* Put simply, air moves down its pressure gradient—that is, it always moves from an area of high pressure to an area of lower pressure. To achieve inspiration, the higher pressure must be outside the body. To achieve expiration, the higher pressure must be inside the body's airways. (P_B, Atmospheric [barometric] pressure; P_A, alveolar pressure—see Figure 24-2.)

BOX 24-1   Gas Laws

A true understanding of respiratory function requires some familiarity with some of what physical scientists call the "gas laws." The gas laws are simply statements of what we have come to understand about the physical nature of gases. The gas laws are based on the concept of an **ideal gas,** that is, a gas whose molecules are so far apart that the molecules rarely collide with one another.

The gas laws are also based on the premise that gas molecules continually collide with the walls of their container and thus produce a force against it called the *gas pressure.* Pressure exerted by a gas depends on several factors. One factor is the frequency of collisions, which is proportional to the concentration of the gas: the higher the gas concentration, the higher the number of collisions with the wall of the container and thus the higher the gas pressure. **Boyle's law** sums up this principle very neatly by stating that a gas's volume is inversely proportional to its pressure (see figure, part *A*). When the volume of a container increases, the pressure of the gas inside it decreases, and when the volume decreases, the gas pressure increases. In this chapter, Boyle's law has been applied to ventilation: when thoracic volume increases, air pressure in the airways decreases (allowing air to move inward), and when thoracic volume decreases, air pressure in the airways increases (allowing air to move outward).

Another factor that affects an ideal gas is its temperature. Temperature is really a measurement of the motion of molecules. Thus, an increase in temperature signals an increase in the average velocity of gas molecules. It follows that all other things remaining the same, an

increase in the temperature of a gas will increase its pressure. However, if the container is expandable, as it is in part *B* of the figure, and thus the pressure is held constant, the volume increases. This principle is summed up in **Charles' law,** which states that volume is directly proportional to temperature ($V \alpha T$) when pressure is held constant (see figure, part *C*). One could extend this notion to state that pressure is proportional to temperature ($P \alpha T$) when volume is held constant. One can assume, then, that during inspiration, air expands in volume as it is warmed by the respiratory mucosa.

Dalton's law takes things a step further by stating the situation when the gas in question is actually a *mixture* of different kinds of gas molecules, as in air (part *E* of the figure). Dalton's law states that the total pressure exerted by a mixture of gases is the sum of the pressure of each individual gas. That is, the collision force created by all of one type of molecule accounts for only a part of the total pressure—the collision forces of all the other types of molecules in the mixture must be included to arrive at the total gas pressure. Dalton's law, also known as the **law of partial pressures,** is used to determine the partial pressure of oxygen (P_{O_2}) in air, for example. Because the partial pressure of a gas is determined by its relative concentration in the mixture of gases, partial pressure values can be used in much the same way as concentration values in determining the direction of net diffusion.

Another gas law, **Henry's law,** describes how the pressure of a gas relates to the concentration of that gas in a liquid solution (part *F* of the figure). If you have a beaker of water surrounded by air, which contains the oxygen, the concentration of oxygen dissolved in the water will be proportional by the partial pressure of oxygen in the air. Henry's law further states that the concentration of the gas in solution is also a function of the gas's **solubility,** or its relative ability to dissolve. Thus, Henry's law states that the concentration of a gas in

a solution depends on the partial pressure of the gas and the solubility of the gas, as long as the temperature remains constant. This principle explains how the plasma concentration of a gas such as oxygen relates to its partial pressure.

Also see Box 24-5, which discusses Fick's law.

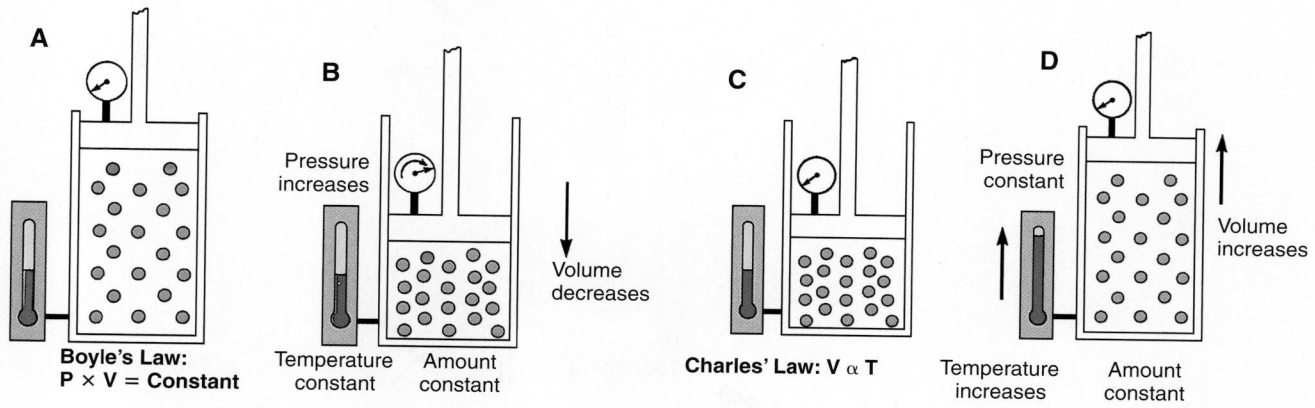

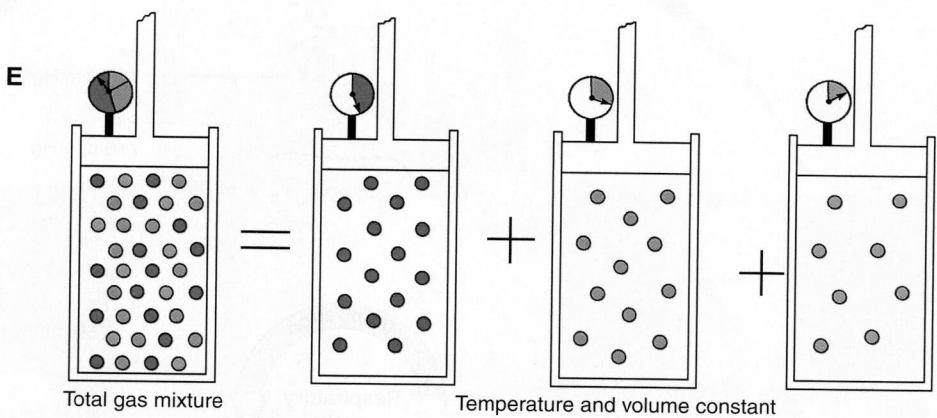

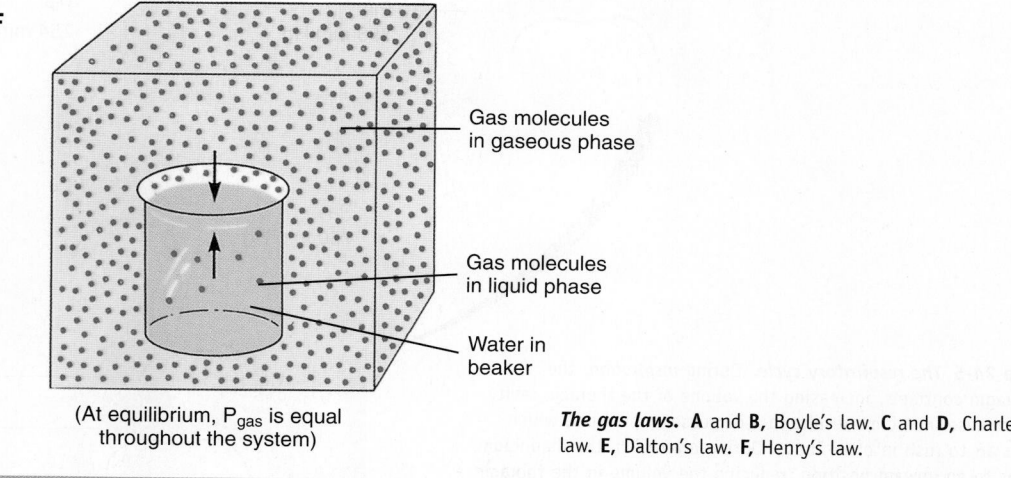

The gas laws. **A** and **B,** Boyle's law. **C** and **D,** Charles' law. **E,** Dalton's law. **F,** Henry's law.

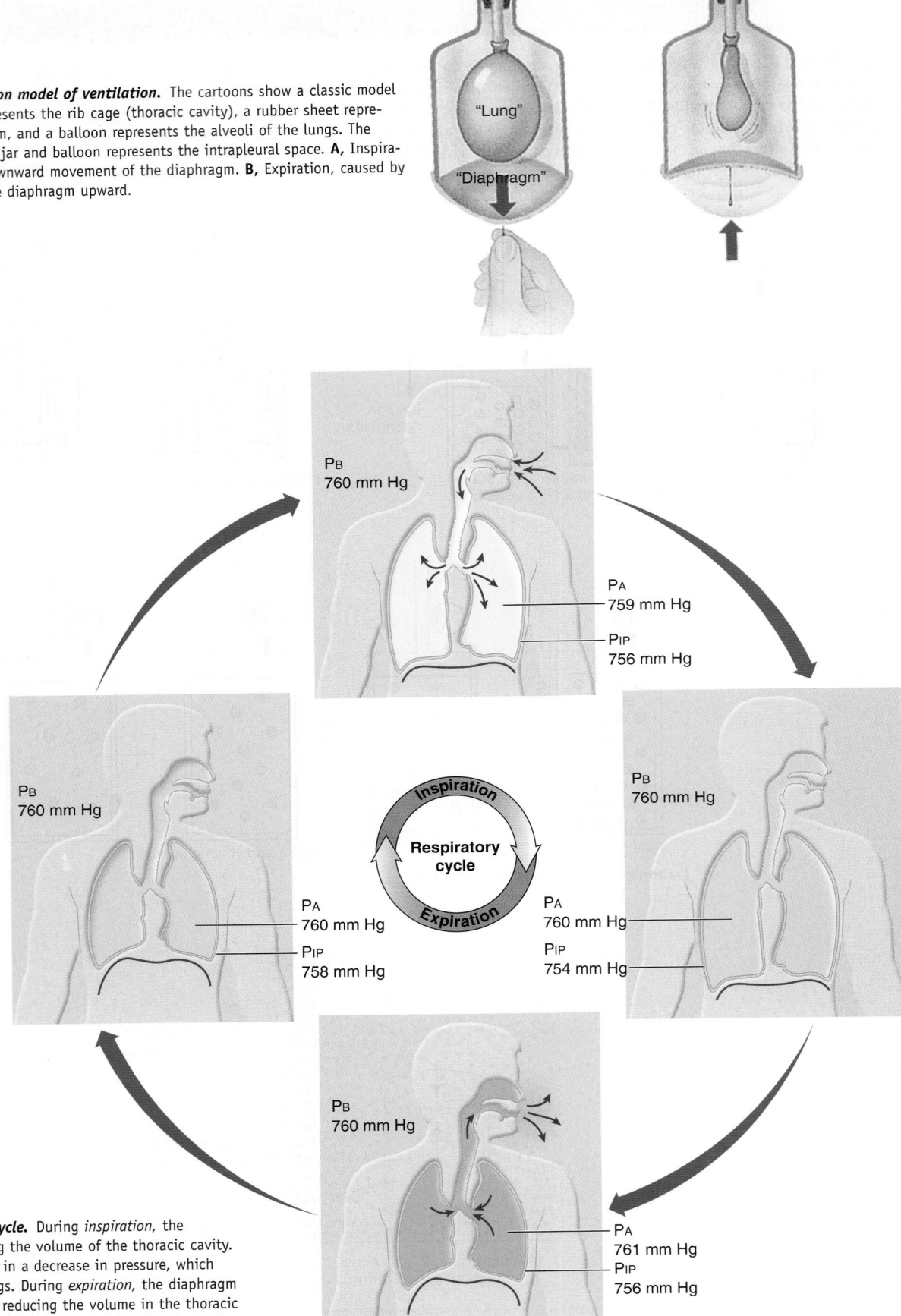

A Air

B

"Lung"

"Diaphragm"

Figure 24-4 *Balloon model of ventilation.* The cartoons show a classic model in which a jar represents the rib cage (thoracic cavity), a rubber sheet represents the diaphragm, and a balloon represents the alveoli of the lungs. The space between the jar and balloon represents the intrapleural space. **A,** Inspiration, caused by downward movement of the diaphragm. **B,** Expiration, caused by elastic recoil of the diaphragm upward.

P_B 760 mm Hg

P_A 759 mm Hg

P_{IP} 756 mm Hg

Inspiration

Respiratory cycle

Expiration

P_B 760 mm Hg

P_A 760 mm Hg

P_{IP} 758 mm Hg

P_B 760 mm Hg

P_A 760 mm Hg

P_{IP} 754 mm Hg

P_B 760 mm Hg

P_A 761 mm Hg

P_{IP} 756 mm Hg

Figure 24-5 *The respiratory cycle.* During *inspiration,* the diaphragm contracts, increasing the volume of the thoracic cavity. This increase in volume results in a decrease in pressure, which causes air to rush into the lungs. During *expiration,* the diaphragm returns to an upward position, reducing the volume in the thoracic cavity. Air pressure thus increases, forcing air out of the lungs. See Table 24-1 for additional details.

Table 24-1	The Respiratory Cycle		
P$_{IP}$	P$_A$	P$_B$	
Inspiration			
758	760	760	The diaphragm is relaxed, putting the thoracic cavity at low volume. At the beginning of inspiration P$_{IP}$ < P$_A$, keeping alveoli open. Since P$_A$ = P$_B$, no air is flowing yet.
756	759	760	The diaphragm contracts, increasing the thoracic volume and reducing P$_{IP}$. A decrease in P$_{IP}$ causes a decrease in P$_A$. Now P$_A$ < P$_B$, and air flows down the pressure gradient (into the lungs).
754	760	760	Eventually, the alveoli fill with air and P$_A$ equilibrates with P$_B$. Inward air flow stops. The cycle is now ready to shift to the expiration phase. Note that P$_{IP}$ is still dropping but P$_A$ has not yet "caught up" with the drop.
Expiration			
754	760	760	As expiration is about to begin, the diaphragm is contracted maximally. Since P$_A$ = P$_B$, there is no air flow.
756	761	760	The diaphragm relaxes, and elastic recoil of the thoracic walls and alveoli increases P$_{IP}$ and P$_A$. Now, P$_A$ > P$_B$. Air moves (outward) down the pressure gradient.
758	760	760	The diaphragm eventually relaxes fully, so the decrease in volume stops. P$_A$ equilibrates with P$_B$, and air flow ceases. The system is now ready for another inspiration phase.

P$_{IP}$ = intrapleural pressure (air pressure in the intrapleural space)
P$_A$ = alveolar pressure (air pressure inside the alveoli)
P$_B$ = atmospheric (barometric) pressure (air pressure of the external environment [atmosphere])
All P values are expressed in *mm Hg* and are examples only.

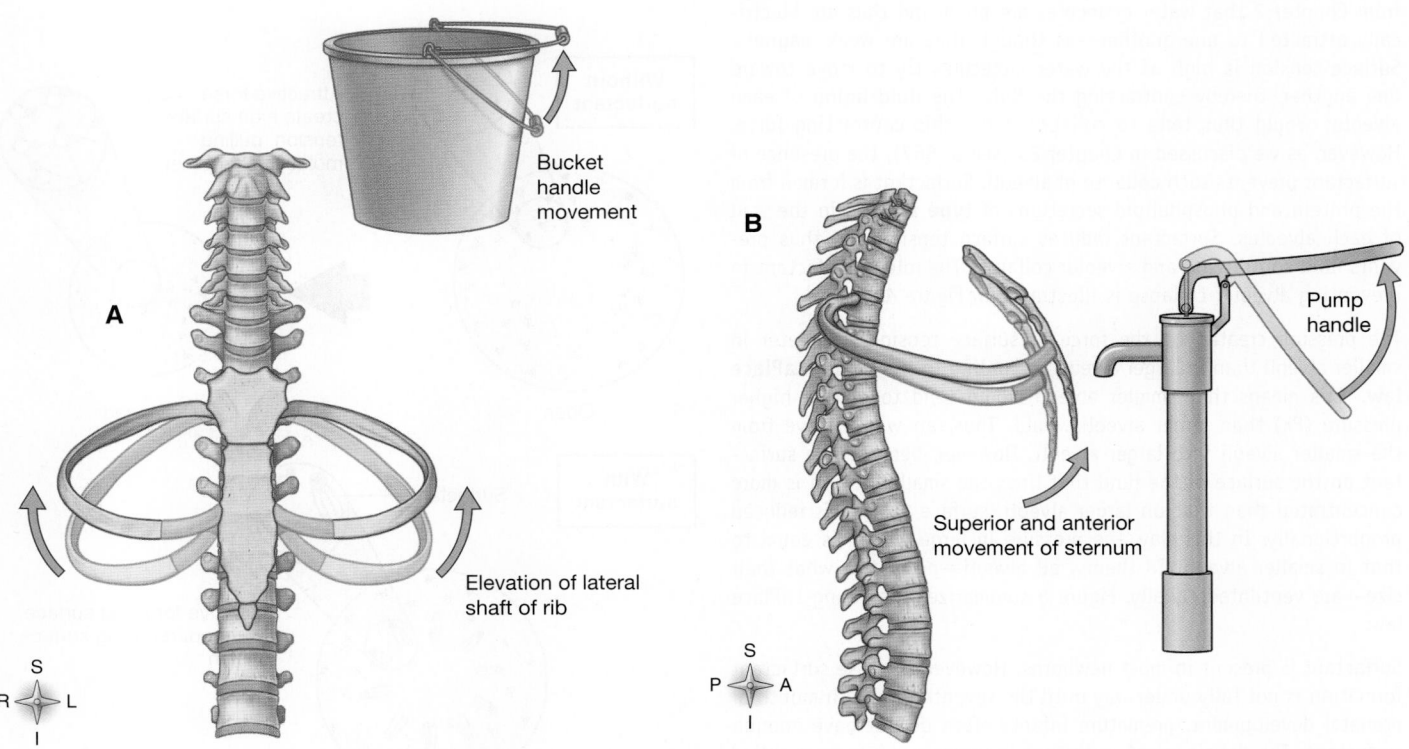

Bucket handle movement

Elevation of lateral shaft of rib

Superior and anterior movement of sternum

Pump handle

Figure 24-6 *Movement of the rib cage during breathing.* **A,** Inspiratory muscles pull the ribs upward and thus outward, as in a bucket handle. **B,** Inspiratory muscles pull the sternum upward and thus outward, as when pulling upward on the handle of a water pump.

At the beginning of each inspiration, intrapleural pressure (PIP) is about 758 mm Hg. Thus, the PIP is about 2 mm Hg less than atmospheric pressure (frequently written −2 mm Hg). During normal quiet inspiration, PIP decreases further to 756 mm Hg (−4 mm Hg) or less. As the thorax enlarges, it pulls the lungs along with it because of cohesion between the moist pleura covering the lungs and the moist pleura lining the thorax. Thus the lungs expand and the pressure in their tubes and alveoli necessarily decreases. Alveolar pressure decreases from an atmospheric level to a subatmospheric level—typically a drop of about 1 to 3 mm Hg. The moment that alveolar pressure becomes less than atmospheric pressure, a pressure gradient exists between the atmosphere and the interior of the lungs. According to the primary principle of ventilation, air moves into the lungs. Eventually, enough air moves out of the lungs to reach a pressure equilibrium between the atmosphere and the alveoli—and the flow of air stops.

The ability of the lungs and thorax to stretch, or **compliance,** is essential to normal respiration. If the compliance of these structures is reduced by injury or disease, inspiration becomes difficult—or even impossible (Box 24-2).

For a summary of the mechanism of inspiration just described, see Figures 24-5 and 24-7.

Expiration

Quiet expiration is ordinarily a passive process that begins when the pressure gradients that resulted in inspiration are reversed. The inspiratory muscles relax, causing a decrease in the size of the thorax and an increase in intrapleural pressure from about 754 mm Hg (−6 mm Hg) before expiration to about 756 mm Hg (−4 mm Hg) or more during respiration. It is important to understand that this pressure between the parietal and visceral pleura is always negative, that is, less than atmospheric pressure

---

BOX 24-2   Surfactant and Lung Compliance

As we have discussed already, inspiration cannot occur without the ability of the lungs and thorax to stretch—a characteristic called *compliance.* Of course, the natural "stretchiness" of the alveolar walls is important in determining lung compliance. Conditions that cause thickening, or fibrosis, of lung tissues reduce the ease of stretch and thus reduce lung compliance. A greater impact on lung compliance is made by **surface tension** in the fluid film that lines the alveoli.

Surface tension in an aqueous (water-based) solution results from the attractive forces between water molecules in the solution. Recall from Chapter 2 that water molecules are polar and thus are electrically attracted to one another—as though they are weak magnets. Surface tension is high as the water molecules try to move toward one another, thereby contracting the fluid. The fluid lining of each alveolus would thus tend to collapse under this contracting force. However, as we discussed in Chapter 23 (see p. 867), the presence of surfactant prevents such collapse of alveoli. Surfactant is formed from the protein and phospholipid secretions of **type II cells** in the wall of each alveolus. Surfactant reduces surface tension and thus prevents fluid contraction and alveolar collapse. The role of surfactant in preventing alveolar collapse is illustrated in Figure A.

The pressure created by the force of surface tension is greater in smaller alveoli than in larger alveoli, according to the **Young-LaPlace law.** This means that smaller alveoli would tend to have a higher pressure (PA) than larger alveoli would. Thus, air would move from the smaller alveoli into larger alveoli. However, because the surfactant on the surface of the fluid that lines the smaller alveoli is more concentrated than that on larger alveoli, surface tension is reduced proportionally. In this way, the pressure in large alveoli is equal to that in smaller alveoli. In theory, all alveoli—no matter what their size—are ventilated equally. Figure B summarizes the Young-LaPlace law.

Surfactant is present in most newborns. However, because surfactant formation is not fully under way until the seventh or eighth month of prenatal development, premature infants often do not have enough surfactant. The deficiency of surfactant in premature infants is called **hyaline membrane disease (HMD).** Because lack of surfactant decreases lung compliance, a premature infant will try to inflate the

alveoli by increasing effort of the inspiratory muscles. Such great effort is needed to maintain normal ventilation that the baby may die of exhaustion. The effects of such alveolar collapse and ventilation difficulty are collectively called **respiratory distress syndrome (RDS).** In infants, it is more specifically called *infant respiratory distress syndrome (IRDS).* See Figure C.

One way to treat IRDS is to use a special type of mechanical respirator with **continuous positive airway pressure (CPAP,** pronounced "SEE-pap"). The respirator artificially inflates the baby's lungs and

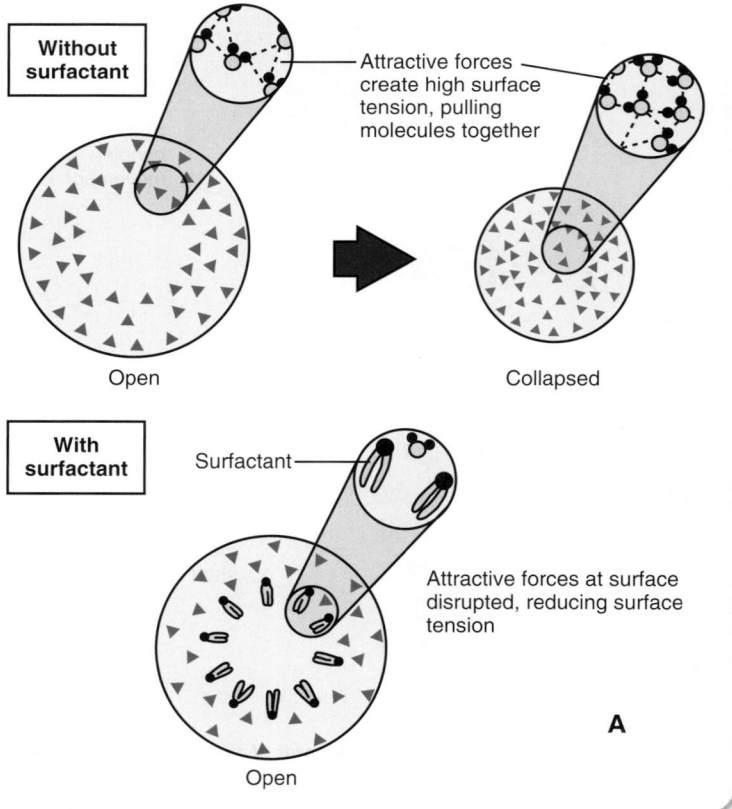

Without surfactant

Attractive forces create high surface tension, pulling molecules together

Open

Collapsed

With surfactant

Surfactant

Attractive forces at surface disrupted, reducing surface tension

Open

A

BOX 24-2 Surfactant and Lung Compliance—cont'd

then maintains enough pressure during expiration to prevent collapse—thus relieving the baby's inspiratory muscles. Synthetic surfactants are also used frequently to prevent or treat IRDS. The surfactant is delivered through a tube directly into the airways—a method called **intratracheal injection.**

Adult respiratory distress syndrome (ARDS) often results from conditions that impair or remove alveolar surfactant. Accidental inhalation of vomit, water, smoke, or chemical fumes can cause ARDS. Edema, resulting from any of various conditions, can also impair surfactant and decrease compliance.

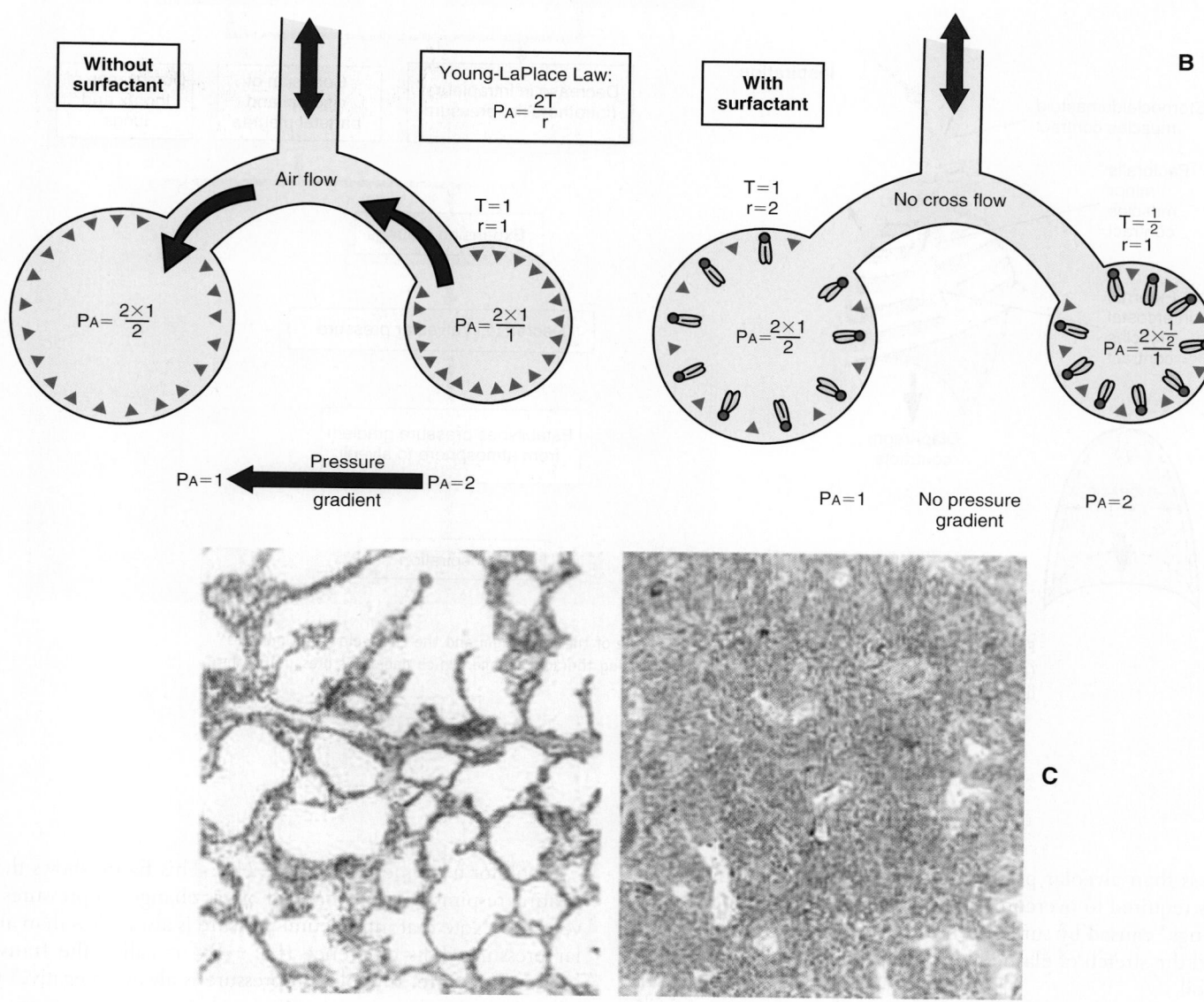

A, Role of surfactant. The surface of the water that lines the small alveoli tends to contract because of its high surface tension, thereby collapsing the entire alveolus. Surfactant disrupts some of the attractive forces and thus reduces surface tension—and the risk of alveolar collapse. *B, Young-LaPlace law.* Also called the *law of LaPlace,* this principle states that P_A is directly proportional to surface tension (T) and inversely proportional to the radius (r) of the alveolus. Without surfactant, the pressure gradient would cause air to flow from the small alveoli to the larger alveoli—thus triggering collapse of the smaller alveoli. When surfactant is present, the concentration of the surfactant is higher as the alveolus gets smaller. Because small alveoli have less surface tension than larger alveoli do (as a result of more concentrated surfactant), the effect of the Young-LaPlace law is counterbalanced. Because P_A thus remains about the same in all alveoli, regardless of size, ventilation is not disrupted. *C, Microscopic effects of RDS.* The light micrograph on the left shows normal lung structure, with many open alveoli. The right image is from an infant who died of RDS. Note the collapse of the alveoli.

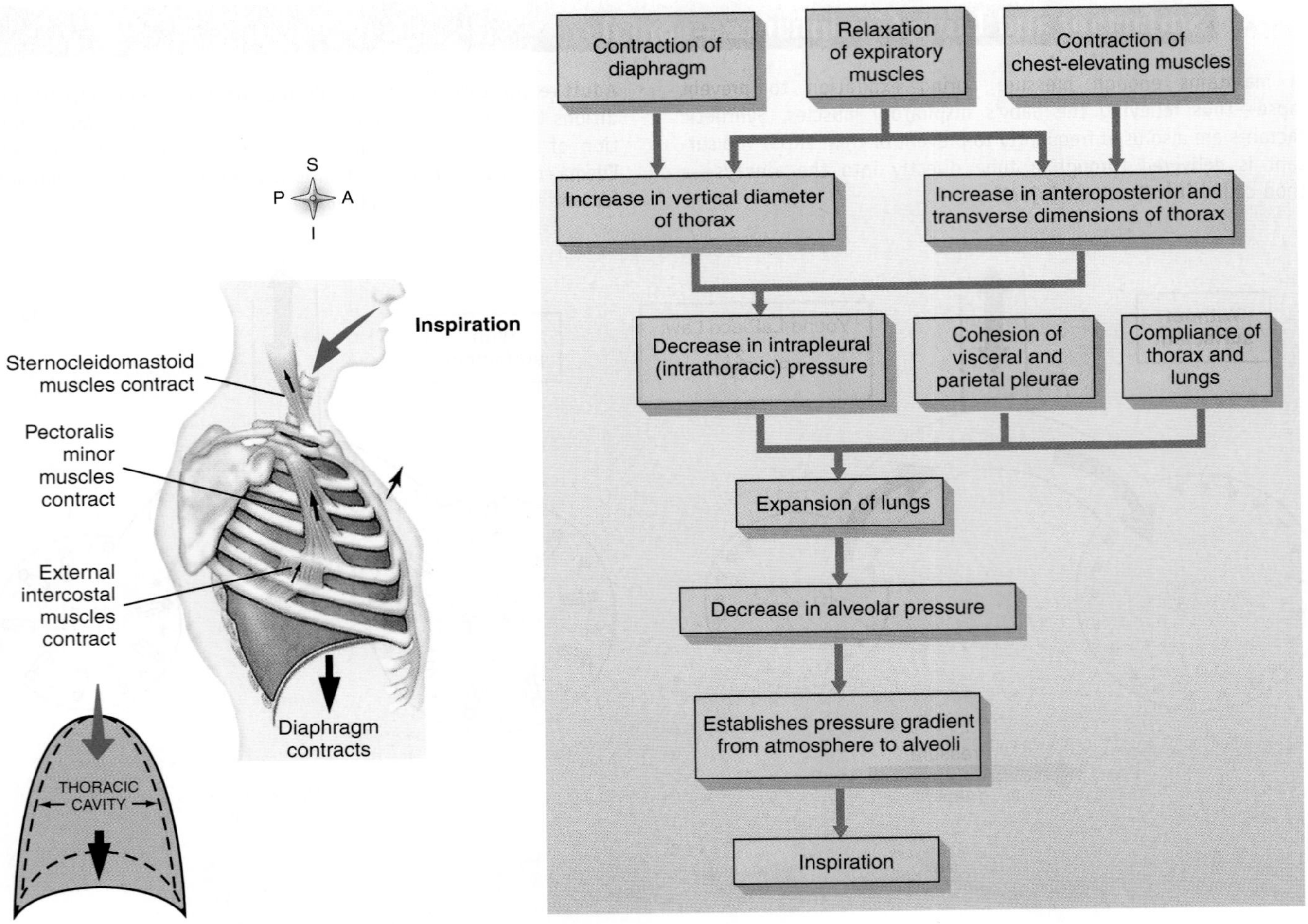

S
P —◆— A
I

Inspiration

Sternocleidomastoid
muscles contract

Pectoralis
minor
muscles
contract

External
intercostal
muscles
contract

Diaphragm
contracts

THORACIC
CAVITY

```
┌──────────────┐  ┌──────────────┐  ┌──────────────┐
│Contraction of│  │ Relaxation   │  │Contraction of│
│  diaphragm   │  │of expiratory │  │chest-elevating│
│              │  │  muscles     │  │   muscles    │
└──────────────┘  └──────────────┘  └──────────────┘
```

Increase in vertical diameter of thorax

Increase in anteroposterior and transverse dimensions of thorax

Decrease in intrapleural (intrathoracic) pressure

Cohesion of visceral and parietal pleurae

Compliance of thorax and lungs

Expansion of lungs

Decrease in alveolar pressure

Establishes pressure gradient from atmosphere to alveoli

Inspiration

Figure 24-7 *Mechanism of inspiration.* Note the role of the diaphragm and the chest-elevating muscles (pectoralis minor and external intercostals) in increasing thoracic volume, which decreases pressure in the lungs and thus draws air inward.

and less than alveolar pressure. The negative intrapleural pressure is required to overcome the so-called "collapse tendency of the lungs" caused by surface tension of the fluid lining the alveoli and the stretch of elastic fibers that are constantly attempting to recoil.

As alveolar pressure increases, a positive-pressure gradient is established from alveoli to atmosphere—and thus expiration occurs as air flows outward through the respiratory passageways. In forced expiration, contraction of the abdominal and internal intercostal muscles can increase alveolar pressure tremendously—creating a very large air pressure gradient.

The tendency of the thorax and lungs to return to their preinspiration volume is a physical phenomenon called **elastic recoil.** If a disease condition reduces the elasticity of pulmonary tissues, expirations must become forced even at rest.

Figures 24-5 and 24-8 summarize the mechanism of expiration just described.

Look for a moment at Figure 24-9. This figure shows the repeating respiratory cycle mapped out as changes in pressures and volumes. Note that intrapleural pressure is always less than alveolar pressure. This difference ($P_{IP} - P_A$) is called the **transpulmonary pressure.** Intrapleural pressure is always "negative" with respect to alveolar pressure. Transpulmonary pressure must be negative to maintain inflation of the lungs, as stated previously.

 QUICK CHECK

1. What is meant by the term *pulmonary ventilation?*
2. What effect does enlargement of the thoracic cavity have on the air pressure inside the lungs?
3. Which requires more expenditure of energy during normal, quiet breathing—inspiration or expiration?

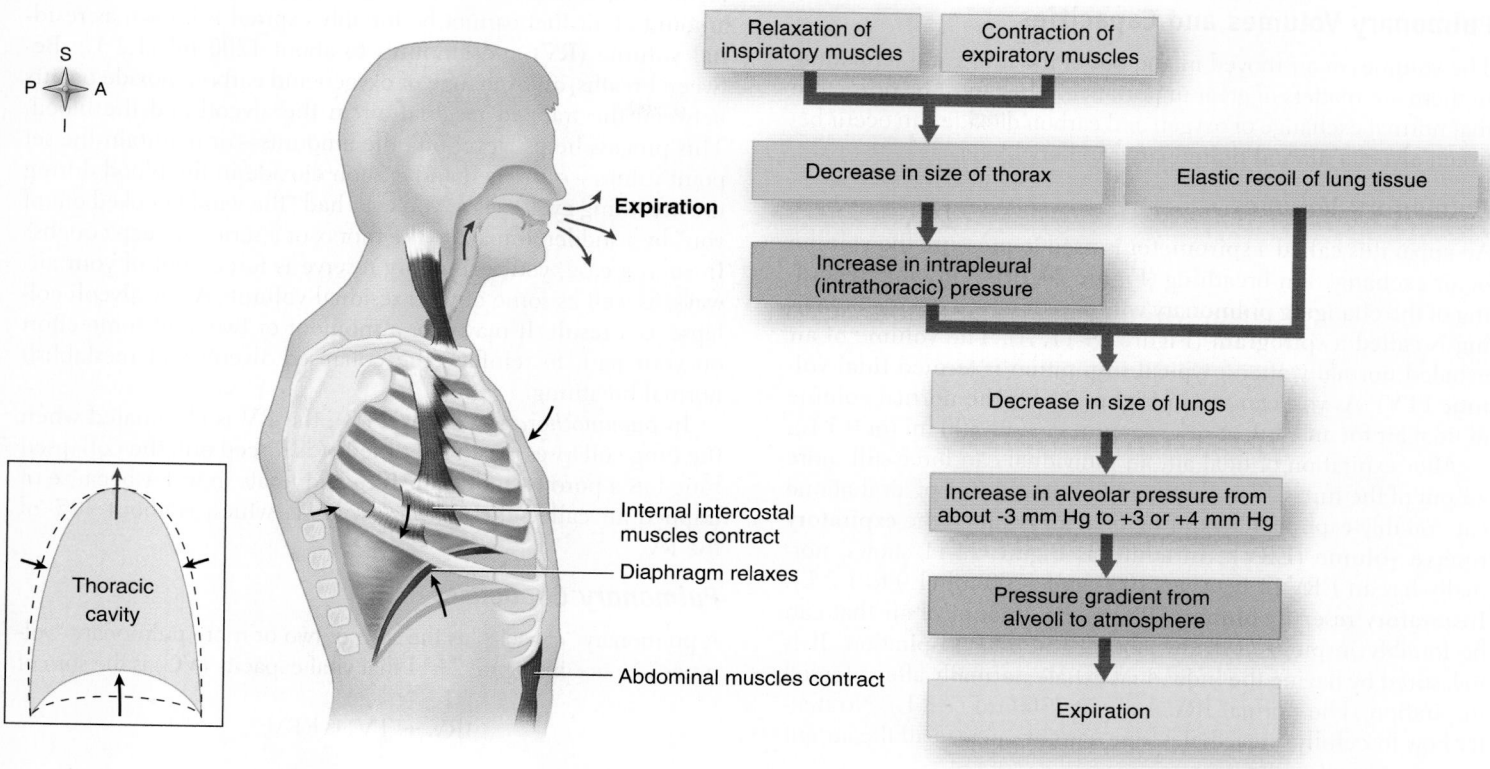

Figure 24-8 *Mechanism of expiration.* Note that relaxation of the diaphragm plus contraction of chest-depressing muscles (internal intercostals) reduces thoracic volume, which increases pressure in the lungs and thus pushes air outward.

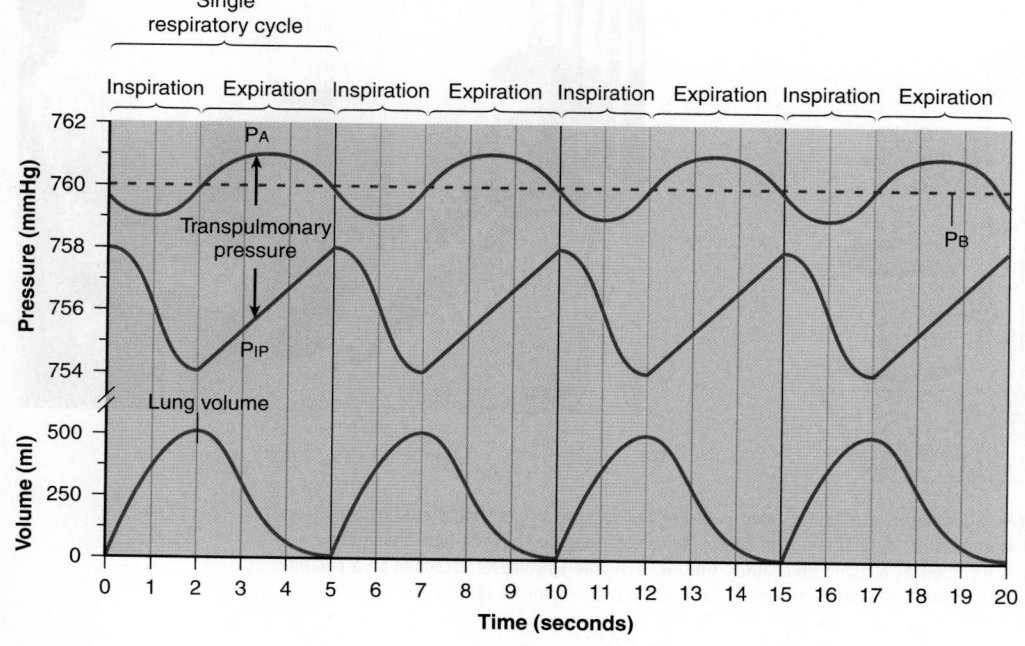

Figure 24-9 *Rhythm of ventilation.* Respiratory cycles repeat continuously in normal, quiet breathing. Notice the rhythmic rise and fall of the intrapleural pressure (P_{IP}) and alveolar pressure (P_A). You can easily see that P_{IP} is always lower than P_A (negative transpulmonary pressure), which helps keep the alveoli inflated. The lowest line shows the change in air volumes during the respiratory cycle.

Pulmonary Volumes and Capacities

The volumes of air moved in and out of the lungs and remaining in them are matters of great importance. They must be normal so that normal exchange of oxygen and carbon dioxide can occur between alveolar air and pulmonary capillary blood.

Pulmonary Volumes

An apparatus called a **spirometer** is used to measure the volume of air exchanged in breathing (Figure 24-10). A graphic recording of the changing pulmonary volumes observed during breathing is called a **spirogram** (Figure 24-11, A). The volume of air exhaled normally after a typical inspiration is termed **tidal volume (TV).** As you can see in Figure 24-11, the normal volume of tidal air for an adult at rest is approximately 500 ml (or 0.5 L).

After expiration of tidal air, an individual can force still more air out of the lungs. The largest additional volume of air that one can forcibly expire after expiring tidal air is called the **expiratory reserve volume (ERV).** An adult, as Figure 24-11 shows, normally has an ERV of between 1000 and 1200 ml (1.0 to 1.2 L). **Inspiratory reserve volume (IRV)** is the amount of air that can be forcibly inspired over and above a normal inspiration. It is measured by having the individual exhale normally after a forced inspiration. The normal IRV is about 3300 ml (3.3 L). No matter how forcefully one exhales, one cannot squeeze all the air out

of the lungs. Some of it remains trapped in the alveoli. This amount of air that cannot be forcibly expired is known as **residual volume (RV)** and amounts to about 1200 ml (1.2 L). Between breaths, an exchange of oxygen and carbon dioxide occurs between the trapped residual air in the alveoli and the blood. This process helps "level off" the amounts—or maintain the set point values—of oxygen and carbon dioxide in the blood during the breathing cycle. Have you ever had "the wind knocked out of you" by a sudden impact to the thorax or a series of deep coughs? In such a case, your expiratory reserve is forced out of your airways, as well as some of your residual volume. A few alveoli collapse as a result. It may take a moment or two, and some effort on your part, to reinflate the collapsed alveoli and reestablish normal breathing.

In *pneumothorax* (see Box 24-3), the RV is eliminated when the lung collapses. Even after the RV is forced out, the collapsed lung has a porous, spongy texture and floats in water because of trapped air called the *minimal volume*, which is about 40% of the RV.

Pulmonary Capacities

A pulmonary "capacity" is the sum of two or more pulmonary "volumes." Notice in Figure 24-11 that **vital capacity (VC)** is the sum of

$$IRV + TV + ERV.$$

A

SIMPLE SPIROMETER

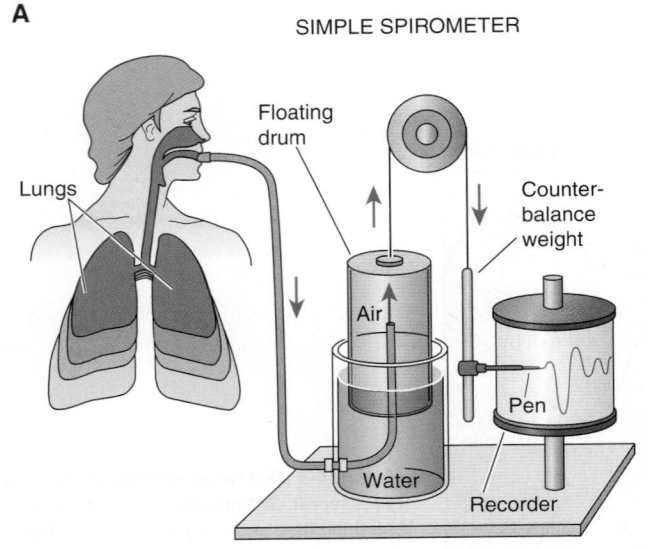

B

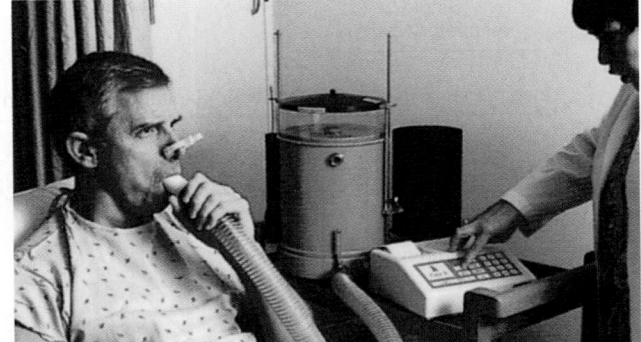

Figure 24-10 *Spirometer.* Spirometers are devices that measure the volume of gas that the lungs inhale and exhale, usually as a function of time. **A,** Diagram of a classic spirometer design showing how the volume of air exhaled and inhaled is recorded as a rising and falling line. **B,** A simple spirometer attached to a computerized recording device. This apparatus is used frequently for routine assessment of ventilation.

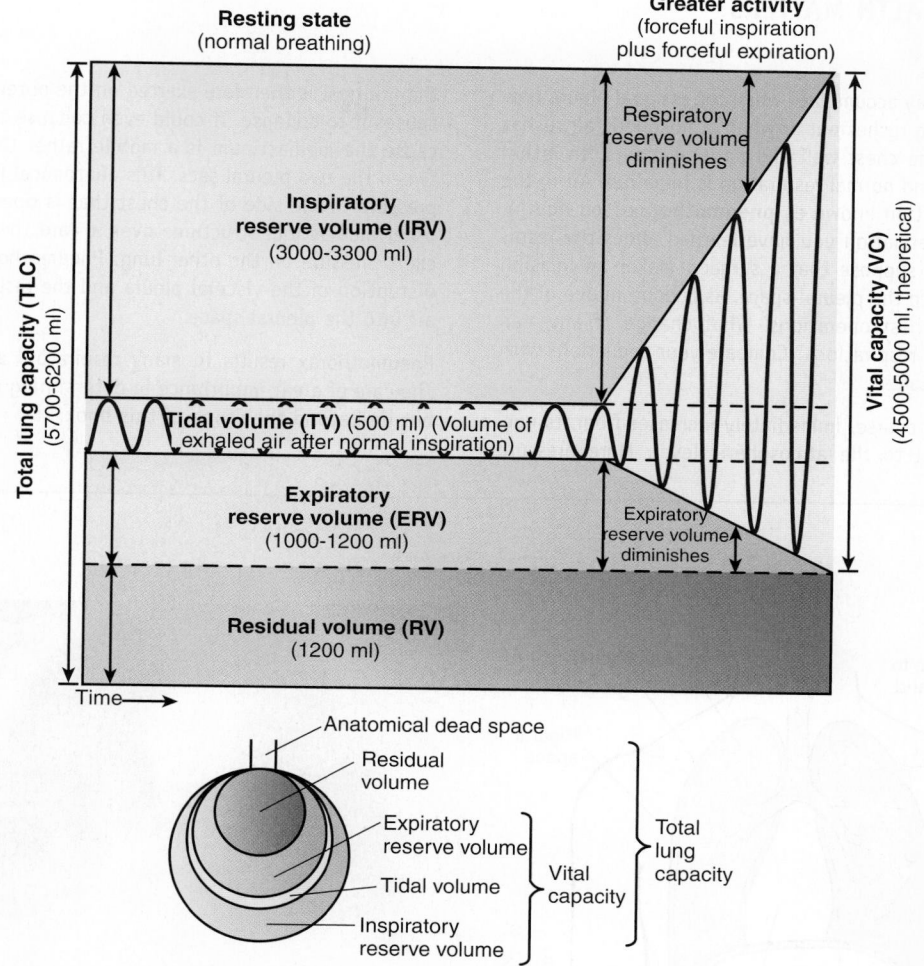

Figure 24-11 *Pulmonary ventilation volumes and capacities.* **A,** Spirogram. **B,** Pulmonary volumes (at rest) represented as relative proportions of an inflated balloon. During normal, quiet respirations, the atmosphere and lungs exchange about 500 ml of air (TV). With forcible inspiration, about 3300 ml more air can be inhaled (IRV). After a normal inspiration and normal expiration, approximately 1000 ml more air can be forcibly expired (ERV). Vital capacity is the amount of air that can be forcibly expired after a maximal inspiration and therefore indicates the largest amount of air that can enter and leave the lungs during respiration. Residual volume is the air that remains trapped in the alveoli.

The vital capacity represents the largest volume of air an individual can move in and out of the lungs. It is determined by measuring the largest possible expiration after the largest possible inspiration. How large a vital capacity a person has depends on many factors—the size of the thoracic cavity, posture, and various other factors. In general, a larger person has a larger vital capacity than a smaller person does. An individual has a larger vital capacity when standing erect than when stooped over or lying down. The volume of blood in the lungs also affects the vital capacity. If the lungs contain more blood than normal, the alveolar air space is encroached on and vital capacity accordingly decreases. This becomes a very important factor in congestive heart disease.

Excess fluid in the pleural or abdominal cavities also decreases vital capacity. So, too, does the disease *emphysema*. In the latter condition, the alveolar walls become stretched—that is, lose their elasticity—and are unable to recoil normally for expiration. This leads to an increased RV. In severe emphysema, the RV may increase so much that the chest occupies the inspiratory position even at rest. Excessive muscular effort is therefore necessary for inspiration, and because of the loss of elasticity of lung tissue, greater effort is required, too, for expiration.

In diagnosing lung disorders a physician may need to know the inspiratory capacity and the functional residual capacity of the patient's lungs. **Inspiratory capacity (IC)** is the maximal amount of

BOX 24-3: HEALTH MATTERS
Pneumothorax

Air in the pleural space may accumulate when the visceral pleura ruptures and air from the lung rushes out or when atmospheric air rushes in through a wound in the chest wall and parietal pleura. In either case, the lung collapses and normal respiration is impaired. Air in the thoracic cavity is a condition known as **pneumothorax** (see figure). To apply some of the information you have learned about the respiratory mechanism, let us suppose that a surgeon makes an incision through the chest wall into the pleural space, as is done in one of the dramatic, modern open-chest operations. What change, if any, can you deduce takes place in respirations? Compare your deductions with those in the next paragraph.

Intrathoracic pressure, of course, immediately increases from its normal subatmospheric level to the atmospheric level. More pressure than normal is therefore exerted on the outer surface of the lung and causes it to collapse. It could even collapse the other lung. Why? Because the mediastinum is a mobile rather than a rigid partition between the two pleural sacs. This anatomical fact allows the increased pressure in the side of the chest that is open to push the heart and other mediastinal structures over toward the intact side, where they exert pressure on the other lung. Pneumothorax can also result from disruption of the visceral pleura and the resulting flow of pulmonary air into the pleural space.

Pneumothorax results in many respiratory and circulatory changes. They are of great importance in determining medical and nursing care but lie beyond the scope of this book.

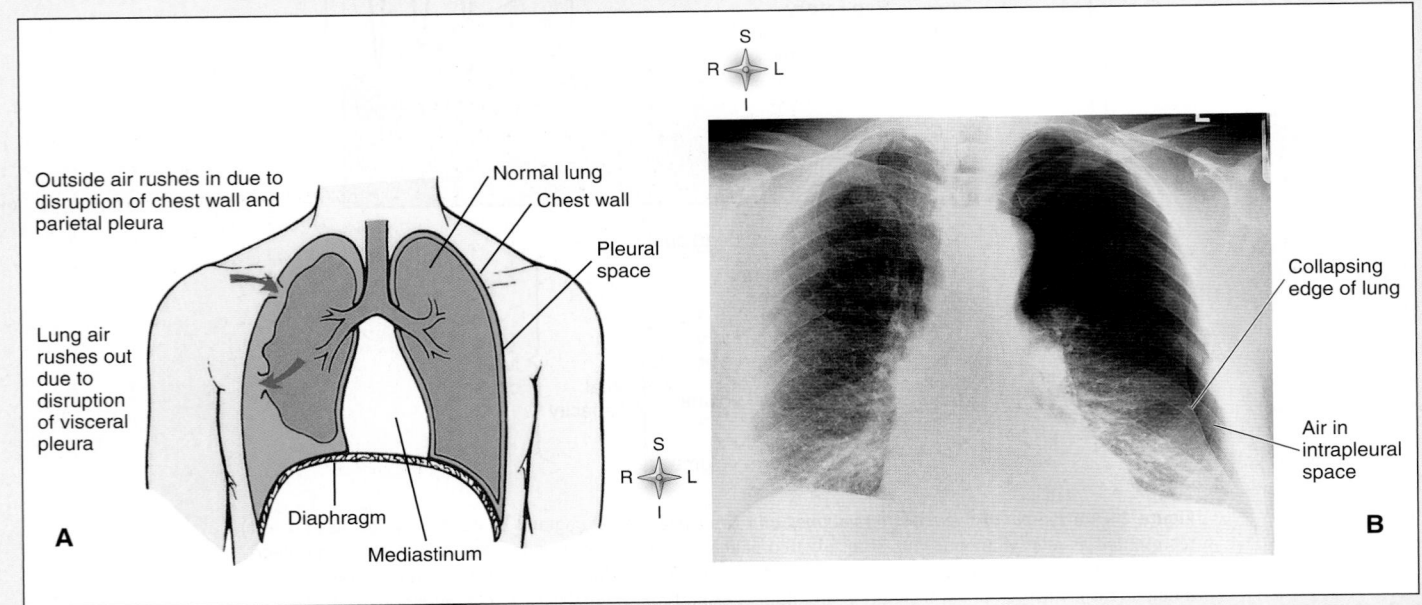

A, Pneumothorax. Diagram showing air entering the thoracic cavity, causing lung collapse. *B, Chest x-ray.* Left-sided pneumothorax showing the left wall of the lung pulling away from the thoracic wall (the air-filled space is slightly darker than the lung in the radiograph).

air an individual can inspire after a normal expiration. From Figure 24-11, you can deduce that

$$IC = TV + IRV.$$

With the volumes given in the figure, how many milliliters is the IC? Check your answer in Table 24-2, which summarizes pulmonary volumes and capacities. **Functional residual capacity (FRC)** is the amount of air left in the lungs at the end of a normal expiration. Therefore, as Figure 24-11 implies,

$$FRC = ERV + RV.$$

With the volumes given, the functional residual capacity is 2200 to 2400 ml (2.2 to 2.4 L). The total volume of air a lung can hold is called the **total lung capacity (TLC).** It is, as Figure 24-11 indicates, the sum of all four lung volumes.

The term **alveolar ventilation** means the volume of inspired air that actually reaches, or "ventilates," the alveoli. Only this volume of air takes part in the exchange of gases between air and blood. (Alveolar air exchanges some of its oxygen for some of the blood's carbon dioxide.) With every breath we take, part of the entering air necessarily fills our air passageways—nose, pharynx, larynx, trachea, and bronchi. This portion of air does

Table 24-2   Pulmonary Volumes and Capacities

VOLUME	DESCRIPTION	TYPICAL VALUE	CAPACITY	FORMULA	TYPICAL VALUE
Tidal volume (TV)	Volume moved into or out of the respiratory tract during a normal respiratory cycle	500 ml (0.5 L)	Vital capacity (VC)	TV + IRV + ERV	4500-5000 ml (4.5-5.0 L)
Inspiratory reserve volume (IRV)	Maximum volume that can be moved into the respiratory tract after a normal inspiration	3000-3300 ml (3.0-3.3 L)	Inspiratory capacity (IC)	TV + IRV	3500-3800 ml (3.5-3.8 L)
Expiratory reserve volume (ERV)	Maximum volume that can be moved out of the respiratory tract after a normal expiration	1000-1200 ml (1.0-1.2 L)	Functional residual capacity (FRC)	ERV + RV	2200-2400 ml (2.2-2.4 L)
Residual volume (RV)	Volume remaining in the respiratory tract after maximum expiration	1200 ml (1.2 L)	Total lung capacity (TLC)	TV + IRV + ERV + RV	5700-6200 ml (5.7-6.2 L)

not descend into any alveoli and therefore cannot take part in gas exchange. In this sense, it is "dead air." Appropriately, the larger air passageways it occupies are said to constitute the **anatomical dead space.** Figure 24-11, *B*, relates the volume of the anatomical dead space to the major pulmonary volumes. In disorders such as **chronic obstructive pulmonary disease (COPD),** some alveoli are prevented from gas exchange and are therefore also "dead space." The anatomical dead space plus any alveolar dead space together make up the **physiological dead space.**

One rule of thumb estimates the volume of air in the anatomical dead space as the same number of milliliters as the individual's weight in pounds. Another generalization says that the anatomical dead space approximates 30% of the TV. TV − dead space volume = alveolar ventilation volume. Suppose you have a normal TV of 500 ml and that 30% of this, or 150 ml, fills the anatomical dead space. The amount of air reaching your alveoli—your alveolar ventilation volume—is then 350 ml per breath, or 70% of your TV.

Emphysema and certain other abnormal conditions, in effect, increase the amount of dead space air or physiological dead space. Consequently, alveolar ventilation decreases, and this in turn decreases the amount of oxygen that can enter blood and the amount of carbon dioxide that can leave it. Inadequate air-blood gas exchange, therefore, is the inevitable result of inadequate alveolar ventilation. Stated differently, the alveoli must be adequately ventilated for an adequate gas exchange to take place in the lungs.

Box 24-4 summarizes some abnormal breathing patterns seen in spirometry.

Pulmonary Air Flow

Various applications of spirometry can be used to generate additional information about air flow in an individual. For example, spirometry can be used to determine pulmonary air flow as the rate of pulmonary ventilation, or **total minute volume** (volume moved per minute). Tidal volume (ml/cycle) multiplied by respiration rate (cycles per minute) yields the total minute volume (ml/min). The total minute volume of a person at rest is about 6000 ml (500 ml/cycle × 12 cycles/min).

Yet another application of spirometry is the **forced expiratory volume (FEV)** test. The FEV test can determine the presence of respiratory obstruction by measuring the volume of air expired per second during forced expiration. The volume forcefully expired during the first second, the FEV_1, is normally about 83% of the vital capacity (Figure 24-12). FEV_2, the total volume expired during the first 2 seconds, is about 94% of the VC. By the end of the third second, FEV_3, 97% of the vital capacity should have been expired. The FEV test is also sometimes called the FVC (forced vital capacity) test.

Some spirometers are capable of producing a graph called the **flow-volume loop.** This type of graph shows a forced expiration (forced vital capacity) as a loop rather than the peaks and valleys of the classic spirogram. In Figure 24-13 you can see that the top portion of the loop represents expiratory air flow (liters per second) along the vertical axis and expiratory volume (liters) along the horizontal axis. The inspiratory air flow and volume are represented by the bottom portion of the loop.

Notice in Figure 24-13 that the top of the flow-volume loop represents the peak expiratory flow, or more simply the *peak flow.* The peak flow is easy to measure even with simple hand-held spirometers. It is no wonder, then, that peak flow measurements are often used at home by asthma patients to keep a diary of air flow function. The flow-volume loop is especially useful in assessing such obstructive disorders because of the characteristic "scooped-out" shape of the expiratory part of the loop. In obstructive disorders, the inspiratory portion of the loop has a normal curve, but is smaller than normal.

QUICK CHECK

4. What is the difference between a pulmonary *volume* and a pulmonary *capacity?*
5. The volume of air that is expired after a normal inspiration during normal, quiet breathing goes by what name?
6. What is meant by the term *vital capacity?*
7. What is the *total minute volume?* How can it be calculated from a spirogram?

BOX 24-4   Types of Breathing

The alternate movement of air into and out of the lungs that we call breathing can occur in distinctive patterns that can be recognized and designated by name (see figure).

Eupnea is the term used to describe normal quiet breathing. During eupnea, the need for oxygen and carbon dioxide exchange is being met and the individual is not usually conscious of the breathing pattern. Ventilation occurs spontaneously at the rate of 12 to 17 breaths per minute.

Hyperpnea means increased breathing that is regulated to meet an increased demand by the body for oxygen. During hyperpnea, there is always an increase in pulmonary ventilation. The hyperpnea caused by exercise may meet the need for increased oxygen by an increase in tidal volume alone or by an increase in both tidal volume and breathing frequency.

Hyperventilation is characterized by an increase in pulmonary ventilation in excess of the need for oxygen. It sometimes results from a conscious voluntary effort preceding exertion or from psychogenic factors (hysterical hyperventilation). **Hypoventilation** is a decrease in pulmonary ventilation that results in elevated blood levels of carbon dioxide.

Dyspnea refers to labored or difficult breathing and is often associated with hypoventilation. A person suffering from dyspnea is aware, or conscious, of the breathing pattern and is generally uncomfortable and in distress. **Orthopnea** refers to dyspnea while lying down. It is relieved by sitting or standing up. This condition is common in patients with heart disease.

Several terms are used to describe the cessation of breathing. **Apnea** refers to the temporary cessation of breathing at the end of a normal expiration. It may occur during sleep or when swallowing. **Apneusis** is the cessation of breathing in the inspiratory position. Failure to resume breathing following a period of apnea, or apneusis, is called **respiratory arrest.**

Cheyne-Stokes respiration is a periodic type of abnormal breathing often seen in terminally ill or brain-damaged patients. It is characterized by cycles of gradually increasing tidal volume for several breaths followed by several breaths with gradually decreasing tidal volume. These cycles repeat in a type of crescendo-decrescendo pattern.

Biot's breathing is characterized by repeated sequences of deep gasps and apnea. This type of abnormal breathing pattern is seen in individuals suffering from increased intracranial pressure.

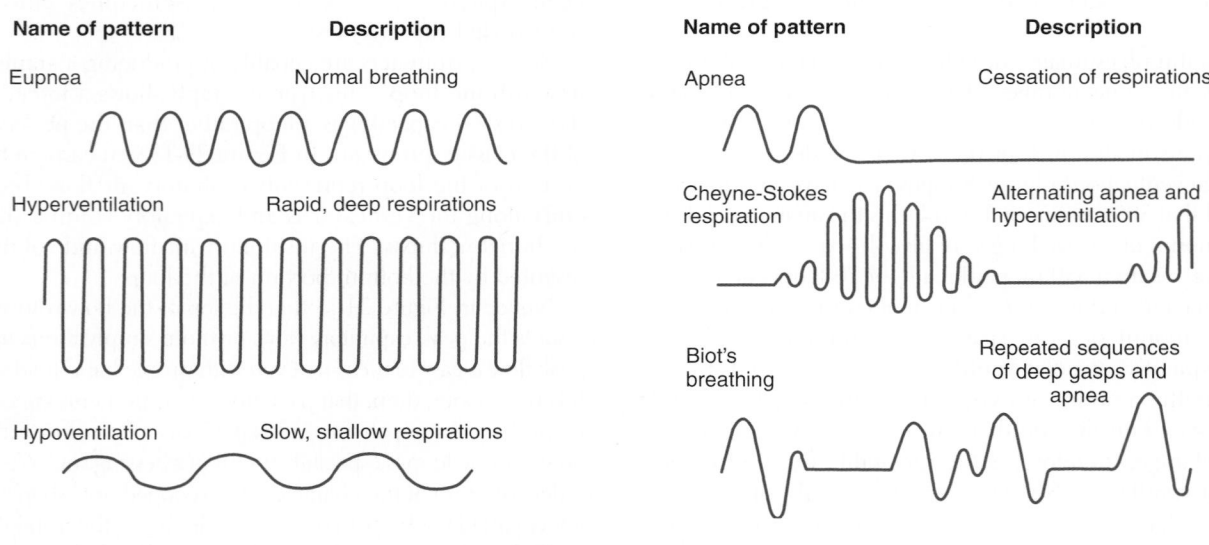

Examples of breathing patterns and spirograms.

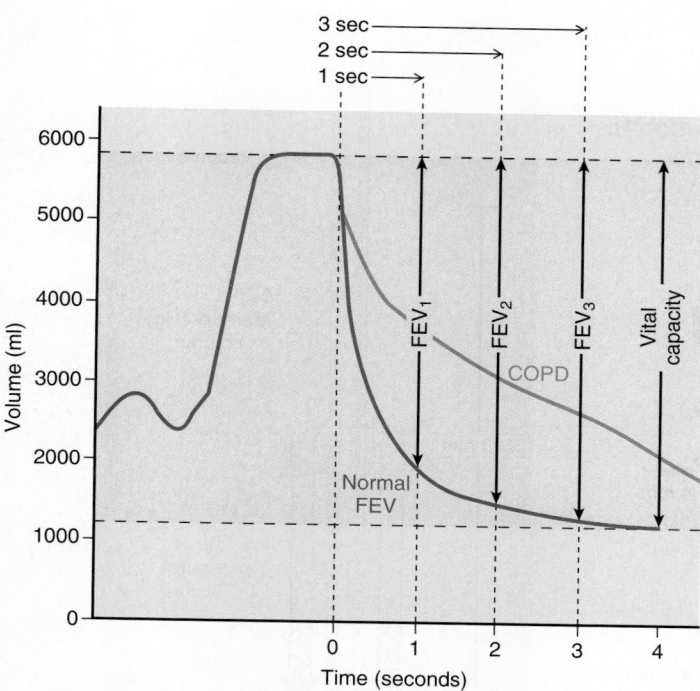

Figure 24-12 *Forced expiratory volume (FEV).* A normal individual forcefully exhales about 83% of the vital capacity (VC) during the first second, 94% at the end of 2 seconds, and 97% by the end of 3 seconds. The red line shows the results from a person with COPD (chronic obstructive pulmonary disease) who cannot forcefully exhale a large percentage of the vital capacity as quickly as a person without pulmonary obstruction.

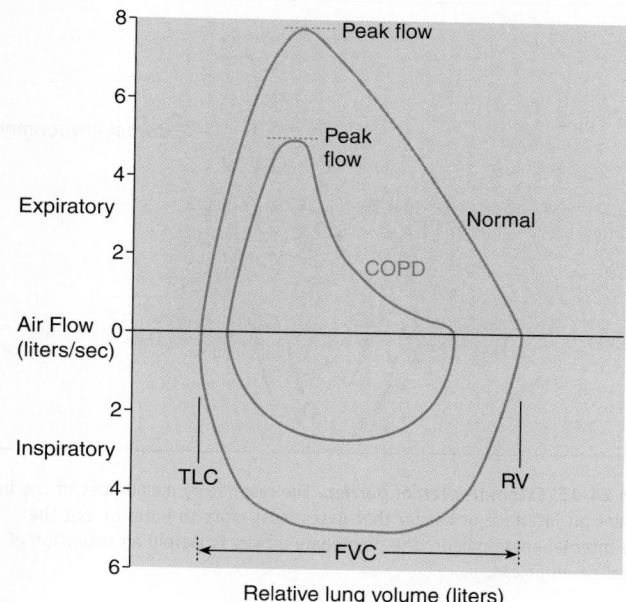

Figure 24-13 *Flow-volume loops.* The top of the loop represents expiratory flow (vertically) and volume (horizontally). The bottom of the loop represents inspiratory flow and volume. Notice that a person with COPD (chronic obstructive pulmonary disease) will produce a smaller loop with a "scooped-out" shape at the end of the expiratory curve.

PULMONARY GAS EXCHANGE
Partial Pressure

Before discussing the exchange of gases across the respiratory membranes, we need to understand the law of partial pressures (Dalton's law). The term **partial pressure** means the pressure exerted by any one gas in a mixture of gases or in a liquid. According to the law of partial pressures, the partial pressure of a gas in a mixture of gases is directly related to the concentration of that gas in the mixture and to the total pressure of the mixture. Figure 24-14 shows how each gas in atmospheric air contributes to the total atmospheric pressure. The partial pressure of each gas is directly related to its concentration in the total mixture. Suppose we apply this principle to compute the partial pressure of oxygen in the atmosphere. The concentration of oxygen in the atmosphere is about 21%, and the total pressure of the atmosphere is 760 mm Hg under standard conditions. Therefore

$$\text{Atmospheric } P_{O_2} = 21\% \times 760 = 159.6 \text{ mm Hg}$$

The symbol used to designate partial pressure is the capital letter P preceding the chemical symbol for the gas. Examples: alveolar air P_{O_2} is about 100 mm Hg, arterial blood P_{O_2} is also about 100 mm Hg, and venous blood P_{O_2} is about 37 mm Hg. The word **tension** is often used as a synonym for the term *partial pressure*—oxygen tension means the same thing as P_{O_2}.

The partial pressure of a gas in a liquid is directly determined by the amount of that gas dissolved in the liquid, which in turn is determined by the partial pressure of the gas in the environment of the liquid. Gas molecules diffuse into a liquid from its environment and dissolve in the liquid until the partial pressure of the gas in solution becomes equal to its partial pressure in the environment of the liquid. Alveolar air constitutes the environment surrounding blood moving through pulmonary capillaries. Standing between the blood and the air are only the very thin alveolar and capillary membranes, and both of these membranes are highly permeable to oxygen and carbon dioxide. By the time blood leaves the pulmonary capillaries as arterial blood, diffusion and approximate equilibration of oxygen and carbon dioxide across the membranes have occurred. Arterial blood P_{O_2} and P_{CO_2} therefore usually equal or very nearly equal alveolar P_{O_2} and P_{CO_2} (Table 24-3).

Exchange of Gases in the Lungs

Exchange of gases in the lungs takes place between alveolar air and blood flowing through lung capillaries. It is important to realize that physiologically speaking, air in the lung is not part of our body. That is, inspired air is not part of the internal environment. As Figure 24-15 shows, the airways are merely inward extensions of the external environment. Before oxygen can enter our internal environment, and before carbon dioxide can leave our internal environment, they must cross the barrier between the external world and the internal world.

Gases move in both directions through the respiratory membrane (see Figure 23-16). Oxygen enters blood from the alveolar air because the P_{O_2} of alveolar air is greater than the P_{O_2} of incoming blood. Another way of saying this is that oxygen diffuses "down" its pressure gradient. Simultaneously, carbon

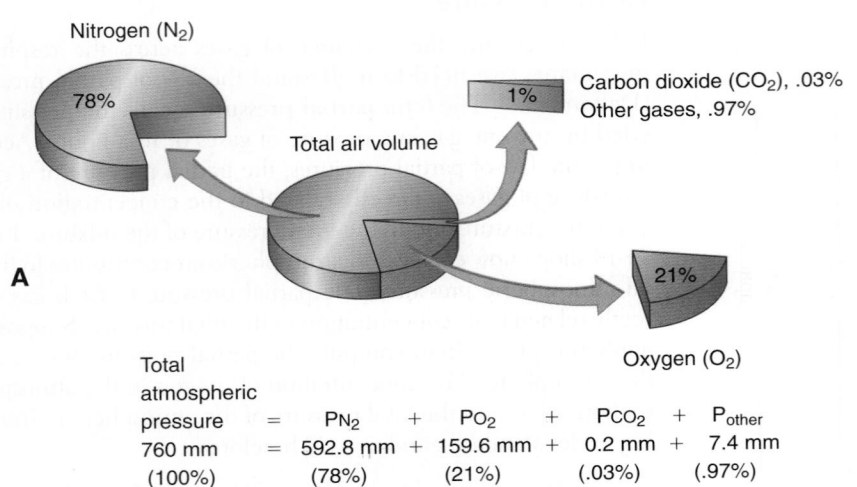

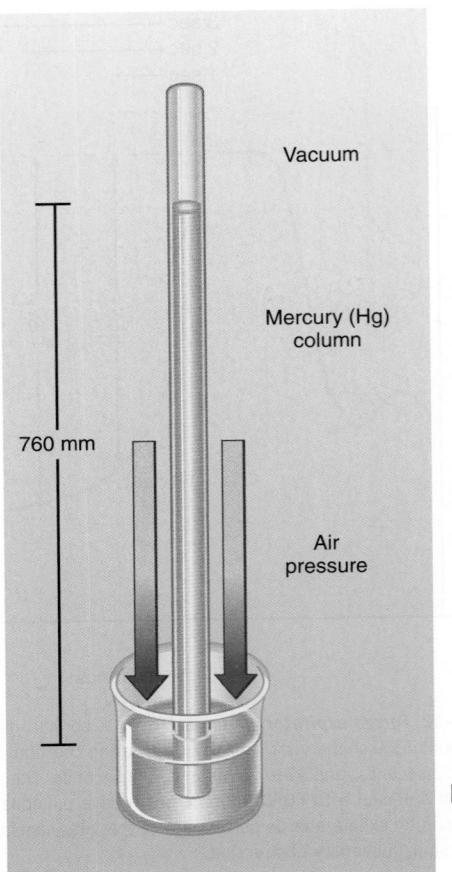

Figure 24-14 *Partial pressure of gases in atmospheric air.* **A,** Composition of dry atmospheric air under standard conditions showing the concentrations of nitrogen, oxygen, carbon dioxide, and other gases. **B,** A mercury barometer. The weight of air pressing down on the surface of the mercury in the open dish pushes the mercury down into the dish and up the tube. The greater the air pressure pushing down on the mercury surface, the farther up the tube the mercury will be forced. Under standard conditions, air pressure causes the mercury column to rise 760 mm. A proportion of this pressure is exerted by each of the gases that comprise air, according to their relative concentrations (see **A**). That is, the total atmospheric air pressure is the sum of the partial pressures of nitrogen, oxygen, carbon dioxide, water vapor, and other gases.

Table 24-3	Oxygen and Carbon Dioxide Pressure Gradients			
	ATMOSPHERE	ALVEOLAR AIR	SYSTEMIC ARTERIAL BLOOD	SYSTEMIC VENOUS BLOOD
P_{O_2}	160*	100	100	40
P_{CO_2}	0.2	40	40	46

*Figures indicate approximate mm Hg pressure under usual conditions.

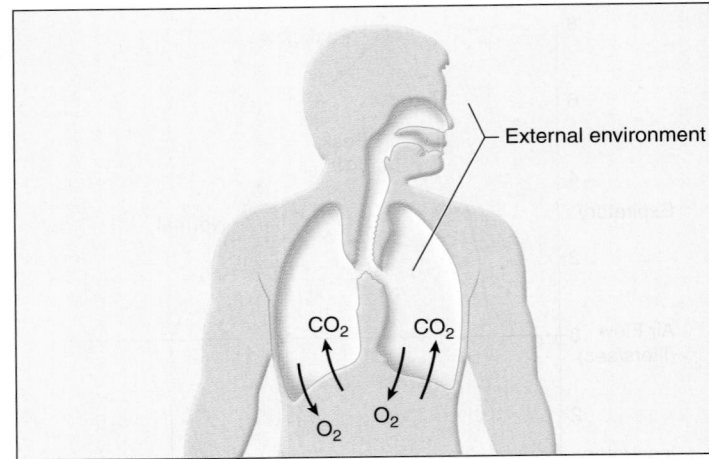

Figure 24-15 *External-internal barrier.* The respiratory membranes of the lung represent an interface or barrier that gases must cross to enter or exit the body's internal environment. The pulmonary airway is merely an extension of the external environment.

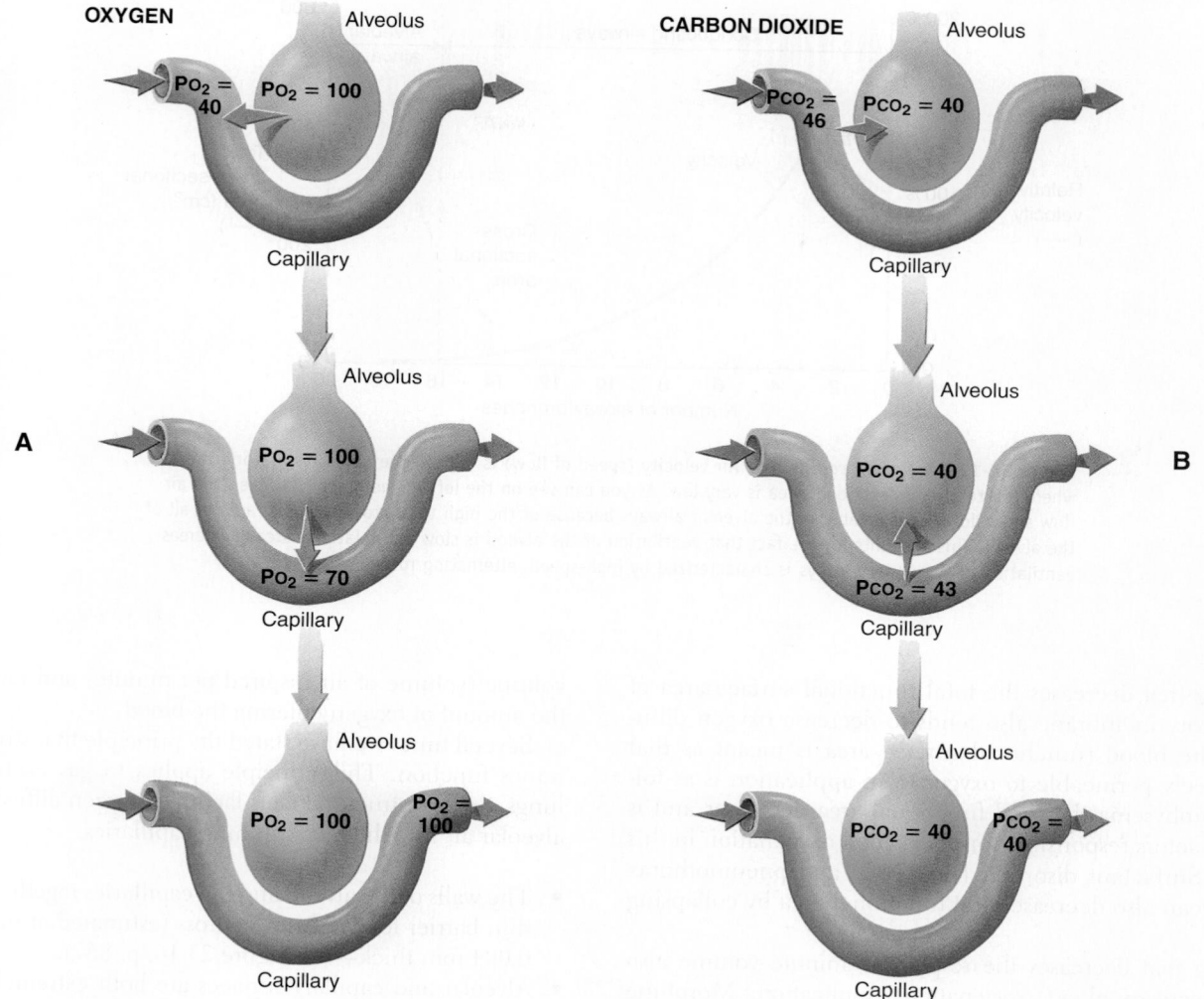

Figure 24-16 *Pulmonary gas exchange.* **A,** As blood enters a pulmonary capillary, oxygen diffuses down its pressure gradient (into the blood). Oxygen continues diffusing into the blood until equilibration has occurred (or until the blood leaves the capillary). **B,** As blood enters a pulmonary capillary, carbon dioxide diffuses down its pressure gradient (out of the blood). As with oxygen, carbon dioxide continues diffusing as long as there is a pressure gradient. P_{O_2} and P_{CO_2} remain relatively constant in a continually ventilated alveolus.

dioxide molecules exit from the blood by diffusing down the carbon dioxide pressure gradient out into the alveolar air. The P_{CO_2} of venous blood is much higher than the P_{CO_2} of alveolar air. This two-way exchange of gases between alveolar air and pulmonary blood converts deoxygenated blood to oxygenated blood (Figure 24-16).

When you look at Figure 24-16, you might wonder why the partial pressures of gases in the alveoli remain constant, whereas the partial pressures of gases in the blood change to equilibrate with alveolar partial pressures. The answer to this question lies in the fact that the alveoli are more or less continually ventilated. That is, there is always new air moving into the alveoli at a relative low, stable velocity (Figure 24-17). Therefore, the average partial pressures of gases in the alveoli as a group are relatively constant.

The amount of oxygen that diffuses into blood each minute depends on several factors, notably the following four:

1. The oxygen pressure gradient between alveolar air and incoming pulmonary blood (alveolar P_{O_2} − blood P_{O_2})
2. The total functional surface area of the respiratory membrane
3. The respiratory minute volume (respiratory rate per minute times volume of air inspired per respiration)
4. Alveolar ventilation (discussed on p. 896)

All four of these factors bear a direct relation to oxygen diffusion. Anything that decreases alveolar P_{O_2}, for instance, tends to decrease the alveolar-blood oxygen pressure gradient and therefore tends to decrease the amount of oxygen entering the blood. An application is as follows: Alveolar air P_{O_2} decreases as altitude increases, and thus less oxygen enters the blood at high altitudes. At a certain high altitude, alveolar air P_{O_2} equals the P_{O_2} of blood entering the pulmonary capillaries. How would this affect oxygen diffusion into blood?

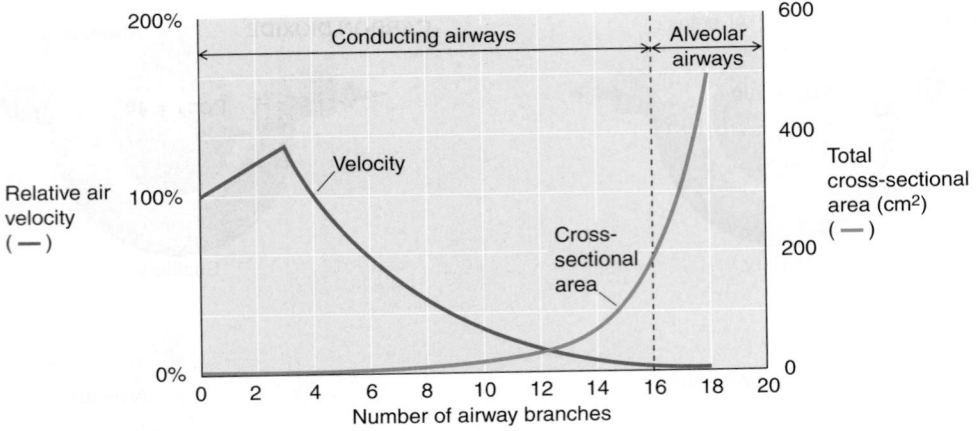

Figure 24-17 *Air flow in capillaries.* Air velocity (speed of flow) is high in the upper respiratory tract, where the total cross-sectional area is very low. As you can see on the left of the graph, however, the air flow slows down considerably in the alveolar airways because of the high total cross-sectional area of all of the alveoli. This accounts for the fact that ventilation of the alveoli is slow and relatively steady whereas ventilation of the upper airways is characterized by high-speed, alternating rushes of air.

Anything that decreases the total functional surface area of the respiratory membrane also tends to decrease oxygen diffusion into the blood (functional surface area is meant as that which is freely permeable to oxygen). An application is as follows: In emphysema the total functional area decreases and is one of the factors responsible for poor blood oxygenation in this condition. Surfactant disorders (Box 24-2) and pneumothorax (Box 24-3) can also decrease total functional area by collapsing alveoli.

Anything that decreases the respiratory minute volume also tends to decrease blood oxygenation. Application: Morphine slows respirations and therefore decreases the respiratory minute

volume (volume of air inspired per minute) and tends to lessen the amount of oxygen entering the blood.

Several times we have stated the principle that structure determines function. This principle applies to gas exchange in the lungs. Several structural facts facilitate oxygen diffusion from the alveolar air into the blood in lung capillaries:

- The walls of the alveoli and the capillaries together form a very thin barrier for the gases to cross (estimated at not more than 0.004 mm thick—see Figure 23-16, p. 867).
- Alveolar and capillary surfaces are both extremely large (See Box 24-5).

BOX 24-5 **Fick's Law**

Fick's law is a principle that describes the diffusion of carbon dioxide (CO_2) and oxygen (O_2) across the respiratory membrane, including the fluid film on the surface of the alveoli. As you can see in the figure, the principle illustrates common sense: each gas diffuses more efficiently (faster) if the surface area (A) is large, if the thickness of the membrane (t) is small, if the solubility of the gas (S) is high, and if the partial pressure (P_{O_2} or P_{CO_2}) gradient is high. Another way of stating Fick's law is that the net gas diffusion rate across a fluid membrane is proportional to the membrane surface area (A), solubility of the gas in the membrane (S), and partial pressure (P) difference—and inversely proportional to the membrane thickness (t).

The human respiratory system takes advantage of this principle by improving what it can in the equation to maximize the rate of gas diffusion. The body builds its respiratory membrane of material with as much solubility to CO_2 and O_2 as possible and as thin as possible. The large number of alveoli in a fractal-like arrangement ensure a very large surface area. And a high partial pressure gradient is maintained across the respiratory membrane.

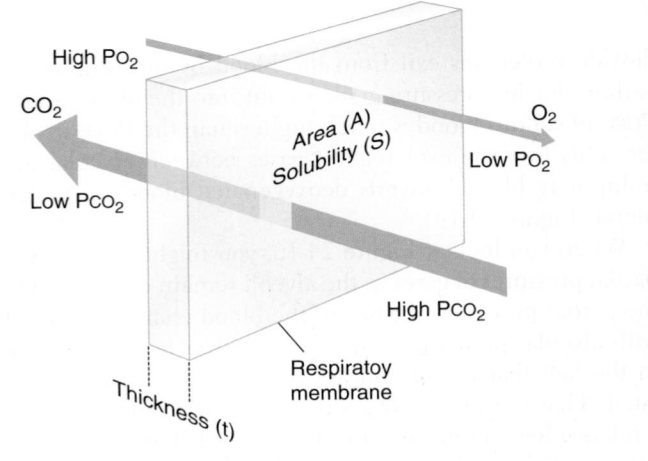

Rate of diffusion of a gas through a membrane. According to Fick's law, the membrane diffusion rate is affected by surface area (A), solubility (S) of the gas, membrane thickness (t), and the partial pressure (P) gradient.

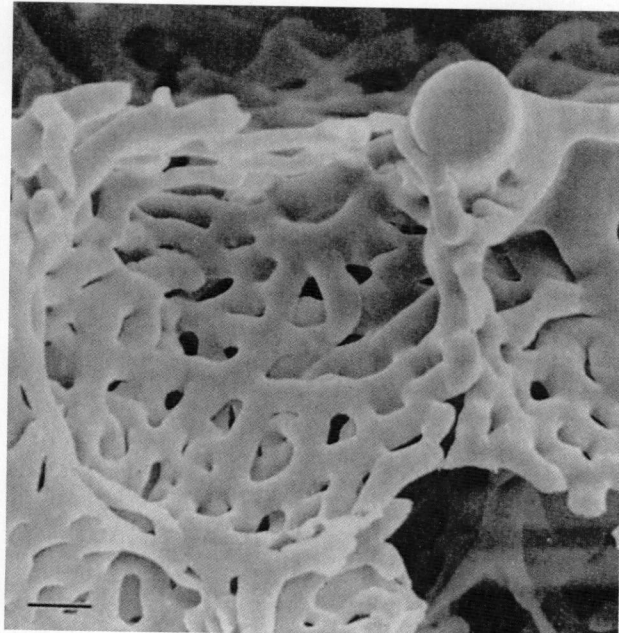

Figure 24-18 *Alveolar blood supply.* Scanning electron micrograph showing the rich blood supply to alveoli (which have been removed). The numerous, narrow branches ensure that each red blood cell is exposed to the alveolar air. (Black bar in lower left represents 10 μm.)

- Lung capillaries accommodate a large amount of blood at one time. The lung capillaries of a small individual—one who has a body surface area of 1.5 square meters—contain about 90 ml of blood at one time under resting conditions (Figure 24-18).
- Blood is distributed through the capillaries in a layer so thin (equal only to the diameter of one red blood cell) that each red blood cell comes close to alveolar air.

 QUICK CHECK

8. How does the partial pressure of a gas relate to its concentration?
9. What determines the direction in which oxygen will diffuse across the respiratory membrane?
10. List two of the four major factors that influence how much oxygen diffuses into pulmonary blood per minute.

HOW BLOOD TRANSPORTS GASES

Blood transports oxygen and carbon dioxide either as solutes or combined with other chemicals. Immediately on entering the blood, both oxygen and carbon dioxide dissolve in the plasma, but because fluids can hold only small amounts of gas in solution, most of the oxygen and carbon dioxide rapidly form a chemical union with some other molecule—such as hemoglobin, a plasma protein, or water. Once gas molecules are bound to another molecule, their plasma concentration decreases and more gas can diffuse into the plasma. In this way, comparatively large volumes of the gases can be transported.

Hemoglobin

Before we begin our discussion of transport of gases in the blood, we will pause and briefly review some facts about **hemoglobin (Hb)** (Figure 24-19). As we outlined in Chapter 17, hemoglobin is a reddish protein pigment found only inside red blood cells.

Hemoglobin is a quaternary protein made of four different polypeptide chains—two alpha chains and two beta chains—each associated with an iron-containing **heme group.** If you look at Figure 24-19, *B*, you will see that an oxygen molecule (O_2) can combine with the iron atom (Fe) in each heme group. Thus, hemoglobin can act as a kind of oxygen sponge that chemically absorbs oxygen molecules from the surrounding solution. Notice also in this figure that carbon dioxide (CO_2) molecules can combine with the amino acids of the alpha and beta polypeptide chains. Thus, hemoglobin can also act as a carbon dioxide sponge and absorb carbon dioxide molecules from a solution. Hemoglobin, then, has exactly the chemical characteristics needed to pick up and transport gases that enter the blood. As you will learn in the following paragraphs, hemoglobin also has the chemical characteristics needed to unload these gases as well. Box 24-6 explains how carbon monoxide interferes with hemoglobin's function.

Transport of Oxygen

Because oxygenated blood has a P_{O_2} of 100 mm Hg, it contains only about 0.3 ml of dissolved O_2 per 100 ml of blood. Many times that amount, however, combines with the hemoglobin in 100 ml of blood to form **oxyhemoglobin.** Because each gram of hemoglobin can unite with 1.34 ml of oxygen, the exact amount of oxygen in blood depends mainly on the amount of hemoglobin present. Normally, 100 ml of blood contains about 15 g of hemoglobin. If 100% of it combines with oxygen, 100 ml of blood will contain 15×1.34, or 20.1 ml, of oxygen in the form of oxyhemoglobin. Figure 24-20 shows how hemoglobin increases the oxygen-carrying capacity of blood.

Perhaps a more common way of expressing blood oxygen content is in terms of volume percent. Normal arterial blood, with a P_{O_2} of 100 mm Hg, contains about 20 vol% O_2 (meaning 20 ml of oxygen in each 100 ml of blood).

Blood that contains more hemoglobin can, of course, transport more oxygen. Thus, blood that contains less hemoglobin can transport less oxygen. Therefore hemoglobin deficiency anemia decreases oxygen transport and may produce marked cellular hypoxia (inadequate oxygen supply).

To combine with hemoglobin, oxygen must, of course, diffuse from plasma into the red blood cells where millions of hemoglobin molecules are located. Several factors influence the rate at which hemoglobin combines with oxygen in lung capillaries. For instance, as the following equation and the **oxygen-hemoglobin dissociation curve** (Figure 24-21) show, an increasing blood P_{O_2} accelerates hemoglobin association with oxygen:

$$Hb + O_2 \xrightarrow{\text{Increasing } P_{O_2}} HbO_2$$

Decreasing P_{O_2}, on the other hand, accelerates oxygen dissociation from oxyhemoglobin, that is, the reverse of the preceding equation. Oxygen associates with hemoglobin rapidly—so rapidly, in fact, that about 97% of the blood's hemoglobin has

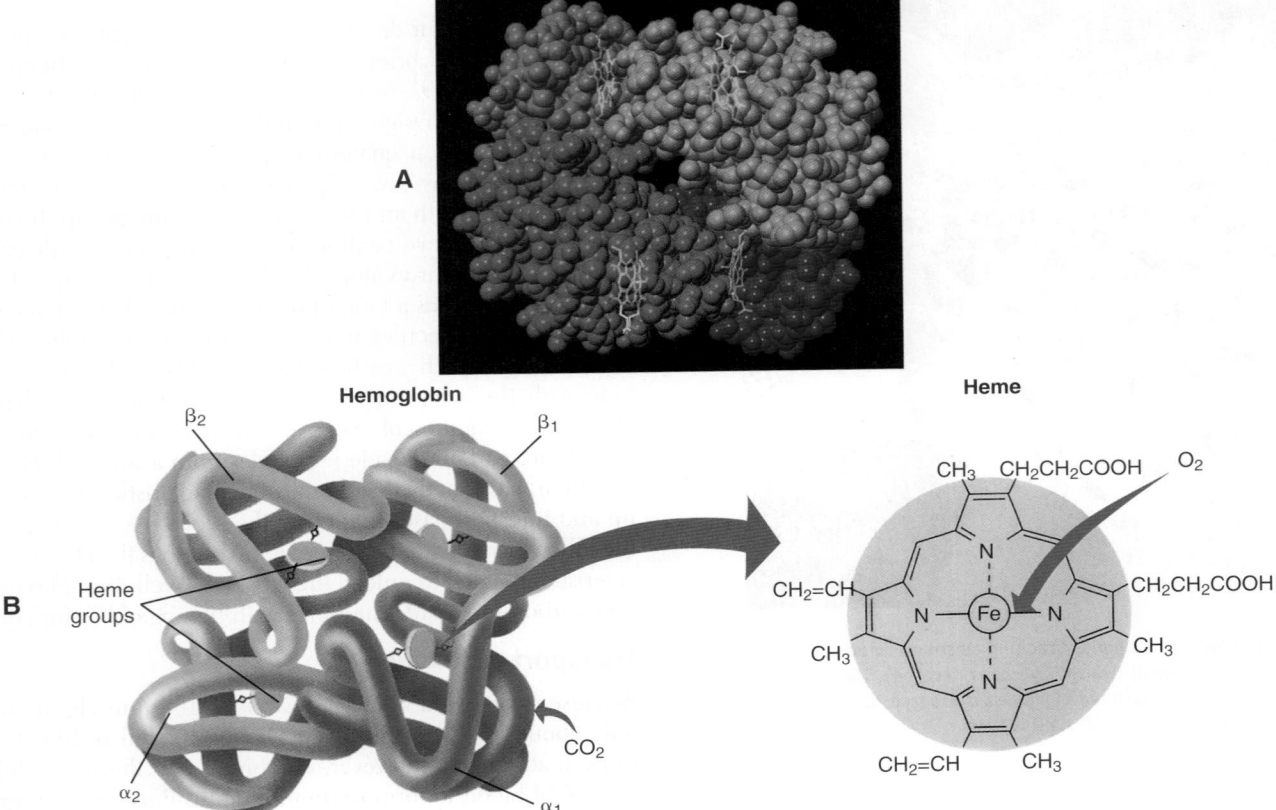

Figure 24-19 *Hemoglobin.* **A,** A space-filling model of the hemoglobin molecule shows four globular proteins fitted together into a thick, doughnut-like structure. **B,** Sketch showing that hemoglobin is a quaternary protein consisting of four different tertiary (folded) polypeptide chains—two alpha (α) chains and two beta (β) chains. Each chain has an associated iron-containing heme group, as seen in detail in the inset. Oxygen (O_2) can bind to the iron (Fe) of the heme group, or carbon dioxide (CO_2) can bind to amine groups of the amino acids in the polypeptide chains.

Carbon Monoxide Poisoning

Gases other than O_2 and CO_2 can bind to the hemoglobin (Hb) molecule. **Carbon monoxide (CO)** is a molecule produced by incomplete combustion in furnaces, engines, and other circumstances. This invisible, odorless gas binds to Hb more than 200 times more strongly than O_2 does. That means that CO "knocks out" O_2 from HbO_2 and forms HbCO. As more and more HbCO is formed, less and less oxygen is being carried by your blood—a life-threatening situation. Because CO binds so strongly, it is hard to remove it from Hb. One strategy to remove CO is to place a person in a pressure chamber where the PO_2 can be driven so high that it "knocks off" the CO from the Hb, allowing O_2 to form HbO_2.

united with oxygen by the time the blood leaves the lung capillaries to return to the heart. In other words, the average oxygen saturation of hemoglobin in oxygenated blood is about 97%.

Summing up, we can say that oxygen travels in two forms: as dissolved O_2 in the plasma and associated with hemoglobin (oxyhemoglobin). Of these two forms of transport, oxyhemoglobin carries the vast majority of the total oxygen transported by the blood. Box 24-7 lists several other oxygen-binding proteins similar to hemoglobin.

Transport of Carbon Dioxide

Carbon dioxide is carried in the blood in several ways, the most important of which are described briefly in the following paragraphs.

Dissolved Carbon Dioxide

A small amount of CO_2 dissolves in plasma and is transported as a solute. About 10% of the total amount of carbon dioxide carried by the blood is carried in the dissolved form. It is this dissolved CO_2 that produces the PCO_2 of blood plasma.

Carbamino Compounds

One fifth to one quarter of the carbon dioxide in blood unites with the NH_2 (amine) groups of the amino acids that make up the polypeptide chains of hemoglobin and various plasma proteins. When carbon dioxide binds to amine groups, it forms *carbamino compounds*. Because hemoglobin is the main protein to combine

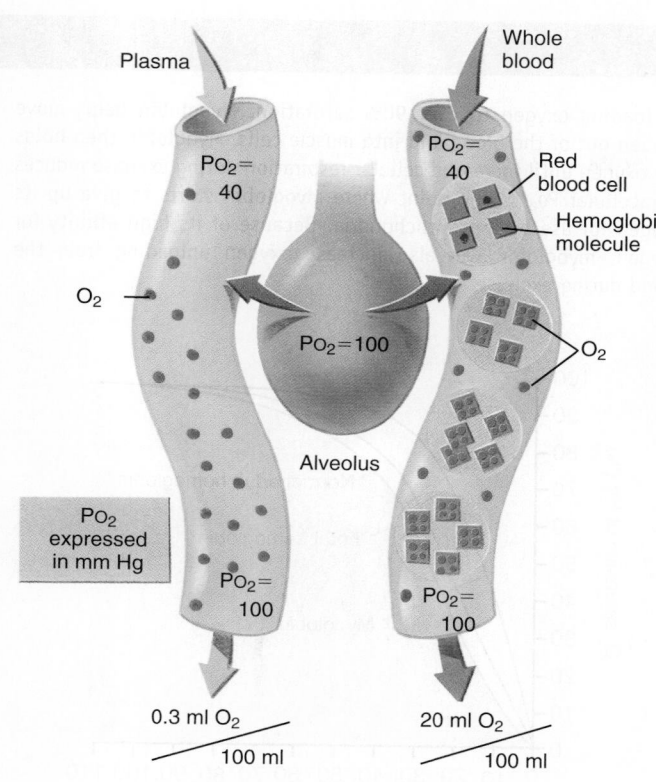

Figure 24-20 *Oxygen-carrying capacity of blood.* If blood consisted only of plasma, the maximum oxygen that could be transported is only about 0.3 ml of O_2 per 100 ml of blood. Because the red blood cells contain hemoglobin molecules, which act as "oxygen sponges," the blood can actually carry up to 20 ml of dissolved O_2 per 100 ml of blood.

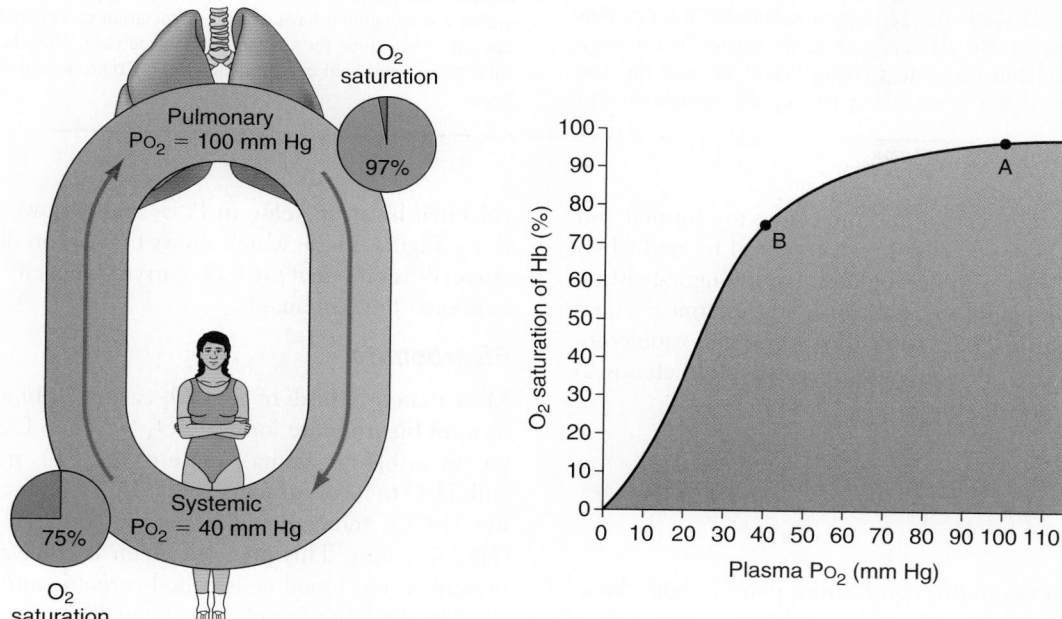

Figure 24-21 *Oxygen-hemoglobin dissociation curve.* The graph represents the relationship between P_{O_2} and O_2 saturation of hemoglobin (Hb-O_2 affinity). The inset shows how the graphed curve relates to oxygen transport by the blood. Notice that at high plasma P_{O_2} values (point *A*), hemoglobin (Hb) is fully loaded with oxygen. At low plasma P_{O_2} values (point *B*), Hb is only partially loaded with oxygen.

BOX 24-7 Oxygen-Binding Proteins

Normal adult hemoglobin (Hb) is obviously an important oxygen-binding protein. Found in great quantity (15 mg Hb/100 ml blood) in red blood cells, hemoglobin acts as an oxygen carrier that picks up oxygen in the pulmonary capillaries and releases oxygen in the systemic capillaries. However, other important oxygen-binding proteins are also important in the bigger picture of human physiology.

One such oxygen-binding protein is **fetal hemoglobin.** Fetal hemoglobin is structurally similar to normal adult hemoglobin, the major difference being that fetal hemoglobin is made up of two alpha chains and two gamma chains rather than two alpha chains and two beta chains (see Figure 24-19). Functionally, fetal hemoglobin is quite different from adult hemoglobin. As the figure indicates, the oxygen–fetal hemoglobin dissociation curve is shifted slightly to the left in comparison to the adult oxygen-hemoglobin dissociation curve. This means that in the placenta (see Figure 18-30, p. 713), fetal hemoglobin can load up with more oxygen at a low partial pressure than adult hemoglobin can. Because the P_{O_2} of maternal blood entering the placenta is only about 32 mm Hg, the high affinity of fetal hemoglobin for oxygen is vital for survival of the fetus.

Another important oxygen-binding protein is **myoglobin.** Recall from Chapter 11 that myoglobin is a protein found in muscle cells. Unlike hemoglobin, which is made up of four polypeptide chains, myoglobin is made up of only one such folded chain—a single alpha chain. Having only one heme group, it can hold only one oxygen molecule. Notice in the figure that the oxygen dissociation curve for myoglobin is shifted far to the left of that for hemoglobin and is also hyperbolic rather than sigmoid (S-shaped). The oxygen-myoglobin dissociation curve shows that myoglobin attracts and holds oxygen much more strongly than hemoglobin does. At a tissue P_{O_2} of 40 mm Hg, the point at which hemoglobin is unloading its oxygen, myoglobin can

be loading oxygen to over 90% saturation. Myoglobin helps move oxygen out of the blood and into muscle cells. Myoglobin then holds the oxygen until increased cellular respiration during exercise reduces intracellular P_{O_2} to the point where myoglobin starts to give up its oxygen for use by the mitochondria. Because of its high affinity for oxygen, myoglobin can also increase oxygen unloading from the blood during exercise.

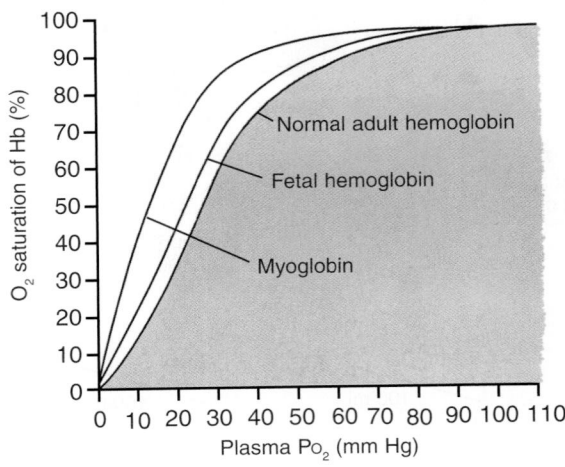

Oxygen dissociation curve for oxygen-binding proteins. Both fetal hemoglobin and myoglobin have oxygen dissociation curves that are shifted to the left of the curve for normal adult hemoglobin. Thus, both of the molecules attract and hold oxygen more strongly than normal adult hemoglobin does.

---

with carbon dioxide, most carbamino molecules are formed and transported in the red blood cells. The compound formed when carbon dioxide combines with hemoglobin has a tongue-twisting name—**carbaminohemoglobin.** The following chemical equation, amplified in Figure 24-22, shows how carbon dioxide combines with amine (NH_2) in hemoglobin's polypeptide chains to produce carbaminohemoglobin ($HbNCOOH^-$) and H^+:

$$Hb-N\begin{smallmatrix}H\\ \\H\end{smallmatrix} + CO_2 \rightleftharpoons Hb-N\begin{smallmatrix}H\\ \\COO^-\end{smallmatrix} + H^+$$

Notice that the arrows in this equilibrium point in both directions. This means that under any given conditions, some carbon dioxide will be associated with hemoglobin and some will not—the reaction is moving in both directions at the same time. The rate of both forward and reverse reactions—carbon dioxide association with and dissociation from hemoglobin—can shift with changes in carbon dioxide concentration. This principle is sometimes called the **rate law** of chemistry. The addition of more carbon dioxide to blood, therefore, will increase the rate of formation of carbaminohemoglobin. Another way to state this principle is to say that the association of carbon dioxide with hemoglobin is ac-

celerated by an increase in P_{CO_2} and is slowed by a decrease in P_{CO_2}. Figure 24-23, which shows the carbon dioxide dissociation curve, illustrates that the CO_2-carrying capacity of blood increases as plasma P_{CO_2} increases.

Bicarbonate

More than two thirds of the CO_2 carried by blood is carried in the form of **bicarbonate** ions (HCO_3^-). When CO_2 dissolves in water (as in blood plasma), some of the CO_2 molecules associate with H_2O to form carbonic acid (H_2CO_3). Once formed, some of the H_2CO_3 molecules dissociate to form H^+ and bicarbonate (HCO_3^-) ions. This process, which is catalyzed by an enzyme present in red blood cells called *carbonic anhydrase*, is summarized by the following chemical equation:

$$CO_2 + H_2O \rightleftharpoons H_2CO_3 \rightleftharpoons H^+ + HCO_3^-$$

Figure 24-24 amplifies this equation. According to the rate law of chemistry we stated earlier, as more CO_2 is added to the plasma, more will be converted to carbonic acid. Because the carbonic anhydrase enzyme in the blood is facilitating the conversion of carbon dioxide and water to carbonic acid, this reaction occurs very rapidly as CO_2 is added to the plasma. Carbonic acid concentration increases as a result, "pulling" the system toward the bicar-

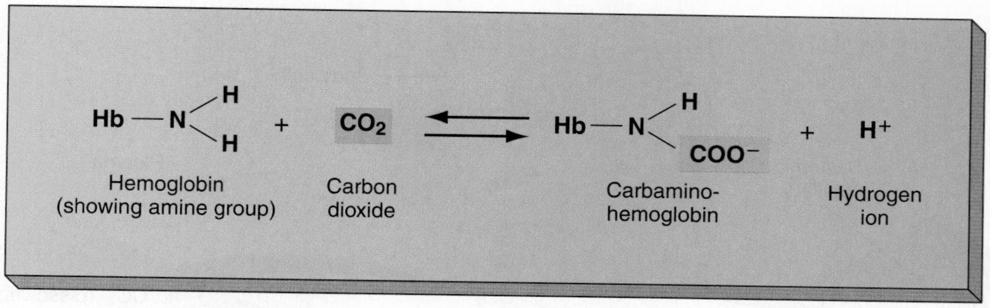

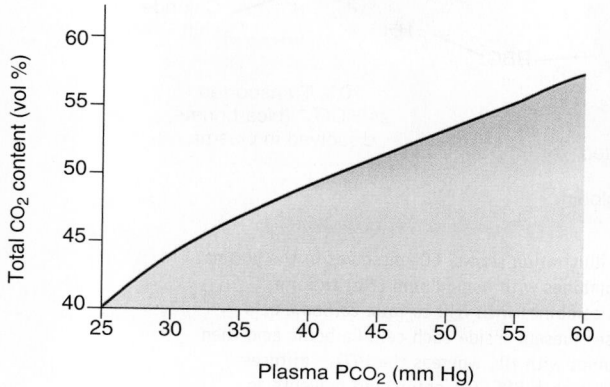

Figure 24-22 *Carbon dioxide–hemoglobin reaction.* Carbon dioxide can bind to an amine group (NH_2) in an amino acid within a hemoglobin (Hb) molecule to form carbaminohemoglobin ($HbNCOOH^-$) and a hydrogen ion. The highlighted areas show where the original carbon dioxide molecule is in each part of the equation.

Figure 24-23 *Carbon dioxide dissociation curve.* The relationship between P_{CO_2} and total CO_2 content (vol%) is graphed as a nearly straight line. Notice that the CO_2-carrying capacity of blood increases as the plasma P_{CO_2} increases.

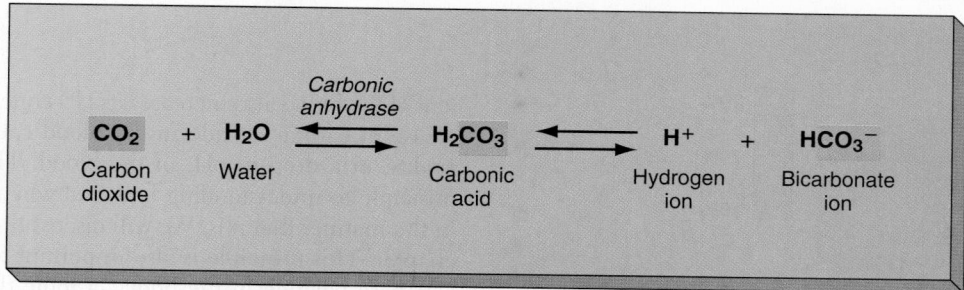

Figure 24-24 *Formation of bicarbonate.* Carbon dioxide can react with water to form carbonic acid, a reaction catalyzed by the RBC enzyme carbonic anhydrase. Carbonic acid then dissociates to form bicarbonate and a hydrogen ion. The highlighted areas show where the original carbon dioxide molecule is in each part of the equation. The double arrows show that each reaction is reversible, the actual rate in each direction governed by the relative concentration of each molecule.

bonate side, thus increasing the rate of bicarbonate formation. The end result is that CO_2 molecules diffusing into plasma will continually be removed from the solution and converted into bicarbonate. This allows room for even more CO_2 to dissolve in the plasma—thus increasing the CO_2-carrying capacity of the blood.

Figure 24-25, which summarizes all three forms of CO_2 transport, shows that once bicarbonate ions are formed, they diffuse down their concentration gradient into the plasma. The exit of this negative ion (HCO_3^-) from the red blood cell is balanced by the inward transport of another negative ion, chloride (Cl^-). This countertransport of negative ions is often called the **chloride shift**.

According to the rate law of chemistry described earlier, when CO_2 is removed from the plasma the entire system, illustrated in

Figures 24-24 and 24-25, shifts in the opposite direction. Thus, the reaction that converts carbonic acid to free CO_2 becomes dominant. The declining concentration of carbonic acid then forces a shift in favor of the conversion of bicarbonate to carbonic acid. In short, CO_2 is unloaded from bicarbonate.

The relative proportions of the three different forms of carbon dioxide carried in the blood are summarized in Figure 24-26.

Carbon Dioxide and pH

You may have noticed by now that when carbon dioxide enters the blood, most of it is converted to carbaminohemoglobin and hydrogen ions (H^+) or to bicarbonate and hydrogen ions. In other words, have you noticed that increasing the carbon dioxide con-

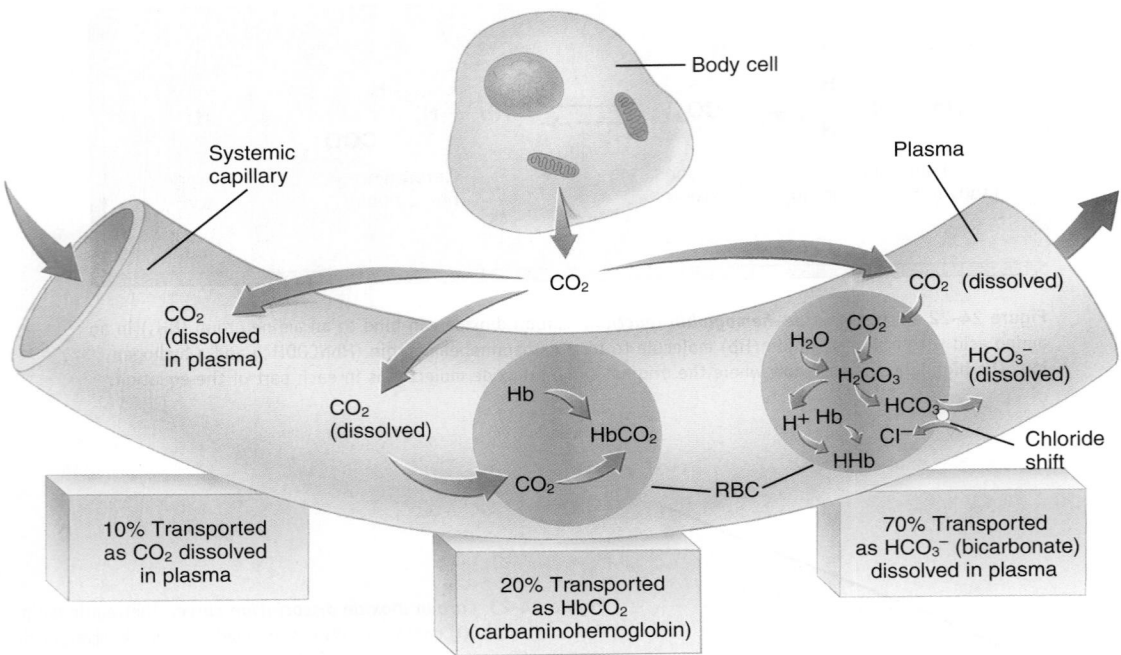

Figure 24-25 *Carbon dioxide transport in the blood.* As the illustration shows, CO_2 dissolves in the plasma. Some of the dissolved CO_2 enters red blood cells (RBCs) and combines with hemoglobin (Hb) to form carbaminohemoglobin ($HbCO_2$). Some of the CO_2 entering RBCs combines with H_2O to form carbonic acid (H_2CO_3), a process facilitated by an enzyme (carbonic anhydrase) present inside each cell. Carbonic acid then dissociates to form H^+ and bicarbonate (HCO_3^-). The H^+ combines with Hb, whereas the HCO_3^- diffuses down its concentration gradient into the plasma. As HCO_3^- leaves each RBC, Cl^- enters and prevents an imbalance in charge—a phenomenon called the *chloride shift*, which is discussed in Chapter 30.

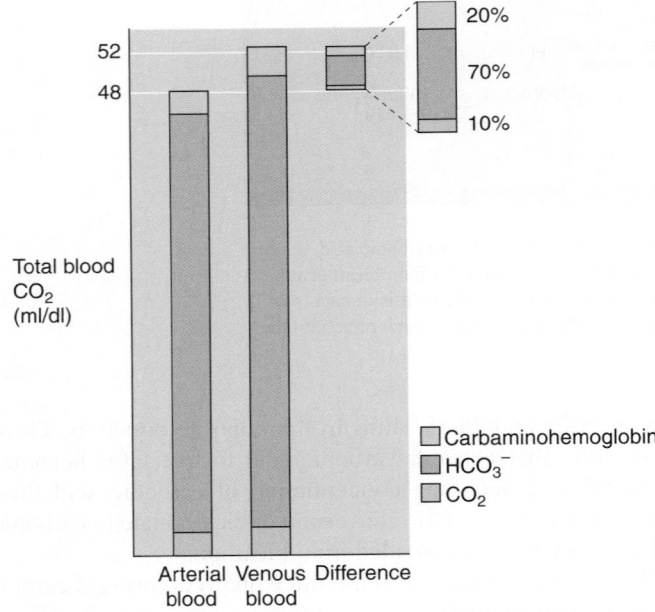

Figure 24-26 *Proportions of carbon dioxide transported in the blood.* This graph shows that systemic venous blood carries more carbon dioxide than systemic arterial blood does. The difference, shown in the upper left, represents the total amount of carbon dioxide loaded into the blood in the systemic tissues. Or it could be viewed as the total amount of carbon dioxide unloaded from the blood in the lungs. Note that most of the carbon dioxide is carried in the form of HCO_3^- (bicarbonate).

tent of the blood also increases its H^+ concentration? Thus, an increase in carbon dioxide in the blood causes an increase in the acidity, or a drop in pH, in the blood. This is a very important principle to understanding how and why respiration is regulated in the manner that it is. We will discuss these matters later in this chapter. This principle is also important to the understanding of acid-base balance in the body—a topic discussed thoroughly in Chapter 30.

QUICK CHECK

11. Most oxygen carried by the blood is transported in what form?
12. Most carbon dioxide carried by the blood is transported in what form?
13. What is oxyhemoglobin? What is carbaminohemoglobin?
14. How does carbon dioxide affect the pH of blood?

SYSTEMIC GAS EXCHANGE

Exchange of gases in tissues takes place between arterial blood flowing through tissue capillaries and cells (Figure 24-27). It occurs because of the principle already noted—that gases move down a gas pressure gradient. More specifically, in the tissue cap-

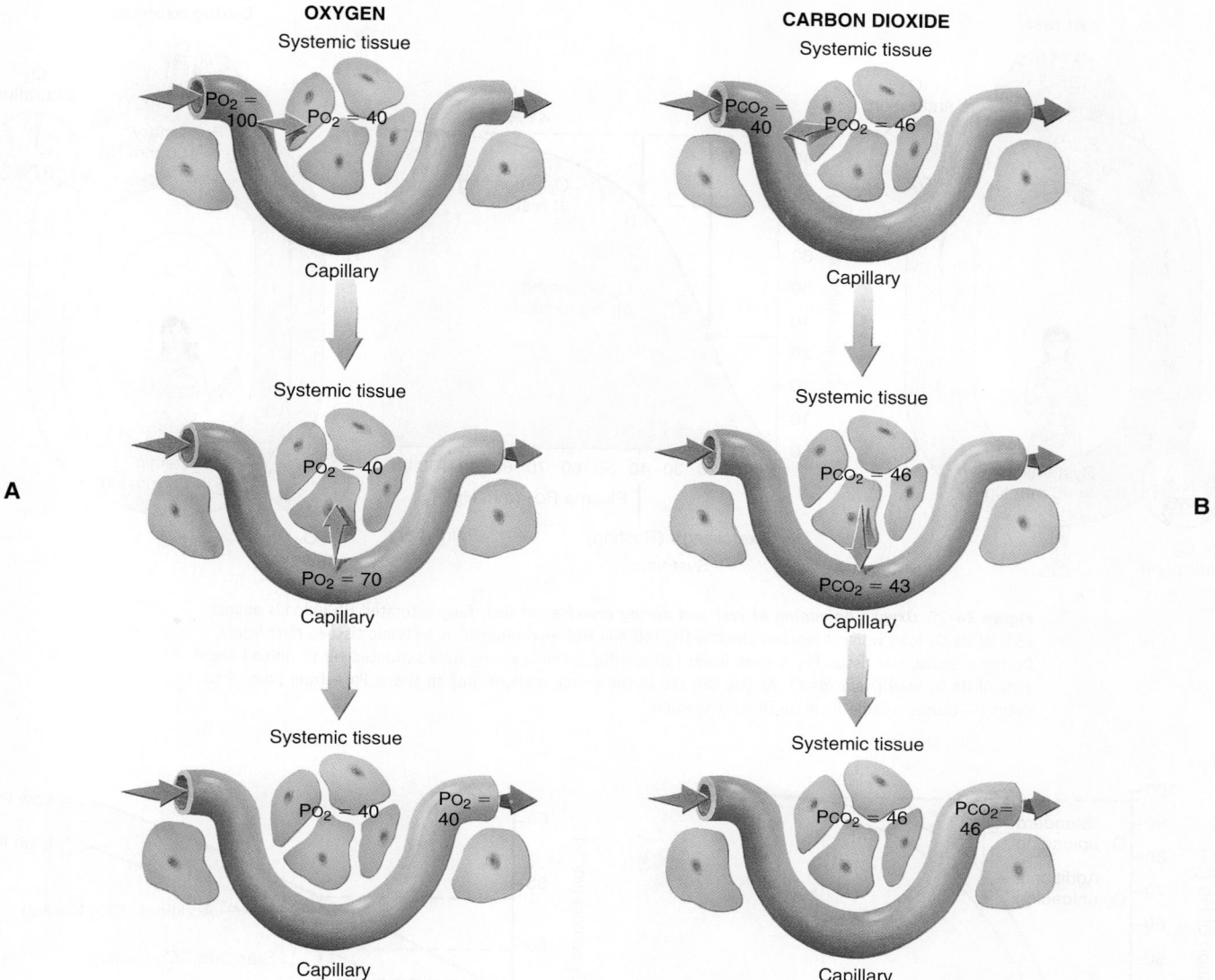

OXYGEN

Systemic tissue

$PO_2 = 100$ $PO_2 = 40$

Capillary

Systemic tissue

$PO_2 = 40$

$PO_2 = 70$

Capillary

Systemic tissue

$PO_2 = 40$ $PO_2 = 40$

Capillary

CARBON DIOXIDE

Systemic tissue

$PCO_2 = 40$ $PCO_2 = 46$

Capillary

Systemic tissue

$PCO_2 = 46$

$PCO_2 = 43$

Capillary

Systemic tissue

$PCO_2 = 46$ $PCO_2 = 46$

Capillary

A

B

Figure 24-27 *Systemic gas exchange.* A, As blood enters a systemic capillary, O_2 diffuses down its pressure gradient (out of the blood). O_2 continues diffusing out of the blood until equilibration has occurred (or until the blood leaves the capillary). **B,** As blood enters a systemic capillary, CO_2 diffuses down its pressure gradient (into the blood). As with O_2, CO_2 continues diffusing as long as there is a pressure gradient.

illaries, oxygen diffuses out of arterial blood because the oxygen pressure gradient favors its outward diffusion (see Figure 24-27). Arterial blood PO_2 is about 100 mm Hg, interstitial fluid PO_2 is considerably lower, and intracellular fluid PO_2 is still lower. Although interstitial fluid and intracellular fluid PO_2 are not definitely established, they are thought to vary considerably—perhaps from around 60 mm Hg down to about 1 mm Hg.

As activity increases in any tissue, its cells necessarily use oxygen more rapidly. This decreases intracellular and interstitial PO_2, which in turn tends to increase the oxygen pressure gradient between blood and tissues and to accelerate oxygen diffusion out of the tissue capillaries. In this way, the rate of oxygen use by cells automatically tends to regulate the rate of oxygen delivery to cells. As dissolved oxygen diffuses out of arterial blood, blood PO_2 decreases, and this accelerates oxyhemoglobin dissociation to re-

lease more oxygen into the plasma for diffusion out to cells, as indicated in Figure 24-28 and the following equation:

$$Hb + O_2 \xrightarrow{\text{Increasing } PO_2} HbO_2$$

Because of oxygen release to tissues from tissue capillary blood, PO_2, oxygen saturation, and total oxygen content are less in venous blood than in arterial blood, as shown in Table 24-4. Carbon dioxide exchange between tissues and blood takes place in the opposite direction from oxygen exchange. Catabolism produces large amounts of CO_2 inside cells. Therefore, intracellular and interstitial PCO_2 are higher than arterial blood PCO_2. This means that the CO_2 pressure gradient causes diffusion of CO_2 from the tissues into the blood flowing along through tissue capillaries (see Figure 24-27). Consequently, the

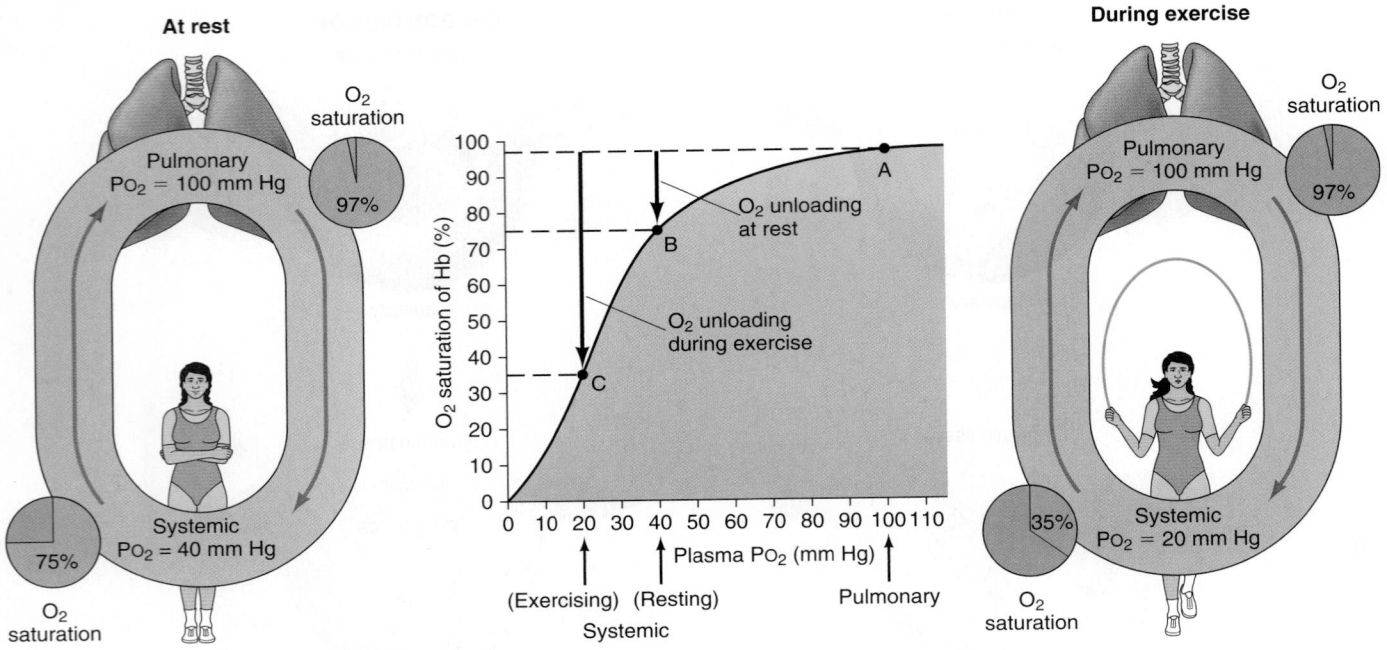

Figure 24-28 *Oxygen unloading at rest and during exercise.* At rest, fully saturated Hb unloads almost 25% of its O_2 load when it reaches the low-P_{O_2} (40 mm Hg) environment in systemic tissues *(left inset)*. During exercise, the tissue P_{O_2} is even lower (20 mm Hg)—thus causing fully saturated Hb to unload about 70% of its O_2 load *(right inset)*. As you can see in the graph, a slight drop in tissue P_{O_2}—from point *B* to point *C*—causes a large increase in O_2 unloading.

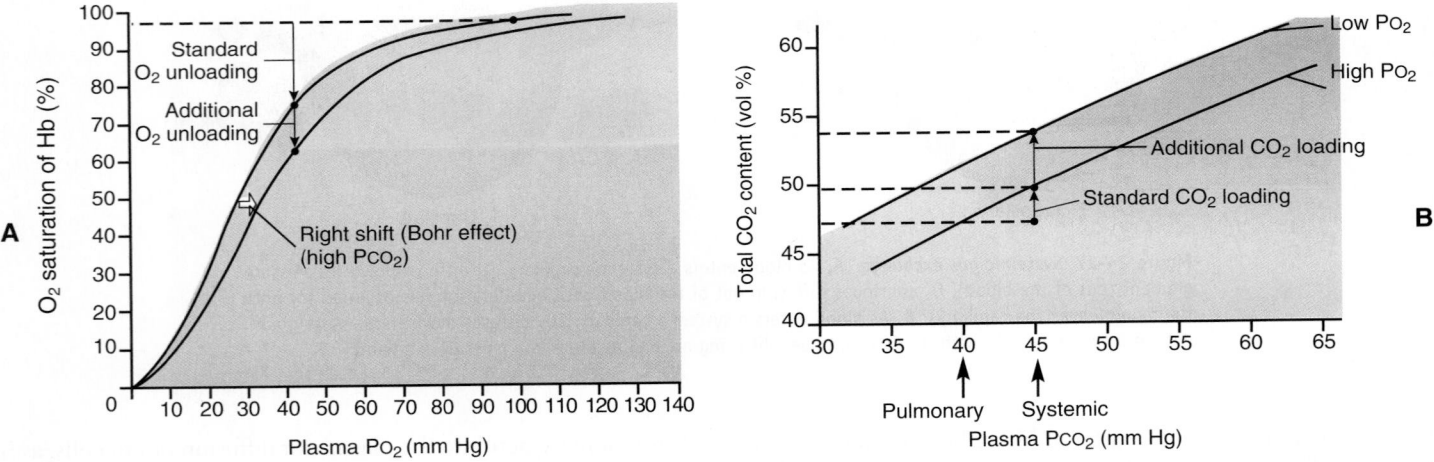

Figure 24-29 *Interaction of P_{O_2} and P_{CO_2} on gas transport by the blood.* **A,** The increased P_{CO_2} in systemic tissues decreases the affinity between Hb and O_2, shown as a right shift of the oxygen-hemoglobin dissociation curve. This phenomenon is known as the *Bohr effect*. A right shift can also be caused by a decrease in plasma pH. **B,** At the same time, the decreased P_{O_2} commonly observed in systemic tissues increases the CO_2 content of the blood, shown as a left shift of the CO_2 dissociation curve. This phenomenon is known as the *Haldane effect.*

P_{CO_2} of blood increases in tissue capillaries from its arterial level of about 40 mm Hg to its venous level of about 46 mm Hg. This increasing P_{CO_2} and decreasing P_{O_2} together produce two effects—they favor oxygen dissociation from oxyhemoglobin and carbon dioxide association with hemoglobin to form carbaminohemoglobin. This reciprocal interrelationship between oxygen and carbon dioxide transport mechanisms is contrasted in Figure 24-29. Note that increased P_{CO_2} decreases the affinity between hemoglobin and oxygen—this is called a "right shift." A right shift of the oxygen-hemoglobin dissociation curve resulting from increased P_{CO_2} is also known as the **Bohr effect,** named for Christian Bohr, who along with other scientists, dis-

Table 24-4	Blood Oxygen	
	SYSTEMIC VENOUS BLOOD	**SYSTEMIC ARTERIAL BLOOD**
P_{O_2}	40 mm Hg	100 mm Hg
Oxygen saturation	75%	97%
Oxygen content	15 ml O_2 per 100 ml blood	20 ml O_2 per 100 ml blood*

*Oxygen use by tissues = difference between the oxygen content of arterial and venous blood (20 − 15) = 5 ml O_2 per 100 ml blood circulated per minute.

covered this phenomenon in 1904. A drop in plasma pH, which normally accompanies an increase in blood P_{CO_2}, also causes a right shift. The **Haldane effect** refers to the increased CO_2 loading caused by a decrease in P_{O_2}. This phenomenon is named for its discoverer John Scott Haldane.

REGULATION OF PULMONARY FUNCTION
Respiratory Control Centers

Various mechanisms operate to maintain relative constancy of the blood P_{O_2} and P_{CO_2}. This homeostasis of blood gases is maintained primarily by means of changes in **ventilation**—the rate and depth of breathing. The main integrators that control the nerves that affect the inspiratory and expiratory muscles are located within the brainstem and are together simply called the *respiratory centers* (Figure 24-30).

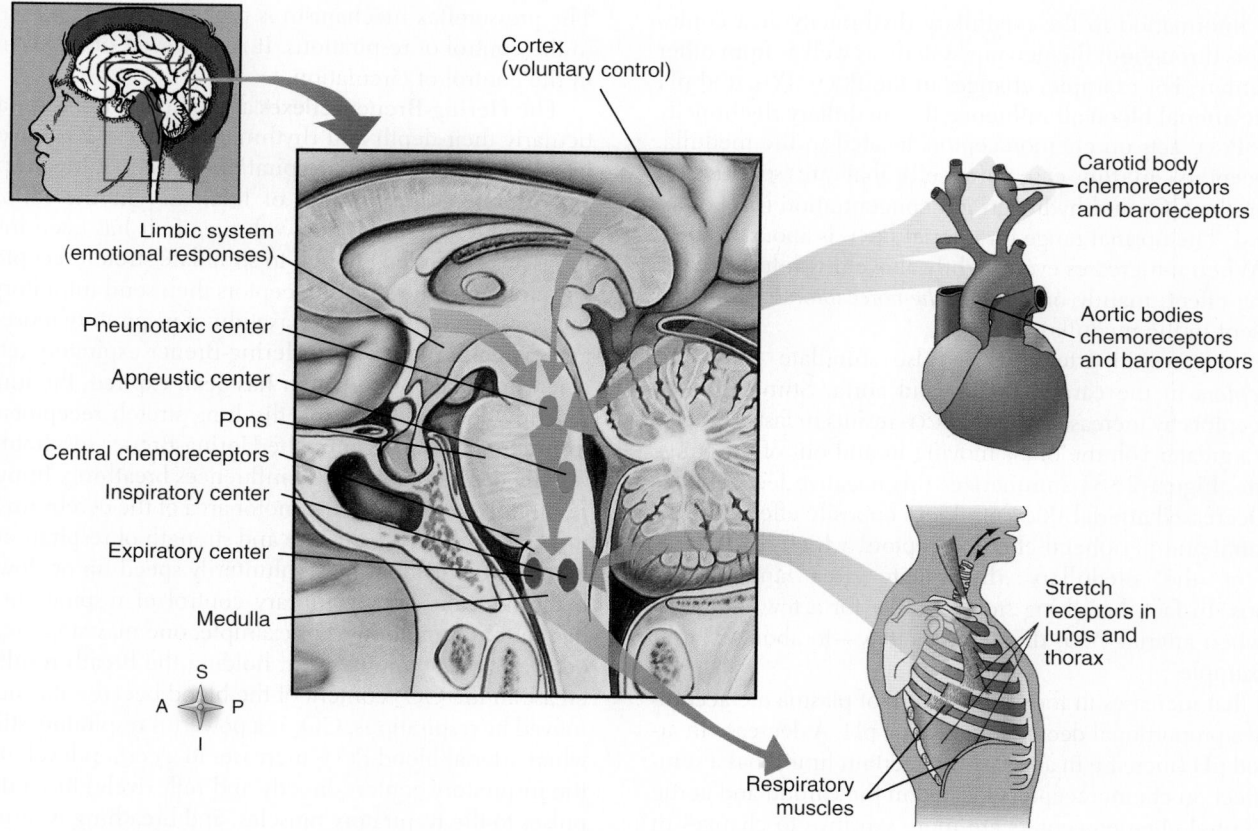

Figure 24-30 *Regulation of breathing.* The inspiratory and expiratory areas of the medulla represent the medullary rhythmicity area. The pneumotaxic center and apneustic center of the pons influence the basic respiratory rhythm by means of neural input to the medullary rhythmicity area. The brainstem also receives input from other parts of the body; information from chemoreceptors, baroreceptors, and stretch receptors can alter the basic breathing pattern, as can emotional (limbic) and sensory input. Despite these subconscious reflexes, the cerebral cortex can override the "automatic" control of breathing to some extent to do such activities as sing or blow up a balloon. Green arrows show flow of information to the respiratory control centers. The purple arrow shows the flow of information from the control centers to the respiratory muscles that drive breathing.

The basic rhythm of the respiratory cycle of inspiration and expiration seems to be generated by the **medullary rhythmicity area.** This area of the medulla consists of two interconnected control centers: the *inspiratory center* and the *expiratory center*. As their names imply, output from the inspiratory center stimulates inspiration, and output from the expiratory center stimulates expiration. Because normal, quiet breathing involves stimulation of inspiratory muscles (mainly the diaphragm) alternating with relaxation of the same muscles, the inspiratory area is thought to act as the primary respiratory pacemaker. The expiratory area seems to be active only when expiratory muscles are needed during forced expiration.

A current hypothesis suggests that the basic breathing rhythm can be altered by different inputs to the medullary rhythmicity area. For example, input from the **apneustic center** in the pons stimulates the inspiratory center to increase the length and depth of inspiration; breathing characterized by abnormally long, deep inspirations is sometimes called "apneustic breathing." The **pneumotaxic center,** also in the pons, normally inhibits both the apneustic center and the inspiratory center. This prevents overinflation of the lungs and thus permits a normal rhythm of breathing.

Factors That Influence Breathing

Feedback information to the medullary rhythmicity area comes from sensors throughout the nervous system, as well as from other control centers. For example, changes in the P_{CO_2}, P_{O_2}, and pH of systemic arterial blood all influence the medullary rhythmicity area. The P_{CO_2} acts on chemoreceptors located in the medulla. Chemoreceptors, in this case, are cells that are sensitive to changes in the CO_2 and hydrogen ion concentration (pH) of arterial blood. The normal range for arterial P_{CO_2} is about 38 to 40 mm Hg. When it increases even slightly above this value, it has a stimulating effect, mainly on *central chemoreceptors* (postulated to be present in the medulla).

Large increases in arterial P_{CO_2} also stimulate *peripheral chemoreceptors* in the carotid bodies and aorta. Stimulation of chemoreceptors by increased arterial P_{CO_2} results in faster breathing, with a greater volume of air moving in and out of the lungs per minute. Figure 24-31 summarizes this negative feedback response. Decreased arterial P_{CO_2} produces opposite effects—it inhibits central and peripheral chemoreceptors, which leads to inhibition of the medullary rhythmicity area and slower respirations. In fact, breathing stops entirely for a few moments (apnea) when arterial P_{CO_2} drops moderately—to about 35 mm Hg, for example.

Recall that increases in the CO_2 content of plasma are accompanied by a proportional decrease in plasma pH. A decrease in arterial blood pH (increase in acid), within certain limits, has a stimulating effect on chemoreceptors located in the carotid and aortic bodies. Central chemoreceptors are more sensitive to changes in pH than are peripheral chemoreceptors. This increased sensitivity results from the fact that fluids of the brain are protected by the BBB (blood-brain barrier) from the buffers present in the blood. Thus, when blood P_{CO_2} increases in the blood, it is partially buffered in the blood—but the CO_2 is *not* buffered in the brain. The brain then senses unbuffered changes in pH.

The role of **arterial blood P_{O_2}** in controlling respirations is not entirely clear. Presumably, it has little influence as long as it stays above a certain level. But neurons of the respiratory centers, like all body cells, require adequate amounts of oxygen to function optimally. Consequently, if they become hypoxic, they become depressed and send fewer impulses to respiratory muscles. Respirations then decrease or fail entirely. This principle has important clinical significance. For example, the respiratory centers cannot respond to stimulation by an increasing blood CO_2 if, at the same time, blood P_{O_2} falls below a critical level—a fact that may become life or death important during anesthesia.

However, a decrease in arterial blood P_{O_2} below 70 mm Hg, but not so low as the critical level, stimulates chemoreceptors in the carotid and aortic bodies and causes reflex stimulation of the inspiratory center. This constitutes an emergency respiratory control mechanism. It does not help regulate respirations under usual conditions when arterial blood P_{O_2} remains considerably higher than 70 mm Hg, which is the level necessary to stimulate the chemoreceptors.

Arterial blood pressure helps control breathing through the respiratory pressoreflex mechanism. A sudden rise in arterial pressure, by acting on aortic and carotid baroreceptors, results in reflex slowing of respirations. A sudden drop in arterial pressure brings about a reflex increase in the rate and depth of respirations. The pressoreflex mechanism is probably not of great importance in the control of respirations. It is, however, of major importance in the control of circulation.

The **Hering-Breuer reflexes** also help control respirations, particularly their depth and rhythmicity. They are believed to regulate the normal depth of respirations (extent of lung expansion)—and therefore the volume of tidal air—in the following way. Presumably, when the tidal volume of air has been inspired, the lungs are expanded enough to stimulate stretch receptors located within them. The stretch receptors then send inhibitory impulses to the inspiratory center, relaxation of inspiratory muscles occurs, and expiration follows the Hering-Breuer expiratory reflex. Then, when the tidal volume of air has been expired, the lungs are sufficiently deflated to inhibit the lung stretch receptors and allow inspiration to start again—the Hering-Breuer inspiratory reflex.

The **cerebral cortex** also influences breathing. Impulses to the respiratory center from the motor area of the cerebrum may either increase or decrease the rate and strength of respirations. In other words, an individual may voluntarily speed up or slow down the breathing rate. This voluntary control of respirations, however, has certain limitations. For example, one may stop breathing and do so for a few minutes, but holding the breath results in an increase in the CO_2 content of the blood because it is not being removed by respirations. CO_2 is a powerful respiratory stimulant. So when arterial blood P_{CO_2} increases to a certain level, it stimulates the inspiratory center (directly and reflexively) to send motor impulses to the respiratory muscles, and breathing is resumed, even though the individual may still will contrarily.

Miscellaneous factors may also influence breathing. Among these are blood temperature and sensory impulses from skin thermal receptors and from superficial or deep pain receptors:

- *Sudden painful stimulation* produces a reflex apnea, but continued painful stimuli cause faster and deeper respirations.
- *Sudden cold stimuli* applied to the skin cause reflex apnea.

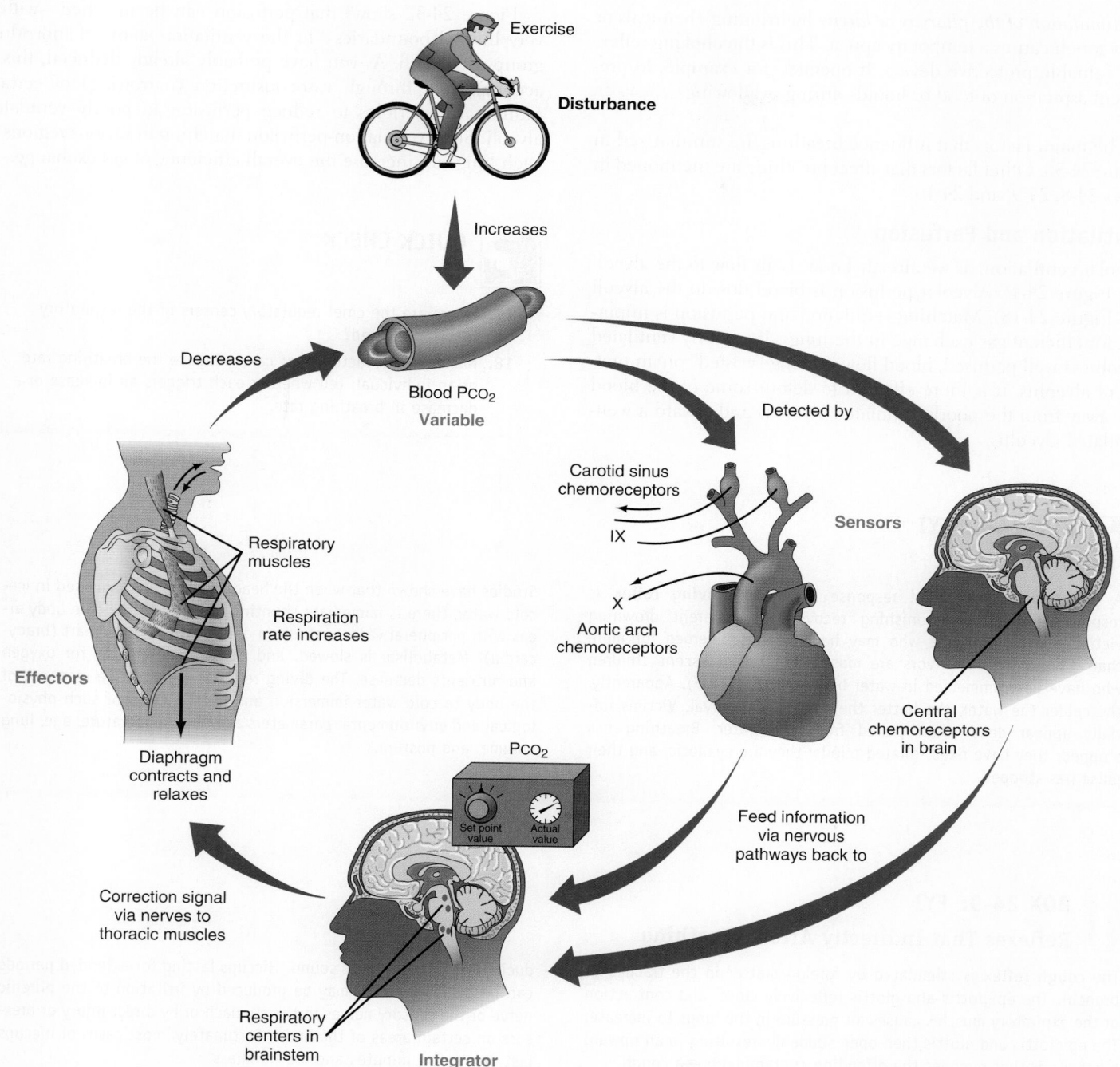

Figure 24-31 *Negative feedback control of respiration.* This diagram summarizes the feedback loop that operates to increase the respiratory rate in response to high plasma P_{CO_2}. Increased cellular respiration during exercise causes a rise in plasma P_{CO_2}—which is detected by central chemoreceptors in the brain and perhaps peripheral chemoreceptors in the carotid sinus and aorta. Feedback information is relayed to integrators in the brainstem that respond to the increase in P_{CO_2} above the set point value by sending nervous correction signals to the respiratory muscles, which act as effectors. The effector muscles increase their alternate contraction and relaxation, thus increasing the rate of respiration. As the respiration rate increases, the rate of CO_2 loss from the body increases and P_{CO_2} drops accordingly. This brings the plasma P_{CO_2} back to its set point value.

- *Stimulation of the pharynx or larynx* by irritating chemicals or by touch causes a temporary apnea. This is the choking reflex, a valuable protective device. It operates, for example, to prevent aspiration of food or liquids during swallowing.

The major factors that influence breathing are summarized in Figure 24-30. Other factors that affect breathing are mentioned in Boxes 24-8, 24-9, and 24-10.

Ventilation and Perfusion

Alveolar ventilation, as we already know, is air flow to the alveoli (see Figure 24-1). Alveolar perfusion is blood flow to the alveoli (see Figure 24-18). Matching ventilation and perfusion is important for efficient gas exchange in the lungs. If a poorly ventilated alveolus is well perfused, blood flow is being "wasted" on an inefficient alveolus. It is more efficient to detour some of the blood flow away from the poorly ventilated alveolus and toward a well-ventilated alveolus.

Figure 24-32 shows that perfusion can be matched—within very limited boundaries—to the ventilation status of individual groups of alveoli. As you have probably already deduced, this is accomplished through vasoconstriction (narrowing) of certain pulmonary arterioles to reduce perfusion to poorly ventilated alveoli. Such ventilation-perfusion matching in various regions of each lung can increase the overall efficiency of gas exchange.

QUICK CHECK

17. Where are the chief regulatory centers of the respiratory function located?
18. Name several factors that can influence the breathing rate of an individual; tell whether each triggers an increase or a decrease in breathing rate.

BOX 24-8: FYI

Diving Reflex

A protective physiological response called the **diving reflex** is responsible for the astonishing recovery of apparent drowning victims—including some who may have been submerged for more than 40 minutes! Survivors are most often preadolescent children who have been immersed in water below 20° C (68° F). Apparently, the colder the water, the better the chance of survival. Victims initially appear dead when pulled from the water. Breathing has stopped; they have fixed, dilated pupils; they are cyanotic; and their pulse has stopped.

Studies have shown that when the head and face are immersed in ice-cold water, there is immediate shunting of blood to the core body areas with peripheral vasoconstriction and slowing of the heart (bradycardia). Metabolism is slowed, and tissue requirements for oxygen and nutrients decrease. The diving reflex is a protective response of the body to cold water immersion and is a function of such physiological and environmental parameters as water temperature, age, lung volume, and posture.

BOX 24-9: FYI

Reflexes That Indirectly Affect Breathing

The **cough reflex** is stimulated by foreign matter in the trachea or bronchi. The epiglottis and glottis reflexively close, and contraction of the expiratory muscles causes air pressure in the lungs to increase. The epiglottis and glottis then open suddenly, resulting in an upward burst of air that removes the offending contaminants—a cough.

The **sneeze reflex** is similar to the cough reflex, except that it is stimulated by contaminants in the nasal cavity. A burst of air is directed through the nose and mouth, forcing the contaminants (and mucus) out of the respiratory tract. Droplets from a sneeze can travel more than 161 km/hr (100 miles/hr) and travel 3 m (12 ft).

The term **hiccup** is used to describe an involuntary, spasmodic contraction of the diaphragm. When such a contraction occurs, generally at the beginning of an inspiration, the glottis suddenly closes, pro-

ducing the characteristic sound. Hiccups lasting for extended periods can be disabling. They may be produced by irritation of the phrenic nerve or the sensory nerves in the stomach or by direct injury or pressure on certain areas of the brain. Fortunately, most cases of hiccups last only a few minutes and are harmless.

A **yawn** is slow, deep inspiration through an unusually widened mouth. Yawns were once thought to be reflexes that increase ventilation when blood oxygen content is low, but newer evidence suggests otherwise. A current theory states that we yawn for the same reason we occasionally stretch—to prepare our muscles and our circulatory system for action. Scientists still have not found the actual physiological mechanism for yawning, however.

BOX 24-10 SPORTS AND FITNESS
Control of Respirations During Exercise

Respirations increase abruptly at the beginning of exercise and decrease even more markedly as it ends. This much is known. The mechanism that accomplishes this increased ventilation rate, however, is not known. It is not identical to the one that produces moderate increases in breathing. Numerous studies have shown that arterial blood P_{CO_2}, P_{O_2}, and pH do not change enough during exercise to produce the degree of hyperpnea (faster, deeper respirations) observed. Presumably, many chemical and nervous factors and temperature changes operate as a complex, but still unknown mechanism for regulating respirations during exercise.

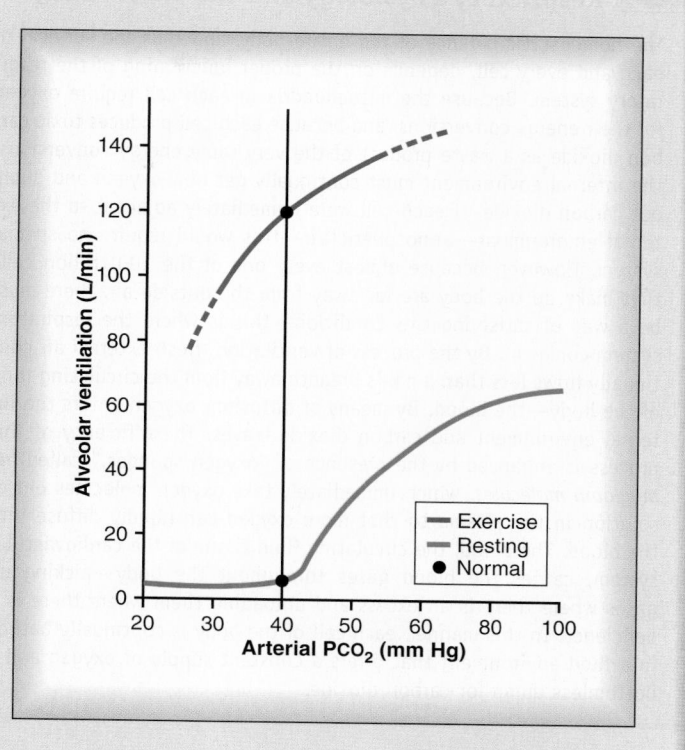

Effects of maximum exercise in an athlete. This graph shows that the breathing rate (vertical axis) is much higher in an athlete exercising maximally than would be expected for any given blood P_{CO_2} (horizontal axis). As you can see at the normal points of a P_{CO_2} of 40 mm Hg, the exercising athlete's breathing (ventilation) rate is 120 L/min. However, at rest the athlete's breathing rate is only about 5 or 6 L/min at the same P_{CO_2}—thus showing that P_{CO_2} is not the major factor causing an increased rate of breathing during exercise.

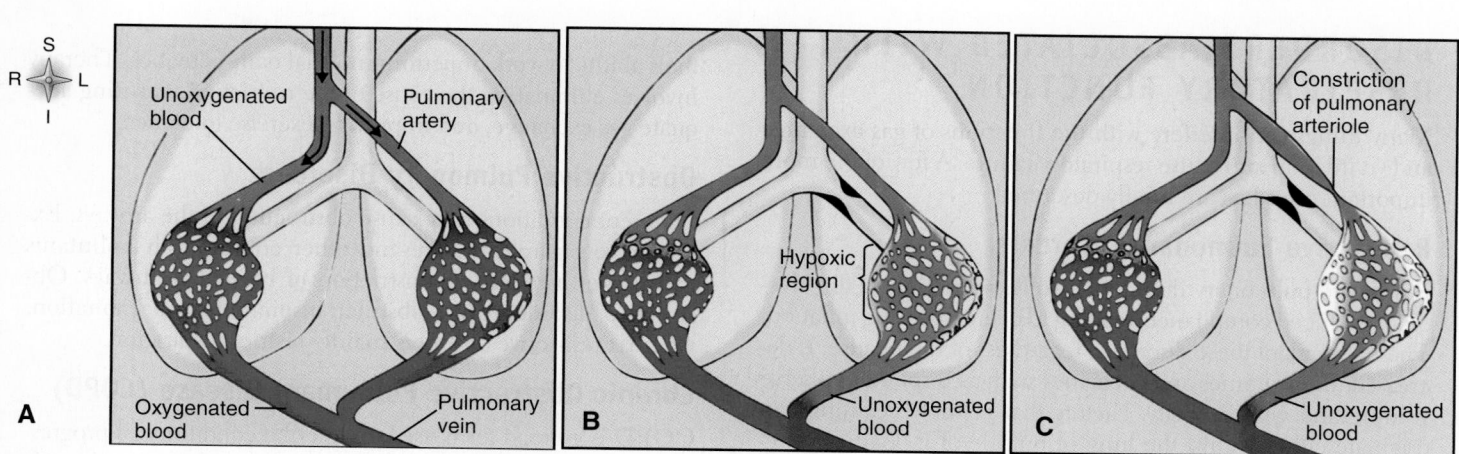

Figure 24-32 *Ventilation and perfusion of the alveoli.* Here, two alveoli represent typical alveoli in the lungs. **A,** Each alveolus is well ventilated with air and well perfused with blood, an efficient combination. **B,** Ventilation to the left alveolus is obstructed, but blood perfusion is unchanged—an inefficient arrangement because blood going to the poorly ventilated alveolus is not being fully oxygenated. **C,** Vasoconstriction of the pulmonary arteriole in the left (poorly ventilated) alveolus reduces blood perfusion—thus efficiently matching the perfusion to the ventilation.

THE BIG PICTURE
Respiratory Physiology and the Whole Body

The homeostatic balance of the entire body, and thus the survival of each and every cell, depends on the proper functioning of the respiratory system. Because the mitochondria in each cell require oxygen for their energy conversions, and because each cell produces toxic carbon dioxide as a waste product of the very same energy conversions, the internal environment must continually get new oxygen and dump out carbon dioxide. If each cell were immediately adjacent to the external environment—atmospheric air—this would require no special system. However, because almost every one of the 100 trillion cells that make up the body are far away from the outside air, there must be a way of satisfying this condition—this is where the respiratory system comes in. By the process of ventilation, fresh external air continually flows less than a hair's breadth away from the circulating fluid of the body—the blood. By means of diffusion, oxygen enters the internal environment and carbon dioxide leaves. The efficiency of this process is enhanced by the presence of "oxygen sponges," called *hemoglobin molecules,* which immediately take oxygen molecules out of solution in the plasma so that more oxygen can rapidly diffuse into the blood. The blood, the circulating fluid tissue of the cardiovascular system, carries the blood gases throughout the body—picking up gases where there is an excess and unloading them where there is a deficiency. In this manner, each cell of the body is continually bathed in a fluid environment that offers a constant supply of oxygen and a bottomless dump for carbon dioxide.

Specific mechanisms involved in respiratory function show the interdependence between body systems observed throughout our study of the human body. For example, without blood and the maintenance of blood flow by the cardiovascular system, blood gases could not be transported between the gas exchange tissues of the lungs and the various systemic tissues of the body. Without regulation by the nervous system, ventilation could not be adjusted to compensate for changes in the oxygen or carbon dioxide content of the internal environment. Without the skeletal muscles of the thorax, the airways could not maintain the flow of fresh air that is so vital to respiratory function. The skeleton itself provides a firm outer housing for the lungs and an arrangement of bones that facilitates the expansion and recoil of the thorax, which is needed to accomplish inspiration and expiration. Without the immune system, pathogens from the external environment could easily colonize the respiratory tract and possibly cause a fatal infection.

Even more subtle interactions between the respiratory and other systems can be found. For example, the language function of the nervous system is limited without the speaking ability provided by the larynx and other structures of the respiratory tract. The homeostasis of pH, which is regulated by a variety of systems, is influenced by the respiratory system's ability to adjust the body's carbon dioxide levels (and thus the levels of carbonic acid).

Mechanisms of Disease

DISORDERS ASSOCIATED WITH RESPIRATORY FUNCTION

Many things can interfere with the functions of gas exchange and ventilation and cause respiratory failure. A few of the more important disorders are briefly described.

Restrictive Pulmonary Disorders

Restrictive pulmonary disorders involve restriction of the alveoli, or reduced compliance, leading to decreased lung inflation. The hallmark of these disorders, regardless of their cause, is decreased lung volumes and capacities such as inspiratory reserve volume and vital capacity. Factors that restrict breathing can originate either within the lung or outside of it. Causes of restrictive lung disorders include alveolar fibrosis (scarring) secondary to occupational exposure to asbestos, toxic fumes, coal dust, or other contaminants; immunological diseases, as in rheumatoid lung; obesity; and metabolic disorders such as uremia. Restriction of breathing can also be caused by pain that accompanies pleurisy (inflammation of the pleurae) or mechanical injuries (such as a fractured or bruised rib). Patients with restrictive lung disease classically experience **dyspnea** (labored breathing) and do not tolerate increased activity, which reduces

their ability to work or perform normal daily activities. Therapy involves eliminating the cause of the restriction, ensuring adequate gas exchange, and improving exercise tolerance.

Obstructive Pulmonary Disorders

Different conditions may cause obstruction of the airways. Exposure to cigarette smoke and other common air pollutants can trigger a reflexive constriction of bronchial airways. Obstructive disorders may obstruct inspiration and expiration, whereas restrictive disorders mainly restrict inspiration.

Chronic Obstructive Pulmonary Disease (COPD)

COPD is a broad term used to describe conditions of progressive irreversible obstruction of expiratory air flow. People with COPD have difficulty breathing, mainly emptying their lungs, and have visibly hyperinflated chests. They have a productive cough and intolerance of activity. The major disorders observed in people with COPD are bronchitis, emphysema, and asthma (see Figure 24-33).

Acute obstruction of the airways, as when a piece of food blocks air flow, requires immediate action to avoid death from suffocation (Box 24-11).

Mechanisms of Disease—cont.

Bronchitis

In chronic **bronchitis,** the person produces excessive tracheo-bronchial secretions that obstruct air flow, and the bronchial mucous glands are enlarged (Figure 24-33, *B*). Risk factors include cigarette smoking (accounting for 80% to 90% of the risk of developing COPD), a normal decline in pulmonary function as a result of age, and environmental exposure to dust and chemicals. With impairment of the alveoli and loss of capillary beds, gas exchange is inefficient, which in turn produces hypoxia.

Emphysema

In **emphysema,** the air spaces distal to the terminal bronchioles are enlarged as a result of damage to lung connective tissue. As the alveoli enlarge, their walls rupture and fuse into large irregular spaces, and gas exchange units are destroyed (Figure 24-33, *D*). Although the etiology is not fully understood, this condition is believed to be caused by proteolytic enzymes that destroy lung tissue. Hypoxia often develops in emphysema victims.

Asthma

Asthma is an obstructive lung disorder characterized by recurring inflammation of mucous membranes and spasms of the smooth muscles in the walls of the bronchial air passages. The inflammation (edema and excessive mucus production) and contractions narrow airways, making breathing difficult (Figure 24-33, *C*). Initial onset of asthma can occur in children or adults. Acute episodes of asthma—so-called "asthma at-

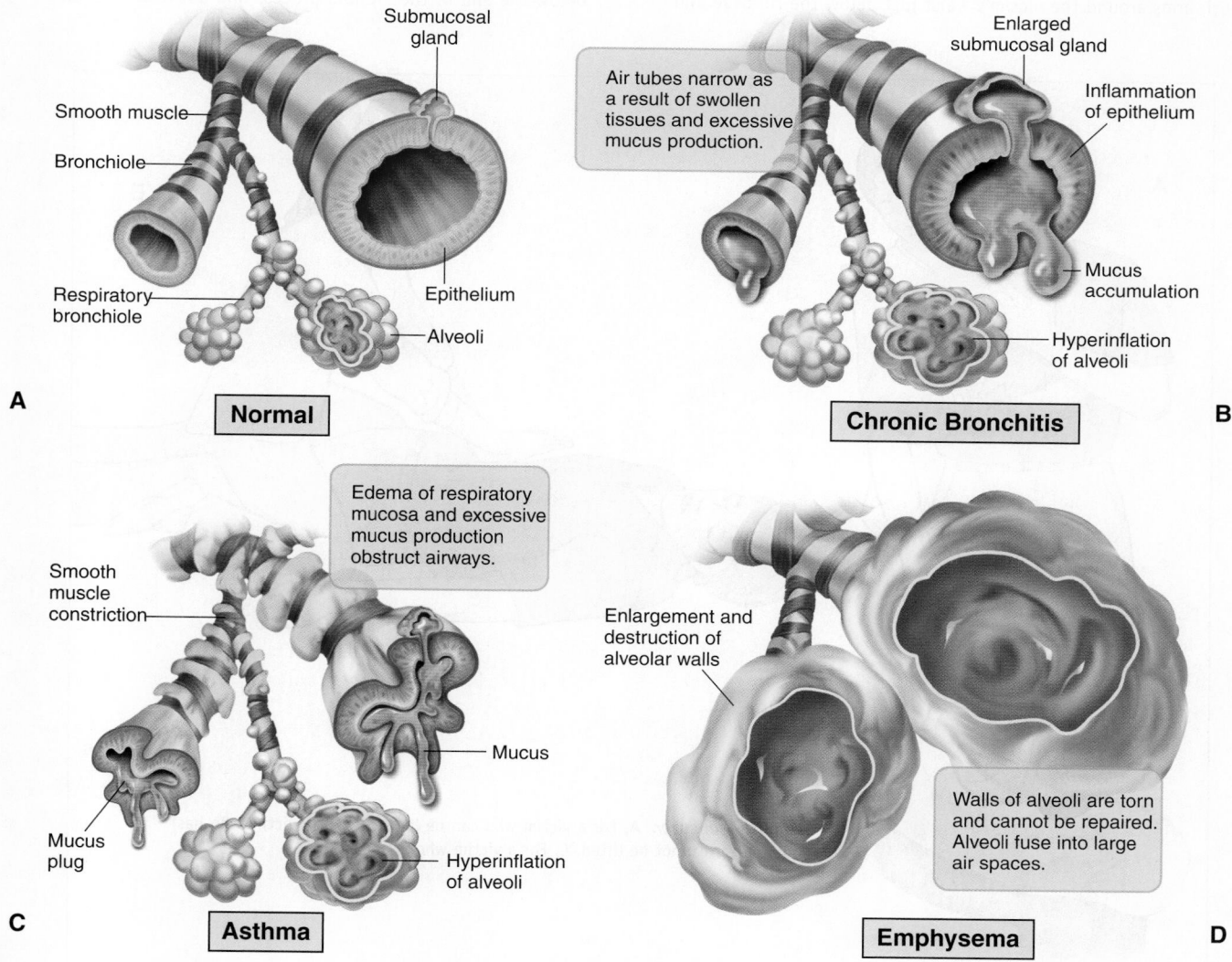

Figure 24-33 *Chronic obstructive pulmonary disease (COPD).* Summary of disorders commonly observed in COPD patients. **A,** Normal airways. **B,** Chronic bronchitis. **C,** Asthma. **D,** Emphysema.

Mechanisms of Disease—cont.

BOX 24-11: HEALTH MATTERS

Heimlich Maneuver

The **Heimlich maneuver** is an effective and often lifesaving technique that can be used to open a windpipe that is suddenly obstructed. The maneuver (see the figure) uses air already present in the lungs to expel the object obstructing the trachea. Most accidental airway obstructions result from pieces of food aspirated during a meal; the condition is sometimes referred to as "cafe coronary." Other objects such as chewing gum or balloons are frequently the cause of obstructions in children. Individuals trained in emergency procedures must be able to tell the difference between airway obstruction and other conditions that produce similar symptoms, such as heart attacks. The key question they must ask the person who appears to be choking is, "Can you talk?" A person with an obstructed airway will not be able to speak, even while conscious. The Heimlich maneuver, if the victim is standing, consists of the rescuer grasping the victim with both arms around the victim's waist just below the rib cage and

above the navel. The rescuer makes a fist with one hand, grasps it with the other, and then delivers an upward thrust against the diaphragm just below the xiphoid process of the sternum. Air trapped in the lungs is compressed, forcing the object that is choking the victim out of the airway. It should be no surprise that the Heimlich maneuver is also called *abdominal thrusts.*

Technique if the Victim Can Be Lifted (Part A of Figure)

1. Rescuer stands behind the victim and wraps both arms around the victim's chest slightly below the rib cage and above the navel. Victim is allowed to fall forward with head, arms, and chest over the rescuer's arms.
2. Rescuer makes a fist with one hand and grasps it with the other hand, pressing thumb side of fist against victim's abdomen just below the end of the xiphoid process and above the navel.

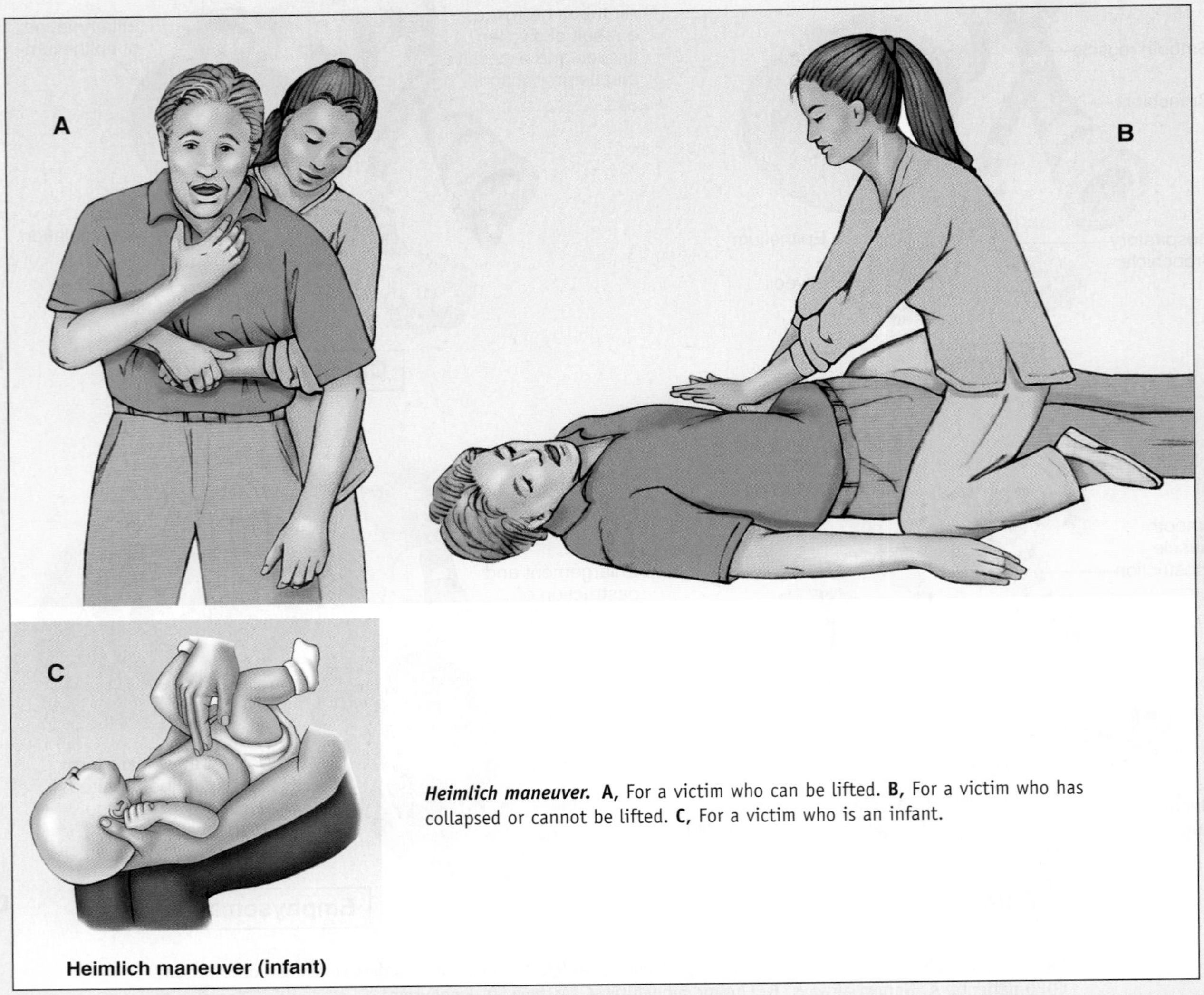

Heimlich maneuver. **A,** For a victim who can be lifted. **B,** For a victim who has collapsed or cannot be lifted. **C,** For a victim who is an infant.

Heimlich maneuver (infant)

Mechanisms of Disease—cont.

BOX 24-11: HEALTH MATTERS
Heimlich Maneuver—cont'd

3. The hands only are used to deliver the upward subdiaphragmatic thrusts. It is performed with sharp flexion of the elbows, in an upward rather than inward direction, and is usually repeated four times. It is very important to not compress the rib cage or actually press on the sternum during the Heimlich maneuver.

Technique if Victim Has Collapsed or Cannot Be Lifted (Part B of Figure)

1. Rescuer places victim on floor face up.
2. Facing victim, rescuer straddles the hips.
3. Rescuer places one hand on top of the other, with the bottom hand on the victim's abdomen slightly above the navel and below the rib cage.

4. Rescuer performs a forceful upward thrust with the heel of the bottom hand, repeating several times if necessary.

Technique if Victim Is an Infant (Part C of Figure)

1. Rescuer cradles infant in one arm, with infant facing the floor.
2. Rescuer delivers five blows to the back, between the shoulder blades.
3. Rescuer then turns the baby over, cradled in one arm (face upward, as shown) and delivers five chest thrusts with the first two fingers (as shown).
4. Rescuer repeats the above sequence as necessary.

tacks"—can be triggered by stress, heavy exercise, infection, or exposure to allergens or other irritants such as dust, vapor, or fumes. Many patients with asthma have a family history of allergies. Dyspnea is the major symptom of asthma, but hyperventilation, headaches, numbness, and nausea can occur. One way to treat asthma is by using inhaled or systemic bronchodilators that reduce muscle spasms and thus open the airways. Other types of treatment involve the use of anti-inflammatory medications or leukotriene modifiers to reduce the inflammation associated with asthma. (Recall from Chap-

ter 21 that leukotrienes are cytokines released by immune cells to regulate the inflammation response.)

Acute respiratory failure can occur when any of the disorders that produce COPD become intense. Heart failure resulting from the pulmonary disease and the vascular resistance that develops with COPD is another possible outcome. Although there is no cure, limiting symptoms can improve quality of life. Bronchodilators and corticosteroids have been used to relieve some of the airway obstruction involved in COPD.

LANGUAGE OF SCIENCE   *(Cont'd from page 883)*

heme group (heem) [*heme* blood]

hemoglobin (hee-moh-GLOH-bin) [*hemo-* blood, *-glob-* ball, *-in* substance]

Henry's law [*William Henry* English chemist]

Hering-Breuer reflexes (HER-ing BROO-er REE-fleks-ez) [*Heinrich E. Hering* German physiologist, *Joseph Breuer* Australian physician, *reflexus* a bending back]

ideal gas

inspiration (in-spih-RAY-shun) [*inspira-* breathe in, *-ation* process of]

law of partial pressures

medullary rhythmicity area (MED-eh-lair-ee rith-MIH-sih-tee) [*medulla-* marrow or middle, *-ary* pertaining to, *rhythm-* rhythm, *-icity* condition of]

myoglobin (my-oh-GLO-bin) [*myo-* muscle, *-glob-* ball, *-in* substance]

oxygen-hemoglobin dissociation curve (AHK-sih-jen hee-moh-GLOH-bin) [*oxy-* sharp, *-gen* generate, *hemo-* blood, *-glob-* ball, *-in* substance, *dis-* reverse, *-socia-* unite, *-ation* process of]

oxyhemoglobin (ahk-see-hee-moh-GLOH-bin) [*oxy-* sharp (oxygen), *-hemo-* blood, *-glob-* ball, *-in* substance]

partial pressure (PAR-shal)

physiological dead space (fiz-ee-oh-LOJ-i-kal ded spays) [*physio-* nature, *-log-* science, *-ical* pertaining to]

pneumotaxic center (noo-moh-TAK-sik) [*pneumo-* wind (breath), *-taxi-* movement or reaction, *-ic* pertaining to]

primary principle of ventilation (ven-tih-LAY-shun) [*prim-* first, *-ary* pertaining to, *principium* foundation, *ventila-* fan or create wind, *-tion* process of]

pulmonary ventilation (PUL-moh-nair-ee ven-tih-LAY-shun) [*pulmon* lungs, *-ary* pertaining to, *ventila-* fan or create wind, *-tion* process of]

rate law

respiratory cycle (RES-pih-rah-tor-ee) [*re-* again, *-spira-* breathe, *-tory* pertaining to]

respiratory physiology (RES-pih-rah-tor-ee fiz-ee-OL-oh-jee) [*re-* again, *-spira-* breathe, *-tory* pertaining to, *physio-* nature, *-logy* science of]

solubility (sol-yoo-BIL-i-tee) [*solubili-* able to dissolve, *-ity* state of]

surface tension

tension

transpulmonary pressure (trans-PUHL-mohn-air-ee) [*trans-* across, *-pulmon-* lungs, *-ary* pertaining to]

type II cells [*II* Roman numeral two]

ventilation (ven-tih-LAY-shun) [*ventila-* fan or create wind, *-tion* process of]

Young-LaPlace law (law of LaPlace) (yung lah-PLAHS) [*Thomas Young* English physician, *Pierre Simon de LaPlace* French physicist]

LANGUAGE OF MEDICINE

adult respiratory distress syndrome (ARDS)
[re- again, -spira- breathe, -tory pertaining to syn-
together, -drome course]

apnea (AP-nee-ah) [a- not, -pne- breathe, -a
condition of]

apneusis (ap-NYOO-sis) [a- not, -pneu- breathe, -sis
condition of]

asthma (AZ-mah) [asthma panting]

Biot's breathing (bee-OHS) [Camille Biot French
physician]

bronchitis (brong-KYE-tis) [bronch- windpipe, -itis
inflammation]

carbon monoxide (CO) (KAR-bon mon-OKS-ide)
[mono- single, -ox- sharp (oxygen), -ide chemical
ending]

carbon monoxide poisoning (KAR-bon mon-OKS-
ide POY-son-ing) [mono- single, -ox sharp (oxygen)]

chest x-ray

Cheyne-Stokes respiration (chain stokes res-pih-
RAY-shun) [John Cheyne Scots physician, William
Stokes Irish physician]

**chronic obstructive pulmonary disease
(COPD)** (KRON-ik ob-STRUK-tiv PUL-moh-nair-ee)
[chron- time, -ic pertaining to, pulmon lung]

continuous positive airway pressure (CPAP)

cough reflex (kof REE-fleks) [reflexus a bending
back]

diving reflex (DYE-ving REE-fleks) [reflexus a
bending back]

dyspnea (DISP-nee-ah) [dys- painful, -pne- breathe,
-a condition of]

emphysema (em-fi-SEE-mah) [em- in, -physema a
blowing or puffing up]

eupnea (YOOP-nee-ah) [eu- easily, -pne- breathe, -a
condition of]

expiratory reserve volume (ERV) (eks-PYE-rah-
tor-ee ree-ZERV VOL-yoom) [ex- out of, -[s]pira-
breathe, -tory pertaining to]

forced expiratory volume (FEV) (eks-PYE-rah-
tor-ee) [ex- out of, -[s]pira- breathe, -tory
pertaining to]

functional residual capacity (FRC)

Heimlich maneuver (HYME-lik mah-NOO-ver)
[Henry J. Heimlich American physician]

hiccup (HIK-up)

hyaline membrane disease (HMD) (HYE-ah-lin)
[hyal- glass, -ine of or similar to]

hyperpnea (hye-PERP-nee-ah) [hyper- excessive,
-pne- breathe, -a condition of]

hyperventilation (hye-per-ven-ti-LAY-shun)
[hyper- excessive, -ventila- fan or create wind, -tion
process of]

hypoventilation (hye-poh-ven-ti-LAY-shun) [hypo-
below, -ventila- fan or create wind, -tion process of]

inspiratory capacity (IC) (in-SPY-rah-tor-ee kah-
PASS-i-tee) [in- in, -spira- breathe, -tory pertaining
to]

inspiratory reserve volume (IRV) (in-SPY-rah-
tor-ee) [in- in, -spira- breathe, -tory pertaining to]

intratracheal injection (in-trah-TRAY-kee-al in-
JEK-shun) [intra- within, -trache rough artery]

orthopnea (or-THOP-nee-ah) [ortho- straight, -pne-
breathe, -a condition of]

pneumothorax (noo-moh-THOH-raks) [pneumo- air
or wind, -thorax chest]

residual volume (RV) (ree-ZID-yoo-al)

respiratory arrest (RES-pih-rah-tor-ee ah-REST)
[re- again, -spira- breathe, -tory pertaining to]

respiratory distress syndrome (RDS) (RES-pih-
rah-tor-ee di-STRESS SIN-drohm) [re- again, -spira-
breathe, -tory pertaining to, syn- together, -drome
course]

sneeze reflex (sneez REE-fleks) [reflexus a bending
back]

spirogram (SPY-roh-gram) [spiro- breath, -gram
drawing or written record]

spirometer (spih-ROM-eh-ter) [spiro- breath, -meter
measurement]

tidal volume (TV) (TYE-dal) [tid- time, -al
pertaining to]

total lung capacity (TLC)

total minute volume

vital capacity (VC) (VYE-tal kah-PASS-i-tee) [vita-
life, -al pertaining to]

yawn

CASE STUDY

Miguel Perez is a 65-year-old farmer who has managed his own large farm for many years. He has smoked two packs of cigarettes a day for 35 years. During the last few years Mr. Perez has experienced slight shortness of breath and a mild cough with activity. He also coughs after arising each morning. Mr. Perez noticed recently that he has difficulty climbing the stairs at home because of fatigue and he must stop at intervals to catch his breath. He has also noticed being short of breath even sometimes at rest. He has lost 10 pounds in the last 2 months. He has been sleeping sitting up or propped by several pillows.

On admission to the emergency room, he is a thin, frail-looking man in acute respiratory distress. He is restless and breathing rapidly. He is sitting on the side of the bed leaning on the bedside table. His heart rate is 120, respirations are 30, and blood pressure is 140/80. Auscultation of the lungs reveals decreased breath sounds with expiratory wheezes. His chest has an increased anteroposterior diameter, giving it a barrel shape, and he is using accessory muscles to breathe. Arterial blood gas results reveal a decreased P_{O_2}, an increased P_{CO_2}, a low pH, and a high bicarbonate level. Pulmonary function tests reveal decreased tidal volume, decreased vital capacity, increased total lung capacity, and a prolonged forced expiratory volume. A **chest x-ray** reveals a flat low diaphragm, hyperinflation of the lungs, clear lung fields, and no evidence of cardiac enlargement. He is started on 2 liters of oxygen by nasal cannula in the emergency room.

1. Which two of the following reasons are the most likely causes of Mr. Perez's dyspnea?

   A. Increased lung compliance and decreased elastic recoil
   B. Decreased elastic recoil and decreased lung compliance
   C. Decreased lung compliance and increased elastic recoil
   D. Increased elastic recoil and increased lung compliance

2. The causes of Mr. Perez's barrel chest include

   A. Excessive secretions
   B. Severe hypoxemia
   C. Hyperinflation of the lungs
   D. Thickening of the bronchial mucosa

3. Retention of carbon dioxide in individuals with chronic obstructive pulmonary disease (COPD), such as Mr. Perez's, is caused by what mechanism?

   A. Alveolar hyperventilation
   B. Dilation of the bronchial tree
   C. Alveolar hypoventilation
   D. Decreased dead space

4. Which of the following best explains Mr. Perez's weight loss?

   A. Cigarette smoking decreases his sense of taste, thus decreasing his appetite
   B. COPD increases the body's need for calories
   C. Cigarette smoking increases the basal metabolic rate leading to weight loss
   D. COPD prevents the body from using nutrients

CHAPTER SUMMARY

RESPIRATORY PHYSIOLOGY (FIGURE 24-1)

A. Respiratory physiology—complex, coordinated processes that help maintain homeostasis
B. Respiratory function includes
   1. External respiration
      a. Pulmonary ventilation (breathing)
      b. Pulmonary gas exchange
   2. Transport of gases by the blood
   3. Internal respiration
      a. Systemic tissue gas exchange
      b. Cellular respiration
   4. Regulation of respiration

PULMONARY VENTILATION

A. Respiratory cycle (ventilation; breathing)
   1. Inspiration—moves air into the lungs
   2. Expiration—moves air out of the lungs
B. Mechanism of pulmonary ventilation
   1. The pulmonary ventilation mechanism must establish two gas pressure gradients (Figures 24-2 and 24-3)
      a. One in which the pressure within the alveoli of the lungs is lower than atmospheric pressure to produce inspiration

      b. One in which the pressure in the alveoli of the lungs is higher than atmospheric pressure to produce expiration
   2. Pressure gradients are established by changes in the size of the thoracic cavity that are produced by contraction and relaxation of muscles (Figures 24-4 and 24-5)
   3. Boyle's law—the volume of gas varies inversely with pressure at a constant temperature
   4. Inspiration—contraction of the diaphragm produces inspiration—as it contracts, it makes the thoracic cavity larger (Figures 24-6 and 24-7)
      a. Expansion of the thorax results in decreased intrapleural pressure (P_{IP}), leading to decreased alveolar pressure (P_A)
      b. Air moves into the lungs when alveolar pressure (P_A) drops below atmospheric pressure (P_B)
      c. Compliance—ability of pulmonary tissues to stretch, thus making inspiration possible
   5. Expiration—a passive process that begins when the inspiratory muscles are relaxed, which decreases the size of the thorax (Figures 24-8 and 24-9)
      a. Increasing thoracic volume increases the intrapleural pressure and thus increases alveolar pressure above the atmospheric pressure

b. Air moves out of the lungs when alveolar pressure exceeds the atmospheric pressure

c. The pressure between parietal and visceral pleura is always less than alveolar pressure and less than atmospheric pressure; the different between P_{IP} and P_A is called transpulmonary pressure

d. Elastic recoil—tendency of pulmonary tissues to return to a smaller size after having been stretched; occurs passively during expiration

C. Pulmonary volumes—the amount of air moved in and out and remaining is important so that a normal exchange of oxygen and carbon dioxide can take place (Figure 24-11)

1. Spirometer—instrument used to measure the volume of air (Figure 24-10)

2. Tidal volume (TV)—amount of air exhaled after normal inspiration

3. Expiratory reserve volume (ERV)—largest volume of additional air that can be forcibly exhaled (between 1.0 and 1.2 liters is normal ERV)

4. Inspiratory reserve volume (IRV)—amount of air that can be forcibly inhaled after normal inspiration (normal IRV is 3.3 liters)

5. Residual volume—amount of air that cannot be forcibly exhaled (1.2 liters)

D. Pulmonary capacities—the sum of two or more pulmonary volumes

1. Vital capacity—the sum of IRV + TV + ERV

2. Minimal volume—the amount of air remaining after RV

3. A person's vital capacity depends on many factors, including the size of the thoracic cavity and posture

4. Functional residual capacity—the amount of air at the end of a normal respiration

5. Total lung capacity—the sum of all four lung volumes—the total amount of air a lung can hold

6. Alveolar ventilation—volume of inspired air that reaches the alveoli

7. Anatomical dead space—air in passageways that do not participate in gas exchange (Figure 24-6)

8. Physiological dead space—anatomical dead space plus the volume of any nonfunctioning alveoli (as in pulmonary disease)

9. Alveoli must be properly ventilated for adequate gas exchange

E. Pulmonary air flow—rates of air flow into/out of the pulmonary airways

1. Total minute volume—volume moved per minute (ml/min)

2. Forced expiratory volume (FEV) or forced vital capacity (FVC)—volume of air expired per second during forced expiration (as a percentage of VC) (Figure 24-12)

3. Flow-volume loop—graph that shows flow (vertically) and volume (horizontally), with the top of the loop representing expiratory flow-volume and the bottom of the loop representing inspiratory flow-volume relationships (Figure 24-13)

PULMONARY GAS EXCHANGE

A. Partial pressure of gases—pressure exerted by a gas in a mixture of gases or a liquid (Figure 24-14)

1. Law of partial pressures (Dalton's law)—the partial pressure of a gas in a mixture of gases is directly related to the concentration of that gas in the mixture and to the total pressure of the mixture

2. Arterial blood P_{O_2} and P_{CO_2} equal alveolar P_{O_2} and P_{CO_2}

B. The exchange of gases in the lungs takes place between alveolar air and blood flowing through lung capillaries (Figures 24-15, 24-16, and 24-17)

1. Four factors determine the amount of oxygen that diffuses into blood

a. The oxygen pressure gradient between alveolar air and blood

b. The total functional surface area of the respiratory membrane

c. The respiratory minute volume

d. Alveolar ventilation

2. Structural facts that facilitate oxygen diffusion from the alveolar air to the blood

a. The fact that the walls of the alveoli and capillaries form only a very thin barrier for gases to cross

b. The fact that alveolar and capillary surfaces are large

c. The fact that the blood is distributed through the capillaries in a thin layer so that each red blood cell comes close to alveolar air (Figure 24-18)

HOW BLOOD TRANSPORTS GASES

A. Oxygen and carbon dioxide are transported as solutes and as parts of molecules of certain chemical compounds

B. Transport of oxygen

1. Hemoglobin is made up of four polypeptide chains (two alpha chains, two beta chains), each with an iron-containing heme group; carbon dioxide can bind to amino acids in the chains and oxygen can bind to iron in the heme groups (Figure 24-19)

2. Oxygenated blood contains about 0.3 ml of dissolved O_2 per 100 ml of blood

3. Hemoglobin increases the oxygen-carrying capacity of blood (Figure 24-20)

4. Oxygen travels in two forms: as dissolved O_2 in plasma and associated with hemoglobin (oxyhemoglobin)

a. Increasing blood P_{O_2} accelerates hemoglobin association with oxygen (Figure 24-21)

b. Oxyhemoglobin carries the majority of the total oxygen transported by blood

C. Transport of carbon dioxide

1. A small amount of CO_2 dissolves in plasma and is transported as a solute (10%)

2. Less than one fourth of blood carbon dioxide combines with NH_2 (amine) groups of hemoglobin and other proteins to form carbaminohemoglobin (20%) (Figure 24-22)

3. Carbon dioxide's association with hemoglobin is accelerated by an increase in blood P_{CO_2} (Figure 24-23)

4. More than two thirds of the carbon dioxide is carried in plasma as bicarbonate ions (70%) (Figures 24-24, 24-25, and 24-26)

SYSTEMIC GAS EXCHANGE

A. The exchange of gases in tissues takes place between arterial blood flowing through tissue capillaries and cells (Figure 24-27)
   1. Oxygen diffuses out of arterial blood because the oxygen pressure gradient favors its outward diffusion
   2. As dissolved oxygen diffuses out of arterial blood, blood P_{O_2} decreases, which accelerates oxyhemoglobin dissociation to release more oxygen to plasma for diffusion to cells (Figure 24-28)
B. Carbon dioxide exchange between tissues and blood takes place in the opposite direction from oxygen exchange
   1. Bohr effect—increased P_{CO_2} decreases the affinity between oxygen and hemoglobin (Figure 24-29, A)
   2. Haldane effect—increased carbon dioxide loading caused by a decrease in P_{O_2} (Figure 24-29, B)

REGULATION OF PULMONARY FUNCTION

A. Respiratory control centers—the main integrators controlling the nerves that affect the inspiratory and expiratory muscles are located in the brainstem (Figure 24-30)
   1. Medullary rhythmicity center—generates the basic rhythm of the respiratory cycle
      a. This area consists of two interconnected control centers
         (1) The inspiratory center stimulates inspiration
         (2) The expiratory center stimulates expiration
   2. The basic breathing rhythm can be altered by different inputs to the medullary rhythmicity center (Figure 24-30)
      a. Input from the apneustic center in the pons stimulates the inspiratory center to increase the length and depth of inspiration
      b. The pneumotaxic center—in the pons—inhibits the apneustic center and inspiratory center to prevent over-inflation of the lungs
B. Factors that influence breathing—sensors from the nervous system provide feedback to the medullary rhythmicity center (Figure 24-31)
   1. Changes in the P_{O_2}, P_{CO_2}, and pH of arterial blood influence the medullary rhythmicity area
      a. P_{CO_2} acts on central chemoreceptors in the medulla—if it increases, the result is faster breathing; if it decreases, the result is slower breathing
      b. A decrease in blood pH stimulates peripheral chemoreceptors in the carotid and aortic bodies and, even more so, stimulates the central chemoreceptors (because they are surrounded by unbuffered fluid)
      c. Arterial blood P_{O_2} presumably has little influence if it stays above a certain level
   2. Arterial blood pressure controls breathing through the respiratory pressoreflex mechanism
   3. Hering-Breuer reflexes help control respirations by regulating depth of respirations and the volume of tidal air
   4. The cerebral cortex influences breathing by increasing or decreasing the rate and strength of respirations
C. Ventilation and perfusion (Figure 24-32)
   1. Alveolar ventilation—air flow to the alveoli
   2. Alveolar perfusion—blood flow to the alveoli

   3. Efficiency of gas exchange can be maintained by limited ability to match perfusion to ventilation—for example, vasoconstricting arterioles that supply poorly ventilated alveoli and allow full blood flow to well-ventilated alveoli

THE BIG PICTURE: RESPIRATORY SYSTEM AND THE WHOLE BODY

A. The internal system must continually get new oxygen and rid itself of carbon dioxide because each cell requires oxygen and produces carbon dioxide as a result of energy conversion
B. Specific mechanisms involved in respiratory function
   1. Blood gases need blood and the cardiovascular system to be transported between gas exchange tissues of the lungs and various systemic tissues of the body
   2. Regulation by the nervous system adjusts ventilation to compensate for changes in oxygen or carbon dioxide in the internal environment
   3. The skeletal muscles of the thorax aid the airways in maintaining the flow of fresh air
   4. The skeleton houses the lungs, and the arrangement of bones facilitates the expansion and recoil of the thorax
   5. The immune system prevents pathogens from colonizing the respiratory tract and causing infection

REVIEW QUESTIONS

1. Define respiratory physiology.
2. What is the main inspiratory muscle?
3. Identify the separate volumes that make up the total lung capacity.
4. Normally, about what percentage of the tidal volume fills the anatomical dead space?
5. Normally, about what percentage of the tidal volume is useful air, that is, ventilates the alveoli?
6. One gram of hemoglobin combines with how many milliliters of oxygen?
7. What factors influence the amount of oxygen that diffuses into the blood from the alveoli?
8. Identify the major factors that influence breathing.
9. Orthopnea is a symptom of what type of disease?
10. Dyspnea is often associated with what type of breathing?
11. Define the diving reflex and explain its physiological importance.
12. Describe four reflexes that indirectly affect breathing.
13. Describe the changes in respirations during a period of exercise.

CRITICAL THINKING QUESTIONS

1. The proper functioning of the respiratory system allows what to occur in the body? What other control system has an impact on this function?
2. Can you identify the various processes that allow the respiratory system to accomplish its function?

3. What is pulmonary ventilation? What evidence can you find to describe whether the lungs are active or passive during this process?
4. How would you compare and contrast inspiration and expiration? Include the importance of elastic recoil and compliance to these processes.
5. How would you summarize the interaction of oxygen and carbon dioxide on gas transport in the blood? Include the Bohr and Haldane effects in your explanation.

6. Suppose your blood has a hemoglobin content of 15 g/100 dl and an oxygen saturation of 97%. How many milliliters of oxygen would 100 ml of your arterial blood contain?
7. Compare and contrast infant and adult respiratory distress syndromes.
8. After strenuous exercise, inexperienced athletes will quite often attempt to recover and resume normal breathing by bending over or sitting down. Using the mechanics of ventilation, how would you modify the recovery practices of these athletes?

CHAPTER 25

Anatomy of the Digestive System

LANGUAGE OF SCIENCE

alimentary canal (al-eh-MEN-tar-ee kah-NAL) [*alimentum* nourishment, *canalis* channel]

anal canal (AY-nal kah-NAL) [*anal* pertaining to the anus, *canalis* channel]

ascending colon (ah-SEND-ing KOH-lon) [*ascendere* to climb, *kolon* large intestine]

bile (byle) [*bilis* bile]

body

cardiac sphincter (KAR-dee-ak SFINGK-ter) [*kardia-* heart, *-ic* pertaining to, *sphingein* to bind]

cheeks

chief cells

colon (KOH-lon)

crown

deciduous teeth (deh-SID-yoo-us) [*decidere* to fall off]

deglutition (deg-loo-TISH-un) [*deglutire* to swallow]

descending colon (dih-SEND-ing KOH-lon) [*descendere* to descend, *kolon* large intestine]

digestion (dye-JES-chun) [*digerere* to break down]

duodenum (doo-oh-DEE-num) [*duodeni* 12 fingers at a time]

endocrine cells (EN-doh-krin) [*endo-* within, *-crine* to secrete, *cella* storeroom]

esophagus (eh-SOF-ah-gus) [*oisophagos* gullet]

fundus (FUN-duss) [*fundus* bottom]

gastrin (GAS-trin) [*gaster* stomach]

gastrointestinal (GI) tract (gas-troh-in-TES-ti-nul) [*gastro-* stomach, *-intestinum-* intestine, *-al* pertaining to, *tractus* trail]

ghrelin (GRAY-lin)

glucagon (GLOO-kah-gon)

greater omentum (oh-MEN-tum) [*great* large, *omentum* entrails]

hard palate (PAL-et) [*palatum* palate]

hepatic (heh-PAT-ik) [*hepar-* liver, *-ic* pertaining to]

ileum (IL-ee-um) [*ileum* groin or flank]

insulin (IN-suh-lin) [*insula* island]

intramural plexus (in-trah-MYOO-ral PLEK-sus) [*intra-* situated within, *-mura* wall, *plexus* network]

Cont'd on p. 956

OVERVIEW OF THE DIGESTIVE SYSTEM

Role of the Digestive System

This chapter deals with the anatomy of the digestive system. The organs of the digestive system together perform a vital function—that of preparing food for absorption and for use by the millions of body cells. Most food when eaten is in a form that cannot reach the cells (because it cannot pass through the intestinal mucosa into the bloodstream), nor could it be used by the cells even if it could reach them. It must therefore be modified in both chemical composition and physical state so that nutrients can be absorbed and used by the body cells. The complete process of altering the physical and chemical composition of ingested food material so that it can be absorbed and used by the body cells is called **diges-tion.** This complex process is the function of both the digestive tract and accessory organs that make up the digestive system.

The first step in the digestive process, called *mechanical diges-tion*, involves physically breaking down ingested food material into smaller pieces. It begins in the mouth and continues with the churning and mixing of food as it passes through the digestive sys-tem. The physical breakdown of ingested food into very small pieces permits effective mixing with digestive secretions. Then, the second phase of digestion, called *chemical digestion*, completes

the breakdown process. It results in the release of nutrient "end products," such as glucose and amino acids, that can be absorbed and enter cells for use as an energy source or for other metabolic functions. It is the digestive process that converts a hamburger and fries lunch into nutrients that can be used at the cellular level.

Controlled movement of ingested food materials through the tract is necessary for normal digestive function. An optimum "rate of passage" of ingested food material permits completion of me-chanical and chemical digestive processes and allows for full ab-sorption of the resulting nutrients. Part of the digestive system, the large intestine, also serves as an organ of elimination. Ingested food material that cannot be put into an absorbable form becomes waste material *(feces)* that is ultimately eliminated from the body. The physiology of the digestive system is discussed in Chapter 26.

Organization of the Digestive System

Organs of Digestion

The main organs of the digestive system (Figure 25-1) form a tube all the way through the ventral cavities of the body. It is open at both ends. This tube is usually referred to as the **alimentary canal** (tract) or, more commonly, the **gastrointestinal (GI) tract.** It is important to realize that ingested food material passing through the lumen of the gastrointestinal tract is actually outside the internal environment

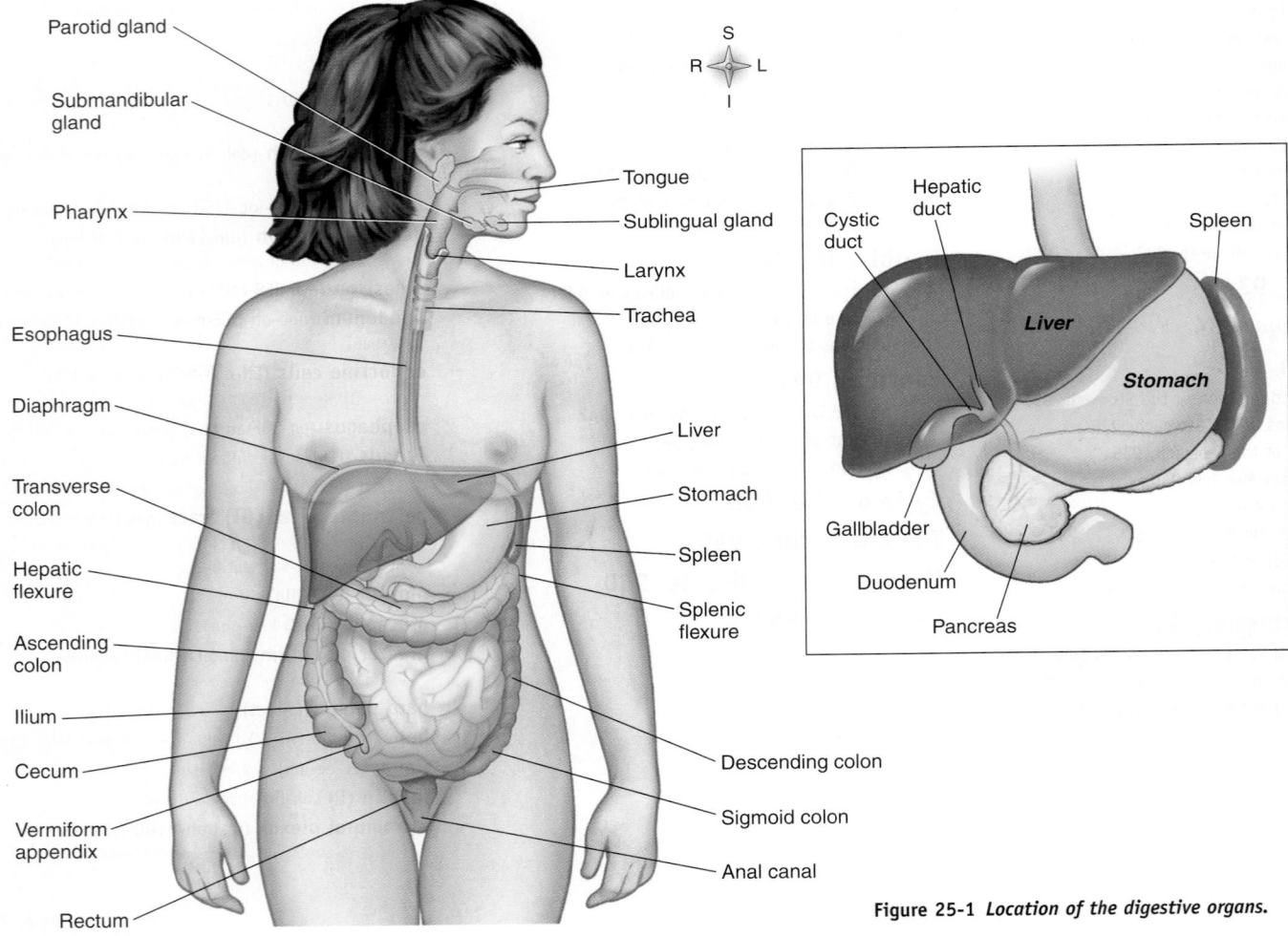

Figure 25-1 *Location of the digestive organs.*

Table 25-1 Organs of the Digestive System

SEGMENTS OF THE GASTROINTESTINAL TRACT	ACCESSORY ORGANS
Mouth	Salivary glands
Oropharynx	Parotid
Esophagus	Submandibular
Stomach	Sublingual
Small intestine	Tongue
Duodenum	Teeth
Jejunum	Liver
Ileum	Gallbladder
Large intestine	Pancreas
Cecum	Vermiform appendix
Colon	
Ascending colon	
Transverse colon	
Descending colon	
Sigmoid colon	
Rectum	
Anal canal	

of the body, even though the tube itself is inside the ventral body cavity. Table 25-1 lists the main organs of the digestive system, that is, the segments of the alimentary canal, and the accessory organs located in the main digestive organs or opening into them. Organs such as the larynx, trachea, diaphragm, and spleen are labeled in Figure 25-1 but are not digestive organs. They are shown to assist in orienting the digestive organs to other important body structures.

Wall of the GI Tract

Layers

The GI tract is essentially a tube with walls fashioned of four layers of tissues: a mucous lining, a submucous coat of connective tissue in which are embedded the main blood vessels of the tract, a muscular layer, and a fibroserous layer (Figure 25-2). Blood vessels and nerves travel through the mesentery to reach the digestive tube throughout most of its length.

Mucosa

The innermost layer of the GI wall—the layer facing the lumen, or open space, of the tube—is called the **mucosa.** Note in Figure 25-2 that the mucosa is made up of three layers—an inner *mucous epithelium,* a layer of loose connective tissue called the *lamina propria,* and a thin layer of smooth muscle called the *muscularis mucosae.*

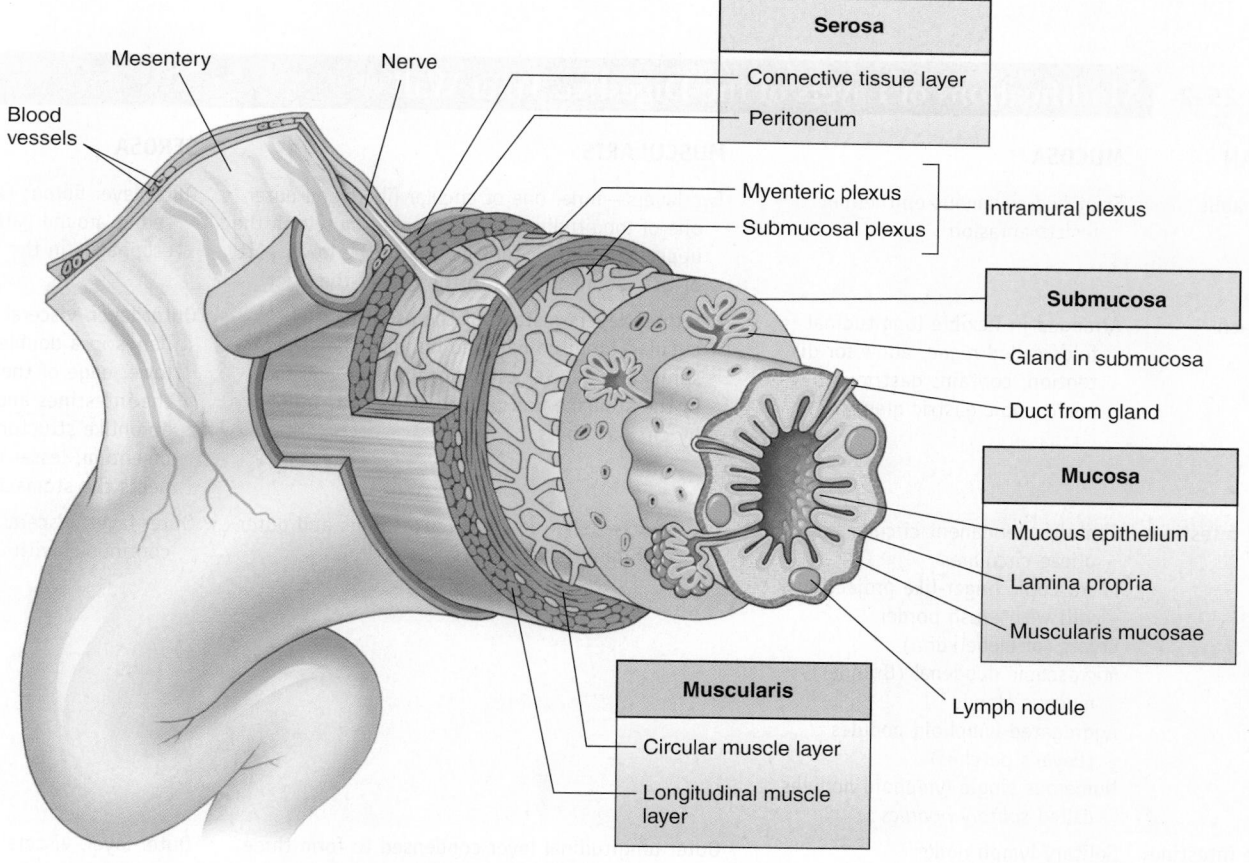

Figure 25-2 *Wall of the GI tract.* The wall of the GI tract is made up of four layers, shown here in a generalized diagram of a segment of the GI tract. Notice that the serosa is continuous with a fold of serous membrane called a *mesentery.* Notice also that digestive glands may empty their products into the lumen of the GI tract by way of ducts.

Submucosa

This layer of the digestive tube is composed of connective tissue and is thicker than the mucosal layer. It contains numerous small glands, blood vessels, and parasympathetic nerves that form the *submucosal plexus* (Meissner plexus).

Muscularis

The muscularis is a thick layer of muscle tissue that wraps around the **submucosa.** This portion of the wall is characterized by an inner layer of circular and an outer layer of longitudinal smooth muscle. Like the submucosa, the muscularis contains nerves organized into a plexus called the *myenteric plexus* (Auerbach plexus). This plexus lies between the two muscle layers. Note in Figure 25-2 that the term **intramural plexus** is used to describe both plexuses. Together they play an important role in the regulation of digestive tract movement and secretion.

Serosa

The **serosa** is the outermost layer of the GI wall and is made up of a connective tissue layer and peritoneum. The term *serosa* is used to describe this layer if it covers segments of the tract that lie within the peritoneal cavity. The serosa is actually the *visceral layer* of the peritoneum—the serous membrane that lines the abdominopelvic cavity and covers its organs. The lining attached to and covering the walls of the abdominopelvic cavity is called the *parietal layer* of the peritoneum. The fold of serous membrane shown in Figure 25-2 that connects the parietal and visceral portions is called a *mesentery*. Notice in Figure 25-2 that exocrine glands empty their secretions into the lumen of the GI tract through ducts.

Modifications of Layers

Although the same four tissue layers form the various organs of the GI tract, their structures vary in different regions of the tube throughout its length. Variations in the epithelial layer of the mucosa, for example, range from stratified layers of squamous cells that provide protection from abrasion in the upper part of the esophagus to the simple columnar epithelium, designed for absorption and secretion, which is found throughout most of the tract. Some of these modifications are illustrated in Figure 25-13 and listed in Table 25-2 and should be referred to when each of these organs is studied in detail.

 QUICK CHECK

1. What is another name for the digestive tract?
2. Can you name the four layers of the digestive tract wall?

Table 25-2	Modifications of Layers of the Digestive Tract Wall		
ORGAN	**MUCOSA**	**MUSCULARIS**	**SEROSA**
Esophagus	Stratified squamous epithelium resists abrasion	Two layers—inner one of circular fibers and outer one of longitudinal fibers; striated muscle in the upper part and smooth muscle in the lower part of the esophagus and in the rest of the tract	Outer layer fibrous (adventitia); serous around part of the esophagus in the thoracic cavity
Stomach	Arranged in flexible longitudinal folds called *rugae;* allow for distention; contains gastric pits with microscopic gastric glands	Has three layers instead of the usual two—circular, longitudinal, and oblique fibers; two sphincters—lower esophageal at the entrance of the stomach and pyloric at its exit, formed by circular fibers	Outer layer, visceral peritoneum; hangs in a double fold from the lower edge of the stomach over the intestines and forms an apronlike structure; greater omentum; lesser omentum connects the stomach to the liver
Small intestine	Contains permanent circular folds, plicae circulares Microscopic finger-like projections, villi with brush border Crypts (of Lieberkühn) Microscopic duodenal (Brunner) mucous glands Aggregated lymphoid nodules (Peyer's patches) Numerous single lymphoid nodules called *solitary nodules*	Two layers—inner one of circular fibers and outer one of longitudinal fibers	Outer layer, visceral peritoneum, continuous with the mesentery
Large intestine	Solitary lymph nodes Intestinal mucous glands Anal columns form in the anal region	Outer longitudinal layer condensed to form three tapelike strips (taeniae coli); small sacs (haustra) give the rest of the wall of the large intestine a puckered appearance; internal anal sphincter formed by circular smooth fibers; external anal sphincter formed by striated fibers	Outer layer, visceral peritoneum, continuous with mesocolon

MOUTH

Structure of the Oral Cavity

The **mouth** is also called the *oral cavity*. The following structures form the oral cavity *(buccal cavity)*: the lips surrounding the orifice of the mouth and forming the anterior boundary of the oral cavity, the cheeks (side walls), the tongue and its muscles (floor), and the hard and soft palate (roof) (Figure 25-3).

Lips

The **lips** are covered externally by skin and internally by mucous membrane that continues into the oral cavity and lines the mouth. The junction between skin and mucous membrane is highly sensitive and easily irritated. The upper lip is marked near the midline by a shallow vertical groove called the *philtrum*, which ends at the junction between skin and mucous membrane in a slight prominence called the *tubercle*. The term *fissure* is often used to describe a cleft or groove between or separating anatomical structures. Therefore, when the lips are closed, the line of contact between them is called the *oral fissure*.

Cheeks

The **cheeks** form the lateral boundaries of the oral cavity. They are continuous with the lips in front and are lined by mucous membrane that is reflected onto the gingiva, or gums, and the soft palate. The cheeks are formed in large part by the buccinator muscle, which is sandwiched with a considerable amount of adipose, or fat, tissue between the outer skin and mucous membrane lining. Numerous small mucus-secreting glands are located between the mucous membrane and the buccinator muscle; their ducts open opposite the last molar teeth.

Hard and Soft Palate

The **hard palate** consists of portions of four bones: two maxillae and two palatines (see Figure 8-5). The **soft palate**, which forms a partition between the mouth and nasopharynx (see Fig-

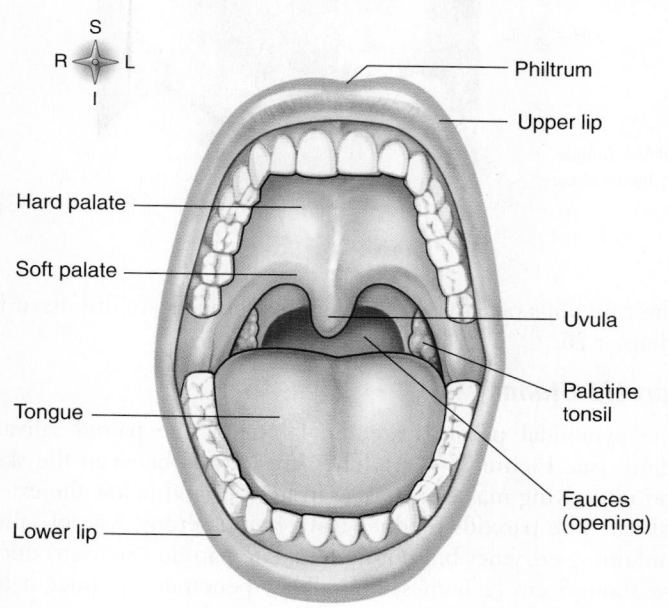

Hard palate

Soft palate

Tongue

Lower lip

Philtrum

Upper lip

Uvula

Palatine tonsil

Fauces (opening)

Figure 25-3 *The oral cavity.*

ure 23-3), is fashioned of muscle arranged in the shape of an arch. The opening in the arch leads from the mouth into the oropharynx and is named the *fauces*. Suspended from the midpoint of the posterior border of the arch is a small cone-shaped process, the *uvula*.

Tongue

The **tongue** is a solid mass of skeletal muscle components (intrinsic muscles) covered by a mucous membrane. The intrinsic muscles of the tongue have, by definition, both their origin and their insertion in the tongue itself. Intrinsic muscles have their fibers oriented in all directions, thus providing a basis for extreme maneuverability. Changes in the size and shape of the tongue caused by intrinsic muscle contraction assist in placement of food material between the teeth during **mastication** (chewing). Extrinsic tongue muscles are those that insert into the tongue but have their origin on some other structure, such as the hyoid bone or one of the bones of the skull. Examples of extrinsic tongue muscles are the genioglossus, which protrudes the tongue, and the hyoglossus, which depresses it (Figure 25-4, *C*). Contraction of the extrinsic muscles is important during **deglutition,** or swallowing, and speech.

Note in Figure 25-4 that the tongue has a blunt *root*, a *tip*, and a central *body*. The upper, or dorsal, surface of the tongue is normally moist, pink, and covered by rough elevations, called *papillae*. The three types of papillae—vallate, fungiform, and filiform—are all located on the sides or upper surface (dorsum) of the tongue. Note in Figure 25-4, *A*, that the large vallate papillae form an inverted **V**-shaped row extending from a median pit named the *foramen cecum* on the posterior part of the tongue. There are 10 to 14 of these large, mushroomlike papillae. You can readily distinguish them if you look at your own tongue. Figure 25-5 shows two micrographs of vallate papillae. Dissolved substances to be tasted enter a moatlike depression surrounding the papillae to contact taste buds located on their lateral surface. Taste buds are also located on the sides of the fungiform papillae, which are found chiefly on the sides and tip of the tongue. The numerous filiform papillae are filamentous and threadlike in appearance. They have a whitish coloration and are distributed over the anterior two thirds of the tongue. Filiform papillae do not contain taste buds.

The *lingual frenulum* (Figure 25-4, *B*) is a fold of mucous membrane in the midline of the undersurface of the tongue that helps anchor the tongue to the floor of the mouth. If the frenulum is too short for freedom of tongue movement—a congenital condition called *ankyloglossia*—the individual is said to be tongue-tied, and speech is faulty (Figure 25-4, photo inset).

Folds of mucous membrane called the *plica fimbriata* (see Figure 25-4, *B*) extend toward the apex of the tongue on either side of the lingual frenulum. The floor of the mouth and undersurface of the tongue are richly supplied with blood vessels. The deep lingual vein can be seen (see Figure 25-4, *B*) shining through the mucous membrane between the lingual frenulum and plica fimbriata. In this region many vessels are extremely superficial and are covered only by a very thin layer of mucosa. Soluble drugs, such as nitroglycerin, are absorbed into the circulation rapidly if placed under the tongue.

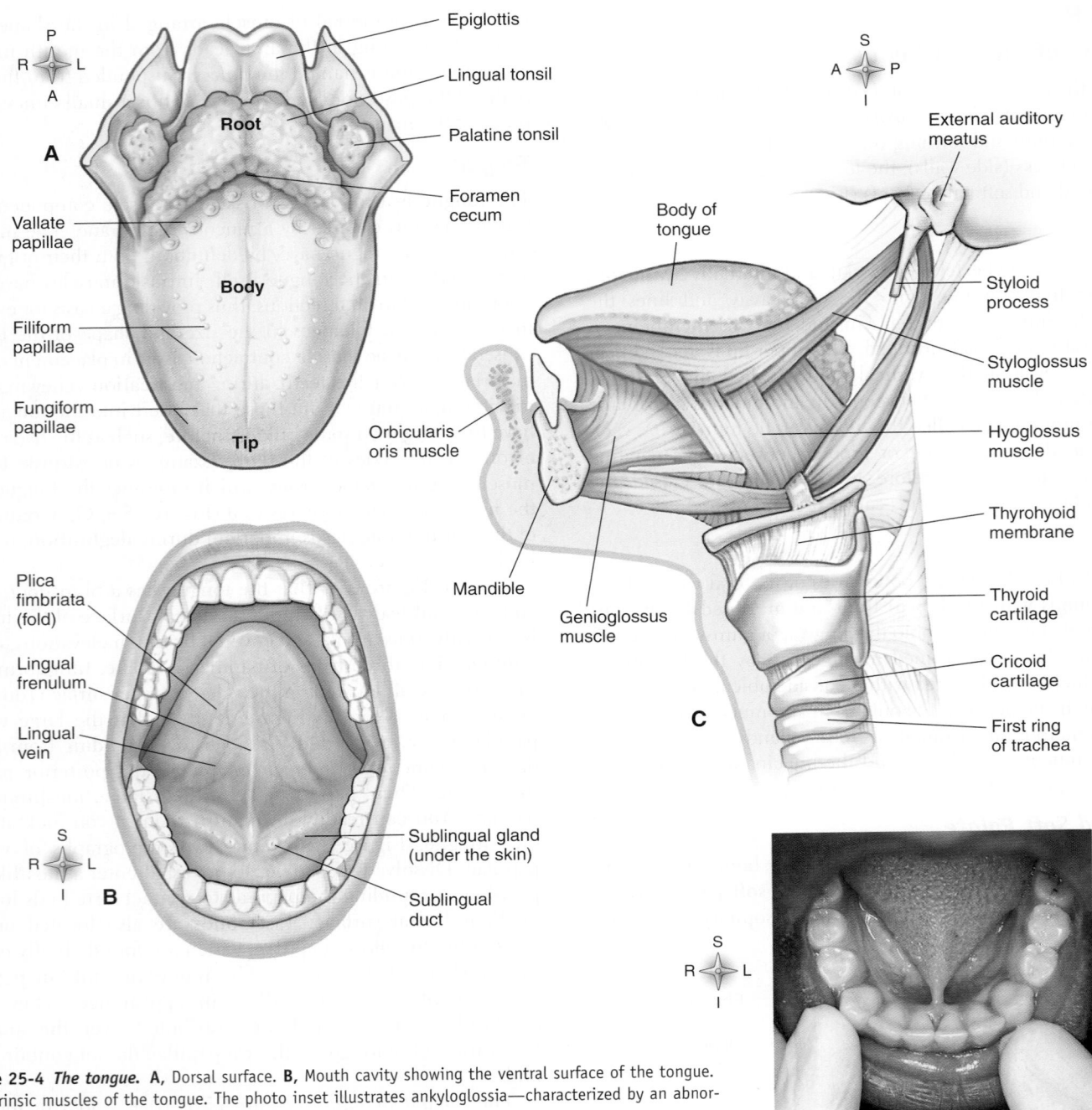

Figure 25-4 *The tongue.* **A,** Dorsal surface. **B,** Mouth cavity showing the ventral surface of the tongue. **C,** Extrinsic muscles of the tongue. The photo inset illustrates ankyloglossia—characterized by an abnormally short lingual frenulum.

Salivary Glands

Three pairs of compound tubuloalveolar glands (Figure 25-6; see Figure 25-1)—the parotid, submandibular, and sublingual glands—secrete a major amount (about 1 liter) of the saliva produced each day. The small glands (buccal glands) that occur in the mucosa lining the cheeks and mouth contribute less than 5% of the total salivary volume. Buccal gland secretion is important, however, to the hygiene and comfort of the mouth tissues. The salivary glands are typical of the accessory glands associated with the digestive system. They are located outside the alimentary canal and convey their exocrine secretions by ducts from the glands into the lumen of the tract (Figure 25-6, A).

The functions of saliva in the digestive process are discussed in Chapter 26.

Parotid Glands

The pyramidal **parotids** are the largest of the paired salivary glands (see Figure 25-6, A). They are located between the skin and underlying masseter muscle in front of and below the external ear. The parotids produce a watery, or serous, type of saliva containing enzymes but not mucus. The parotid (Stensen) ducts are about 5 cm (2 inches) long. They penetrate the buccinator muscle on each side and open into the mouth through the *parotid papilla* opposite the upper second molars. Inflammation of the

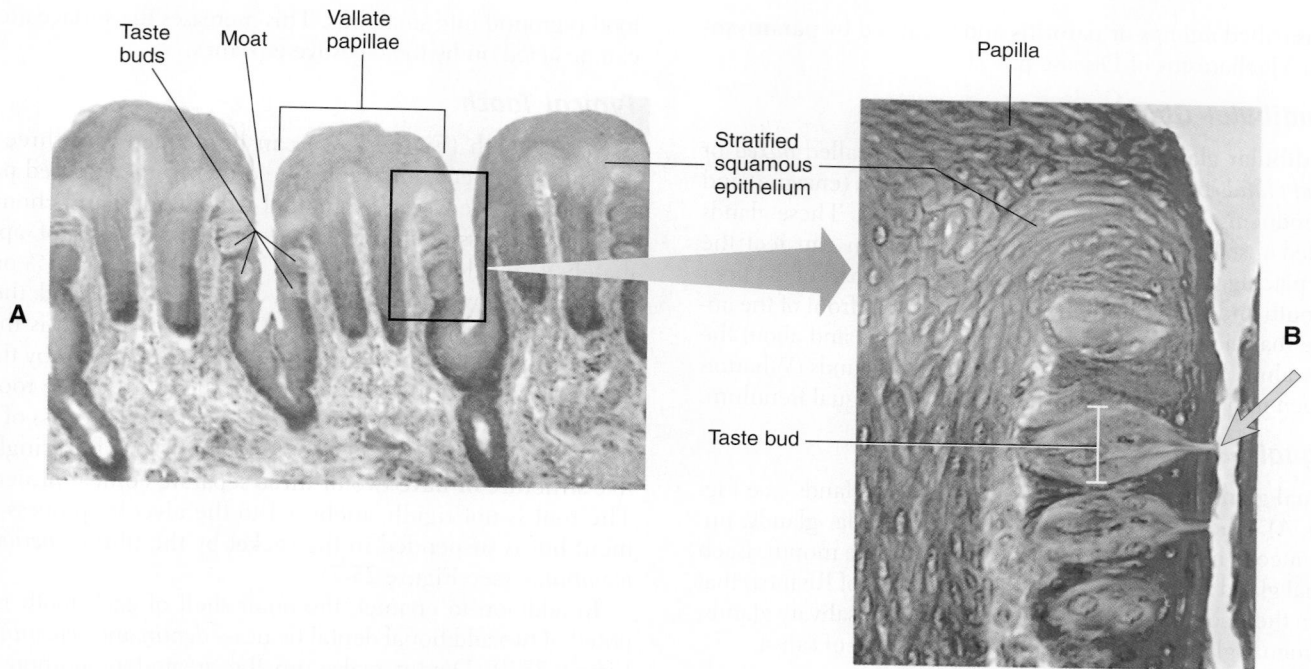

Figure 25-5 *Vallate papillae on the surface of the tongue.* A, Taste buds are located on the lateral surfaces of the papillae. Several taste buds can be seen opening into the moat from the sides of the papillae. (×35.) **B,** Enlargement of the photomicrograph of taste buds in **A.** The arrow points to a pore in the outer surface of the taste bud. (×140.)

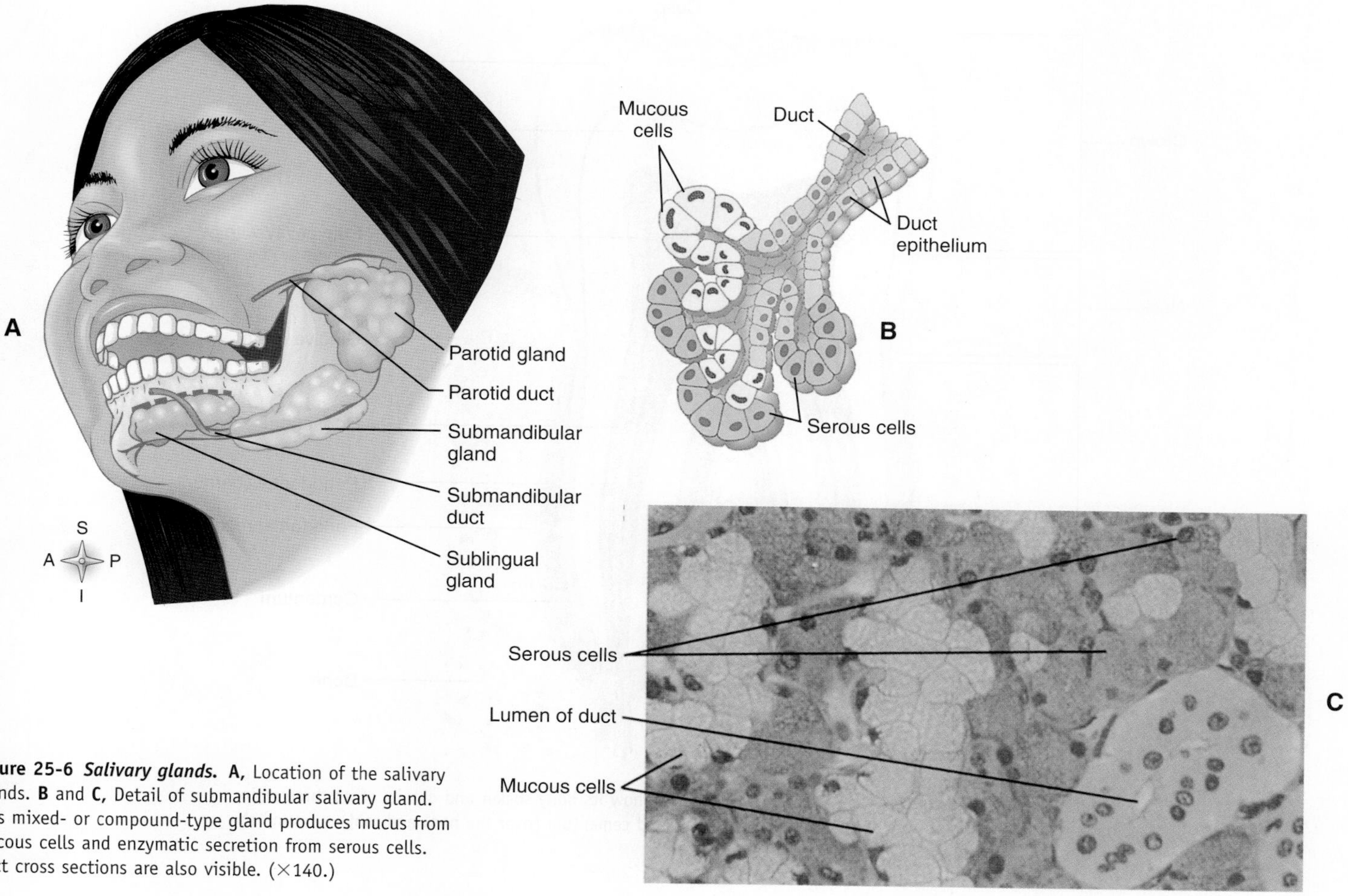

Figure 25-6 *Salivary glands.* A, Location of the salivary glands. **B** and **C,** Detail of submandibular salivary gland. This mixed- or compound-type gland produces mucus from mucous cells and enzymatic secretion from serous cells. Duct cross sections are also visible. (×140.)

parotids is called *mumps* or **parotitis** and is caused by **paramyxovirus** (see Mechanisms of Disease p. 951).

Submandibular Glands

Submandibular glands (see Figure 25-6, A) are called *mixed* or *compound glands* because they contain both serous (enzyme) and mucus-producing elements (Figure 25-6, *B* and *C*). These glands are located just below the mandibular angle. You can feel the gland by placing your index finger on the posterior part of the floor of the mouth and your thumb medial to and just in front of the angle of the mandible. The gland is irregular in form and about the size of a walnut. The ducts of the submandibular glands (Wharton ducts) open into the mouth on either side of the lingual frenulum.

Sublingual Glands

Sublingual glands are the smallest of the salivary glands (see Figure 25-6, A). They lie in front of the submandibular glands, under the mucous membrane covering the floor of the mouth. Each sublingual gland is drained by 8 to 20 ducts (ducts of Rivinus) that open into the floor of the mouth. Unlike the other salivary glands, the sublingual glands produce only a mucous type of saliva.

Teeth

The teeth are the organs of **mastication,** or chewing. They are designed to cut, tear, and grind ingested food so that it can be mixed with saliva and swallowed. During the process of mastication, food is ground into small bits. This increases the surface area that can be acted on by the digestive enzymes.

Typical Tooth

A typical tooth (Figure 25-7) can be divided into three main parts: crown, neck, and root. The **crown** is the exposed portion of a tooth. It is covered by enamel—the hardest and chemically most stable tissue in the body. Enamel consists of approximately 97% calcified (inorganic) material and only 3% organic material and water. It is ideally suited to withstand the very abrasive process of mastication. The *neck* of a tooth is the narrow portion shown in Figure 25-7 that is surrounded by the gingivae, or gums. It joins the crown of the tooth to the root. It is the **root** that fits into the socket of the alveolar process of either the upper or lower jaw. The root of a tooth may be a single peg-like structure or have two or three separate conical projections. The root is not rigidly anchored to the alveolar process by cement but is suspended in the socket by the fibrous *periodontal membrane* (see Figure 25-7).

In addition to enamel, the outer shell of each tooth is composed of two additional dental tissues—*dentin* and *cementum* (see Figure 25-7). Dentin makes up the greatest proportion of the tooth shell. It is covered by enamel in the crown and by cementum in the neck and root area. The dentin contains a **pulp cavity** consisting of connective tissue, blood and lymphatic vessels, and sensory nerves.

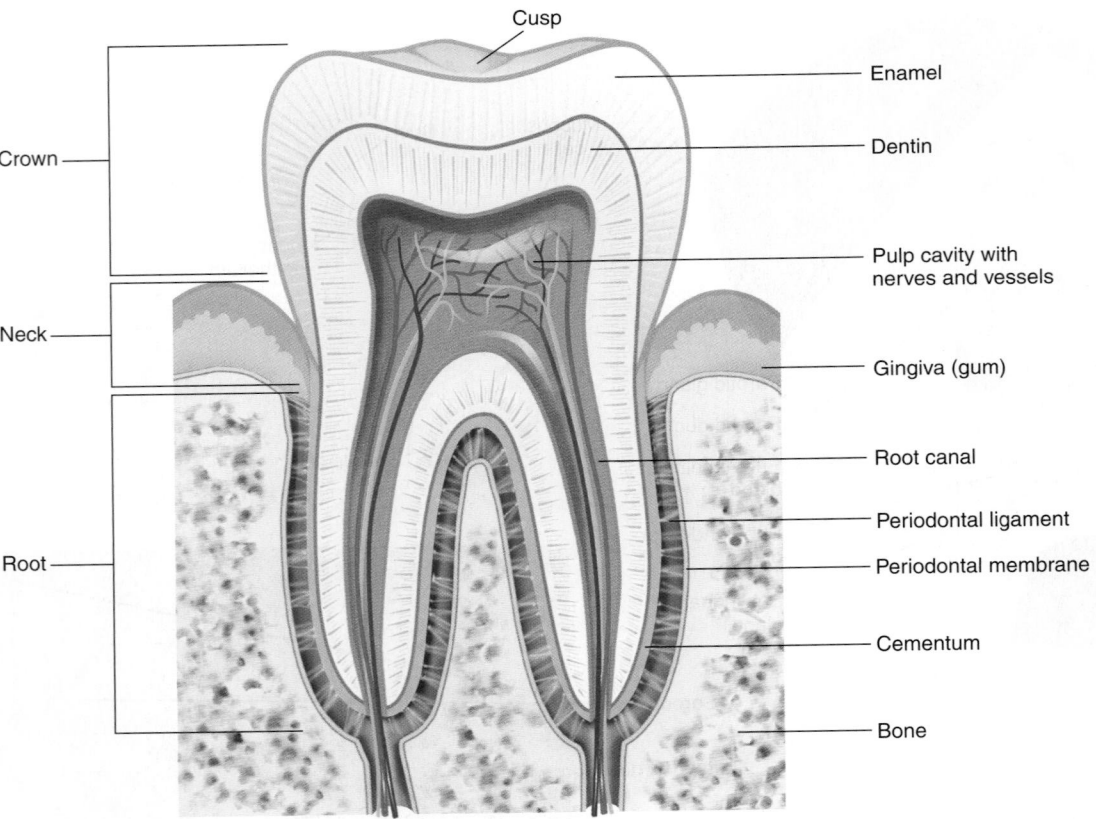

Figure 25-7 *Typical tooth.* A molar tooth sectioned to show its bony socket and details of its three main parts: crown, neck, and root. Enamel (over the crown) and cementum (over the neck and root) surround the dentin layer. The pulp contains nerves and blood vessels.

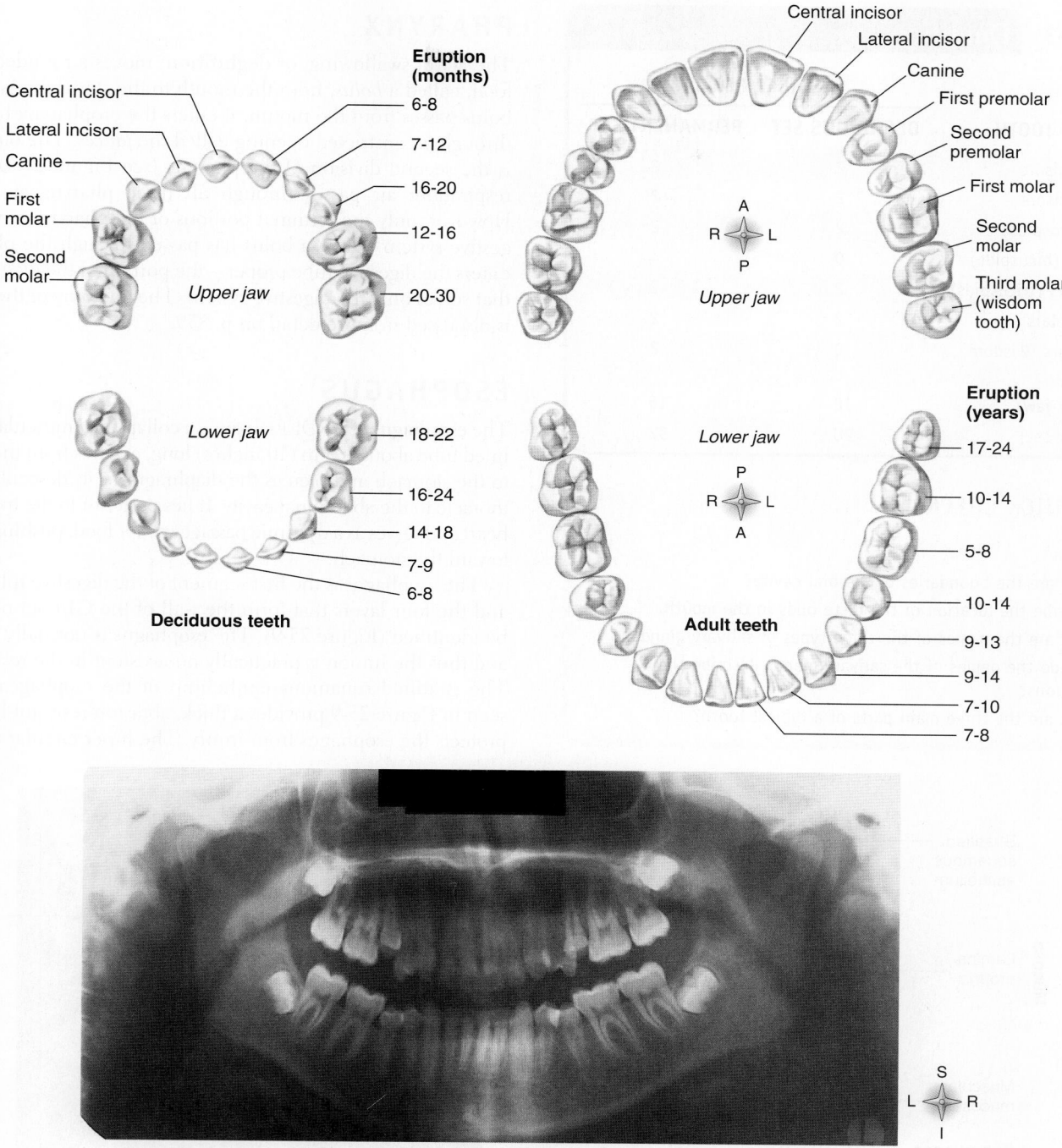

Eruption (months)

Central incisor — 6-8

Lateral incisor — 7-12

Canine — 16-20

First molar — 12-16

Second molar — 20-30

Upper jaw

Central incisor

Lateral incisor

Canine

First premolar

Second premolar

First molar

Second molar

Third molar (wisdom tooth)

A
R — L
P

Upper jaw

Lower jaw — 18-22

16-24

14-18

7-9

6-8

Deciduous teeth

Lower jaw

P
R — L
A

Eruption (years)

17-24

10-14

5-8

10-14

9-13

9-14

7-10

7-8

Adult teeth

S
L — R
I

Figure 25-8 *Deciduous (baby) teeth and adult teeth.* In the deciduous set, there are no premolars and only two pairs of molars in each jaw. Generally, the lower teeth erupt before the corresponding upper teeth. The photo inset is a specialized "Panorex" dental x-ray. It displays the full dentition in a single "flattened-out" image. Can you identify the third molars or wisdom teeth that have not yet erupted?

Types of Teeth

Twenty **deciduous teeth,** or so-called *baby teeth,* appear early in life and are later replaced by 32 **permanent teeth** (Figure 25-8). The names and numbers of teeth in both sets are given in Table 25-3. The first deciduous tooth usually erupts at about 6 months of age. The remainder follow at the rate of 1 or more a month until all 20 have appeared. There is, however, great individual variation in the age at which teeth erupt. Deciduous teeth are generally shed between the ages of 6 and 13 years. The third molars (wisdom teeth) are the last to appear, and usually erupt sometime after 17 years of age.

Table 25-3 Dentition

NAME OF TOOTH	NUMBER PER JAW	
	DECIDUOUS SET	PERMANENT SET
Central incisors	2	2
Lateral incisors	2	2
Canines (cuspids)	2	2
Premolars (bicuspids)	0	4
First molars (tricuspids)	2	2
Second molars	2	2
Third molars (wisdom teeth)	0	2
TOTAL (per jaw)	10	16
TOTAL (per set)	20	32

QUICK CHECK

3. What are the boundaries of the oral cavity?
4. Describe the location of the taste buds in the mouth.
5. What are the names of the three types of salivary glands?
6. How do the names of the salivary glands describe their locations?
7. What are the three main parts of a typical tooth?

PHARYNX

The act of swallowing, or **deglutition,** moves a rounded mass of food, called a *bolus,* from the mouth to the stomach. As the food bolus passes from the mouth, it enters the oropharynx by passing through a constricted opening called the *fauces.* The oropharynx is the second division of the pharynx (see Figure 23-6). During respiration, air passes through all three pharyngeal divisions. However, only the terminal portions of the pharynx serve the digestive system. Once a bolus has passed through the pharynx, it enters the digestive tube proper—the portion of the digestive tract that serves only the digestive system. The anatomy of the pharynx is discussed in more detail on p. 859.

ESOPHAGUS

The **esophagus** (eh-SOFF-ah-gus), a collapsible, muscular, mucus-lined tube about 25 cm (10 inches) long, extends from the pharynx to the stomach and pierces the diaphragm in its descent from the thoracic to the abdominal cavity. It lies posterior to the trachea and heart and serves as a dynamic passageway for food, pushing the food toward the stomach.

The esophagus is the first segment of the digestive tube proper, and the four layers that form the wall of the GI tract organs can be identified (Figure 25-9). The esophagus is normally flattened, and thus the lumen is practically nonexistent in the resting state. The stratified squamous epithelium of the esophageal mucosa seen in Figure 25-9 provides a thick, abrasion-resistant lining that protects the esophagus from injury. The inner circular and outer

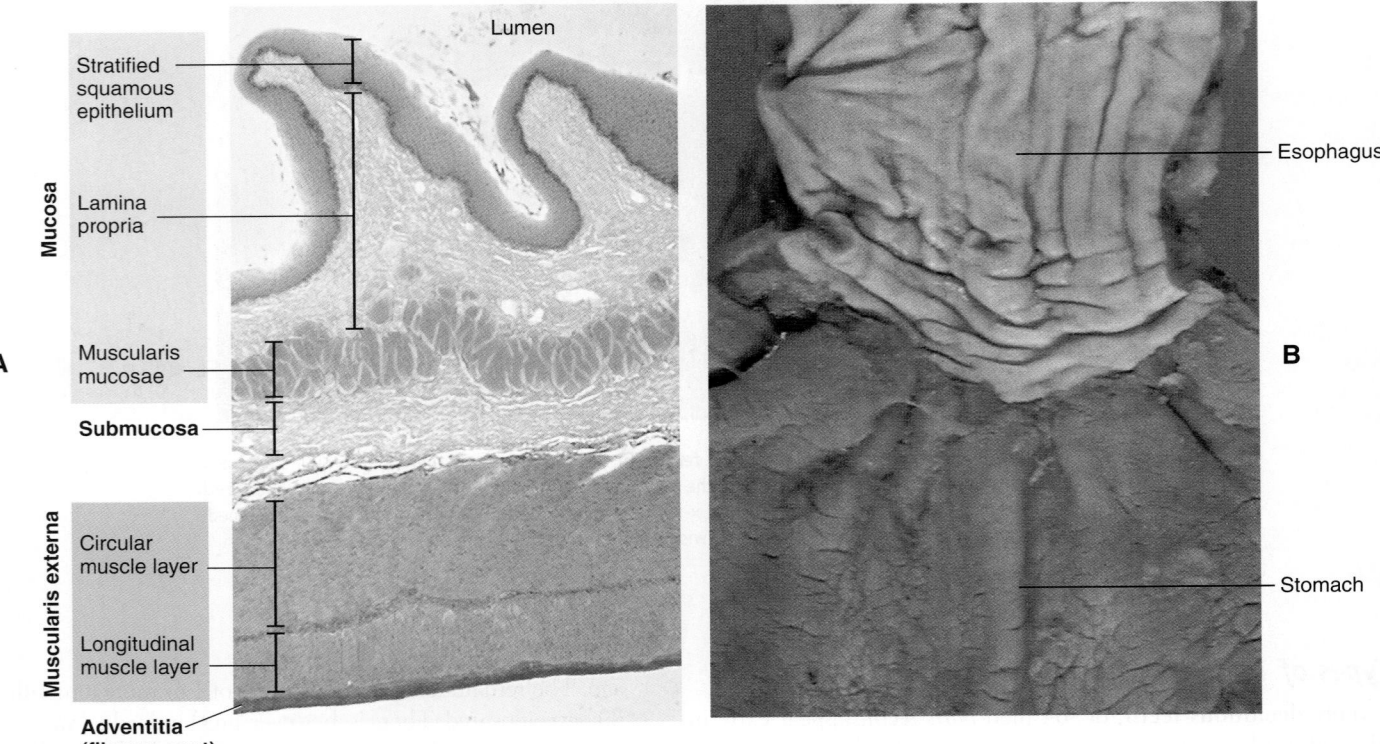

Figure 25-9 *Esophagus.* A, Microscopic appearance of the esophageal wall. The four layers of the GI wall are easily identified (see Table 25-2). **B,** Macroscopic appearance of the junction between the lower part of the esophagus and the stomach—a frequent site of irritation caused by reflux of acidic gastric secretions.

longitudinal layers of the muscular layer are striated (voluntary) in the upper third, mixed (striated and smooth) in the middle third, and smooth (involuntary) in the lower third of the tube.

Each end of the esophagus is guarded by a muscular sphincter. The **upper esophageal sphincter (UES)** helps prevent air from entering during respiration. The **lower esophageal sphincter (LES)**, or **cardiac sphincter,** is located near the *esophageal hiatus*—an opening in the diaphragm located near the junction between the terminal portion of the esophagus and the stomach. The esophageal hiatus in the diaphragm, which permits passage of the esophagus into the abdomen, may become enlarged. Such enlargement may permit bulging of the lower segment of the esophagus and LES and part or even all of the stomach upward through the diaphragm and into the chest. The condition is called a **hiatal** (hye-AYT-al) **hernia.**

Gastroesophageal reflux disease or **GERD** is the term now used to describe the backward flow of stomach acid up through the LES and into the lower part of the esophagus. A potentially serious medical condition, GERD is treated by elimination of the underlying causes, such as a hiatal hernia, by drugs to reduce excess stomach acid, or by surgery to reduce the lumen size or strengthen the LES (see Mechanisms of Disease on p. 953).

The junction between the lower part of the esophagus and stomach (Figure 25-9, *B*) is an important site for a number of pathologic conditions, many associated with repeated exposure to acid gastric secretions. In the last 1- to 1.5-cm-long segment of esophagus below the diaphragm, stratified squamous epithelium is replaced by columnar epithelium. It is this area of transition that is often damaged by exposure to acid and digestive enzymes from the stomach. The area of transition between the lower part of the esophagus and stomach is clearly visible in Figure 25-9, *B*. The stratified squamous epithelial lining of the esophagus appears pale, whereas the columnar gastric epithelium appears brown.

STOMACH

Size and Position of the Stomach

Just below the diaphragm, the digestive tube dilates into an elongated pouchlike structure, the stomach (Figure 25-10), the size of which varies according to several factors, notably the amount of distention. For some time after a meal, the stomach is enlarged because of distention of its walls, but as food leaves, the walls partially collapse, leaving the organ about the size of a large sausage. In adults, the stomach usually holds a volume of up to 1.0 to 1.5 liters.

The stomach lies in the upper part of the abdominal cavity under the liver and diaphragm, with approximately five sixths of its mass to the left of the median line (see Figure 25-1). In other words, it is described as lying in the *epigastrium* and left *hypochondrium* (see Figure 1-14). Its position, however, alters frequently. For example, it is pushed downward with each inspiration and upward with each expiration. When it is greatly distended from an unusually large meal, its size interferes with descent of the diaphragm on inspiration, thereby producing the familiar feeling of dyspnea that accompanies overeating. In this state, the stomach also pushes upward against the heart and may give rise to the sensation that the heart is being crowded.

Divisions of the Stomach

The **fundus, body,** and **pylorus** are the three divisions of the stomach. The fundus is the enlarged portion to the left and above the opening of the esophagus into the stomach. The body is the central part of the stomach, and the pylorus is its lower portion (see Figure 25-10).

Curves of the Stomach

The curve formed by the upper right surface of the stomach is known as the *lesser curvature*; the curve formed by the lower left surface is known as the *greater curvature* (see Figure 25-10).

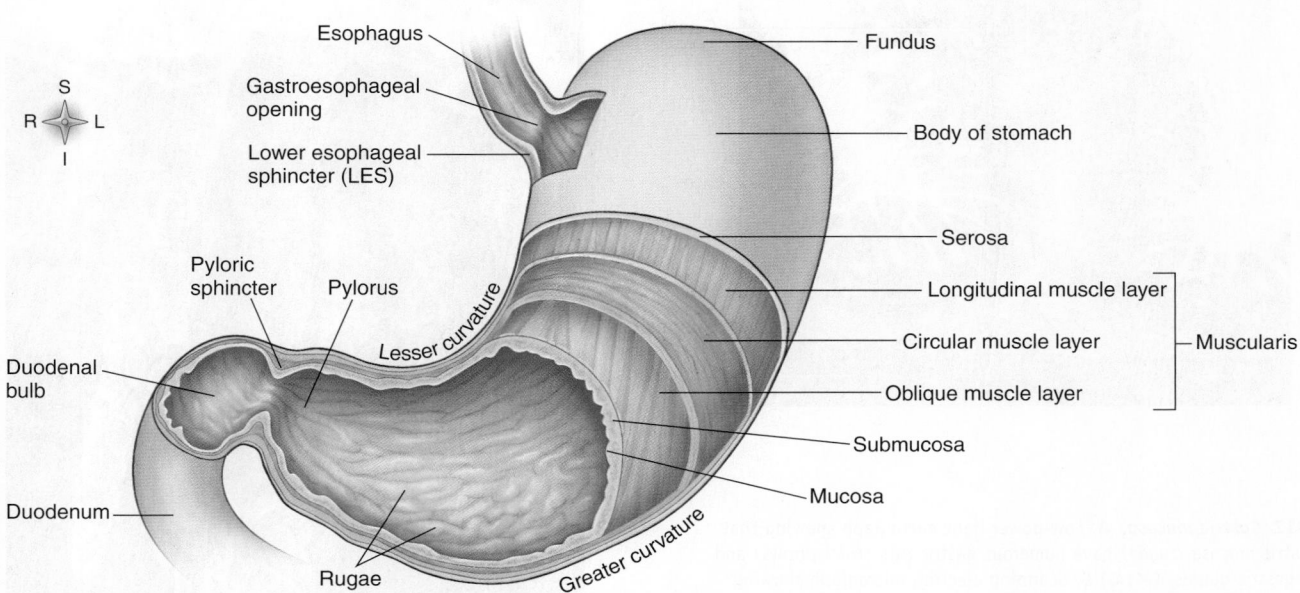

Figure 25-10 *Stomach.* A portion of the anterior wall has been cut away to reveal the muscle layers of the stomach wall. Notice that the mucosa lining the stomach forms folds called *rugae.*

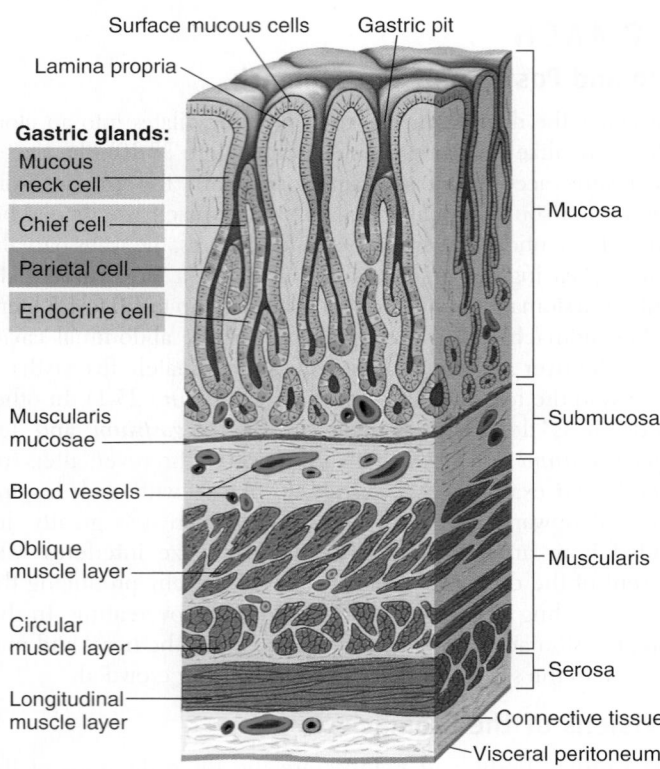

Figure 25-11 *Gastric pits and gastric glands.* Gastric pits are depressions in the epithelial lining of the stomach. At the bottom of each pit is one or more tubular *gastric glands.* Chief cells produce the enzymes of gastric juice, and parietal cells produce stomach acid. Endocrine cells secrete the appetite-boosting hormone ghrelin.

Sphincter Muscles

Sphincter muscles guard both stomach openings. A sphincter muscle consists of circular fibers arranged to form an opening in the center of them (like the hole in a doughnut) when they are relaxed and no opening when they are contracted.

The **lower esophageal sphincter (LES),** or **cardiac sphincter,** controls the opening of the esophagus into the stomach, and the **pyloric sphincter** controls the opening from the pyloric portion of the stomach into the first part of the small intestine (duodenum).

Stomach Wall

Each of the four layers of the stomach wall suits the function of this organ, as summarized in Table 25-2, p. 928 and shown in Figures 25-10 and 25-11. Of particular interest are the modifications to the stomach mucosa and muscularis, both of which are briefly described below.

Gastric Mucosa

The epithelial lining of the stomach is thrown into folds, called *rugae,* and marked by depressions called *gastric pits.* Numerous coiled tubular-type glands, *gastric glands,* are found below the level of the pits, particularly in the fundus and body of the stomach. Figure 25-11 illustrates the anatomical relationship of the gastric pits and gastric glands. The glands secrete most of the gastric juice, a mucous fluid containing digestive enzymes and hydrochloric acid (HCl). Figure 25-12, A, a low-power micrograph of the mucosal lining in the body of the stomach, shows numerous gastric pits and a uniform underlying layer of coiled gastric glands. The mucosal lining is easily differentiated from the deeper submucosal layer in this section. Figure 25-12, B, shows an enlarged view of gastric pits and gastric glands isolated from the submucosa and surrounding tissues.

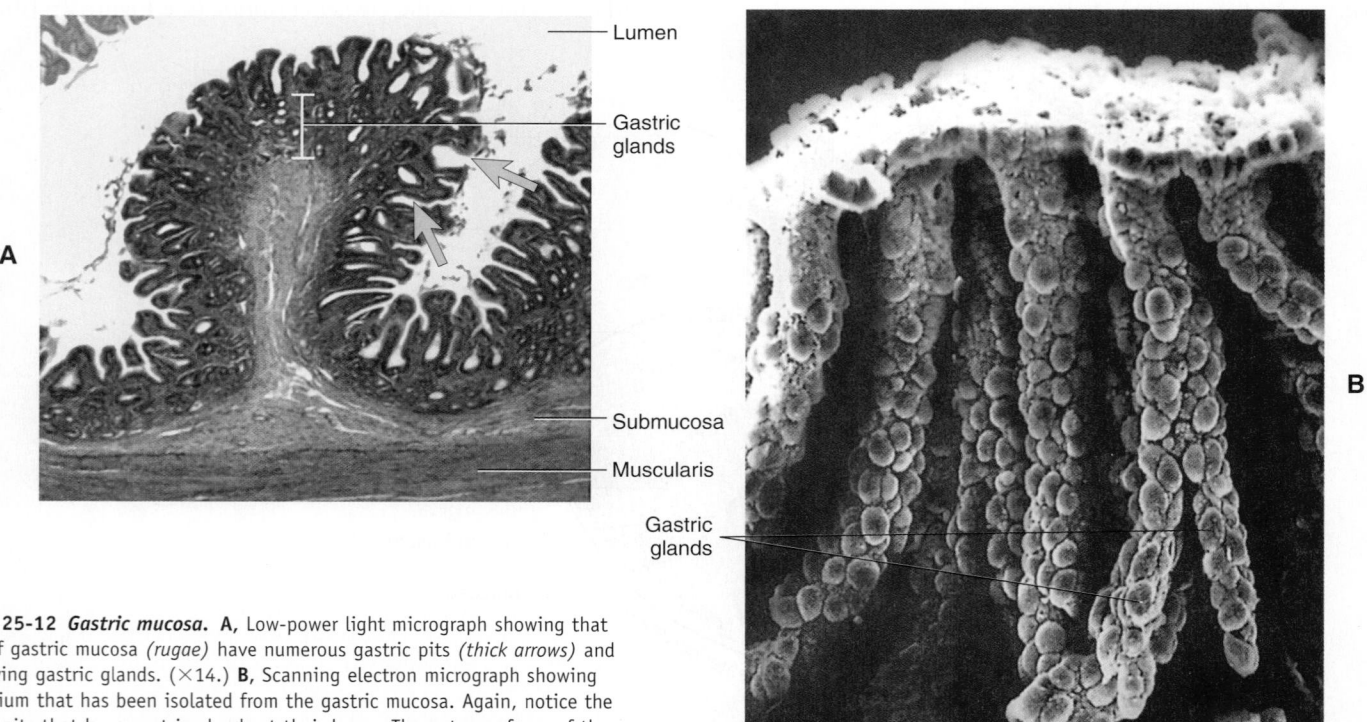

Figure 25-12 *Gastric mucosa.* **A,** Low-power light micrograph showing that folds of gastric mucosa *(rugae)* have numerous gastric pits *(thick arrows)* and underlying gastric glands. (×14.) **B,** Scanning electron micrograph showing epithelium that has been isolated from the gastric mucosa. Again, notice the gastric pits that have gastric glands at their bases. The outer surfaces of the parietal cells are seen as prominent dome-shaped bulges. (×500.)

In addition to the mucus-producing cells that cover the entire surface of the stomach and line the pits, the gastric glands contain three major secretory cells—**chief cells, parietal cells,** and **endocrine cells** (see Figure 25-11). Chief cells (zymogenic cells) secrete the enzymes of gastric juice. Parietal cells secrete hydrochloric acid and are also thought to produce the important substance known as *intrinsic factor.* Intrinsic factor binds to vitamin B_{12} molecules to protect them from digestive juices until they reach the small intestine. Once they reach the small intestine, vitamin B_{12} molecules can be absorbed into the internal environment only if they are bound to intrinsic factor. Endocrine cells secrete *ghrelin*—a hormone that stimulates the hypothalamus to increase appetite—and *gastrin,* which influences digestive functions.

Gastric Muscle

The thick layer of muscle in the stomach wall—the **muscularis**—is made of three distinct sublayers of smooth muscle tissue. As Figure 25-10 shows, there is the usual layer of longitudinal muscles and circular muscles, as well as an additional, underlying oblique layer. The crisscrossing pattern of smooth muscle fibers formed by this arrangement gives the stomach wall the ability to contract strongly at many angles—thus making the mixing action of this organ very efficient.

Functions of the Stomach

The stomach carries on the following functions:

- It serves as a reservoir for food until it can be partially digested and moved farther along the gastrointestinal tract.

- It secretes *gastric juice,* which contains acid and enzymes to aid in the digestion of food.
- Through contractions of its muscular coat, it churns the food, breaking it into small particles and mixing them well with the gastric juice. In time, it moves the gastric contents into the duodenum.
- It secretes *intrinsic factor.*
- It carries on a limited amount of absorption—certain drugs, some water, alcohol, and some short-chain fatty acids found in butter or milk fat.
- It produces the hormones **gastrin,** which helps regulate digestive functions, and **ghrelin,** which increases appetite.
- It helps protect the body by destroying pathogenic bacteria swallowed with food or with mucus from the respiratory tract.

The digestive functions of the stomach are discussed further in Chapter 26.

 QUICK CHECK

8. What is the primary digestive function of the pharynx?
9. Describe the location of the esophagus.
10. What are the three main divisions of the stomach?
11. What are gastric pits?

 BOX 25-1

Diagnostic Study

Upper Gastrointestinal X-Ray Study

An upper GI (UGI) study consists of a series of x-rays of the lower part of the esophagus, stomach, and duodenum, usually with barium sulfate used as the contrast medium. The test is used to detect ulcerations, tumors, inflammation, or anatomical malpositions such as *hiatal hernia* (distention of the diaphragm that allows the stomach to protrude). Obstruction of the upper GI tract is also easily detected.

In this test the patient is asked to drink a flavored drink containing barium sulfate. As the contrast medium travels through the system, the lower portion of the esophagus, gastric wall, pyloric channel, and duodenum are each evaluated for defects. Benign peptic ulcer is a common pathological condition affecting these areas. Tumors, cysts, or enlarged organs near the stomach can also be identified by an anatomical distortion of the outline of the upper GI tract. Can you identify the outline of the stomach in the figure?

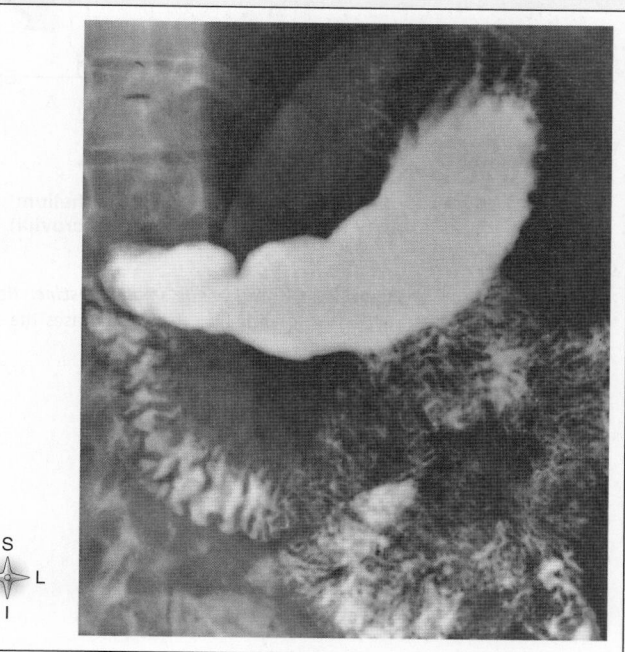

X-ray film of the lower part of the esophagus, stomach, and duodenum.

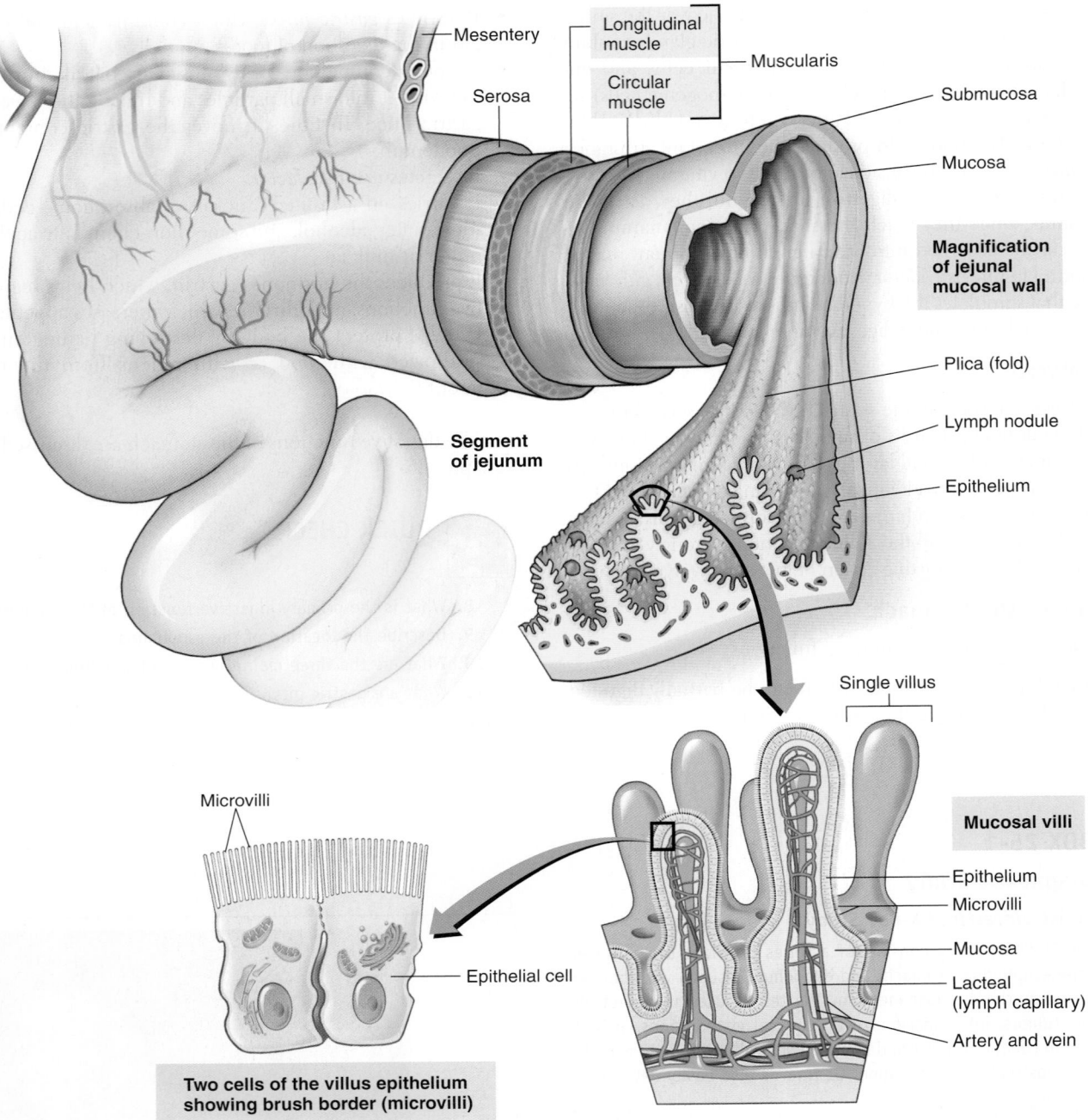

Mesentery

Longitudinal muscle

Circular muscle

Muscularis

Serosa

Submucosa

Mucosa

Magnification of jejunal mucosal wall

Plica (fold)

Lymph nodule

Epithelium

Segment of jejunum

Single villus

Microvilli

Mucosal villi

Epithelium

Microvilli

Mucosa

Lacteal (lymph capillary)

Artery and vein

Epithelial cell

Two cells of the villus epithelium showing brush border (microvilli)

Figure 25-13 *Wall of the small intestine.* Note folds of mucosa are covered with villi and each villus is covered with epithelium, which increases the surface area for absorption of food.

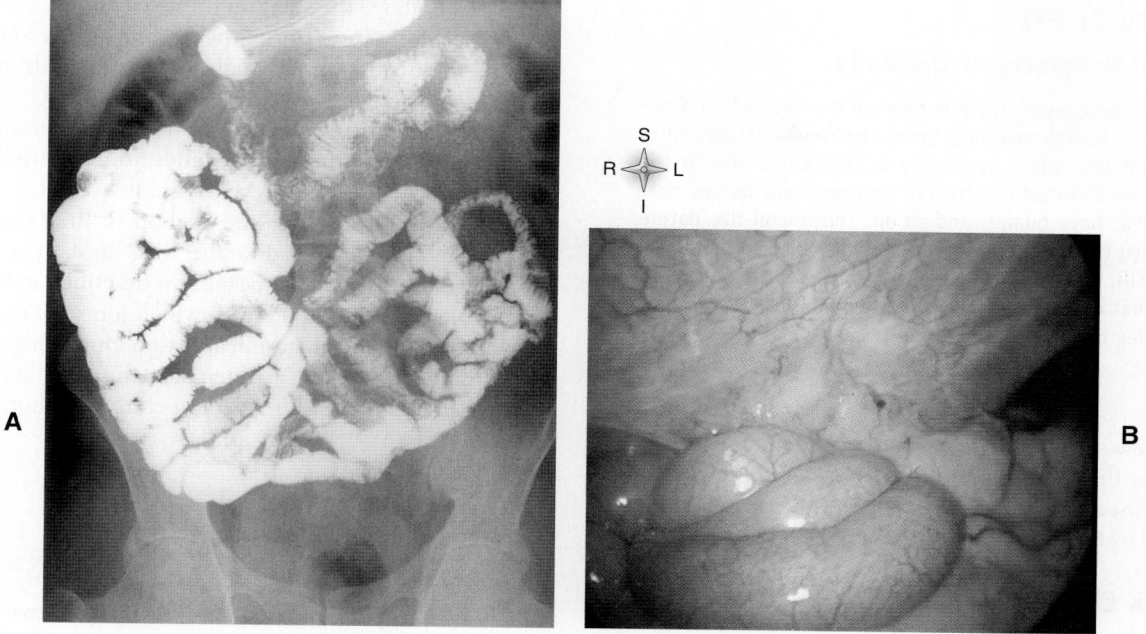

Figure 25-14 *Viewing the small intestine.* **A,** Anteroposterior (AP) x-ray image obtained during a contrast (barium-enhanced) study of the small intestine. The individual is laying supine on the x-ray table. **B,** Laparoscopic view of the small intestine.

SMALL INTESTINE

Size and Position of the Small Intestine

The small intestine is a tube measuring about 2.5 cm (1 inch) in diameter and 6 m (20 feet) in length. Its coiled loops fill most of the abdominal cavity (see Figure 25-1 and Figure 25-14).

Divisions of the Small Intestine

The small intestine consists of three divisions: the duodenum, the jejunum, and the ileum. The **duodenum** is the uppermost division and is the part to which the pyloric end of the stomach attaches. It is about 25 cm (10 inches) long and is shaped roughly like the letter C. The name *duodenum,* meaning "12 fingerbreadths," refers to the short length of this intestinal division. The duodenum becomes **jejunum** at the point where the tube turns abruptly forward and downward. The jejunal portion continues for approximately the next 2.5 m (8 feet), where it becomes the **ileum,** but without any clear line of demarcation between the two divisions. The ileum is about 3.5 m (12 feet) long.

Wall of the Small Intestine

Notice in Figure 25-13 that the intestinal lining has circular *plicae* (folds) that have many tiny projections called **villi.** Villi are important modifications of the mucosal layer of the small intestine. Millions of these projections, each about 1 mm in height, give the intestinal mucosa a velvety appearance. Each **villus** contains an arteriole, venule, and lymph vessel (lacteal) (see Figure 25-13). Epithelial cells on the surface of villi can be seen by microscopy to have a surface resembling a fine brush. This so-called *brush border* is formed by about 1700 ultrafine *microvilli* per cell. Intestinal digestive enzymes are produced in these brush border cells toward the top of the villi. The presence of villi and microvilli

increases the surface area of the small intestine hundreds of times, thus making this organ the main site of digestion and absorption. Mucus-secreting goblet cells are found in large numbers on villi and in crypts (Figure 25-15). Intestinal crypts serve as a site of rapid mitotic cell division. As new cells are produced, older cells are pushed up and out of each crypt, eventually moving to the distal end of a villus, where they are shed. Thus, the intestinal mucosa is

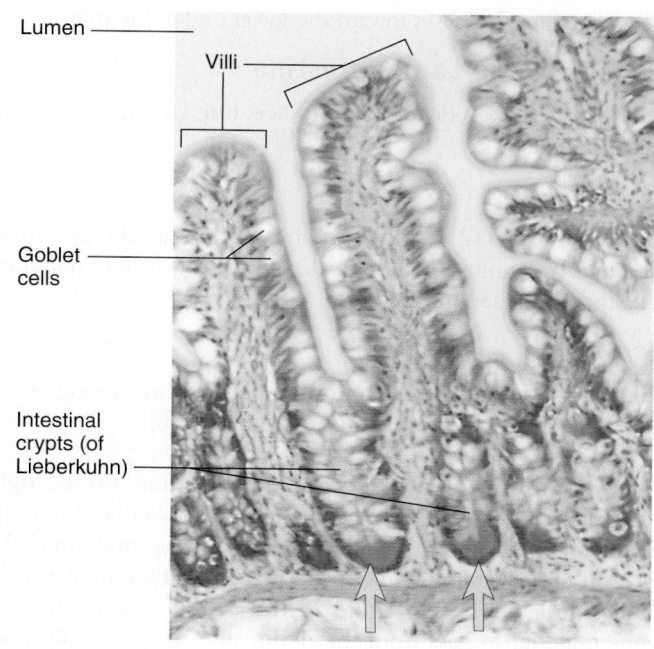

Figure 25-15 *Intestinal villi and crypts.* This micrograph shows a section of the ileum wall with several villi and the intestinal crypts (of Lieberkühn) between them. Cells at the base of each crypt secrete bactericidal enzymes *(arrows)*.

BOX 25-2: FYI
Fractal Geometry of the Body

Biologists are now applying the principles of the new field of **fractal geometry** to human anatomy. Specialists in fractal geometry often study surfaces with a seemingly infinite area, such as the lining of the small intestine. Fractal surfaces have bumps that have bumps that have bumps, and so on. The fractal-like nature of the intestinal lining is represented in Figure 25-13. The plicae (folds) have villi, the villi have microvilli, and even the microvilli have bumps that cannot be seen in the figure. Thus, the absorptive surface area of the small intestine is almost limitless.

continually renewed. At the base of each crypt, secretory cells produce an enzyme that is thought to inhibit bacterial growth in the small intestine. See Table 25-2 and Figure 25-13 for more information on the layers of the small intestine.

QUICK CHECK

12. What are the three main divisions of the small intestine?
13. What are intestinal villi? What is their function?

LARGE INTESTINE
Size of the Large Intestine

The lower part of the alimentary canal bears the name *large intestine* because its diameter is noticeably larger than that of the small intestine. Its length, however, is much less, being about 1.5 to 1.8 m (5 or 6 feet). Its average diameter is approximately 6 cm (2½ inches), but the diameter decreases toward the lower end of the tube.

Divisions of the Large Intestine

The large intestine is divided into the cecum, colon, and rectum (Figure 25-16).

Cecum

The first 5 to 8 cm (2 or 3 inches) of the large intestine is named the *cecum*. It is a blind pouch located in the lower right quadrant of the abdomen (see Figure 25-1).

Colon

The **colon** is divided into the following portions: ascending, transverse, descending, and sigmoid (see Figure 25-16).

- The **ascending colon** lies in a vertical position, on the right side of the abdomen, and extends up to the lower border of the liver. The ileum joins the large intestine at the junction of the cecum and ascending colon, the place of attachment resembling the letter T (see Figure 25-16). The *ileocecal valve* permits material to pass from the ileum into the large intestine, but not usually in the reverse direction.
- The **transverse colon** passes horizontally across the abdomen, below the liver, stomach, and spleen. Note that this part of the

colon is above the small intestine (see Figure 25-1). The transverse colon extends from the *hepatic flexure* to the *splenic flexure*, the two points at which the colon bends on itself to form 90-degree angles.

- The **descending colon** lies in the vertical position, on the left side of the abdomen, and extends from a point below the stomach and spleen to the level of the iliac crest.
- The **sigmoid colon** is the portion of the large intestine that courses downward below the iliac crest. It is called *sigmoid* (meaning "S-shaped") because it describes an S-shaped curve. The lower part of the curve, which joins the rectum, bends toward the left, the anatomical reason for placing a patient on the left side when giving an enema. In this position, gravity aids the flow of enema fluid from the rectum into the sigmoid flexure.

BOX 25-3
Diagnostic Study

Barium Enema Study

The barium enema (BE) study, or lower GI series, consists of a series of x-ray films of the colon that are used to detect and locate polyps, tumors, and diverticula (abnormal "pouches" in the lining of the intestine). Abnormalities in organ position can also be detected.

The test begins with the rectal instillation (enema) of approximately 500 to 1500 ml of fluid containing barium sulfate. The patient is placed in various positions, and the progress of the barium's flow through the intestine is monitored on a fluoroscope. Small polyps and early changes in ulcerative colitis are more easily detected with an **air-contrast barium enema study.** In this study, after the bowel is outlined with a thin coat of barium, air is added to enhance the contrast and outline of small lesions. After the x-ray films are taken, the patient is allowed to expel the barium.

The figure shows a colorized x-ray film of a barium enema produced by a special technique that produces a clear image of the large intestine and its position relative to the skeleton. Can you identify the divisions of the large intestine in the figure? Compare with the plain film x-ray taken after a barium enema shown in Figure 25-16.

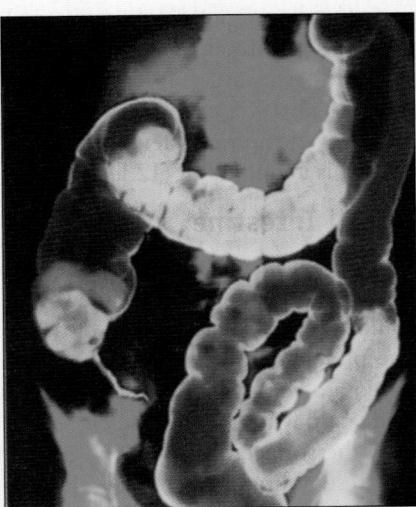

Colorized x-ray film of a barium enema.

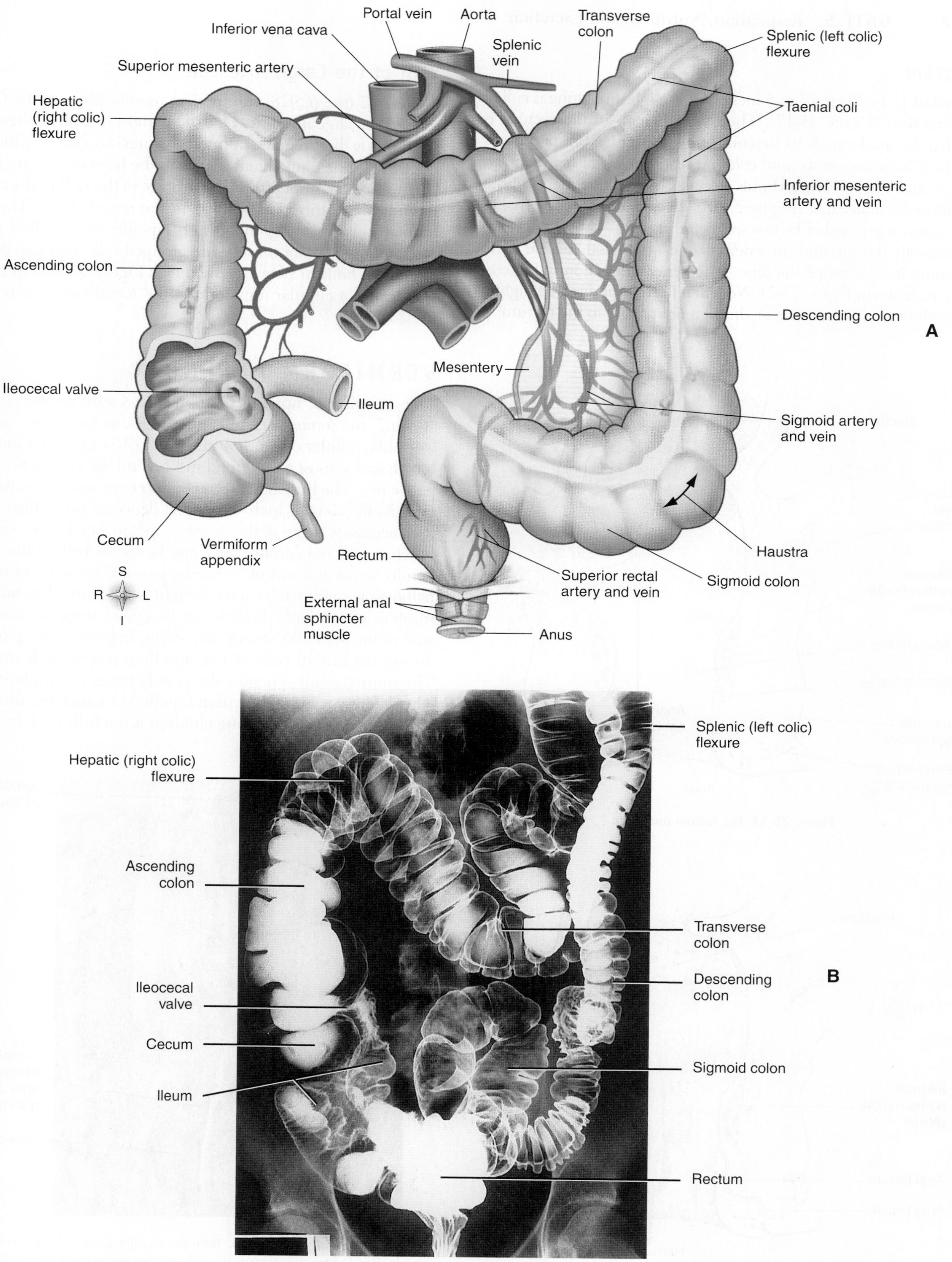

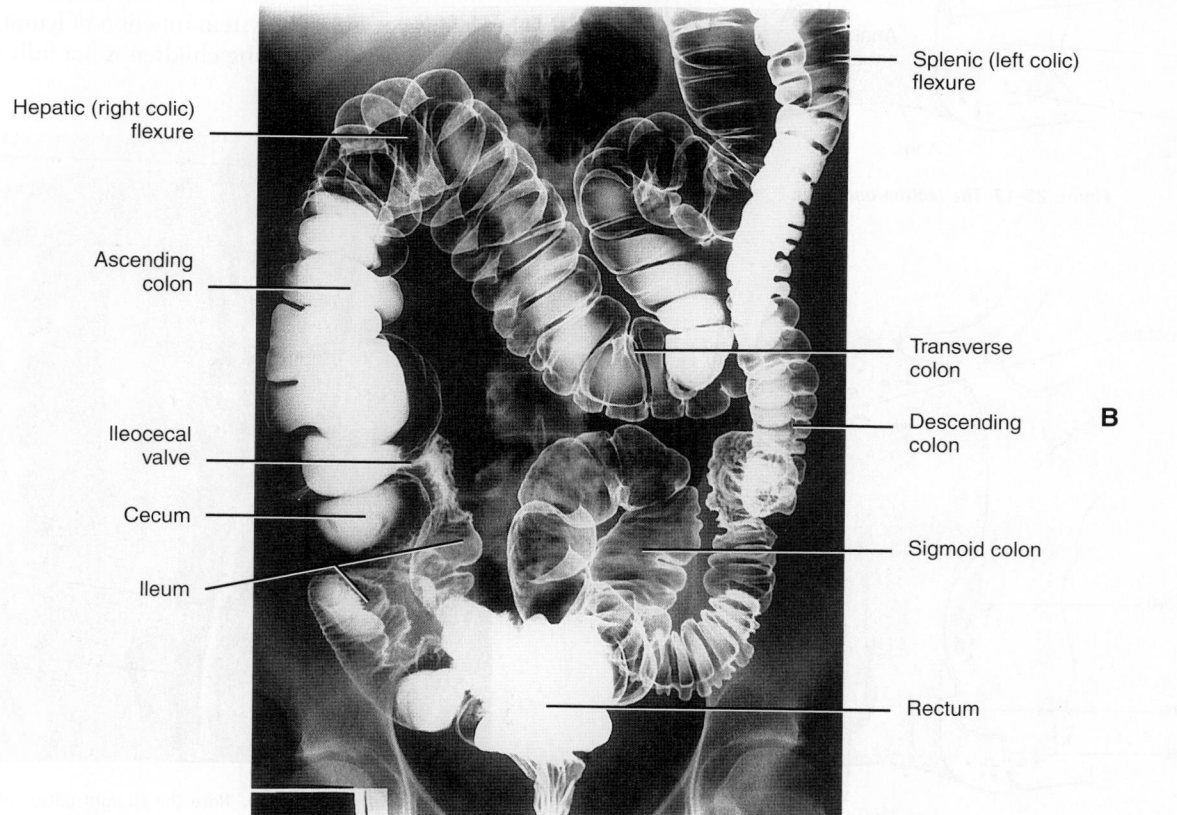

Figure 25-16 *Divisions of the large intestine.* **A,** Illustration showing divisions of the large intestine and adjacent vascular structures. **B,** Normal x-ray film after a double-contrast barium enema (also see Box 25-3).

A

Portal vein
Aorta
Inferior vena cava
Splenic vein
Superior mesenteric artery
Transverse colon
Splenic (left colic) flexure
Hepatic (right colic) flexure
Taenial coli
Inferior mesenteric artery and vein
Ascending colon
Descending colon
Mesentery
Ileocecal valve
Ileum
Sigmoid artery and vein
Cecum
Vermiform appendix
Rectum
Haustra
Superior rectal artery and vein
Sigmoid colon
External anal sphincter muscle
Anus

S
R — L
I

B

Hepatic (right colic) flexure
Splenic (left colic) flexure
Ascending colon
Transverse colon
Ileocecal valve
Descending colon
Cecum
Ileum
Sigmoid colon
Rectum

Rectum

The last 17 to 20 cm (7 or 8 inches) of the intestinal tube is called the **rectum** (Figure 25-17). The terminal inch of the rectum is called the **anal canal**. Its mucous lining is arranged in numerous vertical folds known as *anal columns*, each of which contains an artery and a vein. **Hemorrhoids** (or piles) are enlargements of the veins in the anal canal (Figure 25-18). The opening of the canal to the exterior is guarded by two sphincter muscles—an internal one of smooth muscle and an external one of striated muscle. The opening itself is called the *anus*. The general direction of the rectum is shown in Figure 25-21. Note that the anus is directed slightly posteriorly and is therefore at almost a right angle to the rectum.

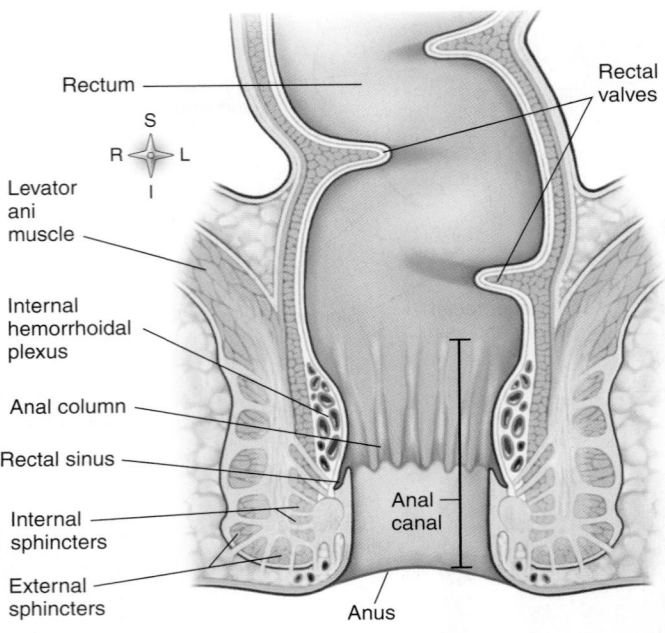

Figure 25-17 *The rectum and anus.*

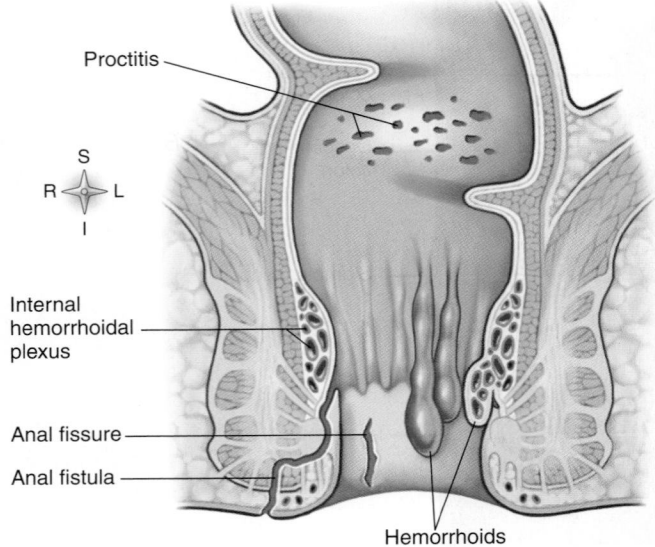

Figure 25-18 *Causes of rectal bleeding.*

Wall of the Large Intestine

Table 25-2 (see p. 928) summarizes modifications of the GI wall seen in the large intestine. One of the most notable of these modifications is the presence of intestinal mucous glands, which produce the lubricating mucus that coats the feces as they are formed (Figure 25-19). Another notable feature of the wall of the colon is the uneven distribution of fibers in the muscle layer. The longitudinal muscles are grouped into tapelike strips called *taeniae coli*, and the circular muscles are grouped into rings that produce pouchlike *haustra* between them (see Figure 25-16). In the rectum, rings of circular muscle form the rectal valves seen in Figure 25-17.

VERMIFORM APPENDIX

The **vermiform appendix** (from the Latin *vermis*, meaning "worm," and *forma*, meaning "shape") is, as the name implies, a wormlike tubular organ. It averages 8 to 10 cm (3 to 4 inches) in length and is most often found just behind the cecum or over the pelvic rim. The lumen of the appendix communicates with the cecum 3 cm (about 1 inch) below the ileocecal valve, thus making it an accessory organ of the digestive system (see Figure 25-16). Its functions are not certain, but some biologists believe that the appendix serves as a sort of "breeding ground" for some of the nonpathogenic intestinal bacteria thought to aid in the digestion or absorption of nutrients. Follicles of lymphoid tissue appear in the wall of the appendix shortly after birth, become more prominent during the first 10 years of life, and then progressively disappear. The normal adult appendix shows only traces of lymphoid tissue. The defense or immune system function of lymphatic tissue present in the appendix of young children is not fully understood.

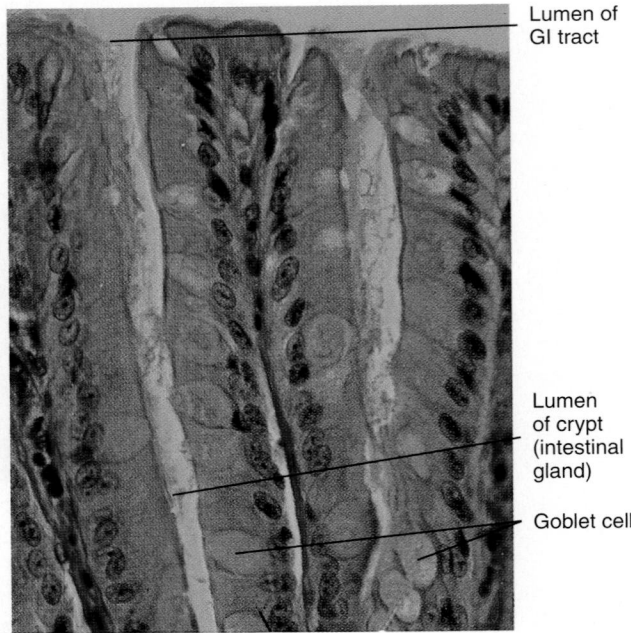

Figure 25-19 *Wall of the colon.* Note the straight nature of the intestinal glands. Many of the columnar epithelial cells are mucus-producing goblet cells. (×140.)

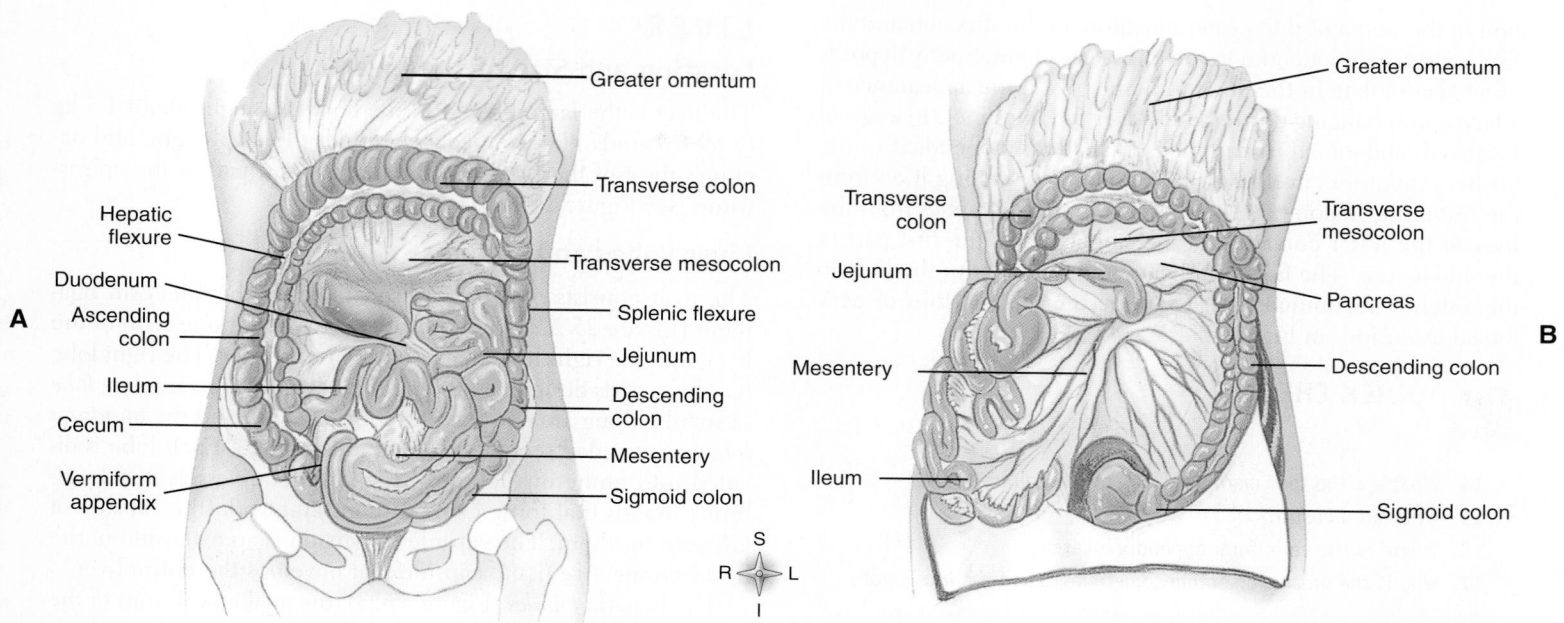

Figure 25-20 *Projections of the peritoneum.* A, Abdominal viscera from the front. The transverse colon and the greater omentum are elevated to reveal the flexures of the colon and the loops of the small intestine. **B,** The transverse colon and greater omentum are raised and the small intestine is pulled to the side to show the transverse mesocolon and mesentery.

Inflammation of the appendix, or **appendicitis,** is a common and potentially very serious medical problem (see Mechanisms of Disease p. 955). The lifetime risk for appendicitis in the United States is 7%. A site on the surface of the anterior abdominal wall is often used to help in the diagnosis of appendicitis and to estimate the location of the appendix internally. It is called *McBurney's point* and is located in the right lower quadrant of the abdomen about a third of the way along a line from the right anterior superior iliac spine to the umbilicus. Extreme sensitivity and pain are common when the abdomen of persons with acute appendicitis is palpated over this point.

PERITONEUM

Now that you have completed the route through the digestive tube, consider for a moment the membrane covering most of these organs and holding them loosely in place. The **peritoneum** is a large, continuous sheet of serous membrane. It lines the walls of the entire abdominal cavity (parietal layer) and also forms the serous outer coat of the organs (visceral layer). In several places the peritoneum forms reflections, or extensions, that bind the abdominal organs together (Figures 25-20 and 25-21). The **mesentery** is a fan-shaped projection of the parietal peritoneum from the lumbar region of the posterior abdominal wall. The attached posterior border of this great fan is just 15 to 20 cm (6 to 8 inches) long, yet the loose outer edge enclosing the jejunum and ileum is 6 m (over 9 feet) long. The mesentery allows free movement of each coil of the intestine and helps prevent strangulation of the long tube. A similar, but less extensive fold of peritoneum, called the **transverse mesocolon,** attaches the transverse colon to the posterior abdominal wall. The **greater omentum** is a continua-

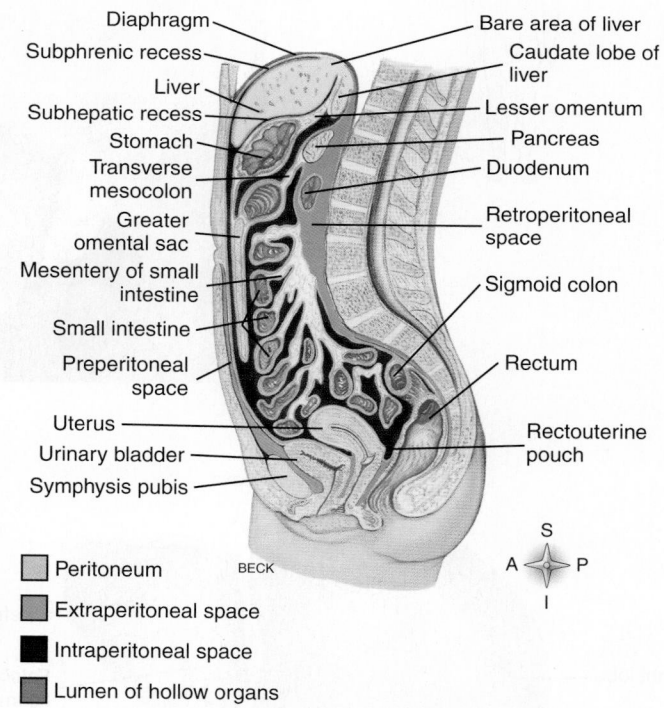

Peritoneum
Extraperitoneal space
Intraperitoneal space
Lumen of hollow organs

Figure 25-21 *Peritoneum.* Sagittal view of the abdomen showing the peritoneum and its reflections. *Intraperitoneal* spaces are shown in black and *extraperitoneal* spaces in green. The portion of the extraperitoneal space along the posterior wall of the abdomen is often called the *retroperitoneal space.*

tion of the serosa of the greater curvature of the stomach and the first part of the duodenum to the transverse colon. Spotty deposits of fat accumulate in the omentum and give it the appearance of a lace apron hanging down loosely over the intestines. In cases of localized abdominal inflammation such as appendicitis, the greater omentum envelops the inflamed area, walling it off from the rest of the abdomen. The **lesser omentum** attaches from the liver to the lesser curvature of the stomach and the first part of the duodenum. The falciform ligament extends from the liver to the anterior abdominal wall. Examine the relationships of peritoneal extensions in Figure 25-21.

QUICK CHECK

14. What are the four main divisions of the colon?
15. What are *haustra*?
16. Where is the vermiform appendix located?
17. Why is the greater omentum sometimes called the *lace apron*?

LIVER

Location and Size of the Liver

The liver is the largest gland in the body. It weighs about 1.5 kg (3 to 4 pounds), lies immediately under the diaphragm, and occupies most of the right hypochondrium and part of the epigastrium (see Figure 25-1).

Liver Lobes and Lobules

The liver consists of two lobes separated by the falciform ligament (Figure 25-22). The **left lobe** forms about one sixth of the liver, and the **right lobe** makes up the remainder. The right lobe has three parts designated the *right lobe proper*, the *caudate lobe* (a small oblong area on the posterior surface), and the *quadrate lobe* (a four-sided section on the undersurface). Each lobe is divided into numerous lobules by small blood vessels and by fibrous strands that form a supporting framework (the capsule of Glisson) for them. The capsule of Glisson is an extension of the heavy connective tissue capsule that envelops the entire liver.

The **hepatic** *lobules* (Figure 25-23), the anatomical units of the liver, are tiny hexagonal or pentagonal cylinders about 2 mm high and 1 mm in diameter. A small branch of the hepatic vein extends through the center of each lobule. Around this central (intralobular)

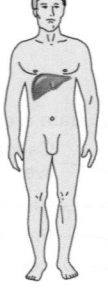

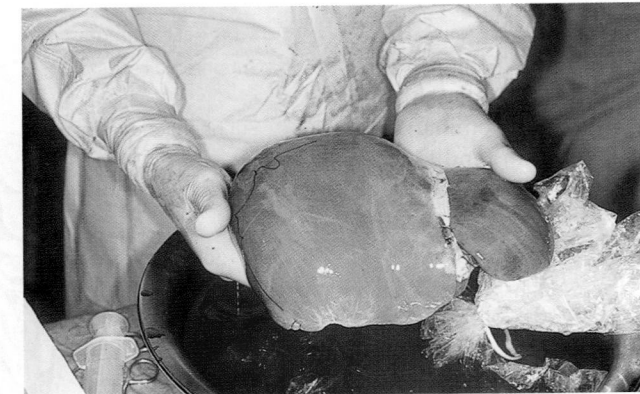

Figure 25-22 *Gross structure of the liver.* **A,** Normal liver prepared for organ transplantation. Diagrams of a normal liver— **B,** anterior view; **C,** inferior view.

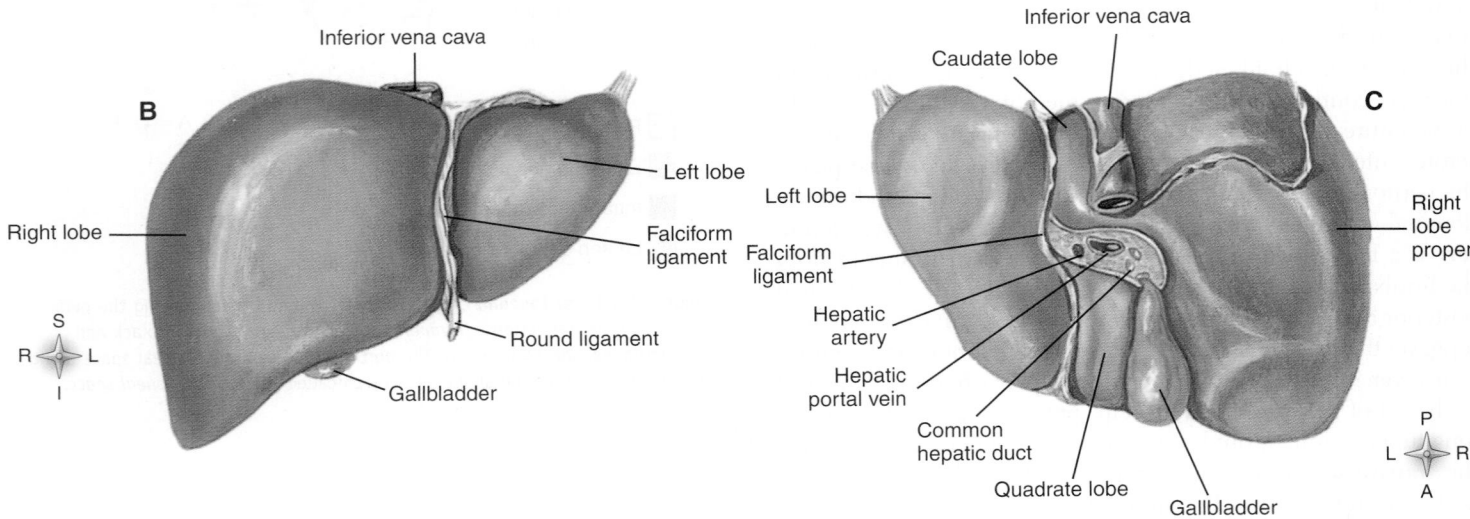

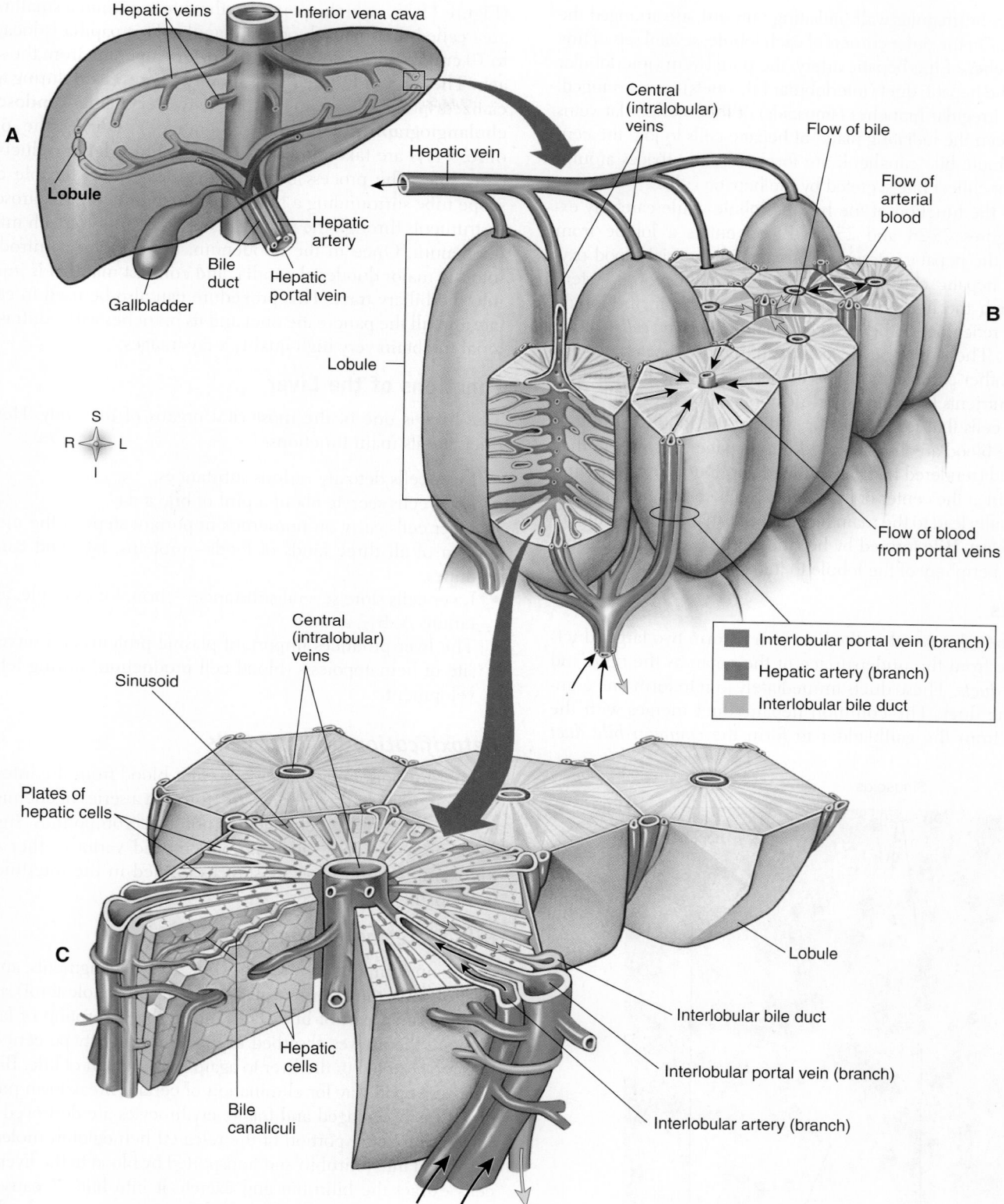

Hepatic veins — Inferior vena cava

A

Lobule

Hepatic artery

Bile duct

Gallbladder

Hepatic portal vein

S
R   L
I

Central (intralobular) veins

Hepatic vein

Flow of bile

Flow of arterial blood

Lobule

B

Flow of blood from portal veins

Central (intralobular) veins

Sinusoid

Plates of hepatic cells

C

Hepatic cells

Bile canaliculi

Lobule

Interlobular bile duct

Interlobular portal vein (branch)

Interlobular artery (branch)

	Interlobular portal vein (branch)
	Hepatic artery (branch)
	Interlobular bile duct

Figure 25-23 *Microscopic structure of the liver.* **A,** This diagram shows the location of liver lobules relative to the overall circulatory scheme of the liver. **B** and **C,** Enlarged views of several lobules show how blood from the hepatic portal veins and hepatic arteries flows through sinusoids and thus past plates of hepatic cells toward a central vein in each lobule. Hepatic cells form bile, which flows through bile canaliculi toward hepatic ducts that eventually drain the bile from the liver.

vein, in plates or irregular walls radiating outward, are arranged the hepatic cells. On the outer corners of each lobule, several sets of tiny tubes—branches of the hepatic artery, the portal vein (interlobular veins), and the hepatic duct (interlobular bile ducts)—are arranged. From these, irregular branches (sinusoids) of the interlobular veins extend between the radiating plates of hepatic cells to join the central vein. Minute bile canaliculi are formed by the spaces around each cell that collect bile secreted by the hepatic cells.

Consider the function of the hepatic lobule while carefully examining Figures 25-23 and 25-24. Blood enters a lobule from branches of the hepatic artery and portal vein. Arterial blood oxygenates the hepatic cells, whereas blood from the portal system passes through the liver for "inspection." Sinusoids in the lobule have many reticuloendothelial cells (mainly *Kupffer cells*) along their lining. These phagocytic cells can remove bacteria, worn RBCs, and other particles from the bloodstream. Ingested vitamins and other nutrients to be stored or metabolized by liver cells enter the hepatic cells that form radiating walls of the lobule. Dissolved toxins in the blood are also absorbed into hepatic cells, where they are detoxified (rendered harmless). Blood continues along the sinusoids to a vein at the center of the lobule. Such central, intralobular veins eventually lead to the main hepatic veins that drain into the inferior vena cava. **Bile** formed by hepatic cells passes through canaliculi to the periphery of the lobule to join small bile ducts.

Bile Ducts

The small bile ducts within the liver join to form two larger ducts that emerge from the undersurface of the organ as the *right* and *left hepatic ducts*. These ducts immediately join to form one common hepatic duct. The common hepatic duct merges with the *cystic duct* from the gallbladder to form the *common bile duct*

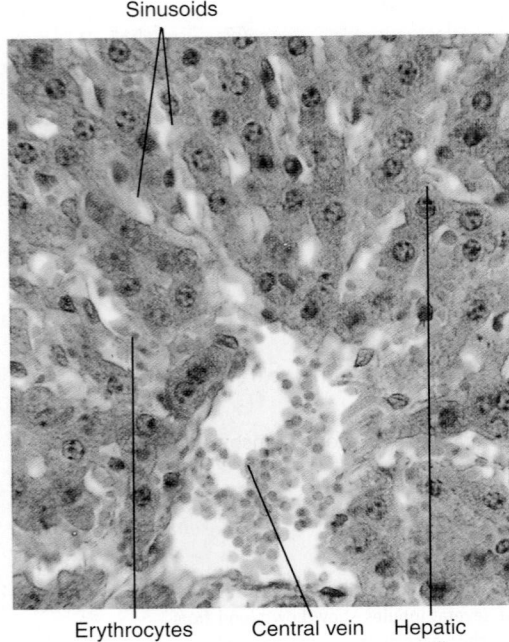

Sinusoids

Erythrocytes   Central vein   Hepatic cell

Figure 25-24 *Liver tissue.* Note the continuity permitting passage of red blood cells from the sinusoids to the central vein. The glandular epithelium (hepatic cells) forms plates between the sinusoids. (Verhoeff stain, ×140.)

(Figure 25-25), which opens into the duodenum in a small raised area called the major duodenal papilla. This papilla is located 7 to 10 cm (2 to 4 inches) below the pyloric opening from the stomach. The inset in Figure 25-25 shows an x-ray taken during a specialized procedure with a tongue-twister name—**endoscopic cholangiography** (koh-lan-jee-OG-rah-fee). During the procedure, x-rays are taken to visualize the gallbladder and ducts that carry bile. The process begins with passage of a flexible endoscope tube surrounding a hollow catheter and other laparoscopic instruments through the mouth, esophagus, and stomach into the duodenum. Once in the duodenum, the catheter is introduced into the major duodenal papilla, and contrast material is injected into the biliary tract. This procedure can also be used to cannulate and fill the pancreatic duct and its branches with contrast material to obtain very high-quality x-ray images.

Functions of the Liver

The liver is one of the most vital organs of the body. Here, in brief, are its main functions:

- Liver cells detoxify various substances.
- Liver cells secrete about a pint of bile a day.
- Liver cells carry on numerous important steps in the metabolism of all three kinds of foods—proteins, fats, and carbohydrates.
- Liver cells store several substances—iron, for example, and vitamins A, B_{12}, and D.
- The liver produces important plasma proteins and serves as a site of hematopoiesis (blood cell production) during fetal development.

Detoxification by Liver Cells

Numerous poisonous substances enter blood from the intestines. They circulate to the liver, where, through a series of chemical reactions, they may be changed to nontoxic compounds. Ingested substances—alcohol, acetaminophen, and various other drugs, for example—and toxic substances formed in the intestines can be detoxified in the liver.

Bile Secretion by the Liver

The main components of bile are bile salts, bile pigments, and cholesterol. Bile salts (formed in the liver from cholesterol) are the most essential part of bile. They aid in the absorption of fats and then are themselves absorbed in the ileum. Eighty percent of bile salts are recycled in the liver to again become part of bile. Bile also serves as a pathway for elimination of certain breakdown products of RBCs. When aged and fragile erythrocytes are destroyed in the spleen, the heme portion of the released hemoglobin molecule is converted into bilirubin and transported by blood to the liver. Liver cells extract the bilirubin and excrete it into bile. Because it secretes bile into ducts, the liver qualifies as an exocrine gland.

Liver Metabolism

Although all liver functions are important for healthy survival, some of its metabolic processes are crucial for survival itself. A fairly detailed description of the role of the liver in metabolism is given in Chapter 27.

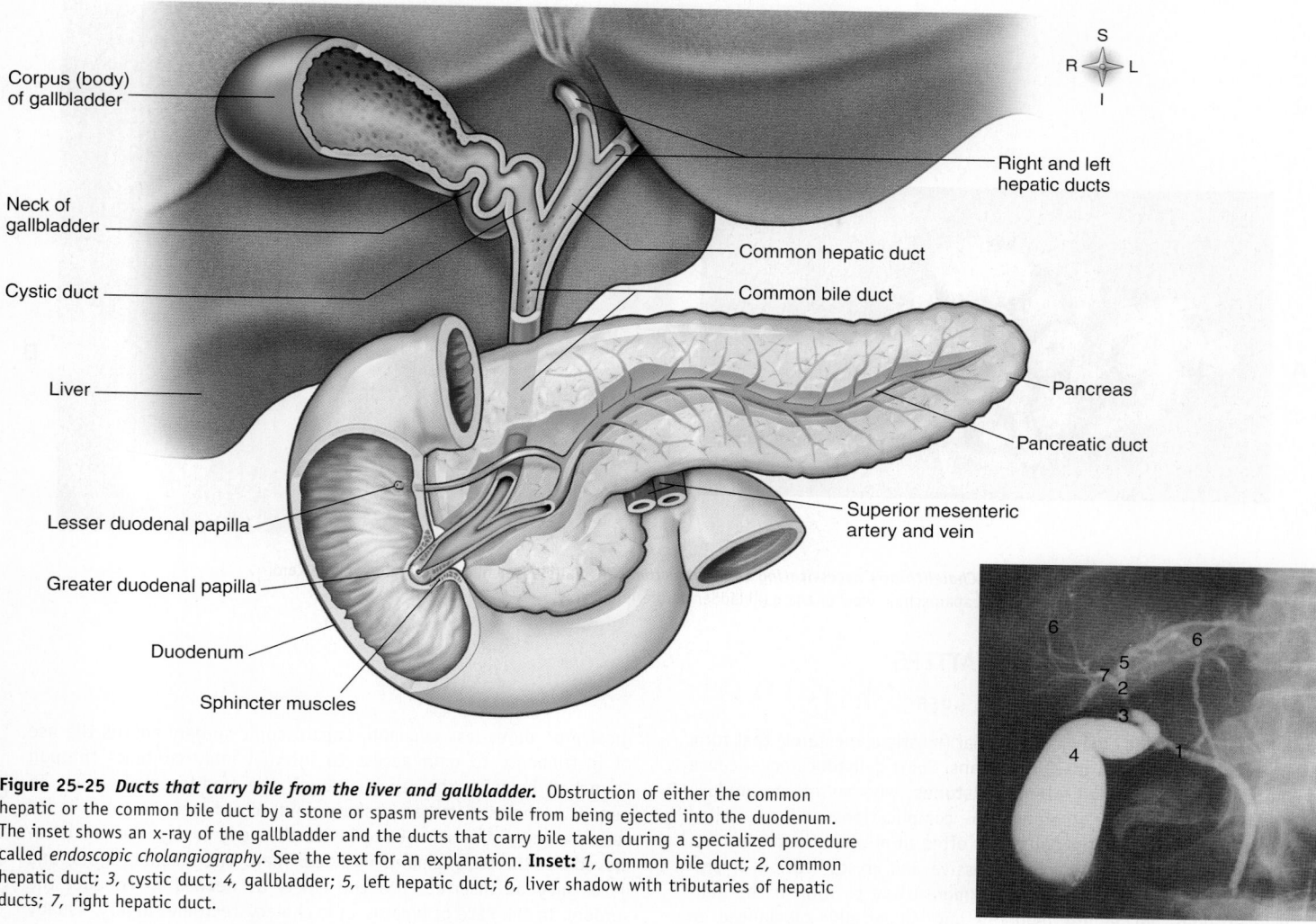

Figure 25-25 *Ducts that carry bile from the liver and gallbladder.* Obstruction of either the common hepatic or the common bile duct by a stone or spasm prevents bile from being ejected into the duodenum. The inset shows an x-ray of the gallbladder and the ducts that carry bile taken during a specialized procedure called *endoscopic cholangiography*. See the text for an explanation. **Inset:** *1,* Common bile duct; *2,* common hepatic duct; *3,* cystic duct; *4,* gallbladder; *5,* left hepatic duct; *6,* liver shadow with tributaries of hepatic ducts; *7,* right hepatic duct.

GALLBLADDER
Size and Location of the Gallbladder

The gallbladder is a pear-shaped sac 7 to 10 cm (3 to 4 inches) long and 3 cm broad at its widest point. (see Figure 25-25). It can hold 30 to 50 ml of bile. The gallbladder lies on the undersurface of the liver and is attached there by areolar connective tissue.

Structure of the Gallbladder

Serous, muscular, and mucous layers compose the wall of the gallbladder. The mucosal lining is arranged in folds called *rugae,* similar in structure to those of the stomach. Inflammation of the gallbladder is called **cholecystitis** (koh-leh-sis-TYE-tis). It is often caused by gallstone formation or **cholelithiasis** (koh-leh-lih-THEE-ah-sis) (Figure 25-26, *A*). Inflammation and stone formation may require surgical removal in a procedure called **cholecystectomy** (koh-leh-sis-TEK-toh-mee). Surgical removal is now commonly performed laparoscopically, a procedure less invasive than traditional surgery (Figure 25-26, *B*). However, efforts to eliminate stones with drugs or nonsurgical methods, such as **ultrasound lithotripsy**, are often the treatment of choice initially.

Functions of the Gallbladder

The gallbladder stores bile that enters it by way of the hepatic and cystic ducts. During this time, the gallbladder concentrates bile five-fold to tenfold. Then later, when digestion occurs in the stomach and intestines, the gallbladder contracts and ejects the concentrated bile into the duodenum. **Jaundice,** a yellow discoloration of the skin and mucosa, results when obstruction of bile flow into the duodenum occurs. Bile is thereby denied its normal exit from the body in feces. Instead, it is absorbed into the blood, and an excess of bile pigments with a yellow hue enters the blood and is deposited in tissues.

PANCREAS
Size and Location of the Pancreas

The pancreas is a grayish pink–colored gland about 12 to 15 cm (6 to 9 inches) long, weighing about 60 g. It resembles a fish with its head and neck in the C-shaped curve of the duodenum, its body extending horizontally behind the stomach, and its tail touching the spleen (Figure 25-27; see Figure 25-1). According to an old anatomical witticism, the "romance of the abdomen" is the pancreas lying "in the arms of the duodenum."

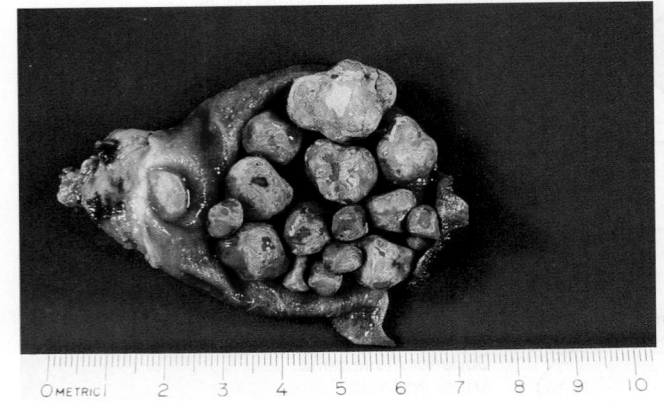
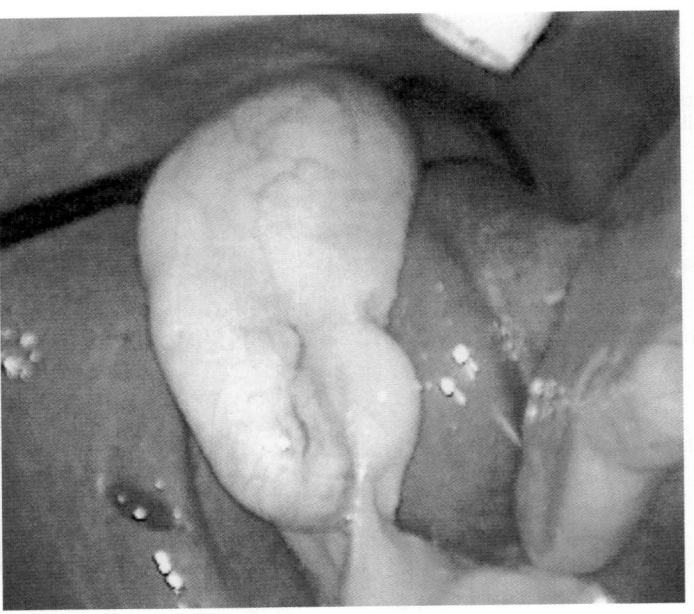

Figure 25-26 *Cholelithiasis necessitating cholecystectomy.* **A,** Gallbladder filled with yellow cholesterol gallstones. **B,** Laparoscopic view of the gallbladder before removal.

BOX 25-4: HEALTH MATTERS

Gallstones and Weight Loss

Gallstones are solid clumps of material (mostly cholesterol) that form in the gallbladder of 1 in 10 Americans. Some gallstones never cause problems and are called **silent gallstones,** whereas others produce painful symptoms or other medical complications and are called **symptomatic gallstones.** Gallstones often form when the cholesterol concentration in bile becomes excessive and crystallization or precipitation occurs. Stones are much more likely to form if the gallbladder does not empty regularly and chemically imbalanced or cholesterol-laden bile remains in the gallbladder for long periods.

The relationship of dieting and weight loss to gallstone formation is currently under intense scrutiny. Physicians have known for years that in severely obese individuals (BMI or body mass index greater than 40), the liver produces higher levels of cholesterol and the risk of gallstones developing is increased. However, only recently have scientists established with certainty that significant and rapid weight loss greatly increases the risk for symptomatic gallstone formation that may require surgery—a procedure called *cholecystectomy* (koh-leh-sis-TEK-toh-mee).

Surgical procedures used for producing weight loss, such as the newer Lap Band Adjustable Gastric Banding System, the more traditional restrictive gastric banding procedure (vertical-banded gastroplasty), or more extensive bypass operations (RGB or Roux-en-Y gastric bypass), almost always result in rapid postsurgical weight loss, but gallstones develop in more than a third of these patients. Unfortunately, individuals who choose nonsurgical approaches to achieve significant and rapid weight loss, such as very-low-calorie or ultra-low-fat diets, also experience higher rates of gallstone formation. In these cases, stone formation is related to imbalances in bile chemistry and delayed emptying or incomplete gallbladder contractions.

If surgery is required for removal of symptomatic gallstones, laparoscopic techniques have now made the need for open abdominal surgical procedures less common. Laparoscopic surgery entails the use of instruments to gain access to internal body contents through "punched holes" rather than through the traditional incision. To punch the holes, a trocar (a sharp pointed rod inside a tube) is inserted through the skin. Once inside the body cavity, the rod is removed but the tube remains in place. Then, instruments, lights, and gases may be inserted into the cavity as needed to conduct the surgery. In the case of laparoscopic cholecystectomy surgery, usually four holes are created. The diagram shows the most common locations for access to the abdominal cavity. Gallstones can sometimes be treated (dissolved) over time or prevented from developing in individuals experiencing rapid weight loss by the oral administration of a naturally occurring bile constituent called *ursodeoxycholic acid* (Actigall).

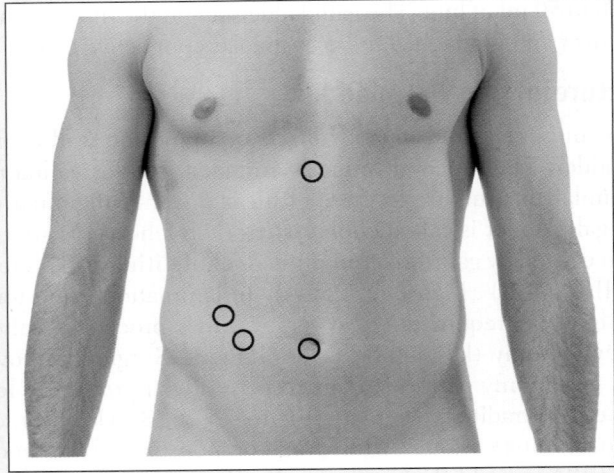

Entry points for laproscopic gallbladder surgery

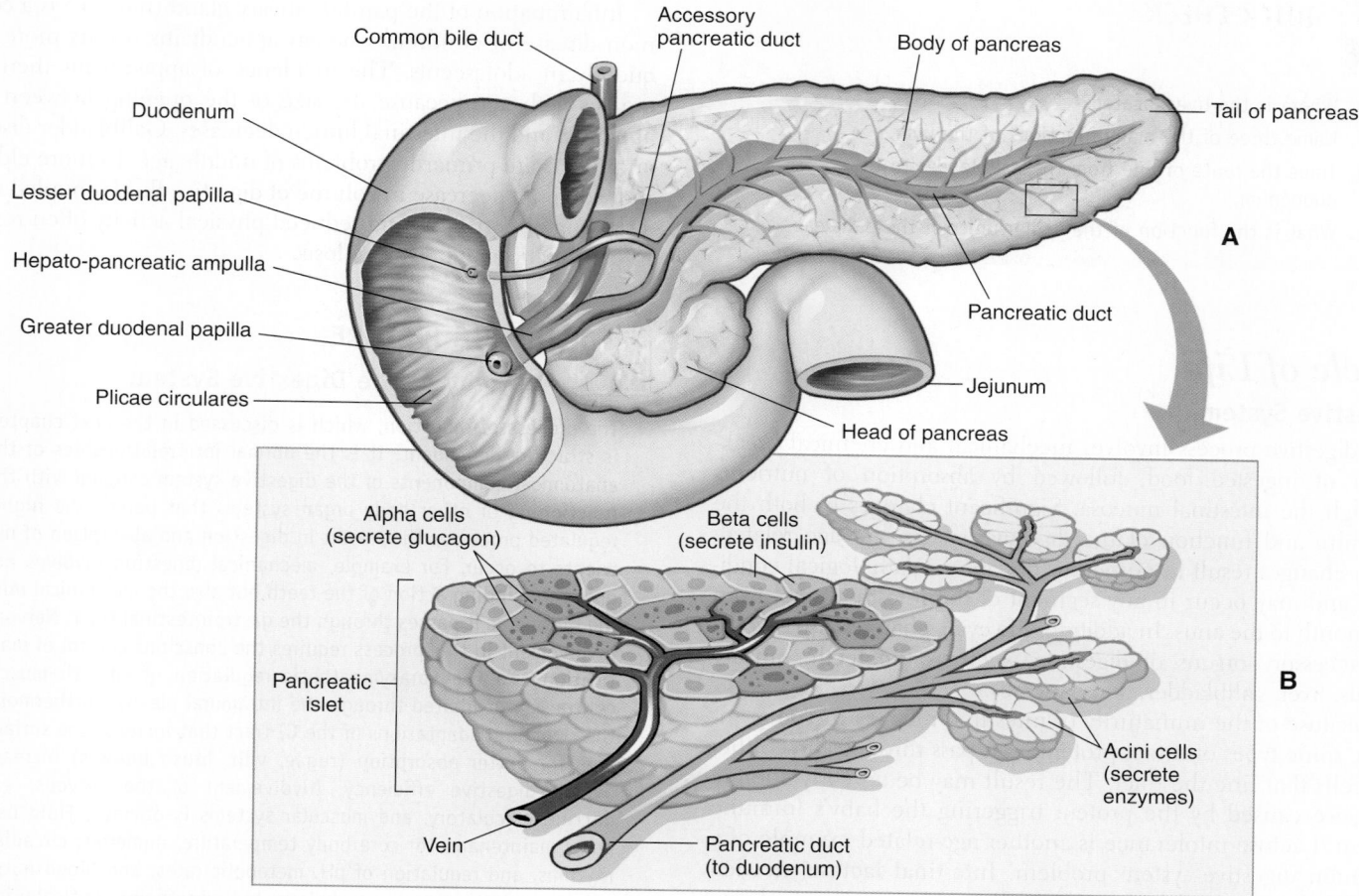

Figure 25-27 *Pancreas.* **A,** Pancreas dissected to show the main and accessory ducts. The main duct may join the common bile duct, as shown here, to enter the duodenum by a single opening at the major duodenal papilla (see Figure 25-25), or the two ducts may have separate openings. The accessory pancreatic duct is usually present and has a separate opening into the duodenum. **B,** Exocrine glandular cells (around small pancreatic ducts) and endocrine glandular cells of the pancreatic islets (adjacent to blood capillaries). Exocrine pancreatic cells secrete pancreatic juice, alpha endocrine cells secrete glucagon, and beta cells secrete insulin.

Structure of the Pancreas

The pancreas is composed of two different types of glandular tissue, one exocrine and one endocrine. Most of the tissue is exocrine, with a compound acinar arrangement. The word *acinar* means that the cells are in a grapelike formation and that they release their secretions into a microscopic duct within each unit (see Figure 25-27, *B*). The word *compound* indicates that the ducts have branches. These tiny ducts unite to form larger ducts that eventually join the main pancreatic duct, which extends throughout the length of the gland from its tail to its head. It empties into the duodenum at the same point as the common bile duct, that is, at the major duodenal papilla. An accessory duct is frequently found extending from the head of the pancreas into the duodenum, about 2 cm above the major papilla (see Figure 25-27).

Embedded between the exocrine units of the pancreas, like so many little islands, lie clusters of endocrine cells called **pancreatic islets** (see Figure 25-27). Although there are about a million of these tiny islands, they constitute only about 2% of the total mass of the pancreas. Special staining techniques have revealed that several kinds of cells—mainly alpha cells and beta cells—make up the islets. They are secreting cells, but their secretion passes into blood capillaries rather than into ducts. Thus, the pancreas is a dual gland—an exocrine, or duct, gland because of the acinar units and an endocrine, or ductless, gland because of the pancreatic islets.

Functions of the Pancreas

- The acinar units of the pancreas secrete the digestive enzymes found in pancreatic juice. Hence the pancreas plays an important part in digestion (Chapter 26).
- Beta cells of the pancreas secrete **insulin,** a hormone that exerts a major control over carbohydrate metabolism (see Figure 25-27).
- Alpha cells secrete **glucagon**. It is interesting to note that glucagon, which is produced so close to where insulin is produced, has a directly opposite effect on carbohydrate metabolism.

QUICK CHECK

18. Where is the liver located?
19. Name three of the many functions of the liver.
20. Trace the route of bile from the gallbladder to the duodenum.
21. What is the function of the acinar units of the pancreas?

Cycle of Life
Digestive System

The digestive process involves mechanical and chemical breakdown of ingested food, followed by absorption of nutrients through the intestinal mucosa. Significant changes in both the structure and function of the digestive system are age related. Such changes result in numerous diseases or pathological conditions and may occur in any segment of the intestinal tract from the mouth to the anus. In addition, life cycle changes also involve the accessory organs of digestion, such as the teeth, salivary glands, liver, gallbladder, and pancreas.

Because of the immaturity of intestinal mucosa in young infants, some types of intact proteins can pass through the epithelial cells that line the tract. The result may be an early allergic response caused by the protein triggering the baby's immune system. Lactose intolerance is another age-related example of a common digestive system problem. Intestinal lactase, needed for the digestion of lactose, or milk sugar, is almost always present at the time of birth. Levels may rapidly diminish in some babies, however, and such individuals soon become unable to digest lactose.

Inflammation of the parotid salivary gland (mumps) is a common disease of children, whereas appendicitis occurs more frequently in adolescents. The incidence of appendicitis then decreases with age because the size of the opening between the appendix and the intestinal lumen decreases. Gallbladder disease and ulcers are primarily problems of middle age. In more elderly individuals, a decrease in volume of digestive fluids coupled with a slowing of peristalsis and reduced physical activity often results in constipation and diverticulosis.

THE BIG PICTURE
Anatomy of the Digestive System

The process of digestion, which is discussed in the next chapter, is structure dependent. It is the normal interrelationships of the anatomical components of the digestive system coupled with the functioning of other body organ systems that permit the highly regulated processes that result in digestion and absorption of nutrients to occur. For example, mechanical digestion involves not only the grinding action of the teeth but also the mechanical mixing of food as it passes through the gastrointestinal tract. Nervous involvement in this process requires the conscious control of mastication and the parasympathetic regulation of smooth muscle contraction mediated through the intramural plexus. Furthermore, the structural adaptations of the GI tract that increase the surface area for better absorption (rugae, villi, brush borders) increase overall digestive efficiency. Involvement of the nervous, endocrine, circulatory, and muscular systems is obvious. Fluid balance, maintenance of core body temperature, numerous circadian rhythms, and regulation of pH, metabolic rates, and blood nutrient concentrations are less obvious, but nonetheless critically important components of the big picture as it relates to the anatomy of the digestive system.

Mechanisms of Disease

DISORDERS OF THE DIGESTIVE SYSTEM

Disorders of the digestive system are described here as well as in the Mechanisms of Disease section at the end of Chapter 26. Disease affects the structure and function of both the primary and accessory organs of digestion and may involve congenital, neural, inflammatory, cancerous, and ulcerative conditions, as well as problems related to secretion, absorption, and motility. Because many clinical manifestations of gastrointestinal tract disorders involve both structural and physiological impairments, students are encouraged to review both disease-related discussions as they complete each chapter.

Diseases of the Salivary Glands

Diseases of the salivary glands, including problems that affect their sympathetic and parasympathetic innervation, may alter both the chemical composition and the amount of saliva produced, about 1 liter per day on average. Inadequate saliva inhibits proper mixing and mastication of food, decreases production of salivary amylase (ptyalin) that initiates digestion in the mouth, and causes an imbalance in salivary pH (normally about 7.4).

Sjögren's Syndrome

Sjögren's (SHOW-grins) **syndrome** is an autoimmune disease in which the body's immune system targets the salivary and tear glands for destruction. The result is a dramatic reduction in both the production of saliva, a condition called **xerostomia** (zer-oh-STOW-mee-ah), and tears. Symptoms of dry mouth and dry eyes, which cause a feeling of irritation and grittiness, becomes progressively worse over time. The syndrome affects more women than men and usually begins around age 50. Among other symptoms, the lack of saliva makes chewing and swallowing difficult and contributes to a higher incidence of

Mechanisms of Disease—cont.

tooth decay. Drugs that stimulate saliva production and the use of artificial tears and saliva help people with Sjögren's syndrome cope with symptoms.

Inflammatory or cancerous conditions of the salivary glands can also affect levels of *immunoglobulin A (IgA)*, which helps reduce the incidence of oral infections after minor injuries and scrapes of the oral mucosa.

Mumps

Mumps is an acute viral disease characterized by swelling and inflammation of the parotid gland *(parotitis)*. Both parotid glands are involved in about 70% of individuals with mumps. It is caused by a paramyxovirus. Initial symptoms include fever, loss of appetite, and a generalized feeling of weakness and discomfort. Within a few hours, swelling of the parotid gland and spasm of the jaw muscles cause pain when the mouth is opened or during chewing movements. As swelling of the parotid gland becomes more pronounced, it extends over the ramus and fills the hollow behind the angle of the mandible to produce the classic "puffy" facial appearance of mumps (Figure 25-28, A). Another helpful diagnostic sign is redness of the parotid papilla on the inside of the cheek opposite the second molar tooth on one or both sides of the upper jaw (Figure 25-28, B). Most of us think of mumps as a childhood disease because it most often affects children between the ages of 5 and 15 years. However, it can occur in adults—often producing a more severe infection. The mumps infection can affect other tissues in addition to the parotid gland, including the joints, pancreas, myocardium, and kidneys. In about 25% of infected men, mumps causes inflammation of the testes, or **orchitis.** Of the 25% of men in whom mumps-related orchitis

develops, only about half experience atrophy of testicular tissue. Furthermore, since the problem is usually unilateral and involves only one testicle, sterility rarely results, although some reduction in fertility may occur.

Disorders of the Mouth and Teeth

Infections, cancer, congenital defects, and other disorders of the mouth and teeth can result in a variety of serious complications. Such conditions may cause pain or damage to the mouth and teeth that makes chewing and swallowing difficult. Mouth infections or cancer may also spread to nearby tissues: the nasal cavity (then on to the sinuses, middle ear, and brain) or pharynx (and on to the esophagus, larynx, and thoracic and other body organs).

Tooth Decay

Tooth decay, or dental caries (KAIR-ees), is a common disease throughout the world. It is a disease of the enamel, dentin, and cementum of teeth that results in the formation of a permanent defect called a *cavity* (Figure 25-29, B). Most people living in the United States, Canada, and Europe are significantly affected by the disease. Decay occurs on tooth surfaces where food debris, acid-secreting bacteria, and plaque accumulate.

If the disease is untreated, tooth decay results in infection, loss of teeth, and inflammation of the soft tissues in the mouth. Bacteria may also invade the paranasal sinuses or extend to the surface of the face and neck, causing serious complications.

Gingivitis (jin-ji-VYE-tis) is the general term for inflammation or infection of the gums. Most cases of gingivitis result from poor oral hygiene—inadequate brushing and no flossing. Gingivitis may also be a complication of other con-

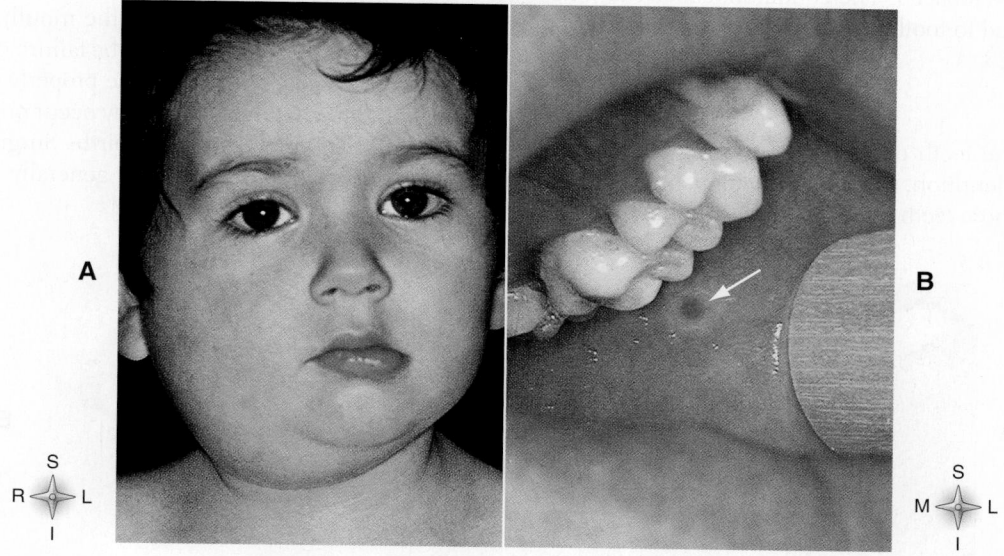

Figure 25-28 *Mumps*. A, This young boy with mumps has unilateral parotid swelling on the right side. **B,** Note the redness of the parotid papilla—an early sign of mumps.

Mechanisms of Disease—cont.

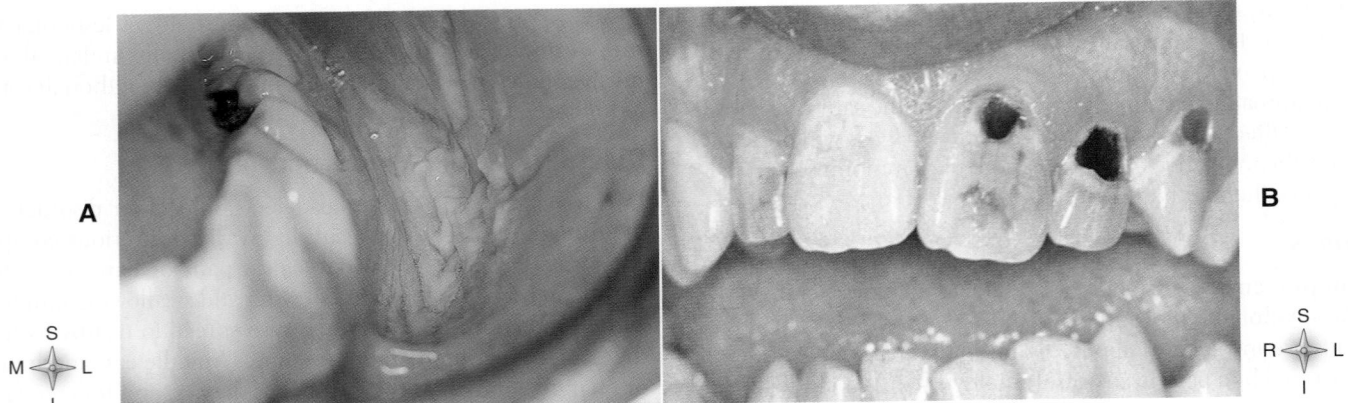

Figure 25-29 *Disorders of the mouth and teeth.* **A,** Leukoplakia (snuff dipper's pouch). This individual has leukoplakia in the area "between the cheek and gum" used for placement of chewing tobacco. **B,** Dental caries. These permanent defects, or cavities, are filled with decayed dental tissues.

ditions such as diabetes mellitus, vitamin deficiency, or pregnancy.

Periodontitis (pair-ee-oh-don-TYE-tis) is inflammation of the periodontal membrane, or *periodontal ligament*, that anchors the tooth to the bone of the jaw. Periodontitis is often a complication of advanced or untreated gingivitis and may spread to the surrounding bony tissue. Destruction of periodontal membrane and bone results in loosening and eventually complete loss of teeth. Periodontitis is the leading cause of tooth loss in adults.

Leukoplakia (loo-koh-PLAY-kee-ah) of the mouth is a precancerous change in the mucous membrane characterized by thickened, white, and slightly raised patches of tissue. Leukoplakia often develops in the fold between the "cheek and gum" in users of smokeless tobacco. The condition, called *snuff dipper's pouch*, may lead to tooth and gum disease, as well as oral cancer (Figure 25-29, A).

Malocclusion

Malocclusion of the teeth occurs when missing teeth create wide spaces in the dentition, when teeth overlap, or when malposition of one or more teeth prevents correct alignment of the maxillary and mandibular dental arches (Figure 25-30, A and B). Malocclusion that results in protrusion of the upper front teeth so that they hang over the lower front teeth is called *overbite* (A), whereas positioning of the lower front teeth outside the upper front teeth is called *underbite* (B).

Dental malocclusion may cause chronic pain and significant problems in functioning of the temporomandibular joint, contribute to the generation of headaches, or complicate routine mastication of food. Fortunately, even severe malocclusion problems can be corrected by the use of braces and other dental appliances. **Orthodontics** (or-thoh-DON-tiks) is the branch of dentistry that deals with the prevention and correction of positioning irregularities of the teeth and malocclusion.

Cleft lip and **cleft palate** (Figure 25-31) are the most common *congenital defects* affecting the mouth. They may occur alone or together and are caused by failure of structures in the upper lip or palate to fuse or close properly during embryonic development. Cleft lip, which may occur on one or both sides, is typically repaired soon after birth. Surgical repair of cleft palate is usually done later—but generally in the first or second year of life.

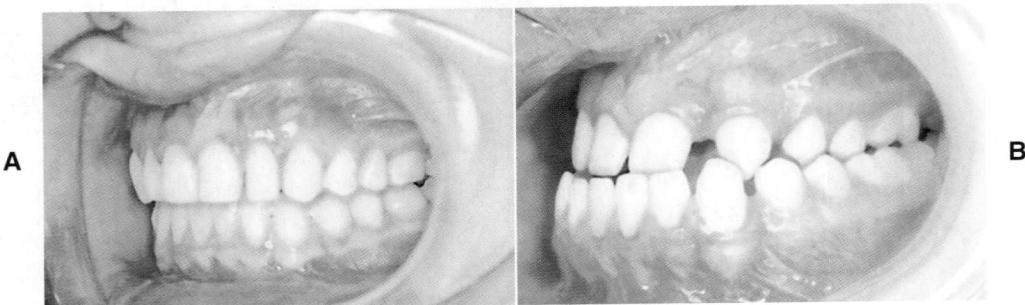

Figure 25-30 *Malocclusion.* **A,** Overbite. **B,** Underbite.

Mechanisms of Disease—cont.

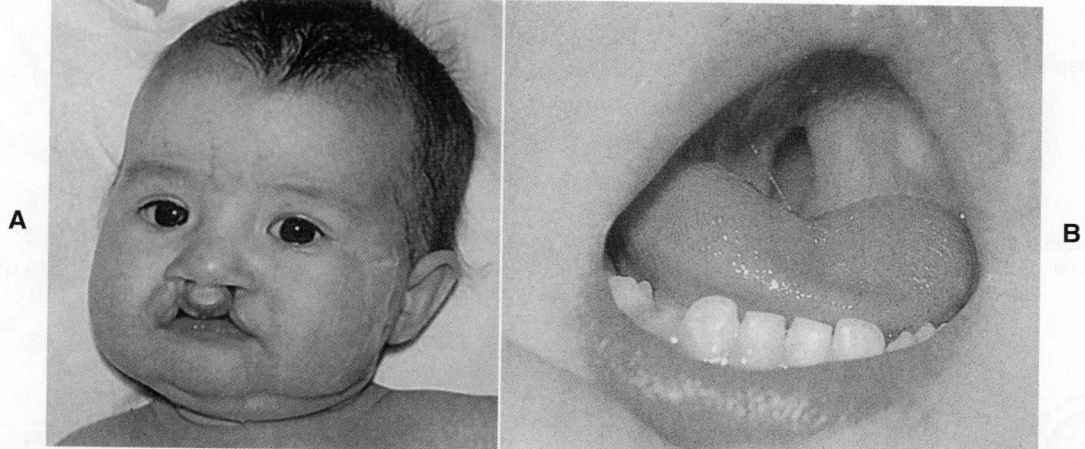

Figure 25-31 *Congenital defects of the mouth.* **A,** Bilateral cleft lip in an infant. **B,** Cleft palate. Modern reconstructive surgery techniques are extremely effective in minimizing the cosmetic, anatomical, and functional problems associated with these defects.

Gastroesophageal Reflux Disease (GERD)

The terms *heartburn* or *acid indigestion* are often used to describe a number of unpleasant symptoms experienced by more than 60 million Americans each month. Backward flow of stomach acid up into the esophagus causes these symptoms, which typically include burning and pressure behind the breastbone. The term **gastroesophageal reflux disease (GERD)** is now used to better describe this common and sometimes serious medical condition.

In its simplest form, GERD symptoms are mild and occur only infrequently (twice a week or less). In these cases, avoiding problem foods or beverages, stopping smoking, or losing weight if needed may solve the problem. Additional treatment with over-the-counter antacids or non–prescription-strength acid-blocking medications called H_2-receptor antagonists (famotidine [Pepcid], others) may also be used. More severe and frequent episodes of GERD can trigger asthma attacks, cause severe chest pain, result in bleeding, or promote a narrowing (stricture) or chronic irritation of the esophagus called **erosive esophagitis** (Figure 25-32, *D*). In these cases, more powerful inhibitors of stomach acid production called proton pump inhibitors (esomeprazole [Nexium], others) may be added to the treatment prescribed. Drugs called promotility agents, which strengthen the lower esophageal sphincter and thus reduce backflow of stomach acid, are also used in moderate to severe cases of GERD.

Two minimally invasive procedures for treating serious cases of GERD are now available (Figure 25-32, *B* and *C*). One, called the **Stretta procedure,** uses radiofrequency energy emitted by a special electrode to produce small burns that tighten the muscular wall of the lower esophageal sphincter and reduce acid reflux from the stomach. The other procedure, called the **Bard endoscopic suturing system,** uses a miniature sewing machine–like device to place two or more stitches in the muscular wall of the lower esophageal sphincter, which are then pulled together to narrow the lumen. In both procedures, which are done on an outpatient basis, a flexible tube called an **esophageal endoscope** is used to insert and then remove the necessary electrode or suturing device required for treatment. As a last resort, a surgical procedure called **fundoplication** is performed to strengthen the sphincter. The procedure involves wrapping a layer of the upper stomach wall around the sphincter and terminal esophagus to lessen the possibility of acid reflux. If GERD is left untreated, serious pathological (precancerous) changes in the esophageal lining may develop—a condition called **Barrett's esophagus.**

Ulcers

An **ulcer** is a craterlike wound or sore in a membrane caused by tissue destruction (Figure 25-33). Current statistics show that about 1 in 10 individuals in the United States will suffer from either a gastric (stomach) or duodenal ulcer in their lifetime. Ulcers cause disintegration, loss, and death of tissue as they erode the layers of the wall of the stomach or duodenum. These craterlike lesions cause gnawing or burning pain and may ultimately result in hemorrhage, perforation, widespread inflammation, scarring, and other very serious medical complications. Usually, perforation does not occur, but small, repeated hemorrhages over long periods can cause anemia.

Two Australian scientists, Dr. Barry Marshall (a microbiologist) and Dr. J. Robin Warren (a pathologist) were awarded the 2005 Nobel Prize in Physiology or Medicine for ulcer research. In 1979 they discovered that infection with a spiral-shaped bacterium called *Helicobacter pylori* (*H. pylori*)—and not excessive acid secretion—was the primary cause of most

Mechanisms of Disease—cont.

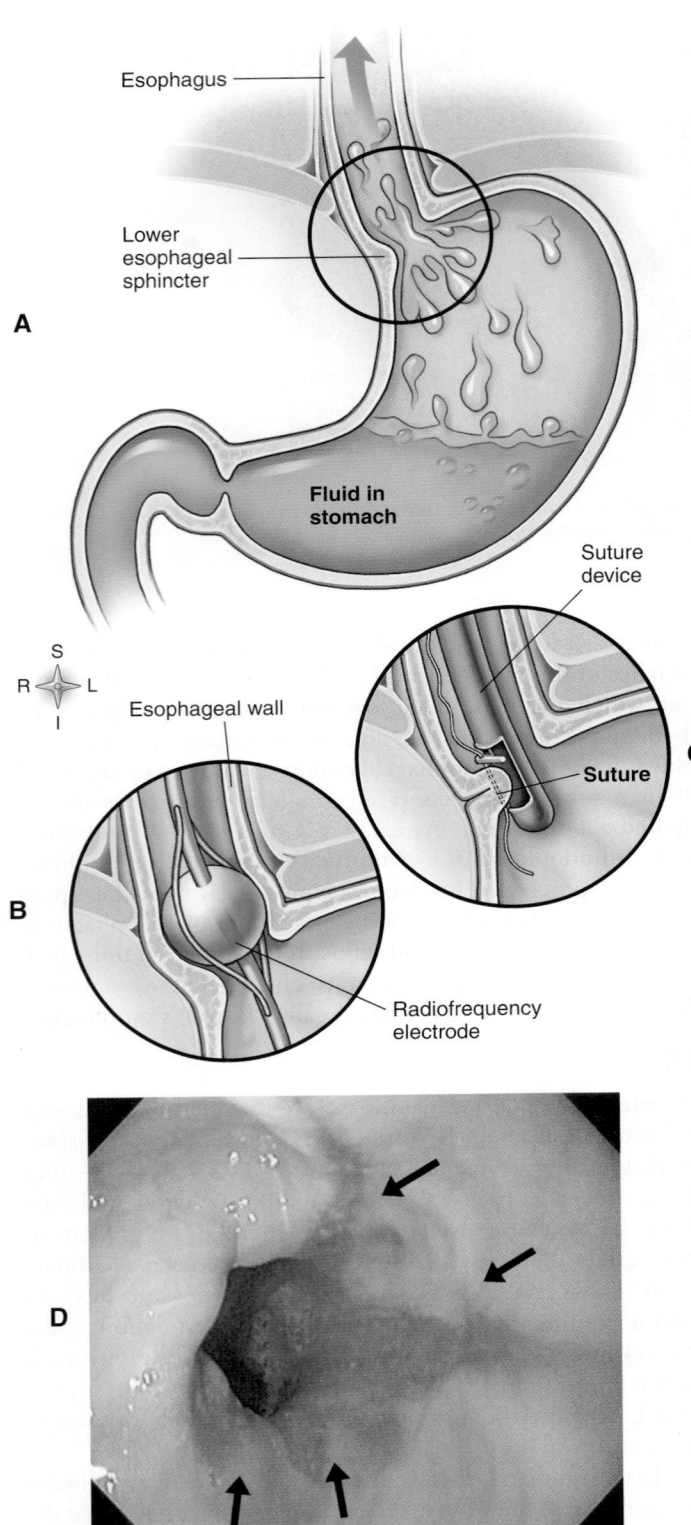

Figure 25-32 *Treating gastroesophageal reflux disease or GERD (see text discussion).* **A,** Reflux of gastric acid up into the esophagus through the lower esophageal sphincter. **B,** Stretta procedure. **C,** Bard endoscopic suturing system. **D,** Endoscopic view of esophageal inflammation *(esophagitis)* caused by "splashing back" of acids from the stomach in a patient with GERD.

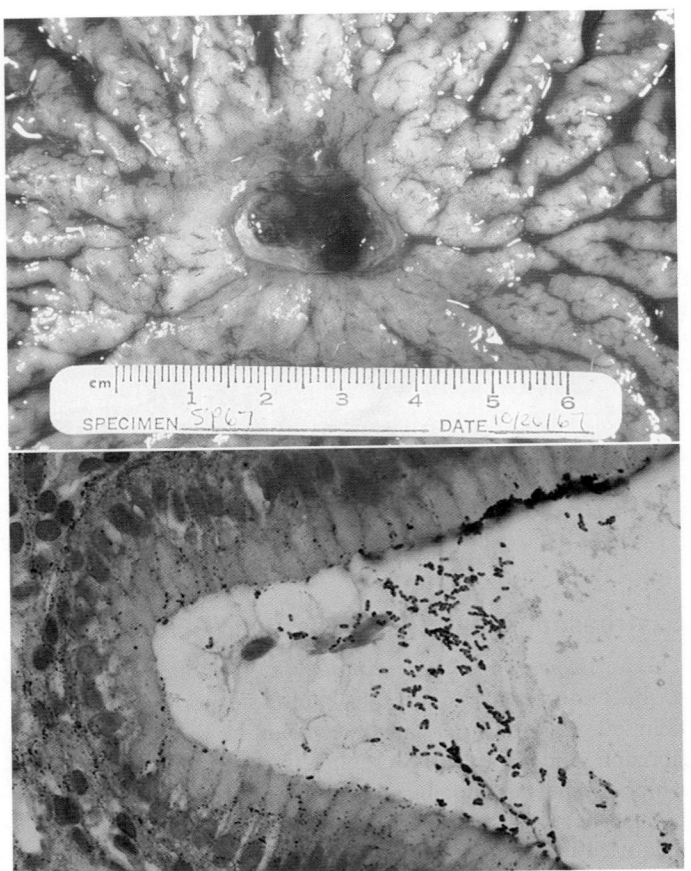

Figure 25-33 A, Gastric ulcer. **B,** *H. pylori* (black particles) infecting the stomach mucosa.

ulcers. The two scientists were quick to report the relationship they observed between inflammation, lack of protective mucus, and tissue erosion around areas of bacterial colonization by *H. pylori* in tissue biopsies obtained from many ulcer patients. However, many in the medical community were initially slow to accept their research results linking *H. pylori* to ulcer development as valid. At the time, it was difficult for many clinicians to accept that ulcers were caused by a bacterium and were, therefore, more like an infectious disease than an illness caused by excess acid. Initial skepticism decreased when Marshall publically swallowed a culture of *H. pylori* and then developed a severe case of gastritis that was successfully treated with antibiotics! Marshall and Warren not only identified a bacterium as the cause of ulcers, their work also suggested that the time honored and traditional use of antacid treatment for ulcers be abandoned and replaced by antibiotic therapy. In awarding the Nobel, their research was described as producing "one of the most radical and important changes in the last 50 years in the perception of a medical condition." What was once considered a radical new explanation for the cause of ulcers is now a proven fact. And, antibiotic treatment is now an important part of the accepted standard of

Mechanisms of Disease—cont.

care for most ulcer patients. *H. pylori* infection in ulcer patients can be diagnosed by biopsy, breath, or blood antibody tests.

Long-term use of certain pain medications such as aspirin and ibuprofen, called **nonsteroidal anti-inflammatory drugs (NSAIDs)**, can also cause ulcers. These drugs interfere with prostaglandins that regulate the mucus lining of the GI tract. NSAID-induced ulcers can be treated by stopping NSAID use and taking acid-reducing drugs until the ulcer heals.

Disorders of the Pyloric Sphincter

The pyloric sphincter is of clinical importance because **pylorospasm** is a fairly common condition in infants. The pyloric fibers do not relax normally to allow food to leave the stomach, and the infant vomits food instead of digesting and absorbing it. The condition is relieved by the administration of a drug that relaxes smooth muscles. Another abnormality of the pyloric sphincter is **pyloric stenosis**, an obstructive narrowing of its opening.

Appendicitis

If the mucous lining of the appendix becomes inflamed, the resulting condition is the well-known affliction **appendicitis**. As you can see in Figure 25-16, the appendix is very close to the rectal wall. For patients with suspected appendicitis, a physician often evaluates the appendix by a digital rectal examination.

The opening between the lumen of the appendix and the cecum is quite large in children and young adults—a fact of clinical significance because food, fecal material, or calcified, stone-like concretions called **appendicoliths** (ah-pen-DIK-oh-liths) may become trapped in the opening, block the lumen, and cause irritation and inflammation resulting in appendicitis. If calcification within the appendix is visible on an x-ray in a patient with pain in the lower right abdominal quadrant, there is an extremely high probability—physicians call it "clinical suspicion"—of acute appendicitis.

The opening between the appendix and the cecum is often completely obliterated in elderly persons, which explains the low incidence of appendicitis in this population.

If infectious material becomes trapped in an inflamed appendix, the appendix may rupture and release the material into the abdominal cavity (Figure 25-34). Infection of the peritoneum and other abdominal organs may result—with sometimes tragic consequences.

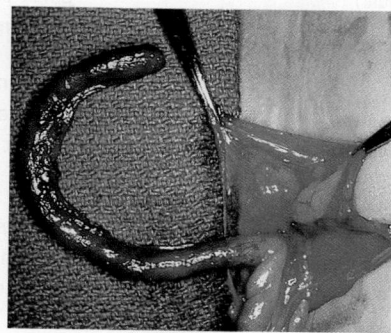

Figure 25-34 *Acute appendicitis.* Note the inflamed tissue surrounding the base of a gangrenous appendix.

Rectal Bleeding

Rectal bleeding is a symptom that should always be investigated. Although it may signal a serious disease problem, such as cancer, most instances of this common problem are indicators of less serious and non–life-threatening conditions (see Figure 25-18).

Hemorrhoids are dilated veins that result from direct irritation or from increases in venous pressure that often accompany pregnancy or result from constipation and the subsequent straining required to pass compact and hardened stools. They most frequently develop near the anal opening or on the wall of the anal canal. Though often painful and irritating, hemorrhoids generally respond readily to treatment and are seldom a serious health concern.

Proctitis, or inflammation of the rectal mucosa, is another frequent cause of rectal bleeding and related symptoms such as mucus discharge or increased frequency of bowel movements. The condition may result from direct irritation or infection. In most cases, proctitis responds quickly to the administration of anti-inflammatory drugs and treatment of the underlying problem.

Anal fissures are generally minor lacerations in the lining of the anus or anal canal that result in rectal bleeding. They are caused by direct irritation—often the result of passing a hardened stool. An **anal fistula** is a more serious problem that may require surgical repair. A fistula is a passageway that often develops between the rectal wall and the skin surrounding the anus. Fistulas occur in Crohn disease, an inflammatory bowel disease.

LANGUAGE OF SCIENCE

(Cont'd from page 925)

jejunum (jeh-JOO-num) [*jejunus* empty]

left lobe

lesser omentum (oh-MEN-tum) [*losian* to lose, *omentum* entrails]

lips

lower esophageal sphincter (LES) (eh-SOF-eh-JEE-ul SFINGK-ter) [*oisophagos-* gullet, *-al* pertaining to, *sphingein* to bind]

mastication (mas-ti-KAY-shun) [*masticare* to chew]

mesentery (MEZ-en-tair-ee) [*mes-* middle or median, *-enteron* intestine]

mouth

mucosa (myoo-KOH-sah) [*mucus* slime]

muscularis (mus-kyoo-LAIR-is) [*musculus* muscle]

pancreatic islets (pan-kree-AT-ik EYE-lets) [*pan-* all, *-kreas-* flesh, *-ic* pertaining to, *islet* island]

parietal cells (pah-RYE-i-tal sells) [*paires-* wall, *-al* pertaining to, *cella* storeroom]

parotids (peh-RAH-tids) [*para-* beside, *-ous* ear]

peritoneum (pair-i-toh-NEE-um) [*peri-* around, *-tenein* to stretch]

permanent teeth [*permanere* to remain]

pulp cavity [*pulpa* flesh, *cavus* hollow]

pyloric sphincter (pye-LOR-ik SFINGK-ter) [*pyle-* gate, *-ouros-* to guard, *-ic* pertaining to, *sphingein* to bind]

pylorus (pye-LOR-us) [*pyle-* gate, *-ouros* guard]

rectum (REK-tum) [*rectus* straight]

right lobe

root

serosa (see-ROH-sah) [*serum* watery fluid]

sigmoid colon (SIG-moyd KOH-lon) [*sigma-* the letter S, *-oid* resembling, *kolon* colon]

soft palate (PAL-et) [*palatum* palate]

sublingual glands (sub-LING-gwall) [*sub-* under, *-lingua* tongue]

submandibular glands (sub-man-DIB-yoo-lar) [*sub-* under, *-mandere* to chew, *glans* acorn]

submucosa (sub-myoo-KOH-sah) [*sub-* under, *-mucus* slime]

tongue

transverse colon (trans-VERS KOH-lon) [*transversus* oblique, *kolon* colon]

transverse mesocolon (trans-VERS MEZ-oh-koh-lon) [*transversus* oblique, *meso-* middle or median, *-kolon* colon]

upper esophageal sphincter (UES) (eh-SOF-ah-JEE-ul SFINGK-ter) [*oisophagos-* gullet, *-al* pertaining to, *sphingein* to bind]

vermiform appendix (VERM-i-form ah-PEN-diks) [*vermis-* worm, *-forma* shape, *appendere* to hang upon]

villus (VIL-us) [*villus* shaggy hair], pleural villi

LANGUAGE OF MEDICINE

air-contrast barium enema study

anal fissures (AY-nal FISH-urs) [*anal* pertaining to the anus, *fissura* cleft]

anal fistula (AY-nal FISS-tyoo-lah) [*anal* pertaining to the anus, *fistula* pipe]

appendicitis (ah-pen-di-SYE-tis) [*appendere-* to hang upon, *-itis* inflammation]

appendicoliths (ah-pen-DIK-oh-liths) [*appendere-* to hang upon, *-lith* a stone]

Bard endoscopic suturing system

Barrett's esophagus (eh-SOF-ah-gus) [*Norman R. Barrett* English surgeon, *oisophagos* gullet]

cholecystectomy (koh-leh-sis-TEK-toh-mee) [*chole-* bile, *-kystis-* bag, *-ectomy* surgical removal]

cholecystitis (koh-leh-sis-TYE-tis) [*chole-* bile, *-kystis-* bag, *-itis* inflammation]

cholelithiasis (koh-leh-lih-THEE-ah-sis) [*chole-* bile, *-lithos-* stone, *-osis* condition]

cleft lip (kleft)

cleft palate (kleft PAL-et)

endoscopic cholangiography (en-doh-SKOP-ik koh-lan-jee-OG-rah-fee) [*endo-* within, *-scope* instrument for observation, *cholangio-* relationship to a bile duct, *-graphy* process of recording]

erosive esophagitis (eh-ROH-siv eh-SOF-ah-jye-tis) [*erodere* to consume, *oisophagos-* gullet, *-itis* inflammation]

esophageal endoscope (eh-sof-ah-JEE-ul EN-doh-skohp) [*oisophagos-* gullet, *-al* pertaining to, *endo-* within, *-scope* instrument for observation]

fractal geometry

fundoplication (fun-doh-plih-KAY-shun) [*fundus-* bottom, *-plicare* to fold]

gastroesophageal reflux disease (GERD) (gas-troh-eh-sof-eh-JEE-all REE-fluks) [*gastro-* stomach, *-oisophagos-* gullet, *-al* pertaining to, *refluere* to flow back, *dis-* reversal, *-aise* ease]

gingivitis (jin-ji-VYE-tis) [*gingiva-* gum, *-itis* inflammation]

hemorrhoids (HEM-eh-royds) [*hemo-* blood vessel, *-rrhois* flow]

hiatal hernia (hye-AY-tal HER-nee-ah) [*haire-* to stand open, *-al* pertaining to, *hernia* rupture]

jaundice (JAWN-dis) [*jaune* yellow]

leukoplakia (loo-koh-PLAY-kee-ah) [*leuko-* white corpuscle, *-plakia* a plate]

malocclusion (mal-oh-CLEW-zhun) [*malus-* bad, *-occudere* to close up]

mumps

nonsteroidal anti-inflammatory drugs (NSAIDs) (non-steh-ROID-al an-tee-in-FLAM-ah-tor-ee) [*non-* not, *-steros-* solid, *-oid-* resembling, *-al* pertaining to, *anti-* against, *-inflammare* to set afire]

orchitis (or-KYE-tis) [*orchi-* testes, *-itis* inflammation]

orthodontics (or-thoh-DON-tiks) [*ortho-* straight or normal, *-odont-* tooth, *-ic* pertaining to]

paramyxovirus (pair-ah-MIK-soh-VYE-rus) [*para-* similar, *-myxa-* mucus, *-virus* poison]

parotitis (pair-oh-TYE-tis)

periodontitis (pair-ee-oh-don-TYE-tis) [*peri-* around, *-odont-* tooth, *-itis* inflammation]

proctitis (prok-TYE-tis) [*proktos-* anus, *itis* inflammation]

pyloric stenosis (pye-LOR-ik steh-NO-sis) [*pyle-* gate, *-ouros-* to guard, *-ic* pertaining to, *stenos-* narrow, *-osis* condition]

pylorospasm (pye-LOHR-oh-spaz-um) [*pyle-* gate, *-ouros* to guard, *-spasmos* twitching or involuntary contraction]

silent gallstones (GAWL-stones)

Sjögren's syndrome (SHOW-grins SIN-drohm) [*Henrik S.C. Sjögren* Swedish ophthalmologist]

Stretta procedure

symptomatic gallstones (simp-toh-MAT-ik) [*symptoma* that which happens]

ulcer (UL-ser) [*ulcus* a sore]

ultrasound lithotripsy (UL-trah-sound LITH-oh-trip-see) [*ultra* beyond or sound, *lithos-* stone, *-tribein* to wear away]

xerostomia (zee-oh-STOH-mee-ah) [*xeros-* dryness, *-stomia* condition of the mouth]

CASE STUDY

Dorothy Wilkerson, 76 years old, comes to the clinic complaining of rectal bleeding. She had this problem 5 years ago and had polyps removed. Since that surgery, she has continued to have an intake high in animal fat and low in fiber foods. She has a family history of polyps and has had ulcerative colitis for more than 15 years.

Ms. Wilkerson is a widow who lives alone. She is 5 feet 3 inches high and weighs 180 pounds. She says that 1 month ago her weight was 215 pounds and she has been steadily losing weight without trying. In addition to rectal bleeding, which is bright red in color, Ms. Wilkerson has noticed an increase in her usual constipation and abdominal pain and fullness. She is very tired and her activity level has decreased. On physical examination, she has guarding and grimaces during abdominal palpation. A mass is palpated in the left lower quadrant and there is tenderness in the right upper quadrant. Fecal material from the rectal examination tests positive for blood. A complete blood count (CBC) reveals anemia. Blood chemistries reveal elevated alkaline phosphatase and bilirubin levels. The results of these tests were all normal when they were performed last, which was when her polyps were removed 5 years ago. She has not been back to the clinic since that time. A barium enema with air contrast is performed and confirms the presence of a rectal tumor with the colon's lumen decreasing in size in this area. A liver scan reveals metastases and obstruction in the liver. A chest x-ray is normal. Sigmoidoscopy and colonoscopy reveal multiple polyps in the rectal portion of the colon and ulcerations along the sigmoid portion of the colon. Biopsy reports reveal adenopolyps and adenocarcinoma of the rectum and ulcerative colitis above the rectal area of the colon. Computed tomography (CT) confirms the existence and extent of the lesions along Ms. Wilkerson's colon.

1. Considering the fact that Ms. Wilkerson evidently has obstruction of the liver, which one of the following bodily functions will *most* likely be affected?

   A. Secretion of glucagon
   B. Secretion of digestive enzymes
   C. Secretion of intrinsic factor
   D. Secretion of bile

2. Ms. Wilkerson's recent weight loss is *best* explained by

   A. Her lack of appetite
   B. The tumor cells competing for nutrients
   C. Her constipation
   D. Liver involvement

3. Ms. Wilkerson's treatment will *most* likely include

   A. Medications to relax smooth muscle
   B. Histamine receptor blockers
   C. Antibiotics
   D. Surgery

4. A barium enema (BE) was performed with air contrast to better detect

   A. The fractal surfaces of the intestinal wall
   B. Tumors and ulcerations of the upper gastrointestinal tract
   C. Small polyps and subtle changes related to ulcerative colitis
   D. Diverticula and abnormalities in organ position of the lower gastrointestinal tract

CHAPTER SUMMARY

OVERVIEW OF THE DIGESTIVE SYSTEM

A. Role of the digestive system
   1. Prepares food for absorption and utilization by all the cells of the body
   2. Food material not absorbed becomes feces that is eliminated
   3. Digestion depends on both endocrine and exocrine secretions and the controlled movement of ingested food materials through the gastrointestinal (GI) tract

B. Organization of the digestive system (Figure 25-1)
   1. Organs of digestion
      a. Main organs of the digestive system form the GI tract that extends through the abdominopelvic cavity
      b. Ingested food material passing through the lumen of the GI tract is outside the internal environment of the body
   2. Wall of the GI tract (Figure 25-2)
      a. Layers—GI tract is made of four layers of tissues: mucosa, submucosa, muscularis, and serosa
      b. Modifications of layers—layers of the GI tract have various modifications to enable it to perform various functions

MOUTH

A. Structure of the oral cavity (buccal cavity) (Figure 25-3)
   1. Lips—covered externally by skin and internally by mucous membrane; the junction between skin and mucous membrane is highly sensitive; when the lips are closed, the line of contact is the oral fissure
   2. Cheeks—lateral boundaries of the oral cavity, continuous with the lips and lined by mucous membrane; formed in large part by the buccinator muscle covered by adipose tissue; contain mucus-secreting glands
   3. Hard and soft palate
      a. Hard palate consists of portions of four bones: two maxillae and two palatines
      b. Soft palate forms the partition between the mouth and nasopharynx and is made of muscle arranged in an arch
      c. Suspended from the midpoint of the posterior border of the arch is the uvula
   4. Tongue—solid mass of skeletal muscle covered by a mucous membrane; extremely maneuverable (Figure 25-4)
      a. Important for mastication and deglutition
      b. Has three parts: root, tip, and body

c. Papillae located on the dorsal surface of the tongue

d. Lingual frenulum anchors the tongue to the floor of the mouth

B. Salivary glands—three pairs of compound tubuloalveolar glands (Figure 25-6) secrete approximately 1 liter of saliva each day; the buccal glands contribute less than 5% of the total salivary volume but provide for hygiene and comfort of oral tissues

1. Parotid glands—largest of the paired salivary glands; produce watery saliva containing enzymes

2. Submandibular glands—compound glands that contain enzyme- and mucus-producing elements

3. Sublingual glands—smallest of the salivary glands; produce a mucous type of saliva

C. Teeth—organs of mastication

1. Typical tooth (Figure 25-7)

a. Crown—exposed portion of a tooth, covered by enamel; ideally suited to withstand abrasion during mastication

b. Neck—narrow portion that joins the crown to the root; surrounded by gingivae

c. Root fits into the socket of the alveolar process and is suspended by a fibrous periodontal membrane

d. Outer shell contains two additional tissues: dentin and cementum

(1) Dentin makes up the greatest portion of the tooth shell; at the crown, covered by enamel, and at the neck and root, covered by cementum

(2) Pulp cavity—located in dentin, contains connective tissue, blood, and lymphatic vessels and sensory nerves

2. Types of teeth (Figure 25-8)

a. Deciduous teeth—20 baby teeth, which appear early in life

b. Permanent teeth—32 teeth, which replace the deciduous teeth

PHARYNX

A. Tube through which a bolus passes when moved from the mouth to the esophagus by the process of deglutition

ESOPHAGUS (FIGURE 25-9)

A. Tube that extends from the pharynx to the stomach

B. First segment of the digestive tube

STOMACH (FIGURE 25-10)

A. Size and position of the stomach

1. Size varies according to factors such as gender and amount of distention

a. When no food is in the stomach, it is about the size of a large sausage

b. In adults, its capacity ranges from 1.0 to 1.5 liters

2. Stomach location: upper part of the abdominal cavity under the liver and diaphragm

B. Divisions of the stomach

1. Fundus—enlarged portion to the left and above the opening of the esophagus into the stomach

2. Body—central portion of the stomach

3. Pylorus—lower part of the stomach

C. Curves of the stomach

1. Lesser curvature—upper right curve of the stomach

2. Greater curvature—lower left curve of the stomach

D. Sphincter muscles—circular fibers arranged so that there is an opening in the center when relaxed and no opening when contracted

1. Lower esophageal sphincter (LES), or cardiac sphincter, controls the opening of the esophagus into the stomach

2. Pyloric sphincter controls the outlet of the pyloric portion of the stomach into the duodenum

E. Stomach wall

1. Gastric mucosa

a. Epithelial lining has rugae marked by gastric pits (Figures 25-11 and 25-12)

b. Gastric glands—found below the level of the pits; secrete most of the gastric juice

c. Chief cells—secretory cells found in the gastric glands; secrete the enzymes of gastric juice

d. Parietal cells—secretory cells found in the gastric glands; secrete hydrochloric acid; thought to produce intrinsic factor needed for vitamin B_{12} absorption

e. Endocrine cells—secrete gastrin and ghrelin

2. Gastric muscularis—thick layer of muscle with three distinct sublayers of smooth muscle tissue arranged in a crisscrossing pattern; this pattern allows the stomach to contract strongly at many angles

F. Functions of the stomach

1. Reservoir for food until it is partially digested and moved further along the GI tract

2. Secretes gastric juice to aid in digestion of food

3. Breaks food into small particles and mixes them with gastric juice

4. Secretes intrinsic factor

5. Limited absorption

6. Produces gastrin and ghrelin

7. Helps protect the body from pathogenic bacteria swallowed with food

SMALL INTESTINE

A. Size and position of the small intestine—tube approximately 2.5 cm in diameter and 6 m in length; coiled loops fill most of the abdominal cavity

B. Divisions of the small intestine

1. Duodenum—uppermost division; approximately 25 cm long, shaped roughly like the letter C

2. Jejunum—approximately 2.5 m long

3. Ileum—approximately 3.5 m long

C. Wall of the small intestine (Figure 25-13)

1. Intestinal lining has plicae with villi

2. Villi—important modifications of the mucosal layer

a. Each villus contains an arteriole, venule, and lacteal vessel

b. Covered by a brush border made up of 1700 ultrafine microvilli per cell

c. Villi and microvilli increase the surface area of the small intestine hundreds of times

LARGE INTESTINE

A. Size of the large intestine—average diameter, 6 cm; length, approximately 1.5 to 1.8 m
B. Divisions of the large intestine (Figure 25-16)
  1. Cecum—first 5 to 8 cm of the large intestine, blind pouch located in the lower right quadrant of the abdomen
  2. Colon
    a. Ascending colon—vertical position on the right side of the abdomen; the ileocecal valve prevents material passing from the large intestine into the ileum
    b. Transverse colon passes horizontally across the abdomen, above the small intestine; extends from the hepatic flexure to the splenic flexure
    c. Descending colon—vertical position on left side of the abdomen
    d. Sigmoid colon joins the descending colon to the rectum
    e. Rectum—last 7 or 8 inches of the intestinal tube; terminal inch is the anal canal with the opening called the *anus* (Figure 25-17)
C. Wall of the large intestine (Figure 25-19)
  1. Intestinal mucous glands produce lubricating mucus that coats feces as they are formed
  2. Uneven distribution of fibers in the muscle coat

VERMIFORM APPENDIX

A. Accessory organ of digestive system; 8 to 10 cm in length; communicates with the cecum

PERITONEUM

A. Large, continuous sheet of serous membrane (Figure 25-20)
B. Made up of parietal and visceral layers
C. Mesentery—projection of the parietal peritoneum; allows free movement of each coil of the intestine and helps prevent strangulation of the long tube (Figure 25-21)
D. Transverse mesocolon—extension of the peritoneum that supports the transverse colon

LIVER

A. Location and size of the liver (Figure 25-22)—largest gland in the body, weighs approximately 1.5 kg; lies under the diaphragm; occupies most of the right hypochondrium and part of the epigastrium
B. Liver lobes and lobules—two lobes separated by the falciform ligament
  1. Left lobe—forms about one sixth of the liver
  2. Right lobe—forms about five sixths of the liver; divides into right lobe proper, caudate lobe, and quadrate lobe
  3. Hepatic lobules—anatomical units of the liver; a small branch of the hepatic vein extends through the center of each lobule (Figure 25-23)
C. Bile ducts (Figure 25-25)
  1. Small bile ducts form right and left hepatic ducts
  2. Right and left hepatic ducts immediately join to form one hepatic duct
  3. Hepatic duct merges with the cystic duct to form the common bile duct, which opens into the duodenum

D. Functions of the liver
  1. Detoxification by liver cells—ingested toxic substances and toxic substances formed in the intestines may be changed to nontoxic substances
  2. Bile secretion by liver—bile salts are formed in the liver from cholesterol and are the most essential part of bile; liver cells secrete approximately 1 pint of bile per day
  3. Liver metabolism carries out numerous important steps in metabolizing proteins, fats, and carbohydrates
  4. Storage of substances such as iron and some vitamins
  5. Production of important plasma proteins

GALLBLADDER (FIGURE 25-25)

A. Size and location of the gallbladder—pear-shaped sac 7 to 10 cm long and 3 cm wide at its broadest point; holds 30 to 50 ml of bile; lies on the undersurface of the liver
B. Structure of the gallbladder—serous, muscular, and mucous layers compose the gallbladder wall; the mucosal lining has rugae
C. Functions of the gallbladder
  1. Storage of bile
  2. Concentration of bile fivefold to tenfold
  3. Ejection of the concentrated bile into the duodenum

PANCREAS

A. Size and location of the pancreas—grayish pink–colored gland; 12 to 15 cm long; weighs approximately 60 g; runs from the duodenum, behind the stomach, to the spleen
B. Structure of the pancreas (Figure 25-27)—composed of endocrine and exocrine glandular tissue
  1. Exocrine portion makes up the majority of the pancreas; has a compound acinar arrangement; tiny ducts unite to form the main pancreatic duct, which empties into the duodenum
  2. Endocrine portion—embedded between exocrine units; called pancreatic islets; constitute only 2% of the total mass of the pancreas; made up of alpha cells and beta cells; pass secretions into capillaries
C. Functions of the pancreas
  1. Acinar units secrete digestive enzymes
  2. Beta cells secrete insulin
  3. Alpha cells secrete glucagon

CYCLE OF LIFE: DIGESTIVE SYSTEM

A. Changes in digestive function and structure are age related
  1. Result in diseases or pathological conditions
  2. May occur in any segment of the intestinal tract
  3. Changes involve accessory organs: teeth, salivary glands, liver, gallbladder, and pancreas
B. Infants—immature intestinal mucosa; intact proteins can pass through epithelial cells lining the tract and trigger an allergic response
C. Lactose intolerance affects infants who lack the enzyme lactase
D. Young age—mumps common in children; appendicitis more common in adolescents and then decreases with advancing age

E. Middle age—ulcers and gallbladder disease common
F. Old age—decreased digestive fluids, slowing of peristalsis, and reduced physical activity lead to constipation and diverticulosis

REVIEW QUESTIONS

1. List the component parts or segments of the GI tract and the accessory organs of digestion.
2. Name and describe the four tissue layers that form the wall of GI tract organs.
3. Identify the structures that form the mouth.
4. Define the following terms associated with the mouth and pharynx: *philtrum, oral fissure, hard and soft palate, fauces, uvula, foramen cecum, lingual frenulum.*
5. What is ankyloglossia?
6. Identify the types of tongue papillae. What is the relationship between papillae and taste buds?
7. List and give the location of the paired salivary glands. Identify by name the ducts that drain the saliva from these glands into the mouth.
8. What type of saliva is produced by the parotid glands? What is meant by the term *mixed* or *compound salivary gland?*
9. Describe a typical tooth. Name the specific types of teeth.
10. Discuss deciduous and permanent teeth.
11. What is meant by the term *deglutition?*
12. List the divisions of the stomach. What is the difference between gastric pits and gastric glands?
13. Identify the three major cell types of the gastric glands. What cell type produces hydrochloric acid? Gastric enzymes? Gastrin? Ghrelin?
14. Describe the seven functions of the stomach.
15. List the divisions of the small intestine from proximal to distal.

16. In what area of the GI tract do you find villi? Haustra? Taenia coli?
17. List the divisions of the large intestine.
18. What is believed to be the function of the vermiform appendix?
19. Discuss the peritoneum and its reflections.
20. Discuss the anatomy of a typical liver lobule.
21. Identify the ducts of the liver and gallbladder.
22. Explain the functions of the gallbladder.
23. Differentiate between endocrine and exocrine functions of the pancreas.
24. Identify the condition that can develop in men who are infected with mumps.
25. What is the difference between dental caries and periodontitis?
26. What is pyloric stenosis?
27. Define cholelithiasis.

CRITICAL THINKING QUESTIONS

1. How would you explain the role of the tongue's intrinsic and extrinsic muscles?
2. How would you describe the unique muscular layer of the esophagus?
3. Increasing the interior surface area of the small intestine allows it to absorb nutrients more efficiently. What examples can you find that add to the interior surface area of the small intestine?
4. Can you explain the x-ray procedure used to diagnose a hiatal hernia? What procedure would be used to diagnose an intestinal diverticulum?
5. If an elderly patient had abdominal pain, why would it be unlikely that it is caused by appendicitis?

CHAPTER 26

Physiology of the Digestive System

LANGUAGE OF SCIENCE

absorption (ab-SORP-shun) [absorbere swallow]
amylases (AM-eh-lays-ez) [amyl- starch, -ase enzyme]
bile (byle)
bile salts (byle)
bilirubin (bil-ih-ROO-bin) [bili- bile, -rub- red, -in substance]
brush border
cephalic phase (seh-FAL-ik fayz) [cephal- head, -ic pertaining to]
chemical digestion
chief cells
cholecystokinin (CCK) (koh-lee-sis-toh-KYE-nin) [chole- bile, -cysto- bladder, -kinin move, -in substance]
chylomicrons (kye-loh-MYE-kronz) [chylo- juice (chyle), -micro- small, -on particle]
chyme (kyme) [chymos juice]
chymotrypsin (kye-moh-TRIP-sin) [chymo- juice, -tryps- rub, -in substance]
contact digestion
defecation (def-eh-KAY-shun) [defeca- to clean the dregs from, -tion process of]
deglutition (deg-loo-TISH-un) [degluti- swallow, -tion process of]
digestion [digest- break down, -tion process of]
elimination (ee-lim-i-NAY-shun) [e- out, -limen- threshold, -ation process of]
emulsified (ee-MULL-seh-fyde) [emuls- to milk out, -i- combining form, -fy process of]
enteric nervous system (ENS) (en-TER-ik) [enter- bowel, -ic pertaining to]
enterogastric reflex (en-ter-oh-GAS-trik) [entero- bowel, -gastr- stomach, -ic pertaining to]
enterokinase (en-ter-oh-KYE-nays) [entero- bowel, -kin- movement, -ase enzyme]
enzymes (EN-zymes) [en- in, -zyme ferment]
feces (FEE-seez) [faex waste matter]

Cont'd on p. 988

961

OVERVIEW OF DIGESTIVE FUNCTION

Now that we are familiar with the structural organization of the digestive system (Chapter 25), we are ready to understand the physiological organization of this system. The primary function of the digestive system is to bring essential nutrients into the internal environment so that they are available to each cell of the body. To accomplish this function, the digestive system uses various mechanisms (Table 26-1). For example, complex foods must first be taken in—a process called **ingestion.** Then, complex nutrients are broken down into simpler nutrients in the process that gives this system its name: **digestion.** To physically break large chunks of food into smaller bits and to move it along the tract, movement (or **motility**) of the gastrointestinal (GI) wall is required. Chemical digestion—that is, breakdown of large molecules into small molecules—requires **secretion** of digestive enzymes into the lumen of the GI tract. After being digested, nutrients are ready for the process of **absorption,** or movement through the GI mucosa into the internal environment. The material that is not absorbed must then be excreted to make room for more material—a process known as **elimination.** Of course, all these activities must be coordinated, which we have already learned is the process of *regulation.* Some of the major digestive processes are summarized in Figure 26-1. Digestive regulation is introduced in Box 26-1.

Note that Figure 26-1 also illustrates the fact that, from a functional perspective, the lumen of the alimentary canal is really a tube-like extension of the external environment that goes right through the middle of the body. Thus, digested materials are not truly "part of the body" until they've been absorbed into the internal environment.

After we have explored the various mechanisms of the digestive process in this chapter, we will be ready for Chapter 27, which discusses the fate of nutrients after they have been absorbed.

DIGESTION

After food is ingested (taken into the mouth), the process of digestion begins immediately. Digestion is the overall name for all the processes that chemically and mechanically break complex foods into simpler nutrients that can be easily absorbed. We begin our discussion with a brief overview of **mechanical digestion,** then later move on to a discussion of **chemical digestion.**

Mechanical Digestion

Mechanical digestion consists of all movement (motility) of the digestive tract that brings about the following:

- Change in the physical state of ingested food from comparatively large solid pieces into minute particles, thereby facilitating chemical digestion
- Churning of the contents of the GI lumen in such a way that they become well mixed with the digestive juices and all parts of them come in contact with the surface of the intestinal mucosa, thereby facilitating absorption
- Propelling the food forward along the digestive tract, finally eliminating the digestive wastes from the body

Mastication

Mechanical digestion begins in the mouth when the particle size of ingested food material is reduced by chewing movements, or **mastication.** The tongue, cheeks, and lips play an important role in keeping food material between the cutting or grinding surfaces of the teeth when chewing. In addition to reducing particle size, chewing movements serve to mix food with saliva in preparation for swallowing.

Deglutition

The process of swallowing, or **deglutition,** may be divided into formation and movement of a food bolus from the mouth to the stomach in three main steps or stages (Figure 26-2):

1. Oral stage (mouth to oropharynx)
2. Pharyngeal stage (oropharynx to esophagus)
3. Esophageal stage (esophagus to stomach)

The first step, which is voluntary and under control of the cerebral cortex, involves the formation of a food bolus to be swallowed on a depression or groove in the middle of the tongue. During the oral stage, the bolus is pressed against the palate by the tongue and

Table 26-1	Primary Mechanisms of the Digestive System

MECHANISM	DESCRIPTION
Ingestion	Process of taking food into the mouth, starting it on its journey through the digestive tract
Digestion	A group of processes that break complex nutrients into simpler ones, thus facilitating their absorption; *mechanical* digestion physically breaks large chunks into small bits; *chemical digestion* breaks molecules apart
Motility	Movement by the muscular components of the digestive tube, including processes of mechanical digestion; examples include *peristalsis* and *segmentation*
Secretion	Release of digestive juices (containing enzymes, acids, bases, mucus, bile, or other products that facilitate digestion); some digestive organs also secrete endocrine hormones that regulate digestion or metabolism of nutrients
Absorption	Movement of digested nutrients through the GI mucosa and into the internal environment
Elimination	Excretion of the residues of the digestive process (feces) from the rectum, through the anus; defecation
Regulation	Coordination of digestive activity (motility, secretion, etc.)

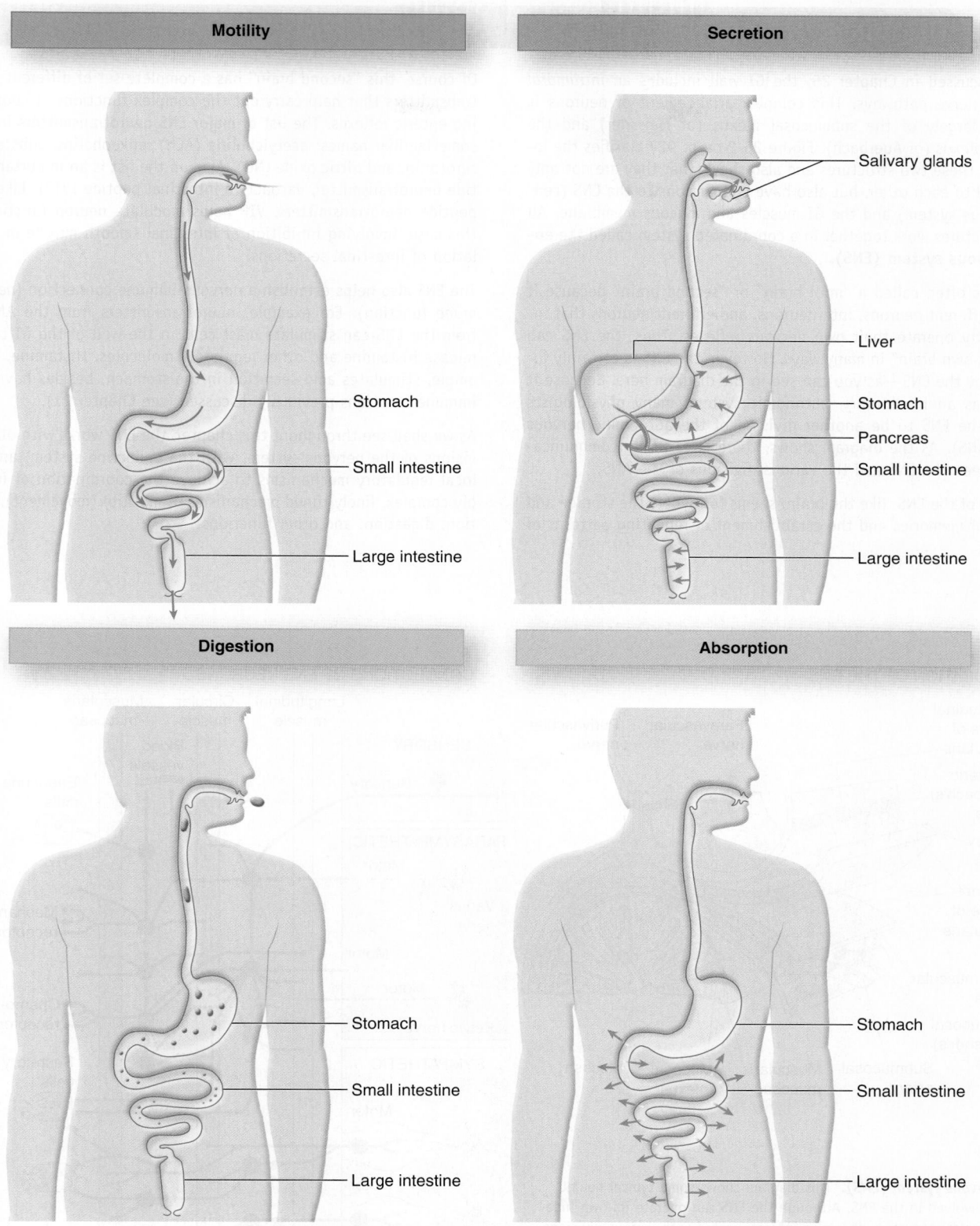

Figure 26-1 *Overview of digestive functions.* Several important digestive functions are summarized in these diagrams. Note that the digestive tract is an extension of the external environment—extending like a tunnel through the body.

BOX 26-1 The Enteric Nervous System

As we discussed in Chapter 25, the GI wall includes an *intramural plexus* of nerve pathways. This complex arrangement of neurons is made up largely of the submucosal plexus (of Meissner) and the myentric plexus (of Auerbach). Figure 25-2 on p. 927 clarifies the locations of these two structures and also shows that they are not only connected to each other, but also have connections to the CNS (central nervous system) and the GI muscles and mucous membrane. All these structures work together in a coordinated system called the **enteric nervous system (ENS).**

The ENS is often called a "mini brain" or "second brain" because it includes afferent neurons, interneurons, and efferent neurons that independently operate their own nervous reflexes. Thus, the ENS can act as "its own brain" in many ways. However, the ENS is certainly influenced by the CNS—as you can see in the diagram here. Because it operates as an involuntary, autonomic system, many physiologists consider the ENS to be another division of the autonomic nervous system (ANS). As the diagram shows, there is certainly communication between the ENS and the various divisions of the ANS.

Operation of the ENS, like the brain, seems to involve the storage and retrieval of memories and the establishment of repeating patterns of response.

Of course, this "second brain" has a complete set of different neurotransmitters that help carry out the complex functions of coordinating enteric reflexes. The list of major ENS neurotransmitters includes some familiar names: acetylcholine (ACh), enkephalins, substance P, serotonin, and nitric oxide (NO). Also on the list is an important peptide neurotransmitter, vasoactive intestinal peptide (VIP). Like other peptide neurotransmitters, VIP helps modulate neuron function—in this case, involving inhibition of intestinal smooth muscle or stimulation of intestinal secretions.

The ENS also helps establish a nervous-immune connection (*neuroimmune* function). For example, neurotransmitters from the ANS and from the ENS can stimulate mast cells in the wall of the GI tract to release histamine and other regulatory molecules. Histamine, for example, stimulates acid secretion in the stomach, besides having the immune functions previously discussed (see Chapter 21).

As we shall see throughout this chapter, the ENS works with other divisions of the nervous system, with the endocrine system, and with local regulatory mechanisms to achieve the coordination of incredibly complex, finely tuned mechanisms of motility (movement), secretion, digestion, and other functions.

LOCATION OF THE ENS

Longitudinal muscle of muscularis
Myenteric (Auerbach's) plexus
Tertiary plexus
Circular muscle of muscularis
Deep muscular plexus
Submucosal (Meissner's) plexus
Paravascular nerve
Perivascular nerve
Submucosal artery
Muscularis mucosae
Mucosal plexus
Mucosa

Enteric nervous system (ENS). This diagram shows some typical neural pathways involved in the ENS. Although the ENS can operate its own independent regulatory system, it is also interconnected with various divisions of the autonomic nervous system (ANS).

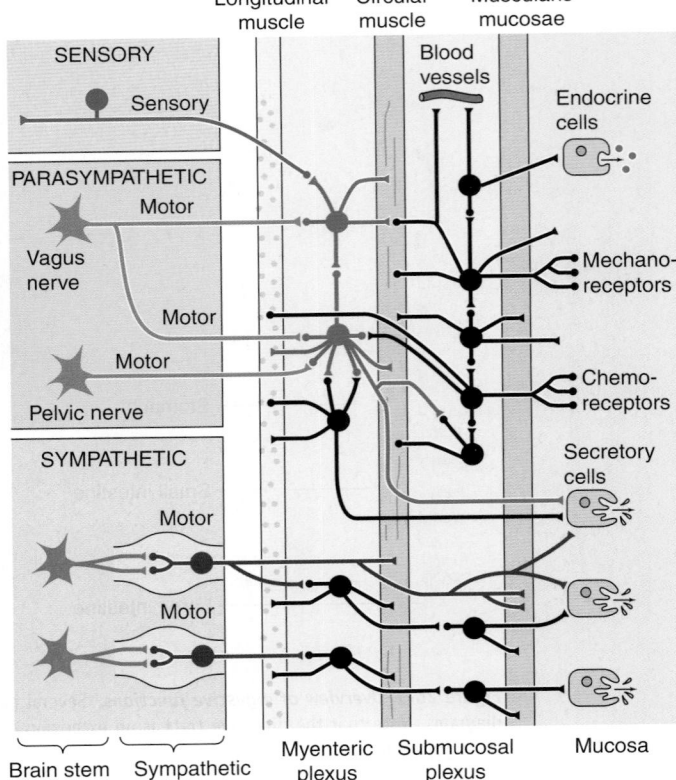

CONNECTIONS OF ENS NEURONS

Longitudinal muscle
Circular muscle
Muscularis mucosae

SENSORY
Sensory
Blood vessels
Endocrine cells

PARASYMPATHETIC
Motor
Vagus nerve
Motor
Mechano-receptors
Motor
Pelvic nerve
Chemo-receptors

SYMPATHETIC
Motor
Secretory cells
Motor

Brain stem or spinal cord
Sympathetic ganglia
Myenteric plexus
Submucosal plexus
Mucosa

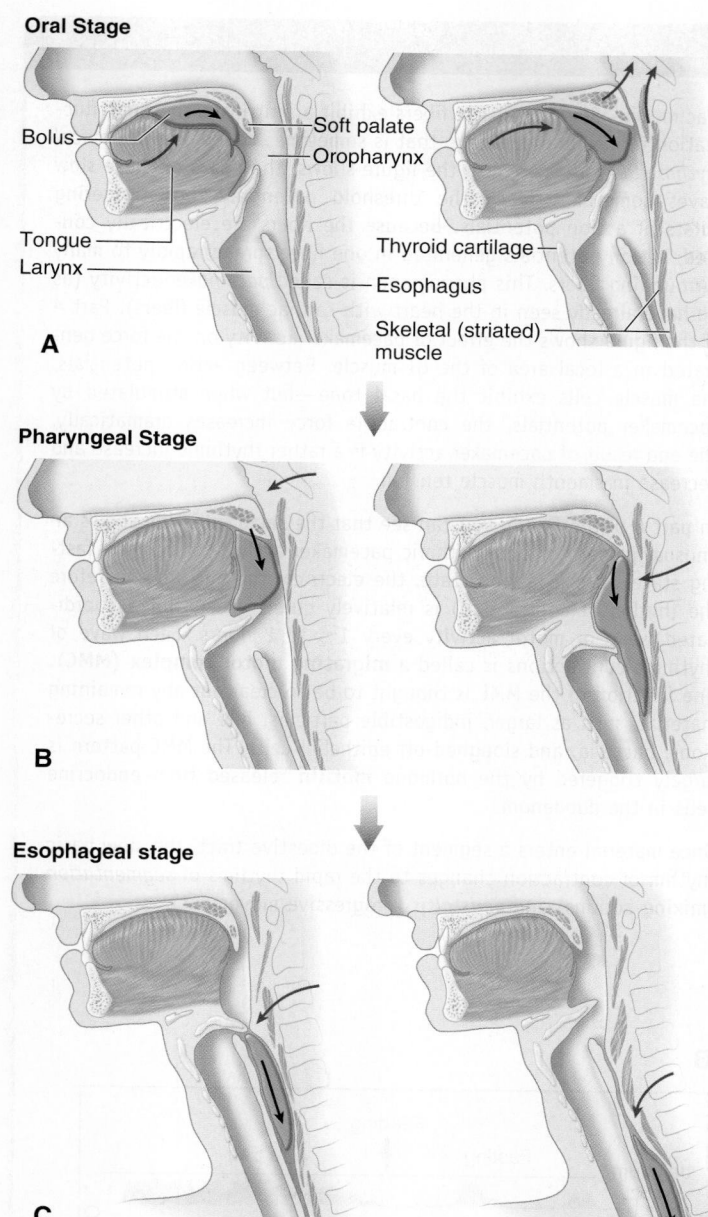

Oral Stage

Bolus — Soft palate
— Oropharynx

Tongue —
Larynx — Thyroid cartilage —
— Esophagus
— Skeletal (striated) muscle

A

Pharyngeal Stage

B

Esophageal stage

C

Figure 26-2 *Deglutition.* **A,** *Oral stage.* During this stage of deglutition (swallowing), a bolus of food is voluntarily formed on the tongue and pushed against the palate and into the oropharynx. Notice that the soft palate acts as a valve that prevents food from entering the nasopharynx. **B,** *Pharyngeal stage.* After the bolus is in the oropharynx, involuntary reflexes push the bolus down toward the esophagus. Notice that upward movement of the larynx and downward movement of the bolus close the epiglottis and thus prevent food from entering the lower respiratory tract. **C,** *Esophageal stage.* Involuntary reflexes of skeletal (striated) and smooth muscle in the wall of the esophagus move the bolus through the esophagus toward the stomach.

then moved back into the oropharynx. The pharyngeal and esophageal stages, both involuntary, consist of movement of food from the pharynx into the esophagus and, finally, into the stomach.

To propel food from the pharynx into the esophagus, three openings must be blocked: mouth, nasopharynx, and larynx. Continued elevation of the tongue seals off the mouth. The soft palate, including the uvula, is elevated and tensed, causing the nasopharynx to be closed off. Food is denied entrance into the larynx by muscle action that causes the epiglottis to block this opening. The mechanism involves raising of the larynx, a process easily noted by palpation of the thyroid cartilage during swallowing. As a result, the bolus slips over the back of the epiglottis to enter the laryngopharynx. Contractions of the pharynx and esophagus compress the bolus into and through the esophageal tube. These steps are involuntary and under control of the *deglutition center* in the medulla. The presence of a bolus stimulates sensory receptors in the mouth and pharynx, thus initiating reflex pharyngeal contractions. Consequently, anesthesia of sensory nerves from the mucosa of the mouth and pharynx by a drug such as procaine makes swallowing difficult or impossible.

Swallowing is a complex process requiring the coordination of many muscles and other structures in the head and neck. Not only does the process occur smoothly, but it must also take place rapidly because respiration is inhibited for the 1 to 3 seconds required for food to clear the pharynx during each swallowing.

Peristalsis and Segmentation

After food enters the lower portion of the esophagus, smooth muscle tissue in the wall of the GI tract takes on primary responsibility for its movement (Box 26-2). The motility produced by smooth muscle is of two main types: peristalsis and segmentation.

Peristalsis is often described as a wavelike ripple of the muscle layer of a hollow organ. The diagram in Figure 26-3 shows step by step how peristalsis occurs. A bolus stretches the GI wall, trigger-

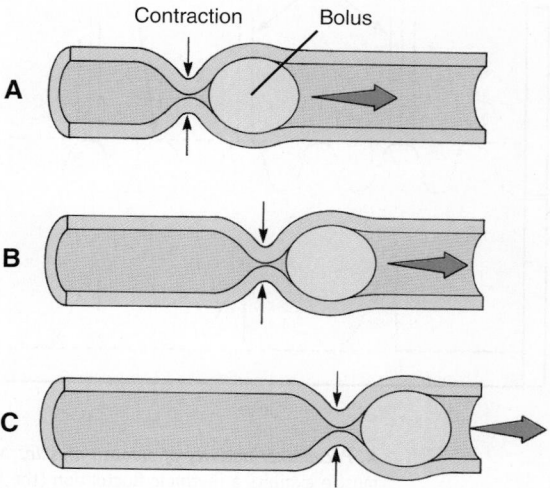

Contraction   Bolus

A

B

C

Figure 26-3 *Peristalsis.* Peristalsis is a progressive type of movement in which material is propelled from point to point along the GI tract. **A,** A ring of contraction occurs where the gastrointestinal (GI) wall is stretched, and the bolus is pushed forward. **B,** The moving bolus triggers a ring of contraction in the next region that pushes the bolus even farther along. **C,** The ring of contraction moves like a wave along the GI tract to push the bolus forward.

BOX 26-2 Smooth Muscle Function in the GI Tract

Smooth muscle tissue in the wall of the GI tract differs from other types of muscle tissue in a number of important ways. For example, smooth muscle is slow compared to skeletal muscle. Because the sliding of myofilaments proceeds at a much slower pace than in skeletal muscle, smooth muscle often exhibits a prolonged contraction phase. Because the same amount of energy is used for slow contraction as for fast contraction, gastrointestinal muscle can sustain tension for long periods without fatigue—exactly what is needed for the long process of digestion.

Another unique functional characteristic of smooth muscle is its ability to maintain *basal tone*. The basal tone is a continuous state of minimal contraction. Continuous tension is maintained in cells that have enough calcium ions free in the sarcoplasm to generate the contraction response. The force of contraction can increase above the basal tone by the effects of an action potential, which cause rapid influx of extracellular calcium.

Gastrointestinal *sphincter* muscles, as you recall from the previous chapter, are ring-like formations of smooth muscle in the GI wall that act as gateways or valves that regulate movement of chyme from one part of tract to the next part. Sphincter muscles generally have a higher basal tone than surrounding smooth muscle does—thus constricting the lumen and "keeping the gate closed." Usually, stimuli in a section preceding the sphincter trigger a decrease in basal tone to allow material to pass through. Some of these reflexes will be discussed later in this chapter.

Most of the smooth muscles in the GI wall are electrically coupled by gap junctions—so-called *single-unit* muscles. Such electrical coupling allows for intrinsic control of smooth muscle contraction, as in cardiac muscle. Smooth muscle fibers exhibit an intrinsic, rhythmic fluctuation in membrane voltage that is sometimes called *basic electrical rhythm (BER)*. As part A of the figure shows, the peaks of these slow waves sometimes reach the threshold potential—thus triggering bursts of action potentials. Because the fibers are electrically coupled, action potentials generated in one fiber spread rapidly to many surrounding fibers. This phenomenon is called *pacemaker* activity (as we have already seen in the heart with cardiac muscle fibers). Part A of the figure shows the effect of pacemaker activity on the force generated in a local area of the GI muscle. Between action potentials, the muscle cells exhibit the basal tone—but when stimulated by pacemaker potentials, the contractile force increases dramatically. The end result of pacemaker activity is a rather rhythmic increase and decrease in smooth muscle tension.

In part B of the figure, you can see that the small intestine shows an unusual pattern of this rhythmic pacemaker activity during the fasting state. In the fasting state, the electrical rhythm (and therefore the rhythm of contraction) is relatively quiet, except for a coordinated wave of motor activity every 1½ to 2 hours. Each wave of rhythmic contractions is called a **migrating motor complex (MMC).** One function of the MMC is thought to be to clear out any remaining material, such as larger, indigestible particles, bile and other secretions, bacteria, and sloughed-off epithelial cells. The MMC pattern is largely triggered by the hormone **motilin** released from endocrine cells in the duodenum.

Once material enters a segment of the digestive tract, the slow basic rhythm of contraction changes to the rapid rhythms of *segmentation* (mixing actions) and *peristalsis* (progressive movement).

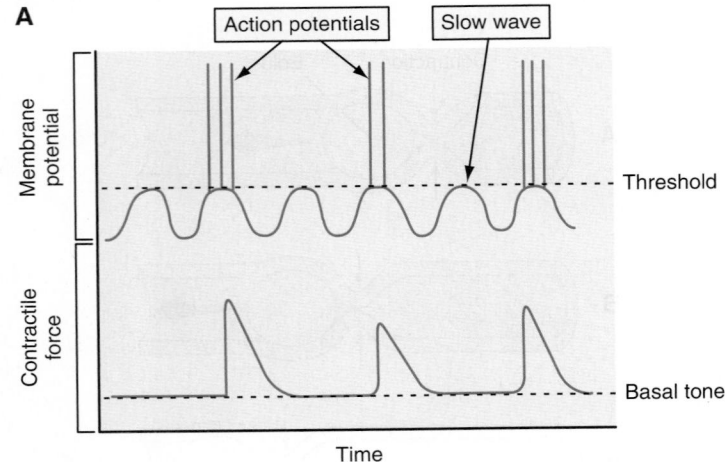

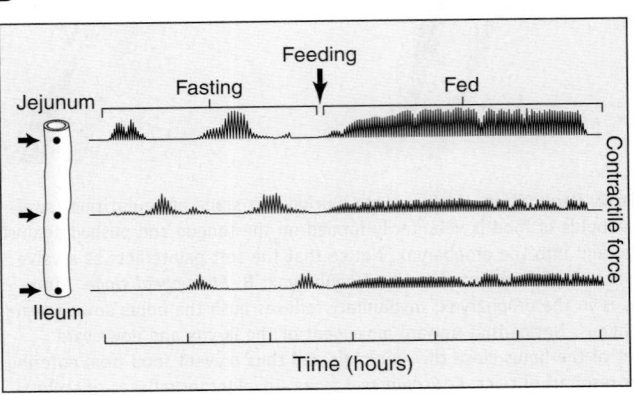

A, Pacemaker activity of smooth muscle. As the upper graph line shows, the membrane potential of smooth muscle exhibits a rhythmic fluctuation (the basic electrical rhythm). Occasionally, the fluctuating membrane potential reaches the threshold—triggering one or more action potentials that spread to surrounding muscle fibers. Smooth muscles are continually partially contracted (the basal tone), but the force of contraction increases dramatically with each burst of action potentials (bottom graph line). The more action potentials that occur in a short series, the stronger the force of contraction. **B, Migrating motor complex (MMC).** In the small intestine, the basal rhythm of smooth muscle contractions during the fasting state is characterized by occasional waves of motor activity called migrating motor complexes (MMCs). These waves of contraction slowly "sweep out" the intestinal lumen between active digestive periods (fed states). During the fed state, the MMC pattern disappears and is replaced by a more rapid, active pattern that includes peristalsis and segmentation.

ing a reflex contraction of circular muscle that pushes the bolus forward. This, in turn, triggers a reflex contraction in that location, pushing the bolus even farther. This process continues as long as the stretch reflex is activated by the presence of food. Peristalsis is a progressive kind of motility, that is, a type of motion that produces forward movement of ingested material along the GI tract. Figure 26-4 shows the effects of peristaltic contractions in the esophagus during deglutition.

Segmentation can be simply described as mixing movement. Segmentation occurs when digestive reflexes cause a forward-and-backward movement within a single region, or segment, of the GI tract (Figure 26-5). Such movement helps mechanically break down food particles, mixes food and digestive juices thoroughly, and brings digested food in contact with intestinal mucosa to facilitate absorption.

Peristalsis and segmentation can occur in an alternating sequence. When this occurs, food is churned and mixed as it slowly progresses along the GI tract.

Regulation of Motility

Gastric Motility

The process of emptying the stomach takes about 2 to 6 hours after a meal, depending on the amount and content of the meal. During its "storage time" in the stomach, food is churned with gastric juices to form a thick, milky material known as **chyme,** which is ejected about every 20 seconds into the duodenum. As you can see in Figure 26-6, while chyme is in the stomach it is continually pushed toward the pyloric sphincter by waves of peristaltic contractions—a

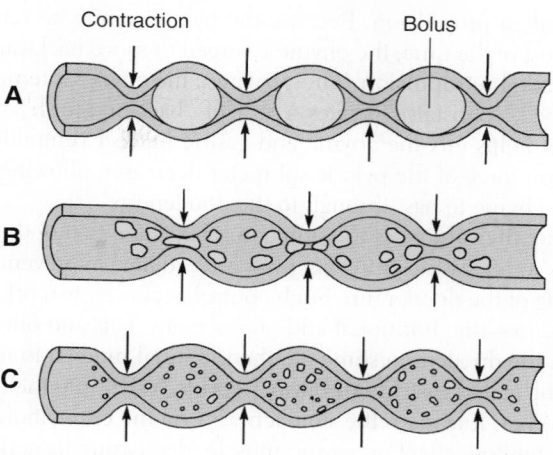

Figure 26-5 *Segmentation.* Segmentation is a back-and-forth action that breaks apart chunks of food and mixes in digestive juices. **A,** Ringlike regions of contraction occur at intervals along the gastrointestinal (GI) tract. **B,** Previously contracted regions relax and adjacent regions now contract, effectively "chopping" the contents of each segment into smaller chunks. **C,** The location of the contracted regions continues to alternate back and forth, chopping and mixing the contents of the GI lumen.

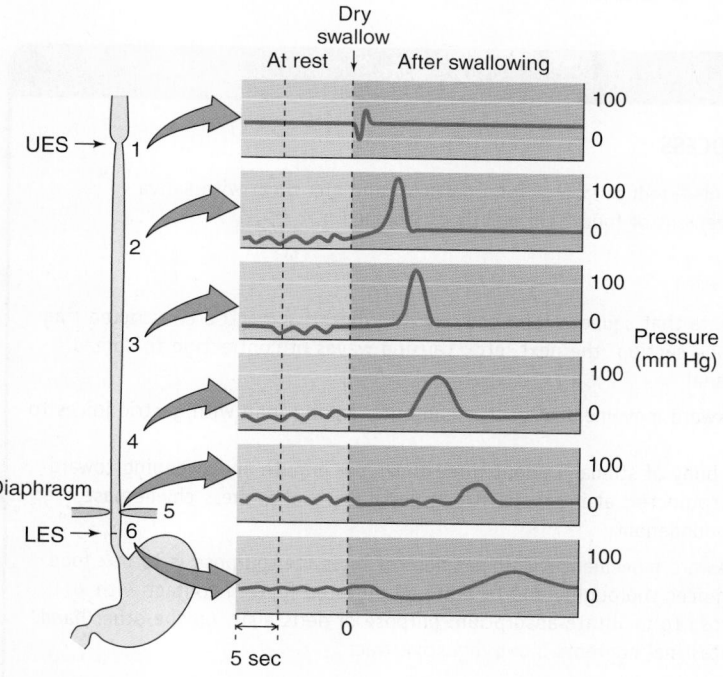

Figure 26-4 *Pressure in the esophagus during swallowing.* The diagram shows the changing pressure (mm Hg) inside the esophagus during the esophageal stage of deglutition (swallowing, see Figure 26-2). Note that an area of high pressure moves progressively down the esophagus—and eventually to the stomach. Note that during the rest period before swallowing, the smooth muscle exhibits a slow rhythm of weak contractions. *UES,* Upper esophageal sphincter; *LES,* lower esophageal [cardiac] sphincter.

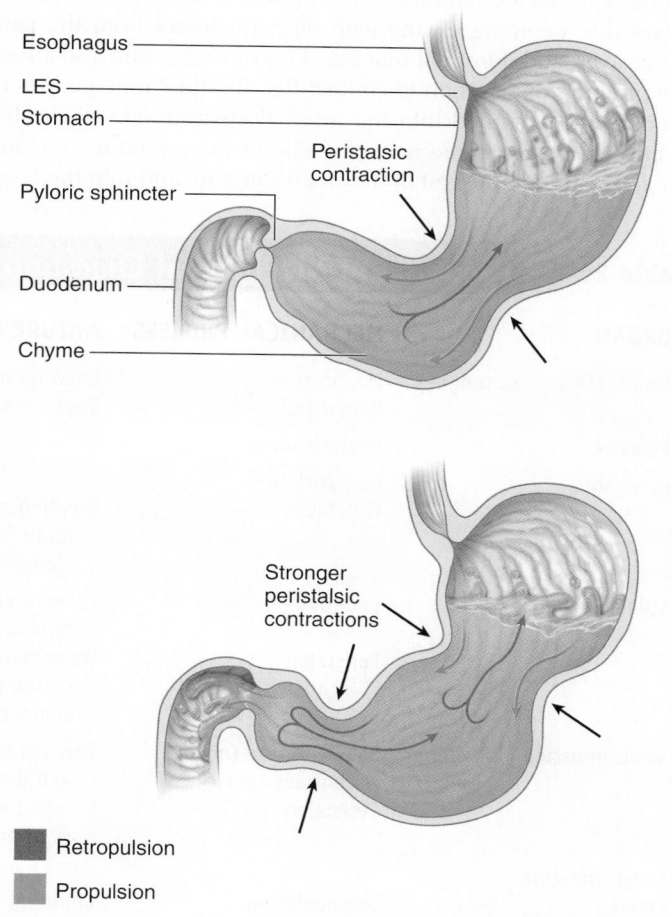

Figure 26-6 *Gastric motility.* Mixing actions in the stomach include both propulsion (forward movement) and retropulsion (backward movement). As peristaltic contractions become stronger, some of the liquid chyme squirts past the pyloric sphincter (which has decreased its muscle tone) and into the duodenum. The stomach continues to mix the chyme as it is gradually released into the duodenum.

process called **propulsion.** Because the pyloric sphincter remains closed most of the time, the chyme is forced to move backward—a process called **retropulsion.** Thus, because the chyme is temporarily "trapped," peristalsis creates a sort of "back-and-forth" movement that helps mix the chyme and gastric juice. Eventually, the contraction force of the pyloric sphincter decreases, allowing a little of the chyme to pass through to the duodenum.

Because the volume of the stomach is large and that of the duodenum is small, gastric emptying must be regulated to prevent overburdening of the duodenum. Such control occurs by two principal mechanisms—one hormonal and one nervous. Fats and other nutrients in the duodenum stimulate the intestinal mucosa to release a hormone called **gastric inhibitory peptide (GIP)** into the bloodstream. When it reaches the stomach wall via the circulation, GIP has an inhibitory effect on gastric muscle, decreasing its peristalsis and thus slowing passage of food into the duodenum. Nervous control results from receptors in the duodenal mucosa that are sensitive to the presence of acid and to distention. Sensory and motor fibers in the vagus nerve then cause a reflex inhibition of gastric peristalsis. This nervous mechanism is known as the **enterogastric reflex.**

Intestinal Motility

Intestinal motility includes both peristaltic contractions and segmentation. Segmentation in the duodenum and upper jejunum mixes the incoming chyme with digestive juices from the pancreas, liver, and intestinal mucosa. This mixing action also allows the products of digestion to contact the intestinal mucosa, where they can be absorbed into the internal environment. Peristalsis continues as the chyme nears the end of the jejunum—moving the food through the rest of the small intestine and into the large intestine. After leaving the stomach, it normally takes about 5 hours for chyme to pass all the way through the small intestine.

Several mechanisms are involved in the control of intestinal motility. Peristalsis is regulated in part by the intrinsic stretch reflexes already described. It is also thought to be stimulated by the hormone **cholecystokinin (CCK),** which is secreted by endocrine cells of the intestinal mucosa when chyme is present.

A list of definitions of the different processes involved in mechanical digestion, with the organs that accomplish them, is presented in Table 26-2.

BOX 26-3 SPORTS AND FITNESS
Exercise and Fluid Uptake

Replacement of fluids lost through sweating during exercise is essential for maintaining homeostasis. Nearly everyone increases the intake of fluids during and after exercise. The main limitation to efficient fluid replacement is how quickly fluid can be absorbed rather than how much a person drinks. Very little water is absorbed until it reaches the intestines, where it is absorbed almost immediately. Thus, the rate of *gastric emptying* into the intestine is critical.

Large volumes of fluid leave the stomach and enter the intestines more rapidly than small volumes do. However, large volumes may be uncomfortable during exercise. Cool fluids (8° to 13° C [46° to 55° F]) empty more quickly than warm fluids. Fluids with a high solute concentration empty slowly and may cause nausea or stomach cramps. Thus, large amounts of cool, dilute, or isotonic fluids are best for replacing fluids quickly during exercise.

Table 26-2	Processes of Mechanical Digestion	
ORGAN	**MECHANICAL PROCESS**	**NATURE OF PROCESS**
Mouth (teeth and tongue)	Mastication Deglutition	Chewing movements—reduce size of food particles and mix them with saliva Swallowing—movement of food from mouth to stomach
Pharynx	Deglutition	
Esophagus	Deglutition Peristalsis	Rippling movements that squeeze food downward in digestive tract; a constricted ring forms first in one section, the next, etc., causing waves of contraction to spread along entire canal
Stomach	Churning	Forward and backward movement of gastric contents, mixing food with gastric juices to form chyme
	Peristalsis	Wave starting in body of stomach about three times per minute and sweeping toward closed pyloric sphincter; at intervals, strong peristaltic waves press chyme past sphincter into duodenum
Small intestine	Segmentation (mixing contractions) Peristalsis	Forward and backward movement within segment of intestine; purpose is to mix food and digestive juices thoroughly and to bring all digested food in contact with intestinal mucosa to facilitate absorption; purpose of peristalsis, on the other hand, is to propel intestinal contents along digestive tract
Large intestine Colon	Segmentation Peristalsis	Churning movements within haustral sacs
Descending colon	Mass peristalsis	Entire contents moved into sigmoid colon and rectum; occurs three or four times a day, usually after a meal
Rectum	Defecation	Emptying of rectum, so-called bowel movement

QUICK CHECK

1. What is meant by the term *motility?*
2. Is deglutition a voluntary or involuntary process?
3. What is the purpose of peristalsis?
4. What triggers the *enterogastric reflex* to inhibit gastric emptying?

Chemical Digestion

Chemical digestion consists of all the changes in chemical composition that foods undergo in their travel through the digestive tract. These changes result from the hydrolysis of foods. **Hydrolysis** is a chemical process in which a compound unites with water and then splits into simpler compounds (see Chapter 2). Numerous enzymes in the various digestive juices catalyze the hydrolysis of foods.

Digestive Enzymes

Overview of Digestive Enzymes

Enzymes were briefly introduced in Chapter 2 and their important functional roles more fully discussed in Chapter 4 (see pp. 114 to 116). Although our interest until now has primarily concerned *intracellular enzymes,* our current discussion will focus on extracellular digestive enzymes. In the following paragraphs we will briefly review enzymes in general and outline some characteristics of *digestive enzymes* in particular.

Recall that enzymes can be defined simply as "organic catalysts"; that is, they are organic compounds (proteins), and they accelerate chemical reactions without appearing in the final products of the reaction.

Recall the two systems used for naming enzymes: the suffix *-ase* is used with the root name of the substance whose chemical reaction is catalyzed (the substrate chemical, that is) or with the word that describes the kind of chemical reaction catalyzed. Thus, according to the first method, lipase is an enzyme that catalyzes a chemical reaction in which a lipid takes part. According to the second method, lipase might also be called a hydrolase because it catalyzes the hydrolysis of lipids. Enzymes investigated before these methods of nomenclature were adopted are still called by older names, such as **pepsin** and **trypsin**—both proteases (protein-digesting enzymes).

Enzymes can be classified as intracellular or extracellular, depending on whether they act within cells or outside them in the surrounding medium. Most enzymes act intracellularly in the body; an important exception is the digestive enzymes. Digestive enzymes are classified as extracellular because they operate in the lumen of the digestive tract, outside any cells of the body. All digestive enzymes are classified chemically as hydrolases because they catalyze the hydrolysis of food molecules—the breakdown of a molecule using water.

Properties of Digestive Enzymes

As with any type of enzyme, digestive enzymes are *specific in their action;* that is, they act only on a specific substrate. This is attributed to a "key-in-a-lock" kind of action, the configuration of the enzyme molecule fitting the configuration of some part of the substrate molecule (Figure 26-7).

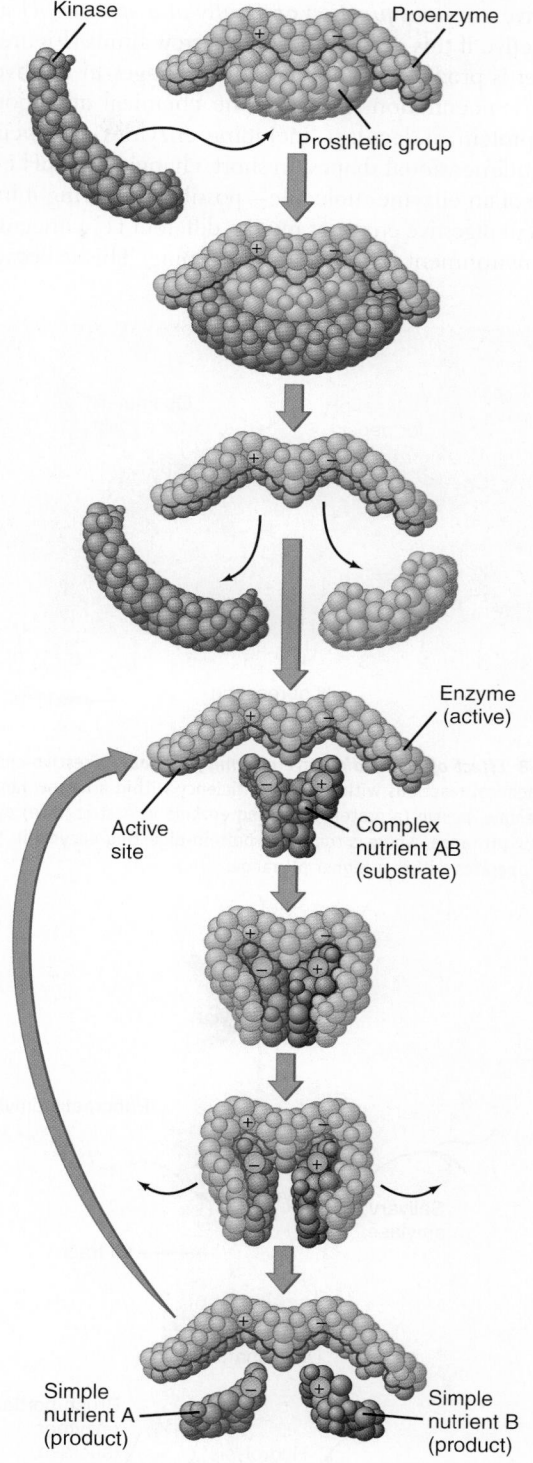

Figure 26-7 *Model of digestive enzyme action.* Enzymes are functional proteins whose molecular shape allows them to catalyze chemical reactions. First, an inactive proenzyme must be altered by an activating enzyme called a *kinase.* The kinase often accomplishes the activation by removing a prosthetic group from the proenzyme, exposing an active site. A complex nutrient molecule *AB* is acted on by the active digestive enzyme, yielding simpler nutrient molecules *A* and *B.* The active enzyme can then digest another complex nutrient molecule.

Digestive enzymes *function optimally at a specific pH* and become inactive if this deviates beyond narrow limits (Figure 26-8). This effect is produced by the fact that changes in the hydrogen ion (H$^+$) concentration influence the chemical attractions that hold all protein molecules—including enzymes—in their complex, multidimensional shapes. In short, changing the pH changes the shape of an enzyme molecule—possibly rendering it inactive.

Different digestive enzymes require different H$^+$ concentrations in their environment for optimal functioning. This is because the

H$^+$ concentration influences the shape of each enzyme molecule. Amylase, the main enzyme in saliva, functions best in the neutral to slightly acid pH range characteristic of saliva. It is gradually inactivated by the marked acidity of gastric juice. In contrast, pepsin, an enzyme in gastric juice, is inactive unless sufficient hydrochloric acid is present. Therefore, in diseases characterized by gastric hypoacidity, dilute hydrochloric acid is given orally before meals.

Most enzymes catalyze a chemical reaction in both directions, the direction and rate of the reaction being governed by the rate law (law of mass action). An accumulation of a product slows the reaction and tends to reverse it. A practical application of this fact is the slowing of digestion when absorption is interfered with and the products of digestion accumulate. However, digestive enzyme reactions do not ordinarily reverse themselves in the digestive tract.

Digestive enzymes are continually being destroyed or eliminated from the body and therefore have to be continually synthesized, even though they are not used up in the reactions they catalyze.

Most digestive enzymes are synthesized and secreted as inactive **proenzymes** (see Figure 26-7). Substances that convert proenzymes to active enzymes are often called *kinases*; for example, enterokinase changes inactive trypsinogen into active trypsin.

Although we eat six main types of chemical substances (carbohydrates, proteins, fats, vitamins, mineral salts, and water), only the first three have to be chemically digested to be absorbed.

Carbohydrate Digestion

Carbohydrates are saccharide compounds. This means that their molecules contain one or more saccharide groups ($C_6H_{10}O_5$). Polysaccharides, notably starches and glycogen, contain many of these groups. Disaccharides (sucrose, lactose, and maltose) contain two of them, and monosaccharides (glucose, fructose, and galactose) contain only one. Polysaccharides are hydrolyzed to disaccharides by

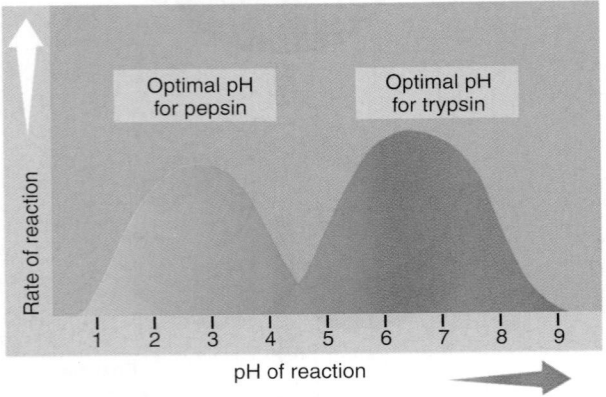

Figure 26-8 *Effect of pH on digestive enzyme function.* Digestive enzymes catalyze chemical reactions with greatest efficiency within a narrow range of pH. For example, *pepsin* (a protein-digesting enzyme in gastric juice) operates within a low pH range, whereas *trypsin* (a protein-digesting enzyme in pancreatic juice) operates within a higher pH range.

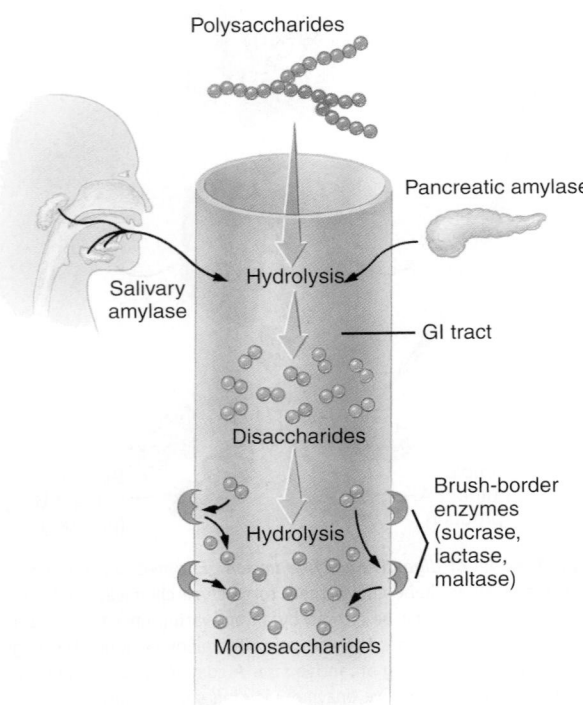

Figure 26-9 *Carbohydrate digestion.* Amylase in saliva and pancreatic juice hydrolyzes polysaccharides into disaccharides. Brush border disaccharidases in the lining of the small intestine then promote hydrolysis of the disaccharides into monosaccharides.

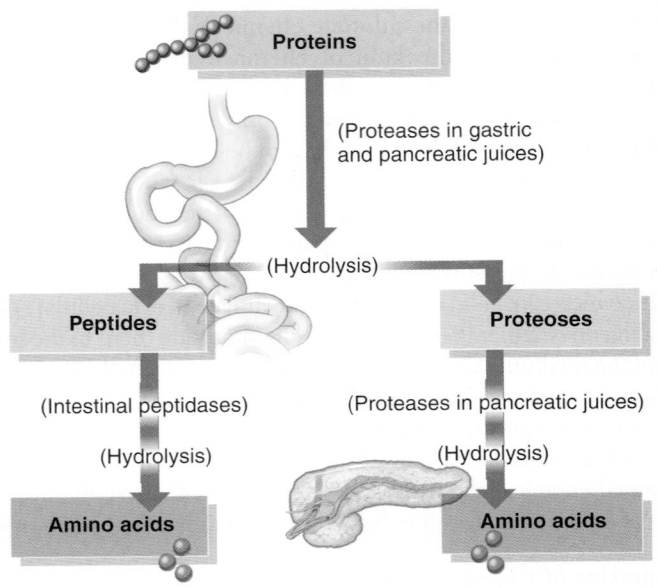

Figure 26-10 *Protein digestion.* Gastric juice protease (pepsin) and pancreatic juice protease (trypsin and chymotrypsin) hydrolyze proteins to proteoses and peptides. Protein digestion is then completed by pancreatic proteases, which hydrolyze proteoses to amino acids, and by intestinal peptidases, which hydrolyze peptides to amino acids.

enzymes known as **amylases,** found in saliva and pancreatic juice (salivary amylase was formerly called ptyalin). The enzymes that catalyze the final steps in carbohydrate digestion are *sucrase, lactase,* and *maltase* (Figure 26-9). These enzymes are located in the cell membrane of epithelial cells covering the villi and, therefore, lining the intestinal lumen. The substrates (disaccharides) bind onto the enzymes at the surface of the brush border, giving the name **contact digestion** to the process. The resulting end products of digestion, mainly glucose, are conveniently located at the site of absorption (and are not floating around somewhere in the lumen).

Protein Digestion

Protein compounds have very large molecules made up of folded or twisted chains of amino acids, often hundreds in number. Enzymes called **proteases** catalyze the hydrolysis of proteins into intermediate compounds, for example, proteoses and peptides, and finally, into amino acids (Figure 26-10). The main proteases are **pepsin** in gastric juice, **trypsin** and **chymotrypsin** in pancreatic juice, and **peptidases** of the intestinal brush border. Each kind of proteases catalyzes the breaking of a specific kind of peptide bond. Because different amino acid combinations within a

---

BOX 26-4 | The Brush Border

The term **brush border** refers to the microvilli on the epithelial mucous cells that line the small intestine, visible in the figures. These microvilli are on the apical surfaces of the epithelial cells—the surfaces that face the interior of the intestinal lumen. Because when viewed under high magnification the microvilli look like the bristles of a brush, the surface of the intestinal mucosa was nicknamed "brush border."

The brush border represents the boundary between the external environment (the lumen of the alimentary canal) and the internal environment of the body. It is across this border that molecules must pass if they are going to be absorbed into the body.

The brush border possesses such an incredibly large surface area because the microvilli increase the apical surface area. There are usually 2000 to 3000 microvilli on each cell! Of course, the presence of intestinal villi, circular folds (plicae circulares), and numerous loops

also adds to the intestinal surface area (see Figure 25-13 on p. 938).

The large surface area of the brush border provides sites for digestive enzymes on the plasma membranes of intestinal cells—the *brush border enzymes.* The efficiency of the last stages of digestion is thus enhanced by having more surface area for more digestive enzymes. Many of the brush border enzymes are on fine, branched filaments that extend from the surface of each microvillus and form a glycoprotein coat on the brush border called the *glycocalyx.* Because each substrate molecule must come in direct contact with a brush border enzyme before it can be digested, the process is called *contact digestion.*

The large surface area also provides more opportunities for the absorption of digested nutrients. More surface area allows for more phospholipid bilayer to absorb fats and more carrier molecules to carry amino acids, peptides, monosaccharides, and other nutrients.

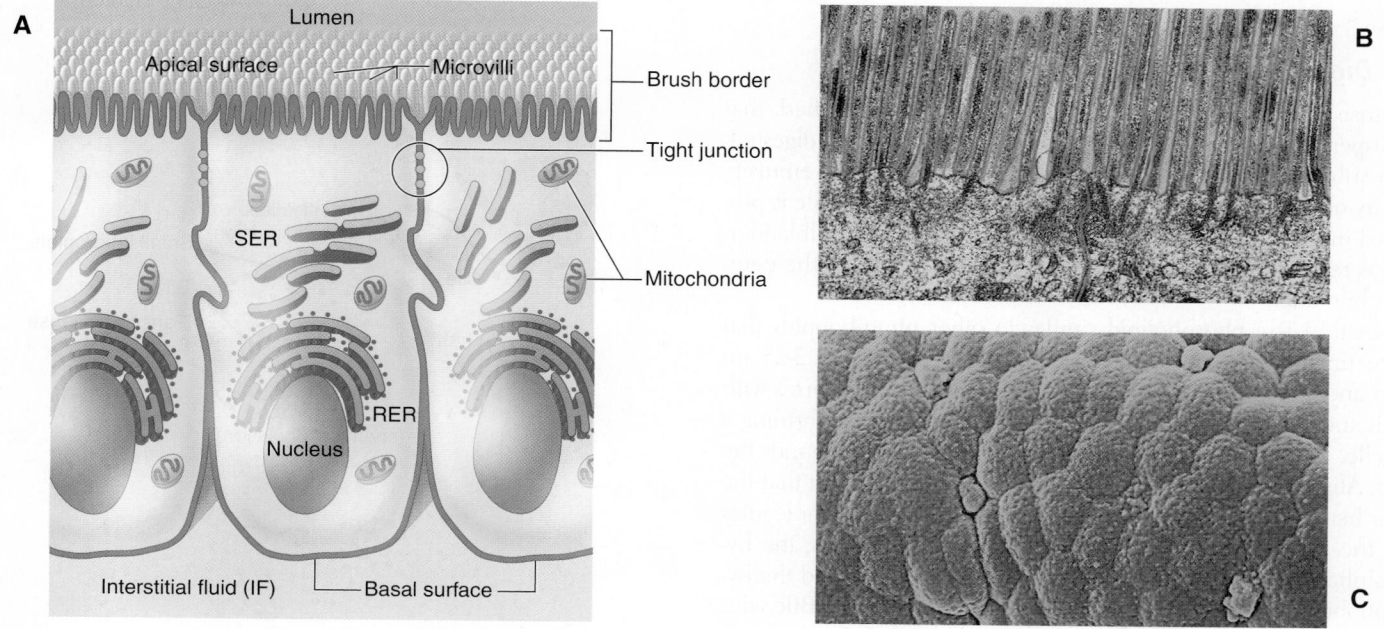

A, *Intestinal epithelium.* The diagram shows intestinal epithelial cells joined by tight junctions and covered with microvilli on their apical (lumen-facing) surfaces. **B, *Cross section of the brush border.*** A transmission electron micrograph shows microvilli (MV), which resemble bristles of a brush. *G,* Glycocalyx containing brush border enzymes. **C, *Surface view of the brush border.*** A scanning electron micrograph of the surface of an intestinal villus shows the epithelial cells joined by tight junctions. Note the brushy appearance of the surface caused by the presence of microvilli. *M,* Cells secreting mucus.

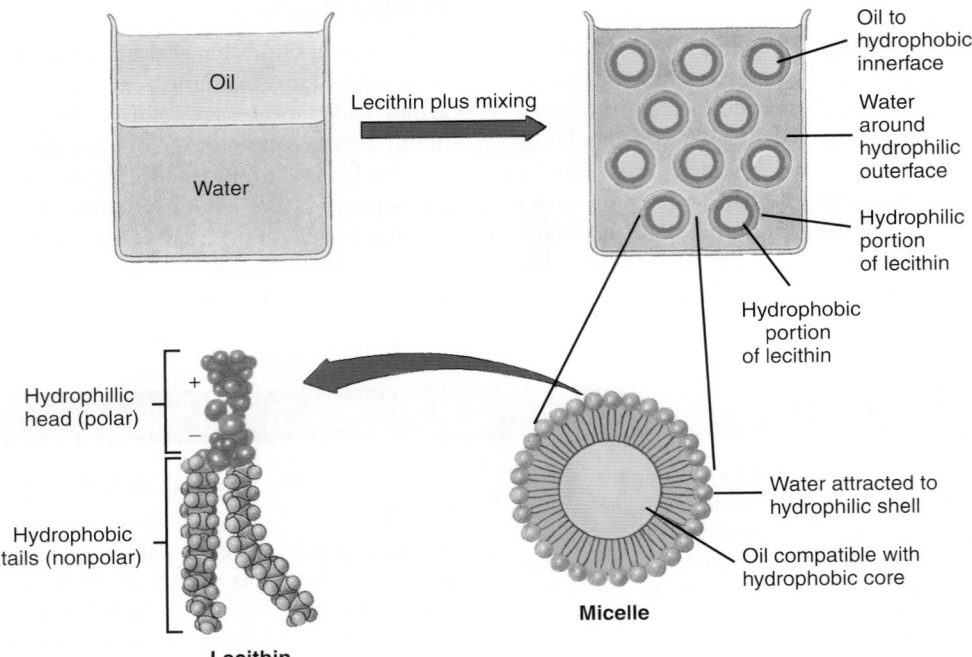

Figure 26-11 *Formation of micelles by lecithin.* Lecithin is a phospholipid in bile that mixes with oil and water to form micelles. Micelles are spherical shells formed by the orientation of lecithin molecules according to their solubility in water and lipids: the polar heads of lecithin face outward (toward water), and the non-polar tails of lecithin face inward (toward lipid). Formation of micelles, a mechanical process called *emulsification,* breaks fat drops into small droplets and makes the droplets water soluble.

protein or polypeptide can have slightly different kinds of peptide bonds holding them together, a whole arsenal of different proteases is needed for efficient protein digestion.

Fat Digestion

Because fats are insoluble in water, they must be **emulsified,** that is, dispersed as very small droplets, before they can be digested. Two substances found in bile, **lecithin** and **bile salts,** emulsify dietary oils and fats in the lumen of the small intestine. Bile is produced in the liver and stored and concentrated in the gallbladder. Bile is released into the lumen of the GI tract by way of the common bile duct.

Lecithin is a phospholipid similar to other phospholipids that make up the bulk of cellular membranes (see Figures 2-25 on p. 85 and 3-3 on p. 64). As Figure 26-11 shows, lecithin mixes with lipids and water, forming tiny spheres called **micelles.** In forming a micelle, lecithin molecules align to form a shell that surrounds the lipid. Alignment of lecithin molecules results from the fact that the polar heads of this molecule are attracted by polar water molecules and the nonpolar tails of lecithin are lipid soluble. Thus, the hydrophilic (polar) heads form the outer face of the shell and the hydrophobic (nonpolar) tails form the inner face of the shell. Bile salts, which are derived from the lipid cholesterol, emulsify fats by forming micelles in the same manner.

The mechanical process of emulsification facilitates chemical digestion of fats by breaking large fat drops into small droplets. This process provides a greater contact area between fat molecules and pancreatic **lipases,** the main fat-digesting enzymes (Figure 26-12).

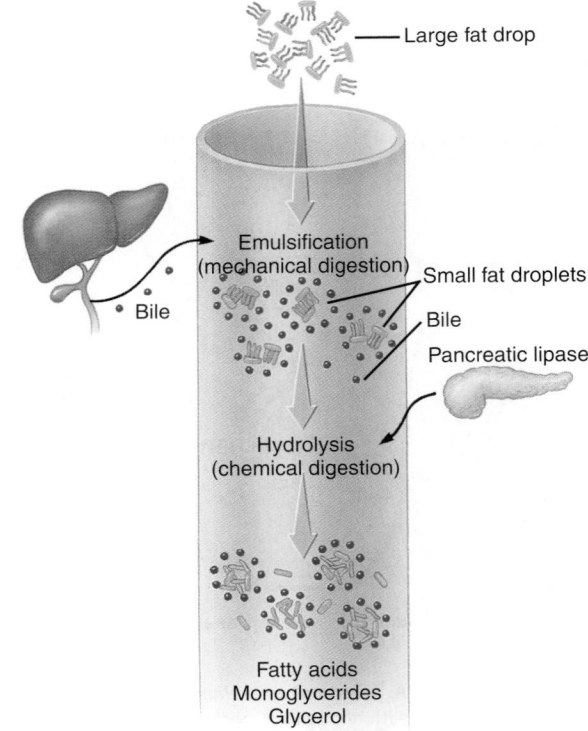

Figure 26-12 *Fat digestion.* Hydrolysis by the enzyme lipase is facilitated by prior emulsion of fats by bile (lecithin and bile salts). Though not pictured, colipase is needed to anchor lipase molecules to the inner face of each micelle.

Table 26-3   Chemical Digestion

DIGESTIVE JUICES AND ENZYMES	SUBSTANCE DIGESTED (OR HYDROLYZED)	RESULTING PRODUCT*
Saliva		
Amylase	Starch (polysaccharide)	Maltose (a double sugar, or disaccharide)
Gastric Juice		
Protease (pepsin) plus hydrochloric acid	Proteins	Partially digested proteins
Pancreatic Juice		
Proteases (e.g., trypsin)†	Proteins (intact or partially digested)	Peptides and **amino acids**
Lipases	Fats emulsified by bile	**Fatty acids, monoglycerides,** and **glycerol**
Amylase	Starch	Maltose
Intestinal Enzymes‡		
Peptidases	Peptides	Amino acids
Sucrase	Sucrose (cane sugar)	**Glucose** and **fructose**§ (simple sugars, or monosaccharides)
Lactase	Lactose (milk sugar)	**Glucose** and **galactose** (simple sugars)
Maltase	Maltose (malt sugar)	**Glucose**

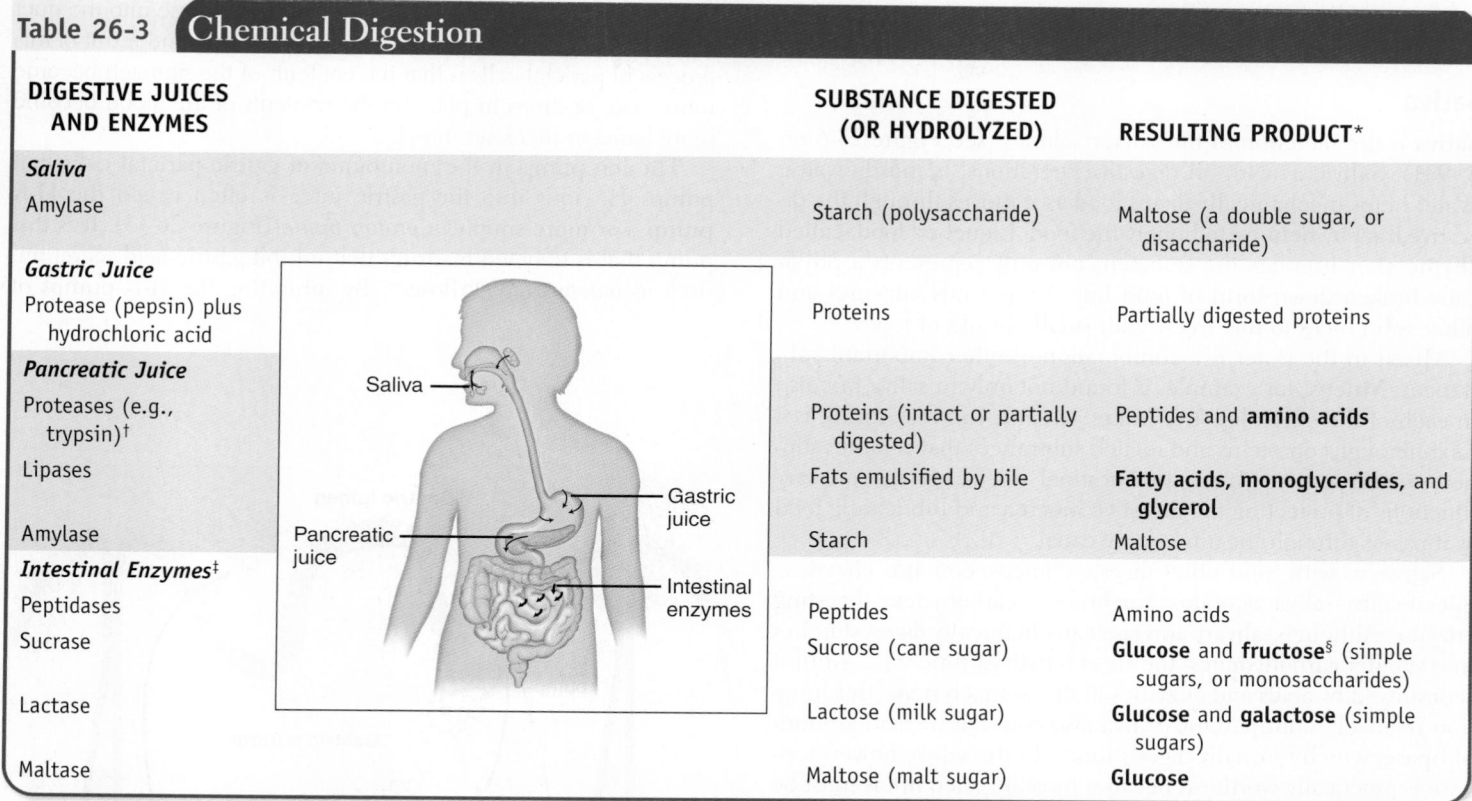

Labels: Saliva; Pancreatic juice; Gastric juice; Intestinal enzymes

*Substances in **boldface type** are end products of digestion (that is, completely digested nutrients ready for absorption).
†Secreted in inactive form (trypsinogen); activated by enterokinase, an enzyme in the intestinal brush border.
‡Brush-border enzymes.
§Glucose is also called *dextrose;* fructose is also called *levulose.*

Triglycerides, important dietary fats, are broken down by lipases to yield fatty acids, monoglycerides, and glycerol molecules. Other lipids are similarly broken down into their respective component chemical groups. For example, a phospholipid molecule can be chemically broken down by a lipase called *phospholipase* to yield one free fatty acid and one lysophosphatide (a phospholipid head with a single fatty acid tail).

The action of lipases is enhanced by a component of pancreatic juice called *colipase* (koh-LYE-payz). Colipase is a coenzyme molecule that anchors a lipase molecule to the inner face of a micelle. This positions the lipase for optimum hydrolysis of lipid molecules within the micelle.

For a summary of chemical digestion, see Table 26-3.

Residues of Digestion

Certain components of food resist digestion and are eliminated from the intestines in the **feces.** Included among these *residues of digestion* are cellulose (a carbohydrate, also known as "dietary fiber") and undigested connective tissue from meat (mostly collagen). These substances remain undigested because humans lack the enzymes required to hydrolyze them. The residues of digestion also include undigested fats. Some fat molecules remain undigested because they have combined with dietary minerals such as calcium and magnesium, which render the fats undigestible. In addition to these wastes, feces consist of bacteria, pigments, water, and mucus.

 QUICK CHECK

5. What type of reaction do all digestive enzymes catalyze?
6. List some factors that alter the shape of an enzyme, thus altering its function.
7. Name the final digestive products of each of the following food molecules:
   a. Protein
   b. Carbohydrate
   c. Triglyceride

SECRETION

Digestive secretion generally refers to the release of various substances from the exocrine glands that serve the digestive system. For example, digestive secretion includes the release of saliva, gastric juice, bile, pancreatic juice, and intestinal juice. In the paragraphs that follow, we will briefly summarize the major digestive juices. Later, we will discuss how secretion is regulated.

Saliva

Saliva is the secretion of the salivary glands (see Figure 25-6 on p. 931). Saliva, as with all digestive secretions, is mostly water. Water helps mechanically digest food as it moves through the digestive tract by helping to liquefy the food. Liquefied food, called **chyme** after it enters the stomach, not only represents a physically broken down form of food but also permits enzymes and other substances to mix freely with small chunks of food.

Mixed in the water is a combination of other important substances. **Mucus,** for example, is found not only in saliva but also in each of the other digestive juices. Mucus, you may recall, is a mixture of glycoproteins and related substances that is rather slippery to the touch. Mucus in intestinal juices has the primary functions of protecting the digestive mucosa and lubricating food as it passes through the alimentary canal.

Saliva, as with most other digestive juices, contains enzymes. Specifically, saliva contains **amylase**—a carbohydrate-digesting enzyme. Although salivary amylase can chemically digest starches into smaller carbohydrates, the short length of time it has until it is destroyed by acids and enzymes in the stomach make this function relatively unimportant. Saliva also contains a small amount of **lipase,** which normally digests lipids. In the saliva, however, lipase is practically worthless because most ingested lipids must be emulsified before lipase can digest them easily.

Saliva also contains a small amount of sodium bicarbonate ($NaHCO_3$). Sodium bicarbonate dissociates in water to form sodium ions (Na^+) and bicarbonate ions (HCO_3^-). You may recall from Chapter 24 that bicarbonate can bind to hydrogen ions, thus taking the hydrogen ions out of solution and causing a decrease in acidity (increase in pH). The slight alkalinity of saliva provides optimum conditions for amylase, which operates at a relatively high pH.

Gastric Juice

Gastric juice is secreted by exocrine *gastric glands*, which have ducts that lead to the gastric lumen by way of the gastric pits (see Figure 25-11 on p. 936). Gastric juice contains not only the basic water and mucus mixture of most other digestive juices but also a unique combination of other substances.

Chief cells in the gastric glands are also called **zymogenic cells** because they secrete the enzymes in gastric juice. The prefix *zymo-* refers to enzymes and *-genic* pertains to making something. Primary among the gastric enzymes is **pepsin,** which is secreted as the inactive proenzyme **pepsinogen.**

Pepsinogen is converted to pepsin by hydrochloric acid (HCl), which is produced by **parietal cells** of the gastric glands. Figure 26-13 shows how carbon dioxide and water form carbonic acid, which then dissociates to form the hydrogen ions needed to actively secrete hydrochloric acid. This is the very same chemical process that was outlined in Chapter 24, where we discussed carbon dioxide transport in the blood (see Figures 24-22 and 24-24, p. 907). Notice here in Figure 26-13 that there is a *chloride shift* similar to that discussed in regard to the respiratory system. In exchange for bicarbonate (also produced by dissociation of carbonic acid), chloride is

shifted into the parietal cell where it can then diffuse into the duct of the gastric gland along with hydrogen ions. The end result of this process in parietal cells is that the contents of the stomach become more acid, or drops in pH, and the contents of the blood become more basic, or increases in pH.

The ion pump in the membrane of gastric parietal cells that pumps H^+ ions into the gastric juice is often called the **H-K pump**—or more simply, a *proton pump* (Figure 26-13). It is this pump that is targeted by drugs that inhibit gastric acid secretion, such as *omeprazole* (Prilosec). By inhibiting the H-K pumps of

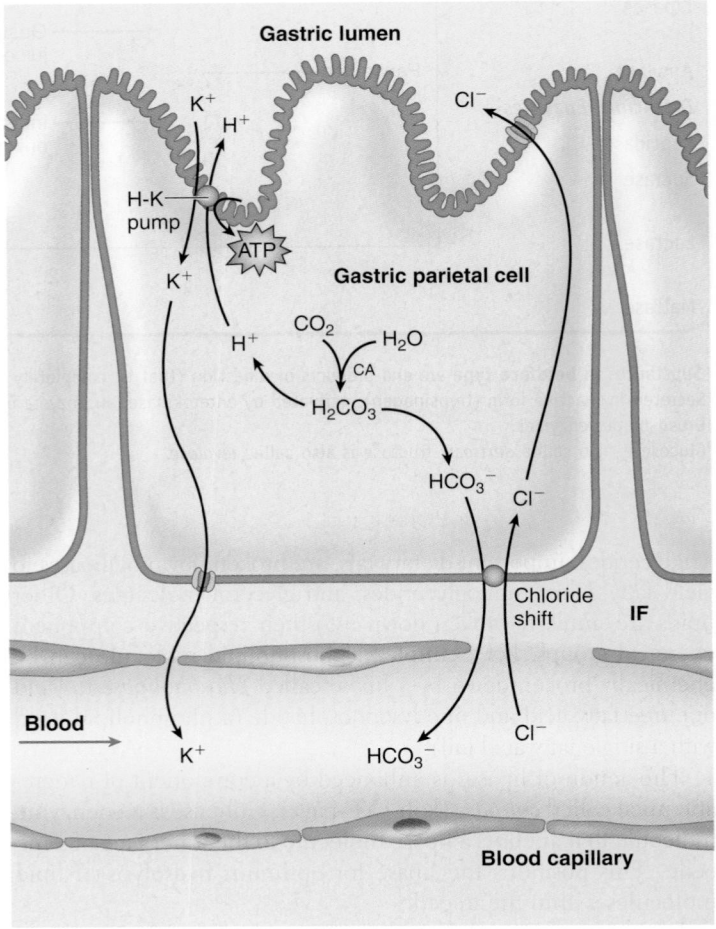

Figure 26-13 *Acid secretion by gastric parietal cells.* In this simplified diagram, you can see that hydrochloric acid (HCl) secretion by gastric parietal cells uses H^+ produced by the dissociation of carbonic acid (H_2CO_3). Recall that carbonic acid is produced by the reaction of water and carbon dioxide—a process enhanced by the enzyme carbonic anhydrase (CA). The chloride (Cl^-) of gastric HCl comes from a chloride shift into the cell in exchange for bicarbonate ions (HCO_3^-) produced by the same dissociation of carbonic acid that yielded the H^+. As Cl^- is shifted into the cell, the intracellular Cl^- concentration rises and produces a concentration gradient with the lumen of the gastric gland—forcing Cl^- to diffuse out of the parietal cell. The net effect is that active pumping of H^+ out of the cell by the H-K pump drives the concurrent shift of Cl^- into the cell from the blood and diffusion of Cl^- out of the cell and into the duct of the gastric gland.

the stomach, these drugs reduce the overall acidity of the stomach contents.

The parietal cells have an interesting and important mechanism involved with secretion of ions. As Figure 26-14, A, shows, when the parietal cell is not actively secreting, it has a relatively small surface area and many internal vesicles. These vesicles have H-K pumps and ion channels and carriers embedded in their membranes. When the parietal cell becomes active, however, these vesicles move quickly to the apical surface (Figure 26-14, B). There, they fuse with the plasma membrane forming more microvilli and increasing the overall surface area by as much as 100 times! The pumps, carriers, and channels then begin the secretion process summarized in Figure 26-13.

Besides secreting acid, parietal cells also produce **intrinsic factor.** Intrinsic factor binds to molecules of vitamin B_{12}, protecting them from the acids and enzymes of the stomach. Intrinsic factor remains attached to B_{12} until it reaches the lower small intestine, where it facilitates the absorption of B_{12} across the intestinal wall (Figure 26-15). Vitamin B_{12}, you may recall, is essential for the production of new red blood cells. In pernicious anemia, caused by insufficient vitamin B_{12} in the body, the stomach may fail to make sufficient intrinsic factor (perhaps because of stomach cancer or ulcers). More frequently, however, an autoimmune mechanism produces antibodies that block the intrinsic factor from binding to vitamin B_{12}.

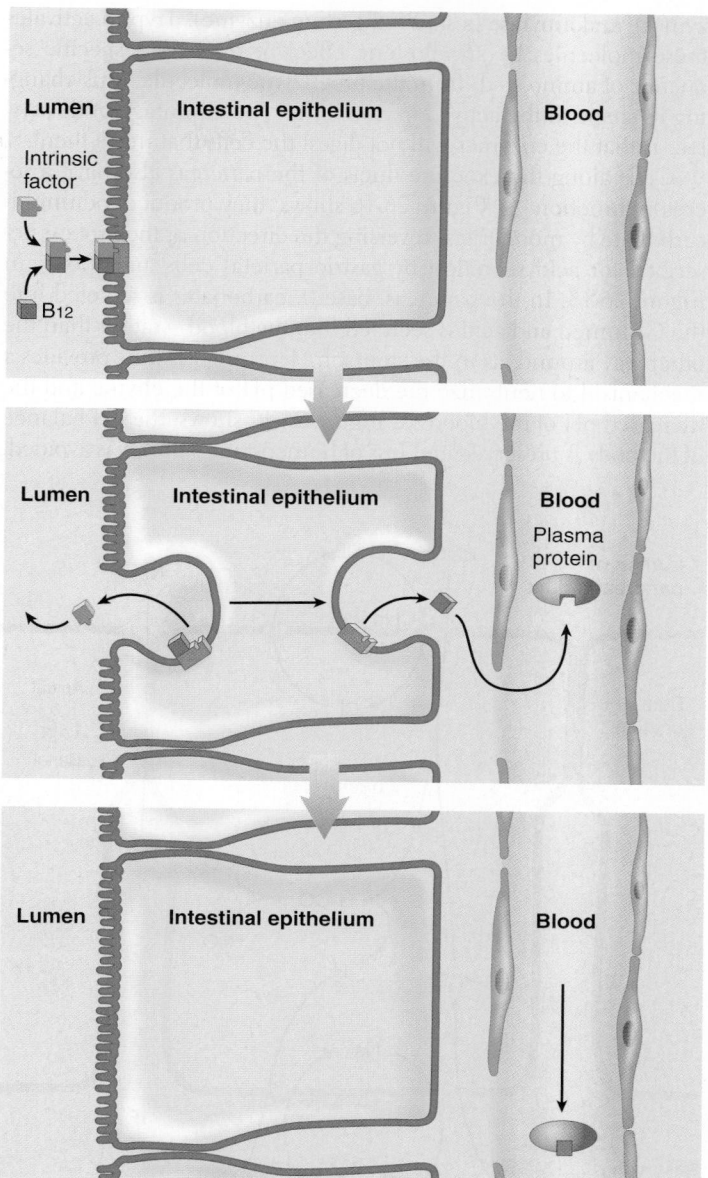

Figure 26-15 *Role of intrinsic factor.* As this series of sketches shows, intrinsic factor secreted from the stomach binds to vitamin B_{12}, which then permits the absorption of vitamin B_{12} into the bloodstream in the intestines.

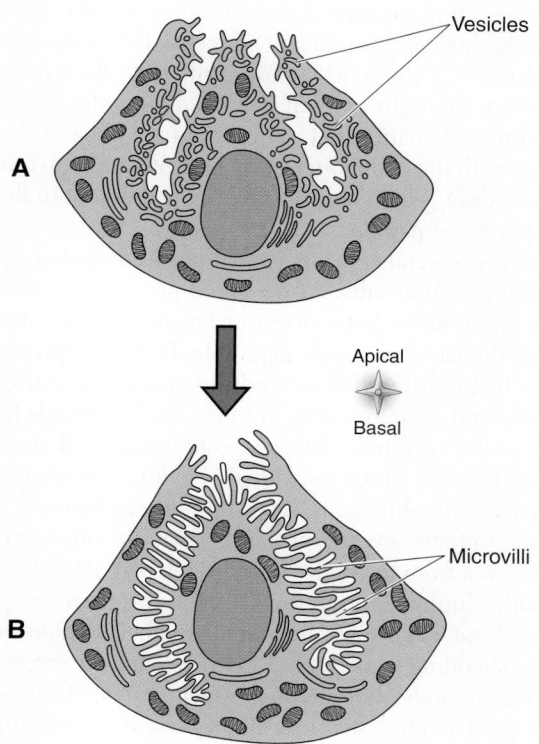

Figure 26-14 *Parietal cell membrane surface.* The surface of the resting parietal cell **(A)** can be enlarged by a factor of 100 when the cell becomes active and vesicles fuse to the apical surface of the cell **(B).** The vesicle membranes contain many ion pumps, carriers, and channels (Figure 26-13) that are thus added to the additional microvilli formed by fusion of the vesicle membrane to the plasma membrane.

Pancreatic Juice

Pancreatic juice is secreted by the exocrine *acinar cells* of the pancreas (see Figure 25-27 on p. 949). As with other digestive secretions, pancreatic juice is mostly water. In addition, pancreatic juice also contains various digestive enzymes. All of these enzymes are secreted as zymogens—inactive proenzymes. For example, protein-digesting **trypsin** is released as the zymogen **trypsinogen,** which is subsequently converted to active trypsin by **enterokinase** in the intestinal lumen. Enterokinase is an activating enzyme bound to the plasma membranes of cells that line the intestinal tract. After it is activated, trypsin can then activate other enzymes such as **chymotrypsin** (and other protein-digesting enzymes), various **lipases** (lipid-digesting enzymes), **nucleases** (RNA- and DNA-digesting enzymes), and **amylase** (a starch-digesting enzyme). Trypsin activates these molecules by an allosteric effect: it removes a specific sequence of amino acids from the proenzyme molecule, thus changing its shape to the active enzyme form. The advantage of this system is that the enzymes will not digest the cells that make them.

Cells along the exocrine ducts of the pancreas also have a secretory function. As Figure 26-16 shows, they produce sodium bicarbonate by more or less reversing the direction of the process described for acid secretion by gastric parietal cells and shown in Figure 26-13. In the pancreas, base (bicarbonate) is secreted into the GI lumen and acid is secreted into the blood—rather than the other way around, as in the stomach. This process thus provides a mechanism to neutralize the decreased pH of the chyme and the increased pH of the blood. As Figure 26-17 shows, the pH balance of the body is preserved and loss of homeostatic stability is avoided.

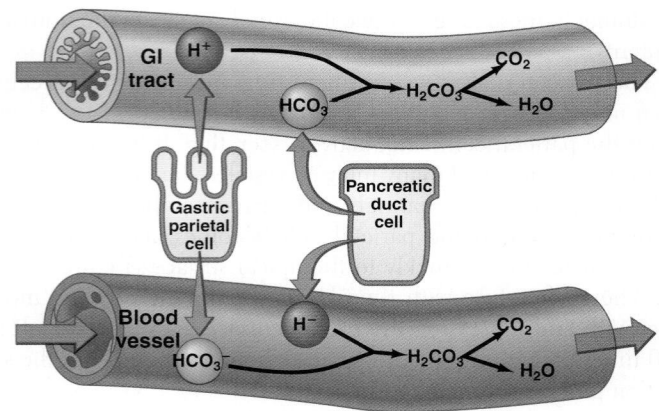

Figure 26-17 *pH balance related to the digestive tract.* Hydrochloric acid (HCl) secretion by gastric parietal cells moves H^+ into the gastrointestinal (GI) lumen, decreasing the pH of chyme. At the same time, gastric parietal cells shift bicarbonate ions (HCO_3^-) into the blood increasing the pH of blood plasma. Eventually, this net loss of acid from the internal environment would cause alkalosis if not for the counterbalancing effects of bicarbonate secretion by the duct cells of the pancreas and other digestive glands such as the liver and intestinal glands. These cells secrete HCO_3^- into the GI lumen, increasing the pH of chyme, at the same time they move H^+ into the blood decreasing the pH of blood back toward normal. This mechanism permits changes in the pH of chyme to facilitate the action of various enzymes without profoundly affecting the pH of the internal environment.

Bile

Bile is an interesting mixture of many different substances that is secreted by the liver and stored and concentrated by the gallbladder. As Figure 25-25 (on p. 947) shows, bile from the liver is conducted through the right and left hepatic ducts, which merge to form a common hepatic duct. The common hepatic duct, in turn, merges with the cystic duct from the gallbladder to form the common bile duct, which delivers bile to the duodenum through the major duodenal papilla.

Bile contains several substances that aid in digestion, specifically **lecithin** and **bile salts.** As we stated previously, both these substances break down large drops of fat into smaller droplets, thus making the fats more easily digestible. Both lecithin and bile salts wrap a hydrophilic shell around the droplets, making them water soluble and therefore able to move freely through the watery chyme in the GI lumen. Bile also contains a small amount of sodium bicarbonate, which as with the sodium bicarbonate secreted by pancreatic duct cells, helps neutralize chyme.

Bile also contains several substances that are ultimately destined for removal from the body by becoming part of the feces that are eventually eliminated from the GI tract. It is therefore proper to state that these waste substances are actually *excretions*, a term that implies shedding waste, rather than *secretions*. Excreted substances in bile include cholesterol, products of detoxification, and bile pigments. The cholesterol in bile represents excess amounts of this lipid that have been picked up from body cells by lipoproteins and delivered to the liver for disposal in bile. Products of detoxification are formed in the liver as it renders toxic molecules harmless, a process called *detoxification*. Bile pigments, chiefly **bilirubin**, are products of hemolysis (the breakdown of old red

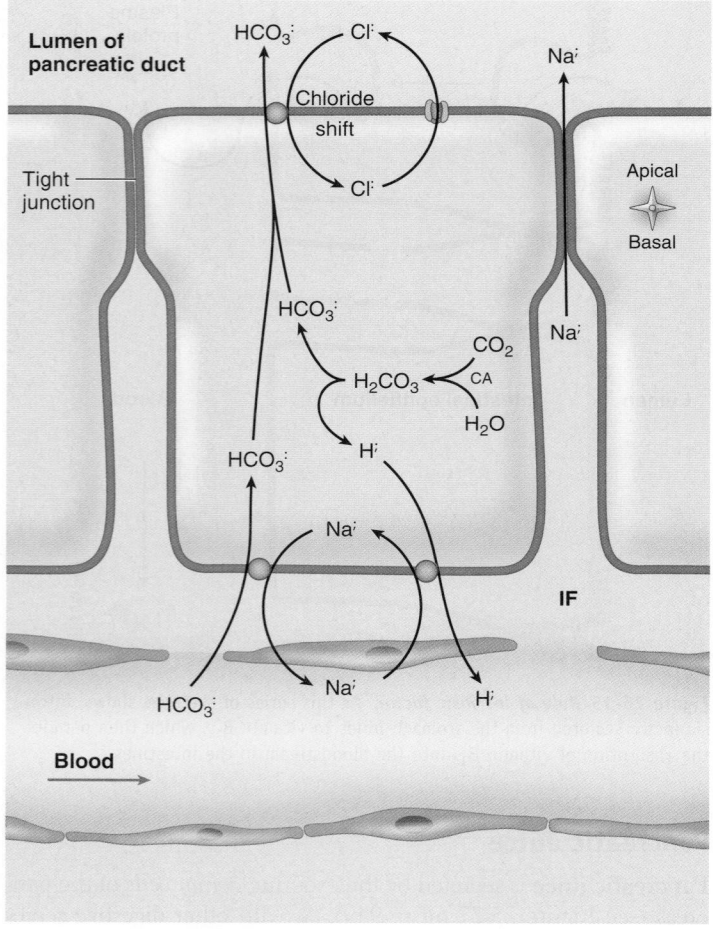

Figure 26-16 *Bicarbonate secretion by pancreatic duct cells.* In this simplified diagram, you can see that bicarbonate (HCO_3^-) secretion by cells of the pancreatic duct uses HCO_3^- produced by the dissociation of carbonic acid (H_2CO_3). Recall that carbonic acid is produced by the reaction of water and carbon dioxide—a process enhanced by the enzyme carbonic anhydrase (CA). Notice that a "reverse" chloride shift occurs as bicarbonate ions (HCO_3^-) are exchanged for Cl^- ions. The outward movement of negative bicarbonate ions into the lumen of the pancreatic ducts creates an electrical gradient that draws positive sodium ions (Na^+) from the IF (interstitial fluid), across the tight junctions, and into the pancreatic juice.

blood cells) by the liver (see Figure 17-8 on p. 654). It is the bile pigments that are eventually eliminated from the GI tract that give feces its characteristic brownish color. Elimination of gray feces, therefore, can ordinarily be interpreted as a sign that bile secretion is abnormally low.

We will discuss many functions of the liver further in the next chapter.

Intestinal Juice

The term **intestinal juice** refers to the sum total of intestinal secretions rather than to a premixed combination of substances that enters the GI lumen by way of a duct. Most intestinal cells produce a water-based solution of sodium bicarbonate. This adds to the buffering effect mentioned in previous paragraphs and illustrated in Figure 26-17. Goblet cells of the intestinal mucosa also produce a watery solution of mucus. Thus, intestinal juice is a slightly basic, mucous solution that buffers and lubricates material in the intestinal lumen.

Intestinal juice is largely a product of the small intestine, but goblet cells in the mucosa of the large intestine produce some lubricating mucus.

Table 26-4 lists various secretions of the digestive tract and their components.

QUICK CHECK

8. Name the primary components of saliva.
9. Name the components of gastric juice. Which type of cell produces each component?
10. Name as many pancreatic enzymes as you can.
11. What is excretion? What components of bile are excretions?

Table 26-4   Digestive Secretions

DIGESTIVE JUICE	SOURCE	SUBSTANCE	FUNCTIONAL ROLE*
Saliva	Salivary glands	Mucus	*Lubricates bolus of food; facilitates mixing of food*
		Amylase	**Enzyme; begins digestion of starches**
		Sodium bicarbonate	Increases pH (for optimum amylase function)
		Water	*Dilutes food and other substances; facilitates mixing*
Gastric juice	Gastric glands	Pepsin	**Enzyme; digests proteins**
		Hydrochloric acid	Denatures proteins; decreases pH (for optimum pepsin function)
		Intrinsic factor	**Protects and allows later absorption of vitamin B_{12}**
		Mucus	*Lubricates chyme; protects stomach lining*
		Water	*Dilutes food and other substances; facilitates mixing*
Pancreatic juice	Pancreas (exocrine portion)	Proteases (trypsin, chymotrypsin, collagenase, elastase, etc.)	**Enzymes; digest proteins and polypeptides**
		Lipases (lipase, phospholipase, etc.)	**Enzymes; digest lipids**
		Colipase	**Coenzyme; helps lipase digest fats**
		Nucleases	**Enzymes; digest nucleic acids** (RNA and DNA)
		Amylase	**Enzyme; digests starches**
		Water	*Dilutes food and other substances; facilitates mixing*
		Mucus	*Lubricates*
		Sodium bicarbonate	**Increases pH** (for optimum enzyme function)
Bile	Liver (stored and concentrated in gallbladder)	Lecithin and bile salts	*Emulsify lipids*
		Sodium bicarbonate	**Increases pH** (for optimum enzyme function)
		Cholesterol	Excess cholesterol from body cells, to be excreted with feces
		Products of detoxification	From detoxification of harmful substances by hepatic cells, to be excreted with feces
		Bile pigments (mainly bilirubin)	Products of breakdown of heme groups during hemolysis, to be excreted with feces
		Mucus	*Lubrication*
		Water	*Dilutes food and other substances; facilitates mixing*
Intestinal juice	Mucosa of small and large intestine	Mucus	*Lubrication*
		Sodium bicarbonate	**Increases pH** (for optimum enzyme function)
		Water	*Small amount to carry mucus and sodium bicarbonate*

*__Boldface type__ indicates a chemical digestive process; *Italic type* indicates a mechanical digestive process.

CONTROL OF DIGESTIVE GLAND SECRETION

Exocrine digestive glands secrete when food is present in the digestive tract or when it is seen, smelled, or imagined. Complicated nervous and hormonal reflex mechanisms control the flow of digestive juices in such a way that they appear in proper amounts when and for as long as needed.

Control of Salivary Secretion

As far as is known, only reflex mechanisms control the secretion of saliva. Chemical, mechanical, olfactory, and visual stimuli initiate afferent impulses to centers in the brainstem that send out efferent impulses to the salivary glands, stimulating them. Chemical and mechanical stimuli come from the presence of food in the mouth. Olfactory and visual stimuli come, of course, from the smell and sight of food.

Control of Gastric Secretion

Stimulation of gastric juice secretion occurs in three phases that are controlled by reflex and chemical mechanisms. Because stimuli that activate these mechanisms arise in the head, stomach, and intestines, the three phases are known as the cephalic, gastric, and intestinal phases, respectively. As you read the description of each phase, glance at the diagrams shown in Figure 26-18.

The **cephalic phase** is also spoken of as the "psychic phase" because psychic (mental) factors activate the mechanism. For example, the sight, smell, taste, or even thought of food that is pleasing to an individual activates control centers in the medulla oblongata from which parasympathetic fibers of the vagus nerve conduct efferent impulses to the gastric glands. Vagal nerve impulses also stimulate the production of **gastrin,** a hormone secreted by endocrine G *cells* in the gastric mucosa. Gastrin stimulates gastric secretion, thus prolonging and enhancing the response.

During the **gastric phase** of gastric secretion, the following chemical control mechanism dominates. Products of protein digestion in foods that have reached the pyloric portion of the stomach stimulate its mucosa to release *gastrin* into the blood in stomach capillaries. When it circulates to the gastric glands, gastrin greatly accelerates their secretion of gastric juice, which has a high pepsinogen and hydrochloric acid content (Table 26-5). Hence, this seems to be a mechanism for ensuring that when food is in the stomach, there will be enough enzymes there to digest it. Gastrin release is also stimulated by distention of the stomach (caused by the presence of food), which activates local and parasympathetic reflexes in the pylorus.

The **intestinal phase** of gastric juice secretion is less clearly understood than the other two phases. Various different mechanisms seem to adjust gastric juice secretion as chyme passes to and through the intestinal tract. Experiments show that gastric secretions are inhibited when chyme containing fats, carbohydrates, and acid (low pH) is present in the duodenum. This probably occurs by means of endocrine reflexes that involve the hormones **gastric inhibitory peptide (GIP), secretin, CCK,** and perhaps several others. These hormones are secreted by endocrine cells in the mucosa of the duodenum. Gastric secretion may also be inhibited by the parasympathetic **enterogastric reflex.** We have discussed how this reflex inhibits gastric motility as food begins to fill the duodenum; now we see that it may inhibit gastric secretion as well.

In summary, we see that the rate of gastric secretion can be adjusted by nervous and endocrine reflex mechanisms in ways that improve the efficiency of the system. Anticipation of swallowing food causes the stomach to prepare itself by increasing its secretion of enzymes and acid. Thus, food enters a stomach already partially filled with gastric juice. The rate of gastric secretion can then be adjusted according to the amount of food present and whether it contains proteins (the only food that can be chemically digested by gastric juice). Gastric secretion—and thus chemical digestion in the stomach—can be slowed when the duodenum becomes full. This prevents the stomach from finishing its task before the small intestine is ready to receive the chyme.

Control of Pancreatic Secretion

Several hormones released by intestinal mucosa are known to stimulate pancreatic secretion. One of these hormones, *secretin*, evokes the production of pancreatic fluid low in enzyme content but high in bicarbonate (HCO_3^-). This alkaline fluid acts to neu-

Table 26-5	Actions of Some Digestive Hormones Summarized	
HORMONE	**SOURCE**	**ACTION**
Gastrin	Formed by gastric mucosa in presence of partially digested proteins, when stimulated by the vagus nerve, or when the stomach is stretched	Stimulate secretion of gastric juice rich in pepsin and hydrochloric acid
Gastric inhibitory peptide (GIP)	Formed by intestinal mucosa in presence of fats and perhaps other nutrients	Inhibits gastric secretion and motility
Secretin	Formed by intestinal mucosa in presence of acid, partially digested proteins, and fats	Inhibits gastric secretion; stimulates secretion of pancreatic juice low in enzymes and high in alkalinity (bicarbonate); stimulates ejection of bile by the gallbladder
Cholecystokinin (CCK)	Formed by intestinal mucosa in presence of fats, partially digested proteins, and acids	Stimulates ejection of bile from gallbladder and secretion of pancreatic juice high in enzymes; opposes the action of gastrin, raising the pH of gastric juice

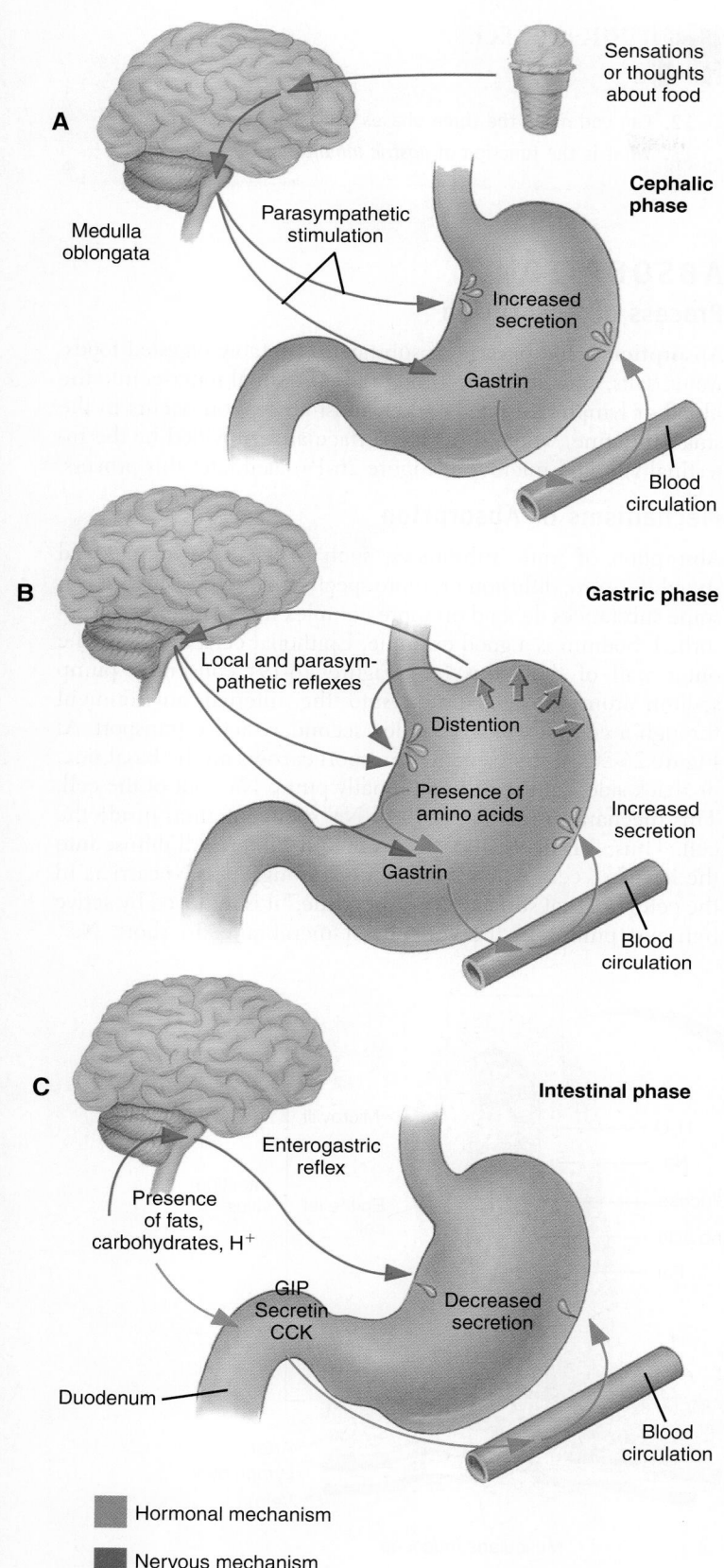

A

Sensations or thoughts about food

Cephalic phase

Medulla oblongata

Parasympathetic stimulation

Increased secretion

Gastrin

Blood circulation

B

Gastric phase

Local and parasympathetic reflexes

Distention

Presence of amino acids

Increased secretion

Gastrin

Blood circulation

C

Intestinal phase

Enterogastric reflex

Presence of fats, carbohydrates, H$^+$

GIP
Secretin
CCK

Decreased secretion

Duodenum

Blood circulation

Hormonal mechanism

Nervous mechanism

Figure 26-18 *Phases of gastric secretion.* **A,** *Cephalic phase.* Sensations of thoughts about food are relayed to the brainstem, where parasympathetic signals to the gastric mucosa are initiated. This directly stimulates gastric juice secretion and also stimulates the release of gastrin, which prolongs and enhances the effect. **B,** *Gastric phase.* The presence of food, specifically the distention it causes, triggers local and parasympathetic nervous reflexes that increase secretion of gastric juice and gastrin (which further amplifies gastric juice secretion). Products of protein digestion can also trigger the gastrin mechanism. **C,** *Intestinal phase.* As food moves into the duodenum, the presence of fats, carbohydrates, and acid stimulates hormonal and nervous reflexes that inhibit stomach activity.

tralize the acid (chyme) entering the duodenum. As you might expect, the presence of acid in the duodenum serves as the most potent stimulator of secretin. (Additional control involving the same hormone is shown by the fact that fats in the duodenum also elicit secretin, which then influences the gallbladder to increase its ejection of the fat emulsifier bile.)

The other intestinal hormone, known as **CCK,** was originally thought to be two separate substances. It has now been identified as one chemical with several important functions: (1) it causes the pancreas to increase exocrine secretions high in enzyme content; (2) it opposes the influence of gastrin on gastric parietal cells, thus inhibiting hydrochloric acid secretion by the stomach; and (3) it stimulates contraction of the gallbladder so that bile can pass into the duodenum.

Control of Bile Secretion

Bile is secreted continually by the liver and is stored in the gallbladder until needed by the duodenum. The hormones secretin and CCK, as described, stimulate ejection of bile from the gallbladder (see Table 26-5).

Control of Intestinal Secretion

Relatively little is known about the regulation of intestinal exocrine secretions. Some evidence suggests that the intestinal mucosa, stimulated by hydrochloric acid and food products, releases hormones into the blood, including *vasoactive intestinal peptide* (VIP), which brings about increased production of intestinal juice. Intestinal secretions contain bicarbonate, which along with pancreatic bicarbonate, neutralizes acid from the stomach. Bicarbonate secretion is regulated by a reflex sensitive to changes in pH of the chyme. Presumably, neural mechanisms also help control the secretion of intestinal juice.

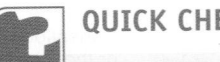

QUICK CHECK

12. Can you name the three phases of gastric secretion?
13. What is the function of *gastric inhibitory peptide*?

ABSORPTION

Process of Absorption

Absorption is the passage of substances (notably digested foods, water, salts, and vitamins) through the intestinal mucosa into the blood or lymph. As stated earlier, most absorption occurs in the small intestine, where the large surface area provided by the intestinal villi and microvilli (Figure 26-19) facilitates this process.

Mechanisms of Absorption

Absorption of some substances, such as water, is simple and straightforward: diffusion or, more specifically, osmosis. However, some substances depend on more complex mechanisms to be absorbed. Sodium is a good example. Epithelial cells that form the outer wall of the villus (see Figure 26-19) constantly pump sodium from the GI lumen into the internal environment through a complex process called secondary active transport. As Figure 26-20, A, shows, active transport carriers on the basal side, or "back side," of the cell continually pump Na^+ out of the cell. This mechanism maintains a low Na^+ concentration inside the cell. Thus, it is likely that Na^+ in the GI lumen will diffuse into the low-Na^+ cell. As Na^+ diffuses in through passive carriers in the cell's luminal surface, or "lumen side," it is removed by active transport pumps in the cell's basal membrane. In short, Na^+

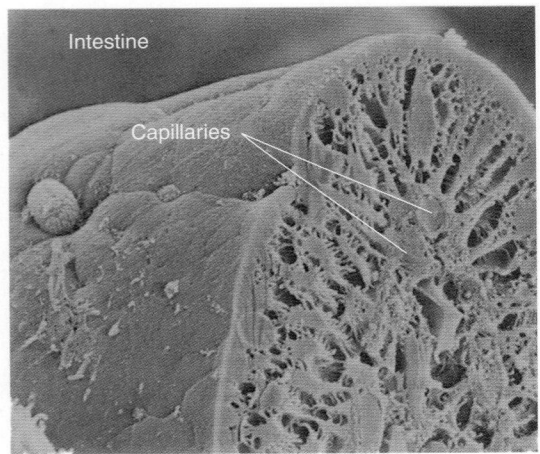

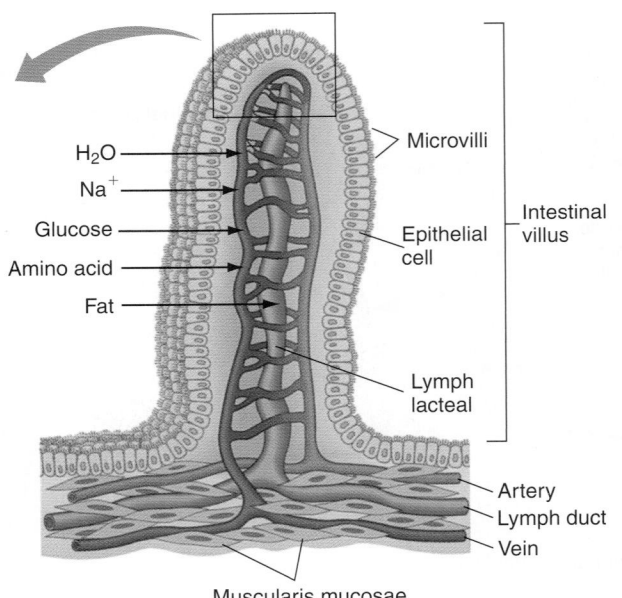

Figure 26-19 *Intestinal villus.* The presence of intestinal villi and microvilli increases the absorptive surface area of the intestinal mucosa. Most absorbed substances enter the blood in intestinal capillaries, with the exception of fat, which enters lymph by way of the intestinal lacteals.

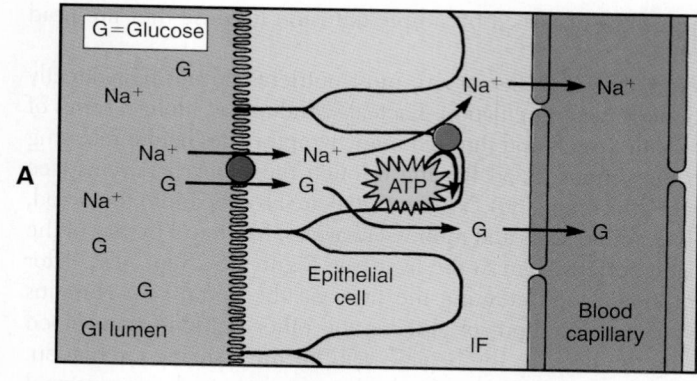

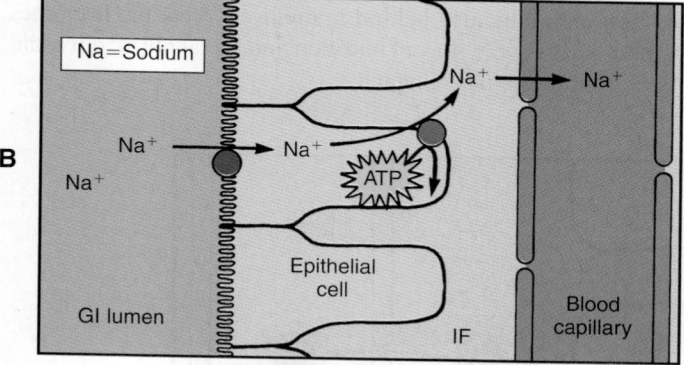

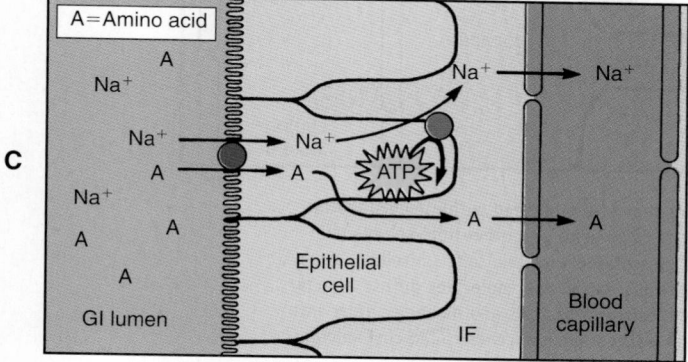

Figure 26-20 *Absorption of glucose, sodium, and amino acids.* Absorption of glucose **(A)**, sodium **(B)**, and amino acids **(C)** is a form of *secondary active transport* because each involves two carriers, one of which is active. The active carrier on the basal side of the epithelial cell maintains a sodium gradient, which facilitates passive transport of sodium, and perhaps another molecule, out of the GI lumen via a passive carrier on the luminal side of the cell.

moves out of the GI lumen only because it is being pumped from the other side of the intestinal cells.

Another good example of a complex transport process is that involving glucose. Though considered an "end product of digestion," glucose is a relatively large molecule and cannot pass freely through the brush border membrane of an intestinal mucosa cell. In addition to physical size, the lipid nature of the cell membrane (see Figure 3-2, p. 84) presents another barrier to glucose absorption. Molecules the size of glucose can pass freely (passively) through the lipid cell barrier only if they are lipid soluble (hy-

drophobic). Because glucose is too large physically and is hydrophilic (water soluble) in nature, it must be transported across the membrane by a carrier to enter the cell. In a process called **sodium cotransport** or coupled transport, carriers that bind both sodium ions and glucose molecules passively transport these molecules *together* out of the GI lumen (Figure 26-20, *B*). However, this is another case of secondary active transport because this movement does not occur without the Na$^+$ concentration gradient maintained by the active transport of Na$^+$ out of the cell's basal membrane. Amino acids and several other compounds are thought to also be absorbed by such a secondary active transport mechanism (see Figure 26-20, *C*).

Other mechanisms of transporting glucose and amino acids have also been proposed. One hypothesis suggests that these compounds are transported by passive carriers on both the apical surfaces (lumen side) and the basal surfaces of the absorptive cells. Another hypothesis suggests that the brush border enzymes also act as carriers. It should also be noted that some short polypeptides can diffuse into absorptive cells where they are hydrolyzed into amino acids that can move into the blood.

Fatty acids and monoglycerides (products of fat digestion) and cholesterol are transported with the aid of lecithin and bile salts from the watery intestinal lumen to absorbing cells on villi. Lecithin and bile salts form microscopic spheres called **micelles,** inside which chemical digestion of fats takes place (see Figure 26-12).

BOX 26-5

Diagnostic Study

Fecal Fat Test

Impaired fat absorption (malabsorption), prevalent in numerous diseases, produces large, greasy, and foul-smelling stools, or **steatorrhea.**

The fecal fat test measures the fat content in the stool. The total output of fecal fat per 24 hours in a 3- to 5-day stool collection provides the most reliable measurement. Each stool specimen throughout the period is collected in a clean, dry container and is sent immediately to the laboratory. The 3- to 5-day collection period is necessary to eliminate daily variations in the amount of fecal fat.

A standard fat content diet is begun 2 or 3 days before collection begins and continues until collection is done. Usually 100 g of fat per day is suggested for adults. In children and infants who cannot eat 100 g of fat per day, a fat retention coefficient is determined by using the formula

$$\frac{\text{Ingested fat} - \text{Fecal fat}}{\text{Ingested fat}} \times 100\%$$

If this fat retention coefficient is lower than 95%, the patient may have steatorrhea.

Analysis of fecal fat is useful in monitoring malabsorption in cystic fibrosis or in any condition characterized by malabsorption, maldigestion, or increased fecal fat.

As Figure 26-21 shows, water-soluble micelles formed in the lumen of the intestine approach the brush border of absorbing cells. There, simple lipid molecules are released to pass through the plasma membrane (its phospholipid bilayer is receptive to lipids) by simple diffusion. After they are inside the cell, fatty acids are rapidly reunited with monoglycerides to form triglycerides (neutral fats). The final step in lipid transport by the intestine is the formation of **chylomicrons,** which are simply another type of micelle. Chylomicrons are formed by the Golgi apparatus of the absorptive cell. The water-soluble chylomicron allows fats to be transported through lymph and into the bloodstream (Table 26-6).

Vitamins A, D, E, and K, known as the "fat-soluble vitamins," also depend on bile salts for their absorption. Some water-soluble vitamins, such as certain of the B group, are small enough to be absorbed by simple diffusion; however, most require carrier-mediated transport. Many drugs (sedatives, analgesics, antibiotics) appear to be absorbed by simple diffusion because they are lipid soluble.

Note that after absorption, most nutrients do not pass directly into the general circulation. Lacteals conduct fats along a series of lymphatic vessels and through many lymph nodes before releasing into the venous blood flowing through the subclavian veins (see Figure 20-2 on p. 780). Nutrients that are absorbed into the blood, such as amino acids and monosaccharides, first travel by way of the hepatic portal system to the liver (see Figure 18-28, p. 712). After absorption, blood entering the liver via the portal vein contains greater concentrations of glucose and other nutrients than blood leaving the liver via the hepatic vein for the systemic circulation. Clearly, the excess of these food substances over and above normal blood levels has remained behind in the liver. What the liver does with them is part of the story of nutrition and metabolism, our topic for discussion in the next chapter.

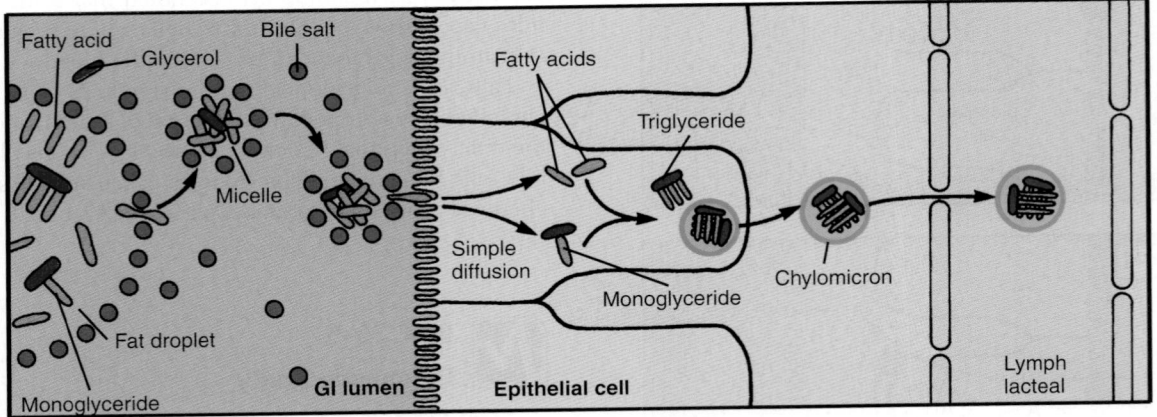

Figure 26-21 *Absorption of fats.* Fats such as triglycerides are chemically digested within emulsified fat droplets, yielding fatty acids, monoglycerides, and glycerol *(left)*. Fatty acids and other lipid-soluble compounds (such as cholesterol) leave the fat droplets in small spheres coated with bile salts (micelles). When a micelle reaches the plasma membrane of an absorptive cell, individual fat-soluble molecules diffuse directly into the cytoplasm. The endoplasmic reticulum of the cell resynthesizes fatty acids and monoglycerides into triglycerides. A Golgi body within the cell packages the fats into water-soluble micelles called chylomicrons, which then exit the absorptive cell by exocytosis and enter a lymphatic lacteal.

Table 26-6	Food Absorption		
FORM ABSORBED	**STRUCTURES INTO WHICH ABSORBED**	**CIRCULATION**	
Protein—as amino acids Perhaps minute quantities of some short-chain polypeptides and whole proteins are absorbed; for example, some antibodies	Blood in intestinal capillaries	Portal vein, liver, hepatic vein, inferior vena cava to heart, etc.	
Carbohydrates—as simple sugars	Same as amino acids	Same as amino acids	
Fats Glycerol and monoglycerides Fatty acids combine with bile salts to form water-soluble substance Some finely emulsified, undigested fats absorbed	Lymph in intestinal lacteals Lymph in intestinal lacteals Small fraction enters intestinal blood capillaries	During absorption, that is, while in epithelial cells of intestinal mucosa, glycerol and fatty acids recombine to form microscopic packages of fats (chylomicrons); lymphatics carry them by way of thoracic duct to left subclavian vein, superior vena cava, heart, etc.; some fats are transported by blood in form of phospholipids or cholesterol esters	

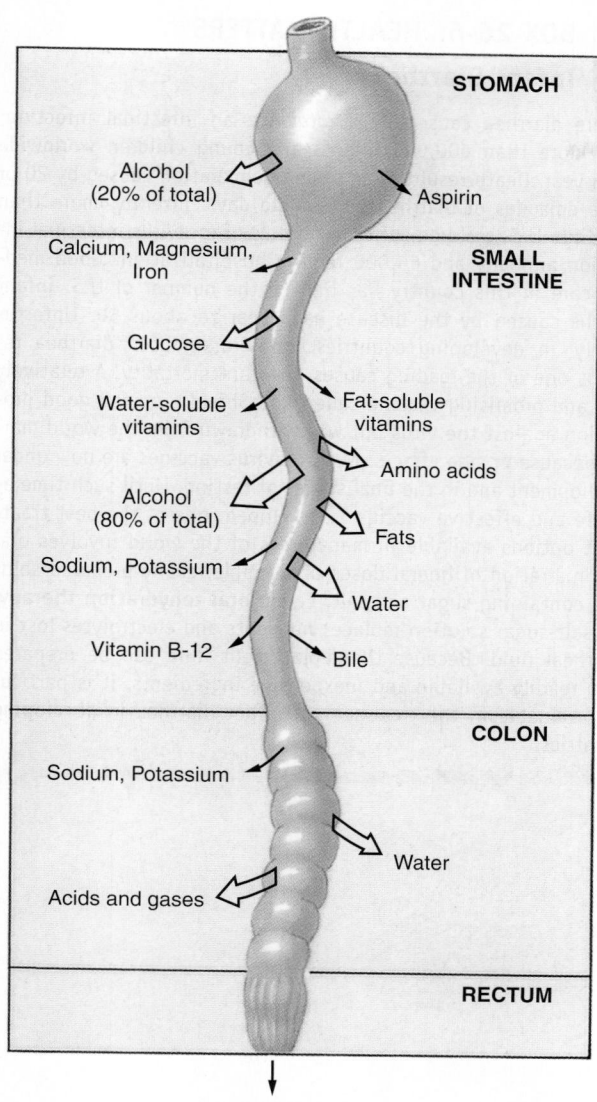

Figure 26-22 *Absorption sites in the digestive tract.* The size of the arrow at each site indicates the relative amount of absorption of a particular substance at that site. Notice that most absorption occurs in the intestines, particularly the small intestine.

Figure 26-22 summarizes the locations where absorption of some important substances takes place. Notice that although the stomach can absorb a small amount of alcohol, almost all absorption takes place in the intestines—particularly the small intestine.

ELIMINATION

The process of **elimination** is simply the expulsion of the residues of digestion—*feces*—from the digestive tract. Formation of feces is the primary function of the colon. The act of expelling feces is called **defecation.**

Defecation is a reflex brought about by stimulation of receptors in the rectal mucosa. Normally, the rectum is empty until mass peristalsis moves fecal matter out of the colon into the rectum. This distends the rectum and produces the desire to defe-

cate. Also, it stimulates colonic peristalsis and initiates reflex relaxation of the internal sphincters of the anus. Voluntary straining efforts and relaxation of the external anal sphincter may then follow as a result of the desire to defecate. Together, these several responses bring about defecation (Figure 26-23). Note that this is a reflex partly under voluntary control. If one voluntarily inhibits

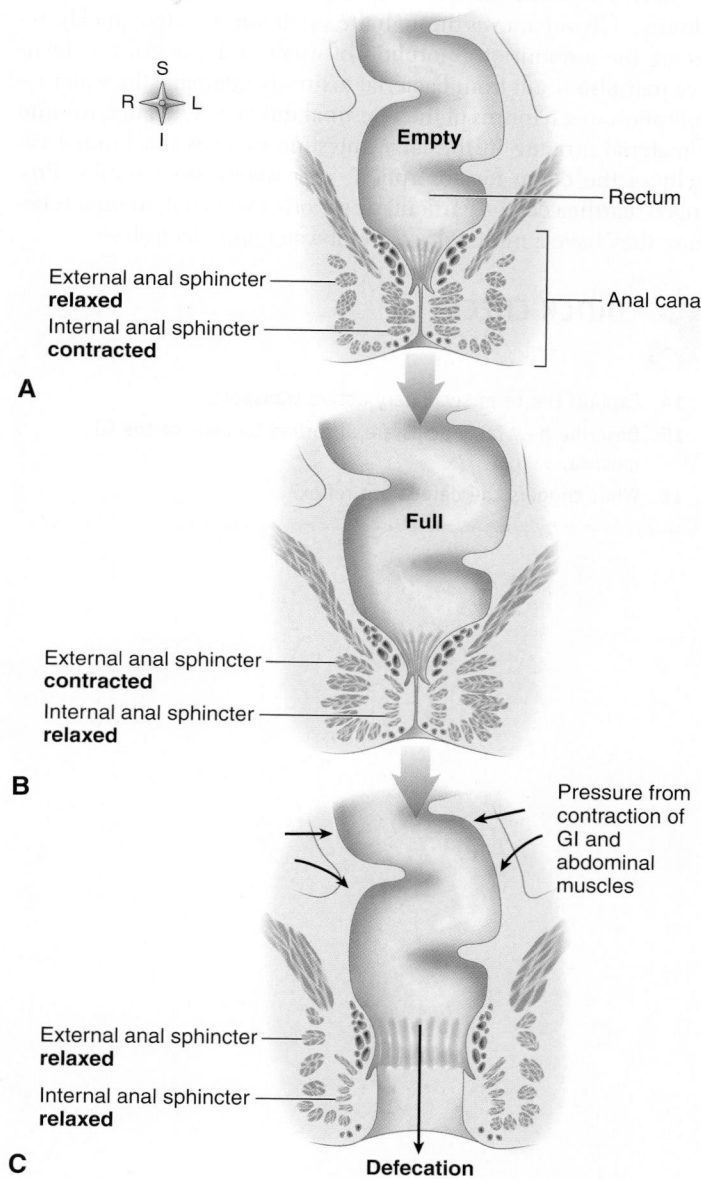

Figure 26-23 *Defecation.* The sketches show the role of the anal sphincters in defecation. **A,** When the rectum is empty (and the rectum is minimally stretched), which is most of the time, thin internal anal sphincters are contracted (closed) and the external sphincters are relaxed (open). **B,** When the rectum becomes full, increased stretch of the rectum triggers the reflexive relaxation of the internal sphincters while at the same time triggering the reflexive contraction of the external sphincters. This shifts the control of the anal opening from an involuntary mode (relaxation of the internal sphincters is involuntary) to a voluntary mode (relaxation of the external sphincters is voluntary). **C,** The conscious decision to defecate causes the external sphincters to relax. This, along with contractions of the colonic smooth muscles and the abdominal skeletal muscles, causes movement of feces out of the alimentary canal.

defecation, the rectal receptors soon become depressed and the urge to defecate does not usually recur until hours later, when mass peristalsis again takes place.

Constipation occurs when the contents of the lower part of the colon and rectum move at a rate that is slower than normal. Extra water is absorbed from the fecal mass, producing a hardened, or constipated, stool.

Diarrhea may occur as a result of increased motility of the small intestine. Chyme moves through the small intestine too quickly, reducing the amount of absorption of water and electrolytes. Diarrhea may also result from bacterial toxins that damage the water reabsorption mechanisms of the intestinal mucosa. The large volume of material arriving in the large intestine exceeds the limited capacity of the colon for absorption, so a watery stool results. Prolonged diarrhea can be particularly serious, even fatal, in infants because they have a minimal reserve of water and electrolytes.

QUICK CHECK

14. Explain the term secondary active transport.
15. Describe how fatty acids are absorbed by cells of the GI mucosa.
16. What triggers the defecation reflex?

BOX 26-6: HEALTH MATTERS
Infant Diarrhea

Severe diarrhea caused by a *rotavirus,* an intestinal infection, kills more than 600,000 infants and young children worldwide each year. Death results from severe dehydration caused by 20 or more episodes of diarrhea in a single day. Currently, more than 3 million U.S. children suffer symptoms of rotavirus intestinal infection annually and 65,000 require hospitalization. Good medical care in this country has limited the number of U.S. infant deaths caused by the disease each year to about 50. Unfortunately, in developing countries, rotavirus-induced diarrhea remains one of the leading causes of infant mortality. A relatively new and promising vaccine called RotaShield provided good protection against the virus but was withdrawn from the world market because of side effects. New rotavirus vaccines are now under development and in the final stages of testing. Until such time as a safe and effective vaccine is developed, one of the best treatment options available in many areas of the world involves oral administration of liberal doses of a simple, easily prepared solution containing sugar and salt. Called **oral rehydration therapy,** the salt-sugar solution replaces nutrients and electrolytes lost in diarrheal fluid. Because the replacement fluid can be prepared from readily available and inexpensive ingredients, it is particularly valuable in the treatment of infant diarrhea in developing countries.

THE BIG PICTURE

Digestion and the Whole Body

The process of digestion, as with any other vital function, provides a means of survival for the entire body and also requires the function of other systems. The digestive system's primary contribution to overall homeostasis is its ability to maintain a constancy of nutrient concentration in the internal environment. It accomplishes this by breaking large, complex nutrients into smaller, simpler nutrients so they can be absorbed (Figure 26-24). The digestive system also provides the means of absorption—the cellular mechanisms that operate in the absorptive cells of the intestinal mucosa. The digestive system also provides some secondary, less vital functions. For example, the teeth and tongue aid the nervous system and respiratory system in producing spoken language. Also, acid in the stomach assists the immune system by destroying potentially harmful bacteria. Some of the various vital and nonvital roles played by the different organs that make up the digestive system are summarized in Figure 26-24.

To accomplish its functions, the digestive system requires functional contributions by other systems of the body. Regulation of digestive motility and secretion requires the active participation of both the nervous system and the endocrine system. The oxygen needed for digestive activity requires the proper functioning of both the respiratory system and the circulatory system. The body's framework (integumentary and skeletal systems) is required to support and protect the digestive organs. The skeletal muscles must function if ingestion, mastication, deglutition, and defecation are to occur normally. As you can see, the digestive system cannot operate alone—nor can any system or organ, for that matter. The body is truly an integrated system, not a collection of independent components.

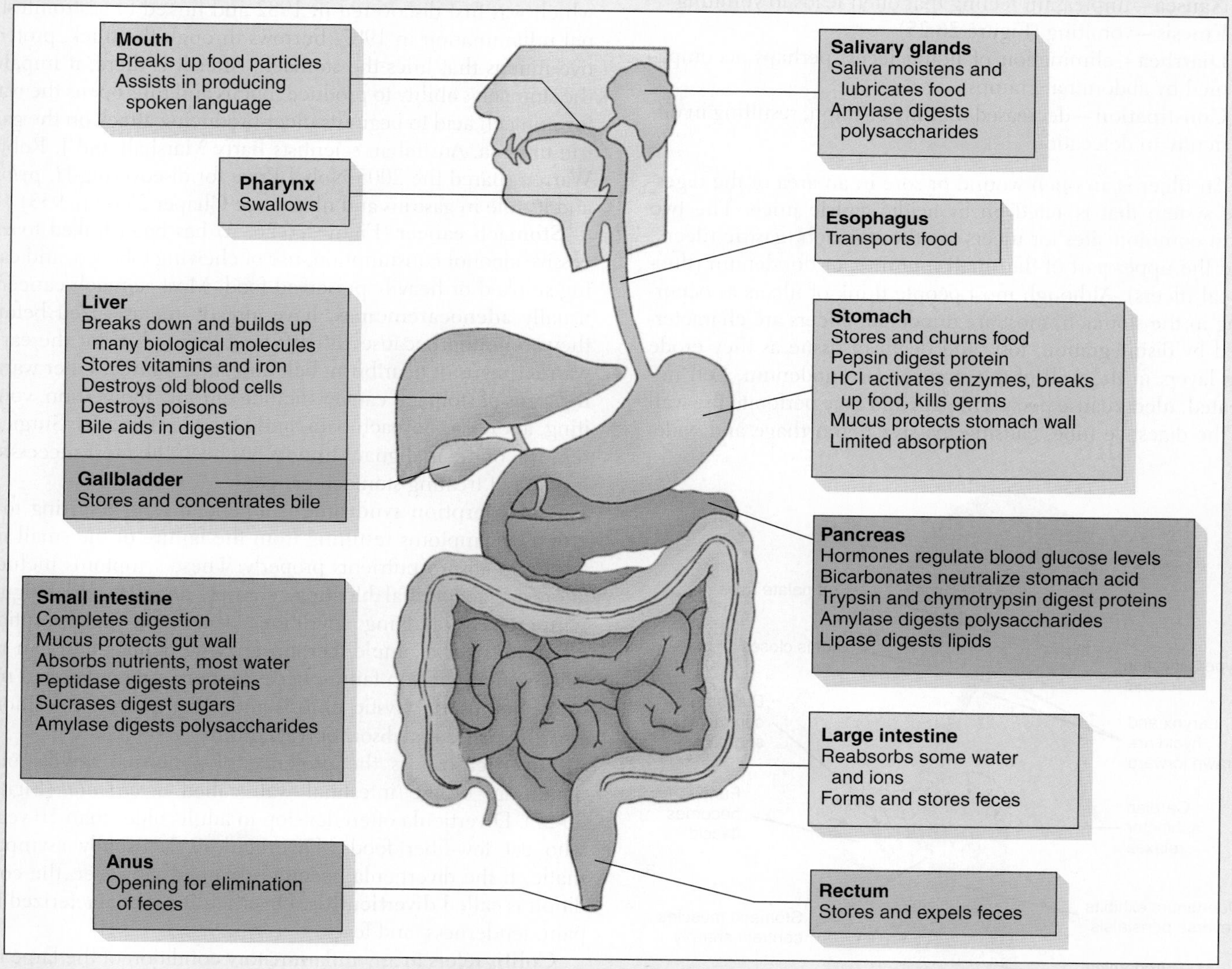

Mouth
Breaks up food particles
Assists in producing
  spoken language

Salivary glands
Saliva moistens and
  lubricates food
Amylase digests
  polysaccharides

Pharynx
Swallows

Esophagus
Transports food

Liver
Breaks down and builds up
  many biological molecules
Stores vitamins and iron
Destroys old blood cells
Destroys poisons
Bile aids in digestion

Stomach
Stores and churns food
Pepsin digest protein
HCl activates enzymes, breaks
  up food, kills germs
Mucus protects stomach wall
Limited absorption

Gallbladder
Stores and concentrates bile

Pancreas
Hormones regulate blood glucose levels
Bicarbonates neutralize stomach acid
Trypsin and chymotrypsin digest proteins
Amylase digests polysaccharides
Lipase digests lipids

Small intestine
Completes digestion
Mucus protects gut wall
Absorbs nutrients, most water
Peptidase digests proteins
Sucrases digest sugars
Amylase digests polysaccharides

Large intestine
Reabsorbs some water
  and ions
Forms and stores feces

Anus
Opening for elimination
  of feces

Rectum
Stores and expels feces

Figure 26-24 *Summary of digestive function.*

Mechanisms of Disease

DISORDERS OF THE DIGESTIVE SYSTEM

Disorders of the GI Tract

Gastroenterology (gas-troh-en-ter-AHL-oh-jee) is the study of the stomach (*gastro-*) and intestines (*entero-*) and their diseases. The gastrointestinal tract is the potential site of numerous diseases and conditions, some of which are briefly described in this section. Many of these disorders, particularly those that primarily affect the stomach or duodenum, are characterized by one or more of the following signs and symptoms:

- **Gastroenteritis**—stomach inflammation (gastritis) and intestinal inflammation (enteritis)
- **Anorexia**—chronic loss of appetite
- **Nausea**—unpleasant feeling that often leads to vomiting
- **Emesis**—vomiting (Figure 26-25)
- **Diarrhea**—elimination of liquid feces, perhaps accompanied by abdominal cramps
- **Constipation**—decreased motility of colon, resulting in difficulty in defecation

An **ulcer** is an open wound or sore in an area of the digestive system that is acted on by acidic gastric juice. The two most common sites for ulcers are the stomach (gastric ulcers) and the upper part of the small intestine, or duodenum (duodenal ulcers). Although most people think of ulcers as occurring in the stomach, most are duodenal. Ulcers are characterized by disintegration, loss, and death of tissue as they erode the layers in the wall of the stomach or duodenum. Left untreated, ulcers cause persistent pain and may perforate the wall of the digestive tube, causing massive hemorrhage and widespread inflammation of the abdominal cavity and its contents. Usually perforation does not occur, but small, repeated hemorrhages over long periods cause anemia.

There is an old saying in medicine: "No acid, no ulcer." Historically, experts believed that too much gastric acid secretion (that is, prolonged hyperacidity) was one of the most important factors in ulcer formation. Hyperacidity is influenced by nervous system factors and by anxiety, other emotional states, and stress.

Therefore, drugs such as omeprazole (Prilosec), ranitidine (Zantac), cimetidine (Tagamet), and other medications that reduce hydrochloric acid formation in the stomach have been widely prescribed for the treatment of ulcers. However, we now know that hyperacidity is only partly to blame for most ulcers. The basic underlying cause of ulcers is a bacterium called *Helicobacter pylori*. This spiral-shaped bacterium, which was first discovered in 1982 and linked to gastrointestinal inflammation in 1987, burrows through the thick, protective mucus that lines the stomach. After it is there, it impairs the stomach's ability to produce mucus and thus opens the way for stomach acid to begin its ulcer-producing attack on the gastric mucosa. Australian scientists Barry Marshall and J. Robin Warren shared the 2005 Nobel Prize for discovering *H. pylori* and its role in gastritis and ulcers (see Chaper 25 on p. 953).

Stomach cancer (Figure 26-26, A) has been linked to excessive alcohol consumption, use of chewing tobacco, and eating smoked or heavily preserved food. Most stomach cancers, usually **adenocarcinomas,** have already metastasized before they are found because patients treat themselves for the early warning signs of heartburn, belching, and nausea. Later warning signs of stomach cancer include chronic indigestion, vomiting, anorexia, stomach pain, and blood in the feces. Surgical removal of the malignant tumors has been the most successful method of treating stomach cancer.

Malabsorption syndrome is a general term referring to a group of symptoms resulting from the failure of the small intestine to absorb nutrients properly. These symptoms include anorexia, abdominal bloating, cramps, anemia, and fatigue. Numerous underlying conditions can cause malabsorption syndrome. For example, certain enzyme deficiencies can result in an absorption failure because there are no digested nutrients to absorb. Cystic fibrosis and other genetic conditions can also cause malabsorption syndrome.

Diverticulosis is the presence of abnormal saclike outpouchings of the intestinal wall called *diverticula* (Figure 26-27). Diverticula often develop in adults older than 50 years who eat low-fiber foods. Diverticulosis is usually asymptomatic. If the diverticula become inflamed, however, the condition is called **diverticulitis.** Diverticulitis is characterized by pain, tenderness, and fever.

Colitis refers to any inflammatory condition of the large intestine. Symptoms of colitis include diarrhea and abdominal cramps or constipation. Some forms of colitis may also produce bleeding and intestinal ulcers. It may also result from an autoimmune disease, as in *ulcerative colitis*. Another type of autoimmune colitis is **Crohn disease,** which often also affects

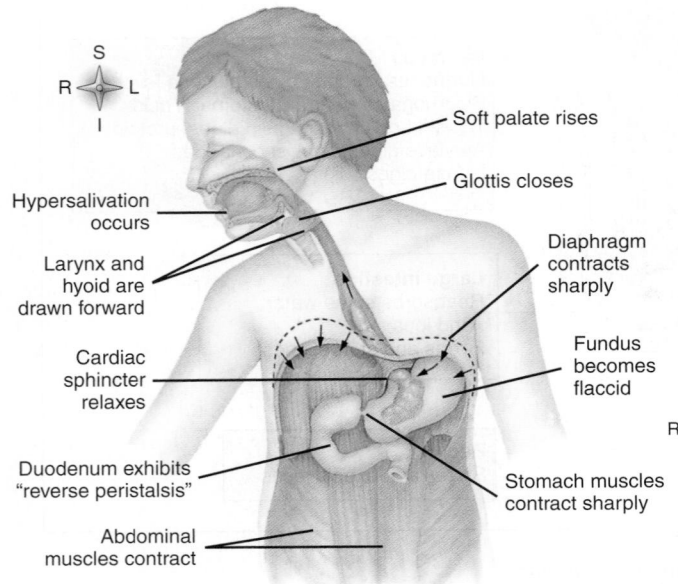

Soft palate rises

Glottis closes

Hypersalivation occurs

Larynx and hyoid are drawn forward

Diaphragm contracts sharply

Cardiac sphincter relaxes

Fundus becomes flaccid

Duodenum exhibits "reverse peristalsis"

Stomach muscles contract sharply

Abdominal muscles contract

Figure 26-25 *Emesis (vomiting).* Summary of components of the vomiting reflex.

Mechanisms of Disease—cont.

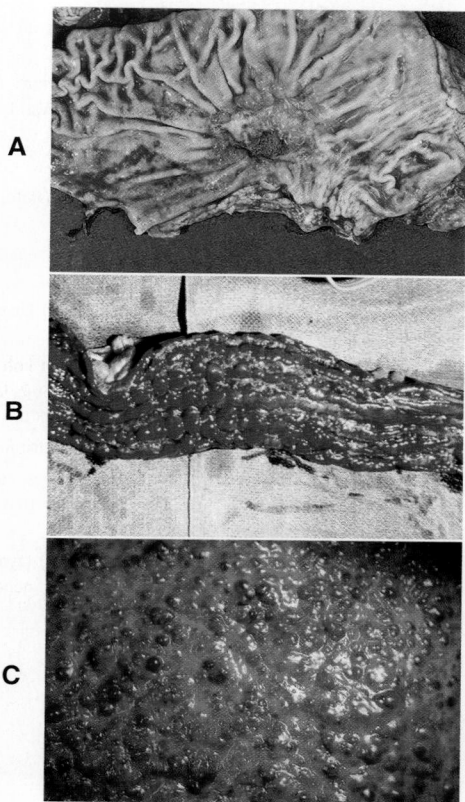

Figure 26-26 *Examples of digestive pathology.* **A,** The abnormal tissue near the center of this opened stomach is gastric carcinoma, or "stomach cancer." Note the distinct, normal rugae (folds) that surround the tumor tissue. **B,** Crohn disease, an autoimmune form of colitis, is characterized by severe inflammation that gives a sort of "cobblestone" appearance to the intestinal lining. **C,** Alcoholic cirrhosis is characterized by hardness of the liver caused by fibrous tissue and by nodules—which can be seen clearly in this photograph of the surface of a cirrhotic liver.

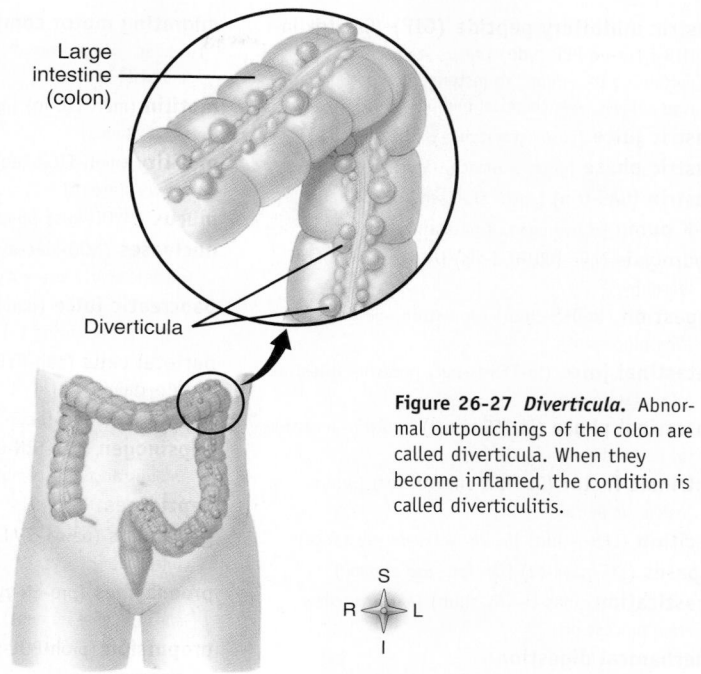

Figure 26-27 *Diverticula.* Abnormal outpouchings of the colon are called diverticula. When they become inflamed, the condition is called diverticulitis.

the small intestine (Figure 26-26, *B*). If more conservative treatments fail, colitis may be corrected by surgical removal of the affected portions of the colon.

Irritable bowel syndrome, or *spastic colon,* is a common chronic noninflammatory condition that is often caused by stress. Irritable bowel syndrome is characterized by diarrhea or constipation with or without pain.

Colorectal cancer is a malignancy, usually an *adenocarcinoma,* of the colon or rectum. Colorectal cancer occurs most frequently after the age of 50, and a low-fiber, high-fat diet and genetic predisposition are known risk factors. Early warning signs of this common type of cancer include changes in bowel habits, fecal blood, rectal bleeding, abdominal pain, unexplained anemia or weight loss, and fatigue.

Disorders of the Liver and Pancreas

Hepatitis is a general term referring to inflammation of the liver. Hepatitis is characterized by jaundice (yellowish discoloration of body tissues), liver enlargement, anorexia, abdominal discomfort, gray-white feces, and dark urine. Various dif-

ferent conditions can produce hepatitis. Alcohol, drugs, or other toxins may cause hepatitis. It may also be a complication of bacterial or viral infection or parasite infestation. *Hepatitis A,* for example, results from infection by the hepatitis A virus. Contaminated food is often a source of infection. Hepatitis A occurs commonly in young people and ranges in severity from mild to life threatening. Another viral hepatitis, *hepatitis B,* is usually more severe. It is also called *serum hepatitis* because it is often transmitted by contaminated blood serum.

Hepatitis, chronic alcohol abuse, malnutrition, or infection may lead to a degenerative liver condition known as **cirrhosis.** The liver's ability to regenerate damaged tissue is well known, but it has its limits. For example, when the toxic effects of alcohol accumulate faster than the liver can regenerate itself, damaged tissue is replaced with fibrous scar tissue instead of normal tissue (Figure 26-26, *C*). *Cirrhosis* is the name given to such degeneration.

Besides the endocrine disorders such as diabetes mellitus discussed in Chapter 16, the pancreas may be involved in numerous other diseases. For example, **pancreatitis,** or inflammation of the pancreas, can be caused by various factors. *Acute pancreatitis* usually results from blockage of the pancreatic duct. The blockage causes pancreatic enzymes to "back up" into the pancreas and digest it. Another condition that blocks the flow of pancreatic enzymes is *cystic fibrosis (CF).* You may recall from Chapter 4 that this inherited disorder disrupts cell transport and causes exocrine glands to produce excessively thick secretions. Thick pancreatic secretions may build up and block pancreatic ducts, disrupting the flow of pancreatic enzymes and damaging the pancreas.

Another serious pancreatic disorder is **pancreatic cancer.** Usually a form of *adenocarcinoma,* pancreatic cancer claims the lives of nearly all its victims within 5 years after diagnosis.

LANGUAGE OF SCIENCE *(Cont'd from page 961)*

gastric inhibitory peptide (GIP) (GAS-trik in-HIB-i-tor-ee PEP-tyde) [*gastr-* stomach, *-ic* pertaining to, *inhibit-* to restrain, *-ory* pertaining to, *pept-* digest, *-ide* chemical ending]

gastric juice [*gastr-* stomach, *-ic* pertaining to]

gastric phase [*gastr-* stomach, *-ic* pertaining to]

gastrin (GAS-trin) [*gastr-* stomach, *-in* substance]

H-K pump [*H* hydrogen, *K* potassium]

hydrolysis (hye-DROHL-i-sis) [*hydro-* water, *-lysis* loosening]

ingestion (in-JES-chun) [*in-* within, *-geste* to carry, *-tion* process of]

intestinal juice (in-TES-ti-nal) [*intestin-* intestine, *-al* pertaining to]

intestinal phase (in-TES-ti-nal) [*intestin-* intestine, *-al* pertaining to]

intrinsic factor (in-TRIN-sik FAK-tor) [*intrins-* inside, *-ic* pertaining to]

lecithin (LES-i-thin) [*lecith-* yolk, *-in* substance]

lipases (LYE-pays-ez) [*lip-* fat, *-ase* enzyme]

mastication (mas-ti-KAY-shun) [*mastica-* chew, *-tion* process of]

mechanical digestion

micelles (my-SELLS) [*mic-* grain, *-elle* small]

migrating motor complex (MMC) [*migra-* to wander, *-ate* process of, *motor* move, *complexus* an embrace]

motilin (moh-TIL-in) [*mot-* move, *-il-* pertaining to, *-in* substance]

motility (moh-TIL-i-tee) [*mot-* move, *-il-* pertaining to, *-ity* state of]

mucus (MYOO-kus) [*mucus* slime]

nucleases (NOO-klee-ay-sez) [*nucle-* nut or kernel (nucleic acid), *-ase* enzyme]

pancreatic juice (pan-kree-AT-ik) [*pan-* all, *-kreat-* flesh, *-ic* pertaining to]

parietal cells (pah-RYE-i-tal) [*pariet-* wall, *-al* pertaining to]

pepsin (PEP-sin) [*peps-* digestion, *-in* substance]

pepsinogen (pep-SIN-oh-jen) [*peps-* digestion, *-in* substance, *-o-* combining form, *-gen* generate]

peptidases

peristalsis (pair-i-STAL-sis) [*peri-* around, *-stalsis* contraction]

proenzymes (pro-EN-zymes) [*pro-* first or in front of, *-en-* in, *-zyme* ferment]

propulsion (proh-PUL-shen) [*propul-* drive forward, *-sion* process of]

proteases (PROH-tee-ay-sez) [*prote-* protein, *-ase* enzyme]

retropulsion (ret-roh-PUL-shen) [*retro-* backward or located behind, *-propul-* drive forward, *-sion* process of]

saliva (sah-LYE-vah)

secretin (seh-KREE-tin) [*secret-* separate, *-in* substance]

secretion (seh-KREE-shun) [*secret-* separate, *-tion* process of]

segmentation (seg-men-TAY-shun) [*segment-* piece out, *-ation* process of]

sodium cotransport (SOH-dee-um koh-TRANS-port) [*sod-* soda, *-ium* chemical ending, *co-* with, *-trans-* across, *-port* carry]

trypsin (TRIP-sin) [*tryps-* rub, *-in* substance]

trypsinogen (trip-SIN-oh-gen) [*tryps-* rub, *-in* substance, *-o-* combining form, *-gen* that which generates]

zymogenic cells (zye-moh-JEN-ik) [*zym-* ferment (enzyme), *-o-* combining form, *-gen* generate, *-ic* pertaining to]

LANGUAGE OF MEDICINE

adenocarcinoma (ad-eh-no-kar-sih-NO-mah) [*adeno-* gland, *-carcin-* cancer, *-oma* tumor]

anorexia (an-oh-REK-see-ah) [*an-* without, *-orexi-* appetite, *-ia* condition]

cirrhosis (sih-ROH-sis) [*cirrhos-* yellow-orange, *-osis* condition]

colitis (koh-LYE-tis) [*col-* colon, *-itis* inflammation]

colorectal cancer (koh-loh-REK-tal KAN-ser) [*colo-* colon, *-rect-* straight, *-al* pertaining to, *cancer* crab]

constipation (kon-sti-PAY-shun) [*constipa-* crowd together, *-ation* process of]

Crohn disease (krohn dih-ZEEZ) [*Burrill B. Crohn* American physician]

diarrhea (dye-ah-REE-ah) [*dia-* through, *-rrhea* flow]

diverticulitis (dye-ver-tik-yoo-LYE-tis) [*diverticul-* turn aside, *-itis* inflammation]

diverticulosis (dye-ver-tik-yoo-LOH-sis) [*diverticul-* turn aside, *-osis* condition]

emesis (EM-eh-sis) [*emesis* vomiting]

gastroenteritis (gas-troh-en-ter-EYE-tis) [*gastro-* stomach, *-enter-* intestine, *-itis* inflammation]

gastroenterology (gas-troh-en-ter-AHL-oh-jee) [*gastro-* stomach, *-entero-* intestine, *-ology* study or science of]

hepatitis (hep-ah-TYE-tis) [*hepat-* liver, *-itis* inflammation]

irritable bowel syndrome (IR-i-tah-bul BOW-uhl SIN-drohm) [*irrita-* to tease, *-ble* able or tending to, *botellus* sausage, *syn-* together, *-drome* course]

malabsorption syndrome (mal-ab-SORP-shun SIN-drohm) [*mal-* poor or bad, *-absorp-* swallow, *-tion* process of, *syn-* together, *-dromos* course]

nausea (NAW-zee-ah) [*nausia* seasickness]

oral rehydration therapy (OR-al ree-hye-DRAY-shun THER-ah-pee) [*oralis* pertaining to the mouth, *re-* back again, *-hydra-* water, *-ation* process of]

pancreatic cancer (pan-kree-AT-ik KAN-ser) [*pan-* all, *-creat-* flesh, *-ic* pertaining to, *cancer* crab]

pancreatitis (pan-kree-ah-TYE-tiss) [*pan-* all, *-creat-* flesh, *-itis* inflammation]

steatorrhea (stee-eh-tah-REE-ah) [*steat-* fat, *-o-* combining form, *-rrhea* flow]

stomach cancer (STUM-ak KAN-ser) [*stomakhos* gullet, *cancer* crab]

ulcer (UL-ser) [*ulcus* a sore]

CASE STUDY

Oscar Perez, age 38, is being seen at the health clinic because of abdominal pain that began 24 hours ago. Mr. Perez describes his pain as sharp and constant, centered in the midepigastric region and slightly over into the upper right quadrant of his abdomen. He denies ever having this severe pain before but has had abdominal pain on and off for the past several months. He has taken Tylenol for abdominal pain in the past, which seems to help a little. Aspirin makes his stomach hurt worse with a kind of "burning" pain. He has not vomited and his stools have not changed. He admits to having stomach bloating after meals, heartburn, belching, and nausea.

His medical history is negative for hospitalizations, surgeries, or trauma. Mr. Perez works as a carpenter and is currently self-employed. His family history is negative for cancer, but his father has non–insulin-dependent diabetes, hypertension, and heart disease. Mr. Perez smokes approximately half a pack of cigarettes per day and has been smoking since the age of 17 years. His alcohol consumption is beer and bourbon, which he consumes mostly on the weekends. He states that he usually drinks until he "gets drunk." He has been drinking since the age of 17 years but has increased his consumption for the past 10 years. Mr. Perez admits the use of illegal intravenous drugs in the past but has been drug free for 12 years. He is married and lives with his wife and two children, ages 17 years and 14 years.

On physical examination, Mr. Perez is noted to be in no acute distress, has normal vital signs, normal hemoglobin and hematocrit, negative urinalysis for bacterial infection, and active bowel sounds in all four abdominal quadrants. On palpation of the abdomen, Mr. Perez complains of pain in the upper epigastric region and right upper quadrant. The physician orders more laboratory blood studies, including liver enzymes, amylase, and a white blood count and differential, and orders gastric radiography (upper GI series) and endoscopy. Mr. Perez is started on ranitidine (Zantac), 150 mg by mouth (PO) in the morning and evening (twice daily [bid]), and is scheduled for a follow-up visit in about 1 week. He is given instructions to call or come in earlier if his pain increases or he notes blood with vomitus or stool.

1. Which of the following could be a possible diagnosis for Mr. Perez's abdominal pain?

   1. Peptic ulcer
   2. Stomach cancer
   3. Liver disease
   4. Malabsorption syndrome
   5. Colitis
   6. Diverticulosis
      A. 1, 2, 3
      B. 4, 5, 6
      C. 1, 2, 5
      D. 2, 3, 6

2. What is the expected action of the medication (Zantac) that was ordered for Mr. Perez?

   A. It will stop the pain.
   B. It will increase the enzyme activity within the intestine.
   C. It will reduce acid formation in the stomach.
   D. It will kill the bacteria in the stomach.

3. The x-ray results are negative, but Mr. Perez's endoscopic biopsy is positive for the presence of *Helicobacter pylori*. He is started on triple therapy with bismuth subsalicylate (Pepto-Bismol), amoxicillin, and metronidazole for a 2-week period. How does the bacterium *Helicobacter pylori* aid in the development of ulcers?

   A. It perforates the wall of the digestive tube, causing massive hemorrhage.
   B. It increases the production of hydrochloric acid within the intestinal tract.
   C. It lowers the acidity level, thus allowing increased gastrointestinal inflammation from amylase.
   D. It produces a phospholipase capable of digesting gastric mucosa and impairs the stomach's ability to produce mucus.

4. A differential diagnosis for Mr. Perez is that of stomach cancer. What factors in his health history or physical examination would alert the practitioner to consider this diagnosis?

   1. Excessive alcohol consumption
   2. Heartburn, belching, nausea
   3. Stomach pain
   4. Intravenous drug use
   5. Occupation
   6. Presence of *Helicobacter pylori*
      A. 1, 3, 4, 5
      B. 2, 3, 5
      C. 1, 2, 3, 6
      D. 2, 3, 4

5. An underlying physiological reason for patients with gastric or pyloric ulcers to often experience abdominal discomfort after eating is

   A. Edematous duodenum resulting from the mechanical pressure caused by the semidigested meal
   B. Increased back-diffusion of ingested acid into the mucosa
   C. Increased gastric concentration in the antral mucosa caused by the ingestion of proteins

6. Which of the following are the most frequent complications of peptic ulcer disease?

   1. Gastric outlet obstruction
   2. Hemorrhage
   3. Anemia
   4. Perforation
      A. 1, 2, 3
      B. 1, 2, 4
      C. 2, 3, 4
      D. 1, 3, 4

CHAPTER SUMMARY

OVERVIEW OF DIGESTIVE FUNCTION

A. Primary function of the digestive system—to bring essential nutrients into the internal environment so that they are available to each cell of the body

B. Mechanisms used to accomplish the primary function of the digestive system (Figure 26-1)
 1. Ingestion—food is taken in
 2. Digestion—breakdown of complex nutrients into simple nutrients
 3. Motility of the GI wall—physically breaks down large chunks of food material and moves food along the tract
 4. Secretion of digestive enzymes allows chemical digestion
 5. Absorption—movement of nutrients through the GI mucosa into the internal environment
 6. Elimination—excretion of material that is not absorbed
 7. Regulation—coordination of the various functions of the digestive system

C. The digestive tract is functionally an extension of the external environment—material does not truly enter the body until it is absorbed into the internal environment

DIGESTION

A. Mechanical digestion—movements of the digestive tract
 1. Change ingested food from large particles into minute particles, facilitating chemical digestion
 2. Churn contents of the GI lumen to mix with digestive juices and come in contact with the surface of the intestinal mucosa, facilitating absorption
 3. Propel food along the alimentary tract, eliminating digestive waste from the body
 4. Mastication—chewing movements
  a. Reduces size of food particles
  b. Mixes food with saliva in preparation for swallowing
 5. Deglutition—process of swallowing; complex process requiring coordinated and rapid movements (Figure 26-2)
  a. Oral stage (mouth to oropharynx)—voluntarily controlled; formation of a food bolus in the middle of the tongue; tongue presses bolus against the palate and food is then moved into the oropharynx
  b. Pharyngeal stage (oropharynx to esophagus)—involuntary movement; to propel bolus from the pharynx to the esophagus, the mouth, nasopharynx, and larynx must be blocked; a combination of contractions and gravity move bolus into esophagus
  c. Esophageal stage (esophagus to stomach)—involuntary movement; contractions and gravity move bolus through esophagus and into stomach
 6. Peristalsis and segmentation—two main types of motility produced by the smooth muscle of the GI tract; can occur together, in an alternating fashion
  a. Peristalsis—wavelike ripple of the muscle layer of a hollow organ; progressive motility that produces forward movement of matter along the GI tract (Figures 26-3 and 26-4)
  b. Segmentation—mixing movement; digestive reflexes cause a forward-and-backward movement with a single segment of the GI tract; helps break down food particles, mixes food and digestive juices, and brings digested food in contact with intestinal mucosa to facilitate absorption (Figure 26-5)
 7. Regulation of motility
  a. Gastric motility—emptying the stomach takes approximately 2 to 6 hours; while in the stomach, food is churned (propulsion and retropulsion) and mixed with gastric juices to form chyme; chyme is ejected about every 20 seconds into the duodenum; gastric emptying is controlled by hormonal and nervous mechanisms (Figure 26-6)
   (1) Hormonal mechanism—fats in duodenum stimulate the release of gastric inhibitory peptide, which acts to decrease peristalsis of gastric muscle and slows passage of chyme into duodenum
   (2) Nervous mechanism—enterogastric reflex; receptors in the duodenal mucosa are sensitive to presence of acid and to distention; impulses over sensory and motor fibers in the vagus nerve cause a reflex inhibition of gastric peristalsis
  b. Intestinal motility includes peristalsis and segmentation
   (1) Segmentation in duodenum and upper jejunum mixes chyme with digestive juices from the pancreas, liver, and intestinal mucosa
   (2) Rate of peristalsis picks up as chyme approaches end of jejunum, moving it through the rest of the small intestine into the large intestine; after leaving stomach, it normally takes approximately 5 hours for chyme to pass all the way through the small intestine
   (3) Peristalsis—regulated in part by intrinsic stretch reflexes; stimulated by CCK

B. Chemical digestion—changes in chemical composition of food as it travels through the digestive tract; these changes are the result of hydrolysis
 1. Digestive enzymes
  a. Extracellular, organic (protein) catalysts
  b. Principles of enzyme action
   (1) Specific in their action (Figure 26-7)
   (2) Function optimally at a specific pH (Figure 26-8)
   (3) Most enzymes catalyze a chemical reaction in both directions
   (4) Enzymes are continually being destroyed or eliminated from the body and must continually be synthesized
   (5) Most digestive enzymes are synthesized as inactive proenzymes

2. Carbohydrate digestion (Figure 26-9)
   a. Carbohydrates are saccharide compounds
   b. Polysaccharides are hydrolyzed by amylases to form disaccharides
   c. Final steps of carbohydrate digestion are catalyzed by sucrase, lactase, and maltase, which are found in the cell membrane of epithelial cells covering the villi that line the intestinal lumen
3. Protein digestion (Figure 26-10)
   a. Protein compounds are made up of twisted chains of amino acids
   b. Proteases catalyze hydrolysis of proteins into intermediate compounds and, finally, into amino acids
   c. Main proteases: pepsin in gastric juice, trypsin in pancreatic juice, peptidases in intestinal brush border
4. Fat digestion (Figure 26-12)
   a. Fats must be emulsified by bile in small intestine before being digested (Figure 26-11)
   b. Pancreatic lipase is the main fat-digesting enzyme
5. Residues of digestion—some compounds of food resist digestion and are eliminated as feces

SECRETION

A. Saliva—secreted by salivary glands
   1. Mucus lubricates food and, with water, facilitates mixing
   2. Amylase is an enzyme that begins digestion of starches; a small amount of salivary lipase is released, function uncertain
   3. Sodium bicarbonate increases the pH for optimum amylase function
B. Gastric juice—secreted by gastric glands
   1. Pepsin (secreted as inactive pepsinogen by chief cells) is a protease that begins the digestion of proteins
   2. Hydrochloric acid (HCl, secreted by parietal cells)
      a. HCl decreases the pH of chyme for activation and optimum function of pepsin (Figure 26-13)
      b. Released actively into the gastric juice by H-K pumps (proton pumps)
      c. Vesicles in the resting parietal cell move to the apical surface when the cell becomes active—thus increasing the surface area for the process of secretion (Figure 26-14)
   3. Intrinsic factor (secreted by parietal cells) protects vitamin B_{12} and later facilitates its absorption (Figure 26-15)
   4. Mucus and water lubricate, protect, and facilitate mixing of chyme
C. Pancreatic juice—secreted by acinar and duct cells of the pancreas
   1. Proteases (e.g., trypsin and chymotrypsin) are enzymes that digest proteins and polypeptides
   2. Lipases are enzymes that digest emulsified fats
   3. Nucleases are enzymes that digest nucleic acids such as DNA and RNA
   4. Amylase is an enzyme that digests starches
   5. Sodium bicarbonate increases the pH for optimum enzyme function; its manufacture also helps restore normal pH of blood (Figures 26-16 and 26-17)

D. Bile—secreted by the liver; stored and concentrated in the gallbladder
   1. Lecithin and bile salts emulsify fats by encasing them in shells to form tiny spheres called micelles
   2. Sodium bicarbonate increases pH for optimum enzyme function
   3. Cholesterol, products of detoxification, and bile pigments (e.g., bilirubin) are waste products excreted by the liver and eventually eliminated in the feces
E. Intestinal juice—secreted by cells of intestinal exocrine cells
   1. Mucus and water lubricate and aid in continued mixing of chyme
   2. Sodium bicarbonate increases pH for optimum enzyme function

CONTROL OF DIGESTIVE GLAND SECRETION

A. Salivary secretion
   1. Only reflex mechanisms control the secretion of saliva
   2. Chemical and mechanical stimuli come from the presence of food in the mouth
   3. Olfactory and visual stimuli come from the smell and sight of food
B. Gastric secretion—three phases (Figure 26-18)
   1. Cephalic phase—"psychic phase," because mental factors activate the mechanism; parasympathetic fibers in branches of the vagus nerve conduct stimulating efferent impulses to the glands; stimulate production of gastrin (by G cells in the stomach)
   2. Gastric phase—when products of protein digestion reach the pyloric portion of the stomach, they stimulate release of gastrin; gastrin accelerates secretion of gastric juice, ensuring enough enzymes are present to digest food
   3. Intestinal phase—various mechanisms seem to adjust gastric secretion as chyme passes to and through the intestinal tract; endocrine reflexes involving gastric inhibitory peptide, secretin, and CCK inhibit gastric secretions
C. Pancreatic secretion stimulated by several hormones released by intestinal mucosa
   1. Secretin evokes production of pancreatic fluid low in enzyme content but high in bicarbonate
   2. CCK—several functions
      a. Causes increased exocrine secretion from the pancreas
      b. Opposes gastrin, thus inhibiting gastric HCl secretion
      c. Stimulates contraction of the gallbladder so that bile is ejected into the duodenum
D. Secretion of bile—bile secreted continually by the liver; secretin and CCK stimulate ejection of bile from the gallbladder
E. Intestinal secretion—little is known how intestinal secretion is regulated; suggested that the intestinal mucosa is stimulated to release hormones that increase the production of intestinal juice

ABSORPTION

A. Process of absorption
   1. Passage of substances through the intestinal mucosa into the blood or lymph (Figure 26-19)
   2. Most absorption occurs in the small intestine

B. Mechanisms of absorption
  1. For some substances such as water, absorption occurs by simple diffusion or osmosis
  2. Other substances are absorbed through more complex mechanisms (Figures 26-20 and 26-21)
    a. Secondary active transport—how sodium is transported
    b. Sodium cotransport (coupled transport)—how glucose is transported
    c. Fatty acids, monoglycerides, and cholesterol are transported with the aid of bile salts from the lumen to absorbing cells of the villi
  3. After food is absorbed, it travels to the liver via the portal system
  4. In summary, most absorption occurs in the small intestine (Figure 26-22)

ELIMINATION

A. Elimination—the expulsion of feces from the digestive tract; act of expelling feces is called *defecation*
B. Defecation occurs as a result of a reflex brought about by stimulation of receptors in the rectal mucosa that is produced when the rectum is distended (Figure 26-23)
C. Constipation—contents of the lower part of the colon and rectum move at a slower than normal rate; extra water is absorbed from the feces, resulting in a hardened stool
D. Diarrhea—result of increased motility of the small intestine, causing decreased absorption of water and electrolytes and a watery stool

THE BIG PICTURE: DIGESTION AND THE WHOLE BODY

A. Primary contribution of the digestive system to overall homeostasis is to provide a constant nutrient concentration in the internal environment
B. Secondary roles of digestive system
  1. Absorption of nutrients
  2. Teeth and tongue, along with respiratory and nervous system, are important in producing spoken language
  3. Gastric acids aid the immune system by destroying potentially harmful bacteria
C. To accomplish its functions, digestive system needs other systems to contribute
  1. Regulation of digestive motility and secretion requires the nervous system and endocrine system
  2. Oxygen for digestive activity is dependent on proper functioning of the respiratory and circulatory systems
  3. Integumentary and skeletal systems support and protect the digestive organs
  4. Muscular system is needed for ingestion, mastication, deglutition, and defecation to occur normally
D. Follow a piece of pizza (containing protein, carbohydrates, and lipids) through the digestive tract, describing what happens to it from the first bite to elimination of its residue.

REVIEW QUESTIONS

1. List four of the mechanical processes that occur during digestion.
2. List the three steps, or stages, in deglutition.
3. Discuss the function of the deglutition center in the medulla.
4. How does gastric inhibitory peptide influence emptying of the stomach? What is the enterogastric reflex?
5. Discuss the functional roles of gastric juice.
6. What is chyme?
7. Describe the classification of digestive enzymes.
8. Discuss three important properties of digestive enzymes.
9. What substances chemically digest proteins? Carbohydrates? Fats?
10. What is meant by the term *cephalic phase of gastric secretion? Gastric phase? Intestinal phase?*
11. What digestive functions does the pancreas perform?
12. Describe the absorption of glucose from the lumen of the small intestine.
13. What vitamins depend on bile salts for their absorption?
14. Name each hormone that controls ejection of bile, stimulation of gastric enzymes, inhibition of gastric emptying, secretion of alkaline fluid from pancreas, and secretion of pancreatic enzymes.
15. Discuss responses that collectively result in defecation.
16. Describe the treatment called *oral rehydration therapy.*

CRITICAL THINKING QUESTIONS

1. In terms of homeostatic balance in the body, what is the function of the digestive system?
2. How would you explain the two types of processes within the digestive system?
3. How would you describe the two types of motility within the intestine?
4. Can you make the distinction between trypsin and pepsin, protein and peptides, and polysaccharides and monosaccharides? Explain why one pair is different from the other two.
5. What is the relationship between hydrochloric acid production by the stomach and bicarbonate production by the pancreas?
6. Because fats are not soluble in water, what process is used by the digestive system in the emulsification of fats?
7. Why do you think emulsification is an example of mechanical digestion rather than chemical digestion?
8. Compare and contrast carbohydrate and protein absorption with fat absorption. Include an explanation of basal active transport.

Nutrition and Metabolism

LANGUAGE OF SCIENCE

aerobic respiration (air-OH-bik res-pi-RAY-shun) [*aero-* air, *-b-* (from *-bio-*) life, *-ic* pertaining to, *re-* again, *-spira-* breathe, *-tion* process of]

amino acids (ah-MEE-no) [*amino* NH_2, *acid* sour]

anabolism (ah-NAB-oh-liz-em) [*anabol-* to build up, *-ism* action]

anaerobic respiration (an-air-OH-bik res-pi-RAY-shun) [*an-* without, *-aero-* air, *-b-* (from *-bio-*) life, *-ic* pertaining to, *re-* again, *-spira-* breathe, *-tion* process of]

anorexigenic (an-oh-rek-sih-JEN-ik) [*anorexi-* without appetite, *-gen-* generate, *-ic* pertaining to]

antioxidant (an-tee-OK-seh-dent) [*anti-* against, *-oxi-* sharp (oxygen), *-ant* agent]

appetite center [*appetere* long for]

ATP synthase (SIN-thays) [*ATP* adenosine triphosphate, *syn-* together, *-ase* enzyme]

assimilation (ah-sim-i-LAY-shun) [*assimila-* make alike, *-tion* process of]

basal metabolic rate (BMR) (BAY-sal met-ah-BAHL-ik) [*bas-* basis, *-al* pertaining to, *metabol-* change, *-ic* pertaining to]

calcitriol (kal-SIT-ree-ol) [*calci-* lime, *-tri-* three, *-ol* alcohol (after *1,25-D3* or *1,25-dihydroxycholecalciferol*)]

calorie (cal) (KAL-or-ee) [*calor-* warmth, *-ie* full of]

catabolism (kah-TAB-oh-liz-em) [*catabol-* to throw down, *-ism* action]

cellulose (SEL-yoo-lohs) [*cell-* storeroom (cell), *-ul-* small, *-ose* carbohydrate]

chylomicrons (kye-loh-MYE-krons) [*chylo-* juice (chyle), *-micro-* small, *-on* particle]

citric acid cycle (SIT-rik) [*citr-* citron tree, *-ic* pertaining to, *acidus* sour, *kyklos* circle]

coenzymes (koh-EN-zymes) [*co-* together, *-en-* in, *-zyme* ferment]

coenzyme A (CoA) (koh-EN-zyme) [*co-* together, *-en-* in, *-zyme* ferment, *A* first letter of Roman alphabet]

deamination (dee-am-i-NAY-shun) [*de-* undo, *-amin-* ammonia, *-ation* process of]

Cont'd on p. 1026

OVERVIEW OF NUTRITION AND METABOLISM

Nutrition and *metabolism* are words that are often used together—but what do they mean? **Nutrition** refers to the foods that we eat and the nutrients they contain. The Council on Food and Nutrition of the American Medical Association defines nutrition broadly as "the science of food; the nutrients and the substances therein; their action, interaction, and balance in relation to health and disease; and the process by which the organism (i.e., body) ingests, digests, absorbs, transports, utilizes, and excretes food substances."

Healthy nutrition requires a balance of different nutrients in healthy amounts. *Malnutrition* is a deficiency or imbalance in the consumption of food, vitamins, and minerals. As a matter of convenient communication, many nutrition experts divide the essential (required) nutrients into two major categories:

1. **Macronutrients.** Macronutrients usually include those nutrients that we need in large amounts, such as carbohydrates, fats, and proteins. Sometimes water is included because we need to ingest a large amount of water each day to remain healthy. Minerals that we need in large quantities to remain in good health are also often included among the macronutrients. For example, sodium, chloride, potassium, calcium, magnesium, and phosphorus are often considered to be macronutrients. This group of minerals can also be more specifically called *macrominerals*. Macronutrients are also sometimes called *bulk nutrients* because we need them in bulk quantities to survive.

2. **Micronutrients.** Micronutrients usually include nutrients that we need in very small amounts, such as vitamins and some minerals. Minerals in this group include iron, iodine, zinc, manganese, cobalt, and a few others. Mineral micronutrients can also be called *microminerals* or *trace elements*.

We have already stated that healthy nutrition requires nutrients in a balance of the proper amounts of various nutrients. What do we mean by "proper balance" of nutrients? That is a puzzle that scientists continue to work on unraveling. The answer to the puzzle will certainly be complicated because we now know that a complex interaction of slight differences in individual genetic codes and individual lifestyles and environments affect how nutrients affect our bodies. Until this puzzle is completely solved, if it ever is, we fortunately have some advice that we can rely on to help us make healthy choices. For example, the United States government makes use of an individually customized food pyramid as a general nutrition guide (Figure 27-1). The Canadian government uses a similar food rainbow to advise eating a healthy, balanced diet (Figure 27-2).

As we proceed through this chapter, we will discuss most of the major nutrients in more detail and their roles in the body.

Metabolism refers to the complex, interactive set of chemical processes that make life possible. A good phrase to remember in connection with the word metabolism is "use of foods" because basically this is what metabolism is—the use the body makes of

Figure 27-1 *United States Food Guide Pyramid.* An abbreviated version of the comprehensive food guide can be found at *www.mypyramid.gov*. The full version includes recommended servings per day and other nutrition advice.

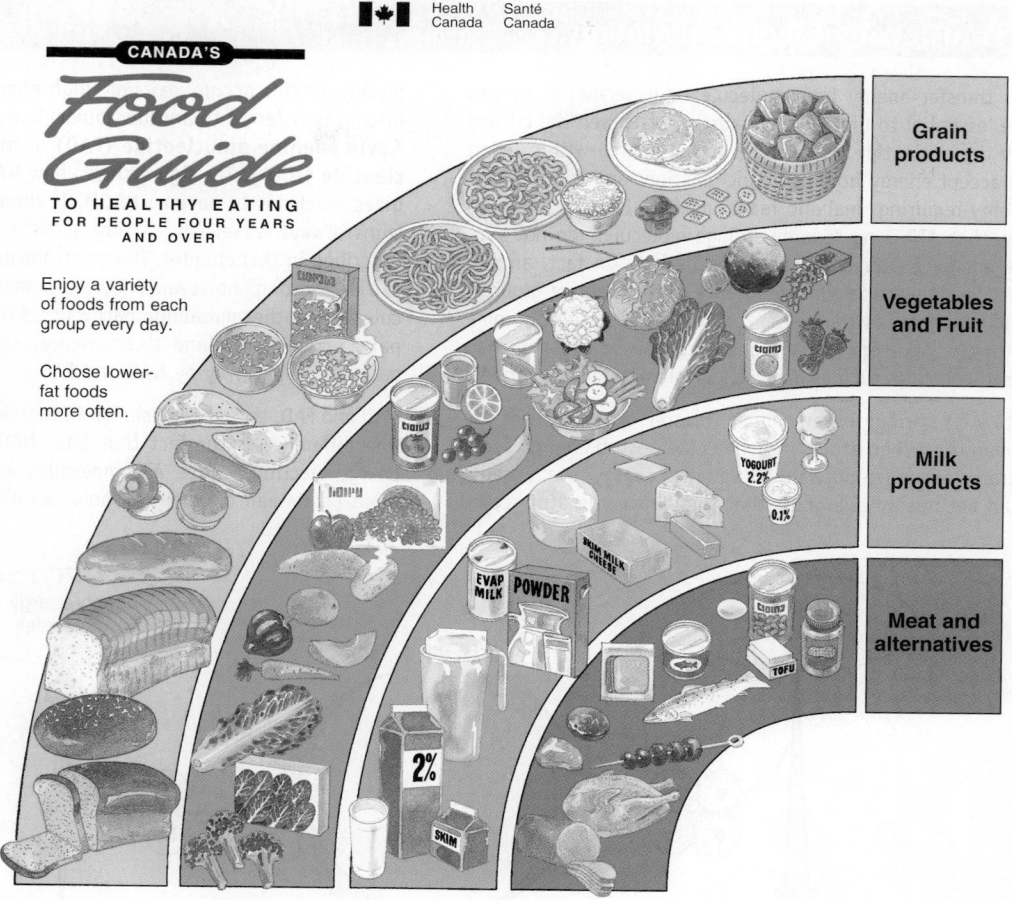

Figure 27-2 *Canadian Food Rainbow.* This is an abbreviated version of the comprehensive food guide found at *www.hc-sc.gc.ca/fn-an/food-guide-aliment/index_e.html.* The full version includes additional nutrition advice.

foods after they have been digested, absorbed, and circulated to cells.

Your body cells use nutrients from food in several ways: as fuel (energy), as material for growth and maintenance, and for regulation of body functions. Before they can be used in these different ways, nutrients have to be assimilated. **Assimilation** occurs when nutrient molecules enter cells and undergo many chemical changes.

Metabolism is a complex process made up of many other processes. Two of the major metabolic processes are **catabolism** and **anabolism.** Each of these processes, in turn, consists of a series of enzyme-catalyzed chemical reactions known as *metabolic pathways.*

Catabolism breaks food molecules down into smaller molecular compounds and, in so doing, releases energy from them. Anabolism does the opposite. It builds nutrient molecules up into larger molecular compounds and, in so doing, uses energy. Catabolism is a decomposition process. Anabolism is a synthesis process. Both catabolism and anabolism take place inside cells. Both processes go on continually and concurrently.

Catabolism releases energy in two forms: heat and chemical energy. The amount of heat generated is relatively large—so large, in fact, that it would hard-boil cells if it were released in one large burst. Fortunately, this does not happen. Catabolism re-

leases heat in frequent, small bursts. Heat is practically useless as an energy source for cells because they cannot use it to do their work. However, this heat is important in maintaining the homeostasis of body temperature. In contrast, chemical energy released by catabolism is more obviously useful. It cannot, however, be used directly for biological reactions. First it must be transferred to the high-energy molecule of adenosine triphosphate (ATP) (Box 27-1).

ATP is one of the most important compounds in the world. Why? It supplies energy directly to the energy-using reactions of all cells in all kinds of living organisms from one-celled plants to trillion-celled humans. ATP functions as the universal biological currency. It pays the energy bills for all cells and is as important in the world of cells as money is in the world of contemporary society.

Look now at Figure 27-3. The structural formula at the top of the diagram shows three phosphate groups attached to the rest of the ATP molecule, two of them by high-energy bonds. Adding water to ATP yields a phosphate group (P), adenosine diphosphate (ADP), and energy, which, as the diagram indicates, is used for anabolism and other cell work. The diagram also shows that P and ADP then use energy released by catabolism to recombine and form ATP. This cycle is called the *ATP/ADP system.*

Metabolism is not identical in all cells. It differs mainly with regard to rate and the kind of products synthesized by anabolism.

BOX 27-1   Transferring Chemical Energy

The ability to transfer energy from molecule to molecule is, as you might imagine, essential to life. We have already discussed the critical role played by the nucleotide ATP in transferring energy within living cells. ATP can accept energy from catabolic reactions and transfer that energy to energy-requiring anabolic reactions (see Figure 27-3). Although we say that ATP is an "energy storage molecule," do not suppose that the energy is stored for very long periods. In fact, an ATP molecule exists for only a brief time before its last phosphate group is broken off and its energy is transferred to another molecule in some metabolic pathway. Long-term storage of energy can be accomplished only by nutrient molecules such as glucose, glycogen, and triglycerides.

In addition to ATP, various other energy transfer molecules are essential to human life. When atoms in a molecule absorb energy, some of their electrons may move outward to a higher energy level (shell). Electrons often become so energized that they leave the atom com-

pletely. As this occurs, pairs of "high-energy" electrons can be picked up and transferred to another molecule by an electron carrier such as **flavin adenine dinucleotide (FAD)** or **nicotinamide adenine dinucleotide (NAD).** The figure shows how NAD$^+$ picks up a pair of energized electrons to become NADH. It should be noted here that electrons always travel with a proton (H$^+$) in the metabolic pathways described in this chapter. The electrons do not stay with the electron carrier for long, however. They are immediately transferred to molecules in another metabolic pathway, as the figure shows. In the cell, pairs of electrons (and their energy) can thus be transferred from pathway to pathway by NAD and FAD.

NAD and FAD, though very similar in function, do have their differences. One difference is the fact that after NAD drops off its pair of high-energy electrons, three ATP molecules are generated and when FAD drops off its pair of electrons, only two ATP molecules are generated.

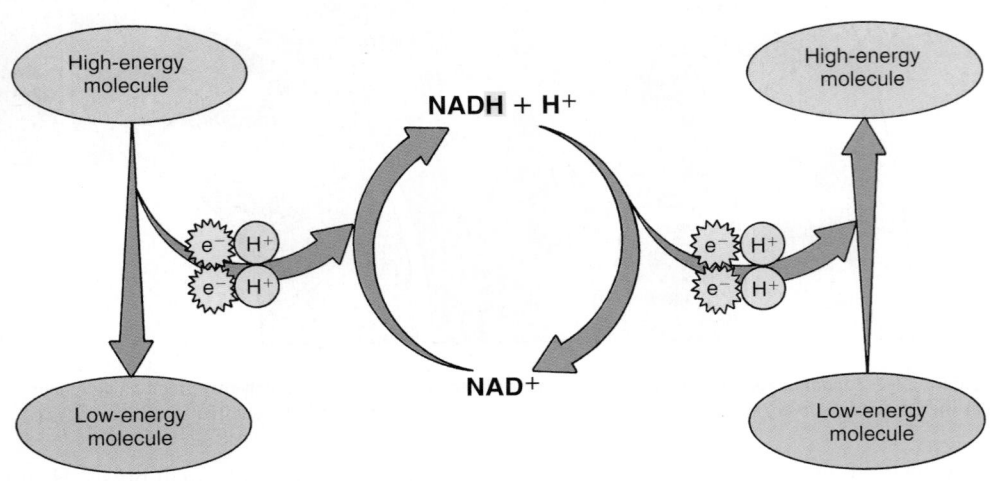

The role of coenzymes in transferring chemical energy.

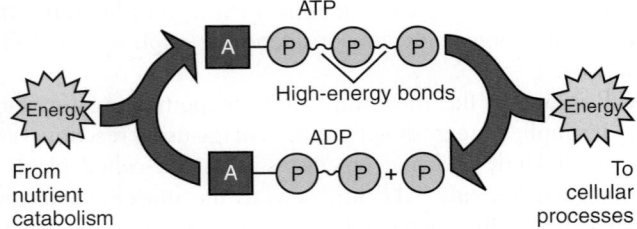

Figure 27-3 *The role of ATP in metabolism.* ATP temporarily stores energy in its last high-energy phosphate bond. When water is added and phosphate breaks free, energy is released to do cellular work. The ADP and phosphate groups that result can be resynthesized into ATP, capturing additional energy from nutrient catabolism. This cycle is called the *ATP/ADP system.*

More active cells have a higher metabolic rate than less active cells do. Anabolism in different kinds of cells produces different compounds. In liver cells, for example, anabolism synthesizes various blood protein compounds. Not so in beta cells of the pancreas. Anabolism here produces a different compound—insulin.

The bulk of this chapter discusses concepts related to the many and varied metabolic pathways of the human body. Our attempt here is to build on what you learned in previous chapters, but not to cover the subject of metabolism in its entirety—if that is even possible. Our goal, therefore, is to underscore the *basic* concepts of nutrition and metabolism. It is important to note that our discussion, as well as the diagrams that accompany the discussion, have been simplified to facilitate understanding of these basic concepts.

CARBOHYDRATES
Dietary Sources of Carbohydrates

Carbohydrates are found in most of the foods that we eat. *Complex carbohydrates*—polysaccharides such as starches in vegetables, grains, and other plant tissues—are broken down into simpler carbohydrates before they are absorbed. **Cellulose,** a major component of most plant tissues, is an important exception to this principle. Because humans do not make enzymes that

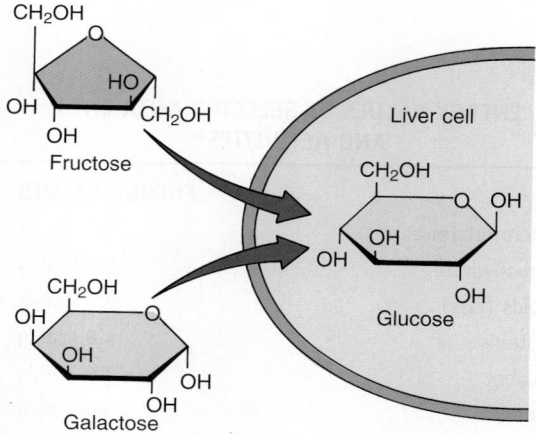

Figure 27-4 *Conversion of monosaccharides.* Monosaccharides *fructose* and *galactose* are usually converted to *glucose* by liver cells. Although simplified in this diagram, conversion to glucose requires several steps. Glucose is the carbohydrate used universally by all cells in the body.

chemically digest this complex carbohydrate, it passes through our system without being broken down. Also called *dietary fiber* or "roughage," cellulose and other indigestible polysaccharides keep chyme thick enough for the digestive system to push it easily. They also help mix chyme, much like the ball inside a can of spray paint. Most biologists believe that a high-fiber diet reduces the risk of many forms of cancer, including colorectal cancer.

Disaccharides such as those in refined sugar must also be chemically digested before they can be absorbed. Monosaccharides in fruits and some "diet foods" are already in an absorbable form, so they can move directly into the internal environment without initially being processed. The monosaccharide *glucose* is the carbohydrate that is most useful to the typical human cell. As Figure 27-4 shows, other important monosaccharides, *fructose* and *galactose*, are usually converted by liver cells into glucose for use by other cells of the body.

 QUICK CHECK

1. Name the two types of metabolism and distinguish between them.
2. Why must energy in nutrient molecules be transferred to ATP?

Carbohydrate Metabolism

The body metabolizes carbohydrates by both catabolic and anabolic processes. Because many human cells use carbohydrates—mainly glucose—as their first or preferred energy fuel, they catabolize most of the carbohydrate absorbed and anabolize a relatively small portion of it. When the amount of glucose entering cells is inadequate for their energy needs, they may make more use of an alternative pathway and catabolize fats or proteins.

We first discussed carbohydrate metabolism in Chapter 4 when we discussed basic concepts of cell metabolism (see pp. 114-116). The subject came up again in Chapter 11 when we discussed energy production in muscle tissue (see pp. 404-407). You may want to flip back to those passages and review the material before proceeding further with this chapter.

As you read through the following sections that briefly describe the process of carbohydrate metabolism, remember the ultimate result of catabolism: the transfer of energy from a nutrient molecule to ATP. It is the continued production of ATP, the energy currency of the cell, that makes nutrient catabolism so incredibly vital to the overall process of life itself.

Glucose Transport and Phosphorylation

Carbohydrate metabolism begins with the movement of glucose through cell membranes. Immediately on reaching the interior of a cell, glucose reacts with ATP to form glucose-6-phosphate. This step, named **glucose phosphorylation,** prepares glucose for further metabolic reactions. **Phosphorylation** (fos-for-i-LAY-shun) is the process of adding a phosphate group to a molecule. In most cells of the body, glucose phosphorylation is an irreversible reaction. However, in a few cells—namely, those of the intestinal mucosa, liver, and renal cortex—glucose phosphorylation is reversible. These cells contain phosphatase, an enzyme that splits phosphate off from glucose-6-phosphate. This reverse glucose phosphorylation reaction forms glucose, which then moves out of the cells into the blood. (Glucose-6-phosphate cannot pass through cell membranes.) Depending on their energy needs of the moment, cells either catabolize (break apart) or anabolize (bind together) glucose-6-phosphate. The measurement of energy is discussed in Box 27-2.

Glycolysis

Glycolysis is the first process of carbohydrate catabolism. It breaks apart one glucose molecule to form two pyruvic acid molecules. (A glucose molecule contains six carbon atoms, and a pyruvic acid molecule contains three carbon atoms. See Figure 27-6.) Glycolysis consists, as Figure 27-5 shows, of a series of chemical reactions. A specific enzyme catalyzes each of these reactions. Probably the most important facts for you to remember about glycolysis are the following:

- Glycolysis occurs in the cytoplasm of all human cells.
- Glycolysis is an *anaerobic* process, that is, it does not use oxygen. It is the only process that provides cells with energy when their oxygen supply is inadequate or even absent.
- Glycolysis breaks the chemical bonds in glucose molecules and thereby releases about 5% of the energy stored in them. Much of the released energy appears as heat, but some of it is transferred to the high-energy bonds of ATP molecules. For every molecule of glucose undergoing glycolysis, a net of two molecules of ATP is formed. About 8 kilocalories (kcal) of energy (under normal physiological conditions) is stored in the high-energy bonds that bind phosphate to ADP to form 1 mole (6.02×10^{23} molecules) of ATP.
- Glycolysis is an essential process because it prepares glucose for the second step in catabolism, namely, the **citric acid cy-**

BOX 27-2: FYI
Measuring Energy

Physiologists studying metabolism must be able to express a quantity of energy in mathematical terms. Internationally, the unit of energy measurement most often used is the **joule (J or j)**. A joule is the amount of work done by a force of 1 newton acting for a distance of 1 meter. A *newton (N)* is the force needed to accelerate 1 kilogram 1 meter per second per second. Most often, nutritionists use a larger version of this unit—the *kilojoule (kJ)*, which is equal to one thousand joules, or the *megajoule (mJ)*, which is equal to 1000 kilojoules.

Although the joule was adopted by the U.S. Bureau of Standards in the 1960s, it is not commonly used in either the U.S. or in Canada. Instead, the currently used unit is the **calorie (cal).** A calorie is the amount of energy needed to raise the temperature of 1 g of water 1° C. Because physiologists often deal with very large amounts of energy, the larger unit, *kilocalorie (kcal)* or *Calorie* (notice the uppercase C), is used. There are 1000 cal in 1 kcal or Calorie.

Nutritionists, who have long preferred to use *Calorie* when they express the amount of energy stored in a food, have now begun to use *kilocalorie* more frequently to avoid confusion. However, sometimes "calorie" is used as the equivalent to "Calorie" or "kilocalorie"—a fact that adds to the confusion over terms.

A simple way to convert kilocalories to kilojoules is kcal × 4.2 = kJ.

Examples of the energy values of the macronutrients and a few examples of physical activity are listed in the table.

ENERGY VALUES OF SELECTED NUTRIENTS AND ACTIVITIES*

	ENERGY VALUES
Macronutrients	
Carbohydrates	4.0 kcal/g
Lipids (fats)	9.0 kcal/g
Proteins	4.0 kcal/g
Alcohol	7.0 kcal/g
Water	0.0 kcal/g
Activities	
Bicycling (pleasure)	−204 kcal/hr
Carpentry	−170 kcal/hr
Cooking	−68 kcal/hr
Hiking	−340 kcal/hr
Housework	−170 kcal/hr
Jogging	−408 kcal/hr
Sexual activity (moderate)	−20 kcal/hr
Sitting (schoolwork)	−34 kcal/hr
Sitting (TV)	0 kcal/hr
Typing	−34 kcal/hr
Walking (2.5 mph)	−136 kcal/hr
Water aerobics	−204 kcal/hr
Yoga	−102 kcal/hr

*Nutrient values are average values of metabolizable energy. Activity values are estimates for additional calories expended (above basal metabolism) in a 150-lb adult. Positive values are energy input into the body; negative values are energy expenditure by body tissues.

cle. Glucose itself cannot enter the cycle but must first be converted to pyruvic acid, then to a compound called *acetyl-CoA* (*coenzyme* A).

Citric Acid Cycle

As stated previously, for every glucose molecule that enters the catabolic pathway described here, two pyruvic acid molecules are produced. Before each pyruvic acid molecule can proceed into the citric acid cycle, it must be converted into an acetyl group (acetate), releasing a carbon dioxide molecule and carried into the citric acid cycle by **coenzyme A (CoA).** Essentially, the citric acid cycle converts the two acetyl molecules to four carbon dioxide and six water molecules. But many chemical reactions intervene. Figure 27-6 shows that one glucose molecule is changed by glycolysis to two pyruvic acid molecules, which, by means of the citric acid cycle, yield six carbon dioxide molecules. Figure 27-7 shows the details of the citric acid cycle.

Glycolysis takes place in the cytoplasm of cells, whereas the citric acid cycle occurs in their mitochondria. Some of the en-

zymes needed for the many steps of the citric acid cycle are dissolved in the matrix of the mitochondrion, and some are attached to the inner membrane of the mitochondrion.

Before it can enter the citric acid cycle, each pyruvic acid molecule combines with coenzyme A (see Figure 27-7) after splitting off CO_2 and a pair of high-energy electrons (with their accompanying protons, H^+) from pyruvic acid, thus forming acetyl-CoA. Coenzyme A then detaches from acetyl-CoA, leaving a two-carbon acetyl group, which enters the citric acid cycle by combining with oxaloacetic acid to form citric acid. This is what gives the citric acid cycle its name. The cycle is also called the **tricarboxylic acid (TCA) cycle** because citric acid is also called *tricarboxylic acid*. For many years this cycle was called the **Krebs cycle** after Sir Hans Krebs, whose brilliant work in discovering this metabolic pathway earned him the 1953 Nobel Prize.

You probably do not need to memorize the names of the intermediate products formed during the citric acid cycle, but notice that all of them are acids. You should note that a little bit of ATP is directly generated by the citric acid cycle (in step 5). In

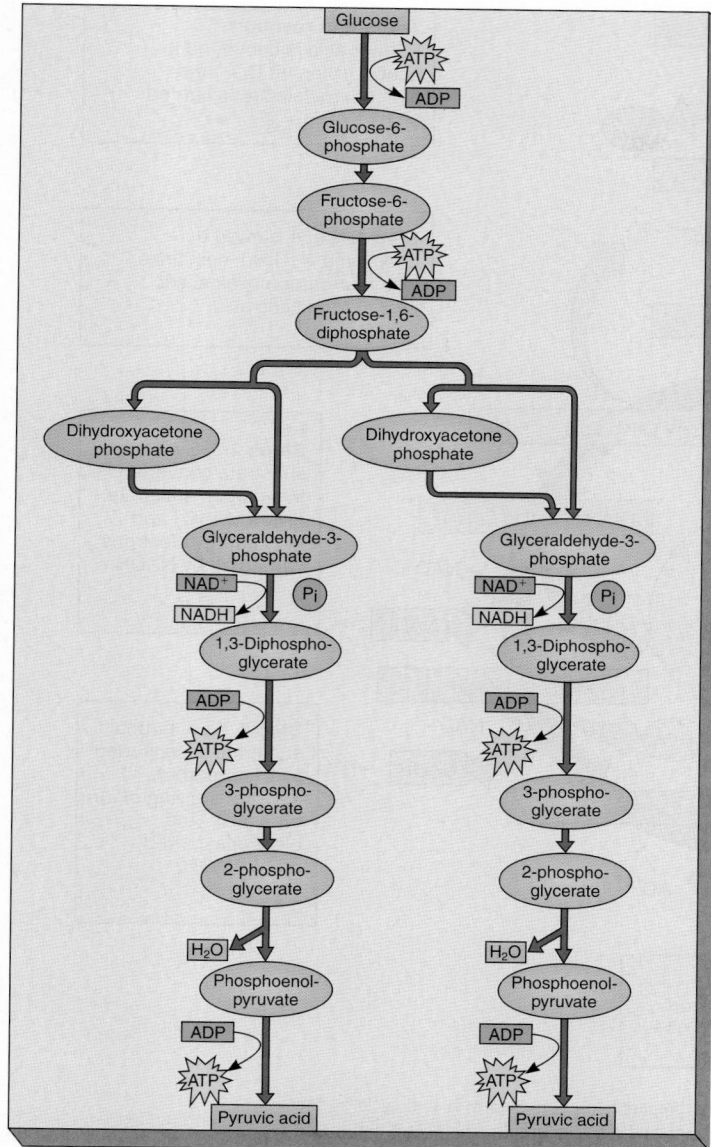

Figure 27-5 *Glycolysis.* The series of enzyme-catalyzed reactions that make up the portion of the catabolic pathway for carbohydrates is called *glycolysis*.

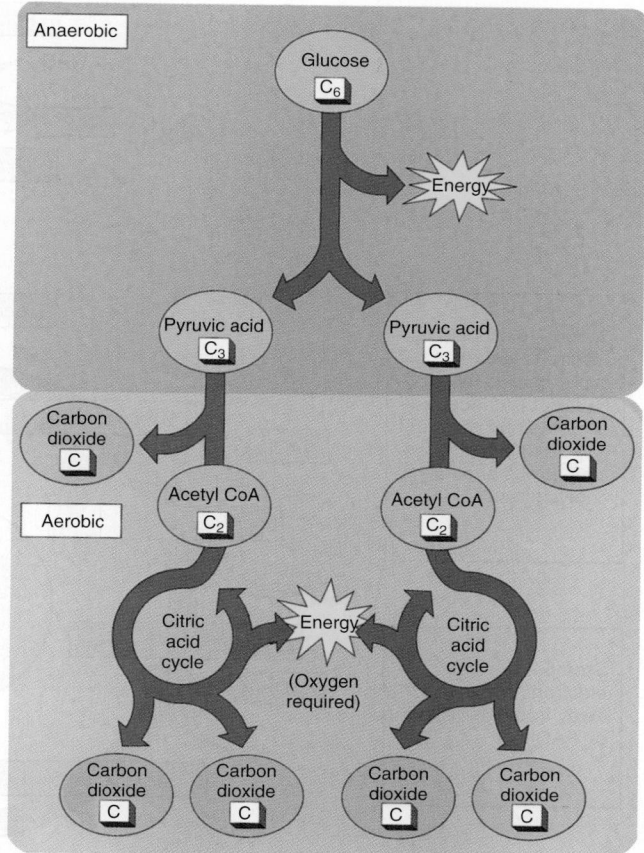

Figure 27-6 *Catabolism of glucose.* Glycolysis splits one molecule of glucose (six carbon atoms) into two molecules of pyruvic acid (three carbon atoms each). The glycolytic pathway does not require oxygen, so it is termed *anaerobic*. A transition reaction removes a carbon dioxide molecule, converting each pyruvic acid molecule into a two-carbon acetyl group that is escorted by coenzyme A (CoA) into the citric acid cycle. There, two more carbon dioxide molecules (one carbon atom each) are released. The carbon and oxygen atoms in the original glucose molecule are thus released as waste products. However, the real metabolic prize is energy, which is released as the molecule is broken down. Because this part of the pathway requires oxygen, it is termed *aerobic*. More detailed depictions of this process are included later in this chapter.

step 5, energy is transferred first to GTP (guanosine triphosphate), a nucleotide similar to ATP, and then finally to ATP.

Observe, too, that for each pyruvic acid molecule entering this pathway, three CO_2 molecules are formed and that certain reactions yield high-energy electrons. Most of the energy leaving the citric acid cycle is in these high-energy electrons. The next section describes how these high-energy electrons are used to generate ATP.

Electron Transport System and Oxidative Phosphorylation

High-energy electrons removed during the citric acid cycle enter a chain of carrier molecules, which is embedded in the inner membrane of mitochondria and is known as the **electron trans-**

port system (ETS). Figure 27-8 shows that high-energy electrons (along with their accompanying protons, H^+) are carried to the electron transport system by NAD and FAD. The electrons quickly move down the chain, from cytochrome to cytochrome, eventually to their final acceptor, oxygen.

As the electrons are transported, some of their energy is used to pump their accompanying protons (H^+) to the intramembrane space between in the inner and outer membranes of the mitochondrion. This creates a concentration gradient of protons, and the intermembrane space thus becomes a virtual reservoir of protons. As with water behind a dam, the reservoir of protons temporarily stores energy. As in a dam possessing water wheels that convert energy, the inner membrane has "proton wheels" built into it—in the form of **ATP synthase.** Figure 27-9 shows how pro-

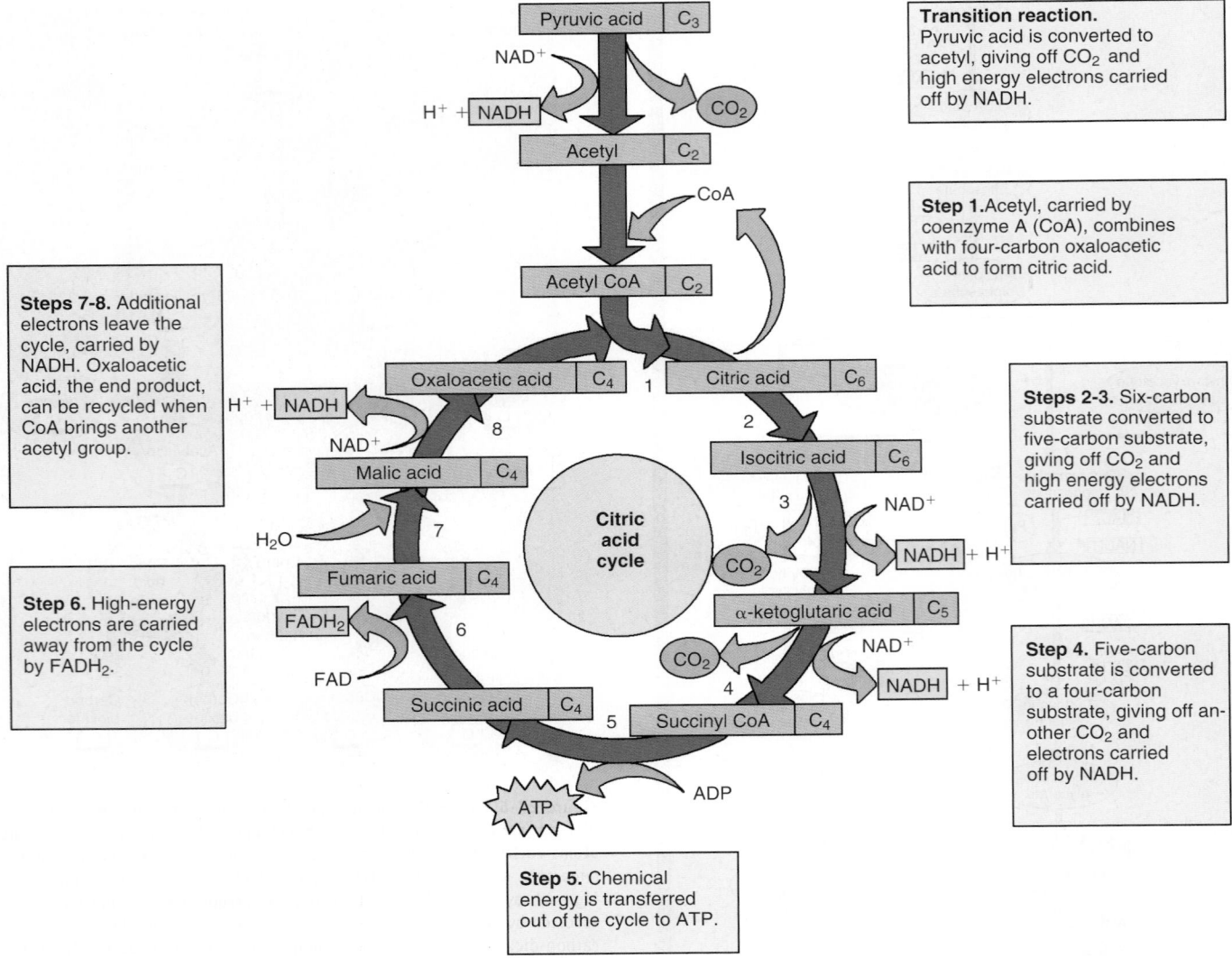

Transition reaction. Pyruvic acid is converted to acetyl, giving off CO_2 and high energy electrons carried off by NADH.

Step 1. Acetyl, carried by coenzyme A (CoA), combines with four-carbon oxaloacetic acid to form citric acid.

Steps 7-8. Additional electrons leave the cycle, carried by NADH. Oxaloacetic acid, the end product, can be recycled when CoA brings another acetyl group.

Steps 2-3. Six-carbon substrate converted to five-carbon substrate, giving off CO_2 and high energy electrons carried off by NADH.

Step 6. High-energy electrons are carried away from the cycle by $FADH_2$.

Step 4. Five-carbon substrate is converted to a four-carbon substrate, giving off another CO_2 and electrons carried off by NADH.

Step 5. Chemical energy is transferred out of the cycle to ATP.

Figure 27-7 *Citric acid cycle.* Each pyruvic acid molecule is prepared to enter the citric acid cycle by the transition reaction, which yields a pair of high-energy electrons and a CO_2 molecule. The acetyl group that is thus formed is picked up by coenzyme A (CoA) and led into the citric acid cycle proper, which is described here as a recurring series of eight steps.

tons move down their concentration gradient and into the ATP synthase structure, turning a molecular wheel that transfers energy by synthesizing ATP from ADP and phosphate.

At this time, the low-energy electrons (e^-) and their protons (H^+) join oxygen, forming water. As you can see, although oxygen is not needed until the very last step of aerobic respiration, its role is vital. Without oxygen to oxidize the hydrogen into water, hydrogen would accumulate and disrupt the functions of the cell.

Oxidative phosphorylation refers to this oxygen-requiring joining of a phosphate group to ADP to form ATP—a reaction whose importance can scarcely be overemphasized.

The arrangement of catabolic "machinery" within the cell, as currently viewed, is shown in Figure 27-10. Glycolytic enzymes in the cytoplasm catalyze the production of pyruvic acid, which diffuses into mitochondria. The enzymes of the citric acid cycle have been localized mostly to the matter (matrix) inside the inner mi-

tochondrial membrane. The high-energy electrons and their accompanying protons are then carried to the cristae of the inner membrane, where the electron transport carriers and mechanism for phosphorylation are found. Because so many of the cell's energy-releasing enzymes are located within the mitochondria, these tiny structures are aptly described as the "power plants" of the cell.

The breakdown of ATP molecules, of course, provides virtually all the energy that does cellular work. Therefore, the process that produces some 90% of the ATP formed during carbohydrate catabolism, namely, oxidative phosphorylation, is the crucial part of catabolism (Figure 27-11). This vital process depends on cells receiving an adequate oxygen supply. Why? Briefly, because only when oxygen is present in cells to serve as the final acceptor of electrons and hydrogen ions can electrons then continue moving down the electron transport chain. If oxygen becomes unavailable, the movement of electrons and hy-

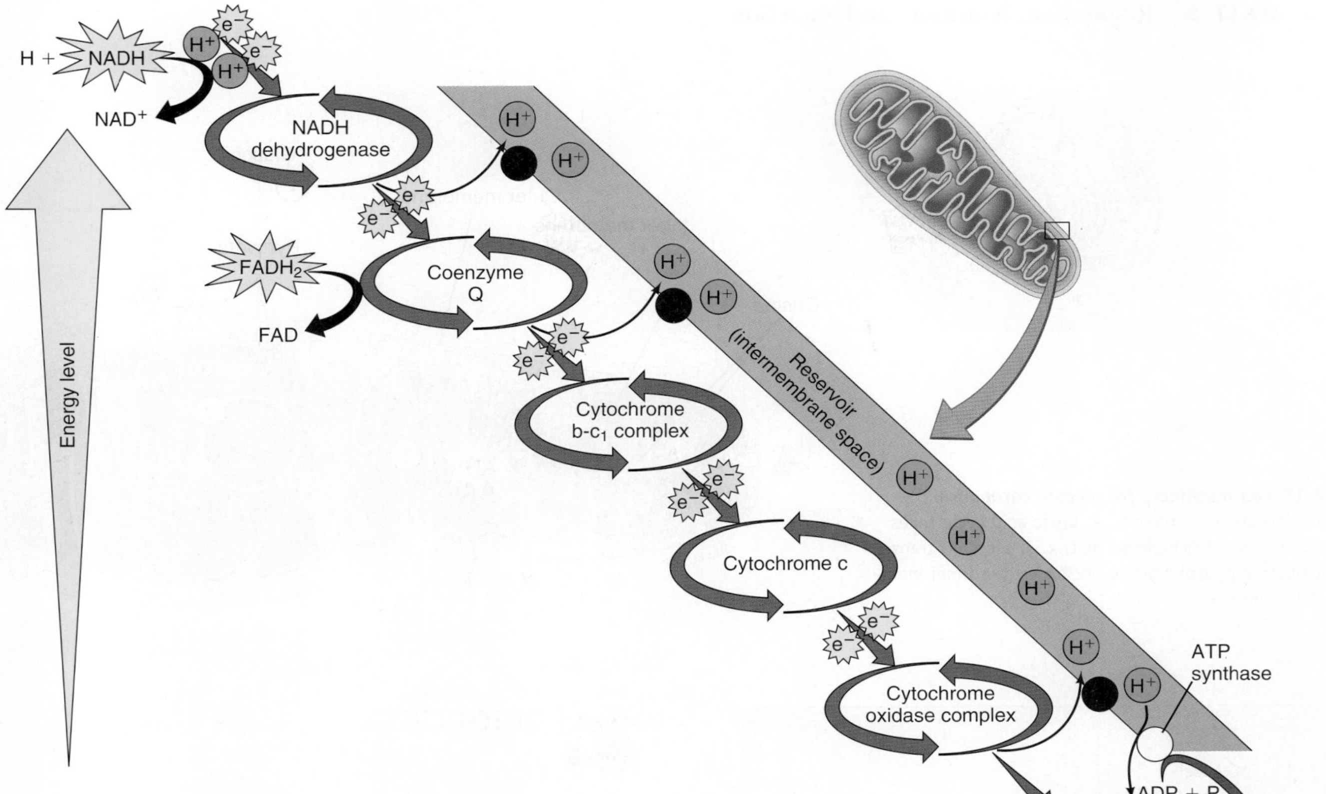

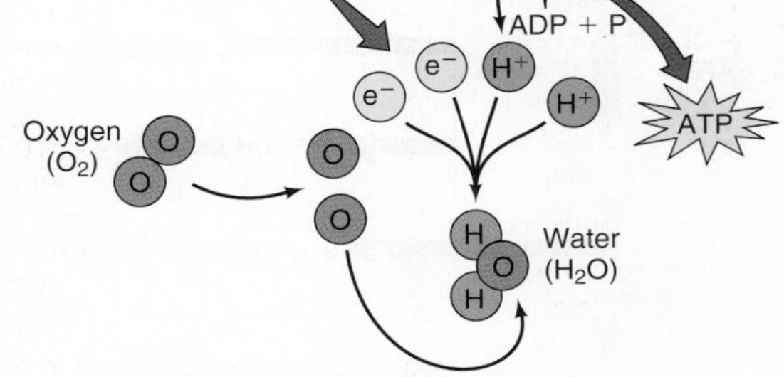

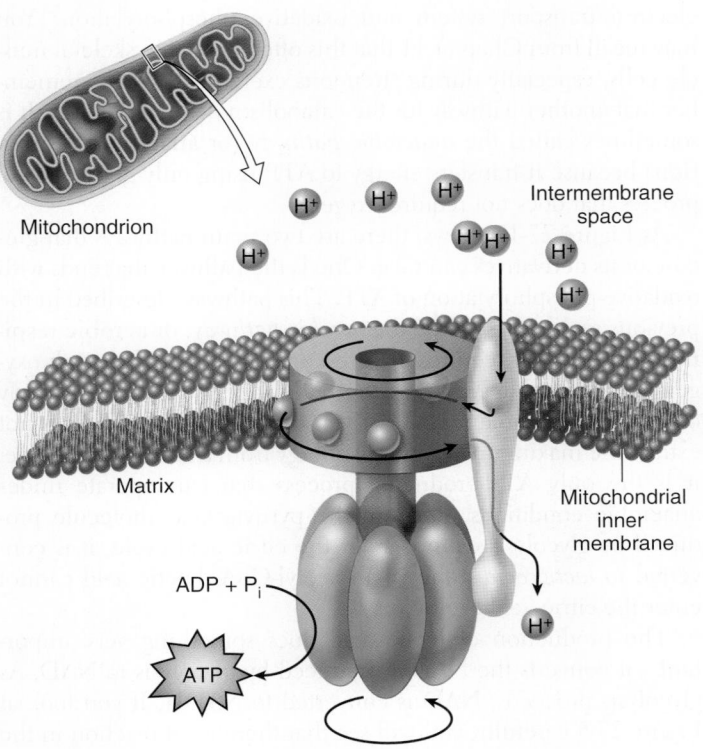

Figure 27-8 *Electron transport system.* Pairs of high-energy electrons and their accompanying protons (H^+) are transferred to the components (cytochromes) of the electron transport system by NAD and FAD. They then jump from cytochrome to cytochrome, losing energy along the way. The energy is used to pump protons *(H^+)* into the compartment between the inner and outer mitochondrial membranes. The diffusion of protons back into the inner compartment drives the phosphorylation of ADP to form ATP (see Figure 27-9). The protons are joined with oxygen and low-energy electrons at the end of the cytochrome chain to form water molecules. This all takes place within each mitochondrion, as the inset shows.

Figure 27-9 *Generation of ATP by ATP synthase.* This simplified model of the proton wheel in the mitochondrial inner membrane shows how protons (H^+) moving down their concentration gradient drive the rotation of a molecular machine. The energy of rotation then phosphorylates (adds phosphate to) ADP to become ATP.

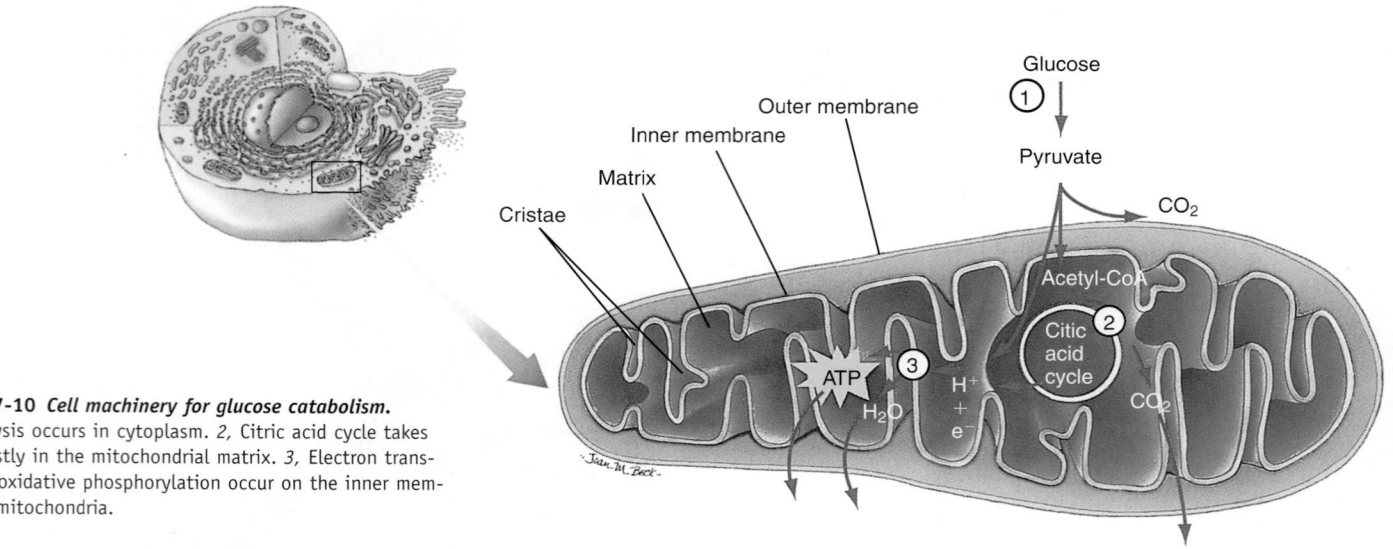

Figure 27-10 *Cell machinery for glucose catabolism.*
1, Glycolysis occurs in cytoplasm. *2,* Citric acid cycle takes place mostly in the mitochondrial matrix. *3,* Electron transport and oxidative phosphorylation occur on the inner membrane of mitochondria.

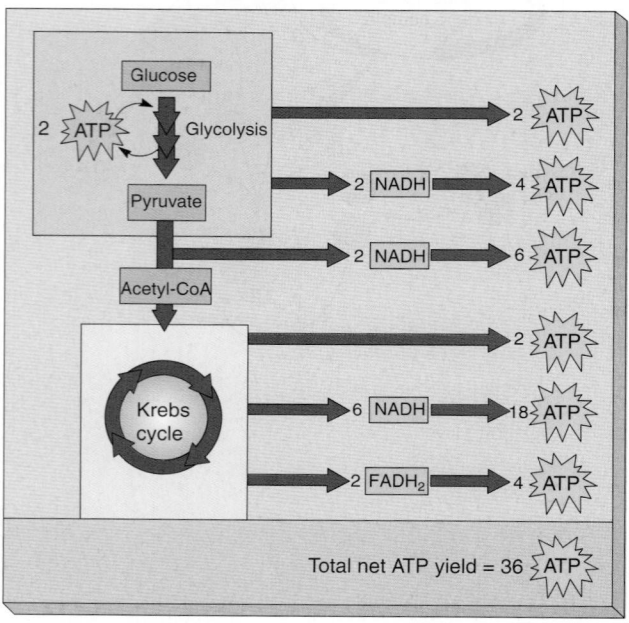

Figure 27-11 *Energy extracted from glucose.* Energy released from the breakdown of glucose is released mostly as heat, but some of it is transferred to a usable form—the high-energy bonds of ATP. In most human cells, one glucose molecule produces enough usable chemical energy to synthesize or "charge up" 36 ATP molecules. Some cells, such as heart and liver cells, shuttle electrons more efficiently and may be able to synthesize up to 38 ATP molecules. This represents an energy conversion efficiency of 38% to 44%, much better than the 20% to 25% typical of most machines.

drogen ions stops. Cessation of ATP formation by oxidative phosphorylation necessarily follows. All too soon, cells have an inadequate energy supply—a lethal condition if it persists for more than a few minutes.

We can summarize the long series of chemical reactions in glucose catabolism with one short equation:

$$C_6H_{12}O_6 + 6\,O_2 \rightarrow 6\,CO_2 + 6\,H_2O + 36\ (\text{or } 38)\ \text{ATP} + \text{Heat}$$

QUICK CHECK

3. What is glycolysis? How much energy is transferred to ATP through this process?
4. What happens to a nutrient molecule as it proceeds through the citric acid cycle?
5. What is the purpose of the electron transport system?

Anaerobic Pathway

What if there is an inadequate amount of oxygen to operate the electron transport system and oxidative phosphorylation? You may recall from Chapter 11 that this often occurs in skeletal muscle cells, especially during strenuous exercise. You may remember that another pathway for the catabolism of glucose exists. It is sometimes called the *anaerobic pathway* (or **anaerobic respiration**) because it transfers energy to ATP using only glycolysis—a process that does not require oxygen.

As Figure 27-12 shows, there are two main pathways that glucose or its derivatives can take. One is the pathway that ends with oxidative phosphorylation of ATP. This pathway, described in the previous sections, is called the *aerobic pathway,* or **aerobic respiration,** because it requires the presence of oxygen. If enough oxygen is not available to operate this pathway, the cell will rely solely on glycolysis to produce ATP. Even though this process does not extract the maximum amount of energy from a glucose molecule, it is the only ATP-producing process that can operate under anaerobic conditions. Because the pyruvic acid molecule produced by glycolysis cannot enter the citric acid cycle, it is converted to *lactic acid* rather than acetyl-CoA. Lactic acid cannot enter the citric acid cycle.

The production of lactic acid does something very important—it converts the NADH produced by glycolysis to NAD. As glycolysis proceeds, NAD is converted to NADH. If you look at Figure 27-5 carefully, you will see that there is no reaction in the glycolytic pathway to turn NADH back into NAD again. This is

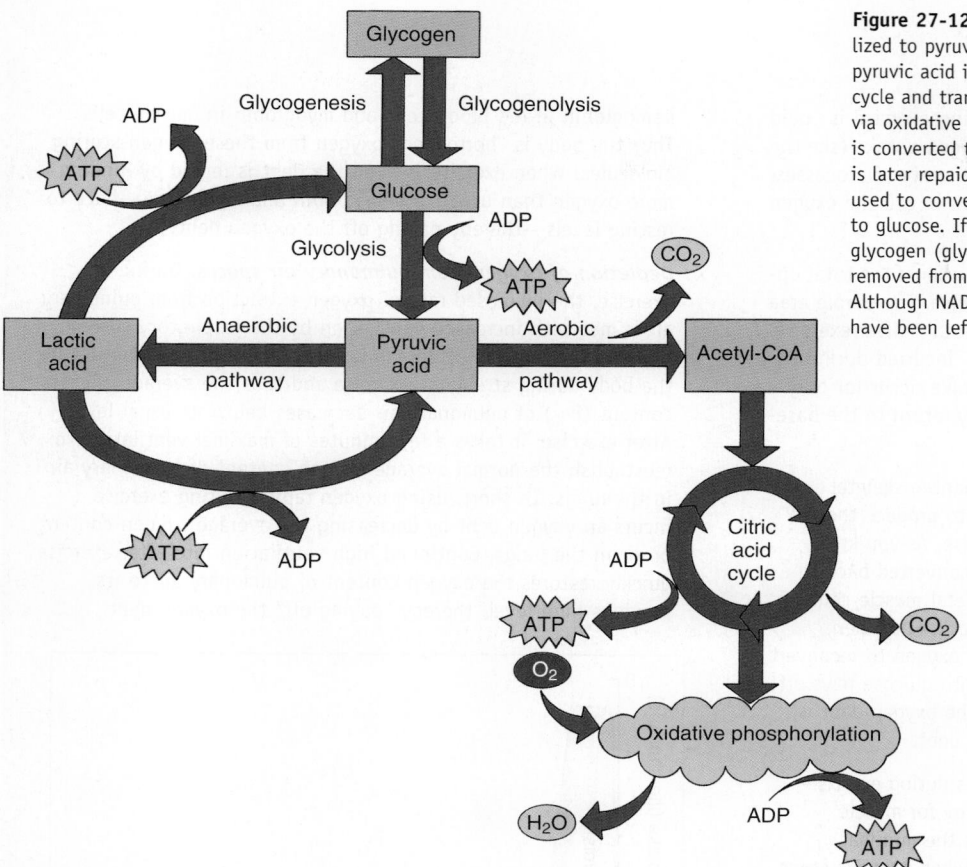

Figure 27-12 *Summary of glucose metabolism.* Glucose is catabolized to pyruvic acid in the process of glycolysis. If oxygen is available, pyruvic acid is converted to acetyl-CoA and then enters the citric acid cycle and transfers energy to the maximum number of ATP molecules via oxidative phosphorylation. If oxygen is not available, pyruvic acid is converted to lactic acid, incurring an oxygen debt. The oxygen debt is later repaid when ATP produced via oxidative phosphorylation is used to convert lactic acid back into pyruvic acid or all the way back to glucose. If there is an excess of glucose, the cell may convert it to glycogen (glycogenesis). Later, individual glucose molecules can be removed from the glycogen chain by the process of glycogenolysis. Although NAD and FAD play important roles in these pathways, they have been left out of this diagram for the sake of simplicity.

BOX 27-3: FYI
Anaerobic Body Cells

Some body cells, specifically the red blood cells (see photo), do not have the cellular organelles and enzymes required to carry out aerobic respiration. These cells must rely solely on glycolysis for their ATP production. Red blood cells produce lactic acid continually, which diffuses into the plasma and is carried to liver cells for conversion back to pyruvic acid and then to glucose and glycogen (see Figure 27-12).

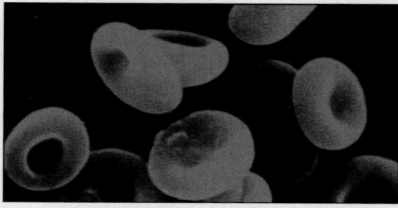

Red blood cells (RBCs). This color-enhanced scanning electron micrograph shows several RBCs, which are unable to perform aerobic respiration because of a lack of mitochondria. They rely solely on anaerobic respiration for their ATP, producing lactic acid as a metabolic waste product.

no problem if the aerobic pathway is in operation—NADH is converted back to NAD in the electron transport system. After a short period of anaerobic respiration, however, glycolysis will turn all of the cell's NAD into NADH and thus glycolysis will have to shut down because of a lack of free NAD. The production of lac-

tic acid solves this biochemical dilemma because this reaction also converts NADH back to NAD.

Once oxygen becomes available again in cells that ordinarily rely on aerobic respiration, some of the lactic acid is converted back to pyruvic acid by the cell (see Figure 27-12). Notice that such reconversion requires energy from ATP. Thus, reconversion in muscle cells, for example, can occur only when phosphorylation has resumed and produced enough ATP to allow reconversion to occur. Once it is converted to pyruvic acid, the molecule can follow the aerobic pathway and become completely catabolized. Much of the lactic acid produced during anaerobic respiration diffuses into the blood and is removed by the liver. Inside liver cells, ATP produced by oxidative respiration is used to convert the lactic acid back into glucose. Because anaerobic respiration requires the later use of ATP molecules produced via oxidative respiration, we can say that an oxygen debt is incurred (Box 27-4). This oxygen debt is repaid when oxygen becomes available to form the extra ATP needed to convert lactic acid back into pyruvic acid or glucose.

Glycogenesis

Imagine what happens in a cell if glucose catabolism is proceeding at maximum rate. What do you do if you see a traffic jam ahead with no possible way to get through it? Probably take an alternative route, if available. Similarly, if glycolytic pathways are "saturated" because of high levels of glucose entering the cell, a "traffic jam" of glucose-6-phosphate will result. Unable to enter glycolysis, glucose-6-phosphate will begin an alternative

BOX 27-4 SPORTS AND FITNESS
The Oxygen Debt

The oxygen debt that skeletal muscles incur during exercise is "paid off" after exercise by a continued high rate of oxygen uptake (see the figure). This oxygen debt actually results from a variety of processes. The primary factors that produce the phenomenon we call the oxygen debt are briefly outlined here:

Exercise and the oxygen debt. The pink area represents the total uptake of oxygen during 4 minutes of maximal exercise. The purple area represents the continued high levels of oxygen uptake after exercise, which constitutes "payment" of the oxygen debt incurred during exercise. Notice that very high rates of oxygen uptake occur for only a few minutes after exercise, but do not completely return to the baseline for quite some time.

1. *Anaerobic respiration.* As described in this chapter, skeletal muscle cells often use the anaerobic pathway to produce the ATP they need for muscle contraction during exercise. As you know, glycolysis produces lactic acid, which is later converted back to glucose in an oxygen-requiring process in skeletal muscle cells, in liver cells, and in other cells. Production of lactic acid, then, incurs an oxygen debt, and subsequent use of oxygen to reconvert lactic acid to pyruvic acid and perhaps then into glucose pays off the oxygen debt. Sometimes this element of the oxygen debt is more specifically called the lactic acid oxygen debt.

2. *Breakdown of ATP.* Use of skeletal muscle cells during exercise requires the breakdown of ATP to provide energy for muscle contraction. Prolonged exercise often depletes the minimal amount of ATP present in a muscle cell. Thus when exercise stops, a great deal of oxygen is required to carry out aerobic respiration at a fast pace so that the ATP can be restored to its original resting level. This takes some doing because ATP is also being used at this time to reconvert lactic acid back into pyruvic acid and perhaps into glucose (see point 1). In short, using up ATP more rapidly than at rest incurs an oxygen debt and subsequently rebuilding the ATP to its set point level pays off the debt.

3. *Extraction of oxygen from hemoglobin and myoglobin.* During prolonged exercise, more oxygen than usual is released from

hemoglobin in red blood cells and myoglobin in muscle cells. Thus the body is "borrowing" oxygen from these oxygen-storing molecules. When exercise is over, the debt is repaid by binding more oxygen than usual to hemoglobin and myoglobin—back to resting levels—thereby paying off the oxygen debt.

4. *Depletion of oxygen from pulmonary air spaces.* During exercise, the increased rate of oxygen extraction from pulmonary air is met with increased ventilation by the respiratory system. However, ventilation often cannot keep pace with oxygen use by the body during strenuous exercise and thus the average oxygen content (P_{O_2}) of pulmonary air decreases below its usual level. After exercise, it takes a few minutes of maximal ventilation to reestablish the normal average oxygen content of pulmonary air in the lungs. In short, using oxygen rapidly during exercise incurs an oxygen debt by decreasing the average oxygen content of air in the lungs. Continued high ventilation rate after exercise quickly restores the oxygen content of pulmonary air to its previous high level, thereby "paying off" the oxygen debt.

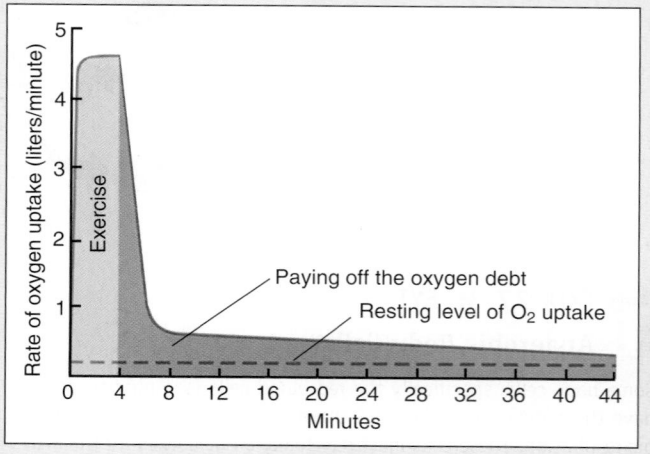

Oxygen uptake during and after exercise.

BOX 27-5: FYI
Glycogenesis

Muscle cells, like liver cells, have a high rate of glycogenesis. Certain cells, however, cannot carry out the process at all. Brain cells, for instance, cannot form or store glycogen. They must depend on the circulation of blood to deliver glucose to them.

route; that is, it will enter the anabolic pathway of glycogen formation. The process of glycogen formation, called **glycogenesis** (see Figure 27-12 and Box 27-5), is a series of chemical reactions in which glucose molecules are joined together to form a structure made of a branched strand of stored glucose "beads" (Figure 27-13).

The process of glycogenesis is part of a homeostatic mechanism that operates when the blood glucose level increases above

the midpoint of its normal range (80 to 100 mg/dl of blood), as indicated in Figure 27-14. Example: Soon after a meal, while glucose is being absorbed rapidly, blood is quickly shunted to the liver via the portal system. Here, a great many glucose molecules leave the blood for storage as glycogen. As a result of glycogenesis, the blood glucose level decreases, ordinarily enough to reestablish its normal level.

Glycogenolysis

Glycogen molecules do not remain in the cell permanently but are eventually broken apart (hydrolyzed). This process of "splitting glycogen" is called **glycogenolysis** (Figure 27-15; see Figure 27-12). It is, in essence, a reversal of glycogenesis. What are the products of glycogenolysis? The answer depends on the cell. Although all cells presumably have the enzymes to break glycogen down to glucose-6-phosphate, only a few cell types (liver, kidney, intestinal mucosa) have the enzyme phos-

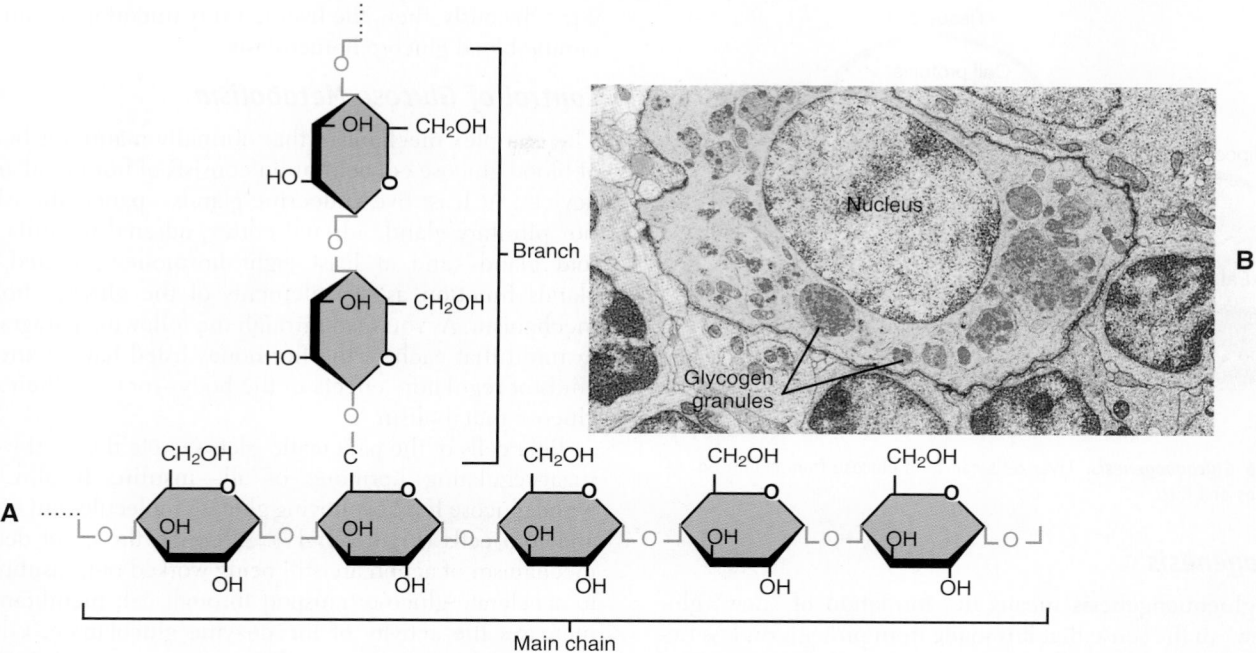

Figure 27-13 *Glycogen.* **A,** A portion of the highly branched strand of glucose subunits called *glycogen.* **B,** Transmission electron micrograph of glycogen granules in a liver cell. Glycogen (a polysaccharide) is the form in which human cells store glucose (a monosaccharide).

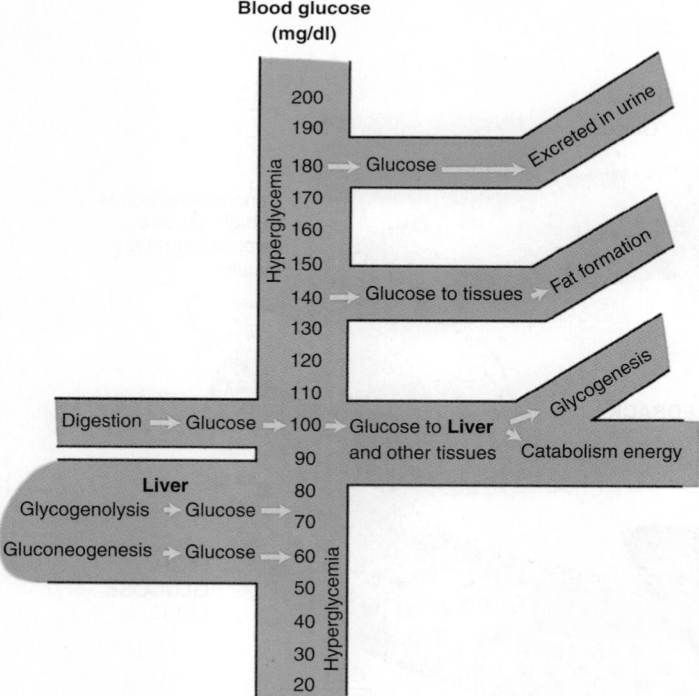

Figure 27-14 *Homeostasis of blood glucose level.* When blood glucose level starts to decrease toward lower normal, liver cells increase the rate at which they convert glycogen, amino acids, and glycerol to glucose (glycogenolysis and gluconeogenesis) and release it into blood. But when blood glucose level increases, liver cells increase the rate at which they remove glucose molecules from blood and convert them to glycogen for storage (glycogenesis). At still higher levels, glucose leaves blood for liver cells to be converted into fat and, at still higher levels, is excreted in the urine.

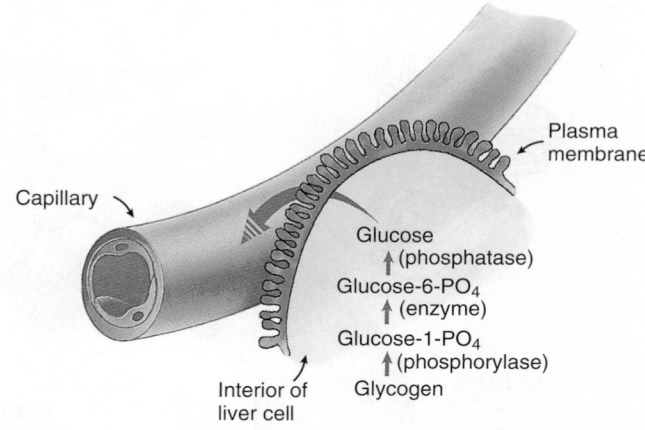

Figure 27-15 *Glycogenolysis in a liver cell.*

phatase, which allows free glucose to form and possibly leave the cell.

So the term *glycogenolysis* means different things in different cells. In muscles, glucose-6-phosphate is the product, which then undergoes glycolysis. But liver glycogenolysis (see Figure 27-15) results in free glucose that can leave the cell and increase the blood glucose level. Accordingly, liver glycogenolysis acts as a part of the homeostatic mechanism to maintain the blood glucose level. Example: A few hours after a meal, when the blood glucose level decreases (see Figure 27-14), liver glycogenolysis accelerates. However, glycogenolysis alone can probably maintain homeostasis of blood glucose concentration for only a few hours because the body can store only small amounts of glycogen.

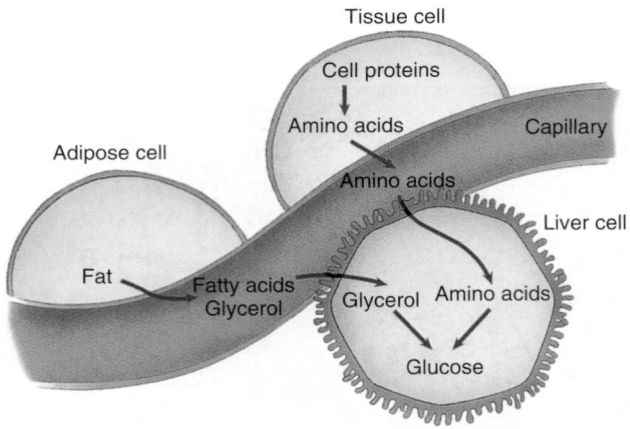

Figure 27-16 *Gluconeogenesis.* Liver cells can form glucose from mobilized tissue proteins and fats.

Gluconeogenesis

Literally, **gluconeogenesis** means the formation of "new" glucose—"new" in the sense that it is made from proteins or, less frequently, from the glycerol of fats—not from carbohydrates. The process occurs chiefly in the liver. It consists of many complex chemical reactions. The new glucose produced from proteins or fats by gluconeogenesis (Figure 27-16) diffuses out of liver cells into the blood. Gluconeogenesis can therefore add glucose to the blood when needed. So, too, can the process of liver glycogenol-

ysis. Obviously, then, the liver is a very important organ for maintaining blood glucose homeostasis.

Control of Glucose Metabolism

The complex mechanism that normally maintains homeostasis of blood glucose concentration consists of hormonal and neural devices. At least five endocrine glands—pancreatic islets, anterior pituitary gland, adrenal cortex, adrenal medulla, and thyroid gland—and at least eight hormones secreted by these glands function as key elements of the glucose homeostatic mechanism. As you read through the following paragraphs, keep in mind that each of the hormones listed have many different kinds of regulatory effects in the body—not just their effects on glucose metabolism.

Beta cells of the pancreatic islets secrete the most well-known sugar-regulating hormone of all—**insulin**. Insulin decreases blood glucose level by moving glucose molecules out of the blood and into cells (Figure 27-17). Although the exact details of its mechanism of action are still being worked out, insulin is known to accelerate glucose transport through cell membranes. It also increases the activity of the enzyme glucokinase. Glucokinase catalyzes glucose phosphorylation, the reaction that must occur before either glycogenesis or glucose catabolism can take place. By applying these facts, you can deduce the main methods by which insulin decreases blood glucose level: increased glycogenesis and increased catabolism of glucose. Figure 27-17 mentions these effects of insulin. By studying this figure, you can also de-

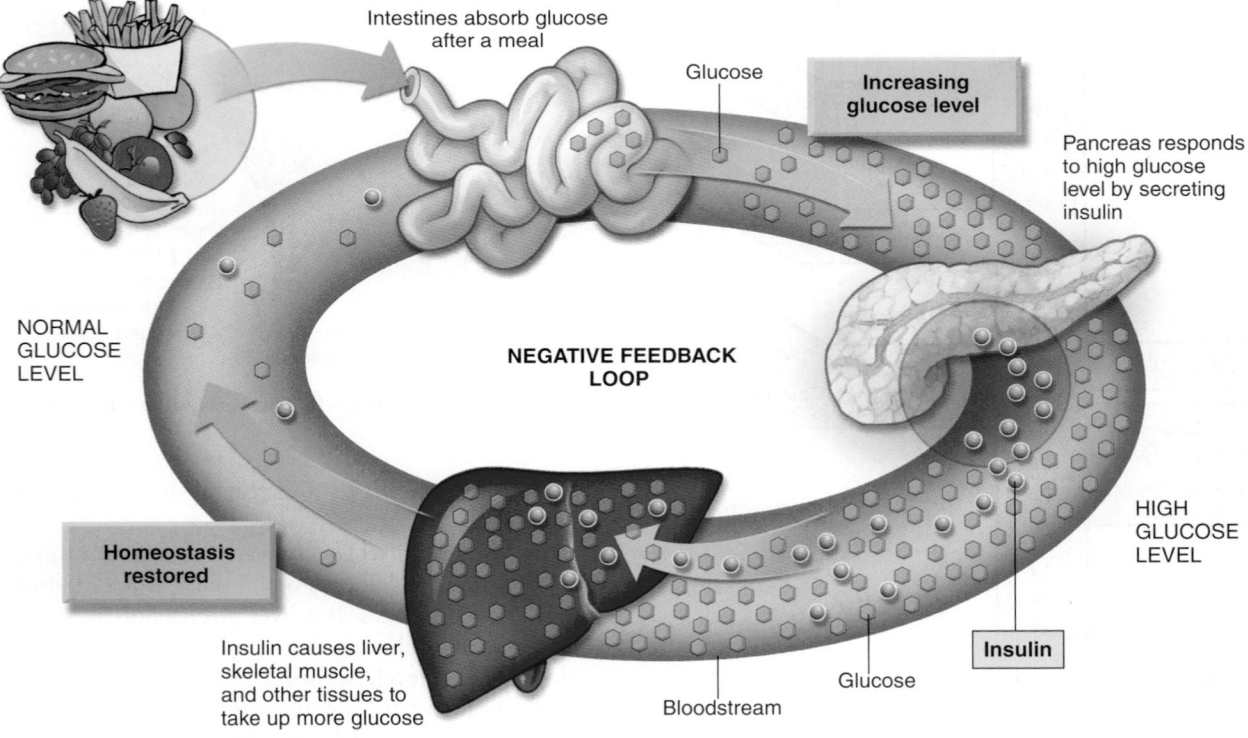

Figure 27-17 *Role of insulin.* Insulin operates in a negative feedback loop that prevents blood glucose concentration from increasing too far above the normal, or set point, level. Insulin promotes uptake of glucose by all cells of the body, enabling them to catabolize and/or store it. The liver and skeletal muscles are especially well adapted for storage of glucose as glycogen. Thus excess glucose is removed from the bloodstream. If the glucose level falls below the set point level (see Figure 27-14), hormones such as glucagon promote the release of glucose from storage into the bloodstream (see Figure 27-18).

duce some of the prominent metabolic defects resulting from insulin deficiency such as in diabetes mellitus. Slow glycogenesis and low glycogen storage, decreased glucose catabolism, and increased blood glucose all result from insulin deficiency.

Alpha cells of the pancreatic islets secrete the sugar-regulating hormone **glucagon.** Whereas insulin tends to decrease the blood glucose level, glucagon tends to increase it. Glucagon increases the activity of the enzyme phosphorylase (see Figure 27-15), and phosphorylase necessarily accelerates liver glycogenolysis and releases more of its product, glucose, into the bloodstream.

Epinephrine is a hormone secreted in large amounts by the adrenal medulla in times of emotional or physical stress. Like glucagon, epinephrine increases phosphorylase activity. This makes glycogenolysis occur at a faster rate. Epinephrine accelerates both liver and muscle glycogenolysis, whereas glucagon accelerates only liver glycogenolysis. Both hormones increase the blood glucose level. Epinephrine is the only hormone whose release into the systemic circulation (and therefore its effects on metabolism) is directly under the control of the nervous system.

Adrenocorticotropic hormone (ACTH) and *glucocorticoids* (e.g., cortisone) are two more hormones that increase blood glucose concentration. ACTH stimulates the adrenal cortex to increase its secretion of glucocorticoids. Glucocorticoids accelerate gluconeogenesis. They do this by mobilizing proteins, that is, the breakdown, or hydrolysis, of tissue proteins to amino acids. More amino acids enter the circulation and are carried to the liver. Liver cells step up their production of "new" glucose from the mobilized amino acids. Glucocorticoids help here, too, by stimulating enzymatic reversal of glycolysis and thus helping the cell in its effort to manufacture more glucose. The glucose streams out of liver cells into the blood and adds to the blood glucose level.

Growth hormone (GH), made by the anterior pituitary, also increases blood glucose level, but by a different mechanism. GH causes a shift from carbohydrate to fat catabolism. It does this by limiting the storage of fat in fat depots. Instead, more fats are mobilized and catabolized. In this way, GH "spares" carbohydrates from catabolism, and the level of glucose in the bloodstream is increased.

Thyroid-stimulating hormone (TSH) from the anterior pituitary gland and its target secretion, *thyroid hormone* (T_3 and T_4), have complex effects on metabolism. Some of these raise, and some lower, the glucose level. One of the effects of thyroid hormone is to accelerate catabolism, and because glucose is the body's "preferred fuel," the result may be a decrease in blood glucose level.

The summary of hormone control shown in Figure 27-18 indicates that most hormones cause the glucose blood level to rise. These hormones are called *hyperglycemic* because they tend to promote a high blood glucose concentration. The one notable exception is insulin, which is *hypoglycemic* (tends to decrease the blood glucose level). See Box 27-6 for more discussion of blood glucose problems.

In looking back over the cells' accomplishments in catabolism, we are struck by a sense of wonder, well expressed by Wayne Becker*: "Respiratory metabolism can be regarded as a marvel of design and engineering. No transistors, no mechanical parts, no noise, no pollution—and all done in units of organization that require an electron microscope to visualize. Yet the process goes on routinely and continuously in almost every living cell with a degree of integration, efficiency, fidelity, and control that we can scarcely understand well enough to appreciate, let alone aspire to reproduce in our test tubes."

*From Becker WM: *Energy and the living cell*, Philadelphia, 1977, JB Lippincott.

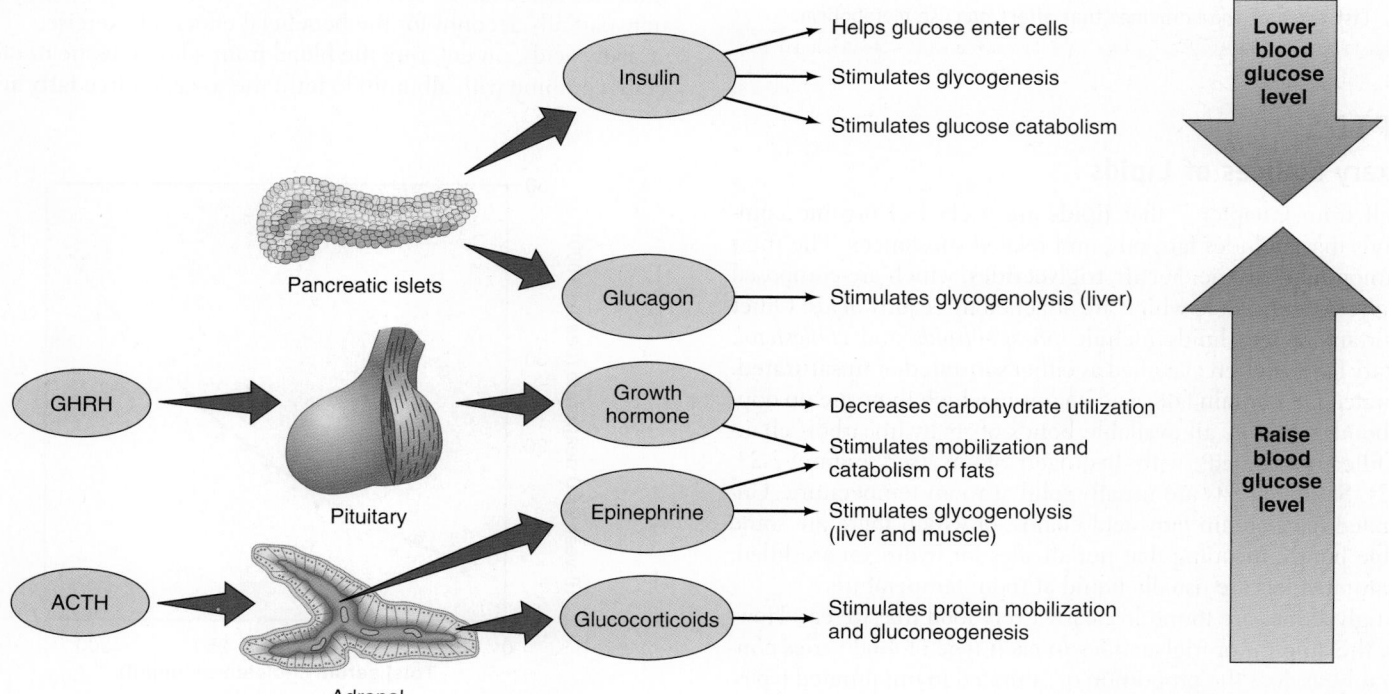

Figure 27-18 *Hormonal control of blood glucose level.* Insulin lowers blood glucose level and is therefore hypoglycemic. Most hormones raise blood glucose level and are called hyperglycemic, or anti-insulin, hormones.

BOX 27-6: FYI
Abnormal Blood Glucose Concentration

The term **hyperglycemia,** which literally means "condition of too much sugar in the blood," is used any time the blood glucose concentration becomes higher than the normal set point level. Hyperglycemia is most frequently associated with untreated diabetes mellitus, but it can occur in newborns when too much intravenous glucose is given or in other situations. If untreated, the excess glucose leaves the blood in the kidney—literally "spilling over" into the urine. This increases the osmotic pressure of urine, drawing an abnormally high amount of water into the urine from the bloodstream. Thus, hyperglycemia causes loss of glucose in the urine and its accompanying loss of water—potentially threatening the fluid balance of the body. Dehydration of this sort can ultimately lead to death.

Hypoglycemia occurs when the blood glucose concentration dips below the normal set point level. Hypoglycemia can occur in various conditions, including starvation, hypersecretion of insulin by the pancreatic islets, or injection of too much insulin. Symptoms of hypoglycemia include weakness, hunger, headache, blurry vision, anxiety, and personality changes—perhaps leading to coma and death if untreated.

QUICK CHECK

6. Why might a cell switch to anaerobic respiration as its major source of usable energy?
7. What is meant by the term *oxygen debt?*
8. Distinguish between glycogenesis and glycogenolysis. Under what circumstances might each occur?
9. List three of the hormones that affect glucose metabolism.

LIPIDS
Dietary Sources of Lipids

Recall from Chapter 2 that **lipids** are a class of organic compounds that includes fats, oils, and related substances. The most common lipids in the diet are **triglycerides,** which are composed of a *glycerol* subunit to which are attached three *fatty acids*. Other important dietary lipids include *phospholipids* and *cholesterol*. Dietary fats are often classified as either **saturated** or **unsaturated.** Saturated fats contain fatty acid chains in which there are no double bonds—that is, all available bonds of its hydrocarbon chain are filled (saturated) with hydrogen atoms (see Figure 2-23, p. 62). Saturated fats are usually solid at room temperature. Unsaturated fats contain fatty acid chains, of which there are some double bonds, meaning that not all sites for hydrogen are filled. Unsaturated fats are usually liquid at room temperature.

Triglycerides are found in nearly every food that we eat. However, the amount of triglycerides in each type of food varies considerably, as does the proportion of saturated to unsaturated types. Phospholipids are also found in nearly all foods because they make up the cellular membranes in and around each cell of all living organisms. Cholesterol, however, is found only in foods of

animal origin. Cholesterol concentration also varies. For example, it is particularly high in liver and the yolks of eggs.

Transport of Lipids

Lipids are transported in blood as chylomicrons, lipoproteins, and free fatty acids. **Chylomicrons** are small fat droplets present in blood after fat absorption. Fatty acids and monoglyceride products of fat digestion combine during absorption to again form fats (triglycerides, or triacylglycerols). These triglycerides plus small amounts of cholesterol and phospholipids compose the chylomicrons. During fat absorption, the so-called *absorptive state*, blood may contain so many of these fat droplets that it appears turbid or even yellowish in color. But during the *postabsorptive state*—usually within about 4 hours after a meal—few, if any chylomicrons remain in the blood. Their contents have moved mostly into adipose tissue cells.

In the postabsorptive state, when chylomicrons are virtually absent from the circulation, some 95% of the lipids in blood are transported in the form of lipoproteins. **Lipoproteins** are produced mainly in the liver, and as their name suggests, they consist of lipids (triglycerides, cholesterol, and phospholipids) and protein. At all times, blood contains three types of lipoproteins, namely, very-low-density lipoproteins, low-density lipoproteins, and high-density lipoproteins. Usually, they are designated by their abbreviations: VLDL, LDL, and HDL. Diets high in saturated fats and cholesterol tend to produce an increase in blood LDL concentration, which in turn is associated with a high incidence of coronary artery disease (CAD) and atherosclerosis (Figure 27-19 and Box 27-7). A high blood HDL concentration, in contrast, is associated with a low incidence of heart disease. One might therefore think of the LDLs as the "bad lipoproteins" and the HDLs as the "good lipoproteins." Considerable evidence indicates that exercise tends to elevate HDL concentration. This may partially account for the beneficial effects of exercise.

Fatty acids, on entering the blood from adipose tissue or other cells, combine with albumin to form the so-called **free fatty acids**

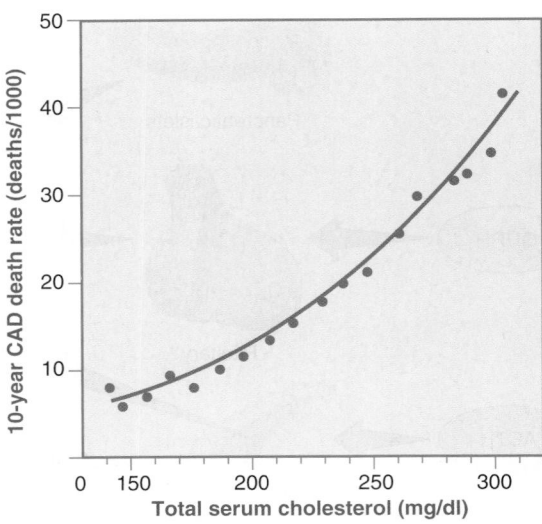

Figure 27-19 *Cholesterol and heart disease.* The graph shows a relationship between the total serum (blood plasma) cholesterol level and coronary artery disease (CAD).

Box 27-7   **Lipoproteins**

As stated in the text, high blood concentrations of low-density lipoproteins (LDLs) are associated with a high risk for atherosclerosis. Atherosclerosis is a form of "hardening of the arteries" that occurs when lipids accumulate in cells lining the blood vessels and promote the development of a plaque that eventually impedes blood flow and may trigger clot formation. Atherosclerosis may also weaken the wall of a blood vessel to the point that it ruptures. In any case, a person with atherosclerosis of the coronary arteries risks a heart attack when blood flow to cardiac muscle is impaired. If vessels in the brain are affected, there is risk of a *cerebrovascular accident (CVA),* or "stroke." Part *A* of the figure is a simplified version of a concept of LDLs function first proposed by Nobel laureates Michael Brown and Joseph Goldstein at the University of Texas. According to their model, LDL delivers cholesterol to cells for use in synthesizing steroid hormones and stabilizing the plasma membrane. Most, if not all, cells have many LDL receptors embedded in the outer surface of their plasma membranes. These receptors attract cholesterol-bearing LDL. Once the LDL molecule binds to the receptor, specific mechanisms operate to release the cholesterol it carries into the cell. Excess cholesterol is stored in droplets near the center of the cell. It seems that, in some individuals at least, cells have so few LDL receptors that they accumulate too much cholesterol in the blood. Some mechanism in endothelial cells moves this excess LDL into the wall of blood vessels.

This has been proposed as a cause for the lipid accumulation characteristic of atherosclerosis.

High blood concentrations of high-density lipoproteins (HDLs) have been associated with a low risk of developing atherosclerosis and its many possible complications. Although the exact details of how this works have yet to be worked out or confirmed, some scientists have made some progress. Jack Oram, a cell biologist working at the University of Washington, has proposed the mechanism illustrated in part *B* of the figure. According to his model, HDL molecules are attracted to HDL receptors embedded in the plasma membranes. Once they bind to their receptors, the cell is stimulated to release some of its cholesterol from storage. The released cholesterol migrates to the plasma membrane, where it may attach to the HDL molecule and be whisked away to the liver for excretion in bile.

Apparently, high blood LDL levels (more than 180 mg LDL per 100 ml of blood) signify that a large amount of cholesterol is being delivered to cells. High blood HDL levels (more than 60 mg HDL per 100 ml of blood) apparently indicate that a large amount of cholesterol is being removed from cells and delivered to the liver for excretion from the body. Currently, researchers are using this information to develop treatments that may prevent—or even cure—atherosclerosis and the disorders it causes.

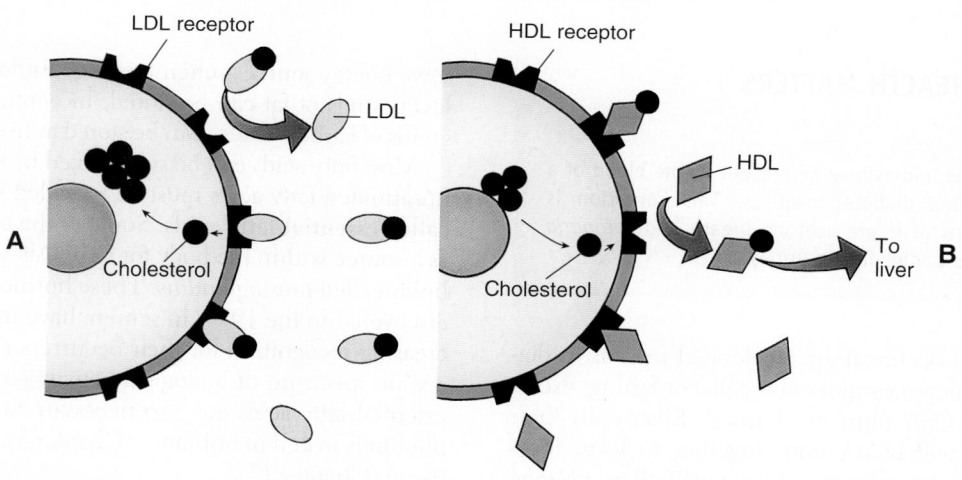

Role of blood lipoproteins. **A,** Simplified diagram of the role of low-density lipoprotein (LDL) in delivering cholesterol to cells. **B,** Proposed role of high-density lipoprotein (HDL) in removing cholesterol from cells.

(FFAs). Fatty acids are transported from cells of one tissue to those of another in the form of free fatty acids. Whenever the rate of fat catabolism increases—as it does in starvation or diabetes—the free fatty acid content of blood increases markedly.

Lipid Metabolism

Lipid Catabolism

Lipid catabolism, like carbohydrate catabolism, consists of several processes. Each of these processes, in turn, consists of a series of chemical reactions. Triglycerides are first hydrolyzed to yield fatty acids and glycerol. Glycerol is then converted to glyceraldehyde-3-phosphate, which enters the glycolysis pathway (see Figure 27-5). Fatty acids, as Figure 27-20 shows, are broken down by a process called *beta-oxidation* into two-carbon pieces, the familiar acetyl-CoA. These molecules are then catabolized via the citric acid cycle. The final process of lipid catabolism therefore consists of the same reactions as carbohydrate catabolism. Catabolism of lipids, however, yields considerably more energy than catabolism of carbohydrates does. Whereas catabolism of 1 g of carbohydrates yields only 4.1 kcal of heat, catabolism of 1 g of fat yields 9 kcal. It is not surprising, then, that lipids are the preferred energy source for muscle tissue.

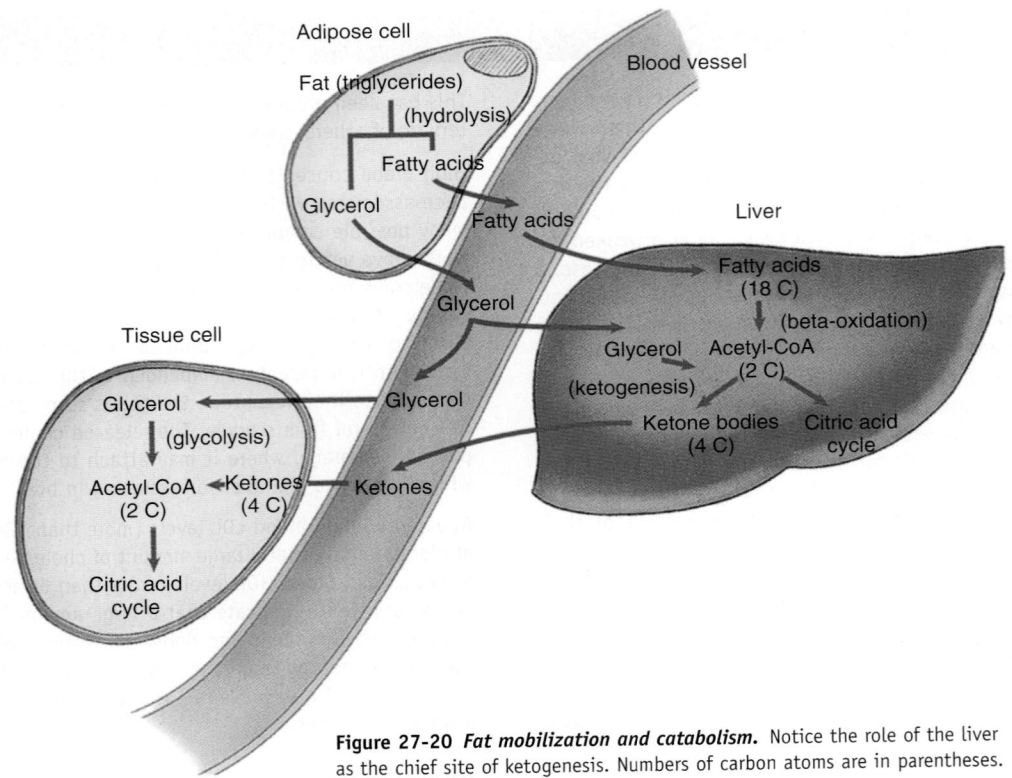

Figure 27-20 *Fat mobilization and catabolism.* Notice the role of the liver as the chief site of ketogenesis. Numbers of carbon atoms are in parentheses.

 BOX 27-8: HEALTH MATTERS

Ketosis

Large amounts of ketone bodies may be present in the blood of a person with uncontrolled diabetes mellitus. This condition is known as **ketosis.** Signs of it are acetone breath and *ketonuria* (high number of ketone bodies in the urine).

When fat catabolism occurs at an accelerated rate, as in diabetes mellitus (when glucose cannot enter cells) or fasting, excessive numbers of acetyl-CoA units are formed. Liver cells then temporarily condense acetyl-CoA units together to form four-carbon acetoacetic acid. Acetoacetic acid is classified as a *ketone body* and can be converted to two other types of ketone bodies, namely, acetone and beta-hydroxybutyric acid—hence the name **ketogenesis** for this process (Box 27-8). Liver cells oxidize a small portion of the ketone bodies for their own energy needs, but most of them are transported by the blood to other tissue cells for the change back to acetyl-CoA and oxidation via the citric acid cycle (see Figure 27-20).

Lipid Anabolism

Lipid anabolism, also called **lipogenesis,** consists of the synthesis of various types of lipids, notably *triglycerides, cholesterol, phospholipids,* and *prostaglandins.* Triglycerides and structural lipids (e.g., phospholipids) are synthesized from fatty acids and glycerol or from excess glucose or amino acids. So it is possible to "get fat" from foods other than fat. Triglycerides are stored mainly in adipose tissue cells. These fat depots constitute the body's largest re-

serve energy source—often too large, unfortunately. Almost limitless pounds of fat can be stored. In contrast, only a few hundred grams of carbohydrates can be stored as liver and muscle glycogen.

Most fatty acids can be synthesized by the body. Certain of the unsaturated fatty acids must be provided by the diet and are thus called **essential fatty acids.** Some of the essential fatty acids serve as a source within the body for synthesis of an important group of lipids called *prostaglandins.* These hormonelike compounds, first discovered in the 1930s in semen, have in recent years gained increasing recognition for their occurrence in various tissues, with a wide spectrum of biological activity (see Chapter 16). Certain essential fatty acids are also necessary to manufacture the phospholipids in cell membranes (Chapter 3) and the myelin in nerve tissue (Chapter 12).

Control of Lipid Metabolism

Lipid metabolism is controlled mainly by the following hormones:

- Insulin
- ACTH
- Growth hormone
- Glucocorticoids

You probably recall from our discussion of these hormones in connection with carbohydrate metabolism that they regulate fat metabolism in such a way that the rate of fat catabolism is inversely related to the rate of carbohydrate catabolism. If some condition such as diabetes mellitus causes carbohydrate catabolism to decrease below energy needs, increased secretion of growth hormone, ACTH, and glucocorticoids soon follows.

BOX 27-9

BOX 27-9 PPARs and Fat Metabolism

A class of proteins called *peroxisome proliferator–activated receptors,* or *PPARs,* may hold some keys to how the body metabolizes fats. PPARs are molecules attached to the DNA molecules of cells. PPARs combine with other proteins to form a complex that regulates genes that determine how the cell takes in and breaks down fat. Initial research shows that chemicals that affect different PPARs can influence fat storage in the body. With more than 60% of the U.S. population overweight and therefore at risk for heart disease, diabetes, cancer, and other health problems, it is no wonder that scientists are scrambling to find out more about how PPARs regulate fat metabolism and how we can influence them to reduce obesity-related health risks.

Table 27-1 Amino Acids

ESSENTIAL (INDISPENSABLE)	NONESSENTIAL (DISPENSABLE)
Histidine*	Alanine
Isoleucine	Arginine
Leucine	Asparagine
Lysine	Aspartic acid
Methionine	Cysteine
Phenylalanine	Glutamic acid
Threonine	Glutamine
Tryptophan	Glycine
Valine	Proline
	Serine
	Tyrosine†

*Essential in infants and, perhaps, adult males.
†Can be synthesized from phenylalanine; therefore nonessential as long as phenylalanine is in the diet.

These hormones, in turn, bring about an increase in fat catabolism. But, when carbohydrate catabolism equals energy needs, fats are not mobilized out of storage and catabolized (Box 27-9). Instead, they are spared and stored in adipose tissue. "Carbohydrates have a 'fat-sparing' effect," so says an old physiological maxim. Or stating this truth more descriptively: "Carbohydrates have a 'fat-storing' effect."

Research has shown that hormonal control of lipid metabolism is very complex—and still not well understood. One intriguing line of research involves the hormone **leptin.** Leptin is secreted by fat-storing cells and seems to regulate *satiety* (feeling of hunger/fullness) and how fat is metabolized. Researchers studying the complex interaction of leptin and other hormones and their receptors hope to find effective treatments for obesity, diabetes, and other fat-storage afflictions.

QUICK CHECK

10. In what forms are lipids transported to cells?
11. How can glycerol and fatty acids enter the citric acid cycle?
12. Which fatty acids cannot be made by the body?

PROTEINS

Sources of Proteins

Recall from Chapter 2 that proteins are very large molecules composed of chemical subunits called **amino acids** (see Figure 2-22, p. 60). Proteins are assembled from a pool of 20 different kinds of amino acids. If any one type of amino acid is deficient, vital proteins cannot be synthesized—a serious health threat. One way your body maintains a constant supply of amino acids is by synthesizing them from other compounds already present in the body. Only about half of the required 20 types of amino acids can be made by the body, however. The remaining types of amino acids must be supplied in the diet. Nutritionists often refer to the amino acids that must be in the diet as essential, or indispensable, amino acids. Table 27-1 lists amino acids according to whether they are considered essential in the diet or nonessential (dispensable) in the diet (synthesized by the body).

BOX 27-10: HEALTH MATTERS
Amino Acids and Disease

Recent research shows that the balance of amino acids circulating in the blood is associated with various diseases. High blood levels of homocysteine, one of several alternate forms of the amino acid cysteine (see Table 27-1), have been linked to heart disease, stroke, and dementias such as Alzheimer disease. Whether such abnormalities in homocysteine levels are the direct cause of these conditions is uncertain. Despite this uncertainty, many physicians recommend lowering abnormally high blood homocysteine levels to reduce the possible risk for these devastating conditions. Homocysteine can be reduced to normally low blood levels when there is adequate vitamin B_6, B_{12}, or B_9 (folic acid) in the diet.

Box 27-10 mentions the link between blood levels of amino acids and disease.

Proteins are obtained in the diet from various sources. Muscle meat and other animal tissues particularly high in proteins contain the essential amino acids. Food from a single plant or other nonanimal source does not usually contain an adequate amount of all the essential amino acids. Therefore, it is important to include meat in the diet or a mixture of different vegetables that provide all the amino acids needed by the body. Plant tissues that are particularly high in protein content include cereal grains, nuts, and legumes such as peas and beans.

Protein Metabolism

In protein metabolism, anabolism is primary and catabolism is secondary. In carbohydrate and fat metabolism, the opposite is true—catabolism is primary and anabolism is secondary. Proteins are primarily tissue-building foods. Carbohydrates and fats are primarily energy-supplying foods.

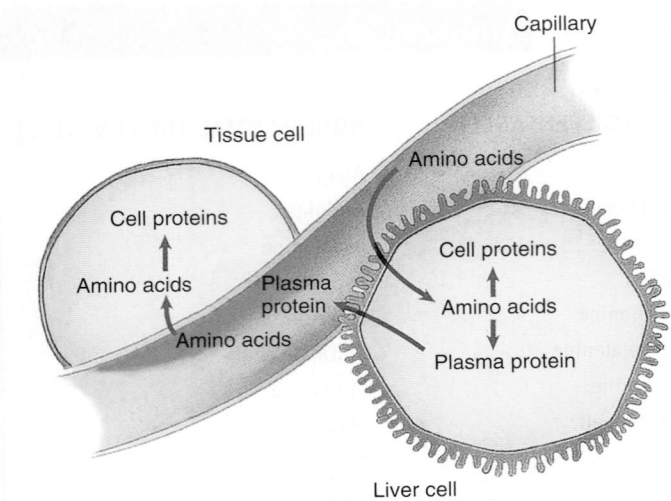

Figure 27-21 *Protein anabolism.* Proteins are synthesized from amino acids transported to each cell by the blood. Notice that the liver cell—like a few other specialized cells in the human body—makes some proteins for export.

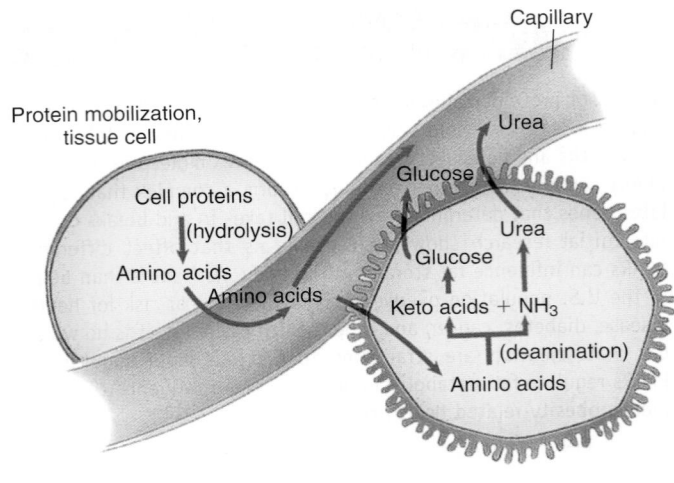

Figure 27-22 *Protein mobilization and catabolism.* Glucocorticoids tend to accelerate these processes and are therefore classified as protein catabolic hormones.

Protein Anabolism

Protein anabolism is the process by which proteins are synthesized by the ribosomes of all cells. The specific mechanisms of protein anabolism were first discussed in Chapter 2 (see pp. 53-59) and further explained in Chapter 4 (see pp. 120-125). We review the basic idea of protein anabolism here.

Every cell synthesizes its own structural proteins and its own enzymes. In addition, many cells, such as liver and glandular cells, synthesize special proteins for export. For example, liver cells manufacture the plasma proteins found in blood (Figure 27-21). The cell's genes, under the influence of signaling mechanisms, determine the specific proteins to be synthesized. Protein anabolism is truly "big business" in the body. Consider, for instance, that protein anabolism constitutes the major process of growth, reproduction, tissue repair, and the replacement of cells destroyed by daily wear and tear. Red blood cell replacement alone runs into millions of cells per second!

Protein Catabolism

The first step in protein catabolism takes place in liver cells. Called **deamination**, it consists of the splitting off of an amino (NH_2) group from an amino acid molecule to form a molecule of ammonia and one of keto acid (e.g., alpha-ketoglutaric acid). Most of the ammonia is converted by liver cells to *urea* and later excreted in the urine. The keto acid may be oxidized via the citric acid cycle (see Figure 27-8) or may be converted to glucose via gluconeogenesis (Figure 27-22) or to fat (lipogenesis). Both protein catabolism and anabolism go on continually. Only their rates differ from time to time. With a protein-deficient diet, for example, protein catabolism exceeds protein anabolism. Various hormones, as we shall see, also influence the rates of protein catabolism and anabolism.

Protein Balance and Nitrogen Balance

Usually a state of **protein balance** exists in the normal healthy adult body; that is, the rate of protein anabolism equals or balances the rate of protein catabolism. When the body is in protein balance, it is also in a state of **nitrogen balance** because the amount of nitrogen taken into the body (in protein foods) equals the amount of nitrogen in protein catabolic waste products excreted in the urine, feces, and sweat.

It is important to realize that there are two kinds of protein, or nitrogen, imbalance. When protein catabolism exceeds protein anabolism, the amount of nitrogen in the urine exceeds the amount of nitrogen in the protein foods ingested. The individual is then said to be in a state of **negative nitrogen balance,** or in a state of "tissue wasting"—because more of the tissue proteins are being catabolized than are being replaced by protein synthesis. Protein-poor diets, starvation, and wasting illnesses, for example, produce a negative nitrogen balance. A **positive nitrogen balance** (nitrogen intake in foods greater than nitrogen output in urine) indicates that protein anabolism is occurring at a faster rate than protein catabolism. A state of positive nitrogen balance therefore characterizes any condition in which large amounts of tissue are being synthesized, such as during growth, pregnancy, and convalescence from an emaciating illness.

Control of Protein Metabolism

Protein metabolism, like that of carbohydrates and fats, is controlled largely by hormones rather than by the nervous system. Growth hormone and the male hormone testosterone both have a stimulating effect on protein synthesis, or anabolism. For this reason, they are referred to as *anabolic* hormones. The protein *catabolic* hormones of greatest consequence are glucocorticoids. They speed up tissue protein mobilization, that is, the hydrolysis of cell proteins to amino acids, their entry into the blood, and their subsequent catabolism (see Figure 27-22). ACTH functions indirectly as a protein catabolic hormone because of its stimulating effect on glucocorticoid secretion.

Thyroid hormone is necessary for and tends to promote protein anabolism and therefore growth when plenty of carbohydrates and fats are available for energy production. On the other

Table 27-2   Metabolism

NUTRIENT	ANABOLISM	CATABOLISM
Carbohydrates	Temporary excess changed into glycogen by liver cells in presence of insulin; stored in liver and skeletal muscles until needed and then changed back to glucose	Oxidized, in presence of insulin, to yield energy (4.1 kcal per g) and wastes (carbon dioxide and water)
	True excess beyond body's energy requirements converted into adipose tissue; stored in various fat depots of body	$C_6H_{12}O_6 \times 6\ O_2 \rightarrow$ Energy $+\ 6\ CO_2 + 6\ H_2O$
Fats	Built into adipose tissue; stored in fat depots of body	Fatty acids $\qquad\qquad$ Glycerol $\downarrow$ (beta-oxidation) $\qquad$ $\downarrow$ (glycolysis) Acetyl-CoA $\rightleftharpoons$ Ketones $\qquad$ Acetyl-CoA $\downarrow$ (tissues; citric acid cycle) Energy (9.3 kcal per g) $+\ CO_2 + H_2O$
Proteins	Synthesized into tissue proteins, blood proteins, enzymes, hormones, etc.	Deaminated by liver, forming ammonia (which is converted to urea) and keto acids (which are either oxidized or changed to glucose or fat)

hand, under different conditions, for example, when the amount of thyroid hormone is excessive or when the energy foods are deficient, this hormone may then promote protein mobilization and catabolism.

Some of the facts about metabolism set forth in the preceding sections are summarized in Table 27-2 and Figure 27-23.

QUICK CHECK

13. What is meant by the term *essential amino acid*?
14. What happens when an amino acid is deaminated?
15. What is the purpose of the process of amino acid deamination?
16. What is meant by the term *nitrogen balance*?

VITAMINS AND MINERALS

One glance at the label of any packaged food product reveals the importance we place on vitamins and minerals. We know that carbohydrates, fats, and proteins are used by our bodies to build important molecules and to provide energy. So why do we need vitamins and minerals?

Vitamins

Vitamins are organic molecules needed in small quantities for normal metabolism throughout the body (Box 27-11). The first vitamins were discovered in the early twentieth century at the University of Wisconsin by Marguerite Davis and Elmer Vernon McCollum. Earlier, vitamins were hypothesized to be amines so they were originally called vitamines, later shortened to vitamins. However, Davis and McCollum discovered two fat-soluble, non-amine substances that fit the proposed function of a vitamin and

BOX 27-11 SPORTS AND FITNESS
Vitamin Supplements for Athletes

Because a deficiency of vitamins (*avitaminosis*) can cause poor athletic performance, many athletes regularly consume vitamin supplements. However, research suggests that vitamin supplementation has little or no effect on a person's athletic performance. A reasonably well balanced diet supplies more than enough vitamins for even the elite athlete. The use of vitamin supplements therefore has fueled controversy among exercise experts. Opponents of vitamin supplements cite the cost and the possibility of liver damage associated with some forms of *hypervitaminosis*, whereas supporters cite the benefit of protecting against vitamin deficiency.

they named these factors fat-soluble A and fat-soluble B. This not only began a system of naming vitamins with letters but also spurred the whole science of nutrition.

Most vitamin molecules attach to enzymes or coenzymes and help them work properly. **Coenzymes** are organic, nonprotein catalysts that often act as "molecule carriers." Many enzymes or coenzymes are totally useless without the appropriate vitamins to attach to them and thus give them the shape that allows them to function properly. For example, coenzyme A (CoA), an important carrier molecule associated with the citric acid cycle, has *pantothenic acid* (vitamin B₅) as one of its major components. Roles of some vitamins in the metabolic pathways described earlier in this chapter are shown in Figure 27-24.

Not all vitamins are involved directly with enzymes and coenzymes. Vitamins A, D, and E play a variety of different, but no less important roles in the chemistry of the body. The form of vitamin A called *retinal*, for example, plays an important role in detecting light in sensory cells of the retina (see Figure 15-24 on p. 578). Vitamin D can be converted to the hormone **calcitriol**, which plays

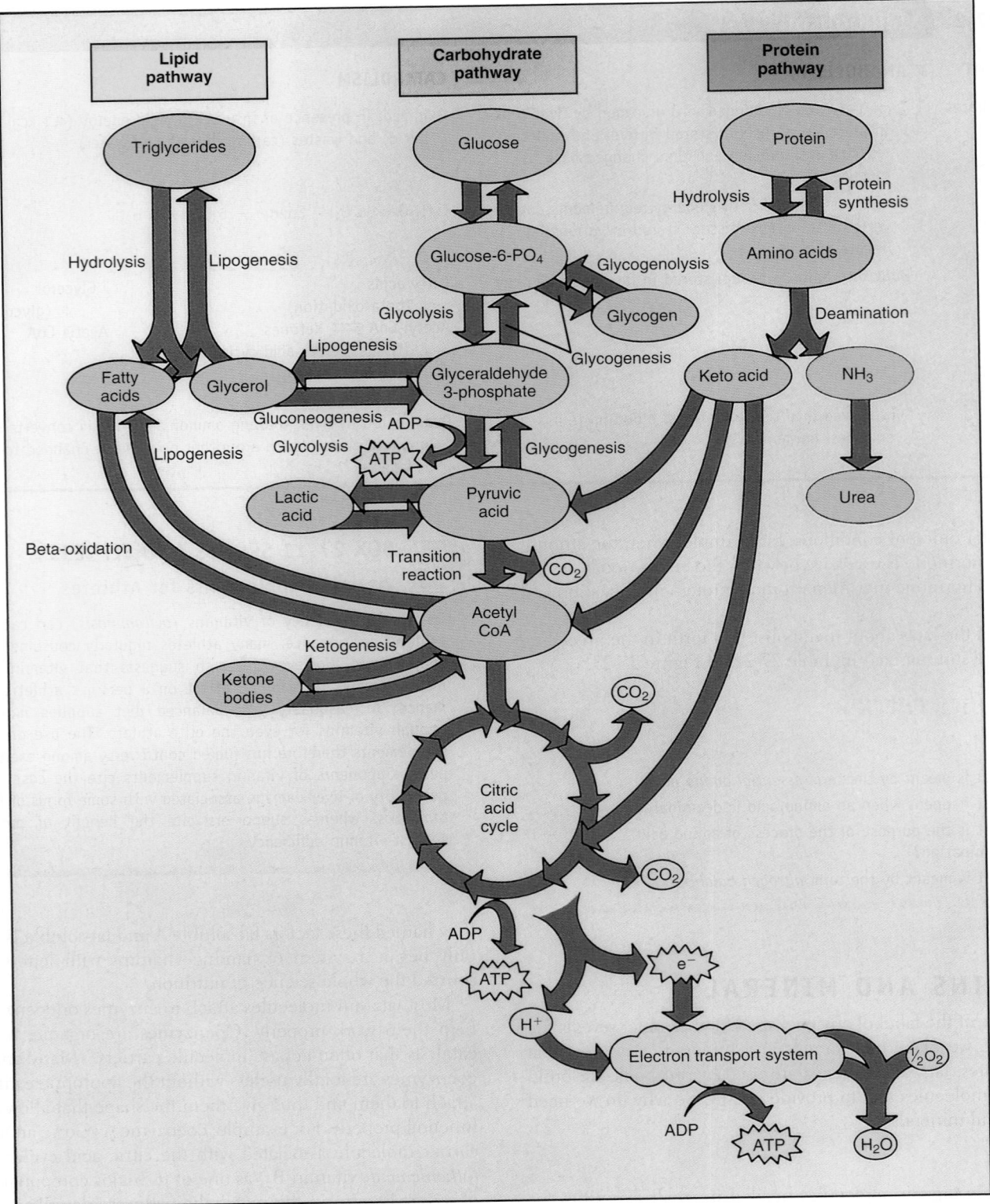

Figure 27-23 *Summary of metabolism.* Notice the central role played by the citric acid cycle and electron transport system. Notice also how different molecules can be converted to forms that may enter other pathways.

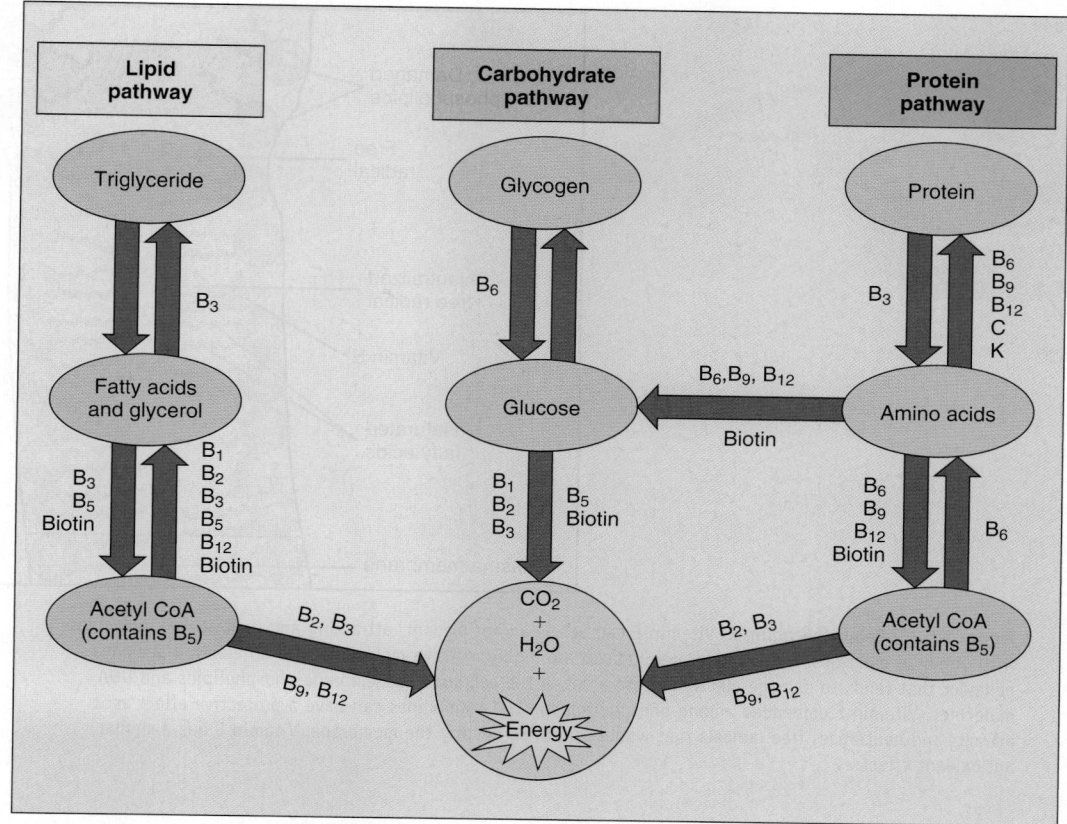

Figure 27-24 *Role of vitamins in metabolic pathways.* Many vitamins attach to enzymes or coenzymes, helping them to work properly. Notice how many different vitamins are needed to run these important metabolic pathways. A deficiency of any one vitamin could disrupt the whole set of pathways. Beta-oxidation of fatty acids also occurs in skeletal muscle and other tissues, too, but it is mainly the liver that releases ketones for use by other cells (see Figure 27-20 on p. 1010).

a role in the regulation of calcium homeostasis in the body. One role of vitamin E is to serve as an **antioxidant** that prevents electron-seeking molecules such as free radicals from damaging electron-dense molecules in the cell membranes and DNA molecules. Figure 27-25 illustrates this function of vitamin E. Vitamin C may also serve as an antioxidant.

All but one vitamin, vitamin D, cannot be made by the body itself. Bacteria living in the colon make two more: vitamin K and biotin. We must eat vitamins, or molecules we can convert into vitamins, in our food to get the rest. The body can store fat-soluble vitamins—A, D, E, and K—in the liver for later use. Because the body cannot store significant amounts of water-soluble vitamins such as B vitamins and vitamin C, they must be continually supplied in the diet. Table 27-3 lists some of the better-known vitamins, their sources and functions, and symptoms of deficiency.

Minerals

Minerals are at least as important as vitamins in our diet. Minerals are inorganic elements or salts that are found naturally in the earth. Like vitamins, mineral ions can attach to enzymes or other organic molecules and help them work. Of course, minerals such as sodium, chloride, and potassium are essential in relatively large amounts for maintaining the fluid/ion composition of the internal fluid environment (see Chapter 29).

Minerals also function in various other vital chemical reactions. For example, sodium, calcium, and other minerals are required for nerve conduction and for contraction in muscle fibers. Without these minerals, the brain, heart, and respiratory tract would cease to function. Iron is needed to manufacture hemoglobin in red blood cells, and iodine is needed to make thyroid hormones T_3 and T_4. Calcium, phosphorus, and magnesium are required to build the strong structural components of the skeleton.

Information about some of the more important minerals is summarized in Table 27-4. Like vitamins, minerals are beneficial only when taken in the proper amounts. Some of the minerals listed in Table 27-4 are required in large amounts and some only in trace amounts. Any intake of minerals beyond or below the recommended amount may become unhealthy—perhaps even life threatening.

Recommended *adequate intakes (AIs)* of minerals can change over the life span. For example, Figure 27-26 shows that calcium intake should increase through childhood and remain high throughout adulthood. However, the same figures shows that actual intake of calcium among females in the U.S. tends to fall short during adulthood—thereby increasing the risk for osteoporosis and other disorders.

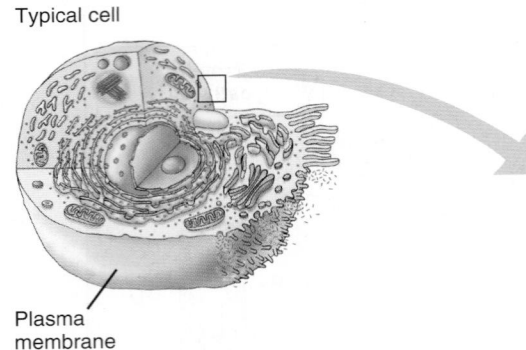

Typical cell

Plasma membrane

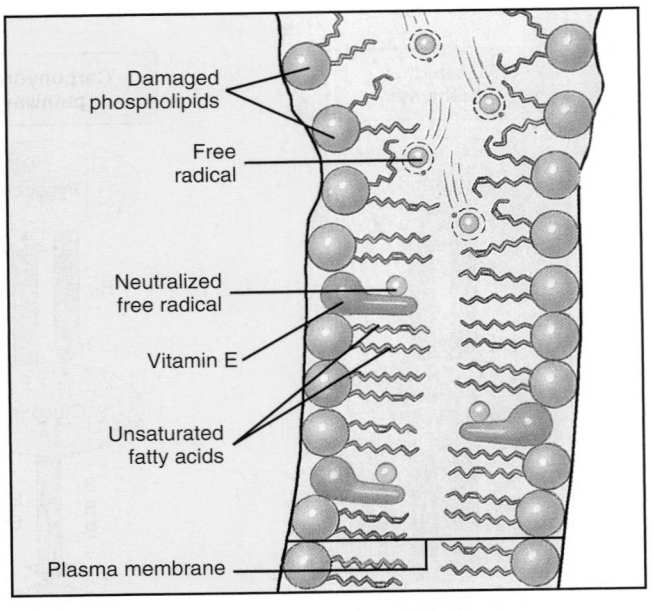

Damaged phospholipids

Free radical

Neutralized free radical

Vitamin E

Unsaturated fatty acids

Plasma membrane

Figure 27-25 *Role of vitamin E.* Vitamin E can act as an antioxidant, attracting and neutralizing molecules with unpaired electrons. For example, *free radicals* are highly reactive molecules with an electron-seeking behavior that tends to damage electron-dense areas of the cell such as membrane phospholipids and DNA molecules. Vitamin E embedded among phospholipids in cell membranes can have a protective effect as it attracts and neutralizes free radicals that would otherwise destroy the membrane. Vitamin C has a similar antioxidant effect.

Table 27-3 Major Vitamins

VITAMIN	DIETARY SOURCE	FUNCTIONS	SYMPTOMS OF DEFICIENCY
Vitamin A	Green and yellow vegetables, dairy products, and liver	Maintains epithelial tissue and produces visual pigments	Night blindness and flaking skin
B-Complex Vitamins			
B_1 (thiamine)	Grains, meat, and legumes	Helps enzymes in the citric acid cycle	Nerve problems (beriberi), heart muscle weakness, and edema
B_2 (riboflavin)	Green vegetables, organ meats, eggs, and dairy products	Aids enzymes in the citric acid cycle	Inflammation of skin and eyes
B_3 (niacin)	Meat and grains	Helps enzymes in the citric acid cycle	Pellagra (scaly dermatitis and mental disturbances) and nervous disorders
B_5 (pantothenic acid)	Organ meat, eggs, and liver	Aids enzymes that connect fat and carbohydrate metabolism	Loss of coordination (rare)
B_6 (pyridoxine)	Vegetables, meats, and grains	Helps enzymes that catabolize amino acids	Convulsions, irritability, and anemia
B_9 (folic acid)	Vegetables	Aids enzymes in amino acid catabolism and blood production	Digestive disorders and anemia
B_{12} (cyanocobalamin)	Meat and dairy products	Involved in blood production and other processes	Pernicious anemia
Biotin (vitamin H)	Vegetables, meat, and eggs	Helps enzymes in amino acid catabolism and fat and glycogen synthesis	Mental and muscle problems (rare)
Vitamin C (ascorbic acid)	Fruits and green vegetables	Helps in manufacture of collagen fibers; antioxidant	Scurvy and degeneration of skin, bone, and blood vessels
Vitamin D (calciferol)	Dairy products and fish liver oil	Aids in calcium absorption	Rickets and skeletal deformity
Vitamin E (tocopherol)	Green vegetables and seeds	Protects cell membranes from being destroyed; antioxidant	Muscle and reproductive disorders (rare)

Table 27-4   Major Minerals

MINERAL	DIETARY SOURCE	FUNCTIONS	SYMPTOMS OF DEFICIENCY
Calcium (Ca)	Dairy products, legumes, and vegetables	Helps blood clotting, bone formation, and nerve and muscle function	Bone degeneration and nerve and muscle malfunction
Chlorine (Cl)	Salty foods	Aids in stomach acid production and acid-base balance	Acid-base imbalance
Cobalt (Co)	Meat	Helps vitamin B_{12} in blood cell production	Pernicious anemia
Copper (Cu)	Seafood, organ meats, and legumes	Involved in extracting energy from the citric acid cycle and in blood production	Fatigue and anemia
Iodine (I)	Seafood and iodized salt	Required for thyroid hormone synthesis	Goiter (thyroid enlargement) and decrease in metabolic rate
Iron (Fe)	Meat, eggs, vegetables, and legumes	Involved in extracting energy from the citric acid cycle and in blood production	Fatigue and anemia
Magnesium (Mg)	Vegetables and grains	Helps many enzymes	Nerve disorders, blood vessel dilation, and heart rhythm problems
Manganese (Mn)	Vegetables, legumes, and grains	Helps many enzymes	Muscle and nerve disorders
Phosphorus (P)	Dairy products and meat	Aids in bone formation and is used to make ATP, DNA, RNA, and phospholipids	Bone degeneration and metabolic problems
Potassium (K)	Seafood, milk, fruit, and meats	Helps muscle and nerve function	Muscle weakness, heart problems, and nerve problems
Sodium (Na)	Salty foods	Aids in muscle and nerve function and fluid balance	Weakness and digestive upset
Zinc (Zn)	Many foods	Helps many enzymes	Inadequate growth

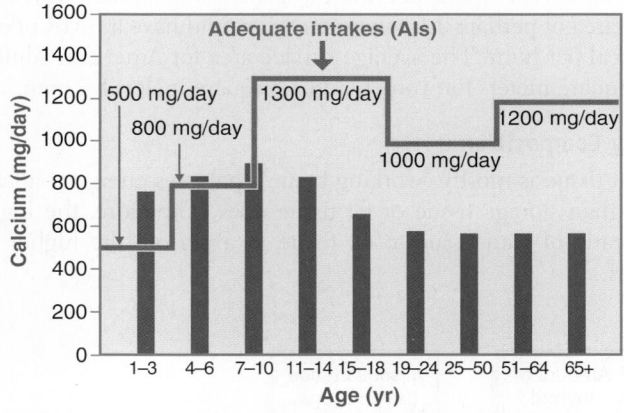

Figure 27-26 *Calcium intake in women.* The chart compares the recommended *adequate intake (AIs)* of calcium for women over the life span with the actual median intake of calcium among females in the United States.

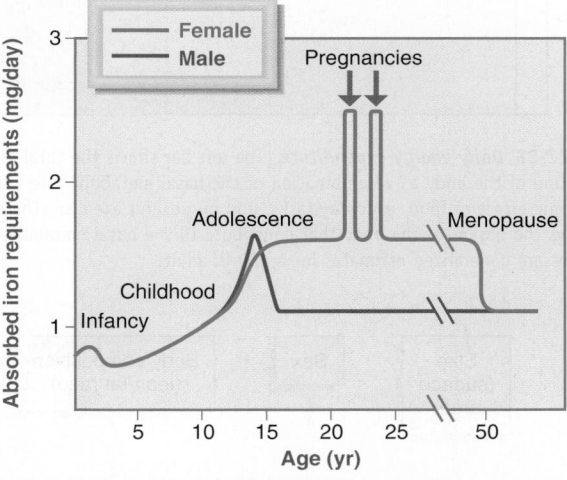

Figure 27-27 *Iron intake requirements.* The chart compares male and female absorbable iron requirements over the life span.

Figure 27-27 shows the requirement for iron over the life span for both men and women. Although both males and females require a large amount of iron during the spurt of growth in the teenage years, the iron requirement remains high only in women during the rest of adulthood. This difference is explained by the fact that adult women must continually replace the iron lost in the menstrual flow. Notice that female iron requirements drop to those of males after menopause. Notice also that the iron requirement peaks during pregnancies—when fetal blood development requires large amounts of iron.

 QUICK CHECK

17. What is a vitamin?

18. List two functions of minerals in the body.

METABOLIC RATES

Metabolic rate means the amount of energy released in the body in a given time by catabolism. It is the energy expended for accomplishing various kinds of work. In short, metabolic rate is the catabolic rate, or the rate of energy release.

Metabolic rates are expressed in either of two ways: (1) in terms of the number of kilocalories of heat energy expended per hour or per day or (2) as normal or as a definite percentage above or below normal.

Basal Metabolic Rate

The **basal metabolic rate (BMR)** is the body's rate of energy expenditure under "basal conditions," namely, when the individual

- Is awake but resting, that is, lying down and, as far as possible, not moving a muscle.
- Is in the postabsorptive state (12 to 18 hours after the last meal).
- Is in a comfortably warm environment (the so-called *thermoneutral zone*, a temperature range at which metabolism is independent of ambient temperature). Note that the BMR is not the minimum metabolic rate. It does not indicate the smallest amount of energy that must be expended to sustain life. It does, however, indicate the smallest amount of energy expenditure that can sustain life and also maintain the waking state and a normal body temperature in a comfortably warm environment.

Factors Influencing Basal Metabolic Rate

The BMR is not identical for all individuals because of the influence of various factors (Figures 27-28 and 27-29), some of which are described in the following paragraphs (Box 27-12).

Size

In computing the BMR, size is usually indicated by the amount of the body's surface area. It is computed from the individual's height and weight. A large individual has the same BMR as a small person per square meter of body surface, if other conditions are equal. However, because a large individual has more square meters of surface area, the BMR is greater than that of a small individual. For example, the BMR for a man in his 20s is about 40 kcal per square meter of body surface per hour (Table 27-5). However, a large man with a body surface area of 1.9 square meters would have a BMR of 76 kcal per hour, whereas a smaller man with a surface area of perhaps 1.6 square meters would have a BMR of only 64 kcal per hour. The average surface area for American adults is 1.6 square meters for women and 1.8 square meters for men.

Body Composition

Lean tissue is mostly "working tissue" that uses energy at a faster rate than storage tissue or fat tissue does. Therefore, the higher the ratio of lean tissue to fat tissue in a person, the higher the BMR.

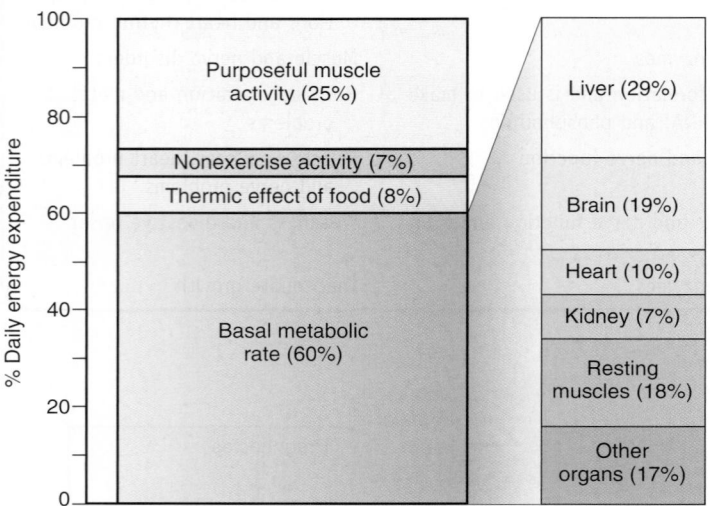

Figure 27-28 *Daily energy expenditure.* The left bar shows the total energy expenditure of the body as a combination of the basal metabolic rate (BMR), the thermic effect of food, everyday tasks, and purposeful exercise. The right bar shows the organs of the body that contribute to the basal metabolic rate. All values are generalized estimates for a 150-lb adult.

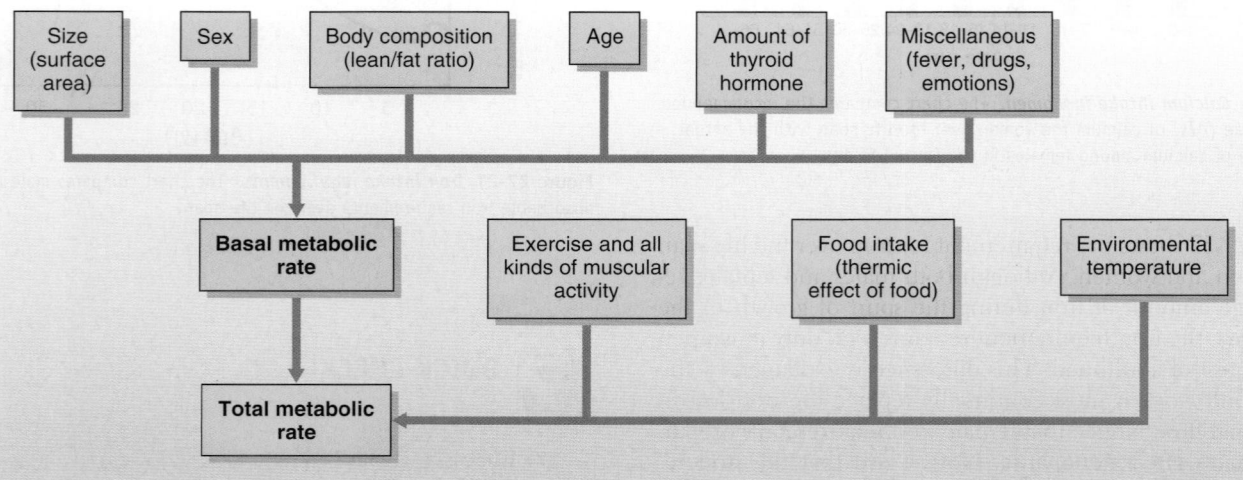

Figure 27-29 *Some factors that determine basal and total metabolic rates.*

BOX 27-12   How BMR Is Determined

BMR can be determined by a method called **indirect calorimetry.** The rationale underlying this method is the fact that BMR (expressed as the number of kilocalories of heat produced per unit of time) can be calculated from the amount of oxygen consumed in a given time (part *B* of the figure). BMR can then be expressed as normal, or as a definite percentage above or below normal, by dividing the actual kilocalorie rate by the known average kilocalorie rate for normal individuals of the same size, sex, and age.

Statistical tables, based on research, list estimated normal BMRs. (If BMR was calculated to be 10% above normal, for example, it would be reported as +10.) Are you curious to know the average BMR for a person of your size, sex, and age? If so, take the following steps:
1. Start with your weight in kilograms and your height in centimeters. (Convert pounds to kilograms by dividing pounds by 2.2. Convert inches to approximate centimeters by multiplying inches by 2.5.) For example, 110 pounds = 50 kg; 5 feet, 3 inches = 158 cm.
2. Convert your weight and height to square meters by using the chart shown here (part *B* of the figure). For example, a weight of 50 kg and height of 158 cm = about 1.5 square meters of body surface area.
3. Find your age and sex in Table 27-5 and then multiply the number of kilocalories per square meter per hour given there by your square meters of surface area and then by 24. For example, the average BMR per day for a 25-year-old woman with a weight of 110 pounds and height of 5 feet, 3 inches = 1332 kcal (37 × 1.5 × 24).

A quick rule of thumb for estimating a young woman's BMR in kcal/hr is to multiply her weight in pounds by 12.

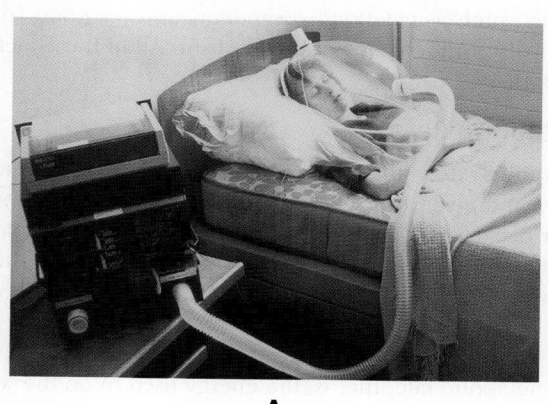

A

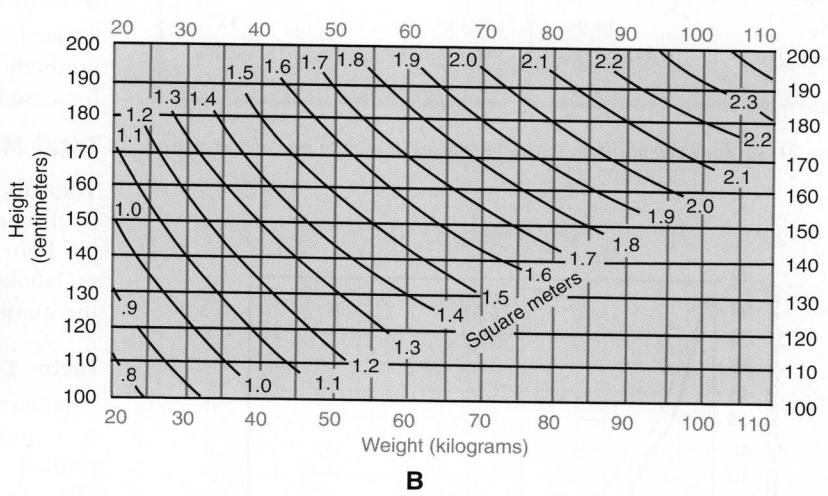

B

A, Measuring BMR. Indirect calorimetry calculates the rate of metabolism by measuring the rate at which oxygen is consumed. *B,* BMR estimate chart.

Table 27-5   Basal Metabolism

AGE (YR)	KILOCALORIES PER HOUR PER SQUARE METER BODY SURFACE	
	MALE	FEMALE
10-12	51.5	50.0
12-14	50.0	46.5
14-16	46.0	43.0
16-18	43.0	40.0
18-20	41.0	38.0
20-30	39.5	37.0
30-40	39.5	36.5
40-50	38.5	36.0
50-60	37.5	35.0
60-70	36.5	34.0

Sex

Men oxidize their food approximately 5% to 7% faster than women do. Therefore their BMRs are about 5% to 7% higher for a given size and age. A man 5 feet, 6 inches tall weighing 140 pounds, for example, has a 5% to 7% higher BMR than a woman of the same height, weight, and age. This gender difference in BMR probably results from the difference in the proportion of body fat determined by sex hormones. Women tend to have a higher percentage of body fat (and thus a lower total lean mass) than men do (Figure 27-30). Fat tissue is less metabolically active than lean tissues such as muscle. Differences in total lean mass not related to gender also affect the BMR. The relationship of BMR to sex and age is illustrated in Figure 27-31.

Age

That the fires of youth burn more brightly than those of age is a physiological and a psychological fact. In general, the younger the individual, the higher the BMR for a given size and sex (see Table 27-5). Exception: the BMR is slightly lower at birth than it

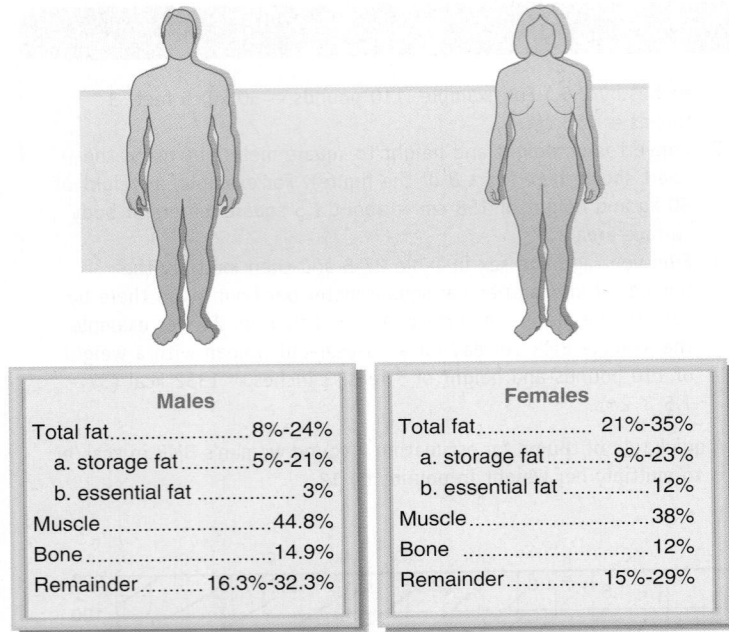

Figure 27-30 *Body composition.* Estimated values in healthy men and women.

Males	
Total fat	8%-24%
a. storage fat	5%-21%
b. essential fat	3%
Muscle	44.8%
Bone	14.9%
Remainder	16.3%-32.3%

Females	
Total fat	21%-35%
a. storage fat	9%-23%
b. essential fat	12%
Muscle	38%
Bone	12%
Remainder	15%-29%

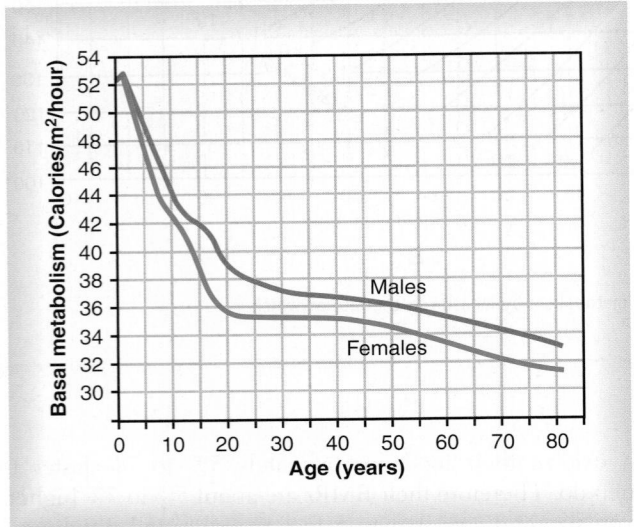

Figure 27-31 *Basal metabolic rates over the life span.* The chart shows normal basal metabolic rates (BMRs) as function of age for each sex.

is a few years later. That is to say, the rate increases slightly during the first 3 to 6 years, then starts to decrease and continues to do so throughout life.

Thyroid Hormone

Thyroid hormone (T_3 and T_4) stimulates basal metabolism. Without a normal amount of this hormone in the blood, a normal BMR cannot be maintained. When an excess of thyroid hormone is secreted, foods are catabolized faster, much as wood in a fireplace is burned faster when the draft is open. Deficient thyroid secretion, on the other hand, slows the rate of metabolism.

Body Temperature

Fever increases the BMR. For every degree Celsius increase in body temperature, metabolism increases about 13%. A decrease in body temperature (hypothermia) has the opposite effect. Metabolism decreases, and because it does, cells use less oxygen than they normally do. This knowledge has been applied clinically by using hypothermia in certain situations—for example, in open-heart surgery. Because circulation is reduced or interrupted during this procedure, oxygen supply necessarily decreases. Cells can tolerate this decreased oxygen supply reasonably well if their oxygen need has also decreased. Induced hypothermia decreases their rate of metabolism and thereby decreases their use of oxygen.

Drugs

Certain drugs, such as caffeine, amphetamine, and lebothyroxine, increase the BMR.

Other Factors

Other factors, such as *emotions*, *pregnancy*, and *lactation* (milk production), also influence basal metabolism. All of these factors increase the BMR.

Total Metabolic Rate

Total metabolic rate is the amount of energy used or expended by the body in a given time. It is often expressed in kilocalories per hour or per day. Most of the factors that determine the total metabolic rate are shown in Figures 27-28 and 27-29. Of these, the main direct determinants are the following:

Factor 1

The basal metabolic rate, that is, the energy used to do the work of maintaining life under the basal conditions previously described. Basal metabolic rate usually constitutes about 55% to 60% of the total metabolic rate.

Factor 2

The energy used to do all kinds of skeletal muscle work—from the simplest activities such as feeding oneself or sitting up in bed to the most strenuous kind of physical labor or exercise.

Factor 3

The *thermic effect* of foods. The metabolic rate increases for several hours after a meal, apparently because of the energy needed for metabolizing foods. Carbohydrates and fats have a thermic effect of about 5%. Proteins have a much higher thermic effect, about 30%. This means that, for 100 kcal of protein, 30 kcal are used for such processes as deamination and oxidation of the protein, leaving just 70 kcal available for other cell work. For this reason, proteins are "worth" fewer calories.

Energy Balance and Body Weight

When we say that the body maintains a state of energy balance, we mean that its energy input equals its energy output. Energy input per day equals the total calories (kilocalories) in the food ingested per day. Energy output equals the total metabolic rate expressed in kilocalories. You may be wondering what energy intake, output, and balance have to do with body weight.

"Everything" would be a fairly good one-word answer. Or, to be somewhat more explicit, the following basic principles describe the relationships between these factors:

- Body weight remains constant (except for possible variations in water content) when the body maintains energy balance—when the total calories in the food ingested equals the total metabolic rate. Example: If you have a total metabolic rate of 2000 kcal per day and if the food you eat per day yields 2000 kcal, your body will be maintaining energy balance and your weight will stay constant.
- Body weight increases when energy input exceeds energy output—when the total calories of food intake per day is greater than the total calories of the metabolic rate. A small amount of the excess energy input is used to synthesize glycogen for storage in the liver and muscles. The rest is used for synthesizing fat and storing it in adipose tissue. If you were to eat 3000 kcal each day for a week and if your total metabolic rate were 2000 kcal per day, you would gain weight. How much you would gain you can discover by doing a little simple arithmetic:

$$
\begin{aligned}
\text{Total energy input for week} &= 21{,}000 \text{ kcal} \\
\text{Total energy output for week} &= \underline{14{,}000 \text{ kcal}} \\
\text{Excess energy input for week} &= 7000 \text{ kcal}
\end{aligned}
$$

Approximately 3500 kcal are used to synthesize 1 pound of adipose tissue. Hence, at the end of this 1 week of "overeating"—of eating 7000 kcal over and above your total metabolic rate—you would have gained about 2 pounds.

- Body weight decreases when energy input is less than energy output—when the total number of calories in the food eaten is less than the total metabolic rate. Suppose you were to eat only 1000 kcal a day for a week and that you have a total metabolic rate of 2000 kcal per day. By the end of the week, your body would have used a total of 14,000 kcal of energy for maintaining life and doing its many kinds of work. All 14,000 kcal of this actual energy expenditure had to come from catabolism of foods because this is the body's only source of energy. Catabolism of ingested food supplied 7000 kcal, and catabolism of stored food supplied the remaining 7000 kcal. That week your body would not have maintained energy balance, nor would it have maintained weight balance. It would have incurred an energy deficit paid out of the energy stored in approximately 2 pounds of body fat. In short, you would have lost about 2 pounds.

Foods are stored primarily as glycogen and fats. Many cells catabolize them preferentially in this order: carbohydrates, then fats. (Skeletal muscle cells, however, seem to put fat first in their order of preference.) If there is no food intake, almost all of the glycogen is estimated to be used up in a matter of 1 or 2 days (Figure 27-32). Then, with no more carbohydrate to act as a fat sparer, fat is catabolized. How long it takes to deplete all of this reserve food depends, of course, on how much adipose tissue the individual has when starting the starvation diet. Finally, with no more fat available, tissue proteins are catabolized (see Figure 27-32). Because significant amounts of protein are not "stored" for use in

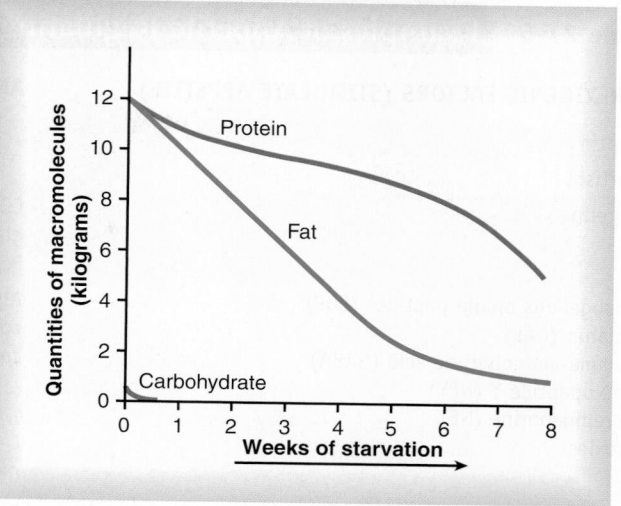

Figure 27-32 *Effects of starvation on the body.* Three major macromolecules serve as primary energy sources: carbohydrates, fats, and proteins. During starvation, the carbohydrate stores (glycogen) are rapidly depleted. However, stored lipids can mobilize and provide much of our energy needs for several weeks. Eventually, lipid stores run low and the body starts using proteins as a major source of energy—causing the breakdown of muscle and other protein-rich tissues. Muscle damage during starvation usually leads to death.

catabolism, this means that important structural and functional proteins are quickly depleted. Thus death soon ensues.

MECHANISMS FOR REGULATING FOOD INTAKE

Mechanisms for regulating food intake are still not clearly established. That the hypothalamus plays a part in these mechanisms, however, seems certain. Numerous studies seem to indicate that a cluster of neurons in the lateral hypothalamus function as an **appetite center**—meaning that impulses from them bring about increased appetite. This effect is often called an **orexigenic** effect (meaning an "appetite-producing" effect).

Additional data suggest that a group of neurons in the ventral medial nucleus of the hypothalamus functions as a **satiety center**—meaning that impulses from these neurons decrease appetite so that we feel sated, or satisfied. This effect is often called an **anorexigenic** effect (meaning "producing an appetite loss" effect).

What acts directly on both of these feeding centers to stimulate or depress them is still being worked out by researchers.

One factor is the temperature of the blood circulating to the hypothalamus. A moderate decrease in blood temperature stimulates the appetite center (and inhibits the satiety center). Result: The individual has an appetite, wants to eat, and probably does. An increase in blood temperature produces the opposite effect, a depressed appetite (**anorexia**). One well-known instance of this effect is the loss of appetite in persons who have a fever.

Another factor is blood glucose concentration and the rate of glucose use. A low blood glucose concentration or low glucose use stimulates the appetite center, whereas a high blood glucose concentration inhibits it. This may be the action of a group of

Table 27-6	Examples of Appetite-Regulating Factors*		
OREXIGENIC FACTORS (STIMULATE APPETITE)	**ANOREXIGENIC FACTORS (INHIBIT APPETITE)**		**SOURCE**
Leptin			Adipose tissue
Cortisol			Adrenal cortex
Ghrelin	CCK glucagon-like peptide-I (GLP-I) Peptide YY_{3-36} (PYY_{3-36})		GI tract
Endogenous opioid peptides (EOP) Galanin (GAL) Gamma-aminobutyric acid (GABA) Neuropeptide Y (NPY) Norepinephrine (NE) Orexins	Alpha-melanocyte–stimulating hormone (α-MSH) cocaine- and amphetamine-regulated transcript (CART) corticotropin-releasing hormone (CRH)		Hypothalamus
	Glucose		Liver
Emotions Environmental stimuli Food sensations Internal stimuli (e.g., blood temperature, glucose) Lifestyle choices and habits	Emotions Environmental stimuli Food sensations Internal stimuli (e.g., blood temperature, glucose) Lifestyle choices and habits		Nervous system†
	Insulin		Pancreas

*Hormones, neurotransmitters, and other factors that affect feeding centers in the hypothalamus.
†Nervous factors not specifically hypothalamic in origin.

BOX 27-13: HEALTH MATTERS
Functional Foods

One of the hottest areas in the field of nutrition today is that of *functional foods*. The food and nutrition board of the Institute of Medicine describes functional foods simply as "foods with ingredients thought to prevent disease." Most functional foods that have been identified so far are plants that contain cancer-fighting chemicals. Cruciferous vegetables (broccoli, cabbage, cauliflower, and their cousins), soybeans, linseed (flax), garlic and its relatives, and citrus fruits are among the better-known functional foods. Cruciferous vegetables, for example, contain chemicals such as indoles, isothiocyanates, ithiolthiones, and chlorophylline that either block carcinogens or suppress tumor growth, or both. Special varieties of some cruciferous vegetables such as broccoli that contain unusually high concentrations of anti-cancer chemicals are now being developed by agricultural geneticists. Lignans, found in linseed and other foods, are antioxidants and may also have cancer-blocking or tumor-suppressing effects. Organosulfur compounds such as diallyl disulfide in garlic, onions, leeks, and shallots may also have anti-cancer effects. Even meat and dairy products, eaten in small amounts, may have anti-cancer properties because they contain the cancer-suppressing chemical known as conjugated linoleic acid. Although the identification of bioactive agents in foods and the beneficial effects they produce are only now being investigated intensely, we know enough to say that a diet high in vegetables, herbs, and fruits—and a high degree of variety—is a wise choice for lowering the risk of developing cancer.

hormones called **orexins** or *hypocretins* produced in the hypothalamus in response to low blood glucose levels.

More recently, a variety of hormones and other regulators have been identified that have a profound impact on the feeding centers of the brain. The hypothalamus itself produces several hormones and neurotransmitters that affect the feeding centers, as you can see in Table 27-6. In addition, appetite-altering factors (hormones and neurotransmitters) are produced in many other organs such as the liver, adipose tissue, pancreas, GI tract, and autonomic nerve pathways (especially the vagal nerve). Of course, factors such as daily eating habits or patterns, emotional responses, the sensations of food, and many others must also be involved in regulating or affecting appetite.

Unquestionably, many factors operate together as a complex mechanism for regulating food intake—a mechanism that is still incompletely understood.

QUICK CHECK

19. Give one of the two ways in which metabolic rates can be expressed.
20. Name three of the factors that influence basal metabolic rate.
21. Distinguish between *basal metabolic rate* and *total metabolic rate*.
22. In which division of the brain would you find the control centers for regulating food intake?

Cycle of Life
Nutrition and Metabolism

The importance of proper nutrition to an individual's well-being begins at the moment of conception and continues until death. In the womb, various nutrients must be obtained from the mother's blood in sufficient quantity to ensure normal growth and development. One critical nutrient during fetal development, infancy, and childhood is protein. Sufficient proteins, containing all the essential amino acids, are required to permit normal development of the nervous system, muscle tissues, and other vital structures. Another critical nutrient during the early years of life is the mineral calcium. Large quantities of calcium are needed by a growing body to maintain normal development of the skeleton and other tissues. In the womb, a steady supply of calcium in the mother's blood is maintained by increased levels of the parathyroid hormone (PTH). Recall from Chapter 16 that PTH increases blood calcium levels by removing it from storage in the bones.

Unless a pregnant woman consumes enough calcium to replace this calcium, she may suffer from the bone-softening effects of calcium deficiency. If proteins, calcium, or other necessary nutrients are in short supply anytime before the beginning of adulthood, the consequences may be permanent. For example, bone deformities resulting from a lack of calcium during childhood could become permanent if not corrected or compensated for before the skeleton ossifies completely.

In late adulthood, the number of food calories needed declines because the metabolic rate declines. This metabolic decline is thought to result largely from age-related changes in the balance of metabolic hormones such as thyroid hormones (T_3 and T_4). Even though the number of required food calories declines, the overall balance of nutrients consumed must be maintained to preserve proper metabolic function. Some nutrients, such as calcium, may be needed in greater quantity in older adults to compensate for (or avoid) age-related bone loss or other conditions.

THE BIG PICTURE
Nutrition, Metabolism, and the Whole Body

Of all the topics we have discussed so far, the topic of nutrition and metabolism has the most easily seen role in the "big picture" of human body function. Each cell in the body cannot remain alive without maintaining the operation of its metabolic pathways. Anabolic pathways are required to build the various structural and functional components of the cells. Catabolic pathways are required to convert energy to a usable form. Catabolic pathways are also needed to degrade large molecules into small subunits that can be used in anabolic pathways. Of course, the basic nutrient molecules—carbohydrates, fats, and proteins of the correct type—must be available to each cell to carry out these metabolic processes. Besides the basic nutrient molecules, cells also require small amounts of specific vitamins and minerals needed to produce the structural and functional components necessary for cellular metabolism.

Various body systems operate to make sure that essential nutrients reach the cells as needed to maintain metabolism in a manner that preserves relative constancy of the internal environment. For example, the nervous, skeletal, and muscular systems help us obtain complex foods from our external environment. The digestive system reduces complex nutrients to simpler, more usable nutrients—then provides the mechanisms that allow us to absorb them into the internal environment. The circulatory system—both the cardiovascular and the lymphatic circulations—transports the absorbed nutrients to the individual cells for immediate use or to the liver or other organs for temporary storage. The endocrine system regulates the balance between immediate use and storage. The respiratory system, working with the cardiovascular system, provides the oxygen needed for oxidative phosphorylation—that is, using the citric acid cycle and electron transport system to transfer energy to ATP. These two systems also provide a mechanism for removing waste carbon dioxide (CO_2) generated by the catabolism of nutrient molecules. Likewise, the urinary system provides a mechanism for removing waste urea generated by protein catabolism. Even the integumentary system becomes involved, by producing vitamin D in the presence of sunlight.

Metabolism, with all the physiological mechanisms that support it, could be described as the essential process of life. It is, after all, the sum total of all the biochemical processes that distinguish a living organism from a nonliving object.

Mechanisms of Disease

METABOLIC AND NUTRITIONAL DISORDERS

Disorders characterized by a disruption or imbalance of normal metabolism can be caused by several different factors. For example, **inborn errors of metabolism** are a group of genetic conditions involving a deficiency or absence of a particular enzyme. Specific enzymes are required by cells to carry out each step of every metabolic reaction. Although an abnormal genetic code may affect the production of only a single enzyme, the resulting abnormal metabolism may have widespread effects. Specific diseases resulting from inborn errors of metabolism, such as **phenylketonuria,** are discussed in Chapter 34.

A number of metabolic disorders are complications of other conditions. For example, you may recall from Chapter 16 that both hyperthyroidism and hypothyroidism have profound ef-

Mechanisms of Disease—cont.

fects on the basal metabolic rate. Diabetes mellitus affects metabolism throughout the body when an insulin deficiency limits the amount of glucose available for use by the cells.

Some metabolic disorders result from normal mechanisms in the body that maintain homeostasis. For example, the body has several mechanisms that maintain a relatively constant level of glucose in the blood—glucose required by cells for life-sustaining catabolism. As mentioned earlier in this chapter, during starvation or in certain eating disorders, these mechanisms are taken to the extreme as they attempt to maintain blood glucose homeostasis. A few of the more well known eating and nutrition disorders are briefly described later in Mechanisms of Disease.

Body Mass Index (BMI)

The **body mass index (BMI)** is an often-used way to assess whether a person's body weight is proportional to height. Thus, the BMI can quickly tell individuals whether they are above or below their ideal weight, and by approximately how much.

To calculate a BMI, simply divide weight (in kilograms) by the square of height (in meters): BMI = kg/m^2. To determine kilogram weight, divide the number of pounds by 2.2. To determine meter height, divide total inches by 39.4.

As discussed in the following sections and illustrated in Figure 27-33, a BMI that is too high or two low is associated with an increased risk of death.

Eating Disorders

Eating disorders have been a part of the medical literature for many years, but there is growing interest and concern as the numbers of reported cases have increased dramatically. The two most common eating disorders are called anorexia nervosa and bulimia. Neither illness is completely understood, and successful treatment is often varied and sometimes controversial.

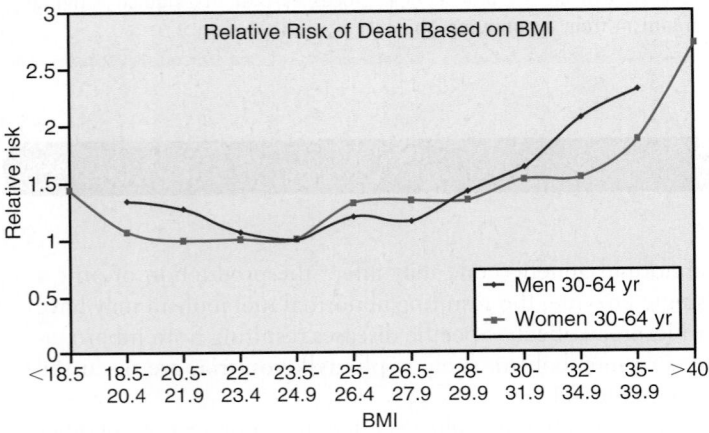

Figure 27-33 *Body mass index and risk of death.* Mortality (death) risk is higher in individuals with a high or low BMI.

Anorexia Nervosa

Anorexia nervosa is primarily a disease of young adults. Most individuals affected are female (90% to 95%) from 12 to 25 years of age. As many as 4% of college-age students suffer from this condition to some degree. These individuals have a disturbed body image, an intense fear of obesity, and diet with a vengeance. They almost always develop unusual eating rituals and scrupulously monitor and restrict their food intake. Anorexic individuals literally starve themselves and, as a result, suffer serious medical complications. The illness is characterized by a 20% to 25% loss of body mass, accompanied by slowed or impaired intellectual functioning. People affected are usually involved in excessive exercise and pursue thinness, regardless of their health. In women, menstruation ceases (amenorrhea), and the basal metabolic rate is decreased as a result of starvation. These individuals suffer from many skin abnormalities and an assortment of psychological, cardiovascular, and hormonal problems. They are at increased risk for sudden death from complications directly related to excessive weight loss and nutritional deficiency. Treatment is directed at resolution of both medical and psychological problems. In addition to psychotherapy and weight stabilization, pharmacological treatment with antidepressants has been used to improve mood and self-image.

Bulimia

Bulimia is an illness characterized by an eating and vomiting, or purging, cycle. It is sometimes referred to as **binge-purge syndrome.** The disease is said to affect about 1% of college-age students. Most people who have bulimia are relatively young, single, white females. The mean age is 25, but the age of patients appears to be increasing. People suffering from bulimia have an uncontrollable urge for food that leads to massive overeating (binging) that is followed by repeated forced vomiting and laxative abuse (purging). Loss of gastric and intestinal contents often leads to serious fluid and electrolyte imbalance. The result is often the development of neurological problems such as convulsions, tetany, and seizures. Vomiting may also cause aspiration pneumonia, erosion of tooth enamel, trauma of the mouth and esophagus, and infection of the salivary glands. The longer the disease is allowed to continue without treatment, the greater the increase in mortality from medical complications. Many people who have bulimia suffer from major depression and have concomitant social problems such as alcohol abuse. About a fourth of people with bulimia are chemically dependent, and many have been victims of sexual abuse. They are especially prone to self-mutilation and suicide attempts. Nutritional counseling, psychotherapy, and treatment with antidepressants help bulimic patients cope with stress and break the binge/purge cycle.

Mechanisms of Disease—cont.

Obesity

Obesity is not an eating disorder itself but may be a symptom of chronic overeating behavior. Like anorexia nervosa and bulimia, eating disorders characterized by chronic overeating usually have an underlying emotional cause. Obesity may also result from metabolic disorders. Obesity is defined as an abnormal increase in the proportion of fat in the body. Usually, a person with a BMI over 30 is considered moderately obese. A person with a BMI over 40 is considered to be extremely obese. Most of the excess fat is stored in the subcutaneous tissue and around the viscera. Obesity is a risk factor in various life-threatening diseases, including many forms of cancer, diabetes, and heart disease (see Figure 27-33).

Nutritional Disorders

Protein-Calorie Malnutrition

Protein-calorie malnutrition (PCM) is an abnormal condition resulting from a deficiency of calories in general and protein in particular. PCM is likely to result from reduced intake of food, but may also be caused by increased nutrient loss or increased use of nutrients by the body. Mild cases occur frequently in illness; as many as one in five patients admitted to the hospital are significantly malnourished. More severe cases of PCM are likely to occur in parts of the world where food, especially protein-rich food, is relatively unavailable. There are two forms of advanced PCM: **marasmus** and **kwashiorkor** (kwah-shee-OR-kor) (Figure 27-34). Marasmus results from an overall lack of calories and proteins, such as when sufficient quantities of food are not available. Marasmus is characterized by progres-

sive wasting of muscle and subcutaneous tissue accompanied by fluid and electrolyte imbalances. Kwashiorkor results from a protein deficiency in the presence of sufficient calories, as when a child is weaned from milk to low-protein foods. At the same time, a child affected with kwashiorkor is also likely to have an underlying infection that further increases calorie and protein needs. Like marasmus, kwashiorkor also causes wasting of tissues, but unlike marasmus, it also causes pronounced ascites (abdominal bloating) and flaking dermatitis. The ascites results from a deficiency of plasma proteins, which changes the osmotic balance of the blood and thus promotes osmosis of water from the blood into the peritoneal space.

Vitamin Disorders

Vitamin deficiency, or **avitaminosis** (ay-vye-tah-mi-NO-sis), can lead to severe metabolic problems. For example, *avitaminosis* C (vitamin C deficiency) can lead to scurvy. Scurvy results from the inability of the body to manufacture and maintain collagen fibers. As you may have gathered from your studies thus far, collagen fibers compose the connective tissues that hold most of the body together. In scurvy, the body literally falls apart in the same way that a neglected house eventually falls apart (Figure 27-35). Other details about scurvy and other types of avitaminosis are given in Table 27-3.

Some forms of **hypervitaminosis**—or vitamin excess—can be just as serious as a deficiency of vitamins. For example, chronic *hypervitaminosis* A can occur if large amounts of vitamin A—more than 10 times the recommended dietary allowance (RDA)—are consumed daily over a period of 3 months or more (Figure 27-36). This condition is first mani-

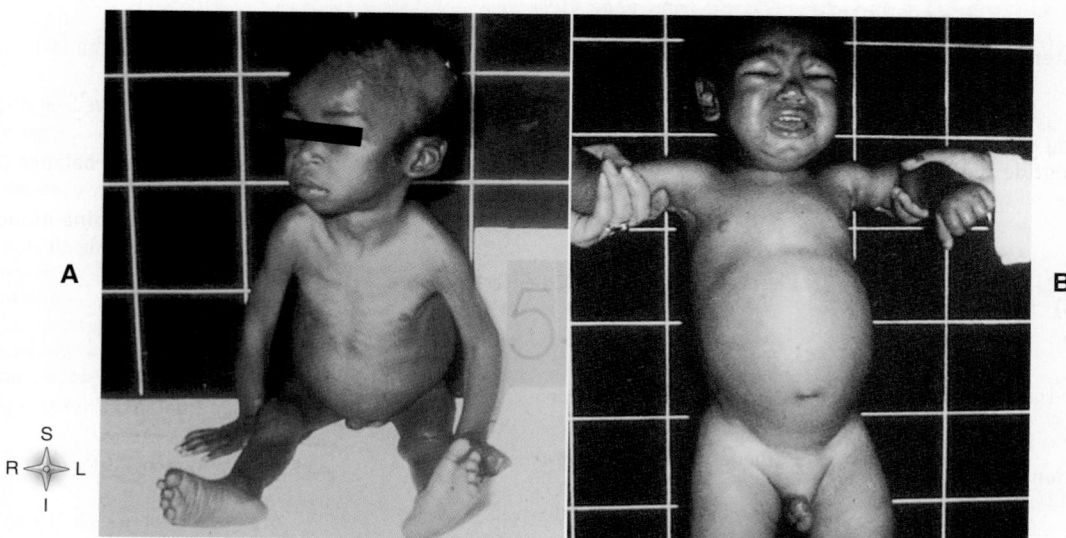

Figure 27-34 *Protein-calorie malnutrition (PCM).* PCM is an abnormal condition resulting from a deficiency of calories in general and protein in particular. Here, two forms of advanced PCM, marasmus **(A)** and kwashiorkor **(B)**, are shown.

Mechanisms of Disease—cont.

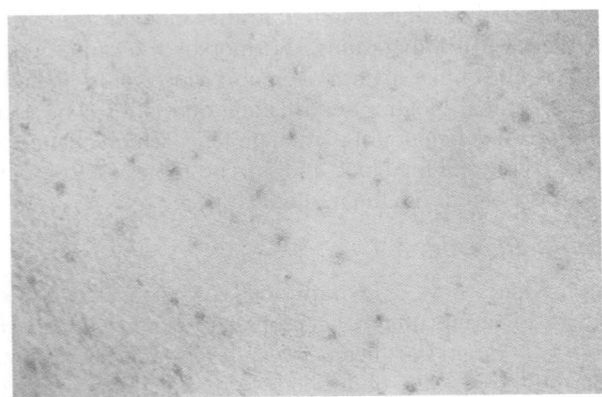

Figure 27-35 *Scurvy.* In scurvy, lack of vitamin C impairs the normal maintenance of collagen-containing connective tissues, causing bleeding and ulceration of the skin, gums, and other tissues, as these lesions on the skin show.

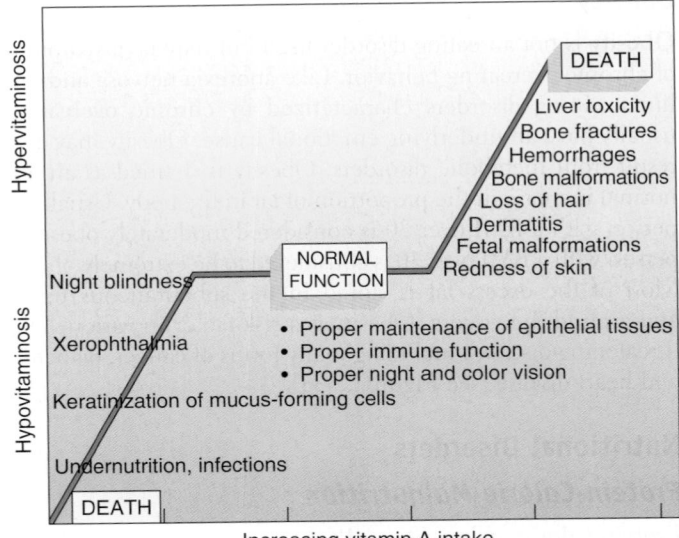

Increasing vitamin A intake

Figure 27-36 *Vitamin A intake.* This chart shows how changing the amount of vitamin A in the diet can lead to hypovitaminosis A or hypervitaminosis A. In extreme, either can lead to death.

fested as dry skin, hair loss, anorexia (appetite loss), and vomiting. However, it may progress to severe headaches and mental disturbances, liver enlargement, and occasionally cirrhosis. Acute hypervitaminosis A, characterized by vomiting, abdominal pain, and headache, can occur if a massive overdose is ingested. Excesses of the fat-soluble vitamins (A, D, E, and K) are generally more serious than excesses of the water-soluble vitamins (B complex and C).

LANGUAGE OF SCIENCE *(Cont'd from page 993)*

electron transport system (ETS) [*elektron* amber, *trans-* across, *-port* carry]

essential fatty acids [*essentia-* quality, *-al* pertaining to, fatty, *acidus* sour]

flavin adenine dinucleotide (FAD) (FLAY-vin AD-en-een dye-NOO-klee-oh-tyde) [*flav-* yellow, *-in* substance, *aden-* gland, *-ine* derived substance, *di-* two, *nucleo-* kernel (nucleus), *-t-* combining form, *-ide* chemical ending]

free fatty acids (FFAs) [*acidus* sour]

glucagon (GLOO-kah-gon) [*gluca-* sweet (glucose), *-agon* lead or bring]

gluconeogenesis (gloo-koh-nee-oh-JEN-eh-sis) [*gluco-* sweet (glucose), *-neo-* new, *-gen-* generate, *-esis* process of]

glucose phosphorylation (GLOO-kohs fos-for-i-LAY-shun) [*gluco-* sweet, *-ose* carbohydrate (sugar), *phos-* light, *-phor-* carry, *-yl-* chemical, *-ation* process of]

glycogenesis (glye-koh-JEN-eh-sis) [*glyco-* sweet, *-gen-* generate, *-esis* process of]

glycogenolysis (glye-koh-jeh-NOL-ih-sis) [*glyco-* glucose, *-gen-* generate, *-o-* combining form, *-lysis* loosening]

glycolysis (glye-KOHL-i-sis) [*glyco-* sweet, *-o-* combining form, *-lysis* loosening]

insulin (IN-suh-lin) [*insul-* island, *-in* substance]

joule (J or j) (jool) [*James Prescott Joule* English physicist]

ketogenesis (kee-toh-JEN-eh-sis) [*keto-* acetone, *-gen-* generate, *-esis* process of]

Krebs cycle [*Sir Hans Adolf Krebs* British biochemist, *kyklos* circle]

leptin (LEP-tin) [*lept-* thin, *-in* substance]

lipids (LIP-ids) [*lip-* fat, *-id* form]

lipogenesis (lip-oh-JEN-eh-sis) [*lipo-* fat, *-gen-* generate, *-esis* process of]

lipoproteins (lip-oh-PROH-teens) [*lipo-* fat]

macronutrients (MAK-roh-NOO-tree-ents) [*macro-* large, *nutri-* nourish, *-ent* agent]

metabolic rate (met-ah-BOL-ik) [*metabol-* change, *-ic* pertaining to]

metabolism (meh-TAB-oh-liz-em) [*metabol-* change, *-ism* process]

micronutrients (MYE-kroh-NOO-tree-ents) [*micro-* small, *nutri-* nourish, *-ent* agent]

negative nitrogen balance (NEG-ah-tiv NYE-troh-jen) [*nitro-* soda, *-gen* generate]

nicotinamide adenine dinucleotide (NAD) (nik-oh-TIN-ah-myde AD-eh-neen dye-NOO-klee-oh-tyde) [*Jacque Nicot* French diplomat (brought tobacco to France), *-in-* substance, *-am-* ammonia, *-ide* chemical ending, *aden-* gland, *-ine* derived substance, *di-* two, *nucleo-* kernel (nucleus), *-t-* combining form, *-ide* chemical ending]

nitrogen balance (NYE-troh-jen) [*nitro-* soda, *-gen* generate]

nutrition (noo-TRIH-shun) [*nutri-* nourish, *-tion* process of]

orexigenic (oh-rek-sih-JEN-ik) [*orexi-* appetite, *-gen-* generate, *-ic* pertaining to]

orexins (oh-REK-sins) [*orexi-* appetite, *-in* substance]

LANGUAGE OF SCIENCE—*cont'd*

oxidative phosphorylation (ahk-si-DAY-tiv fos-for-i-LAY-shun) [*oxi-* sharp (oxygen), *-id-* chemical ending (*-ide*), *-at-* action of (*-ate*), *-ive* pertaining to, *phos-* light, *-phor-* carry, *-yl-* chemical, *-ation* process of]

phosphorylation (fos-for-i-LAY-shun) [*phos-* light, *-phor-* carry, *-yl-* chemical, *-ation* process of]

positive nitrogen balance (POZ-it-iv NYE-troh-jen) [*nitro-* soda, *-gen* generate]

protein balance (PROH-teen)

satiety center (sah-TYE-eh-tee) [*sati-* enough, *-ety* state of, *center*]

saturated (SACH-yoo-ray-ted)

total metabolic rate (met-ah-BOL-ik) [*metabol-* change, *-ic* pertaining to, *rate*]

tricarboxylic acid (TCA) cycle (try-kar-bok-SIL-ik) [*tri-* three, *-carbo-* carbon, *-oxy-* sharp (oxygen), *-yl-* chemical, *-ic* pertaining to, *acidus* sour, *kyklos* circle]

triglycerides (try-GLI-ser-ides) [*tri-* three, *-glycer-* sweet, *-ide* chemical ending]

unsaturated (un-SATCH-yoo-ray-ted)

vitamins (VYE-tah-mins) [*vita-* life, *-amine* ammonia]

LANGUAGE OF MEDICINE

anorexia nervosa (an-oh-REK-see-ah ner-VOH-sah) [*an-* without, *-orexi* appetite, *-ia* condition]

avitaminosis (ay-vye-tah-mi-NO-sis) [*a-* without, *-vita-* life, *-amin-* ammonia, *-osis* condition]

binge-purge syndrome

body mass index (BMI)

bulimia (boo-LEE-mee-ah) [*bu-* ox, *-lim-* hunger, *-ia* condition]

hyperglycemia (hye-per-gly-SEEM-ee-ah) [*hyper-* above, *-glyc-* sweet (glucose), *-emia* blood condition]

hypoglycemia (hye-poh-gly-SEE-mee-ah) [*hypo-* below, *-glyc-* sweet (glucose), *-emia* blood condition]

hypervitaminosis (hye-per-vye-tah-mih-NO-sis) [*hyper-* excessive, *-vita-* life, *-amin* -ammonia, *-osis* condition]

inborn errors of metabolism (meh-TAB-oh-liz-em) [*metabol-* change, *-ism* paction]

indirect calorimetry (in-dir-EKT kal-oh-RIM-eh-tree) [*in-* not, *calori-* warmth, *-metry* process of measuring]

ketosis (kee-TOH-sis) [*keto-* acetone, *-osis* condition of]

kwashiorkor (kwah-shee-OR-kor) [*kwashiorkor* one who is displaced (from the breast)]

marasmus (mah-RAZ-mus) [*marasmos* a wasting]

obesity (oh-BEES-i-tee) [*obesity* fatness, *-ity* state of]

phenylketonuria (fen-il-kee-toh-NOO-ree-ah) [*phen-* shining (phenol), *-yl-* chemical, *-keton-* acetone, *-ur-* urine, *-ia* condition]

protein-calorie malnutrition (PCM) (PROH-teen KAHL-ah-ree mal-noo-TRISH-un) [*mal-* poor, *-nutri-* nourish, *-tion* process of]

CASE STUDY

Georgia Kaplin is a 22-year-old college student with anorexia nervosa. She was diagnosed when she was 15 years old while attending boarding school. She has undergone intensive treatment with recent better control of her disorder. At present she weighs 98 pounds and her height is 5 feet 2 inches. She is fearful of gaining weight. In particular, she is fearful of weighing more than 100 pounds. She has kept her weight over 95 pounds for the last 3 years with close medical follow-up and psychological counseling.

Ms. Kaplan's grades at school are consistently As and Bs, unless she is having difficulty with her condition. She visualizes herself as being overweight, "feels fat," and follows a very strict diet under medical supervision. She shares an apartment with two other roommates who are not aware that she has an eating disorder. She shops for her own food, stores it on a special shelf in the kitchen, and uses a particular shelf in the refrigerator. Her roommates shop, cook, and share meals with each other, but Georgia prefers to eat by herself.

Georgia exercises twice a day. In the morning she jogs with one of her roommates. In the afternoon she works out, concentrating on her abdomen, which she believes is "too fat and sticks out too much." She has not had a menstrual period in more than 3 months. Before that time, her menstrual cycle was irregular and fluctuated with the control of her disease.

On physical examination, Georgia appears pale and thin. Her temperature is 96.2° F (35.7° C), her blood pressure is 64/42, and her pulse is 58 and regular. She states that she has occasional dizziness but, as long as she changes her position slowly from lying to sitting, this dizziness is not usually bothersome.

1. Even though Georgia Kaplan's disease seems to be in control, which one of the following factors indicates a need to monitor her more closely than usual?
   1. A weight of 98 pounds
   2. Fear of gaining weight over 100 pounds
   3. Grades
   4. The fact that her roommates do not know of her disorder
   5. Ritualistic behavior regarding shopping and eating alone
   6. Exercise routine

7. Body image
8. Amenorrhea
9. Dizziness
10. Abnormal temperature, pulse, and blood pressure

    A. 1, 2, 4, and 5
    B. 3, 4, 5, and 10
    C. 4, 5, 6, and 8
    D. 6, 7, 8, and 9

2. The counselor convinces Georgia to share her problem with her roommates and she brings them to her next visit. Discussion with the roommates indicates that Georgia's condition is being complicated by another disorder. She has not gained any weight since the last visit, but her roommates indicate that she is sharing meals with them and progressively eating more food. Which one of the following disorders should be considered?

   A. Phenylketonuria
   B. Obesity
   C. Diabetes mellitus
   D. Bulimia

3. In making nutritional recommendations to Ms. Kaplan, the counselor should stress which of the following factors?

   A. Basic nutrient molecules are necessary for each cell in the body to function.
   B. Body systems will compensate for her inability to eat properly.
   C. Her disorder will not affect other systems in the body.
   D. She will need to take vitamins to counteract all the effects of her disorder.

4. When describing necessary nutrition for Georgia, which one of the following groups of tools is most likely to be used?

   A. Pictures of children with kwashiorkor and marasmus
   B. Recommended daily allowances and the Food Guide Pyramid
   C. Recommended daily allowances and a list of the essential amino acids
   D. The food pyramid and a list of the essential amino acids

CHAPTER SUMMARY

OVERVIEW OF NUTRITION AND METABOLISM

A. Nutrition refers to the food (nutrients) we eat
   1. Malnutrition—a deficiency in the consumption of food, vitamins, and minerals
   2. Categories of nutrients
      a. Macronutrients—nutrients that we need in large amounts (bulk nutrients)
         (1) Macromolecules such as carbohydrates, fats (lipids), proteins
         (2) Water
         (3) Macrominerals—minerals that we need in large quantity; for example, sodium, chloride, calcium
      b. Micronutrients—nutrients we need in very small amounts
         (1) Vitamins
         (2) Microminerals (trace elements)—minerals that are needed only in very small quantities, such as iron, iodine, and zinc
   3. Balance of nutrients is required for good health (Figure 27-1 and 27-2)

B. Metabolism—the use of nutrients—a process made up of many chemical processes (Figure 27-23)
1. Catabolism breaks food down into smaller molecular compounds and releases two forms of energy—heat and chemical energy
2. Anabolism—a synthesis process
3. Both processes take place inside of cells continuously and concurrently
4. Chemical energy released by catabolism must be transferred to ATP, which supplies energy directly to the energy-using reactions of all cells (Figure 27-3)

CARBOHYDRATES

A. Dietary sources of carbohydrates
1. Complex carbohydrates
   a. Polysaccharides—starches; found in vegetables and grains; glycogen is found in meat
   b. Cellulose—a component of most plant tissue; passes through the system without being broken down
   c. Disaccharides—found in refined sugar; must be broken down before they can be absorbed
   d. Monosaccharides—found in fruits; move directly into the internal environment without being processed directly
      (1) Glucose—carbohydrate most useful to the human cell; can be converted from other monosaccharides (Figure 27-4)

B. Carbohydrate metabolism—human cells catabolize most of the carbohydrate absorbed and anabolize a small portion of it
1. Glucose transport and phosphorylation—glucose reacts with ATP to form glucose-6-phosphate; this step prepares glucose for further metabolic reactions
   a. This step is irreversible except in the intestinal mucosa, liver, and kidney tubules
2. Glycolysis—the first process of carbohydrate catabolism; consists of a series of chemical reactions (Figure 27-5)
   a. Glycolysis occurs in the cytoplasm of all human cells
   b. An anaerobic process—the only process that provides cells with energy under conditions of inadequate oxygen
   c. It breaks down chemical bonds in glucose molecules and releases about 5% of the energy stored in them
   d. It prepares glucose for the second step in catabolism—the citric acid cycle
3. Citric acid cycle
   a. Two pyruvic acid molecules from glycolysis are converted to two acetyl molecules in a transition reaction, losing one carbon dioxide molecule per pyruvic acid molecule converted
   b. By end of transition reaction and citric acid cycle, two pyruvic acids have been broken down to six carbon dioxide and six water molecules (Figures 27-6 and 27-7)
   c. Citric acid cycle also called the *tricarboxylic acid (TCA) cycle* because citric acid is also called *tricarboxylic acid*
   d. Citric acid cycle once called Krebs cycle after Sir Hans Krebs, who discovered this process

4. Electron transport system (Figure 27-8)
   a. High-energy electrons (along with their protons) removed during the citric acid cycle enter a chain of molecules embedded in the inner membrane of the mitochondria
   b. As electrons move down the chain, they release small bursts of energy to pump protons between the inner and outer membrane of the mitochondrion
   c. Protons move down their concentration gradient, across the inner membrane, driving ATP synthase (Figure 27-9)
5. Oxidative phosphorylation—the joining of a phosphate group to ADP to form ATP by the action of ATP synthase (Figure 27-10 and 27-11)
6. The anaerobic pathway—a pathway for the catabolism of glucose; transfers energy to ATP using only glycolysis; ultimately ends with the oxidative phosphorylation of ATP (paying the "oxygen debt") (Figure 27-12)
7. Glycogenesis—a series of chemical reactions in which glucose molecules are joined to form a strand of glucose beads; a process that operates when the blood glucose level increases above the midpoint of its normal range (Figures 26-13 and 26-14)
8. Glycogenolysis (Figure 27-15)—the reversal of glycogenesis; it means different things in different cells
9. Gluconeogenesis (Figure 27-16)—the formation of new glucose, which occurs chiefly in the liver
10. Control of glucose metabolism—hormonal and neural devices maintain homeostasis of blood glucose concentration (Figures 27-17 and 27-18)
    a. Insulin—secreted by beta cells to decrease blood glucose level (Figure 27-14)
    b. Glucagon increases the blood glucose level by increasing the activity of the enzyme phosphorylase
    c. Epinephrine—hormone secreted in times of stress; increases phosphorylase activity
    d. Adrenocorticotropic hormone stimulates the adrenal cortex to increase its secretion of glucocorticoids
    e. Glucocorticoids accelerate gluconeogenesis
    f. Growth hormone increases blood glucose level by shifting from carbohydrate to fat catabolism
    g. Thyroid-stimulating hormone has complex effects on metabolism
11. Hormones that cause the blood glucose level to rise are called *hyperglycemic*
12. Insulin is hypoglycemic because it causes the blood glucose level to decrease

LIPIDS

A. Dietary sources of lipids
1. Triglycerides—the most common lipids—composed of a glycerol subunit that is attached to three fatty acids
2. Phospholipids—an important lipid found in all foods
3. Cholesterol—an important lipid found only in animal foods

4. Dietary fats
   a. Saturated fats contain fatty acid chains in which there are no double bonds
   b. Unsaturated fats contain fatty acid chains in which there are some double bonds
B. Transport of lipids—they are transported in blood as chylomicrons, lipoproteins, and fatty acids
   1. In the absorptive state, many chylomicrons are present in the blood
   2. Postabsorptive state—95% of lipids are in the form of lipoproteins
      a. Lipoproteins consist of lipids and protein and are formed in the liver
         (1) Blood contains three types of lipoproteins: very low density, low density, and high density
         (2) Cholesterol lipoproteins associated with heart disease (Figure 27-19)
      b. Fatty acids are transported from the cells of one tissue to the cells of another in the form of free fatty acids
C. Lipid metabolism
   1. Lipid catabolism—triglycerides are hydrolyzed to yield fatty acids and glycerol; glycerol is converted to glyceraldehyde-3-phosphate, which enters the glycolysis pathway; fatty acids are broken down by beta-oxidation and are then catabolized through the citric acid cycle (Figure 27-20)
   2. Lipid anabolism consists of the synthesis of triglycerides, cholesterol, phospholipids, and prostaglandins
   3. Control of lipid metabolism is through the following hormones
      a. Insulin
      b. Growth hormone
      c. ACTH
      d. Glucocorticoids

PROTEINS

A. Sources of proteins
   1. Proteins are assembled from a pool of 20 different amino acids
   2. The body synthesizes amino acids from other compounds in the body
   3. Only about half the necessary types of amino acids can be produced by the body; the rest are supplied through diet—they are found in both meat and vegetables
B. Protein metabolism—anabolism is primary and catabolism is secondary
   1. Protein anabolism—the process by which proteins are synthesized by the ribosomes of the cells (Figure 27-21)
   2. Protein catabolism—deamination takes place in the liver cells and forms an ammonia molecule, which is converted to urea and excreted in urine, and a keto acid molecule, which is oxidized or converted to glucose or fat (Figure 27-22)
   3. Protein balance—the rate of protein anabolism balances the rate of protein catabolism
   4. Nitrogen balance—the amount of nitrogen taken in equals the nitrogen in protein catabolic waste

5. Two kinds of protein or nitrogen imbalance
   a. Negative nitrogen balance—protein catabolism exceeds protein anabolism; more tissue proteins are catabolized than are replaced by protein synthesis
   b. Positive nitrogen balance—protein anabolism exceeds protein catabolism
6. Control of protein metabolism—achieved by hormones

VITAMINS AND MINERALS

A. Vitamins (Table 27-3)—organic molecules necessary for normal metabolism; many attach to enzymes and help them work or have other important biochemical roles (Figures 27-24 and 27-25)
   1. The body does not make most of the necessary vitamins; they must be obtained through diet
      a. The body stores fat-soluble vitamins and does not store water-soluble vitamins
B. Minerals (Table 27-4)—inorganic elements or salts found in the earth
   1. They attach to enzymes and help them work and function in chemical reactions
   2. Essential to the fluid/ion balance of the internal fluid environment
   3. Are involved in many processes in the body such as muscle contraction, nerve function, hardening of bone, etc.
   4. Too large or too small an amount of some minerals may be harmful
   5. Recommended mineral intakes may vary over the life span (Figures 27-26 and 27-27)

METABOLIC RATES

A. Metabolic rate means the amount of energy released by catabolism
B. Metabolic rates are expressed in two ways
   1. The number of kilocalories of heat energy expended per hour or per day
   2. As normal or as a percentage above or below normal
C. Basal metabolic rate—the rate of energy expended under basal conditions
   1. Factors: size, body composition, sex, age, thyroid hormone, body temperature, drugs, other factors (Figures 27-28 through 27-31)
D. Total metabolic rate (Figure 27-28)—the amount of energy used in a given time
   1. Main determinants
      a. The basal metabolic rate
      b. The energy used to do skeletal muscle work
      c. The thermic effect of foods
E. Energy balance and weight—the body maintains a state of energy balance
   1. The body maintains a weight when the total calories in the food ingested equals the total metabolic rate
   2. Body weight increases when energy input exceeds energy output
   3. Body weight decreases when energy output exceeds energy input
   4. In starvation, the carbohydrates are used up first, then fats, then proteins (Figure 27-32)

MECHANISMS FOR REGULATING FOOD INTAKE (TABLE 27-6)

A. The hypothalamus plays a part in food intake
B. Feeding centers in the hypothalamus exert primary control over appetite
  1. Appetite center
    a. Cluster of neurons in the lateral hypothalamus that if stimulated brings about increased appetite
    b. Orexigenic effects—factors that trigger appetite
  2. Satiety center
    a. Group of neurons in the ventral medial nucleus of the hypothalamus that if stimulated brings about decreased appetite
    b. Anorexigenic effects—factors that suppress appetite (anorexia is loss of appetite)

THE BIG PICTURE: NUTRITION, METABOLISM, AND THE WHOLE BODY

A. Every cell in the body needs the maintenance of the metabolic pathways to stay alive
B. Anabolic pathways build the various structural and functional components of the cells
C. Catabolic pathways convert energy to a usable form and degrade large molecules into subunits used in anabolic pathways
D. Cells require appropriate amounts of vitamins and minerals to produce structural and functional components necessary for cellular metabolism
E. Other body mechanisms operate to ensure that nutrients reach the cells

REVIEW QUESTIONS

1. What is metabolism? Nutrition?
2. What two processes make up the process of metabolism?
3. Does the body digest dietary fiber? Why or why not?
4. Briefly describe glycolysis, the first process of carbohydrate catabolism.
5. Where does glycolysis occur?
6. Describe the process of "splitting glycogen."
7. How are dietary fats classified?
8. Explain how lipids are transported in blood.
9. List the hormones involved in the control of lipid metabolism.
10. What are the essential amino acids?
11. What does the term *metabolic rate* mean?
12. List the various factors that influence basal metabolic rate.
13. Describe various factors that influence the amount of food a person eats.
14. Define the term *calorie*.

CRITICAL THINKING QUESTIONS

1. How would you state or interpret, in your own words, the process of carbohydrate catabolism known as the citric acid cycle? Can you draw a picture of what you mean?
2. How would you describe the mitochondria, and why do you think they are referred to as the "power plants" of the cells?
3. Compare anaerobic and aerobic respiration. What important role does lactic acid play in anaerobic respiration?
4. Can you identify and explain the processes and hormones involved in maintaining the homeostatic level of glucose in the blood?
5. State in your own words the process of lipid catabolism. How is it similar to the carbohydrate pathway? How does this process generate ketone bodies?
6. Describe protein catabolism in your own words. How is this process related to a negative nitrogen balance?
7. How would you compare and contrast the functions of proteins, carbohydrates, and fats?
8. What is the function of most vitamins in the body? What examples can you find that do not have this function? What functions do these vitamins have?
9. What is the difference between basal and total metabolic rates?
10. What would be the basal metabolism rate in kilocalories per day of a 22-year-old man who is 6 feet tall and weighs 176 pounds?
11. A man went on a 7-day vacation. Since he planned to be more active than normal, he thought he would not have to be concerned about his diet. During the 7 days, he burned 17,500 kcal. He ate 26,500 kcal. What was the difference in his body weight before and after his vacation?
12. Why do you think high blood concentrations of low-density lipoproteins may lead to atherosclerosis?

Urinary System

LANGUAGE OF SCIENCE

abnormal constituents (ab-NOR-mal kon-STICH-yoo-ents) [*ab-* away from, *-norma* rule, *constituere* to set up]

atrial natriuretic hormone (ANH) (AY-tree-al nay-tree-yoo-RET-ik HOR-mohn) [*atrium* hall, *natrium-* sodium, *-ouron-* urine, *-ic* pertaining to, *hormaein* to set in motion]

Bowman's capsule (BOH-menz KAP-sul) [*William Bowman* English anatomist]

calyx (KAY-liks) [*calyx* cuplike]

collecting duct [*colligere* to gather, *ducere* to lead]

cortical nephrons (KOHR-ti-kal NEF-rons) [*cortic-* cortex, *-al* pertaining to, *nephros* kidney]

countercurrent mechanism [*contra-* against, *-current* flowing, *mechane* machine]

detrusor muscle (dee-TROO-sor) [*detruder* to thrust]

distal tubule (DIS-tall TOO-byool) [*distare* to be distant, *tubulus* little tube]

electrolytes (eh-LEK-troh-lytes) [*elektron-* amber, *-lytos* soluble]

filtration (fil-TRAY-shun) [*filtrare* to strain]

glomerular capsular membrane (gloh-MER-yoo-lar KAP-soo-lahr MEM-brayne) [*glomerulus* small ball, *capsula* little box, *membrana* thin skin]

glomerular filtrate (gloh-MAIR-yoo-lar FIL-trayt) [*glomerulus* small ball, *filtrare* to strain]

glomerulus (gloh-MAIR-yoo-lus) [*glomerulus* small ball]

hilum (HYE-lum) [*hilum* a trifle]

hormones (HOR-mohns) [*hormaein* to set in motion]

juxtaglomerular apparatus (juks-tah-gloh-MER-yoo-lar app-ah-RAT-us) [*juxta-* near, *-glomerulus* small ball, *ad-* toward, *-parare* to make ready]

juxtaglomerular (JG) cells (jux-tah-gloh-MAIR-yoo-lar sells) [*juxta-* near, *-glomerulus* small ball, *cella* storeroom]

juxtamedullary nephrons (jux-tah-MED-oo-lair-ee NEF-ronz) [*juxta-* near, *-medulla* marrow, *nephros* kidney]

Cont'd on p. 1061

OVERVIEW OF THE URINARY SYSTEM

The principal organs of the urinary system are the **kidneys,** which process blood and form urine as a waste to be excreted (removed from the body). The excreted urine travels from the kidneys to the outside of the body via accessory organs: the *ureters, urinary bladder,* and *urethra.*

We often think of the urinary system primarily as a "urine producer," which it certainly is. However, a better image of the system is that of "blood plasma balancer." Each kidney processes incoming blood plasma in ways that allow it to leave the kidney in better condition. The water content is adjusted so that the body does not have too much or too little water to maintain constancy of the internal environment. Likewise, the blood content of important ions such as sodium and potassium is adjusted to match set point levels. Even the pH of the blood can be altered to match the set point level. In these ways, the urinary system regulates the content of blood plasma so that the homeostasis, or "dynamic constancy," of the entire internal fluid environment can be maintained within normal limits.

ANATOMY OF THE URINARY SYSTEM
Gross Structure

Kidney

The kidneys resemble lima beans in shape, that is, roughly oval with a medial indentation (Figure 28-1, *A*). An average-sized kidney measures approximately 11 cm by 7 cm by 3 cm (4.3 by 2.7 by 1.2 inches). The left kidney is often slightly larger than the right. The kidneys lie in a *retroperitoneal* position, meaning posterior to the parietal peritoneum, against the posterior wall of the abdomen (Figure 28-1, *C*). They are located on either side of the vertebral column and extend from the level of the last thoracic vertebra (T12) to just above the third lumbar vertebra (L3). Note in Figure 28-1, *B,* that the superior or upper portions (poles) of both kidneys extend above the level of the twelfth rib and the lower edge of the thoracic parietal pleura. This anatomical relationship has important clinical implications (Box 28-1). Usually the right kidney is a little lower than the left, presumably because the liver takes up some of the space above the right kidney. A heavy cushion of fat normally encases each kidney and holds it in position. Connective tissue, the renal fasciae, anchors the kidneys to surrounding structures and also helps maintain their normal positions.

The medial surface of each kidney has a concave notch called the **hilum.** Structures enter or leave the kidney through this notch. A tough, white fibrous capsule encases each kidney (Figure 28-2).

BOX 28-1: HEALTH MATTERS
Kidney Biopsy

Suspected disease of the kidney, such as renal cancer, often requires a needle **biopsy** to confirm the diagnosis. In these procedures a hollow (biopsy) needle is inserted through the skin surface and then guided into the diseased organ to withdraw a tissue sample for analysis. In a renal biopsy, tissue is removed from the lower rather than the upper or superior pole of the diseased kidney. This avoids the possibility of damage to the pleura caused by the biopsy needle and a resulting pneumothorax.

The coronal section of the right kidney shown in Figure 28-2 portrays the major internal structures of the kidney. Identify the **renal cortex,** or outer region, and the **renal medulla,** or inner region. A dozen or so distinct triangular wedges, the **renal pyramids,** make up much of the medullary tissue. The *base* of each pyramid faces outward, and the narrow *papilla* of each faces toward the hilum. Notice that the cortical tissue dips into the medulla between the pyramids, forming areas known as **renal columns.**

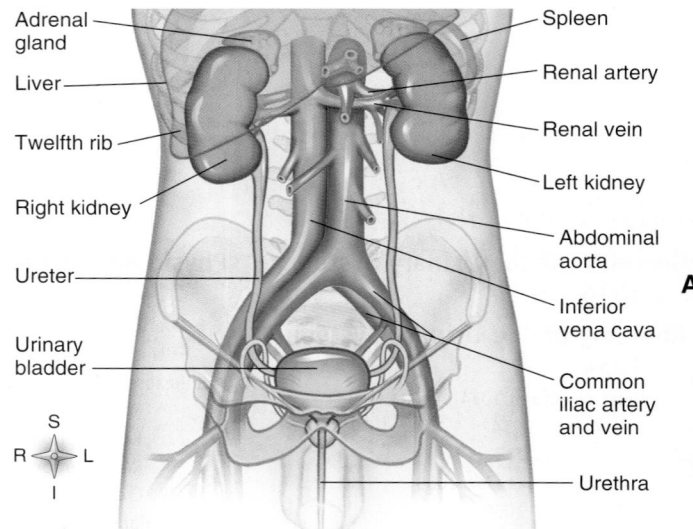

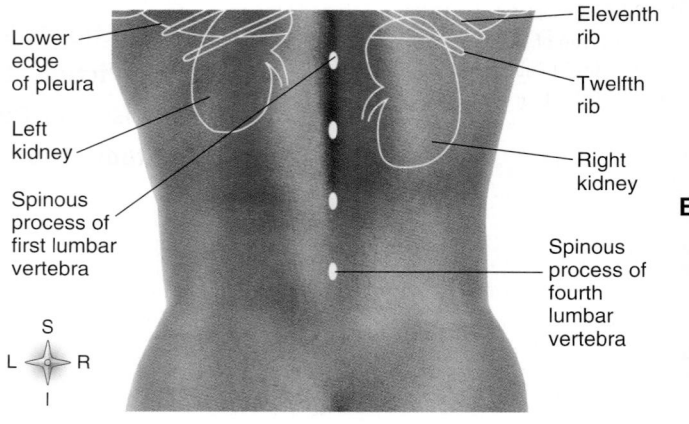

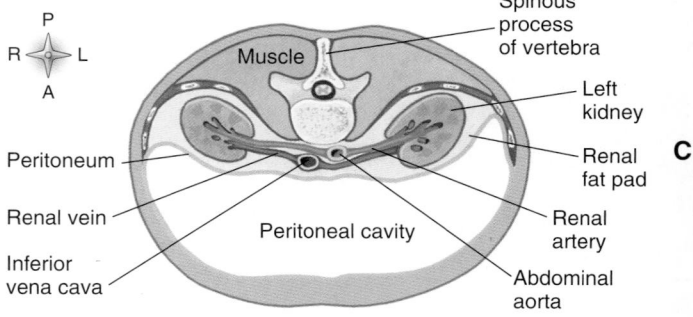

Figure 28-1 *Location of urinary system organs.* **A,** Anterior view of the urinary organs with the peritoneum and visceral organs removed. **B,** Surface markings of the kidneys, eleventh and twelfth ribs, spinous processes of L1 to L4, and lower edge of the pleura (posterior view). **C,** Horizontal (transverse) section of the abdomen showing the retroperitoneal position of the kidneys.

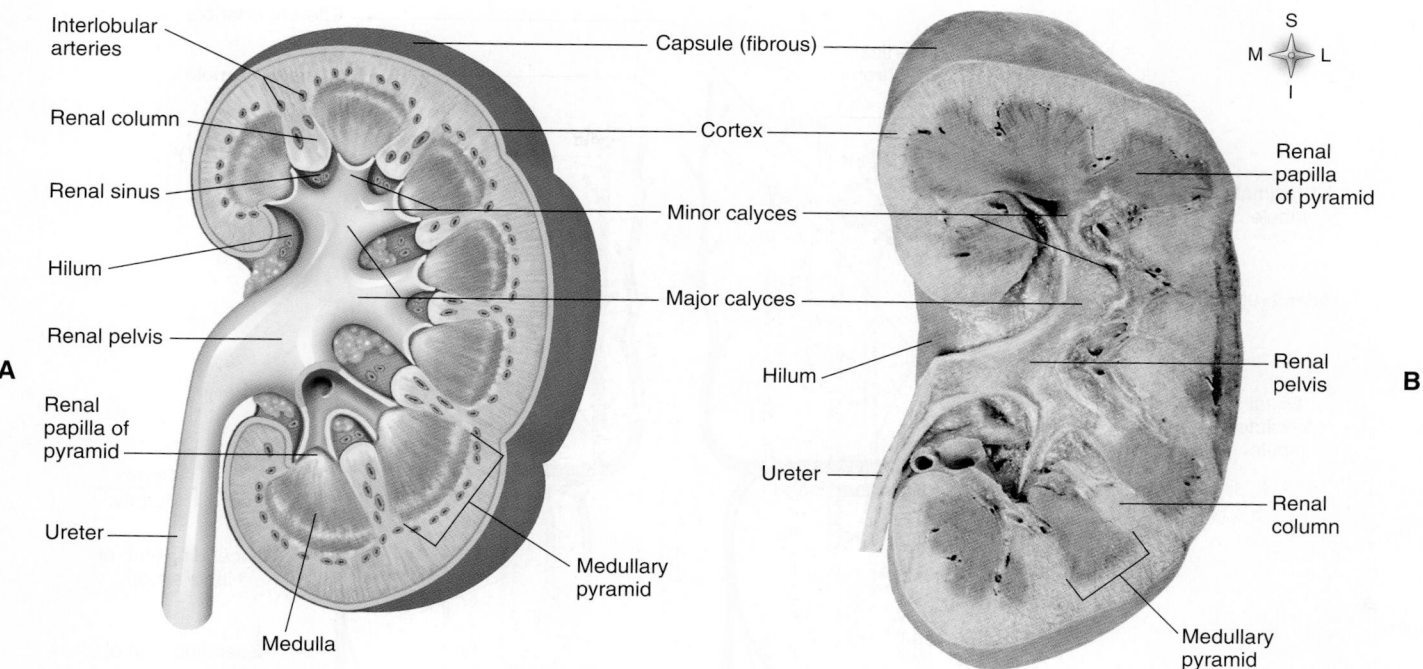

Figure 28-2 *Internal structure of the kidney.* **A,** Coronal section of the right kidney in an artist's rendering. **B,** Photo of a coronal section of a preserved human kidney.

Each renal papilla (point of a pyramid) juts into a cuplike structure called a **calyx.** The calyces are considered the beginnings of the "plumbing system" of the urinary system, for it is here that urine leaving the renal papilla is collected for transport out of the body. The calyces join together to form a large collection reservoir called the **renal pelvis.** The pelvis of the kidney narrows as it exits the hilum to become the ureter.

Blood Vessels of the Kidneys

The kidneys are highly vascular organs (Figure 28-3). Every minute about 1200 ml of blood flows through them. Stated another way, approximately one fifth of all the blood pumped by the heart per minute goes to the kidneys. From this fact one might guess, and correctly so, that the kidneys process the blood in important ways before returning it to the general circulation.

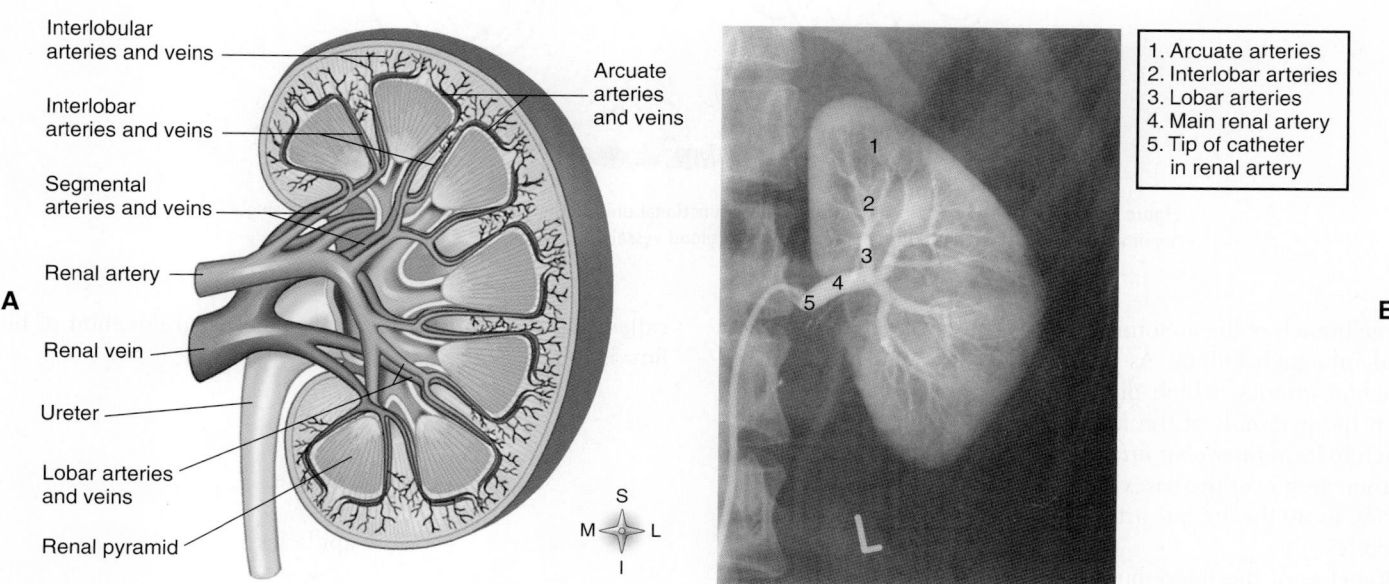

Figure 28-3 *Circulation of blood through the kidney.* **A,** Diagram showing the major arteries and veins of the renal circulation. **B,** Renal arteriogram. Arcuate arteries (1) are seen near the junction of the cortex and medulla, interlobar arteries (2) are present between the medullary pyramids, and lobar arteries (3) are seen branching from the main renal artery (4). Note the tip of the catheter used to inject contrast material (5) into the proximal part of the main renal artery.

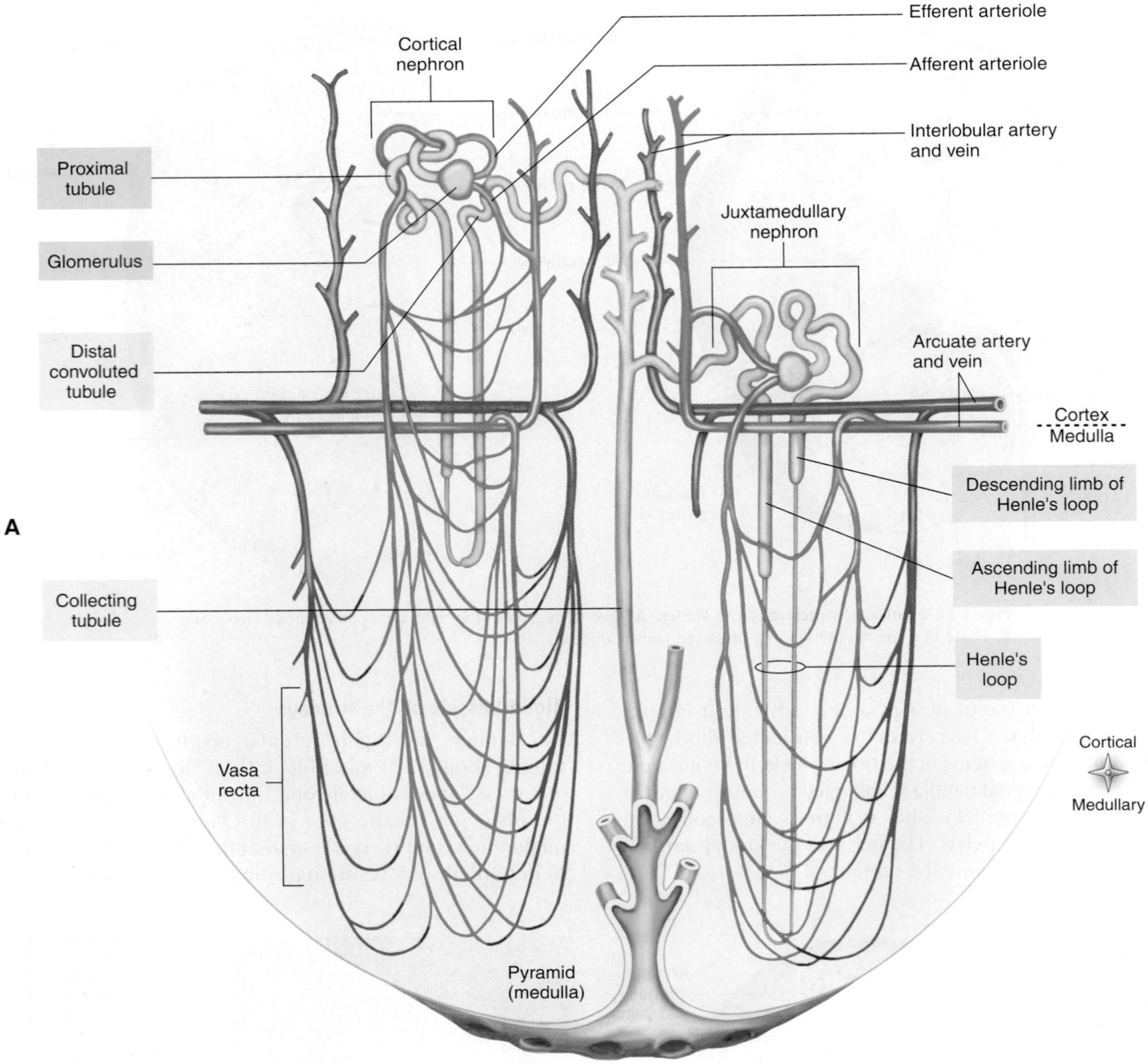

Efferent arteriole

Afferent arteriole

Interlobular artery
and vein

Cortical
nephron

Juxtamedullary
nephron

Proximal
tubule

Glomerulus

Arcuate artery
and vein

Cortex
Medulla

Distal
convoluted
tubule

Descending limb of
Henle's loop

A

Ascending limb of
Henle's loop

Collecting
tubule

Henle's
loop

Cortical

Medullary

Vasa
recta

Pyramid
(medulla)

Figure 28-4 *Nephron.* **A,** The nephron is the basic functional unit of the kidney. This illustration of a single nephron unit also shows the surrounding peritubular blood vessels.

A large branch of the abdominal aorta—the *renal artery*—brings blood into each kidney. As it nears the kidney, it divides into *segmental arteries*, which divide to become *lobar arteries*. Between the pyramids of the kidney's medulla, the lobar arteries branch to form *interlobar arteries* that extend out toward the cortex, then arch over the bases of the pyramids to form the *arcuate arteries.* From the arcuate arteries, *interlobular arteries* penetrate the cortex.

Branches of the interlobular arteries called *afferent arterioles* are shown in Figure 28-4. As you can see in that figure and in Figure 28-6, the afferent arteriole then branches into a tuft-like grouping of five to eight capillaries, the glomerular capillaries,

called the **glomerulus.** Recall that the usual direction of blood flow is

Arteries
↓
Arterioles
↓
Capillaries
↓
Venules
↓
Veins

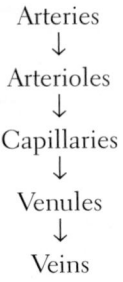

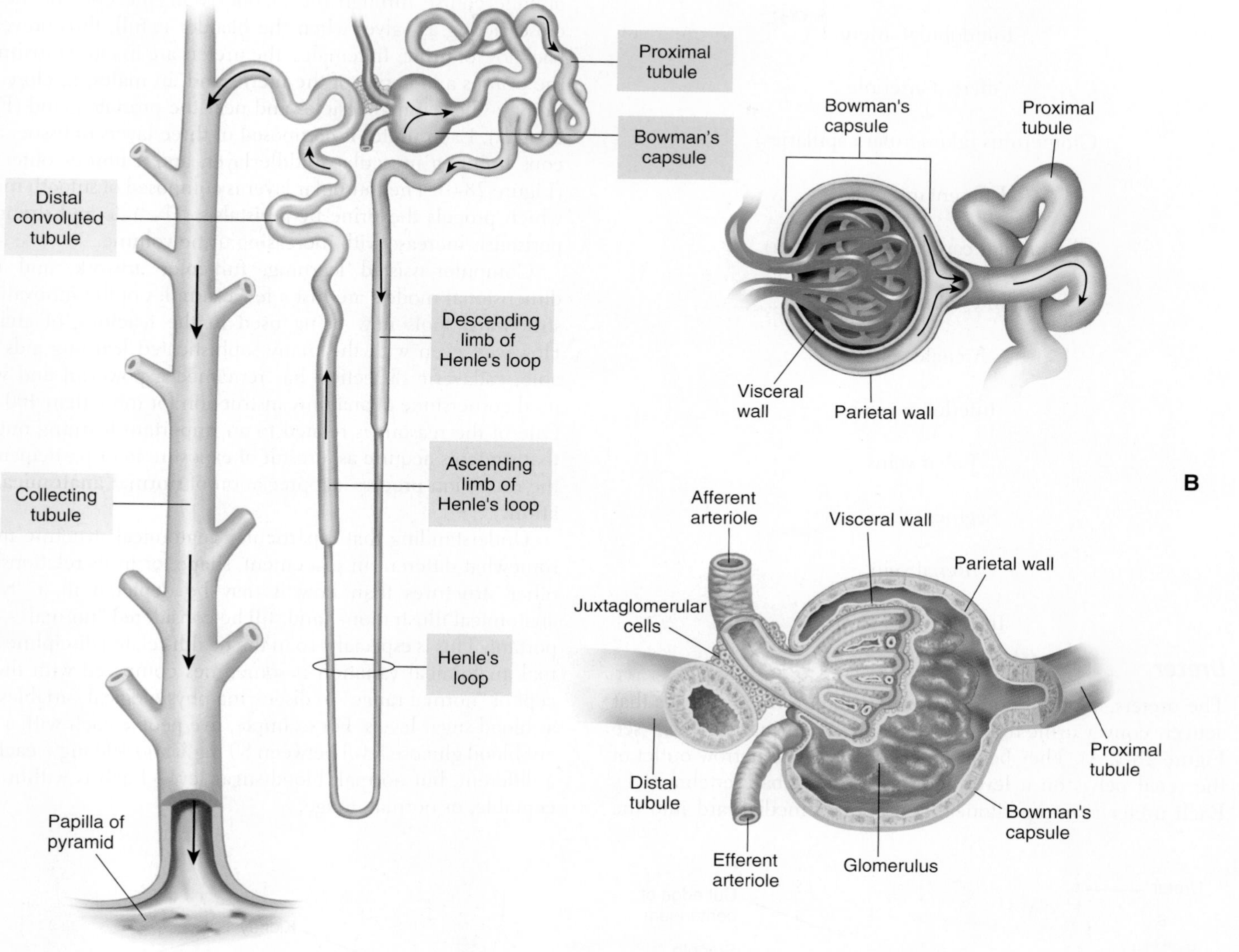

Proximal tubule

Bowman's capsule

Distal convoluted tubule

Descending limb of Henle's loop

Collecting tubule

Ascending limb of Henle's loop

Henle's loop

Papilla of pyramid

Bowman's capsule

Proximal tubule

Visceral wall

Parietal wall

B

Afferent arteriole

Visceral wall

Parietal wall

Juxtaglomerular cells

Distal tubule

Proximal tubule

Bowman's capsule

Efferent arteriole

Glomerulus

Figure 28-4, *cont'd* **Nephron. B,** Components of a nephron showing the direction of fluid flow; schematic of Bowman's capsule showing the parietal and visceral walls, and the relationship of the glomerulus to Bowman's capsule and adjacent structures.

This is not entirely true for the kidney, however. Here, blood leaving the glomerular capillaries flows into *efferent arterioles,* not into venules. From the efferent arterioles, blood moves into the **peritubular capillaries** (meaning "capillaries around the tubules"). This placement of an arteriole (efferent arteriole) between two capillary beds (glomerular capillaries and peritubular capillaries) is unique and found only in the kidney. One portion of the peritubular circulation, the *vasa recta,* loops down into the medulla and back to supply the loop of Henle and collecting ducts; these structures are discussed later in the chapter. Capillary blood then drains in a reverse direction into venules and veins that follow and are named the same as their arterial

counterparts. Ultimately, the renal vein empties into the inferior vena cava. The flow of blood through the kidney tissue follows this pattern:

Abdominal aorta
↓
Renal artery
↓
Segmental arteries
↓
Lobar arteries
↓

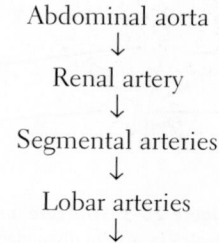

Interlobar arteries
↓
Arcuate arteries
↓
Interlobular artery
↓
Afferent arteriole
↓
Glomerulus (glomerular capillaries)
↓
Efferent arteriole
↓
Peritubular capillaries (vasa recta)
↓
Interlobular veins
↓
Arcuate veins
↓
Interlobar veins
↓
Lobar veins
↓
Segmental veins
↓
Renal vein
↓
Inferior vena cava

Ureter

The ureters, about 28 to 34 cm in length, are the two tubes that actively convey urine from the kidneys to the urinary bladder (see Figure 28-1, A). They begin on each side at the narrow outlet of the renal pelvis on a level with the first lumbar vertebra (L1). Each ureter is retroperitoneal and courses medialward into the pelvis until it reaches the lateral angle of the bladder. It then runs obliquely for about 2 cm through the bladder wall and opens at the lateral angles of the trigone (Figure 28-5, A). Because of its oblique course through the bladder wall, the ends of the tube close, acting as valves when the bladder is full, thus preventing backflow of urine. In females, the ureters are in close proximity to the ovaries and cervix of the uterus and, in males, in close proximity to the seminal vesicles and near the prostate gland (Figure 28-5, B). Each ureter is composed of three layers of tissue: a mucous lining, a muscular middle layer, and a fibrous outer layer (Figure 28-6). The muscular layer is composed of smooth muscle, which propels the urine by peristalsis. The rate and strength of peristalsis increase with increasing urine volume.

Computer-assisted learning, full-color artwork, and three-dimensional models are just a few examples of the innovative instructional tools now being used in the teaching of anatomy. However, even with the many sophisticated learning aids available, cadaveric dissection has remained a powerful and widely used cornerstone of anatomy instruction for more than 400 years. One of the reasons is related to an important learning outcome that students acquire as a result of exposure to or participation in the dissection process—appreciation of **normal anatomical variation.**

Understanding that a particular anatomical structure may be somewhat different in placement, shape, or in its relationship to other structures than how it may be depicted in a "typical" anatomical illustration—and still be considered "normal"—is important. This is especially so in the health-related disciplines. Normal anatomical variation is sometimes compared with the concept of "normal range" in discussing physiological variables, such as blood sugar levels. For example, five people each with a different blood glucose level between 80 mg% and 100 mg% each have a different, but normal, blood sugar level. Each is within an acceptable, or normal, range.

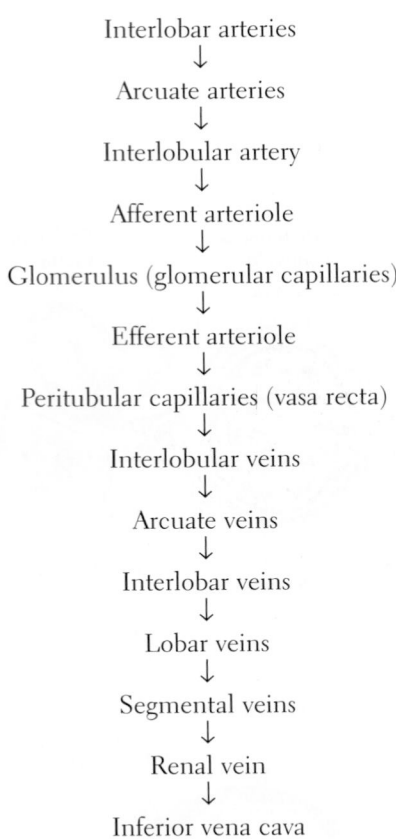

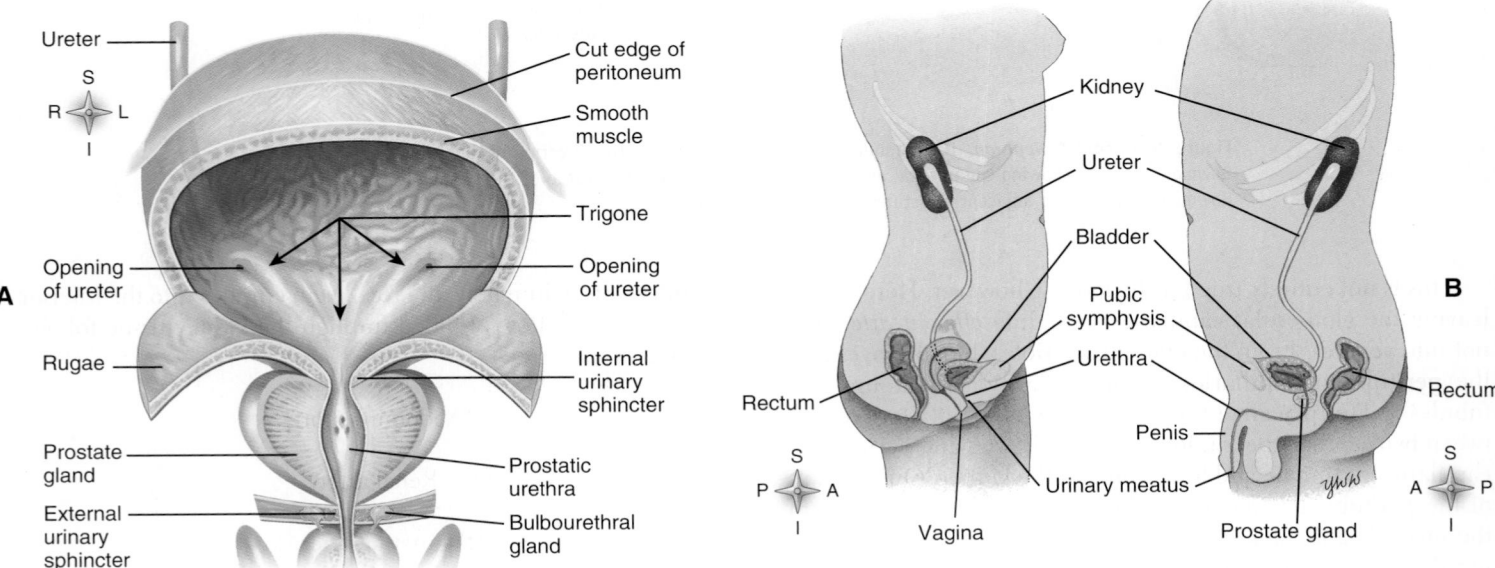

Figure 28-5 *Structure and location of the urinary bladder.* **A,** Frontal view of a dissected urinary bladder (male) in a fully distended position. **B,** Sagittal section of the female urinary system *(left)* and male urinary system *(right),* each showing a partially distended bladder.

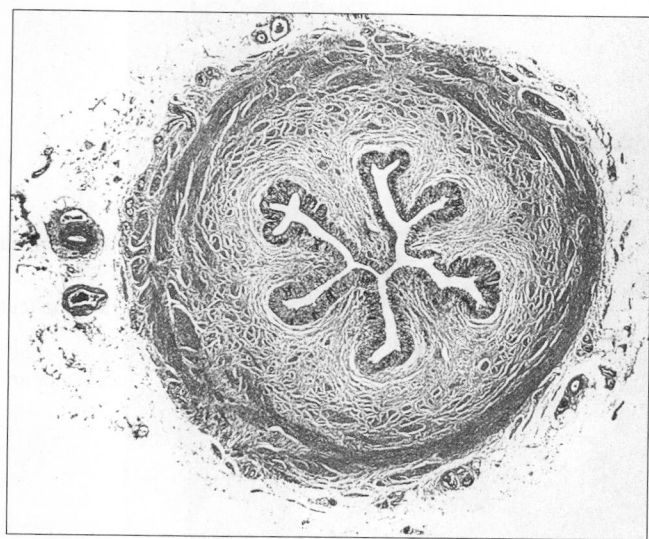

Figure 28-6 *Ureter (cross section).* Low power micrograph. Note the convoluted folds, covered by a specialized mucous lining, that almost fill the lumen. The thick muscular layer is surrounded by a tough fibrous coat.

Look at Figure 28-7. This x-ray, called an *intravenous pyelogram*, is considered normal. It was taken after intravenously administered contrast material had filtered out of the blood and filled the urinary system structures, including the so-called "*cobra head ureters*" seen in this image. The deformed-appearing distal ends of the ureters are dilated as they enter the bladder wall. As a result, they are certainly "atypical" in appearance. However, they are easily recognized as a rather common congenital variant of little clinical significance by radiologists or clinical anatomists, who are able to interpret the image with an appreciation of normal anatomical variation. Such understanding comes with multiple exposure to variability in anatomical dissection specimens or by image interpretation made in the context of experience and clinical findings.

Urinary Bladder

The urinary bladder is a muscular, collapsible bag that is located directly behind the symphysis pubis and in front of the rectum (see Figure 28-5, *B*). It lies below the parietal peritoneum, which covers only its superior surface. The remainder of the bladder surface is covered by a fibrous adventitia. In women it sits on the anterior of the vagina and in front of the uterus, whereas in men, it rests on the prostate.

The wall of the bladder is made mostly of smooth muscle tissue. Often called the **detrusor muscle**, the muscle layer is formed by a network of crisscrossing bundles of smooth muscle fibers. The bundles run in all directions: circular, oblique, and lengthwise. The bladder is lined with mucous transitional epithelium that forms folds called *rugae* (Figure 28-5, *A*). Because of the folds and the extensibility of transitional epithelium, the bladder can distend considerably. There are three openings in the floor of the bladder—two from the ureters and one into the **urethra.** The ureter openings lie at the posterior corners of the triangle-shaped floor (the *trigone*), and the urethral opening lies at the anterior, lower corner.

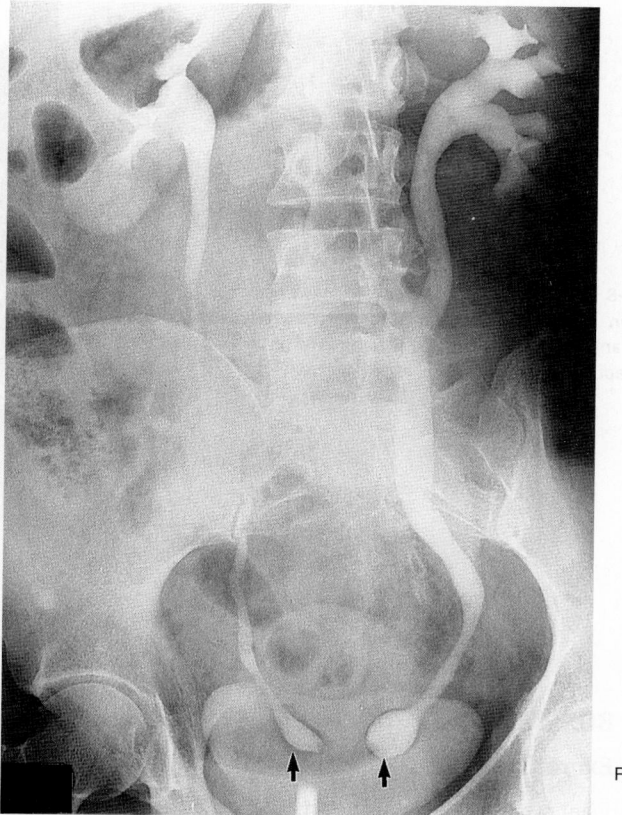

Figure 28-7 *Intravenous pyelogram.* Contrast material fills the urinary system structures. Note the "cobra head" deformity *(arrows)* of the distal ureters as they enter the bladder wall. Such dilations are generally considered a congenital variant of normal with little clinical significance.

The bladder performs two major functions:

1. It serves as a reservoir for urine before it leaves the body.
2. Aided by the urethra, it expels urine from the body.

Urethra

The urethra is a small tube lined with mucous membrane that leads from the floor of the bladder *(trigone)* to the exterior of the body. In females, it lies directly behind the symphysis pubis and anterior to the vagina as it passes through the muscular floor (levator ani muscle) of the pelvis (Figure 28-8). It extends down and forward from the bladder for a distance of about 3 cm (1.2 inches) and ends at the external urinary meatus (Figure 28-5, *B*). The functional importance of the relationship of the urethra and vagina to the muscular pelvic floor, especially after vaginal delivery of a baby, will be discussed in Chapter 32. The male urethra, on the other hand, extends along a winding path for about 20 cm (7.9 inches) (see Figure 28-5, *B*). The male urethra passes through the center of the *prostate gland* just after leaving the bladder. Within the prostate, it is joined by two *ejaculatory ducts*. After leaving the prostate, the urethra extends down, forward, then up to enter the base of the penis. It then travels through the center of the penis and ends as a *urinary meatus* at the tip of the penis.

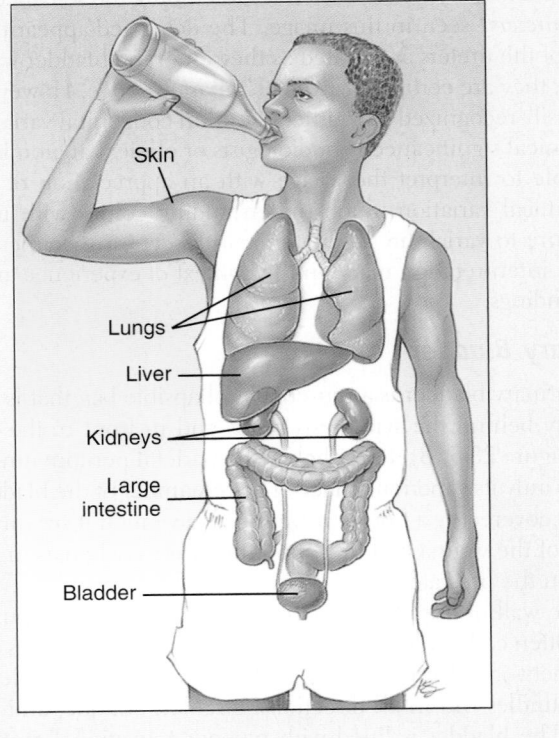

Figure 28-8 *Female urethra piercing the muscular pelvic floor.* The urethra is posterior to the symphysis pubis and anterior to the vagina. (See also text discussion in Chapter 32, p. 1134.)

Labels: Rectum (cut); Vagina (cut); Levator ani; Internal urethral meatus; Urethra; Pubic symphysis

BOX 28-2: FYI

Excretion

The urinary system's chief function is to regulate the volume and composition of body fluids and excrete unwanted material, but it is not the only system in the body that is able to excrete unneeded substances.

The table below compares the excretory functions of several systems. Although all of these systems contribute to the body's effort to remove wastes, only the urinary system can finely adjust the water and electrolyte balance to the degree required for normal homeostasis of body fluids.

SYSTEM	ORGAN	EXCRETION
Urinary	Kidney	Nitrogen compounds Toxins Water Electrolytes
Integumentary	Skin—sweat glands	Nitrogen compounds Electrolytes Water
Respiratory	Lung	Carbon dioxide Water
Digestive	Intestine	Digestive wastes Bile pigments Salts of heavy metals

Labels: Skin; Lungs; Liver; Kidneys; Large intestine; Bladder

Because the male urethra is joined by the ejaculatory ducts, it serves as a pathway for *semen* (fluid containing sperm) as it is ejaculated out of the body through the penis. Thus, we can say that the male urethra is a part of two different systems: the urinary system (when it is used to void urine) and the reproductive system (when it is used to ejaculate semen). Urine is prevented from mixing with semen during ejaculation by a reflex closure of sphincter muscles guarding the bladder's opening. The female urethral tract is separate from the lower reproductive tract (vagina), which lies just behind the urethra (see Figure 28-8).

Micturition

The mechanism for voiding urine begins with voluntary relaxation of the external sphincter muscle of the bladder. In rapid succession, different regions of the detrusor muscle in the bladder wall contract reflexively. This contraction forces urine out of the bladder and through the urethra. Parasympathetic fibers transmit the impulses that cause contractions of the bladder and relaxation of the internal sphincter. Voluntary contraction of the external sphincter to stop voiding is learned. Voluntary control of *micturition* (voiding, or urination) is possible only if the nerves supplying the bladder and urethra, the projection tracts of the CNS, and the motor areas of the brain are all intact and functioning properly. Injury to any of these parts of the nervous system, by a cerebral hemorrhage or a spinal cord injury, for example, results in involuntary emptying of the bladder at intervals. Involuntary micturition is called *incontinence* (see Mechanisms of Disease, p. 1058). In the average bladder, 250 ml of urine causes a moderately distended sensation and therefore the desire to void.

QUICK CHECK

1. Name the accessory organs of the urinary system.
2. What is the general function of the urinary system?
3. Distinguish between the renal cortex and the renal medulla.
4. What are the two major functions of the bladder?

Microscopic Structure of the Nephron

Microscopic functional units, named **nephrons** and numbering about 1.25 million per kidney, make up the bulk of the kidney. The shape of the nephron is unusual, unmistakable, and uniquely suited to its function of blood plasma processing and urine formation (see Figure 28-4). It resembles a tiny funnel with a long, winding stem about 3 cm (1.2 inches) long. About 85% of all nephrons are located almost entirely in the renal cortex and are called **cortical nephrons.** The remainder, called **juxtamedullary** (jux-tah-MED-oo-lair-ee) **nephrons,** lie near the junction (*juxta*) of the cortical and medullary layers but have loops of Henle that dip far into the medulla (see Figure 28-4, *A*). The specialized role of the juxtamedullary nephrons in concentrating urine is discussed later.

As Figure 28-4 shows, each nephron contains the following structures, in the order in which fluid flows through them:

- Renal corpuscle
- Bowman's capsule
- Proximal convoluted tubule
- Loop of Henle
- Distal convoluted tubule
- Collecting duct

As you read the brief description of each of these nephron components, refer often to Figure 28-4, *B*, which shows a schematic diagram of a complete nephron.

Renal Corpuscle

Bowman's Capsule

Bowman's capsule is the cup-shaped mouth of a nephron. It is formed by two layers of epithelial cells with a space, called *Bowman's space,* between them. Fluids, waste products, and electrolytes that pass through the porous glomerular capillaries and enter this space constitute the **glomerular filtrate,** which will be processed in the nephron to form urine. Figure 28-9 shows the relationship between the glomerular tuft of capillaries, formed by the branching of an afferent arteriole, and Bowman's capsule. Formation of a **renal corpuscle,** a glomerulus, and its surrounding Bowman's capsule, is sometimes compared to pushing your fist into the end of an inflated balloon. The mechanism is depicted in the schematic diagrams included in Figure 28-9. Note that as the glomerular tuft of capillaries pushes into the balloon, it becomes surrounded by a double-walled cup with parietal (outer) and visceral (inner) walls similar to Bowman's capsule shown in Figure 28-4, *B*. The parietal, or outer, wall is composed of simple squamous epithelium. It plays no role in the production of glomerular filtrate. The visceral (inner) wall, however, is quite different. It is composed of special epithelial cells called *podocytes* (meaning "cells with feet"). The scanning electron micrograph in Figure 28-11 reveals the odd shapes of podocytes. Notice that the primary branches extending from the cell bodies divide into a network of branches that terminate in little "feet" called *pedicels*. The pedicels are packed so closely together

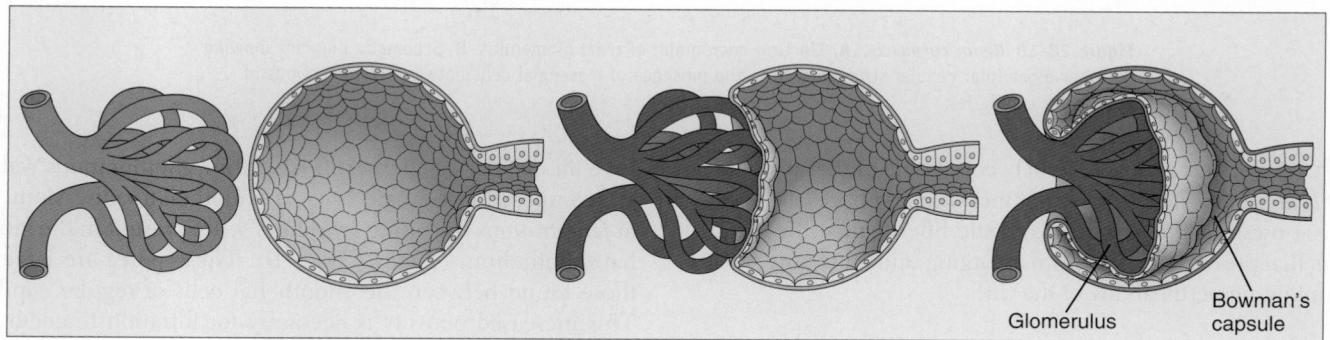

Figure 28-9 *Relationship between the glomerular capillary tuft and Bowman's capsule.* Note that as the glomerular tuft pushes into the inflated balloon-like structure representing Bowman's capsule, the visceral layer of the capsule adheres to the outer surface of the epithelial cells of the glomerular capillaries to become the visceral layer lining the capsular space. The outer layer of cells constitutes the parietal layer of Bowman's capsule. Glomerular filtrate enters the space between the parietal and visceral layers of Bowman's capsule before entering the tubular segments of the nephron.

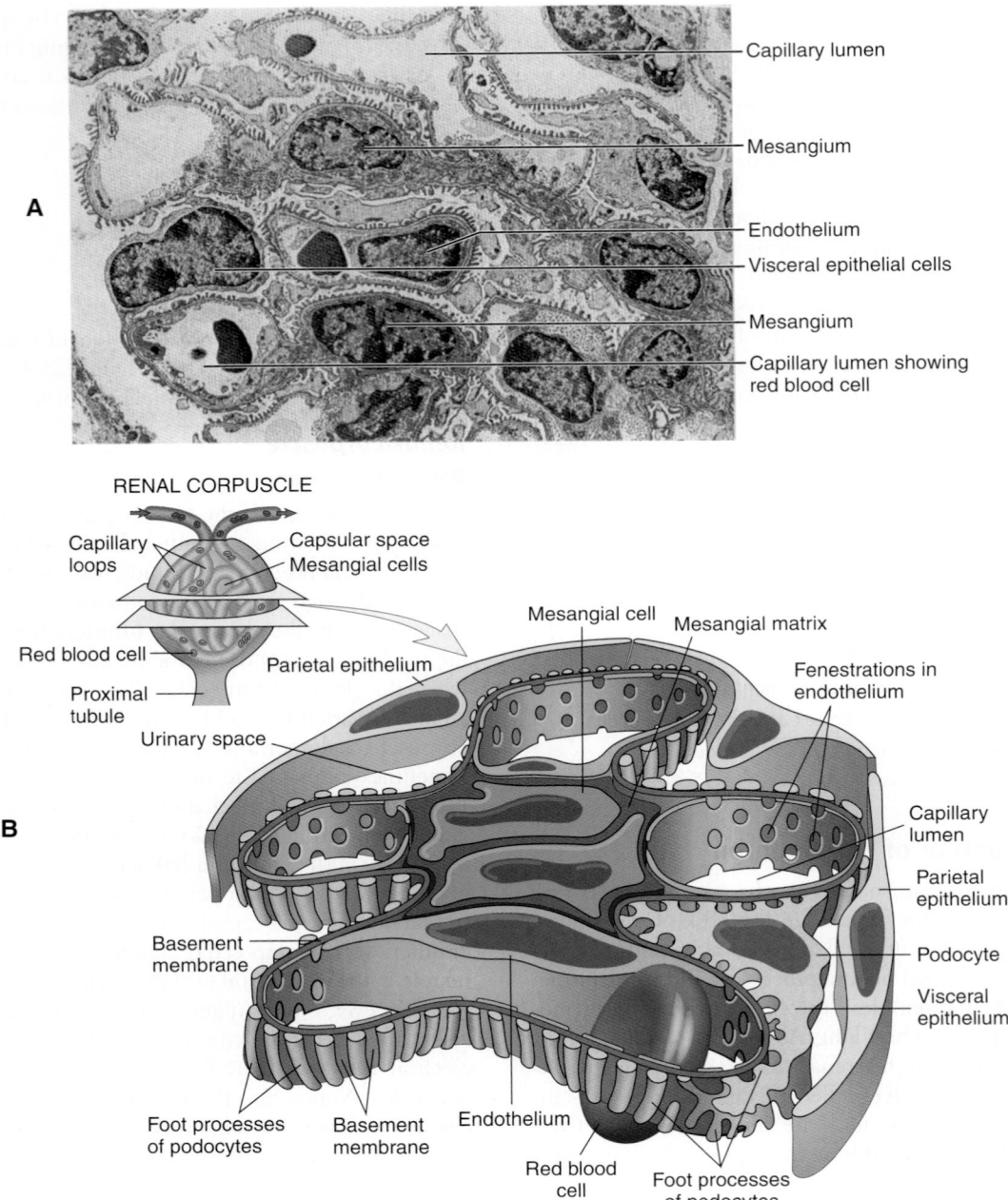

Figure 28-10 *Renal corpuscle.* **A,** Electron micrograph of renal glomerulus. **B,** Schematic diagram showing detail of glomerular cellular structures. Note the presence of mesangial cells between cross sections of glomerular capillaries.

that only narrow slits of space lie between them. These spaces are called *filtration slits*. The slits are not merely open spaces, however. There is a mesh of fine connective tissue fibers called the *slit diaphragm* that prevents the slits from enlarging under pressure while still maintaining permeability of the slit.

Glomerulus

Carefully examine Figure 28-11. Pay particular attention to the structure fitted neatly into Bowman's capsule. It is a network of fine capillaries that has a special name—**glomerulus.** The glomerulus is probably the body's most well known capillary network and is surely one of its most important ones for survival.

Like all capillaries, glomeruli have thin, membranous walls that are composed of a single layer of endothelial cells. Many pores, or *fenestrations* (meaning "windows"), are present in the glomerular endothelium (Figure 28-11, *B*). These pores are larger than those found between the endothelial cells of regular capillaries. This increased porosity is necessary for filtration to occur at the rate required for normal kidney function. The relationship between fenestration size and the rate of glomerular filtration is yet another example of the connection between form and function.

Mesangial cells (Figure 28-10) are unique to renal corpuscles. They are irregular in shape, have numerous cytoplasmic processes, and are scattered in an apparently haphazard way in an extracellular matrix between the twisting glomerular capillaries. Most physiologists believe they serve a support and phagocytic function, similar in some ways to microglia and other types of glia cells found in the nervous system (see Chapter 12, p. 434). However, the presence of myosin-like filaments and angiotensin II receptors in their cytoplasm also suggests a possible role in control of blood flow through the glomerular loop. Mesangial cells are currently the subject of considerable research interest and may well have other important functions or play an important role in human glomerular disease.

Between a glomerulus and its Bowman's capsule lies a basement membrane (basal lamina). It consists of a thin layer of fine fibrils embedded in a matrix of glycoprotein. The visceral layer of Bowman's capsule contacts the basement membrane by means of countless pedicels ("feet") of the podocytes (Figure 28-11, A). The glomerular endothelium, the basement membrane, and the visceral layer of Bowman's capsule constitute the **glomerular capsular membrane,** a structure well suited to its function of filtration (Figure 28-12).

Proximal Tubule

The proximal tubule is the second part of the nephron but the first part of the *renal tubule*. As its name suggests, the proximal tubule is the segment proximal, or nearest, to Bowman's capsule. Because it follows a winding, convoluted course, it is also called the *proximal convoluted tubule*. Its wall consists of one layer of epithelial cells. These cells have a brush border facing the lumen of the tubule. Thousands of microvilli form the brush border and greatly increase its luminal surface area—a structural fact of importance to its function, as we shall see.

Loop of Henle

The **loop of Henle** is the segment of renal tubule just beyond the proximal tubule. It consists of a *descending limb*, a sharp turn, and an *ascending limb*. Note in Figure 28-4, *B*, that the thin segment of the descending limb loops around to become the thick ascending limb. Figure 28-4, *A*, shows a nephron in which the loop of Henle dips deep into the medulla. As noted earlier, a nephron with a loop of Henle that dips far into the medulla is called a *juxtamedullary nephron* (meaning "nephron near the medulla"). Recall that the more superficial *cortical nephrons* constitute about 85% of the total nephron numbers. The length of the loop of Henle is important in the production of highly concentrated or very dilute urine.

Distal Tubule

The **distal tubule,** or *distal convoluted tubule*, is a convoluted portion of the tubule beyond (distal to) the loop of Henle. The **juxtaglomerular apparatus** (meaning "structure near the glomerulus") is found at the point where the afferent arteriole brushes past the distal tubule (see Figures 28-4 and 28-6). This structure is important in maintaining homeostasis of blood flow because it reflexively secretes *renin* when blood pressure in the afferent arteriole drops. Recall from Chapter 19 that renin triggers a mechanism that produces *angiotensin*, a substance that causes vasoconstriction and the resulting increase in blood pressure (see Figure 28-13). The specialized cells of the juxtaglomerular apparatus represent a modification of cells in the walls of both the distal tubule and the afferent arteriole at the point where they touch one another. Large smooth muscle cells in the wall of the afferent arteriole—called **juxtaglomerular (JG) cells**—contain renin granules. These cells are sensitive to increased pressure in the ar-

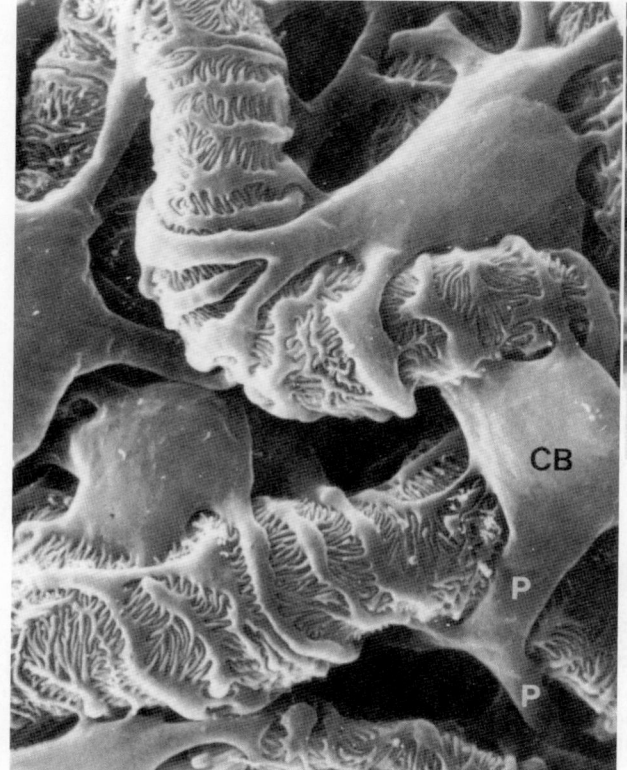

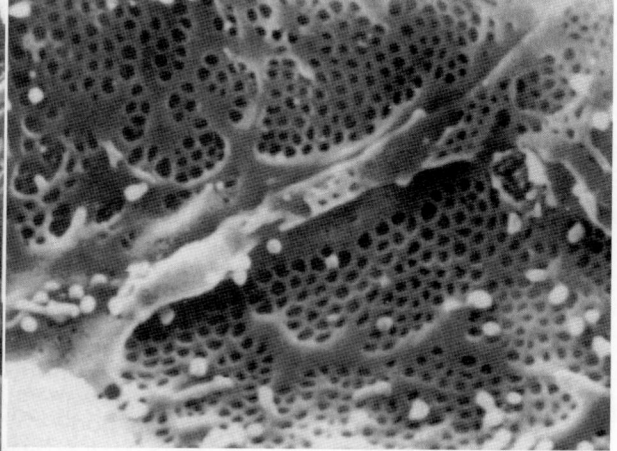

Figure 28-11 *Glomerular capillaries (scanning electron micrographs).* **A,** View of the surface of a glomerular capillary from the vantage point of Bowman's space. A podocyte cell body *(CB)* and podocyte foot processes *(P)* are shown on the outer surface of a capillary endothelial cell. **B,** View of the glomerular capillary wall from the vantage point of the capillary lumen. Note the multiple fenestrations that perforate the endothelial cell walls.

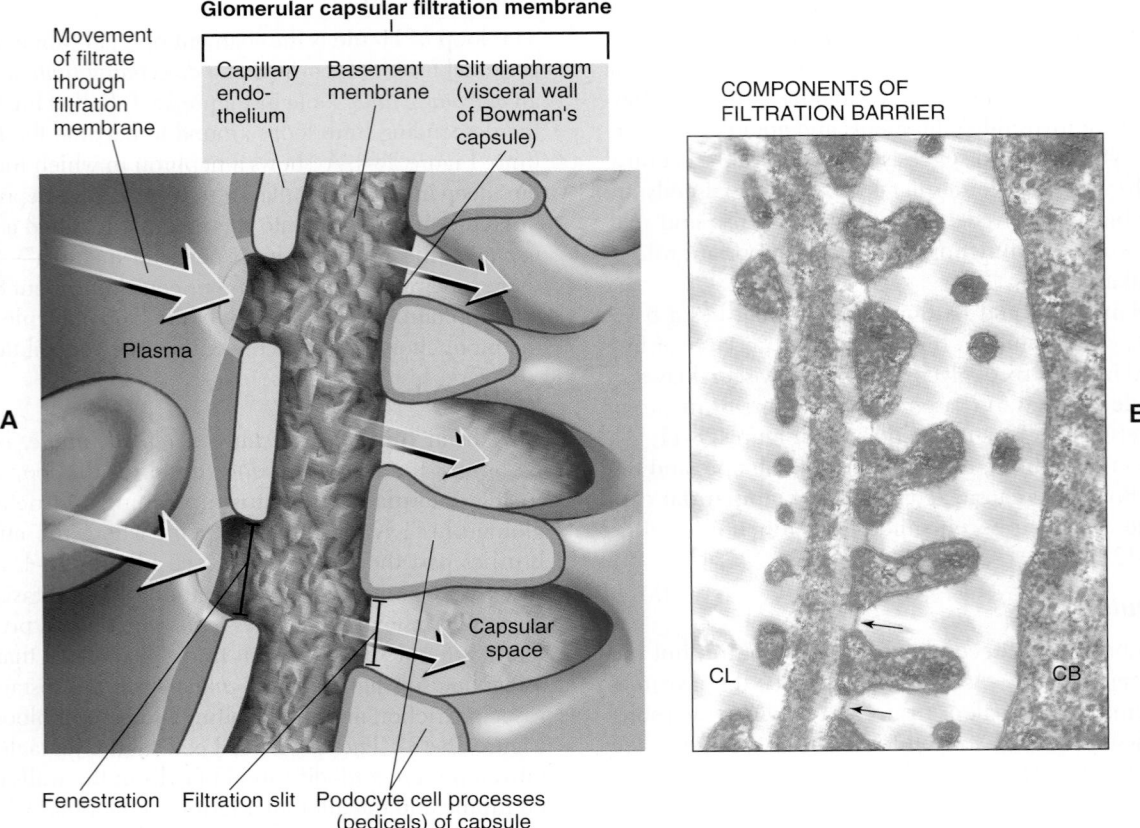

Figure 28-12 *Glomerular capsular membrane*. A, Artist's rendering of a transmission electron micrograph (TEM) showing the filtration membrane formed when footlike extensions (pedicels) of cells forming the visceral wall of Bowman's capsule share a basement membrane with the fenestrated endothelial cells that form the wall of glomerular capillaries. **B,** TEM of the glomerular capsular membrane. Note from left to right, the capillary lumen (CL), fenestrations in the capillary endothelium, and multiple podocyte foot processes separated by slit diaphragms *(arrows)* extending into Bowman's space. A portion of an overarching podocyte cell body (CB) is visible on the right.

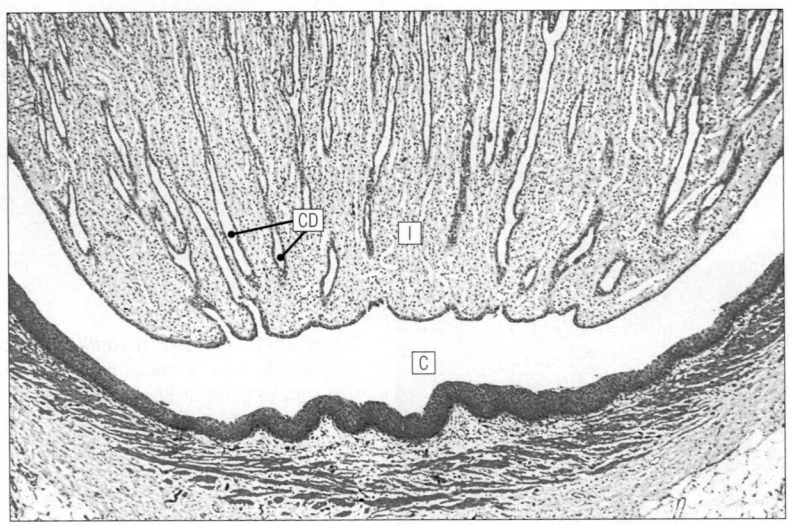

Figure 28-13 *Renal papilla*. Collecting ducts *(CD)* can be seen opening into a calyx *(C)* at the papillary tip. The interstitial tissue *(I)* includes some Henle loops and vasa recta vessels.

teriole and are considered functional *mechanoreceptors*. Modified distal tubule cells in the juxtaglomerular apparatus form a dense, tightly packed structure called the **macula densa.** Cells in the macula densa are *chemoreceptors* that can sense the concentration of solute materials in the fluid passing through the tubule. Acting together, both cell types in the juxtaglomerular apparatus contribute to homeostasis of renal function by influencing the ability of the kidney to produce concentrated urine.

Collecting Duct

The **collecting duct** is a straight tubule joined by the distal tubules of several nephrons. Collecting ducts join larger ducts, and all the larger collecting ducts of one renal pyramid converge to form one tube that opens at a renal papilla into one of the small calyces (Figure 28-13). Bowman's capsules and both convoluted tubules lie in the cortex of the kidney, whereas the loops of Henle and collecting ducts extend into the medulla (see Figure 28-4).

QUICK CHECK

5. Name the segments of the nephron in the order in which fluid flows through them.
6. What characteristics of the glomerular capsular membrane permit filtration?
7. What is the name of the capillary network that surrounds the nephron?

PHYSIOLOGY OF THE URINARY SYSTEM

Overview of Kidney Function

The chief functions of the kidney are to process blood plasma and excrete urine. These functions are vital because they maintain the homeostatic balance of the body. For example, the kidneys are the most important organs in the body for maintaining fluid-electrolyte and acid-base balance. The kidneys do this by varying the amount of water and electrolytes leaving the blood in the urine so that they equal the amounts of these substances entering the blood from various other avenues. Nitrogenous wastes from protein metabolism, notably *urea*, leave the blood by way of the kidneys.

Here are just a few of the blood constituents that cannot be held within their normal concentration ranges if the kidneys fail:

- Sodium
- Potassium
- Chloride
- Nitrogenous wastes (especially urea)

In short, kidney failure means homeostatic failure and, if not relieved, inevitable death.

In addition to processing blood plasma and forming urine, the kidneys also perform other important functions. They influence the rate of secretion of the hormones antidiuretic hormone (ADH) and aldosterone and synthesize the active form of vitamin D, the hormone *erythropoietin*, and certain prostaglandins.

As you already know, the basic functional unit of the kidney is the nephron. It has two main parts—the renal corpuscle and renal tubule—that form urine by means of three processes:

1. **Filtration,** or movement of water and protein-free solutes from plasma in the glomerulus, across the glomerular capsular membrane, and into the capsular space of Bowman's capsule
2. **Tubular reabsorption,** or movement of molecules out of the various segments of the tubule and into the peritubular blood
3. **Tubular secretion,** or movement of molecules out of peritubular blood and into the tubule for excretion

These three mechanisms are used in concert to process blood plasma and form urine. First, a hydrostatic pressure gradient drives the filtration of much of the plasma into the nephron (Figure 28-14). Because the filtrate contains materials that the body must conserve (save), the walls of the tubules start reabsorbing these materials back into the blood. As the filtrate (urine) begins to leave the nephron, the kidney may secrete a few "last minute" items into the urine for excretion. In short, the kidney does not selectively filter out only harmful or excess material. It first filters out much of the plasma, then reabsorbs what should not be "thrown out" before the filtrate reaches the end of the tubule and becomes urine. This mechanism allows very fine adjustments to blood homeostasis, as we shall see. Figure 28-15 shows the amounts of some important molecules that are filtered, then reabsorbed, by the nephron.

Filtration

Filtration, the first step in blood processing, is a physical process that occurs in the kidneys' 2.5 million renal corpuscles (see Figures 28-10 and 28-12). As blood flows through the glomerular

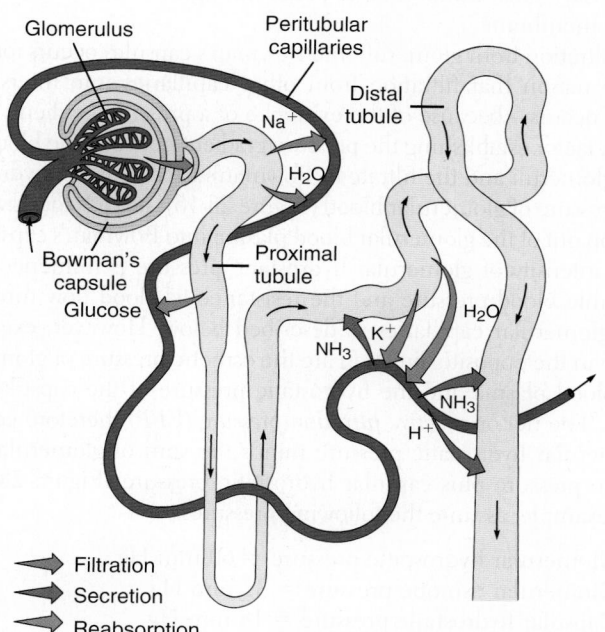

Figure 28-14 *Mechanism of urine formation.* The diagram shows the mechanisms of urine formation—filtration, reabsorption, and secretion—and where they occur in the nephron.

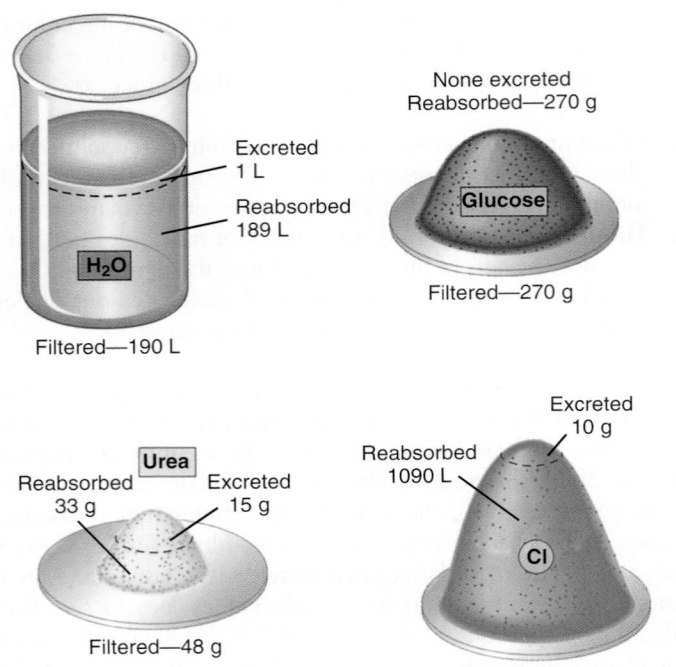

Figure 28-15 *Filtration and reabsorption volumes.* Note the enormous volume of water that is filtered out of glomerular blood per day—180 liters, or many times the total volume of blood in the body. Only a small proportion of this water, however, is excreted into urine. More than 99% of it (179 liters) is reabsorbed into tubular blood.

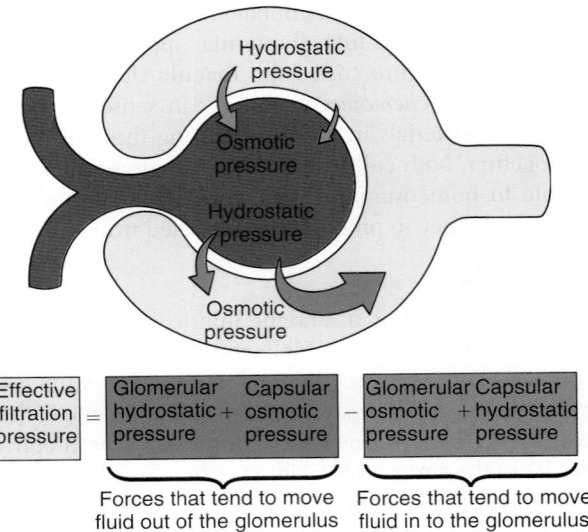

Figure 28-16 *Forces affecting glomerular filtration.* Effective filtration pressure (EFP) is determined by comparing the forces that push fluid into the capillary with those that push it out of the capillary.

capillaries, water and small solutes filter out of the blood into Bowman's capsules. The only blood constituents that do not move out are the blood solids (cells) and most plasma proteins. The result is about 180 liters of glomerular filtrate being formed each day. This filtration takes place through the glomerular capsular membrane.

Filtration from glomeruli into Bowman's capsules occurs for the same reason that filtration from other capillaries into interstitial fluid occurs—because of the existence of a pressure gradient. The main factor establishing the pressure gradient between the blood in the glomeruli and the filtrate in Bowman's capsule is the hydrostatic pressure of glomerular blood (Figure 28-16). It tends to cause filtration out of the glomerular blood plasma into Bowman's capsules. The intensity of glomerular hydrostatic pressure is influenced by systemic blood pressure and the resistance to blood flow through the glomerular capillaries as described below. However, exerting force in the opposite direction are the osmotic pressure of glomerular blood plasma and the hydrostatic pressure of the capsular filtrate. The net or *effective filtration pressure (EFP)* therefore equals glomerular hydrostatic pressure minus the sum of glomerular osmotic pressure plus capsular hydrostatic pressure (Figure 28-16). For example, assume the following pressures:

- Glomerular hydrostatic pressure = 60 mm Hg
- Glomerular osmotic pressure = 32 mm Hg
- Capsular hydrostatic pressure = 18 mm Hg
- Capsular osmotic pressure = negligent (±0 mm Hg)

The EFP (effective filtration pressure) using these figures equals (60 + 0) − (32 + 18), or 10 mm Hg. An effective filtration

pressure of 1 mm Hg, according to some investigators, produces a glomerular filtration rate of 12.5 ml per minute (including both kidneys). With an EFP of 10 mm Hg, the glomerular filtration rate would be 125.0 ml per minute, or about 180 liters in a 24-hour period, a normal rate. Since only about 1.5 liters of urine is excreted each day, more than 99% of the filtrate must be reabsorbed from the tubular segments of the nephron.

Filtration occurs more rapidly out of glomeruli than out of other tissue capillaries. One reason for this is a structural difference between the endothelium of glomeruli and that of tissue capillaries. Glomerular endothelium has many more pores (fenestrations) in it, so it is more permeable than tissue capillary endothelium. Another reason for more rapid glomerular filtration than tissue capillary filtration is that glomerular hydrostatic pressure is higher than tissue capillary pressure. The reason for this, briefly, is that the efferent arteriole has a smaller diameter than the afferent arteriole does. Therefore, it offers more resistance to blood flow out of the glomerulus than venules offer to blood flow out of tissue capillaries.

The glomerular filtration rate (GFR) is directly proportional to the effective filtration pressure (EFP) and can be altered by changes in the diameter of the afferent and efferent arterioles or by changes in the systemic blood pressure (Box 28-3). It can also be altered indirectly by changes in the efficiency of cardiac contraction. Stress may lead to intense sympathetic stimulation of the arterioles with greater constriction of the afferent than the efferent arteriole. Consequently, glomerular hydrostatic pressure falls. In severe stress, it may even drop to a level so low that the EFP falls to zero. No glomerular filtration then occurs. The kidneys "shut down," or in technical language, *renal suppression* occurs.

Glomerular hydrostatic pressure and filtration are directly related to systemic blood pressure. That is, a decrease in blood pressure tends to produce a decrease in both glomerular pressure and the glomerular filtration rate. The converse is also true. However, when arterial pressure increases, a smaller increase in glomerular pressure follows because the afferent arterioles constrict. This de-

Changes in Glomerular Filtration Rate

In some types of kidney disease, the permeability of the glomerular endothelium increases sufficiently to allow plasma proteins to filter out into the capsule. Osmotic pressure then develops in the capsular filtrate. To determine how this affects the effective filtration pressure in the glomeruli, first calculate the normal EFP by using the figures that the table lists as "Normal" in the formula given in Figure 28-16. After you have obtained your answer, apply the formula to the figures given under "Kidney Disease" in the table. You should have found an EFP of 10 mm Hg with the normal figures and an EFP of 15 mm Hg with the kidney disease figures. A change in the glomerular effective filtration pressure produces a similar change in the glomerular filtration rate. Therefore, loss of plasma protein into urine increases not only the EFP but also the glomerular filtration rate.

Intense exercise causes temporary proteinuria in many individuals. Some exercise physiologists believed that intense athletic activities cause kidney damage, but subsequent research has ruled out that explanation. One current hypothesis is that hormonal changes during strenuous exercise increase the permeability of the nephron's filtration membrane, thereby allowing more plasma proteins to enter the filtrate. Some degree of postexercise proteinuria is usually considered normal.

NORMAL AND ABNORMAL PRESSURE IN THE RENAL CORPUSCLE

	HYDROSTATIC PRESSURE	OSMOTIC PRESSURE
Normal		
Glomerular blood	60 mm Hg	32 mm Hg
Capsular filtrate	18 mm Hg	0 mm Hg
Kidney Disease		
Glomerular blood	60 mm Hg	32 mm Hg
Capsular filtrate	18 mm Hg	5 mm Hg

BOX 28-4
Diagnostic Study

Blood Indicators of Renal Dysfunction

Elevated urea levels in blood, as measured in a blood urea nitrogen (BUN) test, was one of the earliest clinical measurements of kidney dysfunction. Elevated BUN levels indicate failure of the kidney to clear urea and, therefore, other substances as well. (**Renal clearance** is the volume of plasma from which a substance is removed from the blood by the kidneys per minute.)

Blood levels of creatinine are also used to test renal function. Creatinine levels in blood seldom change significantly because they are determined by skeletal muscle mass—which seldom changes much. Therefore, an increase in the blood level of serum creatinine is considered to be a reliable indicator of depressed renal function.

creases blood flow into the glomeruli and prevents a marked rise in glomerular pressure or glomerular filtration. For instance, when the mean arterial blood pressure doubles, glomerular filtration reportedly increases only 15% to 20%.

QUICK CHECK

8. What are the three basic processes a nephron uses to form urine?
9. What is the GFR? Why is a high GFR important to kidney function?
10. How does blood pressure affect filtration in the kidney?

Reabsorption

Reabsorption, the second step in urine formation, takes place by means of passive and active transport mechanisms from all parts of the renal tubules. A major portion of water and electrolytes and (normally) all nutrients are, however, reabsorbed from the proximal tubules. The rest of the renal tubule reabsorbs comparatively little of the filtrate. Researchers are still investigating the exact mechanisms of reabsorption in the various segments of the nephron. We have summarized only the essential principles of some of the current concepts in the following paragraphs.

Reabsorption in the Proximal Tubule

As already stated, most of the 180 liters of filtrate that enters the renal tubule from Bowman's capsule each day does not get very far. More than two thirds of it is reabsorbed before it reaches the end of the proximal tubule.

Proximal tubules are thought to reabsorb sodium and other major ions in this manner: sodium ions (Na^+) are actively transported out of the lumen of the tubule and into peritubular blood by the mechanism summarized in Figure 28-17. The microvilli on the luminal surface of each epithelial cell in the tubule wall form a brush border that increases the absorptive surface area of the entire inner face of the tubule. As sodium ions accumulate in the interstitial fluid, the interstitial fluid becomes temporarily positive with respect to the tubule fluid. This electrical gradient (difference in net charge) drives the diffusion of negative ions from the filtrate, into the interstitial fluid and, eventually, into the peritubular blood. In other words, the attraction between negative and positive ions is used to drive the passive transport of chloride (Cl^-), phosphate ($PO_4^=$) and other negative ions out of the tubule.

As the concentration of ions in peritubular blood increases, the blood becomes momentarily hypertonic to the tubule fluid. Through the process of osmosis, water diffuses rapidly from the tubule fluid into peritubular blood, thus making the two fluids isotonic. In short, ion transport out of the proximal tubules causes water osmosis out of them. This is obligatory water reabsorption—obligatory because it is demanded by the principle of osmosis.

Proximal tubules reabsorb nutrients from the tubule fluid, notably glucose and amino acids, into peritubular blood by a special type of active transport mechanism called **sodium cotrans-**

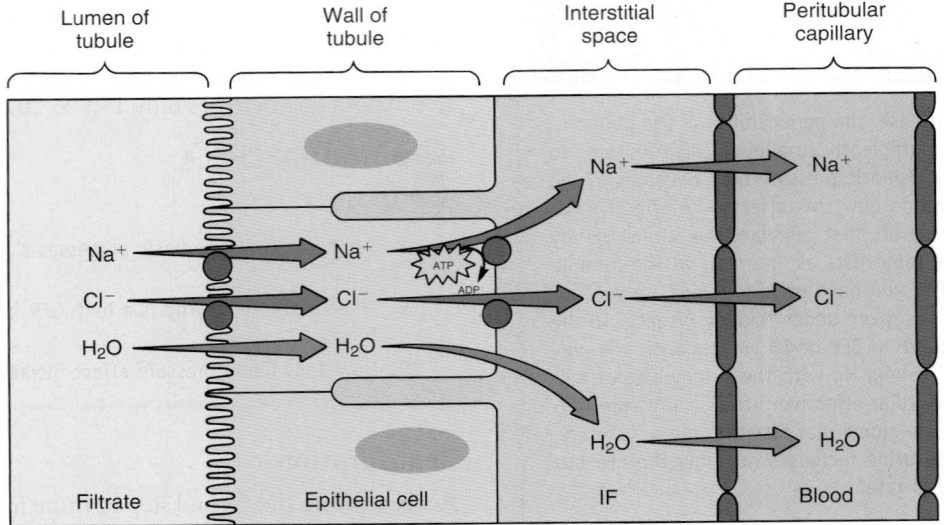

Figure 28-17 *Mechanisms of tubular reabsorption.* Sodium ions (Na^+) are pumped from the tubule cell to interstitial fluid (IF), thereby increasing the interstitial Na^+ concentration to a level that drives diffusion of Na^+ into blood. As Na^+ is pumped out of the cell, more Na^+ passively diffuses in from the filtrate to maintain an equilibrium of concentration. Enough Na^+ moves out of the tubule and into blood that an electrical gradient is established (blood is positive relative to the filtrate). Electrical attraction between oppositely charged particles drives diffusion of negative ions in the filtrate, such as chloride (Cl^-), into blood. As the ion concentration in blood increases, osmosis of water from the tubule occurs. Thus, active transport of sodium creates a situation that promotes passive transport of negative ions and water.

port. Recall from Chapter 26 that, in this mechanism, a carrier molecule in the cell membrane first binds to sodium and glucose (see Figure 26-20, p. 981). The carrier then passively transports both substances through the brush border of a proximal tubule cell into the cell's interior (facilitated diffusion). Sodium moves into the cell because of a concentration gradient maintained by the active transport of sodium out the other side of the cell. Glucose actually moves up its concentration gradient, but no energy is required because it is "riding the coattails" of sodium. Once inside the cell, the substances dissociate from the carrier molecule and diffuse to the far side of the cell. The substances move out of the epithelial cell by different mechanisms. Sodium is transported actively, and glucose is transported passively. Normally, all the glucose that has filtered out of the glomeruli returns to the blood by this sodium cotransport mechanism. Therefore, very little glucose is lost in urine. If, however, the blood glucose level exceeds a threshold amount (about 150 mg/100 ml), not all of the glucose can be reabsorbed. The excess glucose remains in urine. The maximum capacity for moving glucose molecules back into blood is determined by the number of cotransport carriers available.

Urea is a nitrogen-containing waste formed as a result of protein catabolism (see Chapter 27). Actually, toxic ammonia is formed first, but much of it is quickly transformed into the less toxic urea. Urea in the tubule fluid remains in the proximal tubule as sodium, chloride, and water are reabsorbed into blood. Once these materials are gone, a tubule fluid high in urea is left. Because the urea concentration in the tubule is now greater than its concentration in peritubular blood, urea passively diffuses into the blood. About half the urea present in the tubule fluid leaves the proximal tubule this way.

Reabsorption in the proximal tubules can be summarized in the following manner:

1. Sodium is actively transported out of the tubule fluid and into blood.
2. Glucose and amino acids "hitch a ride" with sodium and passively move out of the tubule fluid by means of the sodium cotransport mechanism.
3. Chloride ions passively move into blood plasma because of an imbalance in electrical charges (positive sodium ions have already moved out, thus making the plasma positive and the tubule fluid negative).
4. Movement of sodium and chloride out of the tubule fluid into plasma creates an osmotic imbalance (the blood is hypertonic to the filtrate), so water is obliged by the principle of osmosis to passively move into blood.
5. About half the urea present in the tubule fluid passively moves out of the tubule, with half the urea thus left to move on to the loop of Henle.
6. The total content of the filtrate has been reduced greatly by the time it is ready to leave the proximal tubule. Most of the water and solutes have been recovered by the blood, and only a small volume of fluid is left to continue to the next portion of the tubule, the loop of Henle.

 QUICK CHECK

11. How are NaCl and water reabsorbed in the proximal tubule?
12. What is sodium cotransport?

BOX 28-5: HEALTH MATTERS

Glucose in Urine

Occasionally, the maximum cotransport capacity is greatly reduced, and glucose appears in urine *(glycosuria)*, even though the blood sugar level may be normal. This condition is known as *renal diabetes* or *renal glycosuria*. It is a congenital defect.

Of course, the most common cause of glycosuria is *diabetes mellitus* (see Chapter 16). In this condition, insulin deficiency or target cell dysfunction causes glucose to accumulate in the blood, causing hyperglycemia. The high glucose content of the filtrate formed in the renal corpuscle exceeds the maximum capacity of the cotransport mechanism. Therefore, glycosuria results.

Reabsorption in the Loop of Henle

In juxtamedullary nephrons—those low in the cortex, near the medulla—the loop of Henle and its vasa recta participate in a very unique type of process called a **countercurrent mechanism.** A countercurrent structure is any set of parallel passages in which the contents flow in opposite directions (Figure 28-18). The loop of Henle is a countercurrent structure because the contents of the ascending limb travel in an opposite direction to the flow of urine in the descending limb. The vasa recta also have a countercurrent structure because arterial blood flows down into the medulla and venous blood flows up toward the cortex. The kidney's countercurrent mechanism functions to keep the solute concentration of the medulla extremely high. We will discuss the reason why this is important later. For now, we will briefly discuss how the countercurrent mechanism achieves this goal.

Before we can understand the kidney's countercurrent mechanisms, we must appreciate the histology of the loop of Henle. The descending limb is formed by a much thinner wall than the thick part of the ascending limb (see Figure 28-4). Even more important, the permeability and transport abilities of the two walls are very different. The thin-walled descending limb allows water and urea to diffuse freely into or out of the tubule, depending on their concentration gradients. The thick-walled ascending limb, however, limits the diffusion of most molecules (including water, sodium, chloride, and urea) while actively transporting selected molecules out of the tubule and into the interstitial fluid.

Given the characteristics of each limb, we can see how the system illustrated in Figure 28-19 can develop in the loop of Henle. Look at the ascending limb. You will see that this limb is actively pumping sodium and chloride out of the tubule fluid and into interstitial fluid. This probably occurs by the same mechanism that operates in the proximal tubule. Normally, sodium and chloride ions simply diffuse right back into the tubule fluid to achieve equilibrium. The ascending limb prevents the diffusion of these ions, so they are "trapped" in the interstitial area. Under normal circumstances, water moves from the tubule fluid to the interstitial fluid to achieve osmotic balance. However, the wall of the ascending limb is relatively impermeable to water. In short, salt ions are pumped out of the ascending limb, water is prevented from following osmotically, and thus the tubule fluid develops a low solute concentration (low os-

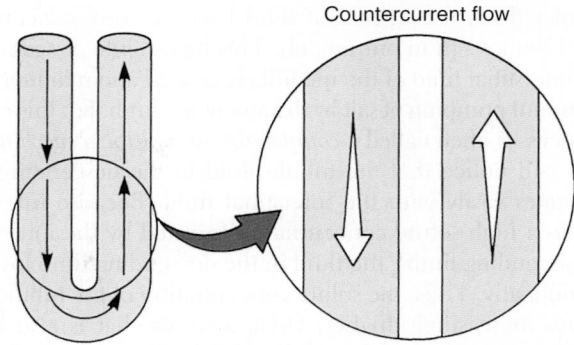

Figure 28-18 *Concept of countercurrent flow.* Countercurrent flow simply refers to flow in opposite directions, as the *inset* shows. Tubule filtrate in the loop of Henle flows in a countercurrent manner, as does blood flowing with the vasa recta of the peritubular capillary network.

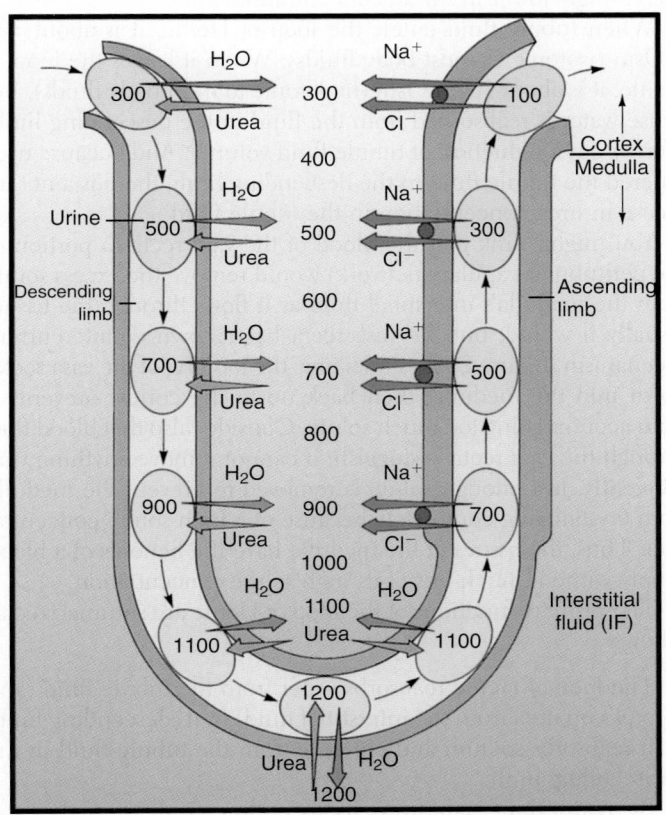

Figure 28-19 *The countercurrent multiplier system in the loop of Henle.* Na^+ and Cl^- are pumped from the ascending limb and moved into interstitial fluid (IF) to maintain high osmolality there. Because the salt content of the medullary IF increases, this is called a "multiplier" mechanism. Ion pumping also lowers the tubule fluid's osmolality by 200 mOsm, so fluid leaving the loop of Henle is only 100 mOsm (hypotonic), as compared with 300 mOsm (isotonic) when it entered the loop. Numbers in the diagram are expressed in milliosmoles (mOsm).

motic pressure) and interstitial fluid develops a high solute concentration (high osmotic pressure).

The ion pumps in the ascending limb can maintain an osmotic difference of 200 mOsm across the wall of the tubule. Notice in Figure 28-19 that the movement of salt out of the tubule at any horizontal level creates a difference of 200 mOsm between the tubule fluid and interstitial fluid. Because salt is continually added to the interstitial fluid, the interstitial fluid becomes very concentrated (up to 1200 mOsm in our model). This high solute concentration of the interstitial fluid of the medulla is created and maintained by the constant pumping of salt by the ascending limb. For this reason, this process is often called a *countercurrent multiplier mechanism.*

You will notice that the tubule fluid in the descending limb equilibrates easily with the interstitial fluid. Because interstitial fluid has a high solute concentration (created by the ion pumps in the ascending limb), the fluid in the descending limb loses water osmotically. Thus, the solute concentration of the tubule fluid becomes increasingly higher. Urea, a solute that is also highly concentrated in the renal medulla, diffuses into the tubule fluid in the descending limb—thereby increasing the solute concentration of the tubule fluid even more. However, as the fluid "rounds the bend" and begins moving into the thick portion of the ascending limb, its Na^+ and Cl^- are removed and it becomes increasingly lower in solute concentration.

When tubule fluid enters the loop of Henle, it is about 300 mOsm (isotonic to most body fluids). When it leaves the loop of Henle, it is about 100 mOsm (hypotonic to most body fluids). Because water is reabsorbed from the fluid in the descending limb, there is a net reduction of tubule fluid volume. And because urea entered the tubule fluid in the descending limb, there is a net increase in urea concentration in the tubule fluid.

You might think that the blood of the vasa recta (a portion of the peritubular capillary network) would remove the excess solute from the medulla's interstitial fluid as it flows through the tissue. Usually it would, but the vasa recta has its own countercurrent mechanism. Figure 28-20 shows how the looping of the vasa recta, down into the medulla, then back up to the cortex, prevents it from accumulating too much solute. Consider also that blood flow through the vasa recta is sluggish; it cannot remove anything very efficiently. Just enough solute is removed to prevent the medulla from crystallizing completely because of a high solute concentration. Thus, the tissues of the medulla have the benefits of a blood supply without much loss of its high solute concentration.

The primary functions of the loop of Henle are summarized as follows:

1. The loop of Henle reabsorbs water from the tubule fluid (and picks up urea from the interstitial fluid) in its descending limb. It reabsorbs sodium and chloride from the tubule fluid in the ascending limb.
2. By reabsorbing salt from its ascending limb, it makes the tubule fluid dilute (hypotonic).
3. Reabsorption of salt in the ascending limb also creates and maintains a high osmotic pressure, or high solute concentration, of the medulla's interstitial fluid.

 QUICK CHECK

13. What is a *countercurrent mechanism*?
14. How does the function of the descending limb of the loop of Henle differ from the function of the thick ascending limb?
15. What is the purpose of the countercurrent multiplier mechanism of the loop of Henle?

Reabsorption in the Distal Tubules and Collecting Ducts

The distal tubule is similar to the proximal tubule in that it also reabsorbs some sodium by active transport, but in much smaller amounts. Left to themselves, the cells that form the distal tubule's walls are relatively impermeable to water. This means that sodium can be removed, but water cannot follow osmotically, so the solute concentration of the tubule fluid continues to decrease. Recall that the tubule fluid is already hypotonic to most body fluids at this point because of the countercurrent system in the loop of Henle.

The cells that form the wall of the collecting duct also prevent water from leaving the filtrate by osmosis. Even though the collecting duct conducts the tubule fluid through the hypertonic medullary region, equilibration does not occur.

Given no other circumstances, the kidney produces and excretes only very dilute (hypotonic) urine (Figure 28-21). This would be catastrophic because the body would soon dehydrate. A regulatory mechanism centered outside the kidney normally prevents excessive loss of water. This mechanism, illustrated in Figure 28-22, involves *antidiuretic hormone (ADH)*, a hormone secreted by the neurohypophysis (posterior pituitary). ADH targets cells of the distal and collecting tubules and causes them to become more permeable to water. When this happens, water is permitted to flow osmotically out of the tubule and into the interstitial fluid, toward an equilibrium. The more ADH present, the more water is allowed out of the tubule, and the closer the tubule

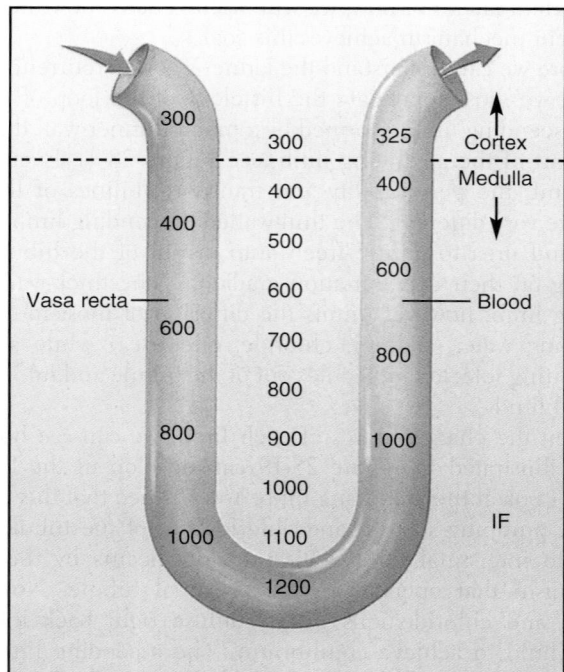

Figure 28-20 *Countercurrent exchange mechanism in the vasa recta.* Because the vasa recta forms a countercurrent loop, blood leaving the capillary bed has only a slightly higher solute content than when it entered. Thus, the high osmolality of medullary tissue fluid is maintained. If peritubular blood instead traveled straight through the tissue, all excess solute in the medulla would be removed, and the osmolality of medullary interstitial fluid would be equivalent to that of the cortex. Numbers in the diagram are expressed in milliosmoles.

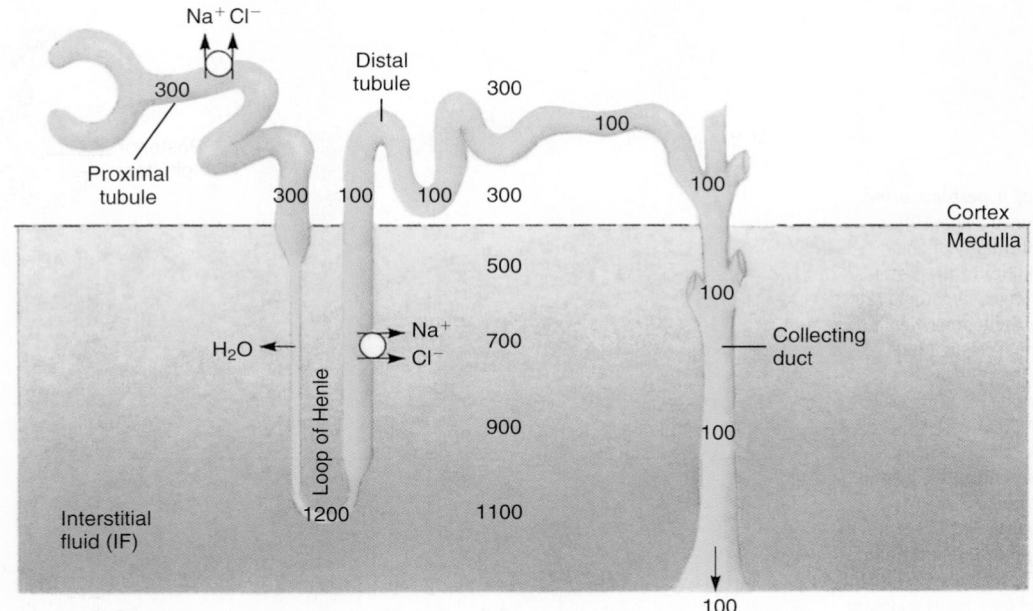

Figure 28-21 *Production of hypotonic urine.* Hypotonic urine is produced by the nephron by the mechanism shown here. The isotonic (300 mOsm) tubule fluid that enters the loop of Henle becomes hypotonic (100 mOsm) by the time it enters the distal tubule. The tubule fluid remains hypotonic as it is conducted out of the kidney because the walls of the distal tubule and collecting duct are impermeable to H_2O, Na^+, and Cl^-. Values are expressed in milliosmoles.

fluid's solute concentration matches that of the surrounding tissue fluid. In this way, the tubule fluid's osmotic pressure could go as high as 1200 mOsm, since the medulla's interstitial fluid can be that high. The solute concentration of the urine excreted depends in large part on the amount of ADH present.

Notice in Figure 28-22 that reabsorption of urea also occurs in the collecting duct when water is reabsorbed under the influence of ADH. As water is reabsorbed from the fluid descending through the collecting duct, the urea concentration of the fluid rises. Because the urea concentration is higher inside the lower part of the collecting duct than it is in the surrounding interstitial fluid, urea diffuses out of the lower collecting duct. The addition of urea to the medullary interstitial fluid assists in maintaining a high solute concentration in the medulla. Less than half the urea that leaves the collecting duct is removed by the vasa recta. The dashed line in Figure 28-22 shows that much of the urea in the medullary interstitial fluid diffuses into the descending limb of the loop of Henle. Thus, urea participates in a sort of countercurrent multiplier mechanism that, together with the countercurrent mechanisms of the loop of Henle and vasa recta, maintains the high osmotic pressure needed to form concentrated urine and therefore avoid dehydration.

Tubular Secretion

In addition to reabsorption, tubule cells also secrete certain substances. Tubular secretion means the movement of substances out of the blood and into tubular fluid. Recall that the descending limb of the loop of Henle removes urea by means of diffusion. The distal and collecting tubules secrete potassium, hydrogen, and ammonium ions. They actively transport potassium ions (K^+) or hydrogen ions (H^+) out of the blood into tubule fluid in exchange for sodium ions (Na^+), which diffuse back into the blood (H^+ transport is discussed further in Chapter 30). Potassium secretion increases when

the blood aldosterone concentration increases. *Aldosterone*, a hormone of the adrenal cortex, targets distal and collecting tubule cells and causes them to increase the activity of the sodium-potassium pumps that move sodium out of the tubule and potassium into the tubule. Hydrogen ion secretion increases when the blood hydrogen ion concentration increases. Ammonium ions are secreted into the tubule fluid by diffusing out of the tubule cells where they are synthesized. Tubule cells also secrete certain drugs, for example, penicillin and para-aminohippuric acid (PAH).

Table 28-1 summarizes the functions of the different parts of the nephron in forming urine.

Regulation of Urine Volume

ADH has a central role in the regulation of urine volume. Control of the solute concentration of urine translates into control of urine volume. If no water is reabsorbed by the distal and collecting tubules, urine volume is relatively high—and water loss from the body is high. As water is reabsorbed under the influence of ADH, the total volume of urine is reduced by the amount of water removed from the tubules. Thus, ADH reduces water loss by the body.

Another hormone that tends to decrease urine volume—and thus conserves water—is aldosterone, a secretion of the adrenal cortex. It increases distal and collecting tubule absorption of sodium, which in turn causes an osmotic imbalance that drives the reabsorption of water from the tubule. Because water reabsorption in the distal and collecting tubule portions requires ADH, the aldosterone mechanism must work in concert with the ADH mechanism if homeostasis of the fluid content in the body is to be maintained. The cooperative roles of ADH and aldosterone in regulating urine volume—and thus regulating fluid balance in the whole body—are summarized in Figure 28-23.

Figure 28-22 *Production of hypertonic urine.*
Hypertonic urine can be formed when ADH is present. ADH, a posterior pituitary hormone, increases the water permeability of the distal tubule and collecting duct. Thus, hypotonic (100 mOsm) tubule fluid leaving the loop of Henle can equilibrate first with the isotonic (300 mOsm) interstitial fluid (IF) of the cortex, then with the increasingly hypertonic (400 to 1200 mOsm) IF of the medulla. As H$_2$O leaves the collecting duct by osmosis, the filtrate becomes more concentrated with the solutes left behind. The concentration gradient causes urea to diffuse into the IF, where some of it is eventually picked up by tubule fluid in the descending limb of the loop of Henle *(dashed line)*. This countercurrent movement of urea helps maintain a high solute concentration in the medulla. Values are expressed in milliosmoles.

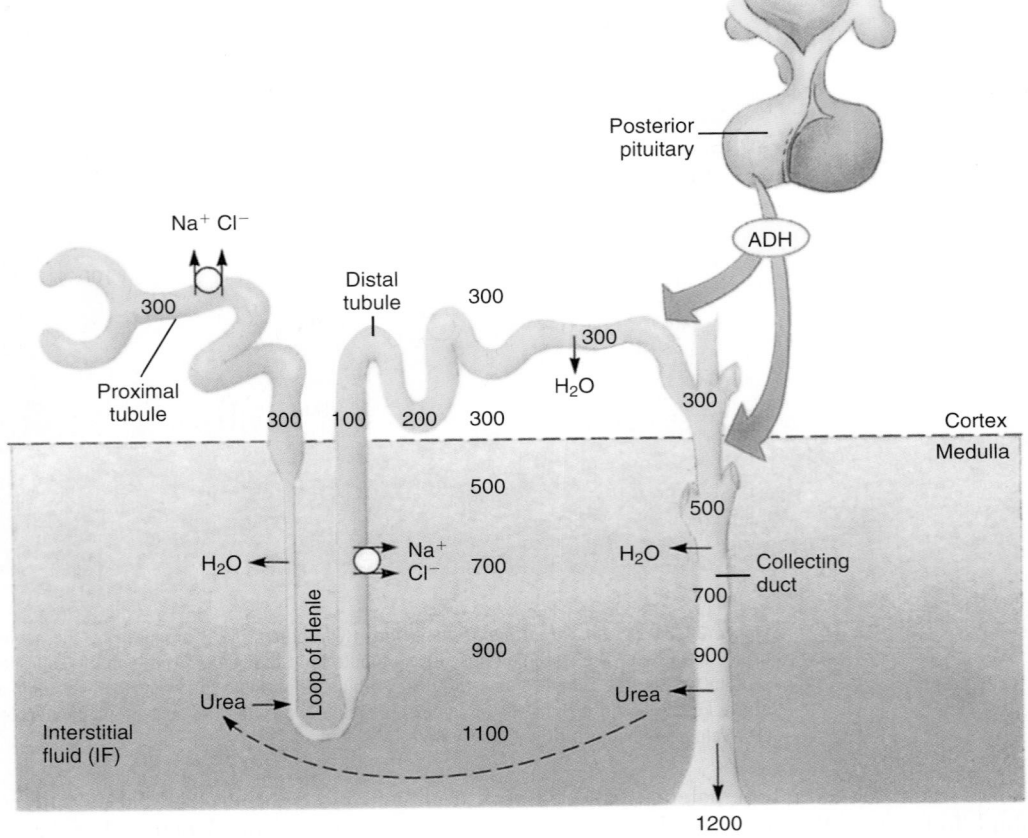

Table 28-1	**Summary of Nephron Function**	
PART OF NEPHRON	**FUNCTION**	**SUBSTANCE MOVED**
Renal corpuscle	Filtration (passive)	Water Smaller solute particles (ions, glucose, etc.)
Proximal tubule	Reabsorption (active)	Active transport: Na$^+$ Cotransport: glucose and amino acids
	Reabsorption (passive)	Diffusion: Cl$^-$, PO$_4^=$, urea, other solutes Osmosis: water
Loop of Henle		
Descending limb	Reabsorption (passive) Secretion (passive)	Osmosis: water Diffusion: urea
Ascending limb	Reabsorption (active) Reabsorption (passive)	Active transport: Na$^+$ Diffusion: Cl$^-$
Distal tubule	Reabsorption (active)	Active transport: Na$^+$
	Reabsorption (passive)	Diffusion: Cl$^-$, other anions Osmosis: water (only in the presence of ADH)
	Secretion (passive)	Diffusion: ammonia
	Secretion (active)	Active transport: K$^+$, H$^+$, some drugs
Collecting duct	Reabsorption (active)	Active transport: Na$^+$
	Reabsorption (passive)	Diffusion: urea Osmosis: water (only in the presence of ADH)
	Secretion (passive)	Diffusion: ammonia
	Secretion (active)	Active transport: K$^+$, H$^+$, some drugs

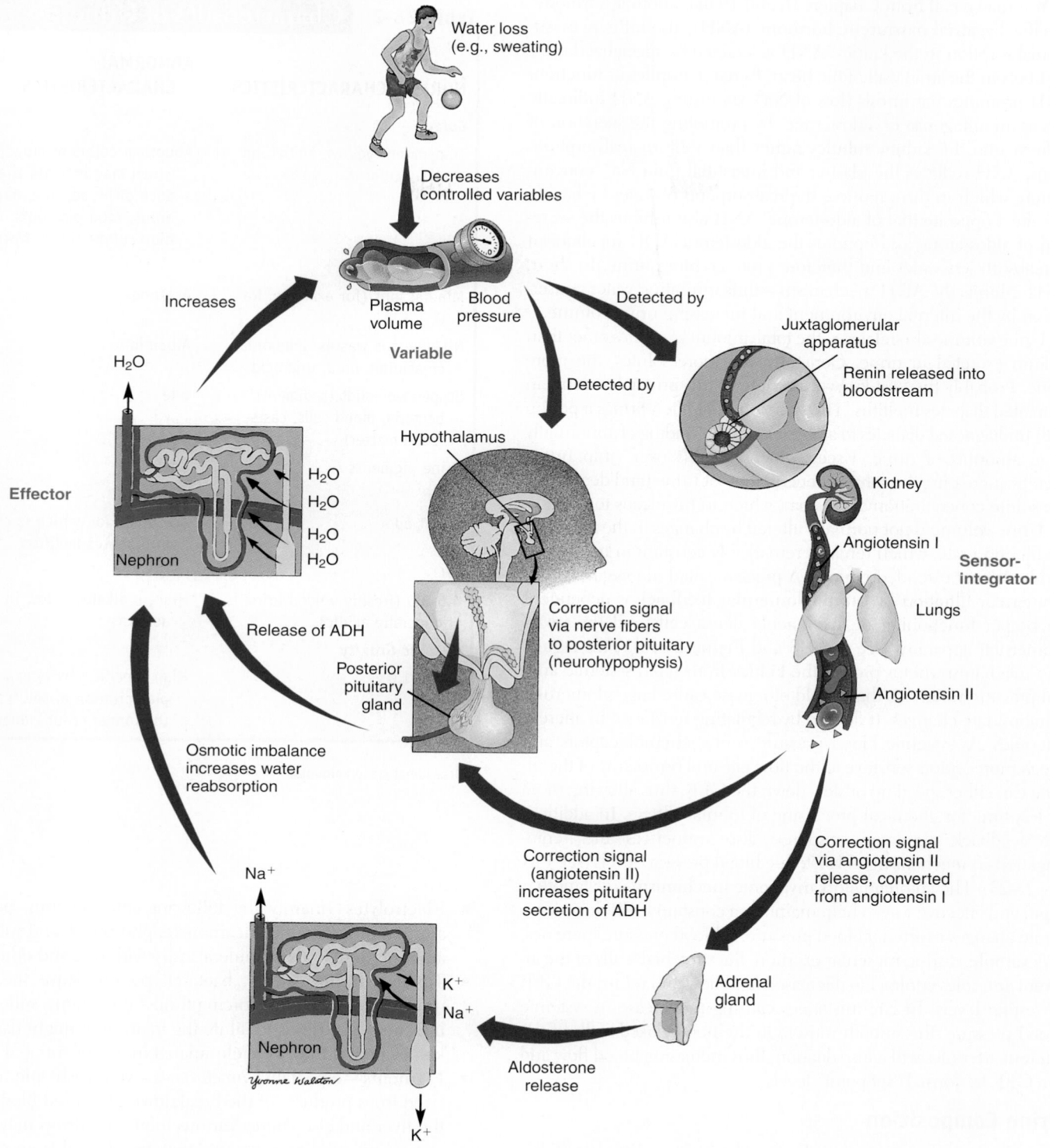

Figure 28-23 *Cooperative roles of ADH and aldosterone in regulating urine and plasma volume.* The drop in blood pressure that accompanies loss of fluid from the internal environment triggers the hypothalamus to rapidly release ADH from the posterior pituitary gland. ADH increases water reabsorption by the kidney by increasing the water permeability of the distal tubules and collecting ducts. The drop in blood pressure is also detected by each nephron's juxtaglomerular apparatus, which responds by secreting renin. Recall from Chapter 16 that renin triggers the formation of angiotensin II, which stimulates the release of aldosterone from the adrenal cortex. Aldosterone then slowly boosts water reabsorption by the kidneys by increasing reabsorption of Na^+. Because angiotensin II also stimulates the secretion of ADH, it serves as an additional link between the ADH and aldosterone mechanisms.

You may recall from Chapters 16 and 19 that another hormone, specifically, **atrial natriuretic hormone (ANH),** also influences water reabsorption in the kidney. ANH is secreted by specialized muscle fibers in the atrial wall of the heart. Its name implies its function: ANH promotes natriuresis (loss of Na^+ via urine). ANH indirectly acts as an antagonist of aldosterone, by promoting the secretion of sodium into the kidney tubules rather than sodium reabsorption. Thus, ANH reduces the plasma and interstitial fluid Na^+ concentration, which in turn, reduces the reabsorption of water by having the effect opposite that of aldosterone. ANH also inhibits the secretion of aldosterone and opposes the aldosterone-ADH mechanism to reabsorb less water and therefore produce more urine. In short, ANH inhibits the ADH mechanism—thus inhibiting water conservation by the internal environment and increasing urine volume.

Urine volume also relates to the total amount of solutes other than sodium excreted in urine. Generally, the more solutes, the more urine. Probably the best known example of this principle occurs in untreated diabetes mellitus. The symptom that often brings a person with undiagnosed diabetes to a physician is the voiding of abnormally large amounts of urine. Excess glucose "spills over" into urine, thereby increasing the solute concentration of urine (and decreasing the solute concentration of plasma), which in turn leads to diuresis.

Urine volume is not normally altered by changes in the glomerular filtration rate, which remains remarkably constant in normal individuals over extended periods. A process called *autoregulation* of glomerular filtration by **tubuloglomerular feedback** is dependent on proper functioning of the macula densa cells and the juxtaglomerular apparatus (see p. 1043 and Figure 28-10). This regulatory mechanism helps protect the kidney from rapid systemic arterial pressure variations that would otherwise cause large glomerular filtration rate changes. It does so by regulating resistance in afferent arterioles. As systemic blood pressure varies, chemoreceptors and mechanoreceptors sensitive to the flow rate and osmolarity of the filtrate can either speed up or slow down the GFR, thus allowing more or less time for chemical processing of tubular filtrate. In addition, the feedback regulatory response also influences the renin-angiotensin mechanism and systemic blood pressure levels (see Figure 28-23). The autoregulatory **myogenic mechanism** is yet another rapid and effective way to help maintain a constant GFR during systemic changes in arterial blood pressure. If blood pressure increases, for example, during muscular exertion, the stretched walls of the afferent arterioles contract to decrease blood flow and return the GFR to resting levels. In circumstances causing a decrease in systemic blood pressure, the smooth muscle in the now relaxed walls of the afferent arterioles will cause dilation, thus increasing blood flow and the GFR to normal "set point" levels.

Urine Composition

The physical characteristics of normal urine are listed in Table 28-2. Notice that normal and abnormal characteristics are listed.

Urine is approximately 95% water, in which are dissolved several kinds of substances; the most important are discussed below:

- **Nitrogenous wastes** from protein catabolism, such as urea (the most abundant solute in urine), uric acid, ammonia, and creatinine.

Table 28-2	Characteristics of Urine	
NORMAL CHARACTERISTICS	**ABNORMAL CHARACTERISTICS**	
Color		
Transparent yellow, amber, or straw color	Abnormal colors or cloudiness, which may indicate the presence of blood, bile, bacteria, drugs, food pigments, or a high solute concentration	
Compounds		
Mineral ions (for example, Na^+, Cl^-, K^+)	Acetone	
Nitrogenous wastes: ammonia, creatinine, urea, uric acid	Albumin	
Suspended solids (sediment)*: bacteria, blood cells, casts (solid matter)	Bile	
Urine pigments	Glucose	
Odor		
Slight odor	Acetone odor, which is common in diabetes mellitus	
pH		
4.6–8.0 (freshly voided urine is generally acidic)	High in alkalosis; low in acidosis	
Specific Gravity		
1.001–1.035	High specific gravity can cause precipitation of solutes and the formation of kidney stones	

*Occasional trace amounts.

- **Electrolytes**—mainly the following ions: sodium, potassium, ammonium, chloride, bicarbonate, phosphate, and sulfate. The amounts and kinds of minerals vary with diet and other factors.
- **Toxins**—during disease, bacterial poisons leave the body in urine. One reason for "forcing fluids" on patients suffering with infectious diseases is to dilute the toxins that might damage the kidney cells if they were eliminated in a concentrated form.
- **Pigments**—especially *urochromes*—yellowish pigments derived from products of the breakdown of old red blood cells in the liver and elsewhere. Various foods and drugs may contain, or be converted to, pigments that are cleared from plasma by the kidneys and are therefore found in the urine.
- **Hormones**—high hormone levels sometimes result in significant amounts of hormone in the filtrate (and therefore in urine).
- **Abnormal constituents** such as blood, glucose, albumin (a plasma protein), casts (chunks of material, such as mucus, that harden inside the urinary passages and then are washed out in urine), or calculi (small stones).

QUICK CHECK

16. Does ADH promote water loss from the internal environment or water conservation by the internal environment?
17. How does aldosterone influence secretion in the kidney tubules?
18. How does aldosterone cause the body to conserve water?
19. What gives urine its characteristic yellowish color?

BOX 28-6: HEALTH MATTERS
Artificial Kidney

The artificial kidney is a mechanical device that uses the principle of dialysis to remove or separate waste products from the blood. In the event of kidney failure, the process, appropriately called **hemodialysis** (Greek *haima*—"blood" and *lysis*—"separate"), is a reprieve from death for the patient. During a hemodialysis treatment, a semipermeable membrane is used to separate large (nondiffusible) particles such as blood cells from small (diffusible) ones such as urea and other wastes. Part *A* of the figure shows blood from the radial artery passing through a porous (semipermeable) cellophane tube that is housed in a tanklike container. The tube is surrounded by a bath, or dialysis solution, containing varying concentrations of electrolytes and other chemicals. The pores in the membrane are small and allow only very small molecules, such as urea, to escape into the surrounding fluid. Larger molecules and blood cells cannot escape and are returned through the tube to reenter the patient via a wrist or leg vein. By constantly replacing the bath so-

lution in the dialysis tank with freshly mixed solution, levels of waste materials can be kept at low levels. As a result, wastes such as urea in the blood rapidly pass into the surrounding wash solution. For a patient with complete kidney failure, two or three hemodialysis treatments a week are required. New dialysis methods are now being developed, and dramatic advances in treatment are expected in the next few years.

Another technique used in the treatment of renal failure is called **continuous ambulatory peritoneal dialysis (CAPD).** In this procedure, 1 to 3 liters of sterile dialysis fluid is introduced directly into the peritoneal cavity through an opening in the abdominal wall (part *B* of the figure). Peritoneal membranes in the abdominal cavity transfer waste products from blood into the dialysis fluid, which is then drained back into a plastic container after about 2 hours. This technique is less expensive than hemodialysis and does not require the use of complex equipment.

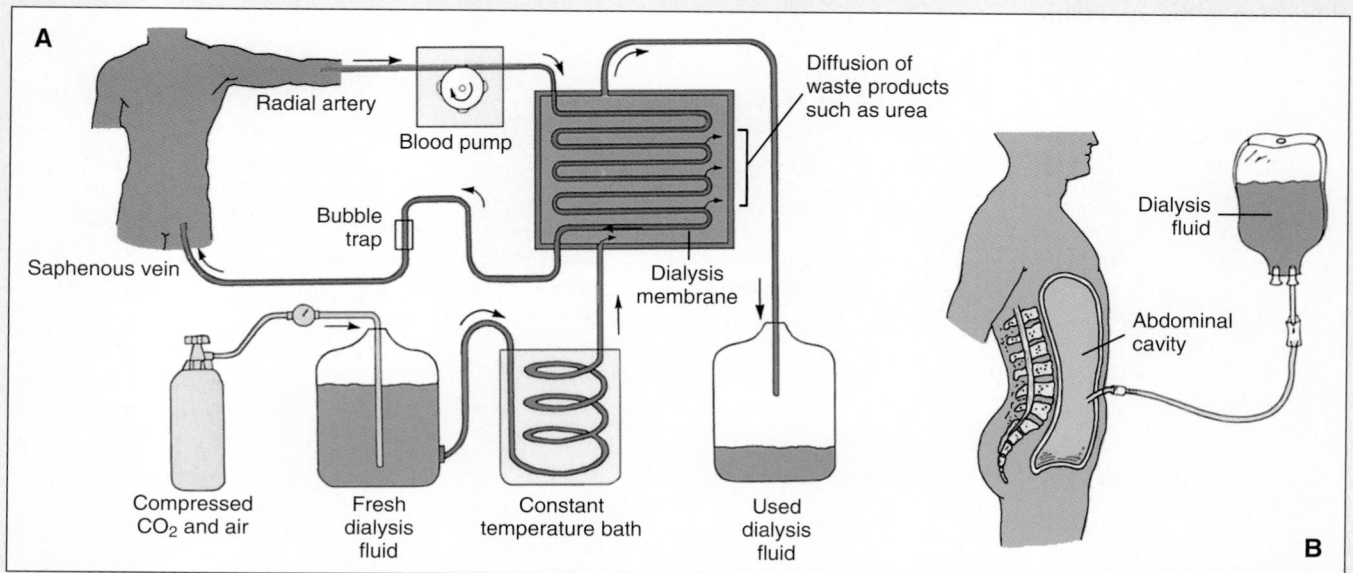

Hemodialysis.

Cycle of Life
Urinary System

The kidney plays a critical role in homeostasis by regulating the levels of many substances in blood. Primary renal functions include filtration, reabsorption, and secretion. All are interrelated by complex control systems involving central nervous system activity and hormonal secretions. More than 1 million nephron units in each kidney serve as the structural framework permitting normal function to occur.

Normally, life cycle changes in kidney structure and function occur only within rather narrow limits. Significant structural changes, such as dramatic decreases in the number of nephron units, almost always indicate serious disease or result from trauma such as crush injuries. Functionally, the kidney is able to operate normally throughout life under a wide array of conditions. If, however, the kidneys cannot cope with extreme conditions, such as water deprivation or disease, death will occur from the buildup of toxins in blood.

Initially, kidney function in a newborn is less efficient than in an older child and adult. As a result, the urine is less concentrated because the regulatory mechanisms required to retain water are not fully operative. Incontinence, or an inability to control urination, is normal in very young children. Reflex emptying occurs when the bladder fills, but normal sphincter activity keeps urine in the bladder until filling occurs. In contrast, many older adults have problems with incontinence because of loss of sphincter tone, or control.

Renal clearance is the ability of the kidneys to clear, or cleanse, the blood of a certain substance in a given unit of time, generally 1 minute. This value for certain substances tends to decrease with advanced age, thus indicating deterioration of kidney function. Changes in the porosity of the filtration membrane also occur in the elderly. Loss of functional nephron units is yet another consequence of aging. It contributes to the gradual decline in renal function in this age group.

THE BIG PICTURE
Urinary System and the Whole Body

As our study of the urinary system has shown us, homeostasis of water and electrolytes in body fluids depends largely on proper functioning of the kidneys. Each nephron within the kidney processes blood plasma in a way that adjusts its content to maintain a dynamic constancy of the internal environment of the body. Without renal processing, blood plasma characteristics would soon move out of their set point range. On the other hand, without the blood pressure generated by cardiovascular mechanisms, the kidney could not filter blood plasma and therefore could not process blood plasma. Thus, the urinary system and the cardiovascular system are interdependent.

Regulation of urinary function, we have seen, is often centered outside the kidney—mainly in the form of endocrine hormone action. Urinary function is also regulated to some extent by nerve reflexes. Thus, both the endocrine system and the nervous system must operate properly to ensure efficient kidney function.

The urinary system also interacts with many other body systems and tissues. For example, the kidneys clear the blood plasma of nitrogenous wastes and excess metabolic acids produced by the chemical activity of nearly every cell in the body. The kidneys also can clear some toxins and other compounds that enter the blood via the digestive tract, skin, or respiratory tract.

In the next chapter, we apply some of what we know about urinary function to a broader study of water and ion homeostasis within the human body's internal environment. After that, we discuss the role of the urinary system and other body systems in maintaining a relatively constant pH in the body's internal environment.

Mechanisms of Disease

You may have experienced the discomfort and pain of a bladder infection or know someone who has. Bladder infection is the most common urinary disorder, but it is not usually serious if promptly treated. There are numerous renal and urinary disorders that are very serious, however. Any disorder that significantly reduces the effectiveness of the kidneys is immediately life threatening. In this section, we will discuss some life-threatening kidney diseases, as well as a few of the less serious, but more common disorders.

Hypertension is the term used to describe elevated blood pressure. Epidemiologists estimate that 25% of the U.S. population and nearly 800 million individuals worldwide are affected. A majority of those with high blood pressure remain asymptomatic and unaware that a problem exists until late in the course of the disease. Unfortunately, the number and severity of health problems caused by hypertension, even in those who have no symptoms, will increase over time and with increasing pressure. The delayed outcomes are often devastating. In most individuals with high blood pressure, no specific pathologic mechanism or precipitating reason can be identified, and the condition is called **essential hypertension** (see Chapter 19, p. 769).

If a condition can be identified that is causing the elevation in blood pressure, the condition is called **secondary hypertension.** As the name implies, it is secondary to an identified body malfunction of some type. So-called *renal hypertension* is a common type of secondary hypertension that may be caused by stenosis (narrowing) of the renal artery, often by the accumulation of atherosclerotic plaque. In these cases, the elevation is secondary to reduced blood flow, resulting in ischemia of kidney tissues. When this occurs, the cells of the juxtaglomerular apparatus secrete renin, which in turn results in angiotensin production and increased blood pressure (see Figures 19-27 and 28-23 for a review of renin-angiotensin reactions). Testing of renal vein blood for increased renin levels is used to confirm the diagnosis, and insertion of a stent into the lumen of the renal artery to increase blood flow to the kidney may be curative.

Obstructive Disorders

Obstructive urinary disorders are abnormalities that interfere with normal urine flow anywhere in the urinary tract. The severity of obstructive disorders depends on where the interference occurs and to what degree the flow of urine is impaired. Obstruction of urine flow usually results in "backing up" of the urine, perhaps all the way to the kidney itself. When urine backs up into the kidney, causing swelling of the renal pelvis and calyces, the condition is called **hydronephrosis** (Figure 28-24). A few of the more important obstructive conditions are summarized here.

Renal Calculi

Renal calculi, or *kidney stones,* are crystallized mineral chunks that develop in the renal pelvis or calyces. Many calculi develop as calcium and other minerals crystallize on the renal papillae, then break off into the urine. Blood uric acid levels become elevated in *gout,* and deposits of uric acid in the kidneys

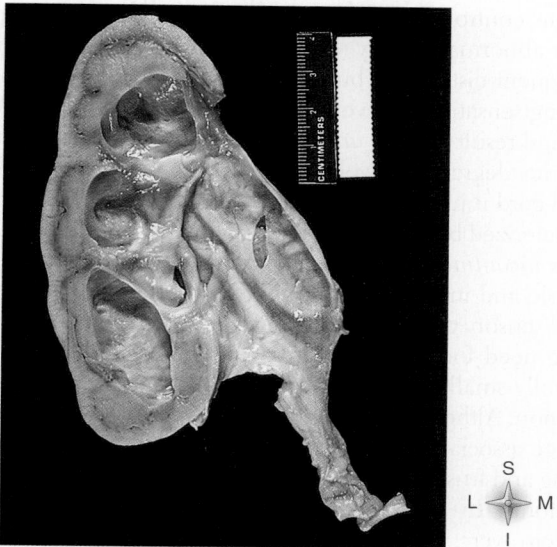

Figure 28-24 *Hydronephrosis.* Note the marked dilation of the renal pelvis and calyces caused by the blockage and "backing up" of urine.

 BOX 28-7: HEALTH MATTERS

Clinical Terms Associated with Urine Abnormalities

Glycosuria or **glucosuria**—Sugar (glucose) in urine
Hematuria—Blood in urine
Pyuria—Pus in urine
Dysuria—Painful urination
Polyuria—Unusually large amounts of urine
Oliguria—Scant urine
Anuria—Absence of urine

produce uric acid stones or calculi. *Staghorn calculi* are large, branched stones that form in the pelvis and branched calyces.

If the stones are small enough, they simply pass through the ureters and urethra and are eventually voided with the urine. Larger stones may obstruct the ureters, causing intense pain called *renal colic* as rhythmic muscle contractions of the ureter attempt to dislodge it. Hydronephrosis may occur if the stone does not move from its obstructing position. In the past, only traditional surgical procedures were effective in removing relatively large stones that formed in the calyces and renal pelvis of the kidney. A technique called **lithotripsy,** which uses an ultrasound generator called a **lithotriptor,** is now used quite frequently to pulverize stones so that they can be flushed out of the urinary tract without surgery.

Neurogenic and Overactive Bladder

Disruption of nervous input to the bladder results in loss of normal control of voiding. The condition is called **neurogenic bladder.** Depending on the nature and severity of the lack of

Mechanisms of Disease—cont.

nervous control, various signs and symptoms of bladder paralysis or abnormal activity result. Involuntary retention of urine, subsequent distention (bulging) of the bladder, and perhaps a burning sensation or fever with chills are common symptoms. The end result is often *urinary incontinence*—leakage of urine or some degree of involuntary urination. Serious stroke or spinal cord injury often results in a type of neurogenic bladder characterized by total loss of normal control of voiding. Called "*reflex incontinence,*" it is characterized by periodic but unpredictable and involuntary urination that occurs in the absence of any sensory warning or awareness. **Overactive bladder** refers to the need for frequent urination. The amounts voided are generally small and feelings of extreme urgency and pain are common. Although serious medical outcomes are rare, incontinence associated with overactive bladder is an often embarrassing and frustrating problem for those who are plagued with symptoms. In the past, treatments for this rather common problem were limited to behavioral techniques and in some cases, surgery. Now, medications are available that reduce the involuntary contractions and incontinence associated with overactive bladder. Commonly used drugs include so-called "anticholinergic" medications such as solifenacin (Vesicare, Detrol, others) and "alpha-adrenergic blocker" drugs that promote smooth muscle relaxation in the bladder and urethra. Examples of this type of drug include alfuzosin and terazosin (Uroxatral, Hytrin, others).

Tumors and Other Obstructions

Tumors of the urinary system typically obstruct urine flow, possibly causing hydronephrosis in one or both kidneys. Most kidney tumors are malignant neoplasms called **renal cell carcinomas.** They usually occur only in one kidney (Figure 28-25).

Bladder cancer occurs about as frequently as renal cancer (each accounts for about 2 in every 100 cancer cases). Renal and bladder cancer have few symptoms early in their development, other than traces of blood in the urine, or **hematuria** (hem-ah-TOO-ree-ah). As the cancer develops, pelvic pain and symptoms of urinary obstruction may occur. Insertion of a **cystoscope** (SIS-toh-skohp) through the urethra and into the bladder permits direct inspection of bladder and other lower urinary tract lesions (Figure 28-26). The hollow tube allows the passage of a light, a viewing lens, and various catheters and operative devices. Figure 28-26, *B*, shows the appearance of a malignant tumor, a transitional cell carcinoma, on the bladder wall before its removal during a surgical cystoscopic procedure.

Various other conditions can obstruct the normal flow of urine. For example, a person with a low proportion of body fat

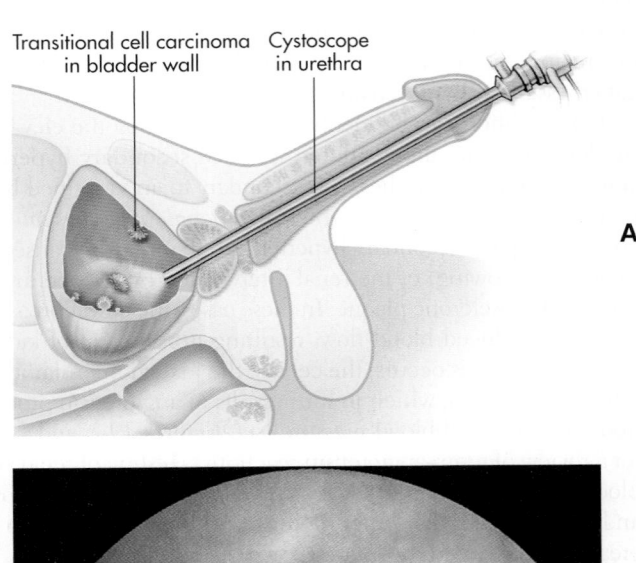

Transitional cell carcinoma in bladder wall   Cystoscope in urethra

A

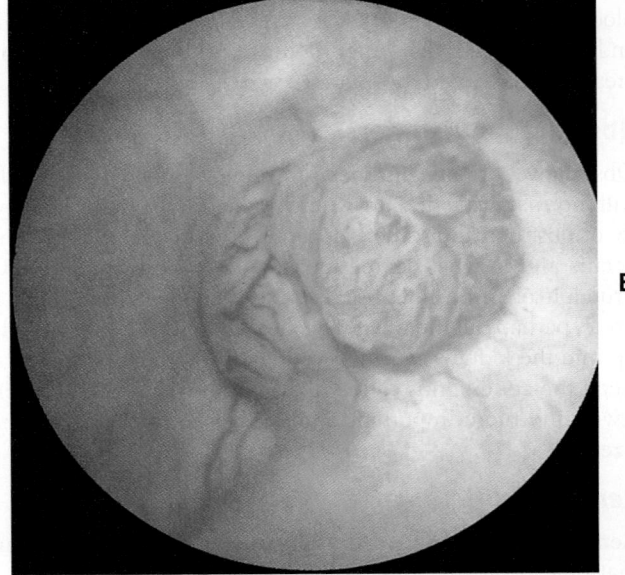

B

Figure 28-26 *Cystoscopic view of bladder cancer.* **A,** Cystoscope in a male bladder. **B,** Cystoscopic view of a cancerous growth (a transitional cell carcinoma) on the bladder wall.

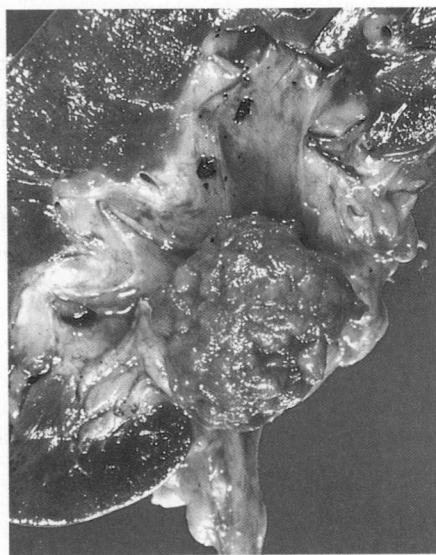

Figure 28-25 *Renal carcinoma.* Surgical specimen. The renal pelvis has been opened to expose the cancerous growth, just proximal to the ureter.

Mechanisms of Disease—cont.

may lack the pad of fat that normally surrounds the kidneys. One or both kidneys may then drop, a condition called **renal ptosis** (TOH-sis). In renal ptosis, the ureters that drain urine out of the kidney may kink and thus obstruct the normal flow of urine. Urinary passages may also be abnormally narrowed through scarring, inflammation, or external pressure—a condition known as a **stricture**.

Urinary Tract Infections

Most *urinary tract infections* (UTIs) are caused by bacteria, usually gram-negative types. UTIs can involve the urethra, bladder, ureter, and/or kidneys. Common types of urinary tract infections are summarized below.

Urethritis is an inflammation of the urethra that commonly results from bacterial infection, often *gonorrhea*. Nongonococcal urethritis is usually caused by a *Chlamydia* infection. Males suffer from urethritis more often than females do.

Cystitis is a term that refers to any inflammation of the bladder. Cystitis commonly occurs as a result of infection but also can accompany calculi, tumors, or other conditions. Bacteria usually enter the bladder through the urethra. Cystitis occurs more frequently in women than in men because the female urethra is shorter and closer to the anus (a source of bacteria) than in males. Bladder infections are characterized by pelvic pain, an urge to urinate frequently, and hematuria. **Interstitial cystitis** is a form of bladder inflammation that occurs without evidence of bacterial infection. It is a persistent and often very painful form of cystitis that is characterized by feelings of urgency, pain on urination, and the appearance of blood in the urine. The bladder wall often shows patches of chronic mucosal ulceration. Anti-inflammatory drugs and supportive treatment of symptoms are often coupled with infusion of sterile fluid to distend the bladder and reduce feelings of urgency. Although the cause is unknown, many physicians believe an autoimmune response is involved. Interstitial cystitis is often associated with lupus erythematosus and other autoimmune disease conditions.

Nephritis is a general term referring to kidney disease, especially inflammatory conditions. **Pyelonephritis** is literally "pelvis nephritis" and refers to inflammation of the renal pelvis and connective tissues of the kidney. As with cystitis, pyelonephritis is usually caused by bacterial infection but can also result from viral infection, mycosis (Figure 28-27), calculi, tumors, pregnancy, and other conditions. The x-ray image shown in Figure 28-27 is called a **retrograde pyelogram**. In this individual, preliminary tests indicated that the right kidney was not functioning. In an attempt to diagnose the cause, a catheter was inserted into the bladder and contrast material was injected into the right ureter to fill the renal pelvis and collecting system on that side. The resulting retrograde pyelogram (image) shows irregular defects in the calyces (arrows) that were found to be filled with fungus. The diagnosis was *renal aspergillosis*—a mycotic infection.

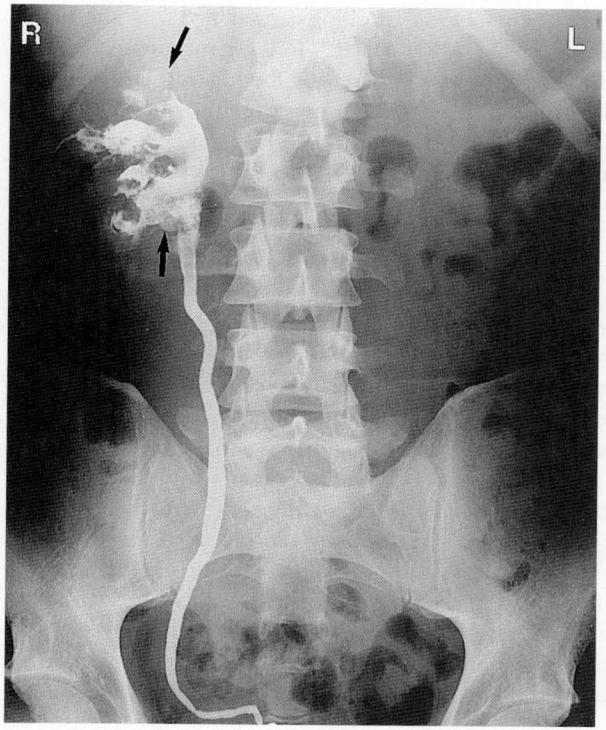

Figure 28-27 *Renal aspergillosis.* A retrograde pyelogram (image) shows irregular defects *(arrows)* caused by the accumulation of fungus balls in the right renal pelvis and calyces (aspergillosis).

Glomerular Disorders

Glomerular disorders, collectively called **glomerulonephritis,** result from damage to the glomerular capsular membrane. This damage can be caused by immune mechanisms, heredity, and other factors. Without successful treatment, glomerular disorders can progress to kidney failure.

Nephrotic syndrome is a collection of signs and symptoms that accompany various glomerular disorders. This syndrome is characterized by the following:

- Proteinuria—presence of proteins (especially *albumin*) in urine. Protein, normally absent from urine, filters through damaged glomerular capsular membranes and is not reabsorbed by the kidney tubules.
- Hypoalbuminemia—low albumin concentration in the blood, resulting from loss of albumin from the blood through holes in the damaged glomeruli. Albumin is the most abundant plasma protein. Because it normally cannot leave the blood vessels, it usually remains as a "permanent" solute in plasma. This keeps the plasma water concentration low and thus prevents the osmosis of large amounts of water out of the blood and into tissue spaces. In hypoalbuminemia, this function is lost and fluid leaks out of the blood vessels and into tissue spaces, thereby causing widespread edema.

Mechanisms of Disease—cont.

- Edema—general tissue swelling caused by the accumulation of fluids in the tissue spaces. The edema associated with nephrotic syndrome is caused by the loss of plasma protein (albumin) and the resulting osmosis of fluid out of the blood.

Acute glomerulonephritis is the most common form of kidney disease. It may be caused by a delayed immune response to streptococcal infection—the same mechanism that causes valve damage in rheumatic heart disease (see Chapter 18). For this reason, it is sometimes called **postinfectious glomerulonephritis.** If antibiotic treatment is not successful, it may progress to a chronic form of glomerulonephritis.

Chronic glomerulonephritis is the general name for various noninfectious glomerular disorders that are characterized by progressive kidney damage leading to renal failure. Immune mechanisms are believed to be the major causes of chronic glomerulonephritis. One immune mechanism involves antigen-antibody complexes that form in the blood when antibodies bind with foreign antigens (or possible self-antigens). These antigen-antibody complexes lodge in the glomerular capsular membrane and trigger an inflammation response. Less commonly, the formation of antibodies that directly attack the glomerular capsular membrane causes chronic glomerulonephritis.

Kidney Failure

Kidney failure, or **renal failure,** is simply failure of the kidney to properly process blood plasma and form urine. Renal failure can be classified as acute or chronic.

Acute renal failure is an abrupt reduction in kidney function that is characterized by oliguria and a sharp rise in nitrogenous compounds in the blood. The concentration of nitrogenous wastes in blood is often assessed by the **blood urea nitrogen (BUN) test**—a high BUN result indicates failure of the kidneys to remove urea from the blood. Acute renal failure can be caused by various factors that alter blood pressure or otherwise affect glomerular filtration. For example, hemorrhage, severe burns, acute glomerulonephritis or pyelonephritis, and obstruction of the lower urinary tract may each progress to kidney failure. If the underlying cause of renal failure is attended to, recovery is usually rapid and complete.

Chronic renal failure is a slow, progressive condition resulting from the gradual loss of nephrons. There are dozens of diseases that may result in the gradual loss of nephron function, including infections, diabetes, glomerulonephritis, tumors, systemic autoimmune disorders, and obstructive disorders. As kidney function is lost, the glomerular filtration rate (GFR) decreases, causing the BUN levels to climb (Figure 28-28). Chronic renal failure can be described as progressing through three stages:

Stage 1. During the first stage, some nephrons are lost but the remaining healthy nephrons compensate by enlarging and

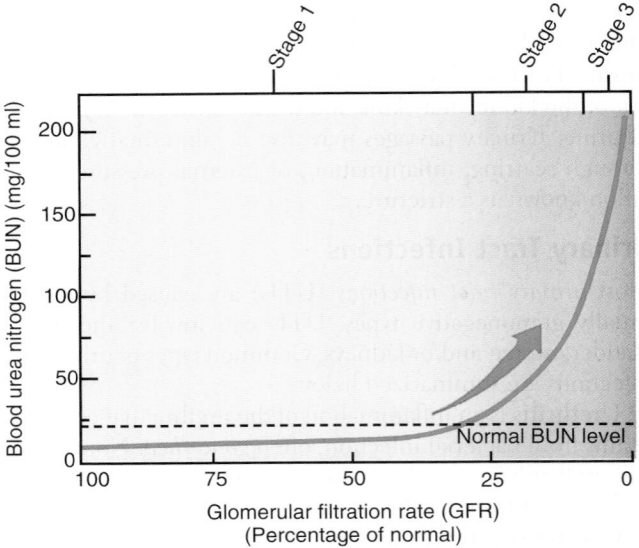

Figure 28-28 *The three stages of chronic renal failure. Stage 1:* As nephrons are lost (indicated by decreasing GFR), the remaining healthy nephrons compensate—keeping BUN values within the normal range. *Stage 2:* As more than 75% of kidney function is lost, BUN levels begin to climb. *Stage 3:* Uremia (elevated BUN) results from massive loss of kidney function.

taking over the function of the lost nephrons. As Figure 28-28 shows, BUN is kept within normal limits even though up to 75% of the nephrons are lost (as indicated by a 75% drop in GFR). This stage is often asymptomatic and may last for years, depending on the underlying cause.

Stage 2. The second stage is often called *renal insufficiency.* It is during this stage that the kidney can no longer adapt to the loss of nephrons. The remaining healthy nephrons cannot handle the urea load, and BUN levels climb dramatically (see Figure 28-28). Because the kidney's ability to concentrate urine is impaired, polyuria and dehydration may occur.

Stage 3. The final stage of chronic renal failure is called **uremia,** or **uremic syndrome.** Uremia literally means "high blood urea" and is characterized by a very high BUN value caused by loss of kidney function (see Figure 28-28). During this stage, a low GFR causes low urine production and oliguria. Because fluids are retained by the body rather than eliminated by the kidneys, edema and hypertension often occur. The uremic syndrome includes a long list of other symptoms caused directly or indirectly by the loss of kidney function. Unless an artificial kidney is used or a new kidney is transplanted, the progressive loss of kidney function will eventually cause death.

LANGUAGE OF SCIENCE

(Cont'd from page 1033)

kidneys

loop of Henle (HEN-lee) [*Friedrich Gustave Henle* German anatomist]

macula densa (MAK-yoo-lah DEN-sah) [*macula* spot, *densus* thick]

mesangial cells (mess-AN-jee-al)

myogenic mechanism (mye-oh-JEN-ik) [*myo-* muscle, *-genic* forming]

nephrons (NEF-ronz) [*nephros* kidney]

nitrogenous wastes (nye-TROJ-i-nes) [*nitron-* soda, *-genous* to orginate from]

peritubular capillaries (pair-ee-TOOB-yoo-lar KAP-i-lair-eez) [*peri-* around, *-tubulus* little tube, *capillaris* hairlike]

pigments [*pigmentum* paint]

reabsorption (ree-ab-SORP-shun) [*re-* back again, *-absorbere* to swallow]

renal clearance (REE-nal) [*ren-* kidney, *-al* pertaining to]

renal columns (REE-nal) [*ren-* kidney, *-al* pertaining to]

renal corpuscle (REE-nal KOR-pus-ul) [*ren-* kidney, *-al* pertaining to, *corpusculum* little body]

renal cortex (REE-nal KOR-teks) [*ren-* kidney, *-al* pertaining to, *cortex* bark]

renal medulla (REE-nal meh-DUL-ah) [*ren-* kidney, *-al* pertaining to, *medulla* marrow]

renal pelvis (REE-nal PEL-vis) [*ren-* kidney, *-al* pertaining to, *pelvis* basin]

renal pyramids (REE-nal PIR-ah-mids) [*ren-* kidney, *-al* pertaining to]

sodium cotransport (SO-dee-um koh-TRANZ-port)

toxins (TOK-sins) [*toxikon* poison]

tubular reabsorption (TOOB-yoo-lar ree-ab-SORP-shun) [*tubulus* little tube, *re-* back again, *-absorbere* to swallow]

tubular secretion (TOOB-yoo-lar seh-KREE-shun) [*tubulus* little tube, *secernere* to separate]

tubuloglomerular feedback (toob-yoo-loh-glow-MER-yoo-lar) [*tubulus-* little tube, *-glomerulus* small ball]

ureters (YOOR-eh-terz) [*ourein* to urinate]

urethra (yoo-REE-thrah) [*ourein* to urinate]

LANGUAGE OF MEDICINE

acute glomerulonephritis (ah-KYOOT gloh-mer-yoo-loh-neh-FRY-tis) [*acutus* sharp, *glomerulus-* small ball, *-nephros-* kidney, *-itis* inflammation]

anuria (ah-NOO-ree-ah) [*a-* not, *-ouron* urine]

biopsy (BYE-op-see) [*bios-* life, *-opsis* view]

blood urea nitrogen (BUN) test (yoo-REE-ah NYE-troh-jen test) [*ouron* urine, *nitron* soda]

chronic glomerulonephritis (KRON-ik gloh-mer-yoo-loh-neh-FRY-tis) [*chronos* time, *glomerulus-* small ball, *-nephros-* kidney, *-itis* inflammation]

chronic renal failure (KRON-ik REE-nal FAIL-yoor) [*chronos* time, *ren-* kidney, *-al* pertaining to]

continuous ambulatory peritoneal dialysis (CAPD) [AM-byoo-lah-tor-ee pair-i-toh-NEE-al dye-AL-i-sis]

cystitis (sis-TYE-tis) [*kystis-* bag, *-itis* inflammation]

cystoscope (SIS-toh-skohp) [*kystis-* bag, *-scope* instrument for observation]

dysuria (dis-YOO-ree-ah) [*dys-* painful, *-ouron* urine]

essential hypertension (ee-SEN-shal hye-per-TEN-shun) [*essentia* quality, *hyper-* excessive, *-tendere* to stretch]

glomerulonephritis (gloh-mer-yoo-loh-neh-FRY-tis) [*glomerulus-* small ball, *-nephros-* kidney, *-itis* inflammation]

glycosuria (glye-koh-SOO-ree-ah) [*glyco-* glucose, *-ouron* urine]

hematuria (hem-ah-TOO-ree-ah) [*hema-* blood, *-ouron* urine]

hemodialysis (hee-moh-dye-AL-i-sis) [*haima-* blood, *-lysis* separate]

hydronephrosis (hye-droh-neh-FROH-sis) [*hydro-* hydrogen, *-nephros-* kidney, *-osis* condition]

hypertension (hye-per-TEN-shun) [*hyper-* excessive, *-tendere* to stretch]

interstitial cystitis (in-ter-STISH-al sis-TYE-tis) [*interstitium* space between, *kystis-* bag, *-itis* inflammation]

lithotripsy (LITH-oh-trip-see) [*litho-* stone, *-tribein* to wear away]

lithotriptor (LITH-oh-trip-tor) [*litho-* stone, *-tribein* to wear away]

nephritis (neh-FRY-tis) [*nephros-* kidney, *-itis* inflammation]

nephrotic syndrome (neh-FROT-ik SIN-drohm) [*nephros-* kidney, *-ic* pertaining to]

neurogenic bladder (noor-oh-JEN-ik BLAD-er) [*neuro-* nerves, *-genic* causing or forming]

normal anatomical variation (an-ah-TOM-i-kal) [*norma* rule, *ana-* up, *-temnein* to cut, *variare* to diversify]

oliguria (ohl-i-GOO-ree-ah) [*olig-* few or little, *-ouron* urine]

overactive bladder

polyuria (pahl-ee-YOO-ree-ah) [*poly-* many, *-ouron* urine]

postinfectious glomerulonephritis (post-in-FEK-shus gloh-mer-yoo-loh-neh-FRY-tis) [*post-* after, *-inficere* to stain, *glomerulus-* little ball, *-nephros-* kidney, *-itis* inflammation]

pyelonephritis (pye-eh-loh-neh-FRY-tis) [*pyelo-* pelvis or kidney, *-nephros-* kidney, *-itis* inflammation]

pyuria (pye-YOO-ree-ah) [*pyon-* pus, *-ouron* urine]

renal calculi (REE-nal KAL-kyoo-lye) [*ren-* kidney, *-al* pertaining to, *calculi* little stone]

renal cell carcinomas (REE-nal cell kar-si-NO-mahs) [*ren-* kidney, *-al* pertaining to, *cella* storeroom, *carcino-* cancer, *-oma* tumor]

renal failure (REE-nal FAIL-yoor) [*ren-* kidney, *-al* pertaining to]

renal ptosis (REE-nal TOH-sis) [*ren-* kidney, *-al* pertaining to, *ptosis* falling]

retrograde pyelogram (RET-roh-grayd PYE-eh-loh-gram) [*retro-* backward, *-gradus* step, *pyelo-* pelvis or kidney, *-gram* drawing or written record]

secondary hypertension (hye-per-TEN-shun) [*secundus* second, *hyper-* excessive, *-tendere* to stretch]

stricture (STRIK-chur) [*stringere* to tighten]

uremia (yoo-REE-mee-ah) [*ouron-* urine, *-emia* blood condition]

uremic syndrome (yoo-REE-mik SIN-drohm) [*ouron-* urine, *-ic* pertaining to]

urethritis (yoo-reh-THRY-tis) [*ourethra-* urethra, *-itis* inflammation]

CASE STUDY

Oscar Brown, age 42, is a farmer from Texas. For the past week he has noted "twinges" in the right lower part of his back. He was awakened this morning by excruciating pain in his back, unresponsive to acetaminophen and the application of a hot water bottle. He spent 4 hours pacing the house, with intermittent bouts of vomiting before going to the hospital emergency department.

In the emergency department, Mr. Brown appears to be in pain and is pacing the floor of the examination room. He reports that his past medical health has been good with no previous hospitalizations, trauma, or surgeries. He does not take any regular prescription medications, does not use tobacco or alcohol, and has an active lifestyle on his farm working 12 to 14 hours a day. He does take over-the-counter medications, acetaminophen (Tylenol) for occasional headache and muscle pain. His family history is negative except that his father and grandfather occasionally "passed blood in their water," but no medical care was ever sought for this problem.

Physical examination reflects a normal temperature, tachycardia at 140 beats per minute, respirations 28, blood pressure of 138/78, cool moist skin, right flank pain, and abdominal pain without rebound

tenderness. Laboratory findings include +1 protein, +3 blood, and a trace of glucose on urine dipstick. Microscopic examination of urine shows red blood cells too numerous to count, a white blood cell count of 8 to 10, bacteria of 1 to 2, and calcium oxalate crystals.

1. The most likely cause of Mr. Brown's pain is

   A. Acute pyelonephritis
   B. Cystitis
   C. Urolithiasis (renal calculi)
   D. Nephrotic syndrome

2. Mr. Brown undergoes intravenous pyelography (IVP), which is an x-ray of the kidneys, ureters, and bladder, and a stone is located blocking the right ureter. This blockage, unless immediately relieved, can lead to which one of the following?

   A. Renal ptosis
   B. Acute glomerulonephritis
   C. Chronic renal failure
   D. Hydronephrosis

CHAPTER SUMMARY

OVERVIEW OF THE URINARY SYSTEM

A. Kidneys—principal organs of the urinary system; accessory organs are the ureters, urinary bladder, and urethra (Figure 28-1)

B. Urinary system—regulates the content of blood plasma to maintain "dynamic constancy" or homeostasis of the internal fluid environment within normal limits

ANATOMY OF THE URINARY SYSTEM

A. Gross structure (Figure 28-2)
   1. Kidney
      a. Shape, size, and location
         (1) Roughly oval with a medial indentation
         (2) Approximately 11 cm by 7 cm by 3 cm
         (3) Left kidney often larger than the right; the right kidney is a little lower
         (4) Located in a retroperitoneal position
         (5) Lie on either side of the vertebral column between T12 and L3
         (6) Superior poles of both kidneys extend above the level of the twelfth rib and the lower edge of the thoracic parietal pleura
         (7) Renal fasciae anchor the kidneys to surrounding structures
         (8) Heavy cushion of fat surrounds each kidney
      b. Internal structures of the kidney
         (1) Cortex and medulla
         (2) Renal pyramids comprise much of the medullary tissue

         (3) Renal columns—where cortical tissue dips into the medulla between the pyramids
         (4) Calyx—cuplike structure at each renal papilla to collect urine; join together to form the renal pelvis
         (5) Renal pelvis narrows as it exits the kidney to become the ureter
      c. Blood vessels of the kidneys—kidneys are highly vascular
   2. Renal artery—large branch of the abdominal aorta; brings blood into each kidney
   3. Interlobular arteries—between the pyramids of the medulla, the renal artery branches; interlobular arteries extend toward the cortex, arch over the bases of the pyramids, and form the arcuate arteries; from the arcuate arteries, the interlobular arteries penetrate the cortex
   4. Pattern of blood flow through the kidneys—Abdominal aorta → renal artery → segmental arteries → lobar arteries → interlobar arteries → arcuate arteries → interlobular artery → afferent arteriole → glomerulus (glomerular capillaries) → efferent arteriole → peritubular capillaries (vasa recta) → interlobular veins → arcuate veins → interlobar veins → lobar veins → segmental veins → renal vein → inferior vena cava (Figure 28-3)
   5. Juxtaglomerular apparatus—located where the afferent arteriole brushes past the distal tubule; important to maintenance of blood flow homeostasis by reflexively secreting renin when blood pressure in the afferent arteriole drops

6. Ureter—tube running from each kidney to the urinary bladder; composed of three layers: mucous lining, muscular middle layer, and fibrous outer layer
7. Urinary bladder (Figure 28-5)
   a. Structure—collapsible bag located behind the symphysis pubis and made mostly of smooth muscle tissue; lining forms rugae; can distend considerably
   b. Functions
      (1) Reservoir for urine before it leaves the body
      (2) Aided by the urethra, it expels urine from the body
   c. Mechanism for voiding
      (1) Voluntary relaxation of the external sphincter muscle
      (2) Regions of the detrusor muscle contract reflexively
      (3) Urine is forced out of the bladder and through the urethra
8. Urethra
   a. Small mucous membrane–lined tube extending from the trigone to the exterior of the body
   b. In females, lies posterior to the symphysis pubis and anterior to the vagina; approximately 3 cm long
   c. In males, after leaving the bladder, passes through the prostate gland where it is joined by two ejaculatory ducts; from the prostate, it extends to the base of the penis, then through the center of the penis, and ends as the urinary meatus; approximately 20 cm long; the male urethra is part of the urinary system, as well as part of the reproductive system

MICROSCOPIC STRUCTURE OF THE NEPHRON

A. Nephrons, the microscopic functional units, make up the bulk of the kidney; those located in the renal cortex are called cortical nephrons; those near the junction of the cortical and medullary layers are called juxtamedullary nephrons; each nephron is made up of various structures (Figure 28-4)
1. Renal corpuscle (Figure 28-10)
2. Bowman's capsule—cup-shaped mouth of the nephron
   a. Formed by parietal and visceral walls with a space between them
   b. Pedicels in the visceral layer are packed closely together to form filtration slits; a slit diaphragm prevents the filtration slits from enlarging under pressure (Figure 28-11 and Figure 28-12)
   c. Glomerulus—network of fine capillaries in Bowman's capsule; together called a renal corpuscle; located in the cortex of the kidney (Figure 28-10)
   d. Basement membrane lies between the glomerulus and Bowman's capsule
   e. Glomerular capsular membrane—formed by glomerular endothelium, basement membrane, and the visceral layer of Bowman's capsule; function is filtration (Figure 28-12)
3. Proximal tubule—first part of the renal tubule nearest to Bowman's capsule; follows a winding, convoluted course; also known as the proximal convoluted tubule

4. Loop of Henle (Figure 28-4)
   a. Renal tubule segment just beyond the proximal tubule
   b. Consists of a thin descending limb, a sharp turn, and a thick ascending limb
   c. Juxtamedullary nephron—a nephron with a loop of Henle that dips far into the medulla
   d. Cortical nephron—a nephron with a loop of Henle that does not dip into the medulla but remains almost entirely within the cortex; constitute about 85% of the total nephrons
5. Distal tubule—convoluted tubule beyond the loop of Henle; also known as the distal convoluted tubule
6. Collecting duct
   a. Straight tubule joined by the distal tubules of several nephrons
   b. Joins larger ducts; the larger collecting ducts of one renal pyramid converge to form one tube that opens at a renal papilla into a calyx

PHYSIOLOGY OF THE URINARY SYSTEM

A. Overview of kidney function
1. Chief functions of the kidney are to process blood and form urine
2. Basic functional unit of the kidney is the nephron; forms urine through three processes (Figure 28-14)
   a. Filtration—movement of water and protein-free solutes from plasma in the glomerulus into the capsular space of Bowman's capsule
   b. Tubular reabsorption—movement of molecules out of the tubule and into peritubular blood
   c. Tubular secretion—movement of molecules out of peritubular blood and into the tubule for excretion
B. Filtration—first step in blood processing; occurs in renal corpuscles
1. From blood in the glomerular capillaries, about 180 liters of water and solutes filter into Bowman's capsule each day; takes place through the glomerular capsular membrane
2. Filtration occurs as a result of a pressure gradient
3. Glomerular capillary filtration occurs rapidly due to the increased number of fenestrations
4. Glomerular hydrostatic pressure and filtration are directly related to systemic blood pressure
C. Reabsorption—second step in urine formation; occurs as a result of passive and active transport mechanisms from all parts of the renal tubules; major portion of reabsorption occurs in the proximal tubules (Figure 28-15)
1. Reabsorption in the proximal tubule—most water and solutes are recovered by the blood, leaving only a small volume of tubule fluid left to move on to the loop of Henle
   a. Sodium—actively transported out of tubule fluid and into blood (Figure 28-17)
   b. Glucose and amino acids—passively transported out of tubule fluid by means of the sodium cotransport mechanism

c. Chloride, phosphate, and bicarbonate ions passively move into blood because of an imbalance in electrical charge

d. Water—movement of sodium and chloride into blood causes an osmotic imbalance, moving water passively into blood

e. Urea—approximately half of urea passively moves out of the tubule with the remaining urea moving on to the loop of Henle

2. Reabsorption in the loop of Henle (Figure 28-19)

a. Water is reabsorbed from the tubule fluid, and urea is picked up from the interstitial fluid in the descending limb

b. Sodium and chloride are reabsorbed from the filtrate in the ascending limb, where the reabsorption of salt makes the tubule fluid dilute and creates and maintains a high osmotic pressure of the medulla's interstitial fluid

D. Reabsorption in the distal tubules and collecting ducts

1. The distal tubule reabsorbs sodium by active transport but in smaller amounts than in the proximal tubule

2. ADH is secreted by the posterior pituitary and targets the cells of distal tubules and collecting ducts to make them more permeable to water

3. With reabsorption of water in the collecting duct, the urea concentration of the tubule fluid increases, which causes urea to diffuse out of the collecting duct into the medullary interstitial fluid

4. Urea participates in a countercurrent multiplier mechanism that along with the countercurrent mechanisms of the loop of Henle and vasa recta, maintains the high osmotic pressure needed to form concentrated urine and avoid dehydration

E. Tubular secretion

1. Tubular secretion—the movement of substances out of the blood and into tubular fluid

2. Descending limb of the loop of Henle secretes urea via diffusion

3. Distal and collecting tubules secrete potassium, hydrogen, and ammonium ions

4. Aldosterone—hormone that targets the cells of the distal and collecting tubule cells; causes increased activity of the sodium-potassium pumps

5. Secretion of hydrogen ions increases with increased blood hydrogen ion concentration

F. Regulation of urine volume (Figure 28-23)

1. ADH influences water reabsorption; as water is reabsorbed, the total volume of urine is reduced by the amount of water removed by the tubules; ADH reduces water loss

2. Aldosterone, secreted by the adrenal cortex, increases distal tubule absorption of sodium, thereby raising the sodium concentration of blood and thus promoting reabsorption of water

3. Atrial natriuretic hormone (ANH), secreted by specialized atrial muscle fibers, promotes loss of sodium via urine; opposes aldosterone, thus causing the kidneys to reabsorb less water and thereby produce more urine

4. Tubuloglomerular feedback mechanism maintains a constant GFR by regulating resistance in afferent arterioles. Protects GFR function from rapid blood pressure variations. Dependent on macula densa cells and the juxtaglomerular apparatus. May influence renin-angiotensin mechanism

5. Myogenic mechanism—rapid and effective regulation of GFR via changes in afferent arteriole smooth muscle contraction and relaxation

6. Urine volume—also related to the total amount of solutes other than sodium excreted in urine; generally, the more solutes, the more urine

G. Urine composition—approximately 95% water with several substances dissolved in it; the most important are

1. Nitrogenous wastes—result of protein metabolism; include urea, uric acid, ammonia, and creatinine

2. Electrolytes—mainly the following ions: sodium, potassium, ammonium, chloride, bicarbonate, phosphate, and sulfate; amounts and kinds of minerals vary with diet and other factors

3. Toxins—during disease, bacterial poisons leave the body in urine

4. Pigments—especially urochromes

5. Hormones—high hormone levels may spill into the filtrate

6. Abnormal constituents—such as blood, glucose, albumin, casts, or calculi

THE BIG PICTURE: URINARY SYSTEM AND THE WHOLE BODY

A. Homeostasis of water and electrolytes in body fluids relies on proper functioning of the kidneys; nephrons process blood to adjust its content to maintain a relatively constant internal environment

B. Urinary and cardiovascular systems are interdependent

C. Endocrine and nervous systems must operate properly to ensure efficient kidney function

REVIEW QUESTIONS

1. List the principal and accessory organs of the urinary system.
2. Name, locate, and give the main function(s) of each organ of the urinary system.
3. Identify the beginnings of the "plumbing system" of the urinary system.
4. How does the mechanism for voiding urine start?
5. The male urethra is part of two different systems. Identify them.
6. Describe the microscopic structure of the kidney.
7. Diagram the flow of blood through the kidney.
8. Define the terms *filtration*, *tubular reabsorption*, and *tubular secretion*.
9. How is effective filtration pressure calculated?
10. Describe the solute concentration of the interstitial fluid of the medulla.

11. What happens to sodium and chloride in the ascending limb of the loop of Henle?
12. What happens to potassium secretion when the blood aldosterone concentration increases?
13. Identify two drugs secreted by tubule cells.
14. What is the normal pH range for freshly voided urine?
15. Identify three body systems in addition to the urinary system that also excrete unneeded substances.
16. Define retention.
17. What is the most common cause of glycosuria?
18. What do elevated BUN levels indicate?
19. Define the term *osmolality*.

CRITICAL THINKING QUESTIONS

1. What would result if the nerves supplying the bladder and urethra were damaged?
2. Can you describe the mechanism of urine formation? How is each step related to the part of the nephron that performs it?
3. If the proximal tubules were unable to transport sodium ions into blood, why would you expect to find high concentrations of both sodium and chloride ions in urine?
4. Why do you think ADH prevents rapid dehydration of the body?
5. How is the function of ANH related to the increase in urine volume?
6. What is the relationship between age and kidney function?

CHAPTER 29

Fluid and Electrolyte Balance

LANGUAGE OF SCIENCE

aldosterone mechanism (al-DOS-ter-own MEK-ah-niz-em)

anions (AN-eye-ons) [*ana-* without, *-ion* electrically charged particle]

blood colloid osmotic pressure (BCOP) (KOL-oyd os-MOT-ik) [*kolla-* glue, *-oid* resembling, *osmos-* impulse, *-ic* pertaining to, *premere* to press]

blood hydrostatic pressure (BHP) (blud hye-droh-STAT-ik PRESS-ur) [*hydro-* water, *-statikos* causing to stand, *premere* to press]

cations (KAT-eye-on) [*kata-* down, *-ion* electrically charged particle]

colloid osmotic pressure (KOL-oyd os-MOT-ik) [*kolla-* glue, *-oid* resembling, *osmos-* impulse, *-ic* pertaining to, *premere* to press]

dehydration (dee-hye-DRAY-shun) [*de-* to remove entirely, *-hydor-* water, *-ation* process]

dissociate (di-SOH-see-ayt) [*dis-* apart, *-sociare* to unite]

electrolytes (eh-LEK-troh-lytes) [*elektron-* amber, *-lytos* soluble]

extracellular (eks-trah-SELL-yoo-lar) [*extra-* outside, *-cella* storeroom]

extracellular fluid (ECF) (eks-trah-SELL-yoo-lar) [*extra-* outside, *-cella* storeroom, *fluere* to flow]

fluid and electrolyte balance (eh-LEK-troh-lyte) [*fluere* to flow, *elektron-* amber, *-lytos* soluble, *bilanx* having two scales]

fluid compartments [*fluere* to flow, *com-* together, *-partiri* to share]

interstitial (in-ter-STISH-al) [*interstitium* space between]

interstitial fluid colloid osmotic pressure (IFCOP) (in-ter-STISH-al KOL-oyd os-MAH-tik) [*interstitium* space between, *fluere* to flow, *kolla-* glue, *-oid* resembling, *osmos-* impulse, *-ic* pertaining to, *premere* to press]

interstitial fluid hydrostatic pressure (IFHP) (in-ter-STISH-al hye-droh-STAT-ik) [*interstitium* space between, *fluere* to flow, *hydro-* water, *-statikos* causing to stand, *premere* to press]

Cont'd on p. 1084

The phrase **fluid and electrolyte balance** implies homeostasis, or constancy, of body fluid and electrolyte levels. It means that both the amount and distribution of body fluids and electrolytes are normal and constant. For homeostasis to be maintained, body "input" of water and electrolytes must be balanced by "output." If water and electrolytes enter the body in excess of requirements, they must be selectively eliminated, and, if excess losses occur, prompt replacement is critical. The volume of fluid and the electrolyte levels inside the cells, in the interstitial spaces, and in the blood vessels all remain relatively constant when a condition of homeostasis exists. Fluid and electrolyte imbalance, then, means that both the total volume of water and the level of electrolytes in the body or the amounts in one or more of its fluid compartments have increased or decreased beyond normal limits.

Table 29-1	Volumes of Body Fluid Compartments*		
BODY FLUID	**INFANT**	**ADULT MALE**	**ADULT FEMALE**
Extracellular fluid			
Plasma	4	4	4
Interstitial fluid	26	16	11
Intracellular fluid	45	40	35
TOTAL	75	60	50

*Percentage of body weight.

INTERRELATIONSHIP OF FLUID AND ELECTROLYTE BALANCE

Several of the basic physical properties of matter discussed in Chapter 2 will help explain the mechanisms of fluid and electrolyte balance. The concept of chemical bonding is a good example. The type of chemical bonds between molecules of certain chemical compounds, such as sodium chloride (NaCl), permits breakup, or dissociation, into separate particles (Na^+ and Cl^-). Recall that such compounds are known as **electrolytes.** The dissociated particles of an electrolyte are called **ions** and carry an electrical charge. Organic substances such as glucose, however, have a type of bond that does not permit the compound to break up, or **dissociate,** in solution. Such compounds are known as **nonelectrolytes.**

Many electrolytes and their dissociated ions are of critical importance in fluid balance. Fluid balance and electrolyte balance are so interdependent that if one deviates from normal, so does the other. A discussion of one therefore necessitates a discussion of the other.

TOTAL BODY WATER

Normal values for total body water expressed as a percentage of total body weight will vary between 45% and 75%. Differences occur because of age, fat content of the body, and sex. In newborn infants, total body water represents about 75% of body weight. This percentage then decreases rapidly during the first 10 years of life. At adolescence, adult values are reached and gender differences, which account for about a 10% variation in body fluid volumes between the sexes, appear. In young, nonobese adults, males weighing 70 kg (154 pounds) will average about 60% of their body weight as water (nearly 40 liters) and females about 50% (Table 29-1). Adipose, or fat, tissue contains the least amount of water of any tissue (including bone) in the body. Therefore, regardless of age, obese individuals, with their high body fat content, have less body water per kilogram of weight than slender people do. In aged individuals of either sex, body water content may decrease to 45% of total body weight. One reason is that old age is often accompanied by a decrease in muscle mass (65% water) and an increase in fat (20% water). In addition, with advancing age the kidneys are less able to produce concentrated urine, and sodium-conserving responses become less effective.

BODY FLUID COMPARTMENTS

Functionally, the total body water can be subdivided into two major **fluid compartments** called the **extracellular** and the **intracellular fluid compartments. Extracellular fluid (ECF)** consists mainly of the *plasma* found in the blood vessels and the *interstitial fluid* that surrounds the cells. In addition, the lymph and so-called *transcellular fluid*—such as cerebrospinal fluid, specialized joint fluids, and humors of the eye—are also considered extracellular fluid. The distribution of body water by compartment is shown in Figure 29-1. **Intracellular fluid (ICF)** refers to the water inside the cells.

Extracellular fluid constitutes the internal environment of the body. It therefore serves the dual vital functions of providing a relatively constant environment for cells and transporting substances to and from them. Intracellular fluid, on the other hand, because it is a solvent, functions to facilitate intracellular chemical reactions that maintain life. When compared according to volume, intracellular fluid is the largest (25 L), plasma the smallest (3 L), and interstitial fluid in between (12 L). Figure 29-2 illustrates the typical normal fluid volumes in a young adult male, and Table 29-1 lists volumes of the body fluid compartments for both sexes as a percentage of body weight.

CHEMICAL CONTENT, DISTRIBUTION, AND MEASUREMENT OF ELECTROLYTES IN BODY FLUIDS

We have defined an electrolyte as a compound that will break up or dissociate into charged particles called ions when placed in solution. Sodium chloride, when dissolved in water, provides a positively charged sodium ion (Na^+) and a negatively charged chloride ion (Cl^-).

If two electrodes charged with a weak current are placed in an electrolyte solution, the ions will move, or migrate, in opposite directions according to their charge. Positive ions such as Na^+ will be attracted to the negative electrode (cathode) and are called **cations.** Negative ions such as Cl^- will migrate to the positive electrode (anode) and are called **anions.** Various anions and

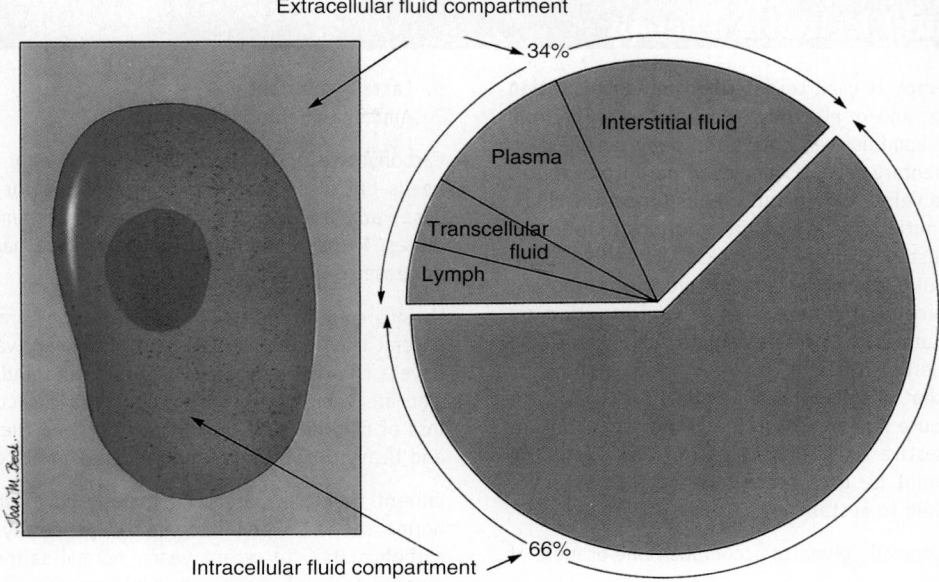

Figure 29-1 *Distribution of total body water.*

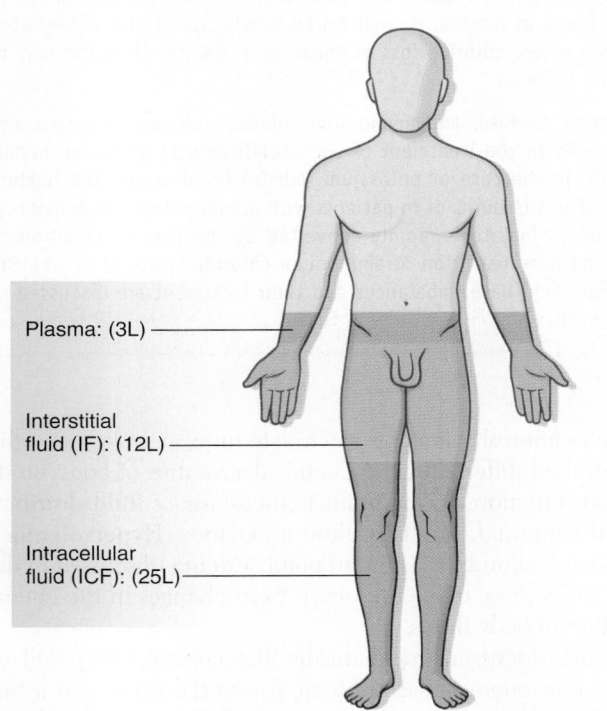

Plasma: (3L)

Interstitial
fluid (IF): (12L)

Intracellular
fluid (ICF): (25L)

Figure 29-2 *Relative volumes of three body fluids.* Values represent fluid distribution in a young adult male.

cations serve critical nutrient or regulatory roles in the body. Important cations include sodium (Na^+), calcium (Ca^{++}), potassium (K^+), and magnesium (Mg^{++}). Important anions include chloride (Cl^-), bicarbonate (HCO_3^-), phosphate ($HPO_4^=$), and many proteins.

The importance of electrolytes in controlling the movement of water between the body fluid compartments is discussed in this chapter. Their role in maintaining acid-base balance is examined in Chapter 30.

Extracellular vs. Intracellular Fluids

Compared chemically, plasma and **interstitial** fluid (the two extracellular fluids) are almost identical. Intracellular fluid, on the other hand, shows striking differences from either of the two extracellular fluids. Let us examine first the chemical structure of plasma and interstitial fluid as shown in Figure 29-3 and Table 29-2.

Perhaps the first difference between the two extracellular fluids that you notice (see Figures 29-3 and 29-4) is that blood contains a slightly larger total of electrolytes (ions) than interstitial fluids do. If you compare the two fluids, ion for ion, you will discover the most important difference between blood plasma and interstitial fluid. Look at the anions (negative ions) in these two extracellular fluids. Note that blood contains an appreciable amount of protein anions. *Interstitial fluid, in contrast, contains hardly any protein anions.* This is the only functionally important difference between blood and interstitial fluid. It exists because the normal capillary membrane is practically impermeable to proteins. Hence, almost all protein anions remain behind in the blood instead of filtering out into the interstitial fluid. Because proteins remain in the blood, certain other differences also exist between blood and interstitial fluid—notably, blood contains more sodium ions and fewer chloride ions than interstitial fluid does.

Extracellular fluids and intracellular fluid are more unlike than alike chemically. Chemical difference predominates between the extracellular and intracellular fluids. Chemical similarity predominates between the two extracellular fluids. Study Figures 29-3 and 29-4 and make some generalizations about the main chemical differences between the extracellular and intracellular fluids. For example: What is the most abundant cation in extracellular fluids? In intracellular fluid? What is the most abun-

BOX 29-1 Fluid and Electrolyte Therapy

The term **parenteral therapy** is used to describe the administration of nutrients, special fluids, and/or electrolytes by injection. The term implies that whatever is administered enters the body by injection and not through the alimentary canal. Examples of **parenteral** routes include **intravenous** (into veins) and **subcutaneous** (under the skin). Significant quantities of nutrient or electrolyte solutions that are injected subcutaneously must be isotonic with plasma or cellular damage will occur. Such solutions may be administered intravenously, however, regardless of tonicity, if correct rates of administration are used. The intravenous route is the preferred route for all fluid and electrolyte solutions. It permits the body to adjust its fluid compartments in the same way that it does after the ordinary intake of water and food. The ideal route for the absorption of nutrients and fluids is, of course, the digestive tract. However, if for any reason the digestive tract route cannot be used, parenteral administration of these substances is required to sustain life.

Parenteral solutions are generally given to accomplish one or more of three primary objectives:
1. To meet current maintenance needs for nutrients, fluids, and electrolytes
2. To replace past losses
3. To replace concurrent losses (additional losses that are in excess of maintenance needs)

Although many different types and combinations of nutrients and electrolytes in solution are available to meet almost every medical need, 85% to 95% of all individuals needing fluid therapy are treated with one or more of the seven basic solutions listed below:
1. Carbohydrate in water
2. Carbohydrate in various strengths of saline
3. Normal saline (0.9% NaCl)
4. Potassium solutions
5. Ringer's solution
6. Lactate solutions
7. Ammonium chloride solutions

Carbohydrate and water solutions not only supply water for body needs but also provide calories required for energy. Dextrose (glucose) and fructose (levulose) are the common parenteral carbohydrates. Perhaps the most frequently used parenteral solution is 5% dextrose in water (D_5W).

Various carbohydrate and saline solutions are also available for parenteral use. Such solutions are of primary value in individuals who have a chloride deficit and ongoing fluid and caloric needs. Patients who are vomiting or undergoing gastric suction that results in the loss of chloride in hydrochloric acid need these solutions. Prolonged and heavy sweating and diarrhea also produce chloride deficits.

Current trends in parenteral therapy have reduced the frequency of normal saline use administered independently of other electrolytes or carbohydrates. In recent years, normal saline as a general purpose electrolyte has been replaced by the use of Ringer's solution, which provides more of the essential electrolytes in physiological proportions. Ringer's solution is often described as normal saline modified by the addition of calcium and potassium in amounts approximating those found in plasma. Normal saline is still useful and widely used in cases where chloride loss is equal to or greater than the loss of sodium, however.

Potassium, lactate, and ammonium chloride solutions are specialty fluids used in the treatment of such conditions as acid-base imbalance or, in the case of potassium, administered during the healing phase of severe burns or in patients with actual potassium deficiency. In acidosis, lactate is rapidly converted by the liver to bicarbonate ions, and administration of ammonium chloride is useful in treating alkalosis. Acid-base imbalances and their treatment are discussed in Chapter 30.

Table 29-2 Electrolyte Composition of Blood Plasma

	CATIONS	ANIONS
	142 mEq Na^+	102 mEq Cl^-
	4 mEq K^+	26 HCO_3^-
	5 mEq Ca^{++}	17 protein
	2 mEq Mg^{++}	6 other
		2 $HPO_4^=$
TOTAL	153 mEq/L plasma	153 mEq/L plasma

dant anion in extracellular fluids? In intracellular fluid? What about the relative concentrations of protein anions in extracellular fluids and intracellular fluid?

The reason we called attention to the chemical structure of the three body fluids is that here, as elsewhere, structure determines function. In this instance the chemical structure of the three fluids helps control water and electrolyte movement between them. Or, phrased differently, the chemical structure of body fluids, if normal, functions to maintain homeostasis of fluid distribution and, if abnormal, results in fluid imbalance. **Hypervolemia** (excess blood volume) is a case in point. **Edema** (discussed in detail on p. 1078), too, frequently stems from changes in the chemical structure of body fluids.

Before discussing mechanisms that control water and electrolyte movement between blood, interstitial fluid, and intracellular fluid, it is important to understand the units used for measuring electrolytes.

Measuring Electrolyte Reactivity

After the important electrolytes and their constituent ions in the body fluid compartments had been established, physiologists needed to measure changes in their levels to understand the mechanisms of fluid balance. To have meaning, measurement units used to report electrolyte levels must be related to actual physiological activity. In the past, only the weight of an electrolyte in a given amount of solution—its *concentration*—was measured. The number of milligrams per 100 ml of solution (mg%) was one of the most

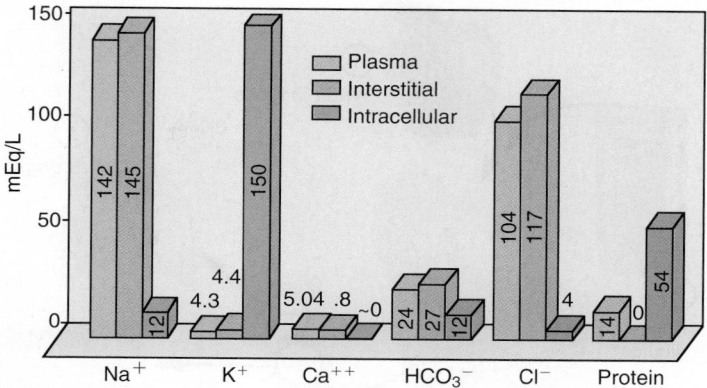

Figure 29-4 *Electrolyte and protein concentrations in body fluid compartments.* This illustration compares individual electrolyte and protein concentration in the three fluid compartments.

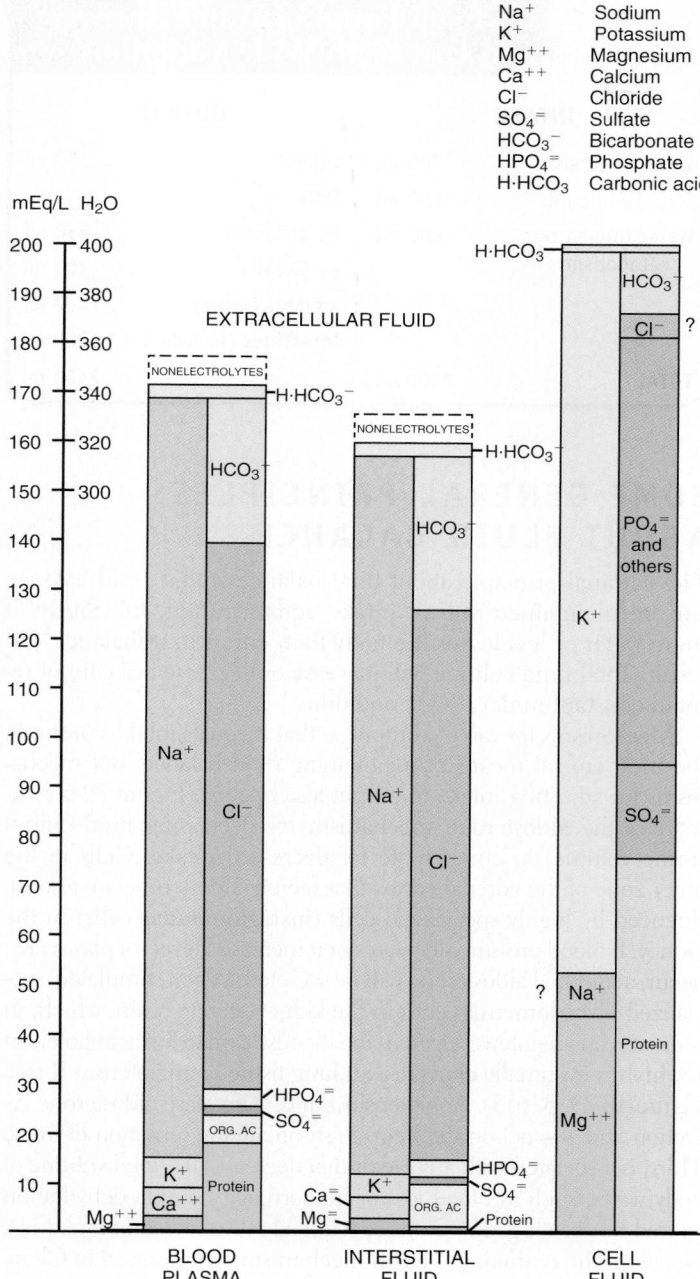

Figure 29-3 *Chief chemical constituents of three fluid compartments.* The column of figures at the left (200, 190, 180, etc.) indicates the amount of cation or anion, whereas the figures on the right (400, 380, 360, etc.) indicate the sum of the cations and anions.

frequently used units of measurement. However, simply reporting the concentration of an important electrolyte such as sodium or calcium in milligrams per 100 ml of blood (mg%) gives no direct information about its chemical combining power or physiological activity in body fluids. The importance of valence and electrovalent or ionic bonding in chemical reactions was discussed in Chapter 2. The reactivity or combining power of an electrolyte depends

not just on the number of molecular particles present but also on the total number of ionic charges (valence). Univalent ions such as sodium (Na^+) carry only a single charge, but the divalent calcium ion (Ca^{++}) carries two units of electrical charge.

The need for a unit of measurement more related to activity has resulted in increasing use of a more meaningful measurement yardstick—the **milliequivalent**. Milliequivalents measure the number of ionic charges or electrovalent bonds in a solution and therefore serve as an accurate measure of the chemical (physiological) combining power, or *reactivity*, of a particular electrolyte solution. The number of milliequivalents of an ion in a liter of solution (mEq/L) can be calculated from its weight in 100 ml (mg%) by using a convenient conversion formula.

Conversion of milligrams per 100 ml (mg%) to milliequivalents per liter (mEq/L):

$$mEq/L = \frac{mg/100\ ml \times 10 \times Valence}{Atomic\ weight}$$

Example: Convert 15.6 mg% K^+ to mEq/L

Atomic weight of K^+ = 39

Valence of K^+ = 1

$$mEq/L = \frac{15.6 \times 10 \times 1}{39} = \frac{156}{39} = 4$$

Therefore, 15.6 mg/100 ml K^+ = 4 mEq/L.

 QUICK CHECK

1. List three important cations and anions that serve critical nutrient or regulatory roles in the body.
2. Name the most abundant chemical constituent in blood plasma, interstitial fluid, and intracellular fluid.
3. Identify the units used to describe electrolyte concentration and electrolyte reactivity.

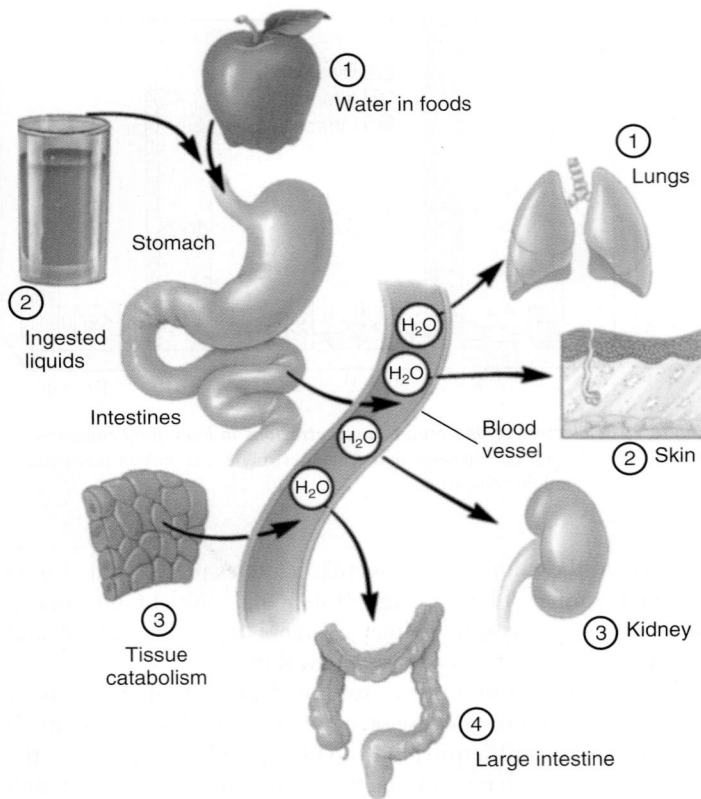

Figure 29-5 *Sources of fluid intake and output.*

Table 29-3	Typical Normal Values (24 Hours) for Each Portal of Water Entry and Exit (with Wide Variations)		
INTAKE		**OUTPUT**	
Water in foods	700 ml	Lungs	350 ml
Ingested liquids	1500 ml	Skin	
Water formed by catabolism	200 ml	By diffusion	350 ml
		By sweat	100 ml
		Kidneys (urine)	1400 ml
		Intestines (in feces)	200 ml
TOTAL	2400 ml		2400 ml

AVENUES BY WHICH WATER ENTERS AND LEAVES THE BODY

Water enters the body, as everyone knows, from the digestive tract—in the liquids one drinks and in the foods one eats (Figure 29-5). But, in addition, and less universally known, water enters the body—that is, is added to its total fluid volume—from its billions of cells. Each cell produces water by catabolizing foods, and this water enters the bloodstream. Water normally leaves the body by four exits: kidneys (urine), lungs (water in expired air), skin (by diffusion and by sweat), and intestines (feces). In accord with the cardinal principle of fluid balance, the total volume of water entering the body normally equals the total volume leaving. In short, fluid intake normally equals fluid output. Figure 29-5 illustrates the portals of water entry and exit, and Table 29-3 gives their normal volumes. These volumes, however, can vary considerably and still be considered normal.

 QUICK CHECK

4. Name the type of chemical compound that breaks up, or dissociates, in solution to form ions.
5. Plasma and interstitial fluid are subdivisions of what major body fluid compartment?
6. List the volumes of body fluid compartments in a young adult female as a percentage of body weight.
7. List the major "portals" of water entry and exit from the body.

SOME GENERAL PRINCIPLES ABOUT FLUID BALANCE

The cardinal principle about fluid balance is this: fluid balance can be maintained only if intake equals output. Obviously, if more water or less leaves the body than enters it, imbalance will result. Total fluid volume will increase or decrease but cannot remain constant under these conditions.

Mechanisms for varying output so that it equals intake constitute the most crucial means of maintaining fluid balance, but mechanisms for adjusting intake to output also operate. Figure 29-6 summarizes the **aldosterone mechanism** for decreasing fluid output (urine volume) to compensate for decreased intake. Cells in the outer zone of the adrenal cortex that secrete aldosterone are also influenced by highly specialized cells (**juxtaglomerular cells**) in the kidney. If blood pressure decreases or if increased levels of plasma K^+ occur, additional aldosterone will be secreted. When stimulated, specialized juxtaglomerular cells in the kidney secrete renin, which, in turn, acts on angiotensinogen in the bloodstream to form angiotensin I, which is eventually converted in lung tissue to angiotensin II (see Figure 16-23, p. 615). Angiotensin I and II increase aldosterone secretion and also act on the brain to stimulate the sensation of thirst. Thirst is associated with any factor that decreases the total volume of body water, such as blood loss or hemorrhage. Simple dehydration caused by sweating also results in reduced saliva secretion and thirst. Details of the **renin-angiotensin mechanism** are discussed in Chapter 16. Figure 29-7 diagrams a postulated mechanism for adjusting intake to compensate for excess output.

Mechanisms for controlling water movement between the fluid compartments of the body constitute the most rapid-acting fluid balance processes. They serve first of all to maintain normal blood volume at the expense of interstitial fluid volume.

MECHANISMS THAT MAINTAIN HOMEOSTASIS OF TOTAL FLUID VOLUME

Under normal conditions, homeostasis of the total volume of water in the body is maintained or restored primarily by mechanisms that adjust output (urine volume) to intake and secondarily by mechanisms that adjust fluid intake.

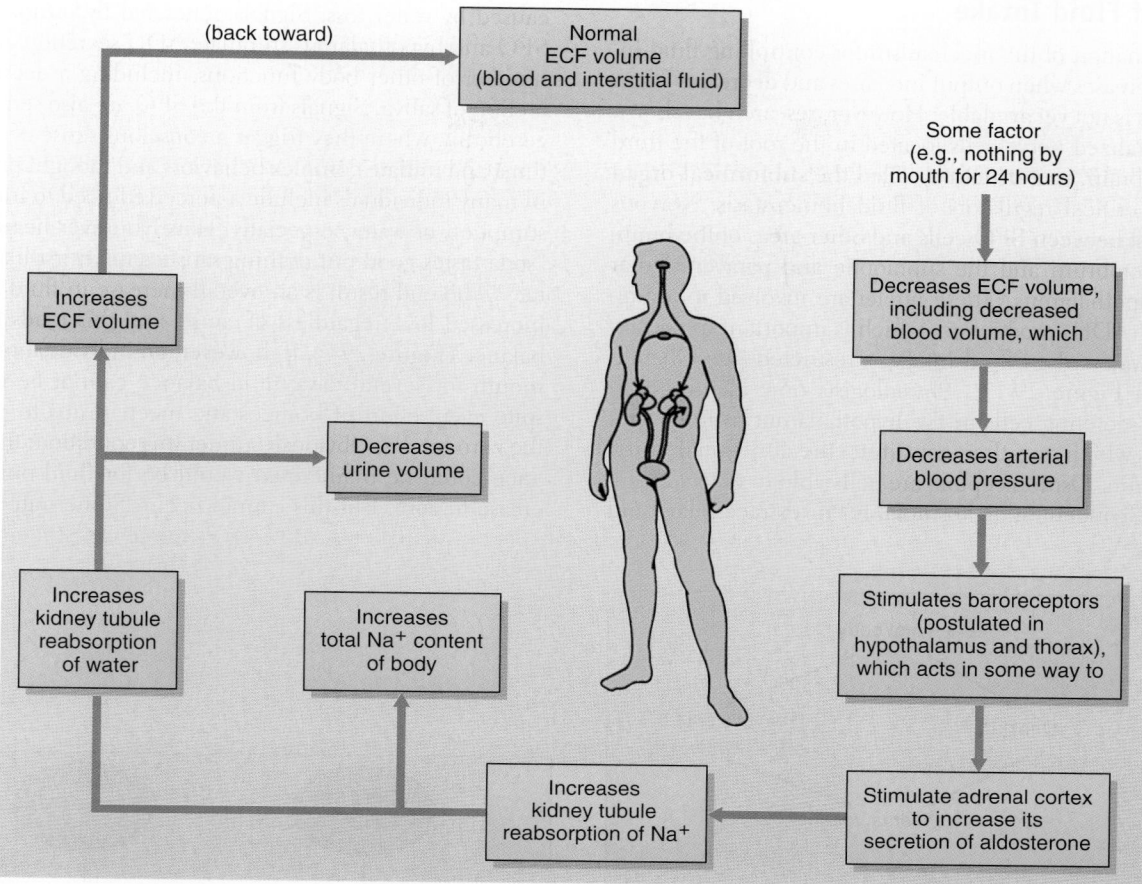

Figure 29-6 *Aldosterone mechanism for ECF homeostasis.* The aldosterone mechanism tends to restore normal extracellular fluid (ECF) volume when it decreases below normal. Excess aldosterone, however, leads to excess ECF volume, that is, excess blood volume (hypervolemia) and excess interstitial fluid volume (edema), as well as an excess of the total Na^+ content of the body. The renin-angiotensin mechanism influencing aldosterone secretion is shown in Figure 16-26.

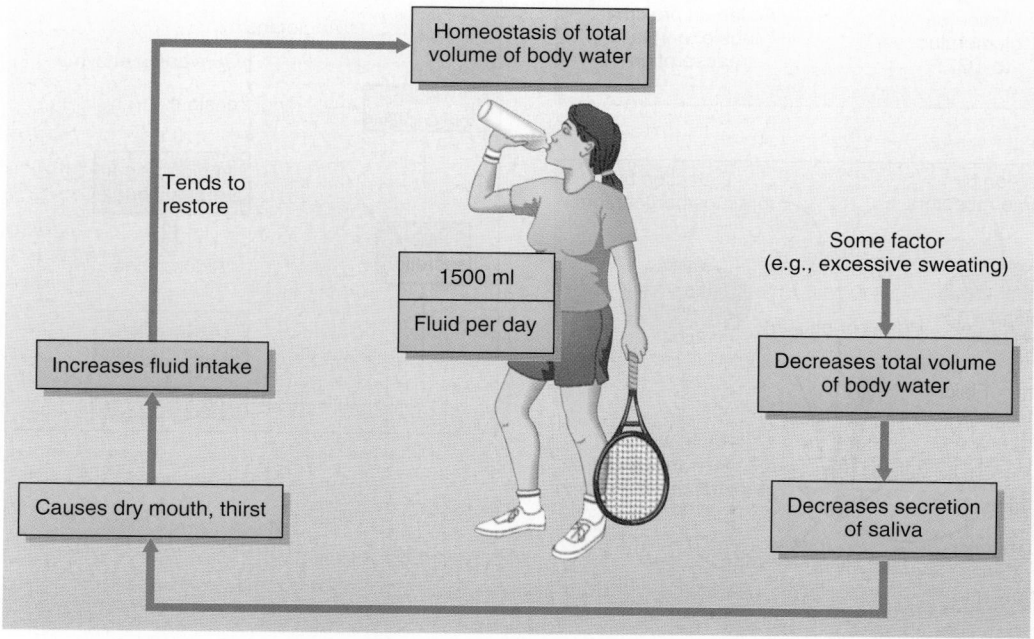

Figure 29-7 *Homeostasis of the total volume of body water.* A basic mechanism for adjusting intake to compensate for excess output of body fluid is diagrammed.

Regulation of Fluid Intake

A detailed explanation of the mechanism for controlling fluid intake so that it increases when output increases and decreases when output decreases is not yet available. However, research has shown that highly specialized nerve cells located in the roof of the third ventricle of the brain, in a structure called the **subfornical organ** or **SFO**, act as critical regulators of fluid homeostasis. Nervous connections exist between SFO cells and other areas of the brain, including the cerebrum and the supraoptic and paraventricular nuclei of the hypothalamus. These nuclei are involved in antidiuretic hormone (ADH) production, which is important in conservation of body water when fluid intake is restricted (see Chapter 16, p. 612) and Figure 29-15). Physiologists now identify both SFO and ADH-secreting cells in the hypothalamus as important **osmoreceptors**, which together constitute the functional **thirst center** of the brain. Osmoreceptors are cells able to detect an increase in solute concentration (osmolality) in extracellular fluid

caused by water loss. Signals generated by osmoreceptors in the SFO and hypothalamus stimulate ADH secretion and also affect a number of other body functions, including a decrease in the secretion of saliva. Signals from the SFO are also sent directly to the cerebrum, where they trigger a conscious sense of dry mouth and thirst and initiate complex behaviors and thought processes, which in many individuals include a perceived need to increase the consumption of water, especially. Have you ever heard someone say "soda tastes good but nothing satisfies my thirst like a glass of water"? The end result is an overall increase in fluid intake to offset increased loss, regardless of cause, and this tends to restore fluid balance (Figure 29-7). If, however, an individual takes nothing by mouth for several days, fluid balance cannot be maintained despite every effort of homeostatic mechanisms to compensate for the zero intake. Obviously, under this condition, the only way balance could be maintained would be for fluid output to also decrease to zero. But this cannot occur. Some output is obligatory.

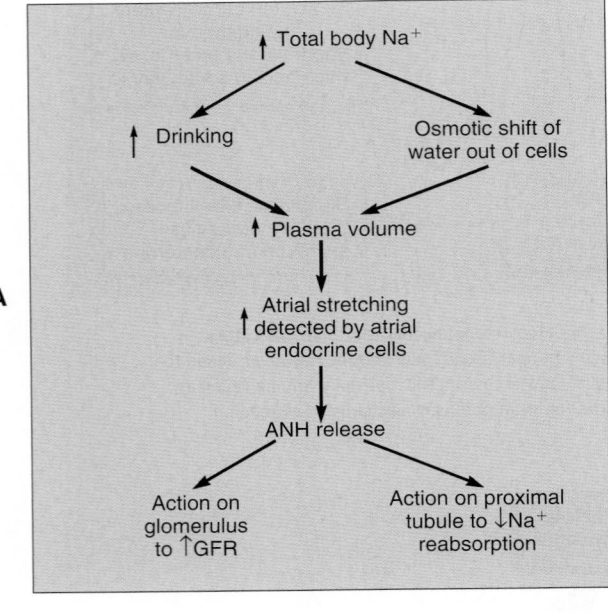

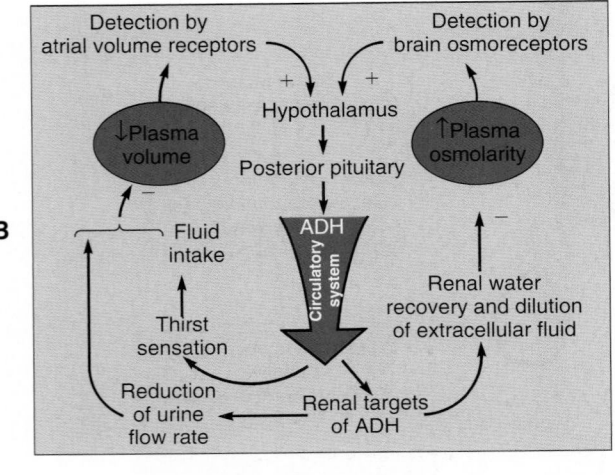

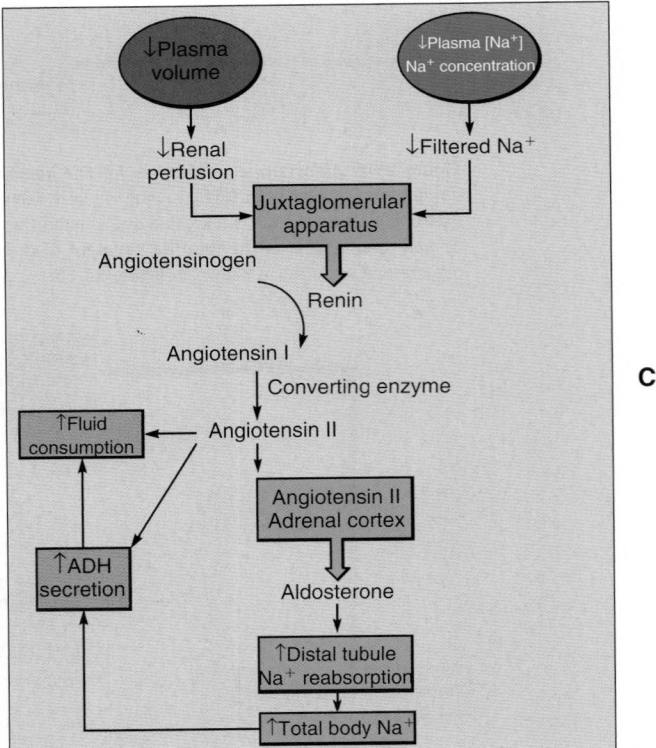

Figure 29-8 *Mechanisms of fluid and electrolyte regulation.* **A,** The ANH system. **B,** The ADH system. **C,** The aldosterone system. *ADH,* Antidiuretic hormone; *ANH,* atrial natriuretic hormone; *GFR,* glomerular filtration rate.

Why? Because as long as respirations continue, some water leaves the body by way of expired air. Also, as long as life continues, an irreducible minimum of water diffuses through the skin.

Regulation of Urine Volume

Two factors together determine urine volume: the glomerular filtration rate and the rate of water reabsorption by the renal tubules. The glomerular filtration rate, except under abnormal conditions, remains fairly constant—hence it does not normally cause urine volume to fluctuate. The rate of tubular reabsorption of water, on the other hand, fluctuates considerably. The rate of tubular reabsorption, therefore, rather than the glomerular filtration rate, normally adjusts urine volume to fluid intake. The amount of antidiuretic hormone (ADH) and aldosterone secreted regulates the amount of water reabsorbed by the kidney tubules (discussed on p. 612; see also Figure 29-6). In other words, urine volume is regulated chiefly by hormones secreted by the posterior lobe of the pituitary gland (ADH) and by the adrenal cortex (aldosterone) and by atrial natriuretic hormone (ANH). Regulation of aldosterone secretion by the renin-angiotensin mechanism has also been discussed.

Although changes in the volume of fluid loss via the skin, the lungs, and the intestines also affect the fluid intake-output ratio, these volumes are not automatically adjusted to intake volume, as is the volume of urine. Figure 29-8 summarizes the fluid and electrolyte regulation mechanisms that involve ADH, aldosterone, and ANH.

Factors That Alter Fluid Loss Under Abnormal Conditions

The rate of respiration and the volume of sweat secreted may greatly alter fluid output under certain abnormal conditions. For example, a patient who hyperventilates for an extended time loses an excessive amount of water via the expired air. If, as frequently happens, the individual also takes in less water by mouth than normal, the fluid output then exceeds intake and a fluid imbalance develops, namely, **dehydration** (that is, a decrease in total body water). The severity of dehydration can be measured by weight loss as a percentage of the normal (hydrated) body weight (Figure 29-9). Symptoms range from simple thirst to muscle weakness and kidney failure. Clinically, dehydration is often detected by loss of skin elasticity or *turgor*. If a fold of skin, when pinched, returns to its origi-

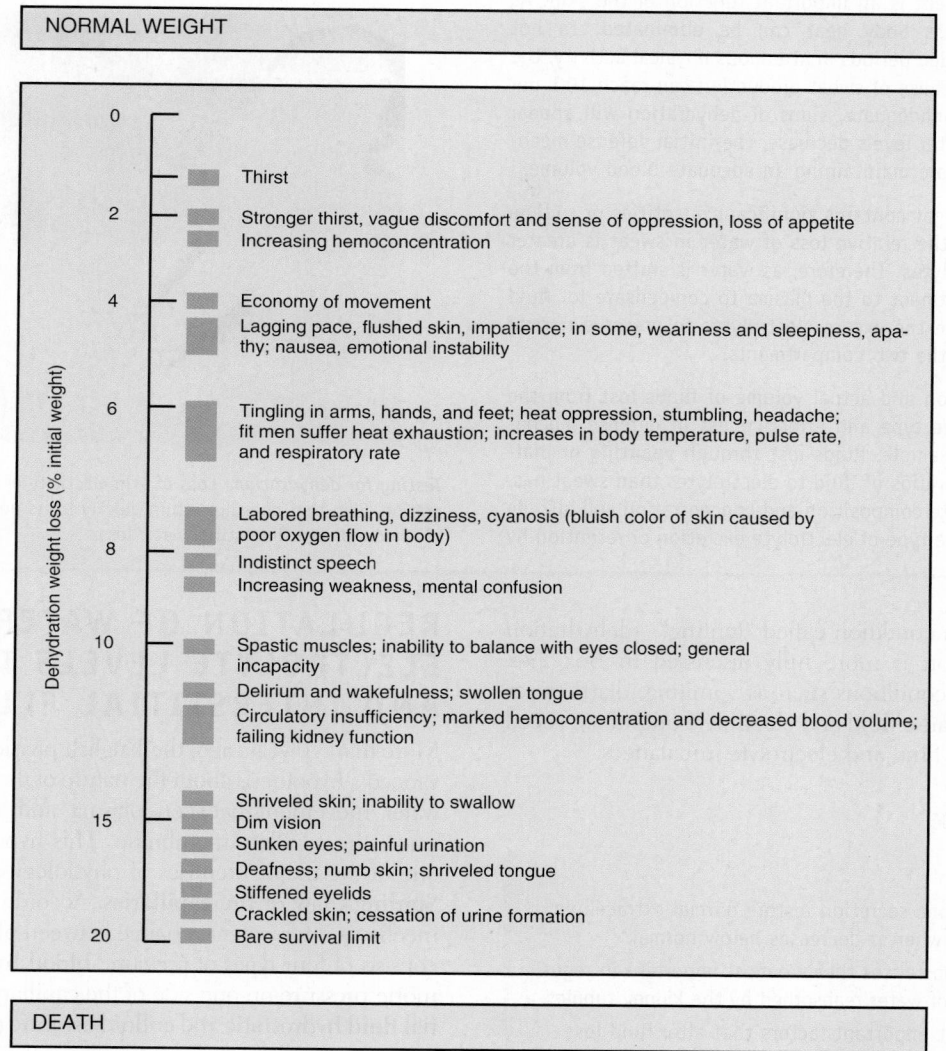

Figure 29-9 *The effects of dehydration.*

BOX 29-2 Dehydration

The term *dehydration* is used to describe the condition that results from excessive loss of body water. Loss of skin elasticity or *turgor* is a sign of dehydration. Water deprivation or loss triggers a complex series of protective responses designed to maintain homeostasis of water and electrolyte levels. Unfortunately, the term *dehydration* is incomplete. It does not, by definition, include the loss of electrolytes. To understand the control mechanisms that ensure fluid and electrolyte balance or properly interpret the clinical signs and symptoms of dehydration in disease states, it is important to realize that in any process of dehydration, water loss is always accompanied by loss of electrolytes. If water intake is reduced to the point of dehydration, a corresponding quantity of electrolytes must be removed to maintain the normal ionic content of body fluids. The same is true in the case of electrolyte loss when an accompanying loss of water must occur to maintain homeostasis of both fluid and electrolyte levels. Understanding the close interrelationships of water and electrolyte loss in dehydration provides the rationale for effective treatment. Water alone is inadequate; treatment of dehydration also requires appropriate electrolyte replacement therapy.

As discussed in Chapter 6, maintaining a constant core body temperature in a hot environment is an important function of the skin. As sweat evaporates, excess body heat can be eliminated. In hot weather or during extended periods of strenuous physical activity, the volume of water lost because of sweat production can reach 10 L per day. If water intake is inadequate, signs of dehydration will appear very rapidly. As body water levels decrease, the initial defense mechanisms are directed toward maintaining an adequate blood volume.

In addition to water, sweat contains significant quantities of sodium and chloride. However, the relative loss of water in sweat is greater than the loss of electrolytes. Therefore, as water is shifted from the interstitial fluid compartment to the plasma to compensate for fluid loss, the kidneys excrete the excess electrolytes to preserve normal ionic concentrations in the two compartments.

The chemical composition and actual volume of fluids lost from the body will also affect the type and effectiveness of defense mechanisms that occur. For example, fluids lost through vomiting or diarrhea will have differing ratios of fluid to electrolytes than sweat has, and the actual electrolyte composition and concentration will also be different. As a result, the type of electrolyte excretion or retention by

the kidneys that will be needed to maintain ionic balance in the fluid compartments will also change.

There is a lag in the volume-electrolyte adjustment mechanism triggered by dehydration. Shifts in fluid occur more quickly between compartments than the adjustment in electrolyte levels. However, if water and electrolyte losses are limited and the interval between loss and replacement is short, the symptoms of dehydration will be mild and transitory.

In severe and prolonged water deprivation or loss, the initial shift of interstitial fluid to plasma will be followed by movement of water from the intracellular compartment as well. Over time, the extracellular and intracellular fluid losses are about equal.

Extracellular (interstitial) water is more "expendable" and quickly accessible as a fluid source to maintain blood volume in the early stages of body fluid loss. It is said to serve as the "first line of defense" against dehydration. As extracellular fluid is depleted, intracellular water must be used to prolong survival time. Ultimately, the volume of extracellular fluid can be reduced by almost 60% and intracellular fluid by 30% before death occurs.

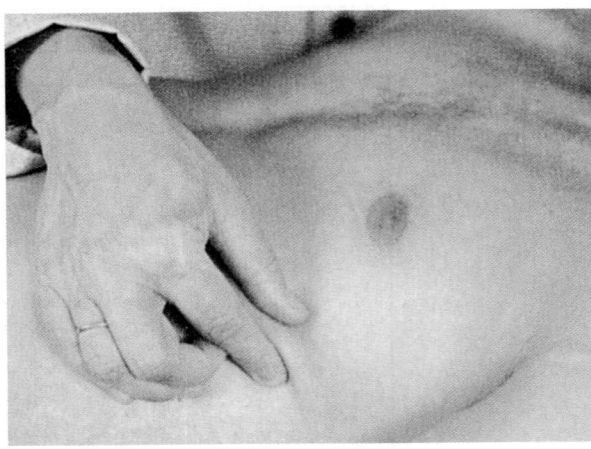

Testing for dehydration. Loss of skin elasticity or turgor is a sign of dehydration. Skin that does not return quickly to its normal shape after being pinched indicates interstitial water loss.

nal shape only slowly—a condition called "tenting"—dehydration is suspected. Dehydration is more fully discussed in Box 29-2 above. Other abnormal conditions such as vomiting, diarrhea, or intestinal drainage also cause fluid and electrolyte output to exceed intake and thus produce fluid and electrolyte imbalances.

QUICK CHECK

8. How does aldosterone secretion restore normal extracellular fluid (ECF) volume when it decreases below normal?

9. Identify the two substances that are most important in regulating the amount of water reabsorbed by the kidney tubules.

10. Name the two most important factors that alter fluid loss under abnormal conditions.

REGULATION OF WATER AND ELECTROLYTE LEVELS IN PLASMA AND INTERSTITIAL FLUID (ECF)

More than 70 years ago, the English physiologist Ernest Starling advanced a hypothesis about the nature of the mechanism that controls water movement between plasma and interstitial fluid—that is, across the capillary membrane. This hypothesis has since become one of the major premises of physiology and is often spoken of as **Starling's law of the capillaries.** According to this law, the control mechanism for water exchange between plasma and interstitial fluid consists of four types of pressure: **blood hydrostatic** and **colloid osmotic pressure** on one side of the capillary membrane and **interstitial fluid hydrostatic** and **colloid osmotic pressure** on the other side.

We are ready now to try to answer the following question: How does the chemical structure of body fluids control water movement between them and thereby control fluid distribution in the body?

According to the physical laws governing filtration and osmosis, **blood hydrostatic pressure (BHP)** tends to force fluid out of capillaries into interstitial fluid (IF), but **blood colloid osmotic pressure (BCOP)** tends to draw it back into them. **Interstitial fluid hydrostatic pressure (IFHP),** in contrast, tends to force fluid out of the interstitial fluid into the capillaries, and **interstitial fluid colloid osmotic pressure (IFCOP)** tends to draw it back out of capillaries. In short, two of these pressures constitute vectors in one direction and two in the opposite direction. This process is similar in many ways to the mechanism responsible for the formation of glomerular filtrate studied in the last chapter. The movement of fluids and electrolytes between plasma and interstitial fluid caused by hydrostatic and colloid osmotic pressure is illustrated in Figure 29-10. Note also in Figure 29-10 that the presence of blind-ended lymphatic vessels in the tissue spaces also serves as a mechanism for drainage of excess interstitial fluid. Osmotic and diffusion forces result in the movement of interstitial fluid and small proteins into the lymphatic system. Eventually, the lymph formed

in this way will enter the circulatory system and become part of the circulating blood volume.

The difference between the two sets of opposing forces obviously represents the net or effective filtration pressure—in other words, the effective force tending to produce the net fluid movement between blood and interstitial fluid. In general terms, therefore, we may state Starling's law of the capillaries this way: the rate and direction of fluid exchange between capillaries and interstitial fluid are determined by the hydrostatic and colloid osmotic pressuresof the two fluids. Or, we may state it more specifically as a formula:

$$(BHP + IFCOP) - (IFHP + BCOP) = EFP*$$

To illustrate operation of Starling's law (see Figure 29-7), let us consider how it controls water exchange at the arterial ends of tissue capillaries. The lower left-hand portion of Figure 29-7 gives

---

*Note that the factors enclosed in the first set of parentheses tend to move fluid out of capillaries and that those in the second set oppose this movement—they tend to move fluid into the capillaries.

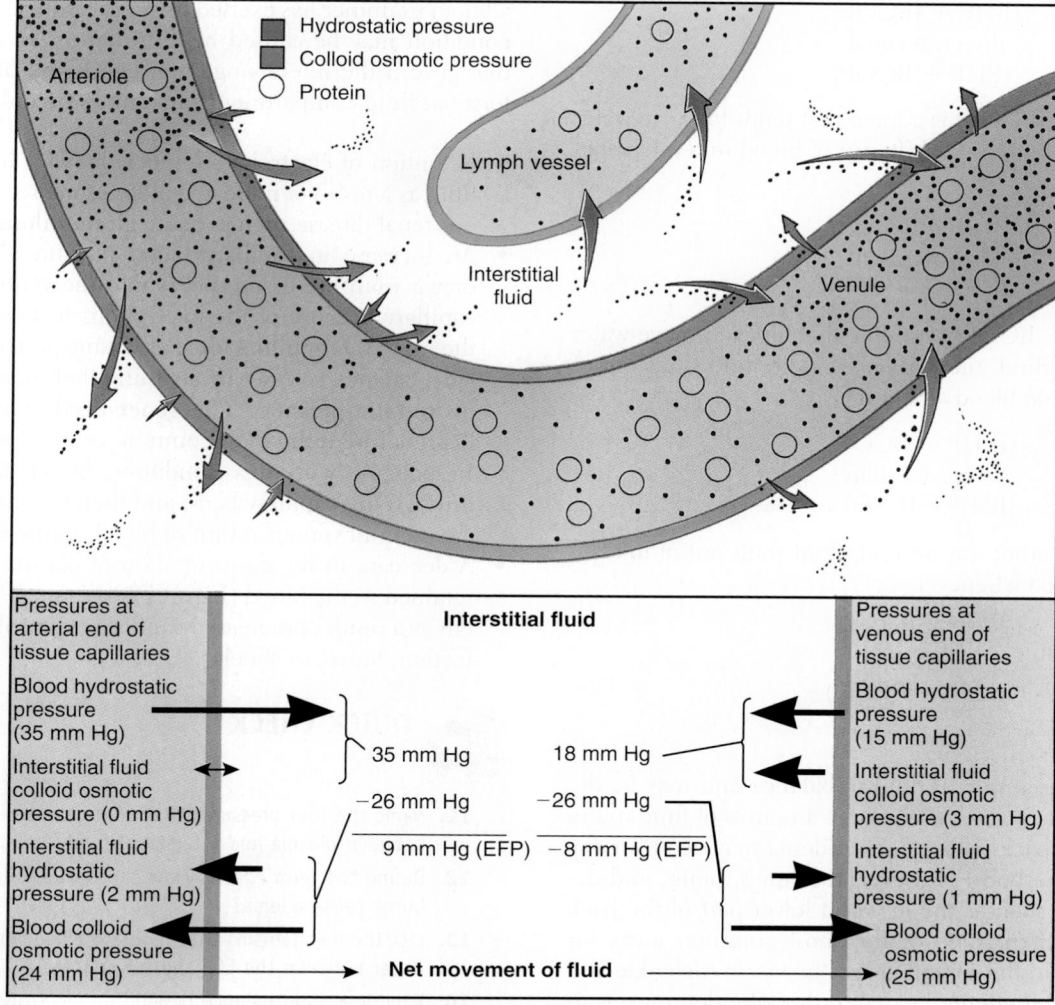

Figure 29-10 *Movement of fluids and electrolytes between plasma and interstitial fluid caused by hydrostatic and colloid osmotic pressure.* *EFP,* Effective filtration pressure. See text for discussion.

normal pressures. Using these figures in Starling's law of the capillaries we get $(35 + 0) - (2 + 24) = 9$ mm Hg net pressure (EFP), which causes water to filter out of blood at the arterial ends of capillaries into interstitial fluid.

The same law operates at the venous end of capillaries (see the lower right-hand portion of Figure 29-7). Again, apply Starling's law of the capillaries. What is the net effective pressure at the venous ends of capillaries? In which direction does it cause water to move? Assuming that the figures given are normal, do you agree that theoretically "the same amount of water returns to the blood at the venous ends of the capillaries as left it from the arterial ends"?

On the basis of our discussion thus far, we can formulate some principles about the transfer of water between blood and interstitial fluid.

1. No net transfer of water occurs between blood and interstitial fluid as long as the effective filtration pressure (EFP) equals 0, that is, when

$$(BHP + IFCOP) = (IFHP + BCOP)$$

2. A net transfer of water, a "fluid shift," occurs between blood and interstitial fluid whenever the EFP does not equal 0, that is, when

$$(BHP + IFCOP)$$
$$\text{does not equal}$$
$$(IFHP + BCOP)$$

3. Because (BHP + IFCOP) is a force that tends to move water out of capillary blood, fluid shifts out of blood into interstitial fluid whenever

$$(BHP + IFCOP)$$
$$\text{is greater than}$$
$$(IFHP + BCOP)$$

4. Because (IFHP + BCOP) is a force that tends to move water out of interstitial fluid into capillary blood, fluid shifts out of interstitial fluid into blood whenever

$$(IFHP + BCOP)$$
$$\text{is greater than}$$
$$(BHP + IFCOP)$$

Or, stated the other way around, fluid shifts out of interstitial fluid into blood whenever

$$(BHP + IFCOP)$$
$$\text{is less than}$$
$$(IFHP + BCOP)$$

Edema

Edema is a classic example of fluid imbalance and may be defined as the presence of abnormally large amounts of fluid in the intercellular tissue spaces of the body. Edema may occur in any organ or tissue of the body. However, the lungs, brain, and dependent body areas such as the legs and lower part of the back are affected most often. One of the most common areas for swelling to occur is in the subcutaneous tissues of the ankle and foot. The term **pitting edema** is used to describe depressions in swollen subcutaneous tissue in this area that do not rapidly refill

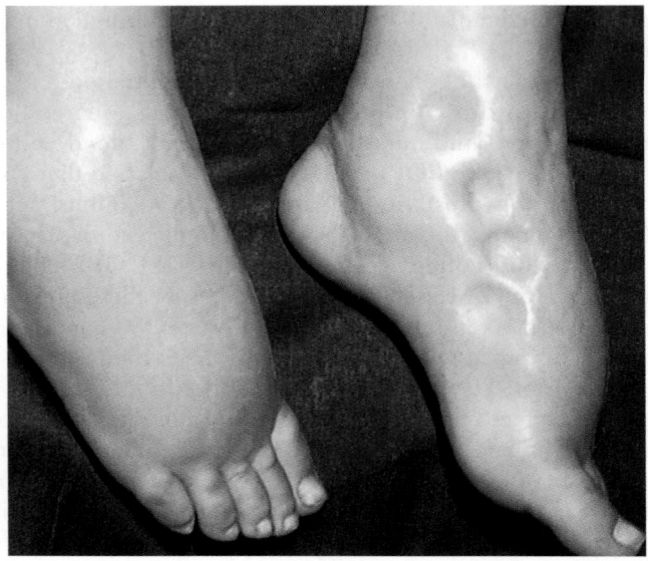

Figure 29-11 *Pitting edema.* Note the finger-shaped depressions that do not rapidly refill after an examiner has exerted pressure.

after an examiner has exerted finger pressure (Figure 29-11). The condition may be caused by disturbances in any of the factors that govern the interchange between blood plasma and the interstitial fluid compartments. Examples include

- Retention of electrolytes (especially Na^+) in the extracellular fluid as a result of increased aldosterone secretion or after serious renal disease such as acute **glomerulonephritis.**
- An increase in capillary blood pressure. Normally, fluid is drawn from the tissue spaces into the venous end of a tissue capillary because of the low venous hydrostatic pressure and the high water-pulling force of plasma proteins (Figure 29-12). This balance is upset by anything that increases the capillary hydrostatic pressure. The generalized venous congestion of heart failure is the most common cause of widespread edema. In patients with this condition, blood cannot flow freely through the capillary beds, and therefore the pressure will increase until venous return of blood improves.
- A decrease in the concentration of plasma proteins normally retained in the blood (Figures 29-12 and 29-13). This may occur as a result of increased capillary permeability caused by infection, burns, or shock.

 QUICK CHECK

11. Name the four pressures that control water exchange between plasma and interstitial fluid.
12. Define the term *edema*. How can tissue inflammation or burns cause edema?
13. List the mechanisms that regulate movement of solutes and water between the ECF and ICF spaces.
14. Why does fluid balance depend on electrolyte balance?

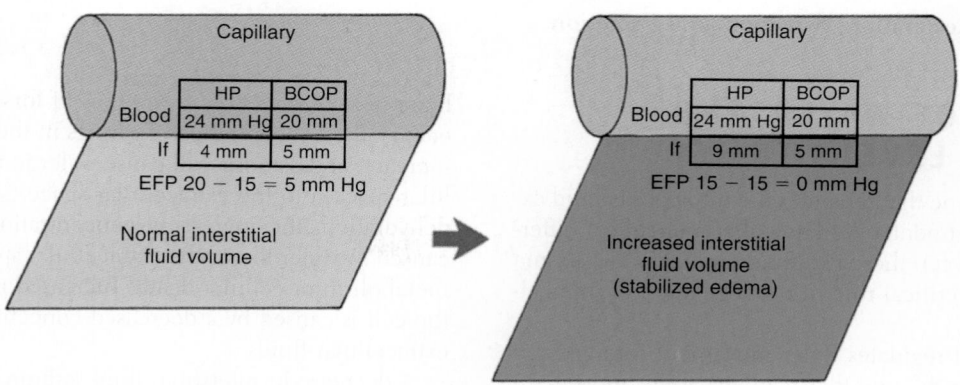

Figure 29-12 *Edema formation.* The mechanism of edema formation can be initiated by a decrease in blood protein concentration and, therefore, a decrease in blood colloid osmotic pressure. In the diagram on the left, blood osmotic pressure has just decreased to 20 from the normal 25 mm Hg. This increases the effective filtration pressure (EFP) to 5 mm Hg from a normal of 0 (see Starling's formula, p. 1076). The EFP of 5 mm Hg causes fluid to shift out of blood into interstitial fluid (IF) until the EFP again equals 0—in this case, when the interstitial fluid volume has increased enough to raise interstitial fluid hydrostatic pressure to 9 mm Hg, as shown in the diagram on the right. At this point a new equilibrium is established, and equal amounts of water once more are exchanged between blood and interstitial fluid. Thus, the increased interstitial fluid volume—that is, the edema—becomes stabilized.

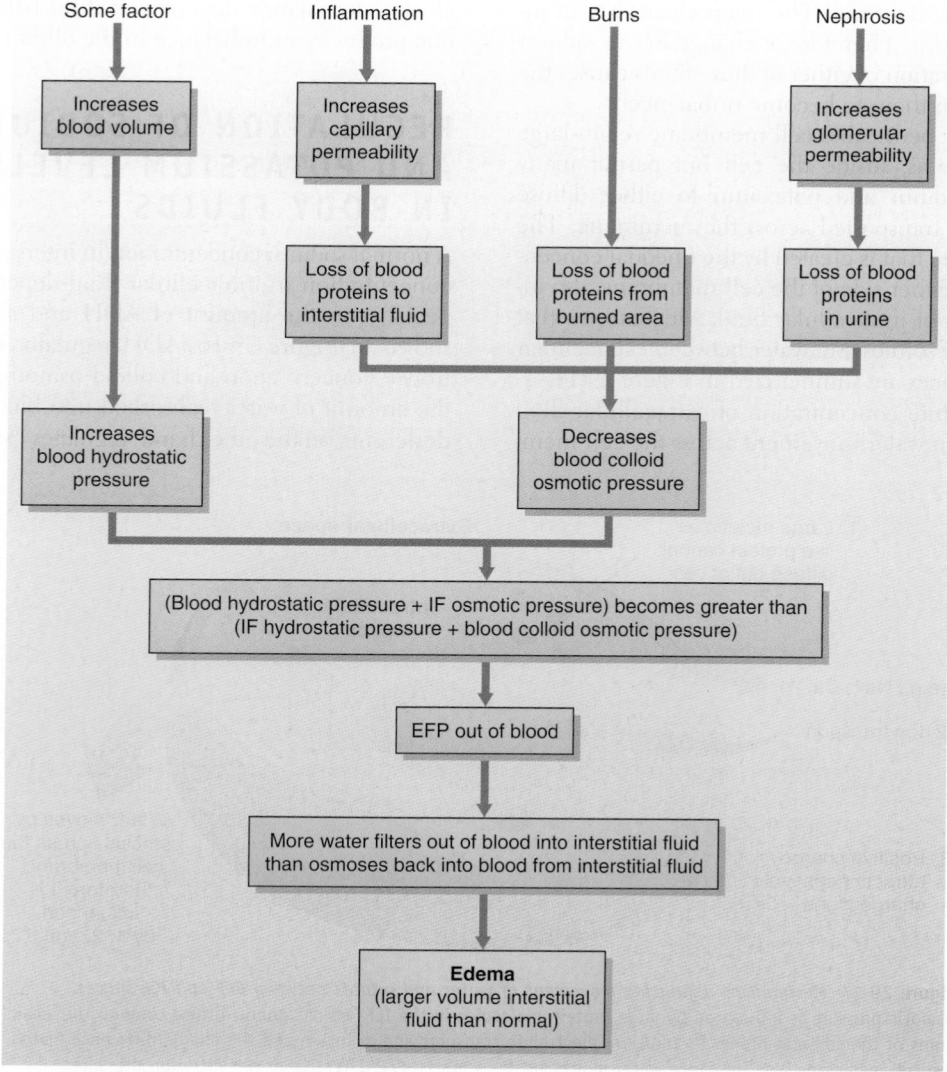

Figure 29-13 *Mechanisms of edema formation in some common conditions.* *EFP,* Effective or net filtration pressure; *IF,* interstitial fluid (see also Figure 29-6).

REGULATION OF WATER AND ELECTROLYTE LEVELS IN ICF

It is the plasma membrane that separates the intracellular and extracellular fluid compartments. We know that a chemical difference predominates between these two fluids, and it is the plasma membrane that plays a critical role in the regulation of intracellular fluid composition.

The mechanism that regulates water movement through cell membranes is similar to the one that regulates water movement through capillary membranes. In other words, interstitial fluid and intracellular fluid hydrostatic and colloid osmotic pressure regulate water transfer between these two fluids. But because the colloid osmotic pressure of interstitial and intracellular fluids varies more than their hydrostatic pressure, their colloid osmotic pressure serves as the chief regulator of water transfer across cell membranes. Their colloid osmotic pressure, in turn, is directly related to the electrolyte concentration gradients—notably sodium and potassium—maintained across cell membranes. As Figures 29-3 and 29-4 show, most of the body sodium is outside the cells. A concentration of 138 to 143 mEq/L makes sodium the chief electrolyte by far in interstitial fluid. The main electrolyte of intracellular fluid is potassium. Therefore, a change in the sodium or the potassium concentration of either of these fluids causes the exchange of fluid between them to become unbalanced.

Pores in the selectively permeable cell membrane retain large molecules, such as proteins, inside the cell but permit many smaller ions such as sodium and potassium to either diffuse through or be selectively transported across the membrane. The electrical charge difference that is created by the unequal concentration of electrolytes on either side of the cell membrane also influences the composition of intracellular fluid. Mechanisms that regulate the movement of solutes and water between extracellular and intracellular fluid spaces are summarized in Figure 29-14.

Any change in the solute concentration of extracellular fluid will have a direct effect on water movement across the cell membrane in one direction or another. If for any reason dehydration occurs, the concentration of solutes in the extracellular fluid will increase, and osmosis will cause water to move from the intracellular space into the extracellular space (see Box 29-2). In severe dehydration, the increasing concentration of intracellular fluid caused by water loss to the extracellular space results in abnormal metabolism or cellular death. Increased movement of water into the cell is caused by a decreased concentration of solutes in the extracellular fluids.

A decrease in interstitial fluid sodium concentration immediately decreases interstitial fluid colloid osmotic pressure, thus making it hypotonic to intracellular fluid colloid osmotic pressure. In other words, a decrease in interstitial fluid sodium concentration establishes a colloid osmotic pressure gradient between interstitial and intracellular fluid. This gradient causes net osmosis to occur out of interstitial fluid into cells. In short, interstitial fluid and intracellular fluid electrolyte concentrations are the main determinants of their colloid osmotic pressure; their colloid osmotic pressure regulates the amount and direction of water transfer between the two fluids, and this regulates their volumes. Hence, fluid balance depends on electrolyte balance. Conversely, electrolyte balance depends on fluid balance. An imbalance in one produces an imbalance in the other (Figure 29-15).

REGULATION OF SODIUM AND POTASSIUM LEVELS IN BODY FLUIDS

A normal sodium concentration in interstitial fluid and potassium concentration in intracellular fluid depend on many factors, especially on the amount of ADH and aldosterone secreted. As shown in Figure 29-16, ADH regulates extracellular fluid electrolyte concentration and colloid osmotic pressure by regulating the amount of water reabsorbed into blood by renal tubules. Aldosterone, on the other hand, regulates extracellular fluid volume

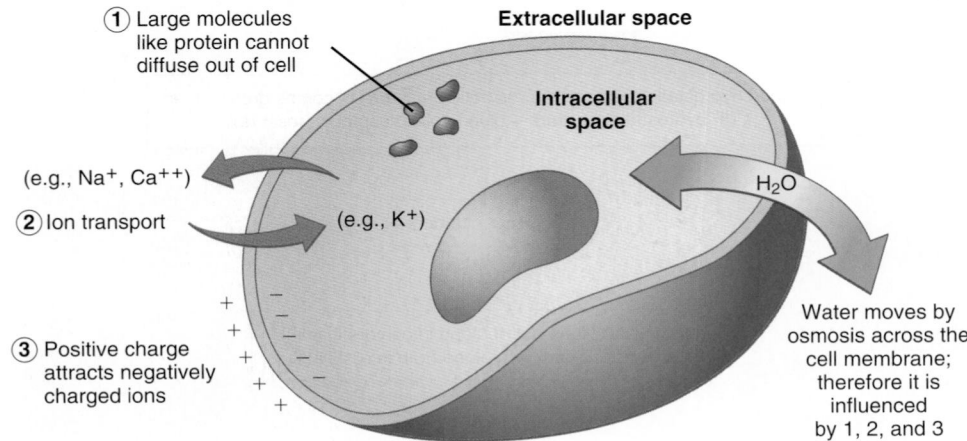

Figure 29-14 *Mechanisms regulating movement of water and solutes between ECF and ICF spaces.*
Osmotic pressure is influenced by large protein molecules in the ICF, which cannot diffuse through the small pores of the cell membrane. In addition, electrolyte transport and diffusion and the charge difference across the cell membrane also influence water movement by osmosis. *EFP,* Effective or net filtration pressure; *IF,* interstitial fluid.

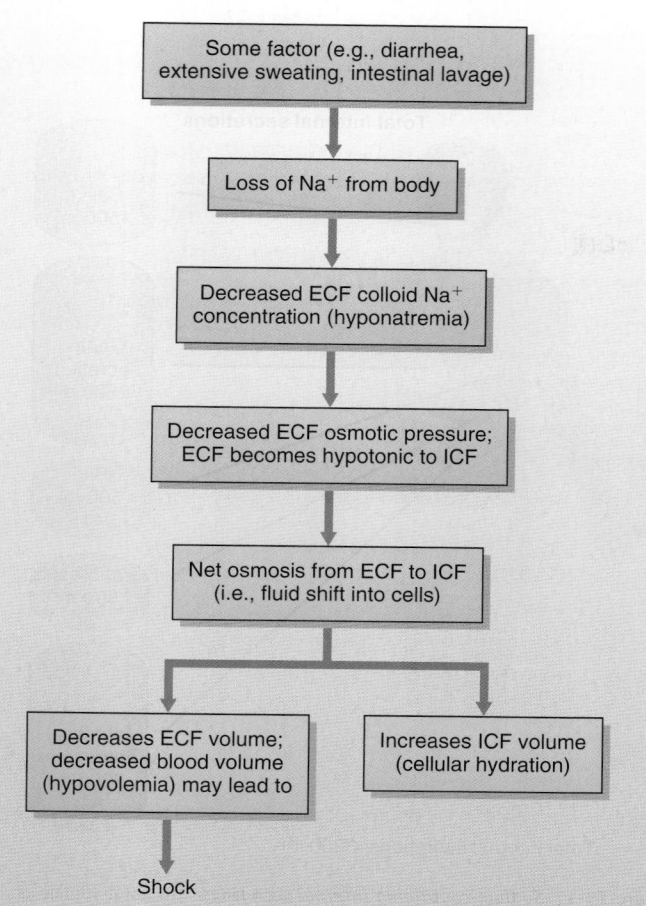

Figure 29-15 *How electrolyte imbalance leads to fluid imbalances.* The schematic uses the example of sodium deficit (hyponatremia) and resulting hypovolemia (cellular hydration). *ECF,* Extracellular fluid; *ICF,* intracellular fluid.

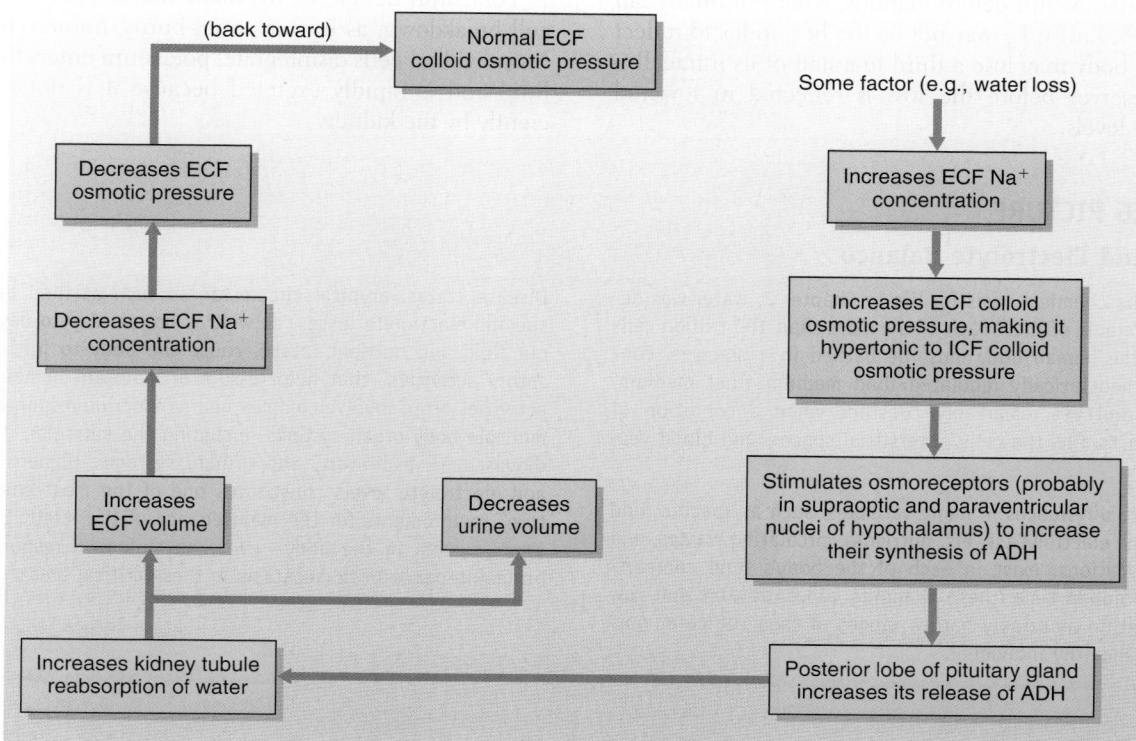

Figure 29-16 *Antidiuretic hormone (ADH) mechanism for ECF homeostasis.* The ADH mechanism helps maintain homeostasis of extracellular fluid (ECF) colloid osmotic pressure by regulating its volume and thereby its electrolyte concentration, that is, mainly ECF Na⁺ concentration. *ICF,* Intracellular fluid.

by regulating the amount of sodium reabsorbed into blood by renal tubules (see Figure 29-6).

If for any reason conservation of body sodium is required, the normal kidney is capable of excreting an essentially sodium-free urine and is therefore considered the chief regulator of sodium levels in body fluids. Sodium lost in sweat can become appreciable with elevated environmental temperatures or fever. However, the thirst that results may lead to replacement of water but not the lost sodium, and because of the increased fluid intake, the remaining sodium pool may be diluted even more. Sodium loss in sweat is not therefore considered a normal means of regulation.

In addition to the well-regulated movement of sodium into and out of the body and between the three primary fluid compartments, there is a continuous movement or circulation of this important electrolyte between a number of internal secretions. More than 8 L of various internal secretions such as saliva, gastric and intestinal secretions, bile, and pancreatic fluid is produced every day (Figure 29-17). The total daily secretion of sodium into these alimentary tract fluids alone will average between 1200 and 1400 mEq. A 70-kg (154-lb) adult has a total body sodium pool of only 2800 to 3000 mEq. Precise regulatory and conservation mechanisms for sodium are required for survival.

Chloride is the most important extracellular anion and is almost always linked to sodium. Generally ingested together, they provide in large part for the isotonicity of extracellular fluid. Chloride ions are usually excreted in the urine as a potassium salt, and therefore chloride deficiency—**hypochloremia**—is often found in cases of potassium loss.

The total body potassium content in the average-sized adult is approximately 4000 mEq. Because the majority of body potassium is intracellular, serum determinations, which normally fall between 4.0 and 5.0 mEq/L, may not be the best index to reflect imbalances. The body may lose a third to a half of its intracellular potassium reserves before the loss is reflected in lowered serum potassium levels.

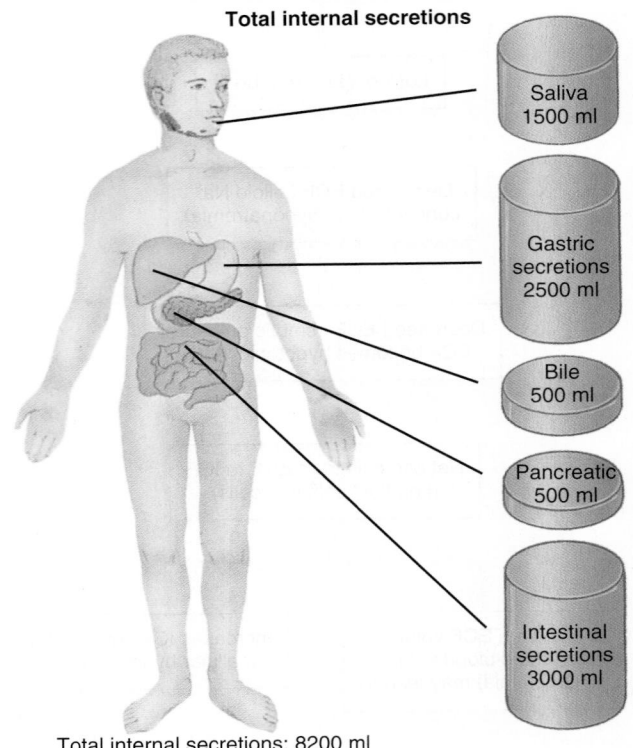

Total internal secretions

Saliva 1500 ml

Gastric secretions 2500 ml

Bile 500 ml

Pancreatic 500 ml

Intestinal secretions 3000 ml

Total internal secretions: 8200 ml

Figure 29-17 *Sodium-containing internal secretions.* The total volume of these secretions may reach 8000 or more milliliters in a 24-hour period.

Potassium deficit, or **hypokalemia**, occurs whenever there is cell breakdown, as in starvation, burns, trauma, or dehydration. As individual cells disintegrate, potassium enters the extracellular fluid and is rapidly excreted because it is not reabsorbed efficiently by the kidney.

THE BIG PICTURE
Fluid and Electrolyte Balance

In discussing the chemical basis for life in Chapter 2, water was described as the "cradle of life." Each of the more than 100 trillion cells that make up the human body must be bathed in a precisely controlled and homeostatically regulated fluid medium. That medium, ever-changing and yet remarkably constant when a condition of homeostasis exists, fills the cells, interstitial spaces, and blood vessels of the body.

Although unique differences in many variables, such as specific fluid volumes, buffers, electrolyte levels, nutrients, circulating wastes, and protein concentrations, exist in each of the body's fluid compartments at any point in time, these changing concentrations and volumes remain within amazingly narrow ranges of their set point normal values in a healthy individual.

Disease states, atypical circumstances such as fluid deprivation or specific electrolyte losses, as well as normal day-to-day variation in our fluid and nutrient intake, cause the body to initiate "compensatory activities" that help restore or maintain homeostasis. These activities often involve complex neuroendocrine responses that affect multiple body organ systems, including the muscular, digestive, cardiovascular, respiratory, and urinary systems. Homeostasis of fluid and electrolyte levels constitutes one of the most crucial "big picture" requirements for the maintenance of life itself. Every cell and organ system in the body—every physiological response—depends on maintenance of homeostasis in these critical areas.

Mechanisms of Disease

FLUID AND ELECTROLYTE DISORDERS

Total body water constitutes 40% to 60% of body weight in adults and is regulated by homeostatic mechanisms of the neuroendocrine system, heart, kidneys, and blood vessels. Normal water losses, known as insensible water losses, occur through expired air from the lungs and from the skin, constituting about 0.4 to 0.5 ml/hr/kg body weight. Abnormally excessive water losses constitute a volume deficit and can lead to a state of dehydration, or **hypovolemia,** in which there is inadequate fluid volume in the extracellular compartment (see Box 29-2). If left untreated, it can result in hypovolemic shock. Many causes of dehydration exist. The most common cause of dehydration is fluid loss from the gastrointestinal tract as a result of vomiting or diarrhea. Dehydration also occurs when an individual fails to take in sufficient oral fluid because of depression, nausea, or oral trauma. *Diaphoresis,* or excessive perspiration, may also cause dehydration, with rapid respirations leading to water vapor losses. Any disorder of the kidneys that increases urine excretion, such as nephritis, can lead to dehydration. Fluid can also shift into a space outside the normal fluid compartments during certain disease states, including ascites, burns, pancreatitis, and traumatic injuries. Regardless of the cause, fluid volume deficits cause low blood pressure and cardiac output, electrolyte disturbances, or acid-base abnormalities. Symptoms include dizziness, lightheadedness, weakness, poor skin turgor, and tachycardia. Restoration of the fluid losses is the main goal of therapy. If the deficit is mild, volume can be replaced orally. If the dehydration is severe, fluid volume is replaced intravenously. Examining the person's weight, skin turgor, blood pressure, and urine output will provide important data on fluid volume stability.

Fluid volume excess, or **hypervolemia,** is an expansion of fluid volume in the body. This can occur if the kidneys retain a large amount of sodium and water, as in congestive heart failure, nephrotic syndrome, renal failure, and liver failure. Manifestations include weight gain (the most consistent sign), edema, dyspnea, tachycardia, and pulmonary congestion. Approaches to this disorder are to treat the original pathological process (i.e., renal failure), monitor the person's weight closely, and use diuretics cautiously to remove the excess fluid (Box 29-3).

Water intoxication may result from the overadministration of water or hypotonic solutions to persons unable to dilute and excrete urine normally. This may occur in patients with kidney insufficiency or abnormal "thirst" mechanisms resulting from neurological disorders. Water content is elevated, and serum sodium levels are diluted. Development of subtle mental changes such as confusion and lethargy occur. If intoxication is severe, stupor, seizures, and coma may result. Correction of the neurological impairment along with water restriction can reverse the symptoms.

BOX 29-3 HEALTH MATTERS
Diuretics

The word **diuretic** is from the Greek word *diouretikos,* meaning "causing urine." By definition, a diuretic drug is a substance that promotes or stimulates the production of urine.

As a group, diuretics are among the most commonly used drugs in medicine. They are used because of their role in influencing water and electrolyte balance, especially sodium, in the body. Diuretics have their effect on tubular function in the nephron, and the differing types of diuretics are often classified according to their major site of action. Examples include (1) *proximal tubule diuretics* such as acetazolamide (Diamox), (2) *loop of Henle diuretics* such as ethacrynic acid (Edecrin) or furosemide (Lasix), and (3) *distal tubule diuretics* such as chlorothiazide (Diuril).

Diuretics can also be classified according to the effect the drug has on the level or concentration of sodium (Na^+), chloride (Cl^-), potassium (K^+), and bicarbonate (HCO_3^-) ions in the tubular fluid.

Nursing implications for caregivers monitoring patients receiving diuretics both in hospitals and in home health care environments include keeping a careful record of fluid intake and output and assessing the patient for signs and symptoms of electrolyte and water imbalance. For example, diuretic-induced dehydration resulting in a loss of only 6% of initial body weight will cause tingling in the extremities, stumbling gait, headache, fever, and an increase in both pulse and respiratory rates.

Disturbances in electrolytes can occur in fluid volume abnormalities and many different disease states. **Hyponatremia** is a condition of decreased serum sodium concentration below the normal range (less than 136 mEq/L) and is usually the result of an excess of water relative to solute. It may also result from excessive losses of sodium. Causes of hyponatremia include skin losses via profuse perspiration, overzealous use of salt-wasting diuretics, adrenal insufficiency, renal or liver failure, low salt intake, or excessive water intake (diluting sodium content). Signs and symptoms include muscle cramps, nausea and vomiting, postural blood pressure changes, poor skin turgor, fatigue, and difficulty breathing. Cerebral swelling can occur in severe cases, causing confusion, hemiparesis (motor weakness on one side of the body), seizures, and coma. Management includes neurological assessment and the administration of sodium orally or intravenously. Water restriction may also suffice.

Hypernatremia is elevation of the serum sodium concentration higher than 145 mEq/L. It is usually indicative of a body water deficit relative to sodium, but can also result from grossly elevated sodium intake. Causes include lack of fluid intake, diarrhea, diabetes insipidus, loss of water via the respiratory tract, heart disease or congestive heart failure, renal failure, or ingestion of salt in abnormal amounts. Signs and symptoms are sim-

Mechanisms of Disease—cont.

ilar to those of dehydration and include thirst, disorientation, lethargy, and seizures. The neurological symptoms are thought to be due to cellular dehydration. Replacement with a hypotonic solution will help lower the sodium level slowly, thereby reducing the risk for cerebral edema.

A common type of electrolyte imbalance is **hypokalemia,** a condition in which potassium is lost from the body, resulting in a serum potassium level below 3.5 mEq/L. Causes include potassium-wasting diuretics, increased urine output with loss of potassium, and vomiting or gastric suctioning without potassium replacement. Hypokalemia can be life threatening and includes manifestations of anorexia, muscle weakness, decreased reflexes, low blood pressure, and cardiac arrhythmias. Potassium can be replaced through the diet, with potassium-rich foods, or intravenously with caution.

The opposite of hypokalemia is **hyperkalemia,** or a serum potassium level above 5.5 mEq/L. This can be even more dangerous than hypokalemia because myocardial muscle can be profoundly affected. The common cause of hyperkalemia is kidney disease, but other factors such as vomiting, diarrhea, potassium-conserving diuretics, extensive tissue damage as in burn or trauma victims, severe infections, and Cushing syndrome are also cited. Notable changes can be seen on the electrocardiogram, such as peaked T waves. Hyperkalemia can induce ventricular dysrhythmias, thereby leading to possible cardiac arrest. Because potassium is a part of neuromuscular functions, the person may experience extremity muscle weakness or failure of the respiratory muscles. Intermittent diarrhea, nausea, and intestinal colic are also manifested. Dietary restriction of potassium is sufficient in mild cases, but emergency intravenous administration of calcium gluconate may be required to correct cardiac symptoms. Also, correction of the underlying condition (i.e., trauma) and dialysis to remove the excess potassium can be instituted to correct severe hyperkalemia.

LANGUAGE OF SCIENCE   (Cont'd from page 1067)

intracellular (in-trah-SELL-yoo-lar) [intra- occurring within, -cella storeroom]

intracellular fluid (ICF) (in-trah-SELL-yoo-lar FLOO-id) [intra- occurring within, -cella storeroom, fluere to flow]

ions (EYE-ons) [ion electrically charged particle]

juxtaglomerular cells (jux-tah-gloh-MER-yoo-lar) [juxta- near, -glomerulus small ball, cella storeroom]

milliequivalent (mil-i-ee-KWIV-ah-lent) [milli- 1/1000 part, -aequus- equal, -valere to be strong]

nonelectrolytes (non-eh-LEK-troh-lytes) [non- not, -elektron- amber, -lytos soluble]

osmoreceptors (os-moh-ree-SEP-tors) [osmos- impulse, -recipere to receive]

parenteral (pah-REN-ter-al) [para- beyond, -enteron bowel]

renin-angiotensin mechanism (REH-nin an-jee-oh-TEN-sin) [ren kidney, angeion- vessel, -tendere to stretch, mechane machine]

Starling's law of the capillaries (STAR-lings) [Ernest H. Starling English physiologist]

subfornical organ (SFO) (sub-FOR-nih-kal) [sub- under, -fornix arch]

thirst center

LANGUAGE OF MEDICINE

diuretic (dye-yoo-RET-ik) [dia- through, -ouron urine]

edema (eh-DEE-mah) [oidema swelling]

glomerulonephritis (gloh-mer-yoo-loh-neh-FRY-tis) [glomerulus- small ball, -nephros- kidney, -itis inflammation]

hyperkalemia (hye-per-kah-LEE-mee-ah) [hyper- excessive, -kalium- potassium, -emia blood condition]

hypernatremia (hye-per-nah-TREE-mee-ah) [hyper- excessive, -natrium- sodium, -emia blood condition]

hypervolemia (hye-per-voh-LEE-mee-ah) [hyper- excessive, -volumen- paper roll, -emia blood condition]

hypochloremia (hye-poh-kloh-REE-mee-ah) [hypo- deficient, -chloros- green, -emia blood condition]

hypokalemia (hye-poh-kah-LEE-mee-ah) [hypo- deficient, -kalium- potassium, -emia blood]

hyponatremia (hye-poh-nah-TREE-mee-ah) [hypo- deficient, -natrium- sodium, -emia blood]

hypovolemia (hye-poh-voh-LEE-mee-ah) [hypo- deficient, -volumen- paper roll, -emia blood]

intravenous injection (in-trah-VEE-nus in-JEK-shun) [intra- occurring within, -vena vein, in- in or within, -jacere to throw]

parenteral therapy (pah-REN-ter-al) [para- supplementary, -enteron bowel]

pitting edema (eh-DEE-mah)

subcutaneous injection (sub-kyoo-TAY-nee-us in-JEK-shun) [sub- under, -cutis skin, in- in or within, -jacere to throw]

CASE STUDY

Brad Chambers, age 28, sustained burns to the upper part of his chest, neck, face, and upper part of his arms when his shirt caught fire while he was trying to start a charcoal fire at a neighborhood bar-becue. One of his neighbors sprayed Brad with a garden hose and took him immediately to the nearby hospital emergency department. On arrival, intravenous fluids are started using a #16 angiocatheter.

1. In severe burn injuries, what is the mechanism responsible for the increased risk of these patients for hypovolemic shock?
   A. Loss of electrolytes through the urinary tract
   B. Retention of electrolytes in the extracellular fluid as a result of increased aldosterone secretion
   C. An increase in capillary blood pressure
   D. An increase in capillary permeability with a resulting decrease in the concentration of plasma proteins

2. Which one of the following is released into the extracellular fluid as a result of cell breakdown that occurs with burn injury?
   A. Sodium
   B. Potassium
   C. Chloride
   D. All of the above

3. Mr. Chambers is stabilized, his burns are cleaned and treated, and he is admitted to the intensive care burn unit. Admitting orders include adding potassium supplements to his current intravenous fluids. The physician orders the potassium in milliequivalents. Which one of the following is the best explanation of what this measurement is?
   A. Milliequivalent is the weight of an electrolyte in a given amount of solution.
   B. Milliequivalent is the percentage of the concentration of that element in a solution.
   C. Milliequivalent measures the number of ionic charges in a solution.
   D. Milliequivalent measures the negative ionic charges obtained within a solution.

4. Hyperkalemia is often an electrolyte imbalance associated with extensive tissue damage. Which of the following can occur as a result of this imbalance?
   A. Peaked T waves seen on the electrocardiogram
   B. Ventricular dysrhythmias
   C. Muscle weakness
   D. All of the above

CHAPTER SUMMARY

INTERRELATIONSHIP OF FLUID AND ELECTROLYTE BALANCE

A. Fluid and electrolyte balance—implies homeostasis
B. Electrolytes have chemical bonds that allow dissociation into ions, which carry an electrical charge; of critical importance in fluid balance
C. Fluid balance and electrolyte balance are interdependent

TOTAL BODY WATER

A. Fluid content of the human body ranges from 40% to 60% of its total weight
B. Fluid content varies according to age, gender, weight, and fat content of the body

BODY FLUID COMPARTMENTS

A. Two major fluid compartments (Figure 29-1)
B. Extracellular fluid (ECF) constitutes the internal environment of the body
   1. Consists mainly of plasma and interstitial fluid
   2. Lymph, cerebrospinal fluid, and specialized joint fluids are considered extracellular
   3. Functions of ECF are to provide a relatively constant environment for cells and transport substances to and from the cells

C. Intracellular fluid (ICF)—water inside the cells
   1. Function is to facilitate intracellular chemical reactions that maintain life
   2. By volume, ICF is the largest body fluid compartment

CHEMICAL CONTENT, DISTRIBUTION, AND MEASUREMENT OF ELECTROLYTES IN BODY FLUID

A. Extracellular vs. intracellular fluids
   1. Plasma and interstitial fluid (ECFs) are almost identical in chemical make-up, with intracellular fluid showing striking differences (Figure 29-3)
   2. Extracellular fluids
      a. Difference between blood and interstitial fluid—blood contains a slightly larger total of ions than interstitial fluid does
      b. Functionally important difference between blood and interstitial fluid is the number of protein anions; blood has an appreciable amount, whereas interstitial fluid has hardly any; because the capillary membrane is practically impermeable to proteins, almost all protein anions remain in the blood
   3. Intracellular fluids—ICF and ECF are more dissimilar than similar
   4. The chemical structure of plasma, interstitial fluid, and intracellular fluid helps control water and electrolyte movement between them

B. Measuring electrolyte concentration—concentration or weight of an electrolyte can be expressed as the number of milligrams per 100 ml of solution (mg%); conversion of mg% to milliequivalents per liter (mEq/L) provides information on actual physiological activity

C. Measuring electrolyte reactivity—milliequivalent—measures the number of ionic charges or electrocovalent bonds in a solution; accurately measures the physiological combining power of an electrolyte solution

AVENUES BY WHICH WATER ENTERS AND LEAVES THE BODY

A. Water enters the body via the digestive tract; water is also added to the total fluid volume from each cell as it catabolizes food, and the resulting water enters the bloodstream (Figure 29-5)

B. Water leaves the body via four exits (Figure 29-5)
1. As urine through the kidney
2. As water in expired air through the lungs
3. As sweat through the skin
4. As feces from the intestine

SOME GENERAL PRINCIPLES ABOUT FLUID BALANCE

A. Cardinal principle of fluid balance: fluid balance can be maintained only if intake equals output

B. Mechanisms are available to adjust output and intake to maintain fluid balance, e.g., aldosterone mechanism (Figure 29-6), renin-angiotensin mechanism (Figure 29-8)

C. Most rapid fluid balance mechanisms are those for controlling water movement between fluid compartments of the body; will maintain normal blood volume at the expense of interstitial fluid volume

MECHANISMS THAT MAINTAIN HOMEOSTASIS OF TOTAL FLUID VOLUME

A. Under normal conditions, homeostasis of total volume of water is maintained or restored primarily by adjusting urine volume and secondarily by fluid intake (Figure 29-7)

B. Regulation of fluid intake—when dehydration begins to develop, salivary secretion decreases, producing the sensation of thirst; increased fluid intake to offset increased output tends to restore fluid balance

C. Regulation of urine volume—two factors determine urine volume
1. Glomerular filtration rate, except under abnormal conditions, remains fairly constant
2. Rate of tubular reabsorption of water fluctuates considerably; normally adjusts urine volume to fluid intake; influenced by amount of antidiuretic hormone and aldosterone
3. Decrease in fluid intake causes osmoreceptors in "thirst center"—wall of third ventricle (subfornical organ) and in supraoptic and paraventricular nuclei of hypothalamus—to increase secretion of ADH (Figure 29-16).

D. Factors that alter fluid loss under abnormal conditions—rate of respiration and volume of sweat secreted may alter fluid output under certain abnormal conditions; vomiting, diarrhea, or intestinal drainage can produce fluid and electrolyte imbalances

REGULATION OF WATER AND ELECTROLYTE LEVELS IN PLASMA AND INTERSTITIAL FLUID (ECF)

A. Law of capillaries—the control mechanism for water exchange between plasma and interstitial fluid consists of four types of pressure: blood hydrostatic and colloid osmotic pressure on one side of the capillary membrane and interstitial fluid hydrostatic and colloid osmotic pressure on the other side; two of the pressures produce a vector in one direction and the other two in the opposite direction
1. Blood hydrostatic pressure (BHP) forces fluid out of capillaries into interstitial fluid (IF)
2. Blood colloid osmotic pressure (BCOP) draws fluid from IF into capillaries
3. Interstitial fluid hydrostatic pressure (IFHP) forces fluid out of IF into capillaries
4. Interstitial fluid colloid osmotic pressure (IFCOP) draws fluid from capillaries to IF

B. The rate and direction of fluid exchange between capillaries and interstitial fluid are determined by the hydrostatic and colloid osmotic pressure of the two fluids (Figure 29-10)

C. Some principles about transfer of water between blood and interstitial fluid
1. No net transfer of water occurs as long as (BHP + IFCOP) = (IFHP + BCOP)
2. A net transfer of fluid occurs when (BHP + IFCOP) ≠ (IFHP + BCOP)
3. Fluid shifts out of blood into interstitial fluid whenever (BHP + IFCOP) > (IFHP + BCOP)
4. Fluid shifts out of interstitial fluid into blood whenever (BHP + IFCOP) < (IFHP + BCOP)

EDEMA

A. Edema—presence of abnormally large amounts of fluid in the intercellular tissue spaces of the body (Figure 29-11)

B. Classic example of fluid imbalance; may be due to
1. Retention of electrolytes in the extracellular fluid
2. Increase in capillary blood pressure
3. Decrease in the concentration of plasma proteins normally retained in the blood (Figure 29-12)

REGULATION OF WATER AND ELECTROLYTE LEVELS IN INTRACELLULAR FLUID

A. Plasma membrane plays critical role in regulating ICF composition

B. IF and ICF hydrostatic and colloid pressure regulates water transfer between ECF and ICF; colloid osmotic pressure is the chief regulator of water transfer across cell membranes, and it is directly related to the electrolyte concentration gradients maintained across cell membranes (Figure 29-14)

REGULATION OF SODIUM AND POTASSIUM LEVELS IN BODY FLUIDS

A. Normal sodium concentration in IF and potassium concentration in ICF depend on various factors, especially the amount of ADH and aldosterone secreted

1. ADH regulates ECF electrolyte concentration and colloid osmotic pressure by regulating amount of water reabsorbed into blood by renal tubules
2. Aldosterone regulates ECF volume by regulating the amount of sodium reabsorbed into blood by renal tubules

B. When conservation of body sodium is required, the kidneys excrete an essentially sodium-free urine; kidneys are considered the chief regulator of sodium levels

C. Chloride—most important extracellular anion and almost always linked to sodium; chloride ions are generally excreted in urine in association with potassium; thus, hypochloremia is often associated with cases of potassium loss

D. Hypokalemia occurs where there is cell breakdown; as cells disintegrate, potassium enters the ECF and is rapidly excreted because it is not reabsorbed efficiently by the kidney

REVIEW QUESTIONS

1. Discuss the changes in total body water content from infancy to adulthood.
2. How does total body water content differ in men and women?
3. List the compartments of extracellular fluid.
4. What are the four exits by which water normally leaves the body?
5. What are the objectives for giving parenteral solutions?
6. What is the cardinal principle about fluid balance?
7. How is urine volume regulated?
8. Define the terms *cation* and *anion*.

9. Are plasma and interstitial fluid chemically similar or different? Explain.
10. Describe the electrolyte composition of blood plasma.
11. Define the term *milliequivalent*. How is it used to measure electrolyte reactivity?
12. What are the four pressures involved in Starling's law?
13. When effective filtration pressure equals 0, what is the net transfer of water between blood and interstitial fluid?
14. When does a "fluid shift" occur between blood and interstitial fluid?
15. Identify various mechanisms that may lead to edema.
16. What role does the plasma membrane play in the regulation of intracellular fluid composition?
17. How does the antidiuretic hormone mechanism maintain homeostasis of extracellular fluid colloid osmotic pressure?
18. In your own words, define dehydration.

CRITICAL THINKING QUESTIONS

1. How would you define fluid and electrolyte balance? Based on what you know, what would be the impact of glucose deficiency on the electrolyte balance of the body?
2. How would you summarize the aldosterone mechanism?
3. Can you make a distinction between anions in the two extracellular fluids? What causes this difference?
4. How would you explain Starling's law of the capillaries?
5. How does hyponatremia lead to fluid imbalance? Can you identify the possible causes?
6. Where are osmoreceptors located, and how are they related to the maintenance of fluid balance?
7. What information would you use to support the view that the homeostasis of fluid and electrolyte levels constitutes one of the most crucial requirements for the maintenance of life itself?

Acid-Base Balance

LANGUAGE OF SCIENCE

acetoacetic acid (ass-ih-toh-ah-SEE-tik ASS-id)
[*acetum*- vinegar, *-ic* pertaining to, *acidus* sour]

acetone (ASS-ah-tohn) [*acetum* vinegar]

acid-forming foods (ASS-id) [*acidus* sour]

acidic ketone bodies (ah-SID-ik KEE-tohn BOD-eez) [*acidus* sour, *keton* form of acetone]

acidity (ah-SID-ih-tee) [*acidus* sour]

acids (ASS-ids) [*acidus* sour]

acidosis (ass-i-DOH-sis) [*acidus*- sour, *-osis* condition]

alkaline (AL-kah-lin) [*al-qaliy* the wood ashes]

alkalinity (al-kah-LIN-ih-tee) [*al-qaliy* the wood ashes]

alkalosis (al-kah-LOH-sis) [*al-qaliy*- the wood ashes, *-osis* condition]

base-forming foods [*basis* foundation]

bases [*basis* foundation]

basic residue (BASE-ick REHZ-ih-doo) [*basis*-foundation, *-ic* pertaining to, *residuum* remainder]

beta-hydroxybutyric acid (BAY-tah-hye-DROK-see-boo-teh-rik ASS-id)

buffers (BUFF-ers) [*buffe* to cushion]

buffer pairs (BUFF-er) [*buffe* to cushion]

calcium (KAL-see-um) [*calx* lime]

carbonic acid (kar-BON-ik ASS-id) [*carbo*- coal, *-ic* pertaining to, *acidus* sour]

carbonic anhydrase (kar-BON-ik an-HYE-drays) [*carbo*- coal, *-ic* pertaining to, *a*- not, *-hydor* water]

chemical buffers (KEM-ih-kal BUFF-erz) [*chemeia* alchemy, *buffe* to cushion]

chloride shift (KLOR-ide shift) [*chloros* green]

chlorine (KLOR-een) [*chloros* green]

compensation (kom-pen-SAY-shun) [*compensare* to balance]

correction

lactic acid (LAK-tik ASS-id) [*lac*- milk, *-ic* pertaining to, *acidus* sour]

magnesium (mag-NEE-see-um) [*magnesium* lodestone]

normal saline

pH (pee-AYCH) [*ph* potential hydrogen]

Cont'd on p. 1102

Acid-base balance is one of the most important of the body's homeostatic mechanisms. The term refers to regulation of hydrogen ion concentration in the body fluids and the study of acid-base physiology is, in a very real sense, the study of the hydrogen ion (H^+). Many of the body's most biologically important molecules contain chemical groups that can either "donate" or "accept" a hydrogen ion (H^+) and thus behave as a weak acid or base. As alterations in a molecule's pH and confirmation change, so does its biological activity. The functional ability of ion channels, membrane receptors, and a variety of enzymes and other important body proteins that influence protein conformation and structure such as hemoglobin and chaperonins (see Chapter 2), is closely dependent on maintaining precise regulation of hydrogen ion concentration. Thus, even slight deviations from the normal pH range in body cells and fluids result in pronounced, systemic, and potentially fatal changes in metabolic activity. For example, activity of the Na-K pump, arguably the most important active transport mechanism in the cell membrane, falls by 50% when pH decreases by approximately 1 pH unit. An even more dramatic effect is seen in the activity of a key enzyme (phosphofructokinase) involved in the breakdown of glucose in the absence of O_2 during anaerobic catabolism (glycolysis). The biological activity of this key enzyme falls by approximately 90% when the pH decreases by only 0.1 unit! Maintaining acid-base balance within narrow and precise ranges is necessary for survival.

MECHANISMS THAT CONTROL pH OF BODY FLUIDS

Meaning of Term pH

Recall from Chapter 2 that water and all water solutions contain hydrogen ions (H^+) and hydroxide ions (OH^-). **pH** is a numerical symbol (0 to 14) used to represent the negative logarithm of the number of hydrogen ions (H^+) present in 1 liter of a solution. In Figure 30-1 the pH value is shown on the right side of the scale and the corresponding logarithmic value is on the left. Take a moment to review the pH unit. **pH** indicates the degree of **acidity** or **alkalinity** of a solution. As the concentration of hydrogen ions increases, the pH goes down and the solution becomes more acid; a decrease in hydrogen ion concentration makes the solution more **alkaline** and the pH goes up. A pH of 7 indicates neutrality (equal amounts of H^+ and OH^-), a pH of less than 7 indicates acidity (more H^+ than OH^-), and a pH greater than 7 indicates alkalinity (more OH^- than H^+). With a pH of about 1, gastric juice is the most acid substance in the body.

With a pH of 7.0, intracellular fluid is essentially neutral. Arterial and venous blood is slightly alkaline because both have a pH slightly higher than 7.0. The slight increase in acidity of venous blood (pH 7.36) compared with arterial blood (pH 7.40) results primarily from carbon dioxide entering venous blood as a waste product of cellular metabolism. Although any pH value above 7.0 is considered chemically basic, in clinical medicine the term **acidosis** is used to describe an arterial blood pH of less than 7.35 and **alkalosis** is used to describe an arterial blood pH greater than 7.45. The lungs remove the equivalent of more than 30 *liters* of carbonic acid each day from the venous blood by elimination of carbon dioxide, and yet 1 liter of venous blood contains only

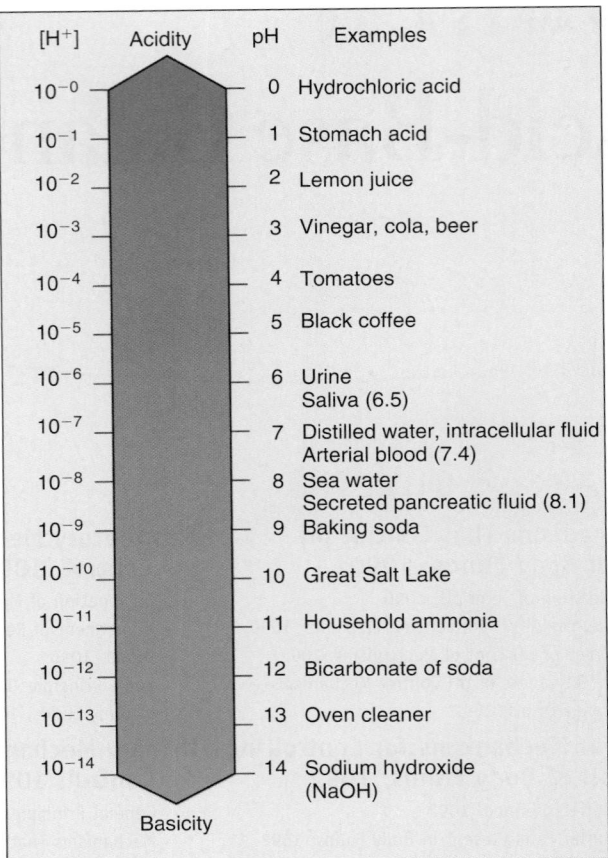

Figure 30-1 *The pH range.* See text for discussion.

about 1/100,000,000 g more hydrogen ions than 1 liter of arterial blood does. What incredible constancy! The pH homeostatic mechanism does indeed control effectively—astonishingly so.

Sources of pH-Influencing Elements

Acids and **bases** continually enter the blood from absorbed foods and from the metabolism of nutrients at the cellular level. Therefore, some kind of mechanism for neutralizing or eliminating these substances is necessary if blood pH is to remain constant. Although both acidic and basic components are important, the homeostasis of body pH depends largely on the control of hydrogen ion concentration in the extracellular fluid. Hydrogen ions are continually entering the body fluids from (1) **carbonic,** (2) **lactic,** (3) **sulfuric,** and (4) **phosphoric acids** and (5) **acidic ketone bodies.**

Carbonic and lactic acids are produced by the aerobic and anaerobic metabolism of glucose, respectively. Sulfuric acid is produced when sulfur-containing amino acids are oxidized, and phosphoric acid accumulates when certain phosphoproteins and ribonucleotides are broken down for energy purposes. Acidic ketone bodies, which include **acetone, acetoacetic acid,** and **beta-hydroxybutyric acid,** accumulate during the incomplete breakdown of fats. Each of these acids contributes hydrogen ions in varying amounts to the extracellular fluid and influences acid-base balance. Toxic accumulation of acidic ketone bodies is a common complication of untreated diabetes mellitus.

BOX 30-1: FYI
Acid-Forming Potential of Foods

Although citrus fruits such as oranges and grapefruit contain organic acids and may have an acid taste, they are not acid forming when metabolized. The acid-forming potential of a food is determined largely by the chloride, sulfur, and phosphorus elements in the "noncombustible" mineral residue, or ash, after metabolism of the food has occurred. Organic acids such as those found in citrus fruits are normally fully oxidized by the cells during metabolism and leave no mineral residue. As a result, they have little influence on acid-base balance.

Table 30-1 pH Control Systems

TYPE	RESPONSE TIME	EXAMPLE
Chemical buffer systems	Immediate	Bicarbonate buffer system
		Phosphate buffer system
		Protein buffer system
Physiological buffer systems	Minutes	Respiratory response system
	Hours	Renal response system

Minerals that remain after food has been metabolized are said to be either **acid** or **base** forming, depending on whether they contribute to formation of an acidic or basic medium when in solution. Acid-forming elements include **chlorine, sulfur,** and **phosphorus**—all are abundant in high-protein foods such as meat, fish, poultry, and eggs. These foods are often designated as **acid-forming foods.** After metabolism is complete, most mixed diets contain a surplus of acid-forming mineral elements that must be continually buffered to maintain acid-base balance. Extremely high-protein diets that produce a predominantly *acid mineral residue* when metabolized may tax the body's ability to remain in acid-base balance if consumed over prolonged periods.

Mineral elements that are alkaline, or basic, in solution include **potassium, calcium, sodium,** and **magnesium.** All these elements are found in fruits and vegetables, which nutritionists often label as **base-forming foods.** The predominantly **basic residue** that results after metabolism of a strict vegetarian diet may also tax the ability of the body to maintain acid-base balance because of a high influx of alkaline components into the extracellular fluid.

Foods containing acids that cannot be metabolized, such as rhubarb (oxalic) and cranberries (benzoic), are said to be direct acid-forming foods, whereas antacids such as sodium bicarbonate and calcium carbonate are examples of direct base-forming substances.

QUICK CHECK

1. Define the term *pH*.
2. Is a solution with a pH above 7 acid or alkaline?
3. Identify three acid-forming and three base-forming elements.
4. Does CO_2 entering venous blood increase or decrease the pH level?

Types of pH Control Mechanisms

The two major types of control systems listed in Table 30-1—*chemical* and *physiological*—operate to maintain the constancy of body pH.

In the discussion that follows, buffer action is defined and the specific types of **chemical** and **physiological buffer systems** are discussed. The rapid-acting **chemical buffers** immediately combine with any added acid or alkali that enters the body fluids and

thus prevent drastic changes in hydrogen ion concentration and pH. As explained later, all **buffers** act to prevent swings in pH even if hydrogen ion concentrations change. If the immediate action of chemical buffers cannot stabilize pH, the **physiological buffers** serve as a secondary defense against harmful shifts in pH of body fluids. pH shifts that are not halted by the immediate effects of chemical buffering cause the respiratory system to respond in 1 to 2 minutes, and changes in the rate and depth of breathing occur. For reasons explained later, such changes in carbon dioxide levels alter hydrogen ion concentration and help stabilize pH. If respiratory mechanisms are unable to stop the pH shift, a more powerful, but slower-acting renal physiological buffer system involving the excretion of either an acid or alkaline urine will be initiated within 24 hours. Collectively, these mechanisms—buffers, respirations, and kidney excretion of acids and bases—might be said to constitute the pH homeostatic mechanism. One example of the relationship between these three mechanisms is shown in Figure 30-2. The equilibrium be-

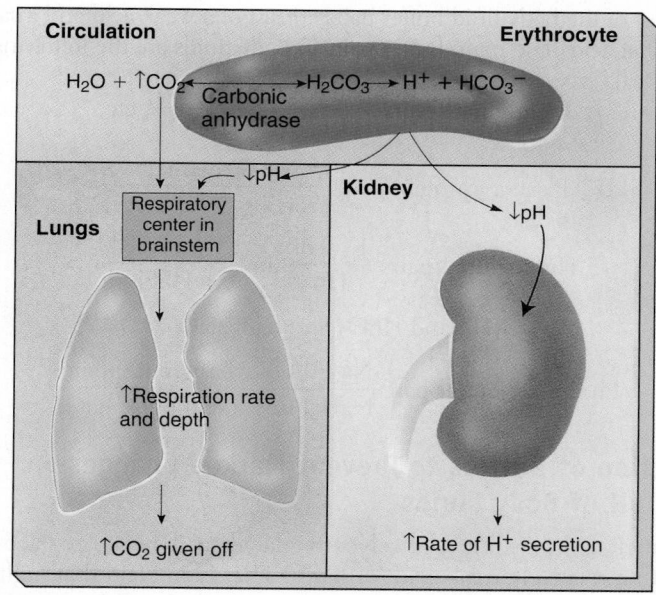

Figure 30-2 *Integration of pH control mechanisms.* Elevated CO_2 levels result in increased formation of carbonic acid in red blood cells. The resulting increase in hydrogen ions, coupled with elevated CO_2 levels, results in an increase in respiratory rate and secretion of hydrogen ions by the kidneys, thus helping regulate the pH of body fluids.

tween hydrogen ions, carbonic acid, bicarbonate, and carbon dioxide within red blood cells illustrates the interrelated nature of the pH control mechanisms. Note that an increase in body carbon dioxide levels results in excess acid formation in red blood cells counteracted by a corresponding increase in elimination of both carbon dioxide by the lungs and excess hydrogen ion in the urine.

Effectiveness of pH Control Mechanisms— Range of pH

The most convincing evidence of the effectiveness of the pH control mechanism is the extremely narrow range of blood pH— normally 7.36 to 7.41.

BUFFER MECHANISMS FOR CONTROLLING pH OF BODY FLUIDS

Buffers Defined

In terms of action, a **buffer** is a substance that prevents marked changes in the pH of a solution when an acid or a base is added to it. Let us suppose that a small amount of a strong acid, hydrochloric acid, is added to a solution that contains a buffer (e.g., blood) and that its pH decreases from 7.41 to 7.27. If the same amount of hydrochloric acid were added to pure water containing no buffers, its pH would decrease much more markedly, from 7 to perhaps 3.4. In both instances, pH decreased after addition of the acid, but much less so with buffers present than without them. Stated another way, buffers do not prevent pH changes, but they do help to minimize them.

In terms of chemical composition, buffers consist of two kinds of substances and are therefore often referred to as **buffer pairs.**

Buffer Pairs Present in Body Fluids

Most of the body fluid buffer pairs consist of a weak acid and a salt of that acid. The main buffer pairs in body fluids are the following:

$$\text{Bicarbonate pairs } \frac{NaHCO_3}{H_2CO_3}, \frac{KHCO_3}{H_2CO_3}, \text{ etc.}$$

$$\text{Plasma} \cdot \text{protein pair } \frac{Na \cdot Proteinate}{Proteins \text{ (weak acids)}}$$

$$\text{Hemoglobin pairs } \frac{K \cdot Hb}{Hb} \text{ and } \frac{K \cdot HbHO_2}{HbO_2}$$

$$(Hb \text{ and } HbO_2 \text{ are weak acids})$$

$$\text{Phosphate buffer pair } \frac{Na_2HPO_4}{NaH_2PO_4} \frac{\text{(basic phosphate)}}{\text{(acid phosphate)}}$$

Action of Buffers to Prevent Marked Changes in pH of Body Fluids

Buffers react with a relatively strong acid (or base) to replace it with a relatively weak acid (or base). That is, an acid that highly dissociates to yield many hydrogen ions is replaced by one that dissociates less highly to yield fewer hydrogen ions. Thus by the buffer reaction, instead of the strong acid remaining in the solution and contributing many hydrogen ions to drastically lower the

pH of the solution, a weaker acid takes its place, contributes fewer additional hydrogen ions to the solution, and thereby lowers its pH only slightly. Because blood contains buffer pairs, its pH fluctuates much less widely than it would without them. In other words, blood buffers constitute one of the devices for preventing marked changes in blood pH.

Let us consider, as a specific example of buffer action, how the sodium bicarbonate ($NaHCO_3$)–carbonic acid (H_2CO_3) system works in the presence of a strong acid or base.

The addition of a strong acid, such as hydrochloric acid (HCl), to the sodium bicarbonate–carbonic acid buffer system would initiate the reaction shown in Figure 30-3. Note how this reaction between HCl and the base bicarbonate ($NaHCO_3$) applies the principle of buffering. As a result of the buffering action of $NaHCO_3$, the weak acid, $H \cdot HCO_3$, replaces the very strong acid HCl, and therefore the hydrogen ion concentration of the blood increases much less than it would have if HCl were not buffered.

If, on the other hand, a strong base such as sodium hydroxide (NaOH) is added to the same buffer system, the reaction shown in Figure 30-4 would take place. The hydrogen ion of $H \cdot HCO_3$, the weak acid of the buffer pair, combines with hydroxide ion (OH^-) of the strong base NaOH to form water. Note what this accomplishes. It decreases the number of hydroxide ions added to the solution, and this in turn prevents the drastic rise in pH that would occur in the absence of buffering.

The principles of buffer action illustrated by the reaction of HCl and NaOH with the sodium bicarbonate buffer pair can be applied equally to the plasma protein, hemoglobin, and phosphate buffer systems.

Carbon dioxide and other acid waste products are being formed continuously as a result of cellular metabolism. The formation of carbonic acid from carbon dioxide and water requires the enzyme **carbonic anhydrase** (see Figure 30-2), which is

BOX 30-2: HEALTH MATTERS
Metabolic Alkalosis Caused by Vomiting

Vomiting, sometimes referred to as **emesis,** is the forcible emptying or expulsion of gastric and occasionally intestinal contents through the mouth. It occurs as a result of many stimuli, including foul odors or tastes, irritation of the stomach or intestinal mucosa, and some vomitive or *emetic* drugs such as *ipecac.* A "vomiting center" in the brain regulates the many coordinated (but primarily involuntary) steps involved (see Figure 26-18, *A*, p. 979). Severe vomiting such as the **pernicious vomiting** of pregnancy or the repeated vomiting associated with pyloric obstruction in infants can be life threatening. One of the most frequent and serious complications of vomiting is metabolic alkalosis. The bicarbonate excess of metabolic alkalosis results because of the massive loss of chloride from the stomach as hydrochloric acid. It is the loss of chloride that causes a compensatory increase of bicarbonate in the extracellular fluid. The result is **metabolic alkalosis.** Therapy includes intravenous administration of chloride-containing solutions such as **normal saline** (0.9% NaCl in water). The chloride ions of the solution replace bicarbonate ions and thus help relieve the bicarbonate excess responsible for the imbalance.

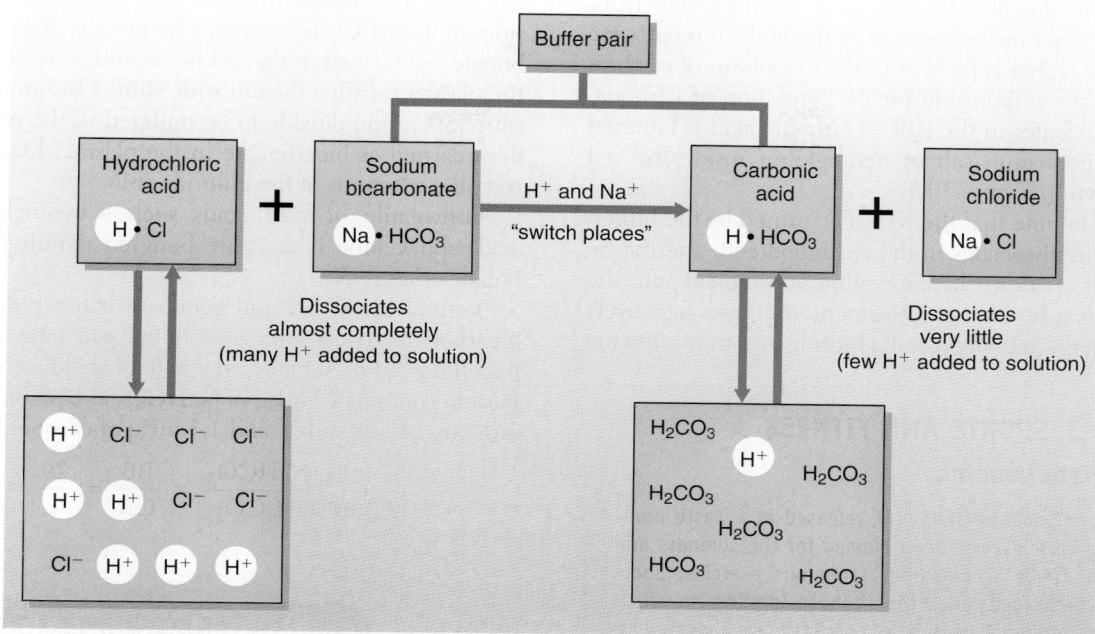

Figure 30-3 *Buffering action of sodium bicarbonate.* Buffering of acid HCl by NaHCO₃. As a result of the buffer action, the strong acid (HCl) is replaced by a weaker acid (H · HCO₃). Note that HCl, as a strong acid, "dissociates" almost completely and releases more H⁺ than H_2CO_3 does. Buffering decreases the number of H⁺ in the system.

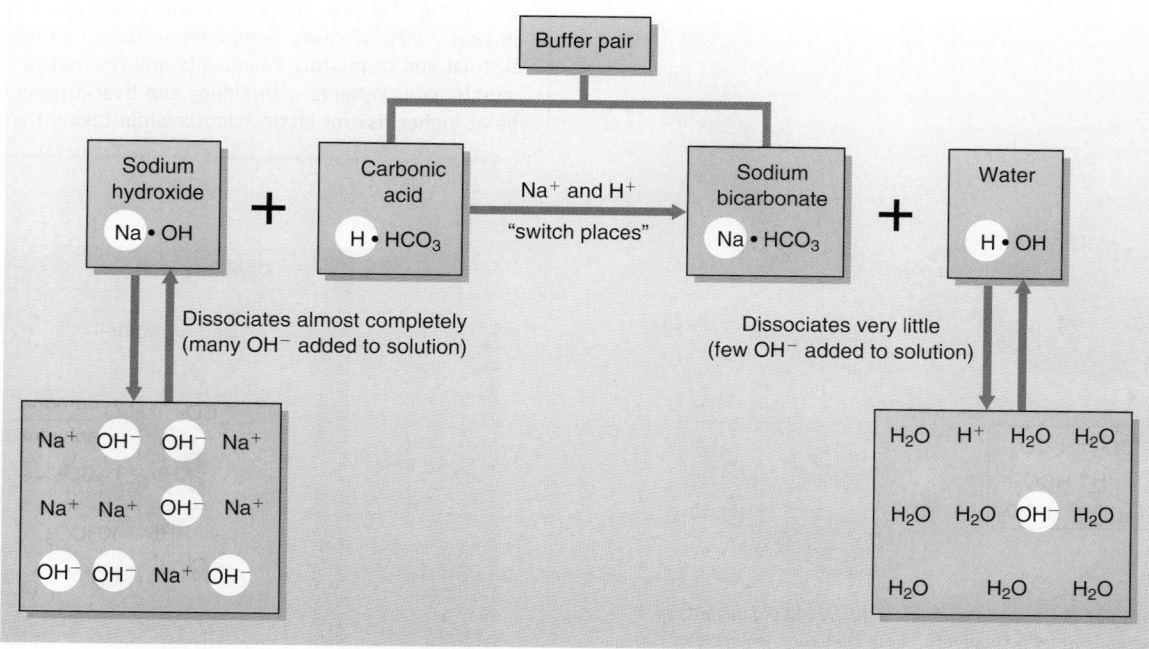

Figure 30-4 *Buffering action of carbonic acid.* Buffering of the base NaOH by H_2CO_3. As a result of buffer action, the strong base (NaOH) is replaced by NaHCO₃ and H_2O. As a strong base, NaOH "dissociates" almost completely and releases large quantities of OH⁻. Dissociation of H_2O is minimal. Buffering decreases the number of OH⁻ ions in the system.

found in the red blood cells. Although there are numerous types of zinc-containing carbonic anhydrases in the body, it is *carbonic anhydrase 1*, or *CA1*, that is present in the cytoplasm of erythrocytes and is primarily responsible for the formation of carbonic acid from CO_2 and water in the RBC. Carbonic acid is buffered primarily by the potassium salt of hemoglobin inside the red blood cell, as shown in Figure 30-5.

It is interesting to note that the $KHCO_3$ formed by the buffering of carbonic acid dissociates in the red blood cell, and the bicarbonate ion diffuses down its concentration gradient into the blood plasma. Because of the movement of these negatively charged ions out of the red blood cell, chloride ions move *into* the

cell from the plasma to maintain the electrical balance on both sides of the RBC membrane. The process of exchanging a bicarbonate ion formed in the red blood cell with a chloride ion from the plasma is called the **chloride shift.** This process makes it possible for carbon dioxide to be buffered in the red blood cell and then carried as bicarbonate in the plasma. Figure 30-6 summarizes the reactions of the chloride shift.

Nonvolatile, or fixed, acids, such as hydrochloric acid, lactic acid, and ketone bodies, are buffered mainly by sodium bicarbonate (Figure 30-7).

Normal blood pH and acid-base balance depend on a base bicarbonate–to–carbonic acid buffer pair ratio of 20:1 in the extracellular fluid. Actually, in a state of acid-base balance, a liter of plasma contains 27 mEq of $NaHCO_3$ as base bicarbonate (BB)—ordinary baking soda—and 1.3 mEq of carbonic acid (CA):

$$\frac{27 \text{ mEq } NaHCO_3}{1.3 \text{ mEq } H_2CO_3} = \frac{BB}{CA} = \frac{20}{1} = \text{pH } 7.4$$

BOX 30-3 SPORTS AND FITNESS

Bicarbonate Loading

The buildup of lactic acid in the blood, released as a waste product from working muscles, has been blamed for the soreness and fatigue that sometimes accompanies strenuous exercise. Some athletes adopt a technique called **bicarbonate loading,** in which large amounts of sodium bicarbonate ($NaHCO_3$) are ingested to counteract the effects of lactic acid buildup. Their theory is that fatigue is avoided because the $NaHCO_3$, a base, buffers the lactic acid. Unfortunately, the excess bicarbonate intake and diarrhea that often result can trigger fluid and electrolyte imbalances. Long-term $NaHCO_3$ abuse can lead to metabolic alkalosis and its disastrous effects.

BOX 30-4   Lactic Acidosis and Metformin

Metformin hydrochloride (Glucophage) is one of the most widely used and effective of the oral antidiabetic drugs. It is used with diet and exercise to lower blood glucose levels in type 2 diabetes mellitus. A rare but very serious complication of metformin therapy is **lactic acidosis.** It is characterized by elevated blood lactate levels, electrolyte disturbances, and decreased blood pH. It is reported in only about 1 in 33,000 patients taking metformin over the course of a year. However, when it does occur it can be fatal in nearly 50% of cases. Symptoms include a variety of gastrointestinal and respiratory complaints and feelings of weakness and muscle pain. Patients with kidney and liver disease are known to be at higher risk for lactic acidosis while taking the drug.

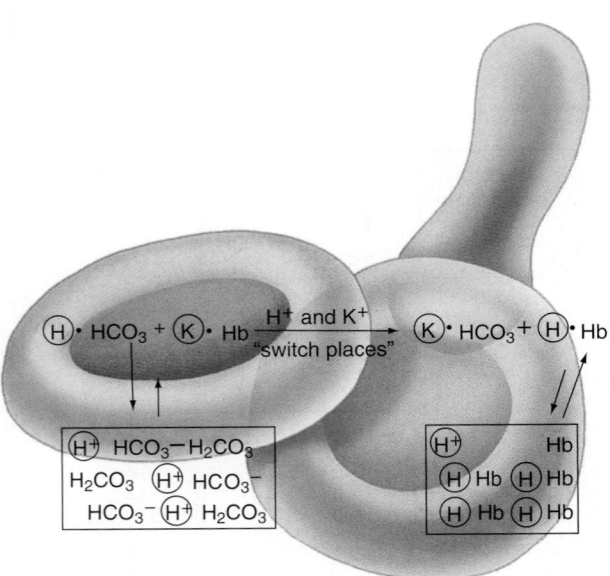

Figure 30-5 *Buffering of the volatile carbonic acid (H · HCO₃) inside red blood cell by potassium salt of hemoglobin.* Note that each molecule of carbonic acid is replaced by a molecule of acid hemoglobin. Because hemoglobin is a weaker acid than carbonic acid, fewer of these hemoglobin molecules dissociate to form hydrogen ions. Hence, fewer hydrogen ions are added to red blood cell intracellular fluid than would be added by unbuffered carbonic acid. Also, because some of the carbonic acid in the red blood cell has come from plasma, fewer hydrogen ions remain in blood than would have if there were no buffering of carbonic acid.

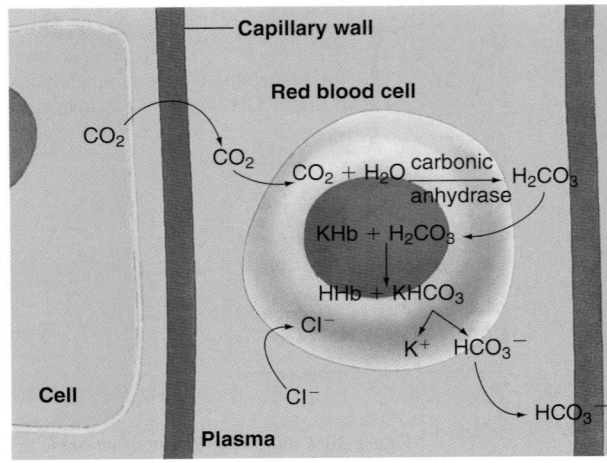

Figure 30-6 *Chloride shift.* The concentration of chloride ions (Cl^-) in red blood cells (RBCs) increases as bicarbonate ions (HCO_3^-) diffuse out of the cell. Bicarbonate ions form as a result of the buffering of carbonic acid by the potassium salt of hemoglobin.

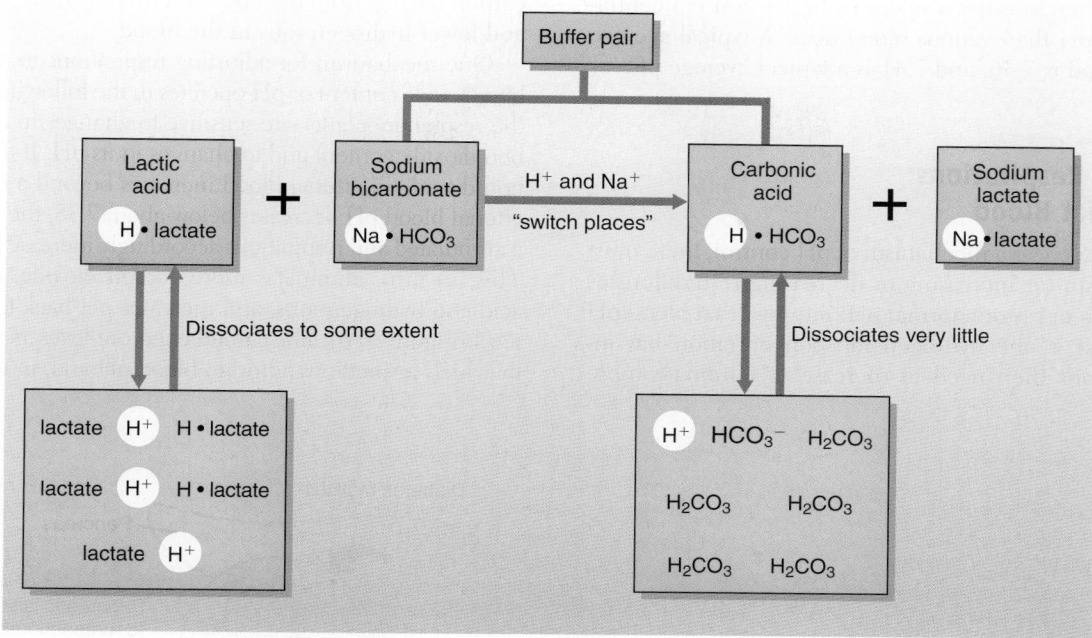

Figure 30-7 *Lactic acid buffered by sodium bicarbonate.* Lactic acid (H · lactate) and other "fixed" acids are buffered by NaHCO₃ in the blood. Carbonic acid (H · HCO₃ or H₂CO₃, a weaker acid than lactic acid) replaces lactic acid. As a result, fewer H⁺ ions are added to blood than would be if lactic acid were not buffered.

The **ratio** of base to acid is critical. If the ratio is maintained, acid-base balance (pH) remains near normal despite changes in the absolute amounts of either component of the buffer pair. This type of adjustment is called **compensation**. For example, a BB/CA ratio of 40:2 or 10:0.5 would result in a compensated state of acid-base balance. However, an increase in the ratio causes an increase in pH (**uncompensated alkalosis**), and a decrease in the ratio causes a decrease in pH (**uncompensated acidosis**). The ability of the body to regulate the amount of either component of the bicarbonate buffer pair—to maintain the correct ratio for acid-base balance—makes this system one of the most important for controlling pH of body fluids. **Correction** of acid-base balance is said to occur when components of the buffer pair return to a normal 20:1 ratio.

Evaluation of the Role of Buffers in pH Control

Buffering alone cannot maintain homeostasis of pH. As we have seen, hydrogen ions are added continually to capillary blood despite buffering. If even a few hydrogen ions were added every time blood circulated and no way were provided for eliminating them, blood hydrogen ion concentration would necessarily increase and thereby decrease blood pH. "Acid blood," in other words, would soon develop. Respiratory and urinary mechanisms must therefore function concurrently with buffers to remove from the blood and from the body the hydrogen ions continually being added to blood. Only then can the body maintain constancy of pH over time.

 QUICK CHECK

5. Define the term *buffer*.
6. Identify the two major types of buffer systems in the body. Which buffer system is the most rapid acting?
7. Use equations to explain the buffering of HCl by sodium bicarbonate and NaOH by carbonic acid.
8. Define the term *chloride shift*.

RESPIRATORY MECHANISMS OF PH CONTROL

Explanation of Mechanisms

Respirations play a vital role in controlling pH. With every expiration, carbon dioxide and water leave the body in the expired air. The carbon dioxide has come from the venous blood—has diffused out of it as it moves through the lung capillaries. Therefore, less carbon dioxide remains in the arterial blood leaving the lung capillaries. The lower P_{CO_2} in arterial blood reduces the amount of carbonic acid and the number of hydrogen ions that can be formed in red blood cells by the following reactions:

$$CO_2 + H_2O \xrightarrow{\text{(carbonic anhydrase)}} H_2CO_3$$

$$H_2CO_3 \rightarrow H^+ + HCO_3^-$$

Arterial blood therefore has a lower hydrogen ion concentration and a higher pH than venous blood does. A typical average pH for venous blood is 7.36, and 7.41 is a typical average pH for arterial blood.

Adjustment of Respirations to pH of Arterial Blood

For respirations to serve as a mechanism of pH control, there must be some mechanism for increasing or decreasing respirations as needed to maintain or restore normal pH. Suppose that blood pH has decreased; that is, the hydrogen ion concentration has increased. Respirations then need to increase to eliminate more carbon dioxide from the body and thereby leave less carbonic acid and fewer hydrogen ions in the blood.

One mechanism for adjusting respirations to arterial blood carbon dioxide content or pH operates in the following way: neurons of the respiratory center are sensitive to changes in arterial blood carbon dioxide content and to changes in its pH. If the amount of carbon dioxide in arterial blood increases beyond a certain level, or if arterial blood pH decreases below about 7.38, the respiratory center is stimulated and respirations accordingly increase in rate and depth. This, in turn, eliminates more carbon dioxide, reduces carbonic acid and hydrogen ions, and increases pH back toward the normal level (Figure 30-8). The carotid chemoreflexes are also mechanisms by which respirations adjust to blood pH and, in turn, adjust pH.

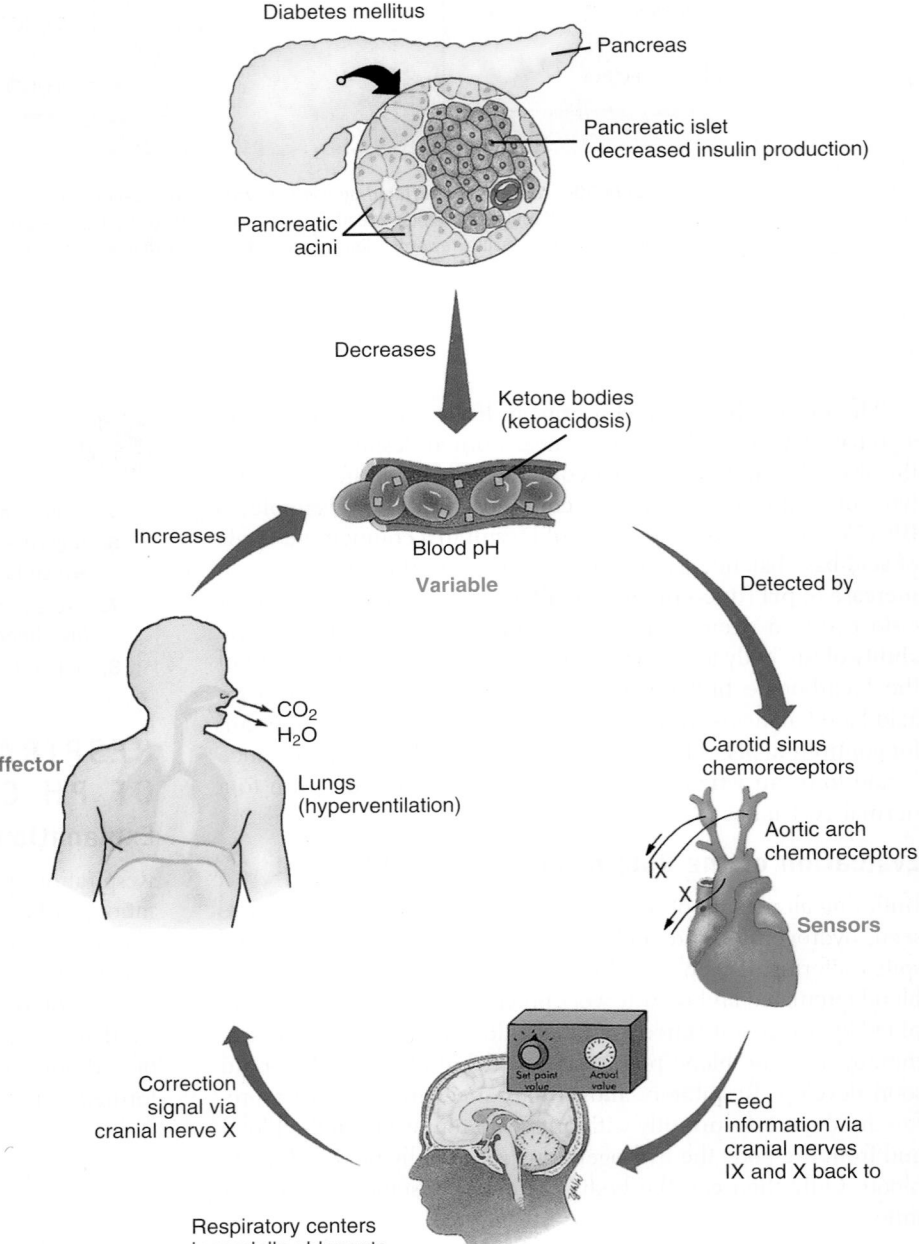

Figure 30-8 *Respiratory mechanism of pH control.* A rise in arterial blood CO_2 content or a drop in its pH (below about 7.38) stimulates respiratory center neurons. Hyperventilation results. Less CO_2 and therefore less carbonic acid and fewer hydrogen ions remain in the blood so that blood pH increases, often reaching the normal level.

Some Principles Relating Respirations and pH of Body Fluids

- A decrease in blood pH below normal (acidosis) tends to stimulate increased respirations (hyperventilation), which tends to increase pH back toward normal. In other words, acidosis causes hyperventilation, which in turn acts as a compensating mechanism for the acidosis.
- Prolonged hyperventilation may increase blood pH enough to produce alkalosis.
- An increase in blood pH above normal (or alkalosis) causes hypoventilation, which serves as a compensating mechanism for the alkalosis by decreasing blood pH back toward normal.
- Prolonged hypoventilation may decrease blood pH enough to produce acidosis.

URINARY MECHANISMS OF PH CONTROL

General Principles About Mechanism

Because the kidneys can excrete varying amounts of acid and base, they, like the lungs, play a vital role in pH control. Kidney tubules, by excreting many or few hydrogen ions in exchange for reabsorbing many or few sodium ions, control urine pH and thereby help control blood pH. If, for example, blood pH decreases below normal, the kidney tubules secrete more hydrogen ions from blood to urine and, in exchange for each hydrogen ion, reabsorb a sodium ion from the urine back into the blood. This, of course, decreases urine pH. But simultaneously—and of far

more importance—it increases blood pH back toward normal. This urinary mechanism of pH control is a device for excreting varying amounts of hydrogen ions from the body to match the amounts entering the blood. It constitutes a much more effective process for adjusting hydrogen output to hydrogen input than does the body's only other mechanism for expelling hydrogen ions, namely, the respiratory mechanisms previously described. But abnormalities of any one of the three pH control mechanisms soon throw the body into a state of acid-base imbalance. Only when all three parts of this complex mechanism—buffering, respirations, and urine secretion—function adequately can acid-base balance be maintained.

Let us turn now to mechanisms that adjust urine pH to counteract changes in blood pH.

Mechanisms That Control Urine pH

A decrease in blood pH accelerates the renal tubule ion exchange mechanisms that acidify urine and conserve blood's base, thereby tending to increase blood pH back to normal. The following paragraphs describe these mechanisms.

- The distal and collecting tubules secrete hydrogen ions into the urine in exchange for basic ions, which they reabsorb. Refer to Figure 30-9 as you read the rest of this paragraph. Note that carbon dioxide diffuses from tubule capillaries into distal tubule cells, where the enzyme carbonic anhydrase accelerates the combining of carbon dioxide with water to form carbonic acid. The carbonic acid dissociates into hydrogen ions and bicarbonate ions. The hydrogen ions then diffuse into the tubular urine, where they displace basic ions (most often sodium)

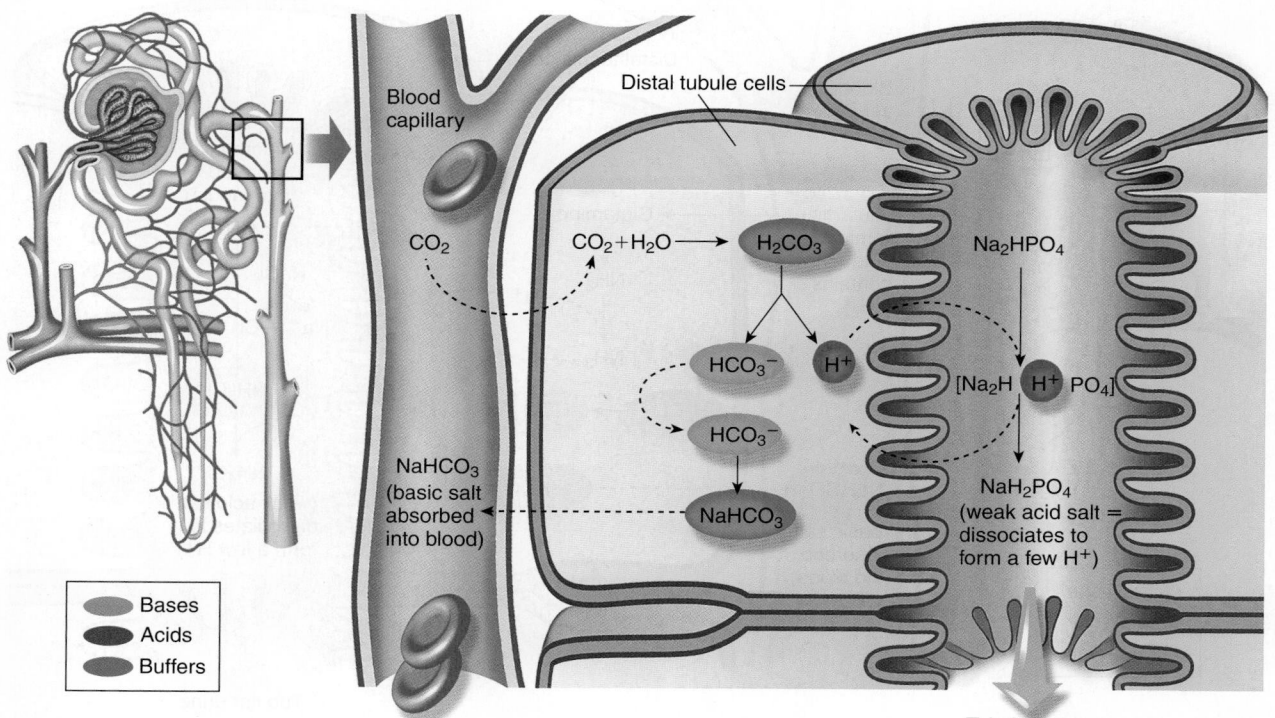

Figure 30-9 *Acidification of urine and conservation of base by distal renal tubule excretion of H$^+$.* See text for discussion of the mechanism.

from a basic salt of a weak acid and thereby change the basic salt to an acid salt or to a weak acid that is eliminated in the urine. While this is happening, the displaced sodium or other basic ion diffuses into a tubule cell. Here, it combines with the bicarbonate ion left over from the carbonic acid dissociation to form sodium bicarbonate. The sodium bicarbonate then diffuses—is reabsorbed—into the blood. Consider the various results of this mechanism. Sodium bicarbonate (or other base bicarbonate) is conserved for the body. Instead of all the basic salts that filter out of glomerular blood leaving the body in the urine, considerable amounts are recovered into peritubular capillary blood. In addition, extra hydrogen ions are added to the urine and thereby eliminated from the body. Both the reabsorption of base bicarbonate into blood and the excretion of hydrogen ions into urine tend to increase the ratio of the bicarbonate buffer pair $B \cdot HCO_3 / H \cdot HCO_3$ (BB/CA) present in blood. This automatically increases blood pH. In short, kidney tubule base bicarbonate reabsorption and hydrogen ion excretion both tend to alkalinize blood by acidifying urine.

- The renal tubules can excrete hydrogen or potassium in exchange for the sodium they reabsorb. Therefore, in general, the more hydrogen ions they excrete, the fewer potassium ions they can excrete. In acidosis, tubule excretion of hydrogen ions increases markedly and potassium ion excretion decreases—an important factor because it may lead to **hyperkalemia** (excessive blood potassium), a condition that can cause heart block and death.

- The distal and collecting tubule cells excrete ammonia into the tubular urine. As Figure 30-10 shows, the ammonia combines with hydrogen to form an ammonium ion. The ammonium ion displaces sodium or some other basic ion from a salt of a fixed (nonvolatile) acid to form an ammonium salt. The basic ion then diffuses back into a tubule cell and combines with a bicarbonate ion to form a basic salt, which in turn diffuses into tubular blood. Thus, like the renal tubules' excretion of hydrogen ions, their excretion of ammonia and its combining with hydrogen to form ammonium ions also tend to increase the blood bicarbonate buffer pair ratio and therefore tend to increase blood pH. Quantitatively, however, ammonium ion excretion is more important than hydrogen ion excretion.

- Renal tubule excretion of hydrogen and ammonia is controlled at least in part by the blood pH level. As indicated in Figure 30-11, a decrease in blood pH accelerates tubule excretion of both hydrogen and ammonia. An increase in blood pH produces the opposite effects.

 QUICK CHECK

9. What is the function of carbonic anhydrase in buffer action?
10. How does respiratory rate affect blood pH levels?
11. List two ways in which acidification of urine occurs.

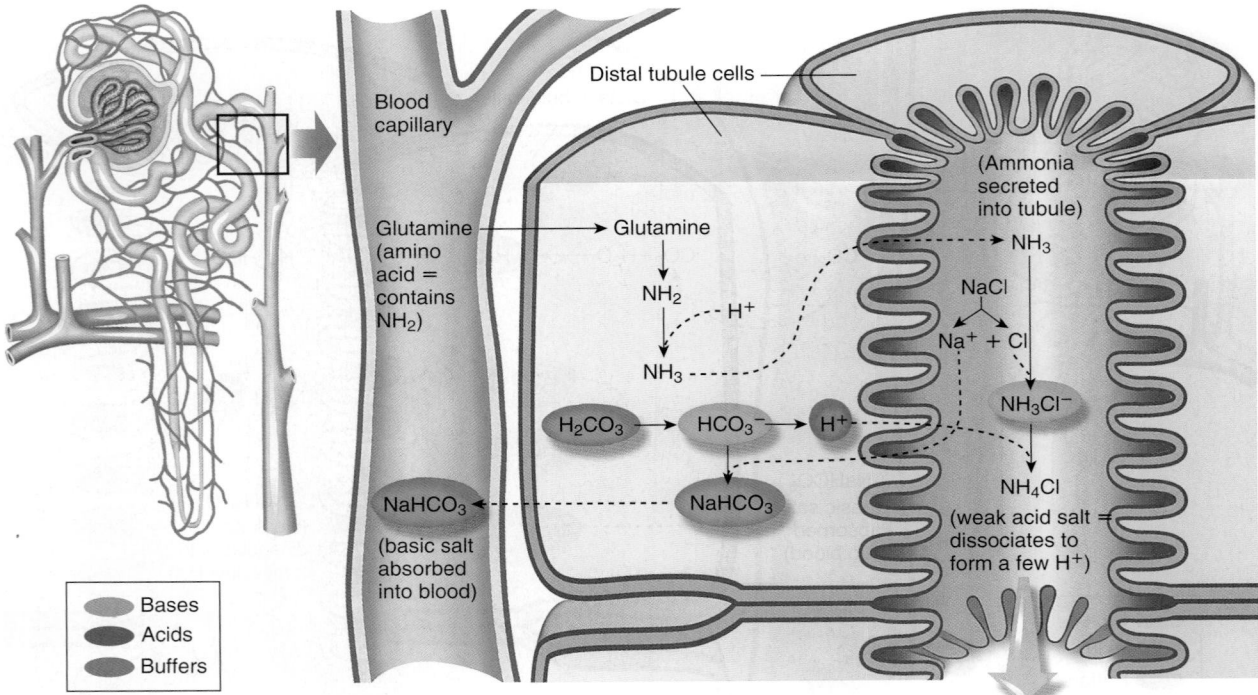

Figure 30-10 *Acidification of urine by tubule excretion of ammonia (NH₃).* An amino acid (glutamine) moves into the tubule cell and loses an amino group (NH₂) to form ammonia, which is secreted into urine. In exchange, the tubule cell reabsorbs a basic salt (mainly NaHCO₃) into blood from urine.

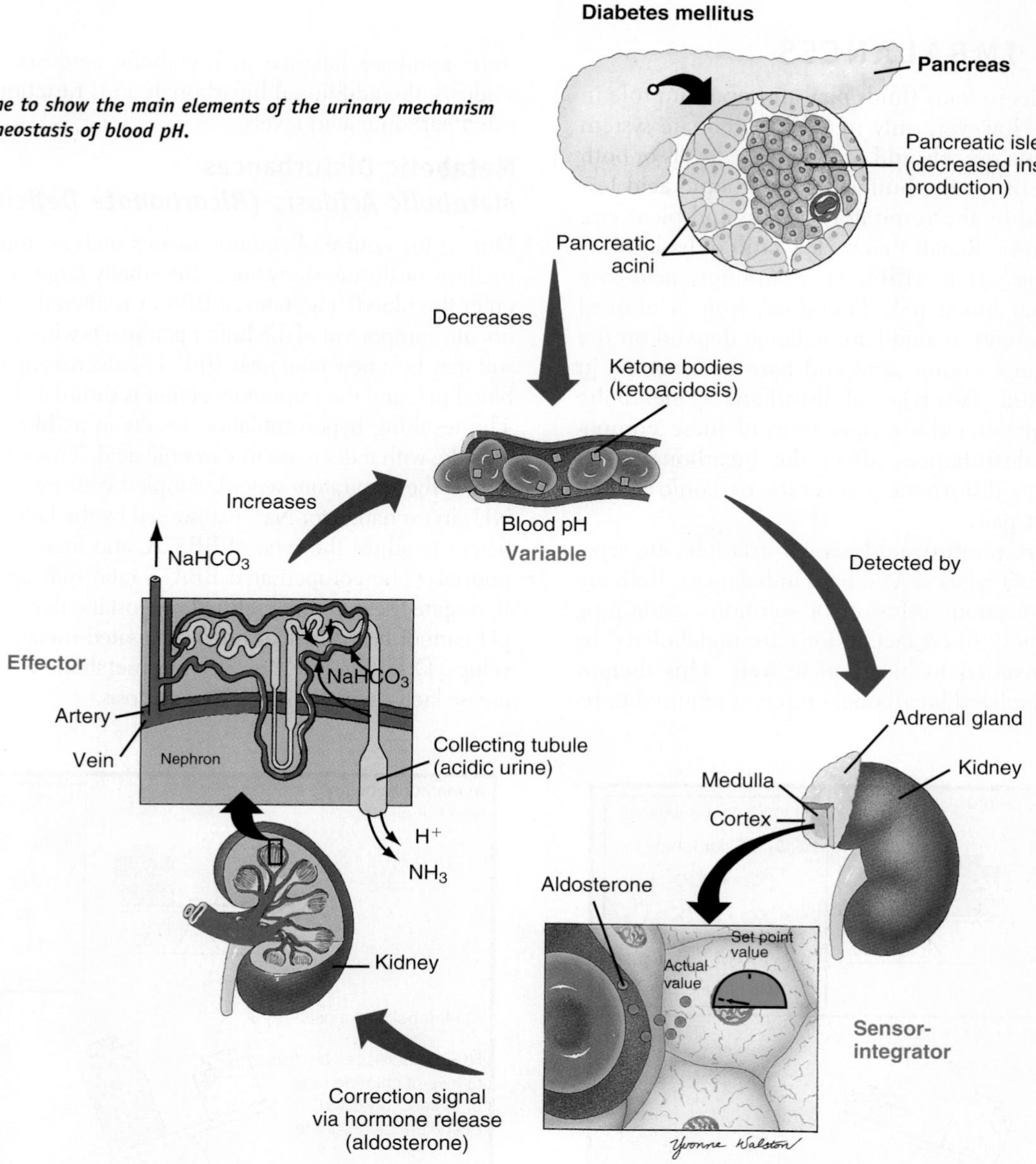

Figure 30-11 *Scheme to show the main elements of the urinary mechanism for maintaining homeostasis of blood pH.*

Diabetes mellitus

Pancreas

Pancreatic islet (decreased insulin production)

Pancreatic acini

Decreases

Ketone bodies (ketoacidosis)

Increases

$NaHCO_3$

Blood pH
Variable

Detected by

Effector

$NaHCO_3$

Artery

Vein

Nephron

Collecting tubule (acidic urine)

H^+

NH_3

Adrenal gland

Medulla

Kidney

Cortex

Aldosterone

Set point value

Actual value

Sensor-integrator

Kidney

Correction signal via hormone release (aldosterone)

Yvonne Walston

THE BIG PICTURE

Acid-Base Balance

In Chapter 1 we discussed the chemical level of organization and its importance in maintaining the complex relationships that exist between the structural and functional units of the body. In Chapter 2 we reviewed the basic principles of chemistry as they apply to the life process. An understanding of these principles is important in the study of every body organ system. They are particularly crucial, however, to an understanding of the homeostatic mechanisms involved in acid-base balance.

Ultimately, all functions occur at the cellular level and, without exception, each vital physiological function depends on the maintenance of an appropriate, stable, and tightly regulated acid-base environment. Regulating chemical reactions at the cellular level permits us to control the flow of energy in the body. We need to control energy flow to accomplish cellular work and to store and transfer energy so that we can meet our immediate and long-term needs. Enzymes are the biological catalysts that permit or assist our cells in

the regulation of all energy-based metabolic reactions required for the maintenance of life. Enzymes involved in metabolic reactions have both optimal pH ranges for maximal activity and limited pH ranges in which activity is maintained. Therefore, anything that disrupts the homeostasis of acid-base balance by disrupting enzyme activity is immediately life threatening because it affects our ability to initiate and regulate the metabolic activity required to sustain life.

Different enzymes require very narrow and often different pH ranges to function. Pepsin, for example, is able to digest proteins optimally in the stomach only if the pH environment remains very acid, whereas trypsin and most other human enzymes work best in a pH range of 6 to 8.

The elaborate and highly sensitive pH control mechanisms intended to provide homeostasis of our acid-base environment are critical for the enzymatic action necessary for healthy metabolism and for life itself.

ACID-BASE IMBALANCES

All of the buffer pairs in body fluids play an important role in acid-base balance. However, only in the bicarbonate system can the body regulate quickly and precisely the levels of both chemical components in the buffer pair. Carbonic acid levels can be regulated by the respiratory system and bicarbonate ion by the kidneys. Recall that a 20:1 ratio of base bicarbonate to carbonic acid (BB/CA) maintains acid-base balance and normal blood pH. Therefore, from a clinical standpoint, disturbances in acid-base balance depend on the relative quantities of carbonic acid and base bicarbonate in the extracellular fluid. Two types of disturbances, metabolic and respiratory, can alter the proper ratio of these components. Metabolic disturbances affect the bicarbonate element, and respiratory disturbances affect the carbonic acid element of the buffer pair.

Metabolic and **respiratory acidosis**, for example, are separate and very different types of acid-base imbalances. Both are treated by the intravenous infusion of solutions containing **sodium lactate**. The infused lactate ions are metabolized by liver cells and converted to bicarbonate ions. This therapy helps replace the depleted bicarbonate reserves required to restore acid-base balance in metabolic acidosis. In respiratory acidosis, the additional bicarbonate ions function to offset elevated carbonic acid levels.

Metabolic Disturbances
Metabolic Acidosis (Bicarbonate Deficit)

During the course of certain diseases such as untreated diabetes mellitus or during starvation, abnormally large amounts of acids enter the blood. The ratio of BB/CA is altered as the base bicarbonate component of the buffer pair reacts with the acids. The result may be a new ratio near 10:1. The decreasing ratio lowers the blood pH, and the respiratory center is stimulated (Figure 30-12). The resulting hyperventilation results in a "blow-off" of carbon dioxide, with a decrease in carbonic acid. This compensatory action of the respiratory system, coupled with excretion of H^+ and NH_3 in exchange for Na^+ reabsorbed by the kidneys, may be sufficient to adjust the *ratio* of BB/CA, and therefore blood pH, to normal. (The compensated BB/CA ratio may approach 10:0.5.) If, despite these compensating homeostatic devices, the ratio and pH cannot be corrected, uncompensated metabolic acidosis develops. Drug use (see Box 30-4) or metabolic conditions that increase lactic acid can also cause acidosis.

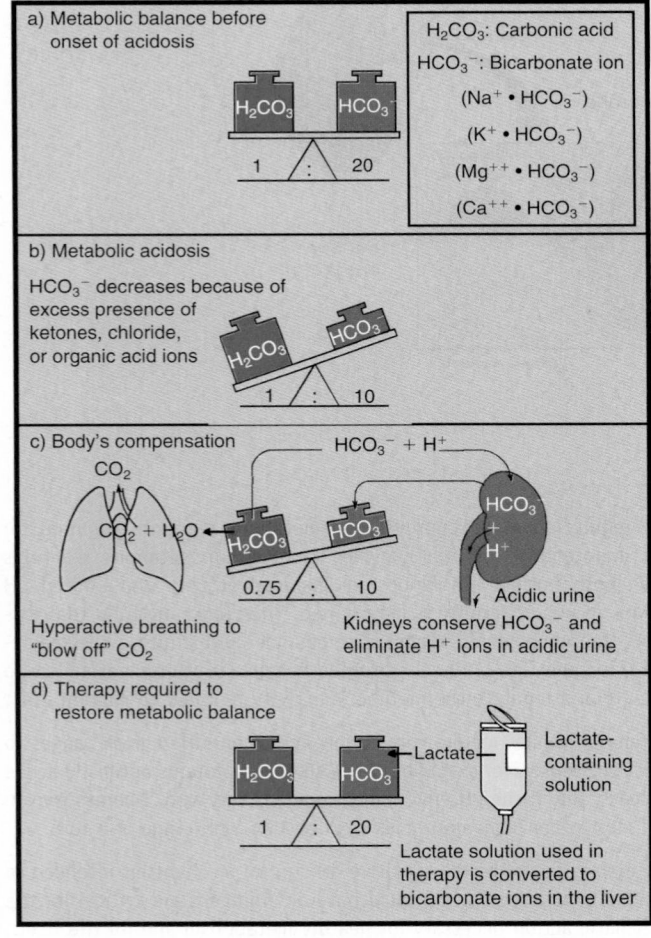

Figure 30-12 *Metabolic acidosis.*

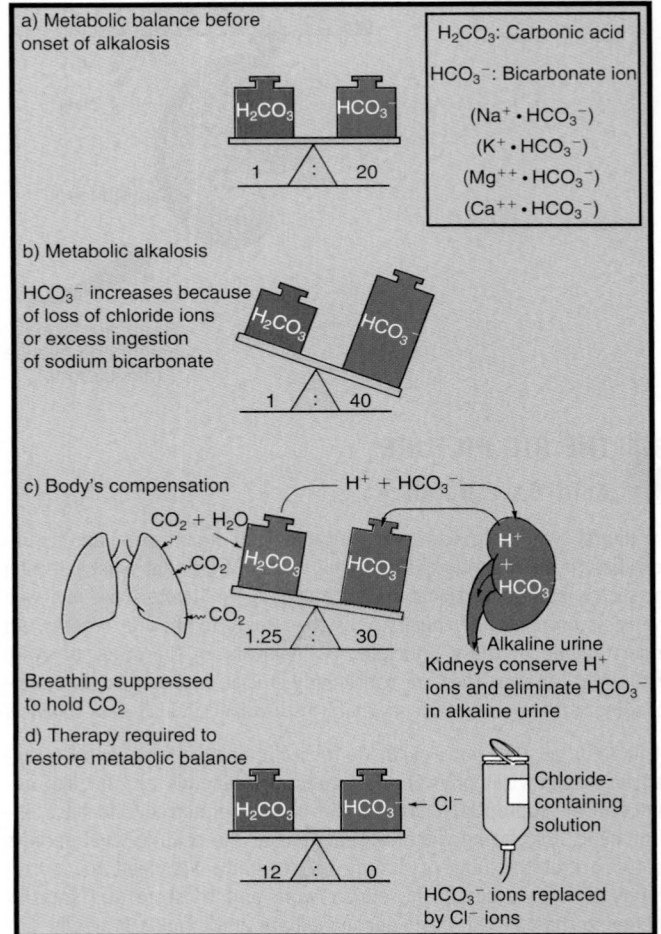

Figure 30-13 *Metabolic alkalosis.*

An increased blood hydrogen ion concentration, that is, decreased blood pH, as we have noted, stimulates the respiratory center. For this reason, hyperventilation is an outstanding clinical sign of acidosis. Increases in hydrogen ion concentration above a certain level depress the central nervous system and therefore produce such symptoms as disorientation and coma. In a terminal illness, death from acidosis is likely to follow coma, whereas death from alkalosis generally follows tetany and convulsions.

Metabolic Alkalosis (Bicarbonate Excess)

Patients suffering from chronic stomach problems such as hyperacidity sometimes ingest large quantities of alkali—often plain baking soda, or sodium bicarbonate—for extended periods. Such improper use of antacids or excessive vomiting (see Box 30-2) can produce metabolic alkalosis. Initially, the condition results in an increase in the BB/CA ratio to perhaps 40:1 (Figure 30-13). Compensatory mechanisms are aimed at increasing carbonic acid and decreasing the bicarbonate load. With breathing suppressed and the kidneys excreting bicarbonate ions, a compensated ratio of 30:1.25 might result. Such a ratio would restore acid-base balance

and blood pH to normal. In uncompensated metabolic alkalosis the ratio, and therefore the pH, remain increased.

Respiratory Disturbances
Respiratory Acidosis (Carbonic Acid Excess)

Clinical conditions such as pneumonia or emphysema tend to cause retention of carbon dioxide in the blood. Also, drug abuse or overdose, such as barbiturate poisoning, suppresses breathing and results in respiratory acidosis (Figure 30-14). The carbonic acid component of the bicarbonate buffer pair increases above normal in respiratory acidosis. Body compensation, if successful, increases the bicarbonate fraction so that a new BB/CA ratio (perhaps 20:2) will return blood pH to normal or near-normal levels.

Respiratory Alkalosis (Carbonic Acid Deficit)

Hyperventilation caused by fever or mental disease (hysteria) can result in excessive loss of carbonic acid and lead to **respiratory alkalosis** (Figure 30-15) with a bicarbonate buffer pair ratio of 20:/0.5. Compensatory mechanisms may adjust the ratio to 10:0.5 and return blood pH to near normal.

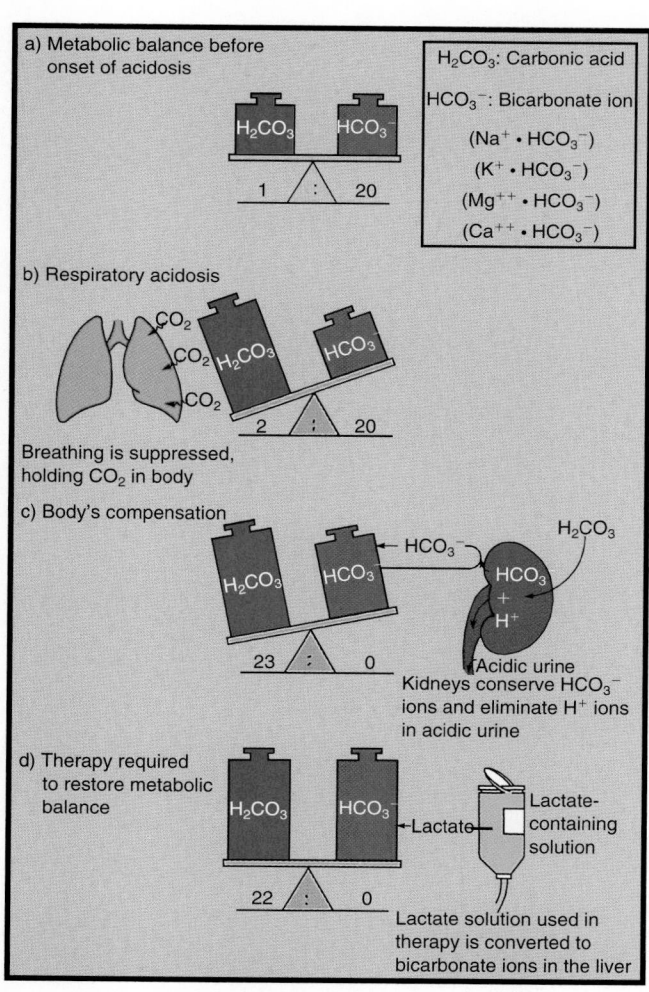

Figure 30-14 *Respiratory acidosis.*

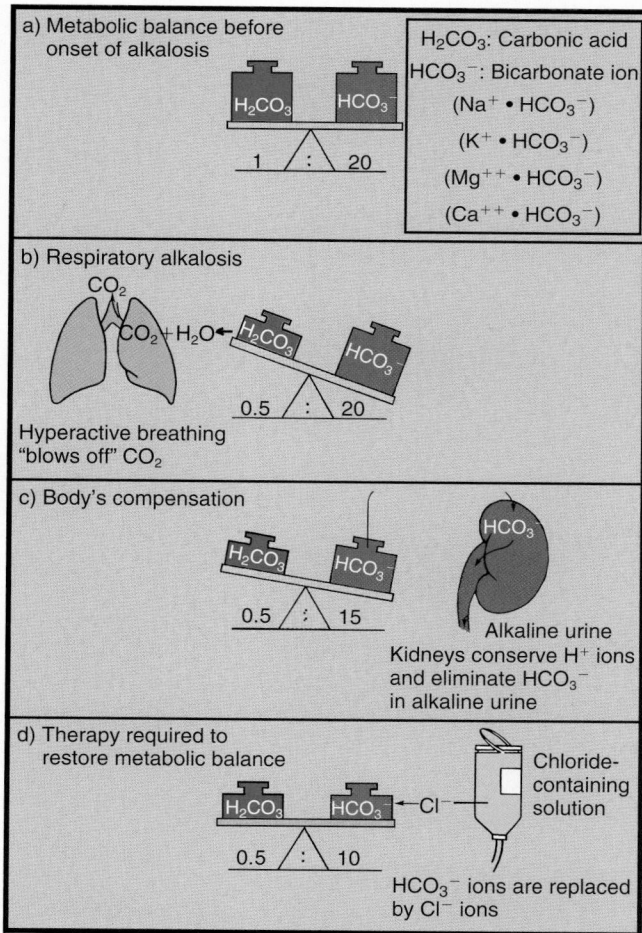

Figure 30-15 *Respiratory alkalosis.*

LANGUAGE OF SCIENCE *(Cont'd from page 1089)*

phosphoric acid (foss-FOR-ik ASS-id) [*phos-* light, *-pherein-* to bear, *-ic* pertaining to, *acidus* sour]

phosphorus

physiological buffers (fiz-ee-o-LOJ-i-kal BUFF-erz) [*physio-* related to nature, *-logos-* science, *-al* pertaining to, *buffe* to cushion]

potassium (poh-TAS-ee-um) [*potasschen* potash]

ratio (RAY-shee-oh) [*ratio* a reckoning]

sodium (SO-dee-um) [*sodo* solid]

sulfur

sulfuric acid (sul-FYOOR-ik ASS-id)

uncompensated acidosis (un-KOM-pen-say-ted ass-ih-DOH-sis) [*un-* not, *-compensare* to balance, *acidus-* sour, *-osis* condition]

uncompensated alkalosis (un-KOM-pen-say-ted al-kah-LOH-sis) [*un-* not, *-compensare* to balance, *al-qaliy-* the wood ashes, *-osis* condition]

LANGUAGE OF MEDICINE

bicarbonate loading (bye-KAR-boh-net LOHD-ing) [*bi-* two, *-carbo* coal]

emesis (EM-eh-sis) [*emesis* to vomit]

hyperkalemia (hye-per-kah-LEE-mee-ah) [*hyper-* excessive, *-kalium-* potassium, *-emia* blood condition]

lactic acidosis (LAK-tik ass-ih-DOH-sis) [*lac-* milk, *-ic* pertaining to, *acidus-* sour, *-osis* condition]

metabolic acidosis (met-ah-BOL-ik ass-ih-DOH-sis) [*metabole* change, *acidus-* sour, *-osis* condition]

metabolic alkalosis (met-ah-BOL-ik al-kah-LOH-sis) [*metabole* change, *al-qaliy-* the wood ashes, *-osis* condition]

pernicious vomiting (per-NISH-us) [*perniciosus* destructive, *vomere* to vomit]

respiratory acidosis (RES-pih-rah-tor-ee ass-ih-DOH-sis) [*respirare* to breathe, *acidus-* sour, *-osis* condition]

respiratory alkalosis (RES-pih-rah-tor-ee al-kah-LOH-sis) [*respirare-* to breathe, *al-galiy* the wood ashes; *-osis* condition]

sodium lactate (SO-dee-um LAK-tayt) [*soda* solid, *lac* milk]

CASE STUDY

David Falcon is a 10-year-old boy who has been admitted to the hospital with type 1 diabetes mellitus. He recently experienced weight loss of 20%, constant hunger, and increased urination. He has felt increasingly sick and began vomiting this morning.

On admission he became less responsive, confused, and nearly unconscious with rapid deep respirations. Physical examination reveals dry skin and dry mucous membranes. His breath has a fruity odor. His initial lab results reveal a blood glucose of 450 mg/dl, positive urine and blood ketones, arterial blood pH of 7.25, P_{CO_2} of 40 mm Hg, and HCO_3^- of 16 mEq/L. Treatment of intravenous fluids with electrolytes and insulin was started and resulted in improvement in David's symptoms.

1. Which of the following responses best explains David's symptoms of increased thirst and increased urination?

   A. Breakdown of fatty acids in the liver
   B. Hyperosmolar body fluids
   C. Loss of potassium in the urine
   D. Gluconeogenesis in the liver

2. David's initial arterial blood gas values represent which of the following conditions?

   A. Compensated metabolic alkalosis
   B. Uncompensated metabolic alkalosis
   C. Uncompensated metabolic acidosis
   D. Compensated metabolic acidosis

3. The body has many ways of responding to alteration in blood pH. David's respiratory pattern is one of these. What is the effect of his rapid deep respirations on blood acid-base balance?

   A. Increases blood pH
   B. Decreases blood pH
   C. Increases blood CO_2
   D. Decreases blood $NaHCO_3$

4. David's vomiting is a symptom of ketone production, and he has ketones in both his blood and urine. Which of the following physiological mechanisms is most likely responsible for his ketosis?

   A. The rate of carbohydrate metabolism has increased.
   B. The rate of protein catabolism has decreased.
   C. The rate of fat catabolism has increased.
   D. The rate of fat catabolism has decreased.

CHAPTER SUMMARY

INTRODUCTION

A. Acid-base balance is one of the most important of the body's homeostatic mechanisms
B. Acid-base balance refers to regulation of hydrogen ion concentration in body fluids
C. Precise regulation of pH at the cellular level is necessary for survival
D. Slight pH changes have dramatic effects on cellular metabolism

MECHANISMS THAT CONTROL pH OF BODY FLUIDS

A. Meaning of pH—negative logarithm of hydrogen ion concentration of a solution (Figure 30-1)
B. Sources of pH-influencing elements
   1. Carbonic acid—formed by aerobic glucose metabolism
   2. Lactic acid—formed by anaerobic glucose metabolism
   3. Sulfuric acid—formed by oxidation of sulfur-containing amino acids
   4. Phosphoric acid—formed in the breakdown of phosphoproteins and ribonucleotides
   5. Acidic ketone bodies—formed in the breakdown of fats
      a. Acetone
      b. Acetoacetic acid
      c. Beta-hydroxybutyric acid
C. Acid-forming potential of foods—determined by chloride, sulfur, and phosphorus content

D. Types of pH control mechanisms
   1. Chemical—rapid-action buffers
      a. Bicarbonate buffer system
      b. Phosphate buffer system
      c. Protein buffer system
   2. Physiological—delayed-action buffers
      a. Respiratory response
      b. Renal response
   3. Summary of pH control mechanisms
      a. Buffers
      b. Respiration
      c. Kidney excretion of acids and bases
E. Effectiveness of pH control mechanisms; range of pH—extremely effective, normally maintain pH within a very narrow range of 7.36 to 7.40

BUFFER MECHANISMS FOR CONTROLLING pH OF BODY FLUIDS

A. Buffers defined
   1. Substances that prevent a marked change in pH of a solution when an acid or base is added to it
   2. Consist of a weak acid (or its acid salt) and a basic salt of that acid
B. Buffer pairs present in body fluids—mainly carbonic acid, proteins, hemoglobin, acid phosphate, and sodium and potassium salts of these weak acids

C. Action of buffers to prevent marked changes in pH of body fluids
 1. Nonvolatile acids, such as hydrochloric acid, lactic acid, and ketone bodies, buffered mainly by sodium bicarbonate
 2. Volatile acids, chiefly carbonic acid, buffered mainly by potassium salts of hemoglobin and oxyhemoglobin (Figure 30-5)
 3. The chloride shift makes it possible for carbonic acid to be buffered in the red blood cell and then carried as bicarbonate in the plasma (Figure 30-6)
 4. Bases buffered mainly by carbonic acid (when homeostasis of pH at 7.4 exists)

$$\text{Ratio } \frac{B \cdot HCO_3}{H_2CO_3} = \frac{20}{1}$$

D. Evaluation of the role of buffers in pH control—cannot maintain normal pH without adequate functioning of the respiratory and urinary pH control mechanisms

RESPIRATORY MECHANISM OF pH CONTROL

A. Explanation of mechanism
 1. Amount of blood carbon dioxide directly relates to the amount of carbonic acid and therefore to the concentration of H^+
 2. With increased respirations, less carbon dioxide remains in blood, hence less carbonic acid and fewer H^+ ions; with decreased respirations, more carbon dioxide remains in blood, hence more carbonic acid and more H^+ ions
B. Adjustment of respirations to pH of arterial blood
C. Some principles relating respirations and pH of body fluids
 1. Acidosis ⟶ hyperventilation
   ↓
   increases elimination of CO_2
   ↓
   decreases blood CO_2
   ↓
   decreases blood H_2CO_3
   ↓
   decreases blood H^+, i.e., increases blood pH
   ↓
   tends to correct acidosis, i.e.,
   to restore normal pH
 2. Prolonged hyperventilation, by decreasing blood H^+ excessively, may produce alkalosis
 3. Alkalosis causes hypoventilation, which tends to correct alkalosis by increasing blood CO_2 and therefore blood H_2CO_3 and H^+
 4. Prolonged hypoventilation, by eliminating too little CO_2, causes an increase in blood H_2CO_3 and consequently in blood H^+, thereby may produce acidosis

URINARY MECHANISMS OF pH CONTROL

A. General principles about mechanism—plays vital role in acid-base balance because kidneys can eliminate more H^+ from the body while reabsorbing more base when pH tends toward the acid side and eliminates fewer H^+ while reabsorbing less base when pH tends toward the alkaline side
B. Mechanisms that control urine pH
 1. Secretion of H^+ into urine—when blood CO_2, H_2CO_3, and H^+ increase above normal, distal tubules secrete more H^+ into urine to displace basic ion (mainly sodium) from a urine salt and then reabsorb sodium into blood in exchange for the H^+ excreted
 2. Secretion of NH_3—when blood hydrogen ion concentration increases, distal tubules secrete more NH_3, which combines with the H^+ of urine to form ammonium ion, which displaces a basic ion (mainly sodium) from a salt; the basic ion is then reabsorbed back into blood in exchange for the ammonium ion excreted

REVIEW QUESTIONS

1. How are carbonic and lactic acid produced?
2. Are fruits and vegetables acid-forming or base-forming foods?
3. Identify several acid-forming elements.
4. What is a physiological buffer?
5. What is the normal range of blood pH?
6. Describe the buffering action of sodium bicarbonate.
7. Identify the main buffer pairs in body fluids.
8. How is the distal renal tubule involved in the acidification of urine and the conservation of base?
9. How can sodium lactate be useful in the treatment of both metabolic and respiratory acidosis?
10. Discuss hyperventilation in relation to the development of an acid-base imbalance.

CRITICAL THINKING QUESTIONS

1. How would you explain pH in terms of the ions involved? What would be the hydrogen ion concentration of a solution with a pH of 4? With a pH of 6?
2. Can you define and summarize the purpose of the chloride shift?
3. What do you predict might happen to the pH if a drug is administered that lowers the $NaHCO_3$ concentration to 23.8 mEq but maintains the H_2CO_3 concentration at 1.3 mEq?
4. Can you explain the role of the respiratory system in maintaining proper blood pH?
5. How would you describe the causes and a possible treatment of the acid-base imbalance that occurs with prolonged vomiting?

Career Choices | Certified Registered Nurse Anesthetist

I practice anesthesiology as a Certified Registered Nurse Anesthetist (CRNA) at Beaumont Hospital in Royal Oak, Michigan. I also have an independent practice in office-based settings for ear, nose, and throat doctors and plastic surgeons. Having researched what a CRNA's job entails, I felt drawn to the career because of the level of independence possible and the wide variety of patients and surgical cases I would encounter.

A trend of major importance in my field is the decentralization of surgical care of low-acuity patients to outpatient/same-day surgical centers. Estimates put the total number of surgical cases performed on an outpatient basis in the United States at 60% to 64%. CRNAs, by virtue of their training and education, are able to serve this population well. An understanding of new medications and their pharmacodynamics and pharmacokinetics is just one part of the requisite body of knowledge that CRNAs use to ensure safe, effective anesthesia care for all of their patients.

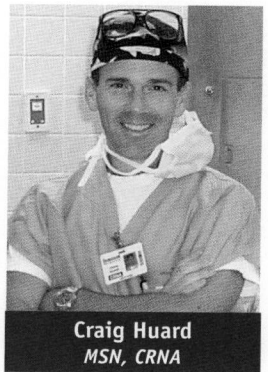

Craig Huard
MSN, CRNA

The most rewarding aspect of my career is the knowledge that I make a real difference in patients' lives. Often a person facing surgery isn't doing so by choice; there is a very high level of stress there. Earning that person's trust in a very short amount of time, seeing them safely through their anesthesia and surgery, and speaking to them afterward brings a sense of completion to my day.

A thorough understanding of A&P is essential to know how and where medications work. It is also essential to know where to place the medications in order to achieve the desired results. I recall studying A&P long into the night, at times wondering if the systems' interplay would ever make sense to me. At the same time, I recall marveling at how it did make sense when the "light" finally came on. I used mnemonics frequently to learn associations, orders, processes, and structures during my study. This is still a useful technique for me.

Reproduction and Development

The chapters of Unit 6 deal with human reproduction, growth, development, genetics, and heredity. The anatomical structures and complex control mechanisms characteristic of the male and the female reproductive systems are intended to ensure survival of the human species. These systems in men and women are adapted structurally and functionally for the specific sequence of events that permit development of sperm or ova, followed by fertilization, normal development, and birth of a baby. Chapter 33 details the developmental changes that occur from fertilization to death. Chapter 34 discusses the scientific study of genetics and heredity.

SEEING THE BIG PICTURE

Male Reproductive System

CHAPTER OUTLINE

LANGUAGE OF SCIENCE

accessory organs (ak-SES-oh-ree OR-gans)
[*accessus* extra, *-organum* instrument]

anal triangle (AY-nal TRY-ang-gul) [*anal* pertaining to the anus, *triangulus* three cornered]

androgens (AN-droh-jens) [*andro-* man, *-gen* that which generates]

androgen-binding protein (ABP) (AN-droh-jen BYND-ing PRO-teen) [*andro-* man, *-gen* that which generates, *proteios* first rank]

blood-testis barrier (blud TES-teez BAYR-ee-er)

body

bulbourethral (BUL-boh-yoo-REE-thral) [*bulbus-* swollen root, *-urethral* pertaining to the urethra]

capacitation (kah-pas-i-TAY-shun)

corpora cavernosa (KOHR-pohr-ah kav-er-NO-sah) [*corpor* body, *caverna* hollow space]

corpus spongiosum (KOHR-pus spun-jee-OH-sum) [*corpus* body, *spongio* like a sponge or related to a sponge]

ejaculation (ee-jak-yoo-LAY-shun) [*ejaculari* to hurl out]

emission (ee-MISH-un) [*emittere* to send forth]

epididymides

epididymis (ep-i-DID-i-miss) [*epi-* on or upon, *-didymos* pair]

erection (ee-REK-shun) [*erigere* to erect]

essential organs (ee-SHEN-shal OR-gans) [*essential* quality, *organum* instrument]

gametes (GAM-eets) [*gamete* marriage partner]

genital ducts (JEN-i-tall) [*genit-* birth or reproduction, *-al* pertaining to]

genitalia (jen-i-TAIL-yah) [*genitalis* belonging to birth]

glans penis (glans PEE-nis) [*glans* acorn, *penis* male sex organ]

gonads (GO-nads) [*gone* semen or seed]

head

interstitial cells (in-ter-STISH-al sells) [*interstitium* space between, *cella* storeroom]

Leydig cells (LYE-dig sells) [*Franz von Leydig* German anatomist]

Cont'd on p. 1124

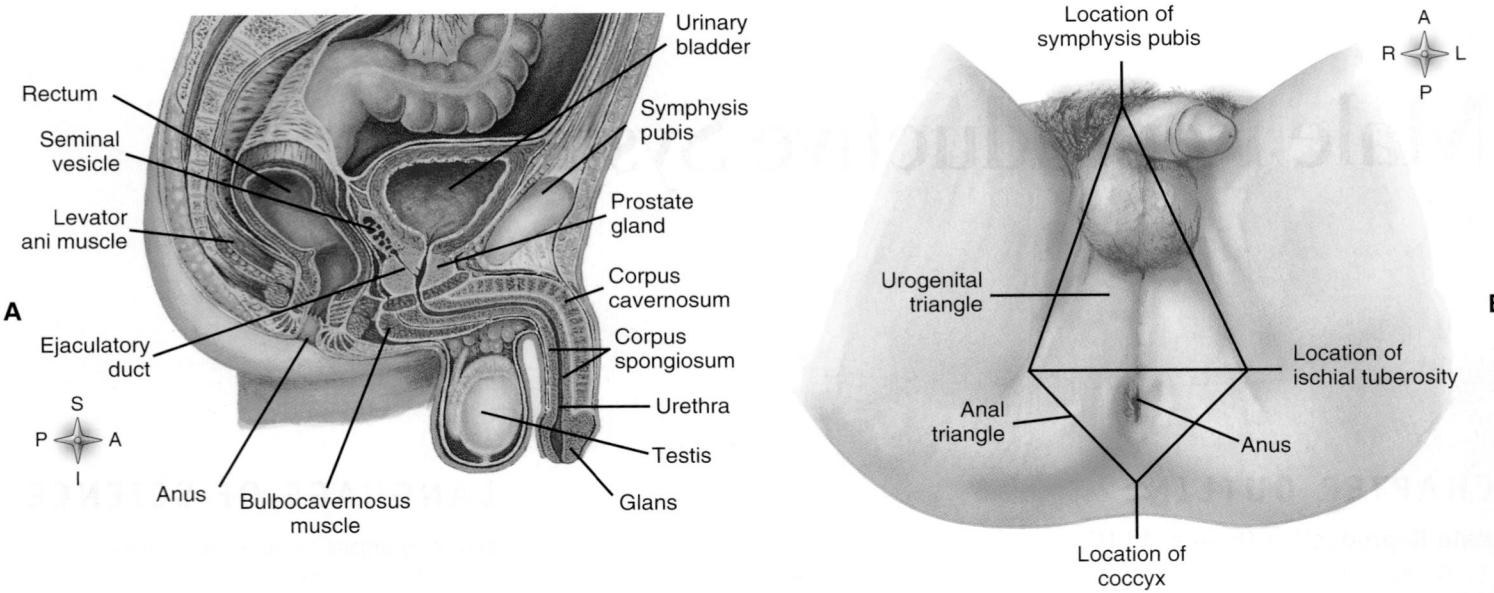

Figure 31-1 *Male reproductive organs.* **A,** Sagittal section of pelvis showing placement of male reproductive organs. **B,** Male perineum showing outline of the urogenital and anal triangles.

The importance of normal reproductive system function is notably different from the end result of "normal function" as measured in any other organ system of the body. The proper functioning of the reproductive system and of its enormously complex control mechanisms ensures survival not of the individual but of the species. In both sexes, organs of the reproductive system are adapted for the specific sequence of functions that are concerned primarily with propagation of the species. Production of hormones that permits development of the secondary sex characteristics occurs as a result of normal reproductive system activity. In humans, sexual maturity and the ability to reproduce occur at puberty. The male reproductive system consists of organs whose functions are to produce, transfer, and ultimately introduce mature sperm into the female reproductive tract, where fertilization can occur. It is the testes that secrete **androgens,** or male sex hormones, notably **testosterone.**

MALE REPRODUCTIVE ORGANS

Organs of the reproductive system (Figure 31-1) may be classified as **essential organs** for the production of **gametes** (sex cells) or as **accessory organs** that play some type of supportive role in the reproductive process.

In both sexes the essential organs of reproduction that produce the gametes, or sex cells (sperm or ova), are called **gonads.** The gonads of the male are the **testes.**

The accessory organs of reproduction in the male include genital ducts, glands, and supporting structures.

Genital ducts convey sperm to the outside of the body. The ducts are a pair of *epididymides* (singular: epididymis), the paired *vasa deferentia* (singular: vas deferens), a pair of *ejaculatory ducts*, and the *urethra*. **Accessory glands** in the reproductive system produce secretions that serve to nourish, transport, and mature sperm. The glands are a pair of *seminal vesicles*, one *prostate*, and a pair of *bulbourethral* (Cowper's) *glands*. **Supporting structures** include the *scrotum*, the *penis*, and a pair of *spermatic cords*.

Perineum

The perineum in the male (Figure 31-1, *B*) is a roughly diamond-shaped area between the thighs. It extends from the symphysis pubis anteriorly to the coccyx posteriorly. Its most lateral boundary on either side is the ischial tuberosity. A line drawn between the two ischial tuberosities divides the area into a larger **urogenital triangle,** which contains the external genitalia (penis and scrotum), and the **anal triangle,** which surrounds the anus.

 QUICK CHECK

1. How is the normal function of the reproductive system different from the end result of "normal function" in other organ systems?
2. Identify the essential and accessory organs of reproduction.
3. Describe the location, shape, and subdivisions of the perineum.

TESTES
Structure and Location

The **testes** are small, ovoid glands that are somewhat flattened from side to side, measure about 4 or 5 cm in length, and weigh 10 to 15 g each. The left testis is generally located about 1 cm lower in the scrotal sac than the right. Both testes are suspended in the pouch by attachment to scrotal tissue and by the spermatic cords (Figure 31-2). Note in Figure 31-2, *B,* that testicular blood vessels, collectively called the *vas afferens,* reach the testes by passing through the spermatic cord. A dense, white, fibrous capsule called the **tunica albuginea** encases each testis and then enters the gland, sending out partitions (septa) that radiate through its interior, dividing it into 200 or more cone-shaped *lobules.*

Each lobule of the testis contains specialized **interstitial cells** (of Leydig) and one to three tiny, coiled **seminiferous**

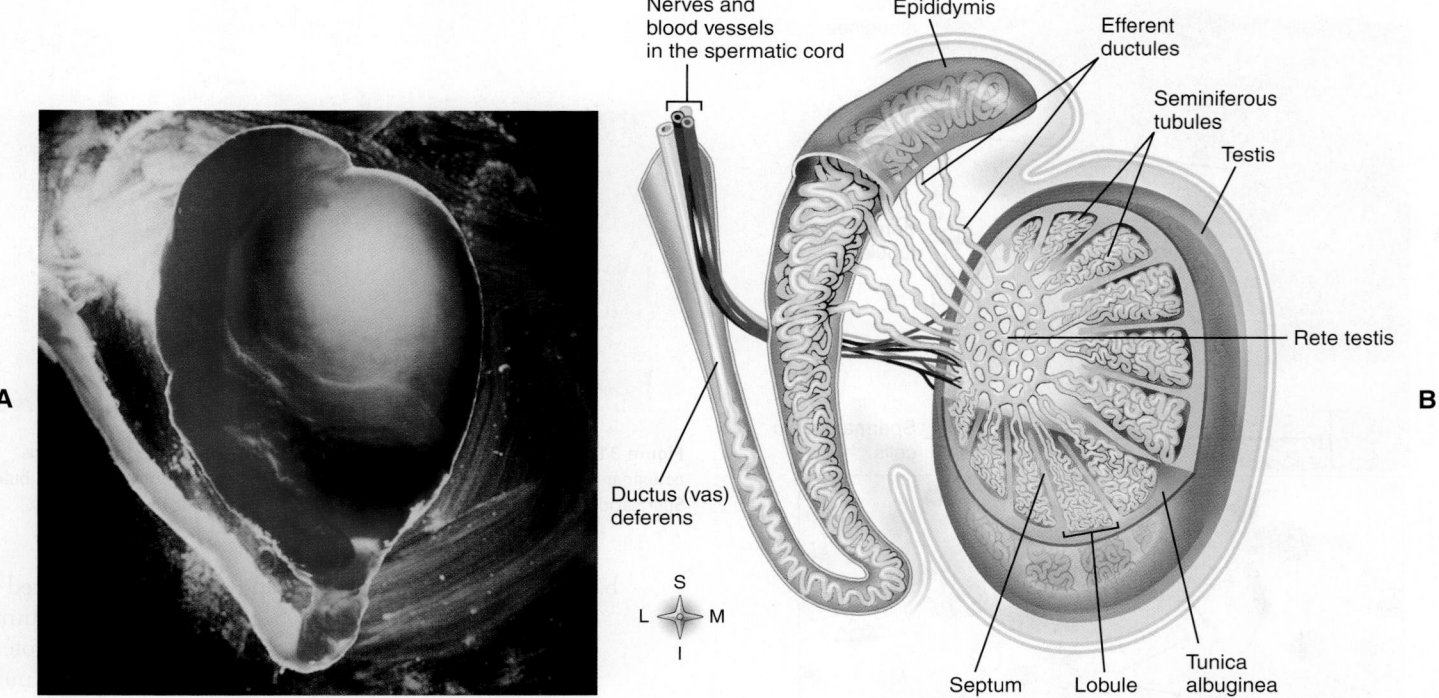

Figure 31-2 *Tubules of the testis and epididymis.* **A,** Transilluminated photograph; the testis is the darker sphere in the center. **B,** Illustration showing epididymis lifted free of testis. The ducts and tubules are exaggerated in size.

tubules, which, if unraveled, would measure about 75 cm (more than 2 feet) in length. The tubules from each lobule come together to form a plexus called the *rete testis.* A series of sperm ducts called *efferent ductules* then drain the rete testis and pierce the tunica albuginea to enter the head of the epididymis (see Figure 31-2).

Microscopic Anatomy of the Testis

Figure 31-3, *A* is a low-power (×70) micrograph of testicular tissue showing a number of cut seminiferous tubules and numerous **interstitial** or **Leydig cells** in the surrounding connective tissue septa. In this figure, maturing sperm appear as dense nuclei with their tails projecting into the lumen of the tubule. The wall of each seminiferous tubule may contain five or more layers of cells. At puberty, when sexual maturity begins, spermatogenic cells in diverse stages of development appear, and the hormone-producing interstitial cells become more prominent in the surrounding septa. Figure 31-3, *B* is a high-power micrograph showing a group of typically round interstitial, or Leydig, cells clustered between seminiferous tubules. A unique cellular feature often visible in Leydig cells is an elongated rectangular-shaped mass called a *Reinke's crystalloid.* Reinke's crystalloids are absent before puberty and then increase in number during the reproductive years into old age. The functional significance of these structures remains uncertain.

Irregular elongated **sustentacular** or **Sertoli cells** have both a supportive and a hormone secretory function important for the developing germ cells. They provide mechanical support

and protection for spermatids attached to their luminal surface and are visible in Figure 31-4 within a section of seminiferous epithelium. They also secrete a hormone, *inhibin,* that plays a role in the rate of maturation and eventual release of mature spermatozoa into the lumen of the seminiferous tubule, thus influencing sperm counts. At sexual maturity, Sertoli cells begin to secrete a specialized protein called **androgen-binding protein** or **ABP** that binds to testosterone and increases its concentration within the seminiferous tubules. Since high concentrations of testosterone are required for normal germ cell maturation, Sertoli cells play an important role in spermatogenesis.

Sertoli cells are columnar in shape and extend from the basement membrane to the luminal surface of the seminiferous tubule (Figures 31-4 and 31-5). Unique *tight junctions* exist between adjacent Sertoli cells and divide the wall of the tubule into two compartments that house either meiotically active cells near the luminal surface or spermatogonia near the basement membrane.

Tight junctions between Sertoli cells form the **blood-testis barrier.** This structure isolates the developing sperm cells, which have active surface antigens different from somatic body cells, from the body immune system. If these antigens were to escape from the tubule epithelium and enter the bloodstream by breaking through the basement membrane, an autoimmune reaction could occur. Note that two meiotic divisions result in a reduction of chromosomes from 46 in the spermatogonia to 23 in the spermatids and mature sperm (see Figure 31-5).

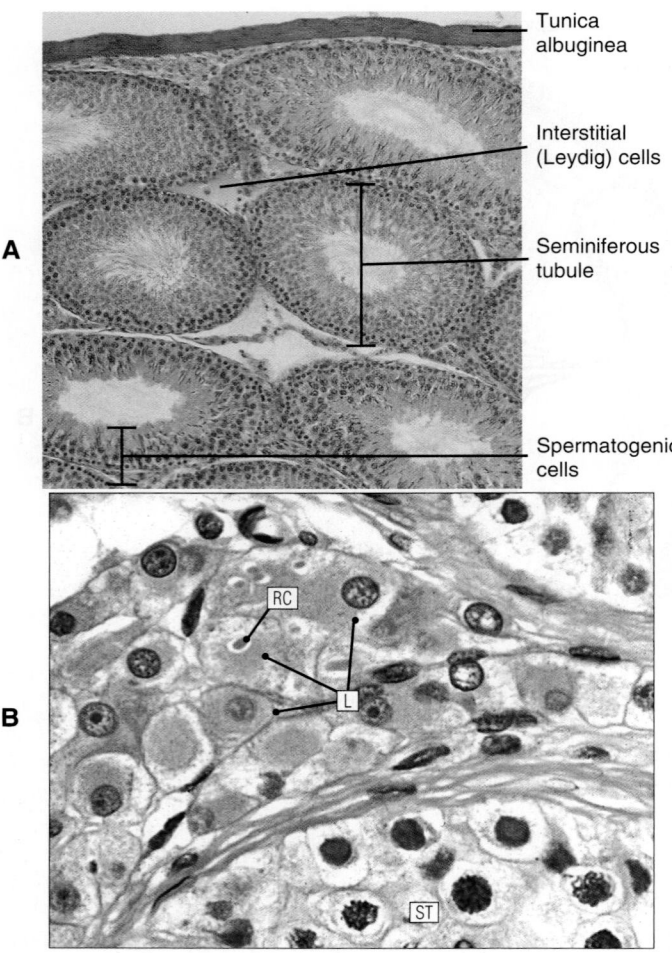

Tunica albuginea

Interstitial (Leydig) cells

Seminiferous tubule

Spermatogenic cells

Figure 31-3 *Testis.* **A,** Low-power view showing several seminiferous tubules surrounded by septa containing interstitial (Leydig) cells. **B,** High-power micrograph showing a cluster of Leydig cells *(L)* between seminiferous tubules *(ST).* Spermatogenic cells can be identified in the wall of a seminiferous tubule. Note the presence of a Reinke's crystalloid *(RC)* in a Leydig cell.

Testes Functions

The testes perform two primary functions: spermatogenesis and secretion of hormones.

1. **Spermatogenesis,** the production of spermatozoa (sperm), the male gametes, or reproductive cells. The seminiferous tubules produce the sperm. Details of spermatogenesis are discussed in Chapter 33. Figure 31-5 shows a section of a seminiferous tubule illustrating the process of meiosis and sperm cell formation.
2. Secretion of hormones—**testosterone** (androgen or masculinizing hormone) by interstitial or Leydig cells and inhibin, by sustentacular or Sertoli cells.
3. Testosterone serves the following general functions:
   a. It promotes "maleness," or development and maintenance of male secondary sex characteristics, accessory organs such as the prostate, seminal vesicles, and adult male sexual behavior.

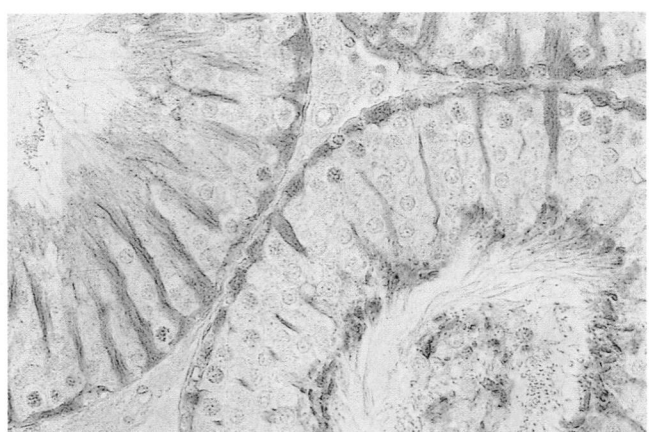

Figure 31-4 *Sertoli cells within seminiferous epithelium.* Sertoli cells are columnar in shape, extending from basement membrane to lumen of the tubule. Spermatids can be seen attached to the luminal surface of the Sertoli cells.

   b. It helps regulate metabolism and is sometimes referred to as "the anabolic hormone" because of its marked stimulating effect on protein anabolism. By stimulating protein anabolism, testosterone promotes growth of skeletal muscles (responsible for greater male muscular development and strength) and growth of bone. Testosterone also promotes closure of the epiphyses (see Chapter 7, p. 231). Early sexual maturation leads to early epiphyseal closure. The converse also holds true: late sexual maturation, delayed epiphyseal closure, and tallness tend to go together.

   c. It plays a part in fluid and electrolyte metabolism. Testosterone has a mild stimulating effect on kidney tubule reabsorption of sodium and water; it also promotes kidney tubule excretion of potassium.

   d. Increased levels decrease hypothalamic secretion of gonadotropin-releasing hormone (GnRH) and lead to reduced secretion of LH and FSH.

The anterior pituitary gland controls the testes by means of its gonadotropic hormones—specifically, FSH and LH. FSH stimulates the seminiferous tubules to produce sperm more rapidly. In the male, LH stimulates interstitial cells to increase their secretion of testosterone.

If the blood concentration of testosterone reaches a high level it will inhibit hypothalamic secretion of gonadotropin-releasing hormone (GnRH). As a result, anterior pituitary secretion of LH will decrease and testosterone levels will return to the normal set point value (Figure 31-6). Increasing blood levels of inhibin will selectively decrease FSH secretion by the anterior pituitary and decrease the rate of sperm production. However, if sperm counts decrease below the normal set point, inhibin secretion will drop, FSH secretion will increase, and sperm numbers will increase to normal levels. Thus a negative feedback mechanism operates between the hypothalamus, the anterior pituitary gland, and the hormone producing cells of the testes—interstitial or Leydig cells producing testosterone and sustentacular or Sertoli cells producing inhibin. The end result is homeostatic control of the full

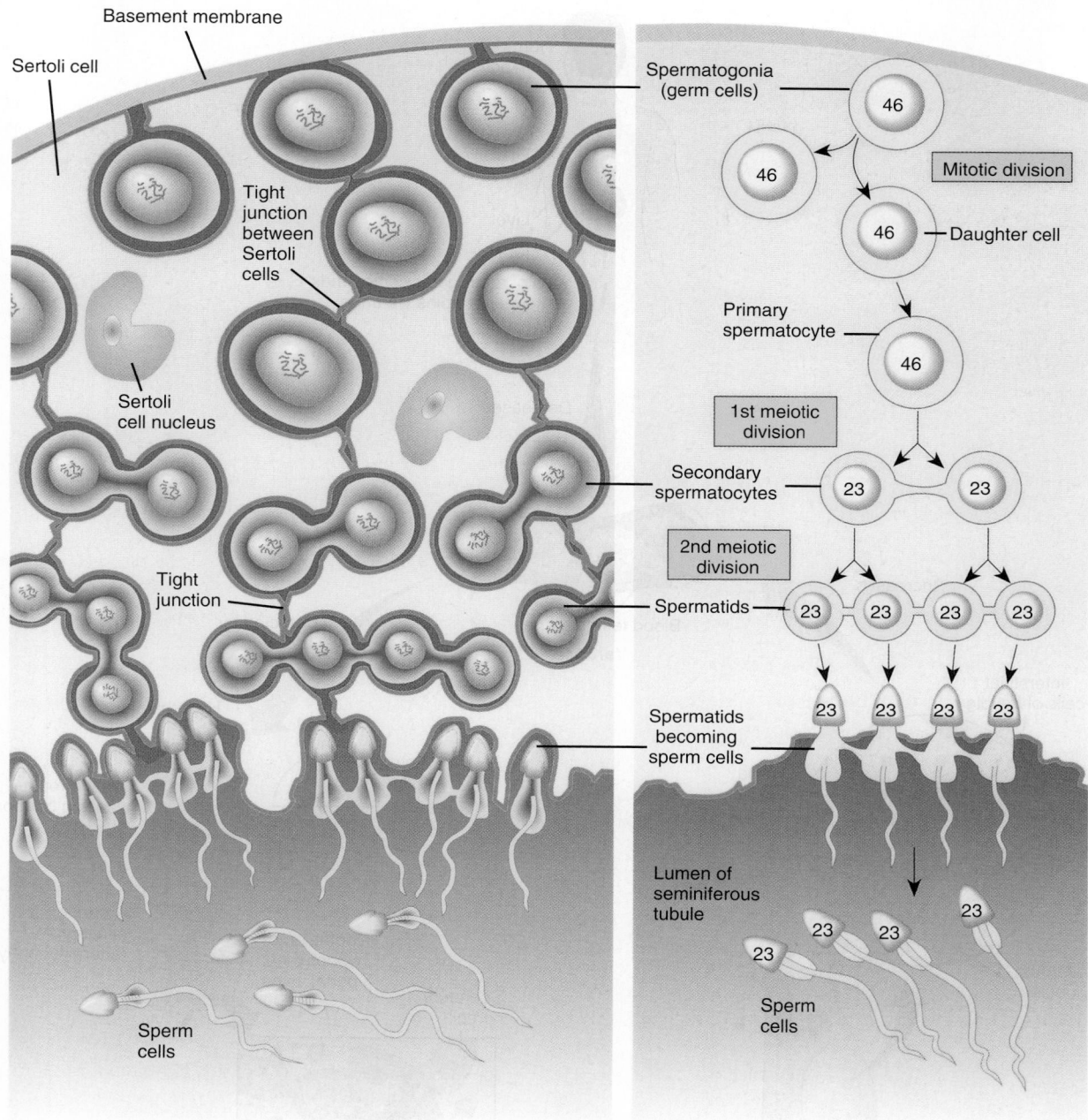

Basement membrane

Sertoli cell

Tight junction between Sertoli cells

Sertoli cell nucleus

Tight junction

Sperm cells

Spermatogonia (germ cells)

Mitotic division

Daughter cell

Primary spermatocyte

1st meiotic division

Secondary spermatocytes

2nd meiotic division

Spermatids

Spermatids becoming sperm cells

Lumen of seminiferous tubule

Sperm cells

Figure 31-5 *Seminiferous tubule.* Section shows process of meiosis and sperm cell formation.

range of effects influenced by testosterone levels and a direct influence on sperm numbers.

Structure of Spermatozoa

The elongated tail-bearing **spermatozoa** seen in the seminiferous tubules (Figure 31-7, *B*) appear fully formed. We know, however, that they undergo a process of "ripening," or maturation, as they pass through the genital ducts before ejaculation. Although anatomically complete and highly motile when ejaculated, sperm must still undergo a complex process called **capacitation** before

they are actually capable of fertilizing an ovum (female gamete). Normally, capacitation occurs in sperm only after they have been introduced into the vagina of the female.

Figure 31-7, *C*, shows the characteristic parts of a spermatozoon: head, middle piece, and elongated, lashlike tail. The head of a spermatozoon is, in essence, a highly compact package of genetic chromatin material, about 5 μm long, covered by a specialized *acrosome* and *acrosomal (head) cap*. The acrosome contains hydrolytic (splitting) enzymes, which are released during capacitation. During the process of capacitation, these specialized acro-

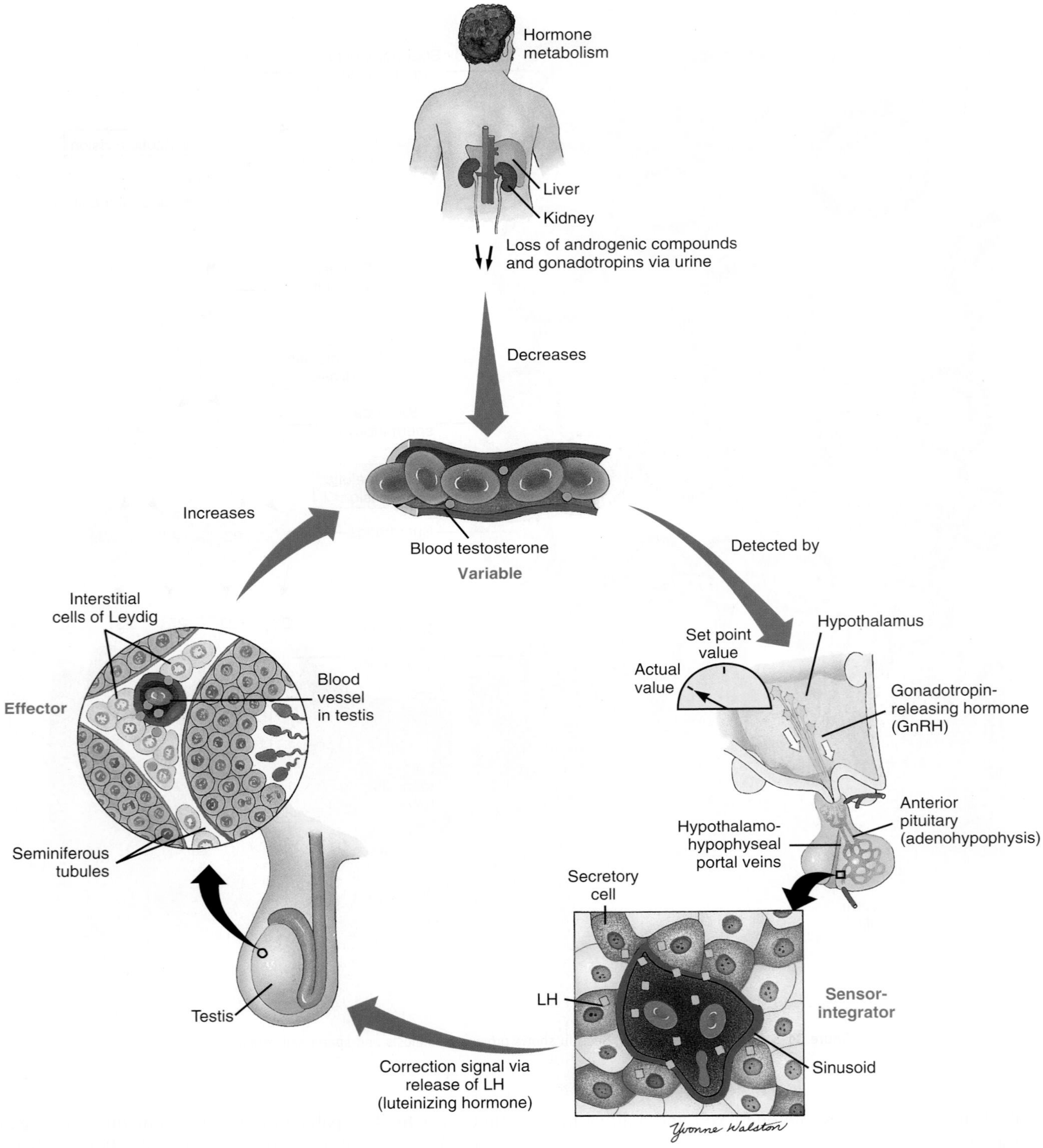

Figure 31-6 *Negative feedback loop controlling testosterone secretion.* Diagram shows the negative feedback mechanism that controls anterior pituitary gland secretion of LH and interstitial cell secretion of testosterone. A similar negative feedback loop exists betweeen inhibin secreting sustentacular (Sertoli) cells in the testis and FSH secreting cells in the anterior pituitary gland. See text for discussion.

BOX 31-1: HEALTH MATTERS
Testicular Self-Examination

Early identification of testicular tumors, the most common neoplasms in young men between 20 and 35 years of age, greatly enhances the potential for successful treatment. If detected early, the cure rate for testicular cancer is approximately 90%.

Monthly **testicular self-examination** is recommended and involves palpating each testis—preferably after a warm shower or bath when the scrotum is relaxed and the testes are descended and accessible. The thumb is placed on the top and the index and middle fingers on the underside of the scrotum to gently roll and palpate each testis. The testes should feel firm, smooth, and rubbery but not hard. The examination may result in some tenderness but should not be painful. The figure shows a large tumor in the left testis. In this case, delay in seeking medical attention was apparently based on the belief that because the growth in testicular size was relatively pain free, it was not serious. Any lump, change in size, or change in texture of a testis discovered during self-examination, regardless of whether painful or not, should be reported to a physician for further assessment.

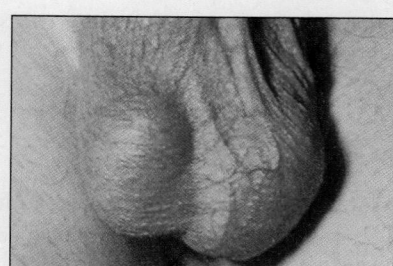

Testicular tumor.

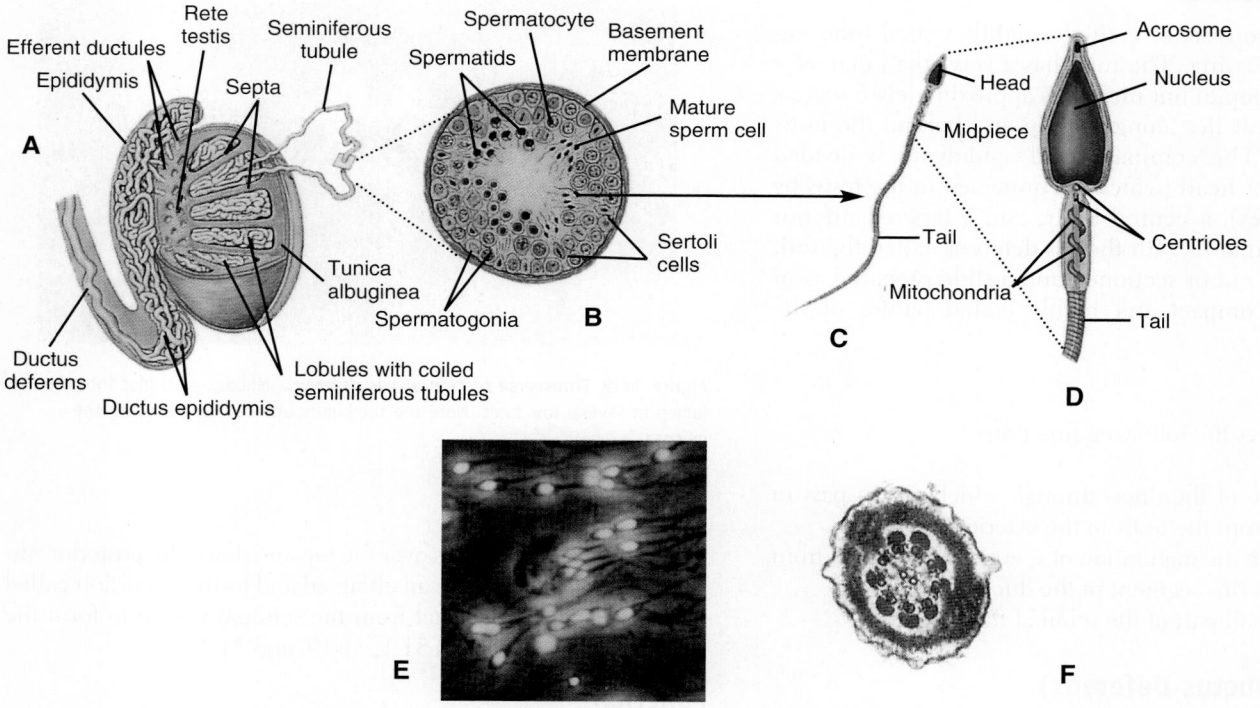

Figure 31-7 *Development and structure of sperm.* A, View of testis and seminiferous tubules. **B,** Spermatid cells. **C,** Mature sperm. **D,** Enlarged view of head and mid piece. **E,** Micrograph of sperm with nuclear material glowing with a fluorescent dye. **F,** Electron micrograph (EM) of a cross section of sperm tail showing nine double microtubules arranged around two single central microtubules.

somal enzymes first break down cervical mucus, allowing sperm to pass into the uterus and uterine tubes. If an ovum is present in the female reproductive tract when **semen** is introduced, continued release of capacitation enzymes assists the sperm cells to digest and penetrate the outer covering of the egg and initiate fertilization. This is the primary reason a high sperm count is essential for male fertility.

The cylindrical middle piece, about 7 μm long, is characterized by a helical arrangement of mitochondria arranged end to end around a central core. It is this mitochondrial sheath that provides energy for sperm locomotion. The tail is divided into a principal piece, about 40 μm long, and a short end piece, 5 to 10 μm in length. If the tail portion of a spermatozoon is cut in cross section and viewed with an electron microscope (Figure 31-7, *F*), its micro structure will appear very similar to other flagella capable of motility (see Chapter 3, p. 96). Note in Figure 31-7, *F* that the central portion of the sectioned sperm tail is a cylinder composed of nine double microtubules arranged around two single microtubules in the center.

QUICK CHECK

4. Describe the location, the size, and the shape of the testes.
5. List the two primary functions of the testes and identify the cell type or structure involved in each function.
6. List the general functions of testosterone.
7. Identify the structural components of a spermatozoon and give the function of each.

REPRODUCTIVE (GENITAL) DUCTS

Epididymis

Structure and Location

Each **epididymis** consists of a single, tightly coiled tube enclosed in a fibrous casing. The tube has a very small diameter (just barely macroscopic) but measures approximately 6 meters (20 feet) in length. It lies along the top and behind the testis (see Figure 31-2). The comma-shaped epididymis is divided into a blunt superior **head** (which is connected to the testis by the efferent ductules), a central **body,** and a tapered inferior portion that is continuous with the vas deferens called the **tail.** If the epididymis is cut or sectioned and a slide prepared as in Figure 31-8, the compact and highly coiled nature of the tubule is apparent.

Functions

The epididymis serves the following functions:

- It serves as one of the ducts through which sperm pass in their journey from the testis to the exterior.
- It contributes to the maturation of sperm, which spend from 1 to 3 weeks in this segment of the duct system.
- It secretes a small part of the seminal fluid (semen).

Vas Deferens (Ductus Deferens)

Structure and Location

The vas deferens, like the epididymis, is a tube. In fact, the duct of the vas is an extension of the tail of the epididymis. The vas deferens has thick, muscular walls (Figure 31-9) and can be palpated in the scrotal sac as a smooth, movable cord. Note in Figure 31-9 that the muscular layer of the vas has three layers: a thick intermediate circular layer of muscle fibers and inner and outer longitudinal layers. The muscular layers of the vas help in propelling sperm through the duct system. The vas deferens from each testis ascends from the scrotum and passes through the inguinal canal as part of the spermatic cord—enclosed by fibrous connective tissue with blood vessels, nerves, and lymphatics—into the abdomi-

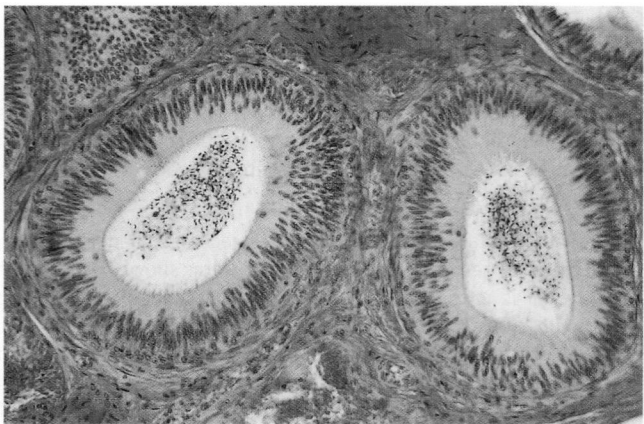

Figure 31-8 *Epididymis section.* Two cross sections of this extensively coiled tubule are visible. Note the presence of spermatozoa within the lumina of the tubules.

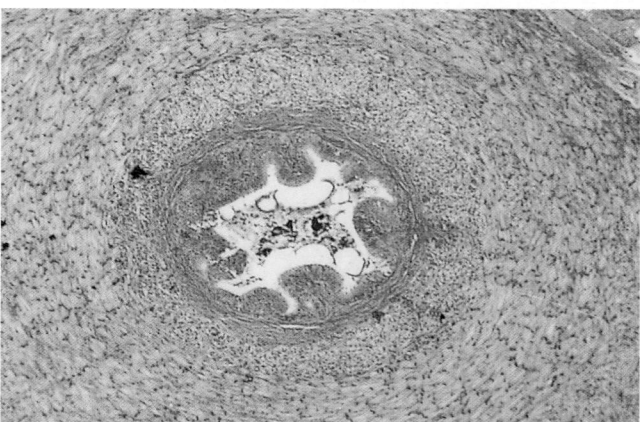

Figure 31-9 *Transverse section of vas deferens.* Mucosa protrudes into the lumen in several low folds. Note the thick muscular coat surrounding the mucosa.

nal cavity. Here it extends over the top and down the posterior surface of the bladder where an enlarged and tortuous portion called the *ampulla* joins the duct from the seminal vesicle to form the ejaculatory duct (Figures 31-1, 31-10, and 31-11).

Function

The vas deferens serves as one of the male genital ducts connecting the epididymis with the ejaculatory duct. Sperm remain in the vas deferens for varying periods of time depending on the degree of sexual activity and frequency of ejaculation. Storage time may exceed 1 month with no loss of fertility.

Ejaculatory Duct

The two ejaculatory ducts are short tubes about 1 cm long that pass through the prostate gland to terminate in the urethra. As Figures 31-10 and 31-11 show, they are formed by the union of the vas deferens distal to the ampulla with the ducts from the seminal vesicles.

BOX 31-2: FYI
Vasectomy

Severing or clamping off the vas deferens—that is, a **vasectomy**, usually done through an incision in the scrotum—makes a man sterile. Why? Because it interrupts the route to the exterior from the epididymis. To leave the body, sperm must journey in succession through the epididymis, vas deferens, ejaculatory duct, and urethra.

Urethra

The urethra in males serves a dual function, which involves both the reproductive system and the urinary system. Refer to Chapter 28, p. 1039, for a discussion of this duct.

ACCESSORY REPRODUCTIVE GLANDS

Seminal Vesicles

Structure and Location

The seminal vesicles are highly convoluted pouches that, when fully extended, are about 15 cm in length. Each is a tubular diverticulum of the vas deferens on one side and is coiled on itself so that it forms a body about 5 to 7 cm in length. The two vasa deferentia lie along the lower part of the posterior surface of the bladder, directly in front of the rectum (see Figures 31-1, 31-10, and 31-11). Figure 31-11 is an isolated cadaver prosection showing the relationships of a number of the male reproductive structures, including the seminal vesicles, when viewed from behind. The highly

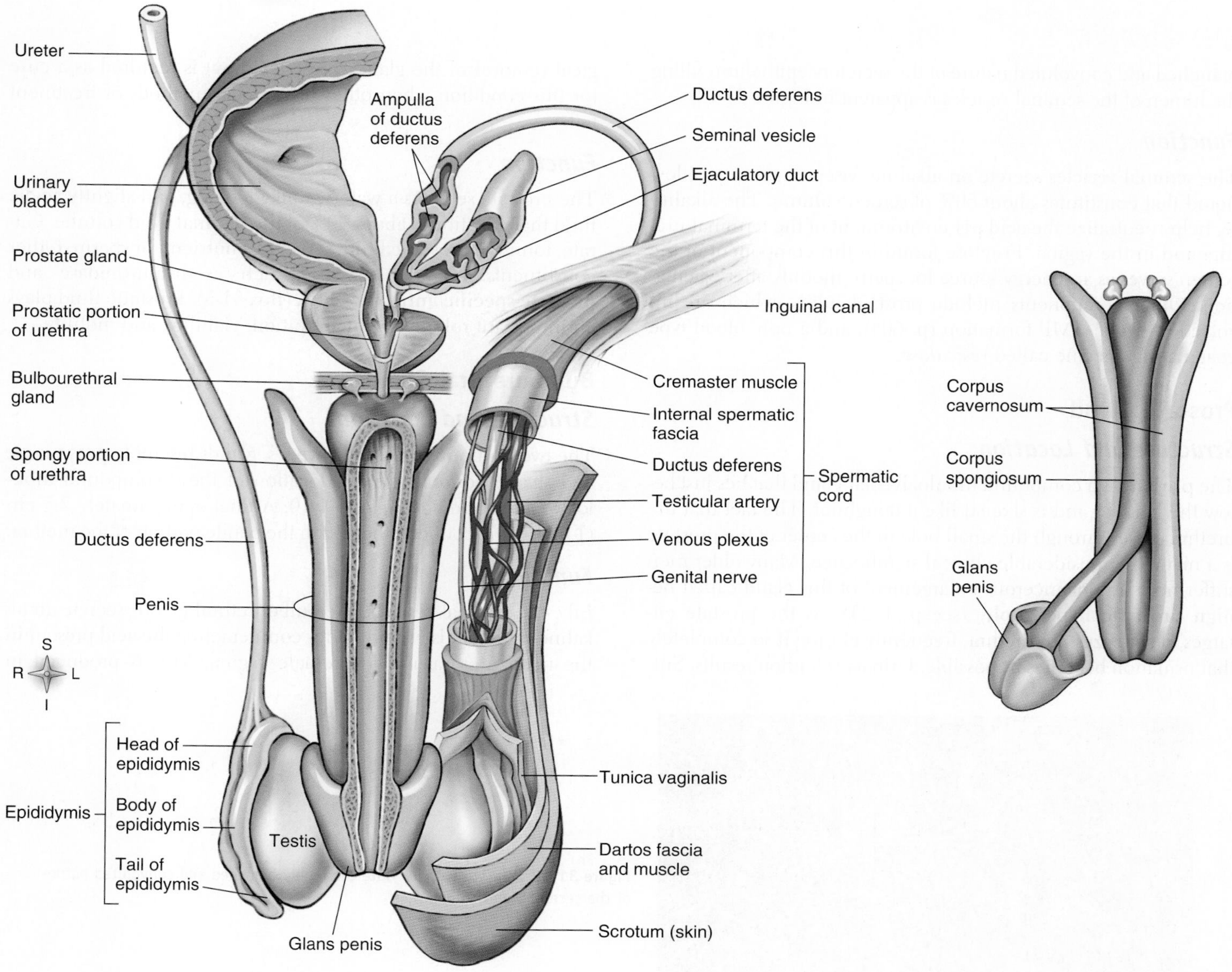

Figure 31-10 *The male reproductive system.* Illustration shows the testes, epididymis, ductus deferens, and glands of the male reproductive system in an isolation/dissection format.

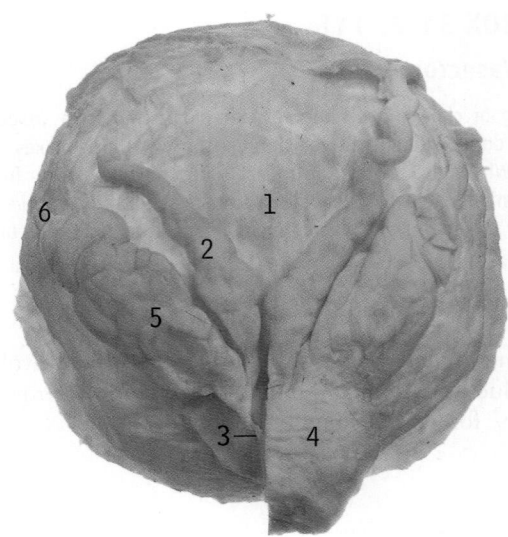

Figure 31-11 *Cadaver prosection.* Photo illustration showing a number of male reproductive structures viewed from behind. The prostate has been sectioned on the left side to reveal the ejaculatory duct. *1,* Base of bladder; *2,* vas deferens; *3,* left ejaculatory duct; *4,* posterior surface of prostate; *5,* seminal vesicle; *6,* ureter.

branched and convoluted nature of the secretory epithelium filling the lumen of the seminal vesicles is apparent in Figure 31-12.

Function

The seminal vesicles secrete an alkaline, viscous, creamy-yellow liquid that constitutes about 60% of semen volume. The alkalinity helps neutralize the acid pH environment of the terminal urethra and in the vagina. Fructose found in this component of the semen serves as an energy source for sperm motility after ejaculation. Other components include prostaglandins, which are involved in cyclic AMP formation (p. 604), and a non–blood type coagulating enzyme called *vesiculase.*

Prostate Gland

Structure and Location

The **prostate** is a compound tubuloalveolar gland that lies just below the bladder and is shaped like a doughnut. The fact that the urethra passes through the small hole in the center of the prostate is a matter of considerable clinical significance. Many older men suffer from a noncancerous enlargement of this gland called **benign prostatic hypertrophy** (see p. 1123). As the prostate enlarges, it squeezes the urethra, frequently closing it so completely that urination becomes impossible. Urinary retention results. Sur-

gical removal of the gland (prostatectomy) is required as a cure for this condition when other less radical methods of treatment fail.

Function

The prostate secretes a watery, milky-looking, and slightly acidic fluid that constitutes about 30% of the seminal fluid volume. Citrate, found in prostatic fluid, serves as a nutrient for sperm. Other constituents include enzymes such as hyaluronidase and **prostate-specific antigen** or **PSA** (Box 31-3). Prostatic fluid plays an important role in sperm activation, viability, and motility.

Bulbourethral Glands

Structure and Location

The two **bulbourethral,** or Cowper's, glands resemble peas in size and shape. You can see the location of these compound tubuloalveolar glands in Figure 31-10. A duct approximately 2.5 cm (1 inch) long connects them with the penile portion of the urethra.

Function

Like the seminal vesicles, the bulbourethral glands secrete an alkaline fluid that is important for counteracting the acid present in the male urethra and the female vagina. Mucus produced in

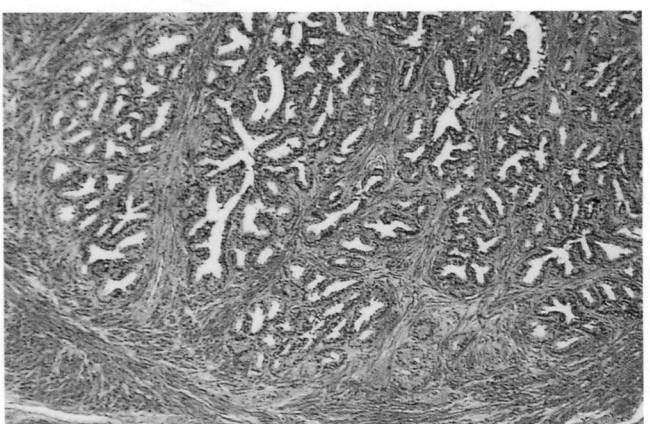

Figure 31-12 *Seminal vesicle.* Note the highly branched and convoluted nature of the secretory epithelium.

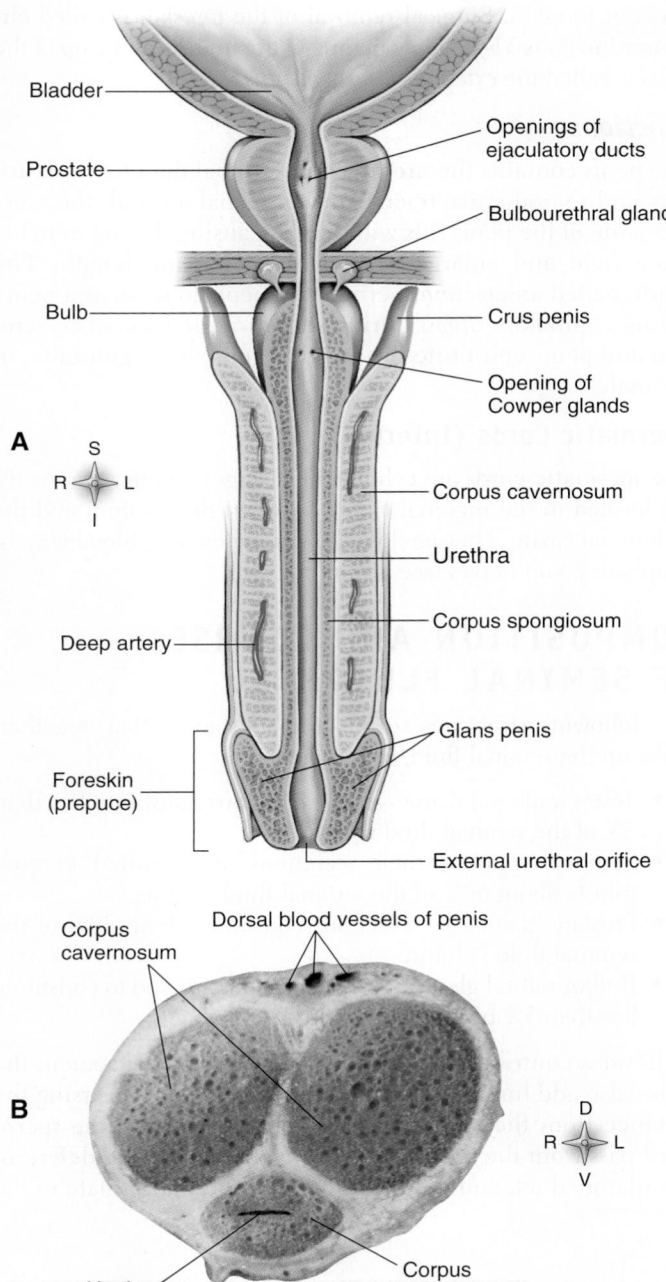

A

S
R ⟷ L
I

Bladder

Openings of
ejaculatory ducts

Prostate

Bulbourethral gland

Bulb

Crus penis

Opening of
Cowper glands

Corpus cavernosum

Urethra

Corpus spongiosum

Deep artery

Glans penis

Foreskin
(prepuce)

External urethral orifice

Corpus
cavernosum

Dorsal blood vessels of penis

B

D
R ⟷ L
V

Urethra

Corpus
spongiosum

Figure 31-13 *The penis.* **A,** In this sagittal section of the penis viewed from above, the urethra is exposed throughout its length and can be seen exiting from the bladder and passing through the prostate gland before entering the penis to end at the external urethral orifice. **B,** Photograph of a cross section of the shaft of the penis showing the three columns of erectile, or cavernous, tissue. Note the urethra within the substance of the corpus spongiosum.

BOX 31-3
Diagnostic Study

Detecting or Predicting the Risk of Developing Prostate Cancer

Many of the 32,000 men who die each year from prostate cancer—the most common nonskin type of cancer in American men—could be saved if the cancer was detected early enough for effective treatment. Several screening tests are available for the detection of prostate cancer once it develops. Cancerous growths in the gland can often be palpated through the wall of the rectum. Unfortunately, by the time prostate cancer can be palpated, it may have spread to other organs. Therefore many cancer experts believe prostate cancer screening, especially in older men, can be more effective if rectal examinations are performed in conjunction with a screening test called the **PSA test.** This test is a type of blood analysis that screens for **prostate-specific antigen,** or **PSA,** a substance often found to be elevated in the blood of men with prostate cancer. A nuclear medicine **bone scan** (see Chapter 7, p. 233) is often used either to exclude metastatic spread of prostate cancer or to locate areas of the body where secondary prostate cancer tumors have already developed. The figure is of a nuclear medicine whole-body bone scan showing numerous metastatic cancerous tumors *(arrows)* throughout the body of an individual judged to have "inoperable" prostate cancer.

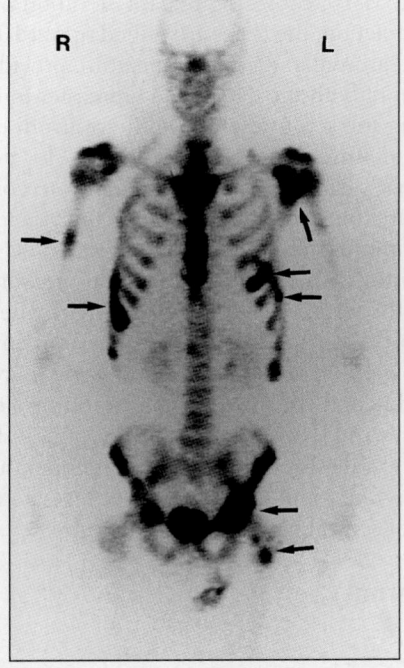

R   L

these glands serves to lubricate the urethra and helps protect sperm from friction damage during ejaculation.

QUICK CHECK

8. List, in sequence, the reproductive ducts sperm pass through from formation to ejaculation.
9. What is the structural relationship between the prostate gland and the urethra?
10. Compare the volume, viscosity, pH, and composition of the secretions produced by the accessory reproductive glands.

SUPPORTING STRUCTURES

Scrotum

The **scrotum** is a skin-covered pouch suspended from the perineal region. Internally, it is divided into two sacs by a septum, each sac containing a testis, epididymis, and lower part of a spermatic cord.

The *dartos* fascia and muscle are located just below the skin of the scrotum (see Figure 31-10). Contraction of the dartos muscle fibers causes slight elevation of the testes and wrinkling of the scrotal pouch. In addition, contraction of the *cremaster muscle*, also seen in Figure 31-10, causes significant elevation of the testes. As a result of contraction, the testes are pulled upward against the perineum. Sexual arousal and cold temperature provide the stimulus for contraction of both the dartos and cremaster muscles.

By elevating the scrotum closer to the warmth of the perineal wall, the temperature of the testes, which are located outside the abdominal cavity, can be maintained at a more constant level in a cold environment. According to one theory, the temperature required for optimum sperm formation is about 3° C below normal body temperature. This is the "functional" reason that justifies placement of the testes outside the body cavity where they are exposed and subject to traumatic injury. In a warm environment the scrotum is longer and its skin loose and wrinkle free, permitting the testes to descend. In the cold, the scrotum elevates and becomes heavily wrinkled, effectively pulling the testes upward toward the body wall. Factors other than temperature, including blood flow dynamics and tissue oxygen levels, are also suggested as "reasons" for scrotal placement of the testes.

Penis (External)

Structure

Three cylindrical masses of erectile, or cavernous, tissue, enclosed in separate fibrous coverings and held together by a covering of skin, compose the **penis** (Figure 31-13). The two larger and uppermost of these cylinders are named the **corpora cavernosa,** whereas the smaller, lower one, which contains the urethra, is called **corpus spongiosum.**

The distal part of the corpus spongiosum overlaps the terminal end of the two corpora cavernosa to form a slightly bulging structure, the **glans penis,** over which the skin is folded doubly to form a more or less loose-fitting, retractable casing known as the **pre-**

puce, or foreskin. Surgical removal of the foreskin is called **circumcision** (Box 31-4). The opening of the urethra at the tip of the glans is called the *external urinary meatus.*

Functions

The penis contains the urethra, the terminal duct for both urinary and reproductive tracts. During sexual arousal, the erectile tissue of the penis fills with blood, causing the organ to become rigid and enlarge in both diameter and length. The result, called an *erection*, permits the penis to serve as a penetrating copulatory organ during sexual intercourse. The scrotum and penis constitute the external genitals, or **genitalia,** of the male.

Spermatic Cords (Internal)

The **spermatic cords** are cylindrical casings of white, fibrous tissue located in the inguinal canals between the scrotum and the abdominal cavity. They enclose the vasa deferentia, blood vessels, lymphatics, and nerves (see Figure 31-10).

COMPOSITION AND COURSE OF SEMINAL FLUID

The following structures secrete the substances that, together, make up the seminal fluid, or **semen:**

- Testes and epididymis—their secretions constitute less than 5% of the seminal fluid volume.
- Seminal vesicles—their secretions are reported to contribute about 60% of the seminal fluid volume.
- Prostate gland—its secretions constitute about 30% of the seminal fluid volume.
- Bulbourethral glands—their secretions are said to constitute less than 5% of the seminal fluid volume.

Besides contributing slightly to the fluid part of semen, the testes also add hundreds of millions of sperm. In traversing the distance from their place of origin to the exterior, the sperm must pass from the testis through the epididymis, vas deferens, ejaculatory duct, and urethra. Note that sperm originate in the

---

BOX 31-4   Circumcision

In the past, medical benefits of routine infant circumcision were thought to include a major reduction in risk for penile cancer and fewer urinary tract infections throughout life. There is little credible research evidence to substantiate such claims. As a result, health care professionals now seriously question whether the procedure has any medical benefits that might outweigh the side effects, which include the potential for urethral injury, infection, hemorrhage, and pain. Elective circumcision performed on newborns is no longer a recommended medical procedure. The decision to circumcise a child is now left to the discretion of the parents and is performed most frequently for religious or cultural reasons. Ritual circumcision is common in approximately one sixth of the global population.

BOX 31-5: HEALTH MATTERS
Hypospadias and Epispadias

The term **hypospadias** describes a congenital condition that is characterized by the opening of the urethral meatus on the underside or ventral surface of the glans or penile shaft (see figure). Rarely the hypospadial opening may appear in the perineal area or be located off center anywhere along the underside of the penile shaft. Surgical correction is performed if the defect is likely to cause urological or reproductive problems. The term **epispadias** refers to a much less common congenital defect that involves the opening of the urethral meatus on the dorsal surface of the glans or penile shaft.

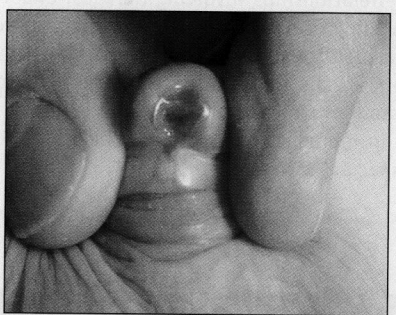

Hypospadias.

BOX 31-6: FYI
Neural Control of the Male Sex Act

Recall that all body functions but one have for their ultimate goal survival of the individual. Only the function of reproduction serves a different, a longer-range, and, no doubt in nature's scheme, a more important purpose—survival of the human species. Male functions in reproduction consist of the production of male sex cells (spermatogenesis, discussed in Chapter 4) and introduction of these cells into the female body (coitus, copulation, or sexual intercourse). For coitus to take place, erection of the penis must first occur, and for sperm to enter the female body, both the sex cells and secretions from the accessory glands must be introduced into the urethra (emission) and semen ejaculated from the penis.

Erection is a parasympathetic reflex initiated mainly by certain tactile, visual, and mental stimuli. It consists of dilation of the arteries and arterioles of the penis, which in turn floods and distends spaces in its erectile tissue and compresses its veins. Therefore more blood enters the penis through the dilated arteries than leaves it through the constricted veins. Hence it becomes larger and rigid, or, in other words, erection occurs.

Emission is the reflex movement of sex cells, or spermatozoa, and secretions from the genital ducts and accessory glands into the prostatic urethra. Once emission has occurred, ejaculation will follow.

Ejaculation of semen is also a reflex response. It is the usual outcome of the same stimuli that initiate erection. Ejaculation and various other responses—notably accelerated heart rate, increased blood pressure, hyperventilation, dilated skin blood vessels, and intense sexual excitement—characterize the male **orgasm,** or sexual climax.

---

testes, glands located outside the body (that is, not within a body cavity), travel inside, and finally are expelled outside.

MALE FERTILITY

Male fertility relates to many factors—most of all to the number of sperm ejaculated but also to their size, shape, and motility. Fertile sperm have a uniform size and shape. They are highly motile. Although only one sperm fertilizes an ovum, millions of sperm seem to be necessary for fertilization to occur. According to one estimate, when the sperm count falls below about 25 million/ml of semen, functional sterility results.

One hypothesis suggested to explain this puzzling fact is this: semen that contains an adequate number of sperm also contains enough hyaluronidase and other hydrolytic enzymes to liquefy the intercellular substance between the cells that encase each ovum. Without this, a single sperm cannot penetrate the layers of cells around the ovum and hence cannot fertilize it. Infertility may also be caused by production of antibodies some men make against their own sperm. This type of sperm destruction or inactivation, called "immune infertility," is caused by an antigen-antibody reaction. A sperm surface protein called fertilization antigen (FA-1) triggers antibody production, which results in infertility.

QUICK CHECK

11. What two structures constitute the external genitalia of the male?
12. What is the function of the dartos and cremaster muscles? How does this function influence fertility?
13. Identify by name the three cylindrical masses of erectile tissue in the penis.
14. What factors influence male fertility?

Cycle of Life
Male Reproductive System

The reproductive system is unlike all other systems of the male body with regard to normal changes that occur throughout the life span. All other systems perform their functions from the time they develop in utero until advanced old age, when degeneration may cause loss of function and, perhaps, death. The male reproductive system, however, does not begin to perform its functions until puberty—usually during the early teenage years. Of course, the biological advantage of this "late start" is that a person does not have the biological, psychological, or social maturity to become a parent before that time.

Initial development of the male reproductive organs begins before birth, when the reproductive tract differentiates into the male form rather than the female form. A couple of months before birth, the immature testes descend behind the parietal peritoneum into the scrotum, guided by the fibrous gubernaculum.

It is not uncommon for them to be late in completing the trip, perhaps not arriving in the scrotum until several weeks after birth. The testes and other reproductive organs remain in an immature form—and thus remain incapable of providing reproductive function—until puberty, when high levels of reproductive hormones stimulate the final stages of their development.

From puberty until advanced old age the male reproductive system continues to operate efficiently enough to permit successful reproduction. A gradual decline in hormone production during late adulthood may decrease sexual desire and fertility to some degree, but a man can usually father a child until the time of death.

THE BIG PICTURE
Male Reproductive System

Propagation of the species is truly a "big picture" outcome related to proper functioning of the reproductive system in both sexes. In the case of at least some of the reproductive structures in males, sexual expression and reproductive system activity and outcomes are also coupled directly with certain urinary system structures and functions.

However, the anatomical structures and functional control mechanisms that result in the production, transfer, and introduction of viable gametes (sperm) into the female reproductive tract are the primary reproductive functions in the male. They depend on complex homeostatic interrelationships involving nervous, endocrine, muscular, reproductive, and circulatory system structures, as well as numerous physiological control mechanisms, including those that affect metabolic, acid-base, and temperature regulation. The chemical, organelle, cell, tissue, organ, and organ system levels of organization are all involved in reproductive functions that affect both the individual and the species.

Mechanisms of Disease

DISORDERS OF THE MALE REPRODUCTIVE SYSTEM

Several disorders of the male reproductive system cause **infertility.** Infertility is an abnormally low ability to reproduce. If there is a complete inability to reproduce, the condition is called **sterility.** Infertility or sterility involves an abnormally reduced capacity to deliver healthy sperm to the female reproductive tract. Reduced reproductive capacity may result from factors such as a decrease in the testes' production of sperm, structural abnormalities in the sperm, or obstruction of the reproductive ducts.

Disorders of the Testes

Disruption of the sperm-producing function of the seminiferous tubules can result in decreased sperm production, a condition called **oligospermia** (ol-i-go-SPER-mee-ah). If the *sperm count* is too low, infertility may result. A large number of sperm is needed to ensure that many sperm will reach the ovum and dissolve its coating—allowing a single sperm to unite with the ovum. Oligospermia can result from factors such as infection, fever, radiation, malnutrition, and high temperature in the testes. In some cases, oligospermia is temporary—as in some acute infections. Oligospermia is a leading cause of infertility. Of course, total absence of sperm production results in sterility.

Early in fetal life the testes are located in the abdominal cavity near the kidneys but normally descend into the scrotum about 2 months before birth. Occasionally a baby is born with undescended testes, a condition called **cryptorchidism** (krip-TOR-ki-dizm), which is readily observed by palpation of the scrotum at delivery (Figure 31-14). The word *cryptorchidism* is from the Greek words *kryptikos* ("hidden") and *orchis* ("testis"). Failure of the testes to descend may be caused by hormonal imbalances in the developing fetus or by a physical deficiency or obstruction. Regardless of cause, in the cryptorchid infant the testes remain "hidden" in the abdominal cavity. Because the higher temperature inside the body cavity inhibits spermatogenesis, measures must be taken to bring the testes down into the scrotum to prevent permanent sterility. Early treatment of this condition by surgery or by injection of testosterone, which stimulates the testes to descend, may result in normal testicular and sexual development.

Most testicular cancers arise from the sperm-producing cells of the seminiferous tubules. Malignancies of the testes are

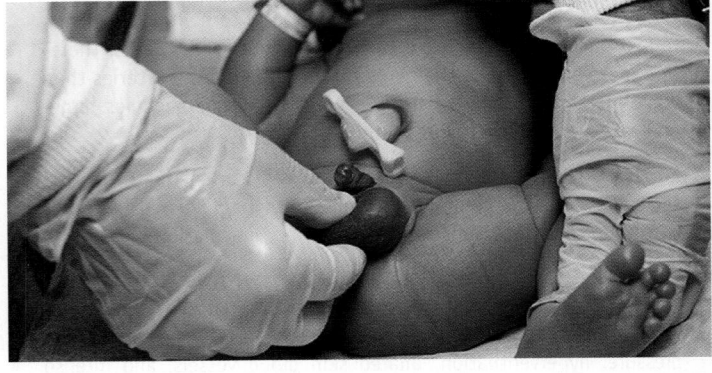

Figure 31-14 *Screening for cryptorchidism in a newborn infant.* Properly descended testes can be palpated easily in the scrotal sac at birth.

Mechanisms of Disease—cont.

most common among men 20 to 35 years old. Besides age, this type of cancer is associated with genetic predisposition, trauma or infection of the testis, and cryptorchidism. Treatment of testicular cancer is most effective when the diagnosis is made early in the development of the tumor. Many physicians encourage male patients to perform regular self-examination of their testes, especially if they are in a high-risk group (see Box 31-1, p. 1115).

Disorders of the Prostate

A noncancerous condition called **benign prostatic hypertrophy** or **BPH** occurs in 75% of men older than 50 years. As the name suggests, the condition is characterized by an enlargement, or hypertrophy, of the prostate gland. As discussed earlier in the chapter, the fact that the urethra passes through the center of the prostate after exiting from the bladder is a matter of considerable clinical significance. As the prostate enlarges, it squeezes and may distort the normal passage of the urethra, frequently closing it so completely that urination becomes very difficult or even impossible. Often the first area of the prostate to enlarge in BPH involves the so-called *periurethral glandular tissue* surrounding the urethra. Tissue near the periphery of the prostate, called *peripheral zone glandular tissue*, may remain normal even though the gland as a whole increases in size. Figure 31-15 shows the appearance of the prostate gland in BPH. Note how the urethra is compressed and its passage through the prostate distorted by pressure from swollen periurethral glandular tissue. Surgical removal of some of the swollen tissue surrounding the urethra in a **transurethral resection** or **TUR** will often be effective in reducing symptoms and improving urine flow rates. In this procedure, a specialized tubelike instrument called a **cystoscope**—which contains operative devices, a lighting system, and viewing lenses, is inserted through the penis and into the prostatic urethra to surgically resect prostatic tissue surrounding the lumen. Alternative procedures using laser therapy, focused ultrasound, or the insertion of urethral stents may also be employed. In severe cases total removal of the gland, a procedure called **prostatectomy** (pros-ta-TEK-toh-mee), may become necessary.

Disorders of the Penis and Scrotum

The penis is subject to numerous sexually transmitted infections, as well as structural abnormalities. One such structural abnormality is **phimosis** (fi-MOH-sis), a condition in which the foreskin fits so tightly over the glans that it cannot retract. The usual treatment for this condition is circumcision—a procedure in which the foreskin is cut along the base of the glans and removed. Severe phimosis can obstruct the flow of urine, possibly causing the death of an infant born with this condition. Milder phimosis can result in accumulation of dirt and organic matter under the foreskin, possibly causing severe infections.

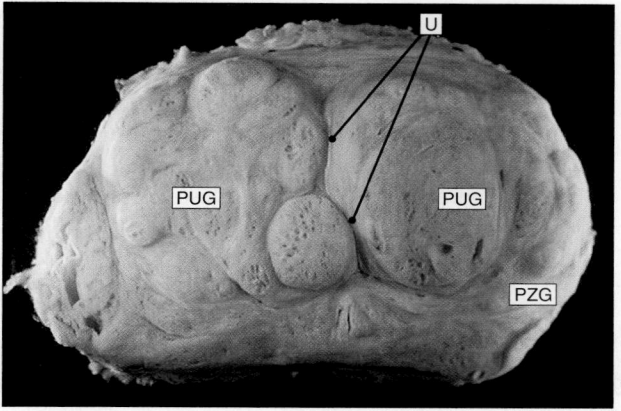

Figure 31-15 *Benign prostatic hypertrophy.* Note (1) hypertrophy of periurethral glandular tissue *(PUG)*, (2) compression and distortion of the urethra *(U)*, (3) normal-appearing peripheral zone glandular tissue *(PZG)*. See text for discussion.

Failure to achieve an erection of the penis adequate enough to permit sexual intercourse is called **impotence** (IM-poh-tense), or **erectile dysfunction (ED)**. ED affects men of all ages but is experienced most often after age 65 years. ED does not affect sperm production but infertility often results because normal intercourse may not be possible. In the past, psychological problems such as anxiety, depression, and stress were often cited as the most important causes of impotence in sexually active men. There is no doubt that such conditions contribute to ED. However, current research suggests that purely psychological problems probably account for far fewer cases of impotence than previously thought. We now know that ED is frequently caused by medical problems related to abnormal vascular or neural control of penile blood flow. Arteriosclerosis, diabetes, alcohol abuse, numerous medications, radiation therapy, tumors, spinal cord trauma, and surgery, especially if pelvic organs such as the prostate are involved, may all cause ED. Treatment options include use of a vacuum pump to pull blood into the penis, surgical implantation of various types of prostheses, or use of drugs that increase blood flow to the spongy cavernous tissue of the penis, causing it to stiffen and become erect. Oral medications are generally preferred by men who do not have medical conditions that preclude their use. As a result of multiple options, even moderate to severe erectile dysfunction occurring in sexually active men can be treated with considerable success.

Swelling of the scrotum can be caused by various conditions. One of the most common causes of scrotal swelling is an accumulation of fluid called **hydrocele** (HYE-dro-seel). Hydroceles may be congenital, resulting from structural abnormalities present at birth. In adults, hydrocele often occurs when fluid produced by the serous membrane lining the scrotum is not absorbed properly. The cause of adult hydrocele is

Mechanisms of Disease—cont.

not always known but, in some cases, it can be linked to trauma or infection.

Swelling of the scrotum may also occur when the intestines push through the weak area of the abdominal wall that separates the abdominopelvic cavity from the scrotum. This condition is a form of **inguinal** (IN-gwi-nal) **hernia** (see discussion of inguinal hernia, p. 420). If the intestines protrude into the

scrotum, the digestive tract may become obstructed—resulting in death. Inguinal hernia often occurs while lifting heavy objects because of the high internal pressure generated by the contraction of abdominal muscles. Inguinal hernia may also be congenital. Small inguinal hernias may be treated with external supports that prevent organs from protruding into the scrotum; more serious hernias must be repaired surgically.

LANGUAGE OF SCIENCE   *(Cont'd from page 1109)*

orgasm (OR-gaz-um) [*orgein* to be lustful]

penis (PEE-nis) [*penis* male sex organ]

prepuce (PREE-pus) [*praeputium* foreskin]

prostate (PROSS-tayt) [*prostates* standing before]

scrotum (SKROH-tum) [*scrotum* bag]

semen (SEE-men) [*semen* seed]

seminal vesicle (SEM-i-nal VES-i-kul) [*semen-* seed, *-al* pertaining to, *vesica* bladder]

seminiferous tubules (seh-mih-NIF-er-us TOOB-yools) [*semen-* seed, *-ferre* to bear]

Sertoli cells (ser-TOH-lee sells) [*Enrico Sertoli* Italian histologist]

spermatic cords (sper-MAT-ik kords) [*sperma-* seed, *-ic* pertaining to, *chorde-* string]

spermatogenesis (sper-mah-toh-JEN-eh-sis) [*sperma-* seed, *-genesis* origin]

spermatozoa (sper-mah-tah-ZOH-ah) [*sperma-* seed, *-zoon* animal]

sustentacular (sus-ten-TAK-yoo-lar) [*sustentare* to support]

tail

testes (TES-teez) [*testiculus* plural of testis]

testosterone (tes-TOS-the-rohn)

tunica albuginea (TOO-nih-kah al-byoo-JIN-ee-ah) [*tunica* tunic, *albus* white]

urogenital triangle (yoor-oh-JEN-i-tal TRY-ang-gul) [*uro-* urine, *-genit-* birth or reproduction, *-al* pertaining to, *triangulus* three cornered]

LANGUAGE OF MEDICINE

benign prostatic hypertrophy (BPH) (be-NYNE pro-STAT-ik hye-PER-troh-fee) [*benignus* kind, *prostatic* pertaining to the prostate, *hyper-* excessive or above, *-trophe* nourishment]

bone scan

circumcision (ser-kum-SIH-zhun) [*circum-* around, *-cadere* to cut]

cryptorchidism (krip-TOR-ki-dizm) [*kryptikos-* hidden, *-orchis-* testis, *-ism* condition of]

cystoscope (SIS-toh-skohp) [*cysto-* bladder, *-scope* an instrument for observation]

epispadias (ep-is-PAY-dee-us) [*epi-* on or above, *-spadon* a rent]

erectile dysfunction (ED) (ee-REK-tyle dis-FUNGK-shun) [*erigere* to erect, *dys-* bad or painful, *-functio* performance]

hydrocele (HYE-dro-seel) [*hydro-* water, *-kele* hernia]

hypospadias (hye-poh-SPAY-dee-us) [*hypo-* under or below, *-spadon* a rent]

impotence (IM-poh-tense) [*im-* not, *-potentia* power]

infertility (in-fer-TIL-i-tee) [*in-* not, *-fertilis* fruitful]

inguinal hernia (IN-gwi-nal HER-nee-ah) [*inguen* groin, *hernia* rupture]

oligospermia (ol-i-go-SPER-mee-ah) [*oligo-* few or little, *-sperma* seed]

phimosis (fi-MOH-sis) [*phimosis* muzzle]

prostatectomy (pros-ta-TEK-toh-mee) [*prostates-* standing before, *-ectomy* surgical removal]

prostate-specific antigen (PSA) (PROSS-tayt speh-CIF-ik AN-tih-jen) [*prostates* standing before, *specific*, *anti-* against, *-gen* that which generates]

PSA test

sterility (steh-RIL-i-tee) [*sterilis* barren]

testicular self-examination

transurethral resection (TUR) (tranz-yoor-REE-thral rih-SEK-shun) [*trans-* across or through, *-ourethra-* urethra, *-al* pertaining to, *re-* again, *-secare* to cut]

vasectomy (va-SEK-toh-mee) [*vas-* vessel, *-ectomy* surgical removal]

CASE STUDY

Jeremy Liebermann, a 10-day-old infant, is brought to the health clinic for a regular follow-up visit. Jeremy was a full-term infant, born by normal vaginal delivery without complications. His hospital stay was uneventful, and he was discharged home with his mother 24 hours after birth. Jeremy is being breast-fed. Jeremy and his parents are Jewish and plan on having Jeremy circumcised according to their religious beliefs.

1. On physical examination of the scrotum, no testis is palpable on the left side. This condition is called:

   A. Hypospadias
   B. Hydrocele
   C. Cryptorchidism
   D. Phimosis

2. Jeremy undergoes surgical correction for his undescended testis and progresses well with no major medical problems into adolescence. He is now 14 years of age and visits the clinic for a physical examination to play football. While reviewing health promotion habits with Jeremy, the nurse practitioner discovers Jeremy does not practice testicular self-examination. When teaching Jeremy how to perform this procedure, which one of the following should be included?

   A. The examination should be done at least twice each month.
   B. The examination should be done after a warm shower or bath when the testes are lower within the scrotum.
   C. The examination is done by placing the first three fingers of one hand under the scrotum while the first two fingers of the other hand are used to gently feel around each testis.
   D. On examination the testes should feel smooth, warm, and hard.

3. Jeremy is married at the age of 24 years and comes to the clinic 2 years later because he and his wife have not been able to conceive a child. He is worried that he might be infertile. Laboratory results confirm that Jeremy has a low sperm count. Jeremy asks "Why is this result important in fertility?" Which of the following would be your *best* response?

   A. A large number of sperm are needed to ensure that many sperm will reach the ovum and dissolve its coating so fertilization can occur.
   B. Oligospermia is only a temporary condition, usually caused by infection, and once identified and treated, normal fertilization will occur.
   C. Low sperm counts will lead to total absence of sperm production and sterility.
   D. Low sperm counts are the result of incomplete "ripening" or maturation of the sperm, and injections of testosterone can correct this condition and enhance fertilization efforts.

4. Jeremy is now 65 years of age and a proud new grandfather. He visits the health clinic at this time with a chief complaint of "blood in his semen." His physician suspects a prostate problem. Which of the following findings would be *most* likely to suggest a diagnosis of prostate cancer?

   1. Decreased prostate-specific antigen
   2. Increased prostate-specific antigen
   3. Prostate on digital rectal examination is large
   4. Prostate on digital rectal examination has a hard nodule on the outer right side surface area
   5. Prostate on digital rectal examination is smooth and rubbery
      A. 1 and 3
      B. 2 and 4
      C. 1 and 4
      D. 2 and 5

CHAPTER SUMMARY

INTRODUCTION

A. Proper functioning of the reproductive system ensures the survival of the species
B. Male reproductive system consists of organs whose functions are to produce, transfer, and introduce mature sperm into the female reproductive tract where fertilization can occur

MALE REPRODUCTIVE ORGANS

A. Classified as essential organs for production of gametes or accessory organs that support the reproductive process (Figure 31-1, *A*)
   1. Essential organs—gonads of the male; testes
   2. Accessory organs of reproduction
      a. Genital ducts convey sperm to outside of body; pair of epididymides, paired vasa deferentia, pair of ejaculatory ducts, and the urethra

      b. Accessory glands produce secretions that nourish, transport, and mature sperm; pair of seminal vesicles, the prostate, and pair of bulbourethral glands
      c. Supporting structures—scrotum, penis, and pair of spermatic cords
B. Perineum—in males, roughly diamond-shaped area between thighs; extends anteriorly from symphysis pubis to coccyx posteriorly; lateral boundary is the ischial tuberosity on either side; divided into the urogenital triangle and the anal triangle (Figure 31-1, *B*)

TESTES

A. Structure and location
   1. Several lobules composed of seminiferous tubules and interstitial cells (of Leydig), separated by septa, encased in fibrous capsule called the tunica albuginea (Figure 31-2)

2. Seminiferous tubules in testis open into a plexus called *rete testis*, which is drained by a series of efferent ductules that emerge from the top of the organ and enter the head of epididymis

3. Located in scrotum, one testis in each of two scrotal compartments

B. Microscopic anatomy

C. Functions
1. Spermatogenesis—formation of mature male gametes (spermatozoa) by seminiferous tubules
2. Secretion of hormone (testosterone) by interstitial cells

D. Structure of spermatozoa (Figure 31-7)—consists of a head (covered by acrosome), neck, mid piece, and tail; tail is divided into a principal piece and a short end piece

REPRODUCTIVE (GENITAL) DUCTS

A. Epididymis
1. Structure and location
   a. Single tightly coiled tube enclosed in fibrous casing (Figure 31-8)
   b. Lies along top and side of each testis
   c. Anatomical divisions include head, body, and tail
2. Functions
   a. Duct for seminal fluid
   b. Also secretes part of seminal fluid
   c. Sperm become capable of motility while they are passing through the epididymis

B. Vas deferens (ductus deferens)
1. Structure and location
   a. Tube, extension of epididymis
   b. Extends through inguinal canal, into abdominal cavity, over top and down posterior surface of bladder
   c. Enlarged terminal portion called ampulla—joins duct of seminal vesicle
2. Function
   a. One of excretory ducts for seminal fluid
   b. Connects epididymis with ejaculatory duct

C. Ejaculatory duct
1. Formed by union of vas deferens with duct from seminal vesicle
2. Passes through prostate gland, terminating in urethra

D. Urethra (see p. 1117)

ACCESSORY REPRODUCTIVE GLANDS

A. Seminal vesicles
1. Structure and location—convoluted pouches about 5 to 7 cm long on posterior surface of bladder
2. Function—secrete the viscous, nutrient-rich part of seminal fluid (60% of semen volume)

B. Prostate gland
1. Structure and location
   a. Doughnut shaped
   b. Encircles urethra just below bladder
2. Function—adds slightly acidic, watery, milky-looking secretion to seminal fluid (30% of semen volume)

C. Bulbourethral glands
1. Structure and location
   a. Small, pea-shaped structures with about 2.5 cm (1 inch) long ducts leading into urethra
   b. Lie below prostate gland
2. Function—secrete alkaline fluid that is part of semen (5% of semen volume)

SUPPORTING STRUCTURES

A. Scrotum
1. Skin-covered pouch suspended from perineal region
2. Divided into two compartments
3. Contains testis, epididymis, and lower part of a spermatic cord
4. Dartos and cremaster muscles elevate the scrotal pouch

B. Penis (Figure 31-13)
1. Structure—composed of three cylindrical masses of erectile tissue, one of which contains urethra
2. Functions—penis contains the urethra, the terminal duct for both urinary and reproductive tracts; during sexual arousal, penis becomes erect, serving as a penetrating copulatory organ during sexual intercourse

C. Spermatic cords (internal)
1. Fibrous cylinders located in inguinal canals
2. Enclose seminal ducts, blood vessels, lymphatics, and nerves

COMPOSITION AND COURSE OF SEMINAL FLUID

A. Consists of secretions from testes, epididymides, seminal vesicles, prostate, and bulbourethral glands
B. Each milliliter contains millions of sperm
C. Passes from testes through epididymis, vas deferens, ejaculatory duct, and urethra

MALE FERTILITY

A. Relates to many factors—number of sperm; size, shape, and motility
B. Infertility may be caused by antibodies some men make against their own sperm

CYCLE OF LIFE: MALE REPRODUCTIVE SYSTEM

A. Reproductive functions begin at time of puberty
B. Development of organs begins before birth; immature testes descend into scrotum before or shortly after birth
C. Puberty—high levels of hormones stimulate final stages of development
D. System operates to permit reproduction until advanced old age
E. Late adulthood—gradual decline in hormone production may decrease sexual appetite and fertility

REVIEW QUESTIONS

1. Name the accessory glands of the male reproductive system.
2. List the genital ducts in the male.
3. List the supporting structures of the male reproductive system.

4. What is the tunica albuginea? How does it aid in dividing the testis into lobules?
5. What are the two primary functions of the testes?
6. What are the general functions of testosterone?
7. Discuss the structure of a mature spermatozoon.
8. What is meant by the term *capacitation?*
9. List the three functions of the epididymis.
10. List the anatomical divisions of the epididymis.
11. Discuss the formation of the ejaculatory ducts.
12. Discuss the type of secretion typical of the prostate gland and seminal vesicles.
13. What and where are the bulbourethral (Cowper's) glands?
14. Describe the structure, location, and function or functions of the scrotum.
15. Name the three cylindrical masses of erectile, or cavernous, tissue in the penis.
16. What and where is the glans penis? The prepuce, or foreskin?
17. What is the spermatic cord? From what does it extend, and what does it contain?
18. Identify and define the male functions in reproduction.

CRITICAL THINKING QUESTIONS

1. How does the function of reproduction differ from all other body functions?
2. Can you identify the functions of the male reproductive system?
3. What is the relationship between the rete testis, seminiferous tubules, and efferent ductules?
4. How is the prostate gland related to the urethra? What problems can result from this relationship?
5. Can you list the structures in the reproductive system that contribute to the formation of seminal fluid?
6. Trace the course of seminal fluid from its formation to ejaculation.
7. What is the chemical in seminal fluid that is important to fertility? What is its function?
8. How is the structure of the spermatozoa related to its function?

CRITICAL THINKING QUESTIONS

Female Reproductive System

CHAPTER OUTLINE

LANGUAGE OF SCIENCE

accessory organs (ak-SES-oh-ree OR-gans)
[*accedere* to approach, *organon* instrument]

ampulla (am-PUL-ah) [*ampulla* flasklike bottle]

anal triangle [*anal* pertaining to the anus, *triangulus* three cornered]

anterior cul-de-sac (an-TEER-ee-or kul-deh-sak)
[*ante-* before, *-prior* foremost, *cul-de-sac* bottom of the bag]

anterior fornix (an-TEER-ee-or FOR-niks)
[*ante-* before, *-prior* foremost, *fornix* arch]

anterior ligament (an-TEER-ee-or LIG-ah-ment)
[*ante-* before, *-prior* foremost, *ligare* to bind]

areola (ah-REE-oh-lah) [*areola* area or space]

body

broad ligaments (LIG-ah-ments) [*ligare* to bind]

cervical canal (SER-vi-kal kah-NAL) [*cervic-* neck, *-al* pertaining to, *canalis* channel]

cervix (SER-viks) [*cervix* neck]

climacteric (klye-MAK-ter-ik) [*klimakter* critical point in human life]

clitoris (KLIT-oh-ris) [*Gk kleitoris*]

corpus albicans (KOHR-pus AL-bi-kans)
[*corpus* body, *albus* white]

corpus luteum (KOHR-pus LOO-tee-um)
[*corpus* body, *luteus* yellow]

cortex (KOHR-teks) [*cortex* bark]

endometrium (en-doh-MEE-tree-um) [*endo-* within, *-metra* womb]

essential organs [*essentil* quality, *organon* instrument]

estrogens (ES-troh-jens) [*oistros-* gadfly, *-gen* that which generates]

estrogenic phase (es-troh-JEN-ik fayz)
[*oistros-* gadfly, *-genic* that which generates]

external os [*externus* outward, *os* bone]

follicle-stimulating hormone (FSH) (FOL-li-kul-STIM-yoo-lay-ting HOR-mohn) [*folliculus* small bag, *stimulare* to incite, *hormaein* to set in motion]

follicular phase (foh-LIK-yoo-lar fayz) [*folliculus* small bag]

Cont'd on p. 1155

In the previous chapter, we discussed the structure and function of the male reproductive system. In this chapter, we discuss the structure and function of the female reproductive system. As you study this chapter, keep in mind that *both systems* must function properly if successful reproduction of offspring is to occur.

OVERVIEW OF THE FEMALE REPRODUCTIVE SYSTEM

Function of the Female Reproductive System

The physiological importance of the female reproductive system is best understood in terms of its final outcome: production of offspring and continued existence of the genetic code. The female reproductive system produces gametes that may unite with a male gamete to form the first cell of the offspring. The function of conception emphasizes the similarity between the male and female reproductive systems. However, this is where the similarity ends. Unlike the male system, the female reproductive system also provides protection and nutrition to the developing offspring for up to several years after conception, as we shall see.

Structural Plan of the Female Reproductive System

So many organs make up the female reproductive system that we need to look first at the structural plan of the system as a whole. As we stated in the previous chapter, reproductive organs can be classified as **essential organs** or **accessory organs,** depending on how directly they are involved in producing offspring.

The essential organs of reproduction in women, the *gonads,* are the paired **ovaries.** The female gametes, or **ova,** are produced by the ovaries.

The accessory organs of reproduction in women consist of the following structures:

- A series of ducts or modified duct structures that extend from near the ovaries to the exterior. This group of organs includes the *uterine tubes, uterus,* and *vagina.* Along with the ovaries, these organs are sometimes collectively called the "internal genitalia."
- The *vulva,* or external reproductive organs. These organs are often called the "external genitalia."
- Additional sex glands, including the *mammary glands,* which have an important reproductive function only in women.

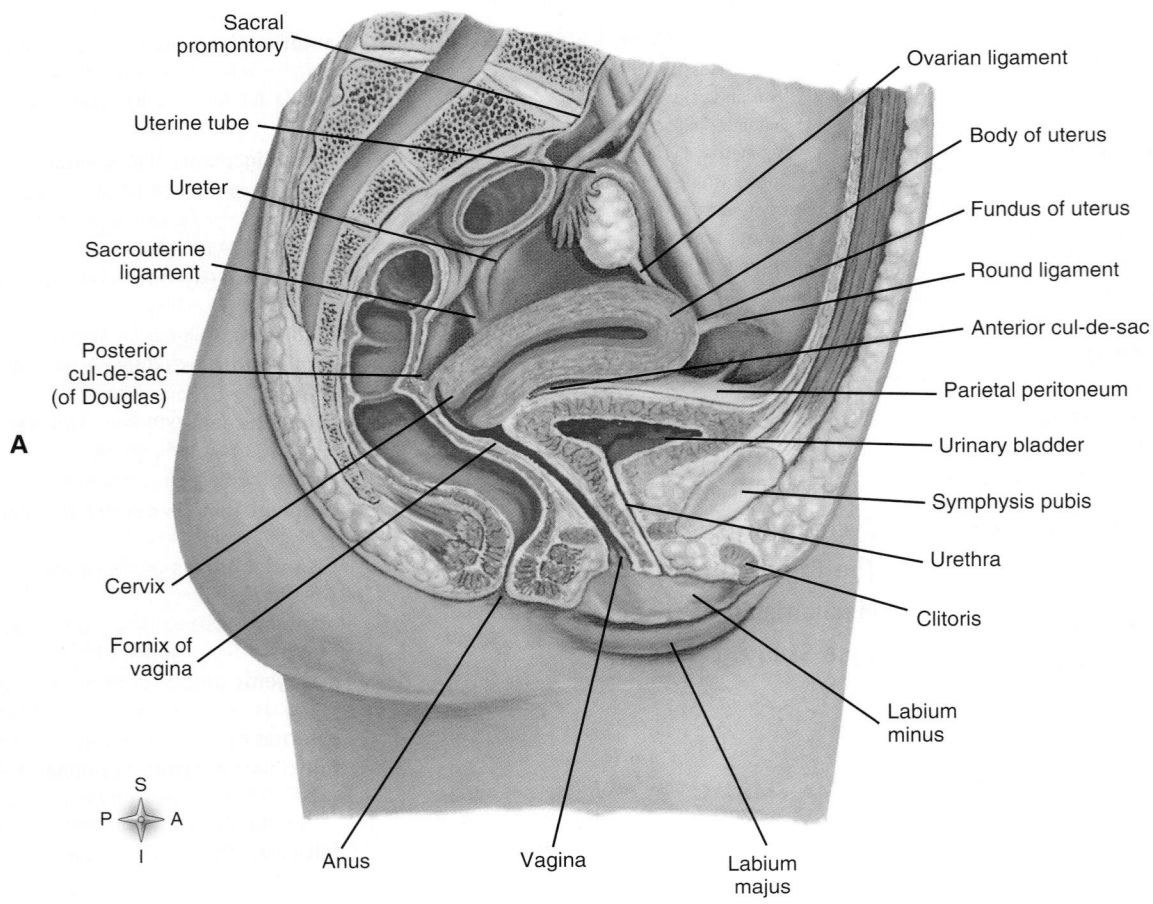

Figure 32-1 *Female reproductive organs.* **A,** Diagram (sagittal section) of pelvis showing location of female reproductive organs.

Many of the essential and accessory organs of the female reproductive system can be seen in Figures 32-1 and 32-2. Refer to these illustrations often as you read about each structure in the pages that follow.

Perineum

The perineum is the skin-covered muscular region between the vaginal orifice and the anus (Figure 32-8). It is a roughly diamond-shaped area between the thighs. The perineum extends from the symphysis pubis anteriorly to the coccyx posteriorly. Its most lateral boundary on either side is the ischial tuberosity. A line drawn between the two ischial tuberosities divides the area into a larger **urogenital triangle,** which contains the external genitalia (labia, vaginal orifice, clitoris) and urinary opening, and the **anal triangle,** which surrounds the anus.

The perineum has great clinical importance because of the danger of its being torn during childbirth. Such tears are often deep, have irregular edges, and extend all the way through the perineum, the muscular **perineal body,** and even through the anal sphincter, resulting in involuntary seepage from the rectum until the laceration is repaired. In addition, injuries to the perineal body can result in partial uterine or vaginal prolapse if this important support structure is weakened. To avoid these possibilities in a woman prone to such injuries, a surgical incision known as an **episiotomy** may be made in the perineum, particularly at the birth of a first baby. In current medical practice, episiotomy procedures are decreasing in frequency and are no longer performed on a routine basis preceding vaginal delivery of a baby.

QUICK CHECK

1. What are the *essential organs* of the female reproductive system?
2. List the *major accessory organs* of the female reproductive system.

OVARIES
Location of the Ovaries

The female gonads, or **ovaries,** are homologous (similar in origin) to the testes in the male. They are nodular glands that after puberty present a puckered, uneven surface, resemble large almonds in size and shape, and are located one on each side of the uterus, below and behind the uterine tubes. Each ovary weighs about 3 g and is attached to the posterior surface of the broad ligament by a structure called the mesovarian ligament, which contains blood vessels and nerves. The ovarian ligament anchors it to the uterus. The distal portion of the uterine tube curves about the ovary in such a way that the finger-like fimbriae at the end of the uterine tube cup over the ovary but do not actually attach to it (see Figure 32-2, A). This fact makes it possible to have a pregnancy begin in the pelvic cavity instead of in the uterus as is normal. Development of the fetus in a location other than the uterus is referred to as an **ectopic pregnancy** (from the Greek *ektopos,* "displaced"). In the cadaver dissection photo of

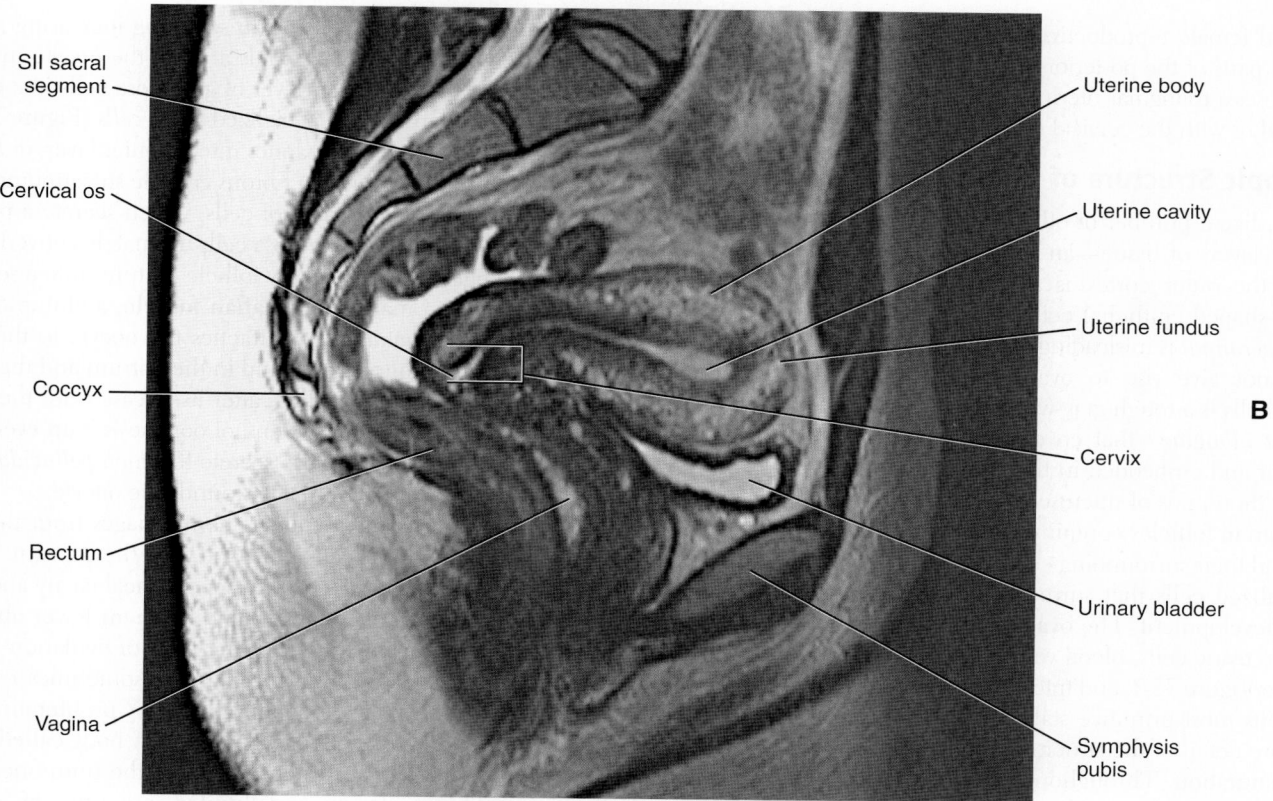

Figure 32-1, cont'd B, MRI scan (sagittal view) of female pelvic viscera.

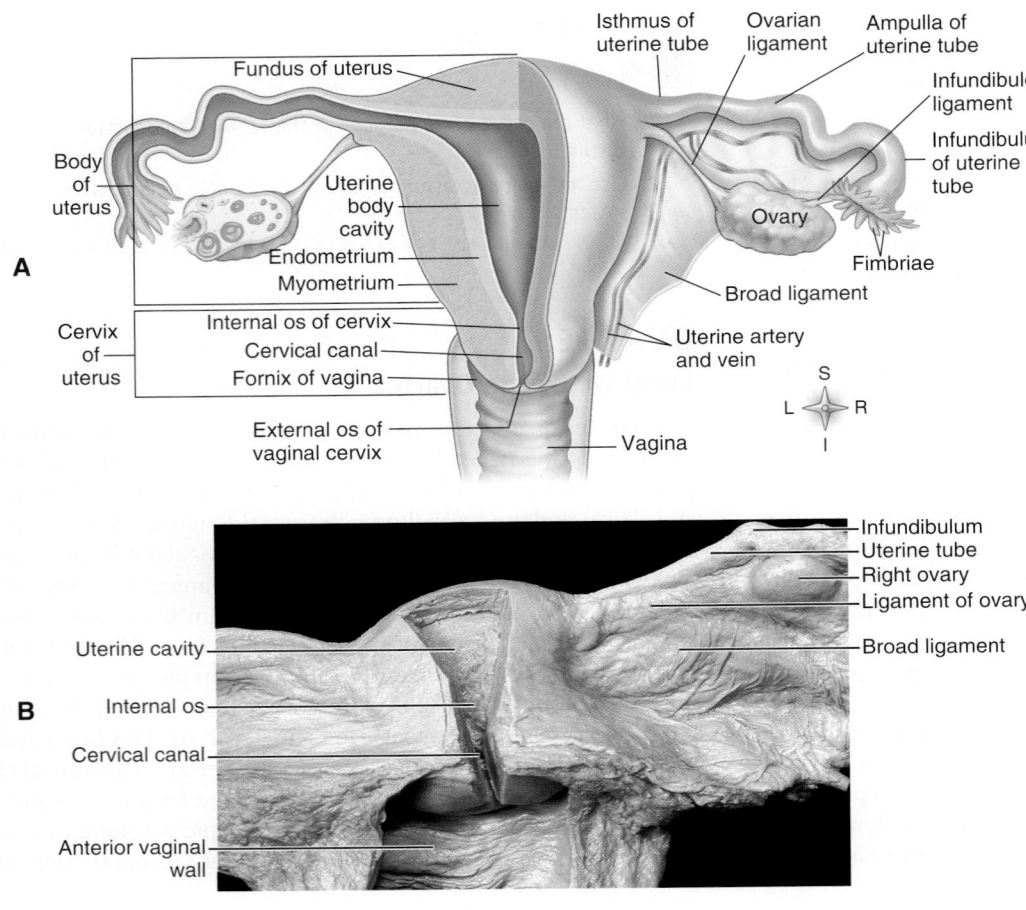

Figure 32-2 *Posterior view of female internal reproductive organs.* **A,** Diagram shows left side of uterus and upper portion of the vagina and the left uterine tube and ovary in a frontal section. The broad ligament has been removed from the posterior surface of the uterus and adjacent structures. **B,** Cadaver dissection showing uterine cavity and cervical canal, exposed by removal of parts of their posterior walls.

the internal female reproductive organs shown in Figure 32-2, *B*, removal of parts of the posterior wall of the body of the uterus and cervix exposes a triangular uterine cavity communicating by way of the internal os with the cervical canal.

Microscopic Structure of the Ovaries

The ovary, like a number of other organs in the body, consists of two major layers of tissue—an outer **cortex** and inner **medulla.** Covering the outer cortex is a surface layer of slightly raised squamous-shaped epithelial cells called the *germinal epithelium.* The term *germinal* is misleading because the epithelial cells of this layer do not give rise to ova. Deep to the surface layer of epithelial cells is a tough gray-white connective tissue layer called the *tunica albuginea* that covers the **ovarian cortex.** Scattered throughout and embedded in the connective tissue matrix of the cortex are thousands of microscopic structures called **ovarian follicles.** Ovarian follicles contain the immature female sex cells, or *oocytes,* and their surrounding cells. After puberty, the oocytes and the specialized cells that surround them are present in varying stages of development. The **ovarian medulla** contains supportive connective tissue cells, blood vessels, nerves, and lymphatics.

Refer to Figure 32-3, and trace the development of a female sex cell from its most primitive state through ovulation. Figure 32-4 shows more detail of the structural differences that appear during follicle maturation. Throughout the process the oocyte grows in size. So, too, does the number of cell layers surrounding it. Initially, the primary follicle is surrounded by a single layer of *granulosa cells.* As maturation proceeds the number of granulosa cell

layers increases and the cells begin secreting increasing amounts of an estrogen-rich fluid that pools around the oocyte in a space called an *antrum.* The outer layer of granulosa cells in a developing follicle condenses into specialized *theca cells* (Figure 32-4, *A*). The theca cell layer soon separates into an outer layer, or *theca externa,* which transforms into a fibrous capsule surrounding the follicle and a *theca interna* layer of cells, which secrete a precursor androgen hormone that granulosa cells ultimately convert into additional estrogen. As the primary follicle matures into a secondary and, eventually, a mature or **graafian follicle,** a clump of granulosa cells called *cumulus cells* attaches the oocyte to the follicle wall when it is surrounded by fluid in the antrum and then covers the mature *ovum,* as it is called, after its release from the follicle. The release of an ovum at the end of oogenesis is an event called **ovulation.** Granulosa cells also secrete the *zona pellucida* (Figure 32-4), a clear gel-like shell that surrounds the oocyte.

When ovulation occurs, blood hemorrhages from the highly vascular theca interna cell layer and fills the antrum. A small quantity of blood may also enter the peritoneal cavity and irritate its pain-sensitive surface, causing the transient lower abdominal pain many women experience at the time of ovulation (see Box 32-3). The blood clot filling the antrum, sometimes called the "corpus hemorrhagicum," is soon replaced by proliferating granulosa and theca interna cells to form a yellow body called the **corpus luteum.** The corpus luteum secretes the hormones progesterone, inhibin, relaxin, and limited amounts of estrogen. Progesterone and inhibin, a peptide hormone, suppress FSH secretion and prevent the continued development of new follicles

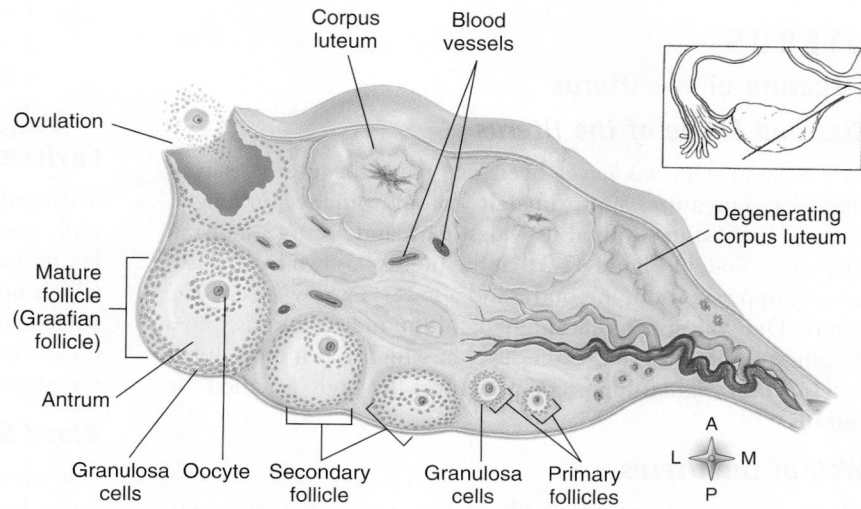

Figure 32-3 *Stages of ovarian follicle development.* Artist's rendition shows the successive stages of ovarian follicle and oocyte development. Begin with the first stage *(primary follicle)* and follow around clockwise to the final stage that is labelled *degenerating corpus luteum*. Remember, however, that all the stages shown occur over time to a *single* follicle, and the presence of all these stages at a single point in time is an artificial construct for learning purposes only.

during the functional life of the corpus luteum. The small amounts of relaxin secreted by the corpus luteum each month help "quiet" or "calm" uterine contractions, thus improving the chances for successful implantation if fertilization should occur. If pregnancy does occur, larger amounts of these hormones continue to be produced by the placenta.

Functions of the Ovaries

Recall that the ovaries are considered to be the *essential* organs of the female reproductive system. This means that it is the ovaries that produce female gametes, or ova. The process that culminates in the release of an ovum is called *oogenesis*, a term that literally means "egg production." As Figure 32-3 shows, ovulation involves the rupture of an ovarian follicle and the subsequent release of fluid and an ovum. The ovum, surrounded by a coat of follicular cells, moves into the uterine tube, where it may draw a sperm cell into it and thus become the first cell of an offspring.

The ovaries are also endocrine organs, secreting the female sex hormones. **Estrogens** (chiefly estradiol and estrone) and **progesterone** are secreted by cells of ovarian tissues. These hormones help regulate reproductive function in the female—making the ovaries even more essential to female reproductive function.

More details of oogenesis and fertilization are discussed in Chapter 33. Further discussion of hormonal regulation of reproductive functions, as well as associated changes within the ovaries, appears later in this chapter.

 QUICK CHECK

3. Briefly describe the location of the ovaries.
4. What are ovarian follicles?
5. List the two major functions of the ovaries.

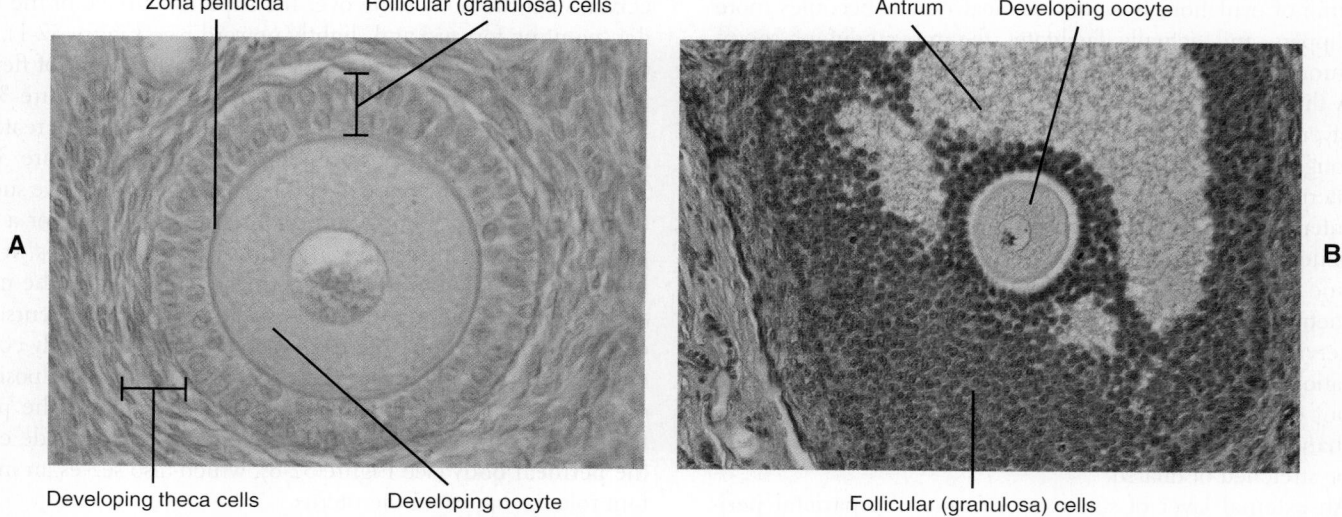

Figure 32-4 *Ovarian follicles.* Female gametes mature within follicles in the outer region of an ovary. Follicles in early stages of development, **A,** and late stages of development, **B,** exhibit a developing oocyte (immature ovum) surrounded by hormone-secreting follicular (granulosa) cells. Notice that the more mature ovarian follicle in **B** has a fluid-filled cavity called the antrum.

UTERUS
Structure of the Uterus
Size and Shape of the Uterus

In a woman who has never been pregnant, the **uterus** is pear shaped and measures approximately 7.5 cm (3 inches) in length, 5 cm (2 inches) in width at its widest part, and 3 cm (1 inch) in thickness. Note in Figure 32-2 that the uterus has two main parts: a wide, upper portion, the **body,** and a lower, narrow "neck," the **cervix.** Did you notice that the body of the uterus rounds into a bulging prominence above the level at which the uterine tubes enter? This bulging upper component of the body is called the **fundus.**

Wall of the Uterus

Three layers compose the walls of the uterus: the inner endometrium, a middle myometrium, and an outer incomplete layer of parietal peritoneum.

1. The lining of mucous membrane, called the **endometrium,** is composed of three layers of tissues: a compact surface layer of partially ciliated, simple columnar epithelium called the *stratum compactum*; a spongy middle, or intermediate, layer of loose connective tissue called the *stratum spongiosum*; and a dense inner layer termed the *stratum basale* that attaches the endometrium to the underlying myometrium. During menstruation and after delivery of a baby, the compact and spongy layers slough off. The endometrium varies in thickness from 0.5 mm just after the menstrual flow to about 5 mm near the end of the endometrial cycle. The endometrium has a rich supply of blood capillaries, as well as numerous exocrine glands that secrete mucus and other substances onto the endometrial surface. The mucous glands in the lining of the cervix produce mucus that changes in consistency during the female reproductive cycle. Most of the time, cervical mucus acts as a barrier to sperm. Around the time of ovulation, however, cervical mucus becomes more slippery and actually facilitates the movement of sperm through the cervix and into the body of the uterus.
2. A thick, middle layer (the **myometrium**) consists of three layers of smooth muscle fibers that extend in all directions, longitudinally, transversely, and obliquely, and give the uterus great strength. The bundles of smooth muscle fibers interlace with elastic and connective tissue components and generally blend into the endometrial lining with no sharp line of demarcation between the two layers. The myometrium is thickest in the fundus and thinnest in the cervix—a good example of the principle of structural adaptation to function. To expel a fetus, that is, move it down and out of the uterus, the fundus must contract more forcibly than the lower part of the uterine wall, and the cervix must be stretched or dilated.
3. An external layer of serous membrane, the **parietal peritoneum,** is incomplete because it covers none of the cervix and only part of the body (all except the lower one fourth of its anterior surface). The fact that the entire uterus is not covered with peritoneum has clinical significance because it makes it possible to perform operations on this organ without the risk of infection that attends cutting through the peritoneum.

Cavities of the Uterus

The cavities of the uterus are small because of the thickness of its walls (see Figure 32-2). The cavity of the body is flat and triangular. Its apex is directed downward and constitutes the **internal os,** which opens into the **cervical canal.** The cervical canal is constricted on its lower end also, forming the **external os,** which opens into the vagina. The uterine tubes open into the cavity of the uterine body at its upper, outer angles.

Blood Supply of the Uterus

The uterus receives a generous supply of blood from uterine arteries, branches of the internal iliac arteries (see Figure 32-2). In addition, blood from the ovarian and vaginal arteries reaches the uterus by anastomosis with the uterine vessels. Tortuous arterial vessels enter the layers of the uterine wall as arterioles and then break up into capillaries between the endometrial glands.

Uterine, ovarian, and vaginal veins return venous blood from the uterus to the internal iliac veins.

Location of the Uterus

The location of the uterus is in the pelvic cavity between the urinary bladder in front and the rectum behind. Age, pregnancy, and distention of related pelvic viscera such as the bladder alter the position of the uterus (see Figure 32-1).

Between birth and puberty the uterus descends gradually from the lower abdomen into the true pelvis. At **menopause** the uterus begins a process of involution that results in a decrease in size and a position deep in the pelvis. Some variation among women in uterine placement within the pelvis is common.

Position of the Uterus

Normally the uterus is said to be *anteflexed* between the body and cervix, with the body lying over the superior surface of the bladder, pointing forward and slightly upward (see Figure 32-1). The cervix points downward and backward from the point of flexion, joining the vagina at approximately a right angle (Figure 32-5). Protrusion of the cervix into the lumen of the vagina creates an **anterior fornix** and a **posterior fornix,** or space (Figure 32-5). These spaces help increase the probability of reproductive success (fertilization) by retaining and pooling seminal fluid for a brief period following intercourse near the external os. This in turn helps increase the numbers of spermatozoa that enter the uterus and, ultimately, the uterine tubes where fertilization occurs. Several ligaments hold the uterus in place but allow its body considerable movement, a characteristic that often leads to malpositions of the organ. Fibers from several muscles that form the pelvic floor (see Figure 10-12, p. 364) converge to form a node called the **perineal body** (see Figure 32-8), which also serves an important role in support of the uterus.

The uterus may lie in any one of several abnormal positions. A common one is *retroflexion*, or backward tilting, of the entire organ. Retroflexion may allow the uterus to prolapse, or descend, into the vaginal canal.

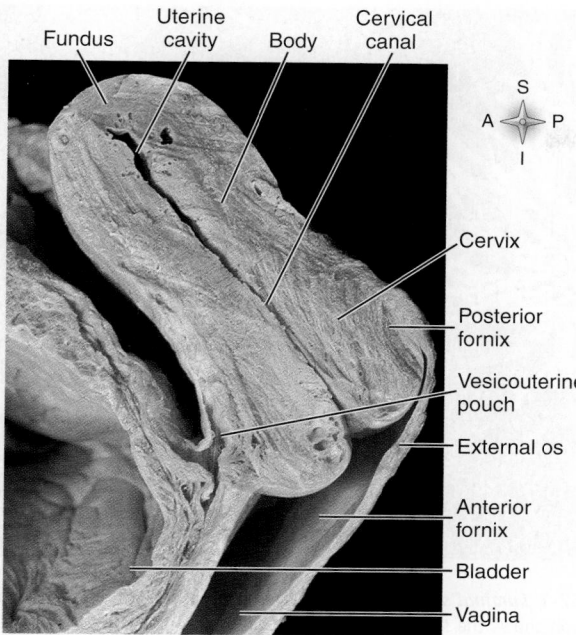

Fundus — Uterine cavity — Body — Cervical canal

S
A P
I

Cervix

Posterior fornix

Vesicouterine pouch

External os

Anterior fornix

Bladder

Vagina

Figure 32-5 *Uterus and vagina in sagittal section.* Note that (1) the body of the uterus is "anteflexed" over the superior surface of the bladder pointing forward and slightly upward, and (2) the cervix protrudes into the vagina at approximately a right angle.

Eight *uterine ligaments* (three pairs, two single ones) hold the uterus in its normal position by anchoring it in the pelvic cavity. These ligaments include the broad (paired), uterosacral (paired), posterior (single), anterior (single), and round (paired) ligaments. Six of these so-called ligaments are actually extensions of the parietal peritoneum in different directions. The round ligaments are fibromuscular cords. Many of the following structures can be identified in Figures 32-1 and 32-2.

- The *two* **broad ligaments** are double folds of parietal peritoneum that form a kind of partition across the pelvic cavity. The uterus is suspended between these two folds.
- The *two* **uterosacral ligaments** are foldlike extensions of the peritoneum from the posterior surface of the uterus to the sacrum, one on each side of the rectum.
- The **posterior ligament** is a fold of peritoneum extending from the posterior surface of the uterus to the rectum. This ligament forms a deep pouch known as the **posterior cul-de-sac (of Douglas),** or **rectouterine pouch,** between the uterus and rectum. Because this is the lowest point in the pelvic cavity, pus collects here in pelvic inflammations. To secure drainage, an incision may be made at the top of the posterior wall of the vagina. The procedure is called a *posterior colpotomy.*
- The **anterior ligament** is the fold of peritoneum formed by the extension of the peritoneum on the anterior surface of the uterus to the posterior surface of the bladder. This fold forms an **anterior cul-de-sac,** or **vesicouterine pouch,** which is less deep than the posterior cul-de-sac (Figure 32-5).
- The *two* **round ligaments** are fibromuscular cords extending from the upper, outer angles of the uterus through the inguinal canals and terminating in the labia majora.

Functions of the Uterus

The uterus, or womb, has many functions important to successful reproductive function. The uterus serves as part of the female reproductive tract, permitting sperm from the male to ascend toward the uterine tubes. If fusion of gametes (*fertilization,* or *conception*) occurs, the developing offspring implants in the endometrial lining of the uterus and continues its development during the term of pregnancy (*gestation*). The tiny endometrial glands are specialized to produce nutrient secretions—sometimes called "uterine milk"—to sustain the developing offspring until a *placenta* can be produced. The placenta is a unique organ that permits the exchange of materials between the offspring's blood and the maternal blood. The rich network of endometrial capillaries promotes efficiency of this exchange function. Rhythmic contractions of the myometrium are inhibited during gestation but are allowed to occur as the time of delivery approaches. Myometrial contractions are the "labor contractions" that help push the offspring out of the mother's body.

If conception or the implantation of the offspring does not occur successfully, the outer layers of the endometrium are shed during *menstruation.* **Menstruation** is a regular event of the female reproductive cycle that permits the endometrium to renew itself in anticipation of conception and implantation during the next cycle. The myometrial contractions seem to aid menstruation by promoting the complete sloughing of the outer endometrial layers. Fatigue of the myometrial muscle tissues may contribute to the abdominal cramping sometimes associated with menstruation.

UTERINE TUBES

The **uterine tubes** are also sometimes called *fallopian tubes,* or *oviducts.*

Location of the Uterine Tubes

The **uterine tubes** are about 10 cm (4 inches) long and are attached to the uterus at its upper outer angles (see Figures 32-1 and 32-2). They lie in the upper free margin of the broad ligaments and extend upward and outward toward the sides of the pelvis and then curve downward and backward.

Structure of the Uterine Tubes

Wall of the Uterine Tubes

The same three layers (mucous, smooth muscle, and serous) of the uterus compose the tubes (Figure 32-6). The mucosal lining of the tubes, however, is directly continuous with the peritoneum lining the pelvic cavity. This fact has great clinical significance because the tubal mucosa is also continuous with that of the uterus and vagina and therefore often becomes infected by gonococci or other organisms introduced into the vagina. Inflammation of the tubes (**salpingitis**) may readily spread to become inflammation of the peritoneum (**peritonitis**), a serious condition. Inflammation of the uterine tubes may also lead to scarring and partial or complete closure of the lumen, even if the original infection is cured with antibiotics (see Mechanisms of Disease, p. 1149). In the male, there is no such direct route by which microorganisms can reach the peritoneum from the exterior.

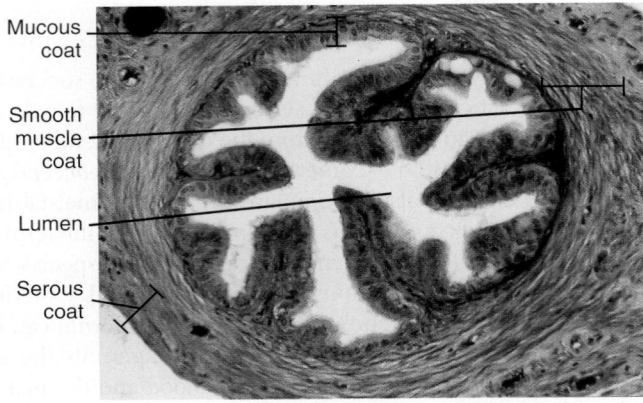

Figure 32-6 *Transverse section of uterine tube.* This micrograph shows a section of the uterine (fallopian) tube in the isthmus region. Notice the highly folded nature of the epithelial, mucous lining of the tube. The folding is even more exaggerated in the ampulla region.

Divisions of the Uterine Tubes

Each uterine tube consists of three divisions (see Figure 32-2):

1. A medial third that extends from the upper outer angle of the uterus called the **isthmus.**
2. An intermediate dilated portion called the **ampulla** that follows a winding path over the ovary.
3. A funnel-shaped terminal component called the **infundibulum** lies just above and extends laterally over the ovary and opens directly into the peritoneal cavity. The open outer margin of the infundibulum resembles a fringe in its irregular outline. The fringelike projections are known as *fimbriae* (see Figure 32-2).

Histology of the Uterine Tubes

Figure 32-6 is a low-power micrograph that illustrates the mucosal lining of the oviduct cut in cross section. These shapes are typical of the appearance of the mucosal lining throughout most of the duct. Note the extensive folds of mucosa that project as shelves into the lumen of the tube. An area of smooth muscle (muscularis layer) can be seen surrounding the mucosa in this section. Cilia, which are important in maintaining currents within the tube to move the ovum toward the uterus, can be seen projecting from the luminal surface in Figure 32-7.

Function of the Uterine Tubes

The uterine tubes are extensions of the uterus that communicate loosely with the ovaries. This arrangement allows an ovum released from the surface of the ovary to be collected by the fimbriae and swept along the uterine tube toward the body of the uterus by ciliary action. The uterine tubes serve as more than mere transport channels, however. The uterine tube is also the site of fertilization. Sperm and ova most often meet, and fertilization occurs, in the ampulla of the uterine tube. A relatively small number of the sperm deposited in the vagina during sexual intercourse move up the uterine tube, where they meet the ovum traveling toward them. It is generally here, within the ampulla of the uterine tube, that the primary function of human

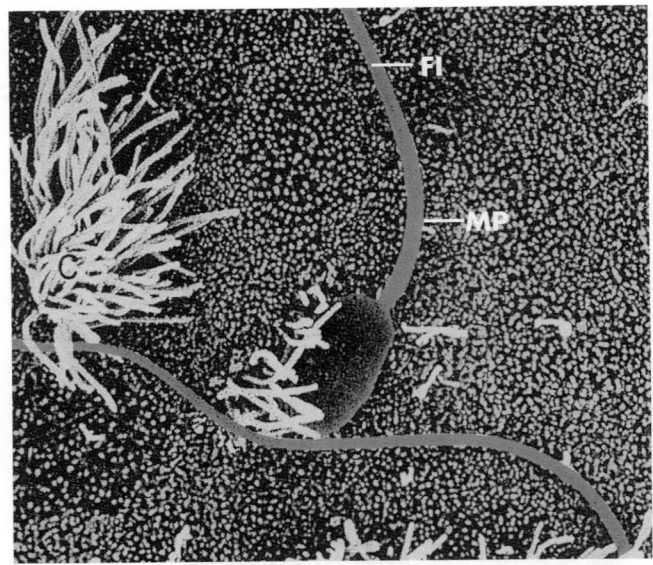

Figure 32-7 *Luminal surface of the uterine tube.* This scanning electron micrograph shows the surface of the uterine tube wall that faces the lumen, as well as some spermatozoa that are present (*Fl*, flagellum; *MP*, mid piece). The smaller, raised projections are microvilli, and the larger projections, found in clumps, are cilia (*C*).

sexual reproduction occurs: recombination of the genetic information from both parents in the first cell of the offspring. Totally blocking the openings into either the distal (abdominal) or proximal (uterine) ends of both uterine tubes, for any reason, results in sterility.

 QUICK CHECK

6. Name the three principal layers of the uterine wall.
7. Describe the anatomical position of the uterus. How is it held in place?
8. List the chief functions of the uterus.
9. What are the functions of the uterine (fallopian) tubes?

VAGINA

Location of the Vagina

The **vagina** is a tubular organ situated between the rectum, which lies posterior to it, and the urethra and bladder, which lie anterior to it (Figure 32-5). It extends upward and backward from its external orifice in the vestibule between the labia minora of the vulva to the cervix (Figures 32-1, 32-2, and 32-8).

The right and left *levator ani* muscles (Figure 10-14, p. 367, and Figure 28-8, p. 1040) unite at the midline to form a hammock-shaped muscular sheet often referred to as the "floor of the pelvis." In the female, the anatomical relationships between the various parts of this supportive muscular sheet and the perineal body, vagina, rectum, and urethra have clinical significance. Fibers of the levator ani insert into and become a part of all these structures. It is not an uncommon occurrence for at least some fibers of the levator

BOX 32-1: FYI
Tubal Ligations

Tubal ligation literally means "tying a tube," and thus this surgical procedure is often referred to as "having one's tubes tied." Tubal ligation involves tying a piece of suture material around each uterine tube in two places, then cutting each tube between these two points (see the figure). Because sperm and eggs are thus blocked from meeting, fertilization and subsequent pregnancy are prevented. For this reason, tubal ligation is also called *surgical sterilization* and is comparable to vasectomy in the male.

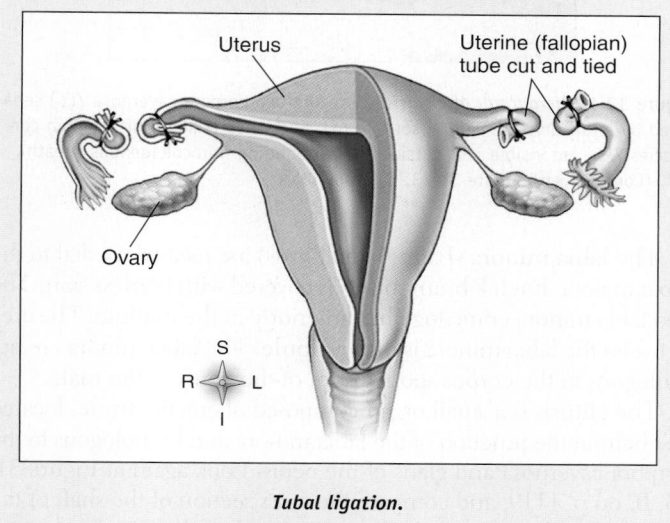

Tubal ligation.

ani to become stretched or damaged in women who have experienced vaginal delivery of a full-term infant. The unfortunate result is often the appearance of a number of troublesome and often chronic symptoms. For example, damage to muscle fibers of the levator ani that constitute a part of the urethral or anal sphincters may result in urinary or fecal *incontinence.* If leakage of urine or fecal material following damage to the sphincter is associated with—or precipitated by, an increase in abdominal pressure caused by events such as coughing, laughing, or lifting weight, the condition is called *stress incontinence.* When muscle fibers in the vagina or perineal body are significantly weakened, the vaginal walls will lose tone and the important role of the perineal body in providing support of pelvic viscera will be affected. The result may involve prolapse of the uterus into the vagina or some degree of rectal prolapse through the anus. In severe cases surgery or other treatments may be required to raise or support the pelvic floor, or repair structural damage. However, in less severe cases where symptoms are limited to occasional leakage of urine, many women benefit from a noninvasive treatment option called **Kegel exercise.** Designed primarily to reduce urinary stress incontinence, Kegel exercises consist of an ongoing exercise program that involves a repetitive series of voluntary contractions of the muscles of the pelvic floor and perineum, similar to that required to stop the flow of urine when voiding. In time, the exercise program will strengthen the urinary sphincter and improve retention of urine.

Structure of the Vagina

The vagina is a collapsible tube about 7 or 8 cm (about 3 inches) long that is capable of great distention. It is composed mainly of smooth muscle and is lined with mucous membrane arranged in rugae. The vaginal mucosa contains numerous tiny exocrine glands that secrete lubricating fluid during the female sexual response. Note that the anterior wall of the vagina is shorter than the posterior wall because of the way the cervix protrudes into the uppermost portion of the tube (see Figures 32-1 and 32-5). In some cases—especially in young girls—a fold of mucous mem-

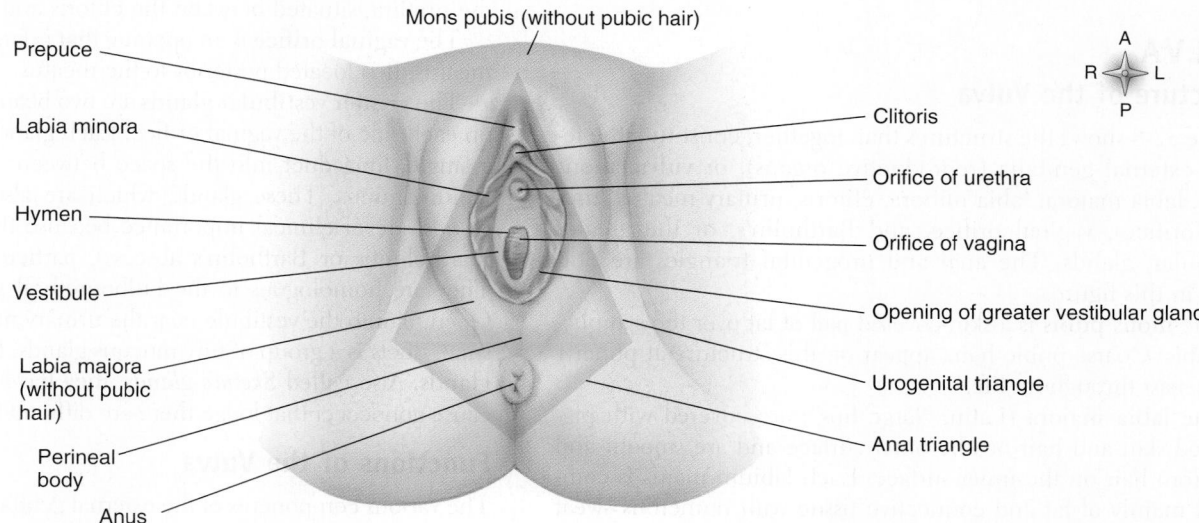

Figure 32-8 *External genitalia of the female showing outline of the urogenital and anal triangles.*

BOX 32-2: FYI

The "G Spot"

In 1950, Dr. Ernest Gräfenberg first described what he called an "erotic zone" about the size of a dime on the anterior wall of the vagina midway between the symphysis pubis and cervix. The area was later named the "Gräfenberg spot," or "G spot." More recently, it has been called the *female prostate.*

The existence of G spot tissue in significant numbers of women continues to be the subject of heated controversy more than 55 years after the term was introduced in the medical community. Although existence of the G spot is supported by some physicians and professionals in the area of human sexuality, most clinicians remain skeptical because of conflicting and inconsistent data.

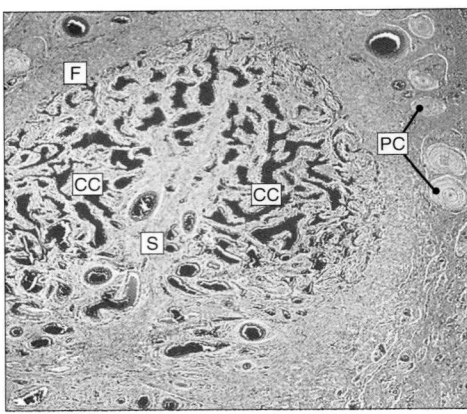

Figure 32-9 *Micrograph of clitoris.* Note the two corpora cavernosa *(CC)* separated by an incomplete central septum *(S).* Sectioned lamellar or pacinian corpuscles *(PC)* are visible just outside the surrounding fibrocollagenous sheath *(F).* (Compare with Figure 31-13, *B,* on p. 1119.)

brane, the **hymen,** forms a border around the external opening of the vagina, partially closing the orifice. Occasionally, this structure completely covers the vaginal outlet, a condition referred to as **imperforate hymen.** Perforation must be performed at puberty before the menstrual flow can escape.

Functions of the Vagina

The vagina is a portion of the female reproductive tract that has several important functions. During sexual intercourse, the lining of the vagina lubricates and stimulates the glans penis, which in turn triggers the ejaculation of semen. Thus the vagina also serves as a receptacle for semen, which often pools in the anterior or posterior fornix of the vagina where it meets the cervix of the uterus (Figure 32-5). Sperm within the semen may move further into the female reproductive tract by "climbing" along fibrous strands of mucus in the cervical canal.

The vagina also serves as the lower portion of the birth canal. At the time of delivery, the offspring is pushed from the body of the uterus, through the cervical canal, and finally through the vagina and out of the mother's body. The placenta, or "afterbirth," is also expelled through the vagina.

Another important function of the vagina is transport of blood and tissue shed from the lining of the uterus during menstruation.

VULVA

Structure of the Vulva

Figure 32-8 shows the structures that, together, constitute the female external genitalia (reproductive organs), or **vulva:** mons pubis, labia majora, labia minora, clitoris, urinary meatus (urethral orifice), vaginal orifice, and Bartholin's, or the greater vestibular, glands. The anal and urogenital triangles are outlined in this figure.

The **mons pubis** is a skin-covered pad of fat over the symphysis pubis. Coarse pubic hairs appear on this structure at puberty and persist throughout life.

The **labia majora** (Latin, "large lips") are covered with pigmented skin and hair on the outer surface and are smooth and free from hair on the inner surface. Each labium majus is composed mainly of fat and connective tissue with numerous sweat and sebaceous glands on the inner surface. The labia majora are homologous to the scrotum in the male.

The **labia minora** (Latin, "small lips") are located medial to the labia majora. Each labium minus is covered with hairless skin. The two labia minora come together anteriorly in the midline. The area between the labia minora is the **vestibule.** The labia minora are homologous to the corpus spongiosum of the penis in the male.

The **clitoris** is a small organ composed of erectile tissue, located just behind the junction of the labia minora and homologous to the corpora cavernosa and glans of the penis. Look again at Figure 31-13, *B,* on p. 1119, and compare the cross section of the shaft of the penis in that illustration with the micrograph of the clitoris shown in Figure 32-9. Absence in the clitoris of a corpus spongiosum with a urethra within its substance is immediately apparent. And, although the two corpora cavernosa are proportionally smaller and separated by a less distinct central septum than in the penis, the homologous relationship that exists between the penis and clitoris is apparent.

An incomplete clitoral *hood* or *prepuce* is located over the superior surface of the clitoris and, on the inferior surface, a thin midline *frenulum* exists that is the female equivalent of the raphe visible on the undersurface of the penile shaft.

The **urinary meatus** (urethral orifice) is the small opening of the urethra, situated between the clitoris and the vaginal orifice.

The **vaginal orifice** is an opening that is larger than the urinary meatus. It is located posterior to the meatus.

The **greater vestibular glands** are two bean-shaped glands, one on each side of the vaginal orifice. Each gland opens by means of a single, long duct into the space between the hymen and the labium minus. These glands, which are also called *Bartholin's glands,* are of clinical importance because they can be infected (bartholinitis or Bartholin's abscess), particularly by gonococci. They are homologous to the bulbourethral glands in the male. Opening into the vestibule near the urinary meatus by way of two small ducts is a group of tiny mucous glands, the **lesser vestibular glands.** Also called *Skene's glands,* they have clinical interest because gonococci that lodge there are difficult to eradicate.

Functions of the Vulva

The various components of the external genitalia of the female operate alone or separately to accomplish several functions important to successful reproduction. The protective features of the mons pubis and labia help prevent injury to the delicate tissues of the clitoris and vestibule. The clitoris becomes erect during sexual stimulation and, like the male glans, possesses a large number of sensory receptors that feed back information to the sexual response areas of the brain. In Figure 32-9, a number of sectioned lamellar or pacinian corpuscles, encapsulated touch and pressure receptors (see Chapter 15), can be seen just outside the fibrocollagenous sheath surrounding the clitoris. Of course, the vaginal orifice serves as the boundary between the internal and external female genitalia.

 QUICK CHECK

10. List three functions of the vagina.
11. What is another name for the external genitalia of the female?
12. How are the clitoris of the female and the glans penis of the male similar in structure and function?

BREASTS

Location and Size of the Breasts

The breasts lie over the pectoral muscles and are attached to them by a layer of connective tissue (fascia) (Figure 32-10). **Estrogens** and **progesterone,** two types of ovarian hormones, control breast development during puberty. Estrogens stimulate growth of the ducts of the mammary glands, whereas progesterone stimu-

lates development of the actual secreting cells. Breast size is determined more by the amount of fat around the glandular tissue than by the amount of glandular tissue itself. Hence the size of the breast is not related to its functional ability.

Structure of the Breasts

Each breast consists of several lobes separated by septa (walls) of connective tissue. Each lobe consists of several lobules, which, in turn, are composed of connective tissues in which are embedded the secreting cells (alveoli) of the gland, arranged in grapelike clusters around the minute ducts. The ducts from the various lobules unite, forming a single lactiferous (milk-carrying) duct for each lobe, or between 15 and 20 in each breast. These main ducts converge toward the nipple, like the spokes of a wheel. They enlarge slightly into small lactiferous sinuses before reaching the nipple (Figure 32-10, A). Each of these main ducts terminates in a tiny opening on the surface of the nipple. A comparatively large amount of adipose tissue is deposited around the surface of the gland, just under the skin, and between the lobes. Suspensory ligaments (of Cooper) throughout the connective tissue of the breast help support the glandular and connective tissues of the entire structure, anchoring them to the coverings of the underlying pectoral muscles.

The nipples are bordered by a circular pigmented area, the **areola** (Figure 32-10, B). It contains numerous sebaceous glands that appear as small nodules under the skin. Sebum produced by these glands helps reduce irritating dryness of the areolar skin associated with nursing. In some lighter-skinned women, the areola

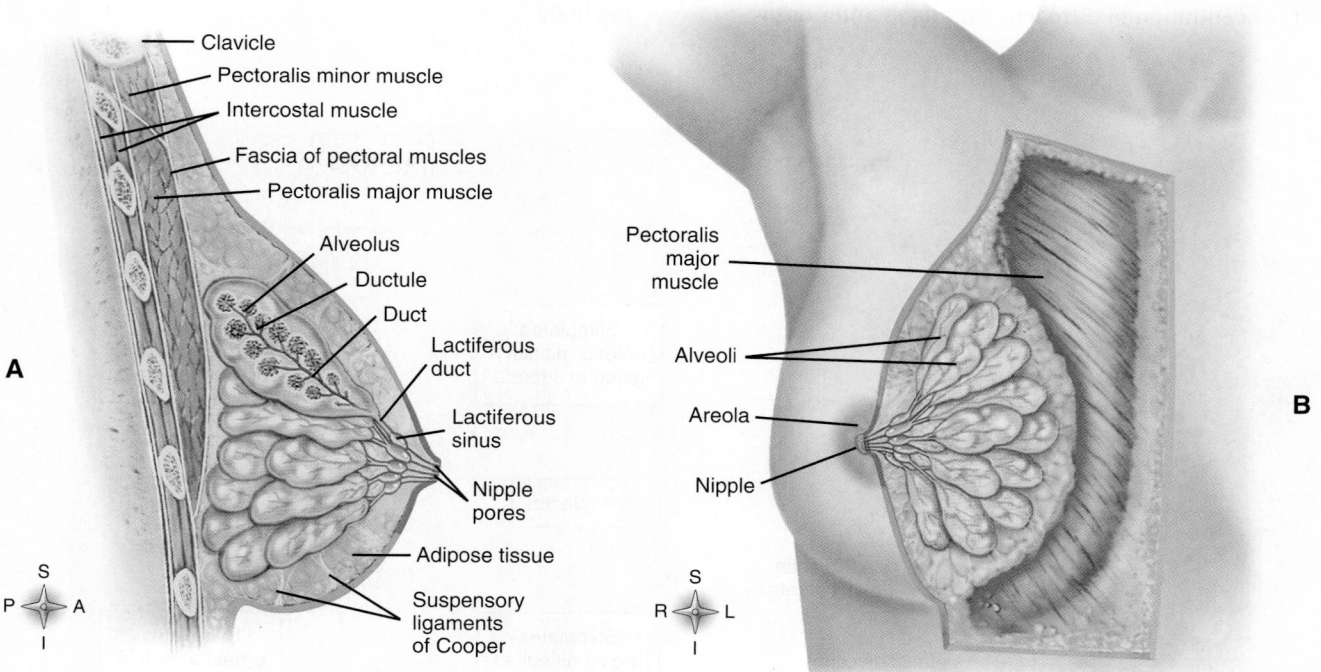

Figure 32-10 *The female breast.* **A,** Sagittal section of a lactating breast. Notice how the glandular structures are anchored to the overlying skin and to the pectoral muscles by the suspensory ligaments of Cooper. Each lobule of glandular tissue is drained by a lactiferous duct that eventually opens through the nipple. **B,** Anterior view of a lactating breast. Overlying skin and connective tissue have been removed from the medial side to show the internal structure of the breast and underlying skeletal muscle. In nonlactating breasts, the glandular tissue is much less prominent, with adipose tissue making up most of each breast.

and nipple change color from pink to brown early in pregnancy—a fact of value in diagnosing a first pregnancy. The color decreases after lactation has ceased but never entirely returns to the original hue. In some darker-skinned women, no noticeable color change in the areola or nipple heralds the first pregnancy.

Knowledge of the lymphatic drainage of the breast is important in clinical medicine because cancerous cells from malignant breast tumors often spread to other areas of the body through the lymphatics. Lymphatic drainage of the breast is presented in Chapter 20.

Function of the Breasts

The function of the mammary glands is lactation, that is, the secretion of milk for the nourishment of newborn infants.

Mechanism Controlling Lactation

Very briefly, **lactation** is controlled as follows:

- The ovarian hormones, estrogens and progesterone, act on the breasts to make them structurally ready to secrete milk. Estrogens promote development of the ducts of the breasts. Progesterone acts on the estrogen-primed breasts to promote completion of the development of the ducts and development of the alveoli, the secreting cells of the breasts. This is an example of hormonal *permissiveness*; estrogen permits progesterone to have its full effect. A high blood concentration of estrogens during pregnancy also inhibits anterior pituitary secretion of prolactin.
- Shedding of the placenta after delivery of the baby cuts off a major source of estrogens. The resulting rapid drop in the blood concentration of estrogens stimulates anterior pi-

tuitary secretion of prolactin. Also, the suckling movements of a nursing baby stimulate anterior pituitary secretion of prolactin and posterior pituitary secretion of oxytocin.
- Prolactin stimulates lactation, that is, stimulates alveoli of the mammary glands to secrete milk. Milk secretion starts about the third or fourth day after delivery of a baby, supplanting a thin, yellowish secretion called *colostrum*. With repeated stimulation by the suckling infant, plus various favorable mental and physical conditions, lactation may continue for extended periods.
- Oxytocin stimulates the alveoli of the breasts to eject milk into the ducts, thereby making it accessible for the infant to remove by suckling.

The mechanism is summarized in Figure 32-11.

The Importance of Lactation

The process of **lactation** plays an important role in the ultimate success of the reproductive system. The biological goal of human reproduction does not lie solely in delivering a healthy infant—the infant must also survive until reproductive age. If a child does not survive to reproduce, then the genetic code cannot be passed on to successive generations and the ultimate goal of reproduction has not been met. Humans and other mammals help ensure the survival of offspring for up to several years by producing nutrient-rich milk. Nursing from the mother's breast provides several advantages for human offspring:

- Milk is a rich source of proteins, fat, calcium, vitamins, and other nutrients in proportions needed by a young, developing body.

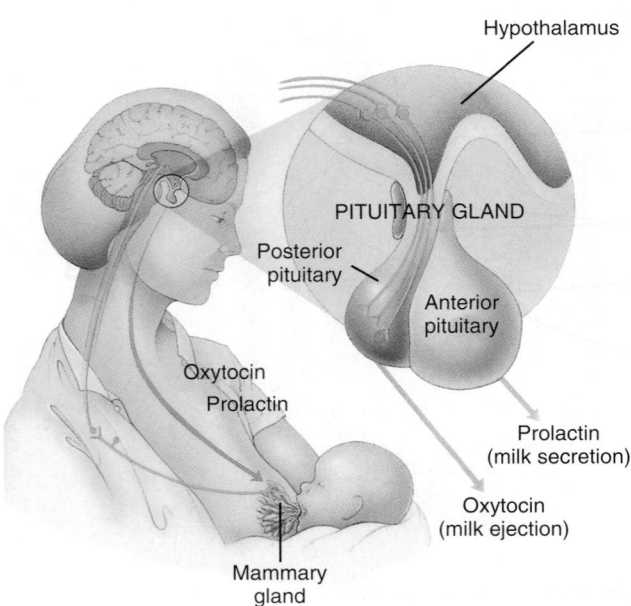

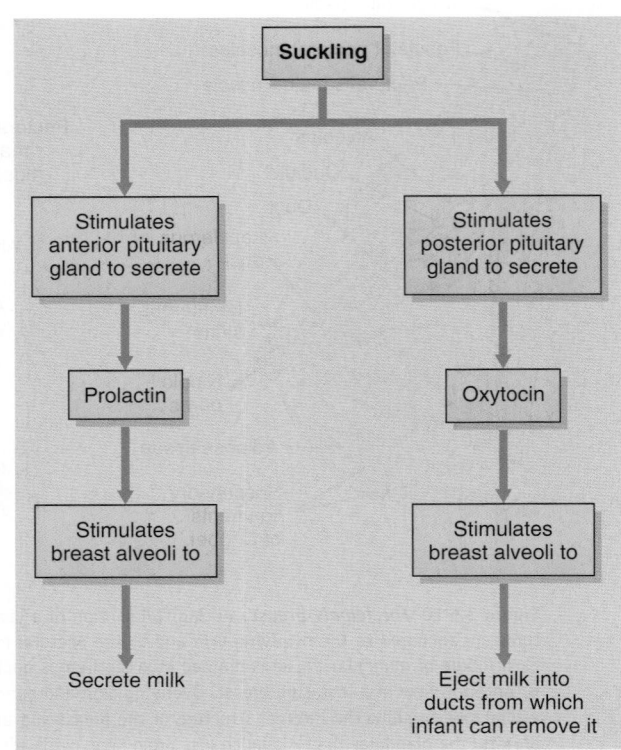

Figure 32-11 *Lactation.* The illustration and accompanying flow chart summarize the mechanisms that control the secretion and ejection of milk.

- Human milk provides passive immunity to the offspring in the form of maternal antibodies present in both colostrum and the milk.
- Nursing appears to enhance the emotional bond between mother and child. Such bonding may foster healthy psychological development in the child and strengthen family relationships that contribute to successful human development.

QUICK CHECK

13. Briefly describe the network of ducts and secreting cells that form the mammary glands.
14. List the hormones that prepare the breast structurally for lactation.
15. Which hormone causes milk to be ejected into the lactiferous ducts?

FEMALE REPRODUCTIVE CYCLES

Recurring Cycles

Many changes recur periodically in the female during the years between the onset of the menses (**menarche**) and their cessation (**menopause,** or **climacteric**). Most obvious, of course, is menstruation—the outward sign of changes in the endometrium. Most women also note periodic changes in their breasts. But these are only two of many changes that occur over and over again at fairly uniform intervals during the approximately 3 decades of female reproductive maturity.

First we shall describe the major cyclical changes, and then we shall discuss the mechanisms that produce them.

Ovarian Cycle

Before a female child is born, precursor cells in her ovarian tissue, called **oogonia,** begin a type of cell division characterized as *meiosis.* Meiotic cell division differs from mitotic cell division in that it reduces the number of chromosomes in the daughter cells by half (recall Chapter 4, pp. 132-134). By the time the child is born, her ovaries contain many primary follicles, each containing an oocyte that has temporarily suspended the meiotic process before it is complete. Once each month, on about the first day of menstruation, the oocytes within several primary follicles resume meiosis. At the same time, the follicular cells surrounding them proliferate and start to secrete estrogens (and tiny amounts of progesterone). Usually, only one of these developing follicles matures and migrates to the surface of the ovary. Just before **ovulation,** the meiosis within the oocyte of the mature follicle halts again. It is this cell, which has not quite completed meiosis, that is expelled from the ruptured wall of the mature follicle during ovulation. Meiosis is completed only when, and if, the head of a sperm cell is later drawn into the ovum during the process of fertilization.

When does ovulation occur? This is a question of great practical importance and one that in the past was given many answers. Today it is known that ovulation usually occurs 14 days before the next menstrual period begins. (Only in a 28-day menstrual cycle is this also 14 days after the beginning of the preceding menstrual cycle, as explained subsequently.)

BOX 32-3: FYI
Mittelschmerz

A few women experience pain within a few hours after ovulation. This is referred to as **mittelschmerz**—German for "middle pain" and generally occurs during days 14 to 16 in a 28-day cycle. It has been ascribed to irritation of the peritoneum by hemorrhage from the ruptured follicle.

Immediately after ovulation, cells of the ruptured follicle enlarge and, because of the appearance of lipoid substances in them, become transformed into a golden-colored body, the **corpus luteum.** The corpus luteum grows for 7 or 8 days. During this time, it secretes progesterone in increasing amounts. Then, provided fertilization of the ovum has not taken place, the size of the corpus luteum and the amount of its secretions gradually diminish. In time, the last components of each nonfunctional corpus luteum are reduced to a white scar called the **corpus albicans,** which moves into the central portion of the ovary and eventually disappears (see Figure 32-3).

Endometrial, or Menstrual, Cycle

During menstruation, bits of the compact and spongy layers of the endometrium slough off, leaving denuded bleeding areas. The dark menstrual discharge generally does not clot and may vary from about 30 to 100 ml, with a majority lost during the first 3 days of the menses. As with the length of the cycle, considerable variation is considered normal. After menstruation, the cells of these layers proliferate, causing the endometrium to reach a thickness of 2 or 3 mm by the time of ovulation. During this period, endometrial glands and arterioles grow longer and more coiled—two factors that also contribute to the thickening of the endometrium. After ovulation, the endometrium grows still thicker, reaching a maximum of about 4 to 6 mm. Most of this increase, however, is believed to be caused by swelling produced by fluid retention rather than by further proliferation of endometrial cells. The increasingly coiled endometrial glands start to secrete their nutrient fluid during the time between ovulation and the next menses. Then, the day before menstruation starts again, a drop in progesterone causes muscle in the walls of the tightly coiled arterioles to constrict, producing endometrial ischemia. This leads to death of the tissue, sloughing, and once again, menstrual bleeding.

The menstrual cycle is customarily divided into phases, named for major events occurring in them: menses, postmenstrual phase, ovulation, and premenstrual phase.

1. The **menses,** or **menstrual period,** occurs on days 1 to 5 of a new cycle. There is some individual variation, however.
2. The **postmenstrual phase** occurs between the end of the menses and ovulation. Therefore it is the **preovulatory phase,** as well as the postmenstrual phase. In a 28-day cycle, it usually includes cycle days 6 to 13 or 14. However, the length of this phase varies more than the others. It lasts longer in long cycles and ends sooner in short cycles. This phase is also called the **estrogenic phase,** or **follicular phase,** because of the high blood estrogen level resulting from secretion by the developing follicle. Increases in estrogen levels cause predictable changes in the appearance,

amount, and consistency of cervical mucus. Collectively, these changes can be used as a fertility sign to predict ovulation (Box 32-4). Increasing estrogen levels cause cervical mucus to become elastic, a property that can be observed by placing the mucus between and then separating two glass microscope slides. The clear, watery cervical mucus found at the time of ovulation will stretch 8 cm or more before the resulting thread will break. This phenomenon is called *spinnbarkeit* (SPIN-bahr-kit). Further, if left to dry on a clean glass slide, cervical mucus produced at or near the time of ovulation will dry in a characteristic feather-like or "fern" pattern. **Proliferative phase** is still another name for it because proliferation of endometrial cells occurs at this time.

3. **Ovulation,** that is, rupture of the mature follicle with expulsion of its ovum into the pelvic cavity (Figure 32-12), occurs frequently on cycle day 14 in a 28-day cycle. However, it occurs on different days in different-length cycles, depending on the length of the preovulatory phase. For example, in a 32-day cycle the preovulatory phase probably lasts until cycle day 18, and ovulation would then occur on cycle day 19 instead of 14. In short, because the majority of women show some month-to-month variation in the length of their cycles, the day of ovulation in a current or future cycle cannot be predicted with accuracy based on the length of previous cycles (see Box 32-4). Typically, there is a decrease in basal body temperature just before ovulation and a rise in temperature at the time of ovulation. This constitutes yet another "fertility sign" (see Figure 32-15).

4. The **premenstrual phase,** or **postovulatory phase,** occurs between ovulation and the onset of the menses. This phase is also called the **luteal phase,** or more simply, the **secretory phase,** because the corpus luteum secretes only during this time. It is also called the **progesterone phase** because the corpus luteum secretes mainly this hormone. The length of the premenstrual phase is fairly constant, lasting usually 14 days—or cycle days 15 to 28 in a 28-day cycle. Differences in length of the total menstrual cycle therefore exist mainly because of differences in duration of the postmenstrual rather than of the premenstrual phase.

Myometrial Cycle

The myometrium contracts mildly but with increasing frequency during the 2 weeks preceding ovulation. Contractions decrease or disappear between ovulation and the next menses, thereby lessening the probability of expulsion of a fertilized ovum that may have implanted in the endometrium.

Gonadotropic Cycle

The adenohypophysis (anterior pituitary gland) secretes two hormones called *gonadotropins* that influence female reproductive cycles. Their names are **follicle-stimulating hormone (FSH)** and **luteinizing hormone (LH).** The amount of each gonadotropin secreted varies with a rhythmic regularity that can be related, as we shall see, to the rhythmic ovarian and uterine changes just described.

BOX 32-4: FYI

Fertility Signs Used in Predicting the Time of Ovulation

Many rhythmic and recurring events that a woman may recognize on almost a monthly schedule during her reproductive years are called **"fertility signs"** and are manifestations of the body changes required for successful reproductive function. They include cyclical changes in the ovaries, in the amount and consistency of the cervical mucus produced during each cycle, in the myometrium, in the vagina, in gonadotropin secretion, in body temperature, and even in mood or "emotional tone." Accurately predicting the time of ovulation in any given menstrual cycle by recognizing one or more of these recurring fertility signs would obviously be of help in either avoiding or achieving conception. However, knowing the length of a previous cycle or series of cycles cannot ensure with any degree of accuracy the time of appearance of other fertility signs in a current cycle or how many days the preovulatory phase will last in the next or some future cycle.

Simply put, prior cycle length is not an accurate fertility sign. This fact accounts for most of the unreliability of the **calendar rhythm method** of fertility planning and its replacement by other more sophisticated **natural family planning** methods that are not based on a knowledge of previous cycle lengths to predict the day of ovulation. Instead, such natural methods base their judgments about fertility at any point in a woman's cycle on other changes. For example, measurement of basal body temperature and recognition of cyclical changes in the amount and consistency of cervical mucus, both of which occur in response to changes in circulating hormones that control ovulation. The time of ovulation can also be approximated by over-the-counter urine tests that detect the high levels of luteinizing hormone (LH) associated with ovulation (LH surge).

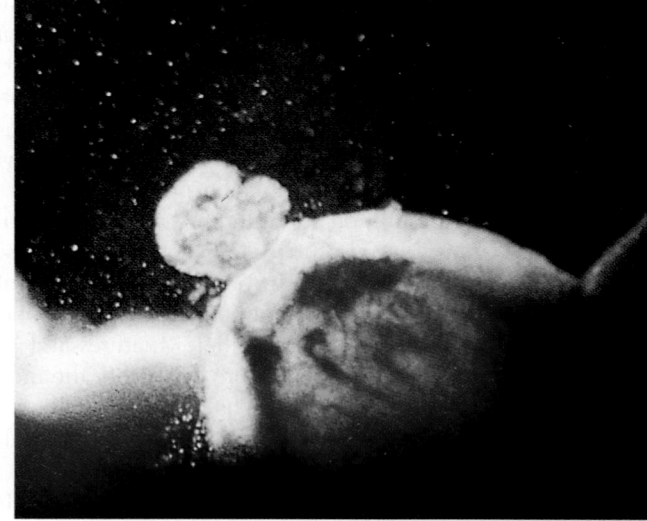

Figure 32-12 *Ovulation.* The rupture of a mature follicle on the surface of an ovary results in the release of an ovum into the pelvic cavity. This process of *ovulation* often occurs on day 14 in a 28-day menstrual cycle, but its exact timing depends on the length of the postmenstrual (preovulatory) phase. Notice in this photograph that the ovum released during ovulation is surrounded by a mass of cells.

QUICK CHECK

16. What is the function of the corpus luteum?
17. How is the corpus luteum formed?
18. What is the difference between the ovarian cycle and the menstrual (endometrial) cycle?
19. Why is the postmenstrual phase of the menstrual cycle sometimes called the *proliferative phase?*

Control of Female Reproductive Cycles

Physiologists agree that hormones play a major role in producing the cyclical changes characteristic in women during the reproductive years. The development of a method called *radioimmunoassay* has made it possible to measure blood levels of gonadotropins. By correlating these with the monthly ovarian and uterine changes, investigators have worked out the main features of the control mechanism.

A brief description follows of the mechanisms that produce cyclical changes in the ovaries and uterus and in the amounts of gonadotropins secreted.

Control of Cyclical Changes in the Ovaries

Cyclical changes in the ovaries result from cyclical changes in the amounts of gonadotropins secreted by the anterior pituitary gland. An increasing FSH blood level has two effects: it stimulates one or more primary follicles and their oocytes to start growing, and it stimulates the follicular cells to secrete **estrogens**. (Developing follicles also secrete very small amounts of progesterone.) Because of the influence of FSH on follicle secretion, the level of estrogens in blood increases gradually for a few days during the postmenstrual phase. Then suddenly, on about the twelfth cycle day, it leaps upward to a maximum peak. Scarcely 12 hours after this "estrogen surge," an "LH surge" occurs and presumably triggers ovulation a day or two later. This hormone surge is the basis of the over-the-counter "ovulation test" (see Box 32-4, p. 1142). The control of cyclical ovarian changes by the gonadotropins FSH and LH is summarized in Figure 32-13. As Figure 32-13 shows, LH brings about the following changes:

1. Completion of growth of the follicle and oocyte maturation with increasing secretion of estrogens before ovulation. LH and FSH act as synergists to produce these effects.
2. Rupturing of the mature follicle with expulsion of its ovum (ovulation). Because of this function, LH is sometimes also called "the ovulating hormone."
3. Formation of a golden body, the corpus luteum, in the ruptured follicle (process called **luteinization**). The name *luteinizing hormone* refers, obviously, to this LH function—a function to which, experiments have shown, FSH also contributes. The corpus luteum functions as a temporary endocrine gland. It secretes only during the luteal (postovulatory, or premenstrual) phase of the menstrual cycle. Its hormones are progestins (the important one of which is progesterone) and also estrogens. The blood level of progesterone rises rapidly after the "LH surge" described earlier. It remains at a high level for about a week, then decreases to a very low

level approximately 3 days before menstruation begins again. This low blood level of progesterone persists during both the menstrual and the postmenstrual phases. What are its sources? Not the corpus luteum, which secretes only during the luteal phase, but the developing follicles and the adrenal cortex. Blood's estrogen content increases during the luteal phase but to a lower level than develops before ovulation.

If pregnancy does not occur, lack of sufficient LH and FSH causes the corpus luteum to regress in about 14 days. The corpus luteum is then replaced by the corpus albicans. Review Figure 32-3, which shows the cyclical changes in the ovarian follicles.

Control of Cyclical Changes in the Uterus

Cyclical changes in the uterus are brought about by changing blood concentrations of estrogens and progesterone. As blood estrogens increase during the postmenstrual phase of the menstrual cycle, they produce the following main changes in the uterus:

- Proliferation of endometrial cells, producing a thickening of the endometrium
- Growth of endometrial glands and of the spiral arteries of the endometrium
- Increase in the water content of the endometrium
- Increased myometrial contractions

Increasing blood progesterone concentration during the premenstrual phase of the menstrual cycle produces progestational changes in the uterus—that is, changes favorable for pregnancy—specifically the following:

- Secretion by endometrial glands, thereby preparing the endometrium for implantation of a fertilized ovum
- Increase in the water content of the endometrium
- Decreased myometrial contractions

As mentioned earlier, low levels of FSH and LH cause regression of the corpus luteum if pregnancy does not occur. This in turn causes a drop in estrogen and progesterone levels, with the result that their maintenance of a thick, vascular endometrium ceases. Thus a drop in estrogen and progesterone levels at the end of the premenstrual phase triggers the endometrial sloughing that characterizes the menstrual phase.

Control of Cyclical Changes in Gonadotropin Secretion

Both negative and positive feedback mechanisms help control anterior pituitary secretion of the gonadotropins FSH and LH. These mechanisms involve the ovaries' secretion of inhibin, estrogens, and progesterone and secretion of releasing hormones by the hypothalamus. Figure 32-14 describes a negative feedback mechanism that controls gonadotropin secretion. Examine it carefully. Note particularly the effects of a sustained high blood concentration of estrogens and progesterone on anterior pituitary gland secretion and the effect of a low blood concentration of FSH on follicular development: essentially, follicles do not mature and ovulation does not occur.

Several observations and animal experiments strongly suggest that sustained high blood levels of estrogens, progesterone, and inhibin decrease pituitary secretion of FSH and LH. These ovar-

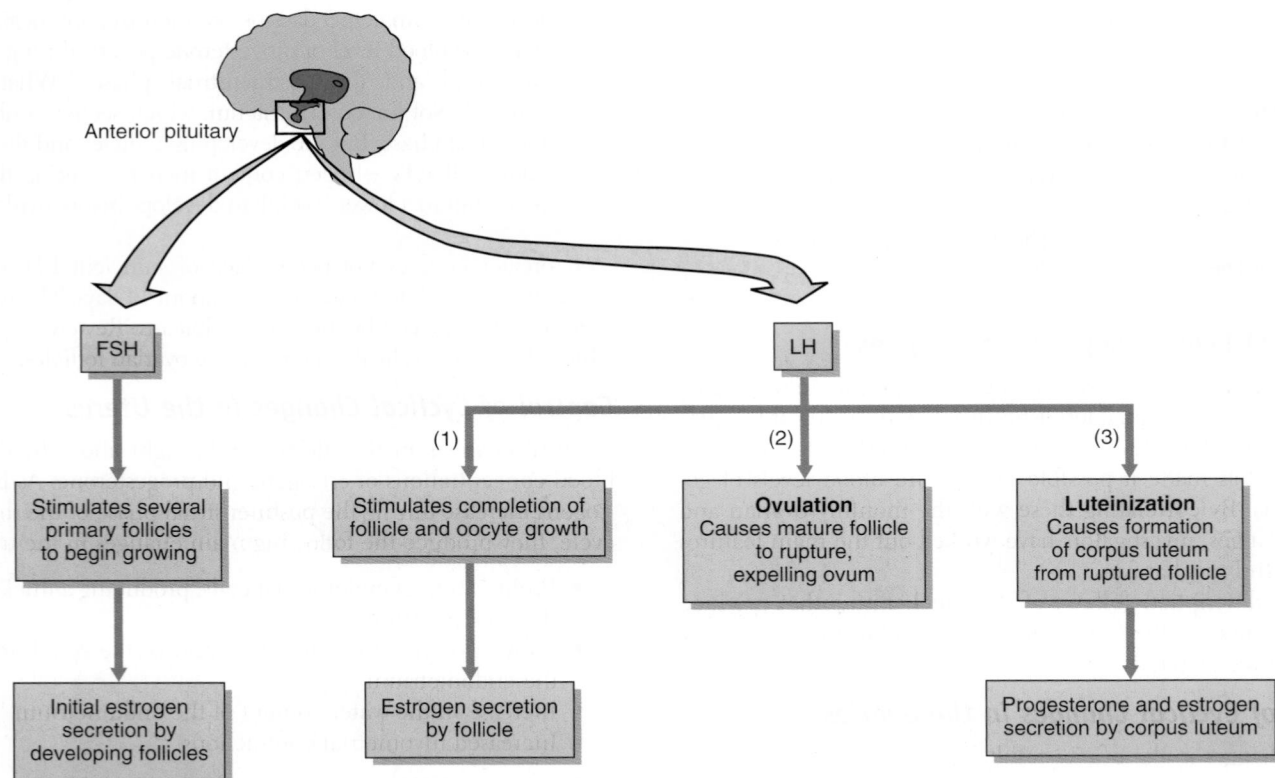

Figure 32-13 *The primary effects of gonadotropins on the ovaries.* Follicle-stimulating hormone *(FSH)* gets its name from the fact that it triggers development of primary ovarian follicles and stimulates follicular cells to secrete estrogens. Luteinizing hormone *(LH)* has several effects on ovaries: *(1)* LH acts as a synergist to FSH to enhance its effects on follicular development and secretion; *(2)* LH presumably triggers ovulation— hence it is called "the ovulating hormone"; and *(3)* LH has a luteinizing effect (for which the hormone was named); recent evidence shows that FSH is also necessary for luteinization.

ian hormones appear to inhibit certain neurons of the hypothalamus (part of the central nervous system) from secreting FSH-releasing and LH-releasing hormones into the hypophyseal portal vessels (see Figure 32-14, on p. 1145). Without the stimulating effects of these releasing hormones, the pituitary's secretion of FSH and LH decreases.

A positive feedback mechanism has also been postulated to control LH secretion. The sudden and marked increase in blood's estrogen content that occurs late in the follicular phase of the menstrual cycle is thought to stimulate the hypothalamus to secrete LH-releasing hormone (LH-RH) into the hypophyseal portal vessels. As its name suggests, this hormone stimulates the release of LH by the anterior pituitary, which in turn accounts for the "LH surge" that triggers ovulation. The fact that a part of the brain—the hypothalamus—secretes FSH- and LH-releasing hormones has interesting implications. This may be the crucial part of the pathway by which changes in a woman's environment or in her emotional state can alter her menstrual cycle. That this occurs is a matter of common observation. Stress, for example—such as intense fear of either becoming or not becoming pregnant—often delays menstruation.

Importance of Female Reproductive Cycles

There are several important functional roles played by the female reproductive cycles. As Figure 32-14 shows, the changes associated with the different cycles are all closely interrelated. The pri-

mary role of the ovarian cycle, for example, is to produce an ovum at regular enough intervals to make reproductive success likely. The ovarian cycle's secondary role is to regulate the endometrial (menstrual) cycle by means of the sex hormones estrogen and progesterone. The role of the endometrial cycle, in turn, is to ensure that the lining of the uterus is suitable for the implantation of an embryo if fertilization of the ovum occurs. The constant renewal of the endometrium makes successful implantation more likely. The cyclical mechanisms of female reproductive function, and the fact that an ovum must unite with a sperm during the first 24 hours or so after ovulation to reach the uterus at the proper stage of development to implant, result in the fact that a woman is fertile only a few days out of each monthly cycle. Human fertility is further limited by the fact that sperm usually cannot survive in the female reproductive tract for more than a few days. Such limited fertility increases the likelihood that conception will occur only when the woman's body is at its reproductive peak.

Infertility and use of Fertility Drugs

Infertility is often defined as failure to conceive after 1 year of regular unprotected intercourse. Infertility may be caused by a wide variety of medical, environmental, and even lifestyle factors, such as smoking or alcohol abuse. Causal factors may be traced to various problems in either the male or female partner, each accounting for about 40% of cases. Of the remaining 20% of affected

Text continued on p. 1147

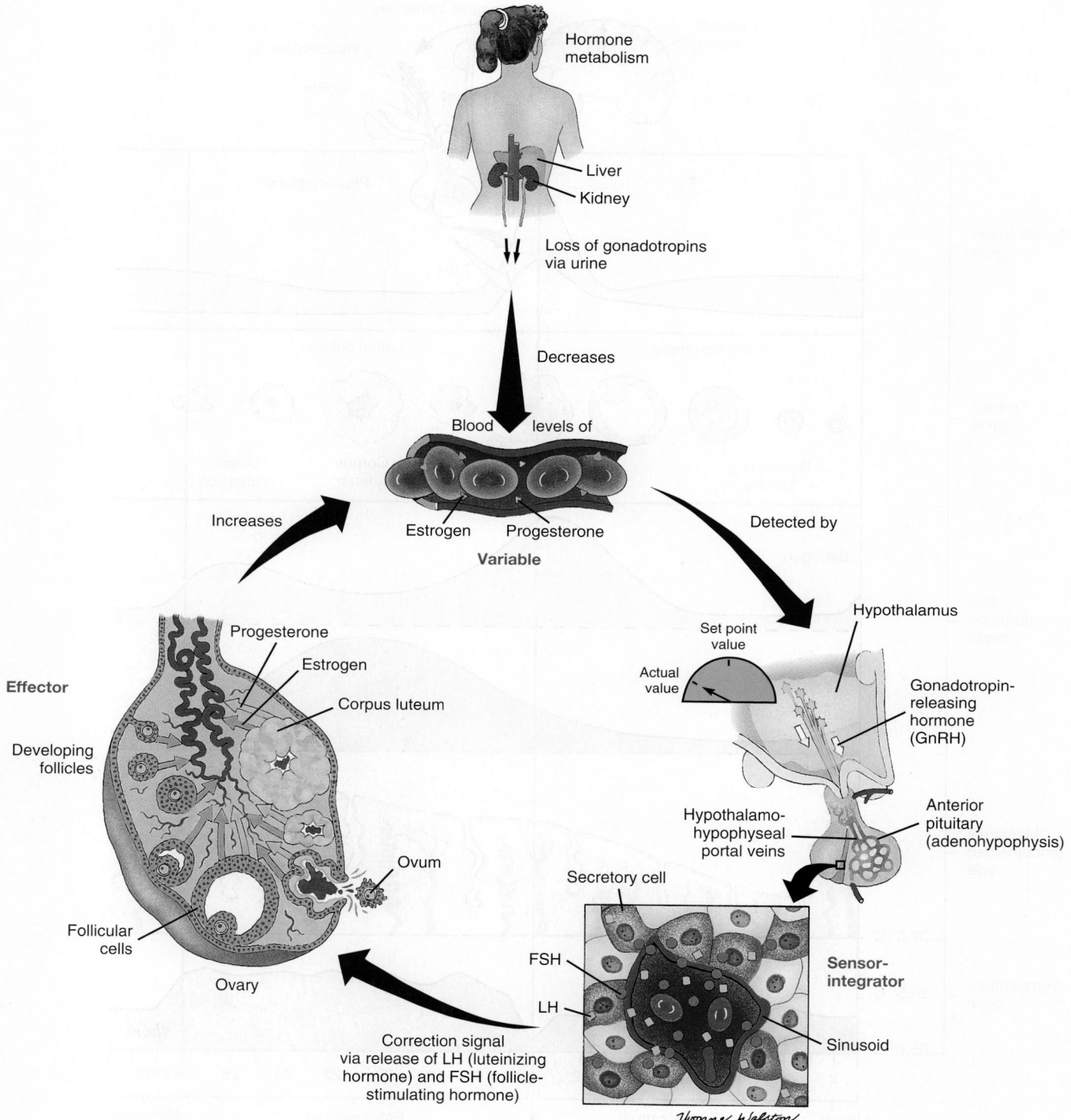

Figure 32-14 *Control of FSH and estrogen secretion.* A negative feedback mechanism controls anterior pituitary secretion of follicle-stimulating hormone (FSH) and ovarian secretion of estrogens. A high blood level of FSH stimulates estrogen secretion, whereas the resulting high estrogen level inhibits FSH secretion. How does this compare with the LH testosterone feedback mechanism in the male? (See Figure 31-6, p. 1114, if you want to check your answer.) According to the diagram, what effect does a high blood concentration of estrogens have on anterior pituitary secretion of FSH? of LH?

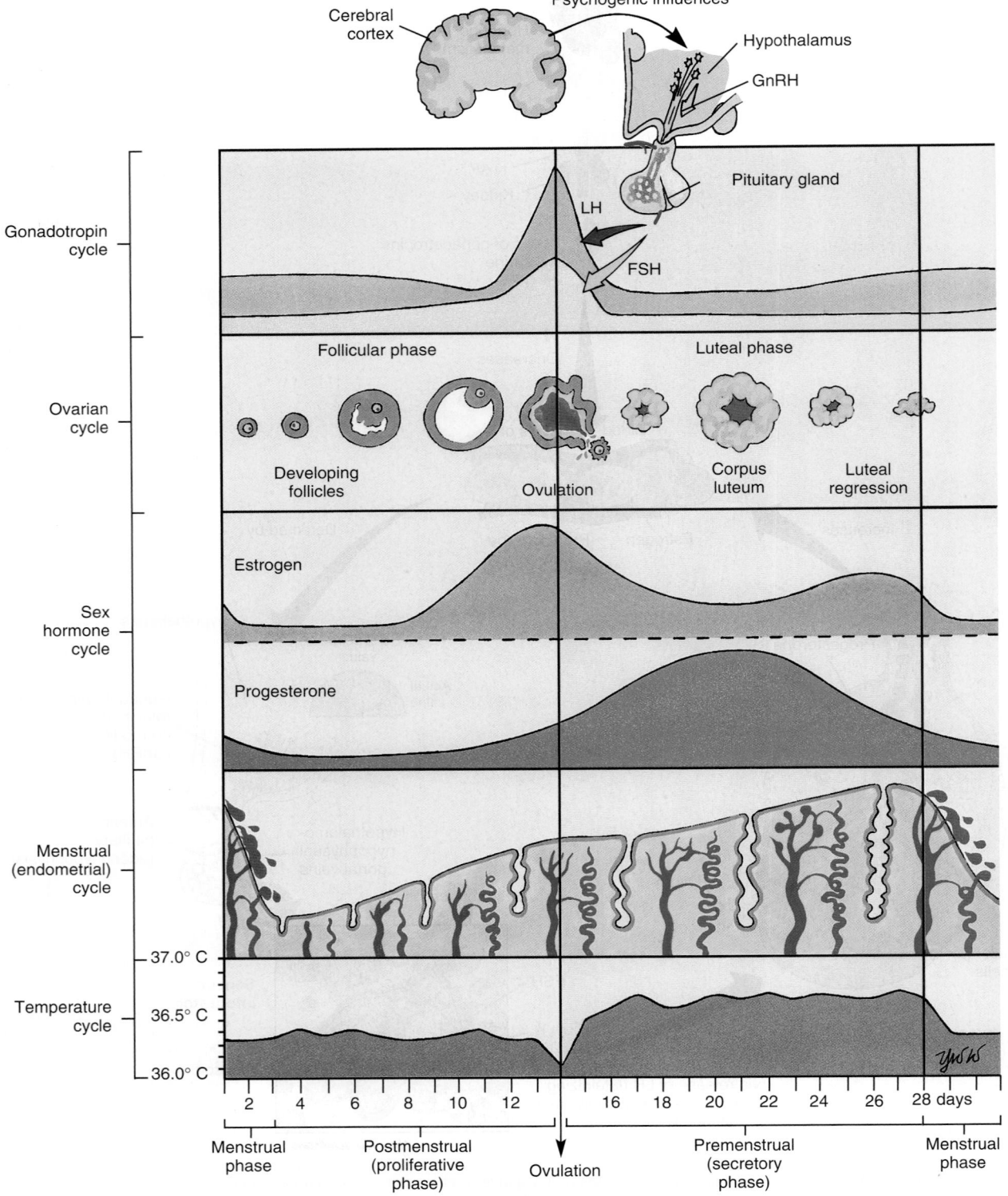

Figure 32-15 *Female reproductive cycles.* This diagram illustrates the interrelationships among the cerebral, hypothalamic, pituitary, ovarian, and uterine functions throughout a standard 28-day menstrual cycle. The variations in basal body temperature are also illustrated.

couples, infertility in about 10% is due to problems shared by both partners and in about 10% the reason is never determined. If testing identifies the female member of the couple as infertile, she joins a subset of about 25% of women in the overall population who will experience some period of infertility during their reproductive years. In many cases, infertility results from a failure to ovulate—often caused by a medical condition such as **polycystic ovary disease** (see Mechanisms of Disease). Significant numbers of infertile women experiencing ovulatory dysfunction desire to become pregnant and, after a sometimes long and complex medical "work-up" and selection process, become candidates to receive so-called **fertility drugs**—either alone or in combination with other "assisted reproductive procedures" such as artificial insemination.

An orally administered fertility drug called *clomiphene* (Clomid, Serophene) is often used to treat women who have anovulatory cycles. It is an antiestrogen agent that competes with estrogen for estrogen-receptor binding sites. By blocking estrogen it acts as an ovulatory stimulant. How? By effectively "tricking" the pituitary gland into producing FSH and LH. The gonadotropin surge causes normal follicle growth and subsequent ovulation. Clomiphene is given in daily doses of 50 or 100 mg for 5 days, generally starting on day 5 of the menstrual cycle. In a successful treatment program, ovulation most often occurs from 5 to 10 days after a course of the drug. The incidence of multiple births is about 5% to 7% (mostly twins), a percentage much lower than what is observed following direct administration (injection) of FSH and LH, where the intent is to produce multiple follicles before inducing ovulation.

Supraovulation, or simultaneous rupture of multiple mature follicles, is an infertility treatment option that may be employed if clomiphene use proves ineffective or if multiple ova are deemed desirable in assisted reproductive procedures such as **in vitro fertilization.** It most frequently involves self-administered injections of either (1) drugs called **menotropins** or (2) genetically developed (recombinant) gonadotropins. Menotropins contain high concentrations of FSH and LH that are extracted from the urine of pregnant women or from postmenopausal women. Recombinant gonadotropins include human menopausal gonadotropins (hMGs) and FSH.

BOX 32-5 SPORTS AND FITNESS
The "Sports Triad" in Elite Female Athletes

A disturbing trend among some elite female athletes involves development of a so-called "triad" of undesirable outcomes. In an attempt to improve performance, these athletes couple severe caloric restriction with overtraining. The result is often a flawed sense of body image that equates thinness with athletic potential. The triad involves (1) *weight loss,* (2) *amenorrhea* (failure to have a menstrual period), and (3) *osteoporosis*. Amenorrhea is caused by a drastic decrease in estrogen secretion as the body attempts to conserve energy by closing down the reproductive function. It is the decrease in estrogen levels that triggers early development of osteoporosis and permanent skeletal damage. The triad of weight loss, amenorrhea, and osteoporosis is often accompanied by development of eating disorders, multiple nutritional deficits, and other serious and potentially fatal outcomes.

Menarche and Menopause

The menstrual flow first occurs (**menarche**) at puberty, at about the age of 13 years, although there is individual variation according to race, nutrition, health, and heredity. Normally, it recurs about every 28 days for about 3 decades, except during pregnancy, and then ceases (**menopause,** or **climacteric**). The average age at which menstruation ceases is reported to have increased markedly—from about age 40 years a few decades ago to between ages 45 and 50 years more recently.

Recall that gonadotropins extracted from the urine of menopausal women (menotropins) are used as fertility drugs. Figure 32-16 shows how the changes just described relate to changes in hormone levels over the life span. Relatively low concentrations of gonadotropins (FSH and LH) sustain a peak of estrogen secretion from menarche to menopause. After menopause, estrogen concentration decreases dramatically—which causes a negative feedback response that increases the gonadotropin levels. Because the follicular cells are no longer sensitive to gonadotropins after menopause, the increased gonadotropin level has no effect on estrogen secretion.

QUICK CHECK

20. A surge in FSH and LH is associated with what major event of the ovarian cycle?
21. How does an increase in estrogen level affect the uterine lining?
22. What is *menopause?* What causes menopause to occur?

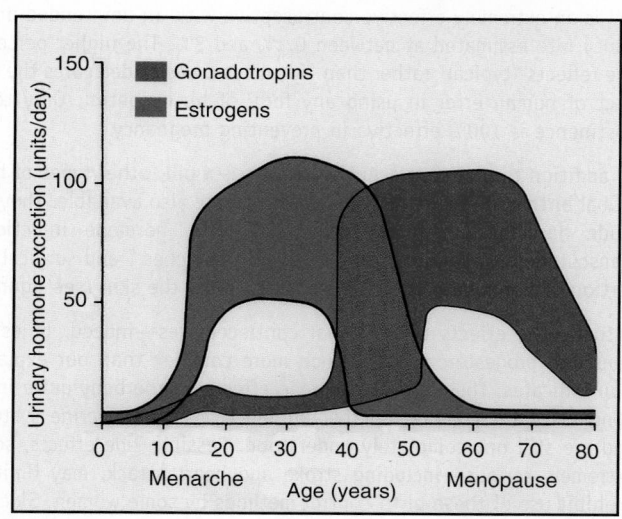

Figure 32-16 *Gonadotropin and estrogen levels over the life span.* This graph shows changes in hormone levels as reflected in urinary excretion rates from birth to advanced old age. Note that an increase in the gonadotropin level at the time of menarche sustains a high but variable level of estrogens until the time of menopause, when ovarian follicular cells cease to respond to gonadotropins. A negative feedback mechanism tries to increase estrogen levels to their former high levels by increasing the gonadotropin levels—a strategy that always fails.

Cycle of Life
Female Reproductive System

As mentioned in Chapter 31, the reproductive system is unlike any other body system with regard to the normal changes that occur during the life span. Unlike other systems, the female reproductive system does not begin to perform its functions until the teenage years (puberty), and unlike the male reproductive system, the female reproductive system ceases its principal functions in middle adulthood.

The female organs begin their initial stages of development in the womb. As a matter of fact, the first stage of meiotic development of all the ova that will ever be produced by a woman is completed by the time she is born. However, full development of the reproductive organs—and the gametes within the ovaries—does not resume until puberty. At puberty, reproductive hormones stimulate the organs of the reproductive tract to become functional and produce a mature ovum one at a time. Reproductive function then continues in a cyclical fashion until **menopause.** Menopause is an event that is usually marked by the passage of at least 1 full year without menstruation. After that time, a woman may continue to enjoy normal sexual activity, but she cannot produce more offspring.

BOX 32-6: HEALTH MATTERS
Methods of Contraception

Hormonal methods of contraception began with establishment of the relationship between sex hormone levels and ovulation. Continuing research in this area led to the development of **oral contraceptives**—often collectively called "the Pill." Numerous oral contraceptive products are now available containing different types, combinations, and dosages of estrogen and progestin products. The so-called "minipill" contains only progestin. Most hormonal contraceptives were developed to prevent pregnancy by initiating negative feedback inhibition of FSH and LH secretion. As a result, mature follicles do not develop, and LH levels required to initiate ovulation do not occur. The next menses, however, does take place if the progesterone and estrogen dosage is stopped in time to allow their blood levels to decrease as they normally do near the end of the cycle to bring on menstruation. For this reason, the Pill can be used to regulate the menstrual cycle, as well as prevent pregnancy. If used correctly and consistently, the pill is an extremely effective contraceptive with an unintended pregnancy rate estimated at between 0.1% and 3%. The higher percentage reflects "typical" rather than "ideal" use, and underscores the impact of human error in using any form of birth control. Only total abstinence is 100% effective in preventing pregnancy.

In addition to oral contraceptives taken as a pill, other types of hormonal birth control "delivery mechanisms" are also available. They include hormone-impregnated vaginal inserts, hormone injections, transcutaneous administration using skin "patches," and surgical insertion of hormone-containing implants under the skin (see figure).

Actually the effects of hormonal contraceptives—indeed, of estrogens and progesterone—are much more complex than our explanation indicates. They have widespread effects on the body quite independent of their action on the reproductive and endocrine systems and are still not completely understood. Possible side effects, some extremely serious—including stroke and heart attack, may limit or prohibit use of these birth control methods by some women. Side effects of hormonal contraceptives are especially troublesome if these products are used for extended periods, by older women, by women who smoke, and in women with blood clotting problems or cardiovascular disease.

In addition to hormonal methods of contraception, many other methods, each with differing rates of effectiveness and unique advantages and disadvantages, are available. For example, *spermicidal methods* involve use of preparations (foams, jellies, and creams) that act to kill sperm, and *mechanical barrier methods* use devices such as condoms, diaphragms, and cervical caps to block sperm from entering the uterus. So-called *behavioral methods* such as tubal ligation (Box 32-1) and male vasectomy (see Box 31-2) result in permanent sterility.

Decision making concerning the use of contraceptive methods to regulate reproductive function involves many complex interactions that are uniquely human—and very personal. The decision to use or avoid contraception—or employ any particular contraceptive method—at any point in time is often influenced by differing medical, social, cultural, ethical, and religious factors as well as by the cost, reliability, safety, or ease of use of a particular method. Informed and thoughtful decision making in this area of human behavior is critically important. It will often be necessary for different individuals to seek out different types of information—from different but credible and knowledgeable sources—in order to make an informed decision that is "right" for that individual. However, seeking counsel and advice from a trusted health care provider early in the process is always recommended.

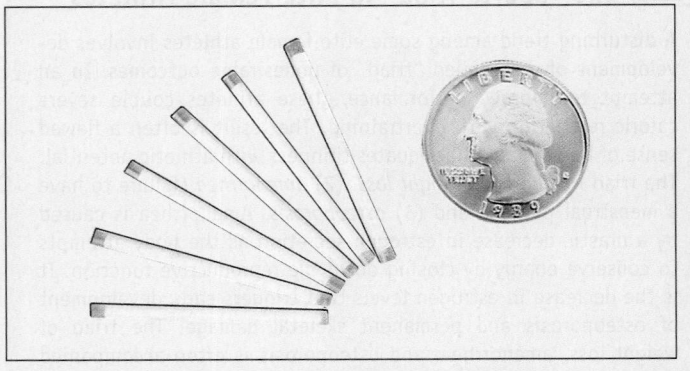

THE BIG PICTURE

Female Reproductive System and the Whole Body

As stated several times in this chapter, the importance of reproductive function lies in the fact that it imparts virtual immortality to our genes. This is important not only to the survival of the human species but also to the survival of life itself. After all, life as we know it could not exist without a genetic code. The "big picture" of human reproduction requires two systems, one reproductive system in each parent. The combined roles of the male and female reproductive systems will be explored as a single topic in the early part of the next chapter.

For now, we will take a closer look at the female reproductive system and its relationships with other systems within a woman's body. As with any system, the female reproductive system cannot function without the maintenance functions of the circulatory, immune, respiratory, digestive, and urinary systems. The female reproductive system

shares a special anatomical relationship with the urinary system. These two systems develop in close proximity to each other and thus share a common structure: the vulva. A special anatomical relationship with the skeletal muscular system is evident in the structure known as the *perineum*. Of course the skeletal and muscular systems both support and protect the internal organs of the female reproductive system. An even more special relationship with the integumentary system should be noted. The breasts, containing the milk-producing mammary glands, are actually modifications of the skin. Although structurally the breasts can be thought of as belonging to the integumentary system, functionally they are best considered as a part of the reproductive system. Nervous and endocrine regulation of female reproductive function has been outlined in this chapter and is explored further in the next chapter.

Mechanisms of Disease

DISORDERS OF THE FEMALE REPRODUCTIVE SYSTEM

Hormonal and Menstrual Disorders

Dysmenorrhea (dis-men-oh-REE-ah), meaning "painful menstruation," is the term used to describe *menstrual cramps*, the painful periods that affect 75% to 80% of women at some time during their reproductive years. For significant numbers of those affected, severe lower abdominal cramping and back pain accompanied by headache, nausea, and vomiting will disrupt their school, work, athletic, or other activities. *Primary dysmenorrhea* is the most common type, occurring primarily in adolescents and young women. Symptoms, which can last from hours to days and vary in severity from cycle to cycle, are caused by an abnormally increased concentration of certain prostaglandins produced by the uterine lining. High concentrations of prostaglandin E_2 (PGE_2) and prostaglandin F_2 (PGF_2) cause painful spasms by decreasing blood flow and oxygen delivery to uterine muscle. Fortunately, primary dysmenorrhea is not associated with pelvic disease, such as an infection or tumor, and can generally be treated effectively with over-the-counter anti-inflammatory and prostaglandin-inhibiting drugs such as ibuprofen and naproxen. In more severe cases a physician may prescribe more powerful anti-inflammatory drugs or certain hormones, including oral contraceptives, to alter menstrual cycle activity or reduce the level or frequency of cyclical uterine contractions. *Secondary dysmenorrhea* refers to menstrual-related pain caused by some type of pelvic pathological condition, including inflammatory conditions and cervical stenosis. Treatment of secondary dysmenorrhea conditions and cervical stenosis involves treating the underlying disorder.

Amenorrhea (ah-men-oh-REE-ah) is the absence of normal menstruation. *Primary amenorrhea* is the failure of menstrual cycles to begin and may be caused by various factors, such as hormone imbalances, genetic disorders, brain lesions, or structural deformities of the reproductive organs. *Secondary amenorrhea* occurs when a woman who has previously menstruated slows to three or fewer cycles per year. Secondary amenorrhea may be a symptom of weight loss, pregnancy, lactation, menopause, or disease of the reproductive organs. Treatment of amenorrhea involves treating the underlying disorder or condition. If amenorrhea occurs as a component of the "sports triad" (see Box 32-5), treatment may become part of extensive and long-term therapy needed to address a number of complex nutritional, hormonal, and self-image issues.

Dysfunctional uterine bleeding (DUB) is irregular or excessive uterine bleeding that most often results from either a structural problem or some type of hormonal imbalance that causes a disruption of blood supply rather than from an infection or other disease condition. Excessive uterine bleeding from any cause can result in life-threatening anemia. Therefore, DUB is a significant medical problem—one that affects nearly 2 million women in the United States each year. To diagnose the cause, a physician may employ specialized ultrasound or x-ray studies, look directly inside the uterus using a telescope-like instrument inserted through the vagina and cervix, or examine tissue obtained by biopsy to exclude cancer.

Transabdominal pelvic ultrasound is perhaps the most widely used and useful imaging technique to look at the uterus and adjacent reproductive structures. During this procedure a transducer is placed on the lower anterior abdominal wall and high-frequency sound waves are transmitted into the pelvic viscera. The returning echoes are then viewed on a screen sim-

Mechanisms of Disease—cont.

ilar to that of a sonar receiver or "fish-finder." The resulting image, a "slice picture," may be enhanced by filling the uterus with saline solution (Figure 32-17). Structural problems, such as growth of a uterine malignancy or benign tumor, may cause DUB by injuring the blood vessels of the uterine wall or lining as the tissue mass grows in size and the normal contours of the organ are distorted. Figure 32-17, A, shows a normal longitudinal pelvic ultrasound image. It was taken "looking down" through the anterior abdominal wall in the mid portion of the pelvis. Note how the bladder was used as a "window" to view the uterus and adjacent reproductive structures below. The bladder, uterus, contours of the uterine cavity, cervix, and vagina are all visible. Figure 32-17, B, shows the appearance of a similar ultrasound image in a patient suffering from DUB caused by a structural problem—abnormal muscular growths called **uterine fibroids** (discussed on p. 1152). The uterus is

enlarged and pushing into the bladder, it appears "lumpy," and the cavity contours are obviously distorted. Surgical removal or treatment to shrink the fibroid growths is usually curative.

If hormonal imbalance is the cause of DUB, it is the excessive growth (hyperplasia) and breakdown of delicate endometrial tissue that results in heavy bleeding. In these cases, treatment generally begins with administration of nonsteroidal anti-inflammatory drugs and hormonal manipulation using low-dose birth control pills. If conservative treatment fails to stop the endometrial lining from hemorrhaging, hysterectomy remains one of the most effective curative options. However, less invasive procedures, including **endometrial ablation** techniques, are now being used more frequently to destroy the endometrial lining and halt or permanently reduce excessive menstrual blood loss in many women suffering from DUB. In **thermal ablation,** a balloon is inserted into the uterus and filled with fluid. A heat probe is then inserted into the balloon and the fluid is heated to a temperature that will destroy the endometrium. In **radiofrequency ablation,** a gold-plated mesh fabric is used to fill the uterine cavity and is then charged with radiofrequency energy that destroys the friable and bleeding endometrial cells. Both procedures carry less risk and have shorter recovery periods than does a hysterectomy.

Premenstrual syndrome (PMS) is a condition that involves a collection of symptoms that regularly occur in many women during the premenstrual phase of their reproductive cycles. Symptoms include headache, irritability, fatigue, nervousness, weight gain, sleep changes, depression, and other problems that are often distressing enough to limit activity and affect personal relationships. Because the cause of PMS is still unclear, current treatments focus on relieving the symptoms.

Infection and Inflammation

Infections of the female reproductive tract are often classified as *exogenous* or *endogenous*. Exogenous infections result from pathogenic organisms transmitted from another person, such as **sexually transmitted diseases (STDs)** (Table 32-1). Endogenous infections result from pathogens that normally inhabit the intestines, vulva, or vagina. You may recall from Chapter 1 that many areas of the body are normally inhabited by pathogenic microorganisms but that they cause infection only when there is a change in conditions or they are moved to a new area.

Pelvic inflammatory disease (PID) occurs as either an acute or chronic inflammatory condition that can be caused by several different pathogens, which usually spread upward from the vagina. PID is a major cause of infertility and sterility and affects more than 800,000 women each year in the United States. It is a common complication following infection by chlamydial and gonococcal STD organisms (see Table 32-1). Infection involving the uterus, uterine tubes, ovaries, and other pelvic organs often results in development of scar tissue and adhesions. Direct laparoscopic examination is often used

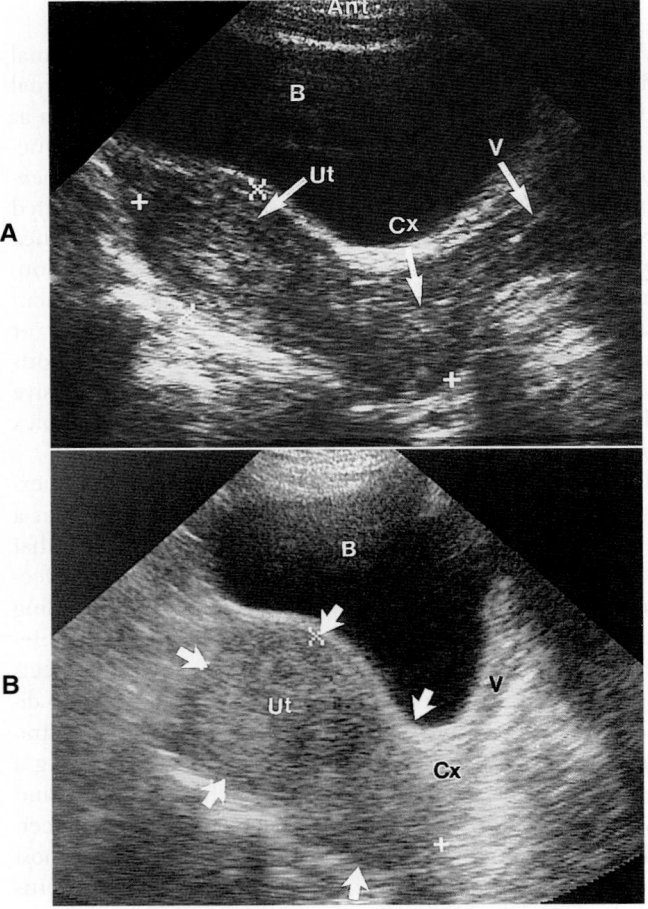

Figure 32-17 *Female pelvic transabdominal ultrasound.* **A,** Normal longitudinal ultrasound image. The bladder *(B)*, uterus *(Ut)*, cervix *(Cx)*, and vagina *(V)*, are visible. **B,** Longitudinal ultrasound image of uterine fibroids causing DUB. Echo pattern shows an enlarged and "lumpy" appearing *(arrows)* uterus *(Ut)* protruding into the bladder *(B)*. The cervix *(Cx)* and vagina *(V)* are visible.

Mechanisms of Disease—cont.

Table 32-1   Examples of Sexually Transmitted Diseases (STDs)

DISEASE	PATHOGEN	DESCRIPTION
Acquired immune deficiency syndrome (AIDS)	*Virus:* Human immunodeficiency virus (HIV)	HIV is transmitted by direct contact of body fluids, often during sexual contact. After an initial infection period that sometimes lasts several years, HIV infection produces the condition known as AIDS. AIDS is characterized by damage to lymphocytes (T cells), resulting in immune system impairment. Death results from secondary infections or tumors.
Candidiasis	*Fungus: Candida albicans*	Yeast infection characterized by a white discharge (leukorrhea), peeling of skin, and bleeding. Although it can occur as an ordinary opportunistic infection, it may be transmitted sexually.
Chancroid	*Bacterium: Haemophilus ducreyi*	Highly contagious STD, characterized by papules on the skin of the genitalia that eventually ulcerate. About 90% of cases are reported by men.
Genital herpes	*Virus:* Herpes simplex virus (HSV)	HSV causes blisters on the skin of the genitalia. The blisters may disappear temporarily but may reappear occasionally, especially as a result of stress.
Genital warts	*Virus:* Human papilloma virus (HPV-6, HPV-7)	Genital warts are nipple-like neoplasms of skin covering the genitalia.
Giardiasis	*Protozoan: Giardia lamblia*	Intestinal infection; may be spread by sexual contact. Symptoms range from mild diarrhea to malabsorption syndrome, with about half of all cases being asymptomatic.
Gonorrhea	*Bacterium: Neisseria gonorrhoeae*	Gonorrhea primarily involves the genital and urinary tracts but can affect throat, conjunctiva, or lower intestines. It may progress to pelvic inflammatory disease (PID).
Hepatitis	*Virus:* Hepatitis B virus (HBV)	This acute-onset liver inflammation may develop into a severe chronic disease, perhaps ending in death.
Lymphogranuloma venereum (LGV)	*Bacterium: Chlamydia trachomatis*	Chronic STD, characterized by genital ulcers, swollen nodes, headache, fever, and muscle pain. *C. trachomatis* infection may cause other syndromes, including conjunctivitis, urogenital infections, and systemic infections. *C. trachomatis* infections constitute the most common STD in the United States.
Scabies	*Animal: Sarcoptes scabiei*	Scabies is caused by infestation by the itch mite, which burrows into the skin to lay eggs. About 2 to 4 months after initial contact, a hypersensitivity reaction occurs, causing a rash along each burrow that itches intensely. Secondary bacterial infection is possible.
Syphilis	*Bacterium: Treponema pallidum*	Although transmitted sexually, syphilis can affect any system. Primary syphilis is characterized by a chancre sore on an exposed area of the skin. If untreated, secondary syphilis may appear 2 months after the chancre disappears. The secondary stage occurs when the spirochete has spread through the body, presenting various symptoms, and is still highly contagious—even through kissing. Tertiary syphilis may appear years later, possibly resulting in death.
Trichomoniasis	*Protozoan: Trichomonas vaginalis*	Urogenital infection, asymptomatic in most women and nearly all men. Vaginitis may occur, characterized by itching or burning and a foul-smelling discharge.

to determine the severity of the PID infection and the reproductive organs involved (Figure 32-18).

PID that results in uterine tube inflammation, a condition called **salpingitis** (sal-pin-JYE-tis) is often characterized by obstruction of the lumen and marked dilation at the end of the tube caused by accumulation of fluid that cannot escape—a condition physicians refer to as **hydrosalpinx** (hye-droh-SAL-pinks) (see Figure 32-18, *B*). Since ultrasound images cannot show if the lumen of the uterine tube is open or obstructed, other imaging techniques are required to make a definitive diagnosis of infertility caused by tubal obstruction. Figure 32-19 shows two specialized x-ray images of the uterus and uterine

tubes called **hysterosalpingograms** (his-ter-oh-sal-PING-go-grams). In the x-ray technique used to produce these images, a cannula is inserted into the cervical canal and contrast material is injected into the uterine cavity and allowed to fill the uterine tubes. In the normal image shown in Figure 32-19, *A*, the contrast material has filled the uterine tubes and flowed out into the peritoneal cavity, demonstrating that the tubes are not obstructed.

Figure 32-19, *B*, is an abnormal hysterosalpingogram of a woman who was treated for salpingitis at an earlier date and later experienced infertility. Note how the contrast material fills and then collects in the dilated portion of the uterine tube

Mechanisms of Disease—cont.

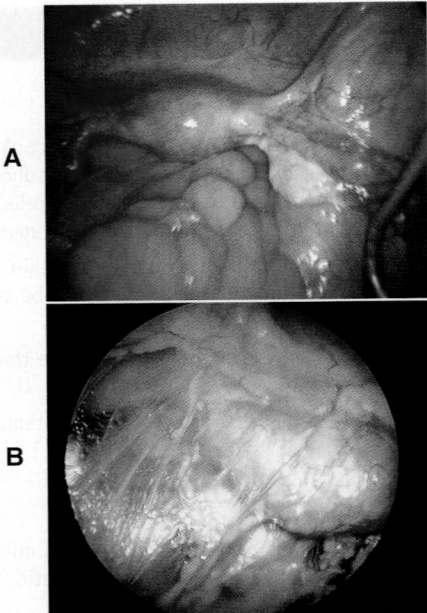

Figure 32-18 *Laparoscopic views of the female pelvis.* **A,** Laparoscopic image shows normal female pelvic reproductive organs. **B,** Laparoscopic view of pelvic inflammatory disease (PID). Note the clear reddish inflammatory membrane that covers and fixes the ovary, uterine tube, and uterus to the surrounding structures. The inflamed and dilated end of the fluid-filled uterine tube (hydrosalpinx) is clearly visible.

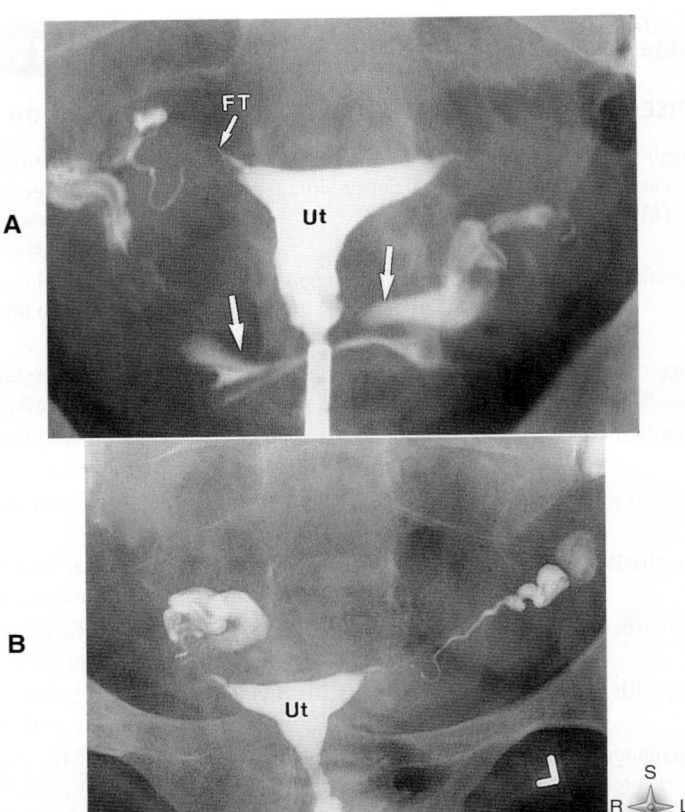

Figure 32-19 *Normal and abnormal hysterosalpingogram.* **A,** Normal image. The uterine cervix is cannulated, and contrast material has filled the uterus *(Ut)* and uterine (fallopian) tubes *(FT)*. *Arrows* indicate contrast material that has flowed out into the peritoneal cavity, indicating that the tubes are not obstructed. **B,** Abnormal image. Contrast material fills the uterine cavity and tubes with no spill into the pelvis. Note the dilated ends of the uterine tubes (hydrosalpinx). Image confirms distal obstruction of the uterine tubes.

(hydrosalpinx) but does not spill into the pelvis. In this case a specialized x-ray image allows the physician to make a definitive diagnosis—infertility caused by uterine tube obstruction—and treat it appropriately.

Although some chlamydial infections may not cause symptoms, most cases of PID are accompanied by fever, pelvic tenderness, and pain. Unfortunately, because of scarring and adhesions, pain may continue even after antibiotic therapy has eliminated the active infection. If left untreated, PID infections may spread to other tissues, including the blood, resulting in septic shock and death.

Vaginitis (vaj-i-NYE-tis) is inflammation or infection of the vaginal lining. Vaginitis most often results from STDs or from a "yeast infection." So-called yeast infections are usually opportunistic infections of the fungus *Candida albicans*, producing **candidiasis** (kan-did-EYE-as-is). Candidiasis infections are characterized by a whitish discharge—a symptom known as **leukorrhea** (loo-koh-REE-ah).

Tumors and Related Conditions

Fibroid, myoma (my-OH-mah), and **fibromyoma** (fye-broh-my-OH-mah) are all terms used to describe benign (non-cancerous) tumors of uterine fibrous or smooth muscle tissue. Individual fibroids may occur, but multiple growths are not unusual. Fibroids are common in women during their reproduc-

tive years and develop most often in the myometrium of the uterine body and rarely in the cervix. The fact that they are seldom seen before puberty, increase in size during pregnancy, and tend to shrink in postmenopausal women suggests that age and estrogen levels may play a role in their development. Fibroids range in size from small asymptomatic nodules to massive tumors that may be painful and exert pressure on other pelvic organs. Growth during pregnancy may result in placental hemorrhage or malpresentation of the fetus, complicating labor and delivery. In addition to pain, symptoms will vary depending on the size and location of the tumor. Even small tumors developing beneath the endometrium can cause severe hemorrhage (DUB). Tumor size, location, and severity of symptoms will determine treatment options. A relatively new technique similar to a heart catheterization, called **uterine artery embolization**, involves snaking a small catheter

Mechanisms of Disease—cont.

through an artery in the groin into the arterial vessel supplying blood to the fibroid. Tiny inert pellets are then injected into the artery blocking the flow of blood. The procedure results in dramatic shrinkage of the treated fibroid and a reduction in symptoms, including hemorrhage. Surgical removal of individual fibroids or, in more severe cases, hysterectomy may be indicated.

Polycystic ovary disease (PCOD) is a condition that affects 3% to 6% of reproductive-age women. It is characterized by ovaries usually twice normal size that are studded with fluid-filled cysts about 0.5 to 1.5 cm in diameter (Figure 32-20). The cysts are found on both ovaries and develop from mature follicles that fail to rupture completely. Corpora lutea are generally absent. Women with PCOD frequently have numerous endocrine abnormalities, including infrequent menstrual cycles and persistent anovulation. Fertility drugs such as clomiphene are often prescribed to induce ovulation.

Ovarian cysts are very common fluid-filled cysts that develop either from follicles that fail to rupture completely *(follicular cysts)* or from corpora lutea that fail to degenerate *(luteal cysts)*. Most women develop a number of these cysts during their reproductive years and their presence does not constitute a diagnosis of polycystic ovary disease. Although ovarian cysts are often multiple, they rarely become dangerous. However, on occasion they may become quite large and painful and be diagnosed by palpation or ultrasonography. Luteal cysts are less common than follicular cysts but tend to cause more symptoms, such as pelvic pain, and menstrual irregularities. Rarely, rupture of a large luteal cyst will result in internal bleeding that requires surgical intervention. The vast majority of all ovarian cysts will disappear within a few months of their appearance, most within 60 days.

Endometriosis (en-doh-mee-tree-OH-sis) is a benign but often painful condition that commonly affects the female reproductive tract. It is characterized by the presence of functioning endometrial tissue outside the uterus. The displaced endometrial tissue is most often found attached to an ovary or to the pelvic or abdominal organs and is occasionally found in other places throughout the body. Just how endometrial tissue spreads from the uterine cavity into the peritoneal cavity is yet to be determined. One theory suggests "retrograde menstruation" or backward movement of some endometrial tissue through the uterine tubes into the peritoneal cavity during the menstrual period. Although rare, the almost bizarre appearance of endometrial tissue in the lungs and lymph nodes has led some researchers to speculate that endometrial tissue "seeds" pass through vascular or lymphatic channels. Regardless of how the movement occurs, the development of endometriosis is an important clinical condition, often causing infertility, dysmenorrhea, and severe pain. Symptoms reflect the fact that displaced endometrial tissue reacts to ovarian hormones in the same way as the normal endometrium—exhibiting a cycle of growth and sloughing off. The disorder afflicts approximately 10% of women, a majority between 30 and 45 years of age.

Breast cancer, often a form of adenocarcinoma, is the most common nonskin malignancy in American women. Although heart disease kills more women than any other cause, breast cancer remains the leading cause of death in women ages 40 to 44 years and of all malignancies, only lung cancer kills more women of all ages. The risk of developing breast cancer increases with increasing age, posing a lifetime risk of one in eight for women who reach advanced old age. Fortunately, treatment of breast cancer is often successful if the cancerous tumor is detected early. Because such tumors are often painless, most physicians recommend monthly breast self-examinations and annual mammograms. Treatments often involve surgery, chemotherapy, and radiation therapy. Surgeries can be very conservative, as in a simple lump removal or **lumpectomy.** If metastasis to surrounding tissue is suspected, a **radical mastectomy** (mas-TEK-toh-mee) may be performed. In this procedure the entire breast, with nearby muscle tissues and lymph nodes, is removed. Just as lumpectomy results in less trauma than radical mastectomy, so-called **limited-field radiation** can provide effective treatment for clearly defined early-stage cancers that have not spread. It does so with shorter treatment cycles and fewer side effects than whole-breast radiation.

In the past, after women had completed their initial treatment for breast cancer they had few options available to lessen the possibility of recurrence. For a number of years the drug **tamoxifen** has been used extensively to prevent the recurrence of breast cancers fueled by estrogen. It does so by blocking the estrogen receptor sites on the cancer cell membrane. Unfortunately, tamoxifen's effectiveness is limited to about 5 years. A relatively new class of drugs called "aromatase inhibitors" (letrozole and others) are now being given to breast cancer patients to prevent recurrence of the disease. Instead of blocking estrogen receptor sites, these drugs block estrogen production. They may replace tamoxifen or be prescribed for use after 5 years of tamoxifen therapy. Other "rational" drugs such as

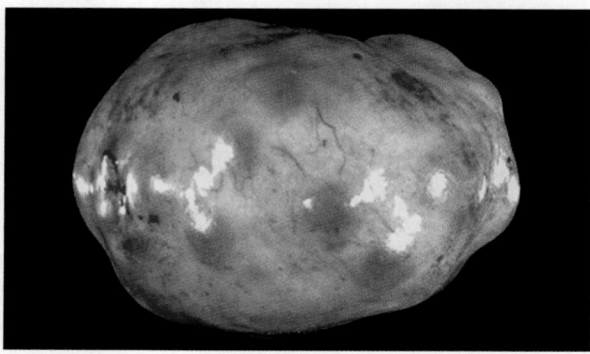

Figure 32-20 *Polycystic ovary disease.* The ovary is studded with fluid-filled cysts developed from follicles that have failed to rupture.

Mechanisms of Disease—cont.

trastuzumab (Herceptin) are now being used to successfully manage recurring forms of breast cancer that in the past were very difficult to treat.

Ovarian cancer is another common malignancy that affects 1 in 70 women in the United States. Usually a type of adenocarcinoma, ovarian cancer is difficult to detect early and is often not diagnosed until it has grown into a large mass. Regular pelvic examinations that include palpation of the ovaries may result in earlier detection. Risk factors for ovarian cancer include age (older than 40 years), infertility, childlessness or few children, a history of miscarriages, and endometriosis. Ovarian cancer is often treated by surgical removal of the ovaries combined with radiation therapy and chemotherapy.

Cancer of the uterus can affect the body of the uterus or cervix. Cancers of the uterine body most often involve the endometrium (endometrial cancer) and mostly affect women beyond childbearing years; a common symptom is postmenopausal uterine bleeding. Risk factors for this type of cancer include obesity, prolonged estrogen therapy, and infertility. Cervical cancer occurs most often in women between the ages of 30 and 50 years. Cervical cancer is often diagnosed early, through screening tests such as the Papanicolaou (pah-peh-ni-koh-LAH-oo) test (Pap smear) (Figure 32-21). In this test, cells swabbed from the cervix are smeared on a glass slide, stained, and examined microscopically to determine whether any abnormalities exist. Because screening tests and other early detection methods have been so successful, the death rates for uterine cancers have dropped dramatically over the last few decades.

Sexually Transmitted Diseases

Sexually transmitted diseases (STDs), or *venereal diseases*, are infections caused by communicable pathogens such as viruses, bacteria, fungi, and protozoans. The factor that links all these diseases and gives this disease category its name is the fact that they can all be transmitted by sexual contact. The term *sexual contact* refers to sexual intercourse in addition to any contact between the genitalia of one person and the body of another person. Diseases classified as STDs can be transmitted sexually but do not have to be. For example, *acquired immune deficiency syndrome (AIDS)* is a viral condition that can be spread through sexual contact but is also spread by transfusion of infected blood and use of contaminated medical instruments such as intravenous needles and syringes. Candidiasis, or yeast infection, is a common opportunistic infection, but it can also be transmitted through sexual contact. Sexually transmitted diseases are the most common of all communicable diseases. Table 32-1 summarizes a few of the principal STDs.

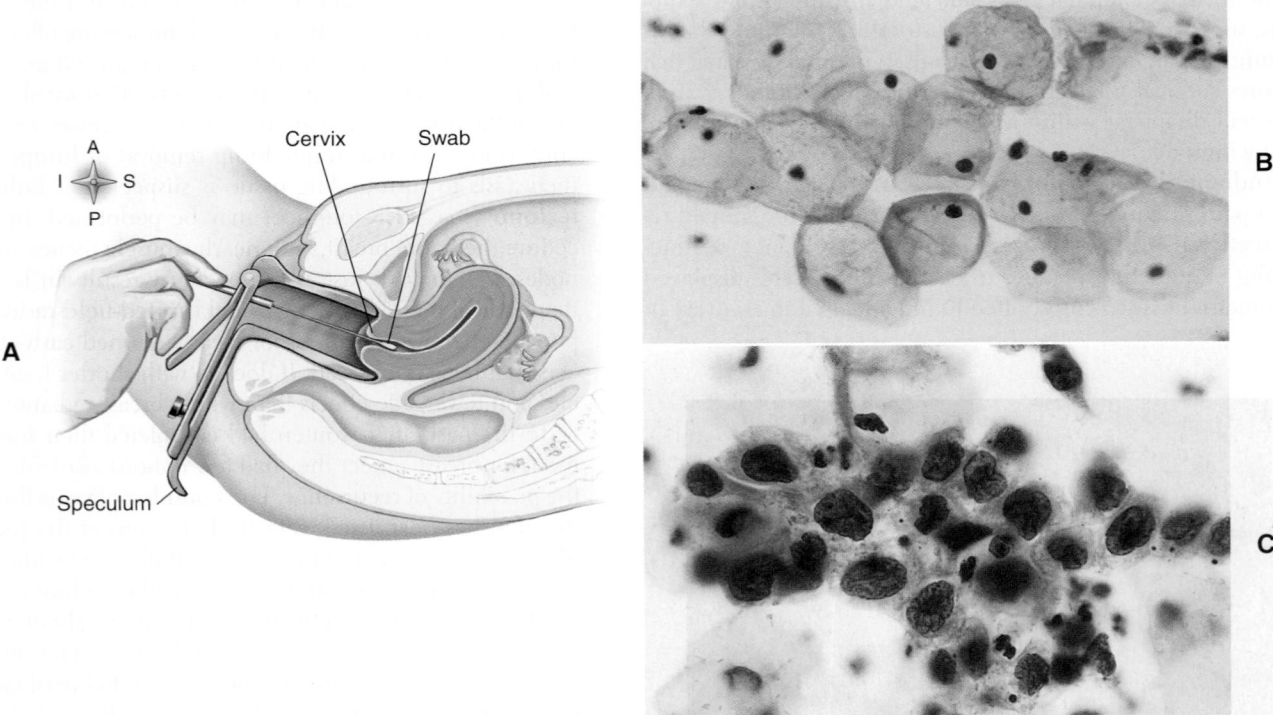

Figure 32-21 *Papanicolaou (Pap) smear.* **A,** Obtaining a Pap smear. **B,** Appearance of normal cervical epithelial cells in Pap smear. **C,** Appearance of cervical cancer cells in Pap smear. Note the reduction in cytoplasm and increased prominence of the nuclei compared with normal epithelial cells.

LANGUAGE OF SCIENCE

(Cont'd from page 1129)

fundus (FUN-duss) [*fundus* bottom]

graafian follicle (GRAH-fee-en FOL-li-kul) [*Reijnier de Graaf* Dutch physician]

greater vestibular glands (ves-TIB-yoo-lar) [*vestibulum* courtyard]

hymen (HYE-men) [*hymen* membrane]

imperforate hymen (im-PER-fah-rayt HYE-men) [*im-* not, *-perforare* to pierce, *hymen* membrane]

infundibulum (in-fun-DIB-yoo-lum) [*infundibulum* funnel]

internal os [*internus* inward, *os* bone]

isthmus (ISS-muss) [*ithmos* a narrow connection or passage]

labia majora (LAY-bee-ah mah-JOH-rah) [*labia* lips, *majora* large]

labia minora (LAY-bee-ah mih-NO-rah) [*labia* lips, *minora* small]

lactation (lak-TAY-shun) [*lact* milk]

lesser vestibular glands (ves-TIB-yoo-lar) [*losian* to lose, *vestibulum* courtyard]

luteal phase (LOO-tee-al fayz) [*lute-* yellow, *-al* pertaining to]

luteinization (loo-tee-in-i-ZAY-shun) [*luteus-* yellow, *-ation* process]

luteinizing hormone (LH) (loo-tee-in-EYE-zing HOR-mohn) [*luteus-* yellow, *-izein* to cause, *hormaein* to set in motion]

medulla (meh-DUL-ah) [*medulla* marrow]

menarche (meh-NAR-kee) [*men-* month, *-archaios* from the beginning]

menopause (MEN-oh-pawz) [*men-* month, *-pausis* to cease]

menses (MEN-seez) [*men* month]

menstrual period (MEN-stroo-al) [*menstrualis* monthly, *peri-* around, *-hodos* way]

menstruation (men-stroo-AY-shun) [*menstruare* to menstruate]

mons pubis (monz PYOO-bis) [*mons* mountain, *pubis* pubes]

myometrium (my-oh-MEE-tree-um) [*myo-* muscle, *-metra* womb]

oogonia (oh-oh-GO-nee-ah) [*oo-* egg, *-gonos* offspring]

ova (OH-vah) [*ova* egg]

ovarian cortex (oh-VAIR-ee-an KOHR-teks) [*ovum* egg, *cortex* bark]

ovarian follicles (oh-VAIR-ee-an FOL-i-kuls) [*ovum* egg, *folliculus* small bag]

ovarian medulla (oh-VAIR-ee-an meh-DUL-ah) [*ovum* egg, *medulla* marrow]

ovaries (OH-var-eez) [*ovum* egg]

ovulation (ov-yoo-LAY-shun) [*ovum-* egg, *-ation* process]

parietal peritoneum (pah-RYE-i-tal pair-i-TOH-nee-um) [*paries* wall, *peri-* around, *-teinein* to stretch]

perineal body (pair-i-NEE-al) [*perineos* perineum]

posterior cul-de-sac (of Douglas) (pohs-TEER-ee-or kul-deh-sak) [*James Douglas* Scottish anatomist]

posterior fornix (pohs-teer-ee-or FOR-niks) [*posterior* behind, *fornix* arch]

posterior ligament (pohs-TEER-ee-or LIG-ah-ment) [*posterior* behind, *ligare* to bind]

postmenstrual phase (post-MEN-stroo-al fayz) [*post-* after, *-menstrualis* monthly]

postovulatory phase (post-ov-yoo-lah-TOR-ee fayz) [*post-* after, *-ovum* egg]

premenstrual phase (pree-MEN-stroo-al fayz) [*pre-* before, *-menstrualis* monthly]

preovulatory phase (pree-ov-yoo-lah-TOR-ee fayz) [*pre-* before, *-ovum* egg]

progesterone (pro-JES-ter-ohn) [*pro-* first, *-gestare* to bear]

progesterone phase (proh-JES-ter-ohn fayz) [*pro-* first, *-gestare* to bear]

proliferative phase (PROH-lif-er-eh-tiv fayz) [*proles-* offspring, *-ferre* to bear]

rectouterine pouch (rek-toh-YOO-ter-in) [*recto-* straight, *-uterus* womb]

round ligaments (LIG-ah-ments) [*ligare* to bind]

secretory phase (SEEK-reh-toh-ree fayz) [*secernere* to separate]

urinary meatus (YOOR-i-nair-ee mee-AY-tus) [*ouron* urine, *meatus* channel]

urogenital triangle (yoor-oh-GEN-i-tal) [*uro-* urine, *-genitalis* fruitful, *triangulus-* three cornered]

uterine tubes (YOO-ter-in toobs) [*uterus* womb]

uteroligaments (yoo-ter-oh-LIG-ah-ments) [*utero-* relating to the uterus, *-ligare* to]

uterosacral ligaments (yoo-ter-oh-SAK-ral LIG-ah-ments) [*utero-* relating to the uterus, *-sacralis* near the sacrum, *ligare* to bind]

uterus (YOO-ter-us) [*uterus* womb]

vagina (vah-JYE-nah) [*vagina* sheath]

vaginal orifice (VAH-ji-nal OR-i-fis) [*vagina-* sheath, *-al* pertaining to, *orificium* opening]

vesicouterine pouch (ves-i-koh-YOO-ter-in) [*vesico-* bladder, *-uterus* womb]

vestibule (VES-ti-byool) [*vestibulum* courtyard]

vulva (VUL-vah) [*vulva* wrapper]

LANGUAGE OF MEDICINE

amenorrhea (ah-men-oh-REE-ah) [*a-* without, *-men-* month, *-rhoia* to flow]

breast cancer [*cancer* crab]

calendar rhythm method

cancer of the uterus

candidiasis (kan-dih-DYE-eh-sis) [*candidus-* white, *-osis* condition]

cervical cancer (SER-vi-kal) [*cervic-* neck, *-al* pertaining to, *cancer* crab]

dysfunctional uterine bleeding (DUB) (dis-FUNK-shun-al YOO-ter-in) [*dys-* difficult, *-functio* performance, *uterus* womb]

dysmenorrhea (dis-men-oh-REE-ah) [*dys-* painful, *-men-* month, *-rhoia* flow]

ectopic pregnancy (ek-TOP-ik) [*ektopos* displaced]

endometrial ablation (en-doh-MEE-tree-al ab-LAY-shun) [*endo-* within, *-metra* womb, *ab-* away from, *-latus* carried away]

endometrial cancer (en-doh-MEE-tree-al KAN-ser) [*endo-* within, *-metra* womb, *cancer-* crab]

endometriosis (en-doh-mee-tree-OH-sis) [*endo-* within, *-metra-* womb, *-osis* condition]

episiotomy (eh-piz-ee-OT-oh-mee) [*episi-* vulva, *-tomy* surgical incision]

fertility drugs (fer-TIL-i-tee) [*fertilis* fruitful]

fertility signs (fer-TIL-i-tee)

fibroid (FYE-broyd) [*fibro-* fiber, *-oid* resembling]

fibromyoma (fye-broh-my-OH-mah) [*fibro-* fiber, *-my-* muscle, *-oma* tumor]

hydrosalpinx (hye-droh-SAL-pinks) [*hydro-* water, *-salpinx* tube]

hysterosalpingograms (his-ter-oh-sal-PING-go-grams) [*hystero-* uterus, *-salpinx-* tube, *-gram* record]

in vitro fertilization (in VEE-troh FER-ti-li-ZAY-shun) [*in-* within, *vitro-* glassware, *fertilis-* fruitful, *-ation* process]

infertility (in-fer-TIL-i-tee) [*in-* not, *-fertilis* fruitful]

Kegel exercise (KEE-gel)

leukorrhea (loo-koh-REE-ah) [*leuko-* white corpuscle, *-rhoia* flow]

limited-field radiation

lumpectomy (lump-EK-toh-mee) [*lump-* mass, *-ectomy* surgical removal]

menotropins (men-oh-TROHP-ins) [*men-* month, *-trepein* to turn]

mittelschmerz (MIT-el-schmertz) [*mittel-* middle, *-schmerz* pain]

myoma (my-OH-mah) [*my-* muscle, *-oma* tumor]

natural family planning

oral contraceptive

ovarian cancer (oh-VAIR-ee-an) [*ovum* egg, *cancer* crab]

ovarian cysts (oh-VAIR-ee-an CISTS) [*ovum* egg, *kystis* bag]

Papanicolaou test (Pap smear) (pah-peh-ni-koh-LAH-oo) [*George N. Papanicolaou* Greek physician]

pelvic inflammatory disease (PID) (PEL-vik in-FLAM-ah-tor-ee) [*pelvis* basin, *inflammare* to set afire, *dis-* without, *-aise* ease]

peritonitis (pair-i-toh-NYE-tis) [*peri-* around, *-tonus-* to stretch, *-itis* inflammation]

polycystic ovary disease (PCOD) (pahl-ee-SIS-tik OH-var-ee) [*poly-* many, *-kystis* bag, *ovum* egg, *dis-* without, *-aise* ease]

premenstrual syndrome (PMS) (pree-MEN-stroo-all SIN-drohm) [*pre-* before, *-menstrualis* monthly]

radical mastectomy (RAD-i-kal mas-TEK-toh-mee) [*radix* root, *mastos-* breast, *-ectomy* surgical removal]

radiofrequency ablation (RAY-dee-oh-FREE-kwen-see ab-LAY-shun) [*radio-* ray, *-frequens* frequent, *ab-* away from, *-lateus* carried away]

salpingitis (sal-pin-JYE-tis) [*salpinx-* tube, *-itis* inflammation]

sexually transmitted diseases (STDs) (SEKS-yoo-al-ee trans-MIH-ted) [*sexus-* sex, *-al* pertaining to, *transmittere* to transmit, *dis-* without, *-aise* ease]

supraovulation (soo-prah-o-vyoo-LAY-shun) [*supra-* above or over, *-ovum-* egg, *-ation* process]

tamoxifen (teh-MOK-seh-fin)

thermal ablation (THUR-mal ab-LAY-shun) [*therme-* heat, *ab-* away from, *-lateus* carried away]

transabdominal pelvic ultrasound (tranz-ab-DOM-i-nal PEL-vik UL-trah-sound) [*trans-* across, *-abdomen* belly, *pelvis* basin, *ultra-* beyond, *-sonus* sound]

uterine artery embolization (YOO-ter-in AR-ter-ee em-boh-lih-ZAY-shun) [*uterus* womb, *arteria-* windpipe, *embolus-* plug, *-ation* process]

uterine fibroids (YOO-ter-in FYE-broyds) [*uterus* womb, *fibra-* fiber, *-oid* resembling]

vaginitis (vaj-i-NYE-tis) [*vagina-* sheath, *-itis* inflammation]

CASE STUDY

Mrs. Beth Calloway, 47 years old, discovered a lump in her right breast. Mrs. Calloway is in good health with no previous hospitalizations or surgeries. She does not smoke or drink, watches the fat content of her diet, and walks about 2 miles each day. Her family history is positive for breast cancer; her mother and maternal grandmother were diagnosed with this disease before 55 years of age.

Mrs. Calloway discovered her breast lump while performing her monthly breast self-examination. She immediately brought it to the attention of her nurse practitioner who, on examination, located a 2- to 3-cm (about 1-inch) mass in the 3 o'clock position of the right breast. This mass felt firm, was well fixed to the chest wall, and was tender to the touch. No skin discoloration, nipple retraction, or drainage was identified. A pea-sized lymph node was located in the right axilla. After examination, the nurse practitioner ordered a mammogram and ultrasound of the breast. Results confirmed a solid mass, 3 cm large, located in the right breast plus a 2-cm–sized lymph node in the right axilla. Mrs. Calloway was immediately referred to a surgeon who ordered a biopsy.

1. Mrs. Calloway is considered to be at high risk for developing breast cancer. Which factor is most positively related to this increased risk profile?

   A. History of breast cancer in maternal side family members
   B. History of cystic breast disease
   C. Early onset of menarche
   D. Trauma related to the birth of her children

2. Which of the following best explains the existence of an enlarged right axillary lymph node in Mrs. Calloway?

   A. The lymph node is the result of an inflammatory reaction that normally occurs with the onset of menses.
   B. The existence of the node is the result of an increased strain on the lymphatic system as a result of cellular degeneration.

   C. The lymph node exists to provide nutrients to the rapidly growing cancer cells.
   D. The lymph node is the result of cancer cells spreading to different tissues within the body.

3. Mrs. Calloway undergoes a modified radical mastectomy and on the third postoperative day, her right arm becomes increasingly swollen and painful. What is the best explanation for the lymphedema?

   A. An electrolyte imbalance is creating an increase in the hydrostatic pressure in the lymphatic system.
   B. A postoperative infection has produced the beginnings of an immune response from the T lymphocytes.
   C. The lymphatic channels are congested with fat particles that are being displaced from the trauma of surgery.
   D. The remaining lymph nodes are inadequate to handle the lymph flow, creating an increase in the hydrostatic pressure.

4. Mrs. Calloway recovers from her surgery and is scheduled to be discharged from the hospital today. The nurse explains to Mrs. Calloway the importance of continuing her monthly breast self-examination and obtaining her yearly pelvic examination and Pap smear test. Which one of the following do you recognize as the rationale for stressing these practices to Mrs. Calloway?

   A. Because Mrs. Calloway had breast cancer, she will be at risk to develop cervical cancer.
   B. Breast cancer often metastasizes to the ovaries and the pelvic examination is an important tool in early identification.
   C. Breast cancer patients will have amenorrhea after surgery and often think they no longer need pelvic and Pap smear tests.
   D. Breast cancer patients are at increased risk for dysfunctional uterine bleeding because of hormonal imbalance caused by the surgery.

CHAPTER SUMMARY

OVERVIEW OF THE FEMALE REPRODUCTIVE SYSTEM

A. Function of the female reproductive system
   1. The function of the female reproductive system is to produce offspring and thereby ensure continuity of the genetic code
   2. It produces eggs, or female gametes, which each may unite with a male gamete to form the first cell of an offspring
   3. It also can provide nutrition and protection to the offspring for up to several years after conception
B. Structural plan of the female reproduction system
   1. Reproductive organs are classified as essential or accessory (Figure 32-1)
      a. Essential organs—gonads are the paired ovaries; gametes are ova produced by the ovaries—the ovaries are also internal genitalia

      b. Accessory organs
         (1) Internal genitalia—uterine tubes, uterus, and vagina— ducts or duct structures that extend from the ovaries to the exterior
         (2) External genitalia—the vulva
         (3) Additional sex glands such as the mammary glands
C. Perineum
   1. The perineum is the skin-covered region between the vaginal orifice and the rectum
   2. This area may be torn during childbirth

OVARIES

A. Location of the ovaries
   1. The ovaries are nodular glands located on each side of the uterus, below and behind the uterine tubes (Figure 32-2)
   2. Ectopic pregnancy—development of the fetus in a place other than the uterus

B. Microscopic structure of the ovaries
   1. The surface of the ovaries is covered by the germinal epithelium
   2. Ovarian follicles contain the developing female sex cells (Figure 32-3)
   3. Ovum—an oocyte released from the ovary
C. Functions of the ovaries
   1. Ovaries produce ova—the female gametes
   2. Oogenesis—process that results in formation of a mature egg (Figure 32-4)
   3. The ovaries are endocrine organs that secrete the female sex hormones (estrogens and progesterone)

UTERUS

A. Structure of the uterus (Figure 32-2)
   1. Size and shape of the uterus
      a. The uterus is pear shaped and has two main parts—the cervix and the body
   2. The wall of the uterus is composed of three layers—the inner endometrium, the middle myometrium, and the outer incomplete layer of parietal peritoneum
   3. Cavities of the uterus—the cavities are small because of the thickness of the uterine walls
      a. The body cavity's apex constitutes the internal os and opens into the cervical canal, which is constricted at its lower end and forms the external os that opens into the vagina
   4. The blood to the uterus is supplied by uterine arteries
B. Location of the uterus
   1. The uterus is located in the pelvic cavity between the urinary bladder and the rectum (Figure 32-1)
   2. The position of the uterus (Figure 32-5) is altered by age, pregnancy, and distention of related pelvic viscera
   3. The uterus descends, between birth and puberty, from the lower abdomen to the true pelvis
   4. The uterus begins to decrease in size at menopause
C. Position of the uterus
   1. Body lies flexed over the bladder
   2. Cervix points downward and backward, joining the vagina at a right angle
   3. Several ligaments hold the uterus in place but allow some movement
D. Functions of the uterus
   1. The uterus is part of the reproductive tract and permits sperm to ascend toward the uterine tubes
   2. If conception occurs, an offspring develops in the uterus
      a. The embryo is supplied with nutrients by endometrial glands until the production of the placenta
      b. The placenta is an organ that permits the exchange of materials between the mother's blood and the fetal blood but keeps the two circulations separate
      c. Myometrial contractions occur during labor and help push the offspring out of the mother's body
   3. If conception does not occur, outer layers of endometrium are shed during menstruation
      a. Menstruation is a cyclical event that allows the endometrium to renew itself

UTERINE TUBES

A. Uterine tubes are also called fallopian tubes, or oviducts
B. Uterine tubes are attached to the uterus at its upper outer angles and extend upward and outward toward the sides of the pelvis
C. Structure of the uterine tubes
   1. Uterine tubes consist of mucous, smooth muscle, and serous lining (Figure 32-6)
   2. Mucosal lining is directly continuous with the peritoneum lining the pelvic cavity
      a. Tubal mucosa is continuous with that of the vagina and uterus, which means it may become infected with organisms introduced into the vagina
   3. Each uterine tube has three divisions: isthmus, ampulla, and infundibulum
D. Function of the uterine tubes
   1. Uterine tubes serve as transport channels for ova and as the site of fertilization

VAGINA

A. The vagina is a tubular organ located between the rectum, urethra, and bladder
B. Structure of the vagina
   1. The vagina is a collapsible tube capable of distention, composed of smooth muscle, and lined with mucous membrane arranged in rugae
   2. The anterior wall is shorter than the posterior wall because the cervix protrudes into its uppermost portion
   3. Hymen—a mucous membrane that typically forms a border around the vagina in young premenstrual girls
C. Functions of the vagina
   1. The lining of the vagina lubricates and stimulates the penis during sexual intercourse and acts as a receptacle for semen
   2. The vagina is the lower portion of the birth canal
   3. The vagina transports tissue and blood shed during menstruation to the exterior

VULVA

A. The vulva consists of the female external genitalia: mons pubis, labia majora, labia minora, clitoris, urinary meatus, vaginal orifice, and greater vestibular glands (Figure 32-8)
B. Functions of the vulva
   1. The mons pubis and labia protect the clitoris and vestibule
   2. The clitoris contains sensory receptors that send information to the sexual response area of the brain
   3. The vaginal orifice is the boundary between the internal and external genitalia

BREASTS

A. Location and size
   1. The breasts lie over the pectoral muscles
   2. Estrogens and progesterone control breast development
   3. Breast size is determined by the amount of fat around glandular tissue (Figure 32-10)

B. Function of the breasts
1. The function of mammary glands is lactation
2. Mechanism of lactation (Figure 32-11)
   a. The ovarian hormones make the breasts structurally ready to produce milk
   b. Shedding of the placenta results in a decrease of estrogens and thus stimulates prolactin
   c. Prolactin stimulates lactation
3. Lactation can provide nutrient-rich milk to offspring for up to several years from birth; some advantages are:
   a. Nutrients
   b. Passive immunity from antibodies present in the colostrum and milk
   c. Emotional bonding between mother and child

FEMALE REPRODUCTIVE CYCLES

A. The female reproductive system has many cyclical changes that start with the beginning of menses
1. Ovarian cycle—ovaries from birth contain oocytes in primary follicles in which the meiotic process has been suspended. At the beginning of menstruation each month, several of the oocytes resume meiosis. Meiosis will stop again just before the cell is released during ovulation (Figure 32-12)
2. Menstrual cycle (endometrial cycle) is divided into four phases
   a. Menses
   b. Postmenstrual phase
   c. Ovulation
   d. Premenstrual phase
3. Myometrial phase
4. Gonadotropic cycle
B. Control of female reproductive cycles
1. Hormones control cyclical changes
2. Cyclical changes in the ovaries result from changes in the gonadotropins secreted by the pituitary gland (Figures 32-13 and 32-14)
3. Cyclical changes in the uterus are caused by changes in estrogens and progesterone (Figure 32-15)
4. Low levels of FSH and LH cause regression of the corpus luteum if pregnancy does not occur. This causes a decrease in estrogen and progesterone, which triggers endometrial sloughing of the menstrual phase
5. Control of cyclical changes in gonadotropin secretion is caused by positive and negative feedback mechanisms and involves estrogens, progesterone, and secretion of releasing hormones by the hypothalamus
C. Importance of the female reproductive cycles
1. The ovarian cycle's primary function is to produce ova at regular intervals
   a. Its secondary function is to regulate the endometrial cycle through estrogen and progesterone
2. The function of the endometrial cycle is to make the uterus suitable for implantation of a new offspring
3. The cyclical nature of the reproductive system and the fact that fertilization will occur within 24 hours after ovulation mean that a woman is only fertile a few days of each month

D. Menstrual flow begins at puberty, and the menstrual cycle continues for about 3 decades

THE BIG PICTURE: THE FEMALE REPRODUCTIVE SYSTEM AND THE WHOLE BODY

A. The female reproductive system shares a special relationship with:
1. The urinary system because of their close proximity and because they share the vulva
2. The skeletal muscles in the perineum
3. The integumentary system because breasts are actually modifications of the skin

REVIEW QUESTIONS

1. Identify the essential and accessory organs in the female reproductive system.
2. Describe the three layers that compose the wall of the uterus.
3. Identify the vessels that supply blood to the uterus.
4. List the eight ligaments that hold the uterus in a normal position.
5. How does the uterus serve as part of the female reproductive tract?
6. What and where are the uterine tubes? Approximately how long are they? What lines the uterine tubes? Their lining is continuous on their distal ends with what? With what on their proximal ends?
7. Trace the development of a female sex cell from its most primitive state through ovulation.
8. What hormones are secreted by the cells in ovarian tissue?
9. Identify all vaginal functions.
10. List all the structures that make up the female external genitalia.
11. Define the term *episiotomy*.
12. Identify the advantages that nursing from the mother's breast provides offspring.
13. What method makes it possible to measure blood levels of gonadotropins?
14. Describe the hormonal changes during menopause.
15. Define the term *mittelschmerz*.

CRITICAL THINKING QUESTIONS

1. Can you name and explain the function of the various hormones that regulate lactation? Where are they produced, and how would you summarize their function and their influence on lactation?
2. Can you list the phases of the menstrual cycle? Which of these phases shows the most variance in length of time? How do the events in each phase contribute to the overall function of the reproductive system?
3. How would you explain the interaction of the hormones that result in ovulation? From what is the name "luteinizing" hormone derived?

4. State in your own words the control of cyclical ovarian changes brought on by FSH and LH.

5. Can you explain, in your own words, the control of cyclical uterine changes brought on by the ovarian hormones? The drop in the level of these hormones triggers what event?

6. Regarding hormone functions, what is the reason contraceptive pills and implants work?

7. How would you correlate the events of the ovarian cycle with the events of the uterine cycle?

8. It is not uncommon for women with eating disorders, such as anorexia, to develop amenorrhea. Explain the link between these two conditions.

9. Female athletes may experience an undesirable condition called amenorrhea. What evidence can you find to support this statement?

CHAPTER 33

Growth and Development

CHAPTER OUTLINE

LANGUAGE OF SCIENCE

adolescence (ad-oh-LESS-ens) [*adolesc-* grow up, *-ence* state of]

adulthood

amniotic cavity (am-nee-OT-ik KAV-i-tee) [*amnio-* fetal membrane, *-t-* combining form, *-ic* pertaining to, *cav* hollow, *-ity* state of]

androgen (AN-droh-jen) [*andro-* male, *-gen* generate]

blastocyst (BLASS-toh-sist) [*blasto-* germ, *-cyst* pouch]

childhood

chorion (KOH-ree-on) [*chorion* skin]

developmental biology

diploid (DIP-loyd) [*diplo-* twofold, *-oid* form of]

ectoderm (EK-toh-derm) [*ecto-* outside, *-derm* skin]

endoderm (EN-doh-derm) [*endo-* inward, *-derm* skin]

fertilization (FER-ti-li-ZAY-shun) [*fertiliz-* fruitful, *-ation* process of]

first polar body [*pol-* pole, *-ar* pertaining to]

fraternal twins

gestation period (jes-TAY-shun) [*gesta-* bear, *-tion* process of]

granulosa cells (gran-yoo-LOH-sah) [*granul-* little grain, *-osa* characterized by, *cella* storeroom]

haploid (HAP-loyd) [*haplo* single, *-oid* form of]

histogenesis (hiss-toh-JEN-eh-sis) [*histo-* tissue, *-gen-* generate, *-esis* process of]

human chorionic gonadotropin (HCG) (kohr-ee-ON-ik go-nah-doh-TROH-pin) [*chorion-* skin, *-ic* pertaining to, *gon-* seed, *-ad-* forming, *-o-* combining form, *-trop-* nourishment, *-in* substance]

identical twins

implantation (im-plan-TAY-shun) [*implanta-* to set into, *-ation* process of]

infancy [*infan* unable to speak, *-cy* state of]

inner cell mass

labor (LAY-bor) [*labor* work]

meiosis (my-OH-sis) [*meiosis* becoming smaller]

mesoderm (MEZ-oh-derm) [*meso-* middle, *-derm* skin]

morula (MOR-yoo-lah) [*morula* little mulberry]

neonatal period (nee-oh-NAY-tal) [*neo-* new, *-nat-* birth, *-al* pertaining to]

Cont'd on p. 1193

Many of your fondest and most vivid memories are probably associated with your birthdays. The day of birth is an important milestone of life. Most people continue to remember their birthday in some special way each year; birthdays are pleasant and convenient reference points to mark periods of transition or change in our lives. The actual day of birth marks the end of one phase of life called the **prenatal period** and the beginning of a second called the **postnatal period.** The prenatal period begins at conception and ends at birth; the postnatal period begins at birth and continues until death. Although important periods in our lives such as childhood and adolescence often are remembered as a series of individual and isolated events, they are in reality part of an ongoing and continuous process. In reviewing the field of *human developmental biology*—study of the many changes that occur during the cycle of life from conception to death—it is often convenient to isolate certain periods such as infancy or old age for study. However, life is not a series of stop-and-start events or individual and isolated periods of time. Instead, it is a biological process that is characterized by continuous modification and change.

This chapter discusses some of the basic concepts of important events and changes that occur in the ongoing development of the individual from conception to death. Study of development during the prenatal period is followed by a review of changes occurring during infancy and adulthood, and, finally, by some of the more important changes that occur in the individual organ systems of the body as a result of aging.

A NEW HUMAN LIFE

Production of Sex Cells

Before a new human life can begin, some preliminary processes must occur. Of utmost importance is the production of mature *gametes*, or sex cells, by each parent. Spermatozoa, gametes of the male parent, are produced by a process called **spermatogenesis.** Ova, gametes of the female parent, are produced by a process called **oogenesis.**

Meiosis

Both types of gamete production require a special form of cell division characterized by **meiosis.** Recall from Chapter 4 that meiosis is the orderly arrangement and distribution of chromosomes that, unlike *mitosis*, reduces the number of chromosomes in each daughter cell to half the number present in the parent cell. The necessity for chromosome reduction as a preliminary step to union of the sex cells is explained by the fact that the cells of each species of living organisms contain a specific number of chromosomes. Human cells, for example, contain 23 pairs or a total of 46 chromosomes. This total of 46 chromosomes per body cell is known as the **diploid** number of chromosomes. Diploid comes from the Greek *diploos*, meaning "twofold." If the male and female cells united without first halving their respective chromosomes, the resulting cell would contain twice as many chromosomes as is normal for human beings. Mature ova and sperm therefore contain only 23 chromosomes, or half as many, as other human cells. This total of 23 chromosomes per sex cell is known as the **haploid** number of chromosomes (from the Greek *haploos*, meaning "single").

Meiotic division consists of two cell divisions that take place one after the other in succession. They are referred to as meiotic division I and meiotic division II, and in both, prophase, metaphase, anaphase, and telophase occur (Figure 33-1). In the interphase that precedes prophase I (of meiotic division I) the same events occur as take place in the interphase preceding mitotic division. Specifically, each DNA molecule replicates and thereby becomes a pair of chromatids, attached to each other only at the centromere. The term *chromosome* applies to any condensed chromatin with its own centromere. For simplicity's sake, in both Figures 33-1 and 33-2, only 4 of the 46 chromosomes are shown. Notice that early in meiosis I, homologous pairs of chromosomes are moved together to form groupings called *tetrads*. During anaphase I the tetrads split (recall that in mitosis it is the chromosomes that split during anaphase). In meiosis I the phenomenon of "*crossing over*" occurs. During crossing over a chromatid segment of each chromosome crosses over and becomes part of the adjacent chromosome in the pair (see Figure 34-4). This is a highly significant event and is discussed in some detail in the next chapter. Because each chromatid segment consists of specific genes, the crossing over of chromatids reshuffles the genes—that is, it transfers some of them from one chromosome to another. This exchange of genetic material can add almost infinite variety to the ultimate genetic makeup of an individual.

Metaphase I follows the last stage of prophase I, and as in mitosis, the chromosomes align themselves along the equator of the spindle fibers, as Figure 33-1 shows. But in anaphase the two chromatids that make up each chromosome do not separate from each other as they do in mitosis to form two new chromosomes out of each original one. In anaphase I, only one of each pair of chromosomes moves to each pole of the parent cell. Therefore, as you can see in Figures 33-1 and 33-2, when the parent cell divides to form two cells, each daughter cell contains two chromosomes or half as many as the parent cell had, although remember, of course, that each chromosome still consists of two sister chromatids joined at the centromere. Thus the daughter cells formed by meiotic division I contain a haploid number of chromosomes, or half as many as the diploid number in the parent cell.

As you can see in Figure 33-1, meiotic division II is essentially the same as mitotic division. In both spermatogenesis and oogenesis the second meiotic division reproduces each of the two cells formed by meiotic division I and so forms four cells, each with the haploid number of chromosomes.

Spermatogenesis

Spermatogenesis is the process by which the primitive sex cells, or **spermatogonia,** already formed in the seminiferous tubules of a newborn baby boy can later become transformed into mature sperm, or *spermatozoa*. Spermatogenesis begins at about the time of puberty and usually continues throughout a man's life. Figure 33-3 shows some of the major steps of spermatogenesis. Trace each step in this diagram with your finger as you read the following paragraph.

Each primary spermatocyte undergoes meiotic division I to form two secondary spermatocytes, each with a haploid number of chromosomes (23). Each secondary spermatocyte undergoes meiotic division II to form a total of four spermatids. Spermatids

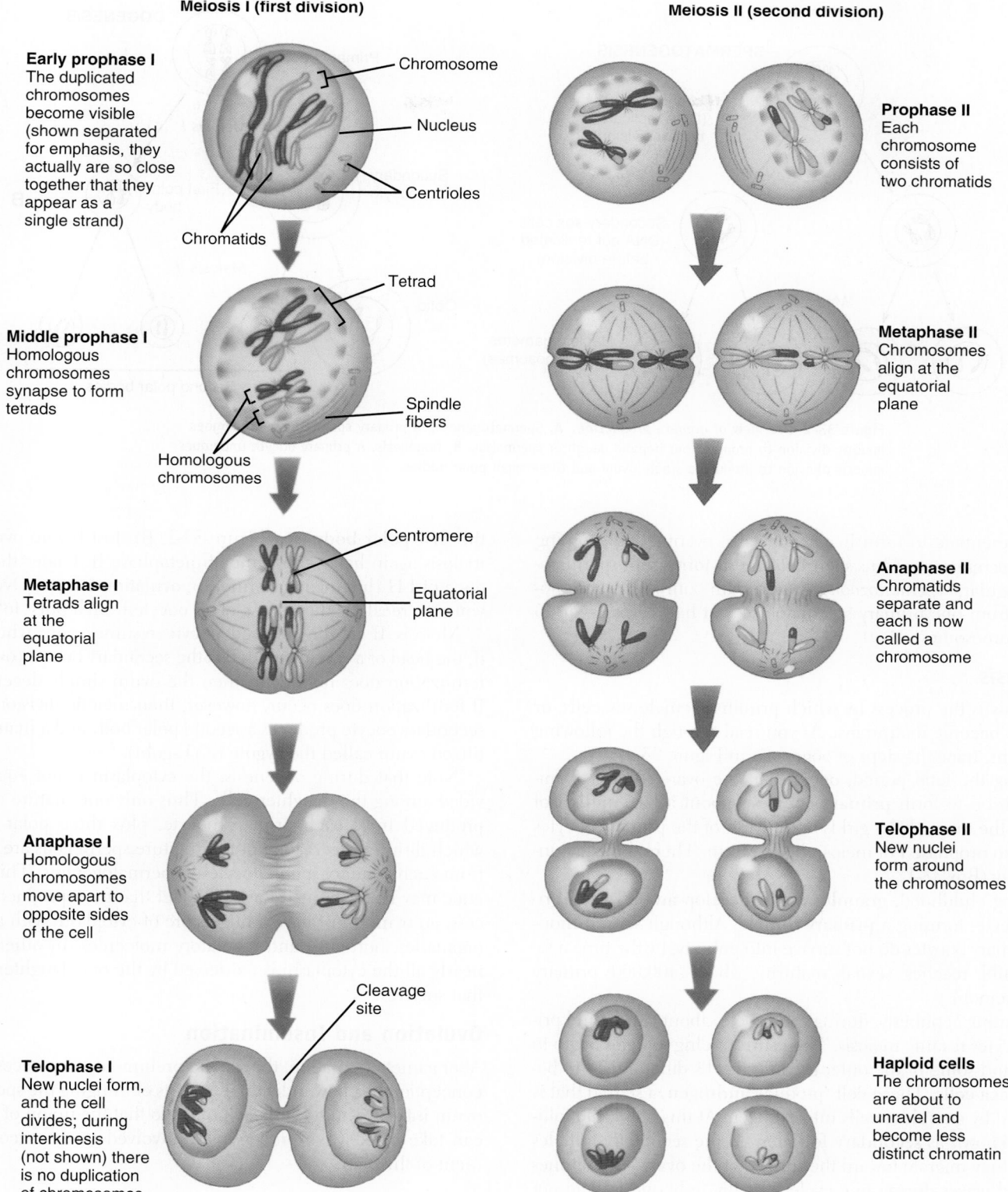

Meiosis I (first division)

Early prophase I
The duplicated chromosomes become visible (shown separated for emphasis, they actually are so close together that they appear as a single strand)

- Chromosome
- Nucleus
- Centrioles
- Chromatids

Middle prophase I
Homologous chromosomes synapse to form tetrads

- Tetrad
- Spindle fibers
- Homologous chromosomes

Metaphase I
Tetrads align at the equatorial plane

- Centromere
- Equatorial plane

Anaphase I
Homologous chromosomes move apart to opposite sides of the cell

Telophase I
New nuclei form, and the cell divides; during interkinesis (not shown) there is no duplication of chromosomes

- Cleavage site

Meiosis II (second division)

Prophase II
Each chromosome consists of two chromatids

Metaphase II
Chromosomes align at the equatorial plane

Anaphase II
Chromatids separate and each is now called a chromosome

Telophase II
New nuclei form around the chromosomes

Haploid cells
The chromosomes are about to unravel and become less distinct chromatin

Figure 33-1 *Meiotic cell division.* Meiosis occurs in a series of two divisions called meiosis I and meiosis II. Notice that four daughter cells, each with the haploid number of chromosomes, are produced from each parent cell that enters meiotic cell division. For simplicity's sake, only four chromosomes are shown in the parent cell instead of the usual 46.

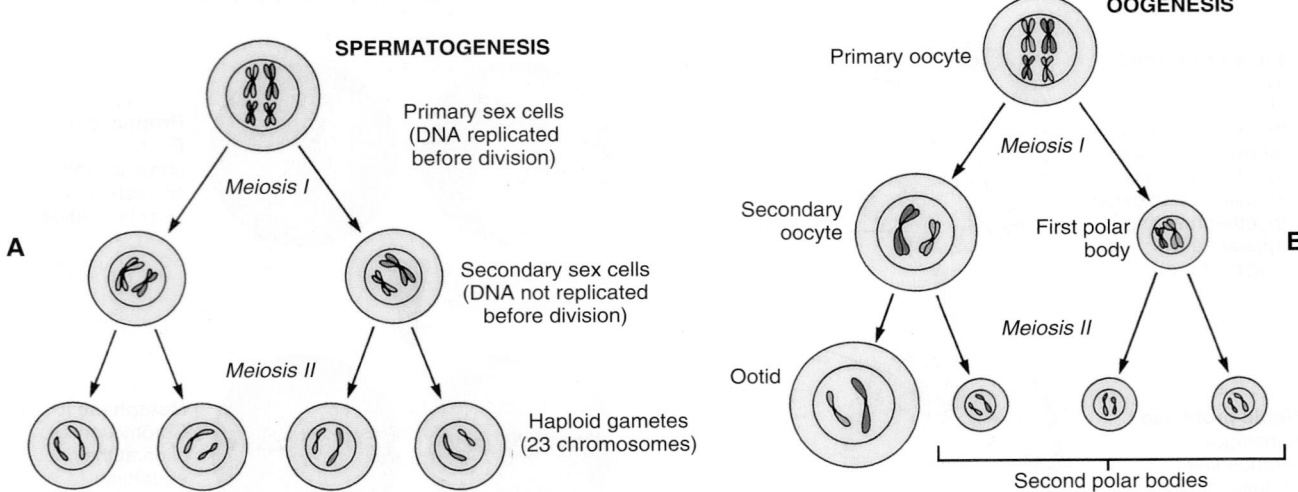

Figure 33-2 *Overview of gamete production.* **A,** Spermatogenesis. A primary spermatocyte undergoes meiotic division to produce four haploid daughter spermatids. **B,** Oogenesis. A primary oocyte undergoes meiotic division to produce a single ovum and three small polar bodies.

then differentiate to form heads and tails, eventually becoming mature *spermatozoa*. Thus spermatogenesis forms four spermatozoa (singular, *spermatozoon*)—each with only 23 chromosomes—from one primary spermatocyte that had 23 pairs, or 46 total chromosomes.

Oogenesis

Oogenesis is the process by which primitive female sex cells, or **oogonia,** become mature ova. As you read through the following paragraphs, trace the steps of oogenesis in Figure 33-4.

During the fetal period, oogonia in the ovaries undergo mitotic division to form **primary oocytes**—about a half million of them by the time a baby girl is born. Most of the primary oocytes develop to prophase I of meiosis before birth. There they stay until puberty (Box 33-1).

During childhood, **granulosa cells** develop around each primary oocyte, forming a **primary follicle.** Although several thousand primary oocytes do not survive into puberty, by the time a female child reaches sexual maturity, about 400,000 primary oocytes remain.

Beginning at puberty, during each cycle, about a thousand primary oocytes resume meiosis. Their surrounding follicles begin to mature and some of the outer granulosa cells differentiate to become **theca cells.** Theca cells produce **androgen,** a steroid that is converted by granulosa cells into estrogen. At this point, the follicles are known as **secondary follicles.** As the secondary follicles mature, they migrate toward the surface of the ovary—sometimes in several waves during one cycle. Usually only one follicle per cycle survives and matures enough to reach the surface of the ovary, where it can be seen as a fluid-filled bump. The fluid-filled space within each mature follicle is called the *antrum.* A mature follicle ready to burst open from the ovary's surface is also called a *graafian follicle.*

By now, meiosis has resumed inside the primary oocyte within the mature follicle. Meiosis I produces a **secondary oocyte** and

the **first polar body** (see Figure 33-2, *B*). Just before ovulation, meiosis again halts—this time at metaphase II. Under the influence of LH (luteinizing hormone), ovulation occurs. Ovulation, you may recall, is the release of an oocyte from a burst follicle.

Meiosis II in the released oocyte resumes only when, and if, the head of a sperm cell enters the secondary oocyte (ovum). If fertilization does not occur, then the ovum simply degenerates. If fertilization does occur, however, then meiotic division of the secondary oocyte produces a second polar body and a mature, fertilized ovum called the **zygote** (ZYE-goht).

Note that during oogenesis, the cytoplasm is not equally divided among the daughter cells. Thus only one mature ovum is produced from each primary oocyte, plus three polar bodies, which disintegrate. A total of four mature sperm cells are formed from each primary spermatocyte in spermatogenesis. This difference may be accounted for by the fact that for reproductive success, an ovum must have a huge store of cytoplasm with all of its organelles, nutrients, and regulatory molecules. In other words, nearly all the cytoplasm is conserved by the one daughter oocyte that survives.

Ovulation and Insemination

After gamete formation, the second preliminary step necessary for conception of a new individual consists of bringing the sperm and ovum into proximity with each other so that the union of the two can take place. Two processes are involved in the accomplishment of this step:

1. Ovulation or expulsion of the mature ovum from the mature ovarian follicle into the abdominopelvic cavity, from which it enters one of the uterine (fallopian) tubes
2. Insemination or expulsion of the seminal fluid from the male urethra into the female vagina. Several million sperm enter the female reproductive tract with each ejaculation of semen. By lashing movements of their flagella-like tails, assisted by

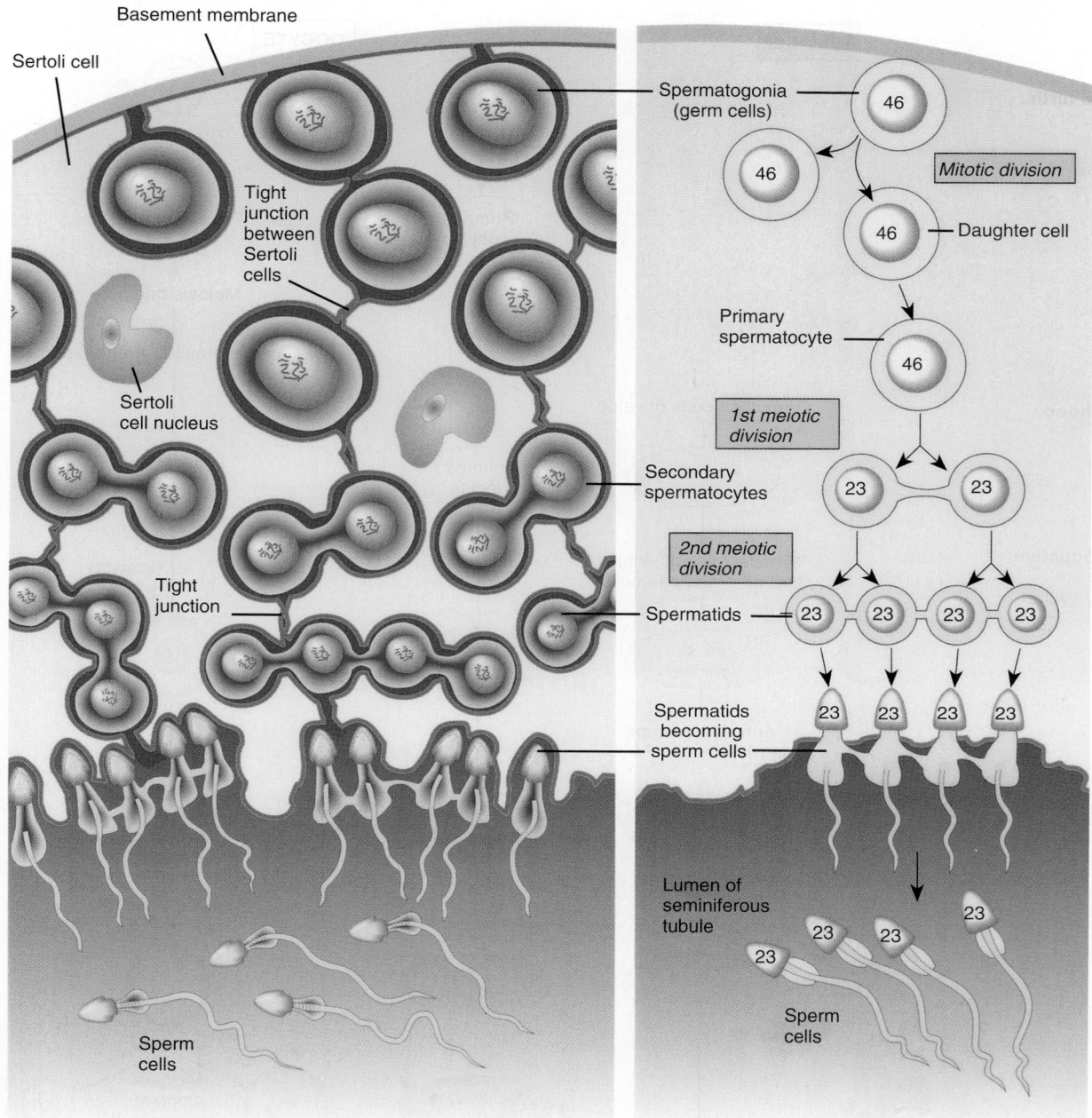

Figure 33-3 *Spermatogenesis.* First, spermatogonia in the outer rim of the seminiferous tubule produce daughter cells by mitotic division. These daughter cells, each with 46 chromosomes, become primary spermatocytes. A primary spermatocyte then undergoes meiotic division I to form two secondary spermatocytes, each with a haploid number of chromosomes (23). Each of the two secondary spermatocytes undergoes meiotic division II to form a total of four spermatids. Spermatids then differentiate to form heads and tails, eventually becoming mature spermatozoa—all with 23 chromosomes. Recall the role of the sustentacular or Sertoli cells, which support the developing male gametes structurally (by physically supporting them) and functionally (by releasing nutrients to them and by secretion of the hormone inhibin).

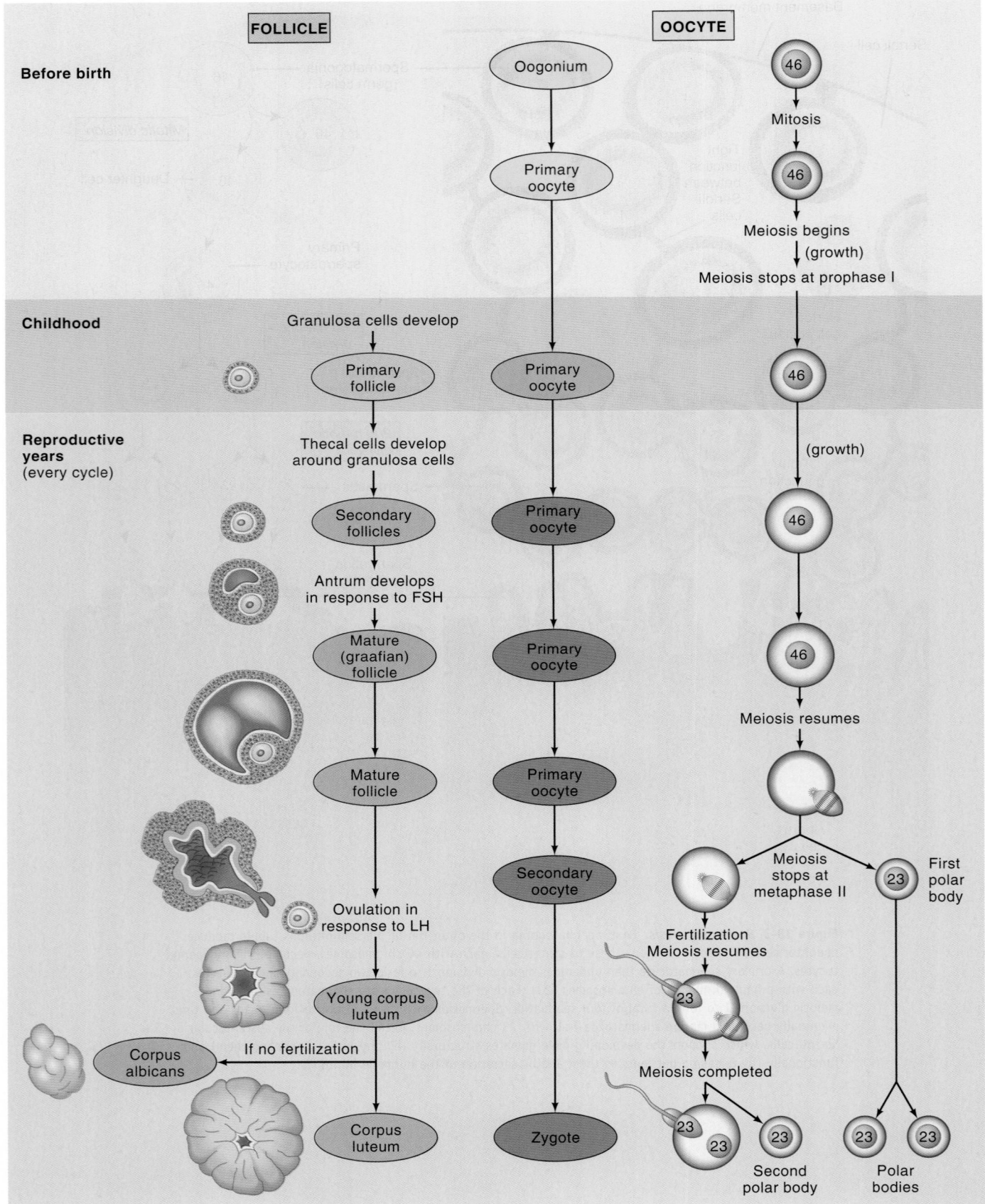

Figure 33-4 *Oogenesis.* Production of a mature ovum (oocyte) and subsequent fertilization are shown on the right as a series of cell divisions and on the left as a series of changes in the ovarian follicle.

BOX 33-1: FYI
Arrest of Oocyte Development

The unique processes that characterize oogenesis have functional advantages that may not be apparent at first glance. For instance, the arrest of oocyte development during prophase I is an excellent example of how the body uses an opportunity to its advantage. Why halt meiosis as it is just beginning? By arresting meiosis at this stage, the oocyte has four copies of the genetic code available (two copies of each chromosome, each of which contains two identical chromatids). With four copies of the DNA code available, the oocyte can quickly synthesize the enormous number of RNA molecules needed for the early stages of human development—the period just after fertilization and before development of the placenta. As soon as enough tRNA, rRNA, and mRNA molecules to regulate the cell through the early developmental stages are made, DNA activity slows and the oocyte waits for the hormonal signal that will stimulate completion of oogenesis. When the hormonal signal, in the form of gonadotropins, arrives, meiosis resumes. Before the cell is released during ovulation, however, the previously transcribed RNA molecules will have directed the synthesis of enough protein to cause a 500-fold increase in the size of the oocyte! The released ovum finally completes meiosis after the sperm DNA has entered it. The fertilized ovum—or **zygote,** as it is now called—has enough cytoplasmic material to sustain it through some of the most critical phases of early human development.

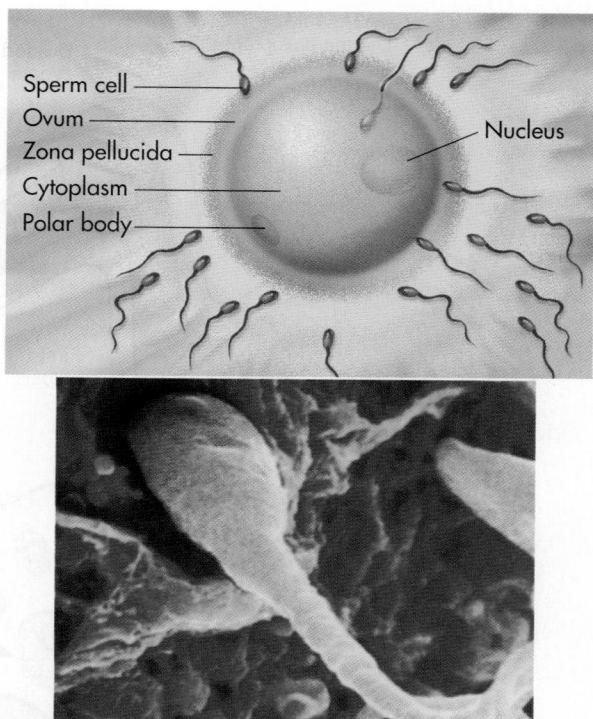

Figure 33-5 *Fertilization.* Fertilization is a specific biological event. It occurs when the male and female sex cells fuse. After union between a sperm cell and the ovum has occurred, the cycle of life begins. The scanning electron micrograph shows spermatozoa attached to the surface of an ovum. Only one will enter the ovum.

various processes in the female reproductive tract, the sperm make their way into the external os of the cervix, through the cervical canal and uterine cavity, and into the uterine (fallopian) tubes.

Fertilization

After ovulation the discharged ovum first enters the abdominopelvic cavity and then soon finds its way into the uterine (fallopian) tubes, where conception, or **fertilization,** may take place (Figure 33-5). Sperm cells "swim" up the uterine tubes toward the ovum. Look at the relationship of the ovary, the uterine tube, and the uterus in Figure 33-6. Recall from Chapter 32 that each uterine tube extends outward from the uterus for about 10 cm (4 inches). It then ends in the abdominal cavity near the ovary, as you can see in Figure 33-6, in an opening surrounded by fringe-like processes, the fimbriae. Sperm cells that are deposited in the vagina must enter and "swim" through the uterus and then move out of the uterine cavity and through the uterine tube to meet the ovum. Fertilization most often occurs in the outer one third of the oviduct, as shown in Figure 33-6. See Box 33-2 for discussion of artificial fertilization outside the body.

The process of sperm movement is assisted by mechanisms within the female reproductive tract. For example, mucous strands in the cervical canal guide the sperm on their way into the uterus. Peristaltic contractions of the female reproductive tract and ciliary movement along the lining of the uterine tubes also assist the movement of sperm. Despite all this, however, only a small fraction of the sperm deposited in the vagina ever reach the

ovum. Only 50 to 100 sperm out of 250 million to 500 million sperm actually reach their target.

The ovum also takes an active role in the process of fertilization. Experiments show that the ovum and its surrounding layers of granulosa cells and a thick jelly-like film called the **zona pellucida** actually attract sperm with special peptides. In addition, receptor molecules on the layers that surround the ovum bind sperm attracted to the area by the peptides. Once bound to a receptor, the sperm releases enzymes from its acrosome that break down the zona pellucida. Once the sperm reaches the surface of the ovum, the two plasma membranes fuse and the nucleus of the sperm moves inside the ovum. In addition to the sperm nucleus, RNA and protein molecules from the sperm cell also enter the egg. The RNA molecules apparently code for proteins needed early in development—thus adding to the ovum's cellular resources. RNA molecules involved in "gene silencing" may also be released into the ovum during fertilization (see Box 4-5 on p. 127).

As soon as the head and neck of one spermatozoon fuse with the ovum (the tail degenerates), complex mechanisms in the egg are activated by sperm proteins to ensure that no more sperm enter. Specifically, sperm proteins trigger an increase in calcium concentration that causes vesicles just inside the ovum's plasma membrane to release enzymes that inactivate the sperm receptors on the zona pellucida. This thick film, then, becomes an impenetrable barrier called the *fertilization membrane.* The 23 chromosomes from the

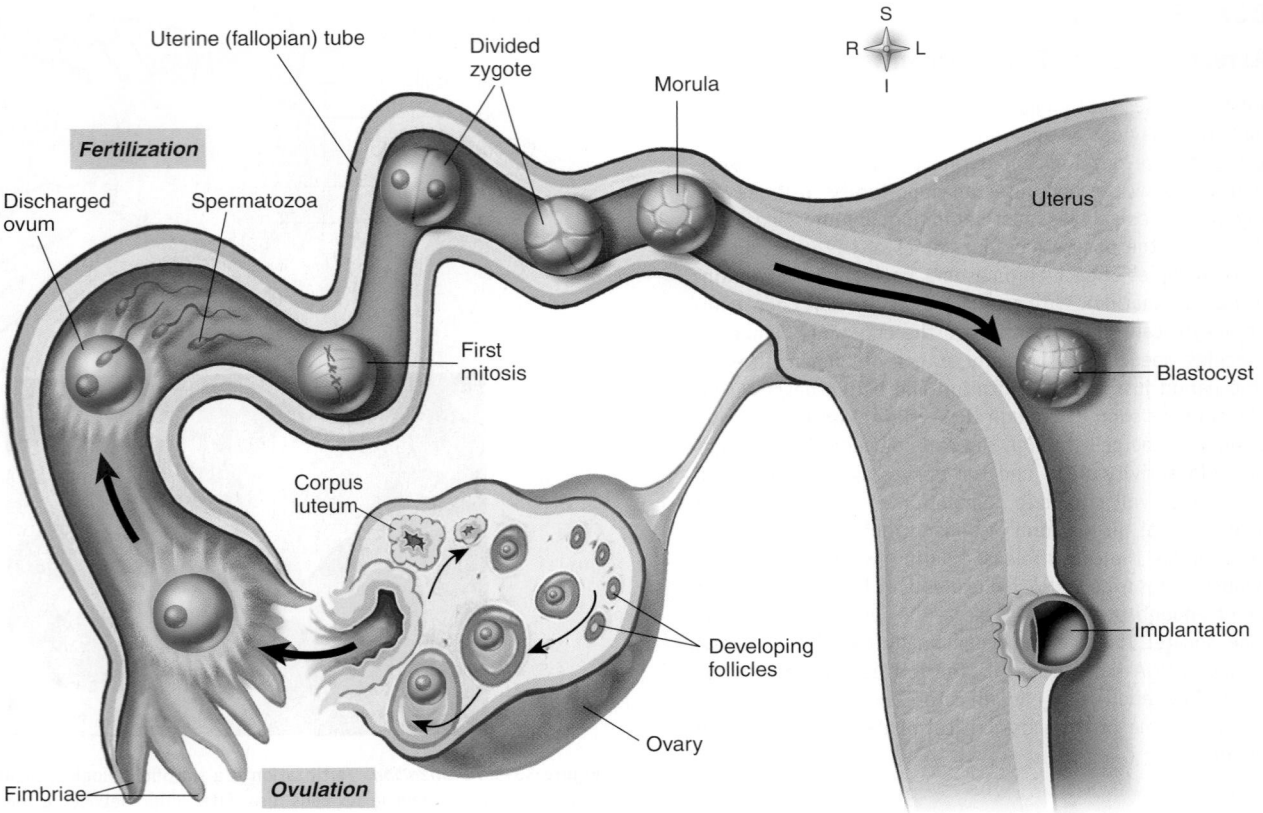

Figure 33-6 *Fertilization and implantation.* At ovulation, an ovum is released from the ovary and begins its journey through the uterine tube. While in the tube, the ovum unites with a sperm to form the single-celled zygote. After a few days of rapid mitotic division, a ball of cells called a *morula* is formed. After the morula develops into a hollow ball called a *blastocyst,* implantation occurs.

BOX 33-2: FYI
In Vitro Fertilization

The Latin term *in vitro* (in VEE-troh) means, literally, "within a glass." In the case of **in vitro fertilization**, it refers to the glass laboratory dish where an ovum and sperm are mixed and where fertilization occurs.

In the classic technique, the ovum is obtained from the mother by first inserting a fiberoptic viewing instrument called a **laparoscope** through a very small incision in the woman's abdomen. Once in the abdominal cavity the device allows the physician to view the ovary and then puncture and "suck up" an ovum from a mature follicle. Over the years refinements to this technique have been made and less invasive procedures, such as the insertion of a needle through the vaginal wall, are currently being used. After about 2½ days' growth in a temperature-controlled environment, the developing zygote, which by then has reached the 8- or 16-cell stage, is returned by the physician to the mother's uterus. If implantation is successful, growth will continue and the subsequent pregnancy will progress. In the most successful fertility clinics in the United States, a normal term birth will occur in about 30% of in vitro fertilization attempts.

sperm nucleus combine with the 23 chromosomes already in the ovum to restore the diploid number of 46 chromosomes.

Inasmuch as the ovum lives only a short time (probably only a day or so) after leaving the ruptured follicle, the fertilization "window" occurs around the time of ovulation. Because sperm may live up to a few days after entering the female tract, sexual intercourse any time from about 3 days before ovulation to a day or so after ovulation may result in fertilization.

The fertilized ovum, or zygote, is genetically complete; it represents the first cell of a genetically new individual. Time, nourishment, and a proper prenatal environment are all that are needed for expression of characteristics such as sex, hair, and skin color that were determined at the time of fertilization.

QUICK CHECK

1. What is the function of meiotic division?
2. How does meiosis differ from mitosis?
3. Where in the female reproductive tract does fertilization usually occur?
4. What is the technical name for the fertilized ovum?

PRENATAL PERIOD

The **prenatal period** of development begins at the time of conception, or fertilization (i.e., at the moment the female ovum and the male sperm cell unite). The period of prenatal development continues until the birth of the child about 39 weeks later. The science of the development of the individual before birth is called **embryology** (em-bree-OL-oh-gee). It is a story of biological marvels, describing the means by which a new human life is begun and the steps by which a single microscopic cell is transformed into a complex human being.

Cleavage and Implantation

As you can see in Figure 33-6, once the zygote is formed it immediately begins to cleave, or divide, and in about 3 days a solid mass of cells called a **morula** (MOR-yoo-lah) is formed. The cells of the morula begin to form an inner cavity as they continue to divide, and by the time the developing embryo reaches the uterus, it is a hollow ball of cells called a **blastocyst** (BLASS-toh-sist). In about 10 days from the time of fertilization the blastocyst is completely implanted in the uterine lining—before nutrients from the mother are available to nourish it. Of course, problems in development or implantation may occur at any stage—resulting in loss of the offspring and termination of the developmental process.

The rapid cell division taking place up to the blastocyst stage occurs with no significant increase in total mass compared with the zygote (Figure 33-7). One of the specializations of the ovum is its incredible store of nutrients that support this embryonic development until **implantation** has occurred.

Note in Figure 33-8 that the blastocyst consists of an outer layer of cells and an **inner cell mass.** The outer wall of the blastocyst is called the **trophoblast** (see Figure 33-8). As the blastocyst develops further, the inner cell mass forms a structure with two cavities called the **yolk sac** and **amniotic** (am-nee-OT-ik) **cavity** (Figures 33-9 and 33-10). The yolk sac is most important in animals such as birds that depend heavily on yolk as a nutrient for the developing embryo. In these animals the yolk sac digests the yolk and provides the resulting nutrients to the embryo. Because the uterine lining provides nutrients to the developing embryo in humans, the function of the yolk sac is not a nutritive one. Instead, it has other functions, including production of blood cells. It is the inner cell mass that eventually forms the tissues of the offspring's body. The trophoblast, on the other hand, forms the support structures described in the following paragraphs. (Box 33-3 discusses the field of *developmental biology.*)

The amniotic cavity becomes a fluid-filled, shock-absorbing sac, sometimes called the "bag of waters," in which the embryo

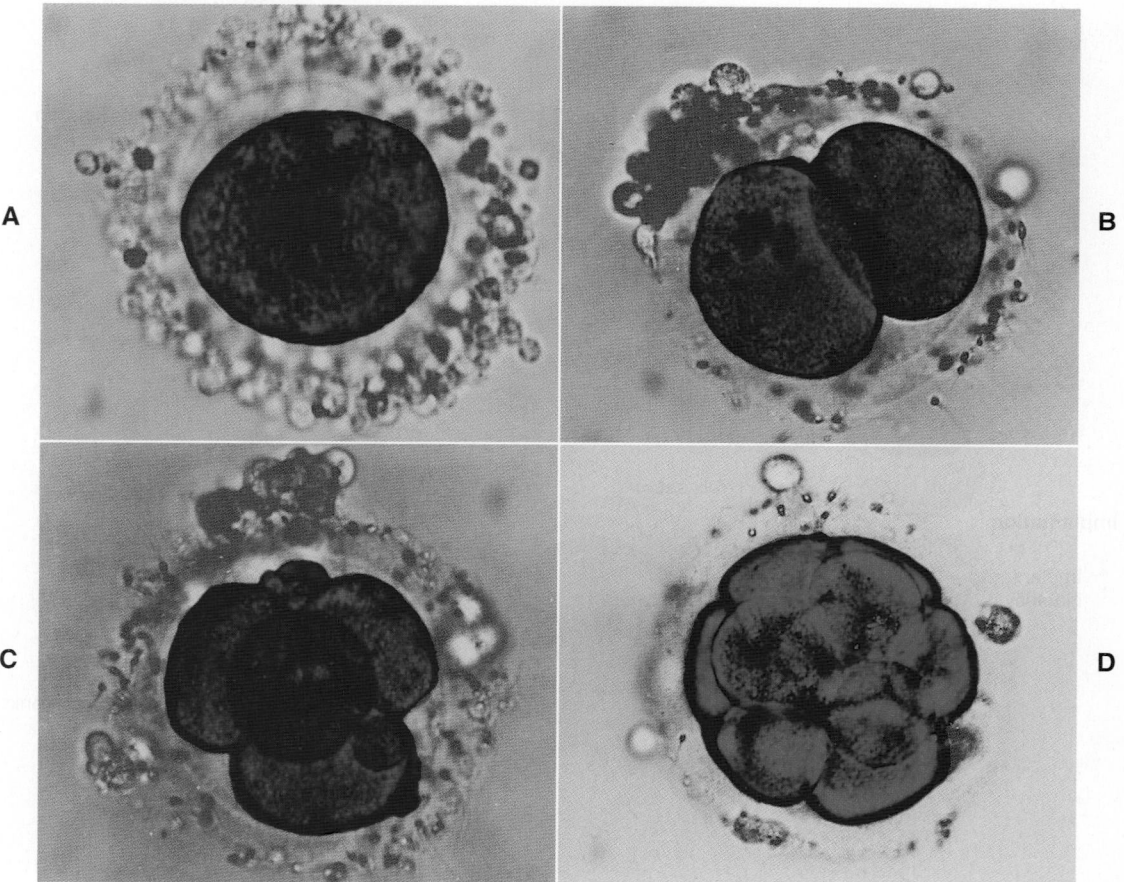

Figure 33-7 *Early stages of human development.* **A,** Fertilized ovum, or zygote. **B** to **D,** Early cell divisions produce more and more cells. The solid mass of cells shown in **D** forms the morula—an early stage in embryonic development.

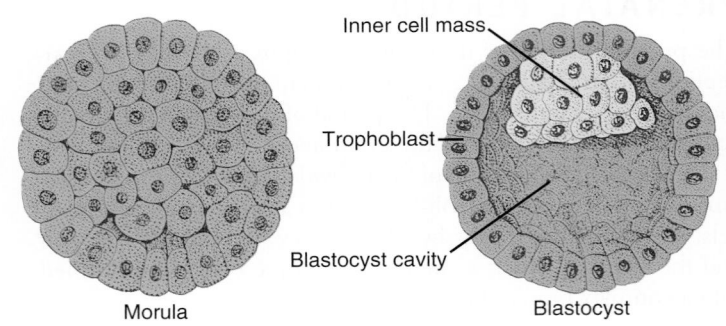

Figure 33-8 *Early stages in the development of the human embryo.* The morula consists of an almost solid spherical mass of cells. The embryo reaches this stage about 3 days after fertilization. The blastocyst (hollowing) stage develops later, before implantation in the uterine lining.

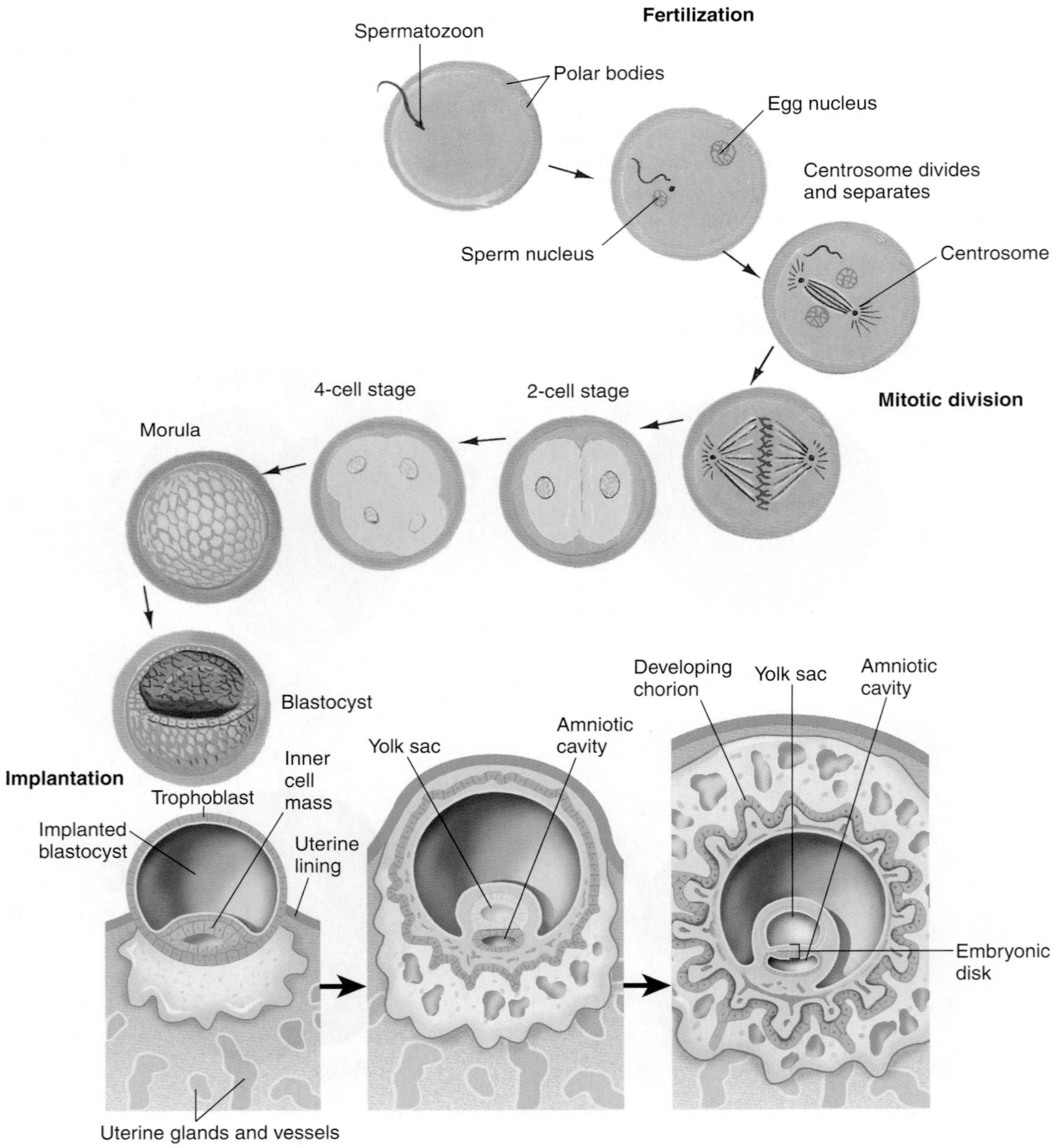

Inner cell mass

Trophoblast

Blastocyst cavity

Morula

Blastocyst

Fertilization

Spermatozoon

Polar bodies

Egg nucleus

Sperm nucleus

Centrosome divides and separates

Centrosome

4-cell stage

2-cell stage

Morula

Mitotic division

Blastocyst

Implantation

Inner cell mass

Trophoblast

Implanted blastocyst

Uterine lining

Yolk sac

Amniotic cavity

Developing chorion

Yolk sac

Amniotic cavity

Embryonic disk

Uterine glands and vessels

Figure 33-9 *Fertilization to implantation and development of the yolk sac.* Rapid growth of uterine glands and vessels covers the developing blastocyst at the time of implantation.

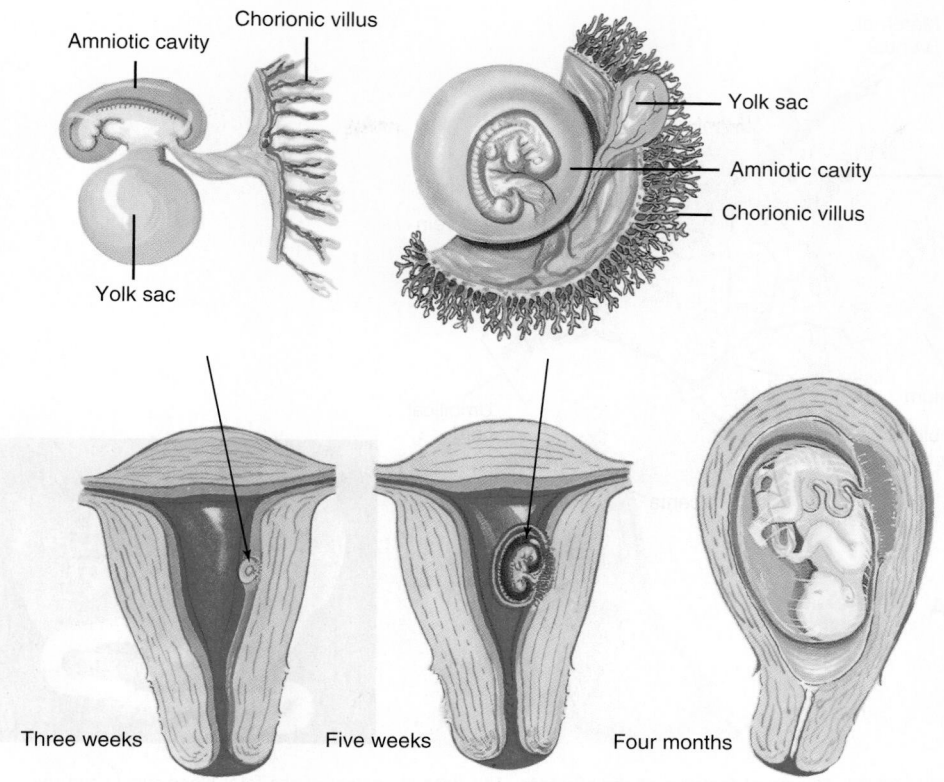

Figure 33-10 *Development of the chorion and amnion.* Development of the chorion and amniotic cavity to 4 months of gestation.

BOX 33-3: FYI

Developmental Biology

Developmental biology is the name given to the branch of life science that studies the process of change over the life cycle. This process of change is called *development*. It is important to realize that in developmental biology, the terms *growth* and *development* do not mean the same thing. Growth is simply an increase in body mass. Development, on the other hand, refers to the complex series of changes that occur at various times of life. Early stages of development—particularly the prenatal stages—are characterized by rapid growth, whereas later stages of development are characterized by little, if any, growth of body tissues.

In this chapter, we briefly discuss various subtopics within the field of human developmental biology. For example, prenatal development is studied by a branch of developmental biology called **embryology.** The biological changes observed during late adulthood are studied by a branch of developmental biology called **gerontology.**

Basic concepts of embryology, gerontology, and other subdisciplines of developmental biology seem to have taken on a greater practical importance during the past few decades than ever before. One reason is the explosion in knowledge of developmental processes and our ability to treat the abnormalities that we can now find. Procedures such as fetal surgery, electrocardiography, and ultrasound permit physicians to diagnose and treat the fetus much like any other patient. Recent discoveries about the processes of aging—as well as the rapidly growing population of aged individuals—have spawned new methods of recognizing and treating physical and psychological problems in the elderly. Another reason developmental biology has taken on great practical importance is that it is a field that serves to unify human biology into a framework that integrates anatomy, physiology, cell biology, molecular biology, medicine, and other disciplines. Thus developmental biology gives us an excellent view of the "big picture" of the human body.

floats during development. The **chorion** (KOH-ree-on), shown in Figures 33-9 to 33-11, develops from the trophoblast to become an important fetal membrane in the **placenta** (plah-SEN-tah). The *chorionic villi* shown in Figures 33-10 and 33-11 are extensions of the blood vessels of the chorion that bring the embryonic circulation to the placenta. The placenta (see Figure 33-11) anchors the developing offspring to the uterus and provides a "bridge" for the exchange of nutrients and waste products between mother and baby.

Placenta

The placenta is a unique and highly specialized structure that has a temporary but very important series of functions during pregnancy. It is composed of tissues from mother and child and func-

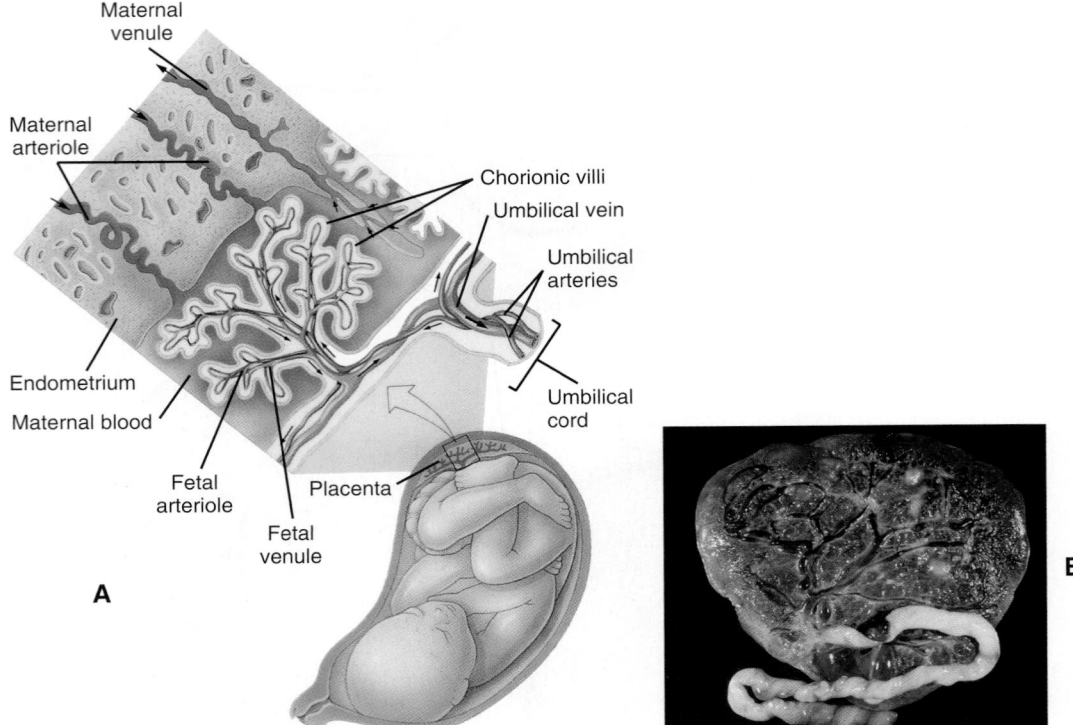

Figure 33-11 *Structural features of the placenta.* The close placement of the fetal blood supply and the maternal blood supply permits diffusion of nutrients and other substances. It also forms a thin barrier to prevent diffusion of most harmful substances. No mixing of fetal and maternal blood occurs. **A,** Diagram showing a cross section of the placental structure. **B,** Photograph of a normal, full-term placenta (fetal side) showing the branching of the placental blood vessels.

tions not only as structural "anchor" and nutritive bridge but also as an excretory, respiratory, and endocrine organ.

Placental tissue normally separates the maternal and fetal blood supplies so that no intermixing occurs. The very thin layer of placental tissue that separates maternal and fetal blood also serves as an effective "barrier" that can protect the developing baby from many harmful substances that may enter the mother's bloodstream. Unfortunately, toxic substances such as alcohol and some infectious organisms may penetrate this protective placental barrier and injure the developing baby (see Box 18-5 on p. 716). The virus responsible for German measles (rubella), for example, can easily pass through the placenta and cause tragic developmental defects in the fetus.

Placental tissue also has important endocrine functions. As Figure 33-12 shows, placental tissue secretes large amounts of **human chorionic gonadotropin (hCG)** early in pregnancy. hCG secretion peaks about 8 or 9 weeks after fertilization, then drops to a continuous low level by about week 16. The function of hCG, as its name implies, is to act as a gonadotropin and stimulate the corpus luteum to continue its secretion of estrogen and progesterone. Recall from Chapter 32 (Figure 32-15 on p. 1146) that reduced levels of the anterior pituitary gonadotropins (FSH and LH) after ovulation normally cause a corresponding reduction in luteal secretion of the estrogen and progesterone needed to sustain the uterine lining. The drop in estrogen and progesterone secretion results from the fact that the FSH and LH needed to maintain the

corpus luteum are now in short supply. To prevent menstruation and to allow successful implantation and development of the offspring, the cells of the trophoblast and, later, the placenta secrete enough hCG to maintain the corpus luteum and thus keep luteal estrogen and progesterone levels high.

As the placenta develops, it begins to secrete its own estrogen and progesterone. As Figure 33-12 shows, as more estrogen and progesterone are secreted from the placenta, a corresponding decrease in hCG secretion produces a drop in luteal secretion of these hormones. After about 3 months, the corpus luteum has degenerated and the placenta has completely taken over the job of secreting the estrogen and progesterone needed to sustain the pregnancy.

Over-the-counter "early pregnancy" tests detect the presence of the hCG that is excreted in the urine during the first couple of months of a pregnancy. Such tests can detect hCG in the urine as early as 1 or 2 days after implantation occurs.

 QUICK CHECK

5. What is a morula? What is a blastocyst?
6. What structures are derived from the trophoblast (outer wall) of the blastocyst?
7. What placental hormone maintains the corpus luteum during the early weeks of pregnancy?

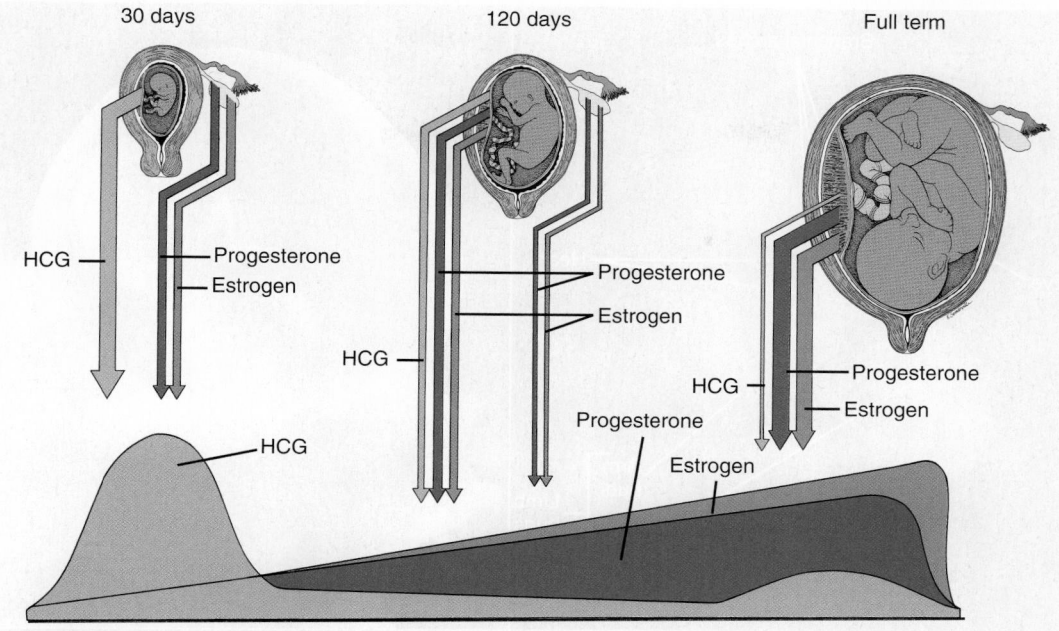

Figure 33-12 *Hormone levels during pregnancy.* Diagram showing the changes that occur in the blood concentration of human chorionic gonadotropin (HCG), estrogen, and progesterone during gestation. Note that high HCG levels produced by placental tissue early in pregnancy maintain estrogen and progesterone secretion by the corpus luteum. This prevents menstruation and promotes maintenance of the uterine lining. As the placenta takes over the job of secreting estrogen and progesterone, HCG levels drop, and the corpus luteum subsequently ceases secreting these hormones.

Periods of Development

The length of pregnancy (about 39 weeks)—called the **gestation period**—is divided into three approximately 3-month segments called *trimesters*. The first trimester extends from the first day of the last menstrual period to the end of the twelfth week. The second trimester extends from the twelfth to the twenty-eighth week of pregnancy. The third trimester extends from the twenty-eighth week of pregnancy until the baby is delivered. Several terms such as *embryo* and *fetus* are used to describe stages of development during the three trimesters of pregnancy.

During the first trimester, or 3 months, of pregnancy, numerous terms are used. Zygote is used to describe the ovum just after fertilization by a sperm cell. After about 3 days of constant cell division, the solid mass of cells, identified earlier as the morula, enters the uterus. Continued development transforms the morula into the hollow blastocyst, which then implants into the uterine wall (see Figure 33-9).

The embryonic phase of development extends from fertilization until the end of week 8 of gestation. During this period in the first trimester, the term *embryo* is used to describe the developing individual. The fetal phase is used to indicate the development extending from week 8 to 39. During this period, the term *embryo* is replaced by the term *fetus*.

By day 35 of gestation (Figure 33-13, *A*), the heart is beating and, although the embryo is only 8 mm (about ⅜ inch) long, the eyes and so-called *limb buds*, which ultimately form the arms and legs, are clearly visible. Figure 33-13, *C*, shows the stage of development at the end of the first trimester of gestation, when the offspring be-

comes known as a fetus. Body size is about 7 to 8 cm (3 inches) long (Figure 33-14). The facial features of the fetus are apparent, the limbs are complete, and sex can be identified. By month 4 (Figure 33-13, *D*), all organ systems are formed and functioning to some extent. Growth of the embryo to 4 months is summarized in the photographs in Figures 33-13 and 33-14; growth to full term is summarized in the graph in Figure 33-15. Box 33-4 discusses methods of diagnosis and treatment in these early developmental stages.

Formation of the Primary Germ Layers

Early in the first trimester of pregnancy, three layers of specialized cells develop that embryologists call the **primary germ layers.** Cells of the *embryonic disk* seen in Figure 33-9 differentiate into distinct types that form each of these three primary germ layers. Cells in each continue to differentiate and thus give rise to the various specific organs and systems of the body, such as the skin, nervous tissue, muscles, or digestive organs (Figure 33-16 and Box 33-5). As new tissues and organs develop, older cells often die through the process of apoptosis (see Chapter 4) and thus make room for newer, more specialized cells. Each primary germ layer is called, respectively, **endoderm** (EN-doh-derm), or inside layer; **ectoderm** (EK-toh-derm), or outside layer; and **mesoderm** (MEZ-oh-derm), or middle layer.

Endoderm

The inner germ layer, or *endoderm*, forms the linings of various tracts, as well as several glands. For example, the lining of the respiratory tract and GI tract, including some of the accessory struc-

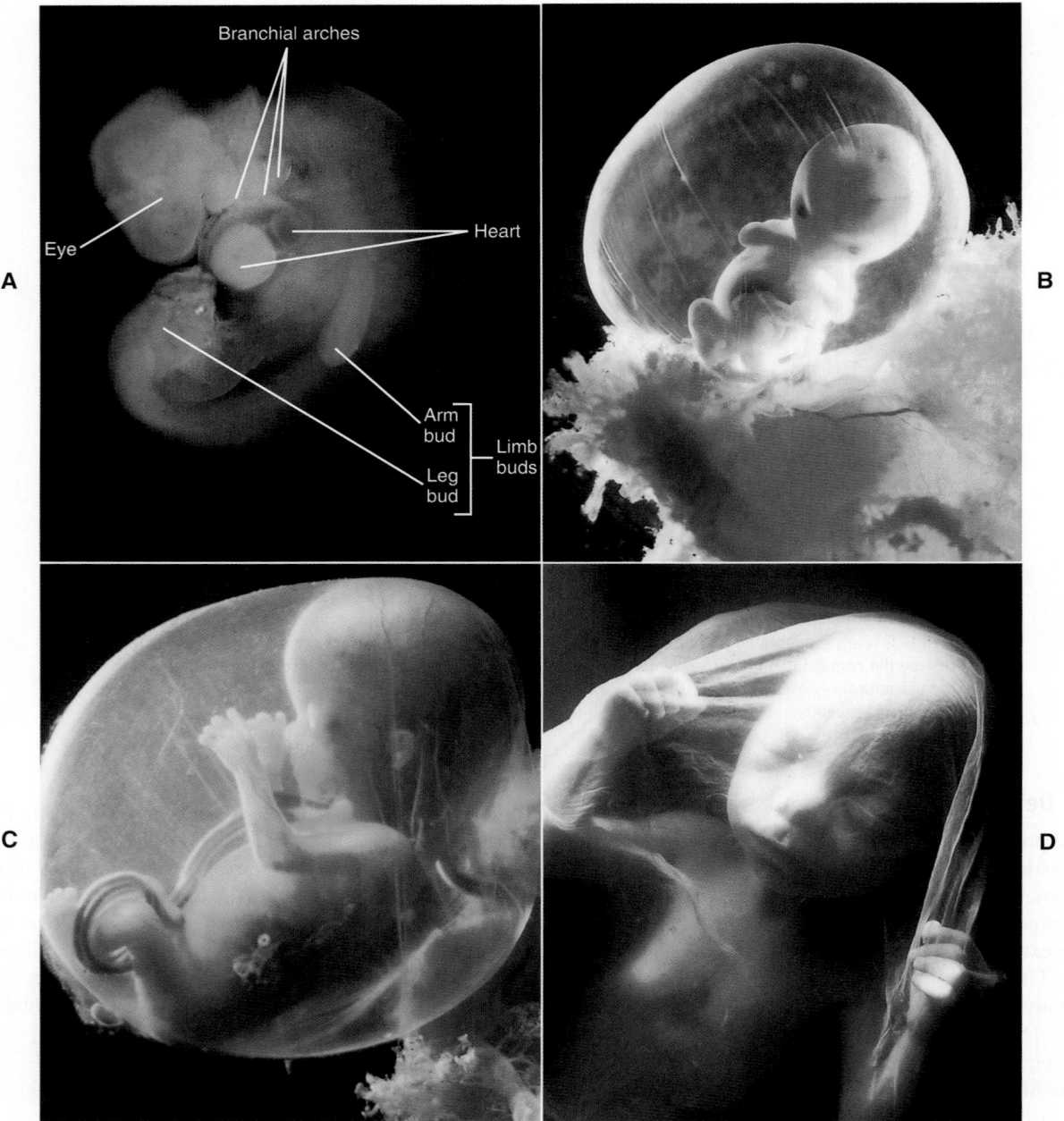

Figure 33-13 *Human embryos and fetuses.* **A**, At 35 days. **B**, At 49 days. **C**, At the end of the first trimester. **D**, At 4 months.

tures such as tonsils, is derived from the endoderm. The linings of the pancreatic ducts, hepatic ducts, and urinary tract also have an endodermal origin. The glandular epithelium of the thymus, thyroid, and parathyroid glands is also derived from the endoderm.

Ectoderm

The outer germ layer, or *ectoderm*, forms many of the structures around the periphery of the body. For example, the epidermis of the skin, enamel of the teeth, and cornea and lens of the eye are derived from the ectoderm. Besides these peripheral structures, various components of the nervous system—including the brain and the spinal cord—also have an ectodermal origin.

Mesoderm

The middle germ layer, or *mesoderm*, forms most of the organs and other structures between those formed by the endoderm and ectoderm. For example, the dermis of the skin, the skeletal muscles and bones, many of the glands of the body, kidneys, gonads,

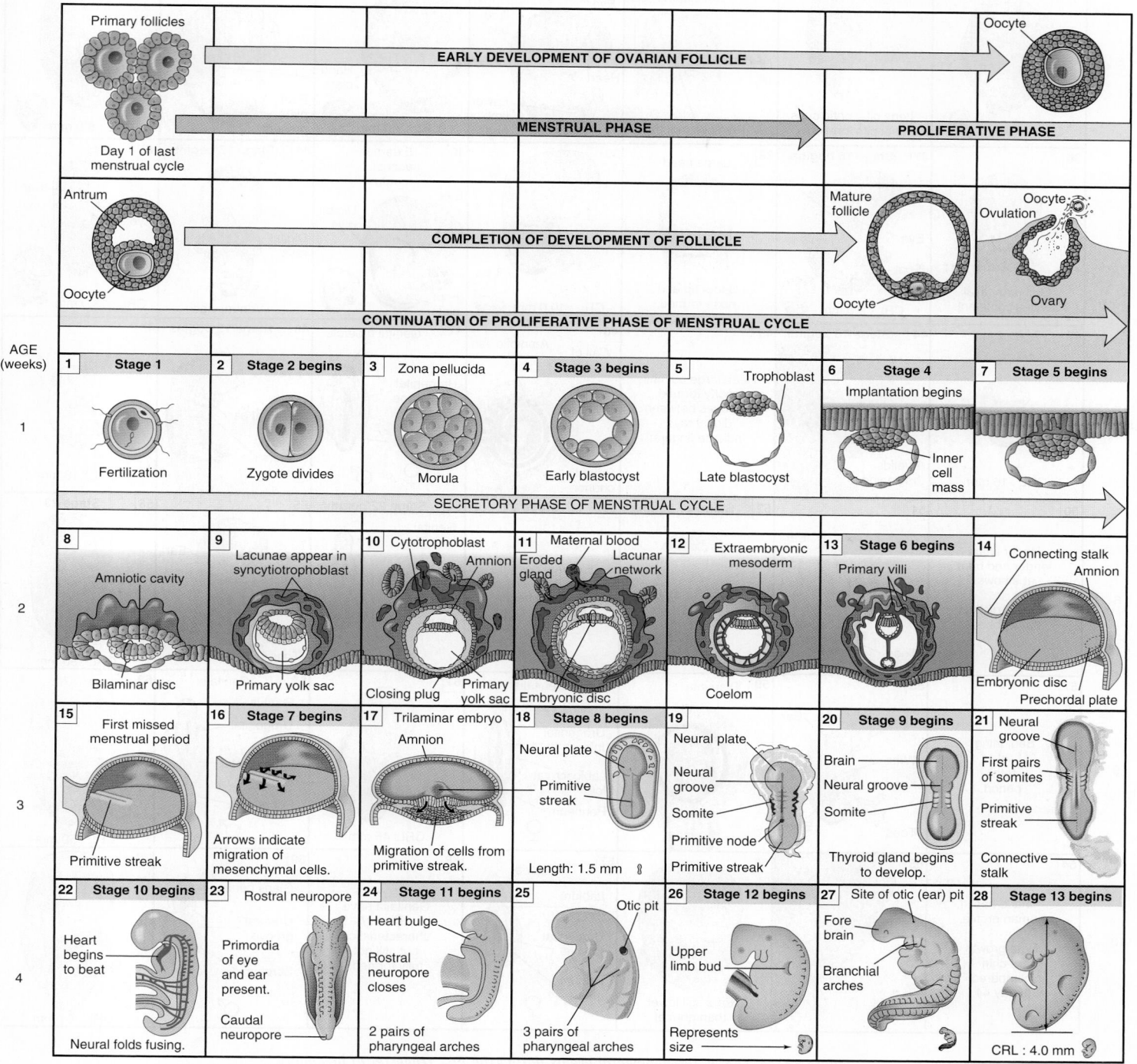

Figure 33-14 *Developmental events during the first trimester of pregnancy.* Each box represents a day, and each row represents a week. Compare this timetable with the preceding figures of this chapter. *CRL,* Crown-to-rump length.

Continued

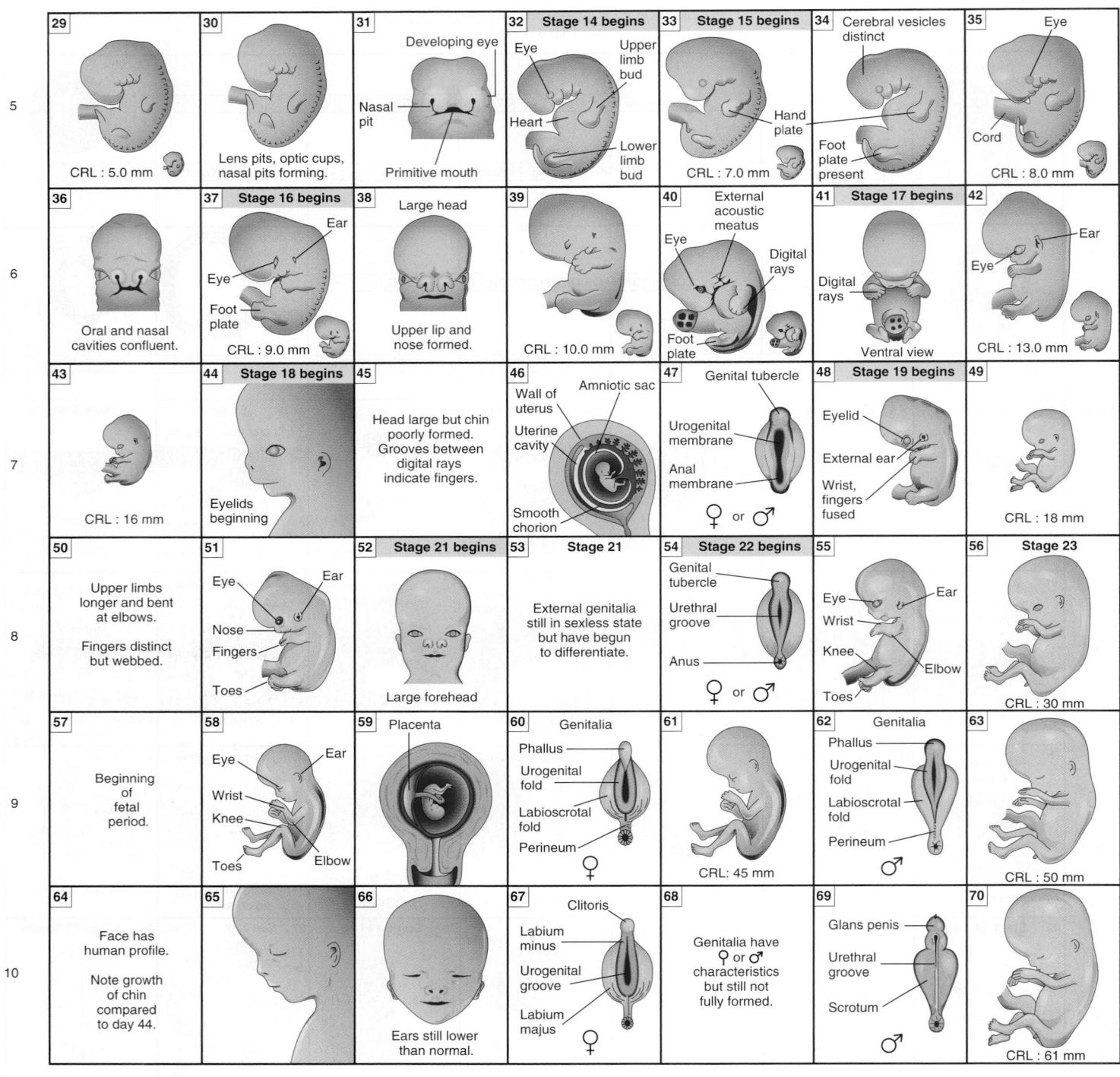

Figure 33-14, cont'd *Developmental events during the first trimester of pregnancy.*

and components of the circulatory system are derived from the mesoderm. Look carefully at Figure 33-16 to discern the logical pattern exhibited by germ layer development and differentiation.

Histogenesis and Organogenesis

The process by which the primary germ layers develop into many different kinds of tissues is called **histogenesis** (hiss-toh-JEN-eh-sis). The way these tissues arrange themselves into organs is called **organogenesis** (or-gah-no-JEN-eh-sis).

The fascinating story of histogenesis and organogenesis in human development is long and complicated; its telling belongs to the science of embryology. However, a brief example of organogenesis that is particularly useful in our current discussion of reproduction and development is the differentiation and development of the sex organs.

As Figure 33-17 shows, the reproductive tracts of both the male and the female begin their development as sets of undifferentiated ducts and gonads. But as embryonic development continues

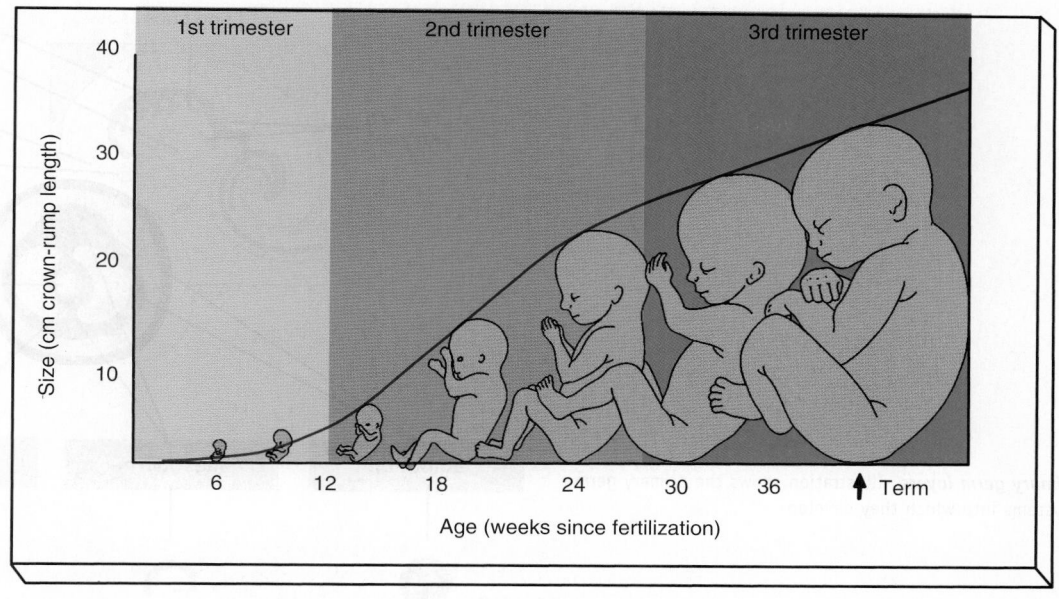

Figure 33-15 *Increase in size during prenatal development.* This graph shows the usual progression in size and body shape during the three trimesters of fetal development.

 BOX 33-4

Diagnostic Study

Antenatal Diagnosis and Treatment

Advances in **antenatal** (from the Latin *ante-,* "before" and *-natus,* "birth") **medicine** now permit extensive diagnosis and treatment of disease in the fetus much like any other patient. This new dimension in medicine began with techniques by which Rh+ babies could be given transfusions before birth.

Current procedures using images provided by ultrasound equipment (see the figure) allow physicians to prepare for and perform, before the birth of a baby, corrective surgical procedures such as bladder repair. These procedures also allow physicians to monitor the progress of other types of treatment on a developing fetus. Part *A* of the figure shows placement of the ultrasound transducer on the abdominal wall. The resulting image (see part *B* of the figure), called an *ultrasonogram,* shows a 20-week-old fetus.

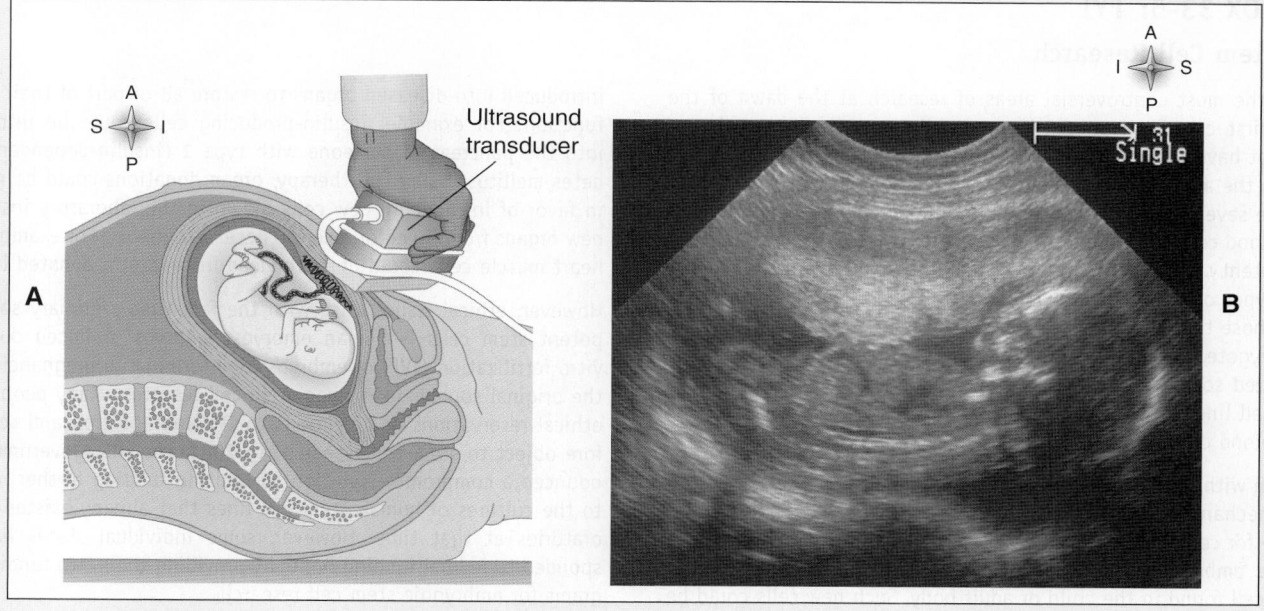

Ultrasonography. **A,** Placement of the ultrasound transducer on the abdominal wall. **B,** Ultrasonogram showing a midsagittal view of a 20-week-old fetus.

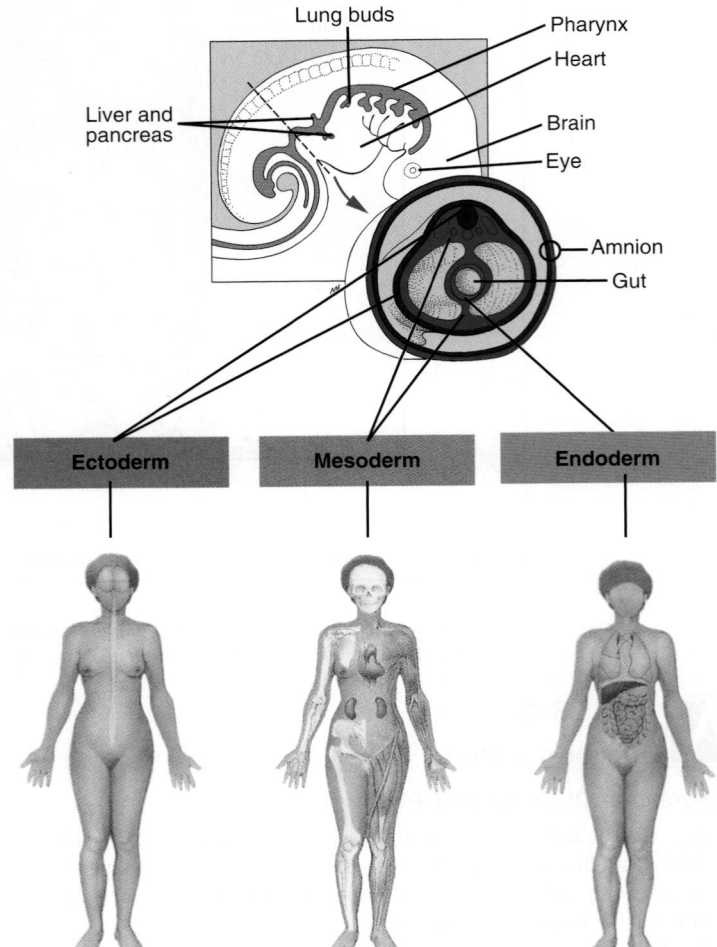

Figure 33-16 *The primary germ layers.* Illustration shows the primary germ layers and the body systems into which they develop.

BOX 33-5: FYI

Stem Cell Research

One of the most controversial areas of research at the dawn of the twenty-first century was human stem cell research. Stem cells are cells that have the ability to divide continually to form new types of cells. In the adult, the stem cells of the red bone marrow are able to generate several different cell lines: red blood cells, several types of white blood cells, and platelets. During embryonic development, even more potent stem cells exist—stem cells that can generate many different types of tissue in the human body. The most potent stem cells of all, those that can generate any and all human tissue types, make up the zygote and morula. By the blastocyst stage, each stem cell has specialized somewhat and can generate only a certain group of different cell lines. By adulthood, only stem cells that produce the different blood cell lines remain.

Research with embryonic stem cells not only helps us understand the basic mechanisms of tissue differentiation, but also holds great promise for *cell therapy,* the use of special cells to treat disease. For example, embryonic stem cells could be used to produce any type of normal cell found in the child or adult body. Such new cells could be introduced into diseased organs to restore all or part of their normal function. For example, insulin-producing cells could be introduced into the pancreas of someone with type 1 (insulin-dependent) diabetes mellitus. Using cell therapy, organ donations could be reduced in favor of introducing new cells grown in the laboratory instead of new organs from living or recently deceased donors. For example, new heart muscle cells could be introduced instead of a donated heart.

However, ethical issues arise with the fact that a primary source of potent stem cells is human embryos. Embryos produced during in vitro fertilization (IVF) or embryos from terminated pregnancies were the original sources for human stem cell research. Many people have ethical reservations or objections to IVF and abortion and so therefore object to stem cell research. In 2001, the U.S. government announced a compromise solution by limiting funding further research to the cultures of human stem cell lines that already existed in laboratories at that time. However, some individual states have responded to federal funding limits by providing their own funding programs for embryonic stem cell research.

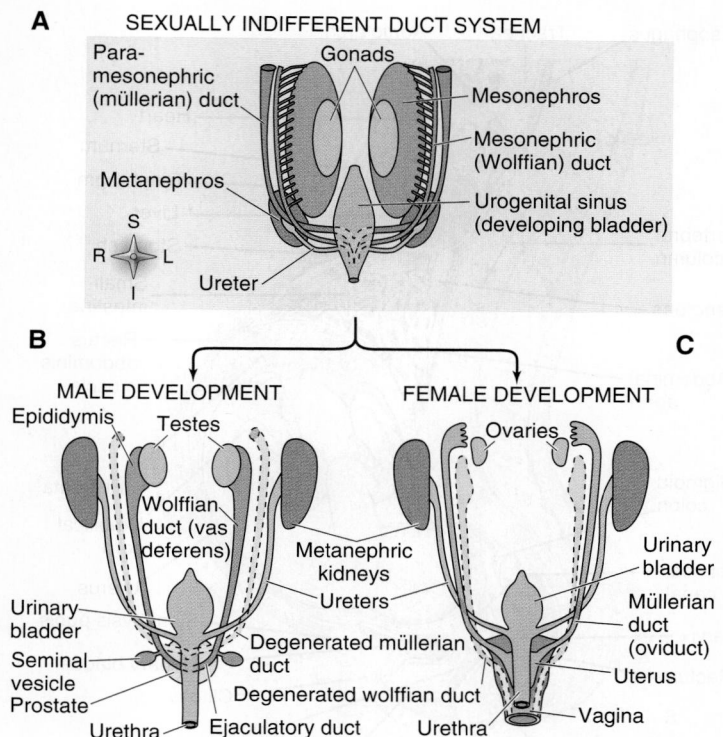

Figure 33-17 *Development of the reproductive tracts.* Both the male and female adult reproductive tracts are similar in their basic outline because they have a shared early development. **A,** Early in embryonic development, an undifferentiated set of gonads and ducts develops in both males and females. **B,** In males, the gonads (now testes) attach to the mesonephric (wolffian) ducts, which develop into the main part of the male reproductive tract. The paramesonephric ducts degenerate in males. **C,** In females, the gonads do not attach directly to a duct. It is the paramesonephric (müllerian) ducts that develop into the female reproductive tract and the mesonephric ducts that degenerate.

in the male, the gonads attach to the *mesonephric (wolffian) duct*—which along with the urethra develops into the male reproductive tract (Figure 33-18, *B*). In the female, by contrast, it is the nearby *paramesonephric (müllerian) duct* that instead develops into a female reproductive tract separate from the urinary tract (Figure 33-18, *C*). Note that the female gonads (ovaries) do not attach to their ducts during development. Figure 33-19 outlines the development of the male and female genitalia. Notice how they develop along slightly different patterns to eventually become distinct types of structures.

A complete outline of the embryonic development of each organ and system is beyond the scope of this book. For the beginning student of anatomy and physiology, it seems sufficient to appreciate that life begins when two sex cells unite to form a single-celled zygote and that the new human body evolves by a series of processes consisting of cell differentiation, multiplication, growth, apoptosis, and rearrangement, all of which take place in a definite, orderly sequence. Development of structure and function go hand in hand, and from 4 months of gestation, when every organ system is in place and functioning to some extent, until term (about 280 days), development of the fetus is mainly a matter of growth.

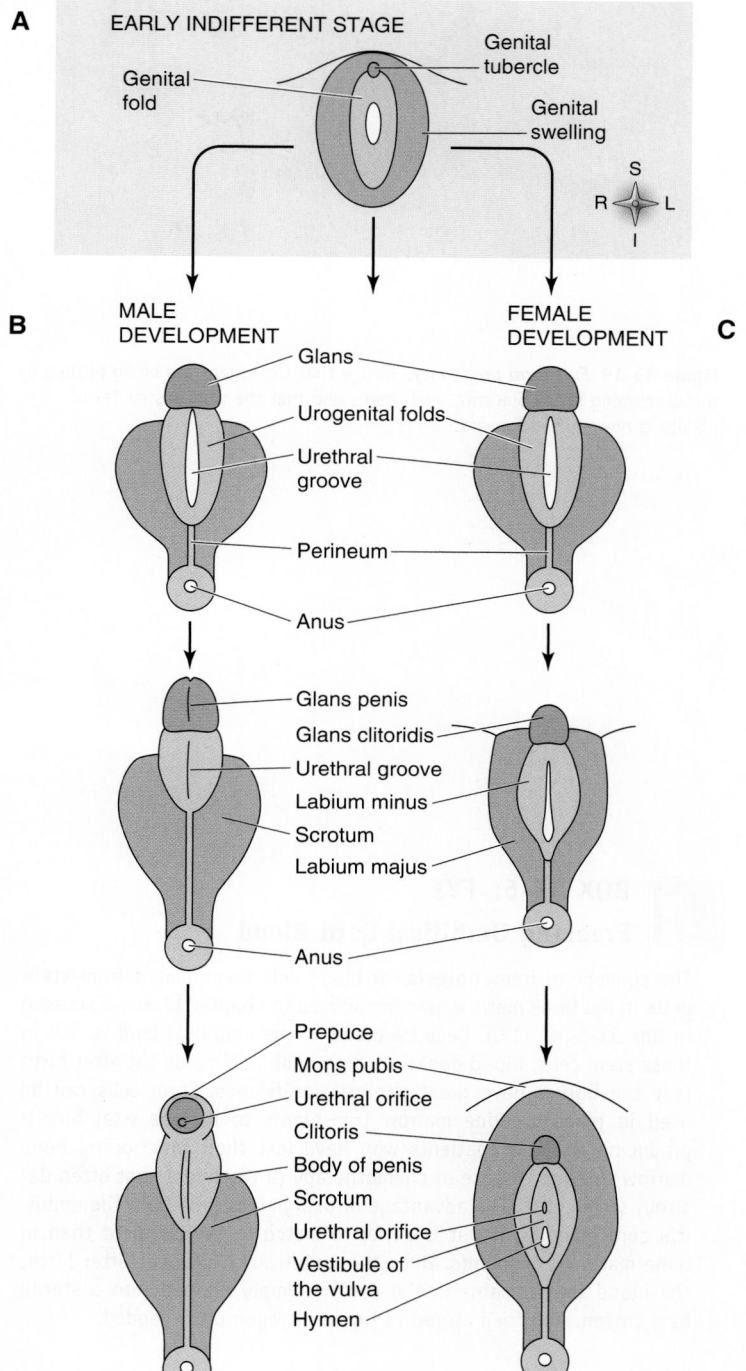

Figure 33-18 *Development of the genitalia.* **A,** In early stages of development, the genitalia are indifferent (not yet distinguishable). **B,** In the male, the genital tubercle eventually becomes the glans of the penis and the folds become the penis shaft and scrotum. **C,** In the female, the genital tubercle becomes the clitoris and the folds become the labia.

Figure 33-19 shows the normal intrauterine position of a fetus just before birth in a full-term pregnancy. The large size of the pregnant uterus toward the end of pregnancy affects the normal function of the mother's body greatly. For example, you might be able to tell from Figure 33-19 that a woman's center of gravity is shifted

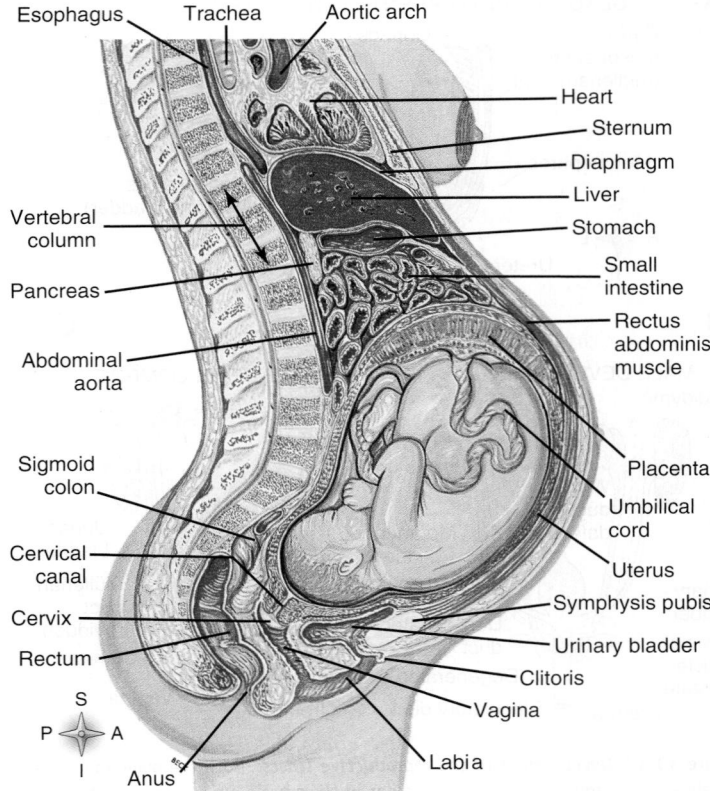

Figure 33-19 *Full-term pregnancy.* Notice that the organs are being pushed by the developing fetus, placenta, and uterus and that the woman's center of gravity is now shifted forward.

BOX 33-6: FYI
Freezing Umbilical Cord Blood

The concept of **hemopoiesis**, or blood cell development from **stem cells** in red bone marrow, was introduced in Chapter 17 and discussed in Box 33-5, p. 1178. Because blood in the umbilical cord is rich in these stem cells, blood donations from umbilical cords cut after birth (see the figure) have great clinical significance. Stem cells can be used in place of bone marrow transplants to replace vital blood-producing tissue in patients who have lost their functioning bone marrow through disease or chemotherapy (a treatment that often destroys stem cells). The advantage of using stem cells from the umbilical cord blood is that it is easier to match to the recipient than in bone marrow transplants. Once the umbilical cord is cut after birth, the blood that remains in the cord is simply drained into a sterile bag, frozen, and then stored in liquid nitrogen until needed.

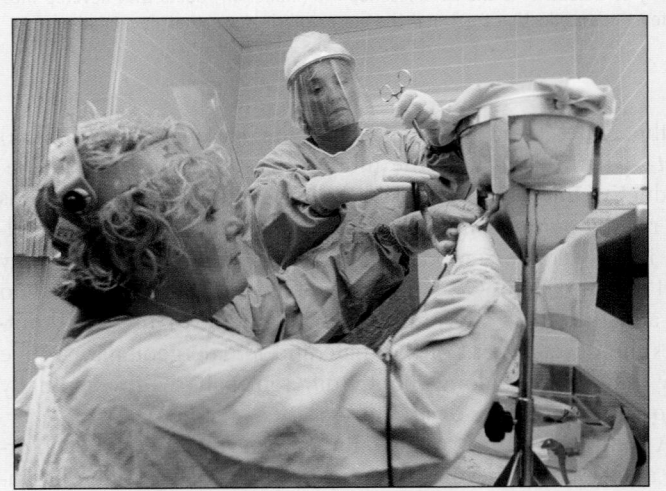

Collecting umbilical cord blood.

forward. This can make walking and other movements of the body difficult—or even hazardous—because the sensory and motor control systems often do not compensate completely for this shift. The pregnant uterus presses on the rectum, sometimes adversely affecting intestinal motility and thus may cause constipation and/or hemorrhoids. Pressure on the bladder reduces its urine-storing capacity, which results in frequent urination. Upward pressure pushes the abdominal organs against the diaphragm, making deep breathing difficult and sometimes causing the stomach to protrude into the thoracic cavity—a condition called *hiatal hernia.*

QUICK CHECK

8. What is a *trimester*?
9. What is the difference between an *embryo* and a *fetus*?
10. Name the three primary germ layers.
11. What is *histogenesis*?

BIRTH, OR PARTURITION

Birth, or **parturition,** is the point of transition between the prenatal and postnatal periods of life. As the fetus signals the end of pregnancy, the uterus becomes "irritable" and, ultimately, muscular contractions begin and cause the cervix to dilate (open),

thus permitting the fetus to move from the uterus through the vagina or "birth canal" to the exterior. The process normally begins with the fetus taking a head-down position fully against the cervix (Figure 33-20, *A*). When contractions occur, the amniotic sac, or "bag of waters," usually ruptures, and labor begins.

Stages of Labor

Labor is the term used to describe the process that results in the birth of the baby. It is divided into three stages (Figure 33-20, *B* to *E*):

1. Stage one—period from onset of uterine contractions until dilation of the cervix is complete
2. Stage two—period from the time of maximal cervical dilation until the baby exits through the vagina

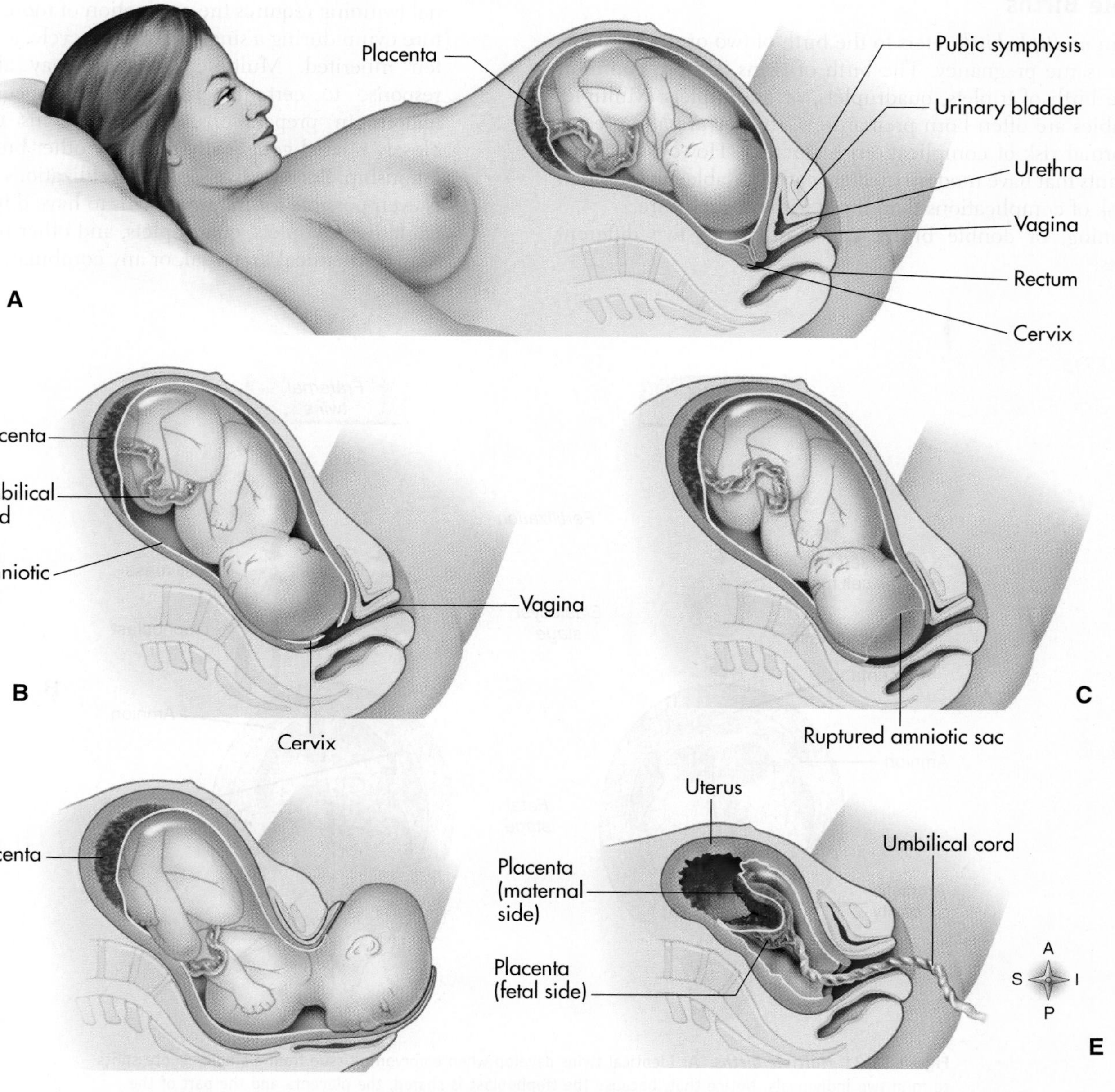

Figure 33-20 *Parturition.* A, The relation of the fetus to the mother. **B,** The fetus moves into the opening of the birth canal, and the cervix begins to dilate. **C,** Dilation of the cervix is complete. **D,** The fetus is expelled from the uterus. **E,** The placenta is expelled.

3. Stage three—process of expulsion of the placenta through the vagina

The time required for normal vaginal birth varies widely and may be influenced by many variables, including whether the woman has previously had a child. In most cases, stage one of labor lasts from 6 to 24 hours, and stage two lasts from a few minutes to an hour. Delivery of the placenta (stage three) is normally within 15 minutes after the birth of the baby.

If abnormal conditions of the mother or fetus (or both) make normal vaginal delivery hazardous or impossible, physicians may suggest a **cesarean** (seh-SAIR-ee-an) **section.** Often called simply a *C-section*, it is a surgical procedure in which the newborn is delivered through an incision in the abdomen and uterine wall.

Multiple Births

The term *multiple birth* refers to the birth of two or more infants from the same pregnancy. The birth of twins is more common than the birth of triplets, quadruplets, or quintuplets. Multiple-birth babies are often born prematurely, so they are at a greater than normal risk of complications in infancy. However, premature infants that have modern medical care available have a much lower risk of complications than those without such care.

Twinning, or double births, can result from two different processes:

1. **Identical twins** result from the splitting of embryonic tissue from the same zygote early in development. One way this happens is that, during the blastocyst stage of development, the inner cell mass divides into two masses. Each inner cell mass thus formed develops into a separate individual. As Figure 33-21, A, shows, identical twins usually share the same placenta but have separate umbilical cords. This is not surprising because in this type of twinning there is a single, shared trophoblast. Because they develop from the same fertilized egg, identical twins have the same genetic code. Despite this, identical twins are not absolutely identical in terms of structure and function. Different environmental factors and personal experiences lead to individuality even in genetically identical twins.

2. **Fraternal twins** result from the fertilization of two different ova by two different spermatozoa (Figure 33-21, B). Fraternal twinning requires the production of more than one mature ovum during a single menstrual cycle, a trait that is often inherited. Multiple ovulation may also occur in response to certain fertility drugs, especially the gonadotropin preparations. Fraternal twins are no more closely related genetically than any other brother-sister relationship. Because two separate fertilizations must occur, it is even possible for fraternal twins to have different biological fathers. Triplets, quadruplets, and other multiple births may be identical, fraternal, or any combination.

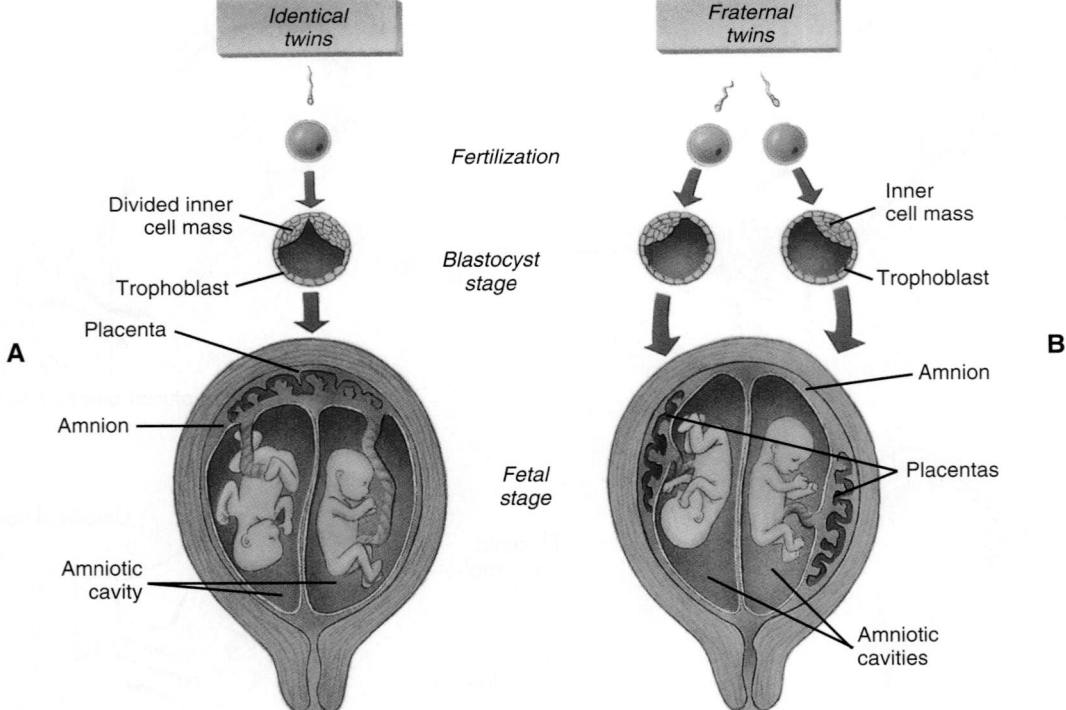

Figure 33-21 *Multiple births.* **A,** Identical twins develop when embryonic tissue from a single zygote splits to form two individuals. Notice that, because the trophoblast is shared, the placenta and the part of the amnion separating the amniotic cavities are shared by the twins. **B,** Fraternal twins develop when two ova are fertilized at the same time, producing two separate zygotes. Notice that each fraternal twin has its own placenta and amnion.

QUICK CHECK

12. Briefly describe the three stages of labor.
13. How does identical twinning differ from fraternal twinning?

POSTNATAL PERIOD

The **postnatal period** begins at birth and lasts until death. It is often divided into major periods for study, but people need to understand and appreciate the fact that growth and development are continuous processes that occur throughout the life cycle. Gradual changes in the physical appearance of the body as a whole and in the relative proportions of the head, trunk, and limbs are quite noticeable between birth and adolescence. Note in Figure 33-22 the obvious changes in the size of bones and in the proportionate sizes between different bones and body areas. The head, for example, be-

comes proportionately smaller. Whereas the infant head is approximately one fourth the total height of the body, the adult head is only about one eighth the total height. The facial bones also show several changes between infancy and adulthood. In an infant the face is one eighth of the skull surface, but in an adult the face is half of the skull surface. Another change in proportion involves the trunk and lower extremities. The legs become proportionately longer and the trunk proportionately shorter. In addition, the thoracic and abdominal contours change from round to elliptical.

Such changes are good examples of the ever-changing and ongoing nature of growth and development. It is unfortunate that many of the changes that occur in the later years of life do not result in an increased function. These degenerative changes are certainly important, however, and will be discussed later in this chapter (see pp. 1187-1190).

The following are the most common postnatal periods: (1) **infancy**, (2) **childhood**, (3) **adolescence** and **adulthood**, and (4) **older adulthood**. Table 33-1 summarizes the projected changes

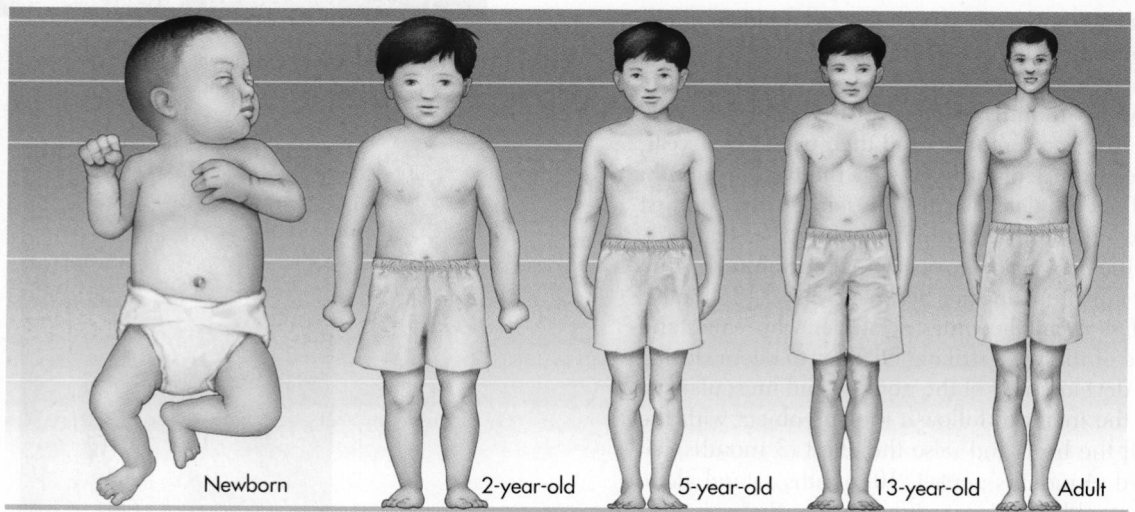

| Newborn | 2-year-old | 5-year-old | 13-year-old | Adult |

Figure 33-22 *Changes in the proportions of body parts from birth to maturity.* Note the dramatic differences in relative head size.

Table 33-1 **U.S. Population—Selected Census Bureau Projections 2000-2050 (numbers in thousands)**

	2000	2010	2020	2030	2040	2050	PERCENT CHANGE
Total Population	282,125	308,936	335,805	363,584	391,946	419,854	49%
Infancy							
0-4 years	19,218	21,426	22,932	24,272	26,299	28,080	46%
Childhood/adolescence							
5-19 years	61,331	61,810	65,955	70,832	75,326	81,067	32%
Adulthood							
20-44 years	104,075	104,444	108,632	114,747	121,659	130,897	26%
45-64 years	62,440	81,012	83,653	82,280	88,611	93,104	49%
Older adulthood							
65-84 years	30,794	34,120	47,363	61,850	64,640	65,844	114%
85+ years	4,267	6,123	7,269	9,603	15,409	20,861	389%

in U.S. population numbers in selected age-groups from 2000 through the year 2050. Notice the proportionally higher rise in older age-groups, particularly those over age 85 years.

Infancy

The period of infancy begins abruptly at birth and lasts about 18 months. The first 4 weeks of infancy are often referred to as the **neonatal period.** Dramatic changes occur at a rapid rate during this short but critical period. **Neonatology** (nee-oh-nay-TOL-oh-jee) is the medical and nursing specialty concerned with the diagnosis and treatment of disorders of the newborn. Advances in this area have resulted in dramatically reduced infant mortality.

Many of the changes that occur in the cardiovascular and respiratory systems at birth are necessary for survival (see Figures 18-31 and 18-32 on pp. 715-716). Whereas the fetus totally depended on the mother for life support, the newborn infant, to survive, must become totally self-supporting in terms of blood circulation and respiration immediately after birth. A baby's first breath is deep and forceful. The stimulus to breathe results primarily from the increasing amounts of carbon dioxide (CO_2) that accumulate in the blood after the umbilical cord is cut shortly after delivery.

Many developmental changes occur between the end of the neonatal period and 18 months of age. Birth weight generally doubles during the first 4 to 6 months and then triples by 1 year. The baby also increases in length by 50% by the twelfth month. The "baby fat" that accumulated under the skin during the first year begins to decrease, and the plump infant becomes leaner.

Early in infancy the baby has only one spinal curvature (Figure 33-23). The lumbar curvature appears between 12 and 18 months, and the once-helpless infant becomes a toddler who can stand (Figure 33-24). One of the most striking changes to occur during infancy is the rapid development of the nervous and muscular systems. This permits the infant to follow a moving object with the eyes (2 months); lift the head and raise the chest (3 months); sit when well supported (4 months); crawl (10 months); stand alone (12 months); and run, although a bit stiffly (18 months).

Childhood

Childhood extends from the end of infancy to sexual maturity, or puberty—12 to 14 years in girls and 14 to 16 years in boys. Overall, growth during early childhood continues at a rather rapid pace, but month-to-month gains become less consistent. By the age of 6 years the child appears more like a preadolescent than an infant or toddler. The child becomes less chubby, the potbelly becomes flatter, and the face loses its babyish look. The nervous and muscular systems continue to develop rapidly during the middle years of childhood; by 10 years of age the child has developed numerous motor and coordination skills.

The *deciduous teeth*, which began to appear at about 6 months of age, are lost during childhood, beginning at about 6 years of age. The *permanent teeth*, with the possible exception of the third molars (wisdom teeth), have all erupted by age 14 years.

Adolescence and Adulthood

The average age range of **adolescence** varies but generally the teenage years (13 to 19) are used as the standard age range. The period is marked by rapid and intense physical growth, which ultimately results in sexual maturity.

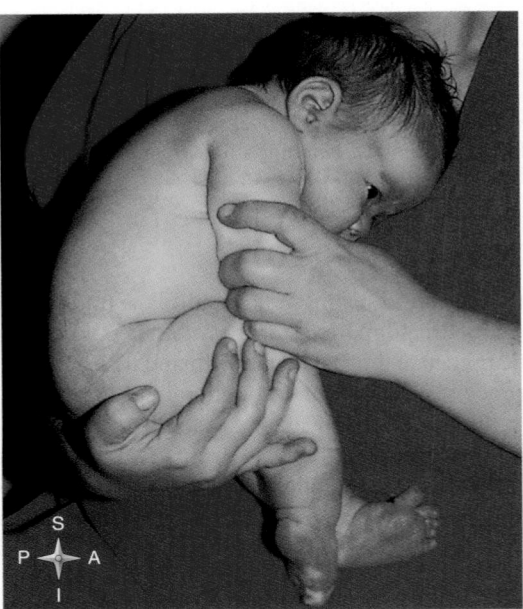

Figure 33-23 *The infant spine.* Photograph showing the normal rounded curvature of the vertebral column in an infant.

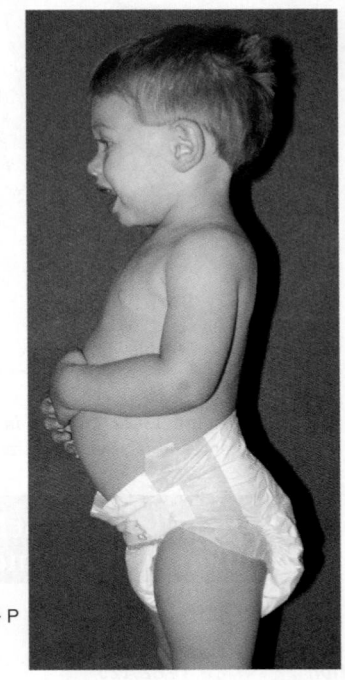

Figure 33-24 *The toddler spine.* Photograph showing the normal curvature of the vertebral column in a toddler. The dark shadow emphasizes the distinct lumbar curvature that develops with the ability to walk (compare with Figure 33-23).

The stage of adolescence in which a person becomes sexually mature is called **puberty.** Many of the developmental changes that occur during this period are controlled by the secretion of gonadotropins (FSH and LH) and sex hormones such as testosterone and estrogen (Figure 33-25). Some of these changes involve development of the gonads themselves and are called **primary sex characteristics.** However, most of the more visible changes involve development of the **secondary sex characteris-**

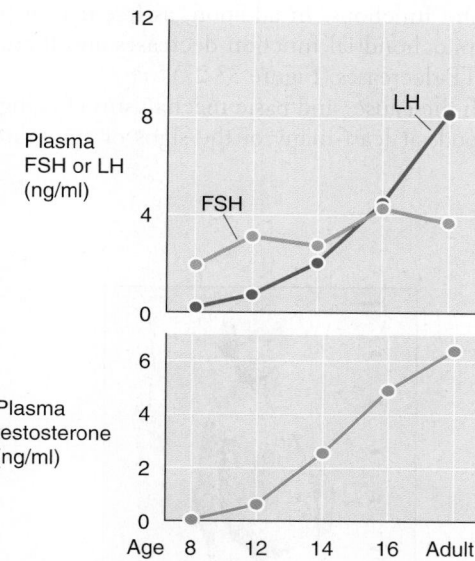

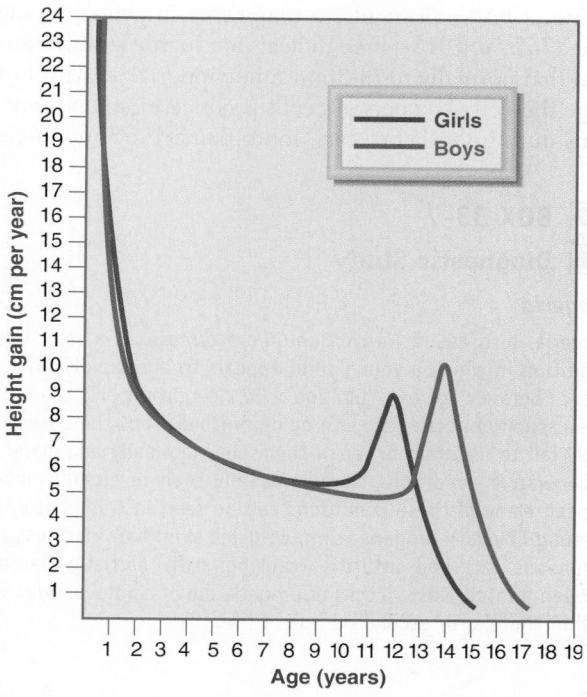

Figure 33-25 *Hormone changes during puberty.* The top graph shows the increase in gonadotropin (LH and FSH) secretion during puberty. The bottom graph shows changes in testosterone levels (in males) during puberty that result from increased gonadotropin levels. Female hormones show a similar change within a monthly cycle of dramatic highs and lows (see Figures 32-14 and 32-15).

Figure 33-26 *Growth in height.* The figure shows typical patterns of gain in height to adulthood for girls and boys. Notice the rapid gain in height during the first few years, a period of slower growth, then another burst of growth during adolescence—finally ending at the beginning of adulthood.

tics such as skeletal changes, fat distribution patterns, growth of pubic and body hair, and growth of the larynx.

Breast development is often the first sign of approaching puberty in girls, beginning about age 9 or 10 years. Most girls begin to menstruate at 12 to 14 years of age. In boys the first sign of puberty is often enlargement of the testes, which begins between 10 and 14 years of age. Both sexes show a spurt in height during adolescence (Figure 33-26). In girls the spurt in height begins between the ages of 10 and 12 years and is nearly complete by 14 or 15 years. In boys the period of rapid growth begins between 12 and 13 years and is generally complete by 16 or 17 years.

Many developmental changes that begin early in childhood are not completed until the early or middle years of **adulthood.** Examples include the maturation of bone, resulting in the full closure of the growth plates, and changes in the size and placement of other body components such as the sinuses. Many body traits do not become apparent for years after birth. Normal balding patterns, for example, are determined at the time of fertilization by heredity but do not appear until maturity. As a rule, adulthood is characterized by the maintenance of existing body tissues. With the passage of years the ongoing effort of maintenance and repair of body tissues becomes more and more difficult. As a result, degeneration begins. This is part of the process of aging, and it culminates in death.

Older Adulthood and Senescence

Most body systems are in peak condition and function at a high level of efficiency during early adulthood. As a person grows older, a gradual but certain decline takes place in the functioning of every major organ system in the body. The study of aging is called **gerontology.** Unfortunately, the mechanisms and causes of aging are not well understood.

Some gerontologists believe that an important aging mechanism is the limit on cell reproduction. Laboratory experiments show that many types of human cells cannot reproduce more than 50 times—thus limiting the maximum life span. Cells die continually in a process called *apoptosis*, no matter what a person's age, but in older adulthood, many dead cells are not replaced—causing degeneration of tissues. Perhaps the cells are not replaced because the surrounding cells have reached their limit of reproduction. Perhaps differences in each adult's aging process result from differences in the reproductive capacity of cells. This mechanism seems to operate in individuals with **progeria** (pro-JEE-ree-ah), a rare, inherited condition in which a person appears to age rapidly (Box 33-7).

Various factors that affect the rates of cell death and cell reproduction have been cited as causes of aging. Some gerontologists believe that nutrition, injury, disease, and other environmental factors affect the aging process. A few have even proposed that aging results from cellular changes caused by slow-acting "aging" viruses found in all living cells. Other gerontologists have proposed that aging is caused by "aging" genes that regulate apoptosis or other cell functions—genes in which aging is "preprogrammed." Yet another proposed cause of aging is autoimmunity. You may recall from Chapter 21 that autoimmunity occurs when the immune system attacks a person's own tissues.

Yet another theory of aging suggests that as a person ages, mitochondria become less able to shuttle adenosine triphosphate (ATP) out to the cytosol. Decreasing ATP availability renders cells less able to perform their functions and even to degenerate—hallmarks of the aging process. One popular theory of aging states that oxygen *free radicals* play a major role in cellular aging (see Figure 27-25, p. 1016). Free radicals are highly reactive forms of oxygen that normally result from mitochondrial activity, but may damage the cell. As a person's cells produce more and more free radicals during the later years, more damage occurs to cellular structures and functions. In addition, as free-radical production increases, mitochondrial function decreases and therefore availability of ATP decreases (Figure 33-27).

Although the causes and basic mechanisms of aging are yet to be understood, at least many of the signs of aging are obvious.

BOX 33-7

Diagnostic Study

Progeria

Progeria, also called *Hutchinson-Gilford disease,* is a rare, inherited condition in which a young child appears to age rapidly. In progeria (*pro-,* "before"; *-geras-,* "old age"; *-ia,* "condition of") the reproductive capacity of cells seems to be diminished. Thus the tissues of the body fail to maintain or repair themselves normally and many of the degenerative conditions more commonly seen in elderly individuals appear. Many of these conditions can be seen in this photograph of a young boy with progeria: thin, wrinkled skin; hair loss; loss of subcutaneous fat; and arthritis (causing stiff, partially flexed, and swollen joints). Most victims of progeria die of cardiovascular disease within the first or second decade of life.

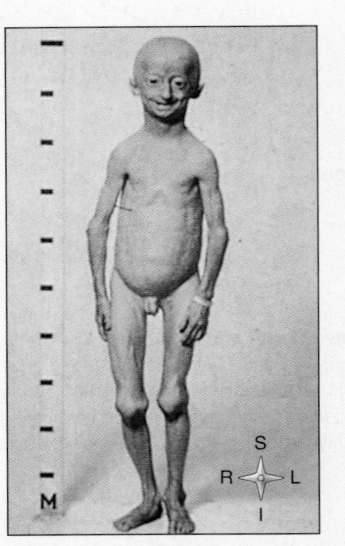

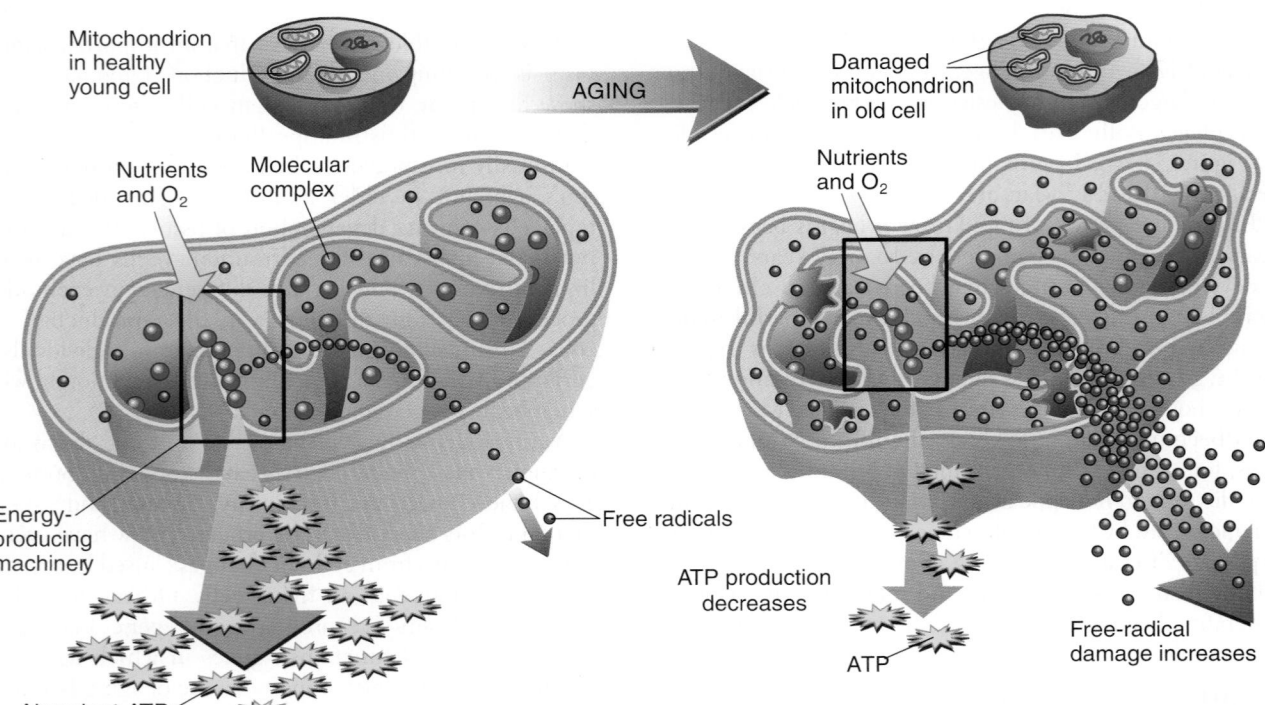

Figure 33-27 *Free-radical theory of aging.* One of many possible mechanisms of the aging processes, free-radical production by cells may increase as a person gets older, increasing the amount of cellular damage. Free radicals are highly reactive forms of oxygen that are normal byproducts of cellular respiration in the mitochondria (shown) and other cell processes. As one ages, the number of free radicals increases as cellular efficiency decreases. Thus, more cellular damage occurs, especially damage to cellular membranes, causing degeneration of the cell.

The remainder of this chapter deals with some of the more common degenerative changes that frequently characterize **senescence** (seh-NES-enz), or older adulthood.

EFFECTS OF AGING

Aging affects each individual in different ways. Environment, genetics, and perhaps even attitude may affect the degree to which the structure and function of a person's body change through older adulthood. Despite advances in understanding how some of the effects of aging can be minimized or even avoided, one cannot completely avoid the fact that body structures degenerate and functions decrease as we get older. Figure 33-28 summarizes a few of the many biological changes that occur by the time we reach late adulthood.

Skeletal System

In older adulthood, bones undergo changes in texture, degree of calcification, and shape. Instead of clean-cut margins, older bones develop indistinct and shaggy-appearing margins with spurs—a process called *lipping*. This type of degenerative change restricts movement because of the piling up of bone tissue around the joints.

With advancing age, changes in calcification of bones reduce the *bone mineral density (BMD)* as you can see in Figure 33-29, A. This may result in reduction of bone size and in bones that are porous and subject to fracture. The lower cervical and thoracic vertebrae are the site of frequent fractures. The result is curvature of the spine and the shortened stature so typical of late adulthood (Figure 33-29, B). The onset of *osteoporosis* and other types of bone loss associated with aging may be avoided—at least to some extent—by maintaining a high BMD through exercise and sufficient calcium in the diet.

Degenerative joint diseases such as **osteoarthritis** (os-tee-oh-ar-THRY-tis) are also common in elderly adults.

Muscular System

Getting older usually involves losing muscle mass. This can begin as early as age 25 years but does not usually reach 10% loss in muscle mass until 50 years of age or so. By age 80 years, many people have lost about 50% of their skeletal muscle mass. Most of the age-related loss of muscle mass is due to a loss of muscle fibers. However, weight training before and during one's later years can increase the mass of the remaining fibers and thus counteract at least some of the reduction in the number of muscle fibers.

Another change in our muscles as we age is that many muscle fibers develop into slower type fibers or into an intermediate form between the "fast type" and "slow type" of muscle fiber. This usually means that the overall ratio of "fast" to "slow" function decreases—that is, our muscle function becomes relatively "slower." As Figure 33-30 shows, the different fiber types also begin to group together in bunches and change their shape slightly. These effects seem unavoidable—even with exercise.

Integumentary System (Skin)

With advancing age the skin becomes dry, thin, and inelastic. It "sags" on the body because of increased wrinkling and skinfolds—as well as a loss of body fat under the skin. Pigmentation changes and the thinning or loss of hair are also common conditions associated with aging.

Urinary System

The number of functioning nephron units in the kidney decreases by almost 50% between the ages of 30 and 75 years. Also, because less blood flows through the kidneys as an individual ages, there is a reduction in overall function and excretory capacity or the ability to produce urine. In the bladder, significant age-related problems often occur because of diminished muscle tone. Muscle atrophy (wasting) in the bladder wall results in decreased capacity and inability to empty, or void, completely.

Respiratory System

In older adulthood the costal cartilages that connect the ribs to the sternum become hardened or calcified. This makes it difficult for the rib cage to expand and contract as it normally does during inspiration and expiration. In time the ribs gradually become "fixed" to the sternum, and chest movements become difficult. When this occurs the rib cage remains in a more expanded position, respiratory efficiency decreases, and a condition called "barrel chest" results. With advancing years a generalized atrophy, or wasting, of muscle tissue takes place as the contractile muscle cells are replaced by connective tissue. This loss of muscle cells decreases the strength of the muscles associated with inspiration and expiration.

Cardiovascular System

Degenerative heart and blood vessel disease is one of the most common and serious effects of aging. Fatty deposits build up in blood vessel walls and narrow the passageway for the movement of blood, much as the buildup of scale in a water pipe decreases flow and pressure. The resulting condition, called **atherosclerosis** (ath-er-oh-skleh-ROH-sis), often leads to eventual blockage of the coronary arteries and a "heart attack." If fatty accumulations or other substances in blood vessels calcify, actual hardening of the arteries, or **arteriosclerosis** (ar-tee-ree-oh-skleh-ROH-sis), occurs. Rupture of a hardened vessel in the brain (stroke) is a frequent cause of serious disability or death in the older adult. **Hypertension (HTN),** or high blood pressure, is also more common.

Special Senses

The sense organs, as a group, all show a gradual decline in performance and capacity as a person ages.

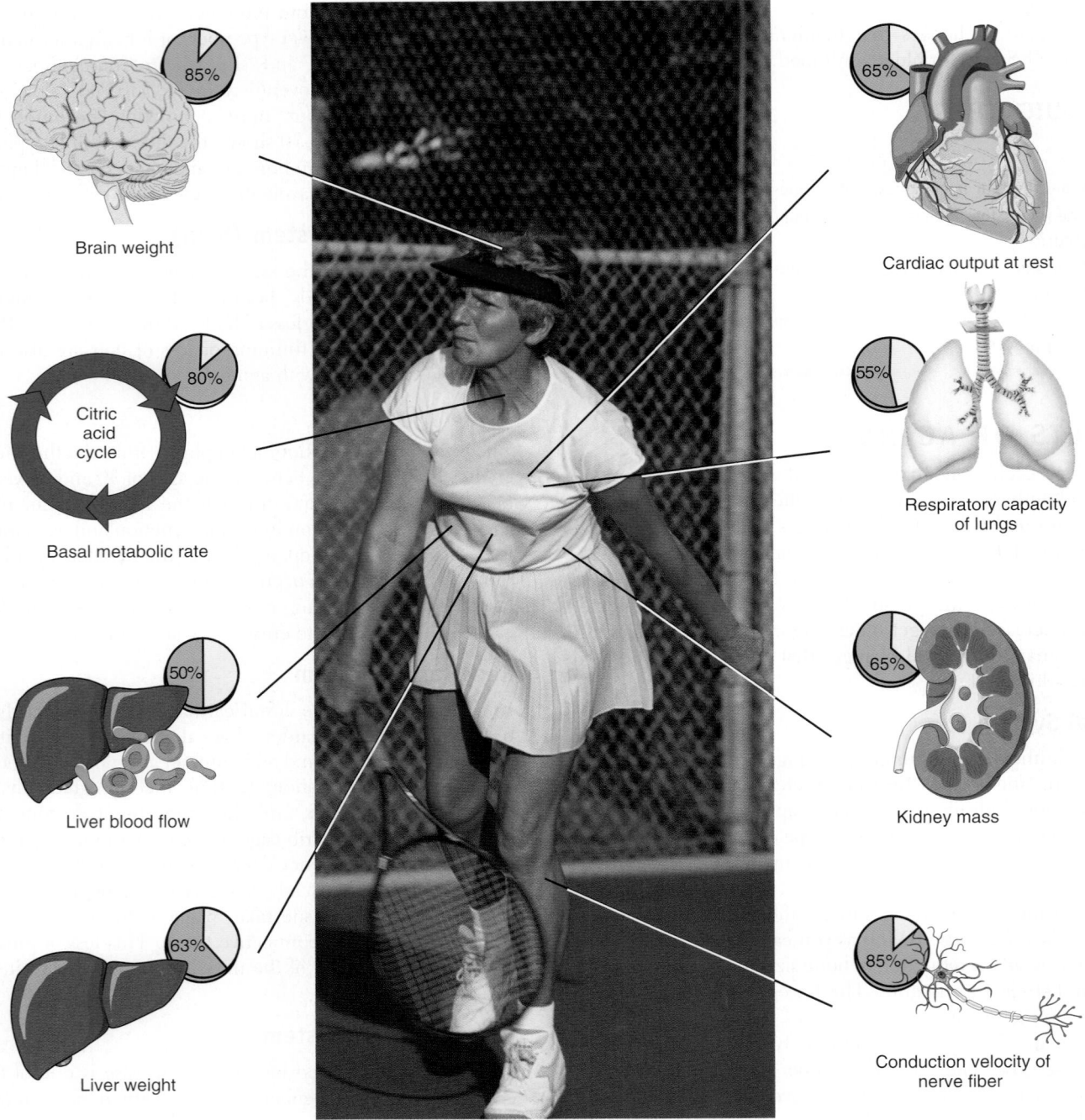

Brain weight

85%

Basal metabolic rate

Citric acid cycle

80%

Liver blood flow

50%

Liver weight

63%

Cardiac output at rest

65%

Respiratory capacity of lungs

55%

Kidney mass

65%

Conduction velocity of nerve fiber

85%

Figure 33-28 *Some biological changes associated with maturity and aging.* Insets show proportion of remaining function in the organs of a person in late adulthood compared with a 20-year-old person.

Most people are farsighted by age 65 years because eye lenses become hardened and lose elasticity; the lenses cannot become curved to accommodate for near vision. This hardening of the lens is called **presbyopia** (pres-bee-OH-pee-ah), which means "old eye." Many individuals first notice the change at about 40 or 45 years of age, when it becomes difficult to do close-up work or read without holding printed material at arm's length. This explains the increased need, with advancing age, for bifocals (glasses that incorporate two lenses) to automatically accommodate for near and distant vision.

Loss of transparency of the lens or its covering capsule is another common age-related eye change. If the lens actually becomes cloudy and significantly impairs vision, it is called a **cataract** and must be removed surgically. The incidence of **glaucoma** (glaw-KOH-mah), the most serious age-related eye disorder, increases with age. Glaucoma causes an increase in

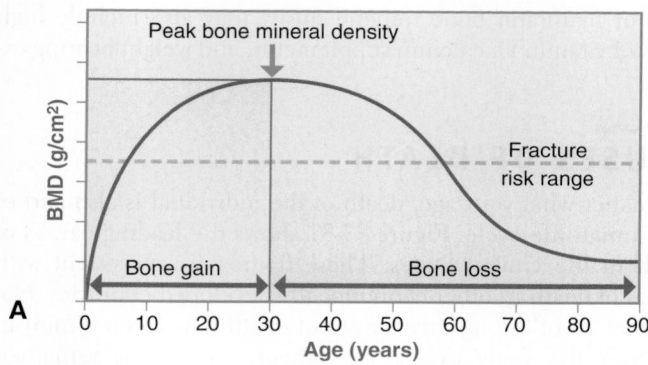

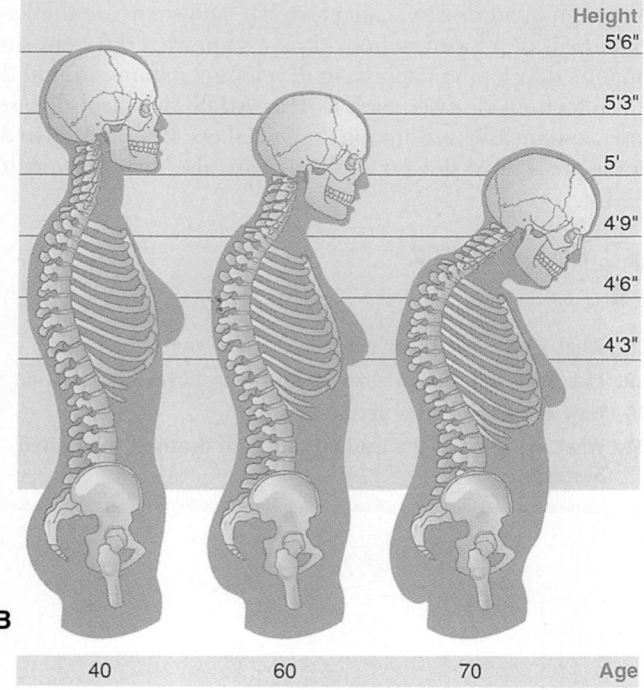

Figure 33-29 *Loss of bone mineral density (BMD) in late adulthood.*
A, Graph showing changes in BMD (bone mineral density) over the life span in females. BMD peaks in young adulthood and decreases in late adulthood to a point where there is an increased risk of bone fracture. Males also have a decrease in BMD beginning around age 50 years, but the decline is more gradual than in women. **B,** The normal spinal curvatures of young adulthood give way to possible changes later in life because of a decrease in BMD.

the pressure within the eyeball and, unless treated, often results in blindness.

In many elderly people a very significant loss of hair cells in the organ of Corti (inner ear) causes a serious decline in the ability to hear certain frequencies. In addition, the eardrum and attached ossicles become more fixed and less able to transmit mechanical sound waves. Some degree of hearing impairment is universally present in the older adult.

The sense of taste is also decreased. Loss of appetite may be caused, at least in part, by the replacement of taste buds with connective tissue cells. Only about 40% of the taste buds present at age 30 years remain fully functional in an individual at age 75 years.

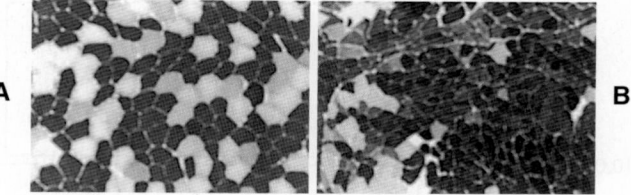

Figure 33-30 *Muscle changes in late adulthood.* Skeletal muscle tissue changes as we age. **A,** Younger muscle shows a typical "checkerboard" pattern of fast and slow fibers (distinguished by stains here). **B,** However, in the elderly the fast and slow fibers tend to group together rather than remain distributed evenly throughout the muscle organ. Note the overall increase in the slower fibers, which appear darker than the other fibers here. Also, the more angular appearance of each muscle fiber cross section in the young changes to a more rounded cross section in the elderly.

BOX 33-8 SPORTS AND FITNESS
Exercise and Aging

A sound exercise program throughout life can reduce some of the common effects of aging. The improved cardiovascular condition associated with aerobic training can prevent or reduce the effects of several age-related circulatory problems. The loss of bone mass that occurs in older adulthood is less troublesome if one enters this period with a higher than average bone density developed through exercising. Finally, the relatively high ratio of body fat to muscle often seen in the elderly can be avoided entirely by a lifelong fitness program.

Interestingly, not everything in the sensory system diminishes with age. Evidence shows that older people are much better than younger people at spotting small movements in a visual scene. This is more a matter of interpreting sensory information in the brain than it is a matter of sensory reception. This should not be surprising, as a growing body of evidence supports our long-standing cultural notion that older adults have a sort of "wisdom" or improved thinking abilities that develop over time.

Reproductive Systems

Although most men and women remain sexually active throughout their later years, mechanisms of sexual response may change, and fertility declines. In men, erection may be more difficult to achieve and maintain, and urgency for sex may decline. In women, lubrication of the vagina may decrease. Although men can continue to produce gametes as they age, women experience a cessation of reproductive cycling between the ages of 45 and 60 years—**menopause.** Menopause results from a decrease in the cyclical production of the primary sex hormones, especially estrogen, with advancing age. The decrease in estrogen accounts for the common symptoms of menopause: cessation of menstrual cycles, hot flashes, and thinning of the vaginal wall. The exact mechanism of hot flashes is not clearly understood, but it is related to the hormonal changes that occur during menopause. Rarely serious, hot flashes usually subside over time. Estrogen therapy may be used to relieve menopause symptoms in some cases.

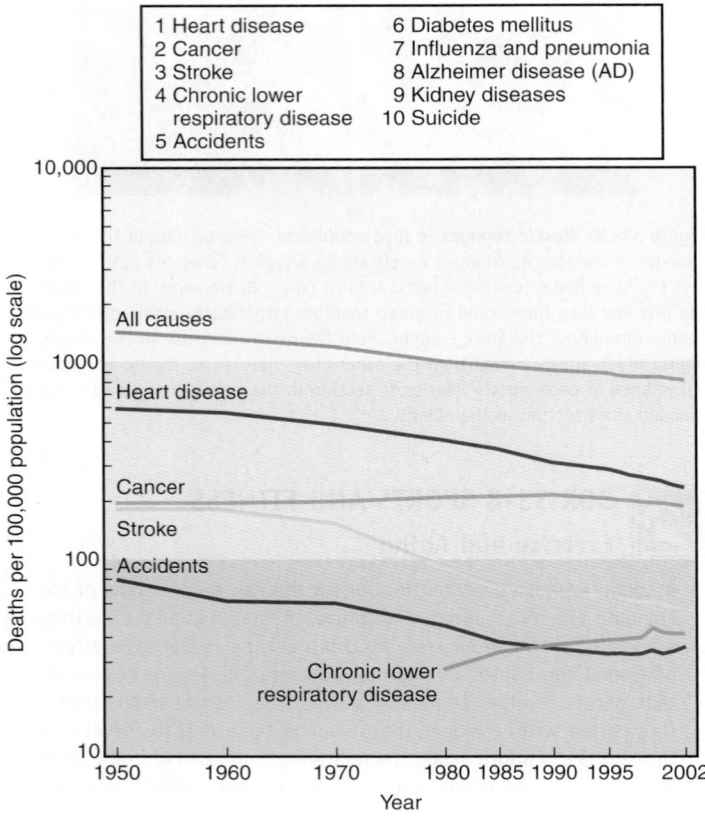

1 Heart disease	6 Diabetes mellitus
2 Cancer	7 Influenza and pneumonia
3 Stroke	8 Alzheimer disease (AD)
4 Chronic lower respiratory disease	9 Kidney diseases
5 Accidents	10 Suicide

Figure 33-31 *Leading causes of death.* The 10 leading causes of death in the U.S. are listed in the inset. The trends in the top five causes of death are graphed out on a logarithmic (powers of 10) scale.

store or maintatin bone mineral dnsity may also include high doses of vitamin D, calcium supplements, and weight bearing exercise.

CAUSES OF DEATH

No matter what your age, death of the individual is also part of the human life cycle. Figure 33-31 shows the leading causes of death in the United States. These figures are consistent with causes of death in other economically developed countries. Notice that all of the top five causes of death have been diminishing over the years except for cancer, which has remained roughly stable.

Although heart disease, cancer, stroke, and so on are the leading killers in developed nations, it is a somewhat different story among the developing nations. In developing nations around the world, infectious diseases such as HIV/AIDS, diarrheal diseases, malaria, and measles are among the top killers. But even so, in developing areas heart disease and stroke are also near the very top of the list.

QUICK CHECK

18. What kinds of changes occur in the skeleton as one ages?
19. List some age-related changes in the cardiovascular system.
20. How can age affect a person's vision?
21. What are some of the leading causes of death in the United States?

The decrease in estrogen levels associated with menopause may also contribute to *osteoporosis*. This condition is characterized by loss of bone mass (see Chapter 7). Osteoporosis is often treated with short term, low dose estrogen replacement therapy (ERT) and nonhormonal bone building drugs. Therapy to re-

THE BIG PICTURE
Growth, Development, and the Whole Body

Understanding basic concepts of human growth and development is essential for understanding the dynamic, ever-changing nature of the human body. It is important to remember that life is a biological process characterized by continuous change in the structure and function of our body.

As we discussed in this and earlier chapters, production of offspring is vital to the survival of the genetic code. Many biologists believe that this gives an important biological meaning to life in general—continued flow of genetic information from one generation to the next. Production of offspring in humans is not successful unless and until a person grows and develops for many years—more than a decade—to the point where viable gametes can be produced. Gametogenesis, then, is the first process necessary to ready the genetic material of the parent to be passed along to the offspring.

Once gametes are available, the sexual union of a male and female may result in fusion of the gametes of the two adults to form the first cell of the offspring. Thus genetic information from two different individuals is now united in a new and unique way in the zygote.

The first cell of the offspring—the zygote—quickly divides again and again to eventually produce a mass of cells that will hopefully implant in the wall of the mother's uterus. Now a process unlike any other we have ever discussed takes over. The offspring and mother biologically connect to each other so that the mother's physiological mechanisms can sustain the offspring until its own systems are developed sufficiently. All the mother's systems alter their function to some degree to maintain the homeostatic balance of both the mother and the offspring.

After many weeks of growth and development—including histogenesis and organogenesis of a full complement of human structures—delivery of the offspring occurs. This event marks the beginning of a continuing series of changes: infancy, childhood, puberty, adolescence, adulthood, senescence, and death.

At some point in adolescence or adulthood many of us have the opportunity to complete the "circle of life" by passing some of the genetic code we received from our parents to yet another human generation.

Mechanisms of Disease

DISORDERS OF PREGNANCY AND EARLY DEVELOPMENT

Implantation Disorders

A pregnancy has the best chance of a successful outcome, the birth of a healthy baby, if the blastocyst is implanted properly in the uterine wall. However, proper implantation does not always occur. Many offspring are lost before implantation occurs, often for unknown reasons. As mentioned in this chapter and the previous chapter, implantation outside the uterus results in an ectopic pregnancy. If the blastocyst implants in a region of endometriosis or normal peritoneal membrane, the pregnancy may be successful if there is room for the developing fetus to grow. Ectopic pregnancies that do succeed must be delivered by C-section rather than by normal vaginal birth. If an ectopic pregnancy occurs in a uterine tube, which cannot stretch to accommodate the developing offspring, the tube may rupture and cause life-threatening hemorrhaging. So-called **tubal pregnancies** are the most common type of ectopic pregnancy.

Occasionally, the blastocyst implants in the uterine wall near the cervix. This in itself may present no problem, but if the placenta grows too close to the cervical opening a condition called **placenta previa** (PREE-vee-ah) results. The normal dilation and softening of the cervix that occur in the third trimester often cause painless bleeding as the placenta near the cervix separates from the uterine wall. The massive blood loss that may result can be life-threatening for both mother and offspring (Figure 33-32, A).

Separation of the placenta from the uterine wall can occur even when implantation takes place in the upper part of the uterus. When this occurs in a pregnancy of 20 weeks or more, the condition is called **abruptio placentae** (ab-RUP-shee-oh plah-SEN-tay). Complete separation of the placenta causes the immediate death of the fetus. The severe hemorrhaging that often results, sometimes hidden in the uterus, may cause circulatory shock and death of the mother within minutes. A cesarean section and perhaps also a hysterectomy must be performed immediately to prevent blood loss and death (Figure 33-32, B).

PIH and Preeclampsia

It is not uncommon for a woman's blood pressure to rise during pregnancy and remain elevated until the end of pregnancy—a condition often called **pregnancy-induced hypertension (PIH)**. In about 6% to 8% of all pregnancies, PIH may progress to a condition called **preeclampsia** (pree-eh-KLAMP-see-ah). Formerly known as *toxemia of pregnancy*, preeclampsia is a serious disorder characterized by the onset of acute hypertension after the twenty-fourth week, accompanied by proteinuria and edema. The causes of PIH and preeclampsia are largely unknown, but intense research efforts have led to the discovery of a gene that regulates how the kidney handles salt and that may also be involved in raising blood pressure during pregnancy. Preeclampsia can result in complications such as abruptio placentae, stroke, hemorrhage, fetal malnutrition, and low birth weight. This condition can progress to **eclampsia**, a life-threatening

Mechanisms of Disease—cont.

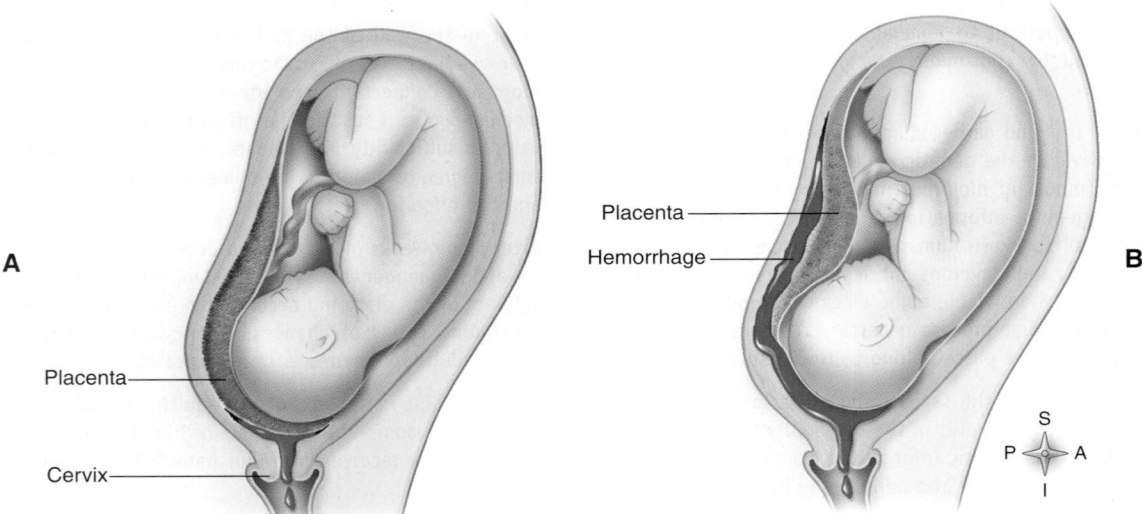

Figure 33-32 *Implantation disorders.* **A,** Placenta previa. **B,** Abruptio placentae.

form of toxemia that causes severe convulsions, coma, kidney failure, and perhaps death of the fetus and mother.

Fetal Death

A **miscarriage** is the loss of an embryo or fetus before the twentieth week (or a fetus weighing less than 500 g, or 1.1 pounds). Technically known as a **spontaneous abortion,** the most common cause of such a loss is a structural or functional defect in the developing offspring. Abnormalities of the mother, such as hypertension, uterine abnormalities, and hormonal imbalances, can also cause spontaneous abortions. After 20 weeks, delivery of a lifeless infant is termed a **stillbirth.**

Birth Defects

Birth defects, also called **congenital abnormalities,** include any structural or functional abnormality present at birth. Congenital defects may be inherited or may be acquired during gestation or delivery. Inherited defects are discussed in the next chapter. Acquired defects result from agents called **teratogens** (TER-ah-toh-jens) that disrupt normal histogenesis and organogenesis. Some teratogens are chemicals such as alcohol, antibiotics, and other drugs. Microorganisms, such as those that cause rubella (a viral infection), can also cross the placental barrier and disrupt normal embryonic development. Radiation and other physical factors can also cause birth defects. Some teratogens are mutagens because they do their damage by changing the genetic code in cells of the developing embryo.

Postpartum Disorders

Puerperal (pyoo-ER-per-al) **fever,** or *child-bed fever,* is a syndrome of postpartum mothers characterized by bacterial infection that progresses to septicemia (blood infection) and possibly death. Until the 1930s, puerperal fever was the leading cause of maternal death—claiming the lives of more than 20% of postpartum women. Modern antiseptic techniques prevent most postpartum infections now. Puerperal infections that do occur are usually treated successfully by an immediate and intensive program of antibiotic therapy.

After a child is born, it needs the nourishment of milk to survive. However, several disorders of lactation (milk production) may occur to prevent a mother from nursing her infant. For example, anemia, malnutrition, emotional stress, and structural abnormalities of the breast can all interfere with normal lactation. **Mastitis** (mas-TYE-tis), or breast inflammation, often caused by infection, can result in lactation problems or production of milk contaminated with pathogenic organisms. In many cultures, the availability of other nursing mothers or breast milk substitutes allows proper nourishment of the infant, even when lactation problems develop. Most breast milk substitutes are formulations of milk from another mammal such as the cow. Infants who lack the enzyme *lactase* may not be able to digest the lactose present in human or animal milk, resulting in a condition called **lactose intolerance.** Infants with lactose intolerance are sometimes given a lactose-free milk substitute made from soy or other plant products.

LANGUAGE OF SCIENCE

(Cont'd from page 1161)

older adulthood

oogenesis (oh-oh-JEN-eh-sis) [*oo-* egg, *-gen-* generate, *-esis* process of]

oogonia (oh-oh-GO-nee-ah) [*oo-* egg, *-gonia* offspring]

organogenesis (or-gah-no-JEN-eh-sis) [*organo-* instrument (organ), *-gen-* generate, *-esis* process of]

parturition (pahr-too-RIH-shun) [*parturi-* desire to bring forth, *-tion* process of]

placenta (plah-SEN-tah) [*placenta* flat cake]

postnatal period (POST-nay-tal) [*post-* after, *-nat-* birth, *-al* pertaining to]

prenatal period (PREE-nay-tal) [*pre-* before, *-nat-* birth, *-al* pertaining to]

primary follicle (FOL-i-kul) [*prim-* first, *-ary* state of, *foli-* bag, *-cle* small]

primary germ layers [*primus* first, *germen* sprout]

primary oocytes (OH-oh-sytes) [*primus* first, *oo-* egg, *-cyte* cell]

primary sex characteristics [*primus* first, *charassein* to engrave]

puberty (PYOO-ber-tee) [*pubertas* age of maturity]

secondary follicles (FOL-i-kuls) [*second* second, *-ary* state of, *foli-* bag, *-cle* small]

secondary oocyte (OH-oh-site) [*second* second, *-ary* state of, *oo-* egg, *-cyte* cell]

secondary sex characteristics

senescence (seh-NES-enz) [*senesc-* grow old, *-ence* state of]

spermatogenesis (sper-mah-toh-JEN-eh-sis) [*spermato-* seed, *-gen-* generate, *-esis* process of]

spermatogonia (sper-mah-toh-GO-nee-ah) [*spermato-* seed, *-gonia* offspring]

theca cells (THEE-kah) [*theca* sheath, *cella* storeroom]

trophoblast (TROH-foh-blast) [*tropho-* nourishment, *-blast* sprout]

yolk sac

zona pellucida (ZOH-nah pah-LOO-sih-dah) [*zona* belt or girdle, *pellucida* transparent]

zygote (ZYE-goht) [*zygotos* yolked]

LANGUAGE OF MEDICINE

abruptio placentae (ab-RUP-shee-oh plah-SEN-tay) [*ab-* away from, *-ruptio* rupture, *placentae* of flat cake]

antenatal medicine (an-tee-NAY-tal) [*ante-* before, *-nat-* birth, *-al* pertaining to]

arteriosclerosis (ar-tee-ree-oh-skleh-ROH-sis) [*arterio-* artery, *-sclero-* harden, *-osis* condition of]

atherosclerosis (ath-er-oh-skleh-ROH-sis) [*athere-* meal, *-sclero-* harden, *-osis* condition of]

cataract (KAT-ah-rakt) [*katarrhakies* waterfall]

cesarean section (seh-SAIR-ee-an SEK-shun) [*cesar* Julius Caesar, *-ean* of]

congenital abnormalities (kon-JEN-i-tall ab-nor-MAL-i-teez) [*congenit* born with, *-al* pertaining to, *ab-* away from, *-normal-* rule, *-ity* state of]

eclampsia (eh-KLAMP-see-ah) [*ec-* out, *-lamp-* shine forth, *-sia* condition]

embryology (em-bree-OL-oh-gee) [*em-* in, *-bryo-* to grow, *-ology* study or science of]

gerontology (jair-on-TAHL-oh-jee) [*geronto-* old age, *-ology* study or science of]

glaucoma (glaw-KOH-mah) [*glauco-* gray, *-oma* growth]

hemopoiesis (hee-moh-poy-EE-sis) [*hemo-* blood or blood vessel, *-poies-* make, *-esis* process of]

hypertension (HTN) (hye-per-TEN-shun) [*hyper-* excessive, *-ten-* stretch or pull, *-sion* state of]

in vitro fertilization (in VEE-troh FER-ti-li-ZAY-shun) [*in vitro* within glass, *fertiliz* fruitful, *-ation* process of]

lactose intolerance (LAK-tohs in-TOL-er-ans) [*lact-* milk, *-ose* carbohydrate (sugar), *in-* not, *-toler-* bear, *-ance* state of]

laparoscope (LAP-ah-roh-skope) [*laparo-* abdomen, *-scope* instrument for observation]

mastitis (mas-TYE-tis) [*mast-* breast, *-itis* inflammation]

menopause (MEN-oh-pawz) [*men-* month, *-o-* combining form, *-pause* cease]

miscarriage

neonatology (nee-oh-nay-TOL-oh-jee) [*neo-* new, *-nat-* born, *-ology* study or science of]

osteoarthritis (os-tee-oh-ar-THRY-tis) [*osteo-* bone, *-arthr-* joint, *-itis* inflammation]

placenta previa (plah-SEN-tah PREE-vee-ah) [*placenta* flat cake, *previa* gone before]

preeclampsia (pree-ee-KLAMP-see-ah) [*pre-* before, *-lamp-* shine forth, *-sia* condition]

pregnancy-induced hypertension (PIH) (PREG-nan-see-in-DOOST hye-per-TEN-shun)

presbyopia (pres-bee-OH-pee-ah) [*presby-* aging, *-opia* visual condition]

progeria (proh-JEER-ee-ah) [*pro-* early, *-ger-* old age, *-ia* condition]

puerperal fever (pyoo-ER-per-al FEE-ver) [*puerp-* childbirth, *-al* pertaining to]

spontaneous abortion (spon-TAY-nee-us ah-BOR-shun) [*ab-* away from, *-or-* be born, *-tion* process of]

stem cells

stillbirth

teratogens (TER-ah-toh-jens) [*terato-* monster, *-gen* generate]

tubal pregnancies (TOO-bal) [*tub-* tube, *-al* pertaining to, *praegnans* pregnant]

ultrasonography (ul-trah-son-OG-rah-fee) [*ultra-* beyond, *-sono-* sound, *-graphy* record]

CASE STUDY

Susan Wallace, 24 years old, has made an appointment with her physician to determine if she might be pregnant. Susan and her husband, Tim, have been trying for almost 1 year to become parents. They have been married for 2 years. Neither have been parents before. Susan took birth control pills for 6 years before she stopped 28 months ago. Both Tim and Susan deny any medical problems. Susan's periods are usually regular, occurring every 26 days. Six weeks have elapsed since Susan has had a menstrual period.

1. While waiting for the results of Susan's pregnancy test, Tim and Susan ask you to explain how fertilization occurs. In giving this explanation, which of the following statements is *true?*

   A. Fertilization occurs in the outer third (ampulla) of the fallopian tube.
   B. Mucous strands in the cervical canal guide the sperm into the uterus.
   C. The ovum attracts the sperm with special peptides.
   D. All the above.

2. During the first couple months of pregnancy, which one of the following hormones is measured in either the serum or urine to confirm pregnancy?

   A. Estrogen
   B. Progesterone
   C. Human chorionic gonadotropin (hCG)
   D. Follicle-stimulating hormone (FSH)

3. Susan is now 5 months pregnant and feeling well. On routine prenatal examination, her physician orders a sonogram to assess fetal development because her fundal height (distance from the top of the symphysis pubis over the curve of the abdomen to the top of the uterine fundus) is slightly larger than expected. It is

determined that Susan has a normally large fetus and that the gestational age may be slightly older than predicted. Additionally, a problem is identified with a low-lying placenta covering the cervical os. What is the correct terminology for this condition?

   A. Ectopic pregnancy
   B. Preeclampsia
   C. Abruptio placentae
   D. Placenta previa

4. It was decided that Susan will need to undergo a cesarean section because of her low-lying placenta, but the remainder of her pregnancy has been uneventful and Susan is now in her last trimester. Which of the following symptoms can Susan expect to develop because of the large size of her uterus?

   A. Her center of gravity will be shifted backward.
   B. Intestinal motility will slow and cause constipation and/or hemorrhoids.
   C. Urine-storing capacity of the bladder will increase.
   D. Displacement of the diaphragm will make deep breathing easier.

5. Susan delivers a normal healthy infant girl that she and Tim name Wendy. Both infant and mother are discharged home with no complications. At the age of 3 months, the pediatrician is testing growth and development of Wendy. Which of the following should Wendy be able to do at this age?

   A. Lift the head and raise the chest
   B. Sit when well supported
   C. Crawl
   D. Drink from a cup held to her mouth

CHAPTER SUMMARY

INTRODUCTION

A. Prenatal period—period beginning with conception and ending at birth
B. Postnatal period—period beginning with birth and continuing until death
C. Human developmental biology—study of changes occurring during the cycles of life from conception to death

A NEW HUMAN LIFE

A. Production of sex cells—spermatozoa are produced by spermatogenesis; ova are produced by oogenesis
   1. Meiosis (Figures 33-1 and 33-2)
      a. Special form of cell division that reduces the number of chromosomes in each daughter cell to one half of those in the parent cell
      b. Mature ova and sperm contain only 23 chromosomes, half as many as other human cells
      c. Meiotic division—two cell divisions that occur one after another in succession

         (1) Meiotic division I and meiotic division II
         (2) Both divisions made up of an interphase, prophase, metaphase, anaphase, and telophase
      d. During prophase I of meiosis, "cross over" occurs where genetic material is "shuffled"
      e. Daughter cells formed by meiotic division I contain a haploid number of chromosomes
      f. Meiotic division II—essentially the same as mitotic division; reproduces each of the two cells formed by meiotic division I and forms four cells, each with the haploid number of chromosomes
   2. Spermatogenesis (Figure 33-3)—process by which primitive male sex cells become transformed into mature sperm; begins at approximately puberty and continues throughout a man's life
      a. Meiotic division I—one primary spermatocyte forms two secondary spermatocytes, each with 23 chromosomes
      b. Meiotic division II—each of the two secondary spermatocytes forms a total of four spermatids

3. Oogenesis (Figure 33-4)—process by which primitive female sex cells become transformed into mature ova
   a. Mitosis—oogonia reproduce to form primary oocytes; most primary oocytes begin meiosis and develop to prophase I before birth; there they stay until puberty
   b. Once during each menstrual cycle, a few primary oocytes resume meiosis and migrate toward the surface of the ovary; usually only one oocyte matures enough for ovulation, and meiosis again halts at metaphase II
   c. Meiosis resumes only if the head of a sperm cell enters the ovum
B. Ovulation and insemination
   1. Ovulation—expulsion of the mature ovum from the mature ovarian follicle, into the abdominopelvic cavity, and then the uterine (fallopian) tube
   2. Insemination—expulsion of seminal fluid from the male into the female vagina; sperm travel through the cervix and uterus and into the uterine (fallopian) tubes
C. Fertilization—also known as *conception* (Figure 33-5)
   1. Most often occurs in the outer one third of the oviduct
   2. Ovum attracts and "traps" sperm with special receptor molecules on its surface
   3. When one spermatozoon enters the ovum, the ovum stops collecting sperm on its surface
   4. The sperm releases its nuclear chromosomes into the ovum; proteins and RNA from the sperm enter the ovum to assist with early development
   5. 23 chromosomes from the sperm head and 23 chromosomes in the ovum make up a total of 46 chromosomes
   6. Zygote—fertilized ovum; genetically complete

PRENATAL PERIOD

A. Begins with conception and continues until the birth of a child
B. Cleavage and implantation (Figure 33-6)—once zygote is formed, it immediately begins to divide
   1. Morula—solid mass of cells formed from zygote; takes approximately 3 days; continues to divide (Figure 33-7)
   2. Blastocyst—by the time the developing embryo reaches the uterus, it has formed a hollow ball of cells, which implants into the uterine lining (Figure 33-8)
   3. Approximately 10 days pass from fertilization until implantation in the uterine lining; ovum has a store of nutrients that support this embryonic development until implantation has occurred
   4. Blastocyst has an outer layer of cells and an inner cell mass
   a. Trophoblast—outer wall of the blastocyst
   b. Inner cell mass—as blastocyst develops, yolk sac and amniotic cavity are formed (Figure 33-9)
     (1) In humans, yolk sac's functions are largely nonnutritive
     (2) Amniotic cavity becomes a fluid-filled, shock-absorbing sac (bag of waters) in which the embryo floats during development (Figure 33-10)
   c. Chorion develops from trophoblast to become an important fetal membrane in the placenta

5. Placenta (Figure 33-11)
   a. Anchors fetus to the uterus and provides a "bridge" for the exchange of nutrients and waste products between mother and baby
   b. Also serves as an excretory, respiratory, and endocrine organ
   c. Placental tissue normally separates maternal and fetal blood supplies
   d. Has important endocrine functions—secretes large amounts of hCG, which stimulate the corpus luteum to continue its secretion of estrogen and progesterone (Figure 33-12)
C. Periods of development (Figures 33-13 to 33-15)
   1. Gestation period—approximately 39 weeks; divided into three 3-month segments called trimesters
   2. Embryonic phase extends from fertilization until the end of week 8 of gestation
   3. Fetal phase—weeks 8 to 39
D. Formation of the primary germ layers
   1. Three layers of specialized cells develop early in the first trimester of pregnancy
   2. Cells of embryonic disk differentiate and form each of the three primary germ layers
   3. Each primary germ layer gives rise to specific organs and systems of the body (Figure 33-16)
   4. Three primary germ layers:
   a. Endoderm—inside layer
   b. Ectoderm—outside layer
   c. Mesoderm—middle layer
E. Histogenesis and organogenesis (Figure 33-16)
   1. Histogenesis—process by which primary germ layers develop into different kinds of tissues
   2. Organogenesis—how tissues arrange themselves into organs
   3. Differentiation and development of the reproductive systems are an example
   a. Reproductive tract (Figure 33-17)
     (1) Gonads attach to mesonephric (wolffian) ducts, which become the male reproductive tract
     (2) Gonads (unattached) and paramesonephric (müllerian) ducts develop into the female reproductive tract
   b. External genitalia (Figure 33-18)
     (1) In the male, the genital tubercle eventually becomes the glans of the penis and the folds become the penis shaft and scrotum
     (2) In the female, the genital tubercle becomes the clitoris and the folds become the labia

BIRTH, OR PARTURITION

A. Transition between prenatal and postnatal periods of life
B. Stages of labor (Figure 33-20)
   1. Stage one—period from onset of uterine contractions until cervical dilation is complete
   2. Stage two—period from maximal cervical dilation until the baby exits through the vagina
   3. Stage three—process of expulsion of the placenta through the vagina

C. Multiple births—birth of two or more infants from the same pregnancy; twins are most common (Figure 33-21)
 1. Identical twins result from the splitting of embryonic tissue from the same zygote early in development
 2. Fraternal twins result from the fertilization of two different ova by two different spermatozoa

POSTNATAL PERIOD

A. Begins at birth and continues until death; commonly divided into a number of periods (Figure 33-22)
B. Infancy begins at birth and lasts until approximately 18 months
 1. Neonatal period—first 4 weeks of infancy; dramatic changes occur at a rapid rate (Figure 33-23)
 2. Changes allow the infant to become totally self-supporting, especially respiratory and cardiovascular systems (Figure 33-24)
C. Childhood extends from end of infancy to sexual maturity, or puberty
 1. Early childhood—growth continues at a rapid pace but slows down
 2. By age 6 years, child looks more like a preadolescent than an infant or toddler
 3. Nervous and muscular systems develop rapidly during the middle years of childhood
 4. Deciduous teeth are lost during childhood, beginning at approximately 6 years of age
 5. Permanent teeth have erupted by age 14 years, except for the third molars (wisdom teeth)
D. Adolescence and adulthood
 1. Adolescence is considered to be the teenage years (from 13 to 19); marked by rapid and intense physical growth, resulting in sexual maturity
  a. Puberty—stage of adolescence during which a person becomes sexually mature
  b. Changes triggered by increases in reproductive hormones (Figure 33-25)
  c. Primary sexual characteristics—maturity of gonads and reproductive tract
  d. Secondary sexual characteristics—fat and hair distribution, skeletal changes, etc. (Figure 33-26)
 2. Adulthood—characterized by maintenance of existing body tissues
E. Older adulthood and senescence
 1. As a person grows older, a gradual decline occurs in every major organ system in the body
 2. Gerontologists theorize a number of different aging mechanisms, all of which may be involved in the processes of aging
  a. Limit on cell reproduction
  b. Environmental factors
  c. Viruses
  d. Aging genes
  e. Degeneration of mitochondria—perhaps progressive damage by oxygen free radicals (Figure 33-27)

EFFECTS OF AGING—COMMON DEGENERATIVE CHANGES THAT FREQUENTLY CHARACTERIZE SENESCENCE (FIGURE 33-28)

A. Skeletal system (Figure 33-29)
 1. Bones decrease in BMD (bone mineral density) and thus change in texture, degree of calcification, and shape
 2. Lipping occurs, which can limit range of motion
 3. Decreased bone size and density lead to increased risk of fracture
 4. Decreased BMD can be avoided (at least partly) by exercise and adequate calcium intake
B. Muscular system (Figure 33-30)
 1. Muscle mass decreases to about 90% by age 50 years and around 50% by age 80 years
 2. The number of muscle fibers decreases as we age but can be offset by an increase in muscle fiber size through exercise
 3. Ratio of "fast" to "slow" functioning in muscle fibers decreases, slowing the function of muscle organs
C. Integumentary system (skin)
 1. Skin becomes dry, thin, and inelastic
 2. Pigmentation changes and thinning hair are common problems associated with aging
D. Urinary system
 1. Number of nephron units in the kidney decreases by almost 50% between the ages of 30 and 75 years
 2. Decreased blood flow through kidneys reduces overall function and excretory capacity
 3. Diminished muscle tone in bladder results in decreased capacity and inability to empty, or void, completely
E. Respiratory system
 1. Costal cartilages become calcified
 2. Respiratory efficiency decreases
 3. Decreased strength of respiratory muscles
F. Cardiovascular system
 1. Degenerative heart and blood vessel disease—one of the most common and serious effects of aging
 2. Atherosclerosis—buildup of fatty deposits on blood vessel walls narrows the passageway for blood
 3. Arteriosclerosis—"hardening" of the arteries
 4. Hypertension—high blood pressure
G. Special senses
 1. Sense organs—gradual decline in performance and capacity with aging
 2. Presbyopia—farsightedness caused by hardening of lens
 3. Cataract—cloudy lens, which impairs vision
 4. Glaucoma—increased pressure within the eyeball; if left untreated, often results in blindness
 5. Decreased hearing
 6. Decreased taste
H. Reproductive systems
 1. Mechanism of sexual response may change
 2. Fertility decreases
 3. In females, menopause occurs between ages 45 and 60 years

CAUSES OF DEATH

A. In developed countries such as the U.S., heart disease, cancer, and stroke (CVA) are among the leading causes of death (Figure 33-31)

B. In developing countries, heart disease and stroke are also leading causes of death, along with infectious diseases such as HIV/AIDS, diarrheal disorders, and malaria

REVIEW QUESTIONS

1. Define the terms developmental biology, growth, and development.
2. Outline the major steps in spermatogenesis. Do the same for oogenesis.
3. Identify the two processes necessary to bring the sperm and ovum into proximity with each other.
4. During fertilization, how does the ovum attract sperm?
5. At what developmental stage does implantation occur?
6. Describe the structural changes between a morula and a blastocyst.
7. How does the placenta develop?
8. What functions does the placenta provide?
9. Outline the hormonal levels of human chorionic gonadotropin (HCG), estrogen, and progesterone at various stages during gestation.
10. During what period is the term embryo replaced by the term fetus?
11. List the various structures derived from each of the three primary germ layers.
12. Explain the three stages of labor.
13. What is the difference between identical and fraternal twins?
14. From birth to maturity, how does the size of the head compare with the rest of the body?
15. What are the time spans of the following postnatal periods: infancy, childhood, adolescence?
16. During what postnatal period do the secondary sex characteristics develop? What initiates this development?
17. What structural changes may result from the changes in bone calcification because of aging?
18. Define the terms atherosclerosis, arteriosclerosis, and hypertension.
19. Identify and describe the most serious age-related eye disorder.

CRITICAL THINKING QUESTIONS

1. How is the process of meiosis different from mitosis?
2. If a diploid cell rather than a haploid cell were used for human reproduction, what would the number of chromosomes per cell be after three generations?
3. How do histogenesis and organogenesis differ? Which of these occurs first in development?
4. Explain the procedure a physician might use if a normal vaginal delivery would be dangerous for the mother or baby.
5. What are the symptoms of the drop in estrogen that occurs during menopause? To what skeletal disorder might this drop be related?
6. Explain the process of in vitro fertilization. What is the probability of the success of this procedure, resulting in a full-term birth?
7. Using physiological principles, explain how a sound exercise program can reduce some of the common effects of aging.

Genetics and Heredity

LANGUAGE OF SCIENCE

autosomes (AW-toh-sohms) [*auto-* self, *-some* body]

carrier

chromatin (KROH-mah-tin) [*chroma* color]

chromosome (KROH-meh-sohm) [*chroma-* color, *-some* body]

codominance (koh-DOM-i-nance) [*co-* together, *-domina-* rule, *-ance* state of]

crossing over

deletion

diploid (DIP-loyd) [*diplo-* twofold, *-oid* form of]

dominant genes

gametes (GAM-eets) [*gamete* marriage partner]

genes [*genein* produce or generate]

gene linkage

genetic mutation (jeh-NET-ik myoo-TAY-shun) [*gene-* produce, *-t-* combining form, *-ic* pertaining to, *muta-* change, *-ation* process of]

genetics (jeh-NET-iks) [*gene-* produce, *-t-* combining form, *-ic* pertaining to]

genome (JEE-nohm) [*gen-* generate (gene), *-ome* entire collection]

genomics (jeh-NO-miks) [*gen-* generate (gene), *-om-* entire collection, *-ic* pertaining to]

genotype (JEN-oh-type) [*geno-* generate (gene), *-type* molded]

haploid (HAP-loyd) [*haplo-* single, *-oid* form of]

heterozygous (het-er-oh-ZYE-gus) [*hetero-* another or different, *-zygo-* yolk, *-ous* characterized by]

histones (HISS-tohns) [*histos* tissue]

homozygous (hoh-moh-ZYE-gus) [*homo-* sameness, *-zygo-* yolk, *-ous* characterized by]

Human Genome Project

insertion

karyotype (KAIR-ee-oh-type) [*karyo-* nucleus, *-type* molded]

meiosis (my-OH-sis) [*meiosis* becoming smaller]

mitochondrial DNA (mDNA, mtDNA) (my-toh-KON-dree-al D N A) [*mito-* threadlike, *-chondri* small grain, *-al* pertaining to]

Cont'd on p. 1223

It seems that today we are hearing more and more about the importance of **genetics,** the scientific study of inheritance, to all fields of human biology—especially anatomy, physiology, and medicine. Popular news magazines are running story after story on the revolution in treating fatal inherited disorders by using something called *gene therapy*. Health and science columns in newspapers keep us informed of the latest discoveries of genes involved with disease, human behavior, and even longevity. Television programs outline the progress of the largest coordinated biological quest that anyone can remember: mapping the entire human genetic code and listing all the proteins encoded there. Clearly, one cannot be informed about human biology today without some knowledge of basic genetics and heredity. In this chapter, we will briefly review the essential concepts of genetics and explain how heredity affects every structure and function in the body.

THE SCIENCE OF GENETICS

History shows that humans have been aware of patterns of inheritance—or *heredity*—for thousands of years, but it was not until the 1860s that the scientific study of these patterns—genetics—was born. At that time, a monk living in Brno, Moravia, was the first to discover the basic mechanism by which traits are transmitted from parents to offspring. That man, Gregor Mendel, proved that independent units (which we now call genes) are responsible for the inheritance of biological traits.

The science of genetics developed from Mendel's quest to explain how normal biological characteristics are inherited. As time went by and more genetic studies were done, it became clear that certain diseases also have a genetic basis. As you may recall from Chapter 1 (p. 30), some diseases are inherited directly. For example, the group of blood-clotting disorders called *hemophilia* is inherited by children from parents who have the genetic code for hemophilia. Directly inherited diseases are often called "hereditary diseases." Other diseases are only partly determined by genetics—that is, they involve genetic risk factors (Chapter 1, p. 30). For example, certain forms of skin cancer are thought to have a genetic basis. A person who inherits the genetic code associated with skin cancer will develop the disease only if the skin is also heavily exposed to the ultraviolet radiation in sunlight.

CHROMOSOMES AND GENES
Mechanism of Gene Function

Mendel proposed that the genetic code is transmitted to offspring in discrete, independent units that we now call **genes.** Recall from Chapters 2 and 4 that each gene is a sequence of nucleotide bases in the deoxyribonucleic acid (DNA) molecule.

Figure 34-1 shows a detailed view of human DNA. Beginning at the left of the diagram, you can see a fully condensed **chromosome** on the left unfold toward the right, where a single double-helix strand of DNA is visible. As the genetic codes of a DNA molecule's genes are being actively transcribed in a cell's nucleus, the DNA is in the threadlike form called **chromatin**. Chromatin, as you can see in Figure 34-1, is actually a thread of DNA wound around little spools made of proteins called **histones**. The chromatin is thus organized into little "thread on spool" subunits called **nucleosomes**.

During cell division, each replicated strand of chromatin coils on itself to form a compact chromosome (Figure 34-1). Each DNA molecule can be called either a chromatin strand or a chromosome, depending on what form it is in. Throughout this chapter we will use the term *chromosome* for DNA, regardless of its actual form, and the term *gene* for each distinct code segment within a DNA molecule.

Each gene in a chromosome contains a genetic code that the cell transcribes to a ribonucleic acid (RNA) molecule (see Box 4-5 on p. 108). Some RNA molecules do not code for polypeptides but have a functional role—for example, ribosomal RNA (rRNA) and transfer RNA (tRNA). A transcribed messenger RNA (mRNA) molecule, however, associates with a ribosome in which the code is translated to form a specific polypeptide molecule. By slight differences in editing of mRNA, one mRNA may perhaps actually produce several specific polypeptides. And the polypeptides may be complete tertiary proteins by themselves—or they may combine with any of several other polypeptides to form several different specific large quaternary proteins (see Figure 2-20 on p. 58).

Many of the protein molecules formed from the polypeptides encoded by genes are *enzymes*, functional proteins that help regulate the various metabolic pathways of the body by catalyzing specific chemical reactions. Because enzymes and other functional proteins such as hemoglobin regulate the biochemistry of the body, they regulate the entire structure and function of the body. Some proteins, such as collagen and keratin, are important structural components of the body—and thus determine important structural characteristics of various body parts.

As you can see, genes determine the structure and function of the human body by producing a set of specific structural proteins, along with many functional proteins and RNA molecules.

The Human Genome

The entire collection of genetic material in each typical cell of the human body is called the **genome** (JEE-nome). The structure of the human genome is summarized in Figure 34-2. The typical human genome includes 46 individual nuclear chromosomes and one mitochondrial chromosome. In 2003, the **Human Genome Project**—a publicly funded, worldwide collaboration to map all the genes in the human genome—was completed. This landmark event coincided exactly with the fiftieth anniversary of the discovery of DNA.

We now know that the human genome contains only about 20,000 to 25,000 genes. This is about one fifth to one quarter of the number originally estimated. Amazingly, it is roughly the same number of genes as in a rat or mouse—and only about one and one-half times as many genes as in a fruit fly!

We also know that less than 2% of the DNA carries protein-coding genes, with most of the rest being "filler" code that is either not used or is edited out of mRNA before it is used to make proteins. The current draft shows us that genes tend to lie in clusters rich in C (cytosine) and G (guanine), separated by long stretches of non-coding DNA rich in A (adenine) and T (thymine). Chromosome 1 has the most genes, with nearly 3000 genes, and the Y chromosome has the fewest, with just over 200 genes. Hundreds of the newly discovered genes in the human genome seem to be bacterial in origin, perhaps inserted there by bacteria in our distant ancestors.

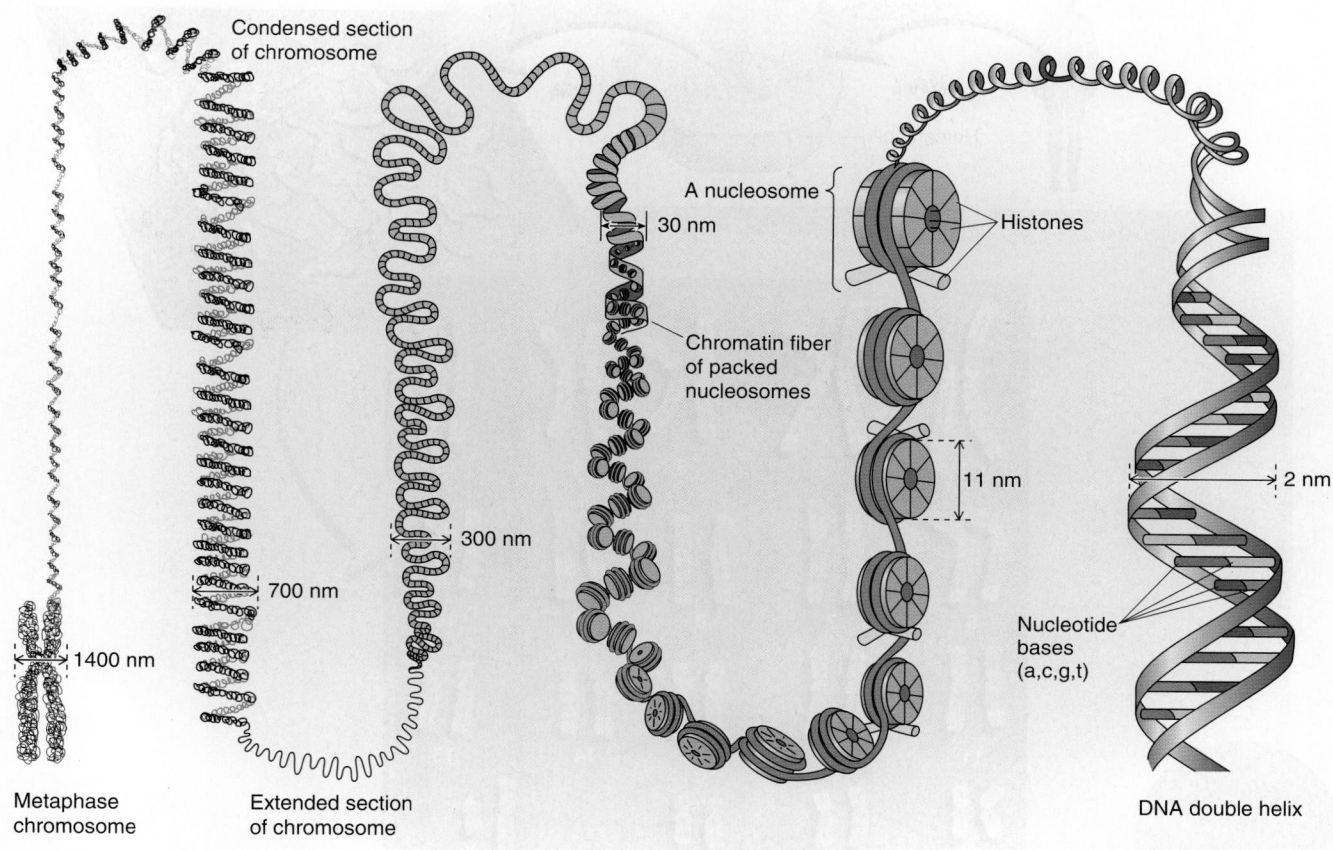

Condensed section of chromosome

A nucleosome

30 nm

Histones

Chromatin fiber of packed nucleosomes

11 nm

2 nm

300 nm

700 nm

1400 nm

Nucleotide bases (a,c,g,t)

Metaphase chromosome

Extended section of chromosome

DNA double helix

Figure 34-1 *Human DNA.* This artist's view of human DNA shows a fully condensed chromosome on the left and unfolding as you look to the right—until you see a single DNA strand (double helix). Note that the scale changes dramatically as you progress from one loop of the diagram to the next.

Although we now have the essential picture of the details of the human genome, much work still lies ahead in the new field of **genomics** (jeh-NOM-iks), the analysis of the genome's code. Besides filling in the remaining details of the rough draft, we have much work to do in discovering all the possible mutations that might exist (see the discussion later in this chapter) and all the proteins encoded by the genes that make up the human genome.

In fact, this quest has generated several other new fields. For example, **transcriptomics** (tran-skript-OME-iks) is the analysis of all the mRNA codes actually transcribed from the human genome—the **transcriptome** (tran-SKRIPT-ome). This field may shed light on which genes are expressed and under what conditions.

A field called **proteomics** (pro-tee-OH-miks) is the analysis of the proteins encoded by the genome. The entire group of proteins encoded by the human genome is called the human **proteome** (PRO-tee-ohm). The ultimate goal of proteomics is to understand the role of each protein in the body. Understanding the roles of every single protein in the body will certainly go a long way in improving our knowledge of the normal function of the body as well as mechanisms for many diseases.

The analysis of the human genome, transcriptome, and proteome has surged forward recently with the widespread use of *RNA interference (RNAi)* techniques. Recall from Box 4-5 (p. 127) that RNAi is a method of silencing particular genes. When harnessed in the laboratory, RNAi can turn off one gene at a time—greatly increasing the chances of figuring out which protein is encoded by that gene and what the function of that protein is.

Information obtained about the human genome can be expressed in a variety of ways. As you can see in Figure 34-2, an **ideogram** (ID-ee-oh-gram), or simple cartoon of a chromosome, is often used in genomics to show the overall physical structure of a chromosome. The constriction in the ideogram shows the relative position of the chromosome's centromere. The shorter segment of the chromosome is called the **p-arm** and the longer segment is called the **q-arm**.

The bands in an ideogram of a chromosome show staining landmarks and help identify the regions of the chromosome. Sometimes physical maps of genes will show exact positions of individual genes on the p-arm and q-arm of a chromosome. A more detailed representation of a gene would show the actual sequence

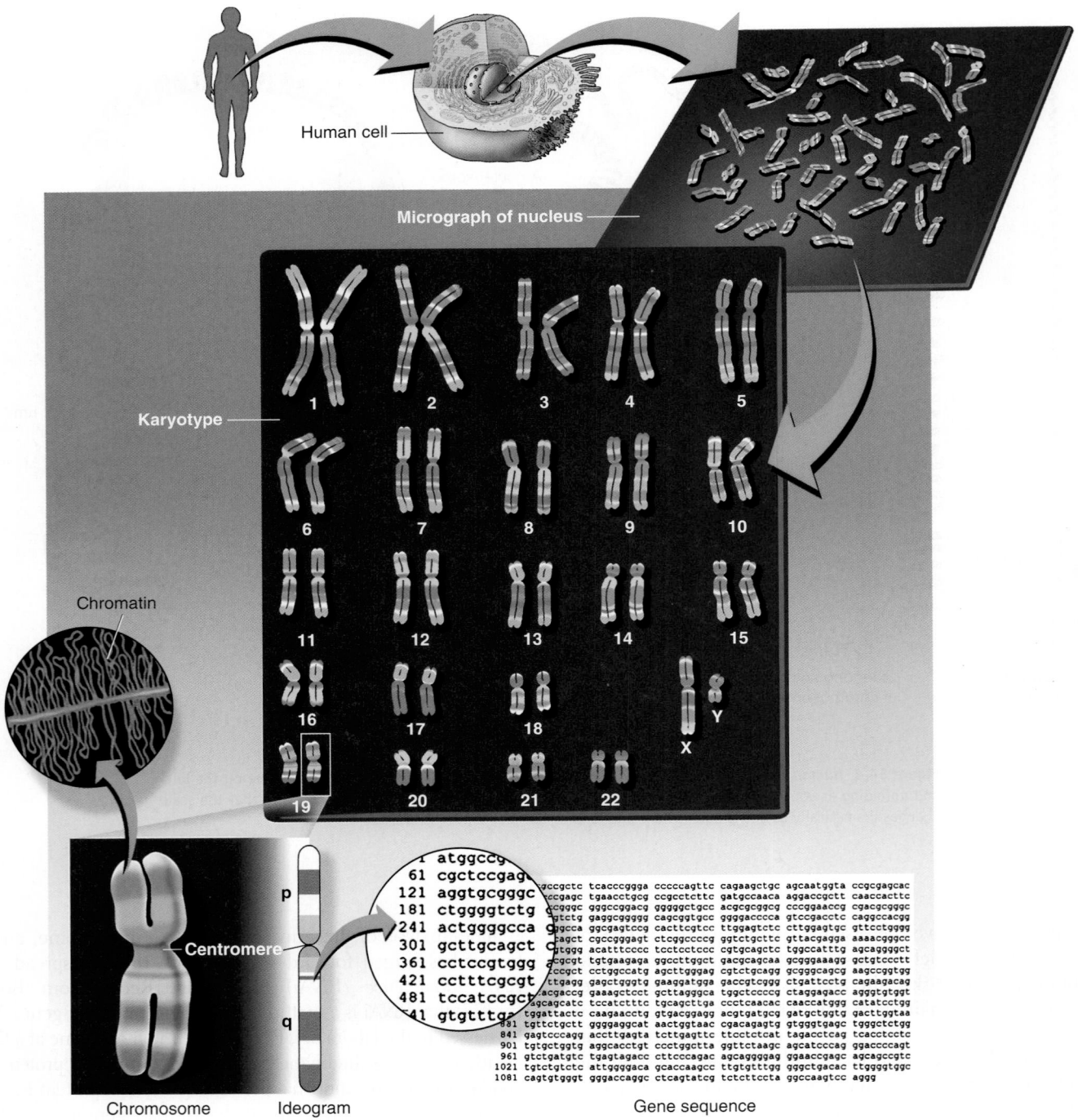

Human cell

Micrograph of nucleus

Karyotype

Chromatin

Chromosome   Ideogram   Gene sequence

Figure 34-2 *Human genome.* A cell taken from the body is stained and photographed. A photograph of nuclear chromosomes is then cut and pasted, arranging each of the 46 chromosomes into numbered pairs of decreasing size to form a chart called the karyotype. Each chromosome is a coiled mass of chromatin (DNA). In this figure, differentially stained bands in each chromosome appear as different, bright colors. Such bands are useful as reference points when identifying the locations of specific genes within a chromosome. The staining bands are also represented on an ideogram, or simple graph, of the chromosome as reference points to locate specific genes. The genes themselves are usually represented as the actual sequence of nucleotide bases, abbreviated as *a, c, g,* and *t*. In this figure, the sequence of one exon (segment) of a gene called GPI from chromosome 19 is shown. Each of these representations can be thought of as a type of "genetic map."

of nucleotide bases, abbreviated *a, c, g,* and *t* for *adenine, cytosine, guanine,* and *thymine,* as shown in Figure 34-2.

Distribution of Chromosomes to Offspring

Meiosis

Each cell of the human body contains 46 chromosomes. The only exceptions to this principle are the **gametes**—male *spermatozoa* and female *ova.* Recall from Chapter 33 that a special form of nuclear division called **meiosis** (see Figure 33-1 on p. 1163) produces gametes with only 23 chromosomes—exactly one half the usual number. This number is called the **haploid** number. This process follows a basic principle of genetics first discovered by Gregor Mendel called the **principle of segregation.** This principle simply states that the two members of a pair of chromosomes separate, or *segregate,* during meiosis.

When a sperm (with its 23 chromosomes) unites with an ovum (with its 23 chromosomes) at conception, a *zygote* with 46 chromosomes is formed. Thus the zygote has the same number of chromosomes (46, the **diploid** number) as each typical body cell in the parents.

As the photograph in Figure 34-2 shows, the 46 human chromosomes can be arranged in 23 pairs according to size. One pair called the **sex chromosomes** may not match, but the remaining 22 pairs of **autosomes** always appear to be nearly identical to each other.

Principle of Independent Assortment

Because one half of an offspring's chromosomes are from the mother and one half are from the father, a unique blend of inherited traits is formed. According to another of Mendel's principles, each chromosome assorts itself independently during meiosis. This **principle of independent assortment** states that as sperm are formed, and chromosome pairs separate (the principle of segregation), the maternal and paternal chromosomes get mixed up and redistribute themselves independently of the other chromosome pairs (Figure 34-3). Thus each sperm is likely to have a *different* set of 23 chromosomes. Because the ova are formed in the same manner, each ovum is likely to be genetically

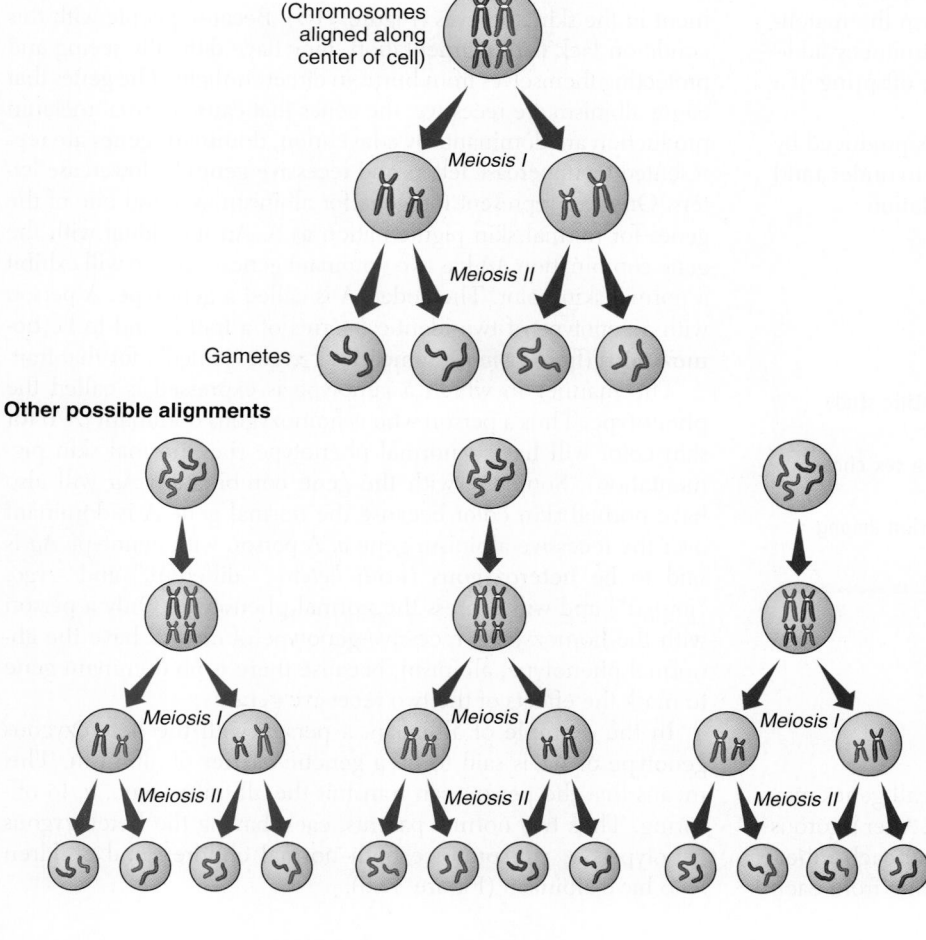

Figure 34-3 *Meiosis and the principle of independent assortment.* In meiosis, a series of two divisions results in the production of gametes with one half the number of chromosomes of the original parent cell. In both meiotic divisions shown here, the original cell has four chromosomes and the gametes each have two chromosomes. During the first division of meiosis, pairs of similar chromosomes line up along the cell's equator for even distribution to daughter cells. Because different pairs assort independently of each other, four (2^2) different alignments of chromosomes can occur. Because human cells have 23 pairs of chromosomes, more than 8 million (2^{23}) different combinations are possible.

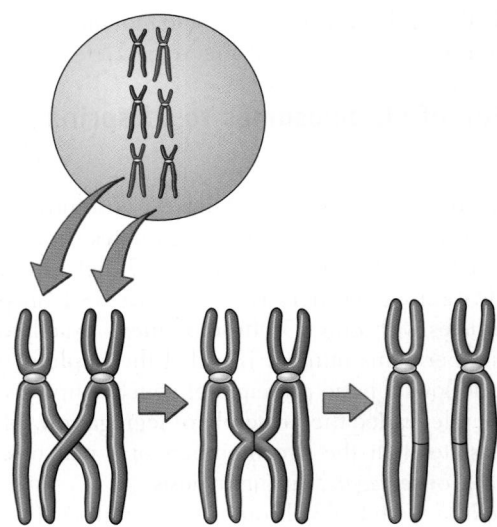

Figure 34-4 *Crossing over.* Genes (or linked groups of genes) from one chromosome are exchanged with matching genes in the other chromosome of a pair during meiosis.

different from the ovum that preceded it. Independent assortment of chromosomes ensures that each offspring from a single set of parents is very likely to be genetically unique—a phenomenon known as *genetic variation.*

According to the principle of **gene linkage**, genes on an individual chromosome tend to stay together. An important application of this principle occurs during one phase of meiosis, when pairs of matching chromosomes line up along the equator of the cell and exchange genes or groups of linked genes with one another. This process is called **crossing over** because genes from a particular location cross over to the same location on the matching chromosome (Figure 34-4). Crossing over introduces additional opportunities for genetic variation among the offspring of a single set of parents.

When one considers the genetic variation that is produced by independent assortment and crossing over, it is easy to understand the tremendous variation seen in the human population.

QUICK CHECK

1. How do genes produce biological traits?
2. Who might be considered the founder of the scientific study of genetics?
3. What is the difference between an autosome and a sex chromosome?
4. List some mechanisms that increase genetic variation among human offspring.

GENE EXPRESSION

Hereditary Traits

Dominant and Recessive Traits

Mendel discovered that the genetic units we now call genes may be expressed differently among individual offspring. After rigorous experimentation with pea plants, he discovered that each inherited trait is controlled by two sets of similar genes, one from each parent. We now know that each autosome in a pair matches its partner in the type of genes it contains. In other words, if one autosome has a gene for hair color, its partner will also have a gene for hair color—in the same location on the autosome. Although both genes specify hair color, they may not specify the *same* hair color. Mendel discovered that some genes are **dominant** and some are **recessive.** A dominant gene is one whose effects are seen and whose effects are capable of masking the effects of a recessive gene for the same trait.

Consider the example of **albinism,** a total lack of melanin pigment in the skin and eyes (Figure 34-5). Because people with this condition lack dark pigmentation, they have difficulty seeing and protecting themselves from burns in direct sunlight. The genes that cause albinism are recessive; the genes that cause normal melanin production are dominant. By convention, **dominant genes** are represented by uppercase letters and **recessive genes** by lowercase letters. One can represent the gene for albinism as *a* and one of the genes for normal skin pigmentation as *A*. An individual with the gene combination *AA* has two dominant genes—and so will exhibit a normal skin color. The code *AA* is called a **genotype.** A person with a genotype of two identical forms of a trait is said to be **homozygous** (from *homo-*, "same," and *-zygo*, "joined") for that trait.

The manner in which a genotype is expressed is called the **phenotype.** Thus a person who is homozygous dominant *(AA)* for skin color will have a normal phenotype (i.e., normal skin pigmentation). Someone with the gene combination *Aa* will also have normal skin color because the normal gene A is dominant over the recessive albinism gene *a*. A person with genotype *Aa* is said to be **heterozygous** (from *hetero-*, "different," and *-zygo*, "joined") and will express the normal phenotype. Only a person with the homozygous recessive genotype of *aa* will have the abnormal phenotype, albinism, because there is no dominant gene to mask the effects of the two recessive genes.

In the example of albinism, a person with the heterozygous genotype of *Aa* is said to be a genetic **carrier** of albinism. This means that the person can transmit the albinism gene, *a*, to offspring. Thus two normal parents, each having the heterozygous genotype *Aa*, can produce both normal children and children who have albinism (Figure 34-6).

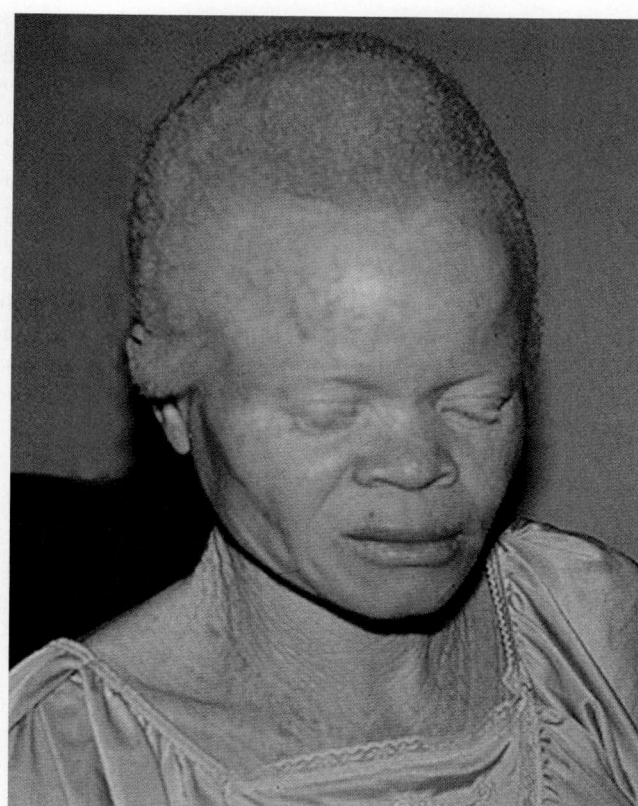

Figure 34-5 *Albinism.* There are several forms of albinism in humans. The type shown here, tyrosinase-negative oculocutaneous albinism, results from the inheritance of two abnormal genes for tyrosinase—the enzymes required to convert tyrosine to melanin pigments. Melanin is normally present in the skin as well as in the eye, where its absence produces vision problems that include sensitivity to light. This African woman would otherwise have dark skin, dark hair, and normal vision. However, as she looks away from the bright camera light as she is photographed, you can see the abnormally light hair and skin typical of this type of albinism.

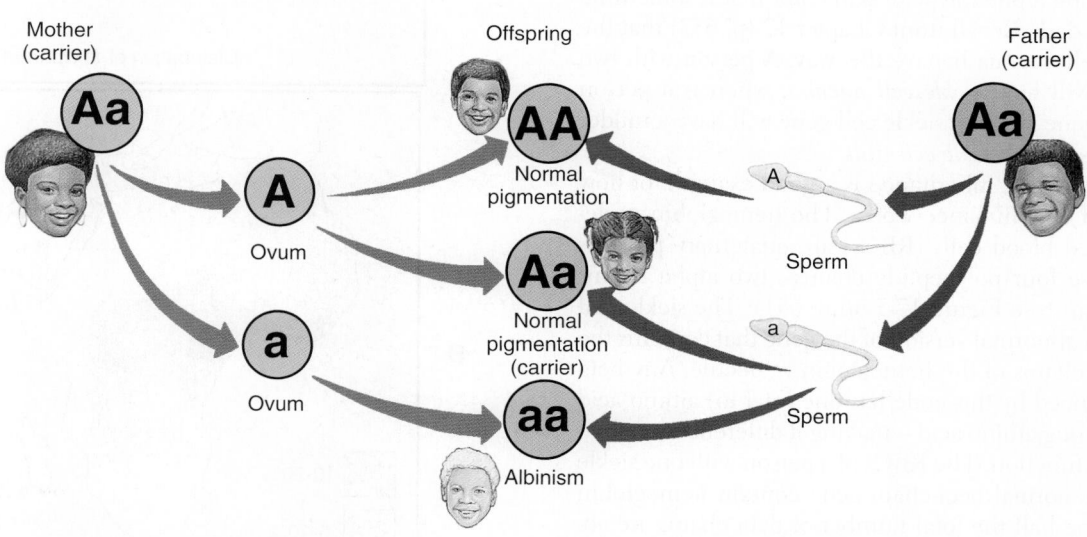

Figure 34-6 *Inheritance of albinism.* Albinism is a recessive trait, producing abnormalities only in those with two recessive genes (a). Presence of the dominant gene (A) prevents albinism.

Polygenic Traits

It is important to note that melanin pigmentation in human skin is actually governed by several different pairs of genes. The gene for the form of albinism discussed here, *tyrosinase-negative oculocutaneous albinism*, involves just one of the gene pairs that regulate skin color. Inherited characteristics, such as skin color and height, which are determined by the combined effect of many different gene pairs, are often called **polygenic** ("many genes") traits to distinguish them from **monogenic**, or single-gene, traits.

Polygenic traits are often hard to study in the phenotype because they are so variable. You can think of a polygenic trait as a "combined trait" because it results from the combined activity of several different genes. Each gene may be dominant or recessive in character. Because each gene is only one of several that govern the combined trait, however, the phenotype may be any of a large number of different variations of the trait. Using skin color as an example, the form of albinism described above involved only one of several genes that govern pigmentation of the skin and eyes. That one gene is critical for all the other genes that govern skin color to have their effects. However, if the dominant form of that critical gene is in place, then variations in any of the other genes that regulate skin color can now exert influence on skin pigmentation.

Codominant Traits

What happens if two different dominant genes occur together? Suppose there is a gene A^1 for light skin and a gene A^2 for dark skin. In a form of dominance called **codominance**, they will simply have equal effects, and a person with the heterozygous genotype A^1A^2 will exhibit a phenotype of skin color that is something between light and dark. Recall from Chapter 17 (p. 653) that the genes for sickle cell anemia behave this way. A person with two sickle cell genes will have *sickle cell anemia*, whereas a person with one normal gene and one sickle cell gene will have a milder form of the disease called *sickle cell trait*.

The case of sickle cell inheritance is a good example of how the mechanism of codominance works. The hemoglobin molecules within all red blood cells (RBCs) are quaternary proteins that each comprise four polypeptide chains—two alpha chains and two beta chains (see Figure 17-5 on p. 651). The sickle cell gene is actually an abnormal version of the gene that contains the code for the beta chains of the hemoglobin molecule. Any beta chain that is produced by this code has one (of 146) amino acid replaced by the wrong amino acid—making it different enough to drastically alter its function. The RBCs of a person with one sickle cell gene and one normal beta-chain gene contain hemoglobin in which about one half the total number of beta chains are abnormal and about one half are normal. The RBCs of a person with two sickle cell genes contain hemoglobin in which all the beta chains are abnormal. Thus in sickle cell trait only some hemoglobin molecules are defective, but in sickle cell anemia *all* of the hemoglobin molecules are defective.

The frequency of occurrence of the abnormal sickle cell gene is an example of an interesting epidemiological phenomenon. Because sickle cell trait provides resistance to the parasite that causes the most deadly form of **malaria** (*Plasmodium falciparum*

malaria), sickle cell disorders persist in areas of the world in which malaria is still prevalent (Figure 34-7). Malaria is a sometimes fatal condition caused by blood cell parasites (*Plasmodium* species) and is characterized by fever, **anemia**, swollen spleen, and possible relapse months or years later. The unique distribution of *P. falciparum* malaria results from the fact that people without sickle cell trait more often die of this condition before producing offspring than those with the malaria-resistant sickle cell trait. Thus the "bad" sickle cell gene is more likely to be transmitted to the next generation than the "good" genes for normal hemoglobin.

The sickle cell/malaria relationship points to an important concept in medical genetics: "disease" genes often provide some biological advantage for a human population in certain circum-

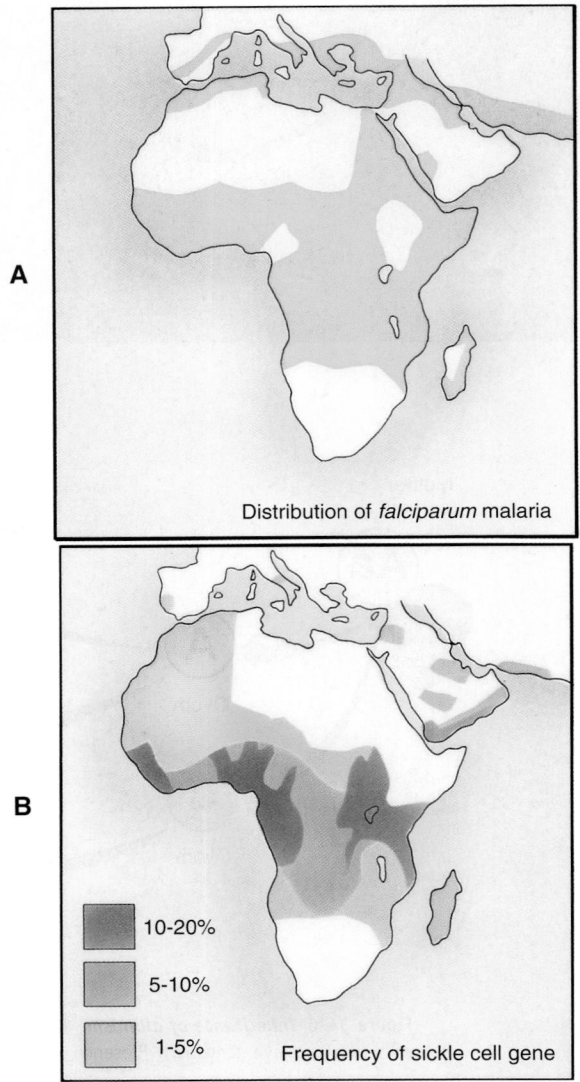

Figure 34-7 *Relationship between the frequency of sickle cell trait and the distribution of malaria.* A, The distribution of the most deadly form of malaria in Africa correlates closely with the frequency of occurrence of the sickle cell gene, **B.** The sickle cell trait provides resistance to malaria, and thus heterozygous individuals are more likely than homozygous individuals to survive and reproduce—spreading the abnormal gene further in the population.

stances. It is only when circumstances change that the gene is seen to do more harm than good. Genes for many other hereditary diseases (e.g., *thalassemia* and *Tay-Sachs disease*) are now known to impart protection against pathogenic conditions in heterozygous individuals.

Sex-Linked Traits

Recall from our earlier discussion that besides the 22 pairs of autosomes, there is one pair of sex chromosomes. Notice in the lower right portion of the photograph in Figure 34-2 that the chromosomes of this pair do not have matching structures. The larger sex chromosome is called the X *chromosome*, and the smaller one is called the Y *chromosome*. The X chromosome is sometimes called the "female chromosome" because it includes genes that determine female sexual characteristics. If a person has only X chromosomes, she is genetically a female. The Y chromosome is often called the "male chromosome" because anyone possessing a Y chromosome is genetically a male. Thus all normal females have the sex chromosome combination XX, and all normal males have the combination XY (Box 34-1). Because men produce both X-bearing and Y-bearing sperm, any two parents can produce male or female children (Figure 34-8).

The large X chromosome contains many genes besides those needed for female sexual traits. Genes for producing certain clot-

ting factors, photopigments in the retina of the eye, and many other proteins are also found on the X chromosome. The tiny Y chromosome, on the other hand, contains few genes other than those that determine male sexual characteristics. Thus both males and females need at least one normal X chromosome—otherwise genes for clotting factors and other essential proteins would be missing. Traits carried on sex chromosomes are called **sex-linked traits.** Some sex-linked traits are called X-*linked traits* because they are determined by genes in the large X chromosome. Other sex-linked traits are called Y-*linked traits* because they are determined by genes in the tiny Y chromosome.

Dominant X-linked traits appear in each person, as one would expect for any dominant trait. In females, recessive X-linked genes are masked by dominant genes in the other X chromosome. Only females with two recessive X-linked genes can exhibit the recessive trait. Since males inherit only one X chromosome (from the mother), the presence of only one recessive X-linked gene is enough to produce the recessive trait. In short, in males there are no matching genes in the Y chromosome to mask recessive genes in the X chromosome. For this reason, X-linked recessive traits appear much more frequently in males than in females.

An example of a recessive X-linked condition is *red-green color blindness*, which involves a deficiency of normal photopigments in the retina (see Chapter 15, p. 579). In this condition, male children of a parent who carries the recessive abnormal gene on an X chromosome may be color blind (Figure 34-9). A female can inherit this form of color blindness only if her father is color blind *and* her mother is either color blind (homozygous recessive) or a color-blindness carrier (heterozygous). The X chromosome has been studied in great detail, and general locations for genes causing dozens of distinct X-linked diseases have been identified (Figure 34-10).

Only one clinically significant Y-linked condition has been identified by geneticists. A missing part of q-arm of the Y chromosome may cause inheritable problems with spermatogenesis—and possible infertility. Such a Y-linked condition may be passed from father to son.

BOX 34-1: FYI

Timing and Sex Determination

Research has shown that X-bearing sperm swim more slowly than Y-bearing sperm. An interesting hypothesis stems from this fact. If insemination occurs on the day of ovulation, Y-bearing sperm, being faster than X-bearing sperm, should reach the ovum first. Therefore a Y-bearing sperm would be more apt to fertilize the ovum. And because Y-bearing sperm produce males, there should be a greater probability of having a baby boy when insemination occurs on the day of ovulation. Statistical evidence supports this view.

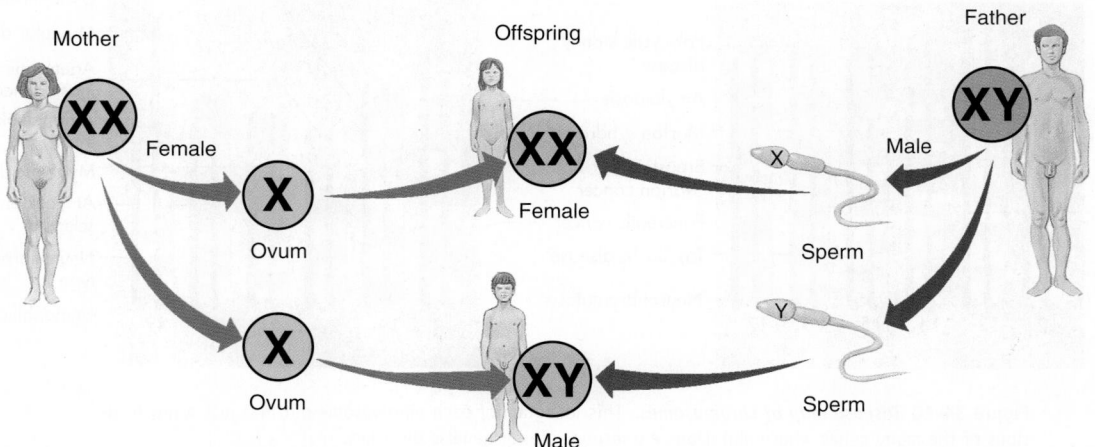

Figure 34-8 *Sex determination.* The presence of the Y chromosome specifies maleness. In the absence of a Y chromosome, an individual develops into a female.

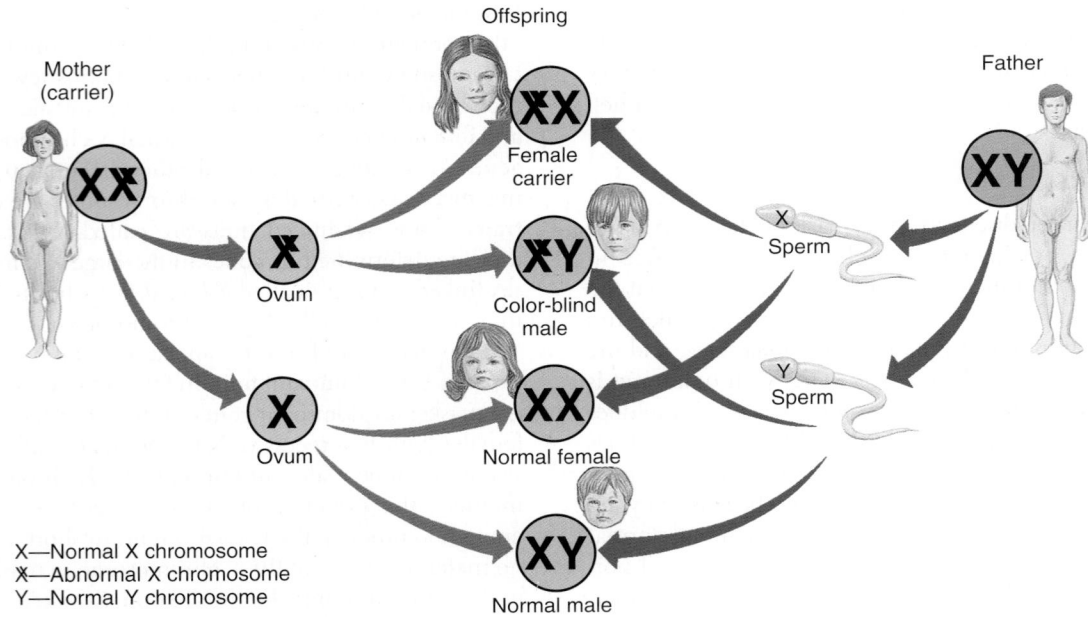

Figure 34-9 *Sex-linked inheritance.* Some forms of color blindness involve recessive X-linked genes. In this case, a female carrier of the abnormal gene can produce male children who are color blind.

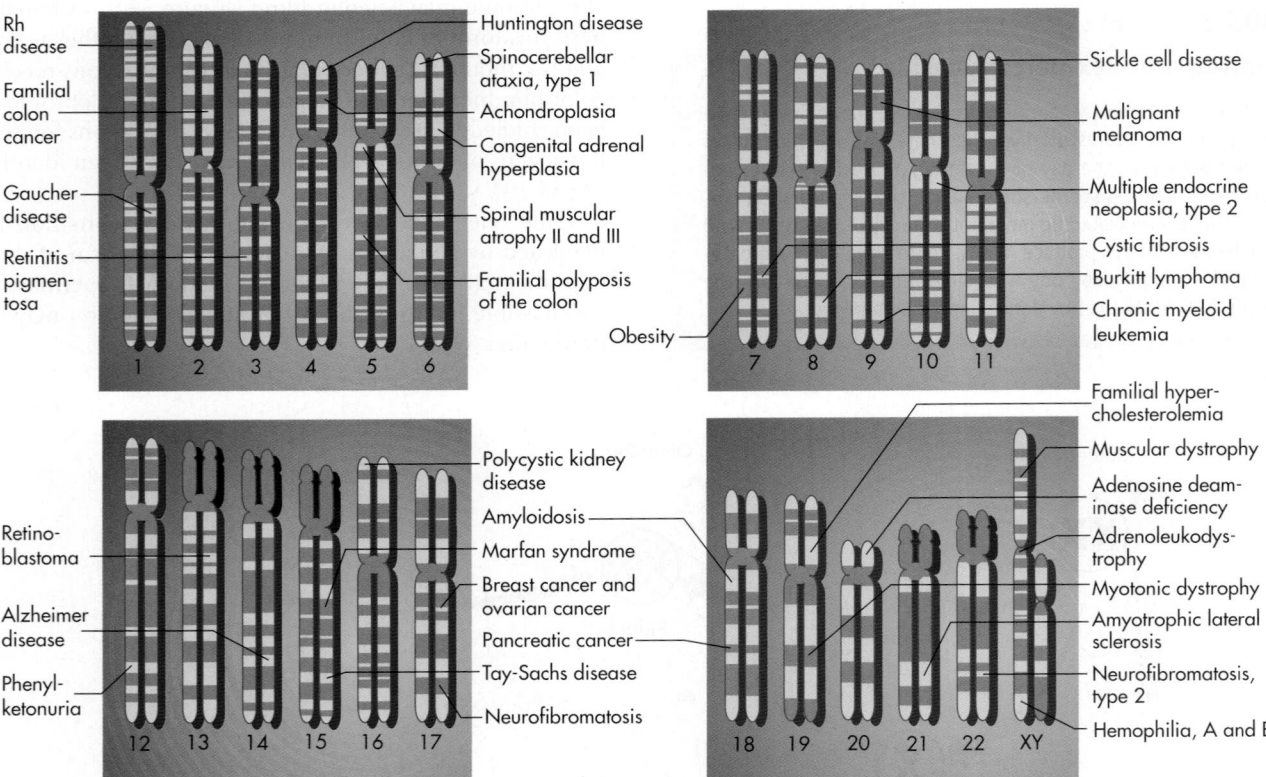

Figure 34-10 *Disease map of chromosomes.* This ideogram of each chromosome outlines just a few locations of the many genes whose mutations are responsible for genetic disorders.

Genetic Mutations

The term *mutation* simply means "change." A **genetic mutation** is a change in an individual's genetic code. Some mutations involve a change in the genetic code within a single gene, perhaps a slight rearrangement of the nucleotide sequence. A mutation called a **deletion** occurs when one or more nucleotide bases in a sequence are missing. An **insertion** mutation occurs when one or more nucleotides appear within the usual sequence of nucleotide bases in a gene. With either type of mutation, the cell cannot read the genetic code normally, and thus the encoded protein cannot be made in its usual form. Other mutations involve damage to a portion of a chromosome or a whole chromosome. For example, a portion of a chromosome may completely break away.

Mutations may occur spontaneously without the influence of factors outside the DNA itself. However, most genetic mutations are believed to be caused by **mutagens**—agents that cause changes in the genetic code by damaging DNA molecules. Genetic mutagens include chemicals, some forms of radiation, and even viruses.

If mutations occur in reproductive cells or their precursors, they may be inherited by offspring. Beneficial mutations allow organisms to adapt to their environments. Because such mutant genes benefit survival, they tend to spread throughout a population over the course of several generations.

Harmful mutations inhibit survival and therefore are not likely to spread widely through the population. Most harmful mutations kill the organism in which they occur or at least prevent successful reproduction—and so are never passed to offspring. Harmful mutations that are recessive, however, may persist at low frequencies in a population indefinitely because they do not cause problems for individuals who inherit just one of these genes. If a harmful dominant mutation is only mildly harmful, it may persist in a population over many generations.

Consider also the case of the mutations that cause sickle cell anemia, thalassemia, and Tay-Sachs disease—the heterozygous genotype produces a phenotype that resists a specific disease, and the homozygous genotype produces a phenotype characterized by a different disease condition.

QUICK CHECK

5. What is a dominant genetic trait? A recessive trait?
6. What is codominance?
7. How can a mutant gene benefit a human population?
8. What is X-linked inheritance?

MEDICAL GENETICS

Mechanisms of Genetic Diseases

As science writer Matt Ridley repeatedly and emphatically stated in his best-selling book *Genome: The Autobiography of a Species in 23 Chapters*, "GENES ARE NOT THERE TO CAUSE DISEASE." Although we often hear of new "disease genes" being discovered—more than 1200 of them since 1981 and the pace is rapidly increasing—the function of these genes is not to cause disease any more than the function of an arm is to cause bone fractures. If you break an arm, a normal bone is broken and fails to serve its usual function. In genetic disorders, a normal gene or chromosome is broken (mutated) and fails to serve its usual function. Such a gene is sometimes called a "disease gene" because when it is broken, it is involved in the mechanism of a particular disease. Keep this simple—but often overlooked—principle in mind as you read the following paragraphs.

Nuclear Inheritance
Single-Gene Diseases

As we just stated, genetic diseases are diseases produced by an abnormality in the genetic code. Many genetic diseases are caused by individual mutant genes in the nuclear DNA that is passed from one generation to the next—making them **single-gene diseases.** In single-gene diseases, the mutant gene may make an abnormal product that causes disease or it may fail to make a product required for normal function. As discussed previously, some disease conditions result from the combined effects of inheritance and environmental factors. Because they are not solely caused by genetic mechanisms, such conditions are not genetic diseases in the usual sense of the word—they are instead said to involve a *genetic predisposition.*

Chromosomal Genetic Diseases

Some genetic diseases do not result from an abnormality in a single gene. Instead, these diseases result from chromosome breakage or from the abnormal presence or absence of entire chromosomes. For example, a condition called **trisomy** may occur where there is a triplet of autosomes rather than a pair. Trisomy results from a mistake in meiosis called **nondisjunction** when a pair of chromosomes fails to separate. This produces a gamete with two autosomes that are "stuck together" instead of the usual one. When this abnormal gamete joins with a normal gamete to form a zygote, the zygote has a triplet of autosomes (Figure 34-11). Trisomy of any autosome pair is usually fatal. However, if trisomy occurs in autosome pair 13, 15, 18, 21, or 22, a person may survive for a time—but with profound developmental defects. **Monosomy,** the presence of only one autosome instead of a pair, may also result from conception involving a gamete produced by nondisjunction (see Figure 34-11). As with trisomy, monosomy may produce life-threatening abnormalities. Because most trisomic and monosomic individuals die before they can reproduce, these conditions are not usually passed from generation to generation. Trisomy and monosomy are congenital conditions that are sometimes referred to as **chromosomal genetic diseases** (Box 34-2).

Mitochondrial Inheritance

Mitochondria are tiny, bacteria-like organelles present in every cell of the body (see Figure 3-11). The major exception, of course, is the red blood cell, which does not reproduce itself. As with a bacterium, each mitochondrion has its own circular DNA molecule, sometimes called **mitochondrial DNA (mDNA** or **mtDNA).** Figure 34-12 shows an ideogram of the structure of a mitochondrial chromosome.

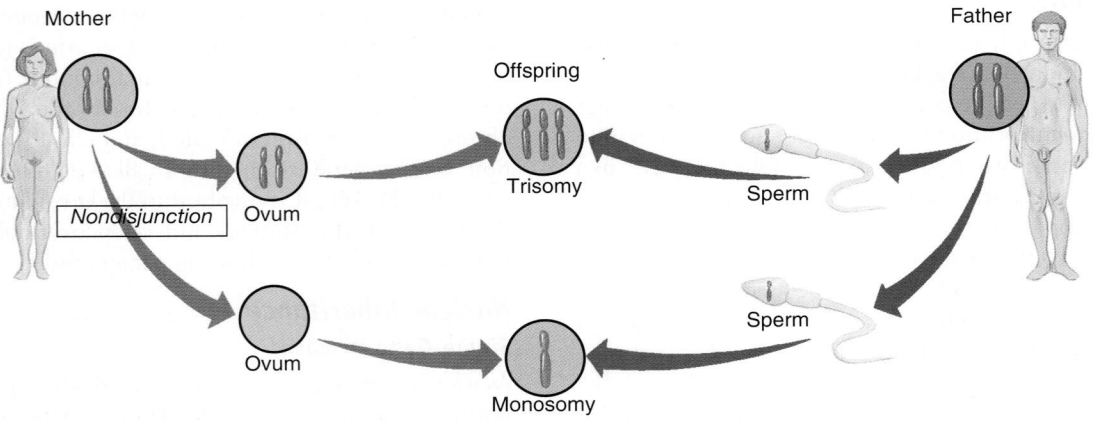

Figure 34-11 *Effects of nondisjunction.* Nondisjunction, failure of a chromosome pair to separate during gamete production, may result in trisomy or monosomy in the offspring.

BOX 34-2: FYI
Congenital Disorders

A congenital disorder is any pathological condition present at birth. As explained in Chapter 33, congenital disorders may have a genetic cause. For example, one form of a facial deformity known as cleft palate is an X-linked inherited condition (see Figure 34-10). However, some congenital disorders are not inherited. For example, fetal alcohol syndrome is a group of congenital deformities that result from exposing a developing fetus to alcohol consumed by the mother (see Chapter 18, p. 716). Thus not all congenital disorders are inherited disorders.

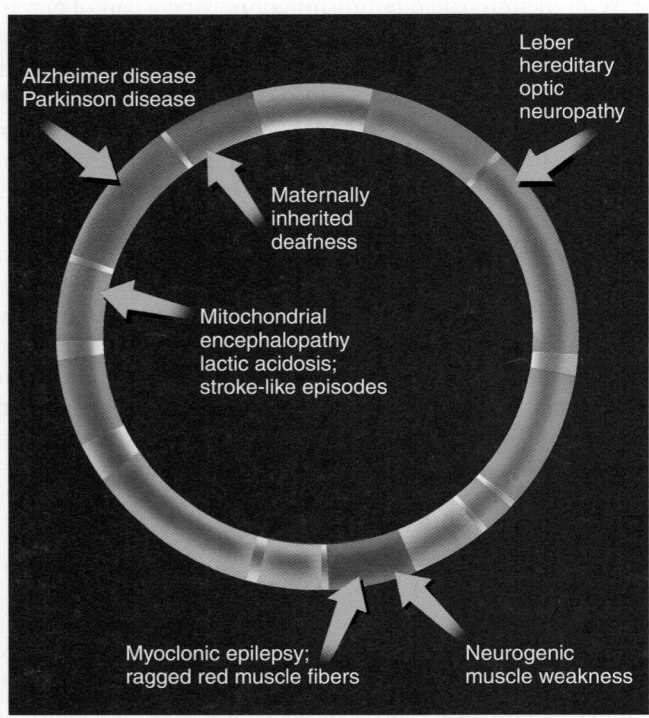

Figure 34-12 *Map of mitochondrial DNA.* Ideogram showing locations of some mtDNA genes associated with various diseases.

Inheritance of mtDNA occurs only through one's mother because the few mitochondria that a sperm may contribute to the ovum during fertilization do not survive. Because mtDNA contains the only genetic code for several important enzymes, it has the potential for carrying mutations that produce disease. *Mitochondrial inheritance* is known to transmit genes for several degenerative nerve and muscle disorders. One such disease is *Leber hereditary optic neuropathy.* In this disease, young adults begin losing their eyesight as the optic nerve degenerates—resulting in total blindness by 30 years of age. Some medical researchers believe that at least some forms of several other diseases are associated with mtDNA mutations. These diseases include **Parkinson disease, Alzheimer disease (AD), diabetes mellitus (DM)** with deafness, and maternally inherited forms of deafness, myopathy, and cardiomyopathy.

QUICK CHECK

9. How are single-gene diseases different from chromosomal conditions?
10. What is *nondisjunction*? How can it cause trisomy?
11. What is mitochondrial inheritance?

Single-Gene Diseases

There are many examples of single-gene diseases. Only a few of the many single-gene diseases are discussed here and summarized in Table 34-1.

Cystic fibrosis (CF), briefly mentioned in Chapter 4 (p. 138), is caused by a recessive gene in chromosome 7 that codes for *CFTR (CF transmembrane conductance regulator).* CFTR normally regulates the transfer of sodium across cell membranes and serves as a chloride channel. When this gene has a deletion of a single codon, the abnormal version of CFTR causes impairment of chloride ion transport across cell membranes. Disruption of chloride transport causes exocrine cells to secrete thick mucus and concentrated sweat. The thickened mucus is especially troublesome in the gastrointestinal (GI) and respiratory tracts, where

Table 34-1   Examples of Genetic Conditions

CHROMOSOME LOCATION	DISEASE	DESCRIPTION
Single-Gene Inheritance (Nuclear DNA)		
DOMINANT		
7, 17	Osteogenesis imperfecta	Group of connective tissue disorders is characterized by imperfect skeletal development that produces brittle bones
17	Multiple neurofibromatosis	Disorder is characterized by multiple, sometimes disfiguring benign tumors of the Schwann cells (neuroglia) that surround nerve fibers
5	Hypercholesterolemia	High blood cholesterol may lead to atherosclerosis and other cardiovascular problems
4	Huntington disease (HD)	Degenerative brain disorder is characterized by chorea (purposeless movements) progressing to severe dementia and death generally by age 55 years
CODOMINANT		
11	Sickle cell anemia / Sickle cell trait	Blood disorder in which abnormal hemoglobin causes red blood cells (RBCs) to deform into a sickle shape; sickle cell anemia is the severe form, and sickle cell trait the milder form
11, 16	Thalassemia	Group of inherited hemoglobin disorders is characterized by production of hypochromic, abnormal RBCs
RECESSIVE (AUTOSOMAL)		
7	Cystic fibrosis (CF)	Condition is characterized by excessive secretion of thick mucus and concentrated sweat, often causing obstruction of gastrointestinal or respiratory ducts
15	Tay-Sachs disease (TSD)	Fatal condition in which abnormal lipids accumulate in the brain and cause tissue damage leading to death by age 4 years
12	Phenylketonuria (PKU)	Excess of phenylketones in the urine is caused by accumulation of phenylalanine in the tissues; it may cause brain injury and death if phenylalanine (amino acid) intake is not restricted
11	Albinism (total)	Lack of the dark brown pigment *melanin* in the skin and eyes results in vision problems and susceptibility to sunburn and skin cancer
20	Severe combined immune deficiency (SCID)	Failure of the lymphocytes to develop properly causes failure of the immune system's defense of the body; it is usually caused by adenosine deaminase (ADA) deficiency
RECESSIVE (X LINKED)		
23 (X)	Hemophilia	Group of blood clotting disorders is caused by failure to form clotting factors VIII, IX, or XI
23 (X)	Duchenne muscular dystrophy (DMD)	Muscle disorder is characterized by progressive atrophy of skeletal muscle without nerve involvement
23 (X)	Red-green color blindness	Inability to distinguish red and green light results from a deficiency of photopigments in the cone cells of the retina
23 (X)	Fragile X syndrome	Mental retardation results from breakage of X chromosome in males
23 (X)	Ocular albinism	Form of albinism in which the pigmented layers of the eyeball lack melanin results in hypersensitivity to light and other problems
23 (X)	Androgen insensitivity	Inherited insensitivity to androgens (steroid sex hormones associated with maleness) results in reduced effects of these hormones
23 (X)	Cleft palate (X-linked form)	One form of a congenital deformity in which the skull fails to develop properly, characterized by a gap in the palate (plate separating mouth from nasal cavity)
23 (X)	Retinitis pigmentosa	Condition causes blindness, characterized by clumps of melanin in retina of eyes
Single-Gene Inheritance (Mitochondrial DNA)		
mtDNA	Leber hereditary optic neuropathy	Optic nerve degeneration in young adults results in total blindness by age 30 years
mtDNA	Parkinson disease	Nervous disorder is characterized by involuntary trembling and muscle rigidity

Continued

Table 34-1	Examples of Genetic Conditions—cont'd	
CHROMOSOME LOCATION	**DISEASE**	**DESCRIPTION**
Chromosomal Abnormalities		
TRISOMY		
21	Down syndrome	Condition is characterized by mental retardation and multiple structural defects
23	Klinefelter syndrome	Condition is caused by the presence of two or more X chromosomes in a male (XXY); characterized by long legs, enlarged breasts, low intelligence, small testes, sterility, and chronic pulmonary disease
MONOSOMY		
23	Turner syndrome	Condition is caused by monosomy of the X chromosome, without a Y chromosome (XO), characterized by immaturity of sex organs (causing sterility), webbed neck, cardiovascular defects, and learning disorders

it can cause obstruction that leads to death. This condition is often treated by the continuous use of drugs and other therapies that relieve the symptoms. CF occurs almost exclusively among Caucasians. The mutation that causes CF is thought to actually protect carriers of the gene from potential deadly cases of diarrhea, as in cholera and serious *Escherichia coli* infections.

Phenylketonuria (PKU) is caused by a recessive gene that fails to produce the enzyme *phenylalanine hydroxylase*. This enzyme is needed to convert the amino acid phenylalanine into another amino acid, tyrosine. Thus phenylalanine absorbed from ingested food accumulates in the body—resulting in the abnormal presence of phenylketone in the urine (hence the name phenylketonuria). A high concentration of phenylalanine destroys brain tissue; babies born with this condition are at risk of progressive mental retardation and, perhaps, death. Many PKU victims are identified at birth by state-mandated screening tests. After being identified, PKU victims are put on diets low in phenylalanine—thus avoiding a toxic accumulation of this amino acid. You may be familiar with the printed warning for people with phenylketonuria commonly seen on products that contain aspartame (NutraSweet) or other substances made from phenylalanine. The mutant PKU gene may have originated among the Celts in western Europe, where it offered protection against the toxic effects of molds growing on grains stored in cold, damp climates.

Tay-Sachs disease (TSD) is a recessive condition involving failure to make a subunit of an essential lipid-processing enzyme, hexosaminidase. Abnormal lipids accumulate in the brain tissue of Tay-Sachs victims, causing severe retardation and death by 4 years of age. There is currently no specific therapy for this condition. TSD is most prevalent among certain Jewish populations. Some epidemiologists believe that this ethnic distribution is related to the hypothesis that heterozygous carriers of the Tay-Sachs gene have a higher than normal resistance to tuberculosis (TB)—a potentially fatal disease that once killed millions in the crowded Jewish ghettos of many large cities. Residents of these TB-infested areas who carried the Tay-Sachs gene apparently survived longer—and reproduced more frequently—than noncarriers.

Osteogenesis imperfecta is a dominant genetic disorder of connective tissues. Its name, which means "imperfect bone formation," describes its chief characteristic. The bones of people with osteogenesis imperfecta do not have normal collagen and thus are often so brittle that the slightest trauma can result in serious fractures. There are different forms of the disease. In its most severe form, it results in fractures of the fetal skeleton in utero—often progressing to death shortly after birth. In a form seen in infancy, this disease is characterized by short, deformed limbs, a thin, enlarged skull, and easily fractured bones (Figure 34-13). In a less severe form, symptoms appear when a child begins to walk and become milder until after puberty—when the symptoms usually disappear.

Multiple neurofibromatosis is a dominant inherited disorder discussed in Chapter 12 (p. 464). This disorder is characterized by multiple, sometimes disfiguring, benign tumors of the glial cells that surround nerve fibers.

Other important inherited disorders include **Duchenne muscular dystrophy (DMD)**, **hypercholesterolemia**, **sickle cell anemia**, **albinism**, certain forms of **hemophilia**, and **Huntington disease (HD)** (Box 34-3). These and other conditions are summarized in Table 34-1.

Chromosomal Diseases

As described earlier, some genetic disorders are not inherited in the usual sense but result from nondisjunction during formation of the gametes. As Figure 34-11 shows, nondisjunction results in gametes that produce either trisomy or monosomy in the cells of offspring. At least 10% of all human sperm and at least 25% of all mature oocytes have extra, missing, or broken chromosomes. Most zygotes and embryos with chromosomal abnormalities do not survive beyond a few days—and thus the mother is not even aware that conception occurred. Of the pregnancies that last long enough to become aware of, between 15% and 20% are spontaneously aborted (miscarried)—with about one half of those having chromosomal abnormalities.

A few of the major chromosomal disorders are summarized here and in Table 34-1.

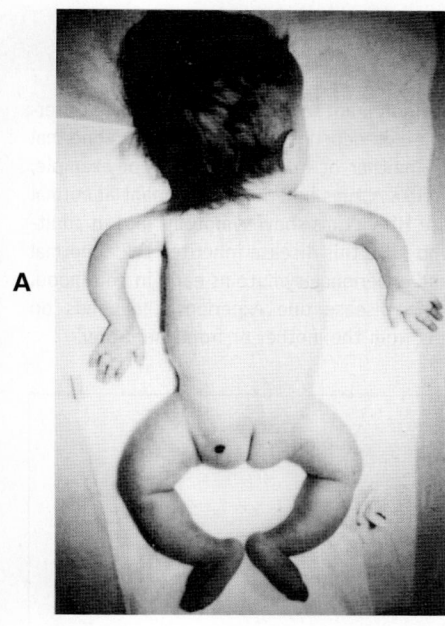

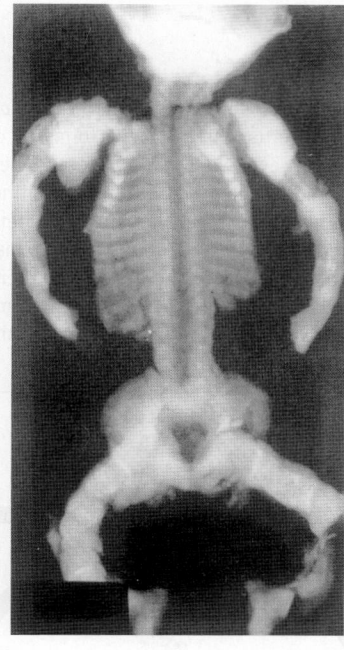

Figure 34-13 *Osteogenesis imperfecta.* **A,** The infantile form of this inherited disease is characterized by imperfect bone development that results in the appearance of this child: curved, brittle bones in the limbs and a thin, enlarged skull. **B,** This radiograph of a fetus with osteogenesis shows the many bone fractures that produce an accordian-like shortening of the limbs.

The most well-known chromosomal disorder is *trisomy 21, which produces a group of symptoms called* **Down syndrome.** As Figure 34-14, *A,* shows, in this condition there is a triplet of chromosomes 21 rather than the usual pair. In the general population, trisomy 21 occurs in only 1 of every 600 or so live births. After age 35 years, however, a mother's chances of producing a trisomic child increase dramatically—to as high as 1 in 80 births by age 40 years. One recent hypothesis that explains this phenomenon states that as a woman ages, her reproductive system becomes less likely to spontaneously abort abnormal blastocysts that have implanted in the endometrium. Thus nondisjunction may occur equally among young and middle-aged women.

Down syndrome results from trisomy 21 and rarely from other genetic abnormalities (which can be inherited). This syndrome is characterized by mental retardation (ranging from mild to severe) and multiple defects that include distinctive facial appearance (Figure 34-14, *B*), enlarged tongue, short hands and feet with stubby digits, congenital heart disease, and susceptibility to acute leukemia. People with Down syndrome have a shorter than average life expectancy but can survive to old age.

Klinefelter syndrome is another genetic disorder resulting from nondisjunction of chromosomes. This disorder occurs in males with a Y chromosome and at least two X chromosomes, typically the XXY genotype. Characteristics of Klinefelter syndrome include long legs, enlarged breasts, low intelligence, small testes, sterility, and chronic pulmonary disease (Figure 34-15).

Turner syndrome, sometimes called *XO syndrome,* occurs in females with a single sex chromosome, X. As with the conditions described previously, the syndrome results from nondisjunction during gamete formation. Turner syndrome is characterized by failure of the ovaries and other sex organs to mature (causing sterility), cardiovascular defects, dwarfism or short stature, a webbed neck, and possible learning disorders (Figure 34-16). Symptoms of Turner syndrome can be reduced by hormone ther-

apy using estrogens and growth hormone. Cardiovascular defects may be repaired surgically.

Genetic Basis of Cancer

Recall from Chapter 5 that some forms of cancer are thought to be caused, at least in part, by abnormal genes called **oncogenes.** Oncogenes are altered (mutated) forms of normal genes.

One hypothesis states that most normal cells contain such cancer-causing genes. However, it is uncertain how these genes become activated and produce cancer. Perhaps oncogenes can transform a cell into a cancer cell only when certain environmental conditions occur. It has also been shown that viruses can transmit oncogenes to human cells.

Another hypothesis states that normal cells contain another class of genes, sometimes called **tumor suppressor genes.** According to this hypothesis, such genes regulate cell division so that it proceeds normally. When a tumor suppressor gene is nonfunctional because of a genetic mutation, it then allows cells to divide abnormally (or fail to die)—possibly producing cancer.

Yet another possible genetic basis for cancer relates to the genes that govern the cell's ability to repair damaged DNA. For example, a rare genetic disorder called **xeroderma pigmentosum** is characterized by the inability of skin cells to repair genetic damage caused by the ultraviolet (UV) radiation in sunlight. Individuals with this condition nearly always develop skin cancer when exposed to direct sunlight. In this condition, the genetic abnormality does not cause skin cancer directly but inhibits the cell's cancer-preventing mechanisms.

These three hypotheses regarding the genetic basis of cancer are not mutually exclusive. They are undoubtedly all important factors in the genesis of cancer. Cancer researchers are now working intensely to determine the exact role various genes play in the development of cancer. The more we understand the genetic basis of cancer, the more likely it is that we will find effective treatments—or even cures.

BOX 34-3: FYI

Maternal and Paternal Genes

According to principles based on Mendel's work, inheritance of a dominant gene for a disease causes that disease regardless of whether the gene originally came from the mother or from the father. However, we know now that maternal genes can be temporarily imprinted or chemically marked differently from paternal genes. Inherited maternal imprints are changed to the paternal form when males produce spermatozoa. Likewise, all paternal imprints inherited by a woman are changed to maternal imprints when she produces ova. The concept of maternal and paternal imprinting is summarized in the figure.

Imprinting explains why some dominant inherited disorders have different forms in different people—it depends on whether the abnormal gene was inherited from the mother or from the father. For example, 90% of Huntington disease (HD) victims inherit the dominant abnormal gene from their fathers. These HD victims show symptoms late in adulthood. The remaining 10% who have this disease inherited the abnormal gene from their mothers—and experience symptoms early in childhood. Thus the form of Huntington disease one experiences depends on whether the abnormal gene is from the mother or from the father.

Ovum

Fertilized ovum

Sperm

Maternal imprint

Paternal imprint

Body cell of female

Inherited imprints affect development of the individual

Body cell of male

Primary oocyte

Original imprints are made

Primary spermatocyte

New imprints are made

New imprints are made

Maternal imprint on all ova

Paternal imprint on all sperm

Maternal imprint

Paternal imprint

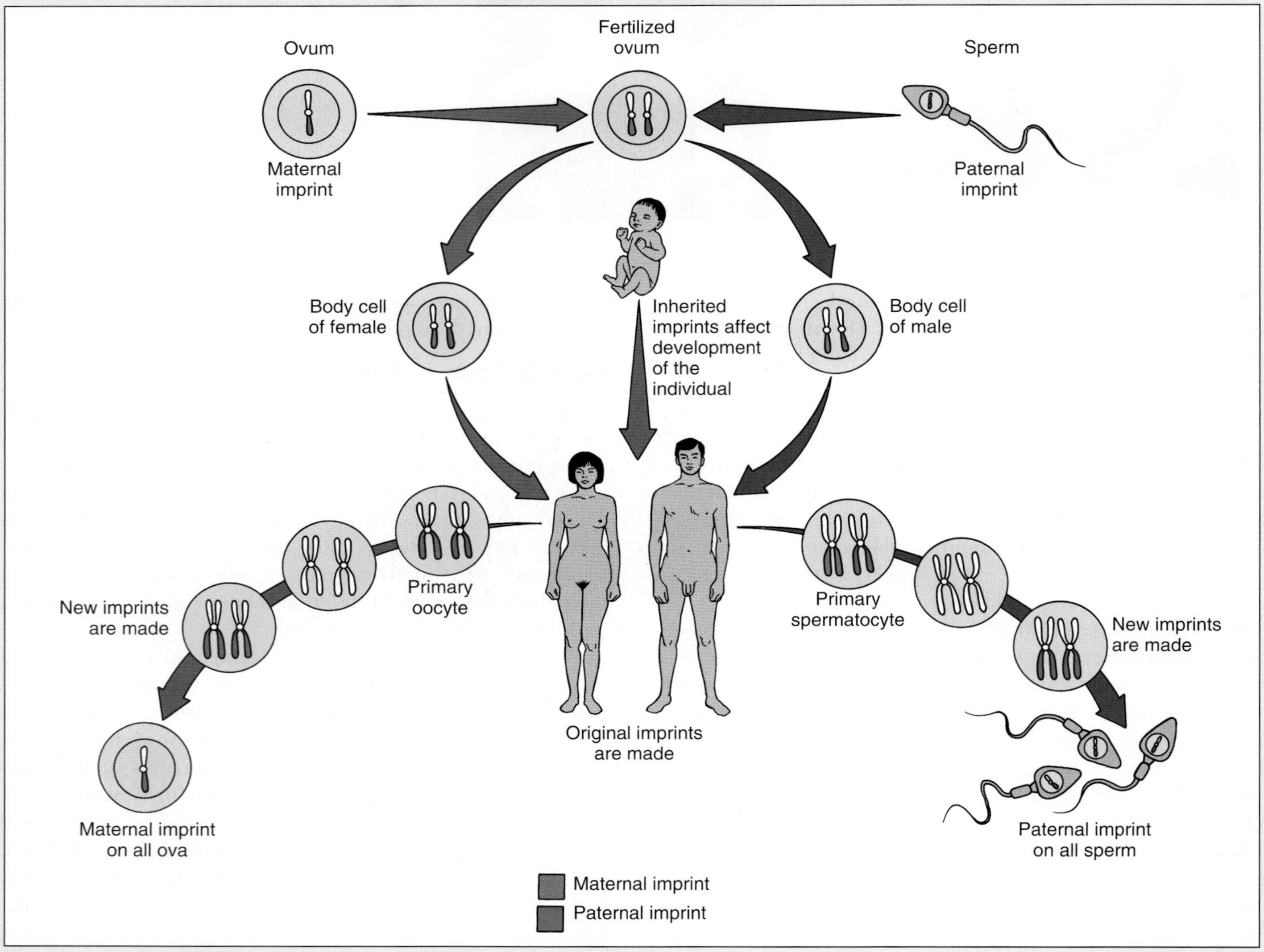

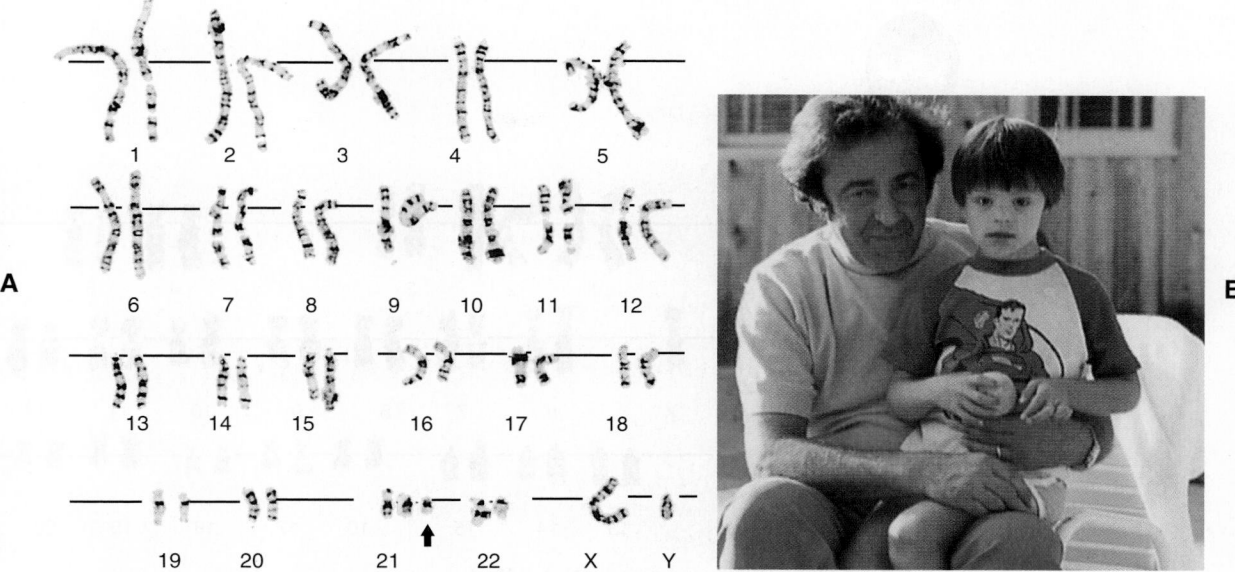

Figure 34-14 *Down syndrome.* A, Down syndrome is usually associated with trisomy of chromosome 21. **B,** A child with Down syndrome sitting on his father's knee. Notice the distinctive anatomical features: "Oriental" folds around the eyes, flattened nose, round face, and small hands with short fingers.

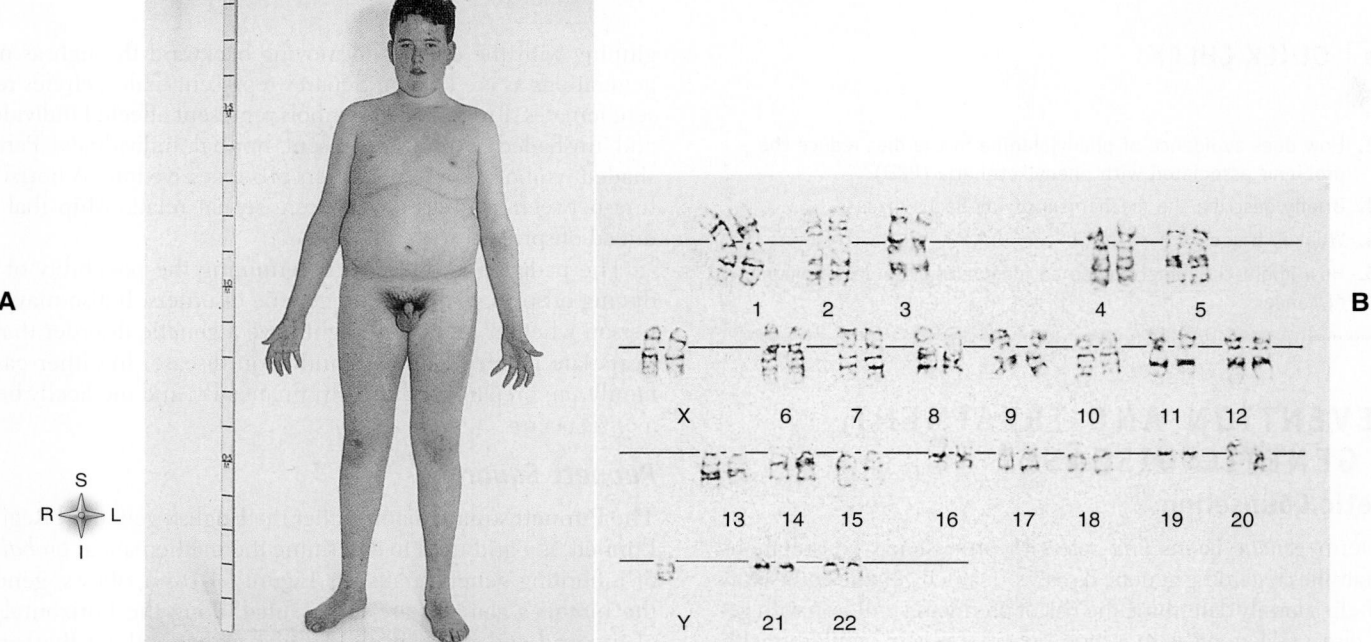

Figure 34-15 *Klinefelter syndrome.* A, This young man exhibits many of the characteristics of Klinefelter syndrome: small testes, some development of the breasts, sparse body hair, and long limbs. **B,** This syndrome results from the presence of two or more X chromosomes with a Y chromosome (genotypes XXY or XXXY, for example).

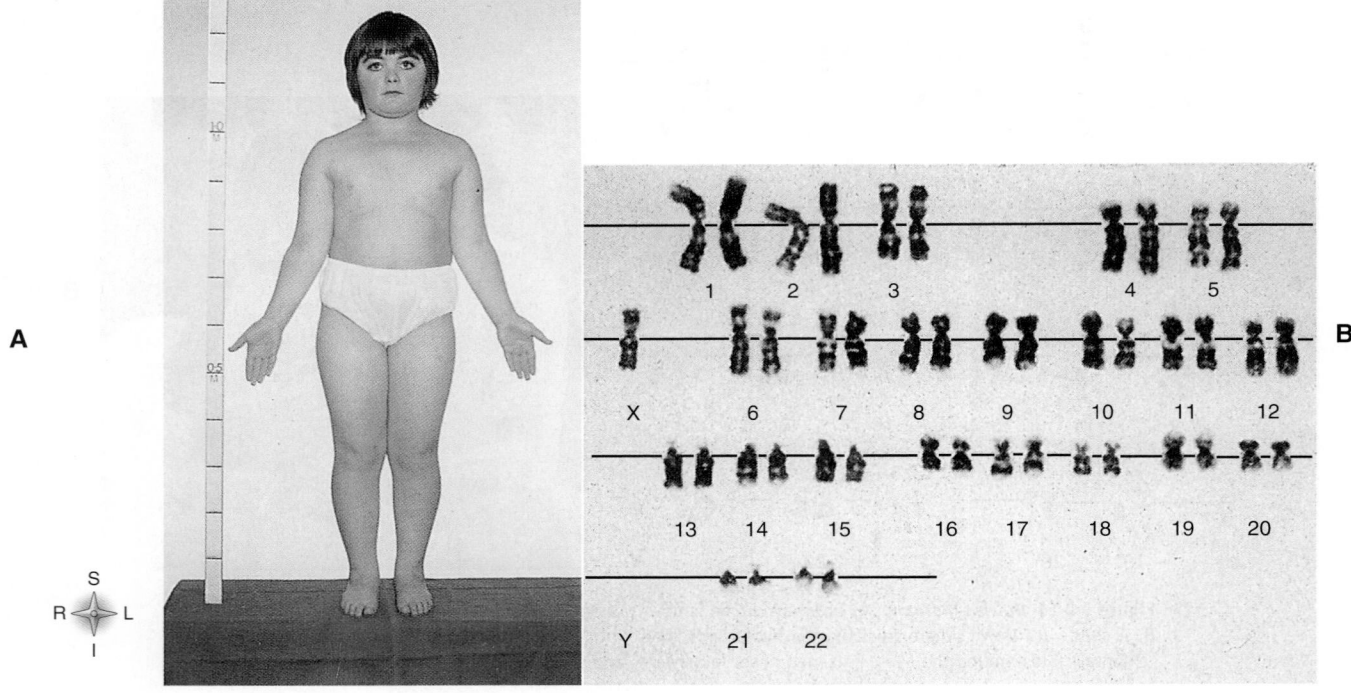

Figure 34-16 *Turner syndrome.* **A,** This woman exhibits many of the characteristics of Turner syndrome, including short stature, webbed neck, and sexual immaturity. **B,** As this karyotype shows, Turner syndrome results from monosomy of sex chromosomes (genotype X0).

QUICK CHECK

12. How does avoidance of phenylalanine in the diet reduce the problems associated with phenylketonuria (PKU)?
13. Briefly describe the mechanism of Tay-Sachs disease.
14. What is trisomy 21?
15. How might the genetic code be involved in the development of cancer?

PREVENTION AND TREATMENT OF GENETIC DISEASES

Genetic Counseling

The term *genetic counseling* refers to professional consultations with families regarding genetic diseases. Trained genetic counselors may help a family determine the risk of producing children with genetic diseases. Parents with a high risk of producing children with genetic disorders may decide to avoid having children of their own. Genetic counselors may also help evaluate whether any offspring already have a genetic disorder and offer advice on treatment or care. A growing list of tools is available to genetic counselors, some of which are described in the following section.

Pedigree

A **pedigree** is a chart that illustrates genetic relationships in a family over several generations (Figure 34-17). Using medical records and family histories, genetic counselors assemble the chart be-

ginning with the client and moving backward through as many generations as are known. Squares represent males; circles represent females. Fully shaded symbols represent affected individuals, and unshaded symbols represent normal individuals. Partially shaded symbols represent carriers of a recessive trait. A horizontal line between symbols designates a sexual relationship that produced offspring.

The pedigree is useful in determining the possibility of producing offspring with certain genetic disorders. It also may tell a person whether he or she might have a genetic disorder that appears late in life, such as Huntington disease. In either case, a family can prepare emotionally, financially, and medically before a crisis occurs.

Punnett Square

The **Punnett square,** named after the English geneticist Reginald Punnett, is a grid used to determine the mathematical *probability* of inheriting genetic traits. As Figure 34-18, A, shows, genes in the mother's gametes are represented along the horizontal axis of the grid and genes in the father's gametes along the vertical axis. The ratio of different gene combinations in the offspring predicts their probability of occurrence in the next generation. Thus offspring produced by two carriers of PKU (a recessive disorder) have a one in four (25%) chance of inheriting this recessive condition (see Figure 34-18, A).

The same grid shows that there is a two in four (50%) chance that a child produced will be a PKU carrier. Figure 34-18, B, however, shows that offspring of a carrier and a noncarrier cannot inherit PKU. What is the chance of an individual offspring being a PKU carrier in this case? The grid in Figure 34-18, C, shows the

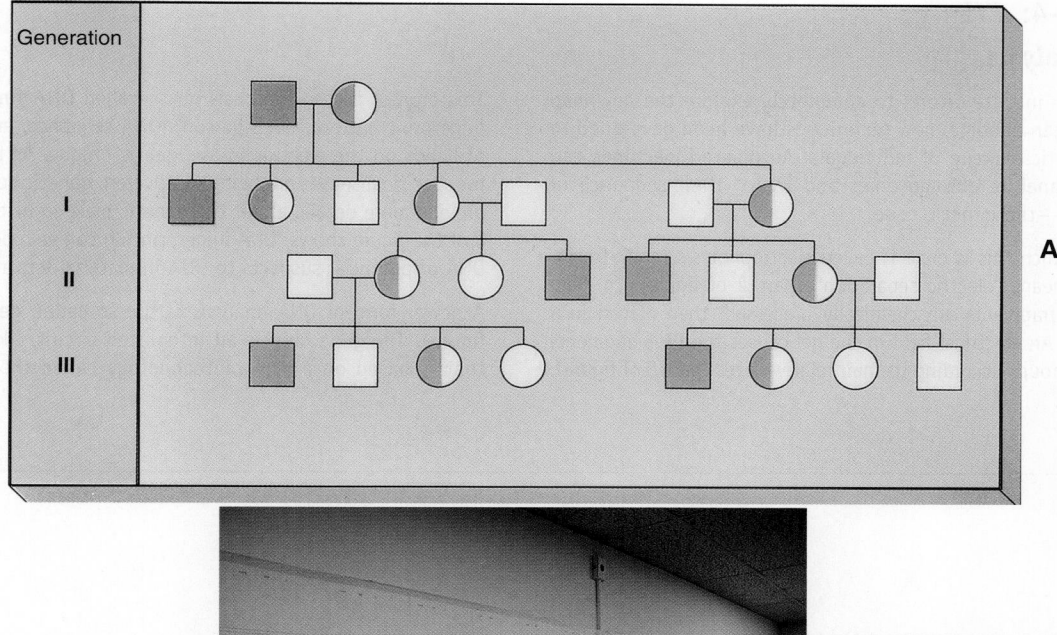

Figure 34-17 *Pedigree.* **A,** Pedigrees chart the genetic history of family lines. Squares represent males, and circles represent females. Fully shaded symbols indicate affected individuals, partly shaded symbols indicate carriers, and unshaded symbols indicate unaffected noncarriers. Roman numerals indicate the order of generations. This pedigree reveals the presence of an X-linked recessive trait. **B,** This huge pedigree compiled by Dr. Nancy Wexler of Columbia University and the Hereditary Disease Foundation traces Huntington disease (HD) over several generations of a large family in Venezuela. Dr. Wexler and other researchers collaborated in using information from this pedigree, along with techniques of molecular biology, to find the gene responsible for this disease and develop a test to detect the presence of the gene before the disease becomes apparent. Dr. Wexler became interested in this disease after her mother's death from HD.

probability of producing an affected offspring when a PKU victim and a PKU carrier have children. Figure 34-18, *D,* shows the genetic probability when a PKU victim and a noncarrier produce children.

Karyotype

Disorders that involve trisomy (extra chromosomes), monosomy (missing chromosomes), and broken chromosomes can be detected after a **karyotype** is produced.

The first step in producing a karyotype is getting a sample of cells from the individual to be tested. This can be done by scraping cells from the lining of the cheek or from a blood sample containing white blood cells (WBCs). Fetal tissue can be collected by **amniocentesis,** a procedure in which fetal cells floating in the amniotic fluid are collected with a syringe (Figure

34-19). **Chorionic villus sampling (CVS)** is a newer procedure in which cells from chorionic villi that surround a young embryo (see Chapter 33, p. 1171) are collected through the opening of the cervix.

Collected cells are grown in a special culture medium and allowed to reproduce. Cells in metaphase (when the chromosomes are most distinct) are stained and photographed using a microscope. The chromosomes are cut out of the photo and pasted on a chart in pairs according to size, as in Figures 34-2 and 34-15, *B.* More recent techniques use digital imaging and computers to automatically generate a karyotype.

Genetic counselors then examine the karyotype, looking for chromosome abnormalities. What chromosome abnormality is visible in Figure 34-15, *B?* Is this a male or female karyotype? Box 34-4 discusses *DNA analysis.*

BOX 34-4: FYI
DNA Analysis

As a result of the intense efforts to completely explore the new map of the entire human genome, new techniques have been developed to analyze the genetic makeup of individuals. Automated machines can now chemically analyze chromosomes and "read" their sequence of nucleotide bases—the genetic code.

One method by which this is done is called electrophoresis (el-ek-tro-fo-REE-sis), which means "electric separation" (part *A* of figure). In electrophoresis, DNA fragments are chemically processed, then placed in a thick fluid or gel. An electrical field in the gel causes the DNA fragments to separate into groups according to their relative sizes (part *B* of figure).

This process is also the basis for so-called **DNA fingerprinting.** Like a fingerprint pattern, each person's DNA sequence, and thus the pattern of bands on the electrophoresis gel, is unique. As the exact sequences for specific diseases are being discovered, genetic counselors are able to provide more details about the genetic makeup of their clients. As part *C* of the figure shows, DNA fingerprinting can also be used to match the DNA of potential suspects to DNA in material left at a crime scene.

A newer form of DNA analysis is the so-called **gene chip** (part *D* of figure). The gene chip is an integrated circuit, like a computer chip, that is based on a DNA biotechnology called the **microarray.** A mi-

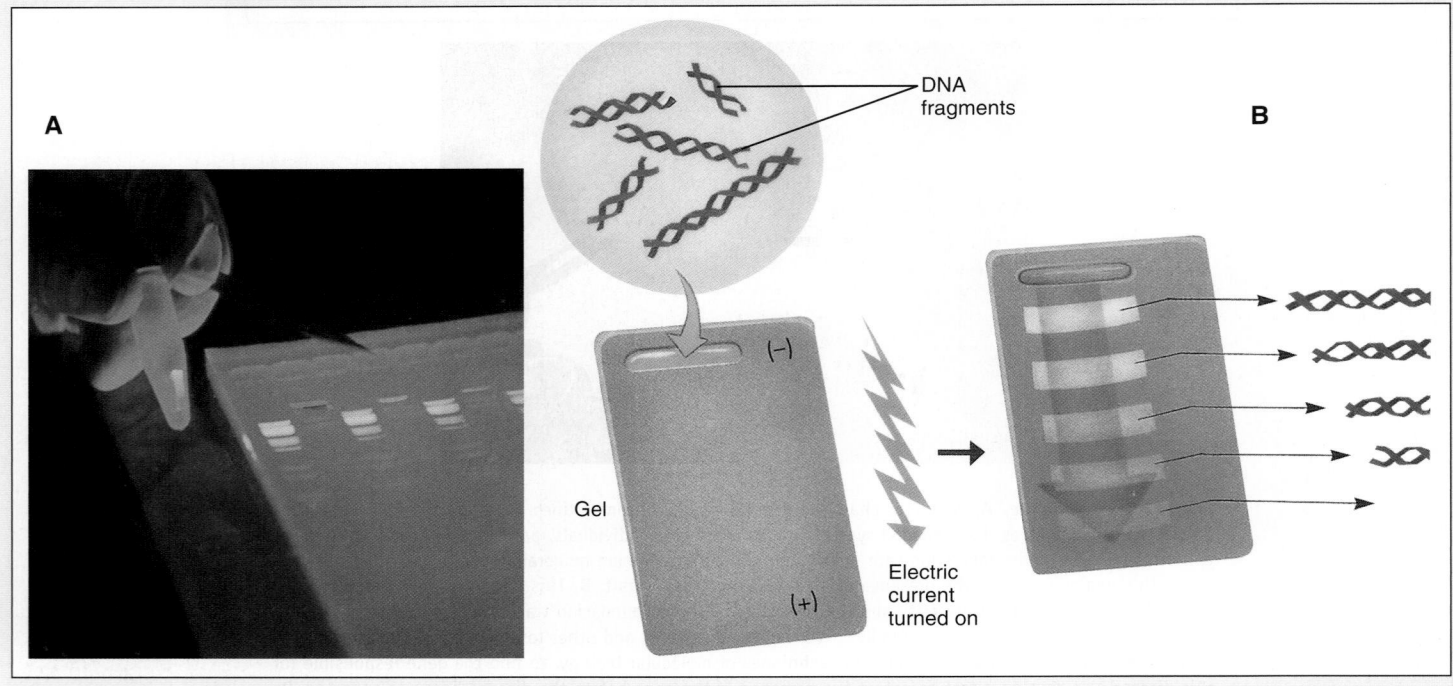

Electrophoresis and gene chip. **A,** Researcher examines bright bands on an electrophoresis gel. **B,** An electric field in the gel causes DNA fragments to separate into groups resembling bands according to their relative molecular sizes.

BOX 34-4: FYI—cont'd
DNA Analysis

croarray is a tiny silicon plate with a grid (array) made up of thousands of tiny wells. Each well contains a short, single-stranded segment of DNA or RNA. Each well contains a different, predetermined gene code segment. Then sample DNA or RNA from a subject is tagged with a fluorescent dye and added to the microarray. When subject DNA or RNA has a complementary code to the short strand in a particular well, it combines and becomes stuck in the well. The well then fluoresces brightly, but other wells where the DNA or RNA did not combine do not fluoresce (look at the background image in part D of figure). The bright spots are read by a scanner and a complete report

is generated. Thus, the presence or absence of certain genes or gene segments can be quickly discovered.

DNA chips like the one pictured may help in treating patients with drug therapies. A physician can screen a patient to see if he or she has the version of a gene that is likely to respond to a drug treatment—or screen out a patient likely to have adverse reactions to the drug. The presence of certain genes may help diagnose specific types of cancer cells. Of course, this technology may also be used to quickly screen for a large number of genetic disorders or disorders with genetic components.

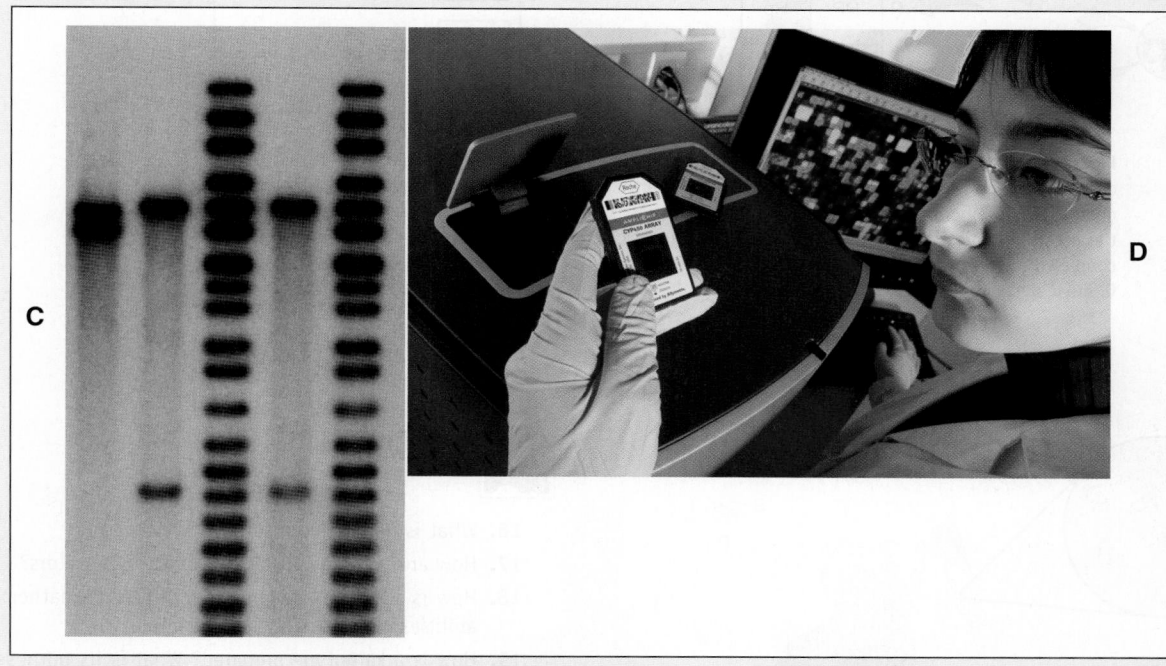

C, When using DNA fingerprinting in criminal justice, DNA from suspects is compared with DNA found at a crime scene. Here an autoradiograph of an electrophoresis gel shows that suspect B's DNA matches the DNA from the crime scene but suspect A's DNA does not. Several different fragments of DNA will be thus compared to reduce the chances of a false match. **D,** This gene chip is a prefabricated microarray that can quickly test for variations in two genes that affect how well a patient is likely to respond to about 25% of the drugs currently available for therapy.

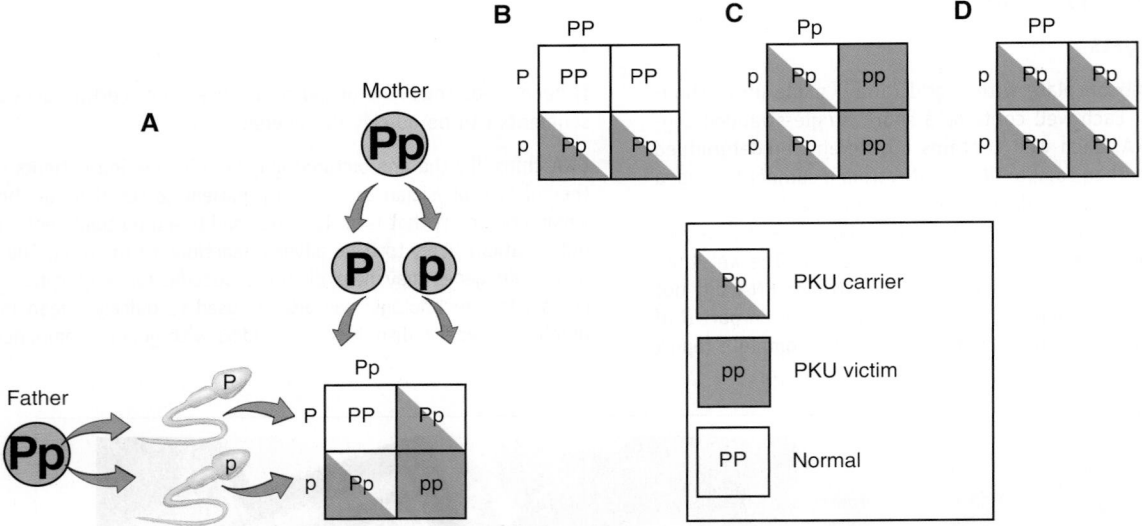

Figure 34-18 *Punnett square.* The Punnett square is a grid used to determine relative probabilities of producing offspring with specific gene combinations. Phenylketonuria (PKU) is a recessive disorder caused by the gene *p*. *P* is the normal gene. **A,** Possible results of cross between two PKU carriers. Because one in four of the offspring represented in the grid have PKU, a genetic counselor would predict a 25% chance that this couple will produce a PKU baby at each birth. **B,** Cross between a PKU carrier and a normal noncarrier. **C,** Cross between a PKU victim and a PKU carrier. **D,** Cross between a PKU victim and a normal noncarrier.

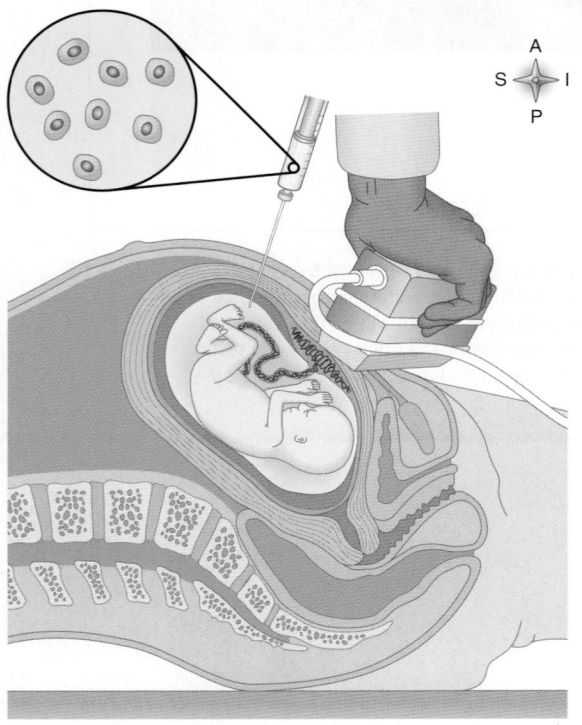

Figure 34-19 *Amniocentesis.* In amniocentesis, a syringe is used to collect amniotic fluid. Ultrasound imaging is used to guide the tip of the syringe needle to prevent damage to the placenta and fetus. Fetal cells in the collected amniotic fluid can then be chemically tested or used to produce a karyotype of the developing baby.

QUICK CHECK

16. What is genetic counseling?
17. How are pedigrees used by genetic counselors?
18. How is a Punnett square used to predict mathematical probabilities of inheriting specific genes?
19. How is a karyotype prepared? What is its purpose?

Treating Genetic Diseases

Introduction to Gene Therapy

Until fairly recently, the only hope of treating any genetic disease was to treat the symptoms. In some diseases, such as PKU, this works well. If PKU victims simply avoid large amounts of phenylalanine in their diets, especially during critical stages of development, severe complications can be avoided. In Klinefelter syndrome and Turner syndrome, hormone therapy and surgery can alleviate some symptoms. However, there are no effective treatments for the majority of genetic disorders. Fortunately, medical science now offers us some hope of treating genetic disorders through **gene therapy.**

In a therapy called **gene replacement,** genes that specify production of abnormal, disease-causing proteins are replaced by normal or "therapeutic" genes. To get the therapeutic genes to cells that need them, researchers are using genetically altered viruses as carriers. Recall that viruses are easily capable of inserting new genes into the human genome. If the therapeutic genes behave as expected, a cure may result. Thus the goal of gene replacement therapy is to genetically alter existing body cells in the hope of eliminating the cause of a genetic disease.

In a therapy called **gene augmentation,** normal genes are introduced with the hope that they will augment (add to) the production of the needed protein. In one form of gene augmentation, virus-altered cells are injected into the blood or implanted under the skin of a patient to produce increased amounts of the missing protein. Another approach is to use bacterial DNA rings called **plasmids** that have been altered by recombinant DNA techniques to carry the therapeutic gene or genes. A more recent approach uses the **human engineered chromosome (HEC).** In the HEC approach, a set of therapeutic genes is incorporated into a separate strand of DNA that is inserted into a cell's nucleus, thus acting like an extra, or forty-seventh, chromosome. Gene augmentation attempts to add genetically altered cells to the body, rather than change existing body cells, as in gene replacement therapy.

RNA interference (RNAi) may also become a weapon against genetic disorders (see Box 4-5 on p. 127). In one proposal, RNAi would be used to silence the abnormal huntingtin gene that causes Huntington disease (HD). In animal studies, RNAi successfully blocked a gene causing high blood cholesterol—without any apparent side effects.

Using Gene Therapy

The use of genetic therapy began in 1990 with a group of young children who had **adenosine deaminase deficiency.** In this rare recessive disorder, the gene for producing the enzyme *adenosine deaminase (ADA)* is missing from both autosomes in pair 20. Deficiency of ADA results in **severe combined immune deficiency (SCID),** making its victims highly susceptible to infection (see Chapter 21, p. 831). As Figure 34-20, A, shows, WBCs from each patient were collected and infected with viruses carrying therapeutic genes. After reproducing a thousand-fold, the genetically altered WBCs were then injected into the patient. Because this

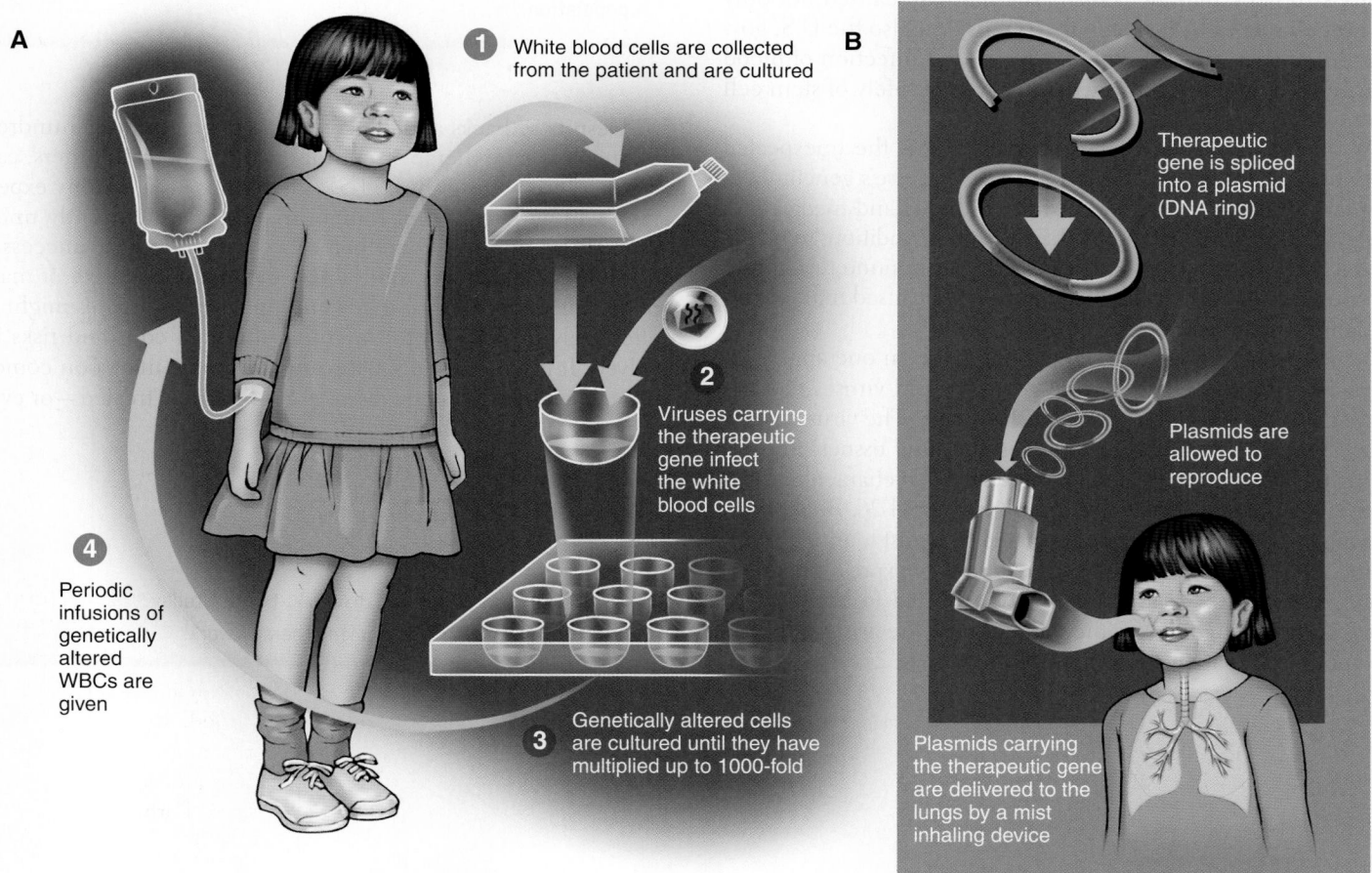

A

1 White blood cells are collected from the patient and are cultured

2 Viruses carrying the therapeutic gene infect the white blood cells

3 Genetically altered cells are cultured until they have multiplied up to 1000-fold

4 Periodic infusions of genetically altered WBCs are given

B

Therapeutic gene is spliced into a plasmid (DNA ring)

Plasmids are allowed to reproduce

Plasmids carrying the therapeutic gene are delivered to the lungs by a mist inhaling device

Figure 34-20 *Gene therapy.* **A,** Gene augmentation therapy used to treat children with severe combined immune deficiency (SCID). White blood cells taken from the patient are infected with viruses carrying the therapeutic *ADA* gene. Virus-altered cells are reproduced and injected into the bloodstream, reducing the effects of SCID. **B,** Gene augmentation therapy in this example uses plasmids containing the therapeutic gene for cystic fibrosis (CF) and delivers them to the lung tissues by means of a common inhaler.

treatment augments cells already present with genetically altered cells, it is a form of gene augmentation therapy.

The first person to receive gene therapy in 1990 for SCID was a girl named Ashanti De Silva, who is now a healthy adolescent. Although the gene therapy seems to have worked in saving her life, she and the others who were part of the initial trials also receive injections of the ADA enzyme encoded by the therapeutic gene. There continues to be debate as to whether success in such cases is from the gene therapy or the enzyme treatments—or both.

In 2000, a group of French researchers were able to successfully alter hemopoietic stem cells (ancestor cells of all new blood cells) in a group of babies with SCID. Because all new WBCs produced in these children have the therapeutic gene, they require no additional enzyme treatments. Although this "cure" was a milestone in gene therapy, researchers are still waiting to see whether the cells will really continue to retain the therapeutic gene as the children get older. Unfortunately, so far three of the children using this method of gene therapy later developed a leukemia-like disorder. Two of the children have already died. Of course, this has been a tragic setback to progress in gene therapy. This particular case caused not only the French researchers to stop their trials, but also the U.S. government to stop gene therapy involving viral infection of blood stem cells. Work is under way to improve the safety of stem cell therapy.

Another tragic setback for gene therapy was the unexpected death of 18-year-old Jesse Gelsinger in 1999. Jesse's genetic liver disorder was being treated successfully with diet and medication. However, Jesse wanted to help others with his condition, so he entered a gene therapy trial. Tragically, a hyperimmune reaction to the virus used to transfer the therapeutic gene caused multiple organ failure—and death.

Gene therapies for CF are also being tried. In one approach, researchers introduce therapeutic genes in cold viruses that are sprayed into the respiratory tract of CF victims. The viruses insert genes for normal chloride channels in the lung tissues, thus reducing the abnormally thick fluid secretions that characterize this disease. In another approach, shown in Figure 34-20, *B*, plasmids carrying the therapeutic gene are carried by a mist to lung tissues. There, the plasmids enter the cells where they express the therapeutic gene. These treatments are not expected to be one-time cures, but with ongoing use they should raise the average life expectancy of CF patients well beyond the current 27 years.

BOX 34-5: FYI
Genes and Longevity

How long can we expect to live? An average life span of 45 to 50 years was the norm in the United States 100 years ago. Today, Americans are approaching 75- to 80-year life spans on average, and the numbers are increasing. Current population studies suggest that centenarians—individuals 100 years of age or older—constitute the fastest growing segment of the U.S. population. Much of this shift has been attributed to environmental factors such as advances in medicine and improved diet. However, genes have a role to play in the aging process as well. Some geneticists believe that as many as 1000 genes will be shown to be involved in the process that we call aging. It is not surprising then that some believe that advances in genetic medicine may help reverse or inhibit some aging effects such as reduced muscle strength, arthritis, some forms of heart disease, and cancer. The discovery of a "longevity gene" on chromosome 4 that is found in many centenarians gives some scientists hope that we can someday actually use a form of gene therapy to lengthen the lives of some individuals. Death is a certainty for all of us, but remarkable and ongoing success in the science of aging and longevity is helping to ensure a longer and healthier life for growing numbers of our population.

Despite setbacks and concerns, there are currently hundreds of ongoing gene therapy trials for diverse genetic disorders, cancer, and even aging (Box 34-5). Thousands of laboratory experiments in anticipation of human trials are also currently under way. Hurdles to overcome before we see widespread success of gene therapies include our lack of detailed knowledge of many of the "disease genes" and how multiple-genes diseases might be effectively treated—not to mention the high costs and risks involved. It is too early to say for sure, but there may soon come a time when many genetic diseases are routinely treated—or even cured—with gene therapy.

QUICK CHECK

20. How are most genetic disorders treated today?
21. How does gene replacement therapy work?

THE BIG PICTURE
Genetics, Heredity, and the Whole Body

The study of human genetics and heredity provides an amazingly clear view of the "big picture" of the structure and function of the human body. In fact, the genetic view extends from the molecular level all the way to the context of the whole human species and beyond.

At the molecular level, the genetic perspective reveals the importance of nucleic acids—especially DNA. DNA molecules serve as tiny data storage devices that contain *all* the information needed to manufacture the molecules needed to build and maintain the human body. Each cell, tissue, organ, and system in the body is built of, and directed by, molecules that have been synthesized with recipes contained in the genes of each of our DNA molecules.

At the molecular level, we see that DNA in concert with various forms of RNA directs the synthesis of structural and functional proteins. Some of the functional proteins, in turn, trigger and direct the synthesis of all other biomolecules that form and regulate the body—lipids, carbohydrates, and so on.

This vast collection of different kinds of molecules also includes the enzymes and other molecules needed to gather and assimilate chemicals from the external environment outside the body. These external chemicals include water, oxygen, vitamins, minerals, sugars, starches, amino acids, fats, and other nutrients. Some of the molecules that ultimately owe their existence in the body to DNA also include those that help us rid the body of wastes such as urea, carbon dioxide, and bile.

If DNA directs all the chemical activity of the body, then it certainly directs cellular metabolism. And, of course, cellular metabolism is the foundation of the function of each tissue, organ, and system in the body. Our study of medical genetics has shown us that even one mistake in one codon of a gene can upset the chemistry of the body enough to shut down an entire system—and thus threaten the survival of the entire body.

LANGUAGE OF SCIENCE   *(Cont'd from page 1199)*

monogenic (mon-oh-JEN-ik) [*mono-* one, *-gen-* generate (gene), *-ic* pertaining to]

monosomy (MON-oh-so-mee) [*mono-* one, *-som-* body (chromosome), *-y* state of]

mutagens (MYOO-tah-jens) [*muta-* to change, *-gen* generate]

nucleosomes (NOO-klee-oh-sohms) [*nucleo-* kernel (nucleus), *-some* body]

p-arm [*p* petite (small), arm]

pedigree (PED-i-gree) [*pied de grue* crane's foot pattern]

phenotype (FEE-no-type) [*pheno-* appear, *-type* molded]

polygenic (pahl-ee-JEN-ik) [*poly-* many, *-gen-* generate, *-ic* pertaining to]

principle of independent assortment

principle of segregation

proteome (PRO-tee-ohm) [*prote-* protein, *-ome* entire collection]

Punnett square (PUN-it) [*Reginald C. Punnett* English geneticist]

q-arm [*q* follows p in Roman alphabet]

recessive genes [*recess* retreat, *-ive* pertaining to, *gene* generate]

sex chromosomes (KROH-moh-sohms) [*chromo-* color, *-some* body]

sex-linked traits

transcriptome (tran-SKRIPT-ome) [*trans-* cross from one to another, *-script* written document, *-ome* entire collection]

trisomy (TRY-so-mee) [*tri-* three, *-som-* body (chromosome), *-y* state of]

LANGUAGE OF MEDICINE

adenosine deaminase deficiency (ah-DEN-oh-seen dee-AM-i-nayse) [*adenosine* blend of adenine and ribose, *de-* remove, *-amin-* ammonia, *-ase* enzyme]

albinism (AL-bi-niz-em) [*alb-* white, *-in-* characterized by, *-ism* state of]

Alzheimer disease (AD) (AHLZ-hye-mer) [*Alois Alzheimer* German neurologist]

amniocentesis (AM-nee-oh-sen-TEE-sis) [*amnio-* birth membrane, *-centesis* a pricking]

anemia (ah-NEE-mee-ah) [*an-* without, *-emia* blood condition]

chorionic villus sampling (CVS) (koh-ree-ON-ik VIL-lus) [*chorion-* protective fetal membrane, *-ic* pertaining to, *villus* shaggy hair]

chromosomal genetic diseases (kroh-moh-SOH-mal jeh-NET-ik) [*chromo-* color, *-soma-* body, *-al* pertaining to, *gene-* generate, *-t-* combining form, *-ic* pertaining to, disease]

cystic fibrosis (CF) (SIS-tik fye-BROH-sis) [*cyst-* sac, *-ic* pertaining to *fibr-* fiber, *-osis* condition]

diabetes mellitus (DM) (dye-ah-BEE-teez MELL-i-tus) [*diabetes* passer through or siphon, *mellitus* honey-sweet]

DNA fingerprinting

Down syndrome (SIN-drohm) [*John L. Down* English physician]

Duchenne muscular dystrophy (DMD) (doo-SHEN MUSS-kyoo-lar DISS-troh-fee) [*Guillaume B. A. Duchenne de Boulogne* French neurologist, *dys-* bad, *-troph-* nourishment, *-y* state of]

gene augmentation (jeen awg-men-TAY-shun)

gene chip (jeen chip)

gene replacement

gene therapy (jeen THER-ah-pee)

hemophilia (hee-moh-FIL-ee-ah) [*hemo-* blood or blood vessels, *-philia* an affinity for]

human engineered chromosome (HEC)

Huntington disease (HD) (HUNT-ing-ton) [*George S. Huntington* American physician]

hypercholesterolemia (hye-per-koh-les-ter-ohl-EE-mee-ah) [*hyper-* excessive, *-chole-* bile, *-stero-* solid, *-ol-* alcohol, *-emia* blood condition]

ideogram (ID-ee-oh-gram) [*ideo-* idea, *-gram* written record]

Klinefelter syndrome (KLINE-fel-ter SIN-drohm) [*Harry F. Klinefelter* American physician]

malaria (mah-LAIR-ee-ah) [*mal-* bad, *-ar-* air, *-ia* condition]

LANGUAGE OF MEDICINE *(Cont'd)*

microarray (mye-kroh-ar-RAY) [*micro-* small, *-array* arrangement in lines]

multiple neurofibromatosis (noo-roh-fye-broh-mah-TOH-sis) [*neuro-* nerve, *-fibr-* fiber, *-oma-* tumor, *-t-* combining form, *-osis* condition]

nondisjunction (non-dis-JUNK-shun) [*non-* not, *-dis-* split in two, *-junction* joint]

oncogenes (ON-koh-jeens) [*onco-* swelling or mass (cancer)]

osteogenesis imperfecta (os-tee-oh-JEN-eh-sis im-per-FEK-tah) [*osteo-* bone, *-gen-* generate, *-esis* process of, *imperfecta* not perfect]

Parkinson disease (PARK-in-son) [*James Parkinson* English physician]

phenylketonuria (PKU) (fen-il-kee-toh-NOO-ree-ah) [*phen-* shining (phenol), *-yl-* chemical, *-keton-* acetone, *-ur-* urine, *-ia* condition]

plasmids (PLAS-mids) [*plasmid* something formed]

proteomics (proh-tee-OH-miks) [*prote-* protein, *-om-* entire collection, *-ic* pertaining to]

severe combined immune deficiency (SCID) (i-MYOON deh-FISH-en-see)

sickle cell anemia (SIK-ul sell ah-NEE-mee-ah) [*sicol* crescent, *cella* storeroom, *an-* without, *-emia* blood condition]

single-gene diseases

Tay-Sachs disease (TSD) (TAY-saks) [*Warren Tay* English ophthalmologist, *Bernard Sachs* American neurologist]

transcriptomics (tran-skript-OHM-iks) [*trans-* cross from one to another, *-script-* written document, *-om-* entire collection, *-ic* pertaining to]

tumor suppressor genes

Turner syndrome (TUR-ner SIN-drohm) [*Harry H. Turner* American endocrinologist]

xeroderma pigmentosum (zeer-oh-DER-mah pig-men-TOH-sum) [*xero-* dry, *-derma* skin, *pigmen-t* paint, *-osum* characterized by]

CASE STUDY

Baby boy Johnson is born to a 34-year-old woman. This is her first pregnancy. During a routine physical examination, the nurse in the delivery room notes that the infant has a small rounded skull with a flat occiput. His inner epicanthal fold and oblique palpebral fissures form an upward outward slant of his eyes. His nose is small with a depressed bridge. His tongue is fissured and protruding. He has short stubby hands and feet with a transverse crease on the palmar side of his hands (simian line), which is characteristic of Down syndrome. He has a heart murmur and subluxation of the left hip.

1. When discussing the infant's disorder with the mother, it would be important to include details about:

   A. Sex-linked traits
   B. Single-gene disorders
   C. Chromosomal disorders
   D. Oncogenes and related disorders

2. Another name for the disorder known as Down syndrome is:

   A. XO syndrome
   B. XXY syndrome
   C. Monosomy 21
   D. Trisomy 21

3. The infant's mother should be aware that her child will most likely have:

   A. Severe mental retardation
   B. Multiple defects
   C. Chronic lymphocytic leukemia
   D. Normal life expectancy

4. Ms. Johnson should also be aware in regard to future attempted pregnancies that Down syndrome is:

   A. A dominant inherited disorder
   B. A recessive inherited disorder
   C. The most well-known chromosomal disorder
   D. More prevalent among certain Jewish populations

CHAPTER SUMMARY

THE SCIENCE OF GENETICS

A. Genetics—scientific study of inheritance; developed to explain how normal biological characteristics are inherited
B. Directly inherited diseases are often called hereditary diseases

CHROMOSOMES AND GENES

A. Mechanism of gene function
   1. Genetic code is transmitted by way of genes, which are segments of DNA
   2. DNA (deoxyribonucleic acid) (Figure 34-1)
      a. Compact form of DNA is a chromosome—which exists only during cell division
      b. Strand form of DNA is chromatin—made up of subunits called nucleosomes, which are like small spools of DNA wound around proteins called histones
   3. Each gene is a sequence of nucleotide bases in the DNA molecule, which the cell transcribes to an RNA molecule
   4. Each mRNA molecule associates with a ribosome, which translates the code to form one or more specific polypeptide molecules
   5. Genes determine the structure and function of the human body by producing a set of specific regulatory RNA and protein molecules—along with specific structural proteins
B. The human genome (Figure 34-2)
   1. Genome—entire set of human chromosomes (46 in nucleus of each cell, 1 mitochondrial chromosome)

   a. Map of the entire human genome (nearly all nucleotides in sequence) was completed in 2003
   b. Contains about 20,000 to 25,000 genes and large amounts of noncoding DNA
   2. Genomics—analysis of the sequence contained in the genome
   3. Transcriptomics—analysis of the mRNA codes actually transcribed from genes in the genome
   4. Proteomics—analysis of the entire group of proteins encoded by the genome and transcriptome, a group of proteins called the human proteome
   5. Genomic information can be expressed in various ways
      a. Ideogram—cartoon of a chromosome showing the centromere as a constriction and the short segment (p-arm) and long segment (q-arm)
      b. Genes are often represented as their actual sequence of nucleotide bases expressed by the letters A, C, G, and T
C. Distribution of chromosomes to offspring
   1. Meiosis (Figure 34-3)
      a. Produces gametes with the haploid number of chromosomes
      b. When a sperm and an ovum unite at conception, they form a zygote with 46 chromosomes
   2. Principle of independent assortment
      a. As sperm and ovum are formed, chromosome pairs separate and the maternal and paternal chromosomes get mixed up and redistributed independently of the other

chromosome pairs, resulting in each gamete having a different set of 23 chromosomes
  b. Genetic variation—independent assortment of chromosomes ensures that each offspring from a single set of parents is genetically unique
  c. Applies to individual genes or groups of genes
  d. Crossing over—during one phase of meiosis, pairs of matching chromosomes line up along the equator and exchange genes with one another (Figure 34-4)
  e. Gene linkage—sometimes an entire group of genes stays together and crosses over as a single unit

GENE EXPRESSION

A. Hereditary traits
  1. Dominant and recessive traits
    a. Each inherited trait is controlled by two sets of similar genes, one from each parent
    b. Each autosome in a pair matches its partner in the type of gene it contains
    c. Different types of genes (Figures 34-5 and 34-6)
      (1) Dominant gene—effects are seen; capable of masking the effects of a recessive gene for the same trait
      (2) Recessive gene—effects are masked by the effects of a dominant gene for the same trait
    d. Genotype—combination of genes within the cells of an individual
      (1) Homozygous—genotype with two identical forms of a gene
      (2) Heterozygous—genotype with two different forms of a gene
    e. Phenotype—manner in which genotype is expressed; how an individual looks because of genotype
    f. Carrier—person who possesses the gene for a recessive trait but does not exhibit the trait
  2. Polygenic traits—when more than one gene is involved in producing a particular trait; a "combined trait" because it results from a combination of genes
  3. Codominant traits—when two different dominant genes occur together, each will have an equal effect
  4. Abnormal "disease" genes that persist in a population often provide some biological advantage, as in the case of the sickle gene that protects against malaria (Figure 34-7)
B. Sex-linked traits (Figures 34-8 and 34-9)
  1. X chromosome—"female chromosome"; larger than Y chromosome; includes genes that determine female sexual characteristics, as well as nonsexual characteristics (Figure 34-10)
  2. Y chromosome—"male chromosome"; smaller than X chromosome; contains few genes other than male sexual characteristics
  3. Sex-linked traits—traits carried on sex chromosomes; also known as X-linked traits
C. Genetic mutations
  1. Mutation—change in the genetic code
    a. Deletion—missing information in the genetic code
    b. Insertion—extra information in the genetic code
    c. Insertions and deletions result in a failure to make the usual protein encoded by a particular gene

  2. Mutations can occur without outside influence
  3. Mutagens—agents that cause most genetic mutations

MEDICAL GENETICS

A. Mechanisms of genetic disorders
  1. Nuclear inheritance
    a. Single-gene diseases—diseases caused by individual mutant genes in nuclear DNA that pass from one generation to the next
    b. Genetic predisposition—disease occurring because of combined effects of inheritance and environmental factors
    c. Chromosomal genetic diseases—congenital conditions such as trisomy and monosomy that produce life-threatening abnormalities; trisomic and monosomic individuals die before they can reproduce (Figure 34-11)
  2. Mitochondrial inheritance
    a. Mitochondrial DNA (mtDNA)—each mitochondrion has its own DNA molecule (Figure 34-12)
    b. Inheritance of mtDNA occurs through one's mother because sperm does not contribute mitochondria to the ovum during fertilization
    c. mtDNA contains the only genetic code for several important enzymes
B. Single-gene diseases
  1. Cystic fibrosis
    a. Results because of recessive genes in chromosome pair 7
    b. Impairment of chloride ion transport across cell membranes causes exocrine cells to secrete thick mucus and concentrated sweat; thickened mucus may obstruct respiratory and gastrointestinal tracts, leading to death
    c. Treatment—use of drugs and other therapies
  2. Phenylketonuria (PKU)
    a. Results from recessive genes that fail to produce phenylalanine hydroxylase
    b. Phenylalanine cannot be converted into tyrosine and thus accumulates
    c. High concentrations of phenylalanine destroy brain tissue
    d. Treatment—diets low in phenylalanine
  3. Tay-Sachs disease
    a. Recessive condition involving failure to make an essential lipid-processing enzyme
    b. Abnormal lipids accumulate in the brain, causing severe retardation and death by 4 years of age
    c. No specific therapy available
  4. Osteogenesis imperfecta (Figure 34-13)
    a. Dominant genetic disorder of connective tissues resulting in imperfect bone formation
    b. Bones are very brittle
  5. Multiple neurofibromatosis
    a. Dominant inherited disorder
    b. Characterized by multiple benign tumors of glial cells that surround nerve fibers

C. Chromosomal diseases—genetic disorders resulting from nondisjunction during formation of the gametes; produce either trisomy or monosomy
   1. Trisomy 21—triplet of chromosomes 21 rather than a pair; characterized by Down syndrome's mental retardation and multiple defects (Figure 34-14)
   2. Klinefelter syndrome occurs in males with a Y chromosome and at least two X chromosomes; characteristics include long legs, enlarged breasts, low intelligence, small testes, sterility, and chronic pulmonary disease (Figure 34-15)
   3. Turner syndrome—XO syndrome; occurs in females with a single X chromosome; characterized by failure of ovaries and other organs to mature, sterility, cardiovascular defects, dwarfism, webbed neck, and learning disorders; symptoms can be reduced by hormone therapy (Figure 34-16)
D. Genetic basis of cancer
   1. Oncogenes—abnormal genes thought to cause some forms of cancer
   2. Tumor suppressor genes regulate cell division so it proceeds normally; when nonfunctional because of a genetic mutation, it allows cells to divide abnormally
   3. Also, genetic abnormalities may inhibit the cell's cancer-preventing mechanisms

PREVENTION AND TREATMENT OF GENETIC DISEASES

A. Genetic counseling—professional consultations with families regarding genetic diseases
   1. Pedigree—chart that illustrates genetic relationships in a family over several generations; helpful in determining the possibility of producing offspring with certain genetic disorders (Figure 34-17)
   2. Punnett square—grid used to determine the mathematical probability of inheriting genetic traits (Figure 34-18)
   3. Karyotype—ordered arrangement of photographs of chromosomes from a single cell; used in genetic counseling to identify chromosomal disorders
B. Treating genetic diseases (Figure 34-20)
   1. Gene therapy involves changing the genetic code of cells to replace normal proteins that are absent in genetic disorders
   2. Gene replacement—abnormal, disease-causing proteins are replaced by "therapeutic" genes; goal is to genetically alter existing body cells in the hope of eliminating the cause of a genetic disease
   3. Gene augmentation—normal genes are introduced to augment the production of the needed protein

REVIEW QUESTIONS

1. Who was the first person to discover the basic mechanism by which traits are transmitted from parents to offspring?
2. Describe albinism in relation to dominance, recessiveness, and genotype.
3. Explain the difference between a genotype and a phenotype.
4. Define codominance, and give an example of a condition that demonstrates codominance.
5. If a certain trait is identified as X-linked recessive, describe the genotype of a female expressing the given trait.
6. Identify several genetic mutagens.
7. How can mutations be beneficial to a species?
8. What role do environmental factors play in relation to certain genetic diseases?
9. Explain the "mistake" in meiosis that results in the condition called trisomy.
10. Describe the genetic inheritance of cystic fibrosis, phenylketonuria (PKU), and Tay-Sachs disease.
11. Identify the chromosomal disorder that involves trisomy 21.
12. Differentiate between oncogenes and tumor suppressor genes.
13. How are gene replacement and gene augmentation therapies used to treat genetic diseases?
14. Define the term *genome*.
15. When is a disorder classified as congenital?

CRITICAL THINKING QUESTIONS

1. Genes regulate protein synthesis. Why is the production of specific proteins so important to the structure and function of the body?
2. Differentiate among the text usage of chromatin, chromosomes, and genes.
3. Explain the processes that increase the variability of the genetic code in the offspring.
4. Explain why the structure of the X and Y chromosomes would predict a greater occurrence of sex-linked disorders in males.
5. Explain what is meant by a pedigree. Explain what is meant by a karyotype. Which of these would be most helpful in the diagnosis of Down syndrome? In the diagnosis of Huntington disease?

Career Choices   Genetic Counselor

Genetic counseling is a health profession that blends the science of medical genetics with the human aspects of counseling families affected or at risk for genetic conditions or predispositions. I work in a private, not-for-profit medical facility. I see prenatal, pediatric, neurology, and cancer patients and their families. For specific details see information from our national support organization at http://www.nsgc.org/CareerInformation.asp#as_a_profession.

I have always had an interest in the medical professions. However, genetic counseling was first introduced to me in an "Ethics of Healthcare" course in my undergraduate studies. After this course sparked my interest, I used job-shadowing opportunities to learn more about what is involved in the field. As I learned more about it, I was convinced that genetic counseling was the profession that I wanted to pursue. I now hold a Master of Science in Human Genetics and Genetic Counseling.

The current trends in my field include positions in research and nontraditional areas, such as positions with insurance companies and clinical laboratories. The Human Genome Project will most likely generate the need for genetic counselors to be integrated into primary care services because we are learning about genetic contributions to many multifactorial conditions (e.g., heart disease, high blood pressure, and diabetes). Counselors can develop specialties in many areas, including prenatal, pediatric, oncology, and neurology clinics.

Christina Zaleski
MS

The variety of patients that I see, the challenges of each case, and the fact that I'm always learning are the greatest rewards of my job. Genetics is a rapidly evolving field. I find personal satisfaction helping patients and families through difficult times by helping to establish a diagnosis, spending time to explain what it may mean for them, making sure they have appropriate follow-up, and helping to find a support group or other resources for them. At times, I spend many hours working with one family, which is different from most other medical fields. We focus on not only the presenting problem, but also the implications to the family as a whole.

Understanding anatomy and physiology is helpful when working with a clinical geneticist to establish a diagnosis. Genetic conditions are often complex, multisystem disorders, and understanding anatomy and physiology is vital. I recommend using repetition, flash cards, and hands-on laboratory activities for studying anatomy and physiology.

GLOSSARY

A

abdominopelvic cavity (ab-DOM-i-no-PEL-vik KAV-i-tee) term used to describe the single cavity containing the abdominal and pelvic organs

abducens nerve (ab-DOO-sens) cranial nerve VI; motor nerve; controls movement of the eye and proprioception

abduction (ab-DUK-shun) moving away from the midline of the body, opposite motion of adduction

abruptio placentae (ab-RUP-shee-oh plah-SEN-tay) separation of normally positioned placenta from the uterine wall; may result in hemorrhage and death of the fetus and/or mother

abscess (AB-ses) cavity, often pus filled, formed by disintegration of tissues

absolute refractory period (AB-so-loot ree-FRAK-toh-ree) time during which the local area of the membrane has surpassed the threshold potential and will not respond to any stimulus

absorption (ab-SORP-shun) passage of a substance through a membrane such as skin or mucosa; often refers to passage of nutrients into blood

accessory gland (ak-SES-oh-ree) a gland that assists organs in accomplishing their functions

accessory nerve (ak-SES-oh-ree) cranial nerve XI; motor nerve

accessory organ ak-SES-oh-ree OR-gan) an organ that assists other organs in accomplishing their functions

accessory spleen (ak-SES-oh-ree) a small version of the spleen commonly found in the mesentery that connects the spleen and stomach (the gastrosplenic ligament); *see* **spleen**

accommodation (ah-kom-oh-DAY-shun) mechanism that allows the normal eye to focus on objects closer than 20 feet

acetylcholine (ACh) (ass-ee-til-KOH-leen) type of neurotransmitter used by motor neurons at neuromuscular junctions to stimulate muscle contraction at or in some autonomic synapses (in ganglionic synapses, at all parasympathetic effectors, and at some sympathetic effectors)

acetylcholinesterase (ass-ee-til-koh-lin-ES-ter-ase) enzyme that rapidly inactivates acetylcholine bound to postsynaptic receptors

acid substance that ionizes in water to release hydrogen ions; substance with a pH of less than 7.0

acidophil (ASS-id-oh-fil) cell that stains easily with acid dyes

acidosis (ass-i-DOH-sis) condition in which there is an excessive proportion of acid in the blood and thus an abnormally low blood pH); opposite of alkalosis

acini (ASS-i-nee) pancreatic cells that secrete serous fluid with digestive enzymes

acne (AK-nee) *see* **acne vulgaris**

acne vulgaris (AK-nee vul-GAR-is) inflammatory skin condition affecting sebaceous gland ducts; occurs most frequently in adolescent years; *see* **comedo**

acoustic nerve (ah-KOOS-tik) cranial nerve VIII; *see* **vestibulocochlear nerve**

acoustic neuroma (ah-KOOS-tik noo-ROH-mah) glial tumor of the Schwann cells surrounding cranial nerve VIII, causing progressive hearing loss and dizziness

acquired immunity (ah-KWYERD i-MYOO-ni-tee) immunity that is obtained after birth through exposure to a specific harmful agent; *see* **adaptive immunity**

acromegaly (ak-roh-MEG-ah-lee) condition caused by hypersecretion of growth hormone after puberty, resulting in enlargement of facial features (for example, jaw, nose), fingers, and toes

acrosome (AK-roh-sohm) specialized structure on the sperm head containing enzymes that break down the covering of the ovum to facilitate conception

actin (AK-tin) contractile protein found in the thin myofilaments of skeletal muscle; *see* **sliding-filament model**

action potential (AK-shun poh-TEN-shal) nerve impulse; membrane potential of an active neuron

active transport (AK-tiv TRANS-port) movement of a substance into or out of a living cell requiring the use of cellular energy

actual osmotic pressure *see* **osmotic pressure**

acuity (ah-KYOO-i-tee) sharpness of visual perception

acute (ah-KYOOT) intense; short in duration, as in acute disease

adaptation (ad-ap-TAY-shun) condition of many sensory receptors in which the magnitude of a receptor potential decreases over a period of time in response to a continuous stimulus

adaptive immunity (ah-DAP-tiv i-MYOO-ni-tee) type of immunity in which specific antigens or particles are recognized, then targeted for destruction; also called *acquired immunity* or *specific immunity*

adduction (ad-DUK-shun) moving toward the midline of the body, opposite motion of abduction

adenine (AD-eh-neen) one of the nitrogenous bases in RNA and DNA

adenocarcinoma (ad-eh-no-kar-si-NO-mah) cancer of glandular epithelium

adenofibroma (ad-eh-no-fye-BROH-mah) benign neoplasm formed in epithelial and connective tissues

adenohypophysis (ad-eh-no-hye-POF-i-sis) anterior pituitary gland, which has the structure of an endocrine gland

adenoid (AD-eh-noyd) literally, glandlike; adenoids, or enlarged pharyngeal tonsils, are paired lymphoid structures in the nasopharynx

adenoma (ad-eh-NO-mah) benign tumor of glandular epithelium

adenosine triphosphate (ATP) (ah-DEN-oh-seen try-FOS-fayt) chemical compound that provides energy for use by body cells

adenyl cyclase (AD-eh-nil SYE-klays) enzyme that promotes ATP change into cyclic AMP

adipose tissue (AD-i-pohs) fat tissue

adolescence (ad-oh-LESS-ens) period between puberty (the onset of reproductive maturity) and adulthood

adrenal (ah-DREE-nal) near the kidney, as in adrenal gland

adrenal cortex (ah-DREE-nal KOR-teks) outer portion of adrenal gland that secretes hormones called *corticoids*

adrenal gland (ah-DREE-nal) endocrine gland that rests on the top of each kidney; made up of cortex and medulla regions

adrenal medulla (ah-DREE-nal meh-DUL-ah) inner portion of adrenal gland that secretes epinephrine and norepinephrine

adrenaline (ah-DREN-ah-len) *see* **epinephrine**

adrenergic (ad-ren-ER-jik) describes axons whose terminals release norepinephrine and epinephrine

adrenocorticotropic hormone (ACTH) (ah-dree-no-kor-teh-koh-TROH-pic HOR-mohn) hormone that stimulates the adrenal cortex to secrete larger amounts of hormones

adulthood period after adolescence

adult respiratory distress syndrome (ARDS) (RES-per-ah-tohr-ee dis-TRESS SIN-drohm) syndrome resulting from impairment or removal of surfactant in the alveoli

aerobic respiration (air-OH-bik res-pi-RAY-shun) catabolic process; the stage of cellular respiration requiring oxygen

aerobic training (air-OH-bik) continuous vigorous exercise requiring the body to increase its consumption of oxygen and develop the muscles' ability to sustain activity over a long time; also known as *endurance training*

afferent division (AF-fer-ent) sensory division (incoming pathways) of the nervous system

afferent impulse (AF-fer-ent IM-pulse) impulse traveling toward the central nervous system

afferent nervous system (AF-fer-ent) subdivision of the peripheral nervous system; consists of all incoming sensory nerves

afferent neuron (AF-fer-ent NOO-ron) neuron that carries impulses toward the central nervous system from the periphery; sensory neuron

agglutinogen (ah-gloo-TIN-oh-jen) substance that stimulates agglutination, particularly of red blood cells; antigens present on red blood cell membranes

agranulocyte (ah-GRAN-yoo-loh-syte) white blood cells without cytoplasmic granules

alarm reaction the initial response to stress

albinism (AL-bi-niz-em) recessive, inherited condition characterized by a lack of the dark brown pigment melanin in the skin, eyes, and hair, resulting in vision problems and susceptibility to sunburn and skin cancer; ocular albinism is a lack of pigment in the layers of the eyeball

albumin (al-BYOO-min) plasma protein that aids in the regulation of the osmotic concentration of the blood

aldosterone (AL-doh-steh-rohn *or* al-DAH-stair-ohn) hormone that stimulates the kidney to retain sodium, ions, and water; only physiologically important mineralocorticoid

aldosteronism (al-doh-STAIR-on-iz-em) hypersecretion of aldosterone

alimentary canal (al-eh-MEN-tar-ee kah-NAL) the digestive tract as a whole

alkaline (AL-kah-lin) substance with a pH of greater than 7.0

alkalosis (al-kah-LOH-sis) condition in which there is an excessive proportion of alkali (base) in the blood; opposite of acidosis

allosteric effector (al-oh-STEER-ik ee-FEKT-or) an agent that alters the function of an enzyme by changing the shape of the enzyme's active site

alpha cell (AL-fah) pancreatic islet cell that secretes glucagon; also called A *cell*

alpha particle (AL-fah PAR-ti-kul) radioactive particle consisting of two protons plus two neutrons

alpha receptor (AL-fah ree-SEP-tor) adrenergic receptor for norepinephrine

alveolar cell (al-VEE-oh-lar) milk-producing cell that releases secretions into ducts of the breast

alveolar duct (al-VEE-oh-lar) airway that branches from the smallest bronchioles; alveolar sacs arise from alveolar ducts

alveolar ventilation (al-VEE-oh-lar ven-ti-LAY-shun) volume of inspired air that actually reaches the alveoli

alveolus (al-VEE-oh-lus) literally, a small cavity; alveoli of lungs are microscopic saclike dilations of terminal bronchioles; gas exchange in the lungs occurs across the membranes of the alveoli

Alzheimer disease (AD) (AHLZ-hye-mer) syndrome characterized by dementia, usually beginning in mid to late adulthood and caused by a combination of factors, including genetic and environmental

amenorrhea (ah-men-oh-REE-ah) absence of normal menstruation

amine (AM-een) organic compound containing nitrogen; neurotransmitter synthesized from amino acid molecules

amino acid (ah-MEE-no) structural units from which proteins are built

amino acid derivative hormone (ah-MEE-no ASS-id deh-RIV-ah-tiv HOR-mohn) category of nonsteroid hormones; each hormone is derived from a single amino acid molecule

amniocentesis (am-nee-oh-sen-TEE-sis) procedure in which a sample of amniotic fluid is removed with a syringe for use in genetic testing, often to produce a karyotype of the fetus

amniotic cavity (am-nee-OT-ik KAV-i-tee) cavity within the blastocyst that eventually becomes a fluid-filled sac in which the embryo will float during development

amphiarthrosis (am-fee-ar-THROH-sis) slightly movable joint such as the one that connects the two pubic bones

ampulla (am-PUL-ah) saclike dilation of a tube or duct; found at end of each semicircular canal, contains crista ampullaris

amygdaloid nucleus (ah-MIG-dah-loyd NOO-klee-us) basal nucleus (region of gray matter of the deep cerebrum) at the tip of the caudate nucleus

amylase (AM-eh-lays) enzyme that digests carbohydrates

anabolic hormone (an-ah-BOL-ik HOR-mohn) hormone that stimulates anabolism in the target organ

anabolism (ah-NAB-oh-liz-em) cells making complex molecules (for example, hormones) from simpler compounds (for example, amino acids); opposite of catabolism

anaerobic (an-air-OH-bik) metabolic process that does not require the presence of oxygen

anaerobic respiration (an-air-OH-bik res-pi-RAY-shun) catabolic process; stage of cellular respiration not requiring oxygen

anal triangle area surrounding anus

anaphase (AN-ah-fayz) latter stage of mitosis; duplicate chromosomes move to poles of dividing cell

anaplasia (an-ah-PLAY-zee-ah) growth of abnormal (undifferentiated) cells, as in a tumor or neoplasm

anastomosis (ah-nas-toh-MOH-sis) connection between vessels that allows collateral circulation

anatomical dead space (an-ah-TOM-i-kal) air passageways that contain air that does not reach the alveoli

anatomical position (an-ah-TOM-i-kal poh-ZISH-un) the standard reference position for the body; gives meaning to directional terms

anatomy (ah-NAT-oh-mee) study of the structure of an organism and the relationships of its parts

anemia (ah-NEE-mee-ah) deficient number of red blood cells, or deficient hemoglobin

anesthesia (an-es-THEE-zhah) state in which a person lacks the feeling of pain

anesthetic (an-es-THET-ik) substance that reduces or eliminates the sensation of pain

aneurysm (AN-yoo-riz-em) abnormal widening of the arterial wall; aneurysms promote formation of thrombi and also tend to burst

angina pectoris (an-JYE-nah PECK-tor-is) severe chest pain resulting when the myocardium is deprived of sufficient oxygen

angiogenesis (an-jee-oh-JEN-es-is) physiological process in which new blood vessels are formed

angioplasty (AN-jee-oh-plas-tee) medical procedure in which vessels occluded by arteriosclerosis are opened (that is, the channel for blood flow is widened)

angiotensin I (an-jee-oh-TEN-sin) formed by conversion of angiotensinogen by renin; causes vasoconstriction and an increase in blood pressure

angiotensin II (an-jee-oh-TEN-sin) formed in the lungs by enzyme conversion of angiotensin I; ultimately stimulates secretion of aldosterone; causes vasoconstriction

angiotensinogen (an-jee-oh-TEN-sin-oh-jen) normal constituent of blood, it is a precursor to angiotensin

anion (AN-eye-on) negatively charged molecule

annulus fibrosis (AN-yoo-lus fye-BROH-sis) tough outer edge of intervertebral disk; surrounds nucleus pulposus

anorexia (an-oh-REK-see-ah) loss of appetite (a symptom, rather than a distinct disorder)

anorexia nervosa (an-oh-REK-see-ah ner-VOH-sah) behavioral eating disorder characterized by chronic refusal to eat, often related to an abnormal fear of becoming obese

antagonism (an-TAG-oh-niz-em) action in which one hormone produces the opposite effect of another hormone

antagonist muscle (an-TAG-oh-nist) muscle that directly opposes prime movers; for example, muscles that flex the upper arm are antagonists to muscles that extend it

anterior (an-TEER-ee-or) front, or ventral; opposite of posterior, or dorsal

anterior cavity (of eye) (an-TEER-ee-or) entire space located in front of the lens of the eye; divided into anterior and posterior chambers

anthrax (AN-thraks) bacterial infection caused by *Bacillus anthracis*, ordinarily affecting herbivores (sheep, cattle, goats, antelope) and often killing them; rarely occurs in humans through accidental or intentional exposure to bacterial spores through inhalation or skin contact; inhalation anthrax is life-threatening but can be treated successfully with drugs; cutaneous anthrax is less serious, characterized by a reddish-brown patch on the skin that ulcerates and then forms a dark, nearly black scab, followed by muscle pain, internal hemorrhage (bleeding), headache, fever, nausea, and vomiting

antiangiogenesis agents (AN-tee-AN-jee-oh-jen-es-is) class of chemotherapy drugs used in cancer treatment that inhibits development of new blood vessels in a tumor; *see* **rational drugs**

antibody (AN-ti-bod-ee) substance produced by the body that destroys or inactivates a specific substance (antigen) that has entered the body

antibody-mediated immunity (AN-ti-bod-ee MEE-dee-ayt-ed i-MYOO-ni-tee) immunity that is produced when antibodies render antigens unable to harm the body

antidepressant (an-tee-deh-PRESS-ant) drug that inhibits feelings of depression or sadness

antidiuresis (an-tee-dye-yoo-REE-sis) opposing the production of a large urine volume

antidiuretic hormone (ADH) (an-tee-dye-yoo-RET-ik HOR-mohn) hormone produced in the posterior pituitary gland to regulate the balance of water in the body by accelerating reabsorption of water in the kidney tubules

antigen (AN-ti-jen) substance, usually a protein fragment, that causes an immune response

antigen-presenting cell (APC) any of a variety of immune cells that present protein fragments (antigens) on their surface and thus allow recognition and reaction by other immune system cells; include macrophages (especially the dendritic cells, or DCs) and B cells

aortic baroreceptor (ay-OR-tik bar-oh-ree-SEP-tor) stretch receptor located in the aorta that is sensitive to changes in blood pressure

aortic body (ay-OR-tik BOD-ee) small cluster of chemosensitive cells that respond to changes in blood levels of carbon dioxide and oxygen

aortic semilunar valve (ay-OR-tic sem-i-LOO-nar) valve between the aorta and left ventricle that prevents blood from flowing back into the ventricle

apatite (AP-ah-tyte) crystals of calcium and phosphate that contribute to the hardness of bone

apex (AY-peks) the point or tip of a structure; *see also* **apical**

aphasia (ah-FAY-zee-ah) language deficit caused by damage in speech centers of the brain

apical (AY-pik-al) pertaining to the apex (tip) of an organ, cell, or other structure; in a cell, often refers to the surface facing the lumen of the organ

apnea (AP-nee-ah) temporary cessation of breathing

apneusis (ap-NYOO-sis) cessation of breathing in the inspiratory position

apneustic center (ap-NYOO-stik) located in the pons; stimulates the inspiratory center to increase the length and depth of inspiration

apocrine glands (AP-oh-krin) glands that collect their secretions near the apex of the cell and then release them by pinching off the distended end; for example, mammary glands

apocrine sweat glands (AP-oh-krin) sweat glands located in the axillae and genital regions; these glands enlarge and begin to function at puberty

aponeurosis (ap-oh-nyoo-ROH-sis) broad, flat sheet of connective tissue

apoptosis (ap-op-TOH-sis) nonpathological, programmed cell death in which specific biochemical steps within the cell lead to the fragmentation of the cell and removal of the pieces by phagocytic cells; occurs when cells are no longer needed and is the normal process by which our tissues remodel themselves

appendicitis (ah-pen-di-SYE-tis) inflammation of the vermiform appendix

appendicular skeleton (ah-pen-DIK-yoo-lar SKEL-eh-ton) bones of the upper and lower extremities of the body

appetite center cluster of neurons in the lateral hypothalamus whose impulses cause an increase in appetite

appositional growth (ap-oh-ZISH-un-al) process by which a flat bone or cartilage grows in size by addition of bony cartilage at its surface

aqueous humor (AY-kwee-us HYOO-mor) watery fluid that fills the anterior chamber of the eye, in front of the lens

arachnoid (ah-RAK-noyd) weblike; particularly the middle layer of the meninges

arbor vitae (AR-bor VI-tay) internal white matter of the cerebellum

areolar (ah-REE-oh-lar) a type of connective tissue consisting of fibers and a variety of cells embedded in a loose matrix of soft, sticky gel

arrector pili muscle (ah-REK-tor PYE-lye) smooth muscles of the skin, attached to hair follicles; when contraction occurs, the hair stands up, resulting in "goose flesh"

arrhythmia (ah-RITH-mee-ah) term referring to any abnormality of cardiac rhythm

arteriole (ar-TEER-ee-ohl) small branch of an artery

arteriosclerosis (ar-tee-ree-oh-skleh-ROH-sis) hardening of arteries; materials such as lipids (as in atherosclerosis) accumulate in arterial walls, often becoming hardened via calcification

artery (AR-ter-ee) vessel carrying blood away from the heart

arthritis (ar-THRY-tis) inflammatory joint disease

articular cartilage (ar-TIK-yoo-lar KAR-ti-lij) layer of hyaline cartilage covering the joint surfaces of epiphyses

articulation (ar-tik-yoo-LAY-shun) joint

artificial immunity (ar-ti-FISH-al i-MYOO-ni-tee) immunization; deliberate exposure to potentially harmful antigens; as opposed to natural immunity

artificial pacemaker (ar-ti-FISH-al PAYS-may-ker) an electrical device that is implanted into the heart to treat problems with heart conduction

ascending tract (ah-SEND-ing) spinal cord tract that conducts impulses up the cord to the brain

ascites (ah-SYE-tees) effusion in the abdominal cavity; abdominal bloating

association tract (ah-so-see-AY-shun) most common cerebral tract; extends from one convolution to another in the same hemisphere

asthma (AZ-mah) obstructive pulmonary disorder characterized by recurring spasms of muscles in bronchial walls accompanied by edema and mucus production, making breathing difficult

astigmatism (ah-STIG-mah-tiz-em) irregular curvature of the cornea or lens that impairs formation of a well-focused image in the eye

astrocyte (ASS-troh-syte) star-shaped neuroglial cell

astrocytoma (ass-troh-sye-TOH-mah) slow-growing neoplasm of astrocyte cells in the central nervous system

atherosclerosis (ath-er-oh-skleh-ROH-sis) type of "hardening of the arteries" in which lipids and other substances build up on the inside wall of blood vessels

atom (AT-om) smallest particle of a chemical element that retains the properties of that element; particles that combine to form molecules (chemical building blocks)

atomic number (ah-TOM-ik) number of protons in an atom of an element

atomic weight (ah-TOM-ik) number of protons plus the number of neutrons in an atom of an element

atrial fibrillation (AY-tree-al fi-bri-LAY-shun) frequent premature contractions of the atrium

atrioventricular (AV) bundle (ay-tree-oh-ven-TRIK-yoo-lar) bundle of specialized cardiac muscle fibers that extend from the AV node to the Purkinje fibers; involved in coordination of heart muscle contraction

atrioventricular (AV) node (ay-tree-oh-ven-TRIK-yoo-lar) a small mass of special cardiac muscle tissue; part of the conduction system of the heart

atrioventricular (AV) valves (ay-tree-oh-ven-TRIK-yoo-lar) two valves that separate the atrial chambers from the ventricles

atrium (AY-tree-um) one of the upper chambers of the heart; receives blood from either the systemic or pulmonary circulation

atrophy (AT-roh-fee) wasting away of tissue; decrease in size of a part; sometimes referred to as *disuse atrophy*

auditory nerve (AW-di-toh-ree) cranial nerve VIII; *see* **vestibulocochlear nerve**

auditory ossicles (AW-di-toh-ree OS-si-kuls) tiny bones in middle ear: malleus, incus, and stapes; they function to amplify sound waves passing from the eardrum to the membranes of inner ear

auditory tube (AW-di-toh-ree) tube that connects the throat with the middle ear and equalizes pressure between the middle ear and the exterior; also known as *eustachian tube*

auricle (AW-ri-kul) part of the ear attached to the side of the head; earlike appendage of each atrium of the heart

autoimmunity (aw-toh-i-MYOO-nit-ee) immune system reaction against self-antigens

autonomic effector (aw-toh-NOM-ik ef-FEK-tor) tissues to which efferent (motor) autonomic neurons conduct impulses

autonomic nervous system (ANS) (aw-toh-NOM-ik) division of the nervous system that monitors and regulates subconscious (involuntary) functions

autopoiesis (aw-toh-poy-EE-sis) concept of self-organization and self-maintenance as a characteristic of all living organisms

autoregulation (aw-toh-reg-yoo-LAY-shun) *see* **intrinsic control**

autosome (AW-toh-sohm) one of the 44 (22 pairs) chromosomes in the human genome besides the two sex chromosomes; means "same body," referring to the fact that each member of a pair of autosomes matches the other in size and other structural features

avascular (ah-VAS-kyoo-lar) free of blood vessels

avitaminosis (ay-vye-tah-mi-NO-sis) general name for any condition resulting from a vitamin deficiency

axial skeleton (AK-see-al SKEL-eh-ton) the bones of the head, neck, and torso

axillary tail of Spence (AK-si-lar-ee tail of Spens) an extension of breast tissue that is in physical contact with several very large lymph nodes

axon (AK-son) in a neuron, the single process that extends from the axon hillock and transmits impulses away from the cell body

axonal transport (AK-soh-nal trans-PORT) process of transporting vesicles, small organelles, and other structures along pathways inside the axon of a neuron

axon collateral (AK-son koh-LAT-er-al) one or more side branches from the axon

axon hillock (AK-son HILL-ok) portion of the cell body from which the axon extends

B

B lymphocyte (B LIM-foh-syte) immune system cell that produces antibodies against specific antigens

bacterium (bak-TEE-ree-um) microorganism capable of causing disease; a primitive, single-celled organism without membranous organelles; plural *bacteria*

ball-and-socket joint spheroid joint, such as shoulder or hip joint; most movable type of joint

balloon angioplasty (AN-jee-oh-plas-tee) medical procedure in which vessels occluded by arteriosclerosis are opened (that is, the channel for blood flow is widened) by use of a balloon at the tip of a catheter (tube) inside the affected vessel; as the balloon is inflated, the blockage is pushed aside; *see* **angioplasty**

baroreflex (bar-oh-REE-fleks) baroreceptors in blood vessels, operating in feedback loops to maintain homeostasis of blood pressure; also called *pressoreflex*

barrel cell organelle that resembles a tiny capsule and is thought to shuttle substances from place to place within a cell; also called *vault*

basal (BAY-sal) pertaining to the base or widest part of an organ or other structure; in a cell, pertains to the surface facing away from the lumen of an organ

basal cell carcinoma (BAY-sal cell kar-si-NO-mah) one of the most common forms of skin cancer; usually occurs on the upper face

basal ganglia (BAY-sal GANG-glee-ah) islands of gray matter located in the cerebrum of the brain that are responsible for automatic movements and postures; now known more properly as *basal nuclei*

basal lamina (BAY-sal LAM-i-nah) glycoprotein material secreted by epithelial cells

basal metabolic rate (BMR) (BAY-sal met-ah-BOL-ik rayt) number of calories of heat that must be produced per hour by catabolism to keep the body alive, awake, and comfortably warm

basal nuclei (BAY-sal NOO-klee-eye) islands of gray matter located in the cerebrum of the brain that are responsible for automatic movements and postures; formerly known as *basal ganglia*; also known as *cerebral nuclei*

base substance that ionizes in water to decrease the number of hydrogen ions; also known as *alkaline*

basement membrane the connective tissue layer of the serous membrane that holds and supports epithelial cells

base pair adenine-thymine or cytosine-guanine; occurs when complementary bases from each helical chain of DNA are held together by hydrogen bonds

basilar membrane (BAYS-i-lar) floor of the cochlear duct, this structure's vibrating response to sound frequencies leads to the transduction of sound waves into nerve impulses

basophil (BAY-soh-fil) white blood cell that stains readily with basic (alkaline) dyes

benign (bee-NYNE) refers to a tumor or neoplasm that does not metastasize (spread to different tissues)

beta blocker (BAY-tah) drug that blocks beta receptors and therefore prevents dilation of blood vessels and increased contraction of heart muscle

beta cell (BAY-tah) pancreatic islet cell that secretes insulin; also called *B cell*

beta particle (BAY-tah) electrons formed in a radioactive atom's nucleus by a neutron breaking down into a proton and an electron

beta receptor (BAY-tah ree-SEP-tor) adrenergic receptor that, when stimulated, causes vessels to dilate and heart muscle to contract faster and stronger

bicarbonate ion (bye-KAR-boh-nayte EYE-on) HCO_3^-; an ion that serves an important role in maintaining normal blood pH

bilateral symmetry (bye-LAT-er-al SIM-eh-tree) concept of the right and left sides of the body being approximate mirror images of each other

biochemistry (bye-oh-KEM-is-tree) science of chemistry of living organisms

biofeedback (bye-oh-FEED-bak) method of learning to consciously control autonomic effectors

biological clock (bye-oh-LOJ-i-kal) internal timing mechanism that governs hunger, sleeping, reproduction, and behavior

biopsy (BYE-op-see) procedure in which living tissue is removed from a patient for laboratory examination, as in determining the presence of cancer cells

bioterrorism (bye-oh-TAIR-or-iz-em) act of using biological weapons such as disease pathogens to spread terror among civilian populations for the purpose of creating fear, panic, and general social disruption

Biot's breathing (bee-OHS) breathing pattern characterized by repeated sequences of deep gasps and apnea due to increased intracranial pressure

bipolar neuron (bye-POH-lar NOO-ron) neuron with only one dendrite and only one axon

blackhead sebum that accumulates, darkens, and enlarges a duct of the sebaceous glands, as in the case of acne; also known as a *comedo*

blastocyst (BLAS-toh-sist) stage of developing embryo that implants in uterine wall; consists of hollow ball of cells plus an inner cell mass

blind spot point on retina where blood vessels and nerves exit the eyeball; "blind" portion of visual field resulting from no photoreceptors present in that portion of retina; also known as *optic disc*

blister (BLIS-ter) fluid-filled skin lesion; *see* **vesicle**

blood-brain barrier structural and functional barrier formed by astrocytes and blood vessel walls in the brain; it prevents some substances from diffusing from the blood into brain tissue

blood coagulation (koh-ag-yoo-LAY-shun) process by which ruptured vessels are plugged up to stop bleeding and prevent loss of precious body fluids

blood serum (SEE-rum) in a sample of blood, the pale yellowish liquid left after a clot forms

blood type one of the four different types of blood; identified by certain antigens on red blood cells (A, B, AB, and O)

body composition percentages of the body made of lean tissue and fat tissue

body planes imagined flat surfaces that cut through the body at any of various angles; *see* **sagittal plane, coronal plane, transverse plane**

Bohr effect (BOR ee-FECT) when increased P_{CO_2} decreases the affinity between hemoglobin and oxygen

bolus (BOW-lus) rounded mass of food that is swallowed

bone highly specialized connective tissue whose matrix is hard and calcified

bone marking specific feature on an individual bone

bone-seeking isotope (EYE-soh-tohp) radioactive element that will substitute for calcium in apatite crystals of bone; causes damage to red marrow and other tissues by radioactive emissions

bovine spongiform encephalopathy (BOH-vyne SPUNJ-i-form en-SEF-al-ah-path-ee) also known as *BSE* or "mad cow disease"; a degenerative disease of the central nervous system caused by prions that convert normal proteins of the nervous system into abnormal proteins, causing loss of nervous system function. The abnormal form of the protein also may be inherited; *see* **prion**

Bowman's capsule (BOH-mens KAP-sul) in the kidney, the cup-shaped top of a nephron that surrounds the glomerulus

brachial plexus (BRAY-kee-al PLEK-sus) plexus located deep in the shoulder that innervates the lower part of the shoulder and the entire arm

bradycardia (bray-dee-KAR-dee-ah) slow heart rhythm (below 50 beats/min)

brain part of central nervous system contained within the cranium; consists of medulla oblongata, cerebellum, and cerebrum

brainstem part of brain containing the midbrain, pons, and medulla oblongata

brain waves fluctuating electrical activity occurring in the brain

bronchial tree (BRONG-kee-al) the trachea, two primary bronchi, and their many branches

bronchiole (BRONG-kee-ol) small branch of a bronchus

bronchitis (brong-KYE-tis) inflammation of the bronchi of the lungs, characterized by edema and excessive mucus production that causes coughing and difficulty in breathing (especially expiration)

brush border lining of small intestine that resembles bristles of a brush; formed by microvilli

buccal cavity (BUK-al KAV-i-tee) oral cavity

buffer (BUFF-er) compound that combines with an acid or with a base to form a weaker acid or base, thereby lessening the change in hydrogen ion concentration that would occur without the buffer; often operates as buffer pairs

buffy coat (BUFF-ee koht) in a centrifuged sample of blood, the thin layer of leukocytes and platelets located at the interface between packed red cells and plasma

bursa (BER-sah) small, cushion-like sacs found between moving body parts, making movement easier

bursitis (ber-SYE-tis) inflammation of a bursa

bypass surgery *see* **coronary bypass surgery**

C

cachexia (kah-KEKS-ee-ah) syndrome involving loss of weight, loss of appetite, and general weakness; usually associated with cancer

cadaver (kah-DAV-er) corpse preserved for anatomical study

calcitonin (CT) (kal-si-TOH-nin) hormone secreted by the thyroid that decreases calcium levels in the blood

calcium pump (KAL-see-um) energy-consuming structure embedded in a cell membrane that moves calcium ions through the membrane against their concentration gradient (that is, from an area of low concentration to an area of high concentration)

callus (KAL-us) bony tissue that forms a sort of collar around the broken ends of fractured bone during the healing process; in the skin, abnormally thick stratum corneum found at points of friction

calorie (C) (KAL-or-ree) heat unit; the amount of heat needed to raise the temperature of 1 gram of water 1° C

Calorie (C) (KAL-or-ree) heat unit; kilocalorie; the amount of heat needed to raise the temperature of 1 kilogram of water 1° C

calyx (KAY-liks) cup-shaped division of the renal pelvis

canaliculus (kan-ah-LIK-yoo-lus) an extremely narrow tubular passage or channel in compact bone; radiates from lacunae and connects with other lacunae and the Haversian canal

canal of Schlemm (kah-NAL of shlem) a ring-shaped venous sinus located deep within the anterior portion of the sclera

cancellous bone (KAN-seh-lus) bone containing trabeculae; also known as *spongy bone*

cancer malignant cellular neoplasm (tumor) that invades other cells and often metastasizes to many parts of the body

capacitation (kah-pass-i-TAY-shun) complex process needed for a mature sperm to become capable of fertilizing an ovum

capillary (KAP-i-lair-ee) tiny vessels that connect arterioles and venules; gas exchange from blood to tissues occurs in capillaries

carbohydrate (kar-boh-HYE-drayt) organic compounds containing carbon, hydrogen, and oxygen in certain specific proportions; for example, sugars, starches, and cellulose

carbonic acid (kar-BON-ik) product of the reaction between carbon dioxide and water

carbonic anhydrase (kar-BON-ik an-HYE-drays) the enzyme that converts carbon dioxide into carbonic acid

carbuncle (KAR-bung-kul) a mass of connected boils, pus-filled lesions associated with hair follicle infections; see **furuncle**

carcinogen (kar-SIN-oh-jen) substance that promotes the development of cancer

carcinoma (kar-si-NO-mah) malignant tumor that arises from epithelial tissue

cardiac cycle (KAR-dee-ak SYE-kul) complete heartbeat consisting of diastole and systole of both atria and both ventricles

cardiac muscle (KAR-dee-ak) specialized muscle that makes up the heart

cardiac output (KAR-dee-ak) volume of blood pumped by one ventricle per minute

cardiac plexus (KAR-dee-ak PLEK-sus) combination of sympathetic and parasympathetic nerves that are located near the aortic arch

cardiac sphincter (KAR-dee-ak SFINGK-ter) a ring of muscle between the stomach and esophagus that prevents food from reentering the esophagus when the stomach contracts

cardiac tamponade (KAR-dee-ak tam-pon-ODD) compression of the heart caused by fluid buildup in the pericardial space, as in pericarditis or mechanical damage to the pericardium

cardiomyopathy (kar-dee-oh-my-OP-ah-thee) general term for disease of the myocardium (heart muscle) causing enlargement

cardiopulmonary resuscitation (CPR) (kar-dee-oh-PUL-moh-nair-ree ree-sus-i-TAY-shun) combined external cardiac (heart) massage and artificial respiration

caries (KAIR-eez) decay of teeth or of bone

carotene (KAR-oh-teen) yellowish pigment; contributes to skin color

carpal tunnel syndrome (KAR-pul TUN-el SIN-drohm) muscle weakness, pain, and tingling in the radial side (thumb side) of the wrist, hand, and fingers, perhaps radiating to the forearm and shoulder; caused by compression of the median nerve within the carpal tunnel (a passage along the ventral concavity of the wrist)

carrier in genetics, a person who possesses the gene for a recessive trait but who does not actually exhibit the trait

carrier-mediated passive transport another name for facilitated diffusion, in which solutes move down their concentration gradient through carrier mechanisms in the membrane wall

cartilage (KAR-ti-lij) a specialized, fibrous connective tissue that has the consistency of a firm plastic or gristle-like gel

catabolism (kah-TAB-oh-liz-em) breakdown of food compounds or cytoplasmic constituents into simpler compounds; opposite of anabolism

catalyst (KAT-ah-list) chemical that speeds up reactions without being changed itself

cataract (KAT-ah-rakt) opacity of the lens of the eye

catecholamines (kat-eh-KOHL-ah-meens) norepinephrine, epinephrine, and dopamine

catheterization (kath-eh-ter-i-ZAY-shun) passage of a flexible tube (catheter) into the body through an exterior opening (as into the bladder through the urethra for the withdrawal of urine)

cation (KAT-eye-on) positively charged particle

cauda equina (KAW-dah eh-KWINE-ah) "horse's tail"; lower end of spinal cord with its attached spinal nerve roots

caveola (kav-ee-OH-la) tiny indentation of a raft in the cell's plasma membrane that traps substances and shuttles them into or through the cell (*plural*, caveolae

celiac ganglion (SEE-lee-ak GANG-glee-on) solar plexus; collateral ganglion that lies just below the diaphragm

cell basic biological and structural unit of the body consisting of a nucleus surrounded by cytoplasm and enclosed by a membrane

cell body the main part of a cell, ordinarily containing the nucleus; in the neuron, the cell body is also called the *soma* or *perikaryon*

cell cycle see **cell life cycle**

cell life cycle the repeating process of a cell's growth and reproduction, producing successive generations of cells and proceeding through the first growth phase (G₁), DNA replication (S), second growth phase (G₂), and mitotic cell division (M); also called *cell cycle*

cell theory concept proposed more than 100 years ago that all living organisms are made up of biological units called *cells*; see **cell**

cellulose (SEL-yoo-lohs) dietary fiber; major component of most plant tissue; nondigestable by humans

central pertaining to the center of the body, as in *central nervous system*

central nervous system (CNS) the brain and spinal cord

central sulcus (fissure of Rolando) (SUL-kus [FISH-ur of roh-LAHN-doh]) groove between frontal and parietal lobes of the cerebrum

centriole (SEN-tree-ol) one of a pair of tiny cylinders in the centrosome of a cell; believed to be involved with the spindle fibers formed during mitosis

centromere (SEN-troh-meer) a beadlike structure that attaches one chromatid to another during the early stages of mitosis

centrosome (SEN-troh-sohm) area of the cytoplasm near the nucleus that coordinates the building and breaking up of microtubules in the cell

cephalic phase (seh-FAL-ik) stage in regulation of stomach secretion in which mental factors stimulate gastric juice secretion

cerebellum (sair-eh-BELL-um) second largest part of the human brain; plays an essential role in the production of normal movements

cerebral cortex (seh-REE-bral KOR-teks) thin layer of gray matter made up of neuron dendrites and cell bodies that compose the surface of the cerebrum

cerebral hemisphere (seh-REE-bral HEM-i-sfeer) either of the two right and left halves of the cerebrum

cerebral nuclei (seh-REE-bral NOO-klee-eye) see **basal nuclei**

cerebrospinal fluid (CSF) (seh-ree-broh-SPY-nal) plasma-like fluid that fills the subarachnoid space in the brain and spinal cord and in the cerebral ventricles

cerebrovascular accident (CVA) (seh-ree-broh-VAS-kyoo-lar) a hemorrhage or cessation of blood flow through cerebral blood vessels resulting in destruction of neurons; commonly called a *stroke*

cerebrum (seh-REE-brum) largest and uppermost part of the human brain that controls consciousness, memory, sensations, emotions, and voluntary movements

cerumen (seh-ROO-men) ear wax

ceruminous gland (seh-ROO-mi-nus) gland that produces a waxy substance called *cerumen* (ear wax)

cervical plexus (SER-vi-kal PLEK-sus) plexus located deep within the neck; innervates muscles and skin of the neck, upper shoulder, and part of the head

cervix (SER-viks) neck; particularly the inferior necklike portion of the uterus

cesarean section (seh-SAIR-ee-an) surgical removal of a fetus through an incision of the skin and uterine wall; also called *C-section*

chaperonins (shap-er-OHN-inz) group of globular proteins that are present in every body cell and di-

rect the intracellular steps required for other proteins to achieve the often twisted and convoluted shape required for them to function

chemical bond energy relationship joining two or more atoms; involves sharing or exchange of electrons

chemical digestion changes in chemical composition of food as it passes through the alimentary canal

chemoreceptors (kee-moh-ree-SEP-tors) receptors that respond to chemicals; responsible for taste and smell and monitoring concentration of specific chemicals in the blood

chemotaxin (kee-moh-TAK-sin) chemotactic factor

chemotaxis (kee-moh-TAK-sis) process by which a substance attracts cells or organisms into (or away from) its vicinity; for example, when inflammation mediators attract white blood cells; see **chemotactic factor**

chemotactic factor (kee-moh-TAK-tik) chemical that attracts a mobile cell such as a neutrophil or macrophage to a specific location, as in an immune response

chemotherapy (kee-moh-THAYR-ah-pee) technique of using chemicals to treat disease (for example, infections, cancer)

Cheyne-Stokes respiration (CSR) (chain-stokes respi-RAY-shun) pattern of breathing associated with critical conditions such as brain injury or drug overdose and characterized by cycles of apnea and hyperventilation

chief cell cells lining the gastric glands of the stomach that secrete pepsinogen and intrinsic factor

childhood period of human development from infancy to puberty

chloride channel (KLOR-ide) a pore in a cell membrane that allows only chloride ions to permeate, or pass through, the membrane

chloride shift (KLOR-ide) diffusion of chloride ions into red blood cells as bicarbonate ions diffuse out; maintains electrical neutrality of red blood cells

cholecystectomy (kol-leh-sis-TEK-toh-mee) surgical removal of the gallbladder

cholecystitis (koh-lee-sis-TYE-tis) inflammation of the gallbladder

cholecystokinin (CCK) (koh-lee-sis-toh-KYE-nin) hormone secreted from the intestinal mucosa of the duodenum that stimulates contraction of the gallbladder, resulting in bile flowing into the duodenum

cholesterol (koh-LESS-ter-ol) steroid lipid found in many body tissues and in animal fats

cholinergic (koh-lin-ER-jik) having to do with acetylcholine

chondral fracture (KON-dral) fracture of an articular (cartilage) surface in a synovial joint

chondrocyte (KON-droh-syte) cartilage cell

chondroitin sulfate (kon-DROY-tin SUHL-fayt) type of proteoglycan that helps thicken and hold together the matrix of connective tissue; see **proteoglycan**

chondroma (kon-DROH-mah) benign tumor of cartilage

chondromalacia patellae (kon-droh-mah-LAY-shee-ah pah-TEL-ah) degenerative process that results in a softening of the articular surface of the patella

chordae tendineae (KOR-dee ten-DIN-ee) string-like structures that attach the AV valves to the wall of the heart

chorion (KOH-ree-on) outermost fetal membrane; contributes to tissues in the placenta

chorionic villi (koh-ree-ON-ik VIL-eye) connection between blood vessels of the chorion and those of the placenta

chorionic villus sampling (koh-ree-ON-ik VIL-lus SAM-pling) procedure in which a tube is inserted through the (uterine) cervical opening and a sample of the chorionic tissue surrounding a developing embryo is removed for karyotyping

choroid (KOH-royd) middle layer of the eyeball; contains a dark pigment to prevent the scattering of incoming light rays

choroid plexus (KOH-royd PLEK-sus) specialized group of capillaries in ventricles of the brain that secrete cerebrospinal fluid

chromatid (KROH-mah-tid) either of the two DNA strands joined by a centromere existing after DNA has replicated (before cell division) but before the centromere has divided

chromatin (KROH-mah-tin) threadlike genetic material in the nucleus

chromosome (KROH-moh-sohm) barlike bodies of chromatin (DNA) that have coiled to form a compact mass during mitosis or meiosis; each chromosome is composed of regions called *genes*, each of which transmits hereditary information

chronic (KRON-ik) long-lasting, as in "chronic disease"

chronic obstructive pulmonary disease (COPD) (KRON-ik ob-STRUK-tiv PUL-moh-nair-ee di-

ZEEZ) general term referring to a group of disorders characterized by progressive, irreversible obstruction of airflow in the lungs; see **asthma, bronchitis, emphysema**

chyle (kyle) milky fluid; the fat-containing lymph in the lymphatics of the intestine

chylomicron (kye-loh-MY-kron) small fat droplet

chyme (kyme) partially digested food mixture leaving the stomach

cilia (SIL-ee-ah) hairlike projections of cells

ciliary body (SIL-ee-air-ee) thickening of the choroid that is located between the anterior margin of the retina and the posterior margin of the iris

ciliary muscle (SIL-ee-air-ee) smooth muscle in the ciliary body of the eye that suspends the lens and functions in accommodation

circulatory system (SER-kyoo-lah-tor-ee) system composed of the heart, blood vessels, and lymphatic vessels; permits transportation of material to and from all the cells of the body

circumcision (ser-kum-SIH-zhun) surgical removal of the foreskin, or prepuce, of the penis

circumduction (ser-kum-DUK-shun) moving a part so its distal end moves in a circle

cirrhosis (sir-ROH-sis) degeneration of liver tissue characterized by the replacement of damaged liver tissue with fibrous or fatty connective tissue

cisterna chyli (sis-TER-nah KYE-lye) an enlarged pouch on the thoracic duct that serves as a storage area for lymph moving toward its point of entry into the venous system

cisternae (sis-TER-nay) tiny, membranous sacs that make up the Golgi apparatus

citric acid cycle (SIT-rik) second series of chemical reactions in the process of glucose metabolism in which carbon dioxide is formed and energy is released; it is an aerobic process; see **Krebs cycle**

cleavage line (KLEEV-ij line) pattern of dense bundles of white collagenous fibers that characterize the reticular layers of dermis; also called *Langer's lines*

cleft palate (kleft PAL-ett) facial deformity that is an X-linked inherited condition; when the palatine bones fail to unite completely

clitoris (KLIT-oh-ris) small, erectile body located within the vestibule of the vagina

clonal selection theory (KLOH-nal) process through which a B or T cell, once sensitized through contact with an antigen, divides rapidly to create a colony of clones that destroys the "selecting" antigen

clone (klohn) any of a family of many identical cells descended from a single "parent" cell

cocaine (koh-KAYN) drug that blocks the uptake of dopamine (neurotransmitter) by neurons, thus inhibiting pain signals or producing a temporary feeling of well-being

cochlea (KOHK-lee-ah) snail shell–like structure in the inner ear that houses the organ of Corti, which is responsible for sense of hearing

cochlear duct (KOHK-lee-ar) membranous tube within the bony cochlea; only part of the internal ear concerned with hearing

cochlear nerve (KOHK-lee-ar) part of vestibulocochlear nerve (cranial nerve VIII); sensory nerve responsible for hearing

codominance (koh-DOM-i-nance) in genetics, a form of dominance in which two dominant versions of a trait are both expressed in the same individual

codon (KOH-don) in RNA, a triplet of three base pairs that codes for a particular amino acid

coenzyme (koh-EN-zyme) organic, nonprotein catalyst that acts as molecule carrier

cofactor (KOH-fak-ter) a nonprotein unit attached to an enzyme molecule than enables the enzyme to function properly

colitis (koh-LYE-tis) any inflammatory condition of the colon and/or rectum

collagen (KAHL-ah-jen) principal organic constituent of connective tissue

collateral ganglion (koh-LAT-er-al GANG-glee-on) sympathetic, prevertebral ganglion; named for nearby blood vessels

collecting duct (koh-LEK-ting dukt) in the kidney, straight tubule joined by the distal tubules of several nephrons

colloid (KOL-oyd) dissolved particles that resist separation from the gas, liquid, or solid medium in which they are dissolved

columnar (koh-LUM-nar) cell classification by shape in which cells are higher than they are wide

coma (KOH-mah) altered state of consciousness from which an individual cannot be aroused

comedo (KOM-ee-doh) inflamed, plugged sebaceous gland duct, common in acne conditions

commisural tract (kom-MIS-yoo-ral) nerve tissue that connects the left and right hemispheres of the brain; see **corpus callosum**

common bile duct (KOM-on byle) duct from the liver that empties into the duodenum; made up of the merging of the hepatic duct with the cystic duct
communicable (kom-MYOO-ni-kah-bil) able to spread from one individual to another
compact bone dense bone; contains structural units called *Haversian systems*
complement any of several proteins normally present in blood serum that when activated kill foreign cells by puncturing them
complementary base pairing bonding purines and pyrimidines in DNA; adenine always binds with thymine, and cytosine always binds with guanine
complete fracture fracture that totally divides a bone into separate pieces
compliance (kahm-PLY-ans) the ease of stretch of a material—as in lung compliance, the ease of stretch of the lung tissues
composite cell (kahm-PAH-zit) artistic representation of a cell that includes features from many different types of cells
compound a chemical combination of two or more elements
compound fracture fracture in which the broken ends of the bone protrude through the skin
computed tomography (kahm-PAYOOT-ed tom-AH-graf-ee) x-ray technique that produces an image representing a detailed cross section of a body structure
concentration gradient (kahn-sen-TRAY-shun GRAY-dee-ent) measurable difference in concentration from one area to another
concentric contractions (kon-SEN-trik) contractions in which the movement results in shortening of the muscle; type of isotonic contraction or dynamic tension
conduction (kon-DUK-shun) in human anatomy, the transfer of heat energy to the skin and then the external environment
conductivity (kon-duk-TIV-uh-tee) ability of living cells and tissues to selectively transmit a wave of excitation from one point to another within the body
condyloid joint (KON-di-loyd) ellipsoidal joint
cone receptor cell located in the retina that is stimulated by bright light
congenital disorder (kon-JEN-i-tall) refers to a condition present at birth; congenital conditions may be inherited or may be acquired in the womb or during delivery
congestive heart failure (CHF) (kon-JES-tiv) left heart failure; inability of the left ventricle to pump effectively, resulting in congestion in the systemic and pulmonary circulations
conjunctiva (kon-junk-TI-vah) mucous membrane that lines the eyelids and covers the sclera (white portion of the eye)
conjunctivitis (kon-junk-ti-VYE-tis) inflammation of the conjunctiva, usually caused by irritation, infection, or allergy
connective tissue (koh-NEK-tiv) most abundant and widely distributed tissue in the body
consciousness state of awareness of one's self and environment and other beings
contact dermatitis (der-mah-TYE-tis) local skin inflammation lasting a few hours or days after being exposed to an antigen
contact digestion when substrates bind onto enzymes located on the surface of the brush border and complete carbohydrate digestion
contractility (kon-trak-TIL-i-tee) ability, as of muscle cell, to contract or shorten to produce movement
contralateral (kon-trah-LAT-er-al) on the opposite side of the body
contralateral reflex arc (kon-trah-LAT-er-al REE-fleks ark) reflex arc whose receptor and effectors are located on opposite sides of the body
contusion (kon-TOO-zhun) a bruise; an injury in which the skin or surface of an organ is not broken but underlying blood vessels may rupture
convection (kon-VEK-shun) in human anatomy, transfer of heat energy to air that is flowing away from the skin
convergence (kon-VER-jens) movement of the two eyeballs inward so that their visual axes come together at the same point on the object viewed; also, when more than one presynaptic axon synapses with a single postsynaptic neuron
convulsion (con-VUL-shun) abnormal, uncoordinated tetanic contractions of varying groups of muscles
cornea (KOR-nee-ah) transparent, anterior portion of the sclera
coronal plane (koh-ROH-nal) frontal plane; divides the body into front and back portions
coronary bypass surgery (KOR-oh-nair-ee) surgery to relieve severely restricted coronary blood flow; veins are taken from other parts of the body and grafted in to bypass the blockage
coronary sinus (KOR-oh-nair-ee SYE-nus) area that receives deoxygenated blood from the coronary veins and empties into the right atrium
cor pulmonale (kohr pul-mah-NAL-ee) failure of the right atrium and ventricle to pump blood effectively, resulting from obstruction of pulmonary blood flow
corpora cavernosa (KOHR-pohr-ah kav-er-NO-sah) two columns of erectile tissue in the shaft of the penis
corpora quadrigemina (KOHR-pohr-ah kwod-ri-JEM-i-nah) midbrain landmark composed of inferior and superior colliculi
corpus albicans (KOHR-pus AL-bi-kans) white scar on ovary that replaces the degenerated corpus luteum
corpus callosum (KOHR-pus kah-LOH-sum) nerve tissue connecting the right and left cerebral hemispheres
corpus luteum (KOHR-pus LOO-tee-um) a hormone-secreting glandular structure formed after ovulation at the site of the ruptured follicle; it secretes chiefly progesterone with some estrogen
corpus spongiosum (KOHR-pus spun-jee-OH-sum) a column of erectile tissue surrounding the urethra in the penis
cortex (KOR-teks) outer part of an internal organ; for example, the outer part of the cerebrum or the outer portion of the kidneys
cortical (KOHR-ti-kal) pertaining to the cortex, or outer area of an organ or structure
corticosteroid (kohr-ti-koh-STER-royd) glucocorticoid secreted by zona fasciculata of the adrenal cortex
corticotroph (kohr-ti-koh-TROHF) cell type of the adenohypophysis (anterior pituitary) that secretes ACTH (adrenocorticotropic hormone) and MSH (melanocyte-stimulating hormone)
cortisol (KOHR-ti-sol) glucocorticoid secreted by zona fasciculata of the adrenal cortex; also known as *hydrocortisone*
cortisone (KOHR-ti-sohn) glucocorticoid secreted by zona fasciculata of the adrenal cortex
costal cartilage (KOS-tal KAR-ti-lij) cartilage that attaches ribs two through ten to the body of the sternum
cotransmission (koh-tranz-MISH-un) theory of efferent autonomic synaptic transmission that states that all or most postganglionic fibers release either norepinephrine or acetylcholine along with NANC (nonadrenergic-noncholinergic) transmitters or modulators and that each substance combines with postsynaptic and/or presynaptic receptors to produce regulatory effects
coumarin (KOO-mar-in) compound that retards blood coagulation; often given as medication to prevent heart attacks
covalent bond (koh-VAYL-ent) chemical bond formed by two atoms sharing one or more pairs of electrons
CPL a submicroscopic vesicle (bubble or membrane) within cells that acts as a "compartment for peptide loading (CPL)," in which proteins are altered in structure
cramps painful muscle spasms (involuntary twitches) that result from irritating stimuli, as in mild inflammation, or from ion imbalances
cranial nerve (KRAY-nee-al) any of twelve pairs of nerves that attach to the undersurface of the brain and conduct impulses between the brain and structures in the head, neck, and thorax
craniosacral division (kray-nee-oh-SAY-kral) parasympathetic division of the autonomic nervous system
cranium (KRAY-nee-um) bony vault, made up of eight bones, that encases the brain
creatine phosphate (CP) (KREE-ah-tin FOS-fayt) substance found in muscle cells and used for the temporary storage of chemical energy to supply ATP and ultimately for muscle contraction
cremaster muscle (kreh-MASS-ter) muscle responsible for elevating the testes
cretinism (KREE-tin-iz-em) dwarfism caused by hyposecretion of the thyroid gland
Creutzfeldt-Jakob Disease (CJD) *see* **variant Creutzfeldt-Jakob Disease (vCJD)**
cribriform plate (KRIB-ri-form) perforated portion of ethmoid bone that separates the nasal and cranial cavities
crista ampullaris (KRIS-tah am-pyoo-LAIR-iss) specialized receptor organ located within the ampulla of the semicircular canals; detects head movements
cristae (KRIS-tee) folds of the inner membrane of a mitochondrion
cross bridge junction of a thick myofilament with a thin myofilament in the myofibril of a muscle fiber, where the head of a myosin molecule in the thick filament binds to the active site of an actin molecule in the thin filament
crossing over phenomenon that occurs during meiosis when pairs of homologous chromosomes synapse and exchange genes
cuboidal (KYOO-boyd-al) cell classification by shape in which cells resemble a cube
cupula (KYOO-pyoo-lah) structure found within the crista ampullaris of the semicircular canal; the gelatinous cap in which the hair cells are embedded
Cushing syndrome (KOOSH-ing SIN-drohm) condition caused by hypersecretion of glucocorticoids from the adrenal cortex
cutaneous membrane (kyoo-TAYN-ee-us) primary organ of the integumentary system; the skin
cuticle (KYOO-ti-kul) skin fold covering the root of the nail
cyanosis (sye-ah-NO-sis) condition of blueness, particularly of the skin, resulting from inadequate oxygenation of the blood
cyclic AMP (SIK-lik A M P) one of several second messengers that delivers information inside the cell and thus regulates the cell's activity
cyclin (SYE-klin) any of several regulatory proteins in the cell that influence the function of activating enzymes called *cyclin-dependent kinases* that drive the cell forward from phase to phase in the cell life cycle and thus regulate the cell's growth and reproduction
cyclin-dependent kinase (CDK) (SYE-klin dee-PEND-ent KI-nays) any of several activating enzymes that drive the cell through the various phases of its life cycle; "cyclin-dependent" refers to the fact that CDK is itself regulated by cellular proteins called *cyclins*
cystic duct (SIS-tik) joins with the common hepatic duct to form the common bile duct
cystic fibrosis (SIS-tik fye-BROH-sis) inherited disease involving abnormal chloride ion (Cl^-) transport; causes secretion of abnormally thick mucus and other problems
cytokinesis (sye-toh-kin-EE-sis) process by which a dividing cell splits its cytoplasm and plasma membrane into two distinct daughter cells; cytokinesis happens along with mitosis (or meiosis) during the cell division process
cytology (sye-TOL-oh-jee) study of cells
cytolysis (sye-TOL-i-sis) literally, "cell bursting"; often occurs when ions and water rush into a cell, causing it to burst
cytoplasm (SYE-toh-plaz-em) gel-like substance of a cell exclusive of the nucleus and plasma membrane; includes organelles (except nucleus) and cytosol (intracellular fluid)
cytoskeleton (sye-toh-SKEL-eh-ton) cell's internal supporting framework
cytosol (SYE-toh-sawl) solution of water and other substances of a cell (outside the nucleus) in which the organelles and cellular inclusions are suspended; it is the liquid portion of the living cell substance known as *cytoplasm*
cytotoxic T cell (sye-toh-TOK-sik T) "cell killing" T cell

D

deamination (dee-am-i-NAY-shun) removal of an amino group from an amino acid to form a molecule of ammonia and one of keto acid; occurs in the liver as first step in protein catabolism
decarboxylase (dee-kar-BOK-si-lays) enzyme that removes carbon dioxide
deciduous teeth (deh-SID-yoo-us) commonly referred to as "baby teeth"; teeth that shed at a certain age before development of permanent teeth
decomposition reaction (dee-KAHM-poh-si-shun) chemical reaction that breaks down a substance into two or more simpler substances
decubitus ulcer (deh-KYOO-bi-tus UL-ser) area of destroyed tissue resulting from inadequate blood supply that often develops when body lies in one position for prolonged periods
deep farther away from the body's surface, as opposed to superficial
defecation (def-eh-KAY-shun) expelling feces from the digestive tract
defibrillation (deh-fib-ri-LAY-shun) application of an electric shock to force cardiac muscle fibers to contract in unison
deglutition (deg-loo-TISH-un) swallowing
dehydration (dee-hye-DRAY-shun) an abnormal loss of fluid from the body's internal environment
dehydration synthesis (dee-hye-DRAY-shun SIN-the-sis) anabolic process by which molecules are joined to form larger molecules
deletion (deh-LEE-shun) genetic mutation that occurs within a DNA molecule when one or more nucleotide bases in a sequence are missing, causing a misreading of the genetic code and a failure to produce a normal protein needed for body function
delta cell pancreatic islet cell that secretes somatostatin; also called *D cell*
denature (deh-NAY-chur) to alter the shape of a protein by a change in pH, heat, or some other manner, to change its chemical properties
dendrite (DEN-dryt) branching or treelike nerve cell process that receives input from other neurons and transmits impulses toward the cell body (or toward the axon in a unipolar neuron)
dentate nuclei (DEN-tayt NOO-klee-eye) paired cerebellar nuclei that are connected by tracts with the thalamus, as well as motor areas of the cerebral cortex
deoxyribonucleic acid (DNA) (dee-ok-see-rye-boh-noo-KLEE-ik) genetic material of the cell that carries the chemical "blueprint" of the body
deoxyribose (dee-ok-see-RYE-bohs) sugar in DNA whose molecules contain only five carbons
depolarization (dee-poh-lar-i-ZAY-shun) electrical activity that triggers a contraction of the heart muscle
dermal papillae (DER-mal pah-PIL-ee) upper region of the dermis that forms part of the dermal-epidermal junction and forms the ridges and grooves of fingerprints
dermatitis (der-mah-TYE-tis) general term referring to any inflammation of the skin
dermatology (der-mah-TOL-oh-jee) study of the integument and its diseases
dermatome (DER-mah-tohm) skin surface areas supplied by a single spinal nerve
dermatosis (der-mah-TOH-sis) general term meaning "skin condition"
dermis (DER-mis) the deeper of the two major layers of the skin, composed of dense fibrous connective tissue interspersed with glands, nerve endings, and blood vessels
descending tract bundle of axons in the spinal cord that conducts impulses down the cord from the brain
desmosomes (DES-moh-sohms) specialized junctions that hold adjacent cells together; consist of dense plate at point of adhesion plus extracellular cementing material
desquamation (des-kwah-MAY-shun) shedding of epithelial elements from the skin surface
detrusor muscle (dee-TROO-sor) smooth muscle tissue making up the wall of the bladder
developmental anatomy study of human growth and development
developmental biology branch of biology that studies the process of change over the life cycle
deviated septum (DEE-vee-ay-ted SEP-tum) abnormal condition in which the nasal septum is far from its normal position, possibly obstructing normal nasal breathing
diabetes insipidus (dye-ah-BEE-teez in-SIP-i-dus) condition resulting from hyposecretion of ADH in which large volumes of urine are formed
diabetes mellitus (dye-ah-BEE-teez MELL-i-tus) condition resulting when the pancreatic islets secrete too little insulin, resulting in increased levels of blood glucose
diad (dye-AD) double structure of T tubules in cardiac muscle fiber
dialysis (dye-AL-i-sis) separation of smaller (diffusible) particles from larger (nondiffusible) particles through a semipermeable membrane
diapedesis (dye-ah-peh-DEE-sis) passage of any formed elements within blood through the vessel wall, as in movement of white cells into an area of injury and infection
diaphragm (DYE-ah-fram) the flat muscular sheet that separates the thorax and abdomen; a major muscle of respiration
diaphysis (dye-AF-i-sis) shaft of a long bone
diarrhea (dye-ah-REE-ah) abnormally frequent defecation of liquid or semi-liquid feces
diarthrosis (dye-ar-THROH-sis) freely movable joint
diastasis (dye-ASS-tah-sis) reduced ventricular filling of the heart
diastole (dye-ASS-toh-lee) relaxation of the heart (especially the ventricles), during which it fills with blood; opposite of systole
diastolic pressure (dye-ah-STOL-ik) blood pressure in arteries during diastole (relaxation) of the heart; clinically more important than systolic pressure
diencephalon (dye-en-SEF-ah-lon) "between" brain; parts of the brain between the cerebral hemispheres and the mesencephalon, or midbrain
differential count (dif-er-EN-shal) percentage of total white blood cell count of the different types of leukocytes
differential white blood cell count (dif-er-EN-shal) percentage enumeration of the different types of leukocytes in a stained blood smear
differentiation (dif-er-EN-shee-AY-shun) process of the development of cell specialization
diffusion (di-FYOO-shun) spreading; natural tendency of small particles to spread out evenly

within any given space; for example, scattering of dissolved particles

digestion breakdown of food materials either mechanically (through chewing) or chemically (via digestive enzymes)

digestive system system composed of mouth, pharynx, esophagus, stomach, small intestine, large intestine, rectum, and anal canal

diploid number (DIP-loyd) normal number of chromosomes per somatic cell (46 in humans)

diplopia (di-PLOH-pee-ah) double vision

dissection (di-SEK-shun) cutting technique used to separate body parts for study

dissociate (di-SOH-see-ayt) when a compound breaks apart in solution forming ions that are surrounded by solvent molecules

distal (DIS-tall) toward the end of a structure; opposite of proximal

distal tubule (DIS-tall TOOB-yool) in the kidney, the part of the tubule distal to the ascending limb of the loop of Henle that terminates in a collecting tubule; also known as *distal convoluted tubule*

disuse atrophy (DIS-yoos AT-roh-fee) loss of muscle tissue mass after a period of few or no contractions, resulting in weakness

divergence (dye-VER-jens) when a single presynaptic axon synapses with more than one different postsynaptic neuron

diverticulitis (dye-ver-tik-yoo-LYE-tis) inflammation of diverticula (abnormal outpouchings) of the large intestine, possibly causing constipation

dominant (DOM-i-nant) in genetics, term referring to genes that have effects that appear in the offspring (dominant forms of a gene are often represented by upper case letters); *see* **recessive**

dopamine (DOH-pah-meen) chemical neurotransmitter

dopaminergic (doh-pah-min-ER-jik) pertaining to a neuron that releases dopamine (neurotransmitter) or to a receptor molecule that is activated by dopamine

dorsal cavity (DOR-sal KAV-i-tee) body cavity that includes the cranial cavity and the spinal cavity; not a standard anatomical term, but used here to help organize the body for the beginning student

dorsal ramus (DOR-sal RAY-mus) branch of spinal nerve that supplies somatic motor and sensory fibers to several smaller nerves

dorsal root ganglion (DOR-sal root GANG-glee-on) small region of gray matter in dorsal nerve root made up of cell bodies of unipolar sensory neurons

dorsiflexion (dor-si-FLEK-shun) when the top of the foot is elevated (brought toward the front of the lower leg) with toes pointing upward

double bond covalent chemical bond in which two pairs of electrons are shared

Down syndrome (down SIN-drohm) group of symptoms usually caused by trisomy of chromosome 21; characterized by mental retardation and multiple structural defects, including facial, skeletal, and cardiovascular abnormalities

Duchenne muscular dystrophy (DMD) (doo-SHEN MUSS-kyoo-lar DISS-troh-fee) common form of muscular dystrophy also called *pseudohypertrophy* (meaning "false muscle growth") because the atrophy of muscle is masked by excessive replacement of muscle by fat and fibrous tissue; characterized by mild leg muscle weakness that progresses rapidly to include the shoulder muscles; caused by mutated gene for dystrophin (needed to hold muscle fiber together during contraction) on the X chromosome, thus making DMD more common in boys than girls

ductus arteriosus (DUK-tus ar-teer-ee-OH-sus) in the developing fetus, it connects the aorta and the pulmonary artery, allowing most blood to bypass the fetus' developing lungs

ductus deferens (DUK-tus DEF-er-enz) vas deferens

ductus venosus (DUK-tus veh-NO-sus) continuation of the umbilical vein that shunts blood returning from the placenta past the fetus' developing liver directly into the inferior vena cava

duodenum (doo-oh-DEE-num) first subdivision of the small intestine; where most chemical digestion occurs

dura mater (DOO-rah MAH-ter) literally "strong" or "hard mother"; outermost layer of the meninges

dural sinuses (DOO-ral SYE-nus-ez) name for large veins of cranial cavity

dwarfism (DWARF-iz-em) condition of abnormally small stature, sometimes resulting from hyposecretion of growth hormone

dynamic equilibrium (dye-NAM-ik ee-kwi-LIB-ree-um) maintaining balance when the head or body is rotated or suddenly moved

dynamic tension (dye-NAM-ik TEN-shun) another name for isotonic contraction

dysplasia (diss-PLAY-zha) abnormal changes in size, shape, and organization of cells in a tissue associated with neoplasms (tumors)

dyspnea (DISP-nee-ah) difficult or labored breathing

dystrophin (DIS-trof-in) protein molecule that forms strands in each skeletal muscle fiber and helps to hold the cytoskeleton to the sarcolemma to keep the muscle fiber from breaking during contractions; normal dystrophin is missing in Duchenne muscular dystrophy (DMD) and related conditions and thus cells break apart more easily, causing the symptoms of DMD

E

eccentric contraction (ek-SENT-rik kon-TRAK-shun) type of isotonic muscle contraction in which muscle lengthens while it is contracting

eccrine sweat gland (EK-rin) water-producing exocrine sweat glands widely dispersed throughout the skin

eclampsia (eh-KLAMP-see-ah) potentially fatal condition associated with toxemia of pregnancy; characterized by convulsions and coma

ectoderm (EK-toh-derm) outermost of the primary germ layers that develops early in the first trimester of pregnancy; gives rise to the skin and the nervous system

ectomorph (EK-toh-morf) thin, lean body type

ectopic pregnancy (ek-TOP-ik) pregnancy in which the fertilized ovum develops in some place other than in the uterus (literally, "out of place")

eczema (EK-zeh-mah) inflammatory skin condition associated with various diseases and characterized by erythema, papules, vesicles, and crusts

edema (eh-DEE-mah) accumulation of fluid in a tissue, as in inflammation; swelling

effector organ, gland, or muscle that responds to a nerve stimulus

efferent division (EF-fer-ent di-VI-shun) the motor division (outgoing pathways) of the nervous system

efferent ductule (EF-fer-ent DUKT-yool) series of sperm ducts that drain the rete testis and pierce the tunica albuginea

efferent nervous system (EF-fer-ent NER-vus SIS-tem) subdivision of the peripheral nervous system (PNS) that consists of all outgoing motor nerves

efferent neuron (EF-fer-ent NOO-ron) neuron that transmits nerve impulses away from the central nervous system to muscles or glands (effectors)

ejaculation (ee-jak-yoo-LAY-shun) sudden discharging of semen from the body

ejaculatory duct (ee-JAK-yoo-lah-toh-ree) duct formed by the joining of the ductus deferens and the duct from the seminal vesicle that allows sperm to enter the urethra

elastic cartilage (eh-LAS-tik KAR-ti-lij) cartilage with elastic, as well as collagenous, fibers; provides elasticity and firmness, as in, for example, the cartilage of the external ear

elastic filament (eh-LAS-tik FIL-ah-ment) in muscle fibers, microscopic protein filaments composed of *titin* (connectin) that anchor the ends of the thick filaments to the Z line and give myofibrils their characteristic elasticity

elastin (eh-LAS-tin) protein found in elastic fiber

electrocardiogram (ECG) (eh-lek-troh-KAR-dee-oh-gram) graphic record of the heart's action potentials

electrocardiograph (eh-lek-troh-KAR-dee-oh-graf) machine that produces electrocardiograms

electroencephalogram (EEG) (eh-lek-troh-en-SEF-loh-gram) graphic representation of voltage changes in brain tissue used to evaluate nerve tissue function

electrolyte (eh-LEK-troh-lyte) substance that dissociates into ions in solution, rendering the solution capable of conducting an electric current

electromyography (eh-lek-troh-my-OG-rah-fee) recording electrical impulses from muscles as they contract

electron (eh-LEK-tron) small, negatively charged subatomic particle

electron microscopy (EM) (eh-LEK-tron my-KRAH-skop-ee) technique of observing small structures by either passing a beam of electrons through a specimen (transmission electron microscopy, TEM) or reflecting a beam of electrons off a specimen (scanning electron microscopy, SEM) and focusing the resulting electron beam(s) to form a magnified image of the specimen

electron transport system (ETS) carrier molecules embedded in the inner membrane of the mitochondria that take high-energy electrons from the citric acid cycle and form water and energy for oxidative phosphorylation

electrophoresis (eh-lek-troh-foh-REE-sis) laboratory procedure in which different types of charged molecules are separated according to molecular weight using a weak electric current

element substance composed of only one type of atom that cannot be broken into simpler constituents by chemical means

elephantiasis (el-eh-fan-TYE-ah-sis) extreme lymphedema (swelling resulting from lymphatic blockage) in the limbs caused by a parasitic worm infestation; so called because the limbs swell to "elephantine proportions"

elimination (eh-lim-i-NAY-shun) defecation

embolus (EM-boh-lus) a moving blood clot circulating in the bloodstream

embryo (EM-bree-oh) animal in early stages of intrauterine development; in humans, the embryonic stage is the first 8 weeks after conception

embryology (em-bree-OL-oh-gee) study of the development of an individual from conception to birth

embryonic disk (em-bree-ON-ik) cells of the early embryo that differentiate into the three primary germ layers

emesis (EM-eh-sis) vomiting

emission (eh-MISH-un) reflex movement of spermatozoa and secretions from genital ducts and accessory glands into prostatic urethra; precedes ejaculation

emmetropic (em-eh-TROHP-ik) relaxed normal eye

emphysema (em-fi-SEE-mah) abnormal condition characterized by trapping of air in alveoli of the lung that causes them to rupture and fuse to other alveoli

endemic (en-DEM-ik) refers to a disease native to a local region of the world

endocardium (en-doh-KAR-dee-um) thin layer of very smooth tissue lining each chamber of the heart

endochondral ossification (en-doh-KON-dral os-i-fi-KAY-shun) process by which bones are formed by replacement of cartilage models

endocrine (EN-doh-krin) secreting into blood or tissue fluid rather than into a duct; opposite of exocrine

endocrine reflex (EN-doh-krin) response that results from feedback loops within the endocrine system

endocrine system (EN-doh-krin) system composed of specialized glands that secrete chemicals known as *hormones* directly into the blood

endocrinology (en-doh-krin-OL-oh-jee) study of the endocrine glands and their hormones

endocytosis (en-doh-sye-TOH-sis) process that allows extracellular material to enter the cell without actually passing through the plasma membrane

endoderm (EN-doh-derm) innermost layer of the primary germ layers that develops early in the first trimester of pregnancy; gives rise to digestive and urinary structures, as well as many other glands and organ parts

endogenous growth (en-DOJ-en-us) *see* **interstitial growth**

endolymph (EN-doh-limf) clear potassium-rich fluid that fills the membranous labyrinth of the inner ear

endometrium (en-doh-MEE-tree-um) mucous membrane lining the uterus

endomorph (EN-doh-morf) body type characterized by excessive fat

endomysium (en-doh-MEE-see-um) delicate connective tissue membrane covering the highly specialized skeletal muscle fibers

endoneurium (en-doh-NOO-ree-um) thin wrapping of fibrous connective tissue that surrounds each axon in a nerve

endoplasm (en-doh-PLAZ-im) material within a cell

endoplasmic reticulum (ER) (en-doh-PLAS-mik reh-TIK-yoo-lum) network of tubules and vesicles in cytoplasm that contributes to cellular protein manufacture (via attached ribosomes) and distribution

endorphin (en-DOR-fin) chemical in central nervous system that influences pain perception; a natural painkiller

endosteum (en-DOS-tee-um) fibrous membrane that lines the medullary cavity of long bones

endothelium (en-doh-THEE-lee-um) squamous epithelial cells that line the inner surface of the entire circulatory system and the vessels of the lymphatic system

endotracheal intubation (en-doh-TRAY-kee-al in-too-BAY-shun) placing a tube in the trachea to ensure an open airway

end-product inhibition (end-PROD-ukt in-hib-ISH-un) process in a biochemical pathway in which the chemical product at the end of the pathway (the end product) becomes an allosteric effector, inhibiting the function of one or more enzymes in the pathway and thus inhibiting further production of the end product

endurance training continuous vigorous exercise requiring the body to increase its consumption of oxygen and develop the muscles' ability to sustain activity over a prolonged period

energy level limited region surrounding the nucleus of an atom at a certain distance containing electrons; also called a *shell*

enkephalin (en-KEF-ah-lin) peptide chemical in the central nervous system that acts as a natural painkiller

enterogastric reflex (en-ter-oh-GAS-trik REE-fleks) nervous reflex causing inhibition of gastric peristalsis in response to the presence of acid and distention of duodenal mucosa; also may inhibit gastric secretion

enterogastrone (en-ter-oh-GAS-trown) hormone theorized to be involved with decreasing gastric peristalsis

enzyme (EN-zyme) biochemical catalyst that allows chemical reactions to take place; functional proteins that regulate various metabolic pathways of the body

eosinophil (ee-oh-SIN-oh-fil) white blood cell, readily stained by eosin (a pinkish acid dye)

ependymal cell (eh-PEN-dih-mal) cells that line the ventricles of the brain and the central canal of the spinal cord

ependymoma (eh-pen-di-MOH-mah) tumor of glial cells called *ependyma* that line fluid spaces of the central nervous system

epicardium (ep-i-KAR-dee-um) inner layer of the pericardium that covers the surface of the heart; it is also called the *visceral pericardium*

epidemic (ep-i-DEM-ik) refers to a disease that occurs in many individuals at the same time

epidemiology (EP-i-dee-mee-OL-oh-jee) study of the occurrence, distribution, and transmission of diseases in human populations

epidermis (ep-i-DER-mis) outermost layer of the skin; sometimes called the "false" skin

epidural space (ep-i-DOO-ral) in the brain, the space above the dura mater

epiglottis (ep-i-GLOT-iss) lidlike cartilage overhanging the entrance to the larynx

epimysium (ep-i-MIS-ee-um) coarse sheet of connective tissue that covers a muscle as a whole

epinephrine (ep-i-NEF-rin) adrenaline; secretion of the adrenal medulla

epineurium (ep-i-NOO-ree-um) tough fibrous sheath that covers the whole nerve

epiphyseal fracture (ep-i-FEEZ-ee-al) when the epiphyseal plate is separated from the epiphysis or diaphysis; this type of fracture can disrupt normal growth of the bone

epiphyseal plate (ep-i-FEEZ-ee-al plate) cartilage plate that is between the epiphysis and the diaphysis and allows growth to occur; sometimes referred to as a *growth plate*

epiphysis (eh-PIF-i-sis) end of a long bone; also, the pineal body of the brain

episiotomy (eh-piz-ee-OT-oh-mee) surgical procedure used during birth to prevent a laceration of the mother's perineum or the vagina

epistaxis (ep-i-STAK-sis) clinical term referring to a bloody nose

epithalamus (ep-i-THAL-ah-mus) small nuclei located outside the thalamus and hypothalamus; considered to be one of the structures of the diencephalon

epithelial membrane (ep-i-THEE-lee-al) membrane composed of epithelial tissue with an underlying layer of specialized connective tissue

epithelial tissue (ep-i-THEE-lee-al) covers the body and its parts; lines various parts of the body; forms continuous sheets that contain no blood vessels; classified according to shape and arrangement

epitope (EP-i-tope) specific portion of an antigen that elicits an immune response

eponym (EP-oh-nim) scientific term based on a person's name, such as *Islets of Langerhans* (equivalent to *pancreatic islets*); eponyms are avoided in modern usage

equatorial plate (eh-kwah-TOH-ree-al) plane at the "equator" of a cell during metaphase where the chromosomes align

equilibration (eh-kwi-li-BRAY-shun) balance between opposing elements

erection (eh-REK-shun) condition of erectile tissue when filled with blood; often refers to the penis' enlargement during sexual arousal

erector spinae (eh-REK-tor SPINE-ee) muscle group in the back consisting of a number of long, thin muscles that travel all the way down the back; the muscles extend (straighten or pull back) the vertebral column and rotate and flex the back laterally

erythema (er-i-THEE-mah) reddening of the skin

erythroblastosis fetalis (eh-rith-roh-blas-TOH-sis feh-TAL-iss) condition of a fetus or infant caused by the mother's Rh antibodies reacting with the baby's Rh-positive red blood cells, characterized by massive agglutination of the blood and resulting in life-threatening circulatory problems for the infant

erythrocytes (eh-RITH-roh-sytes) red blood cells

erythropoiesis (eh-rith-roh-poy-EE-sis) process of red blood cell formation
erythropoietin (eh-rith-roh-POY-eh-tin) glycoprotein secreted to increase red blood cell production in response to oxygen deficiency
esophagus (eh-SOF-ah-gus) muscular, mucus-lined tube that connects the pharynx with the stomach; also known as the *food pipe*
essential fatty acid unsaturated fatty acid that must be provided by the diet; serves as a source within the body for prostaglandin synthesis
essential reproductive organs reproductive organs that must be present for reproduction to occur; the gonads
estrogen (ES-troh-jen) sex hormone secreted by the ovary that causes development and maintenance of female secondary sex characteristics and stimulates growth of the epithelial cells lining the uterus
etiology (ee-tee-OL-oh-jee) theory, or study, of the factors involved in causing a disease
eumelanin (YOO-mel-ah-nin) type of melanin pigment that is dark brown in color
eupnea (YOOP-nee-ah) normal respiration
eustachian tube (yoo-STAY-shun) auditory tube
evaporation in anatomy and physiology, heat lost from the body by vaporization of liquid (sweat) from the skin
exchange reaction chemical reaction that breaks down a compound and then synthesizes a new compound by switching portions of the molecules; for example, AB CD → AD BC
excitability (ek-syte-eh-bill-i-tee) ability of a muscle to be stimulated; also known as *irritability*
excitation (ek-sye-TAY-shun) occurs when a neuron is stimulated and additional Na+ channels open
excitatory neurotransmitter (ek-SYE-tah-toh-ree noo-roh-TRANS-mit-er) neurotransmitter that causes excitation (and thus depolarization) of the postsynaptic neuron
excitatory postsynaptic potential (EPSP) (ek-SYE-tah-toh-ree post-si-NAP-tik poh-TEN-shal) temporary depolarization of postsynaptic membrane following stimulation
excretion (eks-KREE-shun) removal of waste products produced during body functions; occurs by defecation, urination, and respiration and through the skin
exocrine (EKS-oh-krin) secreting into a duct, as in glands that secrete their products via ducts onto a surface or into a cavity; opposite of endocrine
exocytosis (eks-oh-sye-TOH-sis) process that allows large molecules to leave the cell without actually passing through the plasma membrane
exogenous growth (eks-OJ-eh-nus growth) *see* **appositional growth**
experiment test (or series of tests) of a proposed scientific idea or hypothesis; a controlled experiment is one that accounts for, and eliminates, effects of influences other than those being tested
exon (EKS-ahn) segment of a gene in DNA that is used directly for protein synthesis; in the mRNA transcript of a gene, the intervening intron (noncoding) segments are removed and the remaining exon (coding) segments are spliced together to form the final, edited version of the mRNA transcript; *see* **ribonucleic acid (RNA), transcription**
expiration (eks-pih-RAY-shun) exhaling
expiratory center (eks-PYE-rah-tor-ee) one of the two most important respiratory control centers, located in the medulla
expiratory muscles (eks-PYE-rah-tor-ee) muscles that allow more forceful expiration to increase the rate and depth of ventilation; internal intercostals and abdominal muscles
expiratory reserve volume (ERV) (eks-PYE-rah-tor-ee) amount of air that can be forcibly exhaled after expiring the tidal volume (TV)
extensibility (eks-ten-si-BIL-i-tee) ability of a muscle to extend or stretch and return to resting length
extension increasing the angle between two bones at a joint; as opposed to flexion
external auditory meatus (eks-TER-nal AW-di-tor-ee mee-AY-tus) ear canal; a curved tube (approximately 2.5 cm) extending from the auricle into the temporal bone, ending at the tympanic membrane
external ear outer part of the ear: auricle and external auditory canal
external genitalia (JEN-i-tayl-ee-ah) external reproductive organs; penis and scrotum in males; vagina, vulva, and related structures in females
exteroceptor (eks-ter-oh-SEP-tor) somatic sense receptor located on the body surface
extracellular fluid (ECF) (eks-trah-SEL-yoo-lar) liquid found outside of cells, located in two compartments: between cells (interstitial fluid) and in blood (plasma); lymph, cerebrospinal fluid, and joint fluids are also *extracellular fluids*
extracellular matrix (ECM) (eks-trah-SEL-yoo-lar MAY-triks) material between cells in a tissue, made up of water and a variety of proteins
extrapyramidal tract (eks-trah-pi-RAH-mi-dal) motor tract from the brain to the spinal cord anterior horn motor neurons, except for the corticospinal tract
extrinsic control (eks-TRIN-sik) style of physiological regulation in which the control center (regulatory center) is outside, or extrinsic to, the tissue being regulated; for example, the brain's control of a leg muscle or the pituitary gland's regulation of the thyroid gland
extrinsic eye muscle (eks-TRIN-sik) voluntary muscle that attaches the eyeball to the socket and produces movement of the eyeball
extrinsic factor (eks-TRIN-sik) substance secreted in the stomach that allows vitamin B_{12} to be absorbed by the body

F

facial nerve (FAY-shal) cranial nerve VII, mixed nerve
facilitated diffusion (fah-SIL-i-tay-ted di-FYOO-shun) special type of diffusion; when movement of a molecule is made more efficient by action of carrier mechanisms in the plasma membrane
fallopian tubes (fal-LOH-pee-an) uterine tubes; pair of tubes that conduct the ovum from the ovary to the uterus
falx cerebelli (falks ser-eh-BEL-lee) small fold in the dura mater in the posterior cranial fossa
falx cerebri (falks SER-eh-bree) fold in the dura mater that separates the two cerebral hemispheres
farsightedness *see* **hyperopia**
fascicle (FAS-i-kul) small bundle or cluster, especially of groups of skeletal muscle fibers bound together by perimysium
fast muscle fiber white muscle fiber; primarily relies on anaerobic respiration to produce ATP; responds quickly
fatty acid product of fat digestion; building block of fat molecules
fauces (FAW-seez) opening from the mouth into the oropharynx
febrile (FEB-ril) referring to fever
feces (FEE-seez) waste material discharged from the intestines
feedback control loop highly complex and integrated communication control network, classified as negative or positive; negative feedback loops are the most important and numerous homeostatic control mechanisms
feed-forward concept that information may flow ahead to another process to trigger a change in anticipation of an event that will follow
femoral hernia (FEM-or-all HER-nee-ah) rupture of the lower abdominal wall at the femoral ring
fenestrations (fen-es-TRAY-shuns) perforations in the glomerular endothelium
fertilization (FER-ti-li-ZAY-shun) union of an ovum and a sperm; conception
fetal alcohol syndrome (FAS) (FEE-tal AL-koh-hol SIN-drohm) a condition that may cause congenital abnormalities in a baby; it results from a woman consuming alcohol during pregnancy
fetus (FEE-tus) unborn young, especially in the later stages; in human beings, the fetal stage is from the third month of the intrauterine period until birth
fibrillation (fi-bri-LAY-shun) condition in which individual muscle fibers, or small groups of fibers, contract asynchronously (out of time) with other muscle fibers in an organ (especially the heart), producing no effective movement
fibrin (FYE-brin) insoluble protein in clotted blood
fibrinogen (fye-BRIN-oh-jen) soluble blood protein that is converted to insoluble fibrin during clotting
fibrinolysis (fye-brin-OL-i-sis) physiological mechanism that dissolves clots
fibroblast (FYE-broh-blast) connective tissue cell that synthesizes interstitial fibers and gels
fibrocartilage (fye-broh-KAR-ti-lij) cartilage with the greatest number of collagenous fibers; strongest and most durable type of cartilage
fibromyositis (fye-broh-my-oh-SYE-tis) tendon inflammation with myositis (muscle inflammation), as in a charley horse
fibrosarcoma (fye-broh-sar-KOH-mah) cancer of fibrous connective tissue
fibrous connective tissue (FYE-brus koh-NEK-tiv) strong, nonstretchable, white collagen fibers that make up tendons
fight-or-flight reaction changes produced by increased sympathetic impulses allowing the body to deal with any type of stress
filtrate (FIL-trayt) substance remaining in a liquid after it has passed through a filter
filtration (fil-TRAY-shun) movement of water and solutes through a membrane because of a higher hydrostatic pressure on one side
filum terminale (FYE-lum ter-mi-NAL-ee) slender filament formed by the pia mater that blends with the dura mater and then the periosteum of the coccyx
fimbriae (FIM-bree-ee) fringe of tiny finger-like projections around the opening of each fallopian (uterine) tube; the projections help move an ovum into the fallopian tube.
first-degree burn partial-thickness burn, actual tissue destruction is minimal
fissure (FISH-ur) groove
fixator muscle (fik-SAY-tor) muscle that functions as a joint stabilizer
fixed-membrane-receptor model *see* **second-messenger model**
flaccid muscle (FLAK-sid) muscle with less tone than normal
flagellum (flah-JEL-um) single projection extending from the cell surface; only example in human is the "tail" of the male sperm
flavine adenine dinucleotide (FAD) (FLAY-vin AD-eh-neen dye-NOO-klee-oh-tyde) molecule that serves as an electron carrier in the electron transport system
flexion (FLEK-shun) act of bending; decreasing the angle between two bones at the joint; opposite of extension
flow-volume loop a type of spirogram (breathing graph) that shows the flow rate (L/min) of breathing along one axis and volume of breathing (L) along the other axis, thus forming a loop or circle with each respiratory cycle
fluid balance homeostasis of fluids; the volumes of interstitial fluid, intracellular fluid, and plasma and total volume of water remain relatively constant
fluid compartments areas in the body where fluid is located; interstitial fluid, plasma, and intracellular fluid
fluid mosaic model (moh-ZAY-ik) theory of plasma membrane composition in which molecules of the membrane are bound tightly enough to form a continuous layer but loosely enough so molecules can slip past one another
folia (FOH-lee-ah) thin, delicate gyri (raised ridges) of the surface of the cerebellum (singular folium)
follicle (FOL-li-kul) ovarian structure consisting of oocyte surrounded by numerous supporting cells (follicle cells); also, specialized structures required for hair growth; also, a small hollow sphere with a wall of simple cuboidal glandular epithelium, found in thyroid tissue
follicle-stimulating hormone (FSH) (FOL-li-kul STIM-yoo-lay-ting HOR-mohn) hormone present in males and females; in males, FSH stimulates the production of sperm; in females, FSH stimulates the ovarian follicles to mature and follicle cells to secrete estrogen
follicular cell (foh-LIK-yoo-lar) cell that produces thyroid colloid
foramen ovale (foh-RAY-men oh-VAL-ee) in the developing fetus, opening that shunts blood from the right atrium directly into the left atrium, allowing most blood to bypass the baby's developing lungs
forced expiratory volume (EK-spy-rah-tor-ee) maximum volume (mL or L) of air that can be breathed out; also called *forced vital capacity (FVC)*
formed elements red blood cells, white blood cells, and platelets in blood
fornix (FOR-niks) corner of the vagina where it meets the cervix of the uterus
fovea centralis (FOH-vee-ah sen-TRAL-iss) small depression in the macula lutea where cones are most densely packed; vision is sharpest where light rays focus on the fovea
free fatty acid (FFA) fatty acid combined with albumin to be transported by the blood to other cells
free nerve endings specialized receptors in the skin that respond to pain
frontal plane (FRON-tal plane) lengthwise plane running from side to side, dividing the body into anterior and posterior portions
full-thickness burn third-degree burn; skin is severely damaged and nerve endings are destroyed
functional group small cluster of atoms in an organic molecule that gives the molecule particular functional characteristics such as certain chemical binding properties; often represented generically by the letter R
functional protein (FUNK-shun-al PROH-teen) category of proteins that affect the functional operations of a cell ; contrast to *structural protein*
functional residual capacity (FUNK-shun-al reh-ZID-yoo-al kah-PAH-si-tee) amount of air left in the lungs at the end of a normal expiration
fungus (FUNG-gus) organism similar to plants but lacking chlorophyll and capable of producing mycotic (fungal) infections; *plural,* fungi
furuncle (FUR-un-kul) boil; pus-filled cavity formed by some hair follicle infections

G

G protein protein of a cell's plasma membrane involved in transmitting signals from regulatory chemicals (e.g., hormones or neurotransmitters) and their receptor molecules into the target cell
G_0 phase step of the cell life cycle in which a newly formed daughter cell grows in size and then maintains itself for some time, but fails to proceed onward to prepare for later reproduction (cell division); G_0 phase follows the M phase (mitotic division) and represents the cell "opting out" of further reproduction; *see* **cell life cycle**
G_1 phase step of the cell life cycle in which a newly formed daughter cell grows in size, in anticipation of later reproduction (cell division); G_1 phase follows the M phase (mitotic division) and precedes the S phase (DNA replication); *see* **cell life cycle**
G_2 phase step of the cell life cycle in which a cell grows in size after having replicated its DNA and prepares for mitotic reproduction (cell division); G_2 phase follows the S phase (DNA replication) and precedes the M phase (mitotic division); *see* **cell life cycle**
gametes (GAM-eets) sex cells; spermatozoa and ova
gamma ray (GAM-ah) electromagnetic radiation; more penetrating than alpha and beta particles
ganglia (GANG-glee-ah) in peripheral nerves; regions of gray matter made up of unmyelinated fibers
gangrene (GANG-green) tissue death (necrosis) that involves decay of tissue
gap junction cell connection formed when membrane channels of adjacent plasma membranes adhere to each other
gastric juice stomach secretion containing acid and enzymes; aids in the digestion of food
gastrin (GAS-trin) gastrointestinal (GI) hormone that plays an important regulatory role in the digestive process by stimulating gastric secretion
gastroenteritis (gas-troh-en-ter-EYE-tis) inflammation of the stomach and intestines
gastroenterology (gas-troh-en-ter-OL-oh-jee) study of the stomach and intestines and their diseases
gastrointestinal tract (GI tract) (gas-troh-in-TES-ti-nal) alimentary canal; tube formed by the major organs of digestion
gastrulation (gas-troo-LAY-shun) process by which blastocyst cells move and then differentiate into the three primary germ layers
gated channel channel in the plasma membrane that can be opened and closed to alter membrane permeability
gene one of many segments of a chromosome (DNA molecule); each gene contains the genetic code for synthesizing a protein molecule such as an enzyme or hormone
gene linkage when a whole group of genes stay together during the crossing-over process
general adaptation syndrome (GAS) group of changes that make the presence of stress in the body known
gene therapy manipulation of genes to cure genetic problems; most forms of gene therapy have not yet proven to be effective in humans
genetic counseling (jeh-NET-ik) professional consultations with families regarding genetic diseases
genetic mutation (jeh-NET-ik myoo-TAY-shun) change in the genetic material within a genome; may occur spontaneously or as a result of mutagens
genetic predisposition (jeh-NET-ik pree-dis-poh-ZIH-shun) likelihood due to inherited genes of developing a condition even though the condition itself may not be solely caused by genetic mechanisms
genetics (jeh-NET-iks) scientific study of heredity and the genetic code
genitalia (jen-i-TAIL-yah) reproductive organs
genome (JEE-nohm) entire set of chromosomes in a cell; the *human genome* refers to the entire set of human chromosomes
genomics (jeh-NO-miks) field of endeavor involving the analysis of the genetic code contained in the human or another species' genome
genotype (JEN-oh-type) alleles present at one or more specific loci on a chromosome of a given individual; *see* **phenotype**
germinal epithelium (JER-mi-nal ep-i-THEE-lee-um) small epithelial cells that are on the surface of the ovaries
germinal matrix (JER-mih-nal MAY-triks) cap-shaped cluster of cells at the bottom of a hair follicle
gerontology (jair-on-TAHL-oh-jee) study of the aging process
gestation period (jes-TAY-shun) length of pregnancy, approximately 9 months in humans
gigantism (jye-GAN-tiz-em) condition produced by hypersecretion of growth hormone during the early years of life; results in a child who grows to gigantic size
gingivitis (jin-ji-VYE-tis) inflammation of the gum (gingiva), often as a result of poor oral hygiene

glans penis (glans PEE-nis) slightly bulging structure formed by the distal end of the corpus spongiosum; covered by the foreskin in uncircumcised males
glaucoma (glaw-KOH-mah) disorder characterized by elevated pressure in the eye; can lead to permanent blindness
glia (GLEE-ah) nonexcitable supporting cells of nervous tissue; formerly called *neuroglia*
glioblastoma multiforme (glye-oh-blas-TOH-mah mul-ti-FOR-mee) malignant tumor of astrocyte cells of the brain
glioma (glee-OH-mah) any tumor of neuroglia (glial) cells in nerve tissue
glomerular-capsular membrane (gloh-MER-yoo-lar KAP-soo-lar) membrane made up of glomerular endothelium, basement membrane, and visceral layer of Bowman's capsule; function is filtration
glomerulonephritis (gloh-mer-yoo-loh-neh-FRY-tis) inflammatory disease of the glomerular-capsular membranes of the kidney
glomerulus (gloh-MER-yoo-lus) compact cluster, particularly when referring to the tuft of capillaries forming part of the nephron
glossopharyngeal nerve (glos-oh-fah-RIN-jee-al) cranial nerve IX; mixed nerve
glucagon (GLOO-kah-gon) hormone secreted by alpha cells of the pancreatic islets; increases activity of phosphorylase
glucocorticoid (GC) (gloo-koh-KOR-ti-koyd) hormone that influences food metabolism; secreted by the adrenal cortex
gluconeogenesis (gloo-koh-nee-oh-JEN-eh-sis) formulation of glucose or glycogen from protein or fat compounds
glucosamine (gloo-KOHS-ah-meen) component of some proteoglycans that thicken and hold together connective tissues such as cartilage; *see* **proteoglycan**
glucose (GLOO-kohs) monosaccharide, or simple sugar; principal blood sugar used by cells
glucose phosphorylation (GLOO-kohs fos-for-i-LAY-shun) process of converting glucose to glucose-6-phosphate; prepares glucose for further metabolic reactions
glutamate (GLOO-tah-mayt) glutamic acid; amino acid believed to be responsible for up to 75% of the excitatory signals in the brain
glycerol (GLIS-er-ol) product of fat digestion
glycine (GLYE-seen) amino acid; most widely distributed inhibitory neurotransmitter in the spinal cord
glycogen (GLYE-koh-jen) polysaccharide; main carbohydrate stored in animal cells
glycogenesis (glye-koh-JEN-eh-sis) anabolic pathway of glycogen formation; formation of glycogen from glucose or from other monosaccharides, fructose, or galactose
glycogenolysis (glye-koh-jeh-NOL-i-sis) hydrolysis of glycogen to glucose-6-phosphate or to glucose
glycolipid (glye-koh-LIP-id) lipid molecule with attached carbohydrate group
glycolysis (glye-KOHL-i-sis) first series of chemical reactions in carbohydrate catabolism; changes glucose to pyruvic acid in a series of anaerobic reactions
glycophospholipid (glye-koh-fos-foh-LIP-id) molecule that is part sugar and part phospholipid; formed in epidermal cells of the skin to produce a water-proof barrier
glycoprotein (glye-koh-PROH-teen) substance made of molecules that are a combined form of carbohydrate and protein
glycosuria (glye-koh-SOO-ree-ah) glucose in urine; a sign of diabetes mellitus
goblet cell epithelial cell that produces and secretes large amounts of mucus
Golgi apparatus (GOL-jee ap-ah-RAH-tus) organelle consisting of small sacs stacked on one another near the nucleus that makes carbohydrate compounds, combines them with protein molecules, and packages the product for distribution from the cell
Golgi tendon receptors (GOL-jee) sensors that are responsible for proprioception; stimulated by excessive muscle contraction
gomphosis (gom-FOH-sis) fibrous joint where a process is inserted into a socket; for example, the joint between the tooth and mandible
gonadotroph (go-NAD-oh-trohf) cell type of the adenohypophysis (anterior pituitary) that secretes the gonadotropins luteinizing hormone (LH) and follicle-stimulating hormone (FSH)
gonadotropins (go-nah-doh-TROH-pins) hormones (FSH and LSH) produced by the anterior pituitary that stimulate growth and maintenance of the testes or ovaries
gonads (GO-nads) sex glands in which reproductive cells are formed; ovaries in women, testes in men
goniometer (GON-ee-OM-eh-ter) instrument used to measure range of motion (ROM) angles of a joint
gout (gowt) condition characterized by excessive levels of uric acid in the blood that are deposited in the joints
Graafian follicle (GRAF-ee-an FOL-li-kul) a secondary follicle; consists of a mature ovum surrounded by granulosa cells at boundary of fluid-filled antrum
graded potential local potentials that vary according to strength of stimulus and distance on membrane from source
graded strength principle skeletal muscles contract with varying degrees of strength at different times
gradient (GRAY-dee-ent) measurable difference between two points of a given variable such as molecular concentration, pressure, or electrical charge
granulocyte (GRAN-yoo-loh-syte) leukocyte with granules in cytoplasm
Graves disease inherited, possibly immune endocrine disorder characterized by hyperthyroidism accompanied by exophthalmos (protruding eyes)
gray fibers unmyelinated nerve fibers; gray matter
gray matter (gray MAT-ter) type of nerve tissue characterized by a relative lack of myelin and often including many neuron cell bodies and synapses that process information; contrast with *white matter*
gray ramus (gray RAY-muss) short branch by which some postganglionic axons return to a spinal nerve
greater omentum (oh-MEN-tum) pouchlike extension of the visceral peritoneum extending from greater curvature of the stomach to the transverse colon; often called "the lace apron"
greater vestibular glands (ves-TIB-yoo-lar) glands on each side of the vaginal orifice that secrete a lubricating fluid; also called *Bartholin's glands*
greenstick fracture incomplete fracture most commonly seen in children because of resilience of their bones
gross anatomy study of body parts visible to the naked eye
ground substance organic matrix of bone and cartilage
growth normal increase in size or number of cells
growth hormone (GH) (HOR-mohn) hormone secreted by the anterior pituitary gland that controls the rate of skeletal and visceral growth
gustatory (GUS-tah-tor-ee) refers to taste
gustatory cell (GUS-tah-tor-ee) chemoreceptors in tongue that sense taste
gustatory hair (GUS-tah-tor-ee) cilia-like structure projecting from gustatory cells and into taste pores
Guthrie test (GUH-three) blood test to detect phenylketonuria (PKU)
gynecology (gye-neh-KOL-oh-jee) study of the female reproductive system
gyrus (JYE-rus) convoluted ridge, usually refers to rounded elevations of the brain surface

H

hair cells specialized mechanoreceptors in the ear that are responsible for balance and hearing
hair follicle (hair FOL-li-kul) small blind-end tube extending from the dermis through the epidermis that contains the hair root and where hair growth occurs; sebaceous and apocrine skin glands have ducts leading into the follicle
hair papilla (hair pah-PIL-ah) small, cap-shaped cluster of cells located at the base of the follicle where hair growth begins
haploid number (HAP-loyd) halved number of chromosomes in gametes resulting from meiosis; in humans, 23 chromosomes per sex cell
Haversian canal (hah-VER-shun) canal in the Haversian system of bone that runs parallel to the long axis; contains blood vessels and nerves
Haversian system (hah-VER-shun) circular arrangements of calcified matrix and cells that give microscopic bone its characteristic appearance; also called an *osteon*
heart organ of circulatory system that pumps the blood; composed of cardiac muscle tissue
heart block blockage of impulse conduction from atria to ventricles so that the heart beats at a slower rate than normal
heart failure inability of the heart to pump returned blood sufficiently
heart murmur abnormal heart sound that may indicate valvular insufficiency or stenosing of the valve
heat exhaustion (ek-ZAWS-chun) condition caused by fluid loss resulting from activity of thermoregulatory mechanisms in a warm external environment
heat stroke life-threatening condition characterized by high body temperature; failure of thermoregulatory mechanisms to maintain homeostasis in a very warm external environment
helper T cells immune system cells that help B cells differentiate into antibody-secreting plasma cells; also help coordinate cellular immunity through direct contact with other immune cells
hematocrit (hee-MAT-oh-krit) volume percent of blood cells in whole blood; packed cell volume
hematopoiesis (hem-ah-toh-poy-EE-sis) process of blood cell formation
hematopoietic tissue (hee-mah-toh-poy-ET-ik) specialized connective tissue that is responsible for formation of blood cells and lymphatic system cells; in red bone marrow, spleen, tonsils, and lymph nodes
hemisphere (HEM-iss-feer) one half of a generally spherical structure, as in the left and right hemispheres of the cerebellum or cerebrum
hemisphericity (hem-iss-fer-ISS-it-ee) different specialization of function in each hemisphere of an organ, as in the cerebral hemispheres
hemocytoblast (hee-moh-SYE-toh-blast) bone marrow stem cell from which all formed elements of blood arise
hemodialysis (hee-moh-dye-AL-i-sis) therapy involving separation of smaller (diffusible) particles from larger (nondiffusible) particles in blood through a semipermeable membrane, usually employed when a patient's kidneys fail to remove these particles from the blood
hemodynamics (hes-moh-dye-NAM-iks) mechanisms that influence dynamic circulation of blood
hemoglobin (hee-moh-GLOH-bin) iron-containing protein in red blood cells responsible for their oxygen-carrying capacity
hemorrhoid (HEM-eh-royd) varicose vein in the rectum; also called *piles*
hemostasis (hee-moh-STAY-sis) stoppage of blood flow
heparin (HEP-ah-rin) substance obtained from the liver; inhibits blood clotting
hepatic duct (heh-PAT-ik) one of two ducts that drains bile out of the liver
hepatic lobule (heh-PAT-ik LOB-yool) anatomical units of the liver
hepatitis (hep-ah-TYE-tis) inflammation of the liver; may be caused by toxins, viruses (e.g., hepatitis A, hepatitis B), bacteria, or parasites
heredity (heh-RED-i-tee) transmission of characteristics from parents to offspring
Hering-Breuer reflexes (HER-ing BROO-er REE-fleks-ez) control respirations—especially rate and rhythmicity
herniated disk (HER-nee-ayt-ed disk) condition of the vertebral disk when the annulus fibrosis becomes disrupted, allowing the nucleus pulposis to protrude
herpes zoster (HER-peez ZOS-ter) "shingles," viral infection that affects the skin of a single dermatome
heterozygous (het-er-oh-ZYE-gus) genotype with two different forms of a trait
hexose (HEK-sohs) monosaccharide molecule with six carbons
hiatal hernia (hye-AY-tal HER-nee-ah) condition in which a portion of the stomach is pushed through the hiatus (opening) of the diaphragm, often weakening or expanding the cardiac sphincter at the inferior end of the esophagus
hiccup (HIK-up) involuntary spasmodic contraction of the diaphragm
high-density lipoprotein (lip-oh-PROH-teen) a blood lipid fraction (plasma protein) composed of cholesterol, about 50% protein, and triglycerides. Often called "good cholesterol" because it carries cholesterol out of the blood to the liver for elimination
high-energy bond chemical bond that requires an input of energy to form and when broken can result in the transfer of useful energy to cellular processes, as in ATP
hilum (HYE-lum) slit on medial surface of each lung where primary bronchi and pulmonary vessels enter; slit on medial surface of each kidney where blood vessels and other structures enter the kidney
histamine (HIS-tah-meen) inflammatory chemical
histogenesis (hiss-toh-JEN-eh-sis) formation of tissues from primary germ layers of embryo
histology (his-TOL-oh-jee) branch of microscopic anatomy that studies tissues; biology of tissues
histophysiology (his-toh-fiz-ee-OL-oh-jee) study of microscopic anatomy of cells and tissues and how structural information correlates with function
Hodgkin lymphoma (HOJ-kin lim-FOH-mah) type of lymphoma (malignant lymph tumor) characterized by painless swelling of lymph nodes in the neck, progressing to other regions
holocrine gland (HOH-loh-krin) gland that collects secretory product inside its cells, which then rupture completely to release it; for example, sebaceous glands of the skin
homeostasis (hoh-mee-oh-STAY-sis) relative constancy of the normal body's internal (fluid) environment
homeostatic control mechanisms (hoh-mee-oh-STAH-tik) devices for maintaining or restoring homeostasis
homozygous (hoh-moh-ZYE-gus) genotype with two identical forms of a trait
horizontal fissure (FISH-ur) separates the superior lobe of the right lung from the middle lobe
horizontal plane transverse plane; divides the body into superior and inferior portions
hormone (HOR-mohn) substance secreted by an endocrine gland into the bloodstream that acts on a specific target tissue to produce a given response
hormone-receptor complex (HOR-mohn-ree-SEP-tor) when a steroid hormone combines to a receptor site in the target cell and then moves into the nucleus of the target cell
human chorionic gonadotropin (HCG) (kohr-ee-ON-ic go-nah-doh-TROH-pin) hormone secreted early in pregnancy by the placenta that serves to maintain the uterine lining
Human Genome Project (HGP) a worldwide collaborative effort of scientists and others to map out the entire human genome and study the biological, medical, and ethical aspects of their discoveries; the HGP is largely funded by U.S. government sources such as the DOE (Department of Energy) and the NIH (National Institutes of Health); *see* **genome, genomics**
humoral immunity (HYOO-moh-ral i-MYOO-ni-tee) antibody-mediated immunity occurring within blood plasma and other body fluids
huntingtin (HUN-ting-tin) a protein normally present in all cells; an abnormal version of huntingtin is produced in people with Huntington disease (HD), clinging to molecules too tightly in the brain and thus preventing normal function
Huntington disease (HD) (HUN-ting-ton) an inherited disease characterized by chorea (involuntary, purposeless movements) that progresses to severe dementia and death; *see* **huntingtin**
hyaline cartilage (HYE-ah-lin KAR-ti-lij) most common type of cartilage; appears gelatinous and glossy
hyaluronic acid (hye-al-yoo-RAHN-ik) type of proteoglycan that helps thicken and hold together the matrix of connective tissue; *see* **proteoglycan**
human engineered chromosome (HEC) (KROH-moh-sohm) type of gene therapy in which a set of therapeutic genes is incorporated into a separate strand of DNA that is inserted into a cell's nucleus, thus acting like an extra, or forty-seventh, chromosome
hybridoma (hye-brid-OH-mah) hybrid cell formed by fusion of cancerous cell with a lymphocyte to mass produce a specific antibody (monoclonal antibodies)
hydrase (HYE-drays) enzyme that adds water to a molecule without splitting it
hydrocortisone (hye-droh-KOHR-ti-zohn) hormone secreted by the adrenal cortex; cortisol; compound F
hydrogen bond (HYE-droh-jen) weak chemical bond that occurs between the partial positive charge on a hydrogen atom covalently bound to a nitrogen or oxygen atom and the partial negative charge of another polar molecule
hydrogen ion (HYE-droh-jen EYE-on) a proton or a hydrogen atom without its electron; occurs in water and water solutions; produces an acidic solution; has a positive charge (H^+)
hydrolase (HYE-droh-lays) hydrolyzing enzyme
hydrolysis (hye-DROHL-i-sis) chemical process in which a compound is split by addition of H1 and OH− portions of a water molecule
hydrophilic (hye-droh-FIL-ik) "water-loving"
hydrophobic (hye-droh-FOH-bik) "water-fearing"
hydrostatic pressure (hye-droh-STAT-ik PRESH-ur) force of a fluid pushing against some surface
hydrostatic pressure gradient (hye-droh-STAT-ik PRESH-ur GRAY-dee-ent) situation in which a difference in pressure is caused by the force of a fluid as it moves from one point to another
hymen (HYE-men) Greek for "membrane"; mucous membrane that may partially or entirely occlude the vaginal outlet
hypercapnia (hye-per-KAP-nee-ah) excessive carbon dioxide in the blood
hyperextension (hye-per-ek-STEN-shun) stretching an extended part beyond its anatomical position
hyperglycemia (hye-per-glye-SEE-mee-ah) higher than normal blood glucose concentration
hyperkalemia (hye-per-kah-LEE-mee-ah) excessive potassium in the blood
hyperkeratosis (hye-per-ker-ah-TOH-sis) thickening of the horny layer of the skin
hyperopia (hye-per-OH-pee-ah) refractive disorder of the eye caused by a shorter-than-normal eyeball; results in ability to see objects at a distance better than objects nearer; also called *farsightedness*
hyperplasia (hye-per-PLAY-zee-ah) growth of an abnormally large number of cells at a local site, as in a neoplasm or tumor

hyperpnea (hye-PERP-nee-ah) abnormal increase in respiratory rate and depth

hyperpolarization (hye-per-pol-lar-i-ZAY-shun) increase in electrical charges separated by the cell membrane; causes change away from 0 mV

hypersecretion (hye-per-seh-KREE-shun) too much secretion of a substance

hypertension (hye-per-TEN-shun) abnormally high blood pressure

hypertonic (hye-per-TON-ik) solution containing a higher level of salt (NaCl) than is found in a living red blood cell (above 0.9% NaCl); causes cells to shrink

hypertrophy (hye-PER-troh-fee) increased size of an organ or part caused by an increase in the size of its cells

hyperventilation (hye-per-ven-ti-LAY-shun) very rapid, deep respirations

hypervitaminosis (hye-per-vye-tah-mi-NO-sis) general name for any condition resulting from an abnormally high intake of vitamins

hypervolemia (hye-per-voh-LEE-mee-ah) abnormally increased blood volume

hypochloremia (hye-poh-kloh-REE-mee-ah) abnormally low levels of chloride in the blood associated with potassium loss

hypodermis (hye-poh-DER-mis) loose layer of skin, rich in fat and areolar tissue located beneath the dermis

hypoglossal (hye-poh-GLOS-al) under the tongue

hypoglossal nerve (hye-poh-GLOS-al) cranial nerve XII; motor nerve; responsible for tongue movement

hypoglycemia (hye-poh-glye-SEE-mee-ah) lower than normal blood glucose concentration

hypophyseal portal system (hye-poh-FIZ-ee-al) complex of small blood vessels through which releasing hormones travel from the hypothalamus to the pituitary

hypophysis (hye-POF-i-sis) pituitary gland

hypopotassemia (hye-poh-poh-tah-SEE-mee-ah) potassium deficit

hyposecretion (hye-poh-seh-KREE-shun) too little secretion of a substance

hypothalamus (hye-poh-THAL-ah-muss) important autonomic and neuroendocrine control center located inferior to the thalamus in the brain

hypothermia (hye-poh-THER-mee-ah) failure of thermoregulatory mechanisms to maintain homeostasis in a very cold external environment; results in abnormally—sometimes life-threatening—low body temperature

hypothesis (hye-PAHTH-es-iss) idea or scientific concept, usually based on previous ideas or observations, that is proposed as a possible explanation of nature or a natural process; hypotheses (hye-PAHTH-es-eez) undergo intense testing before being accepted widely in the scientific community; *see* **law, science, theory**

hypotonic (hye-poh-TON-ik) adjective used to describe a solution that has a lower potential osmotic pressure than a solution to which it is being compared; a hypotonic solution tends to have low osmolality and thus tends to gain water (by way of osmosis) from the solution to which it is compared

hypoventilation (hye-poh-ven-ti-LAY-shun) slow and shallow respirations

hypoxia (hye-POK-see-ah) deficiency of oxygen in the blood

I

ideogram (ID-ee-oh-gram) a graphic representation of any idea; in genomics, a simple cartoon of a chromosome often used to show the overall physical structure of a chromosome; *see* **chromosome, genomics**

idiopathic (id-ee-oh-PATH-ik) refers to a disease of undetermined cause

ileum (IL-ee-um) distal portion of the small intestine

immune system (i-MYOON SIS-tem) body's defense system against disease

immunization (i-myoo-ni-ZAY-shun) deliberate artificial exposure to disease to produce acquired immunity

immunoglobulin (i-myoo-no-GLOB-yoo-lin) antibodies

immunology (i-myoo-NOL-oh-jee) study of immune system functions and mechanisms

immunotherapy (i-myoo-no-THAYR-ah-pee) therapeutic technique that bolsters a person's immune system in an attempt to control a disease

impermeable (im-PERM-ee-ah-bul) adjective used to describe a membrane that does not allow substances to move through (permeate) it

impermeant (im-PERM-ee-ent) adjective used to describe a substance that is not able to move through (permeate) a membrane

impetigo (im-peh-TYE-go) highly contagious bacterial skin infection that occurs most often in children

impulse (IM-puls) an electrical signal

inborn immunity (IN-born i-MYOO-ni-tee) inherited immunity to disease

inclusion (in-KLOO-zhun) an unidentified particle within a cell or other structure

incomplete tetanus (in-kom-PLEET TET-ah-nus) tetanus with very short periods of relaxation occurring between peaks of muscle tension

incontinence (in-KON-ti-nens) involuntary voiding of urine

incubation (in-kyoo-BAY-shun) early, latent stage of an infection, during which an infection has begun but signs or symptoms have not yet developed

infancy period of human development from birth to about 18 months of age

infant respiratory distress syndrome (IRDS) (IN-fant RES-per-ah-toh-ree di-STRESS SIN-drohm) leading cause of death in premature babies, caused by a lack of surfactant in the alveolar air sacs, inferior lower portion; opposite of superior

inferior lower; opposite of superior

inflammation (in-flah-MAY-shun) group of responses to a tissue irritant marked by signs of redness, heat, swelling, and pain

inflammatory response (in-FLAM-ah-toh-ree) specific process involving tissues and blood vessels in response to injury

infundibulum (in-fun-DIB-yoo-lum) stalk that connects the pituitary gland to the hypothalamus

ingestion (in-JEST-chun) taking in of complex foods, usually by mouth

inguinal hernia (ING-gwi-nal HER-nee-ah) rupture of the lower abdominal wall at the inguinal canal

inherited immunity (in-HAIR-i-ted i-MYOO-ni-tee) inborn immunity; occurs when immune mechanisms are put in place by genetic mechanisms during the early stages of human development

inhibin (in-HIB-in) glycoprotein hormone produced by the ovary to regulate FSH secretion by the anterior pituitary (adenohypophysis)

inhibitory (in-HIB-i-tor-ee) something that slows or stops a process

inhibitory neurotransmitter (in-HIB-i-tor-ee noo-roh-TRANS-mit-er) neurotransmitter that causes hyperpolarization of the postsynaptic membrane, thereby decreasing chances of impulse propagation

inhibitory postsynaptic potential (IPSP) (in-HIB-i-tor-ee post-si-NAP-tik poh-TEN-shal) temporary hyperpolarization that makes the inside of the membrane even more negative than at the resting potential

innate immunity (in-AYT i-MYOO-ni-tee) type of immunity that exists prior to exposure to a specific antigen and can recognize and destroy a variety of harmful agents or conditions; also called *nonspecific immunity*

inorganic compounds (in-or-GAN-ik KOM-powndz) chemical constituents that do not contain both carbon and hydrogen; for example, water, carbon dioxide, and oxygen

insertion (muscle insertion) attachment of a muscle to the bone that it moves when contraction occurs (as distinguished from its origin); (insertion mutation) a type of genetic mutation that occurs when one or more nucleotide bases appear within the usual sequence of nucleotide bases in a gene; in insertion mutations, the cell cannot read the genetic code normally and thus the encoded protein cannot be made in its usual form; *see* **deletion**

inspiratory capacity (IC) (in-SPY-rah-tor-ee kah-PAS-i-tee) maximal amount of air an individual can inspire following a normal expiration

inspiratory center (in-SPY-rah-tor-ee) one of the two most important control centers located in the medulla; the other is the expiratory center

inspiratory reserve volume (IRV) (in-SPY-rah-tor-ee) amount of air that can be forcibly inspired over and above a normal respiration

insulin (IN-suh-lin) hormone secreted by beta cells of the pancreatic islets that increases the uptake of glucose and amino acids by most body cells

integrin (in-TEG-rin) type of protein that acts as a receptor to bind structural proteins in a way that connects a cell's cytoskeleton to surrounding structures such as other cells

integument (in-TEG-yoo-ment) skin; the body's largest organ

integumentary system (in-teg-yoo-MEN-tar-ee) skin and its related structures

intercalated disk (in-TER-kah-lay-ted) cross striations and unique dark bands in cardiac muscle fibers

interferon (in-ter-FEER-on) small protein produced by the immune system that inhibits virus multiplication

interleukin (IL) (in-ter-LOO-kin) any of a class of about a dozen cytokines, regulatory chemicals secreted by leukocytes and other cells, that are involved in regulating a wide variety of immune functions in different cell types; each interleukin is named by the IL acronym with its number, such as IL-1, IL-2, and so on

intermediate fiber (in-ter-MEE-dee-it) muscle fiber exhibiting characteristics between fast and slow fibers

intermediate filament (in-ter-MEE-dee-it FIL-ah-ment) twisted strands of protein, slightly larger than microfilaments that make up part of the cell's internal skeleton (the cytoskeleton)

interneuron (in-ter-NOO-ron) in a three-neuron reflex arc, nerve cell that conducts impulses from a sensory neuron to a motor neuron

interphase (IN-ter-fayz) mitotic phase immediately before visible condensation of the chromosomes during which the DNA of each chromosome replicates itself

interspinales (in-ter-spy-NAH-leez) group of back muscles that connect one vertebra to the next—also helping to extend the back and neck or flex them to the side

interstitial cell (in-ter-STISH-al) small, specialized cells in the testes that secrete the male sex hormone, testosterone

interstitial fluid (IF) (in-ter-STISH-al) fluid located in microscopic spaces between cells

interstitial growth (in-ter-STISH-al) cartilage growth following mitosis and secretion of matrix by chondrocytes; interstitial growth of epiphyseal plate results in growth in length of long bones

intestinal flora (in-TES-ti-nal FLOR-ah) community of various bacterial populations that normally inhabit the colon

intestine (in-TES-tin) part of the digestive tract into which food passes after it leaves the stomach; separated into two segments, the small and the large

intracellular control (in-trah-SELL-yoo-lar) level of homeostatic control of body processes that occurs within cells, as in genetic regulation or enzymatic regulation of the cell

intracellular fluid (ICF) compartment (in-trah-SELL-yoo-lar) fluid located within cells; largest fluid compartment

intramembranous ossification (in-trah-MEM-brah-nus os-i-fi-KAY-shun) process by which most flat bones are formed within connective tissue membranes

intrinsic control (in-TRIN-sik) level of homeostatic control of body processes that occurs within a particular tissue or organ, as when local regulators such as prostaglandins regulate local physiology; may be called *local control* or *autoregulation*

intrinsic eye muscles (in-TRIN-sik eye) involuntary muscles located within the eye; responsible for size of the iris and shape of the lens

intrinsic rhythm (in-TRIN-sik RITH-em) specialized cells in the sinoatrial (SA) node that produce regular impulses without stimulation from the brain or spinal cord

intron (IN-trahn) segment of a gene in a DNA molecule that is not used to code for a protein; introns are removed from mRNA transcripts of the gene and the remaining exons are spliced together to form the final, edited version of the mRNA transcript; *see* **exon, RNA**

involuntary muscle (in-VOL-un-tair-ee) smooth muscle not under conscious control and found in organs such as the stomach, small intestine, and ureters

ion (EYE-on) electrically charged atom or group of atoms

ionic bond (eye-ON-ik) electrocovalent bond; bond formed by transferring of electrons from one atom to another

ionizing radiation (EYE-on-eyz-ing ray-dee-AY-shun) form of radioactivity, or radiation energy, resulting from alpha or beta particles or gamma rays emitted from radioactive atoms scoring direct hits on other atoms and thus ionizing the atoms by knocking electrons out of their outer energy levels; the effect of ionization may injure, kill, or change living cells; *see* **radiation, radiation sickness**

ipsilateral (ip-si-LAT-er-al) on the same side

ipsilateral reflex arc (ip-si-LAT-er-al REE-fleks ark) reflex arc whose receptors and effectors are located on the same side of the body

iris (EYE-ris) colored portion of the eye

irritability (ir-i-tah-BIL-i-tee) ability of a muscle to be stimulated; *see* **excitability**

ischemia (is-KEE-mee-ah) reduced flow of blood to tissue resulting in impairment of cell function

islets of Langerhans (EYE-lets of lahn-GER-hans) pancreatic islets

isometric contraction (eye-soh-MET-rik kon-TRAK-shun) type of muscle contraction in which muscle does not alter the distance between two bones; *see* **isotonic contraction**

isotonic (eye-so-TON-ik) two fluids that have the same potential osmotic pressure

isotonic contraction (eye-so-TON-ik) type of muscle contraction in which the muscle sustains the same tension or pressure and a change in the distance between two bones occurs

isotope (EYE-so-tohp) atoms with the same atomic number but different atomic weights

J

jaundice (JAWN-dis) abnormal yellowing of skin, mucous membranes, and white of eyes

jejunum (jeh-JOO-num) middle third of the small intestine

joint junction between two or more bones; articulation

joint capsule sleevelike extension of the periosteum of each of the bones at an articulation

joint cavity small space between articulating surfaces of the two bones of the joint

juxtaglomerular apparatus (juks-tah-gloh-MER-yoo-lar ap-ah-RAH-tus) in the nephron, the complex of cells from the distal tubule and the afferent arteriole, which helps regulate blood pressure by secreting renin in response to blood pressure changes in the kidney; located near the glomerulus

K

Kaposi sarcoma (KAH-poh-see sar-KOH-mah) rare malignant neoplasm of the skin that often spreads to lymph nodes and internal organs; Kaposi sarcoma is often found in AIDS patients

karyotype (KAIR-ee-oh-type) ordered arrangement of photographs of chromosomes from a single cell used in genetic counseling to identify chromosomal disorders such as trisomy or monosomy

keloid (KEE-lloyd) unusually thick fibrous scar on the skin

keratin (KER-ah-tin) tough, fibrous protein substance in hair, nails, outer skin cells, and horny tissues

keratinization (ker-ah-tin-i-ZAY-shun) process by which cells of the stratum corneum become fitted with keratin and move to the surface

keratinocyte (keh-RAT-i-no-syte) epidermal cell responsible for synthesizing keratin

ketogenesis (kee-toh-JEN-eh-sis) process that produces ketone bodies

ketone body (KEY-tohn) acid produced during fat catabolism

ketosis (key-TOH-sis) large amount of ketone bodies present in the blood of a person with uncontrolled diabetes mellitus

kidney one of the two organs that cleanses the blood of waste products continually produced by metabolism; the kidneys produce urine

killer T cells cytotoxic T cells

kilocalorie (Kcal) (KIL-oh-kal-oh-ree) 1000 calories; *see* **Calorie**

kinase (KYE-nayz) substance that converts proenzymes to active enzymes

kinesthesia (kin-es-THEE-zee-ah) "muscle sense"; that is, sense of position and movement of body parts

Klinefelter syndrome (KLINE-fel-ter SIN-drohm) genetic disorder caused by the presence of two or more X chromosomes in a male (typically trisomy XXY); characterized by long legs, enlarged breasts, low intelligence, small testes, sterility, and chronic pulmonary disease

Krause's end bulb (KROW-sez) skin receptor that detects sensations of touch, low-frequency vibration, and texture differences

Krebs cycle also called *tricarboxylic acid (TCA) cycle*; *see* **citric acid cycle**

Kupffer cell (KOOP-fer) macrophage found in spaces between liver cells

kyphosis (kye-FOH-sis) abnormally exaggerated thoracic curvature of the vertebral column

L

labia majora (LAY-bee-ah mah-JO-rah) "large lips" of the vulva; *singular,* labium majus

labia minora (LAY-bee-ah mi-NO-rah) "small lips" of the vulva; *singular,* labium minus

labor process of expulsion of the fetus and the placenta; childbirth

labyrinth (LAB-i-rinth) bony cavities and membranes of the inner ear

lacrimal apparatus (LAK-ri-mal app-ah-RAT-us) in the eye, the tear (lacrimal) gland plus associated ducts that form tears

lactase (LAK-tayse) enzyme needed to digest lactose

lactation (lak-TAY-shun) milk production

lacteal (LAK-tee-al) lymphatic vessel located in each villus of the intestine; serves to absorb fat materials from chyme passing through the small intestine

lactic acid (LAK-tik) product of anaerobic respiration that accumulates in muscle tissue during exercise and causes a burning sensation

lactose intolerance (LAK-tohs in-TOL-er-ans) condition in which one lacks the enzyme lactase, resulting in an inability to digest lactose (a disaccharide in milk and dairy products)

lactotroph (lak-toh-TROHF) cell type of the adenohypophysis (anterior pituitary) that secretes prolactin

lacuna (lah-KOO-nah) space or cavity; for example, lacunae in bone contain bone cells

lamella (lah-MEL-ah) thin layer, as of bone

Langerhans' cell (LAHNG-er-hanz) cell that plays a limited role in immunological reactions that affect the skin; serves as defense mechanism for the body

Langer's lines (LANG-erz) pattern of dense bundles of white collagenous fibers that characterize the reticular layers of dermis; also called *cleavage lines*

lanugo (lah-NOO-go) extremely fine and soft hair coat on developing fetus

LaPlace, Law of *see* Young-LaPlace Law

laryngitis (lar-in-JYE-tis) inflammation of the mucous tissues of the larynx (voice box)

laryngopharynx (lah-ring-go-FAIR-inks) lowest part of the pharynx

larynx (LAIR-inks) voice box located just below the pharynx; the largest piece of cartilage making up the larynx is the thyroid cartilage, commonly known as the *Adam's apple*

latent period (LAY-tent) period between time a muscle fiber is stimulated and when it contracts; precedes the contraction phase

lateral (LAT-er-al) of or toward the side; opposite of medial

lateral fissure (fissure of Sylvius) (LAT-er-al FISH-ur [FISH-ur of SIL-vi-us]) deep groove between the temporal lobe below and the frontal and parietal lobes of the brain

law scientific idea or explanation that has a very high degree of confidence or certainty after rigorous testing and observation; compare to **hypothesis** and **theory**

length-tension relationship maximum strength that a muscle can develop bears a direct relationship to the initial length of its fibers

lens refracting mechanism of the eye that is located directly behind the pupil

leptin (LEP-tin) a protein hormone produced by fat-storing cells in adipose tissue that plays a role in inhibiting food intake by regulating the satiety center in the hypothalamus; leptin also plays a role in reducing fat storage in nonadipose cells in the liver and skeletal muscles; because of its role in fat storage, it may play a role in future treatments for diabetes, obesity, and other conditions related to fat metabolism; *see* **satiety center**; leptin also helps regulate some immune and neuroendocrine functions and plays a role in development

leukocyte (LOO-koh-syte) white blood cell

leukocytosis (loo-koh-sye-TOH-sis) abnormally high white blood cell numbers in the blood

leukopenia (loo-koh-PEE-nee-ah) abnormally low white blood cell numbers in the blood

levodopa (L-dopa) (LEV-oh-doh-pah) a molecule derived from tyrosine in neurons that is used to produce the neurotransmitter dopamine

ligament (LIG-ah-ment) band of white fibrous tissue connecting bones to other bones

light micrograph (MYK-roh-graf) photographic image of a microscopic structure using a light microscope

light microscope (MYK-roh-skope) magnifying device made of glass lenses that use light transmitted through or reflected from a specimen

limbic system (LIM-bik) parts of the brain involved in emotions and sense of smell; plays key role in coupling sensory inputs to short- and long-term memory; consists of the hippocampus, the hypothalamus, and several other structures

lingual tonsil (LING-gwal TAHN-sil) tonsil located at the base of the tongue

lipid (LIP-id) class of organic compounds that includes fats, oils, and related substances

lipogenesis (lip-oh-JEN-eh-sis) formation of body fat from food sources

lipoma (li-POH-mah) benign tumor of adipose (fat) tissue

lipoprotein (lip-oh-PROH-teen) substance that is part lipid and part protein; produced mainly in the liver

litmus (LIT-mus) pigment used to test for acidity and alkalinity

lobar (LOH-bar) referring to a lobe

lobectomy (loh-BEK-toh-mee) surgical removal of a single lobe of an organ, as in the removal of one lobe of a lung

local potential slight shift from resting membrane potential in a specific region of the plasma membrane

lock-and-key model description of how a specific enzyme will fit into only a specific substrate, like a key fits into a lock

loop of Henle (loop of HEN-lee) extension of the proximal tubule of the nephron

lordosis (lor-DOH-sis) abnormally exaggerated lumbar curvature of the vertebral column

low-density lipoprotein (lip-oh-PROH-teen) a blood lipid fraction (plasma protein) composed of relatively more cholesterol and triglycerides than protein; often call "bad cholesterol" because high levels contribute to atherosclerotic plaque formation and arterial wall damage

lumbar plexus (LUM-bar PLEK-sus) spinal nerve plexus located in the low back

lumbar puncture (LUM-bar) clinical procedure in which some cerebrospinal fluid is withdrawn from the subarachnoid space in the lumbar region of the spinal cord for analysis

lumen (LOO-men) the hollow area of a hollow organ such as the stomach, blood vessel, or urinary bladder; *see also* **luminal**

luminal (LOO-min-al) pertaining to the hollow part, or lumen, of a hollow organ such as the urinary bladder or small intestine; *see also* **lumen**

lunula (LOO-nyoo-lah) crescent-shaped white area under the proximal nail bed

luteinization (loo-tee-in-i-ZAY-shun) formation of a golden body (corpus luteum) in the ruptured follicle

luteinizing hormone (LH) (loo-tee-in-EYE-zing HOR-mohn) in females, acts in conjunction with follicle-stimulating hormone (FSH) to stimulate follicle and ovum maturation, release of estrogen, and ovulation; known as the *ovulating hormone*; in males, causes testes to develop and secrete testosterone

lymph (limf) specialized fluid formed in the tissue spaces that returns excess fluid and protein molecules to the blood via the lymphatic vessels

lymph node (limf node) performs biological filtration of lymph on its way to the circulatory system

lymphangitis (lim-fan-JYE-tis) inflammation of lymph vessels, usually caused by infection, characterized by fine red streaks extending from the site of infection; may progress to septicemia (blood infection)

lymphatic vessels (lim-FAT-ik VES-els) system of blind-ended vessels that collect lymph and deliver it to the circulatory system via the thoracic duct and the right lymphatic duct

lymphedema (lim-fah-DEE-mah) swelling caused by lymphatic vessel blockage

lymphocyte (LIM-foh-syte) one type of white blood cell; *see* B lymphocytes, T lymphocytes

lymphokine (LIM-foh-kyne) chemical compounds released by antigen-bound sensitized T cells

lymphoma (lim-FOH-mah) cancer of lymphatic tissue

lymphotoxin (lim-foh-TAWK-sin) powerful poison that quickly kills any cells it attacks

lysosome (LYE-so-sohm) membranous organelle containing various enzymes that can dissolve most cellular compounds; called *digestive bags* or *suicide bags of cell*

M

M phase (em fayz) step of the cell life cycle in which a cell divides through the process of mitosis (mitotic cell division); M phase follows the G_2 phase (second growth phase) and occurs along with cytokinesis (splitting of the cell's cytoplasm into two daughter cells); *see* cell life cycle

macromolecule (mak-roh-MOL-eh-kyool) large, complex chemical made of combinations of molecules

macrophage (MAK-roh-fayj) phagocytic cell in the immune system

macula (MAK-yoo-lah) highly specialized strip of epithelium in the utricle and saccule; provides information related to head position or acceleration

macula lutea (MAK-yoo-lah LOO-tee-ah) yellowish area near center of the retina where cones are densely distributed

magnetic resonance imaging (MRI) (mag-NET-ik REZ-uh-nens IM-aj-ing) scanning technique that uses a magnetic field to induce tissues to emit radio waves that can be used by computer to construct a sectional view of a patient's body

major histocompatibility complex (MHC) (HIST-oh-kom-PAT-ib-il-it-ee) set of genes in chromosome 6 that all code for antigen-presenting proteins and other immune system proteins; proteins produced by MHC genes in class I and class II also are called *human leukocyte antigens (HLAs)*; MHC proteins present different protein fragments (peptides) at the surface of the cell for possible recognition as either self or nonself antigens by immune system cells

malignant (mah-LIG-nant) refers to a tumor or neoplasm that is capable of metastasizing, or spreading, to new tissues; cancer

maltase (MAL-tays) enzyme that catalyzes the final steps of carbohydrate digestion

mammary gland milk-producing glands of the breasts; classified as external accessory sex organs in females

mammography (mam-OG-rah-fee) x-ray photography of the breast

mandibular nerve (man-DIB-yoo-lar) part of the trigeminal nerve

marrow cavity *see* **medullary cavity**

mast cell immune system cell to which antibodies become attached in early stages of inflammation

mastectomy (mas-TEK-toh-mee) surgical removal of the breast

mastication (mas-ti-KAY-shun) chewing

mastitis (mas-TYE-tis) breast inflammation

mastoiditis (mas-toyd-EYE-tis) inflammation of the air cells within the mastoid portion of the temporal bone; usually caused by infection

matrix (MAY-triks) extracellular substance of a tissue; for example, the matrix of bone is calcified, whereas that of blood is liquid; *see also* **extracellular matrix (ECM)**

matter anything that has mass and occupies space

maxillary nerve (MAK-si-lair-ee nerv) branch of the trigeminal nerve

McBurney's point (mak-BUR-neez) a point on the abdominal wall midway between the right anterior superior iliac spine and the umbilicus; site of pain in acute appendicitis

mechanical digestion process through which food is broken into smaller portions through chewing and movements of the alimentary canal; enables enzymes to act on a larger surface area to accomplish chemical digestion

mechanoreceptor (mek-an-oh-ree-SEP-tor) receptors that respond to physical movement in the environment, such as sound waves; for example, equilibrium and balance sensors in the ears

medial (MEE-dee-al) of or toward the middle; opposite of lateral

medial lemniscal system (MEE-dee-al lem-NIS-kal SIS-tem) posterior white columns of the spinal cord plus the medial lemniscus; functions in touch sensations and conscious proprioception

mediastinum (MEE-dee-ass-TI-num) a portion of the thorax cavity in the middle of the thorax

medulla (meh-DUL-ah) Latin for "marrow"; the inner portion of an organ in contrast to the outer portion, or cortex

medulla oblongata (meh-DUL-ah ob-long-GAH-tah) lowest part of the brainstem; an enlarged extension of the spinal cord; the vital centers are located within this area

medullary (MED-oo-lar-ee) pertaining to the middle or center of an organ or structure

medullary cavity (MED-oo-lair-ee) hollow area inside the diaphysis of the bone that contains yellow bone marrow

medullary rhythmicity area (MED-oo-lair-ee) area in the brainstem that generates the basic rhythm of the respiratory cycle of inspiration and expiration

megakaryocytes (meg-ah-KAIR-ee-oh-sytes) large cells in the spleen that contribute to formation of platelets

meiosis (my-OH-sis) nuclear division in which the number of chromosomes is reduced to half their original number through separation of homologous pairs; produces gametes

Meissner's corpuscle (MYZ-ners KOR-pus-ul) sensory receptor located in the skin close to the surface that detects light touch

melacine (MEL-ah-seen) vaccine made from melanoma cancer cells; used for prevention of recurrence of malignant melanoma following initial treatment and as a less toxic alternative to chemotherapy for treatment of advanced melanoma

melanin (MEL-ah-nin) brown pigment primarily in skin and hair

melanocyte (MEL-ah-no-syte) specialized cell in the stratum basale of the skin that produces melanin

melanocyte-stimulating hormone (MSH) (meh-LAN-oh-syte-STIM-yoo-lay-ting HOR-mohn) hormone secreted by the pituitary that increases production of melanin, leading to darker skin color

melanoma (mel-ah-NO-mah) cancer of pigmented epithelial cells

melatonin (mel-ah-TOH-nin) important hormone produced by the pineal gland; it is believed to regulate onset of puberty and the menstrual cycle; also referred to as the *third eye* because it responds to levels of light and is thought to be involved with the body's internal clock

membrane thin, sheetlike structure

membrane channel pores within the cell membrane through which specific ions or other small, water-soluble molecules can pass

membrane potential difference in electrical charge between inside and outside of the plasma membrane

membrane protein protein embedded in the phospholipid bilayer of the membrane; contributes to cell transport, cell-cell recognition, and other cell functions

membranous organelle (MEM-brah-nus or-gah-NELL) cell organ that is defined or bordered by a membrane

menarche (meh-NAR-kee) first menses occurring at the onset of puberty

meninges (meh-NIN-jeez) fluid-containing membranes surrounding the brain and spinal cord

meningitis (men-in-JYE-tis) inflammation of the meninges caused by various factors, including bacterial infection, mycosis, viral infection, and tumors

meniscus (meh-NIS-kus) articular cartilage disk

menopause (MEN-oh-pawz) termination of menstrual cycles

menses (MEN-seez) periodic shedding of endometrial lining in uterus; occurs in cycles of about 28 days

menstrual cycle (MEN-stroo-al) series of changes that regularly occurs in sexually mature, nonpregnant women that results in shedding of the uterine lining approximately once a month

menstruation (men-stroo-AY-shun) menses; regular event of the female reproductive cycle that allows the endometrium to renew itself

merocrine gland (MER-oh-krin) gland that discharges secretions directly through the cell or plasma membrane

mesentery (MEZ-en-tair-ee) large double fold of peritoneal tissue that anchors the loops of the digestive tract to the posterior wall of the abdominal cavity

mesoderm (MEZ-oh-derm) middle layer of the primary germ layers; gives rise to such structures as muscle, bones, and blood vessels

mesomorph (MEZ-oh-morf) body type characterized by a muscular build

messenger RNA (mRNA) (MESS-en-jer R N A) duplicate copy of a gene sequence on the DNA that passes from the nucleus to the cytoplasm; used by ribosomes to create specific proteins

metabolic acidosis (met-ah-BOL-ik ass-i-DOH-sis) disturbance affecting the bicarbonate element of the bicarbonate-carbonic acid buffer pair in the blood; bicarbonate deficit

metabolic alkalosis (met-ah-BOL-ik al-kah-LOH-sis) disturbance affecting the bicarbonate element of the bicarbonate-carbonic acid buffer pair in the blood; bicarbonate excess

metabolic rate (met-ah-BOL-ik rate) amount of energy released by catabolism in the body over a given period

metabolism (meh-TAB-oh-liz-em) complex, intertwining set of chemical processes by which life is made possible for a living organism; *see* **anabolism, catabolism**

metaphase (MET-ah-fayz) second stage of mitosis, during which the nuclear membrane and nucleolus disappear and the chromosomes align on the equatorial plane

metastasis (meh-TAS-tah-sis) process characteristic of cancer by which malignant tumor cells separate from a primary tumor and then migrate to a new tissue to initiate a secondary tumor

micelles (my-SELLS) droplet of lipid surrounded by bile salts, which makes the lipid temporarily water-soluble

microcephaly (my-kroh-SEF-ah-lee) congenital abnormality in which an infant is born with an abnormally small head

microcirculation (my-kroh-sir-kyoo-LAY-shun) flow of blood through the capillary bed

microfilament (my-kroh-FIL-ah-ment) smallest cell fibers; "cellular muscles"

microglia (my-KROG-lee-ah) type of small neuroglial cell of nerve tissue that serves an immune system function by becoming an active phagocyte when stimulated

microscopic anatomy (my-kroh-SKOP-ik ah-NAT-oh-mee) use of a microscope to study anatomical structures such as cells (cytology) or tissues (histology)

microtubule (my-kroh-TOOB-yool) thick cell fiber (compared to microfilament); hollow tube responsible for movement of substances within the cell or movement of the cell itself

microvillus (my-kroh-VIL-us) brushlike border made up of epithelial cells on each villus in the small intestine; increases the surface area for absorption of nutrients

micturition (mik-too-RISH-un) urination, voiding

midbrain region of the brainstem between the pons and the diencephalon

middle ear tiny and very thin epithelium-lined cavity in the temporal bone that houses the ossicles; in the middle ear, sound waves are amplified

midsagittal plane (mid-SAJ-i-tal plane) cut, or plane, that divides the body or any of its parts into two equal (mirror-image) halves

milliequivalent (mil-i-ee-KWIV-ah-lent) number of ionic charges or electrocovalent bonds in a solution; abbreviated mEq

mineralocorticoid (MC) (min-er-al-oh-KOR-ti-koyd) hormone that influences mineral salt metabolism; secreted by adrenal cortex; aldosterone is the chief mineralocorticoid
minerals inorganic elements or salts occurrring naturally in the earth, many of which are vital to proper functioning of the body and must be obtained in the diet
minute volume volume of blood circulating through the body per minute
miscarriage loss of an embryo or fetus before the twentieth week of pregnancy; after 20 weeks, the event is termed a *stillbirth*
mitochondrial DNA (mDNA) (my-toh-KON-dree-al D N A) DNA specifically in the mitochondrion that has the only genetic code for several important enzymes
mitochondrion (my-toh-KON-dree-on) organelle in which ATP generation occurs; often termed "powerhouse of cell"
mitosis (my-TOH-sis) complex process in which a cell's DNA is replicated and divided equally between two daughter cells
mitral valve (MY-tral) located between the left atrium and ventricle, this valve prevents backflow of blood into the left atrium; named for its resemblance to a bishop's hat (miter); also known as the *bicuspid valve*
mitral valve prolapse (MVP) (MY-tral PROH-laps) condition in which the bicuspid (mitral) valve extends into the left atrium, causing incompetence (leaking) of the valve
mittelschmerz (MIT-el-shmerts) abdominal discomfort experienced by many women at the time of ovulation
mixed nerves nerves with axons of both sensory and motor neurons
mobile-receptor model (MOH-bil ree-SEP-tor) theory that a hormone enters a target cell where it binds to a free (nonstationary) receptor, thereby activating a gene and thus the beginning of mRNA transcription
modiolus (moh-DI-oh-lus) cone-shaped core of bone around which the tube of the cochlea is wound
molar (MOHL-ar) a tricuspid tooth, which is a relatively flat-topped tooth near the posterior of the jaw; *see also* **mole**
mole (mohl) standard unit of measuring amount of a substance; one mole of a substance is made up of 6.02×10^{23} particles of that substance; moles per liter as a concentration of a solute in a solvent is expressed as "molar concentration" or M
molecule (MOL-eh-kyool) formed when two or more atoms join
molecular genetics (moh-LEK-yoo-lar jeh-NET-iks) branch of genetics that studies DNA and RNA and the mechanisms by which they store and transmit hereditary information
monoamine (mon-oh-ah-MEEN) category of small neurotransmitter molecule derived from a single amino acid
monoamine oxidase (MAO) (mon-oh-ah-MEEN OK-si-dase) enzyme located in synaptic knobs of postganglionic neurons; responsible for breaking down neurotransmitters
monoclonal antibody (mon-oh-KLONE-al AN-ti-bod-ee) specific antibody produced from a population of identical cells; *see* **hybridoma**
monocyte (MON-oh-syte) large white blood cell; an agranular leukocyte
monosaccharide (mon-oh-SAK-ah-ride) simple sugar, such as glucose or fructose; building block of carbohydrates
monosomy (MON-oh-so-mee) abnormal genetic condition in which cells have only one chromosome instead of a pair; usually caused by nondisjunction (failure of chromosome pairs to separate) during gamete production
mons pubis (monz PYOO-bis) skin-covered pad of fat over the symphysis pubis in the female
morula (MOR-yoo-lah) solid mass of cells formed by the divisions of a fertilized egg
motility (moh-TIL-i-tee) ability to move spontaneously
motor endplate point at which motor neurons connect to the sarcolemma to form the neuromuscular junction
motor neuron (NOO-ron) transmits nerve impulses from the brain and spinal cord to muscles and glandular epithelial tissues
motor unit functional unit composed of a single motor neuron with the muscle cells it innervates
mucous membrane (MYOO-kus) epithelial membrane that lines body surfaces opening directly to the exterior and secretes mucus
mucus (MYOO-kus) thick, slippery material secreted by mucous membranes that keeps the membrane moist and protected
multifides (mul-TIF-i-deez) group of back muscles that each connect one vertebra to the next, also helping to extend the back and neck or flex them to the side
multiple neurofibromatosis (noo-roh-fye-broh-mah-TOH-sis) disorder characterized by multiple, sometimes disfiguring, benign tumors of the Schwann cells (glia) that surround nerve fibers
multiple sclerosis (MS) (skleh-ROH-sis) most common primary disease of the central nervous system, MS leads to demyelination of nerves, which commonly causes problems with vision, muscle control, and incontinence
multiple wave summation condition in which a series of stimuli come in rapid enough succession that the muscle is not able to relax completely, and sustained or more forceful contraction results
multipolar neuron (mul-ti-POL-ar NOO-ron) neuron with only one axon but several dendrites
multiunit smooth muscle type of smooth muscle tissue composed of many independent single-fiber units that does not usually generate its own impulse but rather responds only to nervous input and is often found in bundles
mumps acute viral disease characterized by swelling of the parotid salivary glands
muscarinic receptor (mus-kah-RIN-ik ree-SEP-tor) type of cholinergic receptor responding to muscarine, as well as acetylcholine
muscle fatigue (fah-TEEG) state of exhaustion produced by strenuous muscular activity
muscle spindle (SPIN-dul) stretch receptor in muscle cells involved in maintaining muscle tone
muscle strain muscle injury resulting from overexertion or trauma and involving overstretching or tearing of muscle fibers
muscle tissue specialized tissue type that produces movement
muscle tone tonic contraction; characteristic of muscle of a normal individual who is awake
muscularis (mus-kyoo-LAIR-is) two layers of muscle surrounding the digestive tube that produce wave-like, rhythmic contractions, called *peristalsis*, that move food material along the digestive tract
mutagen (MYOO-tah-jen) agent capable of causing mutation (alteration) of DNA
mutation (myoo-TAY-shun) change in genetic material within a cell
myalgia (my-AL-jee-ah) muscle pain
myasthenia gravis (my-es-THEE-nee-ah GRAH-vis) chronic autoimmune disease characterized by muscle weakness, especially in the face and throat, caused by immune attacks at the neuromuscular junction
myelin (MY-eh-lin) lipoprotein substance in the myelin sheath around many nerve fibers that contributes to high-speed conductivity of impulses
myelinated fiber (MY-eh-li-nay-ted) axon surrounded by a sheath of myelin formed by Schwann cells (PNS) or oligodendrocytes (CNS)
myelin disorders (MY-eh-lin) any of several disorders characterized by loss or improper development of the myelin sheath that surrounds many axons of the nervous system
myelin sheath (MY-eh-lin sheeth) whitish, fatty sheath surrounding nerve fibers; substance produced by Schwann cells
myeloid tissue (MY-eh-loyd) red bone marrow; specialized type of soft, diffuse connective tissue; the site of hematopoeisis
myocardial infarction (MI) (my-oh-KAR-dee-al in-FARK-shun) death of cardiac muscle cells resulting from inadequate blood supply, as in coronary thrombosis
myocardium (my-oh-KAR-dee-um) muscle of the heart
myofibril (my-oh-FYE-bril) very fine longitudinal fibers found in skeletal muscle cells; composed of thick and thin filaments
myofilament (my-oh-FIL-ah-ment) ultramicroscopic, threadlike structures found in myofibrils; composed of myosin (thick) and actin (thin)
myoglobin (my-oh-GLOH-bin) large protein molecule in the sarcoplasm of muscle cells that attracts oxygen and holds it temporarily
myography (my-OG-rah-fee) procedure in which the contraction of an isolated muscle is recorded
myometrium (my-oh-MEE-tree-um) middle muscle layer in the uterus
myopia (my-OH-pee-ah) refractive disorder of the eye caused by an elongated eyeball; nearsightedness
myopathy (my-OP-ah-thee) general term referring to any muscle disease
myosin (MY-oh-sin) contractile protein found in the thick filaments of skeletal muscle myofilaments
myositis (my-oh-SYE-tis) inflammation of muscle tissue
myxedema (mik-seh-DEE-mah) firm swelling (edema) of the skin caused by deficiency of thyroid hormone in adults

N

nail multilayered, protective structure composed of epithelial cells containing hard keratin; located at the distal ends of fingers and toes
nail body visible part of the nail
nail root part of the nail hidden by the cuticle
nasolacrimal duct (nay-zoh-LAK-rih-mal dukt) tube that extends from lacrimal sac into the inferior meatus of the nose
nasopharynx (nay-zoh-FAIR-inks) uppermost portion of the tube just behind the nasal cavities
natural immunity (i-MYOO-ni-tee) acquired immunity resulting from exposure to disease-causing agents in the course of daily living
natural killer (NK) cells group of lymphocytes that kills many types of tumor cells
nausea (NAW-zee-ah) unpleasant sensation of the gastrointestinal tract that commonly precedes the urge to vomit; upset stomach
near reflex constriction of the pupil for near vision
necrosis (neh-KROH-sis) death of cells in a tissue, often resulting from ischemia (reduced blood flow)
negative feedback feedback control system in which the level of a variable is changed in the direction opposite to that of the initial stimulus
neonatal period (nee-oh-NAY-tal) period of development immediately after birth
neonatology (nee-oh-nay-TOL-oh-jee) diagnosis and treatment of disorders of the newborn
neoplasm (NEE-oh-plaz-em) tumor or abnormal growth; may be benign or malignant
nephron (NEF-ron) anatomical and functional unit of the kidney, consisting of the renal corpuscle and the renal tubule
nerve bundle nerve fibers, plus surrounding connective tissue, located outside the brain or spinal cord
nerve fibers axons of neurons in the nervous system
nerve impulse self-propagating wave of electrical depolarization carries information along nerves; also called *action potential*
nervous system brain, spinal cord, and nerves
nervous tissue specialized tissue type consisting of neurons and glia that provides rapid communication and control of body function
neuralgia (noo-RAL-jee-ah) general term referring to nerve pain
neural networks (NOOR-al NET-works) a network of interconnected neurons in nervous tissue, forming a web of pathways to process information
neurilemma (noo-ri-LEM-mah) nerve sheath
neurilemmocyte (noo-ri-LEM-moh-syte) another name for the Schwann cell; also spelled *neurolemmocyte*
neurobiology (noo-roh-bye-AHL-oh-jee) branch of biology that studies the basic science of the nervous system; compare to neurology
neuroblastoma (noo-roh-blas-TOH-mah) malignant tumor of sympathetic nervous tissue, found mainly in young children
neuroendocrine system (noo-roh-EN-doh-krin SIS-tem) endocrine and nervous systems working in concert to perform communication, integration, and control within the body
neurofibril (noo-roh-FYE-bril) fine strands extending through the cytoplasm of each neuron; formed by cell's cytoskeleton
neuroglia (noo-ROG-lee-ah) nonexcitable supporting cells of nervous tissue; more properly called *glia*
neuroglobin (Ngb) (NOO-roh-gloh-bin) protein in neurons similar to hemoglobin to temporarily store a "back up" supply of oxygen
neurohypophysis (noo-roh-hye-POF-i-sis) posterior pituitary gland
neurology (noo-ROL-oh-jee) branch of medical science that deals with the nervous system, especially its disorders and their treatments; compare to neurobiology
neuroma (noo-ROH-mah) any tumor arising from nervous tissue
neuromodulator (noo-roh-MOD-yoo-lay-tor) "co-transmitter" that regulates the effects of neurotransmitter(s) released along with it
neuromuscular junction (noo-roh-MUSS-kyoo-lar) point of contact between nerve endings and muscle fibers; *see* **motor endplate**
neuron (NOO-ron) nerve cell, including its processes (axons and dendrites)
neuropeptide (noo-roh-PEP-tyde) neurotransmitter with short strands of polypeptides
neurosecretory tissue (noo-roh-SEK-reh-tor-ee) modified neurons that secrete chemical messengers that diffuse into the bloodstream rather than across a synapse; for example, in the hypothalamus
neurotransmitter (noo-roh-TRANS-mit-ter) chemicals by which neurons communicate; the substance is released by a neuron, diffuses across the synapse, and binds to the postsynaptic neuron
neurotrophin (noo-roh-TROHF-in) nerve growth factor
neutron (NOO-tron) neutral subatomic particle located in the nucleus of an atom
neutrophil (NOO-troh-fil) white blood cell that stains readily with neutral dyes
nevus (NEE-vus) small, pigmented benign tumor of the skin; a mole
nicotinamide adenine dinucleotide (NAD) (nik-oh-TIN-ah-myde AD-eh-neen dye-NOO-klee-oh-tyde) molecule that serves as an electron carrier in the electron transport system
Nissl bodies (NISS-ul) found in rough endoplasmic reticulum, provide protein molecules needed for transmission of nerve impulses from one neuron to another; also called *Nissl substance*
nitric oxide (NO) (NYE-trik AWK-side) small gas molecule used as a neurotransmitter that *diffuses backward* from a postsynaptic cell toward the presynaptic neuron, where it has its biochemical effects and gives the opportunity for feedback
nociceptor (no-see-SEP-tor) receptor activated by intense stimuli of any type that results in tissue damage; pain receptor
node of Ranvier (node of rahn-vee-AY) short space in the myelin sheath between adjacent Schwann cells
nonadrenergic-noncholinergic (NANC) transmission (non-AD-ren-er-jik non-KOHL-in-er-jik tranz-MISH-un) neural synaptic transmission in the efferent autonomic pathways that involves transmitters other than the classical autonomic neurotransmitters norepinephrine and acetylcholine; NANC transmitters include GABA, ATP, neuropeptide Y, and others
nondisjunction (non-dis-JUNK-shun) occurs during meiosis when a pair of chromosomes fails to separate
nonelectrolyte (non-ee-LEK-troh-lyte) compound that does not dissociate in solution; for example, glucose
nonmembranous organelle (NON-mem-bran-us or-gah-NELL) cell organ that is not defined or bordered by a membrane
nonpolar (non-POH-lar) describes a covalent chemical bond (or covalently bonded molelcule) in which there is equal sharing of electrons and therefore no distinct areas of electrical charge
nonself concept in immunology that the immune system agents can recocognize certain cell-surface molecules or other molecules as not belonging to that individual and thus being a potential target for destruction or damage by the immune system
nonself antigens (AN-ti-jens) foreign and tumor cell markers that a body's immune cells can recognize as being different from the self; also called *nonself markers*
nonspecific immunity (non-speh-SIF-ik i-MYOO-ni-tee) mechanisms that resist various threatening agents or conditions, not just certain specific agents; *see* **innate immunity**
nonsteroid hormone (non-STAYR-oyd HOR-mohn) hormone synthesized primarily from amino acids rather than from cholesterol
norepinephrine (nor-ep-i-NEF-rin) hormone secreted by adrenal medulla that increases cardiac output; neurotransmitter released by sympathetic nervous system cells
nuclear envelope (NOO-klee-ar) the boundary of a cell's nucleus, made up of a double layer of cellular membrane
nuclear magnetic resonance (NMR) (NOO-klee-ar mag-NET-ik REH-son-ans) magnetic resonance imaging
nuclear pore complex (NPC) (NOO-klee-ar poor KOM-plex) complicated, channel-like structure in the nuclear envelope, the nuclear pore (as it is sometimes called) selectively transports molecules into or out of the nucleus
nuclei (NOO-klee-eye) distinct regions of gray matter in the central nervous system; clusters of neuron cell bodies
nucleic acid (NOO-klay-ic) any of the high-molecular-weight organic compounds composed of nucleotides, a ribose or deoxyribose sugar, and a phosphate group. See as examples **deoxyribonucleic acid (DNA)** and **ribonucleic acid (RNA)**
nucleolus (noo-KLEE-oh-lus) dense, well-defined but membraneless body within the nucleus; critical to protein formation because it "programs" the formation of ribosomes in the nucleus
nucleoplasm (NOO-klee-oh-plaz-im) substance inside a cell's nucleus
nucleotide (NOO-klee-oh-tyde) monomer made up of three types of chemical groups (sugar, phosphate, nitrogen base) that can act alone or to make up a polymer (nucleic acid)
nucleus (NOO-klee-us) membranous organelle that contains most of the genetic material of the cell; also, group of neuron cell bodies in the brain or spinal cord
nucleus pulposus (NOO-klee-us pul-POH-sus) gel-like center of the intervertebral disk; surrounded by annulus fibrosis

O

octet rule (awk-TET) general principle in chemistry whereby atoms usually form bonds in ways that will provide each atom with an outer shell of eight electrons
oculomotor nerve (awk-yoo-loh-MOH-tor) cranial nerve III; motor nerve; controls eye movements
olfactory nerve (ol-FAK-tor-ee) cranial nerve I; sensory nerve; responsible for the sense of smell
oligodendrocyte (ohl-i-go-DEN-droh-syte) small astrocyte with few cell processes; helps to form myelin sheaths around axons within the central nervous system
oligodendroglioma (ohl-i-go-DEN-droh-glye-OH-mah) tumor arising from oligodendrocytes (neuroglia of central nervous tracts)
oncogene (ON-koh-jeen) gene (DNA segment) thought to be responsible for the development of a cancer
oncologist (ong-KOL-oh-jist) cancer specialist
oocyte (OH-oh-syte) developing female sex cell contained within ovarian follicles
oogenesis (oh-oh-JEN-eh-sis) production of female gametes
oogonia (oh-oh-GO-nee-ah) primitive cell from which oocytes derive meiosis
ophthalmic nerve (op-THAL-mik) part of the trigeminal nerve
opsin (OP-sin) protein produced by the breakdown of rhodopsin in rods and cones of the retina; involved in a chemical reaction that initiates an impulse that results in interpretation of light energy as vision
optic chiasma (OP-tik kye-AS-mah) region where right and left optic nerves enter the brain and cross each other, exchanging fibers
optic disc (OP-tic) area in the retina where the optic nerve fibers exit the eye and where therefore there are no rods or cones; also known as a *blind spot*
optic nerve (OP-tik) cranial nerve II; sensory nerve; nerves that conduct visual information to the brain
optic tract (OP-tik) bundles of fibers formed after the optic nerves pass through the optic chiasma
organ group of several tissue types that together perform a special function
organ of Corti (OR-gan of KOR-tee) organ of hearing located in the cochlea; consists of membranes and embedded hair cells capable of converting mechanical motion of sound waves into neural impulses
organelle (or-gah-NELL) any of many cell "organs" or organized structures; for example, a ribosome or mitochondrion
organic (or-GAN-ik) referring to chemicals that contain covalently bound carbon and hydrogen atoms and are involved in metabolic reactions
organism (OR-gah-niz-em) any living entity considered as a whole; may be unicellular (one-celled) or composed of many different cells and body systems working together to maintain life
organogenesis (or-gah-no-JEN-eh-sis) formation of organs from the primary germ layers of the embryo
origin attachment of a muscle to the bone, which does not move when contraction occurs; *see* **insertion**
oropharynx (or-oh-FAIR-inks) portion of the pharynx that is located behind the mouth
orthopnea (or-THOP-nee-ah) dyspnea (difficulty in breathing) that is relieved after moving into an upright or sitting position
osmolality (os-moh-LAL-i-tee) osmotic concentration of a solution; the number of moles of a substance per kilogram times the number of particles into which the solute dissociates
osmoreceptor (os-moh-ree-SEP-tor) special receptors near the supraoptic nucleus that detect decreased osmotic pressure of blood when the body dehydrates
osmosis (os-MOH-sis) movement of a fluid (usually water) through a semipermeable membrane from an area of lesser solute concentration to an area of greater concentration
osmotic pressure (os-MOT-ik) water pressure that develops in a solution across a semipermeable membrane as a result of osmosis
osseous tissue (OS-ee-us) bone tissue
osteitis fibrosa cystica (os-tee-EYE-tis fye-BROH-sah SIS-ti-kah) bone disease caused by hypercalcemia
osteoarthritis (os-tee-oh-ar-THRY-tis) degenerative joint disease; a noninflammatory disorder of a joint characterized by degeneration of articular cartilage
osteoblast (OS-tee-oh-blast) bone-forming cell
osteoclast (OS-tee-oh-klast) bone-absorbing cell
osteocyte (OS-tee-oh-syte) bone cell
osteogenesis (os-tee-oh-JEN-eh-sis) combined action of osteoblasts and osteoclasts to mold bones into adult shape
osteogenesis imperfecta (os-tee-oh-JEN-eh-sis im-per-FEK-tah) dominant, inherited disorder of connective tissue characterized by imperfect skeletal development, resulting in brittle bones
osteoid (OS-tee-oid) organic matrix of bone
osteoma (os-tee-OH-mah) benign bone tumor
osteomyelitis (os-tee-oh-my-eh-LYE-tis) bacterial (usually staphylococcal) infection of bone tissue
osteon (OS-tee-on) *see* **Haversian system**
osteoporosis (os-tee-oh-poh-ROH-sis) bone disorder characterized by loss of minerals and collagen from bone matrix, reducing the volume and strength of skeletal bone
osteosarcoma (os-tee-oh-sar-KOH-mah) bone cancer
otitis media (oh-TYE-tis MEE-dee-ah) a middle ear infection
otoliths (OH-toh-liths) tiny "ear stones" composed of protein and calcium carbonate in the maculae of the ear, which, by responding to gravity and changes in body position, trigger hair cells that initiate impulses resulting in sense of balance
ova (OH-vah) female sex cells (singular, ovum)
ovarian follicle (oh-VAIR-ee-an FOL-i-kul) spherical configuration of cells in the ovary that contains a single oocyte
ovaries (OH-var-ees) female gonads that produce ova (sex cells)
oviducts (OH-vi-dukts) *see* **fallopian tubes, uterine tubes**
ovulation (ov-yoo-LAY-shun) release of an ovum from the ovary at the end of oogenesis
oxidative phosphorylation (ahk-si-DAY-tiv fos-for-i-LAY-shun) reaction that joins a phosphate group to ADP to form ATP
oxygen debt (AHK-si-jen) oxygen required for ATP synthesis to remove excess lactic acid following anaerobic exercise
oxyhemoglobin (ahk-see-hee-moh-GLOH-bin) hemoglobin combined with oxygen
oxytocin (OT) (ahk-see-TOH-sin) hormone secreted by the posterior pituitary gland before and after delivering a baby; thought to initiate and maintain labor, as well as cause the release of breast milk into ducts of the mammary glands

P

pacinian corpuscle (pah-SIN-ee-an KOR-pus-ul) receptor deep in the dermis that detects pressure on the skin surface
packed cell volume (PCV) volume percent of red blood cells in whole blood; *see* **hematocrit**
p-arm the shorter segment of the chromosome, which is divided into two "arms" by the centromere (the longer segment is called the *q-arm*)
palatine tonsil (PAL-ah-tine TAHN-sil) tonsils located behind and below the pillars of the fauces
palpable (PAL-pah-bul) can be identified by touch, such as bony landmarks located beneath the skin
palpebrae (PAL-peh-bree) eyelids
palpebral fissure (PAL-peh-bral FISH-ur) opening between the two eyelids
pancreas (PAN-kree-ass) endocrine gland located in the abdominal cavity; contains pancreatic islets that secrete glucagon and insulin
pancreatic islets of Langerhans (pan-kree-AT-ik EYE-lets) endocrine portion of the pancreas; made up of alpha and beta cells, among others; source of insulin and glucagon
pancreatitis (pan-kree-ah-TYE-tis) inflammation of the pancreas
pandemic (pan-DEM-ik) refers to a disease that affects many people worldwide
Papanicolaou test (pap-peh-ni-koh-LAH-oo) cancer-screening test in which cells swabbed from the uterine cervix are smeared on a glass slide and examined for abnormalities; also called *Pap smear* or *Pap test* [after Dr. George Papanicolaou]
papillae (pah-PIL-ee) small, nipple-shaped elevations
papilloma (pap-i-LOH-mah) benign skin tumor characterized by finger-like projections (for example, a wart)
paralysis (pah-RAL-i-sis) loss of power of motion, especially voluntary motion
parasympathetic division of the autonomic nervous system (ANS) (pair-ah-sim-pah-THET-ik) part of the autonomic nervous system; ganglia are connected to the brainstem and the sacral segments of the spinal cord; controls many autonomic effectors under normal ("rest and repair") conditions
parathyroid gland (pair-ah-THYE-royd) endocrine gland located in the neck on the posterior aspect of the thyroid gland; secretes parathyroid hormone
parathyroid hormone (PTH) (pair-ah-THYE-royd HOR-mohn) hormone released by the parathyroid gland that increases concentration of calcium in the blood
parenteral therapy (pah-REN-ter-al) administration of nutrients, special fluids, and/or electrolytes by injection
parietal (pah-RYE-i-tal) refers to the walls of an organ or cavity
parietal cells (pah-RYE-i-tal) cells located on the basement membrane of gastric glands of the stomach that secrete hydrochloric acid
parietal membrane (pah-RYE-i-tal) portion of the serous membrane that lines the wall of a body cavity
parietal peritoneum (pah-RYE-i-tal pair-i-toh-NEE-um) serous membrane that covers organs of the abdomen, pelvis, and thorax
Parkinson disease (PAR-kin-son) nervous disorder characterized by abnormally low levels of the neurotransmitter dopamine in parts of the brain that control voluntary movement—victims usually exhibit involuntary trembling and muscle rigidity
parotid glands (pah-ROT-id) largest of the paired salivary glands
pars anterior (parz an-TEER-ee-or) major part of the anterior pituitary gland
pars intermedia (PARS in-ter-MEE-dee-ah) forms a small part of the anterior pituitary gland
partial pressure (PAR-shal) pressure exerted by any one gas in a mixture of gases or in a liquid
partial-thickness burn first- or second-degree burn
parturition (pahr-too-RI-shun) act of giving birth
passive transport cellular process in which substances move through a cellular membrane with their own energy supplied directly by the cell or its membrane
pathogen (PATH-oh-jen) microorganism that causes disease
pathogenesis (path-oh-JEN-eh-sis) pattern of a disease's development
pathogenic animal (path-oh-JEN-ik) animal, such as an insect, that causes disease in humans; *see also* **vector -**
pathology (pah-THOL-eh-jee) study of diseased body structures
pathophysiology (path-oh-fiz-ee-OL-oh-jee) study of the underlying physiological aspects of disease
pedicel (PED-i-cel) branches of the podocytes of the Bowman's capsule, packed tightly together with only filtration slits between them
pedigree (PED-i-gree) chart used in genetic counseling to illustrate genetic relationships over several generations
pentose (PEN-tohs) five-carbon sugar such as ribose or deoxyribose
peptide (PEP-tyde) compound made of two or more amino acids connected by peptide bonds (compare to *polypeptide*)
peptide bond (PEP-tyde) bond that forms between the amino group of one amino acid and the carboxyl group of another
peptide hormone (PEP-tyde HOR-mohn) major category of nonsteroid hormones; smaller than protein hormones; made up of a short chain of amino acids; for example, antidiuretic hormone (ADH)
pericardial fluid (pair-i-KAR-dee-al) lubricating fluid secreted by the serous membrane; in pericardial space
pericardial space (pair-i-KAR-dee-al) space between the visceral layer and the parietal layer surrounding the heart, filled with pericardial fluid
pericardium (pair-i-KAR-dee-um) membrane that surrounds the heart
perichondrium (pair-i-KON-dree-um) fibrous covering of cartilage structures
perikaryon (pair-i-KAR-ee-on) cell body
perilymph (PAIR-i-limf) watery fluid that fills the bony labyrinth of the ear
perimysium (pair-i-MEE-see-um) tough, connective tissue surrounding fascicles
perineum (pair-i-NEE-um) area between anus and genitals
perineurium (pair-i-NOO-ree-um) connective tissue that encircles a bundle of nerve fibers within a nerve
periodontitis (pair-ee-oh-don-TYE-tis) inflammation of the periodontal membrane that anchors teeth to jaw bone; common cause of tooth loss among adults
periosteum (pair-ee-OS-tee-um) tough, connective tissue covering the bone
peripheral (peh-RIF-er-al) pertaining to the periphery, or outer boundaries of the body, as in peripheral nervous system
peripheral nervous system (PNS) (peh-RIF-er-al) nerves connecting brain and spinal cord to other parts of the body
peripheral neuropathy (peh-RIF-er-al noo-ROP-eh-thee) general term for any disease or disorder that involves damage to peripheral nerves; also called *peripheral neuritis*, or if many nerves are involved, *polyneuropathy* or *polyneuritis*
peripheral resistance (peh-RIF-er-al) resistance to blood flow caused by friction of blood passing through blood vessels
peristalsis (pair-i-STAL-sis) wavelike, rhythmic contractions of the stomach and intestines that move food material along the digestive tract
peritoneum (pair-i-toh-NEE-um) large, moist, slippery sheet of serous membrane that lines the abdominopelvic cavity (parietal layer) and its organs (visceral layer)
peritonitis (pair-i-toh-NYE-tis) inflammation of the serous membranes in the abdominopelvic cavity; sometimes a serious complication of an infected appendix
peritubular capillary (pair-ee-TOOB-yoo-lar KAP-eh-lar-ee) capillaries around the tubules in the kidney
permanent teeth set of teeth that replaces deciduous teeth
permeable (PERM-ee-ah-bil) adjective used to describe a membrane that allows substances to move through (permeate) it
permeant (PERM-ee-ent) adjective used to describe a substance that is able to move through (permeate) a membrane
permeate (PERM-ee-ayt) to move through or soak through, as water moving through the pores of a fabric or molecules diffusing through the pores of a cellular membrane
peroxisome (per-ahk-si-sohm) organelles that detoxify harmful substances that have entered cells
perspiration (per-spih-RAY-shun) transparent, watery liquid released by glands in the skin that eliminates ammonia and uric acid and helps maintain body temperature; also known as *sweat*
pH (pee-AYCH) units by which acid and base concentrations (relative H^+ ion concentrations) are measured; scale ranges from O (extremely acidic; high H^+ concentration) to 14 (extremely basic, or alkaline; low H^+ concentration)
phagocytes (FAG-oh-sytes) white blood cells that engulf microorganisms and digest them
phagocytosis (fag-oh-sye-TOH-sis) ingestion and digestion of particles by a cell
pharyngeal tonsil (fah-RIN-jee-al TAHN-sil) tonsil located in the nasopharynx on its posterior wall; when enlarged, referred to as *adenoids*
pharyngitis (fair-in-JYE-tis) sore throat; inflammation or infection of the pharynx
phenotype (FEE-no-type) overt, observable expression of a genotype
phenylalanine hydroxylase (fen-il-AL-ah-neen hye-DROK-seh-lays) enzyme needed to convert phenylalanine into tyrosine; its absence causes phenylketonuria
phenylketonuria (PKU) (fen-il-kee-toh-NOO-ree-ah) recessive, inherited condition characterized by excess of phenylketone in the urine, caused by accumulation of phenylalanine (an amino acid) in the tissues; may cause brain injury and death if phenylalanine intake is not managed properly
pheomelanin (fee-oh-MEL-ah-nin) type of melanin pigment that is reddish in color
phlebitis (fleh-BYE-tis) inflammation of a vein
phosphatase (FOS-foh-tayse) enzyme that removes phosphate groups
phospholipid (fos-foh-LIP-id) phosphate-containing fat molecule; an important constituent of cell membranes
phosphorylase (fos-FOR-i-layse) enzyme that adds phosphate groups
photopigment (foh-toh-PIG-ment) chemicals in retinal cells that are sensitive to light
photopupil reflex (foh-toh-PYOO-pul REE-fleks) constriction of the pupil in bright light to protect the retina from too intense or too sudden stimulation; also known as *pupillary light reflex*
photoreceptor (foh-toh-ree-SEP-tor) receptor only in the eye; responds to light stimuli if the intensity is great enough to generate a receptor potential
phrenic nerve (FREN-ik) nerve that stimulates the diaphragm to contract
physiological buffer (fiz-ee-oh-LOJ-i-kal BUF-er) secondary defense against harmful shifts in pH of body fluids; comes into play after the chemical buffer system
physiological dead space (fiz-ee-oh-LOJ-i-kal) the anatomical dead space (pulmonary airway volume outside the alveoli) plus any alveolar dead space (i.e., obstructed or diseased alveoli); *see* **anatomical dead space**
physiological fatigue (fiz-ee-oh-LOJ-i-kal fah-TEEG) fatigue caused by a relative lack of ATP, rendering the myosin cross bridges incapable of producing the force required for further muscle contractions
physiological polycythemia (fiz-ee-oh-LOJ-i-kal pol-ee-sye-THEE-mee-ah) elevated red blood cell numbers and hematocrit values in healthy individuals who live and work in high altitudes
physiology (fiz-ee-OL-oh-jee) scientific study of an organism's body function
physique (fi-ZEEK) body build

pia mater (PEE-ah MAH-ter) vascular innermost covering (meninx) of the brain and spinal cord
pimple (PIM-pul) pustule on the skin, usually associated with acne
pineal body (PIN-ee-al) endocrine gland located in the diencephalon and thought to be involved with regulating the body's biological clock; produces melatonin; also called *pineal gland*
pineal gland (PIN-ee-al gland) *see* **pineal body**
pinna (PIN-nah) flap of the external ear
pinocytosis (pin-oh-sye-TOH-sis) active transport mechanism used to transfer fluids or dissolved substances into cells
pituitary dwarfism (pi-TOO-i-tair-ee DWARF-iz-em) condition resulting from hyposecretion of growth hormone during the growth years
pituitary gland (pi-TOO-i-tair-ee gland) neuroendocrine gland located near base of the brain that has numerous and important regulatory functions; also called the *hypophysis*
placenta (plah-SEN-tah) structure that anchors the developing fetus to the uterus and provides a "bridge" for the exchange of nutrients and waste products between the mother and developing baby
placenta previa (plah-SEN-tah PREE-vee-ah) abnormal condition in which a blastocyst implants in the lower uterus, developing a placenta that approaches or covers the cervical opening; placenta previa involves risk of placental separation and hemorrhage
plantar flexion (PLAN-tar FLEK-shun) the bottom of the foot is directed downward; this motion allows a person to stand on tiptoe
plaque (plak) raised skin lesion greater than 1 cm in diameter
plasma (PLAZ-mah) liquid part of the blood
plasma membrane (PLAZ-mah) membrane that separates the contents of a cell from the tissue fluid, encloses the cytoplasm, and forms the outer boundary of the cell
plasmid (PLAS-mid) small circular ring of bacterial DNA
platelet (PLAYT-let) specialized cell fragments in the blood; thrombocyte; important component in the clotting mechanism
pleura (PLOOR-ah) serous membrane in the thoracic cavity
pleurisy (PLOOR-i-see) inflammation of the pleura
plexus (PLEK-sus) complex network formed by converging and diverging nerves, blood vessels, or lymphatics
pneumonectomy (noo-moh-NEK-toh-mee) surgical procedure in which an entire lung is removed
pneumonia (noo-MOH-nee-ah) abnormal condition characterized by acute inflammation of the lungs in which alveoli and bronchial passages become plugged with thick fluid (exudate)
pneumotaxic center (noo-moh-TAK-sik SEN-ter) group of cells in the pons of the brain that affects the rate of respiration by inhibiting the inspiration center
pneumothorax (noo-moh-THOH-raks) abnormal condition in which air is present in the pleural space surrounding the lung, possibly causing collapse of the lung
podocyte (POD-oh-syte) special epithelial cells making up the visceral layer of the Bowman's capsule
Poiseuille's law (pwah-SWEEZ law) volume of blood circulated per minute is directly related to mean arterial pressure minus central venous pressure and inversely related to resistance
polar molecule (POH-lar MOL-eh-kyool) molecule in which the electrical charge is not evenly distributed, causing one side of the molecule to be more positive or negative than the other
poliomyelitis (poh-lee-oh-my-eh-LYE-tis) viral disease that damages motor nerves, often progressing to paralysis of skeletal muscles
polymer (PAHL-i-mer) large molecule made up of many identical smaller molecules joined together in sequence
polynucleotide (pahl-ee-NOOK-lee-oh-tyde) strand of nucleotides bound to each other by chemical bonds, as in the polynucleotide strand that makes up an mRNA molecule
polypeptide (pahl-ee-PEP-tyde) compound made of many amino acids connected by peptide bonds
polyribosome (PAHL-ee-RYE-boh-sohm) temporary structure formed within a cell by many ribosomes following one another along a strand of mRNA during the process of translation
pons (ponz) part of the brainstem between the medulla oblongata and the midbrain
portal system (POR-tal SIS-tem) arrangement of blood vessels in which blood exiting one tissue is immediately carried to a second tissue before being returned to the heart and lungs for oxygenation and redistribution
positive-feedback mechanism feedback control system that is stimulatory; tends to amplify or reinforce a change in the internal environment
posterior (pohs-TEER-ee-or) located behind; opposite of anterior
posterior cavity (of eye) (pohs-TEER-ee-or KAV-i-tee) all the space posterior to the lens of the eye; contains vitreous humor
posterior chamber (of eye) (pohs-TEER-ee-or CHAM-ber) subdivision of the anterior cavity of the eye; small space behind the iris and anterior to the lens
postganglionic neurons (post-gang-glee-ON-ik NOO-rons) efferent autonomic neurons that conduct nerve impulses from a ganglion to effectors such as cardiac or smooth muscle or glandular epithelial tissue
postnatal period (POST-nay-tal PEER-ee-od) period after birth, ending at death
postsynaptic (post-si-NAP-tik) adjective describing any structure after a synapse (junction) of one neuron to another or any function that occurs after synaptic transmission
postsynaptic neuron (post-si-NAP-tik NOO-ron) in neuron-to-neuron communication, the neuron that receives a stimulus via an adjacent neuron's transmission of neurotransmitters across the synapse
postsynaptic potential (post-si-NAP-tik poh-TEN-shal) local potential produced by opening of ion channels in the postsynaptic membrane
posture (POS-chur) position of the body; often refers to the erect position of the body maintained unconsciously
potassium channel (poh-TASS-ee-um CHAN-el) a pore in a cell membrane that allows only potassium ions to permeate, or pass through, the membrane
potential osmotic pressure (poh-TEN-shal os-MOT-ik PRESH-ur) maximum osmotic pressure that could develop in a solution when it is separated from pure water by a selectively permeable membrane
precapillary sphincter (pree-CAP-pi-lair-ee SFINGK-ter) smooth muscle cells that guard the entrance to the capillary
preeclampsia (pree-ee-KLAMP-see-ah) syndrome of abnormal conditions in pregnancy of uncertain cause; symptoms include hypertension, proteinuria, and edema; also called *toxemia of pregnancy*, it may progress to eclampsia—severe toxemia that may cause death
preganglionic neurons (pree-gang-glee-ON-ik NOO-rons) efferent autonomic neurons that conduct nerve impulses between the spinal cord and a ganglion
pregnancy induced hypertension (PIH) (PREG-nan-see in-DOOST hye-per-TEN-shun) a condition in which a woman's blood pressure rises after the twentieth week of pregnancy and remains elevated until the end of pregnancy; PIH may have a genetic component and may progress to *preeclampsia*
prenatal period (PREE-nay-tal) developmental period after conception until birth
prepuce (PREE-pus) foreskin, especially covering fold of skin over the glans penis
presbyopia (pres-bee-OH-pee-ah) farsightedness of old age
pressoreflex (press-oh-REE-fleks) *see* **baroreflex**
presynaptic (pree-si-NAP-tik) adjective describing any structure before a synapse (junction) of one neuron to another or any function that occurs before synaptic transmission
presynaptic neuron (pree-si-NAP-tik NOO-ron) in neuron-to-neuron communication, the neuron that transmits a signal to an adjacent neuron via release of neurotransmitters that cross the synapse and bind to the postsynaptic neuron
prevertebral ganglion (pree-ver-TEE-bral GANG-glee-on) collateral ganglion; pairs of sympathetic ganglia located a short distance from the spinal cord
primary bronchi (BRONG-kye) the two branches of the trachea that enter the lungs
primary germ layers three layers of specialized cells that give rise to definite structures as the embryo develops; *see* **ectoderm, endoderm, mesoderm**
primary ossification center (os-i-fi-KAY-shun) where a blood vessel enters the cartilage of a developing bone at the midpoint of the diaphysis to initiate bone formation
prime mover main muscle responsible for producing a particular movement
prion (PREE-ahn) a term that is short for "*pro*teinaceous *in*fectious particles," which are proteins that convert normal proteins of the nervous system into abnormal proteins, causing loss of nervous system function; the abnormal form of the protein also may be inherited; a newly discovered type of pathogen, not much is known about how the prion works; *see* **bovine spongiform encephalopathy, variant Creutzfeldt-Jakob Disease (vCJD)**
product molecules or atoms that result from a chemical reaction
proenzyme (proh-EN-zyme) inactive form in which many enzymes are synthesized
progesterone (proh-JES-ter-ohn) steroid hormone produced by the ovaries (particularly the corpus luteum) that helps prepare the uterus for implantation; along with estrogen, helps maintain normal uterine and mammary gland function
prohormone (proh-HOR-mohn) hormone precursor
projection tract (proh-JEK-shun) extension of the sensory spinothalamic tracts and descending corticospinal tracts in the brain
prolactin (PRL) (proh-LAK-tin) hormone secreted by the anterior pituitary gland during pregnancy to stimulate the breast development needed for lactation
prophase (PROH-fayz) first stage of mitosis during which chromosomes become visible
proprioception (proh-pree-oh-SEP-shun) perception of movement and position of the body
proprioceptors (proh-pree-oh-SEP-tors) receptors located in the muscles, tendons, and joints; allow the body to recognize its position
prostaglandin (PG) (pross-tah-GLAN-din) any of a group of naturally occurring lipid-based substances that act in a hormone-like way to affect many body functions, including vasodilation, uterine smooth muscle contraction, and the inflammatory response
prostate gland (PROSS-tayt) lies just below the bladder in the male; secretes a fluid that constitutes about 30% of the seminal fluid volume; helps activate sperm and helps them maintain motility
protein (PROH-teen) large molecules formed by linkage of amino acids by peptide bonds; one of the basic building blocks of the body
proteoglycan (PROH-tee-oh-GLYE-kan) large molecule made up of a protein strand that forms a backbone to which are attached many carbohydrate molecules
proteome (PROH-tee-ohm) the entire group of proteins produced by a cell or by the entire body
proteomics (proh-tee-OH-miks) the endeavor that involves the analysis of the proteins encoded by the genome, with the ultimate goal of understanding the role of each protein in the body
proton (PROH-ton) positively charged subatomic particle
protozoan (proh-toh-ZOH-an) single-celled organism with a nucleus and other membranous organelles that can infect humans; *plural,* protozoa
proximal (PROK-si-mal) next or nearest; located nearest the center of the body or the point of attachment of a structure; opposite of distal
proximal convoluted tubule (PROK-si-mal kon-voh-LOO-ted TOOB-yool) second part of the nephron and the first segment of a renal tubule
psoriasis (so-RYE-ah-sis) chronic, inflammatory skin disorder characterized by cutaneous inflammation and scaly plaques
psychological stressor (sye-koh-LOJ-i-kal STRESS-or) anything an individual perceives as a threat
psychophysiology (sye-koh-fiz-ee-OL-oh-jee) scientific discipline that studies physiological responses of individuals being subjected to psychological stressors
psychosomatic (sye-koh-so-MAT-ik) mind influencing the body
ptosis (TOH-sis) downward displacement of an organ; for example the lowered kidneys of very thin individuals
puerperal fever (pyoo-ER-per-al FEE-ver) condition caused by bacterial infection in a woman after delivery of an infant, possibly progressing to septicemia and death; also called "childbed fever"
pulmonary artery (PUL-moh-nair-ee AR-ter-ee) artery that carries deoxygenated blood from the right ventricle of the heart to the lungs
pulmonary circulation (PUL-moh-nair-ee ser-kyoo-LAY-shun) blood flow from the right ventricle to the lung and returning to the left atrium
pulmonary edema (PUL-moh-nair-ee eh-DEE-mah) congestion of blood in the pulmonary circulation
pulmonary embolism (PUL-moh-nair-ee EM-boh-liz-em) blockage of the pulmonary circulation by a thrombus or other matter; may lead to death if blockage of pulmonary blood flow is significant
pulmonary semilunar valve (PUL-moh-nair-ee sem-i-LOO-nar) valve located at the beginning of the pulmonary artery
pulmonary vein (PUL-moh-nair-ee) any vein that carries oxygenated blood from the lungs to the left atrium
pulmonary ventilation (PUL-moh-nair-ee ven-ti-LAY-shun) breathing; process that moves air in and out of the lungs
pulp cavity cavity in the dentin of a tooth that contains connective tissue, blood and lymphatic vessels, and sensory nerves
pulse point where the pulse can be palpated; that is, wherever an artery lies near the surface and over a bone or other firm background; for example, radial artery
pulse pressure difference between systolic and diastolic pressure
punctae (PUNK-tay) opening into the lacrimal canals located at the inner canthus of the eye
Punnett square (PUN-it) grid used in genetic counseling to determine the probability of inheriting genetic traits
pupil opening in the center of the iris that regulates the amount of light entering the eye
pupillary light reflex (PYOO-pih-lair-ee) photopupil reflex
purine bases (PYOO-reen) one of two types of nitrogenous bases that are vital components of DNA derived from purine; adenine and guanine
Purkinje fibers (pur-KIN-jee) specialized cells located in the walls of the ventricles; relay nerve impulses from the AV node to the ventricles, causing them to contract
pus accumulation of white blood cells, dead bacterial cells, and damaged tissue cells at site of an infection
pustule (PUS-tyool) small, raised skin lesion filled with pus
pyloric sphincter (pye-LOR-ik SFINGK-ter) sphincter that prevents food from leaving the stomach and entering the duodenum
pyramidal tract (pi-RAM-i-dal) fibers that come together in the medulla to form the pyramids; also called *corticospinal tract*
pyramids (PEER-ah-mids) triangular-shaped divisions of the medulla of the kidney
pyrimidine bases (pi-RIM-i-deen) one of two types of nitrogenous bases that are vital components of DNA derived from pyrimidine; cytosine and thymine

Q

q-arm the longer segment of the chromosome, which is divided into two "arms" by the centromere (the shorter segment is called the p-arm)
quaternary (KWAH-ter-nair-ee) fourth or pertaining to the fourth in a series, as in the quaternary structure of a protein, which is the fourth level of complexity in the structure of a protein molecule; a quaternary protein is a protein possessing a fourth level of complexity in its molecular structure

R

radiation (ray-dee-AY-shun) electromagnetic energy, including light, x-rays, heat; in physiology, often refers to flow of excess heat energy away from the body via the blood
radiation sickness (ray-dee-AY-shun SIK-ness) a condition caused by ionizing radiation that can be mild to severe, or even fatal, depending on the level of radiation exposure and the length of time exposed; exposure at lower doses of radiation can result in headache, nausea and vomiting, appetite loss, and diarrhea; exposure to low doses for a longer period of time or a single high-level exposure may cause sterility, damage to fetal development, cancer (including leukemia), cataracts, hair loss (alopecia), and skin damage (radiation dermatitis); *see* **ionizing radiation**
radiation therapy (ray-dee-AY-shun THAYR-ah-pee) treatment often used with cancer in which high-intensity radiation is used to destroy cancer cells; also called *radiotherapy*
radioactive isotope (ray-dee-oh-AK-tiv EYE-so-tope) unstable isotope that spontaneously emits subatomic particles and electromagnetic radiation
radioactivity (ray-dee-oh-ak-TIV-it-ee) the ongoing process of emitting subatomic particles and electromagnetic radiation
radiography (ray-dee-OG-rah-fee) imaging technique using x-rays that pass through certain tissues more easily than others, allowing an image of tissues to form on a photographic plate
radioisotope (ray-dee-oh-EYE-so-tope) an isotope that is unstable and undergoes nuclear breakdown
radiotherapy (ray-dee-oh-THER-ah-pee) radiation therapy
raft also called *lipid raft*, it is a structure made up of groupings of molecules (cholesterol, certain phospholipids, proteins) within a cell membrane that travel together on the surface of the cell, something like a log raft on a lake
ramus (RAY-mus) large branch of a spinal nerve as it emerges from the spinal cavity
rational drugs chemotherapy drugs that target only those specific molecules, enzymes, or receptors unique to cancer cells or tumor growth, thereby affecting only the cancer and sparing normal cells, increasing efficiency and reducing side effects
reabsorption (ree-ab-SORP-shun) process occurring in the nephrons during urine formation whereby

essential materials that were filtered out of the blood in the glomerulus are reabsorbed into the interstitial space

reactant (ree-AK-tant) participant in a chemical reaction that is changed by the reaction

receptor (ree-SEP-tor) portion of a sensory neuron that responds to an external stimulus; also, any molecule on the surface of a cell that binds specifically to other molecules (such as cell markers, hormones, neurotransmitters)

receptor potential (ree-SEP-tor poh-TEN-shal) potential that develops in the receptor's membrane when an adequate stimulus has been received

recessive (ree-SES-iv) in genetics, refers to genes that have effects that do not appear in the offspring when they are masked by a dominant gene (recessive forms of a gene are represented by lowercase letters); *see* **dominant**

recombinant DNA (ree-KOM-bih-nant D N A) joining together of hereditary material into new, biologically functional combinations

recruit to supply with new members, as in stimulating additional muscle fibers to contract simultaneously in a muscle in order to strengthen the contraction force

red fiber muscle fibers that contain large amounts of myoglobin and have a deep red appearance

red marrow bone marrow found in the ends of long bones and in flat bones; so named because of its function in the production of red blood cells

reduction proper alignment of a fractured bone; in chemistry, the gain of one or more electrons by a molecule (as in oxidation-reduction reactions)

reduction division meiosis; when the diploid chromosome number (46) is reduced to the haploid number (23)

referred pain pain from stimulation of nociceptors in deep structures felt on the surface of the body; for example, experiencing pain in the left arm when there is heart pain

reflex automatic involuntary reaction to a stimulus resulting from a nerve impulse passing over a reflex arc

reflex arc impulse conduction route to and from the central nervous system; smallest portion of nervous system that can receive a stimulus and generate a response

refraction (ree-FRAK-shun) bending of a ray of light as it passes from a medium of one density to one of a different density; occurs as light rays pass through the eye

regeneration (ree-jen-er-AY-shun) process of replacing missing tissue with new tissue by means of cell division

relative refractory period (ree-FRAK-tor-ee) in muscle cell contraction, the few milliseconds after the absolute refractory period; time during which the membrane is repolarizing and restoring the resting membrane potential

relaxation state of lessened tension

releasing hormones (ree-LEE-sing HOR-mohns) hormones produced by the hypothalamus that cause the pituitary gland to release its hormones

remission (ree-MISH-un) stage of a disease during which a temporary recovery from symptoms occurs

renal artery (REE-nal AR-ter-ee) large branch of the abdominal aorta that brings blood into each kidney

renal clearance (REE-nal) amount of a substance removed from the blood by the kidneys per minute

renal columns (REE-nal) within the kidneys, the cortical tissue in the medulla between the pyramids

renal corpuscle (REE-nal KOR-pus-ul) within the nephron, the glomerulus plus the Bowman's capsule surrounding it

renal cortex (REE-nal KOR-teks) outer portion of the kidney

renal diabetes (REE-nal dye-ah-BEE-teez) when the maximum cotransport capacity of the kidney is greatly reduced and glucose appears in the urine even though the blood sugar level may be normal; also called *renal glycosuria*

renal glycosuria (REE-nal glye-koh-soo-ree-ah) *see* **renal diabetes**

renal medulla (REE-nal meh-DUL-ah) inner portion of the kidney

renal pelvis (REE-nal PEL-vis) basin-like upper end of the ureter that is located inside the kidney into which the collecting tubules drain

renal pyramids (REE-nal PEER-ah-mids) distinct triangular wedges that make up most of the medullary tissue in the kidney

renal tubule (REE-nal TOOB-yool) a principal part of the nephron, it consists of the proximal tubule, the loop of Henle, and the distal tubule; receives filtrate from Bowman's capsule and transports it to collecting tubules while filtrate is reabsorbed and additional substances are secreted into it in the process of urine formation

renature (ree-NAYT-shur) to reverse the alteration (denaturation) of a protein and thus restore its normal shape and chemical properties

renin (REH-nin) enzyme produced by the juxtaglomerular apparatus of the kidney nephrons that catalyzes the formation of angiotensin, a substance that increases blood pressure; *see* **renin-angiotensin-aldosterone system (RAAS)**

renin-angiotensin-aldosterone system (RAAS) (REH-nun-an-jee-oh-TEN-sin-al-DAH-stair-ohn SIS-tem) causes changes in blood plasma volume mainly by controlling aldosterone secretion

repolarization (ree-poh-lah-ri-ZAY-shun) phase of the action potential in which the membrane potential changes from its maximum degree of depolarization toward the resting state potential

reproduction formation of a new individual; formation of new cells in the body to permit growth

reproductive system system in both sexes that is composed of the gonads, genital ducts, accessory glands, and genitalia

residual volume (RV) (ree-ZID-yoo-al) air that remains in the lungs after the most forceful expiration

resorption (ree-SORP-shun) bone loss resulting from osteoclastic activity

respiration (res-pi-RAY-shun) processes that result in the absorption, transport, and utilization or exchange of respiratory gases between an organism and its environment

respiratory acidosis (RES-pih-rah-tor-ee ass-i-DOH-sis) retention of carbon dioxide in the blood

respiratory alkalosis (RES-pih-rah-tor-ee al-kah-LOH-sis) excessive loss of carbonic acid from the blood

respiratory arrest (RES-pih-rah-tor-ee) cessation of breathing without resumption

respiratory center (RES-pih-rah-tor-ee) center located in the medulla and pons that stimulates muscles of respiration

respiratory cycle (RES-pih-rah-tor-ee) the alternating pattern of inspiration and expiration (inhalation/exhalation) of breathing

respiratory distress syndrome (RES-pih-rah-tor-ee di-STRESS SIN-drohm) difficulty in breathing caused by absence or failure of the surfactant in fluid lining the alveoli of the lung; IRDS is infant respiratory distress syndrome; ARDS is adult respiratory distress syndrome

respiratory membrane (RES-pih-rah-tor-ee) single layer of cells that makes up the wall of the alveoli

respiratory mucosa (RES-pih-rah-tor-ee myoo-KOH-sah) mucus-covered membrane that lines tubes of the respiratory tree

respiratory system (RES-pih-rah-tor-ee SIS-tem) system composed of the nose, pharynx, larynx, trachea, bronchi, and lungs

respiratory tract (RES-pih-rah-tor-ee) organs of the respiratory system, divided into lower and upper respiratory tracts

responsiveness (ree-SPONS-iv-ness) characteristic of life that permits an organism to sense, monitor, and respond to changes in its external environment

resting membrane potential (RMP) membrane potential maintained by a nonconducting neuron's plasma membrane; approximately 70 mV

retention (ree-TEN-shun) inability to void urine even though the bladder contains an excessive amount of urine

reticular activating system (reh-TIK-yoo-lar) network of relays in the brain responsible for maintaining consciousness

reticular formation (reh-TIK-yoo-lar) located in the medulla where bits of gray and white matter mix intricately; this structure is involved in regulating input from sensory neurons, arousal, and motor control

reticulin (reh-TIK-yoo-lin) specialized type of collagen found in reticular fibers

reticulocyte (reh-TIK-yoo-loh-syte) immature red blood cells

reticulocyte count (reh-TIK-yoo-loh-syte) medical procedure to determine the rate of erythropoiesis

reticulum (reh-TIK-yoo-lum) intricate network of fibers

retina (RET-i-nah) innermost layer of the eyeball; contains rods and cones and continues posteriorly with the optic nerve

retrograde signaling (RET-roh-grayd SIG-nah-ling) type of synaptic transmission in which chemical signals are sent from the postsynaptic neuron back to the presynaptic neuron, usually to facilitate or inhibit further presynaptic signals

retroperitoneal (reh-troh-pair-i-toh-NEE-al) area outside of the peritoneum

reversible reaction (ree-VER-sih-bul ree-AK-shun) when the products of a chemical reaction can change back to the original reactants; generally, an equilibrium of products and reactants exists

Rh factor (R h FAK-tor) antigen present on the red blood cells of Rh+ individuals; so named because it was first studied in the rhesus monkey

rheumatic fever (roo-MAT-ik) delayed inflammatory response to streptococcal infection that, if not properly treated, may allow the cardiac valves to become inflamed

rheumatic heart disease (roo-MAT-ik) cardiac damage (especially to the endocardium, including the valves) resulting from a delayed inflammatory response to streptococcal infection; *see* **rheumatic fever**

rhinitis (rye-NYE-tis) inflammation of the nasal mucosa often caused by nasal infections

rhodopsin (roh-DOP-sin) photopigment in rods; *see* **opsin**

ribonucleic acid (RNA) (rye-boh-noo-KLAY-ik) nucleic acid found in both nucleus and cytoplasm of cells; involved in transmission of genetic information from nucleus to cytoplasm and in cytoplasmic assembly of proteins

ribose (RYE-bohse) five-carbon sugar; the sugar in RNA

ribosome (RYE-boh-sohm) organelle in the cytoplasm of cells that synthesizes proteins; sometimes called "protein factory"

ribozyme (RYE-boh-zyme) form of RNA (ribonucleic acid) that acts like an enzyme to promote or regulate chemical reactions in the cell

rickets (RIK-ets) condition primarily seen in infants and children caused by vitamin D deficiency; disease results in soft, pliable bones and skeletal deformities caused by abnormal calcium metabolism

righting reflexes muscular responses to restore the body and its parts to their normal position when they have been displaced

rigor mortis (RIG-or MOR-tis) literally "stiffness of death"; the permanent contraction of muscle tissue after death caused by the depletion of ATP during the actin-myosin reaction, preventing myosin from releasing actin to allow relaxation of the muscle

RNA *see* **ribonucleic acid**

RNA interference (RNAi) a regulatory process of the cell in which a small molecule of dsRNA (double-stranded RNA) called siRNA (small interfering RNA) joins with a RISC (RNA-induced silencing complex) protein structure to break down a specific mRNA (messenger RNA) transcript and thus effectively silence the gene encoded by the mRNA; RNAi is a natural regulatory process thought to be involved with regulating gene expression, as in inhibiting viral infections, but is also used as a research technique to study the human genome

rotation (roh-TAY-shun) joint movement around a longitudinal axis; for example, shaking your head "no"

rotator cuff (ROH-tay-tor) musculotendinous cuff resulting from fusion of the tendons of the supraspinatus, infraspinatus, teres minor, and subscapularis; adds to the stability of the glenohumeral (shoulder) joint

rubrospinal tract (roo-broh-SPY-nal) spinal cord nerves that transmit impulses that coordinate body movements and maintain posture

Ruffini's corpuscle (roo-FEE-neez KOR-pus-ul) receptor that senses deep pressure and continuous touch; located in dermis of the skin

"rule of nines" frequently used method to estimate extent of a burn injury in an adult; the body is divided into areas that are multiples and fractions of 9%

S

S phase step of the cell life cycle in which a growing cell synthesizes a second copy of its nuclear DNA molecules, a process also called *DNA replication*, in anticipation of later reproduction (cell division); S phase follows the G_1 phase (first growth phase) and precedes the G_2 phase (second growth phase); *see* **cell life cycle**

saccule (SAK-yool) part of the membranous labyrinth in the inner ear; contains sensory structure called a *macula*, which functions in equilibrium

sacral plexus (SAY-kral PLEKS-us) plexus formed by fibers from the fourth and fifth lumbar nerves and the first four sacral nerves

sagittal plane (SAJ-i-tal) longitudinal plane that divides the body or a part into left and right sides

salpingitis (sal-pin-JYE-tis) inflammation of the uterine tubes

saltatory conduction (SAL-tah-tor-ee) process in which a nerve impulse travels along a myelinated fiber by jumping from one node of Ranvier to the next

sarcolemma (sar-koh-LEM-ah) plasma membrane of a striated muscle fiber

sarcoma (sar-koh-mah) tumor of muscle tissue

sarcomere (SAR-koh-meer) contractile unit of muscle cells; length of a myofibril between two Z disks

sarcoplasm (SAR-koh-plaz-em) cytoplasm of muscle fibers

sarcoplasmic reticulum (sar-koh-PLAZ-mik reh-TIK-yoo-lum) network of tubules and sacs in muscle cells; similar to endoplasmic reticulum of other cells

satellite cell (SAT-i-lyte) a type of Schwann cell (neuroglial cell) that surrounds the cell bodies of neurons of the peripheral nervous system

satiety center (sah-TYE-eh-tee) cells in the hypothalamus that send impulses to decrease appetite so that an individual feels satisfied

saturated fat (SACH-eh-ray-ted) fats containing triglycerides in which fatty acid chains contain no double bonds (because they are "saturated" with hydrogen atoms)

scala tympani (SKAH-lah TIM-pah-nee) lower portion of the cochlear duct lying below the basilar membrane

scala vestibuli (SKAH-lah ves-TIB-yoo-lye) upper portion of the cochlear duct scar-thickened mass of tissue, usually fibrous connective tissue, that remains after a damaged tissue has been repaired

Schwann cells (shwon *or* shvon) large nucleated cells that form myelin around the axons of neurons; also called *neurilemmocyte*

sciatica (sye-AT-i-kah) neuralgia of the sciatic nerve

science (SYE-ens) field of inquiry that uses rational or logical methods to determine the nature of the universe; biology is the branch of science that specifically studies the living organisms of the universe

sclera (SKLEH-rah) white outer coat of the eyeball

scleroderma (skleer-oh-DER-mah) rare disorder affecting vessels and connective tissue of skin and other tissues, characterized by tissue hardening

scoliosis (skoh-lee-OH-sis) abnormal lateral (side-to-side) curvature of the vertebral column

scrotum (SKROH-tum) pouchlike sac that contains the testes

seasonal affective disorder (SAD) (SEE-son-al ah-FEK-tiv dis-OR-der) mental disorder in which a patient suffers severe depression only in winter; linked to the pineal gland

sebaceous gland (seh-BAY-shus gland) oil-producing glands in the skin

sebum (SEE-bum) secretion of sebaceous glands

second messenger model theory of hormone action in which the hormone binds to receptors of the target cell, which then triggers a second molecule within the cell (such as cyclic AMP) to accomplish its function; also called *fixed-membrane-receptor model*

secondary bronchi (SEK-on-dair-ee BRONG-kye) when the primary bronchus enters the lung on its respective side and immediately splits into smaller bronchioles

secondary ossification center (SEK-on-dair-ee os-i-fi-KAY-shun SEN-ter) growth center located in the epiphyses of long bones; *see* **primary ossification center**

secondary sexual characteristics external physical characteristics of sexual maturity resulting from action of sex hormones; they include male and female patterns of body hair and fat distribution, as well as development of external genitals

second-degree burn burn involving deep epidermal layers of the skin but not causing irreparable damage

second-messenger model model to explain nonsteroid hormone mechanism of action; nonsteroid hormone is "first messenger" acting on cell membrane; intracellular "second messenger"—often cyclic AMP—triggers specific cellular action

secretin (seh-KREE-tin) gastrointestinal hormone; first hormone discovered

secretion (seh-KREE-shun) process by which a substance is released outside the cell

secretory vesicle (SEEK-reh-toh-ree VES-ik-il) bubble made of cellular membrane and containing products of cell metabolism that moves from inside the cell to the plasma membrane, where it breaks open and secretes the products outside of the cell

segmental reflex (seg-MEN-tal REE-fleks) reflex whose mediating reflexes enter and leave the same segment of the spinal cord

segmentation (seg-men-TAY-shun) when digestive reflexes cause a forward-and-backward movement within a single region of the GI tract

seizure (SEE-zhur) *see* **convulsion**

selectively permeable (sel-EK-tiv-lee PERM-ee-ah-bil) adjective used to describe a living membrane that allows only certain substances to move through (permeate) it and only at certain times

self concept in immunology that proposes the immune system agents can recognize certain cell-surface molecules as belonging to that individual, which thus makes them "safe" from destruction or damage from the immune system

self-antigen (self AN-ti-jen) molecule located on the plasma membrane of all body cells that identifies all normal cells of the body for the immune system; also called *self-marker*
semen (SEE-men) ejaculate from the penis that contains spermatozoa plus fluids from the testes, seminal vesicles, bulbourethral glands, and prostate
semicircular canals (sem-i-SIR-kyoo-lar kah-NALS) three tubelike structures located in the inner ear that function in the sense of equilibrium
semilunar valves (sem-i-LOO-nar) valves located between each ventricle and the large artery that carries blood away from it; valves in the veins
seminiferous tubule (seh-mih-NIF-er-us TOOB-yool) long, coiled structure that forms the bulk of the testicular mass and in which spermatozoa develop
senescence (seh-NES-enz) older adulthood; aging
sensation interpretation of sensory nerve impulses by the brain as an awareness of an internal or external event; for example, feeling pain
sensory nerves (SEN-sor-ee) nerves that contain primarily sensory neurons
sensory neurons (SEN-sor-ee NOO-rons) neurons that transmit impulses to the spinal cord and brain from all parts of the body
sensory projection (SEN-sor-ee proh-JEK-shun) brain function that pinpoints the area of the body from which a receptor potential was initiated
sensory receptor (SEN-sor-ee ree-SEP-tor) sense organs in the peripheral nervous system that enable the body to respond to stimuli caused by changes in its internal or external environment
septicemia (sep-ti-SEE-mee-ah) blood poisoning
septum (SEP-tum) a wall that divides two areas; for example, nasal septum
serosa (seh-ROH-sah) outermost covering of the digestive tract; composed of the parietal pleura in the abdominal cavity
serotonin (sair-oh-TOH-nin) neurotransmitter that belongs to a group of compounds called *catecholamines*
serous membrane (SEE-rus) two-layer epithelial membrane that lines body cavities and covers surfaces of organs
serous pericardium (SEER-us per-i-KAR-dee-um) part of pericardial coverings of the heart; made up of a parietal layer and a visceral layer
Sertoli cell (ser-TOH-lee) testis cell found within walls of the seminiferous tubules; provides mechanical support and protection for the developing sperm
serotonin-specific reuptake inhibitors (SSRIs) (sair-oh-TOH-nin speh-SIF-ik ree-UP-tayk in-HIB-it-orz) type of drug that stops the return of serotonin (neurotransmitter) to the presynaptic neuron, thus increasing the baseline amount of serotonin in the synapse and restoring the baseline to a normal serotonin level
sesamoid bones (SES-ah-moyd) small bones found in the substance of tendons
sex chromosomes (KROH-moh-sohms) pair of chromosomes in the human genome that determine gender; normal males have one X chromosome and one Y chromosome (XY); normal females have two X chromosomes (XX)
sex hormones (HOR-mohns) hormones that target reproductive tissue; examples include estrogen, progesterone, and testosterone
sex-linked trait nonsexual, inherited trait governed by genes located in a sex chromosome (X or Y); most known sex-linked traits are X-linked
shoulder girdle clavicle and scapula, which form the only bony attachment of the upper extremity to the trunk
sickle cell anemia (SIK-ul sell ah-NEE-mee-ah) severe, possibly fatal, hereditary disease in which red blood cells become sickle-shaped because of presence of an abnormal type of hemoglobin
sickle cell trait (SIK-ul sell trayt) condition that occurs when only one gene for sickle cell is inherited and only a small amount of abnormal hemoglobin is produced
sign objective deviation from normal that marks the presence of a disease
signal transduction (SIG-nal tranz-DUK-shun) process of changing a signal such as a hormone or neurotransmitter into another form such as enzymatic reaction within the cell receiving the signal (thus the extracellular hormone signal is transduced, or changed, to an intracellular enzymatic signal)
simple diffusion (dih-FYOO-zhun) movement of molecules through a membrane by means of the natural tendency of the molecules to spread and the ability of the spreading molecules to move through, or permeate
simple fracture classification of a broken bone when the skin remains intact
single unit smooth muscle most common type of smooth muscle tissue; smooth muscle tissue in which gap junctions join individual smooth muscle fibers into large, continuous sheets that contract together in an autorhythmic fashion; also called *visceral muscle*
sinoatrial (SA) node (sye-no-AY-tree-al) the heart's pacemaker; where the impulse conduction of the heart normally starts; located in the wall of the right atrium near the opening of the superior vena cava
sinus (SYE-nus) space or cavity inside some cranial bones
sinus arrhythmia (SYE-nus ah-RITH-mee-ah) variation in rhythm of heart rate during the breathing cycle (inspiration and expiration)
sinusitis (sye-nyoo-SYE-tis) sinus infection
skeletal muscle (SKEL-eh-tal) also known as *voluntary* or *striated voluntary muscle*; muscles under willed or voluntary control
skeletal system (SKEL-eh-tal) bones, cartilage, and ligaments that provide the body with a rigid framework for support and protection
sliding-filament model (SLY-ding FIL-ah-ment) theory of muscle contraction in which sliding of thin filaments toward the center of each sarcomere quickly shortens the muscle fiber and thereby the entire muscle
slow muscle fiber red muscle fiber; also called *slow-twitch muscle fiber*
smooth muscle muscles that are not under conscious control; also known as *involuntary* or *visceral muscle*; forms the walls of blood vessels and hollow organs
sodium channel (SO-dee-um CHAN-el) a pore in a cell membrane that allows only sodium ions to permeate, or pass through, the membrane
sodium cotransport (SO-dee-um koh-TRANS-port) complex transport process in which carriers that bind both sodium ions and glucose molecules passively transport the molecules together out of the GI lumen
sodium-potassium pump (SO-dee-um-poh-TAS-ee-um) active transport pump that operates in the plasma membrane of all human cells; transports both sodium ions and potassium ions but in opposite directions and in a 3:2 ratio, thereby maintaining a gradient across the plasma membrane
soft palate (PAL-et) partition between the mouth and nasopharynx
solute (SOL-yoot) dissolved particles in solution
solution (suh-LOO-shun) liquid made up of a mixture of molecule types, usually made of solutes (solids) scattered in a solvent (liquid), such as salt in water
solvent (SOL-vent) liquid portion of a solution in which a solute is dissolved
soma (SO-mah) cell body
somatic (so-MAH-tik) refers to the body or, specifically, all cells of the body except the gametes
somatic motor neuron (so-MAH-tik MOH-tor NOO-ron) motor neuron that stimulates a muscle fiber
somatic nervous system (so-MAH-tik) motor neurons that control voluntary actions of skeletal muscles
somatic reflex (so-MAH-tik) reflexive contraction of skeletal muscles
somatic sense (so-MAH-tik) sense that enables an individual to detect sensations such as pain, temperature, and proprioception
somatic sensory division (so-MAH-tik) division of the nervous system made up of afferent (incoming) pathways from somatic sensory receptors (receptors involved in conscious perception)
somatopsychic (so-mah-toh-SYE-kik) refers to the body affecting the mind; physical disorder that produces mental symptoms; *see* **psychosomatic**
somatostatin (so-mah-toh-STAT-in) hormone produced by delta cells of the pancreas that inhibits secretion of glucagon, insulin, and pancreatic polypeptide
somatotroph (so-mah-toh-TROHF) cell type of the adenohypophysis (anterior pituitary) that secretes growth hormone (somatotropin)
somatotropin (STH) (so-mah-toh-TROH-pin) growth hormone
somatotype (so-MAT-oh-type) classification of body type determined on the basis of certain physical characteristics; *see* **ectomorph, endomorph, mesomorph**
snRNP (snurp) small nuclear ribonucleotide; along with polypeptides, forms subunits of the spliceosome in a cell's nucleus
spatial summation (SPAY-shal sum-MAY-shun) ability of the postsynaptic neuron to add together the inhibitory and stimulatory input received from numerous different presynaptic neurons and produce an action potential based on that collation of information
species resistance (SPEE-shez ree-SIS-tens) genetic characteristics common to all organisms of a particular species that provide natural inborn immunity to certain disease
specific heat the amount of heat required to raise 1 gram of substance 1 degree centigrade
specific immunity (i-MYOON-i-tee) protective mechanisms by which the immune system is able to recognize, remember, and destroy specific types of bacteria or toxins; *see* **adaptive immunity**
spermatic cord (sper-MAT-ik cord) cylindrical casings of white fibrous tissue formed by the ductus deferens and located in the inguinal canal between the scrotum and the abdominal cavity
spermatogenesis (sper-mah-toh-JEN-eh-sis) production of sperm cells
spermatogonia (sper-mah-toh-GO-nee-ah) stem cell population that gives rise to sperm cells
sphygmomanometer (sfig-moh-mah-NOM-eh-ter) device for measuring blood pressure in the arteries of a limb
spinal cord portion of central nervous system that provides two-way conduction from the brain; major reflex center
spinal ganglion (SPY-nal GANG-glee-on) enlarged portion of the dorsal root of the spinal cord, where afferent nerve fibers from sensory receptors synapse with associated sensory neurons on their way to the brain or lower reflex centers
spinal meningitis (men-in-JYE-tis) inflammation of the spinal meninges
spinal nerve nerve that connects the spinal cord to peripheral structures such as the skin and skeletal muscles
spinal reflex reflex arc whose center is located in the spinal cord
spinal tracts white columns of the spinal cord that provide conduction paths to and from the brain; ascending tracts carry information to the brain, whereas descending tracts conduct impulses from the brain
spindle fiber (SPIN-dul FYE-ber) network of tubules formed in the cytoplasm between the centrioles as they are moving away from each other during mitosis
spinothalamic pathway (spy-no-thah-LAM-ik) pathway that conducts impulses that produce sensations of crude touch and pressure
spinothalamic tract (spy-no-thah-LAM-ik trakt) ascending tracts of the spinal cord that convey information about pain, temperature, deep pressure, and coarse touch; these tracts cross over in the spinal cord
spirometer (spih-ROM-eh-ter) instrument used to measure the amount of air exchanged in breathing
spleen largest lymphoid organ; filters blood, destroys worn out red blood cells, salvages iron from hemoglobin, and serves as a blood reservoir
splenomegaly (spleh-no-MEG-ah-lee) condition of enlargement of the spleen
spliceosome (SPLYS-oh-sohm) small structure within the nucleus, about the size of a ribosome and made up of snRNPs (small nuclear ribonucleotides) along with polypeptides; removes introns from mRNA transcript and splices remaining exons into final version of mRNA
spongy bone porous bone in the end of the long bone, which may be filled with marrow
spore (spoor) form assumed by a bacterium that is resistant to heat, drying, and chemicals but can later become active to cause infection
sprain injury to ligamentous joint structures often caused by twisting or wrenching movements
squamous (SKWAY-muss) scalelike
stage of exhaustion third stage of the general adaptation syndrome; when the body can no longer cope or adapt to stressors
stage of resistance second stage of the general adaptation syndrome
staircase phenomenon (STAIR-case feh-NOM-eh-non) *see* **treppe**
Starling's law of the heart principle stating that as myocardial fibers are stretched, the force of contraction is increased
static equilibrium (STAT-ik ee-kwi-LIB-ree-um) sensing the position of the head relative to gravity; compare with dynamic equilibrium
static tension (STAT-ik TEN-shun) another name for isometric (muscle) contraction
stem cell cells that have the ability to maintain a constant population of newly differentiating cells
stenosed (cardiac) valves (steh-NOSD [(KAR-dee-ak)]) valves that are narrower than normal, slowing blood flow from a heart chamber
stereognosis (steh-ree-og-NO-sis) awareness of an object's size, shape, and texture
steroid (STAYR-oyd) any of a class of lipids related to sterols and forming numerous reproductive and adrenal hormones
steroid hormones (STAYR-oyd HOR-mohns) lipid-soluble hormones that pass intact through the cell membrane of the target cell and influence cell activity by acting on specific genes
stillbirth delivery of a dead fetus after the twentieth week of gestation; before 20 weeks it is termed a *spontaneous abortion*
stimulus (STIM-yoo-lus) excitant or irritating agent that induces a response
stimulus-gated channels (STIM-yoo-lus GAY-ted CHAN-nuls) type of cell-membrane channel for the transport of molecules that is controlled by a gate that responds to a stimulus such as a sensory stimulus or chemical (neurotransmitter) stimulus
stomach an expansion of the digestive tract between the esophagus and small intestine, where some protein digestion begins and where food is churned and mixed with gastric juices before entering the small intestine
strabismus (strah-BIS-mus) abnormal condition in which lack of coordination of, or weakness in, the muscles that control the eye causes improper focusing of images on the retina, making depth perception difficult
strain overexertion or trauma-type injury resulting in tearing of skeletal muscle fibers
stratum (STRAH-tum) layer; *plural*, strata
stratum basale (STRAH-tum BAY-sah-lee) "base layer"; deepest layer of the epidermis; cells in this layer are able to reproduce themselves
stratum compactum (STRAH-tum kom-PAKT-um) surface layer of the endometrium in the uterus
stratum corneum (STRH-tum KOR-nee-um) tough outer layer of the epidermis; cells are filled with keratin
stratum germinativum (STRAH-tum JER-mi-nah-tiv-um) term for combined stratum basale and stratum spinosum in the epidermis
stratum granulosum (STRAH-tum gran-yoo-LOH-sum) "granular layer," layer in which the process of keratinization begins
stratum lucidum (STRA-tum loo-SEE-dum) "clear" layer of the epidermis, in thick skin between the stratum granulosum and the stratum corneum
stratum spinosum (STRA-tum spi-NO-sum) "spiny layer"; layer of epidermis that is rich in RNA to aid in protein synthesis required for keratin production
stratum spongiosum (STRA-tum spon-gee-OH-sum) middle layer of the endometrium (of uterus), composed of loose connective tissue
strength training contracting muscles against resistance to enhance muscle hypertrophy; *see* **aerobic training, endurance training**
stress any stimulus that directly or indirectly stimulates neurons of the hypothalamus to release corticotropin-releasing hormone
stress response *see* **stress syndrome**
stress syndrome (stress SIN-drohm) signs and symptoms associated with the body's frequently maladaptive response to stressors; diverse changes initiated by stress
stressor (STRESS-or) any agent or stimulus that produces stress
stretch marks tiny silver-white scars resulting when elastic fibers in the dermis are stretched too much; also called *striae*
stretch reflex automatic response in which the muscle reacts and tries to maintain a constancy of muscle length when a load is applied
striated involuntary muscle (STRYE-ay-ted in-VOL-un-TAIR-ee) cardiac muscle
striated muscle (STRYE-ay-ted) *see* **skeletal muscle**
stroke cerebrovascular accident (CVA); results from ischemia of brain tissue caused by an embolism or ruptured aneurysm in brain blood vessel
stroke volume amount of blood that is ejected from the ventricles of the heart with each beat
structural protein (STRUK-cher-al PROH-teen) any of a category of proteins with the primary function of forming structures of the cell or tissue; contrast with *functional protein*
subarachnoid space (sub-ah-RAK-noyd) within the meninges, space under the arachnoid and outside the pia mater
subareolar plexus (sub-ah-REE-oh-lar PLEKS-us) where the cutaneous plexus and large lymphatics that drain the secretory tissue and ducts of the breast come together
subatomic particles (sub-ah-TOM-ik PAR-ti-kuls) particles that make up atoms; for example, neutrons, protons, and electrons
subdural space (sub-DOO-ral) within the meninges; the space between the dura mater and the arachnoid membrane
sublingual glands (sub-LING-gwal) smallest salivary glands; produce a mucous type of saliva
submandibular glands (sub-man-DIB-yoo-lar) salivary glands located just below the mandibular angle; contain enzyme and mucus-producing elements
submucosa (sub-myoo-KOH-sah) connective tissue layer containing blood vessels and nerves in the wall of the digestive tract

substantia nigra (sub-STAN-shee-ah NYE-grah) midbrain structure that consists of clusters of cell bodies of neurons involved in muscular control
substrate (SUB-strayt) substance on which an enzyme acts
sucrase (soo-krays) enzyme that catalyzes the hydrolysis of sucrose and maltose
sudoriferous glands (soo-doh-RIF-er-us) epidermal sweat glands
sulcus (SUL-kus) furrow, or groove, often associated with the cerebral cortex
superficial (soo-per-FISH-al) near the body surface; opposite of deep
superficial fascia (soo-per-FISH-al FAH-shah) hypodermis; subcutaneous layer beneath the dermis
superficial reflex (soo-per-FISH-al) reflexes elicited by stimulation of receptors located in the skin or mucosa
superior higher; opposite of inferior
suprarenal (soo-prah-REE-nal) situated above a kidney
surface film mixture of residues and secretions from sweat and sebaceous glands with epithelial cells being shed from the epidermis; works as a protective barrier
surfactant (sur-FAK-tant) substance covering the surface of the respiratory membrane inside the alveolus; it reduces surface tension and prevents the alveoli from collapsing
suspensory ligament (sus-PEN-so-ree LIG-ah-ment) fibers attached to the capsule of the lens that help hold it in place
suture (SOO-chur) immovable joint, such as those between the bones of the skull
sweat gland gland in the skin that produces a transparent, watery liquid that eliminates ammonia and uric acid and helps maintain body temperature
sympathetic division (sim-pah-THET-ik) part of the autonomic nervous system; ganglia are connected to the thoracic and lumbar regions of the spinal cord; functions in "fight-or-flight" response
symphysis (SIM-fi-sis) joint characterized by the presence of a pad or disk of fibrocartilage connecting the two bones
symptom (SIMP-tum) subjective deviation from normal that marks the presence of a disease; sometimes used synonymously with *sign*
synapse (SIN-aps) membrane-to-membrane junction between a neuron and another neuron, effector cell, or sensory cell; function to propagate action potential (via neurotransmitters)
synaptic cleft (si-NAP-tik kleft) space between a synaptic knob and the plasma membrane of a postsynaptic neuron
synaptic knob (sih-NAP-tik nob) tiny bulge at the end of a terminal branch of a presynaptic neuron's axon that contains vesicles with neurotransmitters
synarthrosis (sin-ar-THROH-sis) joint in which fibrous connective tissue joins bones and holds them together tightly; commonly called *sutures*
synchondrosis (sin-kon-DROH-sis) joint characterized by the presence of hyaline cartilage between articulating bones
syncytium (sin-SISH-ee-em) continuous, electrically coupled mass of cardiac fibers; allows an efficient, coordinated pumping action
syndesmosis (sin-des-MOH-sis) fibrous joint
syndrome (SIN-drohm) collection of signs or symptoms, usually with a common cause that defines or gives a clear picture of a pathological condition
synergism (SIN-er-jiz-em) with hormones, a combination of hormones has a greater effect on a target than the sum of effects that each would have if acting alone
synergist (SIN-er-jist) muscle that assists a prime mover
synovial fluid (si-NO-vee-al) thick, colorless lubricating fluid secreted by the synovial membrane in synovial joints
synovial membrane (sih-NO-vee-al) connective tissue membrane lining spaces between bones and joints that secretes synovial fluid
synthesis (SIN-the-sis) combination, as in the combination of molecules to form a larger molecule
synthesis reaction (SIN-the-sis ree-AK-shun) reaction that combines two or more reactants to form a more complex structure
system a group of organs that functions as a coordinated team; also called a *body system*
systemic anatomy (sis-TEM-ik ah-NAT-oh-mee) study of anatomy that focuses on learning about body systems
systemic circulation (sis-TEM-ik ser-kyoo-LAY-shun) blood flow from the left ventricle to all parts of the body and back to the right atrium
systole (SIS-toh-lee) contraction of heart muscle
systolic blood pressure (sis-TOL-ik) force with which blood pushes against artery walls when ventricles contract
systolic discharge (sis-TOL-ik) volume of blood pumped by one contraction

T

T cell *see* **T lymphocytes**
T lymphocytes (T LIM-foh-sytes) cells of the immune system that have undergone maturation in the thymus; produce cell-mediated immunity
T tubules (T TOOB-yools) transverse tubules unique to muscle cells; formed by inward extensions of the sarcolemma that allow electrical impulses to move deeper into the cell
tachycardia (tak-eh-KAR-dee-ah) rapid heart rhythm (above 100 beats/min)
target cell cell that, when acted on by a particular hormone, responds because it has receptors to which the hormone can bind
target organs organs that are acted on and respond to a particular hormone
tarsal plate (TAR-sal plate) border of thick connective tissue at the free edge of each eyelid
taste buds chemical receptors in the tongue that generate nerve impulses, resulting in the sense of taste
Tay-Sachs disease (TSD) (tay-SAKS) recessive, inherited condition in which abnormal lipids accumulate in the brain and cause tissue damage that leads to death by age 4 years
tectorial membrane (tek-TOH-ree-al) gelatinous membrane of inner ear that tops hair cells in the organ of Corti
telodendria (tel-oh-DEN-dree-ah) distal tips of axons that form branches; each branch terminates in a synaptic knob
telophase (TEL-oh-fayz) last stage of mitosis
temporal summation (TEM-poh-ral sum-MAY-shun) when synaptic knobs stimulate a postsynaptic neuron in rapid succession and the effects add up over time to produce an action potential (also spatial summation)
tendon bands or cords of fibrous connective tissue that attach a muscle to a bone or other structure
tendon reflex reflex stimulated by tapping on a tendon
tendon sheath (sheeth) tube-shaped structure lined with synovial membrane that encloses certain tendons
teratogen (TER-ah-toh-jen) physical or chemical agent that disrupts normal embryonic development and thus causes congenital defects
terminal hair (TER-mih-nal) coarse pubic and axillary hair that develops at puberty
Terminologia anatomica (ter-mi-no-LOJ-ee-ah an-ah-TOM-ik-ah) literally "Terminology of Anatomy" it is the official list of antatomical terms sponsored by the International Federation of Associations of Anatomists and lists all terms by number and name (in both Latin and English) as well as general principles of usage
testosterone (tes-TOS-teh-rohn) male sex hormone produced by interstitial cells in the testes; the "masculinizing hormone"
tetanus (TET-ah-nus) smooth, sustained muscular contraction caused by high-frequency stimulation
thalamus (THAL-ah-muss) mass of gray matter located in diencephalon just above the hypothalamus; helps produce sensations, associates sensations with emotions, and plays a part in the arousal mechanism
theory (THEE-oh-ree) a scientific idea or explanation that has a reasonably high degree of confidence after rigorous testing or observation; compare to *hypothesis* and *law*
thermoreceptor (ther-moh-ree-SEP-tor) receptors activated by heat or cold
third-degree burn full-thickness burn; most serious of the three degrees of burns
thoracic duct (thoh-RAS-ik) largest lymphatic vessel in the body
thoracolumbar division (thor-rah-koh-LUM-bar) sympathetic division of the autonomic nervous system
thorax (THOR-aks) chest
threshold potential (THRESH-hold poh-TEN-shal) magnitude of voltage across a membrane at which an action potential, or nerve impulse, is produced
threshold stimulus (THRESH-hold STIM-yoo-lus) minimal level of stimulation required to cause a muscle fiber to contract
thrombocytes (THROM-boh-sytes) also called *platelets*; play a role in blood clotting
thrombophlebitis (throm-boh-fleh-BYE-tis) vein inflammation (phlebitis) accompanied by clot formation
thromboxane (throm-BOKS-ayne) prostaglandin-like substance in platelets that plays a role in hemostasis and blood clotting
thrombus (THROM-bus) a nonmoving or fixed blood clot attached to the lining of a vein or artery
thymocytes (THY-moh-sytes) cells in the thymus that develop into T lymphocytes
thymosin (THY-moh-sin) hormone produced by the thymus that is vital to the development and functioning of the body's immune system
thymus (THY-muss) endocrine gland located in the mediastinum; vital part of the body's immune system
thyroid cartilage (THY-royd KAR-tih-lij) largest cartilage of the larynx; Adam's apple
thyroid gland (THY-royd) endocrine gland located in the neck that stores its hormones until needed; thyroid hormones regulate cellular metabolism
thyroid hormone (THY-royd HOR-mohn) hormone that accelerates catabolism of glucose
thyroid-stimulating hormone (TSH) (THY-royd STIM-yoo-lay-ting HOR-mohn) a tropic hormone secreted by the anterior pituitary gland that stimulates the thyroid gland to increase its secretion of thyroid hormone
thyrotroph (THY-roh-trohf) cell type of the adenohypophysis (anterior pituitary) that secretes thyroid-stimulating hormone (TSH)
thyroxine (T_4) (thy-ROK-sin) thyroid hormone that stimulates cellular metabolism
tic douloureux (tik doo-loo-ROO) *see* **trigeminal neuralgia**
tidal volume (TV) (TYE-dal) amount of air breathed in and out with each breath
tight junction connection between cells in which they are joined by "collars" of tightly fused membrane
tissue group of similar cells that performs a common function
TMR *see* **transmyocardial laser revascularization**
tonic contraction (TON-ik) special type of skeletal muscle contraction used to maintain posture
tonicity (toh-NIS-i-tee) potential osmotic pressure
tonsillitis (tahn-sih-LYE-tis) inflammation of the tonsils, usually due to infection
tonsils (TAHN-sils) masses of lymphoid tissue; protect against bacteria; three types: palatine tonsils, located on each side of the throat; pharyngeal tonsils (adenoids), near the posterior opening of the nasal cavity; and lingual tonsils, near the base of the tongue
total lung capacity total volume of air a lung can hold
total metabolic rate (TMR) (met-ah-BOL-ik) total amount of energy used by the body per day
toxin (TOK-sin) poison; chemical that can cause sickness or damage in the body (*tox-* poison, *-in* neutral substance)
trabeculae (trah-BEK-yoo-lee) needle-like threads of spongy bone that surround a network of spaces
trace element elements found in small amounts in the body
trachea (TRAY-kee-ah) windpipe; the tube extending from the larynx to the bronchi
tracheostomy (tray-kee-OS-toh-mee) surgical procedure to cut an opening into the trachea
transcription (trans-KRIP-shun) process in which DNA molecule is used as template to form mRNA
transduction (tranz-DUK-shun) process of changing one form of energy or other physical event to another form, as when sound energy is changed to electrical energy in a microphone
transfer RNA (tRNA) RNA involved with protein synthesis; tRNA molecules carry amino acids to the ribosome for placement in the sequence prescribed by mRNA
translation process in which mRNA is used by ribosomes in the synthesis of a protein
transmyocardial laser revascularization (TMR) (tranz-my-oh-KARD-ee-al LAY-zer ree-VAS-kyoo-lar-i-zay-shun) heart surgery used with patients suffering from severe angina (chest/heart pain) in which a laser beam makes toothpick-sized holes in the myocardium, which stimulates the growth of new blood vessels (angiogenesis) that relieves the angina
transpulmonary pressure (tranz-PUHL-mohn-air-ee) the pressure difference between the alveolar air pressure in the lungs and the fluid pressure in the intraplural space, that is, the pressure difference across the wall of the lung
transverse plane (TRANS-vers) horizontal plane that divides the body or any of its parts into upper and lower parts
treppe (TREP-ee) gradual increase in the extent of muscular contraction following rapidly repeated stimulation; staircase phenomenon
triad (TRY-ad) triplet of tubules; allows an electrical impulse traveling along a T tubule to stimulate the membranes of adjacent sacs of the sarcoplasmic reticulum
tricarboxylic acid cycle (TAC) (try-kar-bok-SIL-ik) aerobic metabolic pathway in which acetyl CoA (from other metabolic pathways) is converted into CO_2 and H_2O with the formation of ATP; also known as *citric acid cycle*, or *Krebs cycle*
tricuspid valve (try-KUS-pid) heart valve located between right atrium and ventricle
trigeminal nerve (try-JEM-i-nal) cranial nerve V; responsible for chewing movements and sensations of the head and face
trigeminal neuralgia (try-JEM-i-nal noo-RAL-jee-ah) pain in one or more (of three) branches of the fifth cranial nerve (trigeminal nerve) that runs along the face; also called *tic douloureux*
triglyceride (try-GLI-ser-ide) lipid that is synthesized from fatty acids and glycerol or from excess glucose or amino acids; stored mainly in adipose tissue cells
triiodothyronine (T_3) (try-eye-oh-doh-THY-roh-neen) thyroid hormone that stimulates cellular metabolism
trimester (try-MES-ter) three-month segments of the gestation period
trisomy (TRY-so-mee) abnormal genetic condition in which cells have three chromosomes (a triplet) where there should be a pair; usually caused by nondisjunction (failure of chromosome pairs to separate) during gamete production
trochlear nerve (TROHK-lee-ar) cranial nerve IV; motor nerve; responsible for eye movements
trophoblast (TROH-foh-blast) outer wall of the blastocyst; contributes in formation of the placenta
tropic hormone (TROH-pik HOR-mohn) hormone that stimulates another endocrine gland to secrete its hormones
tropomyosin (troh-poh-MY-oh-sin) in sliding-filament theory of muscle cell contraction, the long protein molecule that covers the active sites of myosin; myofilament
troponin (troh-POH-nin) in sliding filament theory of muscle cell contraction, the molecules spaced at intervals along the thin filament that block troponin when the myofilament is at rest
true vocal cords lower pair of vocal cords in the larynx, responsible for vocalization
tubal pregnancy (TOO-bal) ectopic pregnancy that occurs in a uterine tube
tuberculosis (TB) (too-ber-kyoo-LOH-sis) chronic bacterial (bacillus) infection of the lungs or other tissues caused by *Mycobacterium tuberculosis* organisms
tumor (TOO-mer) growth of tissues in which cell proliferation is uncontrolled and progressive
tumor suppressor genes (TOO-mer soo-PRESS-or jeens) genes that work against the development of cancerous cells
tunica adventitia (TOO-ni-kah ad-ven-TISH-ah) outermost layer of arteries and veins; made of strong, flexible fibrous connective tissue
tunica albuginea (TOO-ni-kah al-byoo-JIN-ee-ah) tough, whitish membrane that surrounds each testis and enters the gland to divide it into lobules
tunica interna (TOO-ni-kah in-TER-nah) endothelium that lines blood vessels
tunica media (TOO-ni-kah MEE-dee-ah) muscular middle layer in blood vessels; the tunica media of arteries is more muscular than that of veins
Turner syndrome (TUR-ner SIN-drohm) genetic disorder caused by monosomy of the X chromosome (XO) in females; characterized by immaturity of sex organs (causing sterility), webbed neck, cardiovascular defects, and learning disorders
twitch contraction quick jerk of a muscle produced by a single, brief threshold stimulus
tympanic membrane (tim-PAN-ik) eardrum
Type A blood type that has antigen A on red blood cells
Type AB blood type that has both antigen A and antigen B on red blood cells
Type B blood type that has antigen B on red blood cells
Type O blood type that has neither antigen A nor antigen B on red blood cells
tyrosine kinase inhibitors (TYE-roh-seen KIN-ayz in-HIB-it-erz) class of chemotherapy drugs that inhibit tyrosine-activating enzymes in cancer cells

U

ubiquitin (yoo-BIK-wit-in) a class of very small proteins used by the cell to tag proteins for destruction by the proteasome
ulcer (UL-ser) necrotic open sore or lesion
ultrasonography (ul-trah-son-OG-rah-fee) imaging technique in which high-frequency sound waves are reflected off tissue to form an image
umbilical artery (um-BIL-i-kul AR-ter-ee) two small arteries that carry oxygen-poor blood from the developing fetus to the placenta
umbilical cord (um-BIL-i-kul) flexible structure connecting the fetus with the placenta; contains umbilical arteries and vein
umbilical hernia (um-BIL-i-kul HER-nee-ah) rupture of the anterior abdominal wall at the navel
umbilical vein (um-BIL-i-kul) large vein carrying oxygen-rich blood from the placenta to the developing fetus
umbilicus (um-BIL-i-kus) navel
unipolar (pseudounipolar) neurons (yoo-nee-POH-lar [SOO-doh-yoo-nee-POH-lar] NOO-rons) structural category of neurons made up of cells

that appear to have only one extension from the cell body
unmyelinated fiber (un-MY-eh-lin-ay-ted) nerve fiber that does not have a myelin sheath; also referred to as *gray fiber*
unsaturated fat (un-SATCH-yoo-ray-ted) fat-containing fatty acid chains in which there are some double bonds, not all sites for hydrogen are filled; usually liquid at room temperature
upper motor neurons (NOO-rons) neurons whose axons lie in either pyramidal or extrapyramidal tracts
upper respiratory tract (RES-pih-rah-tor-ee) respiratory organs that are not contained within the thorax; includes nasal cavity, pharynx, and associated structures
urea (yoo-REE-ah) nitrogen-containing waste product
uremia (yoo-REE-mee-ah) condition in which blood urea concentration is abnormally elevated, expressed as a high blood urea nitrogen (BUN) value; uremia is often caused by renal failure; also called *uremic syndrome*
ureter (YOOR-eh-ter) long tube that carries urine from kidney to bladder
urethra (yoo-REE-thrah) passageway from bladder to exterior; functions in elimination of urine; in males, also acts as a genital duct that carries sperm to the exterior
urethral stricture (yoo-REE-thral STRIK-chur) narrowing or blockage of the urethra
urethritis (yoo-reh-THRY-tis) inflammation or infection of the urethra
urinary bladder (YOOR-i-nair-ee BLAD-er) collapsible saclike organ that collects urine from the kidneys and stores it before elimination
urinary meatus (YOOR-i-nair-ee mee-AY-tus) external opening of the urethra
urinary system (YOOR-i-nair-ee) system responsible for excreting most liquid wastes from the body
urination (yoor-i-NAY-shun) passage of urine from the body; emptying of the bladder; also called *micturition*
urine (YOOR-in) fluid waste excreted by kidneys
urochrome (YOOR-oh-krohm) pigments from the breakdown of old red blood cells in the liver and elsewhere that are found in the urine
urodynamics (yoo-roh-dye-NAM-iks) force of urine flow in the urinary tract
urticaria (er-ti-KAIR-ee-ah) hives; allergic or hypersensitive response characterized by raised red lesions
uterine tube (YOO-ter-in toob) fallopian tube
uterus (YOO-ter-us) hollow, muscular organ that nourishes developing offspring until birth
utricle (YOO-tri-kul) part of membranous labyrinth of inner ear; involved with equilibrium
uvula (YOO-vyoo-lah) cone-shaped process hanging from the soft palate that helps prevent food and liquid from entering the nasal cavities

V

vaccination (vak-si-NAY-shun) method used to achieve active immunity by triggering the body to form antibodies against specific pathogens
vaccine (VAK-seen) application of killed or attenuated (weakened) pathogens (or portions of pathogens) to a patient to stimulate immunity against that pathogen
vagina (vah-JYE-nah) internal tube from uterus to vulva
vaginal orifice (VAH-ji-nal OR-i-fis) opening of the vagina to the outside of the body
vaginitis (vaj-i-NYE-tis) inflammation of the vagina
vagus nerve (VAY-gus) cranial nerve X; mixed nerve; sensations and movements of organs
variant Creutzfeldt-Jakob Disease (vCJD) (VAR-ee-ant KROYTS-felt YAH-kobe) degenerative disease of the central nervous system caused by prions (*pro*teinaceous *in*fectious particles) that convert normal proteins of the nervous system into abnormal proteins, causing loss of function; *see* **prion**
varicose vein (VAIR-i-kohs) enlarged vein in which blood pools; also called *varix* (*plural,* varices)
varix (VAIR-iks) varicose vein; *plural,* varices
vas deferens (vas DEF-er-enz) testicle duct that extends from the epididymis to the ejaculatory duct; also called *ductus deferens*
vasa vasorum (VAS-ah vah-SOR-um) tiny blood vessels that supply the smooth muscles that surround the walls of larger blood vessels
vasectomy (vah-SEK-toh-mee) surgical severing of the vas deferens to render a male sterile
vasoactive intestinal peptide (VIP) (vay-so-AK-tiv in-TES-ti-nal PEP-tyde) hormone involved with controlling intestinal secretion
vasodilator (vay-so-DYE-lay-tor) drugs that trigger the smooth muscles of arterial walls to relax, causing the arteries to dilate
vasomotor control mechanism (vay-so-MOH-tor kon-trol MEK-ah-niz-em) factors that control changes in the diameter of arterioles
vasomotor pressoreflex (vay-so-MOH-tor press-oh-REE-fleks) reflex that occurs in response to a change in arterial blood pressure
vault *see* **barrel**
vector (VEK-tor) arthropod that carries an infectious pathogen from one organism to another
vein vessel carrying blood from capillaries toward the heart
ventral (VEN-tral) of or near the belly; in humans, front or anterior; opposite of dorsal or posterior
ventral cavity (VEN-tral) body cavity that includes the thoracic cavity and abdominopelvic cavity; not a standard anatomical term, but used here to help organize the body for the beginning student
ventral nerve root (VEN-tral) fibers that carry motor information out of the spinal cord
ventral ramus (VEN-tral RAY-mus) large, complex branch of each spinal nerve
ventral root (VEN-tral) motor branch of a spinal nerve by which it is attached to the spinal cord
ventricle (VEN-tri-kul) a cavity, such as the large, fluid-filled spaces within the brain or the chambers of the heart
ventricular fibrillation (ven-TRIK-yoo-lar fib-ri-LAY-shun) an immediately life-threatening condition caused by the lack of ventricular pumping suddenly stopping flow of blood to vital organs
venule (VEN-yool) small blood vessels that collect blood from capillaries and join to form veins
vermiform appendix (VERM-i-form ah-PEN-diks) tubular structure attached to the cecum
vertebrae (VER-teh-bray) bones that make up the spinal column
vertebral column (ver-TEE-bral KOL-um) the spinal column, made up of a series of separate vertebrae that form a flexible, curved rod; made up of the cervical, thoracic, lumbar, sacral, and coccygeal segments
vertebral foramen (ver-TEE-bral for-AY-men) the central opening in the vertebral column that contains the spinal cord
vertigo (VER-ti-go) abnormal sensation of spinning; dizziness
vesicle (VES-i-kul) any tiny membranous bubble within a cell; clinical term referring to blisters, fluid-filled skin lesions
vestibular membrane (ves-TIB-yoo-lar) roof of the cochlear duct; also called *Reissner's membrane*
vestibular nerve (ves-TIB-yoo-lar nerv) division of the vestibulocochlear nerve (eighth cranial nerve)
vestibule (VES-ti-byool) located in the inner ear; portion adjacent to the oval window between the semicircular canals and the cochlea
vestibulocochlear nerve (ves-TIB-yoo-loh-kok-lee-ar nerv) cranial nerve VIII; sensory nerve; responsible for hearing and equilibrium
villi (VIL-eye) finger-like folds covering the plicae of the small intestines
Vincent infection (VIN-sent in-FEK-shun) bacterial (spirochete) infection of the gum, producing gingivitis; also called *Vincent angina* and *trench mouth*
virion (VYE-ree-on) a complete viral particle (genetic material and protein capsule)
virus (VYE-rus) microscopic, intracellular parasitic entity consisting of a nucleic acid bound by a protein coat and sometimes a lipoprotein envelope
viscera (VISS-er-ah) refers to internal organs
visceral (VISS-er-al) pertaining to the viscera (internal organs); toward or on the internal organs (opposite of parietal)
visceral membrane (VISS-er-al) serous membrane that covers the surface of the viscera
visceral muscle tissue (VISS-er-al) smooth muscle tissue
visceral pericardium (VISS-er-al pair-i-KAR-dee-um) the serous membrane that adheres to and covers the heart
visceral portion (VISS-er-al POR-shun) serous membrane that covers the surfaces of organs in the body cavity
visceral reflex (VISS-er-al) autonomic reflex; contractions of smooth or cardiac muscles or secretion by glands
visceral sensory division (VISS-er-al SEN-sor-ee) division of the nervous system made up of afferent (incoming) pathways from autonomic sensory receptors (receptors involved in subconscious perception) of the internal organs (viscera)
viseroceptor (viss-er-oh-SEP-tor) somatic sense receptors located in the internal visceral organs
vital capacity (VC) (VYE-tal kah-PAS-i-tee) largest amount of air that can be moved in and out of the lungs in one inspiration and expiration
vitamin (VYE-tah-min) organic molecules needed in small quantities to help enzymes operate effectively
vitiligo (vit-i-LYE-go) acquired condition that results in loss of pigment in certain areas of the skin
vitreous humor (VIT-ree-us HYOO-mor) jelly-like fluid in the eye, posterior to the lens
voiding (VOYD-ing) emptying the bladder
volar (VOH-lar) palm of the hand or sole of the foot
Volkmann canals (VOLK-man kah-NALZ) communicating canals between Haversian canals that contain vessels to carry blood from the exterior surface of the bone to the osteons
voltage-gated channels (VOL-tij GAY-ted CHAN-nuls) type of cell-membrane channel for the transport of molecules that is controlled by a gate that responds to a change in voltage (difference in charge across the cell membrane)
voluntary muscle *see* **skeletal muscle**
vulva (VUL-vah) external genitals of the female
vulvitis (vul-VYE-tis) inflammation of the vulva (external female genitals)

W

white fibers muscle fibers containing little myoglobin; also called *fast fibers*; nerve fibers with a thick myelin sheath
white matter nerves covered with white myelin sheath
white ramus (RAY-mus) small branch of myelinated sympathetic preganglionic fibers
windpipe trachea
withdrawal reflex reflex that moves a body part away from an irritating stimulus

X

X chromosome (x KROH-moh-sohm) sex chromosome, which, in females, is paired with another X chromosome; as opposed to a Y chromosome in males
X-linked trait genetic trait associated with the female (X) chromosome

Y

Y chromosome (y KROH-moh-sohm) sex chromosome that contains genes that determine maleness
yeast single-celled fungus (compared to mold, which is a multicellular fungus)
yellow marrow connective tissue rich in fat that is found in the medullary cavity of the long bones of an adult; inactive in red cell production
Y-linked trait genetic trait associated with the male (Y) chromosome
yolk sac in humans, involved with production of blood cells in the developing embryo
Young-LaPlace Law (yung-lah-PLAHS law) principle that states that air pressure (P) in a bubble is inversely proportional to the radius (r) and directly proportional to the surface tension (T) summarized by the equation $P = 2T/r$; named for Englishman Thomas Young and Frenchman Pierre Simon LaPlace; also called *Law of LaPlace*

Z

z disk microscopic structure within the myofibril of a muscle fiber where the thin filaments unite with each other and form a netlike disk; also called the *z line*; serves as boundary of sarcomere unit
zona fasciculata (ZOH-nah fas-sic-yoo-LAY-tah) middle zone of the adrenal cortex that secretes glucocorticoids
zona glomerulosa (ZOH-nah gloh-mair-yoo-LOH-sah) outer zone of the adrenal cortex that secretes mineralocorticoids
zona reticularis (ZOH-nah reh-tik-yoo-LAIR-is) inner zone of the adrenal cortex that secretes small amounts of sex hormones
zone of calcification (zone of kal-si-fi-KAY-shun) deepest layer of the epiphyseal plate; composed of cartilage undergoing rapid calcification
zone of hypertrophy (zone of hye-PER-troh-fee) third layer of the epiphyseal plate; composed of older, enlarged cartilage cells that are undergoing degenerative changes associated with calcium deposition
zone of proliferation (zone of proh-LIF-er-ay-shun) second layer of the epiphyseal plate; composed of cartilage cells undergoing active mitosis
zygomaticus (zye-go-MAT-ik-us) muscle that elevates corners of the mouth and lips; also known as the "smiling muscle"
zygote (ZYE-goht) a fertilized ovum

INDEX